ANALYTE REFERENCE CHART—cont'd

Analyte*	Expected adult reference or therapeutic range			Clinical correlation (page)	Method of analysis (page)
	Conventional units		SI units		
Bence Jones proteins (U)	<60 mg/L		<2.7 μmol/L	730	*(M-691)*
Benzodiazepines (S, P)	See Table 81-8 (p. *M-640*)			740	*(M-627)*
Beta-HCG (beta-human chorionic gonadotropin); *see* Human chorionic gonadotropin					
Beta-hydroxybutyric acid; *see* β-Hydroxybutyric acid					
Beta$_2$-microglobulin; *see* β$_2$-Microglobulin					
Bile acids (S), total	200-2000 ng/mL		—	359	*(M-1098)*
cholic acid	10-100 ng/mL		24-240 nmol/L		
chenodeoxycholic acid	10-80 ng/mL		25-204 nmol/L		
Bilirubin (S), total	<15 mg/L		<26 μmol/L	359, 569	1009
conjugated	<2 mg/L		<3.4 μmol/L		
Free bilirubin (S)	<11.7 μg/L		<20 nmol/L		*(M-1115)*
Reserve bilirubin-binding capacity (S)	>50 mg/L		>87 μmol/L		
Bilirubin amniotic fluid; *see* Amniotic fluid bilirubin					
Blood gases; *see* pH, P_{CO_2}, P_{O_2}				332	*(M-54)*
Breath test $^{14}CO_2$	>2%/hr		—	398	*(M-831)*
Bromide (S)	<1000 μg/mL		<12.5 mmol/L	740	*(M-342)*
BUN (blood urea nitrogen); *see* Urea					
Cadmium (B)	0.1-3.5 μg/L	nonsmokers	0.9-31.1 nmol/L	533	*(M-346)*
	0.3-6.5 μg/L	smokers	2.7-57.8 nmol/L		
(U)	0.3-1.8 μg/L		2.7-16 nmol/L		
Caffeine (S)	5-20 μg/mL	neonates	25-100 μmol/L	795	1116
Calcitonin (P)	<300 pg/mL		<85.7 pmol/L	373, 664	*(M-696)*
Calcium:				373, 664	865
Ionized (S, P, B)	46-53 mg/L		1.14-1.31 mmol/L		*(M-1010)*
Total (S)	85-105 mg/L (atomic absorption)		2.1-2.65 mmol/L		865
Total (U)	<275 mg/24 hr	male	<6.9 mmol/24 hr		
	<250 mg/24 hr	female	<6.2 mmol/24 hr		
Carbamazepine (S)	4-12 mg/L		17-51 μmol/L	594	*(M-642)*
Carbohydrate screen (U)	undetectable			688	*(M-154)*
Carbon dioxide, total (S)	21-31 mEq/L		21-31 mmol/L	313, 332	869
Carboxyhemoglobin (carbon monoxide) (B)	0.42-1.5%	nonsmokers	—	740	*(M-362)*
	2%-10%	smokers	—		
Carcinoembryonic antigens (CEA) (S)	<2.5-4 ng/mL	nonsmokers	—	730	1033
	<10 ng/mL	smokers	—		
Carotenes (S)	0.5-2.5 μg/mL (as total carotenoids)		0.93-4.65 μmol/L	543	*(M-513)*
	0.1-0.85 μg/mL (β-carotene)		0.18-1.58 μmol/L		
Catecholamines (P)	<400 pg/mL hypertensive	norepinephrine	<2.36 pmol/mL	415	*(M-944)*
	<100 pg/mL	epinephrine	<0.55 pmol/mL		
(U)	15-80 μg/24 hr adults	norepinephrine	0.09-0.47 μmol/24 hr		*(M-949)*
	<20 μg/24 hr	epinephrine	<0.11 μmol/24 hr		
	65-400 μg/24 hr	dopamine	0.42-2.6 μmol/24 hr		
Cerebrospinal fluid proteins; *see* Proteins, cerebrospinal fluid					
Ceruloplasmin (S)	150-600 mg/L		1-4 μmol/L	533, 588	*(M-158)*
Chloride (S)	101-111 mEq/L		101-111 mmol/L	313, 332, 346	872
(U)	110-250 mEq/24 hr		110-250 mmol/24 hr		
Sweat	<35 mEq/L		<35 mmol/L		
Cholesterol, total (S, P):	Age, sex, and risk dependent, p. 981			454	974
High-density lipoprotein (HDL) cholesterol (P)	Age and sex dependent, pp. 989 and 990			454	983
Low-density lipoprotein (LDL) cholesterol (P)	Age and sex dependent, pp. 987 and 988				
Cholinesterase (S, amniotic fluid)	Phenotype dependent			594, 740	913
	4.9-12.2 μmol/min/mL		$0.8\text{-}2.0 \times 10^{-4}$ katal/L		
Chromium (U)	<0.5 μg/L		<9.6 nmol/L	533	*(M-520)*
Copper (S)	0.7-1.4 mg/L	males	11.0-22 μmol/L	533	*(M-527)*
	0.8-1.55 mg/L	females	12.6-24.4 μmol/L		
(U)	<30 μg/24 hr		<0.47 μmol/24 hr		
Cortisol (S)	50-230 μg/L	8 A.M.	0.14-0.63 μmol/L	672	*(M-224)*
(U)	4.9-35.3 μg/24 hr		13.5-97.4 nmol/24 hr		
Creatine (S)	2-5.9 mg/L	males	15-44 μmol/L	427	*(M-876)*
	3.9-10 mg/L	females	30-80 μmol/L		
(U)	<45.9 mg/L	males	<0.35 mmol/L		
	<91.8 mg/L	females	<0.7 mmol/L		
Creatine kinase (CK) (S)	<160 U/L	male	$<2.67 \times 10^{-6}$ katal/L	415, 427	917
	<130 U/L	female	$<2.17 \times 10^{-6}$ katal/L		

*B, Whole blood; P, plasma; S, serum; U, urine. Page numbers in parentheses and preceded by *M-* are cross-references to Pesce, AJ, and Kaplan, LA: *Methods in Clinical Chemistry,* St Louis, 1987, CV Mosby Co.

Continued on next page.

ANALYTE REFERENCE CHART—cont'd

Analyte*	Expected adult reference or therapeutic range			Clinical correlation (page)	Method of analysis (page)
	Conventional units		SI units		
Creatine kinase isoenzymes (S)	<4% CK-MB		—	415, 427, 787	922
Creatinine (S)	6.4-10.4 mg/L	males	57-92 μmol/L	346, 427	1015
	5.7-9.2 mg/L	females	50-81 μmol/L		
(U)	1-2 g/day	males	8.8-17.7 mmol/day		
	0.8-1.8 g/day	females	7.1-16 mmol/day		
Creatinine clearance	97-137 mL/min	males	—	346	1015
	88-128 mL/min	females	—		
Cyclosporin (B)	250-1000 ng/mL	RIA	0.21-0.83 μmol/L	795	*(M-427)*
(P)	50-200 ng/mL		0.04-0.16 μmol/L		
Dehydroepiandrosterone (DHEA) (S, P)	3.6-6.3 ng/mL	males	12.5-21.9 nmol/L	650	*(M-232)*
	4.4-6.0 ng/mL	females	15.3-20.8 nmol/L		
Dehydroepiandrosterone sulfate (DHEA-S) (S, P)	1.69-3.15 μg/mL	males	4.61-8.55 μmol/L	650	*(M-232)*
	Varies within menstrual cycle in females				
Delta-aminolevulinic acid; *see* δ-Aminolevulinic acid				496	*(M-1232)*
Digitoxin (S)	9-32 ng/mL		11.8-41.8 nmol/L	415	1088
Digoxin (S)	0.5-2 ng/mL		0.6-2.4 nmol/L	415	1088
Drug screen (U)	Negative		—	740	1091
Electrophoresis; *see* Hemoglobin electrophoresis; Lipoprotein electrophoresis; Protein electrophoresis					
Estriol (serum and urine)	Varies with gestational age			569, 650	944
Ethchlorvynol (S)	<14 mg/L		<96.8 μmol/L	740	*(M-384)*
Ethylene glycol (S)	negative		—	740	*(M-388)*
Fecal electrolytes, fecal osmolality	See Table 106-1 (p. *M-842*)			398	*(M-841)*
Fecal fat absorption	<5% of ingested fat is excreted		—	398	*(M-834)*
Ferritin (S)	20-200 μg/L		45-450 pmol/L	496	*(M-1240)*
α-Fetoprotein; *see* Alpha-fetoprotein				569, 730	*(M-459)*
Folic acid (S)	1.9-14 ng/mL		4.3-31.7 nmol/L	626, 543	1131
Formiminoglutamic acid (FIGLU) (U)	<5.1 mg/L	infants	<29 μmol/L	688	*(M-169)*
	<7.4 mg/24 hr	adults	<42.6 μmol/24 hr		
Galactose-1-phosphate uridylyltransferase (B)	See Table 27-3 (p. *M-179*)		—	688	*(M-173)*
Gamma-glutamyl transferase (S)	8-37 U/L	males	11.7-67 × 10^{-8} katal/L	359	*(M-1120)*
	5-24 U/L	females	6.7-42 × 10^{-8} katal/L		
Gastric fluid analysis	Basal and maximal acid output: see Table 107-3 (p. *M-846*)			398	*(M-843)*
Gentamicin (S)	See under aminoglycosides				
Glucose (S):					
Hexokinase	700-1100 mg/L		3.9-6.1 mmol/L	436	850
Oxygen electrode	650-1100 mg/L		3.6-6.1 mmol/L		
Glycosylated hemoglobin (B)	4.5%-8.5%		—	436	1040
Glycosylated proteins—albumin (S, P)	1.2%-2.3% and 8.4%-15.7% depending on method			436	*(M-118)*
Growth hormone (S)	>10 μg/L in neonates			613	*(M-239)*
	<0.5 μg/L after glucose challenge				
Haptoglobin (S)	600-2700 mg/L		7.1-31.8 μmol/L	512	*(M-1246)*
Hemoglobin (B):					
Hb A_2	1.5%-4%		—	512, 688	1043
Hb F	<0.9%		—	512, 688	1043 *(M-1243)*
Hb S	0		—	512, 688	1043 *(M-1249)*
Hemoglobin electrophoresis (B)	See interpretation		—	512, 688	1043 *(M-1249)*
High-density lipoprotein (HDL) cholesterol (S); *see* Cholesterol					
Homovanillic acid (U)	Age dependent; see Table 92-3 (p. *M-724*)			730	930
Hormone receptors, steroid; *see* Steroid hormone receptors					
Human chorionic gonadotropin (hCG, HCG)				569, 730	938
(S)	nonpregnant <2 mU/mL				
	pregnant >20 mU/mL, varies with gestational age				
(U)	<50 mU/mL				
β-Hydroxybutyric acid (S, P)	<73 mg/L		<0.7 mmol/L	436	*(M-101)*
5-Hydroxyindoleacetic acid (5-HIAA) (U)	1.8-6.0 mg/24 hr		9.4-31.4 μmol/24 hr	719	*(M-714)*
Immunoelectrophoresis (S)	—		—	730, 594	1047
Immunoglobulin (Ig) quantitation (S):	Age dependent, see Table 94-1 (p. *M-735*); adult values below			730, 594	1050
IgA	1.4-3.5 mg/mL		8.7-21.9 μmol/L		
IgD	0-0.14 mg/mL		0-0.76 μmol/L		
IgE	<300 ng/mL		<1.5 nmol/L		
IgG	8.0-16.0 mg/mL		53-106 μmol/L		
IgM	0.5-2.0 mg/mL		0.56-2.2 μmol/L		

**B*, Whole blood; *P*, plasma; *S*, serum; *U*, urine. Page numbers in parentheses and preceded by *M-* are cross-references to Pesce, AJ, and Kaplan, LA: *Methods in Clinical Chemistry*, St Louis, 1987, CV Mosby Co.

Continued on back endpaper.

Clinical Chemistry

Theory, analysis, and correlation

To David
from
Edward E. [illegible]
9/5/93

CLINICAL CHEMISTRY

Theory, analysis, and correlation

LAWRENCE A. KAPLAN, Ph.D.

Medical Research Laboratories,
Cincinnati, Ohio

AMADEO J. PESCE, Ph.D.

Professor, Experimental Medicine,
Pathology, and Laboratory Medicine,
University of Cincinnati Medical Center,
Cincinnati, Ohio

SECOND EDITION

with 95 *contributors*
with 757 *illustrations*

The C. V. Mosby Company

ST. LOUIS • BALTIMORE • PHILADELPHIA • TORONTO 1989

Editor: Stephanie Bircher
Assistant Editor: Anne Gunter
Developmental Editor: Judith E. Kaplan
Production Editors: Sheila Walker, Roger McWilliams, Mary Wright, April Nauman, Celeste Clingan, Jeanne Genz
Book and Cover Design: Gail Morey Hudson
Technical Reviewer: Wendell Davis, Ph.D.

SECOND EDITION

Printed in the United States of America

The C.V. Mosby Company
11830 Westline Industrial Drive, St. Louis, Missouri 63146

Library of Congress Cataloging in Publication Data

Clinical chemistry: theory, analysis, and correlation/[edited by] Lawrence A. Kaplan, Amadeo J. Pesce; with 95 contributors.—2nd ed.

p. cm.
Includes bibliographies and index.
ISBN 0-8016-2704-4
1. Chemistry, Clinical. I. Kaplan, Lawrence A., 1944- II. Pesce, Amadeo J.
[DNLM: 1. Chemistry, Clinical. QY 90 C6415]
RB40.C58 1989
616.07'56—dc19

C/VH/VH 9 8 7 6 5 4 3

Contributors

F. PHILIP ANDERSON, M.S.

Department of Pathology, Medical College of Virginia,
Richmond, Virginia

SUSAN BASSION, Ph.D.

Section Head, Immunology Laboratory,
Department of Pathology, Scott and White Clinic,
Temple, Texas

JOHN D. BAUER, M.D.

Associate Professor, Department of Pathology,
Washington University School of Medicine;
Director of Laboratories, Faith Hospital,
St. Louis, Missouri

GORDON L. BILLS, M.D.

Staff Pathologist, Good Samaritan Hospital, Cincinnati, Ohio

LARRY D. BOWERS, Ph.D.

Associate Professor,
Department of Laboratory Medicine and Pathology,
University of Minnesota, Minneapolis, Minnesota

JOHN M. BREWER, Ph.D.

Professor, Department of Biochemistry,
University of Georgia, Athens, Georgia

MARGE A. BREWSTER, Ph.D.

Professor, Departments of Pathology and Pediatrics,
University of Arkansas for Medical Sciences,
Little Rock, Arkansas

BERNDT B. BRUEGGER, Ph.D.

Research Chemist, BioRad Laboratories, Hercules, California

C. RALPH BUNCHER, Sc.D.

Professor and Director,
Biostatistics and Epidemiology Laboratory,
Department of Environmental Health,
University of Cincinnati, Cincinnati, Ohio

Sr. ELIZABETH ANN BYRNE, M.S., C.L.S.

Assistant Professor, Biology Department,
College of Mount St. Joseph,
Mount St. Joseph, Ohio

R. NEILL CAREY, Ph.D.

Clinical Chemist, Peninsula General Hospital;
Adjunct Associate Clinical Professor,
Department of Medical Technology,
Salisbury State College, Salisbury, Maryland

JOHN F. CHAPMAN, Dr.P.H.

Associate Professor, Department of Pathology,
University of North Carolina at Chapel Hill,
Chapel Hill, North Carolina

I-WEN CHEN, Ph.D.

Professor, Department of Radiology,
Eugene L. Saenger Radioisotope Laboratory,
University of Cincinnati Medical Center, Cincinnati, Ohio

BARBARA J. CLEVELAND, M.P.H., M.T. (ASCP)

Assistant Professor, Department of Clinical Laboratory Sciences,
University of Oklahoma Health Sciences Center,
Oklahoma City, Oklahoma

BRADLEY E. COPELAND, M.D.

Professor, Department of Pathology and Laboratory Medicine,
University of Cincinnati, Cincinnati, Ohio

VICKI N. DAASCH, Ph.D.

Department of Hospital Laboratories,
North Carolina Memorial Hospital,
Chapel Hill, North Carolina

JOSEPH R. DiPERSIO, Ph.D.

Director of Clinical Microbiology,
Department of Laboratory Medicine/Microbiology,
The Christ Hospital, Cincinnati, Ohio

RICHARD F. DODS, Ph.D.

Clinical Laboratory Consultant, Palatine, Illinois

GRAHAM ELLIS, Ph.D.

Associate Professor, Department of Biochemistry,
University of Toronto, Toronto, Ontario, Canada

E. CHRISTIS FARRELL, Jr., Ph.D.

Director, Clinical Chemistry Laboratory, Ohio Valley Hospital,
Steubenville, Ohio

MARIANO FERNANDEZ-ULLOA

Professor of Radiology,
Eugene L. Saenger Radioisotope Laboratory,
University of Cincinnati Medical Center, Cincinnati, Ohio

M. ROY FIRST, M.D.

Professor, Department of Internal Medicine,
Division of Nephrology,
University of Cincinnati Medical Center, Cincinnati, Ohio

CHRISTOPHER S. FRINGS, Ph.D.

Clinical Professor of Pathology,
School of Community and Allied Health,
University of Alabama, Birmingham, Alabama

CARL C. GARBER, Ph.D.

Medical Products Department, E.I. DuPont Company,
Wilmington, Delaware

DAVID L. GARVER, M.D.

Professor, Departments of Psychiatry,
Pharmacology, and Cell Biophysics,
University of Cincinnati Medical Center, Cincinnati, Ohio

JACK GAULDIE, Ph.D.

Associate Professor, Department of Pathology,
McMaster University Medical Center,
Hamilton, Ontario, Canada

STEPHEN M. GENDLER, Ph.D.

Systems Analyst, Ingalls Memorial Hospital, Harvey, Illinois

LEWIS GLASSER, M.D.

Chief, Hematopathology Laboratories; Professor,
Department of Pathology, Arizona Health Sciences Center,
Tucson, Arizona

F. MICHAEL HASSAN, B.S., C.(ASCP)

Assistant Director, Toxicology Laboratory,
Department of Pathology and Laboratory Medicine,
University of Cincinnati Medical Center, Cincinnati, Ohio

WILLIAM R. HEINEMAN, Ph.D.

Distinguished Research Professor, Department of Chemistry,
University of Cincinnati, Cincinnati, Ohio

LINDA A. HEMINGER, B.S.(CNMT)

Department of Radiology,
Eugene L. Saenger Radioisotope Laboratory,
University of Cincinnati Medical Center, Cincinnati, Ohio

DAVID C. HOHNADEL, Ph.D.

Director, Clinical Chemistry,
Department of Laboratory Medicine,
The Christ Hospital, Cincinnati, Ohio

PETER HORSEWOOD, Ph.D.

Assistant Professor, Department of Pathology,
McMaster University Medical Center,
Hamilton, Ontario, Canada

PAUL E. HURTUBISE, Ph.D.

Professor, Department of Pathology and Laboratory Medicine,
University of Cincinnati Medical Center, Cincinnati, Ohio

GAYLE B. JACKSON, M.S.

Director, Hematology/Immunology,
Department of Laboratory Medicine,
The Christ Hospital, Cincinnati, Ohio

STEPHEN N. JOFFE, M.D.

Professor, Department of Surgery,
University of Cincinnati Medical Center, Cincinnati, Ohio

GAIL JONES, M.S., M.T.(ASCP)

School of Medical Technology, Harris Hospital,
Ft. Worth, Texas

STEVEN C. KAZMIERCZAK, Ph.D.

Department of Clinical Pathology and Diagnostic Medicine,
East Carolina University School of Medicine,
Greenville, North Carolina

THADDEUS E. KELLY, M.D., Ph.D.

Professor, Department of Pediatrics,
University of Virginia Medical Center, Charlottesville, Virginia

ELIZABETH J. KICKLIGHTER, M.D.

Department of Internal Medicine,
University of Cincinnati Medical Center, Cincinnati, Ohio

MARY ELLEN KING, Ph.D.

Associate Professor, Department of Pathology,
University of Virginia Commonwealth University,
Medical College of Virginia, Richmond, Virginia

JON R. KIRCHHOFF, Ph.D.

Postdoctoral Fellow, Department of Chemistry,
University of Cincinnati, Cincinnati, Ohio

LEONARD I. KLEINMAN, M.D.

Professor and Director of Newborn Services,
Department of Pediatrics,
State University of New York at Stony Brook School of Medicine,
Stony Brook, New York

ANTHONY KOLLER, Ph.D.

Associate Director, Division of Biochemistry,
Department of Pathology,
Michael Reese Hospital and Medical Center, Chicago, Illinois

WILLIAM J. KORZUN, Ph.D.

Department of Pathology, Medical College of Virginia,
Richmond, Virginia

RICHARD J. KOZERA, M.D.

Professor, Department of Internal Medicine,
Temple School of Medicine, Philadelphia, Pennsylvania

VICKY A. LeGRYS, D.A.

Assistant Professor, Division of Medical Technology,
University of North Carolina at Chapel Hill,
Chapel Hill, North Carolina

BOLESLAW H. LIWNICZ, M.D., Ph.D.

Associate Professor of Pathology and Laboratory Medicine,
University of Cincinnati Medical Center, Cincinnati, Ohio

REGINA G. LIWNICZ, M.D.

University of Cincinnati College of Medicine, Cincinnati, Ohio

JOHN M. LORENZ, M.D.

Assistant Professor, Department of Pediatrics,
State University of New York at Stony Brook School of Medicine,
Stony Brook, New York

JOHN A. LOTT, Ph.D.

Diplomate, American Board of Clinical Chemistry;
Professor of Pathology, Ohio State University Medical Center,
Columbus, Ohio

CRAIG E. LUNTE, Ph.D.

Assistant Professor, Department of Chemistry,
University of Kansas, Lawrence, Kansas

HAROLD MARDER, M.D.

Philadelphia Children's Hospital and Medical Center,
Philadelphia, Pennsylvania

HARRY R. MAXON III, M.D.

Professor of Radiology, Nuclear Medicine;
Associate Professor of Medicine, Endocrinology,
Eugene L. Saenger Radioisotope Laboratory,
University of Cincinnati Medical Center, Cincinnati, Ohio

TIMOTHY G. McMANAMON, Ph.D.

Clinical Chemist, Department of Pathology,
Mercy Hospital Medical Center, Des Moines, Iowa

MICHAEL D.D. McNEELY, M.D.

Director of Laboratories, Island Medical Laboratories,
Victoria, British Columbia, Canada

CHARLES L. MENDENHALL, M.D.

Professor, Department of Internal Medicine,
University of Cincinnati Medical Center, Cincinnati, Ohio

W. GREGORY MILLER, Ph.D.

Associate Professor, Department of Pathology,
Medical College of Virginia, Richmond, Virginia

GERALD MORIARTY, M.D.

Assistant Professor, Department of Neurology,
University of Cincinnati Medical Center, Cincinnati, Ohio

ROBERT L. MURRAY, Ph.D.

Department of Biochemistry, Lutheran General Hospital,
Park Ridge, Illinois

BÉLA NAGY, Ph.D.

Consultant, Cincinnati, Ohio

HERBERT K. NAITO, Ph.D.

Head, Section of Lipids, Nutrition and Metabolic Diseases,
Department of Biochemistry, The Cleveland Clinic Foundation,
Cleveland, Ohio

STEVEN A. NOEL, Ph.D.

Clinical Chemist, Department of Pathology,
Greater Baltimore Medical Center, Baltimore, Maryland

ROBERT J. NORMAN, Ph.D.

Senior Lecturer, Department of Obstetrics and Gynecology,
University of Adelaide, Adelaide, South Australia

RONALD OBERNOLTE, Ph.D.

Clinical Laboratory Consultant, Bass Lake, California

NANCY C. PARKER, M.T.(ASCP), S.C.

Clinical Chemistry Laboratories,
North Carolina Memorial Hospital,
University of North Carolina at Chapel Hill,
Chapel Hill, North Carolina

GERARDO PERROTTA, M.T.(ASCP), S.H.

Assistant Director, Clinical Hematology Laboratory,
Department of Pathology and Laboratory Medicine,
University of Cincinnati Hospital, Cincinnati, Ohio

MICHAEL A. PESCE, Ph.D.

Director, Special Chemistry Laboratory,
Columbia-Presbyterian Medical Center, New York, New York

NATHAN A. PICKARD, Ph.D.

Advanced Chemistry Specialist, E.I. DuPont Company,
Hoffman Estates, Illinois

ALPHONSE POKLIS, Ph.D.

Professor, Department of Pathology;
Director of Clinical Toxicology Laboratory,
Medical College of Virginia, Richmond, Virginia

MORRIS R. PUDEK, Ph.D.

Clinical Chemist, Department of Pathology,
Vancouver General Hospital, Vancouver,
British Columbia, Canada

WOLFGANG A. RITSCHEL, Ph.D.

Professor of Pharmacokinetics and Biopharmaceutics; Head,
Division of Pharmaceutics and Drug Delivery Systems,
University of Cincinnati Medical Center, Cincinnati, Ohio

ALLYN H. RULE, Ph.D.

Associate Professor, Graduate Department of Biology,
Boston College, Boston, Massachusetts

PAUL T. RUSSELL, Ph.D.

Department of Reproductive Endocrinology,
The Christ Hospital, Cincinnati, Ohio

WILLIAM E. SCHREIBER, M.D.

Clinical Assistant Professor, Department of Pathology,
University of British Columbia,
Vancouver, British Columbia, Canada

TIMOTHY J. SCHROEDER, M.S.

Clinical Toxicologist, Department of Pathology and Laboratory
Medicine, University of Cincinnati Medical Center,
Cincinnati, Ohio

ARNOLD L. SCHULTZ, Ph.D.

Associate Professor, Department of Pathology,
University of Colorado Health Sciences Center,
Denver, Colorado

G. BERRY SCHUMANN, M.D.

Professor of Pathology; Director,
Cytopathology and Cytotechnology,
University of Utah College of Medicine,
Salt Lake City, Utah

SUSAN C. SCHWEITZER, M.S., M.T.(ASCP)

Technical Consultant, Department of Pathology,
University of Utah School of Medicine, Salt Lake City, Utah

JOHN E. SHERWIN, Ph.D.

Director of Chemistry, Department of Pathology,
Valley Children's Hospital, Fresno, California

LAWRENCE M. SILVERMAN, Ph.D.

Associate Professor of Pathology and Biochemistry,
University of North Carolina Medical School;
Director, Special Chemistry, North Carolina Memorial Hospital,
Chapel Hill, North Carolina

STEVEN J. SOLDIN, Ph.D.

Department of Laboratory Medicine, Children's Hospital,
National Medical Center, Washington, D.C.

MATTHEW I. SPERLING, B.S.

Research Assistant III,
Eugene L. Saenger Radioisotope Laboratory,
Department of Radiology,
University of Cincinnati Medical Center,
Cincinnati, Ohio

BERNARD E. STATLAND, M.D., Ph.D.

Medical Director and Department Head,
Department of Pathology and Laboratory Medicine,
Methodist Hospital of Indiana, Indianapolis, Indiana

PAULA STEINER, B.S., C.(ASCP)

Vice-President, Operations, Medical Research Laboratories,
Cincinnati, Ohio

GEORGE SULLIVAN, M.T.(ASCP) M.B.A.

Clinical Laboratory Consultant, Loveland, Ohio

JOSEPH SVIRBELY, Ph.D.

Clinical Toxicologist,
Department of Pathology and Laboratory Medicine,
University of Cincinnati Medical Center, Cincinnati, Ohio

M. WILSON TABOR, Ph.D.

Associate Professor of Environmental Health,
Director of Environmental Analytical Chemistry,
Institute of Environmental Health,
University of Cincinnati Medical Center, Cincinnati, Ohio

STEPHAN G. THOMPSON, Ph.D.

Supervisor, Immunodiagnostics, Ames Division,
Miles Laboratories, Inc., Elkhart, Indiana

REGINALD C. TSANG, M.D.

Professor, Department of Pediatrics;
Director, Division of Neonatology,
University of Cincinnati Medical Center, Cincinnati, Ohio

KORY M. WARD, Ph.D.

Instructor, Division of Medical Technology,
School of Allied Medical Professions,
Ohio State University, Columbus, Ohio

ROBERT E. WEESNER, M.D.

Assistant Professor, Department of Internal Medicine,
University of Cincinnati Medical Center;
Director, GI Endoscopy Services,
Veterans Administration Medical Center, Cincinnati, Ohio

DAN WEINER, Ph.D.

Clinical Laboratory Consultant, Cupertino, California

JOHN F. WHEELER

Predoctoral Fellow, Department of Chemistry,
University of Cincinnati, Cincinnati, Ohio

PER WINKEL, M.D., Doc. Med. Sci.

Clinical Chemistry Department,
Rigshospitalet Hospital, Copenhagen, Denmark

LINDA L. WOODARD, M.T.(ASCP)

Clinical Chemistry Laboratories,
North Carolina Memorial Hospital,
Chapel Hill, North Carolina

To our wives

Judith Kaplan and **Anna Pesce**

from whom we took eight years of their lives,

and our mentors

Samuel Natelson and **George F. Grannis**

Foreword

Clinical chemistry is the application of the science of chemistry to the understanding of the human in health and disease. Much of the progress in basic chemistry in the past has been motivated largely by an intense interest in the nature of disease in the human. For example, the alchemist's search for the panacea, to cure all human ills, resulted in the identification of many of the common elements and basic principles of chemistry. However, the term *clinical chemistry,* to designate this specific discipline, did not come into general use until 1949, with the organization of the American Association of Clinical Chemists (now called American Association for Clinical Chemistry) and the subsequent publication of clinical chemistry journals in the United States, Scandinavia, England, Canada, Germany, Italy, and other countries.

Since the early 1950s, substantial progress has taken place in clinical chemistry, not only in highly sophisticated instrumentation, but also in the scope of the science. Instruments not generally available 35 years ago are in common use today in the clinical chemistry laboratory. Examples include automated instrumentation for assaying more than 20 blood components on over 100 small blood specimens per hour, highly sophisticated computers for controlling these devices and analyzing the data, gas chromatography/mass spectrometry assemblies, high-performance liquid chromatography equipment, and atomic absorption and infrared spectrometers. Even the analytical gravimetric balance has been replaced largely by the rapidly weighing electronic balances with digital readout. pH, Pco_2 and Po_2 measurements, which were difficult chores in the fifties, are now carried out routinely with relatively inexpensive instruments, with a degree of accuracy and precision not considered possible 35 years ago. Many components of biological materials that were considered outside the range of capability of the clinical chemistry laboratory are now assayed daily. For example, immunoassay by radiometric, fluorimetric, and enzymatic techniques are performed for the biologically active polypeptides, tumor markers, and other components normally present in minute quantities in the serum such as steroids and iodothyronines. The blood of patients receiving therapeutic drugs is analyzed regularly, even when concentrations are very low. Toxicological sections of the clinical chemistry laboratory have developed extensively, using the modern techniques just mentioned.

With the expansion of the role of the laboratory of clinical chemistry in public health care has come a corresponding increase in their economic importance. Clinical chemistry is now a multibillion-dollar industry. Most major industrial chemical and pharmaceutical complexes are now directly involved in providing instrumentation and supplying reagents for use in the clinical laboratory. Several of these companies operate service laboratories for private physicians and provide reference laboratories or, by contract, operate hospital laboratories. The private-service laboratories have moved from being one- or two-person establishments to national multimillion-dollar enterprises.

The growth of clinical chemistry in the past 25 years has added a major industry to the United States and the world. This growth has been so rapid that adequate time to develop books on the subject, which present all the newer ramifications and could serve as a sophisticated basis for training personnel, has not been available. The text presented by Kaplan and Pesce is designed to fill this void. They accomplished it by assembling the contributions of 95 experts in the field who contributed their talents to this enterprise. In this manner the subject could be addressed in depth and its subject matter kept current.

This book presents the basic analytical principles and integrates them with the newer instrumentation and techniques. The physiology and anatomy of the various organs of the human are then presented, with emphasis on the use of the findings of the clinical chemistry laboratory in the evaluation of their function in disease. Finally, it presents detailed procedures for carrying out the numerous diagnostic tests. In this regard it serves as a detailed laboratory manual.

This text is an excellent source book for those engaged in the practice of clinical chemistry. It is also an excellent manual for the training of medical technologists, clinical chemists, and clinical pathologists. It can serve well the practicing physician who wishes to use the clinical chemistry laboratory's findings in the diagnosis and management of the patient.

Samuel Natelson, Ph.D.
Department of Environmental Medicine,
College of Veterinary Medicine,
University of Tennessee,
Knoxville, Tennessee

Preface to first edition

Clinical Chemistry: Theory, Analysis, and Correlation is designed to serve primarily as a teaching text for medical technologists and medical laboratory technicians. However, this book also contains the depth of information required in a bench reference for clinical chemists and laboratory supervisory personnel.

Clinical chemistry is a multidisciplinary field that draws on diverse disciplines, including pharmacology, toxicology, physiology, immunology, and hematology. The teaching and application of information from these areas to analytical biochemistry is complicated by the volume of knowledge required. We have provided a synthesis of the most relevant and current information from the associated disciplines and thus a complete survey of the clinical chemistry field.

The wide range of information covered in *Clinical Chemistry* includes the use of many terms unfamiliar to students. For example, the term *analyte,* used to describe any substance that can be measured, may not be found in current dictionaries. Therefore definitions of such terms have been included in glossaries in the beginning of each chapter.

The format of *Clinical Chemistry* is designed to separate the large body of facts into three categories:

1. *Laboratory techniques:* the principles of analytical techniques
2. *Pathophysiology:* the ways in which the laboratory data generated are related to disease or organ dysfunction
3. *Methods of analysis:* an in-depth survey of the measurement of commonly analyzed biochemical substances

The chapters within each of the three major sections are organized according to a consistent pattern. Our intent is to facilitate both the presentation of the information by instructors and the retrieval of information by students and clinicians. In addition, to facilitate teaching and learning we have grouped together chapters that explore common themes, such as endocrinology, chromatography, or data processing.

The chapters in Section One, Laboratory Techniques, are directed primarily to baccalaureate-level medical technology students. However, the depth of information provided is intended to be sufficient to meet the requirements of advanced and graduate clinical chemistry students, as well as laboratory supervisors and clinical chemists.

Each of the general information chapters in Laboratory Techniques is organized according to a logical progression of ideas:

1. The scientific principles on which a technique is based
2. The application of a technique to a specific class of analytes or process
3. The limitations, such as sensitivity, specificity, and accuracy of each technique—that is, the factors that govern when a certain technique should or should not be used

Section Two, Pathophysiology, is composed of chapters covering the interpretation of laboratory data obtained from the methods of analysis described in Section Three. To recognize erroneous analytical data, the clinical chemist must have an understanding of pathophysiological mechanisms. It has been our experience that superior medical technologists or technicians are those who have a thorough understanding of pathophysiology and can therefore understand the reason for ordering certain tests and can determine whether the tests ordered are appropriate for the analyte in question. Accordingly, the analytes are discussed in their *usual clinical contexts* by disease or organ dysfunction. For example, rather than providing a chapter on carbohydrates, we wrote the two chapters on diabetes and genetic disease to include discussions on abnormal levels of carbohydrates in the clinical context in which they are most likely to be encountered.

Within each chapter of Pathophysiology, the following organizational plan has been used, where appropriate:

1. A brief description of the anatomy of the organ
2. A discussion of the normal biochemistry and physiology of the organ or system
3. A discussion of pathological conditions, their clinical symptoms, and their effects on function and biochemistry
4. Function and challenge tests that require laboratory analysis
5. A change of analyte with disease section in which the analytes are correlated with various disease states

We have organized the Pathophysiology chapters in this manner to guide the student to a clear comprehension of the biochemical nature of the disease state as compared to the normal, nondiseased state. Thus changes in analyte concentration can be more readily understood. The pathophysiology section therefore highlights the biochemical changes associated with each disease state.

Section Three, Methods of Analysis, is introduced by several chapters that present the following:

1. Classifications of biochemicals by type and a brief description of how biochemical class can affect the laboratory approach of measurements
2. Enzyme analyses, including the variables to be controlled
3. A description of isoenzymes and how they are analyzed
4. A discussion of pharmacokinetic interpretation of data
5. Interferences in spectrophotometric analysis, composed of a discussion of the nature of those interferences and how they can be reduced
6. Additional notes

After these introductory chapters are detailed methods of analysis grouped together as much as possible by common chemical class. The methods cover a wide range of analysis, including toxicology, endocrinology, and the most recent advances in analytical technique. The methods for the analytes included in Section Three are not intended to provide access to all possible methods that may be used in the clinical chemistry laboratory. We have chosen to include the methods that demonstrate most effectively the principles and techniques of analytical chemistry discussed in Section One and that compose those analyses most commonly performed in the clinical chemistry laboratory.

The format used for each analyte is as follows:

1. Appropriate cross-referencing to chapters in Section Two
2. A concise chemical description and classification of the compound. Since *The Merck Index,* ninth edition (tenth for this second edition of *Clinical Chemistry*), can rapidly provide chemical information on a wide variety of biochemicals, whenever possible we have listed *The Merck Index* number for each analyte.
3. A description of the methods, including principles of analyses and the current usage of available procedures
4. A discussion of why one or more procedures is used as a reference method or why it was chosen as the *preferred* method. By comparing the benefits and disadvantages of each method, such as accuracy and precision, the student and clinical chemist can understand more fully the usefulness of each assay.
5. Specimen information, including which type of body fluid is appropriate, specimen stability and preservation, and important interferences
6. A presentation of methods of analysis for the analyte. Since the mid-1970s many analytes have been most commonly measured by automated procedures. In addition, these automated procedures invariably utilize prepackaged reagents available from instrument manufacturers or commercial vendors. Therefore, rather than list a *manual* method for such analytes, we have chosen to *critique* those methods currently available. A comparison of variations of the preferred method to one another, as well as to a reference method (if available), is given with respect to precision, final reagent composition and concentration, and sample fraction (sample volume divided by total volume). This information allows the comparison of current or proposed procedures to the reference or preferred method.

For analytes that are commonly measured by nonautomated, manual procedures, a *complete* description of a working assay has been included. The preferred manual method includes the following:

1. The principle of the reaction or analysis
2. Reagent preparation, stability, and storage
3. The procedure in sufficient detail for ready adaptation by most laboratories
4. Sample calculation as needed
5. Reference ranges using this and other commonly used methods

A concerted effort is being made in clinical chemistry toward the use of SI units (Système International d'Unités) for the reporting of results of laboratory analysis. The SI units are in either moles per unit volume (such as liter) or mass per unit volume (mass per liter) rather than the historical form of mass percent (mass per deciliter). In an attempt to begin this difficult but necessary task, all concentration units are reported as mass per liter or milliliter as well as moles per liter or, for enzymes, in accepted SI units. Because the use of deciliters is still in widespread use, it is our hope that by our expressing mass concentrations per liter, readers will be easily able to convert concentrations to mass per deciliter, and vice versa. Since the current literature uses SI units, the text will allow students to familiarize themselves with these units and enable them to more readily use current journals in their fields.

The amount of material available for these chapters is enormous. To present the necessary information concisely, we have used many figures and summary tables. These figures and tables demonstrate certain patterns of information, allowing greater ease in understanding. For those who desire to have a more profound understanding, numbered references or a general bibliography at the end of each chapter lists appropriate reading material.

The enterprise that produced this text was not undertaken by a single or even several persons. Indeed, the editors of the text have enlisted the aid of scores of gifted clinicians, laboratorians, and educators, and it is to these people that we owe the success of this book.

We would like to give a special note of appreciation to

Drs. David C. Hohnadel, Carl C. Garber, and R. Neill Carey, and to Joseph Svirbely for their help in reviewing and revising several chapters. We would also extend our thanks to the Document Processing Area of the Department of Pathology and Laboratory Medicine, Mary Grannen and Liz Wendelmoot, who typed large portions of the manuscript. Our special thanks also go to Julie Hodde, who directed the artists drawing many illustrations, and to Beverly Etter, Manager of Medical Illustrations of the Biomedical Communications Department, who provided us with superb artistic support.

We also wish to thank our reviewers: Bethany Wise, M.S., M.T. (ASCP), School of Allied Medical Professions, Ohio State University; Edna Mains, B.S., M.T. (ASCP), Medical Technology Program, University of Colorado; Joseph Svirbely, M.S., Department of Environmental Health, University of Cincinnati; Anne Sullivan, M.S., M.T. (ASCP), Department of Medical Technology, University of Vermont; Sharon A. Jackson, M.Ed., M.T. (ASCP), Department of Medical Technology, University of Florida; Janet von Laufen and Barbara Smith Michael, Program in Medical Technology, The University of Texas Health Science Center at Houston; and Lester Hardegree, Jr., M.Ed., M.T. (ASCP), Armstrong State College, Savannah, Georgia. Their suggestions have been invaluable during the development of this book.

We are indebted to Don Ladig of The C.V. Mosby Company, who believed in our ability to deliver this text, and to his associates Rosa Kasper and Cindy Bendet. We would like to thank the Book Editing Department, especially Carl Masthay and Peggy Fagen. Last, we extend our thanks to Dr. Roger D. Smith, Director of the Department of Pathology and Laboratory Medicine, and Dr. Colin Macpherson, who produced the academic environment in which this text could be undertaken.

Lawrence A. Kaplan
Amadeo J. Pesce

Preface

The science of clinical chemistry, as with any other scientific discipline, is not static. Many changes have occurred in the past 5 years as the result of the cumulative advances in technology and our greater understanding of human disease processes.

These changes alone would dictate this revision of *Clinical Chemistry: Theory, Analysis, and Correlation* to provide updated material that reflects recent advances. Every chapter in the second edition has been revised to incorporate new techniques, instrumentation, and diagnostic laboratory tests. For example, Chapter 3, Spectral Techniques, describes the new fluorescent-based analytical assays, and Chapter 14, Automation, reviews the most recent advances in instrumentation, including that used in physicians' offices. New chapters have been added on Human Nutrition (34), Trace Elements (35), and the principles of the new DNA technology (47, Use of Nucleic Acid Probes in the Clinical Laboratory). Certainly this last chapter exemplifies the rapid coming together of advances in technology and biochemistry to create an entirely new approach to disease diagnosis.

Another goal of this edition is to focus the huge amount of material that, by necessity, should be presented to a properly trained health care professional such as a medical technologist. Using the helpful comments of educators of medical technologists, we have tried more than ever to concentrate on general principles of clinical chemistry. Thus Section Three, Methods of Analysis, has been modified to include those analytes that will allow a student to be taught the applications of the principles of analysis described in Section One and those analytes that are frequently measured in the laboratory. The Methods section cannot possibly be all-inclusive. Given the large number of tests now available in clinical laboratories, this would be undesirable in a teaching textbook. For those students who desire a more inclusive review of methods, other resources are available, including *Methods in Clinical Chemistry* by the editors.

In addition to focusing the Methods section, specific chapters have been totally rewritten to present technical information in a form that allows immediate application to laboratory practice. For example, the chapter on computers has been changed to Laboratory Information Systems (Chapter 15). This change focuses attention on the actual application of computers in a laboratory rather than on how computers work. Other chapters, such as Electrochemistry: Principles and Measurements (13); Diabetes Mellitus (29); and Iron, Porphyrin, and Bilirubin Metabolism (32), have been completely rewritten to fit the needs of a teaching text.

As a result of our focusing efforts, the second edition is approximately 20% smaller than the first edition, although it contains much new information. We have retained the overall structure and philosophy of the first edition, as delineated in the preface to that edition (see p. xv), because we have received positive comments from educators, students, and reviewers about the innovative approaches used in the first edition. However, as we already stated, we do not wish to become static and inflexible. We continue to encourage criticisms and suggestions so that we can continue the evolution of this textbook.

This text is designed to serve as an ongoing reference for the student once his or her training is completed. Therefore we have not included material such as case histories or experiments, which are useful only during the student's training. We have placed this material in the students workbook available as a supplement to the first edition. A revised workbook will be available to accompany this text.

We wish to thank Elizabeth Wendelmoot and Georgia Coddington for downloading and retyping the manuscripts. Also, we'd like to thank Dr. Wendell Davis for proofing the manuscripts and Gail Jones and Barbara Cleveland for helping to write the student objectives at the beginning of each chapter. We wish also to note the excellent work of the Medical Center Information and Communications (MCIC) of the University of Cincinnati. We'd like to thank and acknowledge the superb work of Judith E. Kaplan, our medical editor, and Sheila Walker, our copy editor at The C.V. Mosby Company. Lastly, we'd like to thank all our authors, who have once again unselfishly given us their time for the revision of this textbook.

Lawrence A. Kaplan
Amadeo J. Pesce

Contents

CLINICAL CHEMISTRY

Theory, analysis, and correlation

SECTION ONE

Laboratory Techniques

CHAPTER 1 *Basic laboratory principles and calculations*

OBJECTIVES

- Describe the initial setup, operation, and maintenance of each of the following laboratory balances and state when each may be appropriately used in a chemistry laboratory:
  - Trip balance
  - Analytical balance
  - Top-loading balance
- List commonly used laboratory supplies (that is, beakers, pipets, flasks) and describe proper use, quality control, cleaning or maintenance procedures, and advantages or disadvantages of use for each.
- Describe the specifications associated with the quality of reagent chemicals and laboratory water.
- State the laboratory safety regulations related to hazardous situations or use of hazardous chemicals and describe the various types of commonly used laboratory warning signs, interpreting the meaning of each.
- Demonstrate a comprehension of the principles associated with the use of the metric system and reagent and standard curve preparation. Correctly perform mathematical calculations common to clinical laboratories.

KEY TERMS

balances Mechanical or electronic instruments used to measure weight accurately.

beakers Laboratory utensils used to contain liquids or solids.

buret (burette) Laboratory utensil used to deliver a wide range of volumes accurately.

centrifuge Instrument used to separate materials from solution by application of increased gravitational force by rotating or spinning samples rapidly.

chemical purity Degree of purity or homogeneity as designated by various scientific agencies, such as the American Chemical Society and National Bureau of Standards.

desiccant Material used in a desiccator to absorb water from the air.

desiccator Large container used to store material in a water-free environment.

dilution Process of preparing less concentrated solutions from a solution of greater concentration.

Erlenmeyer flask Laboratory utensil used to contain liquids.

funnel Laboratory utensil used to transfer liquids or solids into a container; also used for extraction of liquids.

graduated cylinder Laboratory utensil used to measure a volume of liquid.

metric system A system of measurement of weights, distances, and volumes.

pipet (pipette) Laboratory utensil used to transfer a specific or varying volume of liquid.

Système International d'Unités (SI) An internationally accepted system of measurements.

thermometer A device, physical, electronic, or optical, that is used to measure temperature.

volumetric flask Laboratory utensil used to contain a specific volume of liquid.

water purity Three levels of purity are defined, based on the amount of biological and dissolved organic and inorganic material present in the water.

Part I: Basic laboratory principles

PAULA STEINER

UNITS OF MEASUREMENT

Metric system

The metric system is a system of measurement based on the meter as the unit of length, the gram as the unit of mass, and the liter as the unit of volume. Units are defined as multiples or divisions of these three reference units. Although the metric system has not been widely used in the United States, the International System of Units (Système International d'Unités, or SI), a more coherent system defining seven basic units, has been adopted by the worldwide scientific community.[1,2] Scientific publications and, increasingly, laboratory reports use SI units, and laboratorians must be familiar with conversions between systems. The complete conversion to SI units in the clinical laboratory is a complicated issue because of the complexities of standardizing instrumentation, reprogramming computers, revising report forms, and educating laboratory staff and physicians.[3-5]

SI-derived units

There are seven basic units of the SI system (Table 1-1). Two or more basic units may be combined by multiplication or division to form SI-derived units (Table 1-2). Basic and derived units may be too small or too large for convenient use, and prefixes that form decimal multiples or submultiples of the units are permitted (Table 1-3). A few non-SI units have been retained because of difficulties encountered in converting them to SI units and because of their widespread use. Non-SI units relevant to clinical chemistry and their symbols are time, expressed in minutes (min), hours (h), or days (d), and volume, expressed as liters (L). The General Conference of Weights and Measures (GCPM) has approved l, *l*, or L; however, L is the official abbreviation in the United States and is used in this textbook.

SI units in the clinical laboratory

The SI unit of enzyme activity is defined as the amount of enzyme that will catalyze the transformation of 1 mole of substrate per second in an assay system. The Joint

Table 1-1 Basic quantities and units of the SI (Système International d'Unités)

Quantity	Basic unit	Symbol
Length	Meter	m
Mass	Kilogram	kg
Time	Second	s
Electric current	Ampere	A
Temperature	Kelvin	K
Luminous intensity	Candela	cd
Amount of substance	Mole	mol

Table 1-2 SI-derived units used in medicine

Derived quantity	Derived unit	Symbol
Area	Square meter	m^2
Volume	Cubic meter	m^3
Speed	Meter per second	m/s or $m \cdot s^{-1}$
Substance concentration	Mole per cubic meter	mol/m^3 or $mol \cdot m^{-3}$
Pressure	Pascal	Pa
Work energy or quantity of heat	Joule	J
Celsius temperature	Celsius degree	°C
Activity (radionuclide)	Becquerel	Bq
Power	Watt	W
Electric charge or quantity	Coulomb	C
Electric potential	Volt	V
Resistance		Ω
Conductance	Siemens	S

Commission on Biochemical Nomenclature of the International Union of Biochemistry (IUB) and the International Union of Pure and Applied Chemistry (IUPAC) have recommended that this unit be called the *katal*.* The GCPM has not approved this nomenclature. Whether or not this approval is forthcoming, it seems very likely that the international unit (IU)† will not be replaced by the katal in the near future. Although in many laboratories the IU has replaced traditional units of activity, one IU of activity may differ from the next depending on how the standard conditions of the assay are defined (pH, temperature, type of buffer and its concentration, substrate concentration, activator, and coenzyme concentrations, if required).

The adoption of SI units in the clinical laboratory will change the laboratory report form. Whenever the molecular weight of the analyte is known, its concentration is to be expressed in mol/L or submultiples rather than mass/L. When the molecular weight is unknown, as in specific proteins or their mixtures, concentration should be reported as mass/L. The GCPM does not recommend the use of pH, and hydrogen-ion concentration should be reported as nmol/L. The elimination of the pH scale has not been accepted by most scientific disciplines, and this proposal is still undergoing study.

*The katal (kat) is defined as the amount of enzyme that produces a reaction rate of 1 mole per second in a method of assay.

$$10^{-9} \text{ kat/L (1 nanokat/L)} = 0.06 \text{ IU/L}$$

†In 1964 the IUB defined the international unit as the amount of enzyme that will catalyze the transformation of 1 micromole of substrate per minute under standard conditions.

Table 1-3 SI prefixes

Prefix*	Factor	Symbol	Prefix*	Factor	Symbol
Atto	10^{-18}	a	Deka	10^{1}	da
Femto	10^{-15}	f	Hecto	10^{2}	h
Pico	10^{-12}	p	Kilo	10^{3}	k
Nano	10^{-9}	n	Mega	10^{6}	M
Micro	10^{-6}	μ	Giga	10^{9}	G
Milli	10^{-3}	m	Tera	10^{12}	T
Centi	10^{-2}	c	Peta	10^{15}	P
Deci	10^{-1}	d	Exa	10^{18}	E

*It is recommended that only one prefix be used.

MEASUREMENT OF MASS

Laboratory balances are essential to the operation of the clinical chemistry laboratory. An analytical balance is needed for the preparation of primary standards from the National Bureau of Standards (NBS) standard reference materials and for the gravimetric assay of fecal lipids. The gravimetric calibration of semiautomatic and automatic pipets affords one possible method for checking their precision and accuracy. In the toxicology laboratory, the analytical balance is needed to prepare primary drug standards and for weighing quantities of drugs used in recovery studies. Laboratory balances are mechanical or electronic in design. Both are available in a number of models with different capacities, readabilities, and precision (Table 1-4).

Mechanical balances

Trip balance. The trip balance consists of two pans of equal mass suspended from the ends of a beam that is supported at its center of gravity by a knife-edge fulcrum. The material being weighed is placed in the left pan. The final weight is obtained by adjustment of the position of a rider or small weight on an extension of the beam called the *balance arm bridge*. The balance arm bridge is calibrated in 0.1 g increments up to a maximum of 200 g.

Single-pan balance. The single-pan balance is a modified trip balance with arms of unequal length. The fulcrum is located close to the weighing pan. The right arm of the beam consists of two or three balance arm bridges that support counterbalancing weights. Trip balances are useful for weighing masses quickly when a weight to the nearest 0.1 g is satisfactory, as in the preparation of strong bases or buffers.

Torsion balance. A torsion balance can usually be found in the pharmacy and has application in the chemistry laboratory when a readability is desired intermediate to that of the trip and analytical balances. Torsion balances do not have a knife-edge fulcrum. Metal bands serve to support the weight of the beam. As the balance oscillates, the movement of the pans is restricted by the torque on these bands. Torsion balances cannot support as much weight as the knife-edge balances but can have a readabil-

Table 1-4 Characterization of types of balance in relationship to their operating ranges

Type of balance	Weighing range, g	Scale-reading limit, g	Reproducibility, g
Precision balances			
Electronic	30,000	1	±0.5
	15,000	0.1	±0.05
	3200	0.1	±0.05
	1200	0.01	±0.005
	320	0.001	±0.0005
Mechanical	20,000	2	±1
	10,000	1	±0.5
	5000	0.1	±0.05
	2200	0.01	±0.01
	160	0.001	±0.001
Analytical balances			
Electronic	160	1 mg	±1 mg
	160	0.1 mg	±0.1 mg
	160	0.01 mg	±0.02 mg
Mechanical	160	0.1 mg	±0.05 mg
	160	0.01 mg	±0.01 mg
Microbalances			
Electronic	3.3	10 μg	±10 μg
	3.12	1 μg	±1 μg
	3.10	0.1 μg	±0.3 μg
Mechanical	20	0.001 mg	±0.001 mg

From Richterich R, and Colombo JP: Clinical chemistry, New York, 1981, John Wiley & Sons, Inc.

ity of 2 mg and provide a more accurate, rapid means of weighing small samples of material. By substitution of helical quartz fibers for metal bands, ultramicrotorsion balances can be constructed that will have great accuracy with readabilities of 0.1 to 0.001 microgram. These balances are more likely to be found in a research environment.

Analytical balance. Analytical balances generally have much greater readability, precision, and accuracy than trip or macrotorsion balances. Before 1960, most analytical balances were a two-pan design with a knife-edge fulcrum. The sensitivity of these balances varied inversely with the load. In 1946, Erhard Mettler built an improved analytical balance in Switzerland that was the forerunner of the modern substitution balance. The substitution balance is a single-pan balance with unequal arms. Suspended above the pan are a series of weights that are counterbalanced by a single weight located at the opposite end of the beam (Fig. 1-1). Because the load on either side of the knife-edge is always constant, the sensitivity of a substitution balance does not vary. The material to be weighed is placed inside a tared container or weighing paper on the pan and weights to the nearest 0.1 g are removed from the beam by a dial control lever. Weights less than 100 mg are read from an optical scale attached to the end of the beam opposite the pan. A light source coupled with appropriate lenses and mirrors projects the optical scale (0 to 100 mg) on a screen located in the front of the balance. Weights to the nearest 0.1 or 0.01 mg are read with a vernier. The total weight to the nearest 0.1 g is indicated by a digital register located near the screen. The substitution balance has a number of advantages over the two-pan analytical balance other than constant sensitivity. Errors attributable to unequal arm length are eliminated because both sample and weights are compared on the same arm. The operator does not handle the weights, and the time required for a weighing is reduced to about 30 seconds. Substitution balances are either air or liquid damped.

Top-loading balances. Top-loading balances are modified torsion or substitution balances and are much faster and easier to use than other balances. Weighings can be completed in a few seconds; however, precision is not as good as that obtained with analytical balances and ranges from ± 20 mg to ± 1 mg, depending on the design.

Electronic balances

The electronic balance was introduced in the late 1960s. There are available electronic balances that equal the precision and accuracy of all types of mechanical balances, and these are beginning to replace the latter in research and clinical laboratories. The electronic balance is a single-pan balance that uses an electromagnetic force to counterbalance the load placed on the pan. The pan is attached directly to a coil suspended in the field of a permanent magnet. A current is passed through the coil, producing an electromagnetic force that keeps the pan in a constant position. When a load is placed on the pan, a photoelectric

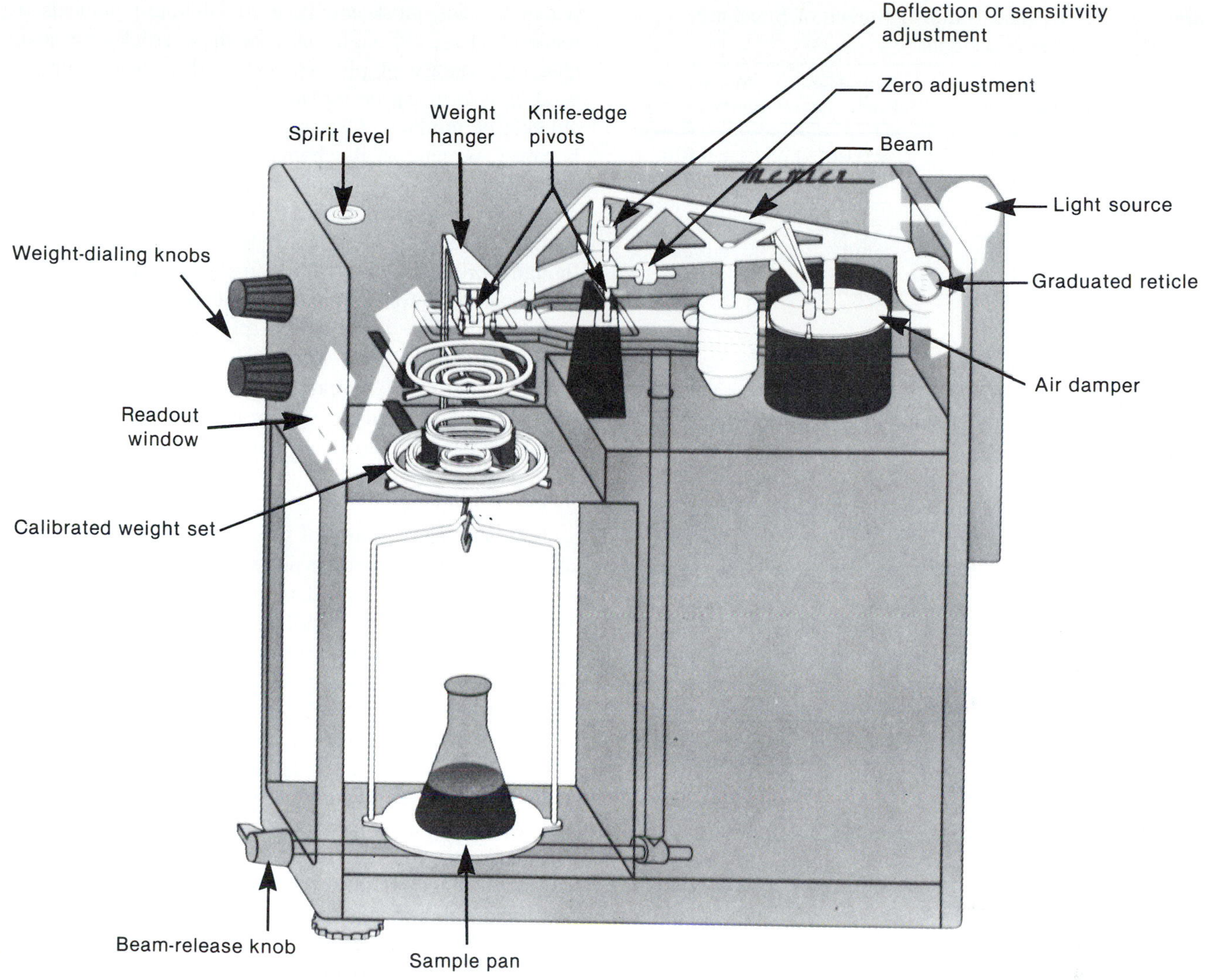

Fig. 1-1 Single-pan balance showing internal design. *(Courtesy The Mettler Instrument Corp, Hightstown, NJ.)*

cell scanning device attached to the lever arm changes position and transmits a current to an amplifier that increases the current flow through the coil and restores the pan to its original position. This current is proportional to the weight of the load on the pan and produces a measurable voltage that is converted by a microprocessor to a numeric display or data output. These balances can be interfaced with data-processing equipment to provide calculations such as weight averaging and statistical analysis of multiple weighings. Electronic balances are either top loading or analytical in design and permit weighings to be made in 5 seconds or less.

Operation and maintenance

Balances should be located in an area of the laboratory that is free of drafts and vibration and has a relatively constant temperature. Some electronic balances have a built-in electronic vibration damper. Excessive vibration can be detected when the variation of the pointer or oscillation of numbers in the last decimal place of the digital display is observed. The analytical balance should be located in a relatively isolated part of the chemistry laboratory away from areas where corrosive chemicals are used or stored. The balance should be placed on a heavy, dedicated weighing table that can be purchased commercially. The balance pan and surrounding area should be kept clean. Chemicals should be weighed on weighing paper, in plastic boats, or in weighing bottles and never placed directly on the pan. Before the balance is used, its level should be checked by adjustment of the foot screws and centering of the bubble in the spirit level. The optical zero should be checked before each weighing.

Table 1-5 Individual National Bureau of Standards tolerances for class S weights

Nominal mass	Individual tolerance (mg)	Maintenance tolerance (mg)
1, 2, 3, 5, 10, 20, 30, 50 mg	±0.014	±0.014
100, 200, 300, 500 mg	±0.025	±0.05
1, 2, 3, 5 g	±0.054	±0.11
10, 20, 30 g	±0.074	±0.148
50 g	±0.12	±0.22
100 g	±0.25	±0.5

Performance

Mechanical and electronic balances that are maintained and used according to the manufacturer's instruction manual may not require service more than once a year, depending on how frequently the balance is used. Most mechanical and electronic analytical balances have internal weights that meet the tolerances for class S weights established by the National Bureau of Standards.[1] The performance and reliability of an analytical balance (often erroneously called "calibration") can be checked if a set of class S weights is available. The College of American Pathologists requires that approved laboratories have access to such weights. A 100 g weight should weigh 100 g ± 0.5 mg. One should check other class S weights to determine whether their apparent weight lies within the NBS maintenance tolerance limits (Table 1-5). If the class S weights weigh greater or less than the maintenance tolerance values, the balance requires service. Adjustments, other than those described in the balance operating instructions, should be performed only by a qualified service technician.

Some electronic top-loading and analytical balances have an internal weight built into the balance for calibration. Information is available concerning the maintenance and performance of balances.[2,6]

LABORATORY SUPPLIES

Laboratory ware or apparatuses are used for three primary functions: storage, measurement, and confinement of reactions. The most widely used laboratory utensils are made of glass, although many of these articles are available in a variety of plastics. Generally, most commonly used glassware can be replaced by the cheaper, more durable plastic products. Glassware is most frequently chosen, however, because of its chemical stability and clarity. For certain specific applications, such as gas chromatography and high-performance liquid chromatography, glass is the preferred material because of its chemical stability. Although borosilicate and aluminosilicate glass are nearly ideal materials for laboratory containers, aqueous solutions, particularly those with a pH greater than 6.0, can attack glassware. Glass also tends to absorb metal ions, possibly altering significantly the concentrations of standard solutions. Good laboratory practice requires that alkaline solutions be stored in plastic containers.[3]

Types of glasses

There are many grades of glassware commercially available that vary in their tensile strength and heat or light resistance. Since soft glass is generally not recommended for general laboratory work, most glass utensils in the chemistry laboratory are constructed from highly resistant borosilicate glass. These are available under the brand names of Kimax (Kimble Glass Company, Vineland, N.J.) and Pyrex (Corning Glass Works, Corning, N.Y.).

Borosilicate glass has a low alkaline-earth content and is free of many contaminants such as heavy metals. This type of glass can be heated to approximately 600° C and will not soften until approximately 820° C. Table 1-6 lists additional types of commonly used glass.

Aluminosilicate glass (Corex, Corning Glass Works, Corning, N.Y.) has all the desirable advantages of borosilicate glass but also much greater mechanical strength. Aluminosilicate glass is more resistant to ordinary breakage, scratching, chipping, and clouding.[3]

Types of plastics

Plastics used for laboratory utensils are constructed from polymerized organic monomers. The properties of the plastics depend on the nature of the monomer and the final

Table 1-6 Types of commonly used glass and their properties

Glass	Properties	Purpose
Kimax or Pyrex	Relatively inert borosilicate glass	All-purpose
Vycor	96% silicate; high resistance to heat and cold shock	For use with temperatures up to 900° C; also has good optical properties
Corning Boron Free Glass	Soft, low-boron glass (<0.2%) highly resistant to alkali; poor heat resistance	For use with highly alkaline solutions
Low-actinic	Glass contains substances giving it an amber or red tint to reduce quantity of light passing through	For use with light-sensitive reagents (such as carotene, bilirubin, or vitamin A)
Corex	Aluminosilicate glass, sixfold to tenfold stronger than conventional borosilicate glass; highly resistant to scratching, alkaline etching, and so on	For glassware to be used in stressful conditions

polymer forms used to prepare the plastic materials. The most commonly used plastics include the polyolefins (polyethylene, polypropylene), polytetrafluoroethylene (Teflon, a fluorinated hydrocarbon produced by du Pont), polystyrene, polycarbonate, and polyvinylchloride.

The polyolefins are noted for their strength and resistance to elevated temperatures. Teflon is almost totally chemically inert and is resistant to a wide range of temperatures. Polycarbonate glassware is very clear and is ideal for graduated cylinders. Teflon, polycarbonate, and

Table 1-7 Chemical resistance and physical properties summary of various plastics*

	Types of resins†									
Classes of substances (20° C)	**LDPE**	**HDPE**	**PP, PA**	**PMP**	**FEP, TFE ETFE**	**PC**	**PSF**	**PVC bottles‡**	**PS**	**Nylon**
Chemical resistance										
Acids, dilute or weak	E	E	E	E	E	E	E	E	E	F
Acids,§ strong and concentrated	E	E	E	E	E	N	G	E	F	N
Alcohols, aliphatic	E	E	E	E	E	G	G	E	E	G
Aldehydes	G	G	G	G	E	F	F	N	N	F
Bases	E	E	E	E	E	N	E	E	E	F
Esters	G	G	G	G	E	N	N	N	N	E
Hydrocarbons, aliphatic	F	G	G	F	E	F	G	E	N	E
Hydrocarbons, aromatic	F	G	F	F	E	N	N	N	N	E
Hydrocarbons, halogenated	N	F	F	N	E	N	N	N	N	G
Ketones	G	G	G	F	E	N	N	N	N	E
Oxidizing agents, strong	F	F	F	F	E	N	G	G	N	N
Physical properties										
Maximum-use temperature (°C)	80	120	135 (PP) 130 (PA)	175	205 (FEP) 150 (ETFE)	135	165	70#	—	—
Brittleness temperature (°C)	−100	−100	0 (PP) −40 (PA)	20	−270 (FEP) −100 (ETFE)	−135	−100	−30	—	—
Sterilization‖										
Autoclaving	No	No	Yes	Yes	Yes	Yes¶	Yes	No#	—	—
Gas	Yes	Yes	Yes	Yes	Yes	Yes	Yes	Yes	—	—
Dry heat	No	No	No	Yes¶	Yes	No	Yes	No	—	—
Chemical	Yes	Yes	Yes	Yes	Yes	Yes	Yes	Yes	—	—

*Modified from 1983–1984 Nalgene Labware Catalog, Nalge Co., Division of Sybron Corp., Rochester, N.Y.

†**Resin codes:** *ETFE,* Tefzel ETFE (ethylene tetrafluoroethylene); *FEP,* Teflon FEP (fluorinated ethylene propylene); *HDPE,* high-density polyethylene; *LDPE,* low-density polyethylene; *PA,* polyallomer; *PC,* polycarbonate; *PMP,* polymethylpentene (''TPX''); *PP,* polypropylene; *PS,* polystyrene; *PSF,* polysulfone; *PVC,* polyvinylchloride; *TFE,* Teflon TFE (tetrafluoroethylene).

Chemical resistance classification: *E,* 30 days of constant exposure cause no damage. Plastic may even tolerate it for years. *G,* Little or no damage after 30 days of constant exposure to the reagent. *F,* Some effect after 7 days of constant exposure to the reagent. Depending on the plastic, the effect may be crazing, cracking, loss of strength, or discoloration. Solvents may cause softening, swelling, and permeation losses with LDPE, HDPE, PP, PA, and PMP. The solvent effects on these five resins are normally reversible; the part will usually return to its normal condition after evaporation. *N,* Not recommended for continuous use. Immediate damage may occur. Depending on the plastic, the effect will be a more severe crazing, cracking, loss of strength, discoloration, deformation, dissolution, or permeation loss.

‡For polyvinyl chloride tubing, see the current Nalgene Labware Catalog.

§Except for oxidizing acids. For oxidizing acids, see ''Oxidizing agents, strong.''

‖Sterilization—*Autoclaving:* Clean and rinse item with distilled water before autoclaving. Certain chemicals that have no appreciable effect on resin at room temperature may cause deterioration at autoclaving temperatures unless removed with distilled water beforehand. *Gas:* Ethylene oxide. *Dry heat:* At 160° C. *Chemical:* Benzalkonium chloride, formalin, ethanol, and so on.

¶Sterilizing reduces mechanical strength. Do not use polycarbonate vessels for vacuum applications if they have been autoclaved.

#Except for the polyvinyl chloride in tubing, which will withstand temperatures to 121° C and can be autoclaved. Refer to ''The Use and Care of Plastic Labware'' in the current Nalgene Labware Catalog for detailed information on sterilization.

Interpretation of chemical resistance. This summary is a general guide only. Because so many factors can affect the chemical resistance of a given product, you should test under your own conditions. If any doubt exists about specific applications of Nalgene products, please contact Technical Service, Nalgene Labware Department, Nalge Company, Box 365, Rochester, NY 14602 or call (716)586-8800. Telex 97-8242.

Effects of chemicals on plastics. Chemicals can affect the strength, flexibility, surface appearance, color, dimensions, or weight of plastics. The basic modes of interaction that cause these changes are (1) chemical attack on the polymer chain, resulting in reduction in physical properties, including oxidation; reaction of functional groups in or on the chain; and depolymerization; (2) physical change, including absorption of solvents, resulting in softening and swelling of the plastic; permeation of solvent through the plastic; dissolution in a solvent; and (3) stress cracking from the interaction of a ''stress-cracking agent'' with molded-in or external stresses. The reactive combination of compounds of two or more classes may cause a synergistic or undesirable chemical effect. Other factors affecting chemical resistance include temperature, pressure, and internal or external stresses (such as centrifugation), length of exposure, and concentration of the chemical. As temperature increases, resistance to attack decreases.

CAUTION. Do not store strong oxidizing agents in plastic labware except that made of Teflon FEP. Prolonged exposure causes embrittlement and failure. Although prolonged storage may not be intended at the time of filling, a forgotten container will fail in time and result in leakage of contents. Do not place plastic labware in a flame or on a hot plate.

some of the polyolefin plastics are autoclavable. Polyvinylchloride plastics are soft and flexible materials used frequently to construct tubing. Table 1-7 reviews some physical properties of commonly encountered plastics and their resistance to a number of chemicals.

Laboratory utensils

Flasks. Some types of flasks used in the laboratory are shown in Fig. 1-2. The most commonly used flask in the clinical chemistry laboratory is the Erlenmeyer flask, which is available in sizes ranging from 25 mL to several liters. Small round-bottom flasks are often used to evaporate samples to dryness.

Beakers. Beakers are wide, straight-sided cylindrical vessels available in a wide range of volumes from 5 mL to several liters.

Flasks and beakers are used for general mixing and preparation of liquid reagents. Because of their narrower mouths, Erlenmeyer flasks contain liquids better than beakers do and thus are more suited for the storage of liquids.

Graduated cylinders. Graduated cylinders are narrow, straight-sided vessels that are used to measure specific volumes (Fig. 1-2 and see below). Graduated cylinders are available in sizes ranging from 10 mL to several liters and

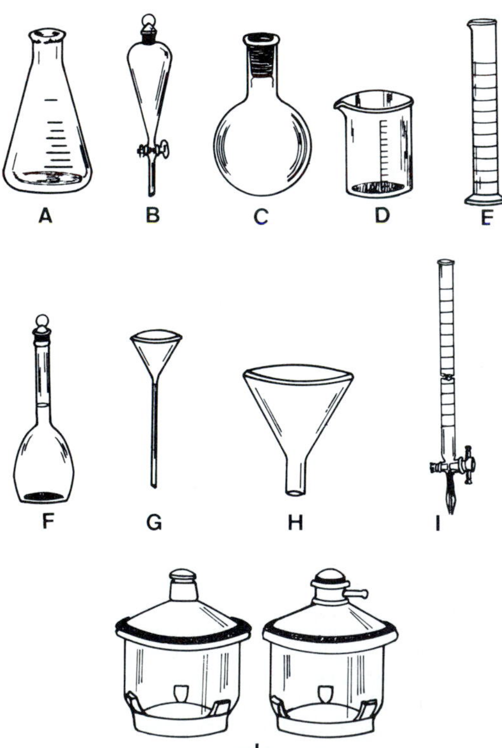

Fig. 1-2 Examples of commonly used laboratory utensils. **A,** Erlenmeyer flask. **B,** Separatory funnel. **C,** Round-bottom flask. **D,** Beaker. **E,** Graduated cylinder. **F,** Volumetric flask. **G,** Long-stem funnel (filtering). **H,** Powder funnel. **I,** Buret. **J,** Desiccators.

may be calibrated *to deliver* (TD) or *to contain* (TC) the volume indicated at specific temperatures. They are graduated into subdivisions of approximately 100 portions of the total volume of the flask.

Volumetric flasks (Fig. 1-2). Volumetric flasks are used to contain (TC) an exact volume when the flask is properly filled to the indicator line at a specified temperature. Volumetric flasks cannot be used to deliver this volume, however.

Burets. Burets are long, graduated tubes with a stopcock at one end (Fig. 1-2). These devices are used to accurately deliver known volumes of liquid into a container. By measuring from graduated line to graduated line, one can deliver fractional volumes (that is, less than 1 mL) of liquid with a high degree of accuracy.

Funnels. Most commonly used funnels are used to transfer liquids or solids into containers. Filtering funnels (Fig. 1-2) are usually 58- or 60-degree–angle funnels with either short or long, thin stems. These funnels are used with filter paper to remove particles from solution. Many funnels have ridges to increase the surface area available for filtering purposes.

Powder funnels (Fig. 1-2) have wide-mouthed stems that allow solids to pass through easily, and thus these funnels are used for transferring solids into a container or flask.

Separatory funnels are constructed with a ground-glass stoppered opening at one end and a stopcock opening at the other end. These devices are used for manual liquid-liquid extractions of relatively large volumes of samples. The lower phase is separated from the upper phase through the stopcock.

Desiccators. Desiccators are used to dry, or keep dry, solid or liquid materials. Desiccators usually have an area at the bottom where a desiccant, or water-absorbing material, is placed (Fig. 1-2). A shelf is placed on top of the desiccant on which the material to be stored can be set. The top of the desiccator has a wide, flat, ground-glass lip that fits snugly against an opposing lip of the bottom part of the desiccator. Stopcock grease is usually placed on the surface of the lips to provide an airtight seal. Many desiccators also have a stopcock outlet on the upper portion to allow the desiccator to be evacuated. A laboratory will often have at least two, sometimes three, desiccators so that material can be stored dry at ambient, 4° and $-20°$ C temperatures.

Some materials used as desiccants are listed in Table 1-8.

Pipets. Most pipets are constructed from glass, although disposable, plastic, serological pipets are available. Pipets are discussed in more detail below.

Glassware washing

It is important that laboratory glassware be thoroughly clean before use for any analytical procedure. Glassware

Table 1-8 Some common drying agents (desiccants)

Desiccant	Properties	Uses
Anhydrous $CaCl_2$	High capacity, slow acting, works well below 30° C	Most conditions, very inexpensive
Anhydrous $MgSO_4$	Neutral, rapid action	Most conditions, inexpensive
Anhydrous Na_2SO_4	Neutral, high capacity, works only below 32° C, slow action	Can remove large volumes of water
Anhydrous $CaSO_4$	Extremely rapid in action, chemically inert, limited capacity to absorb water (6% to 10% weight in water)	More expensive than $MgSO_4$ and Na_2SO_4; sold commercially as Drierite; can be easily regenerated by heating at 230° to 240° C for 3 hours
Al_2O_3 (activated alumina)	Can absorb 15% to 20% of its weight in water	Can be repeatedly reactivated by heating at 175° C for 7 hours

must be free of chemical contamination and completely clean to permit uniform wetting of the surface. Imperfect wetting of the interior surface may result in volume errors because of incomplete drainage of pipets or burets or distortion of the meniscus.

Laboratory utensils are rinsed immediately after use and allowed to presoak in a container filled with 5% bleach. A number of cleaning agents are available for washing laboratory glassware and plasticware. Many detergent-type products are extremely effective, although pipets may require overnight soaking. Automatic washers may require special products. To minimize danger to personnel, it is best to avoid the use of dangerous concentrated acids as cleaning agents on a routine basis.

Adequate rinsing of washed items is essential. After thorough rinsing with tap water, items must be rinsed three to five times with deionized (class I or II) water. When glassware is clean, deionized water drains as a continuous film, whereas unclean vessels will have little drops of water clinging to the surface. After drying, the appearance of dots indicates that glassware is not clean or has been insufficiently rinsed.

Incomplete detergent removal can be detected by rinsing an item with a dilute (20 mg/L) aqueous solution of sodium sulfobromophthalein (Bromsulphalein, BSP) dye. The formation of any pink color indicates detergent residue. A documented routine weekly check for detergent removal is recommended.

For trace metal analysis, plastic containers are used because they are less likely to be contaminated than other containers. To remove the trace levels of metals contained in plastic and to minimize low-level contamination, labware is soaked in 1 M HCl and rinsed in deionized water. For sensitive analytical studies a subsequent soaking period in 1 M HNO_3 (highest purity) and rinsing with deionized water may be necessary. Plastics should be soaked no longer than 8 hours in acids or they may become brittle. Plastic surfaces may be cleaned with alcohol, alkalies, or alcoholic alkalies to remove trace organics that contribute to trace metal absorption.

Chemicals

Chemicals are obtainable in many degrees of purity. In addition, many are analyzed so that the types or amounts of impurities are known. The highest grade or most pure chemicals are obtainable from the National Bureau of Standards. However, only very few such compounds are available for the clinical chemistry laboratory. These are termed *standard, clinical type*.

Additional standards of purity for certain chemicals have been specified by the International Union for Pure and Applied Chemistry (IUPAC). These include atomic-weight standards (grade A); ultimate standards (grade B); primary standards, which are commercially available and have less than 0.002% impurities (grade C); working standards, which are commercially available and have less than 0.05% impurities (grade D); and secondary substances (grade E), which are defined or standardized using a primary standard (grade C) as the reference material.

For ordinary analytical work, chemicals obtainable from various vendors come in many grades. The usual material suitable for analytical work is termed *reagent grade*. These meet specifications defined by the American Chemical Society (ACS). These specifications establish the maximum amounts of various types of impurities that can be present in each chemical. Some manufacturers will sell a certified or very pure material when specifications have not been set by the ACS. Usually these chemicals have the impurities listed and are analyzed by lot; that is, each batch of material has been checked for the most interfering types of impurities. For some chemicals, particularly pharmaceuticals, the specifications of purity are set in *The National Formulary, The United States Pharmacopeia,* and *The Food Chemical Codex*.

Less pure chemicals and reagents are available and are purchased because of their lower cost. Grading of these types of chemicals includes terms such as *practical grade* and *lowest quality technical grade*. These should not be used in analytical work.

For many analyses, particularly those involving spectroscopy and chromatography, it is necessary to purchase chemicals that are more pure than those certified as reagent grade. Such chemicals will specify the maximum amount of impurity or interference. For example, a spectral grade of *n*-butanol will have an absorbance guaranteed to be less than 0.05 at 260 nm when purchased. Compounds with desired purity for ultraviolet and infrared spectroscopy work must be purchased with these specifications. Also available are chromatographic solvents that exceed reagent-grade requirements. Two types of purity analyses are done: one to ensure minimum spectral or detector interference and the other to ensure minimum residual contamination after extraction and evaporation of the solvents in the procedure. For example, pesticide-grade solvents for use in the gas chromatographic analysis of chlorinated organic compounds (including tranquilizers) must have less than 1 part in 100 billion of chlorinated organic impurities. Similar types of high-purity chemicals exist for high-performance liquid chromatography. Most often these chemicals are solvents purified by distillation in glass apparatuses.

Gases, particularly those used for gas chromatography and atomic absorption analysis, must also be of the highest purity. Helium purity must be 99.9999% for use with the gas chromatograph mass spectrometer. Thus careful attention must be paid to the specifications of gases.

A number of chemicals are teratogenic and carcinogenic agents. These are specified to have these properties in the most modern chemical catalogs. They should also be listed in the laboratory safety manual.

VOLUMETRIC MEASUREMENTS

Accurate volumetric measurements and transfers are central to the quantitative analysis of the clinical chemistry laboratory. Such measurements are made by manual pipets, including graduated and volumetric ones, micropipets, dispensing devices, automated pipets, syringes, burets, volumetric flasks, and graduated cylinders. The reliability of these devices can vary considerably, and not all of these devices are suitable for the most accurate work. The maximum allowable error or tolerance limits for these devices are set by the National Bureau of Standards. Class A devices are more accurate, whereas class B ones are less accurate, although they may be as precise. Originally, the class A glassware was defined as NBS certified. Currently, volumetric glassware that meets federal specifications or equivalent is termed *class A*. The College of American Pathologists (CAP) inspection requirements specify that volumetric pipets be of certified accuracy, that is, class A. If not, they must be checked by a gravimetric, colorimetric, or some other verification procedure. In addition, automatic pipets and diluting devices must be checked for accuracy and reproducibility. Volumetric glassware used to make solutions must be of the class A category. Thus for most laboratories the preparation of solutions proceeds with the use of class A glassware. One should remember that accurate delivery of volumes can be accomplished *only* with clean glassware. If liquid does not drain cleanly, that is, if beads of liquid cling to the sides of a pipet, volumetric flask, or buret, one must assume an inaccurate volume. The item of glassware should be set aside and replaced by a clean one.

Manual pipets

There are three types of manual pipets: to contain (TC), to deliver/blow out (TD), and to deliver (TD). *To-contain,* or *rinse-out, pipets* must be refilled or rinsed out with the appropriate solvent after the initial liquid has been drained from the pipet. These pipets contain or hold an exact amount of liquid, which must be completely transferred for accurate measurement. The more common names given to these pipets include the Micro-Folin, dual purpose, Sahli hemoglobin, Kirk Micro, White-Black Lambda, transfer micro, measuring micro, and Lang-Levy. None of these to-contain pipets meets class A specifications.

To-deliver/blow-out pipets are filled and allowed to drain, and the fluid remaining in the tip is blown out. These pipets thus transfer or deliver an exact amount of liquid and are not rinsed out. The common names given to these pipets are Ostwald-Folin, serological, serological long tip, and serological large-tip opening. They can be readily identified by the two frosted bands near the mouthpiece of the pipet (Fig. 1-3). Serological pipets are long glass tubes of uniform diameter. They have volume graduations that extend to the delivery tip of the pipet. Thus the last blown-out drop of liquid is also included in the delivery volume. These pipets come in designs with large-tip openings for delivery of viscous fluids and long tapered tips.

To-deliver pipets are filled and allowed to drain by gravity. To ensure complete draining of the liquid, one must set the flow rates according to maximum NBS flow rates. The pipet must be held vertically, and the tip must be placed against the side of the accepting vessel. Common names given to these pipets include volumetric, Mohr, Mohr long tip, and bacteriological (Fig. 1-3).

An example of a class A volumetric pipet is shown in Fig. 1-4, *B*. It is composed of an open-ended bulb holding the bulk of the liquid, a long glass tube at one end that has the mark (line) to describe the extent to which the pipet is to be filled, and a tapered delivery portion. These pipets hold and deliver only the specific volumes indicated at the upper end of the pipet. After the pipet has completely drained, the amount transferred is equal to the stated value. For each volume desired, a specific size of pipet must be used. In general, the accuracy of these pipets is better than 0.6% and usually on the order of 0.2% (Table 1-9).

Mohr pipets also meet class A standards. These are glass tubes of uniform diameter with a tapered delivery tip. Graduations are made at uniform intervals but well away from the tapered delivery tip (Fig. 1-3). The solution is delivered between the desired markers. The listed accuracy for these pipets is for the *full* volume. If smaller volumes

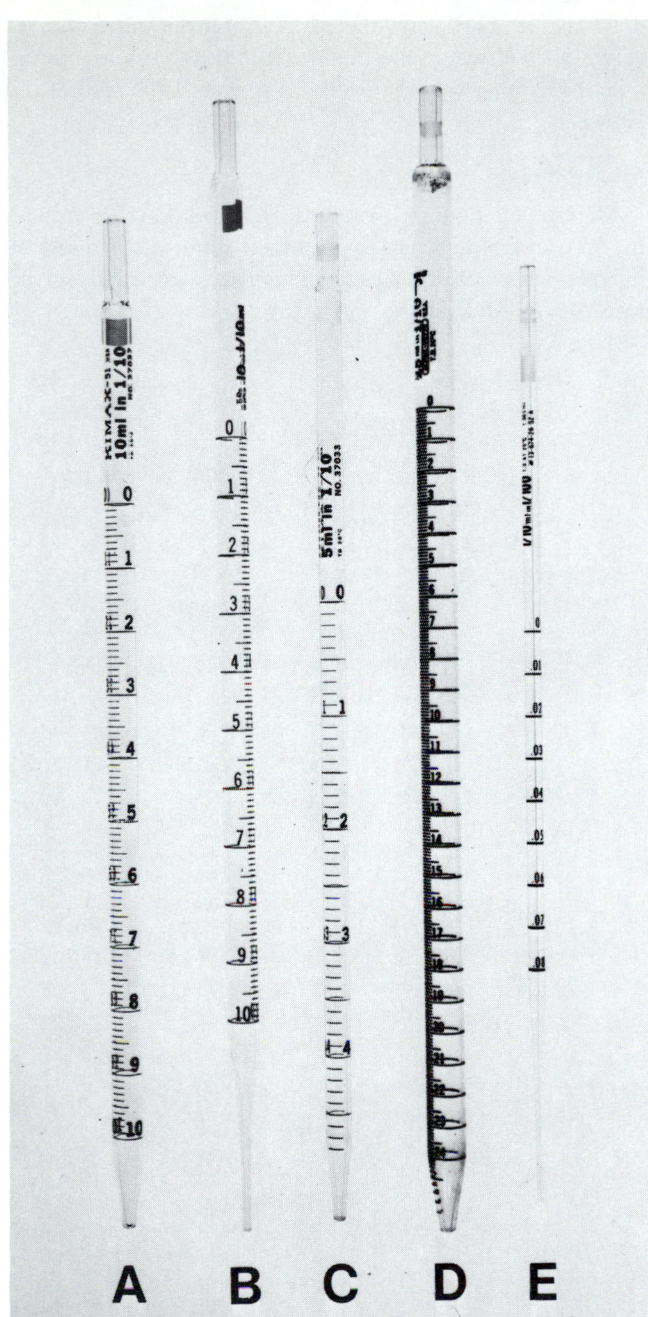

Fig. 1-3 Examples of transfer to-deliver (TD) pipets. **A,** Mohr. **B,** Mohr long tip. **C,** Serological. **D,** Serological large opening. **E,** Serological long tip.

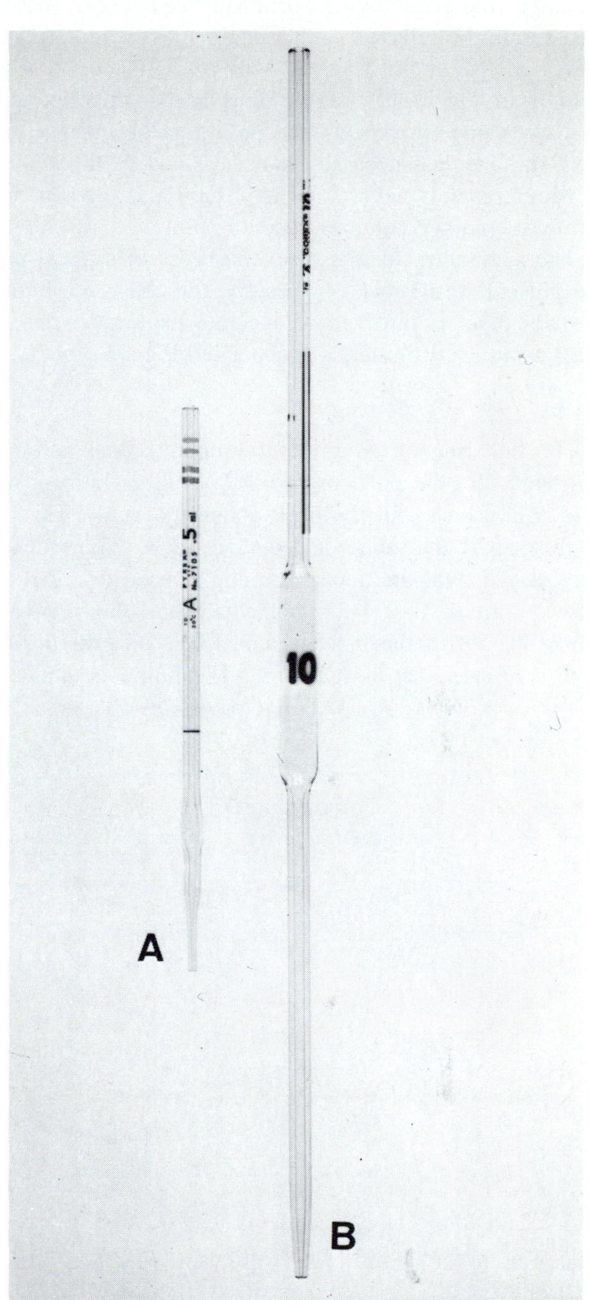

Fig. 1-4 Examples of to-deliver (TD) pipets. **A,** Ostwald-Folin. **B,** Class A volumetric.

Table 1-9 Accuracies of manual pipets

Type of pipet	1.0 mL	5.0 mL	10.0 mL	25.0 mL
NBS standard	—	0.01	0.02	0.025
Class A volumetric	0.006	0.01	0.02	0.03
Mohr	0.01	0.02	0.03	0.10
Mohr long tip	0.02	0.04	0.06	—
Serological	0.01	0.02	0.03	0.10
Serological large opening	0.05	0.10	0.10	0.20
Serological long tip	0.02	0.04	0.06	—

are used, the accuracy proportionally decreases. These pipets should be selected so that the greatest volume is used; therefore maximum accuracy will be achieved. A second version of the Mohr type of pipet is the long tip, which easily fits into small vials. According to the manufacturer, long tips are less accurate than the standard tapered tips. A Mohr pipet is never used as a blow-out type of pipet, but delivers only volumes point to point.

The accuracies for these several types of manual pipets are presented in Table 1-9. Clearly, the class A volumetric transfer pipet is much more accurate than any other. The least accurate is the large-opening serological pipet.

Techniques of pipetting

The first rule of using manual pipets is that nothing is aspirated into the pipet by mouth. A rubber bulb or other device is used to fill the pipet above the mark. The pipet is grasped by the thumb and middle finger. The index finger, placed over the upper opening, is used to control the flow of liquid (Fig. 1-5). With the pipet filled above the mark, the tip of the pipet is wiped free of adhering fluid with a tissue, such as Kimwipe. The liquid is allowed to drain to the mark, and the pipet is transferred to the receiving vessel where the liquid is allowed to drain with the pipet held in a vertical position, tip against the side of the vessel.

Micropipets

Micropipets contain or deliver volumes ranging from 1 to 500 microliters (μL). Another term often used is *lambda pipet* because the term *lambda* was usually given to a volume of 1 μL.

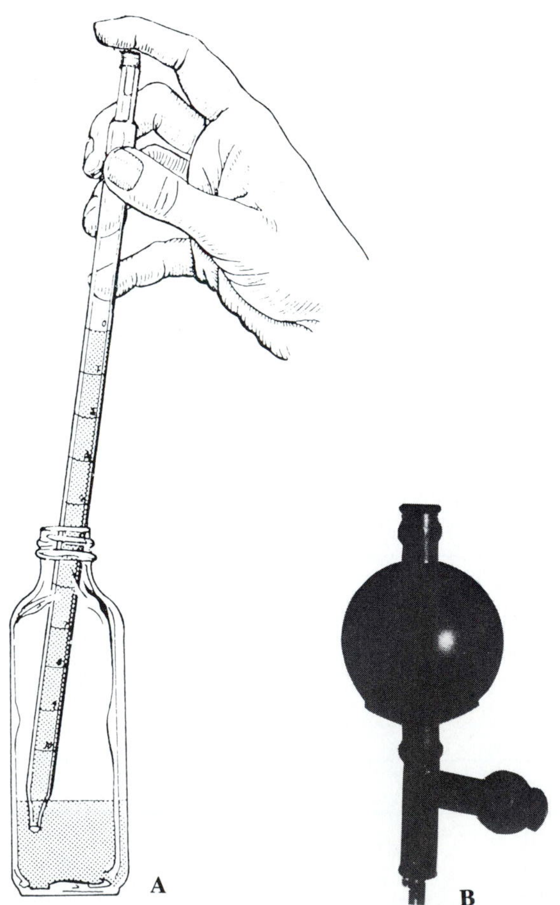

Fig. 1-5 **A,** Proper pipetting technique as described in text. **B,** Example of rubber pipetting bulb used to aspirate sample into pipet.

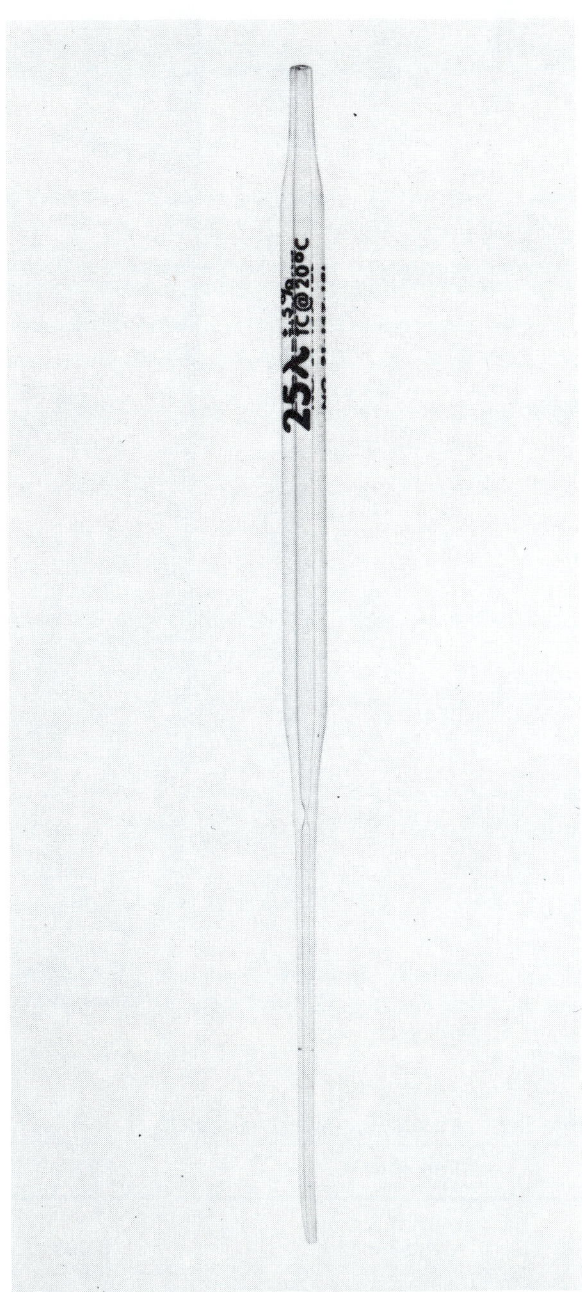

Fig. 1-6 Example of glass micropipet, the Lang-Levy pipet. This pipet is most often used as a to-contain (TC) pipet.

The original, most popular type of micropipet was the Lang-Levy pipet depicted in Fig. 1-6. The pipet has a tapered delivery tip and a constriction at the other end. The pipet is filled to the constriction. These pipets are probably the most accurate ever developed for microvolumes, but because of the difficulties in cleaning and their fragility, they are not commonly used in the clinical laboratory.

An inexpensive, disposable micropipet consists of capillary tubing with a line demarking a specific volume. These are filled to the line by capillary action. The liquid is delivered by positive pressure (blown out) by a medicine dropper or equivalent device. These are usually calibrated "to contain" (TC) and require rinsing to achieve the stated accuracy.

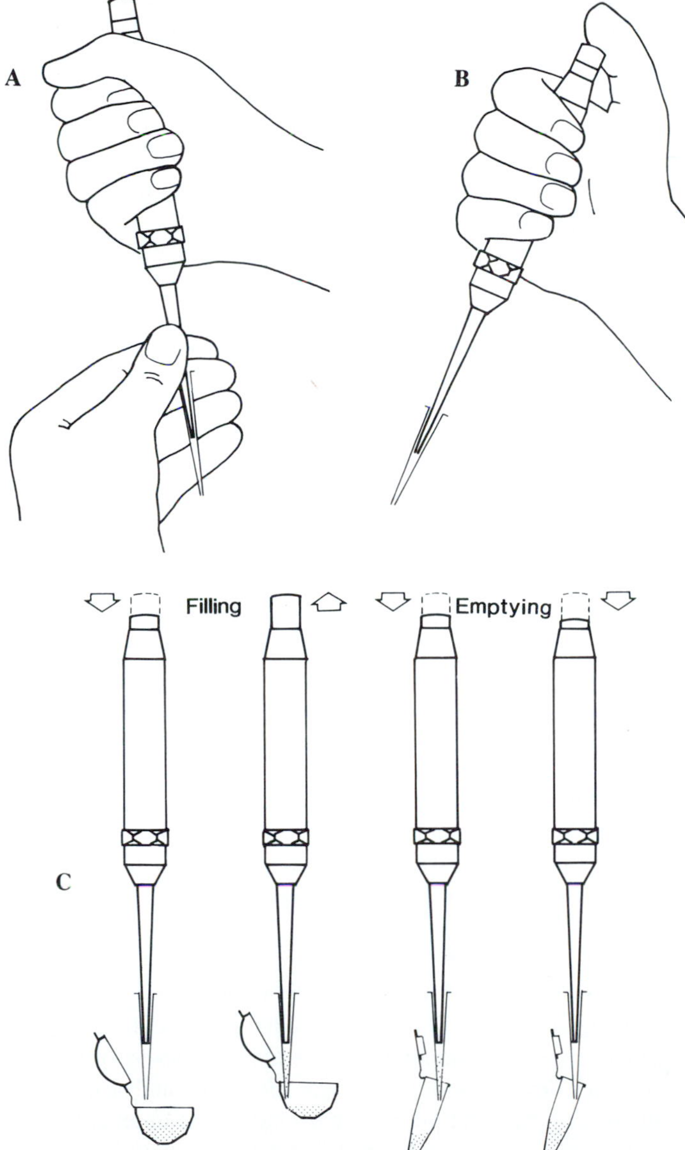

Fig. 1-7 Steps in using Eppindorf type of micropipet. **A,** Attaching proper tip size for range of pipet volume and twisting tip as it is pushed onto pipet to give an airtight, continuous seal. **B,** Holding pipet before use. **C,** Detailed instructions for filling and emptying pipet tip. Follow manufacturer's complete instructions for care and use of micropipets.

The most common type of micropipet used in laboratories was introduced by the Eppendorf Company, and this name has become almost generic for a large number of to-deliver pipets that work by the same principle. These are piston-operated devices. Disposable and exchangeable tips are placed on the barrel of the pipet, and these receive and dispense liquid (Fig. 1-7, *A*). The piston is depressed to a stop position on the device, the tip is placed in the liquid, and then slowly the piston is allowed to rise back to the original position (Fig. 1-7, *C*). This fills the tip with the desired amount of liquid. The tips are not usually wiped because the plastic surface is considered nonwettable, but are drawn along the wall of the vessel from which the fluid is to be taken so that any adhering liquid is removed. The tip is then placed on the wall of the receiving vessel and the piston depressed, first to the stop position, allowing the liquid to drain, and then to a second stop position beyond the first one. This ensures the full dispensing of the fluid. The manufacturer claims to have 99% sample recovery with this device, a standard deviation (for volumes of 10 to 500 μL) of less than 1%, and a reproducibility for multiple pipetting of 0.6% to 0.3%. For volumes under 10 μL the errors become significantly greater.

The pipet tips are disposable so that cleaning is unnecessary. Various types of such tips are available; most are made of polypropylene, since this surface is considered nonwettable. Pipets that automatically eject used pipet tips are available, and these systems allow the user to install a new tip and to remove a used tip without touching it, minimizing analytical contamination problems and biohazard exposure. To achieve the accuracy and precision claimed, disposable tips should be purchased from the pipet manufacturer unless equivalent performance of a substitute can be proven.

For reagents that will react with plastics or for procedures that require the use of reusable tip pipettors, a positive displacement micropipettor, such as the SMI pipettor, is useful. The SMI micropipettor delivers liquid by use of a Teflon-tipped plunger and a glass capillary. Using the same glass capillary, carryover from sample to sample is claimed to be negligible.

Devices with continuously adjustable volumes or multipoint adjustable volumes, which work by the same principles as the Eppendorf-style pipet, are also available. The accuracy claims for volume delivery should be verified at several volumes before use.

Automated dispensers

Automated dispensers are frequently used by the laboratory to add repeatedly a specific volume of reagent or diluent to a solution. Many types of dispensers are available. The Oxford Repipette is a typical example (Fig. 1-

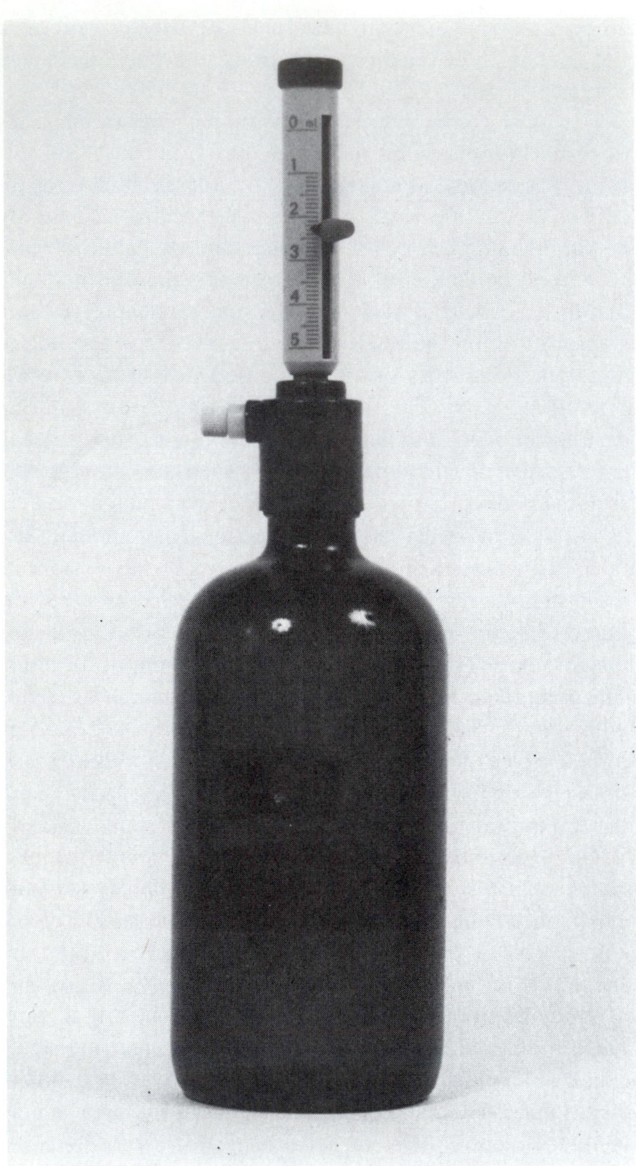

Fig. 1-8 Example of manual dispenser or repipettor.

8). A long tube leading from the dispenser is placed in the reagent bottle. The dispenser is composed of a plunger, valve system, and dispensing tip. Typically, once the dispensing device is primed with liquid, pressing on the plunger dispenses a selected amount of liquid. When the plunger is returned to the original position, usually by spring action, the dispenser chamber is refilled. The manufacturers claim accuracy of 1% and a reproducibility of 0.1% for these devices at full deflection of the plunger. Such devices often require constant cleaning to prevent material from disturbing the piston action and not allowing proper displacement to occur.

Repetitive dispensing pipettors (Eppendorf Repeater, SMI MultiPettor) are useful devices for the serial dispensing of relatively small volumes of the same liquid. The volume dispensed is determined by pipettor setting and by the size of the disposable syringe-type tip, which also acts as the liquid reservoir.

Diluter-dispensers

The diluter-dispenser is often used in automated instruments to prepare a number of samples for analysis. This device pipettes a selected aliquot of sample and diluent into a receiving vessel or instrument. Most of these devices are of the dual piston type (Fig. 1-9). When an electronic signal is given, two pistons are activated. The smaller piston by valvular action aspirates a sample into the pipet tip. The second piston simultaneously fills the second syringe or equivalent device with solvent or diluent. At a second signal the valves are changed, and diluent is sent through the pipet tip. This displaces the sample and rinses the pipet tip. The automatic diluter is now prepared for the next sample. Many ratios of sample to diluent may be selected. Accuracy is to 1% of full syringe volume, and repeatability is on the order of 0.05% of a full-syringe volume.

Syringes

Syringes used for accurate volumetric work are glass cylinders with a precision-bored hole in which is placed a very close fitting plunger. The dispensing tip of the syringe is a metal needle, usually of very fine diameter. Syringes are used in gas and high-performance liquid chromatographic analyses because they can accurately transfer very small volumes of solvent. Such solvents, such as methanol, are usually very volatile and are usually not readily quantitatively transferable by glass pipets. Syringes range in size from 1 to 500 μL. Needles on the end of the syringe are necessary for gas chromatography to pierce the septa of the injection ports. Manufacturers claim accuracy of 1% of the total syringe volume and a repeatability of 1% of the dispensed volume for volumes greater than 5 μL. For those less than 5 μL in size, 2% accuracy is the best that is achievable. Many syringes are not calibrated because internal standards are used in gas and high-performance liquid chromatographic analyses to quantify the samples and correct for transfer errors.

Burets

A buret is one of the most accurate devices for the dispensing of volume. It consists of a long glass cylinder with graduations corresponding to accurate volumes, a stopcock or other device to stop the flow of liquid, and a tapered dispensing tip. Burets meet the class A requirements for volumetric accuracy. A buret is filled with liquid, and an amount of fluid is allowed to flow into a receiving device for waste material. The tip is wiped, and the liquid is allowed to flow into the waste-receiving device with the buret tip on the side of the device until the meniscus reaches a desired graduation. The reading on the buret is recorded. By opening the stopcock, one can allow a specific amount of liquid to flow from the buret into the titrating or receiv-

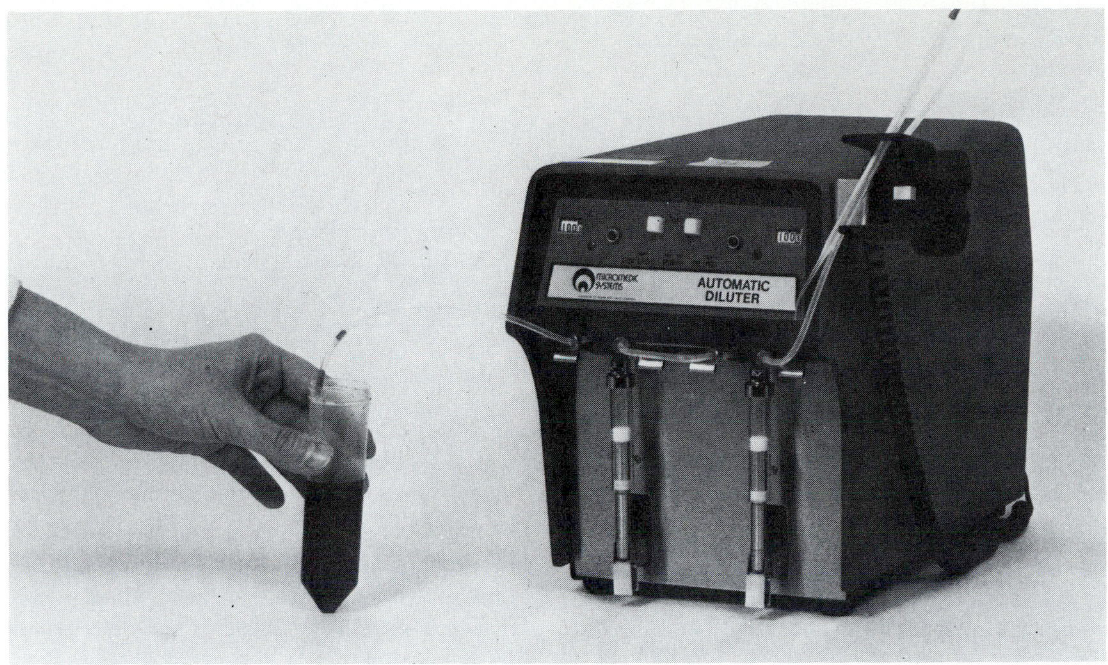

Fig. 1-9 Example of dual-syringe type of diluter-dispenser.

ing vessel. The accuracy requirements of burets to meet NBS standards are presented in Table 1-10.

It is clear from this table that the error of this device is on the order of 0.2% to 0.1%.

Volumetric flasks

Volumetric flasks are essential for the accurate preparation of solutions. Class A specifications are required for such use. The typical volumetric flask consists of a large bulbous lower portion with a flat bottom and a long slender neck. A line or an equivalent mark on the neck describes the position at which the meniscus must be located to achieve the stated volume. The flask is calibrated to contain, *not* deliver, a given volume. Thus it cannot be used as a transfer device. The top of the volumetric flask is capped by a tight-fitting, ground-glass stopper. This allows the flask to be inverted without loss of liquid. Under no circumstances should the flask be heated because this may distort its shape and volume.

To dissolve a solute in a given amount of solvent, several procedures are appropriate. In any event, the solute must be transferred into the volumetric flask and dissolved in a small volume of solvent in the flask. The solute must be completely dissolved before one brings it to final volume because many solutes have appreciable volume changes when placed in solution. Once dissolved, small portions of solvent are added, with swirling, until the meniscus is reached. The volumetric flask is then stoppered and inverted with swirling motions 15 times for 4 to 5 minutes. This should result in uniform mixing.

Volumetric flasks are the most accurate of all devices described. Table 1-11 describes the accuracy of class A volumetric flasks.

One should keep in mind that these figures are accurate only at the temperature specified on the flask. This temperature and other NBS specifications are usually imprinted directly on the glassware (Fig. 1-10).

Graduated cylinders

As their name implies, graduated cylinders have marks that allow for filling to specified volumes. The cylinders are calibrated to deliver or to contain the amount of fluid

Table 1-10 Accuracies of burets

Total volume (mL)	Limit of error (mL)
5	0.01
10	0.02
50	0.05
100	0.10

Table 1-11 Accuracies of volumetric flasks

Capacity (mL)	Limit of error (mL)	Percent error
25	0.03	0.1
50	0.05	0.1
100	0.08	0.08
250	0.11	0.04
500	0.15	0.03
1000	0.30	0.03

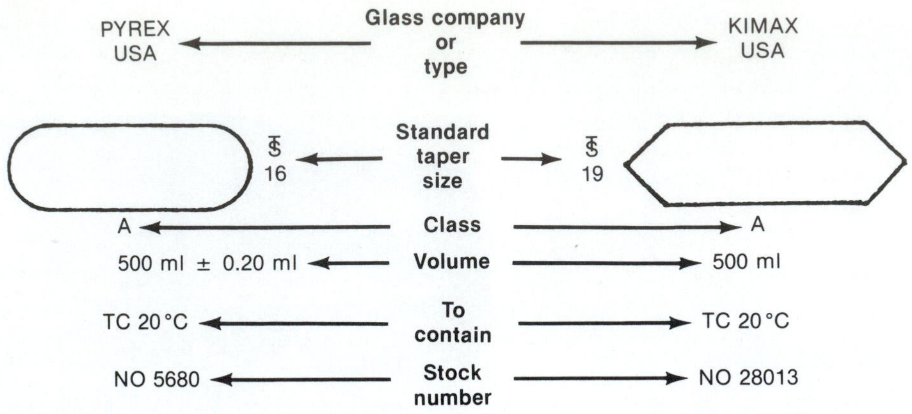

PYREX USA	← Glass company or type →	KIMAX USA
₰ 16	← Standard taper size →	₰ 19
A	← Class →	A
500 ml ± 0.20 ml	← Volume →	500 ml
TC 20 °C	← To contain →	TC 20 °C
NO 5680	← Stock number →	NO 28013

Fig. 1-10 Example of NBS specifications found imprinted on class A volumetric flasks.

described at the marking. They are most often used to deliver fractional volumes of liquid. One can purchase cylinders that meet class A specifications. These specifications and some of the properties are described in Table 1-12.

Table 1-12 Specifications and accuracies of graduated cylinders

Volume (mL)	Subdivision (mL)	Tolerance (mL)	Percent error
10	0.1	0.1	1
25	0.2	0.18	0.9
50	1	0.26	0.5
100	1	0.40	0.4
250	2	0.8	0.4
500	5	1.3	0.26

Quality control procedure for micropipets, dispensers, and diluters

Micropipets. The following protocol establishes a spectrophotometric method for the assessment of accuracy and precision of manual micropipets used in the clinical chemistry laboratory.

Method. Selected manual micropipets with disposable tips are used to deliver measured aliquots of concentrated green dye to class A volumetric flasks. Aliquots are diluted to the mark with doubly distilled water. Most dilutions will employ 10 mL volumetric flasks (refer to Table 1-13 for the appropriate flask size to be used with each pipet). Whenever possible, the same disposable tip should be used with each pipet.

The spectrophotometer is set to wavelength 620 nm at 25° C and zeroed with use of doubly distilled water. Absorbances are measured for each pipet dilution according to standard operating procedure. This yields the value $A_{dilution}$.

Calculations. The delivered volume for each micropipet is calculated from the measured absorbance of each dilution using a conversion factor (CF) based on the mean absorbance value ($A_{1:50\ standard}$) of a manual 1:50 dilution of the stock green food coloring (pipet 1 mL of green dye into a 50 mL volumetric flask; dilute with water up to mark). The conversion factor and pipet volume are calculated according to the following relations:

$$V_{pipet} = A_{dilution} \times CF \qquad \textit{Eq. 1-1}$$

where

$$CF = \frac{V_{flask}(mL) \times 20}{A_{1:50\ standard}} \qquad \textit{Eq. 1-2}$$

V_{flask} is the volume of the receiving flask, and 20 is a volume factor. Equation 1-1 allows the calculation of the actual delivered volume of the micropipet after one measures the absorbance of the diluted solution and multiplies the reading by the concentration factor, CF.

The concentration factor in Equation 1-2 depends only on the volume of the flask (in milliliters) in which the diluted solution is prepared and on the absorbance of the standard, manually diluted 1:50 solution.

Quality control schedule. The accuracy and precision of each manual micropipet should be verified on acquisition and, subsequently, three or four times a year. The mean

Table 1-13 Volumetric dilutions for check of micropipet accuracy

Eppendorf pipet volume (μL)	Volumetric flask (mL)	Dilution
4	10	2500
5	10	2000
10	10	1000
20	10	500
25	10	400
40	10	250
50	10	200
75	10	133.3
100	100	1000
250	100	400
500	100	200

volume should deviate no more than 5% from the labeled volume. The coefficient of variation for repeated pipetting for each pipet should be less than 5% for a minimum of five deliveries. A separate record of accuracy and precision should be maintained for each pipet to facilitate the detection of drifts or shifts in performance. Some micropipets (MLA) permit adjustments to correct detected inaccuracy.

Routine maintenance. The Eppendorf or other piston-type micropipets have a fixed stroke length, which must be maintained. In addition, there are seals to prevent air from leaking into the pipet during its motion. These must be greased to maintain proper operation. Refer to the manufacturer's guidelines for this type of maintenance.

MLA pipets require periodic lubrication and frequent cleaning of the nozzle tip. The nozzle is easily removed by inserting a disposable pipet tip through the hole. The nozzle may be cleaned with a 70% isopropyl alcohol pad and then reinserted. Nozzles that appear worn should be replaced.

Accuracy and precision check for automatic diluters. Weekly, take a stock dye solution of known optical density, prepare a manual dilution similar to the one made by the diluter, and read at a specific wavelength. For example, if the automatic diluter effects a 1:50 dilution, use a 1 mL volumetric pipet to transfer an aliquot of the stock dye solution into a 50 mL volumetric flask. Carefully dilute to 50 mL with water and mix. Using the stock solution, run 20 consecutive dilutions on the automatic diluter. Read all dilutions in a spectrophotometer. Compute $\overline{\chi}$ (mean), SD (standard deviation), and CV (coefficient of variation) of the 20 dilutions. The CV should be equal to or less than 1%. The mean ($\overline{\chi}$) should be within limits established by the laboratory, usually within 2% of the normal dilution. If either the deviation from the mean or the CV is not within limits, check settings and repeat the procedure. Green food dye can be used as stock solution, taking absorbance readings at 620 nm on the spectrophotometer.

Accuracy check for dispensers. Daily, take a clean, unscratched graduated cylinder that can contain several dispensings of the dispenser and use the smallest volume possible. NBS graduated cylinders are available for this purpose. Adjust setting and prime dispenser and ensure that no air bubbles are present. Carefully dispense an aliquot down the side of the cylinder and allow the liquid to drain. Read the volume contained. If the correct volume was delivered, check the maintenance sheet. If it was not, readjust the dispenser settings to correct for the error. Repeat dispensing, calculating the new volume delivered by subtraction. Repeat until correct volume (within an established error, such as 2% to 4%) is delivered.

THERMOMETRY

Types of thermometers

Water baths and metal heating devices must be maintained at relatively constant temperature when kinetic or other temperature-sensitive assays are performed. Liquid-in-glass thermometers, thermistors, and electronic digital thermometers are used to monitor the temperature of heating baths.

The temperature of heating baths should be checked every day and recorded for maintenance and quality control purposes. The accuracy of thermometers that are used to monitor heating baths should be verified at regular intervals, usually 6 months to 1 year. It has been recommended[7] that the thermometer have an accuracy range of one half that of the desired range for the bath. For instance, if the desired accuracy for the heating bath is ± 0.1° C, the thermometer should have a maximum uncertainty of ± 0.05° C.

Calibration of liquid-in-glass thermometers

For calibration standards, one should use thermometers that have been certified by the NBS or calibrated against an NBS-certified thermometer. As part of the NBS Standard Reference Material (SRM) program, certified thermometers are available that can be used to calibrate thermometers in the range of 24° to 38° C.[8] High-quality liquid-in-glass thermometers are available that have ranges greater than the SRM thermometers and have been calibrated against NBS thermometers by the manufacturer. Liquid-in-glass thermometers are available for partial or total immersion. Partial-immersion thermometers are used to measure the temperature of water baths, heating blocks, and ovens. The immersion depth is engraved on the stem and is usually located about 76 mm from the bulb. Total-immersion thermometers are generally used to check refrigerator and freezer temperatures but can be substituted for partial-immersion thermometers if they are verified at the same immersion depth that they will be used in the laboratory.

Verification procedure: noncertified thermometers

1. The mercury column should be checked for a separation or bubbles.
2. The heating bath should be adjusted to the temperature required for the test application.
3. The correction should be determined by comparison of the reference with the noncertified thermometer. The thermometers should be gently tapped before readings are taken so that the mercury column is prevented from sticking.
4. If a total-immersion thermometer is being calibrated for use as a partial-immersion one, it should be immersed in the heating bath to the same depth used for test applications.
5. Electronic thermometers that employ a thermistor probe may be calibrated by use of SRM thermometers or thermometers calibrated against an NBS-certified thermometer.

In the calibration of all types of thermometers, it is important that a liquid bath be used that is maintained at a constant level. It is recommended that the volume of the bath fluid be at least 100 times greater than the volume of the thermometers being calibrated so that a uniform temperature can be maintained throughout the bath.[9] The reference thermometer and thermometer being calibrated should be far enough apart so that the flow of liquid around them is not disturbed. Liquid-in-glass thermometers have a temperature response time of several minutes, whereas thermistors respond in a few milliseconds. This response time must be taken into consideration when taking measurements.

Thermometers differing from the reference thermometer by more than 1° C are discarded or returned to the supplier. Thermometers agreeing with the reference thermometer within 0.1° C may be used for all laboratory purposes, including critical measurements. Thermometers agreeing with the reference thermometer within 0.2 to 1.0° C may be used for general laboratory purposes, such as the monitoring of refrigerators and freezers.

Available from the NBS is a gallium melting point cell that provides a fixed point (melting point of gallium, 29.772° C) for calibrating electronic thermometers that have a thermistor probe. An electronic thermometer calibrated at 29.772° C can be used to standardize liquid-in-glass thermometers in the 20° to 40° C range. The application of the gallium melting point standard for thermometry has been reviewed.[10]

Manual and automated chemistry analyzers are available that use thermistor probes for monitoring the heating-bath temperature. Some of these instruments have flow-through or other cuvette designs that make the quality control measurement of the temperature in the cuvette very difficult. A technique has been devised for monitoring cuvette temperatures by measurement of the change in absorbance of a temperature-sensitive solution.[11] The temperature of the cuvette is obtained when this absorbance is compared with an absorbance-temperature calibration curve. It is claimed that this approach to thermometry is capable of detecting changes of 0.1° C.

CENTRIFUGES AND HEATING BATHS

Types of centrifuges and their uses

Laboratory centrifuges are available in floor or table models with a number of different designs. Centrifuges used most frequently in chemistry have either horizontal or fixed-angle (45- to 52-degree) rotor heads. These centrifuges operate at speeds up to 6000 revolutions per minute (rpm), generating relative centrifugal force (RCF) up to 7300 times the force of gravity (G). Microhematocrit centrifuges for determining red blood cell volumes operate at 11,000 to 15,000 rpm with an RCF up to 14,000 G. Ultracentrifuges are generally used for research; however, a small air-driven ultracentrifuge (Beckman Instruments, Spinco Division, Palo Alto, CA 94304) is available and operates at 90,000 to 100,000 rpm and generates a maximum RCF of 178,000 G. This type of centrifuge has been used to separate chylomicrons from serum, fractionate lipoproteins, perform drug-binding assays, and prepare tissue for steroid hormone receptor assays.

Centrifuges are used in the chemistry laboratory primarily to separate clotted blood or cells from serum or plasma and to clarify body fluids. Although the relative centrifugal force necessary to carry out these separations is not critical, a force of at least 1000 G for 10 minutes will give a good separation. Serum separation tubes are available that contain a silicone gel with a specific gravity intermediate to that of serum and blood coagulum. When these tubes are centrifuged, the gel is displaced up the side of the tube forming a stable, inert barrier between the serum and the clot. The RCF needed to displace the gel is 1000 to 1300 G for 10 minutes. RCF less than 1000 G may result in an incomplete displacement of the gel.

Refrigerated centrifuges have internal refrigeration units that maintain temperatures ranging from −15° to 25° C during centrifugation. This permits centrifugation at higher speeds and protects the specimens from the heat generated by the rotor. The temperature of these centrifuges should be checked daily, and the thermometer checked annually for accuracy.

Maintenance of centrifuges

The speed of a centrifuge is expressed in revolutions per minute (rpm), and the centrifugal force generated is expressed in gravities as relative centrifugal force (RCF). The number of revolutions per minute is related to the relative centrifugal force by the following formula:

$$\text{RCF} = 1.12 \times 10^{-5} \times r \times (\text{rpm})^2$$

r is the radius of the centrifuge expressed in centimeters and is equal to the distance from the center of the centrifuge head to the bottom of the tube holder in the centrifuge bucket. Fig. 1-11 shows a monograph for determining the relative centrifugal force from the number of revolutions per minutes and the radius. The College of American Pathologists (CAP) recommends that the number of revolutions per minute of a centrifuge used in a chemistry laboratory be checked every 3 months. This is most easily accomplished with a photoelectric tachometer or strobe tachometer. Most centrifuges have a hole or plastic window located in the centrifuge lid that permits a stroboscopic check of rpm. Some floor model centrifuges do not have this window but do have a digital tachometer that displays the rpm while the centrifuge is running. The rpm of these centrifuges cannot be checked without taking the lid off the centrifuge or the cowling from around the motor. This creates a hazardous situation; consequently, it is advisable to have factory-trained personnel check these centrifuges.

The CAP also recommends that the timer and speed control be checked on a quarterly basis with any corrections posted near the controls.

Instructions for lubrication, maintenance, and replace-

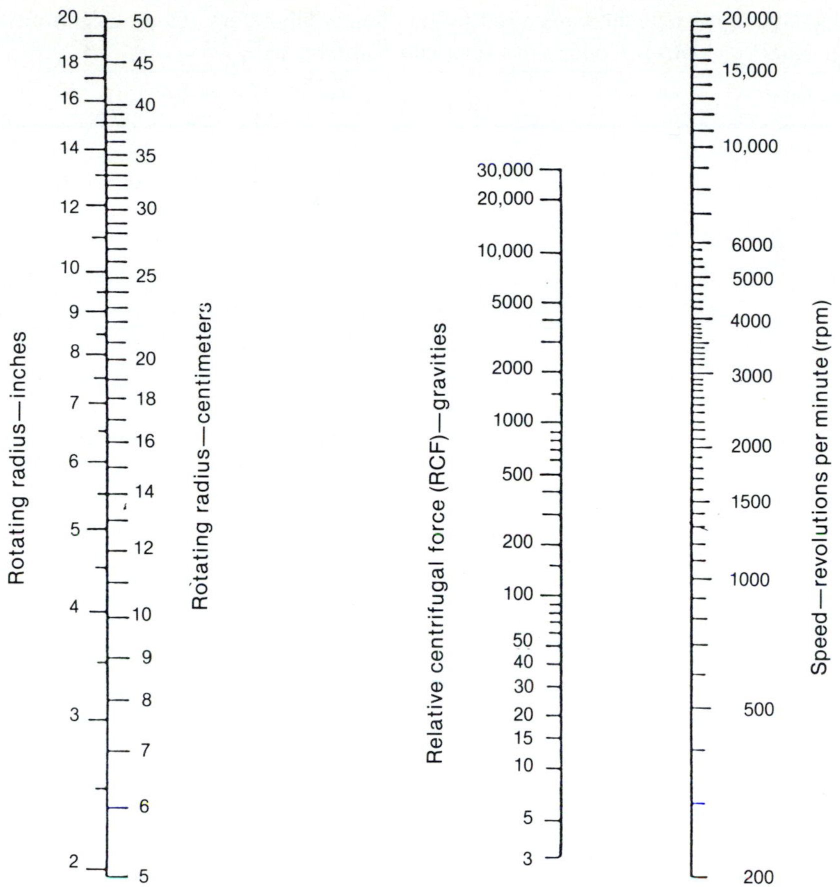

Fig. 1-11 Nomograph for relating relative centrifugal force, RCF, to revolutions per minute, rpm.

ment of brushes are given in the manufacturer's service manual. It is important that the centrifuge cups in the rotor head be balanced before centrifugation. This will permit maximum relative centrifugal force and minimize breakage of tubes and wear on the motor and bearings. Tubes should be capped during centrifugation to prevent release of infectious material inside the centrifuge by aerosol formation. If breakage occurs with release of potentially infectious material, the centrifuge bowl should be cleaned with a germicidal disinfectant, and the rotor head and buckets should be autoclaved. Regardless of manufacturer, drive motor brushes do wear out. If deteriorated brushes are not replaced, the motor may fail and require replacement. Since brushes are considerably less expensive than motors, it is advisable to adhere closely to the recommended frequency of brush inspection. Further information on the maintenance and care of centrifuges is available in the manufacturer's and other manuals.[5,12]

Heating baths and ovens

Laboratory ovens are used to dry chemicals, extracts, media used in electrophoresis, and glassware. For most purposes, temperature control at ±1° C is satisfactory. Thermometers that monitor oven temperature should be checked for accuracy annually. A daily record of the temperature will indicate malfunction of heating elements or thermistor controls and maintain quality control of operation.

Water baths are circulating or noncirculating in design. Noncirculating water baths generally maintain temperature within a range of ±1° C or better. For certain applications, such as enzyme, blood gas, pH, or electrophoretic assays, a narrower range is required (see Chapter 52); these baths have an external or internal circulating pump to maintain thermal equilibrium in the bath. Some of these pumps are coupled with a refrigeration unit to permit temperature control below room temperature. The bath liquid should be type II or type III water. The addition of a bactericidal agent such as a solution of 1:1000 thimerosal (Merthiolate) will reduce the frequency with which the bath and its components must be cleaned.

Metal heating blocks are somewhat less efficient in maintaining a relatively constant temperature and usually operate within ±0.5° C. Blocks that are incorporated into the cuvette compartment of a spectrophotometer will operate with a greater accuracy, usually ±0.2° C or better. The temperature of water or metal heating baths should be recorded daily with a thermometer calibrated against an NBS-certified thermometer.

Table 1-14 Recommendations for reagent water used in the clinical laboratory (Commission on Inspection and Accreditation for Water Quality Control, College of American Pathologists)

Specifications	Type I	Type II	Type III
Resistivity (megohm/centimeter at 25° C)	10 (in-line)	2.0	0.1
Silicate (mg/L, as SiO_2)	0.05	0.10	1.00
pH	NA*	NA	5.0-8.0
Microbiologic content (colony-forming units/mL, maximum)	10	10^3	NA
Particulate matter (>0.2 μm)	<500/L	NA	NA
Organic compounds	Activated carbon treatment	NA	NA

**NA*, Not applicable.

LABORATORY WATER

Significant error can be introduced into clinical laboratory assays by the presence of inorganic or organic impurities in laboratory water. Interference by these impurities may be relatively easy to detect in assays for iron, lead, and trace metals by analysis of a blank, but it is more difficult to recognize where enzymatic, fluorometric, or chromatographic procedures are concerned. The requirement for certification of clinical laboratories by state and federal agencies has resulted in the establishment of well-defined specifications for water purity (Table 1-14).

Three levels of water purity have been recommended by the National Committee for Clinical Laboratory Standards and the College of American Pathologists.[13,14] Type I water should be used in all quantitative chemistry procedures, in the preparation of buffers, controls, and standards, and in electrophoresis, toxicology screening tests, and high-performance liquid chromatography. Type II water is suitable for qualitative chemistry methods and can be used in most procedures carried out in hematology, immunology, and microbiology. Type III water can be used as a water source for producing type II and type I water and for washing and rinsing glassware. The final glassware rinse should be made with either type II or type I water, depending on the use of the glassware.

Storage of water

Type I water will absorb carbon dioxide rapidly and must be used immediately after purification. Type II and type III water can be stored in glass or polyethylene bottles but should be used as soon as possible because of possible contamination with airborne microorganisms. Sterile bottled water is usually equal to or better than type II water and has a shelf life of about 3 years if unopened.

Methods of purifying water

Water originating from rivers, lakes, springs, or wells contains a variety of inorganic, organic, and microbiologic contaminants. The contaminants cannot be completely removed by any single purification system, but a combination of several methods can be used to produce water of type I purity or better (Fig. 1-12).

Distillation. Distillation of water in glass will remove nonvolatile organics, inorganic impurities, and microbiologic organisms. Volatile impurities such as ammonia, carbon dioxide, chlorine, and low-boiling organic compounds will be present in the distillate. Distilled water meets the specifications of type II or type III water.

Deionization. Deionization is accomplished when water is passed through basic and acidic ion-exchange beds or a mixture of the two. This treatment results in the exchange of hydroxyl and hydrogen ions located on the surface of the resin for cation and anion impurities. The resins eventually lose their exchange capacity but can be regenerated by treatment with acid and alkali. Water containing less than 1 million colony-forming units per liter can be deionized to produce type II water. Further treatment with activated charcoal and membrane filtration to remove organic impurities, particulate matter, and microorganisms is necessary to produce type I water.

Reverse osmosis. Reverse osmosis is the passage of water under pressure through a semipermeable membrane made of cellulose acetate, aromatic polyamides, cellulose acetobutyrates, or other materials. This treatment removes approximately 90% of dissolved solids and 98% of organic impurities, insoluble matter, and microbiologic organisms. Reverse osmosis removes about 10% of ionic impurities but does not remove dissolved gases. This method of purification will produce type III water and is frequently used to treat water before its passage through ion-exchange resins.

Filtration. Filtration of water through semipermeable membranes with pore sizes of about 0.2 μm will remove insoluble matter, emulsified solids, pyrogens, and microorganisms. Sterile water can be produced by filtration through membranes with a pore size of 0.3 μm (300 nm) or less.

Adsorption. Organic impurities can be removed from water by adsorption on activated charcoal, clays, silicates, or metal oxides. A combination of deionization, adsorption, and filtration will produce type I water.

Tests for purity of water

Measurement of the conductivity or specific resistance (resistivity) of water is used to indicate the purity of water with regard to inorganic ionized substances. Ion-exchange

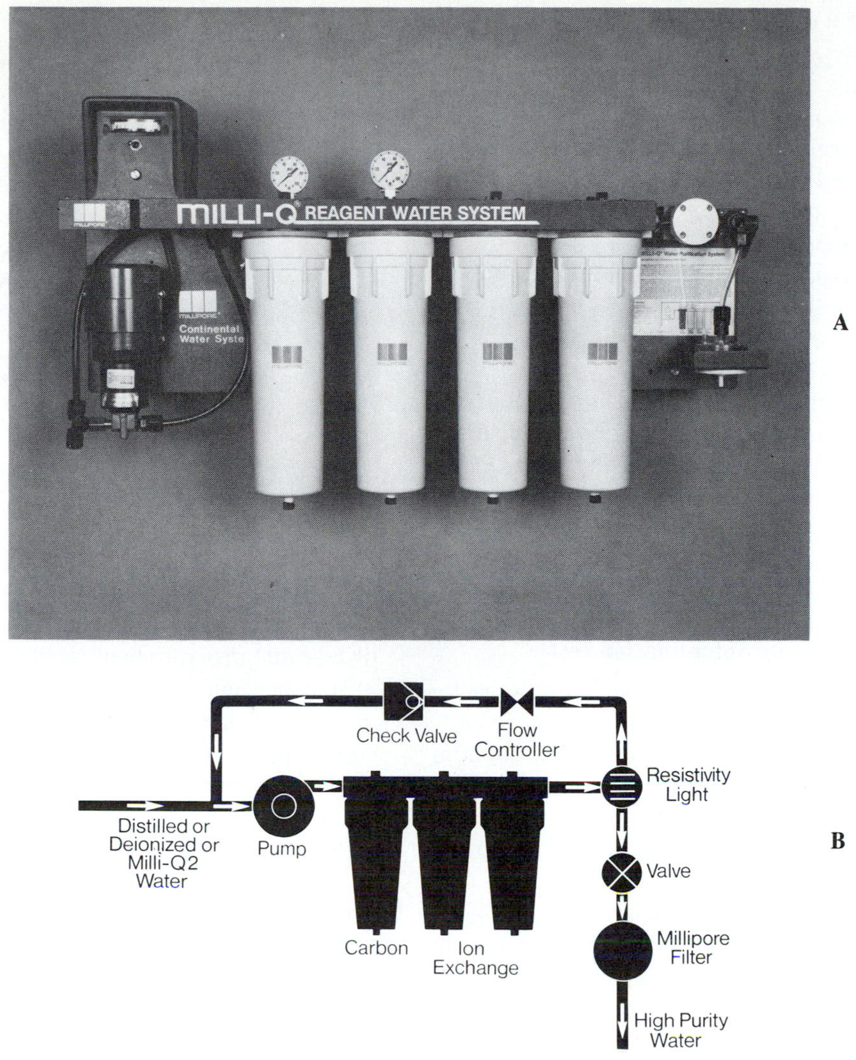

Fig. 1-12 **A,** Type I water purification system. **B,** Scheme of water flow through a type I purification system. *(Courtesy Millipore/Continental Water System, Bedford, ME.)*

tanks usually have a conductivity light located in line at the outlet of the tanks that is adjusted to go out when the resistance of the water goes below 200,000 ohms/cm, indicating that the capacity of the tanks has been exceeded. Systems that supply type I water usually have an ohmmeter located in the line outlet that has a scale graduated in the range of 0.5 to 18 megohms/cm. Type I water should have a resistivity of at least 10 megohms/cm, equivalent to a total dissolved solid concentration of less than one part per million. Procedures for measuring the resistivity and the silicate and microbiologic content of water are available.[13,14]

Water-purity checks

Water-resistance test. This test is a measure of dissolved ionized substances in the water. As the purity of the water increases and the amount of dissolved ionized materials decreases, the ability of the water to conduct an electrical current decreases and the resistance increases.

$$\frac{\text{Purity} \uparrow}{\text{Dissolved ionized material} \downarrow} = \frac{\text{Resistance} \uparrow}{\text{Conductivity} \downarrow}$$

The Beckman Solucomp Ultrapure Water Resistivity Analyzer (Beckman Industrial Corp., 89 Commerce Road, Cedar Grove, NJ 07009) is an example of a portable flowmeter. Only a flow-through conductivity meter can be used to measure the purity of type I water. When the Beckman flowmeter is used according to the manufacturer's instructions, type I water should have a resistance between 15 and 18 megohms/cm. This check is to be performed daily to weekly.

Weekly organic material check

1. Rinse a 1000 mL volumetric flask with distilled, deionized water three times. Collect about 500 mL of this water in the flask.
2. Add 1 mL of concentrated sulfuric acid.

3. Add 0.4 mL of 0.01 N potassium permanganate (see below).
4. Stopper, and mix well by swirling.
5. Let stand for 1 hour at room temperature.
6. If the purple permanganate color does not disappear completely after 1 hour, there is no organic contamination.
7. If the color disappears, this is an indication of the presence of organic contaminants. Notify the supervisor.

Reagents

1. Potassium permanganate, 0.01 N. Dissolve 0.316 g of potassium permanganate in 1 L of boiling distilled water in a 2 L Erlenmeyer flask. Cover with a beaker and simmer for one-half hour. Cool to room temperature. Store in a dark-colored stoppered bottle. Date bottle. Prepare fresh every 6 months.
2. Sulfuric acid, concentrated.

Monthly bacteriological contamination test

1. Obtain a 10 mL sterile culture tube from the bacteriology laboratory.
2. Let approximately 30 to 50 mL of water flow from the deionizer to waste. Quickly open sterile tube and fill halfway with water from deionizer. Close tube and give to bacteriology laboratory for organism check.
3. Record all readings and results on monthly water quality control chart and list any actions taken on water that fails any one of these checks.

SAFETY IN THE CLINICAL LABORATORY*

General considerations

Safety in the clinical laboratory has received increased emphasis in recent years, partly as a result of the manifold regulations of the Occupational Safety and Health Administration (OSHA). Regulations concerning the physical layout of the laboratory and emergency exits will not be discussed here except to note that all personnel should know the exit routes and should be familiar with the locations of the various safety devices—fire extinguishers, emergency showers, eye washers, fire blankets, and other equipment such as respirators and goggles—and with their use and operation. Diagrams that clearly show evacuation routes should be posted in appropriate locations in the laboratory. All employees should be instructed in the location and proper use of safety equipment and protective clothing. (If any information is not understood, laboratory personnel should not hesitate to ask the supervisor for clarification.) Personnel should also recognize the various auditory or visual signals for fire and other emergencies and should fully cooperate in all emergency drills.

*Some material from Bauer JD: Clinical laboratory methods, ed 9, St. Louis, 1982, The CV Mosby Co.

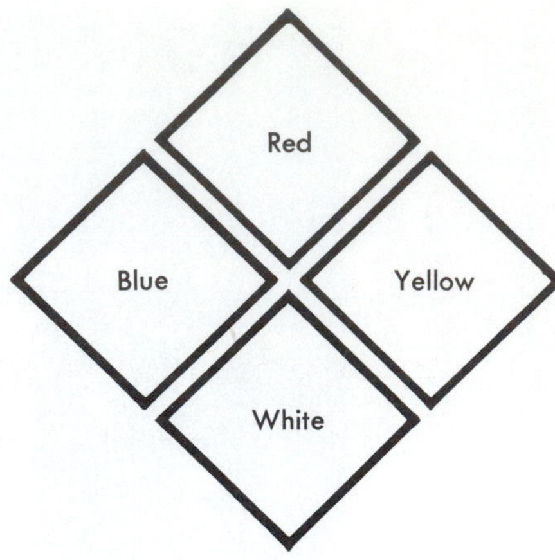

Fig. 1-13 Identification system of the National Fire Protection Association. *(From Bauer JD: Clinical laboratory methods, ed 9, St Louis, 1982, The CV Mosby Co.)*

So that laboratory functions can be performed safely, it is essential that sufficient laboratory space is available and that floors and benches are clean, clutter free, and well maintained. Utilities should be adequate for the laboratory work load, including sufficient sinks, water taps and drains, adequate lighting, sufficient electrical outlets, and proper ventilation and temperature control.

Safety manual. Every laboratory should have a safety manual covering all safety practices and precautions, including regulations concerning the proper use of equipment and the handling of toxic or infectious materials. The material here deals in a general way with factors most common to clinical laboratories.

Warning signs. Consistent warning signs should be used to indicate the presence of hazards. The Hazards Identification System developed by the National Fire Protection Association has been adopted by some laboratories. This system consists of four small, diamond-shaped symbols grouped into a larger diamond shape (Fig. 1-13). The left diamond is blue and is used to identify health hazards. The top diamond is red and indicates a flammability hazard. The diamond on the right is yellow and warns of a reactivity-stability hazard (meaning that materials capable of explosion or violent chemical change are present). The bottom diamond is white and is used to provide special hazard information. It may indicate the presence of radioactivity (Fig. 1-14, *A*), special biohazard (Fig. 1-14, *B*), corrosives, or other dangerous elements (Fig. 1-14, *C*). The severity of each hazard included in the symbol is suggested by numbers from 0 to 4, the greatest degree being 4. This simple system can provide safety information for all personnel and aid in fire fighting as well.

Personal behavior. Smoking, eating, and drinking should be prohibited in all work areas. These activities should be carried out only in designated rest areas completely separated physically from work areas. Foot-operated drinking fountains may be allowed just outside the

Fig. 1-14 Special hazard warnings. *(From Bauer JD: Clinical laboratory methods, ed 9, St Louis, 1982, The CV Mosby Co.)*

work areas. Although the greatest danger from eating or smoking in the laboratory is the possibility of infection from laboratory specimens, there is also the possibility that food or tobacco could contaminate test materials. For the same reasons application of cosmetics should not be done in work areas. No matter what type of work is done in the laboratory, the hands should always be washed before eating, smoking, or leaving the laboratory.

Long hair must be secured so that it will not come in contact with specimens, work areas, reagents, or mediums or become entangled in laboratory equipment. In the microbiology laboratory it may be advisable to use the same precautions that food service workers use in regard to hair containment, since contamination may occur from shedding of long hair and beards.

Material Safety Data Sheets. Manufacturers of chemicals, reagents, and kits provide Material Safety Data Sheets (MSDS) for all products. An MSD sheet contains information about the physical and health hazards of specific chemicals. These documents facilitate compliance with the Federal Occupational Safety and Health Administration Standards and various state and local laws requiring that employers notify employees of known or suspected health hazards. A complete set of MSD sheets for all chemicals and reagents, whether purchased or prepared, must be maintained and must be accessible to all employees and outside contractors working in the laboratory. Training and education regarding the presence of hazards should be conducted annually and should include support staff, secretaries, glassware washers, computer staff, managers, and directors, as well as technicians and technologists.[15]

The Federal Hazard Communication Standard requires that MSD sheets be written in English and contain at least the following information:

1. Name, address, and telephone number of the manufacturer and date of preparation or most recent revision.
2. Chemical identity of all components including common name(s).
3. Toxicity data.
4. Physical and chemical characteristics.
5. Fire and explosion data, including flashpoint.
6. Reactivity data.
7. Health hazard information. This includes routes of entry, acute and chronic hazards, carcinogenicity, symptoms of exposure, conditions aggravated by exposure, and first aid procedures.
8. Precautions for safe handling, including procedures for storage and disposal, as well as spill cleanup.
9. Control measures, including ventilation and protective equipment (goggles, clothing, gloves, respirators, etc.) required.[16]

Volatile flammables

Storage. All refrigerators must be clearly marked on the outside as to the type of material to be placed in them, indicating whether they are explosion proof. (Under no circumstances should food or drinks be placed, even temporarily, in a refrigerator in the work area; these should be left only in designated refrigerators in the rest area.) Volatile solvents should be labeled as to their hazardous qualities and should be kept in explosion-proof refrigerators. These solvents should be kept in these refrigerators only in as small a quantity as is consistent with daily use, preferably not more than 1 week's supply. (There is some question as to the safety of storing flammable solvents such as ethyl ether under refrigeration instead of on an open shelf in a well-ventilated area. Opening a refrigerator door can release evaporation vapors, resulting in ignition and explosion.) Large amounts of such solvents must be kept in a fireproof, specially ventilated area with explosion-proof switches and lighting and a 4-inch sill or ramp at the bottom of the door to prevent the escape of heavier-than-air vapors. Flammable liquids in quantities of more than 2 L should be stored only in clearly labeled, approved red safety storage containers with pressure-relieving, spring-closing covers and flame-arrester screens. Maintaining only the minimum amount of flammables necessary is a good practice.

Handling. All flammable and toxic liquids should be handled in a space with good ventilation, preferably in a fume hood. Evaporation of such liquids must be carried out in a fume hood using a water bath. Extreme caution should be taken with very volatile flammable liquids such as ethyl ether, since the vapors of these liquids can be ignited by any ignition source, even a hot plate or oven. In the evaporation of larger quantities of volatile liquids, glass beads or other antibump granules should be added to prevent bumping and the sudden release of large amounts of vapors. The fume hood function may be checked easily by noting the disposal of smoke or fumes produced after cautiously bringing close together two applicator sticks whose cotton tips have been dipped in strong ammonia and hydrochloric acid, respectively (rinse the tips in running water before discarding).

Compressed gas

Handling and storage. Large cylinders (200 ft^3) of compressed gases are often used in the laboratory. The most commonly used gases are oxygen with acetylene or propane (for atomic absorption or flame spectroscopy), hydrogen or helium (for gas-liquid chromatography), nitrogen, and carbon dioxide. The cylinders are color coded, but the contents must be clearly marked on the cylinder. (Otherwise it should be returned unused to the supplier marked "Contents Unknown.") All cylinders must be securely fastened to the wall or bench or placed in floor holders so that they cannot overturn. A fall can rupture the outlet valve, and the cylinder may then act like a torpedo and inflict serious injury. When transported, the cylinders must always be secured on a dolly or hand truck, and the protective caps must be on when the cylinders are not connected to equipment.

It is advisable to place a large tag on the tank indicating the date that the cylinder was put into use. Empty cylinders must be marked as such. (Note, however, that a small amount of gas must be left in the cylinder and valves closed to prevent negative pressure and drawing in of foreign contaminating materials.) The laboratory should use the smallest cylinder size that is practical. Only the cylinder in use and one spare should be permitted in the immediate laboratory area. Extra cylinders should be stored away from the laboratory in a secure, upright position, preferably in a locked, ventilated, fire-resistant space with the empty cylinders well separated from the full ones.

Precautions. Reduction valves for different types of gases are not interchangeable, and no attempt should be made to connect them by use of adaptors. Laboratory personnel should not attempt to force or free stuck or frozen cylinder valves. When in use, all connections to the cylinders should be tested for leaks with soapy water. Very small leaks of oxygen or nitrogen are of little consequence (except for the waste of gas), but leaks of hydrogen, acetylene, or other flammable gases are not to be tolerated. When one is shutting down flammable gases, the gas is first turned off at the main take-in valve and the gas allowed to burn out; then the reduction valves are closed. A cylinder valve is not turned on unless the reduction valve is off. It should be remembered that propane is heavier than air and that a little leaking gas can flow along the top of the bench to be ignited by a flame elsewhere. Many flame photometers use small cylinders of propane. These are very convenient and usually cause little difficulty. Because of the potential hazards, only properly trained persons should handle compressed gases.

Corrosives

Handling. Caution must be observed when using toxic or corrosive solutions such as those of mercury salts or caustic acids or alkalies. When handling these chemicals, laboratory personnel should wear goggles or a face mask. Strong stirring should be avoided or done in a hood. If considerable stirring is needed, a magnetic stirrer can be used to avoid splashing. In preparing reagents, caution should always be taken to *slowly* add acid to water. It is recommended that laboratories be equipped with automatic eyewash units, either self-contained or attached to faucets, which can direct two streams of water to thoroughly rinse both eyes. If a small amount of any reagent gets into an eye, it should immediately be rinsed well (for at least 15 min) in the eyewash device while one holds the eyelids apart with the fingers and rotates the eyeball. Following treatment for splashes of toxic or corrosive material into the eye, further medical attention should immediately be sought. Contact lenses should not be worn in the laboratory. They prevent proper washing of the eyes in the event a harmful solution does reach them, which could result in serious, permanent damage. Plastic lenses may be damaged by organic vapors.

Spills. All spills should be cleaned up at once. If the material is toxic or corrosive, rubber gloves and a plastic apron should be donned before cleaning up. All potentially infectious materials such as serum or other body fluids should be flooded with a 5% solution of sodium hypochlorite or other disinfectant before being wiped up (wearing gloves); the waste should be disposed of with other infectious materials, not with ordinary wastes. Dilute solutions of acids or bases may be neutralized with solid sodium bicarbonate. For stronger acids, after dilution with water, solid sodium carbonate may be used with caution. For strongly basic solutions, a 1:5 dilution of acetic acid can be used. The "Spill Clean-Up Kit" (Mallinckrodt Chemical Co., St. Louis) is satisfactory for laboratory use. It contains rubber gloves, towels, a scoop, and various chemicals for neutralizing and absorbing acid or alkaline spills, as well as material for absorbing organic solvents. Similar kits can be made with gloves, towel and other tools, sodium bicarbonate, an organic acid such as citric acid, and a quantity of absorbing material such as celite or diatomaceous earth. For such kits only the least expensive grade of technical chemicals need be used. These kits should be kept in strategic places in the laboratory, conspicuously labeled to indicate their use.

Mercury

A particularly troublesome substance to recover is metallic mercury. Although it is not used so much as formerly (with Van Slyke or Natelson apparatus), it may still be used occasionally. Spilled mercury tends to break up into very fine droplets, which are difficult to pick up. Even after collection disposal is a problem, since metallic mercury should not be incinerated or buried. If mercury is used in the laboratory, the "Mercury Absorption/Disposal Kit" (Aldrich Chemical Co., Milwaukee) may be helpful. It contains material that will readily absorb mercury droplets, producing a less toxic substance that may be disposed of by burial. Older benches or floors may contain fine cracks in which minute mercury globules may be lodged. These are very difficult to remove, but the cracks should be cleaned if possible, since mercury is somewhat volatile at room temperature. Rubbing powdered sulfur or sodium polysulfide into the cracks may help to change the mercury to the less volatile sulfide.

Some colorimetric methods, particularly for chloride, use mercury salts in a reagent. The disposal of the large amounts of spent reagent produced in automated methods may be a problem, since the solution should not be disposed of in the sewer. The waste material may be collected in large plastic containers and made slightly acid with acetic acid if necessary; thioacetamide (about 10 g/L) is added. This is stored in a ventilated area (small amounts of hydrogen sulfide may be liberated), and over a period of time the mercury will be precipitated as mercuric sulfide. The supernatant may then be decanted and disposed of in the sewer. The mercuric sulfide may be disposed of by burial.

Azides

Another troublesome chemical, sodium azide, was used as a preservative in the past, although its use has largely been discontinued. Azides form explosive salts with a number of metals such as copper and iron. These salts are readily detonated by mechanical shock. Although the amount used as a preservative is relatively small, continued use with disposal through the sewer can result in a buildup of the metallic salts in the sewer pipes. These salts are extremely explosive; even use of a wrench on such a drain line may result in a violent explosion. It is difficult to remove the azides from the pipes. One suggestion is to close the lower end of a section of pipe and allow a 10% solution of sodium hydroxide to stand at least 16 hours in the pipe; then rinse copiously with water (a minimum of 15 min). It is best to avoid the use of solutions containing azides, particularly since they are also said to be carcinogenic.

Carcinogenic hazards

The use of chemicals that are possibly carcinogenic is another hazard in the laboratory. Many of these are aromatic amines. One, benzidine, has been used frequently in the laboratory for testing for hemoglobin (plasma hemoglobin and occult blood). Generally this can be replaced by the compound 3,3′,5,5′-tetramethylbenzidine dihydrochloride, which is much safer. With proper precautions some potentially carcinogenic compounds can be used occasionally, if necessary. Precautions should include performing the procedure in an isolated area or in a good fume hood, wearing rubber gloves and (if dusty material is handled) a respirator, carefully cleaning up the work area, rinsing the glassware with a strong acid or an organic solvent before placing it in the regular washing cycle, and using disposable equipment as much as possible. No pipetting should be performed by mouth.

Procedures and equipment

Whenever a potentially toxic or dangerous chemical is used in a procedure, the directions in the procedure manual should include warnings of the noxious properties of that chemical. For example, it may seem superfluous to add "Caution: Corrosive" whenever the use of concentrated sulfuric acid is mentioned, or "Caution: Highly Flammable" whenever the use of diethyl ether is mentioned, but these warnings may be helpful to someone who is not familiar with the procedure or to the worker who may become careless through the routine use of a dangerous chemical.

Cracked or chipped glassware should not be used. It is likely to break during use and should be discarded. Flasks or beakers used with corrosive or toxic substances should be rinsed well with water or alcohol (depending on the solubility) before being placed in the collection container for soiled glassware. No pipetting by mouth should be allowed. Although it may seem safe to orally pipet distilled water or saline, it is best to use a pipetting device at all times; thus an attempt to pipet a dangerous substance by mouth is less likely. For simple dilutions with water or saline, the use of a good quality buret with a Teflon stopcock (or automatic dispensing bottle) is often convenient.

Radioactivity precautions

All precautions against eating, drinking, or smoking are particularly applicable to the radioisotope laboratory. Although the amount of radioactivity associated with the kits for radioimmunoassays is small, it can present some hazards. Radioisotopes often used for labeling are ^{125}I and ^{131}I. If either is ingested, the radioactive compound may be broken down in the body and radioactive iodine absorbed and concentrated in the thyroid gland. Thus the thyroid gland can receive a much larger dose of radiation than would be expected from a random distribution in the body. All personnel performing work with radioactive material should wear film badges at all times when in the isotope laboratory.

The radioactive material used in radioimmunoassays is generally so small that it can be harmlessly flushed down the sewer with copious amounts of water. Plastic beads or other resins used to absorb some of the radioactivity should not be incinerated. Burial is satisfactory, since the radioactivity gradually decays. All radioactive material should be stored in special refrigerators or cabinets conspicuously labeled with appropriate signs indicating the presence of radioactive elements. Generally the level of radioactivity used in a radioimmunoassay laboratory that purchases the material in the form of kits and does not prepare its own will not be sufficiently high to require lead shielding for storage.

The procedures using radioactive materials should be carried out in a separate room with a good ventilation system. The door into the room should be conspicuously marked "Radioactive Material—Authorized Personnel Only." All waste containers should be clearly marked as containing radioactive material so that the waste will not be incinerated.

Microbiological hazards

Universal precautions. The presence of the HIV (human immunodeficiency virus) and hepatitis viruses in the blood and body fluids of infected individuals constitutes a hazard to laboratory personnel. Since reliable identification of all patients with HIV or other blood-borne pathogens is not possible, blood and body fluid precautions should be observed for all patients. The Centers for Disease Control "universal precautions" include the following general recommendations:[17]

1. All personnel must routinely use barrier precautions to prevent skin and mucous membrane exposure when contact with blood or body fluids from any patient is expected. Gloves should be worn when handling blood or body fluids or items soiled with blood or fluids and when performing venipuncture. Protective eyewear or face shields should be worn during procedures that are likely to cause splashing to prevent exposure of mucous membranes of the mouth, nose, and eyes to droplets of blood or body fluids. Gowns or aprons should be worn during procedures that are likely to generate splashing.
2. If the hands or skin become contaminated with blood or body fluids, they should be washed immediately and thoroughly. Hands should be washed immediately after gloves are removed.
3. Health care workers should take precautions to avoid injuries from needles, scalpels, and other sharp devices. Needles should not be recapped, bent, broken, or removed from disposable syringes. After use, disposable syringes, needles, scalpel blades, and other sharp items should be placed in puncture-resistant containers such as those manufactured by Sage Products (Cary, IL).
4. Health care workers who have exudative skin lesions or weeping dermatitis should avoid direct patient care and contact with blood or body fluids until the condition is resolved.
5. Pregnant women should be particularly careful to adhere to precautions to minimize the risk of HIV transmission, since the infant is at risk of infection resulting from perinatal transmission.

Precautions for laboratories. Additional precautions are recommended specifically for the clinical laboratory.

1. Specimens of blood and body fluids should be transported in leakproof containers. Care should be taken to avoid contaminating the outside of the container and the accompanying laboratory requisition.
2. Persons processing blood and body fluid specimens (e.g., removing stoppers from evacuated tubes) should wear gloves. Masks and protective eyewear should be worn if mucous membrane contact is anticipated. Gloves should be changed and hands washed after completion of specimen handling.
3. Biological safety cabinets should be used for procedures where there is a high potential for generating droplets, such as blending, sonicating, and vigorous mixing.
4. Mechanical pipetting devices (bulbs, pumps, etc.) should be used for all liquids. Mouth pipetting is forbidden.
5. Use of needles and syringes should be restricted to situations for which there are no alternatives. Used needles should be discarded only in puncture-resistant containers.
6. Laboratory benches should be decontaminated with a chemical germicide (10% household bleach is acceptable) after a spill of blood or body fluids and, routinely, at the end of each work shift.
7. Contaminated materials should be placed in biohazard bags and disposed of in accordance with institutional policies for disposal of infective waste. Infectious waste should either be incinerated or autoclaved before disposal in a sanitary landfill. Blood and liquids may be poured into a drain connected to a sanitary sewer.
8. Instruments and equipment that have been contaminated with blood and body fluids should be decontaminated before repair.
9. All persons should wash their hands after completing laboratory activities and should remove laboratory coats and protective clothing before leaving the laboratory.

Management of exposures. If a laboratory worker is exposed through a needlestick or cut, a mucous membrane exposure (eyes or mouth), or an exposure involving skin contact with large amounts of blood, the source patient should be requested to consent to testing for serological evidence of HIV (as well as hepatitis B). If the patient is positive or refuses testing, the laboratory worker should be tested and evaluated for HIV infection and counseled regarding the risk of infection. Seronegative workers should be retested at 6, 12, and 26 weeks after exposure. These persons are advised to report any acute febrile illness during the first 12 weeks, as such an illness may indicate recent HIV infection. Follow-up beyond 12 weeks generally is not recommended unless the source patient is considered to belong to a high-risk group. It is recommended that seronegative health care workers be retested 6 weeks after the exposure and periodically thereafter to determine whether transmission has occurred. During the follow-up period, particularly the first 6 to 12 weeks after exposure, when most infected persons would be expected to seroconvert (have antiviral antibodies), exposed laboratory workers should follow U.S. Public Health Service recommendations for preventing transmission of HIV.

If a laboratory worker is exposed to infection and the source patient is determined to be seronegative, no further follow-up is necessary unless the source patient is at high risk for developing HIV. If the source patient cannot be identified, decisions concerning follow-up should be individualized. Serologic testing should be readily available to laboratory workers who are concerned that they may have been infected.

If a patient experiences a needlestick, cut, or mucous membrane exposure to the blood or body fluid of a laboratory worker, the patient should be informed and the same exposure management procedure carried out for the source laboratory worker and the exposed patient.

Hepatitis. The presence of hepatitis viruses in blood, tissue, urine, and feces from infected individuals also constitutes a hazard to laboratory personnel. The adoption of ''universal precautions'' provides protection to the laboratory worker from hepatitis as well as HIV.

Vaccines (Hepatovax B, Recombivax HB, Merck & Co., Inc.[18]) are available for immunization against infection caused by the hepatitis B virus, although they do not protect against hepatitis caused by other agents such as hepatitis A virus, non-A, non-B hepatitis viruses, or other viruses affecting the liver. Vaccination is recommended in persons who are at increased risk of infection with hepatitis B, such as laboratory personnel. The vaccine is administered in a series of three doses over a 6-month period. Protective levels of antibodies are induced in 90% to 99% of adults and teenagers receiving the three-dose regimen injected intramuscularly into the deltoid muscle of the arm.

The exposure management procedure should be followed for hepatitis B exposures as well as HIV. Hepatitis B immune globulin should be given as soon as possible and within 24 hours to an individual exposed to hepatitis B.

Electrical equipment

Safeguards. All electrical apparatuses should have three-pronged grounding plugs, and sufficient grounded outlets should be available. The only exceptions may be small items such as clocks, which are totally enclosed in plastic, and microscope lamps, which are similarly enclosed. No attempt should be made to adjust any piece of electrical apparatus while it is plugged in, unless it is absolutely necessary (for example, adjusting the lamp position in a spectrophotometer; this should be done with insulated tools). Under no circumstances should hands be placed inside an electrical instrument with the current on. Rings or jewelry along the forearms should not be worn. Preferably, rubber gloves should be worn for delicate adjustments. All electrical cords should be as short as possible and kept out of any areas where water or other solutions might contact them. Worn electrical cords should be replaced. Personnel should know the locations of the fuse boxes or circuit breakers for the different outlets. If electrical equipment fails to work properly (and particularly if smoke appears or sparks are seen), the apparatus should be disconnected and the maintenance department notified.

Quality control

Safety precautions, like all other aspects of laboratory function, must undergo a sort of ''quality control.'' Inspection of safety equipment operation and availability must be regular. It must be determined whether warning signs are displayed where needed, whether decontamination and disposal practices are adequate, and whether adherence to other safety procedures and rules is being practiced. Maintaining safe working standards may require extra effort, but risk minimization is part of the ''job.''

Part II: Calculations in clinical chemistry

ELIZABETH ANN BYRNE

DILUTION

In several areas of the medical laboratory, one must dilute blood or body fluids to prepare a measurable concentration. Accurate preparation of these dilutions is mandatory for reporting the actual concentrations of body-fluid constituents. Diagnosis, prognosis, and therapy depend on these test results.

Dilution can be defined as expressions of concentrations. Dilutions express the amount, either volume or weight, of a substance in a specified total final volume. A 1:5 dilution contains 1 volume (weight) in a *total* of 5 volumes (weights), that is, 1 volume and 4 volumes.

Expression of a 1:5 dilution can be stated as the common fraction ⅕. This fraction enables one to calculate the actual concentration of a diluted solution.

Example. A 100 mg/mL nitrogen standard is diluted 1:10. The concentration of the resulting solution is $100 \times 1/10 = 10$ mg/mL.

The most commonly used equation for preparing dilutions is

$$V_1 \times C_1 = V_2 \times C_2 \qquad \textit{Eq. 1-3}$$

V_1 is the volume, C_1 is the concentration of solution 1, and V_2 and C_2 are the volume and concentration of the diluted solution. These may be expressed as % (weight/volume, w/v), or molarity, or normality concentration. Similarly V_2 and C_2 are related. This basic equation can also be expressed as

$$\frac{V_1}{V_2} = \frac{C_1}{C_2} \qquad \textit{Eq. 1-4}$$

The most common error in setting up any equation of this type is *not placing* the related volumes or concentrations in the proper place and having the units cancel out, leaving the final, uncanceled units.

One helpful practice for successfully solving laboratory calculations is to label all numbers in any equation with their respective units of measurement. It may take an extra minute but will save many minutes of reviewing calculations when the final result appears illogical or incorrect. A problem that does *not* properly cancel out units cannot be successfully solved. A second helpful practice is to reduce fractions to their least common denominators before one calculates the results.

Example. Prepare 500 mL of 0.5 M NaCl (molecular weight of NaCl = 58.5 g/L).

$$500 \text{ mL} \times \frac{\text{Liter}}{1000 \text{ mL}} \times \frac{0.5 \text{ mol}}{\text{Liter}} \times \frac{58.5 \text{ g}}{\text{mol}} =$$

$$\frac{0.5 \times 58.5 \text{ g}}{2} = \frac{29.25 \text{ g}}{2} = 14.6 \text{ g}$$

14.6 g of NaCl diluted to 500 mL = 0.5 M NaCl

Example. Preparation of 250 mL of 0.1 M HCl from stock 1 M HCl.

Using $C_1 \times V_1 = C_2 \times V_2$
Where V_1 is the unknown $\qquad V_2 = 250$ mL
$C_1 = 1.0$ mol/L $\qquad C_2 = 0.1$ mol/L
$1.0 \text{ mol/L} \times V_1 = 250 \text{ mL} \times 0.1 \text{ mol/L}$
$V_1 = 25$ mL

Measure 25 mL of 1 M HCl; dilute to 250 mL with distilled water. This diluted solution has a 0.1 M HCl concentration. (Mathematical reasoning indicates that a 1:10 dilution of stock 1 M HCl results in a 0.1 M concentration, and 25 mL diluted to 250 ml equals a 1:10 dilution.)

Another application of dilutions

So that a 24-hour urine creatinine concentration could be assayed, the specimen had to be diluted 1:5 before measurement. Calculate the 24-hour excretion if the 24-hour urine volume is 1800 mL and the measured creatinine concentration is 260 mg/L:

Total excretion = Total urine volume × Concentration × Dilution

$$\text{Total excretion} = 1800 \frac{\text{mL}}{24 \text{ hr}} \times 260 \frac{\text{mg}}{\text{L}} \times 5 \times \frac{1 \text{ L}}{1000 \text{ mL}}$$

Total excretion = 2340 mg/24 hr or 2.34 g/24 hr

Exercises

Calculate the concentrations (answers are in the appendix at the end of this chapter).

1. 10 M NaOH, which is diluted 1:20 = ____ M?
2. 2 M HCl, which is diluted 1:5 = ____ M?
3. 1000 mg/L glucose, diluted 1:10 and then 1:2 = ____ mg/L?

Serial dilutions are those in which all the dilutions after the first one are the same. Exceptions to this general description of preparation of serial dilutions are included with certain techniques in serology, such as the antistreptolysin titer.

Serial dilution example. To determine the anti-RH_o (D) titer, serum is diluted 1:5 by addition of 0.2 mL of serum to 0.08 mL of saline solution, in tube 1. Tubes 2 through 8 contain 0.5 mL of saline as diluent. Dilution is performed by transferal of 0.5 mL of tube 1 to tube 2, mixing, and then transferring 0.5 mL through the tubes to tube 8. The concentration of serum in these tubes decreases by a factor of 2 with each dilution: 1:5, 1:10, 1:20, 1:40, 1:80, 1:160, 1:320, and 1:640.

4. For an ABO titer, tube 1 contains 0.9 mL of diluent, tubes 2 to 8 contain 0.5 mL of diluent, 0.1 mL of serum is added to tube 1, and serial dilutions using 0.5 mL are carried out in the remaining tubes. If the last tube showing agglutination with A cells is tube 6, what is the anti-A titer of the serum?
5. All tubes for serial dilution contain 0.5 mL of saline solution, 0.5 mL of serum is added to tube 1, and 0.5 of mL is transferred through the row of tubes. Sheep cells are added to the tubes, and agglutination is demonstrated through tube 7. What is the titer of sheep cell agglutinations?

WEIGHTS AND CONCENTRATIONS

Definitions and examples

Percent concentrations. Percent concentrations are generally expressed as parts of solute per 100 parts of total solution; hence the expression *percent,* or *per one hundred*. The use of percent concentration is derived historically from the early pharmaceutical chemists. Although these terms are still commonly used in the United States, major organizations (AACC, CAP) are attempting to use unified SI units. Concentrations in SI units are described in moles per liter when the molecular weight of the substance is known, or as weight (mass) per milliliter, or weight per liter, when the molecular weight is unknown. Throughout the text SI units are used where possible.

The three basic forms of concentration are as follows:

Weight per unit weight (w/w). Both solute and solvent are weighed, the total equaling 100 g.

Example. 5% w/w of NaCl contains 50 g of NaCl + 950 g of diluent.

Volume per unit volume (v/v). The volume of liquid solute per the total volume of the solute and solvent is expressed.

Example. 1% of HCl (v/v) contains 1 mL of HCl per 100 mL (or 1 dL) of solution.

Weight per unit volume (w/v). The most frequently used expression, concentrations of w/v are reported as grams percent (g%) or g/dL, as well as mg/dL and μg/dL. When percent concentration is expressed without specifying the form, it is assumed to be weight per unit volume. The use of weight percent to describe concentration is being discouraged by professional organizations and, with few exceptions, is not used in this book. With SI units this would be in terms of weight per microliters (μL), milliliters (mL), or liters (L).

Example. To prepare 100 mL of 100 g/L of NaCl, weigh 10 g of NaCl and dilute to volume in a 100 mL volumetric flask.

Molarity. Molarity expresses concentration as the number of moles per liter of solution. One mole is the molecular weight of the substance in grams. A millimole is 1/1000 of a mole. A molar solution contains one gram-molecular weight of a substance per liter.

$$\begin{aligned} 1 \text{ mol (mol)} &= 1000 \text{ mmol} \\ 1 \text{ mmol} &= 1000\ \mu\text{mol} \\ 1\ \mu\text{mol} &= 1000 \text{ nmol} \end{aligned}$$

Examples. 1 M NaOH (molecular weight, or MW = 40.0 g/mol) contains one gram-equivalent molecular weight per liter, or 40 g diluted to 1000 mL with distilled water. A millimolar (1 mM) solution, or 0.001 molar (0.001 M), contains 1 mmol/L. 1 millimole of NaOH is 1/1000 of 40 g, that is, 0.040 g (or 40 mg). When diluted to 1000 mL, the concentration of the solution will be 0.001 M.

Normality. Normality expresses concentration in terms of equivalent weights of substances. Equivalent weights are determined by the valence, which reflects the number of combining or replaceable units. A 1 normal (1 N) solution contains 1 equivalent weight per liter. The equivalent weight of an element or compound is equal to the molecular weight divided by the valence.

Normality and *molarity* relationships can be readily calculated if their definitions are understood.

Examples

1 M HCl = 1 N HCl, since 1 mole of H^+ or Cl^- reacts for every mole of HCl.

1 M H_2SO_4 = 2 N H_2SO_4, since 2 moles of H^+ (that is, equivalents) react for every mole of H_2SO_4.

1 M H_3PO_4 = 3 N H_3PO_4, since 3 moles of H^+ react for every mole of H_3PO_4.

1 M $CaCl_2$ = 2 N $CaCl_2$, since 2 Cl^- can react for every mole of $CaCl_2$.

1 M $CaSO_4$ = 2 N $CaSO_4$, since 2 mole volume electrons are available for reaction with either Ca^{++} or $SO_4^{=}$.

Equivalent weights are known as the number of grams of an element (or compound) that will react with another element (or compound). This so-called law of combining weights is operable for all chemical compounds.

To simplify chemistry procedures and reports, factors can be used to express a quantity of one compound as an equivalent quantity of another compound. This process can be termed *equivalency*.

Example. Calculate the amount of urea if a patient's urea nitrogen level is 800 mg/L. The formula for urea is NH_2—CO—NH_2, and its molecular weight is 60 g/mol. The molecular equivalent weight for nitrogen in the mole is 14 g/mol × 2 molecules = 28. The *urea/nitrogen factor* is determined by the following equation:

$$\frac{\text{MW of urea (60)}}{2 \times \text{MW of nitrogen (28)}} = \frac{x \text{ g of urea}}{1 \text{ g of urea nitrogen}}$$

$$28\,x = 60$$

$$x = 2.14 \text{ (factor)}$$

This factor states that 2.14 g of urea would equivalently represent 1 g of urea nitrogen. So 800 mg/L urea nitrogen × 2.14 equals 1712 mg/L of urea. Laboratory results today are reported as urea nitrogen, since historical methods for this particular test are based on measurement of the urea nitrogen.

Competent laboratory personnel should be able to convert mg/dL to mEq/L. Electrolyte equivalents can be calculated from the equation:

$$\text{mg/dL} \times 10 = \text{mg/L}$$

Since mg/mEq weight is the millimolar weight in milligrams divided by the valence

$$\frac{\text{mg/L}}{\text{mg/mEq}} = \text{mEq/L}$$

or

$$\frac{\text{mg/dL}}{\text{mg/mEq}} \times 10\frac{\text{dL}}{\text{L}} = \text{mEq/L}$$

Example. What is the mEq/L concentration of serum chloride reported as 250 mg/dL? Since the millimolecular weight of chloride is 35.5 (that is, 1 mmol = 35.5 mg), the milliequivalent weight of chloride is

$$\frac{\text{MW}}{\text{Valence}} = \frac{35.5 \text{ g/mole}}{1}$$

$$\frac{250 \text{ mg/dL}}{35.5 \text{ mg/mEq}} \times 10\frac{\text{dL}}{\text{L}} = 70 \text{ mEq/L}$$

Specific gravity. Specific gravity can be used to determine the mass (weight) of solutions. It relates the weight of 1 mL of the solution and the weight of an equal volume of pure water at 4° C (1 g). One practical use of specific gravity is in preparation of dilutions from concentrated commercial acids, the equation being:

Specific gravity × Percent assay =
Grams of compound per milliliter

Example. Concentrated HCl has a specific gravity of 1.25 g/mL and is assayed as being 38% HCl. What is the amount of HCl per milliliter?

$$1.25 \text{ g/mL} \times 0.38 = 0.475 \text{ g of HCl per mL}$$

One common error is neglecting to change the percent assay to its proper decimal; in the above example, 38% = 0.38!

Exercise

6. Using the above example, how many milliliters are needed to prepare 1 L of a 0.1 N HCl solution if the molecular weight of HCl = 36.5?

Water of hydration. Some salts are available in forms both anhydrous (no water) and hydrated (with water molecules). The form of the available salt, including the water of hydration, is listed on the manufacturer's label. To prepare accurate weight concentrations of these salts, calculations must include the molecules of water present in the compound. This is most easily done by calculation of the percentage of the compound that is in the anhydrous form. With this percentage, the weight of the hydrated form can be corrected to that of the anhydrous form.

The advantage of using molar concentrations is that the water of hydration does *not* have to be accounted for in the calculations. For example, 1 mol of $CuSO_4$ = 160 g, and 1 mol of $CuSO_4 \cdot 5\ H_2O$ = 250 g. One gram-equivalent molecular weight of each compound will contain 1 mol of $CuSO_4$; that is:

$$250 \text{ g } CuSO_4 \cdot 5\ H_2O = 1 \text{ mole of } CuSO_4 = 160 \text{ g of } CuSO_4$$

Example. How many grams of $MgCl_2$ are there in 1 g of $MgCl_2 \cdot 3\ H_2O$?

Mg	24	Mg	24
Cl_2	71	Cl_2	71
	95 MW	$3\ H_2O$	54
			149 MW

$$\frac{95}{149} = \frac{x}{1}$$
$$149\,x = 95$$
$$x = 63.7\%$$

One gram of $MgCl_2 \cdot 3\ H_2O$ contains 0.637 g of $MgCl_2$.

Mole fraction. Mole fraction refers to the ratio of the amount of a component to the total mixture of components. Mole fraction is a derived unit that is expressed as either a percent or as a decimal.

Example. What percent Mg is contained in $MgCl_2 \cdot 3\ H_2O$?

$$Mg = 24$$
$$Cl = 35.5$$
$$MgCl_2 \cdot 3\ H_2O = 149$$
$$\frac{24}{149} = 16.1\%$$

Mg is 16.1% of the molecule $MgCl_2 \cdot 3\ H_2O$.

Example. To determine mole percent calcium in calcium carbonate (MW = 100), 1 mole of $CaCO_3$ contains 100 g, comprising 40 (Ca) + 12 (C) + 48 (3 × O), of which 40 is calcium. The mole fraction of calcium in 1 L of 1 mole of calcium carbonate equals 40%.

Example. How much $CuSO_4 \cdot 5\ H_2O$ must be weighed to prepare 1 L of a solution containing 80 mg of $CuSO_4$?

$$\text{Total MW of } CuSO_4 \cdot 5\ H_2O = 250$$
$$\text{MW of } CuSO_4 = 160$$

The proportion of $CuSO_4 \cdot 5\ H_2O$ that is $CuSO_4$ is 160/250 = 0.64. Thus, 1 g of $CuSO_4 \cdot 5\ H_2O$ contains 1 g × 0.64 = 0.64 g of $CuSO_4$. The rest, 0.36 g, is water. Therefore

$$\frac{80}{0.64} = 125 \text{ mg of } CuSO_4 \cdot 5\ H_2O$$

Examples of calculations

a. What is the normality of concentrated HCl that has a specific gravity of 1.19 g/mL and a 38% assay?

$$\text{Specific gravity} \times \text{Percent} = \text{Grams/milliliter}$$
$$1.19 \times 0.38 = 0.452 \text{ g/mL} = 452 \text{ g/L}$$
$$36.5 \text{ g of HCl/L} = 1 \text{ N}$$
$$\frac{452 \text{ g/L}}{36.5 \text{ g/Eq}} = 12.4 \text{ Eq/L}$$

b. If 24.5 g of H_2SO_4 (MW = 98 g/mol) are dissolved in a 1 L solution
(1) What is its molarity?
(2) What is its normality?
(Answer 1) 1 mol of H_2SO_4 = 98 g; therefore 24.5 g equals

$$\frac{1 \text{ mol}}{98 \text{ g}} = \frac{x}{24.5 \text{ g}}$$
$$x = 0.25 \text{ mol}$$
$$0.25 \text{ mol in } 1 \text{ L} = 0.25 \text{ mol/L} = 0.25 \text{ M}$$

(Answer 2) The valence of H_2SO_4 equals 2; therefore the equivalent weight of H_2SO_4 is expressed as follows:

$$\text{Equivalent weight} = \frac{\text{Molecular weight}}{\text{Valence}} = \frac{98 \text{ g}}{2}$$
$$\text{Equivalent weight} = 49 \text{ g}$$

To solve for number of equivalents in 24.5 g

$$\frac{1 \text{ equivalent}}{49 \text{ g}} = \frac{x}{24.5 \text{ g}}$$
$$x = 0.5 \text{ equivalent}$$
$$0.5 \text{ equivalent in } 1 \text{ L} = 0.5 \text{ Eq/L} = 0.5 \text{ N}$$

c. The molecular weight of $CaCO_3$ is 100 g/mol and the atomic weight of calcium is 40 g/mol; how many grams are needed to prepare:
(1) 1 L of 0.1 M $CaCO_3$?
(2) 100 mL of 100 mg/dL of Ca^{++} using $CaCO_3$?
(3) 50 mg/L of $CaCO_3$?
(4) 0.2 mEq/L of Ca^{++} using $CaCO_3$?

(Answer 1) 1 mol of $CaCO_3$ = 100 g; therefore 0.1 mol = 10 g, since 0.1 molar = 0.1 mol/L = 10 g/L

(Answer 2) The percentage weight Ca^{++} in $CaCO_3$ is

$$\frac{\text{Atomic weight } Ca^{++}}{\text{Molecular weight } CaCO_3} = \frac{40}{100} = 40\%$$

In 100 ml, 10 mg of Ca^{++} is needed or

$$\frac{10 \text{ mg}}{\% \text{ Ca in } CaCO_3} = \frac{10 \text{ mg}}{40\%} = \frac{10 \text{ mg}}{0.4} = 25 \text{ mg of } CaCO_3$$

Therefore 25 mg of $CaCO_3$ = 10 mg Ca^{++}

(Answer 3) For 1 L, 50 mg of $CaCo_3$ is needed.

(Answer 4) 1 equivalent weight of $CaCo_3$ equals

$$\frac{\text{Molecular weight of } CaCO_3}{\text{Valence}} = \frac{100 \text{ g/mol}}{2 \text{ equivalents/mol}} = 50 \text{ g/equivalent}$$

To convert to milliequivalents

$$1 \text{ mEq} = \frac{1 \text{ Eq}}{1000}$$

$$\therefore \frac{50 \text{ g}}{1 \text{ Eq}} \times \frac{1 \text{ Eq}}{1000 \text{ mEq}} = 50 \text{ mg/mEq}$$

To calculate amount to prepare 1 L of 0.2 mEq

$$\frac{1 \text{ mEq wt}}{50 \text{ mg}} = \frac{0.2 \text{ mEq wt}}{x}$$

$$x = 10 \text{ mg of } CaCo_3$$

Exercises

7. 3 M $CaCl_2$ (MW 111.1) = ____ N $CaCl_2$

8. 2 H_3PO_4 (MW 98) = ____ M H_3PO_4

9. 2 M H_2SO_4 (MW 98) = ____ N H_2SO_4

10. 250 mL of 5% NaCl contains ____ g of NaCl

11. How much $CuSO_4 \cdot 5\ H_2O$ must be weighed to prepare 100 mL of 5% $CuSO_4$? (MW $CuSO_4$ = 159.61; MW H_2O = 18)

12. What percent of $CuSO_4 \cdot 5\ H_2O$ is water? ____ %

CALCULATIONS BASED ON PHOTOMETRIC MEASUREMENTS (BEER'S LAW)

Refer to Chapter 3 for a description of the relationship between percent transmittance, absorbance, and concentration.

Colorimetry

Colorimetry is the measurement of the kind and amount of light absorbed or transmitted by a solution. These measurements of absorbance or transmittance are logarithmically related. Beer's law (see below) reflects the relationships between the absorbance and concentration of a known standard solution with that of solutions with unknown concentrations, the patients' samples. Beer's law states that the absorbance of a solution is directly related to its concentration. If Beer's law is true, then:

$$C_u = \frac{A_u}{A_s} \times C_s \qquad \textit{Eq. 1-5}$$

C_u and A_u represent concentration and absorbance of the unknown samples, whereas C_s and A_s reflect that of the standard solution. When preparing a colorimetric method for clinical chemistry analysis, one must be sure that Beer's law is followed or this formula cannot be used. In other words, this formula can only be used if the absorbance and concentration are directly related, that is, if the absorbance doubles with a doubling of concentration. A *standard curve* can be used to determine graphically concentration values. Standard-curve preparations are described on p. 36.

Absorbance and transmittance

Absorbance measures the amount of light that is blocked, or absorbed by a solution. Absorbance is also termed "optical density" (OD), a term found in the older literature and not in common use today.

Transmittance measures the amount of light that passes through a solution. The transmittance is usually expressed as a percentage, or *%T*. The %T scale is linear, as noted on a colorimeter readout scale.

As discussed in Chapter 3, the absorbance and percent transmittance are logarithmically related, since absorbance is a logarithmic function. Interconversion of the absorbance and percent transmittance is commonly expressed by the following formula:

$$A = -\log \frac{\%T}{100} \qquad \textit{Eq. 1-6}$$

This equation can be algebraically converted to the following form:

$$A = -(\log \%T - \log 100)$$
$$A = -(\log \%T - 2)$$
$$A = -\log \%T + 2$$
$$A = 2 - \log \%T \qquad \textit{Eq. 1-7}$$

One can obtain absorbance from a hand calculator using this formula by punching in the numbers for %T, converting to the log form, placing a minus sign, and adding 2.

Examples. Determining concentrations using absorbance (A, or OD) readings.

a. If absorbance of an unknown is 0.25 and the concentration of a standard is 4 mg/L with an absorbancy of 0.40, the concentration of the unknown can be calculated using:

$$C_u = \frac{A_u}{A_s} \times C_s$$
$$C_u = \frac{0.25}{0.40} \times 4 \text{ mg/L}$$
$$C_u = 2.5 \text{ mg/L}$$

b. To calculate the concentration of glucose if the following information is known:

$$C_s = 2000 \text{ mg/L}, A_s = 0.40, A_u = 0.25$$

Using the same formula above

$$C_u = \frac{0.25}{0.40} \times 2000 \text{ mg/L}$$
$$C_u = 1250 \text{ mg/L}$$

c. To calculate glucose concentration of unknown (C_u) if the %T is given.
If the 1000 mg/L glucose standard (C_s) reads 49% T, and the unknown reads 55% T, the %T must be converted to absorbance, since absorbance is linearly proportional to concentration:

$$A_s = 2 - \log \%T \qquad A_u = 2 - \log \%T$$
$$A_s = 2 - 1.690 \qquad A_u = 2 - 1.740$$
$$A_s = 0.31 \qquad A_u = 0.26$$
$$49\% \text{ T} = 0.31\ A = A_s$$
$$55\% \text{ T} = 0.26\ A = A_u$$
$$C_s = 1000 \text{ mg/L}, A_u = 0.26, A_s = 0.31$$

Using the formula as in (a) and (b) above:

$$C_u = \frac{0.26}{0.31} \times 1000 \text{ mg/L} = 839 \text{ mg/L}$$

Molar extinction coefficient

Molar extinction coefficients are used in the clinical laboratory to calculate concentrations and activities of enzymes in international units (U) and to determine the purity of dissolved substances. Specific applications are checking standard solutions, such as hemoglobin or bilirubin. The molar extinction coefficient, or molar absorbance coefficient, or ϵ, is defined as the absorbance at a given wavelength of a 1 M solution of the substance in a 1 cm cuvette at 25° C. It is related to absorbance by the formula

$$A = \epsilon cl \qquad \textit{Eq. 1-8}$$

where A = absorbance at a specified wavelength, c = concentration of substance being measured, in moles/L, and l = the path length in cm.

A suitable bilirubin standard, as a 1 M solution in chloroform, would have a theoretical absorbance of 60,700 (mean) ± 800 liters · moles^{-1} · cm^{-1} at 453 nm, when measured in a 1 cm cuvette at 25° C. Logical reasoning suggests that if this standard were diluted to 1:60,700 the absorbance would read 1.

Example. 1 M bilirubin standard is diluted to 1:60,700 and then 1:2, with the final dilution being 1:121,400. The absorbance of this dilution reads 0.495 nm in a 1 cm cuvette. What is the extinction coefficient of this bilirubin standard?

$$\epsilon = \frac{0.495}{1 \text{ mol/L}/121{,}400(1 \text{ cm})}$$
$$\epsilon = 60{,}093 \text{ liters} \cdot \text{mol}^{-1} \cdot \text{cm}^{-1}$$

A major application of ϵ is the measurement of concentrations of substances. If the ϵ of a substance is known and a 1 cm cuvette is used, Beer's law formula is simplified to the following:

$$c = \frac{A}{\epsilon}$$

Example. The ϵ of NADH at 340 nm is 6.22×10^3 liters · mol^{-1} · cm^{-1}. If the absorbance of NADH at 340 nm reads 0.350, what is the concentration?

$$c = \frac{0.350}{6.22 \times 10^3 \text{ L} \cdot \text{mol}^{-1}}$$
$$c = 5.6 \times 10^{-5} \text{ mol/L}$$

Exercises

13. NADH has a molar absorptivity of 3.3×10^3 at a wavelength of 366 nm. Calculate the concentration of a solution that has an A (absorbance) of 0.175 at 366 nm.

14. A chemistry technologist is checking a bilirubin standard. What would be the molar absorptivity of a 1 M solution diluted to 1:60,700 and reading 0.70 in a 7 mm cuvette?

15. The chemistry technologist has a solution of NADH with a concentration of 0.05×10^{-3} mol/L. Calculate the molar absorptivity if it measures 0.300 at 334 nm.

BUFFERS

Buffers resist changes in acidity by forming a weakly ionized acid or base with the added H^+ or OH^- ions. For example, when HCl is added to a solution of Na^+Ac^- (sodium acetate) plus H^+Ac^- (acetic acid), the H^+ of HCl will react with the Ac^- forming more HAc, which is only slightly ionized. The acetate-acetic acid effectively buffers by removing H^+ from the solution.

The Henderson-Hasselbalch equation is used to express acid-base relationships. There are several forms of this equation that will not be delineated at this time but can be used for calculating acid-base problems. The simplest equation is

$$\text{pH} = \text{pK} + \log \frac{\text{Concentration of conjugate base}}{\text{Concentration of weak acid}} \qquad \textit{Eq. 1-9}$$

The pK value depends on a specific set of conditions; these are degree of dissociation, temperature, and pH. The pK for the bicarbonate buffer system in serum or plasma is 6.10 at 37° C. Chemical reference books, such as the *Handbook of Chemistry and Physics,** contain pK values. As capable medical technologists we should grasp the basic calculations of the Henderson-Hasselbalch equations, even though laboratory instruments provide direct "readout" values on patients' acid-base tests.

*Weast RC, editor: Handbook of chemistry and physics, Cleveland, Ohio, CRC Press, yearly updated editions.

Examples

a. Calculate the pH of an acetate buffer composed of 0.20 M sodium and 0.05 M acetic acid. (The pK for acetic acid is 4.76.)

$$\begin{aligned} pH &= pK + \log \frac{[Salt]}{[Acid]} \\ &= 4.76 + \log \frac{0.20}{0.05} \\ &= 4.76 + \log 4 \\ &= 5.3621 \\ pH &= 5.36 \end{aligned}$$

b. Now for a complicated example. Prepare an acetate buffer whose concentration is 0.2 M and has a pH of 5.0. (The pK of acetic is 4.76; the molecular weight of acetic acid is 60 and is 82 for sodium acetate.)

$$pH = 4.76 + \log \frac{[Salt]}{[Acid]}$$

$$\begin{aligned} \log \frac{[Salt]}{[Acid]} &= 5.0 - 4.76 \\ [Salt]/[Acid] &= \text{antilog } 0.24 \\ [Salt]/[Acid] &= 1.7 \end{aligned}$$

The number 1.7 is the ratio of the moles per liter of salt to the moles per liter of acid. Any molar concentrations of salt to acid yielding a ratio of 1.7 will result in a 5.0 pH acetate buffer.

Note. The problem specifies a concentration of 0.2 M solution, or $HAc + Ac^- = 0.2$ M.

If

$$Ac^-/HAc = 1.7$$

then

$$Ac^- = 1.7 \text{ HAc}$$

or

$$HAc + 1.7 \text{ HAc} = 0.2 \text{ M}$$

and

$$\begin{aligned} 2.7 \text{ HAc} &= 0.2 \\ HAc &= 0.074 \text{ mol/L of acid} \\ MW \times M &= g/L \\ 60 \times 0.074 &= 4.44 \text{ g/L of acid needed} \end{aligned}$$

To calculate the weight of salt needed, use:

$$\begin{aligned} \text{Moles/liter of salt} &= \text{Total moles} - \text{Moles/liter of acid} \\ &= 0.2 - 0.074 \\ &= 0.126 \text{ moles of salt/L} \end{aligned}$$

As done for the acid:

$$\begin{aligned} \text{MW of salt} \times M &= \text{Grams of salt/liter} \\ 82 \times 0.126 &= 10.33 \text{ g of salt/L} \end{aligned}$$

When 4.44 g of acid and 10.33 of salt are dissolved in a total volume of 1 L, the resulting buffer concentration is 0.2 M at a pH of 5.0.

ENZYME CALCULATIONS

Expressing enzyme activity in international units has been generally accepted since its recommendation by the International Union of Biochemistry in the early 1960s. One international unit, U, of an enzyme is defined as the amount that will catalyze the transformation of 1 μmole of the substrate per minute under standard conditions. Activity is expressed in terms of enzyme units per liter of serum, or milliunits per milliliter, in the following relationship:

$$1 \text{ U/L} = \mu\text{mol/minute/liter of serum}$$

Explanation of the basic equation for the conversion of absorbance data to international units will not be attempted in this portion of laboratory calculation. Suffice it to state that any change in factors such as temperature or volume must be accounted for in the following basic equation (see Chapter 49):

$$U/L = \frac{\Delta A/\text{min} \times V_t \times 10^6 \ (\mu\text{mol/mol})}{\epsilon \times V_s \times l} \qquad \textit{Eq. 1-10}$$

ΔA/min = Absorbance change per minute
V_t = Total reaction volume including sample, reagent, and diluent
V_s = Serum volume
l = Cuvette path length
ϵ = Extinction coefficient

The factor of 10^6 μmol/mol is added to convert the answer to μmol/min/L (U/L).

Example. What is the lactate dehydrogenase (LD) activity of 0.1 ml of serum + 3 mL of substrate if the NADH being formed showed a 0.002 ΔA/min at 340 nm? ϵ for NADH = $6.22 \times 10^3 \text{ L} \cdot \text{mol}^{-1} \cdot \text{cm}^{-1}$.

Using the above formula:

$$U/L = \frac{0.002 \ (10^6 \ \mu\text{mol/mol}) \ (3.1 \text{ mL})}{1 \text{ min} \left(6.22 \times 10^3 \dfrac{\text{L}}{\text{mol} \cdot \text{cm}}\right) (0.1 \text{ mL}) \ (1 \text{ cm})}$$

$$U/L = 9.9$$

Example. Calculation of international units per liter of alkaline phosphatase activity using *p*-nitrophenol standard requires attention to all factors of the formula:

$$U/L = \frac{\Delta A/\text{min} \times V_t \times 10^6 \ (\mu\text{mol/mol})}{\epsilon \times V_s \times l}$$

If ΔA of sample = 0.070, the ϵ for *p*-nitrophenol is 50,000 L/mol · cm; timing = 15 min; V_t = 5.5 mL; V_s = 0.005 mL.

$$U/L = \frac{0.070 \left(10^6 \dfrac{\mu\text{mol}}{\text{mol}}\right) (5.5 \text{ mL})}{15 \text{ min} \left(50{,}000 \dfrac{\text{L}}{\text{mol} \cdot \text{cm}}\right) (0.005 \text{ mL}) \ (1 \text{ cm})}$$

$$U/L = \frac{0.070 \times 5.5 \times 1000}{50 \times 0.075}$$

$$100 \text{ U/L} = \text{Alkaline phosphatase activity}$$

STANDARD CURVES

Preparation of standard curves on graph paper is an essential way of examining data for validity. Often calculations or computers do not reveal abnormalities of the system, but calculate averages of results. Therefore graphing of data is a very important way to validate assays.

Previously in this chapter, Beer's law was defined as the direct relationship of the absorbance and concentration of a solution. This means that if a 2% solution reads 0.1 *A*, then a 4% concentration will read 0.2 *A*, and an 8% solution will read 0.4 *A*. To repeat, most solutions obey Beer's law; that is, concentration and absorbance are directly proportional only over specified ranges of concentrations.

Graphs

Fig. 1-15. Absorbances of glucose standard concentrations plotted on linear paper result in a straight line, confirming that Beer's law is followed for the concentrations up to 3000 mg/L.

Fig. 1-16. Plotting the %T values of the same glucose concentrations used in Fig. 1-18 on linear paper produces a semicurved line; %T values are *not* linear absorbance versus concentrations. (Recall the logarithmic relationship of absorbance and %T.)

Fig 1-17. Plotting %T values on semilog paper results in a straight line, which can be used to interpolate the concentrations of glucose from %T values.

Exercises (Fig. 1-15 and 1-17)

Find glucose concentrations for the following readings using Fig. 1-15 and 1-17.

16. 0.3 *A* = ___ mg/L
17. 0.39 *A* = ___ mg/L
18. 49% *T* = ___ mg/L
19. 52% *T* = ___ mg/L

Exercises using one known standard value to determine concentrations of unknowns

What are the glucose concentrations of the following patients' samples if the 2000 mg/L standard reads 0.32 *A?*

We can employ either:

$$C_u = \frac{A_u}{A_s} \times C_s$$

or

$$\frac{C_u}{C_s} = \frac{A_u}{A_s}$$

20. 0.22 *A* = ___ mg/L
21. 0.14 *A* = ___ mg/L
22. 0.46 *A* = ___ mg/L

Renal clearance test calculations

Renal clearance tests are used to assess kidney function. Renal clearance is a rate measurement that expresses the volume of blood cleared of the substance being studied (typically creatinine or urea) per unit of time. Therefore the unit for the clearance test is milliliters per minute.

To calculate creatinine clearance, the following information is required:

Serum concentration (S)

Urine concentration (U) (*Caution*: the serum and urine

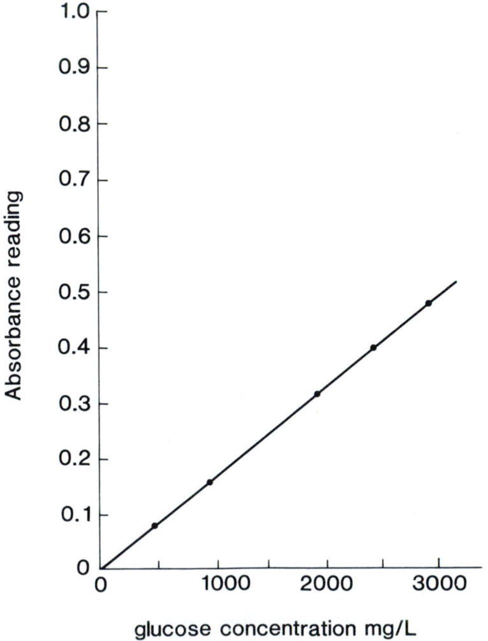

Fig. 1-15 Standard curve for glucose analysis: absorbance versus concentration on linear-linear graph paper.

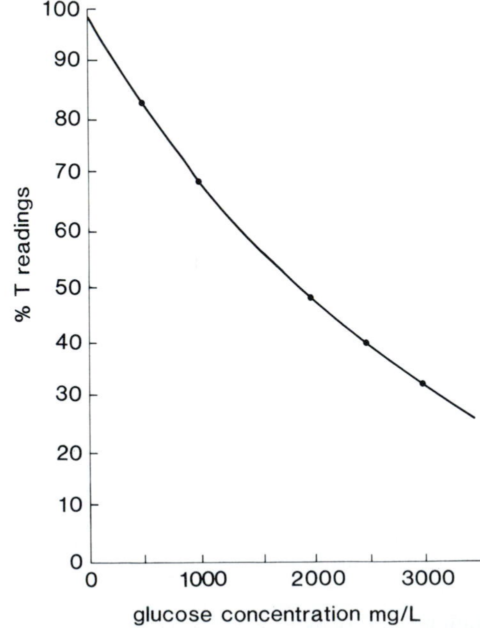

Fig. 1-16 Standard curve for glucose analysis: percent transmittance (%T) versus concentration on linear-linear graph paper.

concentrations may not be in the same units, e.g., mg/L or mg/dL.)

Volume of urine excreted per minute (V) (Volume of urine collected divided by time period in minutes)

$$\frac{\text{Clearance}}{(\text{uncorrected})} = \frac{U \times V}{S}$$

The calculation above does not account for the patient's body surface area. If the physician requests a corrected value, the equation must be multiplied by 1.73/A, where 1.73 equals the average body surface area in square meters and A equals the patient's body surface area.

$$\frac{\text{Clearance}}{(\text{corrected})} = \frac{U \times V}{S} \times \frac{1.73}{A}$$

The patient body surface area is computed from a nomogram, using the patient height and weight or calculated using the following formula:

$$\log A = (0.425 \times \log W) + (0.725 \times \log H) - 2.144$$

where A is the body surface area in square meters; W is the patient's weight in kilograms; and H is the patient's height in centimeters.

Examples

a. Determine the uncorrected creatinine clearance for a patient with a serum creatinine of 25 mg/L. The urine creatinine was 500 mg/L and the urine volume was 312 mL/4 hr.

$$312 \text{ mL}/240 \text{ min} = 1.3 \text{ mL/min}$$

$$C = \frac{U}{S} \times V = \frac{500}{25} \times 1.3 = 26 \text{ mL/min}$$

b. Calculate the corrected creatinine clearance for a child who weighed 22.7 kg, was 95 cm long, and passed 500 mL of urine during a 24-hour period. Serum and urine creatinine values were 0.14 mmol/L and 4.75 mmol/L respectively.

$$\begin{aligned}\log A &= (0.425 \times \log 22.7) \\ &\quad + (0.725 \times \log 95) \\ &\quad - 2.144 \\ \log A &= 0.576 + 1.434 - 2.144 = 0.134 \\ A &= 0.734 \\ 500 \text{ mL}/1440 \text{ min} &= 0.35 \text{ mL/min} \\ C &= \frac{U}{X} \times V \times \frac{1.73}{A} \\ &= \frac{4.75}{0.14} \times 0.35 \times \frac{1.73}{0.734} \\ &= 2.80 \text{ mL/min}\end{aligned}$$

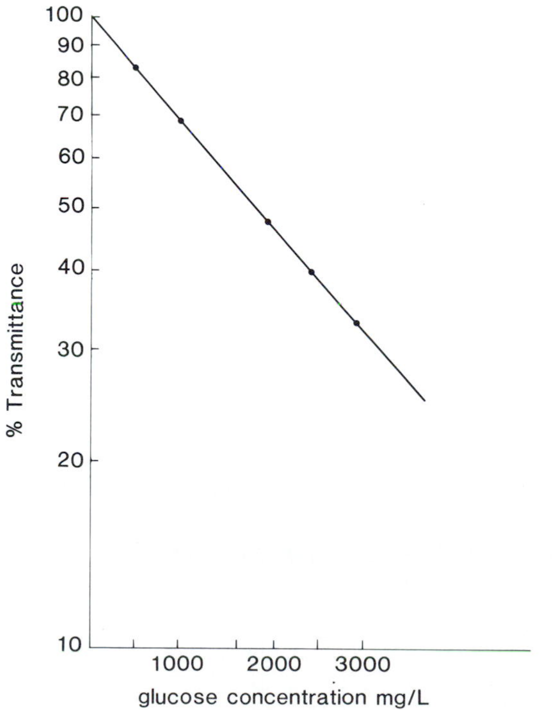

Fig. 1-17 Standard curve for glucose analysis: %T versus concentration on log-linear graph paper.

Timed urine tests

Results of urine tests may be reported in several ways. Values can be reported as concentration units, quantity per total volume, and quantity per unit of time. The clinical usefulness of urine tests is increased when quantitative results are expressed as amount per total volume or amount excreted in a given time period. Good laboratory practice dictates that the urine total volume and the beginning and end time period of collection be recorded.

When urine test results are reported as quantity per total volume, the concentration is measured and the results are corrected as follows:

$$\frac{\text{Amount}}{\text{Total volume}} = \text{Measured concentration} \times \text{Total volume}$$

This calculation requires that both the urine volume and the instrument concentration be reported using the same units.

Examples

a. Calculate the amount of protein in 2400 mL of urine having a concentration of 30 mg/dL.

$$\frac{\text{Amount}}{\text{Total volume}} = \frac{30 \text{ mg}}{100 \text{ mL}} \times 2400 \text{ mL} = 720 \text{ mg}$$

b. Determine the amount of sodium in 1800 mL of urine with a sodium concentration of 35 mEq/L.

$$\frac{\text{Amount}}{\text{Total volume}} = \frac{35 \text{ mEq}}{1000 \text{ mL}} \times 1800 \text{ mL} = 63 \text{ mEq}$$

When results are reported as quantity per time period of collection, the calculation of amount of substance per total volume must be performed first and the result reported for the length of time instead of for total volume.

ADDITIONAL EXERCISES

23. The hemoglobin standard solution contains 200 g/L. What amounts would be used to prepare 6 mL of the following concentrations?

200 g/L? 150 g/L? 100 g/L? 50 g/L?

24. a. What fraction of urea is nitrogen? Urea is $CO(NH_2)_2$.

Atomic weights:
C = 12
O = 16
N = 14
H = 1

b. What percent of urea is nitrogen?

25. There is available concentrated HCl having a 38% assay and a specific gravity of 1.170.
a. What is the weight of HCl present in 1 mL?
b. For the preparation of 100 mL of 10% wt/vol HCl, ____ mL of HCl would be diluted to total volume of ____ mL.

26. Normal saline solution is 0.85% concentration. What is its molarity? (NaCl molecular weight is 58.5)

27. If a protein standard reads 0.48 *A*, and a patient's sample reads 0.36 *A*, what is the patient's protein concentration? Select one of the following answers:
a. Twice the standard concentration
b. Equal to the standard concentration
c. Three fourths of the standard concentration
d. Not enough data for calculation

28. A patient in diabetic coma has high blood glucose levels; so serum from this patient is diluted 1:2 and again 1:2 before it is readable from the glucose chart as 1900 mg/L. What is the actual concentration in (a) mg/L; (b) mg/dL, and (c) g/L?

29. How many mEq/L are there in a solution containing 27.7 mg/dL of potassium? (Atomic weight of K = 39.)

30. How many milliliters of 0.4 N NaOH can be made from 20 mL of 2 N solution?

31. What is the normality of a solution containing 40 mEq per 50 mL?

32. What is the dilution of serum in a tube containing 200 μL of serum, 500 μL of saline, and 300 μL of reagent?

33. Calculate the alkaline phosphatase activity in U/L for the following:

ϵ of standard (*p*-nitrophenol) = 5×10^4 L · mol^{-1} · cm^{-1}
ΔA_{sample} = 0.150
V_{sample} = 0.2 mL
V_{total} = 2.2 mL
Timing = 15 minutes

34. What is the extinction coefficient for the following:

Solution concentration = 1.2 molar
Dilution of solution = 1/121,400
A reading in a 1 cm cuvette = 0.6

35. A 0.01 M Na_2HPO_4 solution needs to be prepared (MW of Na_2HPO_4 = 141.98). Only the hydrated salt, $Na_2HPO_4 \cdot 7\,H_2O$ (MW = 267.98) is available. How many grams will be needed to prepare 250 mL?

36. If a 50 mg/mL solution of Na_2HPO_4 needs to be prepared, how many grams of the hydrated salt need be weighed to make a 1 L solution?

37. A medical technologist desires to prepare 50 mL of a 10 mg/mL solution of NADH (MW = 663.44). To do this accurately, the technologist will first prepare a stock solution containing approximately 50 mg/mL. The absorbance of a 1:1000 dilution of the stock solution will be measured at 340 nm and from the known molar absorbance of NADH at this wavelength (6.22×10^3), the actual concentration will be calculated. A suitable dilution will then be made to prepare the 500 mL of the desired 10 mg/mL solution. Presume that the technologist, following these directions, has prepared a dilution of the stock solution with an absorbance of 0.562. Calculate the concentration of NADH in this stock solution as mmol/L and mg/L. What dilution should be made to prepare the 10 mg/mL of NADH solution?

38. A patient's serum calcium level is 3.5 mEq/L. The expected normal range is 90 to 110 mg/L. Is the patient's calcium level lower than, within, or higher than the expected normal range?

39. A sodium concentration is reported as 3500 mg/L. What is the concentration in mEq/L? (atomic wt. of sodium = eq. wt. = 23 g/mol or 23 mg/mmol)

40. If the cyanmethemoglobin standard, with a concentration of 200 g/L, reads 0.426 *A*, and a patient's blood sample reads 0.297 *A*, what is the concentration of hemoglobin in the sample?

41. A 200 mg/L urea nitrogen standard reads 0.30 *A* and a patient's sample reads 0.40 *A*. The concentration of the standard compared to the patient's level is:
a. Higher
b. Twice as much
c. 3/4 as much
d. 4/3 as much

42. A glucose standard of 2000 mg/L reads 0.4 *A* and a patient's sample reads 1.0 *A*. The technologist should:
a. Report result as 500 mg/dL.
b. Repeat test before reporting.
c. Repeat test on diluted sample.
d. Prepare fresh glucose standard.

APPENDIX: ANSWERS TO PROBLEMS

1. 0.5 M
2. 0.4 M
3. 50 mg/L
4. 1:320
5. 1:128
6. 7.68 mL
7. 6
8. 0.667
9. 4

10. 12.5
11. 7.82
12. 36.05
13. $c = 53 \times 10^{-6}$ mol/L
14. $\epsilon = 60{,}700\ L \cdot mol^{-1} \cdot cm^{-1}$
15. $6.0 \times 10^3\ L \cdot mol^{-1} \cdot cm^{-1}$
16. 1900 mg/L
17. 2400 mg/L
18. 1920 mg/L
19. 1740 mg/L
20. 1375 mg/L
21. 870 mg/L
22. 2870 mg/L
23. 200 g/L = 6 mL + 0 mL of diluent
150 g/L = 4.5 mL + 1.5 mL of diluent
100 g/L = 3.0 mL + 3 mL of diluent
50 g/L = 1.5 mL + 4.5 mL of diluent
24. **a.** 28/60 = 7/15
b. 46.6%
25. **a.** 0.445 g
b. 22.5 mL will be diluted to 100 mL
26. 0.145 mol/L
27. c
28. **a.** 7600 mg/L
b. 760 mg/dL
c. 7.6 g/L
29. 7.1 mEq/L
30. 100 mL
31. 0.8 N
32. 1:5
33. 2.2 U/L
34. 6.07×10^4
35. 0.67 g
36. 94.4 g
37. Concentration of NADH in stock solution is 90.4 mmol/L or 60 mg/mL. Take 8.3 mL of the stock and dilute to 50 mL to prepare the 10 mg/mL of solution.
38. 70 mg/L: lower than the expected range
39. 152 mEq/L
40. 139 g/L
41. 3/4 as much
42. c

REFERENCES

1. Lashor, TW, and Macurdy, LB: Precision laboratory standards of mass and laboratory weights, National Bureau of Standards Circular 547, Washington, DC, 1954, United States Department of Commerce.
2. Laboratory instrument maintenance manual, ed. 3, Skokie, Ill., 1982, College of American Pathologists.
3. Statement from Quadrennial International Symposium on Measurable Properties (Quantities) and Units in Clinical Pathology and Clinical Chemistry, Gaithersburg, Md., August 5 and 6, 1976, Am J Clin Pathol 71:465-468, 1979.
4. The National Committee for Clinical Laboratory Standards Position Paper (PPC-11): Quantities and units (SI), Clin Chem 25:657-658, 1979.
5. Committee on Hospital Care, American Academy of Pediatrics: Metrication and SI units, Pediatrics 65:659-664, 1980.
6. Preventive Maintenance for Precision Balances, Engineering Service, Veterans Administration, Dept. of Medicine and Surgery, G-29, Part 56, May 1974, DM & S Suppl. MP-3.
7. Guide to the selection of accuracy classes of thermistor thermometers and the verification of their accuracy, NCCLS Proposed Standard, PS1-4, Villanova, Pa., 1979, National Committee for Clinical Laboratory Standards.
8. Mangum, BW: Standard reference materials 933 and 934: The National Bureau of Standards' precision thermometers for the clinical laboratory, Clin Chem 20:670-672, 1974.
9. Standard for temperature calibration of water baths, instruments and temperature sensors: NCCLS Approved Standard, AS1-2, Villanova, Pa., 1977, National Committee for Clinical Laboratory Standards.
10. Bowers, GN, Jr, and Inman, SR: The gallium melting-point standard: its application and evaluation for temperature measurement in the clinical laboratory, Clin Chem 23:733-737, 1977.
11. Bowie, L, Esters, F, Bolin, J, and Gochman, N: Development of an aqueous temperature-indicating technique and its application to clinical laboratory instrumentation, Clin Chem 22:449-455, 1976.
12. Preventive Maintenance for Centrifuges, Engineering Service, Veterans Administration, Dept. of Medicine and Surgery, G-29, Part 16, Dec. 1973, DM & S Suppl. MP-3.
13. Specifications for reagent water used in the clinical laboratory, NCCLS Approved Standard, ASC-3, Villanova, Pa., 1980, National Committee for Clinical Laboratory Standards.
14. Commission on Laboratory Inspection and Accreditation, Reagent water specifications, Skokie, Ill., 1978, College of American Pathologists.
15. LeFever, D, and Martin, BG: Hazardous chemicals and employees' right to know, MLO, Aug. 1987, pp. 53-55.
16. Federal Hazard Communication Standard, 29 CFR1910.1200, Occupational Safety and Health Administration, Nov. 25, 1983.
17. U.S. Dept. of Health and Human Services: Recommendations for prevention of HIV transmission in health care settings, MMW Report, Aug. 21, 1987, Vol. 36, No. 2S.
18. Merck Sharp and Dohme: Prescribing information Recombivax HB, West Point, Pa, June 1986, Merck & Co., Inc.

BIBLIOGRAPHY

Alexander, JJ, and Steffel, MJ: Chemistry in the laboratory, New York, 1976, Harcourt Brace Jovanovich, Inc.

Becan-McBride, K: Textbook of clinical laboratory supervision, New York, 1982, Appleton-Century-Crofts.

Brewer, JM, Pesce, AJ, and Ashworth, RB: Experimental techniques in biochemistry, Englewood Cliffs, NJ, 1974, Prentice-Hall, Inc.

Campbell, JB, and Campbell, JB: Laboratory mathematics: medical and biological applications, ed. 3, St. Louis, 1984, The CV Mosby Co.

Christiansen, HE, and Lugibyhl, TT, editors: Suspected carcinogens, Dept. of Health, Education and Welfare, Washington, DC, (NIOSH) 75-188, 1975, US Government Printing Office.

Dharan, M: Total quality control in the clinical laboratory, St. Louis, 1977, The CV Mosby Co.

Diehl, H: Quantitative analysis: elementary principles and practice, Ames, Iowa, 1974, Oakland Street Sciences Press.

Duckworth, JK, and Von Boechman, J: In Commission on Inspection and Accreditation: Clinical Laboratory Improvement seminar, Skokie, Ill., 1980, College of American Pathologists.

Flury, PA: Environmental health and safety in the hospital laboratory, Springfield, Ill., 1978, Charles C Thomas, Publisher.

Halper, HR, and Foster, HS: Laboratory regulation manual, vol. 3, Germantown, Md., 1970, Aspen Systems Corp.

Handbook for analytical quality control in water and wastewater laboratories, United States Environmental Protection Agency, EPA-600/4-79-019, March 1979, Chapter 4.

Ibbot, FA: Micro and ultramicro technic and equipment. In Henry, RJ, Cannon, DC, and Winkleman, JW, editors: Clinical chemistry principles and techniques, ed. 2, Hagerstown, Md., 1974, Harper & Row, Publishers.

Microbiological methods for monitoring the environment, United States Environmental Protection Agency, EPA-600/8-78-017, Dec. 1978, pp 32-37.

Mukherjee, KL: Introductory mathematics for the clinical laboratory, Chicago, 1979, American Society of Clinical Pathologists.

Richterich, R, and Colombo, JP: Clinical chemistry, New York, 1981, John Wiley & Sons, Inc.

CHAPTER 2

Collection and handling of patient specimens

NATHAN A. PICKARD

OBJECTIVES

- Describe appropriate collecting, processing, and storing methods for commonly collected laboratory specimens, summarizing the importance of each method.
- Describe how the following variables relate to the collection of patient specimens, provide examples of the analytes affected and the manner in which the results change:
  - Diurnal variation
  - Posture
  - Stasis
  - Hemolysis
  - Preservation
- List at least six conditions that should be included in laboratory criteria for rejection of blood samples and justify those reasons.
- Describe various types of blood collection tubes, including stopper color, the corresponding anticoagulant, the chemical basis of anticoagulation, the order in which each type should be used, and the clinical applications of each.
- Explain the reasons for, and the advantages or disadvantages of, using the horizontal-head centrifuge and the angle-head centrifuge.

KEY TERMS

accuracy The extent to which a measurement is close to the true value.
adsorption The attachment of one substance to the surface of another.
aerosol A mist produced by the atomization of a liquid.
anaerobic Occurring only in the absence of molecular oxygen.
analyte Any substance that is measured; usually applied to a component of a biological sample.
anticoagulant A substance that can suppress, delay, or prevent coagulation of blood.
calculus An abnormal inorganic mass occurring within the body, usually composed of mineral salts.
chromogen A substance that absorbs light, producing color.
clot Semisolid mass of coagulated blood cells and proteins.
coagulation Process by which blood cells are trapped within a net of precipitated blood proteins to form a clot.
diurnal Occurring during the day.
in vitro Within a glass; observable in a test tube.
in vivo Within the living body.
phlebotomy The opening of a vein for the collection of blood.
plasma The noncellular portion of blood.
relative centrifugal force (RCF) A method of comparing the forces generated by various centrifuges taking into consideration the speed of rotation and radius from the center of rotation.
serum The noncellular portion of blood from which fibrinogen and other clotting proteins have been removed.
stasis Stoppage of the flow of blood.

Quality control in the clinical laboratory begins before the sample is collected from the patient. Accuracy arises from ensuring that the appropriate specimen is collected, the correct collection vessel is used, and the pertinent collection variables are considered. Once the sample is collected and delivered to the laboratory, errors may arise during the period before the analysis of the sample. This chapter reviews these considerations, highlighting those areas where the most common or significant errors may arise.

Phlebotomy is a complex subject, taught as a distinct discipline. The material in this chapter will discuss only those aspects of phlebotomy that affect clinical chemistry analyses. Students are referred to references 1 through 3 for detailed descriptions of phlebotomy technique.

The problem of sample collection and processing is a complex one because there is no simple system that will meet the requirements of all the analytes that the clinical chemistry laboratory measures. Thus the proper method of collection and storage for a particular analyte may not be the correct method for another, and inappropriately collected specimens are a common problem. In addition, it is important to determine the *type* of specimen required for a particular analyte before one begins to obtain a specimen. For example, urine is an appropriate specimen for a toxic-drug screen, whereas a serum sample is not. It is a proper function of the laboratory to provide this information to both physicians and nurses.

The laboratory must communicate to hospital staff and to patients the proper method of collecting and storing urine and stool samples for clinical analysis. Instructions should be provided in writing and attached to all specimen-collection containers. In addition, any instructions or warnings concerning the effects of preservatives, the requirement for dietary restrictions, or other precautions should be clearly labeled.

TYPES OF SPECIMENS

Blood

Blood is, of course, a suspension of cells in a protein-salt matrix. The noncellular portion of blood is a fluid that contains a series of proteins, some of which are involved in the coagulation process. This fluid is called *plasma*. When the coagulation process is allowed to proceed to completion, the noncellular fluid, which can be separated from the clotted material, is termed *serum*.

Blood used for biochemical analysis is collected either from the veins, the arteries, or the capillaries. For most testing, the site of phlebotomy has no analytical or physiological significance, so that venous blood is used because of the ease of collection. For a limited number of analytes such as blood gases and lactic acid, significant differences arise between arterial and venous samples.[4]

Most testing is performed on the serum fraction of blood. The assumption is made that the distribution of constituents between the cellular and extracellular compartments of blood is roughly equal. Therefore one can extrapolate the concentration of an analyte in blood from that measured in serum. This assumption is usually valid.

For some analytes it is necessary to inhibit the blood-clotting process using an anticoagulant. Other analytes require the addition of a preservative for accurate results. A later section will describe the various anticoagulants and preservatives available and their mechanisms of action.

Urine

Urine is commonly analyzed by the clinical chemistry laboratory. Both timed (such as 24-hour) collections, where quantitative analyses of the amount of an analyte excreted in urine over a period of time are desired, and random or spot collections, where the concentration of the analyte in the random sample is desired, are used. Just as in the case of blood samples, there are certain constituents in urine that require stabilization or preservation. This can be accomplished by procedures such as refrigeration or the addition of acid or base.

Other body fluids

The laboratory is also called on to perform a variety of testing on cerebrospinal fluid, amniotic fluid, and other body fluids and drainages. In all instances it is essential to ensure that the proper specimen is collected and contamination is avoided. Simple observation can be used to rule out use of a specimen, such as a bloody spinal tap. In addition, biochemical analyses may be performed on stool samples, gastric aspirates, and renal or biliary calculi.

COLLECTION VESSELS

Syringes

Although syringes are not used as much as they were in the past, they are still required to collect samples for blood gas, spinal fluid, and amniotic fluid analysis. Both glass and plastic syringes may be used for these samples.

Evacuated phlebotomy tubes

Syringe collection has been largely replaced by the use of evacuated blood collection tubes, as exemplified by the Vacutainer system (Becton Dickinson & Co., Paramus, NJ 07652). The tubes may be siliconized to minimize the risk of hemolysis and to prevent the blood from adhering to the wall of the tube (Fig. 2-1). These containers now come presterilized by irradiation and are available in a variety of sizes from 2 to 30 mL, with the 10 mL tube being the most commonly used tube. Table 2-1 lists the different types of phlebotomy tubes available according to the anticoagulants and preservatives present. Also indicated is the mechanism of action, stopper color, and usual application of each type of tube. Specially made needles are used with the Vacutainer systems, with a 20-gauge needle recom-

Table 2-1 Evacuated phlebotomy tubes and agents

Samples	Type of anticoagulant or additive	Stopper color	Chemical basis of anticoagulant or additive	Application	Approximate volume/tube (mL)*
Whole blood	EDTA†	Lavender	Binds calcium	Hematology	7/2
	Na heparin	Brown	Lead free	Lead	7 to 10
Plasma	Na citrate	Blue	Binds calcium	Coagulation	4.5/2.7
	Heparin‡	Green	Inhibits thrombin	Chemistry	10/3.5
	Oxalates	Black	Binds calcium	Coagulation	
Serum	None	Red	None	Chemistry	7/2
	None	Royal blue	Contaminant free	Trace elements	7 to 10
	Serum separator	Gray/red	Gel barrier	Chemistry	6 to 10
	Thrombin	Orange	Increased rate of clotting	Stat. chemistries	10
Antiglycolytic agents					
Serum	Iodoacetate	Gray	Inhibits glyceraldehyde-3-phosphate dehydrogenase	Glucose, lactic acid	7 to 10
Partial plasma	Fluoride/oxalate	Gray	Inhibits enolase	Glucose	7 to 10

*Tubes can come in large or small sizes (first/second filled-tube volumes available).
†Comes as Na^+ or K^+ salt forms.
‡Comes as Na^+, Li^+, or NH_4^+ salt forms.

mended for most patients and the 21- or 22-gauge needles recommended for pediatric patients and those with poor or traumatized veins. The use of needles outside this range increases the likelihood of hemolysis.[5]

The stoppers currently available from manufacturers have, in many cases, been reformulated to exclude TBEP, a common constituent of rubber shown to interfere with therapeutic drug-monitoring results. They are now recommended for most methodologies.[6]

It is the laboratory's responsibility to determine or verify the manufacturer's claim that a specific tube is suitable for a given analysis and that neither the tube itself nor the stopper has an effect on the analyte of interest either by containing a contaminant, adsorbing the substance of interest, altering protein binding, or inhibiting an enzyme reaction. Any effect could depend on the specific procedure that is being used by the laboratory. This is of concern not only for routine biochemical determinations but also for tests for metals, drugs, and hormones.[7] There can be no single comprehensive reference that addresses these problems, and only a review of the literature and experimental study for the analyte of interest by a specific method will minimize the chance of introducing an analytical error or artifact.

Serum separators

Evacuated phlebotomy tubes require very careful separation of plasma or serum from the cellular components of blood. Often, to completely remove contaminating cells, several centrifugation steps are necessary. An improvement over this procedure has been made possible by the introduction of serum-separator materials. These are described later in the section on the serum-separation process.

Other collection containers

Sample collection for patients for whom very little blood volume is available (for example, newborns), or for patients for whom venipuncture is difficult, can be accomplished by use of capillary phlebotomy tubes. These tubes are also used for blood-gas samples and for scalp vein–pH samples. These tubes contain heparin as an anticoagulant. Immediately after the tubes are filled, a piece of metal (''flea'') is inserted into the tube, and the tube is sealed with clay or a plastic cap (Fig. 2-2). A magnet is then used to move the flea back and forth through the tube to achieve proper mixing of the blood and anticoagulant. Immediately before analysis, the flea should be removed.

Special collection containers are also available that facilitate the collection of urine and stool samples, with consideration being taken for the need to add a preservative, when indicated, to stabilize a constituent in the sample. It is also helpful to weigh the container before collection so that minimum manipulation will be required on receipt of the sample in the laboratory. For the determination of bilirubin one may need to provide a light-shielding container, such as a dark glass container or one wrapped in aluminum foil, because of bilirubin's photosensitivity. Patients should also receive clear instructions about the proper method for collecting the sample.

In addition, contamination of the collection vessel can be a problem. Testing for analytes such as heavy metals requires the use of an acid-washed container to avoid contamination of the sample.

COLLECTION VARIABLES

Diurnal variation

Some analytes demonstrate significant diurnal variation, and the time of sample collection must be known for

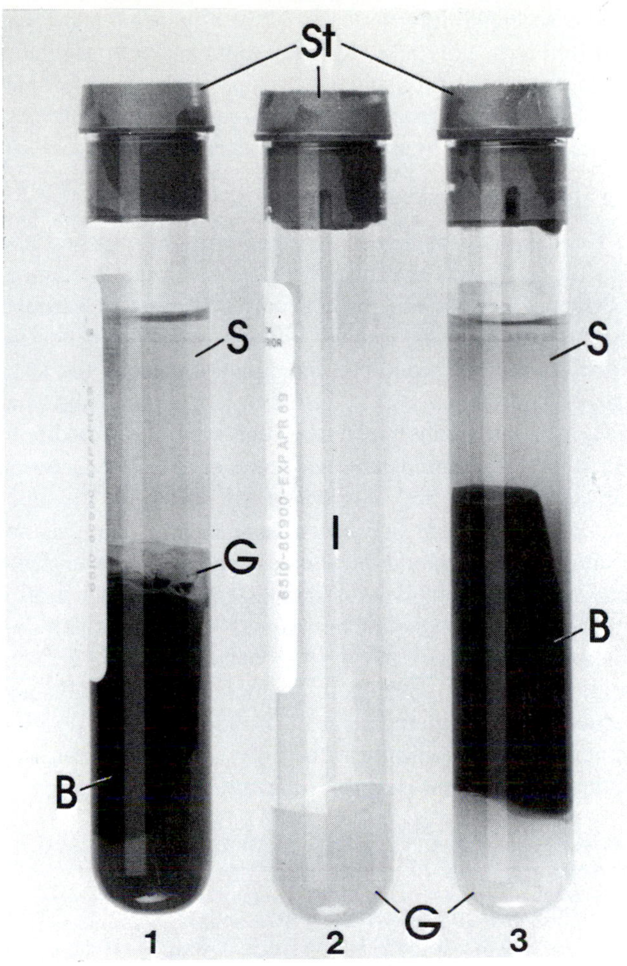

Fig. 2-1 Vacutainer phlebotomy tubes containing barrier gel (red/gray tops). *1,* Tube filled with blood and centrifuged; *2,* unfilled tube; and *3,* tube filled with blood and not centrifuged. Note positions of gel before *(3)* and after centrifugation *(1)*. *B,* Clotted blood; *St,* red/gray stoppers; *G,* barrier gel; *S,* serum.

proper evaluation of the result. Some common examples are listed.

Cortisol	Corticosteroids
Iron	Glucose
Estriol	Triglycerides
Catecholamines	

Within-day variation for these substances may be as much as 30% to 50%. Unless there is a specific contraindication, it is best to collect samples as soon as the patient wakens.

Posture

The posture of the patient at the time of collection can have a significant effect on protein and protein-bound substances in the serum.[8] Serum components known to demonstrate postural variation are the following:

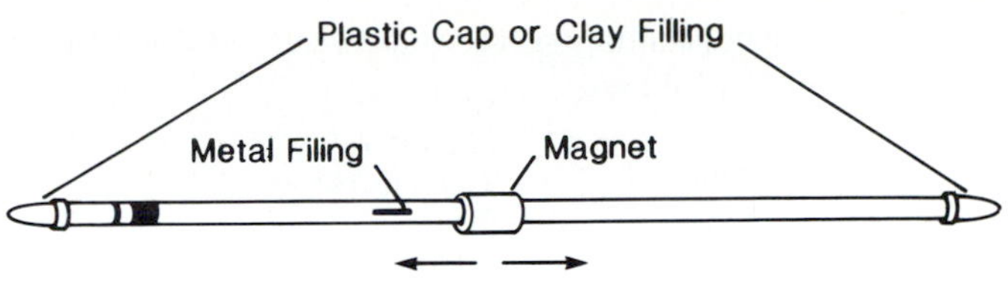

Fig. 2-2 Schematic of heparinized capillary tubes. Magnet is used to move metal filing back and forth through the sealed tube to mix the blood sample with heparin and, later, to remix the sample before analysis.

Total protein	Iron
Albumin	Calcium
Lipids	Enzymes

When patients go from the supine to the standing position, these serum constituents increase their concentration 5% to 15%. This effect is probably attributable to the movement of water out of the intravascular compartment on standing. Again, the only ideal ways to minimize this effect are (1) drawing all blood samples from supine patients and (2) establishing reference ranges using blood drawn from supine, healthy volunteers.

Stasis

Prolonged use of a tourniquet may elevate a number of laboratory results. The application of a tourniquet for an extended period results in a stasis or pooling of blood above the constriction. Tourniquets should always be avoided during the collection of samples for blood-gas analysis or for lactic acid determinations.[5] The resultant slowing of blood flow increases blood pH, decreases Po_2, increases Pco_2, and increases the lactate concentration because lactate is produced by anaerobic metabolism. In addition, prolonged use of a tourniquet will increase the serum concentration of protein and protein-bound substances such as those already described. Significant elevations may be seen with as short as a 3-minute application of a tourniquet.[8]

Hemolysis

During sample collection and until the serum is isolated from the red blood cells, care must be taken to minimize the opportunities for hemolysis. Hemolysis may arise because of the use of too large or too small a needle, moisture in a syringe, vigorous mixing of the blood, rapid expansion of the blood into the tube, or the separation process. Whatever the cause, the net effect is to increase falsely the serum concentration of analytes present in high concentration within the red blood cells.[5] On the other hand, for those substances that exist at lower concentrations in the red cells than outside, hemolysis will result in a dilution effect on the serum constituents. Common analytes whose concentrations are significantly affected by hemolysis are as follows:

Total protein	Triglyceride
Albumin	Norepinephrine
Lipids	Renin
Iron	Aldosterone
Calcium	Potassium
Enzymes	Magnesium
Bilirubin	Inorganic phosphorus
Cholesterol	

This phenomenon does not require visible hemolysis and may result simply from prolonged cell-serum contact before separation. Leaking of potassium and enzymes may occur without the visible leakage of hemoglobin. In addition to the alterations in concentration of certain substances because of hemolysis, the hemoglobin itself may cause a methodological interference. This has been documented in certain procedures for the measurement of albumin, angiotensin II, calcium, carotene, cortisol, iron, insulin, and lipase.[9,10] The effect of chromagens such as hemoglobin on analytic measurement may be reduced in spectrophotometric analysis by a variety of techniques discussed in Chapter 55.

Preservation

Certain measurements, such as analysis of blood-gas parameters and lactate, are affected by red blood cell glycolytic activity. Samples drawn for these analyses must be placed immediately on ice. Analyte loss, leading to erroneous results, will occur within minutes if the sample is not preserved in this manner. Samples for plasma catecholamines, plasma ammonia, and acid phosphatase determinations may also require stabilization by immediate placement in an ice bath until the analysis is performed or the sample processed.

Patient stress

The stress of phlebotomy may also affect laboratory results. Anxiety of the patient may result in changes in catecholamine levels and blood-gas results through direct hormonal effects and hyperventilation. Every effort should be made to calm the patient before collecting the sample. In addition, during the phlebotomy, errors may arise from improper swabbing of the venipuncture site, traumatizing the arm, or drawing too close to an intravenous site.[3]

Phlebotomy tube draw order

Often it is necessary to draw blood into a number of different phlebotomy tubes for the same patient at the same draw time. It is important to prioritize the phlebotomy tubes so that material from one phlebotomy tube does not contaminate the next one. For example, if an EDTA tube (purple top) is used before a regular serum tube (red top), the latter tube could become contaminated with trace amounts of EDTA, which may inhibit certain serum enzymes, such as alkaline phosphatase.

The accepted draw order is to fill *noncoagulated* tubes first (red tops, gray tops, red/gray tops) and anticoagulated tubes next; blue tops and green tops before purple tops. The EDTA tubes are always drawn last to avoid any possible interference by this chelating agent.

TRANSPORTATION

The laboratory must be concerned with the transport history of a specimen, since the transport of a specimen, whether it is delivered by manual or automated carriers, can affect laboratory results. Vibrations can affect the integrity of the red blood cell membranes, causing the leakage of potassium, lactate dehydrogenase, acid phosphatase, aspartate aminotransferase, and so on.[11] In addition, the laboratory should have a record of the time between sample collection and delivery of the sample to the laboratory to ensure that improper transportation delay has not occurred, which could invalidate analysis of an unstable analyte. The time between sample collection and receipt by the laboratory should not exceed 45 minutes. During transport to the laboratory, tubes of blood should be kept in the stopper-up position to promote clot formation and reduce agitation of the sample. In addition, this position minimizes the possibility of hemolysis and contamination with substances released from the stoppers.[12]

LABORATORY CRITERIA FOR UNACCEPTABLE SAMPLES

For the reasons described above, some samples cannot be used for the proper analysis of certain analytes. The clinical laboratory is responsible for establishing criteria for specimen rejection. Documentation of rejected samples should be kept and monitored to establish any patterns that could be corrected. Following are conditions that cause blood specimens to be rejected.

Inadequate sample identification. Each hospital must determine the minimum amount of patient information that must be included on each laboratory slip and specimen. This usually includes the patient's name, address, room, identification number, age, and sex. Phlebotomists must visually and verbally verify the identity of the patient, comparing the name on the wristband, test requisition, and labels. The tube and laboratory slip should be rechecked upon receipt in the laboratory. Differences between the name on the laboratory slip and the name on the sample container are grounds for sample rejection.

Inadequate volume of blood collected into a tube or syringe containing an additive. A specific amount of additive is added to a phlebotomy tube, based on the presumption that the tube will be completely filled with blood. For example, preheparinizing a syringe and then collecting an insufficient volume of blood may cause erroneous results.

Use of an improper collection tube. In general, serum is the preferred sample for most biochemical analyses. Sodium fluoride tubes designed for glucose samples are unsuitable for most other procedures. Chelating agents are unacceptable for many enzymatic methods. Heparin is the anticoagulant least likely to affect clinical chemistry procedures, although this is largely method dependent.

Hemolysis. Visible hemolysis (greater than 200 mg/L of hemoglobin) is unacceptable when one is testing for certain analytes. The degree of interference depends on the extent of hemolysis, analyte concentration, and methodology. The following analytes can be significantly affected:

- Potassium (+)*
- Lactate dehydrogenase (+)
- Aspartate aminotransferase (+)
- Acid phosphatase (+)
- Creatine phosphokinase (+)
- Iron (+)
- Magnesium (+)
- Inorganic phosphorus (+)
- Haptoglobin (−)*
- Bilirubin (−)

Improper transportation. Samples for blood gases, lactic acid, ammonia, and other procedures where there is a significant sample lability must be transported to the laboratory on ice and delivered within a specified time.

Interferents. The presence of potential analytical interferents such as icteric serum, lipemia, turbidity, drugs, dietary components, or isotope exposure that may interfere with a specific analytical procedure is a basis for rejecting a sample.

CHAIN OF CUSTODY INFORMATION

The analysis of specimens whose results will have medicolegal implications (forensic specimens) requires handling in such a manner that the data will be recognized by a court of law. All the processing steps of such specimens, including the collection, transportation, storage, and analytical phases, must be documented to ensure that there was no tampering with the specimen by interested parties, that the specimen belonged to the appropriate individual, and that the results reported are accurate. This is the *chain of custody* document.

Generally all forensic sample collections, such as those intended for blood alcohol analysis or for urine analysis for drugs of abuse, should be witnessed, followed by a foolproof identification of the collected sample. Whenever necessary a *consent form* should be used, signed, and witnessed by the appropriate individuals. Most commonly, this is used in preemployment screening or for a life insurance application. Specimens must have the chain of custody document signed by every individual who has handled the specimens. This is termed a chain of custody. An example of such a chain of custody form is shown in Fig. 2-3. Specimens should be properly stored, usually at 4° C, until time of transportation and must be adequately sealed at all times to ensure their integrity. This is usually done with standard legal-specimen tape, which is readily broken if an attempt is made to tamper with the specimen. When the specimen is kept in storage, it must be in a locked or secure place that can be accessed only by those personnel with a work need. The opening of the specimen must be recorded, witnessed, and the condition of seals recorded. The laboratory's assignment of an identification number must be straightforward and easy to follow through all phases of the testing procedure including instrument logs, labeling of chromatograms, data sheets, and final reports. Quality control data and other information pertaining to the sample must be readily available and easy to trace. If the specimen is to be saved, the same procedure of chain of custody and sealing must be followed.

*"+" denotes a positive interference.
"−" denotes a negative interference.

SAMPLE PROCESSING

Clotting

Blood will normally clot within 20 to 60 minutes; clotting will be faster if the blood collection tube contains a clotting activator and slower if the sample is on ice. Clot formation should be completed before centrifugation, or fibrin formation will continue. This may hamper pipetting or may result in flow problems within the laboratory instrument.

Centrifugation

To separate the liquid portion of the blood from the cellular components, centrifugal force generated by a centrifuge is used. Open tubes of blood should never be centrifuged because an aerosol, produced from the heat and vibration, is generated by the centrifuge. Aerosols will increase the risk of infection to the laboratory staff. Centrifugation of open tubes of blood also may result in evaporation of the sample.

The most common type of centrifuge rotors used in the clinical laboratory are horizontal or fixed angle heads. Other types of centrifuges include ultracentrifuges and special-purpose centrifuges (see Chapter 1).

In the horizontal-head centrifuge the samples are placed in holders that are in a vertical position when the rotor is stationary. As the rotor spins, the holders swing to a horizontal position with respect to the axis of rotation. The angle-head centrifuge has holders in a fixed position, usually at an angle of 52 degrees to the shaft around which it spins. Because of the nature of their construction, horizontal-head centrifuges can attain speeds up to roughly 3000 revolutions per minute (rpm), about 1700 G, without producing excessive heat caused by friction between the head and air. On the other hand, angle-head centrifuges, because of their smoother design, produce less heat and may attain speeds in the range of 7000 rpm (about 9000 G). Cells separated by use of a horizontal rotor may remix

TOXICOLOGY LABORATORY

Chain of Evidence Form

SUBJECT NAME__________ SUBJECT SOCIAL SEC. # __________

DATE/TIME OF COLLECTION__________ COLLECTED BY__________

NUMBER OF SPECIMENS____ TYPE OF SPECIMEN: ___ BLOOD ___ SERUM ___ URINE

WITNESS __________

Sent By Name/Date/Time	Received By Name/Date/Time	Condition of Seals
1.		
2.		
3.		
4.		
5.		

Specimen Opened for Testing Name/Date/Time	Witnessed By Name/Date/Time	Condition of Seals
A. Outside Package 6.		
B. Specimen 7.		

LABORATORY ACCESSION NUMBER: ________

This form must remain with the specimen until line #7 is complete. At that time the form should be turned over to the laboratory supervisor or the designate for filing.

Fig. 2-3 Example of a chain of custody form. *(From Pesce, AJ and Kaplan, LA: Methods in clinical chemistry, St Louis, 1987, The CV Mosby Co.)*

with the serum (plasma) as the tubes drop from the horizontal to vertical position when the centrifuge slows to a stop. This is more of a problem when the centrifuge brake is used than when the head is allowed to gradually coast to a stop. Remixing does not occur when an angle-head rotor is used because the sample cups do not change position on deceleration.[13] When separator or barrier tubes are used, horizontal rotors may give a more uniform barrier between the fluid and cell phases and thus a more discrete separation.

In general, complete separation of the serum from cells is achieved with centrifugation at 1000 to 1200 G for 10 ($\pm$5) minutes.[14] Manufacturer's directions should be followed when using special collection tubes or serum separator devices that may require different conditions.

Most centrifuges display the approximate speed of the centrifuge in revolutions per minute. A more meaningful expression is the RCF, or relative centrifugal force, which takes into account the radius of the centrifugal head and thus the force on the tubes. There are nomograms available that graph the RCF, rpm, and radius in centimeters to allow conversion of the terms (Fig. 1-11, p. 21). Alternatively the RCF may be evaluated by the equation described on p. 20, Chapter 1.

In those cases where a sample is being tested for an unstable analyte (such as lactate), the heat normally generated by the centrifuge would cause sample deterioration. To prevent this from happening, one must use a refrigerated centrifuge to maintain sample integrity.

Serum-separator devices

Serum-separator devices may be broadly separated into two major types, those used during centrifugation and those used after centrifugation.

Devices used during centrifugation. Devices used during centrifugation may be either integrated gel-tube systems or devices inserted into the collection tube just before centrifugation.

Integrated gel-tubes. Integrated gel-tube systems contain a gel that starts at the bottom of the tube. There is no need to remove the stopper before centrifugation, thus saving time and ensuring that aerosol production and evaporation do not occur. During centrifugation, because of its viscosity and density, the gel floats to a position above the cells and below the serum. Depending on the type of gel barrier system used, it is generally safe to store serum on the barrier for up to 48 hours, if not longer, although this should be verified by the laboratory.[15]

Nonintegrated separation systems. There are also devices that can be added to the blood-collection tubes before centrifugation. One such device is a container that is placed on top of the tube, generally also serving as a tube closure. Other separation aids include a variety of nongel devices, which may be beads, crystals, disks, filters, or fiber plugs made of glass, plastic, fiber, or felt. In all cases the centrifugal force (at least 500 G) causes the materials, which are of an intermediate density, to form a barrier between the serum and the cells. Precautions should be taken to avoid hemolysis, evaporation, or aerosols. The barriers formed by these devices are more permeable than those formed by the gel devices, and the serum should be removed from the tube within 1 hour of centrifugation.[16]

Postcentrifugation systems. There are devices that may be added after centrifugation to separate the clot from serum. Generally, they are a plunger-type filter that has a plastic tube with a filter tip at the end in a plastic or rubber base. After centrifugation of the samples these devices are inserted with the filters passing through the serum, stopping just above the surface of the cells. Contact with the cells, which can cause hemolysis, is avoided. The device should then be withdrawn slightly to produce a small air gap below the filter, separating the filter from the cells. This minimizes the potential for leakage of constituents from the cells through the filter into the serum.

SERUM STABILITY AND STORAGE

Stabilization of sample

Certain samples will require stabilization of the analyte if the analysis is not performed immediately. Depending on the analyte, the sample may need refrigeration, freezing, or deep freezing. In addition, some blood and urine tests need acid or alkaline stabilization, which may make that aliquot unsuitable for other testing. Urines for vanillylmandelic acid and catecholamine analysis need acid stabilization, whereas porphyrins are more stable in alkaline urine. See Table 56-1 for stabilization of urine analytes.

Storage

If sample analysis is not going to be completed on the same day as the sample was collected, the sample should be covered and refrigerated or frozen as necessary until it can be analyzed. If it is necessary, the serum should be removed from contact with any serum-separator device or barrier and stabilized.

Evaporation

Excessive exposure of the sample to room air should be minimized to avoid evaporation and the resulting concentration of analytes in the serum. The speed with which a serum evaporates is directly proportional to the room temperature, air flow in the laboratory, duration of exposure, and surface area of the serum exposed to room air.

Greater errors arise when microsample cups, which hold less than 500 mL of serum and have a large surface area, are used.[17] Samples in microcups should probably be analyzed within 10 to 15 minutes after transferal of serum to the cup and should be kept covered as much as possible. In addition, there is a significant loss of carbon dioxide dissolved in the sample to room air, falsely lowering the measured carbon dioxide result.[18] Some investigators have recommended the layering of an organic solvent over the serum to minimize sample evaporation and the loss of carbon dioxide to room air.[19]

Other storage problems

Many other problems may occur if the serum or urine sample is saved for an extended period of time. Medical technologists should be aware of the possibility of bacterial or fungal growth. There has also been a report that placing wood applicator sticks in the serum for extended periods of time may produce analytical artifacts. Potassium, calcium, and glucose concentrations increase after only 3 minutes of contact with wood applicator sticks.[20]

REFERENCES

1. Slockblower, JM, and Blumenfeld, TA: Collection and handling of laboratory specimens: a practical guide, Philadelphia, 1983, JB Lippincott Co.
2. National Committee for Clinical Laboratory Standards: Approved standard procedures for the collection of diagnostic blood specimens by skin puncture, Villanova, Pa, 1982, The Committee.
3. Garza, D, and Becan-McBride, K: Phlebotomy handbook, Norwalk, Conn, 1984, Appleton-Century-Crofts.
4. Cohen, JJ, and Kassirer, JP: Acid/base, Boston, 1982, Little, Brown & Co.
5. Calam, RR: Reviewing the importance of specimen collection, J Am Med Technol 39:297-298, 1977.
6. Kessler, KM, Kewal, J, and Narayanan, S: Effect of blood collection system in percent of free quinidine in serum, Clin Chem 26:1004, 1980.

7. Missen, AW, and Gwyn, SA: Another source of contamination from sample containers, Clin Chem 24:2063, 1978.
8. Statland, BE, Winkle, P, and Bokelund, H: Factors contributing to intra-individual variation of serum constituents. 4. Effects of posture and tourniquet application on variation of serum constituents in healthy subjects, Clin Chem 20:1513-1519, 1974.
9. Frank, JJ, Bermes, EW, Bickel, MJ, and Watkins, BF: Effect of in vitro hemolysis on chemical values for serum, Clin Chem 24:1966-1970, 1978.
10. Young, DS, Pestaner, LC, and Gibberman, V: Effects of drugs on laboratory tests, Clin Chem 21:133D-1334D, 1975.
11. Stige, M, and Jones, JD: Evaluation of pneumatic tube system for delivery of blood specimens, Clin Chem 17:1160-1164, 1971.
12. Pragay, DA, Brinkley, S, Rejent, T, and Gotthelf, J: Vacutainer contamination revisited, Clin Chem 25:2058, 1979.
13. Hicks, R, Schenken, JR, and Steinrauf, MA: Laboratory instrumentation, Hagerstown, Md, 1974, Harper & Row, Publishers.
14. Calam, RR, Benoit, S, and Du Bois, JA: Proposed standards for the handling and processing of blood specimens: National Committee for Clinical Laboratory Standards, 1(16):504-505, 1981.
15. Narayanan, S, et al.: Control of blood collection and processing variables to obtain extended (greater than 5 day) stability in serum constituents, Clin Chem 25:1086, 1979.
16. Seckinger, DL, Antonio Vasquez, D, Rosenthal, PK, and Heller, ZH: Evaluation of a new serum separator, Clin Chem 28:157-159, 1982.
17. Burtis, CA, Begovich, JM, and Watson, JS: Factors influencing evaporation from serum cups, and assessment of their effect on analytical error, Clin Chem 21:1907-1917, 1975.
18. Gambino, SR, and Schreiber, H: The measurement of carbon dioxide concentration with the autoanalyzer: a comparison with three standard methods and a description of a new method for preventing loss of carbon dioxide from open cups, Am J Clin Pathol 45:406-411, 1966.
19. Bandi, ZL: Estimation, prevention and quality control of carbon dioxide loss during aerobic sample processing, Clin Chem 27:1676-1681, 1981.
20. Joseph, TP: Interferences from wood applicator sticks used in serum, Clin Chem 28:544, 1982.

CHAPTER 3

Spectral techniques

CHRISTOPHER S. FRINGS
JACK GAULDIE

OBJECTIVES

- Describe the relationships among wavelength, frequency, energy, and color of the ultraviolet and visible spectra.
- Describe the relationship between percent transmittance (%T) and absorbance (A) and how this relationship affects the color of a solution.
- Describe the Beer-Lambert law and its limitations.
- Illustrate the construction and operation of photometric monochromators and detectors and explain the advantages or disadvantages associated with the use of each in spectral-type instruments. Further describe the principles of spectral isolation and band pass.
- Draw a block diagram of the essential components of the atomic absorption spectrophotometer and the flame emission photometer and state the principle of the operation of each, highlighting similarities and differences. Explain the interferences associated with each.
- Describe how the instrumentation and basic principles of photometry are modified with the applications of turbidity, nephelometry, or fluorometry and identify any unique interferences or sources of error associated with each.

KEY TERMS

absorbance Defined as 2 − log %T, it is directly proportional to concentration of absorbing species if Beer's law is followed.

absorption spectrum The range of electromagnetic energy that is used for spectroanalysis, including both visible light and ultraviolet radiation; also graph of spectrum for a specific compound.

absorptivity Absorbance divided by the product of the concentration of a substance and the sample path length.

angle of detection The angle at which scattered light is measured in nephelometry.

band pass The range of wavelengths that reaches the exit slit of a monochromator; usually referred to as the range of wavelengths transmitted at a point equal to half the peak intensity transmitted.

Beer-Lambert law (most commonly referred to as ***Beer's law***) The concentration of a substance is directly proportional to the amount of radiant energy absorbed.

blank A solution comsisting of all the components including solvents and solutes except the compound to be measured. This solution is used to set I_o, the original light intensity.

cuvette The receptacle in a photometer in which the sample is placed.

electronic transition The change in the orbital position of an electron of an atom or molecule. In the case of the absorption of a photon of light, the electron usually goes from the ground or the lowest energy level to some higher one with a consequent higher energy state (increased energy) of the molecule. Basis of fluorescence phenomena.

emission wavelength The wavelength of light (λ_{em}) that is used to monitor decay of excited molecules into fluorescence; usually refers to the wavelength of output photons measured by a fluorometer.

excitation wavelength The wavelength of radiant energy (λ_{ex}) that is absorbed by a molecule and causes it to be raised to a higher energy state; usually refers to the wavelength of incident energy of a fluorometer.

filter An optical device (usually glass) that allows only a portion of polychromatic, incident light to pass through. The amount of transmitted light is related to the band pass of the filter.

fluorescence The light emitted by an atom or molecule after absorption of a photon. This light is at longer wavelengths (less energy) than the absorbed light and is usually emitted in less than 10^{-8} sec. However, some compounds emit the photon at a slower rate.

grating An optical device consisting of a reflecting, ruled surface that disperses polychromatic light into a uniform, continuous spectrum. Dispersion of light is attributable to interference phenomena at the ruled surface.

hollow-cathode lamp A lamp consisting of a metal cathode and an inert gas. When an electric current is passed through the cathode, the metal is sputtered free and, after colliding with the gas in the lamp, emits a line spectrum of specific wavelengths related to the metal of the cathode.

infrared radiation The region of the electromagnetic spectrum extending from about 780 to 300,000 nm.

internal standard An element or compound added in a known amount to yield a signal against which an instrument or an analyte to be measured can be calibrated.

light scattering The interaction of light with particles that cause the light to be bent away from its original path (cause of turbidity).

line spectrum Discontinuous emission spectrum of elements in which the emitted light bands cover a very narrow (0.1 nm) range of energies.

luminescence Light emitted at low temperatures, often as the result of a chemical reaction *(chemiluminescence)*.

molar absorptivity (ϵ) The absorbance of light, at a specific wavelength, divided by the product of concentration in moles per liter and the sample path length in centimeters. Molar absorptivity is expressed as L/mol·cm.

monochromatic Light of one color (wavelength). In practice this refers to radiant energy composed of a very narrow range of wavelengths.

monochromator Device used to isolate a certain wavelength or range of wavelengths. Usually refers to prisms or grating.

nephelometry A technique that measures the amount of light scattered by particles suspended in a solution.

phosphorescence Similar to fluorescence, the light emitted by an atom or molecule after absorption of a photon. The light is usually emitted at a time greater than 10^{-3} sec after absorption of the photon.

photodetector A device that responds to light (photons) usually in a manner proportional to the number of photons striking its light-sensitive surface. Commonly a current that is proportional to the incident light intensity is generated.

photometer An instrument that measures light intensity; composed of a source of radiant energy, filter for wavelength selection, cuvette holder, detector, and a read-out device.

photon A particle consisting of a discrete packet of radiant energy.

polarized fluorescence The orientation of the emitted fluorescent light. It can be calculated from the polarization formula.

polychromatic Light of many colors (wavelengths), usually referring to white light, or that encompassing a defined portion of the spectrum.

Rayleigh scatter The reflection of light at different angles by particles suspended in a solution. This scattering occurs when the wavelength of light is greater than the size of the particles.

reflectance spectrophotometry A quantitative spectrophotometric technique in which the light reflected from the surface of a colorimetric reaction is used to measure the amount of the reaction product.

refraction A process by which the path of incident light is bent after the light passes obliquely from one medium to another of different density.

refractive index The ratio of the speed of light in two different mediums; usually the reference medium is air.

refractometer An instrument for measuring the refractive index (refractivity) of various substances, especially of solutions.

spectrophotometer An instrument that measures light intensity. It is composed of a source of radiant energy, an entrance slit, a monochromator, exit slit, cuvette holder, detector, and read-out device. Measurements in these instruments can be made over a continuous range of available spectrum.

stray light Radiant energy reaching the detector and consisting of wavelengths other than those defined by the filter or monochromator.

time-delayed fluorescence A technique in which the fluorescence of slowly emitting compounds such as metal chelates is measured. Usually the time between 400 and 1000 msec is monitored.

ultraviolet radiation The region of the electromagnetic spectrum from about 180 to 390 nm.

visible light The radiant energy in the electromagnetic spectrum visible to the human eye (approximately 390 to 780 nm).

wavelength The linear distance traversed by one complete wave cycle of electromagnetic energy.

LIGHT AND MATTER[1-3]

Properties of light and radiant energy

Electromagnetic radiant energy is a form of energy that can be described in terms of its wavelike properties. Electromagnetic waves travel at high velocities and do not require the existence of a supporting medium for propagation.

The wavelength, λ, of a beam of electromagnetic radiant energy is the linear distance traversed by one complete wave cycle and is usually given in nanometers (nm, 10^{-9} meters). The frequency, ν, is the number of cycles occurring per second and is obtained by the relationship

$$\nu = \frac{c}{\lambda}$$

The velocity, c, varies with the medium through which the radiant energy is passing ($c = 3 \times 10^{10}$ cm/sec when measured in a vacuum).

Table 3-1 Electromagnetic spectrum

	Gamma rays	X rays	Ultraviolet (UV)	Visible	Infrared (IR)	Microwaves
Wavelength (nm)*	0.1	1	180	390	780	400×10^3

*This is the wavelength interval where the lowest type of respective radiant energy occurs.

Radiant energy can be shown to behave as if it were composed of discrete packets of energy called *photons*. The energy of a photon is variable and depends on the frequency or wavelength of the radiant energy. The relationship between the energy, E, of a photon and frequency is given by the formula

$$E = h\nu$$

h is Planck's constant and has a numerical value of 6.62×10^{-27} erg·sec. The equivalent expression involving wavelength is

$$E = \frac{hc}{\lambda}$$

This equation shows that shorter wavelengths have a higher energy than longer wavelengths.

The electromagnetic spectrum covers a very large range of wavelengths, as shown in Table 3-1. The areas of the electromagnetic spectrum that are commonly used in the clinical laboratory are the ultraviolet (UV) and visible regions. The visible region is generally specified as the region between 390 and 780 nm, whereas the ultraviolet spectrum usually referred to in the clinical chemistry laboratory falls between 180 and 390 nm. Sunlight or light emitted from a tungsten filament is a mixture of radiant energy of different wavelengths that the eye recognizes as "white." The breakdown of the visible region into color absorbed and color reflected is shown in Table 3-2. If a solution absorbs radiant energy (light) between 400 and 480 nm (blue), it will *transmit* all other colors and appear yellow to the eye. Therefore yellow is the complementary color of blue. If white light is focused on a solution that absorbs energy between 505 and 555 nm (green), the transmitted light and thus the solution will appear purple (blue plus red). If a red light is focused on a red solution, red light will be transmitted because this solution cannot absorb red light. On the other hand, if green light is focused on the red solution, no light is transmitted, since the solution absorbs all light but red. The human eye responds to radiant energy between 390 and 700 nm, but laboratory instrumentation permits measurements at both shorter wavelengths, such as ultraviolet (UV), and longer wavelengths, such as infrared (IR), of the spectrum.

Table 3-2 Colors and complementary colors of visible spectrum*

Wavelength† (nm)	Color absorbed†	Complementary or solution color transmitted
350-430	Violet	Yellow blue
430-475	Blue	Yellow
475-495	Green blue	Orange
495-505	Blue green	Red
505-555	Green	Purple
555-575	Yellow green	Violet
575-600	Yellow	Blue
600-650	Orange	Green blue
650-700	Red	Blue green

*If a solution absorbs light of the color listed in the second column, the observed color of the solution (that is, the transmitted complementary light) is given in the third column.

†Because of the subjective nature of color, the wavelength ranges are only approximations.

Interactions of light with matter

Absorption process. When an atom, ion, or molecule absorbs a photon, the added energy results in an alteration of state, and the species is said to be excited. Excitation may involve any of the following processes:

1. Transition of an electron to a higher energy level
2. A change in the mode of vibration of the molecule's covalent bonds
3. Alteration of its mode of rotation about the covalent bonds

Each of these transitions requires a definite quantity of energy; the probability of occurrence for a particular transition is greatest when the photon absorbed supplies this exact quantity of energy.

The energy requirements for these transitions vary widely. Usually elevation of electrons to higher energy levels requires greater energy absorption than is needed to cause vibrational changes. Rotational alterations usually have the lowest energy requirements. Therefore absorption of energy in the microwave and far infrared regions results in shifts in the rotational energy levels, since the energy of the radiant energy is insufficient to cause other types of transitions. Changes in vibrational levels are caused by absorption in the near infrared and visible regions. Promotion of an electron to a higher energy level occurs after energy absorption in the visible, ultraviolet, and x-ray regions of the spectrum. The energy content of the electrons of covalent bonds varies with the nature of the bonds. The energy of a photon of light needed to excite an electron will therefore vary with the bond, and each type of bond will have its own characteristic pattern of optimum wavelengths of light that can be absorbed by that bond. Table 3-3 gives the electronic absorption bands for a number of organic groups.[4]

Table 3-3 Electron absorption bands for representative chromophores

Chromophore	System	λ_{max}	ϵ_{max}	λ_{max}	ϵ_{max}	λ_{max}	ϵ_{max}
Ether	—O—	185	1000				
Thioether	—S—	194	4600	215	1600		
Amine	$—NH_2$	195	2800				
Thiol	—SH	195	1400				
Disulfide	—S—S—	194	5500	255	400		
Sulfone	$—SO_2—$	180	—				
Ethylene	—C═C—	190	8000				
Ketone	>C═O	195	1000	270-285	18-30		
Esters	—COOR	205	50				
Aldehyde	—CHO	210	strong	280-300	11-18		
Carboxyl	—COOH	200-210	50-70				
Nitro	$—NO_2$	210	strong				
Azo	—N═N—	285-400	3-25				
Nitrate	$—ONO_2$	270 (shoulder)	12				
	$—(C═C)_2—$ (acyclic)	210-230	21,000				
	$—(C═C)_3—$	260	35,000				
	$—(C═C)_5—$	330	118,000				
	C═C—C≡C	219	6500				
Benzene		184	46,700	202	6900	255	170
Anthracene		252	199,000	375	7900		
Quinoline		227	37,000	270	3600	314	2750
Isoquinoline		218	80,000	266	4000	317	3500

From Willard, HH, Merritt, LL, and Dean JA: Instrumental methods of analysis, ed 4, Princeton, NJ, 1965, D Van Nostrand Co, Inc.

The absorption pattern of a complex organic molecule containing tens of thousands of bonds must therefore describe the cumulative sum of the absorption of *all* the individual covalent bonds.

The absorption of radiant energy by a solution can be described by means of a plot of the absorbance as a function of wavelength. This graph is called an *absorption spectrum* (Fig. 3-1). The absorption spectrum reflects the sum of the energy transitions characteristic for a molecule at each wavelength of light. Absorption spectra are often helpful for qualitative identification purposes. This is particularly true for low-energy absorptions such as those found in the infrared region. Irrespective of the amount of energy absorbed, an excited species tends to return spontaneously to its unexcited, or ground, state; in the process it releases energy as kinetic (movement), vibrational, or light (see fluorescence, below) energy.

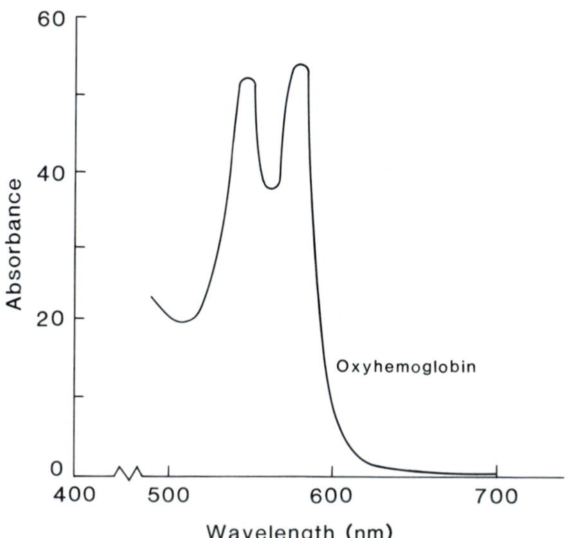

Fig. 3-1 Absorption spectrum of oxyhemoglobin.

Emission process. Some elements and compounds can be excited in such a fashion that when the electrons return from the excited state to the ground state, the energy is dissipated as radiant energy. The radiant energy may consist of one or more than one energy level and therefore may consist of different wavelengths. This principle is used in flame photometry and fluorometric methods and will be further discussed with these topics.

ABSORPTION SPECTROSCOPY

Radiant-energy absorption

Consider a beam of radiant energy with an original intensity, I_o, impinging on and passing through a square cell

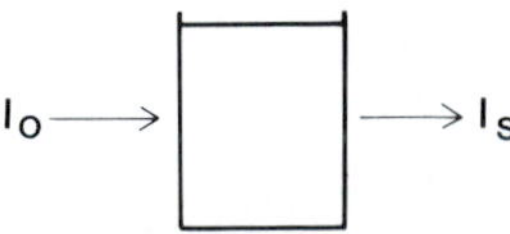

Fig. 3-2 Transmittance of radiant energy through a cuvette. I_o is the incident radiation; I_s is the transmitted radiation.

(whose sides are perpendicular to the beam) containing a solution of a compound that absorbs radiant energy of a certain wavelength (Fig. 3-2). The intensity of the transmitted radiant energy, I_s, will be less than I_o. Some of the incident radiant energy may be reflected by the surface of the cell or absorbed by the cell wall or the solvent. Therefore these factors must be eliminated if one is to consider *only* the absorption of the compound of interest. This is done by using a blank or reference solution containing everything but the compound to be measured. The amount of light passing through the blank solution is set as the new I_o (relative to the reference cell and solution). The transmittance for the compound in solution is defined as the proportion of the incident light that is transmitted:

$$\text{Transmittance} = T = I_s/I_o$$

Usually this ratio is described as a percentage:

$$\text{Percent T} = \%\text{T} = I_s/I_o \times 100\%$$

The concept of transmittance is important because only transmitted light can be measured.

As the concentration of the compound in solution increases, more light is absorbed by the solution and less light is transmitted. Percent T varies inversely and logarithmically with concentration. However, it is more convenient to use absorbance, A, which is directly proportional to concentration. Therefore

$$A = -\log I_s/I_o = -\log T = \log \frac{1}{\text{T}}$$

To convert T to %T, the denominator and numerator are multiplied by 100%:

$$A = \log \frac{1}{T} \times \frac{100\%}{100\%} = \log \frac{100\%}{\%\text{T}}$$

This can be rearranged to

$$A = \log 100\% - \log \%T$$

or

$$A = 2 - \log \%T$$

It is important to remember that absorbance is *not* a directly measurable quantity but can only be obtained by mathematical calculation from transmittance data.

The relationship between absorbance and %T is shown in Fig. 3-3, in which the linear %T scale runs from 0 to 100%, whereas the logarithmic absorbance scale runs from infinity to 0.

Beer-Lambert law

The Beer-Lambert law (most commonly referred to simply as *Beer's law*) states that the concentration of a substance is directly proportional to the amount of radiant energy absorbed or inversely proportional to the logarithm of the transmitted radiant energy. If the concentration of a solution is constant, and the path length through the solution that the light must traverse is doubled, the effect on the absorbance is the same as doubling the concentration, since twice as many absorbing molecules are now present in the radiant energy path. Thus the absorbance is also directly proportional to the path length of the radiant energy through the cell.

The mathematical relationship that connects absorbance of radiant energy, concentration of a solution, and path length is shown by Beer's law:

$$A = abc$$

A is absorbance; a, absorptivity; b, light path of the solution in centimeters; and c, concentration of the substance of interest.

This equation forms the basis of quantitative analysis by absorption photometry or absorption spectroscopy. Absorbance values have no units. The absorptivity is a proportionality constant related to the chemical nature of the solute and has units that are reciprocal of those for b and c.

When c is expressed in moles per liter and b is expressed in centimeters, the symbol ϵ, called the molar absorptivity, is used in place of a and is a constant for a given compound at a given wavelength under specified conditions of solvent, pH, temperature, and so on. It has units of L/mole·cm. The higher the molar absorptivity, the higher the absorbance for the same mass concentration of two compounds. Therefore, in selecting a chromogen for spectrophotometric methods, the chromogen with a higher molar absorptivity will impart a greater sensitivity to the measurement.

Once a chromogen is proved to follow Beer's law at a specific wavelength (that is, a linear plot of A versus c with a zero intercept; Fig. 3-4, A), the concentration of an unknown solution can be determined by measurement of its absorbance and interpolation of its concentration from the graph of the standards. In contrast, when %T is plotted versus concentration (on linear graph paper), a curvilinear relationship is obtained (Fig. 3-4, B). Because of the linear relationship between absorbance and concentration, it is

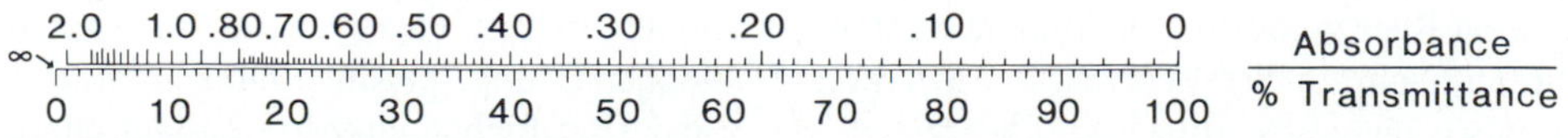

Fig. 3-3 Scale showing relationship between absorbance and percent transmittance.

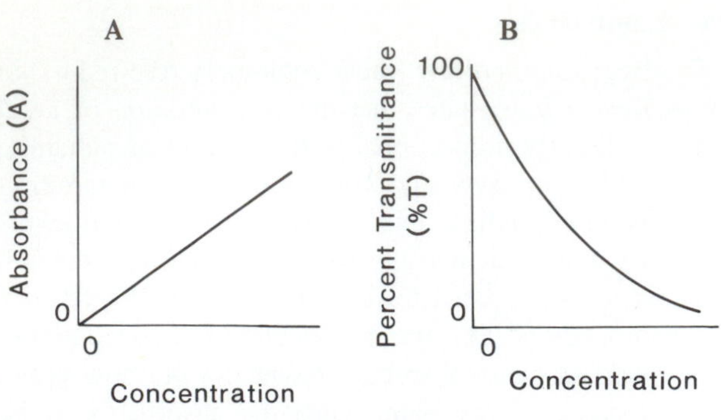

Fig. 3-4 Relationships of absorbance, **A,** and percent transmittance, **B,** to concentration.

possible to relate unknown concentrations to a single standard by a simple proportional equation. Therefore

$$\frac{A_s}{A_u} = \frac{C_s}{C_u}$$

and

$$C_u = \frac{A_u}{A_s} \times C_s$$

Where C_u and C_s are the concentration of the unknown and standard, respectively, and A_u and A_s are the absorbance of the unknown and standard.

The above equation is valid *only* if the chromogen obeys Beer's law and both standard and unknown are measured in the same cell. The concentration range over which a chromogen obeys Beer's law must be determined for each set of analytical conditions.

Beer's law is an ideal mathematical relationship that contains several limitations. Deviations from Beer's law, that is, variations from the linearity of the absorbance versus concentration curve (Fig. 3-5), occur when (1) very elevated concentrations are measured, (2) incident radiant energy is not monochromatic, (3) the solvent absorption is significant compared with the solute absorbance, (4) radiant energy is transmitted by other mechanisms (stray light), and (5) the sides of the cell are not parallel. If two or more chemical species are absorbing the wavelength of incident radiant energy, each with a different absorptivity, Beer's law will not be followed. If the absorbance of a fluorescent solution is being measured, Beer's law may not be followed.

Stray radiation (stray light) is radiant energy that reaches the detector at wavelengths other than those indicated by the monochromator setting. All radiant energy that reaches the detector with or without having passed through the sample will be recorded. Fig. 3-5 shows the effects of stray light on Beer's law. As the amount of stray light increases (or monochromicity decreases), deviation from Beer's law also increases (that is, linearity decreases).

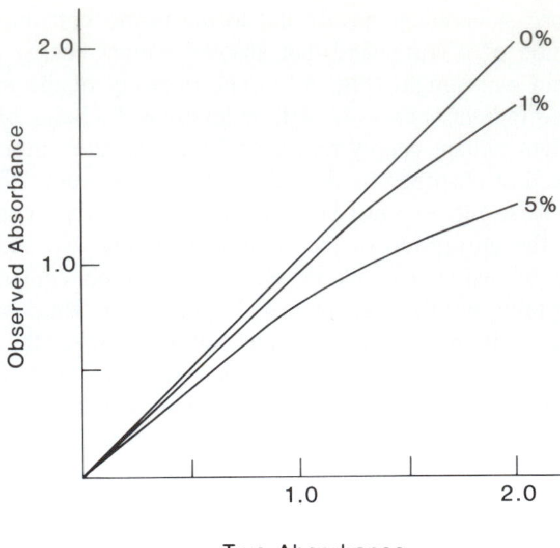

Fig. 3-5 Effect of stray radiation on true absorbance. *(From Frings, CS, and Broussard, LA: Clin Chem 25:1013-1017, 1979.)*

Instrumentation

Single-beam spectrophotometer. The major components of a single-beam spectrophotometer are shown in Fig. 3-6. The apparatus needed can be divided into seven basic components: (1) a stable source of radiant energy; (2) an entrance slit to focus the light; (3) a wavelength selector; (4) an exit slit to focus light; (5) a device to hold the transparent container (cuvette), which contains the solution to be measured; (6) a radiant-energy detector; and (7) a device to read out the electrical signal generated by the detector. If a filter is used as the wavelength selector, monochromatic light at only discrete wavelengths is available, and the instrument is called a *photometer*. If a monochromator is used (that is, a prism or grating, see below) as the wavelength selector, the instrument can provide monochromatic light over a continuous range of wavelengths and is called a *spectrometer* or *spectrophotometer*. Spectrophotometers can be double-beam instruments with two cuvette holders, one for the sample and the other for the blank, or reference sample. Advantages of the double-beam instrument include the capability of making simultaneous corrections for changes in light intensity, grating efficiency, slit width variation, and so on. It is particularly useful for obtaining spectral curves.

Sources of radiant energy. A tungsten-filament lamp is useful as the source of a continuous spectrum of radiant energy from 360 to 950 nm (Fig. 3-7). Tungsten iodide lamps are often used as sources of visible and near-ultraviolet radiant energy. The tungsten halide filaments are longer lasting, produce more light at shorter wavelengths, and emit higher intensity radiant energy than tungsten filaments do.

Fig. 3-6 Components of a spectrophotometer. *A,* Source of radiant energy; *B,* entrance slit; *C,* wavelength selector; *D,* exit slit; *E,* cuvette and cuvette holder; *F,* detector; *G,* read-out device.

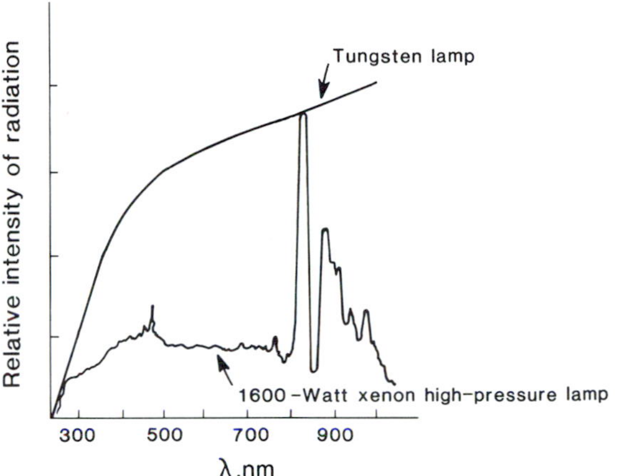

Fig. 3-7 Intensity of radiant energy versus wavelength for a tungsten filament and a 1600 watt xenon light source. Tungsten lamp intensity has been magnified approximately a hundredfold to place it on the same scale as the xenon lamp. *(Modified from Brewer, JM, et al, editors: Experimental techniques in biochemistry, Englewood Cliffs, NJ, 1974, Prentice-Hall Inc.)*

Hydrogen and deuterium discharge lamps emit a continuous spectrum and are used for the ultraviolet region of the spectrum (220 to 360 nm) (Fig. 3-8). The deuterium lamp has more intensity than the hydrogen lamp. Mercury-vapor lamps emit a discontinuous or line spectrum (313, 365, 405, 436, and 546 nm) (Fig. 3-8). This is useful for wavelength calibration purposes but is not used in many spectrophotometers. The mercury lamp is used in photometers or spectrophotometers employed for high-performance liquid chromatography. Recently, light-emitting diodes have been employed as light sources.

It is important to understand that the amount of light emitted from a light source is not constant over a continuous range of wavelengths. Thus a typical lamp has a complex transmittance spectrum with maxima and minima (Figs. 3-7 and 3-8). Lamps of different types, and even of different manufacturers, can vary. Therefore care must be taken in choosing a lamp for a particular analysis, since the amount of light emitted at the desired wavelength may be too little or too much. For example, hydrogen or deuterium lamps, used for ultraviolet analysis, have a maximum output of ultraviolet light in the 250 to 300 nm range. The output of radiant energy at longer wavelengths (greater than 340 nm) is considerably less and can be too weak for many analyses.

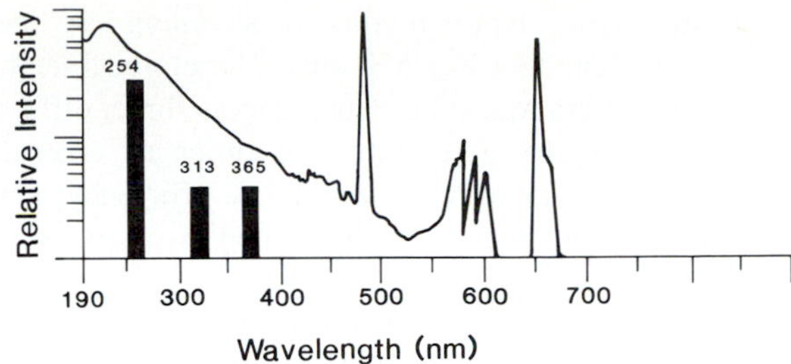

Fig. 3-8 Intensity of radiant energy versus wavelength for a mercury lamp *(solid bars)* and a deuterium lamp *(continuous line)*. For illustrative purposes, the intensity of the mercury emission lines has been reduced several hundredfold, and only those lines (wavelengths are numbers above bars) in the ultraviolet region of the spectra have been depicted.

Wavelength selectors. Isolation of the required wavelength or range of wavelengths can be accomplished by using a filter or monochromator. Filters are the simplest devices, consisting of only a material that selectively transmits the desired wavelengths and absorbs all other wavelengths. In a monochromator, radiant energy from the source lamp is dispersed by a *grating* or *prism* into a spectrum from which the desired wavelength is isolated by mechanical slits.

Filters. There are two types of filters: (1) those with selective transmission characteristics, including glass and Wratten filters, and (2) those based on the principle of interference (interference filters). The Wratten filter consists of colored gelatin between clear glass plates; glass filters are composed of one or more layers of colored glass. Both types of filters transmit more radiant energy in some parts of the spectrum than in others.

Interference filters work on a different principle. The principle is the same as that underlying the play of colors from a soap film, namely, interference. When radiant energy strikes the thin film, some is reflected from the front surface, while some of the radiant energy that penetrates the film is reflected by the surface on the other side. The latter rays of radiant energy have now traveled farther than the first by a distance two times the film thickness. If the two reflected rays are in phase, their resultant intensity is doubled, whereas, if they are out of phase, they destroy each other. Therefore, when white light strikes the film, some reflected wavelengths will be augmented and some destroyed, resulting in colors.

Monochromators. Monochromators can give a much narrower range of wavelength than filters and are easily adjustable over a wide spectral range. The dispersing element may be a prism or a grating.

Dispersion by a prism is nonlinear, becoming less linear at longer wavelengths (over 550 nm). Therefore, to certify wavelength calibration, one must check three different wavelengths. Prisms give only one order of emerging spectrum and thus provide higher optical efficiency, since the entire incident energy is distributed over the single emerging spectrum.

A grating consists of a large number of parallel, equally spaced lines ruled on a surface. Dispersion by a grating is linear; therefore only two different wavelengths must be checked to certify the wavelength accuracy.

Band pass. Except for laser optical devices, the light obtained by a wavelength selector is not truly monochromatic (that is, of a single wavelength) but consists of a range of wavelengths. The degree of monochromicity is defined by the following terms. *Band pass* is that range of wavelengths that passes through the exit slit of the wavelength-selecting device. The *nominal wavelength* of this light beam is the wavelength at which the peak intensity of light occurs. For a wavelength selector such as a filter or a monochromator whose entrance and exit slits are of equal width, the nominal wavelength is the middle wavelength of the emerging spectrum.

The range of wavelengths obtained by a filter producing a symmetrical spectrum is usually noted by its *half-band width* (or *half-band pass*). This describes the wavelengths obtained between the two sides of the transmittance spectrum at a transmittance equal to one half the peak transmittance (Fig. 3-9). For monochromators, the degree of monochromicity is described by the *nominal band width*, which corresponds to those wavelengths that are centered about the peak wavelengths and transmit 75% of the total radiant energy present in the emerging beam of light. For monochromators with variable exit slits, the band pass will also vary.

Slits. There are two types of slits present in monochromators. The first, at the entrance, focuses the light on the grating or prism where it can be dispersed with a minimum of stray light. The second slit, at the exit, determines the band width of light that will be selected from the dispersed spectrum. By increasing the width of the exit slit, the band width of the emerging light is broadened, with a resultant increase in energy intensity but a decrease in spectral purity. In diffraction-grating monochromators, the exit slit may be of fixed width, resulting in a constant band pass. In contrast, prism monochromators have variable exit slits.

The purpose of both slits in filter photometers is to make the light parallel and reduce stray radiation.

Cuvettes. The receptacle in which a sample is placed for spectrophotometric or photometric measurement is called a *cuvette* or *cell*. Glass cuvettes are satisfactory for use in the range of 320 to 950 nm. For measurement below 320 nm it is necessary to use quartz (silica) cells. Such cells can be used at higher wavelengths also. Fig. 3-10 shows the transmission pattern of several types of cuvettes. Cuvettes with a square cross section and with a circular cross section (that is, test tubes) are available. Greater accuracy is achieved by square cuvettes with parallel sides made of *optical glass*. Although cuvettes usually have internal dimensions (that is, path lengths) of 1 cm, cuvettes with other dimensions are available.

Detectors

Barrier layer (photovoltaic) cells. These detectors consist of a plate of copper or iron on which a semiconducting layer of cuprous oxide or selenium is placed. This layer is covered by a light-transmitting layer of metal that serves as a collector electrode. As illumination passes through the transparent electrode to the semiconducting layer, an electron flow is induced in the semiconducting layer, and this flow can be sensed by an ampmeter. These detectors are rugged, relatively inexpensive, and sensitive from the ultraviolet region up to about 1000 nm. No external power is required, and the photocurrent produced is essentially directly proportional to the radiant energy intensity.

Barrier layer cells exhibit the fatigue effect, which

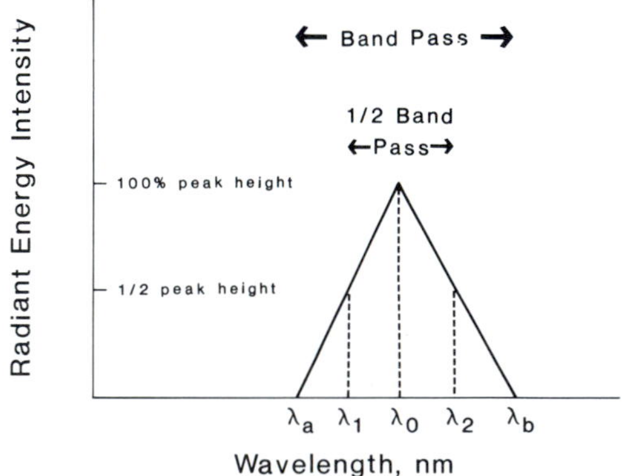

Fig. 3-9 Idealized distribution of radiant energy emerging from exit slit of wavelength selector. For a filter, or a monochromator with entrance and exit slits of equal width, a symmetrical distribution of transmitted energy occurs, as shown.

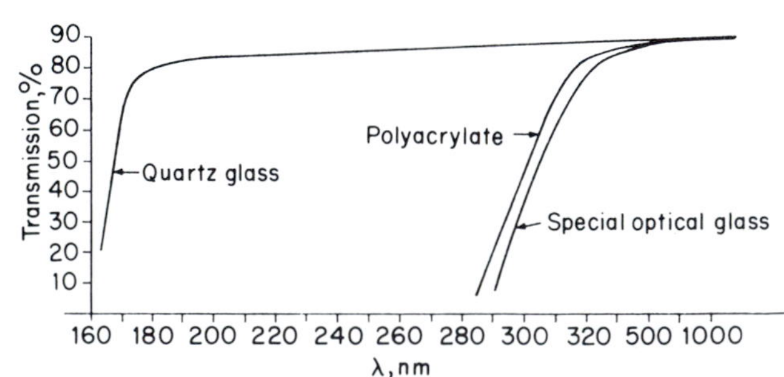

Fig. 3-10 Transmission characteristics of several types of optical materials used for cuvettes. *(From Keller, H: In Richterich, R, and Colombo, JP, editors: Clinical chemistry, New York, 1981, John Wiley & Sons, Inc.)*

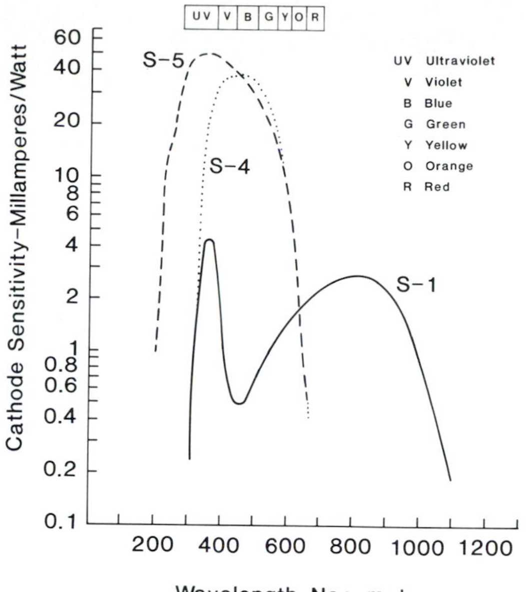

Fig. 3-11 Response of cathode of several photomultiplier tubes to energy of different wavelengths. Sensitivity is expressed as milliamperes of current generated per watt of incident radiation.

means that on illumination, the current rises above the apparent equilibrium value and then gradually decreases. Therefore it is best to wait 30 sec between readings.

Photomultiplier tubes. A photomultiplier tube is an electron tube that is capable of significantly amplifying a current. The cathode is made of a light-sensitive metal that can absorb radiant energy and emit electrons in proportion to the radiant energy that strikes the surface of the light-sensitive metal. These surfaces vary in their response to light of different energies (wavelengths) and so also in the sensitivity of the photomultiplier tube (Fig. 3-11). The electrons produced by the first stage go to a secondary surface, where each electron produces between four and six additional electrons. Each of the electrons from the second stage goes on to another stage, again producing four to six electrons. As many as 15 stages (or dynodes) are present in today's photomultiplier tubes (Fig. 3-12). Photomultiplier tubes have rapid response times, do not show as much fatigue as other detectors, and are very sensitive.

Instrument performance

The sensitivity of response of a spectrophotometer is a combination of lamp output, efficiency of the filter or monochromator in the transmission of light, and response of the photomultiplier. Since these factors are all functions of wavelength, it is clear that the instrument must be reset when one changes wavelengths. This resetting most often takes the form of adjusting the blank solution to read 100% T (zero absorbance) by changing the photomultiplier gain.

A series of recommendations on instrument specifications has been proposed that covers many aspects of instrumentation used for photometric analysis.[9] These specifications are listed in Table 3-4.

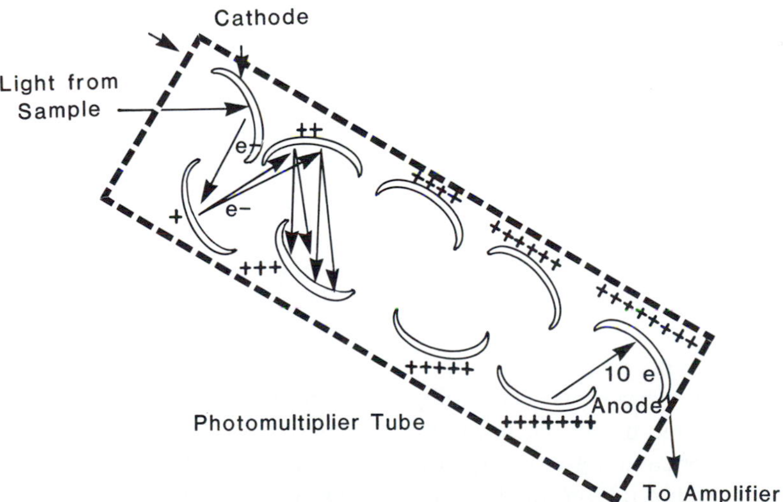

Fig. 3-12 Schematic of photomultiplier tube. Each dynode (electrode used to generate secondary emissions of electrons) is represented by a crescent. Light impinges on cathode and frees an electron. Electron is drawn towards first dynode (stage) by applied voltage. Secondary electrons are released and pass on to successive dynodes, which are at increasingly higher voltages, as depicted by the + symbols. Increasing numbers of secondary electrons are generated at each stage. In this diagram a tenfold amplification of the initial signal is produced at the anode. A photomultiplier tube may increase signal several thousandfold. *(From Simonson, MG: In Kaplan, LA, and Pesce, AJ, editors: Nonisotopic alternatives to radioimmunoassay, New York, 1981, Marcel Dekker, Inc.)*

Table 3-4 Guidelines for photometric enzyme instruments

Parameter	Error or range (95% confidence, ±2 SD)
Carry-over	
Sample to sample	<0.3%
Temperature accuracy	±0.1° C
Equilibration time	20 sec
Sample handling	
Accuracy	1%
Precision	0.5%
Size	50 μL or less
Reagent handling	
Mixing time	≤10 sec
Photometric performance (at a rate of 0.1 *A*/min)	
Initial absorbance 0-1 *A*	<3%
Initial absorbance 1-2 *A*	<5%
Wavelength accuracy	±2 nm
Bandwidth	<8 nm
Wavelength range	Variable
Absorbance range	0-2 *A*
Linearity	<2%
Cell path/placement	<0.6%
Absorbance drift (10 to 60 min)	<2%
Absorbance accuracy	<2%
Absorbance reproducibility	
Low 0-1 *A*	±2%
High 1-2 *A*	±4%

From Instrumentation Guidelines Study Group, Subcommittee on Enzymes: Clin Chem 23:2160-2162, 1977.

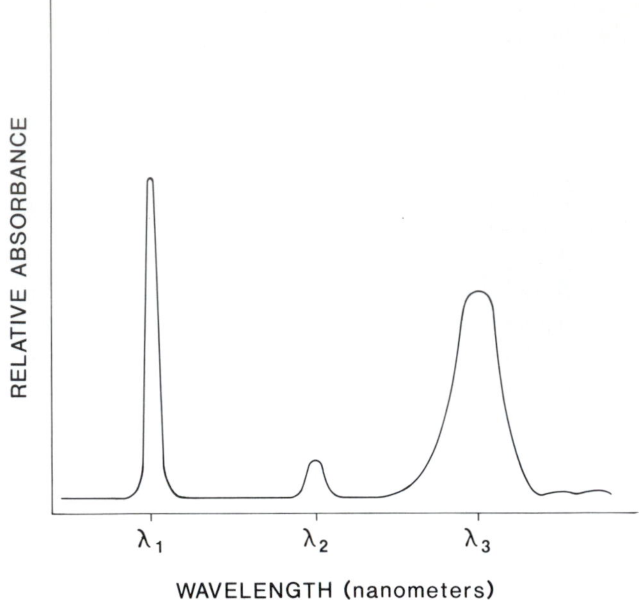

Fig. 3-13 Schematic of idealized absorption spectra. λ_1, λ_2, and λ_3 represent the absorption bands of a chromophore.

Selection of optimum conditions and limitations

When establishing a new spectrophotometric procedure, it is important to record the absorption spectrum of the material that is being measured. This absorption spectrum should be recorded in relation to either water or a reagent blank, depending on the actual method of analysis chosen. Examples of such spectra are presented in the Methods section of this text. This spectrum will help in determining the best wavelength for the spectrophotometric analysis. The optimum wavelength for a specific analysis will depend on several factors, including the absorption maxima of the chromogen, the slope of the absorption peak, and the absorption spectra of possible interfering chromogens.

An example of an absorption spectrum is seen in Fig. 3-13. According to Beer's law, the higher the molar absorptivity, the greater the absorption at a given concentration and wavelength and the higher the sensitivity of the analysis will be. In this spectrum there are three peaks of absorption (highest absorption coefficient): λ_1, λ_2, and λ_3 nm. The absorptivity at λ_2 is too low, and the use of λ_2 can be ruled out immediately.

If an absorption peak is too narrow, as at λ_1, any small error in the setting of the spectrophotometer at this wavelength will result in a large change in absorbance. With spectrophotometers using manually set wavelengths, this can cause large run-to-run imprecision and analytical error. With filter photometers one would require a high-quality, accurate filter to ensure accuracy when monitoring at a narrow absorption peak.

These problems can be avoided by using a wider absorption peak (λ_3). With this absorption peak, small changes in wavelength adjustment will result in small changes in absorptivity, and precision and accuracy will be high.

The sensitivity of many methods may be improved by use of absorption bands at shorter wavelengths (such as the ultraviolet), since very often these are more intense. However, often there is additional nonspecific absorption from buffers or other chemical moieties in the solution at shorter wavelengths. Therefore appropriate blanks must be used to obtain accurate measurements. In some techniques the analyte is purified before analysis, and detection at short wavelengths (ultraviolet) is feasible and provides optimum sensitivity.

Knowledge of the wavelengths at which the commonly interfering chromogens absorb light will also help determine the wavelength of choice. A general rule for selecting the optimum wavelength at which to monitor a spectrophotometric reaction would include three criteria. (1) Choose an absorption peak with the greatest possible molar absorptivity. (2) Choose a relatively broad peak. (3) Choose a peak that is as far as possible from the absorption peaks of commonly interfering chromogens.

Quality control checks of spectrophotometers

Several quality control checks should be performed to certify that spectrophotometers are functioning within specifications. These checks are wavelength accuracy, linearity of detector response, stray radiation (stray light), and photometric accuracy. Details of the spectrophotometer performance checks can be found in references 5 and 10.

Wavelength accuracy. If the wavelength calibration of an instrument changes, the measured absorbance will change. The magnitude of the absorbance error attributable to inaccurate wavelength calibration depends on the relative location of the point on the absorption spectrum of the chromophore to be measured. That is, the absorbance error relative to the wavelength error is greater when the absorbance measurement is on the slope of the absorbance band than when the absorbance measurement is on or near the peak of the absorbance band. Maintenance of wavelength calibration is especially important for analyses such as spectrophotometric enzyme assays.

The most accurate method of checking the wavelength accuracy involves the replacement of the source lamp with a radiant energy source that has strong emission lines at well-defined wavelengths. Useful radiant energy sources are (1) the mercury vapor lamp, which has strong emission lines at 313, 365, 405, 436, and 546 nm, and (2) the deuterium or hydrogen lamp, which has useful emission lines at 486 and 656 nm (Fig. 3-8). Spectrophotometers equipped with a hydrogen or deuterium radiant energy lamp have built-in sources for checking wavelength accuracy.

A second method for checking wavelength calibration involves the use of rare-earth glass filters such as holmium oxide and didymium. Holmium oxide has strong absorption lines at approximately 241, 279, 287, 333, 361, 418, 453, 536, and 636 nm. Didymium has much broader absorption bands at approximately 573, 586, 685, 741, and 803 nm. Because of the possibility of filter deterioration, this wavelength accuracy should be periodically checked.

A third method for checking wavelength calibration involves the use of solutions. A solution of a stable chromogen can be used as a secondary wavelength calibration standard to determine whether the wavelength accuracy of an instrument has changed *after* the wavelength accuracy has been certified by a primary wavelength calibration standard such as a mercury or deuterium lamp. Disadvantages of using chemical solutions for wavelength calibration are that the absorption peaks are generally broad and spectral shifts may result from contamination, aging, or preparation errors.

Irrespective of the method, calibration at two wavelengths is necessary for grating instruments, and calibration at three wavelengths is necessary for prism instruments.

Linearity of detector response. A properly functioning spectrophotometer must exhibit a linear relationship between the radiant energy absorbed and the instrument readout. Instrument linearity is a prerequisite for spectrophotometric accuracy and analytical accuracy. Solid glass filters may be used to check instrument linearity. The most common method for certifying linearity of detector response is through the use of solutions of varying concentrations of a compound known to follow Beer's law. Some compounds used for this purpose are oxyhemoglobin at 415 nm, *p*-nitrophenol at 405 nm, cobalt ammonium sulfate at 512 nm; copper sulfate at 650 nm; and green food coloring at 257, 410, and 630 nm.

The absorbances of solutions containing increasing concentrations of one such compound are plotted against the known concentration. A nonlinear plot of absorbance versus concentration indicates either an error in dilution or an instrument problem. Besides a faulty detector, stray radiation or too wide a slit may cause a nonlinear response.

Stray radiation. An increase in stray radiation is often observed at the extreme ends of the spectral range, where detector response or source energy is at its lowest. Stray radiation usually causes a negative deviation from Beer's law. Methods used to detect stray radiation employ filters or solutions that are highly transmitting over a portion of the spectrum but are essentially opaque below an abrupt "cutoff" wavelength. Several solutions have been used to check for stray radiation, including Li_2CO_3 below 250 nm, NaBr (0.1 mol/L) below 240 nm, and acetone below 320 nm. The exact wavelength at which the cutoff occurs is a function of concentration, cell path length, and temperature; thus the wavelengths reported may vary somewhat. Many filters can detect stray radiation. If solutions or filters that transmit no radiant energy at the measurement wavelength are used, the measured transmittance would be the amount of stray radiation present. Multiplication of this transmittance by 100 would give the percentage of stray radiation. An instrument malfunction is indicated whenever the amount of stray radiation exceeds 1%.

Action taken to eliminate stray radiation includes changing the light source, verifying wavelength calibration, sealing light leaks, realigning instrument components, and cleaning optical surfaces.

Photometric accuracy. When one performs analyses that do not use chemical standards, absorbance accuracy is essential. An absorbance standard should have a constant, stable absorbance at a suitable wavelength that is insensitive to the spectral band width of the instrument and to variations in the geometry of the light beam. Such standards should be easy to use and readily available. The National Bureau of Standards (NBS) has a set of three neutral-density glass filters (SBM 930) that have known absorbances at four wavelengths for each filter. These filters are not completely stable and must be recalibrated by the NBS periodically.

Potassium dichromate solution, cobalt ammonium sulfate solution, and potassium nitrate solution have been used as standards for checking photometric accuracy. Standard solutions are subject to absorbance changes with

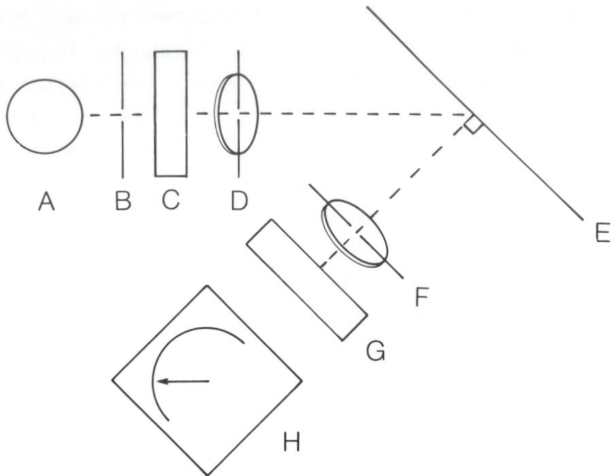

Fig. 3-14 Diagram of reflectance spectrophotometer: *A*, light source; *B*, slit; *C*, filter or wavelength selector; *D*, collimating lens or slit; *E*, test surface; *F*, collimating lens or slit; *G*, detector; *H*, readout device.

time, temperature, and pH, which make them unsuitable as long-term calibration standards for photometric accuracy.

Reflectance spectrophotometry

In reflectance spectrophotometry, a beam of light is directed at a flat surface and the reflected light quantified. The light reflected from the surface is focused onto a photomultiplier tube. The instrumentation can be similar to that of a single-beam filter spectrophotometer (Fig. 3-14). A lamp generates light that passes through a filter and a series of slits and is focused on the test surface. Some of the light incident to a test sample is absorbed by the chromophores on the surface, and the remainder is reflected. (This is analogous to light passing through a solution, for in this case also, some light is absorbed and the remainder passes through.) The reflected light is then passed through a series of slits and lenses and on to a photodetector. The signal is then converted to an appropriate read out. The term *reflection density* is used to describe the absorption of light by chromophores at the surface. Reflection density is related to the intensity of light reflected by the sample.[11] The reflection density, D_R, of the test sample is related to the ratio of the light reflected by the standard reflector (usually a barium sulfate–coated surface), R_O, to the light reflected by the test sample (R_{test}), as described by the equation

$$D_R = \log (R_O/R_{test})$$

This is analogous to the equation that relates %T (and thus absorbance) to incidental transmitted light (see equation, p. 53). In general, the optical properties of different surfaces vary considerably. The optical properties of test paper or plastic strips differ from those of dry film. Therefore, to calibrate an instrument for the measurement of reflection density, one must use a standard with the specific surface employed by the test system. The D_R value in the equation above may be corrected for stray reflectance. A black standard with the same surface characteristics as the test sample can be used to give a value for maximum absorbance. Any reflection read by the instrument under these conditions will be stray reflection. This value can be subtracted from the test value to correct for this variable. The use of reflectance allows quantitative measurement of reactions on surfaces such as a dipstick or dry film.

Disadvantages of the procedure include the problem of standardization. The amount of light reflected and subsequently measured is instrument dependent. The angle at which the reflection is measured, the surface area monitored, and so on are variables. In addition, test surface variations (caused during the manufacturing or handling process) can alter surface reflectance properties.

ATOMIC ABSORPTION[12,13]

Atomic absorption spectrophotometry is widely used in the clinical laboratory for determining calcium, magnesium, lithium, copper, zinc, and other metals.

Principle

Vaporized atoms in the ground state absorb light at very narrowly defined wavelengths. These absorption bands are on the order of 0.001 to 0.01 nm in width, and thus the entire absorption spectrum of atoms is called a *line spectrum*. If these atoms in the vapor state are excited, they can return to the ground state by emitting light of the same discrete wavelengths as the line spectrum. In atomic absorption spectrophotometry the ionic form of the element is not excited in the flame but is dissociated from its chemical bonds and, by attracting free electrons produced by the combustion process, is placed in the atomic ground state. In this form it is capable of absorbing light at the specific wavelengths of its line spectrum.

In atomic absorption a beam of radiant energy containing the line spectrum of the element to be measured is passed through a flame containing the vaporized metal to be determined. The source emitting such radiant energy is called a *hollow-cathode lamp*. The wavelength of the absorbed radiant energy is the same as that which would be emitted if the element were excited. With the aid of a monochromator, the attenuation of one of the wavelengths of the incident light is measured. This attenuation is caused by the photons interacting with ground-state atoms in the flame. Beer's law is valid for relating the concentration of atoms in the flame and transmission or absorption of light. Only a small percentage of atoms in the flame are excited, and most atoms are in a form capable of absorbing radiant energy emitted by the hollow-cathode lamp.

Instrumentation

Fig. 3-15 shows the major components of an atomic absorption spectrophotometer.

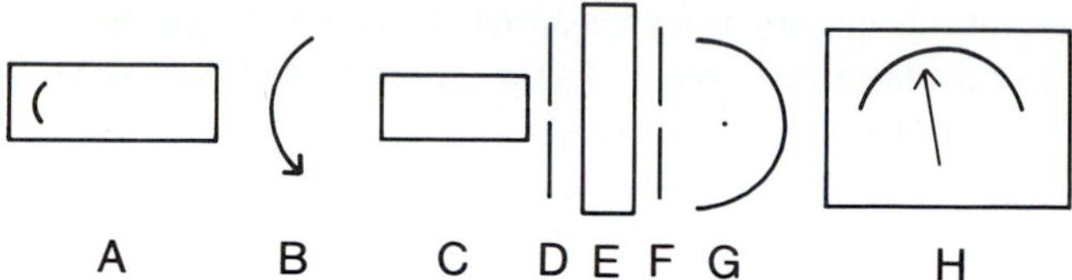

Fig. 3-15 Essential components of atomic absorption spectrophotometer. *A*, Hollow cathode lamp; *B*, chopper; *C*, flame and burner assembly; *D*, entrance slit; *E*, wavelength selector; *F*, exit slit; *G*, detector; *H*, read-out device.

Hollow-cathode lamp. The hollow-cathode lamp is the most practical means of generating a line spectrum of the required spectral purity. The lamps have a hollow or cup-like cathode that is lined with the pure metal of the element to be determined or with an appropriate alloy. A separate lamp is used for each element, except for a few instances in which the cathode can be constructed in such a manner that a single lamp serves for two or three elements (such as calcium and magnesium). The lamp is filled with an inert monatomic gas, usually argon or neon, at low pressure. The inert gas selected for the lamp can vary with the analyte to be measured. For example, lead and iron can be better analyzed with neon-filled lamps, whereas lithium analysis is better performed with argon-filled lamps. For other measured elements, the choice of inert gas is not critical. Quartz or a special glass that allows transmission of the proper wavelength is used as the window. A current is supplied to the cathode, and metal atoms are continually released (sputtered) from the inner surface of the cathode, filling the lamp with an atomic vapor. Atoms in this vapor undergo electronic excitation by collisions with the inert gas, and the resulting excited atoms emit their characteristic radiant energy when returning to the ground-state electron level. This results in a beam of radiant energy with the correct wavelength for absorption by ground-state atoms in the flame.

Burner. In atomic absorption spectrophotometry, the sample solution must be converted into a fine spray or aerosol while it is being introduced into the flame. This process is called *nebulization*. The nebulizer is usually considered part of the burner. Within the flame, solvent evaporates from the aerosol, leaving microscopic particles that disintegrate under the influence of heat to yield atoms. This phenomenon is termed *atomization*. Acetylene is the commonly used fuel in the burner. Temperatures of 2300° C are usually achieved in flame atomic absorption.

Two kinds of burners have been used in most clinical applications. One is the total consumption burner. With this burner, the gases and the sample mix within the flame. The relatively large droplets that are produced in the flame can cause signal noise. The flame in this type of burner can be made hotter, causing molecular dissociations that may be desirable for some chemical systems. The second type of burner is the premix burner (laminar-flow burner). In this burner system, larger droplets from the atomization go to waste and not into the flame, so there is less signal noise. The path length through the premix burner is longer than that of the total consumption burner, which provides greater sensitivity. The flame temperature is not as hot as that of the total consumption burner.

Monochromator and detector. Monochromators (gratings or prisms) and photomultiplier tubes can isolate a pure radiant energy signal and measure the intensity of that signal, respectively. Extraneous radiant energy, both from other wavelengths of the line spectrum and from light generated by the flame, is kept from reaching the photomultiplier tube by the monochromator. The photomultiplier tube converts the radiant energy that was *not* absorbed in the flame into a signal and amplifies this signal to drive a recorder or meter.

Flameless atomic absorption. The purpose of the flame is to convert the sample into an atomic vapor. The flame can be replaced by other atomization processes. One process applicable to mercury analysis uses chemical reactions to convert mercury into an atomic vapor. The sample is decomposed by digestion with acids, then a reducing agent is added to convert mercury to the elemental state, and finally a stream of gas is bubbled through the apparatus pushing mercury vapor into a sealed cell with quartz windows in the optical beam. Absorbance measurements are made at 253.7 nm.

In a more frequently employed atomization technique, the sample is dried on a carbon support platform or tube. The sample is vaporized in an inert atmosphere when an electric current is passed through the support to create an instantaneous temperature sufficient to vaporize the analyte. These atomizers are in the space normally occupied by the flame. The temperatures achieved by flameless atomic absorption (up to 2700° C) are necessary to vaporize heavier metals. Flameless atomic absorption instruments have a greater sensitivity than do flame atomic absorption instruments.

Sources of error

Chemical, ionization, matrix, and burner interferences can occur in atomic absorption measurements. Additional factors that may cause variable behavior from sample to sample or between unknowns and standards include temperature, solvent composition, salt content, viscosity, and surface tension.

Chemical interference. With some elements the presence of certain anions in the sample results in the formation of compounds that are not completely dissociated in the flame. The result is a decrease in the number of ground-state atoms present in the flame. The most common example of chemical interference in atomic absorbance is the formation of a tight complex of calcium with anions, especially phosphate ions. The effect of tightly complexing anions can be minimized or eliminated when lanthanum is added to the sample to displace calcium from the complex. Lanthanum forms a more stable complex with phosphate than calcium does.

Ionization interference. When atoms in the flame become ionized (A^+) instead of remaining in the ground state ($A°$), they will not absorb the incident light. This is termed *ionization interference,* and this effect will result in an apparent decrease in analyte concentration. Ionization interference can be corrected when one adds an excess of a substance that is more easily ionized, thus providing free electrons. The excess free electrons thus shift the reaction

$$A^+ + e^- \rightarrow A°$$

to the formation of ground-state atoms. Ionization interference is minimized by operation of the flame at the lower temperatures of acetylene-air combustion.

Matrix interference. Differences in the matrix between the sample and the standard can result in errors. Protein is sometimes included in the standards when the serum dilution factor is small. The matrix effects are minimized as compositional differences between the standard and the sample become negligible. Calcium standards must contain physiological concentrations of sodium, since sodium will cause a negative interference.

Burner problems. The most critical component in the flame atomic absorption spectrophotometer is the flame and its associated nebulizer. A steady flame is essential, and controlled gas flows are required for both oxidant and fuel. A clean burner head is essential for precise and accurate analysis.

Emission interference. A number of the analyte atoms introduced into the flame will become excited and emit a photon to return to the ground state. The light emitted is, of course, at the same wavelengths as the incident light being measured. The emitted light enhances the signal being received by the photodetector. The increased signal is translated as a decreased absorption and therefore a falsely lower concentration of analyte. This interference can be eliminated by use of a *chopper* (see Fig. 3-15) to create a pulsed beam of incident light from the hollow-cathode lamp. The light caused by emission interference, however, is a constantly produced beam. This steady emission can be electronically differentiated from the pulsed beam of transmitted light and thus eliminated as a source of interference.

FLAME PHOTOMETRY[14,15]

Flame photometry is widely used in the clinical laboratory to determine sodium, potassium, and lithium concentrations in biological fluids.

Principle

Atoms of some metals, when given sufficient heat energy as supplied by a hot flame, will become excited and reemit this energy at wavelengths characteristic for the element as described above. The reactions undergone by ions in the flame are as follows:

$$A^+ + e^- \rightarrow A°$$
$$A° + \text{Heat} \rightarrow A^*$$
$$A^* \rightarrow A° + h\nu$$

A^* represents the excited atom in the flame, $A°$ an atom with ground-state electron energy, and $h\nu$ a photon. Alkali metals are relatively easy to excite in a flame. Lithium produces a red emission; sodium, a yellow emission; and potassium, a red-violet color in a flame. These colors are characteristic of the metal atoms that are present as cations in solution.

The intensity of the characteristic wavelength of radiant energy produced by the atoms in the flame is directly proportional to the number of atoms excited in the flame, which is directly proportional to the concentration of the substance of interest in the sample. The actual number of atoms present in the excited state is a small fraction of the total number of atoms present in the flame.

Instrumentation

Fig. 3-16 shows the major components of a flame photometer. Natural gas, propane, and air are used to produce the flame in most flame photometers encountered in the clinical laboratory. Since it is essential that the flame temperature be held constant, regulators are required to maintain a constant flow of gas. An atomizer is needed to convert the sample into fine droplets before introduction of the sample into the flame. The filter, entrance and exit slits, and detectors have a function similar to those discussed previously for spectrophotometers.

Direct and internal standard instruments

The two types of flame photometers are (1) the direct type and (2) the internal standard type. In the direct type, the intensity of the characteristic spectral emission is used as a measure of concentration. In the internal standard type, the intensity of the spectral emission of the element to be determined is balanced against that of an added element that acts as an internal standard. All flame photometers used in the clinical laboratory are of the internal standard type, and the internal standard is lithium or cesium. The internal standard is normally absent from biological

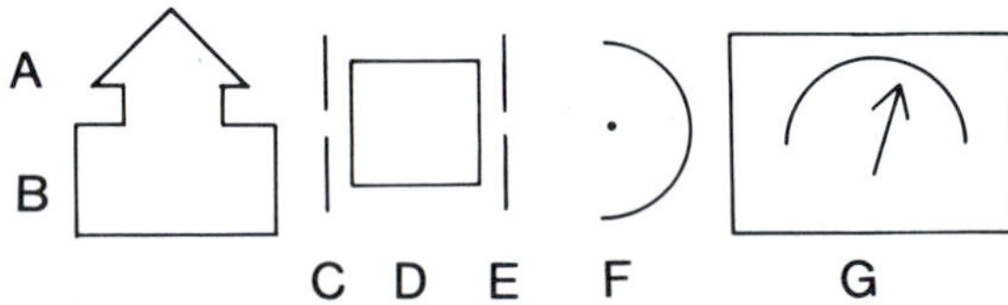

Fig. 3-16 Essential components of a flame photometer. *A,* Flame; *B,* burner head and atomizer; *C,* entrance slit; *D,* wavelength selector (filter); *E,* exit slit; *F,* detector; *G,* read-out device.

fluids and emits light at a wavelength sufficiently removed from that of the elements to be measured (such as sodium and potassium) that it does not interfere with the measurement of these elements. The sample and standards are diluted with a constant concentration of internal standard. The light emission of the internal standard provides a reference signal against which the emission of the other elements can be measured. The flame photometer makes a comparison of the emission of the desired element with the emission of the reference element. Small variations in sample-aspiration rate, atomization rates, solution viscosity, flame temperature, and flame stability will affect the amount of excited atoms present in the flame. In the direct method these variables will affect the intensity of the emission and thus the calculated result. When an internal standard is used, fluctuations of these variables will affect both the internal standard and the element to be measured *simultaneously* and to the same degree. Although the intensity of emitted light for both the internal standard and the analyte will vary, the *ratio* of the two will not. Thus the use of internal standardization allows more precise and accurate determinations.

An internal standard can also be used to minimize the effect of *mutual excitation*. This phenomenon is the result of photons emitted by excited sodium atoms, in turn exciting potassium atoms. This can result in falsely elevated potassium values unless concentrations of sodium and potassium in the standards closely approximate the concentrations of these analytes in unknowns—a difficult achievement for urine specimens. The internal standard (lithium or cesium) acts to absorb the radiation and buffer the potassium atoms from the effect of mutual excitation.

Limitations

Mechanisms that may be responsible for interferences in flame photometry include the following:

1. *Alteration of light emission because of altered flame temperature*. The higher the flame temperature, the less the error that is produced by this phenomenon.
2. *Inadequate selectivity of wavelength*. The magnitude of the background error caused by photoemission by extraneous elements is a function of the filter or monochromator. Accurate selection of the emission lines of sodium or potassium can be achieved with a monochromator; it may or may not be achieved with a filter.
3. *Differences in the viscosities of the standards and samples*. Differences in viscosities can result in differences in aspiration rates. Inclusion of wetting agents in the internal standard used to dilute standards and samples helps to minimize this source of error.
4. *Molecular emission bands*. Certain radicals, such as cyanide, emit spectral bands that overlap spectral lines of metallic ions and therefore produce erroneous enhancement. It is difficult to evaluate the magnitude of error produced as a result of this factor.

FLUOROMETRY[16,17]

Principle

Fluorescence may be considered one of the results of the interactions of light with matter. When light impinges on matter, it can simply pass through, as in a transparent solution, it can be scattered by the interaction, or it can be absorbed. When light is transmitted, there is no loss of energy. When light is scattered, there is no change of energy; the light is the same wavelength before and after it interacts with matter. But when light is absorbed, there is conversion of the light energy into any one of a number of forms, including radiationless transitions (converting the energy into heat) and others, such as fluorescence and phosphorescence, where photons are emitted (Fig. 3-17). Absorption of the light can be used, of course, to determine the concentration of compounds as is done in absorption spectroscopy. If the absorbed light is reemitted, the emitted photons can be used to quantitate the amount of the light-emitting compound (fluor). Quantitation is also possible with use of scattered light, since the amount of

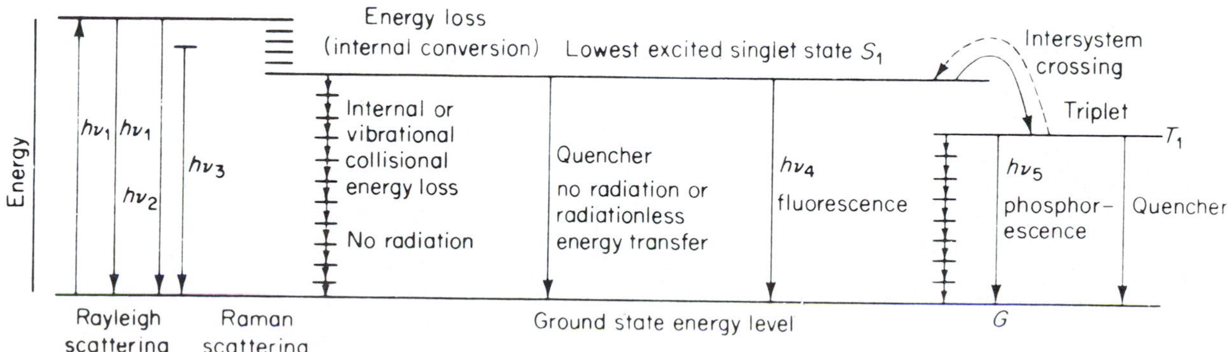

Fig. 3-17 Schematic showing conversion of light energy into different forms of molecular and radiant energy. *(From Pesce, AJ, et al, editors: Fluorescence spectroscopy, New York, 1971, Marcel Dekker, Inc.)*

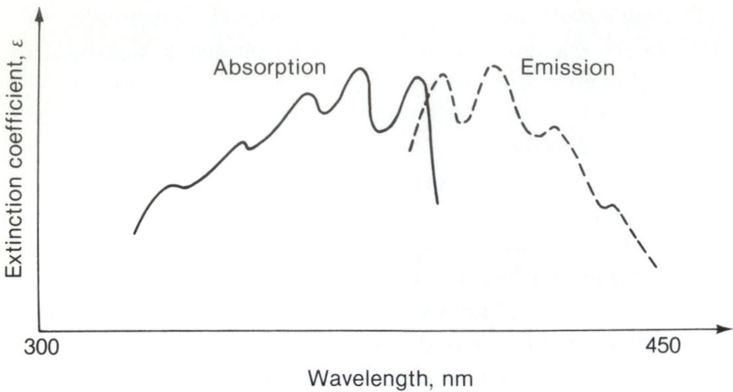

Fig. 3-18 Absorption (excitation) and emission (fluorescence) spectra of a fluorescent compound. *(From Pesce, AJ, et al, editors: Fluorescence spectroscopy, New York, 1971, Marcel Dekker, Inc.)*

scattered light is related to the number and size of particles in solution. Methods that use light scattering are termed *nephelometry* and *turbidity*.

Fluorescent light is the result of the absorbance of a photon of radiant energy by a molecule. Once the molecule absorbs a photon, it has an increased energy level, and because the molecular energy is greater than that of its environment, it will seek to eject the excess energy. When the energy is lost as an ejected photon, the result is fluorescence or phosphorescence emission.

For fluorescence to occur, there has to be a good probability that the energy of the excited state can be converted to the ground state by the ejection of a photon. In general, not all compounds fluoresce; indeed, only a very few fluoresce. Of those that do fluoresce, not every single photon absorbed will be converted to fluorescent light. Some excited compounds will lose energy by radiationless transitions, that is, by transfer of the energy to the solvent. For the same amount of light absorbed, the molecules with higher fluorescence efficiency will have brighter or more intense fluorescence. In solutions, when a molecule returns to the ground state by emitting a photon, there is usually considerably less energy in the emitted photon than was present in the one initially absorbed. In other words, the emitted fluorescent light is at a longer wavelength than the exciting or absorbed radiation (Fig. 3-18).

Instrumentation

The basic components of a fluorometer are similar to those of an absorption spectrophotometer. The major difference is the introduction of a set of filters or a monochromator before and after the cell to isolate the emitted light. A diagram of a fluorometer is presented in Fig. 3-19.

The principal components are the exciting light, filters or monochromators to separate the exciting light from the emitted light, and a sensitive detector. Most often, the measurement of fluorescent light is made at 90 degrees to the exciting light. This is done to maximize the sensitivity of the instrument by minimizing the amount of excitation light that can reach the photodetector. The detector is a photomultiplier or similar device that can quantitate the very small fluorescent light signal and thus achieve the desired level of sensitivity. Because the spectrum of absorption and emission varies from one compound to another, the instrument must be optimized for every analyte measured. This is done by adjusting the exciting wavelength to achieve the maximum absorption of photons, which usually means setting the instrument to the absorption maxima of the compound. By the same token, the wavelength of maximum emission of the fluorescent photons must also be ascertained, and this is the wavelength at which the fluorescence signal is most often recorded.

Limitations

The fluorescence signal of a compound is affected by many variables, including (1) solvent, (2) pH, (3) temperature, (4) absorbance of the solution, and (5) presence of

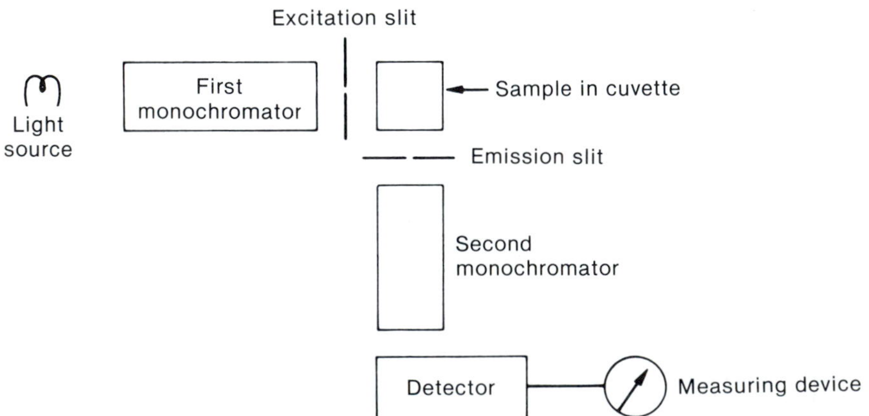

Fig. 3-19 Essential components of fluorometer. *(From Brewer, JM, et al, editors: Experimental techniques in biochemistry, Englewood Cliffs, NJ, 1974, Prentice-Hall, Inc.)*

interfering or specifically quenching compounds. Standardization is not usually done by an absolute procedure as in absorption spectroscopy because the fluorescence varies depending on (1) the intensity of the incident light on the sample, (2) the amount of light intercepted by the detector as controlled by the slits, (3) the band width of light analyzed, and (4) the efficiency of the detector. The quantum yield or efficiency of light emission of a photon is constant if solvent, pH, temperature, and so on are kept constant. But in general the instrument will not be constant on a daily basis. Therefore relative fluorescence yield is used for most measurement. For a reagent blank in a fluorometric assay, only the zero or null fluorescence can be set. There is no equivalent to the 100% scale of absorption. Therefore the electronic signal will vary from instrument to instrument for the same concentration of analyte.

To enable a series of fluorescent standards to form a curve that is linear with concentration, the absorbance of the solutions should not exceed 0.1. Above this absorbance, all portions of the solution are not uniformly illuminated; that is, the initially illuminated layer of the solution will absorb more light than the final layer, and thus the initial layer will fluoresce more than the final layer. With dilute solutions (such as those with absorbance of less than 0.1), this does not occur. However, certain assays employ the quantitative relationship between fluorescence intensity and the amount of light absorbed by a solution whose absorbance is greater than 0.1. Such assays are termed *fluorescence attenuation assays*. In these systems a constant amount of fluorescent dye is placed in each test and control solution. The greater the amount of colored reaction product formed by the analyte, the smaller the amount of light absorbed by the fluorescent dye. Thus the decrease in light passing through the solution results in a proportionate decrease in fluorescence intensity that can be related to the concentration of the analyte.

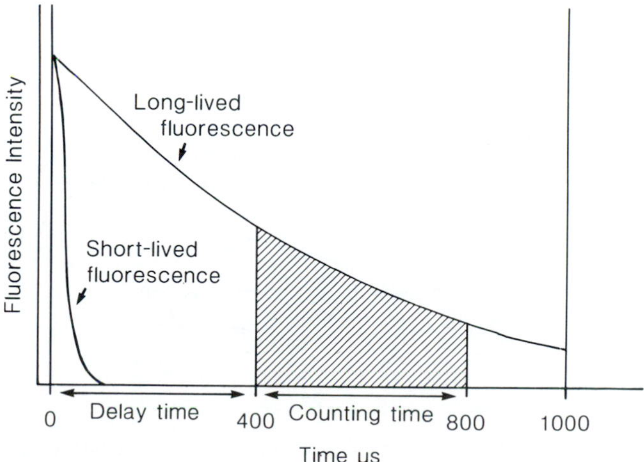

Fig. 3-20 Schema showing difference between short- and long-lived fluorescence.

Time-delayed fluorescence[18]

One approach that can be used for a limited number of analytes to improve the sensitivity of fluorescence techniques is the use of time-resolved fluorescence. The fluorescence emission time of most fluorescence molecules, such as fluoroscein, is in the range of nanoseconds (nsec); that is, the fluorescence signal decays within 100 nsec. Some compounds such as the metal chelate diketone-europium have very long fluorescence lifetimes, 10 to 1000 microseconds (μsec). By measuring the fluorescence after 400 μsec, over a period of an additional 400 μsec the fluorescence intensity of these compounds can be obtained without interference from light scattering or from any other molecules that may fluoresce. Aside from increasing the specificity of analysis, this technique provides greatly increased sensitivity.

Current systems use instruments that illuminate the sample for a period of time, stop the illumination, and measure the emitted fluorescence over a specified time period from 400 to 800 μsec after the illumination (Fig. 3-20).

Although this technique does give enhanced sensitivity, it requires specialized instrumentation. In addition, current systems that use this principle require separation steps, because the chelates cannot be measured in body fluids (see Chapter 10 for more details).

Chemiluminescence[18,19]

A chemiluminescent reaction is any chemical reaction in which one of the products of the reaction is light. The enzyme peroxidase can react with molecules such as luminol (5-amino 2,3-dihydro-1,4-phthalazinedione) to yield light as part of the reaction product. The luminol reaction results in photon emission in the range of 400 to 450 nm. The low photon yield of this reaction has limited its sensitivity and its application. However, by adding enhancer molecules (luciferin, 6-hydroxybenzothiazole) the reaction can be followed for many minutes (30 or more) with a several thousand-fold increase in photon output. The products of these reactions are not known. A partial reaction may be written as follows:

$$2\ H_2O_2 + \text{luminol and enhancer} \xrightarrow{\text{Peroxidase}} 2\ H_2O + h\nu + \text{oxidized luminol}$$

The peroxidase is often part of an enzyme immunoassay system in which the peroxidase is the label. The reaction can be measured by very sensitive photomultiplier tubes. The advantage of this technique is that it can be very sensitive. One molecule of peroxidase can turn over several million molecules of substrate per minute. Thus detection can be more sensitive than that achievable with radioisotopes. A disadvantage of the system is that the reaction is performed in a heterogeneous system in which the peroxidase is attached to a surface. The H_2O_2 and luminol must be in a system free from common biological matrices, such as serum, and therefore a separation step is necessary.

Fluorescence polarization[16,20,21]

Light is considered to be composed of an electronic and a magnetic vector. If the light is polarized, then all the electronic vectors have the same orientation.

When light is absorbed by a molecule it excites a chromophore that has specific orientations, resulting in the transition of an electron to a higher energy level. The excited molecule emits light as the electron (fluorescent oscillator) returns to a lower energy level. Fluorescence polarization measurements require a dye with an electronic orientation such that the emitted light retains the initial orientation of the incident beam (for example, fluorescein). Molecules in solution rotate so that this orientation can be lost. In fluorescence polarization, the dye must be selected so that its molecular rotation is so great between the times of absorption and emission that the molecule becomes randomly oriented during this time. The quantity that is measured is the intensity of the oriented light or the difference between the oriented and unoriented light.

To measure polarized light, several options are available. One approach is to excite the test solution with light polarized in one dimension and record the amount of the emitted fluorescence. The polarizer is placed immediately after the first monochromator as shown in the diagram of a fluorometer given in Fig. 3-19. The solution is then excited by light that is polarized at 90 degrees to the light used in the first excitation, and the emitted fluorescence is recorded. Polarization (P) can be determined from the following equation:

$$P = \frac{I_{vv} - I_{hv}}{I_{vv} + I_{hv}}$$

where I_{vv} equals the signal recorded when the vertically polarized light is used to excite the sample, and I_{hv} is the response when the horizontally polarized light is used to excite the sample. Emitted light is measured from the vertical polarizer at a 90-degree orientation from the incident light using a second polarizer that is placed before the second monochromator. A number of processes affect the final polarization. For many fluorescent dyes, including the most popular one, fluorescein, orientation of light is retained if the molecule is held rigid. When the molecule rotates, polarization is lost.

The process of Brownian motion, leading to rotation of the fluorescent oscillator, is taken advantage of for fluorescence depolarization measurements. The ability of a molecule in solution to rotate partially depends on the viscosity of the solution and on the molecular volume of the molecule. When the viscosity of the solution or the molecular volume increases, the fluorescent molecules rotate slower and the polarization increases. In the case of the fluorescence polarization immunoassay, when the drug fluorescein derivative is bound by antibody, the molecular volume increases. Therefore the polarized fluorescence increases. When the derivative is unbound, the molecular volume is low and the fluorescence polarization is low.

Fluorescence polarization measurements can be made very accurately, and they are less affected by variations in fluorescence intensity than are standard fluorescence measurements (see equation at left). Thus precision on the order of 1% or greater of measurement is readily achieved, which translates into more precise assay measurements. Another advantage of fluorescence polarization is that the technique can be used as a homogeneous assay. Disadvantages include the fact that the technique is limited to assays that can use fluorescent dyes, the instrumentation required for performing fluorescence polarization measurement is very specialized and may only measure fluorescence intensity or polarization, and the system is less flexible than absorption spectroscopy. In addition, when performing fluorescence polarization measurements, it is crucial to control temperature and viscosity.

NEPHELOMETRY AND TURBIDITY[22-24]

Principle

Interaction of light with particles. To understand the principle of nephelometric or turbidimetric assays, we must first examine the concept of light scattering. When a collimated (that is, parallel, nondivergent) beam of light strikes a particle in suspension, some light is reflected, some is scattered, some is absorbed, and some is transmitted. Nephelometry is the measurement of the light scattered by a particulate solution. Turbidity measures light

A SMALL PARTICLES
-LIGHT SCATTERED SYMMETRICALLY
BUT MINIMALLY AT 90° (RAYLEIGH)

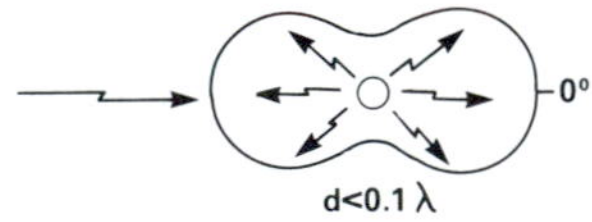

B VERY LARGE PARTICLES
-LIGHT MOSTLY SCATTERED FORWARD
(MIE)

C LARGE PARTICLES
-LIGHT SCATTERED PREFERENTIALLY
FORWARD (RAYLEIGH-DEBYE)

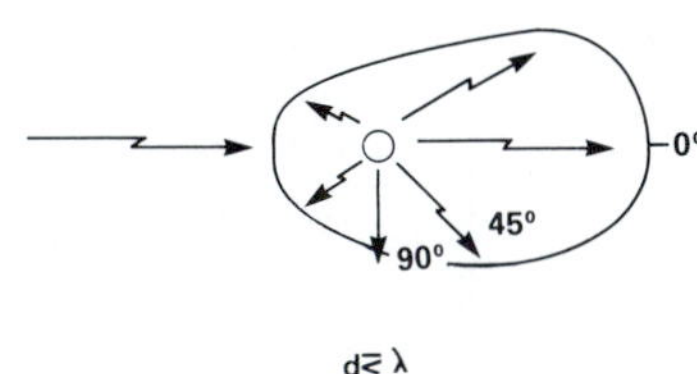

Fig. 3-21 Effect of particle size on scattering of incident light in a homogeneous solution. *(From Gauldie, J: In Kaplan, LA, and Pesce, AJ, editors: Nonisotopic alternatives to radioimmunoassay, New York, 1981, Marcel Dekker, Inc.)*

scattering as a decrease in the light transmitted through the solution.

In considering nephelometry, the question of how light is scattered by a *homogeneous* particle suspension must be examined. Three types of scatter can occur. If the wavelength, λ, of light is much larger than the size of the particle ($d < 0.1\ \lambda$), the light is symmetrically scattered around the particle, with a minimum in the intensity of the scatter occurring at 90 degrees to the incident beam, as described by Rayleigh (Fig. 3-21, *A*).

If the wavelength of the incident light is much smaller than the size of the particle ($d > 10\ \lambda$), most of the light appears to be scattered forward because of destructive out-of-phase backscatter, as described by the Mie theory (Fig. 3-21, *B*).

If, however, the wavelength of light is approximately equal to the size of the particles, more light appears scattered in a forward direction than in a backward direction (Fig. 3-21, *C*), as described by Rayleigh-Debye scatter.[22]

One of the most common uses of light-scattering analyses is the measurement of antigen-antibody reactions. Since most antigen-antibody complex systems are heterogeneous with particle diameters of 250 to 1500 nm and the wavelengths used in most light-scattering analyzers are 320 to 650 nm, the scatter seen is essentially Rayleigh-Debye, with the blank scatter being primarily described by

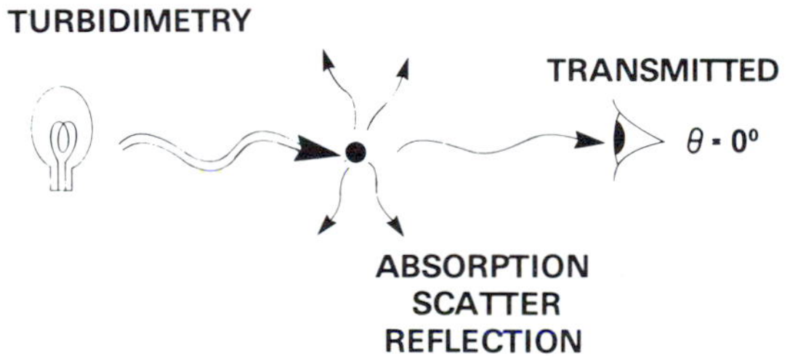

TURBIDIMETRY

TRANSMITTED

$\theta = 0°$

ABSORPTION
SCATTER
REFLECTION

Fig. 3-22 Schematic diagram of turbidity measurements. θ, Angle of detection. *(From Gauldie, J: In Kaplan, LA, and Pesce, AJ, editors: Nonisotopic alternatives to radioimmunoassay, New York, 1981, Marcel Dekker, Inc.)*

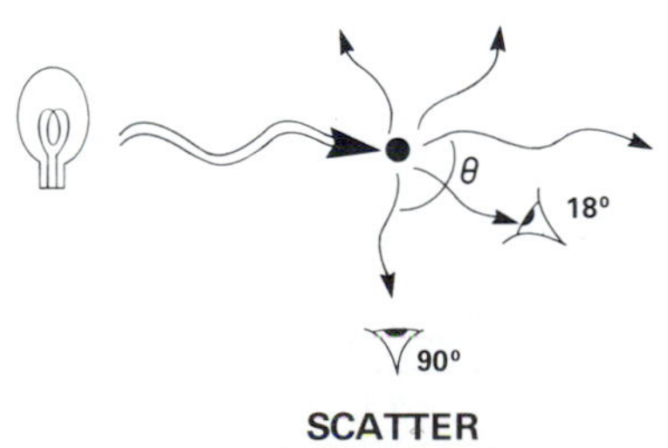

Fig. 3-23 Schematic diagram of nephelometric measurement. θ, Angle of detection. *(From Gauldie, J: In Kaplan, LA, and Pesce, AJ, editors: Nonisotopic alternatives to radioimmunoassay, New York, 1981, Marcel Dekker, Inc.)*

Rayleigh scatter. Thus the ability to detect light scatter in a forward direction ($\theta = 15$ to 90 degrees) would lead to greater sensitivity for nephelometric determinations. Such is the case in the newer rate and laser nephelometers.

Detection of scattered light

Turbidity. Turbidity is a measure of the reduction in the light transmission caused by particle formation, and it quantifies the residual light transmitted (Fig. 3-22). The instrumentation required for turbidity measurements ranges from a simple manual spectrophotometer available in most laboratories to a sophisticated discrete analyzer. Because this technique measures a decrease in a large signal of transmitted light, the sensitivity of turbidimetry is limited primarily by the photometric accuracy and sensitivity of the instrument. Instruments used for turbidimetry can be used for many other assays, such as enzyme assays and those assays based on color development.

Nephelometry. Nephelometry, on the other hand, detects a portion of the light that is scattered at a variety of angles (Fig. 3-23). The sensitivity of this method primarily depends on the absence of blank or background scatter, because the instruments are detecting a small increment of signal at a scatter angle, θ, on a supposedly black or null background. Ideally, no light is detected in the absence of a scattering species, and so subsequent scatter in samples is measured against this black background. The signal is magnified by the use of a photomultiplier, so that the detection range is increased. However, such measurements require the committed use of a nephelometer, which has limited use in other assays.

Instrumentation

Schematic layout of instruments. A schematic layout of the basic components of a nephelometer is shown in Fig. 3-24. Typical systems consist of a light source, a collimating system, a wavelength selector such as a filter (the last two items are unnecessary with laser light sources), a sample cuvette, a stray light trap, and a photodetector.

Light source. Fluoronephelometers such as the Technicon instrument use a medium-pressure mercury-arc lamp as a light source, which serves both for nephelometry and fluorometry. The relatively high intensity light and short-wavelength emission bands make this a good source. Other light sources range from simple low-voltage tungsten-filament lamps and light-emitting diodes to sophisticated low-power lasers. Lasers produce stable, highly collimated, and intense beams of light (typically 1 milliradian divergence) that require no additional optical collimators as do other light sources. In optical systems using laser light, it is easier to reduce stray light, which contributes to background scatter, and to mask the transmitted beam, thus allowing measurement of forward scatter. The increase in light intensity achievable with lasers also results in an improvement of signal-to-noise ratio, but this is limited

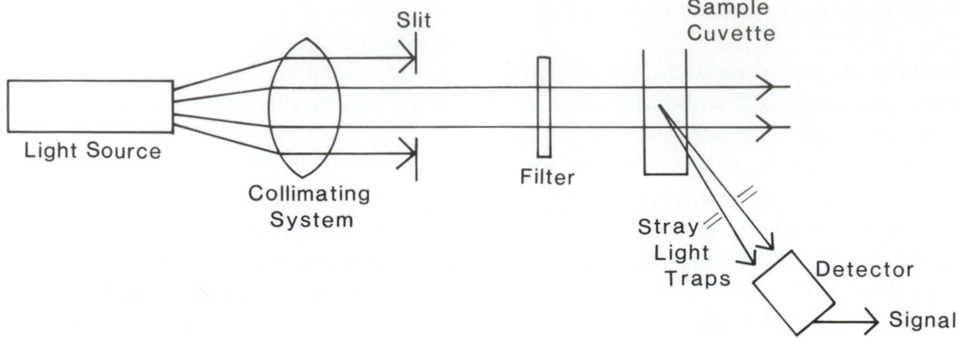

Fig. 3-24 Schematic of basic components of a nephelometer.

somewhat by detector saturation. Disadvantages of laser sources include cost, safety problems, and the restricted availability of limited fixed wavelengths. Since particle size may continually change during the course of reaction analysis, as during immune precipitate formation, light scatter at a single wavelength may change while the average light scatter over a number of wavelengths remains relatively constant. The Beckman Auto-ICS System employs a broad-band filter for selection of a wavelength region from a normal tungsten lamp source to overcome this problem, which is obviously more acute in rate methods when the size of the particle is changing most rapidly.

In all cases, the photodetector system must be matched to the wavelength or wavelengths of the scattered light, which, for nephelometry and turbidimetry, corresponds to the incident light wavelength or wavelengths.

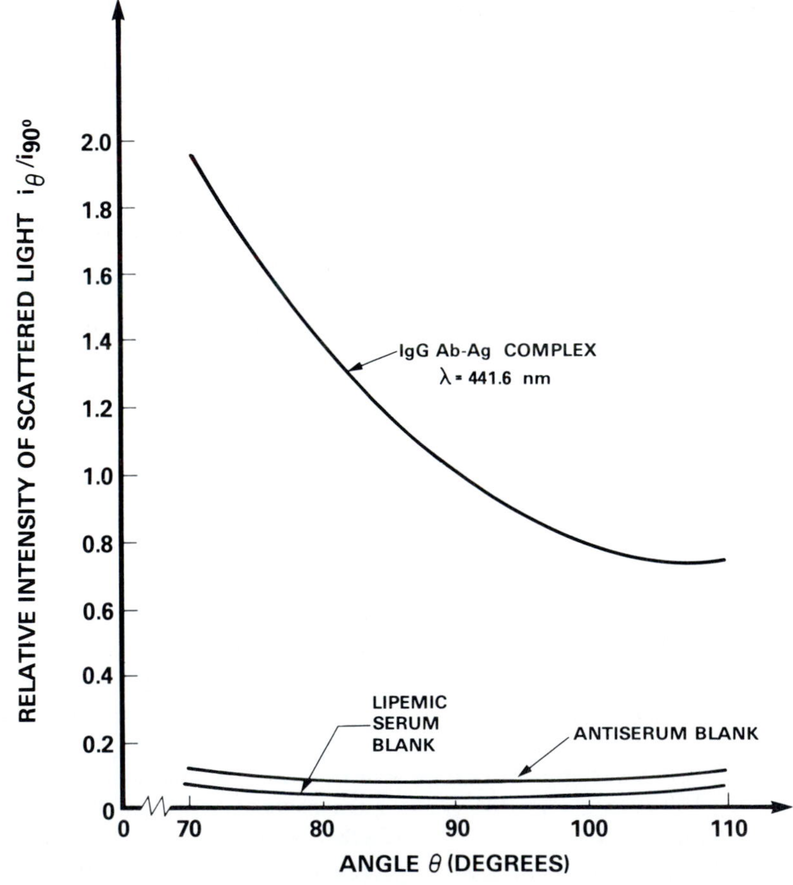

Fig. 3-25 Effect of angle of detection on the apparent light-scattering intensity for IgG/anti-IgG reaction products. Effect on serum blank and lipemic blank. *(From Kusnetz, J, and Mansberg, HP: Optical considerations: nephelometry. In Ritchie, RF, editor: Automated immunoanalysis, part 1, New York, 1978, Marcel Dekker, Inc.)*

Table 3-5 Automated clinical instruments available for light-scattering analysis

	Technicon AIP	Hyland PDQ (OISC 120)	Behring Auto laser LN	Beckman Automated ICS	Reaction rate analyzer*	Centrifugal fast analyzer†
Nephelometer (N) (angle) or turbidometer (T)	N (90°)	N (31°)	N (5°-12°)	N (70°)	T	T
End-point (E) or kinetic (K) analysis of reaction	E + blank	E + blank	E + blank	K	K or E	K or E
Light source (wavelength, nm)	Mercury arc (357 nm)	Laser (632.8 nm)	Laser (632.8 nm)	Tungsten lamp (400-550 nm)	Variable: ultraviolet to 700 nm	Variable: ultraviolet to 700 nm
Samples/hour	120	120	120-240	30-50	20-40	300
Reaction time	10 min	1 hour	1 hour (no PEG)	30 sec	1-2 min (kinetic)	1-2 min (kinetic)
Other type of analysis	+ +	+	+	+	+ + +	+ + + +

Modified from Deverill, I, and Reeves, WG: J Immunol Methods 38:191-204, 1980. PEG, Polyethylene glycol. Plus sign indicates ability of instrument to perform other measurements in addition to light-scattering analysis.
*For example, those by Abbott TDx Turb. (Irving, Tex.)
†For example, Roche (COBAS Bio) (Nutley, N.J.).

Angle of detection. Since particles the size of antigen-antibody complexes appear to scatter light more in the forward direction, there is an increased signal-to-noise ratio as the detector is placed nearer the transmitted path (0 degrees). Fig. 3-25 shows the magnitude of the difference expected in an IgG–anti-IgG system.[25]

The blank signal, described best by Rayleigh scatter (see Fig. 3-21, *A*), is not as affected by an altered angle of detection. Thus, although most early nephelometers detected light scattered at 90 degrees for reasons of manufacturing ease, which limited low-angle measurement capability, the detection of forward light scatter should provide theoretically greater sensitivity. The newer instruments tend to operate with lower detection angles, optimized in many cases to give the highest signal-to-noise ratio for the particular instrument's optics. Several of the systems listed in Table 3-5 detect forward light scatter and possess the increased sensitivity expected. Obviously, detection at 0 degrees is not possible because of the high intensity of the transmitted beam, but some laser-equipped fast analyzers using a mask to block the transmitted beam are able to operate at very low angles. Instruments employing low-angle detectors tend to have greater sensitivity than the 90-degree type of instruments.

Limitations: turbidimetry versus nephelometry

Although the principle of nephelometry—detection of a small signal (amplifiable) on a black background—should lend this method high sensitivity, the sophistication and specifications of the instruments available do not achieve this promise. Turbidimetry—detection of a small decrease in a large signal—should be limited in sensitivity; however, current instruments have excellent discrimination and can quantify small changes in signal, thereby allowing turbidimetric measurements to achieve high sensitivity.

Turbidimetry and nephelometry have similarities to absorption spectrophotometry, and many sources of interference and errors are common to all these systems. Many techniques, discussed in Chapter 55, that can be used to minimize absorption interferences are also applicable to turbidimetry and nephelometry. Because of the uniqueness of nephelometric measurements, especially in the case of antibody-antigen reactions, some specific applications are discussed below.

Endogenous color and choice of wavelength. Basic light-scattering theory predicts that the intensity of scattered light increases as shorter wavelengths of incident light are used. Most immunological assay reactions employ serum protein reactions requiring the choice of a wavelength at which neither the proteins nor colored serum components absorb appreciably. Since proteins absorb strongly below 300 nm and serum has an absorption peak at 400 to 425 nm because of porphyrins, instruments tend to operate in the 320 to 380 or 500 to 650 nm ranges. Reduction of the protein concentration by dilution will decrease background absorption. Most immunochemical reactions measured by nephelometry use high-affinity antibodies that allow for large dilutions of protein and consequent improvement of sensitivity.

Comparison of sensitivity. Sensitivity in nephelometers is largely controlled by the amount of background scatter from sample and reagents. Since background scatter can be high relative to specific scatter, instruments do not reach their full potential of sensitivity. This limitation, coupled with the higher wavelengths generated in laser instruments, accounts for the fact that laser instruments show no great increase in sensitivity over conventional nephelometers.

Sensitivity in turbidimetric measurements depends on the ability of the detector to resolve small changes in light intensity. Using low wavelengths and high-quality spectrophotometers with their high-precision detection systems, sensitivity in turbidimetry is usually adequate for many measurements and in many cases compares well to neph-

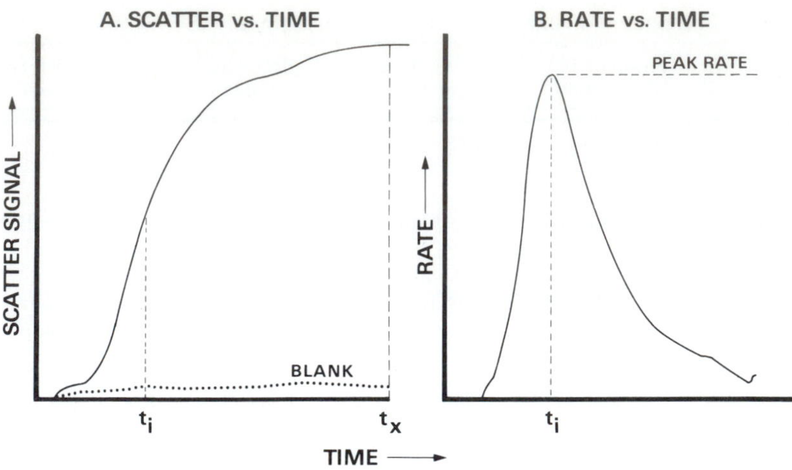

Fig. 3-26 Kinetic analysis of light scattering. *Curve A,* Intensity of scattered light signal versus time; *curve B,* rate of change of scattered light signal versus time.

elometry. Theoretically, with additional refinements, nephelometry ultimately should provide higher sensitivity than turbidimetry.

End-point versus kinetic analysis. Examination of light scattered as a function of time, after there is mixture of an antibody and antigen, shows that after an initial delay there is an almost linear increase in scatter followed by a slower attainment of plateau scatter. The secondary reaction occurs much more slowly than the first because larger particles form and begin to flocculate, and they distort the scatter intensity seen at forward angles. Both turbidity and nephelometry measurements behave in this manner.

There are two basic ways of measuring light scatter caused by this reaction, end-point analysis and rate analysis. End-point analysis requires blank (reagent) determinations and a reasonable amount of elapsed time before final measurement. Fig. 3-26 shows the forward scatter developed at 70 degrees of a rate nephelometric analyzer.[24] Comparing the two graphs, one can see the differences between an end-point analysis (blank value versus reading at $t = x$) and a rate or kinetic analysis (increase in scattered intensity over a set time interval). The kinetic approach, which electronically subtracts any blank signal, does not require a separate reagent blank to be run. Both kinetic and end-point analysis can be applied equally to turbidimetry and nephelometry.

A comparison of several commercially available light-scattering devices and some of their features is given in Table 3-5.

REFRACTIVITY

Principle

When a beam of light impinges on a boundary surface, it can be reflected, absorbed, or, if the material is transparent, pass into the boundary and emerge on the other side. When light passes from one medium into another, the path of the light beam will change direction at the boundary surface if its speed in the second medium is different from that in the first (Fig. 3-27). This bending of light is called *refraction*.[26]

Since the degree of refraction of a light beam depends on the difference in the speed of light between two different mediums, the *ratio* of the two speeds has been expressed as the *index of refraction,* or *refractive index*. The relative ability of a substance to bend light is called *refractivity*. The expression of a refractive index, n, is always relative to air with the convention that n of air = 1. The measurement of the refractive index is the measurement of angles, since the light is bent at an angle proportional to the relationship of n in the medium through which the light is passing:

$$\frac{n}{n_1} = \frac{\sin \theta}{\sin \theta_1}$$

The refractivity of a liquid depends on (1) the wavelength of the incident light, (2) the temperature, (3) the nature of the liquid, and (4) the total mass of solid dissolved in the liquid. If the first three factors are held constant, the refractive index of a solution is a direct measure of the total mass of dissolved solids.

Applications

Refractometry has been applied to the measurement of total serum protein concentration.[27] The assumption of this analysis is that the serum matrix (that is, the concentration of electrolytes and small organic molecules) remains essentially the same from patient to patient. Since the mass of protein is normally so much greater than the mass of other serum constituents, small variations of these other substances have no significant effect on the refractive index of serum. Refractometers are calibrated against "nor-

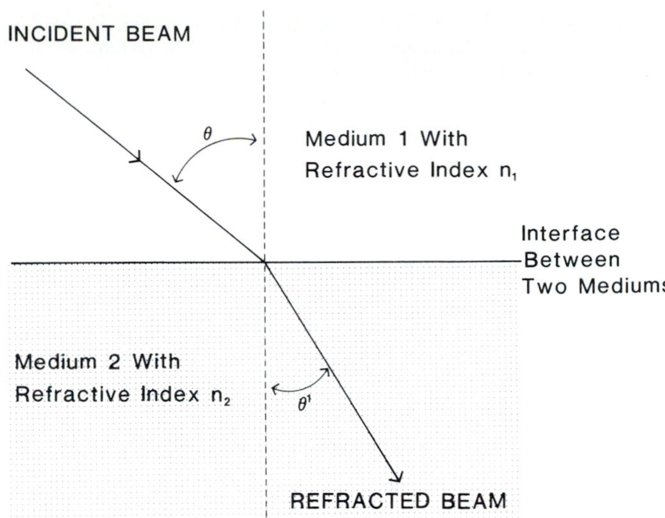

Fig. 3-27 Schematic illustrating bending of light when it passes from a medium of one density into a medium of a different density, with an angle of deflection, θ^1.

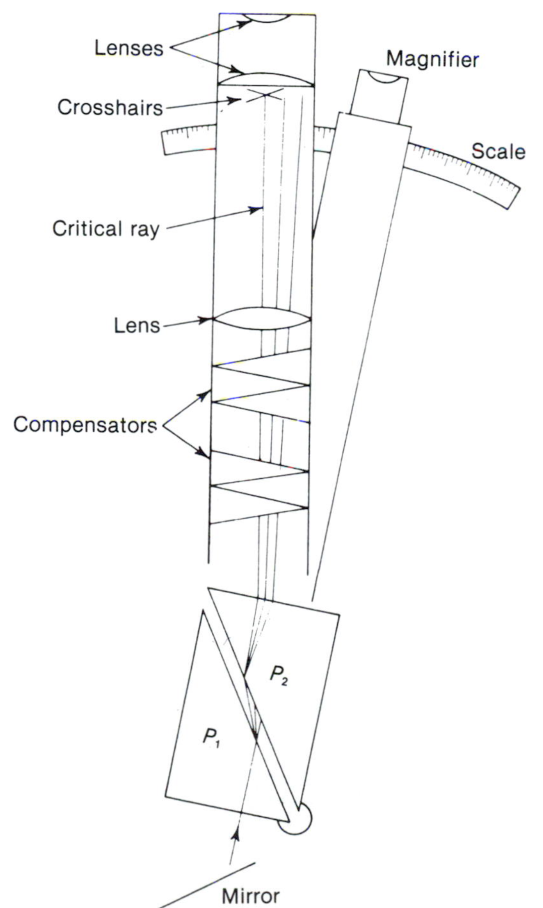

Fig. 3-28 Schematic of an Abbé refractometer. *(From Shugar, GJ, Shugar, RA, and Bauman, L: Chemical technicians' ready reference book, New York, 1973, McGraw-Hill Book Co.)*

mal'' serum, and total protein concentrations are read directly from a scale.

Refractometry is also used to estimate the specific gravity of urine samples. The refractive index is linearly related to the total mass of dissolved solids and thus to specific gravity. This remains valid over most of the range normally encountered for urine (that is, up to 1.035 g/mL).

Interference

When the concentration of small molecular weight compounds or particulate matter greatly increases, positive interference results. This interference occurs in the presence of hyperglycemia, hyperbilirubinemia, azotemia (increased serum urea), lyophilized samples, and hyperlipidemia. Hemolysis will also result in false-positive values for total serum protein.

Instrumentation

Most clinical refractometers are based on the Abbé refractometer (Fig. 3-28), marketed by American Optical Corporation. This refractometer consists of two prisms and a series of lenses. Light passes through the first prism where the light beam is dispersed. The dispersed light passes into and through the thin layer of the liquid sample where it is refracted. The light beam passes through a second prism where the light is again dispersed and on leaving is again refracted. The boundary at the edge of the refracted light beam is aligned perpendicularly to the scale on which serum protein concentrations or specific gravity can be read. The scale for reading serum protein (g/dL or g/L) is established by calibration of the instrument against a ''normal'' serum solution.

This type of refractometer is extraordinarily simple, having no moving or electrical parts. Thus it is easily reproducible, measuring protein with a precision of $\pm 1\%$ and an accuracy of ± 1 g/L. The sample size is on the order of 50 μL.

More complex refractometers are used to monitor column effluents for high-performance liquid chromatography (HPLC) analysis (see Chapter 5).

REFERENCES

1. Richards, WG, and Scott, RR: Structure and spectra of molecules, New York, 1985, John Wiley & Sons, Inc.
2. Clayton, RK: Light and living matter: the physical part, New York, 1970, McGraw-Hill Book Co.
3. Jaffe, HH, and Orchin, M: Theory and applications of ultraviolet spectroscopy, New York, 1966, John Wiley & Sons, Inc.
4. Willard, HH, Merritt, LL, and Dean, JA: Instrumental methods of analysis, ed 4, Princeton, NJ, 1965, D Van Nostrand Co.
5. Frings, CS, and Broussard, LA: Calibration and monitoring of spectrometers and spectrophotometers, Clin Chem 25:1013-1017, 1979.
6. Brewer, JM, Pesce, AJ, and Ashworth, RB, editors: Experimental techniques in biochemistry, Edgewood Cliffs, NJ, 1974, Prentice-Hall, Inc.
7. Keller, H: Optical methods of measurement. In Richterich, R, and Colombo, JP, editors: Clinical chemistry, New York, 1981, John Wiley & Sons, Inc.
8. Simonson, MG: The application of a photon-counting fluorometer for

the immunofluorescent measurement of therapeutic drugs. In Kaplan, LA, and Pesce, AJ, editors: Nonisotopic alternatives to radioimmunoassay, New York, 1981, Marcel Dekker, Inc.
9. Instrumentation Guidelines Study Group, Subcommittee on Enzymes: Clin Chem 23:2160-2162, 1977.
10. Alexander, LR, and Barnhart, ER: Photometric quality assurance instrument check procedures, Atlanta, Ga, 1980, US Department of Health and Human Services, Centers for Disease Control, Bureau of Laboratories.
11. Curme, HG, Columbus, RL, Dappen, GM, et al: Multilayer film elements for clinical analysis: general concepts, Clin Chem 24:1335-1342, 1978.
12. Rubeska, I: Atomic absorption spectroscopy, Cleveland, 1969, Chemical Rubber Co Press.
13. Robinson, JW: Atomic absorption spectroscopy, ed 2, New York, 1973, Marcel Dekker, Inc.
14. Winefordner, JD, editor: Spectrochemical methods of analysis, New York, 1977, John Wiley & Son, Inc, Wiley Interscience.
15. Dvorak, J, Rubeska, I, and Rezak, Z: Flame photometry: laboratory practice, Cleveland, 1971, Chemical Rubber Co Press.
16. Pesce, AJ, Rosen, CG, and Pasby, TL, editors: Fluorescence spectroscopy, New York, 1971, Marcel Dekker, Inc.
17. Wehry, FL, editor: Modern fluorescence spectroscopy, New York, 1976, Plenum Press.
18. Hemmila, I: Fluoroimmunoassays and immunofluorometric assays, Clin Chem 31:359-370, 1985.
19. Enhanced luminescence: a practical immunoassay system, The Medicine Publishing Foundation Symposium Series 18, Oxford, Medicine Publishing Foundation, 1986, pp 1-56.
20. Spencer, RD: Fluorescence polarization. In Kaplan, LA, and Pesce, AJ, editors: Nonisotopic alternatives to radioimmunoassay, New York, 1981, Marcel Dekker, Inc.
21. Jolly, MD, Stroupe, SD, Wang, CJ, et al: Fluorescence polarization immunoassay. I. Monitoring aminoglycoside antibiotics in serum and plasma, Clin Chem 27:1190-1197, 1981.
22. Ritchie, RF, editor: Automated immunoanalysis, parts 1 and 2, New York, 1978, Marcel Dekker, Inc.
23. Deverill, I, and Reeves, WG: Light scattering and absorption developments in immunology, J Immunol Methods 38:191-204, 1980.
24. Gauldie, J: Principles and clinical applications of nephelometry. In Kaplan, LA, and Pesce, AJ, editors: Nonisotopic alternatives to radioimmunoassay, New York, 1981, Marcel Dekker, Inc.
25. Kusnetz, J, and Mansberg, HP: In Ritchie, RF, Automated immunoanalysis, part 1, New York, 1978, Marcel Dekker, Inc.
26. Glover, FA, and Gaulden, JDS: Relationship between refractive index and concentration of solutions, Nature 200:1165-1166, 1963.
27. Rubini, MD, and Wolf, AV: Refractometric determination of total solids and water of serum and urine, J Biol Chem 225:868-876, 1957.

CHAPTER 4

Chromatography: theory and practice

M. WILSON TABOR

OBJECTIVES

- State the general principles of chromatography and describe their application to the divisions and subdivisions of chromatography, supporting each description with an illustration of mechanism.
- State the effect of each of the following chromatographic parameters on the resolution of a chromatographic separation:
  a. Height equivalent to a theoretical plate
  b. Retention time
  c. Mobile phase polarity
  d. R_f
- Describe the separation processes involved with the following types of chromatography and list the class of molecules that can be separated by each type:
  a. Adsorption
  b. Partition
  c. Ion exchange
  d. Gel permeation
- Describe four physiochemical forces that are central to polarity and explain how polarity affects the chromatographic behavior of compounds in normal-phase and reversed-phase chromatography.
- List three basic techniques for sample preparation for chromatography and provide the basis of purification for each technique.

KEY TERMS

adsorption Process whereby one substance adheres to another because of attractive forces between surface atoms of the two substances. (See Fig. 4-13.)

analyte The substance or component in a sample that is being measured.

band A chromatographic zone; that is, a region where the separated substance is concentrated.

capacity factor The ratio of the elution volume of a substance to the void volume in the column. (See Equation 4-6 and Fig. 4-9, *A*.)

chromatogram A series of separated bands or zones detected either visually, as in some paper chromatographic or thin-layer chromatographic separations, or indirectly by a detection system. In the latter case, the detection system usually outputs an electrical signal, which is graphically plotted through time, to display the series of separated bands or zones.

chromatography A method of analysis in which the flow of a mobile phase (gas or liquid) containing the sample promotes the separation of sample components by a differential distribution between this phase and a stationary phase. The stationary phase may be a solid or a liquid coated or bonded on a solid.

dipole The attractive force of compounds having centers of both positive and negative charges that is the result of an unequal sharing of bonding electrons between two elements of the compound with large differences in electronegativity. (See Fig. 4-11, *B*.)

dispersive force The attractive force, sometimes termed *van der Waals' force,* of compounds that results from the induction of a temporary dipole within an individual compound. (See Fig. 4-11, *A*.)

efficiency A measure of chromatographic performance usually related to the sharpness of the peaks in the chromatogram and quantitated by the number, *N,* of theoretical plates of a column. (See Equation 4-3, Figs. 4-6 and 4-7, and Table 4-1.)

electrostatic attraction The attractive force between compounds with formal positive or negative charges. (See Fig. 4-11, *D*.)

eluotropic series Series of solvents or solvent mixtures arranged in the order of their ability to elute a solute from an adsorbent.

elution Removal of a solute from a stationary phase by passage of a suitable mobile phase.

elution volume (v_e) The volume of mobile phase required to elute a solute from a chromatographic column. (See Equation 4-6 and Fig. 4-9, *A*.)

emulsion For a liquid-liquid extraction, a mixture formed when one of the immiscible liquid phases becomes dispersed as fine droplets in the other immiscible liquid phases.

equilibrium concentration distribution coefficient (K_D) The ratio of the concentration of a sample component in one phase to its concentration in a second phase at equilibrium. The two phases may be two immiscible liquids or the mobile phase and the stationary phase. (See Equation 4-1 and Fig. 4-3.)

height equivalent to a theoretical plate (HETP) The number obtained by dividing the column length by the theoretical plate number. (See Equation 4-4 and Fig. 4-7.)

hydrogen bonding The attractive force of compounds where a hydrogen atom covalently linked to an electronegative element, like oxygen, nitrogen, or sulfur, has a large degree of positive character relative to the electronegative atom, thereby causing the compound to possess a large dipole. (See Fig. 4-11, *C*.)

pK_a The pK of an acid is the pH at which it is half dissociated.

partition Process by which a solute is distributed between two immiscible phases. (See Fig. 4-15.)

peak A band or zone in a chromatogram showing a maximum of concentration between two minima.

polarity (P') The attractive forces encompassing the total interaction of solvent molecules with sample molecules and of solvent or sample molecules with the stationary phase.

R_f A ratio used in paper chromatography and thin-layer chromatography that is the distance from the origin to the center of the separated zone divided by the distance from the origin to the solvent front. (See Equation 4-9.)

resolution (R *or* R_s) The degree of separation between two components by chromatography. (See Equations 4-2, 4-8, and 4-10 and Fig. 4-5.)

retention time (t_R) The time that has elapsed from the injection of the sample into the chromatographic system to the recording of the peak maximum of the component in the chromatogram. (See Fig. 4-7.)

selectivity (α) The ratios of the capacity factors for two substances measured under identical chromatographic conditions; sometimes termed *separation factor,* or *chromatographic selectivity*. (See Equation 4-7 and Fig. 4-9, *B*.)

theoretical plate number (N) A number defining the efficiency of the chromatographic column. (See Equation 4-3 and Fig. 4-7.)

void volume (V_o) The interstitial volume of the chromatographic column; that is, the volume of mobile phase imbibed in the pores and around the stationary phase in a column. (See Equation 4-8 and Fig. 4-9, *A*.)

The need for fast, reproducible, and accurate analyses for many classes of analytes present in small amounts is being met today in the clinical laboratory largely as a result of the developments in chromatography during the past two decades. *Chromatography* is a collective term referring to a group of separation processes whereby a mixture of solutes, dissolved in a common solvent, are separated from one another by a differential distribution of the solutes between two phases. One phase, the solvent, is mobile and carries the mixture of solutes through the other phase, the fixed or stationary phase. Chromatographic methods encompass a number of variations in technique in which the mobile phase ranges from liquids to gases and the stationary phase ranges from sheets of cellulose paper to capillary glass tubes as fine as a human hair that are internally coated with a covalently bonded complex or complex organic polymers. A cursory examination of the scientific literature shows that both the numbers of chromatographic methods published and their applications have been growing exponentially.[1-4]

Modern chromatography began in 1906, when Michael S. Tswett detailed his separation of chlorophylls using a column of calcium carbonate (chalk)[5] and introduced a system of nomenclature that is now universally applied to chromatography (''color writing,'' from the Greek words *khramoma, khramomato-,* meaning ''color,'' and *graphe,* meaning ''writing'). This is but the first of the many published accounts of the colorful history of chromatography.[6-10]

BRANCHES OF CHROMATOGRAPHY

Chromatographic methods are generally classified according to the physical state of the solute carrier phase, that is, the mobile phase. These branches are represented in Figs. 4-1 and 4-2 as solution and gas chromatography, referring to the respective liquid and gaseous states of the mobile phase. In Fig. 4-1 these branches are further classified according to how the stationary phase matrix is contained for a particular chromatographic method. For example, solution chromatography is divided into flat and column methods, depending on whether the stationary phase is a thin layer mechanically supported on a sheet or is packed into a column. The flat method of support may involve use of a sheet of paper, such as cellulose, or a thin layer on a mechanical backing, such as glass or plastic.

Column methods are classically referred to as *liquid chromatography*. Furthermore, *column methods* is a phrase generally used to subdivide solution chromatography wherein the stationary phase is packed into a glass or metal tube. However, it is noted that gas chromatography is strictly a column method because a column must be used for containment of the stationary phase (see Chapter 6).

The main divisions of chromatography, based on mobile phase, may also be subdivided according to mechanism of solute interaction with the stationary phase (Fig. 4-2). Two mechanisms, adsorption and partition, are the most commonly encountered for both solution and gas mobile-phase separations. Adsorption chromatography (liquid-solid [L/S] or gas-solid [G/S]) is a process whereby solutes of a sample are separated by their differences in *attraction* to the stationary versus the mobile phase. Partition chromatography (liquid-liquid [L/L] or gas-liquid [G/L]) is a process whereby the solutes of a sample are separated by differences in their *distribution* between two liquid phases (L/L) or between a gas and a liquid phase (G/L). In both

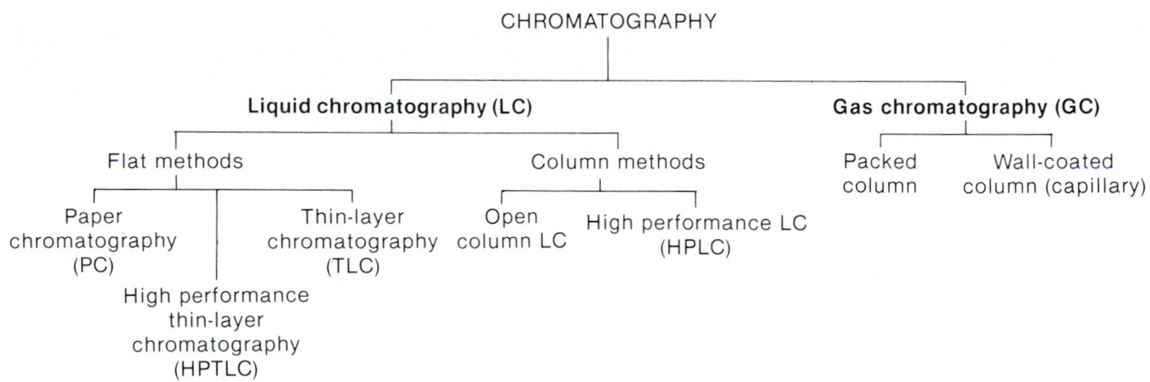

Fig. 4-1 Branches of chromatography according to mobile phase and physical apparatus.

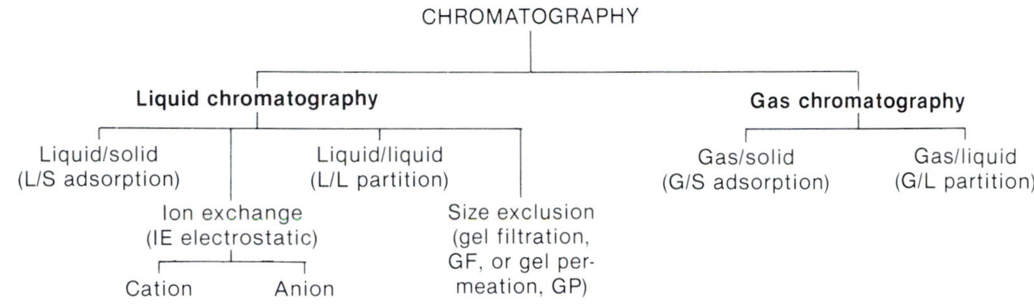

Fig. 4-2 Branches of chromatography according to mechanism of separation on stationary phase.

cases the stationary phase is liquid, and the mobile phase is a liquid or a gas.

Other mechanistic divisions of solution chromatography are ion exchange and gel permeation. Ion-exchange chromatography (IE) uses an insoluble matrix containing covalently linked ionic groups for the stationary phase, which can reversibly exchange either cations or anions with the mobile phase. Gel-filtration (GF) chromatography refers to a stationary phase of a solvent-swollen hydrophilic gel in the form of porous beads that is used with an aqueous-solvent mobile phase. Gel-permeation (GP) chromatography refers to a stationary phase of solvent-swollen hydrophobic gel in the form of porous beads that is used with an organic-solvent mobile phase. Both GF and GP chromatography are sometimes referred to as *molecular exclusion* (or *inclusion*) *chromatography* (see p. 88).

The boundaries between these mechanistically different types of chromatography are not finite, since for some chromatographic separations more than one mechanism may be operating. For example, in GF chromatography adsorptive interactions between the solute molecules and the stationary phase are common, in addition to the prevailing size-exclusion mechanism. More discussion of these mechanisms is presented below, after a brief discussion of chromatographic theory and principles.

GENERAL PRINCIPLES

The theoretical basis of chromatography is well developed, with both solution and gas-phase methods sharing the same foundation.[11] Only a few general concepts of this theory are discussed in this section. For more extensive discussions and additional reference leads, refer to representative books[12-15] and the *Analytical Chemistry* compendium reviews[1-4] on specialized chromatographic techniques.

The separation of a mixture containing two or more components is an operation with the goal of producing fractions, each of which has an increased concentration of one component relative to the other components contained in the original mixture. The physicochemical basis of chromatographic separation techniques is principally distribution equilibrium.[13] Distribution equilibrium refers to the differences in solubility and adsorption of a component in two immiscible phases.

Separation equilibrium can be visualized as the distribution of a solute, *S*, between two immiscible phases, upper phase *(u)* and lower phase *(l)*, at constant temperature and pressure. The ratio of the solute concentrations in the two phases determines the separation, which can be defined by an equilibrium concentration distribution coefficient, K_D, for the molar concentration, C_u and C_l, of solute in the upper and lower phases, respectively:

$$K_D = \frac{C_u}{C_l} \qquad \textit{Eq. 4-1}$$

Molar concentration is defined in the classical sense as the number of moles of solute per unit volume of solvent. The distribution coefficient is sometimes referred to as a *parti-*

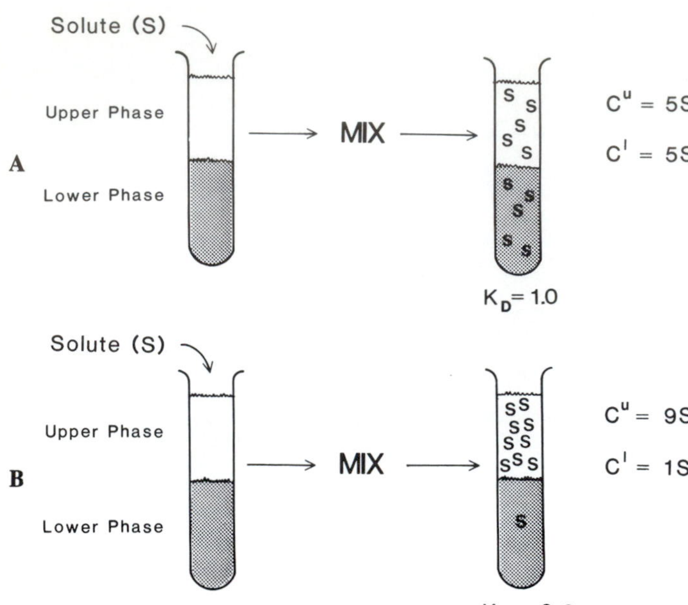

Fig. 4-3 Separation of a solute, *S*, by partition into two different solvent systems. In the first system, **A,** solute has a distribution coefficient, K_D, of 1.0, indicating an equal partitioning between upper and lower phases after mixing. In second system, **B,** solute has a K_D of 9.0, indicating a partitioning of nine parts of the solute in the upper phase and one part of the solute in the lower phase after mixing. C_u, Upper-phase concentration; C_l, lower phase concentration.

tion ratio. In practical terms, a K_D of 1.0 means that 50% of the solute is distributed in the upper phase and 50% is distributed in the lower phase (Fig. 4-3, *A*). Likewise, a K_D of 9.0 means that 90% of the solute is distributed in the upper phase and 10% is in the lower phase (Fig. 4-3, *B*).

For a more generalized application of distribution coefficients to chromatography, let $C_l = C_m$, where C_m refers to the amount of the solute distributed into a unit amount of mobile phase, and let $C_u = C_s$, where C_s refers to the amount of the solute distributed into a unit amount of stationary phase. In this more generalized case, the distribution equilibriums for a solute being separated in a given chromatographic system at constant temperature (that is, isothermal) can be graphically illustrated by a plot of solute concentration in the mobile phase, C_m, versus the concentration in the stationary phase, C_s. This is expressed as an adsorption-distribution isotherm (Fig. 4-4, *A*). The slope of the isotherm is equal to the distribution equilibrium coefficient, K_D.

The resulting shape of a distribution isotherm plot (Fig. 4-4, *A*) depends on several factors. Since solute movement between the mobile phase and the stationary phase is a thermodynamic equilibrium process, the distribution equilibrium coefficient, K_D, is both temperature and pressure dependent. However, the earlier assumption that temperature and pressure are not contributing factors in the situation under discussion is usually made for most routine chromatographic procedures. This assumption and the overall thermodynamic basis of solute-solvent interaction theory[12,14] hold only for dilute solutions of the solute in the mobile phase. This type of situation exists in conventional gas chromatography and high-performance liquid chromatography (HPLC). For these separations, the distribution isotherms approach linearity (Fig. 4-4, *A*), and the solute-concentration profiles resemble a discrete circular spot (Fig. 4-4, *B*) or a symmetric bell-shaped (that is, gaussian) elution peak (Fig. 4-4, *C1*). But it must be noted that one cannot predict behavior of any given sample in any given chromatographic system.

Linearity of distribution isotherms, as in Fig. 4-4, *A1*, is the exception rather than the norm. In most cases, concave or convex distribution isotherms are observed for solute-stationary phase interactions (Fig. 4-4, *A2* and *A3*, respectively). Several factors influence the degree to which nonlinearity is observed. At higher concentrations of solute in the mobile phase, nonlinearity results, since the thermodynamic basis of solute-solvent interactions only holds for dilute solutions. Another factor is the complexity of the sample, that is, the presence of multiple solutes. Interaction of these components with each other will cause a deviation from linearity.

If dilute solutions are used or there are no solute-solute interactions, a Langmuir or convex isotherm (Fig. 4-4, *A2*) may be observed. This isotherm shows that the limited number of adsorptive sites on the stationary phase become occupied by solute with increasing solute concentration, thereby losing their capacity to adsorb in proportion to the overall increase in solute concentration. Therefore the resulting solute-elution profile (Fig. 4-4, *B2* and *C2*) is characterized by a sharp front or leading edge that indicates a high solute concentration. The rear boundary of the solute elution pattern decreases asymmetrically from the peak.

When the solute is poorly adsorbed to the stationary phase, preferring the mobile phase, the anti-Langmuir or concave isotherm (Fig. 4-4, *A3*) is observed. In this situation, the resulting solute-elution profile (Fig. 4-4, *B3* and *C3*) is characterized by sloping (that is, low-solute concentration) front boundaries and sharp (that is, high-solute concentration) rear boundaries. Also, the elution volume for the solute is a function of sample size.

RESOLUTION

The ultimate goal of any given chromatographic technique is to separate the components of a given sample within a reasonable time. The purpose of such a separation is to detect or quantitate a particular component or group of components of interest in pure form. The ability to resolve the components from one another and the degree to which this resolution is accomplished are measures of the adequacy of the chromatographic separation. The question

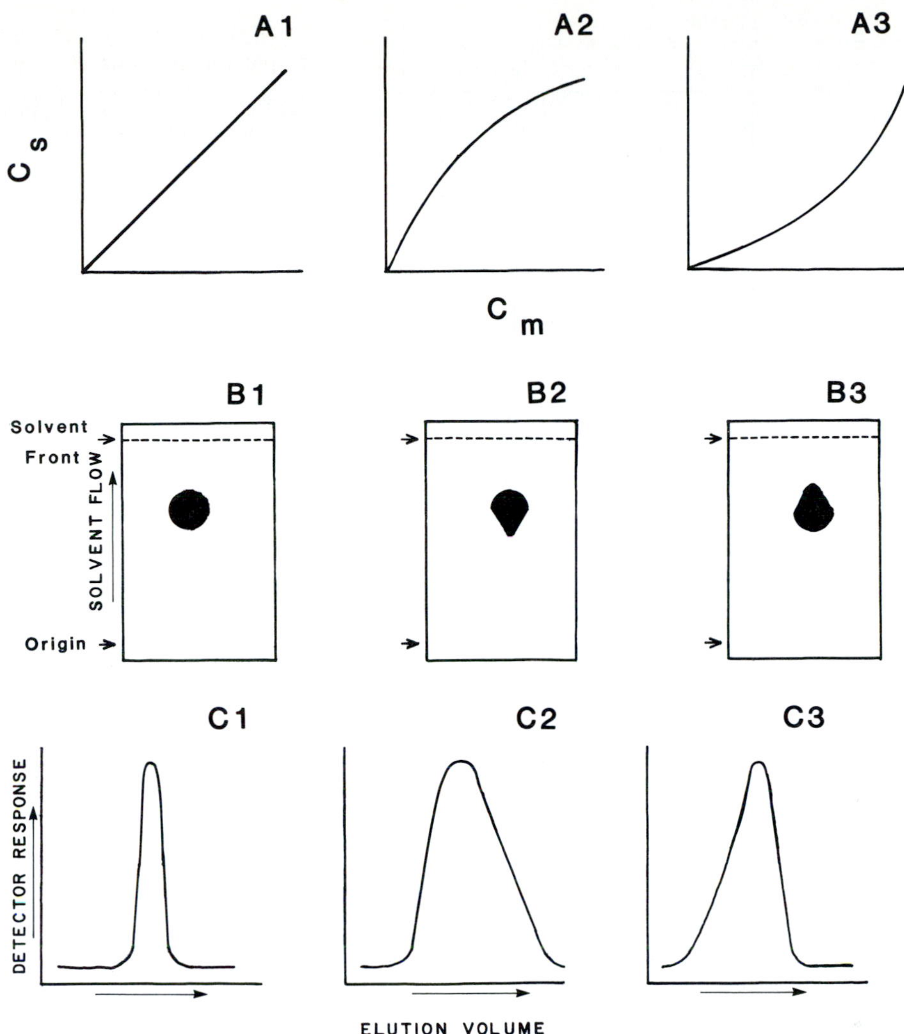

Fig. 4-4 Relationship between equilibrium distribution isotherm *(A1* to *A3)* and solute shape, position, and elution profile in a PC/TLC chromatogram *(B1* to *B3)* or a GC/HPLC chromatogram *(C1* to *C3)*. C_s, Concentration of solute in stationary phase; C_m, concentration of solute in mobile phase.

of what is adequate resolution for a given sample has been detailed by Snyder.[16] This question can be answered by defining the objectives of the chromatographic separation. Generally, objectives for the analyst in a chemical laboratory depend on the following questions. (1) Is a particular substance present in a sample; that is, should a qualitative analysis be followed? (2) How much of a particular substance is present in a sample; that is, should a quantitative analysis be followed?[17] In the following discussion of the theory of resolution, the principal emphasis and corresponding illustrations will be on column techniques, gas or liquid, rather than flat methods.

Note that, by convention, the concentrations of solutes in a chromatographic system are plotted out versus time or distance. The bands or zones of analytes separated are usually termed a *peak*.

An actual chromatographic separation of a three-component mixture by high-performance liquid chromatography is shown in Fig. 4-5, indicating important parameters for assessment of resolution. The quantity, *R*, for any two components is defined as the distance, *d*, between the peak centers of two peaks divided by the average base width, *W*, of the peaks:

$$R = \frac{d_2 - d_1}{1/2(W_1 + W_2)} \qquad \textit{Eq. 4-2}$$

For this calculation, both the distance, *d*, and the peak width, *W*, are measured in the same units, although units of elution volume can be used as well. This method of calculating resolution is based on the assumption that the distribution isotherms for the components being separated approach linearity (Fig. 4-4, *A*) under specified conditions.

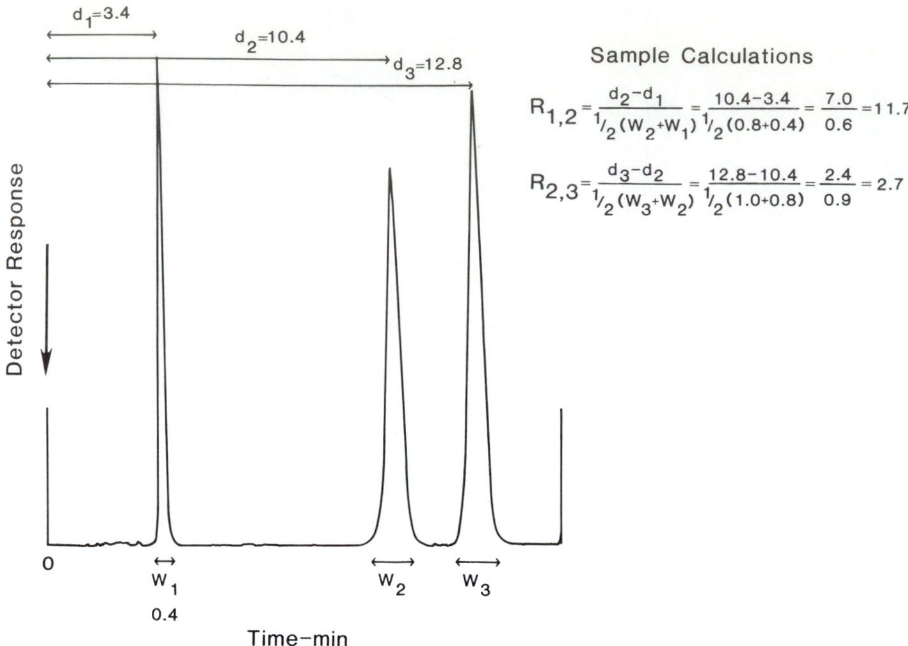

Fig. 4-5 Calculation of resolution of sample components actually separated by HPLC. The distances, d_1 to d_3, are the actual amounts of time from injection (↓) to apex of eluting peak for each component, *1* to *3*, respectively. Peak widths, W_1 to W_3, are measured by triangulation at base of each peak for components *1* to *3*, respectively. Both *d* and *W* must be measured the same way from the time of injection, that is, in units of time (minutes or seconds), length (inches or centimeters), or elution volume (milliliters). Resolution, *R*, is unitless.

The set of specified conditions are mobile phase, stationary phase, and solute-concentration range.

A resolution value of 1.25 or greater is required for good quantitative or qualitative chromatographic analyses. If the resolution is 0.4 or less, the peak shape does not clearly show the presence of two or more components. The actual value of the resolution depends on two factors: (1) width of the peak and the distance between the peak maxima for column separations and (2) diameter of the circular spots and the distance between these spots for flat-method separations.

These determinant factors of resolution also are indicative of the efficiency of the chromatographic process. Efficiency is decreased by the broadening of a solute band as it migrates through the stationary phase. If broadening occurs to any significant extent during the chromatographic process, the resulting peaks will be wide or the resulting spots will be diffuse. The separation of components is then poor, and the sensitivity with which they can be detected is reduced. An example of high- versus low-efficiency of separation is illustrated in Fig. 4-6 for both column and flat-method separations.

Solute-band broadening occurs during the actual chromatographic process and may be described as follows for a column separation. The sample, in a small volume of solvent, is introduced into the mobile phase at a point near the inlet end of the column. Once entering the column, the sample begins to disperse by thermal diffusion processes, which continue as it passes through the column. The longer the time a solute band spends in a column (that is, the longer the retention time), the greater the opportunity for thermal diffusion and the greater the band dispersion. The result is broader but gaussian (symmetrical, as in Fig. 4-4, *C1*) elution peaks for the more retained solutes. This process is similar in flat methods of chromatography, resulting in larger diameter but symmetrical spots, as in Fig. 4-4, *B1*, for the more retained solutes. Several additional factors contribute to solute-band broadening.[18,19] These include nonuniform regions of the stationary-phase, nonuniform particle-size distribution, and nonuniform column packing, which may result in nonuniform passage of solute molecules. In the latter case, some molecules spend more time in the separation system than do others; that is, they have a longer path. These processes result in broader, nongaussian dispersion of solute bands, such as asymmetric eluting peaks or trailing spots, as in Fig. 4-4, *B2* and *C2*, respectively.

Therefore one way to maximize column efficiency (that is, decrease the extent of band broadening) is to use a well-packed chromatography column that contains a stationary-phase packing that is not only small but also uniform in size distribution of particles. Columns meeting

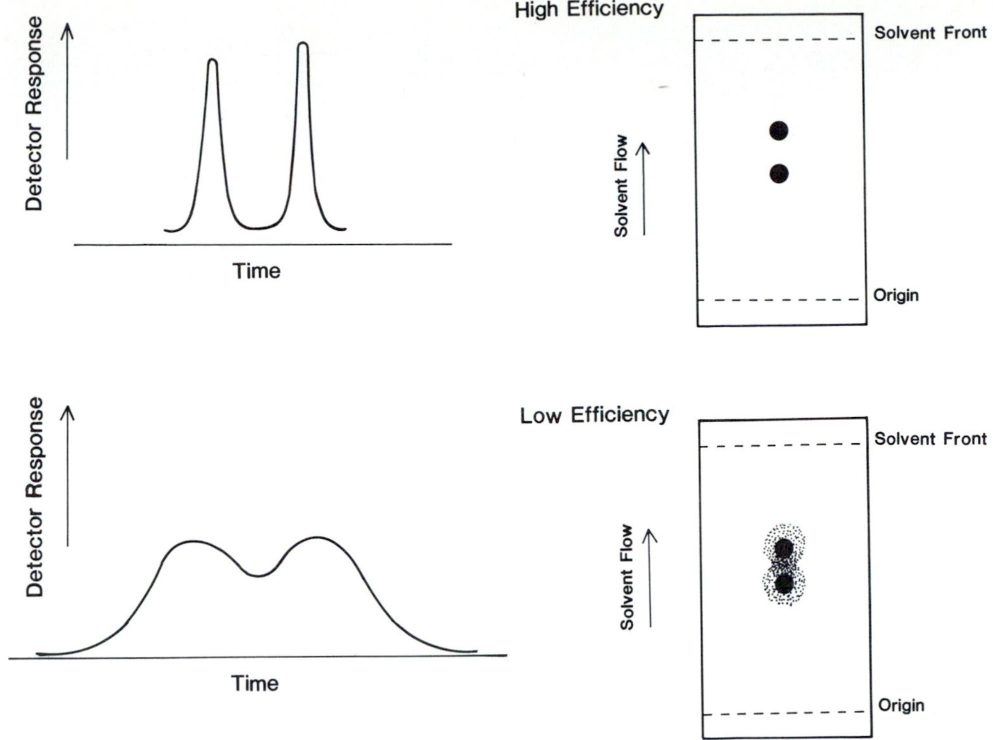

Fig. 4-6 Model chromatograms exemplifying high-efficiency separations and low-efficiency separations.

these criteria are readily obtainable today from numerous commercial sources.

Theoretical plates

A numerical assessment of column efficiency can be obtained by calculating the number of theoretical plates, N, for a given column. A theoretical plate is a microscopic segment of a column where a perfect equilibrium is assumed to exist between the solute in the mobile and stationary phases. The theory originated with Martin and Strain[20] in their mathematical treatment of the chromatographic process. The number of theoretical plates can be calculated for a column directly from the resulting chromatogram (Fig. 4-7) by the following:

$$N = 16 \left(\frac{t_R}{W}\right)^2 \qquad \textit{Eq. 4-3}$$

In this expression, W is the base width of the chromatographic peak, and t_R is the retention time of the solute, that is, the time from introduction of the sample onto the column to the apex of the eluting solute peak. Both t_R and W are measured in the same units—time or distance. (Note that N is dimensionless.) A large number of theoretical plates indicates relatively narrow peaks, that is, an efficient column.

Related to the number of theoretical plates is the column height equivalent to a theoretical plate[20] (HETP), which can be calculated by the following:

$$\text{HETP} = \frac{L}{N} \qquad \textit{Eq. 4-4}$$

In this equation, L equals the column length, usually in millimeters. Maximum column efficiency is obtained when HETP is as small as possible.

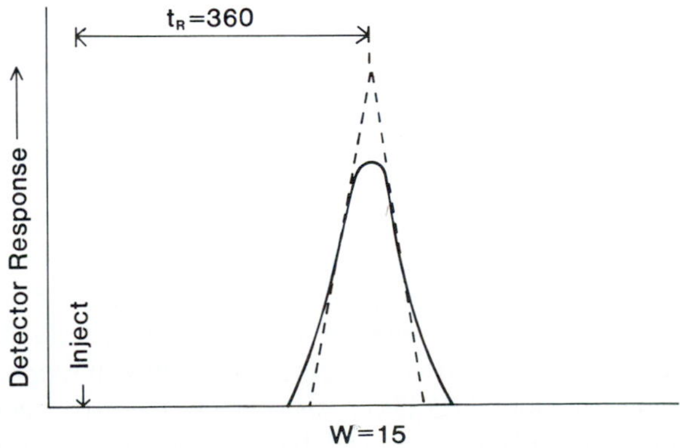

Fig. 4-7 Calculation of number of theoretical plates, N, from an HPLC or gas chromatogram. Elution or retention time, t_R, is measured from time of injection to apex of eluting peak for component. Peak width, W, is measured at the base of the peak by triangulation, as shown. Both are measured in same units.

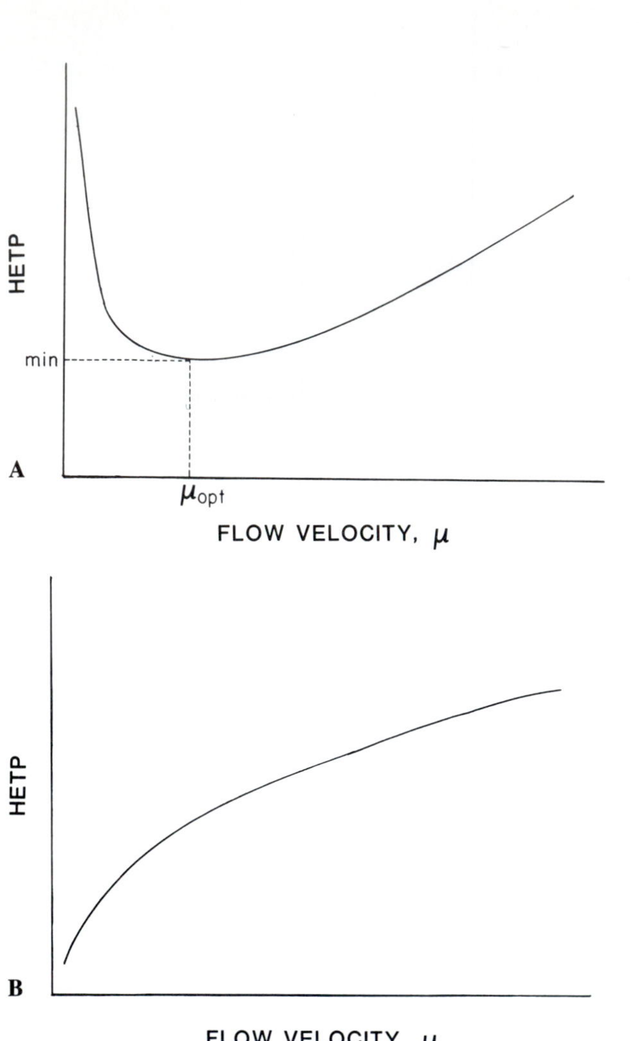

Fig. 4-8 Dependence of theoretical plate height, HETP, on mobile-phase velocity, μ, in gas chromatography, **A,** and in liquid chromatography, **B.**

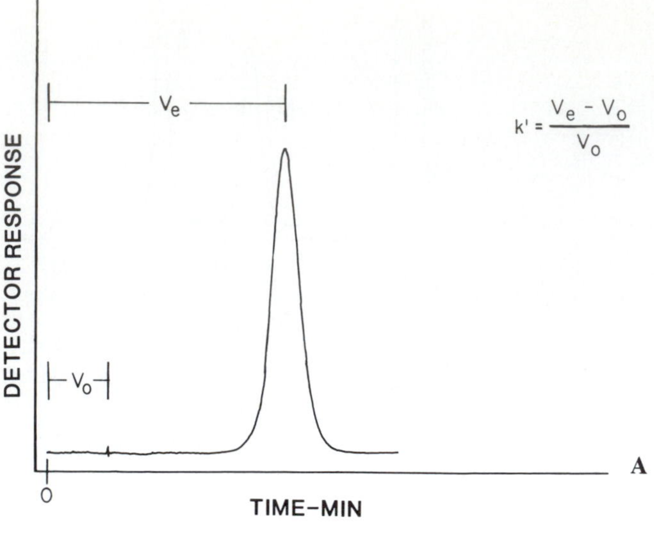

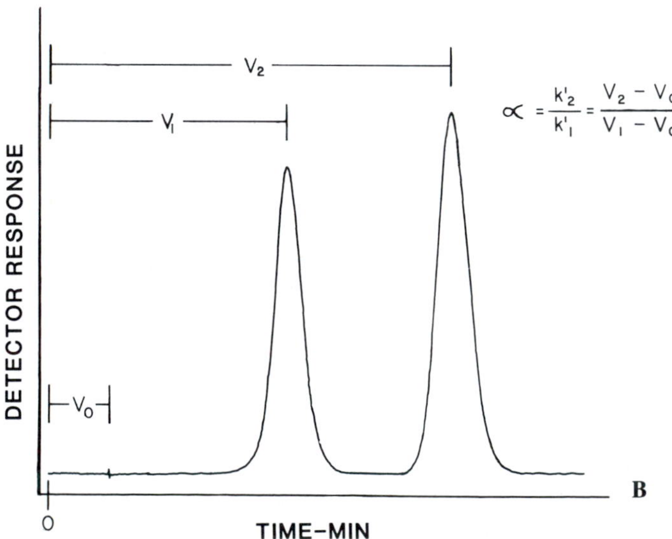

Fig. 4-9 **A,** Calculation of capacity factor, k', from an HPLC chromatogram. Sample was retained by stationary phase and underwent chromatography. **B,** Calculation of selectivity factor, α, from HPLC chromatogram. Solutes *A* and *B* were retained by stationary phase and underwent a chromatographic separation as indicated.

In addition to the factors affecting column efficiency, *N* is affected by the flow rate of the mobile phase. The mobile-phase linear velocity, μ, can be calculated by:

$$\mu = \frac{L}{t_o} \qquad \textit{Eq. 4-5}$$

In this equation, L is column length, and t_o is the time it takes a discrete portion of the solvent to flow through the column. A plot of HETP versus μ shows the experimental relationship of these two variables in gas chromatography (GC) (Fig. 4-8, *A*) and in liquid chromatography (LC) (Fig. 4-8, *B*). At the minimum HETP, the optimum flow velocity, μ_{opt}, is obtained. For gas chromatography (Fig. 4-8, *A*) this plot is the well-known van Deemter plot.[21]

At high mobile-phase velocities, HETP varies linearly with m for gas chromatography (Fig. 4-8, *A*), but in liquid chromatography (Fig. 4-8, *B*) a minimum in HETP is seldom observed, and the HETP tends to level off at high values for m. In practice, GC mobile-phase velocities are usually between one and two orders of magnitude greater than LC. For LC the lower velocities and column efficiencies are necessary because of the active role the mobile phase plays in the chromatographic separation process. Therefore, to improve separation efficiency in LC, one uses smaller stationary-phase particles, usually one tenth to one hundredth the size of those used in GC.

Retention

Resolution depends on factors in addition to theoretical plates. One of these is the ratio of the volumes of mobile and stationary phases in the column (that is, the capacity factor, k'), which can be calculated from the chromatogram (Fig. 4-9, *A*) by the equation

$$k' = \frac{V_e - V_o}{V_o} = \frac{t_R - t_o}{t_o} \quad \textit{Eq. 4-6}$$

The volumes in this equation are the void volume, V_o, of the column, that is, the volume of the mobile phase in the column, and the elution volume, V_e, of a solute retained by the stationary phase and undergoing chromatography. The HPLC chromatogram for Fig. 4-9, *A*, was obtained by injection of a sample containing two solutes, the first of which was not retained by the stationary phase and the second of which was retained, undergoing chromatography. The volumes, V_o and V_e, were then measured from the injection point to the apex of the peak of each component. As indicated above, the capacity factor can be calculated also by the measurement of the times, t_o and t_c, from sample injection to the apex of the peak of the component not retained and of the peak of the component retained, respectively.

Small values of k' indicate that the sample components are little retained by the stationary phase and elute close to the unretained peak. Large values of k' indicate that the sample components are well retained by the stationary phase and that long analysis times are required. For this latter situation, one must remember that solute-band broadening increases with residence time on the column because of an increased diffusion of the solute. Therefore the resulting peaks on elution will be wide and diffuse, decreasing sensitivity and making detection difficult.

The k' value for a particular solute is constant for any given chromatography system at constant mobile-phase compositions and stationary-phase size and composition. Within these limits, the capacity factor varies neither with flow rate of the mobile phase nor with column dimensions, that is, length and diameter.

Selectivity

Another parameter on which resolution depends is the selectivity factor, *alpha* (α), a term that describes the ability to separate two solutes. It is the ratio of the capacity factors for two solutes:

$$\alpha = \frac{k'_2}{k'_1} \quad \textit{Eq. 4-7}$$

The capacity factors for two solutes are determined as shown in Fig. 4-9, *B*, from which the resultant selectivity of the system can be calculated. The selectivity of any given column for the sample is a function of the process of solute exchange between the mobile phase and the stationary phase.[12] Therefore, to affect selectivity, one can change the chemical composition of either the mobile phase or the stationary phase to increase the preference of the solute for one phase or the other. In LC, changes either in the stationary or in the mobile-phases are usually made to improve selectivity. However, in GC, changes in stationary-phase chemistry are used for selectivity improvement.

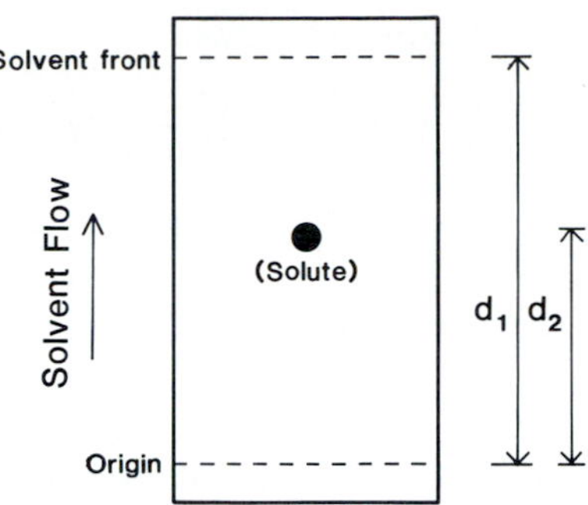

Fig. 4-10 Calculation of R_f (solvent-front ratio) value for a sample component from a paper or thin-layer chromatogram. The distance d_1 from origin to solvent front and distance d_2 from origin to center of spot of separated component are measured, as shown, in terms of same units; that is, length (centimeters or inches). Resulting R_f value, d_2 divided by d_1, is unitless.

Improving peak resolution

General peak resolution. With the definition of the factors affecting resolution, a more fundamental equation can now be written:

$$R = \left(\frac{N^{0.5}}{4}\right)\left(\frac{\alpha - 1}{\alpha}\right)\left(\frac{k'_2}{1 + k'_1}\right) \quad \textit{Eq. 4-8}$$

Therefore the resolution of any two solutes in a given chromatographic system is a function of three factors: (1) theoretical *N*, a column-efficiency factor; (2) a selectivity factor, α; and (3) a capacity factor, k'.

A similar relationship can be derived for flat methods of chromatography.[12] However, a major difference exists between flat and column methods of chromatography. In a column method of chromatography, the solutes in a given sample pass completely through the bed of the stationary phase. But in a flat method of chromatography, the separation process is stopped when the mobile phase has reached the end of the bed of the stationary phase, thereby resulting in the solute bands having migrated through only a portion of the bed (Fig. 4-10). For this type of separation, solute retention is measured in terms of the R_f value, which is the distance, d_2, migrated by the solute divided by the distance, d_1, migrated by the mobile phase, or solvent front. This relationship can be expressed as

$$R_f = \frac{d_2}{d_1} \quad \textit{Eq. 4-9}$$

In consideration of this major feature of flat methods of chromatography, the resolution equation for the separation of two solutes becomes as follows:

$$R = \left(\frac{N^{0.5}}{4}\right)\left(\frac{\alpha - 1}{\alpha}\right)\left(\frac{k'_2}{(1 + k'_2)^{2/3}}\right) \qquad \textit{Eq. 4-10}$$

The terms of this equation are as previously defined.

The question of the practical significance of the resolution equations can now be addressed. Specifically, how do these equations relate resolution to the actual experimental conditions of the chromatographic separation and the physical design of a particular chromatographic device? The three fundamental parameters, N, α, and k', of resolution can be adjusted more or less independently of each other. The dynamics or rates of the various physical processes that occur during the separation determine N.

Variation of N to optimize resolution. Experimentally one can change N by adjusting or varying a variety of parameters or conditions. Remember that a doubling of the value for N will increase the resolution by a factor of 1.4. (Note that in the resolution equation R is proportional to the square root of N.) Experimental parameters that can be changed to optimize N are summarized in Table 4-1.

Variation of capacity factor, k', to optimize resolution. Temperature is one parameter that is varied to effect a change in the capacity factor. Since the solute-stationary phase interactions are temperature dependent, a change in this parameter will affect solute retention. This is especially important in GC, but it is also important to a much lesser extent for other chromatographic techniques.

A second way to vary the capacity factor is by effecting changes in the strength of the stationary phase. For partition chromatography, increasing the percentage of liquid-phase coating on the matrix support will effect an increase in k'. Most GLC analyses are done on columns containing 2% to 10% ratios of liquid phase coating to stationary phase matrix support.[22,23] Since retention time is proportional to the amount of liquid phase present, lower percentage ratios mean lower analysis times. However, higher percentages of stationary phase for partition chromatography mean higher resolution. Therefore the final selection of the percentage of stationary phase must be a compromise between analysis time and resolution.

Table 4-1 Experimental parameters affecting N

Parameters	Direction of change necessary to increase N
Mobile-phase flow rate	Decrease
Column length	Increase
Average particle size of stationary phase	Decrease
Particle-size distribution	Decrease
Volume of sample introduced	Decrease
Viscosity of sample introduced	Increase
Viscosity of mobile phase	Increase
Temperature (as it controls mobile-phase viscosity)	Decrease
Extracolumnar effects (that is, dead volumes, excessive connective tubing)	Decrease

Additionally, one can vary the capacity factor, k', to improve resolution by changing the solvent strength of the mobile phase, but this is applicable only to LC methods (see p. 84). In GC the mobile phase is an inert gas like helium or nitrogen, but in LC the mobile phase is an active component of the chromatographic procedure, ranging from very nonpolar, such as hexane, to polar, such as water, in solvent strength.

For the adsorption and partition modes of LC (Fig. 4-2), the chromatographic process involves a continual distribution of the solute molecules between the mobile and the stationary phases. Any shift in the equilibrium concentration distribution coefficient (K_D) affects the capacity factor; that is, the elution volume per time of a particular solute in the given sample. One way to shift this coefficient is to increase or decrease the polarity of the mobile phase relative to the stationary phase. Numerous examples of mobile-phase alterations could be given. This is a powerful technique for varying the k' of solutes being separated (see p. 103, Chapter 5).

Variation of selectivity to optimize resolution. After optimization of the k' values, the selectivity, or separation factor, α, then can be adjusted to maximize the resolution. In general, the separation factor is varied by changes in the chemistries of the mobile phase, the stationary phase, or the sample. These changes may be accomplished as follows.

Derivatization. The chemical form of the sample can be altered in a variety of ways to affect selectivity. The most common method is derivatization. Knapp[24] has summarized an extensive variety of derivatization reactions for samples to be separated by GC. Derivatization is not only used in GC to improve sample volatility or solute detection limits, but it is also used to improve selectivity.[25] In LC, derivatization techniques are generally used to improve detection,[19,21,26] with the most extensive collection of methods presented in a book by Frei and Lawrence.[27]

Alterations in mobile-phase chemistry. Another method for improving the separation factor in LC is altering the mobile-phase chemistry.[28] Solvent polarity changes can be used to improve the capacity factor, k'. For improving the separation factor, solvents of similar polarity (strength) but different chemical natures are interchanged without affecting the overall polarity of the mobile phase, resulting in changes in the solvent selectivity. For example, the resolution of an HPLC separation initially developed with a mobile phase of methylene chloride/hexane, 50/50 by volume, could be improved by substitution of tetrahydrofuran for methylene chloride without changing the solvent strength.[14,29,30] The reason for improvement in resolution would be that tetrahydrofuran has a quite different selectivity toward certain classes of compounds, such as lipids, than methylene chloride does. In this case, substitution of an ether for an organic halide, the functional groups of the mobile phase have been altered. Changes in solvent functionality without changing solvent strength is a powerful technique for improving selectivity, and thereby resolution, in LC. Depending on the sample and chromatographic and detector requirements, the choice of solvent or solvents that can be used to effect changes in selectivity is extensive, as detailed by Snyder[31] in his compendium of properties of 911 solvents. A further discussion of the properties of solvents is presented in a later section.

Alterations in stationary-phase chemistry. As adjustments in the functionality of the mobile phase are used to improve the separation factor in LC, similar changes in the functionality of the stationary phase are commonly used to improve the separation factor in GC. For example, many polyester stationary phases, such as ethylene glycol succinate, can be replaced by the cyanopropyl silicone stationary phases, such as Silar 10C, allowing for a change in functionality with little or no change in polarity. For relatively nonpolar stationary phases, the hydrocarbon phases, such as Apiezon N, can be readily replaced by the alkyl silicone phases, such as OV-1, for a functionality change. These changes in stationary phase afford an improvement in selectivity, thereby leading to an optimization in resolution. Details as to the initial choice of phase are discussed in the following sections.

POLARITY

In previous sections, the importance of solvent strength and stationary-phase strength were discussed in relationship to their effect on resolution. Both forces affect chromatographic resolution by the capacity factor and the separation factor as described earlier in Equations 4-6 and 4-7. These parameters are linked by the *principle of polarity*.

The role of polarity is central to the interaction of molecules in the liquid or gaseous state and polarity is a major determinant property in the overall chromatographic process. The concept of polarity can be interpreted several ways[19,29,30] but generally is considered to encompass the total interaction of solvent molecules with sample molecules and of solvent or sample molecules with the stationary phase. The physicochemical basis for polarity is the interaction of attractive forces that exist between molecules. These four attractive forces are more specifically referred to as dispersive, dipolar, hydrogen-bonding, and dielectric interactions. As illustrated in Fig. 4-11 and discussed below, all four of these interactions involve the

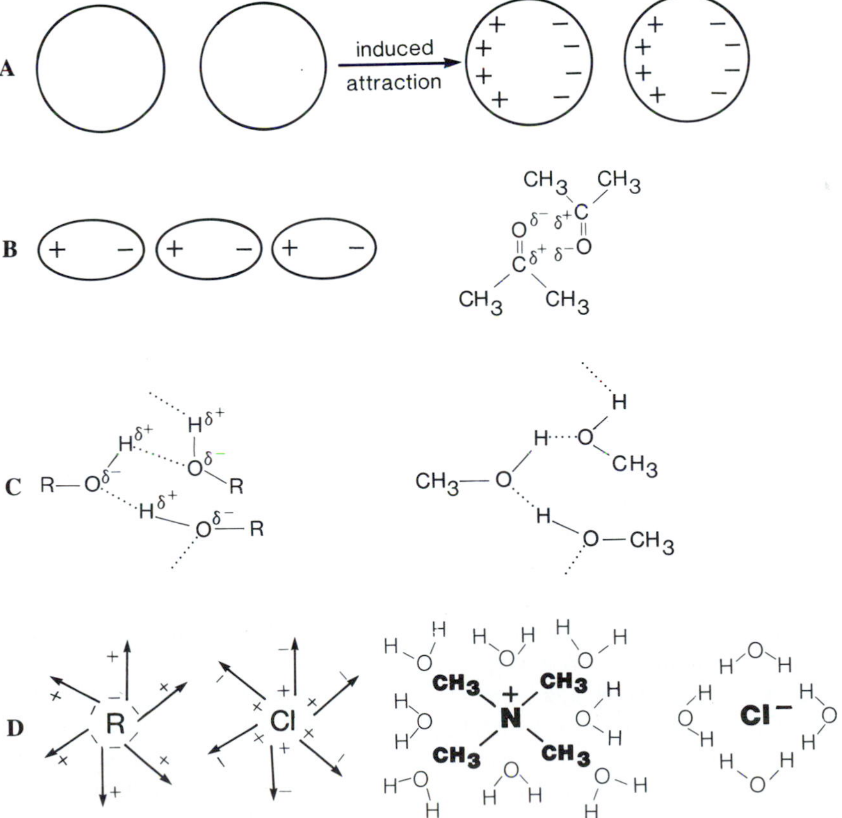

Fig. 4-11 Physicochemical interactions between molecules that constitute concept of polarity: **A,** dispersive or van der Waals' interactions; **B,** dipole interactions; **C,** hydrogen bonding; **D,** electrostatic interactions.

attraction of induced, partial, or formal positive and negative charges.

Dispersion interactions of molecules, sometimes termed *van der Waals' forces,* refer to the induced attraction between two molecules. This temporary separation of opposite charges (dipole) in one molecule induces the polarization of electrons in an adjacent molecule, thereby causing the two molecules to be attracted to each other by electrostatic interactions (Fig. 4-11, *A*). The formation of temporary dipoles in molecules is the physical basis for existence in the liquid state of many compounds composed of elements with small differences in electronegativity, for example, the elements carbon and hydrogen. Generally, dispersive interactions are an important determinant in polarity only when other forces are lacking or when some elemental constituents of molecules are electron-rich species (such as halogens in halohydrocarbons).

Some molecules possess permanent rather than temporary dipoles (Fig. 4-11, *B*). These compounds have centers of positive and negative charge that are the result of an unequal sharing of bonding electrons between two elements with large differences in electronegativity within the same molecule. This overall molecular dipole is enhanced by the presence of elemental nonbonding electron pairs within a compound. Elements such as oxygen, sulfur, halogens, and nitrogen possess nonbonding electrons when covalently linked to other atoms in compounds. The resulting permanent dipole is directional, with one end of the molecule being partially positive while the other is partially negative.

One special category of dipolar molecules is composed of those whose hydrogen is covalently linked to an electronegative element such as oxygen, nitrogen, or sulfur. In these molecules the hydrogen has a large degree of positive character relative to the electronegative atom to which it is bonded. Because of the small size of the hydrogen atom compared with other atoms, this positive end of the dipole can approach close to the negative end of a neighboring dipole. The force of attraction between the two is quite large, about 10 times that of normal dipolar interactions. This special case of dipole-dipole interactions is termed *hydrogen bonding* and is one of the most important types of weak attractive forces. This bonding is illustrated in Fig. 4-11, *C,* for the alcohol methanol.

The fourth type of attractive force is the electrostatic or dielectric interaction. In this case, the solute molecule or the stationary phase is a charged ionic species having either a formal positive or a formal negative charge. A small counterion, such as H^+ or Cl^-, is present but is generally separated from the charged solute or stationary phase because of solvation by the mobile phase (Fig. 4-11, *D*). These ionic species increase the dipolar character of the solvent by an enhancement of polarization. Dielectric interactions are quite strong and favor the dissolution of ionic or ionizable sample molecules in strongly dipolar solvents such as water or methanol.

The polarity of a solute molecule is the result of the four attractive forces described above. The polarity of a molecule will affect its interactions with the mobile and stationary phases. The more polar molecules have this property primarily because of strong dipoles, an ionic character, an ability to form strong hydrogen bonds, or a combination of the three forces. The less polar (nonpolar) molecules have dispersive forces as a primary basis of interaction with a very weak ability to interact through dipolar, hydrogen bonding, or dielectric forces. The practical aspects of these interactive forces and the degree of polarity or nonpolarity form the basis for the mechanisms of chromatography. The role of polarity in these mechanisms is discussed below for each of the three chromatographic constituents—the solute, the solvent or mobile phase, and the stationary phase.

Solvent polarity and solvent strength

Solvent strength in LC is a measure of the ability of the mobile phase to compete with the solute molecules for active sites (that is, interaction or attraction sites) on the stationary phase. When the stationary phase is silica gel, the active sites are the highly polar hydroxyl groups (Si—OH). Therefore, in this case, solvent strength increases with solvent or mobile phase polarity. However, when the stationary phase is nonpolar, such as a polydivinylbenzene (for example, XAD-2), the solvent strength decreases with increases in solvent or mobile phase polarity. The strength of a solvent is directly related to its polarity.

Solvent polarity has been described and quantitated in various ways.[29,30,32-35] The four most common solvent polarity classification schemes are (1) the Hildebrand solubility parameter, which is based on thermodynamic properties of the compound in question[35]; (2) the Rohrschneider polarity scale of P' values, which is based on measurement of solvent properties through the use of model solutes[31]; (3) the elutropic series, which ranks solvents in order of their eluting power for removing solutes from a polar stationary phase such as alumina[36]; and (4) the solvent selectivity grouping of Snyder and Karger, which is a blend of the three previous approaches and incorporates specific solubility parameters for dispersion, dipole, and hydrogen-bonding interactions. A listing of the solvent selectivity groups is given in Table 4-2 along with representative classes of compounds and specific examples for each group.

Solvents within any one group exhibit the same types of attractive forces. For example, the compounds of group 1 (I) are all strong hydrogen-bonding acceptors and weak hydrogen-bonding donors and have intermediate dipole moments. The solvents in group 5 (V) are compounds whose dipole interactions predominate over any hydrogen-bonding interactions. However, the solvents in group 0 are compounds in which only dispersion interactions (that is, temporary dipoles) are the predominate force. These sol-

Table 4-2 Solvent classification by selectivity groups

Solvent group	Representative classes or examples of compound
0	*n*-Alkanes, cyclohexane, saturated fluorochlorohydrocarbons
1 (I)	Aliphatic ethers, trialkylamines
2 (II)	Aliphatic alcohols
3 (III)	Pyridine derivatives, tetrahydrofuran, amides
4 (IV)	Glycols, benzyl alcohol, acetic acid
5 (V)	Methylene chloride, ethylene chloride, bis-2-ethoxyethyl ether
6 (VIA)	Aliphatic ketones and esters, nitriles, dioxane, sulfoxides
7 (VIB)	Nitrocompounds, phenyl alkyl ethers, aromatic hydrocarbons, carbon tetrachloride
8 (VII)	Halobenzenes, diphenyl ether
9 (VIII)	Fluoroalkanols, chloroform, water

vents do not have permanent dipoles, nor do they interact through hydrogen bonding.

The overall degree of the interactive forces of the solvent is quantitated in the polarity index, P.[29,30] A list of solvents commonly used in chromatography is presented in Table 4-3 with values for their polarity indices and their solvent selectivity group classifications. It is noted that solvents within any one group may vary widely in their overall degree of polarity. For example, the solvents listed for group 6 (VIA) vary more than three polarity units. In group 2 (II), the polarity index varies more than two units. Even though these variations are relatively large, one must remember that solvents within the same group exhibit the same kinds of interactions.

This information can be used as a basis for solvent selections and can be used to vary the mobile-phase selectivity, α, to improve resolution. The first important step in solvent selection is to maximize solute solubility in a given solvent. For a solvent to be an effective mobile phase in chromatography, it should dissolve the sample over the range of expected solute concentrations.

The most significant factor affecting solute solubility is solvent polarity. One can vary liquid mobile phase polarity systematically over a wide range of P' values by using a combination of two solvents, since solvent polarities are additive according to:

$$P' \text{ (of combined solvents } A + B) = \phi_A P'_A + \phi_B P'_B \qquad \textit{Eq. 4-11}$$

In this expression, ϕ_A and ϕ_B are the volume fractions of solvents A and B, which have solvent polarities of P'_A and P'_B, respectively. The two solvents should differ in P' values so that maximization of sample solubility can be achieved. It is important that the two solvents are miscible in all proportions. Usually a solvent that is too weak *(A)* to dissolve the sample is mixed with a solvent that is too strong *(B)*. The two solvents are now chosen from the solvent groups (Table 4-3) according to the compatibility of interactive forces with sample solutes and stationary-phase chemical functionalities. For example, to calculate the P' for solvent mixtures of isooctane (solvent A) and chloroform (solvent B), their individual polarity values are obtained from Table 4-3. Any mixtures of these two solvents would range in polarity from −0.4 to 4.3, that is, weakest (100% isooctane) to strongest (100% chloroform). A 50%:50% mixture of these two solvents would have a P' of (0.5 × −0.4) + (0.5 × 4.3), or 1.95. Also, for each 10% change in solvent composition, the polarity of the mixture would change by 0.47 units: 0.1 ($P'_B - P'_A$) or 0.1 (4.3 − [-0.4]) = 0.47.

Table 4-3 Representative solvent polarity indices

Solvent	Polarity index	Solvent group
Isooctane	−0.4	0
Hexane	0.0	0
Carbon tetrachloride	1.7	7
Dibutyl ether	1.7	1
Triethylamine	1.8	1
Toluene	2.3	7
Chlorobenzene	2.7	8
Diphenyl ether	2.8	8
Diethyl ether	2.9	1
Benzene	3.0	7
Ethyl bromide	3.1	6
Methylene chloride	3.4	5
1,2-Dichloroethane	3.7	5
Bis-2-ethoxyethyl ether	3.9	5
n-Propanol	4.1	2
Tetrahydrofuran	4.2	3
Chloroform	4.3	9
Ethyl acetate	4.3	6
Isopropanol	4.3	2
2-Butanone	4.5	6
Dioxane	4.8	6
Ethanol	5.2	2
Nitroethane	5.3	7
Pyridine	5.3	3
Acetone	5.4	6
Benzyl alcohol	5.5	4
Methoxyethanol	5.7	4
Acetic acid	6.2	4
Acetonitrile	6.2	6
Dimethylformamide	6.4	3
Dimethyl sulfoxide	6.5	6
Methanol	6.6	2
Nitromethane	6.8	7
Water	9.0	9

Once a suitable solvent mixture has been found to dissolve the sample, it is then applied to a chromatographic system such as a silica HPLC column. As previously discussed, the retention index, k', for the solutes in the sample is maximized to a value between 2 and 8. This is accomplished by adjusting the solvent polarity by small changes in the relative proportions of the two solvents.

Resolution is then maximized for the solute separation by adjustments in chromatographic selectivity, α. This is done by exchanging one of the mobile-phase solvents for a solvent in another solvent selectivity group (see Table 4-3) but with the *same polarity* of the overall mixture being kept. For example, chloroform, $P' = 4.3$, could be exchanged with isopropanol, $P' = 4.3$, in the previously described isooctane-chloroform mixture. The net effect of this exchange would be to go from a solvent that is a strong hydrogen-bonding donor, chloroform, to a solvent that is both a strong hydrogen-bonding donor and hydrogen-bonding acceptor. Other examples of such changes were given earlier in the discussion on selectivity. Further discussion of solvent selection and optimization are given by Snyder.[19,30]

In principle, this approach is a general guide to solvent choice both for dissolving the sample and for optimizing a mobile phase in LC (HPLC, thin-layer chromatography, and so on). It does, however, require knowledge of the forces involved in the chromatographic separation process. The latter is discussed below in the sections on the role of stationary-phase polarity and the mechanisms of chromatography.

Stationary-phase polarity and selectivity

The stationary phase is the fundamental component of the chromatographic separation (Table 4-4). The role of the stationary phase depends on its selectivity, which, in turn, is determined by the polarity of the phase. The forces constituting stationary-phase polarity are the same interactions responsible for mobile-phase (solvent) polarity: dispersion, dipole, hydrogen bonding, and dielectric.

The relative strength of polarity of stationary phases is more difficult to ascertain than the relative strength of polarity of liquid mobile phases, as previously described. Most attention has been directed to the stationary phases used in GC wherein the two major variables determining the separation are the stationary phase itself and the temperature of the column. In this mode of chromatography, studies of the affinity of solute molecules of varying polarities have led to a classification system for stationary phase according to polarity. Details on the Rohrschneider[37] and McReynolds[38] classifications can be found in Chapter 6.

Table 4-4 Preferred stationary phases

Phase type	Examples
Dimethylsilicone	OV-1, OV-101, SE-30, SP-2100
50% Phenylmethylsilicone	OV-17, SP-2750
Trifluoropropylmethylsilicone	OV-210, SP-2401
Polyethylene glycol	Carbowaxes
Polyesters	DEGS, EGSS-X, EGA
3-Cyanopropylsilicone	Silar 10C, SP-2340, Apolar 10C

From Hawkes, S, Grossman, D, Hartkopf, A, et al: J Chromatogr Sci 13:115-117, 1975.

MECHANISMS OF CHROMATOGRAPHY

The mechanisms by which a chromatographic method can separate sample components are generally based on polar interactions and physical interactions (interactions resulting from the size and shape of the solute molecules). The latter interaction is principally the mechanism for gel permeation chromatography, though molecular size plays a minor role in other modes of chromatography. The other broad mechanistic classes of chromatography are adsorption, partition, and ion exchange. Each is briefly discussed in the following section.

Adsorption

The interactions of solute or mobile phase molecules at the surface of a solid particle form the basis of the adsorption mechanism. There are fundamentally two types of adsorbents, nonpolar and polar. The latter category includes those that are acidic (that is, having electron-accepting surfaces) and those that are basic (that is, having electron-donating surfaces).

The nonpolar adsorbents have limited application in gas-solid chromatography (GSC)[22] (see Chapter 6). The mechanism for adsorption of the solute to nonpolar stationary phases is principally by dispersive interactions. Retention is determined by the adsorption energy and the surface volume of the stationary phase. A decrease in temperature or an increase in pressure increases adsorption. The converse is also true; that is, an increase in desorption can be accomplished by an increase in temperature or a decrease in pressure.

Polar adsorbents are the most widely used stationary phases in liquid-solid chromatography (LSC), with applications in both flat and column methods. Limited applications are found in GSC. The most common stationary phases are silica and alumina for LSC and silica or porous glass and aluminosilicates (zeolites or molecular sieves) for GSC. This latter group also has application in LC for gel permeation chromatography, in which the separation is principally through the mechanism of molecular exclusion. This is discussed later in the section on the mechanism of gel permeation.

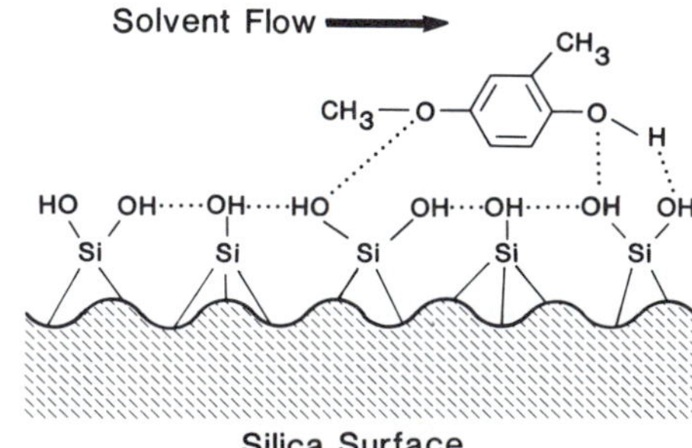

Fig. 4-12 Mechanism of separation of a metabolite of methylanisole by silica gel chromatography. Hydrogen bonds, ...; covalent bonds, ____.

Table 4-5 Selected groups of solutes in order of increased retention in normal-phase and reversed-phase chromatography

Reversed phase	Solute type	Normal phase
Most retained ↓	Fluorocarbons	Least retained ↑
	Saturated hydrocarbons	
	Unsaturated hydrocarbons	
	Halides and esters	
	Aldehydes and ketones	
	Alcohol and thiols	
Least retained	Acids and bases	Most retained

The principal polar adsorbents used in LC are silica and alumina, accounting for more than 95% of the applications in HPLC[19] and thin-layer chromatography (TLC).[15,39,40] Both hydrogen bonding and dipole interactions between the solute and the surface hydroxyl (silica and acid-washed alumina) or the oxygen anionic (base-washed alumina) groups of the stationary phases constitute the mechanism of separation by this method (Fig. 4-12). The number and topographic arrangement of these groups, along with the total surface area, determine the activity and strength of the adsorption. Retention of solutes on these phases increases with increasing polarity of the compound class (Table 4-5). Retention of a solute molecule requires displacement of adsorbed solvent molecules (Fig. 4-13). Adjustments in solvent polarity, as previously described, ultimately determine the strength of adsorption of the solute

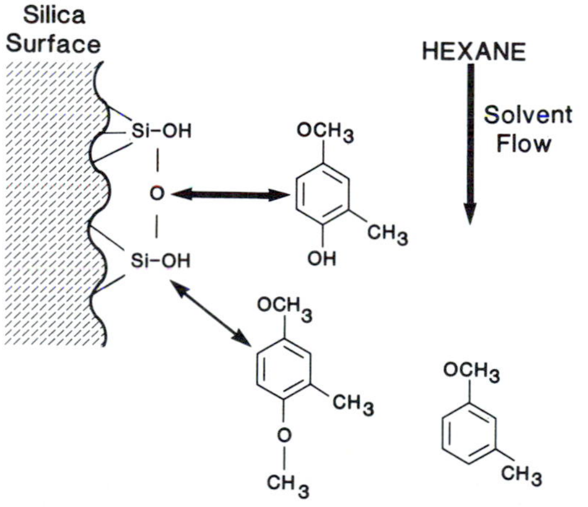

Silica Surface

HEXANE

Solvent Flow

Si-OH

O

Si-OH

OCH_3

CH_3

OH

OCH_3

CH_3

O

CH_3

OCH_3

CH_3

Fig. 4-13 Mechanism of adsorption chromatography by separation of 3-methylanisole and two of its biochemical metabolites. The most polar sample components, such as 3-methyl-4-hydroxyanisole, are retained the most by polar silica gel stationary phase *(heavy arrow)*. Sample components of intermediate polarity, such as 2,5-dimethoxytoluene, are retained to a much lesser degree *(light arrow)*, whereas relatively nonpolar components, such as 3-methylanisole, are not retained and prefer the nonpolar mobile phase, hexane.

to the stationary phase and the retention characteristics of the system.

Adsorption chromatography offers many advantages for use in LC separations. First, an extensive literature is available to the investigator for the separation of many types and classes of compounds by TLC. These methods are readily transferable to adsorption HPLC. Second, the flexibility, speed, and low cost of TLC allow its use in experimental development, particularly for selection of mobile phases. Once the optimum separation has been achieved, the transfer of the method to HPLC is straightforward. Third, TLC has a great value for use in preliminary investigation of samples of unknown constituents, particularly when one considers the advantages noted above. Finally, adsorption chromatography, particularly with silica gel, has been widely used for the separation of drugs in both the HPLC and TLC modes.

Partition

Partition chromatography is based on the separation of solutes by use of differences in their distribution between two immiscible phases. In liquid-liquid chromatography (LLC), the phase support is usually coated with a polar substance (normal phase), with separations accomplished by using an immiscible mobile phase. A normal-phase partition system would be silica coated with a monolayer of water or some other polar liquid and the use of a relatively nonpolar solvent system. Separations in this system are based on solute polarity, with the least polar compounds eluting first and the most polar substances retained the longest (Fig. 4-14). A similar separation system operates in paper chromatography, in which the cellulose is coated with an aqueous monolayer, and immiscible solvents are used as the mobile phase.

In 1969 Halasz and Sebestian[41] introduced a variation in the stationary phase for LLC, in which the silica support was chemically modified to produce a monolayer of a nonpolar organic substituent. These chemically bonded stationary-phase supports are available with a variety of functional groups. The most commonly used bonded phases are hydrocarbon phases such as octadecyl or octyl groups bonded to silica. The organic nature of the bonded phases imparts a nonpolar character to the stationary phase. Therefore the mobile phases commonly used are highly polar, such as water, methanol, or acetonitrile (see Table 4-3). Solutes are separated by their relatively nonpolar character (that is, the most polar eluting first), whereas the nonpolar solutes are retained longer (Fig. 4-15). From this type of separation characteristic, the use of bonded phases in LLC is termed *reversed-phase chromatography*. A further discussion and examples of normal-phase and reversed-phase LC are given in Chapter 5.

Another example of chromatography in which a partition mechanism operates is gas-liquid chromatography (GLC). The forces of interaction between solute molecules

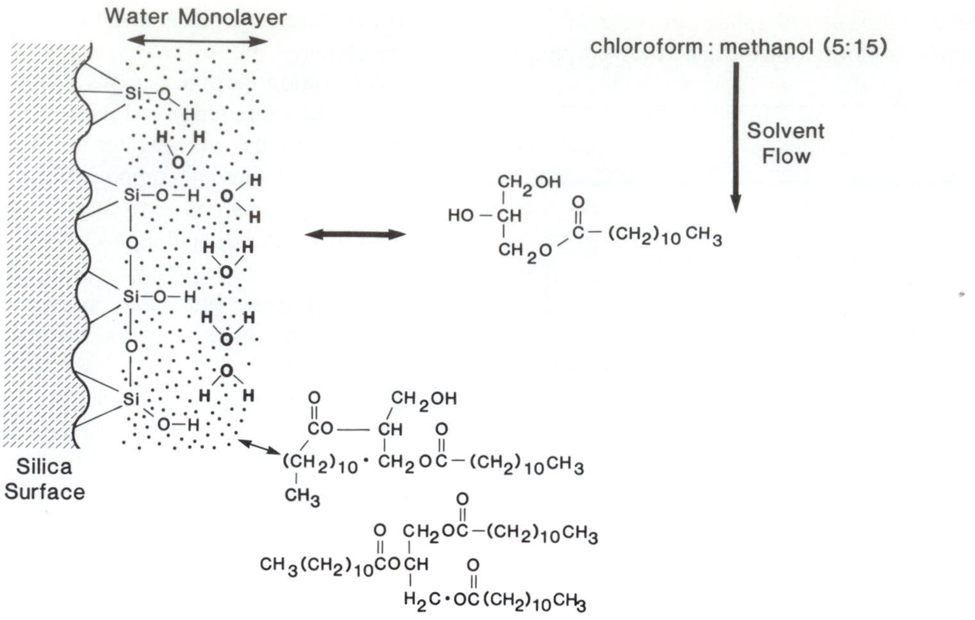

Fig. 4-14 Mechanism of liquid/liquid chromatography by separation of monoglycerides, diglycerides, and triglycerides of lauric acid. Silica gel stationary phase has monolayer of water strongly held by hydrogen bonding. Solute molecules are partitioned between liquid mobile phase (chloroform:methanol) and liquid stationary phase, or water monolayer. Most polar sample components, the monoglycerides, are retained most by polar stationary phase *(heavy arrow)*. Sample components of intermediate polarity, such as the diglycerides, are retained to a much lesser degree *(light arrow)*, whereas relatively nonpolar components, such as the triglycerides, are not retained and prefer relatively nonpolar stationary phase.

and the liquid-coated stationary phase are as previously discussed for LLC. However, for GC, the mobile phase serves as an inert carrier for the sample constituents, whereas in LLC the mobile phase is an active, interacting component in the partition mechanism.

Ion exchange

Ion-exchange chromatography uses stationary phases that possess formal positive or negative charges. The most common retention mechanism is the exchange of sample ions, *A*, and mobile-phase ions, *B*, with the charged groups, *R*, of the stationary phase:

$$A^- + R^+B^- \rightarrow B^- + R^+A^- \quad \text{Anion exchange} \qquad \textit{Eq. 4-12, A}$$

$$A^+ + R^-B^+ \rightarrow B^+ + R^-A^+ \quad \text{Cation exchange} \qquad \textit{Eq. 4-12, B}$$

In the first case, anion exchange is occurring, whereas cation exchange is shown for the second; sample ions compete with mobile-phase ions for ionic sites on the stationary phase. The sample ions that interact weakly with the stationary phase will be retained least, whereas those that interact strongly will be retained the most and will be eluted later. The principal force of these interactions is electrostatic, or the attraction of opposite charges.

To effect a separation of sample constituents, the extent of ionization of sample molecules is controlled by variations in pH of the mobile phase. Since the solutes are predominately weak acids, HA, or weak bases, B, a change in pH will shift the following ionization equilibriums either to the right or to the left:

$$\text{pH}\downarrow \qquad \text{pH}\uparrow$$

$$\text{HA} \rightleftarrows \text{H}^+ + \text{A}^- \qquad \textit{Eq. 4-13, A}$$

$$\text{BH}^+ \rightleftarrows \text{H}^+ + \text{B} \qquad \textit{Eq. 4-13, B}$$

An increase in ionization leads to an increased retention of the sample. Factors other than pH controlling solute retention in ion-exchange chromatography are (1) charge strength of the solute ion, (2) ionic strength of the mobile phase, and (3) charge strength of the counterion on the stationary phase. One can decrease the retardation of solutes by increasing the ionic strength of the mobile phase, decreasing the strength of the counterion, such as use of Na^+ instead of H^+ for cation-exchange phases, or by adjusting the pH of the mobile phase in a manner to decrease dissociation of either the solute, the counterion on the packing, or both.

Gel-permeation (molecular or size exclusion) chromatography

In contrast to the previous mechanisms and modes of chromatography, gel-permeation chromatography (GPC) separation is strictly based on molecular size. The stationary phase for GPC contains pores of a particular average

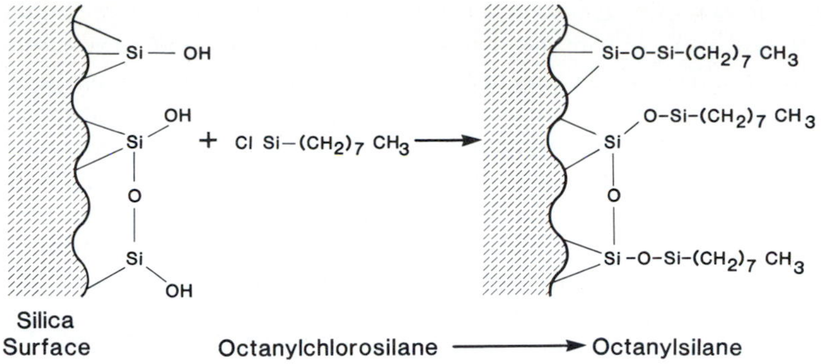

Si — OH
OH
Si
O
Si
OH
+ Cl Si–$(CH_2)_7$ CH_3 →
Si-O-Si-$(CH_2)_7$ CH_3
O-Si-$(CH_2)_7$ CH_3
Si
O
Si -O-Si-$(CH_2)_7$ CH_3
Silica Surface
Octanylchlorosilane → Octanylsilane

Fig. 4-15 Chemical preparation of bonded, stationary phase (reversed phase). Organochlorosilane reacts with nucleophilic hydroxyl (OH) groups of silica gel, forming siloxane covalent bond (Si—O—Si).

size, and if the sample molecules are too large to enter the pores, they are not retained (excluded) by the stationary phase and are eluted from the column first. Small sample molecules permeate deeply into the pores and are retained the longest. They ultimately diffuse from the pores and are swept away by the flow of the mobile phase. Intermediate-sized sample molecules enter the pores to some extent but are not retained as easily as the small sample molecules because they do not penetrate as deeply into the pores. They are eluted from the column in volumes between those needed to elute the largest solute (small V_e) and smallest solutes (large V_e). This mechanism is illustrated in Fig. 4-16.

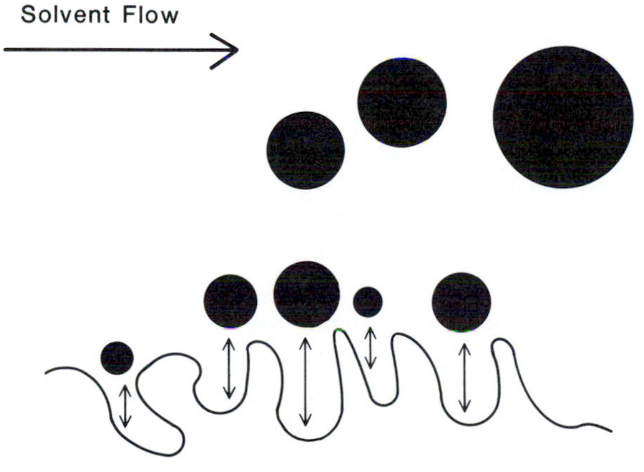

Fig. 4-16 Mechanism of size-exclusion chromatography. Stationary phase, in form of porous beads, contains pores of varying diameter. Mobile phase outside and inside pores is same, except that liquid inside is immobilized. When a sample containing solutes varying from small to large molecules elutes through column, small molecules penetrate all pores and are retained, thus being eluted later than large molecules, which move only in mobile phase. Molecules of intermediate size penetrate only some pores, thereby being retained to a lesser degree than small ones are.

The major advantage of this mode of chromatography is that the LC method can be used to separate virtually any sample, as long as it is soluble in a mobile phase. Additionally, it is applicable to soluble species with an average molecular weight of 50 to more than 10 million. Since molecular size is the property of interest, representative calibration curves should be obtained by use of calibration standards of known molecular weight. Likewise, stationary phase choice is based on the expected molecular weight range of the solute molecules in the sample and compatibility with the mobile phase. A mobile phase for GPC should be chosen first on the basis of sample solubility, and then a compatible stationary phase is selected. Most stationary phases are compatible with aqueous or proton-donating (such as methanol) solvents. However, stationary phases are available that are compatible only with organic solvents. Lists of the available phases for GPC are tabulated in the manufacturers' literature, in reviews,[42] and in books.[19,43]

SAMPLE PREPARATION FOR CHROMATOGRAPHY

Nature of problem

Few chromatographic analyses are conducted on the sample as submitted to the clinical laboratory. For any given sample, the goal of the chromatographic analysis is either a qualitative or a quantitative determination of its components. To achieve this objective, one should separate the components of interest as discrete zones with the same peak or spot distributions and k' or R_f values as the standards under identical chromatographic conditions. However, the complexity of a biological-sample matrix usually renders the chromatographic separation ineffectual by (1) interaction of sample impurities with the stationary phase, causing a reduction in the resolving power of the system; (2) saturation of most chromatographic detector systems, tending to raise the noise level and thereby decreasing sensitivity; (3) interaction of the component of interest with other matrix components, leading to irreproducibility of the separation from sample to sample; and (4) poorly resolved components that interfere with the analysis. To minimize these sample effects in the chromatographic separation, a strategy for separation of the analyte from interfering components (sample cleanup) is required.

Any separation method employed in the laboratory must meet the criteria of yield, separation, capacity, and cost-effectiveness. The advantages of having high yield in any sample manipulation step are obvious, but if recovery is quantitative with little purification, the method is unsatisfactory. The corollary is also true: if the separation is excellent but there is a low yield, the method is of little value. Many separation methods are readily applied on a large scale where large amounts of sample are available, but others are only applicable to small-scale separations. The criterion of cost-effectiveness includes time, equipment, reagents, and labor, which may render a separation method impracticable.

The strategy of preparation for chromatographic analysis should include consideration of whether the objective of the analysis is to qualitatively detect or to quantitate the substance under investigation.[44]

Mechanical methods for initial isolation of analyte

The type of sample matrix received by the analyst in a clinical chemistry laboratory varies from a simple homogeneous-appearing liquid such as perspiration to a complex heterogeneous solid such as feces. However, the most commonly received sample matrices are urine and blood (or plasma). The initial step in analyte preparation for chromatography will vary according to matrix.

Solid samples, such as tissues or feces, are first disrupted or treated for preparation of a homogeneous solution or suspension from which the analyte can be isolated. Homogenization of tissues in a blender such as a Polytron (Brinkmann Instruments, Inc., Westbury, N.Y.) with an appropriate solvent may solubilize the desired analyte. The use of a Potter-Elvehjem tissue grinder is also effective. Tissue can also be extracted in a mortar with a pestle and a small amount of solvent. In addition to these grinding or shearing techniques, solid samples can be disrupted by sonication in solvent or hydrolyzed by acid, base, or enzymes. A procedure for preparing homogeneous powders of feces for subsequent extraction has been developed for the investigation of drug metabolism. It involves grinding the sample with a stainless steel ball mill in the presence of anhydrous sodium sulfate.[45]

Liquid samples may also require an initial treatment for removal of analytes sequestered by matrix components. Mild-base hydrolysis has been used to release sequestered polychlorinated biphenyls from blood lipid components.[46] Whole blood can be diluted with sterile water to disrupt blood cells osmotically before analyte isolation. Another initial treatment applied to blood or urine samples is to remove proteins and other macromolecules through precipitation. Two of the more commonly used protein-precipitating agents are trichloroacetic acid and barium sulfate.

In other mechanical methods of matrix disruption, such as homogenization in a solvent or buffer, centrifugation is commonly employed to remove cell debris, particulate matter, or other large contaminants. An alternative method for the removal of insolubles is filtration, either through an inert material, such as glass wool, or through a nitrocellulose or nylon membrane.

Chromatographic methods for initial isolation of analyte

A common method for the initial isolation of components of interest from aqueous solutions, such as urine or blood, is the use of XAD-2 resin chromatography.[47] This stationary phase of polydivinylbenzene has a large surface area and is of a nonionic character, making it capable of adsorbing many classes of organic compounds from aqueous solution, principally by dispersive and dipole interactions. The adsorbed organics are eluted from the XAD-2 by organic solvents like methanol, acetone, diethyl ether, hexane, methylene chloride, or combinations of these solvents.[48] The XAD-2 method has been applied most often to urine- or blood-screening methods for drugs of abuse and their metabolites, but it can also be applied to isolate trace amounts of compounds.

Another very common chromatographic technique for initial analyte isolation is the use of small columns of silica or of octadecylsilyl-bonded phase.[49,50] The analytes from a relatively large volume of sample are adsorbed from aqueous solution by forces similar to those operating in the XAD-2 procedure. Desorption is accomplished when a small volume of an appropriate solvent is passed through the silica or reversed-phase cartridge; the sample may then be processed for any mode of chromatography. Many additional resins, of the type just described, are currently available for the isolation of compounds of interest to the clinical chemist.[51]

Other types of chromatographic methods have been used in the preparation of samples for analysis. Ion-exchange chromatography has been widely used to isolate charged analytes. Anion exchange, first suggested by Horning and Horning,[52] has been widely used for the isolation of acidic constituents from biological fluids. Although DEAE-Sephadex is the most widely used ion-exchange stationary phase for sample cleanup, other anion exchangers, such as AG1X and Dowex 3, have been used.

Extraction methods for analyte isolation

Liquid-liquid and liquid-solid partition methods have been widely used for both primary and secondary extraction steps in a wide variety of clinical chemistry analyses before the chromatographic quantitation step. The reasons for the use of extraction procedures are numerous, including the isolation of the analyte from large quantities of contaminating materials and its concentration into a small volume of solvent, making detection easier. Liquid-liquid extraction procedures are easily accomplished, usually permitting the workup of multiple samples simultaneously.

The success of an extraction step depends on knowledge of the polarity of the analyte. This information is used to select an extracting solvent that will effectively remove the analyte from the sample. A general rule of solvent selection is that compounds tend to favor solvents having the same polarity interaction forces. It is critical that the chosen solvent be immiscible with the sample matrix.

There are other points to consider in solvent selection. The solvent must be chemically compatible with the analyte; that is, no chemical reaction should be possible between the two. The solvent must be compatible with all subsequent operations after the extraction. For example, a solvent with a high boiling point would be difficult to remove, and so the analyte solution would be difficult to concentrate. The solvent should not introduce any contaminants that would make the analysis difficult. Many laboratory supply companies offer common solvents of high-purity grades, such as (1) *HPLC-grade solvents,* which are compatible with most detector systems and do not contain particulate matter that would foul the HPLC equipment; (2) *pesticide-grade solvents,* which are compatible with electron-capture GC detectors because they do not introduce any contaminating substances; and (3) *lipograde-grade solvents,* which do not contain any greases or other substances that would interfere with the analysis of lipids. These are but a few of the types of quality solvents available. If the solvent is not available in the required purity, a purification of the solvent must be done before use in any sample cleanup procedure. Most commonly, a distillation of the solvent will suffice, but sometimes more extensive purification measures are required. Methods of more rigorous purification procedures for most solvents are described in Weissberger's text.[53] Even with the use of the highest quality solvents commercially available or prior purification of solvents, impurities may still be a problem. The most common contaminant is plasticizers, usually coming from cap liners and other plastic materials.[54] These contaminants, various alkyl phthalates, can interfere with some analyses, particularly when electron-capture gas chromatography is used. Foil-lined screw caps of extraction tubes or sample vials have been shown to be the source of contaminants that interferred with GC.[55] These contamination problems can be eliminated by using screw-cap liners made of Teflon.

Once a decision on the solvents for extraction has been made, the actual operations in extraction must be considered. In general, a repeated series of extractions with smaller volumes of solvent will be more efficient than a single extraction with a large volume. For solid samples, the solvent may be introduced during the mechanical disruption step as previously mentioned. The cycle of grinding, sonication, and so on is repeated several times with several volumes of solvent. However, doing so sometimes does not effectively extract the desired analyte. In this case, the pulverized solid sample may have to be extracted with a Soxhlet extractor or a continuous infusion extractor. Both of these methods are more efficient than manual operations for extracting substances from a solid matrix. However, the requirements for these methods include a reasonably volatile extracting solvent and stability of the analyte at the boiling point of the solvent.

To extract an analyte with ionizable groups, it is best to first solubilize the solid sample in an aqueous solution. The pH of the sample solution is then adjusted below the pK_a of acidic components or above the pK_a of basic components with the addition of acid or base, respectively, to convert the analyte, 95% or greater, into its extractable (nonionized) form. A nomogram relating pK_a values of acids to percent ionization at various pH values has been published by Hopgood.[56] If the pK_a of the analyte is not known, a lowering of the pH of the aqueous solution to a pH of 2.0 by the addition of acid is usually sufficiently low to permit the extraction of most acidic analytes. Likewise, raising the pH to 12 usually is sufficient to permit the extraction of most basic analytes of unknown pK_a.

For liquid-liquid extraction, an increase in the ionic strength of the aqueous layer will enhance the ease of extraction of the analyte, causing it to favor the extracting solvent. An ionic neutral salt, such as sodium chloride or potassium bromide, is commonly used for this purpose.

One of the problems frequently encountered in liquid-liquid extractions is the formation of emulsions, that is, one of the immiscible phases becomes dispersed as fine droplets in the other. To avoid emulsion formation, several precautions can be taken during the actual extraction process: (1) If the two liquid layers have a large contact surface, avoid vigorous mixing of the phases. The use of gentle agitation will accomplish the extraction. (2) Filter all finely divided particulate matter before extraction. (3) Use solvent pairs with large differences in density.

If an emulsion does form, there are several steps that will possibly break it. First, try to get the dispersed droplets to achieve coalescence by mechanically disrupting their surfaces. Stirring with a glass rod or filtration through a loose bed of glass wool sometimes will break it. Second, if the densities of the two solvents are sufficiently different, centrifugation will sometimes effect separation. Third, cooling or freezing the mixture sometimes causes a coalescence of droplets. Fourth, an increase in ionic strength, by the addition of salt or a small amount of an alcohol, such as ethanol or 2-ethyl-1-hexanol, may cause a decrease in the forces stabilizing the emulsion. Fifth, a change in the ratio of the two solvents by addition of more extraction solvent or a partial evaporation of solvent may break the emulsion. Finally, filtration through phase-separation filter paper will break many emulsions commonly encountered.

In most cases, one of these procedures will be successful in breaking the emulsion.

For examples of the use of solvent extraction techniques, see the procedures for the analysis of drugs in Chapters 50, 54, 60, 61, and 66.

Processing of sample extracts

Many analyte extracts are too dilute for direct chromatographic analysis or for derivatization reactions before chromatography and are usually concentrated by evaporation of the extracting solvent.

Any solvent-evaporation procedure must be conducted with care to avoid loss of the analyte. Such a loss of analyte can occur if traces of water are present in the extract. These can be removed by use of an anhydrous salt such as sodium carbonate or sodium sulfate. Alternative desiccating salts such as calcium oxide or magnesium sulfate can also be used. Other purposes for drying an extract may be to conduct subsequently a derivatization procedure, such as acetylation or silylization, or to remove traces of water, which may interfere with the chromatography step.

During concentration of an extract, one must take care to avoid losses of the analyte. The analyte may be lost to the concentration vessel by irreversibly binding to the walls during concentration. This can be avoided by prior silylization of the glassware. Some substances are sufficiently volatile to form azeotrope mixtures with the solvent and be lost during evaporation. To avoid this, many gentle concentration methods or apparatuses are available. Micro-Synder or Kurderna-Danish concentrators evaporate solvent under mild conditions. If the analyte is heat sensitive or sensitive to oxygen, evaporation of the solvent under a stream of purified inert gas, such as nitrogen or argon, can be employed. In this case, one can warm the vessel to a range of 35° to 50° C to expedite the evaporation process. The use of a rotary evaporator under reduced pressure is also a gentle method for solvent evaporation. A comparison of solvent reduction methods has been made by Constable et al.[57] wherein recoveries of analytes varied from 41% to 140%, depending on the method employed. These results emphasize the importance of method validation and the key role that quality assurance samples, such as samples spiked with analyte, play in the use of a specific approach to cleanup.

Another method for concentrating the analyte is the back extraction of the compound of interest from the solvent. For example, MacGee[58] has published a variety of methods whereby the analyte, in the original extracting solvent, is back extracted into a small volume of analyte-derivatizing solvent before gas chromatography. Methods of this type expedite the analysis, since solvent-evaporation steps are not required. Other examples of analyte cleanup procedures for preparing samples for chromatography are detailed in the review by Ko and Petzold[59] and Sunshine.[60] Additional examples are given in Chapters 5 and 6 on HPLC and GC and for individual analytes in Section Three.

REFERENCES

1. Stinshoff, KE, Stein, W, Wood, WG, and Laska, PF: Clinical chemistry, Anal Chem 59:337R-350R, 1987.
2. Clement, RE, Onuska, FI, Eiceman, GA, and Hill, Jr, HH: Gas chromatography, Anal Chem 60:279R-294R, 1988.
3. Barth, HG, Barber, WE, Lochmueller, CH, et al: Column liquid chromatography, Anal Chem 60:387R-435R, 1988.
4. Sherma, J: Thin-layer and paper chromatography, Anal Chem 60:74R-86R, 1988.
5. Tswett, M: Absorption analysis and the chromatographic method: application in the chemistry of chlorophyll, Berichte der deutschen botanischen Gesellschaft 24:384-393, 1906. Translated by Strain, HH, and Sherman, J, in Tswett, M: Absorption analysis and chromatographic methods, J Chem Ed 44:238-242, 1967.
6. Heftman, E: History of chromatography. In Heftman, E, editor: Chromatography: a laboratory handbook of chromatography methods, ed 3, New York, 1975, Van Nostrand Reinhold Co.
7. Strain, HH, and Svec, WA: Differential methods of analysis. In Heftman, E, editor: Chromatography: a laboratory handbook of chromatography methods, ed 3, New York, 1975, Van Nostrand Reinhold Co.
8. Kalasz, H, and Ettre, LS: Chromatography: the state of the art, Wellingborough, UK, 1984, Collet's.
9. Ettre, LS: The development of chromatography, Anal Chem 43:20A-21A, 25A, 27A-31A, 1971.
10. Latiinen, HA, and Ewing, GW, editors: A history of analytical chemistry, Washington, DC, 1977, Analytical Chemistry Division of American Chemical Society.
11. Giddings, JC: Reduced plate height equation: a common link between chromatographic methods, J Chromatogr 13:301-304, 1964.
12. Schoenmakers, OJ: Optimization of chromatographic selectivity: a guide to method development, New York, 1986, Elsevier.
13. Tsuji, K, and Morozowich, W: GLC and HPLC determination of therapeutic agents, part I, New York, 1978, Marcel Dekker, Inc.
14. Snyder, LR: Principles of absorption chromatography, New York, 1968, Marcel Dekker, Inc.
15. Stahl, E: Thin-layer chromatography: a laboratory handbook, New York, 1969, Springer-Verlag.
16. Snyder, LR: A rapid approach to selecting the best experimental conditions for high speed liquid column chromatography. I. Estimating initial sample resolution and the final resolution required by a given problem, J Chromatogr Sci 10:200-212, 1972.
17. Katz, E: Quantitative analysis using chromatographic techniques, New York, 1987, John Wiley & Sons, Inc.
18. Giddings, JC: Non-equilibrium and diffusion: a common basis for theories of chromatography, J Chromatogr 2:44-52, 1959.
19. Snyder, LR, and Kirkland, JJ: Introduction of modern liquid chromatography, ed 2, New York, 1979, John Wiley & Sons, Inc.
20. Martin, AJP, and Synge, RLM: A new form of chromatogram employing two liquid phases. I. A theory of chromatography. II. Application to the micro- determinations of the higher monoamino-acids in proteins, Biochem J 35:1358-1368, 1941.
21. Van Deemter, JJ, Zuiderweg, FJ, and Klinkenberg, A: Longitudinal diffusion and resistance to mass transfer as causes of nonideality in chromatography, Chem Eng Sci 5:271-289, 1956.
22. Grob, RL, editor: Modern practice of gas chromatography, New York, 1977, John Wiley & Sons, Inc.
23. McNair, HM, and Bonelli, EJ: Basic gas chromatography, ed 5, Palo Alto, Calif, 1969, Varian Associates.
24. Knapp, DR: Handbook of analytical derivatization reactions, New York, 1979, John Wiley & Sons, Inc.
25. McMahon, DH: Methods development guidelines for chemical derivatization in gas chromatography, J Chromatogr Sci 23:426-428, 1985.
26. Lawrence, JF: J Chromatogr Sci 23:484-487, 1985.
27. Frei, RW, and Laurence, JF: Chemical derivatization in liquid chromatography, New York, 1977, Elsevier Scientific Publishing Co.
28. West, SD: The prediction of reversed-phase HPCL retention indices

and resolution as a function of solvent strength and selectivity, J Chromatogr Sci 25:122-129, 1987.
29. Keller, RA, and Snyder, LR: Relation between the solubility parameter and the liquid-solid solvent strength parameter, J Chromatogr Sci 9:345-459, 1971.
30. Karger, BL, Snyder, LR, and Eon, C: An expanded solubility parameter treatment for classification and use of chromatographic solvents and absorbents: parameters for dispersion, dipole and hydrogen bonding interactions, J Chromatogr 125:71-88, 1976.
31. Snyder, LR: Solvent selection for separation processes. In Perry, ES, and Weissberger, A, editors: Techniques of chemistry: separation and purification, ed 3, vol 12, New York, 1978, John Wiley & Sons, Inc.
32. Rohrschneider, L: Solvent selection in absorption liquid chromatography, Anal Chem 46:470-473, 1974.
33. Saunders, DL: Solvent selection in absorption liquid chromatography, Anal Chem 46:470-473, 1974.
34. Snyder, LR: Classification of the solvent properties of common liquids, J Chromatogr 92:223-230, 1974.
35. Hildebrand, JH, and Scott, RI: The solubility of non-electrolytes, ed 3, New York, 1964, Dover Publications, Inc.; and Hildebrand, JH, and Scott, RI: Regular solutions, Englewood Cliffs, NJ, 1962, Prentice-Hall, Inc.
36. Trappe, W: Die Trennung von biologischen Fettstoffen aus ihren natürlichen Gemischen durch Anwendung von adsorptionssäulen. I. Mitteilung: Die eluotrope Reihe der Lösungmittel, Biochem Z 305:150-161, 1940.
37. Rohrschneider, L: Eine Methode zur Charakterisierung von gaschromatographischen Trennflussigkeiten, J Chromatogr 22:6-22, 1966.
38. McReynolds, WO: Characterization of some liquid phases, J Chromatogr Sci 8:685-691, 1970.
39. Zlatkin, A, and Kaiser, RE: HPTLC: high performance thin-layer chromatography, Journal of Chromatography Library Series, 9, New York, 1977, Elsevier Scientific Publishing Co.
40. Touchstone, JC, and Dobbins, MF: Practice of thin layer chromatography, New York, 1978, John Wiley & Sons, Inc.
41. Halasz, I, and Sebestian, I: New stationary phase for chromatography, Angew Chem, Int Ed 8:453-454, 1969.
42. Anderson, DMW: Gel permeation chromatography. In Simpson, CF, editor: Practical high performance liquid chromatography, Philadelphia, 1978, Heyden & Sons, Inc.
43. Determann, H: Gel chromatography, ed 2, New York, Berlin, 1969, Springer-Verlag.
44. Tabor, MW: Chemical analysis for assessment and evaluation of human exposure to hazardous anthropogenic compounds, In Abbou, R, editor: Hazardous waste: detection, control, treatment, Amsterdam, 1988, Elsevier Science Publishers.
45. Smith, CC, Khalil, A, and Tabor, MW: Fractionation of urinary and fecal metabolites of the antimalarial drug WR-158,122 following oral doses in rats and rhesus monkey, Toxicologist 3:52, 1983.
46. Que Hee, SS, Ward, JA, Tabor, MW, and Suskind, RR: Screening method for Aroclor 1254 in whole blood, Anal Chem 55:157-160, 1983.
47. Stolman, A, and Pranitis, PA: XAD-2 resin drug extraction methods for biologic samples, Clin Toxicol 10:49-60, 1977.
48. Weissman, N, Lowe, ML, Beattie, JM, and Demetriou, JA: Screening method for detection of drugs of abuse in human urine, Clin Chem 17:875-881, 1971.
49. Shackleton, CHL, and Whitney, JD: Use of Sep-Pak® Cartridges for urinary steroid extraction: evaluation of the method for use prior to gas chromatographic analysis, Clin Chim Acta 107:231-243, 1980.
50. Heikkinen, R, Fotsis, T, and Adlercreutz, H: Reversed-phase C_{18} cartridge for extraction of estrogens from urine and plasma, Clin Chem 27:1186-1189, 1981.
51. Dressler, M: Extraction of trace amounts of organic compounds from water with porous organic polymers, J Chromatogr 165:167-206, 1979.
52. Horning, EC, and Horning, MG: Metabolic profiles: gas-phase methods for analysis of metabolites, Clin Chem 17:802-809, 1971.
53. Riddick, JA, and Bunger, WB: Organic solvents. In Weissberger, A, editor: Techniques of chemistry, vol 2, ed 3, New York, 1978, John Wiley & Sons, Inc.
54. DeZeeuw, RA, Jonkman, JHG, and van Mansvelt, FJW: Plasticizers as contaminants in high purity solvents: a potential source of interference in biological analysis, Anal Biochem 67:339-341, 1975.
55. Denney, DW, and Karsek, FW: Detection and identification of contaminants from foil-lined screw-cap sample vials, J Chromatogr 151:75-80, 1978.
56. Hopgood, MF: Nomogram for calculating percentage ionization of acids and bases, J Chromatogr 47:45-50, 1970.
57. Constable, DJC, Smith, SR, and Tanaka, J: Comparison of solvent reduction methods for concentration of polycyclic aromatic hydrocarbon solutions, Environ Sci Technol 18:975-978, 1984.
58. Kossa, WC, MacGee, J, Ramachandran, S, and Webber, AJ: Pyrolytic methylation/gas chromatography: a short review, J Chromatogr Sci 17:177-187, 1979.
59. Ko, H, and Petzold, EN: Isolation of samples prior to chromatography. In Tsuji, K, and Morozowich, W, editors: GLC and HPLC determination of therapeutic agents, part I, New York, 1978, Marcel Dekker, Inc.
60. Sunshine, I: Manual of analytical toxicology, Boca Raton, Fla, 1971, CRC Press, Inc.

BIBLIOGRAPHY

Braithwaite, A, and Smith, FJ: Chromatographic methods, ed 4, New York, 1985, Chapman & Hall.

Bruner, F: The Science of chromatography, Journal of Chromatography Library Series, 32, New York, 1985, Elsevier Scientific Publishing Co.

Poole, CF, and Schuette, SA: Contemporary practice of chromatography, New York, 1984, Elsevier.

Grob, RL: Modern practices of gas chromatography, New York, 1977, John Wiley & Sons, Inc.

Heftmann, E: Chromatography fundamentals and applications of chromatographic and electrophoretic methods, Journal of Chromatography Library Series, 22, New York, 1983, Elsevier Scientific Publishing Co.

Snyder, LR, and Kirkland, JJ: Introduction of modern liquid chromatography, ed 2, New York, 1979, John Wiley & Sons, Inc.

CHAPTER 5

Liquid chromatography

LARRY D. BOWERS

OBJECTIVES

- Describe the advantage of using peak height or peak area to quantitate an analyte using liquid chromatography.
- Define external standardization, internal standardization, and standard addition; explain how unknown concentrations are determined with each standardization process; state the requirements for internal standard selection and use.
- Differentiate between normal and reversed-phase chromatography; describe bonded reversed phase chromatography, and explain the advantages associated with the use of paired-ion chromatography.
- Diagram the basic column liquid chromatographic system; list four basic components of a high-performance liquid chromatography (HPLC) separation system and state the purpose of each.
- List three types of detector systems used in HPLC and, for each, describe the principle of operation, its use practicality, and the advantages or disadvantages associated with its use.

KEY TERMS

bonded phase A chromatographic support in which the stationary phase is covalently bound to the surface of the support.

effluent Mobile phase that has exited from the column.

eluate A compound or mixture that has been separated in and exited from the column.

eluent Mobile phase.

gradient elution An elution system where the solvent composition is varied during the run.

H Height equivalent to one theoretical plate (HETP).

isocratic elution Elution with a solvent mixture of constant composition

L Length of the chromatographic column, usually in millimeters.

mobile phase The mixture of solvents that is percolated through the column.

μ Solvent velocity in the column ($\mu = L/t_o$).

N Number of theoretical separating plates in a chromatographic column.

normal phase A chromatographic mode in which the mobile phase is less polar than the stationary phase.

permeability A measure of the ease with which the mobile phase can be forced through the column.

R_s Resolution; the degree of separation between two eluates.

reversed phase A chromatographic mode in which the mobile phase is more polar than the stationary phase.

stationary phase The portion of the separation system that is immobilized in the column.

support The particles on which the stationary phase is held.

t_o Time required to elute an unretained substance.

t_R Retention time; the time required to elute a compound from a chromatographic column.

V_o Volume of solvent required to elute an unretained compound; also called "void volume."

V_R Retention volume; the volume of mobile phase required to elute a compound from a chromatographic column at t_R.

Liquid chromatography is a form of separation science in which a liquid mobile phase is percolated through a column or thin layer of particles. Fig. 5-1 shows a schematic diagram of a column chromatograph used in liquid chromatography. The liquid mobile phase is taken from the reservoir and moved through the column, usually by a pump. A method of introducing the sample into the chromatographic system is also required. The most important constituent of a chromatographic instrument is the column. The column is packed with small particles on which specific sites or a layer of solvent (called the *stationary phase*) is held. The differential equilibration of the analytes between the mobile and the stationary phases results in their separation. All chromatographic modes (see Chapter 4) can be used in liquid chromatography. Finally, we can either collect the column effluent for further analysis or analyze the effluent with an on-line detector such as a photometer. Aliquots of the liquid phase can be collected if subsequent analysis is desired. The recording of any parameter that allows the analyte or analytes to be monitored as a function of elution volume or time is called a *chromatogram*.

Liquid chromatography is well suited for use in the clinical laboratory. Since the retention of a compound is determined by equilibria, the position of the peak in the chromatogram (that is, the retention volume) can be helpful in analytical identification. If a substance coelutes with a known compound, it may be the same material; an identical retention, however, does not *prove* identity. In addition, quantitative information can be obtained by measuring the height or area of the peak. Because the components of a mixture are separated, quantitation of several compounds in a single analysis is possible. Such quantitation is useful in the measurement of drugs or intermediates in a metabolic pathway (such as porphyrins). Another advantage of liquid chromatography is that the relatively polar compounds present in body fluids readily dissolve in commonly used mobile-phase solvents. This is in contrast to gas chromatography, which requires volatile analytes. Proteins and peptides are readily separated by liquid chromatography.

Despite the fact that liquid chromatography was discovered before gas chromatography, it was used in only a very small number of analytical applications in the clinical laboratory. Classical liquid chromatography typically required hours or days to complete a separation, whereas gas chromatography required only minutes. With the development of small (10 μ) totally porous particles in the early 1970s, liquid chromatography was able to achieve speed and resolution comparable to packed column gas chromatography (GC). The introduction of covalently bonded stationary phases resulted in the further popularization of high-performance liquid chromatography (HPLC). Developments, such as microprocessor automation, have made both analytical and preparative chromatography easier to perform and more appealing. At present, HPLC is recognized as a true complement to GC in chromatographic analysis.

RESOLUTION, EFFICIENCY, AND SPEED OF ANALYSIS

The object of any chromatographic technique is to separate, or resolve, the species of interest from other compounds of interest or from interferences in the sample matrix. As analysts, we must also be concerned about the speed of the analytical scheme, including any sample preparation steps, the separation itself, and the calculation and reporting of results. Not surprisingly, speed of analysis and resolving power are related. The evolution of HPLC was based on an understanding and optimization of the factors that affect resolution.

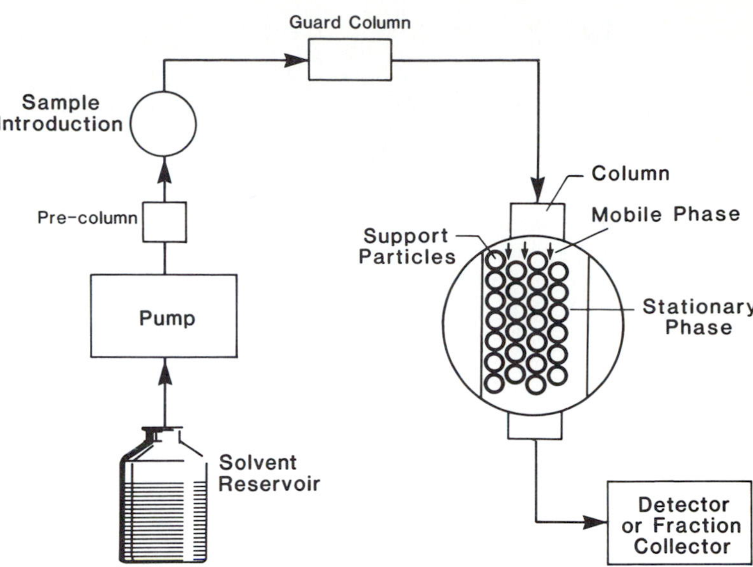

Fig. 5-1 Schematic diagram of a column liquid chromatographic system. *(From Bowers, LD, and Carr, PW: Quantitative Aspects of HPLC Workshop, Minneapolis, Minn, 1983.)*

Resolution

The relative separation of two chromatographic peaks is measured by a parameter known as *resolution,* or, R_s. A further definition of R_s, as well as an example of the calculation of R_s, can be found on p. 78 of Chapter 4. It is dependent on the positions of the centers of the peaks that correspond to each compound and on the width of the peak between the points where it is indistinguishable from the background signal. If the peak tracing reaches the background level before rising for the second peak, base-line resolution has been achieved. If the first peak shown in the bottom panel of Fig. 5-2 were collected up to the minimum in the valley between the two peaks, the compound in that peak would be 100% free of the compound making up the second peak. When the baseline resolution is not achieved (Fig. 5-2, top), some of the compound present in the second peak will be collected along with the compound in the first peak. Table 5-1 shows the relative impurity in the first peak for various resolution values when both peaks are the same size. Snyder and Kirkland have covered this topic more completely.[1] Resolution of peaks will also affect quantitation as discussed later.

Because resolution depends on peak widths, resolution must be controlled by the factors that govern the peak

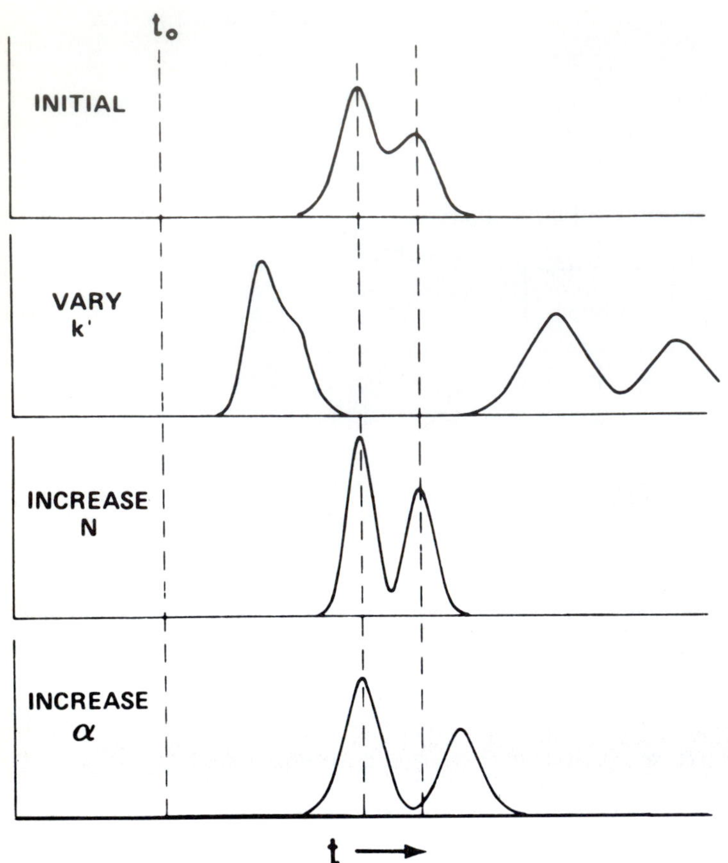

Fig. 5-2 Effect of varying k', N, and α on resolution. *t*, Time. *(From Snyder, LR, and Kirkland, JJ: Introduction to modern liquid chromatography, ed 2, New York, 1979, John Wiley & Sons, Inc.)*

width, namely, efficiency (N, number of theoretical plates) and relative peak retention. Peak retention is determined by the capacity factor (k′) and selectivity (α) as discussed in Chapter 4.

$$R_s = \frac{\sqrt{N}}{4} \frac{(\alpha - 1)}{\alpha} \frac{k'}{1 + K'} \qquad \textit{Eq. 5-1}$$

The effect of changes in each of these parameters on resolution is illustrated in Fig. 5-2.

Chromatographic efficiency (see p. 78, Chapter 4)

Originally this term was used to describe phenomena occurring in the chromatographic column. Thus, one could optimize the efficiency of the column by adjusting the flow rate, increasing the temperature, and so on. The Knox equation as represented by:

$$H = A\mu^{1/3} + \frac{B}{\mu} + C\mu \qquad \textit{Eq. 5-2}$$

relates H, the column height equivalent to a theoretical plate, to the flow velocity, μ. H is defined as the length of the column divided by N. The constants A, B, and C are

Table 5-1 Effect of resolution on various peak parameters*

Resolution	Purity†	% Error in Area‡	% Error in peak height‡
0.6	90	>−25	~15
0.8	95	−10	1
1.0	98	−3	<1
1.25	99.5	<−1	0
1.5	100	0	0

*Assuming a peak-height ratio of 1 to 1 for the two components.
†Purity of major peak, assuming collection is stopped at the lowest point of the valley between the peaks.
‡Error for the smaller peak.

numbers calculated from formulas derived from the mechanisms associated with the movement of a compound through the column. As the diameter of the column packing material (the stationary phase) decreases, H decreases. By decreasing H, N can be increased in the same length of column, resulting in improved separations. Hence, 10, 5, and even 3 μm totally porous stationary-phase particles have been used in HPLC to maximize efficiency. It is doubtful that particles smaller than 3 μm will result in increased efficiency. Modern HPCL columns have about 10,000 theoretical plates.

For the practicing chromatographer, four final points about efficiency are worthy of mention. First, a new column should always be tested upon receipt to be sure that reasonable efficiency is obtained with that column. An initial efficiency value also serves as a bench mark for measuring the decline in column performance as it is used. Second, the column should be tested under the manufacturer's flow and mobile phase conditions. The buyer should be aware that a column tested at 0.1 mL/min to obtain a high number of theoretical plates may not be the best column to use at more practical flow rates. Third, *N* is in reality a measure of system efficiency. Thus a poorly designed detector can make the best column look bad. To achieve the high efficiencies reported by some column manufacturers, the entire system must be optimized. Finally, note that to enhance resolution twofold, one must increase *N* fourfold. Thus, adjustment of *N* is normally used to fine tune a relatively good separation. References 2 to 5 contain a more detailed treatment of system efficiency.

The retention of a compound on a column is often described by using its *capacity factor, k′* (see p. 81, Chapter 4). The capacity factor is a normalizing factor that allows retention on different-sized columns, or columns operated at different flow rates, to be compared because the *k′*, by definition, is related to the equilibrium of the analyte between the mobile and stationary phases. In terms of resolution, *k′* values over five do not increase resolution much and can, in fact, slow analysis and deteriorate the limit of detection. In liquid chromatography, *k′* is adjusted primarily by changes in mobile-phase composition, although it is also inversely related to the temperature.[2,6]

The difference in retention of two compounds as measured by the ratio of their capacity factors is called the *selectivity,* or α (see p. 81, Chapter 4). Note that if α is large, there will be a large difference in the retention volumes of the two compounds, and we can perform the separation with few plates and little column retention. Selectivity is frequently adjusted by changes in solvent composition. One of the major advantages of liquid chromatography is the wide range of selectivity achievable by varying the composition of the mobile phase. For example, an ion-exchange separation depends on the number of charges on the analytes. We can change the selectivity by varying the mobile phase pH, ionic strength, or salt composition (NaCl versus LiCl). Selection of a mobile-phase system requires an understanding of the separation mechanisms and a great deal of experience. A change in the stationary phase can also be used to adjust selectivity because the equilibrium achieved between the stationary and mobile phases is the basis of any chromatographic separation. If the retention of the analytes is adequate ($1.5 \leq k' \leq 6$), one can improve the resolution most readily by changing the selectivity.

Speed of analysis

So far, we have discussed retention only in terms of mobile-phase volume, V_R because (1) the volume of mobile phase used is a direct reflection of cost and (2) changes in the flow rate do not affect the V_R. In contrast, the retention time, t_R, is a function of flow rate and retention volume, that is:

$$t_R = \frac{V_R}{F} \qquad \text{Eq. 5-3}$$

where F is the flow rate in milliliters per minute. For example, if the diameter of the column were doubled, its volume, and thus the retention volume, would increase by a factor of 4. We could keep t_R constant by increasing the flow rate by a factor of 4, but solvent consumption would increase accordingly.

The optimization of analysis speed depends on a number of factors. The most important relationships in achieving a rapid separation are given in the box. Guichon has discussed these factors in great detail.[7] The minimum time possible for a separation is the product of the time required for an analyte to pass one plate and the number of plates required for adequate resolution. The smaller the height of a theoretical plate, H, the faster the separation. Retention time is also increased by the need for more resolution; by small selectivities, α; and by large capacity factors, k'. Again, to obtain the fastest separations, a mobile phase must be selected to maximize α at relatively small values of k'. Under optimal conditions for a separation requiring 3000 plates, an analysis time of 100 seconds is feasible with HPLC. One of the problems in translating theory to practice in the clinical laboratory is that the analyte may have to be separated from an unknown metabolite or interferent, and hence α is not known during the development of the separation. In this case, a slower separation time is acceptable and the selectivity of the detector must be relied upon to indicate a potential problem.

PRACTICAL CONSIDERATIONS IN SPEED OF ANALYSIS

$$t_R = N \cdot \frac{H}{\mu} \cdot (1 + k')$$

$$L = NH$$

$$\Delta P = NH\mu \cdot \frac{\eta}{K_o}$$

μ, Solvent velocity in the column ($= L/t_o$); η, solvent viscosity; K_o, column permeability. (See reference 7 for more detail.) ΔP, Column back-pressure.

QUANTITATION

Approaches

Quantitation in liquid chromatography is achieved when one relates either the peak height or the peak area to the concentration of analyte in the sample. The chromatographic trace is a recording of the concentration of the analyte or analytes as sensed by the detector. At the peak height maximum

$$C_{max} = \frac{C_s V_s}{V_R} \sqrt{\frac{N}{2\pi}} \qquad \text{Eq. 5-4}$$

C_s and V_s are the concentration and volume of sample injected respectively, V_R is the retention volume, and N is the number of theoretical plates. The greater the peak height for any given concentration, the more sensitive the method. Thus minimizing the retention volume and maximizing the number of theoretical plates will result in the most sensitive assay. The factors that decrease the retention volume are a small column void volume and a small capacity factor. Large plate counts can be achieved by low flow rates and small support-particle diameters. Also note that, unlike the sensitivity of gas chromatography, liquid chromatographic sensitivity is improved by injection of larger sample volumes. The peak height can be related to the concentration in the sample by a sensitivity factor, S. The sensitivity factor will change if the retention volume changes, making day-to-day operation without standardization difficult but not impossible.

Example: If an identical sample were injected onto a 4.6 mm internal diameter (i.d.) and a 2.1 mm i.d. column, what would be the relative size of the peaks if all else were unchanged?

$$C_{max} = \frac{1}{V_R}\left[C_sV_s\sqrt{\frac{N}{2\pi}}\right] = \frac{1}{\Phi\pi r^2 L}[\bar{C}] \quad \textit{Eq. 5-5}$$

where Φ is the void fraction, r is the radius, and L is the length (15 cm) of the column; $\bar{c}$ is mass injected into column. If Φ and L are the same for the two columns, then

$$\frac{C_{max}\,(2.1)}{C_{max}\,(4.6)} = \frac{(2.3)^2}{(1.05)^2} = 4.8 \quad \textit{Eq. 5-6}$$

Therefore the peak from the 2.1 mm column would be almost five times as large. This sensitivity advantage is causing increased interest in microbore columns.

How much of a decrease in length would be necessary to observe a similar increase in peak height? What would the decrease in length change that the decrease in radius would not? (k'; α; or N).

The second approach to quantitation of an analyte is measurement of the peak area. Although other methods of integration are available and are described in references 1 to 3, microprocessor-based integrators have become so inexpensive that they are the most reasonable means of measuring peak areas. Most liquid chromatography detectors are concentration dependent; that is, they measure a concentration in the flow cell. (This is in contrast to most GC detectors, which are mass-flow dependent.) The peak area obtained from a concentration-dependent detector is inversely proportional to flow rate. Therefore significant variation in peak area can occur if the flow rate changes during a chromatographic run. In addition, less resolution is required for an equal degree of accuracy when peak heights rather than peak areas are used because the overlap of peaks, even with a resolution of 1.0, affects peak area but does not affect the peak maximum.[1] A rough rule of thumb is: use peak *height* when there are interfering peaks or when maximum accuracy is required, but use peak *area* when precision is the main requirement. For peaks barely above the base-line noise, peak heights should always be used.

Standardization

Standardization in liquid chromatography can be accomplished in any of three ways: external standardization, internal standardization, or standard addition. For external standardization, a calibration curve is constructed from the peak height (or peak area) values obtained with known concentrations of analyte and a constant injection volume. The slope of the curve is the sensitivity factor, S, in peak-height units per concentration unit (such as mm/mM). The concentration of the unknown is then simply its peak height divided by the sensitivity factor. Notice that, if the calibration curve is linear, the sensitivity factor can be obtained from a single standard. It would be important in this case to check several control specimens to verify the validity of the sensitivity factor. The principal sources of error in external calibration are variable losses in the preparative steps before LC analysis and sample injection variability. It is thus important to treat the standards and samples in the same way. It should be possible to achieve 1% precision with external calibration, but up to 5% is commonly observed.

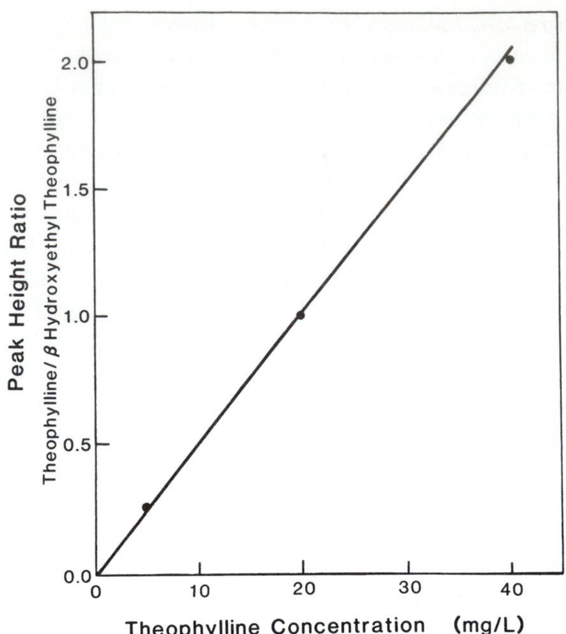

Fig. 5-3 Calibration curve for theophylline using internal standard technique.

Internal standardization uses a compound that is usually structurally similar to the analyte to correct for losses and injection imprecision. The same amount of internal standard is added to each sample before sample pretreatment and chromatography. The calibration curve is then constructed from the ratio of the peak heights (or areas) of the standard and the internal standard at various standard concentrations (Fig. 5-3). Again, unknown concentrations can be measured by using the sensitivity factor or interpolating the value from the calibration curve. It is hoped that any losses or variations that occur will affect the analyte and internal standard equivalently. In practice, this is difficult to achieve. An internal standard can be used not only to improve accuracy and precision but also as a quality control check because its peak height should be the same in all chromatograms. A number of requirements must be met in the selection of an internal standard. These are summarized in the accompanying box. In some cases internal standards do not improve precision and accuracy and may deteriorate precision because of the imprecision involved in measuring two peaks. The precision that can be attained with HPLC has been studied, and sources of error have been analyzed.[8]

The final calibration method, standard addition, requires two analyses to be performed on each sample and thus is

REQUIREMENTS FOR INTERNAL STANDARD (IS) SELECTION AND USE

1. The internal standard must be completely resolved from all peaks in the sample.
2. It must be eluted near the analyte, with $k' \pm 30\%$ being preferable.
3. It must behave similarly to the analyte in pretreatment if losses are to be corrected. This may require more than one internal standard.
4. It must have a peak height or area approximately equal to a standard in the concentration range desired.
5. It must not normally be present in the sample.
6. It should be commercially available in pure form.
7. It should be added as a liquid.

not as popular as the other methods. After a sample is analyzed once, a known amount of the compound of interest is added in a very small amount of liquid so that no dilution occurs, and the sample is reanalyzed. To correct for extraction variability, the addition should be made to the biological fluid and the entire analytical scheme repeated. Quantitative data can be obtained from the ratio of the peak height (or area) before and after addition of the standard. Addition of the standard can also help to verify the identity of the peak because the standard must coelute for the unknown to be the same compound. The converse is not true, however, because more than one compound may elute at the same retention volume.

GENERAL ELUTION PROBLEM

The chromatographer is sometimes required to separate compounds that, though structurally related, behave quite differently in the separation system. In Fig. 5-4, a separation of the bile acid conjugates demonstrates the problem. When the mobile-phase composition is adjusted to achieve resolution of peaks 2 and 3, peak 8 is retained for over 30 minutes. Since there are no other peaks near peak 8, the excessive base line present between peaks 7 and 8 is a waste of valuable analysis time. A mobile phase with a constant composition is referred to as *isocratic*. The alternative to this is to change the mobile-phase composition during the chromatographic run. This can be done as a single change from one mobile phase to another (step gradient) or as a continuous change in any of a variety of shapes (such as linear, segmented linear, or exponential gradient). A complete treatment of gradient elution can be found in reference 9. In brief, the peak retention volume, width, and resolution are determined primarily by the rate of solvent composition change. Thus, many of the concepts discussed earlier are not valid in gradient elution. Any quantitative analyses developed using gradient elution should be carefully documented. In recent years, the reproducibility of HPLC gradients has improved greatly, and accurate quantitative analyses are now possible. In HPLC, the ability to vary mobile-phase composition can be purchased as a part of the solvent-delivery system. It generally increases the cost of the system significantly.

SELECTION OF A CHROMATOGRAPHIC MODE

In Chapter 4 the various mechanisms of chromatography were discussed, including ion exchange, steric exclusion

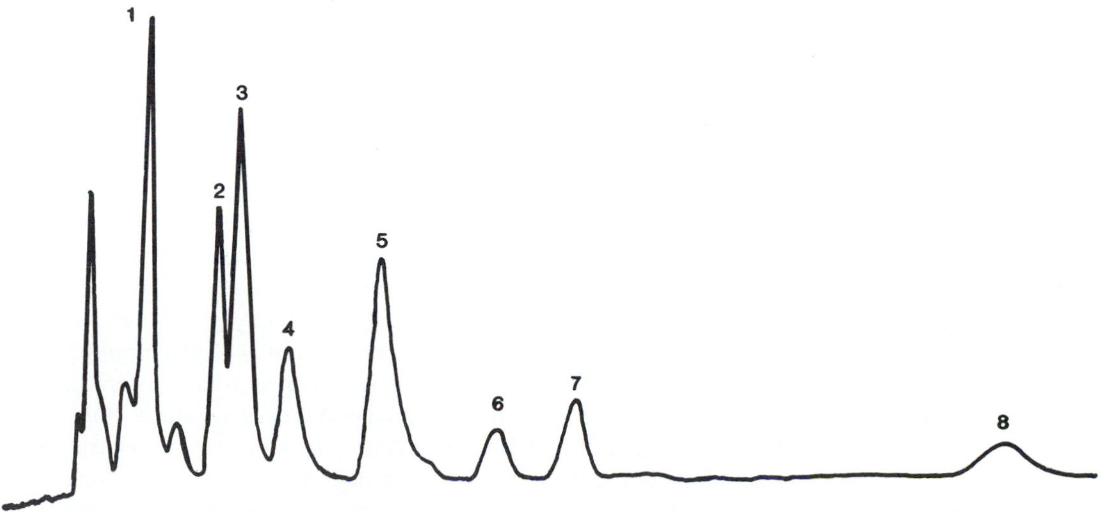

Fig. 5-4 Separation of common human bile acid conjugates by reversed-phase HPLC. Peaks correspond to taurocholate, *1;* taurochenodeoxycholate, *2;* taurodeoxycholate, *3;* taurolithocholate, *4;* glycocholate, *5;* glycochenodeoxycholate, *6;* glycodeoxycholate, *7;* and glycolithocolate, *8. (From Roberts, G, and Bowers, LD, unpublished data.)*

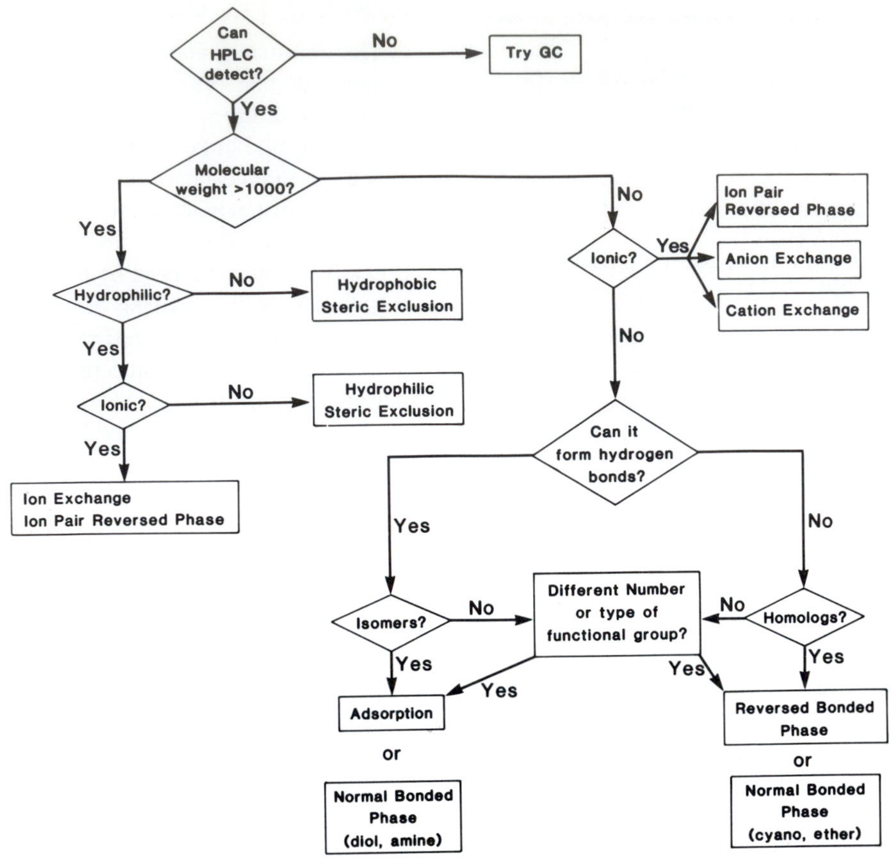

Fig. 5-5 Selection guide for chromatographic modes. *(From Bowers, LD, and Carr, PW: Quantitative Aspects of HPLC Workshop, Minneapolis, Minn, 1983.)*

(gel permeation), adsorption, and partition. As mentioned previously, all these modes are available in liquid chromatography. The selection of the "best" mode is a problem of significant magnitude for the chromatographer. The choice is made based on an understanding of the mechanism of each mode and its strengths and weaknesses, and frequently it is accomplished by experience. Fig. 5-5 illustrates one method of selecting a chromatographic mode using mechanistic considerations.

One of the chromatographer's first considerations is the size of the molecules to be separated. If the molecules are relatively large (> 100,000 MW), *steric exclusion* (SEC) is a logical first choice (see Chapter 4). If, on the other hand, the molecular weights are less than 1000, steric exclusion is probably not the mode of choice.

Another key factor that affects the choice of a separation mode is the existence of an ion or the presence of ionizable groups on the molecule. *Ion exchange* would be a logical choice for molecules that can be charged, regardless of molecular size. Ion exchange is based on the interaction of analyte charges with an oppositely charged group bound to the chromatographic support. One can vary retention by varying pH and ionic strength or by adding organic solvents such as methanol and ethanol. The greater the number of charges on the analyte, the greater the retention. Increasing ionic strength, and with it the number of charged groups (such as Na^+) competing with the analyte for the exchange sites on the support, reduces retention. One limitation of ion-exchange chromatography is that it does not exhibit high theoretical plate counts, and thus separations are relatively slow. Mobile-phase manipulations in bonded-phase partition chromatography called *ion pairing* (see p. 104) can in many cases accomplish the same separation more efficiently.

If the molecule has a molecular weight less than 1000 and is not ionizable, another characteristic of the compound must be used to achieve the separation. One such characteristic is the ability of the compound to form hydrogen bonds. *Adsorption chromatography* is based on the interaction of the analyte with a three-dimensional binding site on the support matrix, which may involve hydrogen bonding. Thus, the structure and polarity of the solute are important in determining retention. For example, it would be relatively easy to separate *p*-dinitrobenzene from *o*-dinitrobenzene using adsorption chromatography because of the structural differences between the two compounds. On the other hand, separating caproic (six carbon atoms) acid from caprylic (eight carbon atoms) acid would be very difficult because the parts of the molecules that interact with the stationary phase are identical.

In adsorption chromatography, compounds are eluted from the column because of competition between the analyte and the solvent for the binding site. A solvent that elutes compounds more rapidly and therefore competes better for the chromatographic sites is called a "strong solvent." Water would be a strong solvent for silica adsorbents because it interacts strongly with the silanol (SiOH) groups responsible for the adsorption mechanism. A solvent that does not compete well for sites is called a "weak solvent." Hexane would be a weak solvent for silica adsorbents. One can adjust solvent strength by using mixtures of solvents. Snyder[10] has developed a scale of solvent strength called the eluotropic series and has extended the solvent strength theory to binary and ternary[11] mixtures. Interestingly, solvent mixtures of the same strength can show differences in selectivity because Snyder's theory considers only the adsorption process, whereas in reality the solute can also interact with the solvent.

One final consideration in adsorption chromatography is the control of the "activity" of the adsorbent. Silica contains several types of silanol groups, which interact differently with various types of compounds. A small amount of water, methanol, acetonitrile, isopropanol, or other strong solvent is used to block the strongest sites and give a more reproducible adsorption surface. This problem has caused some chromatographers to avoid adsorption systems, probably unnecessarily. A separation that involves nonpolar to intermediate polarity compounds that can form hydrogen bonds or that have isomeric components will probably be easily achieved with adsorption chromatography.

The interactions between solute and stationary phase in *partition chromatography* are not nearly so well defined as those for adsorption systems. Any chemical forces that exist between molecules can be used in partition-based separations, including hydrogen bonding, van der Waals' forces, ion-ion interactions, and so on (see Chapter 4). The basis of partition systems is the distribution of a solute between two liquid solvent layers, one stationary and the other mobile. It is essentially analogous to thousands of extractions taking place in a column. In classical partition chromatography, a polar liquid, such as β,β′-oxydipropionitrile (β,β′-ODPN), was coated onto the support particles and an immiscible solvent such as hexane was used as the mobile phase. Since solvent-solute interactions involve the entire molecule, a typical partition system as given here would separate a homologous series (C-6, C-7, and C-8 carboxylic acids) and some positional isomers. A partition system with a polar stationary phase and a nonpolar mobile phase is called *normal phase* because it was the first type of system developed. Later, a system with a nonpolar stationary phase, such as squalane, and a polar mobile phase, such as a water-acetonitrile mixture, was developed and called *reversed phase*. Both types of liquid-liquid partition chromatography had many problems related to the finite solubility of all solvents in each other. For example, the β,β′-ODPN would slowly dissolve in the hexane and slowly change the amount of stationary phase, which in turn would change the retention volumes of the analytes. The temperature had to be very closely controlled, and gradient elution was impossible with partition chromatography. These problems made partition chromatography difficult to use in the clinical laboratory.

All of this changed with the development of chemically bonded stationary phases. These materials are prepared by covalent bonding of an organic moiety onto the surface of the support particle, usually silica. The organic groups include nonpolar functions, such as octadecylsilane (ODS or C-18) or octylsilane (C-8), and polar groups, such as cyanopropyl (CN), aminopropyl (NH_2), or glycidoxypropyl (diol) silanes. The advantages of bonded phases are that (1) polar and ionic compounds are readily separated, (2) the stationary phase does not strip off, (3) gradient elution can be used, and (4) the columns are easy to use and take care of. The main disadvantage of bonded phases is that at pH's below 2 the bonded group is cleaved from the support and at pH's above 8 the silica support particles dissolve. Since most separations for clinical laboratory applications can be performed within the workable pH window, bonded phases are very popular. It has been estimated that 90% of all separations are now performed on reversed-phase (such as C-18) chemically bonded columns. A separate section has been included to discuss the variety of separations feasible with reversed-phase systems.

The use of biochemical interactions, such as the affinity of an enzyme for its substrate or of an antibody for its antigen, is beginning to be recognized in clinical laboratory settings. These interactions have been grouped together as affinity chromatography. The most commonly used affinity technique is the use of phenylboronic acid columns to isolate glycosylated proteins such as glycosylated hemoglobin.

Ion chromatography is also emerging as a potentially useful technique for the clinical laboratory. Ion chromatography is in effect both a separation and detection technique for ionic compounds such as F^-, $H_2PO_4^-$, and CN^- or Pb^{++}, Pt^{++}, and As^{++}. The technique uses very low capacity ion exchange columns (0.1 mEq/g), which are generally not useful for normal ion exchange separations. At present, this is a very specialized technique.

As stated on p. 100, selection of a chromatographic system depends both on an understanding of the strengths and weaknesses of the various separation analysis modes and on the experience of the analyst. It is possible that all separations of interest in a laboratory could be performed on one type of column (such as reversed phase); however, this would require considerable mastery of mobile phase modification and might result in a separation inferior to that obtainable with relative ease on another type of column. In summary, there is no "best" way to solve a chromatographic problem.

BONDED REVERSED-PHASE CHROMATOGRAPHY

The most popular type of bonded-phase support is the octadecylsilane (ODS or C-18) reversed-phase (RP) column packing. The reason for this popularity is the ease with which polar biological molecules are separated. The mobile phase in reversed-phase chromatography is a polar solvent such as water, and therefore the polar or ionic molecules are readily soluble in the mobile phase. The stationary phase is a nonpolar octadecyl-bonded phase. If a molecule has a nonpolar part, as shown for tyrosine in Fig. 5-6, it can interact with the nonpolar stationary phase while the polar part interacts with the polar mobile phase. Molecules with a small nonpolar area do not interact very strongly with the stationary phase, but they spend most of their time in the mobile phase and thus are rapidly eluted from the column. Molecules with a large nonpolar area interact strongly with the stationary phase and may not be eluted with water. In this case, a less polar (stronger) solvent, such as methanol, acetonitrile, or tetrahydrofuran, would be mixed with the water so that the mobile phase can compete with the stationary phase for the nonpolar part of the molecule. The interactions involved are not competition for a specific site as in adsorption chromatography, but rather competition in terms of solubility. Thus nonpolar molecules can be eluted from the column if a sufficient amount of organic solvent is added to the eluent. A detailed discussion of the theory of reversed-phase chromatography can be found in references 12 to 14. In summary, polar molecules are eluted first and nonpolar molecules are eluted last. The selectivity and elution volumes are determined by the mobile-phase conditions, including pH, ionic strength, and percentage of organic solvent.

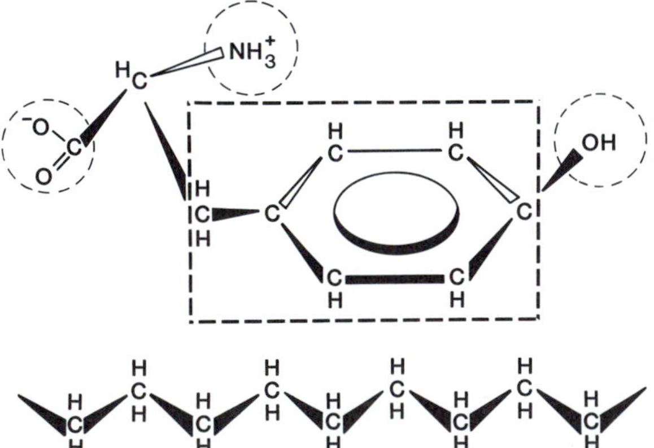

Fig. 5-6 Retention in reversed-phase chromatography is result of interaction of nonpolar portion of compound such as tyrosine *(enclosed in box)* with nonpolar stationary phase. Hydrophilic groups *(circled)* tend to decrease retention.

Stationary-phase considerations

As noted earlier, the separation obtained for any group of analytes is a function of both the stationary phase and the mobile phase. The analyst has the ability to vary mobile-phase conditions but is dependent on manufacturers for the stationary phase, particularly for bonded-phase packings. If one is to do reproducible chromatography, the behavior of the stationary phase toward nonpolar, polar, and ionic compounds must be the same from column to column and lot to lot. In addition, the durability of the column is important, particularly since columns cost $250 to $350 each. Significant improvements in these features have occurred in recent years, but there are some limitations in producing a column with absolutely reproducible reversed-phase column packings.

The preparation of an octadecylsilane (ODS) stationary phase is usually accomplished by reacting the silanol (SiOH) groups on the silica gel with an octadecylsilane such as octadecyldimethylchlorosilane. The resulting surface contains octadecyl groups bound to the surface by siloxone (Si-O-Si) bonds as shown in Fig. 5-7. Most manufacturers of columns use silanes with only one chloride group, and the resulting stationary phase is called *monomeric*. Because of the stereochemistry of the silica gel surface, only about one third of the silanol groups can react with the ODS groups. The remaining silanol groups are polar and can interact with polar analytes, changing the selectivity of the stationary phase. Trimethylchlorosilane can react with about an additional 20% of the surface silanol groups with a resultant increase in the nonpolar character of the support (Fig. 5-7). This process is called *end capping* and is used in many commercially available packing materials. As might be expected, differences in the surface morphology of the silica gel, reaction conditions in the ODS-bonding step, and the presence or absence of end capping make columns purchased from different manufacturers perform differently. In fact, variations from lot to lot of packing material may be quite noticeable in the separations obtained for certain analyses. Thus it is not surprising that in adapting a method to a laboratory one may require significant changes in the mobile phase if a C-18 column from a manufacturer other than that named in the original report is used.[15] Choice of a reversed-phase column still requires trial and error. For the novice, use of the brand of column reported in a publication is probably warranted to obtain acceptable chromatograms in a reasonable amount of time. The situation with respect to columns is improving as a better understanding of the silica backbone is achieved. Several good reviews of columns used in biomedical applications have appeared.[16,17]

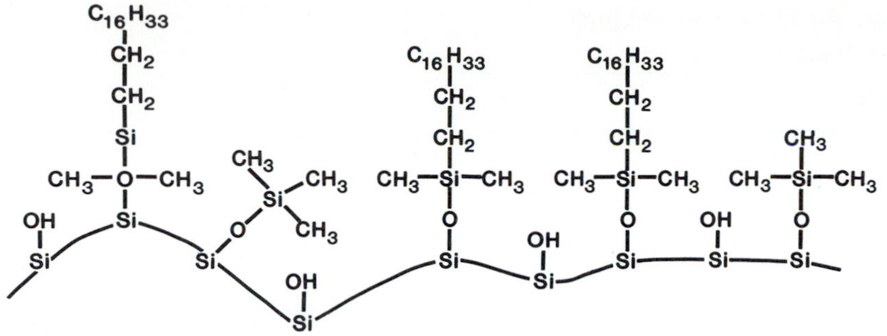

Fig. 5-7 Schematic representation of a silica-based octadecyl reversed-phase support that has been end capped. Note presence of residual silanol groups on surface.

Mobile-phase considerations

The real power in liquid chromatography arises from the fact that changes in mobile-phase composition can have major effects on selectivity and thus on resolution. In reversed-phase systems, two types of changes can be made: (1) changing the type of organic solvent used and (2) additions to the mobile phase that affect its pH, ionic strength, or complexing ability. It has been recognized for some time that the type of organic solvent used has an effect on retention and selectivity.[14] In some cases, a change from methanol to acetonitrile actually alters the elution order of the compounds. Unfortunately at this time the use of solvents to vary selectivity is largely empirical,[18] and the exact role of each solvent in a separation is poorly defined; however, the relative strength of solvents is well defined with solvent strength and, therefore, the ability to elute solutes, increasing in the following order: methanol $<$ DMSO $<$ ethanol $\leq$ acetonitrile $<$ tetrahydrofuran $<$ dioxane $<$ isopropanol. As a rule, when one is adjusting the retention of a solute, a 10% increase in the fraction of organic solvent (for example methanol or acetonitrile) in water causes a twofold or threefold decrease in k' value.

Ion suppression. Mobile-phase modifications that change retention by introducing a second chemical equilibrium process in the mobile phase have been used since the inception of reversed-phase chromatography.[19] The first approach was control of pH to affect retention. If, for example, ascorbic acid was to be separated by HPLC, there would be a strong influence of pH on retention. At pH's above the pK_a of ascorbic acid, the acid would be deprotonated and charged and, therefore, would not partition itself strongly into the nonpolar stationary phase. Retention would be relatively low. On the other hand, at pH's below the pK_a, the acid would be protonated and uncharged and so would be much more strongly retained. In the area about one pH unit on either side of the pK_a, the retention changes rapidly as a function of pH, as shown in Fig. 5-8. The use of pH control to increase retention for acids has been termed *ion suppression*. It is a very useful method of

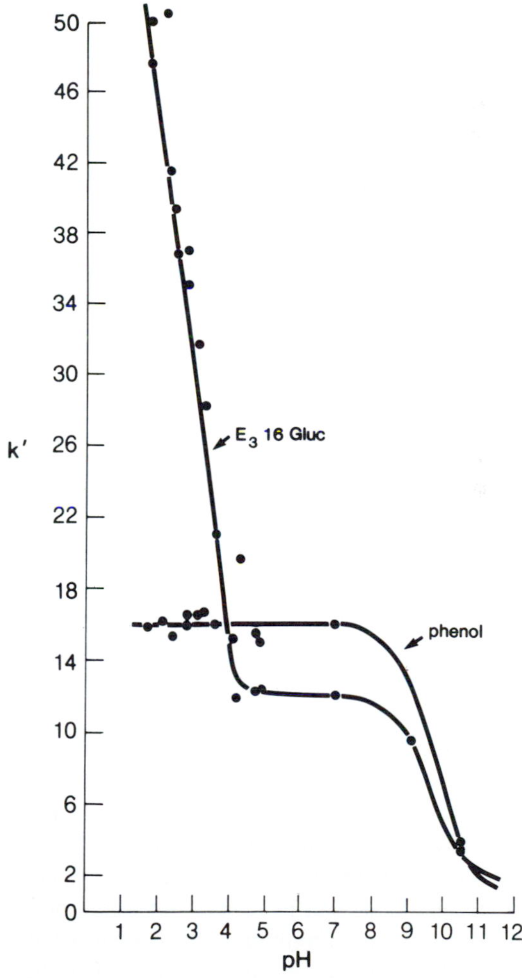

Fig. 5-8 Change in k' as a function of pH for estriol-16α-glucuronide and phenol. Decrease in k' at pH 2 is attributable to ionization of glucuronic acid; decrease at pH 10 is attributable to ionization of phenolic group. *(From Oliphant, C, and Bowers, LD, unpublished data.)*

Table 5-2 Useful buffers for reversed-phase high-performance liquid chromatography

Buffer pair	pK_a
Phosphoric acid/dihydrogen phosphate	2.12
Chloracetic acid/chloracetate	2.87
Succinic acid/monohydrogen succinate	4.23
Acetic acid/acetate	4.77
Piperazine phosphate	5.33
Monohydrogen succinate/succinate	5.65
Tetramethylethylenediamine phosphate	6.13
Dihydrogen phosphate/monohydrogen phosphate	7.20
Tris(hydroxymethyl)-aminomethane	8.19

adjusting retention behavior. Buffers are normally used to control the pH. It is important to remember that an acid-base pair is only a good buffer within one pH unit of the pK_a. Table 5-2 lists some useful buffers for reversed-phase HPLC.

Paired-ion chromatography. Because silica-based reversed-phase materials can only operate in the pH range of 2 to 8, many basic compounds cannot be handled by ion suppression. In these cases, the charge on the analyte can be neutralized by addition to the mobile phase of an oppositely charged ion *(counter ion)* that will form an ion pair with the analyte. This neutral pair will be more strongly retained than the analyte alone. The more hydrophobic the counterion, the greater the increase in retention. For example, the catecholamines, which are weak bases, have their retention increased when ClO_4^-, pentane sulfonate, and octane sulfonate are used as counterions. If equimolar amounts of each were tested, octane sulfonate-modified mobile phases would have the greatest retention because of the nonpolar octyl group. Perchlorate ion-containing mobile phases would have the least increase in retention. Not only the hydrophobicity of the ion-pairing agent, but also its concentration in the mobile phase is important. As the concentration of the counterion is increased from zero, there is a linear relationship between increased retention and concentration as shown in Fig. 5-9. At higher concentrations of counterion, the retention remains constant despite increased concentrations of pairing ion. Thus selection of the type and concentration of pairing ion, along with selection of the pH and of concentration and type of organic modifier or modifiers, allows rather precise adjustment of the retention of compounds of interest. Strong acids can have their retention modified by addition of tetraalkylammonium salts, such as tetraethylammonium perchlorate or amine buffers. The theory associated with ion-pair chromatography (IPC), which is still controversial, is discussed in references 19 to 21. The advantages of ion-pair chromatography are threefold: (1) both ionic and polar compounds can be separated in one chromatographic system, (2) it is more efficient (that is, N is greater) than ion-exchange separations, and (3) the separation can be controlled by mobile-phase modifications. The major disadvantage of IPC is that in some cases the ion-pairing reagents permanently modify the column.

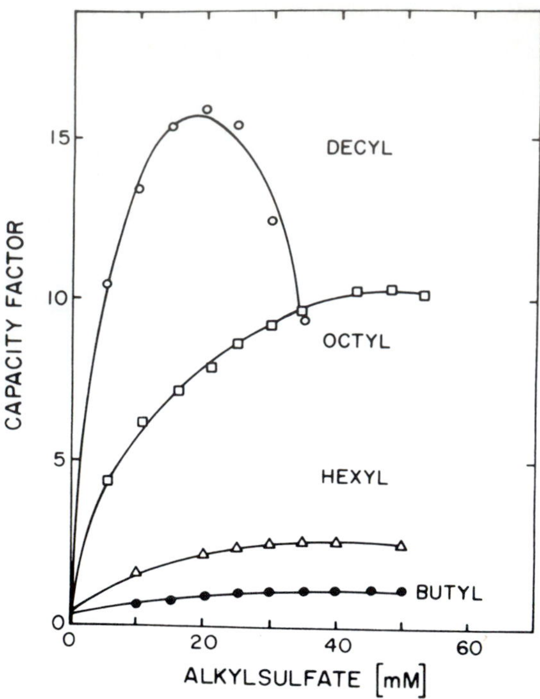

Fig. 5-9 Variation of capacity factor, k', of epinephrine as a function of ion-pairing reagent concentration for several n-alkylsulfonates. *(From Horvath, C, et al: Anal Chem 49:2295, 1977.)*

Other mobile-phase modifiers. It should be apparent that anything added to the mobile phase that makes the analyte less polar will increase its retention on a reversed-phase column. An additive that makes the analyte more polar reduces retention. The addition of silver ions to the mobile phase results in the selective complexation of compounds containing double bonds, making them more hydrophilic. Thus, selectivity between compounds differing only in the number of double bonds (such as prednisolone and cortisol) can be achieved by use of argentation chromatography (that is, addition of Ag^+). The use of cupric, zinc, nickel, and cadmium ions, often in the form of a hydrophobic chelate, has also been reported. For example, the selectivity, α, for the *cis*- and *trans*-dicarboxylic acids fumaric acid and maleic acid, respectively, increased from 0.9 to 2.5 upon addition of a Zn(II)-4-dodecyldiethylenetriamine chelate to the mobile phase. This was the result of the formation of a complex between the dicarboxylic acid and the metal chelate. In summary, anything the chromatographer can do to change the retention of one compound relative to another improves selectivity. The advantage of liquid chromatography is that mobile-phase modifications are relatively easy to make and are limited in selectivity only by the chemical insights of the chromatographer.

INSTRUMENTATION

As mentioned previously, modern HPLC requires relatively sophisticated instrumentation to achieve difficult separations in less than 10 minutes. There are four basic components in the separation system itself: (1) a solvent-delivery system to provide the driving force for the mobile phase, (2) a sample introduction system, (3) the column, and (4) the detector (Fig. 5-1). In addition, a recorder or integrator must be used to display or calculate the results.

Solvent-delivery systems

The most common delivery system is based on the reciprocating piston pump. Other types, including pneumatic amplification, syringe, and diaphragm pumps, are mainly of historical interest and are discussed in references 1 to 3.

In the first pumps used in HPLC, solvent delivery occurred during less than half the cycle time. This meant that flow through the column was erratic. The stoppage of flow and the compression of the solvent that occurred when the pump head refilled also resulted in a signal at the detector, which limited the detection of analyte. In the jargon of chromatography, the stoppage is known as a "pulse." Subsequently, manufacturers have used a variety of designs to minimize pulsation. These range from a very rapid refill stroke relative to the delivery stroke, to two or more pump heads that are out of phase such that one head is always delivering solvent. There are also mechanical devices known as "pulse dampeners" that work quite well under isocratic conditions with all but the most crude pumps. It must be emphasized, however, that although pulses can be minimized in a reciprocating piston pump, they can never be eliminated. An excellent monograph on the care and maintenance of HPLC equipment has been published.[22]

Gradient elution can be attained by mixing the solvents after they have passed through the pumps or by mixing them before they enter the pump. A well-maintained system of either type can function well, and selection of one or the other is based on gradient reproducibility, gradient shapes available, cost, and operator preference.

Sample-introduction systems

The most widely used method of introducing a sample into the chromatographic system is the fixed-loop injection valve. A sample aliquot is loaded into an external loop of stainless steel tubing. The valve is then rotated so that the sample loop is flushed onto the column by the mobile phase from the pump. Returning the valve to the original position allows loading of the next sample. Fixed-loop valves can be used in two ways: partial-fill method and full-loop method. In the latter the entire loop (such as 20 μL) is filled with sample and injected. It is the most precise method. One should recognize, however, that accurate results require flushing of the loop with five to ten loop volumes before loading. In the partial-fill mode, the sample loop is not filled with sample. In this case, the precision of the injection volume is determined by the loading syringe.

In addition to the manual valve injector described above, many automatic sampling devices are now available. These autosamplers allow unattended operation of the HPLC system, making 24-hour-a-day use possible. They are usually quite reliable and precise, due to mechanical advances and computerization. The devices may cost $5,000 to $10,000, but play an important role in busy laboratories.

Columns and connectors

The column is of course the most important part of the separation system. The packing material for columns has been discussed at length. Packing a high-efficiency column is still an art. The column tube itself is usually of No. 316 stainless steel with an inside diameter of between 2.1 and 4.6 mm and a length of 50 to 1500 mm. Column end fittings are required at both ends to connect the column to the other system components and to hold the packing material inside. The frits used should have pores less than one fourth the average diameter of the packing material. In addition to the analytical (or preparative) column, which actually performs the separation, two types of protector columns might be used. A *guard column* (Fig. 5-1) is located between the injector and the analytical column and is 1/15 to 1/25 the volume of the latter. It is packed with a material similar to that of the analytical column. Its function is to collect any particulate matter or any strongly retained components of the sample and therefore to protect the expensive analytical column. A *precolumn* is positioned between the pump and the injection valve. It is always packed with silica, the purpose of which is to saturate the mobile phase with silicate and thus prevent dissolution of the packing material in the analytical column. This has reportedly allowed operation of silica columns at pH's of 10, far beyond the normal pH for dissolution of silica. Rabel[22] has discussed the use of protector columns at length.

The analytical column has been described previously. The high efficiencies obtained with current HPLC columns result from the use of small (3, 5, or 10 mm), totally porous particles. In order to obtain efficient columns and reasonable operating pressures, the range between the largest and smallest particle must be as small as possible. Both spherical and irregularly shaped particles are available. Columns packed with spherical particles seem to give lower operating pressures and are more durable.

In addition to the columns, the connections made between system components are critical because the fittings should introduce as little peak spreading as possible. They should be of the zero dead volume (ZDV) or at least low dead volume (LDV) type. An excellent article on the intricacies of HPLC plumbing has recently been published.[23]

Detection systems

The final component in the chromatographic system is responsible for detecting the compounds as they elute from the column. Ideally, a detector would respond to any compound, would detect picograms or less of the analytes, would be immune to any solvent-related phenomena, and would respond linearly to a wide range of concentrations. Unfortunately, liquid chromatography does not have such a detector. In the clinical laboratory, three types of detectors have been used: photometers, fluorometers, and electrochemical devices. The selection of the appropriate detector will depend on the required selectivity and sensitivity. The performance characteristics of the common LC detectors are summarized in Table 5-3.

Absorbance detection. Ultraviolet and visible wavelength photometers are used in the same manner as other photometers, with quantitation based on Beer's law. The main difference is that, because of the small size of the flow-through cell (5 to 10 μL), intense light sources are required to get a reasonable amount of light through the cell. The most common light source is the mercury-arc lamp, which has an intense emission at 254 nm. A relatively inexpensive dual-beam, fixed-wavelength filter photometer can thus be constructed and is simple, easy to use, and relatively rugged. Absorbance changes as small as 10^{-5} absorbance units can be detected. Fortunately, most compounds that contain an aromatic ring absorb at 254 nm, but the detector is far from universally sensitive. For compounds that do not absorb at 254 nm, there are two alternatives. Most manufacturers offer other lamps or lamp/phosphor systems that have selected emissions in the wavelength range from 280 nm to 660 nm (such as 280, 313, 365, 405, 436, 546, 578, 660 nm). More recently, zinc and cadmium hollow-cathode lamps, which have emissions at 214 and 229 nm, respectively, have been used. The use of a lower wavelength allows detection of more compounds.

The second alternative is a variable-wavelength photometer where the wavelength is chosen by a monochromator from the spectra emitted by a deuterium or tungsten-halide lamp in the ultraviolet or visible region, respectively. The advantage here is that the wavelength of maximum absorption can be selected. On the other hand, because of the diminished intensity of these light sources, the limiting base-line noise is generally five to ten times greater than that obtained when using a fixed-wavelength detector, although recent variable wavelength systems have closed this sensitivity gap. A second advantage is that these detectors can be operated at 190 nm where even sugar moieties such as glucose absorb light. Unfortunately, because almost everything absorbs light below about 205 nm, it is very difficult from a practical viewpoint to use this region of the spectrum. Even the widely used reversed-phase solvents, acetonitrile and methanol, have an absorbance of 1.0 at 190 and 205 nm, respectively. Of course, as they are diluted, the absorbance decreases, but they can absorb a great deal of the available light in the far ultraviolet and thus cause detection-limit problems. One final characteristic of most photometers is that they respond to refractive-index changes that occur because of pressure changes (pulses), temperature changes, or solvent changes (such as a gradient). Well-designed photometers and flow cells, to a large extent, eliminate this problem (which can affect the detection limit of the detector and its utility in gradient work) and are probably worth the greater capital investment.

A recent addition to the detector armamentarium is the diode array detector (DAD). In this detector, polychromatic light is directed through the flow cell and separated into its component wavelengths after passage through the sample. Many wavelengths of light are detected simultaneously in an array of photodiodes. Thus an absorbance spectrum of the compound contained in the chromatographic peak can be obtained "on-the-fly." With recent advances in computer software, this spectral information can be used to detect co-elution of interfering compounds or to identify the compound by comparison of its spectra to spectra in a computerized library. The role of this detector in the clinical laboratory has not been fully determined, but the prospects are exciting.

Fluorescence detection. The technique of fluorescence is based on the ability of a molecule to emit light after it has been excited by light radiation. For a description of fluorescence spectroscopy, refer to Chapter 3. The main differences in fluorometer performance between manufacturers arise from differences in the lamp intensity and the detection efficiency. Commercially available instruments use emission from either deuterium or xenon arc lamps for

Table 5-3 Performance characteristics of commonly used detectors

Ideal detector characteristic	Fixed wavelength	Variable wavelength	Fluorescence	Electrochemical	
				Oxidation	Reduction
Selective?	No	No	Yes	Yes	Yes
Sensitivity to flowrate changes?	Possibly	Possibly	No	Yes	Yes
Limit of detection?	10^{-10} g/L	10^{-9} g/L	10^{-11} g/L	10^{-12} g/L	10^{-9} g/L
Cell volume?	8-10 μL	8-10 μL	20 μL	≤1 μL	
Compatible with gradient?	Yes	Yes	Yes	?	

exciting light. Because the fluorescent intensity is directly proportional to the excitation light intensity, there has been a great deal of interest recently in laser sources. Fluorescence is a highly sensitive detection method for those compounds that fluoresce, but as many molecules do not fluoresce, numerous methods for derivatizing compounds have been developed. Epinephrine and other amine compounds can react with dimethylaminonaphthalenesulfonyl (dansyl) chloride or *o*-phthalaldehyde to produce highly fluorescent compounds.

Electrochemical detection. Although a variety of electrochemical detectors are available, amperometric detectors are the most widely used. In an amperometric cell, the column effluent flows past an electrode to which a voltage has been applied. If the voltage is sufficiently large, the analyte molecules at the interface between the electrode and the solution can either accept electrons and be reduced or give up electrons and be oxidized. The net movement of electrons causes a current to flow, and the current is proportional to the concentration of analyte at the interface, as in the following:

$$i = hnFA(C - C_o) \qquad \textit{Eq. 5-7}$$

where i is the current, h is a measure of the ability of the analyte to reach the electrode surface, n is the number of electrons involved in the oxidation or reduction, F is Faraday's constant, A is the electrode area, and C and C_o are the analyte concentrations in solution and at the electrode surface, respectively. Several detector designs have been developed in an effort to have h, and therefore the current, as large as possible. The two most popular designs, the thin layer or channel cell and the wall jet cell, are shown in Fig. 5-10. Although there are theoretical advantages and disadvantages, these two types of cells are comparable in performance. In addition to the variety of cell designs, many electrode materials have been used, including carbon paste, glassy carbon, gold, and mercury. The different electrode materials have different compatibilities with HPLC solvents and the voltages that can be applied to them. Carbon paste and glassy carbon have a potential range useful for oxidation of compounds such as the catecholamines. Mercury has a range most useful for reductions.

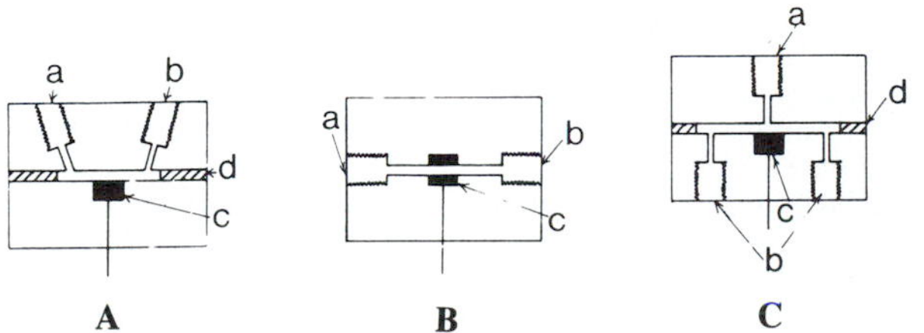

Fig. 5-10 Schematic diagram of thin layer of channel, **A,** tubular, **B,** and wall-jet, **C,** electrochemical flow cell designs. *(From Weber, SG: I & EC Product Research and Development 20:593, 1981.)*

As discussed in Chapter 9, the oxidation or reduction of a compound occurs in a distinct potential range. Thus electrochemical detection can be quite selective. For oxidative (or anodic) processes, amperometric detection is very sensitive. As little as 10 pg of epinephrine can be detected. For cathodic reactions, a number of problems associated with oxygen reduction restrict the limit of detection to the nanogram range. To make the reductive mode more sensitive, all oxygen would have to be removed from the mobile phase. Another advantage of amperometric detectors is the small cell volume, which can be smaller than 1 μL.

Mass spectrometric (MS) detection. Intensive research has gone into the development of a reliable interface between HPLC and a mass spectrometer in recent years. The major problem in developing the interface is the large volume of gas produced from commonly used HPLC liquid flow rates (1 to 2 mL/min). Two recently developed interfaces show great promise. Thermospray is a technique that uses heat to evaporate the mobile phase and provide a source of ions for the mass spectrometer. More recently, particle momentum separators, similar to the jet separators originally developed for GC/MS, have been developed and are commercially available. The next few years should see the common application of HPLC/MS to problems of clinical interest.

SUMMARY

Liquid chromatography has become an integral part of the analytical capability of the clinical laboratory. The evolution of high-performance liquid chromatography was motivated, in part, by the interest expressed by clinical chemists in separating and measuring the concentrations of drugs and their metabolites in body fluids. These early applications took advantage of the speed of HPLC for compounds that were easily measured by the standard detectors (254 nm). It has become apparent, subsequently, that HPLC has several limitations for clinical application: (1) a significant amount of sample clean-up may be required, (2) no sensitive detector responds to all compounds, and (3) there are limited applications of selective detectors that are highly sensitive. The sample preparation problem is being addressed in a number of ways, including off-line preparation with solid-phase extractants such as Bond-Elut and Sep-Paks, or automated systems, such as the du Pont Prep I, and on-line extractions, such as the Technicon Fast-LC system. Although there is nothing at present to fill the universally sensitive detector void, on-line derivatization systems to enhance detection are being investigated. Recently two commercial derivatization systems have been introduced, the Kratos Post Column Reaction System and the Varian System I-III post-column reactor. Refer to Lawrence and Frei's monograph[24] and to Knapp's compilation of methods[25] to appreciate the advantages and disadvantages of derivatization. Current research will have an impact on the limits of detection in HPLC, and before long

Table 5-4 Sample of HPLC usage in the clinical laboratory

Analyte	Sample matrix	Chromatographic mode	Mobile phase*	Detection system†
Antiarrhythmic drugs (procainamide, lidocaine, quinidine, disopyramide, propranolol)	Serum	Reversed phase	I	UV (254), fluorescence
Anticonvulsant drugs (phenobarbital, phenytoin, ethosuximide, primidone, carbamazepine)	Serum	Reversed phase	I	UV (200)
Amino acids	Urine	Reversed phase	G	Fluorescence, dansyl derivatives
Bile acids	Serum	Reversed phase	G	Fluorescence, enzyme reduction of NAD to NADH
Bilirubin	Serum	Reversed phase	I	Vis (403)
Carbohydrates (from fungal infections)	Serum	Normal bonded phase (NH_2)	I	UV (200)
Chloramphenicol	Serum	Reversed phase	I	UV (278)
Cholesterol	Serum	Reversed phase	I	UV (205)
Cortisol	Serum	Reversed phase	I	Fluorescence
Creatinine	Serum	Reversed phase	I	UV (238)
Creatinine kinase isozymes	Serum	Ion exchange	G	Chemiluminescence
1,25-Dihydroxyvitamin D	Plasma	Normal bonded phase (NO_2)	G	CPB
Epinephrine	Plasma, urine	Reversed phase	I	Electrochemical
Estriol	Urine	Reversed phase	I	Fluorescence
Gentamicin	Serum	Reversed phase	I	Fluorescence, *o*-phthalaldehyde derivative
Hemoglobin A_{1c}	Red blood cell hemolysate	Ion exchange	G	Vis (403)
Homovanillic acid	Urine	Reversed phase	I	Fluorescence, *o*-phthalaldehyde derivative
5-Hydroxyindole acetic acid	Urine	Reversed phase	I	Electrochemical
17α-Hydroxyprogesterone	Serum	Reversed phase	I	UV (254)
Metanephrines	Urine	Reversed phase	I	Fluorescence, *o*-phthalaldehyde derivative
Norepinephrine	Plasma, urine	Reversed phase	I	Electrochemical
Porphyrins	Urine	Reversed phase	G	Fluorescence
Prednisone, prednisolone	Serum	Reversed phase	I	UV (254)
Proteins	Serum	Ion exchange	I	UV (254), *p*-bromophenacyl derivative
Serotonin	Serum	Reversed phase	I	Electrochemical
Tricyclic antidepressants	Serum	Adsorption	I	UV (254)
Vitamin A and E	Serum	Reversed phase	I	UV (290)

*Type of elution system used; *I*, isocratic; *G*, gradient.
†*CPB*, Competitive protein binding; *UV*, Ultraviolet photometry; *Vis*, visible light; numbers in parentheses indicate wavelength.

it should be possible to reach the picogram range for a number of compounds of clinical interest. HPLC has come a long way from being a tool limited to the drug analysis laboratory. As summarized in Table 5-4, liquid chromatography has been used in every area of the laboratory for preparative, qualitative identification, and quantitative separations. The table is not meant to be comprehensive, but rather to serve as a representative sampling in terms of equipment requirements and areas of application.

REFERENCES

1. Snyder, LR, and Kirkland, JJ: Introduction to modern liquid chromatography, ed 2, New York, 1979, John Wiley & Sons, Inc.
2. Willard, HH, Merritt, LL, Dean, JA, and Settle, FA: Instrumental methods of analysis, ed 6, New York, 1981, Van Nostrand Reinhold Co.
3. Johnson, EL, and Stevenson, R: Basic liquid chromatography, Palo Alto, Cal, 1978, Varian Associates.
4. Karger, BL, Snyder, LR, and Horvath, C: An introduction to separation science, New York, 1973, John Wiley & Sons, Inc.
5. Giddings, JC: Dynamics of chromatography, New York, 1965, Marcel Dekker, Inc.
6. Majors, RE: In Grushka, E, editor: Bonded stationary phases in chromatography, Ann Arbor, Mich, 1974, Ann Arbor Science Publishers.
7. Guichon, GG: In Horvath, C, editor: High performance liquid chromatography: advances and perspectives, vol 2, New York, 1981, Academic Press, Inc.
8. van der Wal, SJ, and Snyder, LR: Precision of high-performance liquid chromatographic assays with sample pretreatment: error analysis for the Technicon Fast-LC system, Clin Chem 27:1233-1240, 1981.
9. Snyder, LR: Gradient elution. In Horvath, C, editor: High performance liquid chromatography: advances and perspectives, vol 1, New York, 1980, Academic Press, Inc.

10. Snyder, LR: Principles of adsorption chromatography, New York, 1963, Marcel Dekker, Inc.
11. Snyder, LR, Glajch, JL, and Kirkland, JJ: Theoretical basis for a systematic optimization of mobile phase selectivity in liquid-solid chromatography, J Chromatogr 218:299-326, 1981.
12. Horvath, C, Melander, W, and Molnar, I: Solvophobic interactions in liquid chromatography with nonpolar stationary phases, J Chromatogr 125:129-158, 1976.
13. Horvath, C, and Melander, W: Liquid chromatography with hydrocarbonaceous bonded phases: theory and practice of reversed phase chromatography, J Chromatogr Sci 15:393-404, 1977.
14. Atwood, JG, and Goldstein, J: Testing and quality control of a reversed phase column packing, J Chromatogr Sci 18:650-654, 1980.
15. Karger, BL, and Giese, RW: Reversed phase liquid chromatography and its application to biochemistry, Anal Chem 50:1048A-1073A, 1978.
16. Majors, RE: Practical operation of bonded-phase columns in high-performance liquid chromatography. In Horvath, C, editor: High performance liquid chromatography: advances and perspectives, vol 1, New York, 1980, Academic Press, Inc.
17. Glajch, JL, Kirkland, JJ, Squire, KM, and Minor, JM: Optimization of solvent strength and selectivity for reversed-phase chromatography using an interactive mixture-design statistical technique, J Chromatogr 199:57-79, 1980.
18. Karger, BL, LePage, JN, and Tanaka, N: Secondary chemical equilibria in high-performance liquid chromatography. In Horvath, C, editor: High performance liquid chromatography: advances and perspectives, vol 1, New York, 1980, Academic Press, Inc.
19. Horvath, C, Melander, WR, Molnar, I, and Molnar, P: Enhancement of retention by ion-pair formation in liquid chromatography with nonpolar stationary phases, Anal Chem 49:2295-2305, 1977.
20. Melin, AT, Askemark, Y, Wahlund, KG, and Schill, G: Retention behavior of carboxylic acids and their quaternary ammonium ion pairs in reversed phase chromatography with acetonitrile as organic modifier in the mobile phase, Anal Chem 51:976-983, 1979.
21. Walker, JQ, Jackson, MT, Jr, and Maynard, JB: Chromatographic systems: maintenance and troubleshooting, New York, 1957, Academic Press, Inc.
22. Rabel, FM: Use and maintenance of microparticulate high performance liquid chromatography columns, J Chromatogr Sci 18:394-408, 1980.
23. Dolan, J, and Upchurch, P: Interchangeability of HPLC fittings. Upchurch Scientific, Oak Harbor, WA, 1983; also LC Magazine 2:20-21, 1984 and 3:92-95, 1985.
24. Lawrence, JF, and Frei, RW: Chemical derivatization in liquid chromatography, Amsterdam, New York, 1976, Elsevier/North-Holland, Inc.
25. Knapp, DR: Handbook of analytical derivatization reactions, New York, 1979, John Wiley & Sons, Inc.

CHAPTER 6 Gas chromatography

ALPHONSE POKLIS

OBJECTIVES

- Diagram the basic components of a gas chromatographic system, describe the function of each component, and outline the mechanism of gas-liquid and gas-solid chromatography.
- Explain the significance of temperature dependence in gas chromatography, and tell why temperature is the most important single parameter.
- Name four common carrier gases and state the function of a carrier gas, summarizing the criteria for selection of an appropriate carrier gas for chromatographic separation.
- Define derivatization and state why the process is used in gas chromatography.
- Describe the operation of the five types of detectors that may be used in gas chromatography.

KEY TERMS

active sites Places, usually on the stationary phase, that reversibly bind the compound to be separated.

chemical ionization The component molecule to be analyzed is mixed with an ionized gas, such as methane or isobutane. A positive charge is transferred to the molecule, the M + 1 charged molecule and its fragments are separated by the mass spectrometer, and their size and relative abundance are measured. (''M'' is 'mass'.)

corrected retention time The amount of time a compound is retained on a column minus the gas holdup time.

derivative A molecule chemically altered from the original one. Usually in gas chromatography it refers to chemical groups added to increase the volatility of the initial compound.

diffusivity The ability of a molecule to diffuse or spread because of the thermal energy inherent in the molecule.

effective plate (number of effective plates) The number of partitions that are practically available on a column.

electron-capture detector A device that releases beta particles into the carrier gas stream, producing low-energy electrons that are collected on electrodes and measured. This type of detector is very sensitive and specific for compounds with chemical groups of high electronegativity, such as halogens.

electron impact Fragmentation of molecules into specific charged fragments by collision with high-energy electrons.

flame ionization detector A device in which eluted components are mixed with hydrogen and burned in the air to produce flame, which ionizes these components. A pair of electrodes measures the number of ions.

flow regulator A system of valves that is set to yield a desired gas pressure and thus control the rate of gas movement (flow) through a gas chromatographic system.

gas chromatography A physical technique that separates components based on their distribution between a gas and a stationary phase.

gas holdup time The amount of time it takes for the carrier gas to move from the injection port to the detector, analogous to the void volume of liquid chromatography.

gas-liquid chromatography A separation technique in which the stationary phase is a liquid.

gas-solid chromatography A separation technique in which the stationary phase is a solid.

injection port A device usually having a septum and a heating block to volatilize the compounds to be separated. This is placed before the column.

Kovats' index This index relates the logarithm of the retention time of a compound, regardless of its chemical nature, to those of the *n*-paraffins.

liquid phase The nonvolatile fluid that coats the immobile support medium. These fluids have the property of acting as solvents for the compounds to be separated.

mass fragment A degraded portion of a molecule containing one or more charges.

mass spectrometer A device that may be considered a gas chromatographic detector. Compounds are fragmented into specific groups of charged molecules, which are separated into their mass and charge components, and their relative abundance is measured.

mass transfer The movement of mass from one phase to another.

McReynolds' constant A constant describing a system that classifies the stationary phase in terms of its ability to separate various compounds.

negative chemical ionizations Similar to chemical ionization except that the gas is oxygen or hydrogen, which produces primarily negative ions of the form M − 1. (''M'' is 'mass.')

nitrogen-phosphorus detector A device similar to a flame ionization detector but into which alkaline metals are introduced. When nitrogen- or phosphorus-containing compounds are burned, the rate of release of alkaline metal vapor and thus of current flow is increased.

nonvolatile liquid A fluid that does not vaporize or have a form that is readily gaseous in nature.

nonpolar Usually applied to molecules that have a hydrophobic affinity, that is, ''water hating.'' Nonpolar substances tend to dissolve in nonpolar solvents.

overloading When too much of a compound is presented for adsorption by the stationary phase, nonequilibrium between the two phases occurs.

phase ratio The ratio of mobile-phase (gas) volume to stationary-phase (column) volume. Usually indicated by the term ''β.''

plate A chromatographic term that refers to a single partitioning unit of the chromatographic system.

polar Usually applied to molecules that have a hydrophilic affinity, that is, ''water loving.'' Polar substances tend to dissolve in polar solvents.

relative retention time The ratio of the corrected retention time of the reference compound to that of the sample compound.

retention index A system relating the retention time to a standard.

Rohrschneider constant Similar to the McReynolds constant.

selected-ion chromatogram A technique in which only mass fragments of a preselected size are recorded and quantified by the mass spectrometer.

separator A device that removes large portions of the carrier gas and concentrates the solutes before entrance to the mass spectrometer.

septum A device that separates the chromatographic column from the laboratory environment. Usually it is a small disk of silicone rubber through which the solution to be separated is injected into the column.

silanization The chemical process of converting the SiOH moieties of a stationary phase to the ester form.

sorbent A material that has the property of interacting with the compound of interest, usually to make it bind.

thermal compartment The temperature-regulated oven in which the chromatographic column is placed.

thermal-conductivity detector A device that measures the difference between the heat conductivities of the carrier gas and that of the sample-gas effluents. A sample carried in gas increases the heat conductivity.

Van Deemter's equation Relates the HETP (height equivalent to the theoretical plate) to the linear velocity of the carrier gas.

Chromatography is a physical technique that separates two or more compounds based on their distribution between two phases, a stationary and a mobile one. Review Chapter 4 for a description of the basic theory and practice of chromatographic separations. The stationary phase may be a liquid or a solid, and in gas chromatography (GC) the mobile phase is a gas that percolates over the stationary phase. When separation of sample components is accomplished by use of a mobile gas phase and a stationary phase consisting of a thin layer of nonvolatile liquid held on a solid support, the technique is called gas-liquid chromatography (GLC). Gas-solid chromatography (GSC) employs a solid sorbent as the stationary phase. Regardless of the type of mobile or stationary phase, separation is achieved by the difference in partitioning of the various molecules of the sample between the two phases.

Gas chromatographic separation is illustrated by the following example: A sample containing the components to be separated is injected into a heated block in which they are immediately vaporized and swept by a stream of carrier gas through a column of stationary phase. The components are adsorbed onto the stationary phase at the head of the column and then are gradually desorbed by fresh carrier gas. The partitioning between the two phases occurs repeatedly as carrier gas sweeps the components toward the column outlet.

As the components are eluted, they enter a detector where their presence is converted to an electric signal, which is then measured, usually by a strip chart recorder that produces a series of peaks charted versus time (Fig. 6-1). The appearance time, height, width, and area of these chromatogram peaks may be measured to yield valuable qualitative and quantitative data.

MOLECULES THAT CAN BE SEPARATED BY GAS CHROMATOGRAPHY

Theoretically any compound that can be vaporized or converted to a volatile derivative may be analyzed by gas chromatography. Compounds as small as carbon monoxide and methane and as large as 800 daltons have been successfully analyzed. Compounds larger than 800 daltons lack sufficient volatility. Generally a compound must be stable as a vapor to produce a single identifiable chromatographic peak. Unstable compounds may be converted to stable, volatile derivatives. However, if a compound degrades to known products or a consistent number of prod-

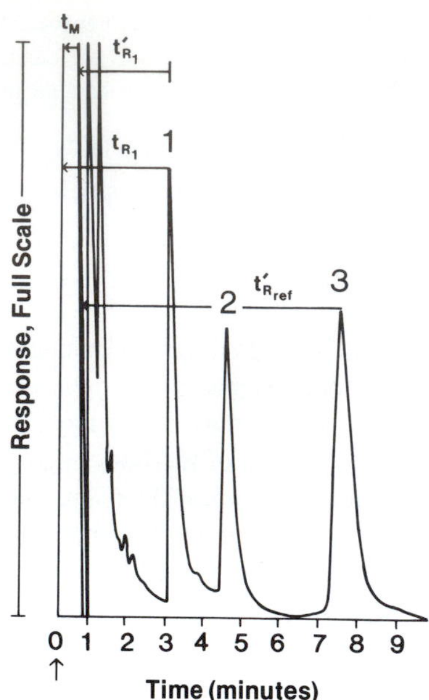

Fig. 6-1 Example of gas chromatogram showing detector response versus time. Vertical arrow (time = 0) indicates time of injection, and initial peak is "dead time," t_M. t_R designates uncorrected retention time, whereas t'_R is corrected retention time. $t'_{R_{ref}}$ is corrected reference time for internal standard or reference compound. Relative retention time for peak 1: $r = t'_{R1}/t'_{R_{ref}}$. *(Modified from Mackell, MA, and Poklis, A: J Chromatogr 235(2):445, 1982.)*

ucts, the resultant pattern of multiple compounds may be used as a means of tentative identification.

Inorganic compounds, or the inorganic salts of organic acids and bases, lack sufficient volatility for gas chromatographic analysis. Thus the technique is generally applied to analysis of organic molecules in their neutral nonionic forms. Before chromatographic analysis, compounds are generally isolated and concentrated by means of solvent extraction and evaporation to dryness. The residues containing the analytes are dissolved in small amounts of volatile organic solvents. The solvent-analyte solution is then chromatographed. The analyte vapor should not interact with the solvent. The solvent should have greater volatility and much less affinity for the stationary phase than the analyte compounds do, thereby eluting far ahead of the analyte and not interfering with the chromatogram.

THEORY OF GAS CHROMATOGRAPHIC SEPARATION

The following is a brief discussion of the basic chromatographic theory as it applies to gas-liquid chromatography. See Chapter 4 for a review of general chromatography theory. For a more complete treatment of the many complex variables that influence gas chromatographic separations, one should read references 1 and 2.

Partition coefficient (K_D)

The general concepts of sample *partitioning* and the *partition coefficient (K_D)* have been described in Chapter 4 (see Equation 4-1 and p. 75). Similarly, definitions of *retention* and *retention time,* and *capacity factor* (or capacity rates) have been described in Chapter 4 on p. 73 and Fig. 6-1. These basic concepts apply to both LC (Chapter 5) and GC. The transit time required for the carrier gas to move from the point of injection to the end of the column is called the "gas holdup time," or "dead time," t_M (Fig. 6-1). It arises from the internal volume of the injector column and detector and is equivalent to the void volume in HPLC. A compound that does not partition into the stationary phase ($K_D = 0$) will be eluted from the column at t_M.

Because the "dead time" is a characteristic of both the particular gas chromatograph used for analysis (injector and detector volumes) and the column volume, it is the same for all sample components and is of no significance in identification. The difference between the uncorrected retention time and the dead time is called the "corrected retention time," t'_R

$$t'_R = t_R - t_m \qquad \textit{Eq. 6-1}$$

Partitioning in GC systems

The K_D, or corrected time, t'_R, of a solute in a specific system must be determined experimentally; it cannot be predicted easily. Both the partition coefficient and retention time of a solute are directly related and depend on the affinity of solute for the stationary phase. The old rule "like dissolves like" offers a simple guide to solute affinity. Polar solutes will have greater partition coefficients, K_D, hence longer retention times on hydrophilic (polar) phases than hydrophobic (nonpolar) stationary phases. Likewise, hydrophobic solutes exhibit greater K_D's and longer retention times on the nonpolar rather than the polar stationary phase. If both a hydrophobic and a polar solute are chromatographed together on a polar stationary phase, the hydrophobic solute will have less affinity for the phase than the polar solute (K_D hydrophobic $<$ K_D polar) and will be eluted before the polar solute ($t'_{R\ \text{hydrophobic}} < t'_{R\ \text{polar}}$).

The K_D of solute is expressed as a concentration (amount/volume) ratio and therefore may be written as

$$K_D = [W_s/V_s][W_m/V_m] = \frac{V_m}{V_s} \cdot \frac{W_s}{W_m}$$

W_S = Weight of the sample in the stationary phase
W_M = Weight of the sample in the mobile phase
V_S = Volume of the stationary phase
V_M = Volume of the mobile phase

The ratio of the mobile-phase (gas) and stationary-phase (column) volumes (V_m/V_s) is called the phase ratio (β). The ratio of sample weights (W_s/W_m) in each phase is equal to the capacity ratio, k. Therefore the partition coefficient can be expressed as

$$K_D = \beta k$$

The phase ratio and capacity ratio are characteristics for a particular column. However, their product, K_D, is independent of the particular column. Therefore the higher the phase ratio (smaller the volume of stationary phase), the smaller the capacity ratio (less time for elution, t_R). In general, the smaller the capacity ratio (shorter t_R), the more difficult it is to achieve a particular separation. The amount of solute analyzed on column without overloading is dependent on the amount of stationary phase that influences β. The greater the amount of stationary phase, the smaller the β and the larger the *k* (greater t_R).

Temperature dependence

Temperature is the most important single parameter in a gas chromatograph separation. This is attributable to the great dependence of the partition coefficient, K_D, on temperature. The inverse relationship between K_D and temperature is given by the equation

$$\log K_D = \frac{\Delta H}{2.3\ R \cdot T_c} + \text{Constant} \qquad \textit{Eq. 6-2}$$

ΔH is the partial molar heat of solution of the solute in the liquid state, *R* is the gas constant, and T_c is the column temperature.

This equation demonstrates that, in a given system, the higher the column temperature, the lower the K_D, which means a lower capacity ratio, *k* (shorter retention time). Therefore the retention times of solute molecules may be readily altered when one changes the column temperatures. Roughly, a 30° C decrease in column temperature will approximately double the retention time. Conversely, a 30° C increase in column temperature will approximately halve the retention time. The influence of temperature on separation is discussed later.

Column performance

The ability of a column to produce optimum separations is measured by two quantities: *efficiency,* which is the ability to produce narrow peaks, and *resolution,* which is the ability to separate two adjacent peaks.

The general concepts of *resolution, theoretical plates,* and *height equivalent to a theoretical plate* (HETP), all used to describe the efficiency of a column have been defined in Chapter 4 (pp. 79 to 81). HETP (or H) uses the uncorrected retention time, t_R. If the corrected retention time, t'_R, is employed instead, the expression *height equivalent to an effective plate,* or HEEP *(h),* is used. The term HETP defines not only the efficiency of the column, but also the overall efficiency of the system because of the gas holdup time of the injector, column, and detector. Because the gas holdup time varies in different instruments, the column efficiency is best expressed by the number of effective plates (HETP or *h*).

The relationship of the HETP to the linear gas velocity, μ, is complex. The factors affecting gas flow are expressed by the Van Deemter's plot as described in Chapter 4, p. 80.

The relationship of plate height to linear velocity, when graphed, corresponds to a hyperbola (see Fig. 4-8). Eventually the curve reaches a minimum, which is the optimum velocity (μ_{opt}) for the smallest plate height. The optimum velocity is ideal for only one compound; however, similar compounds have closely related optimum velocities, and a single flow rate is suitable for their separation. At higher flow rates, the gas sweeps the diffusing molecules from the liquid before all have emerged, thus broadening the peak.

MOBILE-PHASE CONSIDERATIONS

The most commonly used carrier gases are presented in Table 6-1. The carrier gas must be inert so as not to react with the sample components. Large quantities of relatively pure gas must be commercially available because appreciable amounts of carrier gas are used for analysis. Common impurities in carrier gases are moisture, oxygen, and hydrocarbons. Each of these contaminants may adversely affect various detectors, producing unstable recorder baseline or extraneous peaks. In certain situations, carrier-gas impurities may interact with sample components and prevent their analysis. For example, prepurified-grade nitrogen contains up to 20 ppm of oxygen. If high column temperatures are necessary to separate compounds that are readily oxidized, the oxygen impurity in nitrogen carrier gas may degrade the compounds on the column and prevent their detection or may produce multiple extraneous peaks of the degradation products. In such a situation, helium, which contains less oxygen contamination, should replace nitrogen as the carrier gas.

The choice of carrier gas does influence column performance (efficiency and resolution) and time required for analysis (retention time). As presented by the Van Deemter equation, the height equivalent to a theoretical plate (HETP) is related to the linear gas velocity (μ) of the carrier gas. This interaction is highly complex, but the following brief discussion presents the basic effects of carrier gas upon separation.

Table 6-1 Common carrier gases

Gas	Molecular weight	Density (g/L)	Impurities (ppm)
Argon	39.944	1.784	—
Helium	4.007	0.177	Hydrocarbons (1 to 100)
Hydrogen	2.018	0.089	—
Nitrogen	28.014	1.251	Oxygen (20)

In a flowing system (column) where the gas is compressible, the density, pressure, and velocity of the gas are different at each point in the column. The carrier gas is compressible, and the value of the carrier-gas velocity must be corrected to average conditions. The average linear gas velocity ($\bar{\mu}$) is determined by two factors: (1) the time necessary for an unretained solute to pass through the column (dead time, t_M) (p. 112) and (2) the length of the column, L. These factors determine μ by the following equation:

$$\bar{\mu} = L/t_M \qquad \textit{Eq. 6-3}$$

In a given column, the optimum linear gas velocity ($\bar{\mu}_{opt}$) is proportional to the diffusivity of the vapors of the substances being chromatographed. However, the kinetic theory of gases states that the diffusivity in gas is inversely proportional to the square root of the molecular weight or density of the gas (see Table 6-1). Therefore the μ_{opt} will be lower for high-density gases (low solute diffusivity), such as nitrogen and argon, and higher for low-density gases (high solute diffusivity), such as helium and hydrogen. Therefore in the same column different values of μ_{opt} will be obtained for different gases. Fig. 6-2 shows the relationship between the $\bar{\mu}$ of two different carrier gases and the resultant HETP. Nitrogen has a minimum HETP of 0.465 at a μ_{opt} of 7 cm/sec, whereas helium has a minimum HETP of 0.55 at a μ_{opt} of 17.5 cm/sec. This demonstrates that in the same length of column, a high-density gas (nitrogen or argon) will produce better efficiency (more theoretical plates per unit length, lower HETP) than a low-density gas (helium or hydrogen). For a given system, the maximum resolution will be obtained at minimum h, which is determined at the μ_{opt} of the carrier gas. Also, for a given length of column, better resolution will be obtained by the carrier gas producing the lowest h value.

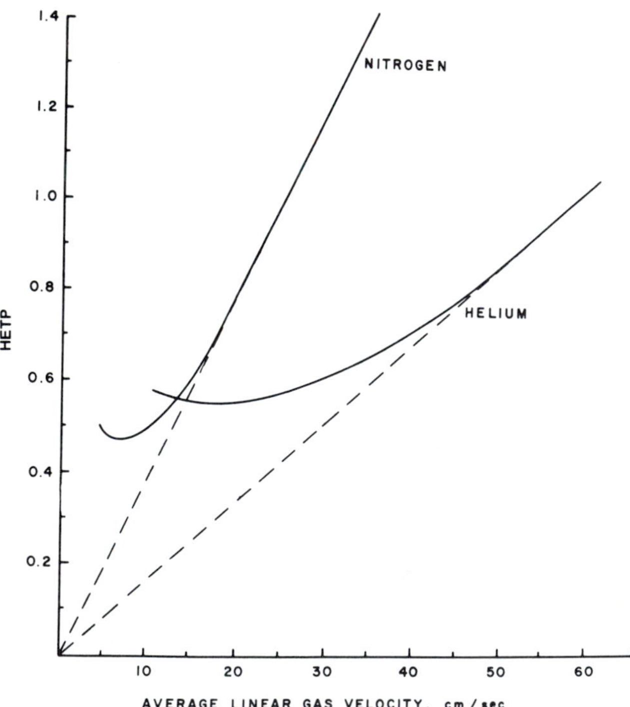

Fig. 6-2 Relationship between HETP (heat equivalent to a theoretical plate) and average linear gas velocity for two different carrier gases: nitrogen and helium. *(From Ettre, LS: Practical gas chromatography, Norwalk, Conn, 1973, Perkin-Elmer Corp.)*

For a low density gas, the diffusivity of the solute and the μ_{opt} in a given column are greater than those of a high-density gas. Therefore, shorter retention times (smaller t_R) are obtained with helium or hydrogen than with nitrogen or argon.

STATIONARY-PHASE CONSIDERATIONS

Gas-solid stationary phases

In gas-solid chromatography (GSC) the column is packed with an adsorptive solid material on which the sample components are partitioned by adsorption on the surface of the solid. This material should possess a large surface area per unit volume to ensure rapid equilibrium between the stationary and gas phases. It should possess uniform particle size and pore structure and be strong enough to resist breakdown during handling and column packing. Theoretically, the smaller the particle size of support, the greater the efficiency of the column. However, the smaller the particles, the greater the resistance to flow and the greater the necessary carrier-gas pressure.

The most common chromatographic solids for adsorption phases are made from diatomaceous earth (kieselguhr). The processed white kieselguhr is sold under many trade names: Chromosorb W, Celite, Gas Chrom, and Anakron. The diatomite may also be crushed, blended, pressed into brick, and processed such that mineral impurities form oxides and silicates, which give the material a pink color. It is marketed as crushed firebrick, or Chromosorb P. This material has greater density and is less fragile than the white material. The pore size of the pink material is only 2 mm compared to 9 mm for the white. Therefore greater efficiency is obtained with the pink material.

Each support possesses individual properties that may enhance or hinder its use for a particular application. The white material is slightly alkaline and will interact with acidic compounds. Its surface, however, is nonadsorptive, a property that favors its application for analysis of polar compounds. The pink material adsorbs polar compounds; thus it is best suited for the separation of nonpolar molecules like hydrocarbons.

Another type of solid stationary phase consists of porous polymer beads, which allow the analyte molecules to partition directly from the gas phase into the amorphous polymer. Porapak, a polymer of ethylvinylbenzene cross-linked with vinylbenzene, is the most popular polymer phase.

Table 6-2 Examples of commonly used stationary phases and their applications

Stationary phase	Structures	Activity	Temperature (°C min/max)	Application	Specific compounds
Silicone OV-1 (100% methyl)	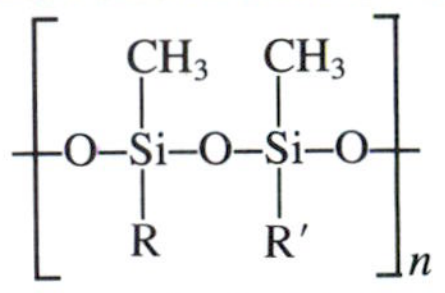 $\left[-O-\underset{R}{\overset{CH_3}{Si}}-O-\underset{R'}{\overset{CH_3}{Si}}-O-\right]_n$ R and R′ = CH_3 in above structure	Nonpolar	100/350	Bacteria, drugs	Fatty acid methyl esters, benzodiazepines
Silicone OV-17 (50% phenyl)	R and R′ = phenyl in above structure	Intermediate polarity	20/350	Drugs, steroids	Tricyclic antidepressants, barbiturates, cholesterol
Silicone OV-210 (50%, 3,3,3-trifluoropropyl)	R and R′ = $-CH_2-CH_2-CF_3$ in above structure	Polar	20/300	Drugs, pesticides	Basic drugs, lindane, aldrin, DDT
Silicone OV-225 (25% cyanopropyl, 25% phenyl)	R = phenyl, R′ = $-CH_2-CH_2-CH_2-CN$ in above structure	Polar	20/275	Steroids	TMS derivatives of 17-ketosteroids
10% Apiezon L 2% KOH	Undefined mixture of high-boiling hydrocarbons	Nonpolar	50/225	Amines	Amphetamine
NPGS (neopentyl glycol succinate)	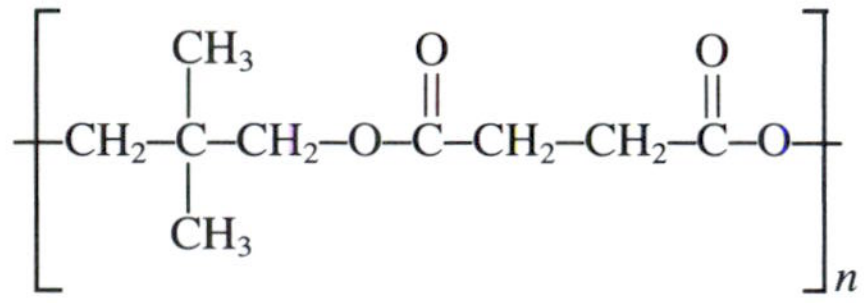 $\left[-CH_2-\underset{CH_3}{\overset{CH_3}{C}}-CH_2-O-\overset{O}{\overset{\Vert}{C}}-CH_2-CH_2-\overset{O}{\overset{\Vert}{C}}-O-\right]_n$		50/240	Volatile fatty acids	Acetic through caproic acids
Carbopack B/5%	$(-CH_2-CH_2-O-)_n$	Polar		Alcohols, aldehydes, ketones	Methanol, ethanol, acetaldhyde acetone
DEGS (diethylene glycol succinate)	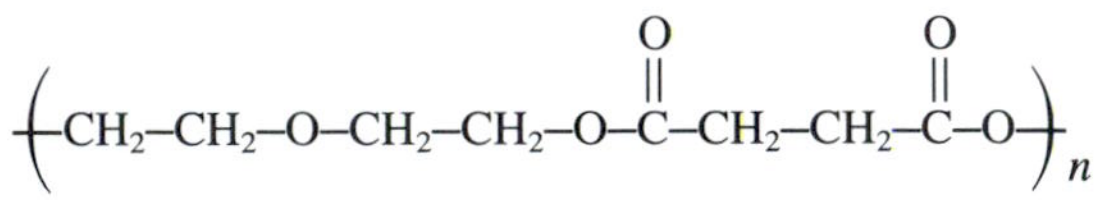 $\left(-CH_2-CH_2-O-CH_2-CH_2-O-\overset{O}{\overset{\Vert}{C}}-CH_2-CH_2-\overset{O}{\overset{\Vert}{C}}-O-\right)_n$	Polar	20/200	Bacteria	Fatty acid methylesters
EGA (ethylene glycol adipate)	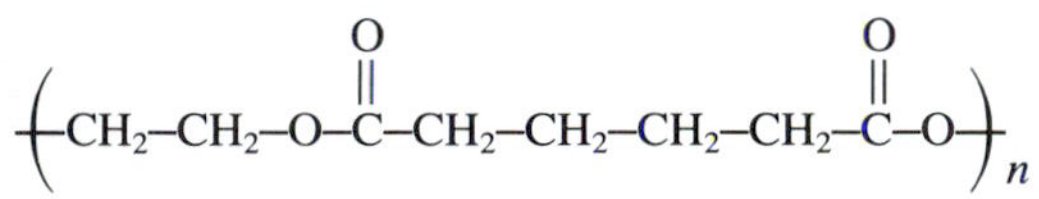 $\left(-CH_2-CH_2-O-\overset{O}{\overset{\Vert}{C}}-CH_2-CH_2-CH_2-CH_2-\overset{O}{\overset{\Vert}{C}}-O-\right)_n$		100/210	Amino acids	NBTFA* derivatives of amino acids
Chromosorb 102 (styrene divinyl benzene polymer)	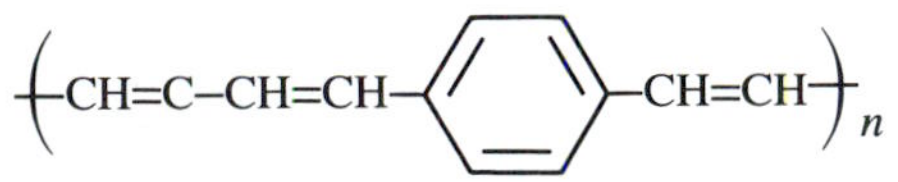 $\left(-CH=C-CH=CH-C_6H_4-CH=CH-\right)_n$		<250° C	Alcohols, aldehydes	Methanol, ethanol, acetaldehyde
Porapak Q (ethylvinyl benzene + divinyl benzene polymer mixture)	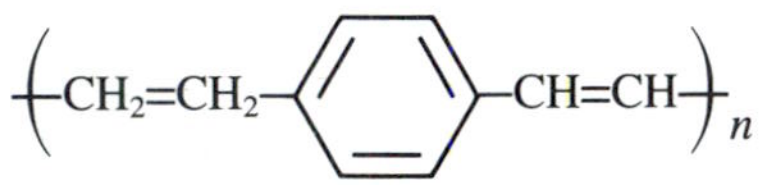 $\left(-CH_2=CH_2-C_6H_4-CH=CH-\right)_n$		<250° C	Low molecular weight	Chlorinated hydrocarbons

**NBTFA*, Nitroblue tetrazolium fatty acid.

The material may be modified by copolymerization with various polar monomers to produce beads of varying polarity. Porapak columns are thermally stable up to 250° C. At temperatures above 250° C the column material will be degraded and eluted, a phenomenon called *column bleed.* These degradation products can be observed by the detector. Water and highly polar molecules are rapidly eluted from the polymer. Porapak is especially useful for baseline separation of aqueous samples containing low molecular weight alcohols, esters, halogens, hydrocarbons, ketones, and mercaptans (Table 6-2).

Gas-liquid-solid supports for GLC

The stationary phase in gas-liquid chromatography (GLC) is a thin film of liquid held on an inert support. In capillary chromatography, the liquid is coated on the walls of the tubing. In packed columns, the liquid is held in a thin-layer film across the surface of an inert support (Fig. 6-3). Many materials that act as stationary phases for GSC are also supports for the liquid phase in GLC. Both the pink and white solid phases described earlier are popular liquid supports. Although the support should be inert and not influence separation, both pink and white materials have *active sites* because of metallic impurities and silanol (-SiOH) and siloxane (SiOSi-) groups, which form hydrogen bonds with polar compounds. This interaction gives rise to distorted (asymmetrical) peaks in the resultant GLC chromatogram. These *active sites* may be removed by acid washing of the mineral impurities from the support and by conversion of the silanol groups to silylesters (silanization) of dimethyl-dichlorosilane or hexamethyldisilazone. Silanization reduces surface activity but also reduces the surface area of the support so that no more than 10% (v/w) of the liquid stationary phase to total column weight may be applied. In certain instances, special additives are mixed with the liquid phase to block the active sites of untreated support material. Two such examples are the incorporation of stearic acid in silicone oil used in separation of fatty acids and the addition of potassium hydroxide to polar liquid phases used to separate amines. Informative data describing preparation, applications, and limitations of support materials are readily available from commercial manufacturers and suppliers.

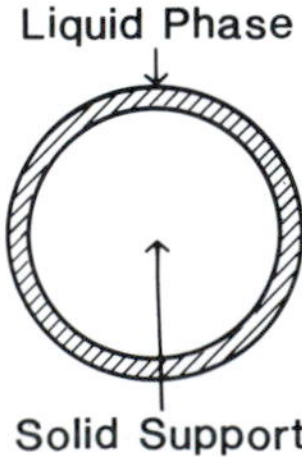

Fig. 6-3 Schematic of solid support particle for gas chromatography with liquid stationary-phase coating.

Liquid phases

The universal popularity of GLC as a separation method is attributable to the large variety of liquid phases with differing solution properties and therefore different affinities for various classes of analytes. The range of liquids used as stationary phases is limited only by their volatility, thermal stability, and ability to wet the support. No single stationary phase will achieve all desired separations. Commercial suppliers typically offer 100 to 200 liquid phases; however, many of these phases are duplicates sold under various trade names or are so similar in character as to have little difference in their separation abilities. In fact, few laboratories require the use of more than a half dozen different liquid phases. Eighty percent of a wide range of organic compounds may be successfully separated using only four to seven phases: OV-101, OV-17, Carbowax 20M, OV-225, DEGS, OV-275, and OV-210.[3,4] Examples of liquid phases, characteristics, and applications are presented in Table 6-2.

Liquid phases may be generally classified into five categories: (1) Nonpolar phases, which are hydrocarbon liquids such as squalane, silicone greases, Apiezon L, and silicone gum rubber. Generally, compounds are eluted from these phases in order of increasing boiling point. (2) Intermediate polarity phases that include polar or polarizable groups attached to a long nonpolar skeleton, such as esters of high molecular weight or alcohols such as diisodecylphthalate. Both polar and nonpolar compounds are separated by these phases, with the more polar ones eluted first. (3) Polar phases, which contain a high concentration of polar groups, such as carbowaxes. These phases differentiate between polar and nonpolar compounds by interacting strongly only with polar compounds, separating these from the earlier eluting, less polar compounds. (4) Hydrogen-bonding phases, which contain many hydrogen atoms readily available for hydrogen bonding, such as glycol phases. Polar compounds have greater affinity for the stationary phase and are eluted more slowly. (5) Special-purpose phases that can be prepared to use a specific chemical interaction between the sample and the stationary phase. An example of a special-purpose phase is silver nitrate dissolved in glycol to enhance separation of unsaturated hydrocarbons by charge-transfer interactions.

Each liquid phase has a specific temperature range for efficient use (Table 6-2). The maximum temperature at which a phase may be used is determined by its volatility. Beyond this temperature the phase is lost because of decomposition or volatilization and is carried into the detector producing extensive background noise (column bleed). A column may be heated above the maximum temperature for brief periods of time as in temperature programming, but the maximum temperature must never be exceeded for isothermal (constant-temperature) analysis. Below the minimum temperature the increased viscosity or solidification of the liquid renders analysis irreproducible.

The amount of stationary phase in the column is expressed in percent by weight of the liquid phase on the support. In general, analytical columns contain 3% to 10% liquid phase. Deviations from these values may occur in specific applications: very low liquid loads for high molecular weight compounds and high loads for small, highly volatile compounds such as hydrocarbons containing one to four carbon atoms. The amount of stationary phase directly affects the sample capacity and efficiency of the column. The greater the amount of liquid phase, the larger the amount of sample that may be chromatographed.

DERIVATIZATION

Often it is desirable to modify a molecule chemically so that a newly formed product has properties that are preferable to its precursors. One may need to derivatize a compound to make it volatile and stable as a gas and thus analyzable by GC. Derivatives are also prepared to achieve increased sensitivity, selectivity, or specificity for a given separation. Derivatives may be eluted from the column sooner, have less tailing, produce sharper peaks, provide stability to thermally labile compounds, and increase resolution. Derivatization involves a chemical reaction between some functional group on the sample molecule (usually a polar group, which reduces volatility or interacts with the stationary phase to increase retention time) and a smaller molecule (derivatizing agent), which forms a new product of increased volatility with a smaller partition coefficient (K_D). The derivatization may be carried out before sample injection or may occur in the injection port of the chromatograph (''on column'' or ''flash derivatization''). A few derivatization techniques are briefly presented, but for a more complete discussion consult the literature.[5,6]

A popular GC derivatization technique is the replacement of an active hydrogen by a trimethylsilyl (TMS) group. The resultant *silyl* derivatives are usually less polar and more volatile and display greater thermal stability than their parent compounds. Silylizing reagents react vigorously with water or alcohol-containing solvents; therefore the conversion reactions are carried out in anhydrous solvents such as acetonitrile or tetrahydrofuran. TMS reagents are flammable, and some are highly corrosive. They should be handled with care.

$$\underset{\text{Alcohol}}{ROH} + \underset{\text{Trimethylchlorsilane}}{(CH_3)_3SiCl} \rightarrow ROSi(CH_3)_3 + HCl$$

Esterification is often used for GC analysis of compounds containing a carboxylic acid group. Methyl esters possess the greatest volatility and hence are most popular. Alkylation reactions with quaternary alkylammonium hydroxides or dimethylformamide-dialkyl acetals have become popular as ''flash-derivatizing'' reagents. Fig. 6-4 presents the derivatization reaction of tetramethylammonium hydroxide and barbiturate drugs. To increase sensitivity, derivatizing reagents that produce halogen- or nitrogen-containing compounds are used with electron-capture detectors.

SELECTION OF A SEPARATION SYSTEM

Choosing the mobile phase

The mobile phase or carrier gas has one major function in gas chromatography: to carry the vaporized sample through the column and into the detector. As previously described, selection of the proper carrier gas is predicated upon three considerations: (1) the operating principles of the detector through which the gas will be continuously flowing, (2) the presence of impurities in the carrier gas, and (3) the desired speed of analysis and performance of the column. Compounds that are negligibly partitioned into a stationary phase cannot be separated from each other. Similarly, compounds with too great an affinity for the stationary phase will have unacceptably long retention times or may be irreversibly retarded.

Stationary-phase selection

Liquid phases with the same physical properties as the sample will retain the sample and generally effect a separation. However, this general rule does not aid in determining which specific stationary phase is potentially the best for a particular separation. Several approaches to the choice of liquid-phase selection for a desired separation are briefly presented.

Many sources of irreproducibility can affect GC analysis. These include variations in assay conditions and variations in stationary-phase packaging (lot to lot, company to company, and so on). To ensure reproducible identification of peaks of interest regardless of exact assay conditions,

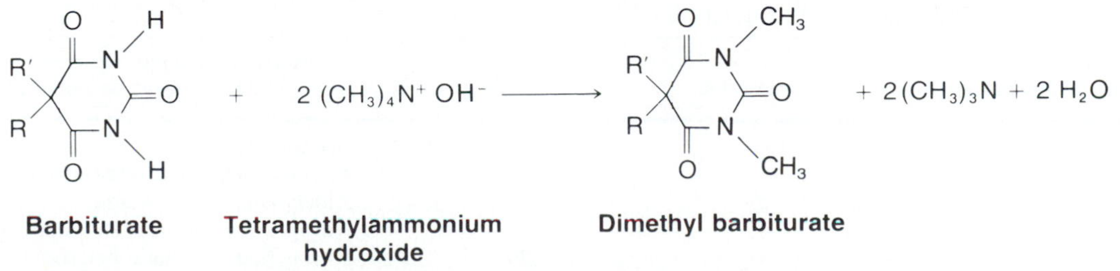

$$\text{Barbiturate} + \underset{\text{Tetramethylammonium hydroxide}}{2\,(CH_3)_4N^+\,OH^-} \longrightarrow \text{Dimethyl barbiturate} + 2(CH_3)_3N + 2\,H_2O$$

Fig. 6-4 Tetramethyl derivatization of barbiturate drugs.

Table 6-3 Data used for calculation of retention index

Compound	Carbon atoms	Symbol	t_R (min)	Log t_R (min)	Retention index (*I*)
Hexane	6	*z*	14.96	1.175	600 (by definition)
Benzene	6	*x*	16.86	1.227	650 (by experiment)
Heptane	7	*z* + 1	19.01	1.279	700 (by definition)
Octane	8	—	24.15	—	800 (by definition)
Nonane	9	—	30.76	—	900 (by definition)

relative retention times are converted to indices or constants. These values can then be used to compare data between analyses, within a laboratory, or between laboratories.

Kovats index. If the components of the sample are known, the most likely stationary phase that will effect a separation may be selected by use of the Kovats retention index.[7,8] Retention indices relate the retention time of a compound, regardless of its chemical nature, to those of the *n*-paraffins (straight-chain hydrocarbons) eluted directly before and after it. Chromatographing the *n*-paraffins on a given column under set conditions yields a nearly linear relationship between the log of their retention time t_R and the number of carbon atoms in each paraffin, as shown in Fig. 6-5. Each *n*-paraffin is given a retention index, *I*, that equals 100 times the number of carbon atoms. The retention index of all other compounds is calculated from the following relationship:

$$I = 100z + 100\,[(\log t_{Rx} - \log t_{Rz})/(\log t_{R(z+1)} - \log t_{Rz})] \qquad \textit{Eq. 6-4}$$

z equals the number of carbon atoms in the unknown compound; t_{Rx} is the retention time of the unknown substance *x;* t_{Rz} is the retention time of the *n*-paraffin eluted immediately before *x;* and $t_{R(z+1)}$ is the retention time of the *n*-paraffin eluted immediately after *x*. For example, assume the data presented in Table 6-3 and graphically represented in Fig. 6-5 were obtained by chromatography from a series of *n*-paraffins and benzene on a given liquid phase. The retention index, *I,* benzene (z = 6) on that liquid phase is calculated to be 650.

$$I = 100(6) + 100(0.052/0.104) = 600 + 50 = 650$$

The calculated *I* applies only to the particular stationary phase and temperature conditions. However, the effects of flow rate of the mobile phase and the percent loading (quantity of liquid phase) will change the retention time proportionally for all *n*-paraffins, and thus the calculated *I* values are unchanged. Extensive lists of retention indices of numerous compounds on liquid phases are available in the *ASTM Gas Chromatography Data Compilation Catalog AMD 25A*[9] and a supplement catalog *AMD 25A S-1*.[10] If several compounds of a different chemical nature are to be separated simultaneously, the compounds of interest are located in the table and their respective *I* values for various stationary phases noted. A difference of at least 30 *I* units between the compounds will indicate that a particular phase will efficiently separate them. The *I* values are based on peak apex only and give no indication of peak width. Therefore *I* values do not indicate the resolution of the compounds. However, since the retention times increase with the retention index, *I,* the retention indices indicate the order in which compounds will be eluted from the column. A recent clinical application of *I* values has been the qualitative identification of drugs on standard liquid phases under both isothermal and temperature-programmed conditions.[11]

Rohrschneider and McReynolds constants. Rohrschneider and McReynolds constants are related systems that classify stationary phases in terms of their separating power.[12,13] The *I* values of a set of reference compounds of varying polarity are determined on the liquid phase being tested and compared against the *I* values for the same compounds obtained on a reference liquid phase. Squalane, a nonpolar liquid, is used as the reference phase. The constants are then calculated as indicated in the following equations:

$$\text{Rohrschneider constant (X)} = 1/100\,(I_{\text{test phase}} - I_{\text{squalene}}) \qquad \textit{Eq. 6-5}$$

$$\text{McReynolds constant (X')} = I_{\text{test phase}} - I_{\text{squalene}} \qquad \textit{Eq. 6-6}$$

Table 6-4 presents data related to five reference compounds used to determine McReynolds constants. Both Rohrschneider and McReynolds constants may be used for two purposes: to select a liquid phase for a particular application and to classify liquid phases as to how similar or different they are in the ability to perform chromatographic

Table 6-4 Reference compounds used to determine McReynolds constants

Reference compound	Abbreviation	*I**	Organic compound expected to display similar behavior on liquid phase
Benzene	*x'*	650	Aromatics, olefins
Butanol	*y'*	590	Alcohols, phenols, weak acids
2-Pentanone	*z'*	627	Aldehydes, esters, ketones
Nitropropane	*u'*	652	Nitrogenous and nitrile compounds
Pyridine	*s'*	699	Nitrogenous aromatic heterocyclics, bases

*Absolute value of retention indices observed on squalane.

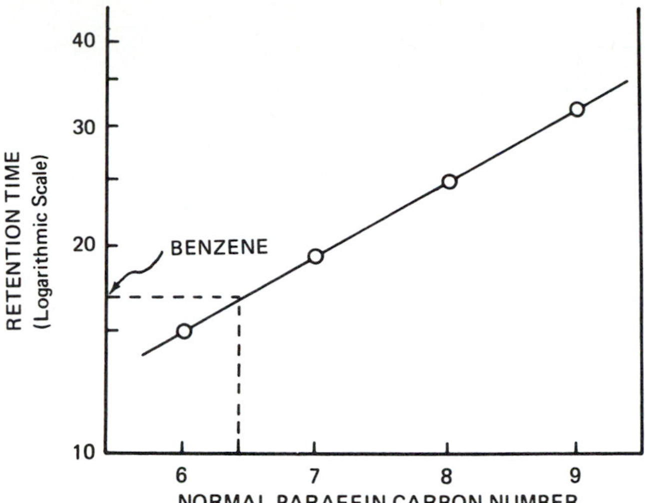

Fig. 6-5 Linear relationship between log of retention time and number of carbon atoms in a series of paraffin hydrocarbons. *(From Rowland, FW: The practice of gas chromatography, Palo Alto, Calif, 1974, Hewlett-Packard Co.)*

separations.[14] An example of the use of McReynolds constants for selection of a liquid phase for a given application would be the separation of saturated and unsaturated fatty acid methyl esters. If one has available liquid phases 1(DEGS) and 2(Carbowax 20M), presented in Table 6-5, an examination of their McReynolds constants permits one to choose the best phase for the separation. For the ability to separate saturated and unsaturated fatty acid esters, one considers the constants x' (olefinic compounds, unsaturated esters) and z' (esters) (Table 6-4). The McReynolds constants for x' and z' (in bold face in Table 6-5) are much higher for DEGS than for Carbowax 20M; therefore DEGS is better suited for the separation of the fatty acid esters.

The characterization of liquid phases as to their ability to perform separations by use of McReynolds constants is demonstrated by examination of phases 3(Emulphor ON-870), 4(Triton X-100), and 5(XE-60) in Table 6-5. All liquid phases are commercially available, and one can determine differences or similarities in separating power as illustrated in Table 6-5. Liquid phases 3 and 4 are almost identical in their ability to separate the reference compounds. Therefore, if used as a liquid phase, they would yield very similar chromatograms. Liquid phase 5 is similar to phases 3 and 4 for constants x' and y'. Therefore separation of compounds characterized by these constants (Table 6-5) on phase 5(XE-60) would be essentially the same on phases 3(Emulphor EN-870) and 4(Triton X-100). The constants z' and u' are higher for phase 5, and the separation of the compounds listed in Table 6-4 for these constants would be better on XE-60 than on the other two phases. However, although keto compounds (constant z') are better separated on phase 5, the separation of alcohols (constant y') would be practically identical on phases 3, 4, and 5. When using packed columns, one finds that differences in McReynolds constants of 20 or less are insignificant. Differences of 100 McReynolds units indicate significantly better separating ability of one phase compared to the other.

Table 6-5 McReynolds constant of various liquid phases

Liquid phase	McReynolds constant*				
	x′	y′	z′	u′	s′
1. DEGS	**496**	746	**590**	837	835
2. Carbowax 20M	**322**	536	**368**	572	510
3. Emulphor ON-870	202	395	251	395	344
4. Triton X-100	203	399	268	402	362
5. XE-60	204	381	340	493	367

*x', Benzene; y', butanol; z', 2-pentanone; u', nitropropane; s', pyridine. For bold-faced type, see text above.

COMPONENTS OF GAS CHROMATOGRAPH

Basically a gas chromatograph consists of six components (Fig. 6-6): (1) a pressurized carrier gas with ancillary pressure and flow regulators, (2) a sample injection port, (3) a column, (4) a detector, (5) an electrometer and signal recorder, and (6) thermostated compartments encasing the column, detector, and injection port.

Carrier gas

The efficiency of a gas chromatograph depends on a constant flow of carrier gas. The carrier gas from a pressurized tank flows through a toggle valve, a flowmeter (range 1 to 1000 L/min), metal restrictors, and a pressure gauge (1 to 4 atmospheres). The flow is adjusted by a needle valve mounted at the base of the flowmeter. The gas moves more slowly at the head of the column than at the outlet because of a pressure drop in the column. Thus the flow rates are measured as the gas exits from the column. This is done with a soap-film flowmeter. A simple sidearm buret with a rubber bulb filled with soap solution is connected to the detector outlet. One determines the flow rate by noting the time required for a film (bubble) to pass between two calibrated volume marks on the buret.

Carrier gas should be inert, dry, and pure. The most common carrier gases are inert, but they may contain contaminants that affect column performance and the response of ionization detectors. Hydrocarbon gases and water are removed from the carrier gas by a molecular sieve trap between the gas cylinder and the chromatograph.

Sample-injection port

Most GC analyses are performed on nonaqueous, liquid samples that are injected by a glass microsyringe. A needle is inserted through a septum into a heated block where the

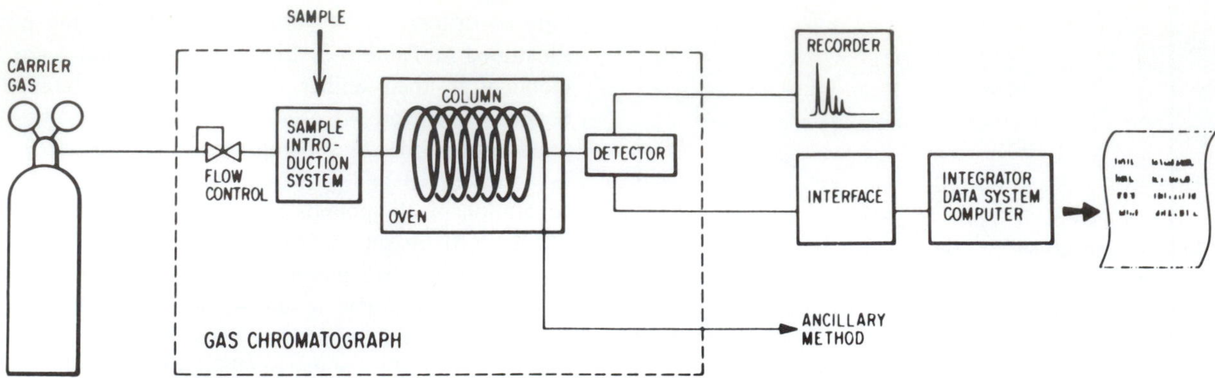

Fig. 6-6 Basic components of a gas chromatographic system. *(From Ettre, LS: Practical gas chromatography, Norwalk, Conn, 1973, Perkin-Elmer Corp.)*

sample is vaporized and swept by carrier gas into the column. The pressure inside the injection port is usually well above atmospheric pressure, and the stream of carrier gas sweeps away the sample and aids in vaporization. Thus a sample may be vaporized at temperatures below its atmospheric boiling point. However, the injection port temperature is usually set at 25° to 50° C higher than the boiling point of the highest boiling components in the sample. This assures that immediate vaporization will occur and that the components will not be diluted by carrier gas and will enter the head of the column as a single band. The time required for vaporization is dependent on the amount and volatility of the sample. Dilute samples vaporize faster than concentrated samples. High-boiling or temperature-sensitive compounds may be diluted with volatile solvents, which lower injection temperatures significantly.

Because heated metal may catalyze the degradation of many biological compounds, many injection ports are equipped with a glass liner or a glass column that extends through the injectors flush to the septum. The latter approach is called "on-column injection." For maximum efficiency it is imperative that the sample be the smallest possible volume (0.5 to 10 μL) consistent with detector sensitivity and be injected as a single, uniform band ("slug injection"). Insertion, injection, and withdrawal of the needle should be performed quickly and smoothly. Gaseous samples are injected by a gas-tight syringe or a calibrated bypass loop. The loop consists of a glass system of three stopcocks, between two of which a standard volume of gas is trapped and introduced into the carrier gas stream when the stopcocks are switched.

A septum separates the chromatographic column from the laboratory environment. Septums are small disks of silicone rubber, and numerous types are available, depending on analytic requirements. Silicone rubber septums may absorb certain types of samples. Special septums (such as Teflon R coated) will alleviate this problem. Low-molecular-weight solvents used in the manufacture of septums may be released as the injection port is heated. This "bleed" of solvent may produce unwarranted peaks (ghost peaks) in chromatograms, and it increases the background level of the detector. Low-bleed septums from which the solvents have been extracted are available. Repeated injections through the septum will gradually destroy its mechanical strength, causing leakage. As a result, the retention time and sensitivity decrease as the carrier gas and part of the sample are released back through the septum into the atmosphere. This problem is easily avoided by regular insertion of new septums.

Various specialized injection systems are commercially available. If large numbers of similar analyses are to be performed, automatic sampling units are commonly used.

Column tubing

The column tubing is a container for the stationary phase (packing material) and directs the carrier gas flow. It should be inert and not affect the separation by reaction with the stationary phase or the sample. Depending on the gas chromatograph employed, the columns may be shaped as a U-tube or coiled in an open spiral or flat pancake shape.

Stainless steel and copper columns are often used for analyses requiring temperatures greater than 250° C. However, for the analysis of drugs, steroids, or other biological compounds, metal columns may absorb these analytes or catalyze their degradation. Therefore glass is the tubing of choice for the majority of clinical analyses. However, glass is fragile and inflexible, and if not properly handled, the columns are easily broken during transport or installation. Recently nickel has been recommended as a substitute for glass. Nickel tubing has been effectively used in the analysis of specific drugs, pesticides, and cholesterol, which previously required glass tubing.[15] However, the application of nickel tubing to the broad range of biological compounds has not yet been established. Until such time, glass tubing should remain the primary support when one is performing a clinical analysis.

Inside column diameters vary from capillary to larger

dimensions. The use of capillary columns coated with a liquid stationary phase is becoming increasingly popular for clinical analyses. Most clinical analyses are currently performed on columns of 2 or 4 mm inner diameter (ID). Columns of 4 mm ID contain four times the stationary phase as 2 mm ID columns of the same length and therefore possess a greater sample capacity. However, the same separation will require higher temperatures and a longer analysis time on the wider column. In addition, columns should be only as long as necessary to effect the desired separation. A short column provides a short analysis time, low temperatures, long column life, and less background in the detector. Columns of 0.7 to 2 m (2 to 6 feet) are sufficient for most chemical separations.

Thermal compartment

Precise control of column temperature is imperative in gas chromatography. The column oven is controlled by a system that is sensitive to changes of 0.01° C and maintains the column temperature to ± 0.1° C of the desired temperature. The column oven, injection block, and detectors should have separate heaters and controls. Analysis may be performed at a constant oven temperature (isothermal), or the temperature may be varied during the analysis (temperature programming). The temperature change during analysis can be programmed to vary with time according to predetermined, reproducible patterns, giving linear, convex, or concave curves when column temperature is plotted against time (Fig. 6-7). Temperature programming is often used in separating a complex mixture, the components of which have widely varying affinity for the stationary phase. Initially the column temperature is set low to permit separation and elution of the compounds with little affinity for the stationary phase. The temperature is then raised to elute compounds of higher stationary-phase affinity. Many chromatographs are equipped with specialized oven controls that uniformly raise the column temperature after each sample injection.

Detectors

As the carrier gas exits from the column, a detector senses the separated components of the sample and provides a corresponding electrical signal. Any physical device that accomplishes this may be used as a detector; however, only a few are commonly used. For proper operation or optimum response, each type of detector requires a specific carrier gas (Table 6-6). The most widely used detectors are discussed in the following section.[16,17]

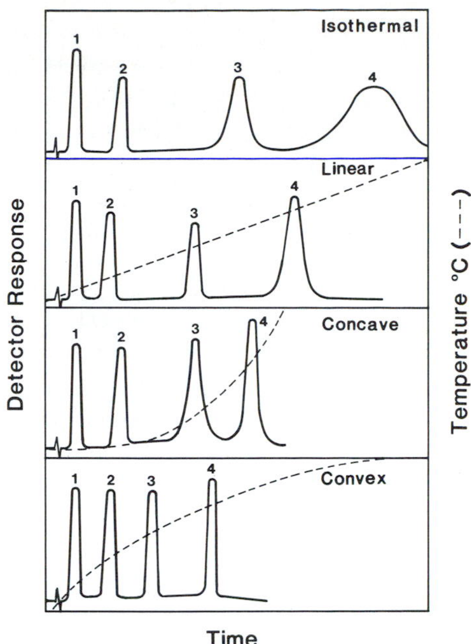

Fig. 6-7 Schematic diagram of theoretical separation of four compounds showing varying elution patterns with different temperature programming.

Thermal-conductivity detector (TCD). A thermal-conductivity detector measures the difference in ability to conduct heat (thermal conductivity) between pure carrier gas and the carrier with sample mixture. A sample carried in the gas increases the thermal conductivity. Usually four heat-sensing elements, thermistors or wires, are mounted in a brass or stainless steel heat sink and connected to form the arms of a Wheatstone bridge (Fig. 6-8). An electric current is passed through the wires composing the bridge. Two filaments in opposite arms of the bridge are cooled by carrier gas (reference), and the other two by the column effluent (sample). The heat lost over both sets of wires is balanced by adjustment of the flow rate of the pure carrier gas. Emerging components from the column increase the rate of cooling of the sample wires because of the increased thermal conductivity of the gas mixture. This changes the electrical resistance of the sample wire pat-

Table 6-6 Detectors and appropriate gases

Detector	Carrier gas	Detector gas
Thermal conductivity (TCD)	Helium, hydrogen	—
Flame ionization (FID)	Helium, nitrogen	Air and hydrogen
Nitrogen-phosphorus (NPD)	Helium, nitrogen	1. Air and hydrogen 2. Air and 8% hydrogen in helium
Electron capture	Nitrogen	5% methane in argon
	5% methane in argon	—

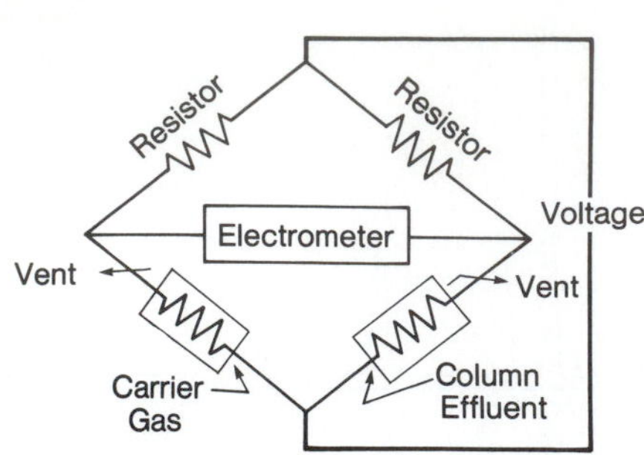

Fig. 6-8 Schematic diagram of thermal conductivity detector. *(From Werner, M, Mohrbacher, RJ, and Riendeau, CJ: In Baer, DM, and Dito, WR: Interpretation of therapeutic drug levels, Chicago, 1981, American Society of Clinical Pathologists.)*

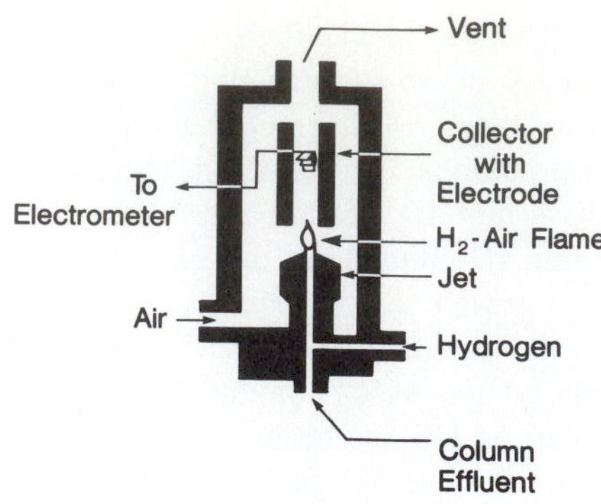

Fig. 6-9 Schematic diagram of flame ionization detector. *(From Werner, M, Mohrbacher, RJ, and Riendeau, CJ: In Baer, DM, and Dito, WR: Interpretation of therapeutic drug levels, Chicago, 1981, American Society of Clinical Pathologists.)*

tern, making the Wheatstone bridge out of balance. This imbalance causes a response on the recorder. Important variables in optimum TCD response are carrier gas, flow rate, filament current, and detector temperature. TCD's lack selectivity because any compound cooling the wire will cause a response. They are not as sensitive as other detectors, with minimum detection ranging from 0.1 to 0.5 mg of analyte per microliter.

Flame-ionization detector (FID). In a flame-ionization detector, eluted components in the carrier gas are mixed with hydrogen and burned in air to produce a very hot flame to ionize organic compounds. A pair of electrodes, charged by a polarizing voltage, collects the ions and generates a current proportional to the number of ions collected. The resultant current is amplified by an electrometer, producing a response on the recorder. The response of an FID is directly proportional to the number of carbons in a molecule bound to hydrogen or other carbon atoms. It is insensitive to water, carbon monoxide, carbon dioxide, and most inorganic compounds. The FID is the most popular detector for the determination of organic compounds. Sensitivity depends on chemical structure; therefore, the detector response must be determined for each compound analyzed. At optimal conditions the minimal detectable quantity of organic compound is 1 ng. A cross section of an FID detector is shown in Fig. 6-9.

Nitrogen-phosphorus detector (NPD). A nitrogen-phosphorus detector is similar to an FID except that ions of an alkali metal (rubidium) are introduced into the hydrogen flame. When a compound containing nitrogen or phosphorus is burned in the flame, the rate of release of alkali metal vapor is increased. The alkali metal vapor readily ionizes in the flame and increases the current flow, which results in enhanced sensitivity for nitrogen and phosphorus. The optimum response is greatly dependent on the flow of hydrogen. The selective interaction of alkali metal ions with these compounds is complex and poorly understood. However, the sensitivity to organonitrogen compounds and lack of response to other organics make the NPD highly advantageous for the analysis of biological samples. At optimum conditions, the minimum detectable quantity of nitrogenous organic compounds is less than 1 ng. A cross section of an NPD detector is presented in Fig. 6-10.

Electron-capture detector (ECD). In an electron-capture detector, a radioactive isotope releases beta particles that collide with the carrier gas molecules, producing many low-energy electrons. The electrons are collected on electrodes and produce a small, measurable, *standing current*. As sample components that contain chemical groups with high electron affinity (electrophilic species), particularly halogen atoms, are eluted from column, they capture the low-energy electrons generated by the isotope to form negatively charged ions. The detector measures the loss of cell current because of the recombination of the electrons. Three techniques are used for the collection of the electrons: (1) direct current (DC), (2) pulsed method, and (3) linear method. In the DC method, a constant voltage is applied to the cell electrodes and the electrons are collected continuously to produce a steady current. The sen-

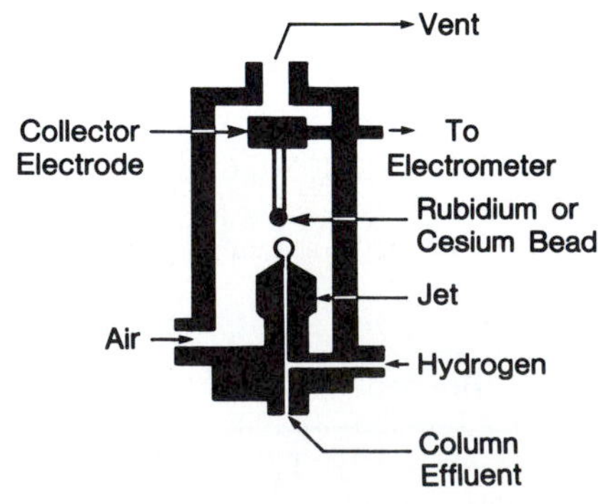

Fig. 6-10 Schematic diagram of alkali-metal flame detector. *(From Werner, M, Mohrbacher, RJ, and Riendeau, CJ: In Baer, DM, and Dito, WR: Interpretation of therapeutic drug levels, Chicago, 1981, American Society of Clinical Pathologists.)*

sitivity of the method is less than that of other ECD methods because both negative ions and free electrons are collected by the electrodes. The reduction in current is smaller than it would be if only free electrons were collected. In the pulsed method, a voltage is applied in continuous pulses of short duration; therefore the heavy negative ions do not have time to respond, and only free electrons are captured. Between pulses, the electron concentration in the detector builds up to levels exceeding those of the DC method. Thus the pulse method has greater sensitivity. The DC and pulsed methods inherently produce a nonlinear response over a wide range of sample concentrations. Such a response is attributable to the finite amount of beta radiation emitted by the detector source per unit time. Because a decrease in current is measured, once a concentration of eluting solute captures a majority of the available low-energy electrons, only small changes in current (detector response) will be observed with increasing concentrations of solute. The linear range is usually 400 to 500 times the detection limit of a solute for a tritium source and 100 times for a nickel-63 source. However, the linearized method uses electronic modifications that operate the detector in a pulsed mode such that constant cell current is produced. The linear range is thus expanded to ranges of 10,000:1 for a nickel-63 source. The sources of beta particles in an ECD are usually tritium or nickel-63. The ECD is the most sensitive detector available, since as little as 1 picogram of halogen-containing compound may be measured. Laboratories using electron-capture detectors must be licensed by the Nuclear Regulatory Commission and are subject to all regulations concerning employee safety and possible environmental contamination set forth by the commission.

Mass spectrometer (MS) as a detector.[18] The mass spectrometer is a specialized chromatographic detector that provides extremely sensitive detection (picogram quantities) and is the ultimate specific identification technique currently available to the analyst. The mass spectrometer is often considered a separate instrument rather than a component of a gas chromatograph because samples may be directly inserted into the instrument (direct-probe technique), or it may be coupled with other chromatographic instruments such as a high-performance liquid chromatograph. The primary difficulty in combining a gas chromatograph with a mass spectrometer is that the GC operates at a positive pressure, whereas the MS operates under a high vacuum. This limitation is overcome by use of large vacuum pumps (700 L/sec) that maintain a vacuum despite the large flow of gas from the GC. Additionally, a *separator* inserted between the effluent of the GC column and the ionization chamber of the MS removes large portions of the carrier gas, concentrating the analyte and reducing the total volume of gas entering the MS. Two types of separators are available, the *membrane* and the *jet*. The membrane separator, made of silicone rubber, is mounted between the effluent of the GC and the vacuum of the MS. The organic solutes are more soluble in the membrane than in the carrier gas and, after being dissolved in the membrane, are pumped away by the vacuum system and transferred into the ion source. The jet separator operates on the principle of differential mobility of sample molecules and carrier gas. The effluent from the GC enters a vacuum through a narrow jet. The core of the molecular beam contains a higher concentration of sample molecules than the perimeter does. A pump removes the lighter gas molecules, whereas the heavier sample molecules in the core are collected by a small orifice and transferred into the MS.

The generation of a spectrum by the mass spectrometer involves three steps: ionization, mass filtration, and detection. At present there are three modes of ionization used in the analysis of biological specimens: (1) electron impact (EI), (2) chemical ionization (CI), and (3) negative chemical ionization (NCI). In the electron-impact mode, the most widely used type of ionization, the sample molecules in the gas phase are bombarded with high-energy electrons (70 eV) so that they produce a charged molecule or shatter the molecule into ionic fragments. The charged fragments are separated and detected according to their atomic masses. The mass spectrum is a display of the different masses of the charged fragments and their relative abundance. The mass spectrum generated by electron impact is characteristic of the molecule analyzed and is often used to establish an unequivocal identification of an unknown drug. The electron-impact spectrum of 3,4-methylenedioxyamphetamine (MDA), a hallucinogenic drug of abuse, isolated from the urine of a fatal overdose case[19] is presented in Fig. 6-11. The identification of MDA is based upon the characteristic mass to charge *(m/e)* fragment pattern and relative ion abundance (height) of these fragments compared to the major ion fragment at ion mass 44 (Table 6-7).

In the chemical ionization technique, the sample molecules are mixed with an ionized gas such as methane or

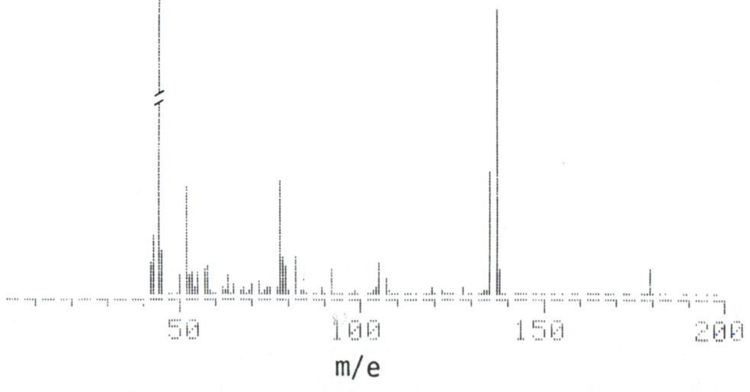

Fig. 6-11 3,4-Methylenedioxyamphetamine (MDA) electron impact spectrum, *m/e*. Ion mass 44 base peak and an ion mass peak at 179. *(From Poklis, A, Mackell, MA, and Drake, WK: J Forensic Sci 24(1):70, 1979.)*

Table 6-7 Ion mass and mass abundance of MDA* standard and MDA extracted from decedent's urine

Ion mass, *m/e*	**Ion abundance of MDA standard**	**Ion abundance of urine compound‡**
44	100.0	100.0
51	12.7	12.1
77	16.4	12.5
78	6.4	4.5
79	6.4	4.5
105	4.3	3.6
135	13.1	13.6
136	35.6	31.0
179	5.2	3.1

From Poklis, A, Mackell, MA, and Drake, WK: J Forensic Sci 24:70-75, 1979.
*3,4-Methylenedioxyamphetamine (MDA).
‡Similarity index, 0.9957; molecular weight, 179.2.

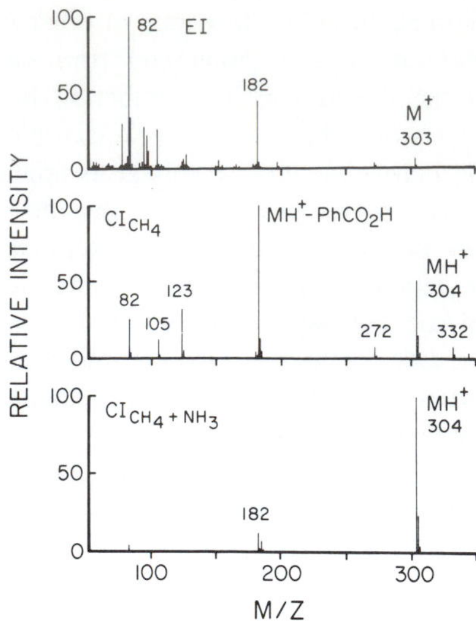

Fig. 6-12 Mass spectra, M/Z, of cocaine. *(From Foltz, RL, Fentiman, AF, and Foltz, RB: GC/MS assays for abused drugs in body fluids, Washington, DC, 1980, US Dept of Health and Human Services.)*

isobutane. A positive charge (proton) is transferred to the sample molecule as a result of the ion-molecule collisions. In the electron beam, methane CH_4 becomes transformed to CH_5

$$CH_4 + \text{electron beam} \rightarrow CH_4{}^{+\cdot}$$
$$CH_4 + CH_4 \cdot \rightarrow CH_5 + \text{Other products}$$

The CH_5 can then interact with the analyte of interest to form a positive M + 1 ion.

$$R + CH_5 \rightarrow RH^+ + CH_4$$

This can then break down to form fragments:

$$RH^+ \rightarrow \text{Fragments}$$

By controlling experimental conditions, one can minimize fragmentation of the sample molecule, and the dominant ion produced equals the molecular weight (M) of sample plus a proton (M + 1, Fig. 6-12). The electron-impact and chemical-ionization mass spectra of cocaine are compared in Fig. 6-12. Unlike the fragmented EI spectrum, the CI spectrum of cocaine obtained using methane and ammonia as reagent gases shows only a single abundant ion corresponding to the protonated molecular ion (*m/z* 304).[20] The CI mass spectrum using methane displays a base peak (fragment of highest abundance) at *m/z* 182, which results from the loss of benzoic acid from the protonated cocaine molecule. In negative chemical ionization, oxygen or hydrogen are mixed with the sample molecules to produce a dominant negative ion of M − 1. The detection of compounds containing halogen (fluorine, chlorine) atoms may be greatly enhanced by negative chemical ionization. The chemical ionization technique determines the molecular weight of the sample and, because of the paucity of the fragment ions, is ideally suited for high-sensitivity, quantitative analysis. Once produced, the ions are directed and separated by mass by a magnetic field (see Chapter 7, p. 138).

Two types of chromatograms may be produced when the MS is used as a GC detector: the *total chromatogram* and the *selective ion chromatogram*. In the total chromatogram, the total ion current of the various compounds in the GC effluent is displayed. The chromatogram has the appearance of any other separation of sample components as produced by the conventional GC detectors, such as the FID or NPD. The mass spectrum of each peak is collected and stored by a computerized data system. The ion or mass spectrum of each separated component may then be produced on a printout of the data system. All ionized compounds that are eluted from the GC are detected. The selective ion chromatogram is produced when a specific mass fragment or ion is monitored. Depending on the capabilities of the instrument, several specific ions may be monitored. Only ionized or fragmented compounds produce the specific charged mass unit and are detected and displayed in the chromatogram. Therefore in GC separation of a complex mixture the total chromatogram will display all components present, whereas selective ion monitoring (SIM) will display only those components that produce a specific chosen ion. The *m/z* 182 ion in the electron-impact spectrum in Fig. 6-12 could be used for selective ion monitoring.

Because of the complexity of operation and the resultant spectral lines, mass spectrometers are connected to computerized systems for operation and data collection. For qualitative analysis, the computer automatically stores spectra of sample compounds and searches a stored spectral library of known compounds for comparison (Table 6-7).

Advances in instrumentation have made widely available relatively inexpensive, bench-top GC/MS instruments. Although they may not have the computer capability of larger GC/MS instruments, their price and versatility have made them popular for drug monitoring and screening for drugs of abuse (see p. 1094 and Chapter 51).

Readout. Strip chart recorders are the most common readout devices in gas chromatography. Recorder sensitivity is usually 1 to 10 mV, with a full-scale response of 1 second or less. Quantitative determinations of separate compounds are performed in two ways: peak-height or peak-area measurements. Both the peak height and peak area of the detector response to the effluent sample are proportional to its concentration. Peak-height measurements are useful in repetitive analyses that are performed by the same operator in a fixed system, that require extensive calibration, or that only partially resolve compounds, making peak-area determinations difficult. In general, peak-area measurements are more precise. Peak area can be determined by manual or automated methods. Electronic integration of the peak area produces both the most precise and the most accurate measurements. Today, detectors may be connected to microprocessor units or to a computerized data system that automatically records the response, identifies the sample components, integrates the signals, performs calculations, stores all data, and prints out the analytical results in final form.

REFERENCES

1. Willett, J, and Kealey, D: Gas chromatography, New York, 1987, John Wiley & Sons, Inc.
2. Grob, RL: Modern practice of gas chromatography, ed 2, New York, 1985, John Wiley & Sons, Inc.
3. Delley, R, and Friedrich, K: System CG72 von bevorzugten Trennflauussig-keiten fur die Gas-chromatographie, Chromatographia 10:593-598, 1971.
4. Hawkes, S, Grossman, D, Hartkopf, A, et al: Preferred stationary liquids for gas chromatography, J Chromatogr Sci 13:115-117, 1975.
5. Siggia, S: Instrumental methods of organic functional group analysis, New York, 1972, John Wiley & Sons, Inc.
6. Ahuja, S: Derivatization in gas chromatography, J Pharm Sci 65:163, 1976.
7. Kovats, E: The Kovats retention index system, Anal Chem 36:31A, 1964.
8. Lorenz, LJ, and Roger, LB: Specification of gas chromatographic behavior using Kovats indices and Rohrschneider constants, Anal Chem 43:1593-1599, 1971.
9. American Society for Testing and Materials: Gas chromatographic data compilation catalog AMD 25A, Philadelphia, Pa, 1967.
10. American Society for Testing and Materials: Gas chromatographic data compilation catalog, suppl 25A S-1, Philadelphia, Pa, 1971.
11. Perrigo, BJ, and Peel, HW: The use of retention indices and temperature-programmed gas chromatography in analytical toxicology, J Chromatogr Sci 19:219-226, 1981.
12. Supina, WR, and Rose, LP: The use of Rohrschneider constants for classification of GLC columns, J Chromatogr Sci 8:217-217, 1970.
13. McReynolds, WO: Characterization of some liquid phases, J Chromatogr Sci 8:685-691, 1970.
14. Ettre, LS: Basic relationships of gas chromatography, ed 2, Norwalk, Conn, 1979, Perkin-Elmer Corp.
15. Fenimore, DC, Whitford, JJ, Davis, CM, and Zlatkis, A: Nickel gas chromatographic columns: an alternative to glass for biological samples, J Chromatogr 140:9-16, 1977.
16. David, DJ: Gas chromatographic detectors, New York, 1974, John Wiley & Sons, Inc.
17. Sevcik, J: Detectors in gas chromatography, New York, 1975, Elsevier/North-Holland, Inc.
18. Message, GM: Practical aspects of gas chromatography: mass spectrometry, New York, 1984, John Wiley & Sons, Inc.
19. Poklis, A, Mackell, MA, and Drake, WK: Fatal intoxication from 3,4-methylenedioxyamphetamine, J Forensic Sci 24:70-75, 1979.
20. Foltz, RL, Fentiman, AF, Jr, and Foltz, RB, editors: GC/MS asays for abused drugs in body fluids, Rockville, Md, 1980, National Institute of Drug Abuse.

CHAPTER 7

Isotopes in clinical chemistry

I-WEN CHEN

OBJECTIVES

- Describe the basic structure of the atom. Explain the significance of the atomic number and atomic mass number as they relate to the existence of isotopes.
- Describe the modes of radioactive decay associated with tritium, carbon-14, and iodine-125 and name the particles emitted.
- Define and explain the importance of decay constants, decay factors, half-life, and specific activity.
- Summarize the operation of scintillation detectors. Diagram and explain the use of the crystal and liquid scintillation counters and describe the operation of each component. Define quenching and outline correction methods.
- Describe the use of isotopes in the clinical laboratories and list radiation safety considerations for monitoring personnel and work areas, contamination control, and waste disposal.

KEY TERMS

becquerel (Bq) Système International d'Unités; (SI) unit of radioactivity corresponding to a decay rate of 1/sec (1 Bq = 1 sec^{-1} = 2.7 × 10^{-9} Ci) (see *curie*).

chemiluminescence Production of light photons by an interaction of the sample material with the solute or solubilizer added to the scintillation solution in liquid scintillation counting.

circuit, anticoincident A circuit used in the pulse-height analyzer of a radioactive-particle counter for setting window width (see *window*). It transmits a pulse arriving at its input from the lower discriminator only if there is no pulse arriving from the upper discriminator at the same time (see *discriminator*).

circuit, coincident A circuit used in a liquid scintillation counter to eliminate the electronic noise. It determines if a pulse from one photomultiplier tube is accompanied by a corresponding pulse from the other within the allowed time interval (see *coincidence resolving time*).

circuit, summation A circuit used in a liquid scintillation counter to sum all coincident pulses to improve counting efficiency for low-energy beta-particle emitters.

coincidence resolving time A time interval within which the output pulses from each photomultiplier tube of a liquid scintillation counter have to arrive at the coincident circuit to be counted.

curie (Ci) Unit of radioactivity. One curie is defined as an activity of a sample decaying at a rate of 3.7 × 10^{10} disintegrations per second (dps).

decay constant A constant unique to each radioactive nuclide (see *nuclide*), representing the proportion of the atoms in a sample of that radionuclide undergoing decay in unit time.

decay factor The fraction of radionuclides remaining after a time, *t*.

discriminator(s) Device(s) used in the pulse-height analyzer of a radioactive-particle counter for setting upper (upper-level discriminator) and lower (lower-level discriminator) voltage limits for counting.

electron capture One mode of radioactive decay in which the neutron-poor nuclides decay by capturing electrons from orbits closest to the nucleus to transform a proton to a neutron. The orbital vacancy created by the electron capture is filled by the electron from a higher orbit, resulting in emission of characteristic x rays.

electron volt (eV) Basic unit of energy commonly used in radiation, defined as the amount of energy acquired by an electron when it is accelerated through an electrical potential of 1 volt.

half-life ($t_{1/2}$) Time required for a given number of radionuclides in the sample to decrease to one half its original value.

isobar Nuclides (see *nuclide*) with the same atomic mass number but different atomic number.

isotope Nuclides (see *nuclide*) with the same atomic number but different atomic mass number.

isotopic abundance Amounts of isotopes present for a given element.

nucleon A collective term for protons and neutrons in the nucleus.

nuclide A nucleus with a particular atomic number and atomic mass number.

radiation absorbed dose (rad) A measure of local energy deposition per unit mass of material irradiated. One rad is equal to 100 ergs of absorbed energy per gram of absorber.

roentgen equivalent, man (rem) That dose of any ionizing radiation causing the same amount of biological injury to human tissue as 1 rad of x, gamma, or beta radiation. In the case of x, gamma, or beta radiation rems are equal to the absorbed dose in rads; in the case of alpha radiation, however, the dose in rems equals the dose in rads multiplied by 20, because only 0.05 rad of alpha radiation is needed to produce the same biological effect as 1 rad of x, gamma, or beta radiation.

roentgen (R) A unit of x rays or gamma rays representing the quantity of ionization produced by photon radiation in a given sample of air. One roentgen equals that quantity of photon radiation capable of producing one electrostatic unit of ions of either sign in 0.001293 g of air.

specific activity Activity of the radionuclide per unit mass of the radioactive sample, expressed as Ci per μg, μCi per μmole, and so on.

specific ionization Number of ion pairs produced per unit path length of ionizing radiation.

transmutation A radioactive decay process that results in a change in nuclear constitution, such as electron capture decay (see *electron capture*).

wavelength shifter The secondary scintillator added to scintillation liquid for shifting the wavelength of light emitted by the primary scintillator for more efficient detection by the photocathodes of photomultiplier tubes in liquid scintillation counting.

window The voltage limit set by the upper-level and lower-level discriminators (see *discriminator*) of the pulse-height analyzer of a radioactive-particle counter for differential counting.

The use of isotopes, both stable and radioactive, has provided a great body of information in the medical sciences, much of which could not have been obtained in any other way. The usefulness of isotopes depends on the fact that isotopes of an element have identical chemical properties but different isotopic properties, such as radioactivity and increased mass, and on the fact that the isotopic properties and chemical properties of an element are independent of each other. Therefore substitution of an atom in the molecule of a substance by other isotopes will not chemically alter that substance, and the isotopic properties of the isotopes incorporated into that substance will remain unchanged. The isotopic properties will make that substance more easily identifiable. For example, thyroxine is a thyroid hormone containing four atoms of iodine. One or all of these iodine atoms may be replaced by radioactive iodine without appreciable alteration of its chemical properties, and the radioiodine atoms incorporated into the thyroxine molecules will maintain their characteristic radioactivity. The radioiodine-labeled thyroxine molecules can be identified and quantified easily by virtue of their radioactivity. Radioiodine-labeled thyroxines are used in various thyroid-function tests.

The application of radioisotopically labeled compounds in clinical chemistry has greatly expanded since the advent of radioimmunoassays, in which the quantity of antigen bound to antibody is determined by measurement of radioactivity (see Chapters 10 and 11). Because of its inherent sensitivity, specificity, and wide applicability, radioimmunoassay has been playing an important role in the diagnosis of various disorders since its discovery in 1956 by Berson and Yalow. In this chapter, some basic principles involved in the measurement of isotopes are discussed.

BASIC STRUCTURE OF AN ATOM

Fundamental particles of an atom

An atom is the smallest unit of matter that still exhibits the chemical properties of an element. The primary building blocks of atoms, the electron, the proton, and the neutron, are termed *elementary particles*. According to the planetary model of the atom developed by Rutherford in 1911, the atom consists of a central, small, positively charged body (the nucleus, composed of protons and neutrons) around which the negatively charged electrons move in defined orbits. Although the Rutherford model is oversimplified, one can use it to explain many atomic phenomena satisfactorily. The planetary model of an atom of carbon is illustrated in Fig. 7-1.

The nucleus of carbon contains six protons and six neutrons. Since complete atoms are electrically neutral, six orbiting electrons are present in the carbon atom to match the six protons in the nucleus. They move around the nucleus in a series of orbits or shells, at varying distances from the nucleus, much as the planets of the solar system travel in different orbits at varying distances from the sun. The orbits, or shells, are called *K, L, M,* and so on, starting from the inner one. Only two electrons can be accommodated in the K shell; the L shell of the carbon atom contains the remaining four.

The physical and chemical differences between the atoms of different elements depend on the number of protons and neutrons contained in an atomic nucleus, which determines both the mass and charge of the nucleus, and the

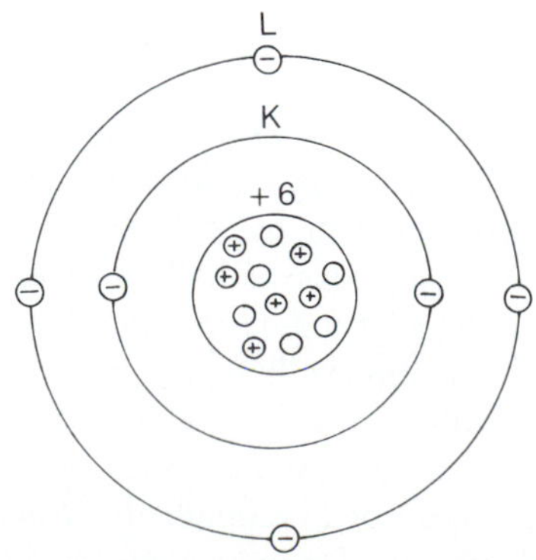

Fig. 7-1 Planetary model of carbon atom.

number and arrangements of the electrons, which determine the chemical properties of elements.

Atomic structure nomenclature

There are several important terms that are helpful in understanding atomic structure:

nucleon A collective term for protons and neutrons in the nucleus.

atomic number, Z The number of protons in the nucleus.

atomic mass number, A The total number of nucleons in the nucleus.

neutron number, N The number of neutrons in the nucleus.

nuclide A nucleus with particular *Z* and *A* numbers.

element, E A nucleus with a given *Z* number.

isotope Nuclides with the same *Z* but different *A* numbers (various nuclear species of the same element).

isobar Nuclides with the same *A* but different *Z* numbers (different elements with the same atomic mass).

The atomic mass number is represented as a left superscript and the atomic number as a left subscript to the chemical symbol. Thus an element, *E*, is written as ${}^{A}_{Z}E$. The most abundant, naturally occurring, stable isotope of carbon has six protons and six neutrons in the nucleus. The atomic number, *Z*, is therefore 6; the atomic mass number, *A*, is 12 ($A = Z + N$); and the whole atom may be written as ${}^{12}_{6}C$. The other naturally occurring but less abundant isotope of carbon is ${}^{13}_{6}C$, which contains seven neutrons in the nucleus. ${}^{12}_{6}C$ and ${}^{13}_{6}C$ are both stable isotopes of carbon, and neither is radioactive. The best-known radioactive isotope of carbon is ${}^{14}_{6}C$, which contains six protons and eight neutrons. Examples of other groups of isotopes of an element commonly used in clinical chemistry are ${}^{1}_{1}H$, ${}^{2}_{1}H$, and ${}^{3}_{1}H$ and ${}^{125}_{53}I$, ${}^{127}_{53}I$, and ${}^{131}_{53}I$. ${}^{1}_{1}H$ is the most abundant naturally occurring isotope of hydrogen and has one proton but no neutron in the nucleus. ${}^{2}_{1}H$ is a stable isotope of hydrogen and is known as *deuterium* because its nucleus contains two nuclear particles, one proton and one neutron. ${}^{3}_{1}H$, called *tritium,* is a radioactive isotope of hydrogen, the nucleus of which is formed by a combination of a proton and two neutrons. All isotopes of hydrogen have a single circling electron and therefore have identical chemical properties; however, their physical properties are different. For example, they will have different boiling and freezing points. The tritium nucleus is unstable and will undergo radioactive transitions to become a different and stable nucleus—the nucleus of helium. The naturally occurring stable isotope of iodine is ${}^{127}_{53}I$. The other two isotopes of iodine mentioned here are radioactive isotopes with different numbers of neutrons in their nucleus, as indicated by their atomic mass numbers. In many cases the atomic number subscript is redundant because the atomic number and the chemical symbol both identify the chemical species. Therefore, except in some equations describing nuclear reactions, the subscript is normally omitted (such as ${}^{14}C$, ${}^{3}H$, and ${}^{125}I$).

PRINCIPLES OF RADIATION AND RADIOACTIVITY

Nuclear radiation

The release of energy or matter during the transformation of an unstable atom to a more stable atom is termed *nuclear radiation*. The numbers and arrangement of protons and neutrons in the nucleus of an atom determine whether the nucleus is stable or unstable.

Nuclear stability. There are favored neutron-to-proton ratios among stable nuclides. The ratio is equal to or close to unity for the light nuclides. When the atomic mass number exceeds 40, no stable nuclides exist with equal numbers of neutrons and protons because, as the number of protons increases, the repulsive coulombic forces between the protons increase at a greater rate than the attractive nuclear force does. Therefore the addition of extra neutrons is necessary to increase the average distance between protons in the nucleus to reduce the coulombic force. For heavy nuclei the neutron-to-proton ratio is 1.5 or greater. For example, the heaviest stable isotope of lead, ${}^{208}Pb$, has a neutron-to-proton ratio of 1.53.

Fig. 7-2 illustrates the relationship between the neutron and proton numbers of the stable nuclides. An imaginary line, called the *line of stability,* represented by a dashed line in the graph, can be obtained from the neutron-proton plot; the stable nuclides are clustered around this line. Nuclides deficient in protons lie below the line of stability and are unstable. Nuclides deficient in neutrons lie above the line and are also unstable. The graph also illustrates the fact that, as nuclides become heavier, more neutrons are required to maintain stability.

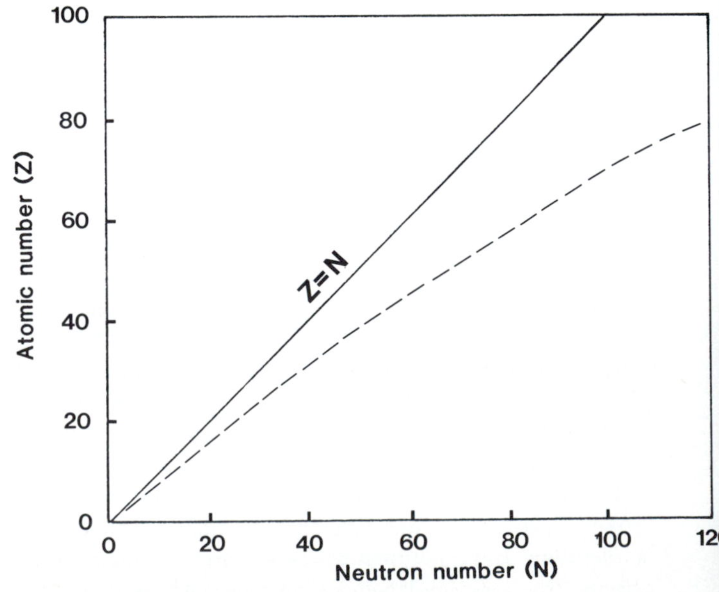

Fig. 7-2 Neutron-proton ratios for stable isotopes, *dashed line*.

In addition to the favored neutron-to-proton ratio, the stable nuclides tend to favor even numbers. For example, 168 out of approximately 280 known stable nuclides have even numbers of both protons and neutrons, reflecting the tendency of nuclides to achieve stable arrangements by pairing up nucleons in the nucleus.

Modes of radioactive decay. Unstable nuclides are generally transformed into stable nuclides by one of the radioactive-decay processes described below.

Decay by alpha-particle emission. An alpha (α) particle consists of two neutrons and two protons and is essentially a helium nucleus. Heavy nuclides that must lose mass to achieve nuclear stability frequently decay by alpha-particle emission, because alpha-particle emission is an effective way to reduce the mass number. The emission of one alpha particle removes two neutrons and two protons from the nucleus, resulting in the reduction of an atomic number by 2 and a mass number by 4. Very heavy radioactive nuclides that decay with alpha-particle emission are of little interest in clinical chemistry. An example of alpha-particle decay is as follows:

$$^{226}_{88}\text{Ra} \rightarrow {}^{222}_{86}\text{Rn} + {}^{4}_{2}\text{He} \text{ (alpha particle)}$$

Decay by beta-particle emission. Beta (β) particles are either negatively charged electrons (negatrons, β^-) or positively charged electrons (positrons, β^+). Proton-deficient nuclides lying below the line of stability, shown in Fig. 7-2, usually decay by negatron emission, because this mode of decay transforms a neutron into a proton, moving the nucleus closer to the line of stability. Neutron-deficient nuclides lying above the line of stability usually decay by positron emission, since this mode transforms a proton into a neutron.

In beta-particle decay processes the mass number does not change because the total number of nucleons in the nucleus remains the same. Such decay processes are known as *isobaric transitions*. However, the atomic number increases by 1 in the negatron emission and decreases by 1 in the positron emission, resulting in a transmutation of elements (conversion of one element to another). Examples of decay by beta emission are as follows:

$$^{3}_{1}\text{H} \rightarrow {}^{3}_{2}\text{He} + \beta^- + \nu \quad \text{(negatron emission)}$$
$$^{11}_{6}\text{C} \rightarrow {}^{11}_{5}\text{B} + \beta^+ + \nu \quad \text{(positron emission)}$$

The neutrino (ν) is a particle with no mass or electrical charge and virtually does not interact with matter. The only practical consequence of its emission from the nucleus is that it carries away some energy released in the decay process.

Decay by electron capture. In addition to the decay by positron emission, the neutron-deficient nuclides may decay by electron capture to transform a proton to a neutron. Thus the electron capture is sometimes called *inverse negatron decay*. It is also an isobaric transition leading to a transmutation of elements. In the electron-capture process the electron is captured from orbits closest to the nucleus, that is, the K and L shells (K and L capture; see Fig. 7-1). The orbital vacancy created by the electron capture is quickly filled by the electron from a higher orbit, resulting in emission of a characteristic x ray.

The daughter nucleus formed by this mode of decay is frequently in an excited or metastable state and may further undergo decay by gamma (γ)-ray emission, as described below.

Decay by gamma-ray emission. In some cases the isobaric transitions previously mentioned (negatron emission, positron emission, electron capture) result in a daughter nucleus in an excited or metastable state, which means that it possesses excess energy above its minimum possible ground-state energy. Such an excited or metastable nuclide decays promptly to a more stable nuclear arrangement by the emission of gamma rays, electromagnetic radiation of very short wavelength:

$$^{125}_{53}\text{I} \xrightarrow[\text{capture}]{\text{electron}} {}^{125}_{52}\text{Te}^{*} \longrightarrow {}^{125}_{52}\text{Te} + \gamma$$

Note that gamma emission is not accompanied by any change in mass number, proton number, or neutron number. This is called an *isomeric transition*.

Radiation energy

In any of the radioactive decay processes mentioned previously, a fixed amount of energy is released with each disintegration. Most or all of the released energy will appear as the kinetic energy of the emitted particles or photons. The basic unit of energy commonly used in radiation is the electron volt (eV). One electron volt is defined as the amount of energy acquired by an electron when it is accelerated through an electrical potential of 1 V. Basic multiples are the kiloelectron volt (keV; 1 keV = 1000 eV) and the megaelectron volt (MeV; 1 MeV = 1000 keV = 1,000,000 eV). In general, the energy of beta particles emitted from radionuclides in clinical use ranges from 18 keV to 3.6 MeV, and that of gamma rays ranges from 27 keV to 2.8 MeV.

Rate of radioactive decay

Decay constant, decay factor, and half-life. Radioactive decay is a spontaneous process; that is, it is not possible to predict when a given radioactive atom will decay, and the probability of decay in a large number of atoms can be given only on a statistical basis. For a sample containing N radioactive nuclei, the number of nuclei decaying at any given moment (dN/dt) can be given by the following:

$$\frac{dN}{dt} = -\lambda N \qquad \textit{Eq. 7-1}$$

*The nuclide is in an excited or metastable state.

In this equation, λ is the decay constant of the radioactive nuclide, and the minus sign indicates that the number of radioactive nuclides is decreasing with time. Each radionuclide has a characteristic decay constant that represents the proportion of the atoms in a sample of that radionuclide undergoing decay per unit time. The *decay constant,* λ, is measured in units of $(\text{time})^{-1}$. Therefore the equivalence $\lambda = 0.05\ \text{sec}^{-1}$ means that, on the average, 5% of the radionuclides are disintegrating per second. On integration of Equation 7-1, we obtain

$$N = N_0 e^{-\lambda t} \qquad \textit{Eq. 7-2}$$

where N_0 is the number of radionuclides present at time $t = 0$, and e is the base of the natural logarithm. Therefore the number of radionuclides remaining after a time, *t(N),* is equal to the number of radionuclides at a time, $t = 0$ (N_0), multiplied by the factor $(e^{-\lambda t})$. This factor is the fraction of radionuclides remaining after a time, *t,* and is termed the *decay factor*. The decay factor, $e^{-\lambda t}$, is an exponential function of time, *t;* that is, a constant fraction of the number of radionuclides present in the sample disappears during a given time interval. A given time interval is customarily expressed as the time required for a given number of radionuclides in the sample to decrease to one half its original value. This time interval is termed the *half-life* $(t_{1/2})$. The half-life of a radionuclide is related to its decay constant as follows:

$$t_{1/2} = 0.693/\lambda \qquad \textit{Eq. 7-3}$$

This equation is derived as follows:

In Eq. 7-2, $N = N_0/2$ after time $t_{1/2}$.
$\therefore N_0/2 = N_0 e^{-\lambda t_{1/2}}$
$\therefore 2 = e^{\lambda t_{1/2}}$
$\therefore \ln 2 = \lambda t_{1/2}$
$\therefore 0.693 = \lambda t_{1/2}$, or $t_{1/2} = 0.693/\lambda$

A plot of the decay factor versus the number of half-lives elapsed, *t,* on a semilogarithmic graph paper gives a straight line, as shown in Fig. 7-3. One can obtain this straight line by connecting two points on the curve, that is, at $t = 0$, decay factor $= 1.0$, and at $t = 2t_{1/2}$, decay factor $= 0.25$. This graph can be used to determine the decay factor of any radionuclide at any given time, provided that the elapsed time is expressed in terms of number of radionuclide half-lives elapsed. For example, the half-life of ^{125}I is in 60 days. To calculate the residual radioactivity at 360 days, the number of half-lives elapsed must first be calculated:

$$\frac{360\ \text{days}}{60\ \text{days/half-life}} = 6\ \text{half-lives}$$

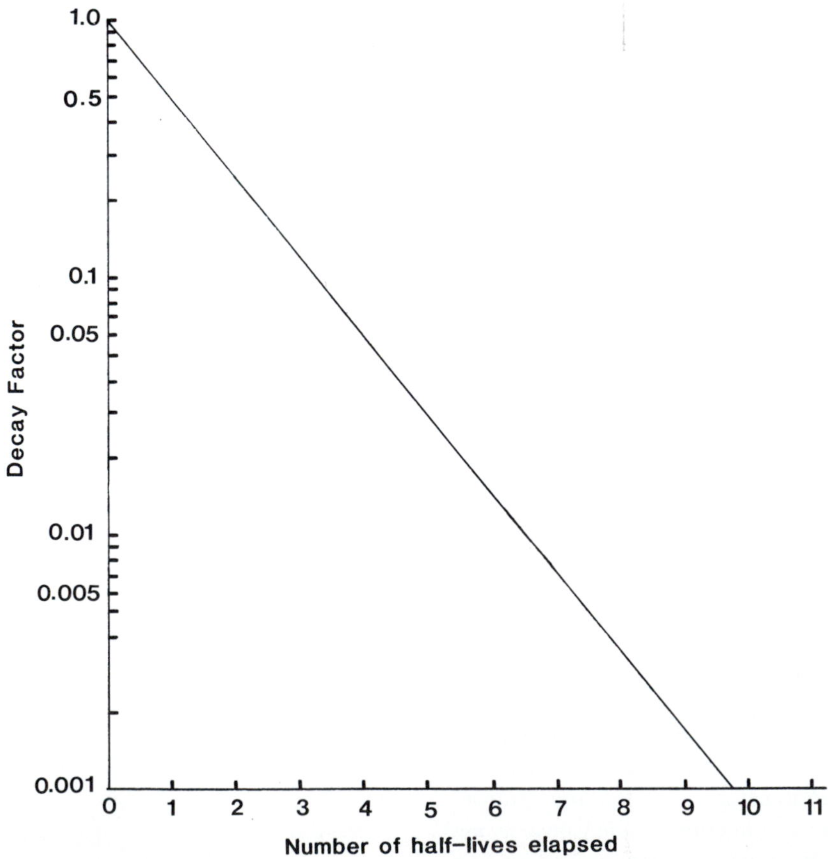

Fig. 7-3 Universal decay curve. Decay factor $(e^{-\lambda t})$ versus number of half-lives elapsed.

Table 7-1 Decay factors for ^{125}I

Days	Days 0	4	8	12	16
0	—	0.955	0.912	0.871	0.831
20	0.794	0.758	0.724	0.691	0.660
40	0.630	0.602	0.574	0.548	0.524
60	0.500	0.477	0.456	0.435	0.416

From Fig. 7-3 one can see that the decay factor at 6 half-lives is approximately 0.016. When this value is multiplied by the initial amount of radioactivity, it gives the residual activity at 360 days.

One can also determine the decay factor from tables of such factors, which are available from most radiopharmaceutical companies and instrument manufacturers. An example of such a table for ^{125}I is Table 7-1. For example, the decay factor for ^{125}I 28 days after the manufacturer's calibration date is 0.724. This number, multiplied by the initial amount of radioactivity, gives the level of radioactivity left at 28 days.

Units of radioactivity. The average rate of decay of a sample (see Equation 7-1), namely, the average number of nuclides disintegrating per second (dps) or per minute (dpm), is the activity of the sample and is used to determine the amount of radioactivity present in the sample.

Radioactivity is measured in curie (Ci) units. One curie is defined as the activity of a sample decaying at a rate of 3.7×10^{10} dps (2.22×10^{12} dpm), which is very close to the activity of 1 g of ^{226}Ra (3.656×10^{10} dps/g). In fact, the curie was originally defined as the activity of 1 g of ^{226}Ra. The basic multiples of the curie are as follows:

$$1 \text{ Ci} = 10^3 \text{ millicurie (mCi)} = 10^6 \text{ microcurie } (\mu\text{Ci}) = 10^9 \text{ nanocurie (nCi)} = 10^{12} \text{ picocurie (pCi)}$$

In clinical chemistry the amounts of radioactivity used are usually in the range of nanocuries to microcuries; occasionally, picocurie quantities are measured. The use of Système International d'Unités (SI) units in radioactivity measurements has been introduced. The basic unit of this system is the becquerel (Bq), in which 1 Bq = 1 dps. Thus

$$1 \ \mu\text{Ci} = 3.7 \times 10^4 \text{ Bq}$$

This system has not gained widespread acceptance in the United States.

In Equation 7-2, N_0 and N are the numbers of radionuclides present at times 0 and t, respectively. These quantities are extremely difficult to measure. However, the effects of the nuclear disintegrations can be measured more easily with use of one of the radioactive detectors described on pp. 133 to 137. In this way the total number of disintegrations per second occurring within the radioactive sample, or the radioactivity, at any given time can be estimated. Since the radioactivity, A, is proportional to the number of atoms N, Equation 7-2 can be written as

$$A = A_0 e^{-\lambda t} \qquad \textit{Eq. 7-4}$$

Therefore the decay constant, the decay factor, and the half-life are also applicable to activity versus time.

It is often necessary to know the specific activity of a radioactive sample. This is the activity of the radionuclide per unit mass of the radioactive sample and thus is measured in microcuries per microgram, microcuries per micromole, and so on or submultiples thereof. The specific activity of the nuclide is inversely proportional to the half-life of the radionuclide and is an important factor in determining the sensitivity of radioimmunoassay. A list of nuclides commonly used in clinical chemistry and some of their radiation characteristics are presented in Table 7-2. The low decay rates of ^3H and ^{14}C with long half-lives give them low specific activity and make them less suitable for radioimmunoassay work. The specific activities of nuclides listed in Table 7-2 are calculated under the assumption that all nuclides present are radioactive (carrier free). This assumption does not always hold true. For example,

Table 7-2 Radiation characteristics of radionuclides commonly used in clinical chemistry

Nuclide	Half-life	Main radiation: Type	Energy (keV)	Specific activity*: mCi/μg	mCi/μmole
^3H	12.3 years	β^-	18	9.7	29
^{14}C	5760 years	β^-	158	0.0044	0.062
^{32}P	14.3 days	β^-	1700	285	9120
^{35}S	87.1 days	β^-	167	42.8	1500
^{51}Cr	27.8 days	EC†	γ320	92	4690
^{59}Fe	45 days	β^-/γ	460/1099	49.1	2900
^{57}Co	270 days	EC†	γ122	8.5	480
^{125}I	60 days	EC†	γ35	17.3	2200
^{131}I	8.1 days	β^-/γ	807/364	123	16,100

*Carrier free.
†Electron capture.

the isotopic abundance of available ^{131}I preparations seldom exceeds 20%; that is, only about 20% of iodine atoms present in the ^{131}I preparation are ^{131}I—the rest are ^{127}I (stable iodine). Therefore the actual specific activity of ^{131}I preparations is only about one fourth of the theoretical specific activity shown in Table 7-2. Similarly, the specific activity of molecules depends on the isotopic abundance of the radionuclides present in the molecules. A thyroxine molecule contains four iodine atoms, and thus the specific radioactivity of radioiodine-labeled thyroxine preparations depends not only on the kind of radioactive iodine used for labeling but also on how many of the stable iodines are replaced by the radioactive iodine.

Properties of radiation and interaction with matter

An understanding of the properties of radiation and the mechanism of the energy loss of radiation as it passes through matter is important, because the operation of every detecting device for any type of radiation depends on one or more of the particular properties of the radiation that is being measured and the interactions of radiation with matter. Further, the safe manipulation of radioactive substances requires a knowledge of the nature of radiation and its ability to penetrate matter. The harmful effects of radiation on tissues are highly dependent on the ability of the radiation to ionize matter and on the energy of the incident radiation.

The interactions of radiation with matter result in the transfer of energy from a radioactive nucleus to the surrounding material. This transfer is accomplished through processes of excitation and ionization; therefore radiation emitted from radionuclides is frequently termed *ionizing radiation*.

Excitation occurs when orbital electrons are perturbed from their normal arrangement by absorbing energy from the incident radiation. Ionization occurs when the energy absorbed is sufficient to cause an orbital electron to be ejected from its orbit, creating an ion pair (a free electron and a positively charged atom or molecule). This ionizing ability of radiation is best expressed by the number of ion pairs produced per unit path length, that is, the specific ionization. Properties of various forms of radiation are presented in Table 7-3.

Particulate radiation. Because of their relatively large mass and double charge, alpha particles produce a great deal of ionization, which causes them to lose their energy quickly in a short distance (high specific ionization; see Table 7-3). Therefore alpha particles are weakly penetrating and can be stopped completely by very thin layers of solid materials. For this reason, they are less hazardous externally. However, if they get into the body, they will irradiate the tissues around them intensely, causing a serious health hazard. Alpha-emitting nuclides are seldom used in medicine.

Beta particles may be negatively or positively charged. There is no known difference between the negatron particle and the electron except for their origin. Positrons are antiparticles of electrons. The energetic negatron and positron particles also lose their energy by excitation and ionization of molecules; but, because of their smaller mass and charge, their specific ionization is not as high as that of alpha particles (see Table 7-3). In general, pure beta particle–emitting nuclides have a continuous spectrum of energy ranging from zero to a maximum of E_{max}. E_{max} is equivalent to the total energy available from the nuclear decay and is characteristic of each radionuclide. Fig. 7-4 shows the beta-particle energy spectrum of ^{14}C.

Electromagnetic radiation. Electromagnetic radiations usually encountered in the field of medicine include gamma-ray and x-ray radiation. Except for possible differences in energy, these photons are indistinguishable and engage in the same type of interactions with matter. Because photons have no mass, are uncharged, and travel with the velocity of light, they might travel through matter for a considerable distance without any interaction and then lose all or most of their energy in a single interaction. Photons can interact with matter in several different ways, depending on their energies and the properties of the material with which they interact.

Photoelectric effect is especially important for photons

Table 7-3 Basic properties of radiation

Radiation	Charge	Energy range	Approximate range of travel in: Air	Approximate range of travel in: Water	Relative specific ionization*
Particles					
α	+2	3-9 MeV	2-8 cm	20-40 μm	2500
β^-	−1	0-3 MeV	0-10 m	0-1 mm	100
β^+	+1	0-3 MeV	0-10 m	0-1 mm	100
Electromagnetic					
X rays	None	1 eV to 100 keV	1 mm to 10 m	1 μm to 1 cm	10
Gamma rays	None	10 keV to 10 MeV	1 cm to 100 m	1 mm to 10 cm	1

*The number of ion pairs produced per unit path length relative to that of gamma rays.

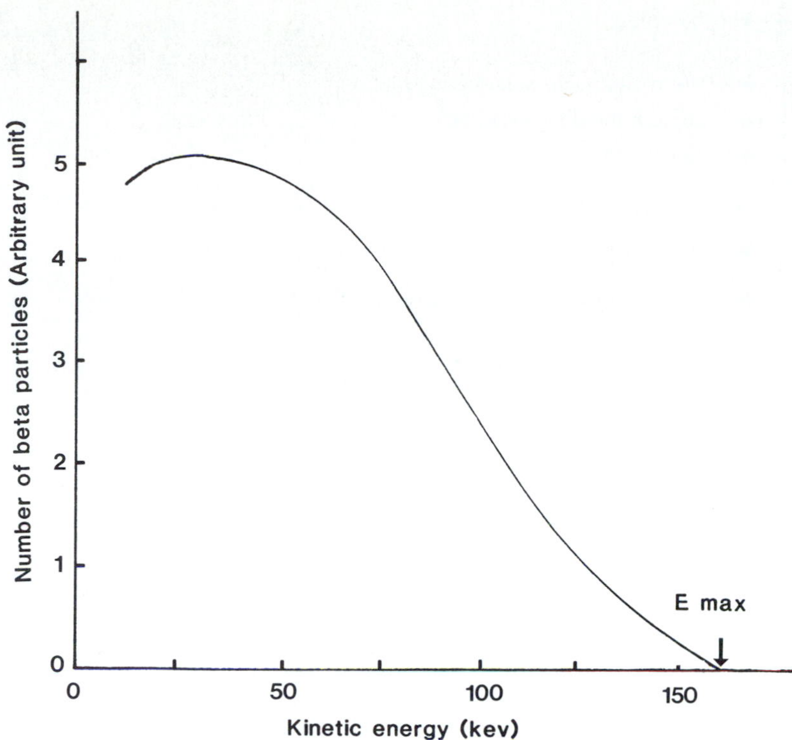

Fig. 7-4 Beta-particle spectrum for ^{14}C.

with low energy (below 0.5 MeV). The photon interacts directly with one of the orbital electrons in matter (a photon-electron interaction), and the entire photon energy is transferred to the electron. Some transferred energy is used to overcome the binding energy of the electron, and the remaining energy is carried by the electron as kinetic energy. The ejected electron (photoelectron) in turn transfers its kinetic energy to many other electrons in its path. The photoelectric effect is especially pronounced if the atomic number of the absorbing material is high.

Compton effect, or *Comptom scattering,* occurs primarily with photons of medium energy (0.5 to 1 MeV). In this process a collision between a photon and an electron results in the transfer of only a portion of the photon energy to the electron. The scattered photon with reduced energy emerges from the site of interaction in a new direction. The ejected electron (Compton electron) and the scattered photon lose more energy by subsequent interactions.

ISOTOPE DETECTION

Radioactive isotopes

Radioactivity measurements depend on the ability of radionuclides to produce ionized or excited atoms within the detector. Two basic types of radiation detectors are in common use: gas ionization and scintillation. The latter is capable of detecting both negatron and gamma radiation and providing information on the type and energy of the radiation and hence is currently the most commonly used detector in the field of medicine. Therefore the following discussion is devoted primarily to scintillation detectors; other detectors are described only briefly.

Ionization of gases. The Geiger-Müller counter and the ionization chamber are examples of gas ionization detectors. In the Geiger-Müller counter the radiation is detected through ionization produced within a suitable gas. Because the ions produced are accelerated by the relatively high voltage applied between the electrodes of the detector, considerable secondary ionization occurs, leading to a large output pulse (electron multiplication). The major advantage of this type of detector, as compared with ionization chambers, is its ability to detect low levels of radiation. The ionization chamber functions on a similar principle. However, because a lower voltage is used, electron multiplication does not occur in the ionization chamber, and the output signal is relatively small. Both detectors are widely used in survey meters for measuring exposure of personnel and locating a spilled radionuclide.

Scintillation detectors. Scintillation counting is based on the principle that a charged particle (alpha or beta) entering the detector, or an electron excited in the detector after an interaction with an incoming photon (gamma ray), will dissipate its energy within the scintillator contained in the detector by various processes of interaction mentioned previously. A portion of the energy absorbed by the scintillator is emitted as photons in the visible or near-ultraviolet region of the electromagnetic spectrum. Scintillators,

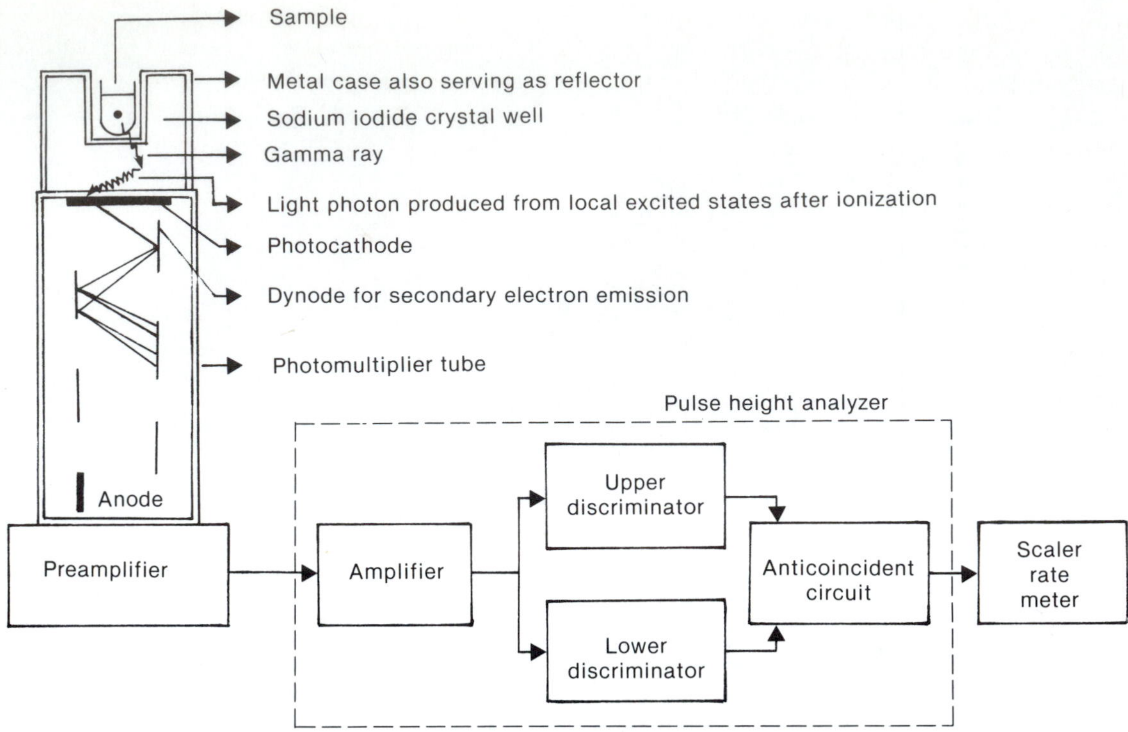

Fig. 7-5 Block diagram showing principal components of typical crystal scintillation counter.

or fluors, are substances capable of converting the kinetic energy of an incoming charged particle or photon into flashes of light (scintillation).

Crystal scintillation detectors. The most commonly used fluor for detecting gamma radiation by scintillation is a single crystal of sodium iodide containing small amounts of thallium (about 1%) as the activator. Fig. 7-5 is a block diagram of the common types of thallium-activated sodium iodide crystal scintillation detectors. The crystal is usually in the shape of a well, and the sample to be counted is allowed to sit in the well. The sodium iodide crystal is very hygroscopic. It is encapsulated in a metal (such as aluminum) can to prevent it from absorbing atmospheric moisture, except for one face (usually the bottom face) of the crystal well, which is covered by a transparent material such as Lucite and is optically coupled to the transparent face of a photomultiplier tube.

A gamma ray emitted from the sample placed in the crystal well is highly penetrating and therefore can pass through the glass or plastic wall of the test tube containing the radioactive sample and enter the crystal. As the gamma ray passes into the crystal, it produces excitation or ionization, and light photons are emitted. About 20 to 30 light photons are produced for each electron volt of energy absorbed. The photons pass through the transparent crystal and strike the photocathode of the photomultiplier tube, causing a release of electrons from the cathode. The energy required to release one photoelectron from the photocathode is about 300 to 2000 eV.

In addition to the conversion of the light photons emitted by the fluor into a pulse of detectable electrons, the photomultiplier also amplifies the minute amount of current produced from the photocathode to a level that can be effectively handled in conventional electronic amplifier circuits. This is achieved by a process of electron multiplication. As illustrated in Fig. 7-5, a series of metal plates, termed *dynodes,* are spaced along the length of the photomultiplier tube. The dynode surface is coated with a material capable of emitting secondary electrons when struck by an accelerated electron. Each dynode is maintained at a potential voltage higher than the preceding one. The initial photoelectrons are accelerated toward the first dynode and strike it to produce secondary electrons, which are then accelerated toward a second dynode. About three or four electrons are released from the dynode for each striking electron. This process is repeated until an amplification of about 10^8 is achieved.

The current output of the photomultiplier tube is amplified, and the resulting voltage pulse is shaped for optimal counting by conventional electronic circuitry such as that shown in Fig. 7-5. The preamplifier reduces the distortion of the electrical signal produced by the photomultiplier tube. The preamplifier output is further amplified by the amplifier to give a voltage of up to 10 V.

The height of the voltage pulse produced by the amplifier is proportional to beta- or gamma-ray energy deposited in the detector. Each radionuclide has a characteristic spectrum of energies (pulse height) such as that shown for

^{14}C in Fig. 7-4. The function of the pulse-height analyzer is to sort out the pulses according to their pulse height and to allow those pulses within a restricted range (the photopeak) to reach the rate meter for counting. This is accomplished by means of *discriminators*. A lower discriminator sets the lower limit, and an upper discriminator sets the upper limit of the energy range to be counted. The lower discriminator excludes all voltage pulses below the lower limit; the upper discriminator excludes voltage pulses above the upper limit. The energy interval represented by the difference between the two discrimination levels is called the *window width*. Only the pulses with energy in the preset discriminator window pass through the *anticoincident circuit* and are counted, because the anticoincident circuit will transmit a pulse arriving at its input from the lower discriminator only if there is no pulse arriving from the upper discriminator at the same time. This mode of operation is termed *differential counting*. The discriminators can also be set to count every pulse that exceeds the setting of the base control. This mode of operation is termed *integral counting*. In some counters the discriminator controls consist of a lower level discriminator, termed the *base control,* and the window-width control. The value of the base control and the window width gives the upper level of the energy window selected.

A scintillation detector equipped with two or more pulse-height analyzers (multichannel analyzers) can be used to simultaneously count two or more radionuclides, either in the same sample or in different samples, provided that there is sufficient energy difference between them so that a certain portion of the energy of one radionuclide can be detected free from the second radionuclide. For example, the major photopeak of ^{125}I occurs at 27 keV and that of ^{131}I at 364 keV. In addition to the 364 keV photopeak, a minor ^{131}I photopeak occurs at 32 keV. For counting a mixture of these two isotopes, one analyzer channel (A) is centered at 27 keV and the other channel (B) at 364 keV, with the window width of about 20 to 40 keV. Channel B gives the true count for ^{131}I, because ^{125}I does not contribute counts to channel B. Counts from channel A, however, represent the sum of the true counts for ^{125}I and the ^{131}I spillover. One can estimate the extent of the ^{131}I spillover by counting the pure ^{131}I standard in both channels.

To save counting time, crystal scintillation counters equipped with up to 48 detectors have been developed in recent years. With such a counter, the total assay counting time can be reduced to as much as one forty-eighth of that required by a single detector instrument. However, it is absolutely necessary to make sure that all detectors in a multidetector instrument perform in an equivalent fashion.

Liquid scintillation detectors. Liquid scintillation detectors are primarily used for counting beta particle–emitting radionuclides such as ^{3}H, ^{14}C, and ^{32}P. Unlike that of gamma photons, the penetration of negatron particles is so short that they cannot penetrate the wall of the sample container for interaction with crystal scintillators. In liquid scintillation counting, the sample is dissolved or suspended in a solution or "cocktail" consisting of a solvent such as toluene, a primary scintillator such as 2,5-diphenyloxazole (PPO), and a secondary scintillator such as 1,4-bis-2(5-phenyloxazolyl)-benzene (POPOP). The beta particles from the radioactive sample dissolved in the scintillation cocktail ionize and excite the molecules of the solvent. The excitation energy is transferred to the primary scintillator, which emits light photons when the excited electrons return to the ground energy level. The wavelength of light emitted by the primary scintillator is frequently too short (about 350 to 400 nm) for efficient detection by the photocathodes of photomultiplier tubes. The secondary scintillator absorbs the photons emitted by the primary scintillator and reemits them at a longer wavelength (about 430 nm). Thus the secondary scintillator is also termed a *wavelength shifter*.

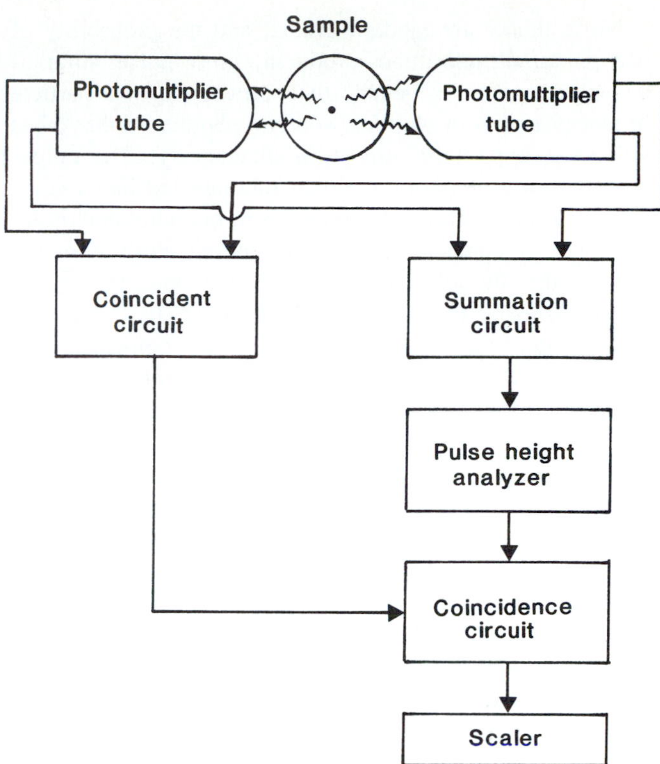

Fig. 7-6 Schematic diagram of liquid scintillation counter.

A typical arrangement of the principal components of a liquid scintillation counter is shown in Fig. 7-6. The light photons produced in the sample vial are detected and amplified by the photomultiplier tubes in the same manner as for the crystal scintillation counter. In the liquid scintillation detector, however, a second photomultiplier tube, a coincident circuit, and a summation circuit are incorporated to eliminate the electronic noise associated with the photomultiplier tube and to improve counting efficiency for low-energy beta-particle emitters.

Noise pulses are random events, and the probability of two photomultiplier tubes producing noise pulses simultaneously is relatively small. In contrast, the beta particle produces a burst of photons, and two photomultiplier tubes will receive photons almost simultaneously. The output pulses from each photomultiplier tube are fed into a coincident circuit to check if a pulse from one photomultiplier tube is accompanied by a corresponding pulse from the other within the allowed time interval (termed the *coincidence resolving time,* usually about 20×10^{-8} second). Pulses within the resolving time produce a coincident signal that is electrically sent to the coincident gate. Most noise pulses do not meet the coincidence resolving time requirement and are excluded. The summation circuit is incorporated to sum all coincident pulses to obtain the true pulse height. The summed coincident pulses are amplified, sorted, and counted in a manner similar to that for the crystal scintillation counter.

In liquid scintillation counting, proper energy transfer cannot occur unless the sample is in contact with the scintillation solution to give a colorless, transparent, homogeneous solution. Some radioactive samples are not soluble in the scintillation solution, and so it may be necessary to add one or more substances to obtain a homogeneous scintillation mixture. Solubilizers such as methylbenzethonium chloride (Hyamine 10 ×) are used to facilitate dissolution of the sample in the scintillation solution, or jelling agents such as aluminum stearate are used to enhance the counting efficiency by stabilizing the sample suspension in liquid scintillators. Many commercial liquid scintillation cocktails of nonpolar mediums (toluene or xylene) contain some type of surfactant such as the Tritons (polyoxyethylene ethers and other surface-active compounds) to maintain aqueous samples in colloidal suspensions so that the aqueous samples can be counted at high efficiency. Nonvolatile, radioactive materials are also counted on solid supports such as filter paper disks or glass fibers immersed in a scintillation solution. The disadvantage of this counting method is the relatively low counting efficiency because of impurity quenching.

Quenching is basically a process that results in the reduction of the overall photon output of the sample. Impurities present in the radioactive sample may compete with the scintillators for energy transfer, that is, the energy is lost to a non-light-producing process. This phenomenon is termed *impurity quenching*. Water in aqueous samples or a support medium such as a filter disk may cause impurity quenching. Colored substances such as hemoglobin may absorb the light photons produced by the scintillation process before they can be detected by the photomultiplier tubes or may change the wavelength of the light photons to a value not suitable for efficient detection by the photocathodes of photomultiplier tubes. Quenching in liquid scintillation counting is detected and corrected by efficiency determination. The efficiency of the measurement is defined as the ratio of the observed counts per minute (cpm) to the absolute units of disintegrations per minute (dpm):

$$\text{Efficiency} = \frac{\text{cpm}}{\text{dpm}}$$

Since the quenching characteristics of each sample are different, the efficiency must be determined for each sample. By knowing the counting rate (cpm) and the counting efficiency of a sample, one can calculate the absolute radioactivity (dpm) of the sample. Several methods for efficiency determination have been developed, but only those most frequently used are discussed.

Internal standardization. Internal standardization, one of the oldest methods, is the most accurate for efficiency determination when properly carried out. In this method the sample is counted before and after the introduction of a calibrated standard of the measured radionuclide. The difference between the count rates before and after the spike, divided by the calibrated activity of the spike in disintegrations per unit time, is termed the *counting efficiency*. The disadvantages of this method are the time-consuming manipulation of the sample and the loss of sample for recount after the introduction of the spike.

Sample-channels ratio. The sample-channels ratio method is based on a downward shift of the pulse-height spectrum of the photon as a result of the quenching-induced decrease in the pulse height of many energetic decays (Fig. 7-7). The degree of the shift is related to the extent of quenching or counting efficiency and is expressed by the change in the ratio of the sample counts obtained from two different discriminator window settings (channels). As shown in Fig. 7-7, one channel is usually set to measure the entire isotope spectrum (L_1 to L_3), and the second channel is restricted to only a portion of the spectrum (L_1 to L_2). In this method, channel ratios (L_1 to L_2) to (L_1 to L_3) of a set of artificially quenched standards of known efficiencies are determined and plotted against counting efficiency to obtain a quench curve. The efficiency of any unknown sample can be determined from its channel ratio and the quench curve. This method requires no additional sample manipulation, unlike internal standardization, and is suitable for handling a large number of samples through automation. This method, however, may result in large errors in highly quenched samples or samples with low count rates.

External standards. Unlike the internal standard method, the known activity in the external standard method is provided by an external source of gamma radiation, such as ^{226}Ra, placed at a fixed position adjacent to the sample vial. The external gamma-ray source generates electrons through the Compton collision process in the scintillation solution. The Compton electrons transfer energy to the solution and cause scintillation in the same way as beta particles do in the scintillation medium. The energy spectrum produced by Compton electrons is also affected

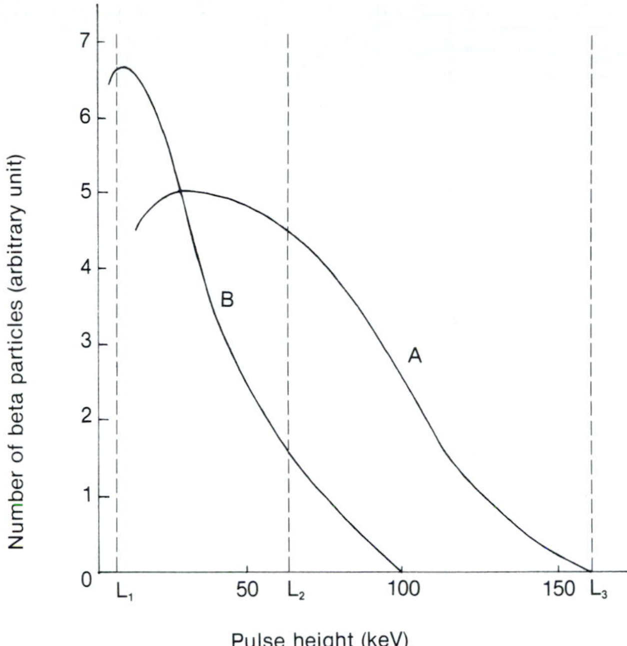

Fig. 7-7 Unquenched, *A*, and quenched, *B*, pulse-height spectra of ^{14}C. L_1 to L_3 denotes discrimination levels.

by the presence of quenching materials, as in a typical beta-particle spectrum. The sample is counted twice, once in the absence of and once in the presence of an external standard. As in the sample-channels ratio method, a set of quenched standards of known efficiencies is used to obtain a correlation curve between the sample-counting efficiency and the count rate of the external standard. The counting efficiency of a sample can be determined from the count rate of the external standard counted with the sample and from the correlation curve. The external standard method has become an integral part of almost all modern liquid scintillation detectors.

Another problem encountered in liquid scintillation counting is *chemiluminescence,* the production of light photons by a chemical reaction between the sample material and the solute or solubilizer added to the scintillation solution. Chemiluminescence gives rise to single photons and can be excluded by the coincident circuit of the liquid scintillation counter. However, when chemiluminescence reactions are of sufficient intensity, non-beta-particle coincident pulses may be generated that interfere with the beta-particle scintillation counting. The chemiluminescent effect will eventually disappear, but this may take several hours or longer, especially at low temperatures. Chemiluminescence can be monitored by repeated counting of the sample. Some modern instruments are capable of automatically monitoring and correcting for the chemiluminescent effects.

As with crystal scintillation counting, mixed-isotope counting is possible with a liquid scintillation counter with multichannel analyzers. A common example of dual-label counting involves a mixture of tritium, with a maximum beta-particle decay of 18.6 keV, and ^{14}C, with a maximum beta-particle decay of 156 keV.

Counting statistics. Since radioactive decay is essentially a random process, it is unlikely that successive measurements on a given sample will result in the same number of counts. However, radioactive decay obeys a *Poisson distribution*. The Poisson distribution density formula can be applied in the calculation of the precision of measurement at a given count rate. If a single measurement of total counts, *N,* is made, then precision of this measurement in terms of the percent coefficient of variation (%CV) can be estimated to be as follows:

$$\%CV = \frac{\sqrt{N}}{N} \times 100\%$$

For example, at a total count of 100, CV = 10%; at 1000, CV = 3.2%; at 10,000, CV = 1%. This approximation is applicable only when the background count is negligible compared with the sample count. When significant background counts are present, the formula for the standard deviation of a difference must be used:

$$\%CV = \frac{\sqrt{N + B}}{S} \times 100\%$$

where *B* is the background count and *S* is the sample count ($N - B$). Thus a total count of 100 in the presence of a background count of 10 gives the following:

$$\%CV = \frac{\sqrt{100 + 10}}{100 - 10} \times 100\% = 11.7\%$$

Precision can be increased by prolonging the counting time, but a 1% CV (that is, 10,000 counts) is satisfactory in most applications.

Stable isotopes

The amounts of stable isotopes present in an element *(isotope abundance)* are determined primarily by a mass spectrometer. The principle of a mass spectrometer is illustrated in Fig. 7-8. A mass spectrometer consists of four principal components: a sample inlet for vaporization of the sample; an ion source for ionization of the vaporized sample by, for example, electron bombardment; a mass separation unit for separation of resulting positively charged ions by means of electrical and magnetic fields according to mass; and a detector-recorder.

In general, radioactive isotopes rather than stable ones are used whenever there is a choice because the radioactivity measurement is usually easier, less expensive, and more sensitive than the measurement of stable isotopes. However, the latest commercially available isotope-abundance mass spectrometers, especially when coupled with the gas chromatograph (gas chromatography–mass spectrometry, or GC-MS) or with another mass spectrometer

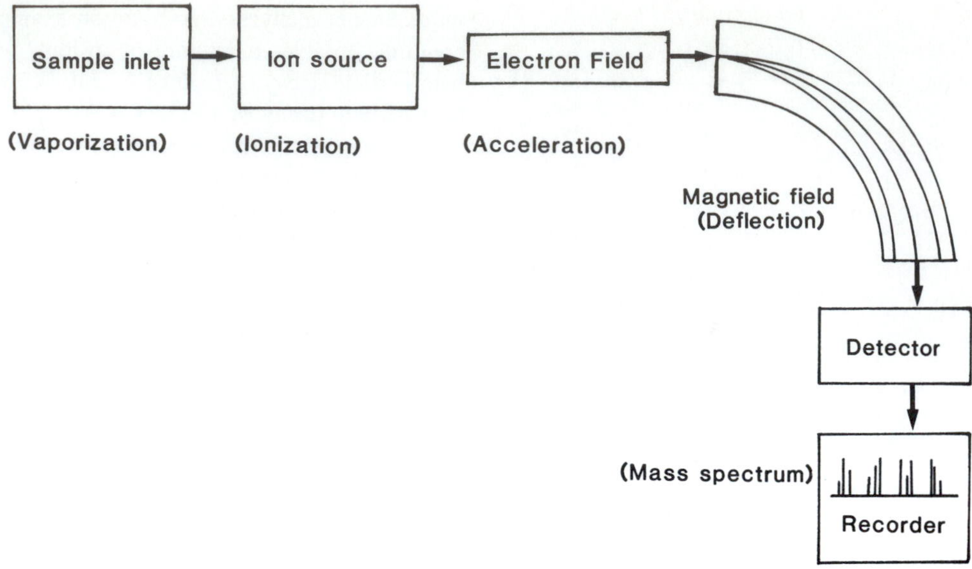

Fig. 7-8 Schematic of a mass spectrometer.

(tandem mass spectrometry, or MS-MS), are accurate, sensitive, specific, and relatively easy to operate and are expected to become as acceptable as scintillation counters in a routine clinical laboratory.

Stable isotopes are nonradioactive and are suitable for use as tracers in humans, especially infants, children, and pregnant women. In the case of nitrogen and oxygen, no suitable radioactive isotopes are available, and the stable isotopes ^{15}N and ^{18}O are used exclusively. In addition to being useful tracers, stable isotopes have also been used in the quantitative analysis of various substances in recent years. For example, isotope-dilution mass spectrometry has been used in definitive methods for calcium, lithium, potassium, chloride, cholesterol, certain steroid hormones, glucose, and other clinically important compounds. GC-MS has been used in the diagnosis of organic acidemia and vitamin B_{12} deficiency.

RADIATION HEALTH SAFETY

Although the quantity of radioactivity handled in the clinical chemistry laboratory is usually very small, a basic knowledge of radiation safety is vital to every laboratory worker who has frequent contact with radioactive substances because the biological effects of long-term exposure to very low doses of ionizing radiation are still largely unknown and may prove to be hazardous to health.

Radionuclides commonly used in the clinical laboratory are either beta-particle emitters, such as ^{14}C and ^{3}H, or gamma-ray emitters, such as ^{125}I and ^{57}Co. Both forms of radiation produce their biological effects by producing ionization and excitation along their paths in the tissue. However, beta radiation is less penetrating than gamma radiation; thus beta-particle emitters are considered to be more hazardous in terms of internal radiation and less hazardous in terms of external radiation than gamma-ray emitters. Therefore the primary concern with beta-particle emitters is to prevent the entry of radioactive materials into the body through inhalation, ingestion, or absorption by the skin; with gamma-ray emitters other factors, such as shielding, exposure time, and exposure distance, are also important when considering radiation safety.

Monitoring

Regular monitoring of both personnel and work areas is an important radiation safety procedure. It is necessary to measure periodically the radiation exposure doses of personnel to ensure that radiation doses received are below the recommended limits. The following three basic units are used to measure radiation exposure and dose.

The roentgen (R) is a unit of x rays or gamma rays and measures the quantity of ionization produced by photon radiation in a given sample of air. One roentgen equals that quantity of photon radiation capable of producing one electrostatic unit of either sign in 0.001293 g of air.

The radiation absorbed dose (rad) is a measure of local energy deposition per unit mass of material irradiated by any ionizing radiation. One rad is equal to 100 ergs of absorbed energy per gram of absorber.

The roentgen equivalent, man (rem) is that dose of ionizing radiation causing the same amount of biological injury to human tissue as 1 rad of x, gamma, or beta radiation. In the case of alpha radiation, the dose in rems equals the dose in rads multiplied by 20 because only 0.05 rad of alpha radiation is needed to produce the same biological effect as 1 rad of x, gamma, or beta radiation. The recommended maximum permissible dose to the whole body is 0.5 rem per year for the general public and 5 rems per year for radiation workers.

Film badges are probably the most commonly used and cost-effective way of monitoring personnel. The photographic film becomes progressively optically dense when exposed to ionizing radiation and thus may be used to monitor the radiation dose received by the wearer. Since most clinical laboratory personnel working with radioimmunoassays routinely handle ^{125}I-labeled compounds, it is advisable to monitor possible accumulation of radioactive iodine in the thyroid glands. Arrangements should be made to have the radioactive content of each worker's thyroid measured at least twice each year or after each radioiodination experiment. It is necessary to keep all records of radiation exposure of all workers handling radioactive materials for at least 5 years. Each laboratory should have a portable radiation detector, such as a portable Geiger-Müller survey meter, to monitor radioactivity in an area in which radioactive materials are routinely handled. Monitoring of beta radiation usually requires taking samples of the work area with swabs and using a liquid scintillation counter to determine the presence of radioactivity.

Contamination control

Internal radiation exposure is controlled only by preventing the entry of radioactive materials into the body. This requires strict adherence to the general rules for radiation safety. No smoking, eating, drinking, applying of cosmetics, or storing of food is allowed in work areas. Mouth pipetting of radioactive materials should never be done. All persons working in radioactive areas must wear the designated protective clothing (a standard laboratory coat is satisfactory in a clinical laboratory involved in radioimmunoassays) and disposable gloves. Radioactive materials must be properly labeled, stored, and used only at specially designated areas. Work involving the possible generation of volatile, radioactive substances, such as radioiodination, should be performed in an exhaust hood. The working surface should be covered by a layer of disposable absorbent material. In addition to the proper operating technique, cleanliness and good housekeeping are essential to prevent and minimize the spread and buildup of contamination.

Persons contaminated by radioactive materials should be quickly decontaminated to prevent the possible transfer of radioactivity to internal organs by absorption through the skin. Facilities for decontamination, such as a shower and an eyewash station, should be available in each laboratory. Absorbent materials should be used to remove spilled radioactive material. The contaminated area should then be scrubbed with soap and water. It is a good practice to cover the contaminated area immediately with a piece of paper to prevent spreading of the radioactivity to other parts of the laboratory.

Waste disposal

The level of radioactivity of the radioactive waste material generated in clinical chemistry laboratories involved in radioimmunoassays is usually very low. The radioactive waste materials in most cases are of such low specific activity or of such a low concentration that they may be disposed of directly by release to air, water, or ground. Radioactive waste disposal is regulated by the Nuclear Regulatory Commission (NRC) of the United States. Recent changes in the disposal regulation adopted by the NRC allow users of most in vitro radioassay kits to dispose of the waste in the same manner as for nonradioactive trash. All radiation labels should be removed and destroyed before the disposal.[4] The NRC proposes to allow a total of 7 Ci of water-soluble radioactivity to be disposed of each year through sewer systems in the following groups:

1 Ci—Undefined waste
1 Ci—Carbon-14
5 Ci—Tritium

Liquid scintillation mediums containing less than 0.5 μCi/g of material may be incinerated. Some states (NRC agreement states) are approved by the NRC to regulate the use, safety, and disposal of radioactive material in the state, provided that the regulations are more restrictive than the NRC regulations. It is important therefore to be familiar with the state regulations on radioactive materials if your laboratory is located in an NRC agreement state.

BIBLIOGRAPHY

Heal, AV: Safety and disposal changes that affect regulations in radioassay labs, Lab World, pp 50-53, Dec 1981.

L'Annunziata, MF: Radionuclide tracers, New York, 1987, Academic Press, Inc.

Peng, CT, Horrocks, DL, and Alpen EL: Liquid scintillation counting: recent applications and development, vol 1, Physical aspects; vol 2, Sample preparation and applications, New York, 1980, Academic Press, Inc.

Rinehart, KL, Jr: Fast atom bombardment mass spectrometry, Science 218:254, 1982.

Sorenson, JA, and Phelps, ME: Physics in nuclear medicine, ed 2, New York, 1987, Grune & Stratton, Inc.

CHAPTER 8 Electrophoresis

JOHN M. BREWER

OBJECTIVES

- State the charge properties of amphoteric substances at acidic, isoelectric, and basic pH. Explain the significance of these properties in electrophoresis.
- Outline the principle of electrophoresis and summarize how electrophoretic separations of molecules are affected by the following techniques:
  - Enhanced resolution
  - Molecular size
  - Molecular size and charge
- Describe the commonly encountered problems associated with electrophoresis and explain probable causes and corrective actions.
- List three common support mediums for electrophoresis and classify commonly used visualization stains according to substance being separated. Describe the physical or chemical reactions that allow a stain to visualize a specific substance.
- Describe the effect(s) on electrophoretic separations of the following parameters:
  - Ionic strength of buffer
  - pH of buffer
  - Electro-osmosis
  - Heating

KEY TERMS

Ampholyte A trade name for a mixture of substances with a range of isoelectric points that have high buffering capacities at their isoelectric points.

amphoteric A substance that can have a positive, zero, or negative charge, depending on conditions.

anion Negatively charged particle or ion.

boundary Edge of a zone, as of a macromolecule solution next to the solvent.

buffer A mixture of proton-donating and proton-accepting substances whose function is to keep the proton concentration (the pH) constant or nearly so. An example is a mixture of acetic acid and sodium acetate.

cation Positively charged particle or ion.

co-ion An ion of the same charge as the one under consideration. Generally a much smaller ion.

conductivity The readiness of a substance to carry a current. In an ionic solution, the sum of the product of the charge concentrations and charge mobilities.

convection Mass or bulk movement of one part of a solution relative to the rest, usually because of density differences.

counterion An ion of opposite charge to the one under consideration. Generally a smaller ion.

disk electrophoresis A stacking or isotachophoretic step followed by zone electrophoresis, usually on a polyacrylamide gel.

discontinuous solvent A solution consisting of at least two separate regions that have different ions in them.

effective mobility The actual mobility of a substance under certain conditions. Generally less than the mobility because of a lower charge or resistance by a supporting medium.

electric field An influence measured in volts (or volts per centimeter) that is manifested by the behavior of a charged particle in it.

electrical neutrality A condition in which total positive charges equal total negative charges.

electroblot Substances separated by electrophoresis are transferred onto a sheet of material using an electric field perpendicular to the original separating field. The material adsorbs and immobilizes the substances in the same pattern as the original separation.

electrodes Substances in contact with a conductor. The substances are connected to a source of an electric field.

electrolyte Ionic substances, usually of low molecular weight, added to provide as constant and uniform an ionic environment for electrophoresis as possible.

electro-osmosis Tendency of a solution to move relative to an adjacent stationary substance when an electric field is applied.

electrophoresis Movement of charged particles because of an external electric field.

frictional coefficient A measure of the resistance a particle offers to movement through a solvent.

gel A network of interacting fibers, or a polymer that is solid but traps large amounts of solvent in pores or channels inside.

ionic strength The sum of the concentrations of all ions in a solution, weighted by the squares of their charges.

isoelectric Condition of zero net charge on an amphoteric substance.

isoelectric focusing The ordering and concentration of substances according to their isoelectric points.

isoelectric point The pH at which a substance has a zero net charge.

isotachophoresis The ordering and concentration of substances of intermediate effective mobilities between an ion of high effective mobility and one of much lower effective mobility, followed by their migration at a uniform velocity.

joule heating Heating of a conductor by the passage of an electrical current.

mobility The velocity a particle or ion attains for a given applied voltage. A relative measure of how quickly an ion moves in an electric field.

molecular sieving Separation of molecules on the basis of their effective sizes.

polyacrylamide Polymer of acrylamide and usually some cross-linking derivative.

polyelectrolyte Substance with many charged or potentially charged groups.

resolving power Ability to separate closely migrating substances.

Sodium dodecyl sulfate (SDS) A detergent and an especially effective protein denaturant.

Southern blot A kind of *electroblot,* in which the substances separated and transferred are nucleic acids.

stacking Ordering or arranging and concentrating macromolecules according to their effective mobilities.

Western blot A kind of *electroblot,* in which the substances separated and transferred are proteins.

zeta potential The potential produced by the effective charge of a macromolecule, usually taken at the boundary between what is moving with the macromolecule and the rest of the solution.

zone A particular region or space within a larger one, generally distinguished by some property, such as its occupancy by a protein.

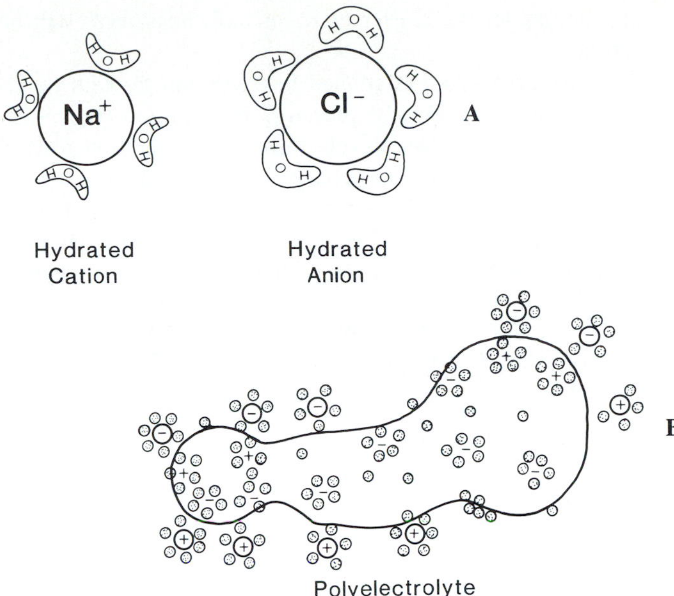

Fig. 8-1 State of charged particles in water solution. **A,** Small ions (Na^+ and Cl^-) with associated water molecules. **B,** Macromolecule with water molecules *(stippled smaller circles)* associated with charged and polar groups. Hydrated co-ions and counterions are also shown as larger circles around plus or minus signs.

Movement of charged particles because of an *external* electrical field is called *electrophoresis.*[1] Since charged molecules can be made to move, different molecules can be separated if they have different velocities in an electric field. Therefore electrophoresis is a separation technique, just as chromatography and ultracentrifugation are, and there are similarities among these techniques.

The movement of a species in aqueous solution is affected by the species' interaction with water molecules, that is, hydration (Fig. 8-1). The more polar a molecule

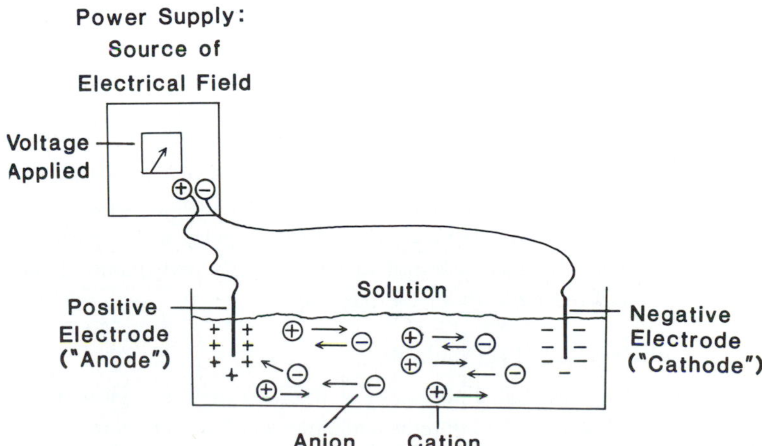

Fig. 8-2 Application of electric field to solution of ions makes ions move.

is, the greater the cluster of water molecules there will be about it.[2]

The electric field is applied to a solution through oppositely charged *electrodes* placed in the solution (Fig. 8-2). A particular ion then travels through the solution toward the electrode of opposite charge. Thus positively charged particles (cations) move to the negatively charged electrode (cathode), while negatively charged particles (anions) migrate to the positively charged electrode (anode).

APPLICATION OF ELECTRIC FIELD TO A SOLUTION CONTAINING A CHARGED PARTICLE

Forces on a particle

The force exerted on a charged particle depends on the electric field, V (in volts or volts per centimeter), and the charge on the particle, Q. The force on the charged particle is the product:

$$F_{elec} = QV \qquad \textit{Eq. 8-1}$$

This electrical force, F_{elec}, when exerted on the particle, will cause it to move. However, a particle moving in a solvent will experience resistance because of the viscosity of the solvent.[1,3] The resistance, which is itself a force, is proportional to the velocity[3]:

$$F_{resistance} = fv \qquad \textit{Eq. 8-2}$$

The proportionality constant, f, is called the *frictional coefficient*.

The frictional coefficient depends on the viscosity of the solvent and the size and shape of the particle. The greater the viscosity, the slower the movement. The bigger or more asymmetrical the particle, the slower its movement through the solvent. The frictional coefficient of a large particle such as a protein is a characteristic property of the particle.

Mobility of a particle

When an electric field is applied to a charged particle, the particle will begin to migrate. The electrophoretic and frictional forces oppose each other, and the particle's velocity increases until the forces are equal ($F_{resistance} = F_{elec}$). The velocity, v, a particle attains for a given electric field, V, is determined by two properties of the particle—its charge and its frictional coefficient. Consequently, the value of v/V is also a characteristic property of the particle and is important enough to be given its own name. It is called the *mobility* of the particle.

Effect of pH on mobility

Each ion has a particular charge and mobility. However, when a solution contains a substance whose pK is near the pH of the solution, that substance exists in both a charged and an uncharged form in the solution. The fraction of species with a charge will depend on the pK of the substance and the pH of the solution. When the pH is equal to the pK_a of a weak acid, only 50% of the particles will be charged. At 1 pH unit below the pK_a, 90% will be uncharged. At 1 pH unit above the pK_a, 90% will be in the charged state. Since the effective (average) charge of a substance varies with the pH, its *effective mobility* also varies with the pH.

This is particularly true for substances such as proteins. Proteins are clearly *amphoteric* substances; that is, they contain acidic and basic groups. Their overall (net) charge will be highly positive at low pH's, zero (isoelectric) at a particular higher pH, and negative at still more alkaline pH's. Since mobility is directly proportional to the magnitude of the charge, the effective mobility of a protein is very much a function of the pH.

The most important practical consequence of this is that electrophoretic solutions must be buffered to maintain a constant pH. The buffer pH is chosen to give an optimum net charge for maximum separation. For proteins, pH's in the 7 to 9 range are generally used. The buffer is used to maintain this pH and thus the net protein charge throughout the electrophoretic process. Buffers are ionic substances themselves and so take part in any electrophoretic process, a fact that must also be considered.

Electrolytes

In electrophoresis, a substance such as a protein is put in one limited region or *zone* of the system and is made to move into another region or zone. Therefore much of the solution will not have protein in it at any particular time. If the protein and its associated counterions are not present to carry the current in a particular region, other ions must be present to carry the current. For this reason it is a common practice to add an excess, usually about 0.1 M, of low-molecular-weight buffer to the solution through which the protein must travel. The buffer and salt ions *(electrolytes)* provide a constant electrical environment so that the overall movement of the protein will be as constant as possible and will be minimally influenced by other protein molecules.

Ion movement and conductivity

In any electrical system, the current produced is proportional to the voltage applied:

$$V = \text{Resistance} \cdot \text{Current} \qquad \textit{Eq. 8-3}$$

or

$$V = \frac{\text{Current}}{\text{Conductivity}} \qquad \textit{Eq. 8-4}$$

In electrophoresis, the current is the flow of ions (in both directions). The conductivity is the sum of the concentrations times the effective mobilities of all ions present. An ion with a higher effective mobility carries a larger fraction of the current.[2]

The voltage, conductivity, and current thus are all related (see Equation 8-4). If the conductivity is increased by an increase in the salt concentration while the current is kept constant, the voltage must decrease. Such a decrease in voltage reduces the electrical force, F_{elec}, on charged particles, slowing the movement of the macromolecules. This increases the time needed for a given separation, and the resolution decreases because of increased diffusion. If the conductivity is increased at a fixed voltage, the current must increase, thus increasing the electric heat (joules) generated by the system, since heating is proportional to the square of the current. Excessive heating produces convective disturbances in the solutions, which distort the electrophoretic patterns and may also denature macromolecules. Since increasing the conductivity at either a fixed voltage or current has deleterious effects, optimum results are achieved when one keeps the concentration of ions and therefore the conductivity at moderate values.

FACTORS AFFECTING MOBILITIES OF MACROMOLECULES

Charge and conformation

The clinical laboratory usually deals with polyelectrolytes, substances with many charged or potentially charged groups on them. The net charge of a polyelectrolyte is determined by the total number of charged groups within the polyelectrolyte and its conformation. The folding, or *conformation,* of a protein exposes sites for any small molecules that it may bind. In some cases electrolyte ions bind strongly and specifically to the macromolecule. For example, bovine serum albumin can bind several chloride ions. This changes the net charge of the macromolecule and therefore its mobility.

Ionic atmosphere and zeta potential

Counterions, ions of opposite charge, naturally tend to hover in the vicinity of the charged groups of macromolecules.[4] However, they do not actually neutralize the charges on the macromolecule but are instead located at a spectrum of distances from the charged groups of the macromolecule, forming a double layer of charge about the macromolecule, called an *ionic atmosphere.*

The macromolecule moves with its entourage of hydration and hydrated counterions (Fig. 8-3). These reduce the effective charge of the macromolecule to a level given by the *zeta potential.* The zeta potential is the potential (voltage) produced by the effective charge of the macromolecule at the *surface of shear.* The surface of shear is the boundary between the entire macromolecular complex in solution (hydration layer and embedded counterions) and the material that is staying behind (the solvent).

Relaxation effect

Because of random thermal motion, electrophoresing macromolecules move irregularly, in jumps, rather than in a continuous straight line. At each jump, the counterion atmosphere is left somewhat behind. The counterions (or their replacements from other parts of the solution) then move to catch up or reposition themselves, but this takes a little time. This is called a *relaxation effect.* It also tends to lower the mobility of the macromolecule, since the retarded or misplaced counterions momentarily produce a charged field that acts in a direction opposite that of the applied field.

Electrophoretic effect

Since ions in water solution are hydrated, the counterions of the electrolyte moving in the opposite direction

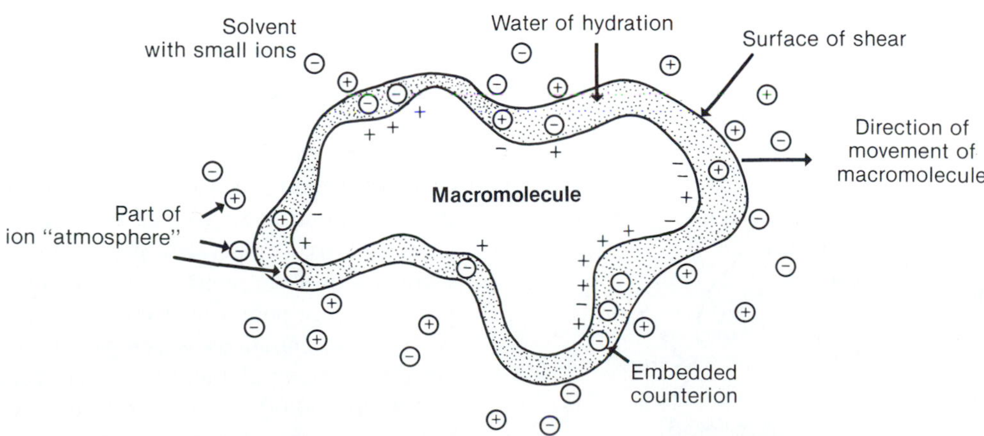

Fig. 8-3 Zeta potential of macromolecule is average effective electric field strength (potential) produced by charges on macromolecule and any charged particles embedded in solvent carried along (water of hydration) with macromolecule. *Stippled area,* Water of hydration. Zeta potential is measured at surface of shear: the boundary between water of hydration and rest of solvent.

carry solvent along with them. The macromolecule is thus moving against a *flow* of solvent, and its mobility is reduced.

SUPPORT MEDIUMS

The goal of electrophoresis is to separate a mixture of macromolecules into completely separate zones. The narrower (thinner) the original zone of a mixture of macromolecules is, the smaller the migration distance necessary to achieve separation will be. Use of narrower zones means that complete separation can be effected in less time. Blurring or remixing of the separated zones as a result of diffusion is also reduced.

The major technical difficulty in using narrow (thin) zones of relatively concentrated macromolecules is a mechanical one. Such zones will be considerably more dense than the solvent; thus the zones would "fall" through the solvent faster than the macromolecules would electrophorese. This is called *bulk flow,* or *convection*. The conventional solution to this problem is to use a supporting medium.

Functional basis

The supporting medium must allow as free a penetration of the material to be separated as possible and yet cut off bulk flow (convection). Most mediums do this by offering a restricted pore size for electrophoretic movement of the macromolecules. A capillary tube has the same effect.

Electro-osmosis

The supporting medium should not interact with the molecules, since this will inhibit or stop migration. The usual interaction problem encountered is not the actual adsorption of the material. More commonly seen are the effects of charged groups attached to the medium that result in *electro-osmosis*. Electro-osmosis is a very general effect.[2] It is more pronounced when charged groups are present in the supporting medium, but it always occurs to some extent. For example, agar, which is often used as a supporting medium in electrophoresis, is a mixture of agarose and agaropectin. The agaropectin has a relatively large number of carboxyl groups in it, which at neutral pH have counterions. If a voltage is applied, the counterions will move, but the carboxyl groups attached to the polysaccharide matrix will not. The counterions carry enough solvent with them to produce a *net* flow of solvent in *one* direction. This is the electro-osmosis effect, sometimes called *endosmosis*.

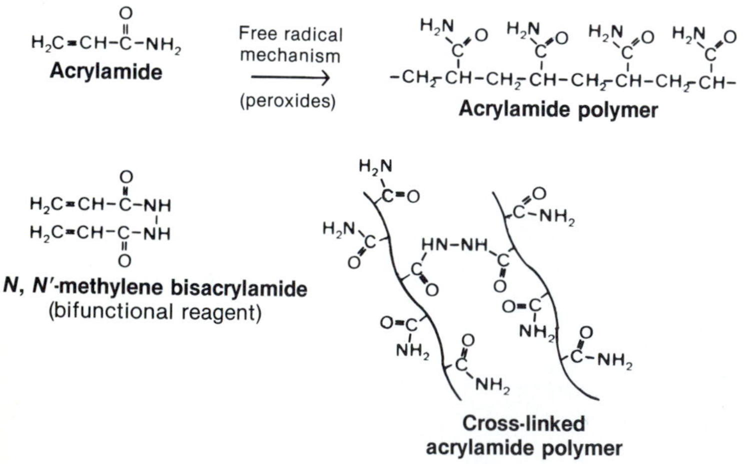

Types of supporting mediums

The supporting medium can be a solution such as a sucrose-density gradient, but, in general, insoluble materials are used. Some are self-supporting, whereas others are mechanically supported by the apparatus. Paper or sheets of plastic such as cellulose acetate have enough mechanical strength to allow electrophoresis on sheets hung or stretched over rods.

Support mediums can also be classified as particulate or continuous. Particulate support mediums include glass beads, Sephadex, and cellulose fibers. Continuous support mediums include polyacrylamide, starch, and agarose gels.

Gels are jellylike solids in which all the solvent is included. Starch gels, for example, are made from starch suspensions that are heated and cooled. The starch fibers interact, tangling with each other but trapping the solvent so that large gaps or pores exist between the fibers. These gaps or pores are available for macromolecular movement. Similar gels can be made from agar, agarose, and some chemical polymers. Gels can also be made by polymerization of acrylamide with a small percentage of a bifunctional acrylamide derivative that cross-links the acrylamide polymers (Fig. 8-4).

Fig. 8-4 Polyacrylamide gels are produced by polymerizing a mixture of acrylamide and a bifunctional (cross-linking) acrylamide derivative. Derivative shown is that in common use.

Molecular sieving

The porosity, or average pore size, of some mediums is fixed, but the porosities of other mediums can be controlled. For example, by changing the gel concentrations of starch or agar, one can vary the pore size. If the average pore size is near the average diameter of the macromolecules that are being electrophoresed, the supporting medium will produce molecular sieving effects.

The average pore size of polyacrylamide gels cast at 5% to 10% concentrations is comparable to the effective diameters of many globular (relatively compact) proteins of 15,000 to 250,000 daltons molecular weight. These gels will filter such solutions, separating macromolecules on the basis of both size and mobility (see Chapter 4, reference 6, and Chapter 15, reference 5). The molecular sieving effects can produce enhanced resolution, that is, narrower zones of macromolecules.

ENHANCED-RESOLUTION TECHNIQUES

Discontinuous buffers

Several electrophoretic techniques employ a system in which different buffer cations or anions are placed along the electrophoretic path on either side of the protein zone. The mobilities are different for each ion. If the ionic species placed before and the species placed after a mixture of high-molecular-weight substances, proteins, for example, have suitable mobilities, the proteins become *stacked* between the two ions. If a series of proteins are electrophoresed in this discontinuous buffer system, then they will arrange themselves according to their effective mobilities and become concentrated, resulting in a much more effective (enhanced) resolution.

There are three major applications of the use of discontinuous systems to produce enhanced resolution: isotachophoresis, isoelectric focusing, and disk electrophoresis. These techniques, briefly described in Table 8-1, are used frequently to analyze proteins.

Separations based on molecular size

By enabling separations on the basis of molecular size as well as molecular charge, the techniques briefly described in Table 8-2 provide enhanced separation ability. They usually employ polyacrylamide or agarose to prepare gels of known pore size. Molecules of smaller molecular size will electrophorese faster than larger molecules carrying similar charges.

SELECTION OF METHODS AND CONDITIONS

Knowledge of the isoelectric point and the molecular weight of the compound to be examined can help determine the optimum conditions for an electrophoretic separation. A buffer should be chosen with a pH that will provide the maximum separation without destroying the properties of the sample. Very acidic or basic conditions pose problems for any system, since an increasing fraction of the current is carried by protons or hydroxyl ions, resulting in poorer separation. A summary of the effects of

Table 8-1 Enhanced-resolution techniques

Technique	Physical basis	Mechanism	Effect	Limitations	Advantages and uses
Isotachophoresis	Electrical neutrality; current in series circuit is constant.[7,8] Lower conductivity solution following higher conductivity solution must move at same velocity and carry same current, so the two solutions experience different voltages and adjust in concentration.[4,7,8]	Solution of lower conductivity containing ion of lower mobility running behind solution of higher conductivity containing higher mobility ion C (e.g., chloride); ion of intermediate mobility (ion D; e.g., protein) sandwiched between; concentration of D increases to carry same current as A and C.	Intermediate mobility ions (e.g., proteins) stack in thin concentrated zones and move in discrete zones.	Ions not separated; resolution not as good as disk electrophoresis	No widespread clinical applications currently
Disk electrophoresis	Effective mobilities of some ions is pH dependent; it is often used with polyacrylamide gels.	As above; then stacked proteins overrun by following ion (ion A) because of pH change; change in pH produced using ion B.	Proteins, now in thin zones, migrate independently.	Ion systems and pH's limited; technically more exacting than ordinary electrophoresis	High sensitivity and resolving power; used extensively to separate proteins, mostly as research tool
Isoelectric focusing[6,7,9,10]	Migration of ions must occur in both directions; amphoteric ions (with both basic and acid groups) have zero effective mobility and zero net charge at their isoelectric point.	Ion movement stops because of zero counterion (ion B) concentration, leaving all ions stacked in pH gradient; leading and trailing ions are an acid and a base.	Proteins in zones at isoelectric (isoionic) points; pH gradient is buffered by carrier ampholytes.	More complicated and exacting than ordinary electrophoresis	High capacity and resolution to 0.01 pH unit possible; primarily a research tool

Table 8-2 Use of supporting mediums in separations

Method	Principle	Effect	Limitations	Advantages and uses
Separations based on molecular size				
Gradient gels[11]	Gels cast with increasing gel concentrations going from origin to end of gel have gradient of pore sizes.	Larger macromolecules move more slowly as they encounter higher gel concentrations; can measure relative sizes and charges.	Difficult to reproduce gels	Research tool
Gels containing denaturants[12,13]	Gels cast with denaturants (e.g., urea of SDS) so that macromolecules migrate in denatured forms; SDS binds uniformly and in large amounts to most proteins.	Proteins in SDS migrate proportionately to lower subunit molecular weights.	Exacting technique; disulfide bonds in proteins must be broken; not all proteins behave normally.	Research tool
Separations based on molecular size and charge				
Gel electrophoresis[11]	Pore size is small enough to restrict diffusion.	Higher resolution	Reproducibility	Better resolution than cellulose acetate; agarose gels widely used mostly research tool
Immunoelectrophoretic methods (see Chapter 10)	Antigen and antibody are brought together using electrophoresis to form precipitate.	Can identify and quantitate specific antigen or antibody	Sensitivity somewhat low	Very widespread (e.g., *rocket immunoelectrophoresis*) clinical use
Two-dimensional electrophoresis[14]	Separation occurs according to one parameter (e.g., isoelectric point) in one dimension (direction), then according to a second parameter (e.g., size) at right angles.	Mixture of proteins spread over a surface; information proportional to square of length of side of surface[14,15]	Exacting technique; difficult to reproduce patterns	High information content; widely used in clinical research

Table 8-3 Common effects of electrophoretic parameters on separation

Parameter	Effect on electrophoresis
pH	Changes charge of analyte and hence effective mobility; can affect structure of analyte, such as denaturing or dissociating a protein.
Ionic strength	Changes voltage or current; increased ionic strength usually reduces migration velocity and increases heating.
Ions present	Can change migration velocity if interaction is strong; can cause tailing of bands.
Current	Too high a current causes overheating.
Voltage	Migration velocity is proportional to voltage.
Temperature	Temperature gradients in support mediums cause bowed bands. Overheating can denature (precipitate) proteins. Lower temperatures reduce diffusion but also reduce migration velocity; there is no effect on resolution.
Time	Separation of bands (resolution) increases linearly with time, but dilution of bands (diffusion) increases with the square root of time.
Medium	Major factors are endosmosis and pore-size effects, which affect migration velocities.

the various parameters on electrophoresis is provided in Table 8-3.

Support mediums

The choice of a supporting medium is based on many considerations. Slabs or flat-surface mediums allow greater amounts of material to be electrophoresed. These mediums are also useful when one is comparing different samples. Use of gel cylinders can provide great sensitivity, whereas thin-layer methods provide even greater sensitivity.

Paper and cellulose acetate. Paper is especially favored for separation of low-molecular-weight substances in specialized biochemical laboratories. The main advantages of these materials are their thinness and mechanical strength. A thinner support means greater sensitivity because less material is needed to produce a detectable spot or zone.

Cellulose acetate is prepared by treating cellulose with acetic anhydride. This puts acetyl groups on the sugar hydroxyl groups. The resulting material is pressed into sheets that are somewhat stronger than paper and a good deal more chemically uniform. Adsorption of material to the sugar hydroxyl groups in paper leads to losses of material and to *tailing* of zones. Cellulose acetate is more inert in this respect, and because of its uniformity and ease of preparation (the strips are merely soaked in electrophoretic buffer so that no air is trapped), it is very widely used in routine clinical work.

Gels. Gels can be cast with varying thicknesses to increase or decrease capacity (the amount of sample). Gels also offer the possibility of molecular sieving effects because their porosity can be controlled by changing their composition. On the other hand, their mechanical strength tends to be low.

Starch and agar. Starch gels are not as extensively used as agarose, acrylamide, or even Sephadex. The starch solution must be heated to 100° C; then degassed, an awkward process; then cast. The starch gels tend to provide greater resolution than agar or agarose does. However, the inconvenience of preparing starch gels has limited their use.

Agar and agarose (agar without the agaropectin) are easier to handle. Because agarose demonstrates a lower electro-osmosis effect and exhibits fewer problems with adsorption, it is preferred over agar. The pore size of agarose is much greater than that of polyacrylamide. This is why agarose is used in most immunoelectrophoretic techniques, since antigen and antibody must be able to migrate through the gel. Another advantage of agarose is that it may be poured after reheating to only about 50° C; thus some proteins, such as antibodies, can be mixed in without denaturing. Even so, the same disadvantages encountered when preparing starch gels are encountered when preparing agar and agarose gels. Precast agarose gels for a variety of separations are available commercially. These gels are used for separation of isoenzymes, hemoglobins, glycoproteins, and so on.

Polyacrylamide. Polyacrylamide gels are less frequently used in clinical laboratories but are a common research tool. They are clear, fairly easy to prepare, and exhibit reasonable mechanical strength over the acrylamide concentration range normally employed for proteins. In addition, they show a low endosmosis effect and have a pore size well suited for the separation of the average proteins, RNA molecules, and smaller restriction fragments of DNA. A major clinical use of polyacrylamide gels is the separation of alkaline phosphatase isoenzymes. However, polyacrylamide gel preparation and casting are somewhat more exacting and time consuming, and reproducibility of gel preparation is difficult to achieve.

Cellulose acetate sheets can be purchased in relatively uniform batches so that results of different electrophoresis experiments are more consistent. After electrophoresis, the strips can be dissolved (in acetone, for example) so that slices containing stained or radioactive material can be more easily quantitated.

On the other hand, resolution with cellulose acetate is not as good as with polyacrylamide. Eight or nine serum protein fractions can be resolved using cellulose acetate, but up to 30 fractions can be detected using disk electrophoresis on polyacrylamide gels.[11] So cellulose acetate is used because it lends itself to fast, easy, reproducible measurements, although of comparatively low resolution.

There is an exception to the requirement that a supporting medium not interact with the material being separated. In the *Western blot* technique,[15] a gel slab containing separated proteins is electrophoresed perpendicularly to the slab, transferring some or all of the separated proteins onto a sheet of material, often nitrocellulose, which adsorbs the proteins.[17] The sheet may be stained for protein or enzyme activity (see below). Usually, however, the sheet is washed with some neutral protein, such as milk protein or bovine serum albumin, to block unoccupied adsorption sites. Then the locations of specific transferred adsorbed proteins are determined, usually by immunochemical assays. The resulting pattern may be compared with the original slab that is stained or examined by autoradiography for all proteins or all labeled proteins.

A major virtue of Western blots is their sensitivity. Adsorption of antibodies takes place at concentrations of antigen and antibody that are far too low for a precipitate to form. As little as 10^{-10} g of protein can be detected and hence identified.[16] Consequently, antigens (and antibodies) in serum that are present at very low concentrations can be detected and even quantitated. This technique is used for detecting acquired immunodeficiency syndrome (AIDS) antigens and antibodies in patients.[15]

Conditions

Horizontal versus vertical position. Electrophoresis can be performed horizontally or vertically; there is no theoretical reason for preferring one or the other procedure. Horizontal electrophoresis places less mechanical stress on the support, whereas vertical electrophoresis supporting mediums are often supported between glass plates.

If horizontal electrophoresis is used and the surface of the medium is open to air, evaporation of the solvent can cause problems. As a result of evaporation, salt concentrations rise, usually unevenly along the support, leading to nonuniform current flow and heating. This can lead to problems of uneven migration, especially at the sides of a horizontal, flat electrophoresis bed. If one buffer reservoir is higher than the other, convective flow of the buffer through the supporting medium can occur. Therefore the electrophoresis apparatus should always be level.

Sample application. Sometimes samples are simply applied to the surface of the supporting medium and allowed to soak in. Sometimes special slots or holes are cast or cut in a gel. Occasionally, the sample may be polymerized into the gel or cast with the gel. Injection of the sample is rarely used, except with isotachophoresis. If electrophoresis takes place in stages (such as in two-dimensional electrophoresis), the gel containing part or all of the sample may be cut out and reattached, sometimes with an agarose "glue," to another gel for the next stage in the separation. Often the sample is layered onto the surface of a gel. The sample is then usually made denser than the solvent by adding sucrose or glycerol. All of the above techniques are equally valid when combined with the appropriate type of assay.

Current and voltage considerations. Electrophoresis can be carried out at constant voltage, constant current, or constant power. Selection of any of these modes often depends on the power supply available. Since diffusion increases with the square root of time, it is best to complete the electrophoresis as quickly as possible. This requires use of the maximum voltage. However, the voltage is always limited by the efficiency of cooling of the apparatus. Some workers claim that temperature gradients of more than 0.1° C across a gel or other support lead to noticeable distortions of macromolecule zones. For many current clinical applications, temperature control does not appear to be necessary, and separations are carried out at ambient temperatures.

However, one recently developed commercial electrophoresis system employs a very efficient cooling system and very thin, preformed agarose gels to perform high-voltage electrophoresis. By using 1000 to 2000 V instead of 100 V, the electrophoresis can be completed in 2 to 3 minutes instead of 20 minutes. This approach maximizes the efficiency of the electrophoretic system in time and resolution.

The conductivity of any electrophoretic system will change with time, because the ionic composition will change as a result of movement (electrophoresis) of the sample along the system. Such changes are minimal in continuous systems, such as high-voltage paper electrophoresis, and application of constant voltage is satisfactory for these systems. For isotachophoresis, a constant velocity of zone migration is desired, and constant current is used. For other electrophoretic systems, heating is usually the limiting factor (see above), and so constant power (wattage) should be used. Disk electrophoresis and isoelectric focusing fall into this category. Pulsed power supplies provide no advantage.[18]

Separation time. In the case of isotachophoresis, the electrophoresis is stopped when the trailing ion emerges. Isoelectric focusing is complete after the gradient is formed and the current has dropped to a stable value. The time to stop a disk or ordinary zone electrophoretic separation is usually indicated by the position of the *tracking dye,* usually when the dye band reaches a predetermined position in the stationary support (typically the end of the support). Dyes that have high mobilities, such as bromphenol blue, are employed. They are usually added to the sample. Since some proteins bind such dyes, their apparent mobilities may be changed.

LOCATING THE ANALYTE

Analysis involves determining where the substances to be identified or quantified are on the support. This can be accomplished by measuring a physical property of a molecule, such as light absorption or refractive index, or by using a chemical reaction such as staining. Generally, measurements of physical properties lack specificity, sensitivity, or resolution. For example, not all proteins absorb strongly at 280 nm.

Staining

Staining often achieves the desired goals of resolution, sensitivity, and specificity (Table 8-4). Since the zones of material broaden by diffusion after electrophoresis is stopped, the first step in the analytical procedure is to eliminate diffusion. One can do this in paper electrophoresis by drying the paper or in autoradiography by freezing the gels.

Proteins

Protein in gels is often denatured (that is, precipitated in the gel matrix) by soaking the gels in dilute acetic acid or more effectively in trichloroacetic acid. Addition of sulfosalicylic acid further improves the denaturing ability of the staining solution. In the Western blot procedure, the proteins are immobilized by adsorption.

Sometimes heat must also be applied to make the proteins insoluble. However, some resist denaturation by all these conventional procedures and remain soluble in the stain. If detergents such as SDS or other soluble agents are present, they will interfere with precipitation. Inclusion of methanol in the acid solutions helps remove such substances before staining.

Choice of stain

Many types of stains are employed, depending on the need. Sometimes it is desirable to stain everything, such as all proteins. A dye called *Stains-All* is suitable.[19] A *silver stain,* which reacts to both proteins and nucleic acids, is an alternative.[20] Proteins, after electrophoresis on cellulose acetate, are most often stained with Ponceau S.[21]

Stains and staining procedures are often specific for one chemical group. The ninhydrin stain for amino groups, often used after paper electrophoresis of peptides or amino acids, is an example of this. Glycoproteins can be treated with periodic acid (for oxidation) and color developed with a dye (fuchsin) in the presence of a reducing agent (sulfite). This periodic acid–Schiff (PAS) stain treatment oxidizes carbohydrate groups to aldehydes, which react with the dye to form a Schiff base. The sulfite reduces the Schiff base, making the stain permanent. There is also a specific stain for phosphoproteins. A fairly complete list of stains is given by Righetti and Drysdale.[21] (See Table 8-4 for a list of commonly used stains.)

Table 8-4 Commonly used stains for various substances

Substance	Stain	Comments
Proteins	Ponceau S*	Less sensitive than amido black, but more specific for proteins Most widely used stain
	Bromphenol blue† Light green SF†	Low sensitivity, but can be used with ampholyte gels
	Coomassie brilliant blue R250*	Can detect less than 0.2 μg of protein; can be used with ampholyte gels
	Silver stain (silver reduced onto oxidized macromolecules)‡	At least 10-50 times more sensitive than others; different proteins give different colors, for unknown reasons
	Stains-All (a cationic carbocyanine dye)§	Generally sensitivity, including phosphoproteins
	Amido black 10B (''buffalo black'')‖	Very sensitive stain
Lipoproteins	Sudan black B¶	
	Oil red O#	
	Coomassie brilliant blue R250#	Used with SDS gels
Glycoproteins	PAS (periodic acid–Schiff)**	Best for neutral glycoproteins; 2-3 μg of carbohydrate detectable
	Stains-All§	Best for sialic acid–rich glycoproteins
Nucleic acids	Stains-All§	Best for RNA, DNA, and mucopolysaccharides
	Silver stain§	
	Ethidium bromide††	Fluorescent bands with DNA; less than 10 ng detectable
Enzymes		
Dehydrogenases	NADH (fluorescence)‡‡	
	Nitroblue tetrazolium chloride‡‡	
Esterases Cholinesterases	Beta-naphthyl esters and tetrazotized o-dianisidine‡‡	
Phosphatases	1-Naphthylphosphate and fast blue B‡‡	

*Righetti, PG, and Drysdale, JW: Isoelectric focusing, New York, 1976, Academic Press, Inc.
†Chapter 19, reference 8.
‡Merril, CR, Goldman, D, Sedman, SA, and Ebert, MH: Science 211:1437, 1981.
§Green, MR, Pastewka, JV, and Peacock, AC: Anal Biochem 56:43, 1973.
‖Wilson, CM: Anal Biochem 96:263, 1979.
¶Swahn, B: Scand J Clin Lab Invest 4:98, 1952.
#Weller, H: Klin Wochenschr 36:563, 1958.
**Matthieu, JM, and Quarles, RH: Anal Biochem 55:313, 1973.
††Brunk, CF, and Simpson, L: Anal Biochem 82:455, 1977.
‡‡Gabriel, O: Methods in enzymology, vol 22, New York, 1971, Academic Press, Inc, p 578.

Once the stain has been introduced, usually by soaking of the support in the stain solution, excess stain must be removed. This can be done electrophoretically or, most commonly, by diffusion. Electrophoretic removal is fast but can result in distortion of the stained zones. Diffusion involves changing the solvent or using a destainer to remove free stain.

Many enzymes are identified by the use of colored or fluorescent substrates or products (zymograms), even in gels or other support mediums. For example, alkaline phosphatase hydrolyzes *para*-nitrophenylphosphate to *para*-nitrophenol, which has a yellow color at pH 8. Soaking a gel in such assay solutions produces colored bands where the enzymes are. If the product of enzymatic activity is of low molecular weight, rapid diffusion of the product may broaden the zone, making location of the enzyme or identification of the isozyme difficult. Products that form polymers or insoluble substances are better, but if this is not possible, a contact print method may be used.[11] A sheet of paper or other material is impregnated with a chromophoric substrate and pressed against the support containing the separated enzyme or enzymes. Often, the support is cut into slices, and the slices are incubated in assay solution. A list of such zymograms is given in the text by Righetti and Drysdale.[21]

Other localization techniques

A common technique for localization of proteins and nucleic acids employs radioactive labels, such as ^{125}I or ^{14}C, incorporated into the macromolecules. Following electrophoresis, a piece of x-ray film is placed in contact with the stationary support in the dark for 12 to 24 hours. After the film is developed, a dark area corresponding to the position of the radioactively labeled macromolecule will be present. This technique, *autoradiography,* is commonly employed in the Western blot technique for proteins

Table 8-5 Commonly encountered problems in electrophoresis

Problem	Likely cause	Corrective action
No migration	Instrument not connected	Check electrical circuits.
	Wrong pH; electrodes connected backwards	Check isoelectric point of protein and pH of buffer; check electrode polarity.
Bowed electrophoretic pattern on edges of support	Overheating or drying out of support	Humidify chamber; check buffer ionic strength.
Tailing of bands	Chemical reaction: subunit dissociation or adsorption to support	Use different support; try different pH.
	Salt in sample	Check sample for salt; dialyze sample against electrophoresis buffer.
	Buffer co-ion effect	Use different buffer co-ion.
Holes in staining pattern	Analyte present in too high a concentration	Apply less concentrated sample.
Very thin, sharp bands	Molecular weight of sample very high for support pore size	Use support with larger pore size.
	Sulfhydryl oxidation and aggregation	Run sample with sulfhydryl reducing compound or at lower pH.
Very slow migration	High molecular weight	Use support with larger pore size.
	Low charge	Change pH so that charge increases.
	Ionic strength too high	Check conductivity; dilute buffer.
	Voltage too low	Increase voltage.
Sample precipitates in support	pH too high or low	Run at different pH.
	Too much heating	Use lower wattage or external cooling.

and also is used in the Southern blot technique for nucleic acids.

Some commonly encountered problems in electrophoresis, their most likely causes, and suggested corrective action are listed in Table 8-5.

CLINICAL APPLICATIONS

The most common uses of electrophoretic techniques in the laboratory today are the following:

1. Specific protein electrophoresis
   a. Quantitative analysis of specific serum protein classes, such as gamma globulins and albumins (see Chapter 63)
   b. Identification and quantitation of hemoglobin and its subclasses (see Chapter 63)
   c. Identification of monoclonal proteins, such as Bence Jones gamma globulins in either serum or urine (see Chapter 63)

Table 8-6 Compounds commonly separated by electrophoresis

Class of compound	Stain	Support medium
Amino acid	Ninhydrin	Paper, cellulose acetate
Serum protein	Ponceau S	Cellulose acetate
	Coomassie blue 250	Polyacrylamide (with or without isoelectric focusing)
Lipoproteins	Oil red O	Agarose
Glycoproteins	PAS	Agarose
Nucleic acids	Ethidium bromide (fluorescent)	Agarose
Hemoglobins	Silver stain	
	o-Dianisidine	Cellulose acetate, agar
	Ferricyanide	
	Peroxide	
Isoenzymes		
Lactate dehydrogenase	Fluorescent NADH or tetrazolium	Agarose
Creatine kinase	Fluorescent NADH or tetrazolium	Agarose
Alkaline phosphatase	1-Naphthylphosphate + fast blue B or 5-bromo-4-chloroindolyl phosphate	Polyacrylamide, cellulose acetate
Immunoglobulins	Coomassie blue 250	Agarose
Specific antigens by immunological electrophoretic techniques (such as Laurell rocket)	Amido black 10B	Agarose

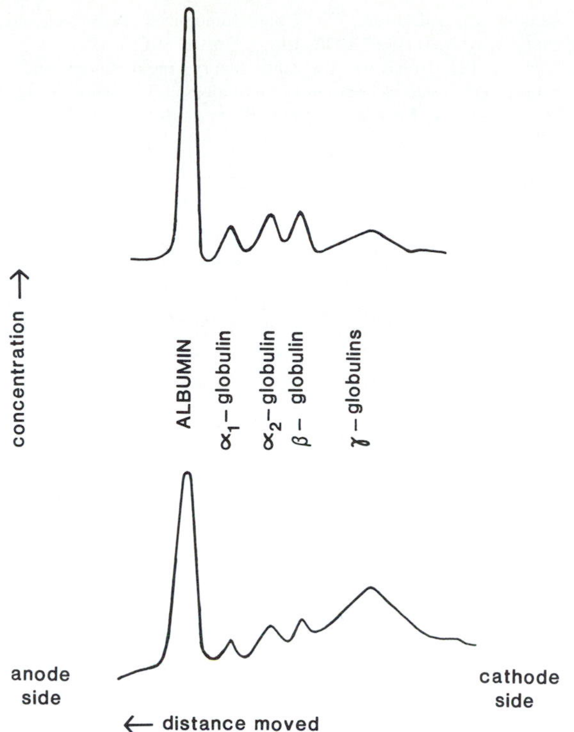

Fig. 8-5 Example of effect of disease, hepatic cirrhosis, on blood serum protein electrophoretic pattern. *Upper profile,* Distribution characteristic of healthy people.

d. Separation and quantitation of major lipoprotein classes (see Chapter 61)

2. Isoenzyme analysis: separation and quantitation of enzymes such as creatine kinase, lactate dehydrogenase, and alkaline phosphatase into their respective molecular subtypes (see Chapters 53 and 59)
3. Immunoelectrophoresis: most often used to determine qualitatively the elevation or deficiency of specific classes of immunoglobulins; also can be used to semiquantitate serum proteins, such as transferrin and complement component C3 (see Chapter 10)
4. Western blot technique to identify a specific protein: used to confirm the presence of antibodies to human immunodeficiency virus (HIV) (AIDS) (see Chapter 10)
5. Southern blot techniques to identify specific nucleic acid sequences (DNA or RNA) (see Chapter 47): used for prenatal diagnosis of inborn errors, diagnosis of viral infections, and identification of risk factors for cancer (see Chapter 49)

The two-dimensional procedures will undoubtedly replace some or all of the preceding procedures, but at this time mostly exploratory work is being done. All the procedures involve measurement of alterations in an electrophoretic pattern compared with a normal control. One can often use these to diagnose specific diseases[14,22] (Fig. 8-5). Nephrotic syndrome, for example, may be accompanied by a decrease of more than 25% in α_2 globulin levels and a decrease of up to 25% in gamma globulin levels because of the loss of low-molecular-weight proteins in the urine. The pattern of decrease or increase in several disease conditions is fairly characteristic, and hence quantitation of stained cellulose acetate strips is useful in diagnosis.

The stained support may be either put through a strip scanner if the support is a strip or put through a modified spectrophotometer if it is not. With due care, the dye absorbance on the support is a good measure of the amount of protein. With cellulose acetate electrophoresis, normal serum is electrophoresed next to the serum to be tested. The plots of absorbance (color) versus distance obtained from a strip scanner (see Fig. 8-5) allow immediate comparison of relative amounts of each class of separated proteins.

Electrophoresis is sometimes used in assays of genetic defects. The two-dimensional techniques now being developed have tremendous potential in that area. Table 8-6 lists compounds normally separated by electrophoresis.

REFERENCES

1. Van Holde, KE: Physical biochemistry, ed 2, Englewood Cliffs, NJ, 1985, Prentice-Hall, Inc.
2. Moore, WJ: Physical chemistry, ed 4, Englewood Cliffs, NJ, 1970, Prentice-Hall, Inc.
3. Tanford, C: Physical chemistry of macromolecules, New York, 1961, Academic Press, Inc.
4. Melvin, M: Electrophoresis, New York, 1987, John Wiley & Sons.
5. Righetti, PG, Van Oss, CJ, and Vanderhoff, JW: Electrokinetic separation methods, New York, 1979, Elsevier/North Holland, Inc.
6. Deyl, Z, Everaerts, FM, Prusik, Z, and Svendsen, PJ: Electrophoresis: a survey of techniques and applications, J Chromatogr 18(series), 1979.
7. Ornstein, L: Disc electrophoresis, Ann NY Acad Sci 121:321, 1964.
8. Brewer, JM, and Ashworth, RB: Disc electrophoresis, J Chem Educ 46:41, 1969.
9. Jovin, TM, Dante, ML, and Chrambach, A: Multiphasic buffer systems output PB 196085 to 196092 and 203016, Springfield, Va, 1970, National Technical Information Service.
10. Fagerhol, MK, and Laurell, CB: The Pi system-inherited variants of serum alpha 1-antitrypsin, Prog Med Genet 7:96, 1970.
11. Andrews, AT: Electrophoresis: theory, techniques, and biochemical and clinical applications, ed 2, New York, 1986, Oxford University Press, Inc.
12. Laemmli, UK: Cleavage of structural proteins during the assembly of the head of bacteriophage T4, Nature 227:680, 1970.
13. Wyckoff, M, Rodbard, D, and Chrambach, A: Polyacrylamide gel electrophoresis in sodium dodecyl sulfate-containing buffers using multiphasic buffer systems, Anal Biochem 78:459, 1977.
14. O'Farrell, PH: High-resolution two-dimensional electrophoresis of proteins, J Biol Chem 250:4007, 1975.
15. Dunn, MJ, editor: Electrophoresis '86, Deerfield Beach, Fla, 1986, VCH Verlagsgesell schaft.
16. Towbin, H, Staehelin, T, and Gordon, J: Proc Nat Acad Sci USA 76:4350, 1979.
17. Lemkin, PF, and Lipkin, LE: GELLAB: a computer system for 2D gel electrophoresis analysis. II. Pairing spots, Comput Biomed Res 14:355, 1981.
18. Allington, RW, Nelson, JW, and Aron, GG: ISCO Applications Research Bulletin, No 18, Lincoln, Neb, 1975, Specialties Co.
19. Green, MR, and Pastewka, JV: Identification of sialic acid-rich glycoproteins on polyacrylamide gels, Anal Biochem 65:66, 1975.
20. Merril, CR, Goldman, D, Sedman, SA, and Ebert, MH: Ultrasensi-

tive stain for proteins in polyacrylamide gels shows regional variation in cerebrospinal fluid proteins, Science 211:1437, 1981.
21. Righetti, PG, and Drysdale, JW: Isoelectric focusing, New York, 1976, Academic Press, Inc.
22. Annino, JS, and Giese, RW: Clinical chemistry, principles and procedures, ed 4, Boston, 1976, Little, Brown & Co.
23. Righetti, PG: Isoelectric focusing: theory, methodology and applications, ed 2, vol 5, Laboratory techniques in biochemistry and molecular biology, New York, 1983, Elsevier Biomedical Press.

CHAPTER 9 Immunological reactions

SUSAN BASSION

OBJECTIVES

- Define antigen and antibody.
- List and explain eight factors affecting antigenicity.
- Describe the composition and structural differences of antibodies. Name five human immunoglobulins and describe the physiological role of each.
- List and explain the forces involved in antigen-antibody reactions and the factors that influence the specificity of immunochemical reactions.
- Outline the mechanism of the following antigen-antibody gel-precipitation reactions and state the principle of each:
  - Double immunodiffusion
  - Radial immunodiffusion

KEY TERMS

affinity Measure of the binding strength of the antibody-antigen reaction.

antibodies Proteins that combine specifically with antigens.

antigenic determinant That portion of an antigen involved in a reaction with an antibody.

antigens Substances that induce an immune response.

avidity Measure of the binding strength of antibodies to multiple antigenic determinants on natural antigens.

B cells B lymphocytes that transform to plasma cells and produce antibodies.

constant region Noncombining region of immunoglobulin polypeptide chains.

cross-reactivity Binding of an antibody to an antigen other than the one initiating the immune response.

Fab fragment Portion of immunoglobulin molecule made by papain degradation and containing the antibody combining site composed of the light chain and a portion of the heavy chain.

Fc fragment Portion of immunoglobulin molecule produced by papain degradation that contains most of the heavy chain, including the complement binding site.

flocculation Precipitation reaction producing large, loosely bound precipitates.

haptens Low-molecular-weight substances that can induce an immune response only when coupled to high-molecular-weight immunogenic molecules.

heavy chain Portion of immunoglobulin molecule consisting of a polypeptide chain of about 50,000 daltons.

hypervariable regions Amino acid sequences in the variable region that have an increased likelihood of variation.

idiotype Portion of immunoglobulin molecule conferring unique character, most often including its binding site.

immunoglobulins (Ig) Proteins with antibody activity.

joining (J) chain Portion of IgM molecule possibly holding structure together.

lattice The cross-linked, three-dimensional structure formed by the reaction of multivalent antigens with antibody.

light chain Portion of an immunoglobulin molecule composed of a polypeptide chain of about 22,000 daltons.

plasma cells Immunoglobulin-producing cells that are the end stage of β cell differentiation.

precipitation A reaction in which antigen and antibody, in proper ratios, produce a large lattice of matrix that becomes insoluble.

secretory piece Polypeptide chain attached to IgA and necessary for secretion into mucosal spaces.

valency The effective number of antigenic determinants on an antigen molecule.

variable region N-terminal portion of immunoglobulin polypeptide chain whose amino acid sequence can change; this region includes the antigen-combining site.

zone of equivalence Region of antibody-antigen reaction in which concentrations of both reactants are equal.

Immunological reactions can occur between two types of substances, *antigens* and *antibodies*. This chapter examines these substances and the interactions between them.

ANTIGENS

Antigens, or *immunogens,* are defined as substances that induce an immune response. The immune response produced may be an antibody (humoral) response or the production of sensitized cells (cellular response). Usually both humoral and cellular responses are stimulated.

Factors affecting antigenicity

Many factors determine the antigenicity of a molecule. The nature and dosage of an antigen, the route of administration, the organism immunized, and the sensitivity of the detection method are important factors in the evaluation of antigenicity. Many other conditions must be satisfied for a molecule to be immunogenic. These conditions are discussed below.

Chemical nature. The first antigens investigated were bacteria and red blood cells. Studies showed that such complex macromolecular structures were composed of many different proteins, carbohydrates, and lipids. Subsequent investigations have proved that immunogens are found in several chemical classes. Proteins, polysaccharides, glycolipids, nucleic acids, and polynucleotides are capable of inducing an immune response.

Size. There is no absolute size requirement, but size is of considerable importance in determining the antigenicity of a molecule. The most potent immunogens are macromolecules with molecular weights greater than 100,000 daltons. The A and B polypeptide chains of insulin (2500 daltons) and of glucagon (3600 daltons) are immunogenic in guinea pigs. Nevertheless, most molecules with molecular weights less than 10,000 daltons are weakly immunogenic, if at all.

Haptens. Substances with low molecular weights can induce an immune response when coupled to higher-molecular-weight immunogenic *carrier* molecules. Such incomplete antigens, or *haptens,* do not elicit an immune response by themselves but do react with antibody. Many low-molecular-weight compounds have been shown to act as haptens, including monosaccharides, lipids, peptides, hormones such as adrenocorticotropic hormone (ACTH) and prostaglandins, toxins such as arsphenamide, and drugs such as barbiturates and sulfonamides.

Complexity. A molecule must exhibit a certain degree of chemical complexity to be antigenic. Synthetic amino acid homopolymers, composed of repeating units of a single amino acid, have been shown to be poor immunogens; copolymers of two or three amino acids are much better immunogens. Increasing immunogenicity follows increasing complexity. For example, the addition of aromatic amino acid residues such as tyrosine to synthetic amino acid copolymers increases their immunogenicity.

Antigenic determinants. The portion of an antigen involved in the reaction with an antibody is called an *antigenic determinant*. An antigen may contain more than one type of antigenic determinant; the number of antigenic determinants per antigen varies with the size and complexity of the molecule (Fig. 9-1). The effective number of anti-

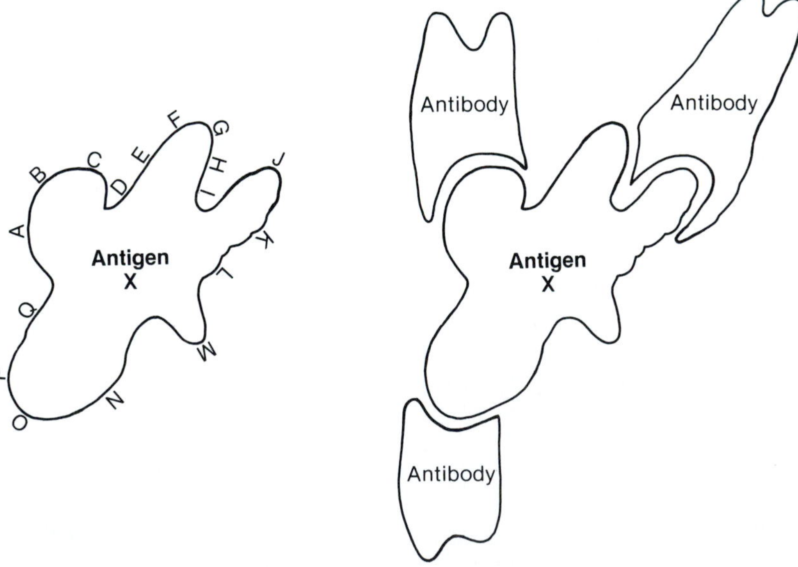

Fig. 9-1 Antigen X contains many different antigenic determinants, designated *A* to *Q* in this schematic representation. Antibody molecules when combined with antigen X bind to different sites. Maximum number of molecules of antibodies bound in this figure is 3; therefore valence is 3.

genic determinants on an antigen is its *valency*. This is the number of antibody molecules that can be bound to an antigen at the same time (see Fig. 9-1). Antibodies recognize the three-dimensional shape of an antigenic determinant (conformational antigenic determinant), as well as the basic amino acid structure (sequential antigenic determinant). An antigenic determinant sometimes comprises as few as four amino acid residues. The combining site of an antibody molecule reacts with an antigenic determinant in the complementary *lock-and-key* manner of protein-enzyme interactions. The affinity of an antibody for an antigenic determinant is directly proportional to the closeness of fit. The capacity of a portion of an antigen to serve as an antigenic determinant is termed its immunopotency.

Conformation and accessibility. The tertiary structure or spatial folding of molecules is a significant factor in their immunogenicity. Antibodies to native proteins do not react with denatured molecules. Antibodies to native proteins are directed primarily to conformational rather than sequential antigenic determinants.

In addition, accessibility or exposure to the environment is an important factor in determining immunogenicity. The terminal side chains of polysaccharides, the portions of a polysaccharide molecule that stick out from the main part of the molecule, are the most immunopotent regions of polysaccharide antigens. Accessibility of an antigen to the environment also depends on the solubility of an antigen in aqueous medium. The more soluble an antigen is, the greater the probability that it will interact with an antibody. The influence of charge on immunogenicity may be a manifestation of accessibility. Charged or hydrophilic residues are more in contact with the environment than hydrophobic residues, which tend to be sequestered in the interior of molecules.

Foreignness. The immune system is capable of distinguishing *self* from *nonself* in such a way that, under normal circumstances, a vigorous immune response is produced only to substances recognized as foreign. The more distant the evolutionary relationship between antigen and host, the more immunogenic the molecule. Thus guinea pig albumin will not evoke an immune response when injected into another guinea pig. The same guinea pig albumin will evoke a strong immune response, however, when injected into a different or more complex (higher) vertebrate, such as a rabbit or a monkey. Similar proteins produced by animals that are closely related in evolutionary terms have remarkably similar amino acid sequences. As animals diverge from each other in evolutionary development, the differences in the amino acid sequence of analogous proteins increase.

Genetics. It has recently been shown that the ability to recognize an antigen and the strength of the immune response produced may be under strict genetic control. Some strains of mice injected with synthetic polypeptides are capable of producing a vigorous immune response. Other mice, with closely related but nonidentical genetic backgrounds, may be poor responders or nonresponders.

• • •

In summary, many factors influence the immunogenicity of an antigen: chemical nature, size, molecular complexity, conformation, accessibility, foreignness, and genetics.

ANTIBODIES

The proteins that combine specifically with antigens are termed *antibodies*. Antibodies are produced by a subset of lymphocytes, called *B lymphocytes,* and by their progeny, *plasma cells*. B lymphocytes, through their production of antibodies, are responsible for the phenomenon of humoral immunity. Proteins with antibody activity are also called *immunoglobulins*. Immunoglobulins are an extremely heterogeneous group of molecules that constitute approximately 20% of the plasma proteins. Immunoglobulins are heterogeneous in their antigen specificity, amino acid sequence, migration within an electrical field, and functions. There may be as many as 10,000 different molecules circulating in the human body that can be classified as immunoglobulins.

Structure

The discovery that electrophoretically homogeneous proteins found in the serum of patients with multiple myeloma were structurally homogeneous and very closely related to normal immunoglobulin was an important advance in the study of immunoglobulin structure. Such myeloma proteins could be isolated in large quantities and chemically characterized. These studies produced an understanding of the precise structure of the antibody molecule.

H and L chains. Antibodies are glycoproteins composed of 82% to 96% polypeptide and 4% to 18% carbohydrate. All immunoglobulin molecules have a common structure of four polypeptide chains. Two identical large, or *heavy,* chains (H chains) and two identical small, or *light,* chains (L chains) are held together by noncovalent forces and covalent interchain disulfide bonds (Fig. 9-2).

The carbohydrate portion of the immunoglobulin molecule is covalently bonded to amino acids in the polypeptide chains. The carbohydrates are usually found bound to the C-terminal half (Fc) of the molecule. Their function is poorly understood. They may be involved in transporting the molecule or protecting it from metabolic degradation.

Fab and Fc fragments. Enzymatic digestion of immunoglobulin molecules has provided further evidence of their structure (see Fig. 9-2). Digestion with papain splits the molecule on the N-terminal side of the disulfide bond, yielding three fragments of approximately equal size. Two of these fragments are identical and retain the antigen-binding capacity associated with an intact immunoglobulin molecule. The fragments with the antibody-combining site *(Fab fragments)* are each composed of an entire light chain

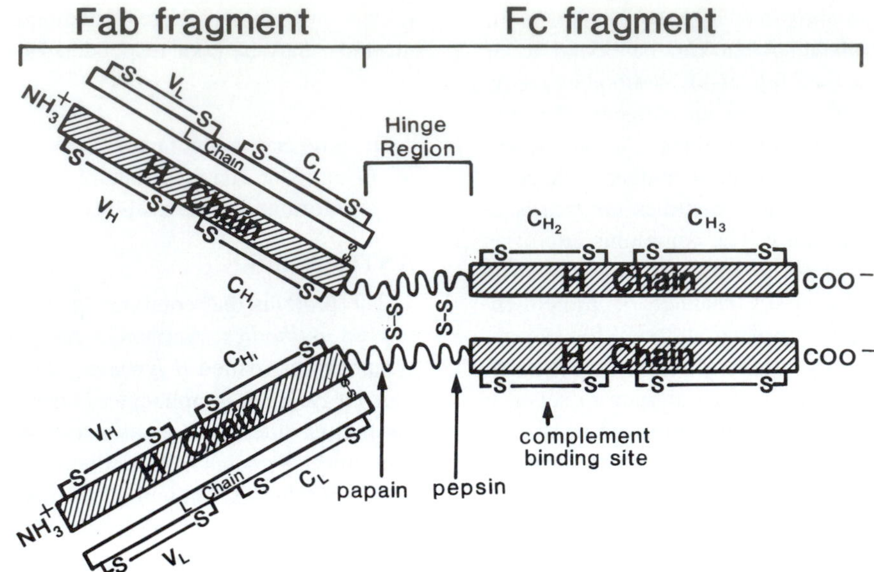

Fig. 9-2 Diagram of IgG molecule (immunoglobulin monomer). *H,* Heavy chain; *L,* light chain; *V,* variable region; *C,* constant region; *S-S,* disulfide bonds. *Arrows,* Papain and pepsin cleavage sites. NH_3^+ indicates N-terminus, and COO^- indicates C-terminus of immunoglobulin.

and a portion of the heavy chain. The third fragment has no antigen-binding activity and is crystallizable *(Fc fragment).* The Fc fragment retains the other biological activities associated with immunoglobulin molecules: interaction with the complement system and binding to tissue. The Fc fragment is composed of the C-terminal half of the heavy chain.

Digestion with pepsin cleaves the antibody molecule on the C-terminal side of the disulfide bond. This digestion results in the Fab'_2 fragment composed of the two Fab fragments linked by the disulfide bond. The remainder of the molecule undergoes extensive degradation.

V and C regions. Each polypeptide chain is composed of *domains,* or peptide sequences of uniform size (100 to 110 amino acid residues), that contain intrachain disulfide bonds. The domain of the N-terminal or antibody-combining site is more variable in its amino acid sequence than the rest of the polypeptide chain and is called the *variable region* (V region). The remainder of the polypeptide chain is composed of domains that are similar in immunoglobulin molecules of the same and other species. These domains are termed *constant regions* (C regions). Light chains are composed of one variable and one constant region (V_L and C_L). Heavy chains are composed of one variable and three or four constant regions (V_H and $C_{H1\text{-}1\text{-}4}$) (see Fig. 9-2).

The specific amino acid sequences of the variable regions of the light and heavy chains of an antibody molecule confer a unique character on the antibody and are termed the *idiotype* of the molecule. The idiotype is determined by the antigenic determinant to which the antibody is directed. The character of the idiotype permits the complementary fit of the antigenic determinant to the antibody-combining site.

Light-chain types. There are two types of light chains found in immunoglobulin molecules, kappa (κ) and lambda (λ). Kappa and lambda light chains differ structurally in the amino acid sequence of their constant regions. A given antibody molecule always has two identical kappa light chains or two identical lambda light chains. An antibody molecule can never have both a kappa and a lambda light chain together. More immunoglobulin molecules have kappa than lambda light chains. In human serum the ratio of kappa to lambda antibody molecules is approximately 2:1.

Heavy-chain types. Five types of heavy chains are distinguished in humans, based on structural differences in the constant regions of the chains. These structural differences permit functional differences. The heavy-chain types are designated gamma (γ), alpha (α), mu (μ), delta (δ), and epsilon (ε). The heavy-chain types vary in molecular weight. The gamma, alpha, and delta heavy chains are composed of three constant regions. The mu and epsilon heavy chains have four constant regions. The heavy-chain type determines the class of immunoglobulin. In humans there are five immunoglobulin classes, corresponding to the five heavy-chain types: immunoglobulin G (IgG), immunoglobulin A (IgA), immunoglobulin M (IgM), immunoglobulin D (IgD), and immunoglobulin E (IgE) (Table 9-1).

In addition, some immunoglobulin classes have subclasses based on additional amino acid differences in their constant regions. IgG has four subclasses, and IgA and IgM have two subclasses each. The biological properties and concentrations of the subclasses may differ.

Table 9-1 Properties of human immunoglobulin classes

Properties	IgG	IgA	IgM	IgD	IgE
Heavy chain	γ	α	μ	δ	ϵ
Subclasses	1-4	1 and 2	1 and 2	None	None
Light chain	κ and λ	κ and λ	κ and λ	κ and λ	κ and λ
Form	Monomer	Monomer and dimer	Pentamer (some monomer may circulate)	Monomer	Monomer
Formula	$\gamma_2\kappa_2$ or $\gamma_2\lambda_2$	$\alpha_2\kappa_2$ or $\alpha_2\lambda_2$	$\mu_{10}\kappa_{10}$ or $\mu_{10}\lambda_{10}$	$\delta_2\kappa_2$ or $\delta_2\lambda_2$	$\epsilon_2\kappa_2$ or $\epsilon_2\lambda_2$
J chain	No	On dimer	On pentamer	No	No
Molecular weight (approximate)	150,000	Monomer 160,000 Dimer 400,000	900,000	180,000	190,000
Complement fixation (classical pathway)	$G_3>G_1>G_2$	No	M_1 and M_2	No	No
Crosses placenta	Yes	No	No	No	No
Concentration in serum	8-16 mg/mL	1.4-3.5 mg/mL	0.5-2 mg/mL	0-0.14 mg/mL	≤300 ng/mL

Immunoglobulin G

IgG molecules are monomers of the basic immunoglobulin subunit. They are composed of two kappa or two lambda light chains and two gamma heavy chains. IgG molecules may therefore be represented as $\gamma_2\lambda_2$ or $\gamma_2\kappa_2$. Approximately 75% of serum immunoglobulin is IgG. The frequency of IgG subclasses varies as follows: IgG_1, 60% to 70%; IgG_2, 14% to 20%; IgG_3, 4% to 8%; and IgG_4, 2% to 6%. There is evidence that antibodies produced to certain antigens may be restricted in their subclasses. Polysaccharide antigens tend to produce IgG_2 and IgG_4 antibodies. Antiviral and antinucleoprotein antibodies are found primarily in the IgG_1 and IgG_3 subclasses.

IgG molecules cross the placenta and are responsible for the immunological defense of the newborn, IgG molecules also fix to the surface of effector cells, which are then capable of antibody-mediated cytotoxic reactions important in protecting the host. IgG molecules bind or "fix" complement, a complex of serum proteins that assists in the lysis of elimination of foreign particles. Complement proteins are bound to the IgG molecule in the midpoint of the heavy chain, near the disulfide bond in the CH_2 domain. This area is called the *hinge region* because the two Fab fragments of the molecule open to expose the complement-binding site (see Fig. 9-2). Molecules in the IgG subclasses differ in their ability to fix complement proteins. IgG_3 is most active, followed by IgG_1, IgG_2, and IgG_4.

Immunoglobulin M

IgM constitutes approximately 10% of serum immunoglobulins and exists primarily as a pentamer of the basic immunoglobulin structure. The five immunoglobulin monomers are held in a circle by disulfide bonds between H chains of the subunits. In addition, the IgM molecule contains a polypeptide chain called the *joining,* or *J,* chain that may help in maintaining its structure. The J chain is a small glycoprotein (MW, 15,000 daltons) that is covalently bonded to the H chains of the molecule.

IgM is the predominant immunoglobulin in the initial immune response to an antigen. It is the most efficient immunoglobulin in fixing complement. This efficiency is a result of its pentameric structure. The polymeric structure of IgM would also be expected to increase its valency. The presence of 10 Fab units conveys on the IgM molecule a theoretical valency of 10. This means that an IgM molecule should be able to bind 10 antigen molecules simultaneously. Although this value has been computed in some experimental systems, it is not normally observed. Steric hindrance may be responsible for this disparity.

Immunoglobulin A

IgA constitutes approximately 15% of the serum immunoglobulin, but it is the predominant immunoglobulin in body secretions such as saliva, tears, sweat, human milk, and colostrum. In serum, IgA exists in both monomeric and polymeric forms. Polymeric serum IgA possesses the J chain. Secretory IgA exists as a dimer of the basic immunoglobulin unit combined with a J chain and an additional polypeptide chain called the *secretory piece*. The secretory piece is bound to dimeric secretory IgA by strong noncovalent linkages. It is important in secretory transport of the molecule and in its protection from proteolytic digestion in the gut. Secretory IgA provides the first line of defense against local infections and is important in the processing of food antigens in the gut.

Immunoglobulin D

IgD is a monomer of the basic immunoglobulin unit and is present in human serum in trace amounts. In addition, it is expressed on lymphocyte cell surface membranes. The main function of IgD is unknown. It may be involved in lymphocyte differentiation.

Immunoglobulin E

IgE is also a monomeric immunoglobulin. It is present in human serum in very low concentrations. IgE, which binds to cells by means of its Fc portion, is responsible for the physiological manifestations of allergy.

ANTIGEN-ANTIBODY REACTIONS

Antigen-antibody reactions were first recognized by bacteriologists who also surmised that such reactions exhibited specificity. Bacteriologists noted that the serum of patients recovering from infectious diseases could agglutinate the organism responsible for their disease but not unrelated organisms. Serum from persons not exposed to the disease or from patients before they contracted the disease could not agglutinate the same organisms. From such evidence, scientists proposed the existence of antibody molecules, the specificity of their interactions with antigens, and the importance of such interactions in host defense.

The following sections consider the forces involved in antigen-antibody binding, the specifity of the reaction, and the mechanism of the reaction.

Binding forces

The strength of the binding of an antigen to an antibody depends on the complementarity of fit of the antigenic determinant to the idiotype of the antibody. It also depends on the sum of weak, noncovalent, intermolecular forces. Such weak, short-range forces can operate between antigen and antibody if their closeness of fit brings them into proximity with one another. The attraction between antigen and antibody molecules involves electrostatic attraction, hydrogen bonding, van der Waals forces, and hydrophobic interactions (see Table 51-2).

In solution at physiological pH, charged polar groups on the amino acid residues of proteins can be strongly attracted to one another. These electrostatic forces are the strongest and most important contributors to noncovalent attraction between antigen and antibody.

Hydrogen bonding between the amino and carboxy groups of peptide bonds also contributes to the attractive forces. Hydrogen bonds are weaker than electrostatic forces, but their numbers make them an important factor.

Van der Waals forces are the weakest forces involved. They can function only within a very small radius because of their low power. The close approximation of antigenic determinant to idiotype induces charge fluctuations within the atoms of the molecules. At very close distances the nucleus of one atom can be attracted to the external orbit electrons of a second atom. These van der Waals forces contribute to binding strength.

The final component of the attractive forces involves hydrophobic bonding between apolar groups in solution. Hydrophobic bonding functions by the exclusion of polar water molecules to bring hydrophobic molecules together. Such interactions also serve to attract polar water molecules to hydrophilic amino acid residues on protein molecules. Antibody molecules have increased numbers of hydrophobic amino acid residues such as alanine, leucine, tyrosine, tryptophan, and methionine in their antibody-combining sites, where they enhance bonding to hydrophobic residues in antigenic determinants.

Antibody affinity

The strength of the binding of a single antigenic determinant to an antibody is a function of the closeness of fit and is called *antibody affinity*. Antibody affinity is an expression of the attraction between molecules of antibody and antigen. It is a function of the sum of the short range, noncovalent, intermolecular forces.

Binding of antigen to antibody is a reversible reaction. The equilibrium of the reaction favors antigen-antibody association if the fit between molecules is good and the forces binding the molecules together are relatively strong and stable. The strength of the association between antigen and antibody is represented by the association constant, which may be derived as follows:

$$Ag + Ab \underset{k_2}{\overset{k_1}{\rightleftharpoons}} Ag \cdot Ab \qquad \textit{Eq. 9-1}$$

$$\frac{[Ag \cdot Ab]}{[Ag][Ab]} = \frac{k_1}{k_2} = K_a \text{ (Association constant)} \qquad \textit{Eq. 9-2}$$

To study these reactions, one places solutions of a small antigen, or hapten, on either side of a semipermeable membrane. As the hapten diffuses across the membrane, the reaction proceeds to the right of Equation 9-1; that is, hapten and antibody associate to form complexes. Eventually, equilibrium is reached. At equilibrium, a constant number of antibody molecules react with hapten to form complexes, and a constant number of antibody molecules react with hapten to form complexes, and a constant number of complexes dissociate to form free hapten and free antibody. One rate constant, k_1, expresses the tendency of the reaction to move toward the right, or the tendency for association. The other rate constant, k_2, expresses the tendency of the reaction to move to the left, or the tendency for dissociation. These reaction-rate constants differ for each antigen and antibody pair. The concentrations of antigen, antibody, and complex at equilibrium are described by Equation 9-2. The equilibrium constant, K_a, expresses the tendency of the reaction to favor association between antigen and antibody.

Heterogeneity of immune response

Analysis of the binding of a simple hapten containing a single antigenic determinant shows variation in binding strength of antibody molecules. Immunization with a single antigenic determinant produces a variety of antibodies with different antibody-combining sites and with a range of antibody affinities. This is termed the *heterogeneity* of the immune response. The immune system responds to three-dimensional antigenic determinants that present more than one area of contact (Fig. 9-3). Antibodies can bind to

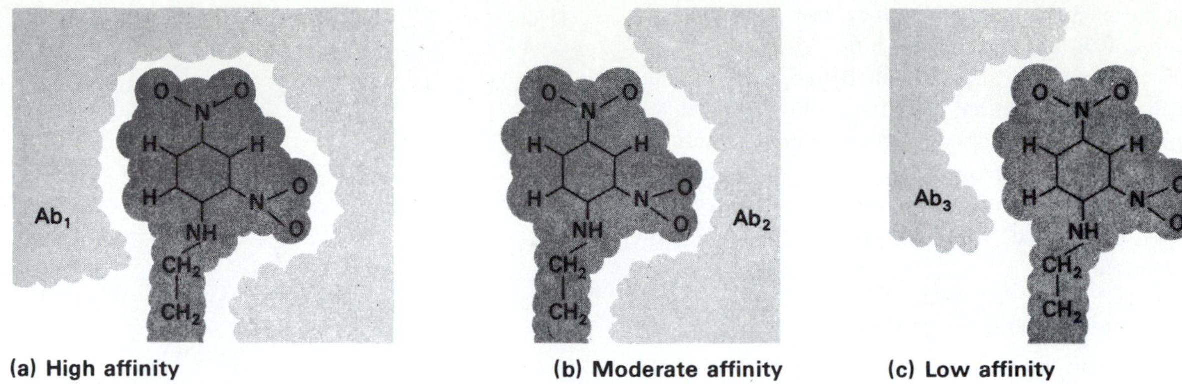

Fig. 9-3 Binding of antibodies present in same antiserum with different affinities to same hapten (dinitrobenzene linked to amino group of lysine). **(a)** Antibody$_1$ fits with nearly whole hapten and is thus of high affinity. **(b)** Antibody$_2$ fits with less of molecule and not so closely and has a moderate binding affinity, whereas **(c)** low affinity antibody$_3$ is complementary in shape to so little of hapten surface that its binding energy is very little above that occurring between completely unrelated proteins. Only a portion of antibody combining site is shown. *(From Roitt, I: Essential immunology, ed 4, Oxford, England, 1980, Blackwell Scientific Publications.)*

different configurations of a single antigenic determinant. Thus different antibody molecules can be produced against the varied configurations of a single determinant. Such different configurations present differing distributions of charged and hydrophobic residues, resulting in varying closeness of fit with an antibody.

Antibody avidity

In natural situations in vivo a variety of antibody molecules are generated in response to a large number of multivalent antigenic stimuli. Thus there are two areas of complexity: (1) multiple antibodies generated to different conformations on a single antigenic determinant and (2) multiple antigenic determinants on a single natural antigen. Both of these generate the *diversity* of the immune response. The measure of binding strength of antibodies to multiple antigenic determinants on natural antigens is termed *avidity*. Avidity is a measure of the stability of the antigen-antibody complex. It is partially dependent on the affinity of each antibody for its complementary antigenic determinant. There is an enhanced effect, however, with multivalent antigens. The sum of the binding is greater than its individual parts. This is caused by the reversible nature of antigen-antibody bonds. It is also caused by the divalent nature of IgG molecules or the multivalent nature of IgM molecules. If the single bond between antigen and antibody dissociates, the antigen escapes. If an antigen has two antigenic determinants, each of which is bound by antibody, the antigen is kept in place until the broken bond re-forms (Fig. 9-4). Thus avidity is a measure of the stability of the multivalent antigen–multivalent antibody complex.

Cross-reactivity

Cross-reactivity of antigen and antibody is a by-product of the heterogeneity of the immune response. As stated previously, immunization with a simple hapten produces a variety of antibodies of differing antibody affinities. Some of these antibodies will combine with chemically related and structurally similar haptens. Cross-reactivity usually involves low-affinity antibodies that exhibit a less precise fit to antigens. The reactivity of an antibody to a different antigen may also indicate that the two antigens in question share a previously unknown but common antigenic determinant. Thus cross-reactivity may result from similar or identical antigenic determinants in different antigens. Such cross-reactivity is often observed with antibodies produced to such drugs as penicillin. The reactivity of an antibody with penicillin may be very high, but metabolic derivatives

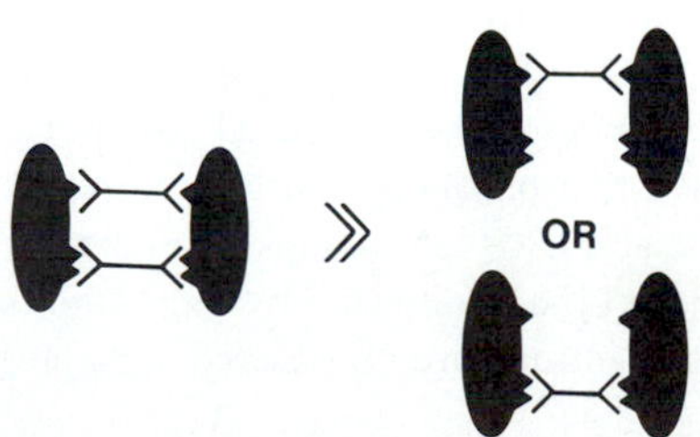

Fig. 9-4 Multivalent bonding of antigen-antibody increases bonding strength. A single bond created by divalent antibody molecules between a single antigenic determinant on two adjacent antigens is much weaker than binding created by two divalent antibodies bound simultaneously to two unique antigenic determinants on two adjacent antigens. Strength and complexity of this multivalent bonding are described by the term *avidity*.

containing the basic drug structure may also react with an antibody produced to the complete drug.

In cases of prolonged antigenic challenge, such as that occurring in natural infection, animals exhibit a natural selection for high-affinity antibodies. As the heterogeneity of the antibody response narrows, the specificity of antigen-antibody reactions increases. This adaptation of the immune response promotes effective protection of the host against infection.

Genetic basis of antibody diversity

The heterogeneity of the immune response or the variety of antibodies produced to a single antigen is known to be genetically determined. The variable (V) regions of light and heavy chains of the antibody molecule encode for antibody specificity. The amino acid sequence of the V region is genetically controlled. Recent evidence suggests that there are positions in the amino acid sequence of the V region that have an even more increased likelihood of amino acid variation. These *hypervariable regions,* scattered throughout the amino acid sequence of the V region, are brought into proximity to each other by the natural folding of the antibody molecule. The natural folding, or tertiary structure, of the antibody molecule is also genetically determined and is also dependent on the amino acid sequence. The sequence dictates possible attraction between polar amino acid residues, as well as the possibility of intrachain disulfide bonds. The approximation of hypervariable regions by the folding of the antibody molecule results in the formation of the antibody-combining site.

ANTIGEN-ANTIBODY PRECIPITATION REACTIONS

The primary reaction of antigen with antibody is usually detected by secondary manifestations of the reation. The nature of the secondary manifestations depends on experimental conditions, the class of antibody involved in the reaction, the number of antigenic determinants on the antigen, and the size and solubility of the antigen. The reaction of antibody with soluble molecules possessing multiple antigenic determinants that permit cross-linking is detected by *precipitation* of the complex out of solution. The term *flocculation* may be used to describe a precipitation reaction that produces a large, loosely bound precipitate. The reaction of antibodies with large, particulate, multivalent antigens is detected by *agglutination* of the antigen. These reactions are considered separately.

Precipitation curve

When a known quantity of antibody is present in solution in a series of tubes to which increasing amounts of antigen are added, precipitation occurs in some of the test tubes. When the amount of precipitate is measured and correlated with the amount of antigen present, one obtains a curve similar to that shown in Fig. 9-5.

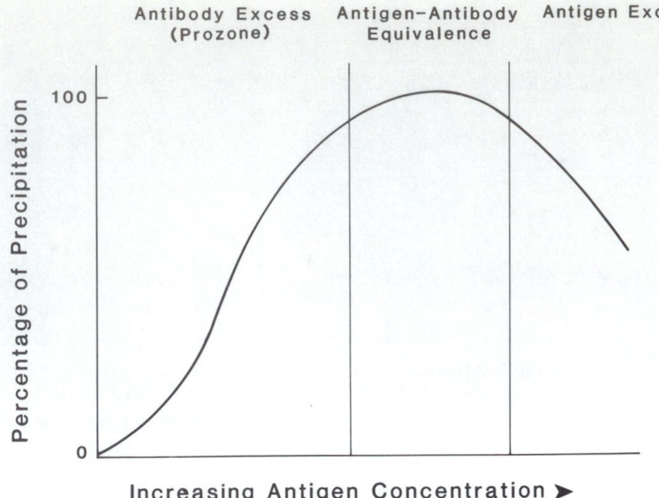

Fig. 9-5 Quantitative precipitin curve in which amount of antibody antigen complex that precipitates is plotted as function of antigen concentration.

In the first phase of the reaction, called the *antibody-excess phase,* no free antigen (an antigen without attached antibody) can be detected in the fluid and essentially no precipitate can be found. Free (unattached) antibody can be detected, however. As increasing amounts of antigen are added, the amount of precipitate detectable at the bottom of the test tubes increases until a point of maximum precipitation is reached. At this point, no free antigen or free antibody can be detected in the fluid. This is called the *zone of equivalence*. As the amount of antigen added continues to increase, the amount of precipitate detected-diminishes. Examination of the fluid phase of the reaction at this time shows no free antibody but increasing amounts of free antigen. This area of the curve is called the *antigen-excess* phase.

Lattice theory

Antigen-antibody complexes precipitate out of solution because of the multivalent nature of both molecules. The reaction of antigens possessing multiple antigenic determinants and antibodies with two (as in IgG) or more (as in IgM) antibody-combining sites produces a lattice of interlocking molecules. Antibody molecules can cross-link antigenic sites on the same or different molecules of antigen. As the size and complexity of the lattice increase, the lattice becomes insoluble and precipitates out of solution (Fig. 9-6).

In the antibody-excess zone, a single molecule of antigen binds to each antibody molecule. The excess of antibody ensures that each molecule of antigen can encounter a free antibody molecule. The absence of cross-linking produces small soluble complexes.

As the antigen concentration increases and the zone of equivalence is entered, complexes of increasing size with

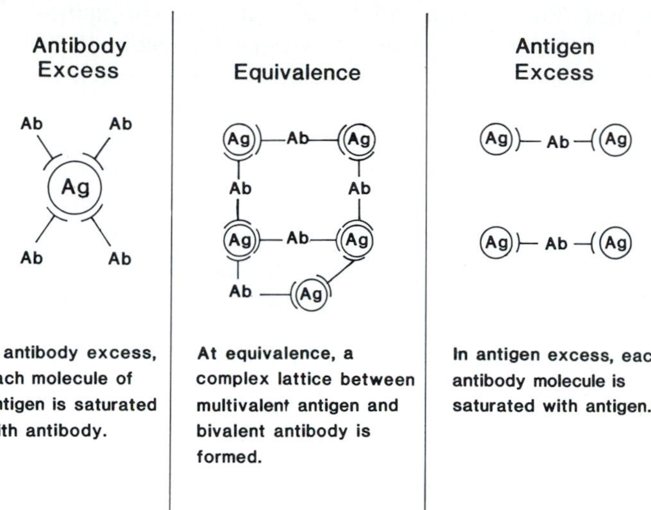
Antibody Excess
Equivalence
Antigen Excess
Ab
Ag
Ab
In antibody excess, each molecule of antigen is saturated with antibody.
At equivalence, a complex lattice between multivalent antigen and bivalent antibody is formed.
In antigen excess, each antibody molecule is saturated with antigen.

Fig. 9-6 Representation of sizes of molecular complexes formed at varying ratios of antigen and antibody.

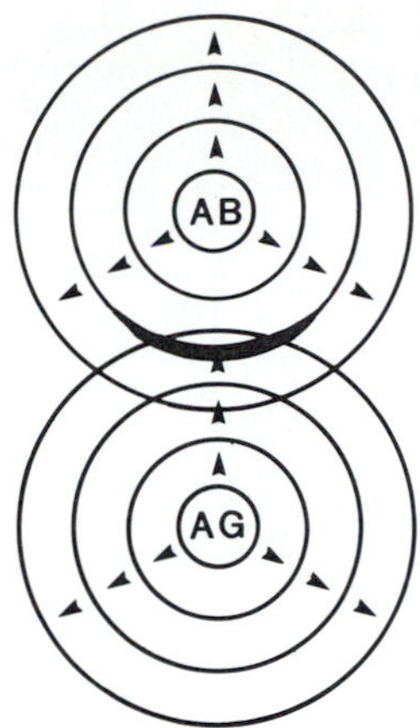
AB
AG

Fig. 9-7 Depiction of protein gradients in radial immunodiffusion. Concentric circles represent decreasing protein concentrations. Both antigen *(AG)* and antibody *(AB)* diffuse radially from application wells. Precipitation, *heavy black arc,* occurs at point of antigen-antibody equivalence. Precipitin line is closer to well of lower concentration and concave toward reagent of higher molecular weight.

increasing levels of cross-linking are formed. Such large, complex lattices precipitate out of solution.

As the antigen concentration continues to increase, the zone of antigen excess is reached. In this portion of the curve, smaller complexes are again seen. The size of the lattice decreases because there is sufficient antigen to permit binding of a free antigen molecule to each antibody-combining site. At high concentrations of antigen, lack of precipitation can result in false-negative results. Obviously, detection of antigen by antibody precipitation requires optimum concentration of both reactants. Formation of lattices best suited for precipitation occurs at an equal equivalent concentration of antigen and antibody or at a slight antigen excess.

Other factors affecting precipitation

The precipitation of antigen-antibody complexes out of solution can be affected by factors other than the ratio of antigen concentration to antibody concentration. Different antibody molecules can precipitate the same antigen to varying degrees. The efficiency of the antibody depends on its affinity and specificity. The charge and shape of the antigen-antibody complex are also important. Highly charged complexes are difficult to precipitate. More energy is required to displace polar water molecules from the complexes and to bring the charged molecules into proximity. The best precipitates involve protein antigens with molecular weights from 40,000 to 160,000 daltons. Proteins in this range are easily cross-linked by multivalent antibody molecules. Polysaccharide antigens, denatured proteins, and viruses produce broader precipitation curves. Their large size sterically hinders precipitation. Precipitation can also be affected by temperature, pH, and ionic concentration. Such factors influence antigen-antibody interactions on a molecular level.

Precipitation reactions in gel

Precipitation reactions are frequently carried out in a gel-support matrix composed of agar or the more purified polysaccharide, agarose. The agar prevents convective mixing of antigen and antibody and thereby ensures establishment of concentration gradients of the two reactants. Precipitation in agar is only a moderately sensitive technique when compared with newer advances, such as radioimmunoassay, but it is widely employed because of its ease and versatility. In addition, precipitation reactions in gels can be modified to permit the study of antigenic relationships among different compounds. The following section discusses two gel-precipitation reactions, double immunodiffusion and radial immunodiffusion.

Double immunodiffusion. In double-immunodiffusion reactions, or *Ouchterlony tests,* agar or agarose is poured onto a solid support, such as a glass slide or petri dish. Wells are then cut into the agar. Antigen and antibody solutions are placed into separate wells. The solutions then diffuse toward one another in the gel in a radial fashion during room-temperature incubation (Fig. 9-7). With diffusion into the agar, the solutions establish concentration gradients that diminish with distance from the well. At the point of antigen-antibody equivalence at the interface of the diffusing fronts, a precipitation line is formed (see Fig. 9-7). The positioning and shape of the line are dictated by the concentration of the reactants and the size of the molecules. The line will be closer to the well with the reactant of lower concentration because the distance traveled is directly proportional to concentration. The rate of diffusion

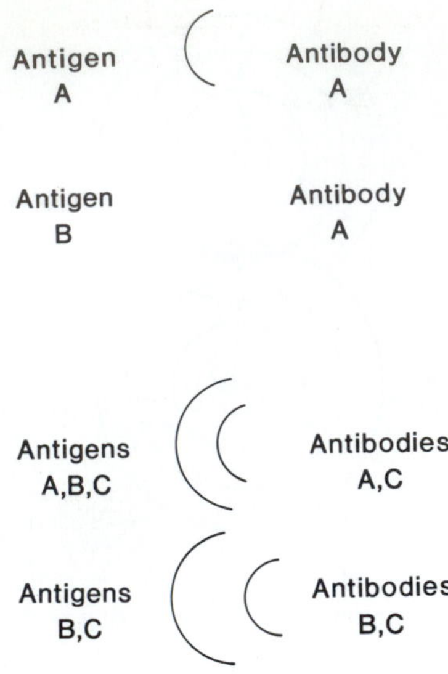

Fig. 9-8 Precipitation occurs at equivalence point of antigen with corresponding antibody. Multiple precipitation lines are seen with multiple antigens and corresponding multiple antibodies.

is also inversely proportional to molecular size. High-molecular-weight compounds, such as IgM, diffuse more slowly than lower-molecular-weight substances, such as IgG. The precipitation line that is formed at the interface of the two-concentration gradients will be concave to the higher-molecular-weight compound, whose diffusion rate is slower. Because precipitation occurs at antigen-antibody equivalence to slight antigen excess, an inappropriate ratio of antigen to antibody will result in failure to form a precipitate.

The presence of different antigenic determinants on the same or different molecules can be detected by the production of more than one precipitation line if the antiserum used contains antibodies against the multiple antigens (Fig. 9-8). Two antigens in the antigen solution will create two independent concentration gradients as they diffuse into the agar. Precipitation will occur at the point of equivalence of each antigen with its corresponding antibody. In this way the components of an antigen mixture can be studied.

Ouchterlony testing also permits analysis of the relationship between two antigenic mixtures. The antibody solution is placed in a center well, which is surrounded by several wells into which antigen solutions are placed. Unrelated antigens that have corresponding antibodies in the antibody solution will form separate precipitation lines corresponding to their distinct components. Radial diffusion from the adjacent wells causes superimposition of the two gradients, but because the concentration gradients of the antigens are independent, two separate and distinct lines are formed (Fig. 9-9). The double spurs formed at the two antigen interfaces are the hallmark of the *reaction of nonidentity*. The two antigens or antigen mixtures have nothing in common in relationship to the antibody solution used.

Two identical antigen solutions that are tested with their common antibody result in a *reaction of identity* (see Fig. 9-9). Precipitation lines form at their separate but adjacent points of equivalence against antibody. Because diffusion through the gel is radial, an area of shared and common antigen concentration between the two adjacent wells is formed. For this reason, there is fusion of the two separate precipitation lines. This fusion or reaction of identity indicates that the antigen solutions are identical with respect to the antibody employed or that the antigen solutions have one anitgenic determinant in common. It does not imply molecular identity. If antigen concentrations in the adjacent wells differ, a common fused precipitation line is still formed at the point of average concentration of the two antigen fronts.

A *reaction of partial identity* is seen when adjacent wells share some but not all the antigens detected by the

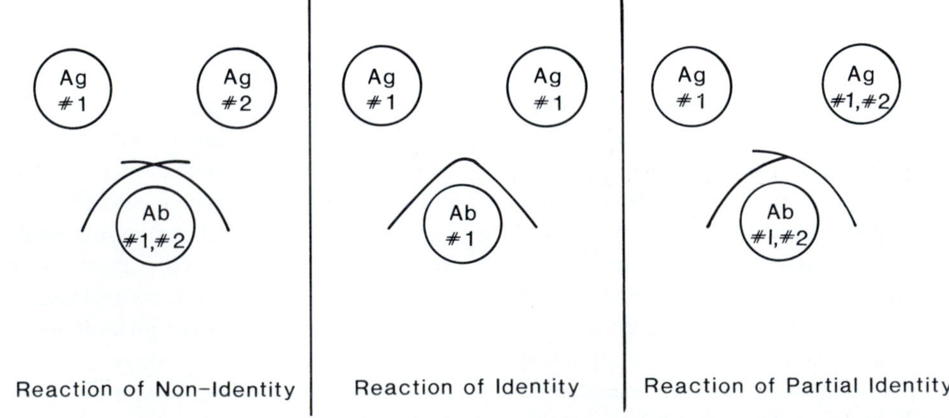

Fig. 9-9 Double immunodiffusion patterns. *Ag*, Antigen; *Ab*, antibody; *heavy line*, precipitin line; *circles*, application wells.

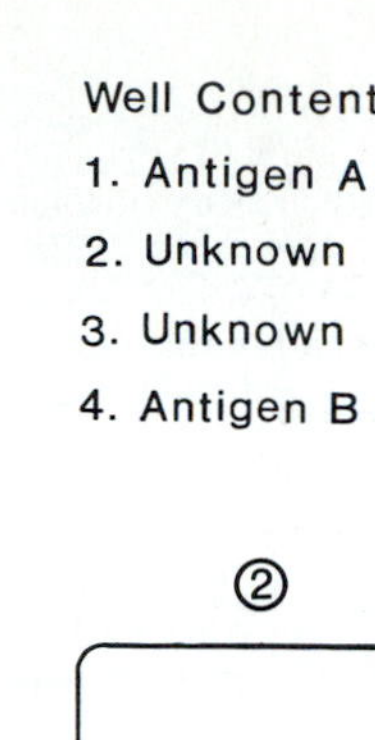

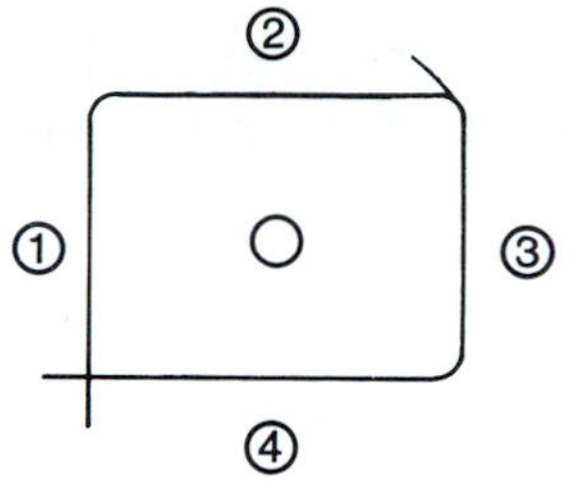

Conclusions:
2 = antigen A
3 = antigen A and antigen B

Well Content
1. Antigen A
2. Unknown
3. Antigen B + unknown
4. Unknown

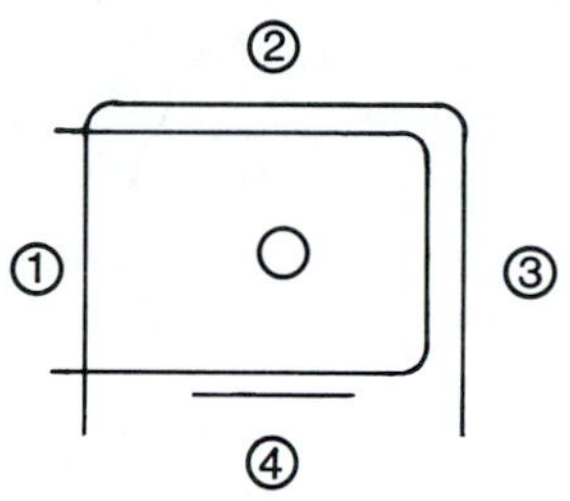

Conclusions:
#2 = antigen A + antigen B
#3 = antigen A + antigen B
#4 = antigen B + antigen C

Well Content
1. Unknown
2. Unknown
3. Antigen A + unknown
4. Antigen B

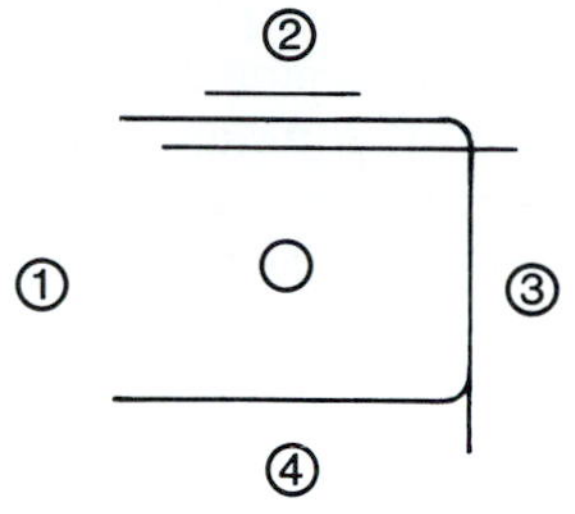

Conclusions:
1 = contains no antigens recognized by antisera or reagent concentrations are not correct
Antigen A and B are the same
#2 = contains antigen A plus antigens C and D
#3 = contains A and E

Fig. 9-10 Interpretations of double-immunodiffusion patterns. Center well contains an antiserum to several possible antigens. Surrounding wells contain test antigens. *Heavy lines,* precipitin reaction.

antibody solution. A reaction of partial identity is a reaction of nonidentity superimposed on a reaction of identity (see Fig. 9-9). A common fused line is formed by reaction of the shared antigen. A second identical equivalence point between the novel antibody and a common antigen creates a superimposed precipitation line that extends into the common area between the adjacent wells. The second antigen creates a concentration gradient that is not contributed to by the adjacent well. The spur indicates that the second antigen solution lacks an antigenic determinant present in the first antigen solution that is recognized by the antibody solution. The spur always points to the well of the antigen that is monospecific with respect to the antibody.

Antigen A diluted 1:2 | Antigen A diluted 1:4 | Antigen A diluted 1:8

Agar with Anti-A

Fig. 9-11 Radial immunodiffusion patterns. Band of precipitation, *stippled area,* extends as a disk from center of each circular well. Area of precipitation is proportional to concentration.

Two double-immunodiffusion patterns that show a complex relationship of antigen and antibody solutions are presented and discussed in Fig. 9-10.

Radial immunodiffusion. Radial immunodiffusion, the *Mancini technique,* is a precipitation reaction carried out by application of antigen solution to a gel that has been impregnated with a monospecific antibody solution. It is

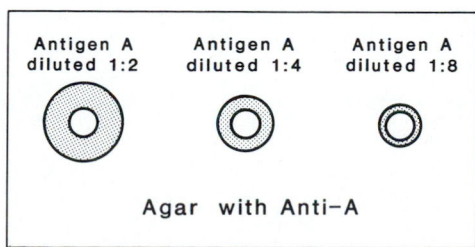

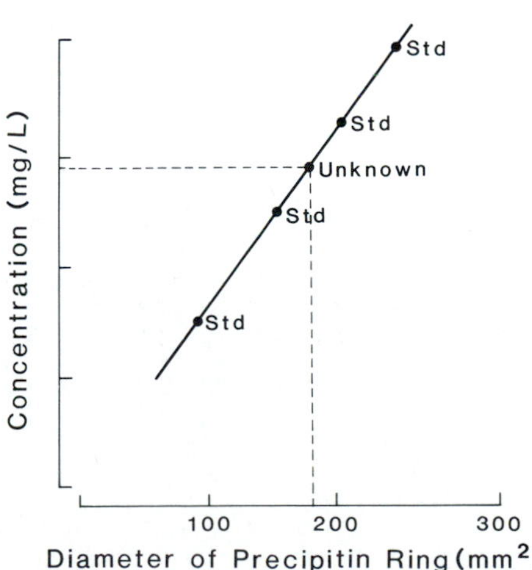

Fig. 9-12 Graph of concentration of antigen expressed as mg/L versus square of diameter of precipitin ring. *Std,* Standard.

an adaptation of a gel-precipitation reaction that permits quantitation of antigen. The antibody-gel solution is applied to a solid support. Wells are then cut into the agar, and dilutions of antigen are placed in the wells. The antigen diffuses out radially into the agar. This diffusion produces a concentration gradient that is inversely proportional to the distance from the well. Antibody concentration within the gel is constant. At the point where antigen and antibody concentrations are equivalent, precipitation occurs. Because diffusion from the well is radial, the precipitation appears as a ring around the well. The square of the diameter of the ring (mm^2) is directly proportional to antigen concentration. The precipitation reaction is not a static but a dynamic one. The precipitation ring first forms close to the well at the initial point of antigen-antibody equivalence. As antigen continues to diffuse from the well, antigen excess causes conversion of the precipitate to soluble complexes that resolubilize and continue to diffuse outward. A new ring is formed at a new point of antigen-antibody equivalence. The final distance traveled at end-point equilibrium is a function of antigen concentration. The thickness of the ring is a function of the final concentration of antigen-antibody complexes at the equivalence point (Fig. 9-11).

With constant sample volume, temperature, pH, incubation time, and antibody concentration, an unknown antigen concentration can be determined. This is accomplished when one compares the square of the diameter of the precipitation ring of the unknown rings obtained by several dilutions of a standard antigen solution. When the concentrations of the diluted standard are plotted against ring area, the concentration of the unknown can be easily determined (Fig. 9-12).

BIBLIOGRAPHY

Antigens

Goodman, JW: Antigenic determinants and antibody combining sites. In Sela, M, editor: The antigens, vol 8, New York, 1975, Academic Press, Inc, pp 127-187.

Williams, CA, and Chase, MW: Methods in immunology and immunochemistry, vol 6, Preparation of antigens and antibodies, New York, 1967, Academic Press, Inc.

Antibodies

Natvig, JB, and Kunkel, HG: Human immunoglobulins: classes, subclasses, genetic variants, and idiotypes, Adv Immunol 16:1-59, 1973.

Nisonoff, A, Hooper, JE, and Spring, S: The antibody molecule, New York, 1975, Academic Press, Inc.

Antigen-antibody interactions

Barrett, JT: Textbook of immunology: an introduction to immunochemistry and immunobiology, ed 5, St Louis, 1988, The CV Mosby Co.

Boguslaski, RC, Maggio, ET, Nakamura, R, editors: Clinical immunochemistry: principles of methods and applications, Boston, 1984, Little, Brown, & Co.

Gill, TJ: Methods for detecting antibody, Immunochemistry 7:887-1000, 1970.

Rose, NR, and Friedman, H, editors: Manual for clinical immunology, ed 2, Washington, DC, 1980, American Society for Microbiology.

Weir, DM, editor: Handbook of experimental immunology, ed 3, Oxford, England, 1978, Blackwell Scientific Publications, Ltd.

Weir, DM, et al, editors: Immunochemistry, ed 4, Oxford, England, 1986, Blackwell Scientific Publications, Ltd.

Williams, CA, and Chase, MW: Methods in immunology and immunochemistry, vol 3, Reactions of antibodies with soluble antigens; vol 4, Agglutination, complement, neutralization and inhibition, New York, 1971, 1977, Academic Press, Inc.

CHAPTER 10

Immunochemical techniques

PAUL E. HURTUBISE
SUSAN BASSION
JACK GAULDIE
PETER HORSEWOOD

OBJECTIVES

- For each of the following techniques, state the principle of the immunoreaction, describe sample requirements and preparation, list common pitfalls, and explain the interpretation of results:
  - Immunoelectrophoresis
  - Counterimmunoelectrophoresis
  - Two-dimensional immunoelectrophoresis
  - Laurell rocket immunoelectrophoresis
  - Immunonephelometry
  - Western blot
- State the sample requirements and preparation and list common pitfalls for the Ouchterlony and radial immunodiffusion techniques.
- Define agglutination and differentiate between direct agglutination, indirect agglutination, and agglutination inhibition reactions.
- Describe the principles of solid-phase, "sandwich" assays, distinguishing between those that measure antigen and those that measure antibody.
- List the labels used for "sandwich" assays, comparing their sensitivity and their ability to be used in heterogeneous and homogeneous assays.

KEY TERMS

agglutination Clumping or aggregating together by specific antibody of particles, such as red blood cells or latex beads to which the specific antigenic determinant is attached.

agglutinin Specific antibody that causes agglutination.

anti-antibody An antibody with specificity for immunoglobulins reacting in test assay systems.

antibody absorption The process of removing or tying up undesired antibody in an antiserum reagent by allowing it to react with undesired antigens.

antibody reagent A high-titer, high-affinity IgG class of antibody prepared in animals.

antigen reagent A stabilized solution containing a known amount of an antigen that is used as a standard.

cold agglutinin An agglutinin that reacts better at temperatures less than body temperature; best reaction is usually at 4° C.

complement A group of serum proteins activated as a result of an antibody-antigen reaction. When the reaction is on the surface of a red blood cell, the activated complement can lyse the cell.

complement fixation A term applied to a set of assays in which complement is activated or "fixed" by a test reaction system.

Coombs' test A type of agglutination reaction. A direct Coombs' test measures the presence of antibody on cells; an indirect test measures its presence in serum.

counterimmunoelectrophoresis An assay in which antigen and antibody migrate toward each other under the influence of an electric field. The presence of antigen is observed by the formation of a precipitin line.

cryoglobulin Protein that precipitates at temperatures less than body temperature; precipitates maximally at 4° C.

fluoroimmunoassay Any immunoprocedure that uses a fluorescent molecule as the indicator label.

hemolysin Anti–sheep red blood cell antibody.

heterogeneous enzyme immunoassay Any technique that uses two phases, usually liquid and solid, to separate reacted from unreacted components.

immobilization The fixation of antigen or antibody onto a solid support such as a plastic tube or microtiter plate.

immunodiffusion Random, spreading movement of antibody or antigen or both in a support medium.

immunoelectrophoresis An immunoprecipitation technique in which antigens are separated from each other by migration in an electric field, followed by reaction with antibody by immunodiffusion.

indicator phase The portion of an immunochemical reaction that can be measured.

inhibition assay A term for those types of immunoassays in which an excess of antigens prevents or inhibits the completion of either the initial or indicator phase of the reaction.

monoclonal antibody A monospecific antibody that is produced by a single plasma cell or a single clone of plasma cells of a lymphocyte myeloma hybrid.

monospecific An antibody that will react with only one type of antigen molecule.

nephelometric assay Measurement of antigen or antibody by determination of the amount or rate of formation of antibody-antigen aggregate by the amount of light scattered.

nephelometric inhibition assay (NINIA) Measurement of haptens by inhibition of formation of antibody-antigen lattice.

nephelometry Measurement of light-scattering properties of large particles (such as antigen-antibody complexes) in solution.

Ouchterlony double diffusion A version of the original gel diffusion technique invented by Oudin in which antigen and antibody in separate wells are allowed to spread (diffuse) toward each other.

polyclonal antibody Heterogeneous antibodies with diverse affinities produced by a large number of plasma cells.

prozone phenomenon Apparently lower reactivity or nonreactivity of antibody solution at low dilutions.

radial immunodiffusion (Mancini technique) Measurement of antigen concentration by allowing antigen to spread (by diffusion) into agarose containing the desired monospecific antibody. The area of the immunoprecipitin ring is proportional to antigen concentration.

rocket (Laurell) immunoelectrophoresis Assay system in which the antigen, under the influence of an electric field, migrates into agarose-containing antibody, with a resultant immunoprecipitation reaction. The precipitin lines appear rocket shaped.

sandwich assay A term applied to a solid-phase immunoassay in which the first layer is immobilized antibody, the second is antigen, and the third is labeled antibody.

specificity Property of an antibody molecule that restricts its reactivity to a defined molecule or group of molecules.

titer Maximum dilution of a specific antibody that gives a measurable reaction with a specific antigen; usually expressed as the reciprocal of that dilution.

Chapter 9 describes the molecular nature of antigens and antibodies, as well as the general characteristics of the antigen-antibody reaction. This chapter deals with many techniques that use the antigen-antibody reaction as the basis to detect, characterize, or quantitate constituents in blood and other body fluids submitted to the laboratory for analysis. These constituents can range from small-molecular-weight drugs and their metabolites to large-molecular-weight proteins, such as IgM and alpha$_2$-macroglobulin. Most frequently, the patient's sample contains the antigen (analyte), and antibody is added as the reagent to detect or measure the antigen. In contrast, in cases of infectious disease, serological determinations, and autoimmune antibody testing, the patient's sample is the source of antibody, and it is the antibody measurement that is clinically important. For these determinations, antigen of known composition is used. The antigen may be soluble or tissue based. However, this latter form of testing is usually performed in the immunology section of the laboratory. Because this chapter is directed primarily at those techniques used in the clinical chemistry laboratory, it will concentrate on the procedures that detect antigen in the patient's sample.

REAGENTS

Antibody as reagent

Reagent antibodies are usually prepared in animals, such as rabbits or goats, by the repeated exposure of the animal to foreign substances. A group of cells are stimulated to respond by producing antibodies. Some of the groups of atoms of the immunizing material are the major determinants of the antigen molecule and cause the production of the largest amount of antibodies; minor determinants, however, also produce antibodies. Since many different antibodies are present that are attributable to the expansion of several classes of antibody-producing cells, the antiserum thus produced is a *polyclonal* reagent antiserum. For example, an antibody against the protein human serum albumin (anti-HSA) is a reagent that has multiple antibodies to antigenic determinants or specific molecular configurations that are characteristic and specific for the surface of HSA.

It is important to demonstrate that this anti-HSA will not react with other serum proteins, such as IgG and transferrin. If this anti-HSA is to be used as a reagent in the clinical laboratory, its specificity (that is, its reactivity with only HSA) must be verified in the same immunological test system used to generate patient results. For every reagent antibody, the specificity of its immunochemical reactivity is the single most important factor in the success or failure of any immunological technique used in the clinical laboratory.

A second-generation technology has been developed using hybridization of a single antibody-producing cell (plasma cell) with tumor cells. These hybrid cells can be isolated and will produce antibody directed to a single antigenic determinant. The antibodies have a single homogeneous primary structure and are called *monoclonal antibodies*. Although monoclonal antibodies have not been made to all antigens, they are expected to play a larger role in the future because they provide a reproducible antibody reagent of known specificity and affinity. Monoclonal antibodies are used in competitive binding assays and in tissue assays to identify specific antigens. In this chapter, antibodies as reagents will be understood to be polyclonal in nature unless otherwise stated.

Selection of antibody as a reagent in an immunological procedure requires information about its characteristics such as its strength (titer), affinity, and specificity. Because not all of the immunoglobulin in the antisera is reactive, the amount of antibody that is available for reactivity in a specific immunological method is termed the "titer of the antibody." The titer is the reciprocal of the maximum dilution of the antibody that gives a detectable reaction for a specific method. The titer of the reagent antibody is often different for each kind of immunological procedure. For example, anti-HSA may react in an immunoprecipitation technique at a maximum dilution of 1:32, but the same antiserum may react at a maximum dilution of 1:6400 in a radioimmunoassay procedure. Occasionally the amount of reagent antibody present may be expressed in weight, that is, milligrams per milliliter. This expression of antibody amount is determined by precipitation techniques and is often helpful in determining the amount of reagent needed. For monoclonal antibodies that are virtually pure, the indicated amount describes the total reactive proteins.

Affinity. Reagent antibodies generally fall into two categories, those of high affinity and those of low affinity. The antibodies in the reagent may be a mixture of both, but one should use reagents where high-affinity antibodies are predominant. This will result in a strong union with the antigen that is not readily reversible and that will not be influenced greatly by alteration of the reaction conditions. Low-affinity antibodies do not bind well with the antigen and can be influenced by temperature, pH, and ionic strength with consequent changes in the reaction, resulting in dissociation of the antibody-antigen complex. Most commercial reagent antibodies are of the high-affinity type. However, if one is preparing reagents, they should be tested to be certain they are of the appropriate, preferably high, affinity.

Specificity. Specificity refers to the ability of the antibody to restrict its reaction to a defined group of molecules. Because these reagents are really a collection of antibodies, they are directed to multiple antigenic determinants on a single antigen and thus could have multiple reactivities. Some antigenic determinants may be shared by different molecules. For example, human serum albumin has many determinants that can also be found on bovine serum albumin and on other animal albumins as well. Therefore human anti-HSA might interact with other animal sera if they are part of the reagents of the reaction. This interaction with other species of albumins could significantly interfere with the results. Other reagent antibodies may react with antigenic determinants that may be common to several molecular forms of plasma proteins. For example, a reagent antibody directed to the IgG molecule should only recognize the IgG molecule, but there may be antibody in the reagent that would also react with light chains of that IgG molecule. Because light chains are common to all the immunoglobulin classes (that is, IgA, IgM, and IgD) the reagent antibody would then react with all immunoglobulin molecules. It would not be appropriate to say that it was recognizing only the IgG molecule. The problem of cross-reactivity with other serum proteins can usually be controlled by a technique termed *antibody absorption,* which binds or removes from the reagent that population of antibody reacting inappropriately with other molecules of the test solution. Absorption is necessary for virtually all antibody reagents of the polyclonal type. It is often accomplished by addition of the undesired reacting antigen or by preparation of pure antibody by affinity columns.

Specificity of the antibody is especially important when one is dealing with reagent antibodies that are directed to drugs or metabolites of these drugs. For example, antibody directed to a small molecule, such as gentamicin, is usually produced by the chemical binding of many of these molecules to a protein or polypeptide backbone, such as poly-L-lysine. This material is subsequently antigenic in the animal, and a good reacting antibody against the small molecule can be produced. Once the reagent antibody is produced, however, problems related to specificity frequently arise. Gentamicin by itself would be considered an antigenic determinant, but there may be close analogs of other molecular configurations that include the polypeptide backbone that could interact with the antibody to gentamicin; tobramycin and kanamycin are examples. This potential problem is uncovered by determination of the interaction of the antibody with as many closely related analogs as possible. Specificity of the reagent antibody is extremely important in the enzyme immunoassays and radioimmunoassays that are most frequently used to measure the presence of small molecules such as drugs and hormones. However, often there is a residual reaction between the reagent antibody and a closely related compound. This reaction between antibody and the undesired antigen is termed cross-reactivity.

There are times, however, when the cross-reactivity with very similar antigenic determinants cannot be avoided. For example, antibody directed to the small molecule trinitrophenol will cross-react with dinitrophenol. To the antibody, these small-molecular-weight entities look very similar. The only way to establish the specificity of the antibody is to determine the relative affinity of the reagent antibody to presumptive cross-reacting molecules at concentrations likely to occur in patients. Often, particularly in the case of antibody reagents used in therapeutic drug monitoring, the degree of cross-reactivity with the metabolites of the drug and with other drugs is given by the manufacturer.

Because reagent antibody is protein, all precautions to prevent denaturation and degradation should be taken. The reagent should be kept free of bacterial contamination and should be stored in the refrigerator (4° C) if it is to be used within several days. Long-term storage usually is adequate at −20° C.

Antigen as analyte

Numerous naturally occurring molecules or antigens that are protein, glycoprotein, or lipoprotein in nature can be detected and measured easily in biological fluids if specific reagent antibodies are available. In addition, many small molecules, such as drugs and hormones, can be measured. The following box lists examples of the large and small molecules that are frequently measured by immunological techniques. To ensure accurate detection and precise measurement of these molecules using immunological techniques, close attention by the technologist to the proper handling and storage of the biological fluid containing these antigens is necessary.

The biological fluids most commonly available to the laboratory for analysis are serum, urine, and cerebrospinal fluid. Antigens present in each of these fluids are subject to degradation depending on (1) the nature of the antigen, (2) its concentration, (3) its susceptibility to various enzymes in the body fluids, and (4) its relative stability at various storage temperatures (such as room temperature, 4° C, −20° C, and −70° C). Each specimen must be stored and handled properly to ensure that the antigen molecule is unaltered and the reagent antibody can react with the appropriate antigenic determinants on the molecule. Stability of antigens must be established for each biological fluid. For example, the C4 component of complement

EXAMPLES OF MOLECULES IN BIOLOGICAL FLUIDS FREQUENTLY MEASURED BY IMMUNOLOGICAL TECHNIQUES

Large molecules	Small molecules
Immunoglobulins (IgG, IgA, IgM, IgD, IgE)	Digoxin and digitonin
Complement components (C3, C4, factor B)	Antibiotics
Coagulation factors (factor VIII, fibrinogen)	Cytotoxic drugs
Lipoproteins	Prostaglandins
Acute-phase proteins (α_1-antitrypsin, C-reactive protein)	Hormones
Albumin	Theophylline
Selected urine and cerebrospinal-fluid proteins	Anticonvulsant drugs
Viral antigens	Antiarrhythmic drugs

of serum is stable and can be measured accurately up to a week after receipt of the serum if the specimen is stored at 4° C before analysis. However, the C4 component of cerebrospinal fluid is very labile and is usually present at very low concentrations. If this kind of sample is stored more than 8 hours at 4° C before analysis, the C4 will have been degraded and will be unmeasurable. Thus spinal fluid must be frozen and stored at −70° C to ensure that the C4 will not be degraded before measurement. Another example is that of antigen denaturation in urine specimens. Because most urine specimens are acidic, immunological measurement of various proteins is often suspect. Proteins are degraded in an acid pH, and many antigenic determinants on these proteins are lost. Beta$_2$-microglobulin is a protein found in both urine and plasma. In urine it is used to estimate renal tubular dysfunction. It is rapidly destroyed if the pH of urine is less than 6.0. Quantitation of specific proteins in urine samples requires immediate neutralization of the acid pH at the time of collection. In contrast, the protein is stable for a week in serum stored at 4° C. The problems associated with specific protein measurement and antigen degradation are not so acute when small molecules are measured, but it is always good laboratory practice to store biological fluids at 4° C if the analysis is to be performed on the same day and in a frozen state if the analysis is to be performed much later.

It should be emphasized that the immunological reactivity of a molecule may not be related to its biological activity. The importance of this distinction is illustrated by the immunological measurement of α_1-antitrypsin and C3, the third component of complement. α_1-Antitrypsin is a potent inhibitor of the proteolytic enzyme trypsin, and its production is under genetic control. In certain individuals genetic variations occur in which the molecule is estimated to be present at normal levels when measured by immunochemical techniques, but the molecule's enzyme-inhibiting capability is greatly impaired. Immunochemically, the genetic variants react as well as the normally functioning molecule does; however, there is a great biological difference. Another example, C3, is a reasonably stable molecule in serum, but in other body fluids, such as joint fluid and cerebrospinal fluid, the C3 molecule may not be present as an intact molecule. When levels of C3 are obtained by immunological methods, they can appear to be normal or increased when, in fact, there is a low level of the complete molecule. This discrepancy is attributable to the reaction of antibody with the breakdown products of C3, which retain the appropriate antigenic determinants. Examples of immunological reactivity without biological activity occur frequently and demonstrate that normal levels of molecules assayed by immunological methods do not necessarily predict normal functional activity.

Qualitative and quantitative measurements of antigen in biological fluids require the use of a highly specific reagent antibody and a known reference standard of antigen. The reactivity of the antibody with the antigen in the patient's biological fluid is compared to the reactivity of the antibody with the standard antigen. For the most part, standards are supplied in immunological test kits. If these test kits are approved by the Food and Drug Administration (FDA), the technologist is reasonably assured that the reagent antibody is detecting the antigen, as stated by the manufacturer. However, it is good practice when using immunological methods to evaluate the test system with reference antigen obtained from other sources. The World Health Organization (WHO) supplies reference antigen for many of the serum proteins as primary standards. Secondary standards have been developed by the College of American Pathologists (CAP) in collaboration with the Centers for Disease Control (CDC) and are easily available. These reference materials, when used in the immunological methods, provide the technologist with a level of confidence that the laboratory is providing valid results on measurement of antigen in biological fluids (Table 10-1).

IMMUNODIFFUSION (OUCHTERLONY)

Immunodiffusion is commonly used to determine if antibody or antigen is present in a test solution. It can also be used to establish if there are changes in antigenic structure or to estimate the purity of either antigen or antibody.

Principles

The principles of this technique are discussed in Chapter 9.

Sample requirements and preparation

Samples must have a suitable concentration of the test antigen or antibody so that an observable reaction takes place. Usually, several dilutions of antibody and antigen must be tested before the desired result is observed. Serum or plasma is often suitable with appropriate dilution. Urine and cerebrospinal fluid often must be concentrated. After samples are concentrated, they must not contain too high a salt concentration or a pH that would prevent the reaction from occurring.

Reagents

If an antibody is to be tested for purity, it is desirable to have the pure antigen and a series of solutions containing antigens capable of reacting with possible contaminating antibodies. In contrast, if an antigen is to be tested for purity, one should use antibodies that can react with possible antigen contaminants.

Instrumentation

Essentially no instrumentation is required, although some laboratories prefer an indirect light apparatus.

Common pitfall

Improper dilution of antigen or antibody may occur.

Table 10-1 Summary of immunological techniques*

Technique	Assay end point	Equipment needed	Assay sensitivity	Time needed for assay results	Common analytes	Comments
Immunodiffusion (Ouchterlony)	Precipitation (qualitative)	Agar template	45 μg/mL	8-72 hours	Bacterial, viral, or fungal antigens	Most frequently used to screen for presence of antigen
Immunoelectrophoresis	Precipitation (qualitative)	Agar template Buffer tanks Electrodes Power-supply unit	500 μg/mL	12-24 hours	Serum, urine, and cerebrospinal-fluid protein	Used to assay complex mixture of analytes in biological fluids
Counterimmunoelectrophoresis	Precipitation (qualitative)	Agar template Buffer tanks Electrodes Power-supply unit	3 μg/mL	2-3 hours	Bacterial, viral, or fungal antigens	Commonly used to screen for antigens associated with infectious agents; more rapid than immunodiffusion
Two-dimensional immunoelectrophoresis	Precipitation (qualitative)	Buffer tanks Cooling blocks Electrodes Power-supply unit	500 μg/mL	8-10 hours	Serum proteins	Research used to examine subtle differences in proteins
Radial immunodiffusion (Mancini)	Precipitation (quantitative) CV, 10%-15%	Calibrated device to measure precipitation diameter Graph paper	50 μg/mL	12-24 hours	Serum and CSF proteins	Most commonly used immunological technique to measure serum and CSF proteins
Laurell rocket immunoelectrophoresis	Precipitation (quantitative) CV, 8%-12%	Buffer tanks Electrodes Power-supply unit Calibrated device to measure precipitated "rocket" height Semilog graph paper	50 μg/mL	4-8 hours	Serum and CSF proteins	More rapid than radial immunodiffusion
Turbidometric	Light-scattering of agregates of antigen-antibody complexes (quantitative) CV, ~8%	Spectrophotometer	50 μg/mL	15 minutes	Serum and CSF proteins	More rapid than radial immunodiffusion; cannot be automated

Immunonephelometry	Light-scattering of aggregates of antigen-antibody complexes (quantitative) CV, 3%-8%	Nephelometer must be equipped with microcomputer if rate of formation of light-scattering aggregates is measured.	1 µg/mL	½-1 hour	*Direct mode:* Serum and CSF proteins *Inhibition mode:* Phenobarbital, theophylline, and phenytoin	Popularly accepted technique to quantitate protein in direct mode; in many laboratories, this technique has replaced radial immunodiffusion
Direct and indirect agglutination	Agglutination of bacteria or red blood cell-containing antigen (semiquantitative)	None	15 µg/mL	1-5 minutes	Antibodies to bacterial antigens (e.g., febrile agglutinins) and red blood cell antigens	Techniques commonly used by serology laboratory and blood bank; not often used in chemistry laboratory
Agglutination inhibition	Inhibition of agglutination (semiquantitative)	None	15 µg/mL	2-5 minutes	Detect antigens such as pregnancy hormones (HCG)	Rapid test procedure often used to screen urine of pregnant women for HCG
Complement fixation	Lysis of red blood cells or inhibition of red blood cell lysis (semiquantitative)	Spectrophotometer Graph paper	10 µg/mL	24 hours	Detect complement-fixing antibodies to bacterial, viral, and fungal antigens	Worldwide, most commonly used serological procedure; sensitivity of assay approaches radioimmunoassay; assay difficult to perform
Enzyme immunoassay (ELISA, sandwich)	Color reaction between enzyme and substrate (quantitative) CV, 8%-15%	Spectrophotometer or microliter plate ELISA reader Graph paper Microtiter plates	1 ng/mL	1-24 hours	Serum proteins (such as IgE) Bacterial, viral, and fungal antigens Antibodies to infectious agents	Excellent assay for screening for presence of small amounts of antigen or antibody
Enzyme immunoassay (competitive binding)	Color reaction between enzyme and substrate (quantitative) CV, 8%-15%	Microtiter plates Spectrophotometer or microtiter plate ELISA reader Graph paper or microcomputer	1 ng/mL	2-4 hours	Small amounts of antigen (such as hormones, drugs, viral antigens)	
Immunoradiometric	Radioisotope decay emission	γ Counter, liquid scintillation counter or microtiter plates	1 ng/mL	1-24 hours	Same as ELISA above	Excellent assay for quantitative measurement of low levels of antigens or antibodies; problem with radioactive wastes
Immunofluorimetric	Fluorescence of dye	Specialized fluorometers	1 ng/mL	1-24 hours	Same as ELISA above	Same as for immunoradiometric but no waste problems

*Abbreviations: *CSF*, cerebrospinal fluid; *CV*, coefficient of variation; *HCG*, human chorionic gonadotropin hormone.

Interpretation of results

Interpretation of results is discussed in Chapter 9.

Limitations

The Ouchterlony technique requires a concentration of antigen greater than 45 μg/mL and thus is not as sensitive as other techniques. It is not as discriminatory as immunoelectrophoresis and resolves only a few antigens compared to other techniques. Large molecules do not diffuse readily into the gel, and the technique resolves these poorly.

IMMUNOELECTROPHORESIS

The immunological technique of immunoelectrophoresis (IEP) is used primarily as a qualitative procedure to evaluate the electrophoretic and immunological characteristics of proteins and glycoproteins in serum, urine, or cerebrospinal fluid. IEP is a more sophisticated technique than serum protein electrophoresis and is used routinely to characterize qualitative abnormalities of specific proteins or to analyze the composition of a complex mixture of proteins in biological fluids. Immunoelectrophoresis is not considered a screening method.

Principles

Immunoelectrophoresis is a two-stage procedure. The first stage involves the separation of the components of antigenic material in biological fluids by use of their differential migration in an electrical field. Generally, serum proteins can be separated into five discrete regions (albumin, α_1, α_2, β, and γ); this stage of the procedure is termed *zone electrophoresis*. The second stage of this technique is the immunological characterization of each of the separated proteins by immunodiffusion procedures. In this process, antibody diffuses into the gel and the antigen-antibody reaction is visualized by precipitation.

In this technique, a solid surface, such as a glass slide, is covered with agarose in a buffered solution (pH 8.2), and the antigen or sample well is cut into the center of the slide. A long, parallel cut in the agarose is made so that antibody in the second stage can diffuse perpendicularly to the separated proteins. Fig. 10-1 illustrates the technique of immunoelectrophoresis. The biological fluid to be analyzed is placed at the application point (antigen well). Because of the alkaline pH of the agarose buffer, most biological molecules assume a net negative charge and in an electrical field will migrate toward the anode. If serum samples are used, proteins are separated into five distinct regions when a potential difference of 3.3 V/cm is established across the agarose plate for 30 to 60 minutes. Albumin travels the farthest toward the anode. After the albumin are the α_1, α_2, β, and γ regions, respectively. After the electrophoresis, antisera are placed in the parallel antibody trough and allowed to diffuse toward the electrophoretically separated proteins. The density, position, and shape of the resultant immunoprecipitin bands are then interpreted as a means of describing each of the precipitated proteins.

Permanent records of the immunoelectrophoretic patterns can be obtained by staining the precipitin bands with protein stains, such as amido black or coomassie blue. Some laboratories obtain permanent records by photo-

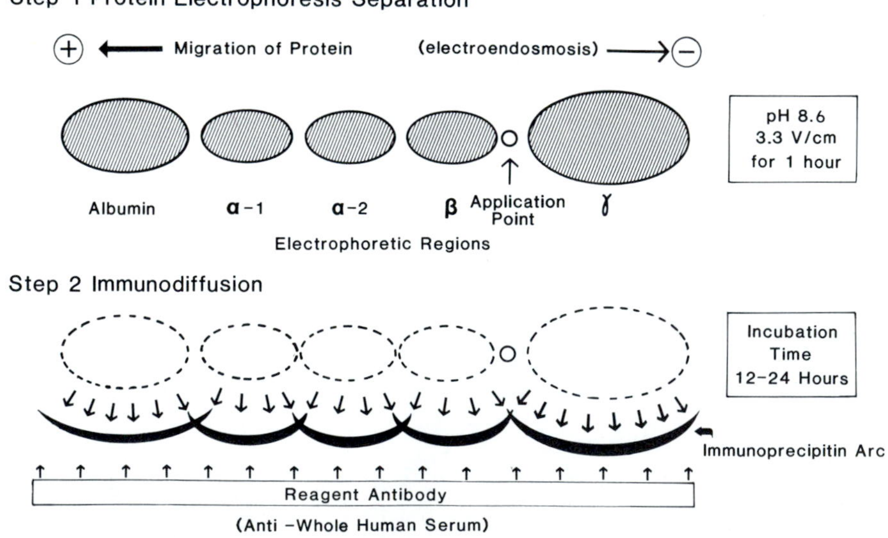

Fig. 10-1 Immunoelectrophoresis technique using normal human serum and anti–whole human serum. Usually many more than five immunoprecipitin arcs are seen.

graphing the immunoprecipitin bands in indirect lighting in lieu of staining.

Sample requirements and preparation

Serum. Most immunoelectrophoretic procedures are established to characterize serum proteins that are present in concentrations greater than 500 μg/mL. The system can be used to detect albumin, the immunoglobulins, and about 30 other serum proteins present in at least these concentrations. No further preparation of the serum specimen is required. The sample should be stored in a frozen state to preserve antigens if the procedure is to be delayed more than a day after receipt of the specimen. Plasma is considered an inappropriate sample for immunoelectrophoresis because of the high concentrations of fibrinogen in an unclotted sample.

Urine. As mentioned earlier, immunoelectrophoretic techniques are used to characterize proteins that are present in concentrations greater than 500 μg/mL. Urine usually contains dilute concentrations of protein and must therefore be concentrated to bring the protein concentrations into the detectable range for the immunoelectrophoretic procedure. Concentration procedures usually involve the use of semipermeable membranes that allow water and salts, but not proteins, to pass through. Methods that concentrate the urinary salts, such as lyophilization, are not acceptable. Many of the urine proteins can be detected by immunoelectrophoretic procedures if the urine is concentrated 50 to 100 times. It is important that urine specimens be properly stored in a frozen state if any delay in immunoelectrophoretic analysis is anticipated.

Cerebrospinal fluid. Cerebrospinal fluid is also a dilute protein solution that requires concentration before immunoelectrophoresis can be performed. The original volume must be reduced 50 to 100 times so that albumin, transferrin, and immunoglobulins, usually the major proteins of clinical significance, can be detected. Again, storage of the sample in a frozen state is important if the analysis is to be delayed more than 1 day.

Reagents

In immunoelectrophoresis, three quality reagents are needed: agarose, buffered solution, and specific reagent antibody. A high-quality agarose, which will serve as the inert support medium through which proteins will migrate, is important. Agarose, unlike agar, does not develop surface changes that can interfere with the migration of proteins during the electrophoresis. The phenomenon of electroendosmosis is less noticeable when agarose is used.

Immunoelectrophoresis is usually carried out in a buffered medium (pH 8.6). Maintaining the alkaline pH of this buffer is important. Barbital or barbituric acid is frequently added to the buffering solution, not only because it has good buffering properties, but also because it is an excellent bacteriostatic reagent. Close attention to the freshness of a buffer solution is necessary each time immunoelectrophoresis is performed.

The final reagent that requires close attention is the reagent antibody used to define the various migrated proteins. This antibody should perform well in an immunoprecipitation reaction and should be titered to the dilution that will give the maximum precipitation band for those detectable proteins that are present in lowest concentration.

Instrumentation

Four specific pieces of equipment are needed for the performance of immunoelectrophoresis: power supply, buffer tanks, proper wicking material, and agar-cutting template. The power supply used in the electrophoretic stage of the procedure should be able to supply a constant voltage over a period of several hours at a rate of 3 to 5 V/cm (Fig. 10-2). Buffer tanks should be able to hold enough buffer to prevent changes in pH during the electrophoresis stage of the technique. Electrodes should be made of platinum because of its high conductance and inertness to chemical degradation. The wicking material should be porous and inert and should allow free passage of buffer from the tank to the agarose-containing plate. The wicking material should have low electrical resistance. Material

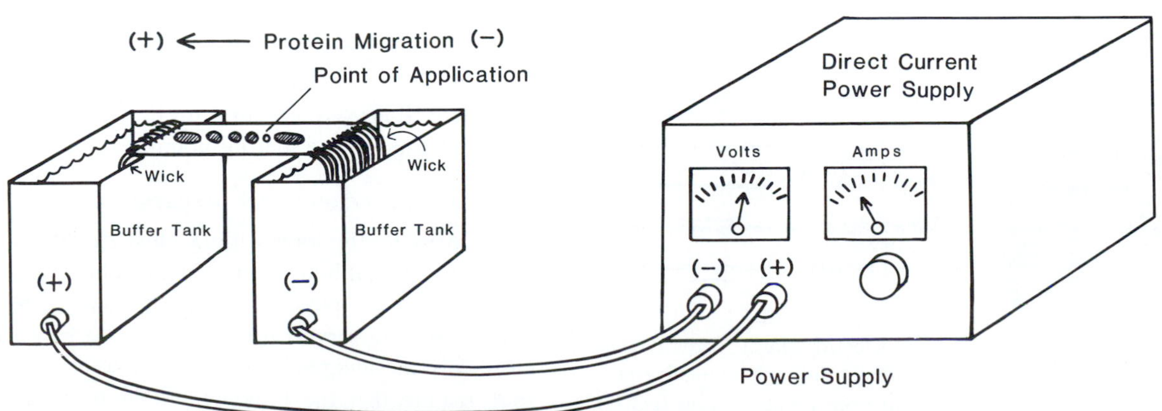

Fig. 10-2 Major equipment components needed to perform immunoelectrophoresis techniques.

that has a high electrical resistance can cause the system to overheat, producing inaccurate results. The agar-cutting template should be built with razor-sharp cutting blades set in a rigid form in such a manner that uniform wells and troughs can be cut into the agarose. Close attention should be paid to maintaining sharp cutting edges in this template.

Common pitfalls

Immunoelectrophoresis is a multicomponent system that requires close attention to each of the major components. Following is a list of common pitfalls.

1. Spent buffer solution, causing improper migration of the proteins during electrophoretic separation. This can cause overheating in the agarose, which may result in some denaturation of the protein and destruction of antigenic determinants. A buffer solution becomes spent when it is used for too many electrophoretic runs.
2. Improper concentration of specimens, resulting in no precipitation, insoluble aggregated proteins, or a concentration too high to observe.
3. Inappropriate attachment of the wick to the agarose plate and buffer tanks, resulting in intermittent or nonuniform flow of electricity across the agarose plate during the electrophoretic procedure. Improper attachment of the wick to the agarose plate is a frequent source of errors in this technique.
4. Dull cutting blades in the template, causing ragged cutting of the agarose plate with consequent artifactual influence on the diffusion properties of the reagent antibody.
5. Improper dilution of the antiserum, usually with loss of precipitation reaction.
6. Improper ionic strength of the buffer. The ionic strength of the buffer solution used should be sufficiently low so that when migration of the antigenic material is induced, the electrical charge is conferred primarily to the antigens rather than to ions of the buffer.
7. Overheating, usually caused by improper wicking, spent buffer solutions, or too long an electrophoretic migration, with a resulting denaturation of the proteins.

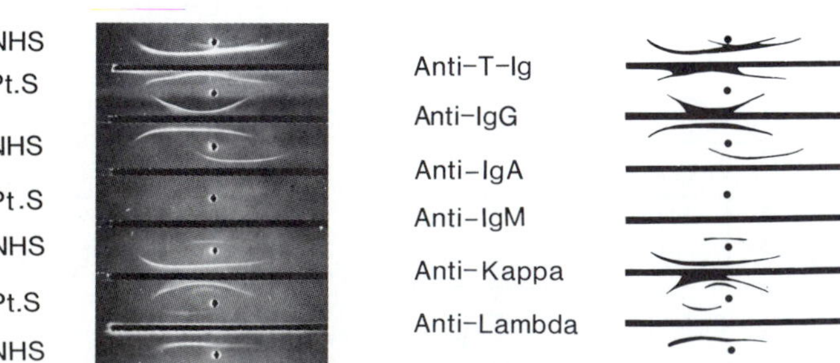

Fig. 10-3 Examples of immunoelectrophoresis patterns. *On left,* Actual gel. *On right,* scheme drawn from photograph. *NHS,* Normal human serum; Pt.S, serum from a patient with IgG(K) myeloma; *Anti-T-Ig,* antisera to all (total) human immunoglobulins; *Anti-IgG,* antisera to human immunoglobulins.

8. Excessive time for the immunodiffusion step, resulting in immunoprecipitation lines in inappropriate positions, making interpretation questionable. Antisera and antigen concentrations should be titrated to a point where the maximum precipitation line can be observed between 14 and 24 hours.
9. Improper staining and photographic procedures.

Interpretation of results

Immunoelectrophoretic analysis of biological fluids requires a firm understanding of the processes of electrophoretic separation of proteins and immunodiffusion analysis. Fig. 10-3 illustrates the use of this technique in the interpretation of normal and abnormal serum. Analysis of these immunoprecipitin bands can provide much useful clinical information. The presence of monoclonal or oligoclonal immunoglobulins, the presence or absence of specific proteins, α_1-antitrypsin, and so on can be used for the diagnosis of disease processes.

COUNTERIMMUNOELECTROPHORESIS

Counterimmunoelectrophoresis (CIE) is also called *double electroimmunodiffusion* (double EID). This technique is frequently used for the detection of single antigens present in a patient sample. It has the advantage of being much more rapid than immunodiffusion techniques.

Principles

In immunodiffusion techniques described earlier, antigen and antibody are allowed to come into contact and precipitate purely by passive diffusion processes. This process can be accelerated if the antigen and the antibody migrate toward each other in an electrical field. Fig. 10-4 illustrates this technique. The antibody-containing well is positioned closest to the positive electrode, whereas the antigen-containing well is positioned toward the negative electrode. If the buffer is selected appropriately, the antibody will migrate toward the negative electrode. The antigen must carry a net negative charge so that it will travel toward the positive electrode. If the electrical current and antigen and antibody concentrations are all in proper proportion, a precipitation line will form halfway between the two wells. Counterimmunoelectrophoresis is a screening procedure. It is used primarily for the detection of bacterial and viral antigens present in biological fluids.

Sample requirements and preparation

Serum is the biological fluid most frequently tested by this technique, although urine and cerebrospinal fluid may also be used. Proper storage of the fluids to preserve the antigen is necessary if the analysis is to be delayed. Occasionally the antigen may be at a low concentration and may require that the biological fluid be concentrated so that the antigen can be detected. This is particularly true of urine and cerebrospinal fluid specimens.

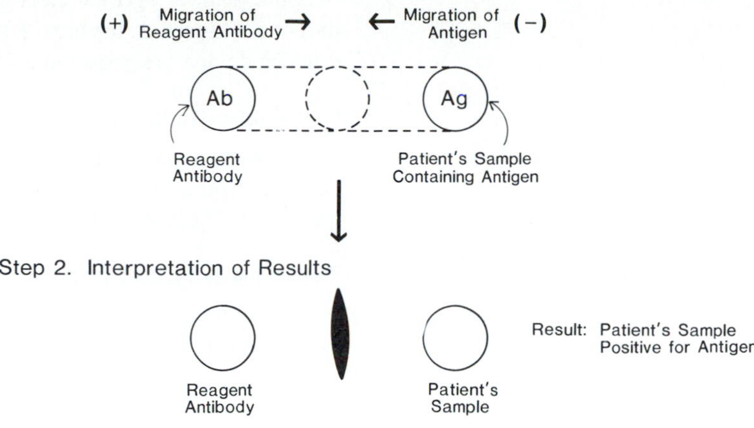

Fig. 10-4 Counterimmunoelectrophoresis.

Reagents

The *reagent antibody* used in counterimmunoelectrophoresis should migrate in an electrical field toward the negative electrode. In general, these antibodies are of the IgG class, since IgM migrates primarily toward the positive electrode. The antibody should also be a good precipitating antibody and have singular specificity for the antigen to be detected.

Instrumentation

All the equipment listed in the immunoelectrophoretic section is required, except that the agar-cutting template configuration is different. This template should be able to cut antigen and antibody wells that are properly spaced in the agarose plate to allow an appropriate antibody-antigen reaction to occur.

Common pitfalls

1. Improper handling and storage of specimens.
2. Antigen to be detected does not migrate toward the positive electrode.
3. Improper distance between the antigen- and antibody-containing wells. If the distance is too great, the antibody-antigen reaction will not occur with sufficient intensity to form an immunoprecipitation line. If the wells are too close, the antibody-antigen reaction may occur within one of the wells and therefore not be visible in the agar between the wells.
4. Improper concentration of specimen.
5. Multiple specificities of the reagent antibody.

Interpretation of results

A test result is considered to be positive when a clear immunoprecipitin band is found between the antibody- and antigen-containing wells. Serial dilution of the test specimen can be used to determine the titration of the antigen present in the test specimen. Counterimmunoelectrophoresis is a screening procedure, and therefore confirmation of positive test results might be considered appropriate.

WESTERN BLOT

The Western blot method is often used in clinical applications to confirm the presence of antibody (such as HIV antibody) to specific antigens after it is detected by the use of a screening technique such as enzyme-linked immunosorbent assay (ELISA). It is also used as a research tool to detect specific antigens.

Principles

This method is a three-stage procedure that uses electrophoresis in the first two stages. An antigen mixture is first electrophoresed in neutral agarose to separate the components by charge/weight differences. After the first stage the agarose film is overlaid with a sheet of nitrocellulose-based filter paper. In a second electrophoretic step, the protein is transferred from the agarose to the nitrocellulose. The nitrocellulose has the property of effectively irreversibly binding the transferred protein. The nitrocellulose sheet is then treated with a protein solution, which reacts with all remaining binding sites, minimizing nonspecific binding in the next step. Next, an antibody solution, usually serum from a patient, is allowed to react with the nitrocellulose sheet. Excess antibody is then removed by washing. The nitrocellulose sheet is incubated with a second labeled antibody that has specificity for the first antibody. The label allows detection of the original antigen-antibody reaction. If the label is an enzyme, such as peroxidase, then the reaction is detected by substrate precipitation, for example, a benzidine dye. A diagrammatic depiction is presented in Fig. 10-5.

A soluble protein solution is the most commonly used antigen preparation.

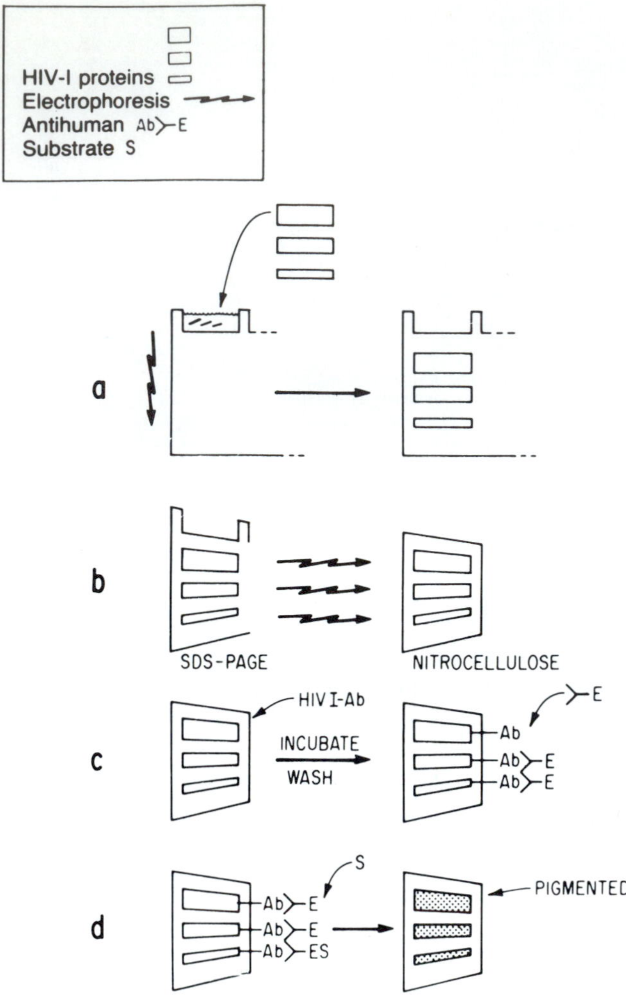

Fig. 10-5 Diagrammatic depiction of enzyme-linked immunoelectrotransfer blot technique (Western blot). **a,** HIV-1 proteins are layered onto an SDS-PAGE, subjected to electrophoresis, and separated according to their molecular weight. **b,** The discrete proteins are then electrophoresed (blotted) to a nitrocellulose matrix and incubated, first with specimen containing HIV-1 antibody (Ab), which binds to the discrete HIV-1 AG bands. **c,** Tagged antihuman AB is then added. The excess is then washed away and substrate is added. **d,** HIV-1 AB directed toward the discrete HIV-1 antigen bands is present, the discrete bands can be visualized as pigmented bands. *(Reprinted from American Clinical Products Review 6(11):16, 1987. Copyright 1987 by International Scientific Communications, Inc.)*

Reagents

The agarose film should contain an appropriate buffer so that adequate separation of the antigen mixture occurs. The nitrocellulose sheet should be able to absorb the proteins.

Reagent antibody. The reagent antibody label must have a high specificity and specific activity so that it can detect only the desired antibody and so that the sensitivity of the assay will be great enough to determine the presence of the reaction.

Instrumentation

A flat plate or vertical electrophoresis apparatus, such as the one described in Fig. 10-2, and an electrophoresis transfer device are required.

Common pitfalls

The assays commonly use peroxidase dye reactions to detect the presence of antibody or antigen. The assays require skilled individuals to read the patterns.

Test sensitivity and interpretation of results

This technique is nearly as sensitive as the enzyme immunoassay technique. However, whereas enzyme immunoassays are employed as quantitative assays, Western blots are used primarily as qualitative techniques; that is, they are used to determine the presence or absence of a particular protein antibody. Thus the most widely used purpose of the technique is confirmation of the results of an enzyme immunoassay. For example, an antigen-antibody reaction detected at a site that corresponds to one or more of the bands of HIV viral protein provides confirmation of the results of other immunoassays.

TWO-DIMENSIONAL IMMUNOELECTROPHORESIS

Two-dimensional immunoelectrophoresis is used as a research technique for the analytical separation of closely related antigens with similarly related electrophoretic properties. It is used in special situations to identify the heterogeneity of certain proteins such as factor VIII (antihemophilic factor) or to distinguish genetic variants of some proteins, such as α_1-antitrypsins. It is a difficult, time-consuming procedure that is not likely to be used in most laboratories.

Principles

This procedure is a two-stage technique that uses electrophoresis in both stages. The antigen mixture is first electrophoresed in a neutral agarose to separate components by charge. After this first stage, the agarose strip containing the separated protein is transferred to a larger plate and abutted to a gel containing antibody. An electric current is applied in a direction perpendicular to the first electrophoretic separation so that the antigens migrate into the antibody-containing agarose, forming immunoprecipitation arcs. This second stage of electrophoresis may take as long as 20 hours. After the reaction, the plate is washed in saline solution to remove excess protein, excess fluid is pressed out with filter paper, and the plate is fixed and stained.

RADIAL IMMUNODIFFUSION

Because of its simplicity and accuracy, radial immunodiffusion is one of the most commonly used techniques for the quantification of antigens.

Principles

The principles of radial immunodiffusion are discussed in Chapter 9.

Sample requirements and preparation

The assay is valid only over a rather narrow range of antigen concentrations. In part, this is dictated by the antibody concentration in the gel and by the ease of diffusion of the antigen into the gel. For some antigens, several dilutions are necessary to achieve the optimum concentration range. Urine and cerebrospinal fluid generally must be concentrated before they can be quantified by this technique. Excess salt must be removed from these specimens if they have been concentrated by lyophilization.

Reagents

Monospecific antibody to the desired antigen is the only crucial reagent.

Instrumentation

A ruler or similar device is needed.

Common pitfalls

If the antiserum measures more than one antigen, a double ring may be seen. Temperature should be kept constant. If the antigen is too large or aggregated, the resulting diffusion pattern will not be quantitative.

Test sensitivity

Test sensitivity depends on antigen size; the greater the size, the poorer the diffusion. For most serum proteins, the assay is sensitive to 50 μg/mL.

LAURELL "ROCKET" IMMUNOELECTROPHORESIS

Laurell immunoelectrophoresis is a quantitative method that is used to measure antigens in biological fluids. It provides data for individual antigens similar to those obtained by radial immunodiffusion. The advantage of this technique over radial immunodiffusion is that the time needed to produce results is 4 to 6 hours rather than 18 to 72 hours. This shortened time is attributable to the use of electrophoresis instead of passive diffusion. The "rocket" technique is usually more expensive to perform than radial immunodiffusion because of the need for electrophoretic equipment and the increased quantity of reagent antibody and agarose used. This technique is best used to measure antigens present in concentrations of 50 to 20,000 mg/mL. Antigens most frequently measured are the serum proteins.

Principles

In this technique the antigen combines with the antibody during electrophoresis of the antigen into an antibody-containing medium such as agarose. The pH of the agarose-antibody medium is usually 8.6, so that the antibody has little or no net negative charge and will not migrate when current is passed through the gel. The antigen to be measured must have a net negative charge so that it will migrate into the gel when the electrophoresis is performed. The agarose-antibody medium is usually prepared with a series of circular wells cut at one end of the gel. Each is filled with a measured amount of antigen (standard) or the test specimen. Electrical current is applied to the gel plate in such a way as to cause migration of the antigen to the center of the plate. As the antigen migrates and combines with the antibody in the gel, cones or rocket-shaped bands of precipitate are formed. Under the influence of the electrical current, the unbound antigen within the rocket-shaped band of precipitate migrates into the precipitate and causes the precipitate to dissolve in antigen excess. The leading edge of the rocket re-forms ahead of the antigen in the direction that the antigen is migrating. The amount of antigen within the leading edge is successively diminished until equivalence of the antigen-antibody complex is reached. At this point, a stable precipitate will form, and the antigen will no longer migrate. One can determine the concentration of antigen in the test sample by taking the length of the rocket from the edge of the circular well to the edge of the precipitin band and comparing it with the lengths of rockets obtained from reference standards run at the same time in the same gel (Fig. 10-6). The length of the rocket in the test sample is proportional to the log concentration of the antigen in the reference standard.

The time of the electrophoresis is important in this technique if accurate measurements are to be obtained. Electrophoresis must continue until the edge of the rocket reaches equilibrium; otherwise, a rounded or blunted apex is observed, which indicates equilibrium has not been reached. The usual electrophoresis time is 4 to 6 hours, and if the proper-shaped rocket fails to form, the test sample should be diluted and run again.

Sample requirements and preparation

As in the other immunological techniques, care must be taken to ensure that the antigen to be measured is not degraded or denatured. These precautions are especially true when one is using the rocket technique to measure complement components, such as C3 and C4, and the immunoglobulin IgM. Breakdown of the protein molecule can cause several peaks to form within the same rocket. The technologist must then determine which peak is the one to measure for accurate quantitation. In some disease states, breakdown of the proteins C3 and C4 occurs in vivo, and the formation of several peaks has clinical significance, but if the test sample has not been properly handled and stored before testing, the significance of the observation of several peaks would be difficult to interpret.

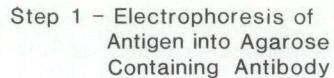

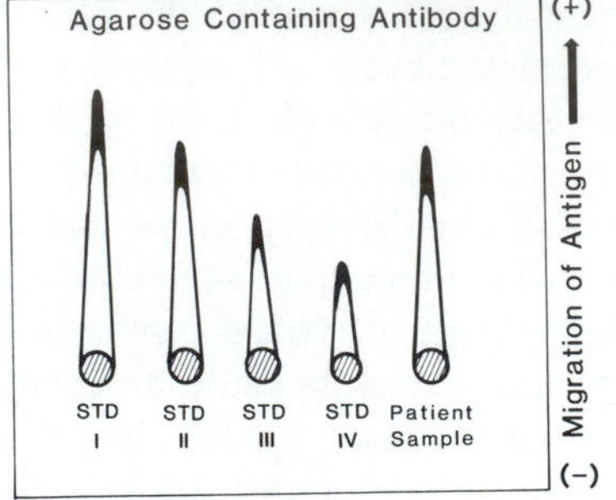

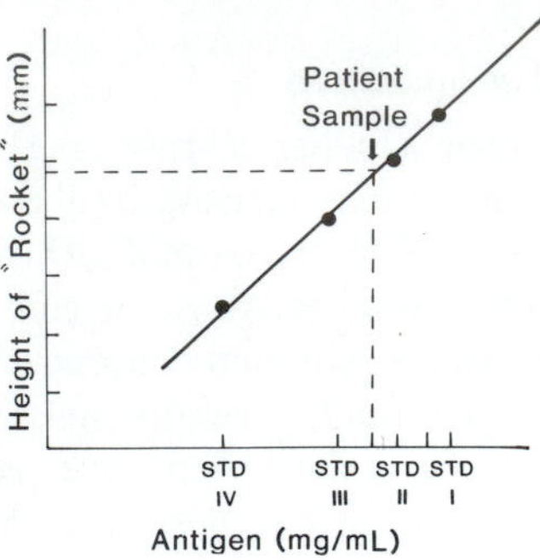

Fig. 10-6 "Rocket" immunoelectrophoresis. *STD,* Standard.

Reagents

Agarose. See p. 173.

Reagent antibody. The reagent antibody should have the same properties as that used in immunoelectrophoresis.

Instrumentation

This technique requires an electrophoresis chamber and an excellent electrical power source. An appropriate cooling apparatus is needed (see "Common pitfalls").

Common pitfalls

1. Improper antigen migration. For an accurate measurement of antigen concentration with this technique, the antigen must be able to assume a net negative charge at a pH of 8.6. If the antigen's isoelectric point is greater than 7.5, the antigen will not migrate adequately into the agarose-antibody medium to obtain peaks of high resolution.

2. Improper wicking. As in all electrophoretic procedures, proper placement of the wicks at the edge of the gel plate such that all of the gel has a constant electrical current is necessary. If the contact point of the wick with the gel is improper, the antigen will not migrate consistently throughout the plate, and false peaks will result.

3. Overheating during electrophoresis. Since the electrophoresis time is prolonged in this technique, close attention to cooling of the gel plate during electrophoresis is essential. Overheating can cause the gel to dry, the antigen to denature, and the antigen to migrate improperly.

4. Improper measurement of peak height.

Test sensitivity and coefficient of variation

The Laurell rocket immunoelectrophoresis technique has a sensitivity of approximately 50 μg/mL for serum proteins found in biological fluids. The upper limits of detection are usually around 20,000 μg/mL for these antigens. Within-run variations are between 5% and 10%, whereas between-run variations, when the same antigen is measured, are 10% to 15%.

IMMUNONEPHELOMETRY

Nephelometry

Principles. When an antibody and an antigen combine in solution, small aggregates that can scatter light quickly form, giving a turbid appearance to the solution. These aggregates then slowly associate to form a larger matrix, which eventually gives rise to the precipitate formed, as seen in immunoprecipitation assays such as double diffusion (Ouchterlony) or radial immunodiffusion (Mancini).

Development of a clinically useful assay was made possible by the observation that for some plasma proteins the intensity of scattered light was a measure of the amount of precipitate formed, as long as the reaction was carried out in antibody excess.

Although the early methods of nephelometric measurement suffered some initial instrument-instability difficulties, the latter half of the 1970s saw the emergence of more stable and reliable instrumentation. This included laser nephelometers, rate or kinetic nephelometers, and light-scattering assays carried out on a number of discrete fast analyzers.

Fast (seconds to minutes) and precise methods for the specific measurement of plasma proteins, such as albumin, immunoglobulins, complement components, and acute-phase reactants, as well as hormones and therapeutic drugs, are now available.

In Chapter 9, the structure of the aggregates created by early lattice formation of immune complexes is shown. These aggregates from primary reactions occur within seconds to minutes, whereas the secondary interactions leading to precipitation occur over a period of hours.

Light-scattering assays measure this early second-order reaction, presumably between antigen and high-affinity antibody, in which there is formed a micelle of protein large enough to scatter light but not large enough to precipitate.

Although the initial reaction of antigen with antibody is fast, the build-up of small light-scattering complexes takes time. This reaction, aggregate formation, can be greatly enhanced with the addition of the water-soluble polymer polyethylene glycol (PEG; MW, 6000) at concentrations of 2% to 4%. The polymer causes a severalfold increase in light scatter while decreasing the reaction time tenfold.

Sample requirements and preparation. The most frequently measured biological fluids are serum and cerebrospinal fluids, although urine may also be measured.

Reagents. Reagent antibody must be of very high titer and affinity and must be specially clarified by microfiltration to minimize scattering.

Instrumentation. Refer to Chapter 3 for a complete description.

Common pitfalls

1. Antibody is not in excess. In these circumstances the amount of precipitate formation will be recorded as a falsely low value.

2. Background scatter is too high. The rate-nephelometric determinations are preferred to end-point determinations because they minimize the background contributions. For example, it is possible to detect a 5% increase in scattering over background using a rate measurement. In contrast, such a difference would not be measurable as an end-point change.

3. Interference is caused by colored solution. End-point methods are most influenced by colored solutions. These absorb the scattered light, tending to yield lower values. However, even kinetic measurements are lower in highly colored solutions.

4. Mixing is insufficient. Because the rate method requires constant agitation to make uniform particles, the mixing efficacy must be constantly monitored.

Limitations. The difficulty of determining whether a given amount of precipitate (light scatter) is occurring in antigen or antibody excess is one that makes all methods suspect when the concentration of molecules, such as immunoglobulins, where one expects fivefold to tenfold variations to occur in pathological conditions, is being measured. Although each instrument manufacturer has devised ways of recognizing antigen excess, the foolproof method requires the laboratorician to carry out the measurements at two different dilutions, or to have available a serum protein electrophoresis of the sample for comparison.

It may be possible, using kinetic assays, to differentiate the early reaction rate seen in antibody excess from that seen in antigen excess, but the microprocessor software needed has not yet been made available.

Light-scattering inhibition immunoassay

Principles. The antibody population that reacts first with antigen represents only a small portion of the total antibody content of a polyclonal antiserum. The concentration of this high-affinity antibody dictates the rapidity and intensity with which the system approaches the optimum scatter for a particular antigen concentration. Addition of an unknown but small amount of antigen to the antibody followed by back-titration of residual antibody by a standard antigen can be used to quantitate very small amounts of antigen and, as has been described for IgE, can allow detection down to 100 ng/mL. This level of protein determination begins to overlap with radioimmunoassay (RIA), although it does not reach the limits of detection of RIA.

Nephelometric-inhibition immunoassay (NINIA) of haptens. If a hormone or drug, such as digoxin, which by itself is nonantigenic, is covalently linked to a large carrier protein, such as keyhole limpet hemocyanin (KLH), the resulting conjugate acts as an antigen and will cause the formation of antibodies that will recognize both KLH and digoxin (hapten) (Fig. 10-7).

If the digoxin is coupled to a different molecule, such as bovine serum albumin (BSA), the conjugate will react with the anti-digoxin-KLH antiserum. Precipitation engendered by the BSA-digoxin (developer) antigen can be attributable only to antibody with antihapten (digoxin) specificity. Precipitation from antihapten activity can be inhibited by addition of the free hapten, and it is on this principle that the nephelometric-inhibition immunoassay is based. The technique was first described for progesterone by Cambiaso et al in 1974, and they referred to the term *NINIA* as an acronym for the approach. With appropriate manipulation of the parameters, one can adjust the number of haptenic groups for adequate precipitation while allowing maximum sensitivity for inhibition by free hapten.

Addition of sample containing free hapten to the reaction mixture neutralizes available antibody and causes a decrease in the signal generated by a standard hapten-conjugate preparation (Fig. 10-8).

Fig. 10-7 Formation of hapten (digoxin) complexes for use as immunizing antigen (keyhole limpet hemocyanin–digoxin) and developer antigen (bovine serum albumin–digoxin) for nephelometric-inhibition immunoassay (NINIA) approach to measurement of small molecules.

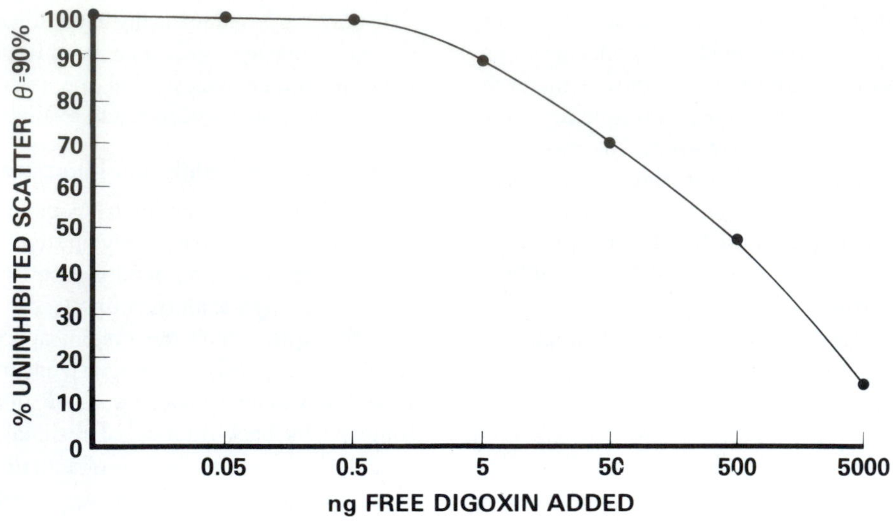

Fig. 10-8 Digoxin standard curve using nephelometric-inhibition immunoassay.

Methods have been developed for rapid analysis of drugs occurring in milligram-per-liter amounts, such as phenytoin, phenobarbital, and theophylline.

Sample requirements and preparation. The most frequently measured biological fluid is serum. The most measured compounds are drugs such as theophylline and phenytoin.

Reagents. The requirements for reagent antibody are similar to those for any nephelometric technique. Reagent antigen is composed of the drug to be monitored covalently bound at several sites to a protein, usually albumin.

Instrumentation. Instrumentation is the same as that used for nephelometry.

Common pitfalls. Pitfalls are similar to those described for nephelometry, except that the reaction is titered to be always in antigen excess.

Additional assay modifications

Particle-enhanced light scattering. Because the amount of light scatter is dependent on the size, amount, and refractive index of the scattering species, an increase in any of these parameters should result in greater sensitivity. This potential is realized when either antigens or antibodies are coupled to various inert carrier particles, but, because of availability and better control of coupling conditions, polystyrene latex beads have become the particles of choice. This type of assay is essentially a type of agglutinating procedure (see discussion of agglutination assays below). The method also offers faster signal generation and greater economy of reagents. The latex-fixation test for detection of rheumatoid factor is the classic example of this type of assay, although it depends on visual observation of agglutination and is thus only semiquantitative. More recently, several particle-enhanced methods have been developed for both direct and inhibition assays, and a wide variety of light-scatter detection techniques have been employed.

Monoclonal antibody reagents. The performance of light-scattering assays depends greatly on the quality of the antiserum used. With conventional polyclonal reagents there is a continual need to monitor and adjust antiserum titer, specificity, and affinity.

Such variability is often overcome with the use of monoclonal antibody preparations from ''immortalized'' hybrid plasmacytoma cell lines. However, unless the antigen has a number of identical antigenic sites, monoclonal antibody cannot cause matrix formation and will cause little or no light scatter. If an appropriate mixture of monoclonal antibodies can be made, complexes will form, causing measurable light scatter. This blending of monoclonal antibodies or the use of monoclonals in particle-enhanced light-scatter assays will ensure constancy of reagent production and give these assays further stability and specificity.

AGGLUTINATION ASSAYS

Agglutination is the clumping and sedimentation of antigen after reaction with antibody. It was first noted when the reaction of bacteria incubated with serum from an infected patient was observed. Observation of the agglutination of red blood cells after incubation with serum led to the discovery of ABO blood groups. Agglutination has been extensively used as a laboratory test because of its ease and versatility. It is, however, only a semiquantitative procedure. Repeat testing shows reproducibility of results only within fourfold dilutions. Agglutinating antibodies *(agglutinins)* may be directed against naturally occurring antigens on the surface of cells *(active* or *direct agglutination)* or against substances that have been applied to the surface of cells or inert particles *(passive* or *indirect agglutination)*.

Principles

Agglutination reactions depend on the formation of antibody bridges by bivalent (IgG) or multivalent (IgM) antibody between antigen particles with multiple antigenic determinants. Large particles, such as red blood cells or bacteria, contain many different antigens, as well as antigens that appear hundreds of times on the cell or particle surface. Thus it is possible for antibody molecules to bind to more than one site on a single particle or to bind to equivalent sites on different particles. Such binding is called *cross-linking*. Antigens with a single antigenic determinant would not permit cross-linking and would therefore not agglutinate. Antibody can bind to antigenic determinants on the same or different particles, creating a high-molecular-weight lattice that clumps together and falls out of suspension. Because of its size and multivalency, IgM is said to be 750 times more efficient at agglutination than IgG. The effectiveness of IgM in agglutination is the reason that IgM molecules are called *complete antibodies*. IgG molecules, because they sometimes cannot be effective at cross-linking, are called *incomplete antibodies*. Agglutination reactions are generally used to detect antibody in serial dilutions of serum containing antibody directed to particulate antigens. *Reverse agglutination* can be used to detect soluble antigen by adsorption of antibody onto cell or particle surfaces. Agglutination reactions are read by the naked eye or with the aid of magnification. Reactions are scored 1+ to 4+ by estimating the extent of agglutination. The titer of the serum is the reciprocal of the highest dilution giving visible (1+) agglutination.

Factors influencing agglutination reaction

Many factors influence agglutination. These include particle charge, antibody type, electrolyte concentration, viscosity of the medium, reactant concentrations, location and concentration of antigenic determinants, and time and temperature of incubation. These factors are considered separately.

Particle charge. Red blood cells, bacteria, and inert particles such as latex have a net negative surface charge, which is called the *zeta potential*. The surface charge of red blood cells is caused by sialic acid residues on the cell membrane. These charges must be overcome to permit the cross-linking that will result in agglutination. Enzyme treatment of red blood cells with papain or ficin makes red cells more easily agglutinable. Such treatment may function by clearing the negative residues from the cell surface to permit more contact between the cells.

Antibody type. IgM antibodies are more efficient at agglutination because their size permits more effective bridging of the gap between cells caused by charge repulsion.

Electrolyte concentration and viscosity. The ionic strength of the medium used for the agglutination reaction can assist in reducing the negative surface charge of particles. This can be accomplished by addition of charged low-molecular-weight molecules, such as albumin, to the medium. The pH of the medium should be near that present in physiological conditions. At neutral pH, high electrolyte concentrations act to neutralize the net negative charge of particles. Increasing the viscosity of the medium with polymerized molecules, such as dextran, also assists in bringing the charged particles together.

Antigenic determinants. As stated earlier, antigens with multiple antigenic determinants are necessary for agglutination. A monovalent antigen would not permit cross-linking. The placement of the antigenic determinants on the particle can also affect agglutinability. Antigenic determinants that are sparsely distributed will not be so easily cross-linked as antigenic determinants that are densely distributed. Antigenic determinants can also be inaccessible to antibody binding because they are buried within cell membranes. IgG antibodies are frequently unable to react with hidden antigenic determinants because such antigens increase the distance that antibody molecules must bridge. Enzyme treatment of red blood cells, in addition to decreasing negative surface charge, may also increase agglutinability of red blood cells by exposing hidden antigenic determinants.

Concentration, temperature, and time of incubation. At higher antigen concentrations, the reaction with antibody will be more rapid. Agitation of the antigen suspension with antibody solution will increase reaction rate by increasing the surface area exposed to antibody. At lower antigen concentrations, reaction time can be shortened by centrifugation, which increases contact between the antigenic particles and the antibody. Concentration of reactants can therefore influence reaction time. Temperature of incubation is also an important variable. Some antigens are bound most readily by antibody at 37° C. These antigens include microbes. Some antigens react optimally with antibody at 4° C. These *cold agglutinins* include antibody to the *i* antigen of red blood cells. Optimum temperature for the agglutination reaction will vary with different antigen-antibody systems. Temperature also affects the behavior of antibodies in vitro.

Direct agglutination

Direct agglutination involves detection of antibody to intrinsic antigens on the surface of red blood cells, to microbes, or to other particulate antigens. The titer of the serum reflects the concentration of the predominant antibody in the serum. Infection by bacteria that contain many different types of antigenic determinants produces antibody against a variety of these different antigens. The titer of the serum will reflect the level of the antibody present in the greatest concentration.

Direct agglutination tests are frequently used in the immunological diagnosis of microbial infections. Early detection of a high titer and documentation of a significant rise in titer are important tools in diagnosis. Antibodies to

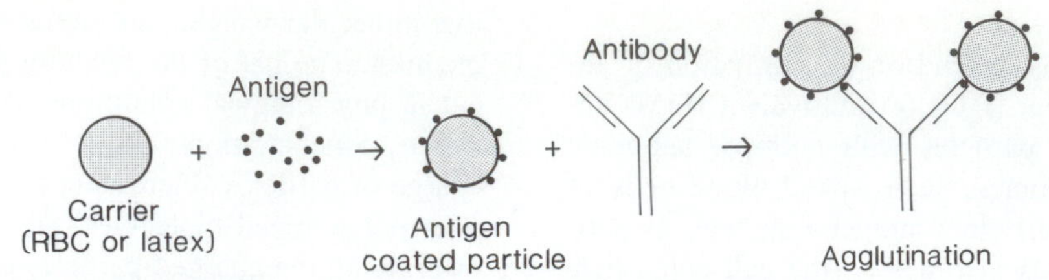

Fig. 10-9 Passive (indirect) agglutination reaction. Antigen is adsorbed onto surface of carrier particle, which is then agglutinated by antigen-specific antibody.

Brucella (brucellosis), *Salmonella* (typhoid fever), and *Proteus* (Rocky Mountain spotted fever) are detected in this way. Direct agglutination tests are also used in the typing of human red blood cells in the blood bank.

Different bacterial antigens may give different patterns of agglutination. Antibodies to bacterial flagella cause cross-linking of the flagella themselves. These antibodies cause formation of a loose, rapidly formed agglutinate. Antibodies to antigens in the body of the bacterium cause cross-linking of the organisms themselves. This results in a granular, compact precipitate that develops more slowly.

Indirect agglutination

Indirect agglutination involves reaction of antibody with antigens that have been passively transferred onto the surface of particles (Fig. 10-9). Red blood cells, usually from human, sheep, or turkey, are employed. Inert particles such as latex (0.81 μm in diameter) and bentonite (clay) are also used. Polysaccharide and some protein antigens, such as albumin and purified protein derivative (PPD), are easily adsorbed onto the particle surface. Other antigens require pretreatment of particles for adsorption. For pretreatment, cells are incubated with tannic acid or chromium chloride, which modifies the cell surface. Treatment probably affects cell surface charge. Antigen can be covalently bound to the cell surface by bifunctional molecules, such as bisdiazobenzidine (BDB) or glutaraldehyde, which bind both to the cell surface and to the antigen. Antigens are often applied to the surface of the cells at the highest concentration that does not produce nonspecific agglutination. This concentration is determined in independent experiments performed before antibody testing is carried out. Because the concentration of antigen can be artificially controlled, the sensitivity of indirect agglutination tests is somewhat greater than that of direct agglutination tests. Indirect agglutination procedures are used in the diagnosis of syphilis. The VDRL (Venereal Disease Research Laboratory) test employs cholesterol crystals coated with cardiolipin antigen. Detection of rheumatoid factor, useful in the diagnosis of rheumatoid arthritis, uses agglutination of latex particles coated with human IgG.

Agglutination inhibition

Agglutination inhibition is an adaptation of the agglutination reaction that permits detection and quantitation of soluble antigen. The concentrations of antibody and particles are carefully controlled to prevent antibody excess. Antibody is incubated with the test antigen solution, and antigen is bound to available antibody-combining sites. The antibody is then added to particulate antigen suspensions. The failure to agglutinate indicates that enough antibody-combining sites have been saturated with a soluble form of the same antigen so that insufficient antibody-binding sites are available for binding and cross-linking of the particulate antigen. Quantitation of the soluble antigen can be accomplished by assessment of the degree of inhibition in serial dilutions of the antigen solution (Fig. 10-10).

Antiglobulin testing

Antiglobulin testing is a modification of the agglutination reaction to permit detection of an incomplete (IgG)

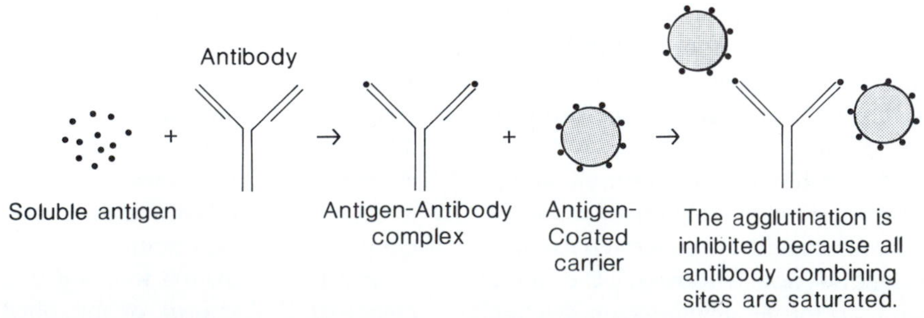

Fig. 10-10 Agglutination-inhibition reaction. Same reaction as that shown in Fig. 10-9 but inhibited by soluble antigen.

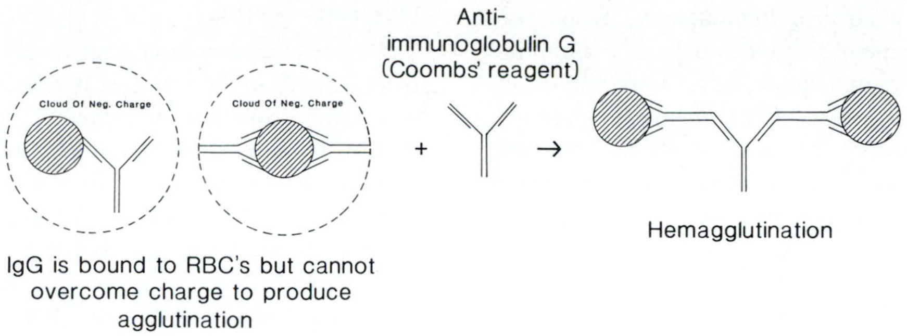

Fig. 10-11 Direct Coombs' test for antibody to red blood cells (RBC).

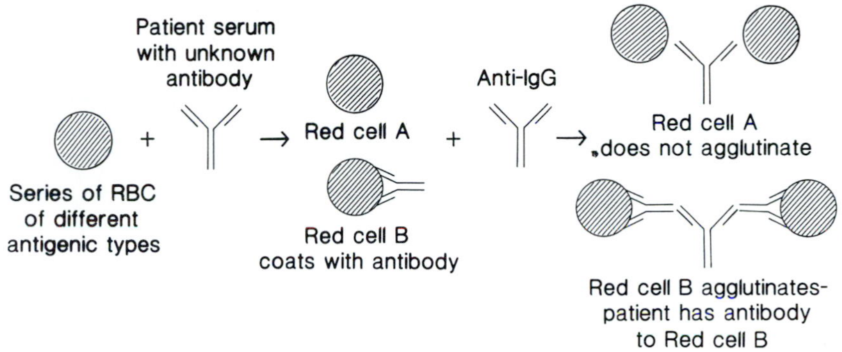

Fig. 10-12 Indirect Coombs' test for antibody.

antibody, which may not produce agglutination even after binding to the particle surface. IgG is less effective at agglutination than IgM because its smaller size is less effective at bridging antigen particles. If an anti-immunoglobulin or *Coombs' reagent* is added to the unagglutinated IgG-coated particles, the Coombs' reagent will bridge the gap between particles by bivalent binding (Fig. 10-11). This enables cross-linking to achieve agglutination. The *direct Coombs' test* is used in blood banks to determine the presence of IgG antibodies to red blood cells.

The *indirect Coombs' test* is a variation of the antiglobulin test and is used to detect free antibody to red blood cells in patient serum (Fig. 10-12). Serum is screened against a panel of red blood cells of known and varied antigenicity. Agglutination of cells to which patient serum and antiglobulin reagent have been added indicates the presence of antibody in patient serum to an antigen present on the agglutinated cells.

Sample requirements

Agglutination reactions can be used to measure components of plasma, serum, or cerebrospinal fluid. Urine must be buffered because of its usual acidity. Care must be taken to consider the possible role of complement in plasma or serum. Specific directions regarding the need for inactivation or dilution to minimize complement should be followed.

Reagents

As cited earlier, particularly for red blood cell agglutination techniques, the factors influencing the reactions are particle charge, type of antibody, electrolyte concentration, and viscosity. Red blood cells, when used fresh, have a shelf life of about 2 weeks. Therefore many manufacturers have developed fixed red blood cells (usually stabilized by tannic acid or glutaraldehyde) or latex beads to overcome the need to prepare the reagents every few weeks.

Instrumentation

The great advantage of this technique is the simplicity of instrumentation. Results can be recorded by eye or with the aid of a mirror or magnifying glass.

Common pitfalls

Antigen excess often results in a prozone phenomenon with false-negative results. Use of expired red blood cells or other reagents can result in lack of agglutination.

Limitations

The technique is only semiquantitative and thus allows estimates of the true value within a dilution factor of two.

COMPLEMENT-FIXATION ASSAYS

The complement-fixation tests are probably the most sensitive of the immunological procedures that were developed early in the history of immunology. Tests more sensitive than complement fixation, such as radioimmunoassay and enzyme immunoassay, have now been developed; but complement-fixation tests are still important, especially in the diagnosis of fungal, viral, and parasitic infections and in the quantitation of functional complement levels (total hemolytic complement) and complement components.

Complement proteins

The term *complement* is used to denote a series of plasma proteins that are activated in sequence after antigen-antibody reactions. Not all classes and subclasses of immunoglobulins are capable of activating complement or can activate it to the same degree (see Table 9-1). The antigen-antibody complex used in complement-fixation tests therefore must involve an antibody that is capable of activating, or *fixing,* complement. As with agglutination reactions, IgM is more efficient in fixing complement than IgG. Because of its pentameric structure, IgM molecules are believed to be a thousandfold more efficient in fixing complement than monomeric IgG molecules when the two immunoglobulins are compared on a molar basis.

The first protein in the complement sequence is bound to a site in the interior of the immunoglobulin molecule in the second constant domain of the heavy chain (C_{H2}). This area is inaccessible on an unreacted or unbound antibody. After interaction and binding with antigen, conformational changes in the immunoglobulin molecule cause the hinge region to open and the complement-binding site to be exposed. The first complement protein binds to this region. Subsequent complement proteins are then activated and can bind to the membrane of the cell to which the antigen-antibody complex is bound. Binding of the complete sequence of nine complement proteins results in small defects in the membranes of the cells. Cytoplasmic cell contents are lost through these holes, and extracellular fluid is admitted. This results in hypotonic swelling of the cell and ends in cell lysis. If the cells used in the test are red blood cells, cell lysis results in release of hemoglobin into the medium. The amount of hemoglobin, which can be quantitated spectrophotometrically, is directly proportional to the amount of complement fixed to the surface of the cells.

One-stage testing

To assess total complement levels (the level of complement activity in the serum that reflects the quantities of all nine major complement proteins) or to assess separately the levels of the nine individual major complement proteins, a one-stage test system is used. A constant volume of red blood cells, usually derived from sheep, is added to a constant amount of anti–sheep red blood cell antibody, or *hemolysin.* A source of complement is added. For total complement measurements, dilutions of the unknown serum are used. For measurement of complement components, serum with an added excess of every complement component but the component to be measured is used. The degree of hemolysis is measured in a spectrophotometer and is directly proportional to the amount of complement available for fixation in the test serum (Fig. 10-13).

Some immunologically mediated diseases cause a decrease in total serum complement levels or in the levels of individual complement components. In addition, hereditary deficiencies of certain components of the complement system have been described.

Two-stage testing

Two-stage complement-fixation testing is used to measure antigen or antibody. In the first reaction, antigen and antibody are incubated with a known amount of complement. In the second stage, the residual complement activity in the solution is determined by an indicator system. The indicator is a suspension of sheep red blood cells that are coated with hemolysin. The degree of hemolysis of the indicator cells is inversely proportional to the amount of complement fixed in the first reaction (Fig. 10-14).

The first reaction can be adapted so that the complement-fixation reaction measures antigen or antibody. To determine antigen levels, one uses a constant volume of antibody. To determine antibody levels, a constant amount of antigen is used.

Sample requirements and preparation

EDTA plasma, serum, and cerebrospinal fluid may be analyzed but certain precautions are necessary. The samples cannot be hemolyzed. For detection of antibody, the endogenous complement of the sample must be inacti-

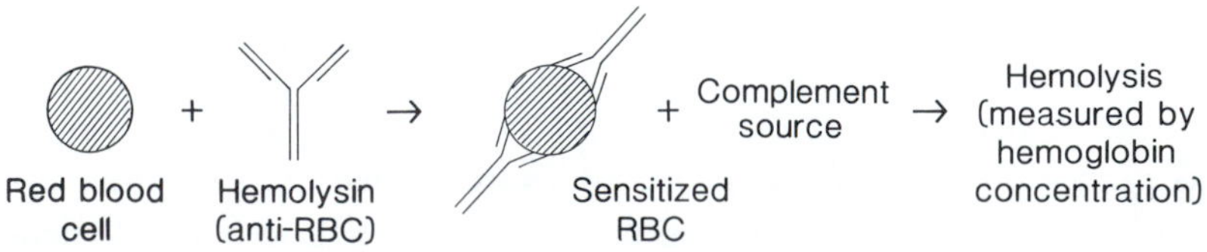

Fig. 10-13 Complement fixation one-stage testing. For measurement of total hemolytic complement, test serum is added as source of complement. For measurement of complement components as complement source, one uses test serums, to which purified complement components are added; for example, to measure complement component 3, one adds all components except C3, in excess to test serum. Therefore reaction is limited only by concentration of C3 in test serum.

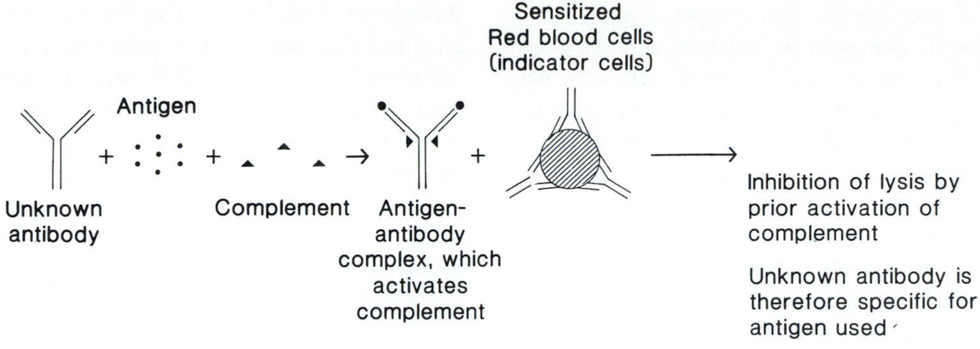

Fig. 10-14 Complement fixation two-stage testing. Antigen or antibody can be measured by holding constant all but the variable to be tested, in this case unknown antibody.

vated. Usually this is accomplished by heating the specimen for 15 to 30 minutes at 56° C. Antigens, if obtained from serum or other fluids, must also be free of endogenous complement.

Reagents

Exogenous complement must be prepared daily and cannot be stored. The complement activity varies significantly from batch to batch and must be standardized daily.

Instrumentation

Only a spectrophotometer is necessary.

Limitations

Complement-fixation tests are sensitive to many variables. They are inhibited by anticomplementary activity (factors that inactivate or interfere with any of the complement proteins) in serum, including factors such as circulating immune (antigen-antibody) complexes, lipemic sera, aggregated immunoglobulins, and heparin.

It is critical to keep all components in the test constant except the one that is to be measured. Red blood cell number, concentration of complement, and concentration of antigen should be rigorously defined for antibody determinations.

Tests are influenced by the instability of some complement proteins, variability in red blood cells, variation between lots of hemolysin, and the narrow range of optimum reactivity for many reagents. The need for fresh reagents and the great variability make this assay one of the most difficult performed by a laboratory.

The Centers for Disease Control (CDC) evaluates complement-fixation reagents and has developed standardized procedures for complement-fixation tests.

INDICATOR-LABELED IMMUNOASSAYS

Indicator-labeled immunoassays use indicator molecules, attached to either the reagent antibody or antigen, to demonstrate that an antibody-antigen reaction has taken place. These indicator molecules can be enzymes, fluorescing molecules, or radioactive compounds. The next section, on immunometric assays, will describe assays that use enzymes or fluorescing molecules as the indicator molecule.

Indicator-labeled immunoassays are very popular in the laboratory today for many reasons. These assays use minimum amounts of reagents, produce test results rapidly, have maximum sensitivity compared to other immunoassays, are reproducible, and are suitable for running large numbers of tests at a single time. Enzyme and fluorescent immunoassays have an additional advantage over radioimmunoassays in that they do not require all the special precautions necessary for the handling of radioisotopes.

Indicator-labeled immunoassays are usually designed to detect the first stage of an antibody-antigen reaction rather than the second stage as described in the immunoprecipitation and agglutination immunoassay section. First-stage detection has several advantages. The rate of reaction is faster, the requirement for electrolytes to support the precipitation phase is obviated, and reagent antibody can be monovalent rather than polyvalent. These types of immunoassays are usually quantitative procedures in the clinical chemistry laboratory and are sensitive to the microgram and nanogram range. Qualitative indicator-labeled immunoassays are very popular in the serology laboratory for the detection of antibodies to infectious organisms and for the characterization of autoimmune antibodies such as antinuclear antibody (ANA) and antithyroid antibodies (ATA).

IMMUNOMETRIC ASSAYS

The heterogeneous immunometric assays are those procedures that require the physical separation of the antibody-antigen complex from the unbound constituents so that measurement of the label associated with either the complex or the unbound constituents can be made. Homogeneous enzyme immunoassays do not require this physical separation for final measurement. Heterogeneous assays can be subdivided into those assays in which the antibody is labeled and those in which the antigen is labeled. One type of assay for antigen involves coating of a surface with antibody, followed by reaction with the test antigen (analyte). The extent of this reaction is assessed by the addition of a second, reaction-labeled antibody. This is frequently referred to as a sandwich type of methodology. A second configuration frequently used is a competitive type in which the antigen is labeled. The labeled antigen competes with unlabeled antigen in the test sample for sites on an antibody that is immobilized on a solid surface, such as polystyrene, latex, or iron.

The appropriate label for an immunochemical reagent must have certain qualities. The label should be inexpensive and should be purified to a high degree. It should also have a specific reactivity associated with it (see Chapter 52 for properties of enzymes). It should not be present in biological fluids in concentrations high enough to interfere with the detection of the antigen. Contaminating label in the biological fluid would increase background and decrease the sensitivity of the assay. This last condition is not essential for heterogeneous assays, but it is crucial for homogeneous assays. Several enzymes, metal chelates, radioisotopes, and fluorophores fulfill most of these requirements and have been successfully used in immunometric assays. Examples of enzymes commonly used are horseradish peroxidase, alkaline phosphatase, and glucose oxidase. Selection of the appropriate enzyme for use as a label for the immunochemical reagents is often empirical, and each has distinct advantages and disadvantages. Selection of radioisotopes is based on the ability of the isotope to label the antigen or antibody, but because of its high specific activity, ^{125}I is often used.

In fluorimmunometric assays, the indicator-reagent antibody is usually conjugated or covalently linked with a fluorochrome molecule. The fluorochrome molecule is a chemical that can absorb electromagnetic energy of short-wavelength light (200 to 400 nm) and then instantaneously emit light at a longer wavelength of the visible spectrum (400 to 700 nm). Intensity of the emitted visible light is the measurable indicator in this assay. The most popular fluorochrome is fluorescein-isothiocyanate, often abbreviated FITC, which can be conjugated easily to free amine groups on the reagent antibody. Other fluorophores include rare earth chelates.

Principles

Sandwich technique (antibody labeled). The sandwich technique can be used to measure either antigen or antibody. Usually there are three layers. For antigen measurement, these are antibody in the first layer, antigen (test sample) in the second layer, and antibody labeled with enzyme, radioisotope, fluorescence dye, or metal chelate in the third layer. For antibody measurements, the three layers are antigen, first layer; antibody (test sample), second layer; and the second labeled antibody, third layer.

For the antigen-measuring system, two different molecules of antibody must bind to the antigen. Thus only large antigens, such as proteins, can be measured by this system. Fig. 10-15 is a schematic of this assay. In the first step, antibody of the desired specificity is immobilized to a solid surface, which may be the wells in a microtiter plate or a plastic test tube. The solid surface is washed to remove all unreacted materials and may then be coated with other material (proteins) to minimize nonspecific reactions with subsequent possible false-positive results. In the second step, the fluid containing the antigen is reacted with the immobilized antibody. All nonreacting material is

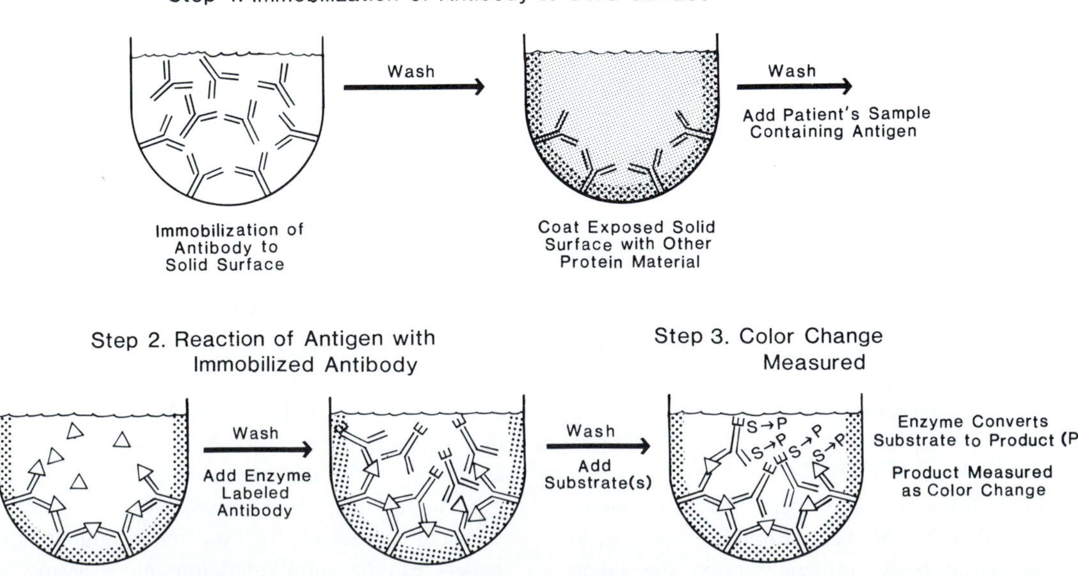

Fig. 10-15 Enzyme immunoassay. Sandwich technique with antibody label.

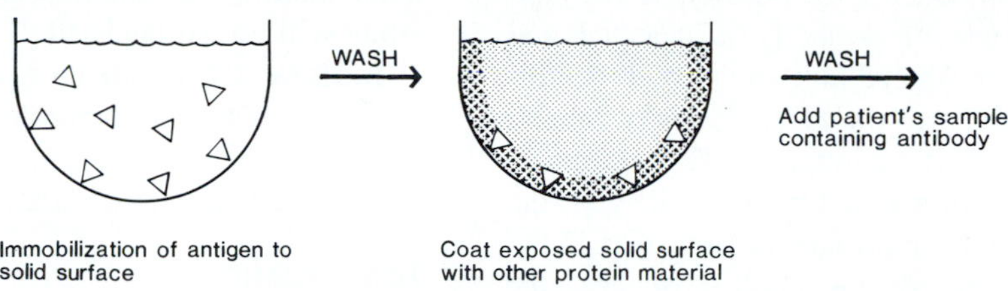

Fig. 10-16 Enzyme immunoassay. Detection of IgE specific for an allergen.

washed away. One should note that, for quantitative results, the amount of antigen added in the test sample should not exceed the antibody-binding capacity. In the third step, the labeled antibody is reacted with the antigen that has now been immobilized by the antibody on the solid phase. All unreacted labeled antibody is then washed away. If enzyme is used as label, substrate with appropriate cofactors is added so that the enzyme on the antibody can convert the substrate to the product. The amount of product is then measured by a color change or color reaction. The intensity of the color is directly proportional to the amount of antigen that has been immobilized on the solid surface by reaction with the antibody. If radioisotope is used, the solid phase can be counted. This format is used most often to detect polyvalent antigens such as various serum proteins found in biological fluids. These polyvalent antigens contain multiple antigenic determinants that can react with the antibody on the solid phase and with the labeled antibody as well.

For an antibody-measuring system, the antigen must be first immobilized on an insoluble matrix, such as a plastic surface or bead. Most often microtiter plates are used. Fig. 10-16 is a schematic of this assay. Immobilization with retention of antigenic reactivity is the first step of this procedure. In the second step, the biological fluid containing presumptive antibody toward the immobilized antigen is

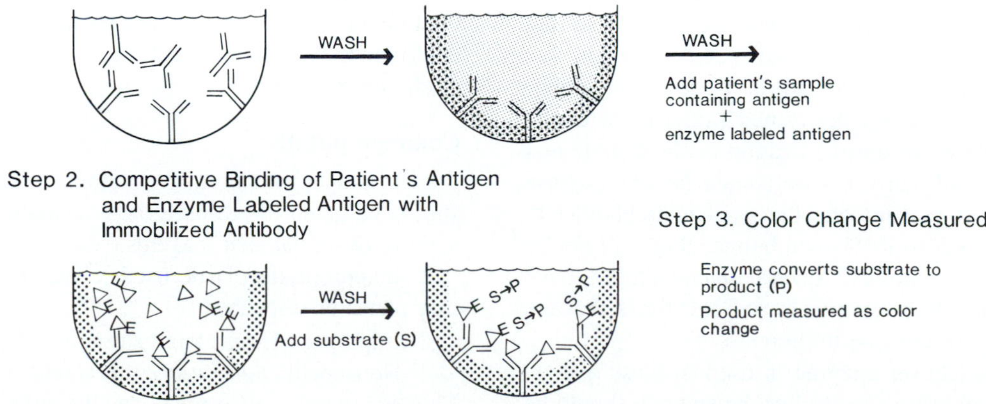

Fig. 10-17 Enzyme immunoassay. Competitive binding.

allowed to react. Any antibody present will be bound to the antigen immobilized on the solid phase. After separation of the unreacted components by washing of the support surface, the presence of antibody is detected and quantitated by addition of labeled anti-antibody that is directed to the class specificity of the antibody under consideration. Finally, the label is quantitated. This format is used to measure IgE in serum specific for an allergen (the so-called RAST test). It is important to note that if the antibody concentration exceeds the number of antigenic sites on the surface, the reaction will not be quantitative. Also the labeled anti-antibody must be present in excess over the test antibody.

Competitive-binding assays. The simplest competitive-binding assay uses labeled antigen (Fig. 10-17). The enzyme-labeled antigen is mixed with the test solution containing an unknown amount of the antigen. The solution containing the labeled and unlabeled antigen is allowed to react with a limited amount of antibody bound to a solid matrix. Unbound antigen (both labeled and unlabeled) is removed by washing, and the amount of labeled antigen is measured by determination of the amount of enzyme bound to the solid surface. This assay is always performed in antigen excess. The concentration of the antigen present in the test sample is inversely proportional to the intensity of the color measured. This format of enzyme immunoassay can be used to detect hapten groups or small molecular antigens found in biological fluids such as drugs and hormones.

Sample requirements and preparation

All the precautions required to preserve antibodies and antigens should be taken when one is using labeled immunoassays. It is particularly important to establish that the antigen or antibody does not bind nonspecifically to the support surface. In some systems where antibody is immobilized or the serum is mixed with coupled antibody, rheumatoid factor may react, yielding false-high values.

Reagents

Radioisotope and enzyme-labeled reagents. Radioisotope and enzyme-labeled reagents must be selected so that the coupled product will have a high specific activity.

Microtiter plates and test tubes. Because the previously described assays are performed primarily with the use of a solid phase (an immobilization of the analyte and reagents into a solid matrix), care should be taken to use the proper type of plastic, one that will hydrophobically attach these reagents so that, once bound, they will not be removed by wash solutions. Several manufacturers have developed plastic microtiter plates and test tubes that are used specifically for enzyme immunoassays.

Substrate. Whichever enzyme is used in these assays, the appropriate substrate specific for the enzyme should be selected to maximize the catalytic activity of the enzyme. Care should be taken to add a sufficient amount of substrate so that if large quantities of enzyme are bound to the solid matrix, full color development can occur and substrate will not be depleted.

Buffers. Close attention should be paid to the ionic strengths, pH, and composition of the buffer used to promote the enzyme's activity. Frequently, specific cations are required for appropriate enzyme activity.

Instrumentation

Measurement of color changes that are a result of enzyme conversion of substrate into product is usually accomplished by use of a spectrophotometer. More recently, the use of microtiter plates, where the volume of fluid in which the color change has occurred is very small (such as 100 to 200 μL), has made it necessary to use special spectrophotometers termed *microtiter readers* or *micro-ELISA readers*. The drawback to the microtiter plate readers, however, is that they are frequently not as sensitive as a standard spectrophotometer. In addition, they are usually end-point readers. However, many have the ability to record the results of a standard 96-well microtiter plate in 1 to 2 minutes.

Instrumentation required for the measurement of radioisotopes is presented in Chapter 7. ^{125}I, the radioisotope most commonly used, requires a γ-counter. Specialized instrumentation is required for the measurement of fluorescent dyes. If metal chelates are employed, techniques such as time-resolved fluorescence may be used (see Chapter 3).

Both the indirect and direct fluoroimmunoassays require a fluorometer or spectrophotofluorometer to obtain accurate reading of the fluorochrome-labeled antibody. The more sensitive the required measurement, the more sophisticated the instrument needed to detect and record the fluorometric signal accurately. Thus for very sensitive measurements, devices such as protein counters are required.

Interfering substances

As indicated earlier, care should be taken to eliminate substances that interfere with the labeled antibody activity, such as antibodies or complement, which might react, binding an artificial matrix, and to eliminate molecules that could interfere with the label, such as enzyme inhibitors or oxidizing reagents.

Common pitfalls

1. Improper storage of immunochemical reagents. Care should be taken to preserve the enzymatic or radioisotopic activity of the labeled reagents.
2. Inappropriate plastic used to manufacture the microtiter plates or test tubes.
3. Improper pH and ionic strengths of buffer.
4. Nonspecific adsorption of reactants to plastic surface. This nonspecific adsorption can be minimized by use of proteinaceous material, such as gelatin or bovine serum

albumin, after initial adsorption of the reagents to the solid phase.

5. Inadequate control of experimental conditions. Factors that must be considered in successful measurement of the enzyme-labeled assays are tight control of temperature, pH, ionic strengths of the buffer, and various cofactor concentrations necessary for the enzyme to convert substrate into product. Finally, since enzymes are protein and can be subject to rapid denaturation under improper incubation conditions, close attention must be paid to preserve the enzyme activity.

6. Substrate depletion. Often, if a high quantity of enzyme-labeled reagents are present on the solid matrix, substrate can be depleted very rapidly. One must take care to have sufficient substrate so that full color development can occur. The upper limits of linearity should be carefully defined.

Test sensitivity and coefficient of variation

Numerous formats are described for the performance of labeled immunoassays. The sensitivity and coefficient of variation of each of these immunoassays depend on the format selected and the instrumentation used to measure the color change. Enzyme immunoassays and radioimmunoassays are generally considered to be sensitive to the nanogram level of detection. Acceptable enzyme immunoassays or radioimmunoassays should have a coefficient of variation of less than 10% in the nanogram range.

Fluoroimmunoassays have a sensitivity for measurement to a lower limit of approximately 100 ng/mL and an upper limit of 200 mg/mL. The between-run variation is between 5% and 10% for the direct assay and usually between 4% and 5%, for the indirect assay. The variation observed within the same run is reported to be 3% to 6%.

SUMMARY

In this chapter, a number of immunological techniques are described and are summarized in Table 10-1. Each of these procedures was developed to meet a specific need to identify or quantitate an antigen present in a patient's sample. The immunoprecipitation techniques are the least sensitive but have the highest degree of specificity in determining the presence of an antigen. Immunonephelometry is the most sensitive of the quantitative assays using precipitation as an end point and is fast becoming the most popular form of assay to quantitate serum and cerebrospinal fluid protein. Recent advances in instrumentation have made this technique the most precise of the immunological techniques for quantitation of antigen, and it is the method of choice over the radial immunodiffusion and rocket immunoelectrophoresis techniques. Immunoelectrophoresis and Western blot immunoelectrophoresis are strictly qualitative tools. The former is commonly employed to assess the quality of an antigen present in a patient's sample, such as polyclonal IgG versus monoclonal IgG or C3 breakdown versus C3 present in native form (the latter for verification of antibody to antigens such as HIV virus).

The agglutination and complement-fixation techniques are procedures used mostly in serology and blood bank laboratories.

Indicator-labeled immunoassays are becoming more popular in the clinical laboratory as quantitative tools because they extend the sensitivity of antigen detection into the range of nanogram quantities. Enzyme immunoassays are the most popular for reasons previously cited. There are many assay formats, and numerous kits are commercially available to measure minute quantities of serum proteins, hormones, and drugs.

Selection of the appropriate immunological technique for detection of an antigen in the laboratory depends on many variables: the technologist's skill, instrumentation, volume of samples to be tested, availability of test format in kit form, quality control reference samples of antigen, ease of the technique, and cost to perform the assay. Whatever assay format is selected, the crucial factors that must always be considered are the specificity of the reagent antibody, the antigen's structure, and sample preservation.

BIBLIOGRAPHY

General

Chan, D, and Perlstein, MT, editors: Immunoassay: a practical guide, Orlando, FL, 1986, Academic Press, Inc.

Roitt, IM, Brostoff, J, and Male, DK: Immunology, London, 1985, Gower Medical Publishing.

Roitt, IM: Essential immunology, ed 6, Oxford, England 1988, Blackwell Scientific Publications.

Rose, NR, Friedman, H, and Fahey, J, editors: Manual of clinical immunology, ed 3, New York, 1986, American Society of Microbiology.

Stites, DP, Stobo, JD, and Wells, JV, editors: Basic and clinical immunology, ed 6, Norwalk, Conn, 1987, Appleton and Lange.

Weir, DM, editor: Handbook of experimental immunology, ed 4, Oxford, England, 1986, Blackwell Scientific Publications.

Immunodiffusion

Crowle, AJ: Immunodiffusion, ed 2, New York, 1973, Academic Press, Inc.

Ouchterlony, O: Handbook of immunodiffusion and immunoelectrophoresis, Ann Arbor, Mich, 1986, Ann Arbor Science Publishers, Inc.

Immunoelectrophoresis

Keren, DF: High resolution electrophoresis and immunofixation, Boston, 1987, Butterworth Publishers.

Penn, GM, and Batya, J: Interpretation of immunophoretic patterns, Chicago, 1978, American Society of Clinical Pathologists.

Ritzmann, SE, and Daniels, JC, editors: Serum protein abnormalities: diagnostic and clinical aspects, New York, 1975, Little, Brown & Co.

Counterimmunoelectrophoresis

Blumberg, BS, Sutnick, AI, and London, WT: Australia antigen as a hepatitis virus, Am J Med 48:1, 1970.

Palmer, DF, Kaufman, L, Kaplan, W, and Cavallaro, JJ: Serodiagnosis of mycotic diseases, Springfield, Ill, 1977, Charles C Thomas, Publishers.

Prince, AM, and Burke, K: Serum hepatitis antigen (SH): rapid selection by high voltage immunoelectrophoresis, Science 169:593, 1970.

Two-dimensional immunoelectrophoresis

Arvan, DA, and Shaw, LM: Two-dimensional immunoelectrophoresis technique and applications in the clinical laboratory, Separation Science 8:123, 1973.

Dunbar, BS: Two-dimensional electrophoresis and immunological techniques, New York, 1987, Plenum Publishing Corp.

Weeke, B: Crossed immunoelectrophoresis. In Axelsen, NH, Kroll, J, and Weeke, B, editor: A manual of quantitative immunoelectrophoresis, Scand J Immunol 2(suppl 1):47, 1973.

Radial immunodiffusion

Fahey, JL, and McKelney, E: Quantitative determination of serum immunoglobulins in antibody-agar plates, J Immunol 94:84, 1965.

Mancini, G, Vaerman, JP, Carbonara, AD, and Heremans, JF: A single radial immunodiffusion method for the immunological quantitation of proteins. In Peeters, H, editor: Protides of the biological fluid, 11th colloquium, Amsterdam, 1964, Elsevier Publishing Co.

Ritzman, SE, Fisher, CL, and Nakamura, RM: Quantitative immunochemical procedures. In Ritzman, SE, and Daniels, JC, editors: Serum protein abnormalities: diagnostic and clinical aspects, Boston, 1975, Little, Brown & Co.

Laurell rocket immunoelectrophoresis

Laurell, CB: Quantitative estimation of proteins by electrophoresis in agarose gel containing antibodies, Ann Biochem 15:45, 1966.

Weeke, B: Rocket immunoelectrophoresis. In Axelsen, NH, Kroll, J, and Weeke, B, editors: A manual of quantitative immunoelectrophoresis, Scand J Immunol 2(suppl 1):37, 1973.

Immunonephelometry

Cambasio, CL, Leek, AE, de Steenwinkel, F, et al: Particle counting immunoassay (PACIA). I. A general method for the determination of antibodies, antigens and haptens, J Immunol Methods 18:33, 1977.

Cambiaso, CL, Riccomi, H, Masson, PL, and Heremans, JF: Automated nephelometric immunoassay. II. Its application to the determination of hapten, J Immunol Methods 5:293, 1974.

Finley, PR, Dye, JA, Williams, J, and Lichti, DA: Rate-nephelometric inhibition immunoassay of phenytoin and phenobarbital, Clin Chem News 27:405, 1981.

Killingsworth, LM, Savory, J, and Teague, PO: Automated immunoprecipitin technique for analysis of the third component of complement (C′3) in human serum, Clin Chem 17:374, 1971.

Nishikawa, T, Kubo, H, and Saito, M: Competitive nephelometric immunoassay methods for antiepileptic drugs in patient blood, J Immunol Methods 29:85, 1979.

Ritchie, RF, Alper, CA, Graves, J, et al: Automated quantitation of proteins in serum and other biological fluids, Am J Clin Pathol 69:151, 1973.

Sternberg, JC: A rate nephelometer for measuring specific proteins by immunoprecipitin reactions, Clin Chem 23:l456, 1977.

Agglutination

Bell, CA, editor: A seminar on antigen-antibody reaction revisited, Washington, DC, 1982, American Association of Blood Banks.

Lennette, EH, Valows, A, Hausler, WJ, Jr, and Truant, JP, editors: Manual of clinical microbiology, ed 3, New York, 1980, American Society for Microbiology.

Williams, CA, and Chase, MW: Methods in immunology and immunochemistry, vol 3, Reactions of antibodies with soluble antigens; vol 4, Agglutination, complement, neutralization and inhibition, New York, 1977, Academic Press, Inc.

Complement fixation

Stansfield, WD: Serology and immunology, New York, 1981, McMillan Publishing Co., Inc.

Fluoroimmunoassays

Hemmila, L: Fluoroimmunoassays and immunofluorometric assays, Clin Chem 31:359-375, 1985.

Nakamura, RM, and Dito, WR: Nonradioisotope immunoassays for therapeutic drug monitoring, Lab Med 11:807, 1980.

Soini, E, and Hemmila, I: Fluoroimmunoassay: present status and key problems, Clin Chem 25:353, 1979.

Enzyme immunoassays

Engvall, E, and Perlmann, P: Enzyme-linked immunosorbent assays (ELISA): quantitative assay of immunoglobin G, Immunochemistry 8:871, 1971.

Kemeny, DM, and Chaldacombe, SJ: Elisa and other solid phase immunoassays, New York, 1988, John Wiley & Sons.

Maggio, ET, editor: Enzyme immunoassay, Cleveland, 1980, CRC Press, Inc.

CHAPTER 11

Competitive-binding assays

STEPHAN G. THOMPSON

OBJECTIVES

- Define and describe competitive-binding assays.
- List three common markers used to label ligands and explain the factors that determine the choice of the ligand label.
- Define cross-reactivity and briefly explain its effect on a competitive-binding assay.
- Describe the process of data reduction and briefly outline the generation of the four most common dose-response curves, given bound counts per minute and nonspecific bound counts per minute.
- Outline the principle of each of the following competitive-binding assays:
  - Radioimmunoassay
  - Enzyme-linked immunoabsorbent assay
  - Enzyme-multiplied immunoassay
  - Fluoresence polarization immunoassay

KEY TERMS

See Chapters 9 and 10 for additional definitions.

competitive-binding assay An analytical procedure based on the reversible binding of a ligand to a binding protein. The ligand competes in proportion to its concentration with a labeled derivative for binding to the limited number of binding sites.

detection limit The smallest concentration of a ligand that can be statistically distinguished from a zero level in an assay. The detection limit is also referred to as the sensitivity of an assay.

enzyme-linked immunosorbent assay (ELISA) A heterogeneous immunoassay that employs an enzyme-labeled ligand and antibody immobilized on a solid phase.

enzyme-multiplied immunoassay technique (EMIT) A homogeneous enzyme immunoassay in which a low-molecular-weight ligand is attached to an enzyme whose activity is inhibited when the conjugate is bound by a specific antibody. Competitive binding of unlabeled ligand to the antibody relieves the inhibition, and a dose-response curve in proportion to the ligand concentration can be generated. This assay does not require a step to separate antibody-bound label and free label.

fluorescence immunoassay (FIA) Competitive-binding assays that employ a fluorophore or fluorogen as the label.

fluorogen A nonfluorescent molecule that becomes fluorescent when modified by a chemical or enzymatic process.

heterogeneous assay A competitive-binding assay in which it is necessary to separate mechanically the protein-bound, labeled ligand from the unbound ligand before measurement of the signal generated by the label.

homogeneous assay Competitive-binding assay in which it is not necessary to separate protein-bound and free ligand fractions because signal of label is modulated by protein binding.

immunoassay Any binding assay in which the binding protein is an antibody.

immunogen A substance that stimulates an antibody response when administered to an appropriate animal. Immunogens include macromolecular antigens and otherwise nonantigenic haptens coupled to a macromolecular carrier.

immunometric assay Competitive and noncompetitive protein-binding assays in which the antibody rather than the ligand is labeled with a radioisotope or other suitable label.

label An atom or molecule attached to either the ligand or binding protein, capable of generating a signal for monitoring the binding reaction.

ligand A molecule or part of a molecule that is reversibly bound by the binding protein in a competitive-binding assay. It usually is the analyte but can also be a cross-reactant.

nonisotopic binding assays Assays with labels other than a radioactive or stable isotope.

prosthetic group Nonprotein components of enzymes and other functional proteins that are required for functional activity of the protein.

sensitivity The degree of response to a change in the ligand concentration in an assay. Sensitivity also refers to and is synonymous with the detection limit.

specificity The degree to which a binding protein binds its particular ligand while not binding structurally similar compounds.

DEFINITION OF COMPETITIVE-BINDING ASSAYS

Competitive-binding assays are based on the noncovalent, reversible binding of a ligand to a specific binding protein. The binding assay is most often described by the following reaction:

$$\text{Ligand} + \text{Binding protein} \rightleftarrows \text{Binding protein: Ligand} \qquad \textit{Eq. 11-1}$$

where the binding protein has a measurable affinity for the ligand that interacts with it. In general, only one binding protein can attach to a small molecule. This reaction can be considered simply as one molecule of ligand reacting with one protein-binding site. The important molecular feature of binding proteins that enables them to be used in quantitative assays is their ability to bind compounds with a high specificity and affinity (see Chapter 9). Examples of specific binding proteins are listed in Table 11-1.

Antibodies that bind small molecules are usually gamma immunoglobulins (IgG) that are produced in animals immune to the specific ligand. The methods of raising such antibodies are discussed in Chapters 9 and 10 and also in more detail in other texts listed in the bibliography at the end of this chapter.

Small molecules that by themselves normally do not provoke an immune response but do elicit antibodies when coupled with larger molecules are termed *haptens*. Because such small molecules are involved in these competitive-binding assays where antibody is used, the ligand may be referred to as a hapten.

The ligand in Equation 11-1 is the analyte to be quantified. Often ligands are drugs (digoxin, theophylline), hormones (cortisol, T_4), or vitamins (B_{12}). For most competitive-binding reactions, the ligand refers to both the analyte and a labeled derivative of the analyte. Both must be able to bind the specific binding protein for the analyte to be measured. The final complex of binding protein and ligand is usually stable and dissociates slowly under favorable circumstances.

Table 11-1 Specific binding proteins present in serum or tissues

Binding protein	Ligand
Antibodies	Varied antigens
Corticosteroid binding globulin (CBG)	Cortisol, corticosterone
Estrogen receptor	Estrogen
Intrinsic factor	Vitamin B_{12}
Thyroglobulin (TBG)	Thyroxine (T_4)
	Triiodothyronine (T_3)
Vitamin D receptor	1-α,25-Dihydroxyvitamin D_3

The reaction described in Equation 11-1 is more complex for large molecules, such as proteins, because these can have many different binding sites (antigenic determinants). Thus the molecule can have one, two, or more binding proteins attached at the same time. In addition, such sites may be quite different from one another so that the population of antibodies generated in response to large molecules is more heterogeneous. The antigenic determinants of high-molecular-weight immunogens and low-molecular-weight haptens are similar when one considers the behavior of antibody-binding reactions. Those reactions involving haptens, however, are much simpler than the interactions of antibodies with high-molecular-weight molecules. Assays other than competitive-binding ones may be used for quantitation of large molecules. These are discussed in Chapter 10.

LAW OF MASS ACTION

The law of mass action describes some aspects of the phenomenon that occurs when molecules bind to one another. This is best illustrated when one examines the concentration of an antibody (Ab) and its ligand (L) or specific binding partner under equilibrium conditions. The bimolecular reaction

$$Ab + L \underset{k_{-1}}{\overset{k_1}{\leftrightarrows}} Ab{:}L \qquad \textit{Eq. 11-2}$$

can be rearranged for calculation of the equilibrium association constant

$$K_a = \frac{k_1}{k_{-1}} = \frac{[Ab{:}L]}{[Ab][L]} \qquad \textit{Eq. 11-3}$$

where k_1 and k_{-1} are the respective rate constants for association and dissociation of the bound complex; $[Ab]$ is the concentration of the unbound or free antibody at equilibrium; $[L]$ is the equilibrium concentration of unbound ligand (a term denoting the antigen, hapten, or other substance); and $[Ab{:}L]$ is the equilibrium concentration of the ligand-antibody complex. K_a, also referred to as the *affinity constant,* is defined in reciprocal molar concentrations (M^{-1}) or liters per mole (L/mol). This is the volume into which a mole of the binding protein can be diluted to yield 50% binding of the ligand. The larger the K_a, the greater the affinity of the antibody for the ligand. It follows that for a constant amount of antibody in the reaction, less ligand is required for a high-affinity antibody to bind 50% of the ligand than is required for 50% binding by a low-affinity antibody. Antiserum produced by an animal that has been immunized with a high-molecular-weight immunogen usually contains a mixture of different populations of antibody (*polyclonal* antisera) to that antigen. These different populations of antibody vary in their ability to bind the ligand (affinity) and in their ability to recognize different sites on the protein's surface.

When the binding protein or antibody is homogeneous, there is only one population of uniform binding sites for a ligand. This usually occurs when the binding reaction employs *monoclonal* antibodies (see Chapter 10). These antibodies are homogeneous, and they have a uniform binding affinity (K_a) and specificity for the antigen.

Low-affinity binding proteins and antibodies typically have association constants in the order of 10^5 to 10^7 L/mol, whereas binders suitable for immunoassays and other competitive-binding assays must have association constants between 10^8 and 10^{11} L/mol. A higher association constant enables one to design assays with sensitivities as low as 10^{-9} to 10^{-12} M, provided the label itself is detectable at these low concentrations.

COMPETITIVE BINDING

The competitive-binding assay can be imagined as the addition of increasing amounts of unlabeled ligand to reaction mixtures containing known, constant amounts of labeled ligand and a specific binding protein. In this case, labeled ligand, L^*, and antibody, Ab, are added together in equimolar amounts. Presuming that all the L^* is bound, the reaction becomes:

$$L^* + Ab \rightleftarrows Ab{:}L^*$$

Two things happen with the addition of increasing amounts of unlabeled ligand (L): (1) unlabeled ligand competes with the labeled ligand for antibody-binding sites, and (2) there is an excess of the total ligand (L and L^*) in solution compared to the number of binding sites. The concentration of antibody-binding sites is therefore limiting with respect to total ligand, thus modifying the preceding reaction as follows:

$$L + AbL^* \rightleftarrows AbL + AbL^* + L^* + L$$

Less L^* is antibody bound (as $Ab{:}L^*$); thus more L^* is free as the amount of L increases. The amount or percentage of L^* in the bound form can be calculated from the amount of L and L^* present. Examples of such calculations are shown in Table 11-2.

Table 11-2 Competive binding response

Moles per liter of reactant × 10^8			
Antibody	**Labeled ligand**	**Unlabeled ligand**	**%Labeled ligand bound**
1	1	0	100
1	1	1	50
1	1	2	33
1	1	3	25
1	1	4	20
1	1	9	10

When the percentage of labeled ligand bound (Table 11-2) is plotted as a function of the concentration of the unlabeled ligand, it yields the dose-response curve shown in Fig. 11-1. The term *dose-response curve* applies to a plot of binding versus increasing amounts of ligand. The curvature of the dose response plot in Fig. 11-1 is attributable to the logarithmic increase in the percentage of L that is

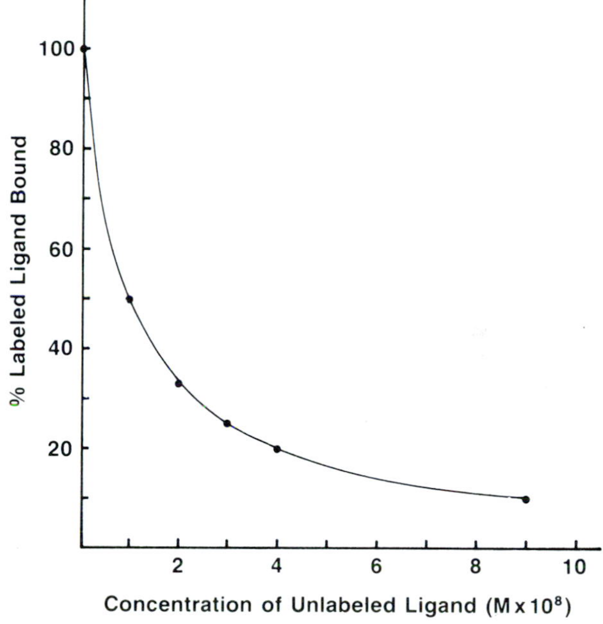

Fig. 11-1 Linear dose-response curve for data shown in Table 11-2.

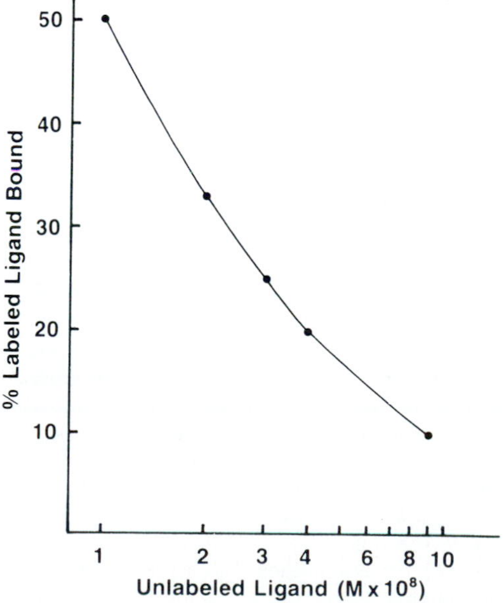

Fig. 11-2 Dose-response curve in which response is plotted as function of log of ligand concentration.

Table 11-3 Techniques to separate bound from free labeled ligand

Technique	Principle
Absorbents	
Nonspecific	Low-molecular-weight ligands such as drugs and steroids are absorbed by particles such as charcoal.
Specific	Antibodies to the ligand or to the ligand-binding antibody are immobilized to a solid matrix such as glass, latex, paper, or plastic. The immobilized antibody-ligand complex is separated from the unbound ligand by decantation, washing, filtration, or diffusion.
Chromatography	The protein-bound ligand moves at a different rate through the chromatographic medium than does the free ligand. This may include multilayer films.
Precipitation	
Ammonium sulfate	The antibody-bound ligand is precipitated by ammonium sulfate (Farr technique).
Double antibody	The antibody-bound ligand is precipitated by the addition of a second antibody specific for the antibody in the antibody-ligand complex.

bound when the concentration of L (dose) in the assay increases arithmetically. Thus the decrease in bound L^* is also logarithmic. Conversion of the concentration of L to a logarithm as shown in Fig. 11-2 makes the relationship become more linear.

SEPARATION TECHNIQUES

Figs. 11-1 and 11-2 show that to derive a dose-response or standard curve one must know the amount of labeled ligand that is antibody bound as a function of the amount of unlabeled ligand added. A variety of techniques have been developed to measure either the bound or free forms of the labeled ligand.

Some of these techniques require that the labeled ligand bound by antibody be physically separated from the free labeled ligand. These assays are called *heterogeneous assays*. Table 11-3 lists some methods used to separate the bound and free labeled ligand. Immunoassay approaches that do not require physical separation of bound and free labeled ligand are called *homogeneous assays*. The signal of the label in a homogeneous assay is altered on binding to the specific binding protein; thus the bound and free labeled ligands can be directly distinguished from one another.

LABELED LIGANDS

Types of labels

Common types of markers used to label ligands include radioisotopes, enzymes, and fluorophores. These can be used in both homogeneous and heterogeneous competitive-binding assays. The type of label, assay to which it is suited, and detection system are presented in Table 11-4.

Factors determining choice of label

Radioisotopes. Radioisotopic labels can be used only with heterogeneous immunoassay systems because binding by antibody does not change the radioactive emission. In general, the desired sensitivity of the assay limits the choice of radioactive labels to certain specific isotopes that have high specific activity, high energy output, manageable half-lives, and ready availability. The radioisotope must be readily incorporated by or coupled to the ligand molecule, and its emission must be easily detected. Isotopes that meet these requirements are listed in Table 11-5. Consideration of all the factors, especially the need for high specific activity and ease of incorporation, has made ^{125}I the label of choice for most radioassays.

Enzymes. Enzymes can be used as labels. Enzymes differ from radioisotopes in that the binding reaction can modify their activity. Again the enzymes must have high specific activity (that is, conversion of many moles of substrate to product per minute per mole of enzyme) and must also be easily coupled to the ligand without losing significant activity. Enzymes that have been used successfully

Table 11-4 Some labels for competitive-binding assays and their means of detection

Label	Assay		Detector
	Homogeneous	Heterogeneous	
Chemiluminescence	Yes	Yes	Photon counter
Enzymes			
Chromogenic substrates	Yes	Yes	Spectrophotometer
Fluorogenic substrates	Yes	Yes	Fluorometer
Enzyme substrates	Yes	Yes	Spectrophotometer, fluorometer
Enzyme prosthetic groups	Yes	Yes	Spectrophotometer, fluorometer
Fluorophores	Yes	Yes	Fluorometer
Metals	No	Yes	Atomic absorption spectrophotometer
Radioactive isotopes	No	Yes	Radioactivity counter

Table 11-5 Radioisotopes used in competitive binding assays

Isotope	Emission	Maximum specific activity* (Ci/g)	Half-life	Counter†
^{3}H	Beta	9.6×10^3	12.3 years	LS
^{14}C	Beta	4.5	5730 years	LS
^{32}P	Beta	2.85×10^5	14.2 days	LS
^{125}I	Gamma	1.74×10^4	60 days	Crystal
^{57}Co	Gamma	8.48×10^3	270 days	Crystal

*The curie (Ci) is a unit of radioactivity equal to 3.7×10^{10} disintegrations per second.
†Beta-particle emitters are counted in liquid scintillation (LS) counters by the release of photons from organic phosphors in solution. Gamma-ray emitters are counted in detectors with a sodium iodide crystal that contains fluor from which photons are released.

Table 11-6 Fluorophores used as labels in competitive-binding assays

Fluorophore	Excitation wavelength (nm)	Emission wavelength (nm)	ε*	Fluorescence quantum yield†
Europium chelate‡	340	613	90,000	>0.95
Fluorescein	490	520	72,000	0.85
Rhodamine	550	585	50,000	0.70
Umbelliferone	380	450	20,000	0.50

*ε is the absorbance of a 1 molar solution through a 1 cm light path.
†The fluorescence quantum yield is relative to the quantum yield of acridine, which is 1.0.
‡Europium: β-diketone chelate.

include peroxidase, alkaline phosphatase, glucose-6-phosphate dehydrogenase, and glucose oxidase.

Fluorophores. Fluorophores used as labels must be able to be attached to the ligand and still fluoresce with a high degree of efficiency. It is also helpful if the absorption (excitation) and emission wavelengths are well separated so that light scatter does not contribute to the fluorescence seen at the emission wavelength. Examples of fluorophores used as labels for competitive-binding assays and their properties are shown in Table 11-6. Of these, fluorescein, europium chelates, and umbelliferone are the most commonly used. Chelates of the rare earth metal europium strongly absorb light and fluoresce with properties that depend on the chelating ligand. The quantum yield (photon output/photon input) is very high while the excitation and emission wavelengths are well separated. Europium and fluorescein have the highest extinction coefficients and quantum yields, but europium has the largest difference between the excitation and emission wavelength (273 nm).

SENSITIVITY

The sensitivity of a binding-reaction assay is a function of the affinity of the binding protein for its ligand. Consequently, for 50% binding to occur, a ligand present at a concentration of 1×10^{-7} M would require an antibody that had a K_a of 10^7 L/mol, whereas a ligand present at a concentration of 1×10^{-10} M would need an antibody that had a K_a of 1×10^{10} L/mol.

Ideally in a competitive-binding assay the binding protein would have the same affinity for both the labeled and unlabeled ligands. Usually this is not the case. In some instances the label or the labeling procedure will alter the immunochemical properties of the ligand to the extent that antibody does not bind it as well as it does the unlabeled compound. The converse is also true: antibodies are often made against haptens in which the chemical bridge used to couple the ligand to the protein carrier is the same bridge used for coupling the ligand to the label. In these cases the antibodies may have higher affinity for the labeled ligand with the chemical bridge than they do for the unmodified ligand. One who designs a competitive-binding assay must consider differences in relative affinity of the antibody for the labeled and unlabeled ligand.

CROSS-REACTIVITY

The specificity of a binding protein for its ligand is measured by its ability to bind only the ligand and not other substances. Cross-reacting molecules are those that are so similar to the ligand in structure that they are capable of being bound by the antibody. The greater the chemical difference between the ligand and a potential cross-reactant, the less likely it is that the cross-reactant will be bound by antibody to the ligand. Differences in antibody binding of ligand and cross-reacting substances are therefore a function of differences in affinity. These differences are reflected by responses to cross-reactants in competitive-binding assays. Table 11-7 presents the relationship between the K_a of the antibody for its ligand and two cross-reac-

Table 11-7 Cross-reactant binding as a function of antibody affinity

Antibody	Bound species	K_a	Concentration (M) required for 50% binding*
1	$Ligand_1$	1×10^8	2×10^{-8}
	$Cross\text{-}reactant_a$	1×10^7	2×10^{-7}
	$Cross\text{-}reactant_b$	5×10^7	4×10^{-8}
2	$Ligand_2$	1×10^{10}	2×10^{-10}
	$Cross\text{-}reactant_c$	2×10^8	1×10^{-8}
	$Cross\text{-}reactant_d$	1×10^6	2×10^{-6}

*When 50% of the ligand or cross-reactant is bound, $B/F = 1$. Since $K_a = \frac{B}{F[Ab]}$; when $B/F = 1$, the $K_a = \frac{1}{[Ab]}$

tants and the concentration of each that is required in the assay to deliver the same binding response. The concentration of $cross\text{-}reactant_a$ required for 50% binding to $antibody_1$ is 10 times the concentration necessary to bind 50% of $ligand_1$. Similarly, 10,000-fold less $ligand_2$ is required to achieve 50% binding to $antibody_2$ than is necessary to bind 50% of $cross\text{-}reactant_d$. Table 11-7 shows that with a lower K_a, more antibody is required to bind 50% of the ligand or cross-reactant, further illustrating the relationship between sensitivity and K_a. The degree to which each cross-reactant in Table 11-7 interferes with the analysis of each ligand depends on the relative concentrations of ligand and cross-reactant in actual samples. For example, cross-reactants c and d would probably *not* interfere in the assay of $ligand_2$, *unless* they were present at concentrations 100-fold or 10,000-fold higher, respectively, than that of $ligand_2$.

Ideally, antibodies or other binding proteins that participate in competitive-binding reactions are very specific for the ligand with essentially no cross-reactivity with closely related molecules. In reality, this is rare, because the antibodies present in a heterogeneous antiserum bind the ligand with different affinities and orientations and are therefore also likely to bind structurally similar molecules. One of the advantages of the use of monoclonal antibodies is the potential for selecting very specific antibodies that have essentially no cross-reactivity with other compounds. Examples of the cross-reactivity of antiserums and a monoclonal antibody are shown in Figs. 11-3 and 11-4. The dose-response curves seen with antiserum, in Fig. 11-3, show that caffeine, which is structurally similar to the antiasthmatic drug theophylline, effectively competes with the label for theophylline binding sites only at much higher concentrations. The degree of caffeine cross-reactivity, determined by the "classical" approach to cross-reactivity, is calculated when one divides the concentration of ligand (in this case theophylline) at 50% of the maximum binding (indicated in Fig. 11-3) by the concentration of cross-reactant (caffeine), which also displaces 50% of the label according to the following equation:

Table 11-8 Caffeine cross-reactivity with theophylline antiserum or monoclonal antibodies

	Concentration (M) when 50% of label is bound		% Cross-reactivity
	Theophylline	Caffeine	
Antiserum	1.29×10^{-7}	1.05×10^{-6}	12.3
Monoclonal antibody	6.15×10^{-8}	5.09×10^{-6}	1.2

$$\frac{\text{[Ligand] at 50\% binding}}{\text{[Cross-reactant] at 50\% binding}} (100\%) = \text{\% cross-reactivity} \qquad \textbf{\textit{Eq. 11-4}}$$

There is 12.3% caffeine cross-reactivity for the antiserum shown in Fig. 11-3, but only 1.2% cross-reactivity shown with the theophylline monoclonal antibody in Fig. 11-4. Table 11-8 summarizes these results. A competitive-binding assay that uses the monoclonal antibody to theophylline is much more specific than one that uses the antiserum shown in Fig. 11-3; consequently the former is less prone to caffeine interference.

DATA REDUCTION

Earlier in this chapter the displacement reaction was described by the equation:

$$L^* + Ab + L \rightleftarrows (Ab{:}L^*) + (Ab{:}L)$$

where *Ab* is the antibody or another binding protein providing a limited number of ligand-binding sites. The concentration of the labeled ligand, L^*, is constant and in excess of the binding-site concentration. The ligand, *L*, will compete with L^* for the available binding sites. As was shown in Table 11-2, the amount of L^* bound by the antibody is inversely proportional to the concentration of *L* in the assay. The dose response is quantitated when one measures either the bound L^* or the free L^* that has been displaced by *L*. Radioimmunoassay (RIA) is a heterogeneous assay that will serve as an example to illustrate the most common methods for generating dose-response curves and reducing data to forms usable for quantifying concentrations of analytes.

The data shown in Table 11-9 were generated by use of a double-antibody RIA for the aminoglycoside antibiotic amikacin, where ^{125}I-amikacin was the labeled ligand. The primary antibody was rabbit antiserum to amikacin. Goat antiserum to rabbit antibodies is included in the assay to precipitate the antibody-bound labeled ligand (^{125}I-amikacin). The precipitate is collected by centrifugation, and the pellet in the bottom of the tube is counted for radioactivity after the supernatant has been removed. The quantity *nonspecifically bound (NSB) counts per minute* refers to the radioactivity nonspecifically bound to the bottom of the tube in the absence of antibody to amikacin.

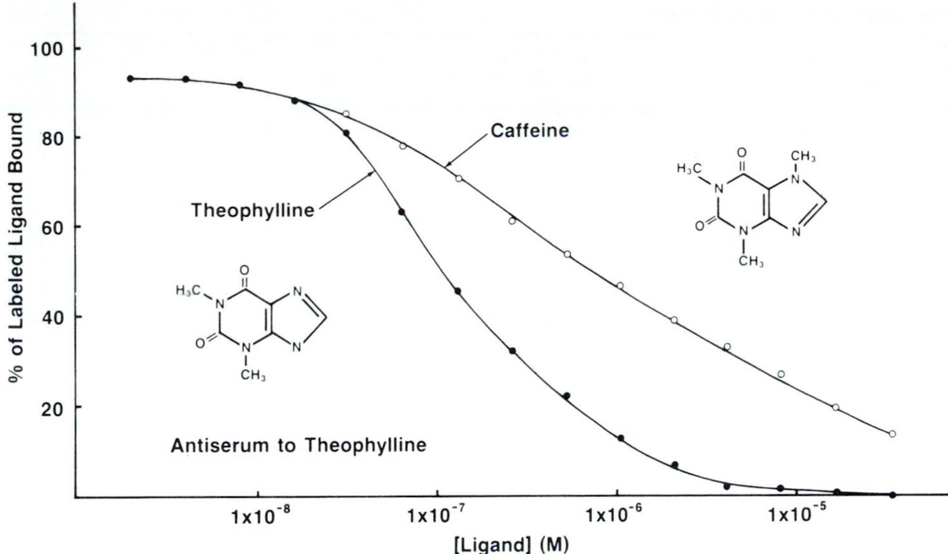

Fig. 11-3 Cross-reactivity of caffeine with an antiserum to antiasthmatic drug theophylline in a substrate-labeled fluorescent immunoassay. Degree of cross-reaction is determined at concentrations of theophylline and caffeine required for 50% of dose response. This is equivalent to 46.5% of bound label. Refer to Table 11-8 for cross-reactivity data.

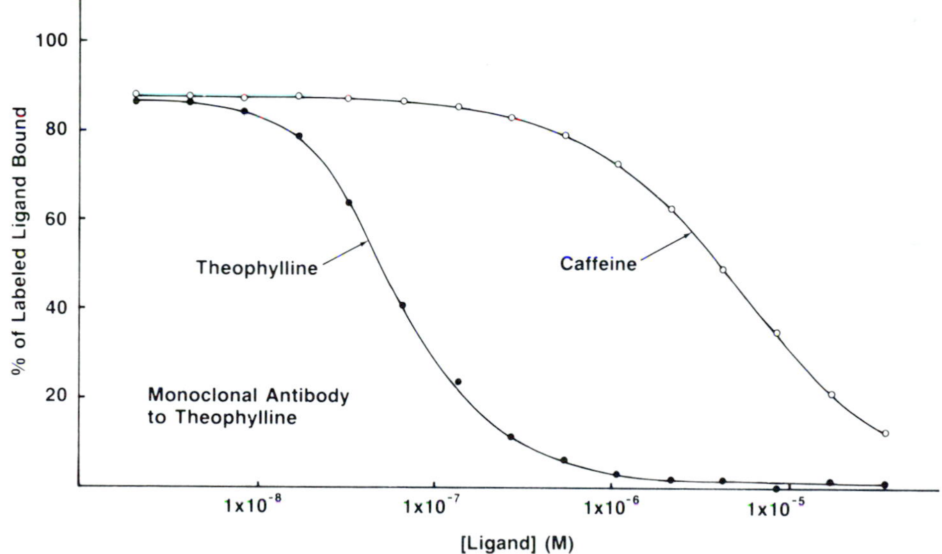

Fig. 11-4 Cross-reactivity of caffeine with a monoclonal antibody to theophylline in substrate-labeled fluorescent immunoassay. Cross-reactivity is determined at 43.2% of bound label.

Table 11-9 shows how the amikacin RIA data are reduced to generate the four most common dose-response curves. In addition to the total bound counts per minute *(TB)*, the *y* axis can be drawn as *%B*, *B/F*, B/B_0, or logit B/B_0. The logit transformation is the following:

$$\text{Logit } (B/B_0) = \ln \frac{B/B_0}{1 - B/B_0} \qquad \textit{Eq. 11-5}$$

The bound counts per minute, *%B*, and *B/F* are never plotted as a function of the arithmetic dose of the ligand concentration because the resulting curves are highly nonlinear (see Fig. 11-1). Instead, they can be plotted versus the log of the amikacin concentration to yield the slightly sigmoid dose-response curve seen in Fig. 11-5. The graph of log dose versus the logit B/B_0 (Fig. 11-6) has been the most accepted empirical method for linearizing competitive protein–binding dose-response curves. Automatic data reduction and processing of the log-logit transformation are easily performed with computerized systems, whereas the plot shown in Fig. 11-5 is constructed either manually or with

Table 11-9 Data for amikacin radioactivity*

Dose of amikacin ($M \times 10^{-8}$)	Total bound cpm (*TB*)	Specifically bound cpm (*B*)†	%*B*‡	*B/F*§	B/B_0‖	Logit B/B_0
0	14019	13588 (B_0)	64.0	1.78	1.00	—
1.07	10694	10264	48.4	0.94	0.76	1.13
2.14	9235	8805	41.5	0.71	0.65	0.61
4.28	7184	6754	31.8	0.47	0.50	−0.01
8.56	5360	4930	23.2	0.30	0.36	−0.56
17.12	3925	3495	16.5	0.20	0.26	−1.06
Unknown 1	8912	8482	40.0	0.67	0.62	0.51
Unknown 2	6910	6480	30.5	0.44	0.48	−0.09
Unknown 3	4340	3910	18.4	0.23	0.29	−0.91

*Total counts per minute (cpm), *T*, of ^{125}I-amikacin in each reaction are 21,225; nonspecifically bound (NSB) counts per minute are 430.
†Specifically bound cpm = *B* = Total bound cpm − NSB.
‡% bound = $(B/T)100$.
§$B/F = \dfrac{B}{TB - B}$
‖ $B_0 = B$ at zero dose of drug.

a calculator. The standard curves demonstrated in Figs. 11-5 and 11-6 can be used for estimation of the concentration of the unknowns listed in Table 11-9. These results are shown in Table 11-10. One can see from Table 11-10 that the interpolated concentrations of the unknown samples are close and relatively independent of the plotting method. Of the three methods, the best linear fitted line over the range of the standard curve is the logit B/B_0 versus log dose. The shallow response seen at higher concentrations of amikacin in the *B/F* versus dose plotting routine is more prone to errors in precision because slight changes in the *B/F* response will cause greater changes in the interpolated dose when the standard curve is shallow (see Fig. 11-1), whereas the same changes in the region of a steep slope (such as seen at lower concentrations) will cause less variation. Since the logit B/B_0 versus log dose plotting method can be routinely linearized for rapid data reduction and processing of the dose responses of unknown concentrations of analyte, this is the most popular method for generating standard curves.

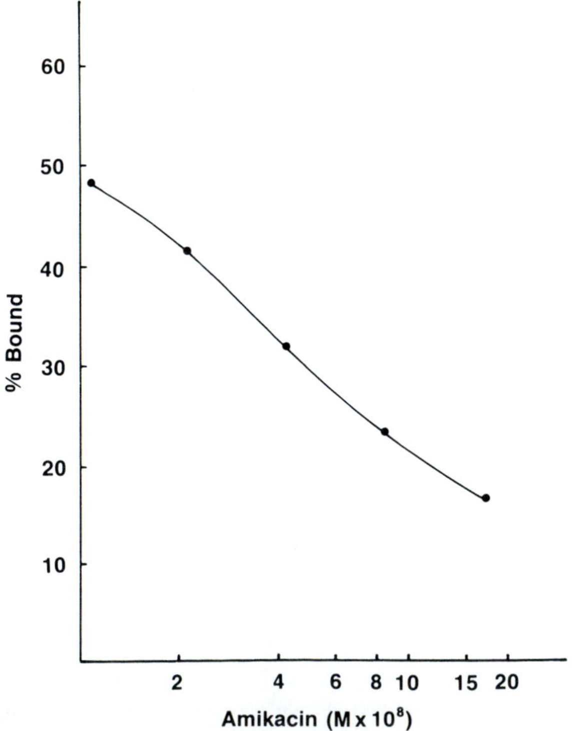

Fig. 11-5 Amikacin radioimmunoassay dose-response curve with percentage of bound amikacin plotted as function of log of amikacin concentration (see Table 11-9).

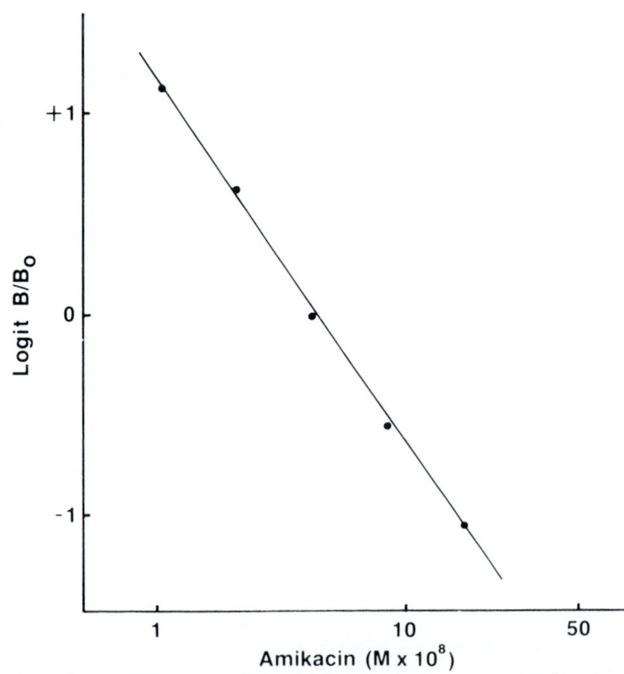

Fig. 11-6 Logit (B/B_0) versus log of amikacin concentration for radioimmunoassay data presented in Table 11-9.

Table 11-10 Concentrations of unknown sample as determined with different data-reduction techniques

	Amikacin (M × 10^{-8})		
Unknown sample	**%*B* versus log dose**	***B*/*F* versus dose**	**Logit B/B_0 versus log dose**
1	2.38	2.40	2.33
2	4.75	4.60	4.93
3	14.06	14.40	13.62

EXAMPLES OF COMPETITIVE-BINDING ASSAYS

Radioimmunoassay

In RIA, the ligand and a constant amount of radioactively labeled ligand compete for a limited number of antibody-binding sites. The concentration of antibody is usually sufficient to bind between 30% and 80% of the labeled material. For example, the antibody to amikacin used in the earlier example bound 64% of the ^{125}I-amikacin in the absence of unlabeled ligand. Addition of unlabeled ligand to the reaction yields a net increase in the total ligand (labeled plus unlabeled), but, because of competition for antibody-binding sites, there will be a decrease in the proportion of labeled ligand that will be bound by the antibody. If one counts the radioactivity bound to the antibody after the separation step, the dose-response curve will have a negative slope like that shown in Fig. 11-7, *A*. As the concentration of unlabeled ligand increases and the antibody-binding sites approach saturation, the slope levels off. When the unbound, radioactively labeled ligand is monitored, the dose-response curve has a positive slope (Fig. 11-7, *B*) with the same shape as in Fig. 11-7, *A*.

Radioimmunoassay is applicable to the measurement of both low-molecular-weight and high-molecular-weight ligands, provided that the labeling procedure or the labeled ligand conjugate itself does not adversely affect the immunoreactivity of the ligand. Some ligands (including proteins) lack the tyrosine residues required for radioactive labeling with the isotopes ^{125}I or ^{131}I. In these cases the ligand must be derivatized with tyrosine before labeling. Derivatization of the ligand with tyrosine, or even nonisotopic labels, could yield a labeled ligand that has lost some or all of its immunoreactivity. Another consequence of conjugation is that some ligands may be unstable when labeled. These problems are sometimes overcome by labeling the antibody rather than the ligand because the major structural differences between IgG molecules are substantially less than the differences between ligands. Losses in immunoreactivity are less likely to occur when the antibody is labeled. IgG has many tyrosines potentially available for labeling with radioactive iodide. The labels are often conjugated to sites on the antibody distal to the binding regions, with such conjugation allowing the antibody to bind the ligand without interference by the label.

Assays based on the use of a labeled antibody are called *immunometric* assays. One that uses a radiolabeled antibody is an immunoradiometric assay (IRMA). With this heterogeneous assay format the ligand in the sample competes with the ligand attached to a solid surface for the binding sites of the labeled antibody (Fig. 11-8). The amount of labeled antibody bound to the solid surface is determined after excess label and sample have been removed. A representative dose-response curve is shown in Fig. 11-8. The immunometric format is considered again in the discussion of ELISA techniques.

Enzyme-linked immunosorbent assay

The enzyme-linked immunosorbent assays (ELISA) are heterogeneous nonisotopic assays that usually have an antibody immobilized onto a solid support (see Table 11-3) and the ligand labeled with the enzyme. Table 11-11 describes some enzymes used for ELISA (or other enzyme immunoassays). These enzymes are useful as labels because they satisfy the following criteria:

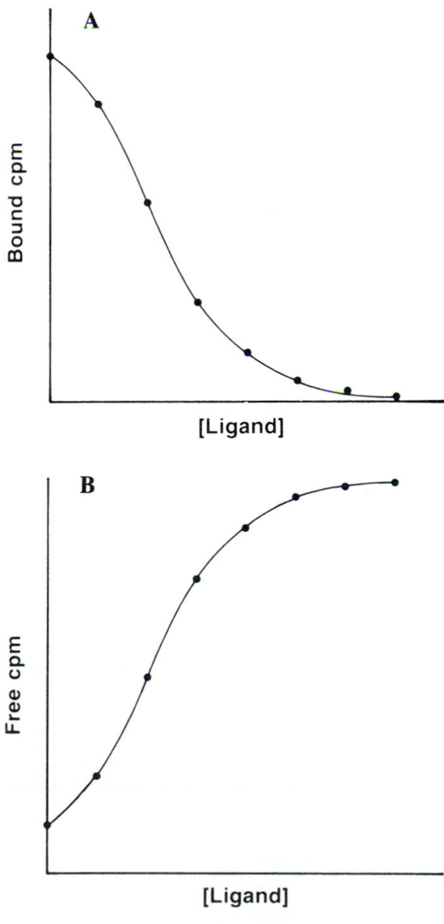

Fig. 11-7 Radioimmunoassay dose-response curves illustrating slopes for antibody-bound, **A,** and free, **B,** radioactivity in counts per minute (cpm) versus dose.

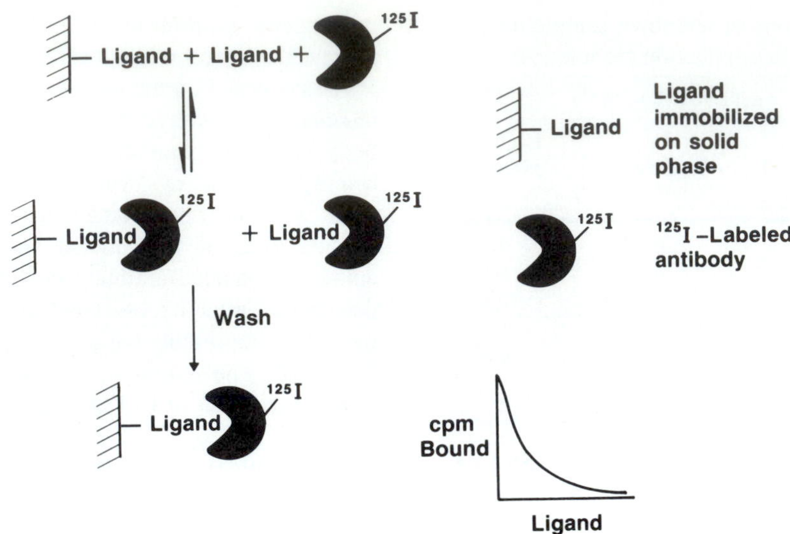

Fig. 11-8 Principle of competitive immunoradiometric assay (IRMA) and typical dose-response curve.

1. The enzymes have a high specific activity. The signal amplification seen with an enzyme is related to the amount of substrate converted to product during the time of incubation. Enzymes with the highest specific activities are potentially capable of giving the greatest amplification. Assays using such enzymes have excellent sensitivity and are able to detect very low concentrations of ligand.

2. The labels are stable during the assay and under refrigerated storage conditions.

3. The enzymes are not present in the biological fluid or tissue sample that is to be analyzed.

4. The enzymes retain most of their activity when attached to the ligand (or antibody).

Alkaline phosphatase and horseradish peroxidase are both available inexpensively in a highly purified form. For this reason and the reasons listed previously, these two enzymes are the enzymes most often used as labels for ELISA. The enzymes listed in Table 11-11 can use either chromogens or fluorogens as substrates, with the respective products yielding an absorbance or fluorescence signal. The advantage of a chromogenic substrate is that its products can be detected visually. Fluorescent products can be detected at 100 to 1000 times lower concentrations than those of chromophores. In addition, the incubation time can be greatly shortened and the sensitivity for the measured ligand is potentially greater when one is using fluorogenic substrates.

Table 11-11 Enzyme labels for immunoassays

Enzyme	Source	Molecular weight	Specific activity (units/mg)*
Alkaline phosphatase	Calf intestine	100,000	400
β-Galactosidase	*Escherichia coli*	540,000	400
Glucose oxidase	*Aspergillus niger*	160,000	200
Glucose-6-phosphate dehydrogenase	*Leuconostoc mesenteroides*	130,000	250
Peroxidase	Horseradish	40,000	900

*A unit of enzyme activity represents the conversion of 1 μmol of enzyme substrate to product per minute.

Many configurations for ELISA have been devised. Some are based on competitive reactions, whereas others are direct immunometric "sandwich" assays. The sandwich ELISA is discussed in Chapter 10. The two basic formats for the competitive assays and the shape of the respective typical dose-response curves describing the signal remaining on the solid phase are shown in Figs. 11-9 and 11-10.

Of the two, the configuration in which the antibody has been immobilized onto the solid surface (Fig. 11-9) has been the most frequently described. This is analogous to the configuration in an RIA because the ligand in the sample competes with the enzyme-labeled ligand for the limited amount of antibody binding sites fixed to the solid phase. After the binding reaction has taken place, the solid phase is washed with buffer to remove the unbound labeled ligand so that it does not contribute to the signal. The amount of enzyme bound to the solid phase is proportional to the absorbance (or fluorescence) of the product formed after the addition of substrate, and it is inversely proportional to the concentration of unlabeled ligand. This method is applicable to both low- and high-molecular-weight analytes.

Instead of the antibody, the ligand can be attached to the solid surface as shown in Fig. 11-10. Only those binding sites in solution not occupied by the ligand from the

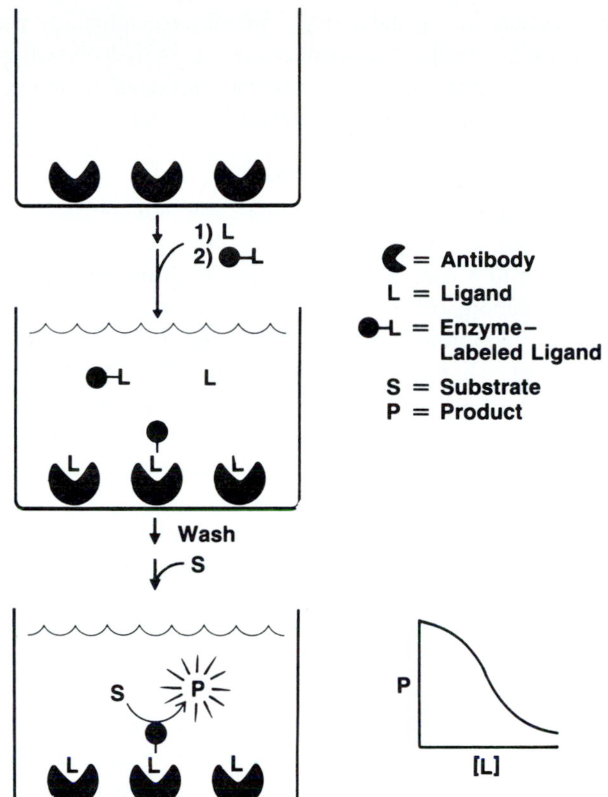

Fig. 11-9 Principles of competitive enzyme-linked immunosorbent assay (ELISA) with ligand labeled with an enzyme and typical dose-response curve.

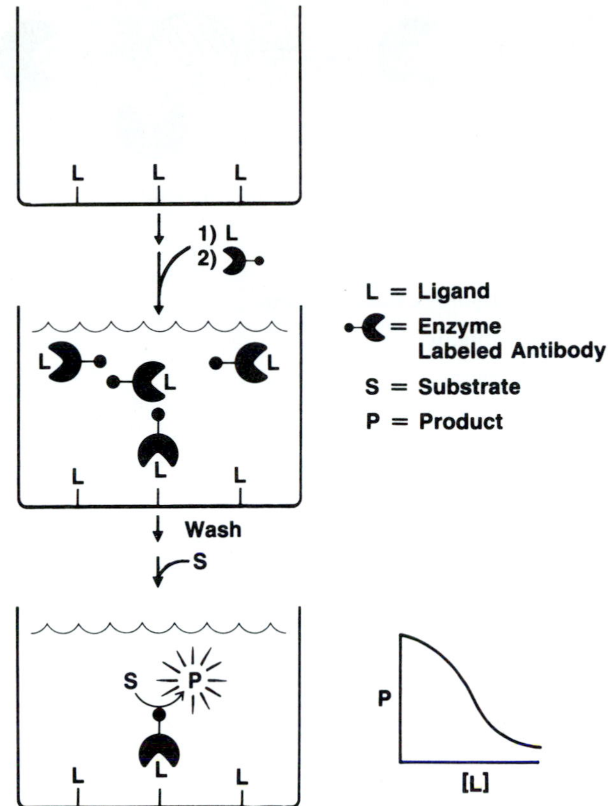

Fig. 11-10 Principles of competitive ELISA where antibody is labeled with enzyme. This is an immunometric assay.

sample will bind the immobilized ligand. The solid-phase ligand:antibody-enzyme complex is washed with buffer before the addition of the substrates. Immunometric ELISAs have the same advantages as IRMAs when compared to an enzyme-labeled ligand ELISA.

Homogeneous enzyme immunoassay

The first homogeneous immunoassay to be described is an enzyme immunoassay for low-molecular-weight ligands. This assay format is known as the *enzyme-multiplied immunoassay technique (EMIT)*. Binding of antibody to the enzyme-labeled ligand changes the enzymatic activity of the label so that antibody-bound enzyme can be distinguished from unbound labeled ligand.

In most instances, binding of antibody to the enzyme-ligand conjugate sterically inhibits the enzyme by limiting the enzyme substrate's access to the catalytic site (Fig. 11-11). A typical enzyme-inhibition profile is shown in Fig. 11-12. Increasing the amounts of gentamicin monoclonal antibody inhibits the activity of the enzyme label, glucose-6-phosphate dehydrogenase (Fig. 11-12). The dose-response curve for the homogeneous gentamicin enzyme immunoassay is shown in Fig. 11-13. The unlabeled ligand competes with enzyme-labeled gentamicin for the limited number of antibody-binding sites. Less labeled gentamicin is therefore bound by the antibody, resulting in an increase in enzymatic activity. Usually this response is reflected by a change in the rate of enzyme activity when two absorbance readings of the reaction are taken 30 seconds apart. The change in absorbance (ΔA) is used to calculate the dose response. The ΔA at zero dose (ΔA_0) is routinely subtracted from all standards, controls, and specimens before the standard curve is plotted and the unknown concentrations are calculated.

On the other hand, the homogeneous enzyme immunoassay for thyroxine (T_4) is based on the antibody-mediated *reversal* of the inhibition of the enzyme malate dehydrogenase by its conjugated ligand, thyroxine. Free thyroxine in the reaction mixture prevents reversal by antibody; consequently malate dehydrogenase activity is reduced as the concentration of thyroxine increases.

Substrate-labeled fluorescence immunoassay

Another homogeneous competitive protein-binding assay is the substrate-labeled fluorescence immunoassay (SLFIA). The assay is based on a label that is a fluorogenic enzyme substrate (β-galactosyl-umbelliferone). When the label is hydrolyzed by a specific enzyme (β-

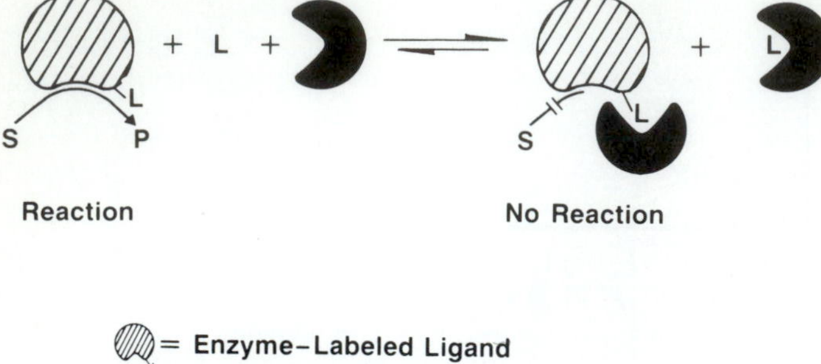

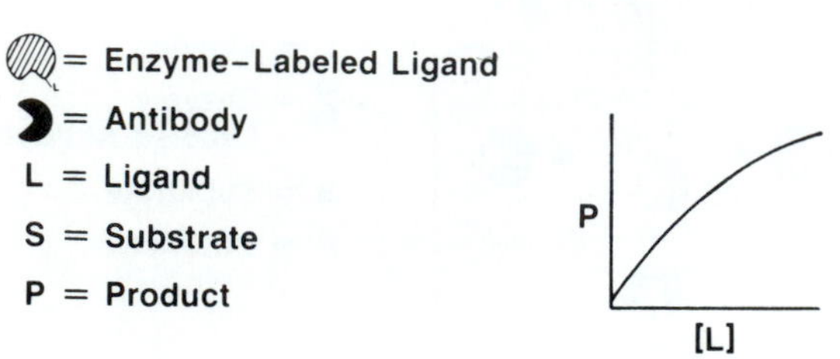

Fig. 11-11 Principle of homogeneous enzyme immunoassay (enzyme-multiplied immunoassay technique, EMIT) and typical dose-response curve.

galactosidase), it yields a fluorescent product (umbelliferone). Binding of the substrate-labeled ligand by specific antibody prevents the enzyme from hydrolyzing the substrate label. Since fluorescence will not be produced by antibody-bound label, the bound label can be distinguished from unbound.

In a competitive-binding reaction assay, the substrate-labeled ligand and sample ligand compete for the limited number of antibody-binding sites as illustrated in Fig. 11-14. As the concentration of ligand in the sample increases, the fraction of β-galactosyl–umbelliferone–ligand not bound by the antibody also increases. It is hydrolyzed by the enzyme, resulting in fluorescence production proportional to the concentration of ligand in the sample.

Time-resolved fluoroimmunoassay

The fluorescence lifetimes of the europium chelates (see Table 11-6) are extremely long—100 μsec and greater (compared to 5 nsec for fluorescein). With the appropriate instrumentation, one is able to measure such fluorescence at times greater than 1 μsec after a pulse of excitation light and distinguish this fluorescence from short-lived background and other fluorescence interferences that have already decayed. This is the basis for a time-resolved fluoroimmunoassay, a heterogeneous assay with inherently low background that is capable of high sensitivity (10^{-12} M). In these assays europium chelate can be used to label either the ligand or the specific antibody. The dose-response curves are similar to those of other heterogeneous, competitive immunoassays.

Fluorescence polarization immunoassay

Fluorescence polarization immunoassay (FPIA) is based on the amount of polarized fluorescent light detected when the fluorophore label is excited with polarized light. The degree of polarization of the emitted fluorescent light depends on the rate of rotation of the fluorophore-ligand conjugate in solution. Small molecules rotate freely, and consequently the fluorescent light emitted by the molecule is depolarized, whereas large molecules, such as proteins,

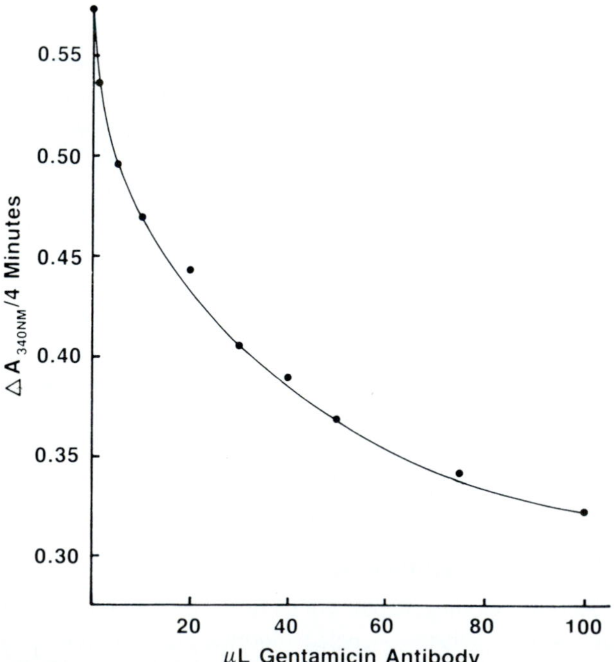

Fig. 11-12 Inhibition of activity of gentamicin–glucose-6-phosphate dehydrogenase conjugate by monoclonal antibody to gentamicin.

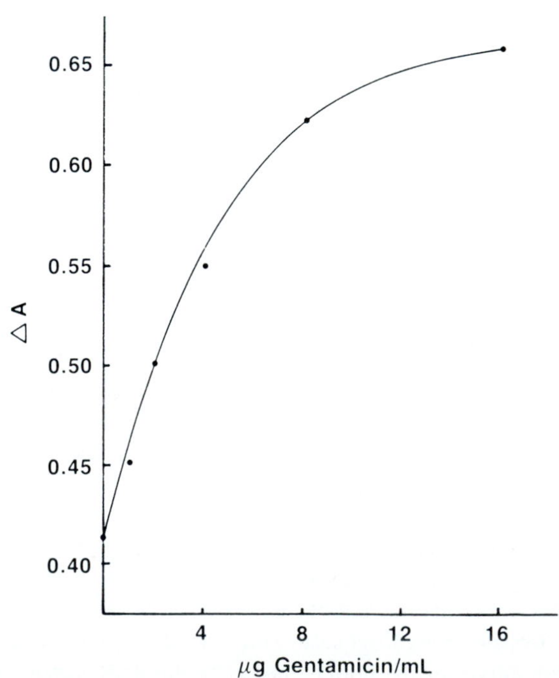

Fig. 11-13 Dose-response curve for homogeneous gentamicin enzyme immunoassay.

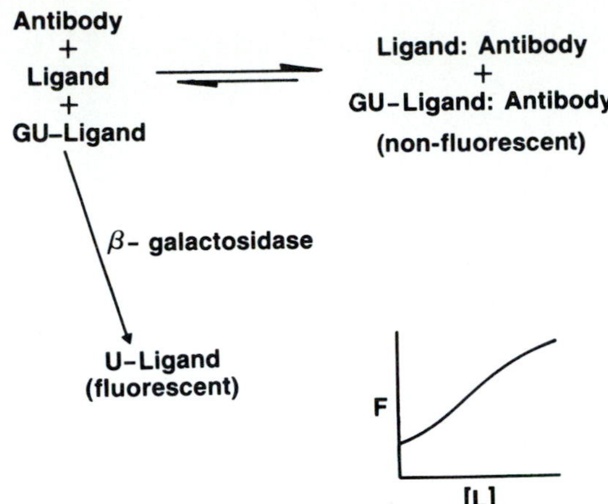

Fig. 11-14 Principle of substrate-labeled fluorescent immunoassay (SLFIA). *GU-Ligand,* β-Galactosyl–umbelliferone-ligand.

rotate more slowly and emit fluorescent light that is polarized. When an antibody binds a low-molecular-weight ligand labeled with a fluorophore (Fig. 11-15), the fluorescence polarization of the labeled ligand is increased because rotation of the labeled ligand-antibody complex is much slower than that of labeled ligands alone. Unlabeled ligand will compete with the labeled ligand for antibody-binding sites so that the amount of polarized fluorescent light resulting from a competitive-binding reaction is inversely proportional to the concentration of unlabeled ligand in the reaction. The principle of the fluorescence polarization immunoassay and a typical dose-response curve are shown in Fig. 11-16.

Apoenzyme reactivation immunoassay system

Apoenzyme reactivation immunoassay system (ARIS) is a homogeneous prosthetic group–labeled immunoassay. Prosthetic groups are nonprotein components of some enzymes that are required for activity. These groups are tightly bound by the enzyme's protein component, the apoenzyme. With some enzymes, the prosthetic group and apoenzyme can be dissociated to render the enzyme inactive; recombining them will regenerate an active enzyme. The prosthetic group–labeled immunoassay depends on the ability of a ligand labeled with the prosthetic group flavin adenine dinucleotide (FAD) to combine with apoglucose oxidase and regenerate glucose oxidase activity. Fig. 11-17 illustrates the principle of the assay. Binding of a constant amount of ligand-FAD by antibody to the ligand prevents the FAD-labeled conjugate from combining with excess apoglucose oxidase in the reaction mixture, thereby reducing the amount of active enzyme regenerated (as monitored by the color produced in the enzyme detection system). Inhibition of reactivation by ligand-specific antibody is reversed in a competitive-binding assay by the addition of unlabeled ligand. The dose-response curve for imipramine ARIS is shown in Fig. 11-18. The absorbance generated in this assay is proportional to the concentration of imipramine in the sample.

ARIS has been applied to dry reagent strip tests for analysis of therapeutic drugs such as theophylline, phenytoin, and phenobarbital. The reactants are impregnated into

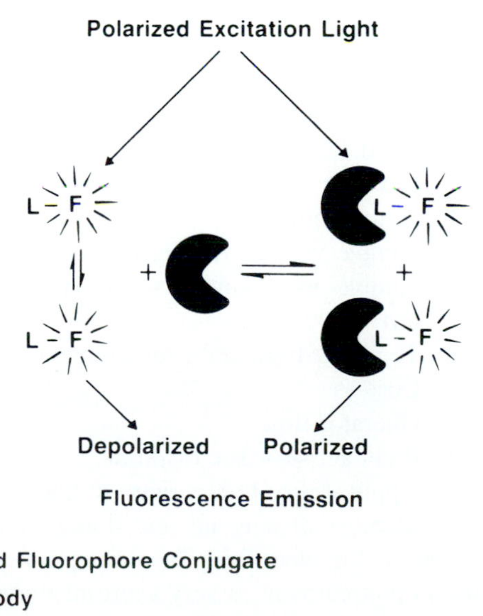

Fig. 11-15 Fluorescence polarization of a fluorophore-labeled ligand free in solution and when bound by an antibody. Unbound fluorescent conjugate rotates in solution so that its fluorescence emission is depolarized, whereas bound label has less rotation, resulting in a polarized fluorescence emission.

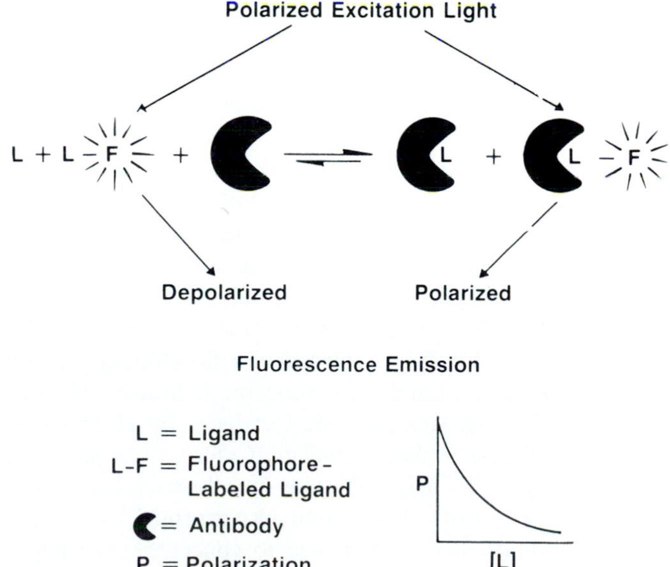

Fig. 11-16 Principle of fluorescence polarization immunoassay (FPIA) and typical dose-response curve.

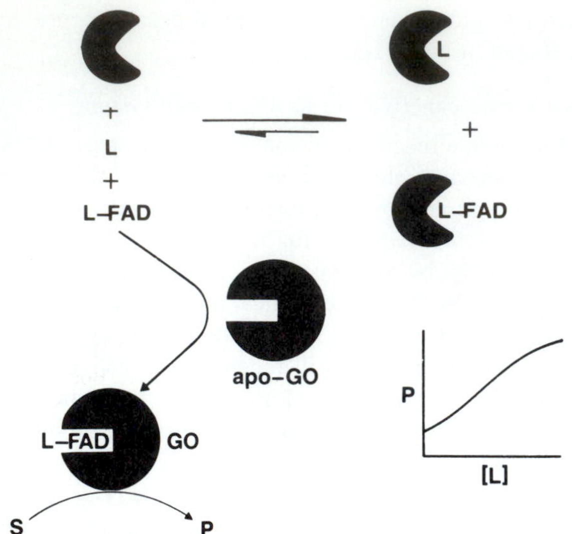

Fig. 11-17 Principle of apoenzyme-reactivation immunoassay system (ARIS) and typical dose-response curve. Ligand, *L*, competes with flavin adenine dinucleotide–labeled ligand *L-FAD*, for antibody-binding sites. Unbound L-FAD is free to react with apo–glucose oxidase *(apo-GO)* to generate an active enzyme.

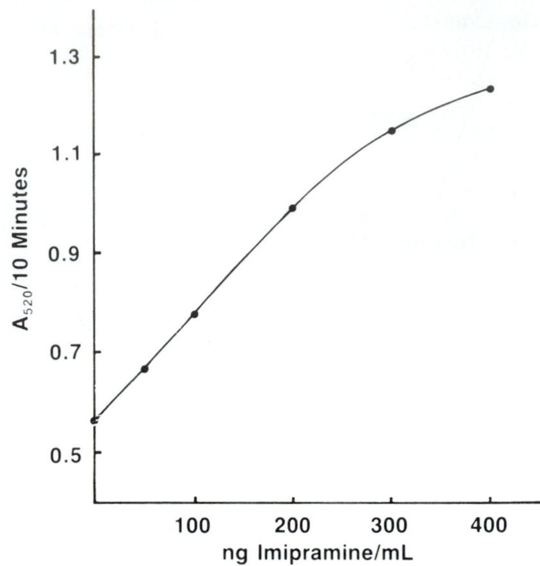

Fig. 11-18 Dose-response curve for imipramine ARIS.

paper that is dried, cut, and affixed to a plastic strip support. The reaction begins when the sample is applied. The response, which is proportional to the drug concentration, is monitored and automatically processed by a reflective photometer. These tests are very convenient and rapid and only require dilution of the serum or plasma specimen.

Enzyme immunochromatography

Immunoassays based on enzyme chromatography do not require instrumentation to obtain a quantitative result because the test can be visually read. Sample is first mixed with an enzyme reagent containing a ligand-peroxidase conjugate and glucose oxidase. A dry paper chromatography strip that has ligand-specific antibody immobilized over its entire surface is then placed in the sample/enzyme reagent mixture so that the liquid reagents can migrate up the strip by capillary action. Since the enzyme conjugate competes with the ligand in the sample for binding to the immobilized antibody, conjugate migration is proportional to ligand concentration. During migration of the reagent mixture, the glucose oxidase is evenly dispersed throughout the strip. After reagent migration and chromatography are complete, the strip is placed in a developing solution containing glucose and a chromogenic indicator, which is oxidized by hydrogen peroxide (produced by glucose oxidase) via the peroxidase conjugate. Color will be produced in the strip to the height that the ligand peroxidase conjugate has migrated. Quantitation is performed by relating the height of color development to concentration using a conversion table. Enzyme immunochromatography tests have been developed for therapeutic drugs such as theophylline, phenytoin, and phenobarbital. These assays can analyze whole blood and do not require instruments for analysis.

ADVANTAGES AND DISADVANTAGES OF DIFFERENT APPROACHES TO COMPETITIVE-BINDING ASSAYS

Before describing some advantages and limitations of the competitive-binding assays described earlier, it is useful to discuss factors to be considered when one designs or chooses a particular assay format. Both the convenience and the performance of an assay must be addressed.

Factors to be considered for convenience are

1. Number of pipetting steps
2. Incubation time
3. Rate or end-point assay
4. Need for additional equipment such as incubators, centrifuges, or specialized detectors
5. Automation
6. Sample volume requirement
7. Sample pretreatment (dilution, extraction, and so on)
8. Temperature-control requirements
9. Cost
10. Operator time
11. Radioactive-waste disposal
12. Applicability to a variety of analytes
13. Stability of reagents and storage temperatures

Obviously, the ideal assay would require very little sample with no pretreatment, a very short incubation time at room temperature, few or no pipetting steps (if automated), and no additional equipment other than the detector or an automated instrument. It would be able to be performed

Table 11-12 Some characteristics of various competitive-binding assays

Immunoassay	Homogeneous or heterogeneous	Detection limit (M)	Amplification	Low- or high-molecular-weight ligands
RIA (radioimmunoassay)	Heterogeneous	1×10^{-12} to 1×10^{-14}	No	Both
ELISA (enzyme-linked immunosorbent assay)	Heterogeneous	1×10^{-11} to 1×10^{-14}	Yes	Both
EMIT (enzyme-multiplied immunoassay technique)	Homogeneous	5×10^{-10}	Yes	Low
SLFIA (substrate-labeled fluorescent immunoassay)	Homogeneous	1×10^{-9}	No	Both
FPIA (fluorescence polarization)	Homogeneous	5×10^{-9}	No	Low
Time-resolved fluorescence IA	Heterogeneous	10^{-12}	No	Both
ARIS (apoenzyme reactivation immunoassay system)	Homogeneous	5×10^{-10}	Yes	Both

quickly, cheaply, with very little operator hands-on time, and it would have a rapid data reduction capability. The reagents would be stable for at least a year at room temperature.

Some of the most important performance factors are as follows:

1. Good assay response over the range of the standard curve.

2. A low background signal (that signal caused by either nonspecific interactions of the reagents in the assay or contributed by other substances in the standards or samples). Background must be kept to a minimum to maximize the slope and range of the standard curves.

3. High antibody affinity. This increases the slope of the dose-response curve and determines the sensitivity or detection limit of an assay (the lowest concentration in the assay that is significantly different from zero). Sensitivity is related to the slope of the dose-response curve, to the experimental error (accuracy and precision of the assay), and to the activity and detectability of the label.

4. Good precision and accuracy. Accurate measurements depend on the accuracy of the values used for the concentrations of standards and on the correct interpolation of the true shape of the dose-response curve between the standards. Obviously, a linear dose response, or a linear transformation of a nonlinear one, enables one to make a reasonably accurate interpolation between the standards.

5. Specificity. This is determined by the ability of the antibody to discriminate between the ligand and similarly structured substances. In most cases, the screening and selection process for suitable antiserums or monoclonal antibodies plays a very important part in determining antibody specificity. This is true for both native ligands (such as proteins) and the synthetic immunogens prepared to produce antibodies to low-molecular-weight substances.

Table 11-12 compares some characteristics of the previously described competitive-binding assays. The heterogeneous assays RIA and ELISA have the greatest potential for sensitivity, with picomolar or less detection limits. In addition to the high specific activities of the radiolabels or enzyme labels, interferences originating either from substances in the sample or from impurities in the reagents are removed by the separation or wash steps. The heterogeneous procedures reduce the background signal so that a greater volume of sample can be used in the assay, thus increasing the sensitivity.

The short half-life of the radioactive labels used in the clinical laboratory (such as ^{125}I) limits the use of RIA reagents to between 6 and 12 weeks. This is a short period compared to reagents for the nonisotope-labeled assays, which often are stable for a year or longer. In addition, radioisotopes may be hazardous and are subject to governmental regulation in their use and disposal.

ELISA and the other nonisotopic immunoassays are neither regulated nor potentially hazardous. Because enzymes are biochemical amplifiers, the systems that employ enzyme labels are capable of producing a greatly amplified signal, depending on the specific activity of the enzyme and the incubation time for conversion of substrate to product. The use of fluorogenic substrates instead of chromogens can enhance the sensitivity of these assays 100 to 1000 times. Their use in the case of ELISA, where the washing steps have removed background fluorescence and other interferences present in the sample or reagents, can be particularly advantageous.

The required washing steps make the ELISA assay less convenient than the homogeneous assays, although instruments are available to automate these steps. ELISA techniques pose a risk of steric hindrance in the ligand-antibody binding reaction because of the introduction of the enzyme label. Coupling of the label to the ligand or binding of the ligand-enzyme conjugate to the antibody could lower the activity of the label, thus reducing the sensitivity of an ELISA. Reproducible attachment of antibody to the solid phase reduces the error often seen in competitive ELISAs and other solid-phase heterogeneous competitive assays.

The homogeneous enzyme immunoassay (enzyme-multiplied immunoassay technique [EMIT]), although limited by the absorbance of the product, is still sensitive to subnanomolar levels of ligand. Like ELISA, the sensitivity may be increased by use of a fluorogenic substrate. EMIT is subject to possible interferences by hemolyzed, lipemic, and icteric samples. Because a rate measurement is used, it is difficult to perform manually but quickly and easily accomplished with automated equipment. However, the assay is not suitable for high-molecular-weight ligands because binding by antibody usually will not cause sufficient enzyme inhibition for unlabeled ligand to provide a dose response.

The fluorescence assays are also sensitive at subnanomolar concentrations of ligand. The homogeneous fluorescence assays avoid errors that can be introduced by the separation steps of heterogeneous systems. Except for perhaps the time-resolved fluoroimmunoassay, which is a heterogeneous assay, these fluorescence assays are prone to interferences by hemolyzed, lipemic, and icteric samples. Other possible sample interferences in homogeneous fluorescence immunoassays include light scattering from lipids and particulates, fluorescence quenching, and background fluorescence from the presence of endogenous fluorophores. The fluorescence polarization immunoassay is sensitive to the depolarized scattered light of particulates in the assay and requires sophisticated instrumentation. Some labeled ligands may be nonspecifically bound by endogenous proteins, resulting in an increased polarization background. Each homogeneous fluorescence assay can be adapted to a heterogeneous assay format.

ARIS has potential for amplification of the signal because an active enzyme is generated in the assay. Unlike other homogeneous enzyme immunoassays, antibody binds a ligand-label conjugate required to generate an active enzyme, rather than binding a ligand coupled to an enzyme that is already active. Therefore ARIS may have lower background and may provide a more sensitive assay. The sensitivity of ARIS can also be improved by the use of a fluorescent readout.

BIBLIOGRAPHY

Ligand binding and competitive-binding assays

Odell, WD, and Franchimont, P, editors: Principles of competitive protein binding assays, ed 2, New York, 1983, John Wiley & Sons, Inc.

Pressman, D, and Grossberg, AL: The structural basis of antibody specificity, Reading, Mass, 1973, WA Benjamin, Inc.

Travis, JC: Fundamentals of RIA and other ligand assays, Anaheim, Calif, 1977, Scientific Newsletter, Inc.

Immunoassays (under a variety of formats)

Albertini, A, and Ekins, R: Monoclonal antibodies and developments in immunoassay. Symposium of the Giovanni Lorenzini Foundation, vol 11, Amsterdam, 1981, Elsevier/North Holland Biomedical Press.

Nakamura, RM, Dito, WR, and Tuckler, ES, III, editors: Immunoassays: clinical laboratory techniques for the 1980's, vol 4. Laboratory and research methods in biology and medicine, New York, 1980, Alan R Liss, Inc.

Radioimmunoassay

Chard, T: An introduction to radioimmunoassay and related techniques, ed 3, New York, 1987, Elsevier Science Publishing Co, Inc.

Theorell, JI, editor: Radioimmunoassay design and quality control. World Federation of Nuclear Medicine and Biology, Oxford, 1983, Pergamon Press, Inc.

Enzyme immunoassay

Maggio, ET, editor: Enzyme-immunoassay, Boca Raton, Fla, 1980, CRC Press, Inc.

Wisdom, GB: Enzyme-immunoassay, Clin Chem 22:1243-1255, 1976.

Substrate-labeled fluorescent immunoassay

Burd, JF, Wong, RC, Feeney, JE, et al: Homogeneous reactant-labeled fluorescent immunoassay for therapeutic drugs exemplified by gentamicin determination in human serum, Clin Chem 23:1402-1408, 1977.

Fluorescence polarization immunoassay

Dandliker, WB, Kelly, RJ, Dandliker, J, et al: Fluorescence polarization immunoassay: theory and experimental method, Immunochemistry 10:215-227, 1973.

Jolley, ME, Sharpe, SD, Wang, C-HJ, et al: Fluorescence polarization immunoassay: monitoring aminoglycoside antibiotics in serum and plasma, Clin Chem 27:1190-1197, 1981.

Time-resolved fluorescence immunoassay

Soini, E, and Kojola, H: Time-resolved fluorometer for Lanthanide chelates—a new generation of nonisotopic immunoassays, Clin Chem 29:65-68, 1983.

Fluorescence immunoassay

Visor, GC, and Shulman, SG: Fluorescence immunoassay, J Pharm Sci 70:469-475, 1981.

Apoenzyme-reactivation immunoassay system (ARIS)

Morris, DL, Ellis, PB, Carrico, RJ, et al: Flavin adenine dinucleotide as a label in homogeneous colorimetric immunoassays, Anal Chem 53:658-665, 1981.

Rupchock, P, Sommer, R, Greenquist, A, et al: Dry reagent strips used for determination of theophylline in serum, Clin Chem 31:737-740, 1985.

Thompson, SG, and Boguslaski, RC: Homogeneous dry reagent immunoassay strips for the determination of therapeutic drugs in human serum or plasma, J Clin Lab Anal 1:293-299, 1987.

Enzyme immunochromatography

Zuk, RF, Ginsburg, VK, Houts, T, et al: Enzyme immunochromatography—a quantitative immunoassay requiring no instrumentation, Clin Chem 31:1144-1150, 1985.

CHAPTER 12

Measurement of colligative properties

LAWRENCE A. KAPLAN

OBJECTIVES

- Identify the relationship between osmosis, osmotic pressure, osmolality, and osmometry.
- List four measurements that depend on colligative properties and describe what happens to each property during sample concentration.
- State the relationship between freezing point depression and moles of particles dissolved per kilogram of water and calculate the osmolality from a given freezing point of a solution.
- Describe the techniques of freezing-point depression and vapor-pressure depression for osmolality determinations.
- State the clinical utility of performing plasma osmolality and urine osmolality determinations. State the clinical situations in which the measurement of the osmolal gap is useful.

KEY TERMS

"acetone" bodies The acetone and other ketones that are present in the serum of patients with diabetic ketoacidosis.

activity The effective concentration of the molecules of a solution.

boiling-point elevation A phenomenon in which addition of solute molecules raises the temperature at which the solution will boil. For water, this is 1.86 centigrade degrees per mole of solute per kilogram.

colligative property A characteristic to which all the molecules of a solution contribute, regardless of their individual composition or nature.

colloid A large molecule, usually in aqueous solution. Normally the term is applied to protein solutions.

colloid osmotic pressure (COP) The osmotic pressure generated by that portion of a solution with high molecular weight (greater than 30,000 daltons).

crystalloids The uncharged solute molecules of a solution.

dew point The temperature at which condensation of water from the vapor phase occurs.

diffusion Mixing or movement of molecules as a result of their random motion.

Donnan effect The distribution of ions caused by having a high molecular weight ion on one side of a semipermeable membrane.

freezing-point depression A phenomenon in which the addition of solute molecules to a solution lowers the temperature at which the solution will freeze.

molality The number of moles of solute per kilogram of water or solvent.

oncotic pressure Another term for colloid osmotic pressure.

osmolal gap The difference between the observed and calculated serum osmolalities. The calculated osmolar values include sodium concentration multiplied by 2, plus glucose and blood urea nitrogen.

osmolality The measurement of the number of moles of particles per kilogram of water.

osmometry The measurement of a colligative property of a solution in which the number of moles of a substrate per unit volume (concentration) are determined.

osmosis Water flow across a semipermeable membrane.

osmotic pressure The hydrostatic pressure required to prevent a change in volume when two solutions of different concentrations are placed on opposite sides of a semipermeable membrane.

plasma expander Usually a high molecular weight dextran that is administered intravenously to increase the oncotic pressure of a patient.

Seebeck effect Voltage difference seen when two ends of a specially made wire are at two different temperatures.

semipermeable membrane A barrier that allows one type of molecule, such as water, to pass but does not allow another type of molecule, such as protein, to pass.

thermistor A device in which the conduction of electrons is temperature dependent. It is derived from the words *thermal resistor*.

thermocouple A device that generates a voltage (Seebeck effect) when the two ends of a wire are at different temperatures.

ultrafiltrate The solution remaining after passage through a semipermeable membrane. Usually it contains only low molecular weight solutes.

vapor pressure depression The phenomenon in which the addition of a solute molecule to a solvent will decrease the amount of solvent in equilibrium between the vapor phase and the liquid phase.

COLLIGATIVE PROPERTIES

Osmosis

Osmosis is neither simply a mixing of two fluids nor simply a diffusion. Diffusion is the mixing of molecules as a result of random motion caused by thermal kinetic energy (brownian motion). For example, if an albumin solution were carefully overlayed with water, the albumin molecules would randomly move back and forth across the original interface boundary. Because there are more albumin molecules in the albumin solution, the odds are great that an albumin molecule will cross into the water side. Thus albumin will diffuse into the water layer until the solutions become homogeneous, that is, until the odds are equal that an albumin molecule will diffuse one way or the other across the original boundary because the concentration is equal on both sides.

The term *osmosis* specifically applies to water flow across a semipermeable membrane such as a cell wall. Although osmosis can occur with any fluid, water is the most important fluid for this discussion. A semipermeable membrane allows some particles (molecules, ions, or aggregates of molecules) to pass through it, but it inhibits the passage of others; hence it is *semi*permeable. The simplest example of a semipermeable membrane is a dialysis membrane, which is usually made of cellophane. It has very small pores through which water and some small molecules and ions pass. However, large molecules, such as proteins, cannot pass through the membrane. To demonstrate, place an albumin solution in a section of dialysis tubing and tie the ends of the tubing. If the tubing is placed in a beaker of water, the albumin molecules cannot move out of the membrane, but water molecules will move in and effect dilution of the albumin. As a result, the tubing will swell as water flows into the albumin solution inside the tubing, increasing the pressure inside the membrane. If the tubing does not burst from this pressure, an equilibrium will be maintained between the water flowing in and the water being forced out by the internal pressure. The hydrostatic pressure built up and maintained by this process is called *osmotic pressure*.

Perhaps the most graphic example of osmosis is lysis of red cells by water. So much water flows into the more concentrated intracellular fluid that the cell swells and bursts. Cells can also shrink if exposed to a fluid of high osmolality. In this case the water in the cell flows out of the cell into the concentrated solution outside. In the laboratory the measurement of mean corpuscular volume (MCV) is affected by this process. If the diluent is not isotonic, that is, of equal osmotic pressure, the cells will swell or shrink, giving an erroneous mean corpuscular volume (MCV) and hematocrit value because the latter is calculated from the MCV.

Osmolality

The term *molarity* is used to characterize concentration, that is, the number of moles of solute per liter of water. *Molality* is the number of moles of solute per kilogram of water. Because a liter of water has a mass of 1 kg, the difference between these two expressions of concentration is usually small. Only for concentrated solutions is the difference appreciable. In practice it is the difference between adding material to a liter of water (molality) and adding water to material to make a liter of solution (molarity).

Molality is the term best suited to osmometry because it gives a simpler theoretical formula for osmotic pressure than molarity does. The term *osmolality* is used to identify the number of moles of particles per kilogram of water.

Because the osmolality of a solution does not depend on the kind of particles but only on the number of particles, it is called a *colligative* property. A solution that is 1 millimolal in sodium chloride is 2 milliosmolal because sodium chloride separates into sodium and chloride ions. Each kind of ion represents a particle that contributes to the osmolality. Furthermore, a 1 millimolal calcium chloride solution is 3 milliosmolal because each molecule ionizes to give one calcium ion and two chloride ions.

Osmometry

Osmometry is the measurement of the concentration, not of a particular molecule, but of molecules and ions in general. In this chapter the clinical importance and use of osmometry are discussed, the techniques for measuring osmolality are reviewed, and examples of instrumentation are provided.

Osmolal gap

There are just a few substances in plasma that contribute significantly to the osmolality, and they are mostly small molecules and ions. For example, plasma usually contains 40 g of albumin per liter, but the number of moles of albumin is very small (only about 50 μmol). In contrast, plasma contains about 150 mmol of sodium ion and 100 mmol of a corresponding anion such as chloride. This is only 5.8 g of sodium chloride per liter. Thus sodium chloride contributes 3000 times more to osmolality than a similar mass of albumin does.

Many formulas have been used to calculate the approximate osmolality of serum or plasma. Most formulas attempt to combine accuracy with simplicity in calculation. A formula that requires measurement of many substances is not very useful clinically. The calculated osmolality can

Table 12-1 Substances affecting plasma osmolality

Substances	Toxic or lethal concentration		Corresponding increase in osmolality (mOsm/kg)
	Historical units (mg/dL)	S.I. units (mmol/L)	
Ethanol	350	80	80
Isopropanol	340	60	60
Methanol	80	24	24
Ethyl ether	180	24	24
Trichloroethane	100	9	9
Acetone*	55	10	10

*Includes other ketones or ketone metabolites.

be compared with the measured osmolality; the difference is called the *osmolal gap*. An abnormal osmolal gap is an important indication of abnormal concentrations of unmeasured substances in the blood. Because the formula predicts the plasma osmolality so well, there is little new information to be gained from *routine* measurements of the osmolality. However, in certain special situations described in the next section, the measurement is informative and worthwhile.

The formulas shown below for the calculation of the osmolality are zero-order "approximations" because they include only the most important contributors to osmolality:

Historical units

$$\text{Calculated osmolality (mOsm/kg)} = 2 \cdot \text{Na (mEq/L)} + \frac{\text{Glucose (mg/dL)}}{18} + \frac{\text{BUN (mg/dL)}}{2.8} \qquad \textit{Eq. 12-1}$$

S.I. units*

$$\text{Calculated osmolality (mOsm/kg)} = 2 \cdot \frac{\text{Na}}{\text{(mmol/L)}} + \frac{\text{Glucose}}{\text{(mmol/L)}} + \frac{\text{BUN}}{\text{(mmol/L)}} \qquad \textit{Eq. 12-2}$$

The S.I. units formula is very straightforward. The factor 2 in both equations counts the cation (sodium) once and the corresponding anion once. Glucose and BUN are undissociated molecules and are counted once each. All other components are ignored. In the historical-units formula the dividing factors represent the respective molecular weights and conversion from deciliters to liters. The calculated osmolality is not corrected for the actual water content of plasma (lipids and protein take up some of the volume) because this correction does not improve the clinical utility of the osmolal gap or of osmolality in general.

Notice that these formulas use molarity rather than molality. This approximation fortuitously compensates for some of the serum components and theoretical corrections that were ignored.

*Système International d'Unités, or simply "metric" units.

The osmolal gap is defined as:

$$\text{Osmolal gap, Osm/kg} = \text{Measured Osm/kg} - \text{Calculated Osm/kg} \qquad \textit{Eq. 12-3}$$

The average osmolal gap is near zero.

CLINICAL USE OF OSMOMETRY

There are two major uses of osmometry: (1) detection of unmeasured substances in the plasma and (2) assessment of renal concentrating ability.

Plasma osmolality

Only a few exogenous substances can be ingested in amounts sufficient to affect the plasma osmolality. Table 12-1 lists the substances and the concentrations necessary to increase the osmolal gap to 10 or more milliosmoles per kilogram. The most common substances are alcohols. If ethanol is measured specifically and does not agree with the osmolal gap within 10 mOsm/kg, with the molar conversion to molal being ignored, there is reason to suspect that another of the substances listed in Table 12-1 is also present. Table 12-1 shows that trichloroethane can be at near-lethal levels in the blood without being readily detected by osmometry.

Although osmometry has long been recommended as a means to detect alcohol, it should be noted that vapor-pressure osmometers are not useful for the detection of alcohol because alcohol forms part of the solvent and thus contributes to its vapor pressure.

An increase in the osmolal gap will also reflect an increase in the anion gap in patients with metabolic imbalance. These changes are caused by the presence of ketone "bodies" (see Chapter 29).

Urine osmolality

Renal concentrating ability is a sensitive measure of kidney function. The urine that is delivered to the bladder is typically one to three times more concentrated than the plasma. A random urine specimen is sufficient to demonstrate the kidney's ability to concentrate urine if the osmolality of the random urine specimen is greater than 600 mOsm/kg. However, if the random urine specimen is di-

Table 12-2 Estimated effect of anticoagulants on osmolality (compared to serum)

Anticoagulant	Full tube (mOsm/kg)	Half-full tube (mOsm/kg)
Heparin	+0	+0
EDTA (disodium salt)	+15	+30
Fluoride-oxalate (sodium fluoride–potassium oxalate)*	+150	+300
Iodoacetic acid (lithium salt)	+5	+10

*This hyperosmolal state accounts for the hemolysis usually observed in the plasma of these samples.

lute, no conclusion about concentrating ability can be made. A definitive follow-up test involves overnight water restriction. After the morning void, at least one urine specimen should exceed 850 mOsm/kg. Patients who are compulsive water drinkers may need continuous observation to assure that no water has been ingested.

The specific gravity as estimated by the refractive index can also be used to measure urine concentration. However, osmometry is less affected by the presence of protein or radiocontrast dyes.

Serum or plasma osmolality

Sample-collection technique is important to obtain a valid specimen for measurement of osmolality. For example, stasis during phlebotomy should be avoided. In addition, the patient's position, supine or upright, will affect the osmolality. Thus a sample from a fasting, hospitalized patient will give the most uniform results, not because lipemia or other effects are avoided but because it is more likely that the patient was supine at the time of the morning phlebotomy rounds.

Serum and heparinized plasma have similar osmolality values. The contribution to the osmolality by fibrinogen in plasma is small, and it is important only in the measurement of colloid osmotic pressure. Freezing-point depression techniques can use whole blood and are not affected by lipemia or hemolysis. Anticoagulants other than heparin increase the measured osmolality. Table 12-2 shows the estimated effect of four anticoagulants. On occasion, the kind of anticoagulant used can be verified by measurement of the osmolality of the plasma.

PRINCIPLES OF MEASUREMENT

Osmolality is a colligative property; thus any of four measurements that depend on colligative properties may be used to measure osmolality. The measurements are (1) osmotic pressure, (2) boiling-point elevation, (3) freezing-point depression, and (4) vapor-pressure depression. Osmotic-pressure measurement has been used only in a special form of osmometry called "colloid osmotic pressure" and is discussed separately in a later section of this chapter. Boiling-point elevation is not useful for clinical samples because proteins will coagulate, causing gross changes in the sample composition. Of the remaining two, freezing-point depression is the most frequently used technique. Vapor-pressure depression as measured by the dew point is more commonly used in pediatric laboratories.

Freezing-point depression

The use of salt to melt ice and snow is a well-known practice. This is an example of freezing-point depression; that is, dissolved salt increases the osmolality, thereby lowering the freezing point of the solution compared to that of the pure solvent (ice or snow). The temperature at which ice and the water solution are in equilibrium is a function of the salt concentration. More precisely, the temperature at equilibrium is a function of the number of particles in solution. The freezing-point temperature is depressed 1.86° C for each mole of particles dissolved per kilogram of water. Because the osmolality of blood is about 0.3 Osm/kg (300 mOsm/kg), the freezing point is −0.62° C. Precise measurement of this temperature requires a sensitive thermometer. A thermistor (thermal resistor) is made from a mixture of transition metal oxides such as manganese, cobalt, and nickel. These materials are semiconductors, and the number of electrons in the conduction band (valence electrons of the metal lattice capable of conducting a current) depends on the temperature. They become better conductors as the temperature rises. The conductance or resistance of the metals can be related to the temperature and hence to the osmolality.

Freezing-point depression is measured as follows:

1. The sample is either cooled by a bath containing an antifreeze solution that is maintained at about −5° C by a conventional refrigerator or by a thermoelectric cooler.
2. The sample is supercooled; that is, its temperature falls below the equilibrium freezing point. This occurs because pure ice crystals are slow to form.
3. The crystallization process is induced by vigorous stirring. Once ice crystals begin to form, additional water molecules are rapidly added to the ice crystals. However, heat is released in the freezing process just as it is absorbed in the melting process. The heat that is released from the formation of the ice crystals raises the temperature of the sample until the rapid freezing stops and an equilibrium temperature is established.
4. The temperature is measured at the plateau, that is, at the temperature at which the heat removed by the cooling bath is matched by the heat released by the freezing process. The temperature at this equilibrium is the freezing point of the solution and is inversely related to osmolality. The plateau's temperature is measured electronically by the thermistor, and the temperature reading is converted to milliosmoles per

kilogram and displayed. At this time, before complete freezing can cause mechanical damage to the thermistor probe, the thermistor is removed from the sample.

Because 1 osmole of solute lowers the freezing point by 1.86° C, osmolality can be calculated directly by the formula:

$$\text{Osmolality (mOsmol/kg)} = \frac{1.86^\circ\ \text{C} \times \text{mOsmol/kg}}{\text{Freezing-point depression}} \quad \textit{Eq. 12-4}$$

However, it is more practical to calibrate the osmometer using saline solutions. Calibration also corrects for systematic or procedural effects, such as the increase in concentration of the sample because of the removal of pure water (as ice) before measurement of the temperature.

A number of factors must be controlled to achieve high precision in freezing-point depression osmometry. These include the bath temperature, fluid composition, and amount of fluid. The fluid composition and volume change as moisture condenses from the room air. The thickness of the sample container and the amount of sample must be standardized. The probe must be rinsed and wiped to minimize carryover from one sample to the next. This is especially important between samples of widely differing osmolality, such as standards and urine samples. Samples can be remeasured, but great care must be taken to warm the sample (for example, by holding the sample cup in one's hand) until *all* of the ice is melted; otherwise the sample will freeze prematurely. Any sample droplets on the side of the cup should be joined with the sample by tipping the cup to coalesce the droplets.

The most common freezing-point osmometers are listed in Table 12-3, along with several key characteristics.

Vapor-pressure depression

Solvent molecules on the surface of a liquid are in constant thermal motion; some of these molecules escape from the surface into the atmosphere above the surface, forming a gaseous vapor phase in equilibrium with the liquid phase. This process is called evaporation. If the liquid contains dissolved solute, some of these solute molecules will occupy the surface layer of the liquid. Generally, a solute molecule will not evaporate but will, by its presence, prevent a solvent molecule from evaporating. As the number of solute molecules increases, the chance that a solvent molecule will evaporate decreases, reducing the vapor phase in equilibrium above the liquid. Thus there is an inverse relationship between the concentration of dissolved solute particles (osmolality) and the vapor pressure above a solution. In vapor-pressure osmometry, the vapor-pressure depression of a solution is compared with that of a standard to determine the osmolality of a solution.

The temperature at which the atmosphere is saturated with solvent can be measured by a thermocouple. A thermocouple generates a voltage (Seebeck effect) between the ends of a wire. The voltage difference between the ends depends on the difference in temperature of the ends.

Thermocouples also exhibit the Peltier effect, which is the opposite of the Seebeck effect. An electrical current through the thermocouple transfers heat from one junction to the other. One junction cools, whereas the other heats. The vapor-pressure osmometer passes an electrical current through the thermocouple in the measurement chamber, causing it to cool. When its temperature falls low enough, water (solvent) begins to condense on it. The electrical current is discontinued, and the thermocouple comes to an equilibrium temperature at which the water condensing on it is matched by the water evaporating from it. This equi-

Table 12-3 Characteristics of clinical osmometers*

Manufacturer	Model	Technique†	Routine sample size (μL)	Precision (%)‡	Measurement time (sec)
Advanced Instrument, Inc.	3D2	FP	200	1.4	90
(Needham Heights, Mass.)	3M0	FP§	20		60
Fiske Associates, Inc.	OS	FP	250	1.8	90
(Needham Heights, Mass.)	One-Ten	FP§	10		60
	OR	FP	50		30
Precision Systems, Inc.	5002	FP	200	1.8	180
(Natick, Mass.)	μOsmette	FP§	50		
Wescor, Inc. (Logan, Utah)	5500	VP	10	2.8	60-90
Instrumentation Lab, Inc. (Lexington, Mass.)	186	COP	300	—	60
Wescor, Inc. (Logan, Utah)	4400	COP	300	—	60-90

*All models are manually loaded with sample and have automated measurement and reporting. Variations in sample size, automated sampling, and printing are available.
†*FP*, Freezing-point depression; *VP*, vapor pressure; *COP*, colloid osmotic pressure.
‡From College of American Pathologists survey for osmolalities in normal range.
§Does not use liquid cooling bath; uses electronic cooling.

librium temperature is measured by the Seebeck voltage, which is linearly related to the osmolality.

Vapor-pressure depression is measured as follows:

1. The sample is sealed in a chamber. The air quickly changes humidity until its humidity is in equilibrium with the sample.
2. The thermocouple cools until its temperature is below the dew point. The electrical current is turned off, and the junction temperature rises as vapor condenses on it.
3. The plateau temperature (the temperature at which an equilibrium exists between condensation and evaporation) is measured.

The vapor pressure of the sample is directly proportional to the thermocouple voltage. Again, it is more practical to calibrate this type of osmometer than to apply theoretical factors. Systematic or procedural effects that must be controlled include the sample volume, the size and composition of the sample absorbent disk, the time delay between sample application and sealing of the chamber, cleanliness of the chamber, and changes in room temperature.

Table 12-3 provides information on the only available clinical vapor pressure osmometer.

COLLOID OSMOTIC PRESSURE

Definitions

Osmotic pressure is a colligative property and hence reflects osmolality. This is strictly true for a semipermeable membrane that is permeable to water only. The difficulty in finding such a membrane has kept the measurement of osmotic pressure from being used as a technique in the assessment of osmolality in clinical samples. However, measurement of the osmolal contribution of a group of molecules responsible for the colloid osmotic pressure (COP) is practical and useful. This property is measured by use of membranes that are permeable to small molecules. Small molecules, less than 30,000 daltons molecular weight, are called *crystalloids* if they are uncharged and *ions* if they are charged. Large molecules are called *colloids*. Hence colloid osmotic pressure measures only the contribution made to osmolality by large, essentially only protein, molecules. An alternate term is the *oncotic* pressure.

Clinical use of colloid osmotic pressure

The major use for the measurement of the COP is detection of conditions leading to pulmonary edema.

In this condition, there is an accumulation of water in the lungs, which interferes with oxygen and carbon dioxide exchange. The actual diagnosis can be obtained from x-ray measurements. Two measurements are needed to predict pulmonary edema: left ventricular heart pressure and COP. As long as the COP is greater than the pulmonary blood pressure (as measured by the "pulmonary artery wedge pressure"), pulmonary edema is unlikely. If heart failure is not present, that is, if the pulmonary blood pressure is normal, then COP measurements *alone* will predict the probability of pulmonary edema.

Knowledge of the albumin or total protein content of the plasma permits the calculation of the COP. However, the formula is inaccurate when used for acutely ill patients (especially patients with heart failure) and for patients who have received dextrans or "plasma expanders." For these groups of patients, measurement of COP is very useful.

Measurement of colloid osmotic pressure

The COP is measured with a microporous filter or membrane that contains pores or channels whose diameters are carefully controlled. Physiological saline is placed in a sealed chamber on one side of the membrane, and the sample is placed on the other. Saline flows into the sample until the back pressure stops further flow. This back pressure, or negative pressure, is sensed by a pressure gauge. In addition to the osmotic pressure, an additional pressure is created by the Donnan effect. This effect arises because at physiological pH most proteins are negatively charged. Because the sample is electrically neutral, there are positive charges equal in number to the negative charges on the proteins. These positive charges are mostly in the form of sodium ions. Charged sodium ions diffuse through the membrane, whereas the corresponding negatively charged proteins do not. This leads to a separation of electrical charges. Because of the charge separation, additional small, negatively charged molecules will be attracted across the membrane. As a result, the number of particles that diffuse will be larger than that resulting from simple osmosis, and the pressure across the membrane will be larger. Because the net charge on proteins changes with pH, the measured COP will also change with pH.

Customarily, COP is reported in millimeters of mercury (mm Hg). In practice a maximum pressure occurs 30 to 90 seconds after the sample is placed into the instrument. This value is chosen because the pressure decays with time as a result of imperfections in the membrane that slowly allow large molecules to diffuse to the saline side, thus reducing the true pressure. Two makers of colloid osmometers are listed in Table 12-3.

BIBLIOGRAPHY

Barlow, WK: Volatiles and osmometry, Clin Chem 22:1231, 1976.

Dorwart, WV, and Chalmers L: Comparison of methods for calculating serum osmolality from chemical concentrations and the prognostic value of such calculations, Clin Chem 21:190-194, 1975.

Eklund, J, Granberg, PO, and Halberg, D: Clinical aspects of body fluid osmolality, Nutr Metab 14(suppl.):74, 1972.

Hendry, EB: The osmotic pressure and chemical composition of human body fluid, Clin Chem 8:241, 1962.

Johnston, RB Jr: Osmolality in serum and urine. In Meites S, editor: Standard methods of clinical chemistry, vol. 5, New York, 1965, Academic Press, Inc., p. 159.

Warhol, RM, Eichenholz, A, and Mulhausen, RO: Osmolality, Arch Intern Med 116:743, 1965.

Wolf, PL, Williams, D, Tsudaka, T, and Acosta, L: Methods and techniques in clinical chemistry, New York, 1972, John Wiley & Sons, Inc.—Interscience.

CHAPTER 13

Electrochemistry: principles and measurements

JON R. KIRCHHOFF
JOHN F. WHEELER
CRAIG E. LUNTE
WILLIAM R. HEINEMAN

OBJECTIVES

- Understand the fundamental differences between potentiometric and voltammetric techniques and understand how each technique is used for clinical measurements.
- Understand the process by which ion-selective electrodes respond to the presence of an analyte.
- Develop a knowledge of the methodology and possible interferences associated with using electrochemistry in the clinical laboratory.
- Understand how various voltammetric and coulometric techniques are used for clinical measurements.

KEY TERMS

activity The effective concentration of a solution species that accounts for interactions with other solution species.

activity coefficient A measure of the degree with which a species interacts with other solution species.

amperometry A controlled-potential technique in which current is measured at a fixed applied potential.

anode The electrode at which oxidation occurs.

auxillary electrode The electrode in the three-electrode electrochemical cell that carries the current to maintain electrolysis at the working electrode.

cathode The electrode at which reduction occurs.

cell potential The quantitative measure of the energy of an electrochemical cell; the difference in electron energy between two electrodes.

charge A quantity of electricity that reflects the total current during a given time: $Q = it$.

conductivity A measure of the relative ability of materials to carry an electrical current.

coulometry A technique in which the charge required to completely electrolyze a sample is measured.

current The rate of charge flow (1 ampere = 1 coulomb/second).

electrolysis A nonspontaneous electrochemical reaction that results from the application of potential to an electrode.

electrolyte solution A solution of ions that provides a conducting medium for electrochemistry.

half-cell potential The quantitative measure of the energy of a half-cell reaction relative to a reference electrode.

half-cell reaction An electrochemical reaction that represents either an oxidation or reduction at one of the electrodes in an electrochemical cell.

hydrodynamic voltammogram A graphical representation of current versus applied potential for a particular electrochemical reaction that occurs in a stirred or flowing solution.

indicator electrode An electrode whose potential varies as the concentration of reactants and products change in solution. This potential is governed by the Nernst equation.

ionic strength (μ) One-half the sum of the concentration (C_i) multiplied by the square of the charge (Z_i) for each ionic species in solution: $\mu = \frac{1}{2}\Sigma_i C_i Z_i^2$.

ionophore A neutral carrier molecule incorporated into an ion-selective electrode to detect a specific ion.

ion-selective electrode An indicator electrode used in potentiometry to respond to specific ions in solution.

limiting current The portion of a hydrodynamic voltammogram where electrolysis is occurring and the current remains constant as a function of increased applied potential.

liquid junction potential A potential that develops at the interface between two nonidentical solutions.

Nernst equation The expression that relates the cell potential to the standard cell potential and the activities of reactants and products in the electrochemical cell.

oxidant In an electrochemical reaction, the species that causes another species to be oxidized and is itself reduced.

oxidation The process whereby a chemical species loses one or more electrons.

polarography Voltammetry performed at a dropping mercury working electrode.

potentiometry The technique where the potential difference between two electrodes is measured under equilibrium conditions.

potentiostat An instrument designed to control the potential of an electrochemical cell.

reductant In an electrochemical reaction, the species that causes another species to be reduced and is itself oxidized.

reduction The process whereby a chemical species gains one or more electrons.

reference electrode An electrode with a stable half-cell potential that is used to measure and control the relative potential of the working electrode.

salt bridge A device that allows ionic movement between compartments of an electrochemical cell to maintain electrical contact and at the same time prevent mixing of the separate solutions.

standard cell potential The electrochemical cell potential measured under standard state conditions.

standard state The condition in which each species is present with unit activity.

stripping voltammetry A voltammetric technique that allows sample preconcentration at the electrode before voltammetric analysis.

voltammetry A technique whereby current is measured as a function of applied potential.

Electrochemistry involves the measurement of electrical signals associated with chemical systems that are incorporated into an electrochemical cell. The cell consists of two or more electrodes that interface a chemical system and an electrical system. The electrical system measures or controls the electrical parameters of voltage and current, which are characteristic of a particular chemical system.

Electroanalytical chemistry makes use of electrochemistry for the purpose of analysis. In this application the magnitude of a voltage or current signal originating from an electrochemical cell is related to the activity or concentration of a particular chemical species in the cell. Excellent detection limits coupled with a wide dynamic range are exhibited by many electroanalytical techniques, with an operating range of 10^{-8} to 10^{-3} M. Measurements can generally be made on very small volumes of sample, that is, in the microliter range. The combination of low detection limits and microliter volume samples allows picomole amounts of analyte to be measured routinely in some instances. Furthermore, electroanalysis lends itself to measurements made in vivo. For example, miniature electrochemical sensors are used to measure pH and Po_2 in the bloodstream of patients with indwelling catheters.

In the clinical laboratory, electroanalysis is routinely used for the determination of many ions, drugs, hormones, metals, and gases. Methods are available for the rapid determination of analytes present at relatively high concentrations, such as blood electrolytes (Na^+, Cl^-, HCO_3^-),

Table 13-1 Electrochemical terms, units, constants, symbols, and conversions

Term	Symbol	Unit or constant	Symbol	Conversion or value
Potential	E	Volt	V	V = J/C
Standard potential	(E^0)			E = i · R
Formal potential	($E^{0'}$)			
Current	i	Ampere	A	A = C/s 1A = 1.05 × 10^{-5}mol of electrons per second
Charge	Q	Coulomb	Q	C = A · s 1C = 1.05 × 10^{-5} mol of electrons
Energy	H	Joules	J	
Resistance	R	Ohm	Ω	
Time	t	Second	s	
Temperature	T	Kelvin	K	
Activity	a	Moles per liter	mol/L	
Concentration	C	Moles per liter (or moles per cubic centimeter)	mol/L (mol/cm^3)	
Area	A	Square centimeters	cm^2	
Diffusion coefficient	D	Square centimeters per second	cm^2/s	
		Gas constant	R	8.31441 J/mol·K
		Faraday constant	F	9.64846 × 10^4C/mol
		Number of electrons in electrode or redox reaction	n	

and analytes present at very low concentrations, such as heavy metals and drug metabolites in blood and urine samples.

The purpose of this chapter is to provide a fundamental background for understanding the electroanalytical techniques found in the clinical laboratory and to illustrate some of the practical applications of electroanalysis. These fundamental electrochemical techniques are divided into three basic categories: potentiometric, voltammetric, and coulometric. Potentiometry is the most widely used clinical application of electrochemistry and involves the measurement of a cell potential under equilibrium conditions. Voltammetry and coulometry are considered dynamic techniques and are based on measurements made on a cell in which electrolysis is occurring. Many common definitions, symbols, and electrochemical nomenclature used in potentiometry, voltammetry, and coulometry are listed in Table 13-1.

POTENTIOMETRIC METHODS

Potentiometric methods are based on the measurement of a potential (voltage) difference between two electrodes immersed in solution under the condition of zero current. The electrodes and the solution constitute an *electrochemical cell*. Each electrode in the electrochemical cell is characterized by a *half-cell reaction* with a corresponding *half-cell potential*. Since no current passes through the cell while the potential is measured, no net electrochemical reaction is occurring; thus a potentiometric technique is an equilibrium method. Potentiometric techniques are important because they can provide accurate measurements of activities, concentrations, and/or activity coefficients of many solution species.*

A typical apparatus for potentiometry is shown in Fig. 13-1. The potential difference between the two electrodes is usually measured with a pH/mV meter. One electrode, the *indicator electrode,* is chosen so that its half-cell potential responds to changes in the activity or concentration of the particular species in solution to be measured. The other electrode is a *reference electrode* whose half-cell potential does not change. It is important to understand that one does *not* measure an individual half-cell potential, but only the potential *difference* between one half-cell and a reference electrode. The most commonly used reference electrodes for potentiometry are the saturated calomel electrode and the silver/silver chloride electrode. The potential of the potentiometric electrochemical cell, E_{cell}, is given by:

$$E_{cell} = E_{ind} - E_{ref} + E_{lj} \qquad \textit{Eq. 13-1}$$

where E_{ind} is the half-cell potential of the indicator electrode, E_{ref} is the half-cell potential of the reference electrode, and E_{lj} is the *liquid junction potential*. The liquid junction potential is the electrical potential that develops at the interface between two liquids as a result of differences in the rates with which ions move from one liquid to the other. For example, the liquid junction potential arises in Fig. 13-1 at the point where the tip of the reference electrode meets the solution. Thus, E_{lj} is a potential that results from differences in charge rather than from an electrochemical reaction at an electrode.

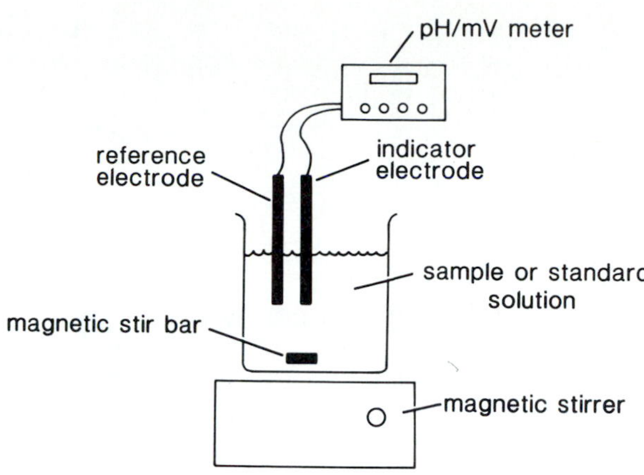

Fig. 13-1 Schematic diagram of apparatus for potentiometry.

Reference electrodes

Since every electrochemical measurement must be made with respect to a reference potential, further discussion is needed with regard to the properties and types of reference electrodes. A reference electrode is an electrochemical half-cell that is used as a fixed reference for the measurement of cell potentials. Ideally, a reference electrode should possess the following characteristics: a stable, easily reproducible half-cell potential; a reversible half-cell reaction; chemical stability of its components; and ease of fabrication and use. Three reference electrodes are discussed below; one is of fundamental significance, and two are of practical importance.

The *standard hydrogen electrode (SHE)* has been chosen as the reference half-cell electrode on which tables of standard electrode potentials are based. In this half-cell, hydrogen gas at a pressure of 1 atmosphere is bubbled over a platinum electrode immersed in acid solution for which the activity of H^+ is unity. The potential of the SHE is defined as 0.0 V at all temperatures, and the potentials of other half-cell couples are referenced to this value. The potentials of other half cells are either negative or positive

*In general, a solution of ions or molecules is characterized by its molar concentration. However, these species can interact with other ions, molecules, or solvent. Depending on the type of interactions that occur, the *effective concentration* of the species may be less than, equal to, or greater than the actual molar concentration of species. The effective concentration is referred to as the *activity* of the species and is related to the molar concentration by an activity coefficient as shown in the equation $a_i = \gamma C_i$, where a_i is the activity, γ is the activity coefficient, and C_i is the molar concentration of the species.

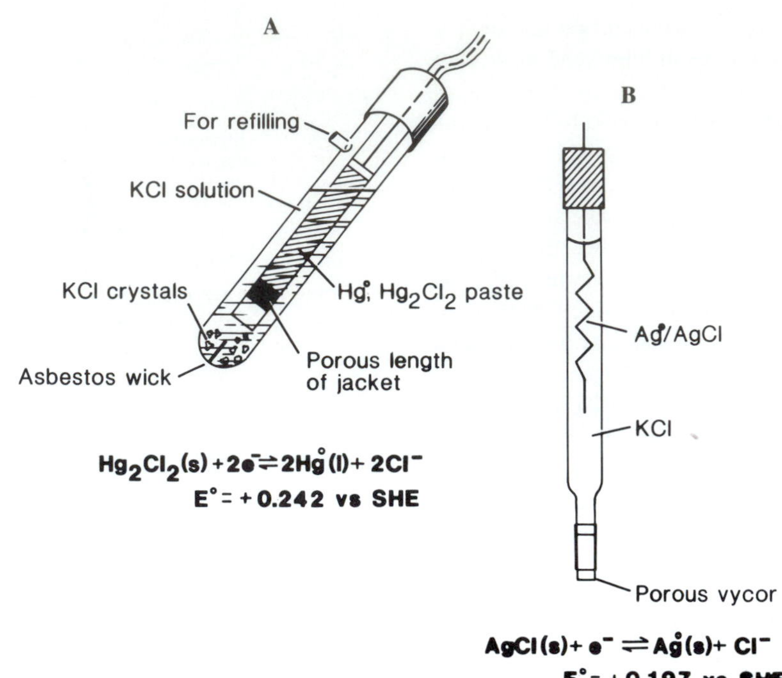

Fig. 13-2 References electrodes. **A,** Saturated calomel electrode, *SCE,* with asbestos wick for salt bridge junction. **B,** Silver/silver chloride electrode, *Ag/AgCl*.

of 0.0 V. Since other reference electrodes are easier to construct and use, the SHE is rarely used in practical applications of electrochemistry.

A commonly used reference electrode is the *saturated calomel electrode (SCE)*. A schematic diagram of a common type of SCE, its half-cell reaction, and standard half-cell potential are shown in Fig. 13-2, *A*. The electrode consists of elemental mercury covered with a thin coating of calomel (Hg_2Cl_2) that is in contact with an aqueous solution saturated with KCl. The potential of the half-cell will be constant so long as the activity of Cl^- does not change. The easiest way to set the activity of Cl^- to a fixed value that is easily verifiable is to saturate the solution with a chloride salt such as KCl. So long as crystals of KCl are present, the experimenter knows that the solution is saturated and that the activity of Cl^- is constant. Note that the other participants in the electrochemical reaction (Hg_2Cl_2 and Hg) are solid and liquid components and, consequently, exhibit unit activity regardless of the amounts present in the cell. Thus the SCE offers the extraordinary convenience of being easily fabricated without the need for accurately preparing the activities of any of the components.

Another commonly used reference electrode is the *silver/silver chloride electrode (Ag/AgCl)*. A representative Ag/AgCl reference electrode is shown in Fig. 13-2, *B*. The electrode is prepared by coating a silver wire with a thin film of AgCl and immersing it in a solution of constant chloride concentration, which fixes the half-cell potential. The Ag/AgCl reference electrode is used routinely, especially as the inner reference electrode in potentiometric membrane electrodes. The SCE and Ag/AgCl electrodes are commercially available or conveniently constructed.

Indicator electrodes

The indicator electrode is the essence of potentiometric analysis. This electrode should interact with the analyte of interest so that E_{ind} reflects the activity of this species in solution and not of other compounds in the sample that might interfere. The relative response of an electrode to one species and not to another species is defined as the *selectivity* of the electrode. The importance of having indicator electrodes that selectively respond to species of analytical significance has stimulated the development of many types of these electrodes.

Ion-selective electrodes. The most common indicator electrode used in clinical chemistry is the *ion-selective electrode* (ISE). The ISE is based on the measurement of a potential that develops across a selective membrane. The response of the electrochemical cell is therefore based on an interaction between the membrane and the analyte that alters the potential across the membrane. The selectivity of the potential response to an analyte depends on the specificity of the membrane interaction for the analyte.

A representative ISE is shown schematically in Fig. 13-3. The electrode consists of a membrane, an internal reference electrolyte of fixed activity, $(a_i)_{internal}$, and an internal reference electrode. The ISE is immersed in a sample solution that contains analyte of some activity, $(a_i)_{sample}$. An external reference electrode is also immersed in this solution. The internal and external reference electrodes constitute the two half-cells of the electrochemical cell. The potential measured by the pH/mV meter (E_{cell}) is equal to the difference in potential between the external ($E_{ref,ext}$) and internal ($E_{ref,int}$) reference electrodes, plus the membrane potential (E_{memb}), plus the liquid junction potential (E_{lj}) that exists at the junction between the external reference electrode and the sample solution.

$$E_{cell} = E_{ref,ext} - E_{ref,int} + E_{memb} + E_{lj} \qquad \textit{Eq. 13-2}$$

If the membrane is permeable to a particular ion (i), a potential develops across the membrane that depends on the ratio of activities of the ion on either side of the membrane. The half-cell potentials of the two reference electrodes are constant, sample solution conditions can be controlled so that E_{lj} is effectively constant, and the composition of the internal solution can be maintained so that $(a_i)_{internal}$ is fixed. Consequently, E_{cell} is described by the Nernst equation:

$$E_{cell} = K + 2.3\,\frac{RT}{zF}\log\,(a_i)_{sample} \qquad \textit{Eq. 13-3}$$

where K represents the constant terms and z is the charge on the analyte ion (cations: $+1$, $+2$, $+3$, etc.; anions: -1, -2, -3, etc.). This logarithmic relationship between cell potential and analyte activity is the basis of the ISE as an analytical device. A plot of E_{cell} versus log a_i for a series of standard solutions should be linear over the working range of the electrode and have a slope of 2.3 RT/zF or 0.0591/z for measurements made at 25° C. Since membranes respond to a certain degree to ions other than the analyte (that is, interferents), a more general expression than Equation 13-3 is needed:

$$E_{cell} = K + 2.3\,\frac{RT}{zF}\log\,[(a_i)_{sample} + k_{ij}a_j^{z/x}] \qquad \textit{Eq. 13-4}$$

where a_j is the activity of the interferent ion *(j)*, x is the charge of the interferent ion, and k_{ij} is the selectivity constant. Small values of k_{ij} are characteristic of electrodes with good selectivity for the analyte, i.

The development of successful ISEs has hinged on the search for membranes that exhibit both sensitivity and selectivity for the analyte of interest. Of the two properties, selectivity is by far the more difficult to achieve. ISEs with selectivity for cations and anions have been developed with three basic types of membranes: liquid and polymer, solid state, and glass. All of these membranes function by selectively incorporating the analyte ion into the membrane and thereby establishing a membrane potential. An ISE membrane must exhibit low solubility in the analyte medium to provide a durable electrode with a stable response. This requirement imposes a severe restriction on the material that can be used for membranes. Also, the membrane must exhibit some electrical conductivity to function in an electrochemical cell. The scope of ISEs has been expanded to include the measurement of gases and neutral organic compounds by combining ISEs with gas-permeable membranes and layers of enzymes, bacteria, and tissues. These general categories of electrodes and specific ISEs are considered in the following sections.

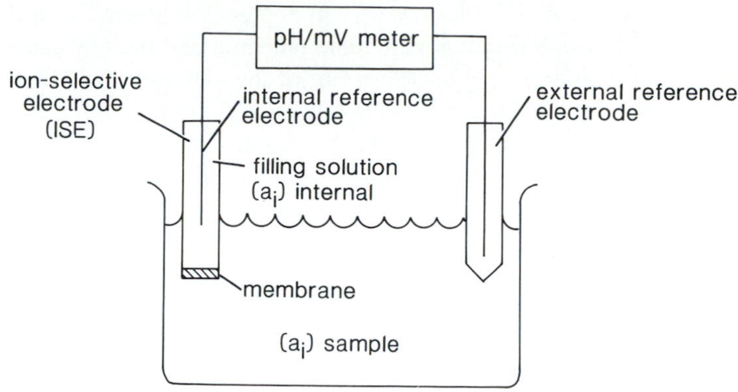

Fig. 13-3 Schematic diagram of an ISE.

Liquid and polymer membrane electrodes. A selective liquid membrane is the basis for a number of excellent ISEs. The liquid consists of a water-insoluble, viscous solvent in which is dissolved an *ionophore,* a hydrophobic organic ion-exchanger or a neutral carrier molecule that reacts selectively with the ion of interest. The liquid is typically soaked into a thin, porous solid membrane such as cellulose acetate, which is then incorporated into the ISE.

Fig. 13-4 shows the schematic diagram of the membrane portion of a liquid membrane ISE and the mechanism whereby an electrode responds to M^+ activity. The liquid membrane is in contact with internal and sample aqueous solutions of analyte M^+. The neutral carrier ionophore (R) reacts with M^+ at each membrane and solution interface and extracts M^+ into the membrane as MR^+. The extraction of M^+ into the membrane generates a positive membrane potential at each interface as a result of the charge difference that occurs when M^+ is extracted into the membrane in the form of MR^+ and the counter anion X^- that remains in the aqueous solution. As the activity of M^+ in solution is increased, the activity of MR^+ in the membrane increases, and the membrane potential increases. This reaction exists at both the outer membrane surface, which is exposed to the sample, and the inner membrane surface, which contacts the inner filling solution of the ISE. The potential of the inner surface of the membrane, $E_{memb(internal)}$, is kept constant by maintaining a constant activity of M^+ in the internal solution. Thus the only potential change measured in the circuit is the potential of the membrane surface contacting the sample, $E_{memb(sample)}$.

The availability of liquid and polymer membrane electrodes for a variety of ions is the result of the development of different neutral carrier ionophores and liquid ion exchangers that react selectively with particular ions. In the surface equilibria shown in Fig. 13-4, any ionic species other than M^+ that react to an appreciable degree with R will also generate a membrane potential and thereby cause an interference. The selectivity of the electrode for M^+ is therefore determined by the relative affinity between R and M^+ and R and the various interferent ions in the sample.

Several ionophores have been found that selectively bind cations to nonaqueous membranes. When it reacts with an ionophore, a cation is essentially inserted in a hydrophobic cavity within the ionophore that allows the cation to exist within a nonaqueous membrane medium. Selectivity for a particular cationic species is controlled by providing an optimum environment in terms of number and position of binding atoms. An excellent example is the antibiotic, valinomycin, which exhibits selectivity for K^+. Fig. 13-5 illustrates the K^+ complex of valinomycin. The hydrated K^+ cation fits into a snug cavity surrounded by oxygen atoms. The electrode exhibits excellent selectivity for K^+ against Na^+ because the smaller Na^+ ion is bound less tightly in the valinomycin cavity. This feature is of considerable practical importance in the clinical determination of K^+ in serum, which contains higher concentrations of Na^+ than K^+. Ionophores for the selective determination of NH_4^+, Ca^{2+}, Na^+, Li^+, and Mg^{2+} have also been incorporated into ISEs. The selective incorporation of these cations occurs by the same principle described for the K^+ electrode. The development of liquid and polymer membrane ISEs has allowed the measurement of ions in samples of diverse origin. The electrodes have been especially successful in clinical laboratories and are now used routinely for measuring Ca^{2+}, K^+, Na^+, and Cl^- in biological fluids. The response characteristics for liquid and polymer membrane ISE systems commonly used in biomedical investigations are shown in Table 13-2.

Electrodes based on ion-pairing reactions have been developed for numerous organic compounds. These electrodes are based on insoluble ion pairs between an ionic form of the organic compound and an ion-pairing reagent. An example is an electrode for the antiepileptic drug, phenytoin (5,5-diphenylhydantoin) based on the ion-pair complex between the 5,5-diphenylhydantoinate anion and the quaternary ammonium cation, tricaprylmethyl ammonium, which is immobilized in polyvinyl chloride.

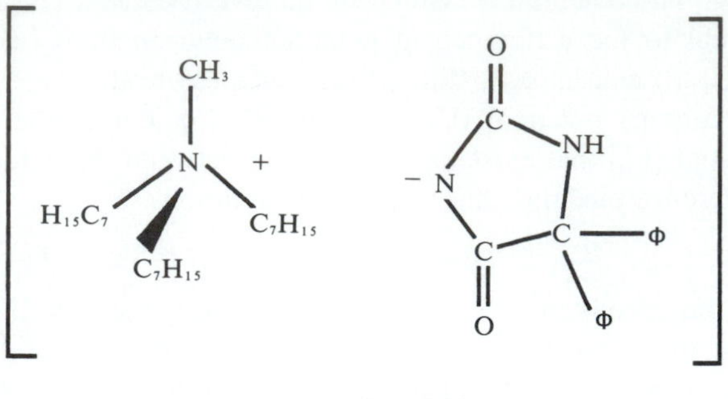

$\Phi = C_6H_5$

The electrode measures phenytoin over the range of 10^{-1} to 10^{-4} mol/L and has a detection limit of 1.5×10^{-5} mol/L. This electrode can be used to determine phenytoin in tablets and capsules. New ion exchangers and neutral carriers are continually being evaluated in an effort to improve selectivity of existing electrodes and to develop electrodes for other ions and molecules.

Solid-state membrane electrodes. Solid-state membranes consist of single crystals or pressed pellets of salts of the ions of interest. The crystal or pellet must have some degree of electrical conductivity and exhibit very low solubility in the solvent in which the electrode is to be used—usually water. An excellent ISE for F^- uses LaF_3 that is doped with Eu^{2+} to provide electrical conductivity. The membrane potential is generated by a selective surface

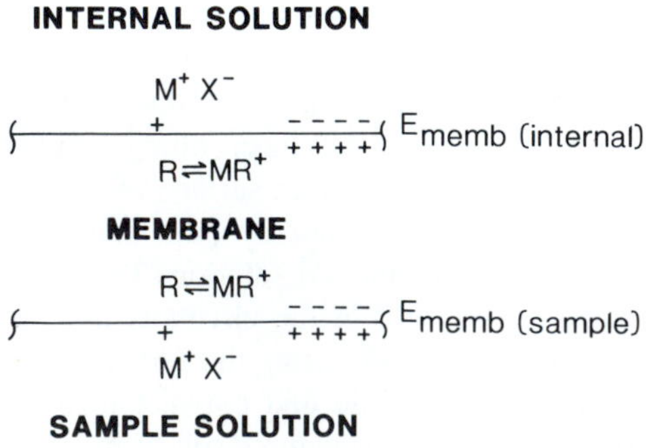

INTERNAL SOLUTION
M^+X^-
$R \rightleftharpoons MR^+$ E_{memb} (internal)
MEMBRANE
$R \rightleftharpoons MR^+$ E_{memb} (sample)
M^+X^-
SAMPLE SOLUTION

Fig. 13-4 Schematic diagram of liquid membrane ISE, where M^+ represents analyte cation, and *R* represents neutral carrier ionophore.

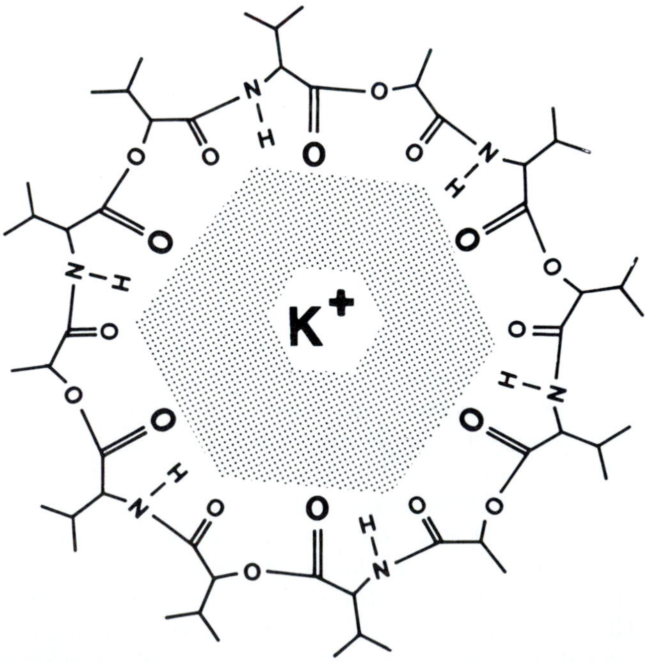

Fig. 13-5 Model of K^+-complex of valinomycin. Stippled region represents volume of hydrated K^+ ion. Bold oxygens are binding atoms.

Table 13-2 ISEs used in clinical chemistry

	Analyte	Membrane composition	Linear response range (mol/L)	Possible interferences
Glass	H^+	72.17% SiO_2, 6.44% CaO, 21.39% Na_2O (mol %)	10^{-12} - 10^{-2}	Na^+
	Na^+	11% Na_2O, 18% Al_2O_3, 71% SiO_2	10^{-6} - 10^{-1}	K^+, Ag^+
Solid state	F^-	LaF_3 crystal	10^{-6} - sat'd	OH^-
Liquid or polymer membrane	Na^+	Na^+ ionophore (ETH 227) 2-nitrophenylocytl ether, sodium tetraphenylborate	10^{-3} - 10^{-1}	Li^+, K^+, Ca^{2+}
	Cl^-	Tri-*n*-octylpropylammonium chloride, decanol	10^{-3} – 10^{-1}	OH^-, Br^-, F^-
	K^+	Valinomycin, dioctyladipate, PVC	3×10^{-5} – 1	NH_4^+
	Ca^{2+}	Ca^{2+} ionophore (ETH 1001), 2-nitrophenyloctyl ether, sodium tetraphenylborate	10^{-7} – 10^{-2}	—
	Ca^{2+}	Calcium di-(*n*-decyl)phosphate, di-(*n*-octylphenyl)phosphonate, PVC	3×10^{-5}– 1	Mg^{2+}
Gas sensors	CO_2	Combination glass pH electrode, 0.01-0.1 M $NaHCO_3$-NaCl filling solution; behind silicone rubber membrane	10^{-4} – 10^{-1}	Organic acids
	NH_3	Combination glass pH electrode, 0.1 M NH_4Cl filling solution; behind porous Teflon gas-permeable membrane	10^{-5} – 5×10^{-2}	Volatile amines

Modified from Meyerhoff, ME, and Opdycke, WN: Ion-selective electrodes. In Spiegel, HE, editor: Advances in Clinical Chemistry, vol 25, New York: Academic Press, Inc, 1986, p 1.

reaction between LaF_3 and F^- in which solution F^- is incorporated into vacancies in the crystal lattice. The selectivity is very good because other anions do not fit well into the crystal structure. The properties of the F^- ISE are shown in Table 13-2. Another clinically important solid-state membrane electrode for determination of Cl^- is based on pressed-pellet membranes of the ionic conductor, Ag_2S, and AgCl. Similar electrodes have also been developed for the detection of Br^-, CN^-, I^-, SCN^-, S^{2-}, Ag^+, Cu^{2+}, Pb^{2+}, and Cd^{2+}.

Glass membrane electrodes. The first and most widely used ISE is the glass membrane electrode for pH measurements. Glasses of certain compositions respond to pH when a membrane potential develops as a result of an ion-exchange mechanism with H^+ that occurs in the thin, hydrated outer layer of the glass membrane that has been soaked in solution. The outstanding properties of the glass pH electrode are attributable to the remarkable selectivity of this surface reaction for H^+.

The basic design of the glass electrode for pH is shown in Fig. 13-6. The electrode consists of a glass or plastic tube with a thin, pH-sensitive glass membrane sealed in the tip. Ordinarily the membrane is only about 50 μm thick and hence is very fragile. The bulb at the end contains an internal solution comprised of 0.1 M HCl into which is dipped a silver wire coated with AgCl, which provides an internal Ag/AgCl reference electrode. This solution also maintains a fixed activity of H^+ to which the internal surface of the membrane is exposed. A shielded cable makes electrical contact between the internal Ag wire and the external pH meter.

The pH response of the glass membrane is determined by the composition of the glass. The glass consists of Na_2O, CaO, and SiO_2. Pure SiO_2 is essentially an insulator that is unresponsive to pH. The addition of Na_2O to the glass formulation disrupts the SiO_2 structure so that nega-

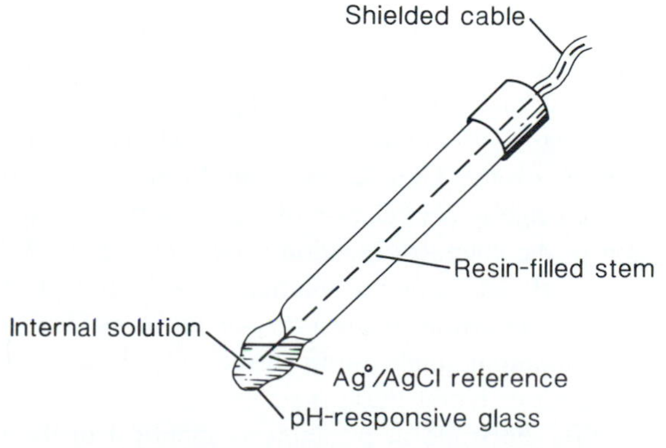

Fig. 13-6 Representative pH electrode. Usually only membrane at tip of electrode bulb is of H^+ ion-selective glass.

tively charged oxide sites (Si-O^-) are paired with Na^+. The mobility of Na^+ in the glass renders the glass membrane slightly conductive to electrical charge. The negative oxide sites serve as ion exchange sites in aqueous solution and provide the basis for pH response. The potential response to pH is extraordinarily accurate over a pH range of 0 to 14. At pH values above about 9 to 10 the electrode exhibits significant response to other monovalent cations such as Na^+ and K^+. This response to alkali cations at high pH is termed the *alkaline error*. It can be minimized by replacing Na_2O and CaO to a certain extent in the glass with Li_2O and BaO.

The membrane response to H^+ is attributed to an ion-exchange process that occurs in the vicinity of the membrane solution interface. On immersion of a dry glass membrane in an aqueous solution, the membrane surface becomes hydrated during the course of a few hours. This uptake of water leads to a gradual dissolving of the glass; this process generally determines the useful lifetime of an electrode. However, surface hydration is essential for electrode function; new electrodes immersed in solution respond poorly until adequately soaked. Soaking establishes a hydrated layer that is only 10^{-4} mm or less in thickness. This hydrated layer then functions as a cation exchange layer in which negatively charged oxygen sites are linked to the glass matrix but in which Na^+ is mobile. Soaking the electrode in acid, for example, results in the replacement of Na^+ with H^+. The membrane response to H^+ can be understood in terms of a surface potential that results from the ion exchange of Na^+ with H^+ in the hydrated gel. Immersion of the electrode membrane in alkaline solution results in exchange of H^+ in the membrane with Na^+ as membrane H^+ moves into solution. The inner membrane potential is held constant by exposure to a fixed activity of H^+ in the internal solution. Glass electrodes for Na^+, Ag^+, and NH_4^+ have been developed by varying the composition of the glass.

Gas-sensing electrodes. Gas-sensing electrodes consist of an ISE in contact with a thin layer of aqueous electrolyte that is confined to the electrode surface by an outer membrane, as shown schematically for a CO_2 electrode in Fig. 13-7. The outer membrane is very thin and is chosen so that it is permeable to the gas of interest; for CO_2, the membrane is made of silicone rubber. This membrane allows CO_2 gas in the sample to pass through the membrane. Dissolution of the CO_2 in the thin layer of electrolyte causes a change in pH because of a shift in the equilibrium position of the chemical reaction shown in Fig. 13-7. The change in pH sensed by the internal ion selective pH electrode is in proportion to the P_{CO_2} of the sample. One of the most important applications of the CO_2 electrode is the measurement of blood P_{CO_2}.

The NH_3 electrode in principle is identical to the CO_2 electrode; here the filling solution is aqueous ammonium chloride. The internal pH electrode senses the change in

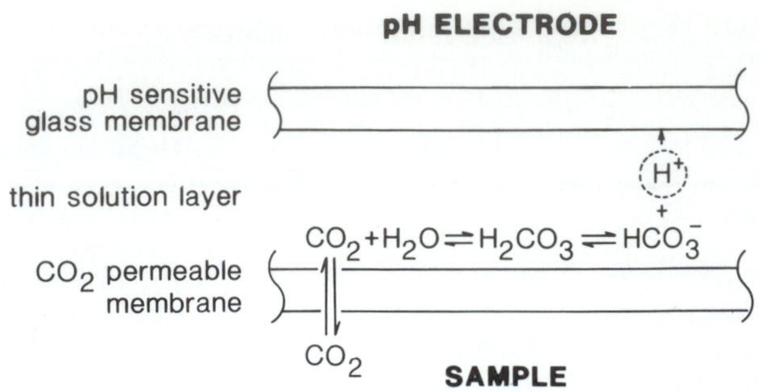

Fig. 13-7 Schematic representation of gas-sensing electrode for CO_2.

pH from the ammonium/ammonia (NH_4^+/NH_3) acid-base equilibrium. The pH change is thus proportional to P_{NH_3} of the sample.

Gas electrodes have been used for other bioanalytical applications such as the measurement of CO_2 in general assays of decarboxylating enzyme activities and the measurement of NH_3 in tissue and serum. Characteristics of the CO_2 and NH_3 electrodes are shown in Table 13-2.

Care and methodology

The care of ISEs is essentially similar for every ion type. Since the sensing tip of the electrodes is made out of extremely fragile and sensitive materials, care must be taken to prevent breakage and maintain the tip in a moist environment. Most commercially available electrodes are supplied with protective coverings to help prevent careless damage to the electrodes while they are not in use. Each electrode is also accompanied by manufacturer's recommendations for specific cleaning and storage requirements. Storage conditions depend on the frequency of use, the type of electrode, and the application. For example, cleaning procedures are different for a pH electrode used in a protein solution and for one used in a solution of inorganic ions. Protein layers are removed by rinsing the electrode with pepsin, bleach, or 0.1 M HCl; whereas an inorganic deposit can be removed with EDTA or acids. After each cleaning procedure, the electrode tip is thoroughly rinsed with distilled water, and the electrode is returned to the storage container.

pH measurements are easily made in the clinical laboratory with a two-point calibration procedure. Standard buffer solutions, which are commercially available, are chosen to bracket the pH of the sample solution. Electrode calibration is always initiated with a pH = 7 standard buffer. The meter is adjusted to read 7.00 after the temperature control has been set to the temperature of the buffer. Subsequently, either an acidic or alkaline buffer is used to complete the calibration, depending on the sample to be measured. Samples can then be measured. Multiple calibrations may be necessary for a large number of samples.

For non-H^+-sensing electrodes a calibration curve that relates the potential difference in mV to concentration is needed. Although ISEs measure analyte activity, the concentration can be related to mV as long as the ionic strength is constant between standards and samples. This is accomplished by addition of a small amount of a solution of high ionic strength to the calibrating standards. This solution must not contain any interfering ions. A potential reading is determined for each standard solution, beginning with the lowest concentration. A plot of log $(C_i)_{standard}$ versus potential is linear for a properly responding ISE. Ion concentrations in samples are then obtained by measurement of the response in mV of the sample and use of the calibration curve. The sample ion concentrations are valid as long as the matrix of the standard solutions is made to closely mimic the samples. Most laboratory electrodes and instruments routinely use a two-point standardization procedure to ensure similar response from analysis to analysis. Other standardization and analysis methods have been developed for various electrodes and applications.

In the clinical laboratory, the analysis of large numbers of samples is a necessity. Thus many ISEs are incorporated into automatic analyzers in a flow-through configuration. This arrangement takes advantage of the rapid response of ISEs by placing them into multi-ion analyzers with large sample throughput capabilities. Calibrants, samples, and rinsing solutions are pumped across the electrode surfaces of the ISEs, which are placed in series. A single reference electrode is used for all ISEs in a system with the exception of the CO_2 sensor, which has its own reference electrode behind the gas-permeable membrane. Calibrants have constant ionic strength that closely matches that of physiological samples to minimize errors that result from differences in liquid junction potentials between samples and standards. It cannot be emphasized enough that the proper care and use of the ISEs in these instruments is essential to ensure accurate and reproducible analyses. This requires a constant monitoring of the performance of both the individual electrodes and the instrument as a whole. The following section will discuss some of the common errors and interferences that can occur with ISE measurements.

Experimental considerations and interferences

Errors in ISE measurement can result for any ion determination if standards and samples are not run at approximately the same temperature, since the Nernst equation is temperature dependent. Perhaps the most important source of error is the response of an ISE to a nonanalyte or interferent ion in the sample. It is therefore important to know the selectivity properties of the electrode being used and to ensure that nonanalyte ions to which the electrode responds are not present in high enough concentrations to constitute an interference. Components in certain samples can also change the sensitivity of an electrode by adsorbing on its surface and thereby blocking access of the analyte to the surface. Such contamination is a problem in samples containing surface-adsorbing species such as proteins. For single electrode determinations in whole blood or serum, techniques are available that isolate the ISE from direct contact with the sample. However, the demand for multiple sample and ion analyses has made single electrode measurements impractical. Modern multi-ion analyzers incorporate a small size–exclusion membrane that protects the ion-selective membrane from the high molecular weight components of biological fluids but allows analyte molecules access to the electrodes.

Although many ISEs are very selective, under certain conditions some ions may interfere and yield erroneous results. Specific examples are listed in Table 13-2 and are briefly discussed below. Detailed descriptions of clinical analyses for many ions can also be found in Chapter 58.

pH measurements have few specific interferences associated with them. The linear response range is from pH 2 to 12. Sensitivity of the glass pH electrode may be reduced at pH values above 10 because of the interference of monovalent cations, especially Na^+. Although monovalent cations can enter and slowly move through the hydrated layer, multivalent cations of 2+ or 3+ charge do not interfere. In solutions of pH less than 1, low water activities also give rise to measurement error.

Na^+ ions are determined by either a glass electrode or a polymer-type liquid membrane electrode. Interferences are minimal because of the high concentration of sodium in biological fluids, especially in blood. The glass electrode exhibits excellent selectivity over K^+ and H^+ because of this fact. Highly acidic urine samples are an exception. The polymer-based ISE, however, can be subject to an interference from Li^+ if a patient is being treated with lithium therapy.

K^+ is usually measured by the valinomycin/polymer electrode described above. Good results are obtained for measurements in blood; however, in undiluted urine samples a negative error may result because of the partitioning of a negatively charged lipophilic component of the urine that is permeable to the polymer. This component can be excluded by using an ISE with a silicone-rubber membrane instead of the polymer. Alternatively, accurate measurements can be obtained by sample dilution.

Determination of Na^+ and K^+ levels in undiluted blood and urine samples requires special attention. Measurements made by the nondilutional ISE method may differ from results obtained by flame atomic emission spectroscopy (FAES). As described in Chapter 58, FAES determines the concentration of ions in the total sample volume. In contrast, the ISE method measures the activity of the ions in the sample. Since the activity can be influenced by the sample environment (for example, protein and water content), deviations may occur between these methods.

Agreement between these methods is usually realized by sample dilution. Methodological differences may be greater in the case of Na^+ determinations because of the higher relative concentration of Na^+ in biological fluids. The influence of physiological effects of Na^+ and K^+ determinations in biological fluids is discussed in further detail in Chapter 58.

Much care must be taken in the determination of Ca^{2+}. Ca^{2+} exists in both the bound form (with proteins or other biological molecules) and the unbound, ionized Ca^{2+} form. The Ca^{2+} ISE is one of the easiest ways to measure ionized Ca^{2+}, since it responds to only the ionized form, which is thought to be the physiologically important form.

The determination of Cl^- in biological fluids by an ISE suffers from frequent fouling of the surface by proteins present in the sample. This problem can be minimized using the semipermeable membrane described above to exclude the large molecules from the electrode surface. Recently a polymer-type liquid membrane electrode that exhibits selectivity for chloride has been developed and used in electrolyte analyzers for biomedical use. The chloride ISE is subject to interference from Br^-, I^-, F^-, CN^-, OH^-, and S^{2-}, but, with the exception of Br^-, these ions are usually not present in physiological samples at a level high enough to interfere.

The F^- ISE exhibits excellent selectivity and suffers only from interference of OH^- at high pH.

ISEs for the measurement of CO_2 are relatively easy to use and interference free. Undiluted blood can be used directly as the sample, and calibration is typically accomplished with 5% and 10% mixtures of CO_2 in an inert gas. Total CO_2 measurements require acidification to convert $CO_3{}^{2-}$ and $HCO_3{}^-$ to CO_2. In this case, calibration is performed with standard $NaHCO_3$ solutions. Response times for total CO_2 measurements are generally longer because of the necessity of establishing equilibrium conditions. $CO_3{}^{2-}$ selective membranes are available in some instruments for total CO_2 measurements. Interferents are mainly organic acids to which the gas membrane is also permeable. The carbonate-selective membranes are subject to interferences from anions such as salicylate, but placement of the ISE behind the silicone-rubber membrane alleviates this problem.

The NH_3 ISE responds very selectively and rapidly to the ammonia concentration in solution. The major drawback is the questionable stability of the membrane and the electrode. Interference can result from nonpolar volatile amines present in the sample. With both the CO_2 and NH_3 electrodes it is important to have a rapidly responding electrode. A decrease in the response time signifies a loss in electrode performance and may require the replacement of the membrane.

VOLTAMMETRIC METHODS

Electrochemical techniques in which a potential is applied to an electrochemical cell and the resulting current from an electrochemical reaction is measured are generally categorized as *voltammetric* methods. Electrochemical cells for voltammetry use a three-electrode configuration. The cell consists of a *working electrode, reference electrode,*, and *auxiliary electrode*. The potential is applied between the working and reference electrode by a *potentiostat;* this applied potential can force changes to occur to any electroactive species at the working electrode surface by *electrolysis*. Electrolysis can occur by a *reduction,* a gain of one or more electrons, or an *oxidation,* a loss of one or more electrons. The current required to sustain the electrolysis at the working electrode and maintain electroneutrality in the cell is provided by the auxiliary electrode. This arrangement prevents the reference electrode from being subjected to large currents that could change its potential. Some voltammetry instrumentation is based on the two-electrode system. Here the auxiliary electrode is absent, and the reference electrode is subjected to the entire cell current.

The basic concept of applying a potential to an electrochemical cell and measuring the current that results from electrolysis can be implemented in numerous ways. Several different techniques have been developed by variations in how the potential is applied and/or how the current is measured. Although the resultant output and the practical applications of these techniques are varied, they all share the common basis of applying a potential, *E,* and measuring a current, *i*, or charge, *Q*. In addition, the solution may be moving or stationary with respect to the working electrode. Voltammetry in an unstirred solution is referred to as *stationary* solution voltammetry. *Hydrodynamic* voltammetry involves the forced movement of solution either through stirring the solution or flowing the solution over the electrode as in *liquid chromatography with electrochemical detection (LCEC)*.

The result of a voltammetric technique is called a *voltammogram* (that is, a current-potential curve). Voltammograms give useful quantitative and qualitative information about the electrochemical reaction. A typical hydrodynamic voltammogram is shown in Fig. 13-8 for the reduction of species *Ox* by one electron to species *Red*.

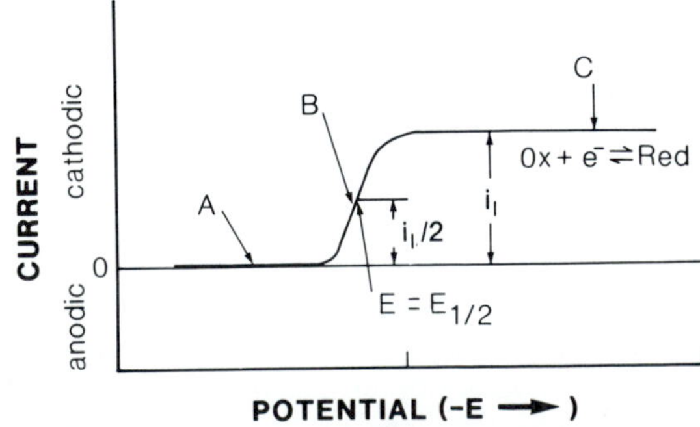

Fig. 13-8 Generalized hydrodynamic voltammogram for reduction of Ox to Red. Potential increases negatively left to right. i_l is limiting current and $E_{1/2}$ is half-wave potential.

As the potential is scanned in the negative direction, the voltammogram can be described by three distinct regions. In region A, the potential applied at the working electrode is insufficient to cause electrolysis to occur; therefore no current is observed. The onset of electrolysis is signaled by the rise in current in region B. The current continues to rise until a maximum value is reached. This is region C, where electrolysis is occurring at the maximum rate possible. The maximum current in region C is defined as the *limiting current* (i_l), and is defined by Equation 13-5:

$$i_l = \frac{nFAD_oC_o}{\gamma} \qquad \textit{Eq. 13-5}$$

where A is the electrode area, D_o is the diffusion coefficient of Ox, C_o is the concentration of Ox, and γ is the diffusion distance. As is illustrated by Equation 13-5, the magnitude of i_l is directly proportional to the concentration of the electrochemically active analyte. Thus voltammetry can be used to quantitatively measure analyte concentration. The practical unit for current is the ampere (A), which is the transfer of one coulomb of charge per second. This corresponds to the passage of 1.05×10^{-5} moles of electrons per second. Since the current involved in most electroanalytical techniques is very small, milliamperes (mA), microamperes (μA), and nanoamperes (nA) are commonly used units.

The *half-wave potential* ($E_{\frac{1}{2}}$) is defined as the potential at one-half the limiting current. $E_{\frac{1}{2}}$ is uniquely characteristic of the species undergoing electrolysis just as the half-cell potential is for the reference electrode, and it can be used for qualitative identification. By convention a reduction is described by a positive or *cathodic* current, whereas an oxidation is described by a negative or *anodic* current. The principles for an oxidation are similar and can be applied for a positive potential scan.

One specific type of voltammetry that is clinically useful is amperometry. Amperometric sensors are devices that measure the current generated at a fixed potential by an electroactive analyte in solution. The potential is set to a value of E where i_l occurs (Fig. 13-8, region *C*), and i_l is then measured for each sample. The current measured is directly proportional to the concentration of species present. Three clinically important amperometric sensors discussed below are the oxygen electrode, the glucose electrode, and LCEC.

Voltammetry electrodes

Working electrodes. Working electrodes have certain properties in common. Good electrical conductance is of foremost importance. Consequently, working electrodes are generally metals or semiconductors. Chemical and electrochemical inertness is important in applications for which the electrode should function simply to transfer electrons to and from species dissolved in solution. This inertness gives a wide potential region with minimum background contributions from electrode and solvent redox properties in which the electrochemistry of the analyte(s) can be easily monitored.

Platinum, gold, mercury, and glassy carbon are commonly used materials for voltammetric electrodes. When used for voltammetry, mercury can be used in the form of a *hanging mercury drop electrode (HMDE)*. To provide the working electrode surface, a reproducible mercury drop can be extruded through a narrow glass capillary by means of a commercially available micrometer syringe. A new drop is formed by simply dislodging the old one and extruding more mercury. The *dropping mercury electrode (DME)* is the working electrode for *polarography*. In this technique, mercury is forced by gravity through a very fine capillary to provide a continuous stream of identical droplets. Each droplet expands, becomes too heavy to be suspended, and breaks loose from the capillary.

Auxiliary electrodes. Auxiliary electrodes are made from any conductive material, typically a piece of platinum wire.

Reference electrodes. The commonly used reference electrodes for voltammetry are the SCE and Ag/AgCl electrodes, which have been previously described in detail.

Oxygen electrode

The oxygen electrode is designed as a complete electrochemical cell. The basic design, which is shown in Fig. 13-9, incorporates a platinum disk as the cathode and an Ag/AgCl electrode as the anode in a buffered electrolyte solution. The electrochemical cell is isolated from the sample by an oxygen-permeable membrane. Oxygen diffuses through the membrane and is reduced electrochemically at the platinum electrode, which is held at a potential that quantitatively reduces oxygen (−0.5 to −0.6 V versus Ag/AgCl).

$$O_2 + 2H^+ + 2e^- \longrightarrow H_2O_2 \qquad \textit{Eq. 13-6}$$

The current generated at the platinum electrode is directly proportional to the concentration (partial pressure) of oxygen dissolved in the sample. As before, the membrane inhibits electrode fouling from serum proteins in blood and also prevents other electroactive substances from being reduced at the electrode. Calibration of the electrode system is done with standard solutions or gases containing known concentrations of oxygen.

Few interferences are associated with the use of the oxygen electrode. Poor response times and variable results may suggest that degradation of the membrane or a change in pH of the buffer solution has occurred. Silver metal may deposit on the platinum cathode and also affect the electrode response. Polishing the electrode with electrode polishing compound will regenerate the platinum surface.

The oxygen electrode is usually incorporated into a blood gas analyzer, which routinely measures oxygen, CO_2, and pH on samples of less than 250 μL of whole blood. Miniaturized O_2 electrodes have been developed that can be used for transcutaneous measurements, elimi-

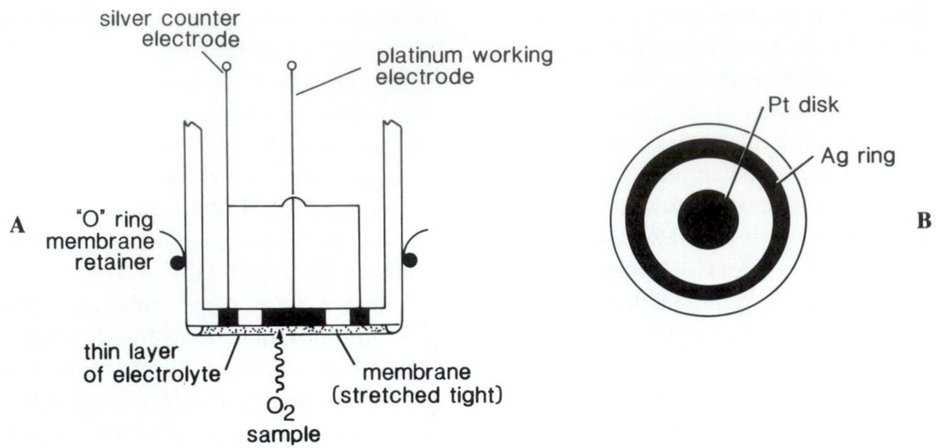

Fig. 13-9 Schematic diagram of oxygen electrode. **A,** Cross-sectional view showing diffusion of O_2 from sample through membrane. **B,** View of electrode assembly from bottom.

nating the need for drawing blood samples. However, the accuracy and response time of these electrodes depend on the physical characteristics of the patient's skin tissue. The oxygen electrode has been used to monitor many reactions that involve consumption of O_2 to measure glucose (glucose oxidase), cholesterol (cholesterol oxidase), and uric acid (uricase).

Glucose electrode

Glucose is another important constituent of serum and plasma that can be measured by an amperometric sensor. A diagram of a typical glucose electrode is shown in Fig. 13-10. The basic design of this electrode uses the enzyme, glucose oxidase, immobilized behind a glucose-selective membrane. Glucose oxidase reacts with the glucose in the sample to generate hydrogen peroxide (H_2O_2) and gluconic acid. The inner membrane is permeable to H_2O_2, which is determined amperometrically by the underlying platinum electrode held at a positive potential sufficient to oxidize H_2O_2 to O_2 (the reverse of the reaction shown in Equation 13-6). The current measured from the H_2O_2 oxidation is directly proportional to the glucose concentration; glucose

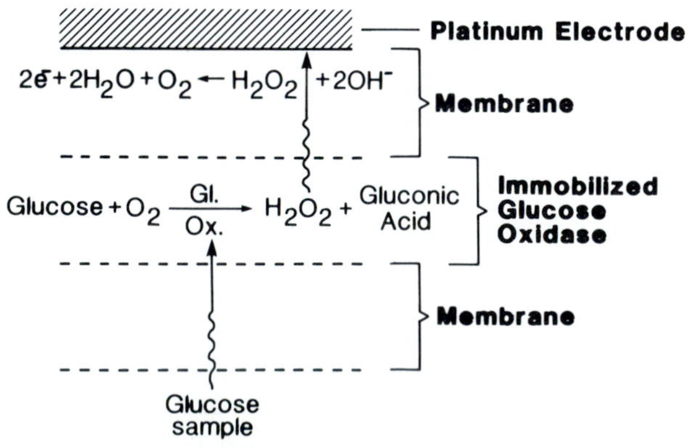

Fig. 13-10 Schematic diagram of glucose electrode.

concentrations have been reported to be quantified in the range of 10^{-7} to 10^{-3} M. Few interferences are noted for this electrode. The inner membrane is impermeable to ascorbic acid, uric acid, and acetaminophen, which are electroactive at positive potentials and may be present in clinical samples. The glucose electrode may be limited by the depletion of O_2 in the internal buffer solution. Current research is involved in constructing future generations of glucose electrodes to overcome this problem. The design for this glucose electrode has been developed by the Yellow Springs Instrument Company (Yellow Springs, OH).

Liquid chromatography with electrochemical detection

LCEC is a hybrid technique that combines chromatography with electrochemistry. As discussed in Chapters 4 and 5, high-performance liquid chromatography provides a means of separating various components of solutions by making use of their chemical affinities to column-packing materials. In LCEC, electroactive materials are detected sequentially as they elute from a chromatographic column and flow across or through the working electrode. Very low detection limits (approximately 1.0 pmol) may be obtained with relatively simple instrumentation. As a result of these features, LCEC has become recognized as a powerful tool for the trace determinations of many clinically important biomolecules. These include several metabolites of the central nervous system that are easily oxidized or reduced at an electrode surface.

To optimize an LCEC determination, it is necessary to consider both chromatographic and electrochemical requirements simultaneously. A primary requirement for electrochemical detection is that the mobile phase has a relatively high conductivity. To this end, buffered mobile phases of moderate ionic strength (0.01 to 0.1 M) are typically used. The vast majority of LCEC applications use reverse-phase, ion-exchange, or ion-pair chromatographic columns for separation (see Chapter 5). Significant amounts (up to 90% v/v) of organic modifiers such as methanol, acetonitrile, and propanol can be added to the aqueous mobile phase to shorten chromatographic retention. Of great advantage in LCEC is the inherent specificity of electrochemical detection, in which only compounds electroactive at the applied potential are detected. In this way many interferences that have similar chromatographic retention times are eliminated. For this reason, sample preparation may be quite facile in comparison with other detection methods available for high-performance liquid chromatography.

Most LCEC applications use a single working electrode in a conventional three-electrode system (see Chapter 5). The choice of working electrode material is an important consideration for LCEC. Carbon paste, a mixture of graphite powder and a coagulant such as paraffin oil, exhibits low background currents at positive potentials and is well suited to oxidative applications. Unfortunately, carbon paste electrodes are incompatible with mobile phases containing significant amounts of organic modifier. Glassy carbon electrodes are compatible with modifiers and may be used at negative potentials. Mercury or mercury–gold amalgam electrodes have the best characteristics when operated at negative potentials, and they are used principally for reductive LCEC. Any of these types of electrode surfaces can become fouled and inefficient when detecting high concentrations of some compounds. Additionally, high concentrations of proteins or other large biomolecules can lead to electrode fouling. This will result in irreproducible data. To alleviate the permanent loss of electrode performance, both physical and chemical cleaning procedures have been developed.

To obtain the best detector sensitivity and detection limits, it is important to know the minimum potential that need be applied to the working electrode for the desired electrochemical reaction to occur. This is most often accomplished by obtaining a hydrodynamic voltammogram (Fig. 13-8) of each component being detected. To generate such a voltammogram, multiple injections are made using various applied detector potentials, and the resulting current is plotted as a function of potential. To achieve maximum sensitivity, the applied potential should be selected so that the observed oxidations or reductions are at the limiting current for all compounds to be detected.

LCEC has become a widely used analytical technique for biomedical analysis. Although space does not permit a complete review of all biochemical applications, this section will consider general classes of compounds detectable by both oxidative and reductive techniques. Table 13-3 lists several compounds of pharmaceutical origin that may be determined using LCEC. Table 13-4 provides the approximate oxidation potentials obtained from hydrodynamic voltammograms that are used to detect some neurochemically important compounds by LCEC.

Oxidative applications. Most phenols are readily oxidized at carbon electrodes. The first and still most common LCEC application is the determination of catecholamines in biological samples. A second major use of LCEC is in the determination of the hydroxyindole metabolites of tryptophan. More recently, an electrochemical enzyme immunoassay has been developed in which phenols are detected as the electroactive product of the enzyme reaction. Aromatic amines are likewise easily oxidized at carbon electrodes and may be studied by LCEC.

Many compounds of biomedical interest are heterocyclic in structure, and electroactive at potentials easily attained using LCEC. Methods for the determination of such heterocycles as ascorbic acid, uric acid, NADH, biotin, and the folates have been developed. Although most amino acids are not electroactive at analytically usable potentials with carbon electrodes, derivatization methods have been developed to produce electroactive products. These methods provide excellent determination of amino acids. Thiols

Table 13-3 Applications of LCEC to drug analysis

Class	Compound
Analgesic	Acetaminophen, codeine, naproxen, phenacetin, salicylic acid, ketobemidone, morphine
Tranquilizer	Diazepam
Anticonvulsant	Nitrazepam
Antibacterial	Amoxicillin
Adrenergic blockers	Labetalol, Mepindolol
Antineoplastic	Methotrexate, procarbazine hydrochloride
Muscle relaxant	Theophylline
Antihypertensive	Sulfinalol hydrochloride
Antitubercular	Rifampin
Antipsychotic	Chlorpromazine
Antiarthritic	Penicillamine

Modified from Lunte, CE, and Heineman, WR: Electrochemical techniques in bioanalysis, Top Curr Chem 143:1-48, 1988.

Table 13-4 Approximate oxidation potentials of neurochemically important compounds

Compound	Oxidation potential
Dopamine	0.3 V
Epinephrine	0.3 V
5-Hydroxyindole acetic acid (5-HIAA)	0.3 V
Norepinephrine	0.3 V
Ascorbic acid	0.3 V
Vanillylmandelic acid (VMA)	0.6 V
Metanephrine	0.6 V
Normetanephrine	0.6 V
3-Methoxy-4-hydroxyphenylglycol (MHPG)	0.7 V
Homovanillic acid (HVA)	0.7 V

Modified from Lunte, CE, and Heineman, WR: Electrochemical techniques in bioanalysis, Top Curr Chem 143: 1-48, 1988.

may also be determined at a gold–mercury amalgam electrode.

Reductive applications. Wide-scale reductive LCEC applications have not been well established in the clinical laboratory because of the difficulties presented by oxygen and trace metal ion interferences at negative potentials. These problems can be overcome with established mobile phase deoxygenation procedures and the use of high purity salts, leading to more clinical applications of reductive LCEC.

Aromatic nitro and nitroso compounds are easily reduced at carbon or mercury electrodes. Other nitro compounds such as nitrate esters, nitroamines, and nitrosamines are also easily reduced. Additionally, several heterocycles of clinical interest may be detected by reductive LCEC, including the K vitamins, the pterins, and several pharmaceuticals.

Anodic stripping voltammetry

Anodic stripping voltammetry is a voltammetric technique that is useful in clinical chemistry for the determination of heavy metals. The determination of Pb^{2+} in biological fluids of patients suspected of having lead poisoning is an example of its use. Stripping voltammetry has the lowest detection limit of the commonly used electroanalytical techniques. Analyte concentrations as low as 10^{-10} M have been determined. The technique consists of two steps. In the first step, analyte is deposited at a mercury electrode by the application of a potential that is sufficient to reduce the species of interest at the working electrode. This step serves to preconcentrate the analyte by electrochemically extracting it into a mercury electrode. It is this preconcentration feature that enables such low concentrations to be reached by stripping voltammetry. In the second step, the deposited analyte is removed, or "stripped," from the electrode by application of increasingly positive potentials, and the resulting current signal is a measure of the amount of species in solution. Since the stripping step gives anodic current (that is, the species is oxidized), the technique is termed anodic stripping voltammetry.

In anodic stripping voltammetry only a fraction of the total analyte is deposited into the mercury electrode by electrolysis during the preconcentration step. Complete deposition of all of the analyte into the electrode is time consuming and generally unnecessary, since adequate concentrations can usually be deposited into the electrode to give a satisfactory stripping signal in much shorter times. Since the deposition is not exhaustive, it is important to deposit the same fraction of analyte for each stripping voltammogram. The parameters of electrode surface area, deposition time, and stirring must be carefully duplicated for all standards and samples. Deposition times vary from 60 seconds to 30 minutes, depending on the analyte concentration, the type of working electrode, and the stripping technique.

Anodic stripping voltammetry has become a useful method in the clinical laboratory for the determination of Pb^{2+} in blood and urine since the development of automated instrumentation by Environmental Science Associates (ESA). ESA also markets a digestion reagent, Metexchange, which frees bound Pb^{2+} from biological components of blood and urine. Detailed procedures for Pb^{2+} determination in blood and urine by anodic stripping voltammetry can be found in the literature.

COULOMETRIC METHODS

Coulometry is a very useful electrochemical method for quantitative analysis. Clinical applications use one form of coulometry that involves the application of a constant current to generate a titrating agent. In principle, the time required to titrate a sample at constant current is measured and is related to the amount of analyte in a sample by Faraday's law (Equation 13-7):

$$Q = it = nFN \qquad \textit{Eq. 13-7}$$

where Q is the charge passed for a finite time (t) at constant current (i), n is the number of electrons involved in the electrochemical reaction, F is Faraday's constant, and N is the number of moles of analyte in the sample. Charge is a quantity of electricity. The unit for charge is the coulomb (C) and corresponds to 1.05×10^{-5} moles of electrons. Since N is measured directly without the need for standards, coulometry is an absolute method and can be used for very precise determinations of analyte.

Titration of chloride

The most common clinical application of coulometry is the determination of Cl^- in serum, plasma, urine, and other body fluids. The analysis of Cl^- takes advantage of the quantitative formation and low solubility of AgCl. Ag^+ ions (the titrating agent) are electrochemically generated at the Ag anode by applying a constant current. Cl^- ions in the sample are rapidly consumed as they react with Ag^+ to form the insoluble AgCl complex. At any point in the titration, the Ag^+ concentration is very low.

$$\text{Anode reaction: } Ag \longrightarrow Ag^+ + e^-$$
$$\text{Solution reaction: } Ag^+ + Cl^- \longrightarrow AgCl\ (s)$$

However, the end point is signaled by a sudden increase in Ag^+ concentration that follows the consumption of all of the Cl^-. A second pair of Ag^+-specific electrodes detects the rise in concentration of silver ions in solution and immediately stops the titration. The amount of Cl^- in the sample is proportional to the amount of Ag^+ ions generated at the anode.

Coulometric determination of chloride is very precise; however, other anions that form insoluble complexes with silver ion can result in Cl^- determinations that are falsely elevated. Poor reproducibility can be a problem at high chloride concentrations because of the large amount of precipitate. Routine cleaning procedures that ensure accurate and reproducible results are discussed in Chapter 58.

BIBLIOGRAPHY

Potentiometric methods

Bates, RG: Determination of pH: theory and practice, New York, 1964, John Wiley & Sons, Inc.

Freiser, H, editor: Ion-selective electrodes in analytical chemistry, vols. I and II, New York, 1978, 1980, Plenum Publishing Corp.

Koryta, J, editor: Medical and biological applications of electrochemical devices, New York, 1980, John Wiley & Sons, Inc.

Meyerhoff, ME, and Opdyke, WN: Ion-selective electrodes. In Spiegel, HE, editor: Advances in clinical chemistry, vol. 25, New York, 1986, Academic Press, Inc., p. 1.

Morf, WR: The principles of ion-selective electrodes and of membrane transport, New York, 1981, American Elsevier.

Simon, W, Ammann, D, Bussmann, W, and Meier, PC: Ion-selective electrodes in biology and medicine. In Laidler, KJ, editor: Frontiers of chemistry, New York, 1982, Pergamon Press, Inc., p. 217.

Solsky, RL: Ion-selective electrodes in biomedical analysis, CRC Crit Rev Anal Chem 14:1, 1982.

Voltammetric methods

Bard, AJ, and Faulkner, LR: Electrochemical methods, New York, 1980, John Wiley & Sons, Inc.

Kissinger, PT, and Heineman, WR, editors: Laboratory techniques in electroanalytical chemistry, New York, 1984, Marcell Dekker.

Lunte, CE, and Heineman, WR: Electrochemical techniques in bioanalysis, Top Curr Chem 143:1, 1988.

Enzyme electrodes

Albery, WJ, and Bartlett, PN: Amperometric enzyme electrodes. Part I, Theory. Part II, Conducting salts as electrode materials for the oxidation of glucose oxidase, J Electroanal Chem 194:211, 1985.

Kessler, M, Clark, LC Jr, Lubbers, DW, et al, editors: Ion and enzyme electrodes in biology and medicine, Baltimore, 1976, University Park Press.

Phenytoin electrode

Cosofret, VV, and Buck RP: A poly (vinyl chloride) membrane electrode for determination of phenytoin in pharmaceutical formulations, J Pharmacol Biomed Anal 4:45, 1986.

CHAPTER 14 Automation

MICHAEL A. PESCE

OBJECTIVES

- Outline the steps in sample processing. Outline the steps in analysis and describe how they may be adapted to automation.
- Describe the proportioning of samples to reagents as accomplished by the following analytical systems:
  a. Dry-film reagent systems
  b. Discrete test analyzers
  c. Continuous-flow systems
- Define sample carryover, identifying systems where it is commonly seen, and suggest at least two ways by which it can be minimized.
- Outline the process of mixing, incubation, and sensing, and describe how they are handled in dry-film reagent systems, continuous-flow systems, and discrete test analyzers.
- List and interpret the eight basic concepts of automation, explaining how the concepts are applied to commonly used automated instruments.

KEY TERMS

analog A measurement derived directly from an instrument's continuous signal (that is, voltage) and usually presented in graphic form.

automation Use of a machine designed to follow repeatedly and automatically a predetermined sequence of individual operations.

bar coding A computer-driven sample-recognition system that identifies both the specimen and the analyses to be performed and relays this information to the automated analyzer.

bulk reagents Those that must be measured before being added to a reaction mixture to attain the desired proportion. Usually a reservoir contains the reagents for more than one analysis.

carryover Contamination of a specimen by the previous one.

centrifugal analyzer Uses centrifugal force to mix the sample aliquot with reagent and to pass the reaction cuvette past a detector.

computation Calculation of a desired result from the signal or readout of an instrument; it can be electronically automated by use of either digital or analog conversion.

continuous flow Instruments that constantly pump reagent and sample through tubing and coils, forming a continuous stream.

dead volume The volume in a sampling container that must be present for proper sample aliquoting but is not consumed.

digital Relating to data available in the form of discrete units or the calculations using such data.

discrete Term applied to instruments that compartmentalize each sample reaction.

dwell time The minimum time required for an instrument to obtain a result as calculated from the initial sampling of the specimen.

flow injection Placement of a sample into the stream of a continuous-flow analyzer.

incubation The time allowed for a chemical reaction or process to proceed.

mixing Process by which individual components of a chemical assay are formed into a homogeneous solution.

proportioning Addition of individual components of a chemical assay in proper ratios or amounts.

readout Visualization of the result of an instrument analysis.

selective instruments An instrument capable of performing multiple tests on a sample but performing only those programmed.

sensor A system or device that monitors changes in the reaction mixture that are related to analyte concentration.

simultaneous analyzers Automated analyzers capable of more than one analysis per sample at the same time.

test repertoire The number of different tests available either at one time or with changing of reagents or instrument components.

throughput The maximum number of individual samples or test analyses that can be practically performed per hour by an assay system, with the required dwell time being taken into account.

unit reagents Premeasured reaction chemicals packaged so that only one package (unit) is used per sample test.

DEFINITIONS

In this chapter the reasons for automation and the ways to achieve it are considered. Examples from the major instrument categories are examined.

In this section the concept of automation is examined in a historical perspective. It is a fact that more tests are being performed per capita every year. This has come about for several reasons: the repertoire of available tests has increased; more patients survive acute illnesses and trauma, thus continuing to need additional tests, and results are generated more rapidly and reported in time to affect medical decisions.

Increases in test work load have been fueled and sustained by improvements in productivity, some of which derive from automation. A major result of automation has been the almost unlimited access to laboratory analyses by physicians. The salutary effect of this phenomenon on patient care and the cost to the health system is still to be determined.

Automation, as applied to clinical chemistry, can be defined as the self-moving or mechanical transfer of a specimen within a complex, industrial assembly to a succession of self-acting machines, each of which completes a specified stage in the total analytical process from crude sample to analytical result. A second definition is the application of fully automated procedures in the efficient performance and control of operations involving a sequence of complex standardized or repetitive processes on a large scale. Automation in the laboratory may be considered in the broader context of the need to obtain information about the state of a patient in a hospital or outpatient setting.

One way to visualize how this information is obtained is shown in the block diagram of Fig. 14-1. The patient is examined by a physician, and there is a decision to obtain specific laboratory information. A specimen is collected, transferred to the laboratory, received, processed by the laboratory, and analyzed to obtain the desired results, which are then transferred back to the requesting physician. In general, laboratory automation has focused on those steps from the reception of the sample in the laboratory to the transfer of information back to the requesting physician. There is, however, increasing interest in automating the process of transferring the sample to the laboratory. This is accomplished by means of vacuum tube systems. Phlebotomy tubes are placed in a cushioned holder that is placed into a *sending station*. The sending station can be programmed to send the holder to a specific *receiving station*. Vacuum tube systems can rapidly (<5 minutes) transport samples with insignificant changes to the samples (that is, hemolysis, gas cavitation in blood gas samples), obviating the need for slower, less reliable human transporters.

Most laboratories have concentrated on purchasing instruments that will automate the steps of analysis. The steps of analysis are outlined in the lower part of Fig. 14-1 and discussed in detail below. Although the term *automation* implies the processing of large numbers of samples, the principles inherent in automation can be applied to the analysis of single samples as well. For example, a blood-gas analyzer is a single-sample analyzer but is commonly highly automated.

The goal of automation is often so desirable that the equipment and its mode of performing certain types of actions often restrict the types of analyses that can be performed. For example, the reference method for glucose analysis requires a protein precipitation followed by the hexokinase reaction. Because of the inability of current technology to automate the precipitation step easily, this method of assay is not available in an automated analyzer.

AUTOMATION OF CHEMICAL ANALYSES

To understand how patient samples are processed by automated procedures, it is necessary to divide the process of analysis, as might be performed in a manual assay, into a series of stages or steps. Commonly the following series of steps is performed during the course of an analysis: (1) obtaining a patient sample in the proper form, (2) mixing an aliquot of the sample with a series of reagents in an

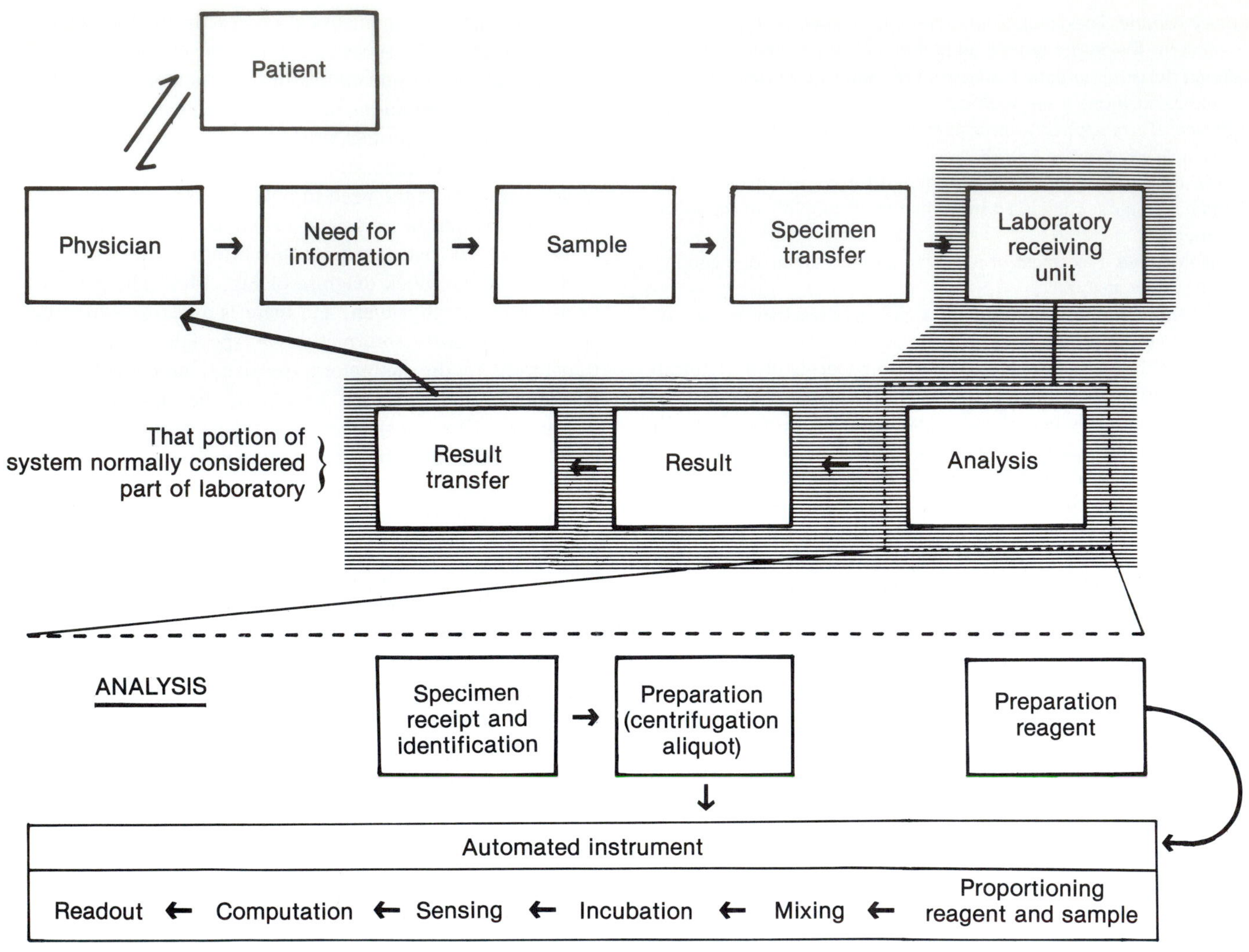

Fig. 14-1 Diagram illustrating specimen and information flow between physician and laboratory. Lower portion of figure outlines steps in sample processing and analysis.

ordered sequence with defined amounts, (3) monitoring or sensing the result of the reaction, (4) quantitating the extent of the reaction, and (5) providing an appropriate readout or permanent record. Automation may be applied to any or all of these steps. The automation of each one of these steps is now discussed in some detail.

Sample preparation and identification

Remarkably little automation has been applied to this step. By and large, the specimens are manually labeled, centrifuged, and divided into aliquots if tests for more than one work station have been requested. The reason for continued manual processing is the need to determine if the sample meets acceptable criteria for analysis (such as use of proper phlebotomy tubes or absence of debris).

The manual performance of these procedures can often lead to errors, which result in the generation of inaccurate patient results. Mislabeling patient samples leads to laboratory information being erroneously transferred. Similarly, the necessity of dividing the sample into aliquots for separate work stations also can lead to improper patient identification.

Various approaches are being tried to avoid or to minimize these manual processing steps. For example, use of computer-based *bar coding* of patient samples allows electronic identification of the sample and the tests requested (see Chapter 15). The use of single instruments capable of measuring at one work station a large number of analytes from an individual specimen has been advocated to delete the need for sample aliquoting. By using whole blood rather than serum as the sample, one can also avoid the centrifugation steps. The type of sample capable of being analyzed by an instrument is usually determined by the chemistry of the analysis rather than the mechanics of automation. Thus most analyzers are capable of processing most biological fluids, such as serum, plasma, urine, and

Table 14-1 Comparison of operational features for several automated analyzers

	RA-1000	SMAC-111	Chem 1	Synchron CX-3	COBAS Mira	ACA IV	Ektachem 700 XR	TDx	Array	Photon Era
Type[a]	D, SL, B	C, S	C, S, SL	D, S, SL	D, SL, B	D, SL, B	D, SL, B	B	B, SL	B
Sample volume (μL)[b]	2-30	443	1-11	120	2-95	10-540	10-11	10-25	20-42	25-200
Minimum volume (μL)[c]	52-80	500	51-61	175	52-145	20-660	50	60-75	170-192	190-1000
Sample identification[d]	None	Keyboard	Keyboard	Keyboard	None	Photo image	Keyboard	None	Keyboard	Keyboard
Reagent[e]	Bulk, A	Bulk, A	Bulk, M	Bulk, M	Bulk, A, M	Unit, M	Unit, M	Unit dose bulk, M, A	Bulk, M	Bulk, M
Proportioning[f]	Vol Add	Flow ratio	Aspiration	Vol Add	Vol Add	Prepackaged, Vol Add				
Minimization of matrix interferences[g]	Dilution	Dialysis	Dilution	Dilution	Dilution	Dilution	Prepackaged, saturation	Vol Add	Vol Add	Vol Add
Mixing	Lateral movement	Coils	Coils	Flow turbulence	Vibration (mechanical)	Vibration (mechanical)	Filtration Diffusion	Dilution Forced turbulence	Dilution Magnetic stirring	Dilution Shaking
Sensing[h]	Spec, electrodes	Spec, electrodes	Spec, electrodes	Spec, electrodes	Spec, electrodes	Spec, electrodes	Reflectance, electrodes	Fluorescence	Nephelometry	Spec
Test repertoire available[i]	14	23	24	8	23	76	34	1	16	1
Total repertoire[j]	40+	23	35	8	29	96	34	56	18	9
Stat capability[k]	Yes	NR	NR	Yes	Yes	Yes	Yes	Yes	Yes	NR
Dwell time (min)[l]	0.5-5	14	17	1-2	2-3	7	3-5	5-20	2	60
Throughput (samples/hr)[m]	100	150	60-360	75	120	50	270	60	40	30[o]
Throughput (tests/hr)[n]	240-720	3450	720-1800	600	120	50	540	60	40-80	30[o]
Readout	Digital	Digital	Digital	Digital	Digital	Digital	Digital	Digital	Digital	Digital

[a]*B*, Batch analyzer, *C*, continous flow; *D*, discrete analyzer; *S*, simultaneous analyzer; *SL*, selective.
[b]Sample volume needed to perform a test (discrete analyzer) or simultaneous profile.
[c]Sample volume plus dead space in sample cup.
[d]Means of identifying sample for final printout.
[e]*A*, Available from alternative sources; *M*, only available from manufacturer.
[f]Manner of adding reagent, diluent, and sample; *Vol Add*, volumetric addition.
[g]Means of minimizing protein interference; *NA*, not applicable.
[h]*Spec*, Photometric; *flame*, flame emission or absorption spectroscopy; *electrodes*, ion-selective electrodes.
[i]The number of tests available at one time, without a change of reagents or instrument module.
[j]Total number of analytes for which reagents are commercially available (as of 1987).
[k]*NR*, Not recommended.
[l]Approximate time between sampling and availability of result.
[m]Number of samples capable of being processed per hour. Data listed are for one simultaneous profile or a *single* test per sample (approximate reaction time for variable analyses = 1 minute) for discrete analyzers.
[n]Calculated by multiplication of maximum number of tests per sample available times number of samples capable of being processed per hour.
[o]Exclusive of offline incubation.

cerebrospinal fluid, but not whole blood. A few instruments, notably those using ion-selective electrodes, are designed for use with whole blood.

Instruments combining several of these system approaches for processing samples are finding their place in the laboratory.

Reagent preparation

Bulk reagents can be manually prepared, although an increasing number are used as concentrates or lyophilates. Thus simple dilution or reconstitution of the reagent with water is all that is required. However, this simple step can lead to analytical errors because of improper dilution or processing. These errors can be avoided by purchasing reagent in a form suitable for instrument use without *any* processing. Unit test-reagent preparation (where sufficient reagent is present for the performance of a single test) has been automated in two ways. The first is the dry-film or impregnated-paper technique. The dry-chemical techniques use either paper or series of thin films impregnated with the desired reagent. The analytical reactions take place when the sample is placed on the dry reagent (see Eastman Kodak Ektachem, Table 14-1 and p. 236). In this type of reagent, preparation consists of wetting the reagent with water, buffer, or sample. The second kind of unit test reagent is a container or test tube containing premeasured liquids or powders to which water, buffer, or sample is added. Unit test reagents tend to be more expensive than bulk reagents, a consideration that can be important for many laboratories.

Proportioning of samples and reagents

Most chemical reactions require the combining of reagent and sample in exact amounts to yield specific, final concentrations of analyte and reagents. Since the reagents, as just described, are prepared in predetermined amounts, the ratio, or proportion, of reagent to sample must be kept constant to achieve reproducible and accurate final reagent concentrations. Thus the addition of sample to reagent is termed *proportioning*. The case of unit test reagents is considered first. In these systems the reagents are already proportioned in the required amounts; therefore only the sample must be proportioned. The dry-film reagent may have the sample added volumetrically or by saturation addition. The latter technique requires some explanation. The film is exposed to an excess of the sample, and the pores of the film allow only a fixed amount of the sample to be absorbed. This fixed amount of sample required to wet the film represents the proportioning mechanism. In some cases of saturation addition, the rate of the diffusion of the sample into the film may also affect the proportioning step. In the case of bulk reagents, proportioning is always accomplished by volumetric addition. There are three automated volumetric dispensing methods in common use. Syringes or volumetric overflow devices are used in discrete test analyzers where sample and reagents are volumetrically added to a test tube or container. The second mechanism is the continuous-flow technique used where sample and reagents are proportioned by their relative flow rates. Typically, peristaltic pumps are used to move the reagents through tubing, and the flow rate is controlled by the cross-sectional area (diameter) of the pump tubing. Usually the sample and reagent streams are allowed to flow continuously through the tubing where mixing and incubation are also accomplished. Additional discussion on this technique will be found in the section on the Technicon Auto-Analyzers. The third type uses electrical valves to control the time the reagents can flow. The flow rate is controlled by the air pressure applied to the reagent container and the flow resistance in the tubing to the reaction vessel.

In almost all systems the sample is introduced into the analyzer with a thin, stainless steel probe. This probe passes into a sample, aspirates a defined quantity of sample, and moves from the sample to dispense the aliquot into an appropriate vessel. Many sample probes have an associated *level sensing device* that permits the tip of the probe to go a specified distance below the level of the sample. Since the same probe is used repeatedly for sequential samples, there is the potential for contamination of a specimen by a preceding one. This is called *sample carryover*. Various techniques have been used to minimize the interaction between samples. These include (1) aspiration of a wash liquid (such as saline solution or water) between sample aspirations and (2) a backflush of the probe. The latter technique has the wash liquid flow through the probe in the direction opposite to the aspiration and then into a waste container.

Measurement of the degree of sample carryover can be achieved by the following procedure. Samples containing low and high levels of analyte are analyzed in the following order: low (L_1); high (H); and a repeat aliquot of the low (L_2). The degree (%) of sample carryover or sample interaction between the high and low concentration samples can be described by the following equation:

$$\text{Percent carryover} = 100\% \, (L_2 - L_1)/H$$

One should determine the percentage of carryover at several levels of analyte to establish those levels of analyte where significant contamination of subsequent samples will occur.

Sample carryover, seen most commonly with continuous-flow analyzers, occurs because the stream exhibits laminar flow. The fluid at the walls of the tubing is stationary, whereas the fluid in the center flows at the highest velocity. Thus an abrupt change between samples will become a gradual change because some of the previous sample is left behind at the walls of the tubing. This material eventually diffuses back into the center of the tube where it is carried onward. Carryover affects the test results by contaminating the current sample with a proportional part of the previous sample. The amount of contamination or carryover that is permitted determines the throughput. If less carryover is permitted, a longer time must be allowed for the previous sample to be flushed out. A longer time spent on each sample reduces the number of samples that are processed per hour.

Almost all continuous-flow analyzers have streams that are *air segmented* to reduce the carryover. Air bubbles are injected into the flowing stream so that the fluid on the walls is forced to move along with the fluid in the center of the tubing. The family of instruments from Technicon Instruments Corporation (Tarrytown, N.Y.) is closely identified with this technique.

Mixing

Those instruments using the dry-film technique mix sample and reagents by the diffusion of sample into the reagents. Most dry-film reagents are premixed during manufacturing, although some are mixed by diffusion, which becomes possible only when the film is wet. The discrete test analyzers can mix the reagent and sample by (1) motion of a test tube or container, (2) stirring by paddle or stick, (3) agitation by air bubbles or ultrasonic waves, or (4) convection resulting from forceful addition of sample into the container. Mixing in continuous-flow analyzers is accomplished by convection in a mixing coil. When the tubing is wrapped in a tight coil, there is a difference between the distance that the fluid on the inside of the flow path within the coil and the fluid on the outside flows. This difference causes the fluid to tumble between the inside and outside of the coil, effecting a mixing process.

Incubation

Automated incubation is merely a delay station where the test mixture is allowed to react. This is performed, in most cases, under conditions of a specified, constant temperature, which is most frequently achieved by heating blocks and air or water baths. These constant temperature devices are monitored electronically by thermocouples. In a continuous-flow system, incubation is accomplished when the length of the tubing is increased. In this case, the delay is the volume of tubing divided by the flow rate. For example, a 4.2 mL incubator with a 1.2 mL/min flow rate would give a 3.5 minute incubation time. Discrete analyzers accomplish incubation by allowing the reaction mixture to dwell in a chamber (test tube or cuvette) for a specified time. A similar approach is used for the dry-film analyzers.

It should be noted that many test methods require an addition of a second reagent, possibly followed by additional incubation. The automated means for doing this are not different from those just discussed.

Sensing

The techniques of automation do not depend on the method of sensing, whether optical, thermal, or electrical. There are two major approaches to automated sensing: in situ and external. The term *in situ* refers to measurement in the vessel where the reaction has taken place, for example, in the reaction cuvette. The term *external* is applied to systems of measurement where the sample is transferred from its original position in the reaction vessel to the sensing device. The dry-film tests are measured in situ by reflectance photometry or by means of integral electrodes (electrometric). Discrete test instruments use both in situ and external sensing mechanisms. The proportional flow instruments use external sensors; this is by definition because it is difficult to define what is meant by in its "original position" for a flowing stream. External sensing generally exposes the sensing chamber to many samples so that care must be taken to eliminate carryover from one sample to the next. Optical or electrode surfaces may also be contaminated by components from the samples. On the other hand, in situ sensing makes special demands on the test chamber or requires an elaborate automated washing procedure. If the test container is disposable, it is impractical to calibrate it for optical or electrical characteristics. Thus such containers must be manufactured with very good reproducibility. However, disposable containers are meant to eliminate the mechanical complexity required to wash and recertify the sensing chamber. Most sensing is done in situ because this approach decreases the mechanical complexity of the instrument.

Chemical reactions can be monitored either at one time point or at many. Commonly, single-point monitoring is used for end-point analysis in which the reaction has gone to completion. Multiple-point monitoring is used for kinetic analysis. Discrete analyzers are easily adapted for multiple time-point monitoring, whereas continuous-flow systems are not.

Computation

Automated computation has taken two forms: *analog* and *digital*. Analog computations use the electrical signal such as a voltage or current from a sensor (such as a phototube) and quantify the signal by comparing it to a reference signal. For example, a "blank" reaction mixture will give a 100% T (blank transmittance), resulting in a certain electronic signal. A test standard will give a lower percentage of T and thus a decreased electronic signal. The analog computer compares the two signals and takes the logarithm of the result. The final result is related to the quantitation of the reaction.

Some reaction signals by their very nature are in a form of discrete numbers. Two examples are individual photon-counting events and counting of radioactive decay. This signal, consisting of a number of individual events, can be monitored by a digital computer that can process the signal. Digital processing is restricted to certain arithmetical functions (such as subtraction or addition) unless the computer is programmable.

For a digital computer to process signals from many types of sensing devices in automated instruments (such as the spectrophotometer and ion-selective electrodes), an analog-to-digital converter is necessary. This converts the voltage or current signal into a digital form, which can be processed by the digital computer.

There are no straightforward rules as to which is the best form of computation. The decision is usually based on economics. However, if any part of the signal processing is done digitally, virtually all the processing is done digitally. Perhaps the major exception is the analog conversion of transmittance to absorbance, which is performed to improve analytical performance.

Readouts

The simplest method that can be used to visualize an instrument readout is the use of light-emitting diodes (LED) or a television monitor (cathode-ray tube, CRT) to express the data in numbers. These devices allow the technologist to review the data before accepting the results. This readout is usually converted to a hard copy such as a paper-tape printout. These data must be manually transferred to laboratory slips or other permanent records. This manual step, of course, is subject to transcription errors.

More sophisticated, automated computer systems collate all the test results for each patient, and the results print directly on the report form. When the results are transferred into a laboratory computer, the instrument readout can be directly interfaced or connected to the computer. Although analog connections are possible, the majority are digital.

CONCEPTS OF AUTOMATION: DEFINITIONS

Test repertoire

Economics suggest that instruments perform more than one kind of test; that is, the investment in automation and the labor in sample loading are not duplicated. Following this logic to an extreme would require that an instrument be capable of performing every conceivable kind of test. This has not been possible. However, the characteristics of the analyses requested are such that six chemistry tests account for 50% of the work load and the next 14 tests account for another 40% of the total work load. Automated instruments have been designed to perform the most frequently ordered tests (electrolytes, BUN, glucose) and those ordered less frequently (drugs, hormones). Automation is usually not essential for rarely ordered tests.

The automation of tests may also be done on the basis of *type* of analysis rather than test volume (number of samples); that is, an automated radioimmunoassay instrument will process RIAs for many different analytes, regardless of numbers of specimens per analysis.

The immediate test repertoire can therefore be defined as the number of tests that can be performed at any one time, whereas the total test repertoire includes the total number of different tests that can possibly be performed on the instrument (that is, by changing reagents and a few components).

Selective

Instruments that are capable of performing multiple tests are selective if the particular tests to be performed on an individual sample can be specified and if no sample and no reagent are consumed by tests that are not requested. For example, the Technicon SMAC is a nonselective instrument because all tests in the immediate repertoire are performed on each sample, regardless of the exact tests requested. A discrete analyzer, such as the ACA, performs only the test requested. Selective analyzers have also been termed *random-access analyzers* for their ability to process different test combinations for individual specimens.

Discrete

Instruments that compartmentalize each sample reaction are discrete analyzers. Typically the sample aliquot and reagent are contained in a single cuvette that is physically separated from all other cuvettes. This contrasts with the continuous-flow instrumentation in which all samples physically flow through the same sample path. Most current instruments are discrete analyzers.

Continuous flow

Instruments that continuously pump reagent through tubing and coils to form a flowing stream and continuously pump sample into that stream are called *continuous-flow analyzers*. The proportioning of sample and reagent is accomplished by control of the respective volumetric flow rates. For example, the sample might be pumped at 0.1 mL/min and the reagent at 0.9 mL/min, giving a 1 to 10 dilution of sample in the final reaction mixture.

Batch analyzer

Instruments that perform the same test simultaneously on all samples presented to it are termed *batch analyzers*. The type of test can vary widely, but usually only a limited number of samples are processed per analysis. Examples of a batch analyzer are the centrifugal analyzers such as the Roche COBAS and Instrumentation Laboratories' Monarch.

Dwell time

The dwell time is the minimum time required to obtain a result after the initial sampling of the specimen. Some instruments can give results in as little as 15 seconds for single tests such as glucose. Commonly, instruments that perform multiple tests on a single sample have longer dwell times, ranging from 60 seconds to 10 minutes. Certain test procedures, such as kinetic analysis for enzyme activity or radioimmunoassays, that require long incubations will, of course, have longer dwell times. Dwell time is extremely important when significant or life-threatening physiological changes can take place rapidly. Thus blood gas determinations (pH, P_{CO_2}, and P_{O_2}) need instruments with dwell times on the order of seconds. On the other hand, a "dwell time" of a week for a vitamin assay would be clinically acceptable.

Throughput

The throughput is the maximum number of samples or tests that can be processed in an hour. For simultaneous analyzers, one can calculate the total test throughput by multiplying the number of samples processed per hour by the number of tests performed on each specimen. For discrete analyzers, the sample throughput will obviously depend on the number of different tests requested on each sample. In addition, the time required per test can vary widely (that is, from less than 30 seconds to approximately 10 minutes). In general, the more tests ordered per sample, the slower the sample throughput on a discrete analyzer. Thus it is more difficult to give a simple, accurate value for the sample throughput for a discrete analyzer. The calculation for throughput does take into account the dwell time; that is, the fact that no results are produced until the dwell time has elapsed. The desired throughput of an instrument is usually matched to the number of samples that need to be processed in a given time period. Generally the throughput of an instrument is chosen so that a maximum number of samples can be processed in a timely fashion. For example, a higher throughput (and more costly) instrument may be required to process samples from a clinic so that results can be made available before the patients return home. In general, an automated analyzer is chosen on the basis of its ability to process the bulk of the routine work load in time for routine clinical decision making.

Stat testing

The word *stat* is an abbreviation of the Latin word *statum,* meaning "immediately." Stat tests, the most widely abused type of testing in a hospital laboratory, account for a large portion of the laboratory work load. Stat specimens must be analyzed before the non-stat samples, resulting in an interruption of the normal work flow of the laboratory. Unfortunately, many stat requests are not ordered because of a medical emergency.

The turnaround time for a stat test should not be greater than 1 hour after the specimen enters the laboratory. However, the total turnaround time can be reduced if the specimen is hand delivered to the laboratory by the housestaff. A "superstat" test, which has turnaround time of less than 10 minutes, can be obtained using whole blood samples for measurement of blood gases, glucose, and electrolytes.

Instruments that are to be used for stat testing need not necessarily have a high throughput but should have a short dwell time. Many stat instruments are dedicated instruments that analyze no more than half a dozen high-frequency tests simultaneously.

Cost

The resources consumed in producing a patient's test result represent the cost of that test. Cost consists of labor (which is the monetary representation of the time spent processing and analyzing the sample), maintenance of the instrument, reagents (which are calculated from the cost of the chemicals used for the test and a proportionate part of those required for instrument start-up), calibration and quality control, consumables (comprising the cost of sample containers, paper, and so on), and capital (which is a proportionate amount of the life of the instrument consumed and hospital overhead including the cost of items such as laboratory slips and maintenance).

COMMON AUTOMATED INSTRUMENTS

This section describes a number of instruments that are common in the hospital laboratory or that illustrate a category of instrument type. A comparison of operational parameters for some automated instruments is shown in Table 14-1. Obviously this table is not meant to be all inclusive but rather to demonstrate the features available for common types of instruments.

Technicon AA, SMA, RA-1000, Chem 1

The Technicon Auto-Analyzers (AA) and the Technicon Sequential Multiple Analyzers (SMA and SMAC, that is, SMA with computer) are a family of instruments that use a method based on a continuous-flow principle invented by Leonard Skeggs in the late 1950s. The first AA processed 20 samples per hour, and the reagent was consumed at 5 mL/min. The SMAs reduced the sample and reagent consumption while simultaneously increasing the throughput. The SMAC can process 150 samples per hour and uses reagent flow rates on the order of 0.75 mL/min. The RA-1000 is a random access discrete analyzer that uses a separate probe for dispensing sample and reagent. The internal and external sides of the probe are coated with a fluorocarbon that prevents contact of the sample with the walls of the probe. Sample is aspirated into the probe together with an air bubble and with a fluorocarbon solution that completely covers the probe tip. Sample, together with the fluorocarbon, is dispensed into the cuvette. Because fluorocarbon is very dense, it will settle to the bottom of the cuvette and will not affect photometric measurements. The fluorocarbon material provides an inert surface that prevents sample-to-sample and reagent-to-reagent carryover. The reactions in the cuvettes are monitored colorimetrically or by UV detection at 30° or 37° C. Between 2 to 30 μL of sample is required, and 240 tests can be determined in 1 hour. The RA-1000 can determine general chemistry tests by end-point, fixed-time, or kinetic assays. Ion-selective electrodes are available for measuring electrolytes. The Chem 1 Analyzer is a continuous-flow system in which a capsule is generated for each test. Two liquid segments, reagent one plus sample and reagent two, are separated from each other by a small air bubble. The capsule is transported through a Teflon tube, the samples and reagents are mixed together, and the absorbance is measured by a series of optical stations. Introduction of the liquid fluorocarbon into the Teflon tube provides an interface between the sample and the air bubble and allows the air bubble to move over the fluorocarbon film assuring a uniform, smooth flow. This eliminates carryover from individual liquid segments. With the Chem 1 system, 1 μL of sample is pipetted with 7 μL of reagent one and 7 μL of reagent two. The reactions are monitored at 30° or 37° C; 720 tests can be performed per hour.

Eastman Kodak Ektachem

The Eastman Kodak Ektachem is a discrete, selective analyzer that uses dry-film technology. The addition of serum sample provides the solvent (water) necessary to rehydrate the dry reagents. The film is composed of multiple layers, some of which serve to ultrafilter the sample (remove protein), whereas others serve to provide reactive reagents. The colored reaction products are measured by reflectance on the side opposite the sample addition. The reflected light is converted into concentration units (of the reaction products) by the Williams-Clapper formula (analogous to Beer's law for converting light transmission into absorbance units) (see p. 60, Chapter 3).

The samples are automatically processed by dispensing 10 μL of serum from the sample cup onto the slide. The slides are automatically dispensed from cartridges, each containing a different method. The slides are incubated while the color forms. Finally the reflectance measurements are performed. The instrument is capable of rate ("kinetic") and enzyme measurement. The system also permits electrometric measurements for the determination of sodium and potassium. The Ektachem uses, by definition, unit reagent. The reagent slides, which must be stored at refrigerated temperature, are available only through the manufacturer. The reagent cost per test is relatively high.

Centrifugal analyzers

There are several manufacturers of these discrete batch analyzers. They were first developed by the Oak Ridge National Laboratory and then adopted by the private sector. The key element is a centrifuge whose rotor contains 16 to 32 cuvettes. The sample and reagent are pipetted into discrete compartments and mixed together by the action of centrifugal force when the rotor is first accelerated. The cuvettes rotate past a stationary light source that permits absorbance readings at a preselected wavelength. These analyzers are useful for rate ("kinetic") measurements and for end-point analyses.

Centrifugal analyzers are batch instruments, usually analyzing only one analyte at a time. Their reagent consumption is small, a feature that makes them economical to operate. Although the theoretical test throughput can be high, in practice it is considerably lower because subsequent runs are often required for reanalysis of samples that give unsatisfactory results, such as for excess enzymatic activity.

Many of these analyzers have automatic sampling devices that take aliquots of the samples and reagents and place them in separate compartments of the rotor. Most sampling devices are separate from the analyzer, although the Roche COBAS has a sampler integrated into the analyzer. Some instruments use disposable rotor cuvettes, whereas others wash the rotor for reuse. Although all the centrifugal analyzers can perform spectrophotometric and turbidimetric analysis, newer instruments (such as the COBAS and Instrumentation Laboratories' Monarch) also have nephelometric and fluorometric capabilities.

The centrifugal analyzers use bulk reagents available from a wide variety of commercial suppliers. The flexibility of these instruments allows for easy adaptation of "in-house" reagents or commercially supplied reagents for routine analysis.

Du Pont ACA, Dimension

The Du Pont ACA IV Clinical Analyzer is a discrete, selective analyzer that can serve as a general chemistry analyzer in small-sized hospitals and a specialty analyzer in medium- and large-sized hospitals. Some 76 test methods are available, ranging from routine and stat chemistries to immunochemistries, therapeutic drug monitoring, and coagulation. It processes 50 tests per hour with a dwell time of 7 minutes. Any combination of tests can be run on any sample at any time. The reagents are prepackaged in plastic packs, which also serve as the reaction cuvettes.

Sample processing begins with automatic sample aspiration. The sample and buffer are then automatically dispensed into the pack. The reagents, which were isolated in separate compartments of the pack, are released and mixed with the sample and buffer. There are stages where two separate reagent additions can take place. The reaction is then measured photometrically when the plastic pack is formed into a 1 cm path-length cuvette. The end-point methods are measured bichromatically, except for several methods that use two packs to permit blank absorbance measurement at the desired wavelength. The rate ("kinetic") and enzyme methods are measured at a single wavelength over a 17-second interval.

The reagent packs are available only from DuPont de Nemours (Wilmington, Del.) at this time and must be stored at refrigerator temperatures. The cost per test is relatively high.

The Dimension is a discrete random-access analyzer that can measure general chemistries, drugs, enzymes, electrolytes, and special chemistry tests. The reagents are bar coded and are contained in a cartridge that has the capability of holding four different reagents for each test. Forty-five cartridges can be placed in the analyzer, and the reagents are stable for 30 days. Depending on the method, each cartridge can hold enough reagent for 15 to 240 tests. For each series of tests performed, the Dimension monitors the reagent inventory, requests new cartridges, and makes the cuvettes. The system dispenses reagent and sample into the cuvette and then mixes them using an ultrasonic probe. Multiple absorbance readings are taken at wavelengths ranging from 293 to 700 nm. After the reaction is complete, the top of the cuvette is sealed, and the cuvette is dropped into a cuvette waste container. With the Dimension, 37 tests are possible with a throughput of 200 chemistry tests and 180 ion-specific electrode tests per hour. The reagent cost is significantly less than that of the ACA, and the sample sizes range from 2 to 44 μL.

Beckman ASTRA, Synchron systems

The ASTRA-4 and ASTRA-8 are discrete, selective analyzers that process the high-volume tests, glucose, creatinine, BUN, Na, K, Cl, and CO_2. The ASTRA-8 will process up to 72 samples per hour and up to 648 tests per hour.

The Synchron CX-3 is a discrete analyzer that can achieve a throughput of 600 tests per hour using 120 μL of sample. The analyzer is a replacement for the ASTRA and is suitable for stat and routine assays. The Synchron CX-4 and CX-5 systems are microprocessor controlled, random access chemistry analyzers that can perform up to 35 different chemistries. The bulk reagents are bar coded. The samples and reagents are automatically pipetted into cuvettes incubated at 30° or 37° C. The reaction carousel has 80 glass cuvettes, which are rotated at 90 rpms, during which time the cuvettes pass through the optics. Ten wavelengths, ranging from 340 to 700 nm are available for analysis. After the reaction is complete, the cuvettes are automatically washed, and a cuvette absorbance check is performed to verify the adequacy of the washing process. The CX-4 and CX-5 systems can determine general chemistries; drugs, by a particle-enhanced turbidimetric system, and specific proteins, by an immunoturbidimetric assay using 3 to 35 μL of sample. The throughput for the CX-4 is 225 tests per hour. The throughput for the CX-5 is 525 tests per hour, with the addition of ion-selective electrodes for measuring electrolytes.

Abbott ABA, Spectrum, TDx

The ABA-100 is a discrete batch analyzer that can process up to 120 tests per hour. The newer version, the VP, is more automated and provides greater throughput than its predecessors. This family of instruments is noted for using bichromatic-photometric measurements. The ABA uses bulk reagents obtainable from a large number of commercial suppliers and from the manufacturer as well.

The Spectrum is a benchtop clinical chemistry analyzer that can be used for random access, tandem access, batch, or stat testing. Tandem access, which is unique to the Spectrum, maximizes the rate at which patient test reports are completed by reviewing the requested test work load and prioritizing the tests in the most efficient order to generate the greatest number of patient reports earlier in the run. The Spectrum uses a polychromatic optical system, which can simulataneously monitor 16 wavelengths for performing monochromatic, bichromatic, or polychromatic analysis. The reagents are bar coded, sample size ranges from 2.5 to 10 μL, and 225 tests can be performed per hour. The Spectrum can determine general chemistries and enzyme assays by photometric analysis and electrolytes with an ion-selective electrode module.

The TDx system was primarily designed for use in the measurement of therapeutic drugs by fluorescent polarization immunoassay (see p. 66, Chapter 3). Up to 20 samples are placed in a 20-position carousel, and these, together with the prepared liquid reagents, which are bar coded, are placed in the analyzer. Depending on the assay, the results are obtained after a dwell time of 5 to 20 minutes. This system is suitable for batch and stat testing. The TDx system can also perform some nondrug chemistries (such as glucose, lactate, and 5-hydroxyindoleacetic acid) by a radiation attenuation procedure, and it can measure specific proteins by nephelometric analysis with a specially designed carousel that contains the light source. The TDx system has unit dose reagents for measuring specific proteins and some drugs. These assays are easy to perform; however, the cost per test is relatively high.

Roche Mira

The Mira is a benchtop random-access discrete analyzer that can be used for stat, profile, or batch testing. Samples are transferred into plastic cups and placed in the sample rack, which can hold up to 30 samples. Twenty-six different profiles can be run. After the appropriate tests are ordered and the reagents loaded, reagent and sample are pipetted into disposable polyacrylic cuvettes. The absorbances are measured using a xenon lamp with five interference filters ranging from 340 to 600 nm. Depending on the tests requested, from 2 to 20 μL of sample are required. With the Mira, up to 23 different tests can be performed, with a throughput of 120 tests per hour. The Mira can determine general chemistries, enzyme assays, and sodium and potassium with an ion-selective electrode module. Syva Co. (Palo Alto, CA) has provided reagents in special cartridges adapted to the Mira for measurement of therapeutic drugs using the enzyme-multiplied immunoassay technique (EMIT) technology.

Hybritech Photon Era

The Photon Era is an automated analyzer designed for measuring Hybritech's Tandem E immunoassays. The function of the sample processer is to pipette calibrators, controls, specimens, and antibody-enzyme conjugate into the reaction tubes containing the antibody-coated beads. The reaction processor is used to wash each bead thoroughly, add substrate, incubate, add quenching solution, and read the absorbance of each reaction. The Photon Era system can measure cancer-specific proteins, hormones, immunoglobulins, and enzymes. It has a throughput of about 30 tests per hour and uses sample volumes ranging from 25 to 200 μL.

Nephelometric analyzers

Nephelometric measurement of specific proteins in serum (see p. 66, Chapter 3) has been automated with the use of the Behring Nephelometer system, the QM300 System from Kallestad, and the Array system from Beckman. Antigen-antibody complex formation increases for a fixed quantity of antibody up to a peak, after which an increase in antigen will yield less complex formation. Therefore it is important that each system be able to monitor antigen excess. With the Kallestad and Beckman systems, antigen excess is checked by addition of more antigen. If further reaction takes place, the antigen is not in excess, and the results are valid. If more antigen is added and an increase in complex formation is not observed, the value is flagged, and the specimen analysis is repeated at a higher dilution.

With the Kallestad and Array systems, calibration curves are stable for 14 days. Temperature is maintained at 25° C for the Kallestad system and 26° C for the Array system. With the Kallestad system, 13 specific proteins can be measured; with the Array system, 18 specific protein assays can be determined. The throughput for the Array system is 80 samples per hour.

With the Behring system, antigen excess is not flagged by addition of more antigen. The reagents are optimized to produce a calibration curve with a wide range of protein values. If the sample values exceed those of the reference curve, they are flagged by the instrument, and the sample must be diluted and repeated. With the Behring system, 34 different tests can be performed with fixed-point, endpoint, and latex-enhanced agglutination reactions. The reagents are bar coded, and the instrument has a throughput of 255 tests per hour.

Hitachi 705

The Hitachi 705 is a discrete, selective analyzer with a throughput of 180 tests per hour. A 16-test repertoire is available at any one time. There is also an ion-selective electrode accessory. A larger instrument, the Hitachi 737, has a throughput of 1500 tests per hour. Both instruments use bulk reagents obtainable from any vendor and can use monochromatic or bichromatic analysis. These analyzers have been designed for microprocessor (microcomputer) control; as a result, the test methods can be readily changed.

Micromedic Concept 4

The Micromedic Concept 4 is a discrete analyzer for RIA. It is based on the use of special test tubes whose inner surfaces have been coated with antibody. The sample and reagent are added to each tube, which can be incubated at room temperature or at 45° C. After an incubation of up to 17 hours, the free ligand (sample analyte and radiolabeled analyte) is aspirated, and the bound ligand attached to the antibody on the test tube is measured in a gamma-ray counter. To increase the throughput, the newer instruments use multiple counting chambers. The system is designed to use racks of 10 tubes each and is capable of processing up to 200 antibody-coated tubes per assay. The precoated tubes are obtainable only from the manufacturer.

Becton Dickinson ARIA II

The Becton Dickinson ARIA II is a continuous-flow analyzer for RIA that sequentially pumps assay mixtures through a reusable chamber containing antibody covalently bound to solid support media. In this system the bound ligand is chemically stripped from the antibody and passed to a flow-through gamma-ray counter. The free ligand can also be counted.

INSTRUMENTS IN PHYSICIANS' OFFICE LABORATORIES

During the past few years, automated and semiautomated analyzers have been introduced for use outside of the traditional hospital clinical laboratory. These instruments have been targeted for use in the physician's office. The criteria a physician uses to choose the ideal instrument depend on many factors: for example, on the type of medical specialty, whether the instrument is to be used for a solo or group practice, whether single test results or chemistry profiling is required, adequacy of turnaround time, and cost of the analyzer and reagents. Ideally, whole blood would be the specimen, and pipetting of sample and reagent would be automated with little or no interaction required from office personnel. Calibration curves should be stable for at least 1 month, and a quality control program and 24-hour service should be available from the instrument manufacturer. The instrument should be interfaced with a computer to provide documentation of laboratory results and a printout of quality control charts. The system should be easy to operate and require little user training because the instrument will usually be operated by personnel with little or no laboratory background. The reagents, dry or wet, should be bar coded, prepackaged, and stable for a least 6 months. Dry chemistry reagents can usually be stored in a small area, which is a significant advantage in an office laboratory.

The number of in-office physician laboratories will increase over the next few years because laboratory data can be obtained while the patient is still in the office, which allows prompt treatment by the physician, leads to decreased anxiety for the patient, and provides an additional source of revenue for the physician. Some of the instruments available for use in physicians' offices are shown in Table 14-2. They have been categorized as either dry or wet chemistry reagent systems.

Table 14-2 Clinical chemistry systems for the physician's office laboratory

Instrument	Manufacturer	Sample type*	Sample volume (μL)	Chemistry	Reagents	Reagents bar coded	Reagent choice	Available assays†	Assay time (min)	Maximum throughput per hr	Output
Seralyzer	Ames	S, P	Prediluted sample, 30	Dry	Strip	No	No	General chemistries, enzymes, TDM	0.5-4	60	Digital display
DT 60 system DT SC DT E	Kodak	S, P	10	Dry	Slide	Yes	No	Enzymes, electrolytes, TDM	5 3 4	65 20 15	Printer
ANALYST	DuPont	S, P	Prediluted sample, 90	Dry	Unitized in rotors	Yes	No	General chemistries, enzymes	10	72	Printer
Reflotron	Boehringer Mannheim	B, S, P	30	Dry	Tab	Yes	No	General chemistries, enzymes	3	20	Digital display
ChemPro	Sentech	B, S, P	150-200	Dry	Cards	Yes	No	Limited general chemistries, electrolytes	1-2	240	Printer
Vision	Abbott	B, S, P	2 drops or 1 capillary tube	Wet	Individual reagent packs	Yes	No	General chemistries, enzymes, TDM limited	8	60	Printer
GEMSTAR	Electro Nucleonics	S, P	10, 50	Wet	Bulk	No	Yes	General chemistries, enzymes	2-15	60	Printer
ASSIST	Technicon	S, P	2.5-20	Wet	Bulk	No	Yes	General chemistries, enzymes	1-4	55	Printer
Uni-Fast	Sclavo	S, P	20-200	Wet	Unitized	No	Yes	General chemistries, enzymes	2-15	30	Printer

**B*, Whole blood; *S*, serum; *P*, plasma.
†*TDM*, Therapeutic drug monitoring.

Dry reagent systems

The Seralyzer system uses test strips consisting of cellulose fibers impregnated with reagents. The reagent strips are stored in vials and are stable for 1 to 4 months after the container is opened. The reaction parameters for each assay are programmed by a dedicated module that is inserted into the analyzer. The sample, which usually requires predilution, is pipetted onto the strip, and the strip is moved into the analyzer. Light from a xenon lamp strikes the strip, and the vertically reflected light is passed through a collimator to the interference filter and is sensed by the sample photodetector. Reflected light is also passed through the reference photodetector. The ratio of the sample to reference reflectance is used to compensate for the drift in electronics and variations in the intensity of the lamp. Profile testing is difficult because each test requires a separate plug-in module.

The Kodak DT system uses a multilayered film technology similar to that of the Ektachem 700 analyzer. Each test slide is bar coded to identify the assay and the lot number of the reagent. The reagents are stable for about 1 year. With a specially designed battery-driven pipet, the sample is placed on the slide and incubated for 5 minutes at 37° C. Light illuminates the slide, and the amount of reflected light is related to the concentration of analyte in the sample. The incubator can hold up to five slides, thus allowing for batch or profile testing of up to five assays. The DT-60 can perform only colorimetric assays. The DT-SC Analyzer, which uses a xenon lamp as a light source with UV or visible wavelength detection, is available to perform enzyme assays and some therapeutic drug testing. The DT-SC analyzer uses the same pipet and basic technology as the DT60. The DT-E module can be obtained for measuring electrolytes.

The Reflotron system uses solid phase reagent tabs containing a magnetic code to identify the tests, reaction parameters, and the calibration curve for each assay. With each reagent lot, the calibration curve is established by the manufacturer. Therefore the laboratory does not have to calibrate the system. The shelf life of each test tab is from 1.5 to 2 years. Blood is placed on a glass fiber pad, and the plasma is separated from the erythrocytes. Plasma is transported away from the erythrocytes by capillary action, and the required sample volume is added to the reagent at a temperature of 37° C. Excess plasma is removed from the reaction area. The concentration of analytes is determined by reflectance photometry. With this system, blood obtained by finger or heel stick can be used for analysis. Since this is a single test system, profile testing is not practical.

The ChemPro system uses individually bar-coded test cards that contain liquid calibrators and an ion-selective electrode system. Each card contains two compartments: one for an aqueous calibrator and one for the sample. After the sample is pipetted onto the card, the calibrators and the sample are transferred by capillary action to ion-selective electrodes where the measurements are made. With this system, the analytical measurements are determined directly on whole blood, and some profile testing is possible.

The ANALYST is a desktop system that consists of specially designed disposable plastic rotors. The rotors contain the bar-coded reagents and are stable for 1 year when stored in a refrigerator. Rotors are available for profile or discrete analysis. Sample, which must be prediluted 10- or 60-fold with pipets supplied by DuPont, is automatically pipetted into the rotor. By centrifugal force the sample is transferred through capillaries into the reaction cuvettes along the rotor's perimeter. These cuvettes contain the reagent tablets. The chemical reactions are monitored at a temperature of 37° C using bichromatic spectrophotometry. After each test profile, the rotor must be discarded whether a complete or partial chemistry profile is performed. With the Analyst, rotors are available for performing a seven- or twelve-test general chemistry profile, a lipid profile, or a discrete analysis of a single analyte. An ion-selective electrode system for measuring sodium and potassium is also available from DuPont.

Wet chemistry systems

The Vision system uses two-dimensional centrifugation with specially designed disposable reagent packs for analysis. The reagent pack is a self-contained unit with a cuvette, liquid reagents, and bar codes to identify the assay. Two drops of sample, either blood, serum, or plasma, are placed into the multichambered reagent pack, and the packs are placed into a 10-position rotor in the analyzer. By centrifugal force, the plasma is separated from the erythrocytes. The packs are rotated at a 90-degree angle, which results in a premeasured amount of sample and reagent being transferred into the cuvette. The reaction is monitored at 37° C, and the absorbances are measured bichromatically. Blood can be obtained by finger or heel stick in a special tube provided by Abbott. The tube is then inserted directly into the reagent pack, eliminating the pipetting of blood from the capillary tube into the analyzer. The reagents are stable for several months, and with this system one to ten analytes can be measured. The Vision system is suitable for batch, profile, or stat assays.

The GEMSTAR is a random-access analyzer that can be used either for batch or profile analysis. Most reagents are supplied in lyophilized form. Reconstituted reagents are usually stable for 5 to 12 days. The GEMSTAR can operate in three modes: the profile, batch, or mix-match mode. In the profile mode, it is possible to run as many as 12 different tests on a single sample. In the batch mode, the same test can be determined on 11 different samples, and in the mix-match mode a profile can be determined on a small batch of samples. The reagents are manually pipetted into disposable cuvettes, and the samples are dispensed into these cuvettes with a special 12-position pipettor system from Electro Nucleonics. The cuvettes are placed in the analyzer and maintained at 37° C, and the absorbance is measured. The STARLYTE, an ion-selective instrument, is available for sodium and potassium determinations.

The ASSIST is a benchtop random-access analyzer. A disk that contains the chemistry programs is loaded into the disk drive. Reagents, which are lyophilized, must be reconstituted and placed in the reagent tray. From 1 to 16 samples are loaded on the sample tray, and the cuvettes, which are disposable, are placed into the analyzer. Samples and reagents are automatically pipetted into the cuvettes and mixed and the absorbances monitored bichromatically. The reagents are stable for 5 days after being reconstituted. A temperature of either 30° or 37° C can be used with this system. The ASSIST analyzer can perform either batch, profile, or stat assays.

The Uni-Fast system uses unitized, lyophilized reagent vials. The reagent is reconstituted with sample and diluent by a pipettor-diluter system that is supplied with the analyzer. The operator aspirates the reaction mixture directly from the vial into a flow-through cuvette where absorbances are measured. For tests that require preincubation, the vial is placed in a separate compartment before the contents are aspirated into the cuvette. Since the Uni-Fast system uses a single reagent vial for each test, it is suitable for stat testing.

TRENDS IN AUTOMATION

One of the most frustrating problems associated with the clinical laboratory is trying to achieve a turnaround time for laboratory test results that meets the real or perceived needs of the clinical staff. A primary cause of turnaround problems is the time required for the specimen to be hand delivered to the laboratory. Depending on where the laboratory is in relationship to the phlebotomist, it can take up to several hours for a specimen to reach the laboratory. In this case, the total time for obtaining the test results will be prolonged, and some of the chemistry results may be inaccurate (for example, glucose levels can be lower than expected). Pneumatic tube systems can be used to automate the transfer of specimens to the laboratory. In this way, the phlebotomist can place all the specimens drawn from a particular area into the tube system and send them to the appropriate laboratory.

When the phlebotomist draws the blood, a bar-coded sticker containing all the patient information and the test requested should be placed on the tube. Once the specimen is in the laboratory, the specimen will be automatically registered by the computer and placed in the appropriate instrument. The laboratory/hospital computer will then inform the computerized instrument of the tests to be performed on that sample. This will eliminate the requisition slip and the registration of patient information by the laboratory clerks.

Whole blood would be the ideal specimen for analysis because it would eliminate the problems associated with centrifugation of blood. Whether blood, serum, or plasma is used in obtaining the test results, the specimen should be sampled directly from the phlebotomy tube. This would eliminate transferring the sample from the tube to the sample cups and potential specimen mixup.

The reagents should be bar coded, prepackaged, and ready for use without the necessity of any interaction by the laboratory personnel. The sample size required for each assay should be small, less than 15 μL, so that assays for the entire hospital population, including pediatric and geriatric patients, can be performed. For example, on a 10 mL sample of blood, it should be possible to perform a chemistry profile, drug analysis, and special chemistry tests. This would reduce the number of blood transfusions necessary because of repeated blood drawing of hospitalized patients.

There is no one ideal instrument for the hospital laboratory. Depending on the hospital size and on the ordering patterns of the staff, several types of instruments can be used. In a large hospital, dedicated instruments should be available for performing the routine profile analyses and the most common stat chemistry assays. Instruments must also be available for determining drugs, special chemistries, and RIAs. By having analyzers perform a wide variety of tests, there may be no need to have separate laboratory sections dedicated to therapeutic drug monitoring, immunochemistry, endocrinology, and other special chemistry tests. In small- to medium-sized hospitals, random access analyzers would be required for the majority of the work load. Any instrument used must be able to generate quality control charts for each assay and should be able to be interfaced to the main laboratory computer to receive ordering information for each sample and to directly transfer laboratory results to the requesting physician.

Many of the routine tests may be shifted out of the hospital laboratory and to the physician's office, operating room, intensive care unit, emergency department, and the patients' homes. "Black box" analyzers will be required in these situations because they will be operated by personnel without laboratory training. Some of these instruments will be portable and battery operated, use whole blood as the sample, and have a built-in quality control system. The test menu for these instruments will include general chemistries, urinalysis, hormones, and some therapeutic drugs. The results will be ready in less than 5 minutes. The cost of these systems will decrease because of the competition between manufacturers to keep instrument and reagent costs down.

BIBLIOGRAPHY

Alpert, N, editor: Chemical Instrument Systems Newsletter, Stamford, CT 06906.

Lifshitz, MS, and DeCresce, RP: An overview of equipment for the physician's office laboratory, Lab Med 17:337-340, 1986.

Pesce, MA: Instrument systems for the physician's office laboratory, J Med Tech 2:566-569, 1985.

Sher, P, editor: New instrument preview section published in *Laboratory Medicine*, Chicago, IL 60612.

Shirey, TL: Development of a layered coating technology for clinical chemistry, Clin Biochem 2:147-155, 1983.

CHAPTER 15

Laboratory information systems

GEORGE SULLIVAN

OBJECTIVES

- Describe the functions of a laboratory information system (LIS) in a clinical laboratory.
- Describe how a technologist interfaces with an LIS's computer terminal to enter or review data.
- Review how patient information is entered into an LIS.
- List how a technologist uses the output function of an LIS.

KEY TERMS

accession number A number assigned by the LIS to a particular patient sample for identification purposes. The number may be reissued after all data on a specimen are obtained.

archiving The process of transferring patient information to a memory storage device that is not immediately accessible to the computer operator.

ASCII Acronym for American Standard Code for Information Interchange. It is the code for numerical values assigned to each character on the keyboard.

bar-code reader An optical reading device that converts a series of black lines into a sequence of numbers or letters for entry into the computer.

command line A series of computer directions entered in a single step. See *string*.

CPU Central processing unit; the part of the computer that controls and performs the execution of instructions (program).

CRT Cathode ray tube; also known as a terminal or video display unit (VDU). A television-like device used to monitor input, output, and system status.

cursor A symbol on the CRT, either flashing or solid, that indicates the user's position on the CRT screen.

data base The systemic storage of patient information that can be accessed by the operator (user).

delta check Process in which the current test value is compared to a previous value or values to detect large changes.

demographics All pertinent patient data, usually collected at admission, including name, age, address, insurance carrier, height, and weight.

field Area provided on the CRT screen for data entry, usually prompted by a cursor.

hardware The physical elements of the computer system, such as CPUs, printers, and terminals.

help text A special aid that displays additional instructions related to a field or screen.

input Any data or instructions placed into the computer system.

interface A communication link that allows the transfer of data between the user and the computer system or between another processor and the computer system.

keyboard A typewriter-like device used to enter information into the system. It is attached to the CRT to allow visual display of the input.

menu A predesigned formatted checklist containing various LIS options.

menu screen A screen displaying several options for selection in a predefined format.

on-line Term signifying the operator's ability to interact with a computer.

order entry A communication tool that enables personnel to enter specific orders and notify the departments without the use of written requisitions.

output Any information generated by a computer as a result of its calculations or processing. Output is either printed, displayed, or transferred to another processor.

printer An output device that produces a typewritten copy of information processed.

program A set of commands or steps that instruct the computer to perform a task.

prompts Instructions that appear on the CRT screen telling the user what to do next or asking for specific information.

RS232C port The Institute for Electrical and Electronics Engineers' standard type of interface in which data are transmitted sequentially along a single wire.

screen The video portion of the CRT that displays the fields to be completed by the user or shows the computer output.

security The ability of LIS to limit access to information. Use of an identification code and secret password or a specific terminal may be required to access data.

software The series of instructions or commands that direct the operation of the information system.

strings A series of commands or alphanumerical codes entered into the computer.

work list A list of patient samples and accession numbers to be processed, generated as LIS output.

BACKGROUND

Purpose

The number of tests performed in the laboratory has increased dramatically in the last 20 years, in part because of the development of new diagnostic tests and the advent of automated analyzers. As a result the amount of patient information generated from these tests has also increased proportionally. The need for assistance in the processing of this data has led to the development and use of a diversity of laboratory information systems (LISs). The LISs integrate sophisticated computers and computer software to serve the specific needs of the laboratory and other components of the health care system.

All medical technologists working in the health care system will have to deal at some interactive level with an LIS. This chapter acquaints technologists with LISs at their interactive level. Since there are so many different systems, this chapter can only give general principles of operation. The term *user* will be employed to describe the technologist who is performing specific tasks or receiving information from the system.

Function

LISs are designed specifically to provide information to various segments of a health care system. These are the laboratories, the physician, and the administrative support services (billing, supplies, patient records). Critical to the successful use of an LIS is the collection of information concerning a specific patient. This information includes requests for specific tests, specimen collection time, and individual patient demographics. After the analyses have been completed, the LIS collects laboratory data from both nonautomated and automated sections of the laboratory. In addition, the LIS must be able to store all the pre-established information in both short-term and long-term memory. The pre-established information includes a library of available tests, test sample requirements, instrument parameters, and quality control data (means, standard deviations, information on report format). All of this pre-established information makes up the *data base* of the system.

A major function of the LIS is to organize the individual bits of information to permit ready access by users. For example, physicians may obtain all the test information generated by different laboratories on one patient in one comprehensive report. Reports of data can be produced either on demand or at prescribed intervals at specific geographical locations, either on paper or by computer terminal (see below). One of the major goals of the LIS is to organize information for ancillary housekeeping purposes for various users of the system. For example, data can be generated that provide billing information for the hospital and lists of available tests for physicians. Special reports can be generated for a laboratory, including lists of samples waiting for analysis, quality control data, and lists of abnormal test results, maintenance records, and statistical reports for review by supervisory personnel. The ancillary reporting functions of an LIS can be tailored to meet the specific needs of the individual laboratory and administrative staff by manipulation of the computer software. A summary of the major goals of an LIS is found in the box below.

Organization

For an LIS to execute the functions listed in the box, it must keep track of all information entered into the system. The LIS achieves this goal by means of a central computer, a number of different input/output devices, and the computer's instruction (software). A patient data file in the memory of the computer coordinates the patient information. Input/output devices allow users to enter or receive information from the central patient data file. Because many of the most important functions of an LIS involve

FUNCTIONS OF AN LIS

- Collection of patient information
  - Demographics
  - Patient location
  - Test ordering
  - Attending physician
- Generation of specimen test results
  - Automated result entry
  - Manual result entry
- Assembling data output
  - Patient report by analyte, specimen, or time
  - Cumulative patient reports
- Producing ancillary reports
  - Work lists of samples to be analyzed
  - Quality control data
  - Billing reports
  - Specifically programmed reports
- Storage of data
  - Short term (for immediate access)
  - Long term (for historical purposes)

input/output functions, interfacing, which allows the functions to proceed, will be discussed next.

INTERFACING

Information exchange

The exchange of information between the computer system and the operator is called *interfacing*. The actual interfacing is achieved by the use of the specific input/output devices discussed below. The interfacing can occur between an instrument and the computer or between a technologist and the computer. The process of information exchange follows a highly structured format that is determined by the computer's software. The format of the information exchange is usually not obvious, and some effort is needed to learn the individual format of each LIS. The term *user friendly* refers to the degree of effort that the system makes to ensure correct and accurate interfacing. Most often the computer information is displayed on a video screen device. This display of computer-generated information is termed a *screen*.

A user-friendly system will have numerous help screens or prompts that explain the options available and make interfacing easier for a new user of the system. A *help screen* allows the user to leave the immediate interactive screen and review the instructions for the options facing the user in the current interactive screen. Many systems permit access to help screens by a certain sequence of commands or by use of special function keys (see below). For example, in some systems pressing the key sequence *control ?* may bring the user to a help screen. The technologist will most likely interface with the LIS to accomplish the following tasks: the ordering of tests and the entry, verification, and transmission of patient results. In addition, the technologist may require pertinent patient information or demographics (for example, age, sex, race, and address) to evaluate test results.

Data input and output

Data input and output can be accomplished by two methods, *command line entry (string entry)* and *menu selection*. In both cases the user is prompted by a cursor on the screen. The cursor can be depicted in many ways, usually as a flashing box.

Command lines are a series of individual computer commands or pieces of data that the user inputs at a keyboard in a single step. For example, the user may want to enter information on the patient identification (account) number and a specific item number representing the test. Instead of entering each piece of information segmentally they are entered together; usually they are "strung" together; the individual command lines are separated from each other by a comma. Using the above example a possible string would be

877653, *, 4814

where the first six digits represent the patient account number, the * represents that the test ordered is stat, and the last four digits represent the test name, in this case an ethanol level.

By entering a series of inputs, all related to each other, as a single computer command, the technologist spends less time at the keyboard, and less computer time is required.

Menus are predesigned checklists containing the various

PATIENT 325. SUPERVISOR ORDER ENTRY
ROOM A1 2106 1

1. *****TOXICOLOGY A-N MENU****	16.
2. ********************************	17. DISOPYRAMIDE (NORPACE)
3.	18. DRUG SCREEN (QUALITATIVE)
4. ACETAMINOPHEN	19.
5. ALCOHOL	20. ETHOSUXIMIDE (ZAROTIN)
6. AMIKACIN	21.
7. AMINOPHYLLINE	22. GENTAMYCIN
8.	23.
9. CARBAMAZEPINE (TEGRETOL)	24. LANOXIN (DIGOXIN)
10.	25. LIDOCAINE
11. DEPEKENE (VALPROIC ACID)	26. LITHIUM
12. DIGITOXIN	27.
13. DIGOXIN	28. MYSOLINE (PRIMIDONE)
14. DILANTIN	29.
15.	30. NORPACE (DISOPYRAMIDE)
OPT ITEM# DESCRIPTION *NO VERIFY*	OPT ITEM# DESCRIPTION

PRIORITY REPEAT THIS SCREEN OPT CHARGE DATE CMD 9-HELP
KEY SELECTION OPTION TO THE LEFT OF ITEMS TO BE ORDERED AND PRESS ENTER.
SELECTION OPTIONS: X—NORMAL P—PRINT LOCAL A—ADDITIONAL INFORMATION

Fig. 15-1 Example of a test request menu. (See text for details.)

options offered by the system. For example, in Fig. 15-1 a list of toxicology tests available for ordering is presented. Movement of the cursor is limited to spaces immediately next to the column numbers. Placement of an *X* at the desired test will instruct the computer to record the test as ordered. Order entry of data or commands may be made by either the command line or menu method. The command line approach is sometimes mixed with a menu entry system so that one can bypass a series of menu choices if one is familiar with the subsequent menu choices. Usually a menu system works well when there are only a few choices or with novice users; experienced users are often impatient flipping through many menus for a few needed answers.

When the technologist has completed the analysis, the result may need to be manually entered into the computer. A cursor on the CRT (video display unit) informs or prompts the user about where on the screen the computer will take input. Areas on the screen that have been predesigned to accept specific data are called *input fields*. These fields can accept numerical or letter data. Alternatively, the result may be uploaded by an instrument directly into the computer. *Up-loading* refers to the transfer of data from an external source such as an analytical instrument with its own processor, to the central processing unit (CPU). *Down-loading* is the reverse of this process; that is, it is the transfer of data (such as specific tests ordered for a specific patient sample) from the computer to the analytical instrument. Typically, the up-loading or down-loading of information does not go directly to or from the instrument but through a small, personal computer.

LIS HARDWARE

The physical or structural components of an LIS are termed the *hardware* of the system. These components include the actual computer (central processor, CPU), the data storage devices (memory), and the peripheral interfacing devices (Fig. 15-2).

Computer

The central component of any computer system is the CPU, or the "brains" of the system. It is in the CPU that the activities determined by the user through given programmed instructions are executed. These instructions are called programs or software (see below). A more complete description of how computers function is beyond the pur-

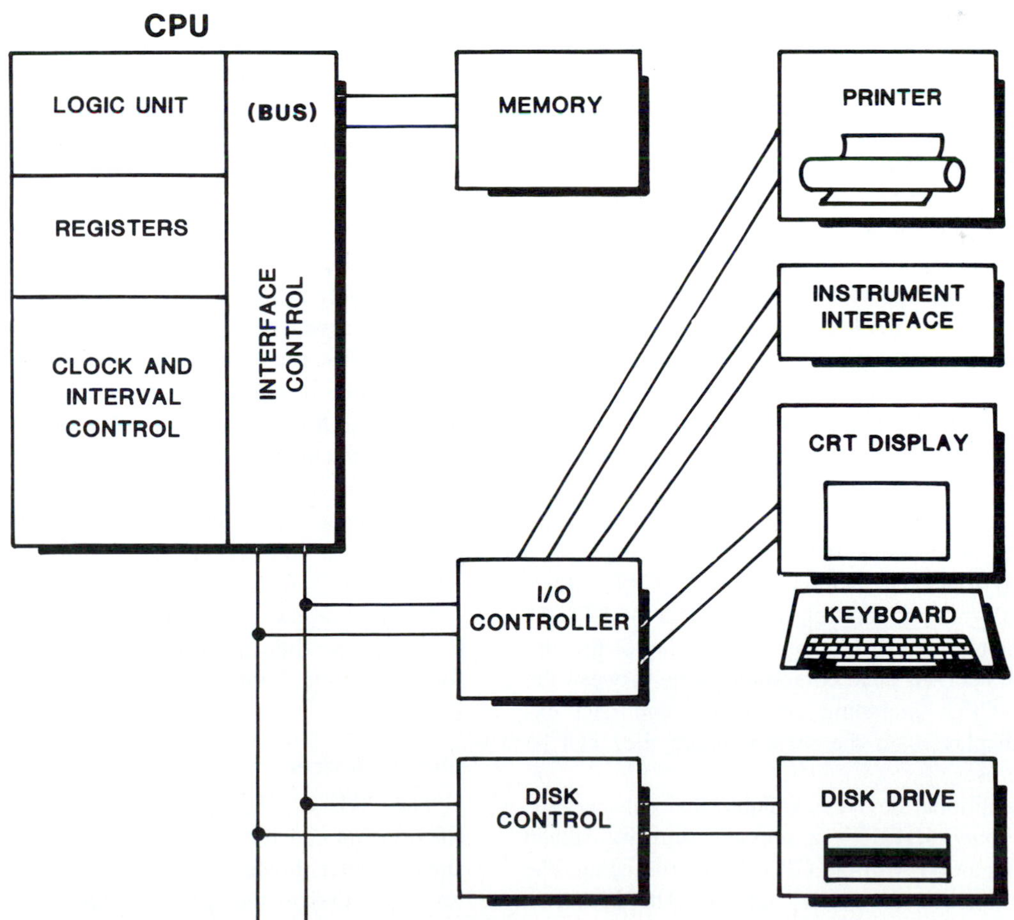

Fig. 15-2 Overall structure of computer with its variety of peripheral units.

view of this book. The interested student is invited to read the review by Hodges listed in the bibliography.

Data storage devices

The second major component of an LIS is the *memory* or *data storage* section. This component contains all the necessary instructions and data needed for operation. In addition, all short-range patient records and laboratory data (for example, quality control data) may also be stored temporarily in the LIS's memory. Devices that store this information include *hard disk* drives, *reel-to-reel* drives, and *magnetic tapes*. The technologist may be more familiar with the "floppy" or flexible disks that are used by a microcomputer for data or program storage. The hard disk, reel-to-reel, and magnetic tape devices can store much larger amounts of information than the internal memory of the computer.

Input devices

The third component of the LIS hardware includes devices by which the CPU and the user communicate. The technologist will need to become familiar with the use of the many different peripheral devices by which information can be entered into the computer. It is through these devices that the technologist will interact with the LIS. Several of these devices can also function as output devices.

Keyboard. The most complex device uses a keyboard. Through the use of standardized codes called ASCII (American Standard Code for Information Interchange), the entire keyboard can be used to enter alphanumerical as well as numerical data into the computer. Alphanumerical input includes both the letters of the alphabet and numbers. Some systems permit input of other symbols such as @, %, &, and #. More complex systems use *function keys*. The function keys automatically enter a series of commands when one key is pressed, for example, F1 = help, F2 = data entry, F3 = page forward, F4 = page backward, F5 = utilities, and F6 = return to start. Such keys reduce the number of keystrokes required to display a help screen, return to the previous screen, enter a different part of the program, or terminate data entry.

Cathode ray tube. The cathode ray tube (CRT) allows exchange of information between the user and the CPU on a specially adapted television tube. The display of this device is termed a *screen*. The communications between the user and the CPU, including commands and data exchange, are displayed on the screen where they can be viewed by the user.

There are additional devices available that enhance this exchange. A *touch screen* is a pressure-sensitive screen that allows interaction with the CPU through a menu. The position of the touch determines choice. The *light pen* (stylus) interacts with a light-sensitive screen to indicate a menu choice. The *mouse* is a manual device that moves a cursor when the mechanism is rolled along a flat surface. It interacts with the computer through a menu.

Fig. 15-3 Example of a bar-code label that can be placed on a phlebotomy tube. The bar code on this label codes for the number at the top of the label.

Bar-code reader. Bar codes, similar to those seen on grocery items, can be printed out by specially designed bar-code printers or can be purchased preprinted according to the laboratory's specifications. The bar codes can be placed on patient samples (phlebotomy tubes or pour-off tubes). The bar-code reader is a device that reads a series of black lines (bars) on a label (Fig. 15-3) and converts this data into a sequence of numbers representing specific information (tests requested, patient sample, accession number, identification of a reagent, blood unit number, etc.). Usually when such data are entered into the CPU, they can be viewed by a technologist on a CRT.

Instrument interfaces. The interfacing of an instrument with the CPU is commonly accomplished by means of a device called an RS232C port. A *port* is a memory location in the CPU that is connected to a series of wires; it permits the main computer to interface with the computer of an instrument. This allows the transfer (up-loading and down-loading) of test requests or result data directly. The RS232C port is the accepted industry standard for serial interface; seven or eight bits of data are transmitted sequentially for each character along a single wire. Supporting software (program instructions) is required to keep track of the transfer process and interpret the data stream. The instrument manufacturers define the specific information or data stream that is sent in this process. For example, the data stream used by the IL Monarch to send a result is as follows: [STX]5,MON 1, 3, 43, 1, 0, 123456789, 0, 0,4.2 [ETX] (Fig 15-4).

Card readers. Card readers are input devices that transform the input information on computer cards, in the form of punched holes or black marks, into computer language.

Output devices

The CRT, RS232C port printer, and other computers on instruments can function as output devices. In each case, the computer directs some data from memory or a storage area on a disk to the specific output device. In the case of the CRT, the processed output data are displayed on the screen. The RS232C port allows electronic information to

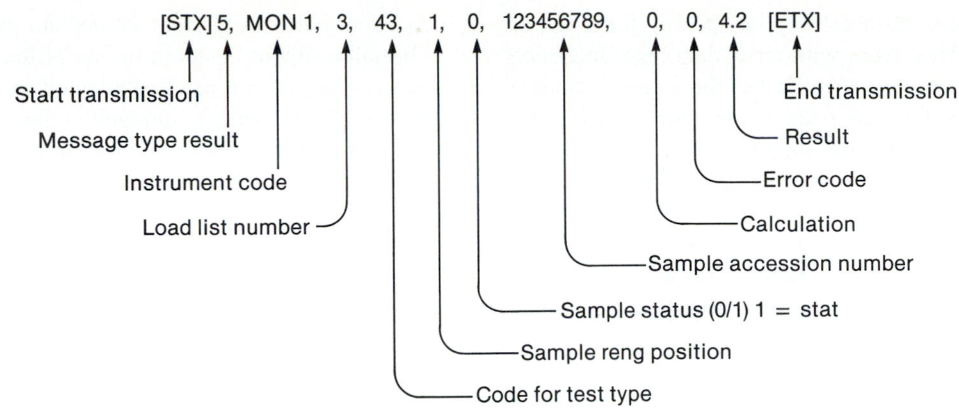

Fig. 15-4 Data stream. (See text for details.)

flow from the LIS to the appropriate device, which may be another computer or processor.

The *printer* is the main output device of the computer. Results printed on paper are the "hard copy" output. The format of the printout is determined by the software (see following discussion). Since this is readily changed, many different formats may be observed by the technologist. The format is often customized to the specifications of a particular laboratory. The printed output containing patient data is placed in the patient's chart and is incorporated into the official legal record of the patient.

SOFTWARE

The complex set of instructions that directs the computer to perform specific tasks in a predetermined order is called the *program*. These instructions direct the collection of data, the assimilation of data, various operations on the data, and the transfer of data throughout the memory of the system. They also contain the information necessary to communicate with input and output devices.

The set of instructions is written in a specific language, which the hardware is designed to accept. Any modifications in the predetermined format of the LIS functions can only be accomplished by changing the program. Although individuals specifically trained to program computers are usually assigned to make the program changes, technologists with sufficient computer experience can often perform this duty.

LABORATORY DATA PROCESSING

Order entry

Patient identification by LIS. Order entry is the process by which requests for laboratory tests, ordered for a specific patient sample, are entered into the LIS. Before order entry, the patient's identification must be established in the computer. In most cases, identification of the patient is accomplished by using information obtained from the patient registration department, usually the hospital information system (HIS). The HIS and LIS are usually interfaced with one another so that the admission, discharge, and transfer information that is entered by the patient registration or admitting departments in the HIS can be used by the LIS.

On admission, the patient is given a patient number. The number of digits varies in different LISs between 6 and 12, and the code sometimes contains a leading alphanumerical character. There are two different general identification processes in use. In the unit record system, a patient is given a unique number the first time he or she is treated by the health care delivery system and maintains that same patient number forever. In some systems the nine-digit Social Security number is used for convenience. Each time the patient returns to the system, a different "encounter number" is assigned to the unique patient number. The encounter number is used to keep track of all services needed or performed for the current visit. In the nonunit record system the patient receives a new patient number for each visit, and no encounter number is needed. In addition, both of these systems may assign other billing numbers for other special purposes. In some cases the laboratory can directly interface with the LIS to enter the patient's demographics (name, address, age, etc.) and receive an assigned patient identification number.

Sample identification by LIS. Test results may come to the attention of the laboratory either by the traditional written request (laboratory slips) or through the computer. If the system employs written requests, then someone in the laboratory must convert the written request into a computer order. Entry of this order into the computer generates an *accession number*. The accession number is specific for that order or that specimen. This is important when multiple determinations of an analyte are ordered for the same patient within a short time, such as blood gases every 15 minutes. Because the orders are separate, each request will generate a separate accession number. In addition to the patient's name and registration account number, the accession number is now a third way to identify a test request and result.

The test accession number may be specific for a test or battery of tests. However, without a date, the accession number may not be very specific for the sample, since a laboratory using a four-digit accession number will start with 0001 and go to 9999 and then begin with 0001 again. A busy laboratory may cycle through 10,000 accession numbers in only a few days. Accession numbers are reissued only when all the test results associated with the sample previously assigned to that number have been entered. Thus there are never two specimens with pending results that have the same accession number.

Some LISs allow order entry directly from the clinical areas if the clinical areas have their own CRTs with a menu-driven program. This system is termed a *ward order entry system*. The advantage of the ward order entry system is that it eliminates both the need for laboratory slips and the intermediate step of order entry by another individual. This can reduce turnaround time and ordering errors.

Data verification

Once the test results have been entered into the LIS, verification of the results is required. This is the most important step in the entire information transfer process. By verifying the data, the technologist asserts that the data, automatically transferred or manually entered, is correct. That is, the result is analytically acceptable and it is properly assigned to the correct patient (identification number) and sample (accession number). The two-step method of entry and verification is required by the Joint Commission on Accreditation of Hospitals and the College of American Pathologists. The safest procedure is one in which the technologist who ran the assay also enters and verifies the results.

Output

The output of the LIS can be in many forms, including paper printouts, electronic information displayed on a CRT, or electronic information transferred to another device (such as a personal computer) or stored in memory devices such as hard disks and magnetic tapes. The technologist will interface with the output at various stages of the testing process. One output function is the generation of printed phlebotomy tube labels at the time of order entry. Another function frequently used by technologists is the generation of a list of test requests with accession numbers. This list defines the work that must be accomplished in a given time period and thus is termed a *work list*. The work list is important in the planning of the day's work for specific areas within a laboratory. LIS output also includes the flagging of critical values and abnormal test results to alert the technologist to transmit critical values to a physician or repeat a test. Some systems will compare a patient request to a previous result obtained on that patient. This is termed a *delta (difference check)*. Unusually different results may alert the technologist to an otherwise undetected error in analysis or a significant change in patient condition that may warrant immediate physician attention.

Initially, output is stored in one of the system's memory devices. This output can be collected over a period of time and then transferred ("dumped") into an output device such as a printer. Usually the format and times of reporting of patient test results are predetermined and carried out by the LIS. A partial exception to this is the selective reporting of data, such as stat broadcasting, where the technologist designates a specific location to which the information is to be transferred. The LIS can produce updated lists of all patient results or summaries of results on individual patients. Most systems do a periodic printing of all data on paper (hard copy). These include various logs, management reports, patient reports, and statistical reports. The patient results are then placed in the appropriate chart. More sophisticated systems may produce an instantaneous update printed at the patient's clinical location. This is particularly useful for the emergency unit and intensive care units.

The output of the computer can be changed to present the data in user-specified formats. As results are entered and verified, an audit trail (a description of what happened to the specimen) can be produced by the system. The computer can retain when the test was ordered, when the specimen was collected, when the test was run, and when the result was entered and verified. In addition, it can record who ordered the specimen (which nurse or clerk), who collected the specimen (which phlebotomist), and who ran the test (which technologist). The name of the verifying technologist also can be entered.

When the results are entered into the computer system either manually or by direct down-loading and then verified as correct, a charge is added to the patient bill. Since the test has its own code number and the patient has an identifying number, the matching of test request and patient takes place rather simply.

Besides the reporting of results, additional information useful to laboratory management can also be generated. Audit trails can be used to determine the volume of testing per time period (work-load recording), the frequency of stat tests by location, the time delay from stat order request to the reporting of a result, or a listing of incomplete, uncollected, or missed requests. These management reports may be helpful in the documentation of the laboratory's participation in the hospital-wide quality assurance program required by accrediting agencies. Many LIS programs have the specific ability to chart and report quality control data. Statistical parameters can be calculated for all tests performed, including standard deviation and coefficient of variation. Charting can include printing Levy-Jennings plots.

MAINTENANCE

Patients and patient data cannot be added to an LIS indefinitely. The time needed to search for data and storage limitations require that, periodically, some accumulated data be removed from the active *on-line* system. Patient data that are removed are called inactive, purged, or archived data, depending on the computer system vendor. In concept, this information is simply removed from immediate access and put in another file. It may still be ''found'' and displayed on the screen or printed, but this requires an additional step, usually performed by LIS personnel. Other maintenance functions also need to be performed periodically. For example, statistical files, quality control files, and work-load files all need to have older data removed. As with patient file maintenance, some maintenance functions are automatic, and some need to be initiated.

The maintenance function of the LIS may have a direct impact on the technologist. Some LISs require that the system become paritally or completely inaccessible during certain maintenance periods. During these times, alternative mechanisms are needed to collect, store, and transmit patient test results.

SECURITY

LISs often interface with other HISs. The data base established when the patient was admitted or registered as an outpatient (often up to 70 questions are asked) is accessed by the LIS for specific data. Other departments also may need to access the main hospital data base or HIS. For this reason a series of LIS programs are necessary to limit a user's access to only those areas or functions that are necessary to accomplish the user's work. The barriers can only be bypassed when the correct password is used. There is at least a two-step security process, with one of the codes being secret and known only to the programmer and the individual user. Usually the degree of access that a user has to the LIS's data base depends on the user's responsibility. For example, a technologist may not be able to enter the data for patient registration but is able to enter results from patient assays. In contrast, a person working in a patient registration area is not able to enter test results but is able to obtain results on past tests.

The security of the computer system is also designed to alert the system controller if tampering occurs. Repeated attempts to gain access into unauthorized areas and other security violations may be countered with automatic shutdown of the terminal.

BIBLIOGRAPHY

Elevitch, FR, and Allen, RD: The ABCs of LIS: computerizing your laboratory information system, Am Soc Clin 1986.

Hodges, DA: Microelectronic memories, Sci Am 237:130-135, 1977.

Nisbet, JA, Owen, DJM, Owen, JA, and Walduck, AG: A comprehensive laboratory computer system working round the clock, Clin Chem Acta 91:89-101, 1979.

Standard Guide for Computer Automation in the Clinical Laboratory, ASTM Manual E792-81, Philadelphia, 1981, American Society for Testing and materials.

Wheeler, LA, and Sheiner, LB: A clinical evaluation of various delta check methods, Clin Chem 27:5-9, 1981.

CHAPTER 16

Laboratory statistics

CARL C. GARBER
R. NEILL CAREY

OBJECTIVES

- Given the appropriate data, construct a frequency and a relative frequency histogram; list the three measures of central tendency, and describe their usefulness when applied to normal and nonnormal distributions.
- Define variance, standard deviation, and coefficient of variation; calculate values for each given appropriate data.
- State the percentage of the population falling within the 68%, 95%, and 99% confidence intervals.
- Explain the differences between parametric and nonparametric statistics.
- Describe the uses of correlation analysis for comparing sets of data.

KEY TERMS

accuracy Estimate of nonrandom, systematic bias between populations of data or between a population of data and the true value.

central tendency The value about which a population is centered. The mean, the median, and the mode are all used to describe the central tendency of a population.

coefficient of variation (CV) A relative standard deviation calculated by multiplying by 100%, the standard deviation divided by the mean.

confidence interval A range around an experimentally determined statistic that has a known probability of including the true parameter.

correlation coefficient A statistic that measures the distribution of data about the estimated linear-regression line.

degrees of freedom The number of independent observations in a data set. It is the number of observations minus the number of restrictions for a set of data.

F test A statistical test used to determine whether there are differences between two variances.

histogram A graphic display of data in which the frequency of a certain value (or range of values) is plotted against a scale of all values.

mean Arithmetic average of a set of data.

median A value or interval of a population occurring in the middle of a population, half of which falls above and half below the median.

mode The value or interval of a population occurring with the greatest frequency.

nonparametric statistics Statistics that do not assume a normal or symmetrical (that is, gaussian) distribution.

null hypothesis The working hypothesis of a statistical test stating that there is no difference between the statistics of two different populations.

parametric statistics Statistics that assume that a population has a symmetrical distribution (such as gaussian or log-normal).

precision Random variation in a population of data.

random error Error that affects reproducibility of a method (precision).

range The difference between the highest and lowest values in a population.

standard deviation Square root of a variance.

statistic A number that describes a property of a set of data or other numbers.

statistics The plural of statistic; also the science that deals with the use and classification of numbers or data.

systematic error Nonrandom error that affects the mean of a population of data and defines the bias between the means of two populations.

t test A statistical test used to determine whether there are differences between two means or between a target value and a calculated mean.

variance A statistic used to describe the distribution or spread of data in a population.

Fig. 16-1 **A,** Histogram (frequency distribution) of glucose results obtained from 20 repetitive measurements of same specimen using bin width 40 mg/L. **B,** With $N = 100$ and bin width = 20 mg/L. **C,** With infinite N and bin width = 10 mg/L.

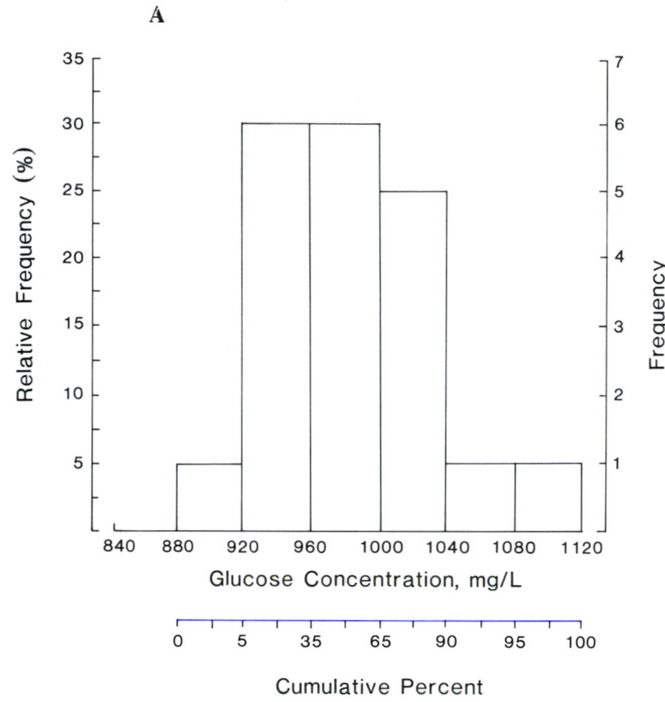

Daniel[1] has defined statistics as ". . . a field of study concerned with 1) the organization and summarization of data, and 2) the drawing of inferences about a body of data when only a part of the data is observed."

A statistic is a number that describes a property of a set of other numbers. A familiar example is the average, or arithmetic mean, of a set of data. Often, only a few simple statistics are required to describe an entire population of data. Thus statistics are the essence that is left after one "boils down" the experimental data. (Notice that *statistic* is singular, *statistics* as a science is singular, and a *group of statistics* is plural.) Statistical descriptions of data sets are useful for several purposes:

1. To assess random variation in a population of data
2. To compare the amounts of random variation among populations of data
3. To test for systematic differences between populations of data
4. To assess the degree of correlation between populations of data

Each of these uses of statistics is briefly described in this chapter.

The importance of statistical methods in clinical chemistry cannot be overemphasized. The reader is encouraged to study statistics textbooks, such as references 1 to 7, to obtain a basic understanding of the intelligent use of statistics.

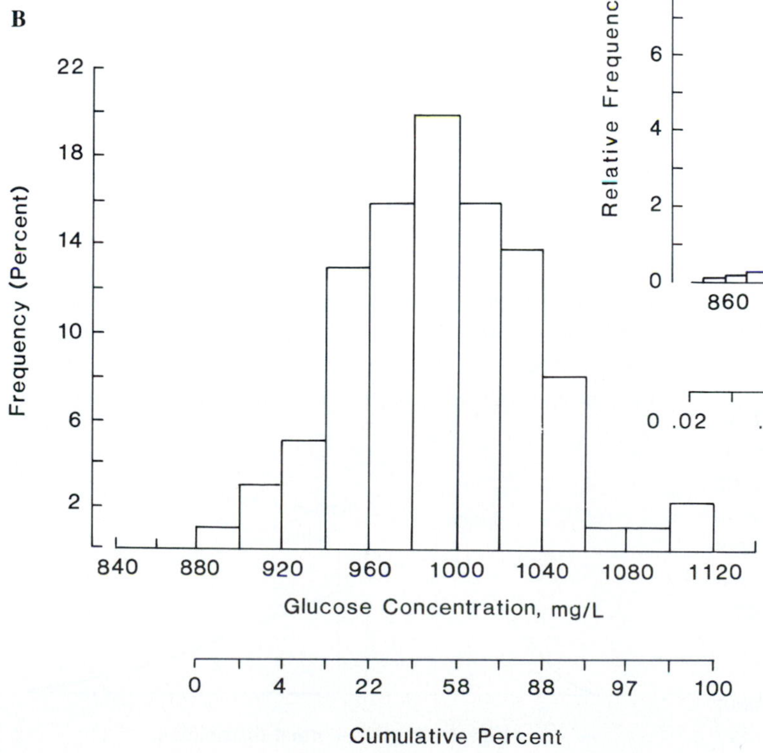

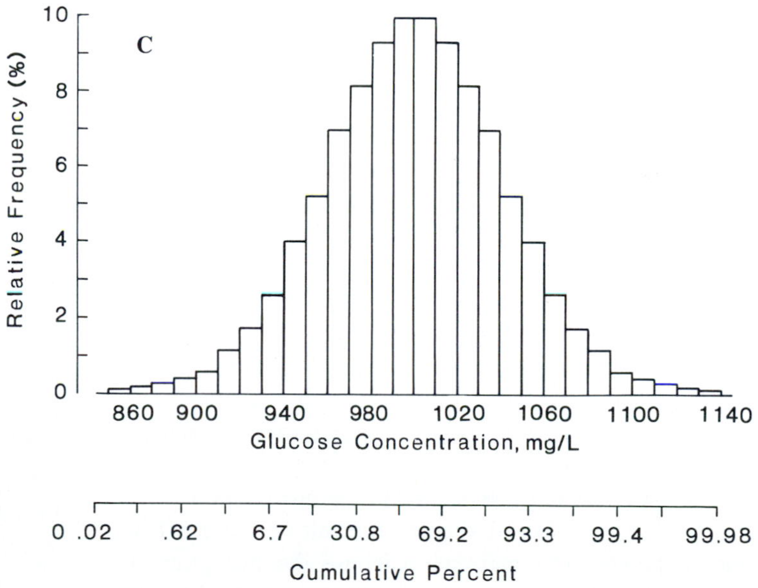

STATISTICAL DESCRIPTIONS OF A POPULATION

No chemical measurement is exact because random variation (random error) is inherent in all laboratory measurements. For example, if the glucose concentration of a single blood specimen is repeatedly measured using the same analyst, method, reagents, and instrument, a population of results (observations) will be obtained. This variation has many causes that one can minimize but not eliminate. Some of these are heterogeneity of the specimen itself with time, variation in the technique of the analyst, heterogeneity of reagents with time, and instrument variation.

Population distributions

Conceptually, perhaps the simplest way to describe a population of data is to construct a *histogram,* also called a *frequency diagram* or *frequency distribution*. A histogram shows the frequency with which a particular value, or range of values, is obtained versus the scale of all values. Fig. 16-1, *A,* is a histogram of the glucose results obtained in the repeated measurement of the blood specimen previously mentioned. The horizontal axis is glucose concentration, divided into small convenient ranges, or *bins*. The vertical axis is the frequency (or relative frequency) with which results from each bin are obtained. When relative frequencies are used, each bin frequency is presented as a percentage of the total number of samples. The histogram's horizontal axis can also represent cumulative percentiles (cumulative percentage of the population up to and including each bin) as well as concentration units. When the number of observations, *N,* is small and the bins are relatively wide, the histogram has a choppy appearance. As *N* increases and the bins are made narrower, the shape of the histogram becomes smoother and the histogram becomes more truly representative of the population. As *N* increases further, the histogram takes on the appearance of a continuous function. In the histogram of hypothetical glucose data in Fig. 16-1, *B,* one can see that the population is centered around 1000 mg/L, few of the observations are less than 920 mg/L, and few are greater than 1080 mg/L. One can also assess the general spread of the data.

If enough data are represented in the histogram and the data are truly random (that is, each result was affected by random processes alone), one can use the histogram to predict the probability of obtaining future results above or below a given value. Fig. 16-1, *C,* shows that there is about a 2.3% chance that a future glucose result from this population will be less than 920 mg/L, and a 2.3% chance that a future glucose result will be greater than 1080 mg/L.

Measures of central tendencies

There are several terms that define the values about which a population is centered. It is customary in everyday life to talk about the "average." The simplest average is the *arithmetic mean,* which one obtains by adding up all the observations and dividing by the number of observations, *N*. The mean, $\bar{x}$, is given by Equation 16-1, where x_i is an individual observation.

$$\bar{x} = \frac{\Sigma x_i}{N} \qquad \textit{Eq. 16-1}$$

The arithmetic mean is representative of the central tendency of the population only if the histogram is symmetrical about the mean, as in Fig. 16-2. Fig. 16-2 also represents the *normal,* or *gaussian, distribution,* which is obtained when repeated measurements of the same analyte are made on the same specimen with only the small, inherent random errors of the method present, as in Fig. 16-1. Many statistical tests assume that data are "normally" distributed. If the entire population of data is used to calculate the mean, this calculated mean is the true mean of the population, indicated by μ. If a subset of the population of data is used, the symbol, $\bar{x}$, denotes that the mean is an estimate of the true mean based on the subset of data (or limited data).

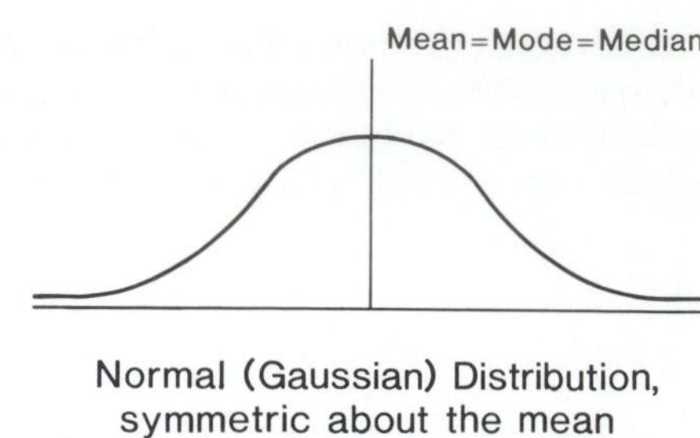

Fig. 16-2 Normal (gaussian) distribution, symmetric about mean.

When a nonsymmetrical distribution is obtained, the arithmetic mean does not describe the center of the distribution satisfactorily. One alternative is to find the most frequently obtained value, called the *mode*. A distribution whose mode differs from the arithmetic mean is shown in Fig. 16-3. Another term describing the center of the distribution is the *median,* which is the value in the middle of the distribution. Half the observations are greater than the median and half are less; the median is the observation at

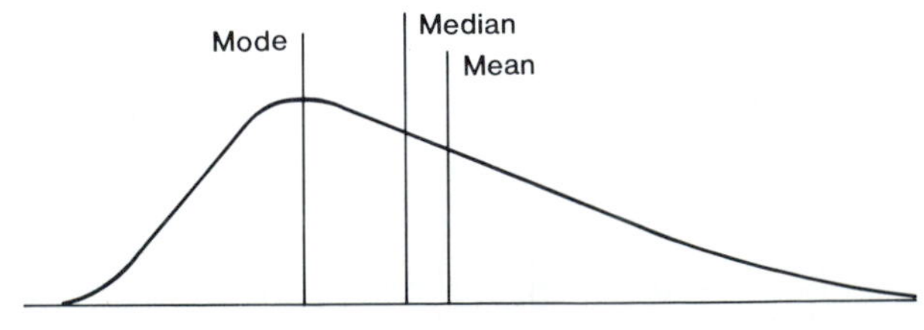

Fig. 16-3 Nonnormal distribution.

the 50th percentile. Mode and median are useful terms to describe distributions that are skewed, for example, with a long tail of higher values, as in Fig. 16-3. An extreme example of the nonsymmetrical distribution, the bimodal distribution, is demonstrated in Fig. 16-4. This distribution would not be expected for repeated testing of the same specimen. Differences between the arithmetic mean, median, and mode alert the analyst to the presence of nonsymmetrical distributions (see p. 263).

Measures of variation

Indicators of the variation of the measurements or the spread of the distribution are given by the range, the variance, and the standard deviation. The range is the differ-

Example 1. Calculation of common statistics from a sample population of repeated glucose measurements.
Construct a frequency histogram of the population below using bins of 20 mg/L. Calculate the arithmetic mean, mode, median, variance, standard deviation, and percent coefficient of variation (%CV).
Population: 1090, 1020, 1040, 970, 1050, 980, 1000, 960, 1030, 970, 960, 970, 970, 1040, 990, 1050, 940, 950, 970, 1000.
Answer:

x_i (mg/L)	$x_i - \bar{x}$	$(x_i - \bar{x})^2$
1090	92.5	8556.25
1020	22.5	506.25
1040	42.5	1806.25
970	−27.5	756.25
1050	52.5	2756.25
980	−17.5	306.25
1000	2.5	6.25
960	−37.5	1406.25
1030	32.5	1056.25
970	−27.5	756.25
970	−27.5	756.25
960	−37.5	1406.25
970	−27.5	756.25
1040	42.5	1806.25
990	− 7.5	56.25
1050	52.5	2756.25
940	−57.5	3306.25
950	−47.5	2256.25
970	−27.5	756.25
1000	2.5	6.25

$\Sigma x_i = 19950$ $\Sigma(x_i - \bar{x})^2 = 31775$

$$\text{Mean} = \bar{x} = \frac{\Sigma x_i}{N} = \frac{19950}{20} = 997.5 \text{ mg/L}$$

Median = 980 mg/L

Mode = 970 mg/L

$$s^2 = \frac{\Sigma(x_i - \bar{x})^2}{N - 1} = \frac{31775}{19} = 1672$$

$s = 40.9$ mg/L

$$\%CV = \frac{s}{\bar{x}} \cdot 100\% = \frac{40.9 \text{ mg/L}}{997.5 \text{ mg/L}} \cdot 100\% = 4.1\%$$

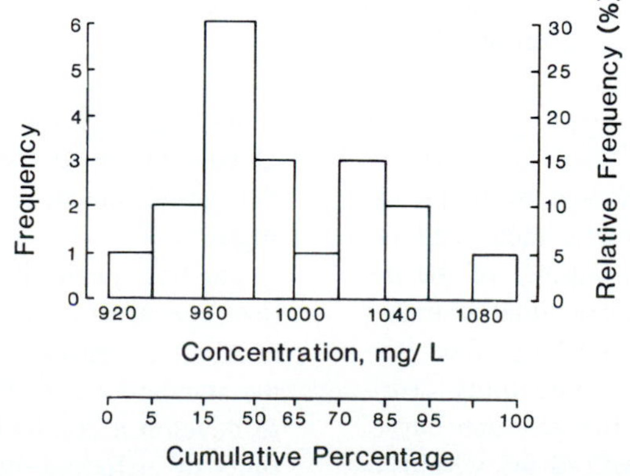

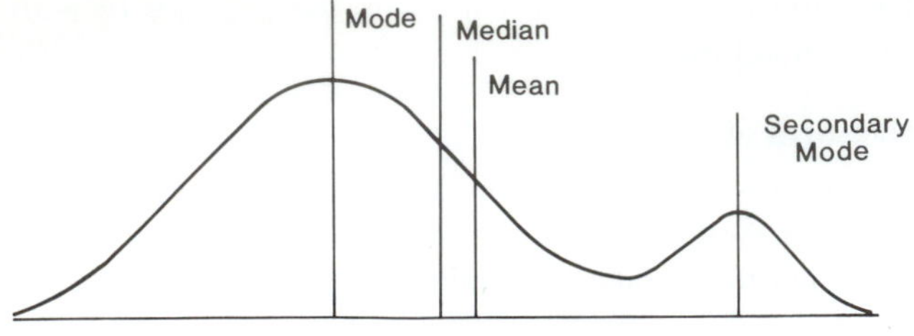

Fig. 16-4 Bimodal distribution.

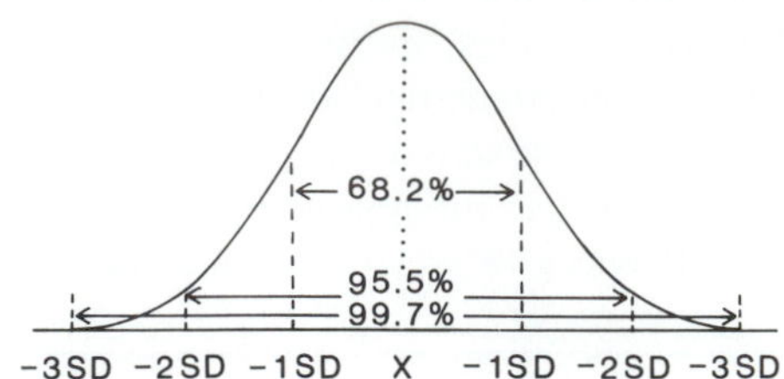

Fig. 16-5 Gaussian curve.

ence between the lowest and highest values in the data set. The *range* is especially useful for indicating the spread of the data when N is small; it makes no assumptions about the shape of the distribution.

One calculates the *variance* by adding the squares of the differences between the individual results and the mean, and dividing by $N - 1$, as in Equation 16-2. The *standard deviation, s,* is the square root of the variance.

$$\text{Variance} = s^2 = \frac{\Sigma(x_i - \bar{x})^2}{N - 1} \qquad \textbf{\textit{Eq. 16-2}}$$

The denominator of Equation 16-2 is $N - 1$ rather than N because there are only $N - 1$ degrees of freedom for the variance, since $\bar{x}$ has been used to calculate the variance. Calculation of mean and standard deviation is demonstrated in Example 1 (p. 253). If the true standard deviation of the population has been calculated from all the data in the population, the Greek sigma, σ, is the symbol for standard deviation. The letter s (or sometimes *SD*) is used when a subset of the population has been used to calculate the standard deviation.

The coefficient of variation, %CV, is simply 100% times the quotient of the standard deviation divided by the mean. The CV is given in units of percentage.

$$\%\text{CV} = 100\%\frac{s}{\bar{x}} \qquad \textbf{\textit{Eq. 16-3}}$$

CVs are often used to express random variation of analytical methods in units independent of analytical methodology. It is often assumed that the CV of an analytical method is independent of concentration; unfortunately this is not the case.

Confidence intervals

If a population is normally distributed, it is completely described by two statistics, the arithmetic mean and the standard deviation. Confidence in the validity of the description increases as N increases. Statistics that assume a gaussian distribution of a population are often called *parametric statistics*. Fig. 16-5 shows the probability distribution for a population described by the gaussian curve. One can see that 68.2% of the observations are within $\pm$ 1.0 s of the mean, 95.5% are within $\pm$ 2.0 s, and 99.7% are within $\pm$ 3.0 s. These intervals containing a stated portion of the data are called *confidence intervals*. They are the basis of statistical quality control "rules" for acceptance and rejection decisions concerning analytical runs. Suppose the same specimen from Example 1 is analyzed for glucose with every batch of patient specimens. Its mean and standard deviation for glucose are well known. Nearly all the data, 99.7%, are within $\pm$ 3 s of the mean (that is, between 875 and 1120 mg/L), so there is only a 0.3% probability of obtaining a result more than $\pm$ 3 s from the mean because of random chance alone. Any result obtained this far from the mean is extremely suspect. Such a result is not likely to be from the same population as the previous data and indicates that the method's performance has changed.

If the measurement of the glucose concentration in Example 1 is inexact, so is the mean of a group of measurements on the specimen, although the mean is more precise than a single measurement. If several means are obtained from different groups of measurements of the same specimen, the individual means are distributed about the grand mean. The random variation in the population of means is described by the standard of error of the mean, $s_{\bar{x}}$, given by Equation 16-4.

$$s_{\bar{x}} = \frac{s}{\sqrt{N}} \qquad \textbf{\textit{Eq. 16-4}}$$

Confidence limits for means of groups of measurements may be set up about the mean using $s_{\bar{x}}$ in a manner analogous to those already described. In the glucose example, if three measurements were made each time, then:

$$s_{\bar{x}} = 40.9/1.732 = 23.6 \text{ mg/L}$$

Thus 99.7% of the means of triplicate determinations of glucose in the specimen will be within 3 $\times$ 23.6 mg/L of the grand mean, or between 926.7 mg/L and 1068.3 mg/L.

The true mean glucose concentration of the example patient specimen cannot be known exactly unless an infinite number of measurements are made. However, one can use the standard error of the experimentally determined mean to develop a confidence interval that has a known probability of including the true mean. The interval is described by Equation 16-5, where μ is the truer mean.

$$\mu = \bar{x} \pm t \cdot s_{\bar{x}} \qquad \textbf{\textit{Eq. 16-5}}$$

Table 16-1 Critical values of *t* for selected probabilities (*p*) and degrees of freedom (*df*)

	Two-sided intervals or tests		
	$p = 0.10$	$p = 0.05$	$p = 0.01$
	One-sided limits or tests		
df	$p = 0.05$	$p = 0.025$	$p = 0.005$
1	6.31	12.70	63.70
2	2.92	4.30	9.92
3	2.35	3.18	5.84
4	2.13	2.78	4.60
5	2.01	2.57	4.03
6	1.94	2.45	3.71
7	1.89	2.36	3.50
8	1.86	2.31	3.36
9	1.83	2.26	3.25
10	1.81	2.23	3.17
12	1.78	2.18	3.05
14	1.76	2.14	2.98
16	1.75	2.12	2.92
18	1.73	2.10	2.88
20	1.72	2.09	2.85
30	1.70	2.04	2.75
40	1.68	2.02	2.70
60	1.67	2.00	2.66
120	1.66	1.98	2.62
∞	1.64	1.96	2.58

Condensed from Davies, OL, and Goldsmith, PL: Statistical methods in research and production, ed 4, New York, 1972, Longman, Inc.

The *t* value is taken from a *t* table (Table 16-1), and depends on the number of degrees of freedom (one less than the sample number, or $N - 1$) and the desired probability, *p*, that the true mean is outside the confidence interval because of chance alone. Thus $p = 0.05$ implies a 95% confidence $[100(1 - p)\%]$ that the interval includes the true mean. The *t* values describe the same probability distribution as is shown in Fig. 16-5. The difference is that in Fig. 16-5 one assumes that the true population mean and standard deviation are known. The *t* table makes allowances for the decreasing confidence in the estimated values of these parameters as *N* decreases. Because a confidence interval is being constructed to include the true mean, the 0.05 probability that the true mean is beyond the calculated limits must be spread over both ends or tails of the distribution (Fig. 16-6). Thus a two-sided, $p = 0.05$, *t* value is used. If one is only interested in stating that the true mean is greater than some limit, a one-sided *t* value would be used. The *t* value for a two-sided interval at $p = 0.05$ is the same as the *t* value for a one-sided limit at $p = 0.025$.

As an example, if a specimen's glucose was measured five times, and the mean and standard deviation of the pentuplicate measurements were 762.7 and 20.3 mg/L, respectively, there is a 95% probability that the true mean is within the interval

$$\mu = 762.7 \pm 2.78 \left[\frac{20.3}{\sqrt{5}}\right]$$

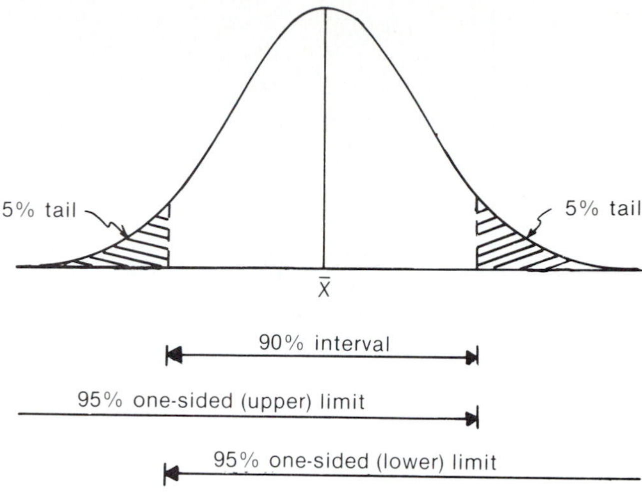

Fig. 16-6 One-sided versus two-sided *t* values.

or μ is between 737.5 and 787.9 mg/L. There is a 95% probability that the true mean exceeds

$$\mu = 762.7 - 2.13 \left[\frac{20.3}{\sqrt{5}}\right]$$

or μ exceeds 743.4 mg/L.

STATISTICAL COMPARISONS OF POPULATIONS

One can use the information in the previous section to define certain analytical parameters used in clinical chemistry. These are the terms accuracy and precision. *Accuracy* is used to describe the ability of an analytical method, after replicate analyses, to obtain the "true" or correct result. The closer the mean of *N* replicate analyses comes to the "true" result, the more accurate a method is. *Precision* defines the reproducibility of a method. After replicate analyses, the narrower the distribution of results, the more precise a method will be.

The definitions of accuracy and precision are demonstrated in Figs. 16-7 and 16-8. Fig. 16-7 shows the results of replicate analyses on an identical sample by three different methods. All three methods have the same mean value and thus the same relative degree of accuracy. However, the distributions of the results by each method differ from each other, with method A having the narrowest distribution (standard deviation) and thus the best precision, whereas method C has the widest distribution of results and the poorest precision. Fig. 16-8 shows two different methods whose distribution of replicate analyses (that is, precisions) are equal but whose means (that is, relative accuracies) are *not* equal. The different means indicate a nonrandom bias between the methods.

Thus the concepts of accuracy (or bias) and precision are independent of each other. This can be conceptualized

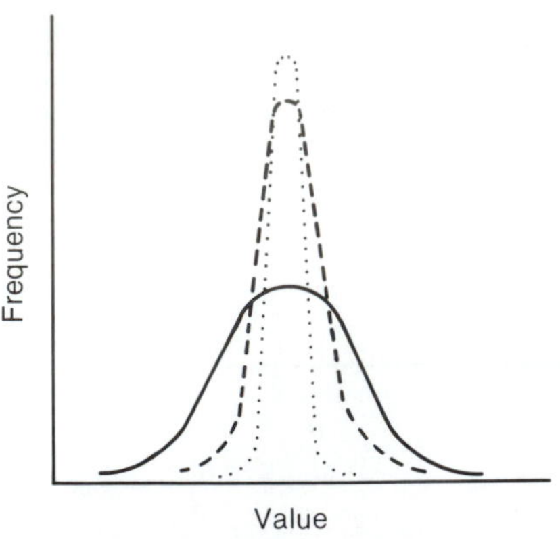

Fig. 16-7 Frequency distributions for three methods having same means, but different distributions, σ. *A* (. . . .) has narrowest distribution, whereas distribution of *C* (——) is wider than that of *B* (– – –).

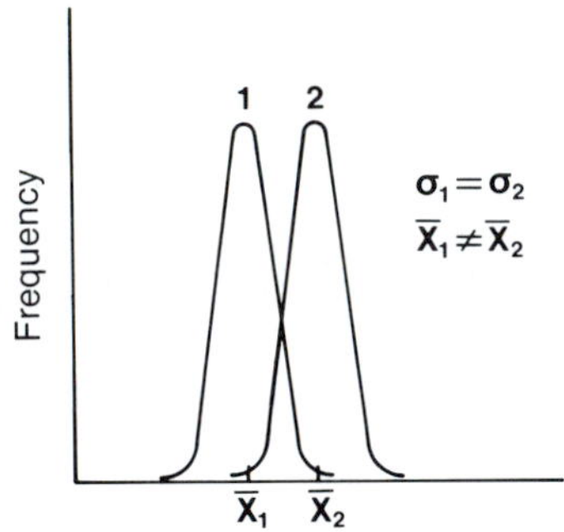

Fig. 16-8 Frequency distributions for replicate analysis by two different methods, *1* and *2*. Both methods are equally precise ($\sigma_1 = \sigma_2$) but are biased from each other ($\bar{X}_1 \neq \bar{X}_2$).

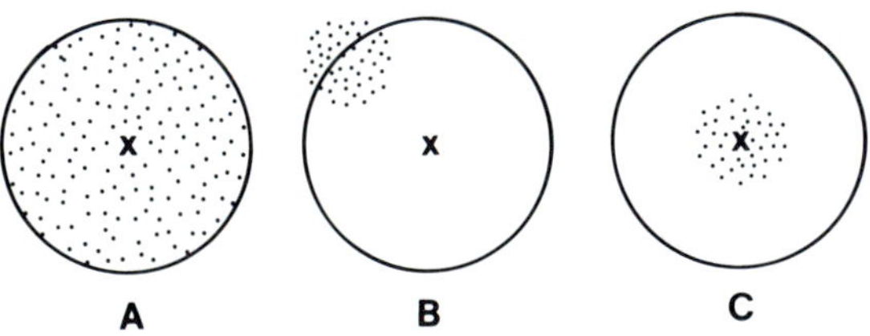

Fig. 16-9 Xs of these targets denote true value for a sample. **A** to **C** denote three replicate analyses by three different methods: **A,** imprecise but accurate; **B,** precise but inaccurate; **C,** accurate and precise.

two-dimensionally by the targets in Fig. 16-9. Repeated attempts to hit the middle of the target ("true" or accurate value) are indicated by dots. A method can be accurate but not very precise, as in Fig. 16-9, *A,* in which the average value falls on the mean of the target. On the other hand, a method can be precise but not very accurate, as shown in Fig. 16-9, *B,* in which the values fall close together but are grouped far from the middle. It is the goal of clinical chemists to design methods that are both accurate and precise (Fig. 16-9, *C*).

When the same specimen is analyzed repeatedly by two different analytical methods, the means and standard deviations calculated for the two methods are almost always different. Unless the specimen is analyzed an infinite number of times by both methods, it is impossible to be absolutely certain that the means or standard deviations really are different. However, one can perform statistical tests on the estimated parameters to determine how likely it is that a significant difference exists between the standard deviations or the means.

Comparison of random variation (precision)

The *F* test is used to determine whether an observed difference in variances is "statistically significant." One performs it by dividing the larger variance (s_l^2) by the smaller variance (s_s^2), as shown in Equation 16-6.

$$F = \frac{s_l^2}{s_s^2} \qquad \textit{Eq. 16-6}$$

The calculated *F* value is compared with a critical *F* value (Table 16-2) to test the *null hypothesis,* which states that there is no difference in the true variances of the two methods. If the calculated *F* value exceeds the critical *F* value for a given probability (such as $p = 0.05$), there is only the given probability, that is, less than 5%, that the calculated *F* value occurred because of chance alone. In this case, one can assume with 95% confidence that there is a statistically significant difference in the variances. Thus one can reject the null hypothesis with 95% confidence and state that the method with the larger estimated variance really is less precise than the method with the smaller estimated variance. If the calculated *F* value does not exceed the critical *F* value, the data do not demonstrate that a real difference exists at the given probability level; that is, there is not a statistically significant difference.

For example, method A has a standard deviation of 30 mg/L ($N = 21$), and method B has a standard deviation of 40 mg/L ($N = 31$). Is method A really more precise?

$$F = \frac{s_l^2}{s_s^2} = \frac{s_B^2}{s_A^2} = \frac{40^2}{30^2} = 1.78$$

From Table 16-2, with 30 degrees of freedom for the numerator and 20 degrees of freedom for the denominator, the critical F value (p = 0.05) is 2.04. The data do not show that the difference in standard deviations is statistically significant.

Comparisons of means (accuracy or bias)

The *t* distribution was used previously to describe intervals that have a stated confidence of containing the true mean. The *t* test is also used to test for statistically significant differences between two experimental means or between an experimental mean and a stated value.

Testing the statistical significance between two means actually involves testing the degree of overlap of their re-

Table 16-2 Critical values of F for $p = 0.05$ and selected degrees of freedom *(df)*

df for denominator	Degrees of freedom for numerator						
	5	10	15	20	30	60	∞
1	230.00	242.00	246.00	248.00	250.00	252.00	254.00
2	19.30	19.40	19.40	19.40	19.50	19.50	19.50
3	9.01	8.79	8.70	8.66	8.62	8.57	8.53
4	6.26	5.96	5.86	5.80	5.75	5.69	5.63
5	5.05	4.74	4.62	4.56	4.50	4.43	4.36
6	4.39	4.06	3.94	3.87	3.81	3.74	3.67
7	3.97	3.64	3.51	3.44	3.38	3.30	3.23
8	3.69	3.35	3.22	3.15	3.08	3.01	2.93
9	3.48	3.14	3.01	2.94	2.86	2.79	2.71
10	3.33	2.98	2.85	2.77	2.70	2.62	2.54
15	2.90	2.54	2.40	2.33	2.25	2.16	2.07
20	2.71	2.35	2.20	2.12	2.04	1.95	1.84
30	2.53	2.16	2.01	1.93	1.84	1.74	1.62
60	2.37	1.99	1.84	1.75	1.65	1.53	1.39
∞	2.21	1.83	1.67	1.57	1.46	1.32	1.00

Modified from Barnett, RN: Clinical laboratory statistics, ed 2, Boston, 1979, Little, Brown & Co.

spective probability distributions. If there is little or no overlap, the populations are demonstrated to be different. If there is significant overlap, one cannot be sure that there is any difference. (For further discussion of hypothesis testing, see references 1 to 5.) A t value is calculated from Equation 16-7, in which subscripts 1 and 2 denote the respective populations.

$$t = \frac{\bar{x}_1 - \bar{x}_2}{\sqrt{\frac{s_1^2}{N_1} + \frac{s_2^2}{N_2}}} \qquad \textit{Eq. 16-7}$$

This t value is compared to the two-sided critical t value for the desired probability level and $(N_1 + N_2 - 2)$ degrees of freedom (if $s_1 \cong s_2$).[1] If the absolute value of the calculated t exceeds the critical t value, the difference between the two means is statistically significant.

For example, assume that a stable control material is included with the patient specimens in each daily batch of glucose measurements. In April the mean was 1110 mg/L, with a standard deviation of 25 mg/L, with $N = 30$. In May the mean was 1090 mg/L, with a standard deviation of 20 mg/L, with $N = 31$. Are the means really different (from two different populations), or can the difference be explained by random variation within a single population?

$$t = \frac{1110 - 1090}{\sqrt{\frac{25^2}{30} + \frac{20^2}{31}}} = 3.44$$

$$df = 30 + 31 - 2 = 59$$

The critical t value for $p = 0.05$ and 59 *df* is 2.00 (see Table 16-1, two-sided interval). Since the calculated t value exceeds the critical t value, one is more than 95% certain of a real difference between the populations.

A special case of comparison of means is the paired-sample t test. It is used to minimize the sources of variation that can cause ambiguity in testing for differences. For example, if glucose method X were being compared to glucose method Y and one compared the means using different random patient specimen populations for each method, the extraneous variation of the populations could mask true differences. To eliminate this variance, the same specimens are analyzed by both methods. This procedure is described in Chapter 19.

Testing the difference between an experimental mean and a stated (given, assigned, known, and so on) value involves testing to see if the stated value is included in the confidence interval around the experimental mean. If it is not, the null hypothesis is rejected, and there appears to be a difference between the stated value and the experimental mean value. Equation 16-8 is used to calculate the t value, in which μ_0 is the stated value.

$$t = \frac{\bar{x} - \mu_0}{s/\sqrt{N}} \qquad \textit{Eq. 16-8}$$

There are $N - 1$ *df*.

Using this example with quality control data for April and May, assume that the glucose concentration in the quality control specimen by the National Bureau of Standards is known to be 1120 mg/L. Is the mean for April (1110 mg/L) significantly different from 1120 mg/L?

$$t = \frac{1110 - 1120}{25/\sqrt{30}} = -2.19$$

The critical t value for $p = 0.05$ and 29 *df* is 2.04. Thus April's mean of 1110 mg/L is statistically significantly different from the assigned glucose concentration of 1120 mg/L.

NONPARAMETRIC STATISTICS

The statistical tests described to this point are termed parametric statistics because they assume a gaussian or other symmetric distribution of the data. Many sampling populations do not meet this criterion, and the analyst needs techniques for describing these populations statistically. This is especially true for distributions of concentrations of some analytes in healthy subjects. The seriousness of this problem depends on the use to be made of the statistics derived from these populations. Statistical procedures involving the dispersion of data in an unknown or nonnormal distribution may require the use of data transformations to yield a normally distributed population (see Chapter 17) or the use of nonparametric statistics.

Nonparametric, or *distribution-free, statistics* require no assumption of the distribution. In this sense, nonparametric statistics may be considered more general. Advantages of nonparametric statistics are that (1) they may be used when the population distribution is unknown, (2) computational procedures are simple and can be performed quickly, and (3) they may be applied to ranked or semiquantitative data and to quantitative data as well.

However, disadvantages in the use of nonparametric statistics are that (1) many data are wasted, and so if the data can be summarized by parametric statistics, this is preferred; (2) some applications may be laborious and tedious for large data sets; and (3) they are less powerful than the corresponding parametric tests. Nonparametric statistical procedures are not commonly used in clinical chemistry. They are described here briefly as an introduction to the concepts.

The simplest nonparametric procedure is to rank the data in order from the lowest (value = 1) to the highest (value = N). The range of the data set is the difference between the lowest and highest value. The middle value (or average of the two central values if N is even) is the median and indicates the central tendency of the data set. A confidence interval of probability, P, may be established for the ranked data by exclusion of the lower $[100(1 - P)/2\%]$ of the data and similarly the upper $[100(1 - P)/2\%]$ of the data. This approach is typically used to determine the limits of reference intervals for those analytes whose distributions are not gaussian. The lower 2.5 percentile and upper 97.5 percentile of the ranked data are usually selected as the lower and upper limits of the reference interval. The central 95% of the data are within the reference interval limits.

A nonparametric test analogous to the t test is the *sign test*, which essentially tests the median rather than the mean of a data set. All the data in a single data set can be compared to some stated (critical) value. Data points higher than the stated value are assigned a plus value (+), lower points are assigned a minus value (−), and zeros are assigned to those values equal to the critical value. The sign test can also be used to compare the results of two methods (A to B). If the B value is higher than A for a given sample, that sample is assigned a plus value. If the B value is less than A, it is assigned a minus value, and zeros are assigned to those samples in which A = B.

If the null hypothesis is true and the median equals the stated value (or method A and method B are equivalent), the probability for plus values is equal to the probability for minus values.

An extension of the sign test is the *Wilcoxon Signed Rank test*. The sign test and the *Wilcoxon Rank Sum Test* are analogous to the parametric paired t test. They are sensitive not only to the sign of the differences but also to the magnitudes of the differences.[5] Sign tests and rank sum tests are described in references 4, 5, and 7.

LINEAR REGRESSION AND CORRELATION

The degree of correlation or association between two variables can be tested by statistical methods. For example, one can determine the degree of correlation between creatinine and pH in urine specimens, or the degree of correlation between measurements of an analyte made by two different methods. Fig. 16-10 and Table 16-3 are examples using blood urea nitrogen (BUN) concentrations obtained from serum samples by an "old" method and a "new" method. The "old" method data are presented as the independent variable, X (that is, the method being used as the accepted or standard method), and the "new" method data are presented as the dependent variable, Y.

Fig. 16-10 appears to disclose a linear relationship between the results obtained by the two methods. It is possible to determine the equation of the line that best fits the

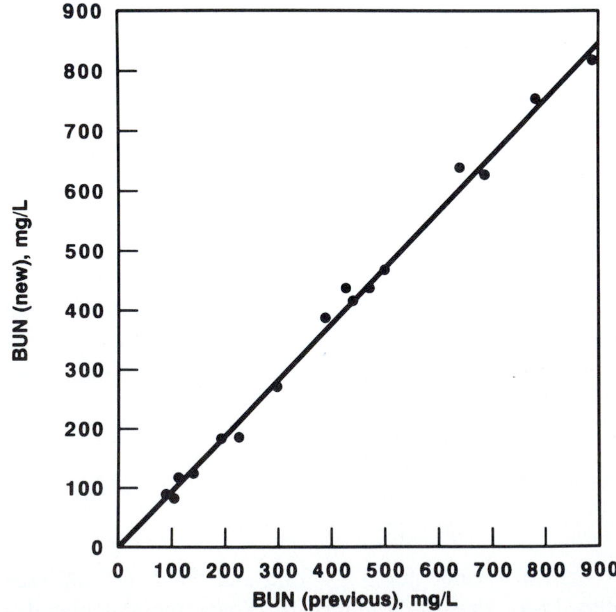

Fig. 16-10 Graph of BUN method comparison data. "New" method *(Y)* versus "old" method *(X)*.

Table 16-3 BUN concentrations (in mg/L) of 16 specimens measured by "old" and "new" methods

Sample number	"Old" (X)	"New" (Y)
1	230	190
2	110	86
3	680	627
4	780	751
5	150	133
6	390	390
7	300	276
8	640	637
9	470	437
10	200	190
11	100	95
12	430	437
13	500	466
14	120	124
15	440	418
16	890	827

data. The general form of this equation, the regression line, is given in equation 16-9:

$$Y = \alpha + \beta x + \epsilon \qquad \textit{Eq. 16-9}$$

where α denotes the true value of the intercept of the line and β, the true value of the slope. The variable x is used to make a prediction of Y and is presumed to be a preset or nonrandom variable. In practice, α and β are unknown and must be estimated as a and b, respectively.

It is unrealistic to assume that every data point will fall exactly on the regression line. A term, ϵ, is included in this model to represent the amount any individual, y, falls off the regression line $\alpha + \beta x$. In practice this term represents all the variability in each measurement caused by unclean glassware, temperature variation, electrical surges, biological variability, nonlinearity, chemical interferences, and so on. In other words, one of the assumptions underlying a regression analysis is that for any fixed, accurately

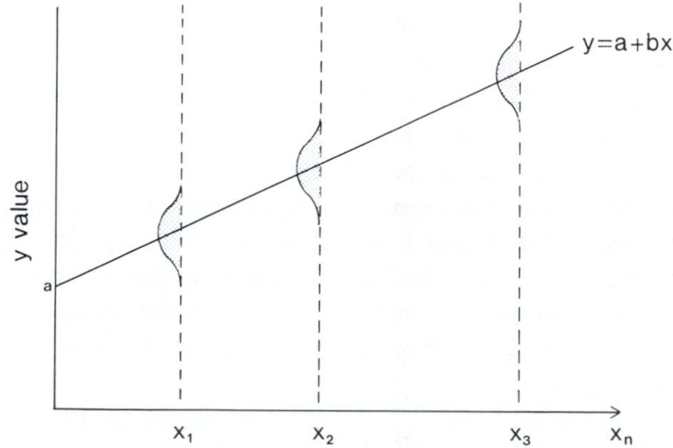

Fig. 16-11 Normal distribution of y values corresponding to a known x value.

known value of x, there is a corresponding normal distribution of y values. This assumption is shown graphically in Fig. 16-11. Notice that the regression line goes through the means of the distributions.

The regression line estimated from limited data is given in Equation 16-10, where Y denotes the predicted mean value of Y for a given value of x.

$$Y = a + bx \qquad \textit{Eq. 16-10}$$

One may determine visually the values for a and b from a scatter plot of the data, aligning a ruler as well as possible along the data. The most rigorous method to determine the regression coefficients is the *method of least squares*. By this approach, the sum of the squares of the difference ($y_i - Y_i$) for each x_i is minimized mathematically, thus the term *least squares*. Formulas for the estimated regression coefficients, a and b, under these conditions are, where N is the number of samples:

$$b = \frac{N\,\Sigma x_i y_i - \Sigma x_i \Sigma y_i}{N\Sigma x_i^2 - (\Sigma x_i)^2} \qquad \textit{Eq. 16-11}$$

and

$$a = \bar{y} - b\bar{x} \qquad \textit{Eq. 16-12}$$

Thus the estimated regression line for the data in Table 16-3 can be written as follows:

$$Y = -4.21 \text{ mg/L} + 0.957x$$

or

$$\text{"New" method} = -4.21 \text{ mg/L} + 0.957 \text{ ("old")}$$

The regression line is graphed in Fig. 16-10. As one can see, very few of the data points fall exactly on the line. As stated before, however, the sum of the squares of these deviations about the regression line is minimized to yield the line of best fit through the data. A measure of the variability of the data points about the line is the estimated standard deviation about the regression line of the differences between the observed and predicted Y values, $s_{y/x}$, otherwise known as the *standard error of the estimate*.

$$s_{y/x} = \sqrt{\frac{\Sigma(y_i - Y_i)^2}{N - 2}} \qquad \textit{Eq. 16-13}$$

The $(N - 2)$ degrees of freedom in the denominator come from the fact that two regression coefficients, a and b, had to be calculated from the data in order to compute Y; thus there are two restrictions placed on the N observations. For the BUN example:

$$s_{y/x} = 17.6 \text{ mg/L}$$

Inference on estimated slope, *b*

A measure of the variability of the estimated slope, b, can be obtained by computation of the standard deviation of b, s_b.

$$s_b = \frac{s_{y/x}}{\sqrt{\Sigma(x_i - \bar{x})^2}} \qquad \textit{Eq. 16-14}$$

In the BUN example, $s_b = 0.018$. It can be shown that a $100(1 - p)\%$ confidence interval for β is as follows:

$$\beta = b \pm t \cdot s_b \qquad \textit{Eq. 16-15}$$

where t is obtained from a two-sided t table for $N - 2$ degrees of freedom and the desired level of significance. Thus a 95% confidence interval for the true slope, β, is given by

$$\beta = 0.957 \pm 2.14\ (0.018)$$

$$\beta = 0.957 \pm 0.039$$

to give a lower limit of 0.918 and an upper limit for the slope of 0.996. The fact that the interval does not bracket the ideal value for the slope of 1.0 implies that the slope of the line is significantly different from 1.0 at the $p = 0.05$ level of significance.

Inference on estimated intercept, *a*

A confidence interval of a can be constructed in a fashion similar to that for b. The estimated standard deviation of a, s_a, is computed as follows:

$$s_a = s_{y/x}\sqrt{\frac{\Sigma x_i^2}{N\ \Sigma(x_i - \bar{x})^2}} \qquad \textit{Eq. 16-16}$$

In the BUN example, $s_a = 8.56$ mg/L. Thus a $100(1 - p)\%$ confidence interval for α is given by:

$$\alpha = a \pm t \cdot s_a \qquad \textit{Eq. 16-17}$$

Substitution of the appropriate values for the BUN example yields an interval from -22.53 to 14.11 mg/L. Since the ideal value for the intercept, zero, is contained within this interval, the true value of the intercept is not significantly different from the ideal value.

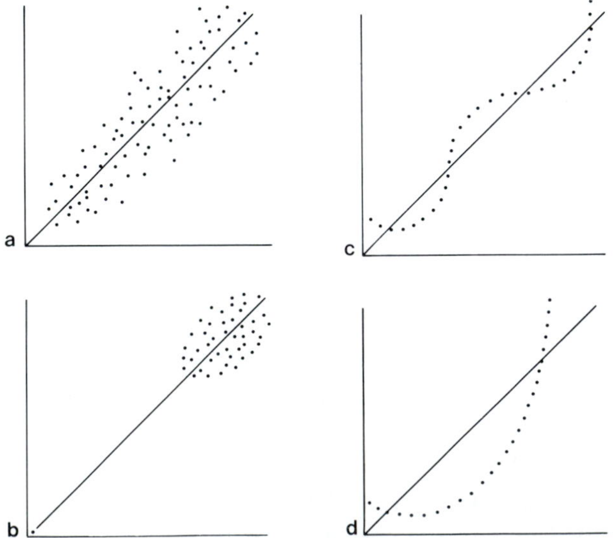

Fig. 16-12 Hypothetical data giving high values for r (estimated Pearson product-moment correlation coefficient).

Correlation and regression

The estimated Pearson product-moment correlation coefficient, r, provides a measure of how closely the data points lie to the estimated regression line. The correlation coefficient can take on value from -1 to $+1$ and will be equal to $+1$ or -1 only if every datum point lies exactly on the regression line. Thus a value of r close to $+1$ or -1 is indicative of a strong (positive or negative) linear relationship between X and Y. The sign of r is the same as the sign of b. The formula for r is given by Equation 16-18.

$$r = \frac{\Sigma[(x_i - \bar{x})(y_i - \bar{y})]}{\sqrt{[\Sigma(x_i - \bar{x})^2][\Sigma(y_i - \bar{y})^2]}} \qquad \textit{Eq. 16-18}$$

Care should be exercised in the interpretation of r. Consider the four hypothetical data sets graphed in Fig. 16-12. Although a high value of r might be obtained for each of the four data sets, only the first one is properly seen as a truly linear relationship between x and y. It is thus recommended that X and Y data always be graphed. In addition, it has been shown that the value of r is sensitive to the scatter of the data and the range of data (see Chapter 19). Although the scatter of the data is inherent in the precision of the method being studied, it is possible to increase the value of r just by extension of the range of data. This is seen in Equation 16-19:

$$r = \frac{1}{s_x}\sqrt{s_x^2 + s_{y/x}^2} \qquad \textit{Eq. 16-19}$$

where s_x is the standard deviation of the x population, indicating the spread in the x data. As s_x becomes very large relative to $s_{y/x}$, r approaches 1.0.

There are statistical approaches available[1,4,7] to determine the significance of the value of r, but the practical meaning of r has in general been somewhat elusive. It appears to be more useful in population studies rather than in comparison of method studies. In any case it is essential to keep in mind that the demonstration of a "relationship" between data sets proves neither cause and effect nor clinically acceptable agreement.

The nonparametric form of the correlation coefficient is the *Spearman rank correlation coefficient*. It tests for the ranking of the X and Y data together. It essentially tests whether the x and y for each datum point are at the same relative positions in their respective distributions. This method ignores their absolute differences from their respective means.[1,7]

In summary, one should be aware of the assumptions made on the properties of data being analyzed by linear regression to avoid inappropriate use and interpretation of this statistical method.

To use linear regression between two data sets, X and Y, it is assumed that:

1. The observed y's are normally distributed to any given x (see Fig. 16-11).
2. The x is known without error.
3. The variance of observed y's for a given x is constant for all x's.
4. The mean y at a given x is a linear function of x.

In general it is not recommended that least-squares analysis be used in situations in which the x value is not known without error. The x values should ideally be predefined or preselected. The application of linear regression analysis to nonpreselected data results in an estimated slope that is biased low (nonaccurate) and an intercept that is biased high. The regression line pivots around the $\bar{x},\bar{y}$ point. It is important to avoid inappropriate application of linear regression to avoid biased results. If the range of data is wide enough, the problem introduced by using imprecise, nonpreselected x values may be minimized. When r is less than 0.99, the regression coefficients may be biased in relation to the true values of the parameters α and β. In such cases other statistical approaches should be used (see Chapter 19).

Prediction of y for a value of x

It is possible to compute a $100(1 - p)\%$ confidence interval for the mean value of Y for a given x (x_0). For example, for samples having a BUN concentration of 270 mg/L by the former method, an approximate confidence interval, CI, for the mean of the corresponding new assay is given in Equation 16-20:

$$\text{CI} = Y(\text{at } x = x_0) \pm t \cdot s_{y/x}\sqrt{\frac{1}{N} + \frac{(x_0 - \bar{x})^2}{\Sigma(x_i - \bar{x})^2}} \qquad \textit{Eq. 16-20}$$

This gives limits extending from 243.3 to 264.9 mg/L for a 95% confidence interval.

A confidence interval for a single assay value (as opposed to the mean of all values) can be computed in a similar fashion by replacement of $1/N$ under the radical with $(1 + 1/N)$. The variability for a single measurement (patient sample) is greater than that for the average; in the BUN example, the 95% confidence interval for a single assay of a sample (x_0 = 270 mg/L) ranges from 214.8 to 293.4 mg/L. A word of caution: If r is less than 0.99 in a method comparison study, the slope will be biased low in relation to the true slope, and the confidence interval as just calculated for Y may not correct for this bias.

Inverse prediction: calibration problem

Consider a situation in which one has fitted a regression line $Y = a + bx$ and is interested in estimating a value, X_0, corresponding to a fixed value, y_0. This situation arises when known corresponding peak heights or areas (y) are measured from a chromatogram. Later a peak height (y_0) is measured for a sample containing an unknown concentration of the analyte. Based on the fitted regression line, it is then possible to obtain an estimate of the corresponding unknown concentration (X_0). One obtains the estimate by solving the equation $y_0 = a + bX_0$ for X_0. Thus Equation 16-9 becomes as follows:

$$X_0 = (y_0 - a)/b \qquad \textit{Eq. 16-21}$$

In the BUN example a y_0 (new method) value of 500 mg/L corresponds to an X_0 value of 526.9 mg/L.

A confidence interval for X_0 can be found in Equations 16-22 and 16-23.[7]

$$\text{CI} = X_0 \pm t \cdot s_{X_0} \qquad \textit{Eq. 16-22}$$

$$\text{CI} = X_0 \pm \frac{t \cdot s_{y/x}}{b}\sqrt{1 + \frac{1}{N} + \frac{(y_0 - \bar{y})^2}{b^2\Sigma(x_i - \bar{x})^2}} \qquad \textit{Eq. 16-23}$$

In the BUN example the 95% confidence interval around an X_0 value of 526.9 mg/L is 486.0 to 567.8 mg/L.

REFERENCES

1. Daniel, WW: Biostatistics: a foundation for analyses in the health sciences, ed 3, New York, 1983, John Wiley & Sons, Inc.
2. Box, GEP, Hunter, WG, and Hunter, JJ: Statistics for experimenters, New York, 1978, John Wiley & Sons, Inc.
3. Draper, N, and Smith, H: Applied regression analysis, ed 2, New York, 1981, John Wiley & Sons, Inc.
4. Davies, OL, and Goldsmith, PL: Statistical methods in research and production, ed 4, New York, 1984, Longman, Inc.
5. Massart, DL, Dijkstra, A, and Kaufman, L: Evaluation and optimization of laboratory methods and analytical procedures, New York, 1978, Elsevier/North Holland, Inc.
6. Barnett, RN: Clinical laboratory statistics, ed 2, Boston, 1979, Little, Brown & Co.
7. Miller, JC, and Miller, JN: Statistics for analytical chemistry, Chichester, West Sussex, England, 1984, Ellis Horwood, Ltd.

CHAPTER 17 Reference values

C. RALPH BUNCHER
DAN WEINER

OBJECTIVES

- Define normality as used in clinical analysis and outline various graphical and statistical techniques for determining data are normally distributed.
- State the purpose of reference ranges and identify primary considerations of the sampling process when establishing a reference range.
- Define clinical sensitivity and specificity, and explain the significance of each in determining reference interval cutoff points.
- Suggest a statistical approach for establishing the reference interval for nongaussian distributions.

KEY TERMS

gaussian A particular symmetrical statistical distribution; also called the *normal distribution*.

log-normal A symmetrical gaussian population distribution obtained by a plot of the logarithm of the data or interval.

log-normal distribution A sample of values with a long tail to the right can often be made to act as a gaussian distribution when one uses the logarithms of values.

negative predictive value The probability that a laboratory result falling within the normal range reflects the true absence of disease; defined as true negatives divided by the sum of true negatives and false negatives.

normal A term with many meanings, including those persons in the nondiseased population and a gaussian distribution (see this chapter for a discussion).

positive predictive value The probability that a laboratory result exceeding the upper limit of normal actually reflects the presence of disease; defined as true positives divided by the sum of true positives and false positives.

predictive value Probability that a laboratory result accurately reflects the true presence or absence of disease. It is dependent on the actual prevalence of the disease.

prevalence The number of persons who have a disease in a given population at any one point in time, or more often the rate of such disease, which is also called the *disease frequency*.

reference range The limits of laboratory values of people without disease; should be further defined by other variables such as gender, age, heritage, and other epidemiological factors.

sensitivity A term used to describe the probability that a laboratory test is positive (that is, greater than the upper limit of normal) in the presence of disease; defined as true positives divided by the sum of true positives and false negatives.

specificity Used to describe the probability that a laboratory test will be negative (that is, within the normal range) in the absence of disease; defined as true negatives divided by the sum of true negatives and false positives.

standard deviation A measure of variability; in the gaussian distribution, two standard deviations above and below the mean encompass the central 95% of the population data, and one standard deviation above and below encompasses 68.3% of the data.

DEFINITION OF NORMAL

This chapter is a discussion of what are considered "normal" values in laboratory medicine and the statistical methods that are used to work with these values. For example, let us assume that some laboratory value, which shall be called the "serum analyte value," was measured. After one has measured the serum analyte value, the next step is to inquire whether this is a "normal" or "abnormal" value. Does the value suggest a healthy person, a condition that can be called normal, or does it indicate that the person's condition is abnormal, or in other words, diseased? If life were simple, the distribution of such values would be as in Fig. 17-1. In this situation each value would be considered to be normal if in the left-hand curve (*A* to *B*) and abnormal if in the right-hand curve (*C* to *D*). If the value were found to be outside the range of normal values, this would be a sign that this person's condition was abnormal. There are almost no laboratory tests that fit this ideal model.

A more common situation is that shown in Fig. 17-2, in which there is some overlap of normal and abnormal values. In this situation there are three choices for each laboratory value: (1) the value may be strictly in the reference range of normal (*E* to *F*), (2) the value may be strictly in the reference range of abnormal (*G* to *H*), or (3) the value may be in the intermediate range (*F* to *G*), in which case only a probability statement can be made classifying that determination as more likely to be normal or more likely to be abnormal. In many cases, as in Fig. 17-3, most results will fall in the area of overlap between the normal and abnormal condition.

At this point, one needs to discuss what is really meant by the word *normal*. Few disease conditions are such that a person's condition can be considered normal at one moment and abnormal at the next. This is especially true of chronic diseases in which persons move slowly from the range of normal into the range of abnormal, often over an interval of many years. Moreover, clinical studies show that the progression is not one of a uniform change from normal to abnormal.

A single laboratory parameter is rarely sufficient to decide whether a person's condition is normal or not. Usually one laboratory test contributes to the diagnosis rather than determines the clinical situation. Moreover, for simplicity, in this discussion only an elevation of serum analyte will show disease, although a lack of a substance or a low concentration often may also indicate disease.

In this text the term *reference range* applies to the values of laboratory tests on analytes, determined on healthy (nondiseased) populations, and is equivalent to the less desirable but frequently used term *normal range*.

The term *normal* has many different meanings in the English language. In one meaning, normal distinguishes a healthy person from the abnormal or unhealthy person. In another meaning, normal means "usual," as in a usual value for a serum cholesterol. Many would say that Americans in general have high levels of serum cholesterol, and in this case even though a value may be considered usual, it is not necessarily the level associated with good health.

In a third meaning, normal is used to describe a set of laboratory results for a test that are distributed in the same manner as a particular bell-shaped statistical distribution (see following discussion), often called the *gaussian distribution*. Although this symmetrical bell-shaped distribution is of great importance in statistics, there is no reason to believe that a particular laboratory analyte has this distribution. In fact, it is clear that most do not. This is another reason for not applying the term *normal* to a set of laboratory values. There are mathematical ways of making the laboratory measurements conform more closely to this distribution, such as use of the logarithm of the concentration rather than the actual measurement. Nevertheless, many statistical determinations depend on this distribution, and despite its name, it is a part of statistical science and should not be confused with the term *healthy*. Some persons consider any symmetrical curve that has a central peak to be a normal distribution. Statisticians reserve this terminology for a particular set of curves with specific mathematical properties.

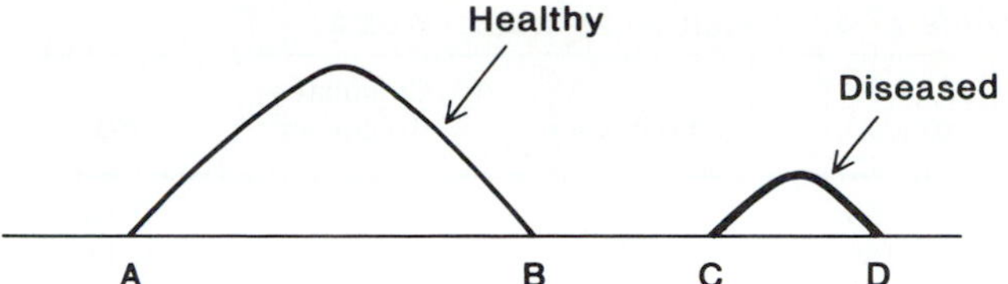

Fig. 17-1 Perfectly separated distributions of healthy and diseased populations. This clear separation does not occur in reality.

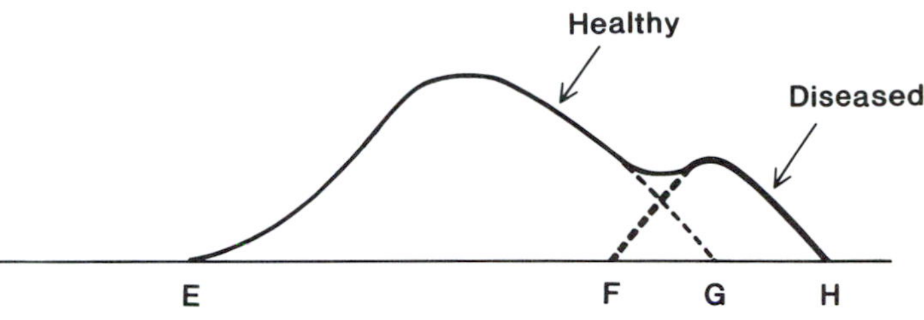

Fig. 17-2 Usual distributions of healthy and diseased populations in which an overlap between the two occur.

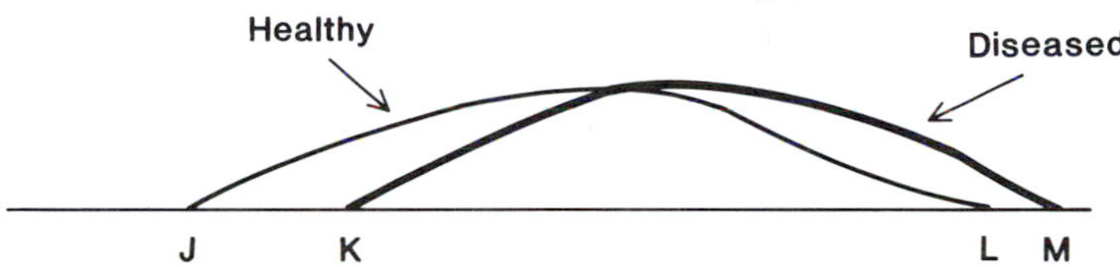

Fig. 17-3 Situation in which degree of overlap between healthy and diseased populations is most common outcome.

ARE DATA NORMALLY DISTRIBUTED?

Probability plot

An important statistical question that must be answered in a clinical laboratory is whether a set of data can be considered to be normally (gaussian) distributed. There are a number of statistical techniques for testing this assumption. One effective and simple method is to graph the data on normal probability paper.

Consider 60 values from parathyroid hormone determinations. It is useful to know whether these data or their logarithms can be considered to be normally distributed. The data are first arranged in increasing numerical order. Then the percentage of the data that are at a given value or less is calculated; this is called the *cumulative proportion of data* and denoted with $F(x)$. The data are shown in Table 17-1.

Next the data are plotted on normal probability paper. This paper is designed and drawn so that data that conform

Table 17-1 Parathyroid hormone data

x(pg/mL)	Frequency	Cumulative frequency	F(x)
≤160	6	6	0.100
180	1	7	0.117
190	1	8	0.133
200	1	9	0.150
230	4	13	0.217
240	1	14	0.233
290	1	15	0.250
320	1	16	0.267
330	3	19	0.317
350	1	20	0.333
360	2	22	0.367
380	5	27	0.450
400	2	29	0.483
410	2	31	0.517
430	1	32	0.533
450	1	33	0.550
470	4	37	0.617
500	1	38	0.633
520	1	39	0.650
530	2	41	0.683
560	1	42	0.700
570	2	44	0.733
580	2	46	0.767
600	1	47	0.783
650	2	49	0.817
670	1	50	0.833
730	3	53	0.883
800	1	54	0.900
920	1	55	0.916
1150	1	56	0.933
1280	1	57	0.950
1370	2	59	0.983
1520	1	60	1.000

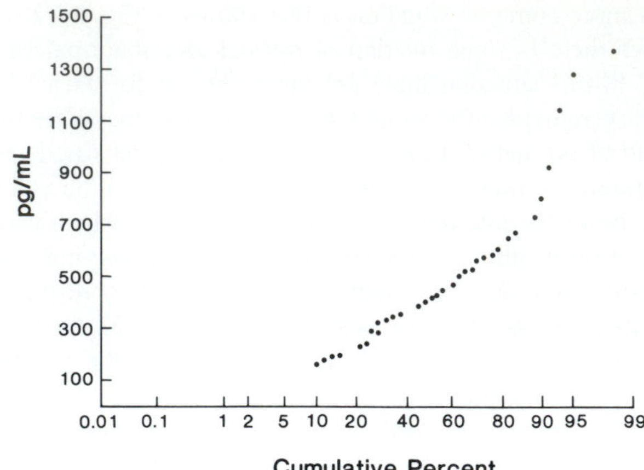

Fig. 17-4 Data for parathyroid hormone (PTH) levels plotted on normal probability paper: concentration of PTH is on *y* axis, and cumulative proportion for each value is on *x* axis.

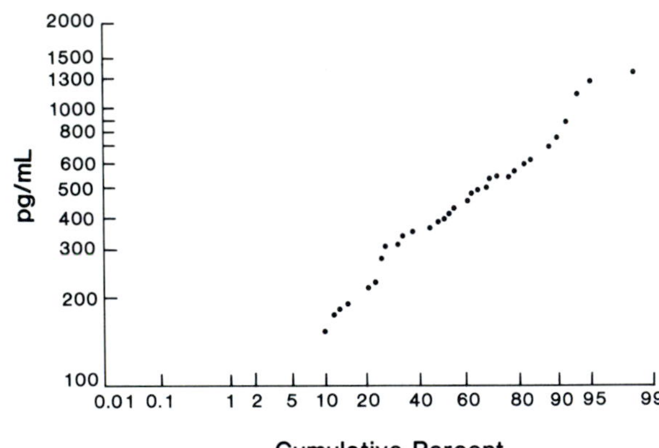

Fig. 17-5 Data for parathyroid hormone (PTH) levels plotted on log-normal probability paper: log of PTH concentration is on *y* axis, and cumulative proportion for each value is on *x* axis.

to the normal distribution will be graphed as a straight line. The horizontal axis is scaled to be a normal (gaussian) distribution, whereas the vertical axis is an ordinary arithmetic scale. We now turn to the data and, for each value of parathyroid hormone, find the point equal to that value (level) on the vertical scale and equal to the cumulative proportion of the whole distribution, *F(x),* on the horizontal scale (plotted as a percentage). These points have been graphed in Fig. 17-4.

A word of caution is necessary at this time. One must be aware that there is neither a 0% nor a 100% on the normal probability paper. Although 99.99% appears to be close to 100%, in actuality, on this type of paper it is not at all close. Therefore one can never plot either a 0% or a 100% point on this type of paper.

One can now look at Fig. 17-4 and decide whether the data appear to be in a straight line. In this instance the data do not lie on a straight line because the points at the higher values bend upward. An alternative would be to take the same data set and graph it on *log-normal* probability paper. On this graph paper the horizontal axis is the same as before, but the vertical axis is drawn on a logarithmic scale. Thus a straight line will indicate a log-normal distribution of the data. This is shown in Fig. 17-5. This simple visual test has shown that these parathyroid hormone data can be easily described by a log-normal statistical distribution. One can then use the mean value and geometric standard deviation to describe the distribution mathematically and to obtain precise percentile values, such as the 2.5 and 97.5 percentiles. These points could be read directly from the graph.

Percentiles

The mean value of a normal distribution is also the 50th percentile, and so the mean value can be read directly from this graph. In the same manner, it is known that from the 16th percentile (approximately) to the 50th percentile is 1 standard deviation, and likewise from the 50th percentile to the 84th percentile is 1 standard deviation. Thus on arithmetic graph paper one can subtract the value corresponding to the 16th percentile from the 84th percentile and divide by 2 to obtain the arithmetic standard deviation. The same procedure can be done with log-normal probability paper, but the standard deviation must be interpreted as a geometric standard deviation. Alternatively one could have subtracted the value corresponding to the 2.5 percentile from the 97.5 percentile and divided this difference by 4 to obtain the standard deviation. If one understands this calculation, it becomes apparent that the scale of the horizontal axis is really in standard deviations.

Occasionally a graph is obtained that appears to consist of two lines joined together, one line in the lower values and a different line in the upper values. This type of result is obtained when a bimodal distribution (two peaks) is graphed and should prompt the investigator to find a way to separate the data into two different normal distributions based on two subpopulations.

Statistical tests for normality

There are numerical tests that one can use to test whether the results from a laboratory determination can be considered to have a gaussian distribution. These include a chi-squared goodness-of-fit test and a Kolmogorov-Smirnov (K-S) nonparametric test of the cumulative distribution.[1] Sometimes, as in our example, a transformation of the data gives a useful fit though the original scale did not. Common transformations include the logarithmic, the square root, and other power functions. A biostatistician can be of help in finding an acceptable transformation.

The *chi-squared* (χ^2) *goodness-of-fit test* is based on the following equation:

$$\chi^2 = \sum_{i=1}^{M} \frac{(O_i - E_i)^2}{E_i}$$

where O_i refers to the observed frequency for the group (bin, cell), E_i is the expected frequency in group i, and M is the total number of groups. If the observed frequencies are in exact agreement with the theoretical distribution, χ^2 will be zero. If the observed frequencies are not in satisfactory agreement with the theoretically expected frequencies, χ^2 will be large. Thus, if the calculated χ^2 is greater than the critical value of χ^2, the null hypothesis (stating that there is no difference between the observed and the theoretical distributions) is rejected. The critical value depends on the degrees of freedom and the level of significance selected.

The chi-squared statistic may be used to test whether a sample distribution fits any type of theoretical distribution, discrete or continuous, including the gaussian distribution.

NUMBER OF SAMPLES NEEDED FOR REFERENCE RANGE

A particularly troubling problem to many people working in clinical laboratories is the question of how many persons must be sampled before a statistically valid reference range is obtained. If this question as stated is brought directly to a statistician, it sounds similar to the question "How high is up?" The training of the statistician requires the laboratory worker to specifiy (1) the size of error that a laboratory is willing to accept, (2) the frequency of various other errors, (3) a difference that is clinically important, and (4) the variability in the measurements. With these pieces of information the statistician can calculate an answer to the question of sample size. Although these types of statistical calculations give mathematically definitive answers, it is possible to provide a simpler response. The classic statistical answer is that the more observations there are, the better. See the following discussion for a specific example.

PROPER SAMPLING OF POPULATION FOR REFERENCE RANGE

When laboratories attempt to establish a reference range, there is a tendency to obtain specimens from persons who are easily available, such as laboratory personnel, nurses, and medical students. These groups are useful in the early stages of testing only. They differ from diseased persons in many ways other than the absence of disease. For example, they tend to be younger, more likely to be more health conscious, and less likely to have other diseases. In addition, these groups may be of a different race than the population to be tested.

A series of recommendations has been developed by the International Federation of Clinical Chemistry (IFCC) on the theory of *reference values*. The first recommendation[2] provides a series of definitions and lists factors to be controlled when one is planning a study to estimate reference values for a procedure. The reference population should be defined in terms of age, gender, race, and possibly genetic background. Other physiological factors that must be defined are phlebotomy specifications (posture, time of day, state of fasting), obesity, smoking, medications, alcohol, and so on.

The clinical situations in which a reference range will be used can affect the actual cutoff point (upper limit of healthy) of a reference range. The first use of a range is to differentiate healthy from nonhealthy persons and determine the specific cause of the illness. The second application of a reference range is the monitoring of patients for progression of disease or for drug therapy after an illness has been identified; in this case the person's laboratory data are compared with previous results. In the case of monitoring a drug level, the renal and hepatic status of the patient can affect the acceptable range of values.

The last purpose of laboratory testing is to determine prognostic or risk factors for a person. A prognostic indicator, such as estrogen receptors for breast cancer, can determine the type of treatment chosen for a patient. In this instance the laboratory data for the person will be compared to a reference range obtained from the population at large, rather than to some specifically defined population.

For most purposes, when the literature suggests that the data or the logarithms of the data have a gaussian distribution, about 40 representative observations will provide sufficient data to obtain reasonable estimates of normal and abnormal levels, even if this does not yield the most precise values for a reference range. If no prior information exists, a larger sample size is essential.[3-5] Clearly a large sample size specifies more precisely the boundary between normal and abnormal. A sample size of 40 can be used by those who want a local reference range and have a limited number of determinations available. A standard reference range brought to the laboratory from outside may not be applicable to local conditions. To determine the reference range if a gaussian distribution is assumed, the mean plus or minus 2 standard deviations can be used to estimate the 2.5 and 97.5 percentiles. These percentiles cover the central 95% of the observations and are usually regarded as the reference range. Alternatively one could obtain nonparametric estimates of these limits. To do this, one ranks the observations and then determines the 2.5 and 97.5 percentiles from these ranked data. Either type of reference range defines ''normal'' as the central 95% of the distribution, automatically excluding 5% (2.5% high and 2.5% low) of the distribution. Thus the implication of either type of reference range is that some ''normal'' but unusual persons will be called ''unhealthy'' because their values are outside the reference range.

One way to test the assumption that 40 samples can give reasonable results is to return to the example of parathyroid hormone data and randomly select 40 values from the 60 values that are presented. One can do this by tossing a die (one of two dice) once for each of the 60 values and using only those results for which the toss results in a 1 or 2 or 3 or 4. This will provide a random sample of about 40 values from the distribution. Then one should reconstruct the two graphs using only the 40 observations. Finally, one can read the mean value and the 2.5 and 97.5 percentiles to see whether they are very different from the results obtained using all 60 observations. Alternatively, one can use a table of random numbers to generalize the sample.

Another test of a reference range based on small numbers of results is provided in Table 17-2. In this table 40 observations are obtained from a known statistical distribution that will have a mean value of 100 and a standard deviation of 20 if the sample size becomes very large. Observations are obtained two at a time, and then the mean value and standard deviation are calculated after each new pair of observations. By the design we know that the main value will be 100 and the standard deviation will be 20, if a very large sample is created. The coefficient of variation will be 20% (= 20/100), and the 95% confidence limits will be 60.8 and 139.2 when the data are numerous. This example shows how rapidly a well-behaved set of data will conform to theory, but it also shows that there will remain some deviation between the theory and the results actually obtained. In this typical situation use of a sample size of 40 was satisfactory but resulted in some error compared to the larger population. The larger the number of observations, the smaller is the deviation from the underlying distribution.

Table 17-2 Random samples from a normal (gaussian) distribution with mean of 100 and standard deviation of 20

First sample	Second sample	Cumulative mean	Cumulative standard deviation
111	94	102.50	12.021
101	92	99.50	8.583
144	141	113.83	23.198
76	70	103.62	27.281
109	100	103.80	24.156
85	103	102.17	22.510
87	105	101.29	21.124
71	101	99.38	21.071
66	121	98.72	22.008
93	103	98.65	20.881
98	67	97.18	20.975
109	97	97.67	20.188
108	80	97.38	19.789
98	103	97.61	19.072
61	96	96.33	19.577
114	86	96.56	19.287
118	103	97.38	19.077
91	97	97.19	18.555
95	106	97.37	18.107
98	110	97.70	17.749

In a real laboratory situation, in addition to the consistent and fixed standard deviation, there are many other sources of variation caused by undefined factors. There are also variations caused by temperature, technician, age of reagents, and characteristics of patients that will change the standard deviation or the mean value in a particular calculation. A small coefficient of variation indicates that the determinations are well-controlled, whereas a larger coefficient of variation suggests the need for a larger sample size.

If the laboratory has many more than 40 points available from which to calculate a reference range, the next suggestion is to try to find out whether there are subgroups in the data. For example, do the men differ from the women in their average values? Do the old differ from the young? Do hypertensive persons differ from normotensive persons? When there are only a small number of observations, it is difficult to discriminate between two subgroups. On the other hand, as the data become more numerous, one can distinguish different subgroups of patients. Similarly, one can compare the determinations carried out by technician A with those done by technician B to see whether there are many differences. One can compare daytime determinations with evening determinations, and so forth.

At this time we should point out that a large number of determinations, such as 200 or more, is very likely to result in some statistically significant differences between subgroups. At this point one must evaluate the practical importance of the differences observed. For example, if men are consistently 10 units higher than women when the average is 100, separate reference ranges will most probably need to be created for men and women. On the other hand, if the difference is only 1%, what does one do? If the sample size is sufficiently large, this 1% difference will eventually be found to be statistically significant. It is a real difference not caused by chance variation, even though very small. Individual variation will be much larger than this average difference between the groups. Therefore this difference is not what we would call "clinically significant," which is to say that there is so much variability anyway that the 1% difference between men and women, even though found consistently, is not of sufficient importance that one should create two separate reference ranges. Rather, in this instance one should use the more simple and efficient procedure of having a single range for all persons. Each laboratory should consult with an appropriate laboratory user to determine what difference can be considered clinically significant.

SENSITIVITY AND SPECIFICITY[6]

Suppose a laboratory has developed a new test for a serum analyte (SA); it may be believed, for example, that diabetic persons will be positive on this test and nondiabetic persons will be negative. Suppose the test is a color reaction, so that only qualitative results are available, either a positive color reaction or a negative reaction. A natural first step is to take a series of diabetic persons and test each for the SA.

In usual terminology the proportion of diabetic (diseased) persons who are positive on this test is called the *sensitivity* of the laboratory test. Using the notation in Table 17-3, this would be given by the fraction $a/(a + c)$. The a represents the number of diseased persons testing positive on this test; the group labeled c consists of persons who are called negative when they should be positive on the test. This group consists of false negatives, and the proportion $c/(a + c)$ is the false-negative rate. Obviously one would want the sensitivity to be as close to 100% as possible.

For comparison one would take a group of healthy persons who are known not to have diabetes and test them on the SA test. The proportion of the nondiabetic persons who are found to be negative on the test, $d/(b + d)$, is the *specificity* of the test. Then there are b false-positive reactions to the test, and $b/(b + d)$ is the false-positive rate. One would want specificity to be as close to 100% as possible.

Table 17-3 Sensitivity and specificity*

	Diseased	Not diseased	Total
Test positive	a	b	$a + b$
Test negative	c	d	$c + d$
Total	$a + c$	$b + d$	N

*Sensitivity = $a/(a + c)$; specificity = $d/(b + d)$.

In general one wishes to have a test that is positive at the earliest possible time when a confirmed diagnosis can lead to useful treatment. Thus a test that was specific for Hodgkin's disease and was diagnostic several months before current tests would be of great value. On the other hand, a perfect blood test that would be diagnostic of all epithelial cancers 3 years before symptoms appeared would produce nothing but frustration because one would still have to wait years to make the specific cancer diagnosis before treatment could be started.

Let us reexamine Table 17-3 to determine how the clinicians will react to the results coming out of the SA test. These results say that $a + b$ persons are positive for the disease, whereas $c + d$ persons are negative. Obviously the clinicians must do additional testing and a clinical workup for the purpose of confirming or denying a diagnosis on the $a + b$ persons who came up as positive in the SA test. An important statistic is the percent *positive accuracy,* or positive predictive value, of the SA tests, which is $a/(a + b)$. If a is much larger than b, the percent positive prediction will be close to 100% and everyone will be happy. On the other hand, if b is larger than a, most persons found to be positive on the test, after an extensive diagnostic workup, will be found to be false positives.

Suppose the SA test is perfected and is tried out on a series of 100 diabetic persons. The clinicians find that 90 of these persons react positively. In another series of 100 other hospital patients confirmed to be free of diabetes, 95 patients are negative on the SA test. Then it can be proudly proclaimed in a medical journal that the sensitivity of the test is 90% and the specificity is 95%, and Table 17-4 can be published. Someone might also calculate that the positive accuracy is 90 (true positives) divided by 95 (all those positive on the test), or 95%, which would imply that the test has excellent sensitivity, accuracy, and specificity. However, there is a basic error in this thinking. In our example half those persons tested were diseased, whereas in ordinary clinical practice perhaps 10% or fewer of the persons undergoing the SA test will be diseased. Table 17-5 shows more reasonable data for this type of test. Maintaining the same sensitivity and specificity, it is now observed that the positive predictive accuracy is two thirds. Recalculating once more shows that if the disease frequency (the prevalence of disease) is only 5%, a major-

Table 17-4 Sensitivity and specificity: disease frequency, 50%

	Diabetic persons	Other persons	Total
Test positive	90	5	95
Test negative	10	95	105
Total	100	100	200

Sensitivity = 90/100 = 90%; specificity = 95/100 = 95%; positive accuracy = 90/95 = 95%.

Table 17-5 Sensitivity and specificity: disease frequency, 10%

	Diabetic persons	Other persons	Total
Test positive	90	45	135
Test negative	10	855	865
Total	100	900	1000

Sensitivity = 90/100 = 90%; specificity = 855/900 = 95%; positive accuracy = 90/135 = 67%.

Table 17-6 Sensitivity and specificity: disease frequency, 5%

	Diabetic persons	Other persons	Total
Test positive	90	95	185
Test negative	10	1805	1815
Total	100	1900	2000

Sensitivity = 90%; specificity = 95%; positive accuracy = 90/185 = 49%.

ity of persons who are found to be positive on the test will be false positives, so the positive predictive accuracy is less than 50% (Table 17-6).

An even more extreme situation exists if one wishes to screen unselected, asymptomatic persons with the SA test rather than selected persons seeking a diagnosis in a hospital. Suppose that one suggests that every person coming into a hospital, or even worse, every person at a local supermarket, be screened with the SA test to see whether they have the disease (diabetes). In this situation let us say that only 1% of the persons tested have the disease. The positive predictive accuracy is then 15%; fully 85% of the people who turn up positive on the SA test will be found after a diagnostic workup to be free from the disease! This example serves to illustrate that the positive predictive accuracy of a test depends on the disease prevalence in the population that is being tested. If the rate of false positives is unacceptable, the specificity of the test must be made greater, for example, more than 98%. Alternatively the test should be applied only to populations in which there is a high prevalence of the disease.

So far we have been discussing only *qualitative tests*

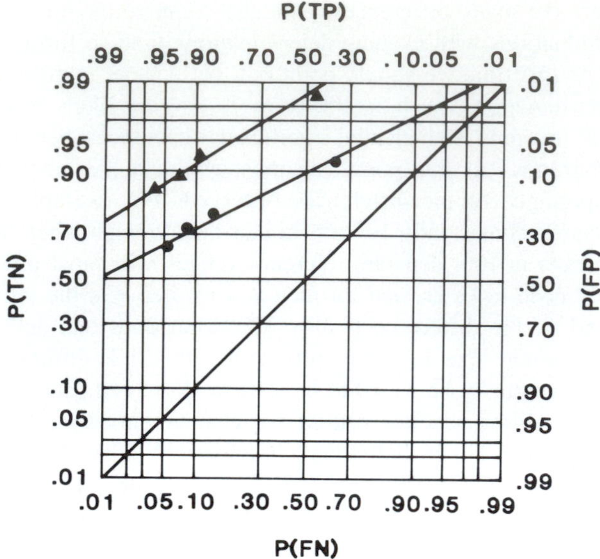

Fig. 17-6 Example of a receiver-operator-characteristic (ROC) curve to determine optimum cutoff point for a test. Horizontal and vertical axes are normal probability scales. Goal is to find a decision point as close to upper left corner as possible. ▲, ROC for more accurate test; ●, ROC for less accurate test; *45-degree diagonal line,* results obtained by coin tossing alone (probability of health, 50%).

rather than those tests that have a numerical measurement, such as quantitative tests. With *quantitative tests* one has the ability to choose different decision points along the quantitative scale. Therefore one can adjust the sensitivity and specificity of a given test. The same is true to a lesser extent with *ordinal tests* (+ versus + + versus + + +; normal-slight-moderate-severe, and so on), in which there is an ordered listing of positivity with three or more levels. By raising the numerical or ordinal level at which one says a person is diseased, one automatically increases the positive predictive accuracy of the test. On the other hand, the price of greater accuracy in the subgroup of those found positive is poorer accuracy in the subgroup of those found negative. Some marginally ill persons will now have laboratory results that fall in the normal interval and will now be declared not diseased.

Choosing the optimum combination of sensitivity and specificity is not easy. Ideally one wishes to have them both as near to 100% as possible. The choice to a great extent depends on the nature of the disease and the nature of the confirmatory diagnostic procedure. When the confirmatory procedures are easy to do, one can accept a greater number of false-positive results. If the confirmatory procedures are more difficult to perform, if they are not conclusive and will have to be repeated, or if they carry an important element of danger, one would like to keep false-positive results to a minimum. Usually one chooses several different levels of a quantitative test and works out the sensitivity and specificity at each of these levels.

One good way of seeing the results of various cutoff levels is to create a *receiver-operator-characteristic (ROC) curve*. An example of such a curve is given in Fig. 17-6. In this type of curve one plots the proportion of false positives (or lack of specificity) against the proportion of true positives (sensitivity). The left vertical axis uses as a scale the probability of obtaining a true-negative $P(\text{TN})$ result. Since the probability of obtaining a false-positive $P(\text{FP})$ result is the complement of the probability of TN, one can also show $P(\text{FP})$ at the same time on the right vertical axis. They are related by the equation

$$1 - P(\text{TN}) = P(\text{FP})$$

The lower horizontal axis is the probability of a false negative (FN), and the upper horizontal axis is the probability of a true positive (TP). They are related by the equation

$$1 - P(\text{FN}) = P(\text{TP})$$

The points (triangles and dots in the figure) for a test are obtained by examination of various levels of the cutoff value that is used to define healthy and diseased persons. Increasing the sensitivity of a test will increase the probability of finding a true positive but will also increase the probability of finding a false positive. The goal is to get as close to the upper left corner as possible. The diagonal shows the results obtained when one tosses a fair coin. Additional information can be found in Swets and Pickett.[7]

REFERENCES

1. Daniel, WW: Biostatistics: a foundation for analysis in the health sciences, ed 4, New York, 1987, John Wiley & Sons, Inc.
2. Grasbeck, R, Siest, G, Wilding P, et al: Provisional recommendation on the theory of reference values (1978). I. The concept of reference values, Clin Chem 25:1506-1508, 1979.
3. Bezemer, PD, Netelenbos, JC, Mulder, C, et al: Determining reference ('normal') limits in medicine: an application, Stat Med 2:191-198, 1983.
4. Dybkaer, R: The theory of reference values. VI. Presentation of observed values related to reference values, Clin Chim Acta 127:441F-448F, 1983.
5. Reed, AH, Henry, RJ, and Mason, WB: Influence of statistical method used on the resulting estimate of normal range, Clin Chem 17:275-284, 1971.
6. Galen, RS, and Gambino, SR: Beyond normality: the predictive value and efficacy of medical diagnoses, New York, 1975, John Wiley & Sons, Inc.
7. Swets, JA, and Pickett, RM: Evaluation of diagnostic systems—methods from signal detection theory, New York, 1982, Academic Press, Inc.

CHAPTER 18

Quality control

BRADLEY E. COPELAND

OBJECTIVES

- Outline an acceptable protocol for initiating new quality control material, including the establishment of an acceptable range.
- State the difference between calibrators and controls, describing the role of each in the clinical laboratory.
- State the purpose of a quality control system in the clinical laboratory and describe the use and intent of the Westgart multirule Shewhart plan.
- Outline the proper procedure for constructing and interpreting a Levy-Jennings plot for ''out of control'' results, trends, and shifts.
- Explain the function of external quality control, contrasting it with use of daily quality control.

KEY TERMS

average difference test A statistical procedure that indicates whether two averages are really different or whether the observed difference is a matter of chance or inherent variability. Also called the *t* test. According to Youden (personal communication), *t* was selected as a nonspecific algebraic term with no acronymic meaning.

control limits Numerical limits within which a control sample must fall to be part of the normal distribution of values, usually target average ± or − 2 USD (usual standard deviation).

external quality control A program in which an external agency provides unknown samples for analysis. The results are submitted for an independent evaluation and are returned to the participant with an evaluation of ''acceptable or ''not acceptable'' performance.

inherent variability The characteristic of repeated measurements on the same material to vary by chance around an average.

internal quality control A program that (1) verifies the validity of laboratory observations and (2) is planned and carried out as part of the daily regular routine within the laboratory.

monthly average The average value of the daily quality control values for a 1-month period.

monthly standard deviation The standard deviation calculated with the daily quality control values for a single month.

out of control The condition in which a quality control sample is outside the numerical control limits. An out-of-control quality control sample must be retested and the validity of the analysis verified.

quality control pool serum A quantity of liquid serum that is preserved in small portions, such as 5 to 10 mL, by freezing or lyophilization and whose constituents remain stable for at least 1 year.

regional quality control A group of 10 to 500 laboratories that jointly purchase a large amount of serum control material so that comparative results can be established.

shift A change in an analytical system that happens abruptly and continues at the new level.

significant difference *Statistically,* a difference that is shown to be outside the expected variability limits; *medically,* a difference that is large enough to influence a medical decision; *operationally,* a statistically significant difference that is not medically significant but one that analysts and supervisors believe to be large enough to require investigation.

standard deviation (SD) An important descriptor of a varying population of measurement data. The average is the central point. Around the average, within ± 1 SD, 67% of the total population fall; within ± 2 SD, 95% of the total population fall; and within ± 3 SD 99.7% (for all practical purposes the total population), fall.

survey specimen A preserved sample similar to a human specimen that is prepared by an independent agency and submitted as an unknown to a group of participating laboratories.

systemic bias *Constant* systematic bias denotes a constant difference between the true value and the observed value, regardless of the concentration level. *Proportional* systemic bias denotes a difference between the true value and the observed value, which changes as the concentration level changes; that is, if the concentration level doubles, the proportional bias doubles. For example, bias at 50 mg = 2 mg; at 100 mg = 4 mg; and at 200 mg = 8 mg.

target average *Temporary*—the average of the 40 initial values collected on a new serum pool. Duplicate values from two vials per day for 10 days. *Final*—the average of the temporary target average, the first month's average, and the second month's average.

trend A gradual change in one direction of the results of repeated analyses of a sample.

usual standard deviation (USD) The average of six monthly standard deviation values. This represents the usual precision capability of an instrument or method.

PURPOSE

The purpose of quality control of analytical testing is to ensure the reliability of each measurement performed on a patient sample. A quality control program is effective if it can be used to ensure that consistently precise and accurate biochemical data have been provided for both short- and long-term medical decision making.[1,2]

Quality control actions should end with decisions based not only on the analytical significance but also on the medical significance of the quality control data. Decisions should be recorded, dated, and signed. The practical decision-making system presented in this chapter has been developed with ease of operation in mind. Additional quality control tools are also presented. Because control samples and patient samples are analyzed by the same system, data from both reflect changes that occur in the system itself. Thus with two sources of data—control samples and patient samples—and several ways of analyzing each set of data, it is possible for the analyst to select an operationally effective combination of quality control tools for each analytical need.

QUALITY CONTROL POOL

Types of material available

The type of quality control material used in the laboratory is based on the laboratory's needs. The daily bench-level work of medical technologists involves determining whether a particular set of patient analyses is valid. Because the quality control material is analyzed along with patient specimens, large amounts (liters) of control material are needed each year. There are currently several sources from which a laboratory can obtain sufficient quantities of quality control material. These are (1) frozen, pooled, patient specimens, (2) commercial lyophilized pool material, and (3) commercial stabilized low-temperature liquid serum pools. Frozen liquid pools have a smaller between-vial variation (as percent coefficient of variation, % CV, 1.3%) than do lyophilized pools (3.0%).[3]

Most often serum or urine pools are used for quality control material, but one can use plasma for specialized purposes. Serum is more frequently used than plasma because it is more readily available and is less likely to have precipitated material. The cost, clarity, stability, validation, and lyophilization error of quality control materials are compared in Table 18-1.

The following statements relate to all quality control serum pools. First, all pooled human material should be free of human immunodeficiency virus (HIV) and hepatitis B virus. The Food and Drug Administration (FDA) requires commercial products to be free of both viruses. Noncommercial frozen pools should not be used if there is evidence of the presence of either virus. Second, all control material requires refrigerator or freezer space for storage of a 1- to 2-year supply. Often commercial distributors will supply quantities on a quarterly basis of a purchased

Table 18-1 Comparison of quality control materials

Criteria	Frozen	Lyophilized	Low-temperature liquid
Cost	Low, if not manipulated* Medium, if manipulated	High	High
Clarity	Clear, if carefully collected	Turbid	Clear
Stability	12 months	18 to 24 months	18 to 24 months
Validation	Compare with accurately measured materials (NBS and CAP†)	Regional and manufacturer's peer group analysis available, or by NBS and CAP	Regional and manufacturer's peer group analysis available
Lyophilization error	Absent	Present	Absent

*That is, if additional analyte is added.
†*NBS*, National Bureau of Standards (now called National Institute of Standards and Technology); *CAP*, College of American Pathologists.

1- or 2-year supply. One would prefer to change control lots only once a year, which is the present practice in most laboratories.

The approach used for urine quality control pools is similar to that used for serum. That is, either pooled, frozen patient specimens or commercially available lyophilized material is most commonly used. For convenience most laboratories use the commercial lyophilized pools.

More than 25 professional groups and manufacturers offer participation in large quality control pool programs in which 100 to 500 laboratories use the same batch of pooled serum. There are both a cost advantage and a scientific advantage when one compares results as a participant in one of these programs.

Target average value of quality control pools

The *target values* of the quality control pool are the estimated concentrations of each analyte within the pool. Each laboratory must establish a target value for each analyte by the regular procedures performed by that laboratory.

When establishing target values for a new lot of quality control material, it is assumed that the analytical systems of the laboratory are performing optimally during the data collection. This is normally ensured by analysis of the new lot of quality control material in parallel with the current quality control program in operation. If the analytical data from the current quality control material indicate satisfactory control performance of the methods, one assumes that the data for the new lot are valid. When setting up a quality control system for the first time, one accepts the current methodology as valid. Future methodology decisions will be based on experience with medical usefulness, significant change limits, and external quality control and accuracy comparisons, as discussed in detail later.

A simple method for establishing target values for quality control pools is described next.

1. Procure a 1-year supply of quality control test material.

2. It is preferable to plan a 6-week lead time to allow for (a) analyses to be run (2 weeks); (b) data to be calculated and evaluated (1 week); (c) methodology decisions to be made (1 week); and (d) 2 weeks for safety because not all planning is perfect. It is also advisable to retain 20 or 30 vials of each expiring pool for use during the subsequent years as reference material for problem solving.

3. Always reconstitute the lyophilized material carefully, following the label directions. Mixing too quickly or too vigorously may interfere with the solubilization of the lyophilized material. This is especially true for enzymes and other proteins that can be denatured and inactivated by such procedures. Date and initial each sample vial on reconstitution. If a frozen liquid pool is used, after thawing, mix six times by inversion because the protein and other compounds become concentrated at the bottom of the tube during freezing.

4. For each constituent, analyze duplicate samples from each of two separate vials for 10 days (20 vials, 40 measurements). An alternative procedure is to reconstitute one vial and analyze in duplicate on each of 20 consecutive days.

5. Calculate average and standard deviation of the 40 analytical values $n = 40$ (Table 18-2).

6. Use the average $\pm$ 2 usual standard deviations as the acceptable control limit for the new lot of quality control material.

7. When a new quality control pool is introduced, the best estimate of the target average value is the preliminary set of 40 values from 20 vials, which are made as already described. This initial average is the *temporary target average (TTA)*. A final target average can be established at the end of the second month. At that time there are three average values (that is, a temporary target value and two monthly averages) that one can average to calculate the final target average (Table 18-2).

Usual standard deviation

Every method has a characteristic inherent variability. The usual standard deviation (USD) is calculated when one averages a series of six monthly standard deviations. This is a valid estimate of the usual day-to-day variability of individual measurements. The USD is used to establish the daily control limits around the *target average (TA)*. The USD is an important performance characteristic of every method (Table 18-2).

The USD has other uses: (1) to establish the statistical significance of the difference between the TA and the monthly average and (2) to establish the statistical significance of the difference between the monthly standard deviation and the USD, as discussed later.

Control limits for each level of control pool

The target average plus and minus two times the usual standard deviation is the control limit for each pool sample. These are 95% confidence limits based on parametric statistics (see Chapter 16).

Introduction of quality control sample into analyte-measuring system

Daily preparation and analysis of quality control samples is a regular responsibility of the analyst. The quality control pools are analyzed as "known" controls by the medical technologists during analysis of patient samples. The frequency of analysis of the quality control material is established by each laboratory for each method. For example, for some large multichannel instruments, controls usually are analyzed every 20 samples. In stable analytical systems the Food and Drug Administration has approved analysis of quality control material once on each 8-hour shift. Automated blood-gas instruments can automatically recalibrate every 60 minutes.

Most laboratories use two different pools, one normal and one abnormal. A *normal* pool contains constituents at concentrations within the nondiseased reference range, whereas an *abnormal* pool contains the analytes at concentrations outside the reference range. For some tests laboratories may employ three pools, low, normal, and high, when medically significant decisions are made at each level. The following discussion assumes that only two pools are used.

The medical technologist must use the data from each quality control analysis to make a decision about the validity of the patient analysis data. This decision should be recorded permanently with date and name of analyst, for example, "in control" or "out of control," plus the initials or signature of the analyst.

Calibration and calibration materials

Controls may not be used as calibrators. Controls and calibrators must be different because each has a separate and important function. Once a material is used to calibrate, it may not be used as a control, and vice versa.

The calibrator has an assigned value that is established by the manufacturer or the user by a reference method. The calibrator is used to standardize the method or instru-

Table 18-2 Example of calculation of temporary target average (TTA), target average (TA), and usual standard deviation (USD) from observed data

Temporary target average (TTA)

Duplicate samples, two vials per day, 10 days; n = 40 measurements of 20 vials per 10 days. The average includes the variability effect of sampling, different vials, and different days. For potassium, mEq/L.

	Vial 1		Vial 2	
Day	**Sample A**	**Sample B**	**Sample A**	**Sample B**
1	6.1	6.1	6.2	5.9
2	6.2	6.2	6.0	6.0
3	5.7	5.8	6.0	6.0
4	5.9	5.8	5.9	5.8
5	6.0	6.0	6.0	6.0
6	5.9	6.0	6.0	6.0
7	5.9	6.0	6.0	6.0
8	5.9	5.8	6.0	5.9
9	6.0	6.1	6.1	6.2
10	6.0	6.1	6.1	6.1

Total, all observations = 239.7 mEq/L
Number of observations = 40

$$\text{Temporary target average} = \frac{\text{Total}}{n} = \frac{239.7}{40} = 5.99 \text{ mEq/L}$$

Final target average

Temporary target average, 5.99 mEq/L
First month, 6.07 mEq/L
Second month, 6.02 mEq/L
Total of averages, 18.75 mEq/L
Number of averages, 3

$$\text{Average of averages} = \frac{\text{Total of avg.}}{\text{Number of avg.}} = \frac{18.75}{3} = 6.05 \text{ mEq/L}$$

This is the *target average,* which replaces the temporary target average in the decision-making process.

Usual standard deviation (USD)

Month	Monthly standard deviation (mEq/L)
April	0.13
May	0.11
June	0.13
July	0.10
August	0.15
September	0.11

Total of monthly SDs, 0.73 mEq/L
Number of monthly SDs, 6

$$\text{Average of monthly SDs, } \frac{\text{Total}}{\text{Number}} = \frac{0.73}{6} = 0.12 \text{ mEq/L}$$

This is the usual standard deviation (USD).

Data from Laboratory Service, Chemistry Section, Veterans Administration Medical Center, Cincinnati.

ment. Differences between an aqueous and serum matrix can affect the ability of a pool to act as a calibrator. These differences can include surface tension, which can affect sample pipetting; the interactions between analytes and proteins; and the effect of the volume fraction occupied by protein, especially lipoproteins, on the actual concentration of certain analytes such as sodium.

Calibrators are usually purchased in lots large enough to last 12 to 18 months. It is recommended that a new lot of calibrator material be tested 6 weeks before use to detect any systematic bias between the calibrator in current use and the new calibrator. Although the FDA requires manufacturers to use reference methods to assign calibrator values, there are frequently significant differences between calibrator lots.

A practical system for new calibrator verification is as follows:

1. Use a 10-day verification period.

2. Each day insert 2 aliquots from one vial of new calibrator as an unknown in the regular daily run (n = 10 vials; 20 values). Calculate the average for each analyte. Compare each average with the value assigned by the manufacturer. This will predict the average change in the quality control pool average value anticipated when the new calibrator is introduced. A change in the quality control pool average greater than 1.0 usual standard deviation is statistically significant, and a decision must be made as to which calibrator value is truly accurate.

3. Each day during this 10-day period reset the instruments using manufacturer's assigned values for the new calibrator.

4. Run 2 aliquots of each quality control pool and four patient samples (patients n = 40).

5. Calculate the average and standard deviation for each quality control pool level measured with the new calibrator (n = 20 values).

6. Compare these averages with current target average for the pool. This step will predict the average values for the quality control pool analytes using the new calibrator.

7. Using the average difference test *(t)*, compare patient sample data obtained using the new calibrator to the data obtained on the same samples using the current calibrator. Calculate a series of 40 differences and the average difference (n = 40).

8. Calculate the standard deviation of the individual differences and then the standard deviation of the average difference ($SD_{avg\ diff}$ = SD_{diff} divided by the square root of n).

9. If the average of the differences is greater than two times the standard deviation of the average difference, there is a statistically significant difference present between the patient values measured in relation to the two calibrators (t = 2.0 or greater, see Chapter 16).

10. Some significant differences may be large enough to be of medical importance, and some may be of operational significance. There are many statistically significant differences that are neither medically nor operationally important; that is, no action is required.

11. Before beginning the use of a new calibrator, a decision should be made as to whether the manufacturer's value for any analyte should be changed. A change would be justified if use of the new calibrator would change the patient values by a medically significant amount. Changes and data supporting these changes should be recorded, dated, and signed by supervisors.

Keep in mind that the overall goal is to maintain a year-in, year-out consistent level of analytical accuracy. The previously outlined verification procedure will identify possible systematic bias caused by analytical bias in the manufacturer's assay procedure.

DAILY DECISIONS

The medical technologist generates a set of quality control data along with each set of patient analyses.

Daily control problems are particularly likely to occur when new batch lots of solution are introduced and immediately after regular maintenance has been done.

It is preferable to check out new batch lots of commercial reagents as soon as they are received rather than just before they are used. It is also good practice to schedule maintenance so that one can establish a test set of controls and run a few patient samples from a previous batch before the next regular daily run is processed. Often maintenance leads to an out-of-control situation. A record of all solution changes, all instrument repairs, and all maintenance procedures is required to be kept to help in subsequent trouble-shooting or in the planning of maintenance (see later section).

It should be emphasized that the frequency of analysis of calibrators and quality control samples must be tailored to the individual instrument as well as to the experience and need of an individual laboratory. Newer instruments can have calibration stabilities of weeks, whereas others have considerably less stability.

Most quality control decisions are made on a daily basis. For some procedures the controls are tested first. Therefore there is an immediate identification of problems, and no patient samples are quantified until the instrument or system is running in control. For other procedures that are run in a batch process, the controls and unknown sample values are only available at the end of the analytical run. Daily bench-level quality control testing can be used to detect only systematic errors and a decrease in precision and not random errors, which occur unpredictably. Random errors are detected when significantly abnormal results are repeated.

Batch analysis decisions using two controls—1:2 SD rule

1. When controls are within ±2 SD. Technologist decision: approve batch and release analyte results.

2. When both controls are outside ±2 SD. Technologist decision: quarantine all results, check the system, and report, if necessary, to supervisor.
3. When one control is outside ±2 SD and the second control is within ±2 SD. Technologist decision: quarantine all results.

The acceptable control range is based on 95% confidence limits (±2 SD). Thus five out of every 100 times a pool is analyzed, results exceeding the 95% limits will be obtained *even though the instrument is functioning optimally*. To ensure that one out-of-control quality control value is not the result of chance, repeat the analysis of both pools. This approach is most feasible for automated or semiautomated analysis. If the repeated controls are within acceptable limits, continue the analysis by releasing all patient results.

If five consecutive quality control values are outside the 1 SD limit and inside the 2 SD limit on either side of the plus or minus side of the average, study the system for a correctable bias. Results can continue to be reported, but prompt review within 2 or 3 days is recommended.

Batch analysis decisions using multirule Shewhart plan[4]

The multiple Shewhart procedure recommended by Westgard et al.[5,6] is as follows: *Decision:* An analytical run is out of control when (1) one control observation exceeds control limits set at ±3 SD from the average (a 1:3 SD rule; that is, one control value outside the 3 SD limit establishes the out-of-control situation); (2) two consecutive control observations exceed control limits set at ±2 SD from the average (a 2:2 SD rule); (3) one control observation exceeds the +2 SD limit and a second control observation exceeds the −2 SD limits (called the range:4 SD rule); (4) four consecutive control observations deviate by more than ± 1 SD from the mean (4:1 SD).

Comparison of the 1:2 SD rule and the multirule Shewhart plan. The Shewhart rules are logical extensions of the 1:2 SD rule, which is recommended here.

Rule 1 (Shewhart). Accept a control value outside 2 SD and inside 3 SD; 1:2 SD rule does not accept values outside 2 SD.

Rule 2 (Shewhart). Accept one control value outside 2 SD, but stop the acceptance when there is a consecutive value outside 2 SD. 1:2 SD stops the acceptance with one value outside 2 SD and inside 3 SD.

Rule 3. The Shewhart range:4 SD would be handled identically by the 1:2 SD rule because both values are outside 2 SD. This rule emphasizes that wide shifts are occurring and that the procedure should be shut down for revalidation.

Rule 4. The 4:1 SD rule detects shifts appropriately and would be part of a daily or weekly 1:2 SD review, depending on how many batches were processed in a day. The Shewhart rule is superior to the four consecutive values above or below the "average rule" that has been suggested by some as an operational out-of-control guideline. Although statistically valid, the latter rule leads to an excessive number of "no action" decisions.

In summary the four Shewhart rules recommended by Westgard et al.[5,6] and the one sample:2 SD rule recommended here as a basic control decision plan illustrate different but comparable decision pathways. The work of Westgard et al.[5-7] using simulated data to predict (1) the probability of rejecting a valid control value and (2) the probability of detecting a statistically significant change in the average represents an interesting exploration of the advantages and disadvantages of multirule control systems.

Each individual laboratory is responsible for defining its system for quality control decisions and is also responsible for making this system a written part of its laboratory manual.

Continuous analysis decisions

A control specimen may be introduced at intervals according to the tendency of the continuous-analysis instrument to drift or according to other factors deemed important. A set of two controls for every 10 samples is rigorous. A set of two controls for every 20 samples is generally acceptable. A calibrating material is inserted after each tenth (rigorous) or at least each twentieth sample. This allows prompt correction of specific operational drift.

The results of all samples analyzed between two sets of *in-control* quality control samples, often termed a *run*, may be reported to the patient's physician.

Stat analysis

For certain procedures each emergency test performed must have a complete set of controls analyzed. For discontinuous stat testing the frequency of control analysis is not dependent on the number of samples analyzed, but on the time between analytical runs. Depending on the stability of the instrument and the nature of the test, it is a reasonable decision to analyze a control sample every 20 to 60 minutes. Other instruments, such as blood-gas instruments, may be automatically recalibrated every hour. Controls must be measured at least at the beginning of every 8-hour shift.

Technologist's response to out-of-control decisions

Once a decision is made to reject a set of analyses by the previous rules, an effort must be made to determine the cause of the improper analysis. Technologists should follow a rational plan for determining the cause of a system failure resulting in out-of-range quality control data. The technologist should begin a check of instrumental systems:

1. Check for calculation error.
2. Repeat standards and controls.
3. Check mechanical and flow systems of the instrument for proper delivery of reagents and sample.

When the source of the trouble is identified, repeat all samples.

4. Prepare and analyze a new dilution of control material.
5. Check to see if sample clots have interfered with the analysis.
6. Check reagent solutions.
7. Check calibration solutions.

If no obvious source of error is identified by this check of systems, release values near the in-control value. Technologists should attempt to trouble-shoot a system and attempt to get it operational by themselves. However, the time spent trouble-shooting before asking for supervisory help must be kept reasonable. Most laboratories have alternative or backup methods of analysis. Therefore a decision must be made about how much time one should spend trouble-shooting before one uses an alternative system. This decision will vary with the type of analyte, the work flow, and the medical situation requiring the result (stat versus routine). Two extreme examples would be a stat potassium analysis versus a 72-hour fecal fat determination. In the case of potassium a delay of more than 30 minutes in providing a backup stat analysis may affect medical decision making, whereas several days' delay in the fecal fat analysis would not be crucial to patient care. These decision should be made in consultation with the supervisory staff.

Thus there is an immediate decision pattern for daily controls that is followed by the individual analyst. When controls are outside the quality control limits, indicating unacceptable performance, one must activate a second level of decision making, which often includes a laboratory supervisor. Again, whenever possible, a structured pattern of immediate response should be planned before the actual problem occurs.

Every quality control decision should be recorded in a permanent record. This is a good practice because it both provides easy follow-up for supervisors and meets College of American Pathologists (CAP), Joint Committee on Accreditation of Hospitals (JCAH), and Medicare requirements for documented control records. The record should state acceptable quality control limits and include a place in which responses to out-of-range quality control values are noted. These response-to-out-of-control notes should

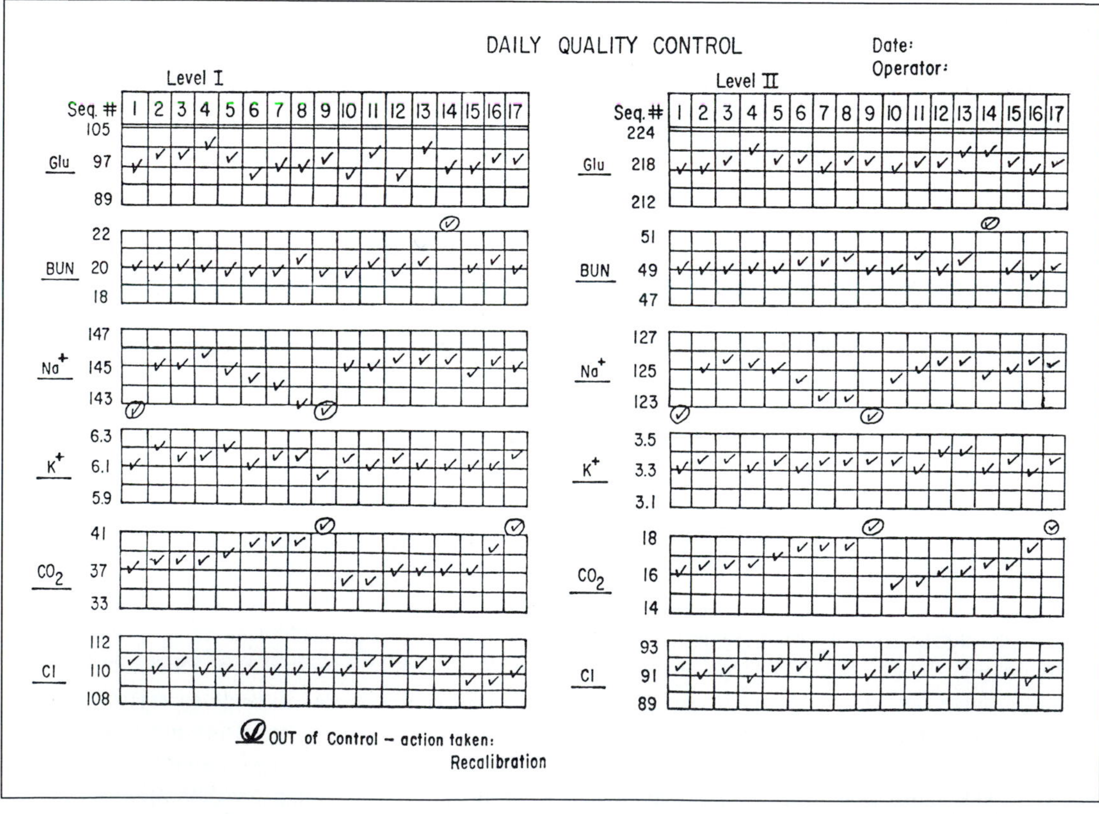

Fig. 18-1 Multiple analyte daily quality control check-off record. As each control is reported, it is quickly logged in on data sheet. Notes of out-of-control values and action taken are included. A daily value for quality control calculation is selected by use of a random number table basis. (Form developed by Rosvoll, RV: In Copeland, BE, Rosvoll, RV, and Casella, JM: Quality control workshop manual, Chicago, 1978, American Society of Clinical Pathologists Commission on Continuing Education.)

include date, analyte, response to correct the problem, and initials of medical technologist[8] (Fig. 18-1). It may be convenient to prepare check-off graphs for each of the multiple constituent levels in an instrument system that provides multiple control values during a single day. This record is useful in predicting the need for maintenance, repair of worn parts, or replacement of deteriorating solutions. Computerized records might obviate the need for written records of daily quality control data, but they will not replace the need for written records of responses to out-of-control problems.

Use of patient data in daily decision making

The results of most patient sample analyses usually fall within the reference (normal) ranges established for each compound. Thus, when one is analyzing a set of patient samples either by continuous or batch operation, the results fall into a familiar pattern, that is, most within the nondiseased-patient reference range and a few elevated and a few low values. Deviations from this pattern of results should alert the medical technologist that a shift in the system may be occurring and that the patient analysis may be invalid. It is unlikely that a consecutive series of two or three patient values will be greatly abnormal. For example, three consecutive patient samples with potassium values above 6 mEq/L should trigger a response to recheck the analyses, even though the controls are within the limits.

An example of this situation is shown in Table 18-3. The data in patient set A show a typical set of results along with acceptable quality control data. In patient set B, however, there is an increased number of abnormal patient results from specimen 4 on, although the quality control data are within acceptable limits. This pattern suggests that the patient samples 4 through 8 are not valid. If sequence B listed in the table is from a continuous operation, one can assume a shift in the calibration at approximately patient sample 4. If the sequence is from a batch operation, one might assume the introduction of some systematic error into the analytical system, such as a pipettor malfunction. In either situation the instrument should be examined by the medical technologist and recalibrated, followed by reanalysis of the patient samples, until the results are consistently repeatable.

Table 18-3 Use of patient data in daily quality control

Sample number	Patient set A	Patient set B
Control I	4.4—in control	4.4—in control
Control II	6.9—in control	6.8—in control
1	3.8	4.1
2	4.6	3.9
3	5.0	5.7
4	4.3	6.1
5	4.2	6.5
6	3.6	5.8
7	4.7	6.4
8	4.0	6.2
9	4.6	5.1
10	3.9	4.7
Control I	4.3—in control	4.6—in control (at 2 SD)
Control II	6.8—in control	7.0—in control (at 2 SD)

Would you wait until the quality control samples after the tenth sample to make a judgment about the system? No. After about the third or fourth patient sample with an extremely high or low value, quality control could be moved ahead and trouble detection should begin. Keep in mind that occasionally by chance a series of specimens from very ill patients may fall in consecutive order. Repeated testing will usually solve this problem.

Daily communication with physicians

When a physician states that he does not believe a particular analytical value, he is exercising an important aspect of his responsibility to the patient, namely, the correlation of all data input into a diagnosis of the patient's condition and plan of treatment. The physician's decision to question the validity of a given measurement is based on his knowledge of the patient's past history, present condition, current therapy, and expected progress. Each laboratory must be ready to confirm any measurement or observation that it has made during the previous week. Each physician must be responsible for immediately alerting the laboratory that a particular value does not fit the clinical picture in his or her opinion. At this point neither the physician nor the laboratory knows whether the values in question are correct or incorrect. A quick confirmation response promotes good patient care and builds confidence in the laboratory.

In the laboratory the most common sources of such problems are errors in data recording, in sample numbering, or in sample labeling. Although not often a laboratory problem, the collection of a sample from the wrong patient does occur. On the other hand, it is my experience that half the questioned reports will be confirmed as valid and will represent specific but unexpected changes in a patient's condition.

WEEKLY REVIEW OF DAY-TO-DAY VALUES

For a weekly review of daily values in the quality control program, it is assumed that the methods of analysis are basically accurate and precise and that each day's quality control data fall within the criteria described previously. However, *short*-term changes in a method can occur, affecting both accuracy and precision. Such variables as new lots of reagent or standards, improper preventive maintenance, and deterioration of instrument, reagents, or standards must be considered as sources of problems. The review of daily quality control data, either on a weekly or monthly basis, is designed to monitor short-term changes in the analytical systems.

Selection of daily quality control value for statistical analysis

How does one select the daily control value for inclusion in long-range data analysis? First, if a single batch is run, there will be only a single value for evaluation. Second, when multiple batches are run or a continuous system is used, there will be multiple quality control analyses. Which value should be selected for the day-to-day analysis? There are two acceptable procedures. A value can be chosen at random from the daily series. One can prepare a random-number series for each month using a random-number table. The standard deviation of these daily values will approximate the inherent day-to-day variability of a single patient analysis. Another acceptable practice is to choose the first quality control analysis of each daily run.

Why select one value from each day? If all the daily results for each day are averaged and the daily average is used to calculate a monthly average and standard deviation, this standard deviation will be the inherent variability of the daily average rather than the inherent variability of a single quality control analysis, which is also the variability of a single patient analysis. Because each patient sample is analyzed only once, the variability of the average of n measurements on the same day is not appropriate as the descriptor of the inherent variability of the single-patient measurement. This explains the preference for the use of a single analysis from each day. This is the estimate of the variability of single measurements on a day-to-day basis. The daily decisions based on analysis of single quality control samples and the daily decisions that the physician makes on the single analysis of the patient sample are related to the variability of the individual measurement on a day-to-day frame of reference.

Alternatively many laboratories average the n values obtained for each quality control pool for each analyte for each day and plot that average on a quality control chart, such as a Levy-Jennings graph. The standard deviation of the analysis obtained in this manner will be less than the standard deviation when one uses only a single, random value. Using this method, narrower control limits are obtained and more statistically significant deviations from the average are observed; however, frequently these are not operationally or medically significant.

Levy-Jennings plots[9]

The data obtained from daily analysis of quality control pools can be plotted to give a visual presentation of the data. The most common visual analysis is the Levy-Jennings plot. The established target average ±2 usual standard deviation is drawn on the y axis, and the days of the month are indicated on the x axis (Fig. 18-2). By using large pieces of paper, one can obtain graphs covering 6 months. Thus cumulative information showing patterns of quality control results can be obtained.

Trends or shifts from the average target value are known as *biases*. Biases can be either positive or negative. Levy-Jennings plots should be observed on a routine basis by supervisory personnel looking for trends or shifts in the data. Normally one notices a trend or a shift within 6 to 7 days after it begins.

An example of a Levy-Jennings plot is shown in Fig. 18-2. If the data slowly deviate up or down from the target value, this is called a *trend* (days 6 to 10, Fig. 18-2). If there is a sudden jump of data points from one average to a new average, this is called a *shift* (days 13 to 20, Fig. 18-2). One can also note changes in the precision of the procedure by observing the scatter or distribution of data points. Thus days 23 to 31 show an obvious decrease in precision, that is, a greater dispersion of data points from that obtained in days 1 to 6. Changes in accuracy and precision can be more formally demonstrated using the procedures described next.

When a systematic bias (that is, accuracy change) or a change of precision is noticed, the supervisory personnel should decide the appropriate action to be taken. This usually includes continuing the analysis while checking reagent, standard, and instrumentation. If the bias becomes severe, so that data points repeatedly exceed the 2 SD

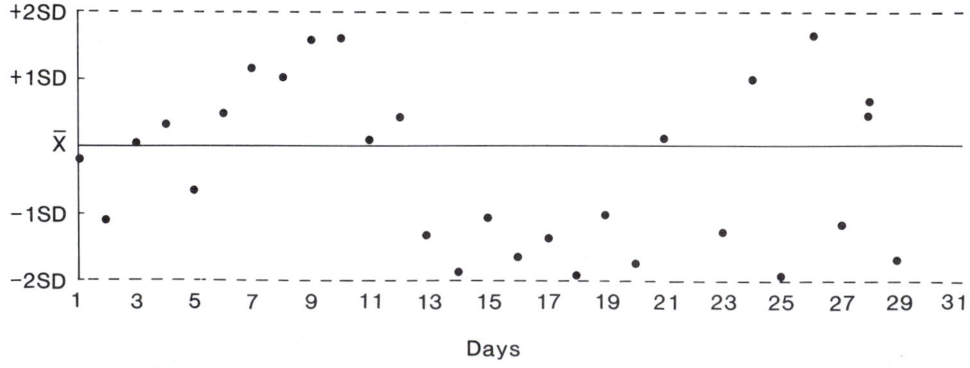

Fig. 18-2 Levy-Jennings plot for recording daily quality control values. Established target average, $\overline{X}$, is given as solid horizontal line. Value of +2 SD and −2 SD are also represented as dashed horizontal lines. Values for each day are plotted sequentially along x axis.

limit, a decision is required to discontinue performing the analysis while seeking the source of the error.

MONTHLY REVIEW OF DAY-TO-DAY VALUES

Before the tenth of the month the average and standard deviations of the past month's daily quality control analyses should be evaluated for change in average from the target average and change in the standard deviation from the usual standard deviation.

Software form for recording monthly quality control data and monthly quality control decisions

Fig. 18-3 is a convenient cumulative form (software) that was developed over 20 years of quality control experience at the New England Baptist Hospital (Boston) and the New England Deaconess Hospital (Boston).[8] Its purpose is to permit easy decision making at the monthly conference and to provide a convenient format for a scanning review of previous months. Two symbols are used in the narrow quality control decision (QCD) columns. A check mark means no statistical difference; an asterisk means there is a statistically significant difference.

When the check mark appears, no added comments are needed. When an asterisk appears, there must be further decisions. These are recorded as follows in the space above the appropriate numbers: *MS*, medically significant; *NMS*, not medically significant; *OS*, operationally significant; *NOS*, not operationally significant; *W*, watch. These decisions are also recorded in a report of the monthly quality control review. In addition, hypotheses about causes of changes and plans to test the hypotheses are recorded. For example, glucose-low bias corresponds to opening a new bottle of calibrator on Feb. 15, 1983. *Action:* Compare a new bottle with a new lot number of calibrator.

An important item to add to the monthly summary is the date when a new lot of calibrator material is introduced.

Method						
Pool						
	TA105[a] mg/dl			4.0[b] USD		
Month	$\bar{X}$	QCD	$\bar{X} - \bar{TA}$	SD	QCD	SD-USD
Initial	105	√		5.0	√	+1.0
Jan.	106	√	+1.0	4.5	√	+0.5
Feb.	104	√	−1.0	3.8	√	−0.2
Mar.	99	*	−6.0[c]	6.3	*	+2.3[d]

Fig. 18-3 Software form for cumulative monthly records, including data and decisions. *a,* Target average (TA) calculated from (1) initial, (2) January, and (3) February. *b,* Usual standard deviation (USD) calculated from previous 6-month experience. *c,* Difference of monthly $\bar{X}$ from TA of greater than 1 USD is statistically significant; it is not medically significant (NMS); it is operationally significant (OS). *d,* Difference greater than 0.5 USD is statistically significant; it is not medically significant (NMS); it is operationally significant (OS). *Decision:* Glucose method needs investigation during April.

Monthly average compared to target average

The monthly average is compared to the target average (TA), which is the best available estimate of the true average value of the quality control sample under the actual conditions of analysis in your laboratory.

The first analysis is to determine if the average for the month changed from the target average. One can do this in the following manner, using Fig. 18-3 as a cumulative record.[8]

Step 1. Record the monthly average in column 1.

Step 2. Subtract the target average from the monthly average. Record the difference in the adjacent column.

Step 3. Make a statistical decision based on the following statistical approximation: *If the difference between the target average and the monthly average is greater than the usual standard deviation, the difference is a statistically significant one at approximately the 5% level of probability*. This means there is a one-twentieth chance of being incorrect in this decision.[10] This is a commonly accepted probability. This decision is based on the average difference test, the *t* test (see earlier discussion).

Necessarily, this decision rule is an approximation.

Decision A. If the difference between the monthly average and the target average is not statistically significant, no action is taken. Place a check mark in column 2, labeled QCD.

Decision B. If the difference is statistically significant, place an asterisk in column 2. In this format it is easy to pick up the significant differences that must be studied. The next action is to make a decision regarding the medical significance of the observed change in the average. In other words, will this change in the observed average cause physicians to misinterpret the tests performed on samples from their patients?

If the answer is yes, there is an immediate emergency situation that must be corrected before further analytical values are sent to physicians.

If the answer is no, place the letters NMS (not medically significant) in the difference box and refer to Decision C.

Decision C. If the difference is statistically significant but is not medically significant, is the difference operationally significant? The persons actually using an instrument can often perceive when the instrument, system, or method is operating properly. At the monthly conference one can arrive at a decision using information from these persons. If the decision is made that the difference is operationally significant, record OS in the difference box. A plan should then be drawn up and carried out to investigate the cause of the observed difference. Since this is not an emergency matter, one can schedule the investigation to be carried out before the next monthly review.

A useful action at this point is to review the daily values of the Levy-Jennings plot for the month under study to see if there is a detectable point at which a shift occurred. Such a point can serve as a focal point for investigating concurrent events such as instrument-maintenance procedures, reagent changes, introduction of new standards, and other possible variables. If a break point is not detectable in the values from the month under study, one may find that consulting the daily values for the previous month may reveal a shift that took place at the end of the previous month. A shift occurring late in the previous month may not be large enough to influence the average value, but when the shift or trend continues, the mean of the following month will be found to be significantly changed.

If the review of the system or instrument shows no obvious reason for the shift, the laboratory may make the decision to ''watch'' the procedure for another month to see whether the change is resolved.

Comparison of monthly standard deviation with usual standard deviation

1. Subtract the usual standard deviation from the current monthly standard deviation. Record in column 5 (SD-USD) of Fig. 18-3.
2. If the difference is greater than one-half the usual standard deviation, the monthly standard deviation is statistically different from the usual standard deviation. The monthly standard deviation decision is based on the following assumptions: $n = 16$ or greater, and the F-ratio test is at a 95% level of certainty.
3. If the difference is not statistically different, place a check mark in the QCD column; if it is statistically different, place an asterisk in the QCD column.
4. At this point the decision can be made about whether the monthly standard deviation has increased to the point at which it is a medically significant change.

One can use the *significant change limit* to evaluate the monthly standard deviation. Table 18-4 lists suggested significant change limits for several analytes. Ask the question: is the significant change limit using the monthly standard deviation useful for medical purposes?

The limits can be estimated by multiplying the USD by 2.8. For example, a usual standard deviation for potassium of 0.07 mEq/L will give a significant change limit of 0.2 mEq/L, which is satisfactory for the required medical decisions. On the other hand, a monthly standard deviation for potassium of 0.2 mEq/L would result in a significant change limit of 0.6 mEq/L. This would require the patient's potassium level to change by more than 0.6 mEq/L before the change in the patient's value would be significant. This would not be satisfactory for medical decision making. The precision needs of physicians are shown in Table 18-4.[2,12] Conservatively a doubling of the precision value may be expected to interfere with medical decision making. Steps to correct a medically significant change in the monthly standard deviation should be taken on an emergency, ''right-now'' basis.

If the decision is that the statistically different monthly standard deviation was not medically significant, an operational decision should be made. An *operationally significant change* requires action, but one can plan it and it need not be taken immediately. One can usually take corrective steps within 10 to 14 days.

Medical importance of a stable usual standard deviation and use of significant change limit

The day-to-day standard deviation of individual measurements is the characteristic frame of reference for medical decisions. The physician assumes that the day-to-day

Table 18-4 Guidelines for usual standard deviation (SD) values

Analyte	Method	Average SD* (normal level)	Estimates of precision needs of physicians (as %CV)*		Significant change limit
			1978[2]	1985[12]	
Glucose	All	31 (500-1000 mg/L)	—	11.2	90 mg/L
Urea	All	6 (100-300 mg/L)	5.0	15.5	20 mg/L
Creatinine	12/60 Technicon	0.7 (9-18 mg/L)	—	13.7	2.1 mg/L
Cholesterol	12/60 Technicon	53 (750-2600 mg/L)	2.0	12.3	150 mg/L
Potassium	12/60 Technicon	0.07 (3.0-4.9 mEq/L)	4.6	4.8	0.2 mEq/L
Sodium	6/60 Technicon	1.3 (133–160 mEq/L)	1.3	2.2	3.9 mEq/L
Calcium	12/60 Technicon	1.6 (70-109 mg/L)	0.6	4.8	4.8 mg/L

Modified from Kurtz, SR, Copeland, BE, and Straumfjord, JV: Am J Clin Pathol 68:463-473, 1977.
*In this table, average SD is equivalent to the usual standard deviation (USD); coefficient of variance (%CV) is SD divided by the mean of the concentration range.

precision is maintained at the same level from month to month and year to year. The physician makes many clinical decisions on the basis of day-to-day differences. The precision of the measurement procedure directly influences the medical interpretation of a day-to-day change. The physician needs to know which changes can be regarded as changes in the patient and which are changes that represent the inherent day-to-day variation in the measurement system.

The *significant change limit* is a decision-making tool that can be given to physicians to help them distinguish day-to-day changes that are caused by the inherent variability of the instruments and methodology, from changes that are caused by changes in the patient. The significant change limit is based on the assumption that the usual standard deviation represents day-to-day method variability. The standard deviation of the difference (SD_{diff}) between two single independent analyses of the same material on different days is related to the usual standard deviation (USD) by the following formula:

$$SD_{diff} = 2(USD)^2 = 1.4 \text{ USD}$$

The 95% confidence limits that define the extent of the inherent method variability are as follows:

$$2\ SD_{diff} = 2.8 \text{ USD}$$

As an approximation, the *significant change limit is three times the usual standard deviation* (Table 18-4).

Changes in patient analyte values smaller than the significant change limit are changes that cannot be differentiated from inherent method variability. Changes greater than the significant change limit represent a real change in the patient. For example, if the usual standard deviation for cholesterol is 50 mg/L, the significant change limit is 150 mg/L. A change from 2000 to 2200 mg/L would exceed the significant change limit and would represent a real change in the patient. A change from 2000 to 1900 mg/L would not exceed the significant change limit and would most likely represent the inherent variability of the method. To facilitate consistent decisions by the attending physician, it is important to maintain a consistent level of precision from month to month and year to year.

Use of patient population average in monthly evaluation

By comparing the results obtained from a large number of patient analyses, one can also evaluate the constancy of a laboratory's analytical values. If the analytical system is constant, the premise is that the distribution of results for a large patient population (most of whom are nondiseased) will be constant. Therefore a statistically significant change in the daily or monthly mean of such a population reflects the introduction of a bias in the method when consecutive comparisons of the average population values are made. This analysis, of course, is invalid if there is a change in the patient population. When this type of data review involves large numbers of data, it usually is performed by computer analysis. A procedure for quality control using the average of patient values can be found in reference 10.

What is acceptable precision?

Often the question is asked, How precise should the measurements be? There are three points to consider: (1) What is the medical use for which these measurements are being made? (2) Will the significant change limit allow the physician to make useful medical decisions as per question 1? and (3) Is the instrument or method being operated with the optimum expertise and the optimum carefulness? If the answers to questions 2 and 3 are yes, the usual standard deviation obtained is valid and performance is acceptable. These two questions emphasize two specific points: first, medical measurements should be medically useful, and, second, each method and instrument should be operated with optimum care.

A summary of the usual standard deviations of frequently performed tests obtained in the university hospitals of the United States is presented in Table 18-4.[12]

As a rule, if a method's day-to-day standard deviation is less than 1.5 times the usual standard deviation indicated in Table 18-4, performance is adequate. If the standard deviation is greater than two times the USD indicated in Table 18-4, the method needs immediate attention. It is specifically recommended that the usual standard deviation for sodium, chloride, and carbon dioxide not exceed 2 mEq/L, and for potassium not exceed 0.2 mEq/L.

It should be understood that in many instances, as the number of persons doing a procedure increases, the standard deviation may also increase. Therefore small laboratories may have an advantage over large laboratories for those tests that are done in a daily batch, since fewer persons contribute to the day-to-day inherent variability. When comparing the standard deviations from different laboratories, never compare the standard deviation of individual values day to day with the standard deviation of a series of daily average values. The standard deviation of a series of averages from multiple daily values will have much less variability than the standard deviation of individual measurements on a day-to-day basis. Compare like with like, or an incorrect decision will be made about the quality of work.

LONG-TERM DECISIONS ON METHODOLOGY RELATED TO ACCURACY AND PRECISION

Consistency goals and methodology evolution

Daily, weekly, monthly, and yearly quality control programs verify that the method in use maintains year-in, year-out consistency, precision, and accuracy. A patient returns to the physician's office or hospital clinic over a period of years. Consistency is important to maintain comparability of measurements so that observed differences represent patient changes and not changes in laboratory systematic bias. Long-range quality control review also allows laboratories to compare current methodology with newly introduced methods.

The technology used in clinical chemistry is not static. New methods of analysis and instrumentation are constantly appearing. The new patient bedside methods[13] and physician office methods[14] are causing considerable discussion[15]; these methods also require strict quality control.[16] After some years of use a new method can be perceived as generally more accurate, more precise, or more cost effective than previous methods and may become accepted as the new state of the art. For example, according to the 1978 CAP quality control summary, 58% of participating laboratories used the diacetylmonoxime method for measurement of urea. In 1988 the percentage of laboratories using that method had decreased to 5%; by then urea was largely being quantitated by enzymatic methods. There have been similar shifts toward different methods for the measurement of glucose, sodium, and potassium. Similarly the precision and sensitivity of many analyte measurements have improved in the past 5 years. Long-range quality control allows a laboratory to compare itself to others and determine if it is using the most accurate and precise methods available.

Although every new method should undergo vigorous validation before introduction, only continual evaluation can determine the long-range validity of a procedure under working conditions. When changing to a new instrument or method, it is important to avoid positive or negative systematic bias changes that may change the reference value ranges.

Another important use of long-term external quality control is to allow the interchangeability of the results from different laboratories if a patient is transferred from one hospital to another. The goal of having a continual and accurate and interchangeable biochemical record of a person can be achieved by laboratories comparing themselves to each other using regional pools and external quality control programs as reference points and by correlation with reference methods and reference laboratories. This goal can be advanced by the increased acceptance of standard procedures as proposed by the NBS, FDA, and NCCLS (National Committee for Clinical Laboratory Standards).

Internal quality control programs for precision and consistency control

The day-to-day internal quality control program is the most useful tool for the maintenance of long-term consistency control and long-term precision control. There are at least 20 to 25 quality control analyses included in each monthly average. This number makes it possible to form statistical decisions that detect small changes before they become operationally significant or medically significant. All regional quality control programs and most commercially available quality control pools offer a computerized reporting system. The monthly reports incorporate the values obtained by all users of the pools. These reports usually list the individual laboratory's monthly mean, standard deviation, and coefficient of variation. Some reports also provide the cumulative mean, standard deviation, and coefficient of variation.

The quality control report is designed so that each laboratory can compare its data to its own peer group, that is, same instrument and method, as well as to other instruments or methods. By comparison of each laboratory's data with peer and other methods, one can detect possible laboratory bias in a method as well as problems in precision.

Medically or operationally significant deviations from the peer mean or standard deviation should be used as an indication of either an improper target value or a true bias in the method. The alerted supervisory staff should look for additional evidence, either continued bias in succeeding months or bias in other external quality control data (see following discussion). A laboratory's standard deviation that is 1.5 times greater than the average standard deviation of a group of regional pool members indicates that the laboratory's procedure has a statistically different precision value.

External quality control programs for accuracy control

To provide independent validation of the internal quality control programs, external surveys have been developed to provide unknown samples for analysis by participating laboratories. Because the analyst has no idea of the target

Table 18-5 External quality control programs

Name	Type of survey	Address
College of American Pathologists	Physician's office General chemistry Enzyme Blood gas Therapeutic drugs Toxicology Instrumentation	7400 North Skokie Blvd. Skokie, IL 60077
American Association of Clinical Chemists	Therapeutic drug monitoring Endocrinology	1725 K St. NW Washington, DC 20006
American Association of Bioanalysis	Chemistry	818 Olive Blvd. Suite 918 St. Louis, MO 63101
Wellmark Diagnostics	Chemistry	650 Woodlawn Rd., West Guelph, Ontario, Canada N1K 1B8

CONSTITUENT UNIT OF MEASURE YOUR LABORATORY'S METHOD/SYSTEM COMPARATIVE METHOD	EVALUATION STATISTICS COMPARATIVE STATISTICS						
	SPECIMEN NUMBER	CODE	YOUR RESULT	MEAN	S.D.	NO. OF LABS	YOUR SDI
CHOLESTEROL (L) MG/DL ENZYMATIC ABBOTT SPECTRUM		14 14	311 266	311.7 262.0	9.6 9.2	174 171	−.1 +.4
ENZYMATIC ALL MULTICON ANALYZERS	C-3 C-4			293.6 248.4	15.8 14.1	4256 4285	+1.1 +1.2

Fig. 18-4 Survey results for cholesterol from the College of American Pathologists (CAP) comprehensive chemistry survey. Survey lists analyte (cholesterol), units of measurement, laboratory's method, and reference method (Enzymatic, All Multicon analyzers). Results of laboratory's analysis of two samples sent by CAP for that survey (C-3 and C-4) are compared to group mean and to reference group.

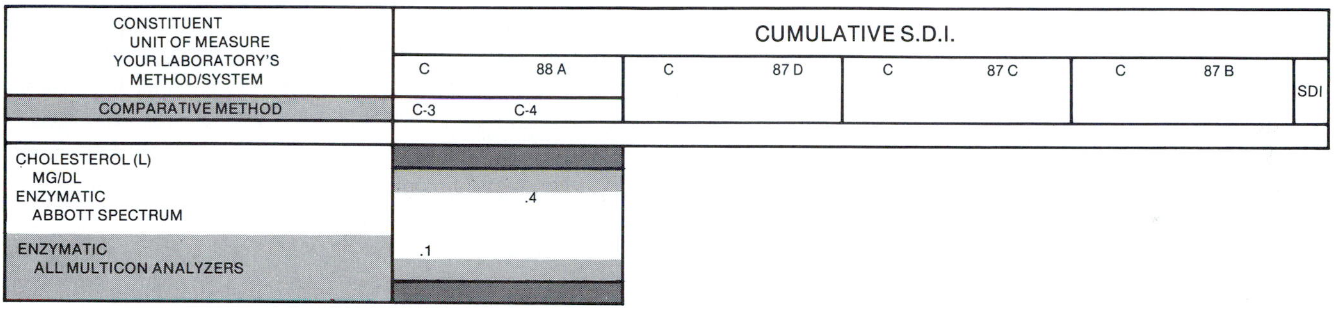

CONSTITUENT UNIT OF MEASURE YOUR LABORATORY'S METHOD/SYSTEM COMPARATIVE METHOD	CUMULATIVE S.D.I.					
	C 88 A C-3	C-4	C 87 D	C 87 C	C 87 B	SDI
CHOLESTEROL (L) MG/DL ENZYMATIC ABBOTT SPECTRUM		.4				
ENZYMATIC ALL MULTICON ANALYZERS	.1					

Fig. 18-5 Graph of CAP survey results from 1 quarter (3 months) for cholesterol. Results are plotted as standard deviation indices (SDIs) from target or average value of other participants using the reporting laboratory's method. Results refer to data in Fig. 18-4, sets C-3 and C-4.

value, operator bias is eliminated. Thus these programs, if properly used, can give a valid estimation of the inherent accuracy of a system. These programs involve the testing of a wide range of analytes, including hormones, drugs, and enzymes, in addition to the usual biochemical analytes in both serum and urine. Some programs that provide laboratories successfully with information on accuracy and precision are listed in Table 18-5.

College of American Pathologists. The CAP Comprehensive Chemistry Survey involves approximately 7000 participating laboratories. This survey covers most common analytes, such as electrolytes, glucose, and urea, and several hormones as well. The results of a typical survey for cholesterol are shown in Fig. 18-4. The report includes the following information: (1) constituent and the reporting laboratory's method and units of measure; (2) comparative method, if available; (3) results submitted by the laboratory ("your result"); these results should have been obtained in a manner identical to that used for patient specimens; that is, the quality control sample should not be analyzed more than once or treated in a more careful fashion; (4) the mean and standard deviation of all participants using a method similar to yours; and (5) your standard-deviation index (SDI), which is calculated when you subtract your result from the group mean and divide it by the group standard deviation. The SDI is thus an indication of distance of the participant value from the group mean value. The SDI is also given in a visual presentation to show the pattern of deviation from the group mean (Fig. 18-5). The previous three survey SDIs are also presented to provide a visual concept of systematic bias development. Information is also provided comparing the submitting laboratory's results to the designated reference method for that analyte. Note that all survey results for the past year are within 1 SDI.

The mean, standard deviation, and coefficient of variation for each method or system are also presented. The summary report for cholesterol for the 1988 sets C-3 and C-4 is shown in Table 18-6. It is important to note the difference in mean and standard deviation values between methods and between the numerous instruments within a method system as well. Each laboratory should compare its results with its closest peer group (same method, same instrument). The difference between the individual laboratory result and average value of the peer or reference group provides an indication of a possible bias. When one

Table 18-6 Example (modified) of an all-participant summary for cholesterol from 1988 College of American Pathologists (CAP) Survey

	Specimen C-3				Specimen C-4			
Cholesterol—mg/dL	**No. of labs**	**Mean**	**SD***	**CV***	**No. of labs**	**Mean**	**SD**	**CV**
All method principles								
All instruments	4485	293.6	16.1	5.5	4515	248.6	14.5	5.8
Enzymatic								
Abbott Spectrum	174	211.7	9.6	3.1	171	262.0	9.2	3.5
Baxter Paramax	220	288.1	11.5	4.0	219	243.3	10.5	4.3
Dupont ACA	482	296.1	8.4	2.9	483	240.0	7.4	3.1
Hitachi 705 (BMD)	207	294.2	9.9	3.4	204	249.5	8.7	3.5
Kodak Ektachem 400, 700	505	282.4	10.4	3.7	502	238.8	8.3	3.5
Technicon RA 1000	250	297.9	11.8	4.0	247	252.6	10.1	4.0
All auto chem instr	4257	293.6	15.8	5.4	4286	248.4	14.1	5.7
All manual chem instr	132	293.0	25.7	8.8	131	251.0	22.2	8.9
All multicon analyzers	4256	293.6	15.8	5.4	4285	248.4	14.1	5.7
Liebermann-Bur w/o ext								
All auto chem instr	41	292.6	18.2	6.2	40	248.4	14.9	6.0
Confirmatory results†		296.7	2.9	—		250.8	2.1	—

**SD,* Standard deviation; *CV,* coefficient of variation.
†Based on CDC referred methods.

is choosing a new method or instrument, this list, showing the standard deviations and average values, may be very helpful in selecting a system that is precise and accurate.

The difference between the mean values of the various groups reflects true methodological biases and not random laboratory error. If a laboratory's result is biased statistically from its peer group, that is, outside ±2 SD, this will be pronounced "not acceptable" performance. When this occurs, the laboratory should review the results for transcription errors,[17] review the quality control for that day, and repeat the analysis on some of the remaining sample or of a follow-up survey-validated sample. The study, conclusions drawn, and decisions made should be recorded on the survey report itself. If repeated survey results for the same constituent show a bias in the nonacceptable range, the method or instrument should be changed. If a peer group is consistently biased from other groups, it is necessary to determine which group is correct.[18] A search of the literature for method comparisons and the use of reference material can often help answer this important question. The CAP survey has been instrumental in clearly showing statistically significant method and instrument biases. Many of these have been corrected by the FDA, Centers for Disease Control (CDC), instrument makers, and reagent suppliers. Proved biases should be corrected by modification of existing procedures or replacement by newer ones. It is important to consider the biases that will affect medical decision making.

There is a difference of opinion among authorities about how frequently external quality control testing should be done. Quarterly (3-month) intervals are considered satisfactory by the secretary of Health and Human Services (HHS), JCAH, CAP, and the American Association of Bioanalysts (AAB).

The quarterly programs conducted by the CAP and the CDC assume that daily, weekly, and monthly internal quality control programs are in regular operation to detect both spurious daily values and shifts and trends of the average. It is also assumed that the external quality control unknown specimens are analyzed when the analytical system is "in control" and therefore is representative of the system.

On the other hand, the American Association for Clinical Chemistry (AACC) Therapeutic Drug Monitoring (TDM) Program (every month) recommends and provides unknown samples more frequently. These programs also assume that daily, weekly, and monthly internal quality control systems are being used (see following section).

In the assessment of accuracy, the regional pool systems conducted by commercial companies and by user groups provide more powerful estimates of interlaboratory accuracy bias (since these comparisons are based on averages of 20 to 25 measurements) than the external monthly or quarterly sample programs, which supply one to three samples and request only one or two measurements per sample. Comparison data with peer groups are published on a monthly or quarterly basis.

In addition, the CAP program provides for purchase of additional lyophilized *survey validated samples,* which are delivered after the survey results have been reported to the individual laboratory. Therefore one can study all methods or instruments that show a systematic bias to identify and correct the source of the bias. This survey-validated serum pool is a very valuable tool for accuracy control.

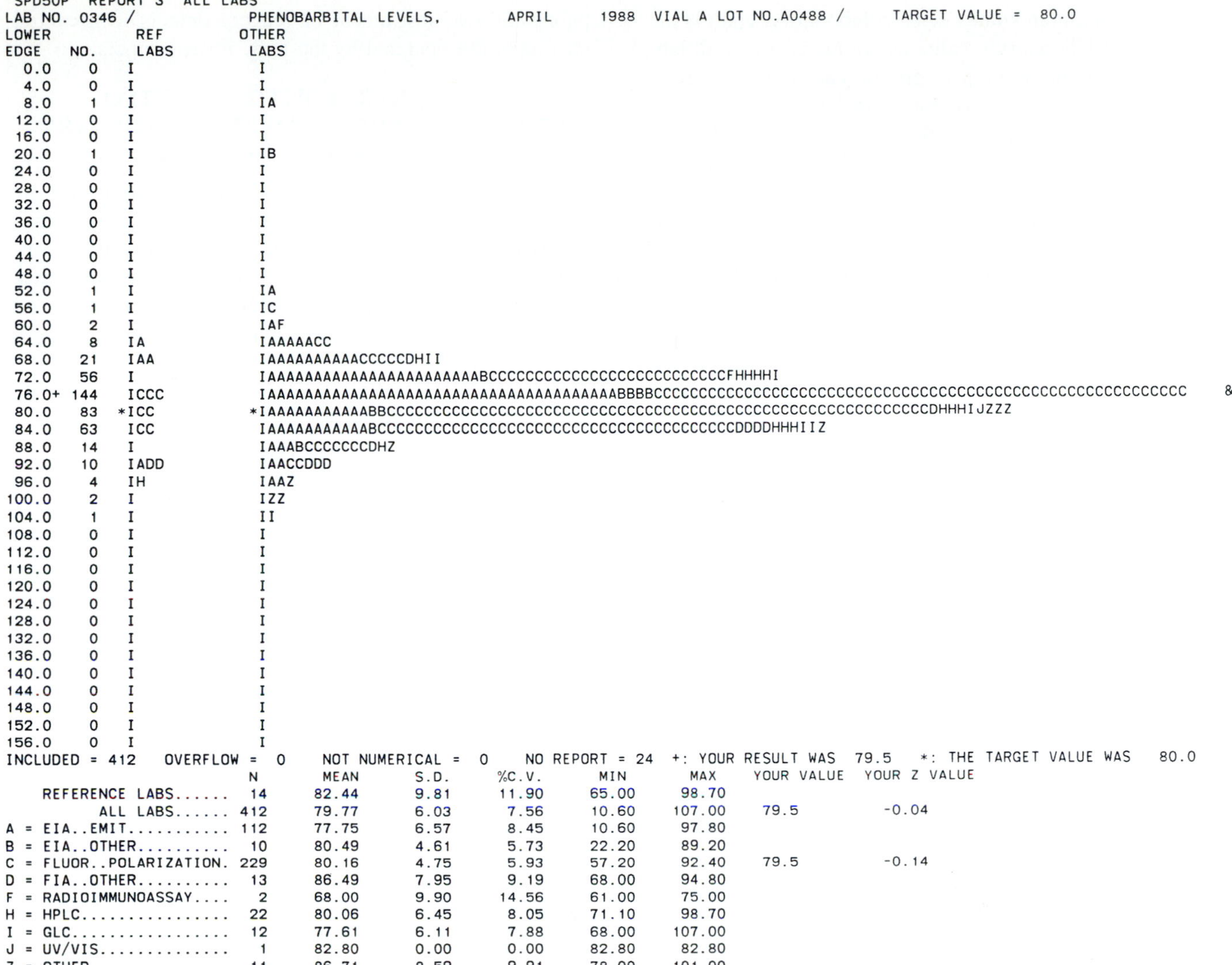

```
SPD50P  REPORT 3  ALL LABS
LAB NO. 0346 /              PHENOBARBITAL LEVELS,       APRIL      1988  VIAL A LOT NO.A0488 /      TARGET VALUE =  80.0
LOWER          REF          OTHER
EDGE    NO.    LABS          LABS
  0.0     0   I              I
  4.0     0   I              I
  8.0     1   I              IA
 12.0     0   I              I
 16.0     0   I              I
 20.0     1   I              IB
 24.0     0   I              I
 28.0     0   I              I
 32.0     0   I              I
 36.0     0   I              I
 40.0     0   I              I
 44.0     0   I              I
 48.0     0   I              I
 52.0     1   I              IA
 56.0     1   I              IC
 60.0     2   I              IAF
 64.0     8   IA             IAAAAACC
 68.0    21   IAA            IAAAAAAAAAAACCCCDHII
 72.0    56   I              IAAAAAAAAAAAAAAAAAAAAAAAABCCCCCCCCCCCCCCCCCCCCCCCCCFHHHHI
 76.0+  144   ICCC           IAAAAAAAAAAAAAAAAAAAAAAAAAAAAAAAAAAAAAAAABBBBCCCCCCCCCCCCCCCCCCCCCCCCCCCCCCCCCCCCCCCCCCCCCCCCCCCCCCCCC     &
 80.0    83  *ICC           *IAAAAAAAAAAABBCCCCCCCCCCCCCCCCCCCCCCCCCCCCCCCCCCCCCCCCCCCCCCCCCCCCCCCCCCCCCCCDHHHIJZZZ
 84.0    63   ICC            IAAAAAAAAAAABCCCCCCCCCCCCCCCCCCCCCCCCCCCCCCCCCCCCCCCDDDDHHHIIZ
 88.0    14   I              IAAABCCCCCCCDHZ
 92.0    10   IADD           IAACCDDD
 96.0     4   IH             IAAZ
100.0     2   I              IZZ
104.0     1   I              II
108.0     0   I              I
112.0     0   I              I
116.0     0   I              I
120.0     0   I              I
124.0     0   I              I
128.0     0   I              I
132.0     0   I              I
136.0     0   I              I
140.0     0   I              I
144.0     0   I              I
148.0     0   I              I
152.0     0   I              I
156.0     0   I              I
INCLUDED = 412   OVERFLOW =  0    NOT NUMERICAL =  0    NO REPORT = 24   +: YOUR RESULT WAS  79.5    *: THE TARGET VALUE WAS   80.0
                              N     MEAN      S.D.     %C.V.      MIN       MAX    YOUR VALUE  YOUR Z VALUE
      REFERENCE LABS......   14    82.44      9.81     11.90     65.00     98.70
            ALL LABS......  412    79.77      6.03      7.56     10.60    107.00    79.5           -0.04
A = EIA..EMIT...........    112    77.75      6.57      8.45     10.60     97.80
B = EIA..OTHER..........     10    80.49      4.61      5.73     22.20     89.20
C = FLUOR..POLARIZATION.    229    80.16      4.75      5.93     57.20     92.40    79.5           -0.14
D = FIA..OTHER..........     13    86.49      7.95      9.19     68.00     94.80
F = RADIOIMMUNOASSAY....      2    68.00      9.90     14.56     61.00     75.00
H = HPLC................     22    80.06      6.45      8.05     71.10     98.70
I = GLC.................     12    77.61      6.11      7.88     68.00    107.00
J = UV/VIS..............      1    82.80      0.00      0.00     82.80     82.80
Z = OTHER...............     11    86.71      8.59      9.91     78.00    101.00
```

Fig. 18-6 Histogram of results of phenobarbital levels from therapeutic drug monitoring (TDM) survey sponsored by American Association for Clinical Chemistry. In this particular survey target value of 79.5 was established when a known amount of drug was added. Values from all reporting laboratories (536) are divided into increments of 4 units (μg/mL), number of laboratories reporting each result, and method used by letter are presented as a frequency histogram. In this particular circumstance, although target value was 80, mean value obtained by all laboratories was 79.8. This laboratory reported a result of 79.5. Since standard deviation was 6.03 for all laboratories, *Z* value was

$$\frac{79.5 - 79.8}{6.03} = -0.04$$

or 0.04 times usual standard deviation. By this laboratory's method (C = fluorometric polarization), however, SD was 4.75; therefore *Z* was

$$\frac{79.5 - 80.10}{4.75} = -0.14$$

which is acceptable, on a comparative basis.

American Association for Clinical Chemistry. The therapeutic drug monitoring (TDM) program sponsored by the AACC is designed specifically to monitor analysis of therapeutic drugs. The TDM program sends out three lyophilized serum specimens 12 times a year. The TDM report includes the following information (Fig. 18-6): (1) presentation of all results for each analyte in a histogram form, which also gives the distribution of data for all instruments and techniques reported; (2) your value and the target value, which is the calculated concentration of the weighed-in drug; (3) the mean, standard deviation, and coefficient of variation of each group, including the minimum and maximum values of each laboratory; and (4) the participating laboratory's value as compared to an all-laboratories value and to its peer group. The statistical deviation of a participating laboratory's result from the overall mean value is designated by the *Z* value, which is similar to the standard-deviation index (SDI) in the CAP survey. In addition, the mean standard deviation and coefficient of variation of reference laboratories and all participants are also presented.

The TDM program is useful because of its large data base (more than 500 participating laboratories) and the frequency of sampling data. This allows rapid accumulation of information covering 12 different time points, which increases the validity of decisions based on an apparent bias. Again the significance of a bias between a participating laboratory and other groups must be determined. Because the TDM program lists an ''accurate weight'' target value, such decisions can be reached rigorously. The AACC quality control program was extended in 1982 to include hormone analysis.

CAP Enzyme Chemistry Survey. Because of the absence of pure preparations of human enzymes, it is difficult to assign absolute values for enzyme concentrations. Measurement of enzymes by their catalytic properties has led to a wide number of assay procedures. This has made it difficult to compare laboratories in terms of accuracy and precision.

A unique approach to this problem has been the CAP Enzyme Chemistry Survey. The participating laboratory receives, on a quarterly basis, a set of three lyophilized samples. This set consists of linearly related pools of serum with increasing amounts of enzymes. Thus laboratories might differ in the absolute values for enzymatic activity, but the relative recovery of activity and the linearity of the recovery should be similar for all laboratories using similar methods and temperatures. This program also gives each participating laboratory an estimate of their short- and long-term precision. This program offers the best means of comparing a laboratory's accuracy and precision for enzyme analysis.

Responsibility for systematic bias

The burden for elimination of systematic bias problems should not rest solely on the shoulders of the user but should also be shared by the manufacturer and distributors of equipment and reagents. Careful perusal of the national surveys in clinical chemistry will quickly reveal instrument systems and reagent systems that show the presence of significant systematic bias in the analyses of the quality control samples. However, Uldall[3] points out the importance of matrix effect on the reliability of analytical results. He indicates that the matrix causing a problem can be in the calibrating standard or in the patient and quality control sample. This information is useful when choosing new equipment. The FDA has developed an excellent reporting system for immediate communication (by a toll-free telephone line) of instrument or reagent defects to a central clearinghouse operated by the U.S. Pharmacopeia.

ACCURACY CONTROL WITH DEFINITIVE METHODS, REFERENCE METHODS, PRIMARY STANDARDS, AND REFERENCE MATERIALS

The control of accuracy is an important ongoing activity that is part of each laboratory's program to maintain a consistent baseline of analytical measurements. Precision control comes first, since accuracy control studies require a knowledge of the inherent variability of the method and the capability of maintaining a consistent average from month to month.

Practical plan for accuracy control

There are several circumstances in which it is necessary to investigate the accuracy of a method: (1) when introducing a new method into the laboratory, (2) when the method in use is questioned because of accumulated external quality control data (see previous section), and (3) when clinicians voice concern about the clinical validity of the results.

There are two approaches to an investigation of the accuracy of a method. The first approach is the parallel analysis of patient samples by two methods: (1) the method under investigation and (2) a reference or definitive method. The reference or definitive method analyses may be performed by another laboratory. Twenty comparisons are sufficient for the initial study. By use of appropriate statistical tests, usually the average difference test (*t* test), one can establish the presence or absence of a statistically significant bias between the methods (see Chapter 19).

The second approach is to analyze a reference material with a known concentration level by the method under investigation. The reference material provides a specific target value for the method under evaluation. Repeated analysis of this material will indicate a bias if one exists. Duplicate values on 2 days with the system in control would be sufficient for calculation of an average value that might indicate a difference (reference value minus the average of the four analyses) greater than 1 USD for the method, suggestive of a true bias.

Definitive and reference methods

A *definitive method*[19] is the scientifically most accurate way to measure a particular chemical substance. This involves instrumentation of the most sophisticated type and a separation procedure that usually purifies the analyte to a greater extent before its concentration is measured. These methods are available in institutions such as the NBS. Because of the complex nature of proteins, peptides, hormones, and enzymes, definitive methods are not available for these analytes at present.

A *reference method* is a method that has been compared to the definitive method and found to give analytical concentration values within ±1% or 2% of the true definite value in the hands of a series of expert laboratories. Thus the reference method has a demonstrated record of transferability and accuracy. The equipment and methodology are such that these methods are usually available in a university hospital–level laboratory. If a definitive method is not available for comparison, the reference method is established by consensus among authorities in the field.

A standard-setting body such as NCCLS or the standards committee of a professional society may make this selection or recommendation, or it may be made by a group of knowledgeable individuals specializing in a given field who form a consensus recommending a reference method.

A *derived* method is any method that has been compared to a reference method and has been shown to give comparable results within a range acceptable to the FDA, which licenses these products, or within a range acceptable to the user. Information on these method comparisons and evaluations are available in the medical literature.

Definitive methods and reference methods are established by groups of interested persons who represent different disciplines and different organizations that include the measurement experts from the NBS, members of methodology and standardization committees of professional organizations, experts from industry, and representative experts from governmental agencies with a legal responsibility, such as the CDC and the FDA. In the United States the NCCLS has been a final meeting place for these expert groups. Publications of the NCCLS represent the best state-of-the-art consensus agreements in the United States. Other similar international consensus organizations are now being formed on the basis of the NCCLS model.

Primary crystalline standards are always required by definitive and reference methods. The NBS provides a number of primary crystalline standards that may be used to prepare primary liquid standards. These include glucose, cholesterol, urea, creatinine, uric acid, sodium, potassium, chloride, phosphorus, calcium, magnesium, lithium, albumin, and hydrogen ion. Aqueous sealed vials prepared from these NBS primary standards are also available from the CAP. When NBS primary standards are used, a method's accuracy may be said to be "traceable to" the NBS primary standard.

Prominent sources for information on reference methods are the U.S. National Bureau of Standards, Analytical Division, Methodology Section, Gaithersburg, Md.; Clinical Chemistry Section, Communicable Disease Center, Atlanta, Ga.; National Committee for Clinical Laboratory Standards, Villanova, Pa.; Standards Committee of the American Association Clinical Chemists; and the Standards Committee of the College of American Pathologists.

Reference materials

There are now available five protein-based, reliable accuracy control materials. Each of these reference materials is useful for investigating the accuracy of a method. Target concentrations assigned by definitive methods are the most accurate values obtainable by state-of-the-art technology and thus are preferable when they are available.

1. National Bureau of Standards Human Serum SRM (909) (Office of Standard Reference Material, Gaithersberg, Md.). The NBS provides a human serum pool assayed by definitive methods for many constituents, including glucose, cholesterol, uric acid, calcium, chloride, potassium, lithium, magnesium, creatinine, sodium, urea, some trace metals, and some enzymes.

In addition, the NBS has also prepared human serum cholesterol reference materials, which have been analyzed by the NBS definitive method.[20] These are the SRM 1951 Cholesterol in Human Serum Frozen (low-, medium-, and high-cholesterol levels) and SRM 1952 Cholesterol in Human Serum Freeze-Dried (low-, medium-, and high-cholesterol levels). These two SRMs can be purchased from the NBS.

2. Seronorm. A commercially available reference pool is sold under the name of Seronorm (Accurate Chemical Company, Westbury, Long Island, N.Y.). This lyophilized serum pool has six constituents assayed by definitive methods and 18 constituents assayed by reference methods. Other analytes are analyzed by derivative methodology. Analyzed values from highly sophisticated laboratories are presented. In addition, separate materials are available for trace metals, proteins, enzymes, lipids, bilirubin, and therapeutic drugs.

3. College of American Pathologists; national survey-validated pools. These pools are from regular CAP Comprehensive Chemistry Surveys and thus have well-documented target values. Since these target values are the mean reported values of all methods, they are not viewed as definitive values but are rather good approximations of the actual concentration. Studies conducted on six vials for seven analytes by the NBS indicate that the definitive method value and the average value of all methods were the same (that is, all within $\pm 2\%$ and a majority within $\pm 1\%$).[21] In addition to the all-methods value, these materials also have estimates for particular methodology, instrument, and reagent combinations. The CAP also has a special collaborative project to provide specific values for serum protein constituents, which has resulted in the CAP-CDC Reference Preparation of Serum Proteins (RPSP). The reference preparation contains albumin, alpha-1-glycoprotein, alpha-1-antitrypsin, alpha-2-macroglobulin, ceruloplasmin, haptoglobin, transferrin, C3 complement, C4 complement, IgG, IgA, and IgM. The assigned values are based on analyses by acknowledged experts in each protein measurement.

4. Some regional pools submit their material for definitive method analysis to the NBS and thus may be used as an accuracy control.

5. In 1987 the CAP released a series of three lyophilized cholesterol reference materials with a pattern of fatty acid esters similar to that found in human serum. The choles-

terol levels in these serum pools have been determined by the NBS method[20] and the CDC definitive method.

Selection of a laboratory for accuracy control assistance

When engaging in an investigation of the accuracy of a method using another laboratory as a reference point, it is important to be completely confident of the quality of its analytical work. One should always obtain information as to the accuracy and precision of the analytical data. It is useful to ask if a definitive or reference method is being used.

How to decide on an accuracy standard

Using the resources just described, a laboratory can choose one or more reference standards. The choice of an accuracy reference point requires an independent decision for each laboratory and for each method. As mentioned previously, within the laboratory consistency from day to day and year to year is the first and most important consideration.

The decision on an accuracy reference point depends on local needs and local opinions. For example, the wishes of the attending physicians should be considered. They may wish to have their values approximate those at a local university hospital or to have results approximate those of a local clinical investigator whose advice is influential in patient care. The opinion of these persons should be included in the discussion of accuracy goals.

YEARLY HEALTHY HUMAN VALUE STUDY

An essential part of a quality control program is the yearly healthy human value study. The average value for each analyte for each year should be compared with the preceding year. If the difference between the average for these 2 years exceeds the usual standard deviation of the procedure, an inaccuracy caused by systematic bias may be found. A practical yearly program is to analyze serum from 12 males and 12 females who are 20 to 30 years old.

Eliminating bias in healthy human values

Each average value and standard deviation includes the inherent variability of the method, as expressed by the usual standard deviation of the method day to day, and also contains any systematic bias present in the system. Therefore it is recommended that the healthy human values study be done each year after about 6 months of experience has been obtained with the quality control pool. In addition, collection of this data should not be done immediately after a new reference calibrator has been introduced, since this is also a time when systematic bias may be introduced. It usually requires 2 or 3 months to be sure that a new calibrator or a new pool is bias free.

Acquisition of a bias-free set of age-related healthy humans

By selecting a series of persons within the same decade (10-year period), one can compare the average values from year to year to detect systemic bias. To provide an under-

Cincinnati University Hospital Clinical Chemistry Laboratory

Maintenance log — IL 813 blood gas Year______

First shift

Required maintenance	Jan	Feb	Mar	Apr	May	June	July	Aug	Sept	Oct	Nov	Dec			
Monthly:															
1. Clean PO_2 cathode and check															
electrolyte solution															
2. Change PO_2 membrane															
Initials of technologist															
Six months:															
1. Change pH tip seal															
2. Change belly band															
Initials of technologist															

Fig. 18-7 Example of preventive maintenance record for a specific instrument.

standing of age-related changes, first review the literature and then compare the decade that the laboratory has studied with the corresponding decade reported in the literature. The difference between the average value of the selected decade and the average reported in the literature for the same decade can be considered to be the systematic bias between the two measurement systems. One may apply this bias to the other decades in the article to obtain the predicted value for different age groups.

Thus, by studying healthy persons in a single decade (I have found the decade from 20 to 30 years of age most convenient), it is possible to make the published literature relevant to the laboratory reference values. Male-female differences can be evaluated by selection of 11 or 12 persons of each sex for the total of 22 or 24 nondiseased persons.

RECORD KEEPING AND SOFTWARE FORM DEVELOPMENT

The best reason for developing convenient and durable forms is to permit a convenient ongoing evaluation of the measurement system. One can study problems and remedy out-of-control situations in a timely fashion. A second reason for good records is found in the requirements of the several national bodies that accredit medical laboratories, such as the Health Care Financing Administration (HCFA) for Medicare, CAP, and JCAH.

These agencies require that daily quality control be run and that records be kept that demonstrate that quality control decisions were made. This requires documentation of quality control decisions on a daily, monthly, and year-end basis. The daily decision is recorded as "in" or "out" of control with signature or initials. With the monthly decisions, a record should be kept and signed by those attending the monthly review conference. The yearly evaluation requires similar documentation.

PREVENTIVE MAINTENANCE

Instrument maintenance principles and documentation

All instrument maintenance should be performed as recommended by the manufacturer's instrument manual, with additional maintenance added by the user as necessary. Any errors in the instrument manual should be brought to the manufacturer's attention and changed with the manufacturer's authorization.

Need to date and sign maintenance records

The maintenance records should be audited once a month for completeness, and instrument down time and actions taken should be noted. The audit should be recorded and signed by those who perform it (Fig. 18-7).

Program of quality control for materials and reagents

All prepared reagents should be dated when received and when put into use. All reagents prepared within the laboratory should have on the label a storage temperature and an outdate period. When using prepared reagents, one should record the lot numbers on the daily work sheet or in a log book so that problems involving reagents can be identified by noting the reagent lot numbers. Reagent lot variation is a frequent source of out-of-control values.

REFERENCES

1. Westgard, JO, and Barry, PL: Cost effective quality control: managing the quality and productivity of analytical processes, Washington, DC, 1986, American Association of Clinical Chemistry.
2. Howanitz, PH, and Howanitz, JH: Laboratory quality assurance, New York, 1987, McGraw Hill Co.
3. Uldall, A: Quality assurance in clinical chemistry, Scand J Clin Lab Invest 47(suppl 187), 1987.
4. Shewhart, WA: Economic control of quality of manufactured products, New York, 1931, D Van Nostrand Co, Inc.
5. Westgard, JO: Better quality control through microcomputers, Diagn Med 61:741, 1982.
6. Westgard, JO, Barry, PL, Hunt, MR, et al: Performance characteristics of rules for internal quality control: probabilities of false rejection and error detection, Clin Chem 27:493-501, 1981.
7. Westgard, JO, Groth, T, Aronsson, T, et al: Performance characteristics of rules for internal quality control: probabilities for false rejection and error detection, Clin Chem 23:1957-1967, 1977.
8. Copeland, BE, Rosvol, RV, and Casella, JM: Quality control workshop manual, Chicago, 1978, American Society of Clinical Pathologists Commission on Continuing Education.
9. Levey, S, and Jennings, ER: The use of control charts in the clinical laboratory, Am J Clin Pathol 20:1059, 1950.
10. Barnett, RN: Clinical laboratory statistics, ed 2, Boston, 1979, Little, Brown & Co.
11. Kurtz, SR, Copeland, BE, and Straumfjord, JV: Guidelines for clinical chemistry quality control based on the long-term experience of 61 university and tertiary care referral hospitals, Am J Clin Pathol 68:463-473, 1977.
12. Skeudzel, LP, Barnett, RN, and Platt, R: Medically useful criteria for analytical performance of laboratory tests, Am J Clin Pathol 83:200-205, 1986.
13. Broughton, PMG, and Buckley, BM: Performance requirements of tests performed nearer the patient, Scand J Clin Lab Invest 47:99-104, 1987.
14. Baer, DM, and Belsey, RE: Quality control in the office laboratory, Clin Lab Med 6:805-813, 1986.
15. Ffyfe, J: Stick testing a cause for concern, Br J Hosp Med 35:196, 1986.
16. Benson, ES: The responsible use of the clinical laboratory, Clin Biochem 19:262-270, 1986.
17. Georges, RJ: Incidence and causes of "wild results," Ann Clin Biochem 24(suppl 2):151, 1987.
18. Ehrmeyer, SS, and Laessig, RH: Interlaboratory profiency testing programs: a computer model to assess their capability, Clin Chem 33:784-787, 1987.
19. Velapoldi, RA, Paule, RC, Schaffer, R, et al: A reference method of the determination of potassium in serum, National Measurement Laboratory, National Bureau of Standards, sp. pub. no. 260-63, Washington, DC, 1979, US Government Printing Office.
20. Schaffer, R, Sniegoski, LT, Welch, MJ, et al: Comparison of two isotope dilution/mass spectrometric methods for determination of total serum cholesterol, Clin Chem 28:5-8, 1982.
21. Gilbert, R: Definitive method comparison with all methods averages from CAP chemistry survey, Am J Clin Pathol 70:450-470, 1978.

CHAPTER 19

Evaluation of methods

R. NEILL CAREY
CARL C. GARBER

OBJECTIVES

- List three purposes of a method evaluation.
- List aspects to consider when selecting a method to evaluate for use in a clinical chemistry laboratory.
- Differentiate between random, constant, proportional, and total error.

KEY TERMS

accuracy The agreement between the mean estimate of a quantity and its true value.[26]

allowable error (E_A) The amount of error that can be tolerated without invalidating the medical usefulness of the analytical result. Allowable error is defined as having a 95% limit of analytical error; only one sample in 20 can have an error greater than this limit.[2]

assigned value The value assigned either arbitrarily (as by convention) or from preliminary evidence (as in the absence of a recognized reference method).[26]

bias A systematic component of analytical error, estimated from a comparison-of-methods experiment.[4] Also known as the difference between two quantities. A measure of inaccuracy.

comparative method The analytical method to which the test method is compared to the comparison-of-methods experiment. This term makes no inference about the quality of the comparative method.[2]

comparison-of-methods experiment An evaluation experiment in which a series of patient samples are analyzed by both the test method and comparative method. The results are assessed to determine whether differences exist between the two methods.[2]

confidence interval The numerical interval that contains, with a specified probability,[4] the population parameter.

constant systematic error (CE) An error that is always in the same direction and magnitude, even as the concentration of analyte changes.[3]

error The difference between a single estimate of a quantity and its true value. If a good estimate of the true value is not available, the difference may have to be expressed as the deviation from an assigned value. Note that errors as defined cannot be classified as "random" or "systematic" without sets of measurements being considered.[26]

ideal value The value of a parameter under conditions of zero error.

imprecision The standard deviation or coefficient of variation of the results in a set of replicate measurements. The mean value and number of replicates must be stated as well as the particular type of imprecision, such as between-laboratory, within-day, or between-day.[26]

inaccuracy The systematic error, estimated from the mean of a set of data relative to the true value or estimated from other approaches, that indicates the difference between the observed value and true or assigned value.

interference The effect of a component on the accuracy of measurement of the desired analyte.[26]

interference experiment An evaluation experiment that estimates the systematic error in a method resulting from interference or lack of specificity.[2]

linear regression An approach to choose a single line through a data set that "best" describes the relation between two subsets or two methods. This mathematical technique minimizes the sum of the squares of the differences between the observed y values and the y values predicted by the regression line for a given x value. One uses this approach assuming that there are no errors in the data by the x method.

medical decision level (X_C) A concentration of analyte at which some medical action is indicated for proper patient care. There may be several medical decision levels for a given analyte.

parameter A number that describes a feature of a population. This is in contrast to a statistic, which is an estimate of a parameter derived from a sample of the population.

precision The agreement between replicate measurements.[26]

proportional systematic error (PE) An error that is always in one direction and whose magnitude is a percentage of the concentration of analyte being measured.[2]

random analytical error (RE) An error, either positive or negative, whose direction and exact magnitude cannot be predicted; imprecision.[2]

recovery experiment An evaluation experiment that estimates proportional systematic error.[2] The amount of analyte recovered is divided by the amount of analyte added to a sample, and the ratio is expressed as the percentage of recovery. The deviation of the percentage of recovery from 100% is the proportional error.

replication experiment An evaluation experiment that estimates random analytical error.[2] Measurements are made on aliquots of a stable sample over specified periods of time, as within a run, within a day, or over a period of days.

sample The appropriately representative part of a specimen used in the analysis. This sample should be called a *test sample* when it is necessary to avoid confusion with the statistical term *random sample from a population*.[26]

standard error of the estimate The standard deviation of the differences, $s_{y/x}$, between the observed y values and the y values predicted by the regression line for a given x. This statistic measures the dispersion or spread of the data around the regression line.

systematic analytical error (SE) An error that is always in one direction; inaccuracy.[2]

test method The method that is chosen for experimental testing or study by means of method evaluation.[2]

total error (TE) A combination of the random and systematic analytical errors that estimates the magnitude of error that might occur in a single measurement.

true value A term considered to have self-evident meaning requiring no definition. In practice the true value is closely approximated by the definitive (method) value and somewhat less closely by the reference (method) value.[26]

variance The square of the standard deviation.

Over the last 15 years the quantitative analytical methods used in clinical laboratories have become more reliable and more standardized. A growing proportion of analytical procedures are supplied by commercial manufacturers. The emphasis of the hospital clinical chemist has shifted away from methods development. Instead he or she selects from among those commercially available methods that suit the laboratory situation best and subsequently evaluates the selected method.

The process of method evaluation has also been evolving.[1] It has been recognized that a method's performance can be objectively judged as acceptable only if its errors are small enough to be acceptable for medical use. The protocols developed by Westgard et al.,[2] the Food and Drug Administration (FDA),[3] and the National Committee for Clinical Laboratory Standards (NCCLS)[4-8] all measure the errors in terms of analyte concentration units and compare them to medically allowable error.

PURPOSE OF METHOD EVALUATION

Laboratory requirements

New analytical methods are usually developed to improve accuracy or precision over existing methods, to allow automation, to reduce reagent or labor cost, or to measure a new analyte. In any case the method's analytical performance in a clinical laboratory setting must be verified experimentally, even if the new method is believed to be an improvement over all previous methods. The extent of the experiments and the interpretation of the data vary, depending on the purpose of the evaluation and who performs the evaluation, but the basic approach and basic experimental design are similar for all evaluations.

The process of evaluating a method is different from the process of routine quality control of a method after it has been introduced into daily use. Routine (daily) quality control (see Chapter 18) is the process by which increases in the size and frequency of the analytical errors of a method in routine use are detected so that the method can be "repaired." Routine quality control detects errors only when they exceed the error that was present in the method when the control ranges were established. The use of routine quality control does not enable the investigator to determine the magnitude of the inherent errors of the method or to decide whether they are acceptable. Method-evaluation experiments are required to assess the inherent analytical errors of the method and relate them to medical requirements.

Manufacturer requirements

When a manufacturer develops a new method and prepares to market it, the manufacturer is required by the FDA to make claims about the analytical performance of the method, specifically about its precision and accuracy.[9] These claims must be supported by experimental method-evaluation data. It is essential that these claims be realistic and conservative. The level of performance of the method in most users' laboratories must be at least as good as that claimed by the manufacturer. However, potential customers often compare claims made by different manufacturers as they select methods, and thus the claims must be competitive. Extensive experimental data will be required for the manufacturer to develop defensible claims. Protocols for manufacturers to follow the experimental testing of methods and statistical treatment of data to produce defensible performance claims have been developed by NCCLS.[4]

Method-evaluation studies are also performed by some other organizations to verify a manufacturer's claims of analytical performance for a method. The FDA, which has indicated its intention to test whether the performance of any commercial method (''in vitro diagnostic device'') meets the claims stated by the manufacturer, has published its testing protocol.[10] NCCLS is also developing testing protocols for claims verification.[5] It is not cost effective to repeat all the experiments performed by the manufacturer to develop the claims. Instead, fewer tests are done, and conservative statistical tests are used to determine whether the actual performance is worse than the claimed performance. The claim will be rejected only if there is a very high probability (such as 95%) that the method's actual performance is worse than its claimed performance.

Most method evaluations are performed by ''users'' (laboratory personnel) in hospitals and commercial laboratories. These evaluations are performed to determine whether the performance of a method meets the requirements for the medical applications intended by the user. The method may be a commercial method, a newly ''home-grown'' method, or a method the user has seen in the literature and is setting up on his or her own laboratory. The user needs to perform the evaluation as efficiently as possible and to determine with a minimum of experimental work whether the method's performance is unacceptable. Thus the user can reject the method without performing all the time-consuming studies that would be required for acceptance.

Medical requirements

The decision to accept or reject a candidate laboratory method should be based on the ability of the method to meet the requirements of the final user, the physician who is interpreting the results of a laboratory test on a patient. The error of the test result is excessive if it causes a misdiagnosis. The greatest chance for misdiagnosis caused by an analytical error in a test result occurs at the concentration at which a medical diagnosis is made; this concentration is termed the *medical decision-level concentration*. For example, a fasting glucose concentration below 500 mg/L may be diagnostic of hypoglycemia.[11] For each decision-level concentration, a performance standard may be formulated, consisting of the decision-level concentration, X_C, and the allowable error, E_A. Allowable error is stated in concentration units so that errors of the test method may be judged by comparison with clinically allowable error. One interprets the method-evaluation data by using the data to estimate the error of the method at the medical decision-level concentration and then comparing this estimate with the allowable error. If the method's error exceeds allowable error, performance is not acceptable. If it is less than allowable error, performance is acceptable.

The amount of error present in the single measurement of an analyte is different each time the analyte is measured because a portion of the error is purely random. Thus the magnitude of error for a measurement on a given patient specimen cannot be known exactly, and one canot predict the absolute maximum error a method could ever make on the analysis of a single patient specimen. However, one can calculate an estimate of the upper limit of the error such that there is only a 5% chance (or less if desired) that the actual error would exceed the upper limit. Similarly, one can define allowable error as a 95% upper limit of error. There will be only a 5% chance that the error might be larger than this defined amount and thus could cause a misdiagnosis.

Exact performance standards (for allowable error) do not exist for most analytes. Performance standards have been proposed for the analytes measured most often, but generally one must use one's own professional judgment and input from clinicians to establish the performance standard for a particular analyte. There are some guidelines for general performance standards; several of these have been combined in Table 19-1. The medical decision-level concentrations (X_C) for Table 19-1 are those suggested by Barnett.[12,13] Tonks[14] proposed that allowable error be either one fourth of the normal range or 10%, whichever is less. Allowable standard deviations and coefficients of variation from all references can be converted into 95% limits of allowable error if they are multiplied by 1.96. (A multiplier of 2 may be used for convenience; see p. 254). For enzymes, the limit is expanded to 20%.[15]

The Aspen Conference[18] sponsored by the American College of Pathologists (CAP) also recommended the use of intraindividual and interindividual biological variations for determining the goals for the precision of a method used for group testing. The analytical coefficient of variation is denoted as CV_A, where

$$CV_A = \tfrac{1}{2}\sqrt{(CV_{intra})^2 + (CV_{inter})^2}$$

and CV_{intra} is the biological variation observed within an individual and CV_{inter} is the biological variation observed between individuals.[18] To enable the physician to monitor intraindividual changes, the method must be even more precise:

$$CV_A = \tfrac{1}{2}\,CV_{intra}$$

Fraser[21,22] combined the criterion for CV_A based on CV_{intra} and pharmacokinetic theory to develop performance standards for therapeutic drug assays (Table 19-2). His proposal supports a total error criterion by stating, ''No bias should exist.''

Solid guidelines for performance exist only for the most

Table 19-1 Recommended medically allowable errors, E_A

Chemistry analyte	Decision level, X_C	Barnett[12,13]	Tonks[14,15]	Gilbert[16]	Cotlove et al.[17]	1976 Aspen Conference[18]		Elion-Gerritzen[19]	Ross[20]
						Group	Individual		
Albumin	35 g/L	5			3	1	1		
Bicarbonate	20 mmol/L	2			1.6				
	30 mmol/L	2			1.6				
Bilirubin	10 mg/L	4	2						
	200 mg/L	30	60						
Calcium	85 mg/L	6	4			15	15		
	110 mg/L	5	5.5	5.5	3.2	2	2	5	
Chloride	90 mmol/L	4	1.8	3.6	1.8	2.7	2		
	110 mmol/L	4	2.2	4.4	1.8	3.3	2.4	4.8	
Cholesterol	2500 mg/L	400	250	400	350	500	100	350	
Creatinine	20 mg/L	3	2	4					
Cortisol	50 μg/L					30	13		
	300 μg/L					180	78		
Globulin	35 g/L	5	50						
Glucose	500 mg/L	100	120		90	31	22		
	1200 mg/L	100	200	130		74	53		
	2000 mg/L							310	
Iron	1000 μg/L			150		360	260		
Magnesium	20 mg/L			3	1.4				
Osmolality	270 mOsm/kg		15						
pH	7.35	0.008							
	7.45	0.012							
P_{CO_2}	35 mm Hg	6							
	50 mm Hg	6							
P_{O_2}	30 mm Hg								12
	80 mm Hg	10							8
	195 mm Hg								31
Phosphorus	45 mg/L	4	4.5	5	4.5	3.1	2.6	7.7 (at 78 mg/L)	
Potassium	3.0 mmol/L	0.5	0.26	0.24	0.28	0.17	0.13	0.47	
	6.0 mmol/L	0.5	0.59	0.5	0.28	0.34	0.26	—	
Protein, total	70 g/L	6	4.9	7	4.4	2.8	2.1	—	
Sodium	130 mmol/L	4.0	2.3	3.8	1.0	1.0	1.0	0.5	
	150 mmol/L	4.0	2.7	4.4	1.0	1.2	1.2	—	
Thyroxine	30 μg/L					4	2	—	
	130 μg/L					18	10	—	
Triglycerides	1600 mg/L	300				1180	420	—	
Urea nitrogen	270 mg/L	40	27	50	30	54	33	50	
Uric acid	60 mg/L	10	6	9	11	6	4		
Enzymes (E_A given as percentage of measured value)									
Acid phosphatase (ACP)			20%						
Alkaline phosphatase (ALKP)			20%			13%	1.8%		
Amylase (AML)			20%						
Aspartate amino transferase (AST, GOT)			10%						
Creatine kinase (CK)						26%	13%		
Lactate dehydrogenase			20%					22.2%	

Table 19-2 Recommended medically allowable errors (E_A) for therapeutic drugs

Drug analyte	Decision level	E_A
Carbamazepine	8 mg/L	1.0
	12 mg/L	1.5
Digoxin	0.8 ng/mL	0.08
	2.0 ng/mL	0.20
Ethosuximide	40 mg/L	4
	100 mg/L	10
Lithium	0.5 mmol/L	0.04
	1.5 mmol/L	0.11
Phenobarbital	15 mg/L	0.7
	40 mg/L	1.8
Primidone	5 mg/L	1.1
	12 mg/L	2.7
Theophylline	10 mg/L	2.8
	20 mg/L	5.6
Valproate	50 mg/L	16
	100 mg/L	33

Data from Fraser, CG: Clinical Chem 33:387-389, 1298-1299, 1987.

frequently measured analytes. It is often observed that the least quantitative aspect of clinical laboratory testing is the medical interpretation of laboratory data. Laboratory workers need advice about medically allowable error from the clinician users of the test in their own hospitals. A continuing dialogue with clinicians helps make their interpretations and expectations more realistic and informs the chemist of the analytical performance needed medically.

Other factors, such as turnaround time, affect the medically allowable error. Clinicians can sometimes accept increased error if turnaround time is fast.

SELECTION OF METHODS

Evaluation of need

The quality of service achievable by a laboratory is determined by selection of personnel, equipment, and analytical methods. The many considerations involved in the process of method selection are shown in Fig. 19-1. Unless this process is well organized, method selection can be a traumatic and costly experience. The box on p. 295 provides a logical sequence to follow in method selection.

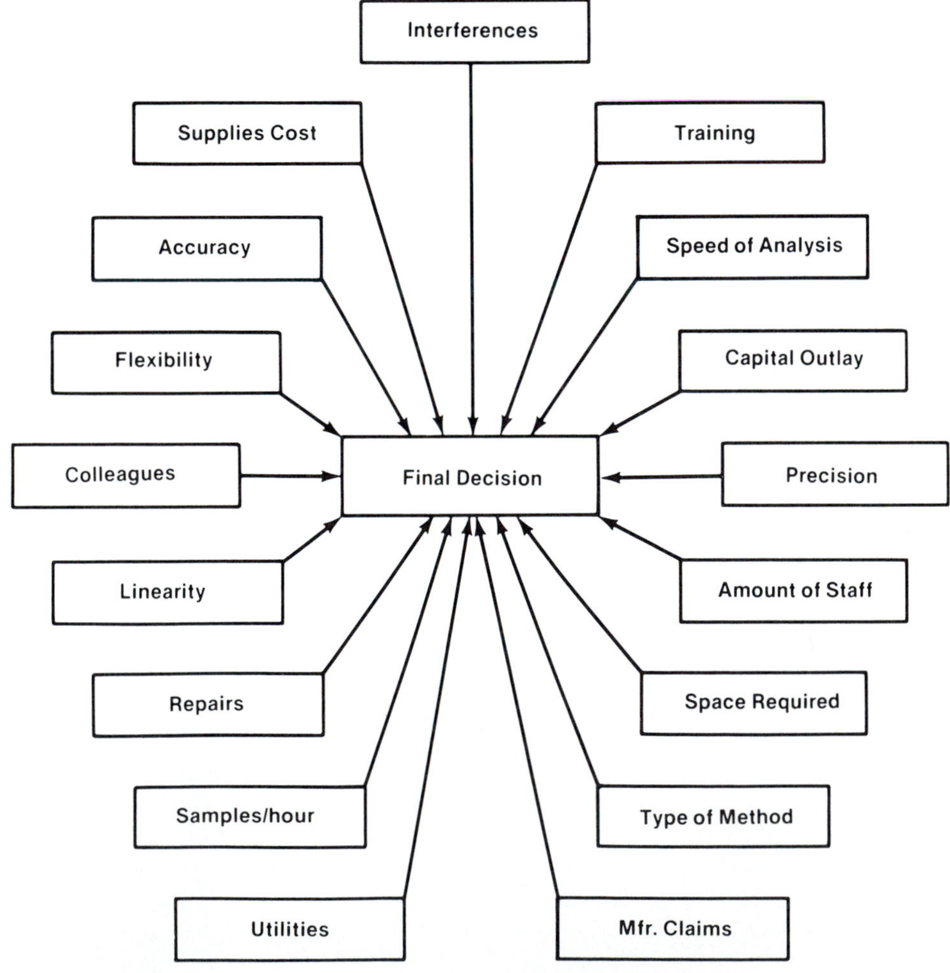

Fig. 19-1 Factors in method selection.

STEPS IN SELECTION PROCESS

Determine need
Define requirements
  Application
  Methodological
  Performance
Review literature
Select candidate method

Often the decision to set up a new method or instrument is based on a medical requirement for a new test provided on site by the laboratory. A change in the methodology of a presently offered test may also be dictated by advances in laboratory practice. For example, in recent years many laboratories have converted their immunoassays from radiolabeled (RIA) to enzyme-labeled (EIA) reagents to reduce or eliminate the special procedures required for managing radioactive materials and to take advantage of the longer shelf lives of EIA reagents. The need for a new method or device may also be dictated by the age and lack of operational reliability of the present analyzer.

Application characteristics

After the need for a new method (or analyzer) has been determined, all the practical features required of the method are defined. These are termed *application characteristics*. They include requirements of sample size, turnaround time, sample throughput rate, specimen type, automated calibration, on-line quality control review, self-diagnostics, laboratory space required, reagent storage facilities required, availability and skill of laboratory staff, time available for training, cost per test, and safety and environmental hazards. Emphasis will be placed on sample size for pediatric applications, on turnaround time and interrupt features for stat applications, and on the sample throughput rate for high-volume screening applications. It is essential that a candidate method meet these fundamental requirements before one considers it any further.

Cost per test has been defined as an application characteristic. This item could be considered separately in light of present emphasis on rising medical costs. The factors affecting the direct cost should be considered when one compares candidate methods. These include the depreciated capital cost, reagent and supplies cost, service and repair cost, and labor cost. Much of this information is available from the manufacturer in terms of initial equipment cost, estimated reagent and supplies consumption and cost, estimated productivity, and service costs (by contract or per visit). Other information, such as the expected workload and anticipated modifications in the productivity based on the internal quality control procedures, are available from within the laboratory. One can use CAP workload recording units to make an initial estimate of the direct labor costs. Users must adjust this information to their own laboratory situation. The example in the box below illustrates how this information may be combined to arrive at a total direct cost per test.

Method characteristics

The next step in the selection process is the definition of ideal methodological characteristics that will enable the

ESTIMATION OF DIRECT COSTS

Laboratory information
  8000 patient tests (samples) per year
  50% estimated test yield (samples, quality control, repeats, dilutions, troubleshooting)
  16,000 total number of assays
  4 CAP units per sample = 64,000 total units
  Five-year depreciation of capital
  $750 setup costs (change laboratory bench)
Manufacturer information
  $11,000: cost of equipment
  $1200: cost of reagents and supplies (for 16,000 assays)
  $1100: service contract (two visits)
  $500: replacement parts
Calculate equipment costs

Capital and setup: $11,750 ÷ 5 yr = $2350/yr ÷ 8000	= $0.294/sample
Service and repair: $1100 + $500 = $1600/hr ÷ 8000	= $0.20/sample
Subtotal	= $0.494/sample

Calculate labor costs (assume $13.50/hour including benefits)

64,000 CAP units ÷ 60 units/hour × $13.50/hour ÷ 8000 samples	= $1.800/sample
Total direct costs	$2.444/sample

selected method to have a good chance for success in the user's laboratory. These characteristics include preferred methodology, which will potentially have the necessary chemical specificity (freedom from interferences) and chemical sensitivity (ability to detect small quantities or small changes in the analyte's concentration). The ability to use primary aqueous standards for calibration (freedom from matrix effects) is also important. The choice of reagents, temperature, reaction time, measurement time, and measurement approach (such as end-point, two-point, or multipoint kinetic methods) are all characteristics of a method and should be defined. A source of recommended principles for clinical chemistry methods has been developed by NCCLS.[23]

Analytical performance characteristics

The method should also be defined in terms of its analytical performance capabilities. Overall goals for analytical performance have been discussed in terms of allowable error based on the medical application of the test. Other aspects of performance that must be defined are working range of the method (linearity), stability of the reagents, ability of the analyzer to detect reagent depletion in the case of enzyme substrates, expected reference range, amount of error caused by interfering substances, precision (within-run, between-run, between-day, and total), and accuracy of the method (determined by comparison of results to those obtained by a reference or standard method). The manufacturer is now required to provide information about precision and accuracy. However, it is essential that the new method be tested to see whether it does perform to manufacturer's specifications. More importantly, the selected method must be evaluated experimentally to determine if the method's actual performance in the user's laboratory is good enough to meet the needs for the medical application in the user's institution. The manufacturer's claimed performance should be considered only as a starting point for determining the actual performance in the user's laboratory setting.

Next, one should review the technical and professional literature to determine what methods are available and to obtain some information about their application and methodological and performance characteristics. It is also useful to confer with colleagues about their experience and recommendations.

The final step in the selection process involves putting all the information together to arrive at a final choice. The use of a rating scheme enables one to obtain a more objective overall rating of the candidate methods.[24,25] The rating scheme can be customized by use of apropriate weighting factors for the characteristics that are more important.

LABORATORY EVALUATION OF A METHOD

A method-evaluation study is not performed to test all methods to determine the method with the smallest error but to determine whether the selected method has acceptably small analytical errors. The process of method evaluation involves estimation of the magnitude of analytical error for a single patient specimen. The laboratory experiments performed to obtain data for estimating the errors are chosen because they give quantitative estimates of random and systematic errors with a minimum of experimental work. The error estimates obtained may be invalid, however, if certain underlying assumptions are not true. These assumptions include operator familiarity with the method's procedure; stability of calibrators, controls, and reagents; and linearity of response throughout the working range.

Familiarization

It is essential that the operators of the method become thoroughly familiar with the details of the method and instrument operation before the collection of any data that will be used to characterize the method's performance. This familiarization period has been addressed by NCCLS[4–8] and may include training by the manufacturer. It should be of sufficient duration so that, at its completion, one can perform all aspects of the method or instrument operation comfortably. Obviously the length of time for device familiarization varies with the complexity of the method or analyzer.

Stability

Verification of the stability of reagents, calibrators, and control materials, especially those prepared in house, can be a lengthy procedure. The matter is simplified considerably for commercially prepared materials. One can use the manufacturer's expiration date during the method evaluation because serious stability problems will be detectable through unacceptable analytical performance of the method. For in-house preparations it is necessary to document these characteristics. One should perform preliminary studies with crossover analyses comparing the results for fresh calibrators and "old" calibrators using fresh reagents to test the stability of calibrators. This should be done several times, and the differences for each specific age of calibrator should be averaged to reduce the effects of different preparations. Similarly, one can test the stability of reagents by periodically (daily, weekly, or monthly, depending on the anticipated decay rate) preparing new reagents and testing them against the older reagents. The older reagents should be stored under specified conditions for the subsequent measurements. One can test the observed differences by use of a *t* test (see Chapter 16).

Linearity

The International Federation of Clinical Chemistry (IFCC) has defined the analytical range in a qualitative sense, stating that it is "the range of concentration or other quantity in the specimen over which the method is applicable without modification.''[26] When the limits of linearity are studied experimentally, the range of concentrations included should at least encompass the limits claimed by the manufacturer. The absolute minimum number of different concentrations that must be measured is three, in a mathematical sense, but at least five different concentrations are recommended, evenly spaced throughout the range of interest. Duplicate measurements should be made on each concentration sample.

Ideally, an initial linearity study should use aqueous standards to identify the capabilities of the method in an ideal specimen matrix. This should be followed by the analysis of an analyte dilution series of samples containing the biological matrix, such as serum or urine. The aqueous and matrix samples will provide important information about the impact of the biological matrix on the method. It may be difficult to prepare specimens in a biological matrix with a range of analyte concentrations from zero to the limit of linearity. For analytes not normally present in the matrix, such as drugs, the analyte is simply added to an analyte-free specimen to obtain the desired maximum concentration, and a dilution series is prepared using analyte-free serum or urine. One can also approximate serum matrices by diluting stock aqueous pools of analyte with human serum albumin, enzyme inactivated serum, or Plasmonate (Cutter Laboratories, Inc., Berkeley, Calif.). One can also use a patient specimen containing the analyte at a concentration known to exceed the linearity of the method and then construct a dilution series using analyte-free materials. The accuracy of the volumetric dilutions is very important, and serial dilutions are not recommended because errors are propagated through the subsequent samples. Rather, each sample should be prepared by direct dilution from the original high sample or pool.

Finally, all the data points should be plotted for visual inspection of linear performance. The actual result of the analysis of each dilution is plotted against the percentage of high pool present in each dilution. The straight portion of the resulting curve represents the linear portion of the assay. In the case of methods with curvilinear response, such as radioimmunoassay procedures, the results obtained from the recommended curve-straightening algorithms should be plotted to show linearity of results.

Random and systematic error

In general, errors that affect the performance of analytical procedures are classified as either random or systematic. Factors contributing to random error are those that affect the reproducibility of the measurement. These include (1) instability of the instrument, (2) variations in the temperature, (3) variations in the reagents and calibrators, (4) variability in handling techniques such as pipetting, mixing, and timing, and (5) variability in operators. These factors superimpose their effects on each other at different times. Some cause rapid fluctuations, and others occur over a longer time. Thus random error has different components of variation that are related to the actual laboratory setting. The *within-run* component of variation (σ_{wr}) is caused by specific steps in the procedure, such as pipetting precision, and short-term variations in the temperature and stability of the instrument. Within-day, *between-run* variation (σ_{br}) is caused by differences in recalibration that occur throughout the day, longer-term variations in the instrument, small changes in the condition of the calibrator and reagents, changes in the condition of the laboratory during the day, and fatigue of the laboratory staff. The *between-day* component of variation (σ_{bd}) is caused by variations in the instrument that occur over days, changes in calibrators and reagents (especially if new vials are opened each day), and changes in staff from day to day. One can combine these components in such a way as to produce an estimate of the total variance of a method (σ_t^2).

$$\sigma_t^2 = \sigma_{wr}^2 + \sigma_{br}^2 + \sigma_{bd}^2$$

Terms used to indicate random error include *precision, imprecision, reproducibility,* and *repeatability*. In each case they refer to the random dispersion of results or measurements around some point of central tendency.

Systematic error describes the error that is consistently low or high. If the error is consistently low or high by the same amount, regardless of the concentration, it is called *constant* systematic error (Fig. 19-2). If the error is consistently low or high by an amount proportional to the concentration of the analyte, it is called *proportional* systematic error.

Factors that contribute to constant systematic error are independent of the analyte concentration, and the magnitude of this error is constant throughout the concentration range of the analyte. Constant systematic error is caused by an interfering substance that gives rise to a false signal. The error can be positive or negative. A reaction between the interfering substance and the reagents caused by a lack of specificity is an example of a constant systematic error.

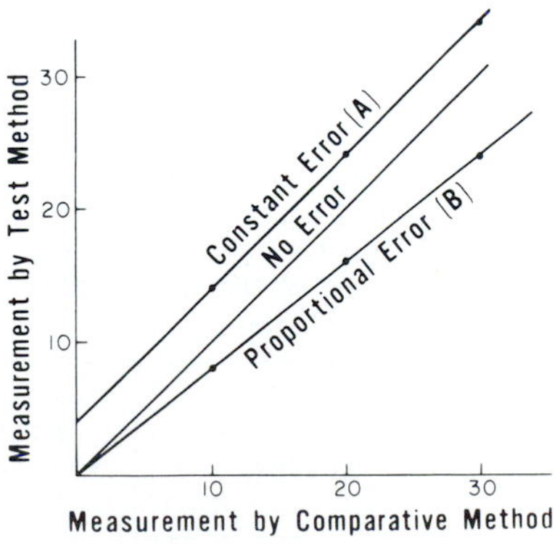

Fig. 19-2 Constant and proportional errors. *(From Westgard, JO, de Vos, DJ, Hunt, MR, et al: Am J Med Technol 44:290, 1978.)*

Another cause of systematic error is an interfering substance that affects the reaction between the analyte and the reagents. This type of error is seen in enzymatic methods using oxidase-peroxidase–coupled reactions in which the hydrogen peroxide intermediate is destroyed by endogenous reducing agents such as ascorbic acid. An interfering substance may also inhibit or destroy the reagent so that it remains in suboptimum amounts for the reaction with the analyte. A nonchemical source of constant systematic error may be caused by improper blanking of the sample or the reagents.

Proportional error may be caused by incorrect assignment of the amount of substance in the calibrator. If the calibrator has more analyte than is labeled, all the unknown determinations will be low in proportion to the amount of error in the calibration, and vice versa. Erroneous calibration is the most frequent cause of proportional error. Proportional error may also be caused by a side reaction for the analyte. The percentage of analyte that undergoes a side reaction will be the percentage of error in the method.

EXPERIMENTS TO ESTIMATE MAGNITUDE OF SPECIFIC ERRORS

In designing experiments that will be used to determine the analytical errors of a method, it is imperative that the experiments be carefully conceived to avoid ambiguous conclusions. The aim of this section is to describe specific experiments that will enable one to estimate the magnitude of a specific error. One can then compare the size of the error to the allowable error to determine the acceptability of the method. This approach is used for all the types of errors described previously. Each type of error is considered individually before combinations of errors are considered. Fig. 19-3 presents an organization of experiments to be performed for specific error determinations arranged in such a way that the easy experiments can be done first. The more extensive (and expensive) final studies are performed only if the errors estimated by these preliminary experiments are acceptable.

TYPE OF ANALYTIC ERROR	EVALUATION EXPERIMENTS	
	PRELIMINARY	FINAL
RANDOM ERROR	REPLICATION WITHIN RUN PURE MATERIALS REAL SAMPLES	REPLICATION RUN TO RUN REAL SAMPLES
CONSTANT ERROR	INTERFERENCE	COMPARISON WITH COMPARATIVE METHOD
PROPORTIONAL ERROR	RECOVERY	

Fig. 19-3 Specific evaluation experiments for estimating specific types of analytical error. *(From Westgard, JO, de Vos, DJ, Hunt, MR, et al: Am J Med Technol 44:290, 1978.)*

Random error estimated from replication studies

The within-run replication experiment is the simplest type of study and should be one of the first performed to assess the performance of a new method. Because it allows assessment of precision over a very short time, the results cannot be extrapolated to indicate long-term performance. The short-term performance must be judged acceptable before one goes to the trouble to study the long-term performance of the method.

The replication study should first be performed with an aqueous solution of calibrator or standard and then repeated with samples whose matrix is as similar as possible to that of the intended patient samples. The concentrations to be studied should be at or near the medical-decision concentrations for the analyte. This is where the laboratory data will be interpreted most critically; thus one must be certain of the method's performance at these concentrations.

An estimate of random error is developed by consideration of repeated analyses of the same specimen. Sixty-eight percent of the results are within ± 1.0 standard deviation of the test mean (SD, or s_{TM} in this chapter), and 95% of the results are within 1.96 SD of the mean. The 95% limit for random error is 1.96 times the standard deviation. (1.96 SD is taken as the 95% limit of random error throughout this chapter. You may wish to round to 2 SD for simplicity, with an accompanying 2% difference from the rigorous estimate of random error. The standard deviation is seldom known accurately to within 2%; thus this rounding would probably not degrade the quality of the estimate of error significantly.) If the estimate of random error is less than allowable error, random error is acceptable. Example calculations are shown on p. 307 and in Chapter 16.

Constant error estimated from interference studies

The interference study measures the constant error caused by the presence of a substance suspected of interfering with the test method. A sample is spiked with a substance suspected of interfering. The volume of this addition should be small, less than 10% of the sample volume, so that the disruption of the matrix is minimal. To compensate for the dilution of the spiked sample, a baseline sample should be prepared by addition of the solvent, used for the interferent, to another aliquot of the sample. The two samples should then be analyzed, at least in duplicate. The difference between the results in the two samples is attributable to an interference caused by the added substance.

A scheme for studying the effects of hemolysis involves taking two blood samples. One is centrifuged and analyzed directly (baseline sample), and the red blood cells in the other blood tube are physically traumatized to rupture the cell membranes to yield an elevated amount of serum hemoglobin. After centrifugation, this hemolyzed sample is analyzed. The difference between the two samples is attributable to the effects of hemolysis. Mild, moderate, or severe hemolysis may be simulated, depending on the volume of red cells traumatized. This approach is more consistent with the actual problems encountered in the laboratory than the approach in which pure hemoglobin is added to a sample. It is not valid if red blood cells contain the analyte.

One may study the effects of lipemia by dividing a lipemic sample into two portions and analyzing one directly while centrifuging the other with an ultrahigh-speed centrifuge to remove the lipoproteins before analysis. The difference in results is attributable to the effects of lipemia. Alternatively, one may also prepare turbid specimens for each decision-level concentration by adding small amounts of Intralipid (Cutter Laboratories, Inc., Berkeley, Calif.) to nonlipemic specimens of appropriate analyte concentrations to obtain slightly, moderately, and grossly appearing lipemia. Baseline concentrations are prepared by addition of equal volumes of water to the original specimens.

Pools with increased amounts of unconjugated bilirubin are produced from a stock solution of bilirubin prepared by the dissolving of pure bilirubin in dimethylsulfoxide to 2500 mg/L. Clear, nonicteric patient sera are spiked to the desired bilirubin concentration. Baseline specimens are prepared as already described.

The choice of substances to be tested is almost infinite. For all absorbance-measuring methods, the effects of hemolysis, icterus, and lipemia should be determined. Other substances that have been reported to affect similar methods should be tested. Pipetting should be (1) precise so that the baseline and spiked samples reflect the same extent of dilution and (2) accurate so that a known amount of interfering substance is added. Again, it is important that the concentration of the analyte in the sample be near the medical-decision levels. A substance that is a possible interferent should be added so that its final concentration is the maximum physiologically expected concentration. If no errors are caused at this high concentration, one can assume that lower concentrations will not adversely affect the performance of the method. If an error is too large at the maximum concentration of interfering substance, it may be appropriate to test the interference at lower concentrations. A slightly icteric sample may be acceptable, but a grossly icteric one may not. It is recommended that these interference studies be conducted on the comparative method (see later discussion) at the same time as a check on the experimental technique.

Calculation of the constant error from the interference experiment data is shown on p. 307. The overall average difference (bias) is called a *constant error (CE)* because it is independent of the analyte concentration. This constant error is compared directly to the allowable error for the appropriate decision level. If the constant error is less than allowable error, the constant error caused by the interference is judged acceptable. This decision is based on clinical limits instead of a statistical test of significance. The standard deviation of the interference values is a measure of the uncertainty of the estimated constant error.

Proportional error estimated from a recovery experiment

Another preliminary study is the recovery experiment. This procedure involves the addition of a known amount of analyte to an aliquot of sample. As in the interference experiment, the sample is divided into two aliquots. One aliquot is spiked with analyte. An equivalent amount of diluent is added to the second; this is the baseline sample. The two samples are then analyzed. The baseline sample provides the original amount of analyte. The difference between the spiked sample and the baseline sample indicates the amount of analyte "recovered." The amount "added" is calculated from the concentration of the stock solution of the analyte and the volume added. The volume of analyte added to the sample should be less than 10% to avoid major disruption of the sample matrix. Pipetting accuracy is critical because the amount of added analyte is calculated from the volume. The concentration of the sample and the amount added should be such that they test the performance of the method near the medical-decision levels of the analyte. In some instances a very small amount of analyte is added to the sample, and the amount recovered is lost in the randomness of the method. Thus it is advisable to make two to four measurements on each sample to reduce the effects of the imprecision of the method. Analysis of these samples with the comparison method is recommended as a check on the experimental technique.

The calculation of recovery is illustrated with an example on p. 308. *Recovery* is defined as the ratio of the amount recovered to the amount added and is given as a percentage. The difference between the calculated percentage of recovery and 100% recovery is the percentage of proportional error. The standard deviation of the percentage of recovery is a measure of the uncertainty of percentage of proportional error. One cannot compare percentage of proportional error directly to allowable error to decide acceptability because percentage of proportional error is not in concentration units. One can convert proportional error to concentration units at the medical-decision level, as shown on p. 308. If the proportional error is less than the allowable error, the proportional error is acceptable. Again, the decision is based on medical requirements rather than statistical tests of significance.

If within-run random error, constant error, and proportional error are acceptable, the day-to-day replication and comparison-of-methods experiments are performed.

FINAL-EVALUATION EXPERIMENTS

The final-evaluation experiments take the most time to perform and potentially yield the most definitive information about the test method's day-to-day performance on real patient specimens.

Between-day replication experiment

The between-day replication experiment is an expansion of the within-run experiment over many days, usually 20. This period must be long enough to allow the random effects occurring over several days to influence the long-term estimate of random error. This experiment and the comparison-of-methods experiment described next are usually combined for better efficiency in the study.

A material known to be stable for the time of the experiment is used, usually a frozen serum or plasma pool or a lyophilized control product. Aliquot-to-aliquot variation of the material must be minimal because it will appear to be day-to-day variance of the test method.

Random error is estimated as 1.96 times the total standard deviation and compared to allowable error, as described previously for the within-run study.

Comparison-of-methods experiment

The comparison-of-methods experiment determines the systematic error of the test method, using real patient specimens. A group of patient specimens are analyzed by both the test method and a comparative method, a method known to be accurate and precise. Systematic differences between the two methods are interpreted as errors of the test method if the results of the comparison method are known to have little or no error (negligible random and systematic errors). Thus the comparative method should be of the highest quality possible so that errors will not be erroneously assigned to the test method.

Methods may be classified in terms of the quality of their performance, definitive, reference, or routine methods. See Chapter 18 for a description of definitive and reference methods.

In practice the comparative method is often the method in routine use in the laboratory and not a method of reference quality. It is useful to see how results from the test method compare with those of the routine method, but differences between the two methods should be interpreted cautiously unless the quality of the comparative method is known to be high.

At least 40 and preferably 100 or more patient specimens should be analyzed. They should include the variety of disease states that will be encountered by the test in routine use. Analyte concentrations of the specimens should be evenly distributed throughout the analytical range; otherwise regression analysis of the comparison data will be inaccurate. Hemolyzed, lipemic, and icteric specimens should be included if not proscribed by the manufacturer of the test method, and if they do not cause errors by the comparative method. If included in the study, they should be identified. Specimens must be carefully selected from the routine workload to be an efficient representation of the patient mix; preanalysis by the routine method is usually necessary.

Specimens are analyzed in duplicate by each method. Results should be examined carefully and plotted daily. Any specimen whose duplicates do not agree closely by either method should be reanalyzed in duplicate by both methods in the next run. Any specimen with large differences between results by two methods should be similarly reanalyzed. If the large difference is confirmed, one should investigate the patient for disease or diseases present and the specimen for other analytes (possible interferents) to determine the cause of the large difference. Immediate follow-up is essential to avoid unanswerable questions about outliers later.

The test and comparative methods should be run at the same time, or as closely in time to each other as possible. If this is not possible, specimens must be stored in a manner that guarantees analyte stability.

The comparison-of-methods experiment is usually combined with the between-day replication experiment. Patient specimens should be evenly spread over at least five runs, and preferably all 20 runs, to ensure that day-to-day effects have a chance to influence the data and to ensure that day-to-day effects are "fully confounded" (in statistical parlance). Both methods must be maintained in acceptable quality control during the period.

***t*-Test statistics: bias, s_d.** The systematic differences between the test and comparative methods are most easily estimated from the comparison-of-methods data by the bias. The *bias* is the difference between the average result by the test method and the average result by the comparative method. Bias can indicate the magnitude of the systematic error between the two methods. (Each patient specimen must be analyzed by both methods for bias to be valid.) Bias is given by Equation 19-1, in which y_i and x_i are the analyte concentrations of the individual specimens by the test method and comparative method, respectively, and N is the number of paired results compared.

$$\text{Bias} = \frac{\Sigma(y_i - x_i)}{N} \qquad \textit{Eq. 19-1}$$

The standard deviation about the bias, called the *standard deviation of the differences,* s_d, is calculated in a manner analogous to the standard deviation from the replication experiment. One may view it as an indicator of the random error between the two methods.

$$s_d = \sqrt{\frac{\Sigma(y_i - x_i - \text{Bias})^2}{N - 1}}$$

The statistical significance of the bias, that is, whether it really differs from zero, or no bias, is determined by use of the t test. A t value is calculated according to the formula

$$t = \frac{\text{Bias}\sqrt{N}}{s_d}$$

The t value is the ratio of a systematic-error term (bias) to a random-error term (s_d). If the bias increases relative to the standard deviation of differences, there is less of a probability that the observed bias is caused by random variations and more of a probability that there really is a systematic difference between the test and comparative-method mean values. For example, in a comparison of glucose methods there were 101 specimens, the bias was 30 mg/L, and the t value was 2.11. The critical t value for $p = 0.05$ and for 100 degrees of freedom (obtainable from a statistics textbook or Chapter 16) is 1.99. (The two-sided critical t value is used because the bias could be either positive or negative.) The calculated t value exceeds the critical t value; therefore a statistically real bias exists between the two methods. (For a more thorough discussion of the t test, see Chapter 16).

The acceptability of the systematic error, as estimated by the bias, is judged by comparison with allowable error. If bias is less than allowable error, the systematic error is acceptable. If bias exceeds allowable error, the systematic error is not acceptable. Decisions about acceptability should never be based on the t value alone. A large bias and large s_d may combine to give an insignificant t value, even though the bias is unacceptably large.

Westgard and Hunt[27] have shown that bias can give inaccurate estimates of systematic error if proportional error is present. Both proportional and constant errors are combined in the bias. Proportional error also increases s_d. Bias should not be used as an estimator of systematic error unless proportional error is absent, or unless the mean analyte concentration as measured by the comparative method is very near the decision-level concentration (X_C) and the data are well distributed around X_C. Otherwise the bias will be weighted toward the side of X_C that has the most samples with large individual biases.

Correlation coefficient. The statistic most frequently cited in reports of comparison-of-methods experiments is the correlation coefficient. It is calculated according to Equation 19-2, in which r is the correlation coefficient. Equations 19-2 and 16-18 are mathematically equivalent; Equation 19-2 is computationally simpler.

$$r = \sqrt{\frac{N\Sigma x_i y_i - \Sigma x_i \Sigma y_i}{[N\Sigma x_i^2 - (\Sigma x_i)^2][N\Sigma y_i^2 - (\Sigma y_i)^2]}} \qquad \textit{Eq. 19-2}$$

The correlation coefficient is the ratio of the covariance (variation in which x and y move proportionally together away from their means) to the total variance of x and y. An r value of zero indicates that there is no correlation between the methods. A value of $+1$ indicates perfect positive correlation.

The correlation coefficient is often considered with the linear regression statistics; it is treated separately here because of its frequent abuse in method evaluation reports. Westgard and Hunt[27] demonstrated that the correlation coefficient is extremely sensitive to the range of analyte concentrations of the patient specimens in the comparison-of-methods experiment. In a comparison of bilirubin methods over a range of 0 to 45 mg/L, a correlation coefficient of 0.950 was obtained. When data pairs with bilirubin concentrations above 15 mg/L were eliminated, the correlation coefficient dropped to 0.773. This is shown in Fig. 19-4.

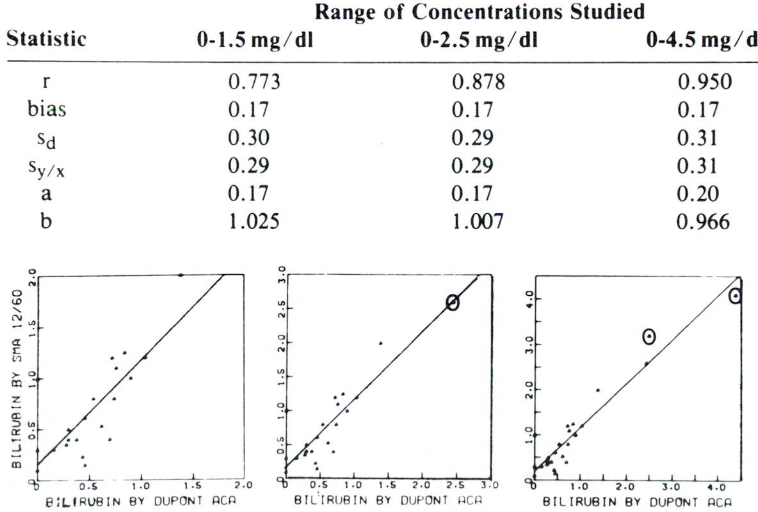

Statistic	Range of Concentrations Studied 0-1.5 mg/dl	0-2.5 mg/dl	0-4.5 mg/dl
r	0.773	0.878	0.950
bias	0.17	0.17	0.17
s_d	0.30	0.29	0.31
$s_{y/x}$	0.29	0.29	0.31
a	0.17	0.17	0.20
b	1.025	1.007	0.966

Fig. 19-4 Effect of range of data on correlation coefficient, *r*. *(From Westgard, JO, de Vos, DJ, Hunt, MR, et al: Am J Med Technol 44:552, 1978.)*

Decisions about the acceptability of the analytical performance of a method should never be based on the value of the correlation coefficient alone.

Linear regression statistics. If the test method and comparative method do correlate with each other, an $X:Y$ plot of results resembles a straight line, which can be described by the linear regression expression

$$Y_i = a + bx_i \qquad \textit{Eq. 19-3}$$

where Y_i is the calculated value on the straight line corresponding to the actual comparative method result, x_i. The proportionality between the methods is given by the slope, b, whose ideal value (no errors) is 1. Constant error is indicated by the y intercept, a. Random error between the methods is indicated by the standard error of the regression, $s_{y/x}$, also called the standard error of estimate and the standard deviation of the residuals. Linear regression statistics are calculated by Equations 19-4 to 19-6.

$$b = \frac{N\Sigma x_i y_i - \Sigma x_i \Sigma y_i}{N\Sigma x_i^2 - (\Sigma x_i)^2} \qquad \textit{Eq. 19-4}$$

$$a = \bar{y} - b\bar{x} \qquad \textit{Eq. 19-5}$$

$$s_{y/x} = \sqrt{\frac{\Sigma(y_i - Y_i)^2}{N - 2}} \qquad \textit{Eq. 19-6}$$

An estimate of systematic error at X_C, the decision-level concentration, may be obtained from the linear regression statistics by substitution of X_C for x_i in Equation 19-3, to calculate Y_C, the concentration the test method would measure for a specimen whose true analyte concentration is X_C. The systematic error, SE, is calculated by subtraction of X_C from this Y_C:

$$\text{SE} = |Y_C - X_C| = |a + bX_C - X_C|$$

This estimate of error will be valid only if the following limitations of linear regression are observed.

The data must be carefully examined for nonlinearity, and the range of data must be limited to the linear range. Nonlinearity at higher concentrations (falling off at high concentrations) will lower the slope, increase the y intercept, and increase $s_{y/x}$.

The data must be carefully examined for outliers. The importance of daily examination and plotting of comparison-of-methods data cannot be overemphasized. Outlier specimens (for which $|y_i - Y_i| > 3.5\ s_{y/x}$) must be detected immediately and reanalyzed by both methods so that the data can correct or confirm the outlier. The linear regression line is "pulled" toward the outlier, with the greatest effects caused by outliers at the extremes of the data. Confirmed outliers should be investigated for their causes. A confirmed outlier really is representative of the true analytical performance of the method. The systematic error of the test method should be calculated both with the outlier included in the data set and with it excluded. If errors are acceptable with the outlier excluded and excessive with it included, extreme caution should be exercised. There are statistical tests for removal of outlier,[28] but no more than one outlier should be excluded in a set of 40 patient comparisons. If more than one outlier is present per 40 patient comparison samples, the test method should be rejected until a cause for the outliers can be found and corrected.

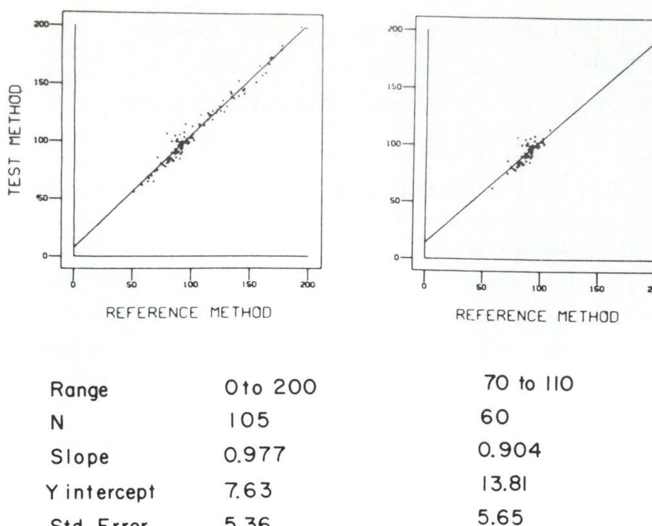

TEST METHOD
REFERENCE METHOD
REFERENCE METHOD
Range 0 to 200 70 to 110
N 105 60
Slope 0.977 0.904
Y intercept 7.63 13.81
Std Error 5.36 5.65

Fig. 19-5 Effect of range of data on linear regression statistics. *(From Westgard, JO, and Hunt, MR: Clin Chem 19:49, 1973.)*

The range of analytical concentrations must be wide. The effects of a narrow range of data on the least-squares statistics are seen in Fig. 19-5. Methods-comparison data often fail to meet one additional assumption of linear regression calculations. This assumption that is violated requires that the x data (comparison) be known without error. Actually, in a methods-comparison experiment, random errors do affect the results of the comparison method. When the range of the data is sufficiently large, the effects of the failure to know the x values without error becomes negligible.[29,30]

Waakers et al.[29] have suggested that the correlation coefficient be used to decide whether the range of data is sufficient for using traditional least-squares calculation. If the correlation coefficient is greater than 0.99, the traditional least-squares approach will calculate a slope whose mathematical error will be less than 1%. If the correlation coefficient is less than 0.99, the slope will be falsely low and the y intercept will be too high. Cornbleet and Gochman[30] have suggested another decision limit. If the ratio of the analytical standard deviation of the comparative method to the standard deviation of the comparison-of-methods experiment specimen population by the comparative method is less than 0.2, the least-squares calculation will be appropriate. If these tests on the data fail, one should use another regression approach, such as those discussed by Cornbleet and Gochman.[30]

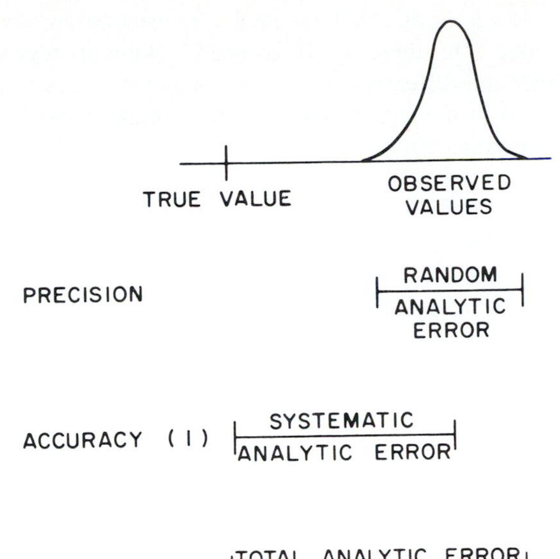

Fig. 19-6 Total analytical error. *(From Westgard, JO, Carey, RN, and Wold, S: Clin Chem 20:825, 1974.)*

Calculation of systematic error by use of linear regression statistics is demonstrated on p. 309.

ESTIMATION OF TOTAL ERROR

Estimates of random error and systematic error are combined to estimate the total error of the test method. This is the most severe criterion for the test method to meet. The rationale for the total error concept is shown in Fig. 19-6. The horizontal line is error in concentration units, and the vertical line is located at the *true-value* concentration, the medical decision-level concentration, or zero error. The vertical distance from the horizontal line represents the probability of obtaining a test method result at any given amount of error (difference from X_C). The bell-shaped curve shows the distribution of test-method data obtained from repeated analyses of a patient specimen whose true analyte concentration is X_C. The distance from the mean of that curve to the true value is the systematic error. The dispersion around the mean of the data is the random error, which has a 95% limit of 1.96 times the standard deviation. There will be (1) instances in which the combined error will be exactly equal to the systematic error, (2) other times when the combined error for a given result will be less than the average systematic error by some amount because of the random error of the method, and (3) other times when the combined error will be greater than the systematic error, again by some amount caused by the random error of the method. The physician has no way of knowing what the various components of error are or when they will cause a larger error. Therefore it is essential to consider the worst-case combination, and define this as total error (TE):

$$TE = RE + SE$$

If total error is less than allowable error, the method's overall performance is acceptable. Calculation of total error is demonstrated on p. 309.

Equations for estimating the magnitudes of the various errors and the criteria for judging their acceptability are summarized in Table 19-3.

CONFIDENCE-INTERVAL CRITERIA FOR JUDGING ANALYTICAL PERFORMANCE

To this point it has been assumed that the error estimated by each of the previous equations is absolutely accurate. However, if the same experiment were to be repeated in as identical manner as possible, a slightly different estimate of error would probably be obtained. Exact measurements of random and systematic errors cannot be obtained from the limited numbers of specimens analyzed in the procedures recommended above.

In the approach developed by Westgard, Carey, and Wold,[31] 95% upper and lower limits of error are calculated. If the 95% upper limit of an error is smaller than the allowable error, one can be at least 95% sure that the method's performance is acceptable. If the 95% lower limit is greater than the allowable error, one can be at least 95% sure the method's performance is not acceptable, and no further testing is indicated. The method should be rejected or modified to improve its analytical performance. When the lower 95% limit is less than allowable error and the 95% upper limit of error exceeds allowable error, one cannot decide whether the method is unacceptable or acceptable, and more data are required to make a definitive decision.

Table 19-3 Point-estimate criteria for acceptable performance

Type of error	Criteria
Random (RE)	$1.96 \cdot s_{TM} < E_A$
Constant (CE)	$\text{Bias} < E_A$
Proportional (PE)	$\frac{\lvert R - 100 \rvert}{100} \cdot X_C < E_A$
Systematic (SE)	If $\overline{X} = X_C$, $\lvert \overline{Y} - \overline{X} \rvert < E_A$ or $\lvert Y_C - X_C \rvert = \lvert a + bX_C - X_C \rvert < E_A$
Total (TE)	$RE + SE = 1.96\ s_{TM} + \lvert a + bX_C - X_C \rvert < E_A$

Calculations of confidence-interval estimates of each type of error are demonstrated on pp. 307 to 310.

Confidence-interval criterion for random error

In the calculation of random error, the true value of the standard deviation is not known. One can estimate upper and lower confidence limits of the standard deviation by multiplying the observed standard deviation by the appropriate one-sided 95% factors. These factors (see Table 19-5) are referenced to $N - 1$ degrees of freedom.

$$s_{TM_u} = s_{TM} \times A_u$$

$$s_{TM_l} = s_{TM} \times A_l$$

where A_u and A_l are the factors for computing the upper and lower one-sided limits of the standard deviation.[32]

The upper confidence limit of random error is 1.96 times the upper confidence limit of the standard deviation, and the lower confidence limit of random error is 1.96 times the lower confidence limit of the standard deviation.

$$RE_u = 1.96 \times s_{TM_u}$$

$$RE_l = 1.96 \times s_{TM_l}$$

Confidence-interval criterion for constant error

Upper and lower limits for constant error are estimated from the interference study data. Again, an upper limit is derived such that there is a 95% probability that the true error is less than that limit.

$$CE_u = CE + \frac{t \cdot s}{\sqrt{N}}$$

and

$$CE_l = CE - \frac{t \cdot s}{\sqrt{N}}$$

The t value is a one-sided, 95% t. Fig. 19-7, *A*, shows the upper 95% limit of constant error, leaving only a 5% chance that the error exceeds this upper limit, CE_u. Similarly, Fig. 19-7, *B*, shows the lower 95% limit of constant error, CE_l. A one-sided t is used only when one is concerned with an upper limit on the constant error without regard for how small the constant error is, and vice versa for a lower limit. (A two-sided t is used to answer the question, "Is there a difference?" without regard to whether the difference is positive or negative, as in the t test used in the interpretation of the comparison-of-methods experiment.)

Confidence-interval criterion for proportional error

A similar approach is used to estimate the upper and lower limits for recovery. A one-sided t value is used in the following equation for $N - 1$ degrees of freedom and a 95% limit.

$$\overline{R}_u = \overline{R} + \frac{t \cdot s}{\sqrt{N}}$$

and

$$\overline{R}_l = \overline{R} - \frac{t \cdot s}{\sqrt{N}}$$

where $\overline{R}$ is the % mean recovery and s is the standard deviation of the recovery (see p. 308). The upper confidence limit of the percentage of proportional error is the confidence limit that is the greater distance from the ideal value of 100%. The lower confidence limit of the percentage of proportional error is the confidence limit that is closer to 100%. These percentages are converted to concentration units at the critical concentration or concentrations to make the comparison with allowable error.

$$PE_u = X_C \left| \frac{\overline{R}_u \text{ or } \overline{R}_l - 100}{100} \right|_u$$

and

$$PE_l = X_C \left| \frac{\overline{R}_l \text{ or } \overline{R}_u - 100}{100} \right|_l$$

Confidence-interval criterion for systematic error

Fig. 19-8 shows the profile of a confidence interval around a least-square regression line. The expression for the limits, w, of this intermal is given as

$$w = t \cdot s_{y/x} \sqrt{\frac{1}{N} + \frac{(X_C - \bar{x})^2}{\Sigma(x_i - \bar{x})^2}} \qquad \textbf{\textit{Eq. 19-7}}$$

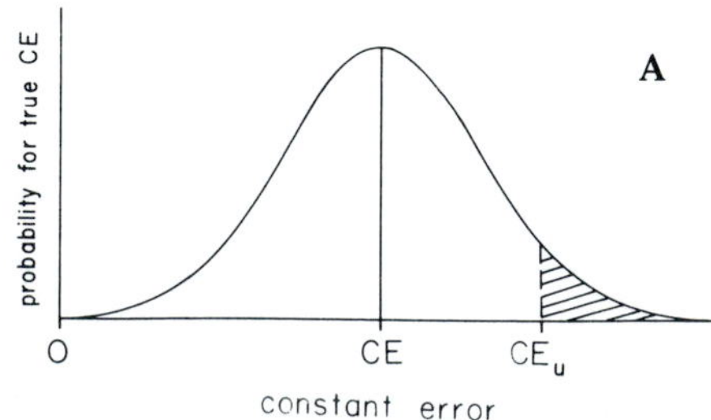

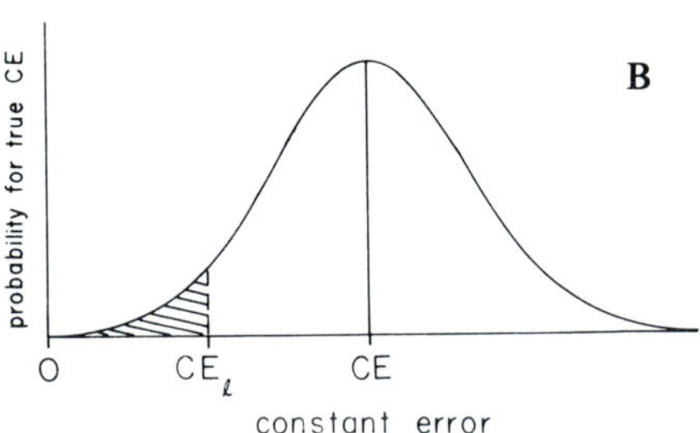

Fig. 19-7 **A,** One-sided 95% upper limit of constant error, CE_u. **B,** One-sided 95% lower limit of constant error, CE_l.

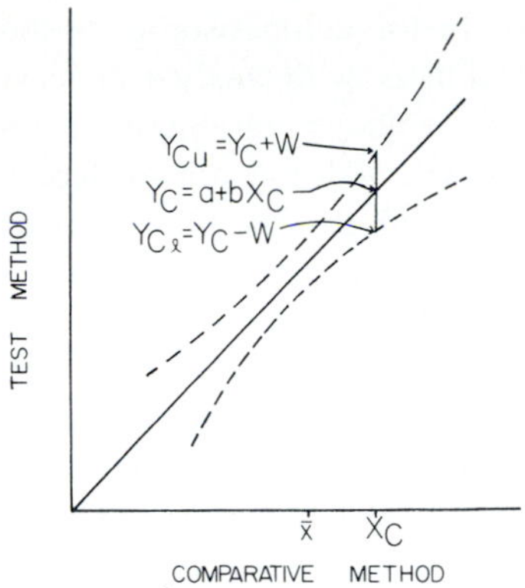

Fig. 19-8 Confidence interval around regression line. *(From Westgard, JO, de Vos, DJ, Hunt, MR, et al: Am J Med Technol 44:727, 1978.)*

This equation is similar to those used to calculate the limits for constant and proportional error in terms of the component $t \cdot s_{y/x}$. The component under the square-root sign becomes $1/N$ if X_C equals the mean of the patient data by the comparative method. As X_C moves away from the mean, the right term begins to contribute to the widening of the limits. One can calculate the denominator of this second term from the standard deviation of the patient population by the comparative method (s_x) as follows:

$$\Sigma(x_i - \bar{x})^2 = s_x^2(N - 1)$$

In this situation, one cannot know exactly what the regression line is, and for a given X_C, the corresponding Y_C could be as large as $(Y_C + w)$, or as small as $(Y_C - w)$. The limit that is farther from the ideal value is used to estimate the upper limit of systematic error, and the limit closer to the ideal value is used to estimate the lower limit of systematic error:

$$SE_u = [(Y_C \pm w) - X_C]_u \qquad \textit{Eq. 19-8}$$

and

$$SE_l = [Y_C \pm w) - X_C]_l \qquad \textit{Eq. 19-9}$$

Equations 19-7 to 19-9 may not provide a valid estimate of the confidence limits of the linear regression line if the precision of the test method is not reasonably constant throughout the concentration range of the patient specimens included in the comparison-of-methods experiment.

Confidence-interval criterion for total error

As described before, total error is the worst-case combination of the random and systematic error. Since both the random and systematic errors have variances included in the equations used to calculate them, they must be combined vectorially. Their variances are combined as shown:

$$TE_u = \sqrt{RE_u^2 + w^2} + SE_u$$

and

$$TE_l = \sqrt{RE_l^2 + w^2} + SE_l$$

If the upper 95% limit of the total error is less than the allowable error, one can be 95% sure that the method performs acceptably. If the lower 95% limit of the total error exceeds allowable error, one can be 95% sure the method does not perform acceptably and should be modified or rejected.

One should note that whenever the ideal value (zero-error condition) is between the upper and lower limits of the estimated error, there is a chance that the true error might be zero. Thus in these situations the lower limit of error is simply zero. The upper limit of error remains as just calculated. This situation can arise for the constant, proportional, or systematic error estimates, but not, of course, for random error. If the lower limit of systematic error is zero, the lower limit of the estimate of total error is equal to the lower limit of the estimate of random error because there may not be a systematic error present.

Equations for calculating confidence intervals of the various errors and criteria for judging their acceptabilities are summarized in Table 19-4.

OTHER EVALUATION PROTOCOLS

Protocols for manufacturers to follow to produce defensible performance claims have been developed by

Table 19-4 Confidence-interval criteria

Error	Acceptable performance	Unacceptable performance
Random (RE)	$1.96\, s_{TM_u} < E_A$	$1.96\, s_{TM_l} > E_A$
Constant (CE)	$\lvert Bias\rvert + \frac{t \cdot s}{\sqrt{N}} < E_A$	$\lvert Bias\rvert - \frac{t \cdot s}{\sqrt{N}} > E_A$*
Proportional (PE)	$\frac{\lvert \bar{R}_u \text{ or } \bar{R}_l - 100\rvert_u}{100} \cdot X_C < E_A$	$\frac{\lvert \bar{R}_l \text{ or } \bar{R}_u - 100\rvert_l}{100} \cdot X_C > E_A$*
Systematic (SE)	$\lvert(a + bX_C \pm w) - X_C\rvert_u < E_A$	$\lvert(a + bX_C \pm w) - X_C\rvert_l > E_A$*
Total error (TE)	$\sqrt{RE_u^2 + w^2} + SE < E_A$	$\sqrt{RE_l^2 + w^2} + SE > E_A$*

*If the ideal value is between the two limits, it is possible that there is no error, and thus the lower limit of error might be zero. In these cases, therefore, CE_l is defined as zero, or PE_l is redefined as zero, or SE_l is redefined as zero, whatever the case might be. If $SE_l = 0$, TE_l becomes RE_l.

NCCLS.[4] These protocols involve more laboratory experimental work than described previously. Claims are stated by statistical tolerance limits:

> For example, one could be 99% confident that 95% of patient samples will agree within the limit of 8 mg/L with the values determined by a selected analytical method. This would be a tolerance limit for total error. A similar tolerance limit statement can be made for imprecision.[4]

Use of a tolerance limit for the upper limit of an error estimate increases the probability that a user who repeats the manufacturer's experiments will obtain an error estimate less than the claimed error.

In the total-error protocol a tolerance limit is developed for the maximum difference between the test and comparative method for 95% of patient specimens analyzed by both methods. This is a larger estimate of total error than the one presented earlier in this chapter because it includes the random error of the comparative method and gives more weight to differences caused by interferents in the patient specimens. The random-error portion of the total-error tolerance limit is derived from $s_{y/x}$ instead of s_{TM}.

The NCCLS has proposed a series of user-oriented protocols. The first, EP5-P, is for evaluation of precision and precision-claims verification.[5] It requires duplicate measurements on at least two different materials in a run, two runs per day, for 20 days. An analysis of variance calculation is used to determine within-run, within-day, and day-to-day components of variance. These are combined to estimate the total standard deviation.

Guideline EP6-P addresses both experimental procedure and data analysis for evaluating the linearity of an assay.[6] Statistical procedures are presented for determining the limit of linearity and for estimating errors caused by nonlinearity at a particular concentration. Another NCCLS guideline, EP7-P,[7] presents two approaches for interference testing in the clinical chemistry laboratory. The first approach is similar to that already discussed in this chapter (pp. 298 and 299). The second approach describes the determination of interferences with increasing concentrations of interferent (dose-response method). This guideline also presents extensive lists of exogenous and endogenous interferents and recommended testing concentrations.

The EP9-P NCCLS guideline[8] addresses key issues in method-comparison studies. Various approaches to estimate the bias between the two methods and its confidence interval are described for use, depending on the nature of the data as determined in the preliminary review.

The FDA has developed a protocol that can be used to verify conformance of in vitro diagnostic devices' performance to that claimed by the manufacturers. Performance is tested separately at 10%, 50%, and 90% of a method's claimed linear range.[10] A chi-square test is used to compare the total standard deviation observed in an analysis of variance experiment to the manufacturer's claimed standard deviation. In the comparison-of-methods experiment 20 patient specimens with analyte concentrations within 10% of each of the above concentrations (60 patient specimens total) are selected and run in duplicate by the test and comparative methods. The bias is calculated for the 20 specimens at each level and is compared to the bias claim by use of a one-sided t test at each level. Since data used to make a decision at each concentration are within 10% of the decision-level concentration, the problems of the bias and t test described earlier in this chapter are eliminated. It is assumed that if bias is within a manufacturer's claim at these three levels, it is also within claimed limits at all intermediate levels. This method does not use data as efficiently as one that uses a linear regression approach, but it avoids the problems caused by violating the underlying assumptions of linear regression.

DISCUSSION

There are situations, as in the study of different enzyme methods, in which suitably close agreement is not expected or possible because of diffferent reaction conditions or different definitions of enzyme units. In these cases, rather than conclude that the method is unacceptable, a new clinical baseline of information is necessary, and a new reference range is needed. Specific disease-related data should be obtained to provide new clinical information for interpretation of test method results.

Evaluation of a method for a ''new'' analyte previously not measured in the user's laboratory is an analogous situation. Since there is no comparative method on site, accurate estimates of systematic error are harder to obtain. Reliance on published evaluation reports increases. The conclusions of these reports must be reviewed cautiously after the analysis of one's own experimental data is completed. If an analyte is not usually measured, emphasis shifts to experiments to estimate specific errors. Accurate recovery studies are essential. The interference studies are expanded to include a broader range of chemicals that could interfere with the measurement reactions. Patient specimens that have been analyzed in another laboratory may be analyzed for comparison purposes, but the reliability of the systematic-error estimate may be decreased by specimen instability and lack of user control of the other laboratory's procedure. However, if the other laboratory is the reference laboratory to which the user has previously referred specimens for measurement of this analyte, the comparison really is being made to present practice.

Smaller laboratories often do not have the resources for exhaustive method-evaluation studies but fortunately are usually not among the first to evaluate a new method. Usually some evaluation reports have been published. Even when a method's performance is well documented by published evaluation studies, the user should still evaluate random error and perform the comparison-of-methods experiment to verify acceptable performance in his or her own laboratory. A reference range study should be performed.

EXAMPLE CALCULATIONS AND DECISION MAKING FOR A PERFORMANCE EVALUATION STUDY FOR GLUCOSE

I. Estimation of random error from replication data

A. Statistics calculations

(y_i = results from the method being tested)

1. Mean:

$$\bar{y} = \frac{\Sigma y_i}{N}$$

2. Standard deviation:

$$s_{TM} = \sqrt{\frac{\Sigma(y_i - \bar{y})^2}{N - 1}}$$

or

$$s_{TM} = \sqrt{\frac{\Sigma y_i^2 - N\bar{y}^2}{N - 1}}$$

3. Coefficient of variation:

$$CV = \frac{s_{TM}}{\bar{y}} \cdot 100\%$$

4. Example: glucose

$$N = 21,\ \Sigma y_i = 24{,}990, \text{ and } \Sigma y_i^2 = 29{,}765{,}480$$
$$\bar{y} = 1190 \text{ mg/L}$$
$$s_{TM} = 37 \text{ mg/L}$$
$$CV = 3.1\%$$

B. Point estimate of random error (RE)

$$RE = 1.96 \cdot s_{TM}$$

If RE < E_A, performance is acceptable.

Example: glucose, E_A = 100 mg/L at X_C = 1200 mg/L
$\bar{x}$ = 1190 mg/L
s_{TM} = 37 mg/L

RE = 1.96 × 37 = 73 mg/L; RE is acceptable.

C. Confidence-interval estimate of random error (RE_u, RE_l)

$$s_{TM_u} = s_{TM} \times (A_{.95}) \ldots \text{(see Table 19-5)}$$
$$= 37 \times 1.358 = 50.2 \text{ mg/L}$$
$$s_{TM_l} = s_{TM} \times (A_{.05})$$
$$= 37 \times 0.7979 = 29.5 \text{ mg/L}$$
$$RE_u = 1.96 \times s_{TM_u} = 98 \text{ mg/L}$$
$$RE_l = 1.96 \times s_{TM_l} = 58 \text{ mg/L}$$

Since $RE_u < E_A$, we are 95% certain that random error is acceptable.

Table 19-5 Factors for computing one-sided confidence limits for standard deviation

Degrees of freedom ($N - 1$)	$A_{.05}$	$A_{.95}$
1	.5103	15.947
5	.6721	2.089
10	.7391	1.593
15	.7747	1.437
20	.7979	1.358
25	.8149	1.308
30	.8279	1.274
40	.8470	1.228
50	.8606	1.199
60	.8710	1.179
70	.8793	1.163
80	.8861	1.151
90	.8919	1.141
100	.8968	1.133

From Natrella, MG: Experimental statistics, National Bureau of Standards Handbook 91, Washington, DC, 1963, US Government Printing Office.

II. Estimation of constant error from an interference study for a glucose method

A. Sample preparation

1. 1.00 mL of serum A + 0.10 mL of water
2. 1.00 mL of serum A + 0.10 mL of 1000 mg/L of creatinine standard
3. 1.00 mL of serum A + 0.10 mL of 3000 mg/L of creatinine standard

B. Results (See table below.)

C. Formulas for calculations

$$\text{Concentration added} = \text{Concentration of standard} \times \frac{\text{Volume standard}}{\text{Total volume}}$$

$$\text{Interference} = \text{Concentration (test)} - \text{Concentration (baseline)}$$

D. Point estimate of constant error (CE)

$$CE = \text{Interference}$$

If CE < E_A performance is acceptable.

Example: For glucose, E_A = 100 mg/L at X_C = 1200 mg/L

In the presence of 91 mg/L of creatinine, CE = 33 mg/L; CE is acceptable.

In the presence of 273 mg/L of creatinine, CE = 110 mg/L; CE is not acceptable.

	Creatinine added (mg/L)	Glucose measured (mg/L)	Interference (mg/L)	Average interference (CE) (mg/L)
1.	—	1200, 1220, 1190	—	—
2.	91	1240, 1240, 1230	+40, +20, +40	+33
3.	273	1310, 1340, 1290	+110, +120, +100	+110

E. Confidence-interval estimate of constant error (CE_u, CE_l)

1. In the presence of 91 mg/L of creatinine, CE = 33 mg/L, s = 11.5 mg/L, N = 3, and t for ($N - 1$), or 2 degrees of freedom, 95% one-sided limit is 2.92.

$$CE_u = CE + t\frac{s}{\sqrt{N}}$$

$$= 33 + \frac{2.92 \times 11.5}{\sqrt{3}} = 52 \text{ mg/L}$$

$$CE_l = CE - t\frac{s}{\sqrt{N}}$$

$$= 33 - \frac{2.92 \times 11.5}{\sqrt{3}} = 14 \text{ mg/L}$$

Since $CE_u < E_A$ (52 < 100 mg/L), the constant error caused by creatinine of 91 mg/L is acceptably small.

2. In the presence of 273 mg/L of creatinine, CE = 110, s = 10 mg/L, N = 3.

$$CE_u = 110 + \frac{2.92 \times 10}{\sqrt{3}} = 127 \text{ mg/L}$$

$$CE_l = 110 - \frac{2.92 \times 10}{\sqrt{3}} = 93 \text{ mg/L}$$

Since $CE_l < E_A < CE_u$, we cannot be 95% sure the method is acceptable or 95% sure the method is not acceptable for glucose analysis in the presence of 273 mg/L of creatinine. More data should be obtained to narrow the confidence limits to one side of E_A or the other.

III. Estimation of proportional error from a recovery study for a glucose method

A. Sample preparation

1. 2.0 mL of serum A + 0.1 mL of water
2. 2.0 mL of serum A + 0.1 mL of 10,000 mg/L of glucose standard
3. 2.0 mL of serum B + 1.0 mL of water
4. 2.0 mL of serum B + 0.1 mL of 10,000 mg/L of glucose standard

B. Results (mg/L) (See table below.)

C. Formulas for calculations

$$\text{Concentration added} = \text{Concentration of standard} \times \frac{\text{Volume of standard}}{\text{Total volume}}$$

$$\text{Concentration recovered} = \text{Concentration (test)} - \text{Concentration (baseline)}$$

$$\%\text{ Recovery} = \frac{\text{Concentration recovered}}{\text{Concentration added}} \times 100$$

D. Point estimate of proportional error (PE)

$$PE(\%) = \bar{R} - 100$$

$$\text{PE (concentration units)} = \left|\frac{\bar{R} - 100}{100}\right| \cdot X_C$$

If PE < E_A, performance is acceptable.
Example: For glucose, E_A = 100 mg/L at X_C = 1200 mg/L

$$\bar{R} = 94.8\%$$

$$PE(\%) = 94.8 - 100 = 5.2\%$$

or

$$PE = \left|\frac{94.8 - 100}{100}\right| \cdot 1200 \text{ mg/L} = 62 \text{ mg/L}$$

PE is acceptable.

E. Confidence-interval estimate of proportional error (PE_u, PE_l)

$\bar{R}$ = 94.8%, s = 3.09%, N = 6, and t for ($N - 1$), or 5 degrees of freedom, and 95% one-sided limit is 2.02.

$$\bar{R}_u = \bar{R} + \frac{t \cdot s}{\sqrt{N}}$$

$$= 94.8 + \frac{2.02 \times 3.09}{\sqrt{6}} = 97.3\%$$

$$\bar{R}_l = \bar{R} - \frac{t \cdot s}{\sqrt{N}}$$

$$= 94.8 - \frac{2.02 \times 3.09}{\sqrt{6}} = 92.3\%$$

The limit that deviates more from the ideal recovery of 100% is used to estimate the upper limit of proportional error, $PE_u\%$.

$$PE_u\% = |92.3 - 100| = 7.7\%$$

and for the lower limit

$$PE_l\% = |97.3 - 100| = 2.7\%$$

To relate $PE_u\%$ or $PE_l\%$ to E_A, convert them to concentration units at X_C.

	Glucose added	Glucose measured	Glucose recovered	Percent of recovery*
1.	—	510, 530, 540	—	—
2.	476	970, 1000, 980	460, 470, 440	96.6%, 98.7%, 92.4%
3.	—	1240, 1200, 1210	—	—
4.	476	1690, 1660, 1640	450, 460, 430	94.5%, 96.6%, 90.3%

*Average recovery ($\bar{R}$) = 94.8%; SD of Rec. (s_R) = 3.09%; SD of Avg. Rec. = 1.26%.

$$PE_u = \frac{PE_u\%}{100} \cdot X_C = \frac{7.7\%}{100} \times 1200 = 92 \text{ mg/L}$$

$$PE_l = \frac{PE_l\%}{100} \cdot X_C = \frac{2.7\%}{100} \times 1200 = 32 \text{ mg/L}$$

Since $PE_u < E_A$, one can be 95% sure proportional error is acceptably small for this glucose method.

IV. Estimation of systematic error from a comparison-of-methods study for glucose

A. In the comparison of an automated glucose oxidase method *(y)* versus the manual glucose national reference method *(x)*, the following statistics were obtained by linear regression analysis:

$Y = 0.973x - 57$ mg/L, $s_{y/x} = 37$ mg/L, $N = 82$, $\bar{x} = 1723$, $\bar{y} = 1619$, $s_x = 571$ mg/L, where s_x is the standard deviation of the x values for the 82 samples, and $r = 0.9941$.

B. Point estimate of systematic error (SE)

1. Consider bias:

$$\text{Bias} = |\bar{y} - \bar{x}| = 104 \text{ mg/L}$$

The bias provides an estimate of systematic error at the mean of the data. However, if $\bar{x}$ is not equal to the X_C of interest, there must be no proportional error between methods for the bias to provide an accurate estimate of systematic error at these other concentrations. If proportional error is present, use linear regression statistics or calculate the bias using only those samples close to X_C.

2. Consider linear regression:

$$SE = |Y_C - X_C|$$

where $Y_C = a + bX_C$

For $X_C = 1200$ mg/L of glucose,

$$Y_C = 0.973 \times 1200 - 57 \text{ mg/L} = 1111 \text{ mg/L}$$
$$SE = |1111 - 1200| = 89 \text{ mg/L}$$

Since $SE < E_A$, systematic error is acceptable.

C. Confidence-interval estimate of systematic error (SE_u, SE_l)

$$Y_{C_u} = Y_C + w$$

$$Y_{C_l} = Y_C - w$$

where

$$w = t \cdot s_{y/x} \sqrt{\frac{1}{N} + \frac{(X_C - \bar{x})^2}{\Sigma(x_i - \bar{x})^2}}$$

where

$$\Sigma(x_i - \bar{x})^2 = s_x^2(N - 1)$$

where w is the width of the confidence interval around the regression line (see Fig. 19-8). The value for t, obtained from a 95% one-sided t table and $N - 2$ degrees of freedom, has the value of 1.66.

$$w = 1.66 \times 37 \sqrt{\frac{1}{82} + \frac{(1200 - 1723)^2}{571^2 \times 81}}$$

$$w = 9 \text{ mg/L}$$

thus

$$Y_{C_u} = 1111 + 9 = 1120 \text{ mg/L}$$

$$Y_{C_l} = 1111 - 9 = 1102 \text{ mg/L}$$

The limit that deviates more from the ideal value for Y_C (ideally, $Y_C = X_C$) will be used to estimate the upper limit of systematic error.

$$SE_u = |(Y_{C_u} \pm w) - X_C|_u = |1102 - 1200| = 98 \text{ mg/L}$$

and

$$SE_l = |(Y_{C_l} \mp w) - X_C|_l = |1120 - 1200| = 80 \text{ mg/L}$$

Since $SE_u < E_A$, we can be 95% sure that the systematic error is acceptable between these methods.

V. Estimation of total error

A. Point estimate of total error (TE)

$$TE = RE + SE \text{ (see Fig. 19-6)} = 1.96\, s_{TM} + |Y_C - X_C|$$

$$TE = 7.3 + 89 \text{ mg/L} = 162 \text{ mg/L}$$

Since $TE > E_A$, the total error of the new glucose method is not acceptable.

B. Confidence-interval estimate of total error (TE_u, TE_l)

$$TE_u = \sqrt{RE_u^2 + w^2} + SE$$

and

$$TE_l = \sqrt{RE_l^2 + w^2} + SE$$

Notice that the variances of the uncertainty in repetitive measurements and the uncertainty in the regression line are added, and the square root of the sum is taken to estimate the overall uncertainty, which is then combined with the point estimate of systematic error to yield the appropriate limit for total error. Thus

$$TE_u = \sqrt{98^2 + 9^2} + 89 = 187 \text{ mg/L}$$

and

$$TE_l = \sqrt{58^2 + 9^2} + 89 = 148 \text{ mg/L}$$

Because $TE_l > E_A$, we are at least 95% sure that the total error is not acceptable. Although both components of error—random and systematic—are fairly large, the errors that are easier to reduce or eliminate are systematic errors. However, in this example, both RE_u and SE_u are almost 100 mg/L. Steps should be taken to reduce both types of error before the automated method is acceptable.

REFERENCES

1. Westgard, JO: Precision and accuracy: concepts and assessment by method evaluation testing, CRC Crit Rev Clin Lab Sci, pp 283-330, 1981, Boca Raton, Fla, CRC Press, Inc.
2. Westgard, JO, de Vos, DJ, Hunt, MR, et al: Concepts and practices in the selection and evaluation of methods, Am J Med Technol; Part I. Background and approach, 44:290-300, 1978; Part II. Experimental procedures, 44:420-430, 1978; Part III. Statistics, 44:552-571, 1978; Part IV. Decision on acceptability, 44:727-742, 1978; Part V. Applications, 44:803-813, 1978.
3. Proposed establishment of product class standard for detection or measurement of glucose, Federal Register 39:126-24136, 1974.
4. National Committee for Clinical Laboratory Standards: NCCLS tentative standards EP2-T to EP4-T, Protocol for establishing performance claims for clinical chemical methods, Instrumentation Evaluation Subcommittee of the Evaluation Protocols Area Committee, Villanova, Pa, 1979.
5. National Committee for Clinical Laboratory Standards: NCCLS proposed guideline EP5-P, Proposed guidelines for user evaluation of precision performance of clinical chemistry devices, Subcommittee for User Evaluation of Precision of the Evaluation Protocols Area Committee, Villanova, Pa, 1982.
6. National Committee for Clinical Laboratory Standards: NCCLS proposed guideline EP6-P, Evaluation of the linearity of quantitative analytical methods, Subcommittee on Linearity Performance of the Evaluation Protocols Area Committee, Villanova, Pa, 1986.
7. National Committee for Clinical Laboratory Standards: NCCLS proposed guideline EP7-P, Interference testing in clinical chemistry, Subcommittee on Interference Testing of the Evaluation Protocols Area Committee, Villanova, Pa, 1986.
8. National Committee for Clinical Laboratory Standards: NCCLS proposed guideline EP9-P, User comparison of quantitative clinical laboratory methods using patient samples, Subcommittee on User Comparison of Methods of the Evaluation Protocols Area Committee, Villanova, Pa, 1986.
9. Labeling requirements and standards development for in-vitro diagnostic products, Federal Register 21 CFR:809-810, 1974.
10. Lee, HT, Daniel, A, and Walker, CD: Conformance test procedures for verifying labeling claims for precision, bias, and interferences in invitro diagnostic devices used for the quantitative measurement of analytes in body fluids, Bureau of Medical Devices Biometrics Report, US Food and Drug Administration pub no 8202, Silver Spring, Md, 1982, US Government Printing Office.
11. Berkow, R, editor: The Merck manual of diagnosis and therapy, ed 13, Rahway, NJ, 1977, Merck Sharp & Dohme Research Laboratories, p 1302.
12. Barnett, RN: Medical significance of laboratory results, Am J Clin Pathol 50:671-676, 1968.
13. Barnett, RN: Analytic goals in clinical chemistry: the pathologist's viewpoint. In Elevitch, FR, editor: Proceedings of the 1976 Aspen Conference on Analytic Goals in Clinical Chemistry, Skokie, Ill, 1977, College of American Pathologists, pp 20-24.
14. Tonks, D: A study of the accuracy and precision of clinical chemistry determinations in 170 Canadian laboratories, Clin Chem 9:217–233, 1963.
15. Tonks, D: A quality control program for quantitative clinical chemistry estimations, Can J Med Technol 30:38-54, 1968.
16. Gilbert, R: Progress and analytic goals in clinical chemistry, Am J Clin Pathol 63:960-973, 1975.
17. Cotlove, E, Harris, E, and Williams, G: Biological and analytic components of variation in long-term studies of serum constituents in normal subjects. III. Physiological and medical implications, Clin Chem 16:1028-1032, 1970.
18. Elevitch, FR, editor: CAP Aspen Conference 1976: analytical goals in clinical chemistry, Skokie, Ill, 1977, College of American Pathologists.
19. Elion-Gerritzen, WE: Analytic precision in clinical chemistry and medical decisions, Am J Clin Pathol 73:183-195, 1980.
20. Ross, JW: Blood gas internal quality control, Pathologist 34:377–379, 1980.
21. Fraser, CG: Desirable standards of performance for therapeutic drug monitoring, Clin Chem 33:387-389, 1987.
22. Fraser, CG: Goals for TDM: a correction, Clin Chem 33:1298-1299, 1987.
23. National Committee for Clinical Laboratory Standards: NCCLS Proposed standard PSC-5, Methodological principles for establishing principal assigned values to calibrators, Villanova, Pa, 1977.
24. Tremblay, MM: Evaluation of instruments in biochemistry, Can J Med Technol 41:65-76, 1979.
25. Shaikh, AH: A systematic procedure for selection of automated instruments in the clinical laboratory, Am J Med Technol 45:710-714, 1979.
26. Buttner, R, Borth, R, Boutwell, JH, et al: Approved recommendation (1978) on quality control in clinical chemistry. Part I. General principles and terminology, Clin Chim Acta 98:129F-143F, 1979.
27. Westgard, JO, and Hunt, MR: Use and interpretation of common statistical tests in method-comparison studies, Clin Chem 19:49–57, 1973.
28. American Society for Testing and Materials: ASTM Standard E178-68: standard recommended practice for dealing with outlying observations, Philadelphia, 1968.
29. Waakers, PJM, Hellendoorn, HBA, Op De Weegh, GJ, and Heerspink, W: Applications of statistics in clinical chemistry: a critical evaluation of regression lines, Clin Chim Acta 64:173-184, 1975.
30. Cornbleet, PJ, and Gochman, N: Incorrect least-squares regression coefficients in method-comparison analysis, Clin Chem 25:432-438, 1979.
31. Westgard, JO, Carey, RN, and Wold, S: Criteria for judging precision and accuracy in method development and evaluation, Clin Chem 20:825-833, 1974.
32. Natrella, MG: Experimental statistics, National Bureau of Standards Handbook 91, Washington, DC, 1963, US Government Printing Office.

SECTION TWO

Pathophysiology

CHAPTER 20

Physiology and pathophysiology of body water and electrolytes

LEONARD I. KLEINMAN
JOHN M. LORENZ

OBJECTIVES

- List the electrolyte composition of the two main compartments of total body water.
- Define anion gap and state its clinical significance; calculate and interpret anion gap results from given data.
- Outline the homeostatic regulation of sodium, potassium, chloride, and body water.
- Define the various states of decreased or increased plasma electrolyte concentrations in terms of an excess or deficit of water or electrolyte.
- List and briefly describe the symptoms and at least two causes or clinical conditions associated with increased and decreased amounts of electrolytes and body water.

KEY TERMS

acidosis Abnormally low body fluid pH. *Respiratory*—caused by an abnormally high P_{CO_2}; *metabolic*—caused by an abnormally low bicarbonate concentration.

active transport The passage of ions or molecules across a cell membrane by an energy-consuming process. This energy is generated by cellular metabolism.

adipsia Absence of thirst.

aldosterone A mineralocorticoid hormone secreted by the adrenal cortex that influences sodium and potassium metabolism.

alkalosis Abnormally high body fluid pH. *Respiratory*—caused by an abnormally low P_{CO_2}; *metabolic*—caused by an abnormally high bicarbonate concentration.

anabolism The process of assimilation of nutritive matter and its conversion into living substance.

angiotensin A vasoconstrictive polypeptide produced by the enzymatic action of renin on angiotensinogen. A converting enzyme from the lung removes two C-terminal amino acids from the inactive decapeptide, angiotensin I, to form the biologically active octapeptide, angiotensin II.

angiotensinogen A serum globulin produced in the liver that is the precursor of angiotensin.

anion An ion that carries a negative charge.

anorexia Diminished appetite for food.

antidiuretic hormone A peptide hormone of the neurohypophysis that acts on the collecting tubule of the kidneys to allow increased water reabsorption and therefore decreased free water excretion by the kidney. Also known as *vasopressin*.

arrhythmia Irregularity of the heartbeat.

ascites The accumulation of fluid in the peritoneal cavity.

asphyxia Interference with oxygen delivery to and carbon dioxide removal from the tissue.

atrial natriuretic factor (ANF) A peptide hormone released by the atria of the heart that increases sodium and water excretion by the kidney.

baroreceptor A nerve ending that responds to change in pressure.

catabolism The breaking down of complex chemical compounds in the body into simpler ones, often with the production of energy.

cation An ion that carries a positive charge.

cirrhosis Progressive disease of the liver characterized by damage to hepatic parenchymal cells.

colloid As used in this chapter, this term applies to the large molecules in the body to which the capillary endothelium and cell membrane are impermeable.

colloid osmotic pressure The effective osmotic pressure of plasma and interstitial fluid across the capillary endothelium, largely the result of the presence of protein.

dehydration Abnormal decrease in total body water (see Table 20-4). *Hypernatremic*—net loss of sodium and water from the body, with net water loss exceeding net sodium loss; *hyponatremic*—net loss of sodium and water from the body, with net sodium loss exceeding net water loss; *normonatremic*—net loss of sodium and water from the body in equal extracellular proportions; *simple*—net loss of body water alone with no net sodium loss.

diabetes insipidus The chronic excretion of very large amounts of hypoosmotic urine caused by inability to concentrate urine because of the lack of antidiuretic hormone (ADH) production, secretion, or effect. *Pituitary*—caused by inadequate ADH synthesis or secretion; *nephrogenic*—caused by unresponsiveness of the renal tubules to ADH.

distension receptor A nerve ending that responds to stretch.

edema An increase in the interstitial fluid volume.

extracellular water (ECW) Water external to cell membranes. *Anatomical*—all body water external to cell membranes; *physiological*—plasma plus body water into which small solutes can freely diffuse; excludes the transcellular portion of the anatomical extracellular water; includes the plasma and interstitial fluid (see Fig. 20-1).

free water Water containing no solute.

Gibbs-Donnan equilibrium The steady-state distribution of permeable ions and transmembrane potential that results across a semipermeable membrane when an impermeant ion exists in unequal amounts on either side of the membrane and solvent movement across the semipermeable membrane is exactly opposed (see Fig. 20-7).

hyperaldosteronism A disorder caused by excessive secretion of aldosterone and characterized by hypokalemic alkalosis, muscular weakness, hypertension, polyuria, polydipsia, and normal or elevated plasma sodium concentration.

hyperchloremia An abnormally high plasma chloride concentration.

hyperkalemia An abnormally high plasma potassium concentration.

hypernatremia An abnormally high plasma sodium concentration.

hyperosmotic Denoting an effective osmotic pressure higher than that of plasma.

hypertonic Denoting a theoretical osmotic pressure higher than that of plasma.

hypochloremia An abnormally low plasma chloride concentration.

hypokalemia An abnormally low plasma potassium concentration.

hyponatremia An abnormally low plasma sodium concentration. *Dilutional*—hyponatremia caused by an excess of water (relative to sodium) in the extracellular compartment.

hypoosmotic Denoting an effective osmotic pressure lower than that of plasma.

hypothalamus Portion of brain beneath the thalamus and connected to the pituitary gland (see Chapter 40).

hypotonic Denoting a theoretical osmotic pressure lower than that of plasma.

hypovolemia An abnormally low blood volume.

insensible water loss Evaporation of water through the skin or from the respiratory tract.

interstitial fluid (ISF) Extravascular, extracellular water (see Fig. 20-2).

intracellular water (ICW) Water inside the cells of the body; water within cell membranes.

ischemia A decrease in arterial blood flow below the minimum level necessary to meet the metabolic demands of the tissue or organ in question.

juxtaglomerular cells Smooth muscle cells that synthesize and store renin and release it in response to decreased renal perfusion pressure, increased sympathetic nerve stimulation of the kidneys, or decreased sodium concentration in fluid in the distal tubule.

macromolecule A molecule of colloidal size, notably proteins, nucleic acids, and polysaccharides.

nephrotoxin A substance that causes dysfunction or death of renal parenchymal cells with some degree of specificity.

obligatory water requirements The minimum volume of water necessary to replace insensible water loss and to excrete the existing renal solute load when the urine is maximally concentrated.

oliguria Abnormally low urine output, that is, less than 400 mL/day in an adult.

osmolarity Osmotic concentration expressed as osmoles or milliosmoles of solute per liter of solvent (see Chapter 12).

osmole The total number of moles of a solute in solution after dissociation.

osmosis The movement of water across a semipermeable membrane from a solution with low solute particle concentration to a solution with high solute particle concentration.

osmotic pressure The force necessary to exactly oppose osmosis into a solution across a semipermeable membrane.

paresthesia An abnormal spontaneous sensation, such as burning, pricking, numbness, and so on.

passive diffusion Movement of ions or solute across a membrane, down electrical or chemical gradients without energy consumption or a carrier process.

plasma The extracellular, intravascular fluid of the body (see Fig. 20-2).

polyanionic Possessing multiple negative charges.

polydipsia Excessive fluid intake secondary to extreme thirst. *Psychogenic*—polydipsia secondary to a psychiatric disorder, without a demonstrable organic lesion.

polyuria Excessive urine output, that is, more than 1 to 2 L/day in the adult.

pseudohyperkalemia Abnormally high plasma potassium concentration in a sample obtained from a patient in the absence of true elevation of plasma potassium concentration in that patient.

renin An enzyme produced, stored, and secreted by the juxtaglomerular cells of the kidney, which acts on circulating angiotensinogen to form angiotensin I.

semipermeable Permeable to certain molecules but not to others; usually permeable to water.

syndrome of inappropriate antidiuretic hormone secretion (SIADH) A constellation of findings, including hypotonicity of the plasma, hyponatremia, and hypertonicity of the urine with continued sodium excretion, that is produced by excessive ADH secretion and improves with water restriction.

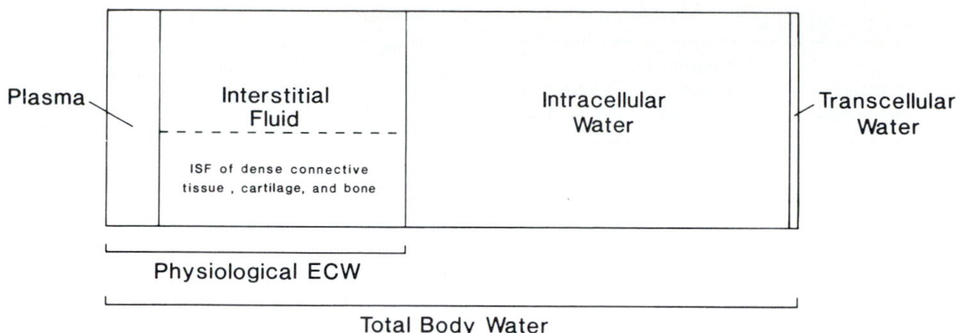

Fig. 20-1 Body water compartments. Note that anatomical extracellular water *(ECW)* includes physiological extracellular water and transcellular water. *ISF,* Interstitial fluid.

total body water All water within the body, both inside and outside the cells, including that contained in the gastrointestinal and genitourinary systems.

transcellular water That portion of extracellular water that is enclosed by an epithelial membrane, whose volume and composition is determined by the cellular activity of that membrane.

transudation The passage of a fluid through a membrane with nearly all the solutes the fluid contains remaining in solution or suspension.

water intoxication An increase in free water in the body; results in dilutional hyponatremia.

Water is the most abundant constituent of the human body, accounting for approximately 60% of the body mass in a normal adult. Water is important not only because of its abundance but also because it is the medium in which body solutes, both organic and inorganic, are dissolved and metabolic reactions take place. The discussion in this chapter focuses on (1) description of the dynamic steady-state compartmentalization of body fluid and its inorganic solutes, (2) physiological mechanisms involved in the maintenance of this compartmentalization, and (3) pathophysiological events that occur during certain clinical states that alter the composition of body fluids.

BODY WATER COMPARTMENTS

Definitions (Figs. 20-1 and 20-2)

Total body water (TBW) includes water both inside and outside of cells and water normally present in the gastrointestinal and genitourinary systems. TBW can be theoretically divided into two main compartments. The *anatomical extracellular water (ECW)* includes all water external to cell membranes and constitutes the medium through which all metabolic exchange occurs. *Intracellular water (ICW)* includes all water within cell membranes and constitutes the medium in which chemical reactions of cell metabolism occur. This compartment is heterogeneous and discontinuous; the interior of each cell is separated from the extracellular water and from the interior of every other cell by the semipermeable cell membrane.

The anatomical ECW is functionally subdivided into *physiological extracellular water* and *transcellular water*. The physiological ECW is the portion of the anatomical ECW whose volume is accessible to direct measurement; it includes *plasma* (intravascular water) and *interstitial fluid* (ISF). The ISF, which includes extravascular, extracellular water into which ions and small molecules diffuse freely from plasma, is the fluid that directly bathes the cells of the body. In addition, there are potential spaces in the body (pericardial, pleural, peritoneal, and synovial) that are normally empty except for a few milliliters of viscous lubricating fluid and are considered to be part of the ISF compartment. Transcellular water includes water in extracellular compartments enclosed by an epithelial membrane, whose volume and composition is determined by the cellular activity of that membrane. These heterogeneous compartments include the aqueous humor in the eye, the cerebrospinal fluid, and water within the gastrointestinal, genitourinary, and nasorespiratory systems. The volume of the transcellular water portion of the anatomical ECW is not included in conventional measurements of extracellular water (see Chapter 38).

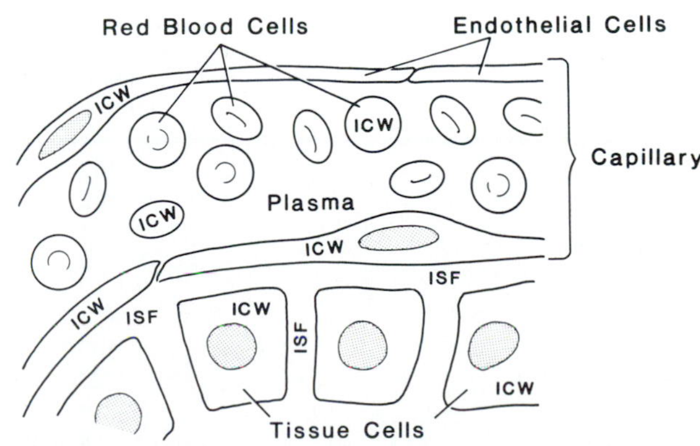

Fig. 20-2 Diagram of plasma, interstitial fluid, *ISF,* and intracellular water, *ICW,* in tissue at microscopic level.

Table 20-1 Compartment volumes

	Percentage of body weight	Percentage of total body water	Volume in 70 kg man
Total body water	60%		42 L
Extracellular water	20%	33%	14 L
Plasma	5%	8%	3.5 L
Interstitial fluid	15%	25%	10.5 L
Intracellular water	40%	67%	28 L

Volume of body water compartments (Table 20-1)

TBW is 65% of body weight in average adult men and 55% of body weight in women. This difference between men and women is largely the result of differences in body fat. As a percentage of total body weight, TBW varies inversely with body fat content, from approximately 70% in very thin persons to 50% in very obese persons.

Physiological ECW volume is approximately 20% of body weight and one third of TBW in the average adult. Unlike TBW, neither physiological nor anatomical ECW volumes can be accurately measured. Plasma volume can be accurately measured and is approximately 5% of body weight. Interstitial fluid volume is calculated as the difference between the ECW and plasma volumes. It is approximately 15% of body weight and one quarter of TBW. Intracellular water is calculated as the difference between the TBW and ECW volumes. It is equal to 40% of body weight and two thirds of TBW in the average adult. ICW volume calculated in this manner includes transcellular water, which has been estimated to be 1% to 3% of body weight.

Maturational changes in body water compartment volumes (Fig. 20-3)

The fraction of body weight that is water and the proportion of TBW that is ECW and ICW do not remain constant during growth. When expressed as a percentage of body weight, TBW gradually decreases during intrauterine gestation and early childhood, reaching a value approximating that in the adult by about 3 years of age. During this time ECW (expressed as a percentage of body weight) decreases and ICW (expressed as a percentage of body weight) increases. Thus ECW becomes a lesser and ICW a greater proportion of TBW. Plasma volume remains constant at 4% to 5% of body weight throughout life. Of course, the absolute volumes of TBW, ECW, ICW, and plasma all increase with growth.

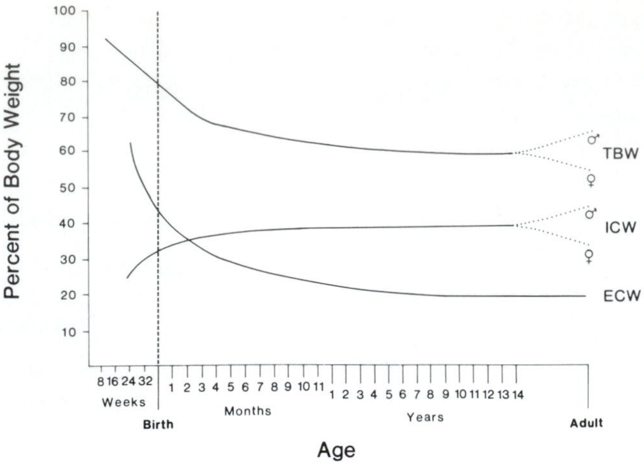

Fig. 20-3 Changes in body water compartments (expressed as a percentage of body weight) with age. *TBW,* Total body water; *ICW,* intracellular water; *ECW,* extracellular water. (Adapted from data of Friis-Hansen, B: Acta Paediatr Scand 46 [suppl 110]:1, 1957.)

Composition of body water compartments (Table 20-2)

Plasma compartment. The plasma compartment is the only compartment whose composition is directly measurable. Notice that the concentration of ions in the plasma is lower than that in *plasma water*. The reason is that plasma is composed of water, ions, and macromolecules. Ions are present only in the water phase. The term *plasma water* is used to indicate this aqueous fraction in distinction to the remainder, which is composed of protein, lipid, and other

Table 20-2 Composition of body compartments

	Plasma (mEq/L)	Plasma water (mEq/L)	Interstitial fluid (mEq/L H_2O)	Intracellular water (mEq/L H_2O)
Cations	153	164.6	153	195
Na^+	142	152.7	145	10
K^+	4	4.3	4	156
Ca^{++}	5	5.4	(2-3)	3.2
Mg^{++}	2	2.2	(1-2)	26
Anions	153	164.6	153	195
Cl^-	103	110.8	116	2
HCO_3^-	28	30.1	31	8
Protein	17	18.3	—	55
Others	5	5.4	(6)	130
Osmolarity (mOsm/L)		296	294.6	294.6
Theoretical osmotic pressure (mm Hg)		5712.8	5685.8	5685.8

macromolecules. The concentration of ions in plasma is lower than that in plasma water because plasma contains *both* the plasma water (in which plasma ions are dissolved) and the macromolecule fraction (in which no ions are dissolved). Plasma water represents only 93% of total plasma volume. Consequently the concentration of ions in plasma is 93% of that in plasma water (see Chapter 58 and p. 885). It should be emphasized that, although the concentration of ions in plasma is that portion conventionally measured and reported, it is the concentration of ions in plasma water that affects the distribution of ions across the capillary endothelium. If there is an abnormally increased amount of macromolecules in the plasma (such as lipids), the measured plasma concentration of ions will be low even though the concentration of ions in plasma water, and the resultant chemical activities of these ions, may be normal. In addition to protein, plasma contains high concentrations of sodium and chloride, moderate concentrations of bicarbonate, and low concentrations of calcium, magnesium, phosphate, sulfate, and organic acids.

The sum of all the charges of positively charged ions (cations) must be equal to the sum of all the charges of negatively charged ions (anions) to maintain electrical neutrality in the plasma. Most often in clinical medicine, however, the plasma concentrations only of sodium, potassium, chloride, and bicarbonate are measured. The sum of these *measured* cations exceeds that of the *measured* anions. Therefore the sum of unmeasured plasma anions must be greater than that of the unmeasured cations. The difference between the sum of measured cations and the sum of measured anions is known as the *anion gap* and is calculated either as [Na] + [K] − [Cl] − [HCO_3] or as [Na] − [Cl] − [HCO_3] (see Chapter 58, p. 859). The latter is frequently used because the plasma potassium concentration is relatively constant and often spuriously elevated (see p. 329). Because total plasma cation concentration

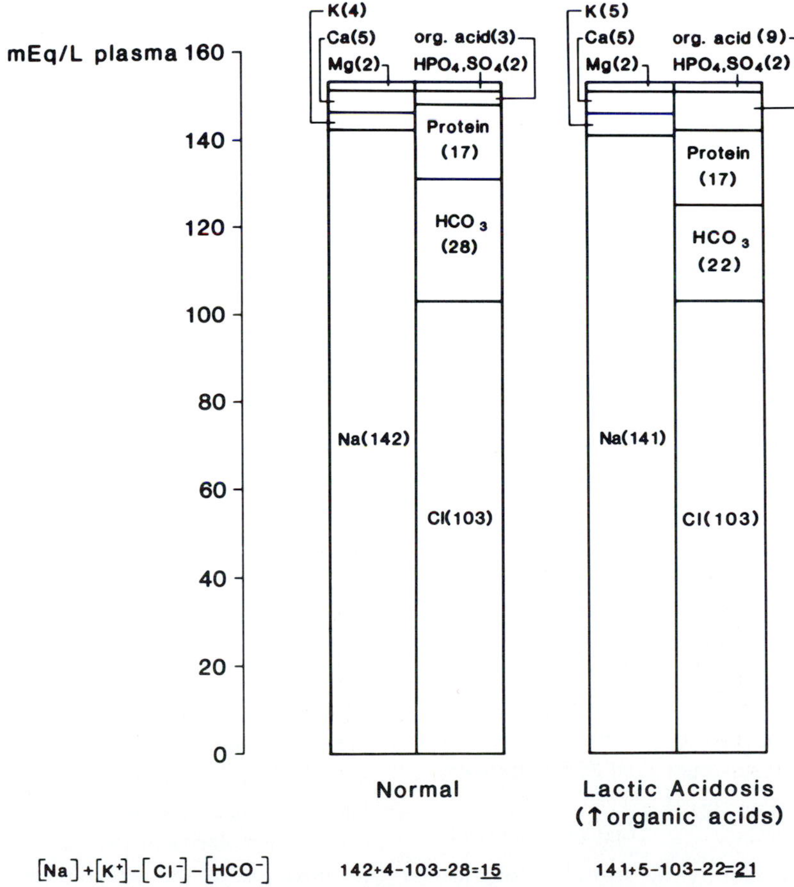

Fig. 20-4 Increased anion gap because of increase in unmeasured anion. *Numbers in parentheses,* Concentration of ions in units of mEq/L plasma. Notice that sum of cations (left-hand side of each bar graph) is always equal to sum of anions (right-hand side of each bar graph), both under normal conditions and in presence of lactic acidosis. Sum of concentrations of unmeasured anions (organic acids, HPO_4, SO_4, and proteins) is larger than sum of concentrations of unmeasured cations (Ca and Mg). During lactic acidosis, difference between unmeasured anions and cations becomes greater because production of lactic acid increases concentration of organic acids.

must equal total plasma anion concentration and decreases in unmeasured cations have little effect in the calculation, an increased anion gap is usually indicative of an increase in the concentration of one or more of the unmeasured anions (Fig. 20-4). A decrease in the anion gap suggests the opposite possibility. The most frequent use of the anion gap clinically is in the differential diagnosis of metabolic acidosis (see Chapter 21).

Interstitial fluid compartment. The interstitial fluid cannot normally be sampled in amounts sufficient for chemical analysis. The major difference between the ISF and plasma is the presence of protein in the plasma and its relative absence in the ISF. Although the concentrations of freely diffusible solute in ISF might be expected to be equal to those in plasma water, this is true only for uncharged solutes. The presence of polyanionic protein molecules in plasma, which cannot cross semipermeable membranes, leads to the Gibbs-Donnan equilibrium (see p. 321). This equilibrium results in plasma water cation concentrations slightly greater than those in ISF and plasma water anion concentrations slightly less than those in ISF. Values for ISF ion concentrations given in Table 20-2 are theoretical approximations based on Gibbs-Donnan equilibrium calculations.

Intracellular water compartment. Solute concentrations in cell water cannot be directly determined. The ICW compartment is heterogeneous; important differences exist in intracellular solute concentrations between different cell types. However, certain features of most cell fluids are quantitatively similar and distinguish ICW from ECW. The major cations of ICW are potassium and magnesium, and the concentration of sodium is always low; the major anions of cell fluids are protein, organic phosphates, and sulfates, whereas chloride and bicarbonate concentrations are low. The profile presented in Table 20-2 is for muscle cells.

Osmotic pressure and osmolarity of body fluids

Osmotic pressure is an important factor determining the distribution of water among the body water compartments. (See Chapter 12 for a description of colligative properties that determine osmotic pressure.) The *theoretical* osmotic pressure (and water attractability) of a solution is proportional to its osmolarity. The theoretical osmotic pressure of a solution at body temperature is calculated as follows:

$$\text{Theoretical osmotic pressure (mm Hg)} = 19.3\ (\text{mm Hg/mOsm/L}) \times \text{Osmolarity (mOsm/L)}$$

Note that the solute permeability of specific biological membranes is not considered in this calculation. The osmolarity and theoretical osmotic pressure of each of the body water compartments are listed in Table 20-2.

Osmotic pressure can be simply seen as the force that tends to move water from dilute solutions to concentrated solutions. When a membrane is permeable to a solute, the solute exerts no osmotic pressure across the membrane—it does not contribute to the *effective* osmotic pressure of the solution. The effective osmotic pressure of a solution thus depends on the total number of solute particles in solution *and* the permeability characteristics of the particular membrane in question. The higher the permeability of a membrane to a solute, the lower is the effective osmotic pressure of a solution of that solute at any given osmolarity. For example, cell membranes are much more permeable to urea than to sodium and chloride. Therefore the effective osmotic pressure of a solution of urea across the cell membrane would be much less than that of a solution of NaCl of the same osmolarity. Measurement of the osmolarity of body compartment water is a measure only of its *theoretical,* not effective, osmotic pressure.

A solution with an *effective* osmotic pressure greater than that of plasma is said to be *hyperosmotic* with respect to plasma. A solution with a *theoretical* osmotic pressure greater than plasma is said to be *hypertonic*. *Hypoosmotic* and *hypotonic* solutions are those with effective and theoretical osmotic pressures, respectively, less than those of plasma.

The effective osmotic pressure of plasma and ISF across the capillary endothelium by which they are separated is referred to as their *colloid* osmotic pressure. The capillary endothelium is freely permeable to most solutes in plasma and ISF. Therefore these solutes contribute to theoretical, but not to effective, osmotic pressure. The capillary endothelium is impermeable to large protein molecules (colloids) under usual circumstances. It is these colloids that are responsible for the effective osmotic pressure of plasma and ISF.

REGULATION OF BODY FLUID COMPARTMENT OSMOLARITY AND VOLUME

Extracellular compartment

Regulation of the ECW osmolarity and volume depends on the independent control of each of these variables by the hypothalamus, the renin-angiotensin-aldosterone system, atrial natriuretic factor, and the kidney.

Water metabolism and hypothalamus (Fig. 20-5). The regulatory centers for water intake and water output are located in separate areas of the hypothalamus in the brain. Neurons in each of these areas respond to increases in ECW osmolarity, to decreases in intravascular volume, and to angiotensin II. Increased ECW osmolarity stimulates these neurons directly by causing them to shrink (increased osmolarity of ISF bathing any cell will cause water to move out of the cell into the ISF—see p. 321). A decrease in intravascular volume causes a reduction in activity of distension receptors located in the atria of the heart, the inferior vena cava, and the pulmonary veins and a reduction in activity of blood pressure receptors in the aorta and the carotid arteries. Relay of this information to the central nervous system stimulates neurons in the water-intake and water-output areas of the hypothalamus. Circulat-

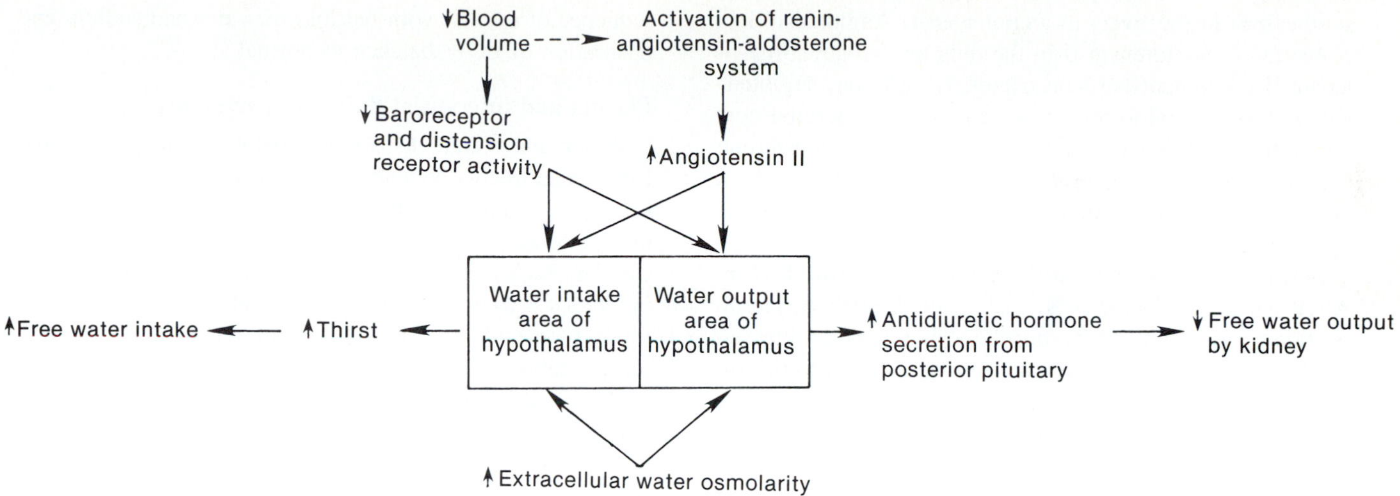

Fig. 20-5 Hypothalamic regulation of water balance.

ing angiotensin II seems to act directly to stimulate neurons located in these water-control areas of the hypothalamus. Stimulation of neurons located in the water-intake area produces the conscious sensation of thirst and thereby stimulates water intake. Stimulation of neurons located in the water-output area results in the release of antidiuretic hormone (ADH) from the posterior pituitary gland. ADH stimulates water reabsorption in the collecting ducts of the kidney, which results in the formation of hypertonic urine and decreased output of free water (water without solute). The integration of all these control mechanisms governing water intake and output ensures maintenance of appropriate water balance.

Water and sodium metabolism and renin-angiotensin-aldosterone system (Fig. 20-6). The renin-angiotensin-aldosterone system functions as a neurohormonal regulating mechanism for body sodium and water content, arterial blood pressure, and potassium balance. Renin is a proteolytic enzyme synthesized, stored, and secreted by cells in the juxtaglomerular bodies of the kidney. Renin secretion is increased by decreased renal perfusion pressure, stimulation of sympathetic nerves to the kidneys, and decreased sodium concentration in the fluid of the distal tubule. Renin converts angiotensinogen (a polypeptide

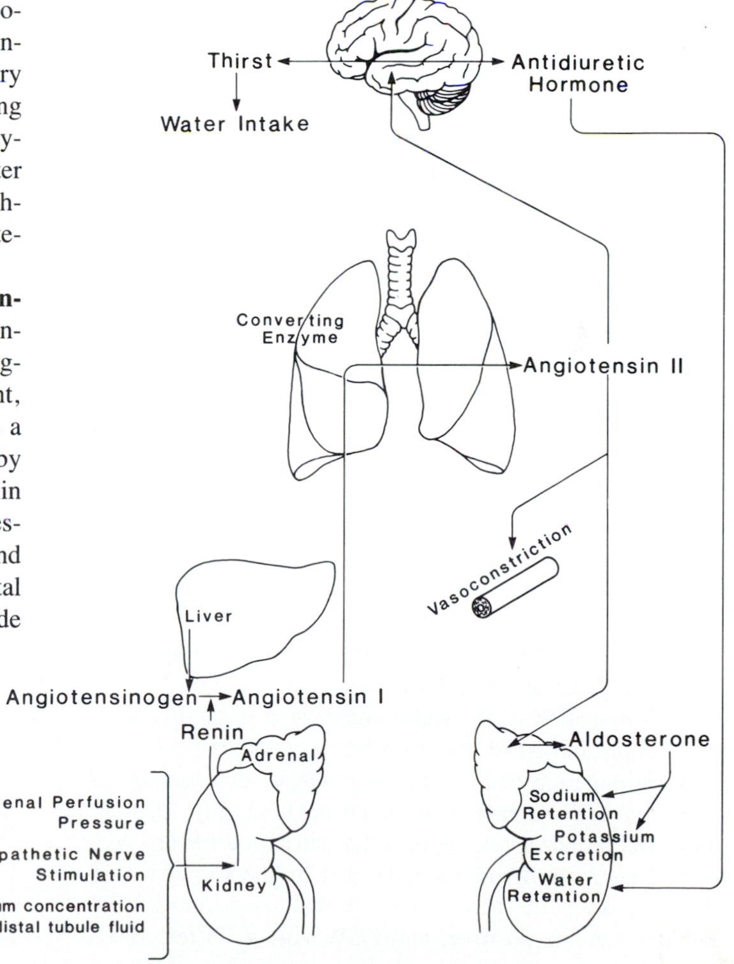

Fig. 20-6 Renin-angiotensin-aldosterone system.

synthesized in the liver) to angiotensin I. Angiotensin I is converted to angiotensin II in the lung and kidney. Angiotensin II is a potent vasoconstrictor. In addition, angiotensin II stimulates aldosterone secretion by the adrenal cortex, thirsting behavior, and ADH secretion. Aldosterone stimulates sodium reabsorption in the distal nephron. As a consequence of this sodium reabsorption, water is retained by the body.

Sodium metabolism and atrial natriuretic factor. Atrial natriuretic factor (ANF, also called atrial natriuretic peptide or atrial natriuretic hormone) is a peptide released by the atria of the heart when they are distended because of expansion of the extracellular space. ANF increases salt and water excretion by the kidney by increasing glomerular filtration rate (GFR) and by inhibiting sodium reabsorption by the renal tubules.

Factors and mechanisms involved in the control of water and sodium excretion by the kidney are discussed in Chapter 22.

Control of extracellular water osmolarity. ECW *osmolarity* is regulated by the hypothalamic control of water intake (regulatory thirst) and renal excretion of free water (see Fig. 20-5). Increased ECW osmolarity stimulates water intake and ADH secretion. ADH secretion decreases renal water excretion. Increased water intake and decreased renal water excretion results in a positive water balance, that is, water gain in excess of water loss. Positive water balance decreases ECW osmolarity to normal. The opposite occurs with decreased ECW osmolarity: thirst and ADH secretion are inhibited. Negative water balance (water loss in excess of water gain) results, and ECW osmolarity is restored to normal.

Control of extracellular water volume. Control of ECW *volume* depends on the integrated control of *water and sodium balance* by the water intake and output areas of the hypothalamus, the renin-angiotensin-aldosterone system, atrial natriuretic factor, and the kidney.

When water and sodium output exceed intake (water and sodium balance are negative), the ECW volume contracts. The associated decrease in plasma volume results in decrease in venous blood return to the heart and decrease in cardiac output. These cardiovascular changes produce the following effects:

1. Stimulation of the water intake area of the hypothalamus and thirst center (see Fig. 20-5)
2. Stimulation of the water output area of the hypothalamus and ADH secretion (see Fig. 20-5)
3. Stimulation of the renin-angiotensin-aldosterone system and increase in angiotensin II (see Fig. 20-6)
4. Inhibition of release of atrial natriuretic factor
5. Retention of sodium and water by the kidney

The net result of these effects is that water and sodium balance become positive, and ECW volume is restored to normal.

Expansion of the ECW volume results in the opposite sequence of events, with net loss of water and sodium and restoration of ECW balance to normal.

Plasma and interstitial fluid compartments

Water and solute distribution between the plasma and ISF compartments depends on an intact capillary endothelial surface and is controlled passively by the interaction of hydrostatic, osmotic, and electrochemical forces. The capillary endothelium functions as a continuous tube, with numerous 4 to 5 nm diameter intercellular channels. It is freely permeable to water and small solutes and relatively impermeable to protein.

Water distribution. Water distribution across the capillary endothelial surface is controlled by the balance of forces that tend to move water from the plasma to the ISF (filtration forces) and forces that tend to move water from the ISF into the plasma (reabsorption forces). The major filtration force is plasma hydrostatic pressure in the capillary. A much weaker filtration force is the ISF colloid osmotic pressure. Since the protein concentration in ISF is negligible, colloid osmotic pressure is low. Another weak filtration force is a small *negative* ISF hydrostatic pressure. The major reabsorption force is the colloid osmotic pressure exerted across the capillary endothelium by plasma proteins. As a broad generalization, plasma hydrostatic pressure (which tends to drive water out of the capillary) exceeds plasma colloid osmotic pressure (which tends to draw water into the capillary) at the arteriolar end of the capillary so that net filtration occurs. As plasma moves along the capillary and filtration occurs, plasma hydrostatic pressure decreases and plasma protein concentration (and therefore plasma colloid osmotic pressure) increases along the course of the capillary so that net reabsorption occurs toward the venous end of the capillary. This is depicted schematically in Fig. 20-7. Overall, filtration exceeds reabsorption; therefore water must be returned to the plasma from the ISF compartment by way of the lymphatic system to prevent edema (defined as an abnormal increase in ISF volume).

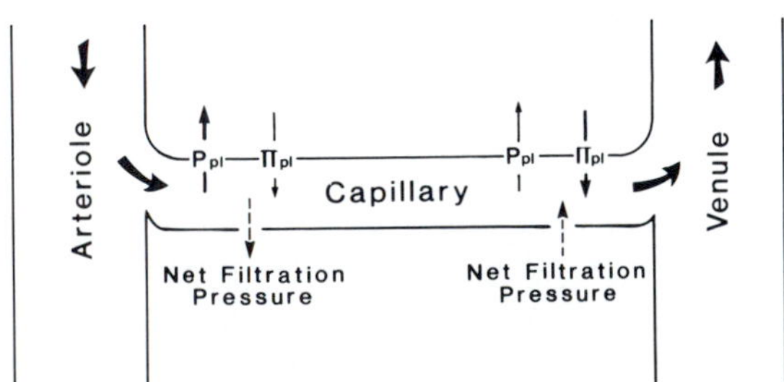

Fig. 20-7 Starling's hypothesis of water distribution between plasma and interstitial fluid compartments. Thickness of arrows representing plasma hydrostatic pressure, P_{pl}, and plasma oncotic pressure, Π_{pl}, indicate their relative magnitudes. *Dashed arrows,* Direction of net filtration pressure.

Solute distribution. The small differences in the concentrations of the various extracellular solutes across the capillary endothelium are the result of the presence of polyanionic protein molecules (that is, having multiple negative charges) in plasma to which the capillary endothelium is relatively impermeable. This results in the Gibbs-Donnan equilibrium (Fig. 20-8): the presence of impermeant polyanionic macromolecules restricted to one side of a membrane permeable to solvent and small ions establishes a characteristic distribution of the permeable ions. At electrochemical equilibrium, the concentrations of *diffusible cations* are slightly *higher* and the concentrations of *diffusible anions* slightly *lower* in the compartment containing the impermeant polyanionic macromolecule.

In the cases of calcium and magnesium, the larger differences between plasma water and ISF concentrations result from the fact that approximately 45% of plasma calcium and 25% of plasma magnesium is protein bound and therefore is nondiffusible.

Intracellular compartment

Water and solute distribution across the cell membrane between the ISF and ICW depends on the integrity of the cell membrane and on osmotic and electrochemical forces; all these factors are sustained by cell metabolism. The cell membrane behaves as though it were an oil film with numerous 0.7 nm diameter pores. This membrane is highly permeable to water but differentially permeable to solutes. The permeability of the cell membrane to a solute is directly related to the lipid solubility of the solute and inversely related to its hydrophilicity (water attractability) and molecular size. Other factors being constant, membrane permeability is greater to anions than to cations.

Cell volume. Cell volume is controlled by ISF osmolarity. Osmolarity inside the cell must equal osmolarity outside the cell because the cell membrane is highly permeable to water and no hydrostatic pressure gradient can be maintained across animal cell membranes. The osmotic content of the intracellular compartment is maintained relatively constant by cell metabolism. Therefore osmotic equilibrium across the cell membrane can be maintained in the face of changes in ISF osmolarity only by the movement of water between the intracellular compartment and interstitial space. A decrease in ISF osmolarity causes movement of water into cells and an increase in intracellular volume. Conversely, an increase in ISF osmolarity causes movement of water out of cells and a decrease in intracellular volume.

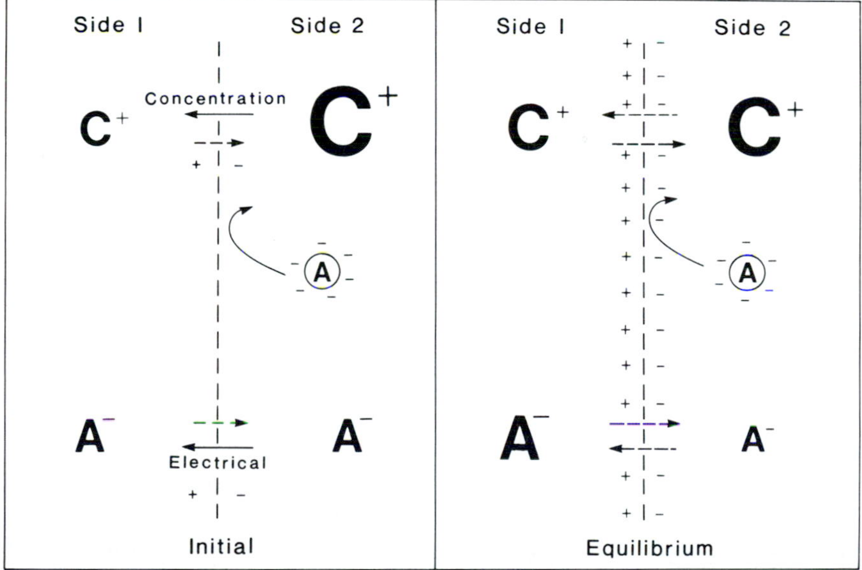

Fig. 20-8 Gibbs-Donnan equilibrium. Distribution of diffusible and nondiffusible ions and development of an electrical potential gradient across a membrane when nondiffusible, polyvalent anion (Ⓐ) with diffusible cation (C^+) is added to one side of membrane in solution of diffusible cation (C^+) and anion (A^-). Initially, diffusible cation moves down its concentration gradient from side 2 to side 1. This movement generates an electrical potential gradient across membrane (side 2 negative with respect to side 1). Diffusible anion moves down this electrical potential gradient from side 2 to side 1. At equilibrium, concentration of diffusible cation will be greater on side 2 than side 1 (as indicated by size of symbols), whereas concentration of diffusible anion will be greater on side 1 than side 2. No *net* movement of diffusible ions occurs across membrane because no net electrochemical gradients exist. Concentration gradient for each ion is balanced by an equal but oppositely directed electrical gradient.

Table 20-3 Water balance in average adult under various conditions

Intake (mL/day)				**Output (mL/day)**			
	Normal	**Hot environment**	**Strenuous work**		**Normal**	**Hot environment**	**Strenuous work**
Drinking water	1200	2200	3400	Urine	1400	1200	500
Water from food	1000	1000	1150	Insensible water			
Water of oxidation	300	300	450	Skin	400	400	400
				Lungs	400	300	600
				Sweat	100	1400	3300
				Stool	200	200	200
Total	2500	3500	5000		2500	3500	5000

Cell solute content. The ionic composition of the intracellular fluid is shown in Table 20-2. This composition is largely the result of an energy-dependent ionic transport pump (Na,K-ATPase) found in the cell membrane that extrudes sodium from the cell in exchange for potassium. In addition, the intracellular solute composition depends on the intracellular production of nonpermeable polyanionic macromolecules. The cellular content of the other, permeable ions results from (1) electrochemical gradients produced by Na-K exchange and nonpermeable intracellular polyanionic macromolecules (Gibbs-Donnan effect), (2) the specific permeability characteristics of the cell membrane to the various ions, and (3) other energy-dependent ion-specific transport pumps. The latter two factors vary from cell type to cell type and are responsible for the differences in ionic content among various cell types.

All factors influencing cell solute content depend on normal cellular metabolism. When cellular metabolism is disrupted, such as during asphyxia, solute and water enter the cell, causing it to swell.

WATER METABOLISM

Water balance

Extracellular water osmolarity is maintained constant at 285 to 298 mOsm/L as a consequence of the dynamic balance between water intake and water excretion, which is controlled by the mechanisms discussed previously. Average daily water turnover in the adult is approximately 2500 mL; however, the range of water turnover possible is great and depends on intake, environment, and activity (Table 20-3).

Under normal conditions approximately one half to two thirds of water intake is in the form of oral fluid intake, and approximately one half to one third is in the form of oral intake of water in food. In addition, a small amount of water (150 to 350 mL/day) is produced by oxidative metabolism. Oral fluid intake is the only source of water that is regulated in response to changes in ECW volume and osmolarity.

Routes of water excretion include urinary water loss, insensible water loss, sensible perspiration (sweating), and gastrointestinal water loss. The kidney is the principal organ regulating the volume and composition of the body fluids. Urine volume varies over a wide range in response to changes in ECW volume and osmolarity. Solute excretion is regulated independently.

Loss of water by diffusion through the skin and through the respiratory tract is known as *insensible water loss* because it is not apparent. It is the only route by which water is lost without solute. Normally, half of insensible water loss occurs through the skin and half through the respiratory tract. The magnitude of cutaneous insensible water loss is a function of body surface area; therefore it is disproportionately greater in infants and children for weight. Insensible water loss varies directly with ambient temperature, body temperature, and activity and inversely with ambient humidity.

Sensible perspiration is negligible in a cool environment but may be quite large with increases in ambient temperature, body temperature, or physical activity. Sodium and chloride are the major ionic components of sweat, but sweat is almost invariably hypotonic to plasma. An increase in ECW osmolarity causes a decrease in the rate of sensible perspiration.

Net water loss from the gastrointestinal tract is normally small, approximately 150 mL/day. However, the *flux* of water and electrolytes between the gastrointestinal tract and ECW compartment is quite large. Therefore, if reabsorption is impaired, water and electrolyte losses from the gastrointestinal tract can be great, as with diarrhea. With the exception of saliva, which is hypotonic, the total solute concentration of most gastrointestinal secretions is similar to ISF.

Disorders of water imbalance

Disorders of water balance (dehydration and overhydration) result from an imbalance of water intake and output or sodium intake and output (Table 20-4).

Dehydration

Deficit of water. *Simple dehydration,* defined as a decrease in total body water with relatively normal total body sodium, may result from failure to replace obligatory water

Table 20-4 Changes in total body water volume and distribution, total body sodium content, and plasma sodium concentration with dehydration and overhydration

	Total body water	Extracellular water	Intracellular water	Total body sodium	Plasma sodium concentration
Dehydration					
Hypernatremic	↓	sl ↓	↓	nl or sl ↓	↑
Normonatremic	↓	↓	nl	↓	nl
Hyponatremic	↓	↓ ↓	↑	↓ ↓	↓
Overhydration					
Water intoxication	↑	↑	↑	nl	↓
Extracellular water volume expansion					
Normonatremic	↑	↑	nl	↑	nl
Hyponatremic	↑	↑	↑	sl ↑	↓

nl, Normal; *sl*, slightly.

losses or failure of the regulatory or effector mechanisms that promote conservation of water by the kidney (see the box). Simple dehydration is by definition associated with hypernatremia and hyperosmolarity because water balance is negative and sodium balance is normal. The increase in ECW osmolarity as water is lost from the body results in movement of water out of the ICW compartment. Therefore simple dehydration results in contraction of both the ECW and ICW compartments (Table 20-4).

Deficit of water and sodium. More often dehydration results from a net negative balance of both water and sodium. In this case water balance may be more negative than, equal to, or less negative than sodium balance (Table 20-4). If water balance is more negative than sodium balance, the result is *hypernatremic* or *hyperosmolar* dehydration; if it is equally negative, *normonatremic* or *isomolar* dehydration results; and if it is less negative, *hyponatremic* or *hyposmolar* dehydration results. Hypernatremic dehydration is most common. Some causes of water and sodium deficits are listed in the box below.

CAUSES OF DEHYDRATION (WATER AND SODIUM DEFICITS)

Hypernatremic dehydration

Water and food deprivation
Excessive sweating*
Osmotic diuresis (with glucosuria)
Diuretic therapy*

Normonatremic dehydration

Vomiting, diarrhea
Replacement of losses in the above conditions with low-sodium liquids

Hyponatremic dehydration

Diuretic therapy†
Excessive sweating
Salt-wasting renal disease
Adrenocortical insufficiency

*If free water intake is inadequate.
†If free water intake is excessive.

The degree of extracellular volume contraction for a given sodium deficit and the associated change in intracellular volume is different for each of these types of dehydration (Table 20-4). The degree of extracellular volume contraction is least with hypernatremic dehydration because the increase in ECW osmolarity causes water to move out of the cell; contraction of ICW volume occurs. Thus the total body water deficit is "shared" by the extracellular and intracellular compartments. The degree of extracellular volume contraction is intermediate with normonatremic dehydration, since no water moves out of or into cells because there is no change in ECW osmolarity. There is also no change in ICW volume. The degree of ECW volume depletion is largest with hyponatremic dehydration because the decrease in ECW osmolarity causes water to move into cells. ICW volume is actually increased.

Symptoms of dehydration. The signs and symptoms of dehydration include thirst, dry mucous membranes, decreased skin turgor, decreased urine output and increased urine osmolarity (except when caused by failure of the kidney to conserve free water), increased blood urea nitrogen, and increased hematocrit. With increasing severity, weakness, lethargy, hypotension, and shock may occur.

Overhydration

Excessive water. *Water intoxication* is defined as an increase in total body water with normal total body sodium. It rarely results from excessive water consumption (polydipsia). More often water intoxication results from impaired renal free water excretion as the result of ADH secretion in excess of that required to maintain normal ECW osmolarity (syndrome of inappropriate ADH secretion, SIADH; see the box on p. 324).

With water intoxication the *dilutional* hyponatremia and hypoosmolarity of the ECW results in water movement

CAUSES OF WATER INTOXICATION

Polydipsia

Psychogenic—secondary to a psychiatric disturbance
Organic—secondary to an anterior thalamic lesion

SIADH

Increased secretion of ADH by hypothalamus secondary to decreased venous return to heart with no decrease in total blood volume
- Asthma
- Pneumothorax
- Bacterial or viral pneumonia
- Positive pressure ventilation
- Chronic obstructive pulmonary disease
- Right-sided heart failure
- Disease of spinal cord or peripheral nerves (Guillain-Barré syndrome, poliomyelitis)

Increased secretion of ADH by hypothalamus in absence of appropriate osmolar or volume stimuli
- Central nervous system disorders (intracranial hemorrhage, hydrocephalus, skull fracture, severe asphyxia, brain tumors, cerebrovascular thrombosis, meningitis, encephalitis, seizures, acute psychoses, and cerebral atrophy)
- Hypothyroidism
- Pain, fear
- Anesthesia or surgical stress
- Drugs such as morphine, barbiturates, cyclophosphamide, vincristine, and carbamazepine

Ectopic, autonomous secretion of ADH

Bronchogenic carcinoma
Adenosarcoma of pancreas
Lymphosarcoma
Duodenal adenocarcinoma
Pulmonary tuberculosis
Pulmonary abscess

into the cells. Therefore water intoxication produces expansion of the ECW and ICW compartments (Table 20-4).

The symptoms of water intoxication are related to the degree and rate of fall in sodium. With an acute fall in serum sodium to 120 to 125 mEq/L, nausea, vomiting, seizures, and coma can occur.

Excessive water and sodium. *Expansion of the extracellular compartment* usually results from retention of sodium *and* water. This occurs with oliguric renal failure, nephrotic syndrome, congestive heart failure, cirrhosis, and primary hyperaldosteronism. In these conditions total body water excess is associated with normal or low serum sodium and osmolarity (Table 20-4). Hypernatremia is rare with water excess. If the serum sodium is normal, the increase in total body water (TBW) will be limited to the ECW. With hyponatremia the increase of TBW will be shared by the ECW and ICW compartments.

SODIUM METABOLISM

Sodium balance

In a normal adult the total body sodium is about 55 mEq/kg of body weight; about 30% is tightly bound in the crystalline structure of bone and thus nonexchangeable. Thus only 40 mEq/kg is exchangeable among the various compartments and accessible to measurement. The exchangeable sodium is distributed primarily in the extracellular space (Fig. 20-9). About 97% to 98% of the exchangeable sodium is found in the ECW space and only 2% to 3% in the ICW space. Approximately 16% of exchangeable sodium is in plasma, 41% is in ISF that is readily accessible to the plasma compartment, 17% is in ISF of dense connective tissue and cartilage, 20% is in ISF of bone, and 3% to 4% in the transcellular water compartment. Total bone sodium (exchangeable plus nonexchangeable) accounts for 40% to 45% of the total body sodium. The concentrations of sodium in the various fluid

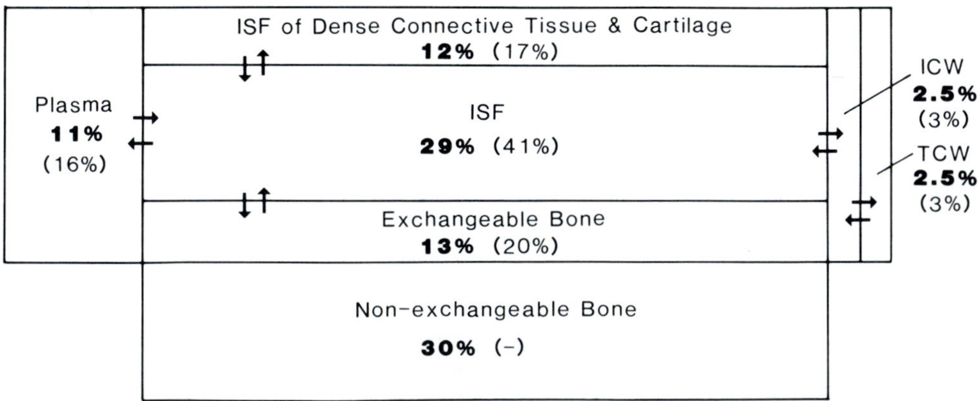

Fig. 20-9 Distribution of sodium among body compartments. *Bold numbers,* Percentages of *total body* sodium in various compartments. *Numbers in parentheses,* Percentages of *exchangeable* sodium in various compartments. *ISF,* Interstitial fluid; *ICW,* intracellular water; *TCW,* transcellular water.

Table 20-5 Electrolyte composition and volume of various gastrointestinal secretions in a normal adult

Fluid	Volume secreted (mL/day)	Electrolyte concentration (mEq/L)			
		Na^+	K^+	Cl^-	HCO_3^-
Gastric juice*	2500	8-120	1-30	8-100	0-20
Bile	700-1000	134-156	4-6	83-110	38
Pancreatic juice	>1000	113-153	2-7	54-95	110
Small bowel	3000	72-120	3.5-7	69-127	30
Ileostomy	100-4000	112-142	4.5-14	43-122	30
Cecostomy	100-300	480-116	11-28	35-70	15
Feces	100	<10	<10	<15	<15

*Electrolyte composition of gastric juice varies, depending on acidity. The higher the acidity, the lower the sodium concentration, the higher the chloride concentration, and the lower the bicarbonate concentration The average sodium concentration is approximately 100 mEq/L. (From Lockwood, JS, and Randall, HT: Bull NY Acad Med 25:228-243, April 1949.)

compartments are depicted in Table 20-2. As discussed previously, the difference in sodium concentration between plasma and ISF is the result of the Gibbs-Donnan equilibrium. The difference in sodium concentration between ISF and ICW is the result of the active transport of sodium out of the cell in exchange for potassium.

The amount of sodium in the body is a reflection of the balance between sodium intake and output. Sodium intake depends on the quantity and type of food intake. Under normal conditions the average adult takes in about 50 to 200 mEq of sodium/day. Sodium output occurs through three primary routes: the gastrointestinal tract, the skin, and the urine.

Under normal circumstances loss of sodium through the gastrointestinal tract is very small. Fecal water excretion is only 100 to 200 mL/day for a normal adult, and fecal sodium excretion only 1 to 2 mEq/day. However, one should bear in mind that although fecal losses of water and electrolytes are normally small, the total volume of gastrointestinal fluid secreted is large, averaging about 8 L/day. Almost all this volume is normally reabsorbed. However, with impaired gastrointestinal reabsorption, losses of water and electrolytes are large. The volume and electrolyte content of various gastrointestinal secretions are shown in Table 20-5. Notice that most of the secretions have sodium contents much higher than that of the feces. Thus with severe diarrhea or with gastric or intestinal drainage tubes, sodium losses through the gastrointestinal tract may exceed 100 mEq/day.

The sodium content of sweat averages about 50 mEq/L but is somewhat variable. The sweat sodium concentration is decreased by aldosterone and increased in cystic fibrosis. The rate of sweat production is highly variable, increasing in hot environments, during exercise, and with fever. Under extreme conditions sweat production can exceed 5 L/day, accounting for a loss of more than 250 mEq of sodium. Under normal conditions, in a cool environment, sodium losses from the skin are small. With extensive burns or exudative skin lesions there is great loss of sodium and water.

The major route of sodium excretion is through the kidney. Furthermore, the urinary excretion of sodium is carefully regulated to maintain body sodium homeostasis, which in turn is critical to control of extracellular volume. The details of the mechanisms and regulation of renal sodium excretion are discussed in Chapter 22.

Sodium is freely filtered by the glomerulus. Approximately 70% of the filtered sodium is reabsorbed by the proximal tubule, about 15% by the loop of Henle, about 5% by the distal convoluted tubule, 5% by the cortical collecting tubule, and another 5% by the medullary collecting duct; thus normally less than 1% of the filtered sodium is excreted.

Disorders of sodium balance

Sodium excess. Sodium accumulates in the body when sodium intake exceeds sodium output because of some abnormality of sodium homeostatic mechanisms. Some major clinical causes of sodium retention are presented in the box.

Since sodium is distributed in the extracellular space, an increase in total body sodium is usually accompanied by an increase in ECW volume. An abnormal increase in ECW volume, particularly an increase in the interstitial space, produces tissue swelling known as *edema*. Thus those clinical conditions associated with sodium retention are frequently characterized by the presence of edema.

CLINICAL CONDITIONS RESULTING IN EXCESS BODY SODIUM

Cardiac failure
Liver disease
Renal disease—nephrotic syndrome
Hyperaldosteronism
Pregnancy

Clinically, edema is characterized by swelling and puffiness of the body.

Congestive heart failure. When the heart begins to fail as a pump, a series of pathophysiological mechanisms occur leading to retention of sodium. The failing heart does not pump as much blood to the kidney, resulting in less sodium filtration, greater reabsorption, and consequently less excretion. The greater venous back pressure generated from the failing heart causes fluid to move from the vascular space to the interstitial space, decreasing the effective plasma volume and cardiac output. These factors stimulate the secretion of angiotensin II, aldosterone, and ADH and decrease release of atrial natriuretic factor. These hormone responses further enhance salt and water retention.

Liver disease. In some liver diseases there is venous obstruction, which results in increased sinusoidal and portal venous pressure. These in turn lead to leakage of fluid out of the vascular space into the peritoneal space (ascites), which lowers the effective plasma volume. The lowered plasma volume leads to salt and water retention by mechanisms similar to those described for heart failure.

Renal disease. If the kidneys are damaged to such a degree that the glomerular filtration rate is greatly reduced and sodium excretion is thereby compromised, sodium retention will occur (see Chapter 22). Sodium retention occurs by another mechanism in the nephrotic syndrome. This syndrome is characterized by proteinuria and decreased serum albumin levels, which result in low plasma colloid osmotic pressure and therefore a shift of fluid from the vascular space to the ISF space. This in turn results in hypovolemia with consequent salt and water retention, as previously discussed.

Pregnancy. The reasons for sodium accumulation during pregnancy are still unclear, but there is no question that most women accumulate between 500 and 800 mEq of sodium during a normal pregnancy. Some suggest that the sodium accumulation may be a resetting of the normal homeostatic mechanism regulating body sodium and water.

Sodium depletion. Sodium depletion occurs when the output of sodium exceeds the intake (see the box). As discussed previously, only small amounts of sodium are lost in the feces under normal conditions. However, under conditions of severe diarrhea or drainage of gastrointestinal secretions, gastrointestinal sodium excretion can be quite large. If this is not replaced by increased intake, sodium depletion will result. Moreover, since the gastrointestinal route may not be available, the intravenous replacement of water and electrolytes may be necessary.

Similarly, losses of sodium through the skin are normally relatively small. However, when the volume of sweat becomes large, when the concentration of sodium in sweat is abnormally high (as with cystic fibrosis), or when there is abnormal exudation of fluid and electrolytes from the surface of the body (as occurs with extensive burns), the amount of sodium lost from the skin may be substantial and sodium depletion may occur.

CLINICAL CONDITIONS THAT CAN RESULT IN DEFICITS OF BODY SODIUM

Gastrointestinal losses—vomiting, diarrhea, fistulas, drainage tubes
Excessive sweating—exercise, fever, hot environment
Renal disease
Adrenal insufficiency—hypoaldosteronism
Diuretic therapy
Osmotic diuresis—diabetes mellitus
Burns
SIADH

When the tubules of the kidney are unable to reabsorb sodium because of disease or hormonal abnormalities, sodium loss can be excessive. For example, aldosterone deficiency, caused by disease of the adrenal gland or abnormalities in the aldosterone-regulating system, leads to decreased reabsorption of sodium in the distal nephron and total body sodium depletion. Inhibition of tubular sodium reabsorption by a diuretic also may lead to body sodium depletion.

In SIADH (see p. 323) there is water retention and hypotonic expansion of the ECW and ICW spaces. This in turn inhibits sodium reabsorption in the proximal nephron, and also perhaps the distal nephron, leading to body salt depletion.

Abnormalities of plasma sodium concentration

Changes in total body sodium are not necessarily associated with similar changes in plasma sodium concentration. That is, with salt retention, plasma sodium concentration is not necessarily increased. In fact, plasma sodium is frequently decreased in sodium-retentive states. Similarly, salt depletion is not necessarily associated with decreased plasma sodium concentrations. Plasma sodium concentration reflects the relative balances of extracellular sodium and water.

Hyponatremia (low serum sodium) occurs when there is a greater excess of extracellular water than of sodium or a greater deficit of sodium than of water. Some causes of hyponatremia are listed in the box on p. 327. Notice that in many cases there is an excess of total body sodium.

The symptoms of hyponatremia depend on the cause, magnitude, and rate of fall in serum sodium. With acute, pronounced hyponatremia caused by water intoxication, nausea, vomiting, seizures, and coma occur. Symptoms are less fulminant with chronic hyponatremia caused by salt depletion in excess of water depletion. With progressively severe degrees of chronic hyponatremia, constant thirst, muscle cramps, nausea, vomiting, abdominal

CLINICAL CONDITIONS ASSOCIATED WITH HYPONATREMIA

Water excess greater than sodium excess
- Heart failure*
- Liver disease*
- Nephrotic syndrome*
- Renal failure*
- Inappropriate ADH secretion
- Psychogenic polydipsia (excessive fluid intake)
- Essential hyponatremia (reset "osmostat")

Sodium deficit greater than water deficit
- Certain gastrointestinal abnormalities—vomiting, diarrhea, fistulas, and intestinal obstruction
- Burns
- Diuretic therapy
- Adrenal insufficiency—hypoaldosteronism

Movement of sodium from extracellular to intracellular water space
- Adrenal insufficiency—hypoaldosteronism
- Sick cell syndrome—shock

Pseudohyponatremia—hyperglycemia, hyperlipidemia, hyperglobulinemia

*Hyponatremia is dilutional, that is, secondary to excessive water retention. Total body sodium may even increase.

cramps, weakness, lethargy, and finally delirium and impaired consciousness occur.

Hypernatremia (high plasma sodium) occurs when there is greater deficit of extracellular water than of sodium. Greater excess of sodium than of water rarely occurs. Causes of hypernatremia are listed in the box below. Notice that in many cases there is actually a deficit of total body sodium.

CLINICAL CONDITIONS ASSOCIATED WITH HYPERNATREMIA

Sodium excess greater than water excess
- Ingestions of large amounts of sodium
- Administration of hypertonic NaCl or $NaHCO_3$
- Primary hyperaldosteronism

Water deficiency greater than sodium deficiency
- Excessive sweating*—exercise, fever, hot environment
- Burns*
- Hyperventilation
- Diabetes insipidus
  - ADH deficiency
  - Nephrogenic—kidney unresponsive to ADH
- Osmotic diuresis*—diabetes, mannitol infusion
- Diminished fluid input—diminished thirst
- Essential hypernatremia—reset "osmostat"
- Certain diarrheal states and vomiting*

*Total body sodium is decreased. Serum sodium concentration is increased because the magnitude of water loss exceeds the magnitude of sodium loss.

Hypernatremia usually occurs as a chronic process secondary to loss of water in excess of sodium. Symptoms are therefore those of dehydration (see p. 322).

POTASSIUM METABOLISM

Potassium balance

Approximately 98% of the total body potassium is found in the ICW space, reaching a concentration there of about 150 to 160 mEq/L. In the ECW space, the concentration of potassium is only 3.5 to 5 mEq/L. Total body potassium in an adult male is about 50 mEq/kg of body weight and is influenced by age, sex, and, very importantly, muscle mass, since most of the body's potassium is contained in muscle.

The concentrations of potassium in the various fluid compartments are listed in Table 20-2. The difference in potassium concentration between plasma and interstitial fluid (ISF) is attributable to the Gibbs-Donnan equilibrium. The difference in potassium concentration in ISF and intracellular fluid is the result of the active transport of potassium into the cell in exchange for sodium. Factors that enhance potassium transport into the cell and thereby increase the ratio of intracellular to extracellular potassium are insulin, aldosterone, alkalosis, and beta-adrenergic stimulation. Factors that decrease potassium transport into the cell or enhance leakage out of the cell include acidosis, alpha-adrenergic stimulation, and tissue hypoxia.

The amount of potassium in the body is a reflection of the balance between potassium intake and output. Potassium intake depends on the quantity and type of food intake. Under normal conditions the average adult takes in about 50 to 100 mEq of potassium/day, about the same amount as sodium. Potassium output occurs through three primary routes: the gastrointestinal tract, the skin, and the urine.

Under normal conditions loss of potassium through the gastrointestinal tract is very small, amounting to less than 5 mEq/day for an adult. The concentration of potassium in the sweat is less than that of sodium, so potassium losses through the skin are usually quite small.

The major means of potassium excretion is by the kidney. The kidney is capable of regulating the excretion of potassium to maintain body potassium homeostasis. The details of the mechanisms of renal potassium excretion are discussed in Chapter 22.

Disorders of potassium balance

Potassium excess. Potassium accumulates in the body when the intake of potassium exceeds output because of some abnormality of the potassium homeostatic mechanisms. Some major conditions causing potassium retention are presented in the box on p. 328. It should be noted that under most conditions the normal kidney is capable of ex-

CAUSES OF POTASSIUM RETENTION

Increased potassium intake
- High-potassium diet
- Oral potassium supplementation
- Intravenous potassium administration
- Potassium penicillin in high doses
- Transfusion of aged blood

Decreased potassium excretion
- Renal failure
- Hypoaldosteronism—adrenal failure
- Diuretics that block distal tubular potassium secretion: triamterene, amiloride, spironolactone
- Primary defects in renal tubular potassium secretion

creting a great deal of potassium, and a high-potassium intake leads to potassium retention only when kidney function is compromised.

Potassium depletion. Potassium depletion occurs when potassium output exceeds intake. As discussed previously, only small amounts of potassium are lost in the feces under normal conditions. As is the case for water and sodium, however, gastrointestinal potassium loss during diarrhea or drainage of gastrointestinal secretions can be quite large (see Table 20-5). Some major clinical conditions causing potassium depletion are presented in the box at right.

Note that alkalosis results in total body potassium depletion. With alkalosis, potassium moves from the extracellular to the intracellular space. In the cells of the distal nephron of the kidney, this increase in intracellular potassium stimulates potassium secretion and therefore increases renal excretion of potassium.

CAUSES OF POTASSIUM DEPLETION

Decreased potassium intake
- Low-potassium diet
- Alcoholism
- Anorexia nervosa

Increased gastrointestinal losses
- Vomiting
- Diarrhea
- Fistulas
- Gastrointestinal drainage tube
- Malabsorption
- Laxative or enema abuse

Increased urinary losses
- Increased aldosterone
  - Primary aldosteronism
  - Adrenal hyperplasia
  - Bartter's syndrome
- Adrenogenital syndrome
- Renal disease
  - Renal tubular acidosis
  - Fanconi syndrome
- Diuretics
  - Thiazides
  - Loop diuretics—ethacrynic acid, furosemide
  - Carbonic anhydrase inhibitors—acetazolamide
- Alkalosis

Abnormalities of plasma potassium concentration

Abnormalities in plasma potassium concentration can occur, not only because of abnormalities in total body potassium, but also because of shifts of potassium between the extracellular and intracellular compartments. Although similar shifts may occur with sodium, the effect of intracellular to extracellular shifting on plasma concentration is

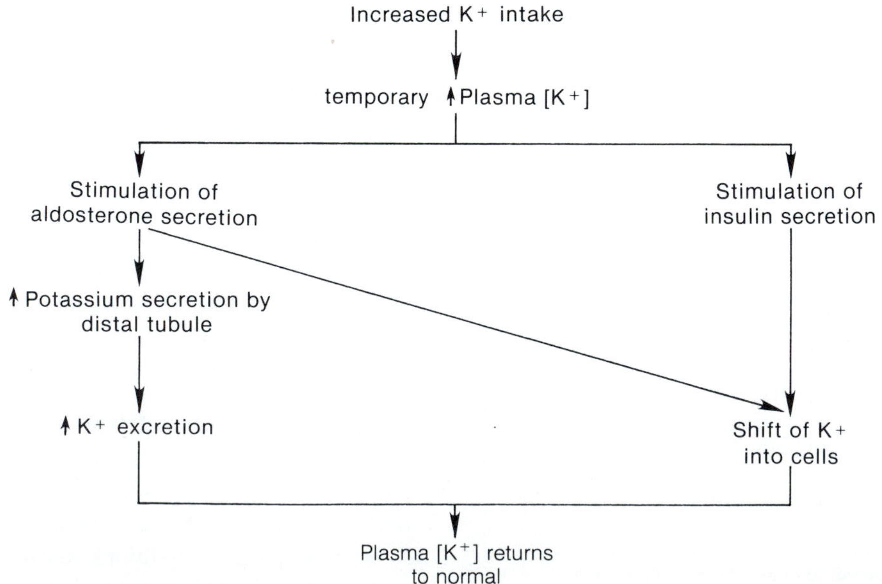

Fig. 20-10 Control of plasma potassium concentration.

CAUSES OF HYPERKALEMIA

Pseudohyperkalemia
- Hemolysis
- Leukocytosis

Intracellular to extracellular shift
- Acidosis*
- Crush injuries
- Tissue hypoxia*
- Insulin deficiency*
- Digitalis overdose*

High potassium intake (see box, p. 328, left)
Decreased potassium excretion (see box, p. 328, left)

*May be associated with total body potassium depletion.

more pronounced for potassium, since 98% of the total potassium is intracellular. For example, if only 2% of the intracellular potassium were to shift to the extracellular space, plasma potassium concentration would double. Fortunately the plasma potassium concentration is held fairly constant despite large fluctuations in potassium intake. This homeostatic mechanism is depicted in Fig. 20-10 for increased potassium intake. The opposite effect occurs with decreased potassium intake.

Hyperkalemia. Clinical conditions associated with elevated plasma potassium are listed in the box. Actual plasma potassium may be normal, but measured plasma potassium may be elevated (pseudohyperkalemia) if the blood sample is hemolyzed or if there is leakage of potassium from white blood cells where there is leukocytosis (elevated white blood cell number). In addition, vigorous arm exercise, tight application of the tourniquet, or squeezing of the area around the venipuncture site may result in cellular potassium release and spurious elevation of plasma potassium concentration.

True hyperkalemia can result from movement of potassium out of the cell into the extracellular water space, increased intake, or decreased output. Hyperkalemia caused by potassium shifts may in fact be associated with total body potassium depletion. This is the case in diabetic ketoacidosis. Hyperkalemia caused by increased intake or decreased output is associated with total body potassium excess.

The clinical signs and symptoms of hyperkalemia include changes in the electrocardiogram, cardiac arrhythmia, muscular weakness, and paresthesias. The greatest danger of hyperkalemia is life-threatening cardiac arrhythmia or arrest.

Hypokalemia. Low plasma potassium concentration (hypokalemia) can be caused by movement of potassium into the cell from the extracellular water space, increased output, or decreased intake (see box). Hypokalemia caused by potassium shifts may in fact be associated with in-

CAUSES OF HYPOKALEMIA

Extracellular to intracellular potassium shift
- Alkalosis
- Increased plasma insulin*
- Diuretic administration

Decreased potassium intake, Increased gastrointestinal losses, Increased urinary losses — See box, p. 328, right

*May be associated with total body potassium excess.

creased total body potassium. Hypokalemia caused by increased excretion or decreased intake is associated with total body potassium depletion.

Signs and symptoms of hypokalemia are numerous and include anorexia, nausea, vomiting, abdominal distension, muscle cramps or tenderness, paresthesias, electrocardiographic changes, arrhythmias, inability to concentrate the urine with resultant polyuria and polydipsia, lethargy, and confusion.

CHLORIDE METABOLISM

Chloride balance

Chloride is the major anion in the ECW space. In a normal adult the total body chloride is about 30 mEq/Kg of body weight. Approximately 88% of chloride is found in the ECW space and 12% in the ICW space. Approximately 14% of total body chloride is in the plasma, 27% in ISF that is readily accessible to plasma, 17% in ISF of dense connective tissue and cartilage, 15% in ISF of bone, and 5% in the transcellular space. The concentrations of chloride in the various fluid compartments are listed in Table 20-2. Note that the concentration of chloride in ISF is greater than that in plasma water, whereas the concentrations of sodium and potassium in ISF are less than that in plasma water. These differences between plasma and ISF are caused by the Gibbs-Donnan equilibrium. Chloride is passively distributed across the cell membrane. The difference in chloride concentration between ISF and ICW is caused by the electrical potential difference across the cell membrane. Since the inside of the cell is negative compared to the outside, the concentration of chloride outside the cell will be higher than that inside.

The amount of chloride in the body is a reflection of the balance between chloride intake and output. Chloride intake depends on the quantity and type of food intake. The chloride content of most foods parallels that of sodium. Under normal conditions the average adult takes in about 50 to 200 mEq of chloride/day. Chloride output occurs by way of three primary routes: the gastrointestinal tract, the skin, and the urinary tract.

Under normal circumstances loss of chloride through the

gastrointestinal tract is very small. Fecal chloride excretion for a normal adult is only 1 to 2 mEq/day. The concentrations of chloride in gastrointestinal secretions are shown in Table 20-5. With severe diarrhea or with gastric or intestinal drainage tubes, chloride by way of through the gastrointestinal tract may exceed 100 mEq/day.

The chloride composition of sweat averages about 40 mEq/L but is somewhat variable. As in the case of sodium, the concentration of chloride in sweat is decreased by aldosterone and increased in cystic fibrosis. Under conditions of excessive sweating, chloride losses through the skin can exceed 200 mEq/day. However, under normal conditions chloride losses through the skin are small.

The major route of chloride excretion is through the kidney. Details of the mechanisms of renal chloride excretion are discussed in Chapter 22.

Disorders of chloride balance

Chloride excess. Chloride accumulates in the body when the intake of chloride exceeds output because of some abnormality of chloride homeostasis mechanism. For the most part, the causes of chloride retention mainly are the same as those of sodium retention. Therefore the pathophysiology of chloride excess is in most cases similar to that of sodium (see box on p. 325). However, there is one clinical condition in which chloride excess may not be associated with sodium excess: certain types of metabolic acidosis. The two major extracellular anions are chloride and bicarbonate. Extracellular bicarbonate is consumed by the reaction with hydrogen ions produced in metabolic acidosis. If no organic anions are produced with the H^+, chloride ions are needed to replace the consumed bicarbonate ions to maintain electrical neutrality. The increase in chloride concentration is caused by the reabsorption of a relatively greater proportion of sodium with chloride than with bicarbonate by the tubules of the kidney.

Chloride depletion. Chloride depletion occurs when the output of chloride exceeds intake. For the most part, the causes of chloride depletion are the same as those of sodium depletion (see box on p. 326). However, in one clinical condition, hypochloremic metabolic alkalosis, there may be chloride depletion without sodium depletion. Hypochloremic metabolic alkalosis may result from loss of chloride in excess of sodium loss, usually from abnormal

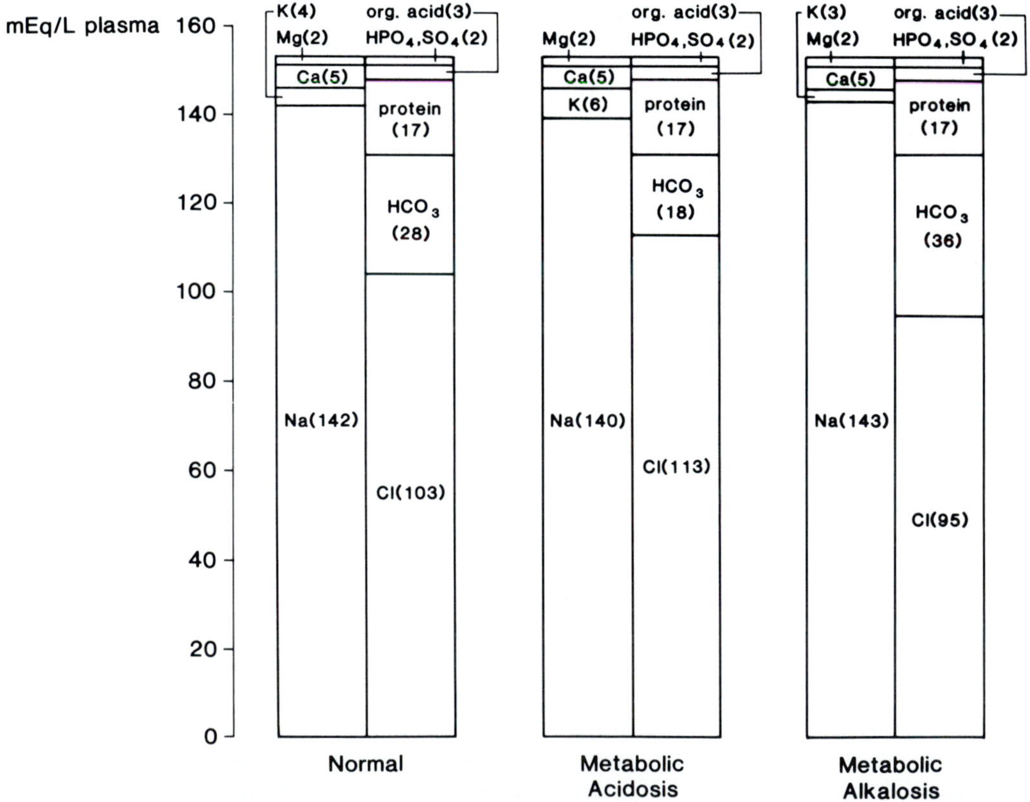

Fig. 20-11 Concentration of electrolytes in plasma (mEq/L) with metabolic acidosis and metabolic alkalosis compared to normal. In example of metabolic acidosis shown, there is no increase in organic acids, only loss of bicarbonate. Metabolic acidosis may be attributable to an increase in organic acids (see Fig. 20-4). In these cases chloride may not be increased. Note that extracellular potassium concentration is elevated in metabolic acidosis and lower in metabolic alkalosis. Under all conditions concentration of anions equals concentration of cations.

loss of gastric fluid. Bicarbonate must be retained to maintain electrical neutrality. Hypochloremia may also be associated with other disorders that involve bicarbonate retention, such as renal compensation for chronic respiratory acidosis (see Chapter 21).

Abnormalities of plasma chloride concentration

As for sodium, changes in total body chloride are not necessarily associated with similar changes in plasma chloride concentration. That is, with body chloride retention the plasma chloride concentration will remain normal if there is a proportional increase in ECW and will decrease if there is a relatively greater increase in ECW. Similarly, plasma chloride concentration may remain normal or even increase with chloride depletion, depending on the concomitant change in ECW.

In most cases the causes of hypochloremia and hyperchloremia are the same as those of hyponatremia and hypernatremia (see p. 327). The major clinical exceptions to the usual parallel changes in plasma sodium and chloride concentrations occur during chronic metabolic acidosis and alkalosis. With metabolic acidosis, hyperchloremia may not be associated with hypernatremia; with metabolic alkalosis, hypochloremia may not be associated with hyponatremia. The reasons for this are those previously discussed for chloride excess and depletion.

Symptoms are not directly attributable to hypochloremia or hyperchloremia. Rather, symptoms that occur in patients with an abnormal serum chloride concentration are caused by the associated abnormality in serum sodium or pH.

A summary of the plasma electrolyte changes in metabolic acidosis and alkalosis is presented in Fig. 20-11.

BIBLIOGRAPHY

Atlas, SA: Atrial natriuretic factor: renal and systemic effects, Hosp Pract 21(7):67, 1986.

Cohen, JJ: Disorders of potassium balance, Hosp Pract 14(1):119, 1979.

Collins, RD: Illustrated manual of fluid and electrolyte disorders, Philadelphia, 1983, JB Lippincott Co.

Hays, RM: Antidiuretic hormone, N Engl J Med 295:659, 1976.

Kokko, JP, and Tannen, RL: Fluids and electrolytes, Philadelphia, 1986, WB Saunders Co.

Levey, M: The pathophysiology of sodium balance, Hosp Pract 13:95, 1978.

Maxwell, MH, Kleeman, CR, and Narins, RA, editors: Clinical disorders of fluid and electrolyte metabolism, ed 4, New York, 1987, McGraw-Hill Book Co.

Oh, MS, and Carroll, HJ: The anion gap, N Engl J Med 297:814, 1977.

CHAPTER 21

Acid-base control and acid-base disorders

JOHN E. SHERWIN
BERNDT B. BRUEGGER

OBJECTIVES

- Outline the blood buffering mechanism of the bicarbonate and hemoglobin buffering systems.
- Explain acid-base balance regulation by the kidney with respect to:
  - Hydrogen ion excretion
  - Bicarbonate ion reaction
  - Sodium-hydrogen exchange
  - Ammonium secretion
- State the Henderson-Hasselbalch equation, identifying the respiratory and metabolic components; relate the equation to acid-base disorders; and calculate the pH given appropriate data.
- For each of the following pathological states, identify the expected pH, Po_2, and Pco_2 values as normal, increased, or decreased, and state the physiological response to the disease states.
  - Drug-induced hyperventilation
  - Acute hyperglycemic ketoacidosis
  - Hypochloremia resulting from persistent vomiting
  - Chronic emphysema
- Define oxygen saturation and P_{50}, and describe the effect of the following on dissociation of oxygen from hemoglobin:
  - 2,3-DPG
  - pH
  - Pco_2

KEY TERMS

acidemia A condition of decreased pH of the blood.

acidosis A pathological condition resulting from accumulation of acid in the blood or loss of base from the blood.

alkalemia A condition of increased pH of the blood.

alkalosis A pathological condition resulting from accumulation of base or loss of acid from the body.

alveoli Small outpouchings of walls of alveolar space through which gas exchange takes place between alveolar air and pulmonary capillary blood.

anion gap The concentration of undetermined anions, calculated as the difference between the measured cations and the measured anions.

apnea Cessation of breathing.

base excess or deficit The difference between the titratable acids and bases of the blood at a pH of 7.4, a Pco_2 of 40 mm Hg, and a temperature of 37° C.

bradycardia Slowing of the heartbeat to less than 60 beats per minute.

buffer A solution consisting of a weak acid and its conjugate base, which resists change in pH on addition of strong acid or base.

carbamino reaction A covalent chemical reaction between CO_2 and the primary amino group ($-NH_2$) of a protein to form a stable, bound form of CO_2.

carbonic anhydrase An enzyme that catalyzes the reaction between CO_2 and water to form carbonic acid (H_2CO_3).

chloride shift Exchange of Cl^- for HCO_3^- in red blood cells in peripheral tissues in response to Pco_2 of blood. The shift reverses in the lungs.

conjugate acid Protonated cationic form of a corresponding weak acid.

Henderson-Hasselbalch equation Describes the relationship between pH, the pK_a of a buffer system, and the ratio of the conjugate base and a weak acid.

hypercapnia A condition of excess carbon dioxide in the blood.

hyperchloremic alkalosis A metabolic alkalosis resulting from increased blood bicarbonate secondary to loss of chloride from the body.

hypoxia A condition of low oxygen content in the blood.

isohydric shift The series of reactions in red blood cells in which CO_2 is taken up and oxygen is released without the production of excess hydrogen ions.

metabolic acidosis Pathological loss of base in the body.

metabolic alkalosis Pathological acumulation of base in the body.

metabolic component The bicarbonate concentration of plasma.

oxygen saturation A term that defines the fraction of total hemoglobin (Hb) in the form of HbO_2 at a defined Po_2. Percentage of saturation = $100(HbO_2)/(HbO_2 + Hb)$.

$\boldsymbol{P_{50}}$ The partial pressure of oxygen at which hemoglobin is half-saturated with bound oxygen.

partial pressure The pressure exerted by a gas, whether it is alone or mixed with other gases. The partial pressure of a gas is denoted by the letter *P* preceding the symbol for that gas; for example, the partial pressure of CO_2 is Pco_2.

pH The negative logarithm of the hydrogen ion concentration.

$\boldsymbol{pK_a}$ The negative logarithm of the dissociation constant of a weak acid; also, the pH of a buffer at which the concentration of a weak acid and its conjugate base are equal. Maximum overall buffering capacity occurs when pK_a = pH.

relative base deficit A term that describes the lowered HCO_3^-/H_2CO_3 ratio caused by an increase in Pco_2. The HCO_3^- (base) is low relative to the Pco_2.

relative base excess Name for the elevated HCO_3^-/H_2CO_3 ratio caused by a decrease in Pco_2. The HCO_3^- (base) is elevated relative to the Pco_2.

respiratory acidosis Pathological retention of CO_2 in the body caused by respiratory change.

respiratory alkalosis Pathological decrease in carbonic acid caused by respiratory change.

respiratory component "αPco_2," or the acid component, which is immediately modified by respiratory status.

surfactant An agent that decreases surface tension. Applies to agents that coat pulmonary alveolar surfaces.

ventilation The exchange of air between the lungs and ambient air.

ACID-BASE CONTROL

Acids and bases

Definitions. Using the simplest definition, an acid is defined as a substance that releases protons or hydrogen ions (H^+), whereas a base is simply defined as a substance that accepts protons or H^+. Both acids and bases are further defined by their degree of affinity for H^+. A strong acid has little affinity for H^+ and so readily dissociates H^+, whereas a weak acid has some affinity for H^+ and thus less readily dissociates H^+. A strong base has a high affinity for H^+; a weak base has low affinity for H^+. If one molecule differs from another by only a proton, the two are called a conjugate acid-base pair. Physiological examples of a weak acid and its conjugate base are carbonic acid (H_2CO_3) and bicarbonate (HCO_3^-). The equilibrium reaction is as follows in Equation 21-1:

$$H_2CO_3 \rightleftarrows H^+ + HCO_3^- \qquad \textit{Eq. 21-1}$$

This equilibrium lies to the right at physiological pH, because bicarbonate, the conjugate base, has a weak affinity for hydrogen ion.

Dietary and metabolic sources of acids and bases. Two types of acids are dealt with in physiological states: fixed acids and volatile acids. Fixed acids are nongaseous acids such as phosphate ($HPO_4^=$) and sulfate (HSO_4^-) or organic acids such as lactic acid, acetoacetic acid, and beta-hydroxybutyric acid. The physiologically important volatile acid is carbonic acid (H_2CO_3). The volatility of carbonic acid arises from its ability to dissociate into water and carbon dioxide (CO_2), which can be relased as a gas, The complete reaction scheme for carbonic acid is as follows:

$$CO_2\ (\text{gas}) \rightleftharpoons CO_2\ (\text{dissolved}) \underset{-H_2O}{\overset{+H_2O}{\rightleftharpoons}} H_2CO_3 \rightleftharpoons H^+ + HCO_3^- \qquad \textit{Eq. 21-2}$$

At one end of the equilibrium is carbon dioxide, which can be considered the anhydrous form of H_2CO_3, and at the other end is HCO_3^-, the conjugate base of H_2CO_3. Although the reaction of CO_2 and water to form H_2CO_3 will occur spontaneously, the enzyme carbonic anhydrase facilitates this reaction in vivo.

Carbohydrates, lipids, and proteins are metabolized by oxidation reactions that generate acids that must be neutralized. Under anaerobic conditions such as those produced by respiratory distress or strenuous exercise, carbohydrates are metabolized to lactic and pyruvic acids, which accumulate until normal oxygenation is achieved. These acids can be further metabolized to the ultimate oxidation product, carbon dioxide, when aerobic metabolism is resumed. Triglycerides are metabolized to fatty acids, which can be further metabolized to ketone bodies (acetoacetic acid and beta-hydroxybutyric acid) under anaerobic conditions. Ultimately these lipid metabolites are further oxidized to carbon dioxide. Proteins are hydrolyzed to amino acids, which are then converted to carbon dioxide. Those proteins composed of sulfur-containing amino acids are catabolized in part to the salt of sulfuric acid. Nucleic acids and some lipids contain phosphorus and are metabolized to salts of phosphoric acid.

pH, hydrogen ion, and buffers. Please review the discussion of pH and buffer calculations in Chapter 1. Remember, rather than describing the concentration of H^+ in moles per liter, the accepted convention is the use of pH, which is defined as the negative logarithm of the concentration of H^+. Normal human whole blood is buffered at a slightly alkaline pH in a range of 7.35 to 7.45, which corresponds to an H^+ concentration of 4.5×10^{-8} M to 3.5 to 10^{-8} M.

Physiological buffers. Buffering capacity depends on the concentration of the buffer and the relationship between the pK_a of the buffer and the desired pH. A buffer is considered most effective within ±2 pH units of its pK_a. It has maximum buffering capacity when its pK_a equals the pH. For maximum blood buffering the pK_a of the buffers should therefore be near physiological pH, that is, pH 7.4. The physiologically important buffers that maintain the narrow pH range observed in the body are hemoglobin, bicarbonate, phosphate, and proteins. Table 21-1 lists the pK_a and concentrations of these buffer systems and their relative buffering capacities.

Table 21-1 Physiologically important buffers and their concentration, pK_a, and buffering capacity

Buffer	pK_a	Concentration (mmol/L)	Relative buffering capacity (mEq/L)
Bicarbonate	6.33	25	1
Hemoglobin	7.2	53	40
Phosphate	6.8	1.2	0.3
Protein	—	—	8

Bicarbonate buffer system. The equation for the bicarbonate-carbonic acid buffer system (Equation 21-1) is as follows:

$$pH = pK_a + \log \frac{[HCO_3^-]}{[H_2CO_3]} \qquad \textit{Eq. 21-3}$$

The measured pK_a is 6.33 at 37° C. Instead of being at the maximum buffer capacity with a 1:1 ratio of HCO_3^- to H_2CO_3, the bicarbonate-carbonic acid buffer system at the blood pH of 7.4 is at a ratio of 20:1

$$pH = pK + \log \frac{[HCO_3^-]}{[H_2CO_3]} = 6.33 + \log 20 = 7.4 \qquad \textit{Eq. 21-4}$$

This 20:1 ratio is maintained primarily by the lungs, which expel CO_2 produced during the metabolism of nutrients.

In Equation 21-3 there are three unknowns: pH, $[HCO_3^-]$, and $[H_2CO_3]$. Although pH is measurable, there is no direct measure of $[HCO_3^-]$ or $[H_2CO_3]$. To use this equation, the term H_2CO_3 is replaced by an analyte that is measurable. The concentration of H_2CO_3 is proportional to the amount of dissolved CO_2 (Equation 21-2). Thus one can replace $[H_2CO_3]$ with the term $\alpha P{CO_2}$ where α (the Bunson coefficient) is the solubility coefficient of CO_2. The pK_a term in the Henderson-Hasselbalch equation is modified to reflect the equilibrium between CO_2 (dissolved) and HCO_3^-. The modified Henderson-Hasselbalch equation describing the equilibrium in Equation 21-2 is as follows:

$$pH = pK_a' + \log \frac{[HCO_3^-]}{\alpha P{CO_2}} \qquad \textit{Eq. 21-5}$$

The apparent pK_a' in human plasma is 6.1 M at 37° C. The solubility coefficient for CO_2 in plasma at 37° C is 0.031 mmol × L^{-1} × mm Hg^{-1}. The concentration of base greatly exceeds that of acid in this plasma buffer system, reflecting the demands put on the body by metabolism, which primarily produces acids. This buffer system is designed to process the primary metabolic waste product, CO_2. The CO_2 component of the buffer system is eliminated by the lungs.

The total CO_2 content ($T{CO_2}$) of plasma is described as

$$T{CO_2} = CO_2 \text{ (dissolved)} + [HCO_3^-] + [H_2CO_3]$$

One can disregard the $[H_2CO_3]$ term because it is so small

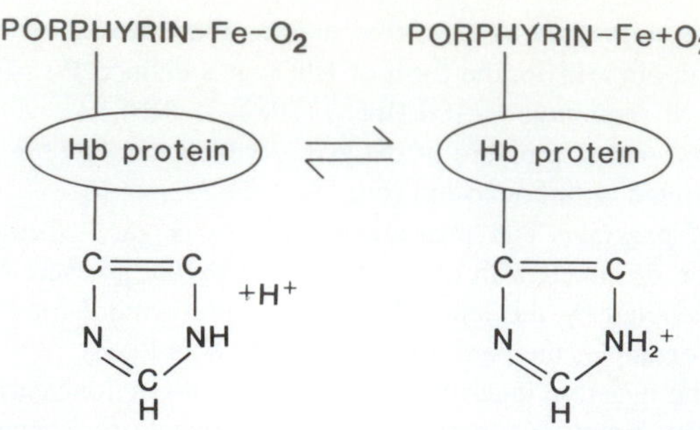

Fig. 21-1 Effect of hemoglobin oxygenation on buffering action of imidazole group of histidine. Binding affects pK_a of imidazole ring making the ring more acidic with release of an H^+.

(one twentieth the $[HCO_3^-]$) and replace CO_2 (dissolved) with the term αCO_2. Thus the equation can be reduced to

$$T{CO_2} = \alpha P{CO_2} + [HCO_3^-] \qquad \textit{Eq. 21-6}$$

Two of the three unknowns are readily determined, allowing calculation of the third (see p. 341).

Hemoglobin. The major buffer of blood is hemoglobin, which is localized in the red blood cells. Hemoglobin (Hb) takes up free H^+ so that the following reaction proceeds to the right:

$$CO_2 + H_2O \rightleftarrows H_2CO_3 \rightleftarrows HCO_3^- + H^+ \searrow HHb^+ + O_2 \;\; (HbO_2) \qquad \textit{Eq. 21-7}$$

Hemoglobin and serum proteins have high concentrations of histidine residues. The imidiazole group of histidine (Fig. 21-1) has a pK_a of approximately 7.3. It is this combination of high concentration and appropriate pK_a that makes hemoglobin the dominant buffering agent of blood at physiological pH. The bulk of the CO_2 formed in peripheral tissues is transported in the plasma portion of blood as HCO_3^- with the H^+ bound to hemoglobin within the erythrocyte (Fig. 21-2). Following exercise, acid wastes are released from tissue, which increases the acidity and $P{CO_2}$ of venous blood. Carbonic acid CO_2 accounts for about 2 mmol/L of CO_2 in venous blood, but accounts for only about 1 mmol/L in arterial blood.

Significant amounts of CO_2 are transported as a protein-bound moiety. CO_2 reacts nonenzymatically with the accessible amino groups of proteins to form a *carbamino* group:

$$O{=}C{=}O + H{-}\underset{H}{N}{-}\text{Protein} \rightarrow {}^-O - \overset{O}{\overset{\|}{C}} - N - \text{Protein} + H^+ \qquad \textit{Eq. 21-8}$$

Carbamino group

Approximately 0.5 mmol/L of CO_2 is transported in this fashion, and there is only a very small arterial-venous difference between arterial and venous concentrations of CO_2. The carbamino reaction with hemoglobin in the red

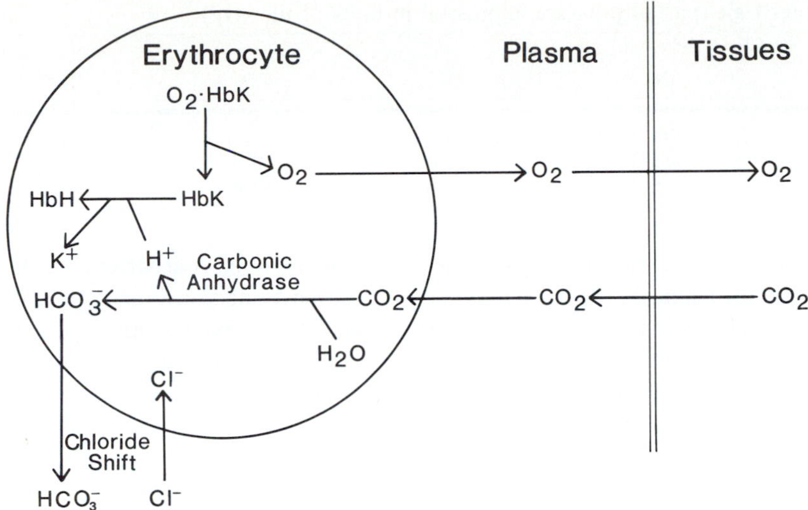

Fig. 21-2 Hemoglobin buffering action in peripheral tissues. HbK is potassium salt of hemoglobin. HbH is protonated form of hemoglobin.

blood cells tends to reduce slightly the buffering capacity of Hb because a proton is released during the reaction.

The observed arteriovenous difference in total CO_2 content is almost entirely the result of formation of bicarbonate in the red blood cell. In the lungs, as deoxygenated hemoglobin becomes oxygenated and the CO_2 is expelled, the H^+ is released from hemoglobin because oxygenated hemoglobin (HbO_2) is a stronger acid than deoxyhemoglobin (HbH). In the lungs then, the reaction in Equation 21-7 proceeds to the left. H^+ is released and reacts with the transported HCO_3^- to form CO_2 that the lungs can now release (Fig. 21-2). The overall equation linking the oxygenation process to buffering is

$$H^+ + HbO_2 \rightleftarrows HbH^+ + O_2 \qquad \textit{Eq. 21-9}$$

The forward reaction occurs in tissues in which there is a relatively high H^+ and a relatively low O_2 concentration, whereas the reverse reaction occurs in the lungs, where the O_2 concentration is relatively high.

Phosphate and proteins. Phosphate is a minor buffering component of the blood, with the following equilibrium reaction occurring:

$$H_2PO_4^- \rightleftarrows H^+ + HPO_4^{-2} \qquad \textit{Eq. 21-10}$$

The pK_a of this reaction is 6.8. Phosphate buffer is an important buffer in urine, which has relatively little protein, hemoglobin, or bicarbonate. The phosphate buffers in the blood are inorganic phosphates, while both inorganic and organic phosphates act as intracellular buffers.

Plasma proteins also act as buffers in the blood. The proteins have prosthetic groups and amino acids such as lysine, arginine, histidine, glutamate, and aspartate that bind or release H^+. This buffering effect is minor compared to the bicarbonate system or hemoglobin system (Table 21-1).

Oxygen and carbon dioxide homeostasis

Partial pressure. In a mixture of gases, the total gas pressure is equal to the sum of the partial pressures of each gas. Blood gas analysis is performed to determine the partial pressures of oxygen and carbon dioxide (PO_2 and PCO_2, respectively). Historically, the units of PO_2 and PCO_2 have been millimeters of mercury (mm Hg), or torr, and these are still used by a majority of laboratories in the United States. The international unit for partial pressure is the *pascal,* or Pa (1 mm Hg = 133.3224 Pa).

Table 21-2 shows the composition of atmospheric air, alveolar air (air inside the lung), and expired air. Humidity makes a substantial contribution to the composition of air in the lungs, thus altering the partial pressures of the other gases. A correction for the gas-volume contributed by water vapor is therefore essential; at 37° C, the PH_2O of blood is approximately 47 mm Hg.

At 37° C and normally encountered atmospheric pressures, the total pressure *P,* is equal to the sum of the partial pressures (PO_2, PCO_2, PN_2, and PH_2O). Table 21-2 shows a total pressure of 760 mm Hg; however extreme examples of lower air pressure (hypobarism) and higher air pressure (hyperbarism) are occasionally encountered, for example, with mountain climbing and deep sea diving. Correction of blood gas data for changes in the total atmospheric pressure is important in these instances, and correction of calibration gases for variations in total pressure is necessary.

Two other terms used in discussing the oxygen content of blood are *oxygen saturation* and P_{50}. Oxygen saturation is the percentage of the total hemoglobin present as oxygenated hemoglobin. The term P_{50} denotes that partial pressure of oxygen at which the hemoglobin is 50% saturated with oxygen.

Table 21-2 Comparison of air (partial pressure expressed in units of mm Hg)

Air	N_2	O_2	CO_2	H_2O	Total pressure
Atmospheric air	598.0	158.0	0.3	3.7	760
Alveolar air	573.0	100.0	40.0	47.0	760
Expired air	566.0	115.0	32.0	47.0	760

Ventilation. Ventilation is differentiated from respiration in that *ventilation* is the mechanical process of moving air in and out of the lungs, and *respiration* is the exchange of gases between the atmosphere and the body. The exchange of gases between air and the capillaries of the pulmonary circulation occurs in the *alveoli*. The normal respiration rate is 13 to 16 times per minute.

The walls of the lung contain elastic connective tissues that would collapse the lung were it not for the surface tension between the surface of the lung and the wall of the thoracic cavity. The surface tension of the inner walls of the alveoli, on the other hand, has a tendency to collapse the alveoli after expiration, when the alveoli are deflated. This surface tension is reduced by the presence of a phospholipid-lipoprotein, *surfactant,* that lines the alveolar walls in a thin film and allows the alveolar walls to be easily reinflated. Premature babies without sufficient surfactant lining the alveolar walls can have respiratory difficulties because of the tendency of alveoli to collapse. It is for this reason that the lecithin/sphingomyelin ratio in amniotic fluid is determined to assess fetal lung development (see Chapter 37).

In the newborn, the lungs are about the same size as the thoracic cavity, and therefore need not stretch to fill the cavity. The thoracic cavity grows faster than the lungs as the child develops, and the surface tension between the lungs and the inner wall of the thoracic cavity prevents the lungs from separating from the thoracic cavity and collapsing. With advanced age, the lungs become less elastic and the chest wall becomes more rigid, reducing the capacity for ventilation.

Gas exchange. Gas transfer in the alveoli is a concentration-dependent phenomenon. Inspired (room) air has a relatively high Po_2 (158 mm Hg) and a low Pco_2 (0.3 mm Hg). Pressures of oxygen and carbon dioxide in capillary blood in the lungs are 50 and 40 mm Hg, respectively. Because the Po_2 of blood is lower than that of inspired air and the Pco_2 of blood is higher than the Pco_2 of room air, the gases diffuse from higher to lower concentration areas. That is, CO_2 gas moves from the capillaries to the alveolar air space, while O_2 moves from the alveoli to the capillaries. Reference values for adult blood gas parameters in arterial and venous blood are sumarized in Table 21-3. Because of its greater water solubility, CO_2 exchanges more rapidly and more efficiently than does O_2. In respiratory acidosis, this phenomenon of differential gas diffusibility can result in low blood Po_2 but relatively normal Pco_2.

Control of ventilation. Ventilatory control regulates the carbohydrate-bicarbonate buffer system, but is in turn regulated by the resulting pH of cerebrospinal fluid and plasma. Control of ventilation is localized in a respiratory center of the brain where chemoreceptors are influenced by the pH of the cerebrospinal fluid. Other chemoreceptors influenced by the changes in pH of arterial blood are located in the carotid and aortic vessels. A rise in Pco_2 of arterial blood will result in a fall in pH. This will in turn stimulate the chemoreceptors, initiating a rise in the respiration rate that will result in the release of more CO_2 from the blood in the lungs.

Acid-base balance

The maintenance of a constant pH is important because changes in pH will alter the functioning of enzymes, the cellular uptake and use of metabolites and minerals, the conformation of biological structural components, and the uptake and release of oxygen.

In the body, physiological buffers act to maintain a constant pH in the following manner. Fixed acids enter into the blood and are immediately neutralized by the bicarbonate buffering system.

$$H^+A^- \text{ [fixed acid] } HCO_3^- \rightleftarrows H_2CO_3 + A^- \text{ [unmeasured anion]}$$
$$\upharpoonleft\downharpoonright$$
$$H_2O + CO_2 \qquad \textit{Eq. 21-11}$$

However, the volatile acid, CO_2, is neutralized by the hemoglobin buffering system, because all the buffering systems are at equilibrium with each other. It is this overall equilibrium that gives the blood the relative buffering capacities described in Table 21-1.

Thus one of the important buffer systems required to maintain the pH of the blood is the carbonic acid–bicarbonate buffer system. Although this system has relatively

Table 21-3 Reference values for adult blood gas parameters in arterial and venous blood

Parameters	Arterial	Venous
pH	7.35-7.45	7.33-7.43
Pco_2	35-45 mm Hg	38-50 mm Hg
Po_2	80-100 mm Hg	30-50 mm Hg
HCO_3^-	22-26 mmol/L	23-27 mmol/L
Total CO_2	23-27 mmol/L	24-28 mmol/L
O_2 saturation	94%-100%	60%-85%
Venous anion gap	5 to 14 mmol/L	
Base excess	−2 to +2 mEq/L	

low buffering capacity (see Table 21-1), it plays a large role in maintaining blood pH because it acts as the immediate buffer when fixed acids enter the blood.

Changes in respiration rate will alter the bicarbonate–carbonic acid ratio and pH. To understand this process, one must reconsider Equations 21-2 and 21-5. A decrease in the ventilation rate will cause a decrease in release of CO_2 from the blood in the lungs. The increased blood CO_2 will result in the formation of more bicarbonate (shifting Equation 21-2 to the right), though the increase in bicarbonate will be less than the increase in P_{CO_2}. Thus there will be a decrease in the bicarbonate-carbonate ratio and a decrease in the pH (see Equation 21-5). If the ventilation rate were to stay constant and the metabolic release of fixed acid were to increase, the same effect would be observed. In this case H^+ reacts with HCO_3^- to form CO_2, which is released in the lungs. There is an immediate decrease in the concentration of bicarbonate with essentially no change in P_{CO_2}, resulting in a decreased bicarbonate/αP_{CO_2} ratio and a decreased pH. If the ventilation rate increases, more CO_2 is released from the blood at the lungs and the bicarbonate-carbonic acid ratio and pH increase. The ventilation rate can range from zero to 15 times normal, allowing a significant degree of regulation of the bicarbonate-carbonic acid ratio. Thus when the rate of ventilation is increased, excess acid in the form of CO_2 is quickly removed. Similarly, when the rate of ventilation is decreased, acid (CO_2) is added to neutralize excess alkaline (HCO_3^-).

The other important blood buffer, hemoglobin, which is vital for the regulation of blood pH, buffers the CO_2 from the tissues. The major function of hemoglobin is the transport of oxygen through the blood to the cells of the body. There is a complex relationship between the degree of oxygenation of hemoglobin and the pH, P_{CO_2}, and total CO_2 (T_{CO_2}) of blood. Oxygenated hemoglobin is a stronger acid than deoxygenated hemoglobin, and therefore in the lungs hemoglobin will release H^+ as it becomes oxygenated (Equations 21-7 and 21-12), decreasing the bicarbonate level and increasing the levels of carbonic acid and its anhydrous form CO_2, and thus increasing the P_{CO_2} of the blood. The rate at which this reaction proceeds is enormously increased by the presence in the red blood cells of the enzyme *carbonic anhydrase*. It is the action of this enzyme that allows the rapid transfer of CO_2 into and out of red blood cells, with consequent buffering by hemoglobin. This process is summarized by the following series of reactions and Fig. 21-3.

$$\text{Gas exchange in the lungs } \underset{\text{(gas)}}{CO_2} \rightleftarrows \underset{\text{(dissolved)}}{dCO_2} \xrightleftharpoons{\text{Carbonic anhydrase}} H_2CO_3 \rightleftarrows H^+ + HCO_3^-$$

$$T_{CO_2} = dCO_2 + HCO_3^- \text{ or } T_{CO_2} = \alpha P_{CO_2} + HCO_3^- \qquad \textit{Eq. 21-12}$$

In the lungs ventilation will eliminate this increased P_{CO_2} by releasing the CO_2 from the blood and thereby return the ratio of bicarbonate to carbonic acid to 20.

Oxygenated hemoglobin is transported in the blood to cells that have relatively low P_{O_2} tension and are releasing metabolic products, such as CO_2 and organic acids, into the blood, thus raising the P_{CO_2} and T_{CO_2} and lowering the pH. The relatively low P_{O_2} causes the dissociation of O_2 from HbO_2 and the consequent delivery of O_2 to the cells. The high CO_2 pressure in the cells drives the CO_2 along a concentration gradient into the red blood cells. Carbonic anhydrase rapidly converts the CO_2 into H^+ and HCO_3^- (see Fig. 21-2). Deoxygenated hemoglobin is a weaker acid than oxygenated hemoglobin. It neutralizes

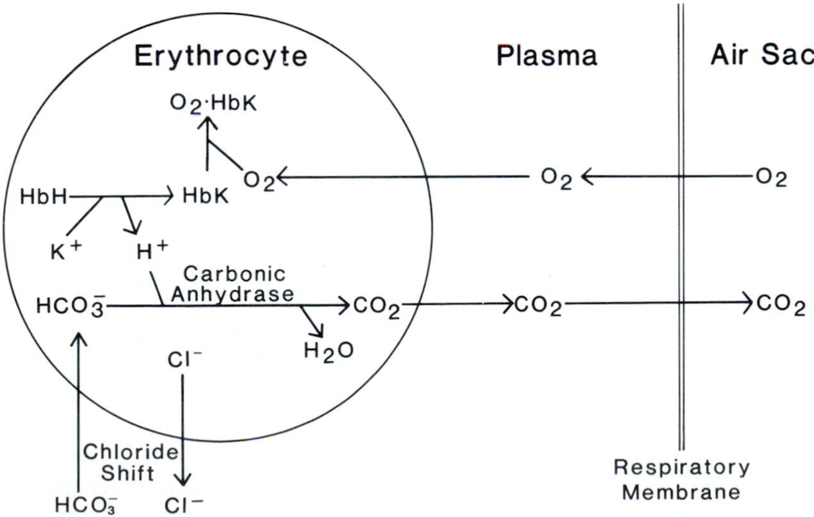

Fig. 21-3 Transfer of CO_2 in lungs from erythrocytes to air sacs. *HbK*, Potassium salt of hemoglobin; *HbH*, protonated form of hemoglobin.

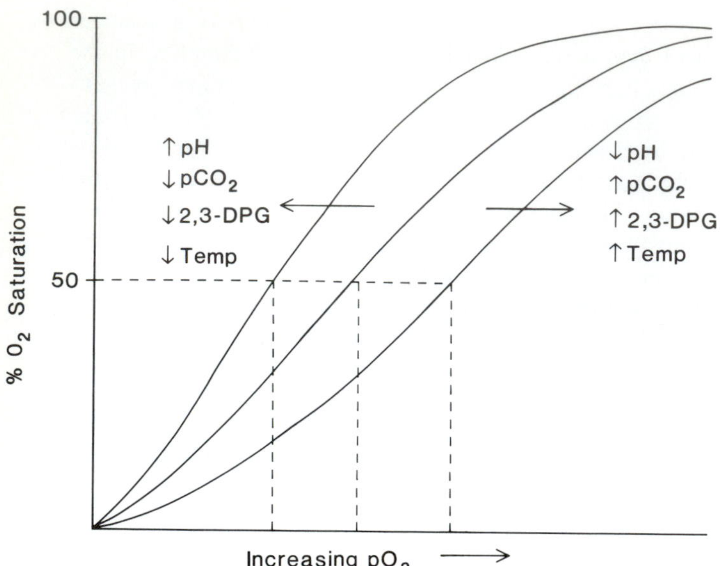

Fig. 21-4 Hemoglobin-oxygen dissociation curves and factors that shift the curve right and left. A shift of curve right or left changes level of Po_2 at which hemoglobin is 50% saturated (P_{50}).

the H^+ to raise the pH and causes the dissociation reaction of carbonic acid to proceed to the right to increase the level of bicarbonate and decrease the Pco_2.

The dissociation of oxygen from hemoglobin as a function of the Pco_2 is shown in Fig. 21-4, which is a graph of the percentage of O_2 saturation of hemoglobin versus Po_2. The sigmoidal shape of the curve indicates that at critical levels of Po_2 near the P_{50} there is a strong increase or decrease in the percentage of O_2 saturation, with a minimal shift in Po_2. In an area of the body in which there is a drop in Po_2 below that P_{50} on the sigmoid curve, the hemoglobin will release a larger portion of O_2 than at a Po_2 level above the P_{50}. Similarly, in areas of high O_2, such as the lungs, the hemoglobin will be essentially saturated with O_2.

One factor that has an effect on the position of the oxygen dissociation curve is 2,3-diphosphoglycerate (2,3-DPG), an intermediate in glycolysis. By interacting with the N-terminal amino groups of the hemoglobin molecule itself, 2,3-DPG induces the release of oxygen from hemoglobin. This is reflected in the shift to the right of the O_2 dissociation curve (Fig. 21-4). With a shift to the right, the critical Po_2 level that causes 50% saturation of hemoglobin by oxygen is increased so that areas of active metabolism, which contain increased 2,3-DPG levels, do not require Po_2 levels as low as those required in areas without increased 2,3-DPG for significant O_2 release from hemoglobin. As is seen in Fig. 21-4, as the hemoglobin-oxygen dissociation curve is shifted to the right, a higher Po_2 is required for the hemoglobin to be 50% saturated. Other metabolites that react with the N-terminal group of the β chain and the ε-amino groups of hemoglobin, causing a shift in the O_2 saturation curve, include glucose (glycohemoglobin) and CO_2.

This type of shift is observed in patients with increased Pco_2 and decreased pH because both will shift the oxygen dissociation curve to the right. This effect of pH and Pco_2 is known as the *Bohr effect*. Even under normal conditions, tissues with active metabolism are increasing the levels of Pco_2 and H^+, causing the oxygen dissociation curve to shift to the right, and thereby lowering the affinity of hemoglobin for oxygen. The net result is an enhanced release of O_2 to the tissues. During exercise, concentrations of H^+, Pco_2, and 2,3-DPG all increase in the areas of the body under stress, thereby shifting the oxygen dissociation curve to the right and enhancing oxygen release to those areas requiring oxygen. (See Chapter 33 for additional detail on 2,3-DPG and the Bohr effect.)

As more oxygen is released in response to increased levels of Pco_2 and H^+, more deoxyhemoglobin, which acts as a buffer, is formed. Increased Pco_2 leads to formation of bicarbonate (HCO_3^-). Some H^+ ions are bound by the deoxygenated hemoglobin, and the rest are buffered by the proteins and phosphate buffer in the plasma. Because all the H^+ formed is buffered, there is essentially no change in the pH. This buffering phenomenon is referred to as the *isohydric shift*. The HCO_3^- formed in the red blood cells as a result of uptake of H^+ by the hemoglobin diffuses out of the cells into the plasma. To preserve the electrical neutrality of the red cell, as HCO_3^- diffuses out of the cell, Cl^- diffuses into the erythrocytes from the plasma. This increase in the erythrocyte Cl^- is termed the *chloride shift*. Thus the plasma chloride concentration of venous blood (where HCO_3^- is formed in red blood cells) is *lower* than that of arterial blood. When CO_2 is expelled from the lungs, Cl^- again shifts out of red blood cells into plasma (see Figs. 21-2 and 21-3). As the polyvalent hemoglobin anions are replaced by the diffusing monovalent chloride anions, the osmolality of the erythrocyte increases, leading to diffusion of water into the erythrocyte, slightly increasing the mean volume of venous red blood cells over the mean cell volume in arterial blood.

Although the intact respiratory system acts as an immediate regulator of the $HCO_3^-/\alpha Pco_2$ system, long-term control is exerted by renal mechanisms (see Chapter 22 for details). The kidneys excrete nonvolatile acids such as sulfuric, hydrochloric, phosphoric, and some of the organic acids into the urine. Hydrogen ions are excreted by the kidneys into the urine and are buffered by HPO_4^{-2} and ammonia, which is derived from deamidation of the amino acid glutamine. Sodium is the cation exchanged for excreted hydrogen ions by the kidney. The kidney also affects the bicarbonate-carbonic acid buffer system by regulating the excretion of bicarbonate (Fig. 21-5). The kidney reabsorbs almost all filtered bicarbonate at plasma bicarbonate concentrations below 25 mEq/L. Only when bicarbonate levels become elevated above 25 mEq/L will bicar-

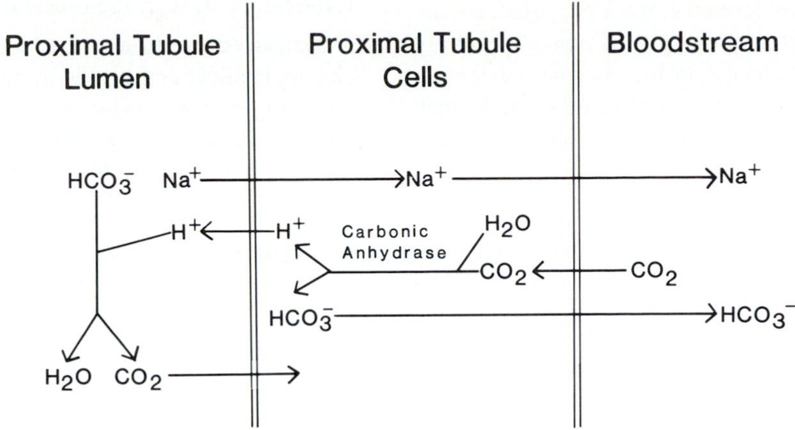

Fig. 21-5 Kidney reabsorption of bicarbonate with excretion of H^+.

bonate be excreted into the urine. The reabsorbed bicarbonate is neutralized by the reabsorbed sodium ions, which have been exchanged for the hydrogen ions excreted in the urine (Fig. 21-5).

ACID-BASE DISORDERS

Definitions

Acid-base disorders are most readily classified in terms of their immediate cause. Thus acidoses and alkaloses are described as of either respiratory or metabolic origin.

These classifications should always be considered in terms of the modified Henderson-Hasselbalch equation:

$$pH = pK'_a + \log \frac{[HCO_3^-]}{\alpha P_{CO_2}} \quad \textit{Eq. 21-13}$$

The term αP_{CO_2} represents the acid component that is directly and immediately modified by respiratory status. Thus the term αP_{CO_2} is called the *respiratory component*. The concentration of bicarbonate is most immediately affected by changes in the hydrogen-ion concentration caused by production of metabolic acids other than CO_2 (that is, fixed acids) and by physiological processes that directly change the concentration of serum bicarbonate. Thus the bicarbonate concentration of plasma is called the *metabolic component* of acid-base status. Acid-base homeostasis is accomplished when one controls one or both of these components. Note that the pH of plasma depends on the *ratio* of the concentration of bicarbonate to αP_{CO_2} rather than on the absolute concentration of these components.

Base-deficient disorders

Any condition associated with a lower-than-normal blood pH is referred to as *acidemia*.

Metabolic acidosis. In base-deficient disorders the pH is below normal. Such disorders occur if metabolic processes result in the accumulation of abnormal amounts of organic acids other than carbonic acid. Examples of acids that accumulate are lactic acid, beta-hydroxybutyric acid, and acetoacetic acid. Metabolic organic acids entering plasma react with plasma bicarbonate to form H_2CO_3, which is immediately converted to CO_2 gas, which is in turn rapidly eliminated from the body by the lungs. The net result is an immediate decrease in bicarbonate concentration with essentially no loss of P_{CO_2}. This leads to a lowered bicarbonate/αP_{CO_2} ratio and a lowered pH, or metabolic acidosis.

In contrast to this accumulation of acids is the pathological loss of base from the body. In severe diarrhea, bicarbonate ion is lost as part of the watery stool, resulting in a base deficit ($\downarrow [HCO_3^-]$, $\downarrow$ bicarbonate/αP_{CO_2} ratio). These types of disorders are termed *metabolic acidoses*.

Respiratory acidosis. A relative base-deficient disorder can result from a decrease in the bicarbonate/carbonic acid ratio as a result of an increase in carbonic acid. This occurs if the lungs are not able to expel CO_2 from the blood. This disorder is termed *respiratory acidosis*. The increase in P_{CO_2} results in an increase in the concentration of bicarbonate as the CO_2 is buffered by hemoglobin. However, the rise in bicarbonate is less than the increase in P_{CO_2}, resulting in a relative base deficit and a decrease in the bicarbonate/αP_{CO_2} ratio, which results in a below-normal serum pH.

Base-excess disorders

Any condition associated with an above normal blood pH is an *alkalemia*.

Metabolic alkalosis. In base-excess disorders, the pH is above normal. If the disorder is caused by an increase in bicarbonate, with little or no change in carbonic acid, the disorder is termed *metabolic alkalosis*. Such a disorder occurs when excess amounts of bicarbonate of soda are ingested or administered or when there is an increased renal reabsorption of bicarbonate, as in hypochloremic alkalosis.

Respiratory alkalosis. If the disorder is caused by a decrease in carbonic acid, as when respiration is overly stimulated, the disorder is termed *respiratory alkalosis*. In this condition, rapid ventilation greatly decreases the P_{CO_2} of blood, with minimal change in bicarbonate concentration. This results in a relative excess of bicarbonate so that the bicarbonate/αP_{CO_2} ratio increases. This increased ratio yields a higher plasma pH.

Laboratory aids in diagnosis of acid-base disorders

Measured parameters. Blood pH is measured with an electrochemical cell that is in contact with the blood sample through an H^+ selective glass membrane (see Chapter 13). To measure P_{CO_2}, one uses a modified pH electrode that has a gas-permeable membrane through which CO_2

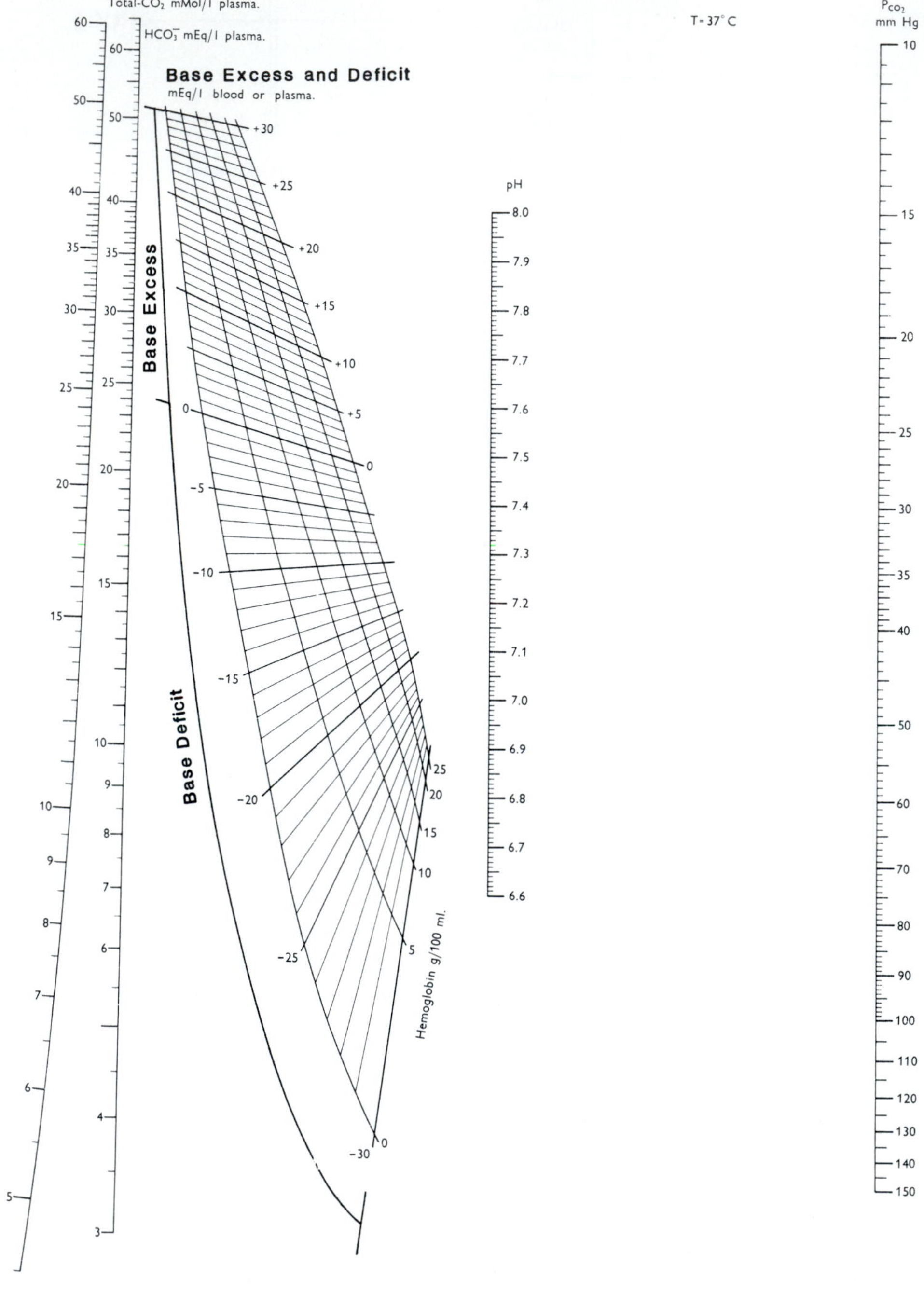

Fig. 21-6 Nomogram of relationship between P_{CO_2}, pH, base excess, hemoglobin, bicarbonate, and total CO_2. A straight line through a value of pH and of P_{CO_2} will connect with a calculated value of HCO_3^- and total CO_2. Base excess or deficit can be derived from that straight line if hemoglobin level is known. (Modified from Radiometer Corp., Copenhagen, Denmark.)

diffuses and alters the pH of a weak bicarbonate buffer solution. The change in pH caused by the diffusion of CO_2 is correlated to the change in the level of CO_2. The oxygen tension of the blood is measured by an electrode that also has a gas-permeable membrane that covers an electrochemical cell. The electrochemical cell has an adjusted cathode potential that is depolarized by the diffusing O_2, causing an increased current to flow that is proportional to the Po_2 level of the blood (see Chapter 13).

Calculated parameters. The remainder of the blood acid-base parameters are not measured but are instead calculated using Equations 21-6 and 21-13. One of the parameters is bicarbonate, which is calculated using the measured parameters pH and Pco_2 in the Henderson-Hasselbalch equation:

$$HCO_3^- = (\alpha Pco_2) \text{ antilog } (pH - pK_a') \quad \textit{Eq. 21-14}$$

Nomograms have also been developed to derive bicarbonate levels from pH and Pco_2 measurements (Fig. 21-6). Bicarbonate levels are useful in the assessment of the degree to which metabolic and renal control are involved in the acid-base status of the patient.

Total CO_2 is measured when a sample of plasma or serum is added to a reaction vessel with sufficient acid to shift the equilibrium of the reaction described by Equation 21-2 far to the left, thereby converting all forms of carbon dioxide into gaseous carbon dioxide. An electrode is then employed to measure the Pco_2 of the acidified plasma in the reaction chamber. Alternatively, the total CO_2 is measured colorimetrically or is measured gasometrically with the Natelson microgasometer. Total CO_2 can also be calculated from the measured parameters of pH and Pco_2 by the following equation:

$$Tco_2 = \alpha Pco_2 + HCO_3^- + H_2CO_3 \quad \textit{Eq. 21-15}$$

The concentration of H_2CO_3 is small enough to be ignored. The bicarbonate concentration is calculated from the measured Pco_2 and pH as described in Equation 21-14. Nomograms have been created to derive Tco_2 from pH and Pco_2 (Fig. 21-6).

Base excess is another calculated parameter that is used to assess the metabolic component of the patient's acid-base disturbance. The term *base excess* is used to describe clinical situations in which there is an excess of bicarbonate (positive base excess) or a deficit of bicarbonate (negative base excess). We use the term *base deficit* for a negative base excess, because this term is a more accurate description of the physiological condition. Base excess in the blood at a pH of 7.40, Pco_2 of 40 mm Hg, a normal hemoglobin of 150 g/L, and a temperature of 37° C is zero. The hemoglobin concentration is important because the blood-buffering capacity is greatly dependent on this quantity. The addition of a base, such as bicarbonate, raises the buffer content of the blood and results in a positive base excess. The loss of base, as occurs in diarrhea or with the addition of acids, lowers the blood buffer content and results in a base deficit. To calculate base excess, the measured parameters, pH and Pco_2, and either an assumed or measured hemoglobin level are used in a relationship developed by Sigaard-Andersen (see Fig. 21-6 and Equation 21-16). The calculation of the base excess or deficit is useful in the management of patients with acid-base disturbances because it permits estimation of the number of milliequivalents of sodium bicarbonate or ammonium chloride that should be administered to correct the patient's pH to normal. In practice the base excess is only a crude estimate because as the patient's condition im-

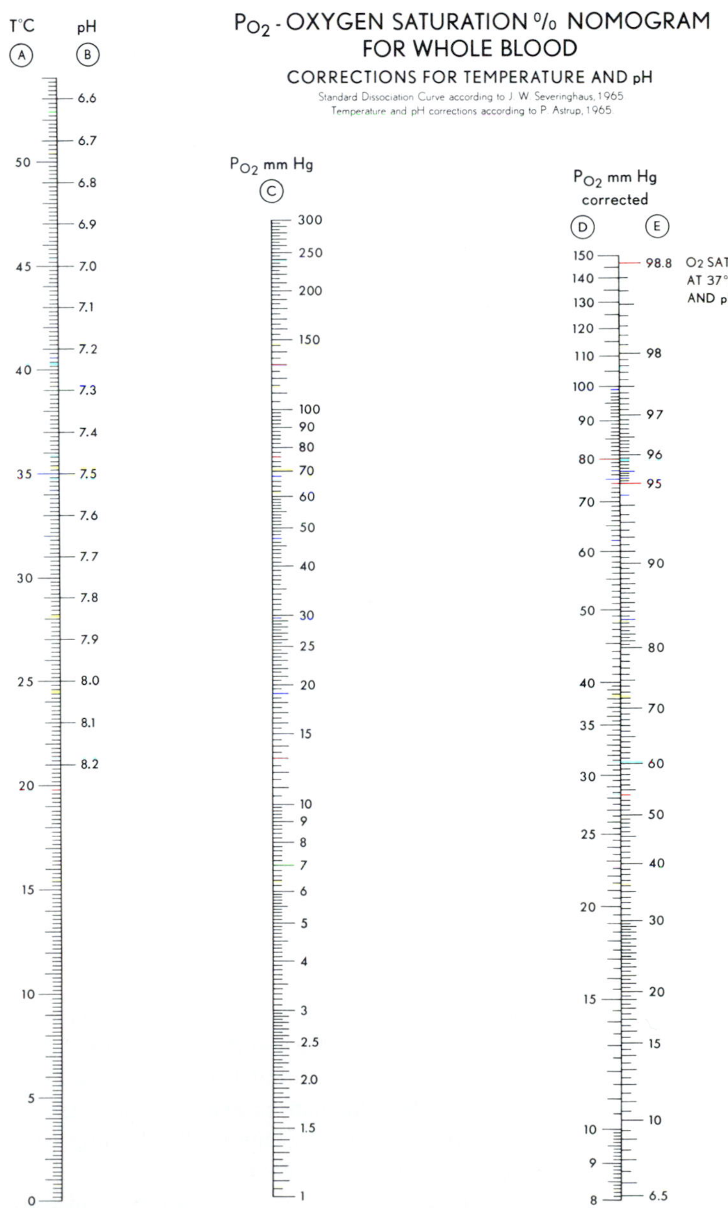

Fig. 21-7 Nomogram of relationship between pH, Po_2, and O_2 saturation. A straight line through a value of pH and of Po_2 will connect with a calculated value of O_2 saturation at 37° C. (Courtesy of Radiometer Corp., Copenhagen, Denmark.)

proves, changes in respiration and metabolism will invalidate the original calculation. It is for this reason that blood gas status is closely monitored by the analysis of sequential blood specimens.

$$\text{Base excess} = (1.0 - 0.0143\ \text{Hgb}) \cdot (HCO_3^-) - (9.5 + 1.63\ \text{Hgb})(7.4\ \text{pH}) - 24 \quad \textit{Eq. 21-16}$$

where Hgb is the hemoglobin concentration in g/dL.

Oxygen saturation indicates the amount of oxygen bound to hemoglobin and is used to determine the effectiveness of respiration or oxygen therapy. Oxygen saturation is calculated by use of the measured parameters of pH and Po_2 and the equation for a normal oxygen dissociation curve. A nomogram to derive O_2 saturation from pH and Po_2 values is presented in Fig. 21-7. Oxygen saturation is also measured directly by use of the difference in the wavelengths of maximum absorbance for oxyhemoglobin and deoxyhemoglobin. This measurement is performed with a co-oximeter. Table 21-3 contains reference values for the calculated blood gas parameters.

Anion gap. If the total measured cations are subtracted from the total measured anions as shown below, the difference is the anion gap, or the amount of unmeasured anions present. Or more simply:

$$\text{Anion gap} = (Na^+) + (K^+) - (Cl^-) - (HCO_3^-) \quad \textit{Eq. 21-17}$$

Usually the only electrolytes measured are sodium, potassium, chloride, and bicarbonate (as total CO_2). However, other anions exist in blood, such as phosphates, ketones, lactic acid, proteins, and sulfates. Because these other anions are not measured while their counter ions are, there is an apparent excess, or gap, of measured cations over measured anions. Increases in the amounts of these unmeasured anions and of the accompanying Na^+ ions will increase the apparent gap. Normally the anion gap averages 12 mEq/L. The anion gap increases with production of organic acids. Diabetic ketoacidosis is the most common cause of an elevated anion gap. If diabetes is ruled out, other causes of the acidosis must be sought, such as lactic acidosis, dehydration, renal tubular acidosis, sepsis, and toxic acidosis.

Acidosis

Metabolic acidosis

Etiology. Increased organic acid production resulting in metabolic acidosis can have a number of causes. Uncontrolled diabetes results in accumulation of acetoacetic and hydroxybutyric acids, which are produced by the excessive oxidation of fatty acids (see Chapter 29). Fasting or fad diets also lead to increased levels of these acids.

Lactic acid increases as a result of increases in anaerobic metabolism caused by strenuous muscular exercise or systemic infections. Lactic acidosis also results from local tissue hypoxia (low tissue Po_2), which is caused by dehydration, poor perfusion as a result of circulatory collapse, or cardiac failure.

Renal tubular acidosis results from a failure of the kidney to acidify the urine by exchanging H^+ for Na^+. This renal insufficiency is acquired as a result of infection or is congenital, as in cases of severe de Toni-Fanconi syndrome. Liver disease that impairs the formation of urea and ammonia will also result in a metabolic acidosis because of retention of H^+.

Salicylate intoxication initially induces respiratory alkalosis because of hyperventilation, which is a result of a stimulatory effect of the drug on the respiratory center. The ingested drug is converted to an acid before excretion, and the large quantities of acid formed ultimately result in metabolic acidosis. Other poisons that are ingested as acids or as compounds that will lead to acid metabolites are methyl alcohol (converted to formic acid), ethylene glycol (converted to oxalic acid), paraldehyde, and ammonium chloride. These compounds initially cause respiratory alkalosis, followed by metabolic acidosis. Infusion of large quantities of isotonic sodium chloride results in a metabolic acidosis because the high sodium load competes with hydrogen ions for renal excretion. Metabolic acidosis is also caused by the ingestion of carbonic anhydrase inhibitors, such as acetazolamide or sulfonamides, which interfere with the formation of bicarbonate in the erythrocyte and the renal tubule cells (see Figs. 21-2, 21-3, and 21-5).

A metabolic acidosis may also be caused by a decreased bicarbonate concentration. Diarrhea and colitis lead to losses of intestinal fluids, which contain high concentrations of bicarbonate. The resultant reduction of the HCO_3^-/H_2CO_3 ratio causes acidemia.

Physiological response. The acidoses just described result from the presentation to the body of an acid load that is compensated for at least in part by retention of bicarbonate. When acidemia occurs as a result of acute metabolic acidosis, the body attempts to correct this acidemia by hyperventilation. Hyperventilation lowers the Pco_2, and to a smaller extent the HCO_3^-, and at least partially increases the ratio of HCO_3^-/H_2CO_3, thereby returning the pH toward normal. This mechanism of correcting the pH during acidosis is known as *compensatory respiratory alkalosis*. The result is a decrease in the Pco_2 and HCO_3^- concentration and a more normal pH.

In a metabolic acidosis that does not involve renal dysfunction, the kidney will excrete organic acids and exchange H^+ for Na^+ in the distal tubule, resulting in a more acid urine. This renal compensatory mechanism becomes effective over a period of time and will eventually normalize both the blood pH and the bicarbonate. This correction can take place only when the underlying cause of the acidosis has been eliminated. Part of the renal response to chronic acidosis is the excretion of ammonia by the renal tubular cells. This excretion of ammonia into the presumptive urine allows additional H^+ to be excreted and thus reduces the H^+ load in blood.

Laboratory findings. The findings seen in metabolic acidosis, summarized in Table 21-4, include a decrease in

Table 21-4 Classes of acid-base disorders with corresponding effects on selected blood-gas parameters

Disorder	pH	P_{CO_2}	HCO_3^-	Base excess
Metabolic acidosis	↓	↓ (N)	↓	↓
Respiratory acidosis	↓	↑	↑ (N)	↑ (N)
Metabolic alkalosis	↑	↑ (N)	↑	↑
Respiratory alkalosis	↑	↓	↓ (N)	↓ (N)

(N), *Initially* normal.

both pH and HCO_3^-. Initially the P_{CO_2} may be normal, but it will decrease as a result of the respiratory response to the acidemia. A base deficit (negative base excess) will also be present.

The laboratory findings in lactic acidosis are those of a metabolic acidosis with a decreased pH, an initially normal P_{CO_2} and P_{O_2}, a decreased O_2 saturation (the hemoglobin saturation curve is shifted to the right), a decreased bicarbonate and total CO_2, an increased anion gap, a negative base excess, and increased potassium and lactic acid concentrations. As the body attempts to correct the metabolic acidosis, the P_{CO_2} decreases.

In cases of toxic drug ingestion, such as methanol, ethylene glycol, or paraldehyde poisoning, the patient develops a metabolic acidosis. Laboratory findings include a decreased pH, an initially normal P_{CO_2} and P_{O_2}, a decreased O_2 saturation, bicarbonate, and total CO_2, an increased anion gap (caused by the ingested poisons or their metabolites), and a negative base excess. As the body attempts to compensate for the acidosis, the P_{CO_2} decreases initially and the bicarbonate slowly increases toward normal as the kidney reabsorbs increasing amounts of bicarbonate.

Treatment. Initially the cause of the acidemia is corrected if possible, for example, by insulin treatment of diabetes. If the pH falls below 7.2, there is sometimes a deleterious effect on the cardiovascular system, and the base deficit may have to be corrected immediately. This is frequently accomplished by administration of bicarbonate, which corrects the base deficit by raising the HCO_3^-/H_2CO_3 ratio. In all cases the cause of the metabolic acidosis must ultimately be corrected.

Respiratory acidosis

Etiology. Respiratory acidoses are caused by disorders that interfere with the normal ability of the lungs to expel CO_2. These disorders include pulmonary edema, bronchoconstriction, pneumonia, emphysema, apnea, and bradycardia. Morphine injection and barbiturate poisoning cause an immediate respiratory depression, resulting in respiratory acidosis.

Respiratory distress syndrome (RDS), which is common in premature infants, results in a respiratory acidosis because the infants lack sufficient levels of surfactant in their lungs to allow the alveoli to expand normally. Normal gas exchange is thus inhibited. Respiratory distress is also seen in some adults who experience systemic shock or oxygen toxicity. Initially, the observed blood gas parameters are decreased pH, increased P_{CO_2}, decreased P_{O_2}, and decreased oxygen saturation. Base excess, bicarbonate, and anion gap are initially within normal limits.

Physiological response. The physiological response to respiratory acidosis includes the increased renal excretion of acids, the retention of sodium and bicarbonate, and if possible, hyperventilation. If a response compensates for the respiratory acidosis and results in its correction, the acidosis is referred to as *compensated respiratory acidosis*. This response may be viewed as the development of a metabolic alkalosis that compensates for the respiratory acidosis. In chronic respiratory acidosis, the pH becomes essentially normal but a base excess remains. When the respiratory disorder is corrected, the normal respiratory response to the acidosis removes the excess CO_2, and a transient metabolic alkalosis may result. Generally this alkalosis does not require treatment.

Laboratory findings. Respiratory disorders frequently lead to an increase in plasma CO_2 concentration with a smaller increase in HCO_3^- and a concurrent decrease in the ratio of HCO_3^-/H_2CO_3, which results in respiratory acidosis. However, as a result of the low oxygen levels in the tissue, a coexisting metabolic lactic acidosis can develop. This metabolic acidosis results in an increased anion gap, and the decrease in bicarbonate results in a negative base excess. Table 21-4 reviews these findings.

As a result of the renal compensatory response, very elevated concentrations of HCO_3^- with almost normal pH are often seen. In chronic respiratory disease the concentration of HCO_3^- and pH are near normal, though the P_{O_2} may be rather depressed.

Medical treatment. Medical treatment is primarily aimed at correction of the respiratory disorder and ventilation of the patient with gases containing higher P_{O_2} and lower P_{CO_2} by use of mechanical respirators. However, initial correction of the acidemia may be achieved by injection of sodium bicarbonate.

Alkalosis

Metabolic alkalosis

Etiology. Occasionally, excessive, chronic ingestion of bicarbonate of soda for gastrointestinal distress results in an increased concentration of blood bicarbonate and a resultant metabolic alkalosis. Similarly, treatment of peptic ulcers with ingestion of large quantities of alkali antacids will also produce metabolic alkalosis. More commonly, metabolic alkalosis arises from the loss of chloride. Prolonged vomiting leads to loss of gastric hydrochloric acid. This in turn raises the pH of the blood, because the loss of the chloride anion during vomiting results in increased renal retention of bicarbonate to counter the sodium reabsorbed by the proximal tubule. This condition is known as *hypochloremic alkalosis*. Corticosteroid administration and diseases such as hyperaldosteronism and Cushing's syndrome, which affect the ability of the kidney to regulate electrolyte balance, will also raise the blood pH. In the distal tubule, Na^+ is retained at the expense of K^+ and H^+. The resultant hypokalemia causes a release of K^+ by cells into the blood and a concurrent balanced movement of H^+ from blood into the cells, thereby leading to a rise in the pH of the blood.

Physiological response. To compensate for the increase in the HCO_3^-/H_2CO_3 ratio during metabolic alkalosis, the respiratory system slows, raising the P_{CO_2} and the bicarbonate concentration of the blood. The P_{CO_2} rises more rapidly than the HCO_3^-, thereby decreasing the pH. This mechanism of readjusting the pH during metabolic alkalosis is termed *compensatory respiratory acidosis*. The result is a more normal pH in the presence of an elevated concentration of HCO_3^-.

If the alkalosis persists, the body will attempt to correct the alkalosis by increasing the renal excretion of the excess bicarbonate unless the proximal tubule of the kidney actually increases the reabsorption of bicarbonate, as in hypokalemia, dehydration, or hypochloremia.

Laboratory findings. During metabolic alkalosis, the ratio of HCO_3^-/H_2CO_3 increases as a result of a rise in the concentration of blood bicarbonate. Because of the physiological response to the alkalemia, additional laboratory findings are an increased P_{CO_2} and an alkaline urine containing titratable bicarbonate (Table 21-4).

Treatment. Treatment of metabolic alkalosis involves administration of NaCl or KCl, depending on the degree of hypokalemia, and perhaps also administration of NH_4Cl if the alkalosis is severe and persistent. The Cl anion of NH_4Cl compensates for the chloride deficit, which may have led to excessive retention of bicarbonate initially. This permits the kidney to begin to excrete the excess bicarbonate and correct the alkalosis.

Respiratory alkalosis

Etiology. Hyperventilation causes respiratory alkalosis. The conditions resulting in hyperventilation include hysteria, excessive crying, pregnancy, salicylate intoxication, impairment of the central nervous system's control of the respiratory system, asthma, fever, pulmonary embolism, and excessive use of a mechanical respirator.

Physiological response. The kidneys respond to the alkalosis by excreting increased amounts of bicarbonate under the conditions of lower P_{CO_2} that occur during respiratory alkalosis. In response to the alkalosis, the proximal tubules of the kidney decrease the reabsorption of bicarbonate. This renal response to respiratory alkalosis is termed *compensatory metabolic acidosis*.

Laboratory findings. Hyperventilation leads to increased loss of CO_2 from the blood at the alveolar surface, which causes the HCO_3^-/H_2CO_3 ratio to increase as carbonic acid is lost. Because of the physiological response to the alkalemia, additional laboratory findings include decreased P_{CO_2} and an alkaline urine containing titratable bicarbonate (Table 21-4).

Treatment. One corrects respiratory alkalosis by lowering the respiration rate with drugs, such as sedatives, or by having the patient breathe air with a higher CO_2 content. One easily accomplishes this by having the patient breathe in a restricted environment, as into a paper bag, which raises the P_{CO_2} of the air and the blood. The increased P_{CO_2} returns the HCO_3^-/H_2CO_3 ratio to normal and corrects the respiratory alkalosis.

CHANGE OF ANALYTE IN DISEASE (Table 21-5)

Diabetic ketoacidosis in patients with uncontrolled diabetes is an example of metabolic acidosis. Laboratory findings include a decreased pH, acidemia, a decreased P_{CO_2}, a normal P_{O_2}, a decreased O_2 saturation, a decreased bicarbonate and total CO_2 (T_{CO_2}), an increased anion gap, a negative base excess, and increased serum potassium, ketones, and lactic acid (caused by the disturbed carbohydrate and fat metabolism).

Emphysema is a disease of impaired respiration that frequently results in respiratory acidosis. Laboratory findings include a decreased pH and P_{O_2}, an increased P_{CO_2} and potassium, and a decreased oxygen saturation. Initially the anion gap, base excess, bicarbonate, and T_{CO_2} are normal. As the body compensates for the acidosis, the bicarbonate and T_{CO_2} rise. As in respiratory distress syndrome, the low P_{O_2} may result in a metabolic acidosis caused by a rise in blood lactate because of increased anaerobic metabolism.

Hemoglobinopathies, such as sickle cell anemia, can lead to unusual oxygen-saturation kinetics caused by the abnormal hemoglobin molecule. Some laboratory findings associated with hemoglobinopathies are decreased oxygen

Table 21-5 Common disorders of acid-base balance and effects on selected blood-gas parameters

Disorder	pH	P_{CO_2}	P_{O_2}	HCO_3^-	Base excess	Anion gap	O_2 saturation
Respiratory distress syndrome	↓	↑	↓	N	N	N	↓
Lactic acidosis	↓	N*	N	↓	↓	↑	↓
Diabetic ketosis	↓	N*	N	↓	↓	↑	↓
Emphysema	↓	↑	↓	N	N	N	↓
Methanol poisoning	↓	N*	N	↓	↓	↑	↓
Renal failure	↓	N*	N	↓	↓	↑	↓

*N, *Initially* normal.

saturation and Po_2 levels, which result in increased anaerobic metabolism and thereby metabolic acidosis. Persistence of the hypoxemia results in decreased bicarbonate and total CO_2, an increased anion gap and blood lactate, and a negative base excess. The respiratory response to this acidosis is hyperventilation, which decreases the Pco_2. The renal response to this acidosis is an increase in the reabsorption of bicarbonate, which tends to return the HCO_3^-/H_2CO_3 ratio to normal.

Renal failure leads to metabolic acidosis with the associated laboratory findings of a decreased pH, initially normal Pco_2 and Po_2, decreased oxygen saturation, increased potassium level, and decreased bicarbonate level and Tco_2. As the anion gap increases because of organic acid production and retention, the base excess becomes negative. The respiratory compensation for the metabolic acidosis will lead eventually to a decrease in Pco_2.

BIBLIOGRAPHY

Arieff, AI, and DeFronzo, RA: Fluid, electrolyte and acid-base disorders, New York, 1985, Churchill Livingstone, Inc.

Beeler, MF: Interpretations in clinical chemistry: self-instructional units, Chicago, 1978, American Society of Clinical Pathologists.

Cohen, JJ, and Kassirer, JP: Acid-base, Boston, 1982, Little, Brown, & Co.

Davenport, HW: The ABC of acid-base chemistry, ed 6, Chicago, 1974, University of Chicago Press.

Harper, WA, Rodwell VW, and Mayes, PA: Review of physiological chemistry, ed 17, Los Altos, Calif, 1979, Lange Medical Publications.

Soloway, HB: How the body maintains acid-base balance, Diagn Med pp 32-41, Feb 1979.

Spearman, CB: Egan's fundamentals of respiratory therapy, St Louis, 1982, The CV Mosby Co.

Spence, AP, and Mason EB: Human anatomy and physiology, Menlo Park, Calif, 1979, The Benjamin-Cummings Publishing Co.

Young, DS, Pestaner LC, and Gibberman V: Effects of disease on clinical laboratory tests, Clin Chem 26(4):supplementary issue, 1980.

CHAPTER 22

Renal function

M. ROY FIRST

OBJECTIVES

- List the five main functions of the kidney, and outline the formation of urine, including the function of the following:
  - Glomerulus
  - Proximal tubule
  - Loop of Henle
  - Distal tubule
  - Collecting duct
- Outline the mechanism by which the kidney conserves protein and describe the metabolic production and renal control of blood urea nitrogen, serum creatinine, and uric acid.
- Given pathological conditions associated with the kidney, state expected abnormal results.
- State the purpose of performing a renal clearance test and the reasons why creatinine is most frequently used for renal clearance testing. State the advantages and disadvantages of using inulin or urea for clearance testing.
- Outline two types of studies used to evaluate renal tubular function.

KEY TERMS

absorption (active and passive) The process of uptake of substance into tissues or cells. Active absorption requires the expenditure of energy to move substances against a concentration gradient, whereas, in the case of passive absorption, substances move from higher to lower concentrations.

aldosterone A steroid hormone produced in the adrenal cortex that acts on the distal tubules to stimulate sodium reabsorption and potassium and hydrogen excretion.

antidiuretic hormone (ADH) Also called vasopressin; a pituitary hormone that acts at the collecting duct to increase absorption of water, resulting in the formation of a more concentrated urine.

anuria A condition in which no urine is formed.

ascending limb The straight portion of the loop of Henle in which the presumptive urine flows up toward the convoluted distal tubule. Osmolality of urine decreases because of loss of chloride (plus Na^+).

Bence Jones protein The light-chain portion of a monoclonal immunoglobulin produced in excess and excreted into urine by patients with multiple myeloma.

bladder A sac used to collect formed urine before voiding.

Bowman's capsule A structure consisting of glomeruli and extended opening of the proximal tubule.

carbonic anhydrase The enzyme at the brush border of the proximal tubule that catalyzes the reaction $H_2O + CO_2 \rightarrow H_2CO_3$.

casts Protein aggregates, outlined in the shape of renal tubules, secreted into the urine.

clearance A theoretical concept expressing that volume of plasma filtered at the glomeruli per unit of time from which an analyte would be completely removed and placed in final urine. It is usually expressed as milliliters of plasma per minute.

collecting tubule The last portion of the nephron, connecting the distal convoluted tubule and the larger collecting ducts, which in turn empty into the ureter. The final concentrating processes under the influence of antidiuretic hormone occur here.

countercurrent mechanism The process by which two streams flowing in opposite directions exchange material. In the kidney, urine and blood form opposing flows, and this mechanism allows reabsorption of substances.

creatinine clearance An estimate of the glomerular filtration rate obtained by measurement of the amount of creatinine in plasma and its rate of excretion into urine.

descending limb The straight portion of the loop of Henle in which forming urine flows down from the convoluted proximal tubule. This portion of the loop of Henle is freely permeable to water, which leaves the presumptive urine.

distal tubule The convoluted tubule connecting the ascending loop of Henle with the collecting tubule. It has secretory and reabsorptive functions as part of the final urine formation and acidification process.

diuretic A drug whose action promotes the increased excretion of salt and water, thus increasing the flow of urine.

filtered load The amount of a substance presented to the tubules for reabsorption.

glomerular filtration rate (GFR) The rate in milliliters per minute at which plasma substances are filtered through the glomeruli into the proximal tubule.

glomerulus Cluster of small blood vessels in the kidney that projects into the expanded end (capsule) of the proximal tubule and functions as a filtering mechanism of the nephron.

hematuria The presence of blood or red blood cells in the urine.

loop of Henle A U-shaped tubule connecting the proximal and distal convoluted tubules. It reduces the volume of tubule fluid.

micturition Urination.

nephron The functional unit of the kidney containing Bowman's capsule, the proximal and convoluted tubules, the ascending and descending limbs of the loop of Henle, and the collecting tubules.

oliguria The formation of small amounts of urine.

proteinuria The presence of protein in urine.

proximal tubule The convoluted tubule beginning at the glomeruli and connecting with the descending loop of Henle. It has secretory and reabsorptive functions as part of the mechanism for urine formation.

pyuria The presence of pus (an inflammation fluid with leukocytes and dead cells) in the urine.

renal cortex The outer part of the kidney that contains mostly glomeruli and convoluted tubules.

renal medulla The inner part of the kidney that contains mostly collecting ducts and the loop of Henle.

renal threshold The plasma concentration of a substance above which it will be present in urine.

renin An enzyme formed by the juxtaglomerular apparatus in the kidney. It converts plasma angiotensinogen to angiotensin I.

specific gravity The ratio of the weight in grams per milliliter of a body fluid compared with water.

titratable acid The combination of hydrogen ion with phosphate present in final urine.

urethra A membranous tube through which urine passes from the bladder to the exterior of the body.

urine The aqueous liquid and dissolved substances excreted by the kidney.

ANATOMY OF KIDNEY

Gross anatomy

The kidneys are paired organs located in the posterior part of the abdomen on either side of the vertebral column. Underneath the capsule of fibrous tissue that encloses the kidney lies the cortex, which contains the glomeruli. The inner portion of the kidney, the medulla, contains the collecting ducts. A vertical section through the kidney is shown on the left-hand side of Fig. 22-1.

The urinary system is illustrated on the right-hand side of Fig. 22-1. The renal pelvis rapidly diminishes in caliber and merges into the ureter. Each ureter descends in the abdomen alongside the vertebral column to join the bladder. The bladder provides temporary storage for urine, which is eventually voided through the urethra to the exterior.

Microscopic anatomy

Each kidney is made up of approximately 1 million functional units, or *nephrons*. The component parts of the nephron are illustrated in Fig. 22-2. The nephron begins with the glomerulus, which is a tuft of capillaries that is formed from the afferent (incoming) arteriole and drained by a smaller efferent (outgoing) arteriole. The glomerulus is surrounded by Bowman's capsule, which is formed by the blind, dilated end of the renal tubule. The proximal convoluted tubule runs a tortuous course through the cortex, entering the medulla and forming first the descending limb of the loop of Henle and then the ascending limb of the loop of Henle. The thick section of the ascending limb of the loop of Henle reenters the cortex, forming the distal convoluted tubule. The merging of two or more distal tubules marks the beginning of a collecting duct. As the collecting duct descends through the cortex and medulla, it receives the effluent from a dozen or more distal tubules. The collecting ducts join and increase in size as they pass down the medulla. The ducts of each pyramid coalesce to form a central duct, which empties through the papilla into a minor calyx, eventually draining into the renal pelvis.

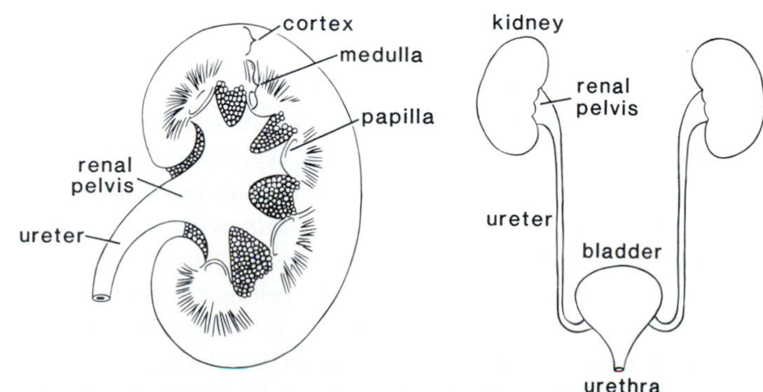

Fig. 22-1 Gross anatomy of kidney and urinary system.

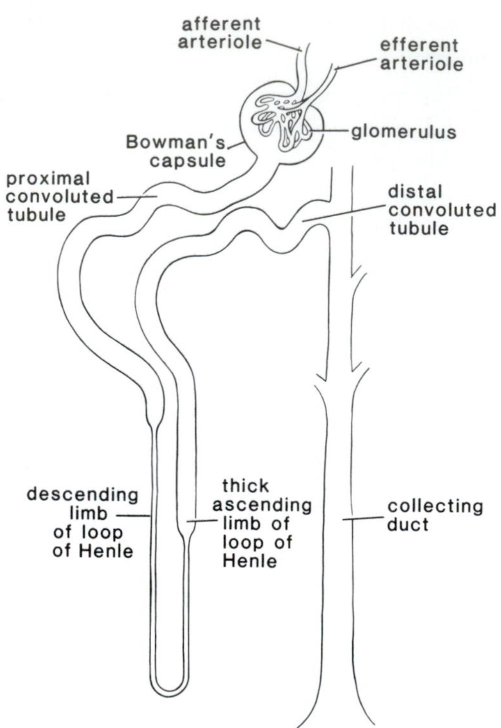

Fig. 22-2 Components of nephron.

RENAL PHYSIOLOGY

The kidney is the chief regulator of all body fluids and is primarily responsible for maintaining homeostasis, or equilibrium of fluid and electrolytes in the body. The kidney has six main functions:

1. Urine formation
2. Regulation of fluid and electrolyte balance
3. Regulation of acid-base balance
4. Excretion of waste products of protein metabolism
5. Hormonal function
6. Protein conservation

The kidney is able to carry out these complex functions because approximately 25% of the volume of blood pumped by the heart into the systemic circulation is circulated through the kidneys; therefore the kidneys, which constitute about 0.5% of total body weight, receive one fourth of the cardiac output.

Urine formation

The removal of potentially toxic waste products is a major function of the kidneys and is accomplished through the formation of urine. The basic processes involved in the formation of urine are *filtration, reabsorption,* and *secretion*. The kidneys filter large volumes of plasma, reabsorb most of what is filtered, and leave behind for elimination from the body a concentrated solution of metabolic wastes called *urine*. In healthy individuals the kidneys, highly sensitive to fluctuations in diet and fluid and electrolyte intake, compensate for any changes by varying the volume and consistency of the urine.

Glomerular filtration. Each minute 1000 to 1500 mL of blood pass through the kidneys. The glomerulus has a semipermeable basement membrane that allows free passage of water and electrolytes but is relatively impermeable to larger molecules. In glomerular capillaries the hydrostatic pressure is approximately three times greater than the pressure in other capillaries. As a result of this high pressure, substances are filtered through the semipermeable membrane into Bowman's capsule at a rate of approximately 130 mL/min; this is known as the *glomerular filtration rate* (GFR). Cells and the large-molecular-size plasma proteins are unable to pass through the semipermeable membrane. Therefore the glomerular filtrate is essentially plasma without the proteins. The GFR is an extremely important parameter in both the study of kidney physiology and the clinical assessment of renal function. In the average healthy person, more than 187,000 mL of filtrate are formed per day. Normal urine output is around 1500 mL per day, which is only about 1% of the amount of filtrate formed; therefore the other 99% must be reabsorbed.

Proximal tubule. The proximal tubular cells perform a variety of physiological tasks. Approximately 80% of salt and water are reabsorbed from the glomerular filtrate in the proximal tubule. All the filtered glucose and most of the filtered amino acids are normally reabsorbed here. Low-molecular-weight proteins, urea, uric acid, bicarbonate, phosphate, chloride, potassium, magnesium, and calcium are reabsorbed to varying extents. A variety of organic acids and bases, as well as hydrogen ions and ammonia, are secreted into the tubular fluid by tubular cells. Under normal conditions, no glucose is excreted in the urine; all that is filtered is reabsorbed. As the plasma concentration of glucose is increased above some critical level, termed the *renal plasma threshold,* the tubular maximum for glucose is exceeded, and glucose appears in the urine. The higher the plasma concentration of glucose, the greater is the quantity excreted in the urine. Renal plasma thresholds also exist for phosphate and bicarbonate ions.

Most of the metabolic energy consumed by the kidney is used to promote *active reabsorption*. Active reabsorption can produce net movement of a substance against a concentration or electrical gradient and therefore requires energy expenditure by the transporting cells. Active reabsorption of glucose, amino acids, small-molecular-weight proteins, uric acid, sodium, potassium, magnesium, calcium, chloride, and bicarbonate is regulated by the kidney according to the levels of these substances in the blood and the body's needs. *Passive reabsorption* occurs when a substance moves by simple diffusion as the result of an electrical or chemical concentration gradient, and no cellular energy is involved in the process. Water, urea, and chloride are reabsorbed in this way.

Tubular secretion, which transports substances into the tubular lumen (that is, in the direction opposite to tubular reabsorption), may also be an active or passive process.

Table 22-1 Filtration, reabsorption, and excretion by kidney

Component	Amount filtered per day	Amount excreted per day	Percentage reabsorbed
Water	180 L	1.5 L	99.2
Sodium	24,000 mEq	100 mEq	99.6
Chloride	20,000 mEq	100 mEq	99.5
Bicarbonate	5000 mEq	2 mEq	99.9
Potassium	700 mEq	50 mEq	92.9
Glucose	180 g	0	100
Albumin	360 mg	18 mg	95.0

Substances that are transported from the blood to the tubules and excreted in the urine include potassium, hydrogen ions, ammonia, uric acid, and certain drugs, such as penicillin. Table 22-1 gives an idea of the magnitude and importance of these reabsorptive mechanisms.

Loop of Henle. The descending limb of the loop of Henle is highly permeable to water. In the medulla, the loop of Henle descends into an environment that is increasingly hypertonic as the papilla is approached. Passive reabsorption of water occurs in response to this osmotic gradient, leaving the presumptive urine highly concentrated at the bottom of the loop. The ascending limb is relatively impermeable to the passage of water but actively reabsorbs sodium and chloride. This segment of the nephron is often called the *diluting segment,* because the removal of salt with little water from the tubular contents lowers the salt and osmotic concentration, in effect diluting the tubular fluid. The ascending thick limb of the loop of Henle transfers sodium chloride actively from its lumen into the interstitial fluid. The tubular fluid in its lumen becomes hypotonic, and the interstitial fluid becomes hypertonic. This phenomenon is known as the *countercurrent mechanism.* A series of successive steps results in sodium chloride being trapped in the interstitial fluid of the medulla. As the isotonic fluid in the descending limb reaches the area into which the ascending limb is pumping out sodium, it becomes slightly hypertonic because of the movement of water into the hypertonic interstitium. The first step repeats itself, and again, as more sodium and chloride are added to the interstitium by the ascending limb, more water is drawn out of the descending limb.

Distal tubule. A small fraction of the filtered sodium, chloride, and water is reabsorbed in the distal tubule. The distal tubule responds to antidiuretic hormone (ADH) so that its water permeability is high in the presence of the hormone and low in its absence. Potassium can be reabsorbed or secreted in the distal tubule. Aldosterone stimulates both sodium reabsorption and potassium secretion in the distal tubule. Hydrogen, ammonia, and uric acid secretion and bicarbonate reabsorption occur, but there is little transport of organic substances. This segment of the nephron has a low permeability to urea.

Collecting duct. ADH controls the water permeability of the collecting tubule throughout its length. In the presence of the hormone, the hypotonic tubular fluid entering the duct loses water. Sodium and chloride are reabsorbed by the collecting tubule, with the transport of sodium stimulated by aldosterone. Potassium, hydrogen, and ammonia are also reabsorbed by the collecting duct. When ADH is present, the rate of water reabsorption exceeds the rate of solute reabsorption, and the concentration of sodium and chloride in the presumptive urine rises. The collecting duct is relatively impermeable to urea.

Regulation of fluid and electrolyte balance

Water. Water is the most abundant component of the human body, constituting approximately 60% of body weight. Body water, and therefore body weight, remains fairly constant from day to day in normal persons, despite wide fluctuations in fluid and salt intake.[1,2] One of the most remarkable properties of the human kidney is its ability to elaborate urine that is either more concentrated or more dilute than the plasma from which it is derived. When the human body needs to conserve water, as in dehydration, the concentrating mechanism operates maximally, and urine osmolality increases to about 1200 mOsm/kg. Conversely, when there is excess water in the body, urine flow increases, and the diluting mechanism reduces urine osmolality to as low as 50 mOsm/kg. The capacity of the kidney to form urine of greatly varying osmolality enables it to regulate the solute concentration and hence the osmolality of body fluids within narrow physiological limits, despite wide fluctuations in intake of salt and water.[1,2] Water balance is controlled primarily through voluntary intake (which is regulated through the thirst center in the hypothalamus) and urinary loss of water. The control of urinary water loss is the major automatic mechanism by which body water is regulated. In dehydrated states the urine is concentrated by reabsorbing water without solute. Conversely, urine is diluted by reabsorbing solute without water.

Sodium. Sodium is the main cation found in extracellular fluid. Sodium is freely filtered through the glomerulus and actively reabsorbed by the tubules. Sodium reabsorption is very important because it affects the regulation of several other electrolytes. Active reabsorption of the sodium ion in the proximal tubule results in passive transport of chloride and bicarbonate as counterions and in passive reabsorption of water. In normal persons, daily urinary sodium excretion fluctuates widely according to the dietary intake, thereby keeping the body sodium content remarkably constant. In the normal person, more than 99% of the filtered load of sodium is reabsorbed by the kidneys. The sodium reabsorption by the nephron is controlled by the renin-angiotensin-aldosterone system (see Chapters 20 and 27).[3,4]

Chloride. The concentration of chloride in the extracellular fluid parallels that of sodium and is influenced by the same factors. However, chloride reabsorption is passive in the proximal tubule and probably active in the distal nephron.

Potassium. Potassium is the chief cation of the intracellular fluid. Maintenance of a normal potassium level is essential to the life of the cells. The normal person maintains potassium balance by excreting daily an amount of potassium equal to the amount ingested minus the small amount eliminated in the feces and sweat. Renal function is the major mechanism by which the body potassium is regulated. Potassium is freely filtered at the glomerulus, and active tubular reabsorption occurs throughout the nephron, except in the descending loop of Henle. Only about 10% of the filtered potassium enters the distal tubule. The distal tubule and collecting ducts are able to both secrete and reabsorb potassium, thus regulating potassium excretion.[5,6] The hormone aldosterone, which stimulates tubular sodium reabsorption, simultaneously enhances potassium secretion in the distal tubule.[6]

Calcium. Calcium is reabsorbed in the proximal tubule under the hormonal influence of parathyroid hormone.[7,8] The maintenance of calcium homeostasis depends on the balance between calcium intake and calcium loss. The body loses calcium in the urine, through the gastrointestinal tract, and in sweat. Calcium balance is achieved largely by the control of calcium absorption rather than by the regulation of calcium excretion. The percentage of ingested calcium absorbed decreases as the dietary calcium content increases, and so the amount absorbed can remain relatively constant. The slight increase in absorption that occurs on a high-calcium diet is reflected in an increased renal excretion.[9]

Phosphorus. Over a wide range of dietary intakes, roughly two thirds of ingested phosphorus is absorbed into the bloodstream. The maintenance of the phosphorus balance is achieved largely through renal excretion.[9] Proximal tubular reabsorption of inorganic phosphate is normally about 90% of the filtered load. Parathyroid hormone depresses the renal tubular reabsorption of inorganic phosphate. In progressive chronic renal failure, there is a progressive increase in the serum phosphorus level.[10]

Magnesium. The filtration of magnesium at the glomerulus and its reabsorption from the proximal tubules parallels that of calcium and is also under the influence of parathyroid hormone. A moderate elevation of the plasma magnesium concentration occurs in patients with advanced chronic renal failure.[11]

Acid-base balance

Each day acid waste products are produced in the body. If they were not disposed of efficiently, they would accumulate and cause cellular damage. Body pH is controlled by three systems: acid-base buffers, the lungs, and the kidneys. (See Chapter 21.)

Excretion of hydrogen ions. In subjects on a normal diet, about 50 to 100 mEq of hydrogen ions are generated each day.[12] To prevent a progressive metabolic acidosis, these hydrogen ions are excreted in the urine. Hydrogen ions are generated in the cells of the proximal and distal tubule and the collecting duct as a result of the formation of carbonic acid by the enzyme carbonic anhydrase (CA). These hydrogen ions are secreted by the cells into the lumen[13]:

$$H_2O + CO_2 \overset{CA}{\rightleftharpoons} H_2CO_3 \rightleftharpoons H^+ + HCO_3^-$$

The kidneys' role in the maintenance of the acid-base balance centers on the generation of bicarbonate. The hydrogen ions are excreted into the urine while newly generated bicarbonate ions pass from the tubular cells into the blood at the same rate as bicarbonate is consumed by the metabolic processes.[14] Four mechanisms exist to handle the hydrogen ions that have been secreted into the tubular fluid.

Reaction with filtered bicarbonate ions. Bicarbonate is completely filterable at the glomerulus. In the tubular lumen, the excreted hydrogen ion combines with the filtered bicarbonate to form carbonic acid, which decomposes to water and carbon dioxide, the latter then diffusing into the cell, where it can be converted by CA to carbonic acid to generate another hydrogen ion. One bicarbonate ion is regenerated for every hydrogen ion that is secreted into the tubular lumen. The bicarbonate ion is reabsorbed into the blood as sodium bicarbonate, thus conserving most of the filtered bicarbonate. The renal threshold for bicarbonate is 28 mM; at a plasma level below this, all filtered bicarbonate is reabsorbed.

Reaction with filtered buffers to form titratable acids. Inorganic monohydrogen phosphate is present in the tubular lumen as the disodium salt. The secreted hydrogen ions react with the filtered phosphate, releasing sodium that combines with the bicarbonate; the sodium bicarbonate is reabsorbed, and dihydrogen phosphate is excreted:

$$Na_2HPO_4 + H^+ \rightarrow NaH_2PO_4 + Na^+$$
$$Na^+ + HCO_3^- \rightarrow NaHCO_3 \text{ (reabsorbed)}$$

Hydrogen ions combine with phosphate to form titratable acid. The rate of excretion of titratable acid is limited by the filtered load of buffer and cannot increase greatly in acidosis. The lowest possible pH of urine is 4.4.

Reaction with secreted ammonia to form ammonium ion. The glomerular filtrate does not contain ammonia. This compound is synthesized in renal tubular cells by deamination of glutamine in the presence of glutaminase.[15] The ammonia diffuses into the tubular fluid, where it reacts with a secreted hydrogen ion to form an ammonium ion. Once again this results in the addition of new bicarbonate to the blood. The most important renal adaptation to acidosis is the increased excretion of ammonium ions[15]:

$$\text{Glutamine} \xrightarrow{\text{Glutaminase}} \text{Glutamic acid} + NH_3$$
$$NH_3 + H^+ \longrightarrow NH_4^+$$

Excretion as free hydrogen ions. Only negligible quantities of hydrogen ions are handled in this way by the kidneys.

Nitrogenous waste excretion

One of the major functions of the kidney is the elimination of nitrogenous products of protein catabolism. The enormous reserves of the kidney for excretion of the products of protein catabolism are indicated by the fact that the blood concentrations of these products are not elevated in renal failure until renal function is reduced to less than one half of normal.[16]

Urea. As amino acids are deaminated, ammonia is produced. The development of toxic levels of ammonia in the blood is prevented by the conversion of ammonia to urea. This takes place in the liver. Urea in the blood is measured as the blood urea nitrogen (BUN). Urea production and the BUN are increased when more amino acids are metabolized in the liver. This can occur with a high-protein diet,[17] tissue breakdown, or decreased protein synthesis. On the other hand, urea production and the BUN are reduced in the presence of a low-protein intake and severe liver disease. Urea production exceeds renal urea excretion in normal persons. The remaining urea is degraded to ammonium ions by intestinal bacteria.[18] Urea is readily filtered, but approximately 40% to 50% of the filtered urea is normally reabsorbed by the proximal tubules. Since a number of factors may influence the BUN level while the GFR remains constant, BUN is a poor indicator of renal function and should not be relied on for that purpose.

Creatinine. Serum creatinine levels and urinary creatinine excretion are a function of muscle mass in normal persons and show little or no response to dietary changes or alteration in electrolyte balance.[19] Creatinine is derived from the nonenzymatic dehydration of creatine in skeletal muscle. The amount of creatine per unit of muscle mass is constant, and thus the rate of spontaneous breakdown of creatine is also constant. As a result, the plasma creatinine concentration is very stable, varying less than 10% per day in serial observations in normal subjects.[20] Since the serum creatinine concentration is a direct reflection of muscle mass, the serum level is higher in males than females. Creatinine is freely filtered at the glomerulus and is not reabsorbed by the tubules. A small amount of the creatinine in the final urine is derived from tubular secretion. Because of these properties of creatinine, the creatinine clearance can be used to estimate the GFR (see p. 354):

```
CH3— NCH2—COOH          CH3— N—CH2
     |                        |    \
     C=NH      ———→           |     C=O + H2O
     |                        |    /
     NH2                   HN=C—NH
   Creatine               Creatinine
```

Uric acid. Uric acid is derived from the oxidation of purine bases. Plasma levels of uric acid are quite variable and are higher in males than in females. Plasma urates are completely filterable, and both proximal tubular resorption and distal tubular secretion occur. With advanced chronic renal failure there is a progressive increase in the plasma uric acid level.

Hormonal function

The kidneys have important metabolic and endocrine functions. The kidney as an endocrine organ will be discussed in this section.[21]

Vitamin D metabolism. In vitamin D metabolism the kidney produces the major biologically active hormone, 1,25-dihydroxycholecalciferol.[22-24] The enzyme responsible for the production of 1,25-dihydroxycholecalciferol is present only in the mitochondria of the renal cortex.[21] (See Chapter 24.)

Renin. The kidney releases renin in response to a decrease in extracellular fluid volume. This results in stimulation of the renin-angiotensin-aldosterone axis, with subsequent sodium and water conservation. (See Chapters 20 and 27).

Erythropoietin. The kidneys play a major role in the production and release of erythropoietin, a hormone that stimulates red blood cell production. The central role of the kidneys in erythropoietin production explains the anemia associated with chronic renal failure.[25]

Protein conservation

Under normal physiological conditions, the kidney helps to maintain the homeostasis of the body's proteins. In humans 180 L of plasma, each containing 70 g of protein, are filtered each day by the glomerulus.[27] Without an efficient conservation mechanism, body protein stores would be depleted very rapidly. Yet normal urine contains less than 200 mg of protein per day, only a minute percentage of the 12,600 g passing through the glomerulus daily.[27] Most of the filtered proteins are absorbed by the proximal tubules and returned to the circulation. Most plasma proteins, except those of very high molecular weight, have been found in the urine. Albumin excretion is less than 20 mg/day.[27] Many proteins of nonserum origin are also found in the urine. One of these, uromucoid or Tamm-Horsfall mucoprotein, is the predominant protein in normal urine, with about 40 mg excreted daily.[28] This high-molecular-weight mucoprotein is excreted by the cells of the distal tubule and collecting ducts. Commercially available dipsticks (Albustix) are in widespread use and are accurate for rapid assessments of urinary protein concentration.

PATHOLOGICAL CONDITIONS OF KIDNEY

There are a number of syndromes that singly or in combination indicate possible renal disease. These have been elegantly described by Coe.[29]

Acute glomerulonephritis

Acute glomerulonephritis is an acute inflammation of the glomeruli, resulting in oliguria, hematuria, increased BUN and serum creatinine levels, decreased GFR, edema formation, and hypertension. The presence of red blood cells in the urine (hematuria) alone is insufficient evidence of acute glomerulonephritis, for blood can originate from elsewhere in the kidney or from the urinary tract. The presence of red blood cell casts in the urine indicates glomerular inflammation and is a finding of great importance. Other abnormalities present in acute nephritis include proteinuria and anemia.

Nephrotic syndrome

The nephrotic syndrome has been classically defined as a clinical entity characterized by massive proteinuria, edema, hypoalbuminemia, hyperlipidemia, and lipiduria.[30] This syndrome, which can have many causes, is characterized by increased glomerular membrane permeability that results in massive proteinuria and excretion of fat bodies. Protein excretion rates are usually greater than 2 to 3 g/day in the absence of a depressed GFR. Hematuria and oliguria may be present. The causes of the nephrotic syndrome are illustrated in the box at left. As a result of the massive loss of serum proteins, primarily albumin, into urine, the plasma protein concentration is decreased, with a concomitant reduction in plasma oncotic pressure. This results in fluid movement from the vascular to interstitial space with consequent edema formation.

CAUSES OF NEPHROTIC SYNDROME

- Associated with various forms of glomerulonephritis
- Associated with generalized disease processes
  - Amyloidosis
  - Carcinoma
  - Systemic lupus erythematosus
  - Diabetic glomerulosclerosis
  - Polyarteritis nodosa
- Associated with mechanical or circulating disorders
  - Renal vein thrombosis
  - Constrictive pericarditis
- Associated with infection
  - Syphilis
  - Malaria
  - Subacute bacterial endocarditis
- Associated with toxins and allergens
  - Penicillamine
  - Gold salts
  - Bee sting
  - Serum sickness
- Miscellaneous
  - Severe preeclampsia
  - Transplant rejection

Tubular disease

In some disorders of renal tubular function, the depressed renal function cannot be explained by the reduction in the GFR. Defects of tubular function may result in depressed secretion or reabsorption of specific biochemicals or impairment of urine concentration and dilution mechanisms. Renal tubular acidosis (RTA) is the most important clinical disorder of tubular function. There are two main types of RTA: (1) proximal RTA, which is a result of reduced proximal tubular bicarbonate reabsorption and which causes *hyperchloremic acidosis,* and (2) distal RTA, in which there is an inability of the tubular cells to create and maintain the usual pH difference between tubular fluid and blood.[31] Failure of either the proximal or distal secretory mechanisms occurs in several disease states. Failure of the proximal tubule to reabsorb bicarbonate causes acidosis because more bicarbonate is passed on to the low-capacity distal mechanism than it can reabsorb. The loss of alkali in the urine causes the blood to become acidotic. Defects in potassium and uric acid secretion may result in elevations of the serum potassium and uric acid levels that cannot be explained by the reduction in the GFR. Reabsorptive disorders of the proximal tubules may result in hypouricemia, hypophosphatemia, aminoaciduria, and renal glucosuria. The Fanconi syndrome is a group of renal defects resulting in glucosuria, aminoaciduria, hypophosphatemia, and renal tubular acidosis. Tubular proteinuria may occur as a result of a tubular defect in the handling of proteins. In tubular proteinuria, less than 2 g/day of protein are excreted. Disorders of urine concentration and dilution occur in all renal disease as the GFR falls appreciably, but occasionally these disorders become extreme and dominate the clinical presentation.[29]

Urinary tract infection

Infection of the urinary tract may occur in the bladder *(cystitis)* or may involve the kidneys *(pyelonephritis).* The presence of a urine bacterial concentration of more than 100,000 colonies/mL is diagnostic of urinary tract infections. In urinary tract infection there is an increased number of white blood cells in the urine. The presence of white blood cell casts indicates pyelonephritis. An increased number of red blood cells may also be present in the urine.

Vascular diseases

Hypertension. Long-standing and severe hypertension can result in progressive renal damage and chronic renal insufficiency (hypertensive nephrosclerosis). On the other hand, hypertension can be caused by the sodium and water retention that occurs in chronic renal failure, acute glomerulonephritis, and the nephrotic syndrome (volume-dependent hypertension), or it can occur as a result of increased renin release from chronically damaged kidneys (renin-dependent hypertension).

Arteriolar disease. Disease of the small arteries of the kidneys (arteritis) may occur in association with generalized disease processes affecting the kidney, such as systemic lupus erythematosus, polyarteritis nodosa, and progressive systemic sclerosis (scleroderma). These diseases may result in the clinical and biochemical abnormalities seen in acute glomerulonephritis, the nephrotic syndrome, or chronic renal insufficiency.

Renal vein thrombosis. Thrombosis of the renal veins results in massive proteinuria and the nephrotic syndrome. Hypertension, edema, hematuria, and impaired renal function may accompany the proteinuria.

Diabetes mellitus

Diabetes mellitus results in a wide variety of abnormalities in kidney function. The early phases of the disease are manifested by the presence of pronounced glucosuria, polyuria, and nocturia as a result of the osmotic diuresis caused by the glucose load. In insulin-dependent diabetes, kidney disease is the leading cause of death. In the juvenile diabetic, overt proteinuria develops approximately 17 years after the diagnosis has been made, hypertension develops 1 to 2 years later, and chronic renal insufficiency is seen after another year.[32] Early in the course of this disease protein excretion, particularly albumin and IgG, is increased. A urinary albumin excretion in the range of 50 to 200 mg/24 hr is usually predictive of diabetic nephropathy.[33]

Urinary tract obstruction

Lower urinary tract obstruction is characterized by residual urine in the bladder after micturition or urinary retention, whereas the presence of upper tract obstruction is established by the demonstration of a dilated collecting system above a constricting lesion.[28] Lower urinary tract obstruction is characterized by a slow urinary stream, difficulty in emptying the bladder, hesitancy in initiating micturition, and dribbling. Chronic renal damage may result from obstruction and incomplete bladder emptying, and symptoms of chronic renal insufficiency may develop. With complete obstruction, oliguria or anuria will occur. Symptoms of urinary tract infection may also be seen. Urinary tract obstruction may occur as a result of congenital disorders of the lower urinary tract, neoplastic lesions (benign prostatic hypertrophy, carcinoma of the prostate or bladder, or lymph nodes compressing the ureters), or acquired disorders (retroperitoneal fibrosis, renal calculi, or urethral strictures).

Renal calculi

Renal calculi, or stones, are seen in combination with renal colic, hematuria, and symptoms of urinary tract infection or obstruction. Kidney stones may form after recurrent urinary tract infections by urease-producing organisms or when the urine is supersaturated by large quantities of calcium, uric acid, cystine, or xanthine.

Acute renal failure

In acute renal failure there is an abrupt deterioration in renal function. Acute renal failure can be classified as follows:

1. Prerenal (occurring before blood reaches the kidney) because of hypovolemia or poor perfusion as a result of cardiovascular failure
2. Renal (occurring in the kidney) because of acute tubular necrosis, which is the most frequently observed cause of acute renal failure, or because of other renal diseases, causing rapid deterioration in renal function, including arterial or venous obstruction
3. Postrenal (occurring after urine leaves kidney) because of obstruction

The causes of acute renal failure are listed in the box. Acute renal failure is usually accompanied by oliguria or anuria; in addition, nonoliguric acute tubular necrosis can occur. Acute renal failure is associated with varying degrees of proteinuria, hematuria, and the presence of red blood cell casts and other casts in the urine. Serum urea nitrogen and creatinine levels increase rapidly, and metabolic acidosis becomes evident. Depending on the cause, acute renal failure can progress to chronic renal failure or can be followed by recovery of renal function. Most patients with acute tubular necrosis recover once the offending cause has been treated or removed.

CAUSES OF ACUTE RENAL FAILURE

- Prerenal
  - Hypovolemia
  - Cardiovascular failure
- Renal
  - Acute tubular necrosis
  - Glomerulonephritis
  - Vasculitis
  - Malignant nephrosclerosis
  - Vascular obstruction
    - Arterial
    - Venous
- Postrenal
  - Obstruction of lower urinary tract
  - Rupture of bladder

Chronic renal failure

Chronic renal failure is a clinical syndrome resulting from the progressive loss of renal function. The symptoms of chronic renal failure result not only from simple excretory failure but also from the onset of regulatory failure, the kidney's failure to regulate certain substances, such as sodium and water; from biosynthetic failure, such as the kidney's inadequate production of erythropoietin, resulting in anemia; and from the excessive production of certain normal substances in response to the chemical derangements that occur in chronic renal failure, such as the excessive production of parathyroid hormone.[34] There are four stages in chronic, progressive renal disease. In the first stage renal function is diminished, but plasma urea and creatinine levels remain normal. At least 50% of normal function must be lost before the concentrations of these chemicals rise above the normal range. The second stage is characterized by mild renal insufficiency. The third stage is the development of frank renal failure with advancing anemia, acidosis, and other clinical and biochemical manifestations. The fourth and final stage is that of uremia, when all the consequences of renal failure become overt.[34] A classification of the causes of chronic renal failure is shown in the box.

CLASSIFICATION OF CAUSES OF CHRONIC RENAL FAILURE

- Primary glomerular diseases
  - Chronic glomerulonephritis of various types
  - Systemic lupus erythematosus
  - Polyarteritis nodosa
- Renal vascular disease
  - Malignant hypertension
  - Renal vein thrombosis
- Inflammatory disease
  - Chronic pyelonephritis
  - Tuberculosis
- Metabolic disease with renal involvement
  - Diabetes mellitus
  - Gout
  - Amyloidosis
- Nephrotoxins
  - Aminoglycosides
  - Analgesic nephropathy
  - Chronic heavy metal poisoning
- Obstructive uropathy
  - Calculi
  - Prostatic hypertrophy
  - Congenital anomalies of lower urinary tract
- Congenital anomalies of kidneys
  - Hypoplastic kidneys
  - Polycystic kidney disease
- Miscellaneous
  - Chronic radiation nephritis
  - Balkan nephropathy

RENAL FUNCTION TESTS

The kidney performs many physiological and excretory functions. By performing a relatively small number of tests, a physician can deduce accurately the functional state of the kidney.[35] The clinician first determines whether any significant impairment of renal function exists and then assesses a particular renal function to make a specific diagnosis.[25] In this section evaluation of glomerular and tubular function and urinalysis will be discussed.

Tests of glomerular function

Glomerular function is most conveniently measured by the creatinine clearance test. Clearance is defined as that volume of plasma from which a measured amount of substance can be completely eliminated into the urine per unit of time. This depends on the plasma concentration of the substance and its excretory rate, which in turn depends on the GFR and renal plasma flow. The creatinine clearance is a renal function test based on the rate of excretion by the kidneys of metabolically produced creatinine. The amount of creatinine produced by endogenous protein metabolism is relatively constant and directly proportional to the body surface area. The amount of creatinine present in the urine depends on renal excretion. Creatinine is freely filtered at the glomerulus and is not reabsorbed by the tubules. The creatinine clearance can therefore be used to estimate the GFR. Generally, a 24-hour urine collection is performed. However, shorter collection periods are acceptable. Precise timing is critical to this test. The bladder is emptied at the beginning of the test period and the urine discarded; all urine passed subsequently during the timed collection is kept in a single container. A sample of blood is drawn during the urine collection period. The creatinine clearance is calculated from the following formula:

$$\text{Creatinine clearance (mL/min)} = UV/P$$

where U is urinary creatinine (mg/L), V is volume of urine (mL/min), and P is plasma creatinine (mg/L). The normal range for creatinine clearance corrected to a surface area of 1.73 m^2 is 90 to 120 mL/min. The creatinine clearance usually parallels true GFR. However, at low filtration rates, creatinine clearance becomes increasingly inaccurate.[36] The creatinine clearance is lower in women, the elderly, and smaller persons.

Inulin clearance is the method of choice when precise determination of the GFR is required.[36] The glomerular capillary wall is freely permeable to inulin, and inulin is not reabsorbed, secreted, or metabolically altered by the renal tubule. The clearance of endogenous creatinine may exceed that of inulin by up to 30% in normal individuals.[36] The main disadvantages in the measurement of inulin clearance are the need for its intravenous administration and the technical difficulty of the analysis.

Urea clearance may also be employed as a measure of the GFR. Urea is freely filtered at the glomerulus, and approximately 40% is reabsorbed in the tubules. Thus urea clearance values will parallel the true GFR.

From a practical point of view, creatinine clearance is used in clinical medicine as an assessment of the GFR. It is important to understand that there is a large margin of reserve in renal function; more than two thirds of the GFR may be lost in the course of chronic renal disease with few clinical symptoms and biochemical abnormalities.[35] For a person whose normal serum creatinine is 7 mg/L, an increase to 14 mg/L, which is still defined as within the normal range for serum creatinine, is indicative of a fall in the GFR to 50% of normal.

Tests of tubular function

Concentration-dilution studies. Assessment of the concentrating and diluting ability of the kidney can provide the most sensitive means of detecting early impairment in renal function, since the ability to concentrate urine and conserve water requires an adequate GFR, renal plasma flow, and tubular mass and healthy tubular cells that are able to pump salt against a sizable electrochemical gradient.[35] The urinary *specific gravity* and *osmolality* are used as measures of the concentrating and diluting ability of the tubules. As long as the urine does not contain appreciable amounts of protein, sugar, or exogenous material such as contrast dye, specific gravity is proportional to osmolality, and a specific gravity of 1.032 will correspond to an osmolality of 1200 mOsm/kg.[35]

Impairment of renal concentrating ability is a relatively early manifestation of chronic renal disease and becomes evident before changes in other function tests appear. However, it is a nonspecific test for reduced renal function, and any disease resulting in chronic renal failure, diabetes insipidus, or the use of diuretics may impair renal concentrating ability. The test is performed after 15 hours of fluid deprivation, and urine is then collected on the hour for 3 hours. Dehydration maximally stimulates endogenous ADH secretion. Under these conditions the urine osmolality should be at least three times that of plasma (286 mOsm/kg). A specific gravity of 1.025 or more or an osmolality of 850 mOsm/kg or above in one of the specimens is accepted as evidence of normal concentrating ability. A patient with completely normal concentrating ability is unlikely to have a serious kidney malfunction of any type.[35] As chronic renal disease progresses, tubular ability to concentrate urine slowly decreases until the urine has the same specific gravity as the plasma ultrafiltrate—1.010. Clinically, the loss of concentrating ability is manifested by nocturia and polyuria.

To test the urinary diluting capacity, one uses the following procedure. The patient empties the bladder and is given 1000 to 1200 mL of water. Urine specimens are then collected every hour for the next 4 hours. Under these circumstances, the urinary specific gravity should fall to 1.005 or less or an osmolality of less than 100 mOsm/kg. In the patient with chronic renal disease who is unable to dilute the urine, there is a danger of fluid overload with this test.

In diabetes insipidus, which can arise from inadequate ADH production or from insensitivity of the renal tubules

Table 22-2 Association of pathological conditions affecting the kidney and clinical and biochemical abnormalities*

	AGN	NS	TD	UTI	HT	RVT	DM	UTO	RC	ARF	CRF
Hypertension	+ +	+	0	0	+ +	±	±	0	0	+	+
Edema	+	+ +	0	0	0	+	+	0	0	+	+
Oliguria or anuria	+	±	0	0	0	±	0	+	0	+	+
Polyuria	0	0	+	0	0	0	+	0	0	0	0
Nocturia	0	±	+	±	0	0	+	±	0	0	+
Frequency	0	0	0	+	0	0	0	±	±	0	0
Loin pain	0	0	0	+	0	+	0	+	+	0	0
Anemia	+	0	0	0	0	0	0	0	0	0	+ +
↑ Blood urea nitrogen	+	0	0	0	±	±	±	±	0	+	+
↑ Serum creatinine	+	–	–	–	±	±	±	±	0	+	+
↓ GFR	+	0	0	0	±	±	±	±	0	+	+
↑ Serum potassium	±	±	0	0	0	0	0	0	0	+	+
↑ Serum phosphorus	±	0	0	0	0	0	0	0	0	+	+
↓ Serum calcium	0	+	0	0	0	0	0	0	0	+	+
↑ Serum uric acid	0	0	+	0	±	0	±	0	±	+	+
Acidosis	0	0	+	0	0	0	0	0	0	+	+
Proteinuria	+	+ + + +	+	±	±	+ +	+	0	0	±	±
Hematuria	+ +	+	±	+	0	+	0	0	+ +	+	±
RBC casts	+	0	0	0	0	0	0	0	0	±	0
Pyuria	±	0	0	+ +	0	0	0	±	±	0	0
WBC casts	0	0	0	+	0	0	0	+	0	0	0
Glucosuria	0	0	+	0	0	0	+ +	0	0	0	0

**AGN*, Acute glomerulonephritis; *ARF*, acute renal failure; *CRF*, chronic renal failure; *DM*, diabetes mellitus; *HT*, hypertension; *NS*, nephrotic syndrome; *RC*, renal calculi; *RVT*, renal vein thrombosis; *TD*, tubular disease; *UTI*, urinary tract infection; *UTO*, urinary tract obstruction. O = absent; ± = variable; + = present. *RBC*, red blood cell; *WBC*, white blood cell.

to ADH, the distal tubular walls are impervious to water. As sodium is reabsorbed, the fluid left behind may be very dilute. In this disease the baseline urine might have a specific gravity of less than 1.005 and an osmolality of 50 mOsm/kg.

Urinalysis

Urinalysis is an indispensable tool for assessing renal disease. It may reveal disease anywhere in the urinary tract. Observations that can be made in the standard urinalysis include the appearance of the specimen, pH, specific gravity, protein semiquantitation, presence or absence of glucose and ketones, and a microscopic examination of the centrifuged urinary specimen. The importance of the urinalysis is indicated in Table 22-2. Microscopic examination of the centrifuged urinary sediment for cells, crystals, and casts should be done on a freshly voided specimen.

Casts are protein conglomerates outlining the shape of the renal tubules in which they were formed. Hyaline casts are composed almost exclusively of protein. Cellular elements may be trapped within hyaline casts, resulting in the formation of granular casts. When there is heavy proteinuria, accumulation of protein within tubular cells leads to fatty degeneration of the cells and desquamation of cells into the urine; these appear in the urine as oval fat bodies. In acute pyelonephritis, white blood cells may aggregate in the tubules to form pus casts. Red blood cell casts are important markers of glomerular inflammation and should be diligently searched for when any form of glomerular nephritis is suspected.

Microscopic examination of the urinary sediment is completed by a search for bacteria and crystals. The presence of crystals in the urine may be a clue to the diagnosis of a specific type of renal calculus. The characteristic urine microscopic findings in healthy individuals and in renal disease are indicated in Table 22-3. (See Chapter 56.)

CHANGE OF ANALYTE IN DISEASE

The changes that occur in analytes have been discussed in the section on pathological conditions of the kidney and summarized in Table 22-2. In this section the following question is examined from a different perspective: What does the finding of a biochemical abnormality or group of abnormalities mean in the diagnosis of the pathological condition in the kidney?

Serum electrolytes (see also Chapter 20)

Sodium. Sodium is the major cation in the extracellular fluid and normally has a serum concentration of 136 to 145 mEq/L. Sodium and its attendant anions are the major contributors to serum osmolality.[37]

Hyponatremia. Hyponatremia with hypoosmolality can occur in renal disease because of an increased extracellular fluid volume resulting from the kidney's inability to excrete water. This state occurs in chronic renal insufficiency and the nephrotic syndrome. Hyponatremia, with decreased extracellular fluid, can be associated with the use of diuretic agents and the syndrome of inappropriate ADH secretion.

Hypernatremia. Hypernatremia, by definition a relative water deficit, can occur in patients with hypotonic fluid loss. Hypernatremia also occurs in diabetes insipidus whenever the oral fluid intake cannot keep pace with the urinary losses.

Chloride. The concentration of chloride in extracellular fluid parallels that of sodium and is influenced by the same factors. Chloride imbalances occur concurrently with sodium imbalances. Hyperchloremia occurs in association with renal tubular acidosis.

Potassium. Potassium is the major cation of intracellular fluid.

Hypokalemia. Hypokalemia is usually associated with overt potassium depletion as a result of excessive losses of potassium-rich fluids. Potassium loss may be renal or extrarenal. Increased renal excretion of potassium occurs with diuretic agents, prolonged use of corticosteroids, primary or secondary aldosteronism, and Cushing's syndrome. Hypokalemia from extrarenal potassium losses usually occurs in the gastrointestinal tract and is seen with prolonged vomiting, diarrhea, fistulas of the intestinal tract, and villous adenomas of the colon.

Hyperkalemia. Hyperkalemia, an acute medical emergency, is usually caused by either increased cellular breakdown exceeding the normal renal excretory capacity or impaired renal excretion.[37] Hyperkalemia may result from (1) increased intake of potassium, as occurs with dietary excess or intravenous potassium administration in the patient with compromised renal function; (2) cellular breakdown, as occurs with extensive burns or rhabdomyolysis (acute muscle necrosis); (3) decreased potassium excretion, as occurs in acute or chronic renal failure, secondary to potassium-sparing diuretics, in adrenal insufficiency, and in

Table 22-3 Characteristic urine microscopic findings in renal disease

Condition	Protein	Red blood cells (per high-power field)	White blood cells (per high-power field)	Bacteria	Casts (per low-power field)
Normal	0-Trace	0-3	0-5	0	Hyaline, occasionally
Glomerulonephritis	$1\text{-}2^+$	>20	0-10	0	Granular red blood cells
Nephrotic syndrome	4^+	0-10	0-5	0	Oval fat bodies; hyaline
Pyelonephritis	$0\text{-}1^+$	0-10	>30	+ +	Granular white blood cells

hypoaldosteronism; or (4) transcellular redistribution of potassium, as occurs with acute acidosis, diabetic ketoacidosis, familial hyperkalemic periodic paralysis, and certain drugs.

Urinary electrolytes

Sodium. Urinary sodium determinations are diagnostically useful in three clinical settings. First, in volume depletion, the measurement of urinary sodium excretion is helpful in determining the route of sodium loss. A low urinary sodium concentration (less than 10 mEq/L) indicates an extrarenal sodium loss, whereas the presence of a high concentration of sodium in the urine indicates renal salt wasting or adrenal insufficiency. Second, in the differential diagnosis of acute renal failure, the urinary sodium excretion will be less than 10 mEq/L in patients with volume depletion who have no intrinsic renal disease and usually more than 30 mEq/L in patients with acute tubular necrosis.[38] In volume depletion a urine-to-plasma osmolality ratio of more than 1.1 and a urine-to-plasma urea ratio of more than 10 is observed, compared with values of less than 1.05 and less than 10, respectively, in acute tubular necrosis.[38] Third, in hyponatremia, a low urinary sodium concentration (less than 10 mEq/L) indicates avid renal sodium retention, which may be attributable to either severe volume depletion or sodium-retaining states seen in cirrhosis, the nephrotic syndrome, and congestive heart failure. When hyponatremia is associated with urinary sodium excretion that equals or exceeds the dietary sodium intake, it is likely that the syndrome of inappropriate ADH secretion is present.[37] In these three situations a random urinary sodium concentration can rapidly supply valuable diagnostic information.

Chloride. The measurement of urinary chloride is of clinical value only in patients with persistent metabolic alkalosis who are not receiving diuretics.[37]

Potassium. Urinary potassium levels are helpful in the evaluation of patients with unexplained hypokalemia.[37] The finding of a urinary potassium concentration of more than 10 mEq/L indicates that the kidney is responsible for the potassium loss, whereas a urinary potassium concentration of less than 10 mEq/L in the presence of hypokalemia strongly suggests that the gastrointestinal tract is the route of potassium loss.

Anion gap

An increased anion gap occurs in renal failure because of the retention of sulfate, phosphate, and organic acid anions. (See Chapters 20 and 21.)

Urea, creatinine, and uric acid

With progressive renal insufficiency there is retention in the blood of urea, creatinine, and uric acid. Normally the ratio of serum urea nitrogen to serum creatinine is between 10:1 and 20:1. In the usual case of renal failure, a similar ratio is seen. Ratios higher than 20:1 occur in disease states of extrarenal origin, such as prerenal azotemia, gastrointestinal bleeding, or excessive protein intake with marginally adequate renal function. On the other hand, urea production and the BUN are reduced in the presence of a low protein intake and in severe liver disease. Uric acid concentration in the blood rises in advanced chronic renal failure, but this rarely results in classical gout.

Calcium and phosphorus

In chronic renal failure there is impaired excretion of phosphate, and progressive hyperphosphatemia occurs. This results in a fall in the plasma calcium concentration (hypocalcemia), giving rise to secondary hyperparathyroidism. The elevated parathyroid hormone level causes calcium resorption from bone, and normocalcemia or hypercalcemia may result. However, hypocalcemia is more prevalent in uremia, both as a result of the reciprocal fall in the plasma calcium concentration as the plasma phosphate level rises and because of reduced calcium absorption in the gut as a result of impaired 1,25-dihydroxycholecalciferol production.[21-24] Hypocalcemia is also present in the nephrotic syndrome, as a result of the hypoalbuminemia. However, the ionized serum calcium level remains normal in this condition.

Proteinuria

Proteinuria may be of two types. In *glomerular proteinuria* large quantities of high-molecular-weight proteins enter the glomerular filtrate and ultimately appear in the urine. Heavy proteinuria (more than 2 g/day) results from increased glomerular permeability, and the protein loss may be great enough to result in the nephrotic syndrome.[30] In *tubular proteinuria* the amount of protein filtered by the glomeruli is not increased, but the low-molecular-weight proteins, which are normally filtered, appear in larger quantities in the final urine because tubular reabsorption is incomplete. Impaired tubular reabsorption of filtered proteins results in modest increases (1 to 3 g/day) in the urinary excretion of low-molecular-weight proteins and albumin.[27] Physiological increases in protein excretion occur during the maintenance of an upright posture, after strenuous exercise, and in normal pregnancy.[27]

Enzymes in urine

Enzymes may appear in urine because of filtration, secretion, or tissue damage.[41] Enzymes of low molecular weight, such as lysozyme and amylase, appear because they are filtered and not completely reabsorbed. High-molecular-weight enzymes, such as lactic dehydrogenase, can be excreted because of parenchymal renal damage.

Hemoglobin and hematocrit

Anemia is a common feature of chronic renal failure, and its severity reflects the extent of renal impairment.[39]

Progressive anemia usually occurs when the GFR falls below 25 mL/min. The anemia of chronic renal failure is attributable to (1) reduced erythropoietin production as renal mass decreases, (2) inhibitors of erythropoiesis present in the serum of the uremic patient, (3) reduced red blood cell survival in advanced renal failure, and (4) iron deficiency caused by blood loss as a result of the hemostatic defect characteristic of renal failure.[40]

REFERENCES

1. Kokko, JP: Renal concentrating and diluting mechanism, Hosp Pract 14:110-116, 1979.
2. Schrier, RW: Renal and electrolyte disorders, ed 3, Boston, 1986, Little, Brown & Co.
3. Peart, WS: Renin-angiotensin system, N Engl J Med 292:302-306, 1975.
4. Rose, BD: Pathophysiology of renal disease, New York, 1981, McGraw-Hill Book Co, Inc, pp 579-585.
5. Suki, WN: Disposition and regulation of body potassium: an overview, Am J Med Sci 272:31-41, 1976.
6. Gonick, HC, Kleeman, CR, Rubini, ME, and Maxwell, MD: Functional impairment in chronic renal disease: studies of potassium excretion, Am J Med Sci 261:281-290, 1971.
7. Sherwood, LM, Mayer, GP, Ramberg, CF, et al: Regulation of parathyroid hormone secretion: proportional control by calcium, lack of effect of phosphate, Endocrinology 83:1043-1051, 1968.
8. Agus, ZS, Gardner, LB, Beck, LH, and Goldberg, M: Effects of parathyroid hormone on renal tubular reabsorption of calcium, sodium, and phosphate, Am J Physiol 224:1143-1148, 1973.
9. Massry, SG, Friedler, RM, and Coburn, JW: Excretion of phosphate and calcium, Arch Intern Med 131:828-859, 1973.
10. Slatopolsky, E, Robson, AM, Elkan, I, and Bricker, NS: Control of phosphate excretion in uremic man, J Clin Invest 47:1865-1874, 1968.
11. Catiguila, SR, Alfrey, AC, Miller, N, and Butkus, D: Total body magnesium excess in chronic renal failure, Lancet 1:1300-1302, 1972.
12. Lennon, EJ, Lemann, J, and Litzow, JR: The effect of diet and stool composition on the net external acid balance of normal subjects, J Clin Invest 45:1601-1607, 1966.
13. Rose, BD: Clinical physiology of acid-base and electrolyte disorders, New York, 1977, McGraw-Hill Book Co, Inc, pp 327-328.
14. Malnic, G, and Giebisch, G: Mechanism of renal hydrogen ion secretion, Kidney Int 1:280-296, 1972.
15. Pitts, RS: Control of renal production of ammonia, Kidney Int 1:297-306, 1972.
16. First, MR: Chronic renal failure, Garden City, NY, 1982, Medical Examining Publishing Co, pp 15-16.
17. Addis, T, Barrett, E, Poo, LJ, and Yuen, DW: The relation between serum urea concentration and the protein consumption of normal individuals, J Clin Invest 26:869-874, 1947.
18. Walser, M: Urea metabolism in chronic renal failure, J Clin Invest 53:1385-1392, 1974.
19. Bleiler, RE, and Schedl, HP: Creatinine excretion: variability and relationship to diet and body size, J Lab Clin Med 59:945-955, 1962.
20. Barrett, E, and Addis, T: The serum creatinine concentration of normal individuals, J Clin Invest 26:875-878, 1947.
21. Stein, JH: Hormones and the kidney, Hosp Pract 14:91-105, 1979.
22. DeLuca, HS: The kidney as an endocrine organ involved in calcium homeostasis, Kidney Int 4:80-88, 1973.
23. Haussler, MR, and McCain, TA: Basic and clinical concepts related to vitamin D metabolism and action, N Engl J Med 297:974-983, 1041-1050, 1977.
24. Lumb, GA, Mawer, EB, and Stanbury, SW: The apparent vitamin D resistance of chronic renal failure: a study of the physiology of vitamin D in man, Am J Med 50:421-441, 1971.
25. Eschbach, JW, Adamson, JW, and Cook, JB: Disorders of red blood cell production in uremia, Arch Intern Med 26:812-815, 1970.
26. Dunn, MJ, and Hood, BL: Prostaglandins and the kidney, Am J Physiol 233:169-184, 1977.
27. Pesce, AJ, and First, MR: Proteinuria: an integrated review, New York, 1979, Marcel Dekker, Inc, pp 80-99.
28. Perlmann, GE, Tamm, I, and Horsfall, FL: An electrophoretic examination of a urinary mucoprotein which reacts with various viruses, J Exp Med 95:99-104, 1952.
29. Coe, FL: Clinical and laboratory assessment of the patient with renal disease. In Brenner, BM, and Rector, FC, editors: The kidney, vol I, Philadelphia, 1981, WB Saunders Co, pp 1135-1180.
30. Pesce, AJ, and First, MR: Proteinuria: an integrated review, New York, 1979, Marcel Dekker, Inc, pp 100-143.
31. Morris, RC: Renal tubular acidosis, N Engl J Med 281:1405, 1969.
32. First, MR, and Pollak, VE: Renal insufficiency in the diabetic patient with heart disease. In Scott, RC, editor: Clinical cardiology and diabetes, vol 3, part 2, Mount Kisco, NY, 1981, Futura Publishing Co, Inc, pp 63-92.
33. Viberti, GC, and Keen, H: The patterns of proteinuria in diabetes mellitus: relevance to pathogenesis and prevention of diabetic nephropathy, Diabetes 33:686-692, 1984.
34. First, MR: Chronic renal failure, Garden City, NY, 1982, Medical Examination Publishing Co, Inc, pp 1-5.
35. Ware, F: Renal function tests: a guide to interpretation, Hosp Med 9:77-92, 1981.
36. Carrie, BJ, Golbetz, HV, Michaels, AS, and Myers, BB: Creatinine: an inadequate filtration marker in glomerular diseases, Am J Med 69:177-182, 1980.
37. Harrington, JT: Evaluation of serum and urinary electrolytes, Hosp Pract 17:28-39, 1982.
38. Oken, DE: On the differential diagnosis of acute renal failure, Am J Med 71:916-920, 1981.
39. Kasanen, A, and Kalliomaki, JL: Correlation of some kidney function tests with hemoglobin in chronic nephropathies, Acta Med Scand 158:213-219, 1957.
40. First, MR: Chronic renal failure, Garden City, NY, 1982, Medical Examination Publishing Co, Inc, pp 17-18.

CHAPTER 23

Liver function

JOHN E. SHERWIN

OBJECTIVES

- Explain the role of the liver in carbohydrate metabolism, nitrogen metabolism, bile pigment formation, and metabolic end-product excretion and detoxification.
- Describe pathological liver conditions and the serum biochemical alterations associated with these diseases.
- Describe the processing of bilirubin by the liver, define jaundice, and describe the various pathological states associated with jaundice.
- List the serum proteins derived from the liver and describe their functions.

KEY TERMS

biliary canaliculi Fine channels running between liver cells.

cirrhosis A liver disorder characterized by loss of normal microscopic architecture with fibrosis. Cirrhosis has a variety of causes, such as obstructive biliary cirrhosis caused by obstruction of major intrahepatic or extrahepatic bile ducts.

Crigler-Najjar syndrome A familial form of nonhemolytic jaundice caused by the absence of glucuronide transferase activity from the liver. Associated with increased serum unconjugated bilirubin, kernicterus, and nervous system disorders.

cytochrome P-450 A series of proteins within cells that are one-electron carriers and whose active centers are heme groups. These are involved in hydroxylation reactions of drugs and other xenobiotics. The 450 refers to the position of the *Soret* absorption band.

detoxification The process of changing the chemical structure of a foreign substance or poison to make it less poisonous or more readily eliminated.

Dubin-Johnson syndrome A familial form of chronic, nonhemolytic jaundice caused by a defect in the hepatic excretion of conjugated bilirubin.

Focal necrosis The death of cells in a small area of tissue.

Gilbert's disease A benign, hereditary form of hyperbilirubinemia and jaundice caused by a defect in the hepatic uptake of unconjugated bilirubin from serum.

gluconeogenesis The formation of glucose from lactate or amino acids by means of the Cori cycle.

glycogenesis The biochemical formation of glycogen from glucose.

glycogenolysis The biochemical degradation of glycogen to form glucose.

hepatitis Inflammation of liver produced by a variety of infections, toxins, and other causes, such as obstruction of the biliary tract in obstructive hepatitis.

hepatobiliary Relating to liver and biliary ducts.

hepatocellular disease Diseases in which the liver cells are destroyed.

hepatocyte A parenchymal liver cell that performs all the functions ascribed to the liver.

jaundice A syndrome characterized by hyperbilirubinemia and deposition of bilirubin pigment in skin and mucosal membranes, giving a yellow appearance to the skin (also called *icterus*).

kernicterus Literally "nuclear jaundice," resulting from deposition of unconjugated bilirubin in nuclei of brain and nerve cells, which causes cell destruction and encephalopathy.

ketone bodies Compounds with carbonyl groups, usually referring to acetoacetic acid, acetone, and beta-hydroxybutyric acid (though the latter compound is not a ketone).

LCAT Lecithin-cholesterol acyltransferase; esterifies cholesterol with fatty acids.

mRNA Messenger ribonucleic acid; is translated into specific proteins.

MSH Melanocyte-stimulating hormone.

neonatal jaundice (physiological jaundice) A disorder of newborns characterized by increased serum levels of unconjugated bilirubin and caused by transient immaturity of liver.

oncofetal proteins Any protein produced by embryological tumors, such as alpha-fetoprotein.

parenchymal cells A general term indicating the functional elements of an organ (see *hepatocyte*).

periportal fibrosis The deposition of fibers or fibrous material in the cells lining the portal blood vessels of the liver.

porphyrias A group of disorders caused by disturbances of porphyrin metabolism characterized by increased formation and excretion of porphyrins and their precursors.

Reye's syndrome Acute, often fatal encephalopathy and fatty degeneration of the liver, seen primarily in children.

Wilson's disease Hepatocellular degeneration, also associated with a change in the iris and lens of the eye and caused by a defect in copper metabolism.

xenobiotics Any organic compound that is foreign to the body, such as drugs and organic poisons.

ANATOMY AND NORMAL FUNCTION OF LIVER

The liver is the largest organ of the body and is responsible for producing most of the endogenous energy sources used by the body. The liver is divided into two primary lobes and is located in the abdominal cavity just below the diaphragm. The two primary cells of the liver are the *hepatocytes* and the *Kupffer cells* (Fig. 23-1). The parenchymal hepatocytes secrete metabolites into veins or the biliary canaliculi; the canaliculi eventually dispense wastes into the bile duct and gallbladder. The hepatocytes are responsible for the metabolic functions of the liver.[1]

The liver is the principal organ for the metabolism of carbohydrates, proteins, lipids, porphyrins, and bile acids. It is capable of synthesizing most body proteins. The liver is also the major site for storage of iron, glycogen, lipids, and vitamins. The liver plays an important role in the detoxification of xenobiotics and excretion of metabolic end products such as bilirubin, ammonia, and urea.

Carbohydrate metabolism and liver function

Polysaccharides are a form of energy storage. The liver is capable of producing glycogen, the principal storage polysaccharide, by glycogenesis and degrading glycogen by glycogenolysis. Whether the glycogen synthetic reaction or the glycogen degradation reaction predominates depends on an individual's metabolic status. Only liver glycogen is available for maintenance of a constant blood glucose, since only the liver (and kidneys) contain the enzyme glucose-6-phosphatase, which converts glucose-6-phosphate to glucose. As a result of the highly branched structure of glycogen, approximately 10% of the glucose of glycogen is available for immediate enzymatic release (see Chapter 29 for details on glucose metabolism). Liver glycogen in the normal adult is not a static storage pool; instead it functions as a source of glucose for the rest of the body, except for muscle. Under conditions of stress, increased body energy requirements must be met by increased glucose use, that is, glycogenolysis and glycolysis. Additional glucose required for the body is provided by increased secretion of glucose by the liver. This is accomplished by increasing the rate of glycogen degradation and

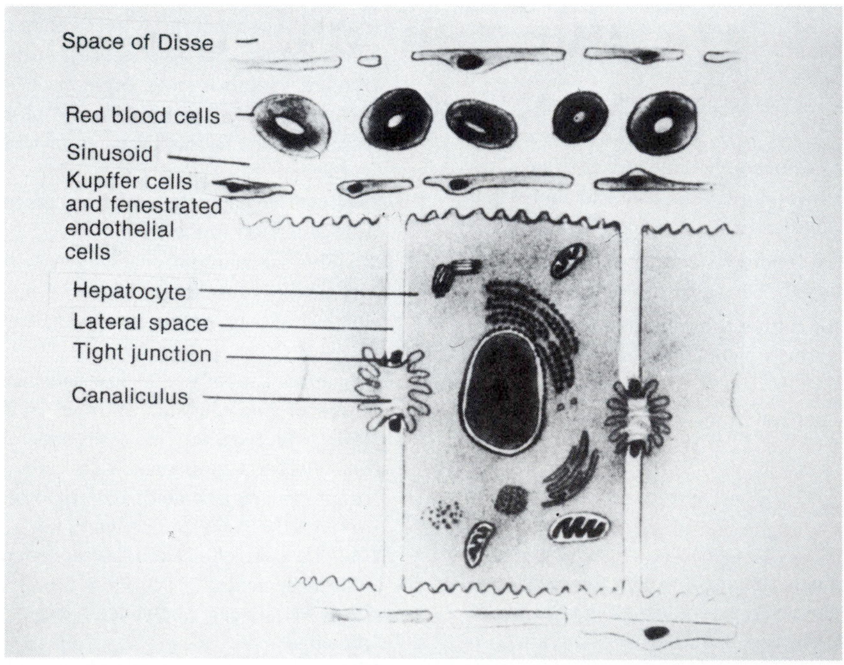

Fig. 23-1 Anatomic relationships of blood, lymphatic, hepatocellular, and biliary systems in liver. *(From Soloway, RD: In Dermers, LM, and Shaw, LM, editors: Evaluation of liver function, Baltimore, 1978, Urban & Schwarzenberg, Inc.)*

the rate of *gluconeogenesis*. Gluconeogenesis is not simply a reversal of glycolysis, since several of the glycolytic enzymes, such as pyruvate kinase, phosphofructokinase, and hexokinase, are not reversible. Lactate and amino acids serve as the precursors for the gluconeogenic pathway. The use of blood lactic acid (of muscle origin) by the liver is an important factor in the clearance of this analyte from serum. Gluconeogenesis is an important source of blood glucose when liver glycogen has been depleted. Liver glycogen depletion will occur within several hours in a fasting person. The net result of these metabolic pathways is to provide a constant supply of glucose to the blood for export to peripheral tissues to meet their energy requirements.

Protein metabolism in liver

Most serum proteins are synthesized in the liver, with two exceptions in the adult: gamma globulin and hemoglobin. In the infant the liver retains the ability to synthesize hemoglobin. As in carbohydrate synthesis, liver function must be extensively impaired before a decrease in protein synthesis can be unequivocally demonstrated. The normal concentrations of the major electrophoretic subgroups of proteins are quite variable and depend at least in part on the person's age.

Many liver enzymes exhibit half-lives of several weeks, though structural proteins are stable nearly indefinitely. Plasma proteins synthesized by the liver exhibit quite varied rates of synthesis and degradation (that is, turnover rates). Under normal conditions, the rate of synthesis of each protein equals its rate of degradation, since its concentration in plasma remains constant. Many proteins synthesized by the liver are excreted into extravascular fluid to carry out specific functions. These functions include nutrition, blood pressure control (oncotic pressure), and transport. Table 23-1 lists some liver proteins found in plasma and some of their properties.

One of the most important serum proteins produced in the liver is albumin. Present in concentrations of 40 to 50 g/L, albumin represents 50% to 60% by weight of all plasma protein. This molecule has an extraordinarily wide range of functions, including nutrition; maintenance of oncotic pressure; and serum transport of Ca^{+2}, unconjugated bilirubin, free fatty acids, drugs, and steroids. Its multifactorial role in human physiology makes it an important analyte in monitoring liver disease.

Table 23-1 Major representative plasma proteins and their properties

Electrophoretic fraction	Protein	Approximate concentration (g/L)	Principal function
α_1	Prealbumin	1	—
	Albumin	40-50	Binds Ca^{+2}, T_4, bilirubin
	Antitrypsin	2-4	Inhibits a number of proteolytic enzymes
	Lipoprotein, high density (HDL)	3-8	Transport of cholesterol from peripheral tissue to liver
	Retinol binding	0.03-0.06	Transport of retinol
α_2	Thyroxine binding	0.01-0.02	Transport of thyroxine
	Haptoglobins (three types)	1-3	Transport of free hemoglobin from destroyed red blood cells
	Lipoprotein, very low density (VLDL)	1.5-2.0	Transport of cholesterol and triglycerides
	Ceruloplasmin	0.2-0.6	Transport of copper, increases use of iron as ferroxidase
	Prothrombin (bovine)	0.1	Proenzyme of thrombin
	Angiotensinogen	—	Precursor of angiotensin
	Erythropoietin	<0.05	Erythropoietic hormone
β_1	Lipoprotein, low density (LDL)	4-10	Transport of cholesterol and other lipids
	Plasminogen	0.3	Profibrinolysin, precursor of fibrinolysin
	Fibrinogen	3	Coagulation factor I
$\beta_1\beta_2$	Complement	0.5-1.8	Lysis of foreign cells
	Transferrin	2-4	Transport of iron
β_2	Glycoproteins	0.3	Unknown
γ	Blood group globulins and immunoglobulins	7-15	Contain various antibodies, blood globulins, complement C_1,C_2, and so on

Modified from Orten, JM, and Neuhaus, OW: Human biochemistry, ed 10, St Louis, 1982, CV Mosby Co, p 440.

Metabolic pools of amino acids are present in the liver. From these pools amino acids are drawn for the synthesis of proteins. When a protein is degraded, the bulk of the constituent amino acids are returned to these intracellular pools. The released amino acids can also be used in gluconeogenesis, transamination, or deamination reactions or be reincorporated into new proteins. Important transamination reactions are catalyzed by the enzymes alanine aminotransferase (ALT or SGPT) and aspartate aminotransferase (AST or SGOT). In the normal person who is in nitrogen equilibrium or negative nitrogen balance, the amino groups of excess amino acids in serum are converted to ammonia or urea for excretion.[2] Some amino acids are also excreted unchanged in the urine. Positive nitrogen balance (that is, insufficient dietary nitrogen) leads to a diminution of amino acid pools and thus decreased urea excretion.

Urea, creatinine, ammonia, and uric acid account for 70% to 75% of the serum nonprotein nitrogen; urea accounts for 60% of the total. Most of the metabolism of nonprotein nitrogen occurs in the liver. Urea is produced in the liver because arginase, the enzyme that converts arginine to urea and ornithine, is present only in the liver. Although blood ammonia concentration is normally quite low, 500 μg/L, ammonia is an important intermediate in amino acid synthesis. Sources of ammonia include hepatic oxidation of glutamate to oxoglutarate, the transamination and oxidative deamination of amino acids and catecholamines, and bacterial breakdown of urea in the gut. The primary mechanisms for the metabolic disposal of ammonia is the synthesis of glutamate, glutamine, and carbamyl phosphate. Carbamyl phosphate may be used to synthesize orotic acid and ultimately pyrimidines for nucleic acids or to synthesize urea, which is the principal pathway for the excretion of excess nitrogen.[2]

Lipid biosynthesis and transport in liver function

Lipids, for the sake of this discussion, include only free fatty acids, triglycerides, glycerophosphatides, sphingolipids, cholesterol, and cholesterol esters. General chemical structures for these lipids are shown in Chapter 51, and their metabolism is discussed in Chapter 30.

Lipids are synthesized in the liver in response to excess carbohydrate intake. The excess carbohydrate is converted to acetyl coenzyme A (acetyl CoA), and a cytoplasmic enzyme system converts it to the fatty acid palmitate, using reduced nicotinamide adenine diphosphonucleotide (NADPH) and adenosine triphosphate (ATP), which are also produced from glucose metabolism.

In the liver, fatty acids are broken down to acetyl CoA, which can then be oxidized to CO_2 by the citric acid cycle. However, a small portion of acetyl CoA is converted to ketone bodies, such as acetoacetate, beta-hydroxybutyrate, and acetone. In normal persons, these products are present in blood to the extent of only about 30 mg/L. In the presence of excess mobilization of fatty acids, as in diabetic ketoacidosis or alcohol intoxication, limiting amounts of NAD and NADP result in an increased hepatic synthesis of the ketone bodies.

Cholesterol is also synthesized in the microsomes of liver from acetyl CoA. Approximately 70% of the total cholesterol in plasma is esterified with fatty acids. The enzyme lecithin cholesterol acyltransferase (LCAT) is also produced by the liver. Bile acids are produced from cholesterol by the cells lining the biliary canaliculi and ductules of the liver.[3] The bile acids are the final excretory metabolites of cholesterol. They also serve as aids in the digestion of dietary lipids (see Chapter 30). Approximately 80% of the available cholesterol is converted into the four major bile acids (Table 23-2). Cholic acid and chenodeoxycholate are the primary bile acids and are present in a fivefold to tenfold excess over the secondary bile acids, deoxycholic acid and lithocholic acid, which are produced by metabolism of the primary hepatic bile acids by intestinal bacteria.

Bile acids are collected in the biliary canaliculi and ductules and stored in the gallbladder. They are then transported to the intestinal lumen, where they emulsify ingested lipids. This emulsification of the lipids permits the intestinal mucosa to digest the lipids and absorb the liberated triglycerides and cholesterol. More than 90% of the secreted bile acids are reabsorbed and returned to the liver through the portal circulation.[3]

Liver function as a storage depot

The liver is an important site for the storage of iron, glycogen, amino acids, and some lipids and vitamins. The adult liver contains about 700 mg of iron. Nutritional iron is absorbed primarily in the intestine. To be absorbed, ferric iron (Fe^{+3}) must be converted to ferrous iron (Fe^{+2}). Immediately after absorption, the Fe^{+2} is reconverted to Fe^{+3} and temporarily stored in the intestinal mucosa as the ferritin complex. The ferritin apoprotein is synthesized in the liver. The iron is then released once more as Fe^{+2} into the plasma and is rapidly oxidized into Fe^{+3} and complexed with transferrin, a hepatic synthesized $alpha_1$ globulin. In the normal adult, transferrin is 25% to 30% saturated with Fe^{+3}. In the liver, transferrin releases iron, and a new ferritin-Fe^{+3} complex is formed. Ferritin is the primary storage form of iron; apoferritin binds iron as a colloidal hydrous ferric oxide. However, a small amount of iron is stored as hemosiderin, which is an insoluble cellular inclusion of Fe^{+3} complexed with ferritin. Hemosiderin granules serve as a storage form for iron when there are insufficient levels of apoferritin. The ratio of iron to protein is much greater in hemosiderin than in ferritin. Additional details of iron metabolism are found in Chapter 32.

Lipid is stored, primarily as triglyceride, in subcutaneous adipose tissue. Lipid is also stored in the liver, where it functions as an energy reservoir. Under normal

Table 23-2 Names and structures of bile acids, their relative contribution to the total bile acid pool, and normal serum concentrations

Name	Structure	Relative content (%) in bile	Normal serum concentration (mmol/L)
Cholylglycine	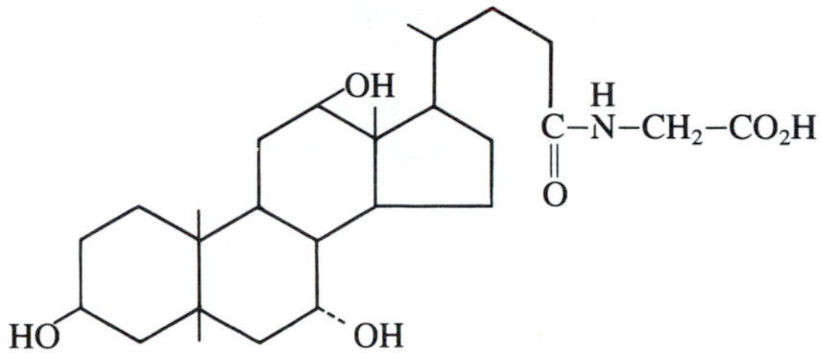 OH; HO; OH; $\overset{H}{C}$–N–CH_2–CO_2H (C=O)	38	0.2-0.9
Deoxycholylglycine	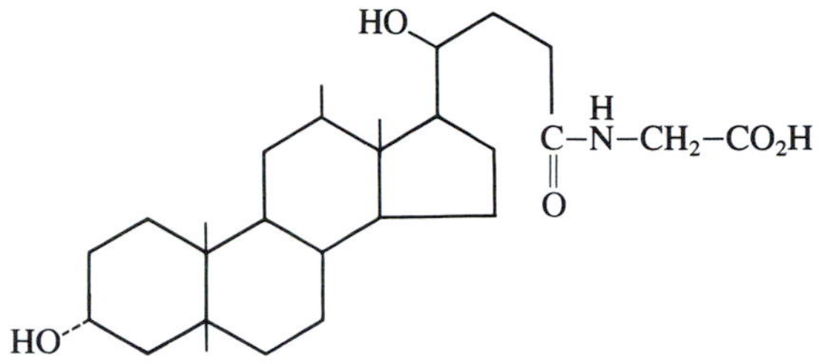 HO; HO; $\overset{H}{C}$–N–CH_2–CO_2H (C=O)	20	0.08-0.7
Lithocholylglycine	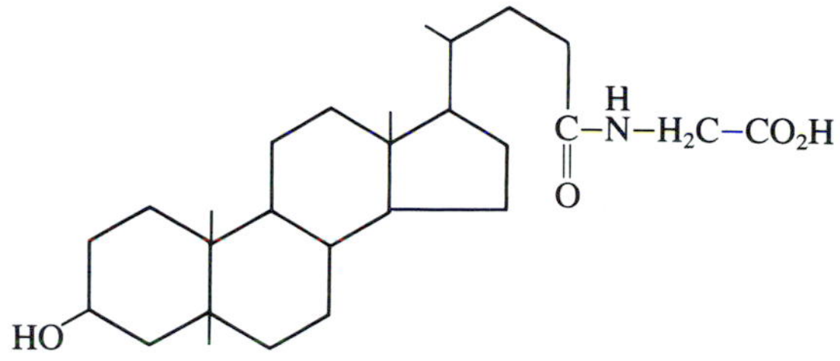 HO; $\overset{H}{C}$–N–H_2C–CO_2H (C=O)	4	0.07-0.3
Chenodeoxycholylglycine	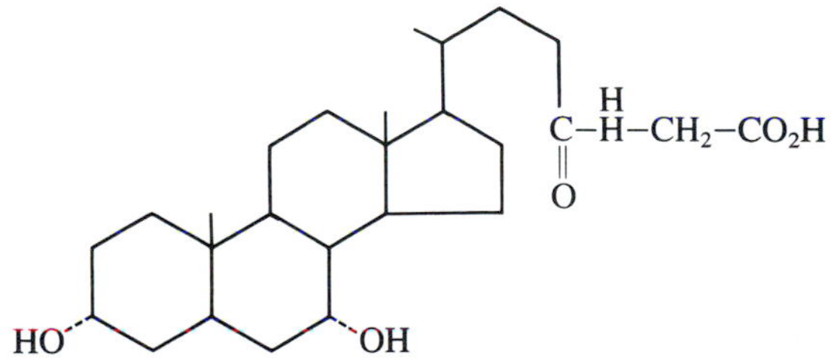 HO; OH; $\overset{H}{C}$–H–CH_2–CO_2H (C=O)	38	0.05-0.2

From Shaw, LM: Lab Management 20:56-63, 1982.

circumstances the liver functions as a temporary storage site for lipids as they are synthesized in the liver or absorbed from the intestine after a meal.

Bile pigment formation

About 126 days after emergence from the reticuloendothelial tissue, the senescent erythrocytes are phagocytized and the hemoglobin is released. The heme portion of hemoglobin is converted to bilirubin, with the release of iron and the globin proteins. The liberated iron is bound by transferrin and returned to the iron stores of the liver or bone marrow; the globin is degraded to its constituent amino acids. The conversion of heme to bilirubin requires 2 to 3 hours (Fig. 23-2). Bilirubin, bound to albumin, is transported from the reticuloendothelial cells to the hepatocytes. Albumin has two types of binding sites for bilirubin. One site has a high affinity constant of about 1×10^8 M, and the other has a weaker binding constant of 1×10^6 M. Bilirubin at the lower affinity binding site is displaced by free fatty acids, barbiturates, and salicylates. Bilirubin bound to albumin is apparently not cytotoxic and remains in solution as it is transported to the liver. A small amount of bilirubin (less than 20 μg/L) is free in the plasma. In the liver, bilirubin is transported through the cellular microvilli of the sinusoids to the hepatocytes. Bilirubin is dissociated from albumin and taken up into the hepatocytes by specific proteins. Within the hepatocyte, bilirubin glucuronide is formed by reaction of bilirubin with uridine diphosphoglucuronate (UDP-glucoronate) in the presence of UDP-glucuronyl transferase. Methodology

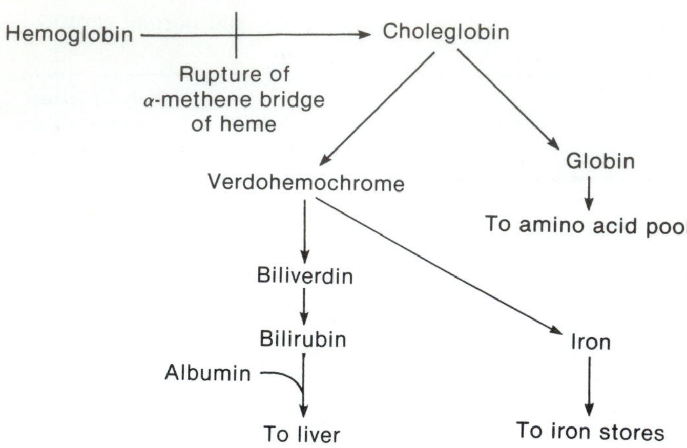

Fig. 23-2 Formation of bilirubin from hemoglobin occurs primarily in reticuloendothelial tissues.

indicates that 8 to 10 mg/L of unconjugated bilirubin is present in normal adult serum and that normal adult serum contains no conjugated bilirubin. Controversy exists about the relative contribution of the monoglucuronide and diglucuronide to fresh bile pigment. It seems likely that the formation of bilirubin diglucuronide represents the normal conjugation reaction, but that in disease states the accumulation of monoglucuronides of bilirubin occurs as a result of accumulation of unconjugated bilirubin in the face of a limited supply of glucuronate.

After formation, the bilirubin-diglucuronide is excreted by the hepatocyte into the biliary canaliculi. Any disease resulting in a decreased secretion of the conjugated bilirubin will result in an increased serum concentration of this analyte. As part of the bile, these water-soluble conjugates are secreted into the lumen of the small intestine. The bilirubin conjugates are hydrolyzed by a beta glucuronidase, and the regenerated bilirubin is converted to *d*-urobilinogen and further reduced to *l*-urobilinogen and *l*-stercobilinogen by the anaerobic bacteria of the intestinal lumen. The urobilinogens are reabsorbed from the intestine and recirculated in the extrahepatic circulation, where they are ultimately excreted in the urine. Stercobilinogen is not reabsorbed from the intestine but is a normal constituent of the feces. These bilinogens oxidize spontaneously in air to the corresponding bilins and thereby contribute to the color of both urine and feces (Fig. 23-3). (See Chapter 32 for additional details on bilirubin metabolism.)

Metabolic end-product excretion and detoxification

Humans have two mechanisms for detoxification of foreign (xenobiotic) materials, such as drugs and poisons, and toxic metabolic products, such as ammonia and bilirubin. The first is to bind the compound reversibly to a protein so that the material is inactivated; thus bilirubin is bound to albumin and lead is bound to hemoglobin. The second mechanism is to modify the compound chemically so that

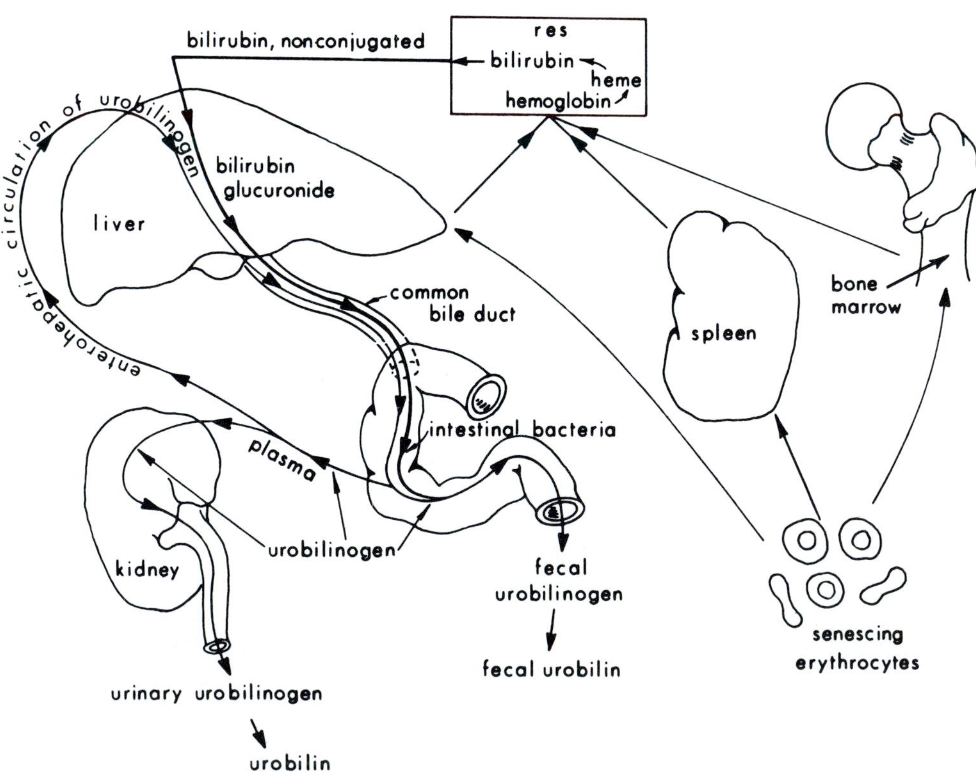

Fig. 23-3 Bilirubin metabolism. *(From Bauer, JD, editor: Clinical laboratory methods, ed 9, St Louis, 1982, The CV Mosby Co.)*

it is readily excreted; thus ammonia is converted to urea and bilirubin is converted to bilirubin-glucuronide.

The inactivation and detoxification of exogenous compounds is usually accomplished by hydroxylation mediated by one of the cytochrome P-450 enzymes or by conjugation of the parent compound or its metabolite with sulfate or carbohydrate. These reactions are localized in the microsomes of the liver. The sulfated compounds are more water soluble than the parent compounds and are excreted directly in the urine. Carbohydrate conjugates are commonly excreted into the intestinal lumen as part of the bile.

LIVER-FUNCTION ALTERATIONS DURING DISEASE

Jaundice

Jaundice is a general condition that results from abnormal metabolism or retention of bilirubin. Jaundice causes a yellow discoloration of the skin, mucous membranes, and sclera. The three principal types of jaundice are prehepatic, hepatic, and posthepatic.

Prehepatic jaundice is the result of acute or chronic hemolytic anemias. Hepatic jaundice includes disorders of bilirubin metabolism and transport defects, such as Crigler-Najjar disease, Dubin-Johnson syndrome, and Gilbert's disease, as well as neonatal physiological jaundice and diseases resulting in hepatocellular injury or destruction.

The specific diseases of bilirubin metabolism each represent a defect in one of the steps in the hepatic processing of serum bilirubin (Fig. 23-4). Thus *Gilbert's disease* is caused by a defect in the transport of bilirubin from plasma albumin into the hepatocyte. Although levels of unconjugated bilirubin are elevated in this familial disorder, levels of conjugated bilirubin are not. Impairment in the conjugation step by UDP-glucuronide caused by a deficiency in the enzyme UDP-glucuronyl transferase will also lead to a large increase in unconjugated bilirubin. When this enzyme deficiency is congenital, it is known as *Crigler-Najjar disease*. However, this impairment is most frequently encountered as *neonatal* or *physiological jaundice*.[4] This enzyme activity is one of the last liver functions to be activated in prenatal life, since unconjugated bilirubin formed in the fetus is cleared by the placenta into maternal blood. However, in premature births, infants are sometimes born without the enzyme activity present. This leads to a rapid buildup of unconjugated bilirubin, which can be life threatening. The unconjugated bilirubin, which is much more lipid soluble than water soluble, readily passes into the brain and nerve cells and is deposited in the nuclei of these cells, resulting in *kernicterus*. Kernicterus often results in cell damage and death. Neonatal jaundice will persist until the enzyme glucuronyl transferase is produced by the newborn's liver.[4] The newborn's blood must be monitored frequently to detect dangerously high levels of unconjugated bilirubin (about 200 mg/L). At this point the infant is treated either with ultraviolet light to destroy the bilirubin as it passes through the capillaries of the skin or with exchange transfusion.[4]

The last step in the hepatic processing of bilirubin is the postconjugation step of excretion of the bilirubin-glucuronide from the hepatic microsomes into the canaliculi. Impairment of this process, called the *Dubin-Johnson syndrome,* results in large increases in the conjugated bilirubin fraction of serum and a urine positive for bilirubin.

Hepatic jaundice also encompasses the disorders characterized by hepatocellular damage or necrosis, such as hepatitis and cirrhosis. Posthepatic jaundice is generally caused by biliary obstructive disease resulting from spasms or strictures of the biliary tract, ductal occlusion by stones, or compression by neoplastic disease. Since the hepatic functions of transport and conjugation of bilirubin are normal in these diseases, the major increase in serum bilirubin involves the conjugated fraction. Unable to be properly excreted by the liver in these disorders, the conjugated bilirubin fraction increases in serum, resulting in the appearance of bilirubin in the urine. The laboratory findings for bilirubin and its metabolites in these diseases are summarized in Table 23-3.

Hepatitis

Hepatitis, a general term meaning inflammation of the liver, is used to describe diseases resulting in hepatocellu-

Table 23-3 Concentrations and changes in concentration of bilirubin and its metabolites in normal persons and those with jaundice

Condition	Serum		Urine		Feces pigment
	Total bilirubin	Conjugated bilirubin	Bilirubin	Urobilinogen	
Normal	2-10 mg/L	0-2 mg/L	Negative	0.5-3.4 mg/day	Brown
Prehepatic jaundice	Increased	Normal	Negative	Increased	Normal
Hepatic jaundice					
Hepatocellular disease	Increased	Increased	Positive	Decreased (normal)	Light brown
Gilbert's disease	Increased	Normal	Negative	Decreased (normal)	Normal
Crigler-Najjar syndrome	Increased	Decreased	Negative	Decreased	Light brown
Dubin-Johnson syndrome	Increased	Increased	Positive	Decreased (normal)	Light brown
Posthepatic obstructive jaundice	Increased	Increased	Positive	Decreased	Light brown

Fig. 23-4 Mechanisms of hyperbilirubinemia. **A,** Normal bilirubin metabolism with hepatocyte uptake of unconjugated bilirubin *(dark arrow)* and microsomal conjugation and excretion of conjugated bilirubin *(striped arrow)*. **B,** Hemolytic jaundice in which increased bilirubin production results in increased excretion of conjugated bilirubin and a rise in excess (exceeding liver capacity) unconjugated bilirubin in blood. **C,** Gilbert's disease in which decreased hepatic uptake results in large increase in blood levels of unconjugated bilirubin. **D,** Physiological jaundice in which microsomal conjugating system is not functional, resulting in a large increase in unconjugated bilirubin. Congenital deficiency is the Crigler-Najjar syndrome. **E,** Dubin-Johnson syndrome in which there is a biochemical defect preventing secretion of conjugated bilirubin, resulting in a backflow into blood. **F,** Intrahepatic or extrahepatic obstruction in which a physical block prevents secretion of conjugated bilirubin. Hepatocellular disease results in a pattern similar to a combination of **C** and **D.** *(From Leevy, CM, editor: Evaluation of liver function, ed 2, Indianapolis, 1974, Lilly Research Laboratories.)*

lar damage. Hepatitis is usually caused by infections or toxic agents. Viral hepatitis is the most common cause of acute hepatocellular disease. Four types of hepatitis virus have been recognized: type A virus; type B virus; non-A, non-B hepatitis virus; and the delta hepatitis virus. The latter virus is defective, requiring the presence of type B virus for infectivity. The clinical symptoms of the four types are similar, and the majority of hepatitis cases are anicteric because, despite the hepatic necrosis, the liver retains sufficient residual functional capacity to handle the bilirubin load from normal hemoglobin turnover. Other viruses known to cause hepatitis include cytomegalovirus, coxsackievirus B, and herpes simplex.

Serum levels of ALT and AST rise rapidly during the early course of hepatitis because of hepatic necrosis. In patients who develop jaundice, the rise in the transaminases precedes the increase in bilirubin, which persists for 2 to 8 weeks. AST is usually more elevated than ALT. Serum alkaline phosphatase and gamma-glutamyl transferase (GGT) are elevated during the early cholestatic portion of the disease and remain elevated until the disease has resolved. Diagnosis of the type of viral hepatitis is accomplished by measurement of the specific hepatitis antigen during the prodromal phase of the illness. Specific antibodies to virus antigens are detectable for several weeks after the antigen is no longer detectable. Chronic hepatitis is generally the result of a persistence of hepatitis B infection. It is associated with elevation of serum bilirubin, minimally but persistently elevated serum transaminases (with ALT now greater than AST) caused by hepatic necrosis, and, occasionally, elevated alkaline phosphatase.

Drug-induced hepatic damage

The most common cause of drug-induced hepatic damage is chronic excessive ingestion of alcohol. Laboratory findings associated with this damage are an elevation of GGT, a mild elevation of the transaminases, an increase in the globulin and a decrease in the albumin fractions of the serum proteins, and a decrease in bromosulfophthalein (BSP) or indocyanine green clearance. (See Chapter 31 for a detailed review of alcoholic liver disease.)

Other classes of drugs that induce hepatic damage include barbiturates, tricyclic antidepressants, and antiepileptics. These drugs typically cause an increase in serum GGT. Drug withdrawal permits liver regeneration. Chemotherapeutic drugs such as vincristine, vinblastine, actinomycin D, and 5-fluorouracil will typically cause an elevation of the serum transaminases and lactate dehydrogenase because of hepatic tissue damage and enzyme release.

Reye's syndrome

Reye's syndrome typically occurs in children between 2 and 13 years of age. The liver has fatty infiltration with necrosis and cholestasis, and encephalopathy occurs because of accumulation of ammonia. Laboratory findings include an elevated blood ammonia, elevated serum transaminases, and a prolonged prothrombin time.

Congenital deficiency syndromes with altered liver function

Porphyrias. The porphyrias are caused by congenital enzyme deficiencies in the pathways leading to the synthesis of the heme moiety of hemoglobin and other heme proteins, such as myoglobin and the cytochromes. Increased excretion of specific porphyrin metabolites, varying with the enzyme defect, is seen. The five types of porphyria are acute intermittent porphyria, acquired congenital hepatic porphyria, erythropoietic porphyria, and erythropoietic protoporphyria. Chapter 32 summarizes porphyrin metabolism and discusses the enzyme defects.

Wilson's disease. Wilson's disease is an autosomal recessive disorder of copper metabolism. The accumulation of copper in the liver results in jaundice followed by liver cirrhosis. Wilson's disease is characterized by a low serum concentration of ceruloplasmin, glycosuria, phosphaturia, aminoaciduria, and an elevated urinary copper concentration with excretion greater than 50 μg/day.

Hemochromatosis. Hemochromatosis is another genetic disorder of metal metabolism. In this disease, iron accumulates in the liver, and the resulting cirrhosis is similar to alcoholic cirrhosis. An elevated serum iron, a low iron-binding capacity, and an elevated serum ferritin value in the presence of normal dietary iron intake are characteristic of this disorder.

Alpha$_1$-antitrypsin deficiency

Alpha$_1$-antitrypsin deficiency is an inborn error of protein metabolism that results in emphysema and liver cirrhosis. Numerous alpha$_1$-antitrypsin phenotypes have been identified carrying varying risks of disease.[5] The tissue damage seen in alpha$_1$-antitrypsin deficiencies may well be caused by hydrolytic damage to structural protein by trypsinlike enzymes that have not been neutralized by alpha$_1$-antitrypsin. Since alpha$_1$-antitrypsin represents approximately 80% of the alpha$_1$ fraction of serum proteins on electrophoresis, a severe deficiency is often diagnosed by the absence of this fraction in the electrophoretogram. Phenotyping of the various alpha$_1$-antitrypsin deficiencies is best done by isoelectric focusing.

Other genetic disorders that are associated with liver disease include the lipid-storage diseases and the glycogen-storage diseases. These are discussed in detail in Chapter 47.

Liver tumors and other hepatic disorders

Congenital hepatic fibrosis, hepatic cysts, and liver abscesses are generally best diagnosed by liver biopsy. However, nonspecific changes in liver enzymes and indocyanine green retention can accompany these disorders.

Table 23-4 Change of serum analyte with disease*

	Alkaline phosphatase	GGT	5′-Nucleotidase	AST	ALT	Bile acids	Albumin	NH_3
Acute hepatitis (viral and so on)	↑	↑	↑	↑↑↑	↑↑	↑↑	N	N, ↑
Alcoholic (drug) hepatitis	N, ↑	↑↑↑	↑	↑	↑	↑	N	N, ↑
Chronic hepatocellular disease	N, ↑	N, ↑	N, ↑	↑	↑	↑	↓	N, ↑
Cirrhosis	N, ↑	N, ↑	N, ↑	N, ↑	N, ↑	↑	↓	N, ↑
Reye's syndrome	N			↑	↑			↑↑
Hepatomas	↑↑	↑↑↑↑	↑↑	↑	↑		N, ↓	N
Cholestatic disease	↑	↑↑	↑↑↑	↑	↑	↑	N	N

**N*, Normal; ↑, elevated; ↓, lowered.

Liver tumors frequently alter liver function as a result of tissue compression during tumor growth and infiltration. This results in an increase in serum alkaline phosphatase, 5′-nucleotidase, and especially GGT. BSP and indocyanine green clearances are frequently prolonged. The demonstration of an elevation of serum alpha-fetoprotein is diagnostic of hepatic tumor in the presence of an abnormal liver scan.

LIVER-FUNCTION TESTS

Liver-function tests are generally used to identify liver disease in the absence of jaundice or when the jaundice is the result of hemolytic disease and the existence of complicating liver disease is suspected. Liver function is tested by injection of a dye intravascularly and observation of the retention of the dye in the serum. It is essential that a dye be chosen that is excreted into the bile rather than filtered by the kidney. Thus dyes such as BSP and indocyanine green are acceptable, whereas phenolsulfonphthalein, which is excreted preferentially in the urine, is not. BSP or indocyanine green retention depends on hepatic blood flow, biliary duct function, and liver cell function. Therefore these tests cannot distinguish between hepatocellular disease and obstructive liver disease. In the normal person the percentage of retention is less than 5% at 45 minutes. An increase in the retention is consistent with liver dysfunction resulting from hepatocellular disease, biliary obstructive disease, and space-filling lesions such as tumors in the liver.

CHANGE OF ANALYTE IN DISEASE

Changes of serum analyte with disease are summarized in Table 23-4.

Enzymes

In this section, six serum enzymes are described and their value in the differential diagnosis of liver disease is examined. The enzymes discussed are alkaline phosphatase, GGT, AST, 5′-nucleotidase, and lactate dehydrogenase. Numerous other enzymes have been identified that are useful in the evaluation of liver function. However, these six enzymes are those generally used. Examples of the change in activity with time for several of these enzymes in hepatic disease are shown in Figs. 23-5 to 23-7.

Alkaline phosphatase (EC 3.1.3.1). Alkaline phosphatase (AP) is actually a group of enzymes that hydrolyze monophosphate esters at an alkaline pH. pH optima of these enzymes are generally about 10. The natural substrates for alkaline phosphatase are not known. The enzyme has been identified in most body tissues and is generally localized in the membranes of cells. Alkaline phosphatase activity is highest in the liver, bone, intestine, kidney, and placenta, and as many as 11 different isozymes of alkaline phosphatase have been identified in serum. Since alkaline phosphatase normally contains significant amounts of sialic acid, it is possible that some of these multiple enzyme forms are the result of different degrees of sialation. The enzyme produced by the placenta is known to have a protein composition different from the other enzyme forms. (See pp.

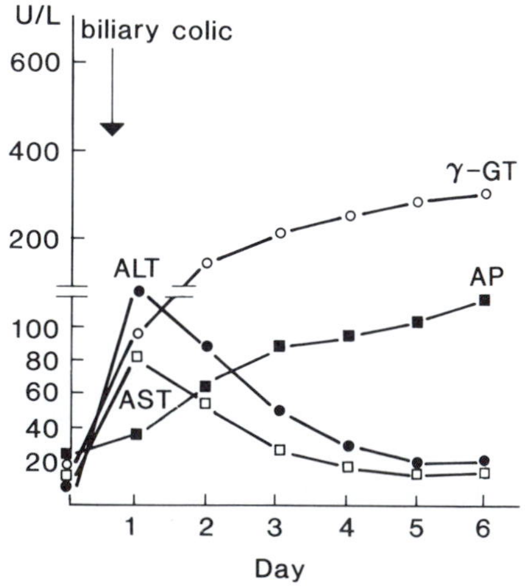

Fig. 23-5 Time course of serum enzyme activities in obstructive jaundice. *(From Schmidt, E, and Schmidt, FW: Brief guide to practical enzyme diagnosis, Houston, 1977, Boehringer-Mannheim Diagnostics.)*

898 and 902 for details on the measurement of this enzyme.)

Measurement of serum alkaline phosphatase is useful in differentiating hepatobiliary disease from osteogenic bone disease. Alkaline phosphatase activity increases greatly (10 times) as a result of membrane-localized enzyme synthesis after extrahepatobiliary obstruction such as cholelithiasis or gallstones. Intrahepatic biliary obstruction is also accompanied by an increased serum alkaline phosphatase activity, but the degree of increase is smaller (two to three times). Liver disease resulting in parenchymal cell necrosis does not elevate serum alkaline phosphatase unless the liver disease is associated with damage to the canaliculi or biliary stasis.

Interpretation of serum alkaline phosphatase measurements is complicated by the fact that enzyme activity can

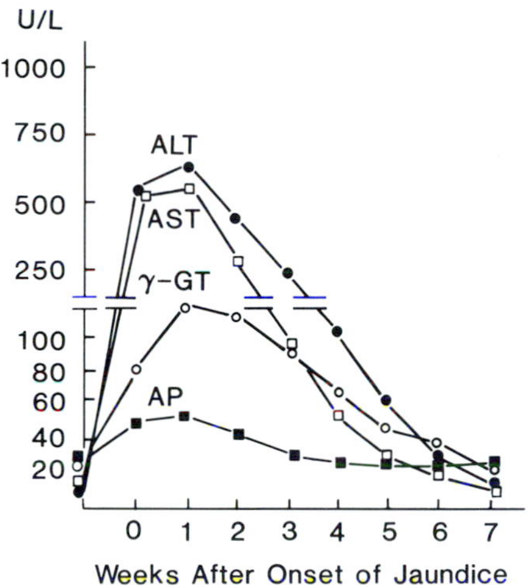

Fig. 23-6 Time course of serum enzyme activities in acute viral hepatitis. *(From Schmidt, E, and Schmidt, FW: Brief guide to practical enzyme diagnosis, Houston, 1977, Boehringer-Mannheim Diagnostics.)*

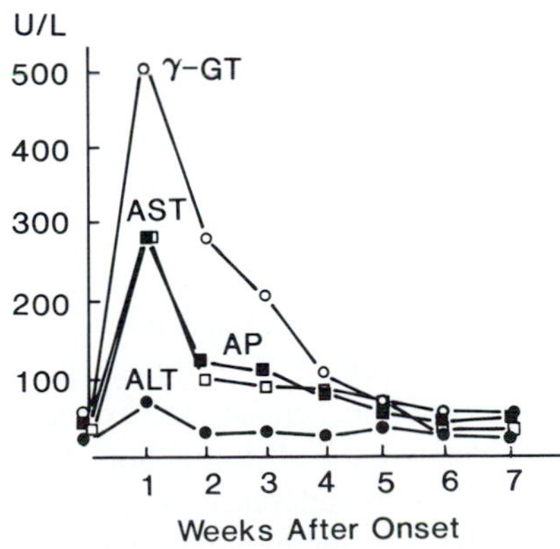

Fig. 23-7 Time course of serum enzyme activities in acute alcoholic hepatitis. *(From Schmidt, E, and Schmidt, FW: Brief guide to practical enzyme diagnosis, Houston, 1977, Boehringer-Mannheim Diagnostics.)*

increase in the absence of liver disease. The most common disorders causing elevation of alkaline phosphatase are bone diseases, such as Paget's disease, rickets, and osteomalacia. Serum alkaline phosphatase activity also rises during puberty because of rapid bone growth and during the third trimester of pregnancy because of alkaline phosphatase released from the placenta.

Gamma-glutamyl transferase (EC 2.3.2.2). GGT (or γ-GT) is a membrane-localized enzyme that plays a major role in glutathione metabolism and resorption of amino acids from the glomerular filtrate. It is also important in the absorption of amino acids from the intestinal lumen. Glutathione (gamma-glutamyl cysteinylglycine) in the presence of GGT and an amino acid or peptide transfers glutamate to the amino acid forming a peptide bond on the gamma-carboxylic acid, thereby forming cysteinylglycine and the corresponding gamma-glutamyl peptide.

Although the GGT activity is highest in renal tissue, serum GGT is generally elevated as a result of liver disease. Serum GGT is elevated earlier than other liver enzymes in diseases such as acute cholecystitis, acute pancreatitis, acute and subacute liver necrosis, and neoplasms of multiple sites at which liver metastases are present. GGT has been advocated as a screening test for alcohol abuse.[6] Since GGT is a hepatic microsomal enzyme, chronic ingestion of alcohol or drugs such as barbiturates, tricyclic antidepressants, and anticonvulsants induces microsomal enzyme production. These drug-induced elevations precede any change in other liver enzymes, and if drug ingestion is stopped at this point, the liver changes are generally reversible. GGT permits differentiation of liver disease from other conditions in which serum alkaline phosphatase is elevated because serum GGT levels are usually normal in Paget's disease, rickets, and osteomalacia and in children and pregnant women without liver disease. Since the prostate contains significant GGT activity, serum activity is higher in healthy men than in women. Serum GGT is most useful in the diagnosis of cholestasis caused by chronic alcohol or drug ingestion, mechanical or viral cholestasis, liver metastases, bone disorders in which alkaline phosphatase is elevated but GGT is normal, and skeletal muscle disorders in which the transaminase AST is elevated but GGT is normal.

5′-Nucleotidase (EC 3.1.3.5). 5′-Nucleotidase (NTD) is a microsomal and cell membrane–localized enzyme that catalyzes the hydrolysis of nucleoside-5′-phosphate esters. The serum enzyme has an apparent pH optimum of 7.5:

$$\text{Nucleoside-5}'\text{-monophosphate} \xrightarrow{\text{NTD}} \text{Nucleoside} + P_i$$

Like GTT, serum NTD is increased in hepatobiliary diseases such as gallstone obstruction of the bile duct, cholestasis, biliary cirrhosis, and obstructive disease caused by neoplastic growth. Serum NTD is not generally elevated in drug-induced liver damage.[7] Therefore it is useful to measure NTD in conjunction with GGT in following the course of chemotherapy for liver neoplasms. Since NTD is not elevated in bone disease, it, like GGT, is useful in differentiating hepatic causes of alkaline phosphatase increase from other causes, such as bone disease, pregnancy, and normal childhood growth.

Lactate dehydrogenase (EC 1.1.1.27). Lactate dehydrogenase (LD) is present in many tissues. LD catalyzes the interconversion of pyruvate and lactate:

$$\text{NAD} + \text{H}^+ + \text{Lactate} \xrightleftharpoons{\text{LD}} \text{Pyruvate} + \text{NADH}^+$$

LD activity is highest in the kidney and heart and lowest in the lung and serum. LD is localized in the cytoplasm of cells and thus is extruded into the serum when cells are damaged or necrotic.

When only a specific organ, such as the liver, is known to be involved, the measurement of total LD can be useful. Total LD is increased in viral or toxic hepatitis, extrahepatic biliary obstruction, acute necrosis of the liver, and cirrhosis of the liver. However, in conditions in which multiple organs are involved, the measurement of total LD is less useful than the measurement of LD isoenzymes. LD_5 and LD_4 account for the primary activity of liver LD, whereas LD_1 and LD_2 account for the predominant activities of heart and kidney LD. Since red blood cells also contain significant LD_1, analysis of hemolyzed serum specimens should be avoided. In these hepatic conditions, LD electrophoresis indicates that the increased total LD is caused by the release of LD_4 and LD_5 into the serum.

Aspartate aminotransferase (EC 2.6.1.1) and alanine aminotransferase (EC 2.6.1.2). The transaminases AST and ALT catalyze the conversion of aspartate and alanine to oxaloacetate and pyruvate, respectively. (See pp. 895 and 909 for discussion of the measurement of the transaminases.) The highest ALT levels are found in liver, whereas AST is present in heart, skeletal muscle, and liver to nearly the same extent. Serum activity of both AST and ALT increases rapidly during onset of viral jaundice and remains elevated for 1 to 2 weeks. In toxic hepatitis, ALT and AST are also elevated, but LD is elevated to an even greater extent as a result of hepatic cell necrosis. Patients with chronic active hepatitis also exhibit increased AST and ALT.

Acute liver necrosis is accompanied by significant increases in the activity of both ALT and AST. The increase in ALT activity is usually greater than the increase in AST activity. In cirrhotic liver disease, serum transaminase activities are generally not elevated above 300 U/L, regardless of the cause of the cirrhotic disease. The elevations of serum ALT and AST seen in Reye's syndrome are directly attributable to hepatic damage, and the increase in ALT is generally greater than the increase in AST. Neoplastic disease also elevates serum transaminase activity.

Measurement of ALT and AST serum levels is valuable in the diagnosis of liver disease. However, these laboratory tests are best used with other enzyme assays, such as LD and creatine kinase, and with other measures of liver and kidney function, such as blood urea, creatinine, ammonia, and bilirubin.[8] This is important in establishing a diagnosis because ALT and AST are present in tissues other than liver, and serum activity of these enzymes can reflect organic disease in tissues other than liver. ALT and AST serum activities are elevated in myocardial infarction, renal infarction, progressive muscular dystrophy, and numerous diseases that only secondarily affect the liver, such as Gaucher's disease, Niemann-Pick disease, infectious mononucleosis, myelocytic leukemia, diabetic ketoacidosis, and hyperthyroidism.

Other hepatic analytes

Bilirubin. Serum bilirubin analysis is helpful in differentiating the cause of jaundice. *Prehepatic jaundice* results in a large increase in unconjugated bilirubin because of the increased release and metabolism of hemoglobin after hemolysis. No increase or only a slight increase in conjugated bilirubin is observed because the transport of bilirubin into the liver and the formation of the glucuronide conjugate become rate limiting. Additionally, because of the increased levels of unconjugated bilirubin excreted by the liver, urinary urobilinogen and fecal urobilin concentrations are elevated, but urinary bilirubin (which is only the freely soluble, conjugated form) is absent. In contrast, *posthepatic obstructive jaundice* is characterized by large increases in serum-conjugated bilirubin. The accumulation of bilirubin in the serum is the result of decreased biliary excretion after the conjugation of bilirubin in the liver rather than the result of an increased bilirubin load caused by hemolysis. Excretion of bilirubin metabolites is low, and urinary bilirubin can usually be demonstrated. *Hepatic jaundice* presents an intermediate pattern wherein both conjugated and unconjugated serum bilirubin are increased to the same degree and conjugated bilirubin is present in the urine. However, the fecal concentration of urobilin is generally decreased.

Cholesterol. Serum cholesterol comprises two forms, free cholesterol and esterified cholesterol. About 70% of the total serum cholesterol is esterified with fatty acids. Since this esterification takes place in the liver, intrahepatic disease or biliary obstruction is characterized by an increase in the free cholesterol and occasionally a shift in the serum free fatty acid profile, though the total cholesterol usually remains unchanged. In chronic disease associated with parenchymal cell destruction, the total cholesterol may fall below the reference range.

Bile acids. Bile acid secretion and production is altered in disease.[3,8] In the healthy adult, serum contains 1 to 2 μg/mL of bile acids. In hepatobiliary disease, serum bile acid concentrations may rise as much as a thousandfold. Other diseases that can cause a significant rise in serum bile acid concentrations are hepatitis, cirrhosis, drug-induced liver disease, and hepatoma. Serum bile acid concentrations are normal in Gilbert's disease, hemochromatosis, and polycystic liver disease. Measurement of serum bile acids is useful in the diagnosis of minimal liver dysfunction when other biochemical parameters are still unchanged.

Triglycerides. Serum triglycerides should be measured in a fasting sample. Increases are relatively nonspecific[8]; liver dysfunction resulting from hepatitis, extrahepatic biliary obstruction, and cirrhosis is associated with an increase in serum triglycerides, but so are such disorders as acute pancreatitis, myocardial infarction, renal failure, gout, pernicious anemia, and diabetes mellitus. Free fatty acids exhibit a similar nonspecificity. They are decreased in chronic hepatitis, chronic renal failure, and cystic fibrosis. Serum free fatty acid concentrations are elevated in Reye's syndrome, hepatic encephalopathy, and chronic active hepatitis, but also in myocardial infarction, acute renal failure, hyperthyroidism, and pheochromocytoma.

Serum proteins in evaluation of liver function. A healthy functioning liver is required for the synthesis of the serum proteins, except for the gamma globulins. The liver has the ability to increase protein output approximately twofold during diseases associated with protein loss. Therefore it is not surprising that total protein measurements are not altered until an extensive impairment of liver function has occurred.

Albumin is decreased in chronic liver disease and is generally accompanied by an increase in the beta and gamma globulins as a result of production of IgG and IgM in chronic active hepatitis and of IgM and IgA in biliary or alcoholic cirrhosis, respectively. Immunoelectrophoresis may facilitate the identification of these subclasses of gamma globulin. However, a decrease in serum albumin is not specific for liver disease, since decreases are also seen in malabsorption, malnutrition, renal disease, alcoholism, and malignant diseases.

The alpha$_1$ fraction of the serum protein globulin is decreased in chronic liver disease, and when this fraction is absent or nearly so, it suggests that alpha$_1$-antitrypsin deficiency may be the cause of the liver disease. Serum alpha$_2$ globulin and beta globulin are increased in obstructive jaundice. Lipoprotein metabolism is altered in liver disease. The increase in alpha$_2$ globulin and beta globulin in obstructive jaundice is largely associated with interferences with normal lipoprotein metabolism. Thus phenotyping of a lipid disorder cannot be performed accurately in the presence of liver disease. Because of the increased serum cholesterol associated with liver disease, the use of high-density lipoprotein cholesterol to assess risk of coronary heart disease is obviated in patients with alcoholic liver disease, biliary obstruction, and acute liver necrosis.

Coagulation factors are produced by the liver and can decrease significantly in the presence of liver disease. Plasma fibrinogen is normally present in a concentration of 2 to 4 g/L. A decrease in plasma fibrinogen is usually an indication of severe liver disease and is associated with decreased concentrations of other clotting factors, most notably prothrombin. Since prothrombin synthesis occurs in the liver and requires the fat-soluble vitamin K, prothrombin time may be increased in biliary obstructive disease, liver cirrhosis or necrosis, hepatic failure, Reye's syndrome, liver abscess, vitamin K deficiency, and hepatitis. The response of the prothrombin time to exogenous vitamin K is useful in differentiating intrahepatic disease from extrahepatic obstructive disease with decreased absorption of vitamin K.

Urea and ammonia in evaluation of liver function. Blood ammonia concentration is higher in infants than adults because the development of the hepatic circulation is completed after birth. Hyperammonemia results infrequently from congenital defects of the urea cycle. The most common of these inborn errors of metabolism is ornithine transcarbamylase deficiency. A much more frequent cause of hyperammonemia in infants is hyperalimentation.[9] Reye's syndrome is frequently diagnosed by an elevated blood ammonia in the absence of any other demonstrable cause.

Adult patients exhibit elevated blood ammonia in the terminal stages of liver cirrhosis, hepatic failure, and acute and subacute liver necrosis. Urinary ammonia excretion is increased in acidosis and decreased in alkalosis, since ammonia salt formation is a significant mechanism for excretion of excess protons. Damage to the renal distal tubules, as occurs in renal failure, glomerulonephritis, hypercorticoidism, and Addison's disease, results in decreased ammonia excretion.

Since urea is synthesized in the liver, liver disease without renal impairment results in a low serum urea nitrogen, though the urea-to-creatinine ratio may remain normal.[10] An elevated serum urea nitrogen does not necessarily imply renal damage, since dehydration may result in a urea nitrogen as high as 600 mg/L, and infants receiving high-protein formula may exhibit a urea nitrogen level of 250 to 300 mg/L. Naturally, renal diseases such as acute glomerulonephritis, chronic nephritis, polycystic kidney, and renal necrosis result in an elevated urea nitrogen.

REFERENCES

1. Johnson, TR: Development of the liver. In Johnson, TR, Moore, WM, and Jefferies, SE, editors: Children are different: developmental physiology, ed 2, Columbus, 1978, Ross Laboratories, pp 155-160.
2. Natelson, S, and Sherwin, JE: Proposed mechanism for urea nitrogen reutilization: relationship between urea and proposed guanidine cycles, Clin Chem 25:1343-1344, 1979.
3. Demers, LM: Serum bile acids in health and in hepatobiliary disease. In Demers, LM, and Shaw, LM, editors: Evaluation of liver function, Baltimore, 1978, Urban & Schwarzenberg, Inc, pp 33-50.
4. Stoner, JW: Neonatal jaundice, Am Fam Physician 24:226-232, 1981.
5. Jeppson, JO, Franzen, B, Cox, DW, et al: Typing of genetic variants

of alpha-1-antitrypsin by electrofocusing, Clin Chem 28:219-225, 1982.
6. Shaw, LM: The GGT assay in chronic alcohol consumption, Lab Management 20:56-63, 1982.
7. Ellis, G, Goldberg, DM, Spooner, RJ, and Ward, AM: Serum enzyme tests in diseases of the liver and biliary tree, Am J Clin Pathol 70:248-258, 1978.
8. Friedman, RB, Anderson, RE, Entine, SM, and Hirschberg, SB: Effects of diseases on clinical laboratory tests, Clin Chem 26:1D-476D, 1980.
9. Lloyd-Still, JD: Disorders of the liver in childhood. In Demers, LM, and Shaw, LM, editors: Evaluation of liver function, Baltimore, 1978, Urban & Schwarzenberg, Inc, pp 171-200.
10. Nanji, AA, and Blank, D: The serum urea nitrogen/creatinine ratio and liver disease, Clin Chem 28:1398-1399, 1982.

BIBLIOGRAPHY

Gitnick, G: Hepatology, New York, 1986, Medical Examination Publishing Co, Inc.

Halsted, JA: The laboratory in clinical chemistry: interpretation and application, Philadelphia, 1976, WB Saunders Co.

Meites, S, editor: Pediatric clinical chemistry: a survey of reference (normal) values, methods, and instrumentation, with commentary, ed 2, Washington, DC, 1981, American Association of Clinical Chemists.

Orton, SM, and Neuhaus, OW: Human biochemistry, ed 10, St. Louis, 1982, CV Mosby Co.

CHAPTER 24 Bone disease

REGINALD C. TSANG
HAROLD MARDER

OBJECTIVES

- Describe the structure and function of normal bone.
- Describe effects of vitamin D metabolites, parathyroid hormone, magnesium, and phosphorus on calcium metabolism and regulation.
- Define osteopenia, osteoporosis, osteomalacia, rickets, osteitis fibrosa, and Paget's disease.
- State expected levels of calcidiol, parathyroid hormone, calcium, phosphorus, magnesium, calcitonin, alkaline phosphatase, and urinary hydroxyproline in disease states.

KEY TERMS

calcidiol Cholecalciferol hydroxylated at the carbon-25 position in the liver.

calcitriol The active vitamin D metabolite; cholecalciferol hydroxylated at both the carbon-1 and carbon-25 positions.

cholecalciferol The parent vitamin D compound.

cortical bone Dense compact bone that provides structural support.

diaphysis Shaft of a long bone.

epiphysis End of a long bone.

metaphysis Region in which diaphysis and epiphysis converge.

osteoblasts Cells that synthesize bone matrix.

osteoclasts Cells that resorb bone.

osteocytes Mature bone cells that have limited function and are encased in bone matrix, the composition of which they help to maintain.

osteoid Bone matrix.

osteomalacia Bone containing normal amounts of osteoid but deficient amounts of mineral.

osteopenia The roentgenographical appearance of subnormally mineralized bone.

osteoporosis A generalized reduction in bone mass involving both mineral and osteoid.

rickets Osteomalacia in childhood.

trabecular bone Interlacing delicate spicules of bone that are predominantly involved in mineral homeostasis.

BONE STRUCTURE AND FUNCTION

Over 90% of the cells in bone are encased in calcified tissue and separated by great distances from a vascular supply. This gives the cells the appearance of inactivity. However, it is now clear that the skeletal system is a dynamic organ. The modern theory of skeletal system function proposes that bone has two interdependent roles: provision of support and maintenance of mineral homeostasis. Both functions are successfully achieved by continuous bone remodeling. Disturbances in the balance and nature of bone formation and resorption produce the common bone diseases.

Bone structure

Macroscopically, the major bones are classified as long bones or flat bones.[1] Long bones are confined to the limbs and consist of a shaft *(diaphysis),* two ends *(epiphyses),* and a region in which the two converge *(metaphysis)* (Fig. 24-1, *A*). Seen in cross section, the diaphysis is lined by dense, compact *(cortical)* bone, whereas the metaphysis contains interlacing bony specules that resemble the structure of a sponge *(trabecular or cancellous bone)* (Fig. 24-

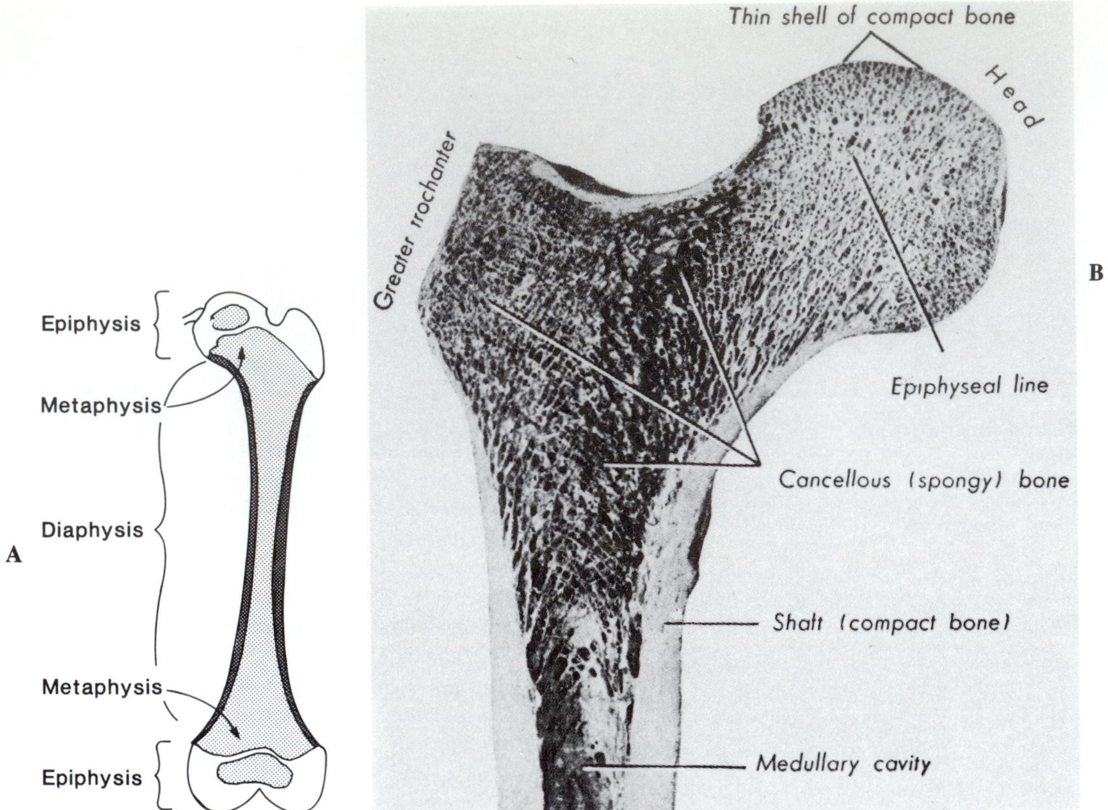

Fig. 24-1 **A,** Parts of a long bone. **B,** Long bone in cross section; note predominance of trabecular, cancellous bone in diaphysis. *(From Copenhaver, WM, Kelly, DE, and Wood, RL, editors: Bailey's textbook of histology, ed 17, Baltimore, 1978, The Williams & Wilkins Co.)*

1, *B*). Flat bones, typified by the bones of the skull, consist of two thin layers of cortical bone that enclose a layer of trabecular bone. Dense cortical bone provides strength needed for structural support, and the spicules of trabecular bone provide a large surface area for bone synthesis and resorption and provide a reservoir of minerals for the maintenance of mineral homeostasis.

Bone contains three major types of mature cells.[2] *Osteoblasts,* found along surfaces of both cortical and trabecular bone, synthesize bone matrix. As osteoblasts become embedded in bone matrix, they differentiate into mature *osteocytes*. Osteocytes synthesize small amounts of matrix continuously to maintain bone integrity, and they are able to resorb bone (osteocytic osteolysis) in exceptional circumstances when normal mineral homeostasis is altered. *Osteoclasts* contain enzymes that demineralize and digest bone matrix. The moth-eaten appearance of bone resorption is evident in histological sections of areas in which osteoclasts are numerous.

Only a small portion of bone is cellular; calcified matrix predominates. This matrix is primarily composed of collagen fibers. Approximately one fourth of the amino acids present in collagen are either proline or hydroxyproline, neither of which is present to any great extent in other tissues. When collagen is metabolized, hydroxyproline-containing oligopeptides are excreted in the urine, and the amount present correlates with the amount of bone turnover. The mineral elements of bone consist mostly of crystals of calcium and phosphate arranged either amorphously or as hydroxyapatite, $Ca_{10}(PO_4)_6(OH)_2$. A wide range of other elements may be present, including sodium, magnesium, copper, zinc, lead, and fluoride.

Bone function

Bone contains 99% of the body's calcium, 85% of the phosphorus, and 66% of the magnesium.[3] Although only a small percentage of these ions may be rapidly mobilized, they provide the initial mineral source when serum concentrations fall. The large surface area and excellent blood supply of trabecular bone permit a quick response to perturbations in plasma mineral concentrations. In contrast, the abundant calcified matrix of cortical bone provides the strength to support body weight. Despite this segregation of structure and function, disturbances in both often coexist. Examples include vitamin D deficiency, which causes both hypocalcemia and easily fractured bone, and immobilization, which causes bone resorption, osteoporosis, and hypercalcemia. Under normal circumstances, such disturbances do not occur because, in bone remodeling that occurs throughout the body, bone formation and bone resorption are "coupled," resulting in equal amounts of bone formation and resorption.[4] Most bone diseases result from alterations in coupling, either anew or secondary to hormonal imbalance, which produce either excessive bone formation or excessive resorption.

Bone remodeling results from both nonhormonal and hormonal stimuli. When subjected to mechanical pressure, bone produces a change in electrical potential, which causes current flow. This flow creates ion fluxes, which, in turn, increase either osteoblastic or osteoclastic activity.

BIOCHEMISTRY AND PHYSIOLOGY

Vitamin D

Attention has been focused on vitamin D since its discovery, in 1925, led to the elimination of the widespread problem of nutritional rickets. Initially considered a vitamin because rachitic patients were cured with oral supplementation of vitamin D, vitamin D is now considered to be a hormone.[5] The major source of vitamin D is not the diet but its production in skin after exposure to sunlight.[6] It is then transported in the bloodstream to the liver and kidneys for activation. It subsequently localizes at sites of activity in intestine and bone because of the presence of specific cellular receptors in these organs. Finally, as in other hormone systems, the plasma level of activated vitamin D is rigidly controlled by feedback regulation. Vitamin D can now be regarded as one of the three major hormones controlling calcium and phosphorus homeostasis and bone mineralization.[5,7]

Biochemistry and metabolism. Vitamin D_3, subsequently referred to as cholecalciferol, is produced in skin through the conversion of 7-dehydrocholesterol by ultraviolet light[5,6] (Fig. 24-2). Cholecalciferol is not the active hormone. Calcitriol (1,25-dihydroxycholecalciferol) was identified (Fig. 24-3) and was shown to act more rapidly and effectively than calcidiol (25-hydroxycholcalciferol) in promoting both intestinal calcium transport and bone mineral release.[5-7] Calcitriol is the most active metabolite of cholecalciferol.

After cholecalciferol is produced in the skin, it is transported to the liver, in which hydroxylation occurs at the C-25 position (Figs. 24-2 and 24-3) to produce calcidiol. Calcidiol is the most abundant vitamin D compound in plasma, with normal concentrations of approximately 30 mg/mL. Production is not under strict metabolic control because the plasma concentration varies with dietary intake[8] and sunlight exposure.[9] Calcidiol is then transported to the kidney, in which hydroxylation occurs at the 1α position to produce calcitriol (Fig. 24-2). Plasma calcitriol concentrations are relatively low (approximately 30 pg/mL). Although normal plasma concentrations vary with age,[10] they are probably under strict feedback control.[11] The details of this control are discussed later. If plasma calcitriol concentrations are sufficient, calcidiol is hydroxylated in the kidney at the C-24 position to yield 24,25-dihydroxycholecalciferol (Fig. 24-2). This last metabolite is currently regarded by most investigators as a waste product of vitamin D metabolism.

Cholecalciferol and its metabolites pass through the circulation attached to vitamin D–binding protein.[12] Certain

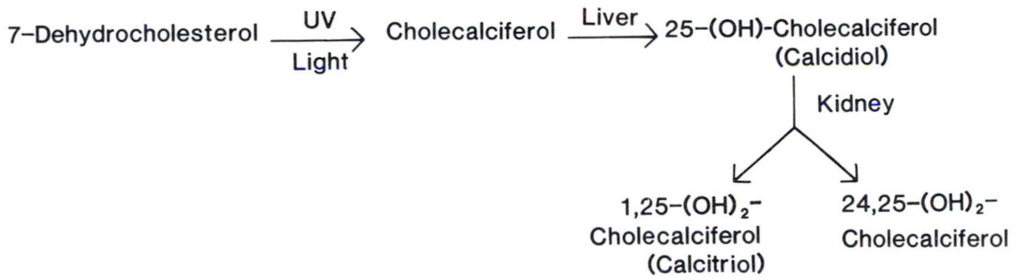

7-Dehydrocholesterol —UV Light→ Cholecalciferol —Liver→ 25-(OH)-Cholecalciferol (Calcidiol)

Kidney

1,25-$(OH)_2$-Cholecalciferol (Calcitriol) | 24,25-$(OH)_2$-Cholecalciferol

Fig. 24-2 Conversion of 7-dehydrocholesterol to activated vitamin D by ultraviolet (UV) light and by liver and kidney.

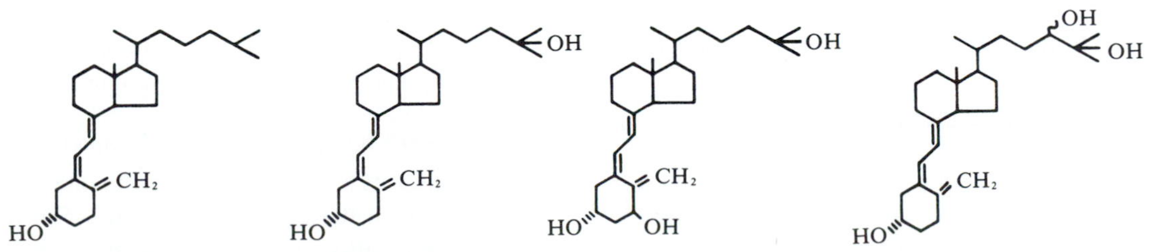

Vitamin D_3 (cholecalciferol) | 25-OH-Vitamin D_3 | 1α, 25-$(OH)_2$ Vitamin D_3 | 24,25-$(OH)_2$ Vitamin D_3

Fig. 24-3 Some common metabolites of cholecalciferol.

TARGET ORGANS FOR CALCITRIOL

Intestine—increased calcium and phosphorus absorption
Bone—enhanced parathyroid hormone–induced bone resorption
Kidney—increased reabsorption of calcium and phosphorus

tissues contain an intracellular protein that serves as a specific receptor for calcitriol. This protein, located in the cytosol of the kidney, intestine, bone, and selected other tissues, functions like other established cellular steroid hormone receptors (see Chapter 40). It facilitates entry of calcitriol into cells and transports the calcitriol to the cell nucleus, where the hormone, in theory, directs protein synthesis to achieve its desired effect.

Mechanisms of action. The function of calcitriol is discussed in many articles[5,7,12] (see box above). The three major target organs of calcitriol are intestine, bone, and kidney. Calcitriol facilitates both calcium and phosphate absorption in the intestine. Phosphate transport accompanies calcium transport, but it is also increased by an unknown, calcium-independent mechanism. Calcitriol works cooperatively with parathyroid hormone to increase bone resorption by incresing osteoclast activity. This may be considered paradoxical because vitamin D is believed to enhance bone mineralization. However, the net effect of the action of calcitriol at bone and intestine is to increase available blood calcium and phosphorus concentrations, which subsequently facilitate mineralization of newly formed bone matrix. Calcitriol increases the renal reabsorption of both calcium and phosphorus, but since 99% of filtered calcium is normally reabsorbed, the overall effect of alterations in plasma calcitriol concentrations on renal calcium reabsorption is small.

Regulation of vitamin D metabolism. The regulation of vitamin D metabolism is easily understood once calcitriol function is known (see box below). Although plasma calcidiol levels are poorly controlled,[10] there appears to be relatively strict control of plasma calcitriol concentrations. Parathyroid hormone (PTH) is the major stimulus for calcitriol formation. PTH administration is now used as a clinical tool to assess the ability of the kidney to produce calcitriol.[13] PTH may stimulate renal 1-hydroxylase, the

PRIMARY STIMULI FOR CALCITRIOL SYNTHESIS

Decreased serum calcium concentration
Increased pathathyroid hormone secretion
Decreased intracellular phosphorus concentration

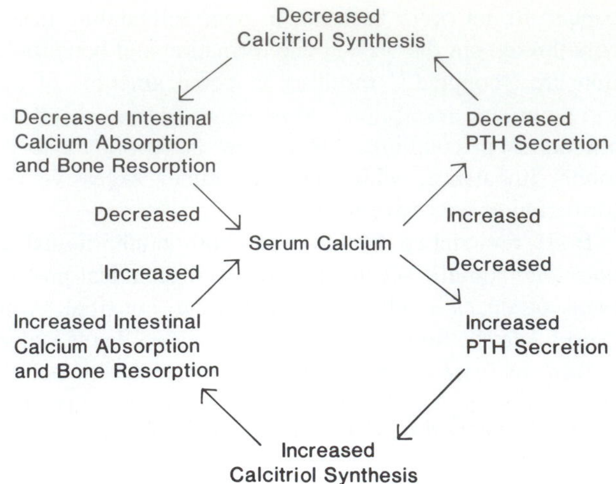

Fig. 24-4 Interrelationships of serum calcium concentrations and parathyroid hormone, *PTH,* and calcitriol.

enzyme that hydroxylates calcidiol at the C-1 position, indirectly by lowering intracellular phosphorus concentrations. Because phosphorus depletion increases calcitriol synthesis in normal or parathyroidectomized animals, decreased intracellular phosphorus levels may be the ultimate common stimulus for calcitriol synthesis.

Understanding the metabolic control of vitamin D allows one to comprehend control of serum calcium and phosphorus concentrations (Fig. 24-4). When the serum calcium concentration falls, PTH is secreted and acutely restores normal serum calcium concentrations by stimulating osteoclasts to resorb bone. Within hours, calcitriol production is increased, which causes enhanced intestinal calcium absorption, which subsequently restores the serum calcium concentration and indirectly the PTH concentration to normal. Elevations in serum calcium concentration produce the opposite effect, that is, a reduction in both serum PTH concentrations and calcitriol synthesis.

Plasma calcitriol concentrations are altered by aging and pregnancy; they are elevated during adolescence and decline in old age.[14] Pregnancy and subsequent lactation are associated with elevated serum estrogen or prolactin concentrations; both hormones increase calcitriol synthesis.[15] It should be noted that the adolescent growth spurt, pregnancy, and lactation all increase requirements for calcium. Thus the elevated serum calcitriol concentrations represent an appropriate response to a physiological need.

Parathyroid hormone (see also Chapter 44)

Biochemistry and metabolism. Parathyroid hormone is synthesized in the parathyroid glands. The precursor hormone for parathyroid hormone is *preproparathyroid hormone*. This precursor is sequentially converted in the gland, first to proparathyroid and then to parathyroid hormone, which is released into the circulation. The N-terminal section of the hormone is the biologically active section, whereas the C-terminal section is inactive. Thus fragments of parathyroid hormone bearing the N terminal generally are active, whereas those bearing the C terminal are inactive.

Mechanisms of action. Parathyroid hormone acts on two major target organs, bone and kidney, to produce three major effects: increase in serum calcium concentrations, decrease in serum phosphorus concentrations, and increase in the active hormonal form of vitamin D (cholecalciferol). In bone, parathyroid hormone predominantly mobilizes calcium and phosphorus to the extracellular fluid, thus raising serum calcium and phosphorus concentrations. At the other target organ, the kidney, parathyroid hormone causes increased calcium retention, increased phosphorus excretion, and increased conversion of 25-hydroxycholecalciferol (calcidiol) to 1,25-dihydroxycholecalciferol (calcitriol). Calcitriol, in turn, as described earlier, predominantly causes increased intestinal calcium and phosphorus absorption. The effects of parathyroid hormone on the kidney are mediated through the formation of cyclic adenosine monophosphate (cAMP), and urinary levels of this substance rise when parathyroid hormone production is increased. The resultant effect of parathyroid hormone on the bone, kidney, and indirectly the intestine is to increase calcium concentrations in the blood. Although phosphorus concentrations may be elevated through parathyroid actions on bone and indirectly the intestine, the effect on increased renal phosphorus excretion overwhelms the other effects and, overall, results in decreased serum phosphorus concentrations (Fig. 24-5).

Regulation of parathyroid hormone production. Parathyroid hormone production is predominantly regulated by the concentration of calcium in the blood bathing the parathyroids. Increased serum calcium concentrations decrease parathyroid hormone production, whereas decreased serum calcium concentrations increase its production. Since parathyroid hormone in turn affects calcium metabolism directly, this inverse relationship of calcium concentrations to parathyroid hormone production is one of the major mechanisms by which calcium is finely regulated in the blood.[16]

Calcitonin

Calcitonin,[16] discovered in 1962, is generally regarded as one of three hormones (along with vitamin D and parathyroid hormone) responsible for the control of calcium and phosphorus homeostasis. Despite great research efforts, a definitive role for calcitonin in calcium homeostasis has not yet been clarified. Neither calcitonin deficiency nor calcitonin excess is clearly associated with bone disease or alteration of serum calcium homeostasis.

Localization, biochemistry, and metabolism. Calcitonin is produced by the parafollicular C cells of the thyroid gland, although the pituitary gland, gastrointestinal

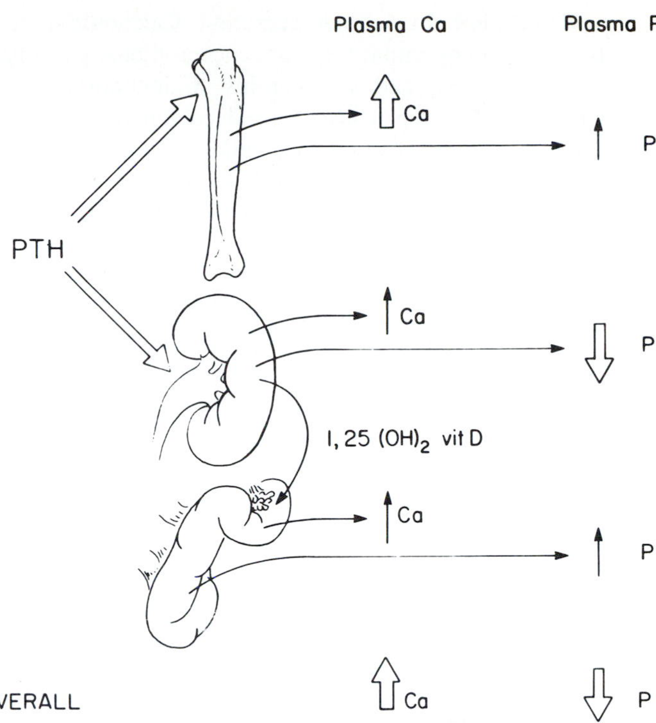

Fig. 24-5 Normal parathyroid hormone physiology. Parathyroid hormone action increases serum calcium concentrations predominantly through its bone and kidney effects but reduces plasma phosphorus concentrations, *P*, through increasing renal phosphorus excretion. *(From Tsang, RC, Noguchi, A, and Steichen, JJ: Pediatr Clin North Am* 26:223, 1979.)

tract, and liver may also produce the hormone.[16] Calcitonin is secreted in a precursor form, with a molecular weight of 15,000 daltons, that is cleaved into the active calcitonin polypeptide, which has a molecular weight of 3500 daltons.[16] Normal serum calcitonin concentrations are less than 100 pg/mL. Calcitonin is rapidly excreted, with a half-life of 10 minutes after intravenous administration.[17] Excretion is predominantly by the kidney, and serum calcitonin concentrations are increased in patients with renal failure.[18]

Biological effects. The biological effects of calcitonin may be divided into those related to calcium and phosphorus homeostasis and those related to gastrointestinal function. Intravenous calcitonin administration causes a prompt decline in the serum calcium concentration. This occurs because of effects of calcitonin on both bone and kidney. Calcitonin alters cell function by increasing intracellular cAMP production.[19] Pharmacological doses of calcitonin inhibit osteoclastic bone resorption,[20] with urinary hydroxyproline excretion decreasing in parallel with the inhibition of bone resorption. Calcitonin also decreases the renal reabsorption of calcium, phosphorus, sodium, potassium, and magnesium.[19] These described effects on both bone and kidney have been produced with pharmacological calcitonin concentrations.

Regulation of calcitonin secretion. Calcitonin secretion is influenced by serum calcium concentrations, gastrointestinal hormones, and sex steroids. Calcitonin release is stimulated by hypercalcemia and inhibited by hypocalcemia.[21] Serum calcitonin concentrations may be higher in pregnant and lactating women than in controls, and men have higher circulating calcitonin concentrations than women.[21]

Magnesium

Magnesium is predominantly an intracellular cation. Sixty percent of the body's magnesium is in bone, and the remainder is equally divided between muscle and other soft tissue. Only 1% of the body's magnesium is in blood. Magnesium is essential for activation of the molecule adenosine triphosphate, which provides the basic energy for cell function. Magnesium also is intimately related to the intracellular synthesis of RNA and DNA.[22]

Metabolism. Magnesium is absorbed through the intestinal tract, with absorption rates ranging from 44% on an ordinary diet to 76% on a low-magnesium diet. There is an efficient conservation of magnesium in the kidney so that in magnesium deficiency extremely low magnesium-excretion rates occur. PTH appears to cause increases in serum magnesium concentrations, possibly by mobilizing magnesium from bone. In acute conditions an increase in the serum magnesium concentration results in suppression of the parathyroids, thus theoretically preventing further parathyroid hormone increase and completing a "feedback" loop for magnesium-parathyroid interrelationships.[22] This beedback mechanism is thus similar to that for calcium-parathyroid interrelationships (see p. 377).

Although acute lowering of serum magnesium appears to increase serum PTH concentrations, chronic magnesium deficiency results in *decreased* release of PTH. In addition to this impairment in parathyroid function, magnesium deficiency can decrease the response of target organs to PTH. Thus magnesium deficiency would lead to hypoparathyroidism and, secondarily, hypocalcemia. Hypocalcemia can also accompany magnesium deficiency because calcium release from bone is decreased. Under normal circumstances, magnesium and calcium undergo an exchange in bone related to their release into the circulation. Lowered magnesium content in bone would result in lowered interchange with calcium and lowered release of calcium from bone.[23]

A PARTIAL DIFFERENTIAL DIAGNOSIS OF OSTEOPENIA

- Osteoporosis
  - Aging (senile)
  - Postmenopausal
  - Juvenile
  - Immobilization
  - Cushing's syndrome
  - Multiple myeloma
  - Leukemia
  - Turner's syndrome
  - Alcoholism
  - Chronic liver disease
- Osteomalacia
  - Vitamin D deficiency
  - Chronic gastrointestinal disease
  - Anticonvulsant induced
  - Vitamin D dependency
  - Vitamin D resistance (hypophosphatemia)
  - Chronic acidosis
  - Fanconi's syndrome
  - Chronic renal failure
  - Phosphorus and calcium deficiency
- Osteitis fibrosa
  - Primary hyperparathyroidism
  - Chronic renal failure
- Paget's disease

BONE DISORDERS

Disorders of calcium, phosphorus, vitamin D, or parathyroid hormone homeostasis frequently produce *osteopenia,* a general term for the x-ray appearance of a subnormal amount of mineralized bone. Many illnesses are associated with osteopenia[24] (see box above). Bone histopathological condition of osteopenia can reveal decreased osteoid (bone matrix) formation, decreased osteoid mineralization, or increased bone resorption.[25] These histological categories correlate with the clinical diagnosis of *osteoporosis, osteomalacia,* or *osteitis fibrosa,* respectively. Osteopenia can result in the crush-fracture syndrome in adults and either fractures or growth failure in children. Trabecular bone is more frequently affected than cortical bone, and so fractures most often occur in vertebrae, the femoral neck, and the distal ends of the long bones, where trabecular bone is abundant. Whereas specific diagnoses of

Table 24-1 Common serum abnormalities associated with metabolic bone disease

Disease	Ca	P	PTH	Alk PO_4	Calcidiol	Calcitriol
Osteoporosis	NL	NL	NL	NL	NL	LO
Osteomalacia	LO	LO	HI	HI	NL or LO	NL or LO
Osteitis fibrosa	HI or NL	LO or HI	HI	HI	NL	HI or LO

NL, Normal; *LO,* decreased; *HI,* increased; *Alk PO_4,* alkaline phosphate.

osteopenic bone are best made histologically, occasionally, characteristic laboratory abnormalities will permit differentiation among osteoporosis, osteomalacia, and osteitis fibrosa (Table 24-1).

Osteoporosis

Osteoporosis is defined histologically as a generalized reduction in bone mass, including both osteoid and mineral. This may be produced by either suppressed bone formation or increased bone resorption. In either case, the "coupling" of bone formation and bone resorption is altered.[26] Over 6 million Americans suffer from acute problems secondary to weakened vertebrae, and approximately 10% of women over 50 years of age have bone loss severe enough to result in fractures.[27] Osteoporosis is frequently diagnosed in older men and women (senile osteoporosis) and postmenopausal women. It occurs only rarely in childhood as an idiopathic illness and can also accompany certain systemic diseases (see box below).

Senile osteoporosis. Progressive bone loss normally occurs during aging. This process begins at 50 years of age in women and 65 to 70 years of age in men and results in a loss of 0.5% of total bone mass per year and approximately 20% in a lifetime.[28] Patients with senile osteoporosis experience accelerated losses of 1% to 2% per year, with symptoms of osteoporosis beginning when 30% of bone mass is lost. It has been suggested that osteoporosis is a natural part of the aging process manifested earlier in those persons who have accrued less skeletal mass during early adult life. The causes of senile osteoporosis are largely unknown. Hormonal alterations that occur during senescence undoubtedly potentiate bone loss (see box). Decreased serum calcitriol concentrations found in elderly persons[29] probably result from a blunted synthetic response to parathyroid hormone.[30] In addition, serum parathyroid hormone concentrations increase[31] and serum calcitonin concentrations decrease[32] with aging. The net effect of these hormonal alterations is diminished intestinal calcium absorption and increased bone resorption.

Postmenopausal osteoporosis. Postmenopausal osteoporosis, which occurs in females at a younger age than does senile osteoporosis, is probably caused by estrogen deficiency. Affected women have diminished intestinal calcium absorption and lower serum calcitriol concentrations than their normal age-matched peers.[29] Although serum PTH concentrations are normal when compared to controls with normal serum calcitriol concentrations, they are low when viewed in the context of calcitriol deficiency.[29] Estrogen supplementation increases intestinal calcium absorption and serum calcitriol and PTH concentrations.[33] These data have been interpreted to indicate that estrogen deficiency produces postmenopausal osteoporosis by causing bone resorption, which releases calcium into the extracellular space and which in turn suppresses PTH secretion, calcitriol synthesis, and intestinal absorption of calcium (Fig. 24-6).

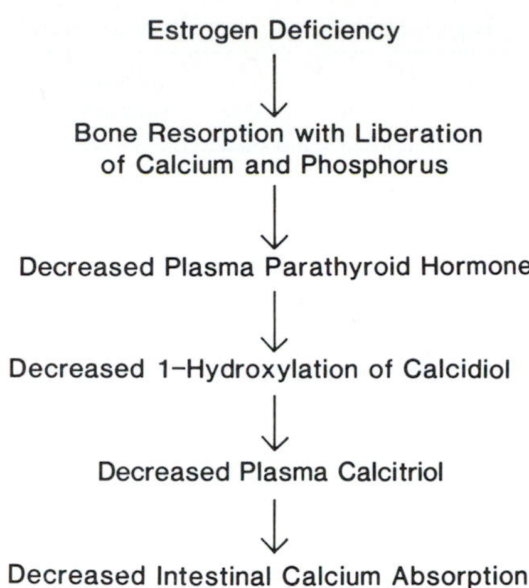

Fig. 24-6 Hypothesized pathogenesis of postmenopausal osteoporosis.

> *CALCIUM-REGULATING HORMONE ABNORMALITIES ASSOCIATED WITH AGING*
>
> Decreased serum calcitriol concentration and calcitriol secretory reserve
> Increased serum parathyroid hormone concentration
> Decreased serum calcitonin concentration

Osteomalacia

Osteomalacia is diagnosed when bone contains normal quantities of osteoid that fail to mineralize. When seen in the growing child, osteomalacia is termed *rickets*. The terms *rickets* and *osteomalacia* are used interchangeably in this chapter.

The major causes of osteomalacia are listed in the box on p. 378 and their associated biochemical abnormalities are summarized in Table 24-2. Historically the most common cause of osteomalacia was *vitamin D deficiency* caused by a combination of insufficient sunlight exposure and inadequate dietary intake of vitamin D–containing foods.[34] Serum calcidiol concentrations, which reflect the adequacy of vitamin D in the body, are low in osteomalacia. Supplementation of foods with vitamin D has virtually eliminated the problem in industrialized countries, but it may still be seen in underdeveloped nations, particularly among dark-skinned people because skin pigment decreases the production of cholecalciferol, which normally occurs after ultraviolet light exposure.

Vitamin D–deficient osteomalacia. Osteomalacia is caused by disturbances in either vitamin D metabolism or the bioavailability of phosphorus. Alterations of vitamin D

Table 24-2 Biochemical abnormalities associated with rickets

	Serum calcium	Serum phosphorus	Parathyroid hormone	Calcidiol	Calcitriol
Vitamin D deficiency	LO	LO	HI	LO	LO or NL
Vitamin D dependency					
I	LO	LO	HI	HI	LO
II	LO	LO	HI	HI	HI
Vitamin D resistance	NL	LO	NL	NL	NL or LO
Dietary phosphorus deficiency	NL	LO	NL	LO	HI

NL, Normal; *LO,* decreased; *HI,* increased.

metabolism range from conditions of insufficient intake or production of cholecalciferol to disturbances in its activation by the liver and kidneys. Generally one can predict the biochemical response to a deficiency of calcitriol (Fig. 24-4 on p. 376). The absorption of intestinal calcium will decrease and produce hypocalcemia. This will stimulate PTH release (secondary hyperparathyroidism), which will mobilize calcium from bone and increase phosphorus excretion by the kidney. Initially, serum calcium concentrations will be maintained at the expense of bone resorption, but as minerals are depleted, hypocalcemia occurs. Hypophosphatemia occurs because of increased urinary phosphorus losses. Thus the characteristic serum abnormalities associated with calcitriol deficiency are hypocalcemia, hypophosphatemia, and hyperparathyroidism (Table 24-2). Osteomalacia also results from phosphorus deficiency. In this situation, low intracellular phosphorus concentrations should stimulate calcitriol synthesis, which will increase both intestinal and renal phosphate absorption. Serum calcium and PTH concentrations should be unaffected (Table 24-2).

Osteomalacia secondary to gastrointestinal disorders. Patients with gastrointestinal disease, particularly those with hepatobiliary disease, often develop osteomalacia. Vitamin D is fat soluble and requires bile acids for absorption (see Chapter 36). Patients with hepatobiliary disease have low serum calcidiol levels that appear to be caused in part by defective intestinal cholecalciferol or ergosterol absorption, impaired calcidiol production by the liver, and enhanced calcitriol metabolism. Osteopenia also may be seen after gastric surgery, although the pathogenesis is not understood.

Osteomalacia secondary to anticonvulsant medication. Rickets may occur in children receiving anticonvulsant medications, such as phenytoin (Dilantin) and phenobarbital, that induce the hepatic microsomal mixed-oxidase enzyme system.[35] This enzyme system, when stimulated, converts calcidiol to polar inactive metabolites, which results in calcidiol deficiency.

Vitamin D–dependent osteomalacia (types I and II). After foods were fortified with vitamin D, it became apparent that normal antirachitic doses of analogs of vitamin D failed to heal the rickets of a small subpopulation of rachitic patients. One group of such patients had the classical signs and symptoms of vitamin D deficiency, including early infantile hypocalcemia, hypophosphatemia, and tetany, but required up to 100 times the normal intake of vitamin D to heal their rickets. This group of patients with *vitamin D–dependent rickets* has recently been subclassified into two groups with distinct pathophysiological bases. Patients with vitamin D–dependent rickets type I have the classical biochemical abnormalities of vitamin D–deficient rickets, but their serum calcidiol concentrations are normal and they lack circulating calcitriol.[36] The presumed defect is an abnormal or absent renal 1α-hydroxylase enzyme. The osteomalacia of these patients heals when physiological doses of calcitriol are administered. Patients with vitamin D–dependent rickets type II have normal calcidiol and calcitriol levels and are resistant to physiological doses of calcitriol, which suggests an end-organ resistance to calcitriol.[37] Both types of vitamin D–dependent rickets are inherited in an autosomal recessive pattern.

Vitamin D–resistant osteomalacia. Patients with *vitamin D–resistant rickets* lack most of the usual biochemical markers associated with rachitic patients. Serum calcium and PTH concentrations are normal, but hypophosphatemia is severe. The major defect in these patients is subnormal reabsorption of phosphorus by the renal tubules.[38] Because low intracellular phosphorus concentrations are a major stimulus for calcitriol synthesis, serum calcitriol concentrations should be elevated in this disorder; however, when measured, serum calcitriol concentrations have been found to be low or low-normal.[36,39] This finding suggests a potential second defect in this condition, that is, dysfunction of the renal 1α-hydroxylase enzyme. Vitamin D–resistant rickets may be inherited in a sex-linked recessive or an autosomal dominant pattern. Traditionally, patients with vitamin D–resistant rickets have been treated with cholecalciferol and phosphate supplements.

Other causes of osteomalacia. Chronic acidosis causes osteomalacia, hypercalciuria, and hyperphosphaturia because of neutralization of acids by bone with subsequent release of bone mineral. Patients with renal Fanconi's syndrome have diminished proximal tubule reabsorption of bicarbonate (resulting in chronic acidosis), phosphorus, glucose, and amino acids. Osteomalacia may be severe because of the chronic acidosis and severe hypophosphatemia. In addition, as part of the proximal tubulopathy, there may be subnormal activity of the renal 1α-hydroxylase enzyme.[40]

Sufficient substrate must be supplied in the diet for proper bone mineralization. Delayed bone mineralization commonly occurs in very low-birth-weight, premature infants who are fed normal infant formulas or breast milk,[41] both of which contain insufficient quantities of calcium and phosphorus for the rapid bone mineralization of premature infants.

Osteitis fibrosa

Osteitis fibrosa is the histopathological bone lesion produced by excessive parathyroid hormone secretion. It is primarily seen in two conditions, primary hyperparathyroidism and chronic renal failure. Bone disease is of lesser significance in primary hyperparathyroidism because surgical removal of the involved parathyroid glands cures the disease. The pathophysiological condition of secondary hyperparathyroidism associated with chronic renal failure is more complex and less amenable to treatment. Thus uremic patients frequently suffer from severe bone disease.

The complex bone abnormality associated with chronic renal failure is termed *renal osteodystrophy*. Two distinct histopathological forms of renal osteodystrophy, osteomalacia and osteitis fibrosa (see box), frequently coexist in the same patient. Osteomalacia is probably caused by decreased synthesis of calcitriol secondary to renal parenchymal disease. Serum concentrations of calcitriol and 24,25-dihydroxyvitamin D are decreased in both children and adults with chronic renal failure, and calcidiol concentrations are normal.[42] The pathogenesis of hyperparathyroidism in patients with renal disease is less clearly understood and may be caused by phosphate retention, decreased calcitriol synthesis, or a combination of the two[43] (Figs. 24-7 and 24-8).

Paget's disease

Paget's disease is a disorder of bone metabolism characterized by increased osteoclastic bone resorption followed by disordered bone formation.[44] The incidence of this disease is difficult to determine because the majority of affected patients are asymptomatic. In autopsy series of persons over 40 years of age, 3% of this group is affected. The cause of the disease is unknown.

The histological pattern of patients with Paget's disease proceeds through three stages. In the early phase of the illness, resorption predominates and the bone marrow is replaced with a highly vascular fibrous connective tissue. In the second phase of the disease, bone formation predominates. The pagetic bone is coarse-fibered, dense trabecular bone. In the final phase the rate of bone resorption declines and continued bone formation produces hard, dense bone. The largest amount of bone resorption that initially occurs produces greatly elevated urinary hydroxyproline concentrations, and the subsequent rapid rate of bone formation results in dramatically elevated serum alkaline phosphatase concentrations. Serum calcium and phosphorus concentrations are normal. However, pathological fractures occur and are treated by immobilization of the patient. Hypercalcemia frequently occurs because immobilization increases the rate of bone resorption. Paget's disease is treated successfully with calcitonin, which inhibits osteoclastic bone resorption.[45]

HISTOPATHOLOGY OF RENAL OSTEODYSTROPHY

Predominant osteitis fibrosa
- Normal serum calcium level—calcitriol responsive
- Pretreatment hypercalcemia—exacerbated by calcitriol

Predominant osteomalacia
- Small amount of fibrosis present—calcitriol responsive
- Pure osteomalacia—hypercalcemia with calcitriol treatment

Mixed osteitis fibrosa and osteomalacia—calcitriol responsive

Mild—calcitriol responsive

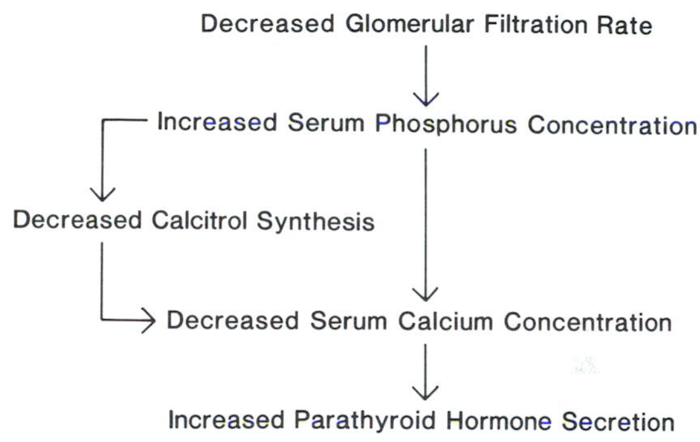

Fig. 24-7 Pathogenetic mechanism of secondary hyperparathyroidism in renal failure according to phosphate theory.

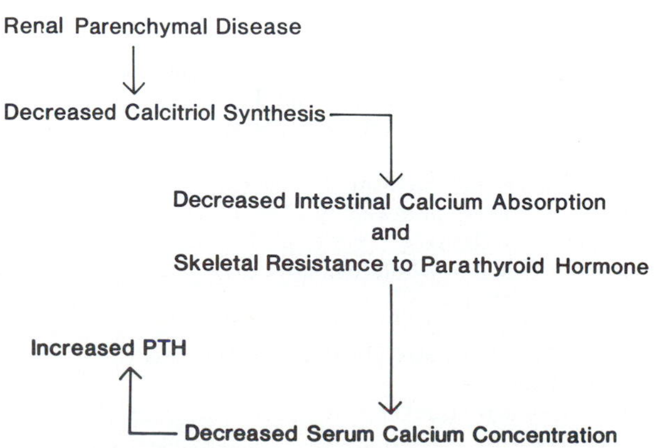

Fig. 24-8 Pathogenetic mechanism of secondary hyperparathyroidism in renal failure according to vitamin D theory.

CHANGE OF ANALYTE IN DISEASE

Vitamin D

Serum concentrations of vitamin D metabolites may be altered in a variety of disease states (see box). Decreased concentrations result from deficient intake, defective metabolic regulation, or increased excretion. Serum calcidiol concentrations are low in patients who have both an insufficient exposure to sunlight and a low intake of foods that contain vitamin D. Patients who receive anticonvulsant drugs convert calcidiol into biologically inactive polar metabolites.[35] Production of calcidiol is impaired in patients with liver disease.[46] Inactive or absent renal 1α-hydroxylase activity and secondary low serum calcitriol concentrations are associated with vitamin D–dependent rickets type I,[36] postmenopausal and senile osteoporosis,[28–33] hypoparathyroidism,[47] pseudohypoparathyroidism,[48a] vitamin D–resistant rickets,[36–39] and chronic renal failure.[48b] Patients with nephrotic syndrome have low serum concentrations of both calcidiol and calcitriol because of urinary losses of both metabolites, as well as the serum protein (vitamin D–binding protein) to which they are attached.[49]

High serum calcidiol concentrations result from either increased exogenous intake or increased endogenous production secondary to an unusually large sunlight exposure. High serum calcitriol concentrations occur in physiological states of increased calcium requirements such as growth[50] and pregnancy and lactation.[51] High serum calcitriol concentrations are also seen in sarcoidosis, in which an extrarenal source of calcitriol production has been implicated, and in hyperparathyroidism, in which the serum concentrations of PTH, a major stimulus for calcitriol production, are elevated.

DISEASES AND CONDITIONS ASSOCIATED WITH CHANGES IN SERUM CONCENTRATIONS OF VITAMIN D METABOLITES

- Calcidiol (25-hydroxycholecalciferol) deficiency
  - Nutritional osteomalacia
  - Anticonvulsant-induced osteomalacia
  - Liver disease
  - Nephrotic syndrome
- Calcitriol (1,25-dihydroxycholecalciferol) deficiency
  - Vitamin D–dependent rickets type I
  - Postmenopausal and senile osteoporosis
  - Hypoparathyroidism
  - Pseudohypoparathyroidism
  - Vitamin D–resistant rickets
  - Nephrotic syndrome
- Calcidiol (25-hydroxycholecalciferol) excess
  - Sunlight exposure
  - Vitamin D intoxication
- Calcitriol (1,25-dihydroxycholecalciferol) excess
  - Childhood
  - Pregnancy and lactation
  - Sarcoidosis
  - Hyperparathyroidism

Parathyroid hormone (see also Chapter 44)

Primary hypoparathyroidism. *Idiopathic hypoparathyroidism* describes the condition of decreased production of PTH. In pseudohypoparathyroidism, production of PTH is intact, but there is target organ resistance to PTH; in other words, PTH, although present, does not exert its physiological actions because the target organs are not responsive. In current terminology, there may be a "receptor defect" for PTH. Another way of describing idiopathic hypoparathyroidism would be *hormone-deficient hypoparathyroidism,* and pseudohypoparathyroidism would be described as *hormone-sufficient, receptor-deficient hypoparathyroidism* (Fig. 24-9). Hypoparathyroidism classically manifests itself with hypocalcemia and hyperphosphatemia, usually in childhood.

Secondary hypoparathyroidism. Hypoparathyroidism may result from other disorders. Inadvertent surgical removal of the parathyroids may occur during thyroidectomy. Since magnesium is important for PTH secretion, magnesium deficiency may result in hypoparathyroidism. An interesting physiological hypoparathyroidism occurs in infants. In utero, calcium is transferred actively across the placenta, and serum calcium concentrations in the fetus are extremely high. These high serum calcium concentrations appear to inhibit fetal parathyroid function. Inhibited parathyroid function persists for a short interval after birth and appears to be the major cause of hypocalcemia in the first 3 days of life, especially in the premature infant.[48a]

The diagnosis of hypoparathyroidism is made from the clinical presentation of lowered serum calcium and elevated serum phosphorus concentrations. PTH concentrations will be low in hypoparathyroidism but elevated in pseudohypoparathyroidism. To further distinguish idiopathic hypoparathyroidism from pseudohypoparathyroidism, PTH infusion is administered. After the infusion, serum calcium and urinary phosphorus and cAMP concentrations are measured. Patients with pseudohypoparathyroidism may have varying "degrees of block" in response to PTH, at the bone site or at various "levels" in the kidney.[52] Patients with hypoparathyroidism are treated with supplements of calcium salts and calcitriol.

Primary hyperparathyroidism. Primary hyperparathyroidism is often described as being related to hyperplasia or adenoma of the parathyroids. In contrast to hypoparathyroidism, which usually begins in childhood, hyperparathyroidism is usually discovered in adulthood. As expected from the physiological action of PTH, excess concentrations of the hormone result in increased serum calcium concentrations and decreased serum phosphorus concentrations. Demineralization occurs as a consequence of the bone lytic action of PTH and is associated with areas of extensive resorption (osteitis fibrosa, see p. 381). Many

Fig. 24-9 In idiopathic hypoparathyroidism, decreased parathyroid hormone results in decreased serum calcium, increased serum phosphorus, and decreased production of 1,25-dihydroxyvitamin D. In pseudohypoparathyroidism, although there is sufficient hormone, target organs are unresponsive and biochemical result is similar. *Inset,* In pseudohypoparathyroidism, resultant low serum calcium concentrations serve as a stimulus to parathyroid hormone production. Since parathyroid glands are intact, in contrast to idiopathic hypoparathyroidism, serum parathyroid hormone concentrations will be *elevated* in an attempt to overcome target organ resistance and rectify hypocalcemia.

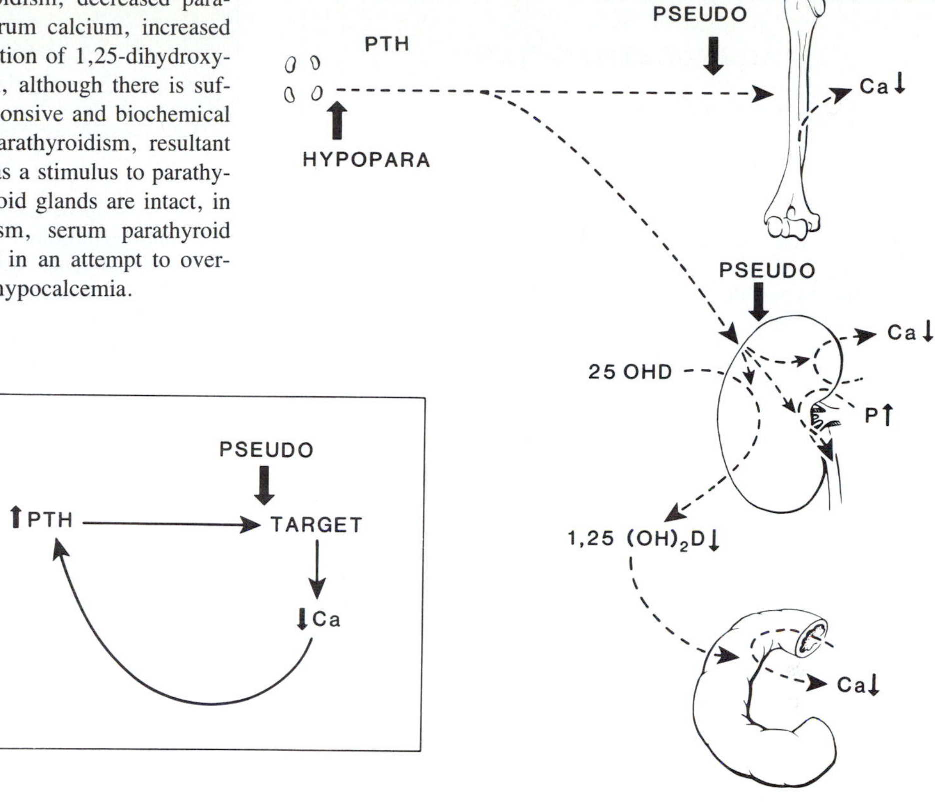

clinical problems are associated with the high serum calcium concentrations. The major organ systems adversely affected by hypercalcemia are the nervous system and the kidney.

Secondary hyperparathyroidism. Conditions that are associated with chronic hypocalcemia will result in chronic stimulation of the parathyroids and secondary hyperparathyroidism. The two major factors resulting in chronic hypocalcemia of nonparathyroid cause are vitamin D–metabolite deficiencies and high phosphorus loads. Any deficiency of vitamin D or its major metabolites will result in decreased intestinal absorption of calcium and hypocalcemia. The initial response to this hypocalcemia will be secondary hyperparathyroidism, which helps maintain serum calcium concentrations in the normal range. High phosphorus loads occur with infusion of phosphorus-containing fluids, ingestion of high phosphorus–containing milk (such as cow's milk given to newborn infants), or retention of phosphorus by failing kidneys. High serum phosphorus concentrations result in a secondary decrease in serum calcium concentrations. With decreased serum calcium concentrations there is compensatory secondary hyperparathyroidism.[53]

The diagnosis of primary hyperparathyroidism is based on the findings of high serum calcium, low serum phosphorus, and high serum PTH concentrations. However, not all hyperparathyroid patients will have increased serum calcium concentrations or increased serum PTH concentrations. Ionized calcium measurements in blood may provide additional diagnostic help because the ionized calcium fraction is the physiologically active calcium.

In secondary hyperparathyroidism, serum calcium concentrations are low or normal because the parathyroid overactivity results from an initial decline in serum calcium. Serum phosphorus concentrations would be low except in situations of phosphorus overload, when they would be high. Serum PTH concentrations should be elevated in secondary hyperparathyroidism.

Calcium

Hypercalcemia. Hypercalcemia resulting from primary hyperparathyroidism has been described earlier. Other causes of hypercalcemia are listed in the box on p. 384.

Endocrine and tumor-related hypercalcemia. Hypercalcemia may occur with disorders of endocrine organs other than the parathyroids. Overproduction of thyroid hormone (thyrotoxicosis) and underproduction of corticosteroids (Addison's disease or abrupt withdrawal of steroid hormones) are associated with hypercalcemia. A wide variety of tumors appear to produce PTH-like substances with osteoclast-stimulatory activity, which results in hypercalcemia (see Chapter 44).

Vitamin A– and vitamin D–related disorders. Excessive intake of vitamin A or vitamin D may result in hypercalcemia. In sarcoidosis, elevated calcitriol concentrations appear to be the cause of the hypercalcemia. Vitamin A in

CAUSES OF HYPERCALCEMIA

Primary hyperparathyroidism
Thyrotoxicosis
Addison's disease
Withdrawal of steroids
Tumors
Vitamin D and vitamin A intoxication
Sarcoidosis
Idiopathic hypercalcemia of infancy
Immobilization
Subcutaneous fat necrosis in infants
Thiazide diuretics
Milk-alkali syndrome
Benign familial hypercalcemia

high doses appears to have a direct effect on bone resorption. Therapy for vitamin D–resistant rickets or hypoparathyroidism with high doses of vitamin D is a common cause of hypercalcemia. Idiopathic hypercalcemia of infants is believed to be related to disordered vitamin D metabolism, possibly increased sensitivity to vitamin D.[54]

Iatrical causes. Immobilization of patients, especially male adolescents, results in rapid mobilization of calcium from bone and resultant hypercalcemia. In infants born after traumatic deliveries, a curious condition of hypercalcemia can occur in association with extensive trauma related to subcutaneous fat necrosis. Use of thiazide diuretics is classically associated with hypercalcemia, presumably because of their augmentation of the calcium-mobilizing effect of parathyroid hormone on bone. Excessive ingestion of milk and alkali in the treatment of peptic ulcer (the milk-alkali syndrome) also results in hypercalcemia.

Familial form. There is a benign form of familial hypercalemia that is inherited as a dominant trait. Mild hypercalcemia (less than 130 mg/L) occurs and apparently is without adverse effects.

Hypocalcemia. The causes of hypocalcemia are currently classified in relation to the major hormone or biochemical involved: vitamin D, PTH, calcitonin, calcium, magnesium, and phosphate (see box at right).

Vitamin D deficiency, which has been covered earlier, occurs as a result of reduced synthesis or intake of the parent vitamin D, altered hepatic metabolism of vitamin D, and decreased renal synthesis of calcitriol, the final active metabolite of vitamin D.

Hypoparathyroidism (primary and secondary) and pseudohypoparathyroidism have been discussed previously. Calcitonin or mithramycin infusions will decrease calcium transport from bone to extracellular space and result in hypocalcemia. Intestinal malabsorption of calcium may lead to hypocalcemia. Acute pancreatitis is associated with fatty acid–calcium complex precipitates in the pancreas and hypocalcemia. Decreased blood ionized calcium occurs with infusion of agents complexing calcium (citrate and acid-citrated blood for transfusion, or EDTA), or alkalosis, which shifts the fraction of calcium that is ionized to that which is protein bound. Hypomagnesemia results in hypocalcemia mostly related to the adverse effect of hypomagnesemia on parathyroid function. Conditions whereby phosphorus concentrations in blood are elevated, as in renal failure (see earilier discussion), phosphate infusion, or infants receiving cow's milk formulas with their high phosphate content, will result in decreased serum calcium concentrations because calcium is shifted from the extracellular space into bone and soft tissues and probably because there is a blunted bone response to the effects of PTH.[55]

CAUSES OF HYPOCALCEMIA

Vitamin D
- Decreased solar exposure and endogenous synthesis
- Decreased intestinal intake (malabsorption, dietary deficiency)
- Altered hepatic metabolism of vitamin D (hepatic disease, anticonvulsants)
- Decreased renal synthesis of calcitriol (vitamin D dependency, renal failure)

Parathyroid
- Hypoparathyroidism (primary and secondary)
- Pseudohypoparathyroidism

Calcitonin
- Calcitonin or mithramycin infusion

Calcium
- Intestinal malabsorption
- Acute pancreatitis
- Infusion of agents complexing calcium
- Alkalosis decreasing ionized calcium

Magnesium
- Magnesium deficiency (see box, p. 386, top right)

Phosphorus
- Renal failure
- Phosphate infusion
- Cow's milk formulas

Changes in parathyroid hormone: vitamin D–axis analytes in hypercalcemia and hypocalcemia. Changes in the serum measurements for phosphorus, the PTH–vitamin D axis, and vitamin D status help in evaluation of the causes of hypercalcemia and hypocalcemia. If the PTH–vitamin D axis is intact, two effects are seen: (1) cAMP production by the kidney is active; renal cAMP is best determined as ''nephrogenous'' cAMP, which takes into account the cAMP not produced in the kidney,[56] and (2) 1,25-dihydroxycholecalciferol (calcitriol) production is also active (Tables 24-3 and 24-4).

Of the *hypercalcemic* disorders listed in Table 24-3, serum phosphorus concentrations are decreased in hyper-

Table 24-3 Parathyroid hormone–vitamin D axis analytes in hypercalcemia

Disorder	Serum phosphorus	Parathyroid hormone (PTH)	Parathyroid hormone–vitamin D axis		Vitamin D status
			Nephrogenous cyclic AMP	Calcitriol (1,25-dihydroxy-cholecalciferol)	Calcidiol (25-hydroxy-cholecalciferol)
Hyperparathyroidism	LO	HI	HI	HI	NL
Vitamin D disorders					
Vitamin D intoxication	NL	LO	LO	HI	HI
High calcitriol in sarcoidosis	NL	LO	LO	NL	NL
Sensitivity to vitamin D: idiopathic hypercalcemia of infancy	NL	LO	LO	NL	NL
Non–parathyroid hormone, non–vitamin D disorders					
Malignancy	NL	LO	HI/LO	LO	NL
Immobilization	NL	LO	LO	LO	NL
Thyrotoxicosis	NL	LO	NL	LO	NL

NL, Normal; *LO,* decreased; *HI,* increased.

Table 24-4 Parathyroid hormone–vitamin D axis analytes in hypocalcemia

Disorder	Serum phosphorus	Parathyroid hormone (PTH)	Parathyroid hormone–vitamin D axis		Vitamin D status
			Nephrogenous cyclic AMP	Calcitriol (1,25-dihydroxy-cholecalciferol)	Calcidiol (25-hydroxy-cholecalciferol)
Parathyroid disorders					
Hypoparathyroidism	HI	LO	LO	LO	NL
Pseudohypoparathyroidism	HI	HI	LO	LO	NL
Vitamin D disorders					
Vitamin D deficiency	LO	HI	HI	NL or LO	LO
Hepatic disease alpha-anticonvulsants	LO	HI	HI	NL or LO	LO
Renal					
Vitamin D–dependent rickets	LO	HI	HI	LO	NL
Osteodystrophy	HI	HI	—	LO	NL
Resistance to 1,25-dihydroxycholecalciferol	LO	HI	HI	HI	NL
Mineral disorders					
Calcium malabsorption	NL	NL or HI	—	—	—
Hypomagnesemia	NL	NL or HI	—	—	NL
High phosphate load	HI	NL or HI	—	—	NL

NL, Normal; *LO,* decreased; *HI,* increased.

parathyroidism because of the phosphaturic effects of PTH; in the remaining hypercalcemic disorders, little effect on serum phosphorus is evident. In hyperparathyroidism, serum PTH concentrations and the PTH–vitamin D axis analytes are increased. In hypercalcemia from other causes, serum PTH is suppressed. In turn the PTH–vitamin D axis may be suppressed, except in sarcoidosis, in which elevation of serum calcitriol concentrations appears to be a primary problem.

Vitamin D status is best assessed through measurement of serum 25-hydroxycholecalciferol (calcidiol) concentrations. Thus serum 25-hydroxycholecalciferol concentrations are elevated in vitamin D intoxication, with or without elevations in the serum 1,25-dihydroxycholecalciferol (calcitriol) concentrations.

In *hypocalcemia* (Table 24-4) related to parathyroid disorders, hyperparathyroidism, or pseudohyperparathyroidism, serum phosphorus is elevated because of decreased urinary phosphorus excretion. The PTH–vitamin D axis is generally hypofunctioning, except in pseudohyperparathyroidism, in which serum PTH concentrations will be elevated because of target organ resistance to the hormone.

In the vitamin D disorders, serum phosphorus concentrations are generally low because one of the major actions of vitamin D is to raise serum phosphorus concentrations. However, in renal osteodystrophy (renal failure) serum phosphorus concentrations are elevated because of decreased renal phosphorus excretion. The PTH–cAMP axis in this circumstance may be increased because of hyperparathyroidism secondary to hypocalcemia; however, serum 1,25-dihydroxycholecalciferol concentrations will remain decreased because of deficiency of vitamin D or blocks in vitamin D metabolism. In the condition of increased resistance to 1,25-dihydroxycholecalciferol, high serum concentrations of the metabolite are found, analogous to elevated PTH concentrations in pseudohyperparathyroidism. Serum 25-hydroxycholecalciferol measurements will be low in vitamin D deficiency or when there is a block in 25-hydroxylation of the vitamin D but normal in vitamin D disorder caused by metabolic blocks beyond the liver step of hydroxylation.

In the mineral disorders causing hypocalcemia, little effect on the PTH–vitamin D axis has been reported. Secondary hyperparathyroidism can be a consequence of hypocalcemia. In hypomagnesemia, however, hypoparathyroidism can occur secondary to magnesium deficiency.

Magnesium

Hypermagnesemia. Excess of magnesium is usually a consequence of increased medicinal intake of magnesium. Magnesium ($MgSO_4$) is used in the treatment of hypertension induced by pregnancy (preeclampsia). The mother will become hypermagnesemic (up to 110 mg/L), as will her infant. Reduced magnesium excretion in severe renal failure may occur, and the use of medicines that contain magnesium (antacids, purgatives) in this situation may result in hypermagnesemia[57] (see box below).

Hypomagnesemia and magnesium deficiency. Severe magnesium deficiency in humans is uncommon, possibly because of the body's highly developed ability to conserve magnesium. Decreased uptake of magnesium caused by gastrointestinal disorders (steatorrhea, malabsorption syndromes, gut resections) can cause magnesium deficiency. Specific intestinal malabsorption of magnesium also occurs and can cause hypomagnesemia in infancy. Protein-calorie malnutrition is often associated with magnesium depletion. Increased urinary magnesium losses may result from generalized renal disease or a specific renal defect in reabsorption of magnesium. Dialysis of patients may result in magnesium depletion if a low–magnesium content dialysate is used. High rates of production of aldosterone (hyperaldosteronism), hyperparathyroidism, and diabetes mellitus cause increased urinary magnesium losses. Alcoholism, intensive diuretic therapy, and treatment with the antibiotic gentamycin also result in increased urinary magnesium losses[58] (see box above).

CAUSES OF HYPERMAGNESEMIA

- Magnesium sulfate therapy
- Magnesium-containing antacids and purgatives
- Renal failure

CAUSES OF HYPOMAGNESEMIA

- Decreased intake of magnesium
  - Steatorrhea
  - Malabsorption syndromes
  - Gut resections
  - Specific intestinal malabsorption of magnesium
  - Protein-calorie malnutrition
- Increased loss of magnesium
  - Renal tubular loss
  - Dialysis with low magnesium dialysate
  - Hyperaldosteronism
  - Hyperparathyroidism
  - Diabetes mellitus
  - Alcoholism
  - Diuretic therapy
  - Aminoglycoside therapy

Magnesium deficiency is often associated with hypocalcemia, and the signs and symptoms of magnesium deficiency normally are the signs of hypocalcemia. Although serum magnesium concentrations can be low, since magnesium is predominantly an intracellular mineral, serum measurements may not reflect intracellular concentrations. Red cell magnesium concentrations have been advocated as a measure of intracellular magnesium status.

Calcitonin

Abnormal serum calcitonin concentrations are rarely found (see following box). Serum measurements are most useful in patients suspected of having medullary thyroid carcinoma, a malignancy of the thyroid C cells. This cancer is frequently seen in different members within families and is often associated with a tendency for other malignancies (termed *multiple endocrine neoplasia syndrome type II*).[59] Serum calcitonin measurements are useful both in the screening of family members who are potentially at risk of developing the disease and in the follow-up examination of previously treated patients suspected of recurrent metastatic disease. Serum calcitonin elevations are produced by a wide variety of other neoplasias, the most frequent one being bronchogenic carcinoma.[60] Because gastrin is a potent stimulus for calcitonin secretion, serum calcitonin concentrations are elevated in Zollinger-Ellison syndrome, a pancreatic tumor of gastrin-secreting cells. Finally, calcitonin excretion is decreased in patients with renal failure, and that decrease results in secondary elevation of serum concentrations of calcitonin in these patients.

Because the thyroid gland is usually the sole source of calcitonin production, athyroid patients lack circulating calcitonin. Calcitonin levels are also decreased in some pa-

DISEASES ASSOCIATED WITH ABNORMAL SERUM CALCITONIN CONCENTRATIONS

Excess
- Medullary thyroid carcinoma
- Bronchogenic carcinoma
- Zollinger-Ellison syndrome
- Renal failure

Deficiency
- Thyroid agenesis
- Thyroidectomy
- Osteoporosis

tients with osteoporosis. This may be caused by altered regulation of calcitonin synthesis or release.[61]

Alkaline phosphatase

In clinical practice, alkaline phosphatase determinations measure a group of enzymes that catalyze the hydrolysis of phosphate esters in an alkaline medium.[62] Alkaline phosphatase is produced by many tissues (see next box), but only the portion produced by bone and liver is detected in serum from healthy persons. The second box below lists causes of abnormal serum bone alkaline phosphatase concentrations. Alkaline phosphatase is produced by osteoblasts and, as previously discussed, lowers bone pyrophosphate levels, which probably facilitates mineralization. Alkaline phosphatase synthesis is deficient in hypophosphatasia, a rare hereditary illness associated with undermineralized bones and pathological fractures, and achondroplasia, an inherited disorder of endochondral bone growth. Production is also decreased with generalized malnutrition or scurvy.

SOURCES OF ALKALINE PHOSPHATASE

Osteoblasts
Bile canalicular cells
Placenta
Leukocytes
Proximal renal tube cells
Active mammary gland

BONE DISEASES ASSOCIATED WITH ABNORMAL SERUM ALKALINE PHOSPHATASE CONCENTRATIONS

Deficiency
- Hypophosphatasia
- Achondroplasia
- Severe malnutrition
- Scurvy

Excess
- Osteoblastic sarcoma
- Osteomalacia
- Paget's disease
- Hyperparathyroidism
- Growing children

CONDITIONS ASSOCIATED WITH ELEVATED URINARY HYDROXYPROLINE CONCENTRATIONS

Paget's disease	Acromegaly
Osteomalacia	Rheumatoid arthritis
Neoplastic bone disease	Osteoporosis
Hyperthyroidism	Aseptic bone necrosis
Osteomyelitis	Chronic renal failure
Burns	

Far more common than decreased concentrations are diseases associated with elevated serum alkaline phosphatase concentrations. Such elevations signify increased osteoblastic activity, as seen in osteoblastic sarcoma, rickets, Paget's disease, and acromegaly. The elevated levels associated with hyperparathyroidism result from secondary bone mineralization rather than PTH-induced osteoclastic activity. One should exercise caution when considering the pathological significance of alkaline phosphatase increases in childhood because growth is an important physiological cause of such elevations.[63] Liver alkaline phosphatase elevations reflect biliary obstruction and do not occur to any great extent with pure hepatocellular disease.

Hydroxyproline

Collagen, which is present predominantly in bone and skin, is the sole source of the amino acid hydroxyproline, which, together with proline, makes up approximately one third of the total amino acid content of collagen. Collagen digestion, associated with either bone or skin breakdown, results in elevated urinary hydroxyproline concentrations (see box above).[64]

REFERENCES

Bone structure and function

1. Shipman, P, Walker, A, and Bichell, D, editors: The human skeleton, Cambridge, MA, 1985, Harvard University Press.
2. Ham, AW, and Cormack, DH: Histology, ed 8, Philadephia, 1979, JB Lippincott Co.
3. Potts, JT, and Deftos, LJ: Parathyroid hormone, calcitonin, vitamin D, bone and bone mineral metabolism. In Bondy, PK, and Rosenberg, LE, editors: Duncan's diseases of metabolism, Philadelphia, 1974, WB Saunders Co.
4. Baylink, DJ, and Lin, CC: The regulation of endosteal bone volume, J Periodontol 50:43-49, 1979.

Biochemistry and physiology

5. DeLuca, HF: The kidney as an endocrine organ for production of 1,25-dihydroxyvitamin D_3, a calcium-mobilizing hormone, N Engl J Med 289:359–365, 1973.
6. Hollick, MF, Frommer, JE, McNeill, SC, et al: Photometabolism of 7-dehydrocholesterol to previtamin D_3 in skin, Biochem Biophys Res Commun 176:107-114, 1977.
7. Avioli, LV: Hormonal aspects of vitamin D metabolism and its clinical implications, Clin Endocrinol Metab 8:547-577, 1979.
8. Haddad, JG, and Stamp, TCB: Circulating 25-hydroxyvitamin D in man, Am J Med 57:57-62, 1974.
9. Avioli, LV, and Haddad, JG: Vitamin D: current concepts, Metabolism 22:507-531, 1973.
10. Chesney, RW, Rosen, JF, Hamstra, AJ, and DeLuca, HF: Serum 1,25-dihydroxyvitamin D levels in normal children and in vitamin D disorders, Am J Dis Child 134:135-139, 1980.
11. Chesney, RW, Rosen, JF, Hamstra, AJ, et al: Absence of seasonal variation in serum concentrations of 1,25-dihydroxyvitamin D despite a rise in 25-hydroxyvitamin D in summer, J Clin Endocrinol Metab 53:139-142, 1981.
12. DeLuca, HF: The vitamin D system in the regulation of calcium and phosphorus metabolism, Nutr Rev 37:161-193, 1979.
13. Eisman, JA, Pounce, RL, Ward, JD, and Moseby, JM: Modulation of plasma 1,25-dihydroxyvitamin D in man by stimulation and suppression tests, Lancet 2:931-933, 1979.
14. Gallagher, JC, Riggs, LB, Eisman, J, et al: Intestinal calcium absorption and serum vitamin D metabolites in normal subjects and osteoporotic patients: effect of age and dietary calcium, J Clin Invest 64:729–736, 1979.
15. Kumar, R, Cohen, WR, Silva, P, and Epstein, FH: Elevated 1,25-dihydroxyvitamin D levels in normal human pregnancy and lactation, J Clin Invest 63:342-344, 1979.
16. Austin, LA, and Heath, H, III: Calcitonin physiology and pathophysiology, N Engl J Med 304:269-278, 1981.
17. Huwler, R, Born, W, Ohnhaus, EE, and Fischer, JA: Plasma kinetics and urinary excretion of exogenous human and salmon calcitonin in man, Am J Physiol 236:15-19, 1979.
18. Ardaillou, R: Kidney and calcitonin, Nephron 15:250-260, 1975.
19. Heersche, JNM, Marcus, R, and Aurbach, GD: Calcitonin and the formation of 3′,5′-AMP in bone and kidney, Endocrinology 94:241-247, 1974.
20. Deftos, LJ: Calcitonin. In Gray, CH, and James, VHT, editors: Hormones and blood, New York, 1979, Academic Press, Inc.
21. Cooper, CW: Recent advances with thyrocalcitonin, Ann Clin Lab Sci 6:119-129, 1976.
22. Alkawa, JK: Magnesium: its biologic significance, CRC series on cations of biological significance, Boca Raton, Fla, 1981, CRC Press, Inc.
23. Tsang, RC: Neonatal magnesium disturbances, Am J Dis Child 124:282, 1972.

Bone disorders

24. Chase, L: Osteopenia, Am J Med 69:915–922, 1980.
25. Parfitt, AM, Oliver, I, and Villanueva, AR: Bone histology in metabolic bone disease: the diagnostic value of bone biopsy, Orthop Clin North Am 10:329-345, 1979.
26. Ivey, JL, and Baylink, DJ: Postmenopausal osteoporosis: proposed roles of defective coupling and estrogen deficiency, Metab Bone Dis Rel Res 3:3-7, 1981.
27. Avioli, LV: Postmenopausal osteoporosis: prevention versus cure, Fed Proc 40:2418-2422, 1981.
28. Wallach, S: Management of osteoporosis, Hosp Pract 13:91-98, 1978.
29. Gallagher, JC, Rigg, BL, Eisman, J, et al: Intestinal calcium absorption and serum vitamin D metabolites in normal subjects and osteoporotic patients: effect of age and dietary calcium, J Clin Invest 64:729-736, 1979.
30. Slovik, DM, Adams, JS, Neer, RM, et al: Deficient production of 1,25-dihydroxyvitamin D in elderly osteoporotic subjects, N Engl J Med 305:372-374, 1981.
31. Gallagher, JC, Riggs, BL, Jerpbak, CM, and Arnaud, CD: Effect of age on serum immunoreactive parathyroid hormone in normal and osteoporotic women, J Lab Clin Med 95:373-385, 1980.
32. Shamonki, IM, Fumar, AM, Tataryn, IV, et al: Age-related changes of calcitonin secretion in females, J Clin Endocrinol Metab 50:437-439, 1980.
33. Gallaher, JC, Riggs, BL, and DeLuca, HF: Effect of estrogen on calcium absorption and serum vitamin D metabolites in postmenopausal osteoporosis, J Clin Endocrinol Metab 51:1359-1364, 1980.
34. Avioli, LV: Hormonal aspects of vitamin D metabolism and its clinical implications, Clin Endocrinol Metab 8:547-577, 1979.
35. Winnacker, JL, Yeager, H, Saunders, JA, et al: Rickets in children receiving anticonvulsant drugs, Am J Dis Child 31:286-290, 1977.
36. Scriver, CR, Reade, TM, DeLuca, HF, and Hamstra, AJ: Serum 1,25-dihydroxyvitamin D levels in normal subjects and in patients with hereditary rickets and bone disease, N Engl J Med 299:976-979, 1978.
37. Brooks, MH, Bell, NH, Love, L, et al: Vitamin D-dependent rickets type II: resistance of target organs to 1,25-dihydroxyvitamin D, N Engl J Med 298:996-999, 1978.
38. Scriver, C: Rickets and the pathogenesis of impaired tubular transport of phosphate and other solutes, Am J Med 57:43-49, 1974.
39. Drezner, MK, Lyles, KW, Haussler, MR, and Harrelson, JM: Evaluation of a role for 1,25-dihydroxyvitamin D_3 in the pathogenesis and treatment of X-linked hypophosphatemic rickets and osteomalacia, J Clin Invest 66:1020-1032, 1980.
40. Chesney, RW, Rosen, JF, Hamstra, AJ, and DeLuca, HF: Serum 1,25-dihydroxyvitamin D levels in normal children and in vitamin D disorders, Am J Dis Child 134:135-139, 1980.
41. Steichen, JJ, Tsang, RC, Greer, FR, et al: Elevated serum 1,25-dihydroxyvitamin D concentrations in rickets of very low-birth-weight infants, J Pediatr 99:293-298, 1981.
42. Chesney, RW, Hamstra, AJ, Mazess, RB, et al: Circulating vitamin D metabolite concentrations in childhood renal diseases, Kidney Int 21:65-69, 1982.
43. Massry, SG, and Ritz, E: The pathogenesis of secondary hyperparathyroidism of renal failure: is there a controversy? Arch Intern Med 138:853-856, 1978.
44. Singer, FR, Schiller, AL, Pyle, EB, and Krane, SM: Paget's disease of bone. In Avioli, LV, and Krane, SM, editors: Metabolic bone disease, vol 2, New York, 1978, Academic Press, Inc.
45. Singer, FR: Huma calcitonin treatment of Paget's disease of bone, Clin Orthop 127:86-93, 1977.

Change of analyte in disease

46. Kooh, SW, Jones, G, Reilly, BJ, and Fraser, D: Pathogenesis of rickets in chronic hepatobiliary disease in children, J Pediatr 94:870-874, 1979.
47. Drezner, MK, Neelon, FA, and Jowsey, J: Hypoparathyroidism: a possible cause of osteomalacia, J Clin Endocrinol Metab 45:114, 1977.

48a. Drezner, MK, Neelon, FA, and Haussler, M: 1,25-Dihydroxycholecalciferol deficiency: the probable cause of hypocalcemia and metabolic bone disease in pseudohypoparathyroidism, J Clin Endocrinol Metab 42:621, 1976.

48b. Juttmann, JR, Buurman, CJ, De Kam, E, et al: Serum concentrations of vitamin D in patients with chronic renal failure: consequences for the treatment with 1-[alpha]-hydroxy derivatives, Clin Endocrinol 14:225-236, 1981.

49. Goldstein, DA, Haldimann, B, Sherman, D, et al: Vitamin D metabolites and calcium metabolism in patients with nephrotic syndrome and normal renal function, J Clin Endocrinol Metab 52:116-121, 1981.
50. Chesney, RW, Rosen, JF, Hamstra AJ, and DeLuca HF: Serum 1,25-dihydroxyvitamin D levels in normal children and in vitamin D disorders, Am J Dis Child 34:135-139, 1980.
51. Kumar, R, Cohen, WR, Silva, P, and Epstein, FH: Elevated 1,25-dihydroxyvitamin D levels in normal pregnancy and lactation, J Clin Invest 63:342-344, 1979.
52. Tsang, RC, and Brown, DR: The parathyroids. In Kelley, V, editor: Practice of pediatrics, vol 1, New York, 1979, Harper & Row, Publishers.
53. Tsang, RC, and Venkataraman, P: Pediatric parathyroid and vitamin D-related disorders. In Kaplan, SA, editor: Clinical pediatric and adolescent endocrinology, Philadelphia, 1982, WB Saunders Co.
54. Taylor, AB, Stern, PH, and Bell, NH: Abnormal regulation of cir-

culating 25-hydroxyvitamin D in the Williams syndrome, N Engl J Med 306:972-975, 1982.
55. Juan, D: Hypocalcemia differential diagnosis and mechanisms, Arch Intern Med 139:1166-1171, 1979.
56. Broadus, AE: Nephrogenous cyclic AMP, Recent Prog Horm Res 37:665, 1981.
57. Tsang, RC: Neonatal magnesium disturbances, Am J Dis Child 124:282, 1972.
58. Alkawa, JK: Magnesium: its biologic significance, CRC series on cations of biological significance, Boca Raton, Fla, 1981, CRC Press, Inc.
59. Grace, K, Spiler, IJ, and Tashjian, AH, Jr: Natural history of familial medullary thyroid carcinoma: effect of a program for early diagnosis, N Engl J Med 229:980-985, 1978.
60. Silva, OL, Brode, LE, and Doppman, JL: Calcitonin as a marker for bronchogenic cancer, Cancer 44:680-684, 1979.
61. McTaggert, H, Ivey, JL, Sisom, K, et al: Deficient calcitonin response to calcium stimulation in post-menopausal osteoporosis, Lancet 1:475-477, 1982.
62. Kaplan, M: Alkaline phosphatase, N Engl J Med 286:200-201, 1972.
63. Root, AW, and Harrison, HE: Recent advances in calcium metabolism, J Pediatr 88:1-18, 1976.
64. Niejadlik, DC: Hydroxyproline, Postgrad Med 51:214-216, 1972.

CHAPTER 25

Pancreatic function

MICHAEL D.D. McNEELY

OBJECTIVES

- Describe the anatomy of the pancreas and its endocrine and exocrine functions.
- Define pancreatitis and distinguish between acute and chronic forms of the disease.
- List pancreatic function tests that may be helpful in diagnosing pancreatic exocrine dysfunction.
- Discuss characteristic elevations of amylase and lipase activity in the diagnosis of pancreatic disease, and list other conditions that may also cause elevation of these enzymes.
- Describe the physiological changes seen in pancreatic exocrine and endocrine function for the following diseases:
  - Adenocarcinoma of the pancreas
  - Endocrine tumors of the pancreas
  - Cystic fibrosis
  - Pancreatic insufficiency

KEY TERMS

acinar From the Latin word *acinus,* which means a berry or grape. In anatomy, the term refers to a small saclike dilation.

ampulla of Vater A muscular sphincter in the duodenum through which the pancreatic and biliary ducts enter the gastrointestinal lumen.

cholangiopancreatography After careful manipulation of a special tube through the duodenum, x-ray contrast material is injected into the bile and pancreatic ducts so that an x-ray visualization can be obtained.

cholecystography A radiological technique in which a patient is given contrast material that will collect in the gallbladder and bile duct. An x-ray examination is then used to observe the anatomy of the biliary tract.

dextran A linear glucose polymer of variable molecular weight in which carbon-1 of each unit is linked to carbon-6 of the next unit.

glucogenesis The biochemical process of glucose synthesis.

glycogenolysis The biochemical process of glycogen breakdown into glucose.

hypoglycemia Low blood glucose concentration (often considered less than 500 mg/L).

islets of Langerhans Clusters of cells within the pancreas that provide its endocrine function.

laparotomy Surgically opening the abdominal cavity.

oncofetal antigen A protein found in fetal and neoplastic tissue.

pancreatic duct A conduit that traverses the pancreas to convey exocrine secretions to the duodenum.

peritonitis Inflammation (usually caused by bacterial infection) of the lining of the abdominal cavity.

proteolytic Having the ability (usually of an enzyme) to break protein molecules into peptides.

sialitis Inflammation of salivary glands or ducts.

vagus nerve Tenth cranial nerve, which carries motor, sensory, and autonomic nerve fibers to neck, thorax, and abdomen.

ANATOMY

The pancreas is a soft, variegated, linear strip of tissue approximately 15 cm long. It lies across the posterior wall of the abdomen and is surrounded on three sides by the duodenum. Its head lies to the right, in association with the midportion of the duodenum, and its body and tail run toward the left to lie under the spleen (Fig. 25-1). It receives a vigorous vascular supply from the aorta. The pancreatic duct runs throughout its length and extends

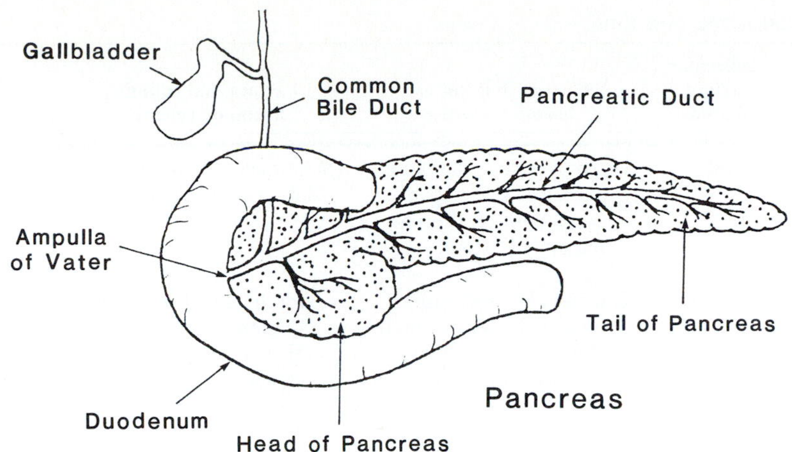

Fig. 25-1 Diagram of pancreas beside duodenum. Pancreatic duct extends throughout organ to convey exocrine enzymes into duodenum. Common bile duct enters duodenum beside or close to pancreatic duct. Endocrine islets are scattered throughout entire organ.

branches throughout the organ. The duct ultimately empties into the duodenum through a muscular sphincter called the *ampulla of Vater*. In some persons this sphincter is shared with the bile duct; in others it is separate but in close association with the bile duct.

The pancreas has two functions. It is an *endocrine* gland that synthesizes the hormones glucagon, insulin, and gastrin. At the same time it serves as an *exocrine* gland by providing digestive enzymes in a bicarbonate fluid to facilitate the duodenal digestion of food.

The endocrine functions are invested in a collection of cells grouped in a characteristic configuration known as the *islets of Langerhans* (Fig. 25-2). These islets contain beta cells, which synthesize insulin; alpha cells, which synthesize glucagon; and delta cells, which produce gastrin.

The exocrine substances are produced by cells in acinar groupings. The secretory regions of these cells are arranged so that the digestive chemicals are emptied into the pancreatic ductal system (Fig. 25-3). The duct epithelium itself contributes to the fluid by secreting both water and electrolytes.

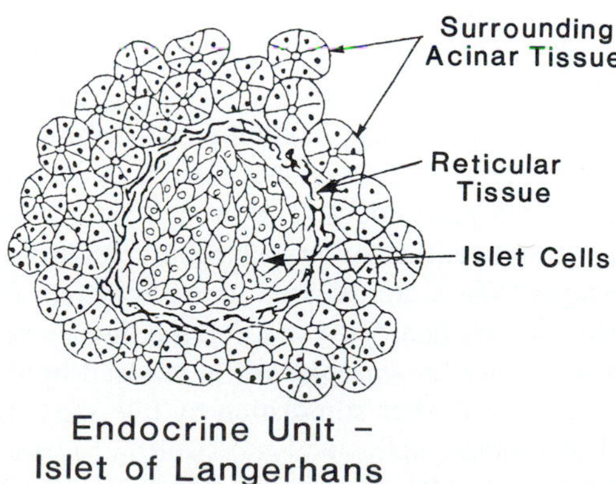

Fig. 25-2 Islet cells are collected together in a cluster separated from acinar tissue by a thin layer of reticular tissue. On routine strains, islet cells appear to be similar but can be distinguished by special staining techniques. Islet cells release their hormones directly into circulation.

ENDOCRINE FUNCTIONS

The endocrine functions of the pancreas are invested in the islets of Langerhans, which are composed of alpha cells, beta cells, and delta cells.[1] The pancreatic endocrine hormones are listed in Table 25-1.

Alpha cells

Alpha cells make up 20% to 30% of all the cells in the islet. Similar cells have been identified in the duodenal mucosa.

The alpha cells are known to produce the hormone glucagon. Glucagon is made up of 29 amino acids in straight chain with a molecular weight of 3485 daltons. A glucagon molecule of gastrointestinal tract mucosal origin is also known to exist. This is immunoreactive material that

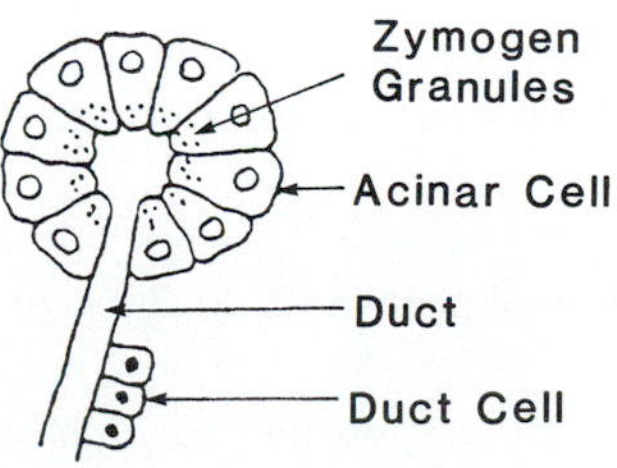

Fig. 25-3 Exocrine acinus terminates with a collection of acinar cells. They contain zymogen granules, which are loaded with proteolytic enzymes. Along wall of duct are located special cells that contribute fluid and bicarbonate.

Table 25-1 Pancreatic endocrine hormones

Cell of origin	Hormone	Molecular weight (daltons)	Factors that initiate hormone release	Factors that inhibit hormone release	Hormone action
Alpha cell	Glucagon	3485	Hypoglycemia, starvation, exercise, stress, pancreozymin, gastroinhibitory polypeptide	Most factors that stimulate insulin release	Increases plasma glucose by stimulating hepatic glycogenolysis and gluconeogenesis and adipose tissue lipolysis
Beta cell	Proinsulin	11,500	Glucose, leucine, arginine, histidine, phenylalanine, sulfonylureas, adrenocorticotropic hormone, glucagon, growth hormone, α-adrenergic blockage, β-adrenergic stimulation, vagal stimulation, secretin, cholecystokinin	α-Adrenergic stimulation, β-adrenergic blockade, thiazide diuretics, phenytoin, diazoxide	Insulin precursor
	Insulin	5734			Stimulates membrane transport, alters membrane-bound enzymes, stimulates protein synthesis, inhibits protein degradation, stimulates messenger RNA synthesis, stimulates DNA synthesis
Delta cell	Gastrin	2098	See Chapters 26 and 42		Role in pancreas not understood
	Somatostatin	1640			Not fully understood
F cell	Pancreatic polypeptide	4226	See Chapters 26 and 44		Not fully understood

has been called *gut glucagon, enteroglucagon,* or *glucagon-like immunoreactive material.*

Insulin-induced hypoglycemia is followed by increases in the concentration of glucagon in the pancreas and peripheral blood. Similarly, starvation, the ingestion of protein, the administration of certain amino acids (arginine or alanine), exercise, and stress also stimulate the secretion of pancreatic glucagon. Pancreozymin and gastroinhibitory polypeptide will increase glucagon. Hyperglycemia will cause suppression of glucagon secretion.

Glucagon stimulates hepatic glycogenolysis, which converts glycogen into glucose. Glucagon also stimulates glucogenesis by promoting the hepatic uptake of amino acids. These mechanisms increase the plasma glucose concentration. Glucagon is also capable of stimulating the release of insulin from the pancreatic beta cells independently of an increase in blood glucose. It is presumed that the physiological role of glucagon is to smooth the metabolic effects of glucose absorption and insulin release.

Beta cells

The beta cells make up 60% to 70% of the pancreatic islet cells.

The beta cells are responsible for synthesizing insulin. First a large, pre-proinsulin molecule is synthesized and stored in beta granules. This large molecule is then enzymatically broken down to liberate active insulin, C-29 peptide, and a variety of intermediate products.

The function of insulin is covered in greater detail in Chapter 29.

Delta cells

The delta cells are minor components (2% to 8%) of the islets. They are also present in the stomach and scattered within the pancreatic ductal epithelium. Delta cells are capable of producing gastrin.

EXOCRINE FUNCTIONS

The exocrine function of the pancreas is to synthesize and convey potent digestive enzymes into the duodenum.[2] The action of these enzymes is discussed in detail in Chapter 26.

The hormone cholecystokinin-pancreozymin is the main stimulus for the release of digestive enzymes from the pancreatic acinar cells.

A number of proteolytic enzymes are formed (Fig. 25-4). These are synthesized as zymogen granules in the acinar cells. When cholecystokinin-pancreozymin or vagal stimulation is received, the granule is released into the lumen of the pancreatic duct. The proteolytic enzyme precursors trypsinogen, chymotrypsinogen, proelastase, and procarboxypeptidase are released in this way into the ductal lumen and are then conveyed into the duodenum. Here an enzyme known as *enterokinase* is released by the duodenal mucosal cell after stimulation by bile salts and proteases. Enterokinase splits the lysine-isoleucine bond in the trypsinogen molecule. The resulting amino acid sequence refolds itself to form an active site. This newly created molecule is known as *trypsin.* Trypsin has the ability to cause catalysis of the trypsinogen. This reaction is very slow and is probably of little physiological consequence.

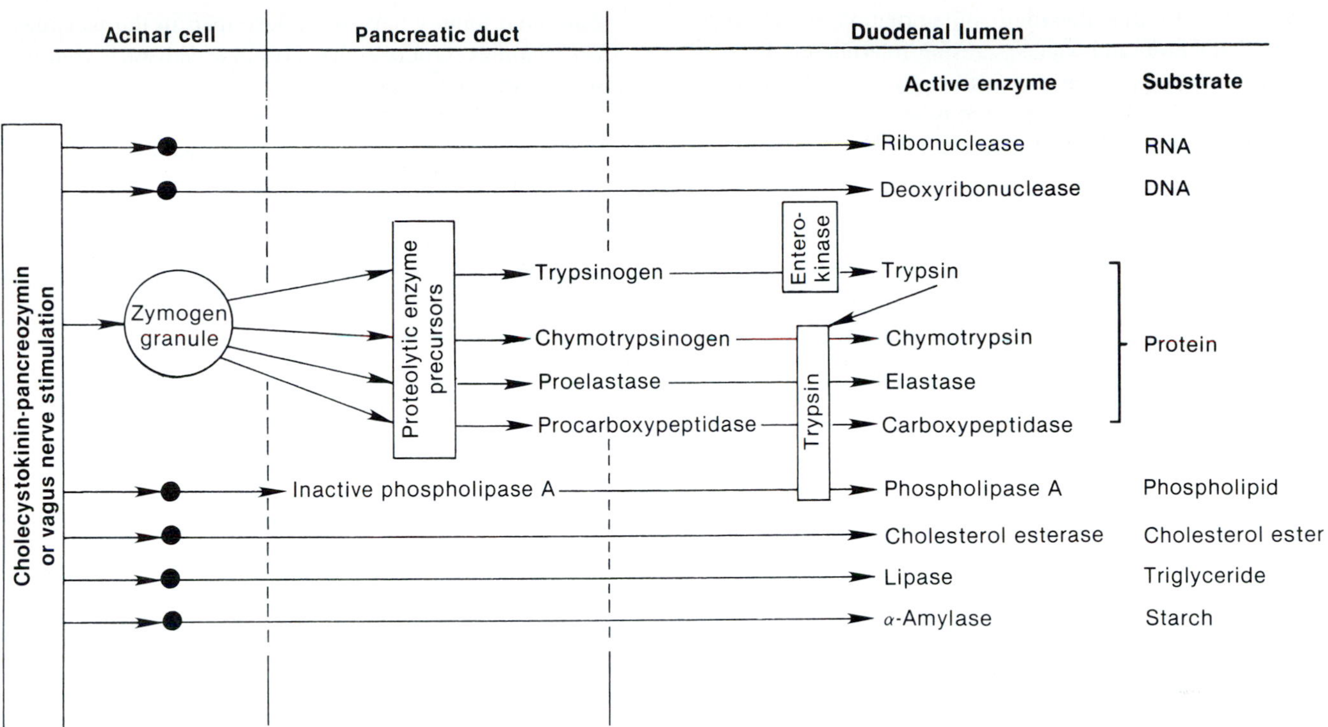

Fig. 25-4 Pancreatic exocrine enzymes.

Trypsin does act on the other zymogens to create the active proteolytic enzymes chymotrypsin, elastase, carboxypeptidase A, and carboxypeptidase B. Each of the proteolytic enzymes has a different function in protein breakdown.

Also secreted into the pancreatic duct are ribonuclease and deoxyribonuclease. These are secreted in the active form.

Alpha amylase is synthesized and released in a similar fashion. It will split starch into limit dextrans and maltose.

Lipids are acted on by the enzymes lipase, phospholipase A, and cholesterol esterase. Phospholipase A requires trypsin activation.

The hormone secretin is elaborated by the S cells of the duodenum and jejunum. It stimulates the pancreatic duct to produce water and bicarbonate. These aid in washing the enzymes into the duodenum and provide the appropriate pH environment for enzyme activity.

PATHOLOGICAL CONDITIONS

Pancreatitis[3-5]

Pancreatitis (inflammation of the pancreas) occurs in two clinical forms: acute and chronic. Two types of acute pancreatitis are recognized.

Acute pancreatitis. The acute edematous variety of pancreatitis occurs in 80% of cases. The remaining 20% of cases are hemorrhagic. Hemorrhagic pancreatitis is much more serious, resulting in a high mortality. It is characterized by parenchymal necrosis and hemorrhage. However, it is usually impossible to distinguish between the two forms when they first occur with severe abdominal pain.

Acute hemorrhagic pancreatitis can cause cardiovascular shock and hemorrhage into the gland. Up to 10 L of fluid may be lost into extracellular spaces in 24 hours. Pancreatic enzymes become activated and cause local damage and microcirculatory effects. Serum calcium may fall precipitously.

Pancreatitis is caused by infections, obstruction to the duct, vascular disorders, toxins, trauma, and metabolic derangements. The common factor is possibly an increase in membrane permeability. Autodigestion of pancreatic cells by proteolytic enzymes is the key event, and it is believed that trypsin activation takes place first. This causes activation of the kinin system, the coagulation cascade, and platelet aggregation. Widespread effects on blood vessels (including those of the lung) contribute to the serious physiological disruption that occurs during pancreatitis.

Current medical management includes replacement of the fluid deficit, relief of pain, and an attempt to suppress pancreatic stimulation by use of nasogastric decompression. Pharmacological intervention is being investigated, but its efficacy is not yet proven. Attention is given to the respiratory system because respiratory failure is a common problem. It is very important to confirm the diagnosis of acute pancreatitis so that treatment can be started early and inappropriate treatment can be avoided.

Chronic pancreatitis. A chronic form of pancreatitis

also occurs. In this disorder, inflammation of the pancreatic tissue ebbs and wanes, causing intermittent discomfort and chronic pancreatic insufficiency. Usually the clinical picture is not dramatic enough to make diagnosis easy. Diagnosis is ultimately confirmed by a variety of visualization techniques, including the barium meal, endoscopic retrograde cholangiopancreatography, two-scale ultrasonography, and angiography.

Cancer of pancreas

Adenocarcinoma. The incidence of adenocarcinoma[6,7] of the pancreas has increased during the past 20 years and now accounts for 5% of all cancer deaths in the United States. The overall 5-year survival rate is less than 1%. The principal reason for this dismal prognosis is the inability to diagnose the disease at an early stage. Approximately 90% of all pancreatic cancers have undergone extension or have metastasized by the time of diagnosis. Most pancreatic cancers are localized in the head of the gland. When the cancer is enlarged in this site, it can obstruct the bile duct and cause obstructive jaundice. Cancers in the body and tail of the pancreas are extremely difficult to diagnose. The most common symptoms are pain, jaundice, and weight loss, which develop insidiously and progressively. The results of laboratory tests are often normal and are never specific. Serum amylase may be elevated, or the patient may be mildly anemic. Pancreatic function tests may reveal abnormalities but are of no diagnostic value. Although a variety of oncofetal (tumor) antigens have been examined with the hope that they may shed some light on this disease, no significant progress has yet been made. Currently the diagnosis is made using direct visualization techniques, radiological studies, and laparotomy.

Endocrine tumors. A group of unusual hormone-synthesizing tumors[8] of the pancreas are recognized. They are derived from cells with the common characteristics of amine-precursor uptake and decarboxylation (APUD).

Insulinomas are the most common of these tumors. They produce a characteristic fasting hypoglycemic syndrome. One can identify them by having the patient fast overnight (some patients may require a fast of up to 72 hours) and discovering hypoglycemia (glucose less than 400 mg/L) and inappropriate plasma insulin concentrations (insulin given as a ratio of microunits per milliliter divided by glucose in milligrams per liter is greater than 0.03).

The glucagonoma is a glucagon-secreting tumor that produces a syndrome of diabetes, characteristic dermatitis, and anemia when it metastasizes. The diagnosis is made when elevated plasma glucagon and plasma insulin are found.

Somatostatinomas do not produce a characteristic syndrome.

Pancreatic polypeptide islet cell tumors have been found in patients with multiple endocrine neoplasia (MEN) I syndrome.

Carcinoid tumors have been described in the pancreas.

Gastrinomas produce the Zollinger-Ellison syndrome, described in Chapter 26.

Vipomas (Verner-Morrison syndrome) cause a voluminous (5 L/day), explosive, watery diarrhea associated with hypokalemia and hypochlorhydria. The syndrome is caused by excess vasoactive intestinal polypeptide.

Pancreatic tumors have been known to produce ectopic adrenocorticotropic hormone (ACTH) and parathyroid hormone.

Pancreatic insufficiency

Pancreatic insufficiency[9] can be caused by repeated attacks of acute pancreatitis or long-standing chronic pancreatitis. In such cases the pancreatic parenchyma is gradually destroyed. When 75% of the gland has been disabled, detectable changes will occur in the digestive process. Because of this tremendous reserve, pancreatic insufficiency becomes apparent only very late in the disease. Diabetes mellitus can also occur because the islets become damaged. Evaluation of pancreatic function is made by stimulation of the pancreas and measurement of the maximum output of pancreatic enzymes. Therapy is directed toward correcting nutritional deficiencies, avoiding alcohol, correcting the diabetes mellitus, and correcting the enzyme defects with oral pancreatic enzymes. The results of pancreatic insufficiency are discussed in Chapter 26.

Cystic fibrosis

Cystic fibrosis[10] is an inherited defect in exocrine gland function in which secretions from the liver, gallbladder, duodenum, pancreas, lungs, salivary glands, and genital tract are abnormal. It is characterized clinically by chronic lung disease and general malnutrition secondary to malabsorption. It may present in the neonatal period as intestinal obstruction with undigested meconium. With modern treatment, the life span of persons with this disease has increased far beyond the 60% mortality before 10 years of age that was usual a decade and a half ago.

A diagnosis of cystic fibrosis requires that at least two of the following four criteria are met: (1) characteristic lung disease, (2) pancreatic insufficiency, (3) increased concentration of sweat chloride, and (4) a positive family history.

The demonstration of an increased sweat chloride is the most characteristic diagnostic test. Sweat-chloride values must exceed 60 mmol/L. Normal values vary with age, and abnormally high values may occasionally be found in other conditions, including glucose-6-phosphatase deficiency, glycogen-storage disease, hypothyroidism, renal insufficiency, and malnutrition. The only acceptable test for measuring sweat chloride is pilocarpine iontophoresis sweat stimulation followed by gravimetric quantitation of the amount of sweat produced and coulometric measurement of the chloride.

Diabetes

Diabetes is a generalized disorder of glucose metabolism that is generally attributed to an insufficiency of insulin. Diabetes is covered in greater detail in Chapter 29.

Hypoglycemia

Pancreatic tumors that produce insulin are one cause of hypoglycemia. This diagnosis is often very difficult to make. It is covered in greater detail in Chapter 29.

PANCREATIC FUNCTION TESTS

Pancreatic assessment[11] has been greatly enhanced by the development of intubation techniques that allow a variety of tubes to be introduced into very specific regions of the gastrointestinal tract.

Secretin-stimulation test

In the secretin-stimulation test[12,13] a double-lumen tube is positioned with fluoroscopic guidance so that one tip is situated in the duodenum and the other is left in the stomach to remove gastric juice. Two basal collections of endogenous duodenal contents are obtained 10 minutes apart. Secretin is then administered intravenously at 1 U/kg of body weight. Secretin administered in this way will stimulate the pancreas to produce water and bicarbonate. For assessment of this response, three 10-minute samples are removed from the duodenum, labeled appropriately, and put on ice.

The volume of all three samples is measured. A normal postsecretin response will be a significant increase in volume (1.5 mL/kg/30 min). Bicarbonate is also measured on all samples; a peak bicarbonate concentration will be at least 90 mmol/L. Low fluid volumes are characteristic of pancreatic duct obstruction. Low bicarbonate concentration suggests chronic pancreatitis. Tumors of the pancreas may also cause decreased flow. Normally the amylase activity will increase, but enzyme behavior is not diagnostically useful.

Pancreozymin-stimulation test[12,13]

Pancreozymin is administered intravenously immediately after the collection of the samples for the secretin-stimulation test. The enzymatic content of the stimulated pancreatic juice is used to assess the function of the pancreas.

Cholecystokinin-stimulation test

Cholecystokinin is occasionally administered along with secretin as a pancreatic function test.

Lundh test

In the Lundh test,[14] trypsin determinations are carried out on duodenal aspirations after a standard meal.

The patient fasts for 12 hours before the test, and medications that affect gastrointestinal motility are withheld. A tube is inserted under fluoroscopic guidance into the third portion of the duodenum. A test meal consisting of 8 g of corn oil, 15 g of casein, and 40 g of glucose in warm water with flavoring is given. The total volume is about 300 mL. The timed meal is taken through a straw over 15 minutes. The patient is then placed on the right side, and the tube is allowed to drain into an iced container. The drainage continues for 2 hours. The volume of material collected is measured, and an aliquot is titrated to determine the hydrogen-ion content. The secretion of hydrogen ion by normal persons is 12 to 20 mmol/min/mL. Those with pancreatic insufficiency secrete less than 10 mmol/min/mL.

An intact duodenal mucosa must be present to mediate the hormone. A vagotomy or duodenal bypass surgery will also diminish the response. The test will be abnormal in 90% of patients with chronic pancreatitis and 79% of those with carcinoma of the pancreas.

CHANGE OF ANALYTE IN DISEASE

(Table 25-2)

Amylase

Amylase[15,16] is the enzyme secreted by the exocrine area of the pancreas for the purpose of breaking down starch into limit dextrans and maltose. Eighty percent of all patients with acute pancreatitis experience an elevation of serum amylase within 24 hours. Amylase is elevated to between 2 and 20 times the upper limit of normal. If the attack is self-limiting, the value will return to baseline within 2 to 7 days. As a result, serum amylase determinations provide a convenient approach to the diagnosis of acute pancreatitis. Hypoglycemia can depress amylase values, and therefore the blood must be drawn before intravenous fluid therapy is initiated. Rarely, acute hemorrhagic pancreatitis will be so severe that all pancreatic tissue is destroyed and the serum amylase level is normal. Seventy percent of all persons receiving morphine will have a raised serum amylase value that is up to three times normal because opiates cause constriction of the sphincter of Oddi and back pressure into the pancreatic duct. In addition to morphine, drugs to be avoided are choline, methacholine, chlorothiazide, pancreozymin, and secretin. Patients with chronic pancreatitis have normal levels of amylase between attacks, and even during attacks amylase elevations may not occur. Carcinoma of the pancreas is not generally associated with a raised amylase level unless the disease has caused some form of destructive lesion. Pancreatic pseudocysts occasionally cause prolonged elevations of serum amylase.

A number of other conditions also cause an increased serum amylase, including mumps, parotitis, sialitis, tuboovarian abscess, sympathetic pleural effusion, intestinal obstruction, abdominal trauma, intra-abdominal surgery, mesenteric thrombosis, peritonitis, cholecystography, common duct obstruction, perforated duodenal or gastric

Table 25-2 Test changes in disease

	Serum amylase	Urine amylase	Serum lipase
Acute pancreatitis	↑ to ↑ ↑ ↑ in 80%	↑ ↑ ↑	↑ to ↑ ↑ in 90%
Opiate administration	↑ in 70%	↑	↑
Cholinergic drugs	↑ many	↑	↑
Chronic pancreatitis	N or ↑	N or ↑	N or ↑
Carcinoma of pancreas	↑ rare	↑ rare	↑ 50%
Pseudocyst	Prolonged ↑	↑	↑
Macroamylasemia	Prolonged ↑	N	↑
Salivary gland disease	↑	↑	N
Tubo-ovarian abscess	↑ or N	↑ or N	N
Sympathetic pleural effusion	↑ or N	↑ or N	N
Intestinal obstruction	↑ or N	↑ or N	N
Abdominal trauma	↑ or N	↑ or N	↑ or N
Mesenteric thrombosis	↑ or N	↑ or N	↑ or N
Peritonitis	↑ or N	↑ or N	↑ or N
Cholecystography	↑ or N	↑ or N	↑ or N
Common duct obstruction	↑ or N	↑ or N	↑ or N
Perforated ulcer	↑ or N	↑ or N	↑ or N
Dissecting aortic aneurysm	↑ or N	↑ or N	↑ or N
Bronchogenic carcinoma	↑ or N	↑ or N	↑ or N
Splenic hemorrhage	↑ or N	↑ or N	↑ or N
Uremia	↑ usually	↓ or N	↑ or N

N, Normal; ↑, elevated; ↓, lowered.

ulcer, dissecting aortic aneurysm, bronchogenic carcinoma, cerebral trauma (rare), splenic hemorrhage, and uremia.

Amylase is a relatively small protein (MW 45,000 daltons) and is therefore filtered readily into the urine. The enzyme can be found in increased concentrations in the urine for longer periods of time than in the serum. An amylase content determined on a 2-hour urine collection is an excellent test for detecting pancreatitis. The test result will be increased by the same conditions that alter the serum amylase level. In addition, urine amylase has been artificially increased by saliva surreptitiously added to the collection vessel.[17]

An interesting condition that has been recognized is macroamylasemia.[18] In this condition, very high levels of serum amylase are found, but pancreatitis is not present. The elevated values persist indefinitely. The amount of amylase found in the urine is low or normal. Macroamylase is a giant molecule that results from the complexing of normal amylase to high-molecular-weight proteins (usually immunoglobulins).

Some recent enthusiasm for the application of the amylase-to-creatinine clearance ratio has been generated on the basis that this procedure gives a more specific indication of pancreatitis than serum or urinary measurements alone. The use of this ratio has not been widely accepted because experience has not proven it to be of any substantial benefit except in highlighting macroamylasemia.[19]

Techniques are available for separating amylase isoenzymes.[20] These methods demonstrate *P* type (from pancreas), *S* type (from salivary glands), and other unusual forms.

Lipase

Serum lipase[21] rises as fast as amylase and remains elevated for a much longer period of time (7 to 10 days). It is not elevated by as many conditions as serum amylase and therefore should be considered much more specific. Unfortunately, analytical difficulties have prevented the widespread use of serum lipase in the routine diagnosis of pancreatitis. The test is elevated in 50% of patients with carcinoma of the pancreas and is occasionally found to be elevated in people with cirrhosis of the liver without pancreatitis. It is elevated in the presence of opiates and cholinergics because they affect the sphincter of Oddi. Serum lipase can be detected in 90% of patients with acute pancreatitis when appropriate analytical techniques are used. Urinary lipase cannot be detected.

Methemalbumin

Detection of methemalbumin[22] in the serum and ascitic fluid of patients with suspected acute pancreatitis is strong evidence that the hemorrhagic variety is present. Methemalbumin is formed by action of pancreatic digestive enzymes on hemoglobin. Other disorders that cause blood loss into the abdominal cavity also result in methemalbumin.

Nonpancreatic enzymes in pancreatic disease

A variety of nonpancreatic serum enzyme changes occur in pancreatic disease. None is specific.

If a pancreatic neoplasm causes obstruction to biliary outflow, typical hepatic enzyme changes will be noted. Early enthusiasm for the use of the serum leucine aminopeptidase determination in the diagnosis of pancreatic tumor has waned because this method has not stood the test of time.

Lactate dehydrogenase increases caused by LD_2 and LD_3 are sometimes caused by malignancy of the pancreas.

In pancreatitis, 5′-nucleotidase, alanine aminotransferase, aspartate aminotransferase, and gamma glutamyl transferase are frequently elevated.

Lactoferrin

It has been reported that the measurement of lactoferrin and trypsin in pancreatic juice can differentiate between patients with pancreatic cancer and chronic pancreatitis.[23] The intubation must be made directly from the pancreatic duct.

Insulin

The primary reason for measuring insulin in serum is the detection of an insulinoma. This tumor will be characterized by an inappropriately increased serum insulin in association with fasting hypoglycemia. Inappropriate means that the insulin (microunits per milliliter) divided by the plasma glucose (milligrams per liter) is greater than 0.03.

Serum insulin also increases in generalized acute illnesses, hyperthyroidism, some diabetic states, acromegaly, adrenal hyperfunction, type IV hyperlipoproteinemia, cirrhosis of the liver, and renal failure. Low levels are found in diabetes mellitus, pheochromocytoma, malnutrition, and cystic fibrosis.

A common dilemma is whether hyperinsulinemia is the result of a tumor or the result of self-administration. Resolution is achieved by measurement of the C-peptide portion of the proinsulin molecule, which is present only with endogenous insulin.

Insulin antibodies

Insulin antibodies are present in most persons who have received pharmaceutical insulin. Some patients have exceedingly high values, which may greatly alter their insulin requirements. It is important to measure both total and unbound antibodies.

REFERENCES

1. Unger, RH: Alpha- and beta-cell interrelationship in health and disease, Metabolism 23:581-593, 1974.
2. Hadorn, B: The exocrine pancreas. In Anderson, CM, and Burke, V, editors: Pediatric gastroenterology, Oxford, England, 1975, Blackwell Scientific Publications, Ltd.
3. Keith, RG, Poncelet P, Mullens, JE, et al: Symposium on pancreatitis, Can J Surg 21:56-74, 1978.
4. Webster, PD, and Spainhour, JB: Pathophysiology and management of acute pancreatitis, Hosp Pract 9:59-66, 1974.
5. Geokas, MC, Van Lancker, JL, Kadell, BM, and Machleder, HI: Acute pancreatitis, Ann Intern Med 76:105-117, 1972.
6. Hermann, RE, and Cooperman, AM: Cancer of the pancreas, N Engl J Med 301:482-485, 1979.
7. Dimagno, EP: Pancreatic cancer: a continuing diagnostic dilemma, Ann Intern Med 90:847-848, 1979.
8. Friesen, SR: Tumors of the endocrine pancreas, N Engl J Med 306:580-590, 1982.
9. Thomson, ABR, Dwyer, J, Gee, M, and Holt, T: Management of pancreatic insufficiency in the adult, Mod Med Canada 34:692-695, 1979.
10. Di Sant'Agnese, PA, and Farrell, PM: Neonatal and general aspects of cystic fibrosis. In Young, DS, and Hicks, JM, editors: The neonate, New York, 1976, John Wiley & Sons, Inc.
11. Malfertheiner, P, and Ditschuneit, H, editors: Diagnostic procedures in pancreatic disease, New York, 1986, Springer-Verlag.
12. Rick, W: Der Secretin-pankreozymin-test in der Diagnostik der Pankreasinsuffizienz, Internist 11:110-117, 1970.
13. Wormsley, KG: Further studies of the response to secretin and pancreozymin in man, Scand J Gastroenterol 6:343-350, 1971.
14. Lundh, G: Pancreatic exocrine function in neoplastic and inflammatory disease: a test of pancreatic function, Gastroenterology 42:275-280, 1962.
15. Webster, PD, and Zieve, L: Alterations in serum content of pancreatic enzymes, N Engl J Med 267:604-607, 654-658, 1962.
16. Salt, WB, II, and Schenker, S: Amylase—its clinical significance: a review of the literature, Medicine (Baltimore) 55:269-289, 1976.
17. Robison, JC, Gitlin, N, Morrelli, HF, and Mann, LJ: P factitious hyperamylasuria: a trap in the diagnosis of pancreatitis, N Engl J Med 306:1211-1212, 1982.
18. Berk, JE, and Fridhandler, L: Advances in the interpretation of hyperamylasemia. In Glass, GBJ, editor: Progress in gastroenterology, New York, 1977, Grune & Stratton, Inc.
19. Warshaw, AL, and Fuller, AF: Specificity of increased renal clearance of amylase in diagnosis of acute pancreatitis, N Engl J Med 292:325-328, 1975.
20. Berk, JE: Amylase in diagnosis of pancreatic disease, Ann Intern Med 88:838-839, 1978.
21. Ticktin, HE, Trujillo, NP, and Evans, PE: Diagnostic value of a new serum lipase method, Gastroenterology 48:12-17, 1965.
22. Geokas, MC, Rinderknecht, H, Walberg, CB, and Weissman, R: Methemalbumin in the diagnosis of acute hemorrhagic pancreatitis, Ann Intern Med 81:483-486, 1974.
23. Fedail, SS, Harvey, RF, and Salmon, PR: Trypsin and lactoferrin levels in pure pancreatic juice in patients with pancreatic disease, Gut 20:983-986, 1979.

CHAPTER 26

Gastrointestinal function and digestive disease

MICHAEL D.D. McNEELY

OBJECTIVES

- Describe the anatomy of the normal digestive tract.
- Outline the digestive and absorptive functions of the normal digestive tract.
- List major pathological conditions of the gastrointestinal tract and the causes of these conditions.
- Describe gastrointestinal function tests and expected test results in disease states.
- State expected results of the following laboratory tests in gastrointestinal disease states: gastrin, carotene, vitamin A, trypsin, fecal fat, fecal occult blood, CEA, 5-HIAA, and urinary oxalate.

KEY TERMS

achlorhydria Literally "without hydrochloric acid." Refers to lack of acid production by the stomach.

anticholinergic A drug that opposes the action of the cholinergic nervous system.

chyme The semisolid end product of gastric action on food. Chyme consists of mucus, gastric secretions, and broken-down food.

gastrointestinal hormones Substances that are produced by gastrointestinal cells and then travel through the bloodstream to act on a separate site. These hormones include cholecystokinin, secretin, glucagon, gastric inhibitory polypeptide, vasoactive intestinal polypeptide, bombesin, somatostatin, motilin, chymodenin, bulbogastrone, entero-oxyntin, and pancreatic polypeptide.

gluten A protein found in wheat and wheat products.

intubation The procedure of introducing a tube-shaped instrument into the body, usually through an anatomical opening, such as the mouth.

pancreatic exocrine enzymes Enzymes required for digestion. Often released in a precursor form. These enzymes include trypsinogen, chymotrypsinogen, proelastase, procarboxypeptidase, ribonuclease, deoxyribonuclease, amylase, lipase, phospholipase A, and cholesterol esterase.

pancreatic hormones Endocrine hormones mainly concerned with carbohydrate intermediary metabolism and including glucagon, insulin, and gastrin.

pyrexia Fever. A body temperature above 37.5° C.

The gastrointestinal tract is a muscular tube lined with epithelial cells and extending 10 m from the mouth to the anus. Along its course its structure is modified to suit particular requirements for the digestion and absorption of food.

The old view of the gastrointestinal tract described it as an inert conduit, across which digested food molecules were allowed to pass into the bloodstream. Today physicians realize that the absorptive surface of the intestine is an extremely elaborate organ covered with minute microvilli that are invested with complex enzyme systems. This intricate microstructure creates an extremely efficient and highly selective absorptive mechanism.[1] In addition, the gastrointestinal tract is controlled by an elaborate hormonal and neural regulatory network.[2]

To complement this enhanced physiological knowledge, new diagnostic techniques, including imaging procedures, fiberoptic intubation, and chemical analyses, have introduced a new era of gastrointestinal diagnoses.

ANATOMY AND GENERAL FUNCTIONS

The gastrointestinal tract has five distinct regions: mouth, stomach, duodenum, jejunum-ileum, and large bowel (Fig. 26-1).

The mouth contains teeth, tongue, salivary glands, and an elaborate swallowing mechanism. It is responsible for tasting, grinding, and lubricating food. The swallowing mechanism propels the food down the esophagus, through the thoracic cavity, and into the stomach.

The stomach is a rough-surfaced, muscular bag coated with a protective mucus layer. The vigorous churning action of the stomach is responsible for the mixing and breakdown of food. Hydrochloric acid and the enzyme pepsin are also secreted and continue the digestive breakdown. These actions convert food into chyme.

The chyme enters the duodenum, into which bile and pancreatic enzymes are secreted. Further enzymatic degradation of the basic food materials takes place in the duodenum and continues as the food material enters the small intestine.

The small intestine is 4 m long. Its absorptive capacity is enhanced by its microvillous substructure. The small intestine consists of two parts—the jejunum proximally and the ileum distally. Here the fragmented food materials are finally broken down and absorbed into the bloodstream.

When the nutrients have been absorbed, the residual matter enters the large intestine, where a process of selective water and electrolyte balance occurs. The digestive process terminates with the formation of feces.

The entire absorptive surface of the gastrointestinal tract is drained by the portal venous blood vessels. These convey the newly absorbed materials directly to the liver so that they may be immediately converted into usable forms.

DIGESTION

Digestion is the chemical process of rendering food into a form that can be absorbed by the body. The digestive process begins in the mouth and is generally completed in the proximal portion of the small intestine. The digestive process for various foods is summarized in Table 26-1.

Digestive action of mouth

Food is tasted in the mouth by the combined action of the taste receptors on the dorsal surface of the tongue and on the palate, pharynx, and tonsils. Four fundamental tastes are recognized: sweet, salt, sour, and bitter. The taste of food also depends on its smell. The integration of these various neural inputs produces the sensation known as *taste*. The taste of a food substance is important to protect us from disagreeable foods, to encourage eating, and to initiate a complex psychoneurogenic reflex that acts on the rest of the gastrointestinal tract.

Part of this reflex stimulates the production of saliva from the three pairs of salivary glands: parotid, mandibular, and sublingual. These glands produce viscid, water-

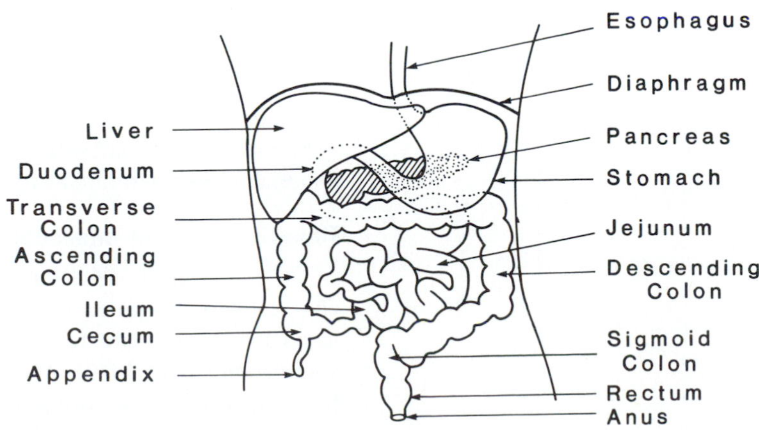

Fig. 26-1 Diagram of gastrointestinal tract.

Table 26-1 Digestion

Food material	Digestive action	End product
Starch	Pancreatic amylase	Disaccharides (mainly maltose)
Disaccharides	Mucosal disaccharidases	Monosaccharides
Monosaccharides	None	
Protein	Gastric hydrochloric acid and pepsin	Partial degradation into large polypeptides
	Pancreatic trypsin, chymotrypsin, and carboxypeptidase	Polypeptides, dipeptides, and amino acids
Long-chain triglycerides	Emulsification with bile, hydrolysis by lipase	Fatty acids and glycerol

based, mucin-containing secretions that act as a lubricant. They also release salivary amylase to initiate the digestion of starch.

Food is masticated by the complex interaction of the teeth, tongue, and mouth. The resulting bolus of food is then propelled to the stomach.

Digestive action of stomach

The stomach is a thin-walled, muscular sac that one can roughly divide into three zones (Fig. 26-2). The very top part of the stomach is known as the *fundus*. The main portion of the stomach is known as the *body*. The outlet of the stomach is known as the *antrum* and is segregated from the duodenum by the pyloric region, which contains a strong, muscular sphincter.

The gastric mucosa is covered with numerous coarse folds known as *rugae*. The rugae assist in mixing food substances during the churning action of the stomach.

The gastric mucosa contains four types of cells. Mucous cells are found throughout the entire stomach and secrete mucus to protect the surface from attack by acid and enzymes. Also found in all parts of the stomach are the surface epithelial cells, which are also capable of secreting mucus but which proliferate rapidly and shed readily, producing a continually viable surface for the stomach. Parietal cells produce hydrochloric acid and intrinsic factor. Chief cells produce the enzyme pepsinogen. These last two cell types are found throughout the body of the stomach.

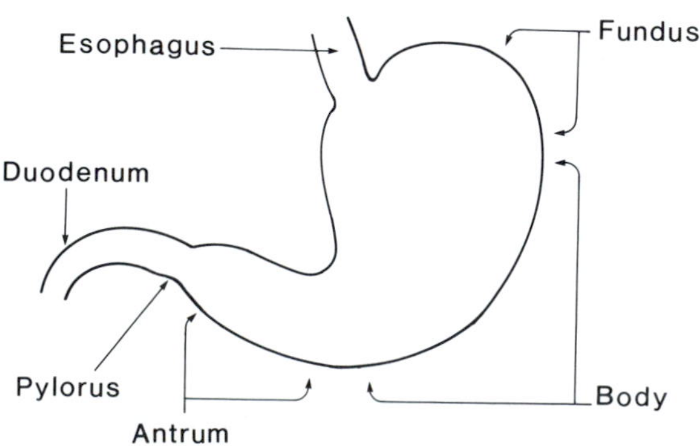

Fig. 26-2 Diagram of stomach.

The antral cells secrete mainly mucus but also some pepsinogen. The hormone gastrin is synthesized and stored in the G cells of the antrum.

There are three phases of gastric activity. The first of these is the cephalic phase, which is initiated by the sight and smell of food. This sensation triggers a direct vagus nerve message from the brain to the stomach to initiate the digestive process.

Next is the gastric phase of digestion, in which a variety of mechanisms interplay to create the digestive milieu (Fig. 26-3):

1. The vagus nerve directly stimulates the parietal cells to release hydrochloric acid.
2. The antral cells are also stimulated by the vagus nerve to secrete gastrin. Gastrin in turn stimulates the parietal cells to produce more hydrochloric acid.
3. Local distension of the gastric antrum also stimulates the production of gastrin and thus the secretion of hydrochloric acid.
4. Cholinergic (vagus) reflexes are further enhanced by distension of the fundus of the stomach.
5. The chief cells contain receptors that respond to the acid environment by secreting the enzyme precursor pepsinogen. Pepsinogen is rapidly converted into its active form (pepsin) at pH 3. Lipase and other enzymes are also liberated, but these enzymes are of little consequence in the digestive process.

The combined action of antral contractions, pyloric sphincter activity, and chemical secretions render the food into a much degraded, mucus-containing solution known as *chyme*.

Stomach activity subsides with time, and the fragmented material is then permitted to pass into the duodenum.

Digestive action of duodenum

The next step in digestion occurs in the duodenum. As chyme enters this portion of the intestine, several gastrointestinal hormones are released by both neural and local stimulation (Table 26-2). These hormones enter the portal blood system and act primarily on various regions of the gastrointestinal tract. See Chapter 42 and reference 3 for a review of the actions of these hormones.

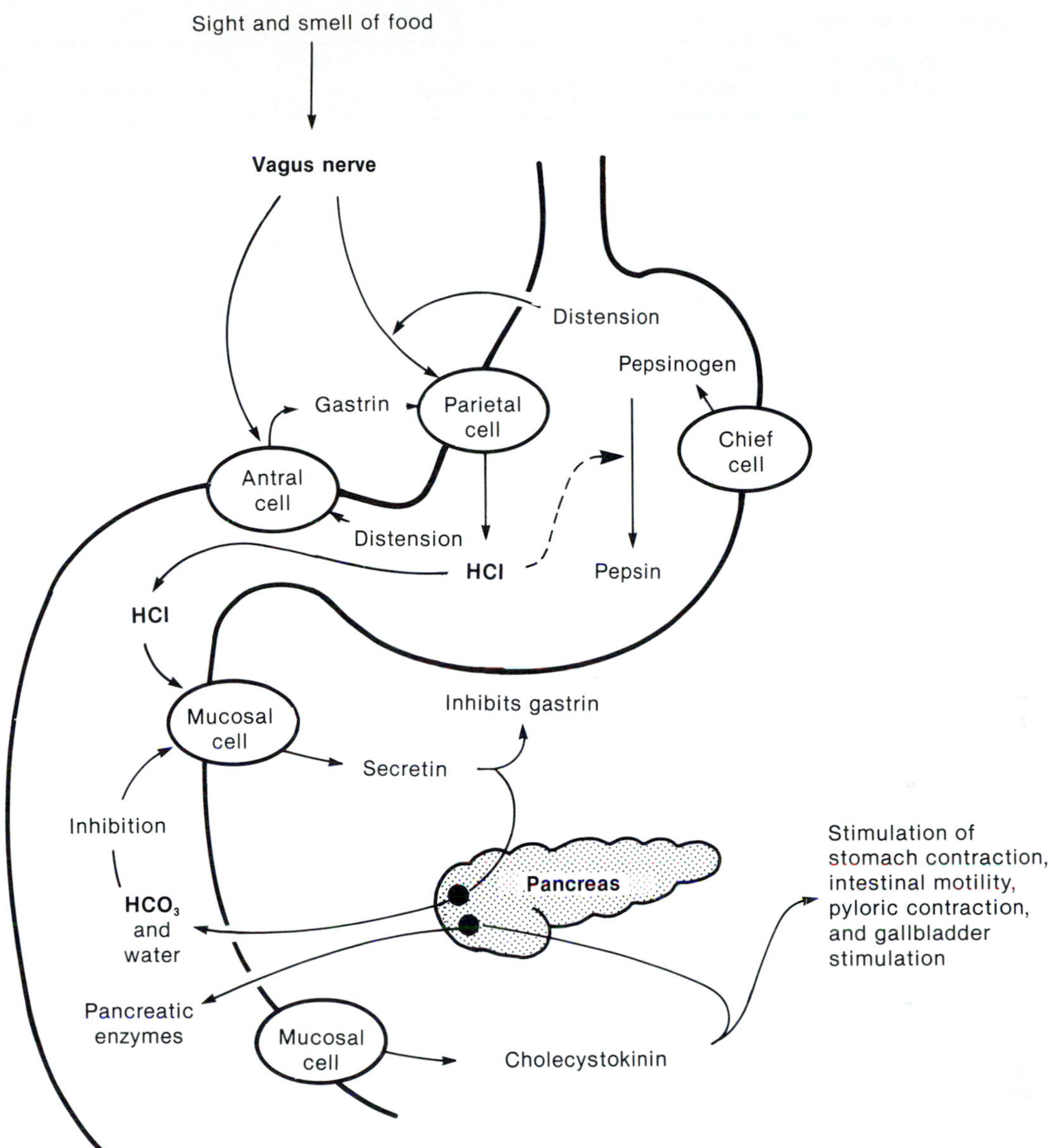

Fig. 26-3 Scheme demonstrating various stimuli of stomach and duodenum.

As a result of intricate hormonal feedback activity, bile salts, bicarbonate, the enzymes amylase and lipase, and a variety of protein-degrading enzymes are secreted into the duodenum. The action of these agents on the primary food substances is now considered. (Additional discussions can be found in Chapter 42.)

Carbohydrate digestion[4]

Carbohydrates are present in the diet as monosaccharides, disaccharides, or complex polysaccharides. Only the polysaccharides require extensive digestion in the duodenum.

Starch is the most common complex polysaccharide. It has a branching structure based on 1,4-carbohydrate or 1,6-carbohydrate linkages. Amylase is capable of hydrolyzing starch into oligosaccharides and ultimately into disaccharides. The dominant disaccharide produced from starch is maltose. Thus as the food leaves the duodenum, monosaccharides and disaccharides from the diet and disaccharides resulting from the action of amylase are passed to the jejunum and ileum, wherein absorption takes place.

Protein digestion[5]

Dietary protein is partially degraded in the stomach by hydrochloric acid and pepsin. In the duodenum, trypsin, chymotrypsin, and carboxypeptidase secreted by the pancreas act on the partially degraded protein to yield polypeptides, dipeptides, and amino acids. These tiny molecules then pass into the ileum and jejunum for assimilation.

Table 26-2 Intestinal hormones

Hormone	Number of amino acids	Source	Stimulating factor	Function
Cholecystokinin	33	Mucosa of upper small intestine	Amino acids, fatty acids, hydrochloric acid, and food in duodenum	Stimulates Pancreatic enzyme secretion Gallbladder contraction Contraction of stomach and pylorus Intestinal motility
Secretin	27	Throughout gut mucosa but concentrated in duodenum	Acid in duodenum	Stimulates pancreatic secretion of water and bicarbonate Stimulates gastric pepsin secretion Stimulates pyloric sphincter Inhibits gastrin secretion and stomach motility
Glucagon	29	Pancreatic and intestinal mucosa	Arginine, alanine, stress	Stimulates gluconeogenesis, raises blood glucose
Gastric inhibitory polypeptide (GIP)	43	Duodenal mucosa	Glucose and fat	Cholecystokinin-like activity
Vasoactive intestinal polypeptide (VIP)	28	Wide distribution throughout gut		Vasodilation and hypotensive effects Inhibits histamine, pentagastrin acid release, and pepsin secretion Stimulates electrolyte and water secretion from pancreas Stimulates bile flow
Bombesin	14			Stimulates pancreatic secretion and gastrin release

Fat digestion[6]

Fat digestion is more complex than digestion of other basic food substances. Most dietary fats are long-chain triglycerides (palmitic, stearic, oleic, and linoleic acids). The stomach decreases the particle size of the fatty substances by its churning action. In the duodenum, fats are emulsified by the detergent action of bile. Emulsification allows the pancreatic enzyme, lipase, to attack the otherwise water-insoluble lipids. Lipase causes stepwise hydrolysis, which first forms a diglyceride, then a monoglyceride, and finally a fatty acid and glycerol (see Chapter 30).

The bile salts, which are so important for fat digestion, are synthesized in the liver from cholesterol and are conjugated to taurine or glycine. The main bile salts are conjugates of cholic and chenodeoxycholic acid.

ABSORPTION

Absorption is the process whereby digested food substances enter the body.[7] Having traversed the duodenum, these digested food substances enter the jejunum and ileum, wherein the final absorption process takes place.

The intestinal mucosa is thrown into many folds, which assist in propelling its contents distally. The surface of each fold is drawn up into fingerlike projections known as villi (Fig. 26-4). Each villus increases the absorptive surface many times. Electron microscopic studies have shown that each villus is covered by hairlike projections known as microvilli. There are 200 million microvilli per centimeter of epithelium. Thus the intestine is given a massive absorptive surface area measuring 500 m^2.

It was once believed that enzymes were secreted by the small intestine to produce a digestive juice known as *succus entericus*. We now know that the main enzymatic action occurs in intimate association with the epithelial surface. Rather than being a purely passive sieve through which food substances are permitted to pass, the intestinal mucosa contains a highly selective mechanism for the absorption of each nutrient.

Although there are regional differences in the ability of the intestine to absorb different food substances, these details are not considered here.

Carbohydrate absorption

The digestive process degrades carbohydrate into monosaccharides and disaccharides. The monosaccharides—glucose, galactose, and fructose—are absorbed by specific active transport mechanisms. That is, they are conveyed from the intestinal lumen into the bloodstream against a concentration gradient. Energy is required for this to occur.

The disaccharides are split into monosaccharides by the enzymatic activity of disaccharidase enzymes located on the microvilli. For example, the milk sugar lactose is split by the enzyme lactase into its component sugars, glucose and galactose. These are then actively absorbed. The disaccharide sucrose is split by the enzyme sucrase into glucose and fructose. Maltose, which is the common product of starch hydrolysis, is split by a surface maltase into two molecules of glucose.

Protein absorption

The digested products of protein are small polypeptides, dipeptides, and amino acids. Dipeptides are absorbed more

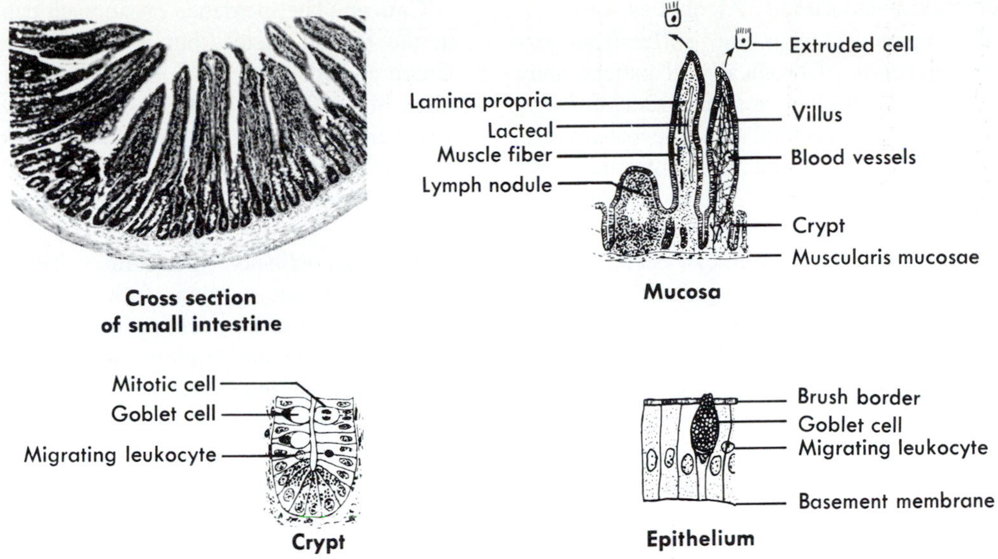

Fig 26-4 Structures of functional components of small intestine. *(From Arey, LB: Human histology: a textbook in outline form, ed 4, Philadelphia, 1974, WB Saunders Co.)*

rapidly than amino acids because of special transport mechanisms. Proteins are not absorbed directly. A very large number of specific absorptive mechanisms designed for various types of amino acids are located in the mucosal surface.

Fat absorption (see also Chapter 30)

The successfully digested fat enters the intestine as a micelle. By diffusion the fatty acids and monoglycerides enter the intestinal epithelial cells, where they then interact with a binding protein. Long-chain fatty acids of 16 to 18 carbons are re-esterified to triglycerides and then bound to apolipoproteins to form chylomicrons. These tiny lipid droplets are released into the lymphatic system and transported to the thoracic duct before entry into the bloodstream. Medium-chain fatty acids (8 to 10 carbons) are not re-esterified and rapidly enter the portal bloodstream bound to albumin.

Vitamins D, E, A, and K[8,9]

Vitamins D, E, A, and K (see also Chapter 36) are not water soluble and must therefore be absorbed with lipids. Thus they depend on normal lipid absorption. Vitamin D absorption is modified by calcium intake and metabolism.

Water and sodium absorption

The control over water absorption is not fully understood, but it is believed that bulk flow after sodium absorption is the mode for water transport in the intestine. Sodium is absorbed by an active-transport mechanism that is linked to the absorption of amino acids, bicarbonate, and glucose.

Calcium

Calcium transport (see also Chapter 24) is under the influence of vitamin D and parathyroid hormone and is regulated by a calcium-binding protein in the mucosal cells.[10]

Iron absorption (see also Chapter 32)

Gastric acid is required to convert iron to the absorbable ferrous form.[11] Iron then enters the mucosal cells and is transported across these cells before being picked up by carrier proteins in the circulation.

Formation of stool

Having passed through the ileum and jejunum, the intestinal contents enter the large bowel. Very little absorption of nutrients occurs in this region. It is here that water is actively absorbed and returned to the circulation. In addition, the balance of electrolytes is regulated. Progressive dehydration of undigested food substances along with the action of the bacteria that normally inhabit the colon leads to the formation of feces.

PATHOLOGICAL CONDITIONS

Malnutrition

Malnutrition is caused by an abnormal food intake. On a worldwide basis, malnutrition is one of the leading causes of death. In North America, overnutrition is the most common malnutrient state. Obesity contributes to the mortality of all diseases but is closely related to diabetes, hypertension, cardiovascular disease, and emotional disorders. Undernutrition may be caused by lack of food, bizarre diets, malabsorption, and hypermetabolic states. There is increasing awareness that many elderly persons in the United States are malnourished. Alcoholics, drug addicts, or mentally impaired persons may suffer from various forms of undernutrition. Chronically ill patients may suffer from anorexia, the loss of appetite. Occasionally, persons will adopt bizarre diets that lack the basic nutritional requirements. Disease of the gastrointestinal tract may prevent nutrients from being absorbed. Malignancy, pyrexia, and endocrine abnormalities may consume food energy at a faster rate than assimilation can occur.[12]

Stomach pathological conditions

Ulcers. The cause of ulcer disease[13] is multifactorial and relates to genetic and psychological makeup. For example, there is increased incidence of ulcer disease in persons with the blood group O. Some workers have claimed that an increased serum pepsinogin occurs in association with a high duodenal ulcer risk.[14] There has been a decrease in the incidence of ulcer disease in the last 20 years, and afflicted persons are now older. Duodenal and gastric ulcers are different disorders.

The pathogenesis of duodenal ulcer has been studied extensively.[15] Patients with duodenal ulcers, as a group, have an increased capacity to secrete acid and pepsin. They have an increased responsiveness to gastrin and increased gastric acid and therefore deliver an increased gastric load to the duodenum.

The pathophysiology of gastric ulceration has not been definitely determined. What seems to be clear is that gastric ulcers are not associated with increased acidity.

An ulcer diagnosis is generally made on clinical grounds, with roentgenographic and endoscopic examinations being of prime importance. The patient's response to therapy is also very helpful. The patient is generally treated with antacids or anticholinergic agents such as cimetidine. Surgery, which usually includes a vagotomy and antral drainage, is generally considered a second line of therapy and is used in cases of repeated ulceration.

After gastric surgery, several physiologically based complications can occur shortly after eating. Some persons with gastric surgery will absorb glucose at an abnormally fast rate. This triggers an extremely rapid release of insulin, which may overshoot and cause hypoglycemia.

Another problem is caused by the rapid entry of osmotically active food particles into the intestine. This causes a fluid and electrolyte shift into the intestine. Hypovolemia and transient hypokalemia cause a generalized nausea and dizziness. The activation of the kallikrein-kinin system has an important but not fully understood role in this dumping syndrome.

Pyloric obstruction. In pyloric obstruction, the outlet of the stomach is constricted by the contraction of an ulcer, malignancy, or congenital abnormality. Obstruction is characterized by vomiting (often projectile), abdominal distention, and loss of hydrochloric acid, leading to severe hypochloremic metabolic alkalosis.

Cancer. The incidence of stomach cancer[16] is declining in the United States, but it remains high in the Soviet Union and in Japan (54% of all cancers). It appears most often in the seventh and eighth decades of life, and the 5-year survival remains at 15%. Over half of all gastric cancers are found in the pylorus or antrum. Surgery in combination with radiotherapy or chemotherapy is used to treat the lesion.

Zollinger-Ellison syndrome. The Zollinger-Ellison syndrome is an extreme form of peptic ulcer disease caused by a gastrin-secreting tumor of the pancreas (gastrinoma).[17] The unrelenting gastrin release stimulates hypersecretion of hydrochloric acid by the stomach.[18] The typical clinical presentation (not seen in all patients) is recurrent peptic ulceration. Seventy-five percent of patients with this syndrome have ulcers in the duodenal bulb or immediate postbulbar area. The tumors are often very small and can be difficult to identify. Sixty percent of tumors metastasize, and multiple tumors are common. Some tumors (10%) arise in the duodenal wall. The excess secretion of hydrochloric acid accounts for most of the clinical manifestations of the syndrome. The large amount of gastric acid entering the duodenum interferes with the digestion of fat and leads to steatorrhea. The prolonged secretion of gastrin causes hypertrophy of the stomach, with parietal cell hyperplasia. The secretion of pancreatic bicarbonate is increased in compensation for the acid load delivered to the duodenum. Often the intestine becomes ulcerated. The proximal intestinal lining often displays abnormal villi, submucosal edema, and hemorrhage. Since gastrin also inhibits salt and water absorption by the intestine, diarrhea will occur in 50% of patients. The diarrhea is enhanced by the very large volumes of gastric contents that are presented to the intestine. The Zollinger-Ellison syndrome is associated with hyperparathyroidism in 20% of patients. Other endocrine abnormalities that appear less commonly include pituitary, adrenal, ovarian, and thyroid tumors. This cluster of endocrine adenomas and carcinomas is known as the *multiple endocrine neoplasia (MEN) syndrome I*. It may occur with autosomal dominant inheritance, as described originally by Werner, or it may occur sporadically.[19]

A fasting serum gastrin concentration four times the upper limit of normal in the absence of achlorhydria or renal failure strongly suggests the Zollinger-Ellison syndrome. This criterion is not met in 40% of cases.

Provocative testing has been used. Serum gastrin is measured after administration of (1) intravenous secretin, 1 to 2 units/kg; (2) intravenous calcium gluconate; or (3) a standard meal. The secretin test, with a postinjection increase of gastrin of 110 pg/mL, is the most reliable.[20,21] A negative secretin response occurs in 5% of patients. Thus the sensitivity of the test is 95% and specificity virtually 100%.

In summary, to diagnose the Zollinger-Ellison syndrome, first, appropriate patients are screened with fasting serum gastrin measurements. Next the secretin test is administered to those patients with fasting serum gastrin values over 100 pg/mL.

Pernicious anemia. Pernicious anemia is a disease that consists of gastric achlorhydria, gastric atrophy, and failure to secrete intrinsic factor. The intrinsic-factor deficiency prohibits absorption of vitamin B_{12}. This leads to the sequelae of mucosal epithelial insufficiency, degeneration of the posterior columns of the spinal cord, and macrocytic anemia. It is covered in greater detail in Chapter 32.

Malabsorption syndromes

In malabsorption syndromes the gastrointestinal tract is impaired so that it cannot absorb a variety of nutrient materials. This is generally the result of a disorder that causes damage to the mucosal lining. Patients suspected of having a generalized malabsorption syndrome should be evaluated with serum iron, vitamin B_{12}, albumin, and calcium determinations. In addition, immunoglobulin determinations can be useful to rule out IgA deficiency, a condition that permits parasitic infestations to occur. The D-xylose absorption test can be used as a screening procedure for generalized malabsorption.

The other category of malabsorption syndrome should in fact be called *maldigestion*. In this group of disorders one of the important factors for the digestive process is in some way impaired. This is most commonly caused by some form of pancreatic insufficiency (see Chapter 25) or surgical procedure. The most common maldigestion syndrome leads to fat malabsorption and steatorrhea.

Steatorrhea. Steatorrhea is a clinical syndrome caused by the malabsorption of dietary fat. The undigested fat travels into the large bowel, and the stools contain an excess amount of lipid and are characteristically pale, bulky, and greasy with a repugnant odor. It is important to distinguish clinically the stools produced in steatorrhea from those produced in diarrhea. When steatorrhea is suspected, testing should be undertaken to estimate the actual amount of fat in the stool. When excess fat has been identified, the specific cause is sought. A deficiency of any factor important for lipid digestion and absorption can cause steatorrhea. Conditions producing steatorrhea include the Zollinger-Ellison syndrome, increased duodenal acid (postgastrectomy syndromes), abnormal bile output, pancreatic insufficiency, intestinal mucosal impairment, and disease of the large bowel that has caused an interruption of bile-salt enterohepatic circulation.

Celiac disease. Celiac disease is an extremely important cause of malabsorption. In this condition persons appear to have an abnormal immunological response to the presence of gluten in the diet. Up to 90% of celiac patients have circulating antibodies to gluten. The response to gluten is shedding of the microvillous mucosal surface of the intestine. This drastically reduces the absorptive surface area and causes malabsorption. Celiac disease may manifest in very subtle ways and may be definitely diagnosed only by the response to a gluten-free diet.[22]

Lactose intolerance and other carbohydrate malabsorption disorders. The most common carbohydrate malabsorption disorder is lactose intolerance.[27] All infants have the intestinal enzyme mechanism necessary to break the milk sugar disaccharide lactose into its components glucose and galactose, thereby allowing absorption to occur. In those population groups who characteristically feed on animal milk throughout life, these enzyme mechanisms persist into adulthood. However, in those groups who are historically not milk drinkers, the enzyme system regresses (African blacks and Orientals). If persons lacking the lactase enzyme ingest milk or milk products, they will fail to split the lactose in the proper fashion. This unabsorbed sugar will create an osmotic force that pulls fluid into the intestinal lumen. This causes cramping, bloating sensations, and diarrhea. Moreover, large-bowel bacteria can metabolize the sugar to produce gas. Although most people with lactose intolerance are aware of their problem and avoid milk products, there are persons with milder forms who experience discomfort in much more subtle ways. The diagnosis of lactose malabsorption is made by use of the lactose tolerance procedures discussed later in this chapter. Malabsorption syndromes of other disaccharides have been reported but are extremely rare. The malabsorption of monosaccharides is seen only in extreme impairment of the mucosal surface.

Carcinoid syndrome[24]

A syndrome consisting of vascular flushing, diarrhea, a carcinoid tumor of the bowel, occasional tricuspid insufficiency, and rarely, pellagra, is called the *carcinoid syndrome*. Carcinoid tumors, which are the most common of small bowel tumors, are located predominantly in the distal ileum. The remainder of the extra-appendicular gastrointestinal carcinoid tumors are found in the rectum and stomach. These tumors metastasize most commonly to the regional lymph nodes, liver, and skeleton. Primary carcinoid tumors of the appendix are common but rarely metastasize, whereas those that arise from other parts of the gastrointestinal tract do metastasize. The tumors produce serotonin and kinins in vast excess. It is these hormonal substances that are responsible for the characteristic clinical syndrome. One can detect the presence of the disorder by measuring serotonin or its metabolites.

Large intestine disease

Diarrhea. Diarrhea[25] is defined as the excessive production of feces, usually as a result of the overabundance of water in the stool. The causes of diarrhea are many. Diarrhea should be clinically distinguished from steatorrhea.

Severe diarrhea causes sodium and water depletion. Potassium is also lost. The acid-base disturbances caused by diarrhea are variable. However, the most common disorder is acidosis, which results from the increased fecal loss of bicarbonate. In chronic, mild diarrhea, hypokalemic alkalosis may be found.

There are three main reasons for diarrhea: solute malabsorption, secretion of fluid into the intestine, and motility disturbance.

Solute malabsorption is caused by the ingestion of poorly absorbed substances, "dumping," or intestinal malabsorption, such as lactose intolerance.

The secretion of fluid occurs in a number of conditions. Passive secretion will occur if the epithelial permeability is increased by obstruction or inflammation. The secretion of anions will occur through the activity of 3′,5′-cyclic adenosine monophosphate as stimulated by cholera toxin, endotoxin, prostaglandins, bile acids, and certain tumor products (such as vasoactive intestinal polypeptide). Another secretory mechanism is the replacement of absorptive epithelium by crypt epithelium (as occurs with viral gastroenteritis).

Motility disturbances are caused by cathartics and nervous tension. These will increase the motility and decrease the transit time and therefore the absorptive efficiency.

Cancer. Malignancies of the colon and rectum[25] account for over half the malignancies of the entire gastrointestinal system. The 5-year cure rate for these lesions runs between 25% and 50%. The cure rate is directly proportional to how early the lesion is detected and is therefore correlated with its proximity to the anus. Thus early detection through screening is the most effective approach to curing this often fatal disorder. Digital and sigmoidoscopic examination of the rectum is supplemented by screening for the presence of occult blood. Roentgenological studies are only useful in screening patients with a high risk of cancer. Colonoscopy is becoming the preferred method of examining high-risk patients.

Blind-loop syndromes. A variety of inflammatory and anatomical disorders of the gastrointestinal tract may cause regional outpouchings to occur in the large bowel.[26] These pockets can trap intestinal material and allow bacterial overgrowth. If this happens, the overabundant bacteria can cause excessive breakdown of bile conjugates. When these materials have been deconjugated, they cannot be reabsorbed by the body and are lost in the feces. This may be the cause of diarrhea. The bile acid breath test can be used in the diagnosis of this condition.

Common bowel disorders. Rarely a chemical diagnostic problem, the most common disorders of the bowel are associated with abnormal motility. Such symptoms as bloating, cramps, and excess flatus production are common clinical problems.

Hyperalimentation

In recent years, techniques for providing nutrition to persons who are otherwise unable to eat[27] have been highly refined. Such techniques employ tube feeding or intravenous alimentation. In the case of those who are undergoing tube feeding, it is important that the material not cause an osmotic load on the gastrointestinal tract and hence produce diarrhea.

The assessment of persons receiving such artificial nutrition is not highly refined. In general, the tests must first allow assessment of whether complications have occurred or whether undernutrition of one or more substances is present. Patients should be monitored carefully by clinical assessment. Biochemical monitoring of the urine is useful and includes measurement of the urine osmolality to evaluate hydration, sodium and potassium measurements to indicate electrolyte load, urea concentration to provide a rough guide to overall nitrogen balance, and ketones and glucose to indicate poor carbohydrate control and caloric loss. Blood analyses are also useful but should be interpreted with caution because the intravenous solution administered at the time of the blood collection can drastically influence the results by affecting lipemia or hyperosmotic forces.

GASTROINTESTINAL FUNCTION TESTS

There are a number of gastrointestinal function tests that have been designed for critical evaluation of various gastrointestinal physiological functions.

Tests of gastric acidity

Tests of gastric acidity are used to screen for the ability of the parietal cells to produce hydrochloric acid. The discovery of achlorhydria (anacidity) is strong evidence for the presence of pernicious anemia and rules out peptic ulcer disease, casting suspicion on gastric ulcer or gastric carcinoma. The presence of acid in the stomach is very strong evidence against pernicious anemia. Acid detection must be carried out to determine the significance of raised serum gastrin determinations.

The only suitable test for the presence of gastric acid is intubation and withdrawal of stomach juice. A pH measurement may then be made directly and should be less than 3. Anacidity is only confirmed by a pH over 6.

A resin-dye exchange method known as the Diagnex Blue test (E.R. Squibb Sons, New York) has been popular in the past. However, because the test produces such a high rate of false-positive and false-negative results, it is not recommended.

Gastric stimulation tests

Gastric analysis. The gastric analysis test involves draining stomach secretions for a baseline period to determine basal or unstimulated acid production. Next, a parietal cell stimulant is administered and gastric juice is collected to evaluate maximum secretory ability. Several different stimuli have been used. The use of a test meal was popular at one time, but it leads to nonreproducible results and interferes with the desired measurements. Histamine has been used.[28] An antihistamine (diphenhydramine, or Benadryl) must be given in advance to ablate the systemic effects of the histamine. Until recently, betazole hydrochloride (Histalog) has been widely used.[28] No antihistamine is needed when this preparation is given. Currently, pentagastrin, the active 5-amino acid portion of gastrin, is the recommended stimulant.[29,30]

Thus the protocol is (1) collect residual gastric fluid from a fasting patient by intermittent suction on a nasogastric tube positioned within the stomach by fluoroscopy, (2) collect basal secretions for four 15-minute periods, (3) administer pentagastrin intramuscularly in a dose of 5 μg/kg of body weight, and (4) collect further stomach secretions for six consecutive 15-minute time periods. All collections are then evaluated for appearance, blood, bile, pH, volume, millimoles of H^+ per liter, millimoles of H^+ per volume, and millimoles of H^+ per hour for each collection.

The pH measurement is useful. Basal pH over 6 is almost certainly caused by anacidity; pH less than 3 indicates normal or excessive parietal cell function; intermediate values (3 to 6 mmol/hour) are not diagnostic and merely indicate a balance between hydrochloric acid and buffer. After stimulation, pH values should fall to less than 2. Failure to do so indicates inadequate parietal cell function, which may be found in pernicious anemia, gastric carcinoma, hypochromic anemia, rheumatoid arthritis, and myxedema. Next, the basal acid output (BAO) should be computed by averaging the millimole-per-hour output for the closest three basal collections. The maximum acid output (MAO) is also calculated as the mean of the two highest, poststimulation values in millimoles per hour.

Normal adult men have a BAO of 2.2 to 2.7 mmol/hour, with 5 mmol/hour being the absolute upper limit.

Table 26-3 Pentagastrin stimulation test

	Basal acid output (BAO) (mmol/hour)	**Maximal acid output (MAO) (mmol/hour)**
Normal adult men		
Under 30 years old	2.2-2.7	14-42
Over 30 years old	2.2-2.7	3-33
Normal adult women	1-1.5	7-20
Zollinger-Ellison syndrome	10-100 (or more)	40%-60% above BAO
Ulcer predisposition		
Likely		>35
Highly likely		>45
Low risk		<11

The MAO for men under 30 years of age is 14 to 42 mmol/hour, and it is 3 to 33 mmol/hour for men over 30 years of age. The values for women are approximately 50% of those for men. Detailed tables of normal values have been published[31] and are reviewed in Table 26-3.

The Zollinger-Ellison syndrome is characterized by high BAO (Table 26-3). The MAO is generally only 40% to 60% higher than the BAO because the stomach is close to maximal stimulation. Indeed, a BAO/MAO greater than 60% is virtually pathognomonic for the Zollinger-Ellison syndrome. If the MAO is over 35 mmol/hour, an ulcer predisposition is likely. In persons over 45 years of age it is highly likely. An MAO less than 11 mmol/hour in a man under 30 years of age suggests a very low peptic ulcer risk.[31] Gastric analysis is used by some surgeons to indicate the nature of surgery to be used for ulcer treatment.

Hollander insulin test. The Hollander insulin test[32] is used to assess whether a surgical vagotomy has successfully denervated the stomach. In this procedure regular insulin (0.15 U/kg of body weight) is administered intravenously to render the patient hypoglycemic (plasma glucose less than 300 mg/L). Vagus stimulation is a normal response to hypoglycemia. Those with an intact gastric vagus will release acid in response to hypoglycemia. A successful denervation will cause an MAO less than 0.05 mmol/hour, and the pH will remain over 3.5. The test is not often performed because clinical evaluation is generally sufficient.

Gastrin stimulation test. The principle of the gastrin stimulation test was described earlier.[20,21]

Blood for serum gastrin is collected from a fasting patient. Secretin (2 units/kg as an intravenous bolus) is then administered and serum gastrin collected at 2, 5, 10, 15, 30, and 60 minutes. An absolute increase of gastrin greater than 110 pg/mL is positive for the Zollinger-Ellison syndrome.

Schilling test

The Schilling test of vitamin B_{12} absorption is an elegant evaluator of both gastric and intestinal function.[33] Radioactively tagged vitamin B_{12} is taken orally by the patient. If absorption is successful, the radioactive material will be excreted in the urine. The vitamin B_{12} is administered with and without intrinsic factor to determine whether that substance is absent. More specifically, the test is carried out in the following way.

The fasting patient is given an oral preparation containing radioactively labeled vitamin B_{12}. One hour later, an intramuscular injection of unlabeled vitamin B_{12} is given. This cold B_{12} saturates the body's binding sites for B_{12}, preventing the labeled oral dose from being stored. Urine is collected for 24 hours. The amount of radioactivity in the urine is determined as a percentage of the original dose.

Table 26-4 Schilling test

Group	Absorption of intrinsic factor–bound vitamin B_{12}	Absorption of unbound vitamin B_{12}	Ratio of bound to unbound
Normal	>15%	>15%	0.5-1.5
Gastric lesion group	>10%	<5%	>2
Pernicious anemia			
Gastrectomy			
Congenital absence of functional intrinsic factor			
Disorders of terminal ileum (see text)	<15%	<15%	0.5

After several days, the test may be repeated. This time, the radioactively labeled vitamin B_{12} is given with intrinsic factor. Again, a 24-hour urine collection is carried out and the amount of vitamin B_{12} excreted is determined. There are three possible results (Table 26-4).

If both stages of the test show greater than 15% excretion, vitamin B_{12} absorption is normal, and any deficiency of serum vitamin B_{12} must be the result of a dietary problem.

If the absorption of vitamin B_{12} is low when it is given alone but normal when it is administered in association with intrinsic factor, one can presume a deficiency of intrinsic factor. Such a situation is found in classic pernicious anemia, gastrectomy, and the rare occurrence of nonfunctional intrinsic factor.

If vitamin B_{12} excretion in the urine is low after administration of both preparations, an absorptive defect of the terminal ileum is presumed to be present. This may be found in tropical sprue, nontropical sprue, regional enteritis, intestinal resection, neoplasm or granuloma, selective ileal vitamin B_{12} malabsorption (Imerslund syndrome) (rare), fish tapeworm *(Diphyllobothrium latum),* bacterial overgrowth, and chronic vitamin B_{12} deficiency.

A two-capsule test has been devised (Dicopac, Amersham Corp. Arlington Heights, Ill.). In this preparation, two forms of vitamin B_{12} are radioactively labeled using different isotopes of cobalt (^{57}Co, ^{58}Co). The advantage of the procedure is that one can give both forms of vitamin B_{12} at essentially the same time. Absorption may then be determined by use of a single 24-hour urine collection. Special counting techniques to subtract the influence of one isotope on the other are required.

Pancreatic challenge tests

The elucidation of malabsorptive disorders depends on the evaluation of pancreatic secretory ability. This is best performed using the pancreatic stimulation tests described in Chapter 25.

Fat absorption tests

The definitive test of fat absorption is the quantitative measurement of fat in timed collections of feces obtained while the patient is maintained on a diet containing a known amount of fat. Because the collection is extremely difficult for the patient, a variety of alternative approaches has been promoted. Unfortunately, none of these entirely replaces the diagnostic ability of the quantitative fecal fat measurement.

Fat screening. Fat screening[34] is carried out first by evaluation of the weight and appearance of the stool. A pale, frothy appearance is virtually diagnostic of excessive fat. More reliable than this is the application of a small amount of the fecal material onto a standard microscopic slide, followed by staining with a fat-specific stain. Trained observers are able to identify excessive fat in 80% of persons with fat malabsorption.

Quantitative fecal fat estimation. Quantitative fecal fat estimation[35] is performed by collecting feces for 3 consecutive days. In the 2 days preceding the collection and during the period of collection, patients must include approximately 100 g of medium-chain triglycerides in their diet.

The easiest way to ensure this intake is to supplement the patient's normal diet with four glasses of whole milk each day and 2 tablespoons of corn oil with each meal.

Feces can be collected in plastic bags. The plastic bags may then be closed with a tin tie and held in a preweighed, 5-gallon paint can. On arrival in the laboratory the can and contents are weighed and the weight of the collection is determined. The chemical analysis is then carried out on a thoroughly mixed aliquot of this 3-day collection.[36]

Serum fat estimations. Another rough screening test for fat absorption is the fat tolerance test.[37] A fasting blood sample is collected. The patient is then given a high-fat meal consisting of 6 ounces of corn oil. Serum is collected at 3 and 6 hours. Estimations of total fat in these samples are then made. The postfat specimens should exhibit a minimum of 50% increase in fat content over the fasting value. This test has poor diagnostic ability.

Isotope tests. Radioactively labeled, medium-chain triglyceride is administered to patients in whom fat malabsorption is suspected.[37] After a suitable time interval, blood is collected and its radioactivity determined. It is assumed that the radioactivity that finds its way into the bloodstream is a result of the successful digestion and absorption of the radioactive fat. Unfortunately, radioactive iodine tags are not suitable for this procedure because they must be linked to unsaturated fats and because the size of the iodine tag gravely distorts the triglyceride molecule, making it susceptible to incidental breakdown. Thus the absorption of the radioactive iodine from these materials might not indicate successful fat abosorption. A ^{14}C-labeled triglyceride has been tested and is measurable in the bloodstream or in the feces.[38]

D-Xylose absorption test[39]

D-Xylose is an aldopentose that is absorbed by the small intestine. It is postulated that D-xylose is absorbed passively; its successful absorption is therefore a reflection of the integrity of the surface area of the small intestine. Once D-xylose is absorbed into the bloodstream there is a small amount of metabolic alteration, but at least 50% is excreted in the urine within the next 24 hours. It has been shown that the amount of D-xylose excreted into the urine over a 5-hour period is closely correlated with the amount of D-xylose absorbed in the gastrointestinal tract.

Several options are available for the performance of this test. The patient is instructed to fast overnight but is encouraged to drink an ample amount of water during this time.

Two doses have been advocated. Most authors suggest that 25 g of D-xylose dissolved in approximately 300 to 500 mL of water is a suitable dose. Smaller subjects are given 1 g/kg of body weight to a maximum of 25 g. We have found that a 5 g dose is adequate.[40] This dose is less likely to cause abdominal cramps and is probably more sensitive because some persons with absorptive defects are capable of absorbing sufficient amounts of the larger dose.

After administration of the sugar, urine is collected for a 5-hour period. A quantitative assay for D-xylose is carried out on this sample. Normally at least 25% of the administered dose will appear in the urine over a 5-hour period if renal function is normal.

For children who cannot be relied on to collect a urinary sample, blood collections at 1 and 2 hours may be substituted. Most persons demonstrate plasma levels greater than 300 mg/L in one of the samples. In children, values above 100 mg/L should be considered within the normal limits.

Low levels of urine or plasma xylose suggest an absorptive defect in the jejunum. Low levels are also seen in ascites, vomiting, delayed gastric emptying, improper urine collection, and high-dose aspirin therapy and with neomycin, colchicine, indomethacin, atropine, and impaired renal function. Normal levels are seen in persons who have absorptive defects occurring in a skip pattern. Such a disease distribution allows a sufficient amount of normal mucosa to remain and absorb an apparently normal amount of D-xylose.

Lactose tolerance test

Some persons do not have a fully developed mucosal lactase enzyme system. This deficiency prevents the normal absorption of lactose, resulting in gastrointestinal bloating, cramps, and diarrhea. The lactose tolerance test is useful for the quick identification of such persons.

In this test, 50 g of lactose dissolved in water is administered orally to the patient, who is observed carefully for the onset of symptoms.

The standard protocol includes the collection of a baseline specimen and 5-, 10-, 30-, 60-, 90-, and 120-minute specimens of blood. The blood sample is analyzed for the presence of glucose. Glucose will be present in the bloodstream if lactose has been successfully cleaved and absorbed. The galactose moiety of the lactose is converted quickly into glucose by the liver. Normal persons will demonstrate a glucose rise of greater than 200 mg/L over the baseline sample. Those with lactase deficiency will exhibit notable abdominal discomfort and will have less than a 100 mg/L increase in the serum glucose concentration.

Recently it has been shown that the most reliable method of determining lactose absorption is the measurement of the amount of hydrogen appearing in exhaled breath after the oral administration of lactose.[41] Lactase-deficient persons will not absorb lactose, and it will find its way into the large bowel, where bacteria will metabolize it. Hydrogen is one of the by-products of this bacterial action. Hydrogen passes quickly into the bloodstream and is removed in the exhaled breath. Special-purpose gas chromatographs can detect the presence of postlactose hydrogen. A normal person allows no lactose to enter the colon and therefore has less than 10 parts per million (ppm) of hydrogen in the exhaled breath. Persons with lactase deficiency demonstrate at least 50 ppm of hydrogen. Intermediate amounts of hydrogen in the breath can be caused by large doses of lactose and are of questionable significance.

The definitive diagnosis is made by tissue enzyme assays carried out on biopsy samples of the intestinal mucosa.

CHANGE OF ANALYTE IN DISEASE (Table 26-5)

Gastrin

The normal fasting serum gastrin concentration[21] is between 30 and 100 pg/mL. Increased serum gastrin may be caused primarily by hyperplasia of the G cells of the antrum, abnormal acid production, such as achlorhydria with compensatory hypergastrinemia or isolated retained antrum after gastrectomy, gastrinomas, and renal disease (see accompanying box).

Low gastrin concentrations are observed in normal persons, in persons with hypothyroidism, and after the administration of oral acid, streptozotocin, and phenformin.

Minimum elevations up to 500 pg are not diagnostic of any specific disorder and are occasionally found in healthy persons. Such values are sometimes produced by the ingestion (or even by the thought or smell) of food, insulin administration, malignant carcinoma of the stomach, pheo-

Table 26-5 Change of analyte and function tests in disease

Disease	Fecal fat	Lactose tolerance	S-carotene S-vitamin A	S-vitamin B_{12} S-folate	Shilling test	D-Xylose absorption	Stool occult blood	Carcinoembryonic antigen	5-Hydroxyindoleacetic acid (5-HIAA)	Pancreatic enzyme testing	Stool examination
Steatorrhea	↑ ↑	N	↓	N, ↓	AB	AB	Neg	N	N	AB	Foul smelling, greasy
Celiac disease	N, ↑	N	N, ↓	N, ↓	N,AB	N,AB	Neg	N	N	N	Variable
Lactose intolerance	N	AB	N	N	N	N	Neg	N	N	N	Loose in association with abdominal cramps
Carcinoid syndrome	N	N	N	N, ↓	N,AB	N	Neg, pos	N, ↑	↑ ↑	N	Loose in association with cutaneous flushing
Functional diarrhea	N	N	N	N	N	N	Neg	N	N	N	Loose
Bowel carcinoma	N	N	N	N, ↓	N	N	Neg or pos	N, ↑	N	N	Change in bowel habits
Inflammatory bowel	N	N	N	N, ↓	N	N	Pos	N, ↑	N	N	Loose, bloody

N, Normal; ↑, elevated; ↓, lowered; *AB*, abnormal; *S*, serum.

chromocytoma, hyperthyroidism, hyperparathyroidism, peptic ulceration, gastritis, cirrhosis of the liver, renal failure, and rheumatoid arthritis. Some cases of Zollinger-Ellison syndrome and pernicious anemia may be found with serum gastrin values in this range, but this is unusual. Values between 500 and 1000 pg/mL are significant. Such results are associated with food ingestion, insulin administration, pheochromocytoma, hyperparathyroidism, renal failure, pernicious anemia, and Zollinger-Ellison syndrome.

Results over 1000 pg/mL are probably caused by either the Zollinger-Ellison syndrome, pernicious anemia, or rarely, parietal cell antibody–positive chronic atrophic gastritis. Most cases of the Zollinger-Ellison syndrome have values greater than 2000 pg/mL.

If the Zollinger-Ellison syndrome is considered, a fasting serum gastrin should be obtained. Results under 100 pg/mL essentially rule out the Zollinger-Ellison syndrome. Results greater than this in persons with normal renal function warrant stimulation testing as described on p. 406.

SERUM GASTRIN CONCENTRATIONS IN DISEASE

Normal (30 to 100 pg/mL)
Low (<30 pg/mL)
  Normal persons, hypothyroidism, administration of oral acid, streptozotocin, phenformin
Minimum elevations (100 to 500 pg/mL)
  Normal persons (occasionally), ingestion of food, insulin administration, malignant carcinoma of the stomach, pheochromocytoma, hyperthyroidism, hyperparathyroidism, peptic ulceration, gastritis, cirrhosis of the liver, renal failure, rheumatoid arthritis, Zollinger-Ellison syndrome (unusual), pernicious anemia (unusual)
Significant increase (500 to 1000 pg/mL)
  Food ingestion, insulin administration, pheochromocytoma, hyperparathyroidism, renal failure, pernicious anemia, Zollinger-Ellison syndrome
Dramatic increase (>1000 pg/mL)
  Zollinger-Ellison syndrome, pernicious anemia, parietal cell antibody–positive chronic atrophic gastritis

Malabsorption testing

Screening approach. Screening for persons with malabsorption syndromes in North American society is best done clinically. Elderly persons are at greatest risk for occult malabsorption. Laboratory screening for malabsorption is not very sensitive; however, measurement of serum albumin, calcium, vitamin B_{12}, and a peripheral smear looking for evidence of macrocytosis and iron-deficiency anemia constitute a reasonable general laboratory screen for malabsorption. If necessary, more specific tests for iron deficiency can be carried out. Measurements of the B vitamins are not generally available but would be particularly useful in elderly persons and in the indigent alcoholic population.

Persons thought to have specific malabsorption syndromes should be tested accordingly. Those with steatorrhea and suspected fat malabsorption should first have their feces examined visually. Next, one should carry out a rapid slide evaluation of a stool sample looking for meat fibers and excess fat. A carotene determination is easily performed and reflects gross abnormalities of fat absorption. A D-xylose absorption test will indicate whether significant generalized absorptive problems are present. Protein malabsorption is difficult to assess biochemically, and only when there is serious amino acid malabsorption will the serum albumin be depressed.

Carotene. Beta-carotene,[42] a naturally occurring pigment, is a common constituent of vegetables and fruit. Collectively, these pigments are called *carotenoids*. The carotenoid chemical structure consists of two 6-member carbon rings joined by a polyunsaturated 18-member carbon chain. Carotenoids are hydrophobic and insoluble in water. Consequently, they must be absorbed into the body in association with fats. Therefore in clinical conditions in which fat malabsorption is impaired, the absorption of carotenoid pigments will be reduced and the serum carotene will be lower than normal. The normal range for serum carotene is 500 to 2500 mg/L.

Serum carotene is decreased in low carotene diets (low vegetable diets), abetalipoproteinemia, Tangier lipoprotein abnormality, liver failure, and 86% of patients with clinically significant fat malabsorption.

A normal serum carotene concentration is found in 14% of patients with fat malabsorption. The concentration of carotene is determined by the balance between the degree of malabsorption and the oral carotene intake.

Vitamin A. Vitamin A[43] is an alcohol derived from beta-carotene by hydrolytic cleavage at the midpoint of the ^{18}C-polyene chain. Vitamin A is found only in animal tissue and animal products such as milk and egg yolk. Because it is chemically similar to carotene, it is also hydrophobic and must be absorbed into the body along with fat. Thus its presence in serum is also a reasonable estimate of the ability of the body to absorb fat. Serum vitamin A concentrations are less dependent on diet than are serum carotene concentrations. Reduced serum vitamin A concentrations are seen in association with fat malabsorption and liver disease.

Because the serum vitamin A determination is significantly more difficult to perform than the serum carotene and because its diagnostic ability is not significantly greater, it has never achieved popular acceptance as a screening test for fat malabsorption.

Trypsin. The measurement of trypsin in stool has been advocated as a screening test of pancreatic insufficiency.[44] Trypsin determination is much more reliable in infants and young children than in adults because there is less colonic degradation of pancreatic enzymes. The simplest test of tryptic activity is the application of a smear of fecal material to a thin film of gelatin. If trypsin is present, an enzymatic breakdown of the film will be seen. The test will be abnormal in all patients with significant pancreatic insufficiency, except in unusual cases of isolated defects of amylase and lipase. Some normal infants, however, will fail to produce sufficient trypsin to degrade the gelatin layer. Although such tests have some validity in screening, a strong clinical suggestion of pancreatic insufficiency warrants specific testing, as outlined in Chapter 25.

Fecal fat determinations.[35] Persons taking a 100 g fat diet will excrete no more than 5 g of fecal fat per day. More than 10 g per day is certain evidence of fat malabsorption.

Failure to adhere to the diet may invalidate the results. Low fat intake will mask minimum fat malabsorption. Grossly excessive fat intake will raise the fecal content above 5 g.

Tests related to specific disorders

Occult blood in stool. Because the survival of persons with cancer of the bowel depends on early diagnosis, reliable methods for detecting carcinoma of the bowel are extremely important.[45] Several color reagents that react to trace amounts of hemoglobin in feces are available. Most rely on the ability of hemoglobin and its derivatives to act as peroxidases and catalyze the reaction between hydrogen peroxide and a chromogenic, organic compound. Benzidine has been used but is carcinogenic and therefore not currently recommended as a laboratory-prepared reagent.

Various commercial systems for stool occult blood measurement are available.[46] These have been evaluated by Morris et al.,[47] who recommend the Hemoccult test (Smith-Kline & French Laboratories, Philadelphia) as the most suitable. In any test of fecal blood loss, a reagent must be suitably sensitive so that it will detect the presence of all gastrointestinal malignancies and specific enough so that investigators are not burdened with large numbers of false-positive reactions.

The Hemoccult test is based on the guaiac-peroxide reaction and is not made positive by insignificant blood loss or meat ingestion. Moreover, it is suitable for patient use because it provides a convenient sample-handling envelope. Those laboratories wishing to use their own reagent system may employ the method of Woodman.[48] Whatever method is employed, the evaluation of occult blood in three separate stool aliquots is recommended for optimum diagnostic detection.

Carcinoembryonic antigen. Carinoembryonic antigen (CEA) is a glycoprotein that is abundant in entodermally derived tissues (gastrointestinal mucosa, pancreas, lung).[49] In the gastrointestinal tract it is located in the glycocalyx of the cell. It is found in fetal tissues at 12 weeks of gestation, and peak levels are observed at 22 weeks. The value decreases near the end of the third trimester. It is not known how CEA enters the circulation. After removal of a CEA-producing tumor, the material disappears within 2 to 4 weeks, most likely because of hepatic breakdown.

Numerous glycoproteins have been found that have immunological cross-reactivity with CEA. They include fetal sulfoglycoprotein, normal colonic antigen (NCA), and normal glycoprotein (NGP). This latter antigen is also known as colonic antigen Be, colonic antigen X, colonic CEA II, and colonic cancer antigen III.

There are three possible clinical uses of CEA: for screening, for diagnosis, and for management of colonic carcinomas.[50]

CEA determination has been advocated as a method of screening for the presence of occult carcinomas of the gastrointestinal tract. Unfortunately, this is a very difficult proposition because CEA is increased in a variety of colonic disorders, some of which predispose to colonic carcinoma. For example, CEA is increased in 20% of colorectal polyps. Thus it is not specific enough to be recommended as a screening procedure.

A variety of studies have been carried out to determine whether the CEA determination is useful to confirm the diagnosis of colonic cancer in individual patients. Unfortunately, the test is not sufficiently reliable to assist in the diagnosis of this potentially fatal disorder. For example, if a person has a 50% chance of colonic cancer, a positive CEA will increase the chance to 80%. Thus the CEA is not recommended as a preoperative indication of whether a malignancy is present. Indeed, the consensus statement of the U.S. Cancer Institute is as follows:

> We cannot recommend, based upon the available data, that CEA be used independently to establish a diagnosis of cancer. However, in a patient with symptoms, a value greater than five to ten times the upper limit of referenced normal should be strongly suggestive for the presence of cancer. . . . Further diagnostic efforts are indicated.[51]

The CEA test plays a well-established role in monitoring the course of disease in persons who have been treated for colonic cancer. For this purpose the CEA should be measured at the time of surgery to establish a baseline level. Six weeks must then be allowed to pass before the next measurement is taken. If the value is elevated, a rising titer over the next few weeks must be noted before residual cancer is considered. If the value is low at the 6-week measurement, one can presume that the original lesion has been removed in its entirety and further measurements at regular intervals should be carried out. Some authors believe that a rising CEA titer requires a second-look operation. However, although some patients will have a rising titer caused by surgically manageable local spread, others will exhibit inoperable metastases.

The CEA concentration approximates the tumor burden and correlates approximately with the clinical course of metastatic disease. Unfortunately, despite many excellent and supportive reports for the use of this test, it has not proved consistently reliable in individual cases.

The normal range for CEA is 0 to 5 ng/mL. However, rising titers significant of disease may be seen within this range. A normal CEA does not exclude an underlying neoplasm. Mild elevations of 5 to 10 ng/mL may be caused by malignancy or by heavy smoking, chronic liver disease, chronic chest disease, inflammatory bowel disease, or chronic renal failure. When the value reaches 10 to 15 ng/mL, malignancy is quite likely because less than 5% of the CEA-inducing, nonmalignant conditions produce results in this range. Values over 15 ng/mL strongly suggest malignancy.

5-Hydroxyindoleacetic acid. The presence of cutaneous flushing and diarrhea is a sufficiently common syndrome to warrant the frequent request for tests that will identify the possibility of the carcinoid syndrome. The most commonly performed biochemical test for this purpose is the 5-hydroxyindoleacetic acid (5-HIAA) determination.[52] This test is performed because the most common biochemical product of this tumor is serotonin, which is formed by the conversion of tryptophan to 5-hydroxytryptamine (serotonin), which is ultimately converted to 5-HIAA.

The amount of 5-HIAA found in the urine is highly method dependent. Many screening procedures for this substance are very nonspecific and should therefore not be used to make diagnoses. An appropriate approach is the use of a screening procedure for all requests for 5-HIAA; those that exhibit an elevated value should be subjected to a somewhat more specific test, such as that of Goldberg.[52]

Normally, up to 15 mg of 5-HIAA is excreted per 24 hours. In the carcinoid syndrome more than 25 mg is usually measured. False-normal results are produced by a number of drugs, including *p*-chlorophenylalanine, corticotropin, ethanol, imipramine, isoniazid, monoamine oxidase (MAO) inhibitors, methenamine, methyldopa, and phenothiazines. Reduction of an elevated value is also seen in renal disease and in phenylketonuria.

In the carcinoid syndrome, results are usually between 25 and 1000 mg/day. False-positive results have been reported in nontropical sprue; intestinal obstruction; pregnancy; sleep deprivation; oat cell carcinoma of the lung; and with ingestion of avocados, bananas, eggplants, pineapples, plums, and walnuts. Drugs that are known to cause an increase in 5-HIAA value are acetanilid, ephedrine, mephenesin, nicotine, phenacetin, phenobarbital, phentolamine, rauwolfia, reserpine, methocarbamol, and glycerol guaiacolate cough medicines.

^{14}C bile acid breath test. The use of breath analysis in gastroenterology has been extensively reviewed by Newman.[53] The most common radioactive breath-testing procedure in current use is the ^{14}C bile acid breath test.[54] In this procedure, glycine-1–^{14}C–cholate is administered orally. If the patient has an intestinal blind loop or other source of bacterial overgrowth, deconjugation of the tracer will occur, allowing ^{14}C to enter the bloodstream. Here it is metabolized to $^{14}CO_2$. Breath CO_2 is trapped in an alkaline solution, and the subsequent detection of radioactivity is a sensitive indication of the presence of bacterial overgrowth.

Tests of celiac disease. The detection of circulating gluten antibodies has been suggested as a means of identifying persons with celiac disease.[55] The test is not sufficiently sensitive or specific to be recommended for routine testing.

IgA. A very good test to perform in persons with chronic undiagnosed diarrhea is the serum concentration of immunoglobulin A (IgA). Approximately 1 in 800 persons suffers from idiopathic deficiency of IgA. Such persons do not have a normal immunodefense of their mucosal membrane system. They are therefore susceptible to infestations of parasites.[56] If a person is discovered to be suffering from a deficiency of IgA, an intestinal intubation with vigorous mucosal biopsy is warranted in search of such parasites as *Giardia lamblia*.

Urine oxalate. The metabolism of oxalate is very complex. A major component of its biochemistry is the enterohepatic circulation of glyoxalate materials secreted in the bile and subsequently reabsorbed by the bowel. In persons who have had gastrointestinal surgery, who have chronic inflammatory disease of the bowel, or who have chronic abnormalities of the gastrointestinal tract, it is possible to have excessive reabsorption of oxalate material, which has been split from the excretory products. Such persons secrete large amounts of oxalate in the urine and are prone to the formation of oxalate renal calculi. Thus the evaluation of urinary oxalate is recommended in anyone who has a chronic anatomical or physiological intestinal abnormality.[57]

Indican. Some intestinal bacteria (including *Escherichia coli* and the *Bacteroides* species) contain the enzyme tryptophanase, which can metabolize dietary tryptophan to indole. This product is then absorbed, converted to indican in the liver, and excreted in the urine.[58] Thus the estimation of indican in a 24-hour specimen of urine has been suggested as a means of diagnosing small bowel bacterial contamination. An increased urinary indican is not always found, however, because the predominant bacteria may not contain the appropriate enzyme. In addition, other small bowel diseases causing malabsorption may encourage indican excretion as a result of absorbed dietary tryptophan. Therefore estimations of urinary indican correlate only roughly with abnormalities in the gastrointestinal tract, and the test is not recommended for regular use in the diagnosis of gastrointestinal pathological conditions. The test is commonly requested by naturopathic physicians who ascribe their own particular interpretation to the results. It is of interest that this was the test that identified the first case of Hartnup's disease.

Serum folate.[59] Folate determinations are widely available and are often used in the assessment of malabsorption. The folate value obtained in red cells reflects a chronic situation, whereas the serum level reflects the day-to-day nutritional impact of folate. Increased serum folate is seen occasionally in pernicious anemia. Decreased serum folate is seen in a series of disorders in which the absorptive capacity of the small intestine is impaired, including Whipple's disease, myelofibrosis, hypothyroidism, celiac sprue, abetalipoproteinemia, vitamin B_6 deficiency, regional enteritis, ileitis, ulcerative colitis, acute and subacute necrosis of the liver, malignancy, and Down's syndrome.

Vitamin B_{12}. The validity of serum vitamin B_{12} determinations has been called to question[60] because many tests of vitamin B_{12} may detect the presence of nonphysiologically active analogs of vitamin B_{12}. Thus they may give false-normal vitamin B_{12} results in B_{12}-deficient persons. Most current methods respect this potential problem.

Low vitamin B_{12} concentrations (less than 150 pg/mL) are found in pernicious anemia, gastric disease, thyroid disease, ileal disease, malabsorption, and vegetarian diets. Increased vitamin B_{12} is seen with vitamin B_{12} administration, liver tissue damage, leukemia, and lymphoma.

REFERENCES

1. Moog, F: The lining of the small intestine, Sci Am 245:154-176, 1981.
2. Track, NS: The gastrointestinal endocrine system, Can Med Assoc J 122:287–291, 1980.
3. Rayford, PL, Miller, TA, and Thompson, JC: Secretin, cholecystokinin and newer gastrointestinal hormones, N Engl J Med 294:1093-1101, 1157-1164, 1976.
4. Dawson, AM: The absorption of disaccharides. In Card, WI, and Creamer, B, editors: Modern trends in gastroenterology, vol 4, London, 1970, Butterworth & Co (Publisher), Ltd.
5. McColl, I, and Sladen, GEC, editors: Intestinal absorption in man, New York, 1975, Academic Press, Inc.
6. Wilson, FA, and Dietschy, JM: Differential diagnostic approach to clinical problems of malabsorption, Gastroenterology 61:911-921, 1971.
7. Ingelfinger, FJ: Gastrointestinal absorption, Nutrition Today, pp 2-10, March 1967.
8. Wiss, O, and Gloor, U: Absorption, distribution, storage and metabolites of vitamin K and related quinones, Vitam Horm 24:576-586, 1966.
9. DeLuca, HF, and Suttie, JW, editors: The fat soluble vitamins, Madison, Wis, 1970, University of Wisconsin Press.
10. Holdsworth, CD: Calcium absorption in man. In McColl, I, and Sladen, GEG, editors: New York, 1975, Academic Press, Inc.
11. Martiner-Torres, C, and Layrisse, M: Nutritional factors in iron-deficiency: food iron absorption, Clin Hematol 2:339-352, 1973.
12. Pearson, WN: Assessment of nutritional status: biochemical methods. In Beaton, GH, and McHenry, EW, editors: New York, 1966, Academic Press, Inc.
13. Mirsky, IA: Physiologic, psychologic, and social determinants in the etiology of duodenal ulcer, AM J Dig Dis 3:285-314, 1958.
14. Grossman, MI: Elevated serum pepsinogen I: a genetic marker for duodenal ulcer disease, N Engl J Med 300:89, 1979.
15. Grossman, MI, Guth, PH, Isenberg, JI, et al: A new look at peptic ulcer, Ann Intern Med 84:57-67, 1976.
16. Shahon, BB, and Horowitz, S: Cancer of the stomach: analysis of 1152 cases, Surgery 39:204-221, 1956.
17. Deveney, CW, Deveney, KS, and Way, LW: The Zollinger-Ellison syndrome—23 years later, Ann Surg 188:384-393, 1978.
18. Walsh, JH, and Grossman, MT: Gastrin, N Engl J Med 292:1324-1334, 1377-1384, 1975.
19. Johnson, GJ, Somerskill, WHK, and Anderson, VE: Clinical and genetic investigation of a large kindred with multiple endocrine adenomatosis, N Engl J Med 277;1379-1386, 1967.

20. Ippoliti, AF: Zollinger-Ellison syndrome: provocative diagnostic tests, Ann Intern Med 87:787-788, 1977.
21. Deveney, CW, Deveney, KS, Jaffe, BM, et al: Use of calcium and secretin in the diagnosis of gastrinoma, Ann Intern Med 87:680-686, 1979.
22. Strober, W: The pathogenesis of gluten-sensitive enteropathy, Ann Intern Med 83:242-256, 1975.
23. Gray, GM: Congenital and adult intestinal lactose deficiency, N Engl J Med 294:1057-1058, 1976.
24. Kuehn, PG, Coley, GM, and Christine, B: Carcinoid syndrome: a study of 16 cases, Hartford Hosp Bull 28:305-307, 1973.
25. Jeejeebhoy, KN: Symposium on diarrhea. I. Definition and mechanisms of diarrhea, Can Med Assoc J 116:737-738, 1977.
26. Gilbertson, VA: The earlier diagnosis of adenocarcinoma of the large intestine, Cancer 27:143-149, 1971.
27. Walravens, PA: Keeping TPN on course with lab monitoring, Diagn Med 4:38-43, 1981.
28. Ward, S, Gillespie, IE, Passaro, ER, et al: Comparison of Histalog and histamine as stimulants for maximal gastric secretions in human subjects and in dogs, Gastroenterology 44:620-626, 1963.
29. Abernethy, RJ, Gillespie, IE, Lowrie, JH, et al: Pentagastrin as a stimulant of maximal gastric acid response in man: a multicentre pilot study, Lancet 1:291-295, 1967.
30. Kirsner, JB, and Ford, H: The gastric secretory response to Histalog: one-hour basal and Histalog secretion in normal persons and in patients with duodenal ulcer and gastric ulcer, J Lab Clin Med 46:307-311, 1955.
31. Blackman, AH, Lambert, DL, Thayer, WR, et al: Computed normal values for peak acid output based on age, sex and body weight, Am J Dig Dis 15:783-789, 1970.
32. McNeely, MDD: Gastrointestinal function. In Sonnenwirth AC, and Jarett, L, editors: Gradwohl's clinical laboratory methods and diagnosis, ed 8, St Louis, 1980, The CV Mosby Co.
33. Herbert, V: Detection of malabsorption of vitamin B_{12} to gastric or intestinal dysfunction, Semin Nucl Med 2:220-234, 1972.
34. Drummey, GD, Benson, JA, and Jones, CM: Microscopical examination of the stool for steatorrhea, N Engl J Med 264:85-87, 1961.
35. Massion, CG, and McNeely, MD: Accurate micromethod for estimation of both medium- and long-chain fatty acids and triglycerides in fecal fat, Clin Chem 19:499-505, 1973.
36. Schwartz, L, Woldow, A, and Dunsmore, R: Determination of fat tolerance in patients with myocardial infarction: method utilizing serum turbidity changes following a fat meal, JAMA 149:364-366, 1952.
37. Silver, S: Radioactive isotopes in medicine and biology, Philadelphia, 1962, Lea & Febiger.
38. Kaihara, S, and Wagner, HN, Jr: Measurement of intestinal fat absorption with carbon-14 labeled tracers, J Lab Clin Med 71:400-411, 1968.
39. Benson, JA, Culver, PJ, Ragland, S, et al: The D-xylose absorption test in malabsorption syndromes, N Engl J Med 256:335-339, 1957.
40. Santini, R, Sheehy, TW, and Martinez-de-Jesus, J: The xylose tolerance test with a five-gram dose, Gastroenterology 40:772-774, 1961.
41. Newcomer AD, McGill, DB, Thomas, PJ, and Hofmann, AF: Prospective comparison of indirect methods for detecting lactase deficiency, N Engl J Med 293:1232-1235, 1975.
42. Onstad, GR, and Zieve, L: Carotene absorption: a screening test for steatorrhea, JAMA 221:677-679, 1972.
43. Parkinson, CE, and Gal, I: Factors affecting the lab management of human serum and liver vitamin A analysis, Clin Chim Acta 40:83-90, 1972.
44. Erlanger, BF, Kokowsky, N, and Cohen, W: The preparation and properties of two new chromogenic substrates of trypsin, Arch Biochem Biophys 95:271-278, 1961.
45. Winawer, SJ, Sherlock, P, Schottenfeld, D, and Miller, DG: Screening for colon cancer, Gastroenterology 70:783-789, 1976.
46. Christensen, F, Anker, N, and Mondrop, M: Blood in feces: a comparison of the sensitivity and reproducibility of five chemical methods, Clin Chim Acta 57:23-27, 1974.
47. Morris, DW, Lee, CS, and Hansell, JR: Presentation at the annual meeting of the American Gastroenterological Society, San Francisco, 1974.
48. MaKarem, A: Hemoglobins, myoglobins and haptoglobins. In Henry, RJ, Cannon, DC, and Winkelman, J, editors: Clinical chemistry principles and technics, New York, 1974, Harper & Row, Publishers.
49. Gold, P, and Freedman, SO: Specific carcinoembryonic antigens of the human digestive system, J Exp Med 122:467-481, 1965.
50. Meeker, WR: The use and abuse of the CEA test in clinical practice, Cancer 41:854-862, 1978.
51. National Institutes of Health Consensus Conference: Carcinoembryonic antigen: its role as a marker in the management of cancer, Br Med J 282:373-375, 1981.
52. Goldberg, H: Specific photometric determination of 5-hydroxyindoleacetic acid in urine, Clin Chem 19:38-44, 1973.
53. Newman, A: Breath-analysis tests in gastroenterology, Gut 15:1-15, 1974.
54. Hofman, AF, and Thomas PJ: Bile acid breath test: extremely simple, moderately useful, Ann Intern Med 79:743-744, 1973.
55. Baker, PG, Barry, RE, and Read, AE: Detection of continuing gluten ingestion in treated coeliac patients, Br Med J 1:486-488, 1975.
56. Jones, EA: Immunoglobulins and the gut, Gut 13:825-835, 1972.
57. Stauffer, JQ, Humphreys, MH, and Weir, GJ: Acquired hyperoxaluria with regional enteritis after ileal resection, Ann Intern Med 79:383-391, 1973.
58. Holdsworth, CD: Intestinal and pancreatic function. In Brown, SS, Mitchell, FL, and Young, DS, editors: Chemical diagnosis of disease, New York, 1980, Elsevier/North Holland, Inc.
59. Rosenberg, IH, and Godwin, HA: The digestion and absorption of dietary folate, Gastroenterology 60:445-463, 1970.
60. Kolhouse, JF, Kondo, H, Allen, NC, et al: Cobalamin analogues are present in human plasma and can mask cobalamin deficiency because current radioisotope dilution assays are not specific for true cobalamin, N Engl J Med 299:785-792, 1978.

CHAPTER 27

Cardiac disease and hypertension

JOHN F. CHAPMAN
LAWRENCE M. SILVERMAN

OBJECTIVES

- Briefly describe myocardial energy metabolism and the important myocardial proteins involved in cell metabolism and function.
- List proteins and enzymes (and their isoenzymes) that are routinely measured in serum to assess myocardial disease, and state the time periods for the expected enzyme elevations.
- List pathological conditions resulting in primary and secondary hypertension.
- Describe the interpretation of renin, catecholamines, BUN, creatinine, potassium, and urine protein measurements in the assessment of hypertension.

KEY TERMS

actin One of the proteins involved in myocardial and arterial smooth muscle contraction.

atherosclerosis ''Hardening of the arteries.'' A process that results in gradual deposition of lipid, fibrin, and calcium in the walls of arteries. The most common cause of death in Western countries.

atrium The chamber of the heart that collects blood from the veins and contracts to expel the blood into one of the two ventricles. There are two atria; the right collects blood from the systemic veins and fills the right ventricle; the left collects blood from the pulmonary veins and fills the left ventricle.

captopril A drug used to treat hypertension. It acts by inhibiting angiotensin-converting enzyme, thereby reducing the concentration of circulating angiotension.

cardiac failure Failure of the heart to maintain the circulation, with resultant accumulation of salt and water and inadequate perfusion of essential organs.

cardiomyopathy A heterogeneous group of disorders that have in common a toxic or genetic insult that affects contracting myocardial cells directly.

Embden-Meyerhof pathway The pathway of anaerobic glycolysis that will convert glucose or glycogen to lactate.

embolus A portion of a blood clot that breaks off from the main clot and travels through the circulation to settle in the lungs (pulmonary embolus) or in one of the systemic arteries (systemic embolus), depending on where the embolus originates.

glycoside A generic term for a group of drugs originally obtained from the foxglove plant. These drugs improve the contractility of the failing heart and slow the rate of ventricular contraction in atrial fibrillation. Digoxin is a member of this family of drugs.

infarction A process of ischemia caused by a loss of blood supply. Thus myocaridal infarction or heart attack occurs when one of the coronary arteries is occluded and the tissue distal to the occlusion dies.

ischemia A reduction of blood supply to tissue that is large enough that the tissue cannot function normally.

Krebs citric acid cycle The pathway of intermediary metabolism that will accept broken-down products from the Embden-Meyerhof pathway and oxidize, decarboxylate, and reduce them, with the production of a relatively large amount of adenosine triphosphate.

myosin One of the contractile proteins involved in cardiac and arterial smooth muscle contraction.

propranolol A beta-blocker drug, that is, a drug that will competitively antagonize catecholamine uptake at beta-receptor sites.

quinidine An antiarrhythmic drug.

renin-angiotensin system A system involved in maintaining the blood pressure.

saralasin The competitive antagonist of angiotensin. It is used intravenously to assess the role of the renin-angiotensin system in maintenance of blood pressure.

technetium-99m ***(^{99m}Tc)*** An isotope injected intravenously to measure left ventricular output or to detect an acute myocardial infarction (*m* means *metastable*).

thallium-201 An isotope injected intravenously to delineate part of the ventricular muscle mass that is ischemic or a scar caused by an old infarction.

ventricles The main pumping chambers of the heart that expel blood into the pulmonary artery and aorta. The left ventricle is more massive and powerful than the right ventricle because the pressure in the systemic circulation is higher than that in the pulmonary circulation; thus a stronger pump is required for the systemic side of the circulation.

Part I: Cardiac disease

ANATOMY AND FUNCTION OF THE HEART

The pumping action of the heart[1] is the prime factor in the maintenance of the body's circulation. The heart is a muscular organ composed of four chambers, the two atria and the two ventricles. The atria collect blood from the systemic circulation and pump it into the ventricles. The right ventricle pumps blood to the lungs for reoxygenation, and the left ventricle pumps blood to the rest of the body, including the heart itself. Cardiac output is determined primarily by the rate at which the heart pumps, by the systemic blood pressure, and by the contractile force developed in the wall of the left ventricle.

Cardiac muscle, which is extremely active, requires large quantities of oxygen for metabolism. Delivery of the oxygen needed to fuel the heart requires a rich capillary bed. Normally, some 75% of the oxygen passing through the heart is extracted from the bloodstream, and changes in metabolic demand, particularly changes in heart rate, are met by alterations in coronary blood flow.

MYOCARDIAL METABOLISM

Energy metabolism

Energy is liberated from various substrates that are used in the heart as fuels by several pathways. These include the Embden-Meyerhof glycolytic pathway, the pentose phosphate shunt, fatty acid oxidation, and the Krebs citric acid cycle (see Chapter 29). The energy produced in the breakdown of substrates is then transported through the electron-transport system of the mitochondria to produce adenosine triphosphate (ATP), the chemical form of stored energy that is used by myocardial tissue to perform work. The heart requires an effective storage method to maintain a reserve of ATP. This is achieved through the synthesis of creatine phosphate, which acts as a backup store for rapid regeneration of ATP when levels fall as a result of increased demand (see p. 431).

The highly aerobic metabolism of the heart allows it to use as fuel many substrates normally present in plasma, and cardiac uptake of most of these substances is proportional to their arterial concentration once certain levels are exceeded. In general terms, the heart uses free fatty acids as its predominant fuel. It also consumes significant quantities of glucose and lactate, as well as lesser amounts of pyruvate, ketone bodies, and amino acids. Most of the energy for cardiac function is obtained from the breakdown of metabolites through the citric acid cycle and oxidative phosphorylation. These enzyme pathways are found principally in the mitochondria, which make up some 35% of the total volume of cardiac muscle.

Enzymes in myocardial cells

Several enzymes found in myocardial tissue[2] are clinically important because detection of their release into the bloodstream can be related to myocardial cell damage and death. Enzymes measured routinely in the clinical laboratory in assessment of myocardial disease are discussed below.

Creatine kinase. The enzyme responsible for production of creatine phosphate is creatine kinase (CK).[2] It has a molecular weight of 85,000 daltons and exists in several isoenzymatic forms (see Chapters 53 and 59). CK is a dimer, composed of the two subunits M (muscle) and B (brain). The three isoenzymes formed from these subunits are found in the cytosol. These isoenzymes are given the abbreviations MM, MB, and BB. If one compares the activity of the enzyme in various tissues, skeletal muscle has by far the highest activity, possessing some 50,000 times the concentration of serum CK. The predominant skeletal isoenzyme is the MM isoenzyme, with only traces of MB and BB isoenzymes. The MB component, however, is increased in certain muscle disorders, particularly the Duchenne type of dystrophy and polymyositis. The skeletal muscle CK activity measured in international units is around 2000 U/g, which compares with the cardiac muscle activity of 500 U/g. In normal serum, at least 95% of the CK present is of the MM type and is probably largely the result of leakage from skeletal muscle, particularly during physical activity. Because of this, serum CK activity in healthy, active persons shows an asymmetrical distribution skewed toward higher values. In addition, values are lower in women than in men and are lower in morning than in evening. Values tend to be lower in hospitalized patients, possibly because bed rest reduces the amount of enzymes released from muscle.

Lactate dehydrogenase. Lactate dehydrogenase (LD)[2] (see Chapters 53 and 59) is a ubiquitous tissue enzyme that catalyzes the reduction of pyruvate to lactate using nicotinamide adenosine diphosphate (NAD). LD has a molecular weight of about 140,000 daltons and is a tetramer composed of four subunits of molecular weight 35,000 daltons each. The subunits, which are of two forms, H (heart) and M (muscle), combine to form the five isoenzymes of LD. The principal isoenzyme in heart (HHHH) is maximally active in the presence of low concentrations of pyruvate but is inhibited by excess pyruvate. In contrast, the major muscle isoenzyme (MMMM) is maximally active in the presence of a higher pyruvate concentration and is less inhibited by excess pyruvate. The heart metabolizes fatty acids and carbohydrates at a fairly constant rate with complete oxidation of pyruvate through the Krebs citric acid cycle. The heart thus has a low tissue concentration of pyruvate and lactate. In contrast, muscle, with sudden demands for increased energy during exercise, has to deal with sudden increases in tissue pyruvate and lactate caused by anaerobic metabolism (see Chapter 53). LD is found in the cytosol of all human cells and thus would have little diagnostic specificity were it not for the presence of isoenzymes that show differences from tissue to tissue. The half-life of LD is different for each isoenzyme; isoenzyme 1 (HHHH) has a half-life of about 100 hours, but isoenzyme 5 (MMMM) has a half-life of only 10 hours. The significance of this is evident in the discussion of the release of enzymes during myocardial infarction.

All five LD isoenzymes occur in all tissues, but relative proportions for each tissue are significantly different (Chapter 53). Heart and red blood cells contain mostly LD_1 and LD_2, whereas skeletal muscle and liver contain LD_5 and to a lesser degree LD_4. Normal serum contains mostly LD_2, with lesser amounts of LD_1 and the other enzymes. If the enzymes are released from cardiac tissue into serum, one frequently sees a change in the ratio of LD_1 to LD_2.

Aspartate aminotransferase. Aspartate aminotransferase (AST; SGOT)[2] (see Chapter 59) catalyzes transfer of an amino group between aspartic acid and pyruvate to form oxaloacetate (alpha-ketoglutarate) and alanine. This ubiquitous enzyme is critical in intermediary metabolism and permits amino acids, aspartic acid, and glutamic acid to be broken down in the Krebs cycle because the reaction is reversible. The enzyme exists in two structurally different forms; one is found principally in the cytoplasm and the other in the mitochondria. It is the cytosol form that is found in serum. Its half-life in serum is probably about 20 hours. Liver has the highest enzyme activity, with some 85 U/g of tissue, whereas heart muscle possesses 75 U/g and skeletal muscle about 50 U/g.

Other important myocardial proteins[2]

Myosin and actin. Myosin and actin proteins form the major part of the contractile apparatus of the myocardial cell. Together they comprise 80% of the muscle cells' protein (see Chapter 28 for details).

Myoglobin. The heart is critically dependent on oxygen, consuming between 6.5 and 10 mL/100 g of tissue per minute at rest. There is a dramatic increase in oxygen requirements with exercise. Normally, some 70% of the oxygen reaching the heart is extracted. The myocardium contains 1.4 mg of myoglobin per gram of tissue. Myoglobin's oxygen dissociation curve is different from that of hemoglobin and thus facilitates entry of oxygen into the cell and is essential for maintaining cardiac function. Myoglobin, with a molecular weight of 17,000 daltons, is small enough to be freely filtered by the glomerulus and is excreted in the urine.

PATHOLOGICAL CONDITIONS

Pathological processes affecting the heart are generally classified into one of the four general categories listed below:

1. Ischemic heart disease
2. Cardiomyopathy
3. Arrhythmias
4. Congenital and valvular heart disease

Ischemic heart disease

Ischemia[1] is a condition in which an organ has an inadequate blood supply to maintain its essential functions. There are many causes of myocardial ischemia, but the most common cause by far is coronary atherosclerosis. This condition can develop over many years, sometimes even beginning in childhood, as the arteries supplying blood to the heart gradually narrow because of deposition of cholesterol and other material in the arterial wall. Coronary vasospasm can also produce ischemia. In coronary vasospasm the arterial wall constricts in an abnormal and prolonged fashion because of supersensitivity to normal vasoconstrictor signals. Other less common causes of myocardial ischemia are inflammation of the coronary arteries, thromboses (blood clots), severe anemia, and severe hypotension.

Effects of occlusion on myocardium. Cessation of blood flow produces a complex series of consequences.

Not only is there serious hypoxia because tissue oxygen concentration drops drastically, but also there is the problem of removal of toxic cellular metabolites from ischemic tissue. Sudden occlusion of the coronary artery leaves myocardial cells with only a few seconds of aerobic metabolism as they rapidly consume the oxygen supplies remaining in the microvasculature. Once the oxygen supply has been exhausted, oxidative phosphorylation is unable to continue. Myocardial metabolism then switches to use of glycogen or glucose in the anaerobic Embden-Meyerhof pathway, instead of the aerobic Krebs cycle. Since the Krebs cycle is unable to function, the end product of anaerobic glucose metabolism, pyruvate, is reduced to lactate, and lactate accumulation is one of the earliest and most dramatic signs of myocardial ischemia.

As ischemia progresses, creatine phosphate reserves are used up, adenosine triphosphate levels fall, and cardiac tissue becomes more acidic as lactate and other acidic intermediates of glycolysis accumulate. Up to 15 to 20 minutes after an ischemic incident, the tissue will recover if it is reperfused. However, after about 20 minutes of occlusion, over 60% of the cellular adenosine triphosphate has been used up and the amount of lactate in wet myocardial tissue is 12 times its normal aerobic level. In addition, all cellular glycogen has been used. Once all the glycogen and creatine phosphate reserves have been used, dramatic ultrastructural changes occur, indicating cell damage. Cell membrane damage is also evident. If, at this point, the obstruction is relieved and the myocardial tissue is reperfused with blood, cell lysis and contracture develop. The integrity of the myocardial cell is rapidly disrupted, and it is unable to tolerate the arrival of fresh blood.

The point when reversible ischemic injury becomes irreversible ischemic injury is the point when the cell is no longer able to maintain membrane integrity. Damage to the cell membrane results in the release of intracellular contents. Of these contents, it is release of enzymes that is particularly significant in the evaluation and confirmation of irreversible ischemic injury. Enzyme release shows irreversible damage to myocytes and indicates that the patient has had a myocardial infarction rather than simply an anginal or transient ischemic episode. The enzymes and proteins that are released after ischemic irreversible injury are soluble enzymes such as creatine kinase (CK), asparatate aminotransferase (AST), lactate dehydrogenase (LD), aldolase, myokinase, alanine aminotransferase (ALT), and myoglobin.

Most enzymes released early in the infarction are cytoplasmic. Mitochondrial enzymes, if soluble, are released also, but there is usually some delay before they appear in plasma. Once membrane damage has occurred, the rate of appearance of an enzyme in the circulation appears to depend on the rate and extent of reperfusion of the damaged myocardium and on the size of the enzyme molecule. The more flow, the better the reperfusion, and the smaller the molecule, the sooner the molecule will be seen in peripheral blood. CK, with a molecular weight of 80,000 daltons, is released before AST, with a molecular weight of 120,000 daltons, which in turn is released before LD, with a molecular weight of 140,000 daltons.

Cardiomyopathy

Cardiomyopathy is a disorder that is characterized by inadequate muscle contraction caused by direct damage to myocardial cells. Although there are some specific forms of cardiomyopathy, most clinical cases of cardiomyopathy are idiopathic; that is, the cause is unclear. Cardiomyopathy manifests as an enlargement of all four chambers of the heart and cardiac failure. The biochemical findings in most cases of cardiomyopathy are nonspecific and reflect the major clinical presentation, that is, cardiac failure. Some cases of cardiomyopathy are sequelae to severe viral infection.

Arrhythmias

Although the structure of the heart as a pump is relatively simple, the regulatory system that balances one side of the heart against the other and regulates cardiac function in relation to the needs of body organs is necessarily much more complex and therefore more at risk of damage.

The anatomical changes that result in cardiac arrhythmias are relatively nonspecific and are frequently more related to the disease process affecting the heart muscle. An important cause of arrhythmias is a myocardial infarction. Whatever the source of the damage, this damage frequently distorts the transmission of the cardiac impulse, producing abnormal and self-sustaining irregular contractile activity. In functional terms the rhythm abnormalities are classified as bradycardias (resulting in heartbeat rates less than 60/min) or as tachycardias (producing heartbeat rates faster than 100/min). The arrhythmias can affect atrial or ventrical contractions and can be acute or chronic. Atrial fibrillation is a fairly common rhythm abnormality in which the atria beat in an irregular, chaotic, and rapid fashion. Chronic arrhythmias are controlled by drugs, whose serum levels should be routinely monitored.

Congenital and valvular heart disease

The multitude of congenital heart abnormalities that have been described will not be discussed. In general terms, all components of the heart can be affected by maldevelopment. Most causes are unknown, but one important exception is rubella infection of the mother during the first trimester of pregnancy.

Of the acquired valvular diseases of the heart, one large group is caused by rheumatic carditis. In susceptible subjects affected by the hemolytic streptococcus, the body develops an immune reaction against myocardial tissue, particularly the valves, which become damaged and deformed.

Table 27-1 Common drugs for treatment of cardiac disease

Name	Class	Mode of action	Comments
Digoxin	Cardiac glycoside	Increases contractility; slows A-V impulse conduction	Narrow therapeutic range; long half-life can produce cumulative toxic effects; can cause arrhythmias
Lidocaine	Antiarrhythmic	Suppresses ventricular abnormalities; attenuates depolarization, automaticity, and excitability	Relatively safe with no cardiac side effects; may cause cerebral toxicity
Quinidine	Antiarrhythmic	Depresses myocardial excitability, conduction velocity, and contractility	Coadministration with digoxin can produe digoxin toxicity; can cause arrhythmias

FUNCTION TESTS

The most important tests for assessing cardiac function are the electrocardiogram (EKG)[3] and myocardial imaging techniques.[1] The EKG involves noninvasive recording of the electrical impulses through the heart and is an effective, but not perfect, means of assessing cardiac rhythm abnormalities. Myocardial imaging techniques (technetium-99m pyrophosphate and thallium 201) are used to assess cardiac output and wall motion abnormalities and to detect nonfunctioning regions of the myocardium caused by infarction.

DRUG THERAPY

Two groups of drugs[4,5] have a direct effect on cardiac tissue and are therefore monitored by the clinical chemistry laboratory. These are the cardiac glycosides and the antiarrhythmic drugs. Table 27-1 lists the most frequently monitored cardiac drugs.

Glycosides

The glycosides increase contractility in the heart and are particularly useful in treating heart failure. The problem with glycosides is that the therapeutic-to-toxic ratio is very low. At toxic dosage levels, they can cause upsets in the electrical activity of the heart and can induce fatal arrhythmias. Since they have long half-lives in the body, they tend to have cumulative effects, necessitating careful monitoring of blood levels to reduce the possibility of toxic side effects.

Lidocaine

Lidocaine is an anesthetic compound that is given intravenously as a bolus or infusion to suppress ventricular irregularities and to prevent the induction of life-threatening arrhythmias, such as ventricular tachycardia. Lidocaine is a particularly attractive drug because it has few side effects and because there is little or no effect on myocardial function and cardiac conduction. Toxicity may be manifested by major convulsive seizures with no prior warning. Accordingly, it becomes important to try to prevent lidocaine toxicity by use of therapeutic monitoring. Toxicity is particularly common in older patients.

Quinidine and other drugs

Quinidine is an antiarrhythmic drug whose major side effect is its capacity to induce severe arrhythmias in its own right. Indeed, many antiarrhythmic drugs have this potential, and for this reason, drug blood levels are usually monitored so that one may use antiarrhythmic drugs as safely as possible. Besides quinidine, other frequently monitored antiarrhythmic drugs are *procainamide* and *disopyramide*.

CHANGE OF ANALYTE IN DISEASE

The analytes that are most frequently used as diagnostic indicators of cardiac disease are the enzymes creatine kinase (CK) and lactate dehydrogenase (LD). Less frequently used are myoglobin and aspartate aminotransferase (AST). These serum markers are used *only* for the diagnosis of acute myocardial infarction (AMI). Currently there are no biochemical markers unique for the other cardiac diseases. This section discusses the clinical value and limitations of these analytes in the assessment of AMI.

Myoglobin

The value of myoglobin assays[6,7] in AMI is limited by myoglobin's early and rather brief appearance in serum after a myocardial infarction and by the presence of large amounts of myoglobin in skeletal muscle, which reduces the specificity of this analyte for myocardial damage.

Myoglobin is rapidly excreted in urine, and, as may be expected, reduced renal clearance results in elevations of serum myoglobin. Practical experience with myoglobin has not led to an increased ability to detect AMI. Thus this analyte is infrequently used to confirm or rule out AMI.

Enzymes and isoenzymes in acute myocardial infarction

After the onset of symptoms in AMI in most patients there is a so-called time window, during which the en-

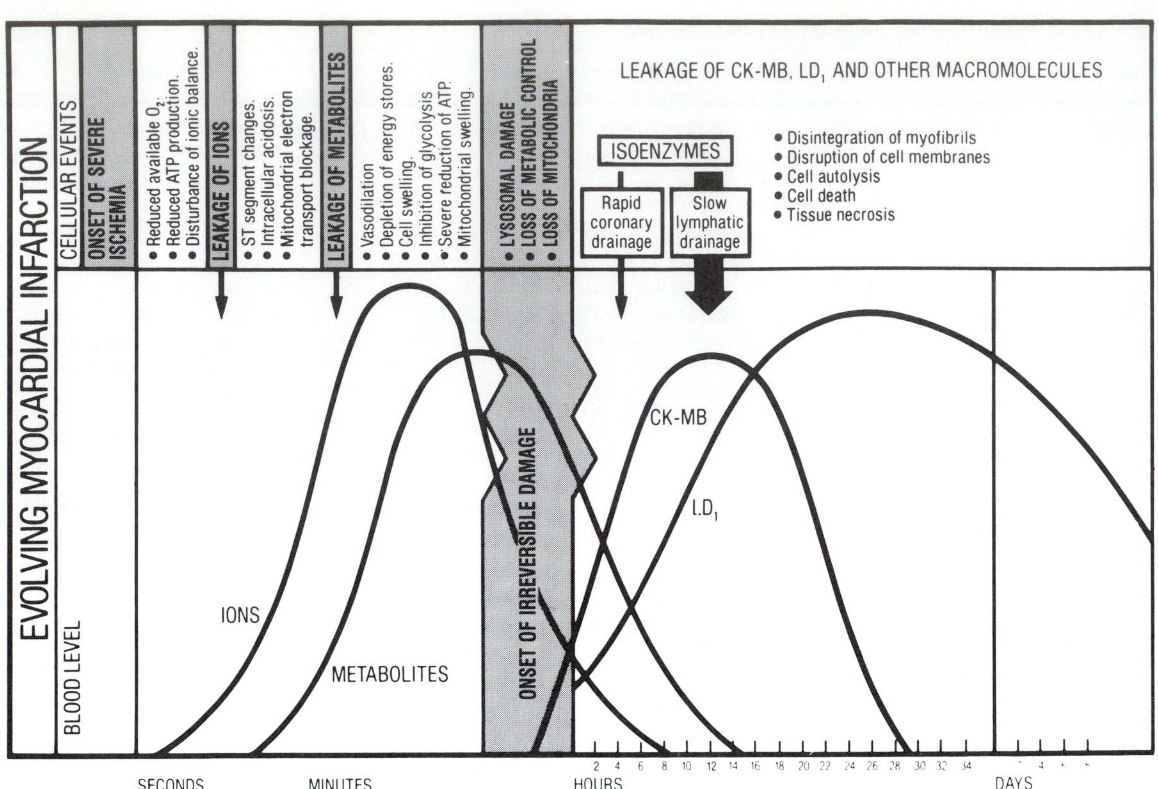

Fig. 27-1 Time course of biochemical events in a typical acute myocardial infarction (AMI). *(Adapted from Usategui-Gomez, M: Lab World 32:49-52, 1981.)*

zymes released from the damaged myocardial tissue are found to be elevated in serum. This temporal relationship is unique for each analyte and varies somewhat between persons, although a typical pattern has been determined (Fig. 27-1). Usually, 4 to 6 hours are required after the onset of chest pain before CK-MB becomes elevated in the serum of patients with AMI.[8] Detection times as short as 2 hours and as long as 15 hours have been reported.[9] This activity peaks at 12 to 24 hours and usually returns to baseline levels within 24 to 48 hours. In a typical time course for CK and LD isoenzymes, CK-MB peaks first, with LD_1 exceeding LD_2 5 to 20 hours later. The average peak times for enzyme release after infarction in a study of 47 patients were 18 hours for CK-MB and total CK, 24 hours for aspartate aminotransferase (AST), and 48 hours for LD.[10] This time course represents the classic temporal sequence of enzyme changes and is often helpful in distinguishing uncomplicated AMI from extension or reinfarction. Whereas this complete pattern is highly sensitive and specific for infarction, slightly different temporal patterns have been reported in up to 25% of patients with AMI.[11] These workers also report that CK-MB levels are higher and persist longer in patients with cardiogenic shock after AMI.

It follows from the above discussion that proper interpretation of enzyme values in the diagnosis of AMI requires that samples be collected at appropriate time intervals. Since the temporal sequence of events is critical in assessment of the biochemical changes after AMI, the laboratory should normally not analyze a single isolated specimen. Most sources recommend that samples be drawn on admission and at 6 hours, 12 hours, and 24 hours when AMI is suspected.[8,9,12] The absolute minimum number of samples recommended for interpretation is two, obtained 12 and 24 hours after the onset of symptoms. Normal results for CK-MB in samples obtained before 12 hours or after 24 hours should not be used to exclude the diagnosis of AMI. A sample collected in this fashion may reflect the period before or after CK-MB elevation occurs in some patients.[13] Note that normal values for total CK should not be used to rule out AMI or to exclude a request for CK-MB analysis. Numerous cases have been documented in which elevated CK-MB levels are accompanied by normal total CK levels during the initial course of an AMI.[13] In most of these cases, however, there is a distinct rise and fall of the total CK values, although none ever exceeds the upper limit of normal. There is now some evidence that three subforms of CK-MM, termed isoforms, may be elevated before CK-MB. Although the pattern is not totally specific for AMI, this approach may be useful in the future as an early indicator of this disease.[14]

For patients who are first seen many hours after the onset of pain, measurement of LD isoenzymes may be helpful, since an increase in LD_1 with concomitant reversal of the normal LD_1/LD_2 ratio, termed the *LD flip,* generally occurs at 12 to 24 hours after infarction and may persist for several days. Elevated AST values can also persist for about 96 hours after infarction and, although less specific than LD_1 elevations, can also provide helpful information when the time of onset is unknown or hospital admission is delayed.[11] When the time of clinical onset is known, normal results from CK-MB, total CK, and LD in properly collected samples are highly reliable for ruling out AMI.[9]

Isoenzymes in conditions other than myocardial infarction

Frequently the pattern of serum enzyme elevations in conditions other than myocardial infarction is different from that seen with myocardial infarction. In these conditions isoenzymes (see boxes) are often chronically elevated. These "false-positive" situations emphasize the importance of testing multiple samples obtained at appropriate intervals. CK isoenzymes undergo fetal developmental expression that is duplicated in certain pathological states. This alteration is most commonly observed in skeletal muscle, in which the adult form of CK is predominantly the MM form. Fetal muscle is predominantly BB until the sixteenth week of gestation,[15] at which time the expression of the gene coding for the M subunit is significantly increased. Diseases of skeletal muscle characterized by blood-perfusion regeneration are often associated with increased levels of the fetal isomeric forms. Thus Duchenne's muscular dystrophy, polymyositis, and some forms of rhabdomyolysis are examples of skeletal muscle diseases in which increased levels of CK-MB are frequently found in serum in proportion to the degree of fiber regeneration.[15] In these chronic diseases, the serum levels of total CK and CK-MB activities do not show the rise and fall characteristics of myocardial infarction (MI), but tend to remain constant over time. However, the levels of CK-MB are frequently in the range associated with myocardial infarction. Since the absolute amount of CK in skeletal muscle is about 5 to 10 times that observed in cardiac tissue, the actual elevations of total CK observed in serum in skeletal muscle abnormalities are frequently dramatically higher than those observed in myocardial infarction. Also, the LD isoenzyme pattern does not reflect myocardial origin.

In addition to the pathological conditions discussed above, elevations of CK isoenzymes, LD isoenzymes, and AST are obviously present in any situation that increases leakage from myocardial cells. Examples include trauma such as surgery or flail chest and diseases involving the myocardium, such as myocarditis, congestive heart failure, and arrhythmias.

Sensitivity and specificity of CK and LD isoenzymes for myocardial infarctions

Clearly, the interpretation of serum enzymes in the diagnosis of AMI becomes obscured by the causes of false-positive and false-negative results. Galen and Gambino[16] have used the predictive value model to express mathematically the clinical usefulness of laboratory tests in var-

Table 27-2 Positive and negative predictive values (PV) for serum CK-MB and LD ratio change ("flip") with varying prevalence of acute myocardial infarction

	Test					
	CK-MB*		LD "flip"†		Both‡	
Prevalence	PV+	PV−	PV+	PV−	PV+	PV−
50%	96	100	100	91	100	96
5%	51	100	100	99.5	100	99.7

*Sensitivity = 100%; specificity = 95%.
†Sensitivity = 90%; specificity = 100%.
‡Sensitivity = 95%; specificity = 100%.

ELEVATED SERUM CK-MB IN CONDITIONS OTHER THAN ACUTE MYOCARDIAL INFARCTION

Severe angina
Chronic atrial fibrillation
Coronary insufficiency
Crush syndrome
Pericarditis
Defibrillation
Insertion of pacemaker
Coronary angiography
Open heart surgery
External cardiac massage or cardiopulmonary resuscitation
Carbon monoxide poisoning
Malignant hyperthermia
Muscular dystrophy, such as Duchenne's syndrome
Poliomyositis
Prostatic surgery or infarction
Dermatomyositis
Reye's syndrome
Malignancy

CAUSES OF INCREASED LD_1/LD_2 RATIO ("FLIP")

Acute myocardial infarction
Acute renal infarction
Hemolysis caused by
- Prosthetic heart valves
- Hemolytic anemias
- Megaloblastic anemias
- Sample during processing

Malignancy

ious patient populations. Table 27-2 lists both positive and negative prediction values of CK-MB and LD ratio change ("flip") for high- and low-prevalence populations. The use of CK-MB in a coronary care population, with a 50% prevalence of disease assumed, clearly increases the predictive values of elevated enzyme levels. In comparison, in a low-prevalence population (5%) in which many conditions outlined in the left box on p. 421 may be seen, the usefulness of this same enzyme test to assess AMI is greatly diminished.

Part II: Hypertension

Hypertension[17,18] is a persistent elevation of blood pressure that continues in the resting (basal) state.

It is now known that treating even mildly hypertensive subjects prolongs life and prevents hypertensive complications such as stroke.[19-21] With the advent of a large number of different therapies, the treatment of hypertension has been revolutionized over the last 25 years and is now available for consideration in a significant proportion, perhaps 10% to 20%, of the American adult population.

REGULATION OF BLOOD PRESSURE

Systemic arterial pressure[22] must be maintained in order for essential organs to function. Maintenance of appropriate vascular pressure is ensured by several control systems. The site of synthesis of these controlling factors and the mechanisms by which these factors control blood pressure are listed in Table 27-3.

Of the many humoral effectors interacting with specific receptors on the arterial smooth cell wall, the most important is angiotensin II (see Chapter 20), which is the most powerful vasopressor agent known to man. Renin, a proteinase, acts on the glycoprotein *angiotensinogen,* produced by the liver, to yield angiotensin I, which in turn is converted by the ubiquitous *angiotensin-converting enzyme* (ACE) to angiotensin II. The renin-angiotensin-aldosterone system is particularly important because many hypertensive persons, particularly young ones and those who have hypertension caused by renal disease, have elevations in the activity of this system. This has both diagnostic and therapeutic possibilities. One can determine if renal disease is the cause of hypertension in some patients by measuring the components of the renin-angiotensin system. More recently, there have become available specific inhibitors of components of the renin-angiotensin system, which offer the chance of treating hypertension in selected subjects with specific therapy.

ROLE OF CATECHOLAMINES IN ARTERIAL TENSION

The two major catecholamine hormones are epinephrine and norepinephrine. Norepinephrine is synthesized primarily at the terminals of the sympathetic nervous system, and epinephrine is synthesized primarily in the adrenal medulla (see Chapter 45). Both hormones, but particularly norepinephrine, elevate blood pressure through their effect on the heart and blood vessels, but there are significant differences in their individual effects. Thus epinephrine vasodilates the arterioles of the skeletal muscles and vasoconstricts those of the skin, mucous membranes, and viscera. Norepinephrine has less effect on cardiac output than epinephrine; its effect is that of general vasoconstriction.

PATHOLOGICAL CONDITIONS[17,18]

Essential or primary hypertension

In most cases of hypertension, generally called *essential hypertension,* the cause is unknown. There are no diagnostic biochemical features associated with this disease. It is, however, very common in developed societies, and its incidence tends to increase with age. Among suggested causal factors are stress and the intake of excessive salt during childhood. However, essential hypertension is most

Table 27-3 Factors that regulate blood pressure

Factor	Site of synthesis	Mechanism and sites of action
Arterial baroreflex activators		
Epinephrine	Adrenal medulla	Vasodilation of arterioles of skeletal muscle; vasoconstriction of arterioles of skin, mucous membranes, and viscera; increases in rate and force of cardiac contraction
Norepinephrine	Terminals of sympathetic nervous system	General vasoconstriction
Body fluid volume regulators		
Antidiuretic hormone (ADH)	Neurohypophysis	Enhanced water reabsorption; increased plasma volume
Aldosterone	Adrenal cortex	Renal tubular sodium reabsorption; increased plasma volume
Renin	Juxtaglomerular cells of kidney	Converts angiotensinogen to angiotensin I
Angiotensin-converting enzyme	Lung	Converts angiotensin I to angiotensin II (potent vasoconstrictor, stimulates aldosterone production)
Vascular autoregulation	Tissue/organ specific	Local mechanisms to maintain constant tissue perfusion

ETIOLOGICAL CLASSIFICATION OF SECONDARY HYPERTENSION

Coarctation of the aorta
Endocrine causes
- Primary aldosteronism
- Pheochromocytoma
- Adrenogenital syndrome
- Cushing's syndrome

Renal causes
- Parenchymal (tumor, cyst, pyelonephritis)
- Renovascular (arteriosclerosis, fibromuscular hyperplasia)

Drug-induced causes
- Oral contraceptives
- Excess licorice ingestion

likely caused by a mosaic of factors, including genetic and environmental components. Idiopathic, or essential, hypertension must be differentiated from secondary hypertension, in which an underlying pathological abnormality can be found to explain the hypertension. The vast majority of persons with hypertension have essential hypertension that is not surgically curable.

Secondary hypertension

Although secondary causes of hypertension amount to only 5% of cases of hypertension, they are important because some forms do offer the possibility of medical or surgical cure. The box above lists some of the common causes of secondary hypertension. Of those listed, renovascular disease, primary aldosteronism, and pheochromocytoma are often curable with surgery. Because of the possibility for surgical cure, it is important to make the correct diagnosis, and laboratory investigation to rule out these diseases is warranted. Table 27-4 lists the laboratory tests that are used for detection and diagnosis of secondary causes of hypertension.

Pheochromocytoma is unique in that hypertension is a major consequence of the condition, whereas in most other causes of secondary hypertension, the hypertension is only one aspect. In pheochromocytoma epinephrine and norepinephrine are produced in excess. The hypertension so produced is usually paroxysmal, manifested initially by devastating headache and palpitations. The pheochromocytoma tumors are usually benign, but sometimes they are malignant (about 10%) and can metastasize to the liver or lung. One problem with this relatively rare condition (about 1 in 1000 cases of secondary hypertension) is that pheochromocytomas can be multiple and recurrence is not unusual.

DRUG THERAPY

There are many drugs[17,18,23] that may be useful for treating hypertensive persons. The most common antihypertensive drugs are listed in Table 27-5. By observation, it has been found that it is best to monitor drug therapy by noting reductions in blood pressure level and the absence of side effects. One can assess the reduction of blood pressure and the absence of side effects most reliably without the use of laboratory procedures.

Diuretics

It is usually found that the reduction in blood pressure with a diuretic is quickly limited, if the dose is increased, by the appearance of side effects, such as prerenal azotemia. Accordingly, clinical practice generally limits the use of diuretics in hypertension to relatively modest doses. Monitoring the serum potassium level is particularly important in the many subjects who are treated with a diuretic for hypertension, since the majority of diuretics have the obverse capability of producing hyperkalemia. In the majority of cases, drastic alterations of potassium level can occur without clinical symptoms, and it becomes important to assess potassium routinely in persons receiving diuretic therapy to avoid the major complication of hypokalemia and hyperkalemia, that is, cardiac arrhythmia.

Renal damage caused by hypertension is often manifest as the development of renal insufficiency, which may be

Table 27-4 Diagnostic laboratory tests for secondary hypertension

Type	Screening tests	Definitive tests
Endocrine disease (see Chapter 45)		
Primary aldosteronism	Plasma electrolytes, urine, Na^+, K^+	Plasma aldosterone (suppressed), plasma renin activity (stimulated)
Pheochromocytoma	Urine metanephrines and vanillylmandelic acid (VMA), plasma catecholamines	Fractionated urine metanephrines, catecholamines; plasma catecholamines
Adrenogenital syndrome	Usually none; history and physical examination findings	11-Desoxycortisol, 17-hydroxyprogesterone, dehydroepiandrosterone
Cushing's syndrome	Plasma, urine electrolytes, CBC, plasma glucose	Plasma cortisol, urine free cortisol
Renal disease (see Chapter 22)		
Renal parenchymal disease, renovascular disease	Urinalysis, renal function tests	Plasma renin activity (peripheral and differential renal vein levels)

Table 27-5 Common antihypertensive drugs

Name	Class	Site of action	Mode of action
Hydrochlorothiazide	Natriuretic diuretic	Kidney, proximal tubules	Blocks resorption of Na^+, K^+, Cl^-
Spironolactone	Potassium-sparing diuretic	Kidney, distal tubule	Blocks distal sodium resorption
Propranolol	Beta-adrenergic receptor blocker	Beta receptors	Blocks action of receptor-stimulating agents
Hydralazine	Peripheral vasodilator	Vascular smooth muscle	Direct relaxation of arteriolar smooth muscle
Methyldopa	Alpha-adrenergic stimulator	Central inhibitory alpha-adrenergic receptors	Metabolite stimulates alpha receptors
Guanethidine	Postganglionic neuronal blocker	Sympathetic neuroeffector junction	Blocks release and distribution of norepinephrine
Captopril	Peptidyldipeptide carboxyhydrolase inhibitor	Renin-angiotensin-aldosterone system	Blocks conversion of angiotensin I to angiotensin II

occult. Consequently, many patients with hypertension have routine serial assessments of renal function, with particular attention being paid to blood urea nitrogen or creatinine levels. A rise in these parameters, even of modest size, is particularly disturbing because it may imply inadequate control of blood pressure. The alternative possibility is that since the therapy being used can in fact reduce renal blood flow, the patient may have developed a degree of prerenal azotemia. This must be guarded against, particularly with diuretic therapy.

Of the many diuretics available, the one most commonly used to treat hypertension is a thiazide diuretic. The fall in blood pressure is initially caused by sodium and water excretion and reduction in the vascular compartment, but a more prolonged effect appears to be the result of a direct relaxant effect on the arteriole.

Thiazides, like most diuretics (except the potassium-retaining ones), may cause potassium loss. In addition, they have certain metabolic complications that must be monitored regularly. These complications include hyperglycemia and occasionally frank diabetes and hyperuricemia. Thus regular monitoring of serum potassium, glucose, and uric acid is necessary.

Potassium-retaining diuretics (such as spironolactone, which antagonizes the action of aldosterone on the renal tubule) are useful in eliminating the need for potassium replacement therapy, but regular monitoring for hyperkalemia is essential. These diuretics should *not* be given with potassium supplements.

Beta blockers

Beta blockers (beta-adrenergic receptor–blocking agents) are a group of drugs that act by blocking the stimulatory input (neural) of the beta-adrenergic receptor sites of muscle cells.

The use of a beta-blocker drug for hypertension is limited by the clinically observed fact that at high dosages, increasing increments of beta blocker produce progressively less satisfactory reductions in blood pressure. It is, of course, possible to measure the serum level of propranolol (the most commonly used beta blocker) and to adjust the dose accordingly, but in practice this is rarely done. Most clinicians gradually increase the dose of propranolol to a level with which they feel comfortable. Then, if adequate control of the blood pressure has not been obtained, they add a second medication, such as a diuretic.

Other drugs

The standard practice after use of a diuretic and beta blocker is to add a vasodilator drug, such as hydralazine, or a centrally active drug, such as methyldopa. The next step would be to use even more potent drugs, such as minoxidil or guanethidine. However, these very potent drugs have an increased incidence of severe side effects.

The capability for assessing the renin-angiotensin system is important, not only for assessing the pathogenesis of hypertension, but also for assessing therapy. Persons with hypertension resulting from the activity of the renin-angiotensin system can be treated with a specific antagonist to angiotensin-converting enzyme, such as captopril, or with a less specific but safer suppressor of renin production, such as a beta blocker. On the other hand, those persons with low plasma renin activities are much more likely to respond to a diuretic or vasodilator drug in the treatment of their hypertension.

CHANGE OF ANALYTE IN DISEASE[24,25]

For the vast majority of hypertensive patients, that is, those with essential hypertension, extensive laboratory evaluation is not warranted. This is because neither standard laboratory tests (ECG, urinalysis, renal function tests) nor more specialized tests (aldosterone, catecholamines, renin) are likely to be very informative. When secondary hypertension is suspected, largely because of historical or clinical information, more extensive laboratory evaluation may be useful.

Renin-angiotensin system

Excess activity of the renin-angiotensin system is a common component in the production and maintenance of hypertension (see Chapters 20 and 45). Measurement of renin activity in plasma is of little value on a casual basis because random samples from normotensive and hypertensive subjects show a wide overlap. The subject should be ambulatory for 2 hours before the test and should take a diuretic to stimulate renin production. Some investigators find renin assays of plasma from the renal vein more valuable than renin measurements in peripheral blood for evaluating renal involvement in hypertension; a unilateral increase in secretion rate is highly suggestive of a renal cause (particularly renal artery stenosis) of the hypertension. These samples have to be obtained directly from the renal veins and require femoral vein catheterization. A ratio of the renin level of a sample obtained from one renal vein to the renin level obtained from the other renal vein of 2:1 (some laboratories use a ratio of $\geq$1.5:1) strongly suggests a renal cause of hypertension and indicates that a surgical approach (unilateral nephrectomy for unilateral parenchymal disease or renal artery surgery for renal artery stenosis) may be curative. With the advent of the saralasin infusion test, it is easier to assess this particular system by competitive antagonism of angiotensin II uptake at arterial receptors. With the patient resting supine, a drop of $\geq$8 mm Hg in the systolic pressure is virtually diagnostic of significant involvement of the renin-angiotensin system in the maintenance of hypertension.

Aldosterone

Patients with primary hyperaldosteronism often, but not always, have hypokalemia. Once diuretics have been eliminated as a possible cause of the low potassium, plasma aldosterone, peripheral renin activity, and 24-hour aldosterone excretion tests should be performed. In patients with bilateral adrenal hyperplasia or an aldosterone-producing adenoma (APA), plasma and urine aldosterone values are typically elevated and plasma renin activity is decreased. Patients with APA typically show little difference in plasma aldosterone values measured from samples taken with the patient in either the supine or standing position. However, patients with bilateral adrenal hyperplasia generally produce higher plasma aldosterone values after 4 hours of erect posture.

Catecholamines

Catecholamine assays have little use in the assessment of most hypertensive persons. In labile hypertension (that is, blood pressure that fluctuates above and below a rather artibrary cutoff level), there is some evidence that excess catecholamine production is a causative factor and, if sustained, may result in permanent hypertension. Catecholamine assays are of particular value if pheochromocytoma is being considered (see Chapter 45) because in this condition hypertension is caused by excessive production of catecholamines (epinephrine, norepinephrine) and their principal metabolites: the metanephrines and vanillylmandelic acid (VMA). Generally, the total urinary metanephrines test on a 24-hour sample is the preferred screening test. Although urinary VMA is usually measured concurrently, it may be reserved for confirmation of an elevated metanephrines result. Fractionation of total urinary metanephrines (metanephrine, normetanephrine) or the measurement of urinary free catecholamines (epinephrine, norepinephrine) is sometimes useful in evaluating the possibility of tumor in extra-adrenal sites. Many patients with adrenal medullary pheochromocytomas produce excessive epinephrine and metanephrine, whereas patients with extra-adrenal tumors tend to produce mainly norepinephrine and normetanephrine.

Renal analytes

A renal profile is a common and regular procedure performed on hypertensive persons. Poorly controlled hypertension frequently damages the kidneys, and renal insufficiency may become manifest only as a progressive rise in blood urea nitrogen (BUN) or creatinine. A rise in BUN, however, occurs late in the disease, by which time a significant amount of renal damage may have occurred. Proteinuria is also a common manifestation of hypertension and is usually measured to monitor the disease. Potassium is an important assay not only because it may point to the underlying pathological condition (such as severe, persistent hypokalemia in Conn's syndrome) but also because hypokalemia is a common problem with diuretic therapy and its development can be so insidious (mild fatigue) that it may eventually present as a life-threatening cardiac arrhythmia. The converse problem, hyperkalemia, may occur with potassium-conserving diuretics, especially if the patient continues to take potassium supplements or if renal insufficiency develops. A creatinine clearance is often useful in separating renal and prerenal azotemia.

REFERENCES

1. Horst, JW, et al, editors: The heart, arteries, and veins, ed 6, New York, 1986, McGraw Hill.
2. Hearse, DJ: Enzymes in cardiology: diagnosis and research, New York, 1979, John Wiley & Sons, Inc.
3. Chou, TC: Electrocardiography in clinical practice, New York, 1979, Grune & Stratton, Inc.
4. Singh, BN, Collett, JT, and Chew, CYB: New perspectives in the pharmacologic therapy of cardiac arrhythmias, Prog Cardiovasc Dis 224:243-301, 1980.
5. Opie, LH, editor: Drugs for the heart, ed 2, Orlando, 1987, Grune & Stratton.
6. Grenadier, E, Keider, S, Hahna, L, et al: The rules of serum myoglobin, total CPK and CK-MB isoenzyme in the acute phase of myocardial infarction, Am Heart J 105:408-416, 1983.
7. Drexel, H, Dworzak, E, Kirchmair, W, et al: Myoglobinuria in the early phase of acute myocardial infarction, Am Heart J 105:642-651, 1983.
8. Guzy, PM: Creatinine phosphokinase-MB (CPK-MB) and the diagnosis of myocardial infarction, West J Med 127:445-460, 1977.
9. Lott, JA, and Stang, JM: Serum enzymes and isoenzymes in the diagnosis and differential diagnosis of myocardial ischemia and necrosis, Clin Chem 26:1241-1250, 1980.
10. Blomberg, DJ, Kimber, WD, and Burker, MD: Creatine kinase iso-

enzymes: predictive value in the early diagnosis of myocardial infarction, Am J Med 59:464-469, 1975.
11. Neufeld, HN, Rabinowitz, B, Clejan, S, et al: Isoenzymes of creatine phosphokinase in acute myocardial infarction, Angiology 28:853-864, 1977.
12. Lee, TH, and Goldman, L: Serum enzyme assays in the diagnosis of acute myocardial infarction, Ann Intern Med 105:221-223, 1986.
13. Irvin, RG, Cobb, FR, and Roe, CR: Acute myocardial infarction and creatine phosphokinase, Arch Intern Med 140:329-334, 1980.
14. Hashimoto, H, Abendschein, DR, Strauss, AW, et al: Early detection of myocardial infarction in conscious dogs by analysis of plasma MM creatine kinase isoforms, Circulation 71:363-368, 1985.
15. Foxall, CD: Changes in creatine kinase and its isoenzymes in human fetal muscle during development, J Neurol Sci 24:483-492, 1975.
16. Galen, RS, and Gambino, SR: Beyond normality, New York, 1975, John Wiley & Sons, Inc.
17. Kaplan, NL: Clinical hypertension, ed 2, Baltimore, 1978, The Williams & Wilkins Co.
18. Marshall, AJ, and Bassett, DW: The hypertensive patient, London, 1980, Pitman Publishers (Pitman Medical).
19. Veterans Administration Cooperative Study Group on Antihypertensive Agents: Effects of treatment on morbidity in hypertension. I. Results in patients with diastolic blood pressures averaging 115 through 129 mmHg, JAMA 202:1020-1034, 1967.
20. Veterans Administration Cooperative Study Group on Antihypertensive Agents: Effects of treatment on morbidity in hypertension. II. Results in patients with diastolic blood pressure averaging 90 through 114 mm Hg, JAMA 213:1143-1152, 1970.
21. Hypertension Detection and Follow-up Program Cooperative Group: Five year findings of the hypertension detection and follow-up program. II. Mortality by race, sex and age, JAMA 242:2572-2577, 1979.
22. Vanhoutte, PM: Vasodilatation, New York, 1981, Raven Press.
23. Drayer, JIM, Lowenthal, DT, and Weber, MA: Drug therapy in hypertension, New York, 1987, Dekker.
24. Tucker, RM: Diagnostic procedures in hypertension. In Duarte, CG, editor: Renal function tests, Boston, 1980, Little Brown & Co.
25. Burke, MD: Hypertension: strategies for laboratory diagnosis, Postgrad Med 67:77-85, 1980.

CHAPTER 28

Muscle disease

BÉLA NAGY

OBJECTIVES

- List the various types of muscles and their functions.
- Describe the normal biochemistry of muscle.
- List enzymes involved in muscle biochemistry, and state the reaction catalyzed by each enzyme.
- Describe the analyte changes observed in pathological conditions of muscle.

KEY TERMS

acetylcholine A quaternary ammonium compound, released at the motor end plate, acetylcholine combines with a receptor on the muscle membrane, thereby transmitting the nerve impulse.

actin The protein component of thin filaments in the sarcomere.

action potential The potential difference between the interior and exterior of the living cell is the membrane potential; brief positive changes in the membrane potential initiating an action are termed *action potential, nerve impulse,* or *nerve spike.*

adenosine deaminase An enzyme in muscle that converts AMP into inosine.

adenosine triphosphate (ATP) A nucleotide compound found in all living cells. It serves as energy storage (fuel) for cell activity.

alanine aminotransferase (ALT, formerly GPT or SGPT) A transaminase found in muscle and other tissues.

aldolase An enzyme of the glycolytic pathway found in muscle and other tissue.

aspartate aminotransferase (AST, formerly GOT or SGOT) A transaminase found in muscle and other tissues.

cardiac muscle Found only in heart, it is cross-striated like skeletal muscle, but the cells are branched and interconnected so that the heart contracts and relaxes as a whole. Cardiac muscle normally is not under voluntary control.

creatine kinase (CK) An enzyme in muscle that catalyzes the transfer of a phosphate group from creatine phosphate to ADP to yield ATP and creatine.

creatine phosphate A creatine–phosphoric acid compound in cells present in large amounts in muscle, another form of energy storage used to rephosphorylate ADP to ATP.

end plate An expansion of the motor nerve terminal, also called a "neuromuscular junction," that contains the presynaptic terminal of the nerve, the synaptic cleft, and the postsynaptic element, which is the skeletal muscle fiber membrane.

glycogen A readily mobilized storage form of glucose. A branched polymer of glucose.

glycolysis Degradation of glucose into three-carbon compounds with a net production of ATP.

involuntary muscles Smooth muscles not under conscious control.

isoenzymes One of the multiple forms in which an enzyme may exist in a single organism, differing chemically and physically but catalyzing the same reaction.

lactate dehydrogenase (LD) An enzyme in muscle and other tissues producing lactate from pyruvate when the amount of oxygen is limiting.

lactic acidosis Lowering of the pH in muscle and serum caused by accumulation of lactic acid.

motor nerve endings Branched or plaquelike endings of axon terminals of voluntary muscles that convey motor impulses.

motor unit The motor nerve fiber together with the muscle fibers that it innervates.

muscular atrophy A motor unit dysfunction, the major cause of which is loss of efferent innervation.

muscular dystrophy A motor unit dysfunction; a hereditary myopathy involving the muscle fibers.

myofibrils The smallest long threads in the muscle fiber, composed of numerous myofilaments.

myogenic Pertaining to muscles, such as cardiac and smooth muscles, that do not require nerves to initiate and maintain contractions.

myoglobin A heme-containing protein of muscle that serves as a reserve supply of oxygen.

myokinase (MK) A muscle enzyme converting two ADP into an ATP and an AMP.

myopathy General motor unit disorder involving the muscle fibers.

myosin The protein that constitutes the thick filaments of the sarcomere.

neurogenic Muscles with well-defined end plates in which contraction is initiated by nerve impulses. If the nerves are cut, the muscles they supply are paralyzed and degenerate.

oxidative phosphorylation An ATP-generating process in which oxygen serves as an ultimate electron acceptor. It occurs in mitochondria and is the major source of ATP generation in aerobic organisms. Also called "respirations."

sarcomere The smallest functional unit of the myofibril, a region between two Z disks from the middle of one I band to the middle of the next I band.

sarcoplasmic reticulum (SR) Well-developed system of tubules around several myofibrils.

skeletal muscle Striated muscles under voluntary control that are attached to bones and usually cross one joint.

sliding filaments The interdigitated thick and thin filaments in the sarcomere. In muscle contraction they slide past each other, shortening the sarcomere without changing their length.

smooth muscle Muscle with narrow, long cells with no cross banding, no end plates, and innervation. It contracts slowly.

tropomyosin Protein component of thin filaments.

troponin Protein component of thin filaments, together with tropomyosin, regulates actin and myosin interactions.

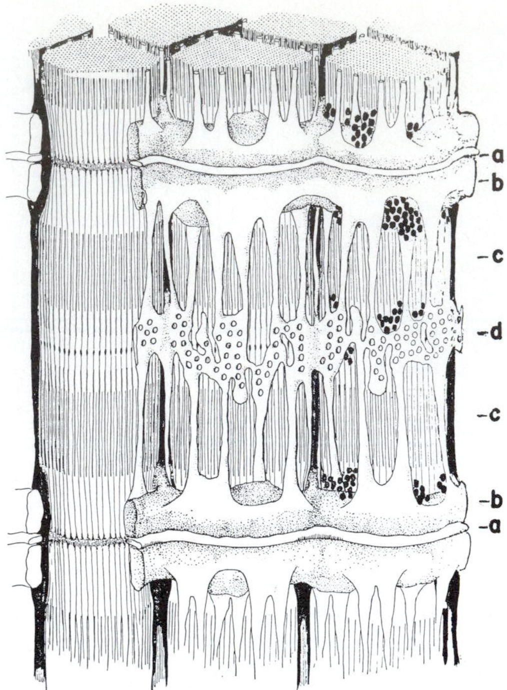

Fig. 28-1 Diagram of skeletal muscle filaments with sarcoplasmic reticulum. Major features of sarcoplasmic reticulum at different levels marked with letters on diagram: *a,* transverse tubules; *b,* terminal cysternae; *c,* longitudinal vesicles; *d,* fenestrated collar at center of A band. Dense granules are glycogen. Transverse tubules and terminal cysternae form the triads at so-called T system. *(From Peachey, LD: J Cell Biol 25:209-231, 1965.)*

ANATOMY OF MUSCLE

There are three distinct groups of muscle: skeletal, cardiac, and smooth. The *skeletal muscle* is the most thoroughly studied. Anatomically, it consists of parallel bundles of unbranched, cylinderically shaped cells or myotubes with multiple nuclei formed as a result of the fusion of many single cells during myogenesis. The fiber-like myotubes, which run the whole length of the muscle, show characteristic cross banding microscopically.

The vertebrate skeletal muscle is made up of these fibers running parallel for the length of the muscle. In each cell or fiber there is a bundle of cylindrical *myofibrils,* which are surrounded by an extensive network of tubular channels, the *sarcoplasmic reticulum* (SR) (Fig. 28-1). Nerve endings are attached to the outer membrane envelope of the cell, the *sarcolemma,* through the motor end plate of an axon. Within the sarcolemma-enclosed space the fibrils are bathed by the intracellular fluid of muscle, the *sarcoplasm.* The fibers or cells of the muscle are linked together by collagenous connective tissue. At the ends of the muscle lie the tendons, which act as the link between the muscle and the skeleton. The muscle then can act on the skeleton as a lever, as either a flexor or an extensor. Muscle contractions are initiated by nerve impulses under voluntary control; therefore skeletal muscles are termed *neurogenic muscles.*

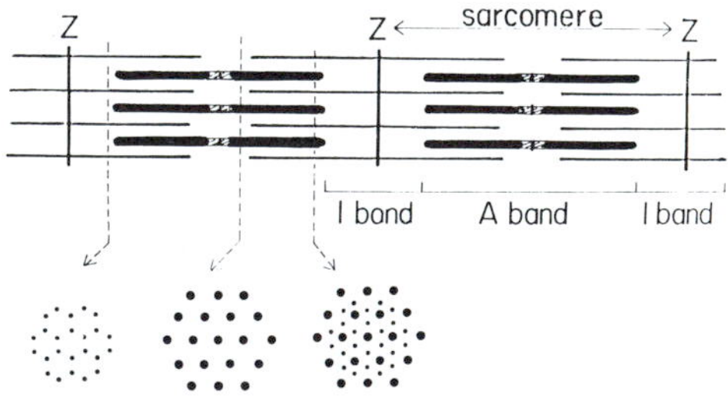

Fig. 28-2 Structure of sarcomere of striated muscle. The sarcomere is the repeating functional unit of muscle. In phase-contrast light microscopy the light I bands alternate with dark A bands. Z line contains material that anchors thin filaments. Transverse sections of sarcomere at different levels are shown in lower part of drawing. At left is a cross section through thin filaments (I band), middle is that of thick filaments (A band), and overlap region cross section is on far right. In overlap array each thick filament is surrounded by six thin filaments. In contraction there is interaction between thin and thick filaments, and they slide past each other to about half the distance of relaxed sarcomere length.

Within the myofibrils the contractile material is organized into repeating units, termed *sarcomeres.* In micro-

Table 28-1 Myofibrillar proteins and their localization in vertebrate skeletal muscle

Proteins	Localization	Amount of protein as percentage of total cellular protein
Contractile proteins		
Myosin	A band	60
Actin	I band	20
Regulatory proteins		
Tropomyosin	I band	3
Troponin	I band	4.5
M protein	M line (M band)	Less than 1
C protein	A band	Less than 1
α-Actinin	Z line, I band	1
Other minor proteins		
Connectin	A and I bands	Less than 1
Z protein	Z line	Less than 1
Desmin	Z line	Less than 1

graphs the sarcomeres show a pattern of alternating light and dark bands, the cross-striation (Fig. 28-2). The band pattern is the result of the alternation of two sets of filamentous contractile proteins with different optical properties, the anisotropic band (A band) and the isotropic band (I band). The contractile material is the protein myosin, in the A band, and actin, in the I band. These and other proteins found in the myofibrillar structure are listed in Table 28-1.

Cardiac muscle is found in the heart. It is also cross-banded, with the same structure as skeletal muscle except that the cells are branched and interconnected. There are no defined nerve end plates in cardiac muscle. Under normal conditions cardiac muscle is not under voluntary control; that is, it does not require nerve signals to initiate and maintain contractions, and it is termed *myogenic*.

Smooth muscle is composed of narrow, fusiform cells with a single nucleus and without obvious cross-banding. It does not have a structurally defined end plate and is not under voluntary control. Smooth muscle is found in walls of tubes or sacs such as blood vessels, the wall of the uterus, the urinary bladder, the intestines, and the bronchioles. It is characterized by slow contraction. It can maintain tension or a given length, without fatigue, at low energy cost.

FUNCTION

Mechanism

The function of the muscles is to produce shortening and force development as a mechanical response to stimulation. This response is translated into a leverage force in the case of skeletal muscle as a result of its attachments to the skeleton.

In cardiac muscle the mechanical force is manifested by the development of pressure within the chambers of the heart, since muscle shortening results in reduced chamber size. Smooth muscle shortening results in the emptying of sacs or cavities, as in the expulsion of urine from the bladder or of a child from the uterus. Usually a considerable force is developed during smooth muscle activity.

Skeletal muscle can shorten to a third of its original length. The sliding filament model is used to explain muscle contraction. The two discontinuous muscle filaments found in the sarcomere interdigitate, and the thick and thin filaments slide past each other during contraction and overlap. This sliding causes the length of the sarcomere to decrease. Adenosine triphosphate (ATP) hydrolysis supplies the immediate energy required for muscle contraction. The amount of ATP hydrolyzed depends on the duration of contraction and on the amount of work done. The ATP hydrolysis sites, which are located at the cross-bridges formed between the interacting thick (myosin) and thin (actin) elements of the sarcomere, are highly active only when myosin and actin interact. Relaxation of the muscle is a passive process. When there is no cross-bridge formation the sarcomere acquires its resting length and the filaments slide back to the less-overlapped position. The same basic principle applies to skeletal, cardiac, and even smooth muscles, although in smooth muscle there are no sarcomeres and the thick and thin filaments are less organized. In smooth muscle the actin filaments are anchored either to the cell membrane or within the cytoplasm to "dense bodies," structures resembling developing Z lines in skeletal muscle.

Acetylcholine as neuromuscular transmitter

Motor nerve endings differ according to the type of muscle fiber they innervate. *Slow-twitch* fibers responding to nerve stimulus with prolonged contraction are innervated by multiple nerve endings. *Fast-twitch* fibers are innervated usually by individual end plates. In both types of motor nerve terminals, acetylcholine is synthesized and stored in vesicles. Nerve impulses release the acetylcholine molecules, giving rise to a complex sequence of events, starting with depolarization of the muscle membrane and leading to contraction of the muscle.

Excitation-contraction coupling

In striated muscle the nerve impulse is transmitted to the membrane of the muscle fiber (sarcolemma) by acetylcholine (Fig. 28-3). Binding of acetylcholine to its receptor site on the cell membrane initiates a change in the voltage potential of the plasma membrane (depolarization). The depolarization spreads as an action potential along the sarcolemma and down to the T tubules around the myofilaments. The response is a massive release of calcium from the sarcoplasmic reticulum. The Ca^{2+} binds to the troponin complex in the thin filaments, initiating interactions with the thick filaments and filamental sliding, that is, contraction. On the right side of Fig. 28-3 is the time scale of these events.

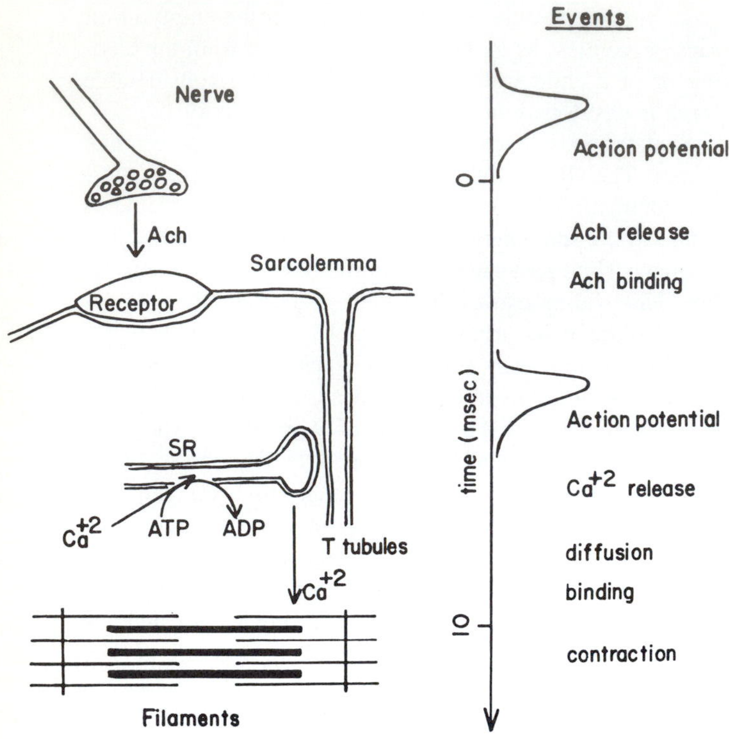

Fig. 28-3 Diagram of nerve impulse and muscle contraction. *Arch,* Acetylcholine; *SR,* sarcoplasmic reticulum.

The released acetylcholine is destroyed by a very fast-acting enzyme, acetylcholinesterase. The level of free Ca^{2+} is decreased by its active accumulation in the sarcoplasmic reticulum, and muscle activity ceases until a new sequence of events occurs.

BIOCHEMISTRY OF NORMAL MUSCLE

Contractile proteins of muscle (Table 28-1)

Myosin is the major protein of skeletal muscle. It is located in the thick filaments (A band) of the sarcomere in the myofibrils. It makes up 60% by weight of the total myofibrillar proteins. Myosin has a molecular mass of about 480,000 daltons and has a long, rodlike shape with a globular head at one end. It consists of two apparently identical polypeptide chains, the heavy chains (H), about 200,000 daltons in mass each, and four smaller peptides, termed light chains (L), ranging in molecular weight from 16,000 to 30,000 daltons. Three light chains are designated as L_1, L_2, and L_3. Each myosin molecule contains two L_2 chains and a pair of either L_1 or L_3 chains, depending on the type of muscle. The functional part of the myosin molecule is located in the globular head, where both the ATPase and the actin-binding sites are located.

Actin is the second most abundant protein in muscle, comprising 20% of the total myofibrillar proteins. Actin, which is a single polypeptide protein, folded into a globular shape, has a molecular mass of 41,785 daltons. It is probably one of the most conserved proteins throughout biological evolution. In skeletal and cardiac muscles a single type of actin, α-actin, is found; slightly modified forms, β-actin and γ-actin, are found in other cells. Actin, a double-beaded, strandlike, linear polymer, interacts with two other proteins, tropomyosin and troponin, and forms the thin filament of the sarcomere. The major functional interaction of actin in the thin filament is with the globular head part of myosin, which stimulates the myosin ATPase activity. The actin-myosin complex is dissociated by a new ATP molecule, and the cycle of actin with myosin interaction is ready to begin again.

Regulatory proteins of muscle

Tropomyosin and troponin are involved with the Ca^{2+}-mediated regulation of actin and myosin interactions. In a noncovalent association with the polymerized actin, they form the thin filaments.

Tropomyosin is a fully α-helical, ropelike protein with two subunits. The molecular mass is about 65,000 daltons for the combined subunits. It makes up about 3% of the total proteins in the myofibril. It forms long end-to-end aggregates, and two of these ropelike strands accupy two grooved sides of the double-stranded actin polymer. Dislocation of these tropomyosin strands, affected by Ca^{2+} binding to troponin from one site to another on actin, is assumed to expose the binding site on actin for myosin.

Troponin attaches at regular intervals on tropomyosin to regulate the binding of tropomyosin on the actin strand. Binding of Ca^{2+} to troponin is a signal for this change. Troponin has three subunits with a combined mass of 70,000 daltons. It comprises about 5.4% of the myofibrillar proteins. The three subunits are troponin C (calcium-binding component), troponin T (tropomyosin-binding component), and troponin I (inhibitory component).

The thin filament, which consists of actin, tropomyosin, and troponin in an approximate molar ratio of 6 : 1 : 1 and comprises about 27.5% of the total proteins of the myofibrils, is localized in the I band of the sarcomere.

Other muscle proteins

Myoglobin is a hemoprotein found in muscle that facilitates the movement of oxygen into muscle cells and provides for local storage of oxygen. It is a single-chain globular molecule with a mass of 17,200 daltons, about one fourth the size of hemoglobin.

Major enzymes in muscle

Creatine kinase (CK) catalyzes the transfer of phosphate from creatine phosphate to adenosine disphosphate; the reversible reaction is

$$\text{Creatine phosphate} + \text{ADP} \underset{}{\overset{\text{CK}}{\rightleftharpoons}} \text{Creatine} + \text{ATP}$$

CK is widely distributed in tissues, where it facilitates the rapid generation of ATP. This is important in contractile and transport systems, where there can be an acute need for ATP. In muscle, CK may represent 10% to 20% weight/volume of cytoplasmic protein. It is a dimeric molecule with a molecular mass of 81,000 daltons. There are three isoenzymes of analytical importance: MM (in muscle), MB (in heart, a hybrid), and BB (in brain). See Table 53-1 for the composition of CK in various tissues.

Aldolase is an enzyme of the lyase group. It catalyzes a reversible cleavage of substrate into two compounds without hydrolysis:

$$\text{Fructose-1,6-diphosphate} \xrightleftharpoons{\text{Aldolase}} \text{Dihydroxyacetone phosphate} + \text{Glyceraldehyde-3-phosphate}$$

Aldolase is a significant enzyme in the glycolytic pathway (see Chapter 29). It is found in every living cell. In tissues where glycolysis supplies a large part of the energy need, the aldolase activity is high. The fructose diphosphate aldolases are present in different isoenzyme forms. The three major forms are aldolase A, predominantly in muscle; aldolase B, in the liver; and aldolase C, in the brain. Each of the three aldolases contain four polypeptide subunits with different amino acid compositions.

Aspartate aminotransferase (AST) catalyzes the transfer of α-amino group from an α-amino acid to an α-keto acid:

$$\begin{matrix}\text{Aspartic acid} + \\ \alpha\text{-Ketoglutaric acid}\end{matrix} \xrightleftharpoons{\text{AST}} \begin{matrix}\text{Oxaloacetic acid} + \\ \text{Glutamic acid}\end{matrix}$$

All tissues contain AST. There are two AST isoenzymes. One is found in the mitochondrial (m-AST), and the other is found in the cytosol (c-AST). The two isoenzymes differ greatly in primary structure.

Alanine aminotransferase (ALT) catalyzes the transfer of the amino group from alanine to α-ketoglutaric acid:

$$\begin{matrix}\text{Alanine} + \\ \alpha\text{-Ketoglutaric acid}\end{matrix} \xrightleftharpoons{\text{ALT}} \begin{matrix}\text{Pyruvic acid} + \\ \text{Glutamic acid}\end{matrix}$$

ALT is also present in all tissues. The appearance in the serum of AST and ALT is a marker of tissue damage. The serum values for ALT are higher in liver damage than in myocardial damage and can be used as a differential indicator.

Lactate dehydrogenase (LD) catalyzes the transfer of two electrons and one hydrogen ion from lactate to nicotinamide dinucleotide (NAD):

$$\text{Pyruvic acid} + \text{NADH} + \text{H}^+ \xrightleftharpoons{\text{LD}} \text{Lactic acid} + \text{NAD}^+$$

Lactate is formed from pyruvate produced by the glycolytic pathway when the amount of cellular oxygen is limiting, as is the case in muscle during intense activity. In tissue damage caused by trauma or disease, LD appears in the serum.

Energetics of muscle contraction

The immediate source of energy in muscle for contraction is ATP:

$$\text{ATP} \longrightarrow \text{ADP} + \text{P}_i + \text{Energy}$$

where P_i is inorganic phosphate. In resting muscle the [ATP]/[ADP] ratio is high and ATP is readily available for the contraction process. This balance is quickly changed during the contraction events. Since these events are rapid, there is a time lapse before appropriate metabolic signals can increase the metabolic rate. To overcome this time discrepancy between immediate energy need and delayed metabolic response, a buffering energy reservoir acts immediately to reestablish the high [ATP]/[ADP] ratio.

Energy reservoir in muscle and the energy buffering system. The energy reservoir in muscle cells consists of the high concentration of creatine phosphate (CP). The immediate effects of the increase in ADP concentration caused by the hydrolysis of ATP during contraction is a disturbance of the equilibrium in the CK-catalyzed reaction described above.

The equilibrium is reestablished by the phosphorylation of ADP to ATP, thus preserving the high [ATP]/[ADP] ratio. To further maintain a high [ATP]/[ADP] ratio, the enzyme *myokinase* (MK) catalyzes the following reaction:

$$2\ \text{ADP} \xrightleftharpoons{\text{MK}} \text{ATP} + \text{AMP}$$

The enzyme *adenosine deaminase* (AD) prevents the accumulation of AMP produced by the myokinase reaction:

$$\text{AMP} \xrightleftharpoons{\text{AD}} \text{IMP} + \text{NH}_3$$

The inosine monophosphate (IMP) generated either returns to the nucleoside pool or is degraded further to uric acid. For a single muscle fiber contraction (twitch) or for short periods of muscle activity the only measurable change in the high-energy phosphate pool is a small change in [CP], and the [ATP]/[ADP] ratio remains unchanged.

Replenishment of ATP in skeletal muscle. Replenishment of ATP in the muscle is accomplished either by oxidative phosphorylation or by glycolysis (see Chapter 29). Human skeletal muscle can contain both red and white fibers, which differ in their metabolic properties. Red, or slow-twitch, fibers are rich in mitochondria and myoglobin. The main metabolic path in these fibers is oxidative phosphorylation. White, or fast-twitch, fibers contain little myoglobin and few mitochondria. In white fibers the main route for energy metabolism is glycolysis. The rates of glycolysis and respiration and thus ATP production depend on the rate of ATP consumption and are regulated by a series of feedback controls. When the ATP/ADP ratio is high, as in the resting state, both the glycolytic and the oxidative paths operate at low rates. During maximum muscle activity when the ATP/ADP ratio is lowered by the rapid consumption of ATP, the rates of activity for the two metabolic paths are increased. At maximum muscle activity the oxygen uptake can increase twentyfold or more in response to the oxygen need for the oxidative process. However, at maximum activity the muscles are still relatively oxygen poor (anoxic) and lactate as the end product of anaerobic muscle metabolism increases in the blood. Lactic acidosis (pH $<$7.4) results from the excessive production of lactic acid and can occur in normal muscles after excessive exercise. The localized acidosis in muscle contributes to fatigue and can result in muscle cramps and pain, especially when it is accompanied by excessive sodium chloride loss.

Fatty acids as energy source. Oxidative phosphorylation by means of the tricarboxylic acid cycle requires either pyruvate derived from glycolysis or fatty acids through oxidation. Both pyruvate and the fatty acid oxidation products enter the cycle as acetyl coenzyme A (acetyl CoA). Fatty acids are oxidized in the mitochondrial matrix. The long-chain fatty acids do not traverse the inner mitochondrial membrane freely. They are carried across the membrane by the carrier molecule *carnitine*. Carnitine is particularly abundant in muscle and stimulates the fatty acid oxidation by mitochondria, especially in red muscle fibers. Three biochemical steps occur before fatty acid utilization in the mitochondria can begin:

1. Activation of fatty acid ($R\text{-}COO^-$):

$$R\text{-}COO^- + CoA + ATP \rightleftharpoons \text{Acyl CoA} + AMP + PP_i$$

CoA is coenzyme A, PP_i is inorganic pyrophosphate, and $R\text{-}COO^-$ is the fatty acid.

2. Acyl (fatty acid) group transfers to the carrier molecule carnitine:

$$\text{Acyl CoA} + \text{Carnitine} \rightleftharpoons \text{CoA} + \text{Acylcarnitine}$$

Acylcarnitine diffuses across the inner mitochondrial membrane. On the matrix side, step 3 occurs:

3. Transacylation to form acyl CoA and free carnitine in the intramitochondrial matrix:

$$\text{Acylcarnitine} + \text{CoA} \rightleftharpoons \text{Carnitine} + \text{Acyl CoA}$$

The transacylation reactions are catalyzed by the enzyme fatty acyl CoA: carnitine fatty acid transferase. The carnitine diffuses to the outer membrane, and steps 1 to 3 are repeated. The mechanism described above is also called the fatty acid shuttle. The fatty acyl CoA becomes the substrate for the fatty acid oxidation system in the inner matrix compartment of the mitochondrion. Defects in the transferase or a deficiency of carnitine impairs the oxidation of fatty acids, causing muscle fatigue and cramps with fasting, exercise, or a high-fat diet.

Glycogen as fuel supply for glycolysis. Glycogen is a stored form of glucose. It is a polymer in which glucose is linked by α-1,4-glycosidic bonds, with branches formed by α-1,6-glycosidic bonds. Glycogen is stored in muscles in the form of granules also containing the enzymes for its synthesis and degradation (see Chapter 29 for details).

Energy metabolism in different fiber types

In skeletal muscle there is a correlation between speed of contraction and the color of the muscle. Red muscles contract more slowly than white muscles. Each muscle type is determined by the proportion of the red and white fibers it contains. There are three general types of muscles: slow-twitch oxidative, fast-twitch oxidative-glycolytic, and fast-switch glycolytic. The classification is based on physiology, histochemistry, and biochemistry (Table 28-2). Many isoenzymes that appear in different forms in different muscle types exhibit specific structural characteristics; the light-chain components of myosin are one example. Fast-twitch muscle myosin contains all three light chains, L_1, L_2, and L_3. Slow-twitch muscle myosin contains only two types of light chains, L_1 and L_2. Differences have been found in the tropomyosin and troponin subunit compositions according to fiber type. The original differentiation based on the color relates to the main source of energy generation: in fast-twitch muscles the ATP is supplied pri-

Table 28-2 Terminology of fiber types of mammalian skeletal muscle

	Contractile muscle type and contraction velocity		
Source	**I Slow-twitch**	**II Fast-twitch**	**III Fast-twitch**
Energy supply	Oxidative	Oxidative-glycolytic	Glycolytic
Color	Red	Intermediate	White
Myoglobin content	High	Intermediate	Low
Creatine kinase activity	High	High	High
Lactate dehydrogenase	Low	Intermediate	High
Aldolase activity	Low	Intermediate	High
Myokinase activity (adenylate kinase)	Low	Intermediate	High
Mg^{2+}-stimulated ATPase (actinomyosin ATPase)	Low	Intermediate	High
Fumarase	High	Intermediate	Low
Citrate synthase	High	Intermediate	Low
Succinate dehydrogenase	High	Intermediate	Low
Myosin stability at pH 9	Low	Intermediate	High
Mitochondrial distribution	High	Intermediate	High
Cytochrome oxidase	High	Intermediate	Low

marily from glycolytic processes, which provide a quick but rapidly exhaustible supply of energy. These muscles are white because of the paucity of myoglobin and mitochondrial cytochromes. Muscles of slow-twitch oxidative or fast-twitch oxidative-glycolytic fibers gain energy primarily through oxidative processes. These muscles are rich in mitochondria and contain significant amounts of myoglobin, which give these muscles a red appearance. The terminology and differences in muscle types are given in Table 28-2.

PATHOLOGICAL CONDITIONS OF MUSCLE

Muscular disorders

The diseases of muscle are all characterized by motor dysfunction. The abnormality is in some part of the motor unit. The motor unit consists of the peripheral neurons whose cell bodies lie in the spinal cord and whose terminations are in skeletal muscles, the motor end plate at the neuromuscular junction, and the 10 to 600 muscle fibers innervated by the neuron. The disorders are conveniently classified according to the part of the motor unit principally affected.

The three major categories of muscle disorders, according to the part of the motor unit affected, are (1) neurogenic muscular atrophies, (2) muscle fiber disorders, and (3) disturbances of the neuromuscular junction.

Neurogenic disorders

Neurogenic muscular atrophies (motor neuron diseases). The muscle weakness and wasting are caused by a loss of efferent innervation as a result of degeneration of some portion of the neuronal network.

Amyotrophic lateral sclerosis and its variants, *progressive spinal muscular atrophy* (Aran-Duchenne muscular atrophy) and *progressive bulbar palsy* (Duchenne's paralysis), are late-onset diseases, generally occurring after 40 years of age.

Infantile spinal muscular atrophy (Werdnig-Hoffman disease) often is fatal before 2 years of age.

Juvenile spinal muscular atrophy (Wohlfart-Kugelberg-Welander disease) is associated with a longer life expectancy than infantile spinal muscular atrophy. Autosomal recessive inheritance is frequently noted in amyotrophic lateral sclerosis and juvenile spinal atrophy.

Anterior root and peripheral nerve involvements

Peroneal muscular atrophy (Charcot-Marie-Tooth disease)

Hypertrophic interstitial neuropathy (Dejerine-Sottas disease) begins usually between childhood and 30 years of age and is slowly progressive.

Acute polyneuropathy (Guillain-Barré syndrome) is a parainfectious and postinfectious disease presumed to be caused by immunological attack on peripheral nerves.

Metabolic neuropathies are seen with metabolic diseases, such as diabetes mellitus, or with malnutrition.

Disorders of muscle fibers

Disorders of muscle fibers, or *myopathies,* are characterized by major defects of the muscle fibers. Certain hereditary progressive myopathies are called, by convention, *muscular dystrophies. Nonhereditary myopathies* can result from inflammation or from an endocrine or a metabolic abnormality.

Muscular dystrophies. Muscular dystrophy is a general name for a group of chronic diseases of muscle characterized by progressive weakness with muscle degeneration but with no evidence of neural degeneration. The dystrophies are inherited diseases with different inheritance patterns. The age of onset, the course of the disease, and the effect on the various fiber types differ among the individual diseases.

Pseudohypertrophic muscular dystrophy (Duchenne muscular dystrophy) is a sex-linked recessive disorder. It affects boys typically between 3 and 7 years of age with steady progression of proximal muscle weakness. First the pelvic girdle and then the shoulder girdle muscles are affected, and most patients are confined to wheelchairs by 10 to 12 years of age. Serum enzymes are greatly elevated in the disease even before symptoms develop; especially noted is the rise in CK. In heterozygous males the CK levels lie between those of affected homozygous males and normal persons. The CK values in heterozygous females and normal persons overlap; unfortunately, only about 70% of heterogyzous females are positively identifiable. Nevertheless, a screening program and genetic counseling have significantly changed the incidence of the disease in the last 15 years, which is fortunate, given the severity of the disease and the absence of effective therapy.

Limb-girdle muscular dystrophy, an inherited autosomal recessive disease, begins later in childhood than Duchenne muscular dystrophy. The involvement of the muscles is similar to that in the Duchenne type of muscular dystrophy, but the progression is less predictable.

Facioscapulohumeral muscular dystrophy (Landouzy-Déjérine muscular dystrophy) involves the facial and shoulder girdle muscles. Its onset is in adolescence, and it progresses throughout life. Life expectancy is normal.

Myotonic muscular dystrophies, or *myotonic myopathies,* are characterized by abnormally slow relaxation of voluntary muscles after a contraction. One form is *Steiner's disease* (myotonia atrophica or myotonic muscular dystrophy). It is an autosomal dominant disorder that can occur at any age and can vary in severity. The muscular weakness is combined with myotonia. Another form is called *myotonia congenita, Thomsen's disease,* or *ataxia muscularis.* It is a rare autosomal dominant myotonia with onset early in life. The muscles become hypertrophied, and muscle stiffness is the most important problem. Therapy can diminish stiffness and cramping, but the weakness is not affected. Exercise is of limited help.

Other rare muscular dystrophies include *benign juvenile*

muscular dystrophy (Becker's muscular dystrophy), which resembles Duchenne's muscular dystrophy in heredity and also in clinical features but is more benign and is not seen as early as Duchenne's dystrophy. *Distal muscular dystrophy* (Gower's muscular dystrophy) has its onset in later years. The progressive weakness begins in the hands and feet and extends proximally. The weakness is only moderate.

Glycogen-storage diseases of muscle. Glycogen-storage disorders are characterized by abnormal accumulation of glycogen in muscle. Inheritance is autosomal recessive. Diagnosis is made by demonstrating absence of the specific enzyme in a biopsy of the muscle (see Chapter 46).

McArdle's disease (glycogen-storage disease, type V) is characterized by clinical symptoms of skeletal muscle weakness, cramps on exercise, and no lactate rise. The missing enzyme is myophosphorylase b. Clinically, this disease can be mild or severe. In healthy persons ischemic exercise results in an acid shift in muscle pH; in patients with McArdle's disease no such shift in pH is present.

Tauri's disease (glycogen-storage disease, type VIII) results in muscle pain, contracture, and exercise intolerance after ischemic work, the same symptoms as are present in type V disease. It is caused by the absence of muscle phosphofructokinase. In both types V and VII there is occasionally myoglobinuria.

Pompe's disease (glycogen-storage disease, type II) is caused by absence of α-1,4-glucosidase and is the most severe form of the glycogen-storage diseases. Besides the symptoms listed above, it can also result in degeneration of anterior horn cells. It may affect the heart, causing cardiomegaly.

Familial periodic paralysis. Familial periodic paralysis is a relatively rare group of inherited disorders. The pattern of inheritance is autosomal dominant. The attach consists of flaccid paralysis that does not affect consciousness and that lasts from an hour to several days. There are three types of familial periodic paralysis, in decreasing order of severity: hypokalemic, normokalemic, and hyperkalemic paralysis. Documentation of serum potassium during an attack is important for the diagnosis.

Endocrine myopathies. Endocrine myopathies are observed in thyroid, parathyroid, and adrenal disorders; in hypercalcemia; and in hypophosphatemia.

Neuromuscular junction disturbances and diseases

Myasthenia gravis. Myasthenia gravis is a causative defect related to impairment of acetylcholine-mediated nerve-impulse transmission to muscle. Recent evidence points to an immunologically induced abnormality in the acetylcholine-receptor protein of the nerve-muscle junction. Antibodies to the receptor protein can often be demonstrated in this disease.

Drug-induced neuromuscular junction block. This nerve block is caused by cholinergic agents, organophosphate insecticides, and most nerve gases, which are cholinesterase inhibitors. These inhibitors prevent the breakdown of acetylcholine and cause persisting nerve depolarization at the neuromuscular junction. Symptoms are myasthenia-like weakness and muscular twitching, myosis, tightness in the chest, gastrointestinal hyperactivity, and increases in bronchial and salivary secretion.

Trauma

Trauma in muscles can be produced in innumerable ways, such as a mechanical blow, heating, freezing, and electrical shock. Extensive necrosis of muscle fibers from any cause results in liberation of muscle enzymes (CK, LD, AST, ALT) and myoglobin into the bloodstream. In all phases of muscle regeneration CK-MM isoenzyme activity decreases and BB and MB isoenzyme activities increase. In later phases of regeneration the MB and MM isoenzymes become active, and finally the MM becomes the dominant isoenzyme. During the early phases of regeneration a sufficiently high MB isoenzyme level appears in serum to interfere with the clinical picture of a myocardial infarction.

CHANGE OF ANALYTE IN DISEASE (Table 28-3)

In general, there is a release of the cellular constituents into circulation when the muscle tissue is damaged. Muscle contains the enzymes aldolase, aspartate transaminase, alanine transaminase, lactate dehydrogenase, and creatine kinase. Myoglobin is also released into peripheral blood following muscle damage. Since the different muscle types have different isoenzyme compositions, the isoenzyme composition can be used as a diagnostic tool in clinical evaluations. For example, CK-MM is virtually the exclusive form of CK found in skeletal muscle, whereas some

Table 28-3 Change of analyte in disease

	Aldolase	Myoglobin	CK	AST	ALT	LD	Lactic acid
Myocardial infarction	—, ↑	↑	↑	↑	↑	↑	—
Skeletal muscle damage	↑	↑	↑	↑	↑	↑	—
Muscular dystrophies	↑	—	↑	—	—	—	—
Fatigue	—	—	—	—	—	—	↑

ALT, Alanine aminotransferase; *AST,* aspartate aminotransferase; *CK,* creatine kinase; *LD,* lactate dehydrogenase.

of the MB form is present in heart tissue. Serum lactic acid elevations occur in cases of shock, muscle fatigue, and tissue hypoxia. For specific diseases, such as myasthenia gravis, specific analytes such as antibodies to the acetylcholine receptor, may be found.

BIBLIOGRAPHY

Bais, R, and Edwards, JB: Creatine kinase, CRC Crit Rev Clin Lab Sci 16:291-335, 1982.

Ebashi, S, Maruyama, K, and Endo, M, editors: Muscle contraction: its regulatory mechanism, New York, 1980, Springer-Verlag New York, Inc.

Flockhart, DA, and Corbin, JD: Regulatory mechanisms in the control of protein kinases, CRC Crit Rev Biochem 16:133-186, 1982.

Foreback, CC, and Chu, J-W: Creatine kinase isoenzymes: electrophoretic and quantitative measurements, CRC Crit Rev Clin Lab Sci 15:187-230, 1981.

Grob, D: Myasthenia gravis: pathophysiology and management, retrospect and prospect, Ann NY Acad Sci 377:xiii-xvi, 1981.

Huxley, HE: Molecular basis of contraction in cross-striated muscles. In Bourne, GH, editor: The structure and function of muscle, vol I, pt I, New York, 1972, Academic Press, Inc.

Korn, ED: Biochemistry of actomyosin-dependent cell motility (a review), Proc Natl Acad Sci USA 75:588-589, 1978.

Korn, ED: Actin polymerization and its regulation by proteins from non-muscle cells, Physiol Rev 62:672-737, 1982.

Kretsinger, RH: Mechanism of selective signaling by calcium, Neurosci Res Prog Bull 19:214-322, 1980.

Morgan, HE, and Wildenthal, K, chairmen: Symposium of the American Physiological Society: protein turnover in heart and skeletal muscle, Fed Proc 39:7-52, 1980.

Nagy, B, and Samaha, FJ: Physiology of normal and diseased muscle. In Frolich, ED, editor: Pathophysiology, Philadelphia, 1983, JB Lippincott Co.

Perry, SV: The regulation of contractile activity in muscle, Biochem Soc Trans 7:593-617, 1979.

Siegel, IM: Muscle and its diseases: an outline primer of basic science and clinical method, Chicago, 1986, Year Book Medical Publishers.

Squire, J: Muscle: design, diversity, and disease, Menlo Park, CA, 1986, Benjamin/Cummings.

Stanbury, JB, Wyngaarden, JB, and Frederickson, DS, editors: The metabolic bases of inherited disease, New York, 1986, McGraw-Hill Book Co.

Zak, R, Martin AF, and Blough, R: Assessment of protein turnover by use of radioisotopic tracers, Physiol Rev 59:407-447, 1979.

CHAPTER 29

Diabetes mellitus

RICHARD F. DODS

OBJECTIVES

- Describe normal glucose homeostasis.
- Differentiate between diabetes mellitus (types I and II), impaired glucose tolerance, and gestational diabetes.
- Describe common acute and chronic complications of diabetes mellitus.
- Lists steps in performing an oral glucose tolerance test.
- State expected levels of the following analytes in controlled diabetes mellitus I and II, ketoacidosis, and hyperosmolar coma: glucose, ketones, pH, bicarbonate, P_{CO_2}, insulin, C-peptide, sodium, potassium, glycohemoglobin, BUN, osmolality, and triglycerides.

KEY TERMS

acromegaly Growth hormone excess in adults. Characterized by enlargement of features such as the head, hands, and feet.

adenosine 3',5'-cyclic monophosphate (cAMP) An organic molecule that is obligatory for the action of enzymes such as protein kinases.

aerobic glycolysis Glycolysis that is linked to the tricarboxylic acid cycle by the presence of oxygen. Aerobic glycolysis produces 36 moles of ATP per mole of glucose.

anaerobic glycolysis Glycolysis that occurs in the absence of oxygen. In this case glycolysis is not linked to the tricarboxylic acid cycle and only 2 moles of ATP are produced.

angiogenesis A complication of diabetes mellitus. Abnormal proliferation of blood vessels in a tissue such as eye lens.

angiopathy A complication of diabetes mellitus. Damage to the basement membranes of blood vessels.

anorexia nervosa Loss of appetite because of psychological reasons. Results in lowered blood insulin.
anoxia Lack of oxygen.
basement membrane A layer of noncellular material that underlies the epithelium.
diabetes insipidus A disorder caused by excessive secretion of arginine vasopressin. As in diabetes mellitus, a characteristic symptom of diabetes insipidus is polyuria.
diabetic ketoacidosis A complication of diabetes mellitus characterized by hyperglycemia, hyperosmolarity, low pH, ketonuria and ketonemia, and lethargy or coma.
disaccharide Two monosaccharides linked by a glycosidic bond.
gestational diabetes Glucose intolerance that occurs in some pregnancies.
glucagon A hormone produced by the alpha cells of the pancreas. Glucagon is primarily involved in energy release.
glucagonoma Excessive glucagon levels caused by a tumor.
gluconeogenesis Production of glucose from lactic acid via pyruvic acid.
glucose A carbon-6 polyhydroxyl aldehyde; primary source of energy in living organisms. Its metabolism produces ATP.
glucosuria Presence of urinary glucose.
glucosylation Reaction in which glucose binds covalently to protein.
glycogen Highly branched, high-molecular-weight polysaccharide composed only of glucose units.
glycogenesis Formation of glycogen from glucose 6-phosphate.
glycolysis Metabolism of glucose 6-phosphate to pyruvic acid or lactic acid.
glycosylation Reaction in which monosaccharides such as glucose, mannose, galactose, xylose, ribose, and fructose covalently bind to proteins.
growth hormone Hormone produced by the anterior part of the pituitary. Also called somatotropin. Raises blood glucose.
hexose monophosphate shunt Metabolic pathway in which glucose 6-phosphate is metabolized to ribose and carbon dioxide.
histocompatibility antigen (human leukocyte antigen [HLA]) Proteins responsible for rejection of tissue transplanted to an individual from another unrelated individual. Specific HLAs are present at a high frequency in persons who develop certain diseases.
hyperglycemia High blood glucose concentrations.
hyperglycemic, hyperosmolar nonketotic coma (HHNC) A complication of diabetes mellitus characterized by hyperglycemia, hyperosmolarity, low pH, normal ketoacid levels, and lethargy or coma.
islet cell antibodies (ICA) Antibodies frequently found in type I diabetics that suggest an autoimmune etiology.
islets of Langerhans Group of cells in the pancreas. The cells are alpha cells, which secrete glucagon, beta cells, which secrete insulin, and delta cells, which secrete somatostatin.
ketonemia Excess of ketones and derived ketoacids in the blood.
ketonuria Excess of ketones and derived ketoacids in the urine.
lactic acidosis Acidosis (low blood pH) caused by excess lactic acid.
lipolysis Hydrolysis of triglycerides to free fatty acids and glycerol.
monosaccharide A polyhydroxyl aldehyde or ketone, such as glucose, fructose, and mannose.
nephropathy A complication of diabetes mellitus. Refers to damage to the glomerulus and capillaries associated with the glomerulus.
neuropathy The most common complication of diabetes mellitus. Refers to reduced motor and sensory nerve conduction velocities caused by axonal degeneration and demyelination.
oxidative phosphorylation The process linking the tricarboxylic acid cycle with ATP formation.
polydipsia Excessive thirst. A symptom of diabetes mellitus.
polyphagia Constant hunger. A symptom of diabetes mellitus.
polysaccharide A carbohydrate composed of more than two monosaccharides linked by glycosidic bonds.
polyuria Excessive urinary output. A symptom of diabetes mellitus.
pre-proinsulin Precursor to proinsulin.
proinsulin Precursor to insulin.
protein kinases Enzymes that phosphorylate other proteins. Some protein kinases depend on AMP for activity.
receptor sites Sites on or in cells where hormones are bound. Hormone binding to the receptor site is the initial step for hormone action.
retinopathy A complication of diabetes mellitus. A disorder of the retina caused in diabetics by cataract formation or proliferation of small blood vessels (angiogenesis).
somatostatin A hormone produced in the delta cells of the pancreas. Inhibits insulin and glucagon secretion.
somatostatinoma A tumor that produces excess quantities of somatostatin, resulting in hyperglycemia.
thyroxine A hormone produced by the thyroid gland. Increases blood glucose levels.
tricarboxylic acid cycle Metabolic pathway that converts glucose 6-phosphate via pyruvic acid to CO_2 and water. When coupled to oxidative phosphorylation, ATP is formed.
uronic acid pathway Converts glucose 6-phosphate to glucuronic acid.

Complications resulting from diabetes mellitus are the third leading cause of death in the United States according to statistics compiled by the National Commission on Diabetes.[1] A recent survey by the National Diabetes Data Group[2] estimates that there are 8 million diabetic Americans, suggesting a prevalence of diabetes of about 6.6%. Approximately half of the diabetics are actually diagnosed. Rates are equal by sex and greater for blacks than whites. The cost of diabetes to the American economy probably exceeds $5 billion annually.

The implications of diabetes extend beyond its direct effects and long-term complications since it has been established that diabetes is a risk factor for coronary heart disease[3] and cerebrovascular disease (stroke).[4] A diabetic has a twofold greater risk of suffering a myocardial infarction than a nondiabetic of the same age and sex.

Recent research has demonstrated that diabetes mellitus is not a single disease but an array of diseases that exhibit a common symptom, inability of the individual to handle glucose (glucose intolerance).

The primary symptoms of diabetes mellitus are polyuria, abnormally high blood and urine glucose levels (hyperglycemia and glucosuria, respectively), excessive thirst (polydipsia), constant hunger (polyphagia), sudden weight loss, and during acute episodes of diabetes mellitus, excessive blood and urinary ketones (ketonemia and ketonuria, respectively). All of these symptoms are the result of the inability to metabolize glucose and the consequences of high glucose levels.

GLUCOSE: PROPERTIES AND METABOLISM

Definition

Carbohydrates are defined as polyhydroxyl aldehydes (aldoses) and ketones (ketoses). Simple carbohydrates such as glucose are also called monosaccharides. Two monosaccharides linked by a bond called a glycosidic bond form a disaccharide. More than two monosaccharides linked by glycosidic bonds form a polysaccharide. Dietary carbohydrates consist of monosaccharides such as glucose, fructose, and galactose; disaccharides such as sucrose, lactose, and mannose; and polysaccharides such as starch. Intestinal enzymes convert disaccharides and polysaccharides to monosaccharides.

Function

The principal biochemical function of glucose is to provide energy for life processes. ATP is the universal energy source for biological reactions. Glucose oxidation by the glycolytic and tricarboxylic acid pathways is the primary source of energy for the biosynthesis of ATP.

Principal glucose metabolic pathways

As shown in Fig. 29-1, glucose is metabolized by five pathways within the cell. Glucose is converted by *glycolysis* to pyruvate and lactate, oxidized to carbon dioxide and water in the *tricarboxylic acid pathway,* converted to glycogen by *glycogenesis,* oxidized to ribose and carbon dioxide in the *hexose monophosphate shunt,* and converted to glucuronic acid in the *uronic acid pathway*. Fig. 29-1 also shows that glucose is rapidly converted within cells to *glucose 6-phosphate,* a major intermediate in the principal pathway of glucose metabolism. The enzyme that catalyzes the phosphorylation of glucose by ATP is hexokinase (or glucokinase in the liver).

Aerobic glycolysis

Glycolysis. Glucose 6-phosphate (G-6P) metabolism by the glycolytic pathway (also called the Embden-Meyerhof cycle) results in the formation of ATP (Fig. 29-2). Glycolysis converts the six-carbon glucose molecule to two molecules of a three-carbon compound called pyruvic acid. This process produces 2 moles of ATP per mole of glucose. An important aspect of glycolysis is the formation of pyruvic acid. In aerobic glycolysis, pyruvate is metabolized further.

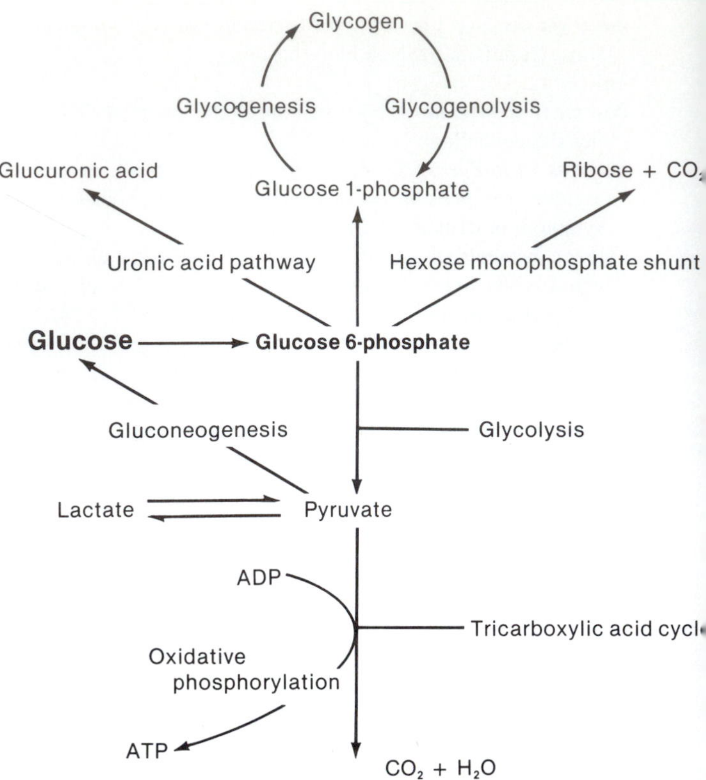

Fig. 29-1 The five principal pathways of glucose metabolism: glycolysis, tricarboxylic acid pathway, glycogenesis, hexose monophosphate shunt, and the uronic acid pathway.

Tricarboxylic acid cycle. Pyruvic acid enters the tricarboxylic acid cycle (citric acid cycle, Krebs cycle) where it is metabolized to carbon dioxide and water. Fig. 29-3 shows the intermediate steps in the tricarboxylic acid cycle and the steps that are used to reduce nicotinamide adenine dinucleotide (NAD) and flavin adenine dinucleotide (FAD) to their corresponding analogs, NADH and $FADH_2$. The tricarboxylic acid cycle does not directly produce ATP, but ATP is produced by the oxidation of NADH and $FADH_2$.

Oxidative phosphorylation. Oxidative phosphorylation is a complex process that involves electron transfer from NADH and $FADH_2$ to a series of compounds, eventually ending up with the reduction of oxygen to yield a water molecule. Thus it is actually the reoxidation of NADH and $FADH_2$, compounds produced by the tricarboxylic acid cycle, that produces ATP. In contrast to glycolysis, which produces 2 moles of ATP per mole of glucose, the tricarboxylic acid cycle linked with oxidative phosphorylation, which takes place in the mitochondria, produces 36 moles of ATP per mole of glucose. The oxidative and ATP synthesis processes are tightly coupled because the availability of ADP controls the rate of oxidation and oxygen availability regulates the rate of phosphorylation.

Anaerobic glycolysis. In fatigued muscle where there is a deficiency of oxygen, or *anoxia,* glucose metabolized by glycolysis to pyruvic acid cannot be further degraded. In-

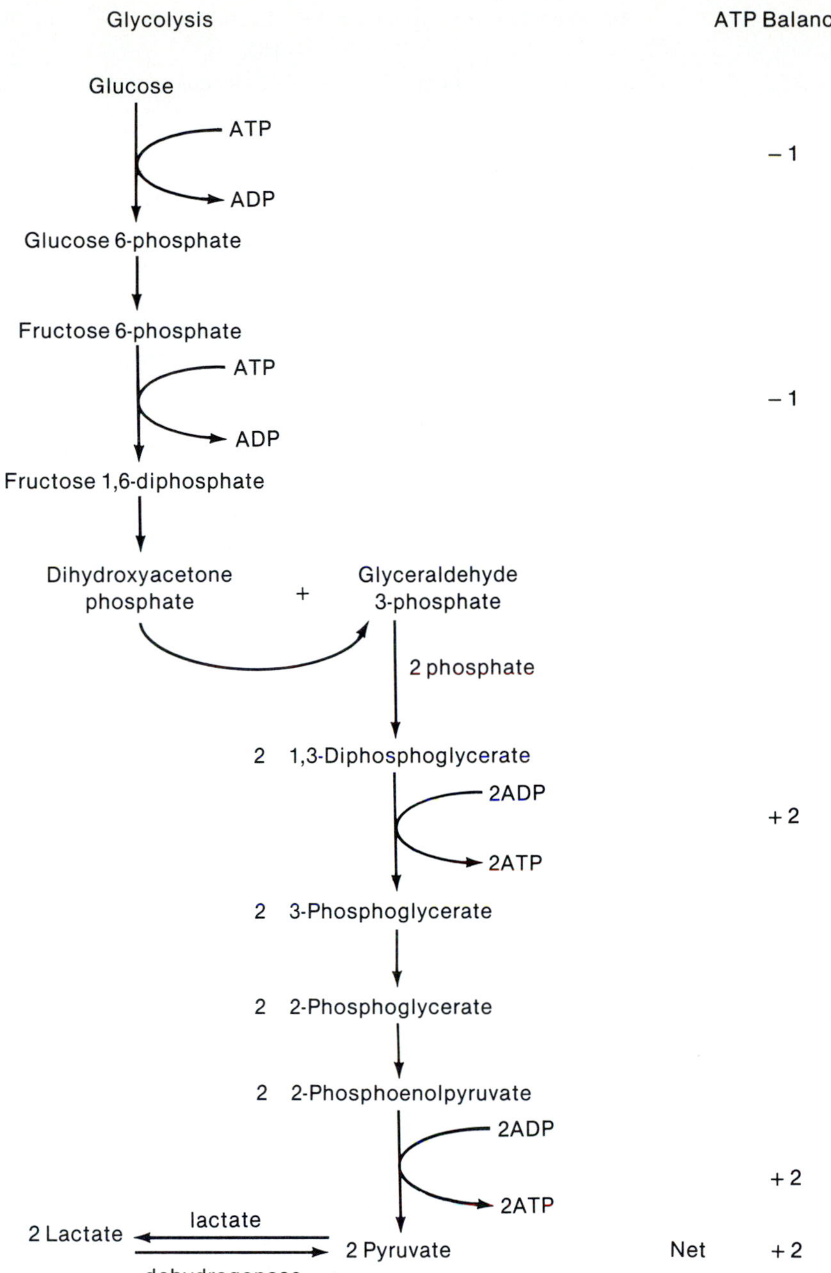

Fig. 29-2 Two stages of glycolysis. First stage proceeds from glucose to formation of 1,3-diphosphoglycerate and consumes 2 moles of ATP. Second stage proceeds from 1,3-diphosphoglycerate to pyruvate and produces 4 moles of ATP. Glycolysis therefore results in net gain of 2 moles of ATP per mole of glucose.

stead (Fig. 29-2) pyruvate is converted by the enzyme lactate dehydrogenase to lactate. This is called anaerobic glycolysis. In contrast to aerobic glycosis, only 2 moles of ATP per mole of glucose are produced by anaerobic glycolysis. Lactic acid produced by anoxic tissue is carried by the circulation to the liver, where it is reconverted to glucose in a process called gluconeogenesis (see below).

Alternate energy sources. As indicated in Fig. 29-3, amino acids and fatty acids also enter the tricarboxylic acid cycle to produce ATP and are therefore alternate sources of energy.

Glycogenesis, glycogenolysis, and gluconeogenesis

Glycogen. Excess glucose is stored in the cells as the polymer glycogen for later energy demands. Glycogen (Fig. 29-4) is a high-molecular-weight polysaccharide composed entirely of glucose units in 1,4-glycosidic bonds with 1,6-branches occurring approximately every 10 units.

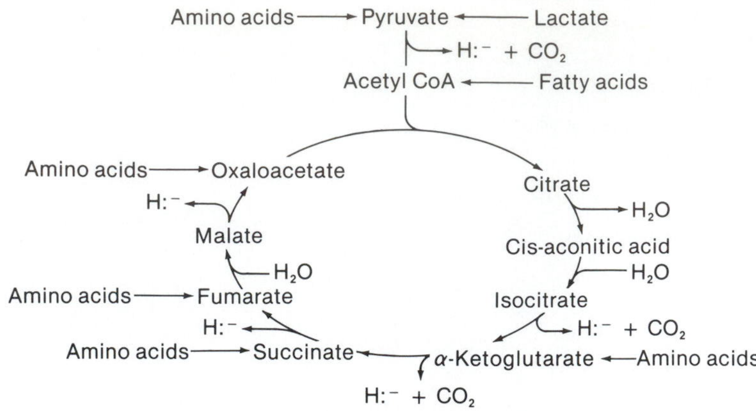

Fig. 29-3 Tricarboxylic acid cycle produces CO_2 and water from pyruvate, fatty acids, and amino acids, which enter the cycle at the points indicated. Hydride ions (H^-) are produced and utilized in oxidative phosphorylation process to produce ATP from ADP and inorganic phosphate.

Glycogen is located in the cytoplasm of liver and muscle cells in granules that contain the enzymes involved in the synthesis (glycogenesis) and hydrolysis (glycogenolysis) of glycogen. Refer again to Fig. 29-2 to see how glucogenesis and glycogenolysis fit into the overall scheme of glucose metabolism. Fig. 29-5 presents a simplified representation of glycogenesis and glycogenolysis.

Glycogenesis. The first step in glycogenesis is conversion of glucose-6-phosphate to glucose-1-phosphate (Fig. 29-5). Reaction of glucose-1-phosphate with uridine 5′-triphosphate (UTP) produces uridine diphosphate glucose (UDP-glucose), a substance that reacts with the glycogen molecule to form 1,4-glucosidic linkages. Synthesis of 1,4-glucosidic bonds is catalyzed by the enzyme *glycogen synthetase*. Glycogen synthetase exists in two forms: the phosphorylated, inactive enzyme form and the dephosphorylated, active enzyme form. Phosphorylation of the inactive enzyme is accomplished by an enzyme called *protein kinase*. Protein kinases are activated by low levels of adenosine 3′,5′-cyclic monophosphate (cAMP). Thus glycogen synthetase activity (and thereby glycogenesis) is regulated by intracellular cAMP levels. Glycogenesis is enhanced by low cAMP levels and inhibited by high cAMP levels; cAMP levels are in turn regulated by insulin.

Branching of glycogen is accomplished by an enzyme called a *branching enzyme*. This branching enzyme hydrolyzes the 1,4-glycosidic bond of glycogen to form five to six glucose unit fragments, which are reattached to the glycogen molecule through 1,6-glycosidic bonds.

Glycogenolysis. Although glycogenolysis (Fig. 29-5) is the opposite of glycogenesis, it does not occur through a

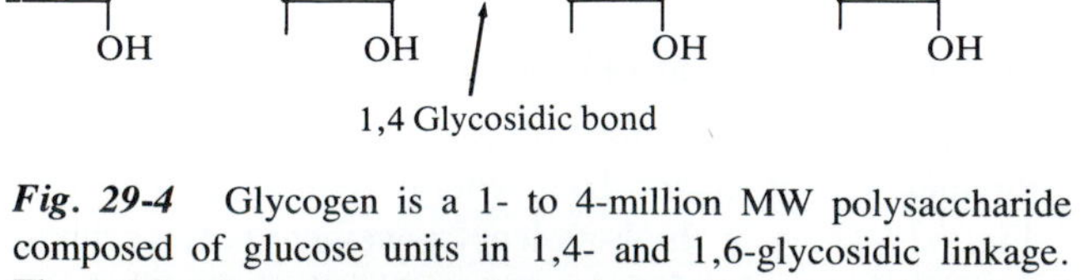

Fig. 29-4 Glycogen is a 1- to 4-million MW polysaccharide composed of glucose units in 1,4- and 1,6-glycosidic linkage. The 1,6-bonds produce branches at intervals of approximately 10 glucose units.

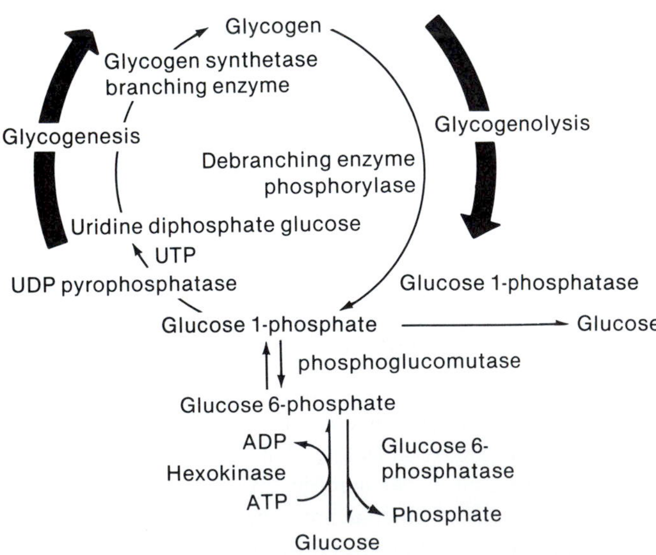

Fig. 29-5 Glycogen, storage molecule for glucose, is synthesized from glucose 1-phosphate by a process called *glycogenesis (left side)*. *Glycogenolysis* releases glucose units from glycogen. Debranching is first step in glycogenolysis *(right side)*.

simple reversal of each step of glycogenesis but by a unique enzyme system. The *debranching enzyme* splits off trisaccharides from branches and reattaches them by 1,4-glycosidic bonds to the ends of the glycogen molecule. *Glycogen phosphorylase* hydrolyzes the 1,4-glycosidic bond producing glucose-1-phosphate:

$$\underset{(n \text{ residues})}{\text{Glycogen}} + P_i \rightarrow \underset{(n-1 \text{ residues})}{\text{Glycogen}} + \text{Glucose-1-phosphate}$$

Like glycogen synthetase, glycogen phosphorylase exists in two forms. The active form, called phosphorylase a, is a tetramer. The inactive form, phosphorylase b, is the dimer. The active tetramer is formed in three steps; phosphorylation of the dimer by a protein kinase called *phosphorylase kinase* followed by binding of the phosphorylated dimer with another phosphorylated dimer. The third and final step in the activation process is the binding of one molecule of pyridoxal phosphate to each subunit of the tetramer. Phosphorylase kinase is activated by cAMP. Note that high cellular cAMP levels favor glycogenolysis over glycogenesis. The activation of glycogen phosphorylase is under hormonal control (see below).

Gluconeogenesis. The steps in gluconeogenesis are shown also in Fig. 29-6. Gluconeogenesis produces glucose-6-phosphate (G-6-P) from amino acids, fatty acids, glycerol, and lactate. Pyruvate is an intermediate in the formation of G-6-P from each of these substances. Glucogenesis is not a simple reversal of glycolysis, although gluconeogenesis does share some of the enzymes of the glycolytic pathway. Glucose is formed only in the liver

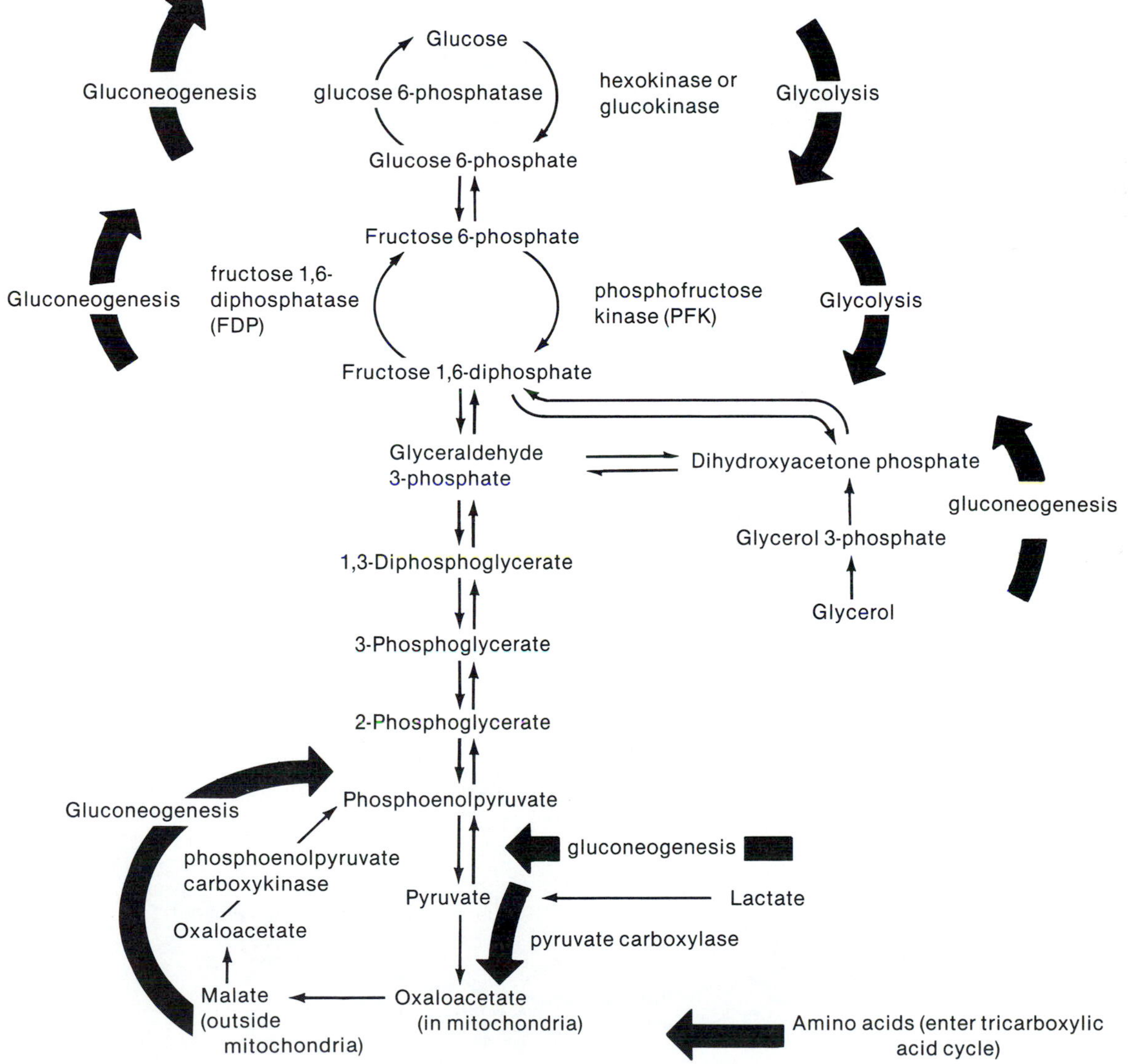

Fig. 29-6 Pathways involved in gluconeogenesis from amino acids, fatty acids, glycerol, and lactate. This pathway shares many of the enzymes of glycolytic and tricarboxylic acid pathways. Gluconeogenesis provides glucose whenever scarcity of glucose occurs and whenever lactate accumulates.

and kidney, which have the enzyme glucose-6-phosphatase, which hydrolyzes G-6-P to glucose. In fact, the liver is the major nondietary source of serum glucose of the body.

Hormone regulation of glucose metabolism

The system for regulating blood glucose levels is designed to achieve two ends. The first is to store glucose in excess of the body's immediate needs in a compact reservoir (glycogen), and the second is to mobilize the stored glucose in order to maintain the blood glucose level. The regulation of blood glucose is essential to keep the brain, whose primary energy source is glucose, supplied with a constant amount of glucose. The role of insulin is to shift extracellular glucose to intracellular storage sites in the form of macromolecules (such as glycogen, fats, and proteins). Thus glucose is stored away in times of plenty for times of need.

In periods of fasting, a series of hyperglycemic agents, in response to low blood glucose, acts on intermediary metabolic pathways to form glucose from storage macromolecules. Thus proteins and glycogen are metabolized to form glucose (gluconeogenesis).

The most important hyperglycemic agents are glucagon, epinephrine, cortisol, thyroxine, growth hormone, and certain intestinal hormones. The behavior of each of these agents in regulating blood glucose is different; whereas insulin promotes anabolic metabolism (synthesis of macro-

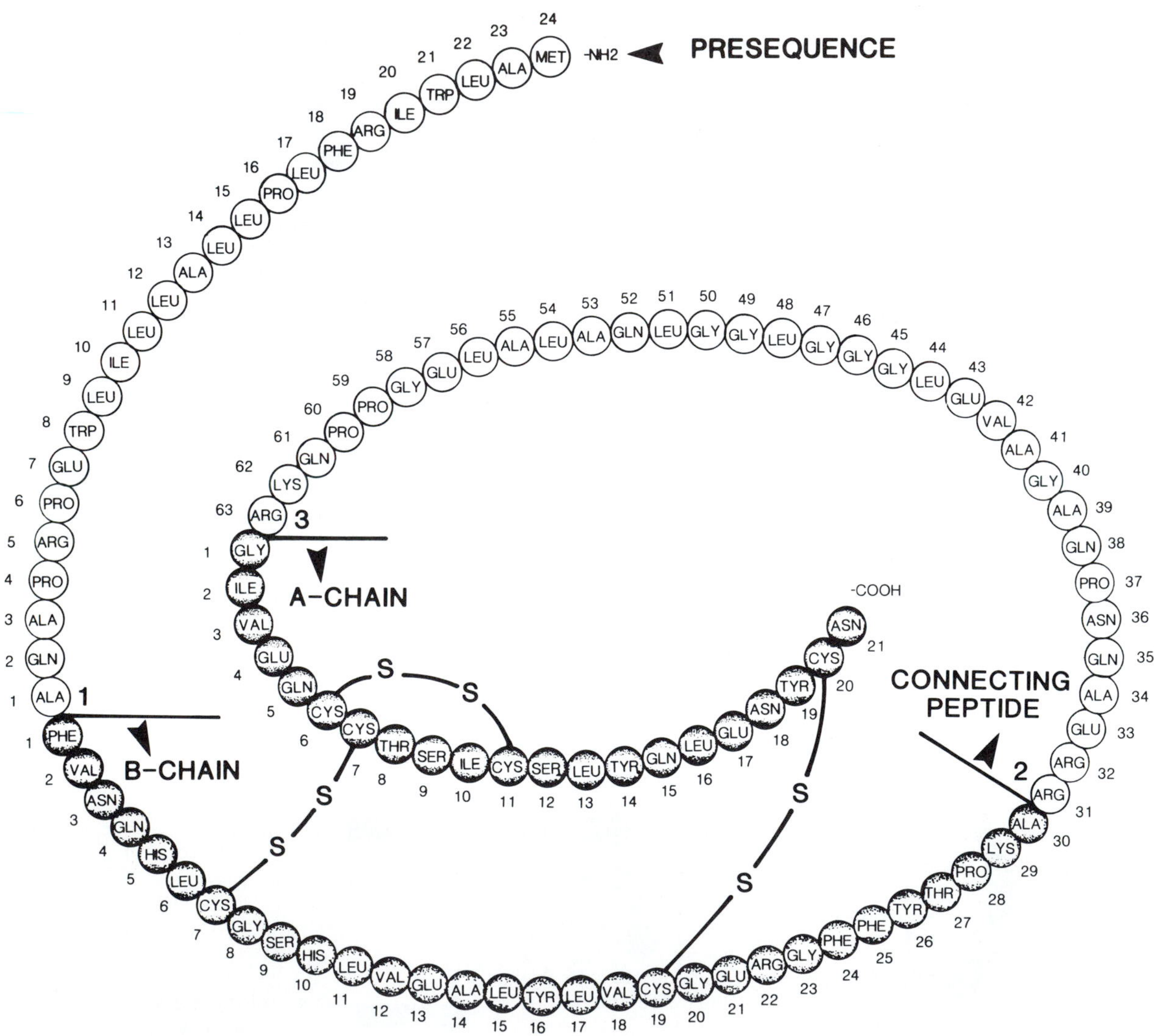

Fig. 29-7 Amino acid sequence of insulin. Amino acid sequence for this figure, a composite of sequences for bovine proinsulin and rat pre-proinsulin, probably does not differ too widely from amino acid sequence of human pre-proinsulin. Enzymatic cleavages of pre-proinsulin *(site 1)* to proinsulin and proinsulin *(sites 2 and 3)* to insulin are described in text.

molecules), these hormones in part induce catabolic metabolism to break down large molecules.

Insulin. Insulin is synthesized in the endocrine pancreas by the beta cells of the islets of Langerhans as a high-molecular-weight precursor called pre-proinsulin.[5] Pre-proinsulin (MW 11,500) is shown in Fig. 29-7. Cleavage at the link marked by the arrow labeled *1* results in the formation of proinsulin (MW 9000).[6] Proinsulin has only 5% of the activity of insulin. The proinsulin molecule consists of the A and B chains of insulin connected by disulfide bonds and by a connective peptide called *C-peptide*. During processing the C-peptide (MW 3000) is removed from the molecule by cleavage at the links marked by arrows *2* and *3*. The resulting insulin molecule (MW 6000) consists of chains A and B connected by two disulfide bonds. This entire process occurs within the beta cell. The initial synthesis of pre-proinsulin occurs at the Golgi apparatus. The molecule is packaged in a vesicle called a beta granule. Cleavage first to proinsulin and next to insulin occurs within the granule. Equal quantities of C-peptide and insulin are released into the circulation when the granule is dissolved at the plasma membrane of the beta cell following neural, dietary, or hormonal stimuli. Only small quantities of proinsulin are found in the circulation.

Glucagon and cortisol. Glucagon is a 3500-dalton polypeptide hormone that is synthesized in the alpha cells of the pancreas.[7] In diabetes mellitus because of insulin deficiency, glucagon levels are elevated and are not suppressed by carbohydrate loading.[8]

Cortisol and the other corticosteroids increase the rate of gluconeogenesis from protein and amino acids, especially in the liver.

Epinephrine. Epinephrine raises glucose levels by inhibiting insulin secretion, stimulating glucagon secretion, stimulating glycogenolysis, and inhibiting gluconeogenesis.

Other hormones. *Growth hormone* and *thyroxine* also act to raise circulating levels of glucose. *Somatostatin* is a polypeptide hormone that is synthesized primarily in the delta cells of the pancreas. Somatostatin inhibits both insulin and glucagon release. *Gastric inhibitory polypeptide* stimulates insulin release. This hormone is located in the intestinal mucosa, and its release is stimulated by glucose and amino acids. Thus food ingestion results in increased levels of circulating insulin.

Opposite actions of insulin and glucagon. Insulin and glucagon have opposing effects. Insulin inhibits proteolysis, lipolysis, and glycogenolysis and stimulates lipid synthesis and glycogenesis in the liver; increases protein synthesis in muscle; and accelerates triglyceride synthesis in fat cells. Insulin acts as the body's only hypoglycemic agent. In contrast, glucagon stimulates lipolysis, ketogenesis, glyconeogenesis, and glycogenolysis. A meal rich in carbohydrates induces insulin secretion and suppresses glucagon release. Hypoglycemia stimulates the release of glucagon. Thus, in general, insulin and glucagon act oppositely to each other, with insulin promoting energy storage and glucagon promoting energy release. The net result of the hypoglycemic agent (insulin) and the hyperglycemic agents is glucose homeostasis.

CLASSIFICATION OF DIABETES MELLITUS

In 1979 the National Diabetes Data Group of the National Institutes of Health developed a classification scheme for diabetes mellitus and other types of glucose intolerance based on current knowledge of the biochemistry of this disease.[9] This new standardized approach to naming diabetic states replaces an inexact nomenclature that had developed over the years which used such terms as prediabetes, latent diabetes, covert diabetes, and juvenile onset and adult-onset diabetes. Table 29-1 summarizes the NIH classification system.

Type I diabetes

Type I, insulin-dependent diabetes mellitus (IDDM), afflicts more than 2 million Americans. Its prevalence is 1 in 300 persons under the age of 20 years, the time when this disease is usually diagnosed. This type of diabetes is caused by insufficient insulin secretion (insulinopenia). Insulin injections are necessary to maintain normal glucose metabolism. Individuals with type I diabetes are especially prone to ketoacidosis. *Ketoacidosis* refers to excessive formation of ketoacids and low blood pH (acidosis). This condition will be discussed in further detail in the section on complications of diabetes. Other complications of type I diabetes include cataracts, diseases of nerves (neuropathy), kidney disease (nephropathy), and blood vessel diseases (angiopathy).

Type II diabetes

Type II diabetes, non-insulin-dependent diabetes mellitus (NIDDM), has no correlation with blood insulin levels. Type II diabetes afflicts nearly 10 million Americans. Onset is usually after the age of 40 years. The type II individual is not dependent on insulin injection, is less prone to ketoacidosis, and is often obese.

Secondary diabetes

Diabetes mellitus caused by other conditions and diseases is called secondary diabetes. Secondary diabetes can be caused by pancreatic disease, acromegaly (growth hormone deficiency), Cushing's syndrome (elevated cortisol), pheochromocytoma (excessive catecholamines), glucagonoma (excessive glucagon because of a tumor), somatostatinoma (excessive somatostatin because of a tumor), primary aldosteronism, severe liver disease, and administration of certain drugs, hormones, and chemicals.

Impaired glucose tolerance

Impaired glucose tolerance (IGT) includes persons who have had an abnormal glucose tolerance test but no frank

Table 29-1 NIH classification of diabetes and other categories of glucose intolerance

Class	Description
Type I diabetes mellitus (insulin-dependent diabetes mellitus [IDDM])	Deficiency of insulin (insulinopenia) Dependence on injected insulin Usually occurs below 40 yr of age Prone to ketoacidosis Prone to diabetic complications Cataracts (six times greater than in nondiabetics) Neuropathy (all show some symptoms; 10% serious) Nephropathy (40%-50% develop renal failure) Angiopathy (high risk for heart attack and stroke)
Type II diabetes mellitus (non-insulin-dependent diabetes mellitus [NIDDM])	Variable levels of insulin Not dependent on exogenous insulin for control of hyperglycemia Usually occurs after 40 yr of age Not prone to ketoacidosis Not prone to diabetic complications
Secondary diabetes mellitus	Diabetes caused by various secondary conditions such as pancreatic disease, acromegaly, Cushing's syndrome, pheochromocytoma, glucagonoma, somatostatinoma, primary aldosteronism, severe liver disease, and certain drugs, chemicals, and hormones
Impaired glucose tolerance	Persons with plasma glucose levels intermediate between upper limit of normal and definitely diabetic levels (1100-1400 mg/L) and persons with an abnormal glucose tolerance test but no frank hyperglycemia
Gestational diabetes mellitus	Diabetes that occurs during pregnancy
Statistical risk classes	
Previous abnormality of glucose tolerance	Previous transient hyperglycemia that occurred either spontaneously or in response to specific stimuli but presently testing normally
Potential abnormality of glucose tolerance	Persons not presently exhibiting any indications of diabetes but at substantially increased risk to develop diabetes in the future; includes monozygotic twin of an NIDDM diabetic; person who has parent, sibling, or offspring who is NIDDM diabetic; obese individuals; members of certain racial or ethnic groups with a high prevalence of diabetes

hyperglycemia. The oral glucose test will be discussed later in this chapter. Estimates of IGT prevalence range from 4.6% to 11.2%.[2]

Gestational diabetes

Gestational diabetes refers to diabetes that occurs during pregnancy. A recent study[10] estimates that 39% of women with gestational diabetes manifest type II diabetes mellitus 20 years after delivery. Screening of pregnant women for gestational diabetes, to prevent perinatal complications associated with maternal hyperglycemia, has become a widespread, accepted practice.

PATHOGENESIS OF DIABETES MELLITUS

Epidemiology

Epidemiologists have studied identical twins and offspring and siblings of diabetics.[11-13] These studies demonstrate clearly that diabetes mellitus develops from a complex interaction between environmental and genetic factors. If the development of diabetes was determined by hereditary factors alone, then the disease should always afflict both identical twins. Three different studies show that when type I diabetes occurs in one twin its subsequent appearance in the other occurs only about 50% of the time. On the other hand, development of type II diabetes in one twin presages its appearance in the other nearly 100% of the time.

Studies of offspring of type II diabetic parents show that diabetes is transmitted to offspring at a frequency of only 6% to 10%. The prevalence of type II diabetes in the general American population is estimated at about 2%. However, a propensity to diabetes exists in the offspring since 25% to 40% of them have abnormal glucose tolerance tests. Similar results have been obtained using siblings of diabetics as subjects.

The principal conclusions derived from these studies are as follows:

1. Offspring and siblings of diabetics are more likely to develop diabetes than those of nondiabetics.
2. The offspring of two diabetics is more likely to develop diabetes than an offspring having only one diabetic parent.
3. The lower than expected frequencies of diabetes in identical twins and offspring and siblings of diabetics suggest the importance of environmental factors in the expression of the genetic component for diabetes.
4. A second event occurring early in life is postulated to trigger type I diabetes in genetically susceptible individuals. This event is likely to be a viral infection or a disturbance in the immune system. Type II diabetes occurs without such an event although its expression is modulated by factors such as obesity.
5. Diabetes is generally transmitted true to type. For example, diabetic siblings and offspring of type I diabetics are usually type I diabetics.

6. Inheritance plays a more important role in development of type II diabetes than type I diabetes.

HLA types

The histocompatibility antigens or human leukocyte antigens (HLAs) are proteins found in nucleated cells. They are responsible for rejection of tissue transplanted to an individual from another unrelated individual. The genes that code for HLAs are on chromosome 6 and have been localized at loci A, B, C, D, and DR. Shortly after their discovery it was shown that certain diseases occurred at a higher frequency in individuals who carried a specific HLA than in those who did not carry that HLA. For example, persons carrying HLA-B27 develop a spine deformity called ankylosing spondylitis 100 times more frequently than persons lacking this HLA.

Type I diabetes has been reported to be associated with the presence of antigens HLA-B8, -BW15, -DR3, and -DR4.[14-17] HLA-B7 is reported to reduce the risk of type I diabetes.[17]

Viruses

Epidemiological studies report a seasonal incidence for type I diabetes[18] and correlate the occurrence of viral infections such as mumps and measles with subsequent development of this type of diabetes.[19,20] Animal studies have strengthened the link between viruses and diabetes by demonstrating that viruses can induce diabetes in some strains of mice.[21,22]

Further evidence for viral infection as a cause of type I diabetes comes from studies with Coxsackie B4 virus. Direct evidence that this virus causes diabetes in humans is derived from the isolation of Coxsackie B4 virus from the pancreas of a child who had developed diabetic ketoacidosis shortly after the onset of a viral infection.[23] The child died of the disease, and autopsy showed extensive beta-cell destruction. Injection of the virus into mice produced diabetes. This experimental evidence strongly links Coxsackie B4 with the onset of diabetes.[24] Since the Coxsackie B4 reports, mumps virus, Coxsackie B1, and rubella reovirus type 3 have been implicated in the transmission of type I diabetes.[25]

An alternative hypothesis proposes an autoimmune basis for the development of diabetes.

Autoantibodies

The production of antibodies to an infective agent is one of the defenses against disease. Occasionally the antibody also attacks nonforeign (host) tissue, resulting in an autoimmune response. Many studies have reported correlations between type I diabetes and autoimmune diseases such as Hashimoto's disease, myasthenia gravis, and thyrotoxicosis.[26] Correlations among HLA-B8, autoimmune disorders, and type I diabetes have also been reported.[27] From these studies a hypothesis for the development of type I diabetes caused by autoimmune disease has emerged. The hypothesis begins with a viral infection, followed by development of antibodies to the viral agent. The antibodies do not distinguish beta-cell protein (probably plasma membrane proteins) from viral protein, and as a result beta-cell damage and eventually diabetes ensue.

Evidence cited in support of this hypothesis includes the isolation of islet cell antibodies (ICAs) from newly diagnosed type I patients that destroy beta islet cells isolated from nondiabetic persons. ICAs are detected in 85% of type I diabetics within 2 years of the onset of the disease. Type II diabetics rarely have ICA.[28]

More recent support for the autoimmune hypothesis derives from experiments that use cyclosporine, an immunosuppressive agent. Administration of cyclosporine causes complete remission of the disease in 50% of type I diabetics.[29]

Insulin metabolism

Diabetes can, but rarely does, result from defects in insulin synthesis, structure, or secretion. Mutations in the structure of proinsulin result in hyperproinsulinemia, which leads to decreased insulin and glucose intolerance.[30] Mutations in insulin itself that produce a biologically ineffective molecule have also been reported.[31,32] Mutations in the proteins that compose the insulin receptor site result in reduced insulin affinity for the site and glucose intolerance[33] (see below).

Receptor site defects

Insulin receptor proteins. Insulin binds reversibly to sites on cell membranes. Insulin-binding sites (called receptor sites) are found only on certain cell types (liver cells, monocytes, adipocytes, and red blood cells). The insulin receptor site is composed of two glycoprotein molecules.[34,35] One subunit is a tyrosine-specific protein kinase.[36] Insulin binding to the receptor site triggers a chain of events resulting in an increase of cell membrane permeability to glucose and amino acids, alteration of enzyme activities, and promotion of protein biosynthesis.

Insulin resistance. In type II diabetes, hyperglycemia is often associated with *hyper*insulinemia. This is in marked contrast to type I diabetes, in which hyperglycemia is always associated with insulin *deficiency*. In fact, whereas type I diabetics depend on insulin to maintain normal blood glucose, type II diabetics respond to relatively high doses of insulin with only small reductions in circulating glucose levels. Type II diabetics are said to be insulin resistant. Insulin resistance in type II diabetes is directly related to a decreased number of insulin receptors.[37] Diabetes caused by decreased numbers of insulin receptor sites is called type A diabetes. Type A diabetes occurs in obese persons. Obese individuals show a significant increase in the numbers of insulin-binding sites and a decrease in symptoms of diabetes when placed on a low-carbohydrate diet.[38] Type A diabetes in obese persons may derive directly from a high-carbohydrate diet rather than obesity per se.

Antibodies to receptor. The presence of circulating antibodies to the insulin receptor has also been reported. Type II diabetes caused by such antibodies to insulin receptor sites is called type B diabetes. Type B diabetics usually have symptoms of autoimmune disorders such as antinuclear antibodies, arthralgia, and enlargement of the parotid gland. Type B diabetes has a lower incidence than type A.

Summary

In summary, it is likely that type I diabetes mellitus is most commonly caused by destruction of islet cells resulting from an autoimmune response to viral infection, whereas most cases of type II diabetes are caused by receptor site defects that either reduce the numbers of insulin-binding sites or affect events after insulin binding. Other etiologies described in this section are probably rare causes of diabetes.

COMPLICATIONS OF DIABETES MELLITUS

The principal complications of diabetes mellitus are retinopathy, neuropathy, angiopathy, nephropathy, susceptibility to infection, hyperlipidemia, ketoacidosis, and hyperglycemic, hyperosmolar nonketotic coma (HHNC). With the singlc exception of HHNC, these diabetic complications occur more frequently for type I diabetics than for type II diabetics.

Retinopathy

Opaque areas in the lens of the eye are called cataracts. Cataract formation is the principal retinopathy of diabetes. Retinopathy is also caused by proliferation of small blood vessels in the lens.

Neuropathy

Neuropathy is the most common complication of diabetes mellitus. It is apparent in about 25% of diabetics and recognized by a variety of symptoms that include pain, numbness, tingling or burning sensations in extremities, dizziness, and double vision. These symptoms are caused by decreased motor and sensory nerve conduction velocities caused by axonal degeneration and demyelination. Secondary manifestations of neuropathy include cardiac failure, excessive sweating, and male impotence.

Angiopathy

Angiopathy refers to damage to linings (basement membranes) of blood vessels. Angiopathy increases the risk of coronary heart disease and stroke and can lead to retinopathy and nephropathy.

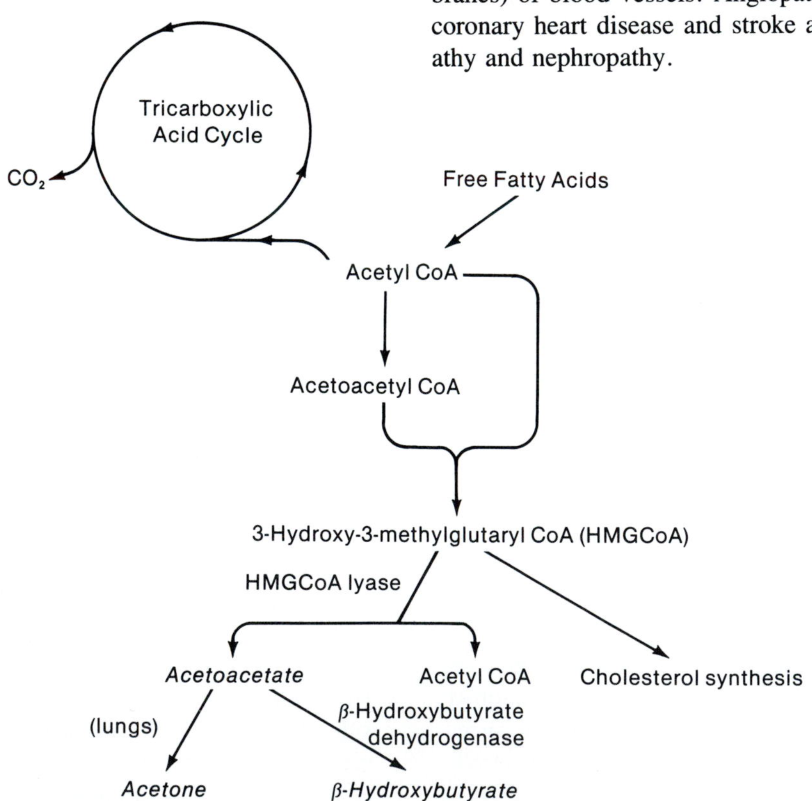

Fig. 29-8 Pathways involved in ketoacid metabolism. Accumulation of ketoacids, acetoacetate, and beta-hydroxybutyrate is a principal feature of diabetic ketoacidosis. Metabolic pathway leading from acetyl CoA to acetoacetate and beta-hydroxybutyrate is accelerated in diabetes because of free fatty acid mobilization.

Nephropathy

Nephropathy refers to damage to the glomerulus (filtering apparatus of the nephron) and capillaries associated with the glomerulus. Capillary damage is caused by angiopathy. The result is a reduction in the filtering capability of the kidneys. Proteinuria is often the first sign of diabetic nephropathy.

Infection

Diabetics are highly susceptible to infection, ulceration, and gangrene (especially in the extremities). Skin disorders are also more common in diabetics than nondiabetics.

Hyperlipidemia and atherosclerosis

High triglyceride and cholesterol levels are often associated with type II diabetes.[39] Increased levels of very low–density lipoprotein (VLDL) have been reported for type II diabetics.[40] High-density lipoprotein (HDL) has been reported to be significantly lower in diabetics than nondiabetics.[41,42] These results are consistent with the higher incidence of coronary heart disease in diabetics,[4] a poor survival rate for diabetics with myocardial infarction,[3] and the inverse relationship between HDL and the risk of coronary heart disease.[43]

Diabetic ketoacidosis

Ketoacid metabolism. As shown in Fig. 29-8, acetyl coenzyme A (acetyl CoA) is at the crossroads of glucose, protein, and lipid metabolism. It either enters the tricarboxylic acid cycle or is metabolized to 3-hydroxy-3-methylglutaryl coenzyme A (HMGCoA). HMGCoA can be metabolized to cholesterol, or it can be converted to acetoacetate. Acetoacetate has two possible fates, spontaneous decarboxylation to acetone (in the lungs) or enzymatic reduction to beta-hydroxybutyrate. Acetoacetate and beta-hydroxybutyrate are commonly called ketoacids, or *ketone bodies*. Ketoacids are normally a source of energy for the brain, kidneys, and skeletal muscle. A considerable quantity of acetoacetate and beta-hydroxybutyrate is excreted by the kidneys with concomitant loss of sodium and potassium. Kidney excretion of sodium and potassium results in the retention of hydrogen ions.

Ketoacids and insulin. In nondiabetics ketoacid formation is a minor pathway. In type I diabetics insulinopenia causes glucose starvation in cells such as muscle and fat cells that depend on insulin for the passage of glucose through the plasma membrane. These cells respond to the insulinopenia by mobilizing fatty acids from triglycerides. Fatty acid degradation increases as it becomes the major source of energy for the cell. Increased fatty acid catabolism produces excessive quantities of acetyl CoA. Although a significant portion of the acetyl CoA is able to enter the tricarboxylic acid cycle to produce energy, an excess quantity of acetyl CoA is metabolized to produce abnormal levels of ketoacids. Increased production of ketoacids elevates blood hydrogen ion concentrations and thereby lowers blood pH (acidosis). This same metabolic pattern occurs in starvation except that hypoglycemia is present instead of hyperglycemia.

Detection. The nitroprusside test (commonly known as Acetest) is useful for the detection of acetoacetic acid (AcAc) in the blood or urine (see p. 856). Nitroprusside does not react with beta-hydroxybutyrate (β-HBA) and reacts only weakly (20%) with acetone. In the early stages of diabetic ketoacidosis, acetoacetate levels are often normal (AcAc:βHB A-1:3) or only mildly elevated while beta-hydroxybutyrate levels are highly elevated (AcAc; rsβHB A-1:30). Under these conditions the nitroprusside test can significantly underestimate the severity of the ketoacidosis. As the ketoacidosis becomes controlled, the beta-hydroxybutyrate is metabolized to acetoacetic acid and the nitroprusside test can become strongly positive.

Diagnosis of ketoacidosis. Blood gases and blood glucose are useful in detecting diabetic ketoacidosis. Low pH, normal P_{CO_2}, low bicarbonate, high anion gap, and high glucose suggest uncompensated ketoacidosis. Low pH, low P_{CO_2}, low bicarbonate, a high anion gap, and high glucose suggest partially compensated diabetic ketoacidosis. The abnormal anion gap is caused by the accumulation of ketoacids (sodium salts).

Lactic acidosis. Lactic acidosis is caused by the accumulation of lactic acid resulting from tissue hypoxia (oxygen deficiency). Like the accumulation of ketoacids, lactate accumulation causes increased blood hydrogen ions and therefore low pH. In the diabetic, lactic acidosis often occurs simultaneously with diabetic keotacidosis, especially if the pH falls below 7.10, if renal insufficiency oc-

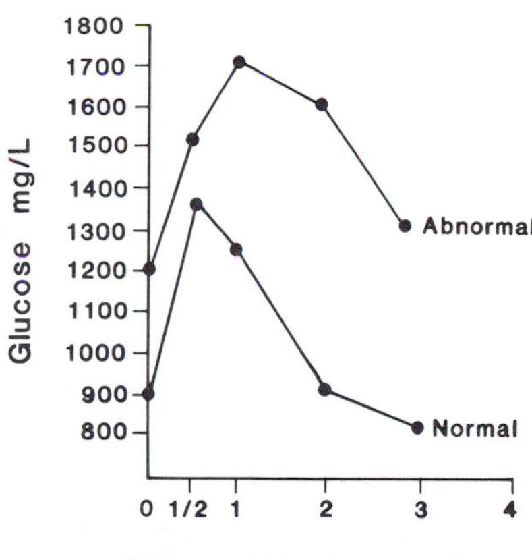

Fig. 29-9 Oral glucose tolerance test (OGTT; see p. 449). Diabetic's response to OGTT is compared to normal response. In diabetics, glucose curve is elevated and delayed. In normal response, peak is reached after 30 minutes and returns to baseline after 2 hours. Type I diabetics produce a nearly flat insulin curve after glucose load. If there is a peak, it occurs late (greater than 1 hour). In type II diabetics, insulin response is often exaggerated, peak is late, and return to baseline is later than 3 hours.

curs, or if certain hypoglycemic agents such as phenformin (DBI) are administered.

Hyperglycemic, hyperosmolar nonketotic coma. Hyperglycemic, hyperosmolar nonketotic coma (HHNC) has been reported with increasing frequency over the past few years. It is characterized by a blood glucose level above 6000 mg/L, normal or slightly low blood pH, serum osmolality above 350 mOsm/kg, normal ketoacid levels, and lethargy or coma. Although diabetic ketoacidosis occurs primarily in type I diabetics, HHNC occurs primarily in type II diabetics. The absence of ketoacids in HHNC is probably caused by the differential sensitivity of lipid and glucose metabolism to insulin. Lipolysis is inhibited by one tenth the insulin level that is required to enhance glucose metabolism.[44] In type I diabetics, insulinopenia enhances lipolysis with resulting accumulation of ketoacids and glucose utilization is blocked, resulting in hyperglycemia. In type II diabetics, although insulin resistance occurs, there is sufficient insulin activity to limit lipolysis and thus ketoacid production but insufficient insulin activity to avoid hyperglycemia. HHNC is often brought on by stressful events and major illness.

PATHOGENESIS OF DIABETIC COMPLICATIONS

Protein glycosylation

Nonenzymatic protein glycosylation commonly occurs in red blood cells, glomeruli, and nerve cells as well as in other tissues. The extent of the glucosylation is proportional to extracellular glucose concentrations. Such glucosylation occurs by the mechanism shown for the glycosylation of hemoglobin (see p. 1040). The carbonyl functional groups of glucose, mannose, galactose, xylose, ribose, and fructose react with free amino groups of proteins to form intermediates called Schiff bases, or adlimines. The amine group that reacts is either an *N*-terminal amino group or a lysine epsilon–amino group. The aldimine subsequently rearranges to form a ketoamine. This rearrangement is called the Amadori rearrangement. The adlimine is labile; it can readily hydrolyze to re-form a free amino group and a carbonyl group. The ketoamine is relatively stable, and its formation is not reversible.

Excessive glycosylation is known to produce significant alterations in a protein's physical and biochemical properties. For example, glycosylation of alpha crystallin, a protein occurring in the lens of the eye, greatly reduces its solubility. Hyperglycemia in rats has been shown to increase alpha-crystallin glycosylation simultaneously with cataract formation.[45] Glycosylation of the basement membrane of blood vessels is known to cause basement membrane thickening similar to that found in most if not all diabetics.[46,47] Functional alterations of immunoglobulin G (IgG) by nonenzymatic glycosylation have been reported[48] to be associated with increased susceptibility to infection.

Based on these and other findings, a hypothesis has been proposed stating that many of the complications of diabetes are caused by glycosylation of specific proteins such as alpha crystallin, IgG, and basement membrane protein.

Sorbitol accumulation

The intracellular accumulation of sorbitol is the basis for another hypothesis designed to explain diabetic complications.[49] Aldose reductase reduces glucose to sorbitol, which in turn is oxidized to fructose by sorbitol dehydrogenase. Sorbitol does not easily cross cell membranes. The removal of sorbitol from the cell depends on its conversion to fructose, which does pass freely through the cell membrane. However, when glucose levels are high, the quantities of sorbitol produced outstrip the cell's ability to convert sorbitol to fructose, resulting in the intracellular accumulation of sorbitol. Intracellular accumulation of sorbitol causes osmotic swelling and injury to cell structures. Only cells that do not depend on insulin for glucose transport across the plasma membrane are affected. These cells include nerve, ocular lens, and glomeruli cells.[50] Supporting this hypothesis are reports of elevations in sorbitol and fructose levels in the nerve and ocular lens cells of diabetics.[51] The strongest support for this hypothesis is derived from studies utilizing aldose reductase inhibitors. The aldose reductase inhibitors, sorbinol and tolrestat, are reported to improve nerve conduction in diabetic rats and diabetic humans.[52,53]

FUNCTION TESTS

Postprandial plasma glucose

Diabetes is more readily detected when the carbohydrate metabolic capacity is tested. This is done by stressing the system with a defined glucose load. Measurement of the rate that the glucose load is cleared from the blood, as compared to the rate of glucose clearance in healthy persons, detects impairment in glucose metabolism. A meal high in carbohydrates is often used as the carbohydrate load, although a 75 g glucose drink is usually preferred over a meal. It is called the postprandial test.

Blood is drawn at 2 hours after ingestion of the meal or glucose drink. Glucose levels above 1400 mg/L are abnormal; levels of 1200 to 1400 mg/L are ambiguous; and levels below 1200 mg/L are normal. The postprandial glucose test, although widely used for detection of diabetes, is highly inaccurate because of several variables that are difficult to control or adjust for. These variables include age, weight, previous diet, activity, illness, medications, time of day that the test is conducted, and actual size of the glucose dose. When a meal is used as the load the effective glucose load depends on the digestion of disaccharides and polysaccharides and their subsequent absorption from the intestinal tract.

O'Sullivan test

The O'Sullivan test is frequently used to detect gestational diabetes. A 50 g load of glucose is used. Blood is drawn at 1 hour. Gestational diabetes is suggested by plasma glucose levels above 1500 mg/L (above 1300 mg/L for whole blood).

Oral glucose tolerance test

The oral glucose tolerance test (OGTT) evaluates glucose clearance from the circulation after glucose loading under defined and controlled conditions. The test has been standardized by the Committee on Statistics of the American Diabetes Association.[54]

Standard conditions call for a minimum carbohydrate intake of 150 g/day for 3 days before the test. Daily carbohydrate intake less than this lowers carbohydrate intolerance. There should be an 8- to 16-hour fast before testing. The patient must be ambulatory since inactivity decreases glucose tolerance. However, exercise and emotional stress should be avoided.

Illness reduces glucose tolerance. Abnormalities of such hormones as thyroxine, growth hormone, cortisol, and catecholamines interfere. Drugs and medications such as oral contraceptives, salicylates, nicotinic acid (found in cigarettes, cigars, pipe tobacco, chewing tobacco), diuretics (including caffeine), and hypoglycemic agents (insulin, sulfonylureas) interfere. Testing time affects the test. The best time to conduct the test is between 7 AM and 12 AM. Evaluation criteria should also be adjusted for age. If adjustments for age are not made, about 80% of persons over 60 years of age will be judged diabetic.[55]

The glucose load should consist of glucose only. Some commercial preparations labeled "100 grams glucose equivalent" contain disaccharides and polysaccharides. The rate that these saccharides are hydrolyzed and absorbed from the intestinal tract varies from person to person. Such a preparation is obviously not desirable for individuals with pancreatic or malabsorptive disorders. The size of the load is 40 g of glucose per square meter of body area. For most subjects 75 g of total glucose is sufficient. The drink can be flavored if caffeine or theophylline is not used.

Blood samples are drawn at fasting, and 1, 2, and 3 hours after ingestion of glucose. Additional samples at ½, 1½, and 2½ hours after glucose ingestion are helpful and sometimes necessary for evaluation of the test.

The OGTT is commonly evaluated by several alternative scoring systems. All use criteria based on the glucose oxidase method for plasma glucose quantification. Plasma glucose values are converted to whole blood values using the following equation:

$$\text{Glucose (whole blood; in mg/L)} = \text{Glucose (plasma; in mg/L)} \times 1.15 + 6\,\text{mg/L}$$

The criteria for the Wilkerson Point System, the Fajans-Conn System, the University Group Diabetes Program (UGDP), and the World Health Organization (WHO) appear in Table 29-2.[56]

The *Wilkerson Point System* uses points as indicated by numbers in parentheses in Table 29-2. Two points or greater suggests diabetes mellitus. A uniform dose of 100 g of glucose is used. The *Fajans-Conn System* judges an individual diabetic if two or more criteria (Table 29-2) are exceeded. A dose of glucose based on 40 g of glucose per square meter of body surface area is employed. WHO suggests impaired glucose tolerance or diabetes when both fasting and 2-hour plasma glucose levels exceed 1390 mg/L (Table 29-2). When the 2-hour value is 1400 to 2000 mg/L, impaired glucose tolerance (IGT) is suggested. When the 2-hour value exceeds 2000 mg/L, diabetes mellitus is suggested. The National Diabetes Data Group has suggested simplifying the evaluation further by eliminating the fasting plasma glucose value. This method employs a uniform dose of 75 g of glucose. UGDP evaluates the oral glucose tolerance test by summing the fasting, 1-, 2-, and 3-hour plasma glucose values (Table 29-2). A sum exceeding 5990 mg/L suggests diabetes mellitus. A glucose load of 30 g per square meter of body surface area is employed.

Shape of the glucose tolerance curve is useful in evaluating OGTT (Fig. 29-9, p. 447). Healthy subjects peak at ½ hour and return to fasting levels at 2 hours. Diabetics peak late (approximately 1 hour) or even show a plateau at 2 to 3 hours and return to baseline after 3 hours.

Insulin determinations performed along with glucose determinations are useful in evaluating the OGTT. Plasma

Table 29-2 Comparison of four criteria for evaluation of oral glucose tolerance test

Time of blood drawing (hr)	Blood glucose levels (mg/L)			
	Wilkerson Point System	Fajans-Conn System	WHO*	UGDP†
Fasting	>1290 (1)	>1840	>1390	Sum
1	>1940 (½)	—	—	Sum
1½	—	>1640	—	—
2	>1390 (½)	>1390	1400-2000 (IGT) >2000 (diabetes)	Sum
3	>1290 (½)	—	—	Sum

*World Health Organization.
†University Group Diabetes Program.

insulin levels after a glucose load differentiate type I from type II diabetes. In nondiabetics, insulin levels peak 1 hour after a glucose load and return to fasting levels at 2 to 3 hours. Type I diabetics respond to a glucose load with little or no insulin increases above fasting levels. Type II diabetics respond to the challenge with an abnormally late and often excessive increase in insulin levels. Type I diabetics often have low fasting insulin levels. Type II diabetics have variable fasting insulin levels.

The OGTT has been criticized.[57,58] Since many of the variables affecting test results are difficult to control, the reproductibility of the test is poor. Different evaluation schemes for the same OGTT often produce different interpretations. In general, the test tends to overdiagnose diabetes. One group of investigators suggests a more conservative evaluation scheme in which glucose levels must exceed 2590 mg/L at 1 hour and 2190 mg/L at 2 hours for the OGTT to be considered abnormal.[59] Others have suggested adding 95 mg/L to the 1-hour glucose value and 53 mg/L to the 2-hour value for each decade after the age of 40 years.[54] The OGTT is best used to assess individuals who have borderline fasting glucose levels or who are at risk for the development of diabetes and to distinguish type I from type II diabetes.

Intravenous glucose tolerance test

The intravenous glucose tolerance test is often used for persons with malabsorptive disorders or previous gastric or intestinal surgery. Glucose is administered intravenously over 30 minutes, using a 20% solution. A glucose load of 0.5 g/kg of body weight is used. Nondiabetics respond with a plasma glucose level of 2000 to 2500 mg/L. Discontinuation of the glucose loading leads to a decrease in plasma glucose levels with fasting levels reached at about 90 minutes. Diabetics demonstrate plasma glucose levels above 2500 mg/L during administration of the load. On discontinuation of the loading, plasma glucose levels of diabetics also return to fasting levels at about 90 minutes. An alternative procedure called the Soskin method uses 50% glucose delivered intravenously within 3 to 5 minutes. The glucose load used is 0.3 g/kg of body weight. Nondiabetics reestablish fasting levels less than 60 minutes after discontinuing the glucose infusion. In diabetics fasting levels are reestablished significantly later than 60 minutes.

CHANGE OF ANALYTE IN DISEASE

The following is a summary of analyte changes in diabetes mellitus. For each analyte, levels in controlled diabetes, diabetic ketoacidosis, and HHNC are compared.

Fasting plasma glucose

Fasting plasma glucose and urinary glucose are the most commonly used markers for diabetes mellitus. In general, repeated fasting plasma glucose levels of 1400 mg/L strongly suggest diabetes provided drugs such as glucocorticoids are not being administered and diseases and conditions such as those listed in the box are not present. Repeated plasma glucose levels of 1150 to 1400 mg/L may indicate the presence of diabetes. Fasting plasma glucose is high only in advanced diabetes.

CONDITIONS AND DISEASES THAT OFTEN CAUSE BOTH HYPERGLYCEMIA AND GLUCOSURIA OR GLYCOSURIA IN ABSENCE OF HYPERGLYCEMIA

Hyperglycemia and glucosuria

Septicemia	Hypercortisolism
Pancreatic cancer	Glucagonoma
Acute pancreatitis	Somatostatinoma
Pheochromocytoma	Primary aldosteronism
Hyperthyroidism	Acute myocardial infarction
Acromegaly	Cerebral hemorrhage

Glucosuria and normal plasma glucose

Pregnancy (renal threshold is reduced)
Vitamin D–resistant rickets
Osteomalacia (proximal tubular malfunction)
Hepatolenticular degeneration

Fasting plasma glucose is directly proportional to the severity of diabetes mellitus. Levels above 1800 mg/L may produce glucosuria. Keotacidosis can occur at almost any level above 1400 mg/L but is more common at levels above 1800 mg/L. HHNC is associated with glucose levels above 6000 mg/L.

Diabetics who are under control exhibit wide variations in their plasma glucose concentrations. Plasma glucose levels in controlled diabetics range during a typical 24-hour period from as low as 250 mg/L to as high as 3250 mg/L. These variations are considerably wider than those of nondiabetics.[60] Wide swings in plasma glucose contribute to the development of diabetic complications. Management of insulin therapy remains a significant challenge for the physician. Excessive quantities of insulin cause insulin-induced hypoglycemia, which often leads to coma. On the other hand, inadequate control of glucose levels causes diabetic complications such as those described earlier. Generally, fasting plasma glucose in diabetics is maintained at normal or slightly above normal concentrations.

Urinary glucose

Urinary glucose is a poor marker for diabetes mellitus. The normal renal threshold for glucose is 1800 mg/L. Blood glucose levels must exceed this value before excessive glucose is apparent in the urine. Further complicating this picture is the fact that the renal threshold in diabetics is often increased to above 3000 mg/L. Some diseases and conditions that produce both hyperglycemia and glucosuria are listed in the preceding box. This box also lists conditions that cause glucosuria in the absence of hyperglycemia.

Glucosylated hemoglobin and plasma albumin

A minor hemoglobin derivative called Hb A_{1c} is produced by glucosylation. Since this reaction is nonenzymatic and since the red cell is completely permeable to glucose, the quantity of Hb A_{1c} formed is directly proportional to the average plasma glucose concentration that the red cell is exposed to during its 120-day life span, that is, the 4 to 6 weeks before sampling. Thus, in long-term hyperglycemia, Hb A_{1c} constitutes a higher percentage of total hemoglobin than in normoglycemia. Transient elevations in plasma glucose only mildly affect Hb A_{1c} levels.

Hb A_1 actually consists of four principal components, called Hb A_{1a1}, Hb A_{1a2}, Hb A_{1b}, and Hb A_{1c}.[61,62] As seen in Table 29-3, each consists of two components, a labile component, which is actually the aldimine, and a stable component, which is actually the ketoamine. For normoglycemic persons, Hb A_{1a1}, Hb A_{1a2}, and Hb A_{1b} constitute 0.4% to 0.8% of the total hemoglobin. Hb A_{1c} constitutes 4% to 5% of total hemoglobin. Total Hb A_1 is normally 5.0 to 7.0%. As shown in Table 29-3, diabetics have total Hb A_1 and Hb A_{1c} percentages that are significantly elevated. The elevations are directly proportional to the long-term degree of hyperglycemia.[63,64] Glucosylated hemoglobins are most useful for monitoring of diabetes; they are not sufficiently sensitive to effectively detect borderline cases of diabetes mellitus.[41]

As stated above, serum albumin is also glucosylated to a degree proportional to plasma glucose levels. The relatively short half-life for albumin of 15 days makes it a good monitor of blood plasma glucose levels.[65]

Insulin

Fasting plasma insulin levels in type I diabetics are usually low. Those of type II diabetics are low only when fasting plasma glucose levels exceed 2500 mg/L. Otherwise, they are normal.[66] A glucose challenge separates type I diabetics from type II diabetics. Glucose loading elicits no significant insulin response for type I diabetics and a delayed, often exaggerated response in type II diabetics.

Table 29-3 Components of glucosylated hemoglobin in diabetics and nondiabetics

Glucosylated hemoglobin fraction	Percent of total hemoglobin Nondiabetics	Diabetics
A_{1a1} (labile + stable)	0.19 ± 0.02	0.20 ± 0.03
A_{1a2} (labile + stable)	0.19 ± 0.4	0.22 ± 0.04
A_{1b} (labile + stable)	0.48 ± 0.15	0.67 ± 0.3
A_{1c} (by high-performance liquid chromatography)		
Labile + stable	3.3 ± 0.3	7.5 ± 2.0
Labile	3.2%	6.9%
Stable	96.8%	93.1%

Adapted from Nathan, DM: Clin Chem 27:1261, 1981; and Bunn, HF, Gabbay, KH, and Gallop, PM: Science 200:21, 1978.

Ketoacids

Significant elevations of acetoacetate and beta hydroxybutyrate cause diabetic ketoacidosis. It is important to measure both blood and urinary ketoacid levels since plasma ketoacid levels can be normal even though urinary ketoacid concentrations are high. This effect is caused by increased urinary ketoacid excretion resulting from renal compensation to low pH. Ketonemia and ketonuria are both absent in HHNC. Controlled diabetics should have both normal plasma and normal urinary ketoacid levels.

Lactic acid

Plasma lactic acid levels are frequently elevated (lactic acidosis) during diabetic ketoacidosis.

Hydrogen ion (pH)

High plasma hydrogen ion concentrations (low pH) occur in diabetic ketoacidosis, ketoacidosis with lactic acidosis, and HHNC. pH levels below 7.00 are associated with a poor prognosis.

Electrolytes

Uncontrolled diabetics can exhibit normal, low, or high plasma sodium levels. Plasma sodium levels in diabetics are influenced by three factors, described next.

Hyperglycemia causes an increase in the osmotic pressure of plasma. As a result water flows from cells to plasma. Plasma substituents are thereby diluted. Thus hyponatremia (low plasma sodium) is common in diabetes. In diabetic ketoacidosis excessive quantities of sodium are excreted in the urine, further lowering plasma sodium levels. However, complicating matters is the preferential excretion of water relative to sodium. This effect often compensates for sodium loss from high plasma glucose levels and ketosis, thus resulting in normal or even elevated plasma sodium levels.

The same factors described above affect plasma potassium levels. However, for potassium, two additional factors are operative. First, insulin causes the transport of intracellular potassium to the plasma. Thus hypokalemia (low plasma potassium) occurs in insulin deficiency (type I diabetes). Second, in acidosis, potassium moves out of cells. Thus, in diabetic ketoacidosis, significant quantities of potassium ion are shifted from cells to plasma. This produces hyperkalemia (high plasma potassium). Type II diabetics normally exhibit hypokalemia or normokalemia.

Plasma bicarbonate levels are normal in controlled diabetes. Ketoacidosis causes low plasma bicarbonate levels. The body responds to ketoacidosis by kidney retention of bicarbonate and rapid, deep respiration called Kussmaul breathing, which removes CO_2. Both of these compensatory mechanisms raise pH. Kussmaul breathing lowers P_{CO_2}. Both plasma bicarbonate and P_{CO_2} are low in diabetic ketoacidosis.

Osmolality

Serum osmolality is increased in both ketoacidosis and HHNC because of the water loss that accompanies glucose excretion. Serum osmolality in HHNC is usually above 350 mOsm/kg, and this is a hallmark of the condition.

Body fluid volume

Renal loss of water in diabetic ketoacidosis produces severe volume depletion, often as much as 6 to 8 L. Patients with HHNC can have fluid deficits greater than 9 L. Low fluid volume (hypovolemia) often coexists with hyponatremia. Insulin therapy restores both fluid volume and plasma sodium to normal.

Anion gap

In ketoacidosis the anion gap is always increased because of the excessive formation of ketoacids. Lactic acidosis further increases the gap because of the high lactate levels.

BUN

BUN levels are increased in both diabetic ketoacidosis and HHNC because of increased protein catabolism and prerenal azotemia secondary to loss of extracellular fluids. Prerenal azotemia refers to increased BUN caused by decreased renal flow. In diabetic ketoacidosis, prerenal azotemia is caused by hypovolemia.

Lipids

Elevated plasma triglyceride, cholesterol, and VLDL are commonly found in diabetics. On the other hand, HDL is usually low.

REFERENCES

1. National Commission on Diabetes: The long range plan to combat diabetes, US Department of Health Education and Welfare, no. 76-1018, National Institutes of Health, 1976.
2. Harris, MR, Hadden WC, Knowler, WC, et al: Prevalence of diabetes and impaired glucose tolerance and plasma levels in the U.S. population aged 20-74 yr, Diabetes 36:523, 1987.
3. Smith, JW, Marcus, FI, Serokman, R, et al: Prognosis of patients with diabetes mellitus after myocardial infarction, Am J Cardiol 54:719, 1984.
4. Oppenheimer, SM, Hoffbrand, BI, Oswald, GA, et al: Diabetes mellitus and early mortality from stroke, Br Med J 291:1014, 1985.
5. Chan, SJ, Keim, P, and Steiner, DF: Cell-free synthesis of rat preproinsulin: characterization and partial amino acid sequence determination, Proc Natl Acad Sci USA 73:1964, 1976.
6. Steiner, DF, and Oyer, PE: The biosynthesis of insulin and a probable precursor of insulin by a human islet cell adenoma, Proc Natl Acad Sci USA 57:473, 1967.
7. Unger, RH, and Orci, L: Glucagon and the A cell, N Engl J Med 304:1518, 1575, 1981.
8. Hartmann, H, Probst, I, Jungermann, K, et al: Inhibition of glycogenolysis and glycogen phosphorylase by insulin and proinsulin in rat hepatocyte cultures, Diabetes 36:551, 1987.
9. National Diabetes Data Group: Classification and diagnosis of diabetes mellitus and other categories of glucose intolerance, Diabetes 28:1039, 1979.
10. O'Sullivan, JB, and Worshop, Y: Subsequent morbidity among gestational diabetes women. In Sutherland, HW, and Stowers, M, editors: Carbohydrate metabolism in pregnancy and the newborn, Edinburgh, 1984, Churchill Livingstone.
11. Tattersall, RB, and Fajans, SS: A difference between the inheritance of classical juvenile-onset type diabetes and maturity onset diabetes of young people, Diabetes 24:44, 1975.
12. Barnett, AH, Eff, C, Lesslie, RDG, et al: Diabetes in identical twins: a study of 200 pairs, Diabetologia 20:87, 1981.
13. Pyke, DA, and Nelson, PG: Diabetes mellitus in identical twins. In Creutsfeldt, W, Kobberling, JV, and Neel, JB, editors: The genetics of diabetes mellitus, New York, 1976, Springer-Verlag New York, Inc.
14. Cudworth, AG, and Woodrow, JC: HLA system and diabetes mellitus. Diabetes 24:345, 1975.
15. Nerup, J, Platz, P, Anderson, OO, et al: HLA antigens and diabetes mellitus, Lancet 2:864, 1974.
16. Koevary, S, et al: Passive transfer of diabetes in the BB/W rat, Science 220:727, 1983.
17. Goldstein, S, and Podolsky, S: The genetics of diabetes mellitus, Med Clin North Am 62:639, 1978.
18. Gamble, DR, and Taylor, KW: Seasonal incidence of diabetes mellitus, Br Med J 3:631, 1969.
19. Hinden, E: Mumps followed by diabetes, Lancet 1:1381, 1962.
20. Johnson, GM, amd Tudor, RB: Diabetes mellitus and congenital rubella infection, Am J Dis Child 120:453, 1970.
21. Onodera, T, Jenson, AB, Yoon, JW, et al: Virus-induced diabetes mellitus: reovirus infection of pancreatic B cells in mice, Science 201:539, 1978.
22. Coleman, TJ, Taylor, KW, and Gamble, DR: The development of diabetes following Coxsackie B virus in mice, Diabetologia 10:755, 1974.
23. Yoon, JW, Onodera, T, Jenson, AB, et al: Virus induced diabetes mellitus. XI. Replication of Coxsackie B3 in human pancreatic beta cell cultures, Diabetes 27:778, 1978.
24. Yoon, JW, Austin, M, Onodera, T, and Notkins, A: Isolation of a virus from the pancreas of a child with diabetic ketoacidosis, N Engl J Med 300:1173, 1979.
25. Craighead, JE: Does insulin dependent diabetes mellitus have a viral etiology? Hum Pathol 10:267, 1979.
26. Powers, AC, and Eisenbarth, GS: Autoimmunity to islet cells in diabetes mellitus, Annu Rev Med 36:533, 1985.
27. Cudworth, AG, and Woodrow, JC: Genetic susceptibility in diabetes mellitus: analysis of HLA association, Br Med J 2:846, 1975.
28. Nerup, J, and Lenmark, A: Autioimmunity in insulin-dependent diabetes mellitus, Am J Med 70:135, 1981.
29. Stiller, CR, Dupre, J, et al: Effects of cyclosporine immunosuppression in insulin-dependent diabetes of recent onset, Science 223:1362, 1984.
30. Kanazaaw, Y, Hayashi, M, Ikeuchi, M, et al: Familial proinsulinemia: a possible cause of abnormal glucose tolerance, Eur J Clin Invest 8:327, 1978.
31. Given, BD, Mako, ME, Tager, H, et al: Circulating insulin with reduced biological activity in a patient with diabetes, N Engl J Med 302:129, 1980.
32. Shoelson, S, Haneda, M, Blix, P, et al: Three mutant insulins in man, Nature 302:540, 1983.
33. Berger, M, and Berchtold, R: Insulin transport and action at target cells. In Marble, A, Krall, LP, Bradley, RE, et al, editors: Joslin's diabetes mellitus, ed 12, Philadelphia, 1985, Lea & Febiger.
34. Kasuga, M, Van Obberghen, E, Yamada, KM, et al: Autoantibodies against the insulin receptor recognize the insulin binding subunits of an oligomeric receptor, Diabetes 30:354, 1981.
35. Van Obberghen, E, Kasuga, M, LeCam, A, et al: Biosynthetic labeling of insulin receptor: studies of subunits in cultured human IM-9 lymphocytes, Proc Natl Acad Sci USA 78:1032, 1981.
36. Roth, RA, and Cassell, DJ: Insulin receptor: evidence that it is a protein kinase, Science 219:299, 1983.
37. Kappy, MS, and Plotnick, L: Erythrocyte insulin binding in obese children and adolescents, J Clin Endocrinol Metab 51:1440, 1980.

38. Bar, RS, Gorden, P, Roth, J, et al: Fluctuations in the affinity and concentration of insulin receptors on circulating monocytes of obese patients, J Clin Invest 58:1123, 1976.
39. Bradley, RF: Cardiovascular disease. In Marble, A, White, P, Bradley, RF, and Krall, LP, editors: Joslin's diabetes mellitus, ed 11, Philadelphia, 1971, Lea & Febiger.
40. Goldberg, RB: Lipid disorders in diabetes, Diabetes Care 4:561, 1981.
41. Dods, RF, and Bolmey, C: Glycosylated hemoglobin assay and oral glucose tolerance test compared for detection of diabetes mellitus, Clin Chem 25:764, 1979.
42. Lopes-Virella, MFL, Stone, PG, and Colwell, JA: Serum high density lipoprotein in diabetic patients, Diabetologia 13:285, 1977.
43. Castelli, WP, Garrison, RJ, Wilson, PWF, et al: Incidence of coronary heart disease and lipoprotein cholesterol levels: the Framingham study, JAMA 256:2835, 1986.
44. Zierler, KL, and Rabinowitz, D: Effect of very small concentrations of insulin on forearm metabolism: persistence of its action on potassium and free fatty acids without its effect on glucose, J Clin Invest 43:950, 1964.
45. Cerami, A, Stevens, VJ, and Monnier, VM: Role of nonenzymatic glycosylation in the development of the sequelae of diabetes mellitus, Metabolism 28:431, 1979.
46. Siperstein, MD, Unger, RH, and Madison, LL: Studies of muscle basement membranes in normal subjects, diabetic and prediabetic patients, J Clin Invest 47:1973, 1968.
47. Butcher, D, Kikkawa, R, Klein, L, et al: Size and weight of glomeruli isolated from human diabetic and nondiabetic kidneys, J Lab Clin Med 89:544, 1977.
48. Kaneshige, H: Nonenzymatic glycosylation of serum IgG and its effect on antibody activity in patients with diabetes mellitus, Diabetes 36:822, 1987.
49. Gabbay, KH: The sorbitol pathway and the complication of diabetes, N Engl J Med 288:831, 1973.
50. Malone, JI, Knox, G, Benford, S, et al: Red cell sorbitol: an indicator of diabetes control, Diabetes 29:861, 1980.
51. Gabbay, KH, Merola, LO, and Field, RA: Sorbitol pathway presence in nerve and cord with substrate accumulation in diabetes, Science 151:209, 1966.
52. Tomlinson, DR, Moriarty, RJ, and Mayer, JH: Prevention and reversal of defective axonal transport and motor nerve conduction velocity in rats with experimental diabetes by treating with the aldose reductase inhibitor sorbinol, Diabetes 33:470, 1984.
53. Notvest, RR, and Inserra, JJ: Tolrestat, an aldose reductase inhibitor, prevents nerve dysfunction in conscious diabetic rats, Diabetes 36:500, 1987.
54. Report on the Committee on Statistics of the American Diabetes Association: Standardization of the oral glucose tolerance test, Diabetes 18:299, 1969.
55. Davidson, MB: The effect of aging on carbohydrate metabolism: a review of the English literature and a practical approach to the diagnosis of diabetes mellitus in the elderly, Metabolism 28:688, 1979.
56. Harris, MI, Hadden, WC, Knowler, WC, et al: International criteria for the diagnosis of diabetes and impaired glucose tolerance, Diabetes Care 8:562, 1985.
57. Siperstein, MD: The glucose tolerance test: a pitfall in the diagnosis of diabetes mellitus, Adv Intern Med 20:297, 1975.
58. Sherwin, RS: Limitations of the oral glucose tolerance test in diagnosis of early diabetes, Primary Care 4:255, 1977.
59. Unger, RH: The standard two hour oral glucose tolerance test in the diagnosis of diabetes mellitus in subjects without fasting hyperglycemia, Ann Intern Med 47:1138, 1957.
60. Mauer, AC: The therapy of diabetes, Am Scientist 67:422, 1979.
61. Schenk, AG, and Schroeder, WA: The relation between the minor components of whole normal human adult hemoglobin as isolated by chromatography and starch block electrophoresis, J Am Chem Soc 83:1472, 1961.
62. Gonen, B, Rochman, H, and Rubenstein, AH: Metabolic control in diabetic patients: assessment by hemoglobin A1 values, Metabolism 28:448, 1979.
63. Koenig, RJ, Peterson, CM, Kilo, C, et al: Hemoglobin A1c as an indicator of the degree of glucose intolerance in diabetes, Diabetes 25:230, 1976.
64. Boden, G, Ravidatt, WM, Gordon, SS, et al: Monitoring metabolic control in diabetic outpatients with gycosylated hemoglobin, Ann Intern Med 92:352, 1980.
65. Guthrow, CE, Morris, MA, Day, JF, et al: Enhanced nonenzymatic glucosylation of serum albumin in diabetes mellitus, Proc Natl Acad Sci USA 76:4528, 1979.
66. Ward, WK, Beard, JC, Halter, JB, et al: Pathophysiology of insulin secretion in non-insulin-dependent diabetes mellitus, Diabetes Care 7:491, 1984.

CHAPTER 30

Disorders of lipid metabolism

HERBERT K. NAITO

OBJECTIVES

- Describe digestion, absorption, and metabolism of cholesterol and triglycerides, including the role of the liver and adipose tissue.
- List the lipoproteins and their apolipoprotein content; state the primary function of each lipoprotein.
- Describe the synthesis and catabolism of HDL, LDL, VLDL, and chylomicrons.
- Discuss each of the six types of hyperlipoproteinemias with respect to lipid and lipoprotein levels, appearance of the specimen, and genetic etiology.
- State the clinical significance of hyperlipidemia.

KEY TERMS

amphipathic (amphi-, "on both sides" + *pathic,* "of feeling"); pertaining to a molecule having two sides with characteristically different properties, or a detergent, which has both a polar (hydrophilic) end and a nonpolar (hydrophobic) end but is long enough so that each end demonstrates its own solubility characteristics.

apolipoprotein The protein component of lipoprotein complexes.

chylomicron Large lipid-protein complexes that are made by the gut, and these molecules serve an important function in the transport of fats (mainly dietary triglycerides).

HDL High-density lipoprotein; this lipid-protein complex is also called *alpha-lipoprotein* and is the most dense of the lipoproteins.

IDL Intermediate-density lipoprotein; this lipid-protein complex has a density between VLDL and LDL and is a product that has a relatively very short half-life and is in the blood in very low concentrations in a normal person. In a type III hyperlipoproteinemic person the IDL concentration in the blood is found to be elevated.

LDL Low-density lipoprotein; this lipid-protein complex is also called *beta-lipoprotein* and is the end product of VLDL catabolism. It is the major carrier of cholesterol.

lipoproteins Lipid (apoprotein)-protein complexes consisting of discrete families of macromolecules with known physical, chemical, and physiological properties.

VLDL Very low–density lipoprotein; also called *pre-beta-lipoprotein*. It is a relatively large lipid-protein complex that transports mainly endogenously synthesized triglycerides.

Part I: Lipids

NORMAL PHYSIOLOGY OF LIPIDS

Lipid composition of foods

The fat found in food is composed mainly of triglycerides, about 98% to 99%, of which 92% to 95% is fatty acid and the remainder is glycerol. The remaining 1% to 2% of the lipids includes cholesterol, phospholipids, diglycerides, monoglycerides, fat-soluble vitamins, steroids, terpenes, and other fats. Most fats are mixtures of triglycerides containing four or five major fatty acids and many more minor or trace constituents. The individual glyceride molecules in most food fats contain both saturated and unsaturated fatty acids. Several polyunsaturated acids (linoleic, linolenic, and arachidonic acids) cannot be synthesized in the animal body and must be provided in the diet. These have been termed *essential fatty acids* (EFAs). The most important EFA is linoleic acid. The primary sources of essential fatty acids in the American diet are vegetable sources.

The small amount of unsaponifiable matter in food fats consists of sterols, fatty alcohols, hydrocarbons, pigments, glycerol esters, and various other compounds. Cholesterol occurs in all animal fats. Most sterols are cholesterol, but depending on the diet, other sterols, such as phytosterols, can make up an appreciable percentage of the total sterols, particularly in people on vegetarian diets. The phytosterols

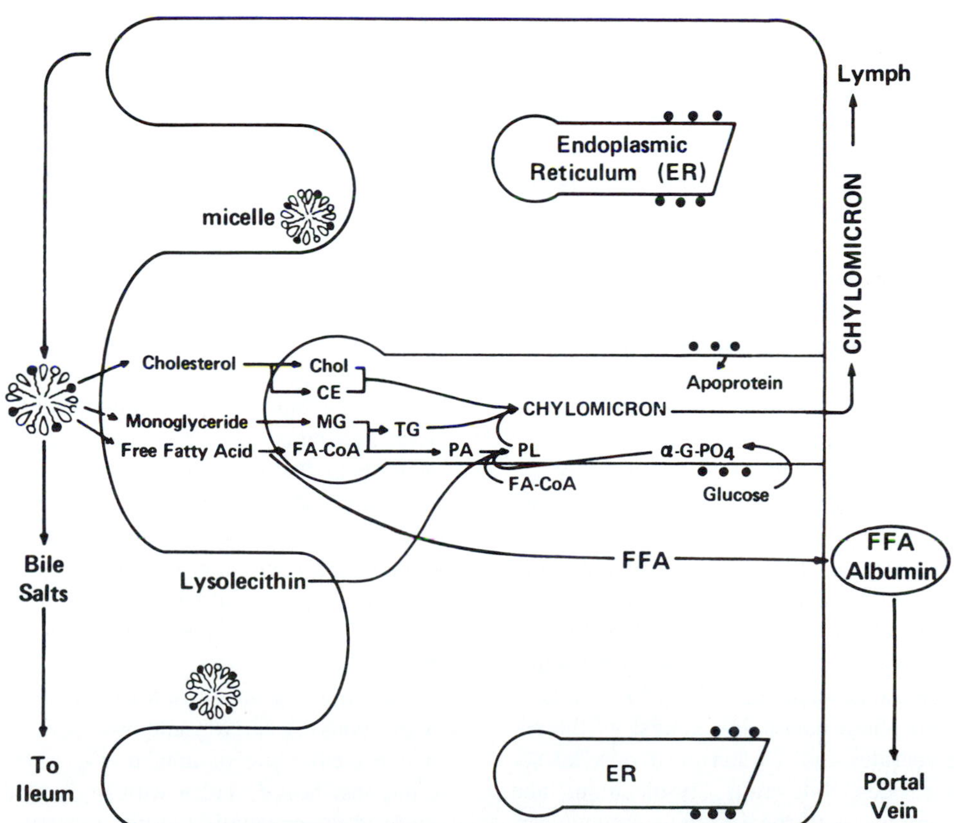

Fig. 30-1 Diagram of micelles along mucosal cell of gastrointestinal tract. After uptake of lipids, they are reassembled with apoproteins in endoplasmic reticulum *(ER)* to form triglyceride-rich particles, chylomicrons, which are carried away by lymph to systemic circulation. Short-chain fatty acids are attached to albumin and are transported through portal vein. *Chol,* Cholesterol; *CE,* cholesterol ester; *MG,* monoglyceride; *FA-CoA,* fatty acid coenzyme A; *PA,* phosphatidic acid; *PL,* phosphatidyl lecithin; *FFA,* free fatty acid; *αG-PO_4,* α-glycerol phosphate; *TG,* triglyceride.

are important because they compete with cholesterol for uptake by the mucosal cell. Thus the more phytosterols consumed, the less dietary cholesterol is absorbed by the mucosal cells of the gut.

Fat digestion, absorption, and metabolism of lipids

Fat absorption occurs in three phases: the *intraluminal*, or digestive, phase, during which the dietary fats are modified both physically and chemically before absorption; the *cellular*, or absorptive, phase, in which the digested material enters the intestinal mucosal cells where it is reassembled into its preabsorptive form; and the *transport* phase, during which the absorbed lipids are carried from the mucosal cell to other tissues through the lymphatics and blood (Fig. 30-1).

Intraluminal phase. Most digestion of food fat is carried on in the intestine through the action of intestinal and pancreatic enzymes (lipases) and bile acids. Bile salts, produced by the liver and stored in the gallbladder as bile, play an important role in this breakdown of food fats because of their *amphipathic* properties. The bile salt molecule is both strongly hydrophilic and strongly hydrophobic. Such molecules tend to arrange themselves on the surface of small triglyceride particles with their hydrophobic end turned inward and the hydrophilic end outward toward the water phase. Because of their surface-active properties, the bile salts emulsify the dietary triglyceride into very small particles with a diameter of approximately 1 μm. The emulsification process thus forms particles that can be readily acted on by digestive enzymes.

The main path of fat digestion progresses from triglycerides to 1,2-diglycerides to 2-monoglycerides and fatty acids. Only a small percentage of the fat is hydrolyzed to free fatty acids (FFAs) and glycerol, perhaps after isomerization to the 1-monoglyceride.

In the intestinal lumen the action of pancreatic lipase on ingested fat results in a complex mixture of triglycerides, diglycerides, monoglycerides, and FFAs. Cholesterol esters are hydrolyzed to free cholesterol and free fatty acid; the reaction is catalyzed by the enzyme cholesterol esterase. In addition, the entry of bile into the duodenum contributes important amounts of bile salts and lecithin; the latter quickly undergoes hydrolysis to lysolecithin. Both of these classes of compounds are essential for solubilization of the lipids in the intestinal contents, which forms a two-phase system—an oil phase containing almost all the triglycerides and diglycerides and a water-clear micellar solution of monoglycerides, bile salts, lysolecithin, and soaps. Bile salts combine with the resulting monoglycerides and fatty acids, forming negatively charged polymolecular aggregates, or micelles, with a diameter in the range of 5 nm. In the conversion of fats from an emulsion phase to a micellar phase, the diameter of the fat-containing particles has been reduced further, approximately 100-fold, and the surface area increased 10,000-fold. The 5 nm micelle now has access to the intramicrovillus spaces (50 to 100 nm) of the intestinal membrane where penetration into the intestinal mucosal cell can take place (Fig. 30-1).

Absorptive phase. The mechanism by which monoglycerides, fatty acids, and other lipids from the intestinal lumen micelles are taken into the mucosal cell is not yet clear. It is generally accepted, however, that the micelle disintegrates on contact with the brush border of microvilli of the mucosal cell membrane, allowing differential uptake of various micelle components (Fig. 30-1).

After monoglycerides and fatty acids enter the endoplasmic reticulum of the mucosal cell, presumably by diffusion, the monoglycerides and fatty acids are re-esterified into triglycerides (Fig. 30-1). There are two main pathways for the resynthesis of triglycerides within the mucosal cell. The monoglyceride pathway is peculiar to the intestinal mucosa and involves the direct acylation of the absorbed monoglyceride from the lumen with activated FFA. The alpha-glycerophosphate pathway present in most tissues involves the acylation of glycerophosphate to form phosphatidic acid, dephosphorylation of the phosphatidic acid to form diglyceride, and further acylation to form triglyceride. Before fatty acids can be utilized for triglyceride resynthesis, they must first be activated by the formation of a coenzyme A (acyl CoA) derivative of the fatty acid. This reaction, which requires adenosine triphosphate (ATP), is catalyzed by the enzyme fatty acid:CoA ligase (AMP). This enzyme has a pronounced specificity for longer-chain fatty acids. Thus long-chain fatty acids appear in thoracic duct lymph transported as triglycerides in the chylomicrons, whereas short- and medium-chain fatty acids are transported bound to albumin in the portal circulation.

Transport phase. Once triglycerides have been resynthesized within the intestinal mucosal cell, they are assembled in the mucosal cell endoplasmic reticulum and the Golgi apparatus into water-soluble macromolecules—chylomicrons. The intestinal lipoproteins leave the mucosal cells presumably by reverse pinocytosis. They first appear in the lymphatic vessels of the abdominal region and later in the systemic circulation. These larger lipid-protein complexes are mixtures of triglycerides, some proteins (as apoproteins), small amounts of cholesterol, and phospholipids.

The bloodstream transports chylomicrons to all tissues in the body, including adipose tissue, which is probably their principal site of uptake (Fig. 30-2). The large chylomicrons, heavily laden with triglyceride (Fig. 30-3), are removed rather rapidly (within minutes). Chylomicrons are normally present in only trace amounts in blood samples taken from individuals after an overnight fast.

Under normal conditions, chylomicron catabolism proceeds in two known phases. In the first, triglycerides are hydrolyzed at extrahepatic tissue sites under the influence of *triglyceride* (or *lipoprotein*) *lipase*. The process results

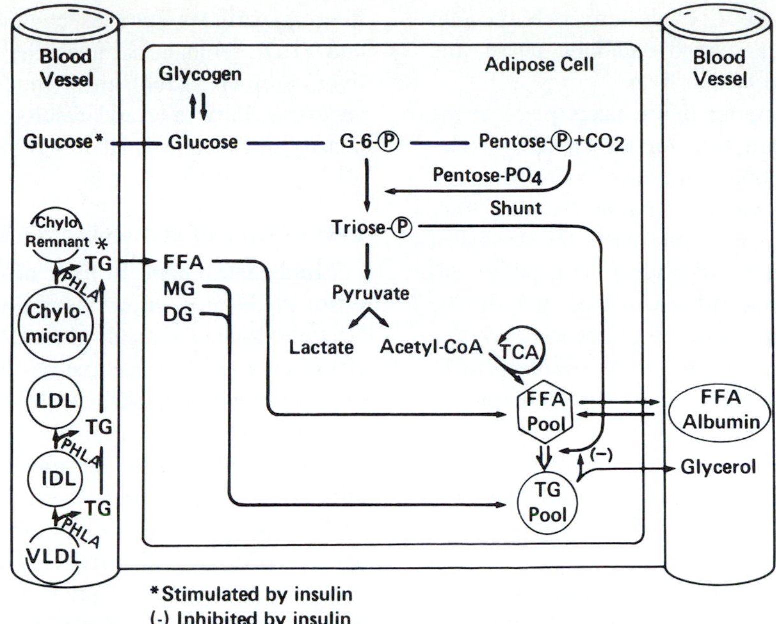

Fig. 30-2 Fat metabolism in adipose cell after meal. This metabolic pathway favors storage of energy (as triglycerides). Insulin promotes entry of glucose, necessary for triglyceride synthesis, and also inhibits hormone-sensitive lipase, the enzyme that promotes triglyceride breakdown to free fatty acids and glycerol. The influx of free fatty acids from chylomicron and very low–density lipoprotein is also an integral part of triglyceride synthesis. *PHLA,* postheparin lipolytic activity; *G-6-P,* glucose-6-phosphate; *TCA,* tricarboxylic acid cycle; *DG,* diglyceride; remainder of abbreviations as in Fig. 30-1 and text.

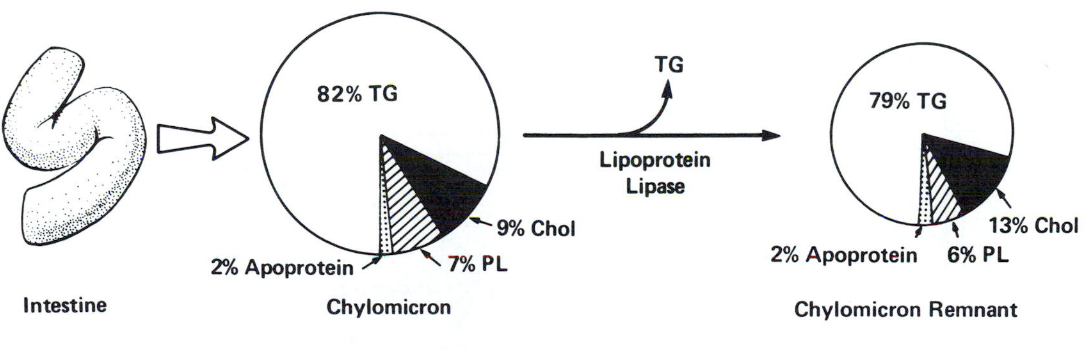

CATABOLIC PATHWAY FOR CHYLOMICRON AND VLDL

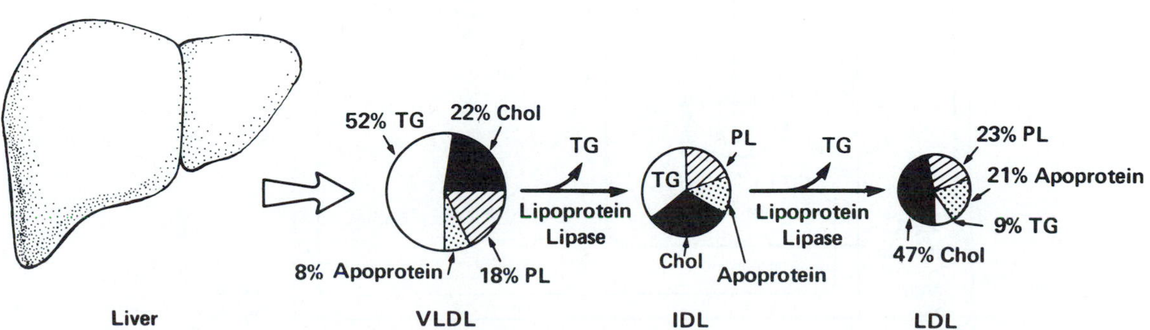

Fig. 30-3 Origin and catabolic pathway of chylomicron and very low–density lipoprotein *(VLDL).* End product of chylomicron is chylomicron remnant; end product of VLDL is LDL. *Chol,* Cholesterol; *PL,* phosphatidyl lecithin; *TG,* triglyceride.

in a relatively triglyceride-poor, cholesterol-rich *remnant* particle (Fig. 30-3). In the second catabolic phase, the remnant particle is removed by the liver.

The hydrolysis by lipoprotein lipase takes place at the luminal surface (bloodstream side) of the capillary endothelium (an intracellular lipoprotein lipase exists but is not involved in metabolism of circulating lipoproteins). Very little triglyceride lipase activity is present in the plasma of normal subjects. The enzyme is bound to the capillary endothelial cells in muscle and adipose tissue and can be released by intravenous administration of heparin. Triglyceride lipase activity is often referred to as *postheparin lipolytic activity* (PHLA). Whereas the adipose tissue enzyme depends on apoprotein C-II and insulin for activation, the hepatic enzyme does not. As a result of the first phase of chylomicron metabolism, unesterified fatty acids are released into the bloodstream, and the diglycerides and monoglycerides are taken up in vacuoles and transported across the capillary wall for hydrolysis.

The fatty acids derived from triglycerides (in chylomicron or VLDL) enter the adipose tissue cell, where they can be stored; the glycerol derived from triglyceride hydrolysis is released into the circulation (Fig. 30-2). The FFAs can also be utilized for energy when needed by all muscular tissue, especially the heart. FFAs are also used for cellular phospholipid synthesis, which leads to membrane formation. EFAs, in addition to their function in membrane synthesis, are the only source for the synthesis of prostaglandins, the ubiquitous local hormones.

The uptake of the triglycerides leads to fewer triglyceride-rich particles such as chylomicron remnant particles and VLDL remnant particles; the former are cleared by the liver. Further catabolism of the VLDL remnant occurs at an extracellular site and results in the formation of low-density lipoprotein (LDL), a cholesterol-rich particle (Fig. 30-3).

Role of liver in metabolism of lipids

Lipid metabolism in liver after meal. Nutrient metabolism in the liver after absorption is summarized in Fig. 30-4. Following a meal, glucose, amino acids, and short-chain fatty acid concentrations rise in portal blood and these nutrients are taken up by hepatocytes. Glucokinase phosphorylates glucose to glucose-6-phosphate. Glycogen is synthesized until the hepatic stores are depleted. If portal hyperglycemia persists, glucose is converted to fatty acid by acetyl CoA. Excess amino acids are deaminated and contribute to the acetyl CoA and pyruvate pool. Some acetyl CoA undergoes further oxidation to carbon dioxide by the tricarboxylic cycle, thus yielding the energy necessary for hepatic functioning. Newly synthesized fatty acids and those produced by elongation of short-chain fatty acids derived from diet are esterified with glycerol, monoglyceride, or diglyceride to form triglyceride.

The newly synthesized triglycerides are coupled with phospholipid, cholesterol, and proteins to form VLDL. These macromolecules are then released into the circulation and transported to adipose tissue.

Hepatic triglyceride synthesis is accelerated when the diet is rich in excess calories. This results in VLDL over-

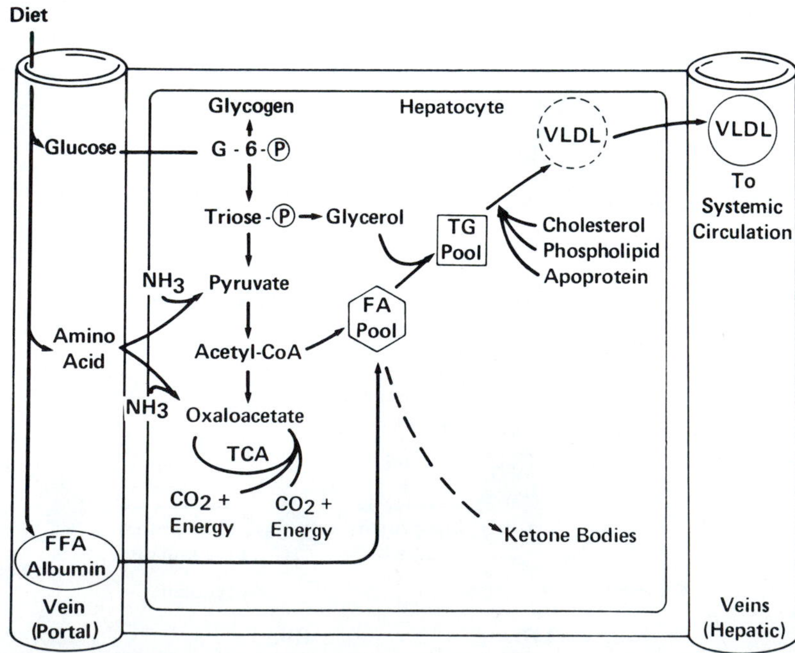

Fig. 30-4 Fat metabolism in liver after meal. Dietary components add to fatty acid pool in liver and ultimately to triglyceride pool. Lipids and apoprotein are packaged in endoplasmic reticulum and released as VLDL in systemic circulation. Abbreviations as in Figs. 30-1 and 30-2.

production, which may explain the occasional transient hypertriglyceridemia observed in normal persons when they consume diets particularly rich in simple sugars.

It is not yet established to what extent chylomicrons and VLDL of intestinal origin released through lymphatic channels into the general circulation are taken up by the liver. These particles could represent yet another source of hepatic triglyceride.

Lipid metabolism in liver during fasting. During fasting, the metabolic pathways outlined in Fig. 30-4 are reversed. Blood glucose concentration falls, and insulin levels are diminished. As a consequence, amino acids are mobilized from muscle, and fatty acids are released from adipose tissue. These molecules are taken up by the liver where gluconeogenesis and ketogenesis predominate. Alanine, the principal gluconeogenic precursor, and other amino acids are incorporated into glucose-6-phosphate (G6P) by deamination and reverse glycolysis. G6P is converted to glucose by the action of glucose-6-phosphatase and then released into the circulation. FFAs derived from adipose tissue are taken up by the liver, and their oxidation to ketone bodies provides energy for gluconeogenesis. FFAs of adipose tissue origin can be esterified to triglycerides, incorporated into hepatic VLDL, and then released into the bloodstream. During periods of stress and in certain metabolic conditions, such as uncontrolled diabetes, FFAs are the principal precursors of hepatic VLDL. Hepatic VLDL triglyceride synthesis is diminished during fasting.

Role of adipose tissue in metabolism of lipids

Lipid metabolism in adipocyte after meal. If the FFAs from dietary sources are not immediately utilized for energy purposes by the peripheral tissues, this concentrated form of energy is stored in the fat cells (adipocytes). After a meal, insulin is secreted by the pancreas into the portal bloodstream. Insulin accelerates glucose entry into adipose cells, where it is metabolized to glycerophosphate and also to fatty acid chains. This is important since adipocytes lack the enzyme glycerokinase, the enzyme needed to re-esterify glycerol, and therefore adipocytes cannot employ dietary glycerol for resynthesis of triglycerides. Thus the uptake of lipoprotein triglycerides depends on the activity of the lipoprotein lipase and the availability of intracellular alpha-glycerophosphate for re-esterification of fatty acids into triglycerides (Fig. 30-2). High levels of insulin accelerate both processes, and after a meal, adipose tissue operates as an effective storage system of alimentary lipids and carbohydrates not immediately used for energy.

Lipid metabolism in adipocyte during fasting. When glucose is in short supply (that is, during fasting), re-esterification (or triglyceride synthesis) is inhibited. Triglycerides in adipose tissue are constantly subjected to the catabolic action of an intracellular enzyme (different from serum lipoprotein lipase) stimulated by hormones such as catecholamines, thyroxine, and glucagon that accelerate the hydrolysis of triglycerides to fatty acid and glycerol. Fatty acids are then released into the circulation where they are bound to albumin (Fig. 30-2).

BIOENERGETICS OF FAT

Energy that the body requires can be derived from foods or the body's own reserves (chiefly fat). Any excess of energy above immediate needs is stored as fat. For complete utilization of the energy from fat, some carbohydrate (glucose) must be utilized simultaneously. If necessary,

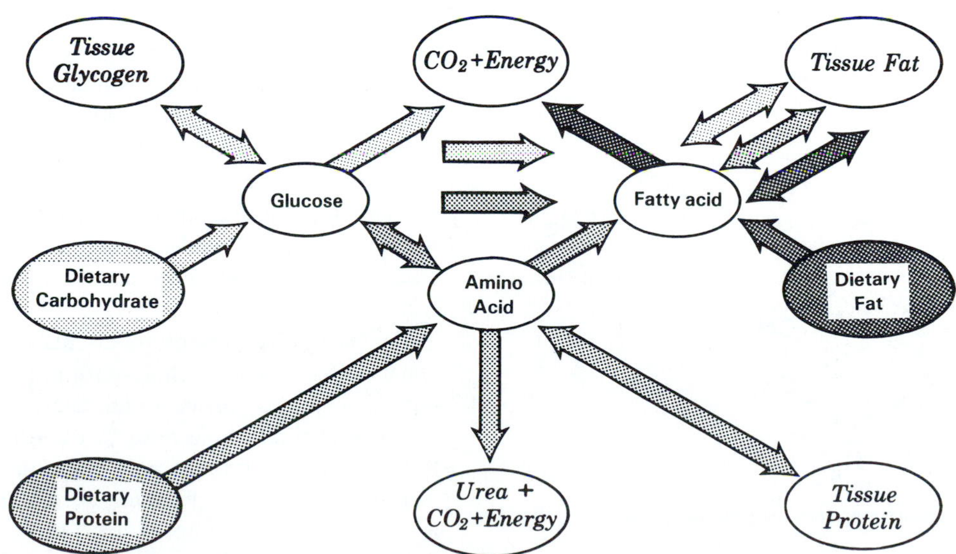

Fig. 30-5 Bioenergetics of fat, carbohydrate, and proteins. Note interrelationship of these three energy compounds.

Table 30-1 Physiological fuel value of major nutrients

	Carbohydrate (Cal/g)	Fat (Cal/g)	Protein (Cal/g)	Ethanol (Cal/g)
Heat of combustion	4.1	9.45	5.65	7.1
Energy from combustion of nitrogen unavailable to body	—	—	1.3	—
Net heat of combustion	4.1	9.45	4.35	7.1
Coefficient of digestibility	0.98	0.95	0.92	Small loss in urine and breath
Physiological fuel value	4.0	9.0	4.0	7.0

the body can synthesize glucose from certain amino acids and from glycogen stored in the liver and muscles. The carbohydrate reserve of hepatic glycogen is sufficient for about 16 hours. It should be evident from Fig. 30-5 that an interrelationship exists among carbohydrates, fats, and proteins in the bioenergetic pathway. One can see from Table 30-1 that fat is an excellent source of energy for physiological needs.

CHOLESTEROL METABOLISM

Biological functions

Cholesterol is a member of a large class of biological compounds called *steroids* that have a similar four-ring structure, a cyclopentanoperhydrophenanthrene ring (Fig. 30-6).

Because of the well-established positive association between plasma cholesterol concentration and coronary heart disease (CHD), we are apt to think of cholesterol as a harmful substance. Contrary to that belief, cholesterol is essential for normal functioning of the organism because it is:

1. An essential structural component of membranes of all animal cells and subcellular particles
2. An obligatory precursor of bile acids
3. A precursor of all steroid hormones, including sex and adrenal hormones

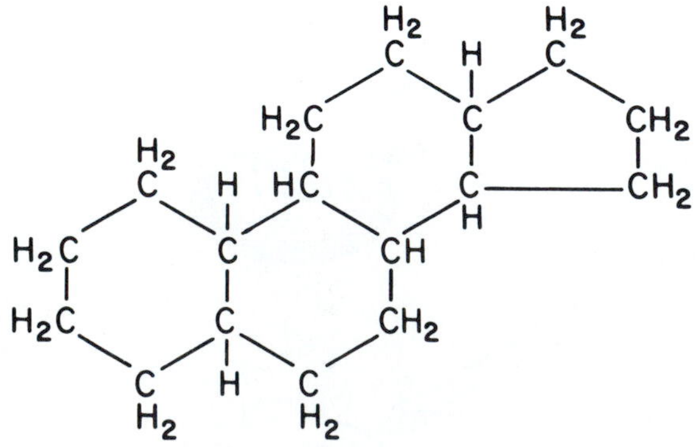

Cyclopentanoperhydrophenanthrene

Fig. 30-6 Chemical structure of cyclopentanoperhydrophenanthrene ring. This common four-ring structure is basic structure of all steroids.

Physiology of cholesterol metabolism

The normal human body contains about 2 g of cholesterol per kilogram of body weight. Much of this is in constant exchange with plasma cholesterol; the turnover rate and the amount of tissue cholesterol that is exchangeable with the plasma cholesterol will vary from one tissue to another.

Because of loss and replacement, about 2% of the body's cholesterol is renewed each day. Since the main channel for outflow from the pool is the gastrointestinal tract, the absolute rate of turnover (in grams per day) can be estimated by measurement of the daily fecal output of bile acids and neutral steroids (cholesterol). Measurements of fecal steroid output indicate probably 1 to 2 g/day for turnover of cholesterol in humans, with excretion of bile acids accounting for about half the total turnover. A schematic representation (Fig. 30-7) illustrates that the concentration of a given cholesterol pool is under the influence of cholesterol input, output, and turnover rates. It should be stressed that because of the continuous cycling of cholesterol into and out of the bloodstream, the plasma cholesterol concentration is not a simple additive function of dietary cholesterol intake and endogenous cholesterol synthesis. Rather, it reflects the rates of synthesis of the cholesterol-carrying lipoproteins and the efficiency of the receptor mechanisms that determine their catabolism. A detailed discussion of the dynamics of lipoprotein concentration can be found in the section on lipoprotein metabolism.

Cholesterol is present in all plasma lipoproteins, but about 60% of the total cholesterol in plasma from a fasting human subject is carried in the LDL. About two thirds of the plasma total cholesterol is esterified with long-chain fatty acids, with linoleic acid being the predominant fatty acid in humans. The cholesteryl esters in the plasma are in a state of constant turnover because of their continual hydrolysis and resynthesis. Hydrolysis of cholesteryl esters takes place in the liver, but synthesis occurs mainly in the plasma by transfer of a fatty acid residue from lecithin to

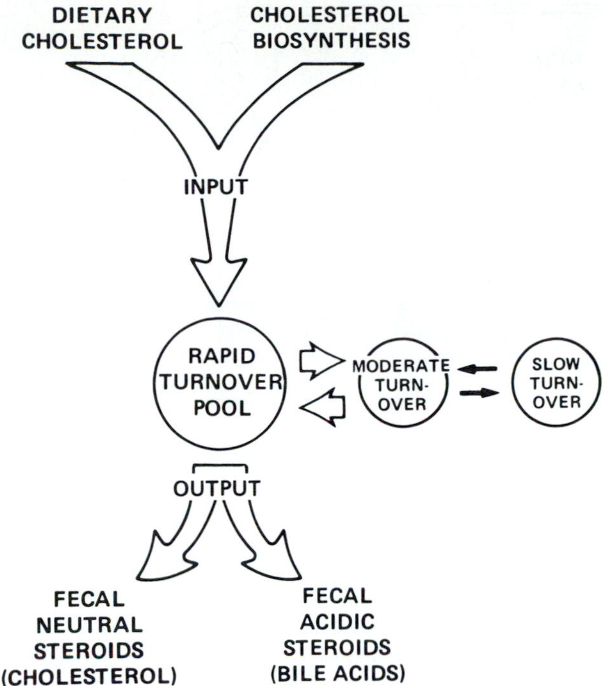

Fig. 30-7 Scheme of dynamics of cholesterol metabolism.

free cholesterol. This reaction is catalyzed by a plasma enyzme known as *lecithin:cholesterol acyltransferase* or (LCAT) (Fig. 30-8). The preferred lipoprotein for human LCAT is high-density lipoprotein (HDL), and it seems likely that the bulk of the esterified cholesterol in the plasma is formed on HDL. The cholesteryl ester then is transferred from HDL to LDL and VLDL, partly in exchange for triglyceride.

It has been suggested that one of the functions of LDL is to transport cholesterol, in esterified form, from the tissues to the liver. In this view, the function of LCAT is to generate esterified cholesterol from the free cholesterol formed in extrahepatic tissues. One might then envisage the following sequence of events (Fig. 30-9). Free cholesterol from peripheral tissues is transferred to HDL; it is then esterified by LCAT, enabling HDL to take up more free cholesterol. The esterified cholesterol formed on HDL is transferred on to LDL and VLDL, where it is incorporated into the nonpolar core of the lipoprotein molecules. LDL, carrying its load of cholesteryl ester, reaches the liver, where the cholesteryl esters are hydrolyzed. The free cholesterol so formed enters the pool of free cholesterol in the hepatocyte. The free cholesterol can leave the pool in the bile or after conversion into bile acids or by reincorporation into plasma lipoprotein (VLDL).

Synthesis

Almost all animal tissues synthesize cholesterol from acetyl CoA. In adults the most actively synthesizing organs are liver and intestinal wall; these two tissues probably supply over 90% of the plasma cholesterol of endogenous origin. Hepatic cholesterologenesis, unlike intestinal cholesterol synthesis, is inhibited by dietary cholesterol. Cholesterol production rate (absorbed cholesterol plus endogenously synthesized cholesterol) amounts to about 1 g/day. In most tissues the rate of synthesis of cholesterol is determined by the capacity of beta-hydroxy-beta-methylglutaryl CoA (HMG-CoA) reductase, the enzyme catalyzing a rate-limiting step in the biosynthetic sequence from acetyl CoA to cholesterol. Although this appears to be the main rate-limiting reaction, there appear to be other sites of suppression in the biosynthetic cholesterol pathway. Hepatic HMG-CoA reductase is subject to induction and repression by several hormones, dietary factors, and drugs. A brief scheme of the control of hepatic cholesterologenesis is shown in Fig. 30-10.

Feedback control of hepatic cholesterologenesis is mediated by cholesterol itself and directly or indirectly by bile acids. Although virtually every cell utilizes cholesterol, the liver is one of the major organs that removes cholesterol from circulation. It does so by two routes: (1) by the excretion of free cholesterol into bile or (2) by the excretion of cholesterol after its conversion to primary bile acids (see Chapter 23). The formation of bile acids is believed to regulate bile acid synthesis by negative product-inhibition feedback.

Absorption

Cholesterol is absorbed from the jejunum in the presence of bile salts. After entering the cells of the intestinal mucosa, it is incorporated into chylomicrons, which enter

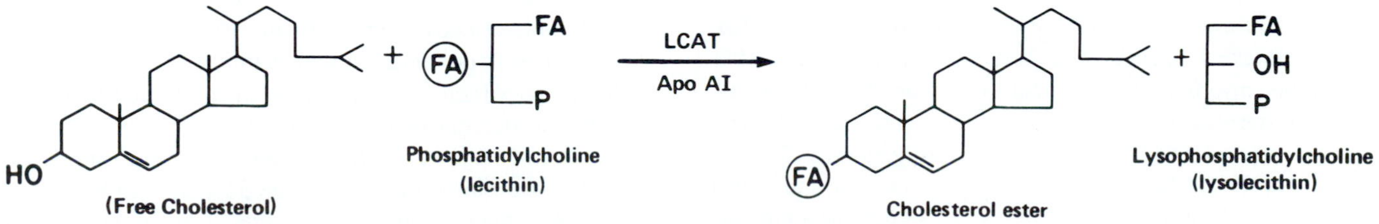

HO (Free Cholesterol) + FA–[FA, P] Phosphatidylcholine (lecithin) —LCAT, Apo AI→ FA– Cholesterol ester + [FA, OH, P] Lysophosphatidylcholine (lysolecithin)

Fig. 30-8 Esterification reaction of free cholesterol; lecithin contributes a free fatty acid to form ester cholesterol. This reaction depends on enzyme lecithin:cholesterol acyltransferase, *LCAT*, and an apolipoprotein, *Apo AI*. *FA*, Fatty acid; *P*, choline.

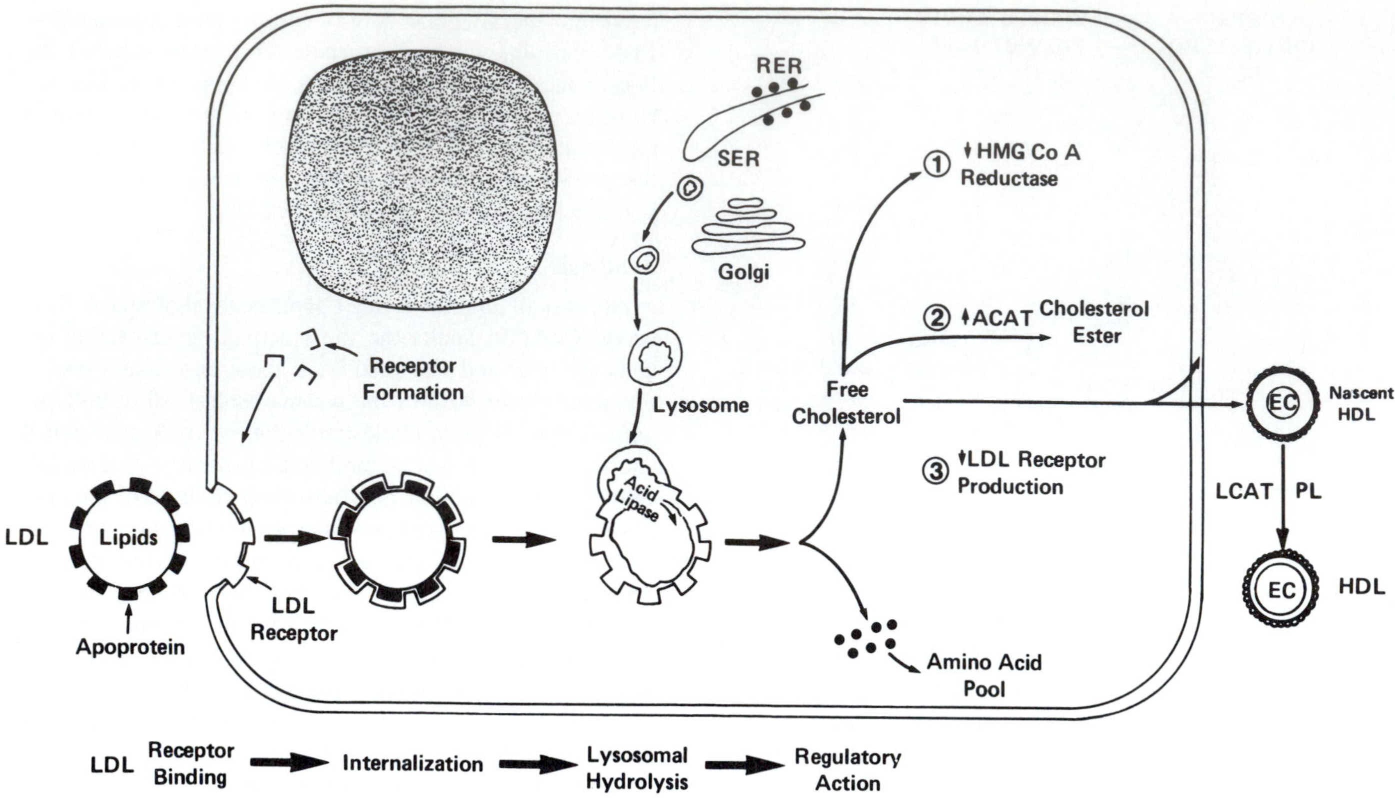

Fig. 30-9 Scheme of LDL uptake and catabolism by a cell. Mechanism not only clears LDL from circulation but also aids in regulation of cholesterol synthesis and storage. High-density lipoprotein, *HDL,* plays an integral role in removing cellular cholesterol, esterifying free cholesterol in blood, and transporting cholesterol to liver for catabolism. *ACAT,* Acetylcholesterol acyltransferase; *HMG CoA reductase,* β-hydroxy-β-methylglutaryl-coenzyme A reductase; *EC,* cholesterol ester; *LCAT,* lecithin:cholesterol acyltransferase; *PL,* phospholipid; *RER,* rough endoplasmic reticulum; *SER,* smooth endoplasmic reticulum.

the blood circulation through the lymphatic system (see Fig. 30-1). The percentage of cholesterol that is absorbed from the diet is self-regulating. Increased levels of triglyceride in the diet tend to promote cholesterol absorption. The size of the bile acid pool appears to have a profound effect on the rate of cholesterol absorption. Diversion of bile acids from the intestinal lumen essentially stops cholesterol absorption, whereas the expansion of the bile acid pool caused by the ingestion of exogenous bile acids results in a greater than normal rate of cholesterol absorption. When the continuous intake of cholesterol is less than about 300 mg/day, more is absorbed (about 40% to 60%). If the intake is increased to 2 to 3 g/day, as little as 10% may be absorbed. In a typical American diet, about 600 mg of steroids are consumed per day, and the coefficient of absorption of dietary cholesterol is generally 25% to 40%. It should be stressed that there is a large individual variation in cholesterol absorption. This could account partially for the difference in individual responsiveness or lack of responsiveness to diet-induced hypercholesterolemia.

Catabolism

In humans, increased absorption of cholesterol is followed by increased excretion of cholesterol from the exchangeable pool. Increased conversion of cholesterol into bile acids is also brought about by interruption of the enterohepatic circulation of bile salts. Bile salts returning to the liver from the intestine repress the formation of an enzyme catalyzing the rate-limiting step in the conversion of cholesterol into bile acids. When bile salts are prevented from returning to the liver, the activity of this enzyme increases and degradation of cholesterol to bile acids is stimulated. This effect may be exploited therapeutically in the treatment of hypercholesterolemia by the use of unabsorbable resins, which bind bile acids in the lumen of the intestine and prevent their return to the liver.

The percentage of neutral sterols and acidic sterols excreted in the bile is on the average 55% and 45%, respectively. The main neutral sterol excreted in the bile is cholesterol.

The mechanisms just stated for excretion of cholesterol via bile acids or cholesterol in the bile depend on the re-

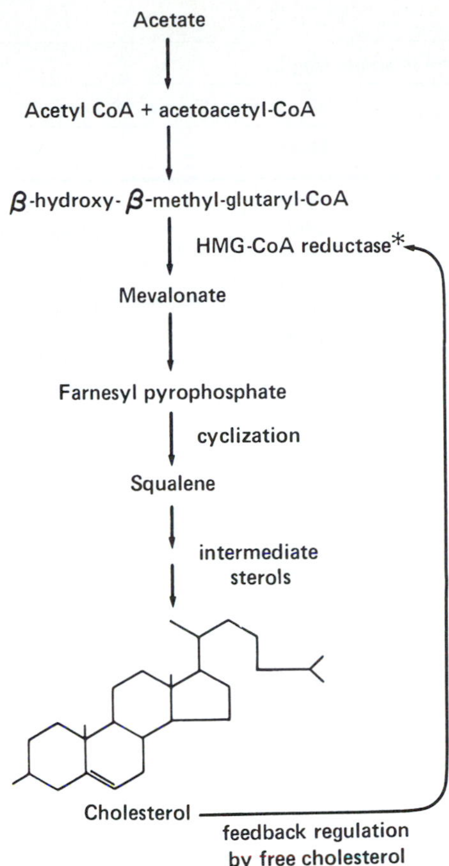

Fig. 30-10 Metabolic pathway of cholesterol synthesis, emphasizing negative feedback end-product inhibition step at β-hydroxy-β-methylglutaryl CoA step with the important enzyme HMG-CoA reductase.

ceptor-mediated activity in the hepatocytes. The hepatocytes have receptor sites that are specific for apoproteins (Apo) B and E. The major function of the liver in lipoprotein clearance is to remove chylomicron remnants (which contain dietary cholesterol) from plasma. Lipoproteins containing Apo E (such as chylomicron remnants and VLDL remnants) and Apo B (such as LDL remnants) are cleared by the same mechanism. However, the Apo-E–containing lipoproteins are cleared with much greater efficiency than the Apo-B–containing lipoproteins. For this reason chylomicron remnants and VLDL remnants (intermediate-density lipoprotein [IDL]) are not measurable in normal individuals (see Fig. 30-3).

The uptake of LDL by the peripheral tissues is also receptor-site dependent (see Fig. 30-9). The binding of LDL to the receptor site followed by internalization and hydrolysis of the LDL leads to free cholesterol in the cell, which then functions (1) as a regulator for the rate of receptor synthesis, (2) as a regulator for cholesterol synthesis by the end-product negative-feedback mechanism, or (3) as a regulator for ACAT (acyl-CoA:cholesterol acyltransferase) activity, which determines how much cholesterol is stored in the cell as cholesteryl oleate, a cholesterol ester. It is believed that one of the factors that causes the efflux of cholesterol from the cell into the blood is the availability of HDL. By this process, the cholesteryl oleate in the cell is hydrolyzed to free cholesterol and fatty acid. The liver, and to some extent the gastrointestinal tract and other organs, such as the adrenal glands and gonadal tissues, take up the HDL and catabolize the HDL to its protein and lipid constituents (including cholesterol).

Normal expected cholesterol values

A problem arises in defining what levels of plasma lipids separate persons with elevated blood fats from the rest of the "normal" population. Before one can address the question of what is adequate dietary or drug therapy for the control of serum lipid and lipoprotein concentrations, one needs first to consider the degree to which blood lipids should be lowered. In other words, at what level should the clinician consider a sample "hyperlipidemic"? Many clinical laboratories and practicing physicians still use normal ranges derived from the central 95% of values when classifying a sample as normal or abnormal. Unfortunately, because of the way we have defined "normal" in the past, many test results do not correlate very well with health-risk conditions on an individual basis. Thus a cholesterol value in the normal range for a given population may not represent a healthy cholesterol level. For example, a cholesterol value of 2500 to 2800 mg/L may be within the 95th percentile of the distribution of an apparently normal male population between 51 and 59 years of age in the United States, but about 40% to 50% of these persons eventually will develop coronary heart disease.

Reference values for serum lipids and lipoproteins that are highly predictive of disease or disease risk, irrespective of the "normal" distribution, should be established.

Cholesterol values depend on many factors. The main variables to consider are age and sex. Table 30-2 gives the distribution of total cholesterol by age and sex for fasting subjects. These values were based on large-scale U.S. population studies (n = 42,653 subjects). Plasma values are lower because of the dilution effect (about 3% to 5%) of anticoagulants such as EDTA, which draw water from the blood cells. Because of a physiological diurnal variation of 5% to 11%, two to three separate cholesterol values should be obtained to determine an individual's true baseline value.

Because of the positive correlation between blood cholesterol concentration and increased risk for CHD, many investigators believe that the average cholesterol concentration for the entire population should be as low as possible. According to clinical data, the individuals with plasma cholesterol values below 1800 mg/L had minimum CHD mortality (about 3.3/1000). However, the relative risk increased by 25% for those with values between 1800

Table 30-2 Distribution of total cholesterol values in U.S. population

	Total cholesterol in serum (mg/L)			
	Male (n = 24,485)		**Female (n = 18,168)**	
Age (yr)	**Mean**	**Range***	**Mean**	**Range***
0 (cord blood)	650	370-960	650	370-960
0-1	1510	1020-1900	1560	1130-2010
2-3	1590	1130-1990	1600	1170-2040
4-5	1620	1170-2180	1630	1200-2020
6-7	1630	1240-2040	1670	1300-2100
8-9	1660	1250-2110	1700	1280-2130
10-11	1650	1290-2100	1660	1300-2090
12-13	1610	1210-2050	1620	1260-2040
14-15	1550	1160-2010	1600	1230-2050
16-17	1540	1150-2040	1610	1230-2060
18-19	1570	1170-2040	1640	1230-2130
20-24	1720	1270-2240	1690	1250-2220
25-29	1870	1360-2510	1760	1310-2290
30-34	1980	1420-2620	1800	1330-2370
35-39	2070	1500-2780	1890	1440-2490
40-44	2130	1550-2760	2000	1510-2600
45-49	2180	1620-2850	2090	1560-2730
50-54	2190	1620-2860	2240	1660-2940
55-59	2200	1600-2850	2380	1770-3100
60-64	1690	1620-2850	2380	1770-3060
65-69	2190	1620-2820	2400	1760-3130
70+	2130	1550-2780	2350	1740-2980

From Cooper, GR: In Faulkner, R, and Meites, S, editors: Selected methods of clinical chemistry, vol 9, Washington, DC, 1982, American Association for Clinical Chemistry.
*Range is 95th and 5th percentiles.

and 2000 mg/L. Between 2000 and 2390 mg/L the relative risk increased about 80%; for values above 2400 mg/L, the relative risk increased almost two and one-half times, or 230%. For individuals with plasma cholesterol concentrations near 2600 mg/L, the relative risk increased by 400%. In simplistic terms, concentrations below 2000 mg/L can be considered to be more ideal than concentrations above 2000 mg/L for the adult population. Using such a definition, this means that about 58% of the adult population has undesirably elevated cholesterol concentrations. The plasma concentration in young American men is essentially within the ideal range. If it did not increase with age, the risk of CHD in the U.S. population would perhaps be much lower than it is now. Again, it should be remembered that the relationship of blood cholesterol concentration to CHD shows no threshold for the disease.

It is well known that as we age we become more susceptible to the atherosclerotic process. It has been calculated that, for an individual with a cholesterol level of 2000 mg/L and no other risk factors, a critical degree of atherosclerosis (greater than 60% stenosis) is reached in many people by the time they reach the age of 70 years. If the same individual had a cholesterol value of 2500 or 3000 mg/L, this degree of coronary artery disease (CAD) would probably be attained by age 60 or 50 years, respectively. This timetable is accelerated when one considers the interaction of risk factors for CHD. With the addition of a CHD risk factor, such as smoking, the critical age is reached by 60 years of age, and by further addition of another CHD risk factor, hypertension, this age drops to 50 years. A plasma cholesterol concentration of 2500 mg/L would move the critical age back to 50 years with one risk factor and to 40 years with two risk factors.

In 1985 the National Institutes of Health recommended new guidelines; the consensus panel recommended treatment of persons with blood total cholesterol and LDL-cholesterol levels above the 75th percentile (upper 25% of values rather than the customary 90th or 95th percentiles).

In 1988 the NIH Adult Treatment Panel of the National Cholesterol Education Program recommended new cutoff levels for cholesterol and LDL cholesterol that should have a larger impact on reducing the morbidity and mortality CHD rates in the United States, where CHD is still a major disease. To simplify the classification system and to make it more convenient to remember the cutoff levels, the panel eliminated the age and sex stratification. These new cutoff levels, shown in Table 30-3, apply to all adults 20 years or older.

All clinical laboratories in the United States should adopt uniform cholesterol cutoff levels for identifying adults at high risk of CHD. The Laboratory Standardization Panel recommended adoption of the cutoff values is-

Table 30-3 Classifications based on total cholesterol and low-density lipoprotein cholesterol

Total cholesterol (mg/L)*	LDL cholesterol (mg/L)
<2000: desirable	<1300: desirable
2000-2390: borderline/high risk	1300-1590: borderline/high risk
≥2400: high risk	≥1600: high risk

From Report of the National Cholesterol Treatment Program Expert Panel on Detection, Evaluation, and Treatment of High Blood Cholesterol in Adults: Arch Intern Med 148:36-39, 1988.
*Milligrams of cholesterol per liter of blood; to convert mg/dL cholesterol to mmole/L, divide by 387 or multiply by 0.002586.

sued by the National Cholesterol Education Program Adult Treatment Panel in January 1988. This required national standardization of cholesterol measurement. In order to use these newly recommended values properly, a laboratory must minimize method-specific biases and achieve adequate precision of cholesterol measurement. Specific attention to method, instrument, and calibration procedures is necessary to minimize method-specific biases.

The 30-year follow-up study carried out as part of the Framingham heart study again shows that cholesterol concentrations are directly related to cardiovascular disease mortality: the overall death rate increased by 5% and cardiovascular disease death by 9% with each 100 mg/L increase. The study suggests that having a very low cholesterol concentration prolongs life.

It should be emphasized that a person's serum or plasma cholesterol concentration is under the influence of several factors.

Genetics. Genetics probably has the most important influence on a person's cholesterol concentration.

Age. Serum cholesterol concentration starts out around 650 mg/L at birth and steadily increases with age.

Sex. Cholesterol concentration in the blood of males is always higher than that in premenopausal females. After menopause, the cholesterol concentration is higher in females than in males. Serum cholesterol levels in males seem to plateau by 50 to 60 years of age.

Diet. Saturated fat in the diet increases serum cholesterol levels, whereas polyunsaturated fat decreases cholesterol concentration. Monounsaturated fats have no apparent effect. Dietary cholesterol appears to elevate serum cholesterol levels. Plant sterols and certain types of fiber tend to decrease serum cholesterol concentration.

Physical activity. Physical activity tends to lower serum total cholesterol. Much of this effect depends on the type, intensity, duration, and frequency of physical activity. Exercise also lowers LDL cholesterol and increases HDL cholesterol concentration.

Hormones. Growth hormone, thyroxine, and glucagon decrease serum cholesterol levels, whereas anabolic steroids and progestins increase cholesterol levels.

Primary disease states. Diabetes mellitus, thyroid dysfunction, obstructive liver disease, acute porphyria, dysgammaglobulinemias, and nephrotic syndrome have an effect on blood cholesterol concentrations.

TRIGLYCERIDE METABOLISM

Biological functions

Triglycerides are the major form of fat found in nature, and their primary function is to provide energy for the cell. One gram of fatty acids liberates about 9 kcal. The human body stores large amounts of fatty acids in ester linkages with glycerol in the adipose tissue. This form of storage of reserve energy is highly efficient because of the magnitude of the energy that is released when fatty acids undergo catabolism (see p. 462 and Table 30-1).

Chemistry

Triglycerides are by far the most abundant subclass of neutral glycerides in nature. Mammalian tissues also contain some diglycerides and monoglycerides, but these occur in trace levels when compared to triglycerides. Most triglyceride molecules in mammalian tissues are mixed glycerides.

Because of their water insolubility, triglycerides are transported in the plasma in combination with other more polar lipids (phospholipids) and proteins, as well as with cholesterol and cholesterol esters, in the complex lipoprotein macromolecules. It appears that the essentially nonpolar triglyceride (and cholesterol ester) is largely in the center of the lipoprotein, with the more polar protein and phospholipid components at the surface, their polar groups directed outward to stabilize the whole structure in the aqueous plasma environment.

Entry of triglyceride into plasma

The concentration of triglyceride in the plasma at any given time is a balance between the rate of entry into the plasma and the rate of removal. A change in concentration may therefore be the result of a change in either or both of these factors. Moreover, a primary change in one may result in a secondary change in the other. Thus perhaps the main problem to be considered in any situation where the plasma triglyceride concentration is abnormally high is whether this is attributable to a rise in the rate of entry or to a fall in the rate of removal of plasma triglycerides.

From intestine. As described earlier, digestion in the intestinal lumen hydrolyzes triglyceride into FFAs and monoglycerides, which are then eventually released into the lymphatics as triglycerides in chylomicrons. These lipid-protein complexes contain about 82% triglyceride, 9% cholesterol (mainly as the ester), 7% phospholipid, and a very small amount (less than 2%) of protein (see Fig. 30-3). Although the amount of protein is small, there is a good deal of evidence that its presence is necessary for the release of the chylomicrons. For example, in abetalipoproteinemia (a genetically determined disease in which apoprotein B cannot be made in the body), triglyceride is not released from the intestinal cells.

An increased influx of chylomicron triglyceride into the plasma occurs after the ingestion of each meal, and this persists for the several hours during which fat is absorbed. It causes an increase in plasma triglyceride concentration, and because the chylomicrons are large enough to scatter light (up to 0.5 μm in diameter), the plasma becomes lactescent (turbid) so as to produce what is commonly called an *alimentary lipemic response*. However, the rise in plasma triglyceride concentration is relatively small when related to the magnitude of the total triglycerides transported by the plasma because the rate of removal of the chylomicron triglyceride also rises rapidly to approximate the increased rate of entry.

Even in starvation or in persons on diets containing little fat, some triglyceride is released from the intestinal cells into the lymphatics, although the amount is naturally much reduced. In such circumstances the lipoprotein complexes themselves are relatively small; on the other hand, when the amount released is large, as during the absorption of a meal containing fat, the complexes increase in size by acquiring a greater load of triglyceride until they become chylomicrons, as these are commonly defined. There may also be an increase in the number of chylomicrons released when a great deal of triglyceride has to be transported from the intestinal cells, but it nevertheless seems likely that much of the rise in output is achieved simply by an increase in the triglyceride loading of each chylomicron unit.

From liver. The liver is the second site of triglyceride release into the plasma. The source of the fatty acids present in the triglyceride entering the blood from this organ depends greatly on the nutritional state. Thus in the fasting state, fatty acids derived from adipose cell triglycerides are taken up by the liver and a portion is re-excreted as VLDL. Following a meal, dietary carbohydrates are taken up by the liver and converted to triglycerides, which are secreted as lipoproteins.

Despite these variations in the source of the triglyceride fatty acids released by the liver, it is important to realize that the release process itself is continuous and that, except during the absorption of dietary fat, the liver is the main contributor of triglyceride to the plasma.

Our knowledge of the process of triglyceride release from the liver indicates that it is probably very similar to triglyceride release from the intestine. Again, as in the intestine, the size of the lipoprotein complexes formed by the liver varies according to the amount of triglyceride that is being released. Thus high rates of release result in large complexes with a high triglyceride load and a correspondingly low density. In fact, the lipoprotein complexes released from the liver under such conditions may reach a size not much below that of the chylomicrons, even though they normally have a somewhat lower triglyceride content and therefore a higher density.

Transport of triglyceride by lipoproteins in plasma

Some lipoproteins present in the plasma are essentially those that have already been described as being produced by the intestine and the liver. These are known either as the VLDL because they are the plasma lipoprotein fraction most readily separated by flotation in the ultracentrifuge or as pre-beta-lipoproteins on the basis of their electrophoretic behavior. During the absorption of dietary fat, chylomicrons are also present, which, because they are of still lower density than the VLDL, are even more readily separated by flotation in the ultracentrifuge.

Although most plasma triglycerides are present in chylomicrons and VLDL, the plasma also contains other triglyceride-carrying lipoproteins. They are called either *low-density lipoproteins* (LDL) or *high-density lipoproteins* (HDL) on the basis of their relative densities and lipid/protein ratios, and they have, respectively, beta-globulin and alpha-globulin electrophoretic mobility. As more knowledge is acquired, it is clear that this relatively simple classification of lipoproteins into four main groups is far from adequate.

Removal of triglyceride from plasma

Since the rate of removal of triglyceride from the bloodstream is normally adjusted quickly to balance its rate of entry, whenever the latter varies, the former will also change.

The rapid removal of triglycerides from chylomicrons and VLDL is a selective process that involves the hydrolysis of the triglyceride moiety by a lipoprotein lipase, as described above. In order to be hydrolyzed, the VLDL and chylomicrons must be trapped or sequestered as they pass through the capillary lumen. This sequestration is of particular interest because it requires an association to take place at the endothelial cell surface between the enzyme and the chylomicrons or VLDL in the bloodstream. In fact, it appears that the association of HDL apoprotein with VLDL and chylomicrons (which occurs mainly after these have been released into the plasma) is a prerequisite for the hydrolysis of the triglyceride fatty acids from the bloodstream.

The action of clearing-factor lipase at the endothelial cell surface not only facilitates the removal of triglyceride fatty acid from the blood but also determines where it is utilized, and this has important consequences. For example, in a state of caloric excess, the proportion of the triglyceride fatty acid in the bloodstream that is in excess of the immediate caloric needs is taken up by adipose tissue. Most fatty acids are reconverted to triglyceride therein and stored. In contrast, in a state of calorie deficit (as during fasting) the tissues derive their energy primarily from the oxidation of unesterified fatty acids, which are mobilized from adipose tissue and carried to the body tissues in the blood. There is still triglyceride in the blood in VLDL under these conditions, but instead of being taken up by adipose tissue for storage, it is now directed away from this tissue and toward muscle to supplement the supply of energy from the mobilized fatty acids. This switch in triglyceride fatty acid uptake is achieved through changes in the activity of clearing-factor lipase in the tissues concerned. Thus fasting results in a fall in the activity of the enzyme in adipose tissue and an increase in its activity in muscle.

Control of intracellular adipose cell lipase

If clearing-factor lipase determines the pattern of triglyceride fatty acid removal from the blood, what regulates the mobilization of triglyceride fatty acids from adipose cells into the circulation? It appears likely that, at least with respect to the enzyme in adipose tissue, the control is humoral. This lipase is concerned with the mobilization of the fatty acids that are stored in adipose tissue and is distinct from the plasma enzyme.

This intracellular adipose triglyceride enzyme is called *hormone-sensitive lipase* because it is converted from an inactive to an active form by epinephrine, norepinephrine, adrenocorticotropin, thyroid-stimulating hormone, and glucagon. Moreover, its activity is promoted by growth hormone. On the other hand, insulin inhibits the activity of this lipase. Unlike the lipoprotein lipase of adipose tissue, hormone-sensitive lipase of other tissue exhibits increased activity during fasting, possibly because of falling insulin levels. It is believed that hormone-sensitive lipase plays an important role in fat mobilization from adipose tissue.

Adipose tissue contains a monoglyceride lipase that vigorously promotes the hydrolysis of beta monoglycerides and, at a slower rate, alpha monoglycerides to glycerol and FFAs. The rate of monoglyceride and diglyceride hydrolysis is much greater than that of triglyceride hydrolysis resulting from the action of hormone-sensitive lipase. However, the activity of the monoglyceride lipase is not influenced by hormones.

Normal expected triglyceride values

Like many cholesterol ranges that represent the "normal" U.S. population, most triglyceride data are obtained from large-scale epidemiological studies. Table 30-4 provides "normal" expected values for both males and females according to age.

Clinical chemistry laboratories usually set their upper limit of normal for fasting plasma triglyceride levels between 1350 and 2000 mg/L based on 95th-percentile determinations for given populations.

Table 30-4 Distribution of triglyceride values in U.S. population

	Total cholesterol in serum (mg/L)			
	Male (n = 24,485)		Female (n = 18,168)	
Age (yr)	**Mean**	**Range***	**Mean**	**Range***
0 (cord blood)	360	100-980	360	310-980
0-1	700	310-820	780	310-980
2-3	550	300-840	660	320-1030
4-5	550	280-910	590	320-970
6-7	560	310-1050	600	330-1070
8-9	580	300-1090	640	340-1100
10-11	590	300-1090	720	370-1240
12-13	720	360-1380	810	420-1380
14-15	760	360-1380	790	400-1370
16-17	800	380-1570	730	400-1200
18-19	850	430-1690	790	410-1350
20-24	1030	450-2070	740	370-1350
25-29	1190	470-2560	770	380-1490
30-34	1320	510-2740	810	400-1550
35-39	1490	550-3310	880	410-1810
40-44	1550	560-3300	1010	460-1970
45-49	1560	590-3370	1080	470-2200
50-54	1560	590-3300	1180	530-2400
55-59	1450	590-2940	1290	560-2700
60-64	1460	590-3000	1310	570-2460
65-69	1410	580-2750	1350	610-2500
70+	1340	590-2660	1350	610-2440

From Cooper, GR: In Faulkner, R, and Meites, S, editors: Selected methods of clinical chemistry, vol 9, Washington, DC, 1982, American Association for Clinical Chemistry.
*Range is 95th and 5th percentiles.

Several highly important decision values exist that should be remembered for specific medical reasons. Perhaps the most useful triglyceride level to remember is above 5000 mg/L, because of the increased risk for acute pancreatitis associated with such levels. In 1984 the National Heart, Lung, and Blood Institute recommended that fasting plasma triglyceride levels equal to 10,000 mg/L be considered to indicate a substantial risk of pancreatitis. Based on this risk and coupled with the large intraindividual variation in repeated triglyceride measurements, a triglyceride level above 5000 mg/L was considered to be abnormal and to warrant the label of significant hypertriglyceridemia.

Part II: Lipoproteins

As discussed earlier, lipids are insoluble in aqueous media, including that of plasma. It is only when the hydrophobic lipids are bound to protein-forming "lipoproteins" that they become soluble in the bloodstream. Because lipoproteins are generally viewed as a class of macromolecules associated with lipid transport, recommendations were made in about 1967 to transfer the diagnostic emphasis from hyperlipidemia to hyperlipoproteinemia.

A lipoprotein can be visualized most simply as a globular structure with an outer solubilizing coat of protein and phospholipid and an inner hydrophobic, neutral core of triglyceride and cholesterol (Fig. 30-11). The protein and phospholipid impart solubility to the otherwise insoluble lipids. The binding of the inner lipid to the phospholipid and protein coat is noncovalent, occurring primarily through hydrogen bonding and van der Waals forces. The protein, free of lipid, is called *apolipoprotein*. Note that the lipids, which are weakly bound to the protein and phospholipid, are bound loosely enough to allow the ready exchange of lipid betweeen the serum lipoproteins themselves, as well as between serum and tissue lipoproteins, yet strongly enough to allow the lipid and protein moieties to be separated in the analytical systems that are used to isolate and classify the lipoproteins.

CLASSIFICATION OF LIPOPROTEINS

Over the years several analytical systems have been used to isolate, separate, and characterize lipoproteins. Most have been based on one or another physiochemical property of the lipid-protein complex. The four most frequently used systems are based on analytical ultracentrifugation, preparative ultracentrifugation, electrophoresis, and precipitation techniques. The most frequently used systems are those based on ultracentrifugation and electrophoresis (Fig. 30-12). With a paper or agarose support medium, electrophoretic patterns show that chylomicrons remain at the origin, while pre-beta-lipoproteins and beta-lipoproteins migrate in the $beta_1$ and $beta_2$-globulin areas, respectively, and alpha-lipoproteins migrate in the $alpha_1$-globulin area.

Using the ultracentrifuge and taking advantage of the fact that lipoproteins are lighter than the other serum proteins, one can separate the lipoproteins into chylomicrons (the lightest lipoproteins, of a density less than plasma), VLDL at a density below 1.006 g/mL (after chylomicron removal), LDL of density 1.006 and 1.063 g/mL, and HDL of density 1.063 to 1.210 g/mL. These lipoprotein classes correlate with electrophoretic patterns; for example, pre-beta-lipoprotein is generally synonymous with VLDL, beta-lipoprotein with LDL, and alpha-lipoprotein with HDL. Table 30-5 and Fig. 30-12 summarize the physical, chemical, and physiological characteristics of the major plasma lipoproteins.

Chylomicrons

Chylomicron is the term originally used to describe the microscopically visible particle appearing in the plasma after a fatty meal. Subsequently, chylomicrons have been

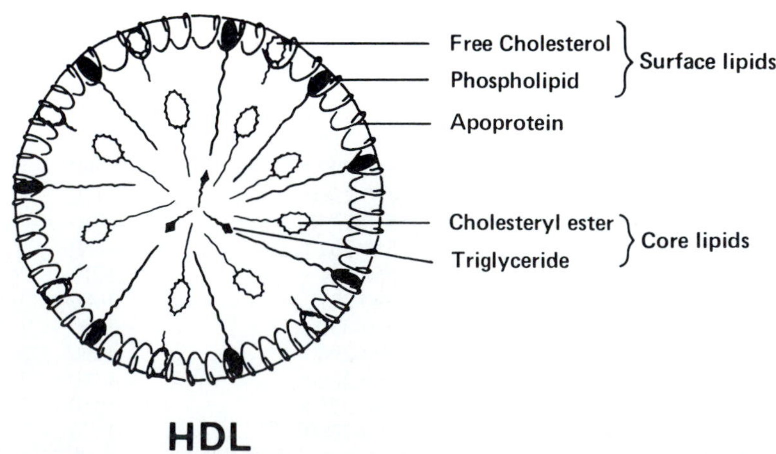

Fig. 30-11 Scheme of a lipoprotein complex showing polar outer surface and a core filled with neutral lipids. *HDL,* High-density lipoprotein.

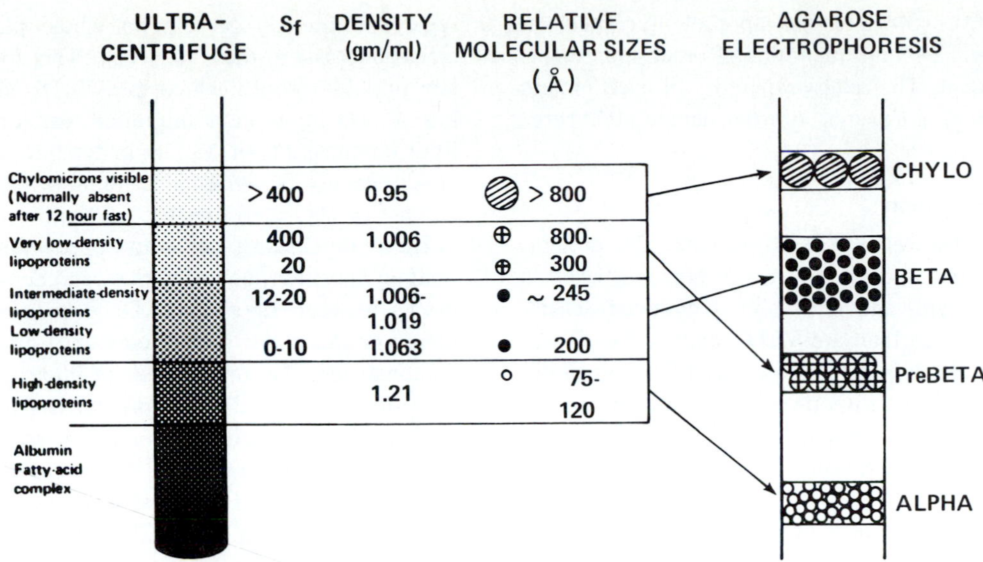

Fig. 30-12 Overview of major types of lipoproteins, showing some basic chemical and physical properties. S_f, Svedberg flotation rate; *chylo,* chylomicrons; *beta,* beta-lipoprotein; *prebeta,* a very-low-density lipoprotein; *alpha,* alpha-lipoprotein.

characterized as triglyceride-rich particles secreted by the intestine, which serve as the major transport form of dietary fat. They contain triglyceride combined with cholesterol, small amounts of phospholipid, and specific apoproteins (Apo B, Apo A, Apo C, and Apo E) (Table 30-5). Most models for chylomicron structure have assumed that the neutral lipids (triglycerides and ester cholesterol) are partially surrounded by an outer shell of phospholipid, free cholesterol, and protein. Under fasting conditions (more than 12 hours after a meal), no chylomicrons are generally found in the blood.

Very low–density lipoprotein

An average preparation of VLDL contains 52% triglyceride, 18% phospholipid, 22% cholesterol, and about 8% protein. Cholesterol and cholesteryl esters occur in a ratio of about 1:1 by weight. Sphingomyelin and phosphatidyl choline are the major phospholipids. The larger the size of a VLDL particle, the greater the proportion of triglycerides and Apo C and the smaller the proportion of phospholipid, Apo B, and other apoproteins. Apo B appears to be present in a constant absolute quantity in all VLDL fractions. Apo B accounts for approximately 30% to 35%, with Apo

Table 30-5 Physical and chemical description of plasma lipoproteins in humans

Feature	Chylomicrons	VLDL	IDL	LDL	HDL
Density (g/mL)	<1.006	<1.006	1.006-1.019	1.019-1.063	1.063-1.21
Electrophoretic mobility	Origin	Pre-beta	Beta	Beta	Alpha
Flotation rate (S_f)	>400	20-400	12-20	0-10	—
Diameter (nm)	80-500	40-80	24.5	20	7.5-12
Lipids (% by weight)	98	92	85	79	50
Cholesterol	9	22	35	47	19
Triglyceride	82	52	20	9	3
Phospholipid	7	18	20	23	28
Apoproteins (% of weight)	2	8	15	21	50
Major	A-I, A-II				A-I, A-II
	B	B	B	B	
	C-I, C-II, C-III	C-I, C-II, C-III			
	E	E	E		
Minor		A-I, A-II		C-I, C-II, C-III	C-I, C-II, C-III
					D
					E

C making up over 50% of the apoprotein content in VLDL. Apo E and varying quantities of other apoproteins may also be present. The relative quantity of each protein varies with the individual and with the degree of hyperlipidemia.

Low-density lipoprotein

LDL contains, by weight, 80% lipid and 20% protein. Consistent with this increased protein content, LDL is smaller (21 to 25 nm) and is of higher hydrated density (1.006 to 1.063 g/mL) than are VLDL and chylomicrons. About 60% of LDL lipid is cholesterol. LDL constitutes 40% to 50% of the plasma lipoprotein mass in humans. Its average concentration in normal adult American males is about 4000 mg/L, and in females it is 3400 mg/L. LDL is the major carrier of cholesterol. Apo B is the major apoprotein of normal LDL, and LDL Apo B represents 90% to 95% of the total plasma Apo B. Experimental evidence suggests that the LDL Apo B in healthy humans is derived almost entirely from VLDL Apo B in plasma. LDL is frequently separated into two classes, LDL_1 (IDL) and LDL_2, on the basis of flotation density. The lower-density fraction, IDL (1.006 to 1.109 g/mL), is more lipid rich than LDL_2, (1.019 to 1.063 g/mL) and probably represents an intermediate in VLDL catabolism (see Fig. 30-3). Thus a comparison of IDL with LDL_2 demonstrates the gradual disappearance of triglyceride and of apoproteins more characteristic of VLDL (Apo C and Apo E) and an enrichment with Apo B and cholesterol ester.

High-density lipoprotein

The HDL macromolecular complex (Fig. 30-11) contains approximately 50% protein and 50% lipid. HDL is the smallest of the lipoproteins (9 to 12 nm) and floats at the highest density (1.063 to 1.21 g/mL) of any of the lipoprotein molecules. The quantitatively most important HDL lipid is phospholipid, although HDL cholesterol is of particular interest. The major phospholipid species is phosphatidyl choline (also known as lecithin), which accounts for 70% to 80% of the total phospholipid. It has an important functional role as a reactant in plasma cholesterol esterification, which is catalyzed by the enzyme lecithin:cholesterol acyltransferase (LCAT).

HDL may be further subfractionated by differential ultracentrifugation into HDL_2 (with a density of 1.063 to 1.110 g/mL) and HDL_3 (1.110 to 1.21 g/mL); the former is present in premenopausal women at about three times its concentration in men. Persons with lower HDL_2 levels are apparently more susceptible to premature CHD.

Other lipoproteins

Floating beta-lipoprotein, or beta-migrating VLDL. This lipoprotein fraction is found in persons with type III hyperlipoproteinemia, or "broad-beta disease" (derived from the broad smear from beta- to pre-beta-lipoprotein regions frequently present on whole plasma lipoprotein electrophoresis in these subjects). This fraction has a density of 1.006 g/mL, which is a VLDL characteristic, but has a beta-lipoprotein migration pattern. The abnormal lipid composition of VLDL in type III hyperlipoproteinemic persons is attributable to a proportionately larger amount of cholesterol in that fraction.

Lp(a) or sinking pre-beta-lipoprotein. Similarities in lipid composition, concentration, and density (1.05 to 1.10 g/mL) between Lp(a) and LDL prevented clear discrimination of these two lipoproteins until immunological tests demonstrated the uniqueness of their protein moieties. Sixty-five percent of Lp(a) protein is Apo B, but another 15% is albumin, and the remainder is an apoprotein unique to Lp(a), called *Apo Lp(a)*. Despite a high frequency in the population, the functional significance of this lipoprotein is uncertain. Recent studies suggest that this lipoprotein is very atherogenic.

Lipoprotein X. Although lipoprotein X has a flotation density similar to that of LDL, the lipid and protein compositions are quite different, and this abnormal lipoprotein migrates electrophoretically differently from LDL. Lipoprotein X is characterized by an unusually high proportion of plasma phospholipid and unesterified cholesterol and by a low protein content consisting of Apo B, Apo C, and albumin. It is found most characteristically in plasma of patients with biliary obstruction.

LIPOPROTEIN METABOLISM

Chylomicrons

As discussed previously, chylomicrons are made exclusively in the intestine and traverse the lymphatic system to the thoracic duct where they then enter the systemic circulation. The major function of the chylomicrons is the transport of dietary or exogenous triglycerides.

It is postulated that the newly synthesized and secreted chylomicron (80 to 500 nm) from the intestinal mucosal cells ultimately picks up Apo C-II from HDL. Apo C-II then catalyzes lipoprotein triglyceride hydrolysis by lipoprotein lipase. The hydrolysis results in the liberation of FFAs and monoglycerides.

As shown in Fig. 30-3, endothelial cell lipoprotein lipase-catalyzed hydrolysis results in progressive triglyceride depletion of the chylomicron molecule resulting in the chylomicron remnant particle. This transformation involves maintenance of lipoprotein structure by simultaneous removal of phospholipid, unesterified cholesterol, and Apo C peptides from the lipoprotein surface to plasma HDL. A reciprocal transfer of cholesterol ester from HDL may occur; Apo D may aid in this transfer process. The circulating chylomicron remnant particle is then released from the capillary wall and cleared from circulation through the liver, where it is metabolized. This particle, now smaller (30 to 80 nm), retains its ester cholesterol and Apo B and Apo E, which play an important role in the uptake of these particles by a high-affinity hepatic receptor uptake mechanism (see Fig. 30-9). When binding occurs, the remnants are immediately internalized by receptor-mediated endocytosis and degraded in hepatic lysosomes.

Very low–density lipoprotein

The major stimulus for hepatic VLDL (30 to 80 nm) synthesis is the demand for triglyceride transport. In the intestine this demand is created by influx of dietary fat, but in the liver the stimulus is the availability of precursors for endogenous triglyceride synthesis. These precursors are nonesterified or free fatty acids (FFAs) and activated glycerol.

After the postprandial rise in chylomicron triglyceride, a secondary rise in triglyceride concentration occurs 4 to 6 hours after a meal. This represents predominantly hepatic VLDL triglyceride synthesized from glucose and chylomicron triglyceride not hydrolyzed in the peripheral tissue. The relative contributions of glucose and dietary fat vary with diet composition. Consumption of a high-carbohydrate diet may lead to a phenomenon known as *carbohydrate-induced hypertriglyceridemia*. With high dietary carbohydrate, glucose influx into the hepatocyte is in excess both of energy demands and of glycogen-storage capacity. This results in the shunting of acetyl CoA into fatty acid synthesis and dihydroxyacetone phosphate into activated glycerol. This phenomenon may not persist in healthy persons, but others may be unusually susceptible to carbohydrate induction of VLDL synthesis. This is the basis for reduction of dietary carbohydrate (simple sugars and alcohol) in the treatment of hypertriglyceridemia, but this approach is not successful if the hypertriglyceridemia has other causes, such as overproduction or a clearance defect.

Hepatic biosynthesis of triglyceride, cholesterol, and phospholipid occurs in the smooth endoplasmic reticulum. The first two lipid components contribute to the lipoprotein core and the last, to the surface coat but, as in the intestine, apoprotein incorporation (especially Apo B) is essential before secretion. Normally, VLDLs represent about 10% to 15% of the total circulating lipoproteins in a normal healthy individual.

VLDL triglycerides are believed to have a fate similar to that of the lipids from chylomicrons (see section on chylomicron catabolism, p. 466). During the catabolism of VLDL, more than 90% of Apo C is transferred to HDL, whereas essentially all the Apo B remains with the original lipoprotein particle (see Fig. 30-3). According to this postulated catabolic pathway, breakdown of VLDL leads to the formation of the cholesterol-rich particle, LDL. HDL plays an important role by serving as an acceptor macromolecule for Apo C and unesterified cholesterol and phospholipids, the excess surface materials from a saturated VLDL. Apo C may recycle from HDL to newly synthesized chylomicrons or VLDL. The half-life of VLDL is 1 to 3 hours.

Intermediate-density lipoprotein

IDL is a transient particle (22 to 28 nm) usually present in very low concentrations in plasma from fasting persons. IDL, as discussed previously, is a lipoprotein derived from VLDL catabolism. The HDL particles interact with the plasma enyzme lecithin:cholesterol acyltransferase (LCAT), which esterifies the excess HDL free cholesterol with fatty acids derived from the 2 position of lecithin, the major phospholipid of plasma. The newly synthesized cholesteryl ester is transferred back to the IDL particles from HDL, apparently through the action of a plasma cholesteryl ester exchange protein (possibly Apo D). The net result of the coupled lipolysis and exchange reactions is the replacement of most of the original triglyceride core of VLDL with ester cholesterol.

After lipolysis, the IDL particles are released from the capillary wall into the circulation. They then undergo a further conversion in which most of the remaining triglycerides are removed and all the apoproteins except Apo B are lost. The resultant particle, which contains almost pure cholesteryl ester in the core and Apo B at the surface, is LDL.

Low-density lipoprotein

As discussed previously, LDL formation occurs primarily from the catabolism of VLDL. In normal healthy persons, LDL cholesterol constitutes about two thirds of the total plasma cholesterol; LDL cholesterol concentration in women is slightly less than that in men (except after menopause). LDL delivers cholesterol to extrahepatic tissues (and to the liver), where it is utilized, deposited, or excreted.

Delivery of the LDL particles to peripheral tissue is accomplished when the LDL binds to high-affinity receptors located in regions of the plasma membrane called *coated pits*. These pits invaginate into the cell and pinch off to form endocytic vesicles that carry the LDL to the lysosomes (Fig. 30-9). Fusion of the vesicle membrane with the lysosomal membrane exposes the LDL to a host of hydrolytic enzymes that degrade the Apo B to amino acids. The cholesteryl esters are hydrolyzed by an acid lipase, and liberated free cholesterol leaves the lysosomes for use in cellular reactions. As a result of this uptake mechanism, extrahepatic cells have low rates of cholesterol synthesis, relying instead on LDL-derived cholesterol. The free cholesterol thus released is used for membrane synthesis and serves to regulate, that is, depress cellular cholesterol synthesis by HMG-CoA reductase. LDL internalization also regulates synthesis of the LDL receptor itself.

Excess cholesterol activates the enzyme acyl-CoA: cholesterol acyltransferase (ACAT), leading to intracellular cholesterol ester storage. Thus the net result of LDL binding and internalization is the reciprocal inhibition and activation of enzymes synthesizing and storing cellular cholesterol and a reduction in the number of receptors available to bind LDL.

The significance of this process for regulation of plasma cholesterol levels in humans is illustrated by patients with the homozygous form of familial hypercholesterolemia. These patients are deficient in LDL receptors and have excessive LDL production and defective LDL catabolism because of an inability of tissues to bind, internalize, degrade, and thus regulate cholesterol synthesis. Except for the receptor-deficient state of familial hypercholesterolemia, however, the role of the LDL receptor in the final control of plasma cholesterol levels is uncertain and is probably only complementary to other regulatory processes. It has been recognized that the specificity of the LDL receptor extends to lipoproteins containing Apo E and Apo B as well. It appears that although extrahepatic receptors take up LDL readily, hepatic receptors take up chylomicron remnants with greater efficiency (about 20 times greater) and LDL with much less efficiency. This difference is probably attributable to the Apo E content of chylomicron remnants, which has a higher affinity than that of Apo B.

In addition to its normal degradation mechanism, the high-affinity LDL receptor pathway, plasma LDL can be degraded by less efficient mechanisms that require high plasma levels to achieve significant rates of removal. One of these mechanisms occurs in scavenger cells (macrophages) of the reticuloendothelial system. When the plasma level of LDL rises, these scavenger cells degrade increasing amounts of LDL. When overloaded with cholesteryl esters, they are converted into foam cells, which are classic components of atherosclerotic plaques. In humans, estimates of the proportion of plasma LDL degraded by the LDL receptor system range from 33% to 66%. The remainder is degraded by the scavenger cell system and perhaps by other mechanisms not yet elucidated.

High-density lipoprotein

Nascent HDL molecules are synthesized in intestinal mucosal cells and in hepatocytes by a process analogous to that of VLDL and chylomicron synthesis. This involves microsomal lipid and protein synthesis followed by secretion. During the synthetic process, phospholipid and free cholesterol are combined with specific apoproteins to form disklike structures that undergo extensive compositional and structural modifications after secretion. The most important of these modifications is the esterification of free cholesterol to form ester cholesterol by an enzymatic reaction catalyzed by LCAT (Fig. 30-8). In humans this is the major source of plasma ester cholesterol. Persons with LCAT deficiency have an accumulation of these ester cholesterol–deficient particles in plasma. This finding possibly indicates that the ester cholesterol formed in the LCAT reaction allows the expansion of the disklike structures to form spheres characteristic of normal plasma HDL (Fig. 30-9). Ester cholesterol thus formed may be transferred to VLDL during catabolism.

The apoprotein profile of nascent HDL is modified concomitantly with changes in lipid content. Apo E is a major component of newly secreted (nascent) HDL, unlike the plasma HDL, which is characterized by a predominance of Apo A with minor contributions by Apo C and Apo E. The functional significance of this modification is not completely understood, but Apo A-I is an activator of LCAT, and its acquisition must facilitate all LCAT reactions. In addition, HDL participates in the regulation of triglyceride catabolism and ester cholesterol formation by providing the respective cofactors, Apo C-II for activation and Apo C-III for inhibition of lipoprotein lipase activity. Also normal HDL may balance LDL transport by mediating cholesterol removal from peripheral sites to degradative and excretory sites. This role of HDL in reverse cholesterol transport may be the basis for the protection afforded by HDL against cardiovascular disease (Fig. 30-9).

The plasma half-life of HDL in normal subjects ranges from 3.3 to 5.8 days. The Apo A-I catabolism and Apo A-II catabolism within HDL are similar. HDL catabolism is enhanced in nephrotic patients but decreased in hypertriglyceridemic subjects, especially those with hyperchylomicronemia. It is also increased in subjects on high-carbohydrate diets and is greatly enhanced in patients with familial HDL deficiency (Tangier disease). It appears that changes in HDL catabolism may play a major role in regulating HDL levels in plasma.

HYPERLIPIDEMIA

By definition, hyperlipidemia is an elevated concentration of lipids in the blood. The major plasma lipids of interest are total cholesterol (free cholesterol + ester cholesterol) and the triglycerides. When one or more of these major classes of plasma lipids is elevated, a condition referred to as *hyperlipidemia* exists. Tables 30-2 and 30-4 provide some guidelines used to define hypercholesterolemia and hypertriglyceridemia.

Cholesterol and triglyceride concentrations can be used to detect hyperlipoproteinemia. Over 95% of persons with hyperlipidemia, as defined previously, have hyperlipoproteinemia. The major exceptions are subjects with excessive amounts of LDL whose plasma cholesterol is kept within normal limits by a concomitant decrease in HDL.

As indicated earlier, the lipid concentrations listed in Tables 30-2 and 30-4 do not indicate healthy values and the NIH suggests that treatment in the United States should be initiated when an adult has a serum cholesterol level above 2000 mg/L. The upper limit of serum triglyceride levels is less clear. Perhaps levels in excess of 5000 mg/L can be considered unhealthy. Dietary therapy should be initiated when the above limits are exceeded in order to prevent or minimize the development of CHD since atherosclerosis seldom occurs with a total cholesterol concentration of less then 1500 mg/L over the life span of a person, unless other risk factors such as genetics, high blood pressure, smoking, and obesity are present.

HYPERLIPOPROTEINEMIA

Hyperlipoproteinemia is an elevation of serum lipoprotein concentrations. With one exception (type III), hyperlipoproteinemia too is defined in quantitative terms. Arbitrary limits of normal lipoprotein concentrations have been established in a manner similar to that used for the establishment of reference range for lipid concentrations.

The quantitation of lipoprotein fractions is usually done by analytical ultracentrifugation. Simpler, but more indirect, procedures can be employed, such as lipoprotein electrophoresis and determination of the cholesterol in the isolated VLDL, LDL, and HDL fractions. Both methods are routinely used in the clinical laboratory.

Table 30-6 Causes of secondary hyperlipoproteinemia

Pattern	Causes
Type I	Insulinopenic diabetes mellitus
	Dysglobulinemia
	Lupus erythematosus
	Pancreatitis
Type II	Nephrotic syndrome
	Hypothyroidism
	Obstructive liver disease
	Porphyria
	Multiple myeloma
	Portal cirrhosis
	Viral hepatitis, acute phase
	Myxedema
	Stress
	Anorexia nervosa
	Idiopathic hypercalcemia
Type III	Hypothyroidism
	Dysgammaglobulinemia
	Myxedema
	Primary biliary cirrhosis
	Diabetic acidosis
Type IV	Diabetes mellitus
	Nephrotic syndrome
	Pregnancy
	Hormone use (oral contraceptives)
	Glycogen-storage disease
	Alcoholism
	Gaucher's disease
	Niemann-Pick disease
	Pancreatitis
	Hypothyroidism
	Dysglobulinemia
Type V	Insulinopenic diabetes mellitus
	Nephrotic syndrome
	Alcoholism
	Myeloma
	Idiopathic hypercalcemia
	Pancreatitis
	Macroglobulinemia
	Diabetes mellitus (insulin independent)

The classification of hyperlipoproteinemia begins with the determination of the type of abnormal lipoprotein pattern. However, other analyses are always necessary; for example:

1. Separation of hyperlipoproteinemia into primary and secondary forms (Table 30-6). The secondary form is caused by another known disease (such as insulinopenic diabetes mellitus, gout, hypothyroidism), which can result in secondary hyperlipoproteinemia manifesting itself in any of the six types of lipoprotein patterns.
2. Differentiation of primary hyperlipoproteinemia into heritable and nonheritable forms.
3. Determination of the relative concentration of the lipoprotein fractions, that is, VLDL cholesterol, LDL cholesterol, and HDL cholesterol.

Nearly all the patients with heritable hyperlipidemia have one of six abnormal lipoprotein patterns. These patterns are summarized in Fig. 30-13, where one can observe that only three of the four lipoprotein families serve as determinants. These three families are (1) chylomicrons, (2) VLDL, and (3) LDL. The original Fredrickson phenotyping system disregarded the importance of HDL and other lipoproteins discussed in this chapter. Because of the fairly recent worldwide epidemiological finding that there is a statistically significant inverse relationship between HDL cholesterol concentration and risk for CHD, laboratories now do LDL cholesterol and HDL cholesterol determinations as part of the overall lipid-lipoprotein profile.

CLASSIFICATION OF HYPERLIPOPROTEINEMIAS

Although the classification of hyperlipoproteinemia (Fig. 30-13) is based on identification of elevated concentration of blood lipids and abnormal lipoprotein patterns, I should emphasize again that each form of dyslipoproteinemia is not a homogeneous entity from a genetic, clinical, or pathological point of view.

The hyperlipoproteinemias are now described in somewhat greater detail, with emphasis on distinctive clinical, diagnostic, genetic, biochemical-pathophysiological, and therapeutic aspects.

Type I hyperlipoproteinemia

Type I hyperlipoproteinemia is characterized by elevated plasma triglyceride concentration, chylomicronemia. LDL and HDL are often low, whereas VLDL may be slightly elevated. Chylomicron removal is reduced, and subjects have deficiencies in lipase activities, such as postheparin-lipolytic activity (PHLA) (specifically extrahepatic, protamine-inactivated lipoprotein lipase). Primary type I hyperlipoproteinemia usually manifests itself early in childhood. Although type I hyperlipoproteinemia is usually primary and familial, the lipoprotein pattern may be pro-

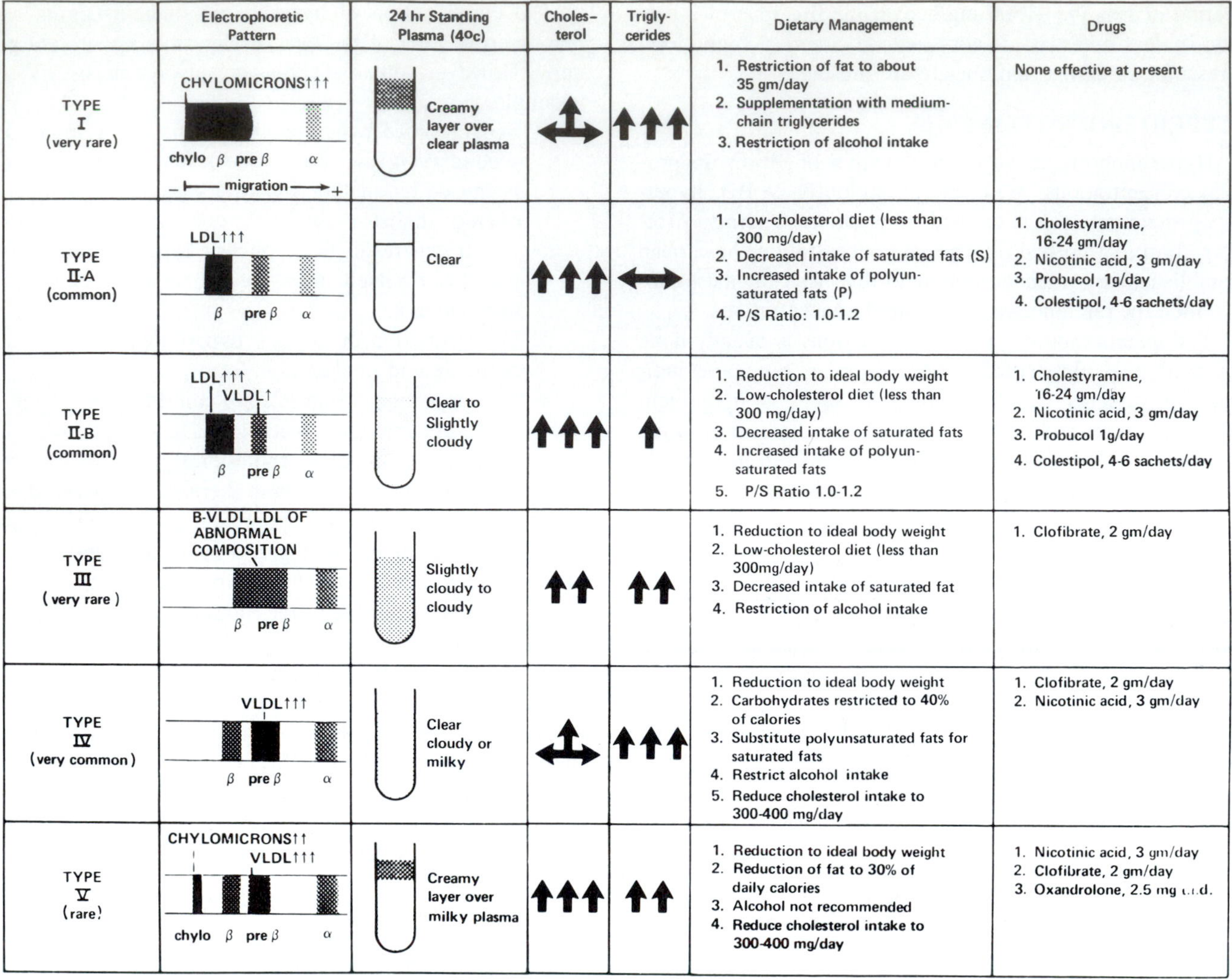

	Electrophoretic Pattern	24 hr Standing Plasma (4oc)	Choles-terol	Trigly-cerides	Dietary Management	Drugs
TYPE I (very rare)	CHYLOMICRONS↑↑↑; chylo β pre β α; − migration → +	Creamy layer over clear plasma	↑↔	↑↑↑	1. Restriction of fat to about 35 gm/day 2. Supplementation with medium-chain triglycerides 3. Restriction of alcohol intake	None effective at present
TYPE II-A (common)	LDL↑↑↑; β pre β α	Clear	↑↑↑	↔	1. Low-cholesterol diet (less than 300 mg/day) 2. Decreased intake of saturated fats (S) 3. Increased intake of polyunsaturated fats (P) 4. P/S Ratio: 1.0-1.2	1. Cholestyramine, 16-24 gm/day 2. Nicotinic acid, 3 gm/day 3. Probucol, 1g/day 4. Colestipol, 4-6 sachets/day
TYPE II-B (common)	LDL↑↑↑; VLDL↑; β pre β α	Clear to Slightly cloudy	↑↑↑	↑	1. Reduction to ideal body weight 2. Low-cholesterol diet (less than 300 mg/day) 3. Decreased intake of saturated fats 4. Increased intake of polyunsaturated fats 5. P/S Ratio 1.0-1.2	1. Cholestyramine, 16-24 gm/day 2. Nicotinic acid, 3 gm/day 3. Probucol 1g/day 4. Colestipol, 4-6 sachets/day
TYPE III (very rare)	B-VLDL,LDL OF ABNORMAL COMPOSITION; β pre β α	Slightly cloudy to cloudy	↑↑	↑↑	1. Reduction to ideal body weight 2. Low-cholesterol diet (less than 300mg/day) 3. Decreased intake of saturated fat 4. Restriction of alcohol intake	1. Clofibrate, 2 gm/day
TYPE IV (very common)	VLDL↑↑↑; β pre β α	Clear cloudy or milky	↑↔	↑↑↑	1. Reduction to ideal body weight 2. Carbohydrates restricted to 40% of calories 3. Substitute polyunsaturated fats for saturated fats 4. Restrict alcohol intake 5. Reduce cholesterol intake to 300-400 mg/day	1. Clofibrate, 2 gm/day 2. Nicotinic acid, 3 gm/day
TYPE V (rare)	CHYLOMICRONS↑↑; VLDL↑↑↑; chylo β pre β α	Creamy layer over milky plasma	↑↑↑	↑↑	1. Reduction to ideal body weight 2. Reduction of fat to 30% of daily calories 3. Alcohol not recommended 4. Reduce cholesterol intake to 300-400 mg/day	1. Nicotinic acid, 3 gm/day 2. Clofibrate, 2 gm/day 3. Oxandrolone, 2.5 mg t.i.d.

Fig. 30-13 Summary of six types of hyperlipoproteinemias. Abbreviations as in Fig. 30-12.

duced by several other disease or metabolic states. Thus secondary type I should be ruled out (Table 30-6).

Once the obvious secondary causes have been ruled out, the following should be checked: (1) presence of eruptive xanthomas, hepatosplenomegaly, lipemia retinalis, abdominal pain, and pancreatitis early in life; (2) intake of drugs that can cause secondary hypertriglyceridemia; (3) presence of reduced plasma levels of triglyceride lipases; (4) reduction in triglyceride levels and disappearance of chylomicronemia on a fat-free diet; and (5) confirmation by family screening of inheritance as an autosomal recessive trait.

Type IIa hyperlipoproteinemia

Type IIa is characterized by elevated concentration of plasma cholesterol with normal triglyceride levels and clear plasma. The lipoprotein pattern is characterized by an elevation of the LDL with normal VLDL. This lipoprotein disorder is recognized as familial hypercholesterolemia, which exhibits the following features: (1) a deficient number of functional LDL receptors in the fibroblast cultures (the pathognomonic feature), (2) an expression of type II pattern in infancy, (3) xanthomatosis in severely affected members, and (4) premature CHD seen by the third and fourth decade.

Secondary type II disorders, such as hypothyroidism, acute intermittent porphyria syndrome, dysgammaglobulinemia, obstructive liver disease, and highly saturated fat and cholesterol diets should be ruled out.

Once secondary hyperlipoproteinemia has been ruled out, the primary disorder can be confirmed by (1) family screening, including children; (2) persistent hypercholesterolemia even after 8 weeks of low cholesterol (less than 300 mg/day), high polyunsaturated fat diet (polyunsaturated fat to saturated fat [P/S] ratio of 1:1.2); and (3) presence of tendinous xanthomas, xanthelasma, and corneal arcus.

Type IIb hyperlipoproteinemia

Another form of familial hyperlipidemia is familial combined hyperlipidemia. The important features include (1) no abnormality in the number of functional LDL receptors in the fibroblast culture, (2) absence of the type II pattern in children, (3) early expression of hypertriglyceridemia, (4) multiple lipoprotein patterns in affected relatives in successive generations, and (5) hypercholesterolemia in the common type II pattern.

This is the most recently described familial hyperlipoproteinemic syndrome and also probably the most common of the primary hyperlipoproteinemias. The characteristic feature of this disorder is a scatter of lipoprotein phenotypes within a family. Most commonly, patients will have an elevation in both LDL and VLDL; however, within a family, hyperbetalipoproteinemia (type IIa) and hyperprebetalipoproteinemia (type IV) will also be found, affecting different persons. In contrast to familial hyperbetalipoproteinemia, subjects with familial combined hyperlipoproteinemia generally do not manifest their disease until adulthood. Clinically, these patients have an increased incidence of CAD. They also are frequently diabetic, have a tendency for hyperuricemia, and show a low incidence of tendinitis and tuberous xanthomas. These features suggest a closer clinical relation to familial hyperprebetalipoproteinemia than to familial hyperbetalipoproteinemia. The mode of inheritance of familial combined hyperlipoproteinemia is still in doubt; however, it is clearly a familial disorder that is seen most commonly with a type IIb pattern.

Remember that subjects with hyperbetalipoproteinemia or hyperprebetalipoproteinemia may also belong within a family group that has familial combined hyperlipoproteinemia.

This lipoprotein disorder is characterized by elevated total cholesterol, LDL, triglycerides, and VLDL with the absence of floating beta-lipoprotein. Any secondary hypercholesterolemia and hypertriglyceridemia should be ruled out before confirmation of the primary type IIb. Family screening is mandatory for recognition of this lipid abnormality. Accurate diagnosis of the type IIb pattern also requires an appreciation of the factors that determine triglyceride levels. Studies in free-living populations in the United States have documented increases in triglyceride levels with age and have indicated that as many as one fourth of middle-aged men exceeded previously published cutoff values for triglyceride levels. Thus, although statistically valid, the critical limits for triglyceride concentrations may not represent physiological limits. Therefore one might expect a greater prevalence of type IIb patterns in older age groups.

The possible effects of dietary carbohydrates should not be overlooked when one is assessing the type IIb pattern. It has been shown that fasting hypertriglyceridemia in patients with type IIb could be attributable to an acute increase in dietary carbohydrates. It appears that triglyceride values greater than 4000 mg/L are rare in patients with type IIb patterns. The few reported cases in the medical literature occurred in postmenopausal women.

Lipoprotein electrophoresis is rarely necessary for the diagnosis of the type II pattern. If the type IIa pattern is present, the elevated cholesterol and the normal triglyceride levels leave little for the lipoprotein electrophoretic pattern to clarify. If the type IIb pattern is present, the decisive diagnostic procedure is differentiation of the type IIb pattern from either the type III or type IV pattern. This task is not easily done with lipoprotein electrophoresis. Also note that the total plasma cholesterol may be normal despite an elevated LDL cholesterol value. This relationship also occurs in familial hypercholesterolemia.

The broad-beta pattern described below is no longer considered unique for type III hyperlipoproteinemia and has been noted in cases homozygous for familial hypercholesterolemia. The presence of a discrete band attributable to high concentrations of Lp(a) or "sinking" pre-beta-lipoprotein can cause diagnostic confusion with the type IIb pattern. When the triglyceride value is normal, the appearance of the IIb pattern of electrophoresis suggests the presence of Lp(a). Thus the electrophoretic pattern is insensitive and is not a highly specific tool for diagnosing type IIb or type III patterns.

Type III hyperlipoproteinemia

Type III hyperlipoproteinemia is characterized by an elevation of both plasma cholesterol and triglyceride concentrations and an abnormal LDL, which floats in the fraction with a density less than 1.006 g/mL. This abnormal LDL often merges with the pre-beta band on electrophoresis to produce a broad-beta band (Fig. 30-13). For accurate diagnosis of type III, an ultracentrifugal study with measurement of cholesterol and triglyceride in the fractions with density below 1.006 g/mL is required to document the presence of the floating beta-lipoprotein.

Measurements of the lipid composition of lipoproteins of a density less than 1.006 appear to offer a more reliable means of identifying type III than is afforded by the detection of the presence of floating beta-lipoproteins by electrophoresis alone. The ratio used clinically is that of the cholesterol content of the VLDL divided by the plasma triglyceride concentration. This ratio appears to be most useful for documenting type III hyperlipoproteinemia when the triglyceride level is at least 1500 mg/L, but it may be subject to error when triglyceride exceeds 10,000 mg/L. It has been suggested that if the VLDL cholesterol/triglyceride ratio is 0.30 or more, the subject may have type III hyperlipoproteinemia. However, when the ratio is less than or equal to 0.25 and not more than 0.29, a diagnosis of possible type III hyperlipoproteinemia should be considered.

The clinical characteristics of subjects with type III vary

widely as a function of age, sex, degree of adiposity, and presence of associated disorders such as hypothyroidism and alcoholism.

The most characteristic xanthoma in subjects with type III is called *xanthoma striatum palmare* (in the literature both ''xanthoma striatum palmaris'' and ''xanthoma striata palmaris'' are improper Latin). In their most subtle form these lesions produce an orange or yellowish discoloration of the palmar creases (xanthochromia striata palmaris), a phenomenon most easily detected in subjects of fair complexion. When more advanced, these lesions may produce planar elevations and even the virtual obliteration of the palmar and digital creases. Raised lesions can occasionally affect the remaining palmar surfaces and in the severe form produce tuberous, incapacitating xanthomas.

Various forms of CHD have been reported in association with type III hyperlipoproteinemia. Hyperlipoproteinemia is readily treated by diet and drugs. The form of cardiovascular disease associated with type III differs significantly from that associated with familial type IIa in that peripheral, and even cerebrovascular, disease appears to be as common as CHD.

Secondary type III hyperlipoproteinemia has been associated with hypothyroidism, gout, and diabetes mellitus and is found in patients with acute renal failure receiving maintenance hemodialysis.

Type IV hyperlipoproteinemia

Type IV hyperlipoproteinemia has also been called *endogenous* or *carbohydrate-induced hyperlipemia*. The latter term is no longer accepted by most workers, since carbohydrate induction of hypertriglyceridemia is also observed in normolipemic individuals. Endogenous hyperlipemia excludes the rare type I hyperlipoproteinemia but includes the uncommon type V as a mixed endogenous and exogenous hyperlipemia. The pure endogenous forms include types IIb, III, and IV.

By definition, type IV hyperlipoproteinemia is an elevation of VLDL (and triglyceride) levels above an arbitrary cutoff point in the absence of either chylomicrons or the abnormal VLDL of type III. LDL levels are normal, and LDL cholesterol measurement discriminates between types IV and IIb.

A tentative diagnosis of type IV may be made if the triglyceride concentration is increased, the total cholesterol is normal or moderately elevated, and the standing plasma reveals no chylomicrons. The biochemical diagnosis is confirmed if electrophoresis reveals a distinct pre-beta-lipoprotein band and the LDL cholesterol is within normal limits. Keep in mind that the presence of a pre-beta band with normal plasma triglyceride levels occurs with ''sinking'' pre-beta-lipoprotein, a triglyceride-poor, apparently normal lipoprotein variant observed in up to 35% of healthy subjects.

Once the biochemical pattern has been confirmed (that is, based on more than one sample under standard conditions), it should be classified according to cause as primary type IV—either familial or sporadic—or secondary type IV (Table 30-6).

The diagnosis of primary familial type IV depends on the following criteria: (1) a type IV pattern, (2) one or more first-degree relatives with type IV, and (3) no close relative with primary type I, III, or V. Other common features are a normal cholesterol level if the triglyceride level is less than 4000 mg/L, a triglyceride level usually below 15,000 mg/L, and a family history of diabetes.

Since a large proportion of our society imbibes alcohol, a brief comment will be made. Although not regarded as a major cause of hyperlipidemia, ethanol is known to cause acute but transient hypertriglyceridemia with an elevation in primarily VLDL, causing mainly type IV hyperlipoproteinemia (sometimes type V hyperlipoproteinemia).

In the hypertriglyceridemia produced by ethanol intake, the following features stand out:

1. The hypertriglyceridemia is usually moderate in extent and limited in duration. The level of triglyceridemia rarely exceeds 10,000 mg/L, and the lipemia peaks in 12 to 14 hours and disappears after 25 to 40 hours. This apparently transient effect appears to hold true, especially for normolipemic individuals.
2. This hypertriglyceridemia is, for the most part, the result of increased VLDL and possibly chylomicrons.
3. The fatty liver associated with alcohol intake plays a vital role in the form and extent of the induced hyperlipoproteinemia, and the resultant changes may be related to the stage of the hepatic damage.
4. The triglyceride concentration in alcoholic hyperlipidemia is intimately related to the quality and quantity of dietary fatty acid intake. It is well known that simultaneous ingestion of ethanol with fat (as in complex meals or singly as corn oil) produces a prolonged and augmented rise in serum triglyceride concentration.

Although it is obvious that hyperlipemia can be produced by an excessive production and release of lipids (hence, lipoproteins) into circulation, by defective removal or clearance of lipids from the blood, or by a combination of these physiological processes, the precise mechanism of ethanol-induced hyperlipemia is still unknown.

Type V hyperlipoproteinemia

Another form of hyperchylomicronemia is the type V hyperlipoproteinemia. It is distinguished by the presence of both elevated VLDL and chylomicrons in the plasma of fasting subjects on a regular diet. This disorder can occur as a primary genetic defect and thus represents a second form of familial hyperchylomicronemia. Triglyceride levels similar to those seen in type I may be observed. As in that disorder, the occurrence of abdominal syndromes, including pancreatitis, and the physical findings of eruptive xanthomas, lipemia retinalis, and hepatosplenomegaly are related to the level of plasma triglyceride. Although the pathophysiology of these manifestations of type V probably does not differ from that observed in type I, a variety of other differences exist in clinical, genetic, metabolic, and biochemical observations.

In sharp contrast to type I hyperlipoproteinemia, most instances of type V occur in adulthood. Full appearance of the abnormality may not occur until the fifth or sixth decade of life, with the females presenting later than the males. Several children with familial type V have been described.

Extremely high triglycerides and hyperchylomicronemia are usually not attributable to primary type V but are found in the setting of several disorders that can lead to secondary type V. These disorders are particularly prone to produce type V if they occur in a patient with primary type IV. For example, pancreatitis may be associated with pronounced hyperchylomicronemia, but later only mild hypertriglyceridemia is found when the patient is re-evaluated under stable conditions. It is also well known that type IV can present as type V in poorly controlled insulin-dependent diabetics and in alcoholics.

Although type V hyperlipoproteinemia is associated with elevated triglycerides (hence, VLDL and chylomicrons), plasma cholesterol concentrations may be slightly to moderately increased. LDL and HDL cholesterols are usually normal to low. The plasma is usually opaque, and a floating cream layer above the turbid plasma may be observed.

When total triglyceride is over 10,000 mg/L, visual appreciation of a discrete floating "creamy" supernate over turbid plasma becomes difficult, and the use of a combination of ultracentrifugation (to remove the chylomicrons) and electrophoresis provides a clear electrophoretic pattern for type V determination. The ultracentrifuge may be used to separate the chylomicrons from the VLDL for individual quantification if needed. Usually, a qualitative assessment of the electrophoresis strip is sufficient to document elevated VLDL and chylomicron lipid as well. Since in practice some overlap of VLDL levels may occur between types I and V, a postheparin lipoprotein lipase measurement should be made as a final characterization. This enzyme should be present in type V subjects. In the absence of a specific assay for the enzyme, a reasonably reliable clue to its presence may be sought by observation of the change in the lipoprotein electrophoresis pattern obtained with a plasma sample drawn 10 minutes after heparin injection (10 U/kg of body weight). The average patient with type V usually has considerably higher triglyceride levels than the patient with type IV. The plasma cholesterol to triglyceride ratio in type V (0.23 ± 0.02) is lower than in type IV (0.86 ± 0.03) because the chylomicrons incorporate less cholesterol than the VLDL does. Major qualitative difficulties can arise in distinguishing type V from type IV hyperlipoproteinemia (endogenous hypertriglyceridemia) because the type V pattern is often transient, with many type V subjects losing their chylomicron band with moderate reductions in triglyceride.

Type V hyperlipoproteinemia is often secondary to a wide variety of diseases, drugs, and dietary habits (see Table 30-6). Since there are many ways of acquiring type V hyperlipoproteinemia, a careful distinction between primary and secondary causes must be made. A routine history of ethanol intake and estrogen or steroid administration, a urinalysis, and the measurement of fasting or 2-hour postprandial blood glucose, liver, thyroid, and renal function tests are all useful in this distinction. Superimposition of poorly controlled diabetes mellitus, alcoholic excess, or estrogens or estrogen-containing oral contraceptives in an individual with pre-existing type IV hyperlipoproteinemia will often produce the type V pattern. In familial type V, particularly with plasma triglycerides above 15,000 mg/L, lipemia retinalis, hepatosplenomegaly, and eruptive xanthomas may be present.

The biochemical defect in type V hyperlipoproteinemia is still not clear. The presence of high levels of chylomicrons in the fasting plasma of a subject whose diet contains a normal or low content of fat clearly indicates that clearance mechanisms are inadequate. A primary defect in removal of plasma triglyceride could also explain the elevated VLDL. That this defect is not simply a variant of type I is established by the easily detectable and often quantitatively normal heparin-releasable plasma PHLA from type V subjects. Thus a different type of problem must lead to this failure in lipoprotein uptake. One possibility is that the synthesis of endogenous triglyceride, and the resulting secretion of VLDL from the liver, may proceed at an abnormally high rate, sufficient to saturate pathways of removal that are shared by chylomicrons, thus leading to an elevation of both lipoproteins. Studies utilizing lipoproteins labeled with radioisotopes have indicated that many patients with endogenous hypertriglyceridemia have elevated synthesis of VLDL triglyceride.

Since types I and V hyperlipoproteinemia are similar in many respects and can at times be hard to differentiate, a summary is warranted.

Familial hyperchylomicronemia may be divided into types I and V hyperlipoproteinemia. Both disorders are manifested by very elevated triglyceride levels and frequently present with eruptive xanthomas, lipemia retinalis, hepatosplenomegaly, and abdominal pain. Type I is caused by a pronounced deficiency of lipoprotein lipase. The triglyceride elevation appears with ingestion of dietary fat and is thus manifested in very young children. Type V may be detected in rare instances in childhood, but the usual presentation is in the adult. It also differs from type I in that lipoprotein lipase is measurable and glucose intolerance and hyperuricemia are commonly associated findings. The only effective treatment for type I is a low-fat diet. Type V is most effectively treated by diet-induced weight loss and will frequently respond to one of the following drugs: nicotinic acid, norethindrone, oxandrolone, benzafibrate, or clofibrate.

TRANSFORMATION OF HYPERLIPIDEMIA TO HYPERLIPOPROTEINEMIA

Limitation of classification of types of hyperlipoproteinemia

The limitations and potentials of the lipoprotein-typing system of Frederickson, Levy, and Lees are well recognized. However, it should be stressed that plasma lipoprotein patterns are not a substitute for an etiological classification of the hyperlipoproteinemias. The approach of Fredrickson, Lees, and Levy is to be regarded as provisional, pending a more fundamental understanding of the causes of the hyperlipidemias.

Using quantitative measurements of cholesterol and triglyceride alone, one can divide patients into three major groups: those with hypercholesterolemia alone, those with hypertriglyceridemia alone, and those with a combination of the two. Subjects with pure hypercholesterolemia usually have the type IIa lipoprotein pattern and those with pure hypertriglyceridemia without chylomicrons, the type IV pattern. Subjects with relatively high cholesterol and triglyceride levels may have lipoprotein patterns of types IIb, III, or IV. In regard to the utility of the lipoprotein-typing system, it is known that there are four areas where the typing system retains general validity:

1. Lipoprotein patterns are useful in sharpening the focus on the diverse metabolic abnormalities that underlie hyperlipidemia.

2. The types of hyperlipoproteinemias so identified are not disease states but represent disorders that similarly affect the concentrations of particular lipoproteins.

3. Each lipoprotein type is often associated with certain distinctive clinical features.

4. Each lipoprotein type, irrespective of cause, is generally more successfully handled by a specific diet and therapeutic approach.

The typing system has four major limitations:

1. For distinction of type III, the diagnosis requires substantiation by determination of the ratio of cholesterol in VLDL divided by the plasma triglyceride as well as electrophoretic confirmation of beta-migrating VLDL. In addition, in subjects with mixed elevations of cholesterol and triglyceride, quantitation of LDL (LDL cholesterol) is important for accurate distinction.

2. The second major area where the typing system has limitations is genetics. No specific type of hyperlipoproteinemia should be considered genotypic; lipid and lipoprotein determinations cannot provide the diagnosis of a specific genetic disorder in a single patient; and there is increasing evidence of strong heterogeneity in lipoprotein patterns in first-degree relatives from families with monogenic familial hyperlipidemia.

3. The third area that the typing system did not cover was the evaluation of alpha-lipoprotein in the classification. It is now known that a low concentration of HDL cholesterol is an independent risk factor for coronary heart disease.

4. Finally, the present electrophoretic systems are limited in their ability to resolve other unusual lipoprotein fractions, such as HDL_c, beta-migrating VLDL, and IDL.

From lipids to lipoproteins: laboratory considerations

In the transformation of hyperlipidemia to hyperlipoproteinemia, lipid analyses alone can be used to determine the lipoprotein disorder with a fair degree of accuracy. If the plasma is clear, the triglyceride level is most likely to be either normal or near normal (less than 2000 mg/L). When triglyceride increases to about 3000 mg/L or higher, the plasma is usually hazy to turbid in appearance and is not translucent enough to allow for clear reading of newsprint through the tube. When plasma triglyceride is over 6000 mg/L, the plasma is usually opaque and milky (lipemic, lactescent). If chylomicrons are present, after several hours at 4° C a thick homogeneous ''cream'' layer may be observed floating at the plasma surface. As summarized in Fig. 30-13, a uniformly opaque plasma sample usually denotes a type IV pattern. An opaque plasma sample with a cream layer on top is usually consistent with the type V pattern. A thick chylomicron cream layer with generally clear plasma infranate is usually consistent with a type I pattern.

In patients with hypercholesterolemia without hypertriglyceridemia, most often with raised LDL levels, the plasma is clear but may have an orange-yellow tint since carotene is carried with LDL. After visual observation, which is ''simple and free,'' the diagnosis of the lipid abnormality can be made in as many as 90% of subjects by quantitation of plasma cholesterol and triglyceride alone.

By analyses of cholesterol and triglyceride, by use of lipoprotein electrophoresis, immunoelectrophoresis, the preparative ultracentrifuge, and differential precipitation of lipoproteins, six lipoprotein phenotypes can be differentiated. The more expensive and cumbersome analytical techniques are often beyond the range of feasibility for small laboratories. Although quantitation of the different lipoprotein classes provides increased information about the clinical and lipoprotein status of the patient, an extensive array of methods is required to facilitate the diagnosis of type III hyperlipoproteinemia and to document the presence or absence of the floating beta-lipoprotein in this disorder.

Most commonly, LDL cholesterol levels are conventionally measured by use of the preparative ultracentrifuge but can conveniently be estimated by the Friedewald formula:

$$\text{LDL cholesterol} = \text{Total cholesterol} - (\text{Triglyceride} + 5 + \text{HDL cholesterol})$$

This estimation requires measurement of HDL cholesterol by various precipitation techniques and is accurate in patients whose triglycerides are less than 4000 mg/L; triglyceride concentrations over 4000 mg/L lead to an inconsistency in the VLDL triglyceride; that is, the cholesterol ratio does not permit division by a fixed number, and the formula cannot be used with great accuracy.

Use of electrophoresis alone, without prior quantitation of cholesterol and triglyceride, is not a logical first step for the following reasons:

1. Except under very controlled conditions, electrophoretic patterns are difficult to quantitate consistently so that they can provide relative or absolute concentrations for the lipoprotein classes. Most of these difficulties relate to differing intensities of dye, application problems, and other variables that are difficult to control routinely in electrophoresis.

2. The ''sinking'' pre-beta-lipoprotein, a ubiquitous lipoprotein, Lp(a), often appears as a pre-beta band on electrophoresis but is not associated with any triglyceride elevation.

Secondary hyperlipoproteinemia

In general, lipoprotein quantification and typing alone will not distinguish the primary from the secondary forms (see Table 30-6). Even the diagnosis of a concurrent disorder that is likely to cause secondary hyperlipoproteinemia does not necessarily establish it as the cause of a patient's hyperlipoproteinemia. Reversal of the lipid abnormality accompanying treatment of the suspected causative disorder, however, is compelling evidence of the secondary nature of the hyperlipoproteinemia. Failure of such reversal to occur implies that the hyperlipoproteinemia may be primary and indicates the need for family screening.

Some disorders that are associated with hyperlipoproteinemia will be obvious from the patient's history and physical examination. Others will require blood or urine tests for diagnosis. If such a screening reveals no abnormalities, it is reasonable to assume that the patient has a primary hyperlipoproteinemia. Whether the hyperlipoproteinemia is established as being familial in origin will depend on the results of family screening.

Other forms of dyslipoproteinemias

In the familial lipoprotein disorders discussed in this chapter, one or more of the four lipoprotein fractions (chylomicron, VLDL, beta-migrating VLDL, and LDL) are present in elevated concentrations.

However, there are other lipoprotein abnormalities in addition to the hyperlipoproteinemias reviewed so far. There are three genetically determined disorders in which one or more of the lipoprotein families are absent from plasma or occur in concentrations that are extremely low. The first of these to be discovered was *abetalipoproteinemia,* in which chylomicrons, VLDL, and LDL are missing. The probable inherited defect is one involving synthesis of the major protein moiety of LDL, Apo B. Another disease is *hypobetalipoproteinemia* (familial LDL deficiency), in which no lipoproteins are missing but LDL concentrations are far below normal. In the third disorder, *familial hypoalphalipoproteinemia,* HDL circulates but contains an abnormal proportion of the two major HDL apolipoproteins. A defect in the synthesis of one Apo A-I is the probable locus affected by this rare mutation. With this abnormal high-density lipoprotein, patients with *Tangier disease* have low plasma HDL concentration, store cholesteryl esters in most parts of the body, and often have neuropathic changes for reasons that are not understood. Their large orange tonsils form an unforgettable part of the syndrome. All these diseases are usually detected initially because of a common manifestation, hypocholesterolemia.

Last, although not classified as a dyslipoproteinemia, *hyperalphalipoproteinemia* is a condition in which the HDL is elevated in the blood beyond two standard deviations for a given age and sex. This condition is under genetic influence, and these persons appear to have a longer life span than those with ''normal'' concentrations of HDL. There are no other clinical symptoms associated with this lipoprotein feature. A more detailed discussion of each of these lipoprotein abnormalities is provided below.

Abetalipoproteinemia. The rare disorder abetalipoproteinemia has five basic features: low plasma LDL concentration, malabsorption of fat, acanthocytosis, retinitis pigmentosa, and ataxic neuropathic disease. These are not specific for this lipoprotein condition. Acanthocytosis may occur in other diseases in which lipoproteins are not deficient. In abetalipoproteinemia LDL is absent, not merely deficient. The plasma cholesterol concentration does not exceed 800 mg/L and is likely to be no higher than 300 mg/L. This is accompanied by concentrations of triglycerides lower than those seen in any other disease, usually less than 200 mg/L. The total phospholipid concentration is also low, that is, less than 1000 mg/L. Both the phospholipid partition and the fatty acid composition of the plasma lipids are abnormal and are frequently reflected in similar abnormalities in erythrocytes and adipose tissue. The gastrointestinal problems of patients with abetalipoproteinemia are stereotypical. Fat malabsorption is present from birth, and the neonatal period is characterized by poor appetite, vomiting, loose voluminous stools, and little weight gain. The neuromuscular manifestations of abetalipoproteinemia are devastating. The cause of the neuromuscular abnormalities in abetalipoproteinemia is obscure. Attention has focused on the abnormal amount of ceroid pigment (lipofuscin) in the cerebellum and other tissues in this disorder.

In abetalipoproteinemia there is a functional disturbance in fat transport. Chylomicrons never enter the plasma, and net transport of endogenous glycerides in VLDL appears to be either absent or sustained at some unchanging minimum level. Heavy feeding of carbohydrate for days, which promotes a brisk rise in the level of plasma glycerides and VLDL in patients with Tangier disease and in nearly all other subjects, fails to do so in patients with abetalipoproteinemia. The defect in abetalipoproteinemia is not known. The most likely one is a failure to synthesize Apo B. The intracellular assembly point of Apo B containing lipoprotein is another possible site of dysfunction, or the defect may possibly lie in the secretion process. Whichever of these may be primary, none of these defects allow one to explain the many secondary manifestations of the disease.

Diagnosis of abetalipoproteinemia can be made in a patient with one of the abnormalities listed at the beginning of this section. The possibility of dysglobulinemia producing antibodies to LDL should also be kept in mind. The single most important laboratory test for screening is the plasma cholesterol determination. The finding of a subnormal value, particularly any concentration below 1000 mg/L, should be followed by a triglyceride determination and lipoprotein electrophoresis. The definitive diagnosis depends on immunochemical demonstration of the absence of LDL Apo B in the plasma. In all patients in whom the diagnosis is made, it is also important to check, in all obligate heterozygotes, the concentration of LDL by immunochemical or ultracentrifugal analysis. In familial hypobetalipoproteinemia, the heterozygote has lower than normal concentrations of LDL and Apo B.

Hypobetalipoproteinemia. Apparently unrelated to abetalipoproteinemia is another genetic disorder, hypobetalipoproteinemia, in which plasma LDL concentrations are about one tenth of normal. This lipoprotein abnormality is inherited as an autosomal dominant trait. The plasma total cholesterol concentrations can be as low as those seen in abetalipoproteinemia. The percentage of cholesterol esterified is normal. The triglyceride levels may be well within the normal range, but sometimes these are in the lower limits of accurate measurement. The phospholipid concentrations may vary from 100 to 1800 mg/L and are usually in the low-normal borderline region in most patients. Vitamin A and E concentrations are normal or low but, if low, are not decreased to the level seen in abetalipoproteinemia. A faint beta-lipoprotein band is visible on electrophoresis. HDL concentrations as measured by precipitation, preparative ultracentrifuge, or analytic ultracentrifuge are normal; VLDL is usually modestly reduced but is present.

LDL is present in serum as measured by immunoprecipitin tests. These measurements have suggested concentrations of LDL that are one eighth to one sixteenth of normal.

Hypoalphalipoproteinemia or analphalipoproteinemia. Tangier disease (familial HDL deficiency) is characterized by severe deficiency or absence of normal HDL in plasma and by the accumulation of cholesteryl esters in many tissues throughout the body, including liver, spleen, lymph nodes, thymus, intestinal mucosa, skin, and cornea. A combination of two features is pathognomonic: a low plasma cholesterol concentration in combination with normal or elevated triglyceride levels and hyperplastic orange-yellow tonsils and adenoid tissue. Some persons may exhibit peripheral neuropathy. The small amounts of HDL in Tangier plasma differ qualitatively and quantitatively from normal HDL, particularly with respect to apolipoprotein content (Apo A-I). The disorder appears to be attributable to an autosomal recessive gene affecting HDL synthesis or catabolism. Heterozygotes in families with known homozygotes can usually be identified by low HDL concentrations (50% below normal); they do not develop neuropathy and cholesteryl ester accumulation. Among the lipoprotein-deficiency states and indeed among all known diseases, the combination of very low cholesterol and elevated triglyceride concentrations gives Tangier disease a unique signature. Some patients may have normal triglyceride levels in the postabsorptive state, however, and may superficially resemble those with LDL deficiency. The plasma total cholesterol level ranges from about 400 to 1250 mg/L, within the range also observed in abetalipoproteinemia and hypobetalipoproteinemia. Individual variation in the plasma triglyceride levels is considerable and is highly contingent on diet. Substitution of carbohydrate for fat often paradoxically lowers the plasma triglyceride concentration in this disorder. The plasma lipoprotein pattern is distinctive: The alpha-lipoprotein band is absent, irrespective of the support medium used. Immunoelectrophoresis may occasionally generate a faint precipitin line of alpha-globulin mobility against anti-HDL antiserum. Most useful for detecting the small amounts of the A apoproteins is immunoanalysis of plasma with specific antiserums to the A-I and A-II apolipoproteins. The reactivity with anti-apo-A-II is generally stronger than with anti-apo-A-I. Estimation of the cholesterol content of the plasma lipoproteins after sequential preparative ultracentrifugation or after ultracentrifugation and selective heparin-manganese precipitation confirms the paucity of HDL.

In addition to HDL absence or deficiency, the following diseases must be excluded:

1. Familial deficiency of LCAT. Here HDL is very low, but the plasma cholesterol level is normal or high and most of the cholesterol is unesterified.

2. Obstructive liver disease, in which the plasma HDL and Apo A may be reduced to levels as low as those seen in Tangier disease. In this disorder the cholesterol level is not low, but high, and most of the cholesterol is not esterified. Appropriate tests of liver function should permit the correct diagnosis.

3. Severe malnutrition or hepatic parenchymal disease in which high-density lipoproteins are decreased; the decrease in cholesterol will also be associated with low triglyceride and LDL levels.

4. Acquired HDL deficiency attributable to dysglobulinemia, including possible development of antibodies to HDL.

5. Other storage diseases associated with foam cells and hepatosplenomegaly. In these conditions HDL levels are higher than those seen in Tangier disease, and typical tonsillar abnormalities are absent.

CLINICAL IMPLICATIONS OF HYPERLIPIDEMIA

Why should there be any concern for hyperlipidemia or hyperlipoproteinemia?

Hyperlipidemia is usually a symptomless biochemical state that, if present for a sufficiently long time, may be associated with the development of atherosclerosis and its complications. Occasionally hyperlipidemia may be associated with specific overt symptoms or signs directly attributable to the presence of hyperlipidemia. Examples are abdominal pain, pancreatitis, and the cutaneous manifestations of hyperlipidemia, such as xanthomas.

Coronary artery disease

CAD is almost always the result of atherosclerosis—hardening of the arteries. Coronary atherosclerosis primarily results from the accumulation of fatty deposits in the walls of coronary arteries, which leads to the formation of fibrous tissue in the vessel wall. CAD is the most common type of heart disease and the leading cause of death in the United States and many other countries. In the United States, an estimated 50% of the adult deaths each year are attributable to CAD.

The exact sequence of events that ultimately leads to severe obstruction of coronary arteries by atheromatous plaques is unknown. Most authorities believe that the process is lifelong, beginning in childhood or early adulthood. Fatty materials, principally cholesterol, move from the bloodstream directly into the lining of the blood vessel and are usually deposited near the beginning of the artery, particularly at the point where it branches. The fatty deposits cause a reaction within the vessel wall, and scarlike fibrous tissue may build up around the deposit, forming a plaque, which can become calcified. Generally, the plaque does not cover the entire circumference of the vessel but only a part of it, protruding into the lumen (the opening) and progressively narrowing it until in some cases total obstruction or occlusion takes place. Significant interference with the blood flow does not occur until well over half the vessel is occluded by a plaque. CAD may be present therefore without evidence of coronary heart disease.

Risk factors associated with coronary artery disease. CAD is much more common among males than among females, affecting five times more males than females under 45 years of age. The sex preference falls off rapidly between 45 and 60 years of age, although males in this group still have about twice as many heart attacks as females do. In the very elderly, the incidence is about the same for both sexes.

PRIMARY AND SECONDARY RISK FACTORS ASSOCIATED WITH CORONARY HEART DISEASE

Primary
- Genetic predisposition for CHD
- Hypertension
- Cigarette smoking
- Elevated total cholesterol (LDL cholesterol)
- Decreased HDL cholesterol

Secondary
- Lack of exercise
- Obesity
- Age
- Male sex
- Stress
- Diabetes mellitus
- Gout and hyperuricemia
- Renal failure patients receiving hemodialysis
- Subjects taking oral contraceptives

Although the basic cause of CAD is unknown, scientists have identified a number of factors associated with a distinct increase in the likelihood that a person will develop a heart attack later in life. These factors, which correlate with the presence of CHD, are spoken of as risk factors. Some risk factors are unavoidable, such as racial and genetic susceptibility, increased prevalence in males, and increased likelihood of having a heart attack as aging occurs. A number of known risk factors are, however, susceptible to modification. Particularly important among these are high blood pressure, cigarette smoking, and elevated serum cholesterol. Approximately 50% of those who have heart attacks are persons who have one or more of these three risk factors. Recent studies show that modification of some of these risk factors may decrease, in part, the original susceptibility to a future heart attack or stroke. A prudent approach therefore appears to be elimination or attenuation of these risk factors whenever they are identified. The box above lists the primary and secondary risk factors associated with coronary heart disease.

Reappraisal of the importance of HDL or HDL-cholesterol levels as predictors of CHD has created much interest. According to the Framingham data, there is a clear gradient of CHD incidence rates in relation to serum HDL-cholesterol concentrations. Persons with levels below 350 mg/L have eight times the CHD rate of persons with HDL-cholesterol levels of 650 mg/L or greater. In 1951 it was first pointed out that a high proportion of serum total cholesterol carried in HDL in babies was similar to that found in animals that appear to have a high resistance to atherosclerotic disease.

These findings tend to support the hypotheses on the atherogenicity of LDL and the seemingly protective effect of HDL. One possible mechanism for the protective role of HDL is that HDL competes with LDL for cell-surface receptors and may not be internalized as extensively as the latter lipoprotein, in effect reducing the cholesterol uptake and its accumulation in the smooth muscle cell. It is also possible that HDL, in concert with phospholipids, may remove cholesterol from cholesterol-laden tissues, such as the arterial intima. Although the mechanism of cholesterol removal from cells is still not known, it has been suggested that lecithin:cholesterol acyltransferase (LCAT) enzyme may play a role in the transport of cholesterol from peripheral tissues to the liver.

Etiology and pathogenesis of coronary artery disease

Initiating factors. The primary physiological event leading to the formation of atherosclerotic plaques is the increased deposition of macromolecules (including cholesterol) within endothelial cells of arteries. The factors that stimulate the deposition are called initiating factors.

Focal, hemodynamically induced endothelial injury with enhanced cell permeability is the probable determinant of the atheromatous process. Additional initiating factors, other than those associated with focal hemodynamics, include the release of platelet constituents, hypertension, carbon monoxide, antigen-antibody complexes, and hyperlipidemia. Cigarette smoking may prove to be an important initiating factor that exerts its effects either directly through an immune mechanism or indirectly through released platelet constituents or carbon monoxide.

Accelerating factors. Factors that influence the nature or rate of lesion development at all subsequent stages are conveniently classified under the generic term *accelerating factors*. Accelerating factors of note include hypercholesterolemia with an associated excess of LDL and disturbances in platelet function, hemostasis, and thrombosis. These accelerating factors, namely LDL and platelet factors, may influence the nature and rate of plaque development by stimulating the proliferation of smooth muscle cells and by stimulating to varying degrees the synthesis by smooth muscle cells of collagen, elastin, and the glycosaminoglycans. In addition, lipoproteins may significantly influence SMC lipid metabolism, including cholesterol synthesis, and the uptake and accumulation of lipids within smooth muscle cells.

Additionally, platelets may directly contribute to plaque growth by serving as components of mural thrombi. LDL may directly induce a spectrum of changes consistent with endothelial injury. In type II hyperlipoproteinemia, LDL has been shown as well to be associated with modified platelet function, which might influence the risk of thrombus formation.

TREATMENT

Control of serum cholesterol concentrations: effect of diet and drugs (Fig. 30-13)

In the treatment of hypercholesterolemia, the first thing that needs to be determined is whether there are any primary diseases that may be contributing to the elevation of the serum cholesterol concentration. Treatment then begins with determining whether the subject is overweight. Reduction to ideal body weight will lower plasma lipids, particularly plasma triglycerides. Further reduction requires special dietary considerations.

Knowing that the quantity and quality of dietary fat influences blood cholesterol concentration, one should reduce saturated fat and cholesterol intake. Cholesterol intake of less than 300 mg/day is recommended. The polyunsaturated fat/saturated fat (P/S) ratio is quantitatively more important than cholesterol consumption in lowering plasma cholesterol concentration.

In the average American diet, the P/S ratio is only about 0.3. For a therapeutic diet to lower plasma cholesterol in a patient with severe type II hyperlipidemia, it is recommended that the ratio be increased to a range of 1 to 1.2. To accomplish this, the patient must consume about 2 tablespoons of polyunsaturated fats or vegetable oil per day and drastically reduce the animal fat intake.

If diet fails to work or has minimum effect on cholesterol reduction in the blood, the next step is drugs. There are no precise criteria for drug therapy in the treatment of hyperlipidemia. Important considerations in prescribing medication are the age of the patient; the severity of the hyperlipidemia; the clinical manifestations, including those of arteriosclerosis; the family history; and the presence of coexisting diseases.

The combination of diet and drug therapy is more effective than drug therapy alone. In providing treatment for the patient, the physician should emphasize that diet and drugs are not curative; however, they can help control hyperlipidemia as long as the patient complies. There are currently about a half-dozen cholesterol-lowering drugs.

These drugs can be divided into several distinct classes.

Bile acid binders. Bile acid binders, including cholestyramine and colestipol, are nonabsorbable anion exchange resins. They interrupt the enterohepatic circulation by binding the bile salts and thus making them unavailable for cholesterol absorption. To compensate for this, the liver cells convert more cholesterol to bile salt, ultimately resulting in a fall in plasma LDL cholesterol concentration. Generally cholesterol will decrease about 25% and LDL-cholesterol will decrease about 35% with a slight increase in HDL-cholesterol.

Nicotinic acid (niacin). Nicotinic acid is a useful reserve drug in the treatment of hypercholesterolemia, hypertriglyceridemia, or mixed hyperlipidemia. It is a potent suppressor of LDL and VLDL levels in most hyperlipidemic states. It is a very effective cholesterol-lowering drug and is especially suitable for mixed or combined hyperlipidemia. Nicotinic acid will generally decrease cholesterol about 25% and triglycerides about 40% to 50%. The severity of side effects accompanying its rapid absorption restricts its wider use.

Probucol. Probucol is poorly absorbed from the gut and becomes concentrated in various lipid-rich tissues. It will predictably lower serum cholesterol levels by approximately 15% to 20%, with negligible influence on triglyceride levels and about a 20% to 25% decrease in HDL cholesterol.

Fibric acid derivatives. These compounds, for example, betafibrate, bind dietary cholesterol and cause a reduction in serum cholesterol.

βHMG CoA reductase inhibitors. These compounds act to suppress endogenous synthesis of cholesterol by inhibiting the enzyme β-hydroxy-β-methyl-glutaryl CoA reductase (βHMG CoA reductase). By decreasing the formation of mevalonic acid (see Fig. 30-10), cholesterol synthesis is also depressed. Examples of this type of inhibitor are lovestatin and synvinalin. Lovestatin has been shown to reduce plasma cholesterol by 30%, slightly reduce triglycerides, and slightly elevate plasma HDL. Data are not currently available to demonstrate either the long-term safety of lovestatin or its ability to lower the incidence of CHD.

Control of serum triglyceride concentrations: effects of diet and drugs

In general, weight reduction and low-fat diets are the most effective means of controlling serum triglyceride levels. In both the obese and hypertriglyceridemic individual, a low-calorie diet is quite successful. In the nonobese hyperlipidemic subject, the choice of the dietary regimen should be guided by the nature of the plasma lipoprotein abnormality. Exogenous fats, whether saturated or not, contribute to the formation of chylomicrons and should be limited in the diet of patients who have an excess of these particles in their fasting plasma (types I and V).

This is the basis for the treatment of familial lipoprotein lipase deficiency (type I). Medium-chain triglycerides, which are absorbed and transported to the liver through the portal system, do not contribute to chylomicron formation and may be used as a substitute for vegetable oil in this disease. The total daily fat intake should be kept between 25 and 35 g/day, and no alcohol should be permitted, since it may promote or enhance pancreatitis.

An excess of simple carbohydrates in the diet usually enhances endogenous hypertriglyceridemia, which is found in types IIb, III, IV, and V. Affected patients should be encouraged to limit their intake of simple sugars. In some cases, a dietary intake reduced in simple sugar, without a reduction in total caloric intake, may completely correct an endogenous hypertriglyceridemia. Alcohol intake should also be limited because it causes acute, transient hypertriglyceridemia.

Drugs are of no value in familial hyperchylomicronemia. In treating familial hypertriglyceridemia (type IV), clofibrate is a first-line drug if the response to the diet has been inadequate. In case of resistance to this medication, one is left with nicotinic acid as a major hypotriglyceridemic agent.

BIBLIOGRAPHY

Christie, WW: Lipid analysis, New York, 1980, Pergamon Press, Inc.

Cooper, GR: Introduction to analysis of cholesterol, triglyceride, and lipoprotein cholesterol: reference values and conditions for analyses. In Faulkner, WR, and Meites, S, editors: Selected methods of clinical chemistry, vol 9. Selected methods for the small clinical chemistry laboratory, Washington, DC, 1982, American Association for Clinical Chemistry.

Current status of blood cholesterol measurement in clinical laboratories in the United States: a report from the Laboratory Standardization Panel of the National Cholesterol Education Program, Clin Chem 34:193-201, 1988.

The Expert Panel: Report of the National Cholesterol Treatment Program expert panel on detection, evaluation, and treatment of high blood cholesterol in adults, Arch Intern Med 148:36-39, 1988.

Fredrickson, DS, and Lees, RS: Familial hyperlipoproteinemia. In Stanbury, JB, Wyngaarden, JB, and Fredrickson, DS, editors: The metabolic basis of inherited disease, ed 3, New York, 1982, McGraw-Hill Book Co.

Frederickson, DS, Levy, RJ, and Lees, RS: Fat transport in lipoproteins: an integrated approach to mechanisms and disorders, N Engl J Med 276:32-44, 94-103, 148-156, 215-224, 273-281, 1967.

Gotto, AM, Jr, editor: Plasma lipoproteins, New York, 1987, Elsevier Science Publishers.

Havel, RJ: Approach to the patient with hyperlipidemia, Med Clin North Am 66(2):319-333, 1982.

Havel, RJ: Familial dysbetalipoproteinemia, Med Clin North Am 66(2):441-454, 1982.

Levy, RJ, Rifkind, BM, Dennis, BH, and Ernst, N, editors: Nutrition, lipids, and coronary heart disease, New York, 1979, Raven Press.

Lewis, LA, editor: Handbook of electrophoresis, vol 3. Lipoprotein methodology and human studies, Boca Raton, Fla, 1983, CRC Press, Inc.

Lewis, LA, and Opplt, JJ, editors: Handbook of electrophoresis, vol 1. Lipoproteinemia: basic principles and concepts; vol. 2: Lipoproteins in disease, Boca Raton, Fla, 1980, CRC Press, Inc.

Miller, ME, and Lewis, B, editors: Lipoproteins, atherosclerosis and coronary heart disease, New York, 1981, American Elsevier Publishing Co, Inc.

Naito, HK, editor: Nutrition and heart disease, New York, 1982, Spectrum Publications, Inc.

Naito, HK: Reducing cardiac deaths with hypolipemic drugs, Postgrad Med 82:102-112, 1987.

National Heart, Lung, and Blood Institute Consensus Conference on Treatment of Hypertriglyceridemia, JAMA 251:1196-1200, 1984.

Rifkind, BM, and Levy, RI, editors: Hyperlipidemia: diagnosis and therapy, New York, 1977, Grune & Stratton.

Schanu, AM, and Spector, AA, editors: Biochemistry and biology of plasma lipoproteins, New York, 1986, Marcel Dekker.

CHAPTER 31

Alcoholism

CHARLES L. MENDENHALL
ROBERT E. WEESNER

OBJECTIVES

- Describe the metabolism of ethanol and the alterations of lipid, protein, and carbohydrate biochemistry that result from excess alcohol consumption.
- List and describe nutritional alterations associated with alcoholism and alcoholic liver disease.
- Describe the systemic diseases associated with alcoholism.
- State expected changes in serum levels of the following laboratory analytes in alcoholic liver disease: aspartate aminotransferase (AST), alanine aminotransferase (ALT), lactate dehydrogenase (LD), gamma-glutamyl transferase (GGT), alkaline phosphatase, 5′-nucleotidase, bilirubin, albumin, globulins, cholesterol, triglycerides, bile acids, and glucose.

KEY TERMS

alcohol withdrawal syndrome The clinical symptoms associated with cessation of alcohol consumption. These may include tremor, hallucinations, autonomic nervous system dysfunction, and seizures.

alcoholic cirrhosis Cirrhosis resulting from chronic excess alcohol consumption. The cirrhotic process in the liver is a pathological process in which progressive injury produces fibrotic bands or scar tissue that entraps liver cells and results in loss of normal microscopic lobular architecture (nodular regeneration).

alcoholic hepatitis Acute toxic liver injury associated with excess ethanol consumption. This is characterized by necrosis, polymorphonuclear inflammation, and in many instances Mallory bodies.

alcoholic ketoacidosis The fall in blood pH (acidosis) sometimes seen in alcoholics and associated with a rise in serum ketone bodies (acetone, beta-hydroxybutyric acid, and acetoacetic acid).

fatty liver Excessive accumulation of fat in the liver parenchymal cells, primarily neutral lipids, triglycerides, and cholesterol. Fatty liver predictably develops after exposure to a variety of hepatotoxins, of which ethyl alcohol is the most common.

fetal alcohol syndrome A group of fetal abnormalities resulting from maternal alcohol consumption during gestation.

hemochromatosis A disorder of iron metabolism characterized by excess iron deposits in tissues, such as those composing the liver, pancreas, and heart, leading to organ injury. The organ injury may manifest itself as cirrhosis, diabetes mellitus, or heart failure.

hepatic fibrosis The deposition of collagen and fibrous tissue in the liver before the development of nodular regeneration and cirrhosis.

kwashiorkor A nutritional disease resulting from protein deprivation. It is characterized by depleted visceral proteins and immunological dysfunction. Total calorie consumption may be deficient, adequate, or even excessive.

Mallory bodies (alcoholic hyalin) An eosinophilic cytoplasmic inclusion that accumulates in the liver cells. It is typically, but not always, associated with acute alcoholic liver injury (alcoholic hepatitis).

marasmus A nutritional disease resulting from calorie deprivation. It is characterized by weight loss and wasting of muscle mass and fat stores.

Wernicke-Korsakoff syndrome A disease of the central nervous system occurring in alcoholics and attributed to thiamin deficiency. Wernicke's encephalopathy consists of ocular disturbances, ataxia, and impaired mental functions. If untreated, it may become Korsakoff's syndrome, which includes memory impairment, confabulations, and deranged perception of time.

Alcoholism represents one of the most serious worldwide socioeconomic and health problems. In the United States, alcohol-related liver disease is the sixth leading cause of death.[1] An alcoholic is a person who consumes an amount of ethanol (ethyl alcohol) capable of producing pathological changes.[2] The amount of ethanol capable of producing disease depends on a variety of factors, including genetic predisposition,[3] malnutrition,[4] and concomitant viral infection of the liver (viral hepatitis).[5] For a susceptible person this may be as low as 35 g/day,[6] which is equivalent to the daily consumption of three cocktails made with 100-proof whiskey. However, for most persons the amount of alcohol necessary to produce disease is in excess of 80 g/day for at least 10 to 15 years. When this amount is converted to the quantity of alcoholic beverages consumed, it represents eight 12-ounce 6% beers, a liter of 12% wine, or a half pint of 80-proof whiskey per day. It should be pointed out that susceptible persons may develop disease with much smaller amounts of ethanol. When the ethanol level is below 7 g/day, however, a pathological condition does not develop even in susceptible persons.[6]

DIAGNOSIS OF ALCOHOLISM

Characteristically alcoholics may attempt to conceal their excessive drinking, so the history of consumption may be unreliable. Verification from the spouse is frequently necessary to determine an accurate drinking history.

Several laboratory tests have been proposed for use as diagnostic tools to confirm the excess consumption of ethanol.[7-9] As few as two drinks per day have resulted in significant increases in mean red cell volume (MCV)[7]; however, the changes are small, and it may be necessary to use baseline predrinking values to recognize the effect of ethanol.

Serum gamma-glutamyl transferase (GGT) has been shown to be readily inducible by a variety of compounds, including ethanol.[9] In the absence of other abnormal laboratory tests for liver disease, elevated levels of GGT may be useful for recognition of the heavy drinker in the predisease state.

BIOCHEMICAL AND METABOLIC ALTERATIONS

Ethanol metabolism

In humans ethanol is primarily an exogenous compound consumed in alcoholic beverages and readily absorbed from the entire gastrointestinal tract. Most of the absorbed ethanol is then degraded by oxidative processes, primarily in the liver, first to acetaldehyde and then to acetate. At least three enzyme systems are capable of ethanol oxidation: (1) alcohol dehydrogenase, (2) microsomal ethanol oxidizing system (MEOS), and (3) catalase. Only 2% to 10% is excreted unoxidized in the urine and lungs.

ADH appears to be the principal pathway for ethanol oxidation. This is especially true for acute intoxication:

$$\text{Ethanol} \xrightarrow[\text{NAD}^+ \quad \text{NADH} + \text{H}^+]{\text{Alcohol dehydrogenase}} \text{Acetaldehyde}$$

The initial step catalyzed by alcohol dehydrogenase appears to be rate limiting for the clearance of ethanol from the blood, but it is not specific for ethanol. Methanol, retinol (vitamin A), and a number of other alcohols are also oxidized by this enzyme. This fact is important clinically in the treatment of acute methanol poisoning.[10] In this instance, ethanol is given clinically to compete with methanol for binding sites on the ADH enzyme; thus oxidation of methanol to the toxic formaldehyde is delayed.

MEOS appears to be a secondary enzyme system for ethanol clearance:

$$\text{Ethanol} \xrightarrow[\text{NADP} + \text{H}^+ \quad \text{NADP} + \text{H}_2\text{O}]{\text{Oxygen} \quad \text{MEOS}} \text{Acetaldehyde}$$

This microsomal enzyme is found in the smooth endoplasmic reticulum (SER) of the hepatocytes. In the chronic alcoholic, there is an increase in MEOS activity.[11] The induction of MEOS activity is associated with increased activity of various other constituents of the SER involved with drug metabolism, such as other enzymes (5′-nucleotidase and GGT), cytochrome P-450 reductase, and cytochrome P-450. These changes have clinical significance because they render the alcoholic more resistant to the effects of many common sedatives and barbiturates, resulting in the necessity for a larger-than-normal dose when sedation is required. If, however, ethanol and barbiturates are consumed simultaneously, competitive enzyme inhibition results in reduced clearance[12] and abnormally high blood levels of barbiturates. Indeed, death has resulted from this interaction.

The role of catalase in the biological oxidation of ethanol is controversial.[13] In vitro in the presence of a peroxide (H_2O_2)-generating system, catalase has been shown to be capable of oxidizing ethanol by the following reaction:

$$\text{Ethanol} + \text{H}_2\text{O}_2 \xrightarrow{\text{Catalase}} 2\text{H}_2\text{O} + \text{Acetaldehyde}$$

It appears that the slow rate at which peroxide can be generated from NADPH oxidase or xanthine oxidase prevents catalase from contributing to more than 2% of the in vivo ethanol oxidation.

The product of all three oxidation pathways is acetaldehyde. It appears that various organ disorders and biochemical alterations are induced by either ethanol or acetaldehyde but not necessarily by both. For example, fetal alcohol syndrome appears to be an ethanol effect independent of acetaldehyde,[14] whereas liver fibrosis and collagen formation are more closely associated with acetaldehyde than with ethanol.[15] The increase in NADH with a concomitant decrease in NAD^+ associated with ADH activity may also produce sequential metabolic changes with clinical consequences. The increased availability of NADH results in increased lactate production, which can result in hyperlactacidemia.[16] The hyperlactacidemia reduces the capacity of the kidney to excrete uric acid, leading to a secondary hyperuricemia.[17] In susceptible persons this may aggravate or precipitate attacks of gout.[18]

Altered lipid, protein, and carbohydrate biochemistry

Lipids accumulate in most tissues in which ethanol is metabolized, and such accumulation results in fatty liver, fatty myocardium, fatty renal tubules, and so on. The mechanism appears to be multifactorial, resulting from both increased lipid accumulation and decreased lipid removal. The concomitant ingestion of ethanol and a fatty meal greatly increases intestinal uptake of fat into chyle[19] accelerates lipid synthesis. The increase in the $NADH/NAD^+$ ratio is associated with a rise in alpha-glycerol phosphate production, which favors hepatic triglyceride formation.[20] Although the secretion of serum lipoproteins is low relative to the lipid load accumulating in the liver, the total amount secreted is increased above normal, and alcoholic hyperlipemia may result.[21] This is a type IV hyperlipemia with increases in both serum triglycerides and cholesterol esters and with a lipoprotein density of less than 1.006 (very-low-density lipoproteins and chylomicron-like particles of hepatic origin).

Enzyme activities related to ethanol oxidation and drug metabolism are frequently altered. These changes may be produced either directly by enzyme induction or suppression or indirectly by the shift in reducing equivalents ($NADH/NAD^+$).

Once synthesized, the cellular release of proteins may be altered. As a consequence of the altered release of protein from the liver cell, one of the earliest liver changes seen in the alcoholic is the accumulation of protein,[22] which occurs concurrently with fat accumulation and contributes to the development of the enlarged liver (hepatomegaly) that is seen in more than 90% of alcoholics with liver disease.[23]

Gluconeogenesis is similarly impaired by ethanol by a variety of mechanisms. Gluconeogenesis is the process whereby glucose is formed from noncarbohydrate sources, that is, glycerol, pyruvate, and a number of amino acids. Because ethanol promotes glycerol-lipid formation and impairs amino acid transport, the availability of these compounds for gluconeogenesis is decreased. The excess NADH from ethanol metabolism favors lactate formation and decreases its availability for gluconeogenesis. Glutamic acid dehydrogenase activity is also diminished by elevated NADH levels, decreasing the availability of alpha-ketoglutarate, which is necessary for the transamination of amino acids before their conversion into glucose.

Typically the storage of glycogen in the liver is also diminished in alcoholics. This results from both poor dietary intake and the liver diseases so frequently associated with chronic alcoholism (fatty liver, alcoholic hepatitis, and alcoholic cirrhosis). The clinical implications of these changes in liver glycogen and impaired gluconeogenesis are discussed later along with insulin and other hormonal involvement in the development of alcoholic hypoglycemia.

In the absence of liver disease, malnutrition, or other predisposing conditions, ethanol itself produces only a slight increase or no change in blood glucose. If it is present, the peak rise in glucose occurs 30 to 45 minutes after ingestion and represents usually no more than a 7% to 10% increase.

NUTRITIONAL ALTERATIONS ASSOCIATED WITH ALCOHOLISM AND ALCOHOLIC LIVER DISEASE

Protein-calorie malnutrition

Alcoholic beverages are high in calories. Each gram of ethanol yields 7 kcal. However, these are low in nutritive value.[24] In fact, chronic alcoholism has long been associated with both malnutrition and liver disease. Malnutrition in the form of protein-calorie deficiency results from a variety of causes: (1) poor dietary intake; (2) malabsorption of consumed nutrients as a result of alcohol-induced gastritis, pancreatitis, or diarrhea; and (3) altered biochemical and physiological processes.

Protein-calorie malnutrition has been classified as marasmus or a kwashiorkor-like disease, depending on whether a caloric deficit (marasmus) or a protein deficit (kwashiorkor-like) predominates[25] (see Chapter 34). In a recent study of 284 alcoholics with alcoholic hepatitis, all patients had some degree of protein-calorie malnutrition, and 80% had features of both marasmus and kwashiorkor-like disease.[4]

Liver injury can develop in the absence of malnutrition. However, the more severe forms are invariably associated with severe malnutrition. Recognition of nutritional deficits appears to be important for the alcoholic with liver disease because correction of these abnormalities by appropriate nutritional therapy has been observed to increase survival and accelerate improvement of the liver injury.[26]

Vitamin abnormalities

Excessive alcohol use commonly leads to vitamin deficiency.[27] Not only is the liver a major storage depot for vitamins but also it converts vitamins into metabolically useful forms (see Chapter 36). Therefore injury to the liver alters vitamin metabolism. Vitamins can be released from necrotic liver cells that are lost from the body and not adequately replaced.[28] Alcohol can increase the body's need for vitamins because of increased nucleic acid synthesis and liver regeneration.[29] Fat-soluble vitamins (vitamins A, D, K, and E) may not be adequately absorbed by the in-

Table 31-1 Vitamins in the alcoholic

Vitamin	Classification of vitamin	Effect of alcohol on intestinal absorption	Incidence (%) of low serum values in alcoholic cirrhosis
A	Fat soluble	Decrease	12
B_1 (thiamin)	Water soluble	Decrease	58
B_2 (riboflavin)	Water soluble	None	6
Nicotinic acid	Water soluble	Not evaluated	33
B_6 (pyridoxine)	Water soluble	None	60
Folic acid	Water soluble	None	78
B_{12} (cyanocobalamin)	Water soluble	Decrease	25
C (ascorbic acid)	Water soluble	Not evaluated	25
D	Fat soluble	Not evaluated	?
E	Fat soluble	Not evaluated	15
K	Fat soluble	Not evaluated	?

testine because of the increased fat loss in the stool seen in some patients with liver disease.[27] In addition, the malnutrition associated with alcohol abuse can affect intestinal absorption.[30] The poor dietary intake of alcoholics can also lead to vitamin deficiency.[31] Because of the many interactions among the liver, alcohol, and vitamins, vitamin replacement plays a major role in the treatment of the alcoholic patient. Table 31-1 gives the incidence of low serum values of vitamins in persons with alcoholic cirrhosis. (See Chapter 36 for a discussion of the biochemistry and physiology of vitamins.)

Vitamin A. Vitamin A is absorbed as retinol, which must be oxidized to retinal with the help of alcohol dehydrogenase to be functional. Alcohol competitively inhibits the conversion of retinol to retinal in the liver.[32]

Vitamin B_1 (thiamin). Thiamin deficiency is encountered frequently in alcoholics.[27] The requirement for thiamin is greatest when carbohydrate is the major source of energy, which is sometimes found in the alcoholic. Thiamin deficiency can produce a neuropsychiatric disorder called *Wernicke-Korsakoff syndrome* in a small number of alcoholics; this syndrome can be genetically determined.[33] There is also increasing evidence that thiamin deficiency may contribute to other forms of brain injury in alcoholics.[32]

Vitamin B_2 (riboflavin). Vitamin B_2 is rarely a problem in the alcoholic.[27]

Nicotinic acid. A severe deficiency of nicotinic acid leading to pellagra can be found in the alcoholic population.[34]

Vitamin B_6 (pyridoxine). In alcoholic hepatitis the serum aspartate aminotransferase (AST, or glutamic-oxaloacetic transaminase [SGOT]) is greater than the serum alanine aminotransferase (ALT, or glutamic-pyruvic transaminase [SGPT]) in 90% of the patients.[35]

Folic acid. Folic acid deficiency is the most common vitamin deficiency among alcoholics.[27] Folate deficiency contributes to the megaloblastic anemia noted in alcoholics.[36] Alcohol does not inhibit intestinal absorption of food-associated folate.[37] Poor dietary intake, increased requirements, and increased excretion are believed to be the major causes of folate deficiency in the alcoholic.[32] In addition, alcohol may directly block folate metabolism.[38]

Vitamin B_{12} (cyanocobalamin). Although alcohol in some way inhibits vitamin B_{12} absorption at the ileum,[32] vitamin B_{12} deficiency is rarely a problem.[27] Because vitamin B_{12} is stored in the liver, acute liver cell necrosis of alcoholic hepatitis may actually produce a noticeable increase in serum B_{12} levels that parallels the severity of the liver injury.

Vitamin C (ascorbic acid). Vitamin C deficiency is not common in alcoholics and does not produce any clear metabolic problems. Low levels in alcoholics have been attributed to poor dietary intake.[39]

Vitamin D. Vitamin D deficiency is not a major problem among alcoholics.[34]

Vitamin E. Chronic alcoholics are susceptible to vitamin E deficiency because of low dietary intake and general malnutrition. Pure vitamin E deficiency is known to produce a testicular lesion, but whether this plays a role in alcoholic hypogonadism remains unclear.[32]

Vitamin K. Because the stores of vitamin K are small, biliary obstruction or severe parenchymal liver disease can produce bleeding abnormalities.[34] Alcoholics often have prolonged prothrombin time.[23] When these patients are given vitamin K parenterally, some show corrected prothrombin time, indicating vitamin K deficiency. Those who do not improve with vitamin K have liver disease too severe to use the vitamin.[34]

Mineral abnormalities (see Chapters 32 and 34)

Iron. The mechanism responsible for the mild to moderate iron accumulation in the liver of some patients with alcoholic liver disease is not clearly understood. First, some alcoholic beverages, especially red wines, contain iron. Making beer in iron kettles has led to hemochromatosis among the South African Bantu.[40] Finally, the effect of alcohol on folate could influence iron absorption. Folate deficiency is associated with ineffective red blood cell formation (erythropoiesis) and iron overloading.[41] Alcoholics with significant iron overload may, in fact, carry the gene for primary idiopathic hemochromatosis.[42] Some alcoholics have low iron levels because of poor dietary intake or chronic blood loss from the intestinal tract (gastritis), peptic ulcer, esophageal and gastric varices, and poor clotting.[34]

Zinc. Zinc, an essential micronutrient element, is important for RNA and DNA synthesis and for a variety of zinc-dependent enzymes, including alcohol dehydrogenase.[32] Vitamin A metabolism can also be altered in zinc deficiency.[43] Alcoholics can have low zinc levels from both reduced oral intake and increased urine loss and may suffer many symptoms seen in zinc-deficient states.[40] Further studies are needed to define the role of zinc in alcohol-related hypogonadism and other related problems.[32]

Lead. The lead content of some wines is high.[32] In addition, when old radiators or lead pipes are incorporated into homemade stills for the production of "moonshine" whiskey, lead poisoning can result.[40] Therefore an effort must be made to detect any signs or symptoms of lead poisoning (see Chapter 35) in alcoholics because of the increased risk.

Magnesium. Chronic alcoholics commonly have magnesium deficiency. This is a result of poor dietary intake, the increased urine loss that is a direct effect of alcohol, starvation ketosis, and vomiting. Magnesium is often low during the alcohol withdrawal syndrome and may contribute to the signs and symptoms of this disorder.[44]

Calcium (see Chapters 24 and 44). Patients with alcoholic liver disease may have low levels of calcium. Alcohol has been found to increase calcium excretion.[45] Patients with cirrhosis have lower calcium absorption, which may be a result of decreased liver hydroxylation of vitamin D.[46] If the alcoholic has fat malabsorption, the calcium in the intestinal lumen may form insoluble soaps with the fats, thus preventing intestinal absorption.

Phosphorus. Hypophosphatemia occurs in about 50% of hospitalized alcoholics.[47] Serum phosphorus is depressed acutely by as much as 10 to 15 mg/L after ingestion of carbohydrates. Alcoholics reach their lowest phosphorus levels on the second to fourth day of hospitalization.[32] The low phosphorus levels in alcoholics may be a result of alcohol itself, poor food intake, diarrhea, vomiting, magnesium deficiency, or use of aluminum-containing antacids.

PATHOLOGY: SYSTEMIC DISEASES OF THE ALCOHOLIC

Liver

Ethyl alcohol is a direct systemic toxin that produces injury to all tissues, depending on the dose and duration of exposure. The degree of injury varies among organ systems. The liver, the organ that is predominantly responsible for ethyl alcohol metabolism, develops the highest incidence and severity of injury.

Three types of pathological liver conditions develop, forming a continuum of disease from very mild reversible changes to life-threatening irreversible disease. The interrelationships among these changes are depicted in Fig. 31-1. The mild form is characterized by fatty infiltration (alcoholic fatty liver) and by some minimum degree of inflammation and necrosis.

In more severe cases of fatty liver, fibrosis may be present, especially around central venous channels. Clinically this is manifested by liver enlargement (hepatomegaly), tenderness over the liver, and anorexia. Laboratory changes are minimum with slight elevation of AST, bilirubin, and alkaline phosphatase levels. When available,

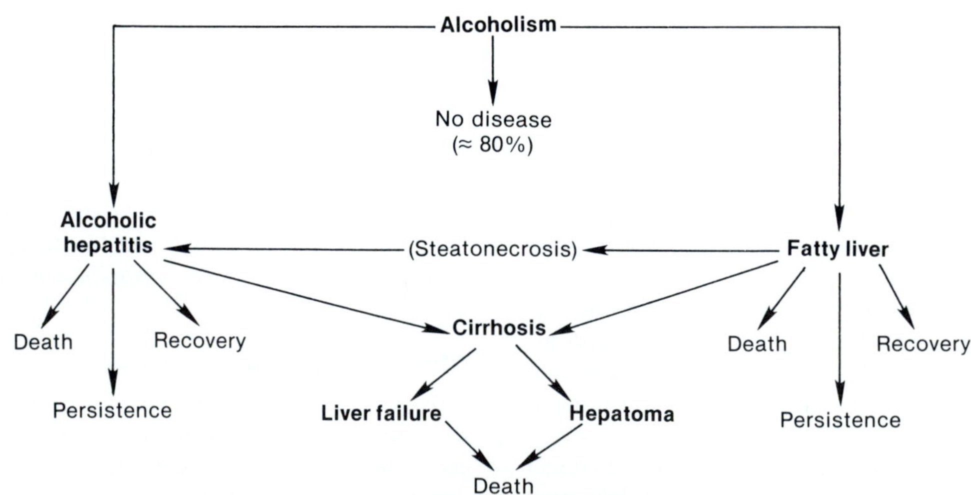

Fig. 31-1 Alcoholic liver disease.

the serum bile acids and dye-clearance tests (indocyanine green and Bromsulphalein) appear to be the most reliable parameters to detect early disease. Alcoholic fatty liver is usually considered a benign, reversible disease. However, sudden deaths have been observed in a small percentage of alcoholics, presumably from fat emboli.

The more severe toxic form of injury, alcoholic hepatitis, is also characterized by fatty liver and hepatomegaly. In addition, inflammation and necrosis are more extensive, and fibrosis is a prominent feature. In the large Veterans Administration Cooperative Study[23] of patients with clinical alcoholic hepatitis, the disease had progressed to cirrhosis in 54.7% of 97 patients in whom histological specimens were available. In a high percentage of patients (76.2%) an irregularly outlined homogeneous meshwork of eosin-staining material could be seen in the cytoplasm of the liver cells (Mallory bodies or alcoholic hyalin). Its presence is not diagnostic for alcoholic hepatitis because it has been seen in a variety of other liver conditions. However, in an alcoholic with liver disease its presence most likely represents alcoholic hepatitis. It has been suggested that alcoholic hyalin results from degenerating microfilaments, which then act as irritants and stimulate fibrosis.[48]

The clinical and laboratory manifestations of alcoholic hepatitis are shown in Table 31-2. None of the symptoms or changes are pathognomonic for this condition.

As the disease progresses, all standard laboratory tests for liver injury become abnormal to varying degrees. Of special note are the serum enzyme changes. Although AST is elevated in more than 75% of even the mild cases, the magnitude of the elevation rarely exceeds 10 times the upper limits of normal. The mean value observed was 84 ± 6 mU/mL. Similarly ALT was only mildly elevated; the ALT value was usually less than 100 mU/mL and almost invariably less than the AST. ALT values of 200 mU/mL or ALT values that are greater than the AST indicate a chronic, persistent hepatitis or chronic, aggressive hepatitis rather than uncomplicated alcoholic hepatitis. In these instances only a liver biopsy can differentiate the cause of liver pathosis.

The bilirubin level and prothrombin time are the tests that most reliably predict the severity and survival prognosis of liver disease. These tests, however, are not sensitive enough to diagnose minimum disease. In very mild cases results may be normal or near normal. However, as the disease progresses, the usefulness of the tests increases. Patients are at high risk when the bilirubin level becomes greater than 200 mg/L and the prothrombin time becomes more than 4 seconds prolonged. In this group mortality exceeds 75%.

The bilirubin and prothrombin time results are used to classify the severity of disease as mild, moderate, or severe.[23] Mild disease is present when the bilirubin is less than 50 mg/L with a normal or only slightly increased pro-

Table 31-2 Initial clinical features and complications of alcoholic hepatitis*[4]

Feature or complication	Severity of disease		
	Mild	**Moderate**	**Severe**
Anorexia†	46.2	63.0	65.7
Weight loss†‡	36.8	27.1	16.2
Fever	18.0	26.2	19.2
Hepatomegaly	85.9	97.1	88.9
Splenomegaly†	24.5	38.6	46.2
Infection	5.2	16.8	8.1
Pancreatitis	13.6	10.3	10.1
Gastrointestinal bleeding	10.4	7.5	14.1
Ascites			
Mild	18.7	19.4	17.2
Moderate	11.0	38.0	45.5
Severe	1.3	20.4	29.3
Combined†	31.0	77.8	92.0
Encephalopathy			
Grade 1	17.3	29.6	25.3
Grade 2	4.5	24.1	41.4
Grades 3 and 4	0.7	3.7	0
Combined†	22.5	57.4	66.7
Probability of surviving 1 yr§	0.91 ± 0.027	0.75 ± 0.045	0.46 ± 0.055

*The diagnoses of the various complicating features and conditions were not defined in the protocol but were determined by the clinical judgment of the participating investigator. Because *n* varied among severity groups (mild, 156; moderate, 108; severe, 99), values are expressed as a percentage of total. Data analysis consisted of an overall chi-squared test, which, when significant at the 0.05 level, was followed by a Bartholomew's test for order. The hypothesized order was $p_{mild} < p_{moderate} < p_{severe}$ except for weight loss where the presence of ascites reversed the hypothesized order.
†$p<0.005$ (Bartholomew's test).
‡The decreasing incidence of weight loss with increasing severity represented an increasing incidence of ascites.
§$p<0.006$. All pairwise comparisons (mild; moderate; severe) determined by normal distribution test statistics.

thrombin time. Moderate disease is present when the bilirubin is equal to or greater than 50 mg/L and the prothrombin time is normal to moderately elevated (less than 4 seconds prolonged). Severe disease is present when the bilirubin exceeds 50 mg/L and the prothrombin time is more than 4 seconds prolonged. As severity increases, the incidence of ascites and encephalopathy also increases, whereas 1-year survival decreases progressively from 91% to 46% (see Table 31-2).

The end stage of chronic alcoholic liver disease is the development of cirrhosis with extensive liver fibrosis, nodular regeneration, distortion of the liver architecture, and ultimately all the clinical complications of chronic liver disease, liver failure, and death.

In some instances cirrhosis may be hard to diagnose clinically, especially when active alcoholic hepatitis with acute necrosis and inflammation are also present. Indeed, of 155 patients recently studied with alcoholic hepatitis without cirrhosis, 13% had grade 2 to 4 hepatic encephalopathy, and 26% had moderate to severe ascites.[49] Only the liver biopsy specimen provided an accurate diagnosis of cirrhosis.

Pancreas

The association between alcohol and pancreatitis is at least as strong as the relationship between alcohol and liver disease is. At postmortem examination 18% to 47% of alcoholics dying of causes other than pancreatitis have some histological evidence of pancreatitis.[50] Alcohol is responsible for at least 30%, in some studies 60% to 90%, of the cases of pancreatitis in the United States.[51] Although there is a wide range of individual variation, the average duration of alcohol consumption before the initial diagnosis of pancreatitis is made is 18 years in men and 11 years in women. This suggests that women have an increased susceptibility. With each approximately 50 g increase in daily alcohol consumption, the risk of developing pancreatitis doubles.[50]

Alcohol stimulates the secretion of at least some acid from the stomach. When acid comes into contact with duodenal mucosa, it stimulates the flow of pancreatic juice.[52] The pancreatic juice induced by alcohol early in the disease process has a high concentration of protein, which precipitates and forms protein plugs. These plugs form within the pancreatic ductules and subsequently obstruct the ductules. This obstruction leads to ductular dilation, ductular proliferation, dilation of acinar tissue, and ductal sclerosis. In time the protein plugs may calcify, giving the x-ray picture of chronic calcific pancreatitis. As scarring of the pancreas continues, the protein or enzyme concentration in the pancreatic juice decreases until not enough enzymes are present for food digestion, and malabsorption occurs.[52,53] For reasons that are not clear, acute painful attacks of pancreatitis characterized by elevations in amylase and lipase may be superimposed on this chronic process.

Because a long interval of steady alcohol consumption is required before the first clinical attack of pancreatitis, most patients develop the first attack of pancreatitis between 30 and 40 years of age.[52] Once pancreatitis has been established, a fairly high percentage of patients who continue to drink have chronic pain.[54] After 80% to 90% of the pancreas has been destroyed, pancreatic insufficiency develops, requiring oral enzyme replacement.[55] In addition to the enzyme-secreting cells being destroyed, insulin- and glucagon-producing cells may also be destroyed. When this occurs, the alcoholic develops diabetes mellitus and the need for insulin. Because the glucagon level is also decreased and is too low to adequately stimulate glucose synthesis, these patients are brittle diabetics and are prone to hypoglycemic attacks.[52] This makes treatment of these patients difficult, especially if alcohol consumption continues.[56]

Less common systemic involvement

Heart. Among alcoholics the incidence of clinically significant heart disease appears to be much less than that of liver or pancreatic disease. It is estimated that 20% to 30% of alcoholics develop liver disease. In two large studies of alcoholics totaling 278 cases at autopsy, 11% to 15% had alcoholic cardiomyopathy.[57] The true incidence of cardiomyopathy is unknown. When it occurs, ethanol produces an acute depressant effect on myocardial function[58] with a progressive cardiac dilation and a low output type of cardiac failure. When death occurs, it is most often a result of gross cardiomegaly and chronic intractable cardiac failure complicated by embolic phenomena.

Central and peripheral nervous system. The most common manifestations of altered brain function in the alcoholic are the symptoms of inebriation, followed in frequency by the withdrawal syndromes of tremor, confusion, hallucinations, delirium tremens, and convulsions. These are usually reversible functional changes. Much less common but frequently irreversible are the metabolic disorders associated with protracted steady drinking and in most instances nutritional deficiencies. These include Wernicke-Korsakoff syndrome, peripheral neuropathy, amblyopia (loss of vision), cerebellar degeneration, cerebral degeneration, central pontine myelinolysis, and progressive dementia with demyelination of the corpus callosum (Marchiafava-Bignami disease). Both clinical and animal studies on mnemonic brain function indicate an ethanol-induced impairment.[59] (Refer to a neurology text for details on each of these syndromes.)

Cancer. Primary liver cancer, a rare tumor, is seen in 5% to 10% of alcoholics with alcoholic cirrhosis.[60] This is believed to be caused by the cirrhotic process. In addition to the hepatomas associated with alcoholic cirrhosis, prospective and retrospective epidemiological studies indicate that chronic alcohol consumption by itself is a cancer hazard.[61] Numerous clinical studies of alcoholics have arrived at this conclusion. Heavy drinkers have an increased incidence of cancer of the mouth, pharynx, larynx, and esophagus.[61] Alcohol consumption may also be related to incidence of cancer of the pancreas, the cardia of the stomach, and the colon.[62] Such epidemiological evidence is most abundant in the case of cancers of the mouth and pharynx, so that a heavy drinker and heavy smoker has a 15 times greater risk of developing oral malignancy than a nondrinker and nonsmoker. It appears that the cancer risk from ethanol is both time and dose related.[62]

Endocrine effects. The endocrine effects of alcohol vary depending on whether the alcohol intake is acute or chronic.[63] Moderate alcohol consumption has little effect on cortisol from the adrenal gland.[64] Alcohol blood levels in excess of 1 g/L, however, may produce an elevated plasma cortisol level. Occasionally chronic alcoholics will show clinical features of Cushing's syndrome and high cortisol levels.[65] Abstention causes the cortisol level to return to normal in 2 to 3 weeks.[63]

Some alcoholics have evidence of pituitary insufficiency. Low blood glucose normally produces a rise in cortisol. This response is decreased or absent in 25% of chronic alcoholics, because the pituitary gland does not secrete enough adrenocorticotropic hormone to stimulate the adrenal gland.[63] Low levels of growth hormone and prolactin after low blood glucose stimulation are further evidence for pituitary insufficiency.[66] The low testosterone levels seen in alcoholics are now believed to be related to pituitary insufficiency.[63]

Hematopoietic effects. The effects of alcohol on the blood and bone marrow are the result of both a direct toxic effect and associated nutritional deficiencies[63] (see Chapters 34 an 36). Not only do alcoholics have low folate levels, but vitamin B_{12} absorption is inhibited by alcohol. Each of these factors alone or in combination can lead to abnormally large red blood cells and megaloblastic anemia.[34,36]

Iron metabolism is affected by alcohol in a number of ways. Alcohol can produce chronic blood loss (ulcers, gastritis, variceal bleeding) leading to iron-deficiency anemia.[34] An increase in serum and tissue iron can also be seen with excess alcohol consumption and causes abnormal iron use within the bone marrow. Finally, alcohol excess can cause acute hemolytic anemia.

Inhibition of the bone marrow by alcohol can result in a low number of circulating white blood cells[67] and platelets.[68] Alcohol also can impair the function of white blood cells,[69] and alcoholic cirrhosis impairs platelet function,[70] which may lead to increased susceptibility to infection and bleeding. Vitamin K deficiency, seen in alcoholics, causes a prolonged prothrombin time because of the lack of vitamin K–dependent clotting factors normally produced in the liver. When liver disease is severe, inadequate clotting factors are produced even when vitamin K is present.

Immune system. Ethanol is known to affect the immune system. Increased circulating B-cell antibodies have been identified in most patients with alcoholic liver disease[23] (see discussion below). Specific antibodies against alcoholic hyalin have been identified in patients with alcoholic hepatitis,[71] which may explain the frequent elevation in IgA.[72] The primary changes have been observed in the cell-mediated immune system. Total numbers of T cells are decreased[73]; their ability to synthesize DNA is impaired (lymphocyte transformation)[74]; their response to antigens and mitogens is suppressed[74]; and their cytotoxic properties are increased.[75] These are manifested clinically by a low lymphocyte count, anergy to skin tests,[4] altered response to vaccinations,[23,76] and increased susceptibility to infections.

These immunological alterations may be partly responsible for some of the liver cell injury and progression of the disease seen in patients with alcoholic hepatitis.[23,75]

CHANGE OF ANALYTE IN DISEASE (Table 31-3)

Enzymes

Increases in serum enzyme levels in liver disease may result from leakage of enzymes into the serum because of damaged cell membranes (AST, ALT, and lactate dehydrogenase [LD]) or from increased enzyme production (alkaline phosphatase and GGT).

Aspartate aminotransferase and alanine aminotransferase. Serum AST and ALT are two serum enzymes that are found in the liver and readily leak from the cells during necrosis and cell injury. Although alcoholic liver injury is characterized by liver cell necrosis and inflammation, the increase in these enzymes is minimum to moderate, with the rise in AST almost always exceeding that observed in ALT. Abnormalities in AST are common and occur early, but the magnitude of the change may not parallel the clinical severity of the liver injury.[23] Explanations for this minimum response to injury are inadequate. Because pyridoxal phosphate is required for transamination reactions, deficiency in pyridoxine may in some alcoholics contribute to the diminished response.

Alkaline phosphatase. In alcoholic liver disease the increase in alkaline phosphatase activity tends to parallel the changes seen in bilirubin.

5′-Nucleotidase. 5′-Nucleotidase is not as sensitive to liver injury as alkaline phosphatase. It is measured along with alkaline phosphatase to determine whether a rise in the latter enzyme is the result of hepatobiliary disease or bone disease.

Gamma-glutamyl transferase. Increases in serum levels of GGT have been reported in alcoholics with minimum or no liver disease. This has been attributed to microsomal enzyme induction rather than liver injury[77] (see p. 485).

Bilirubin

Although alcohol does not directly alter bilirubin metabolism, the liver injury produced by alcohol does. Jaundice, the clinical sign of elevated bilirubin, is seen in about 60% of patients with alcoholic hepatitis.[23] Bilirubin elevation,

Table 31-3 Prevalence of laboratory changes in alcoholic hepatitis[31]

Laboratory test	Direction of analyte change	Incidence of abnormal results, % (ULN × X̄ of group)*		
		Group I	Group II	Group III
Hemoglobin	Decrease	74 (0.70)	85 (0.66)	93 (0.60)
Hematocrit	Decrease	76 (0.80)	83 (0.76)	95 (0.70)
MCV	Increase	73 (1.1)	83 (1.1)	90 (1.1)
	Decrease	2	8	0
WBC	Increase	13	7 (1.1)	54 (1.24)
	Decrease	11 (0.8)	9	8
AST (SGOT)	Increase	79 (2.1)	98 (3.1)	91 (2.5)
ALT (SGPT)	Increase	62 (1.9)	73 (1.9)	56 (1.9)
Total bilirubin	Increase	53 (1.6)	100 (13.5)	100 (18.7)
Alkaline phosphatase	Increase	67 (1.4)	100 (2.3)	88 (1.9)
Prothrombin time	Increase	65	90	100
BUN	Increase	3	17 (0.7)	36 (1.4)
	Decrease	3 (0.5)	0	0
Creatinine	Increase	2 (0.60)	17 (0.76)	29 (1.35)
	Decrease			
Albumin	Decrease	36 (0.74)	90 (0.54)	96 (0.48)
Cholylglycine	Increase	85 (8.2)	100 (25.2)	100 (15.1)
Immunoglobulins				
IgG	Increase	49 (1.1)	72 (1.3)	83 (1.6)
IgA	Increase	84 (1.7)	98 (2.6)	93 (2.8)
IgM	Decrease	17 (0.66)	26 (0.73)	43 (0.95)

*Data are expressed as the change in analyte, and the incidence of abnormal occurrence is expressed as the percentage of total and the magnitude of the change (times upper limit of normal [UNL] for mean); $n = 89$ in group I, 58 in group II, and 37 in group III.

along with a prolonged prothrombin time unresponsive to vitamin K, and depressed serum albumin correlate well with the severity of alcoholic hepatitis. In some cases of alcoholic hepatitis the bilirubin will rise along with the alkaline phosphatase level so that common bile duct obstruction is suspected.[24] Because the patient with alcoholic hepatitis does not tolerate surgery well, a diagnostic procedure such as endoscopic or percutaneous cholangiography should be done to confirm the diagnosis of extrahepatic obstruction before surgery.

Proteins

Albumin. Many tests used to evaluate liver disease measure alterations in biochemical functions of the liver. Because serum albumin is synthesized and secreted by the liver, its serum concentration has been used as a test of liver function. Serum levels of albumin are depressed in all forms of liver disease, including those associated with alcoholism. It is not uncommon to find values below 20 g/L in severe disease.[23] Of 111 patients with severe alcoholic hepatitis in the Veterans Administration Cooperative Study,[23] 18% had values in this range, whereas only 1% had values within the normal range. In patients with alcoholic liver disease the decline in serum albumin may not be related only to the decreased synthesis that is caused by liver disease. Many of these patients have moderate to severe malnutrition; thus the decreased albumin level may reflect decreased visceral proteins resulting from poor dietary protein intake (kwashiorkor). Further, many of these patients have ascites with expanded extravascular pools. The low serum albumin in these persons may represent a shift of albumin from the intravascular to the expanded extravascular space.

Globulins. Changes in serum globulins are not truly a test of liver function, because the gamma globulins that change during liver disease are produced by lymphocytes and plasma cells in response to antigenic stimulation. This is more a test of immune changes in liver disease. Serum globulins usually rise as the albumin falls. The increase in many instances may be disproportionately high. Of the three types of gamma globulins (IgG, IgA, and IgM), IgA increases most readily in alcoholic liver disease. It is elevated in 90% of cases, with a mean increase of 118%.[23] IgM increases the least and is elevated in 25% of cases. IgG is intermediate at 64% with a mean increase of 25%. The mechanism for the increased globulins is not well understood. The IgG increase seems to result from failure of the damaged liver to remove intestinal bacterial antigens, particularly *Escherichia coli*.[78]

The IgA elevation is more difficult to explain. The changes show no correlation with intrahepatic shunting and are independent of IgG. Some have suggested that they represent antibodies to alcoholic hyalin.[72]

Lipids

Triglycerides and cholesterol. Triglyceride and cholesterol changes in serum are frequently observed and are discussed on p. 486. These changes are not commonly used in the diagnosis of alcoholic liver disease.

Bile acids. Because the liver is responsible for almost all phases of bile acid metabolism, it is not surprising that bile acids are very sensitive indicators of even minimum liver injury. In alcoholic liver disease, serum bile acids, especially cholic acid conjugates, have been correlated with biopsy findings.[79] Fatty liver alone was not associated with elevated levels, whereas alcoholic hepatitis and cirrhosis were accompanied by significant increases in serum levels. In the Veterans Administration Cooperative Study[23] on alcoholic hepatitis, cholylglycine was abnormal more often than any other parameter of liver injury. These observations indicate that cholylglycine can be useful in detecting the presence of early injury.

Although results of most laboratory tests used in the diagnosis of liver injury are abnormal to varying degrees in alcoholic liver disease, none of these tests is pathognomonic. In many instances only a liver biopsy can establish the diagnosis with certainty.

Carbohydrates

Hyperglycemia. Only rarely has alcohol been responsible for producing gross glucose intolerance or frank diabetes.[56] Hepatocellular damage, regardless of its cause, is an important cause of glucose intolerance and may play a major role in the hyperglycemia found in some alcoholics.[80] Alcohol-induced elevation in cortisol, especially when sufficient to produce pseudo-Cushing's syndrome, is another contributing factor in alcoholic hyperglycemia.[81]

Hypoglycemia. Although hypoglycemia may occur in severe liver disease with fulminant hepatic failure, it may also occur in association with alcohol use and a relatively normal liver.[63] Hypoglycemia may be the cause of sudden death among alcoholics, with an 11% mortality in adults and a 25% mortality in children.[82] This form of hypoglycemia (alcohol-induced fasting hypoglycemia) occurs in chronically malnourished alcoholics or when moderate to large amounts of alcohol are consumed after a 6- to 36-hour fast.[56] The patients are stuporous or comatose, smell of alcohol, and are often hypothermic. The diagnosis is made on the clinical findings of hypoglycemia and elevated blood alcohol. Lactic acidosis is not uncommon.[63] Plasma insulin levels are low and glucagon levels high.[83] Although growth hormone and cortisol levels are elevated, the elevation is still less than expected for the severity of the hypoglycemia.[83] The mechanism for the hypoglycemia is probably multifactorial, with alcohol inhibition of gluconeogenesis playing a major role.[84] Other contributing factors include a relatively mild adrenocortical insufficiency and a defect in growth hormone secretion.[63]

Another form of hypoglycemia occurs when alcohol is consumed with carbohydrates (alcohol-induced reactive hypoglycemia).[56] Alcohol has the capacity to potentiate the insulin-stimulating properties of glucose.[85] Therefore alcohol consumed at a meal may potentiate insulin secretion and lead to nocturnal hypoglycemia. Thus large amounts of sweetened coffee used to sober up an alcoholic could cause a rapid and severe reactive hypoglycemia that could contribute to an accident during the drive home.

Alcohol can potentiate the action of hypoglycemic medications, especially insulin (alcohol potentiation of drug-induced hypoglycemia).[56] Although alcohol inhibits the liver's ability to mobilize glucose from glycogen (gluconeogenesis), the exact mechanism for hypoglycemia associated with hypoglycemic drugs is unknown. There is also slender evidence that reactive hypoglycemia may occur more frequently in chronic alcoholics than in controls (''essential'' reactive hypoglycemia in alcoholics).[56]

Alcoholic ketoacidosis. Ketoacidosis is an uncommon condition occurring in nondiabetic alcoholics.[63] Patients often have abdominal pain, vomiting, and a history of no recent food intake. The patients are acidotic but conscious, with high levels of serum ketones, salt and water depletion, and normal glucose levels.[86] The mechanism for alcoholic ketosis is uncertain.[63] Serum insulin concentrations are low with high cortisol levels and mild elevations of growth hormone. Cortisol may be ketogenic when insulin levels are low, and alcohol may be directly converted to ketone bodies through acetate production.[63]

TESTS OF LIVER FUNCTION

Indocyanine green dye excretion

Compared with dyes no longer in common use, such as bromosulphothalein (Bromsulphalein [BSP]), indocyanine green is free of side effects; as with BSP, it is rapidly excreted by the liver. Unlike BSP and bilirubin, indocyanine green is not conjugated by the liver before excretion into the bile. This dye has the advantage of being monitored accurately and continuously with a dichromatic ear densitometer; thus one can eliminate venous blood sampling.[87] Disappearance of the dye is believed to be related to hepatic blood flow. Although the test is not used routinely, it is probably the current dye excretion test of choice.

REFERENCES

1. US Bureau of the Census: Statistical abstract of the United States, 1975, Washington, DC, 1975, US Government Printing Office.
2. Criteria Committee, National Council on Alcoholism: Criteria for the diagnosis of alcoholism, Ann Intern Med 77:249-258, 1972.
3. Bailey, RJ, Krasner, N, Feddleston, ALW, et al: Histocompatibility antigens, autoantibodies, and immunoglobulins in alcoholic liver disease, Br Med J 2:727-729, 1976.
4. Mendenhall, CL, Anderson, S, Weesner, RE, et al: Protein-calorie malnutrition associated with alcoholic hepatitis, Am J Med 76:221-222, 1984.
5. Hall, P, editor: Alcoholic liver disease: pathobiology, epidemiology, and clinical aspects, New York, 1985, John Wiley & Sons, Inc.
6. Rydberg, U, and Skerfuing, S: Toxicity of ethanol: a tentative risk evaluation. In Gross, EM, editor: Alcohol intoxication and withdrawal, New York, 1977, Plenum Publishing Corp.
7. Kristensson, H, Trell, E, Eriksson, S, et al: Serum-γ-glutamyltransferase in alcoholism, Lancet 1:609, 1977.
8. Lieber, CS: Pathogenesis and early diagnosis of alcoholic liver injury, N Engl J Med 298:888-893, 1978.
9. Dragosics, B, Ferenci, P, Pesendorfer, F, et al: Gamma-glutyamyl-

transpeptidase (GGTP): its relationship to other enzymes for diagnosis of liver disease. In Popper, H, and Schaffner, F, editors: Progress in liver disease, New York, 1976, Grune & Stratton, Inc.
10. Mendenhall, CL, and Weesner, RE: Alcohols and glycols. In Hanenson, IB, editor: Quick reference to clinical toxicology, Philadelphia, 1980, JB Lippincott Co.
11. Lieber, CS, and DeCarli, LM: Effect of drug administration on the activity of the hepatic microsomal ethanol oxidizing system, Life Sci 9:267-276, 1970.
12. Lieber, CS, and DeCarli, LM: The role of the hepatic microsomal ethanol oxidizing system (MEOS) for ethanol metabolism in vivo, J Pharmacol Exp Ther 181:279-287, 1972.
13. Freytmans, E, and Leighton, F: Effects of pyrazole and 3-amino-1,2,4-triazole on methanol and ethanol metabolism by the rat, Biochem Pharmacol 22:349-360, 1973.
14. Mathinos, PR: Determination of the proximal teratogen of the fetal alcohol syndrome in CD/Mice, doctoral dissertation, Cincinnati, 1982, University of Cincinnati.
15. Mendenhall, CL, and Kromme, C: Acetaldehyde as a stimulus to liver collagen synthesis in the alcoholic (abstract), Paper presented at the thirty-second annual meeting of the American Association for the Study of Liver Diseases, Chicago, Nov. 1981, 1:531, 1981.
16. Lieber, CS, Jones, DP, Losowsky, MS, et al: Interrelation of uric acid and ethanol metabolism in man, J Clin Invest 41:1863-1970, 1962.
17. Olin, JS, Devenyi, P, and Weldon, KL: Uric acid in alcoholics, Q J Stud Alcohol 34:1202-1207, 1973.
18. Newcombe, DS: Ethanol metabolism and uric acid, Metabolism 21:1193-1203, 1972.
19. Mendenhall, CL, Greenberger, PA, Greenberger, JC, et al: Dietary lipid assimilation after acute ethanol ingestion in the rat, Am J Physiol 22:377-382, 1974.
20. Nikkila, EA, and Ojala, K: Role of hepatic L-α-glycerophosphate and triglyceride synthesis in production of fatty liver by ethanol, Proc Soc Exp Biol Med 113:814-817, 1963.
21. Losowsky, MS, Jones, DP, Davidson, CS, et al: Studies of alcoholic hyperlipemia and its mechanism, Am J Med 35:794-803, 1963.
22. Baraona, E, Leo, M, Borowsky, SA, et al: Alcoholic hepatomegaly: accumulation of protein in the liver, Science 190:794-795, 1975.
23. Mendenhall, CL, and the Cooperative Study Group on Alcoholic Hepatitis: Pathogenesis, diagnosis, and treatment of alcoholic hepatitis, Clin Gastroenterol 10:417-441, 1981.
24. Pirola, RC, and Lieber, CS: The energy cost of the metabolism in drugs, including ethanol, Pharmacology 7:185-196, 1972.
25. Blackburn, GL, Bristrian, BR, Maini, BS, et al: Nutritional and metabolic assessment of the hospitalized patient, J Parent Ent Nutr 1:11-22, 1977.
26. Nosrallah, SM, and Galambos, JT: Amino acid therapy of alcoholic hepatitis, Lancet 2:1276-1277, 1980.
27. Leevy, CM, Thompson, A, and Baker, H: Vitamins and liver injury, Am J Clin Nutr 23:493-498, 1970.
28. Frank, O, Baker, H, and Leevy, CM: Vitamin binding capacity of experimentally injured liver, Nature 203:302-303, 1964.
29. Leevy, CM, ten Hove, W, Frank, O, et al: Folic acid deficiencies and hepatic DNA synthesis, Proc Soc Exp Biol Med 117:746-748, 1964.
30. Halsted, CH, Robles, EA, and Mezey, E: Decreased jejunal uptake of labeled folic acid (^{3}H]-PGH) in alcoholic patients: roles of alcohol and nutrition, N Engl J Med 285:701-706, 1971.
31. Leevy, CM, Cardi, L, Frank, O, et al: Incidence and significance of hypovitaminemia in a randomly selected municipal hospital population, Am J Clin Nutr 17:259-271, 1965.
32. Thomson, AD, and Majumdor, SK: The influence of ethanol on intestinal absorption and utilization of nutrients, Clin Gastroenterol 10:263-293, 1981.
33. Blass, JP, and Gibson, GE: Abnormality of a thiamine-requiring enzyme in patients with Wernicke-Korsakoff syndrome, N Engl J Med 297:1367-1370, 1977.
34. McIntyre, N, and Morgan, MY: Nutritional aspects of liver disease. In Wright, R, Alberti, KGMM, Karran, S, et al, editors: Liver and biliary disease, Philadelphia, 1979, WB Saunders Co.
35. Cohen, JA, and Kaplan, MM: The SGOT/SGPT ratio: an indicator of alcoholic liver disease, Dig Dis Sci 24:835-838, 1979.
36. Herbert, V, Zalusky, R, and Davidson, CS: Correlation of folate deficiency with alcoholism and associated macrocytosis, anemia, and liver disease, Ann Intern Med 58:977-988, 1963.
37. Baker, H, Frank, O, Zetterman, R, et al: Inability of chronic alcoholics with liver disease to use food as a source of folates, thiamin and vitamin B_6, Am J Clin Nutr 28:1377-1380, 1975.
38. Sullivan, LW, and Herbert, V: Suppression of hematopoiesis by ethanol, J Clin Invest 43:2048-2062, 1964.
39. Beattie, AD, and Sherlock, S: Ascorbic acid deficiency in liver disease, Gut 17:571-575, 1976.
40. Flink, EB: Mineral metabolism in alcoholism. In Kissin, F, and Begleiter, H, editors: The biology of alcoholism, vol I, Biochemistry, New York, 1971, Plenum Publishing Corp.
41. Celada, A, Rudolph, H, and Donath, A: Effect of experimental chronic alcohol ingestion and folic acid deficiencies on iron absorption, Blood 54:906-915, 1979.
42. Powell, LW: The role of alcoholism in hepatic iron storage disease, Ann NY Acad Sci 252:124-134, 1979.
43. Smith, JC, McDaniel, EG, Fan, FF, et al: Zinc: a trace element essential in vitamin A metabolism, Science 181:954-955, 1973.
44. Wolfe, SM, and Victor, M: The relationship of hypomagnesemia to alcohol withdrawal seizures and delirium tremens, Ann NY Acad Sci 162:973-984, 1969.
45. Kalbfleisch, JM, Lindeman, RD, Ginn, HE, et al: Effects of ethanol administration on urinary excretion of magnesium and other electrolytes in alcoholic and normal subjects, J Clin Invest 42:1471-1475, 1963.
46. Jung, RT, Davie, M, Chalmers, JO, et al: Abnormal vitamin D metabolism in cirrhosis, Gut 19:290-293, 1978.
47. Knochel, JP: The pathophysiology and clinical characteristics of severe hypophosphatemia, Arch Intern Med 137:203-220, 1977.
48. Popper, H: The problem of hepatitis, Am J Gastroenterol 55:335-346, 1971.
49. Mendenhall, CL: Unpublished data presented at the fifth study group meeting, Veterans Administration Cooperative Study No 119 on Alcoholic Hepatitis, Chicago, 1982.
50. Durbec, JP, and Sarles, H: Multicenter survey of the etiology of pancreatic diseases: the relationship between the relative risk of developing chronic pancreatitis and alcohol, protein, and lipid consumption, Digestion 18:337-350, 1970.
51. Camerson, JL, Zuidema, GD, and Margolis, S: A pathogenesis for alcoholic pancreatitis, Surgery 77:754-763, 1975.
52. Banks, PA, editor: Pancreatitis, ed 1, New York, 1979, Plenum Publishing Corp.
53. Sarles, H: Chronic calcifying pancreatitis—chronic alcoholic pancreatitis, Gastroenterology 66:604-616, 1974.
54. Amman, RW, Largiades, F, and Akovbiantz, A: Pain relief by surgery in chronic pancreatitis? Relationship between pain relief, pancreatic dysfunction, and alcohol withdrawal, Scand J Gastroenterol 14:209-215, 1979.
55. DiMagno, EP, Go, VLW, and Summerskill, WHJ: Relationship between pancreatic enzyme outputs and malabsorption in severe pancreatic insufficiency, N Engl J Med 288:813-815, 1973.
56, Marks, V: Alcohol and carbohydrate metabolism, Clin Endocrinol Metab 7:333-349, 1978.
57. Schnek, EA, and Cohen, J: The heart in chronic alcoholism: clinical and pathologic findings, Pathol Microbiol 35:96-104, 1970.
58. Fisher, VJ, and Kavaler, F: The action of ethanol upon the contractility of normal ventricular myocardium. In Rothschild, MA, Oratz, M, and Schreiber, SS, editors: Alcohol and abnormal protein biosynthesis, Elmsford, NY, 1975, Pergamon Press, Inc.
59. Tamarin, JS, Weiner, S, Poppen, R, et al: Alcohol and memory, Am J Psychol 127:1659-1667, 1971.
60. Sherlock, S: Alcohol and the liver: treatment, early recognition. In Sherlock, S, editor: Diseases of the liver and biliary system, ed 6, London, 1981, Blackwell Scientific Publications, Ltd.
61. Keller, AZ, and Terris, M: The association of alcohol and tobacco with cancer of the mouth and pharynx, Am J Public Health 55:1578-1585, 1965.
62. Williams, RR, and Horm, JW: Association of cancer sites with tobacco and alcohol consumption and socioeconomic status of patients: interview study from the Third National Cancer Survey, J Natl Cancer Inst 58:525-547, 1977.

63. Johnston, DG, and Alberti, KGMM: The liver and the endocrine system. In Wright, R, Alberti, KGMM, and Karran, S, et al, editors: Liver and biliary disease, Philadelphia, 1979, WB Saunders Co.
64. Jenkins, JS, and Connolly, J: Adrenocortical response to ethanol in man, Br Med J 2:804-805, 1968.
65. Merry, J, and Marks, V: Hypothalamic-pituitary-adrenal function in chronic alcoholics. In Cross, MM, editor: Alcohol intoxication and withdrawal: experimental studies, Advances in experimental medicine and biology, New York, 1973, Plenum Publishing Corp.
66. Chalmers, RJ, Bennie, EH, Johnson, RH, et al: Growth hormone, prolactin and corticosteroid responses to insulin hypoglycaemia in alcoholics, Br Med J 2:745-748, 1978.
67. Liu, YK: Leukopenia in alcoholics, Am J Med 54:605-610, 1973.
68. Cowan, DH, and Hines, JD: Thrombocytopaenia of severe alcoholism, Ann Intern Med 74:37-43, 1971.
69. McFarland, W, and Leibre, EP: Abnormal leukocyte response in alcoholism, Ann Intern Med 59:865-877, 1963.
70. Thomas, DP, Ream, VJ, and Stuart, RK: Platelet aggregation in patients with Laennec's cirrhosis of the liver, N Engl J Med 276:1344-1348, 1967.
71. Chen, T, Kanagasundaram, N, Kakumu, S, et al: Serum autoantibodies to alcoholic hyalin in alcoholic hepatitis (abstract), Gastroenterology A13:813, 1975.
72. Zinneman, HH: Autoimmune phenomena in alcoholic cirrhosis, Am J Dig Dis 20:337-345, 1970.
73. Bernstein, IM, Webster, KH, Williams, RC, Jr, et al: Reduction in circulating T lymphocytes in alcoholic liver disease, Lancet 2:488-490, 1974.
74. Hsu, CCS, and Leevy, CM: Inhibition of PHA-stimulated lymphocyte transformation by plasma from patients with advanced alcoholic cirrhosis, Clin Exp Immunol 8:749-760, 1971.
75. Kakumu, S, and Leevy, CM: Lymphocyte cytotoxicity in alcoholic hepatitis, Gastroenterology 72:524-526, 1977.
76. Smith, WI, Jr, VanThiel, DH, Whiteside, T, et al: Altered immunity in male patients with alcoholic liver disease: evidence for defective immune regulation, Alcohol Clin Exp Res 4:199-206, 1980.
77. Rosalki, SB: Gamma-glutamyl transpeptidase, Adv Clin Chem 17:53-62, 1975.
78. Pomier-Layrargues, G, Huet, PM, Richer, G, et al: Hyperglobulinemia in alcoholic cirrhosis: relationship with portal hypertension and intrahepatic portal-systemic shunting as assessed by Kupffer cell uptake, Dig Dis Sci 25:489-493, 1980.
79. Milstein, HJ, Bloomer, JR, and Klatskin, G: Serum bile acids in alcoholic liver disease, Am J Dig Dis 21:281-295, 1976.
80. Lundquist, GAR: Glucose tolerance in alcoholism, Br J Addict 61:51-55, 1965.
81. Rees, LH, Besser, GM, Joffcoate, WJ, et al: Alcohol-induced pseudo-Cushing's syndrome, Lancet 1:726-728, 1977.
82. Madison, LL, Lochner, A, and Wulff, J: Ethanol induced hypoglycemia. II. Mechanism of suppression of hepatic gluconeogenesis, Diabetes 16:252-258, 1967.
83. Joffe, BI, Seftel, HC, and Van As, M: Hormonal responses in ethanol-induced hypoglycaemia, J Stud Alcohol 36:550-554, 1975.
84. Arky, RA, and Freinkel, N: Alcohol hypoglycemia. V. Alcohol infusion to test gluconeogenesis in starvation, with specific reference to obesity, N Engl J Med 274:426-433, 1966.
85. O'Keefe, SJ, and Marks, V: Lunchtime gin and tonic, a cause of reactive hypoglycaemia, Lancet 1:1286-1288, 1977.
86. Cooperman, MT, Davidoff, F, Spark, R, et al: Clinical studies of alcoholic ketoacidosis, Diabetes 23:433-439, 1974.
87. Leevy, CM, Smith, F, Longueville, J, et al: Indocyanine green clearance as a test for hepatic function: evaluation by dichromatic ear densitometry, JAMA 200:236-240, 1967.

CHAPTER 32

Iron, porphyrin, and bilirubin metabolism

WILLIAM E. SCHREIBER

OBJECTIVES

- List the physiological functions of iron, and describe its absorption from the gut and its transport in the body.
- Describe the pathological conditions leading to iron deficiency and overload, and describe changes in the following analytes in iron deficiency anemia, anemia of chronic infection, thalassemia, hemochromatosis, and lead poisoning: ferritin, serum iron, iron-binding capacity, and free erythrocyte protoporphyrin.
- Define porphyria and distinguish between primary and secondary porphyrias.
- List porphyrin analytes and RBC enzymes that will be elevated in each of the primary and secondary porphyrias.
- Outline the formation and catabolism of bilirubin.

KEY TERMS

ampulla of Vater The junction of the common bile duct and pancreatic duct proximal to their opening into the duodenum.

anemia A reduction in the quantity of hemoglobin or number of red cells in blood.

bile The yellow-green fluid secreted by the liver and poured into the duodenum via the bile ducts.

bile canaliculi The fine tubular canals running between liver cells into which bile is discharged.

chelate A chemical compound in which a metallic ion is firmly bound to a chelating molecule.

cholestasis The stoppage of the flow of bile.

cirrhosis A liver disease characterized by the loss of the normal microscopic architecture, with fibrosis and nodular regeneration.

erythropoiesis The production of erythrocytes.

hemolytic anemia Anemia caused by shortened survival of mature red blood cells.
hypochromic Referring to an abnormal decrease in the hemoglobin content of erythrocytes.
hypoplasia The underdevelopment of an organ or tissue.
MCH Mean corpuscular hemoglobin, the average amount of hemoglobin per red blood cell.
MCHC Mean corpuscular hemoglobin concentration, the average concentration of hemoglobin per red blood cell.
MCV Mean corpuscular volume, the average red blood cell volume.
megaloblastic anemia Anemia characterized by large red cell precursors in he bone marrow.
microcytic Referring to erythrocytes that are smaller than normal.
palpitation The subjective sensation of a rapid or irregular heartbeat.
parenchyma The functional tissue of an organ (excluding the fibrous framework).
pernicious anemia A megaloblastic anemia caused by failure to absorb vitamin B_{12}.
phagocyte Any cell that ingests microorganisms or other cells and foreign particles.
photoisomers Isomers produced on exposure to light.
photosensitivity Abnormal reactivity of the skin to sunlight.
porphyrinemia The presence of excess porphyrin in blood.
porphyrinuria The presence of excess porphyrin in urine.
reticuloendothelial system A functional system composed of highly phagocytic cells having both endothelial and reticular attributes, located in blood vessels, lymph nodes, liver, spleen, bone marrow, and other tissues.
sideroblastic anemia A heterogenous group of anemias in which iron stores of the reticuloendothelial tissues are increased and bone marrow normoblasts contain iron (sideroblasts).
tachycardia Rapid heart rate.
thalassemia A heterogenous group of hereditary hemolytic anemias that have a decreased rate of synthesis of one or more hemoglobin polypeptide chains.

Part I: Iron metabolism

FUNCTION

Iron is one of the most abundant elements on earth, yet only trace amounts are present in living cells. Nevertheless, it is essential for the existence of all plants and animals. In humans, most of the iron is contained within the porphyrin ring of heme (Fig. 32-1), in proteins such as hemoglobin, myoglobin, catalase, peroxidases, and cytochromes. There are also iron-sulfur proteins, such as NADH dehydrogenase and succinate dehydrogenase, in which iron is present in clusters with inorganic sulfur. In all of these systems, it is the ability of iron to interact reversibly with oxygen and to function in electron transfer reactions that makes it biologically indispensable.

An average adult male has about 4 g of body iron. About 65% to 70% of the total is found in hemoglobin,

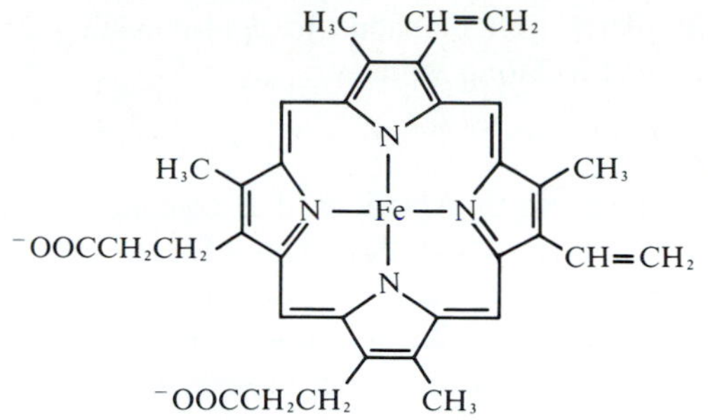

H_3C $CH{=}CH_2$ H_3C CH_3 N Fe N N N $^-OOCCH_2CH_2$ $CH{=}CH_2$ $^-OOCCH_2CH_2$ CH_3

Fig. 32-1 Structure of heme, an organometallic complex of protoporphyrin IX and Fe^{2+}.

4% in myoglobin, and less than 1% in other iron-containing enzymes and proteins. The remaining 25% to 30% represents the storage pool of iron. By comparison, the average adult woman has only about 3 g of iron in her body. This difference is mainly due to the much smaller iron reserves in women. There is also less hemoglobin iron because women have a slightly lower hemoglobin concentration in blood than do men. Iron distribution is summarized in Table 32-1.

METABOLISM

Daily requirements for iron vary depending on the person's age, sex, and physiological status. Although iron is not excreted in the conventional sense, there is a daily loss of about 1 mg because of the normal shedding of intestinal mucosal cells and skin epithelial cells and the loss of small numbers of erythrocytes in urine and feces. Therefore an iron intake of 1 mg per day is sufficient for men and postmenopausal women. However, the blood loss of each menstrual cycle drains 20 to 40 mg of iron, so women in their reproductive years need 1.5 to 2 mg of iron per day. The diversion of iron to the growing fetus during pregnancy, blood loss at delivery, and subsequent breast feeding of the infant consume 900 mg of iron on average. This

Table 32-1 Iron distribution and function in a normal adult

Compound	Function	Iron (mg)
Hemoglobin	O_2 transport, blood	2500
Myglobin	O_2 storage, muscle	300
Enzymes		
Catalase	H_2O_2 decomposition	
Peroxidases	Oxidation	
Cytochromes	Electron transfer	
Iron-sulfur*	Electron transfer	
Transferrin*	Iron transport	4
Ferritin* and hemosiderin*	Iron storage	1000 (men) 100-400 (women)

Modified from Schreiber, WE: Medical aspects of biochemistry, Boston, 1984, Little, Brown & Co.
*Nonheme iron compounds.

increases daily iron demands to approximately 3 mg in pregnant and lactating women.

Absorption

An average North American diet contains between 10 and 20 mg of iron per day. Only 5% to 10% of this amount is absorbed, mainly in the duodenum and upper small intestine. Most dietary iron is in the ferric (Fe^{3+}) state, which is poorly absorbed. Gastric secretions and hydrochloric acid reduce ferric iron to the absorbable ferrous (Fe^{2+}) form. Iron absorption is also enhanced in the presence of ascorbic acid, sugars, amino acids, and other compounds that form soluble iron chelates. Substances that form insoluble complexes with iron, such as phosphates (in eggs, cheese, and milk), oxalates and phytates (in vegetables), and tannates (in tea), decrease iron absorption. Heme iron, which comes mainly from meat and fish, is processed differently. After it is released from the surrounding polypeptide chain, heme is absorbed intact by the mucosal cell, the porphyrin ring is split, and iron is liberated. This process is more efficient than the absorption of nonheme iron and is not affected by dietary factors.

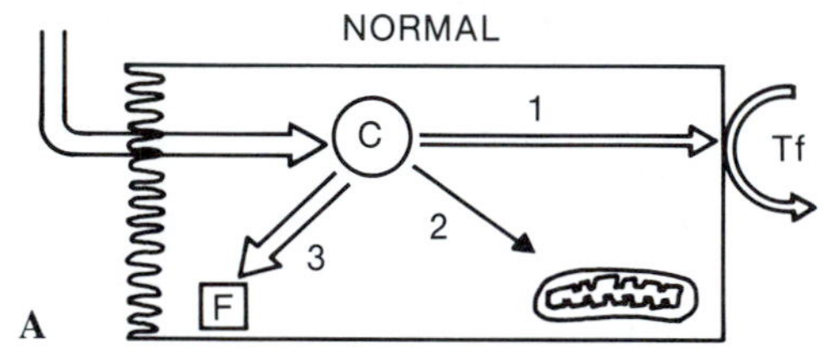

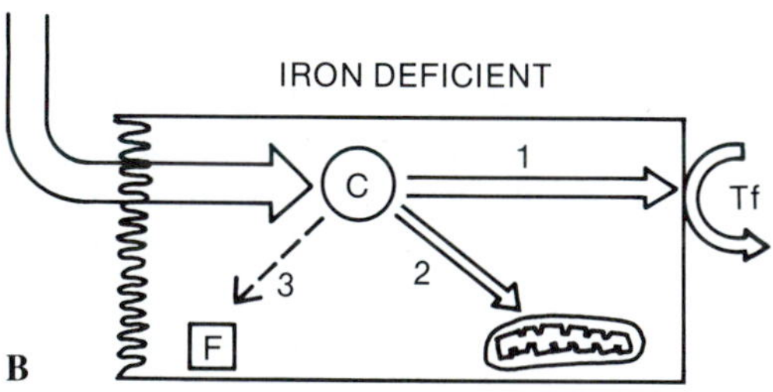

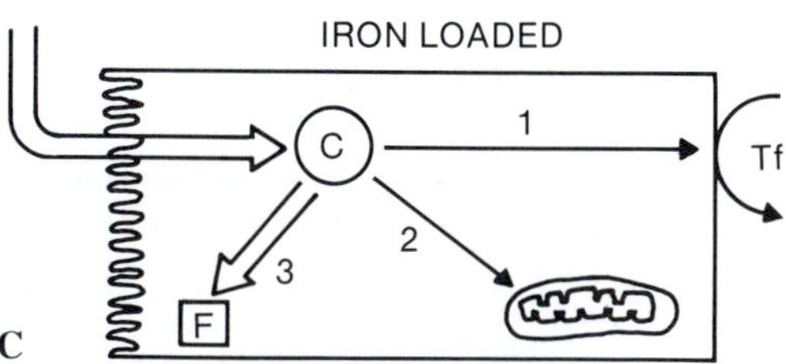

Fig. 32-2 Iron uptake and disposition within the small intestinal epithelial cell. **A,** Normal iron status. **B,** Iron deficiency: increased uptake and transfer of iron to plasma *(1)* and mitochondria *(2)*, with decreased transfer to ferritin *(3)*. **C,** Iron loaded: normal uptake with decreased transfer to plasma *(1)* and increased transfer to ferritin *(3)*. *C,* Intracellular iron carrier; *F,* ferritin; *Tf,* transferrin. *(From Jacobs, A: Clin Haematol 2:323, 1973.)*

Since iron loss is a continuous and largely unregulated process, the body's iron supply is controlled by changes in absorption. The intestinal cells take in a sizable proportion of iron presented to the luminal surface, considerably more than the 5% to 10% that eventually will enter the body's iron pool. Within the intestinal cell, iron is transported by a carrier protein to mitochondria for heme synthesis or to ferritin for storage (Fig. 32-2, *A*). Alternatively, iron may be transferred directly into plasma and join the circulating iron pool. Stored iron can subsequently be mobilized as necessary for transport into plasma. However, the iron in mitochondria and ferritin is lost when the mucosal cells are shed. New cells take their place, and the cycle of iron build-up starts again.

The control of iron transfer from the intestinal mucosa to the plasma is not entirely understood. However, ferritin synthesis is stimulated by iron, so an increase in body iron leads to higher levels of ferritin within intestinal mucosal cells. This favors trapping of iron within the cell, rather than transfer of iron to plasma (Fig. 32-2, *C*). Conversely, a decrease in cellular ferritin (because of iron deficiency) would promote movement of iron directly into plasma (Fig. 32-2, *B*). This explains the increased efficiency of iron absorption (10% to 20%) when iron reserves are depleted.

Red cell turnover

Absorbed iron represents only a fraction of what is required for heme synthesis. Most of this iron (20 to 25 mg/day) comes from the destruction of old erythrocytes by phagocytic cells. Within these cells, heme oxygenase breaks open the porphyrin ring to release iron. Most of the iron is taken up by plasma transferrin, which then carries it to the bone marrow for hemoglobin synthesis. Small amounts of iron are also delivered to other tissues or are conserved within the storage compartment. In this manner, the reticuloendothelial system continuously recycles used iron from old red cells into new ones.

Transport

Transferrin, a single-chain polypeptide with a molecular weight of 79,500 daltons, is the transport protein for iron in blood. Each transferrin molecule has two binding sites for ferric iron, and these sites are normally 20% to 50% saturated. The need for a specific carrier protein derives from the toxicity and relative insolubility of free iron; virtually all of the plasma iron is protein bound. At any given moment, plasma transferrin carries about 3 to 4 mg of iron. However, at least 10 times this amount of iron is transported through plasma daily, so the circulating iron pool is in constant equilibrium with other body pools of iron. Transferrin delivers iron to cells with specific surface receptors for this protein.

Storage

Iron is stored in tissues in either of two forms, ferritin or hemosiderin. Ferritin consists of a multisubunit protein shell, known as apoferritin, surrounding a core of up to 4500 iron atoms. Ferritin takes up iron as Fe^{2+}, oxidizes it to Fe^{3+}, and then deposits it within the iron core as a hydrous ferric oxide–phosphate complex. When iron is released, it is reduced back to Fe^{2+}. Ferritin is present in nearly all cells (especially hepatocytes) and is a readily mobilized form of storage iron. It serves to package and isolate iron atoms from the intracellular environment, thus preventing any toxic action on cell constituents. Hemosiderin is an insoluble complex derived from ferritin that has lost some of its surface protein and become aggregated. It is present in granules 1 to 2 μm in diameter and is visible by light microscopy after staining with Prussian blue. Hemosiderin has a higher iron concentration than does ferritin, but it releases iron more slowly.

About one third of body iron is stored in the liver (see Chapter 23), one third in the bone marrow, and the remainder in the spleen and other tissues.

PATHOLOGICAL CONDITIONS

Iron deficiency

Iron occupies a central position in hemoglobin synthesis and erythropoiesis. One might expect a deficiency of iron to interfere with this process and, in time, to lead to anemia. In fact, iron-deficiency anemia is the most common type of anemia in humans. In the United States, about 2% of adult men and about 20% of women of child-bearing age (up to 50% of pregnant women) are deficient in iron.

Blood loss is the most common cause of iron deficiency in adults. The high prevalence of iron deficiency among women is due to the obligatory blood loss of each menstrual cycle. Bleeding from the gastrointestinal tract (for example, becuse of peptic ulcer, diverticulosis, or malignancy) is the primary cause of iron deficiency in men. Increased demand for iron in infants and young children, adolescents, and pregnant women may also lead to an iron-deficiency state. This is frequently compounded by inadequate intake of iron because of an iron-poor diet of cereals and milk or junk food. Impaired absoption of iron following partial or total gastrectomy and in patients with chronic diarrhea and/or malabsorption may also cause depletion of iron reserves.

The signs and symptoms of iron deficiency are largely attributable to the anemia it produces. Patients may complain of weakness, fatigue, dizziness, and palpitations. Nonspecific symptoms such as nausea, anorexia, constipation, and menstrual irregularities also may accompany iron deficiency. Pallor and tachycardia are the usual findings on physical examination. Some individuals develop pica, a craving for unnatural articles of food such as clay or starch.

Iron-deficiency anemia is one of several anemias in which the red blood cells become hypochromic and microcytic. A complete blood count reveals a drop in hemoglobin, and all of the red blood cell indices (mean corpuscular volume [MCV], mean corpuscular hemoglobin [MCH], and mean corpuscular hemoglobin concentration [MCHC]) are decreased. The peripheral blood smear shows erythrocytes that on average are smaller and paler than normal. No stainable iron is visible in the bone marrow.

Laboratory tests of iron status are especially useful in distinguishing iron deficiency from other causes of hypochromic, microcytic anemia. The concentration of serum iron decreases, while the total iron-binding capacity (TIBC), which measures the capacity of transferrin for iron, increases. The transferrin saturation, which is equal to iron concentration divided by TIBC, is well below its normal value. A decrease in serum ferritin, which is a reflection of body iron stores, is the single most reliable indicator of iron deficiency. Free erythrocyte protoporphyrin is increased, but it is not specific for iron deficiency and is best used as a screening test.

Iron overload

Hemochromatosis. Hemochromatosis is a hereditary disorder characterized by a progressive increase in body iron stores leading to organ impairment and damage. Inheritance is autosomal recessive. Among populations of Northern European descent, 10% of people carry the gene, and 0.3% are homozygotes. For reasons that are not clearly understood, only a fraction of homozygotes develop the full-blown disease. Men are affected five to ten times more frequently than women because of the protective effect of menstrual blood loss and pregnancy. The age of onset of symptoms is usually 40 years or older.

Patients with hemochromatosis absorb about 4 mg of iron per day, even on a normal diet. The mechanism of enhanced iron absorption remains unknown. Under normal conditions, excess iron is processed by cells of the reticuloendothelial system. However, in hemochromatosis, iron is deposited directly in parenchymal cells of the liver, pancreas, heart, and other organs. After accumulating for years, the excessive amounts of intracellular iron lead to tissue injury and, ultimately, organ failure. At this stage, the amount of storage iron may exceed 20 g.

Signs and symptoms are related to the involved organ systems. The liver is nearly always enlarged and in time may become cirrhotic. This predisposes patients to an unusually high risk of hepatocellular carcinoma. About two thirds of patients develop diabetes mellitus; both a genetic predisposition to diabetes and direct injury to the pancreas appear to play a role in its development. An increase in skin pigmentation is found in about 90% of patients. It is caused mainly by increased melanin production, but iron deposition in skin may contribute as well to the characteristic bronze color. Cardiac involvement may take the form of congestive heart failure or arrhythmias. Testicular atrophy and loss of libido are caused by a drop in the production of gonadotropins by the impaired hypothalamic-pituitary axis. Arthritis also occurs in one fourth to one half of patients.

In hemochromatosis, the serum iron concentration increases and the TIBC decreases, the opposite of the changes seen in iron deficiency. The transferrin saturation is much higher than normal and is a particularly sensitive index of iron overload. Serum ferritin concentration is usually increased early in the course of disease, before signs and symptoms become apparent. The definitive test for hemochromatosis is liver biopsy because it allows for the direct measurement of hepatic iron as well as microscopic evaluation of tissue damage and iron stores.

Acquired hemochromatosis. Iron overload can also be an acquired disorder. At first, excess iron is deposited in reticuloendothelial cells of the liver, spleen, and bone marrow, rather than directly in parenchymal cells. This initial stage of iron overload is referred to as *hemosiderosis,* and tissues remain anatomically and functionally normal. As the iron load increases, its distribution pattern changes, and iron is deposited in the parenchymal cells of the liver, pancreas, heart, and other organs. The clinical picture then resembles the hereditary form of hemochromatosis.

Acquired hemochromatosis may be a complication of chronic anemias such as thalassemia and sideroblastic anemia. Not only is iron absorption increased in these disorders, but patients are frequently treated with multiple blood transfusions. Under the right circumstances, excessive chronic iron ingestion may lead to iron overload. The Bantu of South Africa, who drink home-brewed alcoholic beverages with a very high iron content, have a high incidence of the disease. Use of medicinal iron supplements is rarely, if ever, by itself a cause of hemochromatosis. Alcoholics with chronic liver disease (for example, cirrhosis) may also develop an increase in tissue iron stores. However, alcoholics with massive iron overload probably have the genetic form of the disease.

CHANGE OF ANALYTE IN DISEASE

There are three iron compartments, accounting for more than 90% of total body iron, that the clinical laboratory can measure. The largest of these is the iron contained in hemoglobin, whose concentration is determined as part of a complete blood count. Next largest is the storage compartment, and serum ferritin levels are proportional to the size of this pool. Finally, circulating iron is evaluated by measuring the serum concentrations of iron and its transport protein, transferrin. This combination of hematological and biochemical studies enables one to identify disorders of iron metabolism (Table 32-2). However, the results must be interpreted in light of the patient's age, sex, physiological state, and clinical presentation to make an accurate diagnosis.

Hematological studies

A complete blood count gives the number of erythrocytes per liter, hemoglobin concentration, hematocrit, and red blood cell indices. The World Health Organization criteria for anemia, expressed as hemoglobin concentration, are below 130 g/L for men, below 120 g/L for menstruating women, and below 110 g/L for pregnant women. Iron-deficiency anemia is a hypochromic, microcytic anemia, so cell size (MCV) and hemoglobin content (MCH) are reduced, and the concentration of hemoglobin per cell (MCHC) is also down. The peripheral blood smear shows a wide variation in the size and hemoglobin content of erythrocytes, with a large proportion of cells that are smaller and paler than normal. One must remember that these are the hemotological findings in clear-cut iron deficiency. At an early stage of iron depletion, both hemoglobin concentration and red cell indices remain normal. Moreover, hypochromic, microcytic anemia is characteristic of thalassemia, sideroblastic anemia, and the anemia of chronic disease as well as iron deficiency. Red blood cell parameters thus define the presence or absence of anemia and its morphological character, but other studies are required to identify the cause of anemia.

Table 32-2 Laboratory measurements of iron status*

Disease	Serum iron (μg/L)	TIBC (μg/L)	Transferrin saturation (%)	Serum ferritin (μg/L)	Free erythrocyte protoporphyrin (μg/L cells)	Tissue iron stores
Normal	500-1600	2500-4000	20-55	15-200	170-770	N
Storage iron depletion, no anemia	N	N	N	↓	N	↓
Iron-deficiency anemia	↓	↑	↓	↓	↑	↓
Anemia of chronic disease	↓	↓	↓	N or ↑	↑	N or ↑
Thalassemia	↑	↓	↑	↑	N	↑
Sideroblastic anemia	↑	N	↑	↑	N or ↑	↑
Hemochromatosis	↑	↓	↑	↑	N	↑

**N*, Normal; ↓, decreased; ↑, increased.

Erythrocyte studies do not contribute to the diagnosis of hemochromatosis.

Serum iron

The circulating iron pool turns over 10 to 20 times per day, so a typical iron atom spends no longer than 2 hours in plasma. Large changes (20% or greater) in serum iron concentration may occur suddenly, even in healthy people, because of momentary imbalances in iron inflow and outflow. There is also a diurnal variation, with a fall in iron concentration in the evening; significant day-to-day variations occur as well. All of these factors limit the diagnostic usefulness of serum iron measurements.

The reference interval for men is 500 to 1600 μg/L and for women is 400 to 1500 μg/L. Serum iron is decreased in iron deficiency, in the anemia of chronic disease, in malignancies, in inflammation (for example, infections, myocardial infarction, after surgery), following recent blood loss, and during menstruation. Serum iron concentrations are increased in red cell disorders with ineffective erythropoiesis or hemolysis, such as megaloblastic anemia, thalassemia, and sideroblastic anemia. Bone marrow hypoplasia, viral hepatitis, acute iron poisoning, and hemochromatosis also cause an increase in serum iron concentration. Serum iron values should always be interpreted in combination with TIBC and transferrin saturation.

Total iron-binding capacity and transferrin saturation

The TIBC measures the maximum amount of iron that serum proteins can bind. This is an indirect way of assessing transferrin levels. The serum iron concentration divided by TIBC then gives the transferrin saturation. The reference interval for TIBC is 2500 to 4000 μg/L, and transferrin saturation is normally 20% to 55%.

TIBC is increased in iron deficiency, hepatitis, pregnancy, and women taking oral contraceptives. TIBC is decreased in malignancies, nephrosis, inflammation, chronic disease, starvation, megaloblastic and hemolytic anemias, and hemochromatosis. The combination of low serum iron and high TIBC in iron deficiency causes a low transferrin saturation, but pregnancy and chronic disease may also be causes. High transferrin saturation is characteristic of iron overload and is the most sensitive test for diagnosing hemochromatosis. Thalassemia, sideroblastic anemia, and acute iron poisoning also cause the transferrin saturation to increase.

Serum ferritin

The small amounts of ferritin that normally circulate in plasma are in equilibrium with tissue iron stores. The serum ferritin concentration is therefore an especially useful indicator of total body iron status. The reference interval is 15 to 200 μg/L in men and 12 to 150 μg/L in women.

A low serum ferritin concentration is diagnostic of iron deficiency. Ferritin levels drop early in the development of iron deficiency, before serum iron and transferrin saturation become abnormally low. An increase in serum ferritin is seen in iron overload, before the development of signs and symptoms of hemochromatosis. However, the release of ferritin from damaged tissues in hepatitis, acute inflammatory conditions, and a variety of tumors also dramatically increases the serum ferritin level. Increases in serum ferritin must therefore be interpreted in light of the patient's overall health.

Free erythrocyte protoporphyrin

In the course of heme synthesis (see p. 502), small numbers of protoporphyrin molecules escape from the pathway and do not complex with Fe^{2+}. Most of these molecules bind Zn^{2+} instead to produce zinc protoporphyrin, which then attaches to a heme site of hemoglobin and circulates in the mature erythrocyte. Assays of red cell porphyrins have traditionally involved an acid extraction step, which removes zinc and leaves behind the metal-free compound. Thus the term *free erythrocyte protoporphyrin* (FEP) is actually a misnomer. Fortunately, it does indicate the amount of nonheme protoporphyrin in red cells, which is a clinically useful measurement. The reference interval is 170 to 770 μg/L of cells.

A decrease in available iron increases the formation of zinc protoporphyrin and FEP. Both iron deficiency and chronic disease, in which iron use is impaired, will increase FEP. Lead poisoning interferes with the final step in heme synthesis and may produce large increases in FEP. Protoporphyria, a hereditary deficiency of ferrochelatase, is associated with very high FEP values.

The FEP assay is most commonly used as a screening test for iron deficiency and lead poisoning. When performed with a hematofluorometer, the test is rapid, technically simple, and reproducible and requires only a drop of blood. An increased FEP should be followed by more specific tests of iron status.

Part II: Heme synthesis and the porphyrias

STRUCTURE AND FUNCTION

Porphyrin molecules contain a nucleus of four pyrrole rings joined into a macrocyclic compound by methenyl (=CH—) bridges (Fig. 32-3). The extended network of alternating single and double bonds causes porphyrins to absorb visible light; it is this group that imparts a red color to hemoglobin. Porphyrins also fluoresce a reddish-pink color under long-wavelength ultraviolet light, a property that is very useful when detecting and measuring porphyrins in body fluids. Another unique property of porphyrins is the arrangement of four nitrogen atoms at the center of the ring, enabling porphyrins to chelate metal atoms. From a physiological standpoint, iron is the most important metal that complexes with porphyrins.

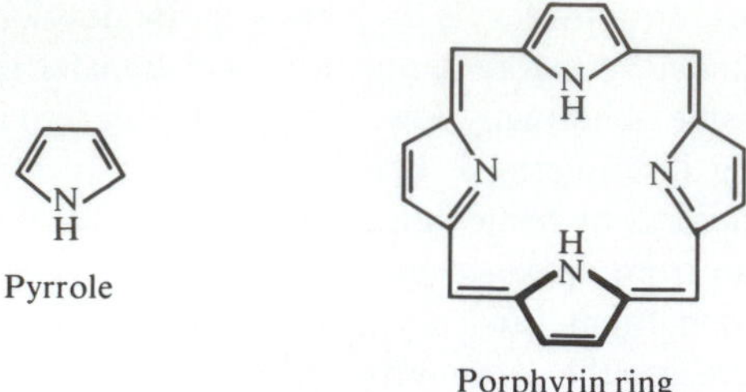

Fig. 32-3 Structures of pyrrole and the porphyrin ring. Structure in boldface is basic pyrrole ring.

Differences in porphyrin structure depend on the type and position of side chains located at the corners of the pyrrole rings. In humans, there are three major porphyrins: (1) uroporphyrin (URO), (2) coproporphyrin (COPRO), and (3) protoporphyrin (PROTO) (see Figs. 32-4 and 32-5). URO has four propionate and four acetate side chains, whereas COPRO has four propionate and four methyl side chains. These groups may be arranged in four different structural configurations, of which the type III isomer is normally produced. PROTO has two propionate, two vinyl, and four methyl groups that can be arranged in any of 15 different configurations. However, only the type IX isomer is produced by the body.

Free porphyrins are by-products of the heme biosynthetic pathway and have no biological function on their own. Heme, the iron chelate of protoporphyrin, is the prosthetic group for a number of proteins and enzymes (see above and Table 32-1). Trace amounts of zinc protoporphyrin also occur naturally, but no physiological role has been attached to this compound.

BIOCHEMISTRY

Heme synthesis takes place in all cells but occurs to the greatest extent in the bone marrow (red cell precursors) and liver. It is helpful to think of the pathway in two halves: (1) formation of the porphyrin ring by repeated condensations of precursors (see Fig. 32-4) and (2) modification of the side chains and insertion of iron (see Fig. 32-5). This arbitrary division simplifies a long and seemingly complex series of reactions.

Synthetic pathway

The synthetic pathway begins with the condensation of succinyl CoA and glycine (activated by pyridoxal phosphate) to form δ-aminolevulinic acid (ALA). This reaction, catalyzed by ALA synthase, is the rate-limiting step in heme synthesis. Two ALAs then condense to form porphobilinogen (PBG), a pyrrole with propionate and acetate side chains at its corners. Next, four PBG molecules condense in head-to-tail fashion to form a linear tetrapyrrole that cyclizes spontaneously. This is a critical step in that one of the pyrrole units must change its orientation of propionate and acetate side chains to produce the type III isomer of uroporphyrinogen. Porphobilinogen deaminase catalyzes the condensation reaction, but uroporphyrinogen III cosynthase performs the isomerization of one pyrrole unit.

At this point the basic ring structure is in place. Modification of the side chains begins with the decarboxylation of the four acetate groups to form coproporphyrinogen III. Two propionate groups are then decarboxylated and dehydrogenated to vinyl groups, producing protoporphyrinogen IX. The bridging carbon atoms are then oxidized from methylene (—CH_2—) to methenyl (=CH—) to yield protoporphyrin IX. In the final step, iron (Fe^{2+}) is inserted into the protoporphyrin ring to produce heme.

Points of interest

Several aspects of porphyrin synthesis deserve a closer look. The pathway is controlled primarily by changes in the activity of ALA synthase, the first and rate-limiting enzyme. Excess heme inhibits the activity of ALA synthase, whereas a deficit of heme stimulates the enzyme. Therefore abnormalities in the production or degradation of heme will affect the number of molecules entering the pathway. Two enzymes, ALA dehydratase and ferrochelatase, are inhibited by lead, and this has implications for the build-up of certain metabolites in lead poisoning. Another interesting enzyme is uroporphyrinogen III cosynthase, which produces the type III isomer or uroporphyrinogen at the ring cyclization step. In the absence of this enzyme, only the type I isomer, which is not a precursor of heme, is formed.

Note also that the latter half of the pathway contains porphyrinogen intermediates. Porphyrinogens differ from porphyrins in that the bridging carbon atoms are fully reduced and all four nitrogen atoms are protonated. There is no network of alternating single and double bonds, so these compounds are colorless and nonfluorescent. Porphyrinogens that are not used by the regular pathway spontaneously and irreversibly oxidize to the corresponding porphyrin. That is why URO, COPRO, and PROTO (not the porphyrinogens) are the major excretion forms.

Finally, the porphyrin pathway begins and ends in mitochondria, but four of the intervening steps take place in the cytosol. The intracellular distribution of enzymes is shown in Fig. 32-6. Since erythrocytes lose their mitochondria as they mature, only half of these enzymes can be assayed in circulating red cells.

PATHOLOGICAL CONDITIONS

What would happen if the enzymes involved in heme synthesis did not function properly? The answer to that question can be found in the study of the porphyrias, a group of genetically determined disorders of heme synthesis. All but one of the porphyrias are inherited as an autosomal dominant trait. Since the patient has only one gene producing a functional enzyme (instead of two), there is about 50% of normal enzyme activity. This partial defect

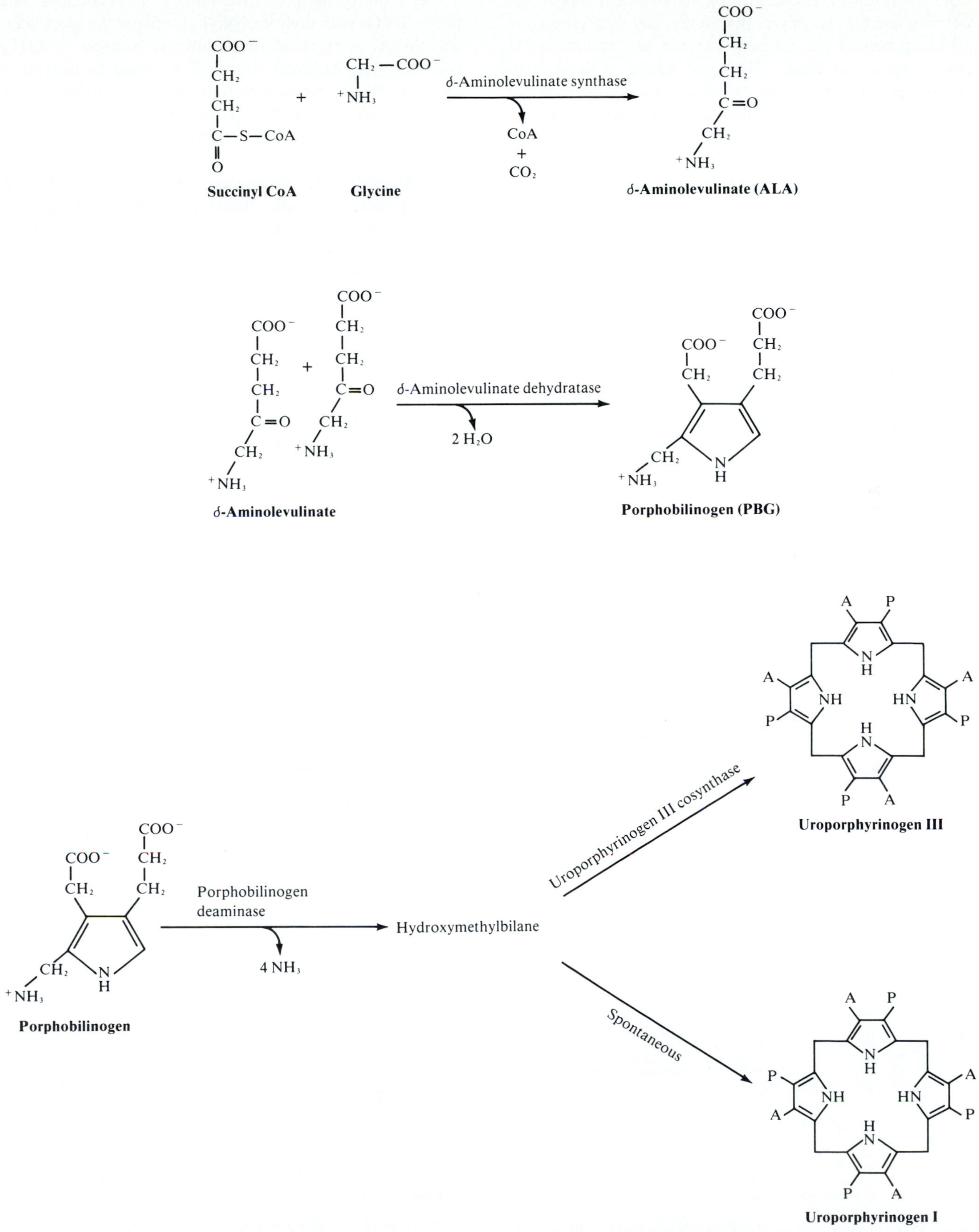

COO^-–CH_2–CH_2–C(=O)–S–CoA + CH_2–COO^- / $^+NH_3$ —δ-Aminolevulinate synthase→ COO^-–CH_2–CH_2–C(=O)–CH_2–$^+NH_3$ (CoA + CO_2 released)

Succinyl CoA **Glycine** **δ-Aminolevulinate (ALA)**

2 δ-Aminolevulinate —δ-Aminolevulinate dehydratase→ Porphobilinogen (2 H_2O released)

δ-Aminolevulinate **Porphobilinogen (PBG)**

Porphobilinogen —Porphobilinogen deaminase→ Hydroxymethylbilane (4 NH_3 released)

Hydroxymethylbilane —Uroporphyrinogen III cosynthase→ **Uroporphyrinogen III**

Hydroxymethylbilane —Spontaneous→ **Uroporphyrinogen I**

Fig. 32-4 Initial steps in porphyrin synthesis. *A*, Acetate; *P*, propionate.

does not result in a deficiency of heme, so patients do not develop anemia. However, porphyrins and their precursors build up behind the deficient enzyme and accumulate in body tissues and fluids. The toxic nature of these compounds produces the signs and symptoms characteristic of each porphyria. The excretion of excess porphyrins and their precursors is the basis for diagnosing these disorders.

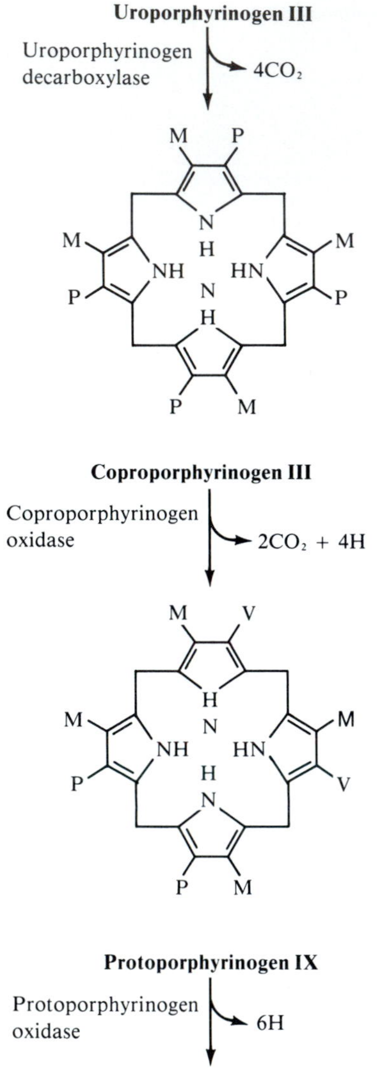

Uroporphyrinogen III

Uroporphyrinogen decarboxylase → $4CO_2$

Coproporphyrinogen III

Coproporphyrinogen oxidase → $2CO_2$ + 4H

Protoporphyrinogen IX

Protoporphyrinogen oxidase → 6H

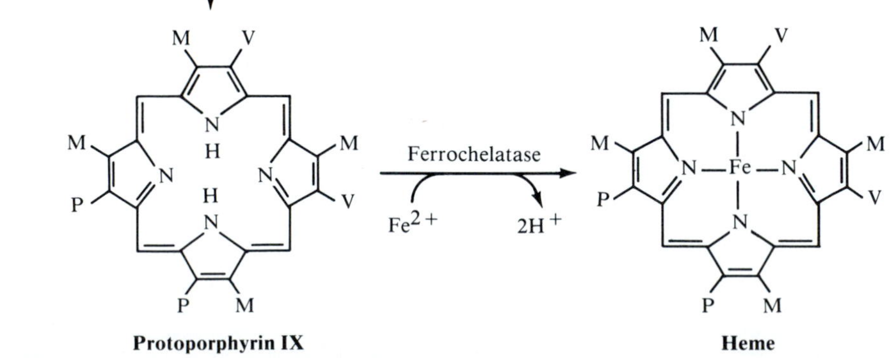

Protoporphyrin IX

Ferrochelatase (Fe^{2+} → $2H^+$)

Heme

Fig. 32-5 Latter half of the heme biosynthetic pathway. *P,* Propionate; *M,* methyl; *V,* vinyl.

The route of porphyrin excretion is a function of solubility. URO, with eight carboxyl groups, is the most water soluble and is excreted almost entirely in urine. PROTO, with only two carboxyl groups, is excreted exclusively in feces. COPRO, which has four carboxyl groups, is excreted by either route. The prophyrin precursors ALA and PBG are sufficiently water soluble to be eliminated in urine.

Traditionally the porphyrias have been classified as erythropoietic or hepatic, based on the site of overproduction of the porphyrins and their precursors. A more useful approach is the classification of the porphyrias by signs and symptoms (neurological versus cutaneous) because this allows one to think in terms of clinical presentation.

Neurological porphyrias

There are three porphyrias characterized by acute attacks of abdominal pain, constipation, neuromuscular signs and symptoms, and psychotic behavior. The acute attacks, which may last from days to weeks, are accompanied by an increase in the excretion of ALA and PBG in urine. Despite the association of porphyrin precursors with the attacks, their biochemical basis is not well understood. A number of people who have the enzymatic defect for one of these porphyrias do not develop acute attacks and are said to have *latent porphyria*. Each enzyme defect is inherited as an autosomal dominant trait.

The signs and symptoms of the neurological porphyrias usually begin during adolescence or later and affect women more often than men. Abdominal pain is the most constant finding and is frequently accompanied by constipation, nausea, and vomiting. Abnormalities of the autonomic nervous system include tachycardia, hypertension, sweating, and urinary retention. Peripheral neuropathy may involve pain in the extremities, areas of reduced or altered sensation, muscle weakness, and paralysis. There may also be seizures, coma, temporary loss of vision, and inappropriate secretion of antidiuretic hormone. Some patients say they cannot think clearly, and they may have a history of nervousness, emotional instability, or psychotic

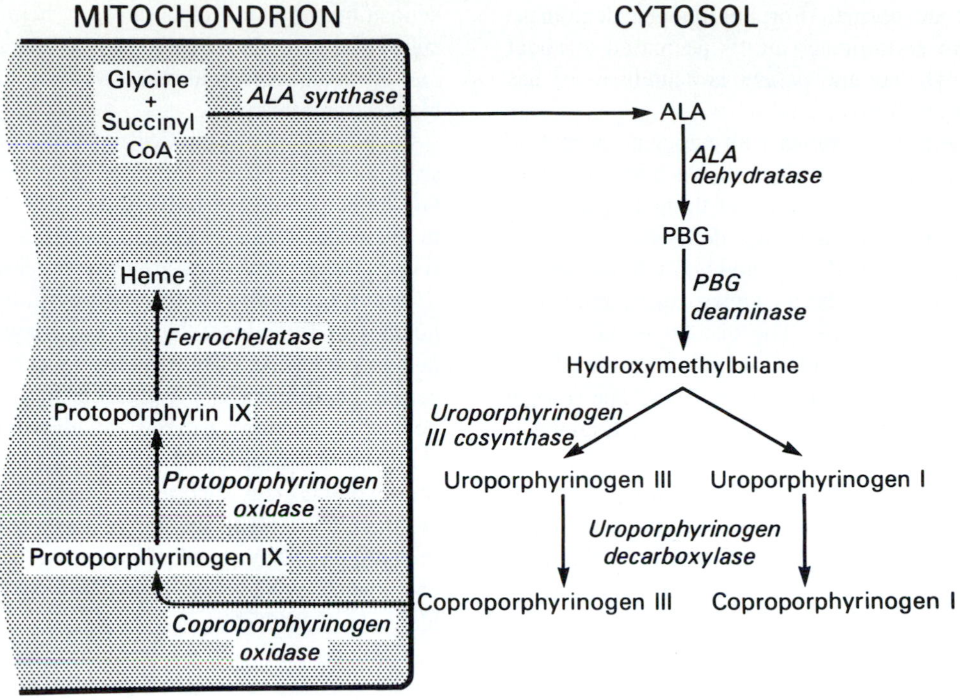

Fig. 32-6 Distribution of the porphyrin pathway between mitochondria and cytosol.

behavior. Attacks may occur spontaneously or may be precipitated by exposure to alcohol, toxins (for example, lead), hormones (for example, estrogens), and a wide variety of drugs. Between attacks, the signs and symptoms of porphyria are usually absent.

The unique features of each neurological porphyria are reviewed below and in Table 32-3.

Acute intermittent porphyria. Acute intermittent porphyria is the most common of the neurological porphyrias. Patients with this disease have a partial deficiency of porphobilinogen deaminase, the enzyme that joins four PBG molecules to form uroporphyrinogen. The defect causes ALA and PBG to accumulate behind the enzyme block, while the drop in the production of heme induces the activity of ALA synthase. Thus ALA and PBG are excreted in the largest amounts in this porphyria. Since the defect does not involve the porphyrinogen portion of the pathway, porphyrins are not produced in excess, and photosensitivity does not occur.

The major diagnostic laboratory finding is an increase in urine ALA and PBG during the acute attacks. However, between attacks their levels may revert to normal. Small amounts of URO may be produced in urine by the nonenzymatic condensation and cyclization of PBG molecules.

Table 32-3 Biochemical and clinical features of the porphyrias

	Acute intermittent porphyria	Variegate porphyria	Coproporphyria	Congenital erythropoietic porphyria	Protoporphyria	Porphyria cutanea tarda
Enzyme defect	Porphobilinogen deaminase	Protoporphyrinogen oxidase	Coproporphyrinogen oxidase	Uroporphyrinogen III cosynthase	Ferrochelatase	Uroporphyrinogen decarboxylase
Inheritance	Autosomal dominant	Autosomal dominant	Autosomal dominant	Autosomal recessive	Autosomal dominant	Autosomal dominant
Signs and symptoms						
Abdominal pain and neurological symptoms	Yes	Yes	Yes	No	No	No
Photosensitivity and cutaneous lesions	No	Yes	Yes	Yes	Yes	Yes
Metabolic expression	Liver	Liver	Liver	Erythroid cells	Erythroid cells and liver	Liver

Fecal porphyrins are normal. Porphobilinogen deaminase may be assayed in erythrocytes and is decreased to about 50% of normal, whether the patient is acutely ill or has the unexpected latent form.

Variegate porphyria. Patients with variegate porphyria suffer from both acute neurological attacks and sensitivity of the skin to sunlight and mechanical trauma. The enzymatic defect appears to be a partial deficiency of protoporphyrinogen oxidase. PROTO, and to a lesser extent COPRO, accumulate in the body, giving rise to photosensitivity and cutaneous lesions. The disease is most common among South African whites and has been traced to a couple who emigrated from Holland in 1688. The disease also has a royal past since King George III of England is thought to have had variegate porphyria.

The finding of ALA and PBG in urine during acute attacks establishes the presence of a neurological porphyria. Variegate porphyria is distinguished from the other two neurological porphyrias by the increased excretion of PROTO, as well as some COPRO, in feces.

Coproporphyria. A partial deficiency of coproporphyrinogen oxidase in this disease causes COPRO to accumulate. Photosensitivity occurs in about one third of patients, less than in variegate porphyria. Increased amounts of COPRO are excreted in feces and urine. By contrast, urinary levels of ALA and PBG are increased only during acute atttacks.

Cutaneous porphyrias

The three cutaneous porphyrias have in common an excess of porphyrins in body tissues, including skin. These molecules absorb light at about 400 nm, which includes a higher energy state. It is thought that this energy may be transferred to molecular oxygen, generating an excited oxygen species that can attack cellular constituents. These events at the molecular and cellular level ultimately translate into photosensitivity and skin lesions.

Unlike the neurological porphyrias, ALA and PBG are not excreted in excess, and no neurological signs or symptoms are present. However, one must remember that two of the neurological porphyrias may also have skin manifestations. Each of the cutaneous porphyrias is briefly discussed below and reviewed in Table 32-3.

Congenital erythropoietic porphyria. Congenital erythropoietic porphyria is the rarest and most severe of the porphyrias. Deficiency of uroporphyrinogen III cosynthase limits production of the type III isomer. In its place, large amounts of the type I isomer series are produced and are eventually oxidized to form URO I and COPRO I. The disease normally manifests in early childhood with extreme photosensitivity. Light-exposed areas of the skin become scarred and, as patients grow older, extensive scarring and mutilation of the fingers, nose, and ears may occur. A unique finding of the disease is *erythrodontia,* the reddish-brown staining of teeth caused by porphyrin deposition. Patients also develop hemolytic anemia and enlargement of the spleen. This is the only porphyria inherited as an autosomal recessive trait and the one with the bleakest prognosis.

Patients excrete a pink or red urine because of the massive amounts of URO and COPRO present. Red cells contain large amounts of URO and COPRO and can be shown to fluoresce when examined under a fluorescent microscope. Fecal porphyrins are also increased.

Protoporphyria. In protoporphyria there is a partial deficiency of ferrochelatase, the last enzyme in the synthetic pathway for heme. The resulting accumulation of PROTO causes photosensitivity beginning in childhood. After exposure to the sun, patients develop burning and itching followed by swelling and redness of the involved area. Sun-exposed areas such as the backs of the hands and face are mainly affected, but skin changes are mild and scarring uncommon. A minority of patients also develop liver disease because the liver is involved in excreting the large amounts of PROTO.

The concentration of free erythrocyte protoporphyrin (FEP) is greatly elevated. Fecal PROTO is also increased; however, urine porphyrins and their precursors are normal.

Porphyria cutanea tarda. Porphyria cutanea tarda is probably the most common of all porphyrias. There is a partial deficiency of uroporphyrinogen decarboxylase, and patients with active disease excrete large amounts of URO and a 7-carboxyl porphyrin in urine. Signs and symptoms consist of photosensitivity, hyperpigmentation or hypopigmentation, mechanical fragility of skin, and blister formation. Porphyria cutanea tarda does not usually appear until adulthood. Moreover, the disease remains dormant until some form of liver dysfunction develops, such as an overload of hepatic iron or alcoholic liver disease. Estrogen therapy may also activate the disease. A toxic porphyria resembling this disease may also occur in people exposed to hexachlorobenzene or other polychlorinated hydrocarbons.

High levels of URO and especially the 7-carboxyl porphyrin in urine are characteristic of porphyria cutanea tarda. Fecal porphyrins are usually normal, but the presence of an unusual porphyrin, isocoproporphyrin, is distinctive.

Secondary disorders of porphyrin metabolism

Alterations in porphyrin metabolism and excretion can occur in situations other than the porphyrias. Several common disorders are described below.

Lead poisoning. Lead poisoning may occur in young children who eat chips of paint containing lead or in adults who are exposed to lead compounds in an industrial setting or who drink ''moonshine'' whiskey distilled in lead-containing equipment. The signs and symptoms include abdominal pain and neurological abnormalities that may mimic an acute attack or porphyria. Lead inhibits two enzymes in the porphyrin pathway, ALA dehydratase and ferrochelatase. Consequently, there is an increase in urinary ALA (but not PBG) and in the erythrocyte concentration of zinc protoporphyrin. In chronic lead poisoning, urinary COPRO is also increased. Although these findings are typical of lead poisoning, the diagnosis rests on the measurement of lead levels in whole blood.

Iron deficiency. Patients with iron deficiency have an imbalance between protoporphyrin, which is produced in normal amounts, and iron, which is not readily available for heme synthesis. As a result, zinc protoporphyrin accumulates in red cells to above normal levels. Because it is such a widespread condition, iron deficiency is the most common cause of porphyrinemia. Other conditions with a decrease in available iron (malignancies, infections, chronic disease) also produce porphyrinemia. Measurement of FEP (that is, zinc protoporphyrin) is a useful screening test for iron deficiency. Of course, this diagnosis is confirmed by studies of serum iron, iron-binding capacity, and ferritin.

Coproporphyrinuria. An isolated increase in urinary COPRO is the most common abnormal finding in screening urine for porphyrins. Although it may indicate a porphyria, it is much more often caused by problems unrelated to heme synthesis, such as liver disease, acute illness, or exposure to toxic compounds. A small, isolated increase in urinary COPRO is therefore usually a nonspecific finding of limited diagnostic value.

CHANGE OF ANALYTE IN DISEASE

The diagnosis of porphyrin disorders is satisfying from a biochemical viewpoint because it is based on the measurement of intermediates in the heme synthetic pathway. The three types of samples analyzed are urine, feces, and whole blood. The laboratory work-up of suspected porphyria depends on the clinical presentation; however, collection of random urine and stool specimens is sufficient for screening purposes. As a rule, positive screening tests are followed with quantitative measurements and, in the case of porphyrins, identification of the elevated porphyrin. The laboratory diagnosis of porphyria is summarized in Table 32-4.

Porphobilinogen

Urinary PBG is elevated in acute intermittent porphyria, variegate porphyria, and coproporphyria. The screening test for PBG is positive during the acute attacks but may be negative between attacks. A positive screening test is confirmed with a 24-hour urine collection for quantitative PBG. The reference interval is 0 to 2 mg/day.

δ-Aminolevulinic acid

Urinary ALA parallels the increase in PBG in the three neurological porphyrias. It is also elevated in lead poisoning and is therefore a less specific indicator of porphyria than PBG. Measurements are made on 24-hour urine collections, and the reference interval is 1.3 to 7.0 mg/day.

Urinary porphyrins

Screening tests for urinary porphyrins are usually positive in all of the porphyrias except protoporphyria and possibly acute intermittent porphyria. Positive urine screens are followed by identification of the porphyrin (URO and/or COPRO) that is elevated. A slight to moderate elevation of COPRO is seen in liver disease, lead poisoning, alcohol ingestion, and acute illness. Larger elevations, especially in URO, require quantitative analysis of a 24-hour urine collection. Reference intervals are less than 40 μg/day for URO and less than 235 μg/day for COPRO.

Patients with congenital erythropoietic porphyria excrete huge amounts of URO and COPRO. In porphyria cutanea tarda, excretion of URO and a distinctive 7-carboxyl por-

Table 32-4 Laboratory diagnosis of the porphyrias*

Porphyria	Urine ALA and PBG†	Urine porphyrins	Fecal porphyrins	Blood porphyrins
Neurological				
Acute intermittent porphyria	↑↑	URO ↑	N	N
Variegate porphyria	↑↑	COPRO ↑	PROTO ↑↑ COPRO ↑	N
Coproporphyria	↑↑	COPRO ↑	COPRO ↑↑ PROTO ↑	N
Cutaneous				
Congenital erythropoietic porphyria	N	URO ↑↑ COPRO ↑↑	COPRO ↑ PROTO ↑	URO ↑↑ COPRO ↑↑
Protoporphyria	N	N	PROTO ↑	PROTO ↑↑
Porphyria cutanea tarda	N	URO ↑↑ COPRO ↑	N	N

**N*, normal; ↑, increased; ↑↑, greatly increased.
†May be increased only during acute attack.

phyrin, with lesser amounts of 6-carboxyl, 5-carboxyl, and coproporphyrin, is seen and depends on the current activity of the disease. Variegate porphyria and coproporphyria are characterized by an excess of COPRO in urine. Patients with acute intermittent porphyria sometimes excrete excess URO because of the nonenzymatic condensation of PBG molecules.

Fecal porphyrins

A positive screening test for fecal porphyrins is typical of four of the porphyrias and is often followed by quantitation of COPRO and PROTO. This is especially important in differentiating variegate porphyria (PROTO greater than COPRO) from coproporphyria (COPRO greater than PROTO). An increase in fecal PROTO is also seen in protoporphyria, whereas urine studies in these patients are normal. The reference intervals for fecal porphyrins are less than 30 μg/g dry weight for COPRO and less than 60 μg/g dry weight for PROTO.

Blood porphyrins

Blood porphyrins are increased in congenital erythropoietic porphyria (URO and COPRO) and protoporphyria (PROTO). Screening is done by the usual solvent extraction methods or by fluorescence microscopy of an unfixed, unstained blood smear. The free ethrytocyte protoporphyrin assay gives quantitative results for blood porphyrins. As previously mentioned, FEP is increased in iron deficiency and lead poisoning as well as protoporphyria. The concentration of blood porphryins is normally less than 500 μg/L erythrocytes.

Red cell enzymes

Only one enzyme, porphobilinogen deaminase, is routinely measured by clinical laboratories. Its activity is decreased to about 50% of normal in all individuals with acute intermittent porphyria, whether the disease is latent or in an acute phase.

Part III: Bilirubin

FORMATION AND STRUCTURE

Heme is degraded in cells of the reticuloendothelial system (Fig. 32-7). Heme oxygenase, assisted by O_2 and NADPH, opens the protoporphyrin ring to release iron, carbon monoxide, and a liner tetrapyrrole, biliverdin. Iron is recycled for future heme synthesis, and carbon monoxide is excreted by the lungs. However, biliverdin is metabolized one step further; the double bond at the center of the molecule is reduced to form bilirubin.* This yellow-

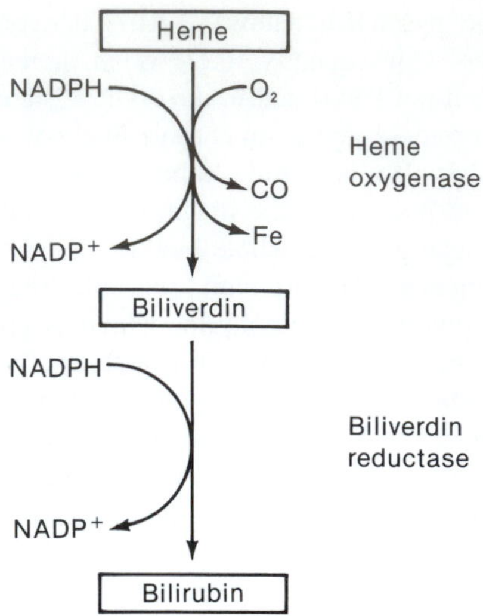

Fig. 32-7 Formation of bilirubin from heme.

orange pigment is the major waste product of heme metabolism.

As a result of the chemical structure of bilirubin, the molecule is very hydrophobic in nature and thus highly insoluble in aqueous solutions. The instability of the molecule in the presence of light is the basis for treating jaundiced infants with phototherapy, since the photoisomers of bilirubin are more polar than the parent compound and can be more easily excreted directly into bile.

METABOLISM

Production

The breakdown of heme-containing proteins generates about 250 to 300 mg of bilirubin per day. Approximately 80% to 85% of bilirubin is derived from the hemoglobin in aged erythrocytes. Most erythrocytes are destroyed within reticuloendothelial cells (extravascular hemolysis), mainly in the spleen. A small percentage of erythrocytes is destroyed within the circulation as well (intravascular hemolysis). The remaining 15% to 20% comes from the destruction of red cell precursors in the bone marrow (ineffective erythropoiesis) and from the turnover of hemoproteins in nonerythroid tissues.

Transport

As bilirubin is released into plasma, it binds very tightly to albumin. Each albumin molecule has one high-affinity binding site for bilirubin and two sites of weaker affinity. Transport of bilirubin in the bound state prevents it from crossing cell membranes and entering tissues, where it would exert toxic effects. Certain anionic drugs, such as sulfonamide antibiotics and salicylates, compete for the bilirubin binds sites on albumin. This is not usually a problem in adults, but infants have a lower binding capacity for bilirubin. In the presence of competing substances,

*Bilirubin is more properly referred to as bilirubin IXα, since it is derived from the type IX isomer of protoporphyrin and ring cleavage takes place at the α-methene bridge. For convenience, the term *IXα* will be omitted in this section.

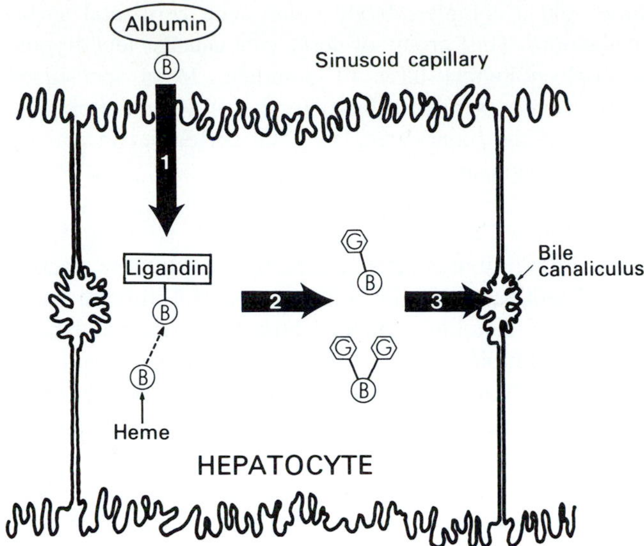

Fig. 32-8 Hepatic metabolism of bilirubin. The indicated steps are *(1)* uptake, *(2)* conjugation, and *(3)* excretion. A small amount of bilirubin is produced by breakdown of heme-containing proteins within hepatocytes. *B,* Bilirubin; *G,* glucuronic acid.

bilirubin may be displaced from albumin, cross the blood-brain barrier, and enter brain cells to cause kernicterus.

In addition to the usual reversible binding, bilirubin may also become covalently bound to albumin in patients with impaired hepatic excretion of bilirubin. This delta bilirubin continues to circulate with its albumin carrier until the albumin itself is removed from the circulation. Because it is attached to a water-soluble protein, delta bilirubin reacts in conventional assays as if it were conjugated (that is, "direct") bilirubin.

Hepatic uptake, conjugation, and excretion

During passage through the microcirculation of the liver, albumin releases bilirubin to hepatocytes (Fig. 32-8). After bilirubin diffuses across the cell membrane, it binds to two cytosolic proteins, ligandin and Z protein. Ligandin is the more important binding protein, and a variety of organic anions bind to it. The cytosolic binding proteins do not appear to be involved in bilirubin uptake. Their function is probably the prevention of the diffusion of bilirubin out of the hepatocyte or into other cellular compartments.

In the endoplasmic reticulum, glucuronic acid residues are added to one or both of the propionate side chains of bilirubin. The reaction is catalyzed by bilirubin UDP–glycuronyl transferase, and UDP–glucuronic acid is the carbohydrate donor. The addition of one or two sugar residues increases the water solubility of bilirubin so that it can be excreted.

Excretion into bile is the rate-limiting step in the hepatic metabolism and bilirubin. It is an energy-dependent process in which bilirubin is transported against a concentration gradient. About 85% to 90% of bilirubin appears in bile as the diglucuronide and about 10% to 15% appears as monoglucuronide. Small amounts of other bilirubin-sugar conjugates (glucosides, xylosides), as well as free bilirubin (photoisomers), are also found in bile.

Intestinal transit

Following excretion into bile, conjugated bilirubin passes through the hepatic and common bile ducts and into the intestinal lumen. There, a portion of the glucuronic acid residues is enzymatically released by beta-glucuronidase. In the distal small intestine and colon, anaerobic bacteria reduce bilirubin to a variety of compounds known collectively as urobilinogens. Unlike the polar bilirubin glucuronide, urobilinogens may be reabsorbed by the intestine and returned via the portal circulation to the liver (enterohepatic circulation). More than 90% of these recirculated urobilinogens is taken up by liver cells and reexcreted into bile; the remainder is filtered by the kidneys

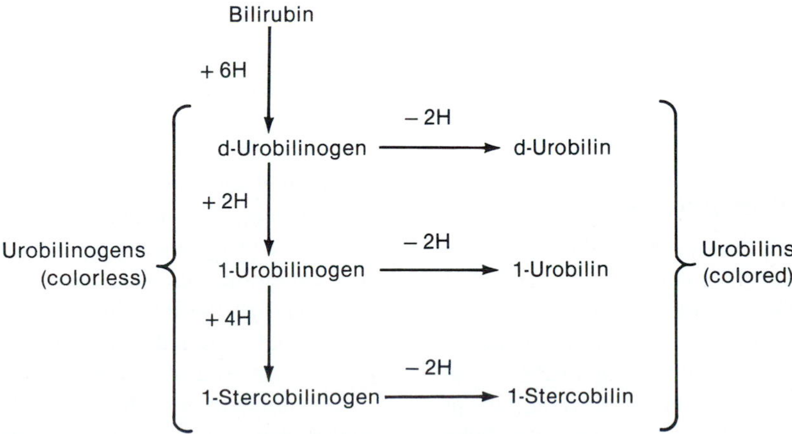

Fig. 32-9 Conversion of bilirubin to urobilinogens in the intestine. It is not known whether glucuronic acid is hydrolyzed before or after the reduction steps.

and excreted in urine. Urobilinogens are colorless, but in the presence of air they oxidize spontaneously to the corresponding urobilins (Fig. 32-9). These colored compounds contribute to the normal color of feces and urine.

PATHOLOGICAL CONDITIONS: HYPERBILIRUBINEMIA

Diseases or conditions that interfere with bilirubin metabolism may cause a rise in its serum concentration. Jaundice, a yellowish discoloration of the sclera and skin, appears when serum bilirubin reaches about 25 mg/L. By itself, hyperbilirubinemia is usually not a threat to health, since adequate mechanisms exist for binding and detoxifying this compound. However, it indicates an abnormality in the production or subsequent metabolism of bilirubin.

Based on the foregoing discussion of normal bilirubin metabolism, there are five processes that lead to hyperbilirubinemia and jaundice:

1. Overproduction
2. Impaired uptake by liver cells
3. Defects in the conjugation reaction
4. Reduced excretion into bile
5. Obstruction to the flow of bile

The first three mechanisms cause an increase in unconjugated serum bilirubin, and the latter two mechanisms lead to the presence of a large component of conjugated bilirubin in serum as well. Since the last four mechanisms are related to liver disease and have been fully discussed in Chapter 23, only the first mechanism is discussed below.

Overproduction

Increased production of bilirubin is usually the result of accelerated red cell breakdown. When the red of hemolysis exceeds the liver's capacity to clear bilirubin from blood, the patient develops hyperbilirubinemia. Serum bilirubin does not usually exceed 50 mg/L in hemolytic states, and it is almost entirely unconjugated. In chronic hemolytic anemias such as sickle cell disease or hereditary spherocytosis, prolonged overproduction of bilirubin may lead to the formation of bilirubin (pigment) gallstones. Ineffective erythropoiesis, seen in thalassemia, pernicious anemia, and other disorders, causes a similar increase in bilirubin production.

CHANGE OF ANALYTE IN DISEASE

An increase in serum bilirubin accompanies a wide variety of pathological states. The majority of diseases or conditions causing jaundice originate in the liver and biliary tree; a smaller percentage are the result of hematological disorders. The clinical laboratory can measure both conjugated and unconjugated fractions of bilirubin, an important first step in diagnosing the cause of jaundice. Other first-line laboratory investigations include aspartate and alanine aminotransferase, alkaline phosphatase, prothrombin time, and a complete blood count with peripheral smear evaluation. This group of tests can usually identify the pathophysiological basis of jaundice. More specialized tests, such as serologicaltests for hepatitis or radiological studies of the biliary tree, are then performed to make a specific diagnosis.

Bilirubin

Serum bilirubin is most commonly measured by reaction with diazotized sulfanilic acid. Conjugated bilirubin reacts in aqueous solution (''direct'' bilirubin), whereas unconjugated bilirubin does not. Addition of an accelerator (alcohol or caffeine) that disrupts internal hydrogen bonding allows the unconjugated fraction to react as well, providing a value for total bilirubin. The unconjugated or ''indirect'' bilirubin is obtained by subtracting direct from total bilirubin. Reference intervals are 2 to 10 mg/L for total bilirubin and 0 to 2 mg/L for direct bilirubin.

Urine bilirubin is derived from the conjugated bilirubin in blood, a portion of which is filtered by the kidneys and excreted in urine. One can easily test for urine bilirubin with commercial dipstick methods, and it is normally undetectable. The presence of bilirubin in urine indicates an elevation in the conjugated fraction of serum bilirubin.

Delta bilirubin

The albumin-bound fraction of serum bilirubin (delta bilirubin) is included in the conjugated fraction of bilirubin when determined by conventional methods. Special techniques, such as high-performance liquid chromatography or a commercial thin-film system developed by Kodak, make it possible to measure delta bilirubin individually. Because it is covalently bound to albumin, delta bilirubin continues to circulate in blood for a week or more after urine bilirubin has disappeared. Clinical studies show that, during recovery from hepatocellular jaundice, the percentage of delta bilirubin may increase in 80% to 90% of total bilirubin. This suggests a role for delta bilirubin in monitoring the recovery phase of hepatic disease.

Urobilinogen

The normal excretion of urobilinogen in urine is 1 to 4 mg/day. Overproduction of bilirubin (for example, hemolytic anemia) increases the amount of urobilinogen formed in the intestine and therefore the amount that is reabsorbed and excreted into urine. Hepatocellular disease may also increase urinary urobilinogen by interfering with its uptake and excretion into bile. Processes that reduce the flow of bilirubin into the intestine (for example, common bile duct obstruction) limit the formation of urobilinogen and thus the amount of urobilinogen in urine.

Urobilinogen may be detected in urine with a commercial dipstick, and an increase provides evidence of a hemolytic cause of jaundice. However, urobilinogen is of little use in the diagnosis of liver disease.

BIBLIOGRAPHY

Iron metabolism

Bothwell, TH, Charlton, RW, Cook JD, and Finch, CA: Iron metabolism in man, Oxford, England, 1979, Blackwell Scientific Publications, Inc.

Bothwell, TH, Charlton, RW, and Motulsky, AG: Idiopathic hemochromatosis. In Stanbury, JB, et al, editors: The metabolic basis of inherited disease, ed 5, New York, 1983, McGraw-Hill, Inc.

Cavill, I, Jacobs, A, and Worwood, M: Diagnostic methods for iron status, Ann Clin Biochem 23:168-171, 1986.

Cook, JD: Clinical evaluation of iron deficiency, Semin Hematol 19:6-18, 1982.

Crosby, WH: Hemochromatosis: current concepts and management, Hosp Pract 22:173–192, 1987.

Fairbanks, VF, and Beutler, E: Iron metabolism. In Williams, WJ, et al, editors: Hematology, ed 3, New York, 1983, McGraw-Hill, Inc.

Fairbanks, VF, and Beutler, E: Iron deficiency. In Williams, WJ, et al, editors: Hematology, ed 3, New York, 1983, McGraw-Hill, Inc.

Finch, CA, and Huebers, H: Perspectives in iron metabolism, N Engl J Med 306:1520-1528, 1982.

Jacobs, A, and Worwood, M, editors: Iron in biochemistry and medicine, II, London, 1980, Academic Press.

Powell, LW, and Isselbacher, KJ: Hemochromatosis. In Braunwald, E, et al, editors: Harrison's principles of internal medicine, ed 11, New York, 1987, McGraw-Hill, Inc.

Schafer, AI, and Bunn, HF: Anemias of iron deficiency and iron overload. In Braunwald, E, et al, editors: Harrison's principles of internal medicine, ed 11, New York, 1987, McGraw-Hill, Inc.

Heme synthesis and the porphyrias

Goldberg, A, and Moore, MR, editors: The porphyrias, Clin Hematol 9:225-451, 1980.

Hindmarsh, JT: The porphyrias: recent advances, Clin Chem 32:1255-1263, 1986.

Kappas, A, Sassa, S, and Anderson, KE: The porphyrias. In Stanbury, JB, et al, editors: The metabolic basis of inherited disease, ed 5, New York, 1983, McGraw-Hill, Inc.

Meyer, UA: Porphyrias. In Braunwald, E, et al, editors: Harrison's principles of internal medicine, ed 11, New York, 1987, McGraw-Hill, Inc.

Schmid, R, editor: The hepatic porphyrias, Semin Liver Dis 2:87-176, 1982.

With, TK: Diagnostic tests for porphyria, Lab Med 11:446-454, 1980.

Bilirubin

Billing, BH: Bilirubin metabolism. In Schiff, L, and Schiff, ER, editors: Diseases of the liver, ed 6, Philadelphia, 1987, JB Lippincott Co.

Isselbacher, KJ: Disturbances of bilirubin metabolism. In Braunwald, E, et al, editors: Harrison's principles of internal medicine, ed 11, New York, 1987, McGraw-Hill, Inc.

Isselbacher, KJ: Jaundice and hepatomegaly. In Braunwald, E, et al, editors: Harrison's principles of internal medicine, ed 11, New York, 1987, McGraw-Hill, Inc.

Ostrow, JD, editor: Bile pigments and jaundice: molecular, metabolic, and medical aspects, New York, 1986, Marcel Dekker, Inc.

Weiss, JS, Gautam, A, Lauff, JJ, et al: The clinical importance of a protein-bound fraction of serum bilirubin in patients with hyperbilirubinemia, N Engl J Med 309:147-150, 1983.

Wolkoff, AW, Chowdhury, JR, and Arias, IM: Hereditary jaundice and disorders of bilirubin metabolism. In Stanbury, JB, et al, editors: The metabolic basis of inherited disease, ed 5, New York, 1983, McGraw-Hill, Inc.

Wu, T-W: Delta-bilirubin: the state of the art and future prospects. In Goldberg, DM, and Walker, WHC, editors: Clinical biochemistry reviews, vol 1, New York, 1987, Pergamon Press, Inc.

Zimmerman, HJ, and Deschner, KW: Differential diagnosis of jaundice, Hosp Pract 22:99-122, 1987.

CHAPTER 33

Hemoglobin

JOHN D. BAUER

OBJECTIVES

- Describe the structure and function of hemoglobin.
- List and define the derivatives of hemoglobin.
- Define thalassemia, and describe the molecular and clinical aspects of alpha and beta thalassemias.
- Define the molecular and clinical aspects of sickle cell anemias.

KEY TERMS

anemia Reduction in hemoglobin or number of red cells.

anoxia Lack of oxygen in tissues.

cyanosis Bluish color caused by lack of oxygen.

erythrocytosis Above-normal increase in number of red cells (polycythemia).

erythropoietin Red cell formation–stimulating hormone.

Heinz body Precipitated hemoglobin in red blood cells.

heme An iron-containing porphyrin (protoporphyrin), which with a protein (globin) composes hemoglobin.

hemoglobin (Hb) An iron-containing respiratory pigment of red blood cells.

hemolysis Destruction of red cells resulting in release of hemoglobin.

hemosiderosis Accumulation of hemosiderin in various organs.

hepatosplenomegaly Increased size of liver and spleen.

hypochromia Reduction of hemoglobin in red cells.

hypoxia Reduced oxygen level in tissues.

myoglobin Oxygen-carrying protein in muscle.

normochromia Normal amount of hemoglobin in red cells.

oxygenation Combination with oxygen.

Plasmodium A blood parasite and causative organism of malaria.

reticulocyte Young red cell containing a network of precipitated organelles visualized when treated with vital dyes.

reticuloendothelial system Collective term for phagocytic cells in tissues, blood vessels, spleen, liver, bone marrow, and lymph nodes.

STRUCTURE AND FUNCTION OF HEMOGLOBIN

Structure

Hemoglobin (Hb) is the red-pigmented, oxygen-carrying protein of red blood cells of vertebrates. It is a globoid tetramer (mw 68,000 daltons) consisting of two pairs of unlike polypeptide chains (Fig. 33-1). Each chain carries an iron-containing porphyrin derivative called *heme,* a ferroprotoporphyrin IX, in which one iron atom is deposited in the center of the porphyrin ring (see Chapter 32). The polypeptide chains (minus heme) are collectively called the *globin moiety* of hemoglobin. Each polypeptide chain is designated by a Greek letter: α, β, γ, δ, ϵ, and ζ. These chains have quite similar three-dimensional structures. The *primary structure,* the amino acid sequence, has been known since the 1960s and demonstrates a closer relationship between β, γ, δ, and ϵ chains than between ζ and α chains. The α chain contains 141 amino acid residues; and the δ, ϵ, and β chains each contain 146 amino acid residues. The amino acid sequence is genetically controlled, and the substitution of one single amino acid can be responsible for at times fatal pathophysiological changes.[1]

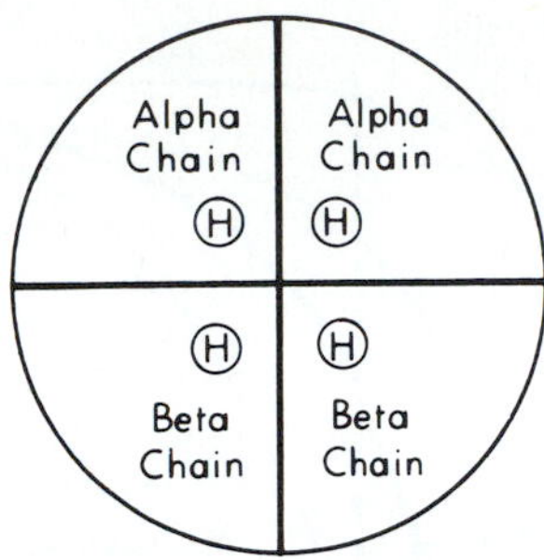

Fig. 33-1 Diagram of structure of Hb A molecule. Four heme groups are attached to one globin molecule, which consists of four polypeptide chains, two of which have an identical amino acid sequence of one type (α chain) and the other two an identical amino acid sequence of another type (β chain). Each polypeptide chain is conjugated to one heme moiety. *H,* Heme *(From Bauer, JD: Clinical laboratory methods, ed 9, St Louis, 1982, The CV Mosby Co.)*

Three fourths of each polypeptide chain is arranged in coiled, α-helical structures. There is further folding of the helices into a specific irregular pretzel configuration, the three-dimensional *tertiary structure*.[2] The heme molecule lies in a pocket between two helices, deeply embedded in the hydrophobic interior of the protein. The complete hemoglobin tetramer forms a globular, doughnutlike molecule that has a central water-filled cavity. The central cavity allows the entrance of 2,3-diphosphoglycerate (2,3-DPG) and salts. The relationship of the subunits to each other in the hemoglobin tetramer is the *quarternary structure* (Fig. 33-2).

Globin chain synthesis and genetics

Differences among the various normal human hemoglobins are based on the variations in the amino acid sequences (the primary structure) of six different globin chains (α, β, γ, δ, ϵ, and ζ), the synthesis of which is controlled by at least six structural genes. The β, δ, and probably the ζ and ϵ chains are coded by one pair of genes, whereas the γ and α chains are coded by two pairs of genes.[2] The α genes are on chromosome 16 and the non-α genes (γ, δ, and β) are closely linked on chromosome 11. The positions of the ϵ and ζ gene loci are not known. Mutation in any of the steps can cause decreased globin synthesis, as demonstrated in the various forms of thalassemia. The ontogenetic sequence of hemoglobin production at different stages of development is also under genetic control, which requires suppression or activation by separate chromosomes or by the gene itself.[3] The α chains are formed early in embryonic life, and once released from the ribosomes they spontaneously, quickly, and sequentially (during the fetal period) combine with the

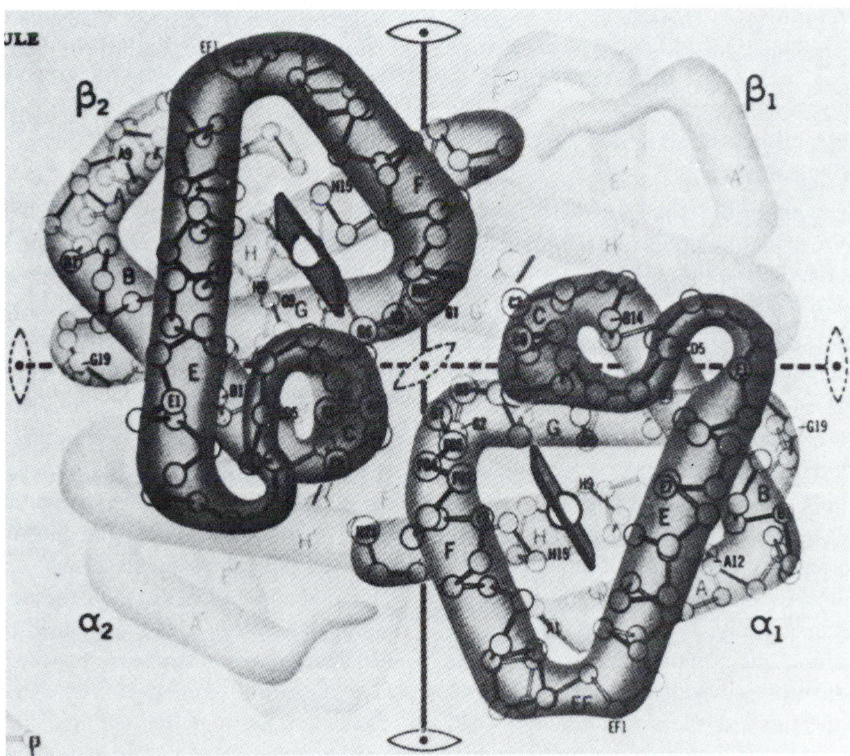

Fig. 33-2 Quaternary structure of hemoglobin. The α_1 and β_2 chains are in foreground, and $\alpha_1\beta_2$ contact is at center. *(From Dickerson, R.E., and Geis, I.: The structure and action of proteins, Menlo Park, Calif., 1969, Benjamin/Cummings.)*

ε, γ, β, and δ chains to form the embryonic (ε), fetal (γ), A_2 (δ), and adult hemoglobins (β). Two single peptide chains of the same type combine to form dimers, and the dimers of different types then combine to form the various tetramers.[4] The proportions and the rate of formation of the dimers conform to the various percentages of the normal hemoglobins, but the factors involved in precipitating the switch from intrauterine to adult hemoglobins are unknown.[5]

NORMAL BIOCHEMISTRY

Assembly of hemoglobin

Adult hemoglobin is a tetrameric protein having two α and two β polypeptide chains. Both are synthesized in equal amounts, although there may be an excess of α chains[6] in the cytoplasm of young red cells. The assembly process starts with the release of α and β chains from the ribosomes into the cytoplasm. They immediately incorporate heme (see Chapter 32 for discussion on the synthesis of heme) and form monomer combinations and dimer aggregates followed by the synthesis of tetramers.[7] In hemoglobinopathies the concentrations of the two like chains (such as βA and βS) may differ, even though their rates of synthesis are the same. There is evidence that the relative rates of assembly (rather than synthesis) of the two hemoglobins (Hb A and Hb S) from their subunits may differ because of difference in affinities of βA and βS for α chains. The α chains, if in short supply, prefer to combine with normal β chains rather than with the variant chains,[8] and the excess variant chains are then removed by proteolysis.

Relationship between function and structure of hemoglobin (see also Chapter 21)

Hemoglobin and oxygen: the oxygen-dissociation curve. The four subunits of hemoglobin each contains a heme moiety deep in the pocket of the globin chains, leaving one edge of the heme exposed to receive the oxygen. Each of the four heme iron atoms can bind reversibly one oxygen molecule. Because the iron stays in the ferrous form, the reaction is an oxygenation, not an oxidation.

To fulfill its function as respiratory pigment, hemoglobin must specifically bind with high affinity to large quantities of oxygen, transport them, and unload them at the relatively high oxygen tension of tissues. Approximately 1.34 mL of oxygen is bound by each gram of hemoglobin. The tetrameric structure of hemoglobin is responsible for its unique oxygen-binding capacity and renders it physiologically superior to single hemoglobin subunits or to myoglobin. The heme iron has six valence bonds, four of which are occupied by the four pyrrole rings of heme. The fifth iron valency bond attaches heme to globin, leaving the sixth iron valency bond available for a reversible combination with oxygen or other ligands.[9]

The affinity of hemoglobin for oxygen depends on the partial pressure of oxygen (Po_2). A plot of the oxygen content (percent oxygen saturation) against Po_2 in case of myoglobin or hemoglobin subunits results in a hyperbolic oxygen-dissociation curve, but a similar plot using hemoglobin gives a sigmoid curve (Fig. 33-3). The hyperbolic curve indicates appreciable release of oxygen at very low partial pressures only, whereas the sigmoid curve indicates release of oxygen much earlier at relatively high oxygen tension to allow adequate oxygenation of tissues. Its sigmoid shape is related to the fact that oxygenation of one heme group increases the oxygen affinity of the others, a phenomenon called the *heme-heme interaction* (or subunit cooperativity), which is responsible for the physiologically efficient uptake and release of oxygen. There is a progressive change in oxygen affinity as each heme molecule becomes oxygenated; the affinity for oxygen is low at first but increases as each heme molecule takes up oxygen. In the case of myoglobin, the hyperbolic dissociation curve indicates that each molecule is oxygenated independently. In the lung, at a Po_2 of about 95 mm Hg, arterial blood

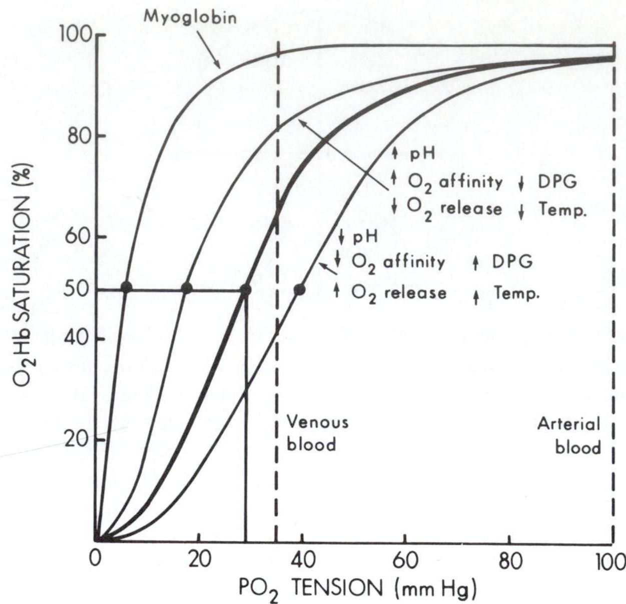

Fig. 33-3 Oxygen-dissociation curves of normal human hemoglobin. *Heavy middle line,* Dissociation curve of normal adult blood (temperature 37° C, pH 7.4, Pco_2 35 mm Hg). *Dots,* P_{50} values, partial pressure of oxygen (27 mm Hg) at which hemoglobin solution is 50% oxyhemoglobin and 50% deoxyhemoglobin. If temperature increases, pH decreases, or carbon dioxide tension (Pco_2) increases, the curve shifts to right. This shift increases release of oxygen from hemoglobin at given oxygen tension by decreasing its oxygen affinity. If temperature decreases, pH rises, or carbon dioxide tension decreases, oxygen-dissociation curve moves to left. This shift increases oxygen-binding capacity of hemoglobin at given oxygen tension; thus there is a decrease in oxygen release. *(From Bauer, JD: Clinical laboratory methods, ed 9, St Louis, 1982, The CV Mosby Co.)*

becomes 97% saturated with oxygen and carries 20 volumes of oxygen per equivalent blood volume. In the capillary bed, venous blood at a Po_2 tension of about 40 mm Hg is still about 75% saturated with oxygen but is nevertheless able to give up 46 volumes of oxygen per 1000 mL of blood. The 75% of hemoglobin returned to the lung in oxygenated form establishes a large reservoir for improved oxygen delivery to tissues.

The position of the oxygen-dissociation curve is determined by a number of factors (see following discussion) affecting the affinity of hemoglobin for oxygen. It is conventionally indexed by the P_{50} value, the Po_2 at which the hemoglobin is 50% saturated with O_2, which normally occurs at a Po_2 of 27 mm Hg. The higher the P_{50}, the lower the affinity of hemoglobin for oxygen. A decreased P_{50} indicates a shift to the left, an increased oxygen affinity of hemoglobin, and an impaired oxygen release to tissues. It is seen (1) in high concentration of Hb F, the γ chain of which binds 2,3-DPG poorly; (2) in hemoglobin ligands, such as methemoglobin and carboxyhemoglobin; (3) in certain hemoglobin variants, such as Hb Ranier; and (4) after massive transfusions with DPG-depleted blood (Fig. 33-3). A shift to the right indicates a decreased oxygen affinity, which eases the delivery of oxygen to tissues. It is seen in various types of hypoxia, such as high altitude, severe anemia, and heart and lung disease.[10]

Oxygen affinity and transport. Oxygen affinity and transport depend not only on Po_2 (see previous discussion of oxygen-dissociation curve) but also on temperature, pH (Bohr effect), and 2,3-DPG concentration.

Bohr effect. The Bohr effect expresses the fact that the oxygen affinity of hemoglobin varies with the pH. In the physiological pH range, the affinity of hemoglobin for oxygen decreases as the acidity increases and the dissociation curve shifts to the right. The Bohr effect aids in the transfer of oxygen (and carbon dioxide) in the acid milieu of tissues in which carbon dioxide and acid metabolites accumulate. Deoxyhemoglobin binds H^+ more actively than does oxyhemoglobin because the latter is a stronger acid. The preferential binding of H^+ to deoxyhemoglobin, the concomitant linkage to 2,3-DPG, and the increased CO_2 tension at tissue levels shift the dissociation curve to the right, decreasing the affinity of hemoglobin to oxygen. The uptake of hydrogen ions by hemoglobin aids in the buffering of the acid metabolites (see Chapter 21) and in the stabilization of deoxyhemoglobin.

2,3-Diphosphoglycerate (2,3-DPG). Unlike other tissues that contain only trace amounts of 2,3-DPG, red cells contain large quantities of the organic ester, the concentration of which is equimolar with that of deoxyhemoglobin. Of the factors that affect oxygen release from hemoglobin (temperature, pH, Po_2, Pco_2, and 2,3-DPG), 2,3-DPG is the most important.[11] It is the most abundant glycolytic intermediate in red cells (see Chapter 29). 2,3-DPG combines with deoxyhemoglobin, and, by reducing the affinity of hemoglobin for oxygen, it shifts the dissociation curve to the right. Functionally, the 2,3-DPG concentration is closely related to the efficiency of oxygen transport and depends on the red cell pH and Po_2. When the pH drops, as in acidosis, the oxygen-dissociation curve moves to the right, but the resulting inhibition of 2,3-DPG corrects the shift by an equal change to the left. An elevated red cell pH pushes the dissociation curve to the left, but the rising 2,3-DPG concentration shifts it to the right, returning it to the base position. The common denominator of the DPG-Bohr effect is the rate of glycolysis,[12] which is stimulated by alkalosis and suppressed by acidosis[13] because the former stimulates phosphofructokinase activity and the latter suppresses it.[14]

Hemoglobin and its derivatives

In addition to oxyhemoglobin and deoxyhemoglobin discussed previously and in Chapter 21, other chemically modified forms of hemoglobin exist.

Carboxyhemoglobin. Carbon monoxide (CO) is a ligand that, like oxygen, binds reversibly to the ferrous ion of hemoglobin; however, it forms a toxic compound carboxyhemoglobin (CO-Hb). It also binds other heme-containing proteins, such as myoglobin, cytochrome P-450, and cytochrome oxidase.[15] The affinity of hemoglobin for CO is 218 times greater than that for oxygen.[16] Because both CO and O_2 compete for the same heme-binding sites, CO displaces O_2, reduces the concentration of oxyhemoglobin, and prevents the formation of deoxyhemoglobin. At a CO concentration of 0.1% in the inhaled air, more than 50% of the hemoglobin is not available for O_2 transport. CO combines with hemoglobin more slowly than oxygen, but the union is much firmer and the release of CO is 10,000 times slower than the release of oxygen from O_2-Hb. In the presence of CO, even the oxyhemoglobin dissociates more slowly because the iron atoms not bound to CO have a higher affinity for O_2, causing the oxygen-dissociation curve to shift to the left. CO binds to heme in a manner similar to oxygen.

Methemoglobin. Methemoglobin (Met-Hb) is oxidized hemoglobin (oxy- or deoxy- forms) in which the ferrous ion of hemoglobin has been oxidized to the ferric state to form ferrihemoglobin. Methemoglobin cannot bind oxygen reversibly and is unable to act as an oxygen carrier. If present in high enough concentrations (over 30% of total hemoglobin), it is responsible for hypoxia and cyanosis (methemoglobinemia).

Normally, after prolonged standing (auto-oxidation), oxyhemoglobin turns brown because of methemoglobin formation, which is also responsible for the brown color of blood in acid urine.

Physiologically, methemoglobin is continuously being formed within erythrocytes because of spontaneous oxidation of hemoglobin, but it is prevented from accumulating within the red cells by the reduction of oxidized heme by a number of enzyme systems that restrict its normal concentration to less than 1% of total hemoglobin.

Hemichromes. When the oxidation of hemoglobin to methemoglobin continues, methemoglobin is converted to compounds called hemichromes, which ultimately precipitate to form *Heinz bodies,* which are responsible for the lysis of the affected red cells.[17] Hemichromes are greenish ferric compounds with a characteristic absorption spectrum. The steps leading to cell lysis are as follows:

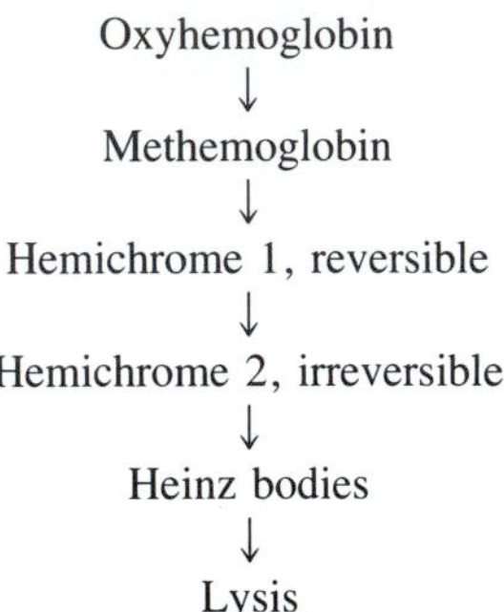

Sulfhemoglobin. Sulfhemoglobin (S-Hb) is a stable compound resulting from the linkage of sulfur to hemoglobin. The toxic effects of certain drugs on hemoglobin not only lead to the formation of methemoglobin but also to concomitant S-Hb production.[18] Sulfhemoglobinemia appears in some persons after exposure to sulfonamides, phenacetin, acetanilid,[19] and trinitrotoluene (TNT). The structure of S-Hb is unknown, but sulfur is probably linked to heme.[20] The S-Hb complex is stable and irreversible (thus differing from the reversible Met-Hb) and does not disappear from the circulation until the involved red cells complete their life cycle. Sulfhemoglobinemia produces anoxia and cyanosis, which clinically are indistinguishable from the anoxia and cyanosis of methemoglobinemia.

Normal human hemoglobins

Molecular structure. Adult and fetal hemoglobins contain α subunits linked to β, γ, or δ chains. In the embryo the ζ and ϵ chains are their precursors (Table 33-1). Hemoglobin is a mixture of variants, some genetically controlled and some acquired. The genetically controlled variants differ from each other in the structure of their globin chains (Table 33-1).

Table 33-1 Normal human hemoglobins

Designation	Tetrameric structure	Hemolysate (%)	
		Adult	Newborn
Adult			
Hb A	$\alpha_2\beta_2$	95-98	20-30
Hb A_2	$\alpha_2\delta_2$	2-3	0.2
Fetal			
Hb F	$\alpha_2\gamma_2$	>1	80
Embryonic			
Gower I	$\zeta_2\epsilon_2$	0	0
Gower II	$\alpha_2\epsilon_2$	0	0
Hb Portland	$\zeta_2\gamma_2$	0	0

Hemoglobin A ($\alpha_2\beta_2$). Hemoglobin A makes up the major portion (95% to 98%) of the adult hemolysate. Small amounts of Hb A are produced in the last 6 weeks of fetal life (see Chapter 37), along with the predominant Hb F (Fig. 33-4). During the ensuing 6 to 12 months of the infant's life, Hb A reaches adult concentrations and is accompanied by only traces of Hb F and Hb A_2. One pair of polypeptide chains of Hb A is composed of α chains, and the other is composed of β chains, so that the complete molecule has the $\alpha_2\beta_2$ formula.

Hemoglobin A_{1c}. Postsynthetic, nonenzymatic glycosylation of the terminal amino group (valine) of the β chain is responsible for the production of Hb A_{1c}. Variations of this hemoglobin are the adduct of glucose-6-phosphate[21] or of fructose-1,6-diphosphate to the β chain.[22] See Chapter 29 for further discussion of glycosylated hemoglobins.

Hemoglobin A_2. Hb A_2 contains a pair of α chains and a pair of δ chains, so that its molecular structure is $\alpha_2\delta_2$. It is a minor component of hemoglobin that makes its first appearance close to term (0.2% of cord blood hemolysate) and remains at low concentration (2.5%) throughout adult life. Its exact function is unknown, but it is probably similar to that of Hb A.[23] Its concentration is increased in β-thalassemia and in β-chain unstable hemoglobins. Normal or decreased values are seen in α-thalassemias and in hemoglobin Lepore heterozygotes.[24] In hemoglobin Lepore homozygotes, Hb A_2 is absent because there is no δ-chain synthesis.[25] Hb A_2 is decreased in iron deficiency and in lead poisoning.

Fetal hemoglobin (Hb F). Hemoglobin F is the major hemoglobin of fetal life, preceded by the embryonic hemoglobins Gower I, Gower II, and Portland. Its molecular formula is $\alpha_2\gamma_2$. It is a mixture of two molecular species in which the γ chains have either glycine ($^G\gamma$) or alanine ($^A\gamma$) at position 136.[26] At birth the Hb F $^G\gamma/^A\gamma$ ratio is about 3:1, whereas in the normal adult hemoglobins the $^G\gamma/^A\gamma$ ratio of the small amount of Hb F (less than 1%) is 2:3. In the first months of fetal life, Hb F competes with the Gower hemoglobins, which are replaced by Hb F at the end of the second month, at which time the fetal hemoglobin concentration is about 90%. It remains at this level until just before term. At birth the red blood cells contain about 70% to 90% Hb F, although higher concentrations have been reported. After birth, Hb F decreases rapidly to about 50% to 70% at the end of the first month, to 25% to 60% at the end of the second month, and to 10% to 30% at the end of the third month. After 6 months and in the first year of life, the Hb F concentration falls from 8% to 2%; in the second year it falls to 1.8%; and in the third year it falls to 1.0%; it finally levels off to the adult level of less than 0.4%,[27] a level not detectable by routine laboratory methods.

The following characteristics and functions of Hb F:

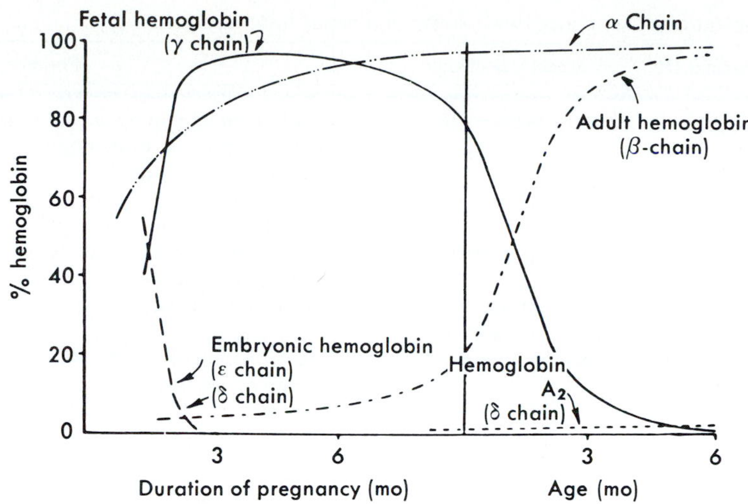

Fig. 33-4 Developmental changes in human hemoglobins. α-Chain synthesis begins well before third month of fetal life and continues into adult life. γ-Chain synthesis replaces embryonic ϵ-chain at about third month, and at about same time β-chain synthesis begins. *(From Bauer, JD: Clinical laboratory methods, ed 9, St Louis, 1982, The CV Mosby Co.)*

1. Electrophoretically it is slower than Hb A.
2. It resists alkali denaturation, a feature that is the basis of the Singer test for Hb F.[28]
3. It is twice as resistant to acid elution as Hb A, a characteristic that forms the basis of the Kleihauer elution technique.[29]
4. It is oxidized to Met-Hb twice as fast as Hb A, so that the newborn readily develops cyanosis as a result of methemoglobin concentration of 15 to 20 g/L blood in response to oxidant chemicals and drugs (see the discussion of methemoglobin).
5. It has a higher oxygen affinity than Hb A, since it binds 2,3-DPG to a lesser degree than adult hemoglobin because of its γ chain. This characteristic allows oxygen transport across the placental villi, despite their low oxygen concentration (80%).[30]

Hb F shows three cellular distribution patterns:

1. Hb A and Hb F are seen in strictly *separate cell populations:* This distribution pattern is seen in fetal-maternal hemorrhage if the mother's blood is examined or in maternal-fetal hemorrhage if the infant's blood is examined.
2. *Even distribution* of Hb F and Hb A within red cells: This distribution is seen in hereditary persistence of Hb F.
3. *Uneven distribution* of Hb A and Hb F in the red cells: This pattern is seen in thalassmeia, S-S disease, Fanconi's anemia, and hereditary spherocytosis.

Embryonal hemoglobins. The embryonal hemoglobin is a mixture of hemoglobins Gower I ($\zeta_2\epsilon_2$), Gower II ($\alpha_2\epsilon_2$), Portland ($\zeta_2\gamma_2$), and Hb F ($\alpha_2\gamma_2$). The first two hemoglobins are not detectable after the first months of gestation,[27] whereas small amounts of Hb Portland may persist into fetal life.[31] The embryonic hemoglobins are able to combine with oxygen at the low oxygen tension and low pH of interstitial fluid. They are detectable in red cells by a modification of the Kleihauer method for Hb F.

PATHOLOGY

Hemoglobinopathies

The inherited disorders of hemoglobin, the hemoglobinopathies, are genetic disorders of structure and synthesis of one or more of the globin polypeptide chains. Inherited disorders of the same moiety are not known. One can divide hemoglobinopathies into a number of overlapping groups[32]: (1) the structural hemoglobin variants that involve substitution, addition, or deletion of one or more amino acids of the globin chain; (2) the thalassemias, a group of disorders in which there is a quantitative defect in globin chain production; (3) combinations of types 1 and 2 that result in complex hemoglobinopathies; and (4) hereditary persistence of fetal hemoglobin, an asymptomatic disorder.

Structural hemoglobin variants

Nomenclature. Hemoglobin variants are assigned letters of the alphabet (such as *S, C,* and *E),* places of discovery (such as Hb D-Los Angeles, Hb-Köln, and Hb C-Harlem), and names of families in which they were first discovered (such as Hb Lepore). A more logical nomenclature is based on the position of the amino acid substitution, such as Hb S = Hb $\alpha_2\beta_2^{6val}$. There are about 500 variant hemoglobins identified at present (Table 33-2).

Excellent tables dealing with various aspects of hemoglobin variants, such as nomenclature, molecular structure,

Table 33-2 Clinical manifestations associated with some abnormal hemoglobins

Disorder	Abnormal Hb	Structural change	Comments
Hemolytic anemia	H	$alpha_2beta_2 \rightarrow beta_4$	Unstable hemoglobin occurring in some forms of alpha-thalassemia; precipitation of hemoglobin and hemolysis are accelerated by certain drugs
	S	beta 6 glu → val	Forms molecular aggregates when deoxygenated, producing sickle cell anemia in homozygotes
	C	beta 6 glu → lys	Low solubility lessens plasticity of red cells, causing hemolytic anemia in homozygotes
	D_{Punjab}	beta 121 glu → gln	Mechanism unknown
	E	beta 26 glu → lys	
	Zurich	beta 63 his → arg	Unstable hemoglobin precipitated by certain drugs, producing hemolytic anemia in heterozygotes
	Köln	beta 98 val → met	
	Sydney	beta 67 val → ala	
	Santa Ana	beta 88 leu → pro	
	Philly	beta 35 tyr → phe	Unstable hemoglobin causes congenital nonspherocytic hemolytic anemia in heterozygotes; precipitated hemoglobin tends to form inclusion bodies within red cells, under certain conditions
	Gun Hill	beta deletion of 5 residues between 90 and 96	
Cyanosis caused by methemoglobinemia	M_{Boston}	alpha 58 his → tyr	
	M_{Iwate}	alpha 87 his → tyr	Methemoglobin causes cyanosis in heterozygotes; some also have evidence of hemolytic anemia
	$M_{Hyde\ Park}$	beta 92 his → tyr	
Cyanosis caused by increased deoxyhemoglobin	Kansas	beta 102 asn → thr	Decreased oxygen affinity of hemoglobin causes cyanosis in heterozygotes
Polycythemia	$J_{Capetown}$	alpha 92 arg → gln	
	Chesapeake	alpha 92 arg → leu	Increased oxygen affinity of hemoglobin hinders release of oxygen to tissues, causing compensatory polycythemia in heterozygotes
	Rainier	beta 145 try → cys	
Hydrops fetalis	Bart's	$alpha_2gamma_2 \rightarrow gamma_4$	Unstable hemoglobin with high oxygen affinity occurring in high concentration in stillborn fetuses with homozygous alpha-thalassemia

From Schmidt, RM, and Brosious, EM: Basic laboratory methods of hemoglobinopathy detection, Atlanta, 1978, Centers for Disease Control.

clinical manifestations, and electrophoretic mobility are found in reference 33.

Classification. Hemoglobin variants are classified according to (1) the molecular mechanisms responsible for the abnormal hemoglobin (or thalassemia), (2) clinical and functional manifestations, and (3) their electrophoretic behavior.

Molecular mechanisms responsible for structural hemoglobin variants. Five basic types of structural changes are responsible for most hemoglobin variants: (1) amino acid substitution, (2) deletions and insertions, (3) unequal crossing-over (fusion genes), (4) chain elongation, and (5) frame shift variance.[34]

Clinical consequences of structural alterations of hemoglobin molecule

Structural alterations of the hemoglobin molecule are responsible for a wide range of clinical manifestations. Most mutations are asymptomatic because they do not interfere with hemoglobin function. Others produce disease because they affect the stability, shape, or function of the hemoglobin molecule. A person homozygous for an abnormal hemoglobin may have striking clinical manifestations (such as sickle cell anemia—Hb S), whereas a person heterozygous for the abnormal hemoglobin (Hb A–Hb S) may be asymptomatic. Even in the homozygous state, some hemoglobin variants (Hb C, Hb D, Hb E) produce only mild symptoms, whereas others are responsible for almost specific pathophysiological changes, such as cyanosis and erythrocytosis.

The clinical disorders can be grouped as follows[34] (Table 33-2):

1. *Hemolytic anemias*. Hb S and Hb C alter the shape and deformity of the red cells. Intraerythrocytic crystals may be formed. Unstable hemoglobins are responsible for intraerythrocytic Heinz body inclusions. The affected cells are prematurely destroyed in the spleen; thus their life span is greatly shortened.
2. *Cyanosis*. M hemoglobins are responsible for methemoglobinemia caused by amino acid substitution near the heme pocket.
3. *Erythrocytosis*. Amino acid substitution causes high oxygen affinity and tissue hypoxia, with the latter

being responsible for erythropoietin stimulation. Hb Chesapeake, Hb Rainier, and Hb Ypsilanti fall in this group.

4. *Hypochromic anemias*. The mutation reduces the hemoglobin output. Examples are Hb Lepore and Hb Constant Spring.

Sickling disorders: sickle hemoglobin

Sickling disorders embrace the homozygous form of the sickle cell gene (sickle cell anemia), the heterozygous form of the sickle cell gene (sickle cell trait), and the combination with other structural hemoglobin variants or with various types of thalassemia.

Sickle hemoglobin (Hb S) is the result of the substitution of valine for the normally occurring glutamine residue at the 6 position of the β chain, and so its structural formula is Hb $\beta^{6(A_3)glu\text{-}val}$. The result of this substitution is the formation of sickled cells by deoxygenated Hb S.

Hb S is not the only hemoglobin that sickles, but it is the most important one. In America and Africa, Hb S is the most common hemoglobin variant, with an incidence of approximately 8% in American blacks (heterozygous form) and 30% in African blacks. The mutation probably originated in Central Africa and spread to countries bordering the Mediterranean Sea so that it can also be found in nonblack inhabitants of these areas. In Africa the high frequency of the sickle cell gene has persisted because heterozygotes for Hb S are somewhat protected from malaria because the *Plasmodium* spp. fails to grow in Hb S and also leads to premature destruction of the affected cell, the Hb S of which sickles readily because of the intraerythrocytic acidosis.[35]

Molecular mechanism of sickling. Deoxygenated Hb S tetramers aggregate to form rodlike fibers that align and are responsible for an intraerythrocytic phase change from solute to gel, which is designated as *gelation*.[36] The critical process, the gel formation, is favored by (1) a high concentration of deoxyhemoglobin S, (2) appropriate temperature (37° C is optimum), (3) acidosis (optimum pH of 6.8), (4) a high concentration of 2,3-DPG, (5) low ionic strength, and (6) a low state of oxygenation.[36] Sickling is reduced by a lowered concentration of Hb S, by a lowered mean corpuscular hemoglobin concentration (MCHC), by the presence of other hemoglobins (Hb A, Hb F, and Hb C), and by increased oxygen tension. All these points are clinically significant (Table 33-3). The admixture of Hb D and Hb O is responsible for a clinically more severe disease.[37]

Pathophysiology of sickle cell disease. Hb S is inherited as an autosomal codominant trait. The sickle-shaped cells temporarily or permanently block microcirculation, and the resulting stases lead to hypoxia and ischemic infarcts of various organs.[36] The vascular endothelial lining, damaged by lack of oxygen, attracts platelets, which initiate the process of diffuse intravascular coagulation (DIC).[38]

Table 33-3 Varying clinical severity of the different sickle syndromes

Genotype	% of hemoglobin S	% of non-S hemoglobin	Clinical severity
SA	30-40	60-70 (A)	0
SF*	70	30 (F)	0
SS	80-90	5-15 (F)	+ +/+ + + +
S-thalassemia	80	20 (A + F)	+/+ + +
SC	50	50 (C)	+/+ + +
SO.SD	30-40	60-70 (O.D)	+ +/+ + + +

From Bunn HF: Sickle cell anemia and other hemoglobinopathies. In Beck, WS, editor: Hematology, Cambridge, MA, 1981, MIT Press.
*Double heterozygous state for hemoglobin S and hereditary persistence of fetal hemoglobin.

Sickled cells lack the deformability of normal cells because the polymerization of deoxyhemoglobin S produces a membrane defect that greatly shortens the life span of the affected cells. The ensuing hemolytic anemia is augmented by the inability of the bone marrow to respond adequately to the anemia because of ineffective erythropoiesis. The hemolysis is responsible for hyperbilirubinemia, reticulocytosis, bone marrow erythroid hyperplasia, gallbladder pigment stones, and osteoporosis as a result of the expanding bone marrow.

Sickle cells exhibit oxygen-transport abnormalities.[39] In sickled cells the oxygen-dissociation curve is shifted to the right. The resulting decreased oxygen affinity favors the release of oxygen at higher oxygen tensions but also supports the formation of deoxyhemoglobin and sickling.[32] The shift to the right of the oxygen equilibrium is caused by an elevated 2,3-DPG concentration and by an Hb S polymerization–mediated effect.

Sickle cell trait (Hb AS). About 8% of American blacks (30% of Central African blacks) are heterozygous for Hb S because they inherit Hb S gene (β^S) from one parent and normal Hb A (β^A) from the other.

Persons with sickle cell trait are usually asymptomatic and have a normal hemogram and red cell survival. The demonstration of Hb S is of no clinical significance but should suggest genetic counseling. There are rare reports of sickling complications in AS patients. They include (1) spontaneous hematuria in about 3% of patients and more frequently hyposthenuria because of the impairment of the concentrating power of the kidneys, both signs pointing to sickling within the vessels of the medulla; (2) rupture of the infarcted spleen; (3) sudden death after strenuous exercise at high altitude; (4) sickling crisis in flight; and (5) rarely, proliferative retinopathy.

Sickle cell anemia (Hb SS). Sickle cell anemia is a chronic, moderate-to-severe hemolytic anemia in a person homozygous for Hb S, having inherited the Hb S gene from both parents, a fact that may have to be ascertained to differentiate sickle cell anemia from sickle cell–β thalassemia or Hb S–hereditary persistence of Hb F. The disease, a β-chain variant, is not evident at birth and does not manifest itself until the γ chains of the newborn are replaced by β^S chains after 3 to 6 months of life.

The clinical severity of Hb SS disease varies from patient to patient. The pathophysiological consequences of Hb SS can be summarized under (1) the hemolytic anemia and (2) the sickle cell crisis, both the result of the polymerization of Hb S within the red cells. Variations in the clinical presentation of Hb SS disease are the result of the influence of genes other than the β^S genes that affect hemoglobin formation.[40]

The hemoglobin values hover around 70 to 80 g/L accompanied by a greatly elevated reticulocytosis (10%). The hemoglobin electrophoretogram shows absence of Hb A (no β^A chains), 80% to 95% Hb S, 2% to 4% Hb A_2, and 2% to 20% Hb F. Three outstanding biochemical findings should be mentioned: hyperuricemia in patients with altered tubular function; reduced zinc levels in plasma, red cells, and hair; and high LD levels in patients in crisis.

Sickle cell–Hb C disease. Sickle cell–Hb C disease has a relatively high incidence (1:833 births among blacks in the United States) because Hb S and Hb C are the most common hemoglobin variants.[41] The patient inherits one hemoglobinopathy from each parent, and the resulting disease is a mild-to-moderate hemolytic anemia associated with the same vaso-occlusive complications seen in Hb SS disease. These complications, however, usually occur with a lower frequency. Since the genes are allelic β-chain mutations, no normal β chains are formed and Hb A is absent.

The smear of the peripheral blood shows many target cells, rare sickle cells, and red cells with straight or curved hemoglobin crystals. The hemoglobin values range from 100 to 130 g/L. The reticulocyte count varies from 3% to 10%. The alkaline cellulose acetate electrophoretogram (pH 8.6) shows only Hb C and Hb S bands. Hb A is absent. Hb A_2 is normal, but its band is hidden by the Hb C band. Hb F may be slightly elevated (1% to 2%). At alkaline pH, Hb O has the same mobility as Hb C but can be separated by citrate agar electrophoresis. Family studies confirm the diagnosis.

Hb C trait and disease

Hb C trait (Hb AC). Hb AC disease affects about 3% of American blacks. It is asymptomatic and the peripheral smear is normal, except for a few target cells. Electrophoresis patterns show about 30% to 40% Hb C, 50% to 60% Hb A, 3% to 4% Hb A_2, and 1% Hb F.

Hemoglobin C disease (Hb CC). The homozygous form is rare, occurring in 1 out of 10,000 American blacks,[42] and is asymptomatic.

Unstable hemoglobin disorders

The unstable hemoglobins are structural variants of Hb A, in which the mutant hemoglobin is less stable than normal hemoglobin. There are about eight unstable hemoglobins described, representing the largest single group of hemoglobin variants.[43] The molecular distortion responsible for unstable hemoglobins produces a series of pathophysiological effects that one can evaluate by laboratory methods, although they are not equally expressed by all unstable hemoglobins. They are (1) hemolytic anemia, (2) increased methemoglobin and sulfhemoglobin production, (3) hemichrome formation, (4) inclusion (Heinz) body formation, (5) altered oxygen dissociation, (6) drug sensitivity, (7) altered electrophoretic mobility (rare), (8) altered response to hemoglobin stability tests, and (9) passage of dark urine. The deeply pigmented urine is caused by mesobilifuscin, a dipyrrole, derived from the catabolism of Heinz bodies or free heme.[44]

Thalassemias

Definitions. Thalassemias are inherited hemoglobinopathies resulting from a decreased rate or production of one or more of the globin chains of hemoglobin.[32,45] They are quantitative hemoglobinopathies that differ from the qualitative hemoglobinopathies by the fact that the structure of the affected globin chain (or chains) is normal but its synthesis is reduced or absent. The decreased hemoglobin synthesis results in decreased red cell hemoglobin, hypochromia, microcytosis, and a variable hemolytic component.

Defective synthesis of one set of globin chains results in excess production of the unaffected pair,[32] which precipitates in the red cells in the form of inclusion bodies, resulting in hemolysis (imbalanced globin-chain synthesis).

Classification. The classification of thalassemias as thalassemia major, intermedia, minor, and minima describes the clinical severity of the disorder and disregards the genetic makeup. The preferred genetic classification is based on the particular deficient polypeptide chain. In α thalassemia the synthesis of α chains is diminished, in β thalassemia the synthesis of β chains is diminished, and so on.

MAIN FORMS OF THALASSEMIAS

- α Thalassemias
  - α Thalassemia 1 (α^0 Thal)
  - α Thalassemia 2 (α^+ Thal)
  - Hb Constant Spring
- β Thalassemias
  - β^+ Thalassemia
  - β^+ Thalassemia, high F
  - β^0 Thalassemia
  - $(\delta\beta)^0$ Thalassemia
  - $(\delta\beta)^+$ Thalassemia
- Hb Lepore
- Hereditary persistence of fetal hemoglobin (HPFH)

Thalassemias involving γ, ϵ, or ζ genes may lead to fetal or embryonic death.[45] A thalassemia-like condition that is asymptomatic is the hereditary persistence of fetal hemoglobin (HPFH). The main forms of thalassemia are classified in the accompanying box.

The inheritance of thalassemia is autosomal and is similar to that of Hb S. From the clinical viewpoint it is recessive because the heterozygous form is asymptomatic. Similar to Hb S (β^S) gene, the thalassemia gene may express itself in homozygous, heterozygous, and doubly heterozygous states.

The clinical spectrum varies from normal to a severe, life-threatening condition and can include growth retardation, hepatosplenomegaly, bone overgrowth, bone pain, and jaundice.

Alpha thalassemias. The alpha (α) thalassemias are a group of genetic disorders that result in defective α-chain synthesis.[46] Diminished α-chain synthesis depresses Hb A, Hb F, and Hb A_2 because they contain α chains, and it leads to excess β and γ chains, which polymerize to the tetrameric forms γ_4 (Hb Barts) and β_4 (Hb H). The presence of these hemoglobins is the hallmark of α thalassemia. There are at least three α-thalassemia genes[32]: alpha thalassemia^0 (α thalassemia 1), in which no α chains are produced, alpha thalassemia$^+$ (α thalassemia 2), in which the α-chain production is reduced, and Hb Constant Spring (Hb CS), in which the α-chain production is thalassemia-like reduced.[47] For a summary of laboratory findings on α thalassemia see Table 33-4.

Beta thalassemia. The beta (β) thalassemias are a group of genetic disorders that result in diminished (β^+ thalassemia) or absent (β^0 thalassemia) β-chain synthesis. There are at least two forms of β^+ thalassemia, one with about one-half (β^+) and one with two-thirds (β^{++}) normal β-chain production. On a clinical basis the severe homozygous disorders of both types of β thalassemia have been called *thalassemia major,* and the heterozygous carrier states have been called *thalassmemia minor* and *minima*. The designation *thalassemia intermedia* describes clinical manifestations of a form of β thalassemia more severe than the trait and milder than the homozygous form.

Beta thalassemias are widely distributed throughout the world but occur most frequently in the Mediterranean populations; they also occur in Southeast Asia, the Middle East, India, and Pakistan. In Greeks and in American blacks the β^+ thalassemia is most common, whereas in Italy the β^0 is predominant.

Like α thalassemia, β thalassemia is transmitted as a mendelian autosomal recessive characteristic. The output of β chains is reduced or absent because of a defect in transcription of the β-thalassemia genes. In β^0 thalassemia homozygotes no β-chain synthesis occurs, whereas in β^+ and β^{++} thalassemia homozygotes, β-chain synthesis occurs at a low rate.

Heterozygous β thalassemia, whether β^+ or β^0, is an asymptomatic disorder that may or may not be associated with a mild degree of anemia.[48] It is the most commonly found thalassemic disease in North America.[46] Characteristically there is a slight-to-moderate erythrocytosis of poorly hemoglobinized cells. The mean corpuscular hemoglobin (MCH) and the mean corpuscular volume (MCV) are always strikingly decreased (MCH 20 pg and MCV between 55 and 56 fL). The MCH is variable. MCV values greater than 75 fL rule out thalassemia.[47] Free red

Table 33-4 Laboratory findings in α thalassemias

Genotype	Anemia (hypochromic)	Hb types	α-Chain deletion
α Thalassemia 1 trait	±	Birth: Hb Barts 5%-10% Hb CS 1%-2% Adult: Hb A,A_2,F	2
α Thalassemia 1/α thalassemia 1 (hydrops)	+ + +	Birth: Hb Barts 80% Traces of Hb H and Portland Adult: Not compatible with life	4
α Thalassemia 2/trait	±	Birth: Hb Barts 1%-2% Hb CS 1%-2% Adult: Hb A,A_2,F	1
α Thalassemia 1/α thalassemia 2 (Hb H disease)	± (Inclusions)	Birth: Hb Barts 1%-15% Hb H 4%-30% Adult: Hb A,A_2,F Hb H 8%-10%	3
α Thalassemia 1/Hb CS (Hb H/CS)	+ + (Inclusions)	Birth: Hb Barts Hb H, Hb, CS Adult: Hb H Hb A,A_2,F,CS	2 plus α-chain termination mutation
α Thalassemia 2/Hb CS	+	Birth: Hb Barts Adult: Hb A, CS	1 plus α-chain termination mutation
Hb CS/Hb CS	+	Birth: Hb Barts Adult: Hb A,A_2,F,CS	α-chain termination mutation

cell protoporphyrin is elevated, a feature distinguishing β-thalassemia trait from iron deficiency.[48] Pregnancy may lead to severe anemia in patients with thalassemia trait.[49] The hemoglobin electrophoretogram shows a slightly elevated Hb F (1% to 7%) in 50% of cases and a diagnostic elevation of Hb A_2 (3.5% to 7.5%).[46] Distribution of Hb F within the red cells demonstrated by the acid elution technique reveals a heterogeneous pattern.

Homozygous β^0 thalassemia leads to complete suppression of β-chain synthesis and to complete absence of Hb A. It is the cause of a severe lethal transfusion-dependent hemolytic anemia accompanied by characteristic clinical and hematological findings. Homozygous β^+ thalassemia is a heterozygous disorder that on the basis of the amount of Hb A synthesized is divided into three main types. Type 1, in which 5% to 15% Hb A is synthesized, is the Mediterranean and Oriental form characterized by a severe transfusion-dependent anemia. Type 2, of African background, has 20% to 30% Hb A and is responsible for a milder disease. Type 3 leads to a mild form of thalassemia intermedia.[32]

Heterozygous HPFH. Hereditary persistence of fetal hemoglobin (HPFH) is a benign condition characterized by the continuous synthesis of Hb F into adult life. It is considered to be a form of δβ thalassemia[50] because the persistence of the γ-chain synthesis compensates for the deficient δ- and β-chain production. HPFH can be broadly classified according to the distribution pattern in the red cells into pancellular and heterocellular groups.[51] In the first group, Hb F is uniformly distributed among the red cells, whereas in the latter it shows a heterogeneous pattern.

The hemoglobin electrophoretogram reveals about 25% Hb F, which by the Kleihauer method shows a pancellular distribution. Hb A_2 concentration is lower than normal (average value of 1.6%),[8] and the remaining hemoglobin mass is Hb A.

Methemoglobinemia

Methemoglobinemia is classified into acquired and hereditary forms.

Acquired methemoglobinemia. The acquired form is caused by various drugs and chemicals (or their metabolites) that are able to oxidize hemoglobin at such a rate as to overwhelm the reductive capacity of the red cells.[16] These substances include nitrites, nitrates, aniline dyes, sulfonamides, and phenacetin. Exposure to toxins may be responsible for very high levels of acute methemoglobinemia.

Hereditary methemoglobinemia. The hereditary forms are subdivided into two types. One is associated with Hb M, variant hemoglobins in which the amino acid substitutions in the α or β chains stabilize methemoglobin and render it resistant to reduction. The other is caused by deficiency in the methemoglobin-reducing ability of the red cells.

Four metabolic pathways for the reduction of the Met-Hb to hemoglobin are available[52]: (1) the NADH methemoglobin reductase pathway, (2) the reverse (NADPH) methemoglobin reductase pathway, (3) the reduction by ascorbic acid, and (4) the reduction by reduced glutathione.

NADH reductase (or diaphorase or NADH cytochrome b_5 reductase) catalyzes the reduction of methemoglobin to hemoglobin, using NADH generated in the anaerobic glycolytic pathway of Embden-Meyerhoff. Methemoglobinemia caused by deficiency of NADH methemoglobin reductase is an inherited disease transmitted as an autosomal recessive trait.[2]

The M hemoglobins—like all hemoglobin variants—show a recessive inheritance pattern and unlike many hemoglobinopathies do not produce hemolytic anemias. If only the heterozygous form is encountered—the homozygous form being incompatible with life—only one pair of globin chains (either α or β) is affected and the methemoglobin level does not exceed 25% to 30%. If α chains are involved, the cyanosis may be present at birth, whereas β-chain substitutions are responsible for cyanosis in the later months[53] because of the later appearance of these chains.

Clinically, patients with hereditary methemoglobinemia have erythrocytosis and slate gray cyanosis from birth, which is not associated with cardiopulmonary disease.[2] Methemoglobin concentrations of 10% to 20% of total hemoglobin produce cyanosis but no other ill effects. Methemoglobin concentrations of 30% may be responsible for headache and dyspnea, and concentrations of 70% and over may be fatal. A low incidence of mental retardation and early death can be observed in cases of methemoglobinemia caused by NADH methemoglobin reductase deficiency, a disease that lends itself to prenatal diagnosis.[53]

CHANGE OF ANALYTE IN DISEASE

Interpretations of hemoglobin values

Normal mean values for hemoglobin in adults are 151 g/L for men and 135 g/L for women. The values for newborns, infants, and children up to puberty differ (Table 33-5).

Hemoglobin values are influenced by physiological variations and by pathological processes. The physiological variations include age, sex, physical exercise, posture, dehydration, and altitude. The influence exerted by age is readily apparent in Table 33-5. The high birth level falls rapidly during the first year of life and then gradually rises to the adult level attained around puberty. Sex differences make their appearance at puberty and fade away in old age. Estrogen can influence the lower levels in women although they are not corrected by oral contraceptives.[54] Strenous exercise raises the hemoglobin level, probably through fluid loss, and a transient increase is also experienced after one changes from the recumbent to the stand-

Table 33-5 Reference values for hemoglobin in "apparently healthy" subjects, white and black

Subjects	Hgb (g/L)
Adult men	151 (139-163)
Adult women	135 (120-150)
Boys	
Birth	200 (185-215)
1 mo	170 (155-185)
3 mo	150 (135-165)
6 mo	140 (130-160)
9 mo	130 (120-140)
1 yr	121 (100-140)
2 yr	123 (105-142)
4 yr	126 (112-143)
8 yr	134 (120-148)
14 yr	140 (125-150)
Girls	
Birth	195 (180-210)
1 mo	170 (158-189)
3 mo	148 (133-164)
6 mo	138 (128-148)
9 mo	128 (117-139)
1 yr	122 (100-140)
2 yr	122 (105-142)
4 yr	127 (113-142)
8 yr	130 (115-145)
14 yr	132 (116-148)

From Miale, JB: Laboratory medicine: hematology, ed 6, St Louis, 1982, The CV Mosby Co.

ing position. Dehydration is responsible for a rise in hemoglobin concentration of such magnitude as to mask a significant anemia. High altitude is responsible for increased hemoglobin levels because of the erythropoietin-stimulating effect of hypoxia.

Pathological levels are encountered in anemias and in polycythemias. The overall reduction in the hemoglobin level in the peripheral circulation is one of the main parameters used in the diagnosis of all types of anemia. Because of the physiological variations and the wide normal and usual ranges of hemoglobin concentrations, other factors must also be considered in the diagnosis of anemia. There are three main causes of anemia: impaired production of red cells by the bone marrow, excessive blood loss and impaired delivery of red cells to the peripheral blood, or a combination of these factors. Decreased erythropoiesis is seen in iron deficiency and in impaired synthesis of heme, globin, and DNA (B_{12}, folate deficiencies) and in impaired bone marrow production. Excessive blood loss may be caused by hemorrhage, hemolysis, abnormal splenic sequestration, parasites, toxins, and trauma to red cells.

Increased hemoglobin values are encountered in polycythemia vera, erythrocytosis, dehydration, newborns, acquired or congenital cyanosis, chronic heart and lung disease, high altitude, renal cysts, and a number of erythropoietin-producing tumors.

Hb F

Wood,[55] using fluorescent anti–Hb F antibodies, identified 0.2% to 7% of adult red cells as Hb F–containing cells; these represent a specific group of red blood cells called *F cells*. In normal adults the concentration of F cells is fairly constant, but there are both genetic and acquired hematological conditions in which their concentration is increased. The genetic disorders include thalassemias (β and δβ), hereditary persistence of Hb F, sickle cell anemia, and unstable β-chain variants. The acquired conditions include pregnancy at about midterm, recovery from bone marrow depression,[56] leukemias (highest values in Philadelphia chromosome-negative juvenile chronic myelocytic leukemia),[57] thyrotoxicosis, and hepatoma.[58]

Hb A_{1c}

Hb A_{1c} levels depend on the time-integrated blood levels of glucose. See Chapters 29 and 63 (p. 1040) for a more detailed discussion. In hemolytic anemias the Hb A_{1c} level is low because the red cell life span is shortened by lysis.[59]

CO-Hb

Some carboxyhemoglobin is produced endogenously as 1 mole of (CO) is generated by the degradation of 1 mole of heme to bilirubin (see Chapter 32). Although this endogenous production of carboxyhemoglobin can present a hazard when exhaled air is concentrated in ill-ventilated small spaces, the exogenous generation of CO from combustion of organic material in confined spaces is more likely to cause intoxication. Exogenous CO is derived from the exhaust of automobiles and from industrial pollutants such as coal gas, charcoal burning, and tobacco smoke. Under the most favorable conditions the CO-Hb concentration in blood is 0.2% to 0.8%. It is never zero because of its endogenous generation and can physiologically be elevated in hemolytic anemias.[60] In smokers the level may vary from 4% to 20%. In workers who have a greater exposure to CO, the average level may be 10%. Because of its firm liaison to hemoglobin, long exposure to even low CO concentrations can lead to toxic accumulations to which the most oxygen-dependent organs, such as brain and heart, are most susceptible. Mild symptoms such as slight headaches and slight dyspnea on exertion can occur at levels of 10% to 15% saturation. At levels of 20% to 30%, the headaches will be more severe and will be accompanied by fatigue and impaired vision and judgment. Levels of more than 50% cause increasingly severe symptoms, coma, and convulsions; and levels of 60% and over are usually fatal, although death has occurred at levels as low as 20%. The half-life of elimination of CO is about 4 hours for a person breathing atmospheric air, but in smokers the level may remain high.[61] Chronic exposure to CO may be responsible for a relative polycythemia.[62]

Oxygen saturation

Clinically oxygen saturation is used as an indicator of tissue hypoxia or hyperoxia. Tissue hypoxia is produced by decreased oxygen content of inspired air, as in high altitude, or by decreased alveolar-capillary oxgyen exchange in the lungs, as in pulmonary fibrosis, emphysema, and chronic heart disease with right-to-left shunt. Tissue hypoxia is also produced (1) by a defect in the erythrocytic oxygen transport as in severe anemia; (2) when hemoglobin ligands are present that prevent oxygen binding, such as carboxyhemoglobin, sulfhemoglobin, and methemoglobin; (3) in hemoglobinopathies; (4) in inappropriate concentrations of erythrocytic 2,3-DPG; and (5) in intraerythrocytic enzyme deficiencies. Therapeutic hyperoxia must be carefully monitored because of the danger of oxgyen toxicity. In newborns it can be responsible for retrolental fibroplasia and in adults for hyaline membrane disease of the lungs (adult respiratory distress syndrome).

2,3-Diphosphoglycerate (2,3-DPG)

DPG and hemoglobin interaction in hypoxia is related to the increased intraerythrocytic deoxyhemoglobin, which binds large amounts of DPG. This binding results in a feedback mechanism that stimulates glycolysis and DPG synthesis. Increased deoxyhemoglobin concentrations raise the pH, which in turn also stimulates the synthesis of DPG. Carrell and Lehmann[11] emphasize that there is a reciprocal relationship between hemoglobin and DPG concentrations. Pyruvate kinase deficiency leads to a buildup of DPG, decreased oxygen affinity, and a low hemoglobin concentration. Hexokinase deficiency leads to a decrease in DPG and to a compensatory erythropoietic response with increased hemoglobin values.[11]

Hemoglobin electrophoresis. The Hb F concentration can vary from 10% to 90%, and the Hb F distribution in the red cells as determined by the Kleihauer technique can be heterogeneous (see also Chapter 63, p. 1043.) The Hb A_2 concentration may vary from 1.4% to 20%, embracing low, normal, elevated, and very high values. In homozygous β thalassemia, even if the A_2 value is low or normal in the patient, it will be high in both parents. If the Hb A_2 level in the patient is expressed as a percentage of the total hemoglobin, it spans the previously mentioned spectrum from low to high; but if it is expressed in relation to the Hb A value only, the ratio is decreased in all cases of β-thalassemia, that is, an A/A_2 ratio of about 10:1 as compared to the normal A/A_2 ratio of about 40:1. If the Hb A_2 level is greatly increased, Hb F is normal or only slightly elevated, and vice versa. In B^0 thalassemia there is a total absence of Hb A, and so the patient's hemoglobin consists only of Hb F and Hb A_2, whereas in β^+ thalassemia diminished amounts of Hb A are found (5% to 20%). In both thalassemias, free α-chains may be seen close to the application point of the electrophoretogram at alkaline pH. The severe reduction in β-globin chains leads to a β/α ratio of less than 0.25 to 0.3.[48]

Other related biochemical findings. Because of the hemolytic component of the anemia, serum unconjugated bilirubin levels are elevated and haptoglobin is absent. Serum aspartate aminotransferase (AST), lactic dehydrogenase (LD), and erythropoietin concentrations are also raised. The erythropoietin elevation is responsible for the 20% to 30% increase in erythropoietic marrow and is a result of the anemia and the high oxygen affinity of Hb F, which further increases the tissue anoxia. The liver involvement (transfusion hemosiderosis) can lead to a bleeding tendency. Gross examination of the urine may show the brown color of dipyrroles, caused by excessive intramedullary hemolysis of normoblasts.

REFERENCES

1. Winslow, RM, and Anderson, WF: The hemoglobinopathies. In Stanbury, JB, Wyngaarden, JB, and Fredrickson DS, editors: The metabolic basis of inherited disease, ed 5, New York, 1982, McGraw-Hill Book Co.
2. Perutz, MF: Sci Am 239:92, 1978.
3. Gaul, A, and Axel, R: Proc Natl Acad Sci USA 72:3966, 1976.
4. Lodish, HF: Annu Rev Biochem 45:39, 1976.
5. Weatherall, DJ: J Clin Pathol 8(suppl):1, 1974.
6. Clegg, JB, and Weatherall, DJ: Nature 240:190, 1972.
7. McDonald, MJ, Shaeffer, JR, Turei, SM, and Bunn, HF: Subunit assembly of hemoglobin A&S. In Brewer, GJ, editor: The red cell, Fifth Ann Arbor Conference, New York, 1981, Alan R Liss, Inc.
8. Shaeffer, JR: J Biol Chem 255:2322, 1980.
9. Roughton, FJW: Transport of oxygen and carbon dioxide. In Senn, WO, and Rahn, H, editors: Handbook of physiology, respiration, Washington, DC, 1964, American Physiological Society.
10. Bunn, F, and Forget, BG: Hemoglobin: molecular, genetic, and clinical aspects, Philadelphia, 1986, WB Saunders Co.
11. Carrell, RW, and Lehmann, H: Abnormal hemoglobins. In Brown, SS, Mitchell FL, and Young, DS, editors: Chemical diagnosis of disease, Amsterdam, 1979, Elsevier North-Holland Biomedical Press.
12. Rorth, M: Ser Haematol 5:1, 1972.
13. Astrup, P: N Engl J Med 283:202, 1970.
14. Rabitzis, ET, and Mills, GC: Biochem Biophys Acta 141:439, 1967.
15. Urbanetti, JS: Carbon monoxide poisoning. In Wallach, DFH, editor: The function of the red blood cells: erythrocyte pathobiology, New York, 1981, Alan R Liss, Inc.
16. Jaffe, ER: Methemoglobin pathophysiology. In Wallach, DFM, editor: The function of the red blood cells: erythrocyte pathobiology, New York, 1981, Alan R Liss, Inc.
17. Carrell RW, Winterbourn, CC, and Rachmilewitz, EA: Br J Haematol 30:250, 1975.
18. Finch, CA: N Engl J Med 239:470, 1948.
19. Reynolds, TB, and Ware, A: JAMA 149:1538, 1952.
20. Berzofsky, JA, Peisach, J, and Blumberg, WF: J Biol Chem 246:3367, 1971.
21. Haney, EN, and Bunn, HF: Proc Natl Acad Sci USA 73:3534, 1976.
22. McDonald, MJ, Shapiro, R, Bleichman, M, et al: J Biol Chem 253:2327, 1978.
23. Bunn, HF, and Biehl, RW: J Clin Invest 49:1088, 1970.
24. Efremov, GD, Miadenovisky, B, Petkov, G, et al: New Istanbul Contrib Clin Sci 12:211, 1978.
25. Roberts, AV, Weatherall, DJ, and Clegg, JB: Biochem Biophys Res Commun 47:81, 1972.
26. Schroeder, WA, Huisman, THJ, Shelton, R, et al: Proc Natl Acad Sci USA 60:537, 1968.
27. Huehns, ER, and Beaven, GH: Developmental changes in human hemoglobin. In Benson, P, editor: The biochemistry of development, London, 1971, Spastics International Medical Publications, William Heinemann Medical Books.
28. Singer K, Chernoff, AJ, and Singer, L: Blood 6:413, 1951.
29. Kleihauer, E: Beihefte Arch Kinderheilkd 53(Suppl), 1966.
30. Perutz, MF, and Lehmann, H: Nature (London) 219:90, 1968.

31. Pataryas, HA, and Stamatoyannopoulos, G: Blood 39:688, 1972.
32. Weatherall, DJ, and Clegg, JB: The thalassemia syndrome, Oxford, England, 1981, Blackwell Scientific Publications, Ltd.
33. Fairbanks, VF: Nomenclature and taxonomy of hemoglobin variants. In Fairbanks, VF, editor: Hemoglobinopathies and thalassemias, New York, 1980, Brian C Decker, Publisher.
34. Weatherall, DJ: Abnormal hemoglobins and thalassemias. In Hoffbrand, AV, Brain, MC, and Hirsh, J, editors: Recent advances in hematology, London, 1977, Churchill Livingstone.
35. Briehl, RW: Physical chemical properties of sickle cell hemoglobin. In Wallach, DFH, editor: The function of red blood cells: erythrocyte pathobiology, New York, 1981, Alan R Liss, Inc.
36. Bunn, HF: Hemoglobin II, sickle cell anemia and other hemoglobinopathies. In Beck WS, editor: Hematology, Cambridge, MA, 1981, MIT Press.
37. Milner, PF, Miller, C, Gray, R, et al: N Engl J Med 283:1417, 1970.
38. Hibbel, RP: J Clin Invest 65:154, 1980.
39. Nagel, RL, and Bookchin, RM: Oxygen transport and the sickle cell. In Wallach, DFH, editor: The function of red blood cells: erythrocyte pathobiology, New York, 1981, Alan R Liss, Inc.
40. Milner, PF: The clinical effects of Hb S, an overview. In Wallach, DFH, editor: The function of red blood cells: erythrocyte pathobiology, New York, 1981, Alan R Liss, Inc.
41. Matulsky, AF: N Engl J Med 288:31, 1973.
42. Schneider, RG: J Lab Clin Med 44:133, 1954.
43. Milner, PF, and Wrightstone, RN: The unstable hemoglobins: a review. In Wallach, DFH, editor: The function of red blood cells: erythrocyte pathobiology, New York, 1981, Alan R Liss, Inc.
44. Kreimer-Birnbaum, M, Pinkerton, PH, Bannerman, RM, et al: Br Med J 2:396, 1966.
45. Fairbanks, VF: Thalassemias and hereditary persistence of fetal hemoglobins (HPFH). In Fairbanks, VF, editor: Hemoglobinopathies and thalassemias, laboratory methods and case studies, New York, 1980, Brian C Decker, Publisher.
46. Miale, JB: Laboratory medicine hematology, ed 6, St Louis, 1982, The CV Mosby Co.
47. Worwood, M, Edwards, A, and Jacobs, A: Nature 229:409, 1971.
48. Okene-Frempong, K, and Schwartz, E: Pediatr Clin North Am 27:403, 1980.
49. Sicuranza, BJ, Tisdall, LH, Saveck, R, et al: NY State J Med 78:1691, 1978.
50. Charache, S, Clegg, JB, and Weatherall, DJ: Br J Haematol 34:527, 1976.
51. Boyer, SH, Margolet, L, Boyer, ML, et al: Am J Hum Genet 29:256, 1977.
52. Jaffe, ER: Clin Haematol 10:99, 1981.
53. Jaffe, ER: Clin. Haematol. **12:**99, 1981.
54. Burton, JL: Lancet 1:978, 1967.
55. Wood, WG, Stamatoyannopoulos, G, Lim, G, et al: Blood 46:671, 1975.
56. Dover, GJ, Boyer, SH, and Zinkham, WH: J Clin Invest 63:173, 1979.
57. Hardisty, RM, Speed, DE, and Till, M: Br J Haematol 10:551, 1964.
58. Weatherall, DJ, Pembrey, ME, and Pritchard, J: Clin Haematol 3:467, 1974.
59. Bunn, HF, Haney, DN, Kain, S, et al: J Clin Invest 57:1652, 1976.
60. Landau, SA, and Winchell, HS: Blood 36:642, 1970.
61. Astrup, P: JAMA 230:1064, 1974.
62. Smith, JR, and Landau, SA: N Engl J Med 298:6, 1978.

CHAPTER 34

Human nutrition

LAWRENCE A. KAPLAN

OBJECTIVES

- Discuss the importance of individual nutrient classes in human metabolism.
- Define kwashiorkor and marasmus and explain the difference between them.
- List the populations at risk for nutritional deficiencies, and list some of the nutrients that are used to monitor nutritional deficiencies.
- Discuss the relationship between excessive alcohol consumption and nutritional disease states.

KEY TERMS

basal metabolism Those energy-requiring metabolic processes, such as sleep, that occur during a resting state.

Calorie The amount of food (dietary component) containing the energy-producing value of one large calorie (amount of heat required to raise the temperature of 1 kg of water 1° C).

diet All the components that an organism consumes; includes foods (nutrients) and water.

essential amino acids Those amino acids that the human body is unable to synthesize at all or in amounts capable of sustaining life.

essential nutrients Nutrients that cannot be synthesized at all or can only be partially synthesized by an organism. These include vitamins, elements, and certain amino acids and fatty acids.

kwashiorkor A malnourished state caused by a diet deficient in protein.

malnutrition A suboptimum nutrition resulting from an inadequate or unbalanced intake of nutrients or an impaired ability to use nutrients. Can lead to pathological conditions.

marasmus A malnourished state caused by an overall deficiency in caloric (food) intake.

nitrogen balance The algebraic sum of daily intake of nitrogenous compounds (primarily proteins and amino acids) and of nitrogen excretion. Used to estimate changes in body protein stores.

nutrient Any chemical that is used by an organism to provide energy or cellular components.

nutrition The science that studies the processes by which nutrients are made available to an organism.

recommended dietary allowance (RDA) Suggested nutrient intake needed to maintain a healthy state. The allowances are intended to provide for individual variations among normal persons as they live in the United States under usual environmental stresses. The RDAs do not represent either a minimum requirement or an optimum level of intake for every individual but rather are intakes of nutrients that meet the nutritional needs of the majority of healthy persons.

total metabolism The sum of all the metabolic processes needed for the continuation of life.

Nutrients are the necessary chemical substances required by organisms for growth and the maintenance of life. *Nutrition* is the clinical science that deals with the physiological functions that allow nutrients to be made available to an organism. The physiological functions include the digestion of dietary foodstuffs, intestinal absorption of digested nutrients, conversion of absorbed nutrients to active or usable molecules, the actual metabolic reactions that employ the active molecules, and the eventual catabolism and elimination of the molecules. From the digested nutrients, an organism derives the energy needed for building and maintaining tissues, as well as the actual building blocks needed for construction of macromolecules.

Table 34-1 lists the basic classes of nutrients required by humans. The table also lists the chapter that discusses the metabolism and functions of these nutrients. Many of these chapters also review the body systems (such as bone and electrolyte and water balance) that involve these nutrients.

Table 34-1 Basic classes of nutrients

Nutrient	Chapter discussed
Carbohydrate	Diabetes, 29
Lipids	Lipid, 30
Proteins	Liver, 23
Inorganic elements	
Na, K, Cl	Electrolytes and water balance, 20
Ca, Mg, inorganic phosphorus	Bone, 24
Fe^{++} (Fe^{+++})	Iron, porphyrins, bilirubin, 32
Trace minerals	Trace elements, 35
Vitamins	Vitamins, 36
Water	Renal, 22 Electrolytes and water balance, 20

ENERGY AND NUTRITIONAL NEEDS IN NORMAL HUMAN METABOLISM

A primary goal of human nutrition is the provision of sufficient energy to maintain life. Most of the nutrient intake serves to supply the energy needs of an organism. The energy needs of humans have been defined as the *calories* required for *total metabolism* and *basal metabolism*. Basal metabolism, occurring in the resting, postnutrient absorptive state, is needed for cellular ion transport and basic body functions, such as brain and heart activity, respiration, and maintenance of body temperature. Basal metabolism for adult humans uses approximately 1600 to 2000 Calories/day for men and 1400 to 1700 Calories/day for women. Energy requirements for total metabolism vary with the size of the individual and the types of activities performed. In addition, the daily energy needs vary with gender (greater for males), age (diminishing with age), and general health and nutritional status of the individual.

For most individuals living in the highly industrialized, wealthy Western countries, the major nutritional problem is one of chronic, excessive nutrient intake. Food intake in excess of that necessary to support an individual's total energy metabolism will result in conversion of excess calories to a stored energy form, primarily fat. Excess body weight (fat) can lead to the development of chronic disease states, such as hypertension, maturity-onset diabetes, and cardiovascular disease, and eventually to decreased longevity.

The 1988 U.S. surgeon general's report on nutrition and health clearly defined the problem of excess nutrient intake. The report described significant relationships between dietary influences and 4 of the 10 leading causes of death in the United States: heart disease, cancer, stroke, and atherosclerosis. Decreased consumption of total dietary fat to below 30% of total calories, decreased consumption of saturated fats, and increased consumption of dietary fibers and complex carbohydrates were all part of the recommendations of the surgeon general's report. The report also recognized recent epidemiological evidence that dietary carotenoids and vitamin A have a protective effect against certain types of epithelial cancers, especially those of the oral cavity, bladder, and lung.

On the other hand, if caloric intake is below the minimum number of calories, weight loss will result. Stored energy, in the form of fat, carbohydrate, and protein, will be mobilized. Prolonged weight loss can lead to an increased susceptibility to infectious diseases, to an increased frequency of ketosis and acidosis, and to decreased longevity.

Obviously, one of the goals of proper energy nutrition is the consumption of sufficient food to provide energy for the maintenance of an appropriate body weight.

ROLE OF INDIVIDUAL NUTRIENT CLASSES IN NORMAL HUMAN METABOLISM

Carbohydrates

Carbohydrates serve as the principal energy source of the body, supplying 50% to 60% of the body's energy requirements. Dietary carbohydrates can be divided into two classes, *complex carbohydrates* and *simple carbohydrates* (natural and refined sugars). The complex carbohydrates, primarily starches, fiber, and dextrose, supply 50% to 60% of dietary carbohydrates. A diet containing a higher proportion of complex carbohydrates is believed to lower the incidence of certain disease states, including maturity-onset diabetes, hypertension, and cardiovascular disease. The digestion and metabolism of carbohydrates are reviewed in Chapters 26 and 29, respectively.

Of course, consuming excessive carbohydrates (especially simple sugars, such as glucose) can lead to an overweight condition and its associated problems. However, insufficient dietary carbohydrate can lead to mobilization of stored nutrients and to ketosis (see Chapter 29), loss of body electrolytes, and dehydration.

Undigestible dietary carbohydrates (for example, cellulose, gums, and lignin) provide bulk and fiber to diets and aid the digestive process. In addition, recent epidemiological studies have associated increased dietary fiber with both a decreased risk for certain cancers (for example, breast and colon) and lowered serum cholesterol.

Lipids

Dietary fats are high energy–yielding nutrients that provide 35% to 45% of the typical caloric intake of Americans. Aside from satisfying metabolic energy needs, dietary fats provide essential fatty acids and facilitate the uptake of the fat-soluble vitamins (see Chapter 36). Diets low in fat, especially those low in saturated fat and cholesterol, are associated with lowered risk of cardiovascular disease. Recent epidemiological evidence also suggests that a decrease in fat consumption may reduce the risk for certain cancers, especially breast and colon cancers. Chapter 30 reviews the metabolism of lipids and their relationship to disease.

The Food and Nutrition Board of the National Research Council as well as the 1988 surgeon general's report recommends that total fat intake should constitute no more than 30% of the total caloric intake and that a *minimum* of 10% of dietary calories should be provided by polyunsaturated fatty acids.

Proteins

Dietary protein serves primarily as a source of amino acids and nitrogen, the building blocks for tissue formation and maintenance. Although the body can synthesize most amino acids, nine amino acids cannot be synthesized de novo and must be obtained from dietary sources. These amino acids, termed *essential* amino acids, are listed in the accompanying box.

The ability of a food source to provide the necessary amounts of total and essential amino acids varies greatly. For example, certain foods such as meat, fish, or poultry contain greater amounts of essential amino acids and therefore can better serve as sources of these amino acids. This is termed the *biological value of proteins*. In general, animal proteins have a higher biological value than do plant proteins.

The *nitrogen balance* of an individual is used to determine if that individual is receiving sufficient dietary amino acids or protein since almost all dietary and excretory nitrogen is derived from these nutrients. Dietary nitrogen intake can be estimated from the diet record. The urine creatinine and urea levels can be used to estimate the nitrogen output of an individual. A stable, healthy individual is in nitrogen balance, and the sum of nitrogen intake equals the sum of the nitrogen output in the urine and stool:

$$\text{Nitrogen (N) balance} = \text{Food N} - (\text{Fecal N} + \text{Urine N})$$

A positive nitrogen balance is associated with a net gain of body protein and general good health, and a negative nitrogen balance is associated with a relative deficiency of nitrogen (protein/amino acids). Ill patients, such as cancer patients and surgical patients, should have their nitrogen balance assessed to ensure that dietary nitrogen intake is sufficient.

ESSENTIAL AMINO ACIDS

Histidine (for children and infants, for growth)
Isoleucine
Leucine
Lysine
Methionine
Phenylalanine
Threonine
Tryptophan
Valine

The requirements for dietary protein will vary greatly, depending on an individual's needs. The recommended dietary allowance (RDA) of protein for infants is 2.0 to 2.2 g/kg of body weight, much higher than that recommended for adults; and a woman's protein needs increase by 50% during pregnancy. Table 34-2 lists the RDA for protein for a number of populations. Note that the average RDA for protein is 0.8 g of mixed proteins per kilogram of body weight per day.

An excess of dietary protein will not result in increased growth or improved health in normal individuals. In fact, an excess of protein can lead to various pathological conditions. However, a long-term deficiency of protein can result in a malnutrition state called *kwashiorkor*. First used in Africa, this term is now used to define a protein-deficient, but not necessarily a calorie-deficient, condition. The disease typically occurs in children, often after weaning, when they begin to consume a high-starch diet. In adults kwashiorkor-like conditions occur after chronic protein deprivation. Children with this disease have edema, multiple skin lesions and rashes, diarrhea, and hair changes; ultimately they die of the diseases. Kwashiorkor is a major scourge in many parts of Africa and the poorer countries of the world.

Inorganic elements

The National Research Council's RDAs for minerals present in relatively high concentrations (and thus not termed *trace* elements [discussed in next section]) are pre-

Table 34-2 Recommended daily dietary allowances for protein*

	Age (yr)	Weight (kg)	Weight (lb)	Height (cm)	Height (in)	Protein (g)
Infants	0.0-0.5	6	13	60	24	kg × 2.2
	0.5-1.0	9	20	71	28	kg × 2.0
Children	1-3	13	29	90	35	23
	4-6	20	44	112	44	30
	7-10	28	62	132	52	34
Males	11-14	45	99	157	62	45
	15-18	66	145	176	69	56
	19-22	70	154	177	70	56
	23-50	70	154	178	70	56
	51+	70	154	178	70	56
Females	11-14	46	101	157	62	46
	15-18	55	120	163	64	46
	19-22	55	120	163	64	44
	23-50	55	120	163	64	44
	51+	55	120	163	64	44
Pregnant						+30
Lactating						+20

From Recommended dietary allowances, ed 9, Washington, DC, 1980, National Research Council.
*Revised 1980 by the Food and Nutrition Board of the National Academy of Sciences–National Research Council. See chapter glossary for definition of recommended dietary allowances.

Table 34-3 Recommended daily dietary allowances for minerals*

	Age (yr)	Weight (kg)	Weight (lb)	Height (cm)	Height (in)	Minerals: Calcium (mg)	Minerals: Phosphorus (mg)	Minerals: Magnesium (mg)
Infants	0.0-0.5	6	13	60	24	360	240	50
	0.5-1.0	9	20	71	28	540	360	70
Children	1-3	13	29	90	35	800	800	150
	4-6	20	44	112	44	800	800	200
	7-10	28	62	132	52	800	800	250
Males	11-14	45	99	157	62	1200	1200	350
	15-18	66	145	176	69	1200	1200	400
	19-22	70	154	177	70	800	800	350
	23-50	70	154	178	70	800	800	350
	51+	70	154	178	70	800	800	350
Females	11-14	46	101	157	62	1200	1200	300
	15-18	55	120	163	64	1200	1200	300
	19-22	55	120	163	64	800	800	300
	23-50	55	120	163	64	800	800	300
	51+	55	120	163	64	800	800	300
Pregnant						+400	+400	+150
Lactating						+400	+400	+150

From Recommended dietary allowances, ed 9, Washington, DC, 1980, National Research Council.
*Revised 1980 by the Food and Nutrition Board of the National Academy of Sciences–National Research Council. See chapter glossary for definition of recommended daily allowance.

sented in Table 34-3. Normal, balanced diets will usually provide sufficient amounts of these elements. Certain non-disease conditions, however, such as pregnancy and lactation, may require increased dietary intake. Often an imbalance of these elements leading to body deficiencies or excesses occurs as a result of disease states; for example, malabsorption syndromes can lead to a deficiency of many inorganic elements. Discussions on the metabolism of calcium, magnesium, and phosphorus can be found in Chapter 24; the metabolism of sodium, potassium, and chloride can be found in Chapter 20; and a discussion of iron metabolism can be found in Chapter 32.

Trace elements

Trace elements are those elements that are present in human tissues in very small amounts, typically in the range of micrograms per gram of tissue. A discussion of the role of these micronutrients in human metabolism is found in Chapter 35.

Vitamins

Vitamins are a series of chemically diverse compounds that are required in a large number of biochemical reactions but that cannot be synthesized by humans. Chapter 36 discusses the role of vitamins in normal human metabolism and pathological conditions resulting from deficiencies or excesses of vitamins.

NUTRITIONAL CONCERNS IN DISEASE

Clinical nutrition probably received more attention from American physicians in the early quarter of this century than in the middle of the century. In large part, the decreased interest in nutritional matters was the result of our country's success in achieving a generally well-nourished population. The identification, characterization, and synthesis of many vitamins, as well as the elucidation of the role of vitamins and minerals in human physiology, coupled with the large-scale production and thus inexpensive availability of these nutrients, have almost obliterated most of the classical nutrition-deficiency states in the United States. However, many of these deficiency diseases are still common in many less developed countries, which represent a substantial portion of humanity.

In the last decade, however, there has been an increased awareness, both by the public and by the medical community, that certain populations are at risk for *malnutrition*. Malnutrition is defined here as a level of intake of a specific nutrient, or nutrients in general, that can result in a pathological state. Although the following discussion reviews deficiency states, pathological conditions can also result from an excess of nutrients, as discussed earlier.

The populations that can be considered at risk for malnutrition are usually seen in a hospital setting. They include cancer patients, burn patients, the geriatric population, low–birth weight newborns, and surgical (especially gastrointestinal) patients. These patients are at risk for a suboptimum nutritional state because of their inability to ingest nutrients by mouth and because of the increased nutrient requirement that results from the disease state. Instead, they may receive a substantial portion (if not all) of their nutrients parenterally, that is, by intravenous infusion (see box). These patients may need to be monitored at ap-

CLINICAL CONDITIONS THAT COMMONLY NECESSITATE TOTAL PARENTERAL NUTRITION

- Bowel obstruction
  - Prolonged ileus
  - Adhesions
  - Strictures
  - Neoplasia
  - Congenital atresia syndrome
  - Pseudo-obstruction
  - Gastroschisis or wound dehiscence
- Severe inflammation or loss of bowel surface
  - Radiation enteritis
  - Regional enteritis/ulcerative colitis
  - Intractable diarrhea of infancy
  - Short bowel from resections, high fistulas
- Metabolic conditions associated with persistent vomiting or anorexia
  - Cancer chemotherapy
  - Hypertension
  - Anorexia nervosa
  - Severe depression
  - Chronic renal failure

From Howard, L, and Meguid, MM: Nutritional assessment in total parenteral nutrition, Clin Lab Med 1:611-630, 1981.

propriate intervals to assess their nutritional state (Table 34-4). The assessment might include trace metals (such as zinc, copper, and chromium) and vitamins such as E and B_{12}. Certainly, patients who are at risk for malnutrition and who also show signs of failure to thrive, weight loss, immunoincompetence (as demonstrated by infections resistant to usual therapy), or hyperglycemia should be investigated for specific trace-metal or vitamin malnutrition. A decision tree that can be used to determine when to investigate such patients is shown in Fig. 34-1.

People who live substantially below a country's poverty line and do not receive sufficient caloric sustenance are also at increased risk for malnutrition. Protein-calorie malnutrition can be classified as *marasmus* or as a kwashiorkor-like disease, depending on whether a caloric deficit (marasmus) or a protein deficit (kwashiorkor-like) predominates. Marasmus is associated with weight loss, loss of depot fat, and decreased skeletal muscle mass. In the more advanced disease, the immune response is also impaired. A kwashiorkor-like syndrome caused by protein depletion may not be associated with weight loss or the other findings of marasmus. However, visceral proteins, such as albumin, transferrin, and retinol-binding protein, are more noticeably depleted, and the immune system impaired. A caloric malnutrition (marasmus) or a protein-deficient malnutrition (kwashiorkor) can certainly be identified on clinical grounds; however, identification may require measurements of serum amino acids, specific serum proteins, total serum protein, and plasma lipids and estimation of immune competence.

The specific serum proteins most frequently measured for nutritional evaluation are transferrin, retinol-binding protein, prealbumin, and albumin. The half-lives of retinol-binding protein and prealbumin are fairly short (approximately 46 and 8 hours, respectively), are sensitive indicators of the presence of protein malnutrition, and can be used to monitor patient response to nutritional therapy. Albumin, with a half-life of approximately 12 days, is a more specific marker for protein malnutrition. Serum albumin levels below 28 g/L and serum prealbumin levels below 110 mg/L are associated with substantial protein malnutrition. Of course, all these serum proteins can be decreased in liver disease in the absence of malnutrition. Certainly any person who demonstrates specific signs of vitamin or trace-metal nutritional deficiency should be investigated for malnutrition for that specific analyte.

Table 34-4 Chronological occurrence of complications from total parenteral nutrition (TPN)

	First 48 hr	First 2 wk	Three mo onward
Mechanical	Complications from catheter insertion Cephalad displacement Pneumothorax Hemothorax Detachment of line at catheter hub with blood loss or air embolism	Catheter coming out of vein, more common if Silastic Detachment of line at catheter hub with blood loss or air embolism	Detachment of line at catheter hub with blood loss or air embolism
Metabolic	Hyperglycemia Hypophosphatemia Hypokalemia	Hyperosmolar nonketotic hyperglycemic coma Hyponatremia Hypomagnesemia (↓ K, ↓ Ca) Acid-base imbalance	Essential fatty acid deficiency Zinc, copper, chromium, selenium, molybdenum deficiency Iron deficiency Vitamin deficiency TPN metabolic bone disease TPN liver disease
Infectious		Catheter-induced sepsis	Catheter-induced sepsis Tunnel infections

From Howard, L, and Meguid, MM: Clin Lab Med 1:611–630, 1981.

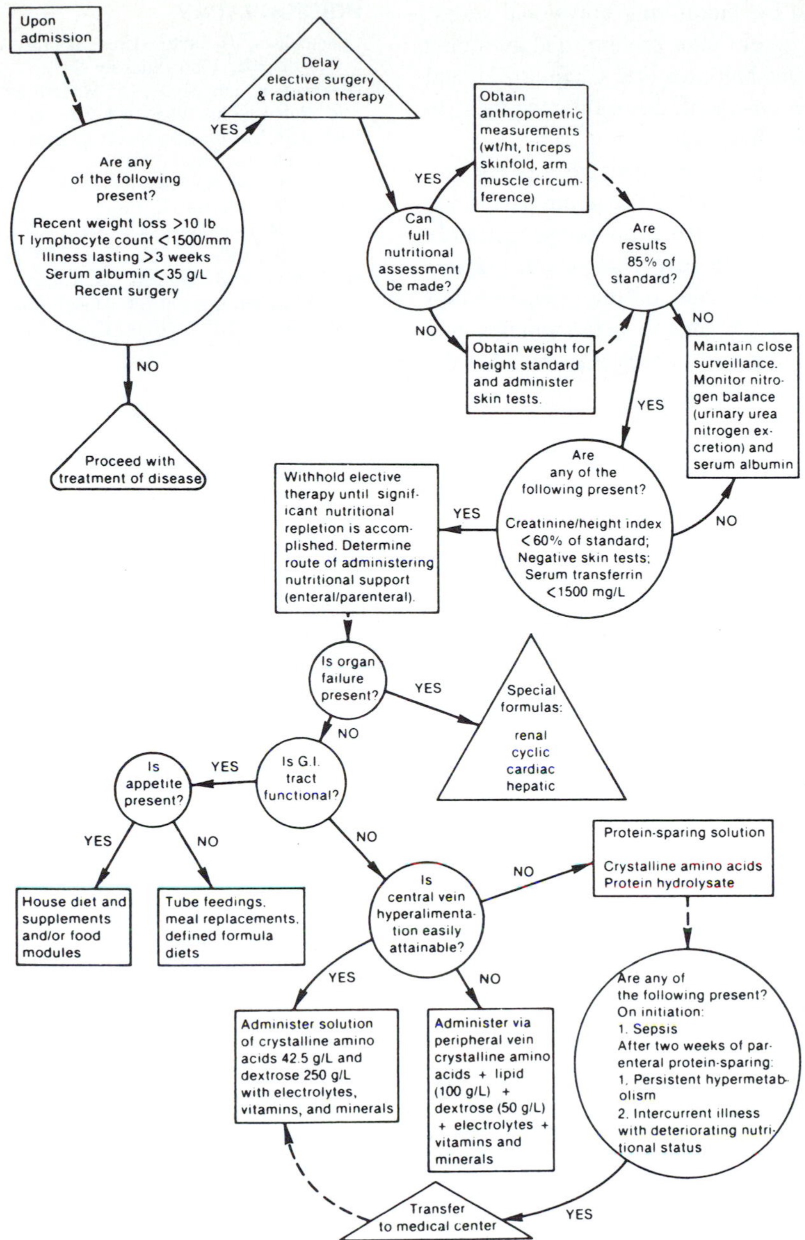

Fig. 34-1 Decision tree for nutritional support. *(From Blackburn, GL, and Harvey, KB: In Garry, PJ, editor: Human nutrition: clinical and biochemical aspects, Washington, DC, 1981, American Association for Clinical Chemistry.)*

NUTRITIONAL ALTERATIONS ASSOCIATED WITH ALCOHOLISM AND ALCOHOLIC LIVER DISEASE

Chronic alcoholism has long been associated with both malnutrition and liver disease. Malnutrition in the form of protein-calorie deficiency results from a variety of causes: (1) poor dietary intake; (2) malabsorption of consumed nutrients as a result of alcohol-induced gastritis, pancreatitis, or diarrhea; and (3) altered biochemical and physiological processes. In a study of 284 alcoholics with alcoholic hepatitis, all patients had some degree of protein-calorie malnutrition, and 80% had features of both marasmus and kwashiorkor-like disease.

The role of malnutrition in the development of liver injury is still controversial. Certainly liver injury can develop in the absence of malnutrition. However, the more severe forms are invariably associated with severe malnutrition. Recognition of nutritional deficits appears to be important for the alcoholic with liver disease because correction of these abnormalities by appropriate nutritional therapy has been observed to increase survival and accelerate improvement of the liver injury.

Many substances used for monitoring nutritional states, such as serum albumin, serum total protein, and ammonia (see Chapter 23); creatinine and urea (see Chapter 22); and lipids (see Chapter 30) are available for routine analysis in most laboratories or larger hospitals.

With the ready availability of inexpensive vitamin and trace-metal preparations, more studies are being performed to evaluate the toxic effects of these nutrients. Certainly vitamins A and E can show toxic effects at elevated doses. In addition, epidemiological studies on the effects of biochemicals, such as vitamins A and E, beta-carotene, and selenium, on the development of cancer may broaden the future role of these micronutrients.

BIBLIOGRAPHY

Cunningham, JJ, editor: Contemporary clinical nutrition, a conspectus, Philadelphia, 1986, Stickley Co.

Goodhart, R, and Shills, M: Modern nutrition in health and disease, ed 7, Philadelphia, 1985, Lea & Febiger.

Lang, CE: Nutritional support in critical care, Rockville, Md, 1987, Aspen Publishers, Inc.

Orten JM, and Neuhaus, OW: Human biochemistry, ed 10, St Louis, 1982, The CV Mosby Co.

Reddy, BS, and Cohen, LA: Diet, nutrition and cancer: a critical evaluation, Boca Raton, Fla, 1986, CRC Press.

Surgeon General's Report on Nutrition and Health, 1988. (Available from Superintendent of Documents, Government Printing Office, Washington DC 20402-9325.)

CHAPTER 35 Trace elements

LAWRENCE A. KAPLAN

Nutritional trace elements
- **Chromium**
- **Cobalt**
- **Copper**
- **Fluoride**
- **Manganese**
- **Molybdenum**
- **Iodine**
- **Selenium**
- **Zinc**

Nonnutritional toxic trace elements
- **Aluminum**
- **Arsenic**
- **Cadmium**
- **Lead**
- **Mercury**

OBJECTIVES

- List the nutritive trace metals and describe their primary biochemical role.
- Briefly list the clinical symptoms associated with deficiencies of cobalt, copper, fluoride, selenium, and zinc.
- List metals that are toxic to humans at trace levels.

KEY TERMS

caries The pathological process of localized destruction of tooth tissue by microorganisms.

micronutrients Essential food material that is required or is present in the body, in very small amounts. Includes some metals and vitamins.

toxic trace elements Elements naturally found in the environment that are antagonistic to biochemical processes. When present in high enough levels in human tissues, they can be toxic and eventually fatal.

trace elements Elements present in the body at very low amounts (micrograms to milligrams) that are essential for certain biochemical processes.

Those elements found in body tissues at levels in the micrograms per gram of tissue or less are designated essential "trace" elements. These elements, also termed *micronutrients,* can be subdivided into three groups:

1. Those for which a definite biological role is recognized and recommended dietary allowances (RDAs) have been determined. The elements in this group are zinc and iodine. The RDAs for these elements are listed in Table 35-1.
2. Those for which there is reasonable evidence of an essential role in human metabolism but for which an RDA has not been established. The elements in this group include copper, manganese, fluoride, chromium, cobalt, selenium, and molybdenum. The estimated safe and adequate dietary intakes of these elements are also listed in Table 35-1.
3. Those for which there is little or no evidence of an essential role in human metabolism but which are present at levels that suggest a possible role. These include nickel, silicon, tin, and vanadium. The elements in this group will not be discussed.

In addition to these essential micronutrients, certain elements with no known purpose can be found in human tissue and in fact are known to cause pathological conditions in humans. Since these elements are present in trace amounts and are known to cause disease states, they will be discussed in this chapter.

NUTRITIONAL TRACE ELEMENTS

Chromium (Cr)

Biochemistry. Despite a quarter of a century of research into the role of chromium in carbohydrate metabolism and in the mechanisms of insulin action, the precise function of chromium in human physiology remains ill defined and obscure.[1] Nevertheless, a human nutritional requirement for trivalent chromium has been accepted by the National Research Council.[2] Although inorganic chromium compounds appear to have minimum biological activity, they are converted in vivo (see next section) to an organochromium complex, termed the *glucose tolerance factor*.[3] This factor is believed to enhance the action of insulin, perhaps by facilitating its reaction with its receptor site.[4]

Table 35-1 RDAs and safe and adequate daily dietary intakes of selected nutritional trace elements*

	Age (yr)	Weight (kg)	Weight (lb)	Height (cm)	Height (in)	Trace elements† Copper (mg)	Manganese (mg)	Fluoride (mg)	Chromium (mg)	Selenium (mg)	Molybdenum (mg)	Zinc (mg)	Iodine (μg)
Infants	0.0-0.5	6	13	60	24	0.5-0.7	0.5-0.7	0.1-0.5	0.01-0.04	0.01-0.04	0.03-0.06	3	40
	0.5-1.0	9	20	71	28	0.7-1.0	0.7-1.0	0.2-1.0	0.02-0.06	0.02-0.06	0.04-0.08	5	50
Children	1-3	13	29	90	35	1.0-1.5	1.0-1.5	0.5-1.5	0.02-0.08	0.02-0.08	0.05-0.1	10	70
	4-6	20	44	112	44	1.5-2.0	1.5-2.0	1.0-2.5	0.03-0.12	0.03-0.12	0.06-0.15	10	90
	7-10	28	62	132	52	2.0-2.5	2.0-3.0	1.5-2.5	0.05-0.2	0.05-0.2	0.1 -0.3	10	120
	11+					2.0-3.0	2.5-5.0	1.5-2.5	0.05-0.2	0.05-0.2	0.15-0.5		
Adults (all ages and genders)						2.0-3.0	2.5-5.0	1.5-4.0	0.05-0.2	0.05-0.2	0.15-0.5		
Males	11-14	45	99	157	62							15	150
	15-18	66	145	176	69							15	150
	19-22	70	154	177	70							15	150
	23-50	70	154	178	70							15	150
	51+	70	154	178	70							15	150
Females	11-14	46	101	157	62							15	150
	15-18	55	120	163	64							15	150
	19-22	55	120	163	64							15	150
	23-50	55	120	163	64							15	150
	51+	55	120	163	64							15	150
Pregnant												+5	+25
Lactating												+10	+50

From Recommended dietary allowances, ed 9, Washington, DC, 1980, National Research Council (revised 1980 by the Food and Nutrition Board of the National Academy of Sciences–National Research Council). (See p. 526, Chapter 34, for definition of RDA.)

*RDAs are for zinc and iodine. Because insufficient information is available on which to estimate RDAs for the other trace elements, only recommended intakes are provided.

†Since the toxic levels for many trace elements may be only several times usual intakes, the upper levels for the trace elements given in this table should not be habitually exceeded.

Pathophysiology. Although it has been suggested that urinary chromium excretion might serve as a means of assessing the nutritional status of persons with respect to chromium,[5] no direct relationship has been established between levels of urine chromium and glucose intolerance. However, patients at risk for chromium deficiency, such as those treated with chronic parenteral nutrition, can develop symptoms of glucose intolerance and weight loss that are reversed with supplementary chromium.[6]

The best current understanding of chromium metabolism is that organic complexes may arise from the direct interaction of aqueous Cr(III) salts with oxygen, nitrogen, and perhaps SH ligands to form the stable Cr(III)-coordination complexes that only very slowly (days) allow further ligand exchange to take place. Cr(VI), the most common form found in nature, is apparently readily reduced to Cr(III) in many biological systems. Thus Cr(VI), actively or passively transported through cell membranes, can be reduced to Cr(III) and then, as Cr(III), can immediately react with available ligands within the cell. These ligands may be proteins, nucleic acids, or small metabolites, which then can be further metabolized as a consequence of the complex formation.

The form in which chromium exists in serum and urine is not known with any certainty. Most of the urinary chromium is present as organic complexes with a molecular weight greater than 1000 daltons. These complexes are heterogenous in size and most likely represent a limited number of peptides that are filtered by the renal glomeruli or produced by renal tubular metabolism. Dietary changes can affect the excretion rates of chromium.

Food sources. Brewer's yeast, molasses, meat products, cheeses, and whole grains serve as the best food sources for chromium.

Toxicity. Chromium has toxic properties that most frequently manifest themselves after occupational exposure. The hexavalent form of chromium, present in chromates, has strong oxidizing properties and is much more toxic than the trivalent form. Breathing contaminated industrial air or contact exposure to chromate dust or other forms of chromium pollution is known to result in inflammation and necrosis of the skin and nasal passages, allergic contact dermatitis, and lung cancer. Oral ingestion can result in damage to the gastrointestinal tract, circulatory shock, and renal failure. Treatment usually consists in removal of the person from the work site or location of exposure. Subacute toxicity and chromium accumulation in the tissues of rats have been reported with chronic exposure to as little as 5 mg of Cr(VI) per liter of drinking water. Toxicity of Cr(III) in experimental animals has been observed when the compounds are given through parenteral administration.

Reference range. Approximately 97.5% of healthy, fasting, nonchromium-supplemented subjects have a urine chromium concentration less than 0.5 μg/L (9.6 nmol/L).[7] The estimated safe and adequate dietary levels for chromium are listed in Table 35-1.

Cobalt (Co)[8,9]

Biochemistry. The only known function of cobalt is that it is an integral part of vitamin B_{12}. Cobalt fits into the corrin ring of vitamin B_{12} and is absolutely necessary for biological activity of this vitamin. See Chapter 36 for details on the functions of vitamin B_{12}. Dietary cobalt is poorly absorbed and is stored in the liver, probably as vitamin B_{12}. Cobalt is excreted in bile.

Pathophysiology. A cobalt deficiency is accompanied by all the signs and symptoms of a vitamin B_{12} deficiency. The most important symptom is anemia. See Chapter 36 for details.

Food sources. Liver and pancreas are good sources of cobalt.

Toxicity. An excess of cobalt can lead to polycythemia.

Reference range. No RDA for cobalt has yet been established.

Copper (Cu)

Biochemistry. Copper is an essential element in biological systems. A 70 kg human adult body contains approximately 80 mg of copper,[10] one third in muscle and the remainder in other tissues and fluids.[11] Copper, in the divalent state (Cu^{2+}), has the capacity to form complexes with many proteins. In a large number of cuproproteins in mammals, copper is part of the molecule and hence is present as a fixed proportion of the molecular structure. These metalloproteins form an important group of oxidase enzymes and include ceruloplasmin (ferroxidase), superoxide dismutase, cytochrome oxidase, lysyl oxidase, dopamine beta-hydroxylase, tyrosinase, uricase, spermine oxidase, benzylamine oxidase, diamine oxidase, and tryptophan-2,3 dioxygenase (tryptophan pyrrolase).[12] Although not an enzyme, the metalloprotein metallothionein binds a number of heavy metals, including copper.[13] Copper also complexes readily with L-amino acids, which facilitate its absorption from the stomach and duodenum.[14] The importance of copper in the efficient use of iron makes it essential in hemoglobin synthesis[15] (see next section).

Approximately 56% of dietary copper is absorbed in the adult.[16] The main route of excretion of copper is in the bile,[10] with less than 50 μg/day excreted in urine.

In plasma, copper is present in two forms. Approximately 90% of the total plasma copper is firmly bound to the α_2-globulin ceruloplasmin,[17] which is an oxidase involved in the binding of iron to transferrin in the plasma and also in the utilization of iron.[18] Approximately 10% of plasma copper is loosely bound to albumin, which acts as the plasma carrier of copper.[19] A very small fraction of the plasma copper is complexed with amino acids[20] or present in other copper enzymes.

Pathophysiology. Copper deficiency in the human infant produces three distinct syndromes. The first is characterized by symptoms of anemia, hypoproteinemia, and low serum iron and copper levels, which are corrected by combined iron and copper supplementation.[21] A second syndrome occurs in malnourished infants receiving high-calorie, low-copper, rehabilitation diets, usually by hyperalimentation. Pronounced neutropenia, anemia, diarrhea, bone changes, and hypocupremia in these infants respond to copper supplementation.[22]

The third syndrome is known as Menkes' or kinky hair syndrome. This syndrome is caused by X-linked genetic defect in copper absorption from the intestinal mucosa that results in low blood, liver, and hair copper levels; progressive mental deterioration; and defective keratinization of hair.[23]

Wilson's disease is an inborn error of copper metabolism inherited as an autosomal recessive characteristic in either sex in humans. This disease is distinguished by excessive accumulation of copper in the liver, brain, cornea, and kidneys, with low levels of ceruloplasmin and high nonceruloplasmin copper in serum.[24] The effectiveness of therapeutic administration of chelating agents in Wilson's disease can be monitored by measurement of copper excretion in the urine.[25]

Food source. Shellfish, liver, kidneys, egg yolk, and some legumes are rich in copper. There are significant amounts of copper in most other food sources.

Toxicity. Copper toxicity is a potential complication in patients on chronic hemodialysis.[26]

Reference ranges. It is generally agreed that there is a gender difference in serum copper levels. The range of copper levels in females is 0.80 to 1.55 mg/L (12.6 to 24.4 μmol/L), and the slightly lower range in males is 0.70 to 1.40 mg/L (11.0 to 22.0 μmol/L). In newborn infants, levels of 0.12 to 0.67 mg/L (1.9 to 10.5 μmol/L) have been reported, rising to a range of 0.27 to 1.52 mg/L (4.2 to 23.9 μmol/L) in children 3 to 10 years old.[27] Elevated serum copper levels, over 2.0 mg/L (31.5 μmol/L), can result from the use of oral contraceptives and estrogen therapy and are usual during the third trimester of pregnancy. Serum copper levels are elevated in a large number of chronic and acute illnesses. Among these are Hodgkin's disease, leukemia and many other malignancies, megaloblastic and aplastic anemia, hemochromatosis, rheumatic fever, major and minor thalassemia, trauma, and collagen diseases. Serum levels are also elevated in some patients undergoing renal dialysis.[28]

Low serum copper levels are found in patients with Wilson's disease, Menkes' or kinky hair syndrome, some iron-deficiency anemias, protein malnutrition, chronic ischemic heart disease, and in burn patients.

Red blood cell copper levels are relatively constant at approximately 900 μg/L of packed cells.[10]

The normal daily urine excretion of copper is reported

to be 2 to 37 μg. An upper limit of normal of approximately 30 μg/day (0.47 μmol/day) has been suggested.[29] The estimated safe and adequate dietary intake of copper is listed in Table 35-1.

Fluoride (F^-)[30,31]

Biochemistry. The importance of fluoride lies in the body's ability to incorporate the anion into the calcium-phosphate crystalline structure of bone. The resulting mineral, fluoroapatite, is a stronger structural material, that possesses anticaries properties.

There is also evidence that fluoride has a direct effect on calcium and phosphate metabolism and may reduce symptoms of osteoporosis. It also is known to inhibit several enzymes, especially enolase. This inhibition is sometimes used clinically to preserve blood by inhibiting glycolysis and preventing changes in blood glucose and lactate levels.

Pathophysiology. Trace amounts of fluoride are known to produce tooth enamel that has increased resistance to bacterial degradation, resulting in caries-resistant teeth. This resistance may be the result of the fluoride-containing crystalline structure, a direct inhibition of bacterial degradative processes, or both. Whatever the cause, the caries-resistance effect of fluoride has prompted many communities to add fluoride to their drinking water when the natural fluoride level is below that concentration which promotes caries resistance (1 mg/L). When this fluoride level is achieved, the incidence of dental caries is reduced by over 50%. Although not quite as effective as fluoride in water, fluoride added to table salt also results in a significant decrease (40% to 50%) in the incidence of dental caries.

Food source. Fish is the primary food source for fluoride, although drinking water can also be an important natural source, depending on the locale. For those areas with low natural levels of fluoride in the drinking water, supplementation, either in the water or in table salt, can be effective.

Toxicity. In contrast to the caries resistance imparted by relatively low levels of fluoride (1 mg/L) in drinking water, excessive amounts of fluoride ingested during active tooth calcification (for example, greater than 10 mg/L) can result in *dental fluorosis*. This condition results in teeth with a patchy, dull white, even chalky-looking appearance. A brown mottled appearance can also occur. Instead of an increased structural strength, the teeth pit readily and have increased fragility.

Reference ranges. The estimated safe and adequate dietary intake levels of fluoride are listed in Table 35-1.

Manganese (Mn)[32]

Biochemistry. Manganese is a cofactor for a number of enzymes including arginase, cholinesterase, phosphoglucomutase, pyruvate carboxylase, mitochondrial superoxide dismutase, as well as several phosphatases, peptidases, and glycosyltransferases. In certain instances the specificity for Mn^{2+} is not very high; Mn^{2+} may be replaced by Co^{2+} or Mg^{2+}. Manganese also functions with vitamin K in the formation of prothrombin. Most manganese is probably stored in the liver, and excess manganese is excreted in the feces.

Pathophysiology. Manganese deficiency has never been reported in humans, so the clinical appearance of a putative deficiency is unknown. However, Mn^{++} deficiencies have been produced experimentally in animals, and the symptoms seen in these studies may have parallels in humans. Rats fed a manganese-deficient diet develop sterility, males show testicular degeneration, and females demonstrate inability to suckle young. In chicks, a manganese-deficient diet results in disorders of bone growth.

Food source. Liver and kidney are the primary meat sources of manganese. Wheat germ, legumes, and nuts are the best vegetable sources of manganese.

Toxicity. Excess manganese can interfere with the absorption of dietary iron. If the excess continues, it may result in an iron-deficiency anemia.

Reference ranges. The estimated safe and adequate dietary intake of manganese is found in Table 35-1.

Molybdenum (Mo)[33]

Biochemistry and pathophysiology. No deficiency has been described in humans or other animals. Molybdenum is an essential trace element for plants and serves as a cofactor for xanthine and aldehyde oxidases. Dietary molybdenum is readily absorbed by the intestine and is excreted in urine and bile.

Food source. Liver and kidney are good meat sources of molybdenum; whole grains, legumes, and leafy vegetables serve as vegetable sources.

Reference ranges. The estimated safe and adequate range of molybdenum dietary intake is listed in Table 35-1.

Iodine

Iodine is important in human metabolism because of its role in the formation of the active thyroid hormones, thyroxine (T_4) and triiodothyronine (T_3). Chapter 41 discusses the metabolism of iodine in detail, as well as the metabolic role of the thyroid hormones.

Food sources. The primary sources of dietary iodine are seafood and drinking water. However, the amounts of iodine in drinking water can vary greatly from one location to another. For that reason, approximately half of the table salt in the United States is iodized, that is, contains sodium iodide.

Selenium (Se)

Biochemistry. The only known biochemical role of selenium in humans is to serve as an essential constituent of the enzyme *glutathione peroxidase* (glutathione:H_2O_2 oxidoreductase, E.C. 1.11.1.9).[34] Selenium is covalently linked to cysteine residues in the protein as selenocysteine. This enzyme is found in the cytoplasm and mitochondria of liver, erythrocytes, platelets, and other tissues. Along with vitamin E, selenium–glutathione peroxidase acts as part of the antioxidant defense system of cells, protecting lipids in cell membranes from the destructive effects of H_2O_2 and superoxide (O_2^-) generated by excess oxygen. Because the antioxidant role of selenium parallels that of vitamin E, the amount of selenium needed as a nutrient in animals is inversely proportional to the dietary levels of vitamin E, and selenium-deficiency diseases will respond to tocopherol treatment.[35]

Pathophysiology. Selenium deficiency in animals can lead to dysfunction of a number of organs including the brain, cardiovascular system, liver, and muscles and can affect fetal outcome.[35] Chronic selenium deficiency has been epidemiologically associated wtih certain types of cancer,[36] whereas selenium levels greater than those needed for nutrition have been shown to inhibit carcinogenesis in experimental models.[37]

Chronic selenium deficiency in humans is also associated with cardiovascular disease[38] and endemic cardiomyopathy.[39] The latter is seen in certain rural areas of China where it is known as *Keshan disease*. Severely low blood selenium levels (17 μg/L) have been found in this population.

Blood and urine selenium determinations have two diagnostic uses: nutritional and toxicological. Lower whole blood selenium values are expected in patients receiving hyperalimentation[34] and in most cancer patients. The latter observation may be attributable to a nutritional effect related to the anorexia and cachexia of cancer. Because the liver is important in selenium metabolism, diseases that severely affect the liver, such as cirrhosis and hepatitis, will also cause low selenium values. Because the majority of selenium in the blood is attached to the important selenoenzyme glutathione peroxidase, nutritional selenium status can also be determined by measurement of plasma or platelet gluthathione peroxidase.[35]

Food sources. Liver and kidney are rich in selenium. Grains and vegetables are good sources of selenium, but the selenium content of these foods varies greatly with the local soil content of this metal.

Toxicity. Selenium toxicity in humans is rare. In animals, excessive selenium intake results in *alkali disease,* characterized by liver and neuromuscular disorders that are usually fatal. Because of physiological regulation, blood selenium measurements are not useful for assessment of selenium toxicity. Urinary selenium measurements are much better indicators of selenium toxicity. The greatest level of selenium found in body fluids was 4900 μg of selenium per liter of urine. This was measured in a selenium worker who had inhaled selenium dust. From these studies, Glover[40] has suggested that in public, rural, and industrial health situations, the selenium urinary level should be below 100 μg/L.

Reference range. The whole blood selenium normal values range from 100 to 190 μg/L (1.3 to 2.4 μmol/L). The normal range depends on the amounts of environmental selenium in an area and therefore should be determined in each area. A lower selenium normal range is expected in the low selenium areas, whereas a higher selenium normal range would be expected in the high selenium areas.

Glover[40] has reported an average urinary selenium level of 84 μg/L.

The estimated safe and adequate dietary intake of selenium is listed in Table 35-1.

Zinc (Zn)

Biochemistry. Zinc, an important nutritive factor, is a cofactor for many metalloenzymes. Over 90 metalloenzymes require zinc for proper function, including red blood cell carbonic anhydrase, alkaline phosphatase, lactic and alcohol dehydrogenases, and many enzymes involved in RNA and DNA synthesis (such as DNA- and RNA-dependent RNA polymerases).[41] Because it is required by many of the enzymes needed for DNA and RNA synthesis, zinc is necessary for the growth and division of cells, especially during stages of life with high rates of growth, that is, infancy, adolescence, and pregnancy. Zinc is also included in the native structures of insulins. Cancer and burn patients are at high risk for zinc deficiency. Zinc deficiency is also associated with renal disease, gastrointestinal disease, and alcohol abuse.

The rapidly dividing cells of the immune system are sensitive to zinc deficiencies. The role of zinc in the development and maintenance of a normally functioning immune system has been well established.[42] In patients with severe zinc deficiency, there can be underdevelopment of the thymus, decreased response of lymphocytes to specific stimulators (phytahemagglutin [PHA]), and decreased hyperimmune response. Immunoglobin levels can be normal or decreased.

Pathophysiology. Zinc deficiency has many causes, but malnutrition and malabsorption are the most common. Pregnant women are at risk for suboptimum levels of zinc, which can adversely affect fetal outcome.[43] Patients receiving chronic total parenteral nutrition are also at risk unless they are given supplemental zinc.[42] If no supplemental zinc is given, plasma zinc levels can rapidly decrease until subnormal levels are reached and clinical signs associated with zinc deficiency can be seen.[44] The most important clinical syndrome of zinc deficiency is acrodermatitis enteropathica. The symptoms associated with this disease are those seen, with varying degrees of severity, in many zinc-deficiency states: skin lesions on body extremities or orifices, diarrhea and anorexia, loss of hair, severe growth retardation, extreme irritability, and increased susceptibility to infection.[42] In 1973, Moynahan and Barnes[45] studied a patient with acrodermatosis and noted a very low serum zinc level (190 μg/L). Symptoms seen in milder zinc-deficiency states include mental lethargy, growth retardation, poor appetite, male hypogonadism, and skin changes.[44]

Although serum or plasma has been the usual specimen for assessment of zinc status, urine[46] and leukocyte zinc levels[47] have also been used. In a study of experimental zinc deficiency, urinary zinc levels changed most quickly in response to a decrease in dietary zinc.[46] Leukocyte levels may best reflect total body levels of zinc and thus may be used to assess chronic zinc deficiency.

Food sources. Meat, seafood, and eggs are good dietary sources of zinc. Although vegetables and whole grains (especially wheat germ) contain significant amounts of zinc, the zinc in these foods has a low bioavailability, especially if high levels of phytates are present. Milk is an important source of zinc; colostrum is especially rich in zinc.

Toxicity. Zinc toxicity is very rare.

Reference ranges

Serum. 654 to 1150 μg/L (10.0 to 17.6 μmol/L).

Urine. 150 to 1200 μg/24 hours (2.3 to 18.3 μmol/24 hours).

Cerebrospinal fluid. 9.2 ± 2.0 (S.D.) μg/L (0.14 ± 0.03 μmol/L).

The RDA for zinc is listed in Table 35-1.

NONNUTRITIONAL TOXIC TRACE ELEMENTS

Aluminum (Al)

Biochemical basis for toxicity. The clinical syndrome resulting from aluminum toxicity has been recognized for over half a century. In the last decade this syndrome has become a major area of concern because of the increased numbers of people with end-stage renal disease who are being treated by dialysis,[48] and the patients who are being treated chronically with aluminum-contaminated parenteral nutrition fluids.

The major source of aluminum in end-stage renal failure is the oral administration of aluminum hydroxide gels used to treat the hyperphosphatemia of renal failure. Although more than 98% of the oral aluminum passes through the gastrointestinal tract, some is absorbed into plasma where it is bound primarily to transferrin.[49] Some of the plasma aluminum equilibrates with tissues and is deposited primarily in bone and nerve (especially brain) tissue. The exact biochemical mechanism of the toxic effect of aluminum on cells is not known. Aluminum is excreted almost exclusively in urine.

Clinical signs and symptoms. Chronic aluminum gel therapy can lead to a progressive, fatal neurological syndrome. The onset of the syndrome is manifested by a speech disturbance characterized by hesitancy and stuttering, followed by dysarthria, dyspraxia, and dysphasia.

Table 35-2 Reference ranges for aluminum

Serum concentration, μg/L (μmol/L)	Comment
0-10 (0-0.37)	Normal serum aluminum concentration
16-60 (0.59-2.22)	Aluminum concentrations in this range not considered toxic
61-100 (2.26-3.37)	Aluminum concentrations in this range indicative of increased aluminum uptake; there is a need to identify sources before problems from aluminum toxicity occur
101-200 (3.74-7.4)	Concentrations >100 μg/L indicative of a potential clinical problem of aluminum toxicity; patient should be under close surveillance for indications of toxicity
>200 (>7.4)	Aluminum concentrations >200 μg/L generally lead to clinical symptoms

Later the patients develop myoclonic movement, seizure activity, a progressive global dementia, and parietal lobe signs such as constructional apraxia or directional disorientation. The syndrome progresses to a stage where the patients cannot perform any purposeful movements and eventually may die. This neurological disorder is now recognized as a definite syndrome and has been named *dialysis encephalopathy* or *dialysis dementia*.[48,50]

In addition to the neurological problems associated with oral aluminum therapy, other long-term complications may include bone disease, such as vitamin D–resistant, aluminum-induced osteomalacia and microcytic, hypochromic anemia.

Treatment. Once the presence of aluminum toxicity has been established, the patient can be treated with the chelating agent desferrioxamine.[50] The desferrioxamine binds aluminum from tissue deposits and is then removed by dialysis. After successful desferrioxamine treatment, serum aluminum levels rise transiently as aluminum is removed from tissues and then fall, usually to well below the pretreatment levels. Reversal of symptoms of aluminum toxicity has been noted in patients treated with desferrioxamine. Such therapy may need to be continued for years.

Reference ranges. Table 35-2 gives reference ranges for aluminum. The aluminum content of dialysis fluid should be less than 15 μg/L.[48]

Arsenic (As)[51-53]

Biochemical basis for toxicity. All the biochemical actions of arsenic are attributed to the trivalent forms of arsenic. These compounds can covalently react with sulfhydryl groups (SH) of enzymes (E), inactivating these groups:

$$R{-}As{=}O + 2SH{-}E \rightarrow R{-}As\begin{matrix} \diagup S{-}E \\ \diagdown S{-}E \end{matrix} + H_2O$$

Since sulfhydryl groups are often required for certain enzyme activity, arsenocompounds inactivate these enzymes, leading to eventual cell death.

Clinical signs and symptoms. Arsenic trioxide and the derivative arsenites and arsenates are the principal causes of occupational and environmental poisoning. Bone marrow depression is observed in chronic arsenic poisoning. Leukopenia and normochromic anemia predominate and are believed to reflect such bone marrow depression.

When a large quantity of a soluble arsenic compound is ingested, especially on an empty stomach, death may occur within a few hours. This results from acute poisoning of the myocardium, which may or may not be associated with brainstem medullary failure. A less dramatic result occurs with ingestion of smaller amounts of arsenic. Usually swallowing of arsenic is painless, but occasionally some epigastric pain is experienced. Shortly after ingestion, vomiting and diarrhea will occur. During the next several days there may be inflammation of the conjunctival and respiratory mucous membranes, epistaxis, transient jaundice, cardiomyopathy, erythematous or vesicular rashes, and sweating. Hematological, renal, or pancreatic dysfunction may be observed. Symptoms of neuropathy become noticeable 1 to 2 weeks later and typically appear as numbness with tingling and paresthesia in the extremities.

Arsine gas was implicated in several hundred poisoning cases, many of which (25% to 30%) proved to be fatal. Generally a delay of 2 to 24 hours occurs before onset of symptoms. Characteristic symptoms are abdominal pain, nausea, vomiting, hematuria, and jaundice. Clinical laboratory abnormalities of greatest signficance involve anemia and in vivo hemolysis. Blood hemoglobin concentrations less than 100 g/L are often observed, and plasma hemoglobin levels over 2 g/L have been detected. Proteinuria and methemoglobinemia are frequently observed to cause renal failure and death.

The symptoms of chronic arsenic exposure are exfoliation and pigmentation of the skin, neurological symptoms, polyneuritis, altered hematopoiesis, and degeneration of the liver and kidneys.

The main risks of cancer involving arsenic are skin cancers caused by arsenic from environmental sources such as water and cancers of the respiratory tract caused by arsenic from industrial sources such as plants and mines where arsenic is present at high levels.

Treatment. Treatment of acute poisoning has often involved intramuscular injection of dimercaprol (British Anti-Lewisite, BAL), continued for several days.

Treatment proposed for severe poisoning includes exchange transfusion and hemodialysis. Since there is prolonged anemia in these cases, hemodialysis is preferred to peritoneal dialysis. Administration of BAL has not been demonstrated to be truly effective because it offers no protection against red blood cell destruction.

Reference range. Normal urine levels range from 0 to 50 μg (0 to 0.667 μmol)/24 hr.

Cadmium (Cd)

Biochemical basis for toxicity. Cadmium is toxic to virtually every system of the body[54-57] and has been implicated in renal disease, prostatic carcinoma, hypertension, anemia, itai-itai (''hurts-hurts'') disease, and other conditions.

There seems to be no special mechanism for the absorption of cadmium from the gut; estimates of 1% to 5% absorption of ingested cadmium have been made. After absorption or injection, the metal is transported quite rapidly by the blood to the liver, kidney, intestinal mucosa, and other tissues. There is some accumulation in the red blood cells, but blood concentrations, as compared to those in the kidney and liver, are generally quite low. Cadmium is bound chiefly to the low-molecular-weight protein metallothionein in all tissues, and it has been suggested that it is in this form that it is transported. The synthesis of this binding protein is apparently induced by cadmium. After long-term exposure, the metal accumulates in the kidneys, and it has been suggested that the cadmium metallothionein complex is itself toxic to the kidneys. There will also be significant concentrations of the metal in the liver.

Cadmium is excreted primarily in the urine. Urine excretion is slow; concentrations in nonoccupationally exposed persons are quite low, below 2 μg of cadmium per gram of creatinine. Because of its firm binding in the tissues, cadmium has a long biological half-life, estimated at between 20 and 30 years. The half-life of cadmium in the blood has been estimated at 2.5 months; therefore, blood levels are useful only as indicators of recent exposure.

Clinical signs and symptoms. Acute poisoning may occur from inhalation of freshly generated cadmium vapors, causing severe pulmonary and bronchial irritation, nausea, and vomiting. Repeated acute episodes may result in lung emphysema, and death may occur from pulmonary edema during an acute attack.

Ingestion of contaminated food or beverages or excess intake in workers exposed to cadmium dust may lead to an *acute* attack with gastrointestinal tract irritation, nausea, vomiting, abdominal cramps, and diarrhea. A single dose exceeding 300 mg can be fatal, but with lower doses recovery is usually complete within a few days.

Chronic poisoning can affect various organs of the body, but it is apparent[56] that the critical organ is the kidney. The earliest indications of toxicity may be low-molecular-weight proteinuria associated with glucosuria, aminoaciduria, impaired acid secretion, and other indications of renal disease. The critical level of cadmium in the kidney cortex is believed to be 200 to 250 μg/g of wet weight of tissue although Cohn et al.[58] suggest a critical range of 300 to 400 μg/g (wet weight) for the renal cortex, based on neutron activation analyses. Chronic poisoning can also affect the lungs (that is, produce emphysema and pulmonary fibrosis) and the bones (that is, itai-itai disease). It has been suggested that chronic cadmium poisoning can lead to cancer and hypertension, but such effects have not been proven.

Medical surveillance of workers exposed to cadmium should include blood and urine cadmium analyses, standard hematological and chemistry profiles, urine protein estimations (total protein and β_2-microglobulin), and routine urinalysis with microscopic examination. Also, a chest roentgenogram and lung-function tests should be performed. For details refer to the excellent pamphlet by Lauwerys.[56]

Treatment. Treatment of cadmium poisoning can only be supportive. The contaminating source must, of course, be discovered and removed. In industrial situations, an affected person should be moved to an area free of cadmium contamination before critical levels are reached. The effects of a combined treatment with diethylthiocarbamate and diethylenetriamine pentaacetate have been studied,[59] but the overall effect appears to be a redistribution of the metal in the various tissues rather than a significant removal of the metal from the body.

Reference ranges. Although estimation of the cadmium content of the kidney cortex in a person is useful when one is determining if the proposed critical concentration of 200 μg/g of wet tissue weight has been realized,[60] it is not possible, of course, to make this measurement routinely. Therefore one must rely on known relationships among blood, urine, and kidney cortex concentrations to predict the severity of renal disease caused by cadmium. For example, it is generally agreed that a urine cadmium level of about 10μg/g of creatinine corresponds to a renal cortex concentration of 200 μg of cadmium per gram of tissue. A World Health Organization (WHO) study group in 1980[61] agreed that the urine cadmium concentration in a person should *not* be allowed to reach 10 μg/g of creatinine, and the group assigned a critical level for blood cadmium of 10 μg/L when the exposure is chronic and long term.

Stoeppler[62] stated in 1983 that the probable mean, normal values for cadmium in body fluids of healthy, nonexposed adults are as follows:

1. Whole blood in nonsmokers: ≤1 μg/L
2. Whole blood in smokers: ≤1.5 μg/L
3. Serum in nonsmokers: ≤0.05 μg/L
4. Serum in smokers: ≤0.1 μg/L
5. Urine in nonsmokers: ≤1 μg/L

Mean reference (normal) levels for urinary cadmium are probably less than 1 μg of cadmium per gram of creatinine.

It is an established fact that tissue levels of cadmium in animals and humans increase with age. Thus, when one is preparing multilevel tissue reference standards; tissues are simply selected from animals of different ages. In humans, smoking increases tissue levels of cadmium.

Reference ranges will vary in different parts of the world depending on the cadmium content of the food, air, water, and soil and the varying cadmium content of tobacco smoked in the different areas.

Lead (Pb)

Biochemical basis of toxicity. Although only 8% to 12% of orally ingested lead is absorbed by the small intestines, the toxic effect of lead is a major problem in many societies. Lead, like many other heavy metals, can react with sulfhydryl groups in enzymes, thereby inactivating them. Most important of these enzymes are the δ-aminolevulinic acid (δ-ALA) dehydratase and δ-ALA synthetase. This leads to the important hematological manifestations of lead poisoning detailed in Chapter 32.

Clinical signs and symptoms. The pathological effects of lead have been recognized for centuries. Many physiological systems, including renal, nervous, reproductive, endocrine, immune, and hemopoietic, are affected. These effects have been the subject of numerous reviews and research publications.[63]

Exposure to lead, either acute or chronic, presents a variety of signs, symptoms, and chemical evidence. The exposed persons may be asymptomatic or symptomatic. Lane et al.[64] listed the following signs and symptoms for exposed adults. Mild symptoms include tiredness, lack of energy, constipation, slight abdominal pain and discomfort, anorexia, altered sleep, irritability, anemia, pallor, and less frequently diarrhea and nausea. Formation and precipitation of lead sulfide may be evidenced as a blue-black ''lead line'' near the gingival margin of the teeth. Severe symptoms and signs include colic, reduction of muscle power, muscle tenderness, paresthesia, and symptoms or signs of neuropathy or encephalopathy. As compiled by the CDC[65] the symptoms most frequently seen in children are irritability, vomiting, abdominal pain, ataxia, anorexia, behavioral changes, speech disturbances, seizures, intercurrent fever, and dehydration. The symptoms for children, listed in descending order of frequency of presentation by Sachs et al.,[66] are drowsiness, irritability, vomiting, gastrointestinal symptoms, ataxia, stupor, and fatigue.

A number of reviews of the clinical presentation of lead exposure have been published, including an excellent one by Granick et al.[67] Other biochemical parameters indicative of lead exposure include δ-ALA, δ-ALA dehydratase, and porphyrins.

Treatment. Treatment of lead poisoning consists of supportive measures and the use of chelating agents, such as EDTA, dimercaprol, and penicillamine.

Reference ranges. Two classes of persons must be considered in regard to lead exposure: (1) those exposed occupationally and (2) those exposed environmentally. The maximum allowable blood level for lead in workers has been set at 600 μg/L (2.9 μmol/L),[68] a level at which health effects are observed. The typical range for healthy nonexposed adults is 100 to 200 μg/L (483 to 965 nmol/L), with lower values for children.[68,69] Blood lead and urine lead concentrations for both children and adults have been reported to be about 10% to 20% higher in males than in females.[70]

Mercury (Hg)

Biochemical basis of toxicity. Like cadmium and lead, mercury compounds readily react covalently with sulfhydryl groups in proteins, resulting in inhibition of functional activity. Enzymes containing sulfhydryl groups will lose this enzymatic activity. Both inorganic mercury and organic mercury compounds are potent toxic compounds.[71]

Clinical signs and symptoms. The most frequent causes of exposure to toxic levels of mercury are those related to the industrial use of this metal, that is, accidental or chronic exposure in the workplace. Exposure to mercury vapor produces toxic effects because of the accumulation of mercury in the brain. Neurological signs displayed may include increased excitability, severe behavioral and personality changes, insomnia, and loss of memory.[72]

Gastrointestinal disturbance and renal damage are likely after acute exposure to mercuric salts. A 1 g dose of mercuric salt is considered lethal in humans.

Chronic mercury poisoning results in the atrophy and degeneration of the sensory cerebral cortex. Hearing and visual impairment may also result.

Treatment. Treatment for severe mercury poisoning should include hemodialysis and administration of mercury-chelating agents (cysteine, *N*-acetyl-penicillamine). In less severe cases, administration of the chelating agents may be effective to mobilize the mercury.

Reference range. Normal urine levels of inorganic mercury for nonexposed subjects range from 0 to 20 μg/24 hr (0 to 0.1 μmol/24 hr).

REFERENCES

Chromium

1. Shapcott, D, and Hubert, J, editors: Chromium in nutrition and metabolism, Dev Nutr Metab 2:1-339, 1979.
2. National Academy of Sciences, Recommended Dietary Allowances, ed 9, Washington, DC, 1980, National Academy of Sciences.
3. Anderson, RA: Nutritional role of chromium, Sci Total Environ 17:13-29, 1981.
4. Hambridge, KM: Chromium nutrition in man, Am J Clin Nutr 27:505-514, 1974.
5. Uusitupa, MKJ, Kumpulainen, JT, Voutilainen, E, et al: Effect of inorganic chromium supplementation on glucose tolerance, insulin response, and serum lipids in noninsulin-dependent diabetics, Am J Clin Nutr 38:404-410, 1983.
6. Freund, A, Atamian, S, and Fischer, JE: Chromium deficiency during total parenteral nutrition, JAMA 241:496-498, 1979.
7. Anderson, RA, Polansky, MM, Bryden, NA, et al: Urinary chromium excretion of human subjects: effect of chromium supplementation and glucose loading, Am J Clin Nutr 36:1184-1193, 1982.

Cobalt

8. Mertz, W: The essential trace elements, Science 213:1332, 1981.
9. Herbert, V: Drugs effective in iron deficiency anemia and other hypochromic anemias. I. In Goodman, LS, and Gilman, A, editors: The pharmacological basis of therapeutics, ed 6, New York, 1987, Macmillan Publishing Company.

Copper

10. Cartwright, GE, and Wintrobe, MM: Copper metabolism in normal subjects, Am J Clin Nutr 14:244-232, 1964.
11. Tipton, IH, and Cook, MJ: Trace elements in human tissue. II. Adult subjects from the United States, Health Phys 9:103-145, 1963.
12. Vallee, BL, and Wacker, WEC: In Neurath, H, editor: The proteins: composition, structure, and function, ed 2, vol V, Metalloproteins, New York, 1965, Academic Press, Inc.
13. Lerch, K: Copper metallothionein, a copper binding protein from *Neurospora crassa*, Nature 284:368-370, 1980.
14. Crampton, RF, Matthews, DM, and Poisner, R: Observations on the mechanisms of absorption of copper by the small intestine, J Physiol (London) 178:111-126, 1965.
15. Hart, EB, Steenbock, H, Waddell, J, and Elvehjam, CA: Iron in nutrition. VII. Copper as a supplement to iron for hemoglobin building in the rat, J Biol Chem 77:797-812, 1928.
16. Strickland, GT, Beckner, WM, and Lew, ML: Absorption of copper in homozygotes and heterozygotes for Wilson's disease and controls: isotope tracer studies with ^{67}Cu and ^{64}Cu, Clin Sci 43:617-625, 1972.
17. Delves, HT: The micro-determination of copper in plasma protein fraction, Clin Chim Acta 71:495-500, 1976.
18. Osaki, S, Johnson, DA, and Frieden, EJ: The possible significance of the ferrous oxidase activity of ceruloplasmin in normal human serum, J Biol Chem 241:2746-2751, 1966.
19. Gubler, CJ, Lahey, ME, Cartwright, GE, and Wintrobe, MM: Studies in copper metabolism. IX. The transportation of copper in blood, J Clin Invest 32:405-414, 1953.
20. Neuman, PZ, and Sass-Kortsak, A: The state of copper in human serum: evidence for amino-acid-bound fraction, J Clin Invest 46:646-648, 1967.
21. Schubert, WH, and Lahey, ME: Copper and protein depletion complicating hypoferric anemia of infancy, Pediatrics 24:710-735, 1959.
22. Cordano, A, Placko, RP, and Graham, GG: Hypocupremia and neutropenia in copper deficiency, Blood 28:280-283, 1966.
23. Sass-Kortsak, A, and Bern, AG: Hereditary diseases of copper metabolism. In Standbury, JB, and Frederickson, DS, editors: The metabolic basis of inherited disease, New York, 1978, McGraw-Hill Book Co.
24. Scheinberg, IH, and Sternlieb, I: Wilson's disease, Annu Rev Med 16:119-134, 1965.
25. Deiss, A, Lynch, RE, Lee, GR, and Cartwright, GE: Long term therapy of Wilson's disease, Am J Intern Med 75:57-65, 1971.
26. Blomfield, J, Dixon, SR, and McCredie, DA: Potential hepatotoxicity of copper in recurrent hemodialysis, Arch Intern Med 128:555-560, 1971.
27. Scheinberg, IH, Cook, CD, and Murphy, JA: The concentration of copper and ceruloplasmin in maternal and infant plasma at delivery, J Clin Invest 33:963, 1954.
28. Cartwright, GE, Gubler, CJ, and Wintrobe, MM: Studies on copper metabolism. XI. Copper and iron metabolism in the nephrotic syndrome, J Clin Invest 33:685-698, 1954.
29. Lyneger, GV, Kollmer, WE, and Bowert, HJM: The elemental composition of human tissues and body fluids, Weinheim, 1978, Verlag Chemic.

Fluoride

30. Shaw, JH, and Sweeney, EA: Nutrition in relation to dental medicine. In Goodhart, RS, and Shils, ME, editors: Modern nutrition in health and disease, Philadelphia, 1980, Lea & Febiger.
31. Loesche, WJ: Topical fluorides as an antibacterial agent, J Prev Dent 4:21-25, 1977.

Manganese and molybdenum

32. Doisy, EA, Jr: Effect of deficiency in manganese upon plasma levels of clotting proteins in man. In Huekstra, WG, Suttie, HE, Ganther, HE, and Mertz, W, editors: Trace elements in animals, ed 2, Baltimore, 1974, University Park Press.
33. Paige, DM, editor: Manual of clinical nutrition, Pleasantville, NJ, 1983, Nutrition Publications, Inc.

Selenium

34. Lane, HW, Barroso, AO, Englert, D, et al: Selenium status of seven chronic intravenous hyperalimentation patients, J Parenteral Enteral Nutr 6:426-431, 1982.
35. Combs, GF, Jr, and Combs, SB: The nutritional biochemistry of selenium, Annu Rev Nutr 4:257-280, 1984.
36. Schrauzer, GN, White, DA, and Schneider, CJ: Cancer mortality studies. II. Statistical association with dietary selenium intakes, Bioinorganic Chem 7:23-34, 1977.
37. Shamberger, RJ, and Frost, DV: Possible protective effect of selenium against human cancer, Can Med Assoc J 104:682, 1969.
38. Shamberger, RJ, Wills, CE, and McCormack, LJ: Selenium and heart disease. II. Blood selenium and heart mortality in 19 states. In Hemphill, DD, editor: Trace substances in environmental health, vol 12, Columbia, MO, 1978, University of Missouri Press.
39. Chen, X, Chen, X, Yang, GQ, et al: Relation of selenium to the occurence of Keshan disease. In Spallholz, JE, Martin, JL, and Ganther, HE, editors: Selenium in biology and medicine, Westport, CT, 1981, AVI Publishing Co.
40. Glover, JR: Selenium in human urine: a tentative maximum allowable concentration for industrial and rural populations, Ann Occup Hyg 10:3-14, 1967.

Zinc

41. Riordan, JF: Biochemistry of zinc, Med Clin North Am 60:661-674, 1976.
42. Hansen, MA, Fernandes, G, and Good, RA: Nutrition and immunity: the influence of diet on autoimmunity and the role of zinc in the immune response, Annu Rev Nutr 2:151-177, 1982.
43. Mukherjee, MD, Sanstead, HH, Ratnaparkhi, MV, et al: Maternal zinc, iron, folic acid and protein nutriture and outcome of human pregnancy, Am Soc Clin Nutr 40:496-507, 1984.
44. Prasad, AS: Clinical manifestations of zinc deficiency, Annu Rev Nutr 5:341-363, 1985.
45. Moynahan, EJ, and Barnes, PM: Zinc deficiency and a synthetic diet for lactose tolerance, Lancet 1:676, 1973.
46. Baer, MT, and King, JC: Tissue zinc levels and zinc excretion during experimental zinc depletion in young men, Am J Clin Nutr 39:556-570, 1984.
47. Patrick, J, and Dervish, C: Leukocyte zinc in the assessment of zinc status, CRC Crit Rev Clin Lab Sci 20:95-114, 1984.

Aluminum

48. Wills, MR, and Savory, J: Aluminum poisoning, dialysis encephalopathy, osteomalacia, and anaemia, Lancet 1:29-34, 1983.
49. Bertholf, RL: Aluminum binding to plasma proteins: a study of the speciation of aluminum in the sera of healthy, uremic, and Alzheimer's diseased individuals, doctoral dissertation, Department of Biochemistry, Charlottesville, 1985, University of Virginia.
50. Berthold, RL, Roman, JM, Brown, S, et al: Aluminum hyroxide-induced osteomalacia, encephalopathy and hyperaluminemia in CAPD: treatment with desferrioxamine, Peritoneal Dial Bull 4:30-32, 1984.

Arsenic

51. Fowler, BA, Ishinishi, N, Tsuchiya, K, and Vahter, M: Arsenic. In Friberg, L, Nordberg, GF, and Vouk, VB, editors: Handbook of the toxicology of metals, Amsterdam, 1979, Elsevier/North Holland Biomedical Press.
52. Kaye, S: Handbook of emergency toxicology, ed 3, Springfield, IL, 1972, Charles C Thomas, Publisher.
53. Savory, J, and Sedor, JA: Arsenic poisoning. In Brown, SS, editor: Clinical chemistry and chemical toxicology of metals, Amsterdam, 1977, Elsevier/North Holland Biomedical Press.

Cadmium

54. Roels, H, Lauwerys, R, Bucket, JP, et al: Significance of urinary metallothionein in workers exposed to cadmium, Int Arch Occup Environ Health 52:159-166, 1983.
55. Piscator, M: Role of cadmium in carcinogenesis with special reference to cancer of the prostate, Environ Health Perspect 40:107-120, 1981.
56. Lauwerys, RR: Health maintenance of workers exposed to cadmium: a guide for physicians, produced for the Cadmium Council, 2392 Madison Ave, New York, NY 10017.
57. Tohyama, C, Shaikh, ZA, Nogawa, K, et al: Urinary metallothionein as a new index of renal dysfunction in "itai-itai" disease patients and other Japanese women environmentally exposed to cadmium, Arch Toxicol 50:159-166, 1982.
58. Cohn, SH, Ellis, KJ, Vartsky, D, and Wielopolski, L: Applications of nuclear technologies for *in vivo* elemental analysis, Neuro Toxicology 4:34-47, 1983.
59. Gale, GR, Atkins, LM, Walker, EM, and Smith, AB: Effects of combined treatment with diethyldithiocarbamate and diethylenetriaminepentaacetate on organ distribution and excretion of cadmium, Ann Clin Lab Sci 13:425-431, 1983.
60. Roels, H, Lauwerys, R, Bucket, JP, et al: In vivo measurement of liver and kidney cadmium in workers exposed to this metal: its significance with respect to cadmium in blood and urine, Environ Res 26:217-240, 1981.
61. Recommended health-based limits in occupational exposure to heavy metals, Technical Report Series 647, Geneva, 1980, Word Health Organization.
62. Stoeppler, M: Analytical aspects of sample collection, sample storage, and sample treatment. In Bratter, P, and Schramel, P, editors: Trace element analytical chemistry in medicine and biology, vol 2. Berlin, 1983, Walter de Gruyter & Co.

Lead

63. Damstra, T: Toxicological properties of lead, Environ Health Perspect 19:297-307, 1977.
64. Lane, RE, et al: Diagnosis of inorganic lead poisoning: a statement, Br Med J 4:501, 1968.
65. Centers for Disease Control: Increased lead absorption and lead poisoning in young children, J Pediatr 87:824, 1975.
66. Sachs, MK, et al: Ambulatory treatment of lead poisoning: report of 1,155 cases, Pediatrics 46:389, 1970.
67. Granick, JL, Sassa, S, and Kappas, A: Some biochemical and clinical aspects of lead intoxication, Adv Clin Chem 20:287-339, 1978.
68. Criteria for a recommended standard . . . occupational exposure to inorganic lead, revised criteria, Washington, DC, 1978, NIOSH, DHEW Publication No. 78-158, Superintendent of Documents, US Government Printing Office.
69. Tsuchiya, K: Lead. In Friberg, L, Nordberg, GF, and Vouk, VB, editors: Handbook on the toxicology of metals, Amsterdam, 1979, Elsevier/North-Holland Biomedical Press.
70. Airborne lead in perspective. Report of the Committee on Biological Effects of Atmospheric Pollutants of the National Research Council, Washington, DC, 1972, National Academy of Sciences, Printing and Publishing Offices.

Mercury

71. Clarkson, TW: Mercury poisoning. In Brown, SS, editor: Clinical chemistry and chemical toxicology of metals, Amsterdam, 1977, Elsevier/North Holland Biomedical Press.
72. Berlin, M: Mercury, In Friberg, L, Nordberg, GF, and Vouk, VB, editors: Handbook on the toxicology of metals, Amsterdam, 1979, Elsevier/North Holland Biomedical Press.

CHAPTER 36

Vitamins

MARGE A. BREWSTER

Fat-soluble vitamins
- **Vitamin A**
- **Vitamin E**
- **Vitamin K**
- **Vitamin D**

Water-soluble vitamins
- **Ascorbic acid (vitamin C)**
- **Riboflavin**
- **Pyridoxine**
- **Niacin**
- **Thiamin**
- **Biotin**
- **Pantothenic acid**
- **Vitamin B_{12}**
- **Folic acid**

OBJECTIVES

- Define vitamin.
- List the fat-soluble vitamins, their functions, and conditions that result from a deficiency.
- List water-soluble vitamins, their functions, and conditions that result from a deficiency.
- Describe the functions of vitamin B_{12} and folic acid and describe disease conditions that are a result of deficiencies of these vitamins.

KEY TERMS

angular stomatitis Inflammation at the corner of the mouth.

anorexia Appetite loss.

antioxidant A substance that protects against oxidation, usually by being readily oxidized itself.

aphonia Loss of voice; motions of crying but little sound.

avidin A glycoprotein in raw egg white with strong affinity for biotin.

Batten's disease A progressive childhood encephalopathy with disturbed metabolism of polyunsaturated fatty acids.

carotenoids Compounds structurally similar to β-carotene (provitamin A) occurring naturally in vegetables and pigmented fruits.

cholelithiasis Presence of gallstones.

CNS Central nervous system.

coagulopathy A disorder of coagulation.

creatinuria Excess excretion of creatine in the urine.

cyanosis Blue skin color caused by lack of oxygen.

dry beriberi Thiamin deficiency resulting in poor appetite, fatigue, and peripheral neuritis.

dyspnea Difficult or painful breathing.

ecchymoses Skin discolorations caused by oozing of blood into tissues.

fibrinolysis The process of dissolution of fibrin clots.

flavins Riboflavin, flavin adenine dinucleotide (FAD), and flavin mononucleotide (FMN).

glossitis Smooth tongue.

hydrolases Enzymes that cleave ester bonds by addition of water.

hyperuricemia Increased concentration of serum uric acid.

ischemia Lack of adequate blood supply to tissues.

megaloblastic anemia Anemia in which marrow and blood cells are large and have multilobed nuclei.

osteoblasts Cells that form bone.

osteoclasts Cells that degrade bone.

ozone O_3; increased in air in areas with heavy automobile traffic.

pancytopenia Decrease in all blood cells and platelets.

paresthesias Abnormal sensations such as burning, prickling, or tingling.

pellagra Niacin deficiency resulting in diarrhea, dementia, and dermatitis.

pernicious anemia Anemia (no longer considered pernicious) caused by antibodies interfering with vitamin B_{12} absorption.

photophobia Abnormal sensitivity to light.

PUFA Polyunsaturated fatty acids.

pyrexia Fever.

pyridine nucleotides NAD, NADH, NADP, NADPH.

RDA Recommended Dietary Allowances, quantity of vitamin recommended for daily ingestion to meet essential needs of a healthy person.

rebound scurvy Symptoms of scurvy occurring on sudden withdrawal of a megadosage of ascorbic acid.

rickets Muscle hypotonia and skeletal deformities in children.

scurvy Ascorbic acid deficiency characterized by swollen gums with loss of teeth, skin lesions, and pain and weakness in the lower extremities.

steatorrhea Excessive lipid in feces.

subclinical vitamin deficiency Chemical indices of vitamin status indicate deficiency, but there is no *apparent* clinical symptom.

thioester Ester bond involving —SH.

vitamins Small-molecular-weight components required for metabolic activity that must be supplied by dietary intake.

Wernicke-Korsakoff syndrome Neurological/behavioral alterations seen in some patients with thiamin deficiency.
wet beriberi Thiamin deficiency resulting in edema and cardiac failure.
xanthomatosis Presence of yellow skin deposits.

The vitamins comprise a group of small-molecular-weight compounds with diverse functions in biological tissues. Primarily they serve as cofactors of numerous enzymatic reactions. Without these cofactors a wide range of enzymes that play critical roles in cellular metabolism become inactive. The term *vitamin* is derived from early evidence that one such compound, thiamin, was an amine vital for health. These compounds or their biologically inactive precursors must be obtained, at least partially, from food sources or, in some instances, from intestinal bacterial synthesis. Inadequacies in supply (resulting from inadequate diet or inadequate intestinal absorption) are termed *vitamin deficiencies*. Abnormalities of metabolism requiring an abnormally high supply of one of these cofactors may be termed *vitamin insufficiencies* or *vitamin dependencies,* depending on the level of supply demanded for physiological function. Variabilities in clinical expression of vitamin abnormalities result from differences in specific etiology and degree and duration of vitamin inadequacy. Additional factors contributing to this variability include the genetic constitution of the individual, simultaneous presence of multiple nutritional insufficiencies, or increased metabolic demands imposed by conditions such as infection, pregnancy, or cancer. We still have limited knowledge of human mechanisms of absorption, transport, storage, and metabolism of vitamins, and we have much to learn with regard to the effects of diseases, therapies, and exogenous agents on these processes.

Because of the pervasive action of vitamins in metabolic processes, the clinical symptoms of vitamin deficiencies are usually nonspecific and vague, often delaying a definitive diagnosis. This is especially true in the early stages of vitamin deficiency or in mild chronic deficiency states; also vitamin deficiency may be complicated by other simultaneous processes. A combination of dietary history, physical examination, and biochemical measurements is often required to diagnose a vitamin deficiency. In addi-

Table 36-1 Vitamin functions and symptoms of deficiency or toxicity

Vitamin	Function	Clinical deficiency	Toxicity
A	Vision, growth, reproduction, mucus secretion, immune responses	Night blindness, growth retardation, appetite loss, reduced taste, recurrent infections, dermatitis, dry mucous membranes Late—bone growth failure, aspermatogenesis, xerophthalmia (dry, thickened, lusterless eyeballs), blindness	*Acute*—raised intracranial pressure and skin desquamation *Chronic*—liver damage, skin changes, and exostoses
E	Antioxidant (membrane stability)	Mild hemolytic anemia, ataxia, loss of tendon reflexes, pigmentary retinopathy	Creatinuria, decreased platelet aggregation, impaired wound healing, antiinflammatory activity, hepatomegaly, impaired fibrinolysis, potentiation of vitamin K deficiency, coagulopathy
K	Coagulation (gamma—carboxylation of inactive clotting factors—prothrombin, factors II, IX, and X)	Hemorrhage (ranging from easy bruising to massive ecchymoses, mucous membrane hemorrhage, or post-traumatic bleeding)	Adults—cardiac and pulmonary signs Newborns—hemolytic anemia
D	Bone calcification	Children—rickets Adults—osteomalacia	Anorexia, vomiting, headache, drowsiness, and diarrhea
C	Collagen formation, catecholamine synthesis, cholesterol catabolism, antioxidant	Early—weakness, lassitude irritability, vague aches and pains Late—scurvy (hemorrhages into skin, alimentary and urinary tracts, other tissues; osteoporotic bones, defective tooth formation, anemia, pyrexia, delayed wound healing)	Increased excretion of oxalate and urate, diarrhea, dyspepsia

Table 36-1 Vitamin functions and symptoms of deficiency or toxicity—cont'd

Vitamin	Function	Clinical deficiency	Toxicity
Riboflavin	Oxidative enzymatic reactions	Angular stomatitis (mouth lesions), glossitis (smooth tongue), photophobia, blepharospasm (eyelid spasm), conjunctival congestion and other ocular changes, dermatological changes, neurological alterations (behavior changes, decreased hand grip strength, burning feet in adults, retarded intellectual development and EEG changes in children), and hematological dyscrasia (anemia and reticulocytopenia)	Low toxicity
Pyridoxine	Enzyme systems involving amino acid transaminases, phosphorylases, racemases, decarboxylases, deaminases	Infants—irritability, seizures, anemia, vomiting, weakness, ataxia, abdominal pain Adults—facial seborrhea	Usually low systemic toxicity. Reduced milk production? Sensory neuropathy?
Niacin	Oxidation—reduction (as pyridine nucleotides NAD and NADP)	Early—lassitude, anorexia, weakness, digestive disturbances, anxiety, irritability, and depression Late—pellagra (dermatitis, mucous membrane inflammation, weight loss, disorientation, delirium, dementia)	Cutaneous flushing, gastric irritation, mild liver dysfunction, jaundice, hyperuricemia, impaired glucose tolerance
Thiamin	Decarboxylations, ketol formation	Infants—dyspnea and cyanosis, diarrhea, vomiting, wasting, aphonia Adults—"dry beriberi" (poor appetite, fatigue, peripheral neuritis) or "wet beriberi" (edema and cardiac failure), Wernicke-Korsakoff syndrome (intelligence disturbance, apathy, ataxia, double vision, nystagmus, drooping eyelids, loss of recent memory)	Anxiety, headache, convulsions, weakness, trembling, neuromuscular collapse
Biotin	Coenzyme for CO_2 carboxylation reactions and for carboxyl group exchange	Dermatitis progressing to mental and neurological changes, nausea, anorexia, peripheral vasoconstriction, or coronary ischemia in some cases	None described
Pantothenate	Acyl-group transfer reactions (as part of coenzyme A and acyl carrier protein)	Never spontaneously seen—with chemical agonist: apathy, depression, increased infection, paresthesias (burning sensations), muscle weakness	None described
B_{12}	Myelin formation, methionine synthesis, folate interconversions, and DNA synthesis	Early—cognitive impairment? Late—megaloblastic anemias, neurological abnormalities (paresthesias progressing to spastic ataxia)	Infrequent adverse reactions are mostly allergic (possibly related to contaminants or preservatives)
Folate	One-carbon transfers	Megaloblastic anemia; organic mental changes?	Few reports—mostly allergic reactions

Compiled from references 5-11.

Table 36-2 1980 Recommended Dietary Allowances[1]

Age (yr)	Vitamin A (μg RE*)	Vitamin E (mg α-TE†)	Vitamin D (μg‡)	Vitamin C (μg)	Thiamin (mg)	Riboflavin (mg)	Niacin (mg NE§)
Infants							
0-6 months	420	3	10	35	0.3	0.4	6
6-12 months	400	4	10	35	0.5	0.6	8
Children							
1-3	400	5	10	45	0.7	0.8	9
4-6	500	6	10	45	0.9	1.0	11
7-10	700	7	10	45	1.2	1.4	16
Males							
11-14	1000	8	10	50	1.4	1.6	18
15-18	1000	10	10	60	1.4	1.7	18
19-22	1000	10	7.5	60	1.5	1.7	19
23-50	1000	10	5	60	1.4	1.6	18
50+	1000	10	5	60	1.2	1.4	16
Females							
11-14	800	8	10	50	1.1	1.3	15
15-18	800	8	10	60	1.1	1.3	14
19-22	800	8	7.5	60	1.1	1.3	14
23-50	800	8	5	60	1.0	1.2	13
50+	800	8	5	60	1.0	1.2	13
Pregnant	200‖	2‖	5‖	20‖	0.4‖	0.3‖	2‖
Lactating	400‖	3‖	5‖	40‖	0.5‖	0.5‖	5‖

*One retinol equivalent (RE) = 1 μg of retinol of 6 μg of β-carotene; 3 μg of retinol = 10 U.
†One mg of α-tocopherol equivalent (TE) = 1 mg of *d*-α-tocopherol.
‡As cholecalciferol; 10 μg of cholecalciferol = 400 U.
§One mg of niacin equivalent (NE) = 1 mg of niacin or 60 mg of dietary tryptophan.
‖RDA needed in addition to normal requirements for females.

tion, therapeutic trial may also be required for a definitive diagnosis. Because vitamin metabolism is complex and interactive, vitamin supplementation without proper studies may lead to other nutrient deficiencies, toxicities, or an irreversible state of untreated deficiency with symptoms masked by inappropriate therapy. Vitamin functions and the usual symptoms seen in deficient and toxic states are listed in Table 36-1.

Suspicion of dietary deficiency of a vitamin arises primarily from knowledge of dietary sources and dietary practices likely to provide inadequate intake or absorption. Recommended Dietary Allowances (RDA) are defined by the Food and Nutrition Board[1] as the levels of intake of essential nutrients considered, on the basis of available scientific knowledge, to be adequate to meet the nutritional needs of practically all "healthy persons." These levels are defined from information on the daily intake requirements needed to avoid deficiency symptoms and to maintain specific functions. RDA values are generally set high enough to meet the needs of 97.5% of the population; they are sometimes even higher if the nutrient is poorly absorbed or insufficiently used. Naito[2] states:

> It is important to emphasize that RDA do not take into consideration the losses of nutrients during the processing and preparation of foods, nor are the RDA designed to cover increased needs resulting from severe stress, disease, or trauma above the usual minor infections and stresses of everyday life. The processing losses must be allowed for separately in planning a food supply, and the increased needs represent clinical problems that must be given consideration.

These RDA are summarized in Table 36-2.

Biochemical indices of vitamin status become abnormal before development of obvious clinical changes, thus allowing detection of a vitamin deficiency at a subclinical or nonclassical stage. Chemical determination of human vitamin status has been approached in the following ways:

1. Measurement of the active cofactor(s) or precursor(s) in biological fluids or blood cells
2. Measurement of urinary metabolite(s) of the vitamin
3. Measurement of a biochemical function requiring the vitamin (such as enzymatic activity) with and without in vitro addition of the cofactor form
4. Measurement of urinary excretion of vitamin or metabolite(s) after a test load of the vitamin

Table 36-3 Concentration or excretion rates associated with classical vitamin deficiency symptoms*†

Vitamin	Chemical value
A	<0.1 mg/L of plasma retinol
E	<5.0 mg/L of plasma α-tocopherol
K	Plasma prothrombin time > normal
D	See Chapter 25
Ascorbic acid (C)	<2.4 mg/L of serum ascorbate
	<3 mg/L of whole blood ascorbate
	<80 mg/L of leukocyte ascorbate
Riboflavin (B_2)	>1.4 AC‡ of erythrocytic glutathione reductase
	<0.1 mg of riboflavin/L of erythrocytes
	<0.12 mg of urinary riboflavin per day
	≤0.08 mg of urinary riboflavin per gram of creatinine
Pyridoxine (B_6)	≥1.5 AC of erythrocytic AST
	≥1.25 AC of erythrocytic ALT
	<0.8 mg of urinary 4-pyridoxic acid per day
	>25 mg of urinary xanthurenic acid per day
Thiamin	>1.25 AC of erythrocyte transketolase
	<0.1 mg of urinary thiamin per day
Niacin	≥1 urinary ratio (α-pyridone/*N'*-methylnicotinamide)
Biotin‡	<0.7 μg/L of whole blood?
	<15 μg of urinary biotin per day?
Pantothenic acid§	<1.0 mg/L of whole blood pantothenate
	<1.0 mg of urinary pantothenate per day
B_{12}	<150 ng/L of serum vitamin B_{12}
	≥24 mg of urinary methylmalonic acid per day
Folate	<140 μg/L of erythrocyte folate
	<3.0 μg/L of serum folate
	≥30 mg of urinary N^5-formiminoglutamic acid (FIGLU) per 8 hours

*These are general guidelines with normal values dependent on age and methodology used.
†Compiled from references 4, 5, and 36.
‡*AC*, Activity coefficient; ratio of activities with and without added cofactor.
§Deficient values for biotin and pantothenate are not well established.

5. Measurement of urinary metabolites of a substance, the metabolism of which requires the vitamin, after administering a test load of the substance

However, reduced serum concentrations of a vitamin do not always indicate a deficiency that interrupts cellular function; on the other hand, normal values do not always reflect adequate function. Interpretation of chemical values must be done with caution and with knowledge of the physiological and methodological factors that can confound a diagnosis.

Table 36-3 lists representative biochemical data that are usually associated with classical deficiency symptoms. These values are, of course, somewhat affected by the age of the patient and by laboratory methodology. References 3 through 9 provide information about laboratory assessment methods[3,4] and method selection,[4,5] toxicological effects,[5] drug interactions,[4,5] environmental effects,[6] and metabolism[7-9] of the vitamins.

FAT-SOLUBLE VITAMINS

Because the fat-soluble vitamins (A, E, K, and D) are absorbed as part of the chylomicron complex (see Chapter 30), their absorption depends on the presence of adequate bile and pancreatic secretions and on healthy bowel mucosa as well. Therefore chronic malabsorptive states are often associated with a deficiency of one or more of these vitamins (see Chapters 25 and 26). The malabsorptive states include biliary tract disease, pancreatic disease, fistula, small bowel obstruction, and alcoholic liver disease (see Chapter 31). Deficiency of this class of vitamins generally develops slowly as stored supplies of vitamins are depleted. Vitamin A can be stored in liver parenchymal cells for a year or longer, and vitamin E can be stored in body fat for several months. Paradoxically, although they are fat-soluble, vitamins K and D appear to be stored only for days or weeks.

Vitamin A

First described in 1909 and found to prevent night blindness in 1925, vitamin A is now known to be made up of three biologically active forms: retinol, retinal, and retinoic acid. These major vitamin A compounds all contain a trimethyl-cyclohexenyl group and an all-*trans* polyene chain with four double bonds (Fig. 36-1). These compounds are derived directly from dietary sources, primarily as retinyl esters, or from metabolism of dietary carotenoids (provitamin A), primarily β-carotene. Major dietary sources of these compounds include animal products (vitamin A) and pigmented fruits and vegetables (carotenoids). Each of these compounds is soluble in organic solvents, with retinoic acid being more polar than the others. Oxidation of retinol or retinal by peripheral cells is irreversible; thus neither retinoic acid nor retinal is metabolically converted to retinol.

Metabolism. Enzymes of the small intestinal mucosa convert dietary β-carotene and retinal to the predominant form of vitamin A, retinol (Fig. 36-2). The retinyl esters of dietary animal products are cleaved to retinol by pancreatic and mucosal hydrolases (vitamin A esterases). Once in the mucosal cell retinol is reesterified, forming retinyl esters (primarily retinyl palmitate) that are transported in lymph chylomicrons to the systemic circulation. After the chylomicrons release their triglycerides to adipose tissue, the retinyl esters are transported to the liver where they are stored associated with lipid droplets in hepatocytes. The more polar retinoic acid does not require this lipoprotein transport route but is directly absorbed into the portal circulation. However, this form is not stored in liver but is excreted through bile as a glucuronide conjugate.

When body demands require mobilization of hepatic vi-

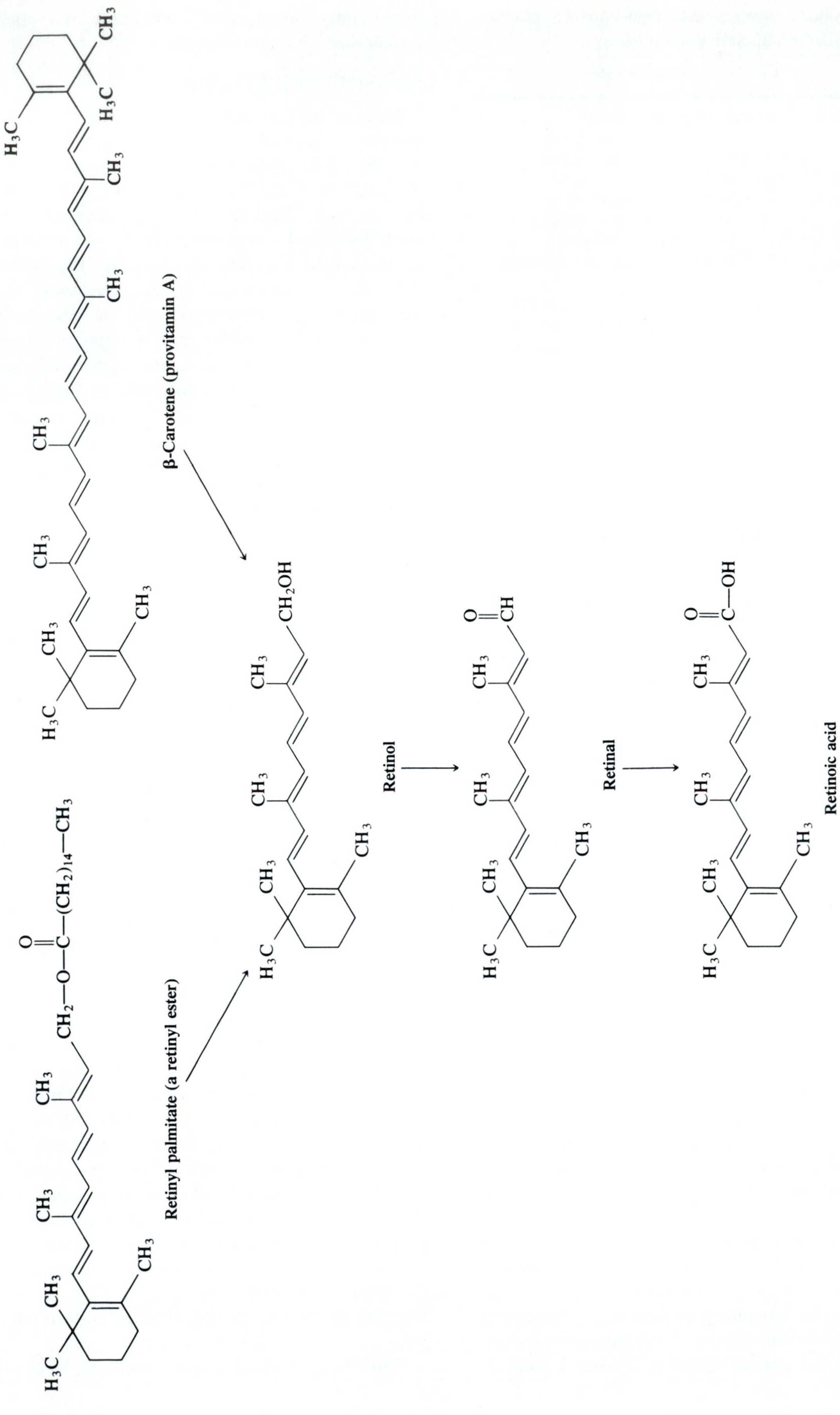

Retinyl palmitate (a retinyl ester)

β-Carotene (provitamin A)

Retinol

Retinal

Retinoic acid

Fig. 36-1 Structures of vitamin A (retinol) with its precursors and metabolites.

tamin A, the stored retinyl palmitate is hydrolyzed and the free retinol combines with retinol-binding protein (RBP). RBP-retinol is then secreted into the circulation, where it complexes with prealbumin. This large complex then circulates to target tissues that have specific receptor sites for RBP, and retinol is transferred intracellularly to another specific binding protein termed *cytosol-retinol–binding protein* (CRBP). CRBP-retinol presumably transports the retinol to its functional site within the cell. Retinol metabolism is shown in Fig. 36-2.

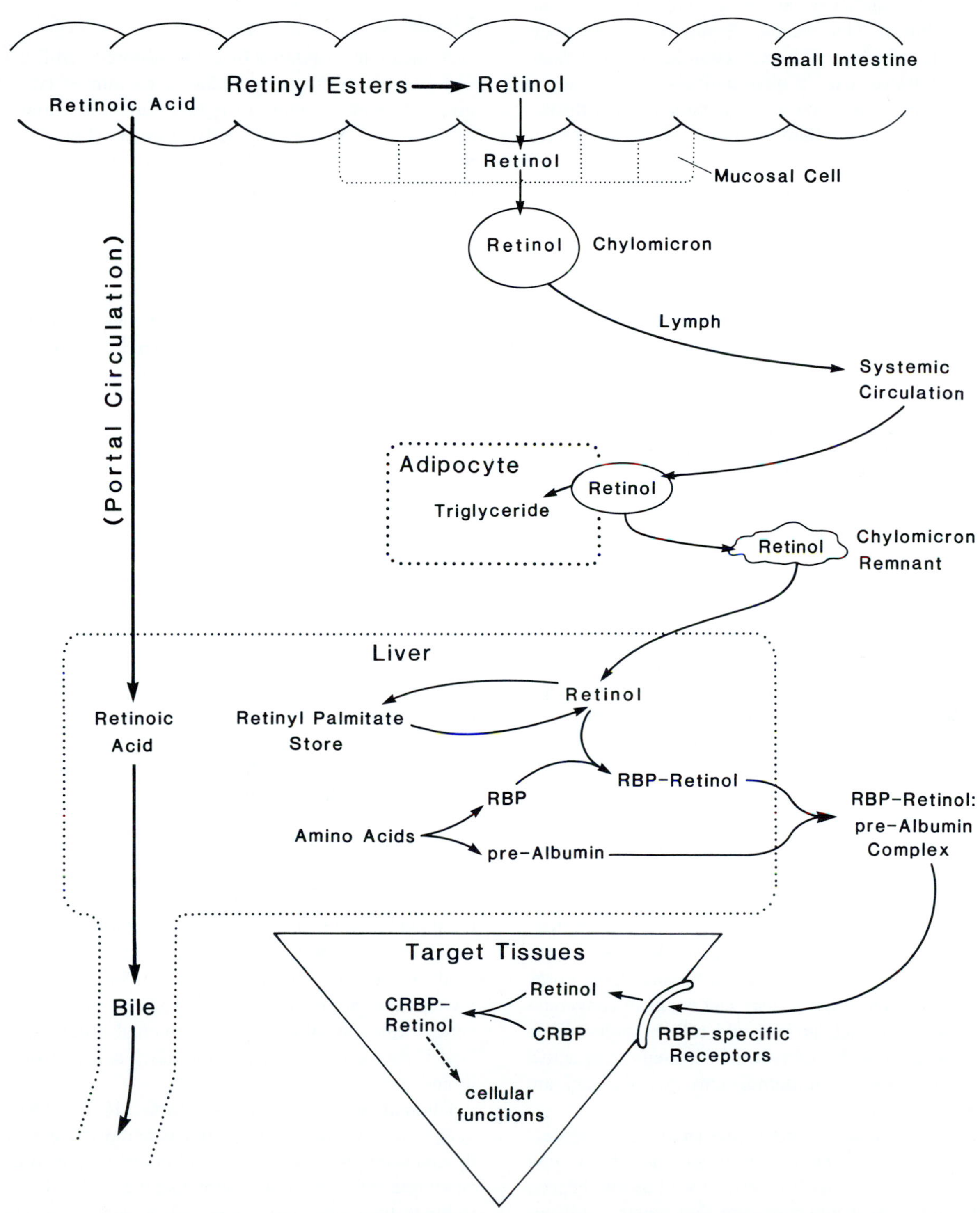

Fig. 36-2 Retinol metabolism. *RBP*, Retinol-binding protein; *CRBP*, cytoplasmic retinol-binding protein.

Function (see Table 36-1). The only clearly defined physiological role for retinol is its role in vision. Retinol is oxidized in the rods of the eye to retinal, which, when complexed with opsin, forms rhodopsin, allowing dim-light vision. In vitamin A deficiency states epithelial cells (cells in the outer skin layers and cells in the lining of the gastrointestinal, respiratory, and urogenital tracts) become dry and keratinized. Thus vitamin A may help to maintain epithelial cells, which provide protection against infectious organisms. Retinoic acid, a quantitatively minor component of vitamin A, is known to function in growth and maintenance of the epithelium but not in vision and reproduction.

Clinical and chemical deficiency signs. Clinical signs of vitamin A deficiency can be separated into early and late signs (Table 36-1). The chemical sign of deficiency is reduction in plasma vitamin A. Generally retinol values below 0.1 mg/L are associated with clinical symptoms, and values above 0.2 mg/L are not.[3,10] Vitamin A itself is not excreted in human urine. Although several metabolites are excreted, they do not seem to reflect the tissue status of vitamin A.

Pathophysiology. Because retinol and RBP are secreted from liver as a 1 : 1 complex, low plasma concentrations of both are seen in vitamin A deficiency. Adequate concentration of plasma retinol usually indicates dietary and tissue adequacy, but low concentrations *do not* always indicate dietary deficiency. Factors that reduce hepatic synthesis of RBP, or secretion of the RBP-retinol complex, lower plasma concentrations of retinol and RBP, even though dietary intake and the hepatic retinol store are adequate. These states are primarily recognized by the absence of an increase in plasma retinol after oral therapy with vitamin A and include protein-calorie malnutrition, liver disease, zinc deficiency, and cystic fibrosis.

Pathophysiological conditions that can result in increased retinol and RBP include chronic renal disease and use of oral contraceptives.

Toxicity and therapeutic uses. Retinoids have been used therapeutically to prevent noise damage to auditory function and to treat a variety of skin disorders. One retinoid, 13-cis-retinoic acid (Accutane) has become the therapy of choice for several forms of acne. Accutane, however, must be used with great caution because large doses of retinoid taken during pregnancy can result in congenital malformations. The anticarcinogenic role of high-dosage retinoids has been well established in animals, and active clinical trials to test for antitumor activity in humans are in progress.

Vitamin A deficiency is often associated with disorders of fat absorption.[10-12] Premature infants are born with lower serum retinol and RBP levels, as well as low hepatic stores of retinol. Sick newborns are thus treated with vitamin A as a preventive measure. With excess intake of vitamin A, liver storage capacity can be exceeded, resulting in circulation of free retinol and retinyl esters in lipoproteins. In this instance retinyl ester concentration in plasma becomes quite high. A test of fat malabsorption involves an oral loading dose of vitamin A. With adequate absorption, plasma retinol is unchanged and retinyl palmitate is elevated within 4 hours.

Vitamin E

A factor in vegetable oils that restored fertility to rats was isolated in the early 1920s as vitamin E; later it was given the generic name *tocopherol* and was shown to include several biologically active isomers (Fig. 36-3). The word *tocopherol* is of Greek derivation, meaning an oil that ''brings forth in childbirth,'' but the fertility role of these compounds is still questionable. α-Tocopherol is the predominant isomer in plasma and is the most potent isomer by current biological assays. Whether the tocopherol isomers have separate physiological effects is unknown.

Dietary sources of tocopherols include vegetable oils, fresh leafy vegetables, egg yolk, legumes, peanuts, and margarine. Diets suspect for vitamin E deficiency are those low in vegetable oils or fresh green vegetables or those high in unsaturated fats.

Metabolism. The absorption, transport, storage, and metabolism of tocopherol are only partially understood. Absorption is believed to be associated with intestinal fat absorption. Approximately 40% of ingested tocopherol is absorbed; the percentage is affected by the amount and by the degree of unsaturation of dietary fat, as well as by the isomer type. The physiological requirement for vitamin E increases with increasing polyunsaturated fatty acids (PUFA) in the diet. Absorbed vitamin E is first associated with circulating chylomicrons and very low–density lipoproteins (VLDL) and later with other lipoproteins. Vitamin E is predominantly found in adipose tissue, although increased dietary α-tocopheryl acetate is reflected by increased concentrations in all animal tissues, including plasma, erythrocytes, and platelets.

Function. Vitamin E functions as an antioxidant, protecting unsaturated lipids from peroxidation (cleavage of fatty acids at unsaturated sites by oxygen addition across the double bond and formation of free radicals).[13] The role of vitamin E in protecting the erythrocyte membrane from oxidant stress is presently the major documented role of vitamin E in human physiology. There is evidence for preventive roles of vitamin E in retrolental fibroplasia, intraventricular hemorrhage, and mortality of small premature infants.[8]

Clinical deficiency signs (Table 36-1). The major symptom of vitamin E deficiency is hemolytic anemia. Although such vitamin E use is still controversial, premature newborns are commonly supplemented with vitamin E to stabilize their red blood cells and prevent hemolytic anemia. Patients with conditions resulting in fat malabsorption, especially cystic fibrosis and abetalipoproteinemia,

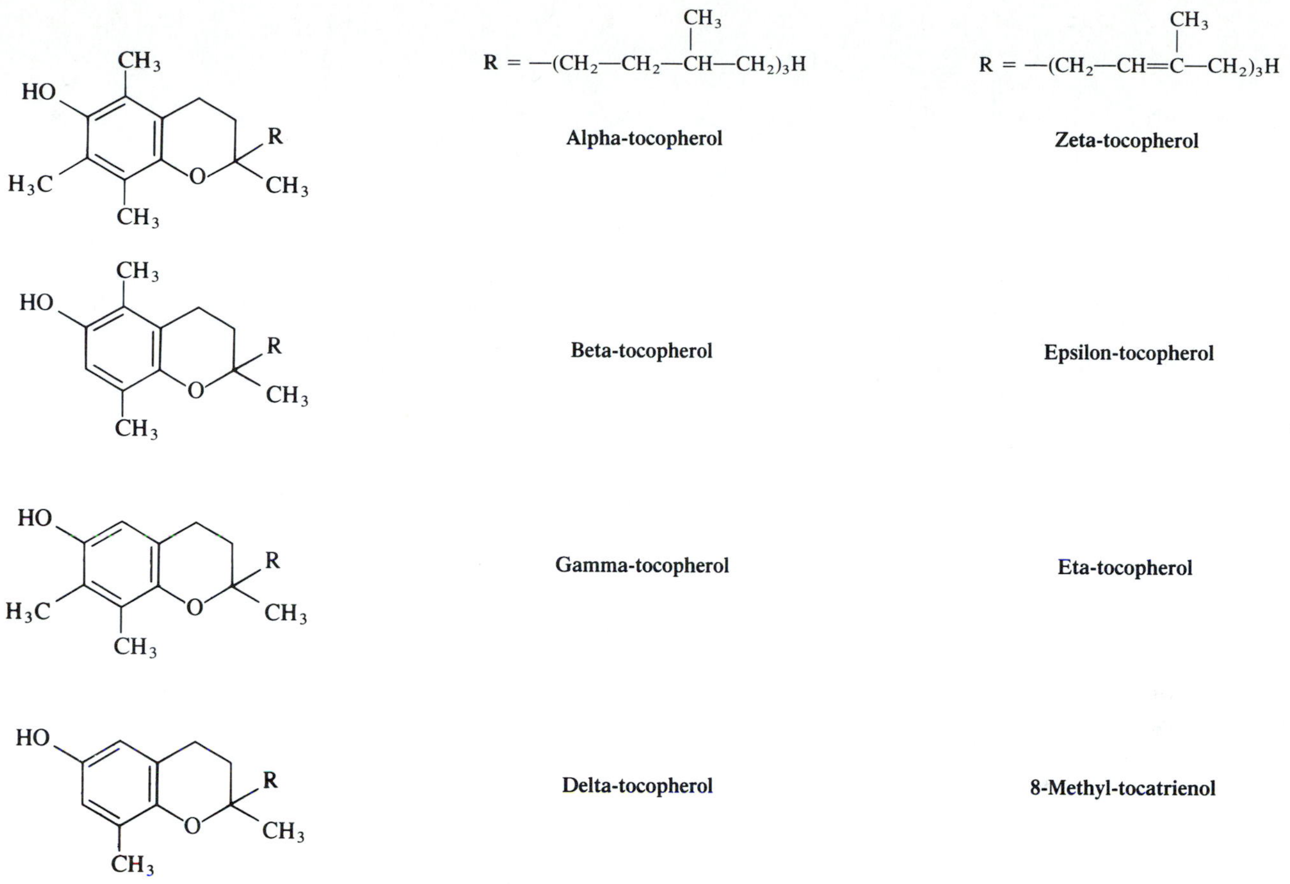

$R = -(CH_2-CH_2-CH(CH_3)-CH_2)_3H$

$R = -(CH_2-CH=C(CH_3)-CH_2)_3H$

Alpha-tocopherol **Zeta-tocopherol**

Beta-tocopherol **Epsilon-tocopherol**

Gamma-tocopherol **Eta-tocopherol**

Delta-tocopherol **8-Methyl-tocatrienol**

Fig. 36-3 Vitamin E isomers.

are very suspect for vitamin E deficiency; a relationship has been recognized between vitamin E deficiency and progressive loss of neurological function in infants and children with chronic cholestasis.[14]

Chemical deficiency signs. Plasma concentrations of α-tocopherol below 5 mg/L are associated with increased erythrocyte hemolysis in the presence of hydrogen peroxide and are thus designated "deficient."[8,10] There is a strong correlation between plasma α-tocopherol and plasma lipids, suggesting that plasma concentrations should be interpreted relative to plasma lipid levels; 0.8 mg of α-tocopherol per gram of total plasma lipids appears to indicate adequate levels of vitamin E in infants. Elevation of plasma total lipids above 15 g/L can apparently shift erythrocyte α-tocopherol to plasma, potentially altering erythrocyte susceptibility despite "adequate" plasma concentrations of α-tocopherol in hyperlipidemic states.[15,16]

Pathophysiology. At the present time assessment of vitamin E status is primarily indicated in newborns, in persons with potential fat malabsorption states, and in persons receiving synthetic diets.

Elevated values of serum vitamin E have been reported during pregnancy and in patients with Batten's disease (a progressive childhood encephalopathy with disturbed PUFA metabolism). Decreased serum values have been reported in patients with grand mal seizures[5,17] and in persons exposed to nonsymptomatic doses of organophosphates.[6]

Toxicity. Toxicity may result from chronic voluntary overdoses. Premature infants receiving vitamin E sufficient to sustain serum levels above 30 mg/L have an increased incidence of sepsis and necrotizing enterocolitis. Patients receiving synthetic diets should be monitored to avoid vitamin E toxicity (Table 36-1).

Vitamin K

Experiments in the mid-1930s led to the discovery of the antihemorrhagic factor later called *vitamin K* (from German, *Koagulation*). Purification efforts revealed several quinone-containing compounds possessing this antihemorrhagic activity, and the term *vitamin K* is now used as a generic descriptor for menadione and derivatives exhibiting this activity. There are a large number of these compounds that are related to those shown in Fig. 36-4 by number and substituents of polyisoprenoid side chains and degree of unsaturation.[18]

The K vitamins are unstable in acidic or alkaline conditions and are readily oxidized. Major dietary sources are

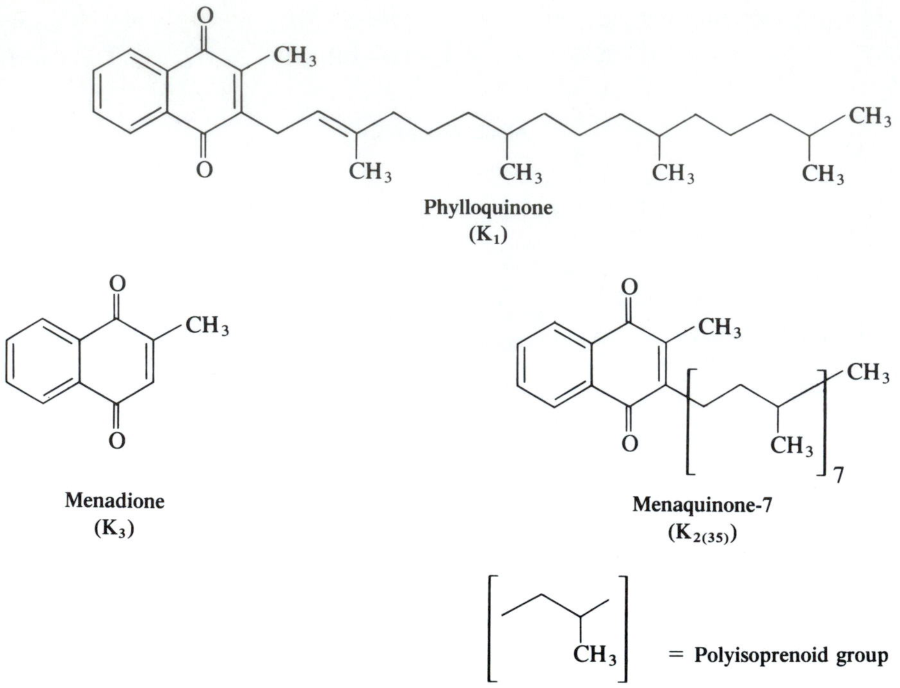

Fig. 36-4 Structures of vitamin K forms.

cabbage, cauliflower, spinach and other leafy vegetables, pork, liver, soybeans, and vegetable oils. Uncomplicated dietary deficiency is considered rare in healthy children and adults, but a study of elderly persons revealed deficiencies in a large percentage that were correctable by oral administration of vitamin K.[19]

Metabolism. In infants vitamin K is absorbed in the colon, where bacterial synthesis is the major source of this vitamin. Older children and adults absorb dietary vitamin K in the upper small intestine, where the contribution of intestinal bacteria is insignificant. Absorption of vitamin K in the intestines is chylomicron mediated, and vitamin K malabsorption states include cystic fibrosis, biliary atresia, cholelithiasis and obstructive jaundice resulting from hemolytic anemias, as well as other disorders leading to dysfunction of the upper small intestine. Depletion of body stores sufficient to manifest deficiency usually requires 3 weeks.

Function. First shown to be involved in prothrombin synthesis, vitamin K has also been found to be required for the synthesis of other active clotting factors (factors VII, IX, and X). Vitamin K–dependent γ-carboxylation of glutamic acid residues of inactive precursor proteins occurs in the liver.

Clinical deficiency signs (Table 36-1). The clinical manifestation of a vitamin K deficiency is hemorrhage as a result of reduction in prothrombin and other clotting factors.

Chemical deficiency signs. Direct measurement of vitamin K in blood is not usually performed for adults. Determination of prothrombin time (velocity of clotting after addition of thromboplastin and calcium to citrated plasma) is an excellent index of prothrombin adequacy. This time is prolonged in deficiency of vitamin K and also in liver diseases characterized by decreased synthesis of prothrombin. Deficiency of vitamin K also results in prolongation of the partial thromboplastin time, but thrombin time is normal. Clinical improvement of these clotting times is noted within a few hours of vitamin K administration. Immunological measurement of acarboxy-prothrombin (protein induced by vitamin K absence; PIVKA-II) is useful in the detection of vitamin K deficiency in neonates and appears to have greater sensitivity than do the coagulation tests.[20]

Pathophysiology. The vitamin K malabsorptive states are described above. The prothrombin reduction seen in patients with alcoholic fatty liver is corrected by administration of vitamin K only if there is associated malnutrition or malabsorption. Paradoxically, administration of vitamin K to patients with preexisting subclinical hepatic dysfunction often results in prolongation of a previously normal prothrombin time. Vitamin K action is antagonized by coumarin or indanedione anticoagulants. Oral contraceptive use increases the levels of prothrombin and factors VII, IX, and X and apparently reduces the requirement for vitamin K.

Gastrointestinal infection associated with diarrhea, combined with prolonged antibiotic therapy and non-vitamin K–supplemented milk-substitute formulas, has led to hemorrhagic symptoms in infants. Recognition of this combination of causes of vitamin K deficiency has led to supplementation of formulas and additional vitamin K administration when appropriate, making this circumstance rare.

Vitamin D

Since 1822, it has been known that rickets (muscle hypotonia and skeletal deformities) could be cured with cod liver oil. This antirachitic factor was identified as vitamin D a century later and soon was shown to have multiple forms. These antirachitic compounds are now collectively termed *vitamin D*. The naturally occurring fish oil vitamin is cholecalciferol (vitamin D_3); it is produced in the skin from ultraviolet activation of 7-dehydrocholesterol. It is now known that vitamin D_3 is a prohormone that is converted in the liver to 25-hydroxycholecalciferol (calcidiol), which is further hydroxylated in the kidney to form the hormone 1,25-dihydroxycholecalciferol (calcitriol). Calciferol (vitamin D_2), which is the primary dietary form (the form used in food fortification), is similarly hydroxylated (see Chapter 24).

Major dietary sources include irradiated foods and commercially prepared milk. Small amounts occur in butter, egg yolk, liver, salmon, sardines, and tuna fish.

Physiological actions, regulation, and assessment of the hormone forms of vitamin D are discussed in Chapter 24. Deficiency results in rickets (children) or osteomalacia (adults), both of which are forms of abnormal bone synthesis. Excess dietary intake is common in the United States, creating concern as to the hypervitaminosis D role in development of arteriosclerosis.[21]

WATER-SOLUBLE VITAMINS

The nine water-soluble vitamins are absorbed without the involvement of fat absorption, and excess vitamin is almost immediately excreted in the urine. Thiamin is stored for only a few weeks, and most other B-complex vitamins and ascorbic acid are stored for less than 2 months. Development of deficiency can therefore be quite rapid. Vitamin B_{12}, although a water-soluble vitamin, is stored in the liver for several years. The water-soluble vitamins function as coenzymes, and all except ascorbic acid and biotin must be metabolically converted to active forms.

Ascorbic acid (vitamin C)

The symptom cluster known as *scurvy* (swollen gums with loss of teeth, skin lesions, and pain and weakness in the lower extremities) was clearly described during the Crusades and became commonplace when long sea voyages began. In 1747, a naval surgeon experimenting with diets to cure scurvy identified the efficacy of citrus fruits. This antiscurvy agent, vitamin C, was isolated in 1932 and later given the name *ascorbic acid* (from its antiscorbutic effect). The structures of this water-soluble vitamin and its oxidized form, dehydroascorbate, are shown in Fig. 36-5. By virtue of its ene-diol group, ascorbate is a very strong reducing compound. Although plants and most animals can synthesize this vitamin, humans cannot, and so dietary ingestion is essential.

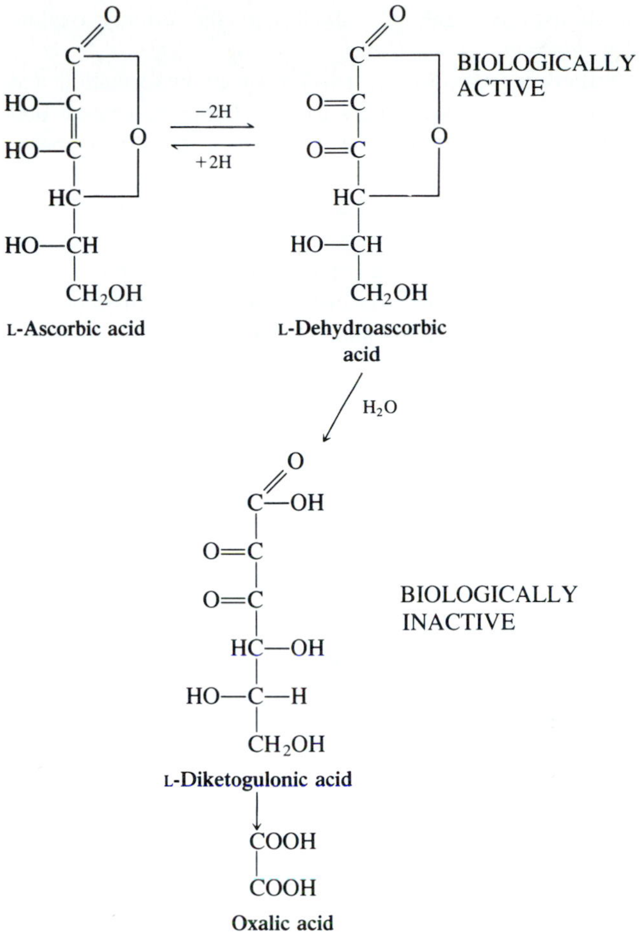

L-Ascorbic acid ⇌ (−2H / +2H) L-Dehydroascorbic acid (BIOLOGICALLY ACTIVE) → (H_2O) L-Diketogulonic acid (BIOLOGICALLY INACTIVE) → Oxalic acid (COOH–COOH)

Fig. 36-5 Structures of ascorbic acid and metabolites.

Major dietary sources include fruits (especially citrus) and vegetables (tomatoes, green peppers, cabbage, leafy greens, potatoes). Since ascorbate is labile to heat and oxygen, fresh and uncooked foods are highest in ascorbate content. Dietary deficiency is highly unlikely in a person who eats one daily serving of a fresh fruit or vegetable but is commonly seen in infants exclusively receiving cow's milk.

Metabolism. Ascorbate is absorbed in the small intestine. It is widely distributed in tissues (most concentrated in the adrenal cortex and pituitary) and passes the placenta readily. The normal body store requires several months to deplete before the appearance of symptoms of scurvy. Cerebrospinal fluid (CSF) concentrations are higher than those in plasma. Excess ascorbate and a metabolite, oxalate, are readily eliminated in urine. Generally ascorbate metabolism accounts for about half the urinary oxalate (Fig. 36-5).

Function. Ascorbic acid is important in formation and stabilization of collagen by hydroxylation of proline and lysine for cross-linking and in the conversion of tyrosine to catecholamines (by dopamine β-hydroxylase). In addition, there is evidence for ascorbate's involvement in hydroxylation reactions (including conversion of cholesterol to bile acids, metabolism of foreign organic compounds, and steroid synthesis) and protection against lipid peroxidation. It enhances absorption of iron and may be involved in heme synthesis.[22,23]

Clinical and chemical deficiency signs. Clinical signs of ascorbic acid deficiency are listed in Table 36-1. Clinical signs of scurvy have been associated with serum ascorbate values below 2.4 mg/L or whole blood ascorbate levels below 3 mg/L. Transient low values, however, do not necessarily reflect a tissue deficiency. Increased ascorbate intake raises serum ascorbate levels to a maximum of about 14 mg/L, at which time renal clearance rises sharply. Leukocyte ascorbate is considered to reflect tissue stores more closely but is technically more difficult to assay. Measurement of urinary ascorbate is not recommended for status assessment because it reflects recent intake and has numerous analytical difficulties. Drugs known to increase the urinary excretion of ascorbate include aspirin, aminopyrine, barbiturates, hydantoins, and paraldehyde.

Pathophysiology. Ascorbic acid requirements are increased in chronic illness, during pregnancy, and during oral contraceptive use. Deficiency has been observed in infants receiving unsupplemented cow's milk and in infants receiving breast milk from deficient mothers. There is controversy as to whether the transient tyrosinemia observed in many infants is a harmless condition or one needing therapy with ascorbate to avoid impaired mental development.

Toxicity and therapeutic uses. The usefulness of ascorbate in preventing colds and cancers has not been scientifically supported (except possibly for gastric and esophageal cancers by prevention of nitrosamine formation)[24] but has led to the common practice of megadose intakes of vitamin C. Although there is little toxicity of this ascorbate, excess intake may interfere with metabolism of vitamin B_{12} and with drug actions (aminosalicylic acid, tricyclic antidepressants, and anticoagulants). A sudden discontinuation of megadose intake can result in a rebound scurvy.

Riboflavin

The yellowish fluorescent pigment in milk, eggs, yeast, and liver discovered in the nineteenth century (named *flavin* from the Latin for "yellow"), the "yellow enzyme" found in yeast, and the yellow heat-stable animal growth factor crystallized from milk in the 1930s were finally realized to be the same substance—a vitamin consisting of a pigment, flavin, attached to the carbohydrate D-ribitol—riboflavin. Its structure and that of its two cofactor-active forms, riboflavin 5′-phosphate (flavin mononucleotide; FMN) and flavin adenine dinucleotide (FAD) are shown in Fig. 36-6. FAD, the most water-soluble of these three flavins, exhibits orange fluorescence, whereas the other two fluoresce greenish yellow. Aqueous solutions of flavins are stable to heat and oxidizing agents, but the riboflavin in milk is dramatically reduced on exposure to sunlight.

Foods high in riboflavin include milk, liver, eggs, meat, and some green leafy vegetables. Intestinal organisms synthesize riboflavin, but the distal location of these bacteria may preclude absorption.

Metabolism. Consumed protein complexes of FAD and FMN release these flavins on gastric acidification and proteolysis. After dephosphorylation, they are absorbed in the proximal small intestine and rephosphorylated; they then appear in the circulation weakly bound to albumin and other plasma proteins, including specific riboflavin-binding proteins. Most flavins are quickly taken up by the liver and the kidneys, although measurable amounts are found in most tissues. The three flavin forms can be interconverted enzymatically. FAD is the predominant tissue form. Small amounts of flavin are excreted in bile, feces, sweat, and breast milk; urinary flavin excretion is greater. Flavin is excreted in urine mostly in the free riboflavin form; the amount excreted depends on tissue stores and on the amount ingested. Only a small amount of the ingested riboflavin is metabolized by the body. Riboflavin, but not FMN or FAD, is the moiety transferred from plasma into brain. The flavins in blood are present primarily in the erythrocytes and function in the two FAD enzymes, glutathione reductase and methemoglobin reductase. The fetus obtains free riboflavin derived from maternal erythrocytic FAD through the placenta.

Function. Flavoprotein enzyme systems contain FAD or FMN as prosthetic groups. Most flavoproteins catalyze removal of either a hydride ion or a pair of hydrogen atoms from the substrate. Riboflavin has key roles in respiratory enzymes such as D-amino acid oxidase, pyruvate dehydrogenase, xanthine oxidase, glutathione reductase, and NADH dehydrogenase. In addition, flavins are involved in the metabolism of iron, pyridoxine, and folate; they are also involved in protection against peroxidation, in metabolism of xenobiotics, and in superoxide generation by granulocytes.[25]

Clinical and chemical deficiency signs. Clinical signs of riboflavin deficiency are listed in Table 36-1. Erythrocytic concentrations of riboflavin, FMN, and FAD are more sensitive indices of riboflavin status than flavin measurements in urine or plasma are, but these blood indices are altered only late in the progression of deficiency. A functional approach to assessment of riboflavin status in-

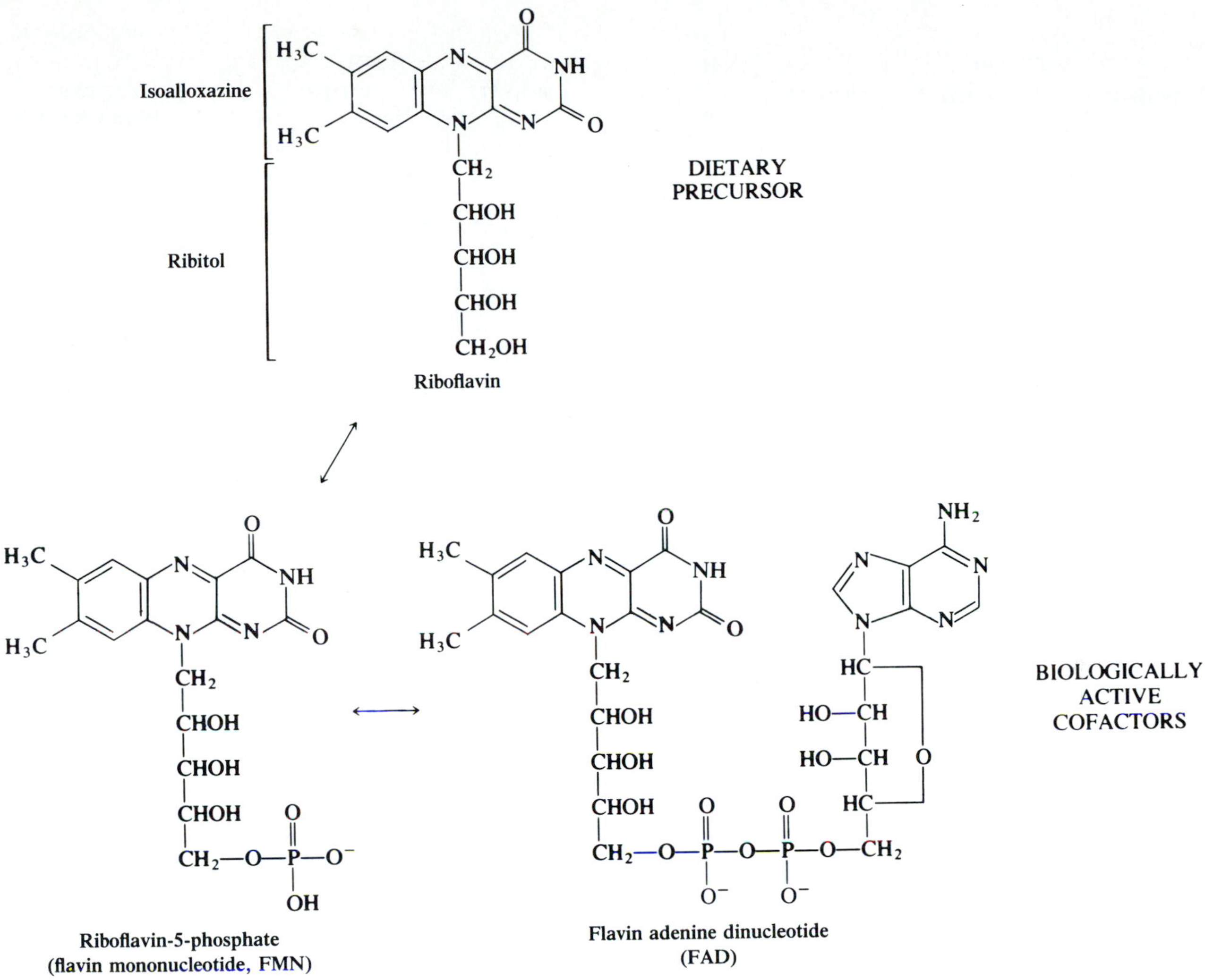

Isoalloxazine

Ribitol

Riboflavin

DIETARY PRECURSOR

Riboflavin-5-phosphate (flavin mononucleotide, FMN)

Flavin adenine dinucleotide (FAD)

BIOLOGICALLY ACTIVE COFACTORS

Fig. 36-6 Riboflavin and its active cofactor forms.

volves measurement of the increase in erythrocytic glutathione reductase (EGR) activity after in vitro addition of FAD (EGR index). This EGR index plateaus rapidly and does not continue to increase as deficiency progresses. In subjects deficient in glucose-6-phosphate dehydrogenase (G6PD) the EGR index is misleading because the EGR does not lose its coenzyme even in severe riboflavin deficiency.[26]

Pathophysiology. Ariboflavinosis is most commonly encountered when intake is inadequate as a result of poverty. It often develops as a consequence of pregnancy, especially in the pregnant adolescent. American surveys have indicated inadequate riboflavin nutrition in up to 10% of children, up to 47% of teenagers, and up to 32% of geriatric subjects. Ariboflavinosis can occur as a result of disease states, including prolonged febrile illness, malignancy, hyperthyroidism, cardiac failure, diabetes mellitus, and gastrointestinal diseases. The malnutrition associated with alcoholism affects riboflavin and other nutrients as well. Riboflavin decomposition is accelerated by phototherapy for neonatal jaundice. Negative nitrogen balance, seen in all catabolic states (including stress, physical exertion, fasting, prolonged bed rest), results in increased urinary riboflavin excretion. Riboflavin deficiency often occurs concomitantly with deficiencies of other B vitamins. Additionally, riboflavin interacts in metabolic processes involving other vitamins. These interactions can therefore result in a mixed clinical picture and may explain how therapy with one vitamin may improve symptoms of another vitamin deficiency.

Phenothiazines, oral contraceptives, and hormonal imbalance can impair riboflavin use; prednisolone reportedly interferes with conversion to coenzymes. Adriamycin-induced cardiac and skeletal myopathy may relate to its binding of riboflavin.[27]

Pyridoxine

Vitamin B_6 is known chemically as pyridoxine. Pyridoxal and pyridoxamine were first identified with vitamin activity; then their phosphorylated forms were recognized

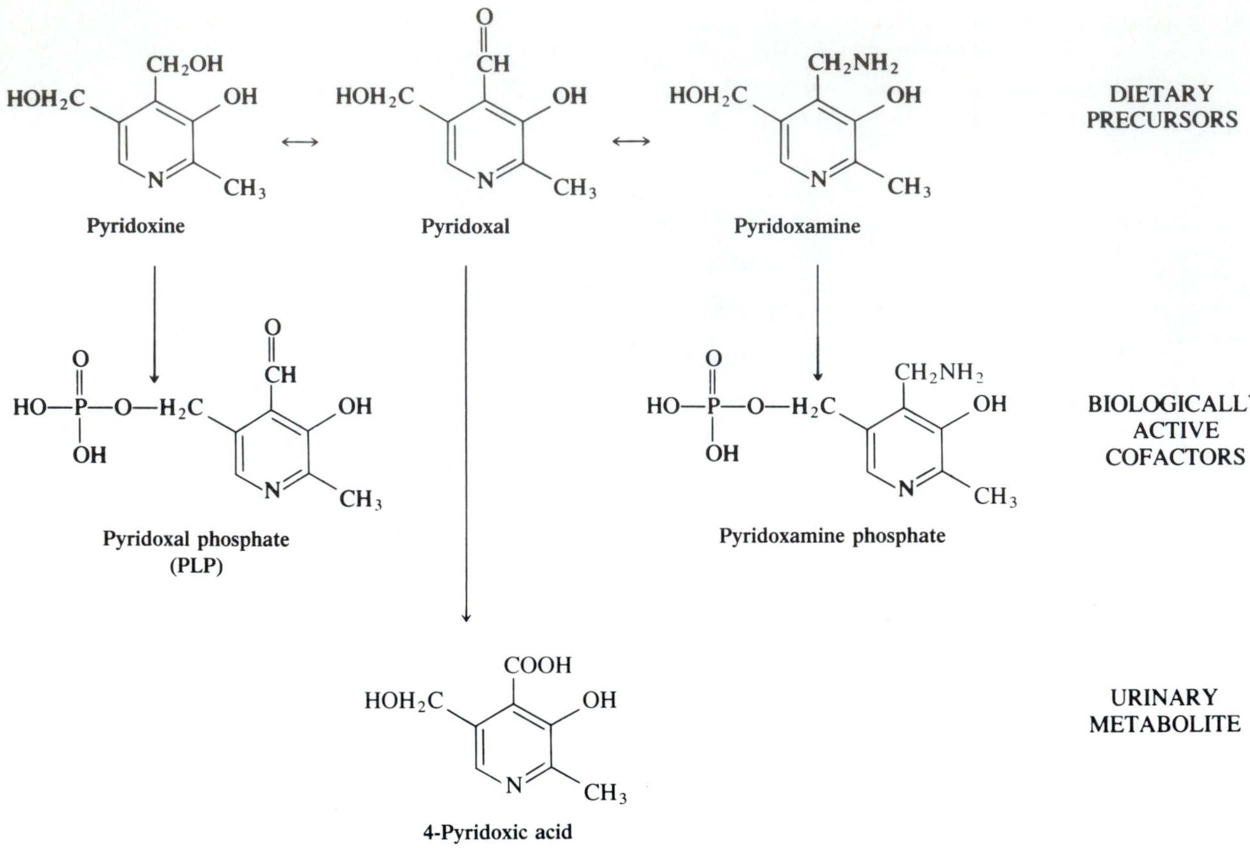

Fig. 36-7 Vitamin B_6 forms and major metabolites. *PLP*, Pyridoxal phosphate.

as active cofactors. Thus the term *vitamin B_6* now refers to the family of compounds structurally related to pyridoxal phosphate (Fig. 36-7). Pyridoxine occurs mainly in plants, whereas pyridoxal and pyridoxamine are present mainly in animal products. These three pyridine derivatives are metabolically interconverted.

Major dietary sources of vitamin B_6 are meat, poultry, fish, potatoes, and vegetables; dairy products and grains contribute lesser amounts. The predominant food form is pyridoxal phosphate, which is readily lost in food processing.

Metabolism. Pyridoxine does not bind to plasma proteins; pyridoxal and pyridoxal phosphate bind mainly to albumin. Erythrocytes rapidly take up pyridoxine, convert it to pyridoxal phosphate and pyridoxal, and then release pyridoxal into plasma. Metabolism of pyridoxal appears to occur primarily in the liver with formation of 4-pyridoxic acid, which is excreted in urine (Fig. 36-7). Pyridoxal phosphate synthesis depends on a flavin enzyme, which interrelates riboflavin and pyridoxine nutrition. Vitamin B_6 concentration is high in the brain and the CSF; the nonphosphorylated forms best enter the CSF, choroid plexus, and brain.

Function. Almost all pyridoxal phosphate–catalyzed reactions are concerned with transformations of amino acids; the major exception is provided by the phosphorylases. The pyridoxine cofactor forms act in over 60 different enzyme systems catalyzing a variety of reaction types, including the transaminases AST and ALT. The best-known functions of the pyridoxine cofactors are their roles in the conversion of tryptophan to 5-hydroxytryptamine (serotonin) and in the separate pathway of tryptophan to nicotinic acid ribonucleotide (the "niacin pathway"), both shown in Fig. 36-8.

Clinical and chemical deficiency signs. Clinical signs of pyridoxine deficiency for infants and adults are listed in Table 36-1. Chemical indices of pyridoxine depletion include reduction in plasma and erythrocyte concentrations of pyridoxine or pyridoxal phosphate. Urinary pyridoxine (normally representing less than 10% of the pyridoxine intake) and pyridoxic acid, the major urinary metabolite, are also reduced. An oral tryptophan load given to persons suspected of being deficient in pyridoxine results in excretion of several tryptophan metabolites in higher amounts than usual, xanthurenic acid being the one most commonly measured (Fig. 36-8). The involvement of other metabolic and hormonal factors in this pathway necessitates cautious interpretation of the tryptophan challenge test. The tissue status of pyridoxal phosphate can be assessed by measurement of the increment of erythrocytic aspartate (or alanine)

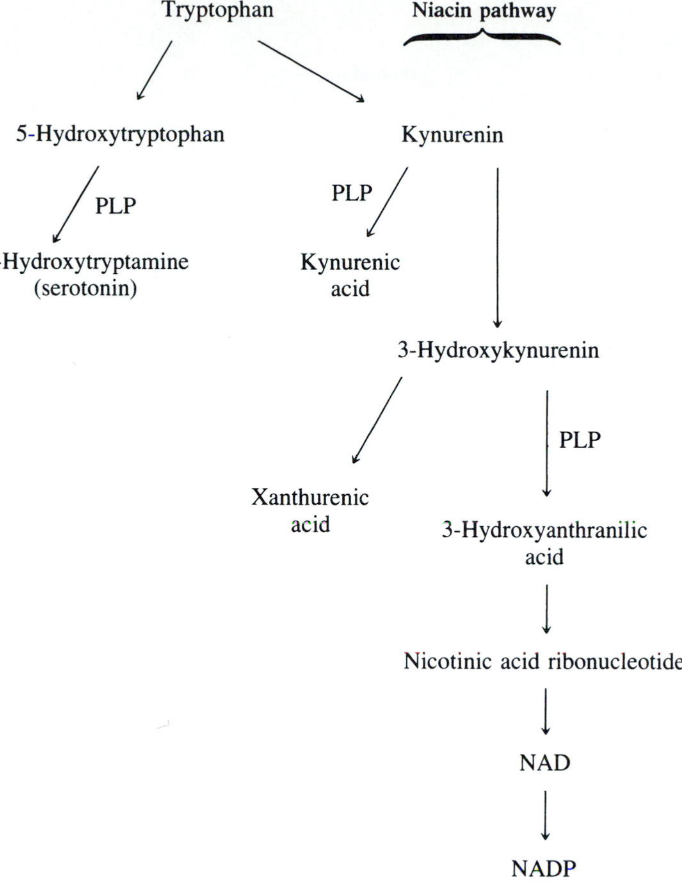

Fig. 36-8 Role of vitamin B_6 in tryptophan metabolism. *NAD,* Nicotinamide adenine dinucleotide; *NADP,* nicotinamide adenine dinucleotide phosphate; *PLP,* pyridoxal phosphate.

aminotransferase (AST or ALT, respectively) after in vitro addition of the pyridoxal phosphate cofactor. Elevation in the ratio of activity plus or minus pyridoxal phosphate (the EAST index) is suggestive of inadequate tissue stores. Pyridoxine-depleted subjects may require several months of repletion to increase enzymatic activity and reduce the stimulation index. It has been suggested that urinary pyridoxic acid and erythrocyte enzyme stimulation both be measured to evaluate short-term and long-term pyridoxine status, respectively. The newer methods of plasma pyridoxal phosphate determination may prove to be equal to (or better than) the widely accepted EAST index.

Pathophysiology. Conditions associated with low pyridoxine indices include celiac disease, acute alcoholism, psychoses such as paranoia and schizophrenia, epilepsy, ulcerative colitis, renal calculi, and lactation. Although low plasma pyridoxal-5-phosphate levels are seen during the acute phase of myocardial infarction, B_6 deficiency does not appear to be a risk factor for ischemic heart disease.[28] Pyridoxine requirements increase during pregnancy as a result of fetal demand and hormonal induction of maternal enzymes, which increases maternal requirements. Pyridoxine inadequacy during pregnancy has been linked to suboptimal birth outcomes (infants with low Apgar scores and low birth weight).[29]

Drugs known to antagonize pyridoxine include isonicotinic acid hydrazide (isoniazid, INH), steroids, and penicillamine. Oral contraceptives lower indices of pyridoxine status; depressive mood changes in patients receiving these agents may relate to pyridoxine's role in serotonin synthesis. Pyridoxine deficiency enhances susceptibility to carbon disulfide toxicity. The vitamin B_6 compounds are of low systemic toxicity, and no teratogenic effect has been detected.

Niacin

Over 200 years ago pellagra (from the Italian, meaning "rough skin"), which is associated with diarrhea, dementia, dermatitis, and death (the "four D's"), was attributed to poor diet. From 1910 to 1935 pellagra was the worst nutritional disease outbreak in U.S. history, with over 150,000 cases reported annually, most of whom were poor persons in the South. In 1912 nicotinic acid was extracted from rice polishings and was claimed to have vitamin-like effects, but it was not until 1935 that nicotinic acid (also called "niacin") was shown to cure black-tongue in dogs (a disease similar to pellagra in humans). Thus the responsible nutrient was identified as niacin, and its deficiency was associated with diets high in corn.

Niacin is a simple derivative of pyridine and is extremely stable. Moderately resistant to heat, acid, and alkali, niacin is related chemically to nicotine but has very different physiological properties. The active cofactor forms of NAD and NADP (Fig. 36-9) derived from niacin can also be synthesized from liver tryptophan (Fig. 36-8) so that sufficient dietary tryptophan can abolish the requirement for niacin.

Meats and grains are major sources of niacin, and many manufactured food products are supplemented with this vitamin, especially those made from cereals. Dietary deficiency is most likely in persons with diets consisting primarily of corn (such as a diet of pork fat and hominy grits or sorghum). Corn is especially poor in tryptophan, and part of the niacin in corn is not absorbed. The high concentrations of leucine in cereals somehow interfere with the niacin pathway and with conversion of niacin to its cofactor forms. The niacin equivalent of tryptophan is approximately 60 mg of tryptophan equaling 1 mg of niacin. Most diets have $\geq$600 mg of tryptophan, but diets low in protein have much less.

Metabolism. Both niacin and nicotinamide are readily absorbed in the gut. Niacin is transported in blood mainly in erythrocytes. There is little storage of niacin in the body, and urine contains nicotinamide and other metabolites of niacin (Fig. 36-9). Plasma nicotinamide readily enters the CSF, whereas niacin does not. Brain tissue does not express the niacin pathway of tryptophan and so must

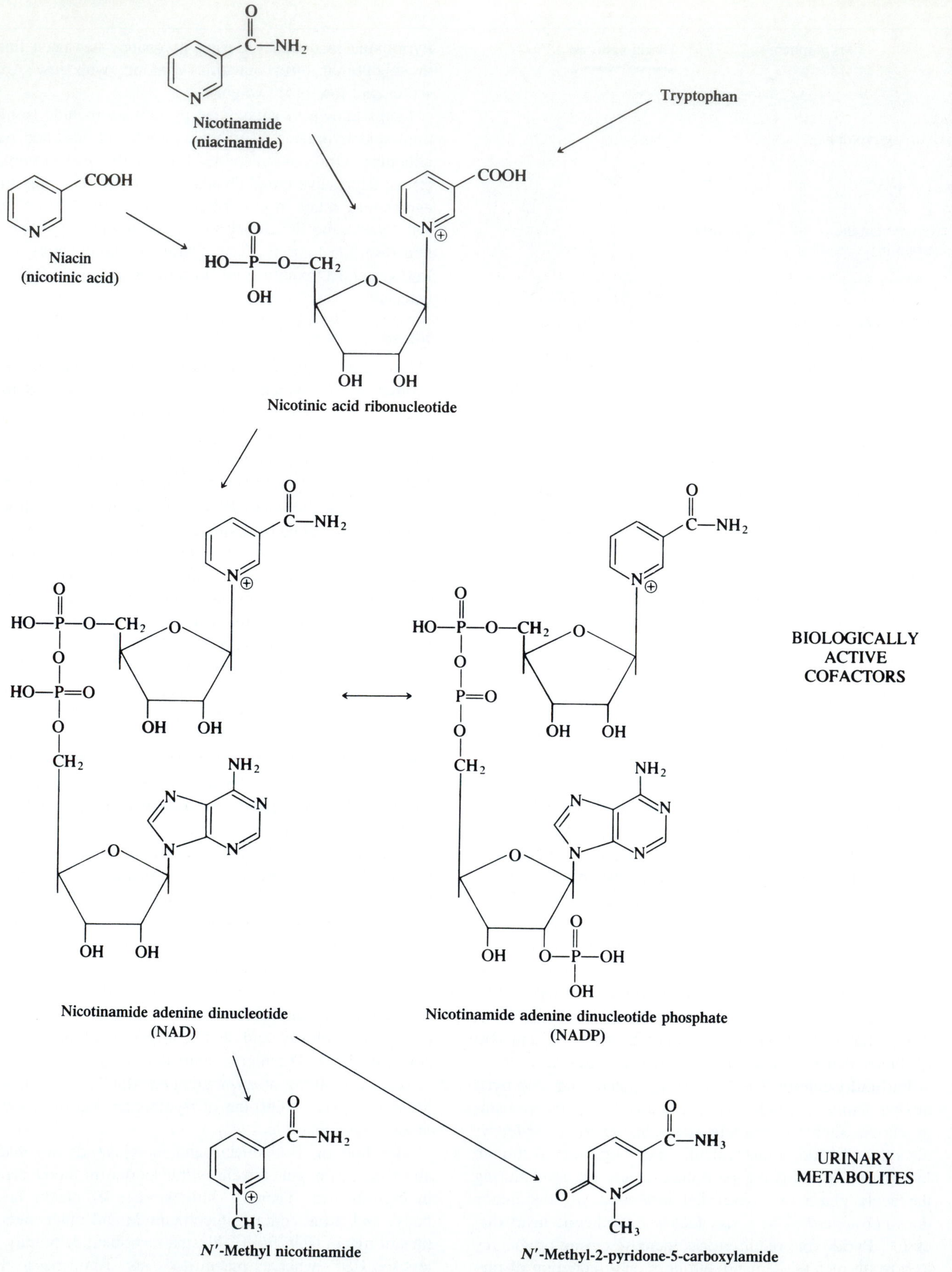

Fig. 36-9 Cofactor forms and metabolites derived from niacin or tryptophan.

use plasma nicotinamide from the diet or dephosphorylated forms of the cofactors.

Function. NAD and NADP are involved in a large number of oxidation-reduction reactions catalyzed by dehydrogenases including alcohol, glutamate, glucose-6-phosphatase dehydrogenase and glycerolphosphate dehydrogenase. Reduction yields the dihydronicotinamide (NADH or NADPH) that has a strong absorption at 340 nm, a feature widely used in assays of pyridine nucleotide–dependent enzymes (see Chapter 52).

Clinical and chemical deficiency signs. Clinical signs of niacin deficiency are listed in Table 36-1. Chemical measures of niacin status primarily involve the two major urinary metabolites *N'*-methylnicotinamide and *N'*methyl-2-pyridone-5-carboxylamide. Ratio of 2-pyridone compound to *N'*-methylnicotinamide is reduced in niacin deficiency; reduction in the individual metabolites is also seen. These metabolites are normally present in plasma also. Only a small percentage of administered niacin is excreted as niacin or as nicotinamide. Measurement of niacin or its nucleotides in blood is generally not considered a reliable index of niacin status, although erythrocytes of deficient persons have lower levels of NAD and NADP and higher amounts of nicotinamide ribonucleotide.

As in the case of pyridoxine, niacin deficiency results in greater susceptibility to toxicity of carbon disulfide. Oral contraceptives stimulate the niacin pathway of tryptophan. Niacin and nicotinamide are widely used in megadoses in the treatment of sprue, psychiatric conditions such as schizophrenia, and a wide range of circulatory disorders including hypertension, cerebral thrombosis, and intermittent claudication (limping). Niacin and its analogs are used in treatment of some hyperlipidemic states to reduce heart and vascular effects.[30] Nicotinic acid is one of several agents known to elevate HDL cholesterol; whether this therapy can retard or reverse the progression of atherosclerosis is not yet clear.[31] Nicotinic acid–induced hyperbilirubinemia response is of diagnostic value in patients with Gilbert's syndrome.[32]

Thiamin

Beriberi, a polyneuritis disease, was first believed to be caused by a toxin that was neutralized by rice husks. This disease was later shown to be a nutrient deficiency caused by removal of an essential factor from rice as it was polished. This antiberiberi factor was finally isolated and crystallized in 1925; then its structure was shown to be a substituted pyrimidine linked by a methylene group to a substituted thiazole (Fig. 36-10). Thus the name reflects its components of amine and sulfur (thia-) groups.

Thiamin is easily destroyed in alkaline media. It resists temperatures up to 100° C but is destroyed at greater temperatures, a fact significant for fried foods or those cooked under pressure. Highly water soluble, thiamine is easily leached out of foodstuffs being washed or boiled.

Sources of thiamin include yeast, wheat, whole grain, and enriched breads and cereals, nuts, peas, potatoes, and most vegetables. Dietary deficiency is suspect in persons with a diet primarily consisting of polished rice; alcoholics; persons with anorexia, vomiting, or diarrhea; and postoperative patients. Deficiency of thiamin is common in the elderly.[33]

Metabolism. Dietary thiamin is absorbed in the intestine by a carrier-mediated process that is saturated at an oral intake of about 10 mg. Blood thiamin appears in the CSF and brain to a small degree; the phosphorylated forms are found even less commonly. Thiamin is excreted unchanged or after cleavage between the ring systems by intestinal microorganisms. Thiamin pyrophosphate (TPP) is the predominant moiety in tissues, whereas the major form in plasma is thiamin. Erythrocytes contain TPP at concentrations about fivefold higher than those found in plasma. Liver, heart, and brain have higher concentrations than do muscle and other organs.

Function. In its TPP cofactor form, thiamin catalyzes the decarboxylation of α-keto acids (pyruvate and α-ketoglutarate), the oxidative decarboxylation by α-keto acid dehydrogenases, and the formation of ketols. Thiamin pyrophosphate functions in major carbohydrate pathways and also functions in metabolism of branched-chain amino acids. Thiamin triphosphate (TTP) may be the thiamin derivative released from nerves after electrical stimulation and as such may play a role in sodium-ion conductance. There is some evidence for decreased brain TTP and an inhibitor of its synthesis (demonstrable in tissues and urine) in patients with Leigh's encephalopathy.

Clinical deficiency signs. Table 36-1 lists the major clinical signs of thiamin deficiency. The Wernicke-Korsakoff syndrome responds to thiamin therapy, and there is evidence for an abnormality of neurotransmitter metabolism, perhaps involving TTP. These patients typically accumulate excessive amounts of pyruvate and lactate in physiological fluids.[34]

Genetic variations in TPP-dependent enzymes modify the effect of dietary thiamin deficiency. Although most patients with thiamin deficiency do not develop Wernicke-Korsakoff syndrome, mild deficiency leads to impairments in higher integrative functions (including memory). More severe deficiency leads either to "dry" or "wet" beriberi, but seldom do both occur together.

Chemical deficiency signs. Chemical indices of thiamin deficiency commonly used are reduction in urinary thiamin, reduction in erythrocyte transketolase (ETK) activity, and stimulation of ETK by in vitro TPP. Prolonged deficiency results in decreased synthesis of the ETK apoenzyme so that the ETK stimulation test may underestimate the magnitude of deficiency. There is also evidence of reduction of ETK in undernutrition, diabetes, and liver disease without a TPP stimulation effect. As is possible with all proteins, genetic heterogeneity of ETK has been dem-

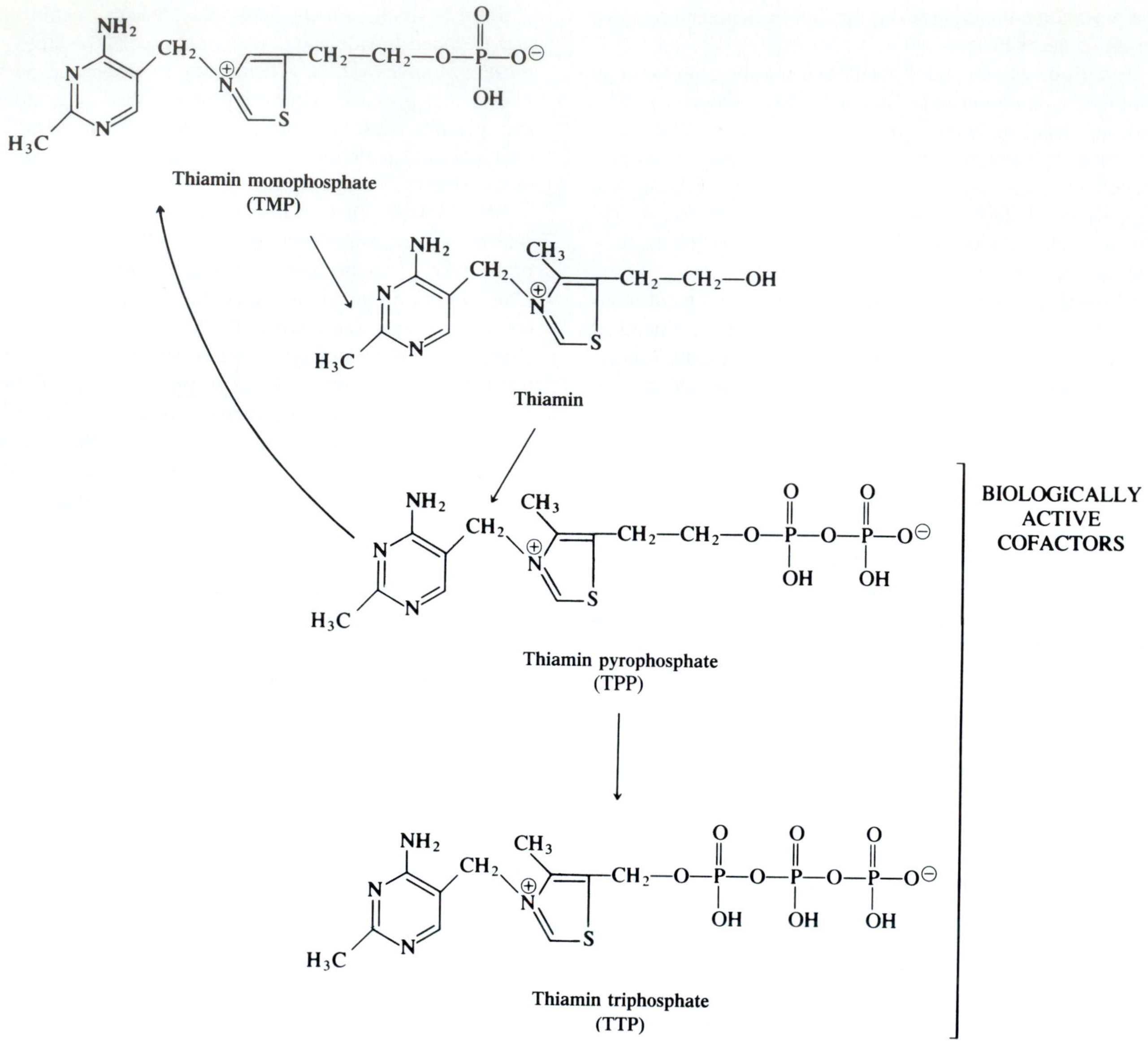

Thiamin monophosphate (TMP)

Thiamin

Thiamin pyrophosphate (TPP)

Thiamin triphosphate (TTP)

BIOLOGICALLY ACTIVE COFACTORS

Fig. 36-10 Thiamin and its cofactor forms.

onstrated in humans. Correlation between ETK stimulation and dietary thiamin or clinical signs is not always seen. There are conflicting reports as to the usefulness of blood thiamin levels; this is possibly related to low concentrations and measurement difficulties.

Pathophysiology. Populations of Southeast Asia who eat foods rich in antithiamin substances commonly develop beriberi. Milder forms of thiamin deficiency are common among pregnant women, elderly persons, and alcoholics. Magnesium deficiency (common in alcoholics) impairs thiamin use. Oral contraceptive use may induce deficiency. The total body store of thiamin is only 30 mg; 30 times the daily requirement. Thiamin deficiency symptoms can occur after about 1 month of a thiamin-deficient diet. Thiamin replacement is often warranted in persons with persistent vomiting or prolonged gastric aspiration, and in those who go on long fasts. Patients with typical sporadic amyotrophic lateral sclerosis (ALS) reportedly have a very high incidence of decreased CSF thiaminemonophosphate with inversion of the thiamin/TMP ratio.[35]

Biotin

Growth factors separately discovered under the names of bios II, coenzyme R, and biotin were noted to be similar; other research demonstrated the symptom complex produced in rats by feeding them raw egg white was corrected by cooking the eggs or by addition of other foods presumably containing "vitamin H," or "protective factor X." The linkage between vitamin H and biotin was made and the clinical role of biotin was established when human

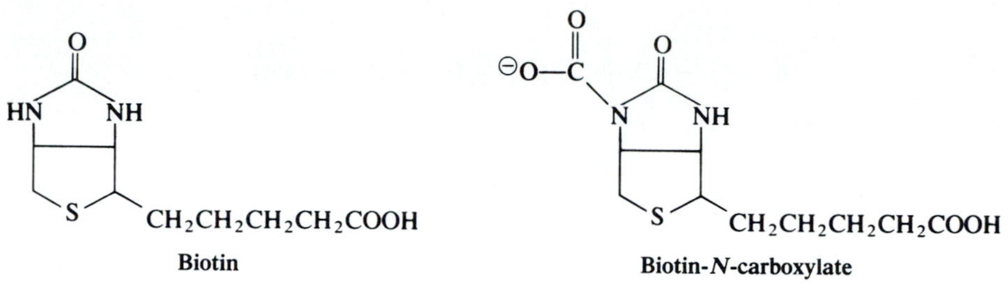

Biotin

Biotin-*N*-carboxylate

Fig. 36-11 Biotin and its active form.

volunteers ingested large amounts of raw egg white and confirmed the "egg white injury" of animals that is correctable with biotin (structure in Fig. 36-11).

Numerous foods contain biotin, although no one food is especially rich (up to 2 mg/100 g). Dietary intake is low in the neonatal period despite the fact that concentrations increase as newborns switch from colostrum to mature breast milk. Enteric bacteria are known to synthesize biotin, although the relative contribution of this source is unclear.

Metabolism. Biotin is absorbed in the proximal half of the small intestine and circulates in blood largely bound to plasma proteins. Avidin, a glycoprotein of egg albumin, has very high affinity for biotin, thus explaining the biotin deficiency resulting from raw egg white ingestion.

Function. Biotin is a coenzyme for carboxylation and carboxyl exchange reactions. Important enzymes include acetyl CoA, propionyl CoA, and pyruvate carboxylases, as well as methylmalonyl-oxaloacetic transcarboxylase.

Clinical deficiency signs. Reported dietary deficiency cases are rare, but each has a history of raw eggs as a large dietary component for months to years. Table 36-1 lists the major signs of biotin deficiency.

Chemical deficiency signs. Dietary deficiency is accompanied by decreased urinary and plasma[36] biotin and increased urinary organic acids, indicating functional deficiency of β-methylcrotonyl CoA carboxylase and propionyl CoA carboxylase. Genetic alterations in these carboxylases may result in biotin-dependent states that cause metabolic acidosis and require pharmacological biotin doses; these enzyme deficiencies are confirmed in leukocytes. Genetic deficiency of biotinidase, which can be detected by a blood spot assay, is treated with biotin.[37]

Pathophysiology. With proper modern nutrition biotin deficiencies are extremely rare. However, biotin deficiency might be suspected in patients receiving long-term total parenteral nutrition.[15,34] No biotin toxicity has been described.[38]

Pantothenic acid

A growth factor occurring in all types of animal and plant tissue was first designated *vitamin B_3* and later named *pantothenic acid* (from Greek, meaning 'from everywhere'). This factor was chemically identified in 1938, and its deficiency was linked to the "burning feet syndrome" in 1949.

Dietary sources include liver and other organ meats, milk, eggs, peanuts, legumes, mushrooms, salmon, and whole grains. Approximately 50% of pantothenate in food is available for absorption. Pantothenate is unstable to acid, alkali, heat, and some salts. Intestinal microorganisms are a source of B_3 for animals, possibly including humans.

Metabolism. As shown in Fig. 36-12, pantothenate is metabolically converted to 4′-phosphopantetheine, which becomes covalently bound to either serum acyl carrier protein (ACP) or to coenzyme A. Little is known of pantothenate metabolism. Urinary excretion of pantothenate correlates well with intake; no saturation is seen with intakes of 10 mg/day. Free pantothenate is the major form in both urine and serum, whereas coenzyme A is the major erythrocytic form. In contrast to other B vitamins, pantothenate tissue repletion is gradual. Tissues known to contain pantothenate are liver, adrenal glands, brain, kidneys, and heart.

Function. Coenzyme A is a highly important acyl-group transfer coenzyme that is involved in a large number of reactions of a great variety of reaction types. Acyl derivatives of coenzyme A are first formed (by thioester linkage), followed by transfer of the acyl group to an acceptor molecule.

Clinical and chemical deficiency signs. No clear-cut case of a pantothenate deficiency has been reported; Table 36-1 lists clinical signs observed in experimentally induced deficiencies. Whole blood pantothenate of less than 1000 mg/L and urinary excretion of less than 10 mg/day are regarded as indicative of deficiency. Most past measurements have used biological assays for free pantothenate, releasing pantothenate from coenzyme A by multiple enzymatic treatments.

Pathophysiology. Low urinary excretion and reduced blood levels of pantothenate have been reported in patients with chronic malnutrition, acute alcoholism, and acute rheumatism. Patients with circulatory and cardiovascular diseases and those with peptic ulcers have reduced circulating pantothenic acid; chronic alcoholics have increased excretion, an indication of impaired use. There is evidence that the increased occurrence of hypertension in some Jap-

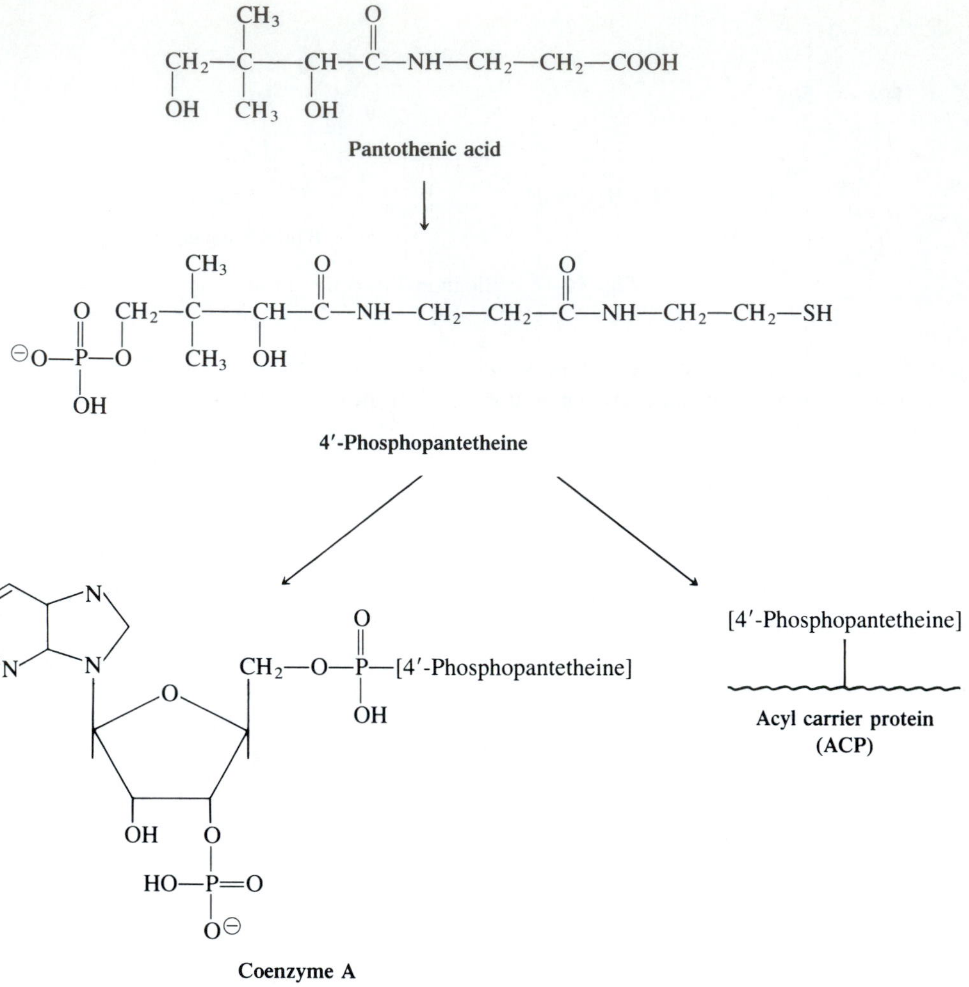

Pantothenic acid

4′-Phosphopantetheine

[4′-Phosphopantetheine]

Acyl carrier protein (ACP)

Coenzyme A

Fig. 36-12 Pantothenic acid and its active cofactors.

anese populations may be related to panthothenate deficiency.[39] Pantothenate has been given postoperatively to stimulate the gastrointestinal tract and also to treat streptomycin-induced neuropathy. No toxicity is known.

Vitamin B_{12}

Pernicious anemia was described over 100 years ago and was recognized as a disease amenable to liver extract therapy. The antipernicious anemia factor (extrinsic factor), finally isolated, crystallized, and characterized between 1948 and 1956, is now known as "vitamin B_{12}." In the early 1930s a similar entity, "pernicious anemia of pregnancy," was shown to be different in that it did not respond to liver extract therapy but did respond to an autolyzed yeast substance. Purification of this substance led to its identification as pteroylglutamic acid, more commonly known as "folic acid." Thus similar clinical symptoms were found to result from deficiencies of two totally different structures (Figs. 36-13 and 36-14), neither of which could replace the other. The clinical similarities and the metabolic interactions of vitamin B_{12} and folic acid usually dictate their simultaneous assessment.[40]

Vitamin B_{12} bears a corrin ring (containing pyrroles similar to porphyrin) linked to a central cobalt atom. The pyrroles are almost saturated with side chains (methyl, acetamide, or propionamide groups). Different corrinoid compounds, or cobalamins, are distinguished by the substituent linked to the cobalt with methylcobalamin and 5′-deoxyadenosylcobalamin being the two known coenzyme forms.

Dietary sources of vitamin B_{12} are of animal origin (meat, eggs, milk) except for the plant comfrey. Total vegetarian diets are therefore likely settings for deficiency. Animals derive their vitamin B_{12} from intestinal microbial synthesis. The average daily diet contains 3 to 30 μg, of vitamin B_{12} of which 1 to 5 μg is absorbed. The frequency of dietary deficiency increases with age, occurring in over 0.5% of persons over age 60.

Metabolism. Most vitamin B_{12} absorption is through a complex with intrinsic factor (IF), a protein secreted by gastric parietal cells (Fig. 36-15). This IF-B_{12} complex binds with specific ileal receptors. ''Blocking'' IF-antibodies prevent binding of vitamin B_{12} to IF, and ''binding'' antibodies can combine with either free IF or the IF-B_{12} complex, thus preventing attachment of the complex to ileal receptors and intestinal uptake of the vitamin. Parietal cell antibodies have also been identified as a cause of pernicious anemia.

Released from the IF complex within the mucosal cell, vitamin B_{12} circulates in plasma bound to specific transport proteins and is deposited in liver, bone marrow, and other tissues. There is a significant enterohepatic circulation of vitamin B_{12}. As a result of this biliary reabsorption, 10 to 12 years are required for a strict vegetarian to become clinically deficient. A person with normal B_{12} stores but lacking IF requires less time (1 to 4 years) for deficiency to become evident.

Transcobalamin II (TC II) appears to be the major serum protein transporting exogenous vitamin B_{12} to tissues.

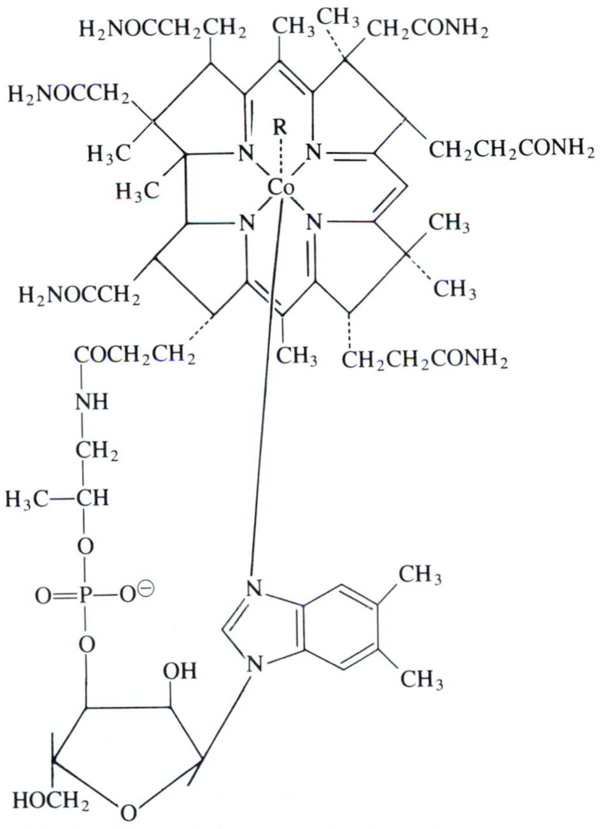

	R		
(Me-B_{12})	CH_3	**Methylcobalamin**	ACTIVE COFACTOR FORMS
(Ado-B_{12})	5′-Deoxy-adenosine	**Deoxyadenosyl-cobalamin**	
(OH-B_{12})	OH	**Hydroxocobalamin**	DIETARY FORMS AND THERAPY
(CN-B_{12})	CN	**Cyanocobalamin**	

Fig. 36-13 Vitamin B_{12} forms.

Cobalophilin (previously R, or rapidly migrating, binding protein, or TC I) transports endogenous B_{12} and is the binder of food cobalamins. Saliva, breast milk, and granulocytes contain large amounts of this binding protein compared with relatively small amounts of cobalamin. Plasma contains both types of transport protein and the three forms of vitamin B_{12} (hydroxycobalamin, methylcobalamin, and deoxyadenosylcobalamin). Absorption and transport of this vitamin are illustrated in Fig. 36-15.

Functions. The adenosyl cobalamin form (Ado-cbl) is a cofactor for odd-numbered chain fatty acid metabolism (myelin sheath formation); deficiency results in accumulation of the intermediate methylmalonyl CoA and excretion in urine of methylmalonic acid. The methylcobalamin form (Me-cbl) is required for methyl group transfer reactions; lack of it results in blockage of DNA synthesis and megaloblastic anemia. Dietary deficiency of the precursor of these two coenzyme forms results in deficiencies of both active cobalamins. Genetic deficiencies of enzymes in either pathway are known. In the few case reports of inherited deficiency of TCII, symptoms of pancytopenia and failure to thrive develop within a few months of birth. Several large studies have shown a racial difference in serum vitamin B_{12} levels, with blacks maintaining higher B_{12} levels than whites.[41]

Clinical and chemical deficiency signs. Table 36-1 lists signs of vitamin B_{12} deficiency. Diagnostic tests for vitamin B_{12} deficiency include measurement of serum B_{12} by microbiologic or radioligand assay methods, measurement of urinary or serum methylmalonic acid, and the Schilling test (see p. 407).

Early serum B_{12} competitive binding methods used B_{12} binders with variable purity and binding specificity, yielding unreliable results; there are numerous reports of cobalamin deficiency in patients with normal serum cobalamin concentrations. GC/MS assay of urinary methylmalonic acid may be more sensitive than serum cobalamin radioimmunoassay with purified IF.[42] Improved methodology provides more accurate clinical information.

Definition of the cause of vitamin B_{12} deficiency may require additional assays including tests for specific antibodies, the deoxyuridine suppression test, and assessment of the transport proteins.

The deoxyuridine suppression test is based on the fact that preincubation of normal bone marrow with an appropriate concentration of deoxyuridine severely suppresses

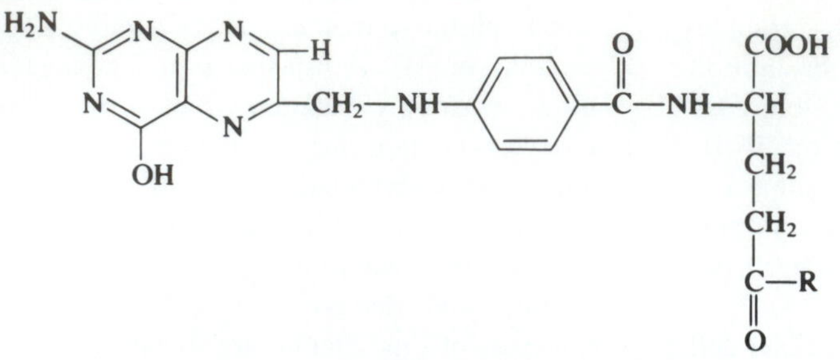

Folic acid
(pteroylglutamic acid)

Monoglutamate: R = —OH
Polyglutamates: R =

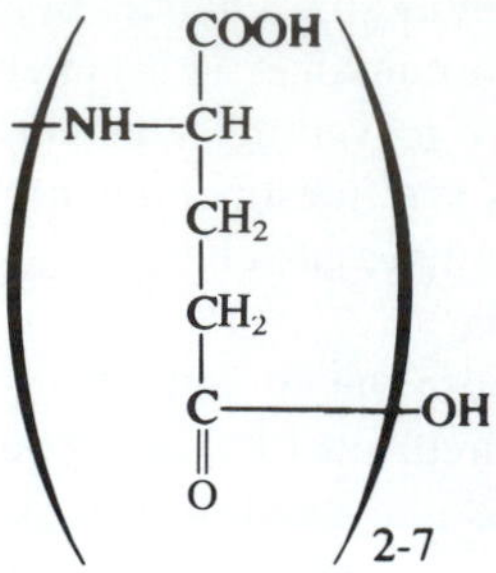

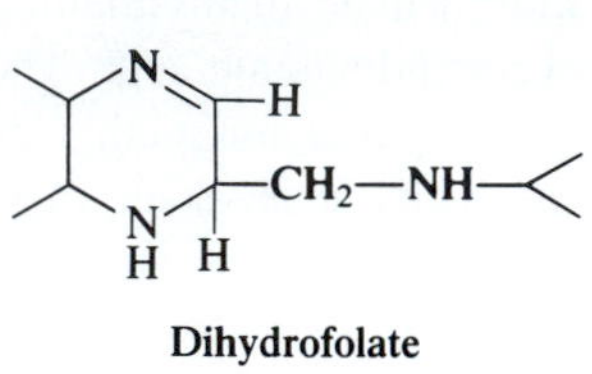

Dihydrofolate
(DHF)

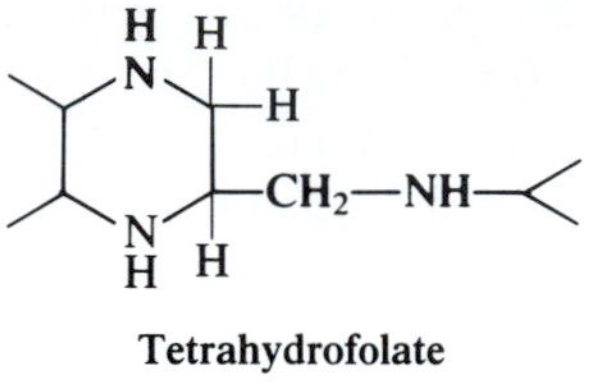

Tetrahydrofolate
(THF)

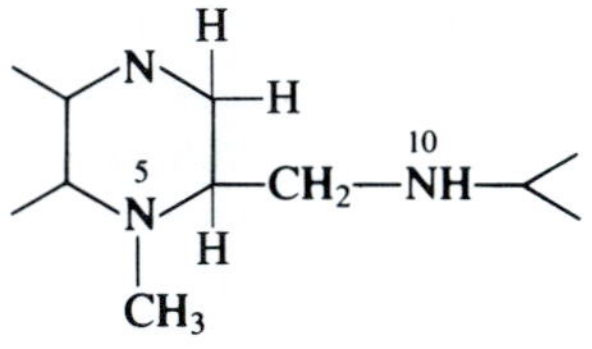

N-5-Methyltetrahydrofolate
(5-Me-THF)

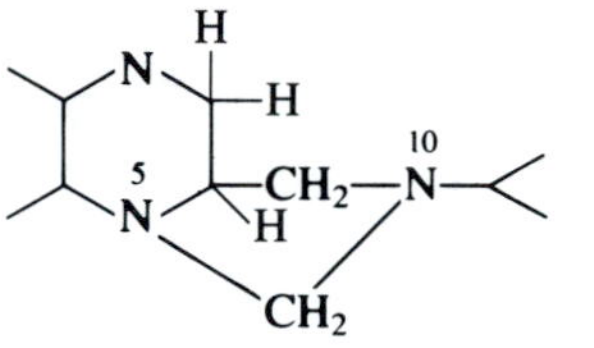

5,10-Methylenetetrahydrofolate
(5,10-CH_2-THF)

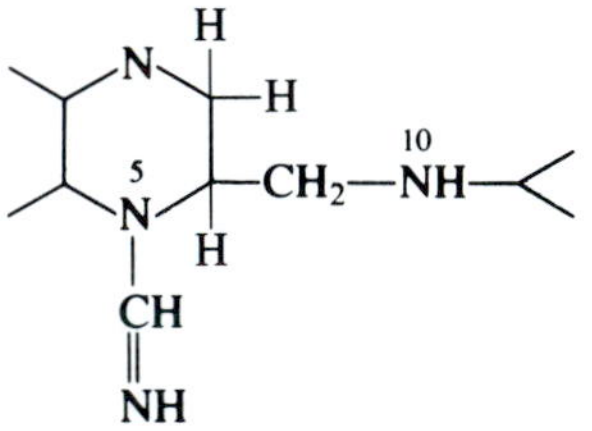

N-5-Formiminotetrahydrofolate
(FI-THF)

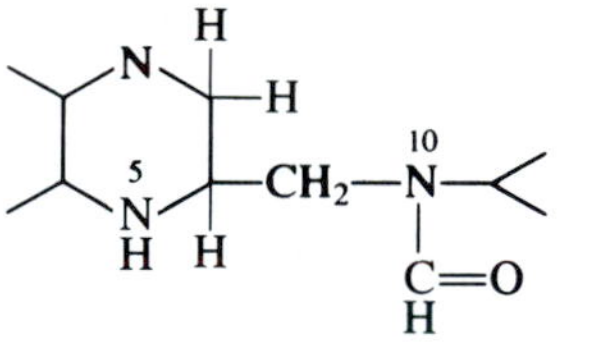

N-10-Formyltetrahydrofolate
(F-THF, folinic acid, leukovorin)

Fig. 36-14 Structures of folic acid forms.

the subsequent incorporation of tritiated thymidine into DNA. This suppression is subnormal with bone marrow from patients deficient in either vitamin B_{12} or folate and is correctable in vitro by addition of the appropriate deficient vitamin. This test will also show abnormal results in patients with megaloblastic changes resulting from neoplastic, chemotherapeutic, or other agents interfering with DNA synthesis.

Pathophysiology. Inadequate secretion of IF may accompany lesions of the gastric mucosa, gastric atrophy, gastrectomy, iron deficiency, and some endocrine disorders. The IF-B_{12} complex may be inadequately formed in pancreatic insufficiency because there is insufficient pancreatic protease activity to split the dietary vitamin B_{12} from cobalophilin in the duodenum. The IF-B_{12} complex may be inadequately absorbed in ileal malfunction (sprue, enteritis, ileal resection, neoplasias, granulomas, and so on). The term *pernicious anemia* is now most commonly applied to vitamin B_{12} deficiency resulting from lack of IF. Antibodies to IF and to parietal cells are common in pernicious anemia patients, in their healthy relatives, and in patients with other autoimmune disorders. Blocking antibodies can result in normal or high serum B_{12} levels in patients with pernicious anemia. Numerous drugs induce vitamin B_{12} malabsorption, and excessive ascorbate intake may convert vitamin B_{12} to analog, blocking forms. Cobalophilin is elevated in patients with chronic myeloproliferative disorders, polycythemia vera, and various neoplasms.

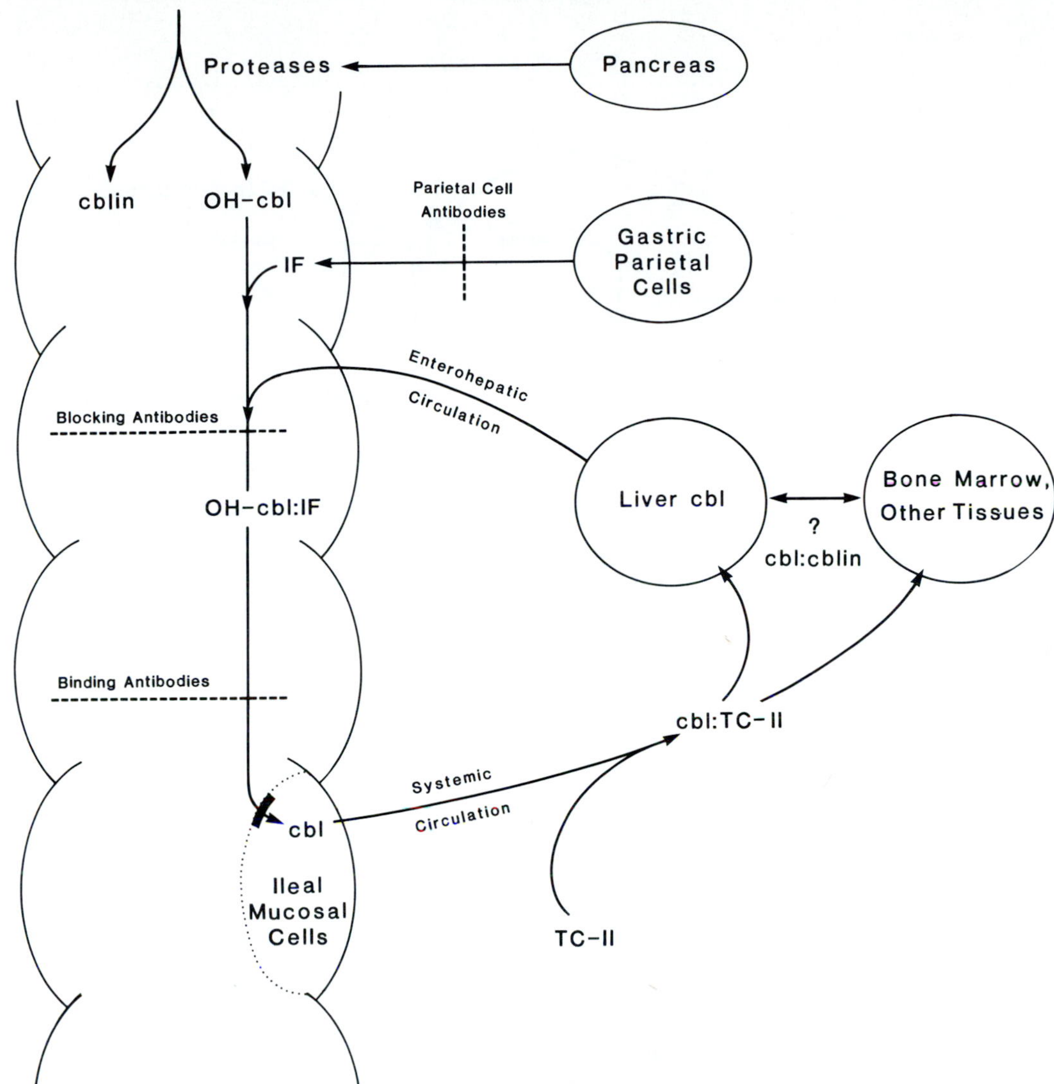

Fig. 36-15 Absorption of dietary vitamin B_{12} and its transport to storage sites. *cblin,* Cobalophilin; *cbl,* cobalamin; *OH-cbl,* hydroxycobalamin; *IF,* intrinsic factor; *TC-11,* transcobalamin II.

Folic acid

Structural relatives of pteroylglutamic acid (folic acid) are the metabolically active compounds usually referred to as "folates" (Fig. 36-14). Up to eight glutamate residues may be found in these naturally occurring compounds.

Food folates are primarily found in green and leafy vegetables, fruits, organ meats, and yeast. Excessive boiling of foods and use of large quantities of water result in folate destruction. The average American diet may be inadequate in folate for adolescents and for pregnant or lactating females.

Metabolism. The naturally occurring folate polyglutamates are hydrolyzed to monoglutamate forms before absorption (which occurs primarily in the proximal jejunum) by the intestinal mucosal cells. After this, folate enters the liver through the portal circulation. The liver converts some of these folate monoglutamates to polyglutamates, which are presumably then stored; another fraction of the folate is excreted in bile as N^5-methyltetrahydrofolate (MeTHF), which is reabsorbed and is the major circulating form of folate. Serum folate (MeTHF), in the monoglutamate form, readily enters the choroid plexus and the CSF. Folic acid, on the other hand, is readily transported from CSF to plasma. A folate-binding protein has been identified in the choroid plexus, probably accounting for the high CSF/plasma ratio. The CSF form is mainly MeTHF; brain folates are predominately polyglutamate forms of dihydrofolate (DHF). Folate catabolism involves cleavage

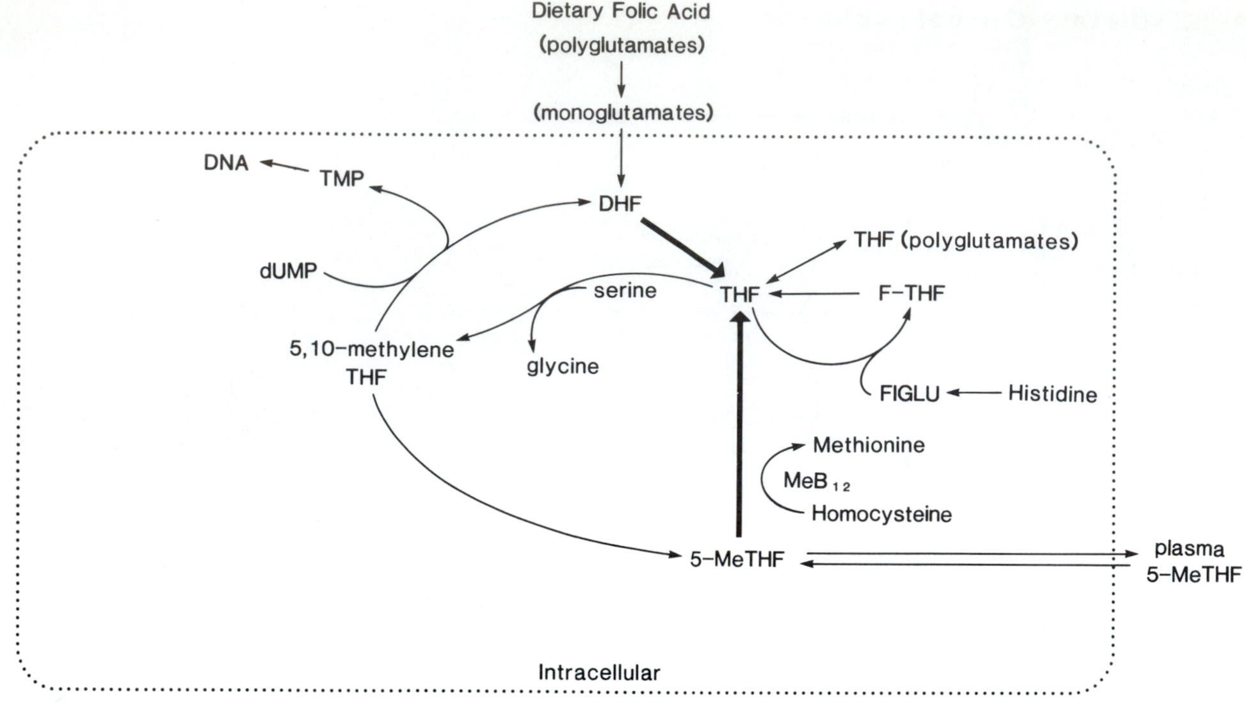

Fig. 36-16 One-carbon transfer using folic acid forms as cofactors. *DHF,* Dihydrofolate; *DNA,* deoxyribonucleic acid; *dUMP,* deoxyuridine monophosphate; *FIGLU,* formimino-L-glutaric acid; *F-THF,* folinic acid; *5-MeTHF, N*-5-methyltetrahydrofolate; *MeB*$_{12}$, methyl cobalamin; *THF,* tetrahydrofolate; *TMP,* thymidine monophosphate.

of the pterin ring, followed by acetylation to form the excreted product, *p*-acetamidobenzoylglutamic acid.

Function. Folate (MeTHF) is a cofactor for enzymatic reactions involving one-carbon transfers. After cellular uptake, the MeTHF is converted to THF while transferring a carbon to homocysteine to yield methionine. As mentioned previously, this reaction requires vitamin B_{12}. In the absence of vitamin B_{12}, the folate is essentially trapped in the MeTHF form, making it unavailable for other reactions, including the synthesis of thymine for DNA (Fig. 36-16).

Clinical and chemical deficiency signs. The major clinical symptom of folate deficiency is megaloblastic anemia. Chemical indices of deficiency are, in order of occurrence, low serum folate (see p. 545), hypersegmentation of neutrophils, high urinary FIGLU (a histidine metabolite accumulating in absence of folate), low erythrocyte folate, macro-ovalocytosis, megaloblastic marrow, and finally anemia. The deoxyuridine suppression test, discussed with vitamin B_{12}, is also an index of folate status. Serum folate, although an early index of deficiency, can frequently be low despite normal tissue stores. Hypersegmentation of neutrophils may not be seen in folate deficiency of pregnant women. The urinary FIGLU test requires ingestion of an oral load of histidine, followed by a timed urine collection, and can be abnormally high in deficiencies of either folate or vitamin B_{12}. Because most folate storage occurs after the vitamin B_{12}-dependent step, erythrocyte folate can also be reduced in deficiency of either B_{12} or folate. Despite this overlap, erythrocyte folate concentration is currently accepted as the best laboratory index of folate deficiency.

Pathophysiology. Folate requirement is increased during pregnancy and especially during lactation. The increase during lactation partially results from the presence in milk of high-affinity folate binders. Other instances of increased folate requirement include hemolytic anemias, iron deficiency, prematurity, and multiple myeloma. Patients receiving dialysis treatment rapidly lose folate. Folate deficiency because of malabsorption can occur in sprue, celiac disease, inflammatory bowel diseases, cardiac failure, and systemic bacterial infections. Both acidic and alkaline conditions can interfere with intestinal transport; the organic anions that accumulate in uremia also impair folate use. Genetic alterations in most of the folate-interconverting enzymes have been reported. Several mentally retarded children have been described who are unable to transfer folates from blood into CSF; this is probably a result of abnormal folate transport-binding systems in the central nervous system. Folate deficiency of dietary origin commonly occurs in the elderly.[43]

Phenytoin (Dilantin) therapy accelerates folate excretion

and interferes with folate absorption and metabolism. Alcohol interferes with folate's enterohepatic circulation, whereas the chemotherapeutic agent, methotrexate, inhibits the enzyme dihydrofolate reductase. Although decreased serum folate can occur with use of oral contraceptives (cycle-day dependent), this is not believed to cause a functional deficit unless some other problem is also present. There has, however, been concern expressed that there may be a relationship between oral contraceptive–induced cervical folate deficiency and cervical carcinoma. A specific folate-binding protein, possibly of granulocytic source, has been reported in the serum of a number of women who were receiving oral contraceptives or who were pregnant.

The therapeutic form of folate is 5-formyl-THF (also known as leucovorin, citrovorum factor, or folinic acid). This form of folate can bypass MeTHF and enter the cycles of folate's one-carbon transfer reactions (Fig. 36-16). This feature is useful in the ''leucovorin rescue'' of cancer patients given high-dose methotrexate therapy with toxic levels of methotrexate.

REFERENCES

1. Food and Nutrition Board: Recommended Dietary Allowances, ed 9, Washington, DC, 1980, National Academy of Science.
2. Naito, HK: Role of vitamin C in health and disease. In Brewster, MA, and Naito, HK, editors: Nutritional elements and clinical biochemistry, New York, 1980, Plenum Publishing Corp, pp 69-116.
3. Pesce, AJ and Kaplan, LA: Methods in clinical chemistry, St Louis, 1987, The CV Mosby Co.
4. Labbe, RF, editor: Clinics in laboratory medicine, vol 1, Laboratory assessment of nutritional status, Philadelphia, 1981, WB Saunders Co.
5. Briggs, MH, editor: Vitamins in human biology and medicine, Boca Raton, Fla, 1981, CRC Press, Inc.
6. Calabrese, EJ: Nutrition and environmental health, vol 1, The vitamins, New York, 1980, John Wiley & Sons, Inc.
7. Brewster, MA, and Naito, HK, editors: Nutritional elements and clinical biochemistry, New York, 1980, Plenum Publishing Corp.
8. Pereira, GR, and Zucker, A: Nutritional deficiencies in the neonate, Clin Perinatol 13:175-189, 1986.
9. Rosenthal, MJ, and Goodwin, JS: Cognitive effects of nutritional deficiency, Adv Nutr Res 7:71-100, 1985.

Vitamin A

10. Garry, PJ: Vitamin A. In Labbe, RF, editor: Clinics in laboratory medicine, vol 1, Laboratory assessment of nutritional status, Philadelphia, 1981, WB Saunders Co.
11. Shamberger, RJ: Vitamin A alterations in disease. In Brewster, MA, and Naito, HK, editors: Nutritional elements and clinical biochemistry, New York, 1980, Plenum Publishing Corp, pp 117-130.
12. Sklan, D: Vitamin A in human nutrition, Prog Food Nutr Sci 11:39-55, 1987.

Vitamin E

13. Bland, J: Lipid antioxidant nutrition. In Brewster, MA, and Naito, HK, editors: Nutritional elements and clinical biochemistry, New York, 1980, Plenum Publishing Corp, pp 139-167.
14. Sokol, RJ, et al: Frequency and clinical progression of the vitamin E deficiency neurologic disorder in children with prolonged neonatal cholestasis, Am J Dis Child 139:1211-1215, 1985.
15. Farrell, P.M., and Bieri, J.G.: Megavitamin E. supplementation in man, Am. J.Clin. Nutr. **28**: 1381, 1975.
16. Bieri, JG, Evarts, RP, and Thorp, S: Factors affecting the exchange of tocopherol between red cells and plasma, Am J Clin Nutr 30:686, 1977.
17. Ogumekan, AO: Vitamin E deficiency and seizures in animals and man, Can J Neurol Sci 6:43-45, 1979.

Vitamin K

18. Suttie, JW: Role of vitamin K in the synthesis of clotting factors. In Draper, HH, editor: Advances in nutritional research, vol 1, New York, 1977, Plenum Publishing Corp.
19. Hazell, K, and Baloch, KH: Vitamin K deficiency in the elderly, Gerontol Clin 12:, 1970.
20. Motohara, K, Endo, F, and Matsuda, I: Screening for late neonatal vitamin K deficiency by acarboxyprothrombin in dried blood spots, Arch Dis Child 62:370-375, 1987.

Vitamin D

21. Taylor, CB, and Peng, S: Vitamin D—its excessive use in the U.S.A. In Brewster, MA, and Naito, HK, editors: Nutritional elements and clinical biochemistry, New York, 1980, Plenum Publishing Corp, pp 132-138.

Ascorbic acid

22. Sauberlich, HE: Ascorbic acid. In Labbe, RF, editor: Clinics in laboratory medicine, vol 1, Laboratory assessment of nutritional status, Philadelphia, 1981, WB Saunders Co.
23. Englard, S, and Seifter, S: The biochemical functions of ascorbic acid, Ann Rev Nutr 6:265-304, 1986.
24. Mirvish, SS: Effects of vitamins C and E on *N*-nitroso compound formation, carcinogenesis and cancer, Cancer 58:1842-1850, 1986.

Riboflavin

25. Komindr, S, and Michaels, GE: Clinical significance of riboflavin deficiency. In Brewster, MA, and Naito, HK, editors: Nutritional elements and clinical biochemistry, New York, 1980, Plenum Publishing Corp, pp 685-698.
26. Bates, CJ: Human riboflavin requirements and metabolic consequences of deficiency in men and animals, World Rev Nutr Diet 50:215-266, 1987.
27. Pinto, J, Raiczyk, GB, Huang, YP, and Rivlin, RS: New approaches to the possible prevention of side effects of chemotherapy by nutrition, Cancer 58:1911-1924, 1986.

Pyridoxin

28. Vermaak, WJ, Bernard, HC, Potgieter, GM, and Theran, HD: Vitamin B_6 and coronary artery disease: epidemiological observations and case studies, Atherosclerosis 63:235-238, 1987.
29. Shuster, K, Bailey, LB, and Madan, CS: Vitamin B_6 status of low-income adolescent and adult pregnant women and the condition of their infants at birth, Am J Clin Nutr 34:1731, 1981.

Niacin

30. Wahlqvist, ML: Effects on plasma cholesterol of nicotinic acid and its analogues. In Briggs, MH, editor: Vitamins in human biology and medicine, Boca Raton, Fla, 1981, CRC Press, Inc, pp. 81-84.
31. Glueck, CJ: Nonpharmacologic and pharmacologic alteration of high-density lipoprotein cholesterol: therapeutic approaches to prevention of atherosclerosis, Am Heart J 110:1107-1115, 1985.
32. Gentile, S, et al: Comparison of nicotinic acid—and caloric restriction-induced hyperbilirubinaemia in the diagnosis of Gilbert's syndrome, J Hepatol 1:537-543, 1985.

Thiamin

33. Flint, DM, and Prinsley, DM: Vitamin status of the elderly. In Briggs, MH, editor: Vitamins in human biology and medicine, Boca Raton, Fla, 1981, CRC Press, Inc, pp 65-80.
34. Blass, JP: Thiamin and the Wernicke-Korsakoff syndrome. In Briggs, MH, editor: Vitamins in human biology and medicine, Boca Raton, Fla, 1981, CRC Press, Inc, pp 107-136.
35. Poloni, M, Mazzarello, P, Patrini, C, and Pinelli, P: Inversion of T/TMP ratio in ALS: a specific finding? Ital J Neurol Sci 7:333-335, 1986.

Biotin

36. Roth, KS, Allen, L, Yang, W, et al: Serum and urinary biotin levels during treatment of holocarboxylase synthetase deficiency, Clin Chim Acta 109:337, 1981.

37. Wolf, B, et al: Biotinidase deficiency: initial clinical features and rapid diagnosis, Ann Neurol 18:614-617, 1985.
38. Roth, KS: Biotin in clinical medicine: a review, Am J Clin Nutr 34:1967, 1981.

Pantothenic acid

39. Schwabedal, PE, Pietrzik, K, and Wittkowski, W: Pantothenic acid deficiency as a factor contributing to the development of hypertension, Cardiology 71:(Suppl 1):187-189, 1985.

Vitamin B_{12} and folic acid

40. Steinkamp, R.C.: Vitamin B_{12} and folic acid: clinical and pathophysiological considerations. In Brewster, MA, and Naito, HK, editors: Nutritional elements and clinical biochemistry, New York, 1980, Plenum Publishing Corp, pp 169-240.
41. Saxena, S, and Carmel, R: Racial differences in vitamin B_{12} levels in the United States, Am J Clin Pathol 88:85-87, 1987.
42. Ho, CH, Chang, HC, and Yeh, SF: Quantitation of urinary methylmalonic acid by gas chromatography mass spectrometry and its clinical applications, Eur J Hematol 38:80-84, 1987.
43. Infante-Rivard, C, Krieger, M, Bascon-Barre, M, and Rivard, SE: Folate deficiency among institutionalized elderly: public health impact, J Am Geriatr Soc 34:211-214, 1986.

CHAPTER 37

Pregnancy and fetal function

PAUL T. RUSSELL

OBJECTIVES

- Describe amniotic fluid including its formation, functions, and normal constituents.
- Describe normal and pathological maternal biochemical changes that occur during pregnancy.
- Describe fetal biochemical changes that occur during prenatal development.
- Describe pathological conditions associated with pregnancy or the perinatal period and list expected levels of significant laboratory analytes.

KEY TERMS

anencephaly A defective development of the brain wherein the cerebral and cerebellar hemispheres are absent.

blastocyst An early stage in embryonic development characterized by a fluid-filled cavity within the cell mass covered by the trophoblast.

hemopoietic Related to the process of formation and development of the various types of blood cells.

hydramnios The presence of an excessive amount of amniotic fluid.

keratinization The process in skin development and differentiation whereby keratin, a proteinaceous substance, is produced.

meningomyelocele A protrusion of the membranes and spinal cord resulting from a defect in the vertebral column.

spina bifida A developmental abnormality characterized by defective closure of the spinal cord.

transudation The passage of a substance through a membrane as a result of a difference in hydrostatic pressure.

trophoblast The cell layer covering the blastocyst, which erodes the inner lining of the uterus during the process of implantation. Trophoblastic cells do not become part of the embryo itself but contribute to the formation of the placenta.

ANATOMICAL AND PHYSIOLOGICAL INTERACTION OF MOTHER AND FETUS

Placenta and fetal membranes

In the human the fertilized egg enters the uterus 3 days after ovulation and implants itself into the uterine lining 2 to 5 days later. After implantation a series of complex and remarkable relationships unfold that establish a supportive environment for the developing embryo and later the fetus.

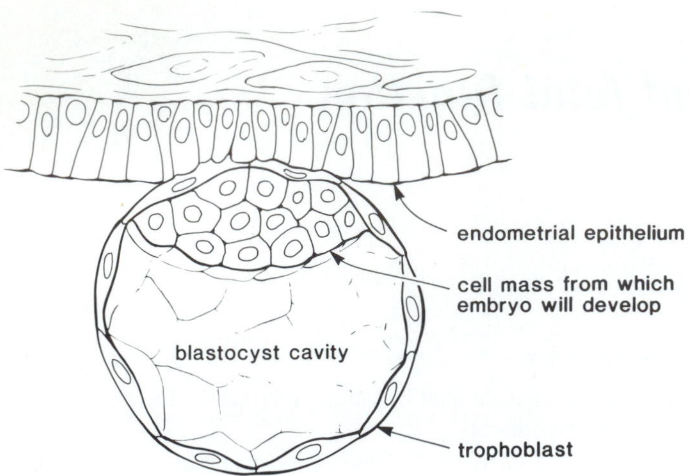

Fig. 37-1 Attachment of blastocyst to endometrial wall.

The capsulelike wall of the blastocyst, the trophoblast, both secretes hormones and establishes an intimate nutritive relationship with the uterus (Fig. 37-1). The invasion of the endometrium by the trophoblast is not a one-sided phenomenon because the maternal tissues also adapt to the new relations and demands. Both sets of coordinated changes lead to the production of a specialized organ, the placenta.

As the demands for nourishment and oxygen by the rapidly growing embryo increase, the trophoblast increases its surface area by forming villi. The surface area of these villi is enormous, and from the villi the fetal circulation of the placenta is established. An innermost membrane, the amnion, immediately surrounds the embryo and is fluid filled. This fluid, referred to as the *liquor amnii* or *amniotic fluid,* bathes the fetus, thereby preventing desiccation, and buffers the fetus against physical shocks.

Fetal growth and nutrition

The fetus grows and develops, using the nutrients provided by maternal blood. The rate of fetal growth is shown in Table 37-1. Gestation is calculated from the last menstrual period. Unlike the *infant* whose diet is a complex mixture of carbohydrates, fats, and proteins, the *fetus* has a diet consisting largely of glucose and sufficient quantities of amino acids to satisfy the nitrogen requirements of protein synthesis. Also included are small amounts of materials, such as fatty acids, vitamins, and minerals, that are essential to normal growth and function. Glucose provides the energy needed for the synthesis of tissues.

The source of glucose for the fetus is maternal blood. The amount of glucose available to the fetus depends on the concentration of glucose in maternal blood, which is regulated by the action of a number of hormones, including insulin and the placental hormone, human chorionic somatomammotropin (also called *human placental lactogen,* or HPL).

Table 37-1 Correlation between length of gestation and embryonic and fetal weight

	Gestation (weeks)	Weight (g)	Weight at 40 weeks (%)
Embryo	7	0.07	0.002
	8	0.22	0.006
	10	3.5	0.1
	12	11	0.3
	14	33	1
	16	80	2
	18	170	5
	20	316	9
	22	460	14
	24	630	19
	26	823	24
Fetus	28	1045	31
	30	1323	39
	32	1680	49
	34	2074	61
	36	2478	73
	38	2914	86
	40	3405	100

Glucose passes readily from the maternal to the fetal circulation by means of ''facilitated diffusion,'' crossing the placenta at a faster rate than would be expected on physiological grounds alone. A rise in the glucose concentration in maternal blood is followed rapidly by a comparable increase in its concentration in fetal blood. The two levels do not become equal, however, and a concentration gradient from mother to fetus is always present. In addition, a significant proportion of glucose is consumed by the placenta to meet its own energy requirements.

Since glucose levels in the fetus mirror those in the mother, there is neither need nor opportunity for the fetus to regulate its own blood glucose concentrations. Mechanisms for doing so do develop during the fetal period but are largely dormant until birth, when the supply of glucose through the placenta ends abruptly. Nevertheless, two important processes are active from an early stage. The first is the storage of glucose as glycogen or fat to provide for the metabolic needs of the newborn until feeding begins. The second is the control of the rate at which glucose is used by the growing tissues, and this is attributable primarily to the action of insulin secreted by the fetal pancreas beginning at 8 to 9 weeks' gestational age.[1]

Role of placenta in gas exchange

To meet its metabolic needs, the fetus is completely dependent on a continuous delivery of oxygen across the placenta. Transplacental exchanges, including those of gases, depend on both perfusion and permeability. Placental perfusion is a composite of uterine and umbilical blood flows, whereas permeability is a characteristic of the placental membrane. Under normal conditions with a well-oxygenated mother, oxygen transport across the placenta is primarily regulated by blood flow. To meet the increased demands of the growing fetus, uterine blood flow and placental membrane permeability increase during gestation. In the presence of maternal hypoxia, oxygen transport across the placenta becomes limited primarily by the permeability of the placental membrane.

Carbon dioxide rapidly diffuses across the placenta in either direction. This allows the fetus to maintain a normal PCO_2 and the mother to eliminate carbon dioxide. The placenta has limited permeability to bicarbonate ions. The placenta, therefore, allows the fetus to dispose of carbon dioxide while protecting it from maternal metabolic acidosis.[2]

Formation of amniotic fluid

Amniotic fluid volume at any point in time is the result of a dynamic balance between production and removal.[3] Amniotic fluid originates from multiple sources, and the relative contribution from each source varies, depending on the stage of fetal development. The multiple sources include the placenta, fetal kidneys, skin, membranes, lungs, and intestine. In the first half of pregnancy, amniotic fluid forms as a transudate from the skin surface of the fetus. The composition of amniotic fluid is similar to that of extracellular fluid, and the amniotic fluid should be considered as an extension of the fetal extracellular fluid space.[4] Later in pregnancy, fetal kidneys and lungs assume the major role in amniotic fluid formation, and its volume now depends on a balance between fetal urination and volume of amniotic fluid that is swallowed.[5]

Fluid moving from the trachea and pharynx into the esophagus can enter the amniotic cavity. This provides an explanation for the appearance of pulmonary surfactant in amniotic fluid. Respiratory movements by the fetus readily mix fluid with surfactant because the movements produce a tidal volume exchange (in and out) of about 600 to 800 mL/day[6] through the fetal lungs throughout the third trimester.

Amniotic fluid disappearance is in part effected by fetal swallowing. It is estimated that between 200 and 450 mL of amniotic fluid per day flows out from the amniotic cavity by this route, accounting for about half of the daily urine production of the fetus. Since the amniotic cavity gains a fluid volume of no more than 10 mL/day in the third trimester (the total solute concentration always remains in the normal range), a sizable quantity of urine must be reabsorbed by other pathways.[7]

Amniotic fluid volume increases throughout pregnancy until it reaches a maximum volume at about 36 weeks of gestation. Normal variations in volume throughout pregnancy are large (Table 37-2).

Table 37-2 Amniotic fluid volume in normal pregnancy

Gestational age (weeks)	Volume of fluid (mL)
12	5-200
14	50-200
16	150-300
18	200-400
20	225-775
22	300-500
24	500-675
26	500-700
28	500-875
30	400-1300
32	400-1375
34	500-1350
36	525-1500
38	300-1525
40	325-1450
42	600

BIOCHEMISTRY OF AMNIOTIC FLUID

Excellent monographs on the subject of amniotic fluid are available that contain compendiums of the biochemical constituents of amniotic fluid.[8,9]

Water, electrolytes, and nitrogenous products

Because of the shift in the source of amniotic fluid that occurs about midway through pregnancy, constituents of amniotic fluid also change during gestation. Before keratinization of the skin, amniotic fluid can result as a transudation from the surface of the fetus. After keratinization and with progressive development of the renal system, fetal urine makes a more prominent contribution to the amniotic fluid compartment. The biochemical composition of amniotic fluid therefore reflects the routes of formation of the fluid and is related to the developmental stage of the fetus.

Amniotic fluid is isotonic during early pregnancy but by term becomes moderately hypotonic (mean total solute concentration, 255 mOsm/kg of water) compared with fetal and maternal plasma. This changing concentration of amniotic fluid reflects the maturation of fetal renal function.[10] The osmotic and oncotic pressures in fetal and maternal tissues cause the transfer of water from mother to fetus to amniotic fluid, and then back to the mother.[11] It has been calculated that the net transfer of water from mother to fetus reaches 20 to 25 mL/day in late pregnancy.[12]

Amniotic fluid concentrations of nitrogenous products of metabolism, creatinine, urea, and uric acid increase toward term (see below) and are manyfold lower than concentrations found in maternal urine. The composition difference between amniotic fluid and maternal urine is readily measurable and can be used to determine whether a sample obtained from vaginal leakage or an errant amniocentesis is amniotic fluid.[13]

Proteins

Proteins derived from many sources have been identified in amniotic fluid. Under normal conditions, amniotic fluid proteins of fetal origin come from the skin, the urinary and gastrointestinal tracts (urine and meconium, respectively), and the respiratory tract. Proteins from the respiratory tract are often part of the proteolipid product secreted by type II epithelial cells of the lung. These products function as components of the lung surfactant system.[14] Proteins of maternal origin can enter amniotic fluid by transudation across the amnion. Under abnormal circumstances, unusual avenues for protein exchange can occur, such as central neural tube development defects, which increase the alpha-fetoprotein levels in amniotic fluid.

Over 50 enzymes have been identified in amniotic fluid,[8,9] but the origin of many of these enzymes and their significance in the fluid are not understood. The enzymes fall into two categories, those whose activity is maximally expressed early in pregnancy (12 to 20 weeks) and those active at the latter stage of pregnancy (35 to 40 weeks). Some enzymes of fetal origin are used in the diagnosis of particular inborn errors of metabolism.

Hormones

Examples of the hormones that have been identified in amniotic fluid are listed in the box below. This list includes steroid, polypeptide, and amino acid–derived hormones such as the catecholamines. Although many of these hormones are products of urinary or biliary excretion from the fetus, a few have clinical usefulness and are discussed later. A more extensive list is available.[8,9]

HORMONES IDENTIFIED IN AMNIOTIC FLUID

Protein and polypeptide
- Insulin
- Growth hormone
- Somatomedin
- Somatostatin
- Prolactin
- Follicle-stimulating hormone
- Luteinizing hormone
- Human chorionic gonadotropin
- Adrenocorticotropic hormone
- Endorphin
- Human placental lactogen
- Oxytocin
- Relaxin
- Thyrotropin
- Thyroxine
- Renin
- Angiotensin

Steroids
- Estriol
- Estradiol
- Estrone

Prostaglandins
- E_2
- $F_{2\alpha}$

MATERNAL BIOCHEMICAL CHANGES DURING PREGNANCY

Human chorionic gonadotropin (HCG)

The urine and serum of pregnant women contain high concentrations of human chorionic gonadotropin (HCG or hCG) produced by the trophoblast, which provide the basis of tests for the diagnosis of pregnancy. Specific and sensitive analytical methods for the β-chain subunit of HCG permit the detection of pregnancy as early as 8 days after ovulation (1 day after implantation). HCG concentrations climb early in pregnancy, reaching a maximum by 8 to 10 weeks of gestation (Fig. 37-2).

HCG is one of a family of closely related glycoprotein hormones that regulates reproductive and metabolic functions. This family includes follicle-stimulating hormone (FSH), luteinizing hormone (LH), and thyroid-stimulating hormone (TSH). These hormones are composed of two polypeptide subunits referred to as alpha (α)- and beta (β)-chains and each contains carbohydrate; thus the term *glycoprotein* is used. The α-chains of HCG, LH, FSH, and TSH are similar in their amino acid sequences, a finding that accounts for the immunological similarity of these protein hormones.[15]

The β-chains of the glycoprotein hormones each have a unique amino acid sequence, which gives these hormones their biological specificity. HCG has amino acid sequences in both the α- and β-chains that are virtually identical to those in LH, differing to a significant degree only in carbohydrate content, with HCG containing about 30% carbohydrate by weight and LH approximately half that amount. This similarity allows HCG to be used instead of LH for such purposes as ovulation induction. The carbohydrate content difference perceptibly influences the metabolic patterns of these hormones; thus the biological half-life of HCG is about 6 hours, whereas the biological half-life of plasma LH is 12 to 45 minutes.[16] These differences in metabolic clearance correlate well with our understanding of the functions of these hormones, with LH regulating the complicated process of ovulation and subsequent formation of the corpus luteum and HCG providing continued stimulation of the corpus luteum to ensure uninterrupted progesterone production until the placenta can provide sufficient progesterone to maintain the pregnancy.

Estrogens

Nearly a half century ago it was recognized that a substantial increase in estrogen excretion accompanied pregnancy. The predominant estrogen was identified as estriol, not the usual ovarian estrogens, estradiol or estrone (Fig. 37-3). It was not until many years later that the unique relationships were elucidated that described the pathway

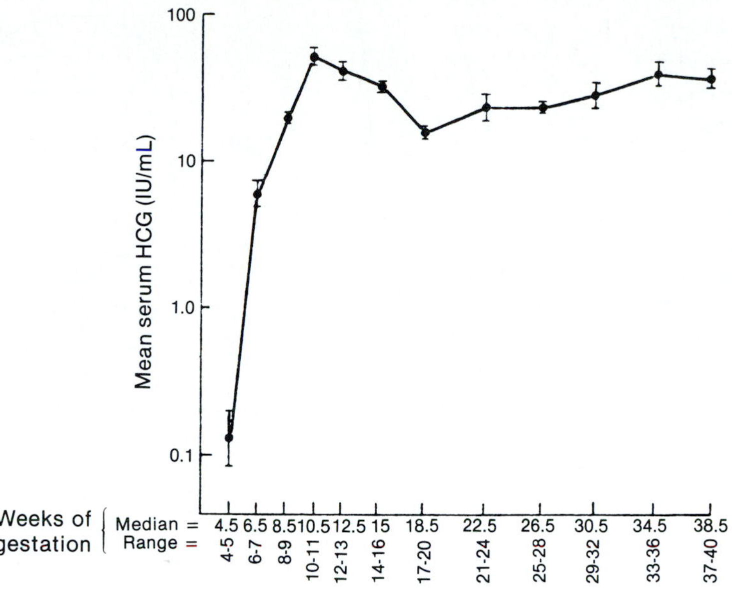

Fig. 37-2 Serum chorionic gonadotropin (HCG) concentrations during pregnancy. *(From Goldstein DP, Aon T, Taymor MT, et al: Am J Obstet Gynecol 102:110-114, 1968.)*

for estriol formation.[17] Estriol formation during pregnancy occurs only in the higher primates, and the obligatory interaction of fetus and placenta required for the formation of this "estrogen of pregnancy" provides the basis for one of the few biochemical tests available to monitor fetal well-being.

Estrogen formation proceeds in an obligatory sequence of reactions that converts cholesterol to progestins, then to androgens, and then to estrogens (refer to Chapter 43). In the ovary, this sequence occurs solely within that organ. In pregnancy, for this sequence to occur, a complementary relationship between the placenta, the fetal adrenal cortex, and the fetal liver leads to estriol formation. This unique relationship is referred to as the *fetoplacental unit*[18] (Fig. 37-4).

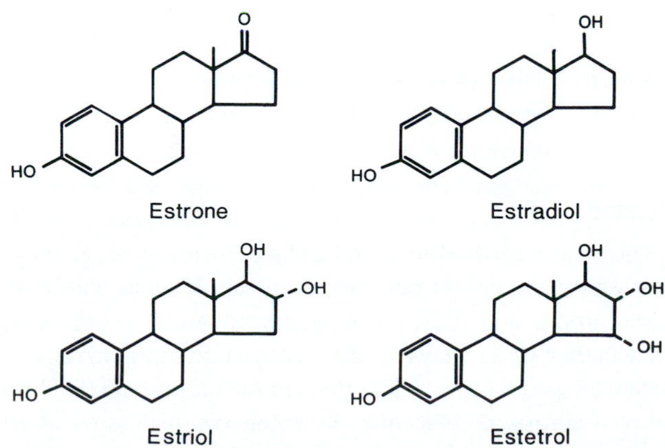

Fig. 37-3 Structures of estrogens. *Dashed lines,* α-Stereoconfigurations; *solid lines,* β-stereoconfigurations of hydroxyl groups.

The fetal adrenal gland synthesizes dehydroepiandrosterone (DHEA), which is converted to 16α-hydroxy-DHEA by the fetal liver. The 16α-hydroxy-DHEA is converted to estriol by placental aromatizing enzymes. In addition, the placenta possesses a very active sulfatase. Sulfation-desulfation of estriol appears to be integrally involved in the transfer of steroids from the placenta, since, in conditions where the placental sulfatase is absent, low maternal estriol levels are characteristically observed.[19]

Estriol comprises over 90% of the known maternal estrogens of pregnancy. It is metabolized to both sulfate and glucuronide conjugated forms by maternal liver. These conjugates are the primary excretory forms of estriol. Concentrations in maternal serum increase with advancing gestation and reach nearly 40 ng/mL at term.[20] Fig. 37-5 indicates the normal increase in plasma estriol and a pattern seen with a diabetic patient, with fetal death, and with growth retardation. Estriol is also found in amniotic fluid.[21]

The functional role for estriol in pregnancy has prompted much speculation. In many biological test systems, estriol is a weak estrogen, demonstrating only one hundredth the potency of estradiol and one tenth the potency of estrone per unit weight. However, estriol can be demonstrated to be equipotent to estradiol in its ability to promote uteroplacental blood flow. For this reason, its role in pregnancy may be to ensure optimum blood flow in the gravid uterus.[22]

More recently, it has been suggested that levels of the estrogen *estetrol* (Fig. 37-3) offer more information than estriol levels about the status of the fetus in utero.[23] Este-

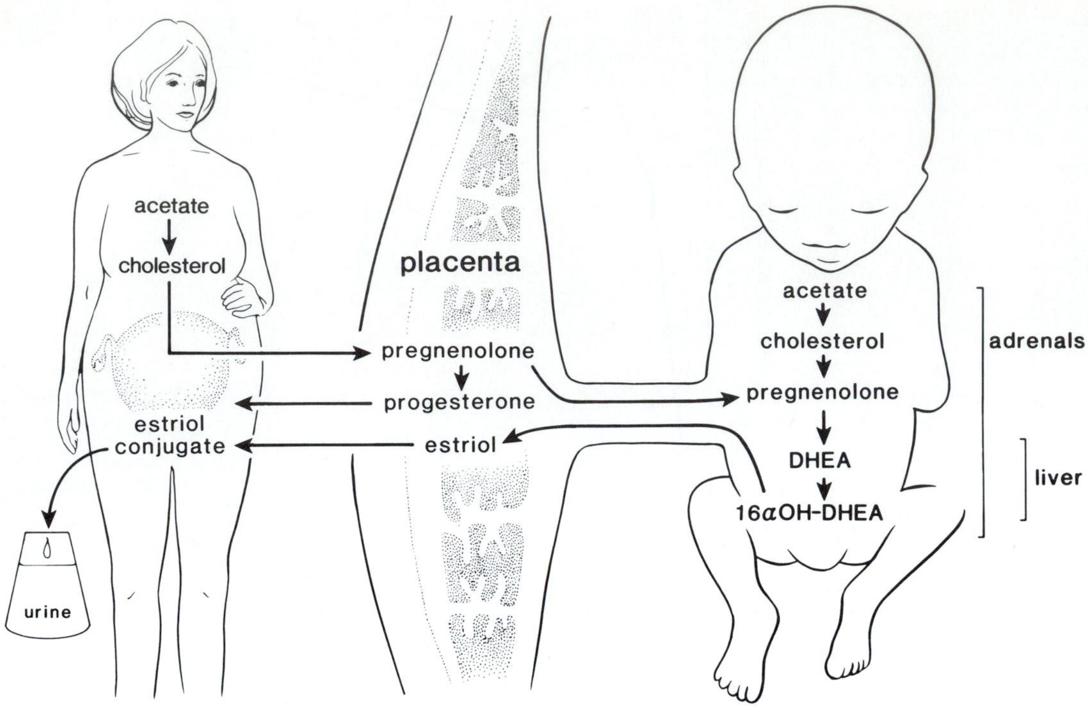

Fig. 37-4 Scheme of fetoplacental unit. *DHEA,* Dehydroepiandrosterone; *16α-OH-DHEA,* 16α-hydroxydehydroepiandrosterone.

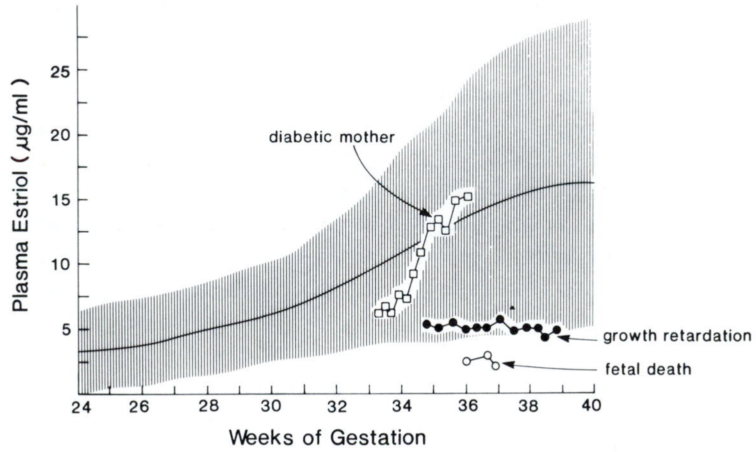

Fig. 37-5 Mean *(solid line)* and estimated 5th and 95th percentiles *(shaded area)* for plasma unconjugated estriol during normal pregnancy. Estriol patterns from three actual pregnancy conditions are shown.

trol (E_4) is not metabolized further as estriol is in the maternal liver. In addition, E_4 levels specifically reflect fetal liver activity, whereas E_3 levels depend on the fetal adrenal glands. Once E_4 is formed by fetal liver, it enters the maternal circulation and is excreted in urine as E_4-glucuronide. The presumed advantage of E_4 measurement is based on the belief that this estrogen reflects fetal activity more directly. Clinical evaluations of fetal well-being have not shown a clear advantage of estetrol over estriol, however.[24]

Thyroid

The maternal thyroid gland enlarges during pregnancy. The reason for this is not entirely clear. Human chorionic gonadotropin and TSH have nearly identical α-subunits, but whether HCG plays a role in thyroid enlargement is a matter of speculation.[25] The thyroid hormones and TSH do not pass across the placenta. An extensive discussion of all facets of thyroid and parathyroid function in pregnancy can be found elsewhere[25] and in Chapters 41 and 44.

Increased estrogen production during pregnancy induces

the synthesis of thyroxin-binding globulin (TBG) by maternal liver.[25,26] The concentration of TBG is doubled by the end of the first trimester, and it remains elevated throughout the rest of pregnancy. Because of the increased number of binding sites available for thyroid hormones, the total amount of both thyroxine (T_4) and triiodothyronine (T_3) is increased in the blood; this increase is greater for T_4 than for T_3, which is generally only raised in the last trimester. Pregnant women are generally euthyroid, although hyperthyroxinemic.

Serum lipids

Hyperlipidemia develops with normal pregnancy. Total serum lipid concentration rises progressively from the end of the first trimester and is 40% to 50% above nonpregnant levels at term.[27] All components of the serum lipids are increased, but the triglyceride fraction shows the largest proportionate rise.[28]

Serum proteins and liver function

The total concentration of serum proteins decreases by about 1 g/L during pregnancy. Most of the decrease occurs during the first trimester. The decrease is mainly in serum albumin.[29] $Alpha_1$, $alpha_2$, and beta globulins rise slowly and progressively. Gamma globulin probably decreases slightly. The IgG component, which is the major immunoglobulin transferred to the fetus, falls progressively.[30]

Throughout pregnancy fibrinogen increases progressively, and values at term are 30% to 50% above nonpregnant levels. Clotting factors VII, VIII, IX, and X are also increased,[31] whereas prothrombin and factors V and XII are reduced.[32] Alterations that occur in the levels of clotting factors and plasminogen are probably brought about by estrogen action on the liver.

Under the influence of increased estrogens, the maternal liver increases the synthesis of transcortin (corticoid-binding globulin, CBG). This results in total cortisol levels increasing during pregnancy, almost doubling by late pregnancy. However, the levels of free, active cortisol are normal.

A number of liver function tests change as a consequence of normal pregnancy. For example, nonspecific alkaline phosphatase activity in serum nearly doubles during normal pregnancy and can reach levels that would be considered abnormal in the nonpregnant woman. Much of this increase is attributable to isoenzymes of this enzyme originating from the placenta.

Glucosuria

Glucosuria is common in healthy, pregnant women. Glucose excretion rises very early in pregnancy, reaching a peak between 8 and 11 weeks of gestation.[33] The degree of glucosuria varies thereafter. The cardinal feature of the glucosuria of pregnancy is a conspicuous variability both from day to day and during the course of a day. The cause of glucosuria in pregnancy appears to be the reduced efficiency of the kidneys to reabsorb glucose.[34]

From these comments it can be seen that pregnancy is potentially diabetogenic. Diabetes mellitus may be aggravated by pregnancy, and clinical diabetes may appear in some women only during pregnancy. The renal processing of glucose during pregnancy is of particular interest because of the frequent appearance of clinical glycosuria and the necessity to differentiate this "renal glycosuria" from that of pregnancy-aggravated diabetes mellitus. Because of the many variables associated with an altered renal physiology in pregnancy, pregnant diabetic women should be monitored by blood sugar levels because urine testing can yield misleading values. Screening for gestational diabetes by a glucose challenge test has become routine. It is important to maintain normal blood glucose levels to prevent perinatal morbidity associated with gestational hyperglycemia.

FETAL BIOCHEMICAL CHANGES DURING PRENATAL DEVELOPMENT

Liver function

Fetal liver contains a large number of hematopoietic cells that disappear after birth. In very early fetal life, the liver is the major blood-forming organ, but by 22 to 24 weeks of gestation the bone marrow has assumed the major responsibility for the formation of blood. Because of widely varying amounts of the different cell types in the newborn liver, enzyme-activity patterns are markedly different from the adult.

The fetal yolk sac and, later, the fetal liver produce alpha-fetoprotein (AFP), which is released into the fetal circulation. It passes from the bloodstream by way of the urine to amniotic fluid. It is normally removed from amniotic fluid by fetal swallowing. Alpha-fetoprotein appears in maternal serum throughout gestation (Fig. 37-6), where it can be easily measured to screen for neural tube defects.

The fetus' need for increased quantities of amino acids for protein synthesis is satisfied by placental transport. This is an active process against a concentration gradient that depends on placental blood flow and, to a lesser degree, on the concentration of amino acids in the maternal plasma.[35]

Because maturation of the fetal liver is not complete by the time of birth, some jaundice regularly occurs in virtually every newborn during the first week of life. This is known as *physiological jaundice*. The yellow pigmentation, or jaundice, is caused by bilirubin that comes from the normal destruction of red blood cells. However, since the immature fetal liver has not developed its full capability to clear bilirubin from the blood, jaundice occurs.

Renal function

The primary function of the mammalian kidney is to maintain water and electrolyte homeostasis. This is accom-

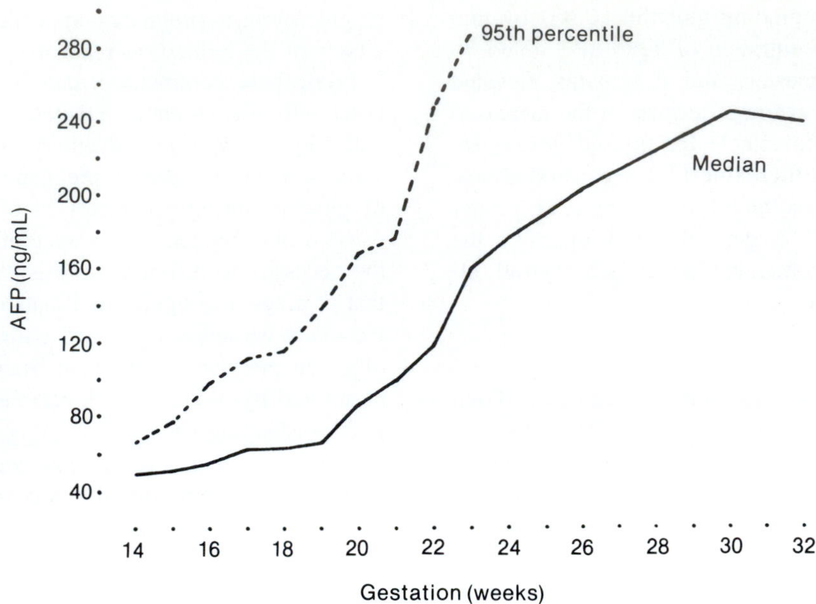

Fig. 37-6 Median and 95th percentile of alpha-fetoprotein, *AFP*, in maternal serum. *(From Crandall BF: In Kirkpatrick AM, and Nakamura RM, editors: Alpha-fetoprotein: laboratory procedures and clinical applications, New York, 1981, Masson Publishing Co.)*

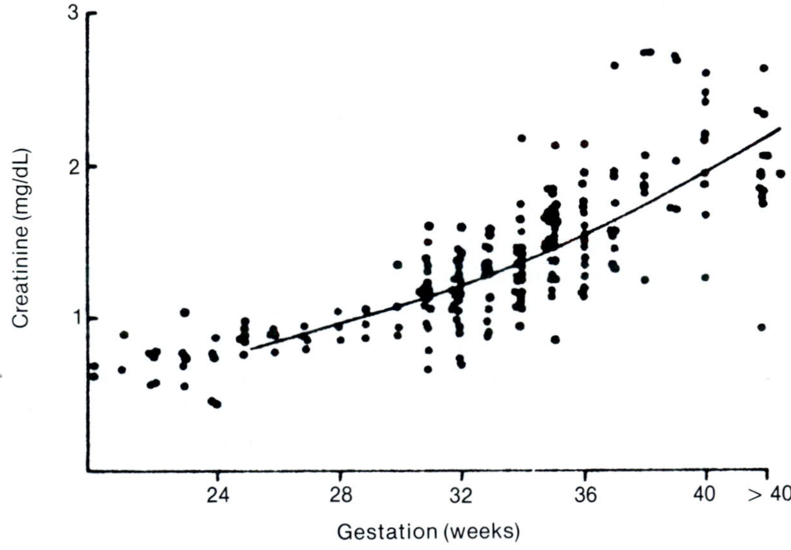

Fig. 37-7 Distribution and regression curve of amniotic fluid creatinine concentration in milligrams per deciliter. *(From Lind T: In Fairweather DVI, and Eskes TKAB, editors: Amniotic fluid: research and clinical application, ed 2, Amsterdam, 1978, Excerpta Medica.)*

plished by selective excretion or retention of water and solutes as conditions dictate. In the fetus, body water and electrolyte balance are maintained largely by the placenta. For this reason fetuses without functional kidneys often show no water or electrolyte abnormalities. Thus the full development of mature renal function occurs when it is needed—after birth.[36]

Organic nitrogenous compounds such as urea, uric acid, and creatinine (Fig. 37-7) gradually increase in concentration in amniotic fluid as the renal system of the fetus matures. In early pregnancy these compounds are present in amniotic fluid in concentrations similar to concentrations in maternal and fetal blood. Concentrations increase gradually to become significantly higher than levels in maternal or fetal blood. A sharp rise in creatinine at about the thirty-seventh week of gestation elevates the amniotic fluid concentrations of urea and creatinine to levels two to three times higher than those of normal serum.[37]

Table 37-3 Composition (by weight percent) of pulmonary surfactant[39]

Constituent	Percent
Phospholipids	85
Saturated phosphatidylcholine	60
Unsaturated phosphatidylcholine	20
Phosphatidylglycerol	8
Phosphatidylinisitol	2
Phosphatidylethanolamine	5
Sphingomyelin	2
Others	3
Neutral lipids and cholesterol	5
Proteins	10

Lung development

At birth there is an abrupt physiological transition that requires the newborn infant to assume vital functions that were previously handled by the maternal circulation. The lung is shifted from a fluid-filled organ to a gas-exchange system in a few brief minutes. This functional transition is possible only if sufficient maturation of the fetal lung has occurred during development. The lungs should have developed alveoli with adequate surface area for gas exchange; also sufficient surfactant must be available to sustain the ventilatory movements needed for pulmonary function. These processes are highly organized and are coordinated by the timing of anatomical and biochemical events.

Surfactant is composed principally of phospholipid containing highly saturated fatty acid moieties.[38,39] Table 37-3 lists its composition. In addition to the highly unusual saturated lecithins, other important constituents of the surface-active system include phosphatidyl glycerol[39,40] and "surfactant apoprotein."[14,41] Gluck and Kulovich[42] were the first to correlate phospholipid concentrations in amniotic fluid to the functional status of the fetal lung. The lecithin/sphingomyelin ratio (L/S) (Fig. 37-8) is used widely for that purpose. Detailed accounts of the function and composition of pulmonary surfactant[43-45] and the accepted nomenclature have been published.

Surfactant facilitates pulmonary function in at least two ways: it maintains alveolar stability by preventing collapse of the terminal respiratory tree, and it reduces the pressure that is needed to distend the lungs in the initial phase of inspiration. Infants who develop the respiratory distress syndrome have a higher surface tension at the alveolar air-liquid interface as a result of a pulmonary surfactant deficiency. Infants with RDS require increased respiratory effort. Clinical management is aimed at maintaining the infant in an oxygen-rich environment and keeping the alveoli open artificially during spontaneous or mechanically assisted breathing until such time as surfactant synthesis is adequate for unassisted respiration.[43]

Surfactant is formed in the large alveolar epithelial cells

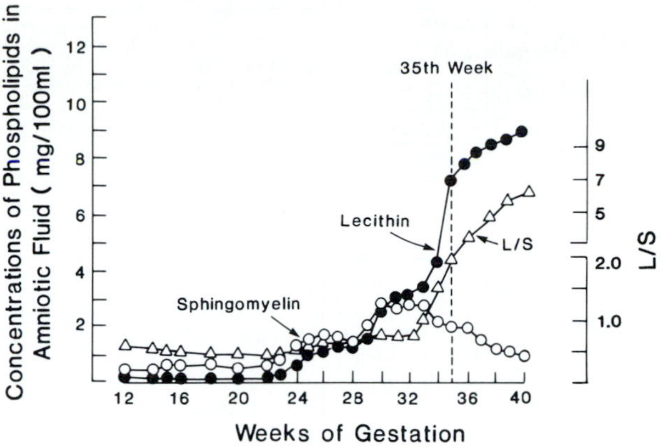

Fig. 37-8 Lecithin, sphingomyelin, and lecithin/sphingomyelin ratios in amniotic fluid during normal pregnancy. *(Adapted from Gluck L, and Kulovich MV: Am J Obstet Gynecol 115:539-546, 1973.)*

known as type II pneumocytes, which comprise about 10% of the cells of the lung. Synthesis of the phospholipid component takes place by the CDP-choline pathway.[46] Cortisol stimulates the process of pulmonary maturation, and exogenously administered synthetic corticoids have been used clinically to hasten pulmonary maturation under circumstances where delivery of a preterm infant is imminent.[47]

Hemoglobin

Embryos have a hemoglobin that is unique to the embryonic stage of development, which is replaced during fetal life by "fetal hemoglobin" (Hb F) and finally by adult hemoglobin (Hb A).[48] The pattern of hemoglobins formed during development is presented in Fig. 37-9. Fetal

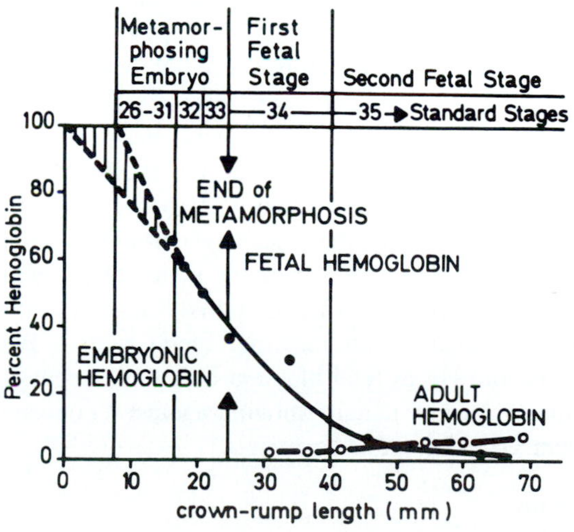

Fig. 37-9 Relationship between hemoglobin types and developmental stages in early human life. *Dashed lines and hatched area,* Expected development. *(From Kleihauer E: In Stave U, editor: Perinatal physiology, New York, 1978, Plenum Publishing Corp.)*

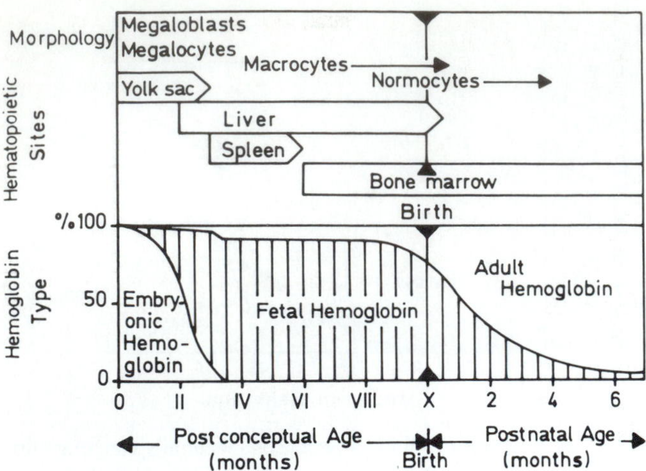

Fig. 37-10 Developmental changes in hematopoietic sites, red blood cell morphology, and hemoglobin types. *(From Kleihauer E: In Stave U, editor: Perinatal physiology, New York, 1978, Plenum Publishing Corp.)*

hemoglobin has been found to comprise 34% of the total in an embryo just less than 7 weeks of age.[49] By approximately 10 weeks of gestation, the embryonic hemoglobins decrease to 10% of the total hemoglobin present. Before the end of the first trimester (less than 12 weeks) the Hb F has increased to approximately 90% of the total, with the remaining percentage constituted by Hb A. From this point on Hb F remains constant until about the thirty-sixth week of gestation when there is a decline. The decline is primarily caused by an increase in Hb A synthesis rather than by a decrease in Hb F. Sharp increases in Hb A are seen in reticulocytes and erythrocytes by birth. The developmental changes in hematopoietic sites, the red cell morphology, and the hemoglobin types are shown in Fig. 37-10.

The physiological differences in the hemoglobins are summarized by Kleihauer[48] and include differences in affinities for oxygen; resistance to acid, base, and heat denaturation; and electrophoretic and chromatographic properties. The higher affinity that fetal hemoglobin has for oxygen is reflected in the fetal oxyhemoglobin saturation curve, which lies to the left of the maternal oxyhemoglobin curves under standard conditions.[50] In fact, when oxygen is diffusing from maternal blood with a P_{CO_2} of about 34 mm Hg to fetal blood at a P_{CO_2} of about 30 mm Hg, the oxygen is actually moving against a concentration gradient.[51]

Bilirubin

Erythrocyte destruction precedes the formation of bilirubin. Biliverdin is the principal and initial degradation product of hemoglobin and an important intermediate pigment in the formation of bilirubin. These relationships are presented in simplified terms:

$$\text{Hemoglobin} \xrightarrow[\text{Globin}]{} \text{Biliverdin} \xrightarrow{\text{Biliverdin reductase}} \text{Bilirubin}$$

Plasma concentrations of bilirubin are usually low in the fetal circulation except in unusual circumstances, such as in severe erythroblastosis fetalis. Even in circumstances where the rapid breakdown of erythrocytes leads to accelerated bilirubin production, as in severe maternal-fetal blood group incompatibility, cord blood bilirubin rarely exceeds 50 mg/L. This fact attests to the rapid, efficient transfer of this pigment across the placenta and the equally efficient disposal of fetal bilirubin by the mother.

After birth, the newborn loses the placental mechanism for bilirubin removal. As a result, there is a modest accumulation of unconjugated bilirubin in the plasma. The jaundice resulting from this change in physiological circumstance is related to the limited uptake, conjugation, and excretion of bilirubin by the immature liver. The degree of neonatal jaundice occurring at birth depends on the maturity and health of the fetal liver at birth. A discussion of the transport and liver metabolism of bilirubin can be found in Chapters 23 and 32.

PATHOLOGICAL CONDITIONS ASSOCIATED WITH PREGNANCY AND PERINATAL PERIOD

Placental disorders

Adequate exchange across the placenta between the maternal and fetal circulations is essential for normal fetal growth and metabolism.[51] Less than optimum quantities of nutrient result in small-for-gestational-age (SGA) fetuses, whereas excessive quantities of nutrient, as with maternal diabetes, result in large-for-gestational-age (LGA) infants.

Few pathological conditions involving the placenta exist where the monitoring of chemicals by the laboratory is useful. These instances, when they arise, usually take advantage of the unique hormone production by the placenta. One such example is the hydatidiform mole. Molar tissue is a developmental anomaly of the placenta that has the potential for malignant growth. It is the most common lesion antecedent to choriocarcinoma. Since the mole is trophoblastic tissue, HCG is produced, resulting in a positive pregnancy test. If serum or urinary HCG levels exceed values typical of specific times in pregnancy, a mole may be suspected. However, because of the variations in gonadotropin values possible for normal pregnancy, no single value can be established as the borderline between normal and abnormal. Because highly sensitive and specific methods are available for monitoring serum HCG, this hormone is useful in monitoring the response of hydatidiform moles to therapy.

Human placental lactogen (HPL) is produced by the placenta, and since its pattern of production is proportional to the size (mass) of the placenta (Fig. 37-11), this hormone can be used to gain insight into the growth of the placenta.

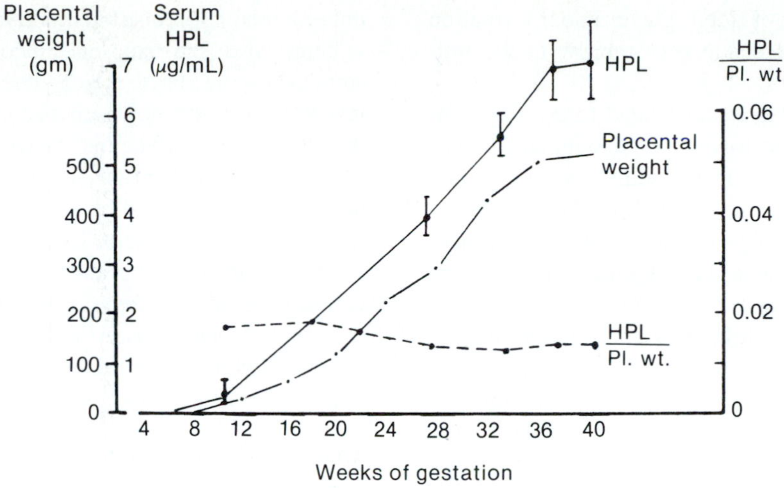

Fig. 37-11 Concentrations of human placental lactogen, *HPL*, during normal pregnancy. *(From Selenkow HA, Saxena SM, Dana CL, and Emerson K, Jr: In Pecile A, and Fenzi P, editors: The foeto-placental unit, Amsterdam, 1969, Excerpta Medica.)*

Fetal immaturity

The last of the organ systems to mature sufficiently to support extrauterine life is the lungs. Fetal "immaturity" generally implies a failure of the pulmonary system to sustain life. The respiratory distress syndrome (RDS) occurs most often when insufficient lung surfactant is present.[52] In most cases the deficiency in surfactant occurs as a consequence of the inability to synthesize or secrete surfactant in amounts sufficient for normal neonatal respiration.

When using methods of the type originally proposed by Gluck,[53] L/S values of 2 or greater have been found to correlate with fetal lung maturity. Before 34 weeks of gestation, lecithin and sphingomyelin are present in amniotic fluid in approximately equal amounts (Fig. 37-8), but at about 34 weeks, the concentration of lecithin begins to rise. It is possible to demonstrate (reference 54, for example) that when the concentration of lecithin in amniotic fluid becomes at least twice that of the sphingomyelin, the likelihood of respiratory distress after delivery is minimum. However, because of early reports that a greater risk of the RDS is associated with the diabetic pregnancy,[55,56] values of 2.5 or greater have often been used for these pregnancies. In addition, because the L/S method is less successful in predicting fetal lung maturity in diabetic pregnancies, tests for other surface-active lipids such as phosphatidylglycerol (PG), and more recently proteins associated with the surfactant complex,[57,58] have been developed and used either with the L/S ratio or solely as independent tests.

Therapy for newborns who have the RDS is basically supportive.[43] Thermoregulation and assisted ventilation are two primary measures. Because of inadequate ventilation, acid-base disturbances are monitored by use of pH, P_{CO_2}, and bicarbonate values. It is also customary to maintain a venous hematocrit of 40% to 45% during the acute phase of the respiratory distress syndrome to support an adequate oxygen-carrying capacity. Experience with the administration of aerosolized phospholipids to infants with the RDS is limited because of the problems associated with duplicating the composition of natural surfactant.[14]

Fetal hemolytic disorders (Rh problems)

Hepatic excretory capacity does not become fully mature until nearly 4 weeks post partum in full-term human infants. The hepatic processing of bilirubin therefore falls short of maximum before that time. When uptake and conjugation of bilirubin are forced to operate at rates exceeding the capacity of the liver to excrete the quantity formed, as in infants with severe hemolytic disease, conjugated bilirubin accumulates in liver and serum.

Erythroblastosis fetalis, caused by the incompatibility of an Rh-negative mother and an Rh-positive father, is a common cause of rapid red blood cell destruction. If the fetus is also Rh-positive, fetal cells entering the maternal circulation may elicit an antibody response to the Rh blood factor. The maternal antibodies (IgG) cross the placenta into the fetus where they destroy fetal red cells.

The fetus normally generates approximately 35 mg of bilirubin from the catabolism of 1 g of hemoglobin. The high maternal-to-fetal plasma protein gradient facilitates rapid transplacental extraction of unconjugated fetal bilirubin and at the same time suppresses glucuronide conjugation by the fetal liver. The transferred bilirubin is so efficiently conjugated and excreted by the mother that it is uncommon for the neonate to have an elevated cord blood bilirubin level. However, in severe erythroblastosis, particularly if coupled with placental deterioration, unconjugated and blood bilirubin levels can run as high as 80 mg/L. Fetuses receiving intrauterine transfusions are often born with high levels of conjugated bilirubin, probably arising from stimulation of fetal glucuronide formation coupled with decreased placental permeability to the bilirubin glucuronide.

Also unique to the newborn and related to developmental immaturity is the tissue toxicity of unconjugated bilirubin, especially to the brain. In the adult, serum bilirubin elevations are viewed as an important clinical or laboratory sign of disease or altered physiological state. In the neonate, hyperbilirubinemia has a dual significance as a clinical sign and also as a toxin.

Conjugated bilirubin is highly water soluble and is therefore readily excreted in fluids. Unconjugated or indirect bilirubin, on the other hand, is insoluble in aqueous solution but highly soluble in lipids. Under normal circumstances, unconjugated bilirubin is bound to plasma albumin, and this binding prevents the entrance of free or unbound indirect bilirubin into the lipid-rich central nervous system. When the albumin-binding capacity is exceeded, unbound, unconjugated bilirubin readily passes into the central nervous system cells. Unconjugated bilirubin is toxic to the central nervous system and causes necrosis, a pathological process referred to as "kernicterus." Surviving infants may have mental retardation, hearing deficits, or cerebral palsy. Many affected infants, particularly those of low birth weight, may have no neonatal symptoms but later in childhood can develop hearing deficits, perceptual handicaps, and hyperkinesis.

Usually there is no detectable bilirubin in amniotic fluid when fetus and mother are normal.[59] However, a neonate who demonstrates significant elevations of unconjugated bilirubin in serum frequently also passes bilirubin into its amniotic fluid. The route by which bilirubin is transferred into amniotic fluid from the fetus is unclear.

Maternal diabetes

There is an increased association of intrauterine deaths, congenital malformations, and perinatal mortality and morbidity in fetuses of diabetic women. For this reason, the pregnant diabetic woman is monitored closely throughout the course of her pregnancy. The fetal pancreas does not ameliorate maternal diabetes, since insulin does not cross the placenta.[60,61] Exogenous insulin therapy and diet are necessary therefore for management of the mother's insulin-deficient state.

As placental size increases, increasing amounts of HPL and other factors are produced that modify or oppose the action of insulin. HPL may be the cause of the increased insulin requirements of pregnancy.[62] Opposing the effect of HPL, maternal pancreatic β-cell hyperplasia frequently occurs with pregnancy, increasing the production of maternal insulin. Thus, normal insulin-glucose homeostasis is maintained when there is an increased glucose demand by the fetus.

When insufficient insulin is available to maintain normal glucose homeostasis, maternal hyperglycemia results. This in turn increases the blood glucose of the fetus, which produces a fetal hyperinsulinemia. At delivery, when the fetus is deprived of the maternal glucose source, the excess insulin in the newborn rapidly decreases the blood glucose levels so that the newborn becomes hypoglycemic. Life-threatening hypoglycemic crises are frequently encountered in untreated infants of diabetic mothers. When this happens, glucose must be administered to the newborn until proper glucose-insulin balance can be achieved.

In pregnancies complicated by diabetes, a significant increase (five or six times) in the incidence of the RDS has been noted. The reason surfactant assessment in uncontrolled diabetic pregnancies is not reliable for predicting fetal lung status is unknown.[58]

Toxemia (hypertensive disease of pregnancy)

Toxemia of pregnancy is characterized by hypertension (blood pressure over 140/90), edema, and proteinuria. Toxemia is subdivided into preeclampsia and eclampsia. Eclampsia is characterized by convulsions and is believed to be the sequela to the preeclamptic state.[63]

It is believed that the precipitating cause of pregnancy hypertension is a compromised uteroplacental blood flow. Management includes bed rest, dietary restriction of salt, and the carefully controlled use of magnesium sulfate to blunt exaggerated neuronal reflex activities. The only satisfactory "cure" for toxemia is delivery, and for this reason information about the pulmonary status of the fetus is of considerable importance. The frequent association of toxemia with diabetes and with vascular disease provides a basis for the correlation of this condition with small-for-gestational-age fetuses. Routine tests for serum creatinine and urine protein are the most common laboratory means for monitoring the toxemic pregnancy.

Spina bifida and anencephaly

Spina bifida and anencephaly are relatively common neural tube defects (NTD)[64] that constitute a large portion of the serious congenital malformations in man. In spina bifida there is a midline defect of the spine that results in a protrusion of the meninges or spinal cord or other neural elements. In anencephaly the brain is a disorganized mass of neural tissues and the cranial vault is absent.

It is possible to identify women who are at increased risk for carrying a fetus with an NTD by measuring maternal serum AFP levels (Chapter 47). Population *screening* for NTD, long accepted in Great Britain, is becoming more widely accepted in the United States. Prenatal *diagnosis* of NTDs by measuring alpha-fetoprotein and fetal-specific acetylcholinesterase levels in amniotic fluid has become routine in recent years. In addition, anencephaly specifically affects estriol formation. The absence of fetal pituitary function and hence fetal adrenal hypoactivity because of reduced ACTH results in very low rates of production of DHEA (dehydroepiandrosterone) from the fetal adrenal glands. Since this androgen is a precursor to estriol, estriol levels are characteristically low.

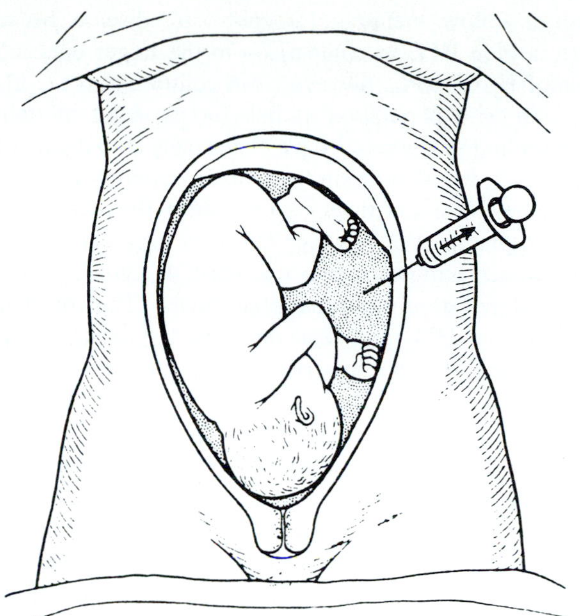

Fig. 37-12 Diagram demonstrating amniocentesis. *(From Queenan JT: Clin Obstet Gynecol 9:491-507, 1966.)*

Tubal pregnancy

Fertilization normally occurs in the fallopian tubes. Under usual circumstances, the fertilized egg migrates down the tube and enters and implants in the uterus. Occasionally, implantation takes place outside the uterus, most commonly in the fallopian tube itself. Any implantation outside the uterus is called an *ectopic pregnancy*, and an implantation that occurs in the tube is called a *tubal pregnancy*. An ectopic pregnancy can be caused by endocrine imbalances, residual effects stemming from tubal infections, or retrograde movement of the embryo from the uterus to the fallopian tube.[65]

Clinical symptoms include lower abdominal pain and vaginal bleeding. Amenorrhea is not a characteristic feature. Before rupture of the fallopian tube occurs, tubal pregnancies usually give a positive pregnancy test. However, since compromised placentas (that is, those in the process of abruption, degeneration, or penetration of the tubal wall) cannot produce chorionic gonadotropin in usual quantities, the tests may turn negative and become misleading. The presence of fetal erythrocytes in the maternal circulation might assist in the diagnosis of cases where the usual tests fail to establish the presence of a pregnancy.[66]

CHANGE OF ANALYTE IN DISEASE

Over the past two decades significant advances elucidating the course of growth and development of the fetus have been made. Part of this newfound knowledge has been possible through the development and application of improved analytical techniques and the improved safety of amniocentesis through ultrasound. Application of much of this knowledge has dramatically altered the course of management of problem or "high-risk" pregnancies (see references 67 and 68 for a discussion of these techniques).

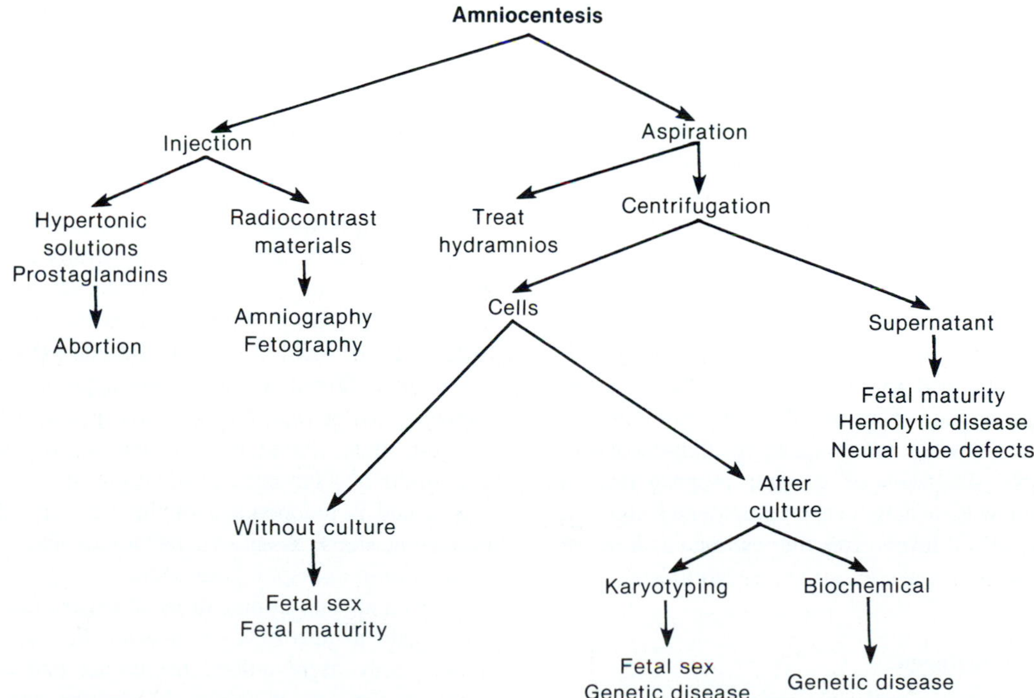

Fig. 37-13 Clinical applications of amniocentesis. *(From Pritchard JA, and MacDonald PC: In Pritchard JA, and MacDonald PC, editors: Williams obstetrics and gynecology, ed 15, New York, 1976, Appleton-Century-Crofts.)*

Fig. 37-12 presents a representation of the technique of amniocentesis, and Fig. 37-13 indicates the multifaceted array of its applications.

Three clinical problem areas have been the primary beneficiaries of amniocentesis: (1) the management of Rh-antigen incompatibility of mother and fetus, (2) the identification of the earliest possible time in pregnancy that delivery can be performed with minimal risk of prematurity, and (3) the identification of potential developmental or genetic disorders. Genetic disorders are discussed in Chapter 46.

Human chorionic gonadotropin (HCG)

HCG is used to identify and follow both trophoblastic disease and the course of normal pregnancy. In a normal pregnancy, urinary levels of HCG rise to a range of 20,000 to 100,000 U/day and then decrease to a range of 4000 to 11,000 U/day later in the pregnancy. Pregnancy can be detected by analysis of urinary HCG about a week after a missed period. By assays of plasma HCG, particularly with sensitive and specific methods such as those for the HCG β-chains, pregnancy can be detected as early as a few days after conception. In cases of hydatidiform mole, urinary HCG titers rise to over 300,000 U/day. After molar evacuation, these values drop within 1 month, and in about 90% of cases HCG is not detectable by urinary assay after 3 months. In cases where trophoblastic tissue remains, such as retained choriocarcinoma, values remain elevated, and serial assays of urine or serum HCG are of great value in determining the results of treatment, usually chemotherapy.

On a molecular basis, HCG shows about 1/4000 the thyrotropic activity of pituitary TSH.[69] If HCG levels are very high, thyroid-stimulating activity is possible. For this reason, the levels of HCG attained in molar pregnancies are believed to be the reason hyperthyroidism is often associated with molar pregnancies.[69,70] If HCG values exceed 100,000 U/day in urine or 300 U/mL in serum, the presence of hyperthyroidism and hydatidiform mole should be suspected.

HCG levels are often low for gestational age with ectopic, including tubal, pregnancies. Positive pregnancy tests have been obtained in only 50% of ectopic pregnancies. A negative test therefore does not exclude ectopic pregnancy. Newer, more sensitive immunoassays may improve the diagnosis of ectopic pregnancies. In addition, women with a diagnosis of threatened abortion who have low HCG levels for the estimated time of gestation have been shown to go on to complete abortion.

Human placental lactogen

HPL (Fig. 37-11) is not entirely satisfactory as a clinical measurement of placental function because there is a wide range of normal values as gestation advances. HPL is not used to follow metastatic trophoblastic disease because there is little HPL in comparison to the larger concentrations of HCG. HPL, however, can contribute to the identification of a population at risk for perinatal mortality. Spellacy and coworkers[71] have extensively investigated the use of HPL for identifying the fetus in jeopardy and found that values were less than 4 μg/mL after 39 weeks of gestation in pregnancies in which the fetus subsequently died. HPL values almost always exceeded 4 μg/mL after 30 weeks of gestation in normal pregnancies. The correlation between low HPL and fetal death was especially strong when severe maternal hypertension was present.

Estriol

Low serum or urinary estriol levels, or, more importantly, declining estriol levels, carry an unfavorable prognostic significance[72] (see Fig. 37-5, showing serial estriol values during various pregnancies). As a rough guideline, a decline of 30% to 50% from the mean of 3 previous days indicates probable impending danger to the fetus.[73] Because of the pronounced diurnal variation in estrogen formation, maintaining a consistent time of day for sampling (particularly when measuring estrogen levels in urine) is important. Since the precursors of estriol are androgens produced by the fetal adrenal gland, drugs such as synthetic corticoids that can cross the placenta and suppress ACTH secretion depress estriol levels in both blood and urine.

Serum or urinary estriol levels that are greater than the 95th percentile should suggest the possibility of twins. Estriol values are good predictors of impending fetal death in hypertensive disease, renal disease, and diabetes, particularly when there is a vascular component. Conditions when chronically low values are seen include toxemia, anencephaly, and placental sulfatase deficiency. Estriols are not helpful in monitoring erythroblastosis fetalis.

Phospholipids

Lecithin (phosphatidyl choline) is the most abundant of the phospholipids identified in amniotic fluid. Even though fetal maturity can be assessed by determination of the concentration of phospholipids in amniotic fluid,[74,75] the use of the ratio of lecithin to sphingomyelin (L/S) has become the more accepted method. This approach offers the advantages that the L/S ratio is independent of the volume of amniotic fluid (found to vary widely even within acceptable limits of what are considered to be "normal pregnancies") and is independent of the recovery of lipid during extraction steps, assuming that losses are comparable for both lecithin and sphingomyelin.

In a composite compilation of experience with the L/S ratio, with a ratio of 2 or greater, the risk of respiratory distress was slight unless the mother had diabetes. With diabetes, the risk of respiratory distress and in turn death from respiratory distress was much greater even though the L/S ratio was greater than 2. If the L/S ratio was 1.5 to 2, respiratory distress was found in 40%; if the L/S ratio was below 1.5, respiratory distress was found in 73% of the infants. Although 73% of infants did develop respiratory distress when the L/S ratio was below 1.5, it proved fatal in only 14% of the infants (Table 37-4). At times, obviously the risk to the fetus from a hostile environment will be greater than the risk of death from respiratory distress.

Gluck and Kulovich[42] and others[76] have reported that in certain pathological pregnancy conditions pulmonary maturation appears to be accelerated, whereas in others it is delayed (Table 37-5). Diseases in which fetal lung maturation may be delayed include diabetes mellitus and hemolytic disease in the fetus. Maternal hypertension and premature rupture of the membranes with delayed delivery have been reported to hasten maturation through surfactant production by the fetal lung.[42,76]

False mature L/S values have been reported after intrauterine transfusions,[77] and blood-contaminated amniotic fluid may give falsely lowered values.[78] Also the RDS can develop in an asphyxiated neonate despite a mature L/S value.[54]

Although phosphatidylglycerol (PG) represents only a minor fraction of the phospholipids of surfactant (Table 37-3), it contributes significantly to the functional properties of surfactant. For this reason PG, isolated from amniotic fluid, can carry prognostic value for fetal lung maturity. Functional lung maturity is associated with measurable quantities of PG,[79-81] but the absence of PG does not necessarily mean that the RDS is inevitable (Table 37-6). Since PG appears to be but a very small constituent of blood,[79] the measurement of PG is especially valuable at times when fetal lung status must be predicted from blood-contaminated amniotic fluid samples. The measurement of PG is considered important also when one is evaluating the maturity of fetuses of diabetic mothers in whom the L/S ratio may be less reliable. When amniotic fluid obtained from leakage into the vagina is the most readily accessible sample, PG is the only phospholipid that should be measured. If a proper vaginal pool sample is obtained, false-positive results are rare, but there can be high rates of false-negative results. Technologists concerned that a vaginal pool sample is heavily contaminated with maternal urine should measure urea levels in the sample to help determine the nature of the sample.

Surface tension

In an attempt to reduce the time and effort necessary for obtaining the L/S ratio, the foam-stability test, or so-called shake test, was introduced by Clements and associates.[82] The foam-stability test depends on the ability of surfactant to generate stable bubbles in the presence of ethanol. When the shake test was positive within 72 hours of parturition, less than 1% of infants developed RDS. About 50% of infants with a negative test developed RDS.[83] Particular care must be taken to ensure that glassware for the test is clean and that the reagents and amniotic fluid samples are not contaminated.

Table 37-4 Relationship of lecithin/sphingomyelin ratio to development of respiratory distress

L/S ratio	Number of infants	Respiratory distress (%)	
		All cases	Deaths
<1.5	162	73	14
1.5-2.0	223	40	4
>2.0	1596	2.2	0.1
>2.5	543	0.9	0

From Harvey D, Parkinson CE, and Campbell S: Lancet 1:42, 1975.

Table 37-5 Fetal lung maturation in relation to disorders

Accelerated fetal lung maturation with:	Delayed fetal lung maturation with:
Maternal hypertension	Diabetes mellitus, classes A, B, C
Renal	Hydrops fetalis
Cardiovascular	Infections
Severe, "chronic" toxemia	Chorioamnionitis
Hemoglobinopathy—sickle cell disease	Fetal infections, viral
Narcotics addiction (heroin, morphine)	Placental conditions
Diabetes mellitus, classes D, E, F	Chronic abruptio placentae
	Prolonged rupture of membranes

From Gluck L: In Lung maturation and the prevention of hyaline membrane disease, Proceedings of the 7th Ross Conference on Pediatric Research, Columbus, Ohio, 1976, Ross Laboratories.

Table 37-6 Phosphatidylglycerol (PG) with L/S ratio less than 2

	Respiratory distress syndrome	
	No RDS (%)	RDS (%)
PG present	91.6	8.4
PG absent	20.0	80.0

From Bent AE, Gray JH, Luther ER, et al: Am J Obstet Gynecol 139:259-263, 1981.

Bilirubin (ΔA_{450})

A maternal anti-Rh antibody titer (indirect Coombs' test) of 1 to 16 or more in most cases warrants amniocentesis and appropriately timed measurements of bilirubin pigment in amniotic fluid. The absorbance of bilirubin, when measured in a continuously recording spectrophotometer, is demonstrable as a "hump" or inflection with maximum absorbance at 450 nm (Fig. 37-14). The mag-

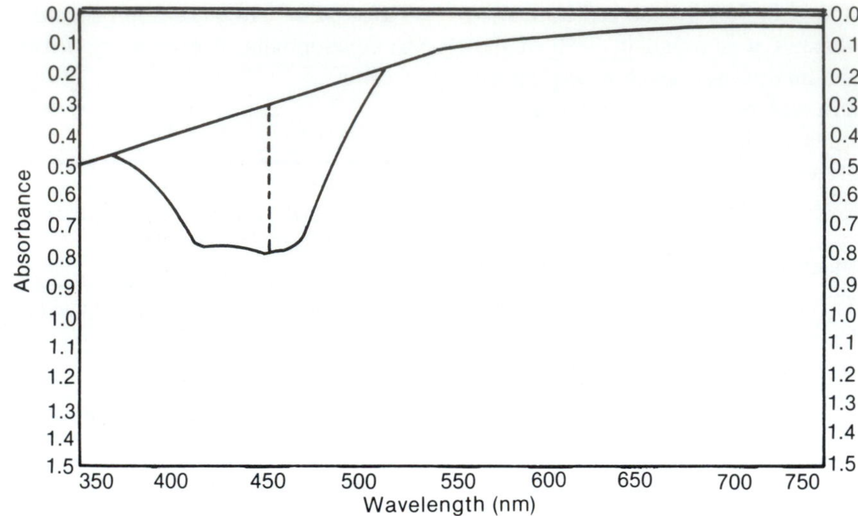

Fig. 37-14 Spectrum of bilirubin. *Dashed line,* Absorbance at 450 nm. *(From Queenan JT: Clin Obstet Gynecol 14:505-536, 1971.)*

nitude of the increase in optical density above the baseline value (ΔOD_{450}) usually but not always correlates well for any gestational age with the intensity of the hemolytic disease.

Liley[84] has developed a graph that reasonably predicts the severity of hemolytic disease and recommends clinical management (Fig. 37-15). The higher the ΔOD_{450}, the more severe is the hemolytic disease relative to the gestational age of the pregnancy. In general, a decreasing amniotic fluid bilirubin trend is a good prognostic indicator that a fetus will survive, whereas a horizontal or rising

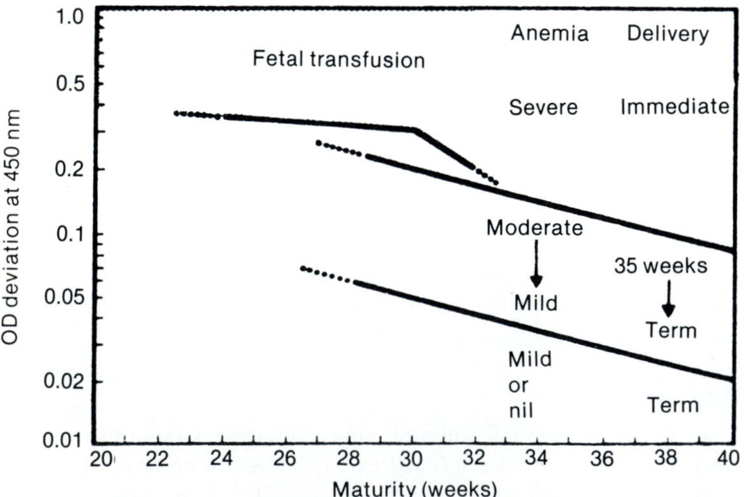

Fig. 37-15 Relationship of absorbance at 450 nm, gestational age of amniotic fluid associated with fetal anemia, and suggested clinical management. *OD,* Optical density (absorbance). *(From Liley AW: Am J Obstet Gynecol 86:485-494, 1963.)*

bilirubin level indicates that there is severe erythroblastosis fetalis. The main source of error is the fetus with polyhydramnios because it can cause a false low-bilirubin determination.

In severe erythroblastotic disease, bilirubin levels of 45 to 60 mg/L and occasionally as high as 80 mg/L are seen.[85] These levels are important in deciding whether to use immediate or delayed exchange transfusions. In neonates receiving intrauterine transfusions, it is not uncommon to see significantly elevated levels of direct bilirubin.[86]

Glucose

After delivery, the glucose concentration in offspring of diabetic mothers declines rapidly below that observed in normal infants. Blood glucose levels of these infants must be carefully monitored to see if life-threatening hypoglycemia occurs. Approximately 60% of babies from insulin-dependent mothers have glucose concentrations below 300 mg/L in the first 6 hours of life.[87]

Renal function tests

During normal pregnancy, renal blood flow and the glomerular filtration rate are significantly increased above nonpregnant levels. With the development of pregnancy-induced hypertension, renal perfusion and glomerular filtration are reduced. Most often, therefore, the creatinine or urea concentration in plasma is not appreciably elevated. The plasma uric acid concentration is much more commonly elevated, especially in women with more severe renal disease. The elevation is a result primarily of decreased renal clearance of uric acid by the kidney, a decrease that exceeds the reduction in glomerular filtration rate and creatinine clearance.[88]

Alpha-fetoprotein (AFP)

When anencephaly or meningomyelocele has occurred in a preceding pregnancy, the risk of recurrence of either defect is approximately 5%. This is detected when one finds elevated AFP in the amniotic fluid. If the AFP is not elevated, anencephaly can be reliably excluded, but only about 90% of the pregnancies with meningomyelocele give positive results.[64]

Elevations of maternal serum AFP (reported as *M*ultiples *o*f the gestational *M*edian level, or MoM) indicate a woman with an elevated risk for carrying a fetus with a neural tube defect (NTD). With correct dating of gestational age and exclusion of uterine pregnancy, an elevated maternal serum AFP level (>2.5 MoM) can detect 80% of open spina bifida cases and 95% of anencephaly cases. Final diagnosis must be made by a combination of ultrasound examination of the fetus and measurement of amniotic fluid AFP levels (see Chapter 46).

Magnesium

Women with toxemia of pregnancy or premature labor are often treated with high levels of $MgSO_4$. These women, usually under hospital care, must be closely monitored for excessive hypermagnesemia (>80 mg/L).

REFERENCES

1. Grillo, RAI, and Shima, K: Insulin content and enzyme histochemistry of the human foetal pancreatic islet, J Endocrinol 36:151-158, 1966.
2. Hytten, FE: Placental transfer. In Hytten, F, and Chamberlain, G, editors: Clinical physiology in obstetrics, Oxford, 1980, Blackwell Scientific Publications, p 475.
3. Ryan, KJ: Hormones and the placenta, Am J Obstet Gynecol 84(Part 2):1695-1713, 1962.
4. Lind, T, Kendall, A, and Hytten, FE: The role of the fetus in the formation of amniotic fluid, J Obstet Gynaecol Br Commonw 79:289-298, 1972.
5. Abramovich, DR: The volume of amniotic fluid and its regulating factors. In Fairweather, DVI, and Eskes, TKAB, editors: Amniotic fluid: research and clinical application, ed 2, Amsterdam, 1978, Excerpta Medica, p 31.
6. Duenhoelter, JH, and Pritchard, JA: Fetal respiration: quantitative measurements of amniotic fluid inspired near term by human and rhesus fetuses, Am J Obstet Gynecol 125:306-309, 1976.
7. Seeds, AE: Current concepts of amniotic fluid dynamics, Am J Obstet Gynecol 138:575-586, 1980.
8. Sandler, M, editor: Amniotic fluid and its clinical significance, New York, 1981, Marcel Dekker, Inc.
9. Fairweather, DVI, and Eskes, TKAB, editors: Amniotic fluid: research and clinical applications, Amsterdam, 1978, Excerpta Medica.
10. Lind, T, Billewicz, WZ, and Cheyne, GA: Composition of amniotic fluid and maternal blood through pregnancy, J Obstet Gynecol Br Commonw 78:505-512, 1971.
11. Hutchinson, DL, Gray, MJ, Plentl, AA, et al: The role of the fetus in the water exchange of the amniotic fluid of normal and hydramniotic patients, J Clin Invest 38:971-980, 1959.
12. Hytten, FE: Water transfer. In Chamberlain, GVP, and Wilkinson, AW, editors: Placental transfer, Tunbridge Wells, 1979, Pitman Medica, p 90.
13. Mithell, G, Field, B, and Kerr, C: Biochemical differences between amniotic fluid and maternal urine, Br Med J 1(6064):811-812, 1977.
14. Weaver, TE: Pulmonary surfactant-associated proteins, Gen Pharmacol 18:1-8, 1987.
15. Pierce, JG, and Parsons, TF: Glycoprotein hormones: structure and function, Annu Rev Biochem 50:465-495, 1980.
16. Catt, KJ, and Pierce, JG: Gonadotropic hormones of the adenohypophysis (FSH, LH and prolactin). In Yen, SSC, and Jaffe, RB, editors: Reproductive endocrinology, Philadelphia, 1978, WB Saunders Co, pp 34-62.
17. Diczfalusy, E: Endocrine fuctions of the human fetus and placenta, Am J Obstet Gynecol 119:419-433, 1974.
18. Eberlein, WR: The fetal adrenal cortex. In Christy, NP, editor: The human adrenal cortex, New York, 1971, Harper & Row, Publishers, Inc, pp 317-327.
19. France, JT, Seddon, RJ, and Liggins, GC: A study of a pregnancy with low estrogen production due to placental sulfatase deficiency, J Clin Endocrinol Metab 36:1-9, 1973.
20. Goebelsmann, U: The uses of oestriol as a monitoring tool, Clin Obstet Gynecol 6:223-241, 1979.
21. Schindler, AE, and Siiteri, PK: Isolation and quantitation of steroids from normal human amniotic fluids, J Clin Endocrinol Metab 28:1189-1198, 1968.
22. Resnick, R, Killam, AP, Battaglia, FC, et al: The stimulation of uterine blood flow by various estrogens, Endocrinology 94:1192-1196, 1974.
23. Kundu, N, Carmody, PJ, Didolkar, SM, and Petersen, LP: Sequential determination of serum human placental lactogen, estriol, and esterol for assessment of fetal morbidity, Obstet Gynecol 52:513-520, 1978.
24. Notation, AD, and Tagatz, GE: Unconjugated estriol and 15α-hydroxyestriol in complicated pregnancies, Am J Obstet Gynecol 128:747-756, 1977.
25. Mestman, JH: Thyroid and parathyroid diseases in pregnancy. In Quilligan, EJ, and Kretchmer, N, editors: Fetal and maternal medicine, New York, 1980, John Wiley & Sons, Inc, pp 49-531.
26. Oppenheimer, JH: Role of plasma proteins in the binding, distribution and metabolism of the thyroid hormones, N Engl J Med 8:1153-1162, 1968.
27. DeAlvarez, RR, Gaiser, DF, Simkins, DN, et al: Serial studies of serum lipids in normal human pregnancy, Am J Obstet Gynecol 77:743-759, 1959.
28. Peters, JP, Heineman, M, and Man, EB: The lipids of the serum in pregnancy, J Clin Invest 30:388-394, 1951.
29. Robertson, EG: Increased electrolyte fragility in association with osmotic changes in pregnancy serum, J Reprod Fertil 16:323-325, 1968.
30. Studd, JW, and Wood, S: Serum and urinary proteins in pregnancy. In Wynn, RM, editor: Obstetrics and gynecology annual, New York, 1976, Appleton-Century-Crofts, pp 103-111.
31. Todd, ME, Thompson, JH, Bowie, EJW, and Owen, CA: Changes in blood coagulation during pregnancy, Mayo Clin Proc 40:370-383, 1965.
32. Talbert, LM, and Langdell, RD: Normal values of certain factors in the blood clotting mechanism in pregnancy, Am J Obstet Gynecol 90:44-50, 1964.
33. Davison, JM: Changes in renal function and other aspects of hemostasis in early pregnancy, J Obstet Gynaecol Br Commonw 81:1003, 1974.
34. Davison, JM: The urinary system. In Hytten, F, and Chamberlain, G, editors: Clinical physiology in obstetrics, Oxford, 1980, Blackwell Scientific Publications, p 306.
35. Levy, HL, and Montag, PP: Free amino acids in human amniotic fluid: a quantitative study by ion-exchange chromatography, Pediatr Res 3:113-120, 1969.
36. Kleinman, LI: The kidney. In Stave, U, editor: Perinatal physiology, New York, 1978, Plenum Publishing Corp, pp 589-616.
37. Van Geuns, HJ, and Van Kessel, H: Creatinine in amniotic fluid and fetal renal function. In Fairweather, DVI, and Eskes, TKAB, editors: Amniotic fluid: research and clinical application, ed 2, Amsterdam, 1978, Excerpta Medica, pp 81-92.
38. King, RJ, and Clements, JA: Surface active materials from dog lung. II. Composition and physiological correlations, Am J Physiol 223:715-726, 1972.
39. Jobe, A, and Ikegami, M: Surfactant for the treatment of respiratory disease syndrome, Am Rev Respir Dis 136:1256-1275, 1987.
40. Godinez, RI, Sanders, RL, and Longmore, WJ: Phosphatidylglycerol

in rat lung. II. Comparison of occurrence, composition, and metabolism in surfactant and residual fractions, Biochemistry 14:835-840, 1975.
41. Gikas, EG, King, RJ, Mescher, EJ, et al: Radioimmunoassay of pulmonary surface-active material in the tracheal fluid of the fetal lamb, Am Rev Respir Dis 115:587-593, 1977.
42. Gluck, L, and Kulovich, MV: Lecithin/sphingomyelin ratios in amniotic fluid in normal and abnormal pregnancies, Am J Obstet Gynecol 115:539-546, 1973.
43. Martin, RJ, Fanaroff, AA, and Skalina, MEL: The respiratory system. In Fanaroff, AA, and Martin, RJ, editors: Behrman's neonatal-perinatal medicine, St Louis, 1983, The CV Mosby Co, pp 432-437.
44. Farrell, PM, and Avery, ME: Hyaline membrane disease, Am Rev Respir Dis 111:657-688, 1975.
45. Farrell, PM, and Kotas, RV: The prevention of hyaline membrane disease: new concepts and approaches to therapy, Adv Pediatr 23:213-269, 1976.
46. Blackburn, WR, Travers, H, and Potter, DM: The role of the pituitary-adrenal axes in lung differentiation, Lab Invest 26:306-318, 1972.
47. Liggins, GC, and Howie, RN: A controlled trial of antepartum glucocorticoid treatment for prevention of the respiratory distress syndrome in premature infants, Pediatrics 50:515-525, 1972.
48. Kleihauer, E: The hemoglobins. In Stave, U, editor: Perinatal physiology, New York, 1978, Plenum Publishing Corp, pp 215-239.
49. Hecht, F, Jones, RT, and Koler, RD: Newborn infants with Hb Portland 1, an indicator of α-chain deficiency, Ann Hum Genet 31:215-218, 1967.
50. Hellegers, AE, and Schruefer, JJP: Nomograms and empirical equations relating oxygen tension, percentage saturation, and pH in maternal and fetal blood, Am J Obstet Gynecol 81:377-384, 1961.
51. Longo, LD: Disorders of placental transfer. In Assali, NS, editor: Pathophysiology of gestation, vol 2, New York, 1972, Academic Press Inc, pp 1-76.
52. Avery, ME, and Mead, J: Surface properties in relation to atelectasis and hyaline membrane disease, Am J Dis Child 97:517-523, 1959.
53. Gluck, L, Kulovich, MV, and Borer, RC: Estimates of fetal lung maturity, Clin Perinatol 1:125-139, 1974.
54. Donald, IR, Freeman, RK, Goebelsmann, U, et al: Clinical experience with the amniotic fluid lecithin/sphingomyelin ratio, Am J Obstet Gynecol 115:547-552, 1973.
55. Robert, MF, Neff, RK, Hubbell, JP, et al: Association between maternal diabetes and the respiratory-distress syndrome in the newborn, N Engl J Med 294:357-360, 1976.
56. Myers, JL, Harrell, MJP, and Mill, FL: Fetal maturity: biochemical analyses of amniotic fluid, Am J Obstet Gynecol 21:961-967, 1975.
57. Katyal, SL, Amenta, JS, and Singh, G: Deficient lung surfactant apoproteins in amniotic fluid with mature phospholipid profile from diabetic pregnancies, Am J Obstet Gynecol 148:48-53, 1984.
58. McMahan, MJ, Mimouni, F, Miodovnik, M, et al: Surfactant associated protein (SAP-35) in amniotic fluid from diabetic and nondiabetic pregnancies, Obstet Gynecol 69:2-5, 1987.
59. Liley, AW: The administration of blood transfusions to the foetus in utero, Triangle 7:184-189, 1966.
60. Freinkel, N, and Goodner, CJ: Carbohydrate metabolism in pregnancy. I. The metabolism of insulin by human placental tissue, J Clin Invest 39:116-131, 1960.
61. Posner, BI: Insulin metabolizing enzyme activities in human placental tissue, Diabetes 22:552-563, 1973.
62. Klopper, A: Placental metabolism. In Hytten, F, and Chamberlain, G, editors: Clinical physiology in obstetrics, Oxford, 1980, Blackwell Scientific Publications, p 453.
63. Chesley, LC, editor: Hypertensive disorders in pregnancy, New York, 1978, Appleton-Century-Crofts.
64. Kirkpatrick, AM, and Nakamura, RM: Alpha-fetoprotein: laboratory proceedings and clinical application, New York, 1981, Masson Publishing Co.
65. Martinez, F, and Trounson: An analysis of factors associated with ectopic pregnancy in a human in vitro fertilization program, Fertil Steril 45:79-87, 1986.
66. Wingate, MB, Iffy, L, Kelly, JV, and Birnbaum, S: Diseases specific to pregnancy. In Romney, SL, Gray, MJ, Little, AB, et al, editors: Gynecology and obstetrics: the health care of women, New York, 1975, McGraw-Hill Book Co, Inc, p 721.
67. Bennett, MJ: Amniocentesis. In Sandler, M, editor: Amniotic fluid and its clinical significance, New York, 1981, Marcel Dekker, Inc, pp 27-36.
68. Root, AW, and Reiter, EO: Human perinatal endocrinology. In Quilligan, EJ, and Kretchmer, N, editors: Fetal and maternal medicine, New York, 1980, John Wiley & Sons, Inc, pp 15-58.
69. Hershman, JM: Hyperthyroidism induced by trophoblastic thyrotropin, Mayo Clin Proc 47:913-918, 1972.
70. Nisula, BC, and Ketelslegers, J: Thyroid-stimulating activity and chorionic gonadotropin, J Clin Invest 54:494-499, 1974.
71. Spellacy, WN, Toeh, E, Buhi, W, et al: Value of human chorionic somatomammotropin in managing high risk pregnancies, J Obstet Gynecol 109:588-598, 1974.
72. Little, B, and Billar, RB: Endocrine disorders. In Romney, SL, Gray, MJ, Little, AB, et al, editors: Gynecology and obstetrics: the health care of women, New York, 1975, McGraw-Hill Book Co, Inc, p 398.
73. Dumont, M, Cohen, M, Cohen, H, and Bertrand, J: Diagnosis of fetal distress in utero by determination of plasma levels of nonconjugated estrogens, Rev Fr Gynecol Obstet 70:171-178, 1975.
74. Arvidson, G, Ekeland, H, and Asted, B: Phospholipid composition of human amniotic fluid during gestation and at term, Acta Obstet Gynecol Scand 51:71-75, 1972.
75. Falconer, GF, Hodge, JS, and Gadd, RL: Influence of amniotic fluid volume on lecithin estimation in prediction of respiratory distress, Br Med J 2:689-691, 1973.
76. Bauer, CR, Stern, L, and Colle, E: Prolonged rupture of membranes associated with a decreased incidence of respiratory distress syndrome, Pediatrics 53:7-12, 1974.
77. Lemons, JA, and Jaffe, RB: Amniotic fluid lecithin/sphingomyelin ratio in the diagnosis of hyaline membrane disease, Am J Obstet Gynecol 115:233-237, 1973.
78. Gibbons, JM, Huntley, TE, Joachim, E, et al: Amniotic fluid analysis for fetal maturity: effect of maternal blood contamination, Obstet Gynecol 39:631, 1972.
79. Bent, AE, Gray, JH, Luther, ER, et al: Phosphatidylglycerol determination on amniotic fluid 10,000 × g pellet in the prediction of fetal lung maturity, Am J Obstet Gynecol 39:250-263, 1981.
80. Bent, AE, Gray, JH, Luther, ER, et al: Assessment of fetal lung maturity: relationship of gestational age and pregnancy complications to phosphatidylglycerol levels, Am J Obstet Gynecol 139:664-669, 1981.
81. Painter, DC: Simultaneous measurement of lecithin, sphingomyelin, phosphatidylglycerol, phosphatidylinositol, phosphatidylethanolamine, and phosphatidylserine in amniotic fluid, Clin Chem 26(Part 2):1147-1151, 1980.
82. Clements, JA, Platzker, ACG, Tierney, DF, et al: Assessment of the risk of respiratory distress syndrome by a rapid test for surfactant in amniotic fluid, N Engl J Med 286:1077-1081, 1972.
83. Harvey, DR, and Parkinson, CE: In Barson, AJ, editor: Laboratory diagnosis of fetal disease, London, 1981, John Wright PSG, Inc, pp 267-298.
84. Liley, AW: Amniocentesis and amniography in hemolytic disease. In Greenhill, JP, editor: Yearbook of obstetrics and gynecology, 1964-1965 series, Chicago, 1964, Year Book Medical Publishers, pp 256.
85. Bellingham, AJ: Hemoglobins with altered oxygen affinity, Br Med Bull 32:234-238, 1976.
86. Benz, EJ, and Forget, BG: The biosynthesis of hemoglobin, Semin Hematol 11:463-523, 1974.
87. Pennoyer, MM, and Hartman, AF, Sr: Management of infants born of diabetic mothers, Postgrad Med 18:199-206, 1955.
88. Pritchard, JA, and MacDonald, PC: Hypertensive disorders in pregnancy. In Williams Obstetrics, ed 15, New York, 1976, Appleton-Century-Crofts, p 556.

CHAPTER 38 Extravascular biological fluids

LEWIS GLASSER

OBJECTIVES

- Define serous fluid and describe the formation of serous fluids.
- Differentiate transudate and exudate based on cause, mode of formation, and protein and enzyme levels.
- List analytes of serous fluids which may be markers of disease conditions.
- Define synovial fluid and describe a normal synovial fluid.
- Describe changes in synovial fluid in pathological conditions.

KEY TERMS

arthritis Inflammation of a joint.
ascites Serous fluid in the peritoneal cavity.
chyle Fatty lymph fluid originating from the intestinal lymphatics. It is milky white in appearance.
chylous effusion Chyle in a body cavity.
colloid osmotic pressure of plasma The difference in osmotic pressure between the plasma and interstitial fluid that drives water into the bloodstream from the interstitial spaces.
effusion Pathological accumulation of fluid in a body cavity.
empyema Usually pus in the pleural cavity.
epicardium The visceral layer of pericardium.
exudate The accumulation of a fluid with a high concentration of protein in a body cavity caused by increased capillary permeability.
gout An inflammatory arthritis of the joint secondary to crystallization of monosodium urate in the joint.
hemothorax Blood in the pleural cavity secondary to rupture of the blood vessels.
hyaluronic acid A high molecular weight polymer made up of repeating units of the disaccharide *N*-acetylglucosamine and glucuronic acid.
hydrostatic pressure The lateral pressure of water within a vessel that tends to drive fluid out of the capillaries into the interstitial space.
joint An articulation between bones.
neuroarthropathy Disease of a joint secondary to a disease of the nervous system.
osmotic pressure The force with which a solvent passes through a semipermeable membrane.
osteoarthritis A degenerative form of arthritis that is primarily a disease of the bones with joint involvement.
osteochondromatosis A joint disease characterized by the development of cartilaginous nodules in the synovial tissues.
paracentesis Aspiration of fluid from a body space.
parapneumonic effusion An accumulation of fluid in the pleural space secondary to pneumonia.
parietal The wall of a cavity.
pericardium The sac enclosing the heart.
peritoneum The serous membrane lining the abdominal cavity and the organs of the abdominal cavity.
permeable Allowing the passage of fluids through a membrane.
pigmented villonodular synovitis A disease of the joints of unknown cause characterized by fingerlike proliferative growths of the synovial tissue with hemosiderin deposition within the synovial tissue.
pleura The serous membrane lining the inner surface of the thorax, the diaphragm, and the outer surface of the lungs.
pseudogout An inflammatory arthritis of the joint secondary to crystals of calcium pyrophosphate.
psoriatic arthritis A chronic destructive joint disease that occurs in some patients with the skin disease psoriasis.
Reiter's syndrome A syndrome of unknown cause characterized by inflammation of the joints, urethra, and conjunctivae.
rheumatoid arthritis A chronic progressive inflammatory disease of unknown cause involving multiple joints.
serous fluid Fluid having the characteristics of serum.
synovial fluid Joint fluid.
systemic lupus erythematosus A multisystem disease of an autoimmune cause involving mostly the skin, kidneys, joints, and serosal membranes.
thoracentesis Removal of fluid from the pleural cavity.
transudate The accumulation in a body cavity of a fluid having a low concentration of protein.

SEROUS FLUIDS

In this chapter the term *serous fluids* is restricted to pleural, pericardial, or peritoneal fluid. The word *serous* is derived from *serum* and accurately expresses the derivation of the body fluids from plasma. Body fluids are designated by a variety of medical terms. *Pleural fluid* (thoracic or chest fluid) is obtained by surgical puncture of the chest wall *(thoracentesis)*. *Empyema* refers to pus in the pleural cavity. Peritoneal fluid is frequently designated by the nonanatomical term *ascitic fluid*. *Ascites* is derived from the Latin word for "bag" and describes the bloated abdomen of the patient afflicted with a massive accumulation of peritoneal fluid. *Paracentesis* means aspiration of fluid from a cavity, and *abdominal paracentesis fluid* is synonymous with *peritoneal fluid*. Whole blood in the body cavities is designated with the prefix "hemo," as in *hemothorax*. A *chylous effusion* refers to the accumulation of lymph (chyle) in the body cavity.

Formation

Normal formation. Each body cavity is lined by a thin serosal membrane. The lining of the body wall is the parietal membrane, and the lining of the organs is the visceral membrane. The two membranes are continuous, and the space between them is the body cavity (Fig. 38-1). The serosal membrane is composed of a thin layer of connective tissue containing numerous capillaries and lymphatics and a superficial layer of flattened mesothelial cells.

Serous fluid is an ultrafiltrate of plasma derived from the rich capillary network in the serosal membrane. Its formation is similar to the production of extravascular interstitial fluid anywhere in the body. Three factors are important: hydrostatic pressure, colloid osmotic pressure, and capillary permeability. Hydrostatic pressure drives fluid out of the capillaries and into the body cavities. Impermeable protein molecules in the plasma exert a force that counteracts the hydrostatic pressure and causes capillaries to absorb fluid. This force is called the colloid osmotic pressure (COP) and is proportional to the molar concentration of protein. Lymphatics also play an important role in the absorption of water, protein, and particulate matter from the extravascular space. In the thoracic (chest) cavity, fluid is formed at the parietal pleura because the high hydrostatic pressure of the systemic circulation exceeds the colloid osmotic pressure, and fluid is reabsorbed at the visceral pleura, where capillary colloid osmotic pressure exceeds the low hydrostatic pressure of the pulmonary circulation (Fig. 38-2). Normally there is less than 10 mL of fluid in each pleural cavity, 20 to 50 mL in the pericardial sac, and less than 100 mL in the peritoneal cavity.[1]

Abnormal formation. Effusions will form when the normal physiological mechanisms responsible for the formation or absorption of serosal fluid are impaired. Thus fluid will accumulate if capillary permeability increases, hydrostatic pressure increases, colloid osmotic pressure decreases, or lymphatic drainage is obstructed. Hydrostatic pressure is increased in congestive heart failure, a frequent cause of effusions. Hypoproteinemia decreases the colloid osmotic pressure. A decreased plasma protein can be secondary to decreased synthesis or increased loss of protein. Albumin synthesized in the liver is the most important protein in the maintenance of colloid osmotic pressure. Diseases of the liver may impair albumin synthesis; the liver disease most frequently associated with hypoproteinemia and effusions is cirrhosis. Hypoalbuminemia is also caused by an increased loss of serum protein, which occurs in the nephrotic syndrome. Capillary permeability increases if the pleural surfaces are inflamed. Increased capillary permea-

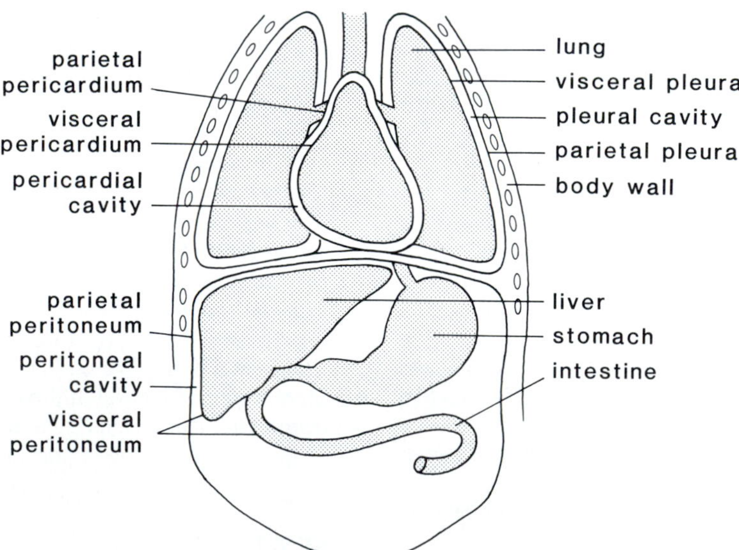

Fig. 38-1 Relationships of serous membranes, body cavities, and viscera. Heart is enclosed in pericardial sac. Outer layer of pericardium is called "parietal pericardium." Lining the exterior surface of heart is visceral pericardium, which is also called "epicardium." Parietal peritoneum lines wall of abdominal cavity. Visceral peritoneum invests stomach, liver, and intestines. Peritoneal cavity is space between two layers of peritoneum.

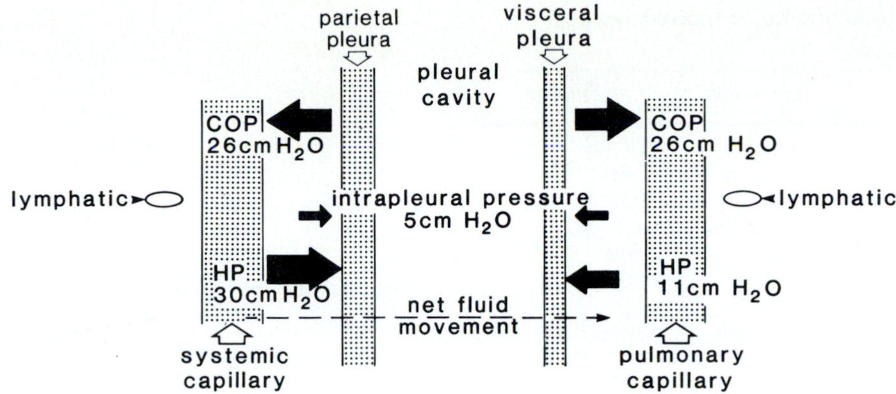

Fig. 38-2 Pleural fluid is formed at parietal pleura as net forces for flow of fluid out of systemic capillaries exceed net colloid osmotic pressures. Movement of fluid is toward visceral pleura where net colloid osmotic pressure exceeds outward forces because of low hydrostatic pressure in pulmonary capillaries. Lymphatics play a role in absorption of water, protein, and particulate matter. *COP*, Colloid osmotic pressure; *HP*, hydrostatic pressure.

Table 38-1 Causes of effusions

Cause	Finding	Pathogenesis
Transudates		
Congestive heart failure	↑ HP	Systematic and pulmonary venous hypertension
Hepatic cirrhosis	↑ HP	Portal and inferior vena cava hypertension
	↓ COP	Hypoalbuminemia
Nephrotic syndrome	↓ COP	Hypoalbuminemia
Exudates		
Pancreatitis	↑ CP	Inflammation secondary to chemical injury
Bile peritonitis	↑ CP	Inflammation secondary to chemical injury
Rheumatoid disease	↑ CP	Inflammation of serosa
Systemic lupus erythematosus	↑ CP	Inflammation of serosa
Infections (bacterial, tuberculosis, fungal, viral)	↑ CP	Inflammation secondary to microorganisms
Infarction (myocardial, pulmonary)	↑ CP	Inflammation secondary to extension of process to serosal surface
Neoplasms	↑ CP	Increased permeability of capillaries supplying tumor implants; pleuritis secondary to obstructive pneumonitis
	↓ LyD	Lymphatic obstruction secondary to lymph node infiltration
Chyle		
Trauma Surgery Neoplasms Idiopathic	↓ LyD	Disruption of lymphatic ducts

HP, Hydrostatic pressure; *COP*, colloid osmotic pressure; *CP*, capillary permeability; *LyD*, lymphatic drainage.

bility also results in loss of protein from the vascular space, and so the physical forces that lead to excess fluid formation are accentuated. Conditions causing an increase in capillary permeability include inflammatory diseases, infection, and metastatic tumors. If the lymphatics are obstructed, a protein-rich fluid will accumulate. Neoplasms of the lymph nodes frequently produce pleural effusions. The causes of effusions and the underlying pathogenesis are listed in Table 38-1.

Change of analyte in disease

Transudates and exudates. Serous effusions are designated as transudates or exudates, depending on the protein content of the fluid. The distinction is important because transudates are noninflammatory fluids caused by disturbances of hydrostatic or colloid osmotic pressure, whereas exudates are inflammatory fluids caused by increased capillary permeability secondary to diseases that directly involve the surfaces of the body cavities. The separation of transudates and exudates involves the use of arbitrary medical decision levels that have been empirically determined. The higher the protein content, the more likely it is that the fluid is caused by a process that alters the capillary permeability and involves the surfaces of the body cavity. Measuring the specific gravity will indirectly measure the protein concentration. Pleural fluids are classified as exudates if the specific gravity is greater than 1.015 g/mL or the total protein is 30 g/L or greater. Measurement of total protein is preferable to measurement of specific gravity. The separation of exudates and transudates in pleural fluids is even more precise if the fluid protein is compared to the serum total protein. Dividing the concentration of the protein in the fluid by the concentration of the protein in the serum gives a ratio. A ratio of 0.5 or greater is indicative of an exudate.[2] The distinction

Table 38-2 Diagnostic criteria of transudates and exudates in pleural fluid

Test	Transudate	Exudate
Appearance	Clear	Cloudy
Fibrinogen	No clot	Clots
Specific gravity	<1.015	≥1.015
Total protein	<30 g/L	≥30 g/L
Total protein (fluid/serum)	<0.5	≥0.5
Lactate dehydrogenase (fluid/serum)	<0.6	≥0.6
Glucose	~ serum	Often <600 mg/L

is further improved if a large protein molecule such as lactate dehydrogenase (LD) is used as a marker of capillary permeability. Pleural fluid to serum LD ratios of 0.6 or greater are diagnostic of exudates.[2] The differences between transudates and exudates in pleural effusions are summarized in Table 38-2. Different cutoff values are used for peritoneal fluid. Protein levels greater than 25 g/L classify the fluid as an exudate.[1] Use of the difference between the serum and peritoneal fluid albumin concentrations provides significantly better discrimination between transudative and exudative ascites than use of the total protein levels; differences less than 11 g/L correlate with malignant effusions.[3,4]

Glucose. Pleural fluid glucose concentrations are similar to plasma levels in normal fluids and transudates. Glucose is decreased in exudative processes such as bacterial infection, tuberculosis, neoplasia, and rheumatoid disease.[5,6] A pleural fluid glucose less than 600 mg/L or a difference between the plasma and fluid glucose of greater than 300 mg/L is clinically significant. Only low glucose levels are diagnostically useful, and the various diseases associated with low levels are also associated with normal values. Any etiological diagnosis on the basis of a low glucose level alone is unreliable. Two mechanisms are operative in producing low values. One is increased glucose use, and the second is a relative block in the transport of glucose from the blood to the fluid. The latter occurs in rheumatoid effusions.[7] Interpretation of low glucose concentrations in peritoneal and pericardial fluid is similar to that in pleural fluid.

pH. Measurement of pleural fluid pH is clinically useful in the management of patients with parapneumonic effusions. Patients with pneumonia develop effusions because the infectious process extends to the visceral pleura, causing exudation of fluid into the pleural space. Complications of parapneumonic effusions include loculation and pus in the pleural cavity. Fluids are divided into potentially benign and complicated effusions on the basis of the pH. Fluids with a pH greater than 7.30 resolve spontaneously, whereas a pH less than 7.20 is an indication for tube drainage.[8,9] A cautionary note: the specimen must be collected anaerobically in a heparinized syringe, stored on ice, and measured at 37° C. Measurement of peritoneal or pericardial fluid pH is of no clinical value.

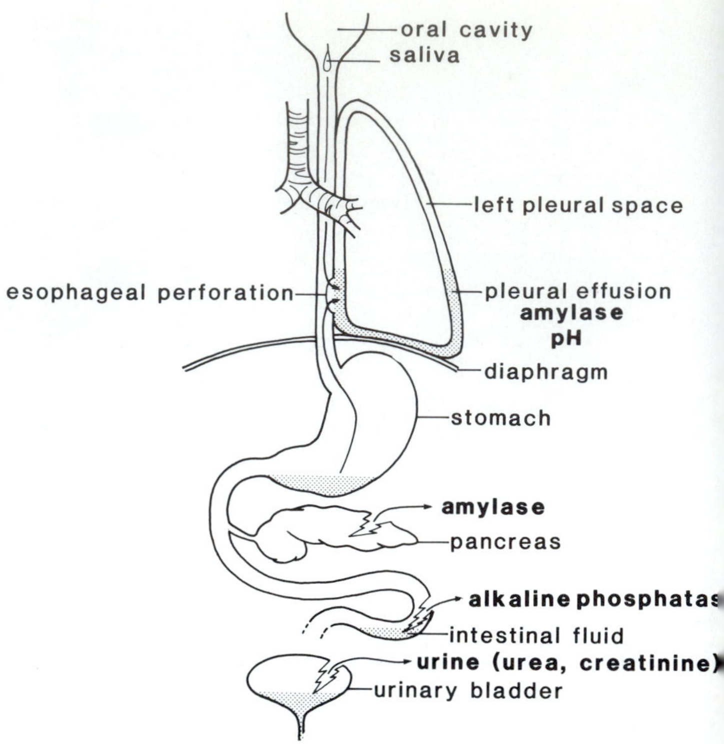

Fig. 38-3 Chemical determinations of body fluids as markers for specific organ involvement.

Lipid. Chyle is a milky white emulsion of fatty lymph fluid originating from the intestinal lymphatics. The accumulation of chyle in the pleural space is rare. Even less frequent is chyle accumulation in the peritoneal or pericardial cavities. Chylous fluid accumulates because of the disruption of the thoracic duct. Chylomicrons found on lipoprotein analysis are the best evidence for a chylous effusion. Triglyceride values above 1100 mg/L are highly suggestive of chylous effusion.[10] Cholesterol values do not distinguish between chylous and nonchylous effusions.

Analytes as markers for organs. Chemical substances can serve as markers for the specific organ involved in the pathogenesis of the effusion. The rationale for these tests is easily understood when one considers the anatomical location of the viscera and normal biochemistry (Fig. 38-3). Analytes that have been used as markers include amylase, pH, alkaline phosphatase, urea nitrogen, and creatinine. Pleural effusions accompany most cases of esophageal rupture. The perforation allows secretions from both the oral cavity and stomach to contaminate the effusion fluid. Pleural fluid amylase levels can be elevated, and the levels are higher than the serum amylase. Electrophoretic studies indicate that the amylase is from the saliva.[11] Another indicator of esophageal perforation is the pH of the pleural fluid. Normal gastric juice has a pH below 3.5. Leakage of gastric contents through the esophageal tear will acidify the pleural fluid.[12,13] A pH below 6.0 is clinically significant. The measurement may be performed at the bedside using pH reagent paper.

Amylase is a well-accepted marker of pancreatic disease. In acute pancreatitis, amylase-rich fluid seeps into the peripancreatic tissue and causes a chemical peritonitis, with the formation of small amounts of peritoneal fluid in most cases. One study reports fluid amylase levels of 27,800 ± 7560 U/L (15,030 ± 4086 Somogyi units/dL).[14] The fluid levels are higher and persist longer than the corresponding blood amylase levels.[15] *Pancreatic ascites* is the chronic accumulation of massive amounts of fluid in association with pancreatitis. It is not certain whether the fluid represents leakage of pancreatic secretions from ruptured ducts or exudation of fluid from the serosal surfaces secondary to chemical irritation.[16] In pancreatic ascites, peritoneal fluid amylase ranges from 680 to 129,500 U/L (370 to 70,000 Somogyi units/dL).[17] Pleural effusions are present in 5% to 15% of cases of pancreatitis. Increased amylase levels are caused by transdiaphragmatic lymphatic drainage or seepage of the enzyme across the diaphragm.[17] In a rare case pleural fluid is present because of a direct communication between the pleural and peritoneal cavities.[18]

Alkaline phosphatase has been shown to be a marker enzyme for pathological processes of the small intestine. The source of the enzyme can be leakage of alkaline phosphatase-rich fluid from the intestinal contents or extravasation from the wall of the intestine.[19] The enzyme is elevated in peritoneal serous effusions in association with intestinal perforation and infarctions of the small bowel and in peritoneal blood in patients with traumatic injuries of the small intestine.[20] The values are higher than corresponding peripheral blood levels.

Both urea nitrogen and creatinine are helpful in the differential diagnosis of a ruptured urinary bladder after abdominal trauma. Extravasated urine will have high levels of both urea nitrogen and creatinine. The former is freely diffusible and will also elevate the blood urea nitrogen; however, the peritoneum is relatively impermeable to creatinine and blood levels will not increase. In uncomplicated serous effusions the fluid urea nitrogen and creatinine are low. If the physician inadvertently aspirates urine from the bladder, both urea nitrogen and creatinine will be high but their concentrations in the blood will be normal.

SYNOVIAL FLUID

Joints are articulations between bones. Freely movable joints are composed of hyaline articular cartilage and a fibrous capsule lined on its inner surface by a membrane (Fig. 38-4). Synovial fluid fills the joint cavity and acts as a lubricant, keeping to a minimum the friction between bones during movement or weight bearing. The fluid also provides the sole nutrition for cartilage. Synovial fluid enters the cartilage by diffusion and by a spongelike effect when the cartilage is compressed and relaxed. The term *synovium* is derived from the Greek *syn* ("with") and the Latin *ovum* ("egg"), suggesting the fluid's resemblance to raw egg white.

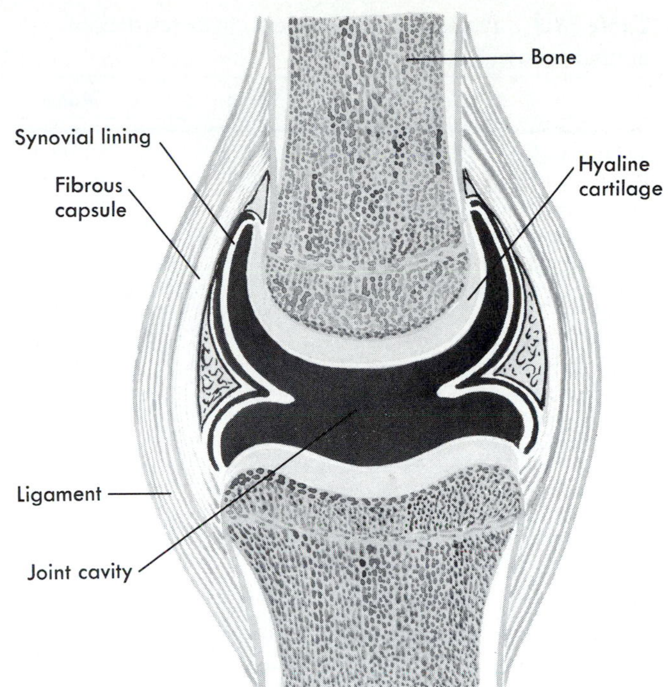

Fig. 38-4 Diagram of normal synovial joint. *(From Beck, EW: Mosby's atlas of functional human anatomy, St Louis, 1982, The CV Mosby Co.)*

Synovial fluid is a dialysate of plasma mixed with hyaluronic acid. Ultrafiltration in the rich vascular network in the synovial tissue produces this fluid, whereas hyaluronic acid, a mucoprotein, is secreted into the dialysate by synovial cells.

Normal synovial fluid

The fluid volume in the normal joint depends on the size of the structure. The knee joint usually contains 0.1 to 3.5 mL of fluid.[21] Normal synovial fluid is rarely obtained. If it is obtained, the fluid is clear or pale yellow with a specific gravity close to plasma but a high viscosity relative to water because of the protein-polysaccharide complex, hyaluronic acid. This composes 99% of the mucoproteins present in the fluid and is a long-chain, high-molecular-weight polymer made up of repeating units of acetyl glucosamine and glucuronic acid. The molecule is destroyed in inflammatory states by hyaluronidase, an enzyme contained in the neutrophils. When this occurs, the fluid viscosity decreases significantly, giving the clinician a bedside test for the presence of inflammatory fluid.

Synovial fluid protein concentrations are related to their molecular weights because of the dialytic effect in the synovium. Thus albumin is present in relatively higher

Table 38-3 Physical and chemical characteristics of normal synovial fluid

	Mean	Range
Volume (mL)*	1.1	0.13-3.5
Relative viscosity at 38° C	235	5.7-1160
Hyaluronic acid (g/L)	3600	1700-4050
Total protein (g/L)	17.2	10.7-21.3
Immunoglobulins (mg/L)		
IgG	4530	330-8500
IgA	740	270-1770
IgM	370	0-840
Fibrinogen (mg/L)	0	0
Complement (CH_{50} U/mL)	20†	16-25
Glucose (mg/L)	‡	650-1200
Uric acid (mg/L)	‡	25-72

*Knee.
†Values are approximately 10% of plasma values.
‡Fasting values are similar to plasma values.

concentrations than the higher molecular weight globulins are. Fibrinogen is not present because of its high molecular weight, and so normal synovial fluid does not coagulate. Glucose and uric acid diffuse freely into the synovial fluid and in the fasting state are concentrated as much as in plasma.

The characteristics of normal synovial fluid are summarized in Table 38-3.[22-25]

Change of analyte in disease

The physical and chemical changes that occur in the synovial fluid during disease reflect basic pathological processes occurring in the joint; therefore selected testing of chemical constituents is useful in predicting the underlying disease. A pathological classification of synovial fluids with the diseases associated with each category is summarized in Table 38-4. The laboratory tests discussed in this section include viscosity, fibrinogen, total protein, complement, glucose, and uric acid.

Clinically there is no need for a sophisticated measurement of viscosity. Instead, one measures it at the time of aspiration by placing a finger at the tip of the syringe and stringing out the fluid. Noninflammatory fluids will "string out" longer than 4 cm. One can also drip fluid off the needle and syringe, observing it string out if it is a noninflammatory fluid and drip like water if it is the result of inflammation. The depolymerization of the hyaluronic acid by neutrophil hyaluronidase decreases the viscosity in inflammatory disease.

The mucin clot test *(Ropes' test)* is seldom advocated but is frequently described in the literature. This test reflects the degree of hyaluronate polymerization. One performs it by dropping fluid into a dilute acid solution and determining the quality of the clot. Although the test is often mentioned without critical comment in discussions of synovial fluid, the quality of the information obtained by use of this test is inferior to that obtained by use of other procedures, and the test should be considered obsolete.

The practical importance of fibrinogen is not in its measurement but in its presence. Normal synovial fluid contains no fibrinogen, but because inflammatory synovitis permits the passage of large molecular weight proteins into the fluid, fibrinogen can be present and spontaneous clotting can occur. Thus anticoagulants are necessary when specimens are collected for microscopic and bacteriological examination.

Unlike serous fluids, the synovial fluid total protein concentration is not used to distinguish noninflammatory from inflammatory fluids, because the leukocyte count is used to make that distinction. Thus total protein is not included in the routine examination of synovial fluids; however, its measurement can be helpful in interpreting complement levels.[26] Complement proteins normally have low values in synovial fluid compared to serum.[27-29] In systemic inflammatory conditions, complement behaves as an acute-phase reactant and there is hypercomplementemia. In some conditions, such as Reiter's disease, the joint fluid comple-

Table 38-4 Pathological classification of synovial fluids

Test	Noninflammatory	Inflammatory	Septic	Hemorrhagic
Volume (mL)	>3.5	>3.5	>3.5	>3.5
Color	Yellow	Yellow-white	Yellow-green	Red-brown
Viscosity	High	Low	Low	Low
Leukocytes (cells/μL)	200-2000	2000-100,000	10,000->100,000	>500
Neutrophils (%)	<25	>50	>75	>25
Glucose (mg/L)	~ serum	>250 mg/L lower than serum	>250 mg/L lower than serum	~ serum
Culture	Negative	Negative	Positive	Negative
Diseases	Osteoarthritis Osteochondritis dissecans Osteochondromatosis Traumatic arthritis Neuroarthropathy	Gout Pseudogout Psoriatic arthritis Reiter's syndrome Rheumatoid arthritis Systemic lupus erythematosus	Bacterial infection Fungal infection Tuberculous infection	Hemophilia Trauma Pigmented villonodular synovitis

ment has been reported to be even higher than that in the serum.

In systemic immune complex diseases such as systemic lupus erythematosus (SLE), complement is consumed widely and can be low in both the serum and synovial fluid. In other diseases, such as rheumatoid arthritis and viral synovitis, complement is consumed locally in the synovia while serum levels are usually normal or high.

C_3 and C_4 are more stable than CH_{50}, which will be falsely low if the fluid sits out at room temperatures. A decreased C_4 level suggests classical pathway activation by immune complexes and is more sensitive than C_3 or CH_{50}. The latter two are low in both classical or alternative pathway activation of complement and, if low, suggest more profound systemic activity. The proper approach to interpreting complement levels in synovial fluid is controversial. For practical purposes compare synovial fluid and serum complement levels and consider synovial fluid complement low if it is less than 30% of serum levels. However, in SLE and other severe immune complex diseases both levels may be low. In such a situation one can compare the synovial fluid and serum complement levels with total protein in each fluid.

Interpretation of synovial fluid glucose levels requires knowledge of the patient's simultaneous serum glucose. This is best done in the fasting state, but such preparation is not always clinically feasible. In the ideal situation after an 8-hour fast, the difference between serum and synovial fluid is less than 100 mg/L; levels 250 mg/L or more below the serum level suggest inflammation, and differences greater than 400 mg/L suggest sepsis. In the nonfasting state, synovial fluid glucose levels less than half of serum levels should definitely arouse suspicion of a septic process. Rarely, such findings are noted also in rheumatoid arthritis effusions. Lactic acid has been advocated in the diagnosis of septic arthritis; however, this test has not gained general acceptance.[30,31]

Serum uric acid levels are important in the diagnosis of gout. The synovial fluid uric acid level is similar to that of serum. Its measurement in synovial fluid is of no diagnostic value[1]; however, formation of monosodium urate crystals and their identification by polarized light microscopy in synovial fluid is the cardinal diagnostic feature of gouty arthritis.

REFERENCES

1. Krieg, AF: Cerebrospinal fluid and other body fluids. In Henry, JB, editor: Clinical diagnosis and management by laboratory methods, Philadelphia, 1979, WB Saunders Co, pp 657-667.
2. Light, RW, MacGregor, MI, Luchsinger, PC, and Ball, WC, Jr: Pleural effusions: the diagnostic separation of transudates and exudates, Ann Intern Med 77:507-513, 1972.
3. Pare, P, Talbot, J, and Hoefs, JC: Serum-ascites albumin concentration gradient: a physiologic approach to the differential diagnosis of ascites, Gastroenterology 85:240-244, 1983.
4. Rector, WG, Jr, and Reynolds, TB: Superiority of the serum-ascites albumin difference over the ascites total protein concentration in the separation of "transudative" and "exudative" ascites, Am J Med 77:83-85, 1984.
5. Light, RW, and Ball, WC, Jr: Glucose and amylase in pleural effusions, JAMA 225:257-260, 1973.
6. Carr, DT, and Mayne, JG: Pleurisy with effusion in rheumatoid arthritis, with reference to the low concentration of glucose in pleural fluid, Am Rev Respir Dis 85:345-350, 1962.
7. Dodson, WH, and Hollingsworth, JW: Pleural effusion in rheumatoid arthritis: impaired transport of glucose, N Engl J Med 275:1337-1342, 1966.
8. Potts, DE, Levin, DC, and Sahn, SA: Pleural fluid pH in parapneumonic effusions, Chest 70:328-331, 1976.
9. Light, RW: Management of parapneumonic effusions, Chest 70:325-326, 1976.
10. Staats, BA, Ellefson, RD, Budahn, LL, et al: The lipoprotein profile of chylous and nonchylous pleural effusions, Mayo Clin Proc 55:700-704, 1980.
11. Sherr, HP, Light, RW, Merson, MH, et al: Origin of pleural fluid amylase in esophageal rupture, Ann Intern Med 76:985-986, 1972.
12. Dye, RA, and Lafaret, EG: Esophageal rupture: diagnosis by pleural fluid pH, Chest 66:454-456, 1974.
13. Abbott, OA, Mansor, KA, and Logan, WD: Atraumatic so-called spontaneous rupture of the esophagus, J Thorac Cardiovasc Surg 59:67-82, 1970.
14. Geokas, MC, Olsen, H, Carmack, C, and Rinderknecht, H: Studies on the ascites and pleural effusion in acute pancreatitis, Gastroenterology 58:950, 1970.
15. Keith, LM, Zollinger, RM, and McCleery, RS: Peritoneal fluid amylase determinations as an aid in diagnosis of acute pancreatitis, Arch Surg 61:930-936, 1950.
16. Donowitz, M, Kerstein, MD, and Spiro, HM: Pancreatic ascites, Medicine 53:183-195, 1974.
17. Salt, WB, and Schenker, S: Amylase—its clinical significance: a review of the literature, Medicine 55:269-289, 1976.
18. Goldman, M, Goldman, G, and Fleischner, FG: Pleural fluid amylase in acute pancreatitis, N Engl J Med 266:715-718, 1962.
19. Lee, YN: Alkaline phosphatase in intestinal perforation, JAMA 208:361, 1969.
20. Delany, HM, Moss, CM, and Carnevale, N: The use of enzyme analysis of peritoneal blood in the clinical assessment of abdominal organ injury, Surg Gynecol Obstet 42:161-167, 1976.
21. Ropes, MW, Rossmeisl, EC, and Bauer, W: The origin and nature of normal human synovial fluid, J Clin Invest 19:795-799, 1940.
22. Hamerman, D, and Schuster, H: Hyaluronate in normal synovial fluid, J Clin Invest 37:57-64, 1958.
23. Hoeprich, PD, and Ward, JR: The fluids of the parenteral body cavities, New York, 1959, Grune & Stratton, Inc.
24. Pekin, TJ, and Zvaifler, NJ: Hemolytic complement in synovial fluid, J Clin Invest 43:1372-1382, 1964.
25. Pruzanski, W, Russell, ML, Gordon, DA, and Ofryzlo, MA: Serum and synovial fluid proteins in rheumatoid arthritis and degenerative joint diseases, Am J Med Sci 265:483-490, 1973.
26. Bunch, TW, Hunder, GG, McDuffie, FC, et al: Synovial fluid complement determination as a diagnostic aid in inflammatory joint disease, Mayo Clin Proc 49:715-720, 1974.
27. Fostiropoulos, G, Austen, KF, and Block, KJ: Total hemolytic complement (CH_{50}) and second component of complement (C′2hu) activity in serum and synovial fluid, Arthritis Rheum 8:219-232, 1965.
28. Lundh, B, Hedberg, H, and Laurell, AB: Studies of the third component of complement in synovial fluid from arthritis patients. I. Immunochemical quantitation and relation to total complement, Clin Exp Immunol 6:407-411, 1970.
29. Ruddy, S, and Austen, KF: The complement system in rheumatoid synovitis. I. An analysis of complement component activities in rheumatoid synovial fluids, Arthritis Rheum 6:713-722, 1970.
30. Brook, I, Reza, MJ, Bricknell, KS, et al: Synovial fluid lactic acid: a diagnostic aid in septic arthritis, Arthritis Rheum 21:774-779, 1978.
31. Borenstein, DG, Gibbs, CA, and Jacobs, RP: Gas-liquid chromatographic analysis of synovial fluid: succinic acid and lactic acid as markers of septic arthritis, Arthritis Rheum 25:947-953, 1982.

CHAPTER 39

Central nervous system

GERALD MORIARTY

OBJECTIVES

- Describe the formation of cerebrospinal fluid.
- List the major characteristics of normal cerebrospinal fluid.
- Describe the common pathological conditions affecting the central nervous system.
- State expected levels of proteins, glucose, and lactic acid in pathological conditions.

KEY TERMS

anticonvulsant drugs Drugs given therapeutically to prevent seizures of various types.

arachnoid membrane The middle membrane of the three membranes covering the brain.

blood-brain barrier The barrier between the brain and its membranes and extravascular fluid that maintains a specified composition of cerebrospinal fluid.

cerebrospinal fluid (CSF) Clear, colorless fluid contained within the four ventricles of the brain, the subarachnoid space, and the spinal cord.

choroid plexus Vascular folds in the pia membrane of the third, fourth, and lateral ventricles that synthesize cerebrospinal fluid.

coma A state of unconsciousness from which patients cannot be aroused, even by the strongest stimuli.

dura The outermost of the three membranes covering the brain; it is the toughest and most fibrous of the three.

epilepsy Transient disturbances of brain function involving impairment of consciousness, motor function, and other neurological functions of the brain.

IgG index The ratio (CSF IgG $\times$ serum albumin)/(serum IgG $\times$ CSF albumin) used as an indicator of the source of elevated cerebrospinal fluid protein, that is, within the central nervous system versus leakage across the blood-brain barrier.

meninges The three membranes covering the brain and spinal cord; dura, arachnoid, and pia.

meningitis Inflammation of the meninges, often caused by viral or bacterial infections.

pia The innermost of the three membranes, directly covering the brain and spinal cord.

seizure Sudden attack of epilepsy.

stroke Sudden onset of symptoms caused by acute ischemia in the brain resulting from hemorrhage, embolism, and so on; it is evidenced by loss of neurological functions.

subarachnoid space The space between the arachnoid and pia membranes.

ventricles Four cavities within the brain filled with cerebrospinal fluid and lined by the pia and the choroid plexus.

xanthochromia A yellow coloring to the cerebrospinal fluid caused by the presence of breakdown products of hemoglobin.

BASIC NEUROANATOMY

The central nervous system (CNS) consists of the brain and spinal cord. The brain includes the two cerebral hemispheres, which are roughly mirror images of one another; the brainstem, a narrow structure through which all the pathways entering and leaving the two hemispheres must pass, which contains the centers that control breathing, heart rate, eye movements, and many other critical functions; and the cerebellum, a rounded structure about the size of a baseball that helps control movement and balance (Fig. 39-1). The cerebellum is attached to the back of the brainstem and is located just beneath the cerebral hemi-

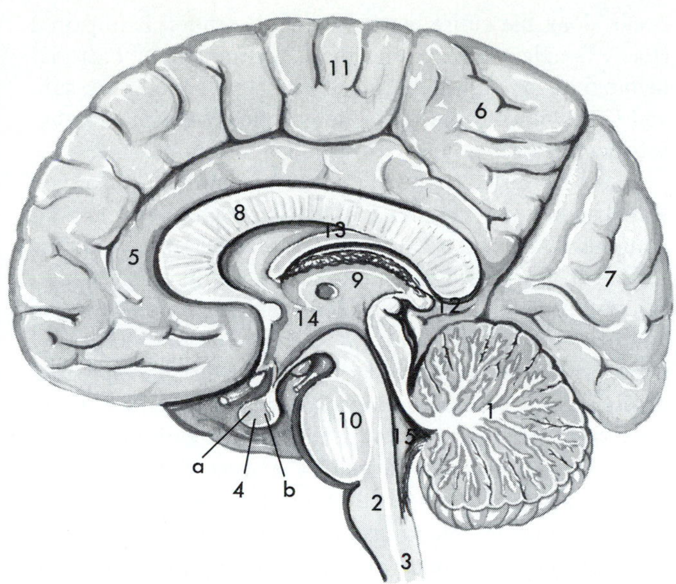

Fig. 39-1 Scheme of functional or motor-control areas of brain (right hemisphere, medial view). *1,* Cerebellum; *2,* medulla oblongata; *3,* spinal cord; *4,* pituitary gland, *a,* anterior lobe, *b,* posterior lobe; *5,* frontal lobe; *6,* parietal lobe; *7,* occipital lobe; *8,* corpus callosum; *9,* thalamus; *10,* pons; *11,* cerebrum; *12,* pineal body; *13,* fornix; *14,* third ventricle; *15,* fourth ventricle. *(From Beck, EW: Mosby's atlas of functional human anatomy, St Louis, 1982, The CV Mosby Co.)*

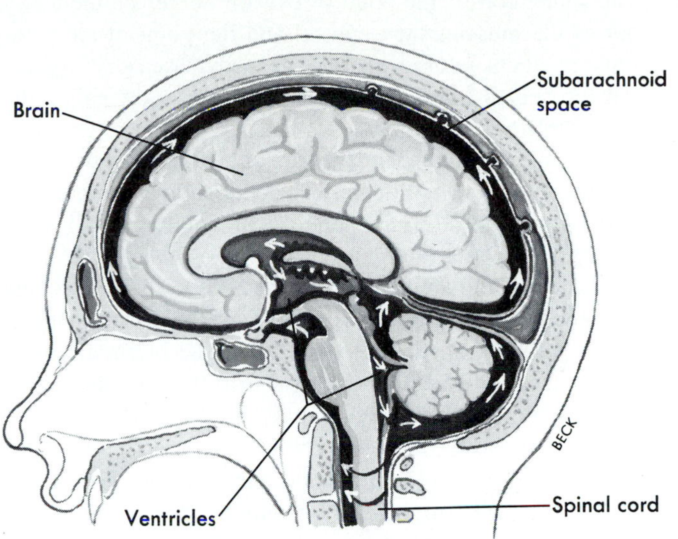

Fig. 39-2 Scheme of brain showing relationships of ventricles and subarachnoid space with rest of brain. *(From Beck, EW: Mosby's atlas of functional human anatomy, St Louis, 1982, The CV Mosby Co.)*

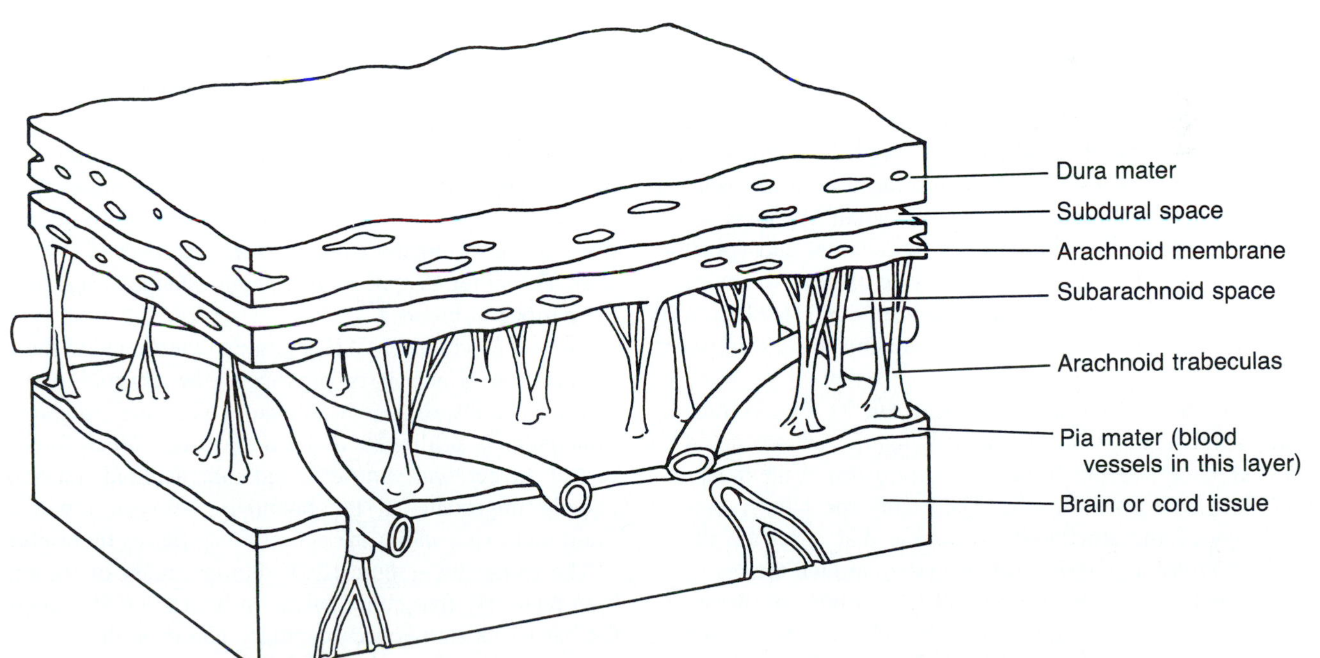

Fig. 39-3 Scheme of meninges. Arrangement may be compared with an underground parking garage. Dura and arachnoid form roof with pia membrane as floor. CSF flows in subarachnoid space. *(From Prezbindowski, KS: Guide to learning anatomy and physiology, St Louis, 1982, The CV Mosby Co.)*

spheres. The lower brainstem flows into the spinal cord. The spinal cord is the point of exit for nerves on their way out to the muscles they control and the point of entry for sensory fibers returning from the body's sensory organs. All the nerves outside the CNS are collectively called the *peripheral nervous system*.

The two cerebral hemispheres are built around a connecting system of hollow spaces called the *ventricular system*. The *ventricles* are filled with cerebrospinal fluid (CSF) (Fig. 39-2).

The brain and spinal cord are both covered by a double membrane called the *meninges* (Fig. 39-3). Its inner membrane, called the *pia,* lies directly on the brain. Its outer membrane, the *arachnoid,* lies next to the outermost covering of the brain and spinal cord, the *dura*. The dura is a tough, nonelastic membrane that essentially wraps the brain and spinal cord in a nondistensible sac. The brain, blood, and CSF are thus sealed within a space whose volume is fixed. The space between the pia and the arachnoid is called the *subarachnoid space* and communicates directly with the ventricular system.

PHYSIOLOGY AND BIOCHEMISTRY

Formation of cerebrospinal fluid

The ventricular system and the subarachnoid space are filled with CSF. The total volume of CSF in adults is about 150 mL, including 30 mL within the cerebral ventricles and 120 mL in the subarachnoid space (including its spinal segment). CSF is constantly produced and reabsorbed at a rate of approximately 500 mL/day (0.35 mL/min). This means the total amount of CSF is replaced every 4 to 6 hours.

CSF is produced in the ventricles by a specialized spongelike structure called the *choroid plexus*. Beginning in the lateral ventricles, where it is formed, it circulates into the third ventricle and then into the fourth ventricle. It leaves the fourth ventricle by three small openings, or foramina, to circulate through the intracranial and spinal subarachnoid spaces. Circulation may be blocked in any of the ventricles or at the foramina between them, leading to an *obstructive hydrocephalus* (accumulation of fluid in the brain).

If new CSF is constantly produced but its total volume remains constant, it has to be leaving the subarachnoid space in some fashion. In fact, it is *absorbed* at the arachnoid villi and *granulations*. These are specialized outpouchings of the arachnoid membrane that penetrate the dura and protrude into the venous system around the brain. These small structures are actually labyrinths of folded tubes whose walls separate the subarachnoid space from the venous space and allow fluid to flow in only one direction by their valvelike action. These arachnoid villi and granulations are scattered along the entire inner table of the skull and down the spinal canal to the points at which the spinal nerves exit the dura. Thus CSF reabsorption can occur along the entire neuroaxis. If absorption is impaired (for example, after meningeal inflammation, bacterial meningitis, or subarachnoid hemorrhage), CNS pressure and CSF volume both rise; this is called a *communicating hydrocephalus*.

Factors that determine the rate at which CSF is formed and absorbed are complex and not completely understood. Any increase in the size of one component (that is, brain, CSF, or blood) leads to a sharp increase in pressure within the system unless there is a corresponding decrease in the volume of one of the other two components. For example, if the total volume of CSF increases (through increased production or decreased reabsorption), there is less room for blood, and the pressure in the system and around the blood vessels increases overall. The same is true if brain volume increases through swelling: there is more pressure and less room for blood and CSF. If volume is added to the dural space, the brain may suffer from direct effects of the abnormally high pressure or perhaps from having its blood flow decreased or through a variety of other mechanisms.

Blood-brain barrier

In 1885 Paul Ehrlich discovered that if an aniline dye is injected intravenously it will stain the dura and many body tissues but fail to stain the brain or CSF. But if dye is injected directly into the subarachnoid space (by the intracisternal route), it will diffuse through the CSF and stain brain tissue. In 1921 Stern and Gautier used the term *blood-brain barrier* to refer to a physiological barrier separating brain and CSF from substances borne in the blood. As described later, the blood-brain barrier allows brain and CSF composition to be maintained at levels quite different from those of blood with respect to proteins, ions, and other molecular elements. The blood-brain barrier is extremely important in clinical practice. It determines the access of antibiotics to the brain and meninges and contributes to the exquisite control exercised over the brain's chemical milieu despite simultaneous changes occurring in the peripheral blood.

The composition of CSF is largely determined at the cell surfaces on which it is produced (the choroid plexus); where it is absorbed (the arachnoid villi and pacchionian granulations); at the arachnoid membrane, which lines the rest of the cerebral ventricles and subarachnoid space; and equally important, at the boundary between blood and brain that exists along the cerebral capillaries themselves.

The extracellular fluid (ECF) compartment of the brain is in relatively free communication with the CSF, whereas the barrier exists between capillary blood on the one hand and the ECF compartment and CSF on the other.

Factors that significantly influence the access of substances to brain and CSF include molecular weight, protein binding, and lipid solubility. With molecular weight, entry is inversely related to size, hence the 1:200 CSF-to-plasma ratio of albumin (molecular weight, 69,000 daltons). Drugs that are highly protein bound enter the CSF much less readily than unbound smaller molecular weight substances. For example, phenytoin is 95% protein bound and 5% free in blood. Only 5% of the total measured blood phenytoin level is easily able to enter the CNS, and only that 5% is "active." In certain cases a clinician may need to know both the free and bound fractions to determine the level of active drug at a given dosage. Calcium, magnesium, and metabolites such as bilirubin are also highly protein bound and thus relatively restricted from CSF. Highly lipid-soluble substances such as carbon monoxide and neuroactive drugs and alcohol readily enter the CNS. Substances that are highly ionized at physiological pH are relatively excluded. Highly polar substances, such as some amino acids, enter slowly and require an active-transport mechanism.

Bicarbonate equilibrates slowly between CSF and blood, whereas the opposite is true of carbon dioxide. Thus, if bicarbonate is infused intravenously, serum pH rises but CSF pH drops for a period until equilibration occurs. This happens because systemic alkalosis is accompanied by respiratory slowing and CO_2 retention. Increasing CO_2 immediately enters the CNS while CSF bicarbonate is still low, and since pH is determined by the ratio of bicarbonate to CO_2, CSF pH drops paradoxically. Thus in cases of established systemic acidosis, such as diabetic ketoacidosis, the rapid venous infusion of bicarbonate can lead to coma.

The blood-brain barrier is readily permeable to water but not to electrolytes. The major cations, sodium and potassium, require hours to reach equilibrium with CSF after changes in peripheral blood. Changes in blood osmolality are followed by parallel CSF changes after a lag time of a few hours.

In the case of drugs, their pK_a, or *ionic dissociation constant,* is important in determining how readily they cross the blood-brain barrier. pK_a refers to the pH at which 50% of a compound is ionized. A nonionized drug is relatively lipid soluble and so more freely enters the CNS. The polar ionized fraction is relatively excluded. The degree of ionization of a drug at any pH (for example, physiological pH of 7.4 for blood, 7.32 for CSF) is given by the Henderson-Hasselbalch equation:

$$\text{Weak-acid pH} = pK_a + \log\frac{[\text{Salt}]}{[\text{Acid}]}$$

$$\text{Weak-base pH} = 14 - pK_a - \log\frac{[\text{Salt}]}{[\text{Base}]}$$

Thus phenytoin, with a pK_a equal to 8.3, is almost entirely un-ionized at pH 7.4, and so its steady-state concentration in brain and CSF is determined by the free or non-protein bound fraction in blood.

Stated somewhat differently, weak acids are relatively excluded from acidic compartments (those with low pH). Since phenobarbital is a weak acid, barbiturate intoxication can be more of a problem in a setting of systemic acidosis, since that condition favors its entry into the CNS with consequently higher CNS concentrations.

Although these remarks illustrate the importance of the blood-brain barrier, they only begin to suggest the complexity of factors that determine the access of substances to the brain. In practice, blood levels of some compounds and even their CSF levels may give little indication of their concentration or activity within actual neural tissue. For example, it is not uncommon to see a delirium caused by lithium intoxication continue to worsen into coma and even death after serum levels of lithium have begun to drop. Thus one must consult available literature for details about individual substances and remain aware that frequently in clinical medicine a discrepancy remains between what theory predicts and what actually occurs.

Finally, the characteristics of the blood-brain barrier can be dramatically altered by disease states. Penicillin, an acidic substance, is normally excluded from the CNS after parenteral injection, and yet it is an effective agent for treating meningitis. The reason is that meningeal inflammation alters (damages) the blood-brain barrier, allowing greater access of drugs, such as penicillin, that normally would not reach infected tissue.

Some specific factors that alter permeability of the blood-brain barrier are as follows:

1. Inflammation can increase the ease of entry into the nervous system of macromolecules such as albumin and penicillin.
2. Neovascularity, in association, for example, with tumor, trauma, or ischemia, alters the blood-brain barrier. This may be caused by defects in the new vessels or by their immaturity.
3. Toxins can change blood-brain barrier characteristics, and some agents used in radiographic studies (such as iodopyracet [Diodrast] and diatrizoate meglumine [Hypaque]) increase its permeability by direct toxic effects. When they are injected in hyperosmolar concentrations, the effect is greater.
4. Adrenal steroids and thyroid hormone actually help stabilize the integrity of the blood-brain barrier. Destabilization may contribute to the elevated CSF protein frequently seen in hypothyroid states.
5. Finally, the blood-brain barrier of the immature nervous system is more permeable to a variety of substances. For infants below the age of 6 months, CSF protein is normally as high as 1000 mg/L.

Functions of cerebrospinal fluid

Why should the brain be suspended in and bathed by this distinctive fluid? First, CSF provides mechanical support to the brain, and a "floating brain" weighs less than if it were simply resting on the bony table of the skull. Second, CSF probably functions to help remove metabolic products or waste from the brain, a function that is poorly understood but probably important in both normal and diseased states. Third, there is some evidence that CSF transports biologically active compounds that may function as chemical messengers. Finally, it plays an important role in maintaining the chemical environment of the brain. Although its communication with the plasma compartment is tightly regulated, CSF seems to be in relatively free communication with the brain's extracellular fluid compartment, which aids brain cells themselves. The following section examines the composition of CSF more closely.

CHARACTERISTICS OF NORMAL SPINAL FLUID

Total volume: 150 mL
Color: colorless, like water
Transparency: clear, like water
Osmolarity at 37° C: 281 mOsm/L
Specific gravity: 1.006 to 1.008
Acid-base balance:

pH	7.31
Pco_2	47.9 mm Hg
HCO_3^-	22.9 mEq/L

Sodium: 138 to 150 mEq/L
Potassium: 2.7 to 3.9 mEq/L
Chloride: 116 to 127 mEq/L
Calcium: 2 to 2.5 mEq/L (40 to 50 mg/L)
Magnesium: 2 to 2.5 mEq/L (24.4 to 30.5 mg/L)
Lactic acid: 1.1 to 2.8 mmol/L
Lactic acid dehydrogenase: Absolute activity depends on method; approximately 10% of serum value
Glucose: 450 to 800 mg/L
Proteins: 200 to 400 mg/L
At different levels of spinal tap:

Lumbar	200 to 400 mg/L
Cisternal	150 to 250 mg/L
Ventricular	150 to 100 mg/L

Normal values in children:

Up to 6 days of age	700 mg/L
Up to 4 years of age	244 mg/L

Electrophoretic separation of spinal fluid proteins (% of total protein concentration):

Prealbumin	2% to 7%
Albumin	56% to 76%
α_1-globulin	2% to 7%
α_2-globulin	3.5 to 12%
β- and γ-globulin	8% to 18%
γ-globulin	7% to 12%
IgG	10 to 40 mg/L
IgA	0 to 0.2 mg/L
IgM	0 to 0.6 mg/L
κ/λ ratio	1

Erythrocyte count:

Newborn	0 to 675/mm^3
Adult	0 to 10/mm^3

Leukocyte count:

<1 year of age	0 to 30/mm^3
1 to 4 years of age	0 to 20/mm^3
5 years of age to puberty	0 to 10/mm^3
Adult	0 to 5/mm^3

Composition of cerebrospinal fluid (see box at left)

The ionic and molecular composition of CSF differs from that of plasma for some components and is the same for others. Changes in serum sodium are followed by corresponding changes in CSF sodium so that after a lag time of about 1 hour sodium values are nearly the same. However, CSF potassium is lower than plasma potassium, and furthermore potassium is maintained within a very narrow concentration range in CSF despite wide fluctuations in plasma values. Active transport in and out of the CSF space appears to be largely responsible for maintaining these differences. Chloride and magnesium are somewhat higher in CSF than in plasma, and bicarbonate is somewhat lower.

CSF glucose normally ranges from 450 to 800 mg/L, that is, between 60% and 80% of the blood glucose concentration after equilibration. Blood and CSF glucose equilibrate only after a lag period of about 4 hours, so that CSF glucose at a given time reflects blood glucose levels during the past 4 hours. When a lumbar spinal puncture (LP) is performed and CSF glucose is to be determined, a simultaneous sample of peripheral blood must also be drawn. When glucose determination is critical (as it may be, for example, in the diagnosis of bacterial or carcinomatous meningitis), the LP and blood glucose should be obtained only after the patient has fasted for at least 4 hours. CSF glucose is altered by certain disease processes, as is discussed later. Equilibrated CSF glucose is definitely abnormal when it is less than 40% of the simultaneous blood glucose value; values less than 400 to 450 mg/L are almost always abnormal.

One should also be aware that the expected ratio of CSF to blood glucose (60% to 80%) falls as blood glucose rises. That is, one would expect a CSF-to-blood ratio of 0.5 when blood glucose values reach 5000 mg/L and a ratio of 0.4 if blood glucose reaches 7000 mg/L.

Proteins found in the CSF ordinarily originate from serum and reach the CSF space by pinocytosis across the capillary endothelium. The normal ratio of serum to CSF protein is 200:1 (with serum equal to 70 g/L and CSF equal to 350 mg/L).

Brain metabolism

The brain's metabolic rate is one of the highest of any of the body's organs, whether one is awake or asleep. But unlike most other organs, which store and reserve some supplies of energy to sustain themselves, the brain has almost no energy reserve. It depends entirely on an uninterrupted supply of glucose and oxygen delivered by peripheral blood. The brain uses glucose almost exclusively to supply its energy needs. To get an idea just how hungry the brain is and how dependent on a constant, swift flow of blood, consider that under resting conditions total cerebral blood flow equals 15% to 20% of cardiac output (or about 500 mL per 100 g of brain per minute). The percentage of cardiac output flowing to the adult brain is much higher than that in infants and young children. Although *total* cerebral blood flow remains remarkably constant, discrete areas within the brain show striking variability. Gray matter (predominantly composed of cell bodies) enjoys a flow of blood three to four times greater than that of white matter (predominantly composed of the fiber connections between nerve cells). Gray matter lies on the surface of the brain (the *cortex*) and is also arranged in some nuclear structures deep inside it. The rest of the brain is composed of white matter connections. Moreover, *regional* blood flow is known to vary during performance of certain tasks, with regional flow increasing in the appropriate areas during tasks such as hand movement, speaking, or mental problem solving. Blood flow is also altered in response to disease states, as in stroke.

The brain depends exclusively on respiration and glycolysis to supply its energy needs. It needs energy for several critical reasons: to maintain membrane potentials, to continue transmitting impulses, and continuously to synthesize new structural elements (since in any living organ, structural components are being constantly broken down and replaced). Available energy is packaged in high-energy phosphate bonds, and the compound that contains the bonds is ATP. Glycolysis and respiration produce energy or ATP as described in Chapters 28 and 29.

Respiration provides 38 mol of ATP per mol of glucose compared with only 2 mol of ATP derived from glycolysis. It is uncertain how much, if anything, glycolysis contributes to the brain's energy metabolism under normal conditions (very low levels of lactate are found in the brain under normal conditions, and lactate is derived from glycolysis).

It is clear that glucose is the predominant substrate for cerebral metabolism, with 0.31 μmol (5.5 mg) of glucose used per minute per each 100 g of brain tissue. During starvation or in states of ketoacidosis, ketone bodies can be used as metabolic substrates and can provide up to 30% of the brain's oxidative metabolic needs. Under ordinary circumstances, however, potential substrates other than glucose, such as fatty acids, have their access to the brain narrowly limited by the blood-brain barrier. Glucose, on the other hand, moves readily across the blood-brain barrier by a process of facilitated *transport*. Once available, about 35% of the entering glucose is rapidly metabolized to CO_2, and the remainder is used for synthetic and structural needs by being incorporated into amino acids, proteins, and lipids.

PATHOLOGICAL CONDITIONS

Damage to *discrete* (focal) areas of the brain or spinal cord produce predictable circumscribed signs and symptoms (for example, paralysis of an arm, leg, or side of the body, loss of ability to speak or comprehend spoken language, incoordination, and so on). *Diffuse* impairment of cerebral tissue, on the other hand, leads to its own characteristic clinical picture. Failure of various intellectual functions such as attention, concentration, judgment, memory, problem-solving ability, and insight are early findings with mild diffuse disease. Other symptoms include changes in alertness beginning with clouding of consciousness and proceeding to drowsiness, stupor, and coma. Seizures can accompany both diffuse and focal damage.

Various disease states tend to produce either focal or diffuse brain damage, so that the pattern of deficits is often helpful to the clinician in working backward toward a specific diagnosis. Some examples of conditions that cause focal damage are stroke caused by arterial occlusion or hemorrhage; trauma; cerebral abscess; and tumors. Many of these conditions also cause changes in the CSF by damaging the blood-brain barrier (elevating protein) and stim-

Table 39-1 Causes of coma and altered mental states

Type	Cause	Laboratory findings
Metabolic	Alcoholism	Increased blood ethanol, metabolic acidosis, and ketosis
	Hyperosmolar coma	Blood glucose ≥10,000 mg/L, no ketosis, dehydration
	Diabetic ketoacidosis	Increased blood glucose, ketosis, acidosis, dehydration
	Metabolic acidosis of other origin	Decreased pH, HCO_3^-; increased lactic acid
	Hypoglycemia	Decreased blood glucose (<500 mg/L)
	Hypercalcemia or hypocalcemia	Changes in calcium levels; hypomagnesemia can be found with hypocalcemia
	Drugs	Presence of any of a number of drugs, often at very high levels
Systemic metabolic diseases	Hepatic coma	Increase in blood ammonia, increased liver function tests
	Uremic coma	Increased serum urea, creatinine with metabolic acidosis
	Ischemic, cardiac, pulmonary	Tissue hypoxia, lactic acidosis
Encephalopathy	Epilepsy	Subtherapeutic levels of antiepileptic drugs
	Intracranial hemorrhage	Bloody spinal tap
Trauma	—	None
Infectious	Bacterial, viral	Decreased CSF glucose, increased protein
Psychiatric	—	None

ulating inflammatory changes (with leukocytosis), tissue necrosis (elevating CSF protein and cell count), or shedding of tumor cells in cytological specimens.

Examples of conditions associated with generalized cerebral dysfunction (encephalopathic states) are anoxia, generalized ischemia, hypoglycemia, sepsis, thyroid abnormalities, disseminated intravascular coagulation, and the entire group of toxic and metabolic derangements. The diagnosis of these states often rests on laboratory findings.

The clinical and pathological changes commonly found in a number of conditions are briefly discussed. Not all the neurological diseases known to produce changes in laboratory values are discussed, and none are presented in depth.

Coma

Coma is most simply defined as a state of unconsciousness from which the patient cannot be aroused. A coma is but one aspect of altered states of consciousness that can be present in patients. *Confusion* is the least altered state, in which there is disorientation with respect to time, associated drowsiness, and altered attention span. *Stupor* is a state in which the patient is unresponsive but can be aroused back to a near-normal state with appropriate stimuli.

A patient with an altered mental state, as seen acutely in an emergency room, must first be given any life support necessary to maintain vital functions, such as ventilation. The next step is to determine the underlying cause of the altered mental status. Readily treatable causes can be treated if they exist, such as administration of dextrose to relieve coma caused by severe hypoglycemia. Table 39-1 lists the most important causes of coma and altered mental states, which include metabolic, structural, or infectious causes. Many of these causes are described in detail below.

Intracranial bleeding

Bleeding from a vessel on the surface of the brain, such as an arterial aneurysm, pours blood between the brain's surface and the pia and arachnoid and is called a *subarachnoid hemorrhage*. Blood thus mingles with CSF, and red blood cells appear in it. Furthermore, because blood is an extremely irritating substance when it escapes from its usual vascular channels, it may provoke an inflammatory response in the meninges called a *chemical meningitis*. That is, leukocytes are shed into the CSF by the irritated meninges; and since meninges are pain sensitive, a subarachnoid hemorrhage is very painful (patients complain of the worst headache in their lives).

Infectious and inflammatory disease

Both the bacterial or viral organism and the intracranial structures they invade help determine CSF changes seen in the infectious process. The CSF parameters that reflect CNS invasion by an infectious agent are white cell count and differential, glucose, and to a lesser extent protein concentration.

Meningitis is an inflammation of the meninges leading to several clinical patterns. Bacterial or *purulent* meningitis is associated with a CSF polymorphonuclear (PMN) leukocytosis, with counts ranging from a very few to many hundreds—even frank pus-containing countless PMNs. The glucose levels may be depressed to strikingly low values, and protein may be elevated. Later in the course of illness lymphocytes can become prominent or dominant, especially if the infection has been partially treated with antibiotics. Partial treatment of bacterial meningitis obscures the CSF findings, making diagnosis difficult, but fails to eradicate the disease—a potentially treacherous circumstance.

Viral meningitides cause a predominantly or exclusively lymphocytic leukocytosis. Glucose usually remains within the range of normal, and protein is usually normal or only slightly elevated.

Fungi also invade the CNS and may cause no change in CSF other than a lymphocytosis and increased protein. Glucose usually remains normal.

Bacteria should be searched for on Gram stain and culture. Viruses can be sought with appropriate serological tests or even culture, and fungi can be found with culture or immunological procedures as well as with appropriate staining. (India ink, for example, may reveal *Cryptococcus* species if carefully performed.)

The CSF findings in syphilis depend on the stage of illness, disease activity, and whether previous treatment was given in adequate amounts. The subject should be reviewed in a more detailed text.

Organisms can invade the brain substance, in which case the term *encephalitis* is used. CSF findings are comparable to findings in meningitis or may be quite minimum.

Abscess formation in the brain may produce no CSF changes even though a potentially deadly infection is rampant.

Finally, any of the above conditions can and frequently do lead to increased CSF pressure. This is especially true of meningitides, which obstruct the usual flow of CSF, causing an obstruction hydrocephalus. Meningitis can also impair CSF absorption, causing a communicating hydrocephalus.

Ischemia

Although immensely dependent on glucose, the brain's own reserves of that substance are tiny. It contains about 1 mmol of free glucose per kilogram and 3 mmol of glycogen per kilogram (70% of which can be immediately converted to glucose). *With oxygen available* and the brain's metabolic needs being supplied by respiration using only glucose stored in the brain itself, those stores could supply the brain's needs for only 2 to 3 minutes. If *both* glucose and oxygen are cut off—as in cardiac arrest—glycolysis would become the dominant source of energy. In that relatively inefficient state, glucose stores in the brain could support its energy metabolism for only about 14 seconds. When energy metabolism ceases, the integrity of cellular membranes essentially fails, potassium begins leaking from the cells, osmotic balance is lost, fluid rushes into the damaged cells, and within seconds cells begin to die.

The term *ischemia* can be defined as inadequate blood flow to a tissue. In the brain, ischemia can be present for many reasons. It does not distinguish between decreased perfusion and total cessation of flow, situations that have different implications for tissue viability. For example, if the heart stops, total cerebral blood flow ceases; if the blood pressure drops low enough, the flow of blood becomes inadequate; and if a single large vessel such as the carotid artery becomes narrowed, too little blood passes through. If a cerebral blood vessel becomes occluded by an embolus or an atherosclerotic plaque, the tissue that it irrigates becomes ischemic. Other blood vessels usually try to supply the area and make up the difference, but if the area of brain supplied by an occluded vessel cannot be supplied with blood from surrounding vessels, that area dies. The event is called a cerebral *infarction* or a *stroke*.

Stroke. Consider what happens to an area of brain that becomes severely ischemic during a stroke. The drop in blood flow is accompanied by a rapid rise in tissue osmolality to a level in the range of 600 mOsm, a change that leads to severe osmotic cellular damage. As a result, the involved area of brain (whether the whole brain or a small area) begins to swell. Swelling leads to rising intracranial pressure, which in turn can further reduce cerebral perfusion. If total intracranial pressure exceeds the systemic arterial pressure, cerebral blood flow stops. In the presence of inadequate oxygen and glucose, tissue respiration ceases and glycolysis becomes the dominant source of energy for a short period. During that time, glycolysis produces high levels of lactic acid, which of course accumulates rapidly because, with blood flow stopped, it cannot be removed. High levels of tissue lactic acid and the accumulation of other possibly toxic products of failing energy metabolism cannot be removed and so add to the damage being done. Cellular membrane integrity is lost; the cells swell and rapidly die.

Clinically, the functions served by the infarcted region are damaged. For example, if the area controlling strength and movement in one extremity or one side of the body is infarcted, that extremity or side becomes weak or paralyzed. If the area subserving speech is damaged, the patient may lose the ability to talk or to comprehend what is heard. It is important to note that the dying or dead area of brain releases protein into the CSF. Also, because blood vessels in the area are also damaged by ischemia, some bleeding may occur. Only a few cells may appear in the CSF, an indication that only a little blood has escaped from the area, or the CSF may become frankly bloody if an actual hemorrhage occurs in the damaged area. Finally, as the brain begins to clear away the damaged tissue, some white cells may appear in the CSF. In summary, then, after a stroke that has not involved significant hemorrhage, one can expect to find the CSF either normal or more likely containing an elevated protein level, a few red cells, and possibly scattered white cells.

Diffuse cerebral ischemia and hypoxia. If total cerebral circulation stops, as in cardiac arrest, consciousness is lost within 6 to 8 seconds. On the other hand, if oxygen supply becomes inadequate but circulation continues, the clinical result is usually a feeling of light-headedness followed by mental confusion in mild cases, proceeding to loss of consciousness, seizures, and coma with moderate to severe hypoxia. Precipitating events include pulmonary edema, carbon monoxide poisoning, pulmonary embolism, strangulation, respiratory failure during mechanical ventilation, and exposure to ambient air at high altitudes.

Failure to restore cerebral circulation and oxygenation within 4 to 5 minutes after their total cessation may result in cell death and irreversible damage. The precise metabolic events occurring in brain tissue subjected to hypoxia, anoxia, ischemia, hypoglycemia, and so on in various combinations are incompletely understood.

Clinically, the onset of acute generalized cerebral ischemia, hypoxia, or severe hypoglycemia leads to signs and symptoms of diffuse *cerebral impairment* as can occur in toxic, metabolic diseases.

Chronic hypoxia

Anemia, congestive heart failure, pulmonary disease, and changes in hemoglobin that interfere with oxygen binding (for example, carbon monoxide poisoning, methemoglobinemia) are examples of conditions associated with chronic insufficiency of cerebral perfusion or oxygenation. Lethargy, confusion, disorientation, and failing memory and judgment are early signs in such instances. Seizures, myoclonus (that is, a sudden single jerk of a large muscle group or even the whole extremity), focal neurological deficits, and finally stupor and coma may all occur when the defects become severe or other disease is superimposed on the state of chronic insufficiency. A blood Po_2 less than 40 mm Hg is necessary to produce significant cerebral symptoms if cerebral blood flow is adequate. Similarly, with oxygenation available, preserved cerebral blood flow must fall to 20 mL/100 g/min before loss of consciousness occurs.

Epilepsy and other seizure disorders

A seizure is a paroxysmal, unregulated burst of electrical firing at some point on the cerebral cortex. If the region of abnormal discharge is small and remains confined, it is called a *focal seizure*. The signs and symptoms of such a focal seizure depend on the site of the electrically discharging tissue.

During such an event the patient's arms, legs, or face may twitch, or the patient may experience a period of confusion. On the other hand, the region of abnormal discharging may not remain confined; it can become generalized to large areas of cerebral cortex and perhaps even involve structures deep inside the brain. This generalized seizure, if it is mild, can cause a lapse of consciousness or a *grand mal,* or major motor, seizure with loss of consciousness, violent shaking, and often loss of urinary continence.

The cause of seizures in most instances is not known. Seizures occur in association with other diseases: birth injuries, brain tumors, strokes, metabolic abnormalities such as hypoglycemia or heavy alcohol use (especially around the time when the patient stops heavy drinking), head trauma, infections such as meningitis or encephalitis, ingestion of toxins such as illegal drugs, or even medications. Each person who comes to a physician with seizures has to be evaluated carefully to see if there is some underlying disease or provocation. Often none is found.

Some persons continue to have seizures throughout their lives and are said to have a *seizure disorder,* or epilepsy. Most people with seizure disorders are perfectly normal in every other way. Others have associated brain damage or disease.

Persons with a seizure disorder are placed on anticonvulsant drugs usually for months or even for the rest of their lives in an attempt to stop the seizures entirely or at least reduce their frequency. In other cases, in which some underlying problem such as an infection or metabolic encephalopathy can be found and corrected, anticonvulsant drugs may be needed only during the active phase of the illness.

Intoxication with drugs and poisons

Many drugs and poisons affect the nervous system directly, producing confusion, drowsiness, stupor, coma, seizures, or psychotic states. Drugs may also cause respiratory depression, alter the systemic metabolic balance, or otherwise indirectly damage the nervous system. In many cases, the differential diagnosis of these states requires laboratory confirmation of the presence of an offending drug or toxin. When specific drugs are known or suspected to be available to the patient, the search is simplified. Often, however, a *toxic screen* for common substances is necessary because the circumstances surrounding an ingestion are unclear.

Physical findings may raise suspicion of a certain class of drugs; for example, small pupils suggest opiates and widely dilated pupils suggest drugs with atropine-like effects such as tricyclic antidepressants or amphetamines with adrenergic actions. Unfortunately, many cases of intoxication involve multiple substances, whether "street" drugs or medications.

Effects of metabolic diseases

Hypoglycemia. The earlier discussion of cerebral metabolism made clear the importance of an uninterrupted supply of glucose to the brain. At a blood glucose of 300 mg/L or lower, consciousness is impaired and then lost. At levels of 200 to 300 mg/L, oxygen use is reduced by 25%, and at 100 to 200 mg/L, oxygen use drops by half. Prolonged hypoglycemia eventually leads to energy failure.

The most common causes of hypoglycemia are insulin overdose and iatrogenic hyperinsulinemia, but it also occurs in alcoholics, with various pancreatic and hepatic diseases, and occasionally with insulin-secreting tumors.

Blood glucose levels approaching dangerous levels (300 to 400 mg/L) should be corrected quickly; since one cannot know, given a single sample, whether glucose levels are rising, falling, or remaining steady, dangerously low levels must be reported without delay. Prolonged hypoglycemia can lead to permanent and profound dementia if the patient survives.

Hyperglycemia. Diabetic ketoacidosis also causes impaired consciousness. The associated hyperosmolar state usually seen with blood glucose levels of more than 10,000 mg/L is almost certainly a primary cause, but the effects of dehydration, systemic acidosis, and perhaps the local accumulation of hydrogen ion in brain tissue may also contribute. There is little correlation, however, between acidosis and coma in this condition.

Liver failure. Signs of diffuse cerebral dysfunction ranging from mild encephalopathic features such as quiet apathetic delirium to seizures and coma can accompany liver disease. Examples are cirrhosis or the side effects of *portal systemic shunts*.

Several possible toxic agents have been proposed as contributing to brain dysfunction in hepatic disease, but none fully explains the findings. Blood *ammonia* levels rise as the urea-producing system of the liver fails to detoxify nitrogenous products from the gut. Although it is clear that elevated ammonia levels do not fully explain hepatic encephalopathy, clinicians often follow blood ammonia levels to monitor the progress of liver-induced encephalopathy. When the diagnosis is in doubt, highly elevated CSF α-ketoglutamate levels are highly indicative of liver-induced encephalopathy.

In liver failure, routine CSF examination is usually normal, though some bilirubin may be found if the serum bilirubin level is high. Serum bilirubin levels of 40 to 60 mg/L are necessary before that substance appears in the spinal fluid. In practice, when bilirubin does appear in the CSF, serum bilirubin will usually be in the 100 to 150 mg/L range or greater.

Severe liver damage can also result in the loss of glycogen stores so that hypoglycemia occurs. Finally, hepatic encephalopathy is almost invariably accompanied by hyperventilation, with a resulting lowering of arterial Pco_2 (an accompanying metabolic alkalosis may mask the increased respiratory drive, however). Since an encephalopathy *not* accompanied by respiratory alkalosis (or metabolic alkalosis) is almost certainly not caused by liver failure, arterial gases may be important diagnostically.

Renal disease. Kidney failure leads to the uremic state and an associated encephalopathy characterized by confusion, delirium, and finally coma. Seizures and myoclonic jerks are common. A restless agitation is perhaps more characteristic of rapidly developing renal failure, and a quiet, drowsy, confused state is more usual in chronic states. As with liver failure, the precise mechanisms of CNS impairment are incompletely understood. An altered blood-brain barrier, which allows entry of substances normally excluded from the CNS (such as certain dialyzable molecules), can be crucial. In any case, the magnitude of azotemia (elevation of BUN) is not well correlated with neurological symptoms, and to some extent the rapidity of onset of the uremic state rather than the absolute rise of BUN seems more likely to determine the severity of symptoms. Urea itself is not the offending agent, since urea infusions do not lead to the typical uremic encephalopathy.

The CSF may show changes indicative of an aseptic meningitis with lymphocytes, PMNs (generally less than $200/mm^3$), and moderately elevated protein (rarely greater than 1000 mg/L). Systemic acidosis is generally not reflected in the CSF. Finally, some element of water intoxication may be present when serum osmolality is less than 260 mOsm/L.

Dialysis itself may cause neurological abnormalities. Because the blood-brain barrier is only slowly permeable to urea and other *osmogens,* as the serum osmolality falls during dialysis the brain becomes hyperosmolar relative to blood and water enters the CNS. If the dialysis is too rapid, the resultant CNS water intoxication can lead to its own encephalopathy. Although serious changes are uncommon and occur in less than 5% of patients, as many as half of all patients on dialysis experience at least mild dialysis disequilibrium symptoms.

Pulmonary disease. The lungs perform two critical functions of gas exchange: they allow oxygen from ambient air to diffuse rapidly into the blood and allow carbon dioxide to diffuse rapidly from the blood into the lungs and out into the ambient air. Some neurological consequences of hypoxia and anoxia have previously been considered in this chapter. But rising carbon dioxide levels (so-called hypercarbia) have their own important impact on cerebral function.

The rapidity of onset of respiratory insufficiency is a significant factor determining its impact on brain function. For example, patients with chronic lung disease who habitually run Pco_2 levels of 60 mm Hg may show no cerebral symptoms or only the slightest dulling of intellectual processes. The same level of hypercarbia developing rapidly can render a patient stuporous. The degree of CSF acidosis accompanying elevated Pco_2 is probably an important determinant of impairment. Severe hypercarbia acts as an anesthetic; it has little effect on O_2 consumption but leads to decreases of 40% to 50% in glucose use.

Many diseases cause CO_2 narcosis, but they fall into three general categories: (1) those involving intrinsic damage to alveolar surfaces leading to impaired gas exchange, such as emphysema; (2) those that impair the rate or depth of respiration, such as neuromuscular disease, poisoning with respiratory depressants such as barbiturates, brainstem damage, and many others; and (3) those that impede or block air flow, such as strangulation, aspiration, or increased dead space.

The symptoms of CO_2 narcosis or encephalopathy usually begin with dull headache and progress to mental confusion, drowsiness, stupor, and finally coma.

Endocrine diseases

Diabetes. Diabetes mellitus can affect the nervous system by a number of paths. Some examples in this chapter that have already been discussed are hyperosmolar states, diabetic ketoacidosis, hypoglycemia usually caused by insulin overdose or relative fasting, and complications of the treatment of ketoacidosis (that is, CNS acidosis resulting from too rapid correction of systemic acidosis) together with CNS hyperosmolarity leading to an influx of fluid into the CNS and consequent cerebral edema. The latter also results from a too rapid correction of systemic hyperosmolarity.

Adrenal disease. Inadequate release of cortisol affects the brain in ways that are complex and not well understood. In chronic untreated states, apathy, depression, fatigue, and even mild delirium are not uncommon. Stupor and coma usually occur only when there is an abrupt severe worsening of chronic illness, the so-called addisonian crisis. Several metabolic derangements—hyponatremia, hyperkalemia, hypoglycemia, and hypotension—occur and add their own insult to the brain.

Excess glucocorticoid products and the administration of steroid medications are associated in some patients with disturbances of mood (depression, elevation, or hypomania), mild confusion, delusions, hallucinations, impaired insight, and grossly inappropriate behavior.

Thyroid disease. The thyroid hormones thyroxine (T_4) and triiodothyronine (T_3) have important effects on brain function. Both hyperthyroidism (thyrotoxicosis) and hypothyroidism (myxedema) predictably affect the brain.

Hypothyroidism. In the fetus and during infancy, hypothyroidism can cause irreversible brain damage and profound mental retardation (cretinism) unless the condition is corrected without delay. Chronic thyroid insufficiency is associated with depression or lability of mood, listlessness, confusion, and sometimes psychosis (that is, delusions and hallucinations usually of a depressive or persecutory nature). Peripheral neuropathy and unsteady gait related to impaired cerebellar functions also occur together with abnormal deep tendon reflexes. For obscure reasons, elevated CSF protein is a common finding in hypothyroidism.

So-called myxedema coma accompanied by decreased body temperature, slowed respiration, and hypometabolism usually occurs in a setting of chronic hypothyroidism on which some acute event is superimposed, such as infection, surgery, trauma, or congestive heart failure. Because myxedema coma is rapidly fatal, correct diagnosis and prompt treatment are essential.

Thyrotoxicosis. Signs of hypermetabolism distinguish the state of thyroid excess that is associated with disturbances in thinking and emotion.

CHANGE OF ANALYTE IN DISEASE (Table 39-2)

Appearance of cerebrospinal fluid

CSF is normally crystal clear and free from all pigmentation, that is, ''clear and colorless.'' One should examine it in a glass tube while comparing it with a tube of water. Both are held in white light against a pure white background. It is best to look down the long axis of the tube. At least 1 ml of fluid should be observed.

A red cell count of 500/mm^3 gives a pink or yellow tinge to the fluid. White counts of 200/mm^3 will give a slightly cloudy appearance. Xanthochromia (a yellow tinge) will appear when blood has been mixed with CSF. This yellowing does not occur immediately but requires from 2 to 4 hours. Since blood in the CSF may represent subarachnoid bleeding or may be simply the result of a traumatic tap (a common occurrence), the CSF sample should be centrifuged immediately. If this is done promptly, bleeding that is caused by a traumatic tap should produce no xanthochromia. Xanthochromia suggests that the bleeding occurred at least 2 to 4 hours before observation of the sample. It may be helpful to know that even after a subarachnoid hemorrhage, CSF may not become xanthochromic until 2 to 4 hours have passed. As many as 10% of patients with subarachnoid hemorrhage actually have clear CSF at 12 hours, but beyond that time 100% will show xanthochromia if the sample is examined carefully.

Protein will also give a slightly xanthochromic appearance when present in concentrations greater than 1500 mg/L. Hemoglobin from hemolyzed red cells will appear in CSF after about 10 hours. When a patient is jaundiced (that is, when serum bilirubin is elevated, as it may be in liver failure), bilirubin may enter the CSF. However, this requires serum bilirubin levels of at least 100 to 150 mg/L before CSF xanthochromia is found.

Proteins of cerebrospinal fluid

In diseased states the local production or modification of proteins within the CNS may lead to diagnostically useful changes in CSF protein patterns. In general, diseases that interrupt the integrity of the capillary endothelial barrier lead to an increase in total CSF protein. Examples are brain tumor, purulent (bacterial) meningitis, cerebral infarction, and trauma. The elevation in total CSF protein that often occurs in hypothyroid states is poorly understood but probably reflects changes in the blood-brain barrier.

Immunoelectrophoresis allows further fractionation of CSF protein constituents. The major *immunoglobulins* in CSF are IgG, IgA, and IgM (with only trace amounts of IgD and IgE). Of all these, IgG is quantitatively the most important. It is often useful to know whether an elevated IgG value is caused by local production of that immunoglobulin within the CNS (as may be the case in some demyelinating diseases such as multiple sclerosis) or by the leakage of 1 IgG across a damaged blood-brain barrier (as in some infections). Since the normal serum IgG is 15% to 18% of total serum protein and normal CSF IgG is 5% to 12%, the ratio of IgG to total protein is sometimes used to estimate the source of IgG elevation. That is, if the ratio

Table 39-2 Change of analyte in CNS diseases

	Analyte in CNS					
Disease	**Glucose**	**Total protein**	**IgG**	**IgG index**	**Xanthochromia**	**Lactic acid**
Stroke (cerebral infarction)	N	↑	N	↓	N, ↑	N, ↑
Hemorrhage	N	N, ↑ ↑	N	N	↑ ↑	N
Epilepsy	N	N	N	N	N	N
CNS tumor	N, ↓	↑	N, ↑	↓	N, ↑	N, ↑
Infection						
Fungal, bacterial	↓	↑	↑	↑	N	↑
Viral	N	N	↑	↑	N	N
Coma	↑ ↑ hyperosmolar ↓ hypoglycemia	↑ (trauma)	N	N	N, ↑ (trauma) N	N
Meningitis, viral	N	N, ↑	N, ↑	↑	N	N

N, Little or no change; ↑, increase; ↑ ↑, large increase; ↓, decrease.

in a sample more nearly approximates the ratio ordinarily found in serum, one tends to suspect that the IgG has been somehow transferred into the CSF from serum. But this is a crude and not especially reliable estimate. Another measure currently popular is the IgG-albumin index. The formula for determining it is as follows:

$$\text{IgG index} = \frac{\text{IgG (CSF)} \times \text{Albumin (serum)}}{\text{IgG (serum)} \times \text{Albumin (CSF)}}$$

The upper limit of normal for this index must be determined for each laboratory, but generally it ranges between 0.25 and 0.85. The IgG index is elevated in diseases in which there is increased CNS IgG production and an intact blood-brain barrier (as in infections). The IgG index is decreased when the blood-brain barrier is compromised, allowing serum proteins to cross into the CSF (as in strokes, some tumors, and some forms of meningitis).

Myelin basic protein has aroused interest as a potential indicator of demyelination. Myelin, which is a complex substance that surrounds many CNS axons like the insulation on a wire cable, is necessary for normal conduction of nerve impulses down the axon. *Demyelinating diseases* are a group of disorders in which the primary insult is some form of damage to the myelin coating of CNS axons. The resulting abnormal propagation of impulses in the brain and spinal cord leads to a variety of clinical signs and symptoms. Myelin basic protein (MBP) is one constituent of normal myelin. It has been found to be elevated in a variety of conditions involving myelin damage: multiple sclerosis is an important example but not the only disease associated with elevated myelin basic protein. In other words, MBP is a very nonspecific indicator of disease; because it lacks specificity, it is being used less commonly in clinical neurology.

Gamma globulin synthesis in the central nervous system

Although elevations of CSF gamma globulin may be caused by changes in serum proteins (for example, small-molecular-weight Bence Jones proteins in multiple myeloma cross the blood-brain barrier and appear in the gamma fraction of CSF proteins), there is evidence that local CNS immunoglobulin production occurs in many diseases. Examples include multiple sclerosis, subacute sclerosing panencephalitis (a devastating process of myelin damage that occurs in children and young adults in association with greatly elevated CSF measles titers), many chronic and acute infections (neurosyphilis, tuberculous meningitis, abscess, viral meningoencephalitis, and sarcoidosis), and some brain tumors. As a practical point, in those settings in which CSF total protein rises as a result of increased permeability of the blood-brain barrier, the addition of serum protein (which normally contains 15% to 18% gamma globulin) raises the CSF gamma fraction. It thus becomes difficult to estimate the upper limit of normal for gamma globulin as a percentage of total protein when total protein is significantly elevated. In any case, when CSF gamma globulin is elevated, a clinician may order a simultaneous serum protein electrophoresis to help determine the source of the increased CSF gamma fraction.

Oligoclonal bands

The gamma fraction is composed of a variety of proteins. Agarose gel electrophoresis performed on concentrated CSF demonstrates elevation of a population of proteins within the gamma range that all share the same electrophoretic mobility. The population of gamma proteins that separates as *oligoclonal bands* is believed to derive from a few groups of immunocompetent cells or *clones*. The appearance of oligoclonal bands has been reported in 79% to 90% of patients with multiple sclerosis and in a variety of CNS inflammatory conditions. Interestingly, this change in the composition of the gamma fraction may occur without any increase in the total gamma globulin concentration.

A final practical point should be made about protein determinations in CSF. When there is blood in the CSF from a traumatic tap or bleeding within the nervous system, one expects that the blood will elevate the CSF protein value. One can still determine the CSF protein level by correcting for the amount of blood present. Simply allow 10 mg/L of protein for every 1000 red blood cells/mm^3. For example, if the red cell count is 10,000 in the CSF sample and its total protein is 1000 mg/L, the corrected total protein equals 900 mg/L. The cell count should be performed as rapidly as possible after lumbar puncture, preferably in the first half hour and certainly not later than 2 hours, since hemolysis will be occurring after that time.

Glucose in cerebrospinal fluid

Determination of CSF glucose helps distinguish bacterial from viral meningitis; the glucose value is often quite *low* (less than 40% to 45% of simultaneously analyzed, equilibrated serum glucose) in bacterial meningitis and tuberculous meningitis and is generally normal in viral disease. Carcinomatous meningitis (widespread infiltration of the meninges by tumor cells) also drives CSF glucose values below the normal range.

Anticonvulsant drugs

Fortunately, a number of effective anticonvulsant drugs are available: phenobarbital, phenytoin, valproic acid, primidone, and carbamazepine, which are excellent for a wide range of seizure types; diazepam (Valium) and lorazepam, which are used acutely only to interrupt a seizure or flurry of seizures; ethosuximide used for petit mal epilepsy; and clonazepam used primarily for myoclonic seizures. Table 39-3 lists some optimum therapeutic ranges of commonly used anticonvulsant drugs.

Table 39-3 Commonly used anticonvulsant drugs

Drug	Optimum therapeutic range (μg/mL)
Phenytoin	10-20
Phenobarbital	15-40
Carbamazepine	4-8
Ethosuximide	40-100
Valproic acid	50-100
Primidone	5-12

In general, a physician begins therapy with one anticonvulsant drug believed likely to be effective and increases the dosage until either adequate control of the seizures is achieved or a toxic side effect that cannot be overcome occurs. The dosage is then lowered, a second drug is added, and the process is repeated. (A hypersensitivity reaction such as rash, great suppression of white count, or evidence of actual liver damage requires that the drug be stopped.)

Although the patient's clinical response—control of seizures, failure to control, or development of side effects—is the principal concern, monitoring serum levels of anticonvulsant drugs is quite useful. For example, failure to control seizures is often traced to inadequate serum levels, and that inadequacy is in turn traced to poor compliance, poor absorption of the drug, or too rapid metabolism of it. One can adjust dosage accordingly. Symptoms can be related to drug toxicity, or toxicity may be ruled out as the explanation. Finally, monitoring of anticonvulsant levels gives important information about drug kinetics and interactions. Periodic measurements of anticonvulsant serum levels is a routine part of the management of seizures.

BIBLIOGRAPHY

Adams, RD, and Victor, M: Principles of neurology, New York, 1977, McGraw-Hill Book Co.

Fishman, RA: Cerebrospinal fluid in diseases of the nervous system, Philadelphia, 1980, WB Saunders Co.

Plum, F, and Posner, JB: The diagnosis of stupor and coma, ed 3, Philadelphia, 1980, FA Davis Co.

Glasser, L: Tapping the wealth of information in CSF, Diagn Med, pp 23-33, Jan-Feb 1981.

Killingsworth, LM, Cooney, SK, Tyllia, MM, and Killingsworth, CE: Deciphering cerebrospinal fluid patterns, Diagn Med, pp 23-29, March-April 1980.

CHAPTER 40 Endocrine function

BOLESLAW H. LIWNICZ
REGINA G. LIWNICZ

OBJECTIVES

- Define hormone.
- List and describe the three main classes of hormones, including chemical composition, transport, and mechanisms of action.
- Differentiate between negative and positive feedback systems in the regulation of hormone levels.
- List the hormones or factors secreted by the hypothalamus and pituitary and describe their actions and regulation.
- Describe pathological conditions that may result in hypersecretion or hyposecretion of the hormones listed above.

KEY TERMS

acromegaly A pathological condition caused by the hypersecretion of growth hormone in adults.

activation The change in the tertiary structure of a cytoplasmic, steroid hormone receptor induced by the binding of a specific hormone to the receptor; allows translocation of the complex to the nucleus.

adenohypophysis The anterior lobe of the pituitary gland that secretes trophic hormones.

amenorrhea The absence or abnormal stoppage of menstrual flow.

cyclic AMP Cyclic 3′,5′-adenosine monophosphate; acts as a second messenger, relaying the message of a protein hormone to the nucleus where specific gene stimulation occurs, leading to the final biochemical change or changes specific for that hormone.

diabetes insipidus A disorder caused by a deficient quantity of antidiuretic hormone manifested by excessive urine secretion.

endocrine gland A ductless gland that secretes hormones directly into the bloodstream.

feedback loop A loop modulating an interaction between two endocrine organs.

galactorrhea Persistent secretion of milk, irrespective of nursing.

gigantism A pathological condition caused by the hypersecretion of growth hormone in children.

hormone A chemical used to transmit messages to a distant target organ to elicit a specific response or action.

hormone transport The manner in which a hormone is carried in the blood, that is, protein bound or free.

hypothalamic nucleus An anatomical structure composed of neurons with a similar function.

hypothalamus The portion of the brain involved in the endocrine and autonomic functions.

immunocytochemical stains A cytological method for detecting hormones based on labeled antibodies.

neurohypophysis The part of the pituitary gland that is an extension of the central nervous system.

panhypopituitarism Deficient secretion of all the pituitary hormones.

pituicyte A cell of the neurohypophysis.

pituitary pericapillary space The intermediate space between the secretory cell and the capillary sinusoid.

pituitary portal system The vascular system connecting the hypothalamus with the adenohypophysis.

pituitary stalk The structure connecting the pituitary gland with the hypothalamus.

receptor Specific cellular protein that must first bind a hormone before a cellular response can be elicited; located in either the cell membrane or the cytoplasm.

release-inhibiting factors Peptides secreted by the hypothalamus that inhibit secretions by the adenohypophysis.

releasing factors Peptides secreted by the hypothalamus that stimulate secretions by the adenohypophysis.

sexual precocity The unusually early development of primary and secondary sex characteristics.

target endocrine organ An endocrine gland whose function is controlled by the pituitary gland.

translocation The process by which a steroid hormone-cytoplasmic receptor complex moves from the cytoplasm into the nucleus, where it can activate specific genes.

trophic hormones Hormones secreted by the adenohypophysis that stimulate the target organs. The suffix *-tropic* means "turning" or "changing"; *trophic* means "nurturing" or "causing growth"; both are used in regard to hormones.

Part I: General aspects of the endocrine system

INTRODUCTION AND DESCRIPTION OF ENDOCRINE FUNCTION

Information is transmitted throughout the body by the nervous and the endocrine systems. In the nervous system messages are carried between peripheral tissue and the brain by means of a nerve network, whereas in the endocrine system messages are carried from one cell to another by the bloodstream through chemical substances called *hormones*. These hormones are chemical messengers produced by specialized cells in endocrine or ductless glands. Hormones are released directly into the blood for transport to a distant site, called the *target tissue,* where they bring about their specific action. Hormones possess a high degree of structural specificity, and any alteration in their molecular composition results in a dramatic change in their physiological activity.

Hormones can be divided into three chemical classes: (1) the protein hormones, (2) the aromatic amines, and (3) the steroid hormones. The hormones listed in Table 40-1 differ from one another in structure, chemical composition, transport, metabolism, and mechanisms of action. The names and locations of the various endocrine glands in the body are shown in Fig. 40-1.

Among the endocrine glands, the most important are the anterior pituitary and the hypothalamus. These two glands function as the primary regulators of the entire endocrine system. In response to external and internal stimuli, the hypothalamus secretes a number of small neuropeptide hormones, called *releasing hormones,* into the portal circulation of the pituitary gland. These hormones are characterized by a rapid onset of action and a brief biological half-life. They regulate production and release of the anterior pituitary hormones that in turn affect most of the endocrine glands.

HORMONE TRANSPORT

Once the hormone is released by the endocrine gland, it is transported to the site of action by the bloodstream. The water-soluble protein hormones are transported in the unbound state in the bloodstream. The steroid and thyroid hormones, which are insoluble in water, are rendered soluble when they are bound to carrier proteins such as albumin, sex hormone–binding protein, and transcortin or

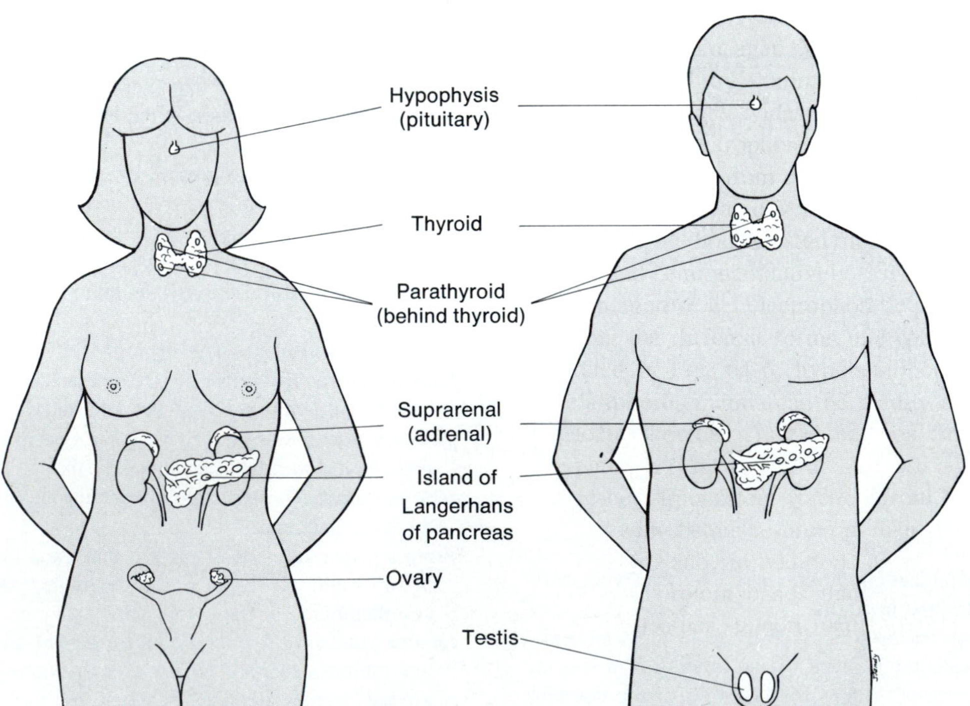

Fig. 40-1 Location of endocrine glands. *(From Toporek, M: Basic chemistry of life, St Louis, 1980, The CV Mosby Co.)*

Table 40-1 Summary of endocrine hormones

Hormones	Target tissues	Chemical nature	Primary actions
Steroid hormones			
Androgens (male sex hormones)			
Testosterone (and other androgens)	Peripheral tissues, testis	Steroid	Stimulates development of secondary sex characteristics in men, protein anabolic effect
Estrogens (female sex hormones)			
Estrone, estradiol	Peripheral tissues, ovary	Steroid	Stimulates development of secondary sex characteristics in women, protein anabolic effect
Progesterone	Uterus	Steroid	Prepares uterus for implantation of ovum
Adrenal cortex hormones			
11-Deoxycortisol	Renal distal tubule, large intestine	Steroid	Maintains electrolyte and water balance
Cortisone, or hydrocortisone (cortisol)	Muscle, liver, adipose cells	Steroid	Stimulates gluconeogenesis from amino acids, antiinsulin effects on glucose and fat metabolism
Aldosterone	Renal distal tubule, large intestines	Steroid	Regulates retention of sodium ions, excretion of potassium ions
Amino acid–derived hormones			
Adrenal medulla hormones			
Epinephrine Norepinephrine	Liver, adipose cells	Phenolic amines	Stimulates glycogenolysis, hypertensive effect
Thyroid hormones	Most tissues	Amino acid (iodinated): thyroxine or derivatives	Regulates rate of metabolism, increases serum glucose
Protein and polypeptide hormones			
Anterior pituitary			
Thyroid-stimulating hormone (TSH), thyrotropin	Thyroid	Protein	Stimulates development and secretion of thyroid gland
Adrenocorticotropic hormone (ACTH)	Adrenal cortex	Protein	Stimulates growth and secretion of adrenal cortex
Follicle-stimulating hormone (FSH)	Ovarian follicle (women) Seminiferous tubules (men)	Protein	Stimulates growth of follicles and production of estrogen in women, formation of spermatozoa in men
Luteinizing hormone (LH)	Corpus luteum (women) Testes (men)	Protein	Triggers formation of corpus luteum and production of progesterone in women, production of androgens by interstitial cells in men
Prolactin (lactogenic hormone, LTH)	Milk glands	Protein	Initiates lactation
Growth hormone (GH)	Most tissues	Protein	Stimulates growth (also affects fat and carbohydrate metabolism)
Posterior pituitary hormones			
Vasopressin (antidiuretic hormone, ADH)	Renal collecting ducts, bladder	Peptide	Stimulates reabsorption of water in kidney tubule
Oxytocin (Pitocin)	Uterus, mammary glands	Peptide	Contracts uterus
Calcitonin (thyroid gland)	Bone	Peptide	Lowers serum calcium
Parathyroid hormone	Bone, small intestines, kidney	Protein	Regulates blood calcium
Pancreatic hormones			
Insulin	Most cells	Peptide	Facilitates carbohydrate catabolism
Glucagon	Liver	Peptide	Raises blood glucose by hepatic glycogenolysis
Gastrointestinal hormones			
Secretin	Pancreas, gallbladder	Peptide	Stimulates flow of pancreatic juice and bile to a much smaller extent
Cholecystokinin-pancreozymin	Gallbladder, pancreas	Peptide	Contraction of gallbladder; stimulates secretion of pancreatic enzymes
Gastrin	Stomach	Peptide	Stimulates secretion of gastric juice (HCl)
Renal erythropoietin	Bone	Peptide	Stimulates red cell formation
Hypothalamic factors	Anterior pituitary	Peptide	Stimulates or inhibits release of corresponding trophic hormones

From Toporek, M: Basic chemistry of life, St Louis, 1980, The CV Mosby Co.

Table 40-2 Serum carrier proteins for hormones

Hormone	Protein (in order of importance)
Testosterone	Sex hormone–binding protein (SHBG) Albumin
Estradiol	Albumin Sex hormone–binding protein
Progesterone	Corticosteroid-binding globulin (CBG)
Cortisol	Corticosteroid-binding globulin
Thyroxine (T_4)	Thyroxine-binding globulin (TBG) Thyroxine-binding prealbumin (TBPA) Albumin

cortisol-binding globulin (Table 40-2). The aromatic amines are also transported in the blood attached to the serum proteins, thyroxine-binding globulin and thyroxine-binding prealbumin.

MECHANISMS OF REGULATORY CONTROL OF HORMONE LEVELS

Secretion and release of a hormone by an endocrine gland is not a steady and continuous process but is a dynamic event varying in response to endogenous and exogenous stimuli. There are two regulatory mechanisms that govern the activity of an endocrine gland. One is the feedback mechanism, which directly controls the secretory activity of the endocrine gland, and the other is a series of interacting regulatory mechanisms of the hypothalamus-pituitary-target gland system, which maintains the integrity of the entire endocrine system.

The primary mechanism with which the body maintains the levels of circulating hormones within set physiological

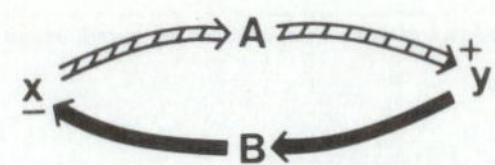

Fig. 40-2 Scheme of feedback loop system in which gland *x* releases product *A*, which has a positive feedback (+) on gland *y* to produce substance *B*. Increased concentrations of *B* feedback on gland *x* to inhibit (−) the release of product *A*.

limits is the feedback mechanism. There are two general types of feedback systems operative in biological systems. These are the positive feedback system, which is rarely seen, and the more common negative feedback system. In a typical endocrine system an increase in one variable, for example, A, causes an *increase* in another variable, B. In a negative feedback system, an increase in B causes a *decrease* in A. These two systems can combine into one continuous loop. For example, compound A causes an increase in the synthesis of B, which causes a decrease in the synthesis of A, thus tending to oppose the primary change (A) (Fig. 40-2). This combined system is used for the hypothalamic-pituitary-target gland control system.

Control of endocrine gland function is based on the use of the feedback principle. This control is exerted either directly on the gland or by the hypothalamic pituitary axis. An example of feedback control exerted directly on the endocrine gland is the regulation of extracellular fluid (ECF) calcium by parathyroid hormone (PTH). Rising levels of PTH increase the concentration of calcium in ECF through its effect on bone, kidney, and gut (positive feedback). The rising levels of calcium in ECF cause the parathyroid to suppress secretion of PTH (negative feedback).

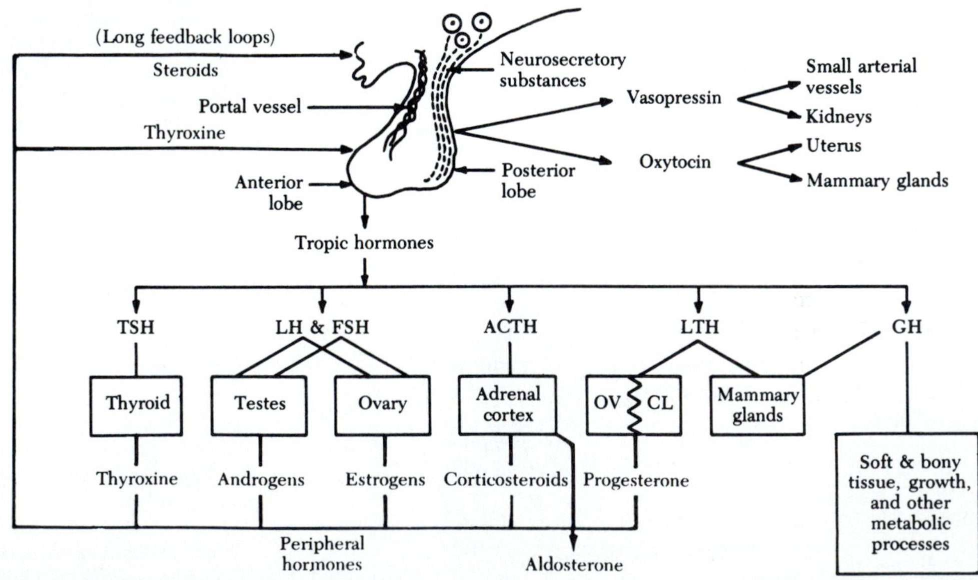

Fig. 40-3 Scheme of relationships between hormones of anterior and posterior lobes of pituitary gland and their target structures. *CL*, Corpus luteum; *OV*, ovary. *(From Orten, JM, and Neuhaus, OW: Human biochemistry, ed 10, St Louis, 1982, The CV Mosby Co.)*

The functional relationship among the hypothalamus, pituitary, and an endocrine gland is discussed in greater detail in Part II of this chapter. Fig. 40-3 illustrates the relationships between the pituitary gland hormones and their target organs.

MECHANISMS OF HORMONE ACTION

Steroid and protein hormones act as messengers, transporting information to all parts of the body through the bloodstream. The hormones, however, circulate in extremely low concentrations. Blood levels of these hormones range from 10^{-10} to 10^{-12} M, whereas glucose circulates in the blood at the level of 10^{-2} M and sodium at the level of 10^{-1} M. It is therefore imperative that these chemical messengers be selectively picked up only by the appropriate target cells and that the message they carry be accepted, processed, and translated into the appropriate response. This highly selective function is carried out by means of hormone receptors located only in the target cells, which thus determine end-organ specificity.

Receptors are cellular proteins with the unique ability to bind to a single type of hormone molecule. For example, estrogen receptors will bind to estrogens and not to androgens or progesterones. This selectivity of receptor binding is caused by their specific primary, secondary, and tertiary molecular structures. Alterations in receptor structures can compromise the specificity, resulting in a physiological disorder. For example, the lysine-lysine-arginine-arginine sequence of a receptor protein is extremely important in the binding of adrenocorticotropic hormone (ACTH) to the adrenal cell. Receptors of adrenal tumorlike cells apparently lose this amino acid sequence and in turn the receptor specificity, thereby permitting binding by thyroid-stimulating hormone (TSH), follicle-stimulating hormone (FSH), or luteinizing hormone (LH) to this site.

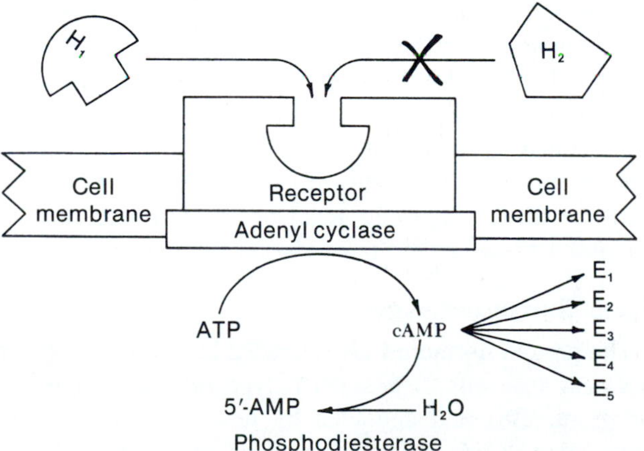

Fig. 40-4 Concept of hormone action by way of second messenger, cyclic AMP. H_1 is hormone recognized by specific membrane receptor site, whereas H_2 is a hormone that is not sensed in the environment. E_1 to E_5, Various metabolic effects of increased cAMP. *(From Orten, JM, and Neuhaus, OW: Human biochemistry, ed 10, St Louis, 1982, The CV Mosby Co.)*

Mechanism of action of protein hormones

The mechanism of action of protein hormones differs considerably from that of steroid hormones. Protein hormones, which are unable to pass through the plasma membrane because of their molecular composition, attach themselves to receptors on the external surface membrane of the target cell (Fig. 40-4). Once bound to the cell membrane, the receptor-hormone complex stimulates the membrane-bound enzyme adenyl cyclase, which activates the conversion of adenosine triphosphate (ATP) to cyclic 3′,5′-adenosine monophosphate (cAMP). The cAMP functions as a second messenger, relaying the message of the protein hormone (first messenger) to initiate biochemical processes inside the cell:

$$\text{Endocrine gland} \xrightarrow[\text{hormone}]{\text{First messenger}}$$

$$\text{Target cell} \xrightarrow[\text{(cAMP)}]{\text{Second messenger}} \text{Metabolic effects}$$

The cAMP directly transmits the message, probably by activation of the enzyme called protein kinase (PK), which in turn activates other regulatory intracellular enzymes. Protein kinase, which is present in both the cytoplasm and the nucleus, acts as a link between cAMP and the activation of the biochemical pathway by bringing about an array of enzymatic reactions involving phosphorylase, esterase, and desmolase, leading to the final regulatory effect of the hormone.

Mechanism of action of steroid hormones

The iodinated thyroid hormones and all steroid hormones act on a wide variety of target cells to alter their function. The steroid hormone enters the target cell by diffusion through the cell membrane and is quickly bound by the specific receptor in the cytoplasm. After hormone binding the receptors are believed to undergo conformational changes called *transformation,* a steroid-induced effect that allows the passage of the steroid-receptor (SR) complex through the nuclear membrane. *Translocation* is the passage of the SR complex from the cytoplasm through the nuclear membrane into the nucleus after transformation.

Within the nucleus the SR complex binds to nonhistone protein of chromatin. This interaction leads to the formation of new messenger RNA (mRNA). The mRNA then migrates into the cytoplasm, where it attaches to ribosomes. The mRNA is translated to affect the formation of proteins that eventually bring about specific changes in the target cell.

Fig. 40-5 summarizes the most generally accepted mechanism of steroid hormone action, though significant differences do exist between hormones. The various steps involved in the mechanism of the steroid hormone action are (1) passage through the target cell membrane, (2) binding to a cytoplasmic receptor, (3) activation of the cytoplasmic receptor, (4) transfer of the activated cytoplasmic

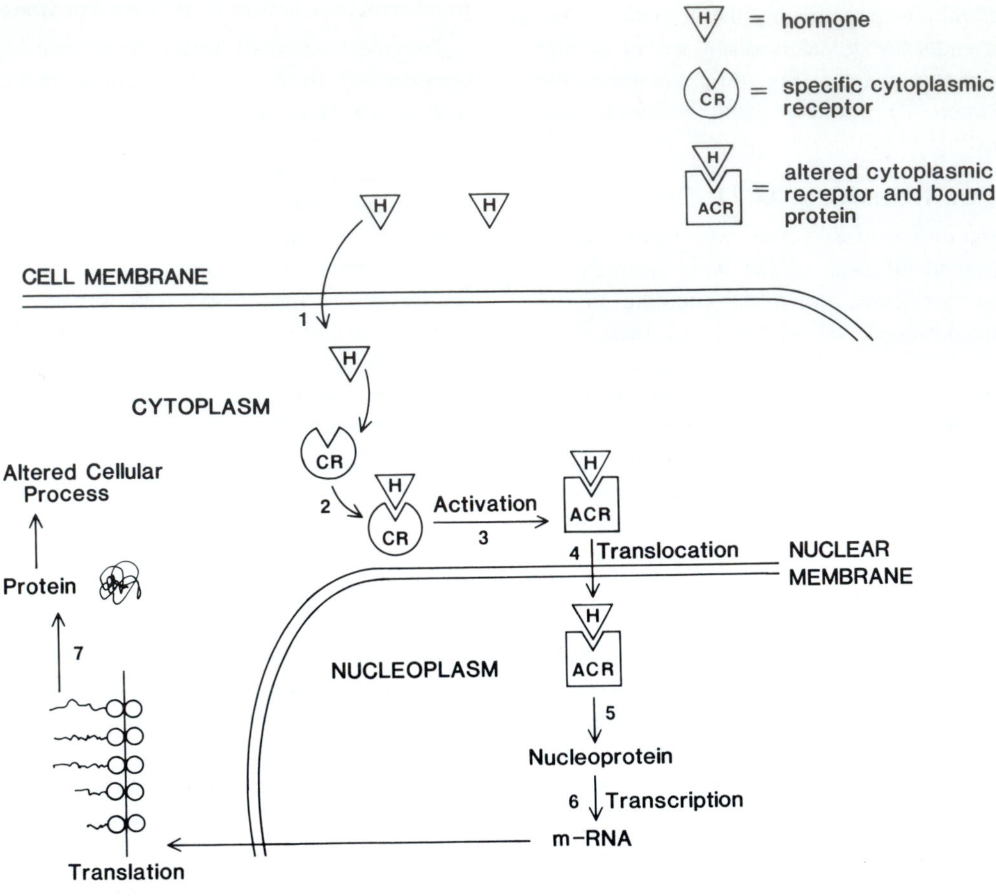

Fig. 40-5 Summary of steps of mechanism of action of steroid hormone on a target cell. Steps 1 to 7 are listed in text.

receptor (ACR) complex across the nuclear membrane to the nucleus, (5) binding of the ACR complex to specific sites in nuclear chromatin DNA, (6) synthesis of new RNA, and (7) synthesis of proteins resulting in modified cellular function.

The unbound or free steroid hormone, which is a low-molecular-weight (less than 500 daltons) and fat-soluble compound, diffuses freely across the plasma membrane into the cytoplasm of any tissue cell. The responsiveness of a cell to the hormone depends on the presence of an appropriate intracellular receptor protein. Steroid hormone receptors are proteins with molecular weights of about 200,000 daltons. There are three types of steroid hormone receptors, each binding one of the three functional classes of steroid hormones: estrogens, androgens, and progesterones. These receptors have been identified in target cells for estrogens, progesterone, androgens, cortisol, aldosterone, and even the sterol 1,25-dihydroxyvitamin D_3.

The concentration of receptors in the tissue determines its responsiveness. Many tissues that have few or no receptors are not responsive to that hormone. Testicular feminization, a syndrome of congenital androgen insensitivity, is an example of absence of androgen receptors in the gonad. Variation in receptor concentration rather than total absence may explain some problems of inadequate hormone response.

Testosterone unlike other steroid hormones is not a potent hormone but rather a prehormone. After its entry into the cytoplasm, it does not directly bind with the androgen-receptor sites. Instead, it is first metabolized by the membrane-bound enzyme 5′Δ-reductase, which converts it to a more potent androgen, dihydrotestosterone (DHT). DHT binds to the cytoplasmic receptor to form the receptor complex for subsequent transmission to the nucleus.

Catabolism of hormones

The protein hormones are degraded by proteases present in serum and cells. The small polypeptides are filtered by the glomerulus and degraded by tubular cells. The aromatic amino acids are catabolized by deiodination or oxidation to inactive forms.

The protein-bound steroid hormones are converted to more water-soluble, inactive forms by hydroxylation, oxidation, or reaction with glucuronic acid or sulfate to form conjugates of these compounds, which are excreted by the kidney. Peripherally, the active steroids are oxidized to

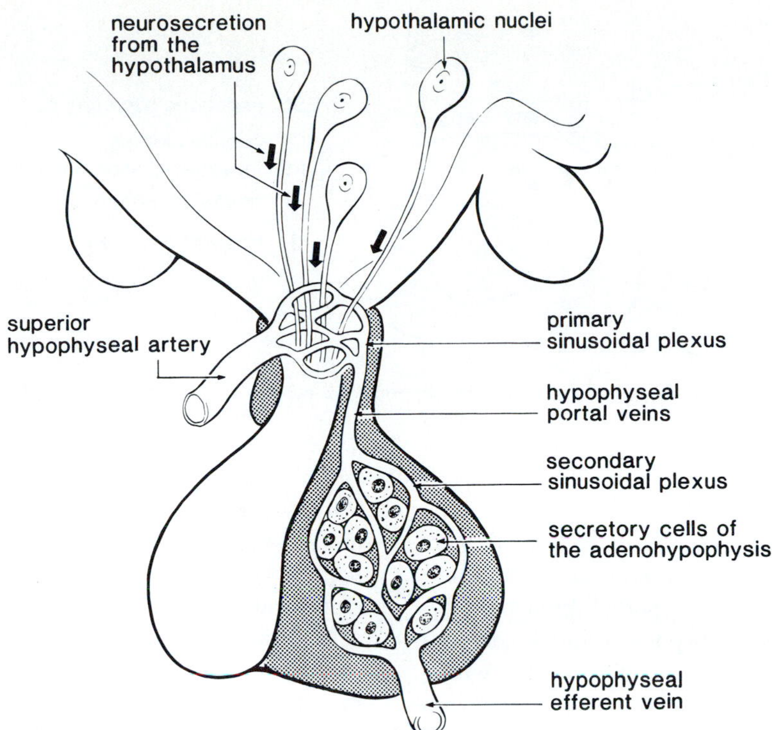

Fig. 40-6 Sinusoid portal system of pituitary gland.

less active hormones, which are then converted by the liver to inactive forms.

Part II: Hypothalamopituitary system

PITUITARY GLAND

The pituitary gland is composed of two major parts: the anterior lobe (adenohypophysis) and the posterior lobe (neurohypophysis). The major bulk of the pituitary gland consists of the adenohypophysis.

Adenohypophysis

There is no direct arterial supply to the adenohypophysis. The superior hypophyseal arteries form a capillary bed that extends upward into the hypothalamus and drains into portal veins of the adenohypophysis, which in turn give rise to a second capillary bed within the adenohypophysis (Fig. 40-6). The sinusoid portal system connects the hypothalamus with the adenohypophysis. The hypothalamic hormones, which modulate the secretion of trophic hormones of the adenohypophysis, are released into this system.

After routine hematoxylin and eosin staining, the adenohypophysis shows three cell populations: acidophils, staining red; basophils, staining blue; and chromophobes, not staining. Table 40-3 lists hormones secreted by each of these cell types as determined by immunocytochemical staining with labeled antihormone antibodies.

The basic functional unit of the adenohypophysis is a secretory cell with surrounding sinusoidal capillaries. The secretory cells release hormone to the pericapillary space, from which it freely diffuses through the endothelial pores into the lumen of the sinusoidal capillaries (bloodstream). For transport of the hypothalamic releasing factors into the secretory cells, the same route is used, but the direction is reversed (Fig. 40-7).

Neurohypophysis

The neurohypophysis does not synthesize any known hormone but serves as an endocrine storehouse. The two

Table 40-3 Cell types and hormones of the adenohypophysis

Hematoxylin and eosin	Trophic hormones
Acidophilic cells	Prolactin (PRL) Growth hormone (GH)
Basophilic cells	Follicle-stimulating hormone (FSH) Luteinizing hormone (LH) Thyroid-stimulating hormone (TSH)
Chromophobic cells	Adrenocorticotropic hormone (ACTH) Prolactin (PRL) Growth hormone (GH) Follicle-stimulating hormone (FSH) Luteinizing hormone (LH) Thyroid-stimulating hormone (TSH) No hormone

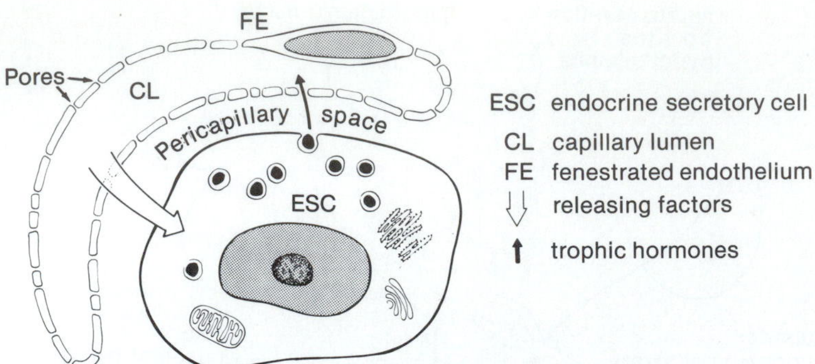

Fig. 40-7 Basic functional unit of adenohypophysis (secretory cell, pericapillary space, and sinusoid capillary).

hormones stored in the neurohypophysis, antidiuretic hormone (ADH, vasopressin) and oxytocin, are synthesized in the hypothalamic nuclei and transported through the pituitary stalk into the neurohypophysis. From the neurohypophysis they are either secreted directly into the bloodstream or stored in extracellular globules for later secretion (Fig. 40-8).

HYPOTHALAMUS

The hypothalamus is a portion of the central nervous system. It lies at the base of the brain beneath the thalamus and is connected to the pituitary gland below by the pituitary stalk. The hypothalamic neurons form many nuclear groups that either have neuroendocrine function or control the autonomic nervous system. The nuclei of the autonomic system control appetite, sexual drive, body temperature, and water and calorie balance. Small neurosecretory cells forming the parvicellular nuclei secrete releasing or release-inhibiting neuropeptides, which are transported to the adenohypophysis by the pituitary portal system. Large neurosecretory cells with long processes form the paraventricular and supraoptic nuclei. The processes of these cells traverse the pituitary stalk and enter the neurohypophysis, where they release their hormones (ADH and oxytocin, see above).

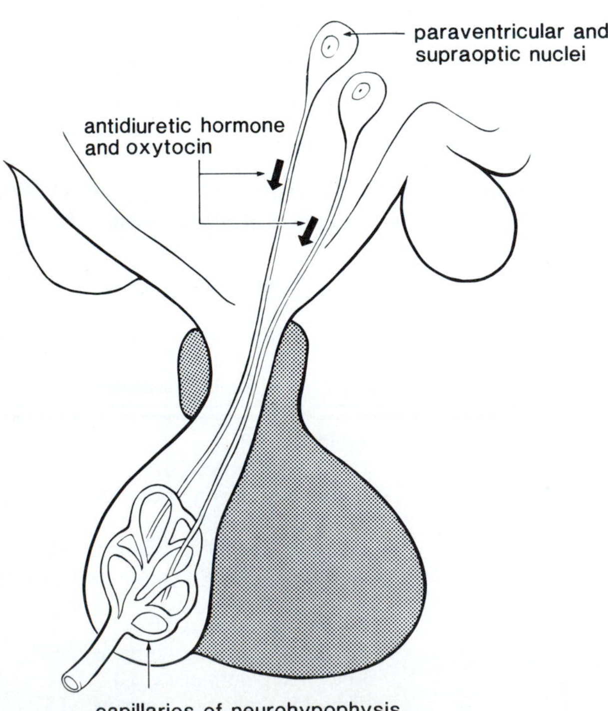

Fig. 40-8 Secretory pathway of neurohypophysis.

FUNCTION OF HYPOTHALAMOPITUITARY SYSTEM

The adenohypophysis secretes trophic hormones that stimulate the release of secretions from target endocrine glands or directly stimulate the metabolism and growth of tissues. The main trophic hormones are growth hormone (GH), prolactin (PRL), FSH, LH, TSH, and ACTH (Table 40-4). The secretion of these trophic hormones depends on (1) the blood level of the hormone secreted by the target endocrine organ, (2) the metabolic state of the patient, and (3) the control exerted by the central nervous system through secretion of hypothalmic factors that stimulate or inhibit the release of pituitary trophic hormones (Table 40-5).

The interaction of these components is complex, and the action of the hypothalamus-pituitary-target endocrine gland-end cell system is explained by the concept of *feedback* loop. The feedback loop is composed of two endocrine organs, the pituitary and the target endocrine gland. The pituitary secretes the trophic hormone, which stimulates the target endocrine gland to secrete its hormone. The hormone of the target endocrine gland in turn affects the secretion of the pituitary gland. The hormone secreted by the target endocrine gland can inhibit the secretory function of the pituitary, lowering the level of trophic hormones (negative feedback), or it can stimulate the pituitary to secrete more of its trophic hormone (positive feedback).

Table 40-4 Action of hormones of pituitary gland

Trophic hormone	Structure (molecular weight in daltons)	Target tissue	Hormone of target endocrine organ
GH	Protein, MW 27,000	Liver, kidney, soft tissue, bone	Somatomedin
PRL	Protein, MW 25,000	Breast	None
FSH	Glycoprotein, MW 30,000 (α and β subunits)	Ovary, testis	Estradiol
LH	Glycoprotein, MW 26,000 (α and β subunits)	Ovary, testis	Progesterone, testosterone
TSH	Glycoprotein, MW 25,000 (α and β subunits)	Thyroid	Thyroxine, triiodothyronine
ACTH	Peptide, MW 45,000	Adrenal cortex	Cortisol
ADH	Cyclic octapeptide, MW 1000	Collecting ducts of renal nephrons	None
Oxytocin	Cyclic octapeptide, MW 1000	Uterus, breast	None

Table 40-5 Hormones or factors of hypothalamus acting on adenohypophysis

Hormone or factor	Abbreviation	Effect on adenohypophysis: Stimulation	Effect on adenohypophysis: Inhibition
Thyrotropin-releasing hormone	TRH	TSH PRL (may not be physiological) GH FSH (in men may not be physiological)	
Luteinizing hormone-releasing hormone or gonadotropin-releasing hormone	LHRH or GnRH	FSH LH	
Prolactin-releasing factor	PRF	PRL	
Growth hormone-releasing factor	GHRF	GH	
Corticotropin-releasing factor	CRF	ACTH	
Growth hormone release-inhibiting hormone (somatostatin)	GHRIH (SRIF)		GH TSH
Prolactin release-inhibiting factor	PIF		PRL

For example, negative feedback exists between the pituitary and the thyroid gland. The pituitary secretes TSH, which stimulates the secretion of thyroxin by the thyroid. The elevation of the thyroxin serum level has an inhibitory effect on the secretion of TSH by the pituitary (Fig. 40-9). This mechanism protects the organism from excessive thyroxin secretion, which would elevate the body's metabolism to a damaging level. An example of positive feedback is the interaction between the secretion of FSH and LH by the pituitary and the secretion of estradiol by the graafian follicle of the ovary. At a threshold plasma level, estradiol stimulates the secretion of FSH and LH. FSH in turn further stimulates the secretion of estradiol. Potentially this could raise the blood level of estradiol dangerously high.

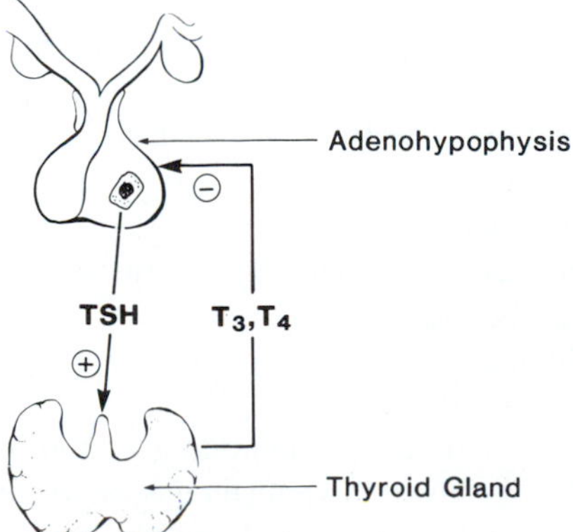

Fig. 40-9 Negative feedback loop between pituitary gland and thyroid gland. T_3, Triiodothyronine; T_4, thyroxine; ⊕, positive effect; ⊖, negative feedback.

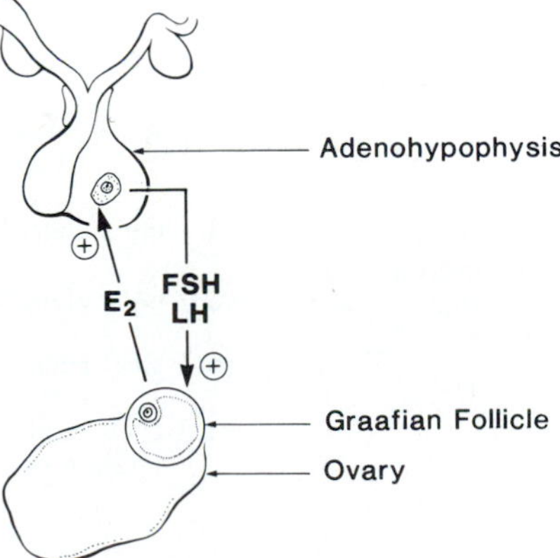

Fig. 40-10 Positive feedback loop between pituitary gland and graafian follicle of ovary. This functional unit is unique for a living organism. E_2, Estradiol; *FSH,* follicle-stimulating hormone; *LH,* luteinizing hormone; ⊕, positive feedback.

However, when the graafian follicle reaches maturity, a synergistic action of FSH and LH results in its rupture and the destruction of the cells secreting estradiol (Fig. 40-10). The negative feedback preserves the homeostasis of the organism, whereas the positive feedback occurs only in systems in which a transient acceleration of growth is needed.

The above examples describe only the interaction between the pituitary and endocrine target organ. One can further expand the feedback model to include the hypothalamus by adding a feedback loop between the hypothalamus and the target endocrine organ. This feedback loop incorporates a releasing factor secreted by the hypothalamus, which stimulates the release of a corresponding trophic hormone from the pituitary, which in turn stimulates the secretion of the hormone of the endocrine target organ. This latter hormone stimulates the hypothalamus to secrete an inhibitory releasing factor. This type of feedback is called indirect inhibitory feedback because it incorporates an intermediate organ, the pituitary. It also differs from the direct negative feedback in that it not only inhibits the secretion of the releasing factor but can also stimulate the secretion of a release-inhibiting factor. An example of this system is the hypothalamic-pituitary-thyroid axis (Fig. 40-11).

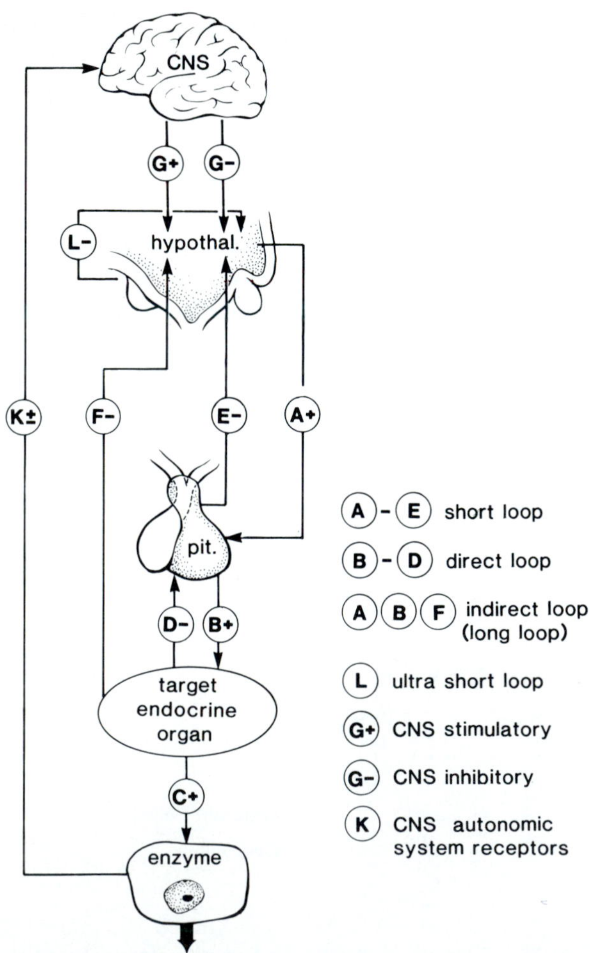

Fig. 40-11 Types of endocrine feedbacks. +, Positive effect; −, negative feedback.

Two trophic hormones, GH and PRL, directly stimulate the target cells without the intermediate involvement of a target endocrine organ (see Table 40-4). PRL directly influences secretions of the breast (lactation), and GH promotes growth in soft tissue, cartilage, and bones. The four other adenohypophyseal hormones act through the stimulation of intermediary target endocrine organs. TSH stimulates the thyroid gland to synthesize and secrete its hormones: triiodothyronine (T_3) and thyroxine (T_4) (see Table 40-4). FSH in women stimulates development of follicles of the ovary and in men spermatogenesis in the testes. LH in women precipitates ovulation by acting synergistically with FSH and later maintains the secretory function of corpus luteum. In men, this hormone is called *interstitial cell–stimulating hormone* (ICSH) and acts primarily by stimulating the interstitial Leydig cells in the testes to secrete testosterone. ACTH stimulates the two inner zones of the cortex of the adrenal glands, the zonae fasciculata and reticularis, which secrete the glucocorticoid hormones and a small quantity of sex hormones.

The neurohypophysis stores and releases two hormones made by the hypothalamic neurons: antidiuretic hormone (ADH), which acts directly on the collecting ducts of the renal nephrons increasing the reabsorption of water, and oxytocin, which in women stimulates the contractions of the muscles of the uterus and the lactating mammary glands.

PATHOLOGICAL CONDITIONS

Adenohypophysis

The clinical manifestations of disorders of the adenohypophysis can be the result of hormone hypersecretion or hyposecretion. Hypersecretion usually involves only one hormone but is not uncommonly associated with hyposecretion of another trophic hormone. For example, hypersecretion of PRL can be associated with hyposecretion of FSH. On the other hand, hyposecretion usually involves more than one trophic hormone.

Hypersecretion. The following factors can cause pituitary hypersecretion:

A. Primary factors
   1. Pituitary adenomas, benign tumors of the pituitary gland
   2. Pituitary hyperplasia
B. Secondary factors
   1. A lack of release-inhibiting factors caused by damage to the hypothalamus or to the hypothalamopituitary tract; known to cause the hypersecretion of PRL
   2. Extrapituitary (ectopic) secretion of trophic hormones by carcinomas (Table 40-6)
   3. Hyposecretion of endocrine target organs

Table 40-6 Ectopic pituitary hormones produced by nonendocrine tumors

Hormone	Tumor
GH	Poorly differentiated lung carcinoma
PRL	Oat cell carcinoma of lung
	Renal carcinoma
ACTH	Oat cell carcinoma of lung
	Carcinoids of lung, thymus, gut, and pancreas
	Medullary carcinoma of thyroid
	Pheochromocytoma
	Neuroblastoma
ADH	Oat cell carcinoma of lung
Oxytocin	Oat cell carcinoma of lung

Growth hormone. The effect of hypersecretion of GH depends on the age of the patient. In adults, it leads to acromegaly, which is the progressive enlargement of the acral (distal) parts of the extremities. In children, the hypersecretion of GH causes gigantism, characterized by excessive growth; a height of more than 8 feet can be reached. The accelerated growth is associated with osteoporosis and muscle weakness. Acromegaly and gigantism are usually caused by pituitary adenomas secreting GH and compressing the adjacent tissues of the pituitary gland, causing hyposecretion of other trophic hormones.

Prolactin. Hypersecretion of PRL causes galactorrhea (lactation). This can be associated with infertility and amenorrhea in women (Forbes-Albright syndrome) and with decreased libido and impotence in men. PRL hypersecretion is usually caused by a pituitary adenoma.

Adrenocorticotropic hormone. Hypersecretion of ACTH by the pituitary is usually caused by pituitary microadenomas or hyperplasia and is called *Cushing's disease*. The symptoms of Cushing's disease are caused by the enhanced quantities of cortisol and androgens secreted by the adrenal cortex. The resulting clinical picture of *Cushing's syndrome* is described in Chapter 45. In about 75% of cases Cushing's syndrome is caused by excessive production of ACTH. The remaining cases are caused by the primary diseases of the adrenal gland.

Thyroid-stimulating hormone. The clinical syndrome that results from tissue exposure to high levels of thyroid hormones is called thyrotoxicosis. In a great majority of cases, it is caused by hyperactivity of thyroid gland rather than by hypersecretion of TSH (see Chapter 41). The rare case of TSH hypersecretion is usually associated with acromegaly.

Gonadotropins. Sexual precocity associated with increased gonadotropin secretion (FSH and LH) is usually induced by cerebral tumors in the region of the hypothalamus.

Hyposecretion. Pituitary hyposecretion can be seen as hyposecretion of one hormone, hyposecretion of a group of hormones, or the total hyposecretion of all hormones; the latter condition is called *panhypopituitarism*. A single hormone hyposecretion is commonly the result of lesions within the hypothalamus. It can also be caused by an uncontrollable elevation of release-inhibiting factors or a lack of releasing factors. In cases of pituitary adenomas, hypersecretion of one hormone, the one made by the tumor is commonly accompanied by hyposecretion of the remaining pituitary hormones because of the destruction of the pituitary gland by the compressing tumor. Adenomas can destroy up to three fourths of the pituitary mass without any evident clinical symptoms of hypopituitarism, and the patient can survive a destruction of up to 90% of the pituitary mass. However, the pituitary does not regenerate to a significant degree; therefore a massive loss of cells of the pituitary has to be treated with administration of the pituitary hormones. Factors that can destroy the pituitary gland are listed in Table 40-7. The clinical symptoms observed in hyposecretion do not depend on the causative factor but merely reflect the lack of trophic hormones.

Table 40-7 Etiological factors in hypopituitarism

Factors within sella	Factors outside sella (hypothalamus)
Tumors (pituitary adenomas and craniopharyngiomas, gliomas, meningiomas, or metastatic tumors)	Tumors (craniopharyngioma, gliomas, germinomas, teratomas, or metastatic tumors)
Infections (tuberculosis, other basal meningitides, syphilis, encephalitis)	Infections
Granulomas (sarcoidosis, Hand-Schüller-Christian disease)	Granulomas
Vascular (postpartum necrosis, hypertensive hemorrhages, infarcts, diabetes mellitus, cranial arteritis, vascular malformations)	Vascular
Iatrogenic (surgical hypophysectomy, irradiation, prolonged therapy with hormones)	Trauma
Trauma	
Amyloidosis	
Hemosiderosis	

Depending on the location of the lesion, a different set of clinical symptoms will occur. In intrasellar lesions symptoms are usually caused by the deficiency of all the pituitary hormones. Hypothalamic lesions can result in the hyposecretion of a single hormone and can also affect centers of autonomic functions, such as centers of appetite, thirst, or sexual drive. Therefore a patient can have hypogonadism caused by the lack of gonadotropins associated with obesity as a result of excessive appetite. An increase in size of an intrasellar tumor can lead to a compression of the optic chiasm, which manifests itself in a lack of peripheral vision.

Growth hormone. Deficiency of GH in children leads to pituitary dwarfism. The child is small but proportionally

Table 40-8 Range of normal levels of pituitary hormones in blood

Hormone	Conditions	Value	Hormone	Conditions	Value
ACTH	8:00 AM	40-120 ng/L	LH	Men	1.5-10 U/L
	8:00 PM	<5-50 ng/L		Women	
				Early follicular	2.5-15 U/L
				Midfollicular	Up to 20 U/L
				Mid cycle	5-70 U/L
				Luteal	Up to 13 U/L
FSH	Children 1 year of age to puberty	>2.5 U/L	PRL		Up to 800 mU/L
			TSH		Up to 5 mU/L
	Adult men	1-7 U/L	GH	Hypoglycemia	>0.5 mU/L
	Adult women				
	Premenopausal	1-10 U/L			
	Midcycle peak	6-25 U/L			
	Postmenopausal	30-120 U/L			

Table 40-9 Tests of hypothalamopituitary function

Hormones	Test	Procedure	Normal response	Pituitary lesion	Hypothalamic lesion
GH, ACTH, PRL	Insulin tolerance	0.1-0.3 U/kg of weight intravenously	Increase in GH (above 40 mU/L), ACTH, PRL, corticosteroids (above 550 nmol/L)	Panhypopituitarism: no increase in GH, ACTH, PRL, and corticosteroids Partial hypopituitarism: no increase in GH	No response of ACTH and corticosteroids with normal response to vasopressin
GH, ACTH, PRL	Glucagon (used when insulin is contraindicated)	1 mg subcutaneously	Increase in GH (men at least 14 mU/L, women 20 mU/L), corticosteroids (above 550 nmol/L)	As above	As above
GH	Arginine	30 g intravenously (children 1.5 g/kg)	Increase (at least 14 mU/L)	Hypopituitarism: no increase	
GH in children	Bovril	20 g/1.5 m^2 body surface orally	Increase (above 20 mU/L)	As above	
GH	Oral glucose tolerance	50 g orally	Decrease (4 mU/L)	Hypersecretion (acromegaly and gigantism): no decrease	
LH/FSH	Clomiphene stimulation	50 mg daily for 5 days	Increase in LH and FSH above laboratory norms	Hypopituitarism: no increase in LH and FSH	No increase in LH and FSH
ADH	Water deprivation	No fluids for 8 hours	Urine 600 mOsm/kg or above, plasma not above 300 mOsm/kg, urine flow rate less than 0.5 mL/min		Hyposecretion of ADH (diabetes insipidus): urine less than 270 mOsm/kg, plasma above 300 mOsm/kg, urine volume not reduced
GH, ACTH, PRL, TSH, LH, FSH	Combined pituitary function test	Insulin 0.1-0.3 U/kg of body weight: TRH 200 μg; LHRH 100 μg intravenous	To insulin—increase in GH (above 400 mU/L), PRL (usually masked by response to TRH), and corticosteroids (above 250 nmol/L) To TRH—increase in TSH (3-15 mU/L) and PRL (1800 mU/L) To LHRH—increase in LH (15-42 U/L) and FSH	Panhypopituitarism: no increase in GH, ACTH, PRL, corticosteroids, TSH, LH, and FSF Hypopituitarism: no increase in individual hormones	Increase in PRL, TSH, LH, and FSH

built. This condition is usually not recognized until after the first year of life.

Prolactin. The only known clinical effect of the deficiency of PRL is the lack of lactation in postpartal women.

Adrenocorticotropic hormone. The clinical picture is essentially similar to changes seen in Addison's disease (see Chapter 45), but there is no hyperpigmentation and only slight impairment in aldosterone secretion.

Thyroid-stimulating hormone. TSH deficiency may be difficult to differentiate from a primary defect of the thyroid gland. Usually there is evidence of other trophic hormone deficiencies, for example, gonadal deficiency.

Gonadotropins. Gonadotropins are usually the first trophic hormones to be affected in lesions of adenohypophysis. The child is often taller than usual, and at adolescence secondary sex characteristics fail to develop. In adults, secondary sex characteristics show signs of atrophy, with scanty or absent pubic and axillary hair, infertility, and possible loss of libido (see Chapter 43).

Neurohypophysis

Hypersecretion of antidiuretic hormone (vasopressin). An excessive secretion of ADH, called the syndrome of inappropriate ADH secretion (SIADH), can occur in a wide variety of clinical conditions, such as meningitis, brain abscess, head injury, cerebrovascular thrombosis, tuberculosis, hypoadrenalism, hypothyroidism, and liver cirrhosis. This syndrome is associated with hyponatremia and urine that is hypertonic relative to plasma in spite of normal renal and adrenal function. The clinical symptoms include weakness, malaise, and poor mental status, ultimately progressing to convulsions and coma.

Hyposecretion of antidiuretic hormone (vasopressin). The clinical entity associated with the hyposecretion of ADH is diabetes insipidus. It can be caused by lesions directly involving the neurohypophysis, hypothalamus, or hypothalamopituitary tract. The clinical manifestations are a sudden onset of insatiable thirst, polydipsia (excessive drinking), and intense polyuria (large urine volume). The greatest danger in the rapid onset of diabetes insipidus is dehydration. The urine concentration is usually on the order of 100 mOsm/kg. Plasma urea and electrolytes and full blood count show the expected changes of dehydration.

FUNCTION TESTS

To assess the pituitary-hypothalamic function it is essential to separate the function of the hypothalamopituitary system from the secretory function of the target endocrine organ. This is done using sensitive immunoassays for specific trophic hormones and challenge tests. The challenge tests also allow for the separation of endocrine abnormalities caused by pituitary lesions from those caused by hypothalamic lesions. Table 40-8 gives the range of normal values of pituitary hormones in blood, and Table 40-9 lists examples of challenge tests.

BIBLIOGRAPHY

General aspects of hormone physiology

Chan, L, and O'Malley, BW: Mechanism of action of the sex steroid hormones, N Engl J Med 294:1322-1326, 1372-1379, 1976.

Felig, P, et al: Endocrinology and metabolism, ed 2, New York, 1987, McGraw-Hill Book Co.

Norman, AW, and Litwack, G: Hormones, Orlando, FL, 1987, Academic Press, Inc.

Robison, GW, Butcher, RW, and Sutherland, EW: Cyclic AMP, New York, 1971, Academic Press, Inc.

Sutherland, EW: Studies on the mechanism of hormone action, Science 177:401, 1972.

Hypothalamopituitary system

Alsever, RN, and Gotlin, RW: Handbook of endocrine tests in adults and children, Chicago, 1978, Year Book Medical Publishers.

Beardwell, C, and Robertson, GL, editors: The pituitary, London, 1981, Butterworth & Co, Publishers, Ltd, pp 47-75, 175-210, 238-264.

Hall, R, Anderson, J, Smart, GA, and Besser, M: Fundamentals of clinical endocrinology, Tunbridge Wells, England, 1980, Pitman Medical Publishers, pp 1-74.

Jeffcoate, SL: Efficiency and effectiveness in the endocrine laboratory, New York, 1981, Academic Press, Inc.

Mazzaferri, EL: Textbook of endocrinology, ed 3, New Hyde Park, NY, 1986, Medical Examination Publishing Co.

McCain, SM, editor: Endocrine physiology II, Int Rev Physiol 24:97-156, 1981.

Tolis, G, Martin, JB, Labrie, F, and Naftolin, F, editors: Clinical neuroendocrinology: a pathophysiological approach, New York, 1979, Raven Press, pp 1-46, 89-128, 366-384.

CHAPTER 41

Thyroid

MARIANO FERNANDEZ-ULLOA
HARRY R. MAXON III

OBJECTIVES

- Describe the synthesis, transport, function, and regulation of thyroid hormones.
- Describe pathological conditions resulting in hyperthyroidism or hypothyroidism.
- State the importance of the analysis of the free hormone levels of T_4 and T_3 and the meaning of the free thyroxine index.
- Describe the thyroid function tests and which analytes are used.
- State the diagnostic value of the following laboratory tests in assessing thyroid conditions: T_4, T_3, FT_4, T_3RU, FTI, and TSH.

KEY TERMS

acromegaly Enlargement of the extremities (especially hands and feet) caused by an increased secretion of pituitary growth hormone.

adrenergic activity Metabolic effects caused or mimicked by the hormone epinephrine (adrenaline).

anabolic agents Compounds (usually androgens) that promote increase or construction of new tissue.

androgen Natural or synthetic substance (usually hormones) that produces masculinizing effects.

aplasia Lack of development of any organ.

C cells Calcitonin-secreting cells of the thyroid.

calorigenesis Production of heat and energy.

cretinism Hypothyroid condition caused by congenital lack of thyroid hormone secretion and characterized by impaired physical and mental development.

DIT Diiodotyrosine.

dysgenesis Abnormal or defective development.

dysplasia Abnormal development of an organ resulting in alteration of configuration, size, or cell organization.

endocytosis The uptake by a cell of material from the environment by invagination of its plasma membrane.

factitial Produced by artificial means; unintentionally produced.

follicular cells Thyroid hormone-producing cells arranged in spherical vesicle-like units.

goiter Enlargement of the thyroid gland; multinodular-enlarged thyroid containing numerous superficial and deep indurations.

Graves' disease Immune disorder caused by antibodies binding to thyroid-stimulating hormone (TSH) receptors, resulting in an unregulated increase in thyroid hormone production and release.

Hashimoto's thyroiditis Inflammatory process of the thyroid caused by a derangement of the immune system, which may or may not lead to abnormal thyroid function.

hyperthyroidism Metabolic and clinical state caused by an increase in circulating active thyroid hormone.

hypothyroidism Metabolic and clinical state caused by decreased levels of circulating active thyroid hormone or increased tissue resistance; *primary*—decreased thyroid function caused by disease of the thyroid gland; *secondary*—decreased thyroid function caused by disease of the pituitary gland; *tertiary*—decreased thyroid function caused by disease of the hypothalamus.

iatrogenic Any adverse condition resulting from the actions of a physician.

interstitium Pertaining to the intercellular spaces of a tissue.

iodine trapping The ability of the thyroid gland to sequester the iodine against a concentration gradient.

medullary carcinoma (thyroid) Cancer of the C cells.

MIT Monoiodotyrosine.

monodeiodination Loss or removal of a single iodine atom.

myxedema Advanced hypothyroid state characterized clinically by distinctive external appearance: pallor, skin edema (swelling) of face and hands, apathy, and so on.

organogenesis Incorporation of ionic form of iodine into the molecular structure of tyrosine.

parafollicular C cells Calcitonin-secreting cells located between follicles.

parenchyma A group of basic morphological and functional cellular units that constitute any organ.

prohormone A compound requiring chemical transformation to become an active hormone.

protein-bound iodine (PBI) The total amount of iodine bound to proteins in the plasma.

resin uptake test Measurement of the number of available binding sites of plasma thyroid hormone–transporting proteins.

solitary nodule Localized enlargement of a portion of the thyroid gland.

struma ovarii Rare teratoid tumor of the ovary composed almost entirely of thyroid tissue.

T_3 Thyroid hormone with iodine atoms in positions 3, 5, and 3′ (triiodotyrosine).

T_4 Thyroid hormone with iodine atoms in positions 3, 5, 3′, and 5′ (tetraiodotyrosine).

rT_3 Triiodotyrosine with iodine in positions 3, 3′, and 5′.

thyroglobulin A glycoprotein of molecular weight 660,000 daltons produced by the follicular cells and containing the precursors of T_3 and T_4.

thyroid colloid The material found within the follicles of the thyroid and containing thyroglobulin and thyroid hormone.

thyroiditis A general term for inflammation of the thyroid gland.

thyrotoxicosis Condition caused by excess thyroid hormone secretion, often used as synonym for hyperthyroidism.

thyroxine-binding globulin (TBG) A glycoprotein of alpha-mobility that transports thyroid hormone in the blood.

TRH (thyrotropin-releasing hormone) A tripeptide that promotes release of TSH.

trophic action Stimulation of cell reproduction and enlargement.

trophoblastic tumor Tumor originating from extraembryonal cells of ectodermic nature located in the blastocyst.

TSH (thyroid-stimulating hormone, thyrotropin) The glycoprotein composed of alpha and beta subunits released from the pituitary that promotes thyroid hormone production and release.

TSI Thyroid-stimulating immunoglobulin.

Wolff-Chaikoff effect Decreased formation and release of thyroid hormone in the presence of excess iodine.

ANATOMY

The thyroid is formed of two lobes, one on either side of the neck. The thyroid gland consists of two types of cells, follicular and parafollicular (Fig. 41-1). Follicular cells are arranged spherically in a single layer. Follicular cells have an apical end facing the center of the follicle and a basal end facing the interstitium, which contains the blood supply and parafollicular cells. Follicular cells produce thyroid hormone, which is then stored in the central portion of the spherical follicle in a material called "colloid." Parafollicular cells are located in the interstitium between follicles and secrete the hormone calcitonin. For this reason they are called "C cells."

THYROID PHYSIOLOGY

The thyroid gland has as its main function the production and secretion of metabolically active hormones that are essential for the regulation of various metabolic functions. Thyroid hormones are produced within the cells of the follicles from the amino acid tyrosine and the halogen element iodine. The two most important thyroid hormones are thyroxine (T_4), which contains four iodine atoms, and triiodothyronine (T_3), which contains three iodine atoms (Fig. 41-2).

Thyroid hormones are released from the colloid and secreted into circulation in response to stimulation of the thyroid gland by the pituitary hormone called *thyroid-stimulating hormone* (TSH).

Metabolism of iodine and thyroid hormone synthesis

Iodine is a natural component of many foods and in the United States is provided in adequate amounts by a well-balanced diet. Extra amounts of iodine are currently provided by the ingestion of iodine-enriched foods and numerous "vitamin pills."

The daily intake of iodine varies widely in different parts of the world. In the United States iodine intake ranges from 250 to 700 mg or more daily. In countries such as Japan, intake may reach several milligrams per day; whereas in some areas of Africa, South America, Asia, and Europe daily intake may be as low as 50 μg.

Under physiological conditions, iodine is absorbed in the small bowel and then enters either the excretory or metabolic pathways (Fig. 41-3). Between 60% and 80% of the ingested iodine is excreted by the kidneys. Small

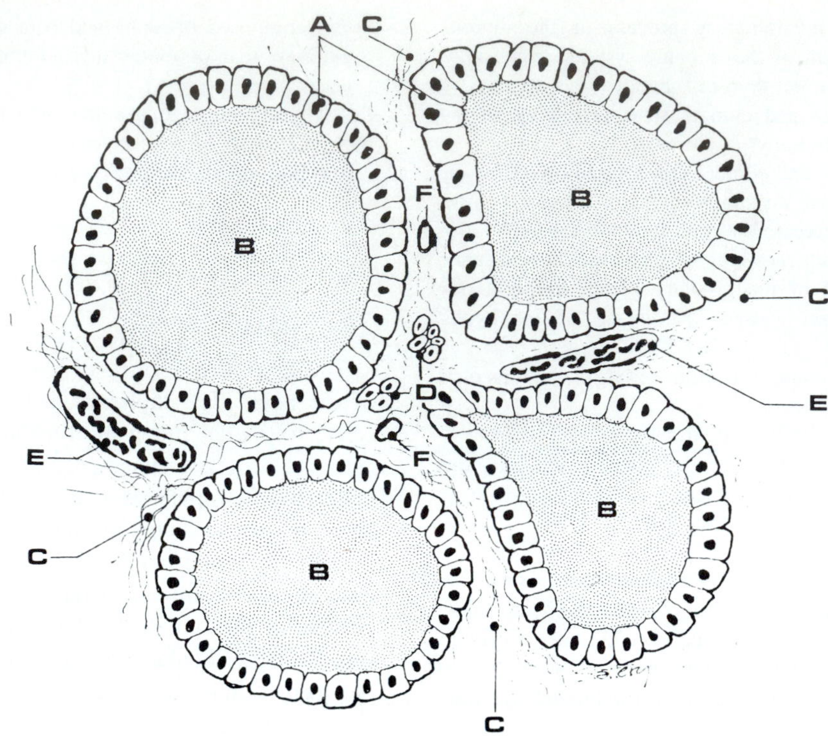

Fig. 41-1 Thyroid gland structure consists of follicular cells *(A)*, enclosing colloid *(B)*, and parafollicular "C" cells *(D)*, in the interstitium *(C)*. *E*, Venule; *F*, capillary.

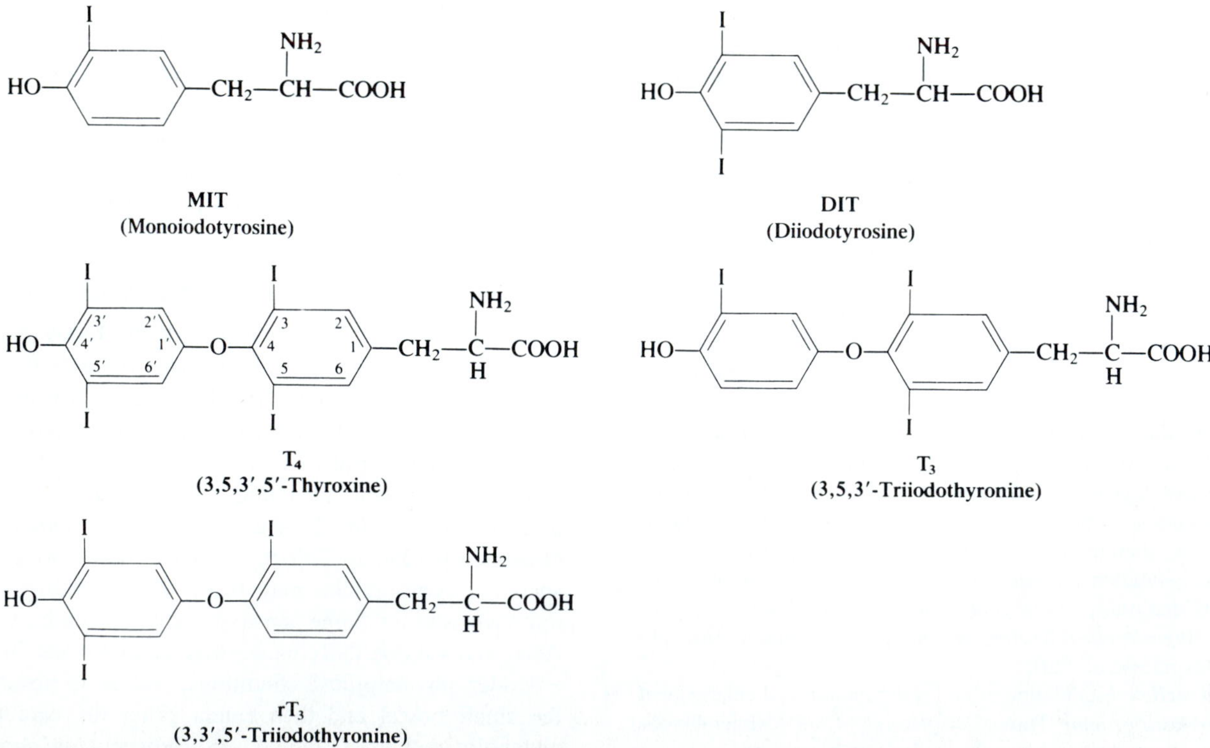

Fig. 41-2 Chemical structure of thyroid hormones and iodinated precursors and metabolites.

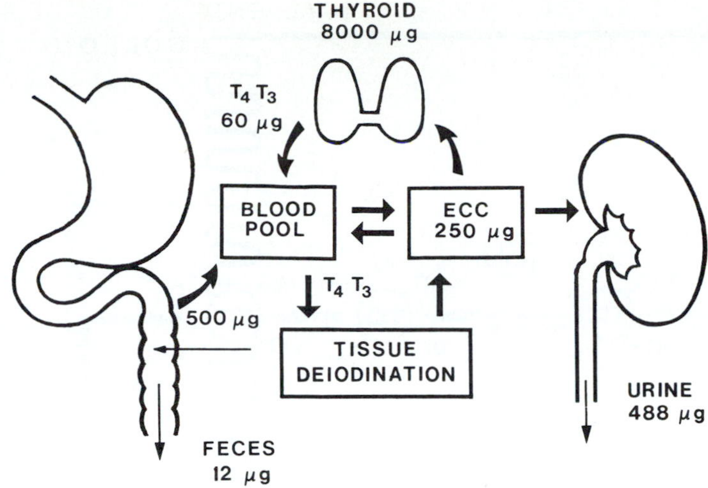

Fig. 41-3 Iodine metabolic pathway in a 24-hour period. *ECC,* Extracellular compartment.

amounts are excreted through the intestinal route. Fecal excretion is derived for the most part from hormones degraded by the liver and excreted into the bowel by the biliary tract. The remainder of the iodine is distributed into the extracellular and thyroid compartments. The intrathyroid iodine compartment contains about 90% of the total body iodine and can amount to as much as 6000 to 12,000 μg. The extracellular compartment contains most other iodine, except for a small but important amount within cells. The metabolism of iodine is closely related to the process of thyroid hormonogenesis.

Classically, intrathyroidal iodine metabolism has been divided into the following stages: (1) iodine trapping or uptake of iodine by the follicular cells, (2) organification, (3) coupling, (4) storage, and (5) secretion (Fig. 41-4). During trapping, thyroid cells concentrate iodine against high chemical and electrical gradients that require an active energy-dependent mechanism at the level of the cell membrane.[1]

This trapping mechanism for iodine has been exploited for many years in clinical tests to assess thyroid function. Radioactive iodine is given orally to patients, and its degree of concentration in the thyroid is subsequently measured. Various physiological and pharmacological factors influence trapping. The most important factor is thyroid-stimulating hormone (TSH), which stimulates trapping of iodine. Iodine excess inhibits the transport of iodine; iodine deficiency stimulates it.

The second step of intrathyroidal iodine metabolism is organification by means of which iodine is incorporated into thyroid hormone.[2] The iodine used comes from trapping and from the intrathyroid deiodination of stored thyroid hormone precursors. Iodine in the thyroid is oxidized in the presence of a peroxidase enzyme into a reactive form that combines with the protein thyroglobulin.[2]

Thyroglobulin is a glycoprotein with a molecular weight of about 660,000 daltons. Thyroglobulin serves as a preformed matrix containing 115 tyrosyl groups to which reactive iodine is attached to form residues of monoiodotyrosine (MIT), diiodotyrosine (DIT), triiodothyronine (T_3), and thyroxine (T_4). After their formation, coupling of MIT and DIT takes place to form intrathyroglobulin T_3 and T_4. The thyroglobulin is then released from the cells into the colloid of the follicle where it remains stored. The storage function of thyroglobulin provides a constant pool of thyroid hormone.

When TSH stimulates the thyroid, a series of cellular and biochemical changes takes place in the gland, all directed toward the synthesis and release of thyroid hormone.[3] Initially, droplets of colloid are engulfed by the colloid cells and digested by proteases releasing MIT, DIT, T_3, and T_4 from the thyroglobulin matrix. Intracellular MIT and DIT are immediately deiodinated, and their iodines are reused in subsequent thyroid hormone synthesis. The released T_3 and T_4 are resistant to intrathyroid deiodination and are secreted as active hormones. The daily secretion of thyroid hormone includes about 90 mg of T_4 and about 7 mg of T_3.[4] Small amounts of reverse T_3 (rT_3) are also secreted by the thyroid.

Transport of thyroid hormones

On their release into the bloodstream, circulating free T_3 and T_4 hormones enter body cells, where they become part of an intracellular pool of hormone and exert their metabolic effects. The T_3 and T_4 are also transported to portions of the brain and pituitary gland, where they participate in the regulation of pituitary TSH secretion. In the bloodstream the thyroid hormones are transported in two forms, protein bound and free. The free hormone form, which is metabolically active, represents only a small fraction (less than 1%) of the total plasma thyroid hormone content. The bound hormone is metabolically inactive and serves as a stable reservoir of hormone that, by virtue of its size and state of equilibrium with the free hormone compartment, maintains a constant supply of hormone available to tissues.

The binding of thyroid hormones to proteins in the blood is accomplished by three plasma proteins.[5] The most important is the thyroxine-binding globulin (TBG), which is a glycoprotein synthesized in the liver. The second most important is thyroxine-binding prealbumin (TBPA) with a molecular weight of 50,000 daltons. The third transporting protein is albumin.

The role of each of these binding proteins in the transport of T_3 and T_4 depends on their relative affinities for each of the thyroid hormones and on their relative concentrations in plasma. Almost all the circulating T_4 (99.95%) is bound to these plasma proteins. Under physiological circumstances, TBG transports 70% to 75% of total T_4, and TBPA and albumin transport 15% to 20% and 10% of T_4, respectively. Most (99.7%) plasma T_3 circulates in the

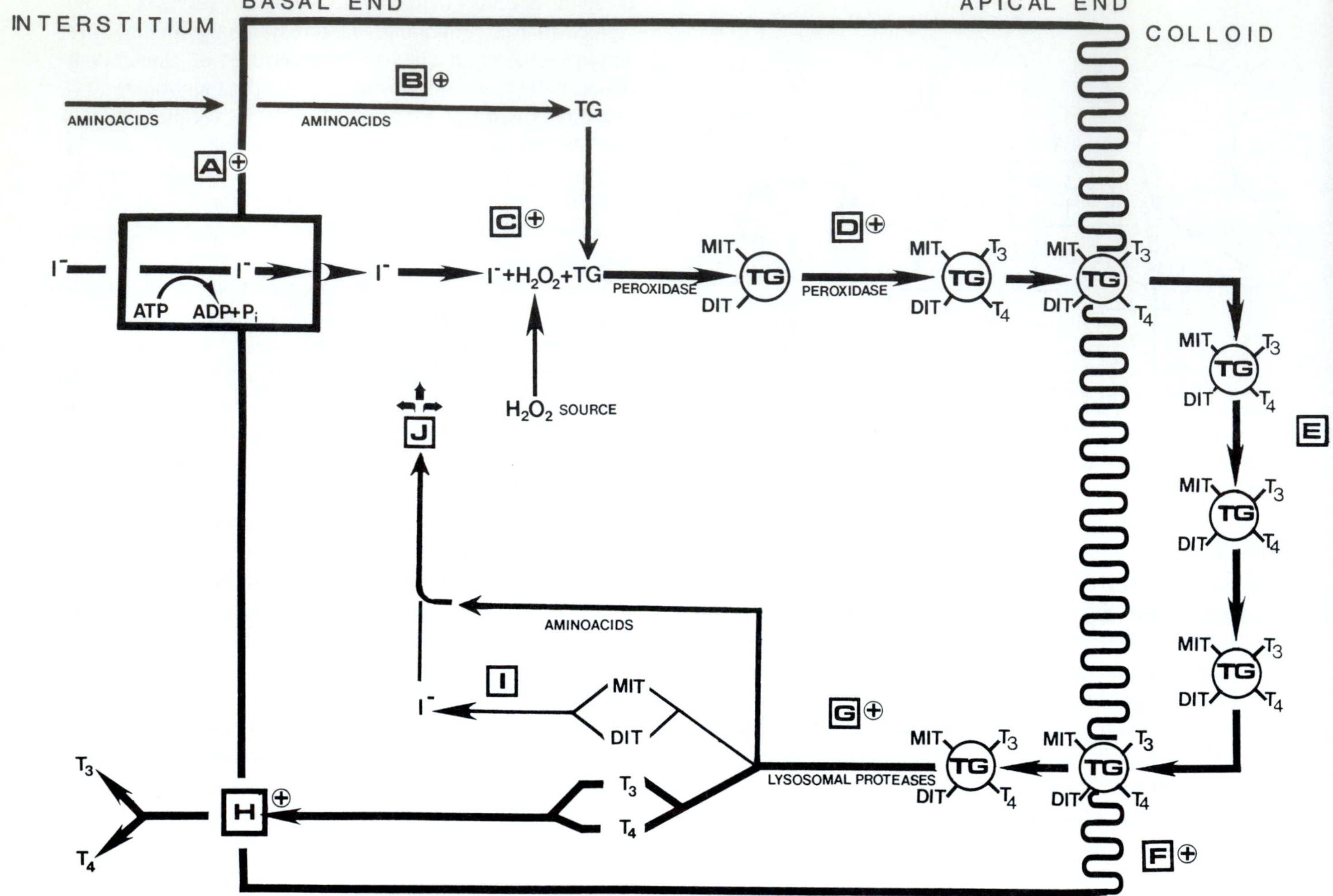

Fig. 41-4 Thyroid cell. Schema depicting stages of thyroid hormonogenesis and intrathyroidal iodine metabolism. *A,* Iodine transport; *B,* thyroglobulin (TG) synthesis; *C,* iodine organification; *D,* intrathyroglobulin coupling; *E,* storage; *F,* endocytosis; *G,* hydrolysis; *H,* hormone secretion; *I,* intrathyroidal deiodination; *J,* recycling; steps influenced by the thyroid-stimulating hormone (TSH) are indicated by the symbol ⊕.

bound form. The affinity of TBG for T_3 is lower than its affinity for T_4, and binding of T_3 to TBPA is negligible. T_3 is bound to albumin. Abnormalities of the binding proteins may result in abnormal total (bound) hormone concentrations in the blood with normal amounts of free hormone. Since both T_4 and T_3 are bound mainly to TBG, changes of TBG levels affect total serum T_4 and T_3 levels. Changes in TBG concentrations in blood have an indirect effect on the negative feedback mechanisms of thyroid hormone regulation (Table 41-1). An increase in TBG will result in an increased binding of free thyroid hormone with a subsequent decrease in circulating free hormone. The decrease in FTH immediately triggers the secretion of TSH, which results in increased thyroid hormone production. All these changes are transient, and a new equilibrium with preservation of the normal thyroid status is quickly reached. Decreases of TBG will cause the opposite biological changes.

Metabolism of thyroid hormones

Circulating T_3 and T_4 are either incorporated into the intracellular pool, where they undergo partial transformations and exert their metabolic effects, or are degraded and eliminated by excretory organs.

All circulating T_4 originates in the thyroid gland, which secretes 80 to 100 mg of T_4 per day. The more metabolically important T_3 originates from both direct thyroid secretion (about 20%) and peripheral conversion of T_4 to T_3 by monodeiodination (about 80%).[4] The total daily production of T_3 from all sources ranges from 22 to 47 μg.[4] Although T_3 has been established as the main active hormone, the role of T_4 as a hormone with direct biological

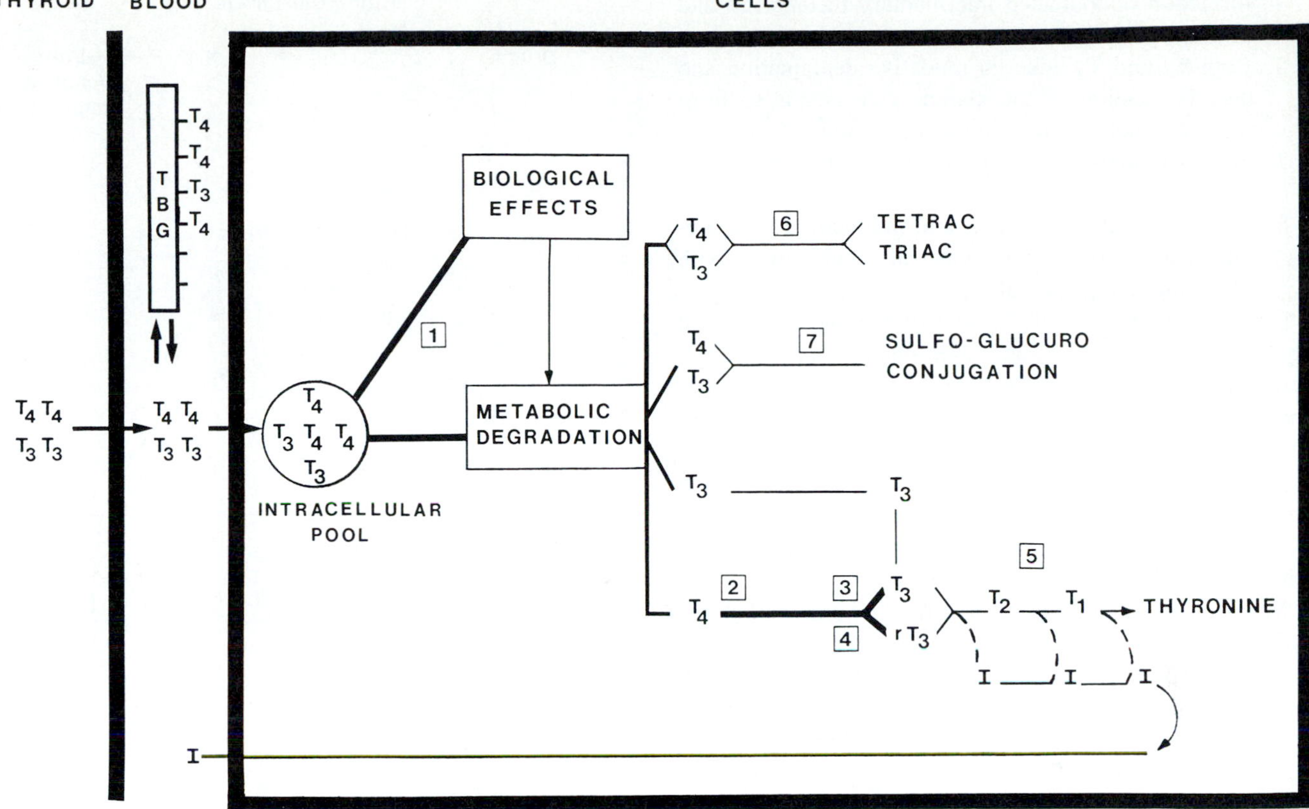

Fig. 41-5 Metabolic pathways of thyroid hormone. *1,* Biological effects through binding to intracellular receptors; *2,* main deiodinative pathway for T_4; *3,* conversion of T_4 into T_3; *4,* conversion of T_4 into rT_3; *5,* serial deiodinations of T_3 and rT_3; *6,* deamination and decarboxylation pathway; *7,* conjugative pathway.

activity has been questioned, and some workers consider T_4 to be a prohormone. However, T_4 is known to have some direct biological activity, albeit less than T_3.

The thyroid hormones are metabolized through deiodinative and nondeiodinative mechanisms. The following are some of the most important metabolic steps (Fig. 41-5):

1. Both T_4 and T_3 exert their biological effects by binding to specific cellular receptors and are subsequently degraded through successive deiodinations.
2. Deiodination accounts for 80% to 85% of the metabolism of T_4 and T_3.[6]
3. About 35% to 50% of the T_4 undergoing deiodination is converted to T_3.[6,7]
4. About 50% to 65% of the deiodinated T_4 is converted into rT_3.[7]
5. Most of the T_4, T_3, and rT_3 are metabolized through a chain of successive deiodinations, resulting in the

Table 41-1 Physiological relationship between thyroxine-binding globulin (TBG) and serum thyroid hormone concentrations

Biological modulators	Initial biochemical changes	Intermediate biological response	Final equilibrium conditions
Increased TBG Pregnancy Oral contraceptives	Increased TBG levels, decreased saturation Augmented binding of hormone (T_4, T_3) Decreased free T_4, T_3	Decreased negative feedback mechanism Increased serum TSH Increased T_4 and T_3 production	Increased TBG Elevated serum T_4 and T_3 levels Normal TBG saturation Normal free T_4 and T_3
Decreased TBG Androgens Malnutrition Liver disease	Decreased TBG, increased saturation Diminished binding of hormone (T_4, T_3) Increased free T_4 and T_3	Increased negative feedback mechanism Decreased serum TSH Decreased T_4 and T_3 production	Decreased TBG Decreased T_4 and T_3 Normal TBG saturation Normal free T_4 and T_3 serum

formation of iodinated intermediary metabolites and ultimately thyronine.

6. Both T_4 and T_3 undergo oxidative deamination and decarboxylation of the alanine side chains to form the acetic acid analogs tetrac and triac.[4]
7. Small amounts of free T_4 are eliminated in the bile and urine.
8. Small amounts of T_3, rT_3, and indirectly T_4 are metabolized through processes of conjugation with glucuronic acid and sulfate.

Conversion of T_4 to T_3 is one of the most important metabolic pathways of T_4 and takes place in many tissues, particularly the liver and the kidney.[4] In T_3, iodine atoms are located at positions 3 and 5 of the inner (nonphenolic) ring and 3′ of the outer (phenolic) ring (Fig. 41-2). In the case of rT_3, iodine atoms are found in position 3 of the inner ring and positions 3′ and 5′ of the outer ring. Almost all rT_3 derives from T_4 monodeiodination[6] and is essentially inert without metabolic activity.

Mechanisms of action of the thyroid hormones at the cellular level have been the focus of intensive research. The thyroid hormone initially appears both to bind to cell membrane receptors and to cross the cell membrane by direct diffusion. Once inside the cell, the thyroid hormone binds to cytosol and nuclear receptor sites. The latter results in formation of messenger ribonucleic acid (mRNA), which in turn directs the synthesis of proteins and enzymes responsible for metabolic functions.[8] Other postulated mechanisms of action at the cellular level are (1) mitochondrial activation, (2) stimulation of Na^+-K^+ adenosinetriphosphatase (ATPase) activity,[9] (3) stimulation of cell membrane functions, probably through a specific receptor, and (4) interaction with the adrenergic system.

Control and regulation of thyroid function

Hypothalamic-pituitary-thyroid axis (HPTA). The HPTA comprises a group of physiologically interrelated neuroendocrine and endocrine organs that regulate and control the secretion of thyroid hormone (Fig. 41-6). The ultimate effector in this axis is the thyroid gland, which is directly in charge of producing, storing, and secreting the hormones thyroxine and triiodothyronine. The hypothalamus, located in the brain, acts as an important regulating organ (see Chapter 40).

Thyrotropin-releasing hormone (TRH) is a tripeptide produced in the hypothalamus and secreted into the venous system. The TRH attaches to receptor sites in the pituitary, where it causes increased production and secretion of thyroid-stimulating hormone (TSH). TSH is a glycopeptide structurally composed of two subunits, alpha and beta. The beta subunit confers on TSH the specific physiological properties that differentiate it from other pituitary glycopeptides. TSH is released from the pituitary into the bloodstream. At the thyroid, TSH attaches to specific cell receptors and exerts two main actions. The first action, a trophic action, is the stimulation of cell reproduction and hypertrophy. The second effect is the stimulation of production and secretion of thyroid hormone by the thyroid cell.

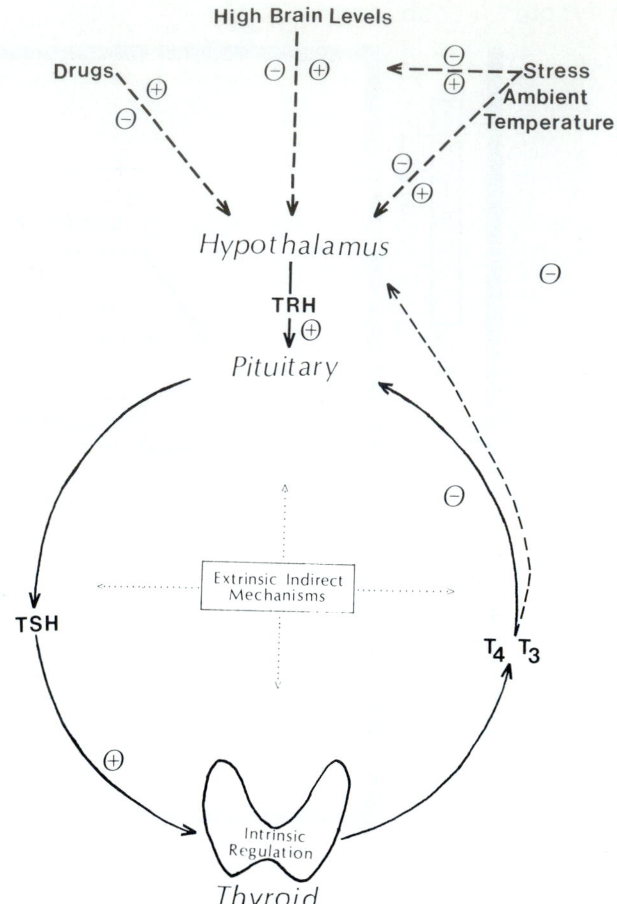

Fig. 41-6 Hypothalamic-pituitary-thyroid axis (HPTA). Stimulatory, ⊕, or inhibitory, ⊖, effect of agent.

In normal persons an increase in blood TRH levels will affect the blood levels of TSH and thyroid hormone (Fig. 41-7). After the intravenous administration of synthetic TRH, blood levels of TSH begin to increase within 10 minutes, reach a maximum at 15 to 45 minutes, and return to normal base levels in 1 to 4 hours.[10]

Elevations of TSH after the administration of TRH result in increases in serum T_4 and T_3. The initial increases in T_3 are higher (75% increase over basal levels) than those in T_4 (15% to 50% over basal levels). The T_3 and T_4 levels subsequently drop slowly. The TSH (and occasionally T_3 and T_4) responses to intravenous administrations of synthetic TRH are currently used for diagnostic purposes.

Negative-feedback system on pituitary secretion of TSH. Increased levels of thyroid hormones in blood inhibit TSH secretion by the pituitary (negative feedback). The mechanism of this inhibitory effect appears to involve circulating free T_3 and T_4. The pituitary gland also is able

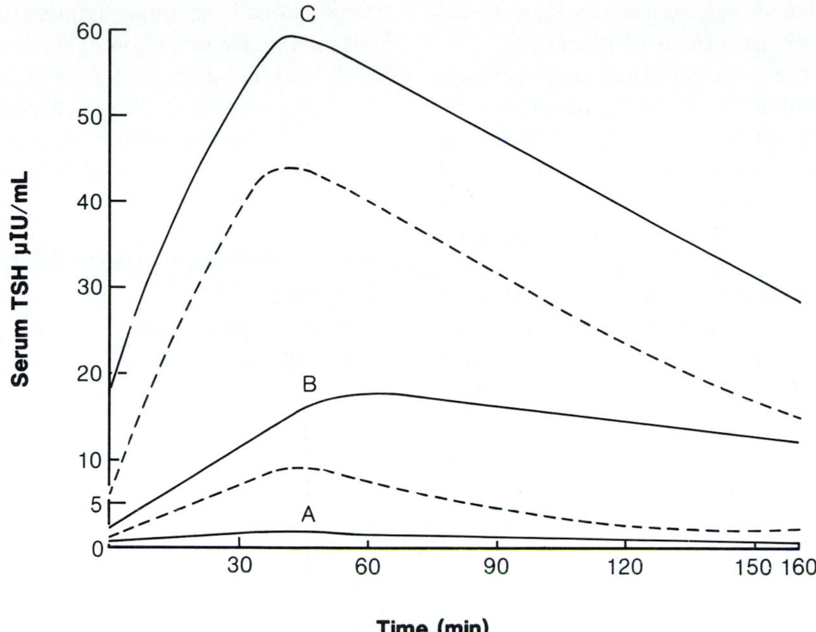

Fig. 41-7 Changes in serum TSH levels in response to TRH (given at 0 min). The upper and lower ranges of normal responses are shown by the dashed lines. Serum TSH changes following TRH challenge in pituitary (A), hypothalamic (B), and primary hypothyroid (C), diseases are also shown.

to convert intrapituitary T_4 directly to T_3. This locally generated T_3 then acts with circulating T_3 on the pituitary gland cells to inhibit the secretion of TSH in response to TRH. Conversely, decreased levels of thyroid hormone result in increased secretion of TRH and TSH.

Other factors affecting thyroid function. Thyroid function is finely regulated by extrinsic and intrinsic mechanisms. The extrinsic direct mechanism is represented by the HPTA (hypothalamic-pituitary-thyroid axis). The extrinsic indirect mechanisms encompass a host of collateral factors of neurogenic, metabolic, and pharmacological nature that exert inhibitory and stimulatory effects centrally at all levels of the HPTA and peripherally on the metabolism of the thyroid hormones. The intrinsic mechanisms are those taking place within the thyroid cells. They are concerned with the maintenance of adequate amounts of

Table 41-2 Effects of various drugs on thyroid function[13-16]

Mechanism of action	Drug	Effects
Hypothalamic stimulation	Amphetamine	Increased T_4, T_3, and free T_4
Inhibition of TSH secretion	Dopamine	Decreased TSH
	Levodopa	Blunted TSH response to TRH
	Glucocorticoids	
	Bromocriptin	
Enhanced TSH secretion	Metoclopramide	Increased TSH response to TRH
Blocking of iodine trapping	Perchlorate	Block production T_4 and T_3
	Thiocyanates	
	Nitroprusside	
Inhibition of organification	Propylthiouracil	Decreased T_4 and T_3 production
	Methimazole	
	Sulfonylureas	
	6-mercaptopurine	
Inhibition of release of T_4 and T_3	Lithium	Mild transient decrease of T_4, T_3
		Induce goiter
Decreased conversion of T_4 and T_3	Propylthiouracil	Decreased T_3
	β-adrenergic blockers	
	Amiodarone	
Increased degradation	Phenobarbital	Slight decrease of T_4 and T_3 and increase in TSH
Modulation of excess of β-adrenergic receptors	β-adrenergic blockers	Blunted adrenergic related effects of T_3 and T_4

intrathyroid hormone. These mechanisms are closely dependent on the availability and effects of iodine.

When the thyroid is exposed to rapid and large increases of iodine, there is an inhibition of the formation and release of thyroid hormone known as the "Wolff-Chaikoff effect."[11] If the thyroid continues to be exposed to increased concentrations of iodine, the inhibitory effects of excess iodine on hormone formation and release decrease and eventually cease.[11]

Effects of drugs and other compounds on thyroid function. Various drugs and substances may interfere with thyroid function. Their pharmacological effects may take place at one or several levels (Table 41-2). For example, amiodarone is an antiarrhythmic drug that causes inhibition of T_4 monodeiodination. This results in elevations of free and total T_4, decreased T_3 and increased rT_3.[12] Increases in TSH, as well as clinical hypothyroidism and hyperthyroidism, are also observed. Various other drugs affect TBG levels and thus total T_3 and T_4 determinations (see accompanying box).

CAUSES OF ABNORMALITIES IN THYROXINE-BINDING GLOBULIN (TBG)

Quantitative

- Increased TBG serum levels
  - Pregnancy
  - Estrogen therapy
  - Oral contraceptives
  - Perphenazine
  - Acute hepatitis
  - Hypothyroidism
  - Neonatal period
  - Acute intermittent porphyria
  - Genetic TBG excess
- Decreased TBG serum levels
  - Androgens
  - Anabolic agents
  - Cirrhosis
  - Acute illness
  - Surgical stress
  - Severe chronic illness
  - Severe hypoproteinemia
  - Nephrotic syndrome
  - Hyperthyroidism
  - Corticosteroid therapy
  - Active acromegaly
  - Klinefelter's syndrome
  - Cushing's syndrome
  - Down's syndrome
  - Type III hyperlipidemia
  - Chronic metabolic acidosis
  - Genetic deficiency

Qualitative

- Genetic
  - Genetically determined increase in binding affinity
  - Genetically determined decrease in binding affinity
- Drugs competing with T_4 and T_3 for TBG-binding sites
  - Phenytoin (diphenylhydantoin, Dilantin)
  - Dicumarol
  - Heparin
  - Atromid S
  - Aspirin
  - Phenylbutazone

Interrelationships between thyroid gland and other functions

Various physiological factors indirectly influence thyroid function and the effects of thyroid hormones. The effects of gender and sex hormones on thyroid function are variable. Thyroid disease is known to predominate in females. Estrogens cause an increase in TBG and, therefore, total serum T_3 and T_4 content, whereas androgenic hormones have the opposite effect. On the other hand, states of decreased (hypothyroidism) and increased (hyperthyroidism) thyroid function are associated with alterations of the reproductive system, abnormalities of the menstrual cycle, delayed or precocious puberty, infertility, and growth retardation.

The newborn has lower levels of T_3 and higher levels of T_4 and rT_3 than adults.[17] Immediately after birth there is an increase of TSH, with corresponding elevations of T_3 and T_4 that reach their zenith the second day after birth and then gradually decrease toward normal adult levels by the end of the first year of life. There is also a slow progressive decrease of T_3 levels in adults during the aging process.

Striking changes of thyroid function occur during pregnancy (see also Chapter 37). The normal placenta produces significant amounts of nonthyroid hormones (especially estrogens), which stimulate the formation of increased TBG levels. Both total T_4 and, to a lesser extent, total T_3 levels increase during pregnancy because of these increases in TBG concentration.[18]

In fasting states, changes in levels of thyroid hormone in serum occur within 24 to 48 hours. Initially serum T_3 decreases rapidly and subsequently, if fasting is maintained, more slowly. The decrease of T_3 levels is caused by decreased peripheral conversion of T_4 to T_3. Concomitantly there is an increase in rT_3 levels. Decreased caloric intake also results in a decline in thyroxine-binding prealbumin (TBPA), probably caused by decreased hepatic production.

General metabolic and physiological effects of thyroid hormone

Increases and decreases in hormone production will result in states of hyperfunction (hyperthyroidism) and hypofunction (hypothyroidism), respectively. The clinical presentation of hyperthyroidism or hypothyroidism reflects the basic effects of thyroid hormone on various organs and metabolic systems.

Table 41-3 Basic physiological effects of thyroid hormone and their relationship with syndromes of thyroid dysfunction

System	Thyroid hormone effects	Usual symptoms	
		Hyperthyroidism	**Hypothyroidism**
Metabolic	Increased calorigenesis and O_2 consumption Increased heat dissipation Increased protein catabolism Increased glucose absorption and production (gluconeogenesis) Increased glucose use	Heat intolerance Flushed skin Increased perspiration Increased appetite and food ingestion Muscle wasting and weakness Weight loss	Cold intolerance Dry and pale skin Decreased appetite and food ingestion Generalized weakness Weight gain
Cardiovascular	Increased adrenergic activity and sensitivity Increased heart rate Increased myocardial contractility Increased cardiac output Increased blood volume	Palpitations Fast heart rate (tachycardia) Increased blood pressure, mainly systolic Bouncy, hyperdynamic arterial pulses Shortness of breath	Slow heart rate (bradycardia) Low blood pressure Heart failure Heart enlargement
Central nervous	Increased adrenergic activity and sensitivity	Restlessness, hypermotility Nervousness Emotional lability Fatigue Exaggerated reflexes	Apathy Mental sluggishness Depressed reflexes Mental retardation
Gastrointestinal (GI)	Increased motility	Hyperdefecation	Constipation

Table 41-3 depicts some general effects that thyroid hormone has on various metabolic functions and systems. The effects of thyroid hormone can be classified according to their clinical expression into (1) general metabolic effects, (2) growth and maturation effects, and (3) organ-specific effects. Generally speaking, both intermediary metabolic pathways and specific metabolic pathways are stimulated by the thyroid hormone resulting in increased oxygen consumption and calorigenesis. Thyroid hormone has important actions as a promoter of cell differentiation, growth, and maturation. Deficiency of thyroid hormone in early life results in severe impairment of physical growth, maturation, and brain development. The third class of thyroid hormonal effects represents a number of direct effects exerted on specific organs and systems.

Finally, increased thyroid function is associated with an elevated catecholamine (hyperadrenergic) state important in the genesis of some symptoms of hyperthyroidism.

Decreased thyroid hormone results in deposition of mucoproteins and mucopolysaccharides in various tissues such as the skin, muscles, and heart. This phenomenon partly explains the thick skin, muscle weakness, enlarged heart, and symptoms of heart failure seen in hypothyroidism.

Thyroid function in nonthyroid disease (NTD)

There is a group of conditions that do not directly alter the output of hormone by the thyroid but rather interfere with the normal transport and metabolism of thyroid hormones. These conditions, characterized by abnormalities in thyroxine-binding globulin (see box, p. 628), are associated with debilitating diseases of various etiologies. These conditions produce changes in standard blood tests of thyroid function but are generally not accompanied by true thyroid dysfunction. The disorders are of clinical importance because they may mimic true thyroid disease or confound the diagnosis of concomitant thyroid dysfunction.

Malnutrition and renal and hepatic dysfunction are the three most important nonthyroid factors leading to abnormalities of thyroid function tests. States of malnutrition are often seen in chronic and debilitating diseases such as congestive heart failure, diabetes mellitus, chronic obstructive lung disease, cancer, and postsurgical states. The most conspicuous physiological abnormality found in all these entities is the impaired peripheral conversion of T_4 to T_3.[19]

The liver has important actions in the metabolism of thyroid hormones. These include production of thyroxine-binding proteins, conversion of T_4 to T_3 or rT_3, and removal of thyroid hormones. Alterations in these functions can occur in liver disease, resulting in abnormal test results in the absence of thyroid disease.

The kidneys have two main functions in relation to the thyroid: The first concerns iodine metabolism, since the kidneys represent a major pathway of iodine elimination; the second is the prevention of excessive losses of thyroxine-binding globulins in the urine. Kidney disease can result in either decreased iodine clearance and excretion, with secondary increases in blood and interstitial pools of iodine as seen in renal failure, or augmented loss of thyroxine-binding proteins as seen in the nephrotic syndrome. Finally, although less important than that of the liver, the kidney has the metabolic function of converting T_4 to T_3.

PATHOLOGICAL CONDITIONS

Hyperthyroidism

Causes. Hyperthyroidism refers to the clinical syndrome caused by an excess of circulating active thyroid hormone. Hyperthyroidism is caused by a number of pathological conditions that have as a common denominator an increase in circulating thyroid hormone (see accompanying box).

Graves' disease is caused by an immunological disorder in which serum autoantibodies bind to TSH receptors in the thyroid cell and stimulate the production and release of thyroid hormone. These antibodies compose a heterogeneous group of serum immunoglobulins that belong to the IgG fraction and are generically termed *thyroid-stimulating immunoglobulins* (TSI)[20] because of their ability to stimulate function in thyroid cells. The long-acting thyroid stimulator (LATS) antibody was the first TSI described and is present in about 60% of patients with Graves' disease.

Graves' disease is characterized clinically by the presence of diffuse goiter (enlarged thyroid gland), symptoms and signs of hyperthyroidism (Table 41-3), ophthalmopathy, and, occasionally, pretibial edema. Graves' ophthalmopathy is characterized by myxedematous infiltration of the tissues and muscles of the orbit, resulting in protrusion of the eyes and ocular muscle dysfunction. The ophthalmopathy may exist without accompanying thyroid hyperfunction.

Toxic multinodular goiter is a frequent cause of hyperthyroidism and usually appears in patients with preexisting nodular goiters. Portions of the thyroid gland are no longer under normal feedback control and secrete excess amounts of thyroid hormone. This condition occurs more frequently in elderly patients and is not accompanied by ophthalmopathy or pretibial edema.

Thyroid adenomas are benign tumors that do not respond to normal control mechanisms and can occasionally produce excess thyroid hormone. In addition to clinical signs and laboratory findings of thyroid hyperfunction, patients with adenomas have thyroid nodules that concentrate radioactive iodine avidly (hot nodules). It should be emphasized that most adenomas do not cause thyroid hyperfunction. Thyroid cancer is a rare cause of hyperthyroidism.

Thyroiditis is a general term used to describe an inflammation of the thyroid gland. All forms of thyroiditis can potentially cause hyperthyroidism because large quantities of hormone can be released from the inflamed and disrupted follicles. Subacute thyroiditis (SAT) and its variant, painless thyroiditis, are considered to be caused by a viral infection of the gland and have two phases. The early phase is characterized by active inflammation of the thyroid, which results in enlargement and tenderness of the gland and clinical and laboratory findings of thyroid hyperfunction caused by release of hormones.[21,22] During the late phase, recuperation takes place and function usually returns to normal. Some patients may evolve through an intermediary state of hypothyroidism. Chronic lymphocytic thyroiditis (Hashimoto's thyroiditis) is occasionally associated with an overactive thyroid state but more often results in hypothyroidism.

A sudden release of hormone may be seen after irradiation of the thyroid gland (radiation thyroiditis). Thyroid hyperfunction caused by this radiation thyroiditis as suggested by elevations of serum T_3 and T_4 is usually mild and self-limited.

Exogenous hyperthyroidism is caused by the administration of excessive thyroid hormone by the physician (iatrogenic) or as a result of surreptitious intake of thyroid hormone by patients (factitious).

Tumors originating from the trophoblast, or outer cellular layer of the forming embryo, can secrete large amounts of human chorionic gonadotropin (HCG). This hormone has been found to have a weak thyroid-stimulating action. As a result, such tumors may cause an increased secretion of thyroid hormones. Hyperthyroidism caused by pituitary tumors that secrete high levels of TSH is a rare occurrence.

CAUSES OF HYPERTHYROIDISM

Primary hyperthyroidism

- Primary thyroid abnormalities
  - Toxic multinodular goiter
  - Thyroid adenoma
  - Thyroid carcinoma
  - Struma ovarii

Secondary hyperthyroidism

- Endogenous: increased serum levels of TSH or thyroid-stimulating substances, resulting in thyroid hyperactivity:
  - Graves' disease
  - Neonatal hyperthyroidism
  - Pituitary tumors
  - Trophoblastic tumors
    - Hydatidiform mole
    - Choriocarcinoma
    - Embryonal carcinoma of testes
- Exogenous
  - Iatrogenic
  - Factitious hyperthyroidism

Thyroiditis

- Subacute thyroiditis
- Lymphocytic (Hashimoto's) thyroiditis
- Radiation

Table 41-4 Hyperthyroidism: laboratory findings in various clinical conditions

Clinical entity	T_4	T_3	FT_4	T_3RU	FT_4I	TSH	TRH stimulation	TSI	Thyroid ^{123}I uptake
Graves' disease	↑	↑	↑	↑	↑	↓,U	Blunted	+	↑
Euthyroid Graves' disease	N	N	N	N	N	N	Blunted, N	+	N
Toxic multinodular goiter	↑	↑	↑	↑	↑	↓,U	Blunted	−	↑,N
Toxic adenoma	↑	↑	↑	↑	↑	↓,U	Blunted	−	↑,N
T_3 toxicosis	N	↑	N	N, ↑	N	↓,U	Blunted	+, −	N, ↑
Hyperthyroidism in pregnancy	↑	↑	↑	N, ↓	↑	↓,U	Blunted	+, −	*
Neonatal hyperthyroidism	↑	↑	↑	↑	↑	↓,N	Blunted	+	*
Subacute thyroiditis	↑,N	↑,N	↑,N	↑,N	↑,N	N, ↓,U	Blunted, N	−	↓,N
Exogenous hyperthyroidism with T_4	↑	↑	↑	↑	↑	↓,U	Blunted	−	↓
Trophoblastic tumors	↑	↑	↑	↑	↑	↑,N, ↓	Blunted	−	↓,N, ↑
Pituitary TSH-secreting tumors	↑	↑	↑	↑	↑	↑	N, ↑	−	↑
Pseudohyperthyroidism	↑	↑,N	↑,N	↑,N	↑,N	↑,N	N	−	N or ↑

FT_4, Free thyroxine; *FT_4I,* free thyroxine index; *N,* normal; *TRH,* thyrotropin-releasing hormone; *T_3RU,* triiodothyronine resin-uptake test; *TSH,* thyroid-stimulating hormone; *TSI,* thyroid-stimulating immunoglobulins; *U,* undetectable.
↑, Elevated; ↓, decreased; +, present; −, absent; *, test contraindicated or not recommended.

Laboratory findings. Graves' disease is the classic example of hyperthyroidism. Its laboratory abnormalities (Table 41-4) include (1) elevation of thyroid hormones in serum, (2) decreased serum levels of TSH, and (3) blunted responses to TRH. In addition, patients with Graves' disease may have elevated serum levels of thyroglobulin and thyroid-stimulating immunoglobulin. In general there is a disproportionately higher rate of production of T_3 in Graves' disease.

A syndrome has been described with symptoms of hyperthyroidism and normal total T_4 and free T_4 concentrations, normal or mildly elevated thyroidal radioactive iodine uptake, and elevated T_3 levels in blood (T_3 thyrotoxicosis). This entity may present as part of Graves' disease but also has been observed in patients with toxic nodular goiter, after iodine ingestion in patients with previous iodine deficiency, and with recurrent hyperthyroidism after treatment with radioactive iodine, thyroid blocking agents, or surgery.

A less frequently found entity, T_4 toxicosis, refers to a condition in which the T_3 is normal or only mildly elevated and T_4 is quite elevated. This condition is seen most often in patients with an alteration in conversion of T_4 to T_3 as a result of chronic debilitating diseases or in patients acutely exposed to large amounts of iodine (for example, x-ray contrast media).

Neonatal hyperthyroidism refers to a state of increased thyroid function seen in newborn infants of mothers whose serum may contain thyroid-stimulating immunoglobulins (TSI). This condition is postulated to be caused by transplacental transfer of maternal TSI.[23] Its course is benign with spontaneous remission.

Treatment. The treatment of hyperthyroidism is aimed at either eliminating excess functioning thyroid tissue (surgery or radioiodine) or blocking the production of hormone (thyroid-blocking drugs). All these methods are currently used, and the choice depends on the cause of the hyperthyroidism, special clinical situations, and the physician's personal preferences.

Hypothyroidism

Causes. A clinical state of hypothyroidism develops whenever insufficient amounts of thyroid hormone are available to tissues. By and large the most common group of entities causing hypothyroidism are those that involve the thyroid gland itself (see accompanying box).

Hashimoto's thyroiditis is probably the single most common cause of hypothyroidism. It is believed to result from a derangement of the cell-mediated and humoral compartments of the immune system.[24]

The most important characteristics of Hashimoto's thyroiditis are the presence of an organification defect and lymphocytic infiltration of the gland with concomitant loss of thyroid tissue. These elements of the disease are reflected clinically by the presence of an enlarged thyroid gland, thyroid hypofunction, and the presence in serum of antithyroid antibodies.

Hypothyroidism can result from various other conditions. It often follows surgical thyroidectomy or therapy with radioiodine for treatment of Graves' disease. Devel-

CAUSES OF HYPOTHYROIDISM

Primary thyroid dysfunction

Parenchymal damage
- Thyroiditis
  - Chronic lymphocytic (Hashimoto's) thyroiditis
  - Subacute thyroiditis
- Therapeutic ablation
  - After ^{131}I therapy
  - After surgery
- Thyroid dysgenesis
  - Aplasia
  - Dysplasia
- Thyroid infiltration
  - Tumors
- Abnormal hormonogenesis
  - Iodine deficiency
  - Iodine excess
  - Thyroid-blocking drugs
  - Congenital and acquired defects of hormone synthesis and thyroglobulin metabolism defects

Pituitary hypothyroidism (TSH deficiency)

Hypothalamic hypothyroidism (TRH deficiency)

Reduced peripheral response to thyroid hormone

opmental abnormalities and tumors and other infiltrative disorders that displace and destroy thyroid tissue can occasionally cause hypothyroidism. Congenital defects in hormonogenesis and the effect of drugs can also result in hypothyroidism. Pituitary and hypothalamic disease are rare conditions that occasionally lead to hypothyroidism caused by inadequate TRH or TSH secretion.

Laboratory findings. Regardless of the cause, the laboratory findings in hypothyroidism are characterized by decreased total serum T_4, T_3, FT_4, FT_4I, and T_3RU. The degree of abnormality of these tests varies widely, depending on the cause of hypothyroidism and stage of the disease.

Some distinctive laboratory findings can be found in Table 41-5. For instance, hypothyroidism caused by intrinsic thyroid disease (primary hypothyroidism) will display elevated TSH levels and exaggerated TSH responses to TRH stimulation. Hypothyroidism caused by pituitary (secondary) and hypothalamic (tertiary) disease exhibits low or borderline normal levels of TSH and TRH test responses as noted earlier.

Hypothyroidism in the neonatal period can lead to severe growth and maturation retardation of the central nervous system (cretinism). Routine screening of neonates for hypothyroidism is an important tool for the early diagnosis of hypothyroidism and a significant step in preventive pediatric medicine.[25]

Treatment. The treatment of hypothyroidism consists in thyroid hormone replacement given orally, which reverses the abnormal laboratory findings and clinical symptoms and signs, provided that there are no abnormalities in the transport and peripheral use of thyroid hormones.

Goiter

Goiter refers to an enlargement of the thyroid gland. Diffuse goiter is characterized by uniform enlargement of the thyroid gland. In multinodular goiter the thyroid gland is enlarged in a nonuniform fashion, resulting in nodules located both superficially and deep within the gland.

Solitary nodule

A solitary nodule refers to the presence of a solitary localized enlargement of a portion of the thyroid gland. Although most of these nodules represent benign conditions, such as cysts, localized hemorrhages, focal thyroiditis, and adenomas, they may also represent malignant tumors of the thyroid. External radiation therapy used in the past for acne and enlarged tonsils and other conditions of the head, neck, and chest has been associated with an increased risk of the subsequent development of both thyroid cancer and benign thyroid neoplasms.[26]

Thyroid cancer

The most common primary malignant thyroid tumors originate from the epithelial cells of the follicles. Patients with thyroid cancer usually do not display significant abnormalities of thyroid function. These tumors often release thyroglobulin into the circulation, where it may be followed as a tumor marker. Since other thyroid diseases can

Table 41-5 Laboratory findings in hypothyroidism

Type	T_4	T_3	T_3RU	FT_4I	TSH	TRH stimulation*
Primary	↓	↓, N	↓, N	↓	↑	↑
Secondary	↓	↓	↓	↓	↓, N	↓
Tertiary	↓	↓	↓	↓	↓, N	N
Peripheral unresponsiveness	↑	↑	N	↑ or N	↑ or N	N or ↑

*Assessed by response of serum TSH to TRH administration.
N, Normal; ↑, elevated; ↓, decreased. See Table 41-4 for abbreviations.

also result in increased levels of thyroglobulin, this parameter is most useful for following activity of the disease rather than for specific diagnostic purposes.

A different type of thyroid cancer originates from the parafollicular C cells of the thyroid and is termed *medullary carcinoma.* These tumors secrete calcitonin, a hormone that lowers blood calcium, which is a useful tumor marker for diagnosing and following these patients.

Pseudohyperthyroidism

Cases of abnormally high T_4 and T_3 hormone levels and normal levels of TBG have been described in patients with otherwise normal thyroid function. This condition is probably caused by decreased affinity and responsiveness of tissue receptors for T_3 and T_4. Laboratory findings in these cases demonstrate elevated T_4 and free T_4 levels with normal levels of T_3.

TESTS OF THYROID FUNCTION

The iodine-concentrating property of the thyroid is used to estimate thyroid function and to obtain functional anatomical images of the gland. For these studies tracer amounts of a radioiodine (^{123}I or ^{131}I) are administered to the patient. Gamma rays emitted by the radioiodine concentrated in the thyroid are detected by specially designed imaging and counting devices and transformed into thyroid radioiodine uptakes and images. The more desirable radioisotope is ^{123}I because it produces better images and delivers lower radiation doses to the patient's thyroid. Serum samples for hormone radioassays that are obtained after the administration of radioactive materials to the patient must be checked for residual radioactivity before the assay is started or spurious results may be obtained.

Thyroid iodine uptake

This parameter measures the percentage of an administered dose of radioiodine that concentrates in the gland by the trapping and organification mechanisms. This measurement may be obtained at various intervals after radioiodine administration. Theoretically, the degree of thyroid uptake in the thyroid at a given time reflects thyroid hormone synthesis and secretion and therefore reflects the functional status of the thyroid. Low thyroid uptakes are found not only in hypothyroidism but also in certain conditions actually associated with thyroid hyperfunction, including (1) thyrotoxicosis factitia caused by exogenous thyroid hormone, (2) subacute thyroiditis, (3) iodine-induced hyperthyroidism, and (4) certain forms of chronic thyroiditis.

TRH stimulation test

The TRH stimulation test takes advantage of the interrelationship between the TRH and TSH secretions. Normally, after the intravenous administration of TRH, there is an increase of TSH levels in blood that in turn elicits an elevation of T_3 and T_4 serum levels.[11,27] Various factors influence TSH responses to TRH stimulation. Abnormal responses to TRH stimulation are of two types. The first type is characterized by lower-than-normal responses of serum TSH to TRH stimulation. In these cases the TSH response to TRH is said to be blunted. The second type of abnormal response consists of higher-than-normal increments of serum TSH after TRH stimulation. Responses of TSH to TRH in various clinical conditions are depicted in Fig. 41-7.

The main clinical applications of the TRH stimulation test are (1) diagnosis of subclinical and early biochemical hyperthyroidism, (2) evaluation of patients with ophthalmopathy without overt hyperthyroidism, and (3) diagnosis of hypothalamic and pituitary hypothyroidism (see also Chapter 48, for differentiation of types of depressive disorders).

TSH stimulation test

Administration of exogenous TSH will stimulate all phases of thyroid function, which will be reflected by increases in the radioiodine thyroid uptake and blood levels of thyroid hormone. The TSH stimulation test is performed by monitoring of the thyroidal radioiodine uptake before (baseline) and after the administration of bovine TSH for 3 consecutive days. Normally there is an increase of serum T_4 and radioiodine uptake to more than 1.5 times the baseline value in response to TSH administration. The chief use of this test has been to differentiate primary hypothyroidism from secondary or tertiary hypothyroidism, and it has been replaced largely by TSH determinations in serum and by the TRH stimulation test.

Triiodothyronine (T_3) or Cytomel suppression test

This test consists of a determination of 24-hour thyroid radioiodine uptake before (baseline) and then after the oral administration of T_3 for 7 to 10 days.[28] Images and thyroid uptake determinations are obtained simultaneously. The purpose of this test is to establish the presence of thyroid tissue that has become autonomous and unresponsive to TSH changes. Normally the administration of thyroid hormone (T_3 or T_4) will shut off the secretion of TSH by the pituitary gland. The subsequent absence of TSH results in diminished thyroid concentration of radioiodine. A thyroid gland that is overactive is often autonomous, and so suppression of TSH production by administration of exogenous thyroid hormone will not be followed by corresponding declines of thyroid radioiodine uptake. Normally a drop in thyroid uptake (suppression) to 30% to 50% of the baseline value occurs after T_3 administration.

Perchlorate discharge test

The perchlorate discharge test detects defects of iodine organification present in conditions such as Hashimoto's thyroiditis and congenital goiters. It is also used to determine the degree of organification defect caused by certain

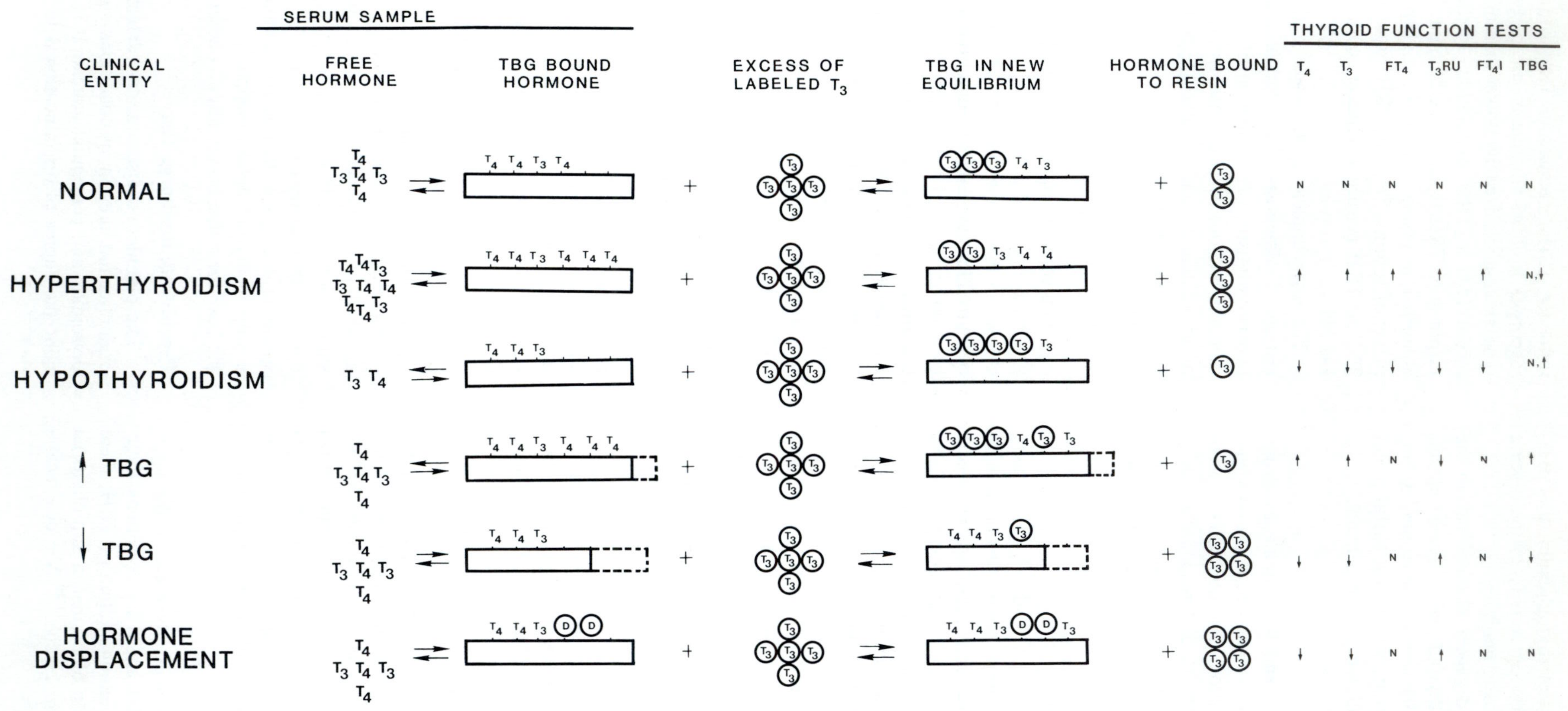

SERUM SAMPLE: FREE HORMONE; TBG BOUND HORMONE

EXCESS OF LABELED T_3

TBG IN NEW EQUILIBRIUM

HORMONE BOUND TO RESIN

THYROID FUNCTION TESTS

CLINICAL ENTITY	T_4	T_3	FT_4	T_3RU	FT_4I	TBG
NORMAL	N	N	N	N	N	N
HYPERTHYROIDISM	↑	↑	↑	↑	↑	N,↓
HYPOTHYROIDISM	↓	↓	↓	↓	↓	N,↑
↑ TBG	↑	↑	N	↓	N	↑
↓ TBG	↓	↓	N	↑	N	↓
HORMONE DISPLACEMENT	↓	↓	N	↑	N	N

Fig. 41-8 Interrelationships between serum TBG (thyroxine-binding globulin), the T_3RU, and other thyroid function tests. *D*, Drugs occupying binding sites on TBG; *FT_4I*, free thyroxine hormone index.

thyroid-blocking drugs used for treatment of hyperthyroidism to assess their therapeutic effects.

CHANGE OF ANALYTE IN DISEASE

Thyroid function and its alterations caused by disease can be assessed by determination of various analytes in blood. Tables 41-4 to 41-6 summarize the changes of some analytes in various disease states.

Protein-bound iodine (PBI)

Historically PBI was used to indirectly assess the concentrations of thyroid hormone in the blood. This test measures iodine contained in thyroid hormones, as well as iodine of nonhormonal origin, which is bound to proteins. For this reason, the PBI measurement has been abandoned as an indirect indicator of T_4 levels in blood. The PBI may be useful to detect increases of endogenous non-T_4 iodoproteins that may be released into the serum in conditions such as Hashimoto's and subacute thyroiditis.

Serum T_4

Elevations in total serum T_4 can occur as a result of increased hormone synthesis, increased hormone release from the thyroid cells, and increased binding capacity of plasma proteins, especially TBG. Increased hormone secretion is most frequently seen in states of hyperthyroidism. Causes of hyperthyroidism are listed in the box on p. 630.

Increased T_4 release also occurs in subacute thyroiditis and Hashimoto's thyroiditis and after radiation. Increased levels of serum TBG from various causes (see box, p. 628) produce elevations of total serum T_4. These conditions are not accompanied by hyperthyroidism.

Serum T_3

In general, elevations of serum T_3 parallel those of T_4 and are found in most states of hyperthyroidism. In addition, isolated elevations of T_3 are found in T_3 thyrotoxicosis. This entity is found in Graves' disease, toxic multinodular goiter, and toxic adenoma and in patients with recurrent hyperthyroidism after treatment with antithyroid drugs, radioactive iodine, and surgery. Increased serum concentration or binding capacity of TBG also results in elevations of T_3.

Resin-uptake test (T_3RU or T_4RU)

Thyroid hormones in the blood are distributed in two compartments: protein-bound and unbound or free compartments (Fig. 41-8). Variations in measurable total thyroid hormone in blood can result from changes in the binding-protein levels. Hypothyroidism and hyperthyroidism will only occur if a net persistent decrease or increase of free unbound thyroid hormone exists in the blood. Blood level determinations of T_4 and T_3 are clinically more meaningful if the functional levels of thyroid hormone-binding protein in blood are known. Thus the resin-uptake test does not measure a specific analyte per se but the functional state (ability to bind hormone) of an analyte (such as TBG).

The resin-uptake test reveals the number of free available binding sites in the plasma thyroid hormone–transporting proteins, especially TBG, since this is the most important transporter of thyroid hormones. TBG concentration is most commonly assessed by the resin-uptake test. Samples of patient serum are mixed with radiolabeled thyroid hormone (T_3 or T_4). The amount of thyroid hormone that remains unbound in the mixture is inversely proportional to the number of available binding sites (Fig. 41-8). This unbound hormone is separated from the mixture when one adds relatively low affinity binding resin to the system (Fig. 41-8). The resin is then separated from the serum, and the amount of uptake of radioactive hormone by the resin is determined. In cases of increased TBG in plasma, although free thyroid hormone levels are normal, the total T_4 is increased, since the total number of binding sites in TBG is increased. Labeled T_4 or T_3 will find an increased pool of TBG (available binding sites), and less labeled hormone will remain free to bind resin. Therefore the resin uptake is low. The opposite changes occur in states of decreased TBG. Conditions that cause abnormalities of the binding sites (TBG) are listed in Table 41-4. Certain drugs compete with thyroid hormone for the TBG-binding sites. This phenomenon is reflected by normal free T_4, low total T_4, and high resin uptake values.

In hyperthyroidism, free thyroid hormone in plasma increases. This results in complete or almost complete saturation of binding sites in TBG, which in turn will result in fewer molecules of labeled hormone bound to TBG and increased amounts of free-labeled hormone (increased resin uptake).

Free thyroxine hormone index (FT_4I)

It is the free hormone that induces metabolic and biological effects in target cells. The free T_4 index (FT_4I) indirectly assesses the level of free T_4 in blood and adjusts for most interferences caused by binding-protein abnormalities.[29] The FT_4I is determined from total T_4 and resin-uptake values obtained on the same sample. The FT_4I is calculated as follows:

$$\frac{\text{Total serum } T_4 \times \text{Patient's resin uptake}}{100}$$

As expected, most alterations of binding proteins produce reciprocal changes in resin uptake and total serum T_4 or T_3, resulting in normal values of FT_4I, whereas true alterations of free thyroid hormone content cause concordant unidirectional changes of FT_4I. Total T_4 and FT_4I are still considered the most practical and valuable tests in the initial evaluation of thyroid status (Fig. 41-8). The FT_4I is elevated in hyperthyroidism (Table 41-4) and is decreased in states of hypothyroidism (Table 41-5).

The free T_3 index is calculated in a similar fashion as the FT_4I from the total serum T_3.[30] It has similar applications and significance as the FT_4I. The FT_3I may be helpful to exclude T_3 toxicosis in some patients taking oral contraceptives in whom isolated serum T_3 increases without corresponding elevations of T_4. In general, the FT_3I offers no advantages to the FT_4I and is used less frequently in clinical practice.

Free T_4 and free T_3

Serum FT_4 correlates very well with secretion rates of T_4 and its metabolism. FT_4 and FT_3 levels tend to parallel changes in total T_4 and total T_3 concentrations. Measurement of the free hormone levels is useful in routine clinical practice and in instances in which other test results are borderline or conflicting. Elevations and decreases of free T_4 and free T_3 theoretically are true reflections of hyperthyroidism and hypothyroidism. However, the results are very dependent on the analytical method used to measure "free" hormone and clinically have been less useful than originally hoped.

Thyroxine-binding globulin (TBG)

Many diseases produce alterations of TBG (see box, p. 628) and other thyroid hormone-transporting proteins in plasma. Changes in concentrations of such proteins affect the total plasma concentrations of T_4 and T_3. Since most of the T_4 and T_3 is bound to TBG, changes in this protein level are the most important clinically.

Quantitative abnormalities are those characterized by absolute increases or decreases of TBG levels in plasma. Consequently, the total amount of thyroid hormone transported in plasma will increase (increased TBG) or decrease (decreased TBG).

Qualitative abnormalities of TBG refer to those stemming from alterations of the binding affinity rather than from absolute changes of TBG amounts in plasma. The binding affinity of the TBG may be increased or decreased in intrinsic TBG defects. Various compounds and drugs may strongly bind to TBG and result in displacement of thyroid hormone from binding sites. The result is decreased total serum T_4 and T_3 (Fig. 41-8).

Serum thyroid-stimulating hormone (TSH)

The principal use of thyroid-stimulating hormone (TSH) determinations in serum is the diagnosis of hypothyroidism. Patients with untreated hypothyroidism stemming from intrinsic thyroid defects (primary hypothyroidism), regardless of the cause, have elevated serum levels of TSH. Patients with hypothyroidism caused by pituitary lesions (secondary) or hypothalamic (tertiary) lesions have normal or low TSH levels. Differentiation between secondary and tertiary hypothyroidism is accomplished by use of the TRH stimulation test.

Most patients with hyperthyroidism have low or undetectable TSH levels in serum, reflecting the inhibitory effects of high levels of circulating thyroid hormone on the hypothalamic-pituitary axis.

Immunoglobulins

It has been suggested that immunological abnormalities play an important role in thyroid pathology.[20] Antibodies against components of thyroid cells are found in many patients with thyroid disease and in older individuals without apparent thyroid disease. High levels of antibodies against thyroglobulin or thyroid microsomes frequently are found in patients with Hashimoto's thyroiditis. High titers of thyroid-stimulating immunoglobulins may be found in Graves' disease.

Laboratory findings in nonthyroid disease (NTD)

As previously mentioned, patients with NTD may develop abnormalities of thyroid tests. These changes can mimic thyroid disease or modify and confound laboratory findings in patients with thyroid disease. Chronic diseases that alter laboratory tests of thyroid function are more frequent than thyroid disease. NTD includes patients with general chronic debilitating conditions and patients with specific conditions, namely, alcoholic cirrhosis, hepatitis, and renal failure. Table 41-6 depicts the most important laboratory findings in several nonthyroid diseases.

As a group, euthyroid patients with NTD have either

Table 41-6 Thyroid-function tests in nonthyroid diseases (NTD)*

	T_4	T_3RU	FT_4I	Free T_4	Free T_3	T_3	rT_3	TSH	TBG
General NTD†	N, ↓	↑, N	↓, N	N, ↑	↓, N	↓, N	↑, N	N	↓, N
Acute hepatitis	N, ↑	↓	N, ↓	N, ↑	N, ↓	↓, N	↑	N	↑
Chronic active hepatitis	N, ↑	↓	N, ↓	N, ↑	N	N	↑	N	↑
Alcoholic cirrhosis	N, ↓	↑, N	N, ↑	N, ↑	↓	↓	↑	N	↓, N
Renal failure	N, ↓	N	N, ↓	N, ↓	↓	↓	N	N	N

*Normal thyroid function in patients with very complex problems; findings in individual patients may vary widely.
†Includes all patients with NTD.
N, Normal; ↑, augmented, ↓, diminished.

normal or low total T_4, normal or low FT_4I, normal or low total T_3 and free T_3, normal or high rT_3, normal or high T_3RU, and normal or slightly high TSH levels in serum.[31-33] The free T_4 index (FT_4I), which is often a reliable indicator of thyroid function, is usually normal but may also be low in patients with NTD.

Low total T_4, low FT_4I, and low or subnormal T_3 levels may suggest the presence of hypothyroidism in patients with NTD. The presence of high serum levels of rT_3 is useful in excluding hypothyroidism in this circumstance.[31] TSH concentrations are usually normal.

Liver disease is characterized by abnormalities of protein synthesis, including the thyroid-binding proteins, and by alterations of T_4 metabolism. Alcoholic liver cirrhosis is accompanied by decreased binding capacity of thyroid hormones, resulting in high T_3RU[34] (Table 41-6). Total T_4 is usually normal or low,[34] and free T_4 is slightly elevated. There is depressed conversion of T_4 to T_3 with decreased serum levels of T_3 and occasionally also the free T_3.

Acute hepatitis is characterized by an increase of the serum TBG[35] (Table 41-6), which results in increased levels of T_4 and decreased T_3RU. In general, chronic active hepatitis also produces the same changes in blood as those seen in acute hepatitis. These include (1) lower than expected T_4, (2) low free T_4 index, (3) decreased free T_4, (4) normal serum TSH, and (5) abnormally increased TSH response to TRH.[36]

In patients with renal failure the most common findings are decreased total and free T_3[37] in the serum, caused by diminished peripheral conversion of T_4 to T_3. Serum thyroxine-binding protein, T_3RU, and total T_4 usually are normal. However, the total T_4 and free T_4 may be slightly low.[38] Diminished total T_4 can result from renal failure with severe catabolic states, probably caused by decreased thyroxine-binding proteins in serum. Concentration of TSH in serum is normal.

Association of NTD with hyperthyroidism is found occasionally. In these situations there is a disproportionate increase of T_4 in serum with only modestly elevated or normal T_3 levels. This is caused by a decreased peripheral conversion of T_4 to T_3 concomitant with the increased T_4 production.

Laboratory findings after therapeutic interventions for thyroid disease

The two most frequently found clinical situations that require closely monitored treatment are hyperthyroidism and hypothyroidism.

Graves' disease, when successfully treated surgically, results in normalization of T_4, T_3, T_3RU, RT_4I, and TSH. In patients treated with antithyroid drugs, such as propylthiouracil (PTU), improvement of thyroid function tests may be occasionally seen as soon as 2 weeks after initiation of treatment. Because of the blocking effect of PTU on conversion of T_4 to T_3, a disproportionately rapid decrease in T_3 may be seen in some cases.

Therapy with ^{131}I is now used more frequently to treat hyperthyroidism. Laboratory tests show a decrease of T_4, T_3, T_3RU, and FT_4I approximately 6 to 10 weeks after treatment. Residual radioactivity in the serum must be considered when one is using radioassay methods within 90 days of therapy.

Patients treated for thyrotoxicosis by any of the methods may follow one of three courses. Ideally normal thyroid function ensues and is reflected by normal laboratory values. Occasionally patients remain hyperthyroid and abnormal laboratory tests persist. More often patients become hypothyroid with low T_4, T_3RU, and FT_4I values. The earliest indication of the ensuing hypothyroidism is a depressed T_4 followed by an elevation of TSH levels as the hypothalamic-pituitary axis recovers. The TSH may remain elevated with normal T_4 levels, a condition known as biochemical hypothyroidism. Patients with treated Graves' disease may develop an ongoing hyperthyroid status with normal or subnormal T_4 values but elevated T_3 levels.

The treatment for hypothyroidism is thyroid hormone replacement. During the monitoring of thyroid hormone replacement it is very important to keep in mind the effect that changes in TBG levels produce on thyroid hormone levels. Laboratory tests should therefore include those that indirectly or directly assess TBG levels, such as the T_3RU, FT_4I, and TBG determinations. When preparations containing T_4 and T_3 are used, normalization of both T_3 and T_4 values is expected. When preparations containing only T_3, such as Cytomel, are used, T_3 normalizes or becomes elevated but T_4 remains low. With pure T_4 preparations the T_4 may be mildly elevated, but T_3 values are usually normal. Normalization of T_4 and T_3 values is accompanied by a corresponding decline in TSH levels. Monitoring of patients on T_4 replacement must take into consideration the fluctuation of serum levels caused by the oral administration of the preparation. Significant increases of T_4 from baseline levels can be observed 2 to 10 hours post ingestion.[39] More meaningful interpretation of serum T_4 levels is accomplished if this potential problem is avoided.

ACKNOWLEDGMENT

We wish to acknowledge Lisa A. Fooks, Ruth M. McDevitt, R.N., B.S.N., and Patrick Slattery for their assistance in the compilation of this chapter and preparation of illustrations.

REFERENCES

1. Wolff, J: Transport of iodide and other anions in the thyroid gland, Physiol Rev 44:45-79, 1964.
2. DeGroot, LJ, and Niepomniszcze, H: Biosynthesis of thyroid hormone: basic and clinical aspects, Metabolism 26:665-718, 1977.
3. Dumont, JE: The action of thyrotropin on thyroid metabolism, Vitam Horm 29:287-412, 1971.
4. Chopra, IJ, Solomon, DH, Chopra, U, et al: Pathways of metabolism of thyroid hormones, Recent Prog Horm Res 34:521-567, 1978.
5. Woeber, KA, and Ingbar, SH: The interactions of the thyroid hormones with binding protein. In Greer, MA, and Solomon, DH, edi-

tors: Thyroid, American handbook of physiology, vol 3, Washington, DC, 1973, American Physiological Society.
6. Chopra, IJ: An assessment of daily production and significance of thyroidal 3,3′,5′-triiodothyronine (reverse T3) in man, J Clin Invest 58:32-40, 1976.
7. Schimmel, M, and Utiger, RD: Thyroidal and peripheral production of thyroid hormones, Ann Intern Med 87:760-768, 1977.
8. Oppenheimer, JH, and Samuels, HH, editors: Molecular basis of thyroid hormone action, New York, 1983, Academic Press, Inc.
9. Edelman, IS, and Ismail-Beigi, F: Thyroid thermogenesis and active sodium transport, Recent Prog Horm Res 30:235-257, 1974.
10. Erfurth, EM, Norden, NE, Hedner, P, et al: Normal reference interval for thyrotropin response to thyroliberin: dependence on age, sex, free thyroxine index and basal concentrations of thyrotropin, Clin Chem 30:196-199, 1984.
11. Wolff, J: Iodide goiter and pharmacologic effects of excess iodide, Am J Med 47:101-124, 1969.
12. Mason, JW: Amiodarone, N Engl J Med 316:455-466, 1987.
13. Kaplan, MM: Clinical and laboratory assessment of thyroid abnormalities, Med Clin North Am 69:863-880, 1985.
14. Oppenheimer, JH, Schwartz, HL, and Surks, MI: Propylthiouracil inhibits the conversion of L-thyroxine to L-triiodothyronine: an explanation of the antithyroxine effects of propylthiouracil and evidence supporting the concept that triiodothyronine is the active thyroid hormone, J Clin Invest 51:2493-2497, 1972.
15. Singer, I, and Rotenberg, D: Mechanisms of lithium action, N Engl J Med 289:254-260, 1973.
16. Cavalieri, RR, Gavin, LA, Wallace, A, et al: Serum T_4, free T_4, T_3 and rT_3 in diphenylhydantoin-treated patients, Metabolism 29:1161-1165, 1979.
17. Abuid, J, Stinson, DA, and Larsen, PR: Serum triiodothyronine and thyroxine in the neonate and the acute increases in these hormones following delivery, J Clin Invest 52:1195-1199, 1973.
18. Selenkow, HA, Birnbaum, MD, and Hollander, CS: Thyroid function and dysfunction during pregnancy, Clin Obstet Gynecol 16:66-68, 1973.
19. Portnay, GI, O'Brian, JT, Bush J, et al: The effect of starvation on the concentration and binding of thyroxine and triiodothyronine in serum and on the response to TRH, J Clin Endocrinol Metab 39:191-194, 1974.
20. McKenzie, JM, Zakarija, M, and Sato, A: Humoral immunity in Graves' disease, Clin Endocrinol Metab 7:31-45, 1978.
21. Christiansen, NJB, Sierboek-Nielson, K, Hansen, JEM, and Christiansen, LK: Serum thyroxine in the early phase of subacute thyroiditis, Acta Endocrinol 64:359-363, 1970.
22. Dorfman, SG, Cooperman, MT, Nelson RL, et al: Painless thyroiditis and transient hyperthyroidism without goiter, Ann Intern Med 86:24-28, 1977.
23. McKenzie, JM, and Zakarija, M: Pathogenesis of neonatal Graves' disease, J Endocrinol Invest 2:183-189, 1978.
24. Brown, J, Solomon, DH, Beall, GN, et al: Autoimmune thyroid disease—Graves' and Hashimoto's, Ann Intern Med 88:379-391, 1978.
25. Dussault, JH, Coulombe, P, Laberge, C, et al: Preliminary report on a mass screening program for neonatal hypothyroidism, J Pediatr 86:670-674, 1975.
26. Maxon, HR, III, Saenger, EL, Thomas, SR, et al: Clinically important radiation-associated thyroid disease: a controlled study, JAMA 44:1802-1805, 1980.
27. Snyder, PJ, and Utiger, RD: Response to thyrotropin-releasing hormone (TRH) in normal man, J Clin Endocrinol Metab 34:380-385, 1972.
28. Werner, SC, and Spooner, M: A new and simple test for hyperthyroidism employing L-triiodothyronine and the 24 hour ^{131}I uptake method, Bull NY Acad Med 31:139-145, 1955.
29. Stein, RB, and Price, L: Evaluation of adjusted total thyroxine (free thyroxine index) as a measure of thyroid function, J Clin Endocrinol Metab 34:225-228, 1972.
30. Sawin, CT, Chopra, D, and Albano, J: The free triiodothyronine (T_3) index, Ann Intern Med 88:474-477, 1978.
31. Chopra, IJ, Solomon, DH, Hepner, GW, et al: Misleadingly low free thyroxine index and usefulness of reverse triiodothyronine measurement in nonthyroidal illnesses, Ann Intern Med 90:905-912, 1979.
32. Chopra, IJ, Solomon, DH, Chopra, U, et al: Alterations in circulating thyroid hormones and thyrotropin in hepatic cirrhosis: evidence for euthyroidism despite subnormal serum triiodothyronine, J Clin Endocrinol Metab 39:501-511, 1974.
33. Grenn, JRB: Thyroid function and thyroid regulation in euthyroid men with chronic liver disease: evidence of multiple abnormalities, Clin Endocrinol 7:453-461, 1977.
34. Inada, M, and Sterling, K: Thyroxine turnover and transport in Laennec's cirrhosis of the liver, J Clin Invest 46:1275-1282, 1967.
35. Tabei, A, and Shimoda, S: Increased TBG-T_4-binding capacity in acute hepatitis, Folia Endocrinol 49:1025-1033, 1973.
36. Schussler, GC, Schaffner, F, and Korn, F: Increased serum thyroid hormone binding and decreased free hormone in chronic active liver disease, N Engl J Med 299:510-515, 1978.
37. Spector, DA, Davis, PJ, Helderman, JH, et al: Thyroid function and metabolic state in chronic renal failure, Ann Intern Med 85:724-730, 1976.
38. Lim, VS, Fang, VS, Katz, A, et al: Thyroid dysfunction in chronic renal failure, J Clin Invest 60:522-523, 1977.
39. Maxon, H, Volle, C, Hertzberg, V, et al: Variation in serum thyroxine concentrations with time after oral replacement dose, Clin Nucl Med 12:369-370, 1987.

CHAPTER 42 Gastrointestinal hormones

STEPHEN N. JOFFE

OBJECTIVES

- Describe the brain-gut axis.
- Describe the difference among endocrine, paracrine, and neurocrine hormones.
- In general terms define gastrointestinal hormones and describe their role in gastrointestinal function.
- State the site of origin, the action, and the clinical significance of the following hormones: gastrin, CCK-PZ, secretin, VIP, PP, GIP, motilin, enteroglucagon, and somatostatin.

KEY TERMS

APUD cells Acronym for cytochemical properties of an endocrine cell that has an amine content, an amine precursor uptake, and decarboxylation.

brain-gut axis Similar peptides found in the gut, nerves, and central nervous system.

GEP hormones Gastroenteropancreatic hormones.

neurocrine hormones Peptides released from nerve synapses of the autonomous system.

paracrine hormones Peptides released into extracellular space where they can exert their effect on adjacent cells.

triple hormone control Endocrine, paracrine (locally acting), and neurocrine, including neurotransmission.

ANATOMY

Gastrointestinal hormones are a heterogeneous group of biologically active substances that are involved in the regulation of gastrointestional function. They are secreted by endocrine cells so widely distributed throughout the gut mucosa that the gut has been labeled the largest endocrine organ in the body.[1] In addition, a significant number of these biologically active peptides are present in the nerves of the gastrointestinal tract[2] and in the central nervous system[3] (Table 42-1). Some of these peptides function as classical hormones, whereas others are paracrine or neurocrine in their modes of action (Fig. 42-1).

Endocrine cells may either release the contents of their storage granule (hormones) into the bloodstream where they become classical hormones acting at distant sites, or they may act locally in paracrine roles. The *paracrine* method of transmission involves the release of peptides into the extracellular space where they can interact with adjacent cells. As a result, peptides may spill over into the lumen of the gastrointestinal tract ("lumones").

A third group of hormones are classified as *neurocrine* hormones. These gut peptides do not appear to be distributed by the circulatory system. They have extremely short biological half-lives after injection and have multiple actions on many different organs. The neurocrine peptides

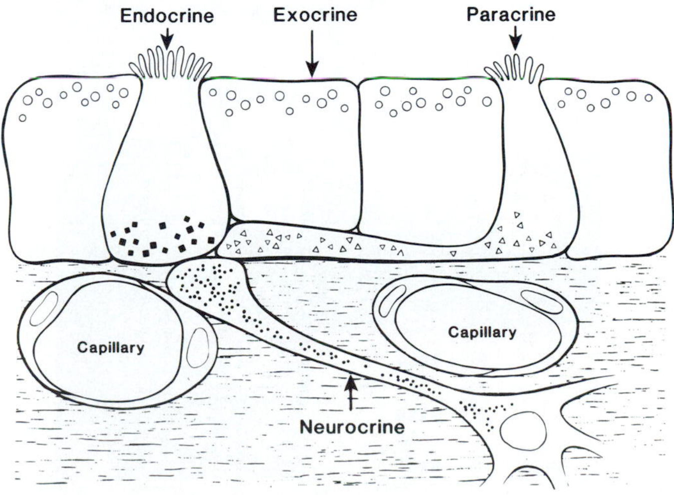

Fig. 42-1 Schematic representation of endocrine, paracrine, and neurocrine systems.

Table 42-1 Human gut hormones

Classical hormones	Candidate hormones (paracrine)	Biologically active	Immunoreactive
Gastrin*	Motilin*	Enterogastrone	Bombesin-like*
Cholecystokinin*	Enkephalin*	Antral chalone	Eledoisin-like
Secretin	Enteroglucagon	Bulbogastrone	Thyrotropin releasing hormone-like*
Gastric inhibitory polypeptide (GIP)	Vasoactive intestinal polypeptide (VIP)*	Gastrone	
Insulin	Urogastrone	Villikinin	
Glucagon	Neurotensin*	Vagogastrone	
	Substance P*	Entero-oxyntin	
	Endorphin	Coherin	
	Somatostatin	Enterocrinin	
	Pancreatic peptide (PP)		
	Chymodenin		

*Present in both brain and gut.

may be released from nerve synapses in a manner analogous to the release of catecholamines or acetylcholine, but, instead of acting locally as neurotransmitters, they are released into the general circulation and have neuroendocrine functions.

The gastrointestinal hormones of neuroectodermal origin can be classified under their common cytochemical characteristics. The initials "APUD" are used to designate a group of cells that have an *a*mine content, amine *p*recursor *u*ptake, and *d*ecarboxylation. These cells produce a number of amino acid or peptide hormones. The APUD cells are scattered throughout the gut and nervous system[4] (Fig. 42-2). APUD cells can be divided into two groups: The first group includes those cells that secrete polypeptides, and the second group includes those cells whose polypeptide product (if there is one) has not been identified. The first group includes most endocrine cells of the alimentary tract, including the islets of Langerhans and cells of the anterior pituitary, pineal, and thyroid glands (Table 42-2). The second group includes cells in many parts of the body; the most important of these are probably the cells in the adrenal medulla.

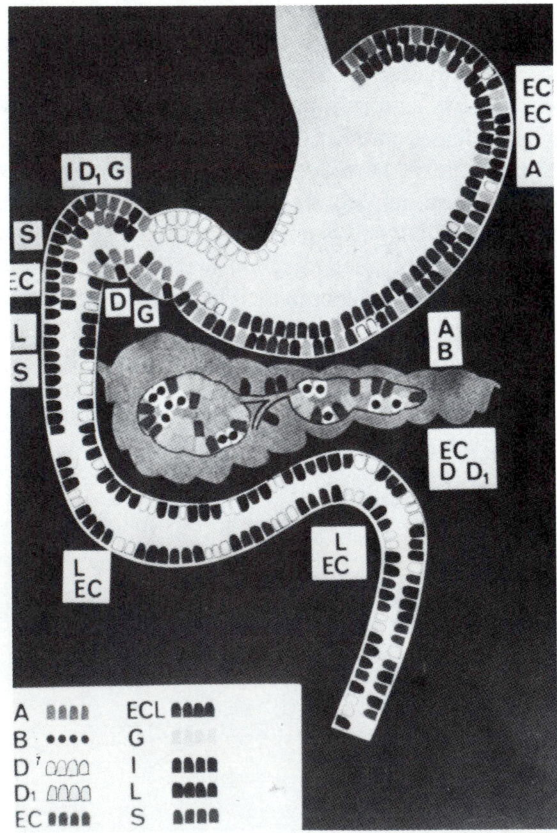

Fig. 42-2 Distribution of APUD cells in stomach, small bowel, and pancreas.

Brain-gut axis

A considerable number of hormones originally described in the gut are also found in nerves and the central nervous system. Conversely, certain brain peptides have also been identified in the gut (Table 42-2).[5] It has been suggested that the nervous system exists in three functional units: autonomic, somatic, and neuroendocrine.[6] The knowledge that gut hormones are distributed not only in endocrine cells but also in peripheral and central nerves has established the fact that these peptides function not only as hormones but also as neurotransmitters.[7]

Gut hormone structure

The two main families of gut hormones are the gastrin and the secretin families. The *gastrin* family consists primarily of gastrin and cholecystokinin, but, in addition, motilin and enkephalin share a number of structural identities. The *secretin* group includes secretin, gastric inhibitory polypeptide (GIP), vasoactive intestinal polypeptide (VIP), glucagon, and bombesin.[8]

Gut hormones exhibit molecular heterogeneity. In macroheterogeneity there are a number of different hormone forms that differ considerably; in microheterogeneity the

Table 42-2 APUD cells with known polypeptide products with related APUDomas and syndromes

Organ	Cell	Polypeptide	Amine	APUDoma	Syndrome
Alimentary tract					
Islets of Langerhans	B (β)	Insulin	Dopamine or 5-HT*	Hyperplasia, Adenoma, Carcinoma	Hypoglycemia
	A (α)	Glucagon			Diabetes, dermatitis
	D ($\beta_1\alpha_1$)	Somatostatin			—
		? Gastrin			Zollinger-Ellison
Stomach	G	Gastrin	—	Hyperplasia, Carcinoma	Zollinger-Ellison
	AL (A-like)	Enteroglucagon	—	—	—
	D	Somatostatin	—	—	—
Duodenum and small intestine	S	Secretin	—	—	—
	D_1	Gastric inhibitory polypeptide (GIP)	—	—	—
	EC	Motilin ? Substance P	5-HT	Carcinoid	Malignant carcinoid
	D	Somatostatin	—	—	—
Large intestine	D	Vasoactive intestinal polypeptide (VIP)	—	—	—
	EC	Enteroglucagon	—	—	—
Other sites					
Anterior pituitary	c (corticotroph)	ACTH†	Tryptamine?	Hyperplasia, Adenoma, Carcinoma	Cushing's
	m (melanotroph)	Melanocyte-stimulating hormone (MSH)	Tryptamine?		Pigmentation
	s (somatotroph)	Growth hormone (GH)			Acromegaly or gigantism
	l (lactotroph)	Prolactin			Forbes-Albright
Pineal	P	Melatonin	5-HT	Pinealoma	Hypogonadism
Thyroid	C	Calcitonin	5-HT	Medullary carcinoma	Medullary carcinoma

*5-Hydroxytryptamine.
†Adrenocorticotropic hormone.

differences involve one amino acid residue or only a side chain. Thus gastrin has been found to circulate as four components—I, II, III, and IV—and each of these components has various forms. This molecular heterogeneity has also been described for cholecystokinin and glucagon, and it presumably exists in many of the gut hormones. One molecular form probably represents the principal hormone, and the other forms are either biosynthetic precursors or postsecretory degradation products.[9,10]

The particular amino acid sequences are important because they confer a specific biological activity to the peptide and also determine the potency of action on a particular organ. In the gastrin-cholecystokinin (CCK) family, the COOH terminal pentapeptide amide sequence (Gly-Trp-Met-Asp-Phe-NH_2) is the main site of similarity and is a biologically active site for the two molecules. It is therefore not surprising that the two hormones share similar biological activity. The NH_2-terminal part of the molecule, however, influences potency for different targets. Thus CCK acts mainly on the gallbladder and pancreas, whereas gastrin is a more potent stimulus for gastric parietal cells. In addition, the localization of the sulfated tyrosyl residue is important both in determining the potency of the peptide and in influencing whether its action will be gastrinlike or CCK-like. If the sulfated residue is moved only one position, the hormone activity can be changed from gastrinlike to CCK-like. If the sulfate radical is removed, the hormone molecule becomes far less potent. The effects of amino acid changes on activity are not so clear in the secretin family, but all share common actions such as inhibition of acid secretion, stimulation of insulin release, and hepatic glycogenolysis. The main sites of origin and action of the gut hormones are listed in Table 42-3.

PHYSIOLOGY

An in-depth discussion of the digestive process is found in Chapter 26. Gut hormones are regarded as regulators of digestion and absorption. They are released in response to the presence of nutrients in the lumen of the gastrointestinal tract and provide acid, bicarbonate, and enzymes for digestion of food. Once the nutrients enter the blood, the pancreatic metabolic hormones are released. This link has led to the term *enteroinsular axis*. Different steps of digestion, absorption, and storage can be both stimulated and inhibited by different gastroenteropancreatic (GEP) peptides. This interplay prevents an overreaction and produces a self-limiting negative feedback control.

Somatostatin is a unique peptide that inhibits most of

Table 42-3 Main sites of origin and action of gut hormones

Tissue	Hormone	Action
Stomach	Gastrin	Stimulation of gastric acid; gastric mucosal growth
Pancreas	Pancreatic polypeptide	Inhibition of gallbladder contraction; inhibition of pancreatic exocrine secretion
Upper small intestine	Secretin	Stimulation of pancreatic bicarbonate
	Cholecystokinin	Gallbladder contraction; stimulation of pancreatic enzyme secretion; stimulation of pancreatic growth
	Motilin	Stimulation of upper GI motor activity
	GIP	Enhancement of insulin release
Lower small intestine	Enteroglucagon	Inhibition of intestinal transit; enhancement of mucosal growth
Colon	Neurotensin	?
	Enteroglucagon	Inhibition of intestinal transit; enhancement of mucosal growth

the gastrointestinal secretory and motor functions. It slows down the entry of nutrients and thus prevents an overflow of metabolic pathways. Its release is stimulated by nutrients, acid, and several gut hormones.

Each gastrointestinal function has several agonists and antagonists. Final control thus depends on a fine balance of numerous influences. In the case of gastric acid secretion, at least 21 different factors appear important in its normal control. Motilin, gastrin, VIP, and glucagon are the major hormones involved in the control of secretion, absorption, motility, and growth in the stomach and intestine. Secretin, CCK, VIP, and pancreatic polypeptide control exocrine pancreatic function, whereas insulin and GIP are involved in the enteroinsular axis. Insulin, glucagon, and somatostatin are primarily involved in the metabolism of carbohydrate, fats, and protein. Substance P, VIP, and the enkephalins have a major neurotransmitter involvement in the central, peripheral, and autonomic nervous systems.

CHARACTERISTICS OF GASTROENTEROPANCREATIC (GEP) HORMONES AND POLYPEPTIDES

Gastrin

Chemistry. Gastrin exists in multiple molecular forms. The four main components, gastrin I, II, III, and IV, differ in the number of amino acids in the polypeptide chain. Component II is gastrin 34 (G 34) or big gastrin; component III is gastrin 17 (G 17); and component IV is gastrin 14 (G 14). The numbers identify the number of amino acids present. A molecule designated "big, big gastrin" probably does not exist except as an artifact in radioimmunoassay or in certain tumors that may produce macromolecules.[10]

Distribution and biosynthesis. The cellular origin of gastrin is the G cell of the gut. These cells extend into the gut lumen, where they terminate in a tuft of microvilli. Because luminal stimuli regulate gastrin release, it is possible that the microvilli bear receptors for these active luminal factors. The secretory granules that contain gastrin are located in the basal parts of the cells, which allows direct secretion of the hormone into the bloodstream. Gastrin is produced mainly by the antral G cells and to a lesser extent by G cells of the proximal duodenum.[11] The biosynthesis of gastrin probably proceeds through a large precursor that is converted first to G 34 and then to G 17.

Biological effects and control. Gastrin stimulates acid secretion, gastric motility, and the growth of fundic small bowel and mucosa.[12] Secretion of gastrin is mediated by luminal, nervous, and bloodborne stimuli. The principal luminal stimuli are the amino acid products of protein digestion.[13] Other luminal stimuli include calcium, alcohol, and pH. In normal subjects gastrin release is dependent on intragastric acidity. At pH 5.5 the rate of gastrin release is three times as high as at pH 2.5. The inhibition effect of acid provides a feedback mechanism for autoregulation of gastrin release. Carbohydrates and fats have little effect on gastrin release. Distension of the antral pouch has long been recognized to release gastrin, and both cholinergic nerve stimulation and noncholinergic mechanisms also play an important role. The feedback regulation of gastrin release is mediated by antral acidification and by neural and hormonal regulatory mechanisms generally characterized by the concept of an enterogastric axis.[14] Several peptides known to inhibit the release of gastrin include secretin, GIP, VIP, glucagon, calcitonin, and somatostatin. Peptides may also modulate gastrin release by paracrine or local action.

In normal subjects a meal such as eggs and toast stimulates an increase in total serum gastrin, which reaches a peak in about 30 to 60 minutes and then declines to basal over the next 1 to 2 hours. Gel filtration analysis indicates that the postprandial serum increases are almost entirely a result of G 17 and G 34. Maximal concentrations of G 17 are achieved 20 minutes after the meal, whereas a peak concentration of G 34 is reached at 45 to 60 minutes.[12] Total serum gastrin concentration is not a reliable indicator of circulating biological activity, however, and it seems likely that G 17 accounts for nearly all the circulating gastrin that is biologically active after a normal meal.[14]

The kidney, liver, and small intestine are important sites of gastric degradation. The smaller COOH terminal fragments of gastrin, such as the pentapeptide, may be almost completely cleared by the liver on a single pass.[14] The normal fasting gastrin concentration is usually less than 25 pmol/L or 50 pg/mL (human G 17 equivalents).

Pathology

Gastrinomas or Zollinger-Ellison syndrome. Plasma gastrin is particularly important in the diagnosis of pancreatic endocrine tumors (gastrinomas). The Zollinger-Ellison syndrome is characterized by peptic ulceration, gastric acid hypersecretion, and a non-beta cell islet cell tumor of the pancreas.[15] These tumors produce large quantities of gastrin, hence the descriptive name "gastrinoma." Basal serum gastrin concentrations above 500 pmol/L (1000 pg/mL) accompanied by gastric acid hypersecretion are virtually diagnostic for gastrinoma. Patients with achlorhydria (decreased HCl output) may have serum gastrin in the gastrinoma range,[16] but can be readily distinguished on the basis of acid secretory tests. A very small number of patients may have hypergastrinemia and hyperchlorhydria resulting from an isolated antrum from previous surgery. Many, perhaps most, gastrinomas are malignant and, at the time of discovery, have metastasized to other parts of the body.

It is now recognized that five types of APUDomas (tumors of the APUD system) may produce excessive gastrin and cause a variant of the Zollinger-Ellison syndrome. Those recognized include G cell hyperplasia or carcinoma in the stomach; islet cell tumor (adenoma or carcinoma) or D-cell hyperplasia in the pancreas; and carcinoma or adenoma in the duodenum or splenic hilum.[17]

In patients with renal failure, elevated serum gastrin concentrations return to normal after renal transplantation.[18] Serum gastrin concentrations are also raised in patients after massive small bowel resection as a result of either decreased gastrin clearance or removal of an intestinal factor that normally suppresses gastrin release.

Peptic ulcer. The role of gastrin in the pathogenesis of peptic ulceration remains uncertain. It has been established, however, that patients with duodenal ulcers tend to have higher-than-normal basal and maximal acid secretion; they are also more sensitive than normal to exogenous gastrin.[19,20] As in normal subjects, G 17 and G 34 account for most of the increase in gastrin after a meal, with a G 34 accounting for the greater increase. Because G 34 is a weaker stimulant of acid secretion than is G 17, it is not yet certain whether there is a significant elevation of biologically active gastrin in duodenal ulcer patients.

Patients with gastric ulcer have elevations in both basal and postprandial serum gastrin concentrations. These patients have a lower basal and maximal acid output than normal, and it seems likely that the higher serum gastrin is the result of a decreased inhibition. After a meal, the peak increase in G 14 is similar to normal, but the increase in G 34 is about four times higher than normal, which is similar to levels seen in duodenal ulcer patients.[12]

Function tests. Many gastrinoma patients have serum gastrin levels that overlap with the upper end of the range for normal subjects and patients with ordinary peptic ulceration. In 15 patients with gastrinoma, 6 were found to have serum gastrin concentrations less than 500 pg/mL.[21] One or more provocation tests are therefore needed to establish the presence of gastrinoma in such patients. The most widely used tests are the *standard test meal,* the *secretin test,* and the *calcium challenge.*[22] The secretion of gastrin in a patient with a gastrinoma caused by a pancreatic tumor is usually unaltered by food. Secretin does increase in response to food in a patient with antral G-cell hyperplasia. Rapid intravenous injection of secretin (2 to 4 clinical units/kg) stimulates a pronounced increase of gastrin in about 80% of all gastrinoma patients with peaks at 2½ to 10 minutes that reach 100% above the basal level.

After a calcium infusion, the serum gastrin level noticeably increases in patients with gastrinoma.[23] In one study all patients responded with a twofold to fiftyfold increase in plasma gastrin. Calcium, however, may provoke a modest rise in serum gastrin levels in normal subjects and in patients with duodenal ulcers, but these responses usually fall short of those seen in gastrinoma patients. There is an association between gastrinoma and hyperparathyroidism such that the hypercalcemia may exacerbate through the release of gastrin from tumors.

Cholecystokinin-pancreozymin (CCK-PZ)

Chemistry. It is now well recognized that cholecystokinin (CCK) and pancreozymin (PZ) are a single substance.[24] The molecule is a basic peptide of 33 amino acid residues. Based on the length of the polypeptide, it is convenient to refer to the cholecystokinins as CCK 33, CCK 39, and CCK 8. These amino acid residue forms may be sulfated or desulfated. Sulfation is particularly important for conferring biological activity, though it is possible that unsulfated forms may have an alternative physiological role as neurotransmitters. The carboxyl-terminal octapeptide portion of CCK is at least tenfold more potent than the intact hormone.[25]

Site and mode of synthesis. CCK is found in the brain and in the K cells of the upper small intestinal mucosa, which correspond to the I cell previously identified by electron microscopy. It is probable that CCK 33 and 39 are biosynthetic precursors of the smaller CCK-like peptides, such as CCK 8, but they may result from cleavage of larger precursors.

Biological effects. The finding of CCK-like peptides in the central nervous system and in the gut suggests that these peptides function as both neurotransmitters and hormones. The physiological role of CCK is related to the regulation of motility of the gallbladder and intestine and the regulation of secretion by the pancreas. Physiological actions of CCK in the pancreas include the stimulation of enzymes, the potentiation of the action of secretin, and stimulation of the growth of the pancreas. CCK also affects smooth muscle other than the gallbladder. It appears to inhibit gastric emptying, relaxes the sphincter of Oddi, and stimulates the small intestine to contract. In addition, it has been reported that CCK stimulates the secretion of pepsin, Brunner's glands, and hepatic biocarbonate and water. Other activities probably relate to its gastrin homologs and include the weak stimulation of gastric acid secretion and antagonism of secretin-stimulated bicarbonate secretion.[26] Its function in the central nervous system is not yet clear, but it has been suggested that it plays a role in the satiety (state of satisfied appetite) mechanisms.[27]

Release of CCK 33 has been studied for the most part by biological methods that involve monitoring gallbladder contraction and pancreatic enzyme secretion in response to the instillation of food into the small intestine. As in the case of gastrin, pure undigested protein does not stimulate hormone release, but after partial proteolysis the mixture of polypeptides and amino acids is a strong stimulus.[28] Fatty acids with chains longer than nine carbons also stimulate CCK release.

Fasting values of CCK concentration in normal humans vary from 26 pg/mL (equivalent to porcine CCK 33) to much higher values. Estimates of CCK release after stimulation by feeding also vary widely. Standard meals can increase CCK from 200 to 300 pg/mL.[29] The half-life of exogenous CCK is reported to be 2½ minutes, and the expected increase of plasma CCK during an infusion of 250 pg/kg/hour would be about 100 pmol/L (400 pg/mL).

Pathology. Plasma CCK is reported to increase in patients with celiac disease and pancreatic exocrine insufficiency.[30]

Secretin

Chemistry. Secretin isolated from hog duodenum is a basic peptide of 27 amino acid residues with strong similarities in sequence to glucagon. Early studies indicate that the intact peptide is needed for full activity and that there is no minimum active fragment comparable to that of gastrin or CCK. Several experiments suggested that the NH_2-terminal sequence of secretin is required for biological activity and that affinity for the receptor is increased by the COOH-terminal portion of the molecule.[31]

Site and mode of synthesis. Secretin is predominantly located in the S cells of the mucosa of the duodenum and jejunum.[32] The frequency of S cells in the jejunum is about one third of that in the duodenum, but the greater length of the jejunum relative to the duodenum means that the bulk of intestinal secretin is jejunal in origin.

Biological effects. Secretin inhibits smooth muscle contraction, decreases gastric acid secretion, lowers the lower esophageal sphincter pressure, and stimulates pancreatic growth. In addition to these effects, it stimulates water and bicarbonate secretion from the liver and Brunner's glands and augments gallbladder contraction. It has a synergistic action with CCK in the stimulation of gallbladder contraction and pancreatic enzyme secretion.[29] The primary physiological role of secretin appears to be the modulation of pancreatic bicarbonate secretion.

A principal stimulus for secretin release is stomach acid, but in the adult jejunum there is seldom if ever likely to be sufficient acid to liberate secretin.[33] Fatty acids with 10 or more carbons in the chain weakly stimulate secretin release, but undigested fat has no activity. The threshold for release of secretin is a pH of 4.5, and below this the amount of secretin released is proportionate to the load of acid entering the small intestine.[29] The optimum conditions for secretin release, that is, pH less than 4.5, are normally obtained only in the first few centimeters of duodenum. The rise in plasma secretin therefore is very modest after a normal meal. The secretion of bicarbonate from the pancreas closely matches the entry of acid into the duodenum. There is little doubt therefore that gastric acid stimulates pancreatic bicarbonate output. It seems likely that the increase in plasma secretin after a normal meal is by itself insufficient to stimulate the pancreas, but its action is strongly potentiated by CCK.

The concentrations of secretin in plasma from fasting humans vary from 0.6 to 50 pg/mL (0.2 to 20 pmol/L).[34] After a meal, there is a small but measurable increase in normal subjects, and increases of about 2 pmol/L to a peak of 5 to 6 pmol/L have been shown.[34] Plasma concentrations rise rapidly after stimulation by acid and reach a peak about 3 minutes before declining to basal levels in about 60 minutes, suggesting a half-life of about 3 minutes.

Pathology. Because acid is the main stimulus for secretin release, it is not surprising that secretin concentrations are elevated in patients with gastric acid hypersecretion (such as gastrinomas).[35] Other conditions characterized by elevated plasma secretin have not yet been described. In patients with celiac disease, an increased number of S cells was found by immunocytochemistry, and the granule content appeared higher than normal, possibly as a result of failure of hormone release. Evidence has suggested decreased secretion of secretin in patients with celiac disease.[34]

Secretin is clinically useful in confirming the diagnosis of gastrinoma. For reasons as yet unclear, the administration of secretin as a bolus intravenously (2 clinical units/kg) provokes a rapid rise in plasma gastrin in patients with gastrinoma. It has little or no effect on the plasma gastrin levels when hypergastrinemia results from other causes.[36]

Vasoactive intestinal polypeptide (VIP)

Chemistry and site of synthesis. VIP has 28 amino acids and a molecular weight of 3320 daltons. VIP has been shown to be present in both endocrine cells and nerves of the gut and central nervous system[37]; there are a number of molecular forms of VIP.

Biological effects. VIP has a wide spectrum of activities, including inhibition of gastric acid secretion, stimulation of insulin release, stimulation of pancreatic water and bicarbonate secretion, stimulation of hepatic glycogenolysis, stimulation of intestinal fluid and electrolyte secretion, stimulation of mucosal cAMP, relaxation of smooth muscle to cause vasodilation, and a positive inotropic action on the heart.[38] Its exact mode of action is uncertain, but it probably functions primarily as a paracrine neurotransmitter substance rather than as a circulating hormone.[39,40] There is no significant release of VIP into the blood after a meal. By radioimmunoassay, the mean serum concentration of fasting plasma VIP is found to be 1.5 pmol/L, but there is a skewed distribution so that the upper end of the normal range is 21 pmol/L, with a mean normal of 2.1 pmol/L.[38,39]

Pathology. Abnormally high tissue levels of VIP have been found in Crohn's disease, whereas in the aganglionic colon segment of patients with Hirschsprung's disease VIP levels have been reported to be extremely low.[8] The most important application of the measurement of VIP is that elevated levels are found in the plasma in some patients with watery diarrhea syndrome (Verner-Morrison syndrome). Verner and Morrison in 1958 reported on two patients with pancreatic non-beta islet cell tumor causing watery diarrhea without peptic ulceration.[41] Subsequently it was named "WDHA" after the initial letters of its main characteristics: *w*atery *d*iarrhea, *h*ypokalemia, and hypo- or *a*chlorhydria. The syndrome is rare; about one tenth as common as Zollinger-Ellison syndrome. It is sometimes part of multiple endocrine neoplasia.

In patients with WDHA syndrome, a non-beta islet cell tumor (vipoma) of the pancreas is usually present (D_1 cells); about half of these tumors are malignant. Tumors can occur elsewhere (such as bronchial, probably oat cell, carcinoma, or retroperitoneal neuroblastoma) and secrete VIP. In large doses, VIP causes vasodilatation with facial flushing, increases intestinal blood flow, induces watery diarrhea, and inhibits gastric secretion. The diarrhea is explosive, consists of up to 30 stools per day, and does not respond to simple measures. This causes a profound hypokalemia (1 to 3 mmol/L), but the associated hypercalcemia is unexplained.[42]

The diagnosis is made by elimination of the common causes of watery diarrhea and hypokalemia. Gastric secretion tests often show the presence of hypochlorhydria, and the diagnosis is confirmed by measurements of elevated blood levels of immunoreactive VIP.[43] A characteristic tumor blush of dilated vessels caused by the local action of VIP is found on selective arteriography. The pseudo-Verner-Morrison syndrome refers to patients with watery diarrhea, hypokalemia, and achlorhydria who have normal serum VIP levels and no obvious tumor. The electrolytes and dehydration must be corrected preoperatively, and if the diarrhea is severe and continuous, steroids (prednisone 60 mg/day) or somatostatin may be given.[44]

Plasma VIP measurement is a very useful screening test for the detection of vipomas and is an effective tumor marker to detect occult metastases.

Pancreatic polypeptide (PP)

Chemistry. PP contains 36 amino acids arranged in a linear sequence. The entire biological activity is displayed by the last 6 amino acids,[45] which do not differ among pig, bovine, and human PP.

Site of synthesis. PP is found almost entirely in the pancreas, where it is distributed in the D_2 cells of the islets, but it is also scattered throughout the acinar cells of the exocrine pancreas and the walls of the pancreatic ducts. Most of the PP cells are in the head of the pancreas.

Biological effects. Pharmacological studies show that PP opposes the effects of cholecystokinin.[45] It therefore inhibits the pancreatic output of enzymes and gallbladder contraction. At somewhat higher doses it has a biphasic effect, initially stimulating and then later inhibiting pancreatic bicarbonate output. Gastric acid output is stimulated, whereas motor activity of the upper intestine is inhibited.[46] PP may play a role in the modulation of insulin and glucagon secretion by the islets.

Fasting human PP concentrations have been variously reported as 20 to 80 pmol/L.[47] Plasma PP levels increase with age. There is a significant tenfold biphasic rise occurring after a meal, with peak values achieved at 30 minutes. The main constituents of the meal that are responsible for the PP release appear to be fat and protein. PP levels are undetectable in patients after pancreatectomy. Postprandial rise reflects signals from the gut to the pancreas. Changes in plasma concentration of amino acids, fat, or glucose do not affect PP release directly. During insulin-induced hypoglycemia, a large rise in PP concentration is seen, which is abolished by vagotomy, suggesting an important vagal influence. PP is liberated into the blood in large quantities after a meal by mechanisms that appear to be dependent on vagal nerve stimulation.[48]

Pathology. Considerably elevated plasma PP levels have been documented in a high percentage of patients with vipomas, insulinomas, gastrinomas, and glucagonomas.[49,50] Although some of these tumors contained PP cells, in others only PP cell hyperplasia in the rest of the pancreas was evident. This suggests that elevated PP levels might serve as biochemical markers for pancreatic endocrine tumors. A number of patients with diarrhea were found to have pancreatic endocrine tumors composed of only PP cells (PPoma). Pure pancreatic polypeptide APUDomas (PPomas) are rare, though elevated plasma PP levels are found in one third of pancreatic APUDomas. Basal and meal-stimulated serum PP levels can be decreased in pancreatitis.[51]

Gastric inhibitory polypeptide (GIP)

Chemistry and site of synthesis. Gastric inhibitory peptide is a straight-chain polypeptide of 43 amino acids with a molecular weight of 5105 daltons. GIP shows considerable sequence homology with the other gastrointestinal hormones.[52] In mammals there is no evidence of any interspecies difference in amino acid sequence. GIP is found in the K cells of the jejunal mucosa and to a lesser extent in the duodenum and ileum.[52]

Biological effects. Reported fasting human plasma levels of GIP range from 15 to 100 pmol/L. There is a relatively prolonged rise in GIP after a meal.[53] Infusion of GIP in human volunteers has demonstrated a considerable enhancement of glucose-induced insulin release at a blood level near the physiological range.[54] Little effect on insulin is seen, however, when subjects are studied in the fasted state. Thus GIP has been renamed "*g*lucose-dependent *i*nsulin-releasing *p*eptide" and has been proposed as the main hormone enhancing insulin release when nutrients are taken orally. GIP thus may be the major component of the enteroinsular axis.[55]

Pharmacological studies have shown that GIP has the capacity to inhibit gastric acid and pepsin secretion and intestinal motility. The physiological role of GIP in the control of acid secretion is at present controversial. GIP is also a potent stimulant of small intestine fluid and electrolyte secretion.

Motilin

Chemistry and site of synthesis. Motilin is a straight-chain peptide of 22 amino acids with a molecular weight of 2700 daltons. Motilin is found in the enterochromaffin (EC) cells of the small intestine, predominantly in the upper portion, though it may also be present in the central nervous system.[56]

Biological activity. Motilin is released into the plasma after a meal, particularly if the fat content is high. In humans duodenal acidification causes an increase in plasma motilin. Because it is released by a physiological stimulus (a meal), it has been suggested that motilin plays a role in the modulation of gastrointestinal function.

The pharmacological actions of motilin include stimulation of small intestine and gastric motility, increase in lower esophageal sphincter pressure, and stimulation of pepsin output. Recent studies have indicated that motilin accelerates gastric emptying in humans and that it also may be responsible for emptying the small intestine between meals.

Fasting plasma motilin concentrations in humans show a skewed distribution, with the upper limits of normal at about 300 pmol/L.[57] Motilin circulates in only one molecular form, and no disease has yet been recognized in which plasma motilin levels are abnormal.

Glucagon-like immunoreactive peptides

Chemistry and site of synthesis. Several molecular forms of this family of hormones have been identified, each probably with a separate physiological role, although only one cell type containing glucagon-like immunoreactivities is found in the intestine.[58] Nomenclature is thus still confusing, and names such as GLI, N-GLI, gut glucagon, or enteroglucagon are used. It has been suggested that enteroglucagon contains the entire amino acid sequence of pancreatic glucagon with an additional N-terminal sequence.[59] The full amino acid sequence has not yet been determined. The pancreatic alpha cells produce a polypeptide of 29 amino acids with a molecular weight of 3485 daltons called *pancreatic glucagon,* which shows numerous amino acid homologies with secretin; consequently many of the biological functions are similar to those of VIP, GIP, and secretin.

Biological activity. Glucagon and the glucagon-like immunoactive peptides have a number of biological actions, which include relaxation of smooth muscle, inhibition of pancreatic enzyme secretion, inhibition of gastric acid secretion, stimulation of intestinal fluid and electrolyte secretion, and stimulation of cardiac output. Pancreatic glucagon is secreted primarily in response to hypoglycemia and is important in the mobilization of hepatic glycogen stores and carbohydrate homeostasis.[60] The inhibitory action of glucagon on gastrointestinal motility has enabled it to be of clinical use to endoscopists and radiologists. Fasting human plasma enteroglucagon levels have been reported to be about 20 pmol/L, rising after a normal meal to 40 pmol/L. Low enteroglucagon levels are seen after starvation, and high enteroglucagon levels are associated with hyperphagia (overeating). After partial gut resection, enteroglucagon rises rather than falls, and very high enteroglucagon levels are found in patients after jejunoileal bypass. Similarly, in states of mucosal damage often associated with malabsorption (such as celiac disease), enteroglucagon levels are elevated.

Pathology. A glucagon-secreting tumor of the pancreas is associated with diabetes, skin rash, and a high concentration of plasma glucagon.[61] The diagnosis is easily confirmed by radioimmunoassay measurement of elevated plasma glucagon levels.[62] An arginine provocation test has been reported to be of some use in the diagnosis of cases with equivocal plasma glucagon levels.

Somatostatin

Chemistry and site of synthesis. Somatostatin is a peptide of 14 amino acids with a molecular weight of 1640 daltons, bridged across at the third and fourteenth positions by a cystine link. Two forms of biologically active somatostatin are somatostatin 14 and somatostatin 28.

Certain analogs are relatively selective in their action, and the analog $DES^{1,2,4,5,12,13}$ D $Tryp^{8}$ is almost as potent as the parent molecule, despite missing 6 of the 14 amino acids. (The superscripts represent amino acid positions.)

Somatostatin is found throughout the brain, but the highest concentrations are found in the hypothalamus, particularly in the median eminence. In the gut the greatest amounts are found in the antrum of the stomach, in the upper small intestine, and in the pancreas. Somatostatin is localized in the mucosa of the alimentary tract, in the D cells of the islets of Langerhans, and in a small number of fine nerve fibers.[63]

Biological activity. Somatostatin acts as an inhibitory hormone throughout the body. Somatostatin inhibits gastric emptying, pepsin secretion, gallbladder contraction, and bile and pancreatic enzyme secretion.[64] Because somatostatin has so many effects, it is probable that it functions as a local hormone and part of the paracrine system.

Pharmacological studies have demonstrated that somatostatin is one of the most potent inhibitors of endocrine secretions known.[65] It has now been shown that it not only completely inhibits the release of growth hormone but that it also inhibits the release of thyroid-stimulating hormone, insulin, glucagon, pancreatic polypeptide, gastrin, secretin, motilin, enteroglucagon, and neurotensin.[66] In addition, it can inhibit the effect of these hormones on their target tissues and thus will prevent acid secretion during pentagastrin infusion and pancreatic bicarbonate secretion during secretin infusion.[64]

Pathology. A number of somatostatin-producing tumors of the pancreas have recently been described and are known as somatostatinomas.[67] These tumors produce an ill-defined clinical syndrome that includes diarrhea, gallbladder disease, and diabetes. In patients with other pancreatic endocrine tumors, somatostatin cell hyperplasia is often present in the surrounding pancreatic tissue. The reason for this is not clear, but it has been suggested that it represents an attempt by the pancreas to decrease tumor hormone production. Because somatostatin is a potent inhibitor of peptide release in both physiological and pathological situations, its use as a therapeutic agent in the management of tumors would be appropriate. The administration of long-acting somatostatin analogs has resulted in inhibition of peptide secretion in a number of patients with gastrinomas, glucagonomas, and VIPomas, with amelioration of the clinical symptoms.[68] Somatostatin has also recently been implicated in the etiology of peptic ulceration.

Miscellaneous peptides

With the advent of improved methodology and increased interest in the gut as an endocrine organ, a large number of substances have been postulated and, in some cases, identified (see Table 42-1). The exact function in the gut of many of these substances is still not fully determined; however, their presence in gut endocrine cells or nervous system tissue requires explanation.[7] Terms such as *incretin* or *enterogastrone* have been used to postulate the existence of gut peptides that respectively modulate insulin release or inhibit gastric acid secretion. In many cases, such peptides are known only in crude biological extracts, and in other cases the term is used merely to describe the function rather than a known agent.

Peptides found in both gut and nervous system (brain-gut axis). Several of these peptides have been discussed previously; the list includes gastrin, cholecystokinin, VIP, motilin, somatostatin, neurotensin, bombesin, substance P, and enkephalin. The most recently identified peptides in this group are neurotensin, substance P, and the endogenous opiate group.

Neurotensin. Neurotensin is a peptide of 13 amino acids localized primarily in the mucosa of the ileum in specific endocrine cells called *N cells*.[69] A number of molecular forms of neurotensin exist in both plasma and tissue. Neurotensin appears to be released independently of the vagus, but its exact physiological function is not yet known.

Substance P. This decapeptide has a molecular weight of 1348 daltons.[70] It has been demonstrated in the brain, intestinal nerve plexus, and enterochromaffin cells of the gut mucosa. The pharmacological actions of substance P include stimulation of intestinal motility and salivary secretion and increase in esophageal sphincter pressure.[71] Elevated plasma levels have been found in patients with carcinoid syndrome. In Hirschsprung's disease, substance P levels are extremely low in the aganglionic segments.

Endogenous opiates. This group consists of two main subgroups: the pentapeptide enkephalins and the endorphins.[72] The two enkephalins differ in structure by only one amino acid. They are found in high concentrations both in the brain and in the G cells and in the myenteric plexus of the antrum and duodenum, where they probably have a paracrine function. Morphine and methionine enkephalin are partial agonists of gastric acid secretion, and their action can be blocked by naloxone, atropine, or metiamide.[73] Evidence suggests that this group of substances may play a role in the regulation of gastric acid secretion and gut motility.

Peptides related to acid secretion. Human *urogastrone* is a polypeptide of 53 amino acids with a molecular weight of about 6000 daltons, although larger molecular forms occur.[74] Urogastrone has a structure and function similar to those of epidermal growth factor. Both peptides are powerful inhibitors of gastric acid secretion and stimulate epithelial cell proliferation.

Peptides related to pancreatointestinal function. A number of these peptides have been tentatively identified. They include chymodenin, coherin, duocrinin, enterocrinin, and villikinin. Further work is required to confirm both their existence and action.

Other tumors

Mixed pancreatic islet cell tumors. Mixed tumors are of clinical importance because the spectrum of symptoms may change with time, and metastases may contain only some of the primary tumor cell types. Because raised plasma concentration of hormones may originate either from the primary or metastatic tumor tissue or from hyperplastic extratumoral cells, immunocytochemical studies are necessary. Ideally the plasma concentration of as many hormones and polypeptides as possible should be measured preoperatively and postoperatively to confirm removal of all tumor tissue. APUDomas that secrete a combination of several polypeptides, such as ACTH, MSH, and gastrin, have been described.[75]

Multiple endocrine neoplasia. Multiple endocrine neoplasia describes a group of syndromes, often familial, in which two or more endocrine glands undergo hyperplasia or tumor formation in the same individual, either at the same time or consecutively. The hyperfunctioning glands secrete their normal major hormones (orthoendocrine syndromes) and abnormal hormones (paraendocrine syndromes). There are two main types of multiple endocrine neoplasia syndromes.

Multiple endocrine neoplasia type 1 (MEN I or MEA I), or Werner's syndrome. This syndrome manifests from the second decade to old age with an equal sex distribution. The areas involved in order of frequency are parathyroids (88%), pancreatic islets (81%), anterior pituitary (65%), adrenal cortex (38%), and thyroid follicular cells (19%).

Changes in the thyroid vary, and occasionally carcinoid tumors are present in the lungs, pancreas, or intestines. Peptic ulceration, especially duodenal, is a common feature in some families and may affect more than half the patients. These may form part of the Zollinger-Ellison syndrome as a result of an associated pancreatic gastrinoma or hyperparathyroidism. Many of these lesions are APUDomas, and the syndrome may result from a widespread dysplasia of the APUD cells.

Multiple endocrine neoplasia type II (MEN II OR MEA II), or Sipple's syndrome. This syndrome is usually inherited, affects sexes equally, and may manifest itself from the first decade onward, most commonly from 20 to 40 years. Three forms of the syndrome are recognized. The most common tumor is medullary carcinoma of the thyroid with a pheochromocytoma in one or both adrenal glands. Hyperparathyroidism resulting from hyperplasia or adenoma may be present and is probably caused by hypercalcemia. Another variant is a medullary carcinoma of thyroid and pheochromocytoma with multiple small subcutaneous and submucous neuromas of the eyelids, tongue, and buccal mucosa with a diffuse hypertrophy of the lips. These lesions are present from birth. In the last type, all features are present together with autonomic ganglioneuromatosis and various other congenital abnormalities.

REFERENCES

1. Pearse, AGE, Polak, JM, and Bloom SR: The newer gut hormones, Gastroenterology 72:746, 1977.
2. Polak, JM, and Bloom, SR: Neuropeptides of the gut, World J Surg 3:393, 1979.
3. Rehfeld, JF: Gastrointestinal hormones. In Crane, RK, editor: International review of physiology: gastrointestinal physiology III, vol 19, Baltimore, 1979, University Park Press.
4. Pearse, AGE: The cytochemistry and ultrastructure of polypeptide hormone-producing cells of the APUD series and the embryologic, physiologic, and pathologic implications of the concept, J Histochem Cytochem 17:303, 1969.
5. Rehfeld, JE: Immunochemical studies in cholecystokinin. II. Distribution and molecular heterogeneity in the central nervous system and small intestine of man and hog, J Biol Chem 253:4022, 1978.
6. Pearse, AGE, and Takor, T: Neuroendocrine embryology and the APUD concept, Clin Endocrinol 5(suppl):229S, 1976.
7. Grossman, MI: Neural and hormonal regulation of gastrointestinal function: an overview, Annu Rev Physiol 41:27, 1979.
8. Bloom, SR, and Polak, JM: In Glass, B, editor: Progress in gastroenterology, New York, 1977, Grune & Stratton, Inc.
9. Rehfeld, JF: Heterogeneity of gastrointestinal hormones, World J Surg 3:415, 1979.
10. Rehfeld, JF, Schwartz, TW, and Stadil, F: Immunochemical studies on macromolecular gastrins: evidence that ''big big gastrin'' in blood and mucosa artifactual—but truly present in some large gastrinomas, Gastroenterology 72:469, 1977.
11. Larsson, LI: Peptides of the gastrin cell. In Rehfeld, JH, and Amdrup, E, editors: Gastrin and the vagus, New York, 1979, Academic Press, Inc.
12. Dockray GJ: Gastrin overview. In Bloom, SR, editor: Gut hormones, Edinburgh, 1978, Churchill Livingstone.
13. Strunz, UT, Walsh, JH, and Grossman, MI: Stimulation of gastrin release in dogs by individual amino acids (40072), Proc Soc Exp Biol Med 157:440, 1978.
14. McGuigan, JE: Gastrointestinal hormones, Annu Rev Med 29:307, 1978.
15. Zollinger, RM, and Ellison, EH: Primary peptic ulceration of the jejunum associated with islet cell tumors of the pancreas, Ann Surg 142:709, 1955.
16. Yalow, RS, and Berson, SA: Radioimmunoassay of gastrin, Gastroenterology 58:1, 1970.
17. Welbourn, RB, and Joffe, SN: The apudomas. In Taylow, S, editor: Recent advances in surgery, Edinburgh, 1977, No. 9, Churchill Livingstone.
18. Hamsky, J, King, RW, and Holdsworth, S: In Thompson, JC, editor: Gastrointestinal hormones, Austin, 1978, University of Texas Press.
19. Walsh, JH, and Grossman, MI: Gastrin, N Engl J Med 292:3324, 1975.
20. Isenberg, JI, Grossman, MI, Maxwell, V, et al: Increased sensitivity to stimulation of acid secretion by pentagastrin in duodenal ulcer, J Clin Invest 55:330, 1975.
21. Creutzfeldt, RA, and Frerichs, H: Insulinomas and gastrinomas. In Bloom, SR, editor: Gut hormones, Edinburgh, 1978, Churchill Livingstone.
22. Primrose, JN, Ratcliffe, JG, and Joffe, SN: Assessment of secretion provocation test in the diagnosis of gastrinoma, Br J Surg 67:744, 1980.
23. Passaro, E, Basso, N, and Walsh, JH: Calcium challenge in the Zollinger-Ellison syndrome, Surgery 72:60, 1972.
24. Jorpes, JE, and Mutt, V: Secretin, cholecystokinin, pancreozymin, and gastrin, New York, 1973, Springer-Verlag.
25. Lin, TM: Actions of gastrointestinal hormones and related peptides on the motor function of the biliary tract, Gastroenterology 69:1006, 1975.
26. Brooks, AM, and Grossman, MI: Effect of secretin and cholecystokinin on pentagastrin stimulated gastric secretion in man, Gastroenterology 59:114, 1970.
27. Sturdevant, RAL, and Goetz, H: Cholecystokinin both stimulates and inhibits human food intake, Nature 261:713, 1976.
28. Meyer, JH, and Kelly, GA: Canine pancreatic responses to intestinally perfused proteins and protein digests, Am J Physiol 231:682, 1976.
29. Rayford, PL, Miller, TA, and Thompson, JC: Secretin, cholecystokinin, and newer gastrointestinal hormones, N Engl J Med 294:1093, 1976.
30. Harvey, RF, Dowsett, L, Hartog, M, et al: A radioimmunoassay for cholecystokinin-pancreozymin, Lancet 2:826, 1973.

31. Robberecht, P, Conlon, TP, and Gardner, JD: Interaction of porcine vasoactive intestinal peptide with dispersed pancreatic acinar cells from the guinea pig, J Biol 251:4653, 1976.
32. Polak, JM, Pearse, AGE, Joffe, SN, et al: Quantification of secretin release by acid using immunocytochemical and radioimmunoassay, Experientia 31:462, 1975.
33. Schaffalitzky de Muckadell, OB, Fahrenkrug, J, Watt-Boolsen, S, et al: Pancreatic response and plasma secretin concentration during infusion of low dose secretin in man, Scand J Gastroenterol 13:305, 1978.
34. Hacki, WH, Greenberg, GR, and Bloom, SR: In Bloom, SR, editor: Gut hormones, Edinburgh, 1978, Churchill Livingstone.
35. Straus, E, and Yallow, RS: Hypersecretinemia associated with marked basal hyperchlorhydria in man and dog, Gastroenterology 72:992, 1977.
36. Deveney, CW, Deveney, KS, Jaffe, BM, et al: Use of calcium and secretin in the diagnosis of gastrinoma, Ann Intern Med 87:680, 1977.
37. Bryant, MG, Bloom, SR, Polak, JM, et al: Possible dual role for vasoactive intestinal peptide as gastrointestinal hormone and neurotransmitter substance, Lancet 1:991, 1976.
38. Said, SI: Vasoactive intestinal peptide (VIP) current status. In Thompson, JC, editor: Gastrointestinal hormones, Austin, 1975, University of Texas Press.
39. Fahrenkrug, J: Vasoactive intestinal polypeptide: measurement distribution and putative neurotransmitter function, Digestion 19:149, 1979.
40. Fahrenkrug, J, Schaffalitzky de Muckadell, OB, and Holst, JJ: Plasma secretin concentration in anaesthetized pigs after intraduodenal glucose, fat, amino acids or meals with various pH, Scand J Gastroenterol 12:273, 1977.
41. Verner, JV, Jr, and Morrison, AB: Islet cell tumor and a syndrome of refractory watery diarrhea and hypokalemia, Am J Med 25:374, 1958.
42. Bloom, SR, Polak, JM, and Pearse, AGE: Vasoactive intestinal peptide and watery diarrhea syndrome, Lancet 2:14, 1973.
43. Ebeid, AM, Murray, PD, and Fischer, JE: Vasoactive intestinal peptide and watery diarrhea syndrome, Ann Surg 187:411, 1978.
44. Devine, BL, Carmichael, A, Russell, RI, et al: Cyclical release of vasoactive intestinal polypeptide (VIP) from a pancreatic islet cell apudoma, Postgrad Med 54:566, 1978.
45. Lin, TM, Evans, DC, Chance, RE, et al: Bovine pancreatic polypeptide: action on gastric and pancreatic secretion in dogs, Am J Physiol 232:E311, 1977.
46. Parks, DL, Gingerich, RL, Jaffe, BM, et al: Role of pancreatic polypeptide in canine and gastric acid secretion, Am J Physiol 236:E488, 1979.
47. Floyd, JC: Human pancreatic polypeptide, Clin Endocrinol Metab 8:379, 1979.
48. Modlin, IM, Lamers, CB, and Jaffe, BM: Evidence for cholinergic dependence of pancreatic polypeptide (PP) release by bombesin—a possible application, Surgery 8:75, 1980.
49. Polak, JM, Bloom, SR, Adrian, IE, et al: Pancreatic polypeptide in insulinomas, gastrinomas, VIPomas, and glucagonomas, Lancet 1:328, 1976.
50. Schwartz, TW: Pancreatic polypeptide and endocrine tumors of the pancreas, Scand J Gastroenterol 14(suppl 53):93, 1979.
51. Stern, I, Robet-Thomson, IC, and Hansky, J: Correlation between pancreatic polypeptide response to secetin v. ERCP., chronic pancreatitis, Gut 23:235, 1982.
52. Brown, JC, Pederson, RA, Jorpes, E, et al: Preparation of a highly active enterogastrone, Can J Physiol Pharmacol 47:113, 1969.
53. Anderson, D, Elahi, D, Brown, JC, et al: Oral glucose augmentation of insulin secretion, J Clin Invest 62:152, 1978.
54. Brown, JC: Physiology and pathophysiology of GIP, Adv Exp Med Biol 106:169, 1978.
55. Brown, JC, and Otto, SC: GIP and the entero-insular axis, Clin Endocrinol Metab 8:365, 1979.
56. Polak, JM, Heitz, P, and Pearse, AGE: Differential localization of substance P and motilin, Scand J Gastroenterol 11(suppl 39):39, 1976.
57. Bloom, SR, Mitznegg, P, and Bryant, MG: Measurement of human plasma motilin, Scand J Gastroenterol 11(suppl 39):47, 1976.
58. Grimelius, L, Polar, JM, Solcia, E, et al: Gut hormones, Edinburgh, 1978, Churchill Livingstone.
59. Moody, AJ, Sundby, F, Jacobsen, H, et al: The tissue distribution and plasma levels of glicentin (gut GLI-1), Scand J Gastroenterol 13(suppl 49):127, 1978.
60. Moody, AJ, Frandsen, EK, Jacobsen, H, et al: Heterogeneity of gut glucagon-like immunoreactivity (GLI). In Foa, PP, Bajaj, JS, and Foa, NL, editors: Glucagon: its role in physiology and clinical medicine, Amsterdam, 1978, Excerpta Medica.
61. Mallinson, CN, Bloom, SR, Warin, AP, et al: A glucagonoma syndrome, Lancet 2:1, 1974.
62. Hardy, JD, and Doolittle, RD: Zollinger-Ellison syndrome, Ann Surg 185:661, 1977.
63. Polak, JM, Pearse, AGE, Grimelius, L, et al: Growth hormone release-inhibiting hormone in gastrointestinal and pancreatic D cells, Lancet 1:1220, 1975.
64. Creutzfeldt, W, and Arnold, R: Somatostatin and the stomach: exocrine and endocrine aspects: First International Somatostatin Symposium, Freiburg, Germany, 1977, Metabolism 27(9 suppl 1):1309, 1978.
65. Raptis, S, and Gerich, JE: Metabolism 27(suppl 1): 1129, 1978.
66. Vale, W, Rivier, C, and Brown, M: Regulatory peptides of the hypothalamus, Annu Rev Physiol 38:473, 1977.
67. Krejs, GJ, Orci, L, Conlon, JM, et al: Somatostatinoma syndrome: biochemical morphology and clinical features, N Engl J Med 301(6):285, 1979.
68. Bloom, SR: New specific long-acting somatostatin analogues in the treatment of pancreatic endocrine tumors, Gut 19:446, 1978.
69. Carraway, R, and Leeman, S: The isolation of a new hypotensive peptide, neurotensin, from bovine hypothalami, J Biol Chem 24(19):6854, 1973.
70. von Euler, US, and Pernow, B, editors: Substance P: proceedings of Nobel Symposium, New York, 1978, Raven Press.
71. Skrabanek, P, and Powell, D: Substance P, vol 1, Edinburgh, 1977, Eden Press.
72. Ambinder, RF, and Schuster, MM: Endorphins: new gut peptides with a familiar face, Gastroenterology 77(5):1132, 1979.
73. Konturek, SJ, Tasler, J, Cieszkowski, M, et al: Comparison of methionine-enkephalin and morphine in the stimulation of gastric acid secretion in the dog, Gastroenterology 78:294, 1980.
74. Gregory, H: The identification of urogastrone in serum, saliva, and gastric juice, Gastroenterology 77(2):313, 1979.
75. Joffe, SM, Elias, E, Rehfeld, JF, et al: Clinically silent gross hypergastrinaemia from a multiple hormone-secreting pancreatic apudoma, Br J Surg 65:277, 1978.

CHAPTER 43 The gonads

ELIZABETH J. KICKLIGHTER
ROBERT J. NORMAN

OBJECTIVES

- Describe the roles of LHRH, LH, and FSH in the production of steroid hormones
- Discuss the menstrual cycle, including the effects of specific hormones and changes in hormonal levels during menstruation.
- Outline the pathways of biosynthesis, transport, and catabolism of estrogens, androgens, progesterone, and testosterone.
- Describe pathological conditions of the gonads and the expected levels of the following analytes in these conditions: LH, FSH, androgens, estrogens, and progesterone.

KEY TERMS

amenorrhea Absence or abnormal cessation of menstruation.

androgens Sex steroid hormones, responsible for the development of the male secondary sex characteristics. The major androgens are testosterone, dihydrotestosterone, and androstenedione.

anorchia Congenital absence of the testis.

endometrium Inner layer of the uterus, the thickness and structure of which vary with the phase of the menstrual cycle.

epididymis Elongated cordlike structure along the posterior portion of the testis that contains ducts capable of storing spermatozoa.

estrogens Sex steroid hormones responsible for the development of the female secondary sex characteristics. The major estrogens include estradiol, estrone, and estriol.

fallopian tube Long, slender, hollow tube that extends from the upper lateral portion of the uterus to the ovary on the same side.

Frommel-Chiari syndrome Postpartum condition possibly caused by hypothalamic dysfunction; characterized by atrophy of the uterus, persistent lactation, galactorrhea, prolonged amenorrhea, and low levels of estrogens and gonadotropins.

gametogenesis The development of male and female sex cells, or gametes.

germinal aplasia Absence of the primordial germ cells within the gonads.

gonadotropins Hormonal substances having affinity for or a stimulating effect on the gonads.

graafian follicle Ovarian structure consisting of the ovum and its encasing cells.

granulosa cells Modified epithelial cells that surround the ovum and secrete the follicular fluid. After ovulation, they are transformed into glandular cells of the corpus luteum.

gynecomastia Benign glandular enlargement of the mammary glands in males.

hirsutism More hair than is cosmetically acceptable for women in a given culture.

Kallmann's syndrome Condition characterized by the absence of the sense of smell because of agenesis of the olfactory bulbs and by secondary hypogonadism because of the lack of LHRH.

Klinefelter's syndrome Condition characterized by small testes with fibrosis of seminiferous tubules, by impairment of function and clumping of Leydig cells, and by increased gonadotropins. It is believed to be caused by the presence of an extra sex (X) chromosome.

Laurence-Moon-Bardel-Biedl syndrome Condition characterized by obesity, hypogenitalism, and mental deficiency.

Leydig cells Interstitial cells of the testes that produce the male sex steroids.

menopause (climacteric) Syndrome of endocrine and somatic changes occurring at the termination of the reproductive period in the adult woman.

menstruation Cyclical, physiological uterine bleeding that normally occurs, usually at approximately 28-day intervals, during the reproductive period of the female in the absence of pregnancy.

myometrium Smooth muscle portion of the uterus that forms the main portion of the uterus.

ovum Female reproductive cell produced in the ovary.

polycystic ovary syndrome (Stein-Leventhal syndrome) Condition associated with bilateral polycystic ovaries and characterized by secondary amenorrhea and anovulation.

pseudohermaphroditism Condition in which the gonads are of one sex, but contradictions exist in the morphological expression of the sex, for example, a genetic and gonadal female with partial masculinization.

Sertoli cells Elongated cells in the tubules of the testes that provide support, protection, and apparent nutrition of spermatids until they become transformed into mature spermatozoa.

Sheehan's syndrome Postpartum pituitary necrosis.

spermatogonia Undifferentiated germ cell of a male originating in the seminal tubule.

Turner's syndrome Condition characterized by a female phenotype, short stature, undifferentiated (streak) gonads, and variable abnormalities including webbing of the neck and cardiac defects. It is associated with an absence of the second (X) chromosome.

virilism Development of male secondary sexual characteristics in a female, including enlargement of the clitoris, growth of facial and body hair, stimulation of sebaceous glands (often with acne), and deepening of the voice.

All reproductive processes are regulated closely by a complicated series of interrelated systems involving the hypothalamus, the pituitary gland, and the male and female gonads. Contrary to popular belief, both men and women secrete the same sex steroid hormones, that is, androgens, estrogens, and progesterone. The difference in steroid hormone secretion between the sexes is not of an absolute or all-or-none type, but rather a matter of degree. Testosterone is the major sex steroid in men, whereas estradiol is the primary sex hormone in women.

GENDER DIFFERENTIATION AND ANATOMY OF REPRODUCTIVE SYSTEM

Gender differentiation

Chromosomal gender is determined by the type of sex chromosomes present in the cells, two X chromosomes, resulting in a genetic female, and one X and one Y, resulting in a genetic male. The presence of the Y chromosome is a prerequisite for the development of the male gonad. Male *gonadal gender* is determined by the production of a specific protein coded for by the Y chromosome. In the presence of this factor testes develop, while in its absence ovaries develop. The development of testes in turn gives rise to testosterone secretion during fetal life and the development of male internal and external genitalia. Lack of fetal testosterone secretion will always result in the establishment of the female gonad. If testosterone production fails, the embryonic gonads and internal and external genitalia will develop as female, regardless of the chromosomal gender. In both sexes the gonads have a dual function—the production of germ cells (gametogenesis) and the secretion of sex hormones.

Anatomy of female reproductive system

The female organs of reproduction are classified as external or internal structures. The external organs and the vagina serve for copulation, whereas the internal organs located within the pelvis provide for development and birth of the fetus. The external organs are the vulva, mons pubis, the labia (majora and minora), the clitoris, and the vestibule of the vagina. The mammary glands are accessories to the reproductive system.

The internal organs of the female reproductive tract consist of two ovaries, two fallopian tubes, the uterus, and the vagina. Unlike the male reproductive system, the internal female reproductive organs show regular cyclical changes that may be regarded as periodic preparations for fertilization and pregnancy.

The ovaries are located on either side of the uterus and serve the dual purpose of production of the ova and secretion of the female sex hormones—the estrogens and progesterone (Fig. 43-1). Every female at birth has about 400,000 microscopically small, immature follicles in the ovaries, each containing an immature ovum. During the reproductive life, however, only 300 to 400 follicles reach maturity. The mature follicle is composed of three layers of cells—the *theca externa,* the *theca interna,* and the *granulosa cells.* The cells of the theca interna are the primary source of estrogens.

The follicle that ruptures at the time of release of an ovum (ovulation) promptly fills with blood, forming what is sometimes called a *corpus hemorrhagicum.* The granulosa and theca cells of the follicle's lining promptly begin to proliferate, and the clotted blood is rapidly replaced with yellowish, lipid-rich luteal cells, forming the corpus luteum. The luteal cells secrete estrogens and progesterone. If pregnancy occurs, the corpus luteum persists and there are usually no more menstrual periods until after delivery. If there is no pregnancy, the corpus luteum begins to degenerate about 4 days before the next menses and is eventually replaced by scar tissue to form a corpus albicans.

The fallopian tubes, also known as *oviducts* or *uterine tubes,* lead from the vicinity of the ovaries to the uterus. They allow transport of the sperm from the uterine cavity

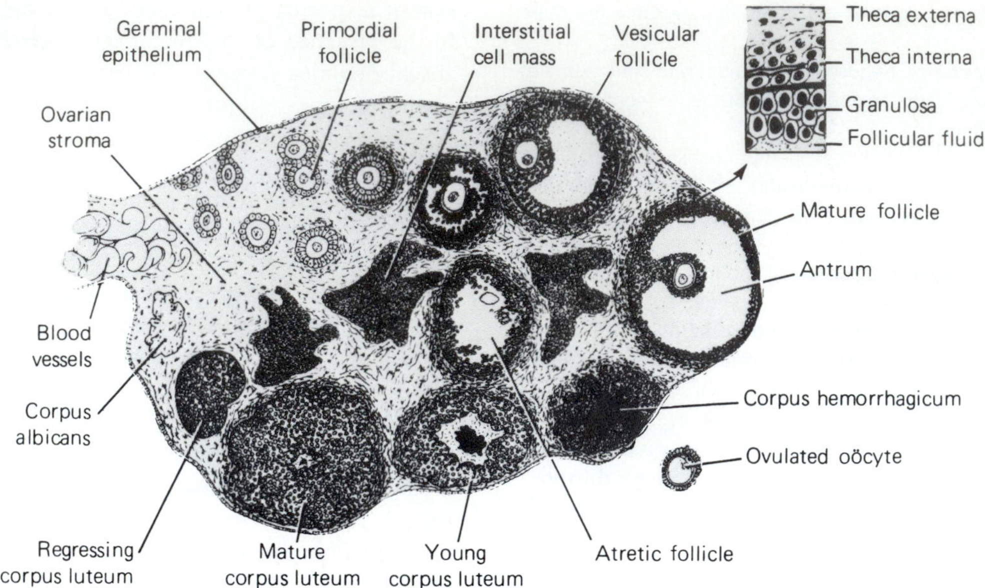

Fig. 43-1 Diagram of ovary, showing sequential development of a follicle and formation of corpus luteum. Section of wall of a mature follicle is enlarged at upper right. *(From Gorbman A, and Bern HA: A textbook of comparative endocrinology, New York, 1962, John Wiley & Sons.)*

and the fertilized egg to the uterus. Fertilization takes place in the fallopian tubes.

The uterus is a thick-walled muscular chamber connecting the fallopian tubes and the vagina in which the fetus develops until the time of delivery. The muscular layer of the uterus is called the *myometrium*. The innermost layer is the *endometrium*. The endometrium undergoes cyclical changes, and during menstruation the endometrium is shed.

Anatomy of male reproductive system

The male reproductive system consists of two testes producing sperm and androgens, a penis, and a system of exocrine glands whose secretions constitute the bulk of seminal fluid in which sperm are conveyed. These glands are the two bulbourethral glands (Cowper's glands); two seminal vesicles; the prostate gland; and a system of bilateral ducts through which sperm pass outside of the body, namely, the epididymis, the vas deferens, and the ejaculatory duct.

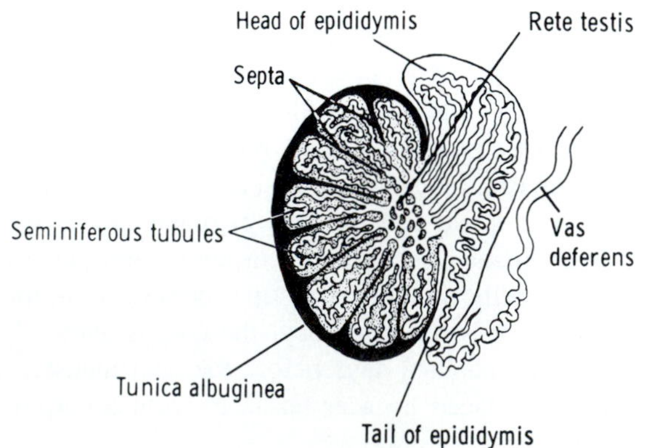

Fig. 43-2 Anatomy of testis. *(From Ganong WF: Review of medical physiology, ed 11, Los Altos, Calif, 1983, Lange Medical Publications.)*

The normal mature testis (Fig. 43-2) contains approximately 250 pyramidal lobules of seminiferous tubules, which are separated by the fibrous septa. The tubules compose over 85% of the volume of the testis. Surrounding the central lumen of the seminiferous tubule is a structured epithelium containing Sertoli cells and spermatogenic cells. The interstitial tissue between the tubules contains Leydig cells, which are responsible for the production of testicular androgens.

PHYSIOLOGY OF SEX STEROID HORMONES

General physiology

The interrelationships between the hypothalamus, pituitary, and gonads are extremely complex. The hypothalamic decapeptide, luteinizing hormone-releasing hormone (LHRH) (also known as gonadotropin–releasing hormone [GnRH]), enhances the release of both pituitary gonadotropins, luteinizing hormone (LH) and follicle-stimulating hormone (FSH), affecting LH more than FSH.

The pituitary gonadotropins, FSH and LH, play an important role by influencing different functions of both the male and female internal sex organs. In women, FSH induces the growth of the follicles of the ovary and causes them to secrete estrogens, whereas LH acts in the final stages of follicular growth to bring about ovulation (or release of the ovum from the follicle) and acts as a luteal cell stimulator, inducing progesterone production. On the other hand, FSH in men brings about spermatogenesis, whereas LH stimulates testosterone production. In other words, the same hormones can act on different target tissues to produce different biological effects. This is also true for the effects the primary sex hormones, testosterone and estradiol, have on their target tissues.

Levels of both gonadotropins (LH and FSH) vary with age and sex and, in the female, with the phase of the menstrual cycle. In addition, both hormones are secreted in rhythmic pulses several times each hour.

Feedback of the various sex steroids at the pituitary and hypothalamic levels is complicated, because several hormones may exert additive effects. The effects of various sex steroids may be positive under one set of circumstances and negative with another. For example, estrogen in low levels has a negative feedback effect on the pituitary, whereas estrogen in high levels has a positive effect. These effects vary at different times during the menstrual cycle (for example, pituitary responsiveness to LHRH depends on the phase of the menstrual cycle).

Estradiol is the primary secretory product of the ovary. Estrone, another important estrogen, is the product of both ovarian secretion and the peripheral (nongonadal) conversion of prohormones of adrenal or ovarian origin. Progesterone is produced by the corpus luteum and is present in significant amounts only after ovulation.

Testosterone is the primary secretory hormone of the testis. In females, testosterone is derived from peripheral conversion of prohormones, mainly androstenedione.

Sex steroid hormones in both sexes during adolescence stimulate characteristic spurts of growth and union of epiphyses with bone shafts, which terminates future growth. In addition, they promote protein synthesis. Although androgens affect the skeletal muscle, giving a muscular appearance in men, estrogens more strongly affect smooth muscle. The distribution of subcutaneous fat and smooth character of the skin in women are estrogen dependent. Androgens induce secondary sex characteristics typical of the male, whereas estrogens stimulate typical female secondary sex characteristics and the growth and development of the female reproductive organs. Estrogens, together with progesterone and other pituitary factors, stimulate the growth and development of the mammary glands. Progesterone may be regarded as the hormone of the mother-to-be, and both estrogen and progesterone bring about characteristic developments during the menstrual cycle.

Menstrual cycle

The physiology of reproduction in the female is dominated by the events in the menstrual cycle. Menstruation is a regular cyclic shedding of the surface layer of the endometrium along with blood; these pass out of the body in the menstrual flow. These periodic uterine events are the result of a cyclic supply and withdrawal of hormonal support and are well coordinated with the morphologic and hormonal changes taking place in the ovary. The hormonal control of the menstrual cycle involves a complex interplay between the hypothalamus; the anterior pituitary hormones FSH and LH; and the ovarian hormones, estradiol and progesterone (Fig. 43-3). The average menstrual cycle, the interval between successive menstruations, is 28 days, starting from the first days of menstruation. The average menstruation period, the period of active uterine bleeding, is 4 days. The length of the menstrual cycle can vary among individual women from 21 to 42 days, and the menstrual period can last from 3 to 7 days.

Ovarian events during menstrual cycle. The ovarian events during the normal menstrual cycle include both morphological and hormonal changes. Under the influence of FSH, several follicles start to develop at the beginning of the cycle, and only one will mature in about 10 to 12 days. The one developing follicle that matures and ruptures, releasing the ovum, is called the *graafian follicle,* and the process of rupture to release the ovum is called *ovulation*. Ovulation presumably takes place on day 14 of the 28-day cycle. This cycle continues until the ovarian function ceases 30 to 35 years after puberty (menopause). The ruptured follicle undergoes morphological changes to become a wrinkled, yellow structure called the *corpus luteum,* which becomes a new endocrine organ. At about day 25 the corpus luteum begins to shrink and degenerate if conception has not occurred. This degeneration is soon followed (within 2 to 3 days) by menstruation, which heralds the beginning of the next menstrual cycle. When the follicles are the dominant ovarian structures, the ovary is said to be in the follicular phase; this occurs during the first half of the cycle. When the corpus luteum is the dominant ovarian structure, the ovary is said to be in the luteal phase of the cycle; this occurs during the second half of the cycle.

Changes in ovarian morphology such as the graafian follicle development and corpus luteum formation correlate well with the changes in ovarian steroid hormone production. The most important ovarian hormones (those which effect the uterine changes during the menstrual cycle) are estradiol and progesterone. Estradiol alone dominates the events during the follicular phase, and progesterone, combined with estradiol, dominates the events during the luteal phase. Thus the ovary plays a key endocrine role. The morphological and hormonal function of the ovary depends solely on the appropriately timed release of both FSH and LH by the pituitary in response to hypothalamic gonadotropin-releasing hormone of the hypothalamus.

Hormonal changes during menstrual cycle. Early in the cycle, when the levels of estrogen and progesterone are relatively constant and low, the FSH levels are rising and

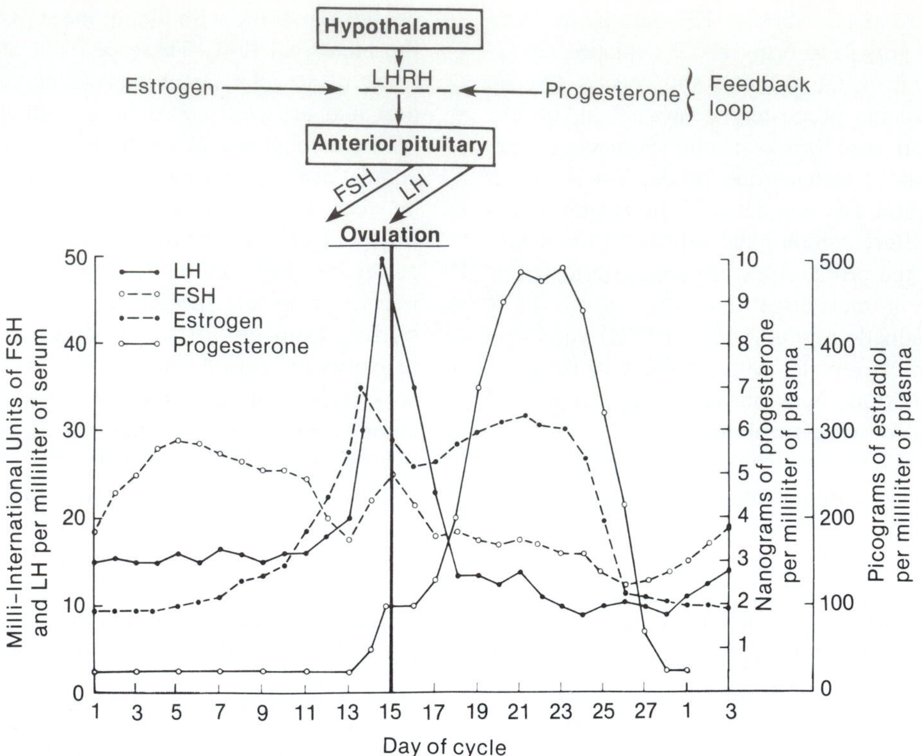

Fig. 43-3 Changes in blood levels of various hormones throughout menstrual cycle. Also shown is hypothalamic-pituitary axis and feedback exerted by estrogens and progesterone. (Pituitary actually produces only one hormone, gonadotropin-releasing hormone, also known as LHRH, because it stimulates secretion of LH more than FSH. A separate FSHRH has never been identified.) *(Modified from Taymor ML, Berger MJ, Thompson IE, and Karamo KS: Am J Obstet Gynecol 114:445, 1972.)*

high and the LH levels are low (Fig. 43-3). These high levels of FSH stimulate follicular growth and the release of estrogens, particularly estradiol. By days 7 to 8, the estradiol level is rising at a rapid rate, reaching its first peak before ovulation. The rising levels of estradiol result in a negative feedback to the hypothalamus and pituitary glands, causing a fall in FSH levels because of the inhibitory action of estradiol on FSH release. Concurrently, the rise in estradiol triggers a rapid rise in the LH (positive-feedback effect). Estradiol reaches a maximum on the day before the LH peak. During the midcycle there is a peak rise of LH, which leads to maturation and rupture of the graafian follicle, releasing the ovum (ovulation) 16 to 24 hours after the LH peak. Before the LH surge and before ovulation, the estradiol level drops considerably and then rises again after ovulation. The ruptured follicle becomes the corpus luteum. Progesterone is released by the corpus luteum and progesterone levels begin to rise, causing an inhibition of the secretion of LH. A sharp increase in progesterone follows, reaching a maximum in about 8 or 9 days after the LH peak (days 23 to 25 of the cycle). As estradiol and progesterone increase, FSH and LH decline throughout the luteal phase. As the corpus luteum regresses, the levels of both estradiol and progesterone begin to diminish. The removal of the inhibitory effect of these two compounds results in the increase of FSH, which stimulates the growth of a new crop of follicles in the ovary. During the menstruation phase, estradiol, progesterone, and LH are at a relatively constant but low level, whereas FSH is the only hormone present at elevated and rising levels.

Effect of ovarian hormones on uterus (endometrium). As a physiological result of the ovarian hormones, estradiol and progesterone, definite changes occur in the endometrium in preparation to receive and implant a fertilized ovum. The rising levels of estradiol stimulate the reconstruction of the endometrium, blood vessels, and secretory glands of the uterus. This change in the endometrial growth is called the ''proliferative phase'' and corresponds with the follicular phase of the ovary. The newly formed glands begin to release glandular substances. Hence this phase is called the *secretory phase* of the endometrium, and it corresponds to the luteal phase of the ovary. During days 1 to 5 of the cycle, enough estrogen is secreted by the follicles to start the deep layer of the endometrium growing, but it is not enough to support the thickened se-

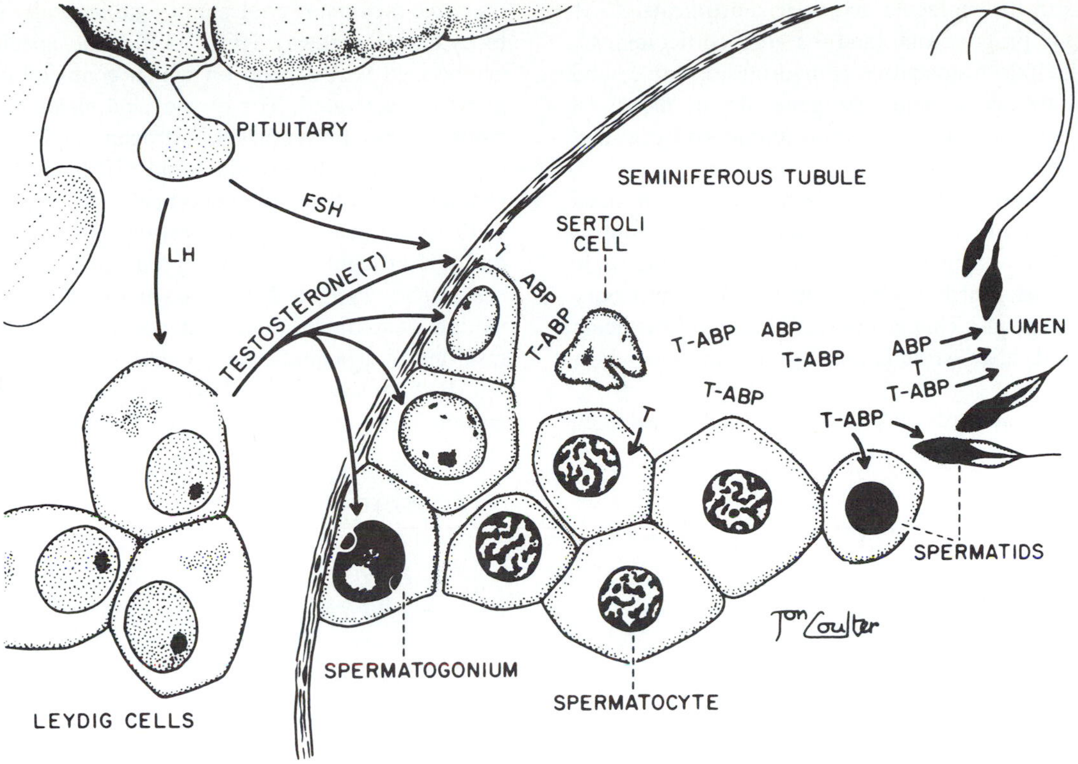

Fig. 43-4 Relationship of gonadotropins and testosterone to tubular function. *(From Felig P, et al: Endocrinology and metabolism, New York, 1981, McGraw-Hill Book Co.)*

cretory endometrium of the previous cycle. As a result, this well-developed surface of the endometrium sloughs off as the estradiol levels decrease. This is menstruation. Thus menstruation occurs during the later stages of the proliferative phase of the endometrium because of withdrawal of hormonal support. If a fertilized egg is successfully implanted in the enriched endometrium, the implantation causes additional estrogens to be produced; thus menstruation is prevented and the development of a pregnancy is begun (see Chapter 37).

Hormonal regulation in the male

The testes, like the ovaries, produce both steroid hormones and gametes. These two activities, androgen synthesis and sperm production (spermatogenesis), are carried out by two anatomically different structures. Androgen synthesis occurs in the Leydig cells, and spermatogenesis takes place in the seminiferous tubules (Fig. 43-4). The anterior pituitary hormones control both of these functions.

FSH stimulates spermatogenesis, and LH (called *interstitial cell–stimulating hormone* [ICSH] in the male) stimulates testosterone production. LH binding in the testis occurs on Leydig cell membranes located in the interstitial tissue of the testis, stimulating cell synthesis of testosterone.

FSH receptors in the testis are located in the Sertoli cells of the seminiferous tubules. The primary effect of FSH is the stimulation of protein synthesis, especially androgen-binding protein (ABP). Testosterone acts first on spermatogonia and then stimulates primary spermatocytes to complete meiotic division and to form secondary spermatocytes and young spermatids. FSH then induces maturation of late spermatids to spermatozoa.

Increased levels of serum testosterone feed back on the hypothalamic-pituitary axis to modulate the amounts of LH being released. These events proceed normally at a rather fixed rate during the adult life.

BIOSYNTHESIS OF STEROID HORMONES

The two trophic hormones produced by the anterior pituitary that have a direct effect on gonadal function are FSH and LH. FSH and LH are glycoproteins composed of two polypeptide subunits: the alpha-subunit is common to FSH, LH, human chorionic gonadotropin, and thyroid-stimulating hormones; the beta-subunit confers biological specificity. The pituitary hormones are transported by the systemic circulation to the target organs.

Under control of the anterior pituitary hormones, the ovaries, the testes, and the adrenals secrete a group of steroid hormones. The steroids are characterized by the presence of a basic ring structure—the cyclopentanoperhydrophenanthrene nucleus (see Fig. 30-6). The steroid hormones are classified as estrogens, androgens, progestogens, and corticosteroids, depending on the number of carbon atoms in the molecule and their substituents. C-21 steroids include progesterone (and the glucocorticoids), C-19 steroids include testosterone, dihydrotestosterone, and the adrenal androgens, and estrogens are in the C-18 group. The basic structures of testosterone and estradiol are depicted in Fig. 43-5.

Biologically active steroid hormones are synthesized through a series of enzymatic reactions starting with a common precursor, cholesterol, which serves as the basic skeleton for all steroid hormones. The enzymes necessary to bring about specific changes in the cholesterol molecule are synthesized after stimulation by the pituitary hormones.

Thus all classical steroid hormones are synthesized from the same precursor, cholesterol, and basically use the same biosynthetic pathway (Fig. 43-5). The specific hormone synthesized will depend on which genes of the endocrine gland are activated. The ovaries and testes have the enzymatic pattern to synthesize estrogens and androgens, but they lack the 21-hydroxylase and 11β-hydroxylase system, characteristic of the adrenocortical cell. The conversion of C-19, androgens, to C-18, estrogens (estrone and estradiol), is brought about by a group of enzymes called the *aromatizing system*. These enzymes catalyze C-19 hydroxylation, loss of the carbon at position 19, and reduction, resulting in a phenolic or aromatic ring; hence the process is called *aromatization*. This process is important for the production of ovarian hormones. The testes differ from the

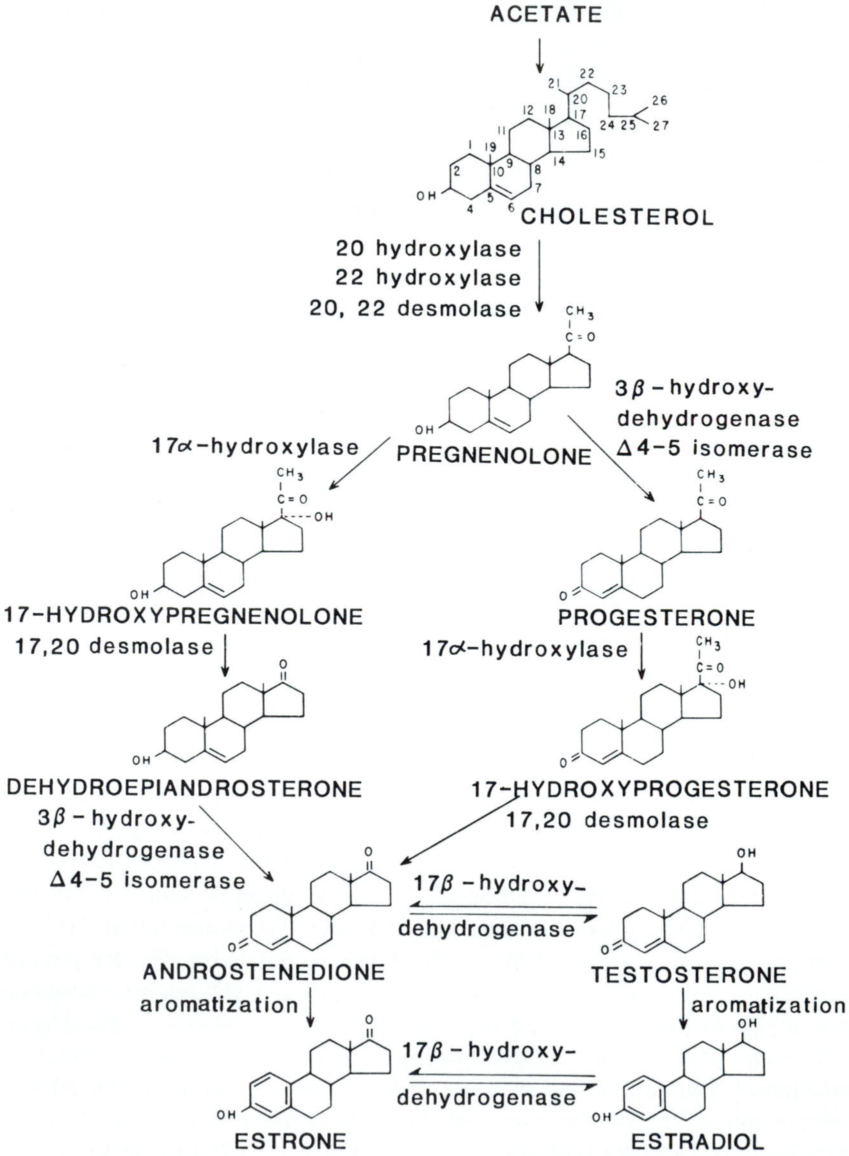

Fig. 43-5 Pathways of sex steroidogenesis. *Capitals,* Metabolic products; *lower-case lettering,* enzymes in metabolic pathway. *(From Felig P, et al: Endocrinology and metabolism, New York, 1981, McGraw-Hill Book Co.)*

ovaries enzymatically by the functional absence of the aromatizing enzyme system. Under normal circumstances, the adrenals do not appear to contain desmolase and the aromatizing system and hence do not synthesize biologically active androgens and estrogens. Some adrenal tumors do produce testosterone or estrogens, or both, indicating the activation and presence of the desmolase and aromatizing systems. Thus all the steroid-producing glands have the ability to synthesize other hormones in addition to their own; however, the synthesis of these enzymes is repressed during normal cell activity.

Ovarian steroid hormones

The ovary produces estrogens, androgens, and progesterone, following the pathways of steroid biosynthesis outlined by Fig. 43-5. Estradiol is the chief secretory product of the maturing follicles, and progesterone is the chief secretory product of the corpus luteum.

The predominance of each of these steroids at certain points during the menstrual cycle has been explained by the "two-cell" hypothesis. The two-cell hypothesis presents a logical explanation of ovarian steroidogenesis (Fig. 43-6). It proposes that LH acts on the theca cells of the ovary to produce androgens. These androgens are transferred from theca cells into granulosa cells where, before ovulation, under the influence of FSH, the androgens are converted to estrogens by the enzyme aromatase. After ovulation, the granulosa cells secrete progesterone and estrogen directly into the bloodstream under the influence of LH.

Estrone, the other major biologically important estrogen, is derived from peripheral conversion of prohormones, particularly androstenedione, from interconversion of estradiol or from secretion by the ovary. In the postmenopausal woman, estrone is the dominant plasma estrogen.

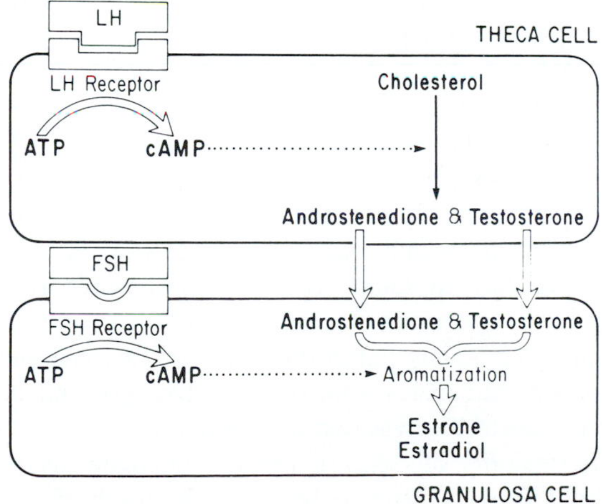

Fig. 43-6 "Two-cell hypothesis" of ovarian estrogen production. Luteinizing hormone stimulates androgen production in thecal cells. Androgens then diffuse into granulosa layer, where they are aromatized into estrogens, in an action specifically stimulated by FSH. *(From Felig P, et al: Endocrinology and metabolism, New York, 1981, McGraw-Hill Book Co.)*

Very little estriol is produced by nonpregnant females. Among the androgens secreted by the ovaries, androstenedione is the chief secretory androgen.

Testicular steroid hormones

The testis in adult men produces large amounts of testosterone in the Leydig cells (about 5 to 12 mg/day) plus a very small amount of other steroids, including estradiol, androstenedione, and dehydroepiandrosterone (Fig. 43-5). Secretion of testosterone by the testes is episodic; a circadian pattern can be demonstrated, with a maximum in the early morning (about 7:00 AM) and a minimum about 13 hours later.

Peripheral conversion of adrenal androstenedione accounts for approximately 5% of testosterone. The major adrenal androgen dehydroepiandrosterone has weak biological potency.

TRANSPORT AND METABOLISM

Transport

Steroid hormones are water-insoluble compounds that are transported in the bloodstream to their site of action by binding to proteins in the plasma. Circulating testosterone and estradiol are largely bound to sex hormone-binding globulin (SHBG), though there is some binding to albumin. Progesterone is preferentially bound to corticosteroid-binding globulin (CBG). The binding proteins are synthesized in the liver.

The binding of a hormone to its carrier protein is a reversible process resulting in an equilibrium between bound and free fractions. It is important to understand that, as for all nonprotein hormones, the *free* fraction is the only biologically active hormone. The percentage of total serum estradiol that is found in the free fraction is 3%, whereas 2% of serum testosterone is not bound to serum protein. The significance of this equilibrium is that the plasma level of active, free testosterone becomes a variable fraction of total amount of testosterone, depending on the level of SHBG. The free testosterone is therefore a better index of androgenicity than the total serum testosterone.

Testosterone and dihydrotestosterone bind more strongly to SHBG than estradiol does. Albumin, on the other hand, binds estrogen more strongly than it binds testosterone. The concentration of SHBG is decreased by administration of testosterone or growth hormone and is increased by estrogen or thyroid hormone.

When the binding capacity of the serum increases as levels of binding protein increase, removal of steroid from plasma by peripheral tissue and subsequent metabolism are reduced. For example, in a patient with hyperthyroidism, serum SHBG levels are increased. This increased binding capacity reduces the level of free testosterone, which in turn stimulates the production of additional testosterone. As a result, the total plasma level of testosterone increases; however, the newly formed testosterone also binds in large part to SHBG. At a certain point the new levels of SHBG and testosterone reach concentrations where the free testosterone level is in the desired range.

Steroid catabolism

The catabolic reactions are mostly reductive and occur mainly, though not exclusively, in the liver. The steroid hormones are converted by the liver to inactive compounds by a group of hepatic enzymes such as hydrogenases, dehydrogenases, and hydroxylases. The metabolic products of these reactions are a series of more polar steroids; however, because the products are still not very hydrophilic, they remain bound to serum proteins and hence cannot be readily filtered by the kidneys. To aid their renal excretion, these metabolites are converted to still more polar metabolites—largely glucuronide and sulfate conjugates. The conjugation is carried out in the liver by specific enzymes, sulfokinase catalyzing the conversion of hydroxysteroids to corresponding sulfates and glucuronyl transferase catalyzing the conversion of hydroxysteroids to corresponding glucuronides. Most steroids are excreted largely as glucuronides and filtered readily by the kidney, except for dehydroepiandrosterone (DHEA), which is largely excreted as a sulfate. The sulfate conjugates are cleared much more slowly by the kidneys.

It should be pointed out that about two thirds of the urinary 17-ketosteroids are of adrenal origin, and one third is of testicular origin. Testosterone itself is also excreted directly into the urine conjugated with glucuronic acid, where its level correlates with the production level of testosterone.

Estrogen sulfates are known to circulate in the blood as biologically active compounds. Although urine represents the major (50% to 70%) route of elimination of estrogen metabolites, considerable amounts of their metabolites do appear in the feces.

Route of excretion of steroid metabolites. A great proportion of steroid metabolites can enter the intestinal tract but are reabsorbed through the portal circulation and are recirculated through the liver. After conjugation by the liver, the metabolites finally appear in the urine. Some metabolites, especially the less polar steroids, are excreted in the bile. The phenolic steroids, such as estrogen, are secreted to a considerable extent into the bile and appear in the feces. The metabolites of androgens and corticosteroids, on the other hand, largely appear in the urine.

PATHOLOGICAL CONDITIONS

Abnormalities of testicular function

Abnormalities in testicular function can be divided into two categories: those resulting in decreased androgen production, or hypogonadism, and those resulting in excessive androgen production, or hypergonadism. Either disorder may be attributable to a primary dysfunction of the testes, or may be secondary to a derangement of the pituitary-hypothalamic axis.

Hypogonadism. The clinical picture of hypogonadism is directly related to the time of development of androgen deficiency. In prepubertal hypogonadism, absence of androgen production by the testis is associated with persistent infantile genitalia, a barely palpable prostate, female distribution of pubic and axillary hair, a partial to total lack of facial and body hair, and the absence of a deepening voice. Although the growth rate is decreased, growth continues with the absence of epiphyseal closure, eventually resulting in tall stature with long arms and legs. Prepubertal hypogonadism is usually inapparent until the adolescent period, when normal adolescent development, including genital and secondary sexual changes, does not occur.

Postpubertal hypogonadism results in minimal changes. In young males, there is usually diminished beard growth and thinning axillary and other body hair. The prostate atrophies, and sexual desire and performance decrease. The genitalia may decrease somewhat in size. In older men, none of these changes may be noted. Beard and body hair growth usually persist, and there may be no noticeable change in libido or sexual function.

Primary hypogonadism. Because of the lack of androgenic feedback on the pituitary-hypothalamic axis, primary hypogonadism is manifested by increased serum and urine gonadotropins and by decreased serum androgen levels and decreased urinary 17-ketosteroid levels. The testicular abnormality may be secondary to a developmental abnormality, such as a genetic or embryological defect, or it may occur at any time later in life. Developmental abnormalities account for most prepubertal cases of hypogonadism. Many syndromes are included in these disorders. The majority are very rare, including Reifenstein's syndrome (male pseudohermaphroditism), male Turner's syndrome, germinal aplasia, and anorchia (congenital absence of testes). Probably the most common syndrome resulting in a developmental abnormality is Klinefelter's syndrome. These patients possess an extra sex chromosome (X) resulting in a karyotype consisting of 44 autosomes plus two X chromosomes and one Y chromosome. Klinefelter's syndrome has an incidence of approximately 1 in 400 males. The spectrum of clinical manifestations is broad, including obviously feminized males on one end and normally virilized males with only microscopic and biochemically detected disease on the other end.

The majority of cases of primary hypogonadism manifesting themselves after puberty are the result of testicular infections, trauma, irradiation, a tumor that has replaced the testicular parenchyma, or surgical or accidental castration. Other rare congenital disorders may manifest themselves as a primary hypogonadal state after puberty. These include such conditions as myotonia dystrophica and cystic fibrosis.

Another condition that may be considered primary hypogonadism is that of Leydig cell failure occurring as a result of male climacteric. Population studies in males indicate total serum testosterone levels remain stable until about 70 years of age, after which they begin to decline. LH and FSH levels begin a slight but definite rise at this time.

Secondary hypogonadism. Secondary hypogonadism results from failure of the pituitary to produce LH (ICHS) and FSH. This is usually the result of primary hypopituitarism, though rarely this condition may be secondary to a defect within the hypothalamus resulting in a failure to release LHRH. This results in an absence of positive feedback on the pituitary gland and inadequate synthesis of FSH and LH. Kallmann's syndrome is a rare genetic condition in which a lack of LHRH leads to secondary hypogonadism.

In rare cases of primary hypopituitarism, isolated deficiencies of gonadotropic hormones have been described. In most cases, however, there is an associated loss of other pituitary hormones resulting in decreased thryoid, adrenal, and gonadal function at all ages. This panhypopituitarism may be idiopathic, part of a congenital disorder, or secondary to a neurohypophysial lesion such as a neoplasm, cyst, or granulomatous process. When there is a progressive loss of the pituitary function because of a neurohypophysial lesion, a decrease in gonadotropins is at times the first deficiency observed and the patient may present with isolated hypogonadism.

The absence of serum and urinary gonadotropins after the age of adolescence in patients with diminished gonadal function is diagnostic of secondary hypogonadism. Studies of growth, thyroid, adrenal, and antidiuretic hormones may reveal clinically unsuspected deficiencies in other pituitary hormones.

Hypergonadism. Hypergonadism may occur as a primary process because of excessive androgen production from a testicular tumor (Leydig cell or interstitial cell carcinoma). Hypergonadism may also occur secondary to altered pituitary-hypothalamic axis function with increased LH/FSH secretion. Primary hypergonadism is noted for high serum androgen levels, high urinary 17-ketosteroids, and low serum gonadotropins. Secondary hypergonadism is differentiated by elevated androgens and their urinary metabolites, and elevated gonadotropins.

The production of excessive quantities of androgenic hormones in adult males results in little if any morphological change. However, excessive production in children results in precocious puberty. This usually manifests itself by early onset of puberty with enlargement of genitalia, appearance of pubic or axillary hair, deepening of the voice, and the beginning of acne. Sexual precocity secondary to increased gonadotropins in males, although uncommon, it often familial in contrast to secondary precocious puberty occurring in females. When precocious puberty occurs in males without a family history, it is almost always associated with a space-occupying lesion in the region of the third ventricle of the brain. The continued production of androgens at a level that is excessive for the age of the patient results in accelerated skeletal growth followed by early closure of the epiphysis, resulting in an adult who is frequently shorter than his contemporaries.

Premature puberty in the male may also result from excessive adrenal androgens. The pattern of growth and development is similar to that produced by increased testicular steroids; however, the testes remain prepubertal in size.

Abnormalities of ovarian function

Endocrine disorders of the ovary may also be classified as "hypo-" or "hyper-"; primary ovarian in origin or secondary to disturbances of hypothalamic-pituitary secretion; and congenital or acquired. Clinical manifestations may be very subtle because the ovary is mainly an organ of reproduction. Its disturbances may be picked up only in a work-up for infertility (see below). Other ovarian disorders are manifest clinically by precocious or delayed puberty.

Ovarian hypofunction. As in the male, the symptoms of gonadal hypofunction depend on whether the condition manifests itself before or after puberty. Ovarian hypofunction that develops in the prepubertal period will manifest itself clinically as delayed or absent menarche or primary amenorrhea. Ovarian hypofunction developing after puberty may manifest itself as secondary amenorrhea.

Primary ovarian hypofunction. Because of the lack of estrogenic feedback on the hypothalamic-pituitary axis, primary ovarian hypofunction is characterized by increased levels of gonadotropins in association with decreased estrogen levels. The two most common causes of primary ovarian failure are Turner's syndrome and the female climacteric, or menopause.

Turner's syndrome is a genetic disorder characterized by 45 chromosomes, consisting of 44 normal autosomal chromosomes with only one X chromosome. In this condition, the primordial germ cells appear to degenerate for obscure reasons without undergoing transformation into premordial follicles, resulting in primitive or "streak" gonads incapable of ovulation or estrogen secretion. The presence of other congenital abnormalities such as short stature, webbing of the neck, short metacarpals, and coarctation of the aorta often suggest the diagnosis during the prepubertal years. Cyclic replacement hormonal therapy is indicated to prevent osteoporosis, premature aging of the skin, and other consequences of estrogen deficiency.

Menopause is a form of primary ovarian hypofunction that is experienced by all women who survive to middle life. The current median age of onset of menopause is 50 years. In the majority of instances, no explanation can be found. It appears to reflect the progressive depletion of oocytes and the loss of their inductive effect upon the granulosa and thecal cells. With a diminution of the steroid feedback influence on the hypothalamus or pituitary, the gonadotropin titer rises and the release of FSH and LH is no longer coordinated. Termination of menstrual function before 40 years of age may be considered premature. Symptoms associated with menopause can be variable. The most common is the occurrence of "hot flashes," seen in 85% of the patients. In patients in whom prior cancer of the breast or uterus has been excluded, estrogen therapy is usually given to relieve hot flashes and delay the onset of osteoporosis.

Secondary ovarian hypofunction. Secondary ovarian hypofunction is characterized by decreased estrogen and progesterone levels in association with decreased gonadotropin levels. It may be attributable to hypothalamic, pituitary, or constitutional disturbances. Hypothalamic disorders that result in hypofunction of the ovary include the Laurence-Moon-Bardet-Biedl syndrome, which is a congenital disorder, and the Chiari-Frommel syndrome, which occurs after childbirth. Emotional strain appears to alter the afferent neural input into the hypothalamic centers controlling the secretion of LHRH and is perhaps the commonest cause of secondary amenorrhea, apart from pregnancy.

Tumors of the pituitary and necrosis resulting from postpartum hemorrhages (Sheehan's syndrome) are the most frequent pituitary lesions resulting in ovarian insufficiency. Granulomatous diseases may also result in diminished pituitary function.

Severe constitutional illnesses such as congenital heart disease, chronic renal disease, or rheumatoid arthritis may cause amenorrhea and sterility. Dieting with too rapid weight loss, anorexia nervosa, poorly controlled diabetes mellitus, and hyperthyroidism may also impair reproductive function. The exact mechanism by which these different disorders affect the hypothalamic-pituitary-ovarian axis is not known.

Ovarian hyperfunction

Primary ovarian hyperfunction. The main cause of primary ovarian hyperfunction is estrogen-secreting tumors. Granulosa and thecal cell tumors are the most common of the estrogen-producing tumors. Approximately 5% arise before puberty, 55% during the period of reproductive life, and 40% after menopause. The symptoms depend on the age of the patient. Precocious puberty and intermittent uterine bleeding result from tumors during the premenarchal years. Irregular uterine bleeding frequently alternating with periods of amenorrhea is common during the active reproductive life. Uterine bleeding is the characteristic manifestation of tumors during the postmenopausal years. Primary ovarian hyperfunction results in decreased levels of FSH and LH because of increased negative feedback on the hypothalamic-pituitary axis.

Secondary ovarian hyperfunction. Secondary ovarian hyperfunction is characterized by increased levels of gonadotropins resulting in increased estrogen secretion. Sexual precocity from pituitary stimulation of ovarian function is generally idiopathic in origin. Gonadotropin secretion can usually be reduced, with regression of the secondary sexual characteristics, by intramuscular injections of progesterone. An unusual form of precocious puberty is associated with hypothyroidism, where the ovary has increased sensitivity to endogenous gonadotropins. The precocity can be reversed by treatment of the hypothyroidism. These observations contrast with the effects of hypothyroidism on adult women who commonly experience a failure of ovulation.

Miscellaneous conditions

Hirsutism. Hirsutism is usually defined as more hair than is cosmetically acceptable to a woman living in a certain culture. Many women with increased hair growth are not found to have a gross hormonal abnormality. Increased body hair in women is commonly attributable to familial traits. Idiopathic hirsutism and polycystic ovary syndrome make up the largest group of the remaining women. Simple increase in hair may be caused by androgen production from any source, or by an increase in the sensitivity of the hair follicles to androgens, or by other abnormalities.

The hormonal abnormalities leading to hirsutism are heterogeneous. Excess androgens may be produced by the ovaries, by the adrenal glands, or by peripheral conversion of prohormone, or by all these sources. There is considerable controversy at present whether ovarian, adrenal, or mixed disorders predominate.

On the other hand, virilism (frontal balding, hirsutism, deepening of the voice, and acne) is almost always associated with excess testosterone, usually from an ovarian source, such as the rare testosterone-secreting tumors of the ovary. An arrhenoblastoma is the most common masculinizing tumor of the ovary.

Polycystic ovary syndrome (Stein-Leventhal syndrome). The Stein-Leventhal syndrome is characterized by bilateral polycystic ovaries in association with infertility and anovulatory menstrual irregularities. It may present initially with oligomenorrhea, hirsutism, or both, in varying degrees. This disorder is one of the most confusing of all endocrine diseases and thus far has lacked a concrete definition. According to some, the diagnosis requires the presence of bilaterally enlarged ovaries, characteristically with multiple bluish follicular cysts. Numerous patients have been described who cannot be differentiated from this group on clinical grounds but whose ovaries are normal or only slightly enlarged. The anatomical findings, the laboratory abnormalities, and the clinical presentation of these women span a wide spectrum from near normal to extremes. Obesity is common but is not a criteria for diagnosis. Plasma testosterone levels (total) and urinary 17-ketosteroid excretion are usually at or slightly above the upper limits of normal. LH levels may be elevated, with normal or low levels of FSH. A ratio of LH to FSH over 1.5 would support the diagnosis of polycystic ovary syndrome.

INFERTILITY AND ITS THERAPY

Infertility affects at least 10% of couples and is a result of abnormalities in the male in 40% of cases, the female in 40% of cases, and unexplained causes in 20% of cases. In the male, azospermia or oligospermia are usually not associated with any endocrine causes and testosterone levels are normal. A small proportion of patients have a mild type of androgen resistance affecting spermatogenesis. In secondary hypogonadism, exogenous FSH and LH in the form of human menopausal gonadotropin (hMG) may be used to treat infertility. In the female, anovulation is frequent and is easily treated with antiestrogen, clomiphene citrate, or hMG. Ovulation during treatment can be monitored by ultrasound or by measurement of the luteal phase progesterone secreted by the corpus luteum. Hyperprolactinemia as a cause of anovulation must be considered. Infertility in some women may result from the presence of antibodies against sperm.

Obtaining ova for in vitro fertilization requires injection of hMG with or without the use of clomiphene citrate. Recruitment of several follicles leads to secretion of high levels of estrogen, and these are usually measured in the plasma as part of the monitoring of therapy.

FUNCTION TESTS

The gonad is a part of the hypothalamic-pituitary-gonadal axis (control system), which produces a hormone, which in turn interacts with the target tissue (compliance system). As a result of this complex system, the cause of an endocrine disorder may rest within any one of the components of the control system or the compliance system. Differential diagnosis involving a series of stimulation and suppression tests may be necessary to establish the proper diagnosis of endocrine disorders.

Measurement of the hormonal status before stimulation is usually extremely valuable, with FSH, LH, prolactin, estradiol (female), and testosterone (male) providing much information. The following are provocative tests of hormonal function.

Gonadotropin-releasing hormone test

This assessment is performed using the synthetic GnRH decapeptide, usually in a dose of 100 μg given intravenously. Two specimens of blood are obtained before injection and then every 15 minutes for 1 hour thereafter. The serum is analyzed for FSH and LH concentrations, with peaks in LH and FSH occurring within 30 and 45 minutes of injection. This rise of FSH and LH in normal persons indicates a normally responsive pituitary gland (before puberty, response is less pronounced) and excludes hypopituitarism as a cause of hypogonadism. In postpubertal patients with hypogonadotropic hypogonadism, a depressed or absent response to GnRH is evidence that there may be hypopituitarism. The possibility that the abnormality may result from a hypothalamic problem may be confirmed by the clomiphene citrate test.

Clomiphene citrate test

The clomiphene citrate (Clomid) test is another means of assessing the hypothalamic-pituitary-gonadal axis. Clomiphene citrate, a nonsteroidal agent, probably acts by competitive inhibition of estradiol at the hypothalamic receptor level. As a result of the decreased binding of estradiol to receptor sites, the hypothalamus response by increasing the output of GnRH, resulting in increased secretion of FSH and LH. The rise in gonadotropins thus depends on an intact hypothalamic-pituitary axis.

Blood is obtained before the administration of the drug; 50 to 100 mg of drug is given for 5 days (day 2 to 6 of the female menstrual cycle, anytime for a male). Blood levels of FSH and LH are measured regularly over the next 10 days. A positive response is indicated by an approximate doubling of the LH concentration. A diminished response to clomiphene with a good response to GnRH indicates that a hypothalamic lesion is a likely basis for the secondary hypogonadism.

Human chorionic gonadotropin stimulation test

The human chorionic gonadotropin (hCG) stimulation test is used to test testicular reserves of testosterone. Injection of hCG mimics the effect of LH by stimulating testosterone production. It is employed to establish the presence of active tissue in the investigation of delayed puberty and to characterize the testicular function of patients with hypogonadotropic hypogonadism. In the usual test procedure, 3000 U/m^2 is given intramuscularly on the first and third days of the test, and blood is drawn to monitor the rise in testosterone. A normal response is a rise of two to three times the basal levels of testosterone.

Dexamethasone suppression test

The dexamethasone suppression test is used to suppress the adrenocorticotropic hormone stimulation of the adrenals to investigate the cause of increased androgens in hirsutism. Used as an overnight dose of 1 mg of dexamethasone or 0.5 mg 6 hourly for 2 days it may also be used to exclude Cushing's syndrome (see Chapter 45). A longer course (0.5 mg 6 hourly for 14 days) is used for suppression of androgens.

Progestogen withdrawal test

The progestogen withdrawal test is used as an indirect measure of endogenous estradiol levels. The principle of the test is that progestogen withdrawal will only produce uterine bleeding if there has been prior estrogen priming of the endometrium. Therefore sufficient concentration of circulating estradiol must have been present to induce synthesis of progesterone receptors. The test is performed by administration of a progestogen (norethisterone acetate or medroxyprogesterone acetate) orally daily for 5 days. After administration of the drug is stopped, vaginal bleeding should be reported. There is no measurement of hormones in this test.

Table 43-1 Change of analyte in disease

Disease	Analyte			
	FSH	**LH**	**Estradiol**	**Testosterone**
Female				
Primary ovarian failure (including menopause)	↑	↑	↓	—
Secondary ovarian failure	↓	↓	↓	—
Feminizing tumors	↓	↓	↑	
Masculinizing tumors	↓	↓	—	↑
Exogenous gonadotropins or gonadotropin-producing tumor (rare)	↑	↑	↑	—
Polycystic ovary syndrome	N to ↓	↑	—	↑
Male				
Primary testicular failure	↑	↑	—	↓
Secondary testicular failure	↓	↓	—	↓
Primary testicular overactivity (e.g., tumor, testotoxicosis)	↓	↓	—	↑
Secondary testicular overactivity (e.g., precocious puberty)	↑	↑	— —	↑

↑, Increased; ↓, decreased; N, normal.

CHANGE OF ANALYTE IN DISEASE (Table 43-1)

Gonadotropins (LH and FSH)

Measurement of serum LH and FSH concentrations provides important indices of hypothalamic-pituitary axis function and helps distinguish primary from secondary gonadal dysfunction syndromes. Pulsatile fluctuations of LH and FSH are commonly noticed in the female throughout the reproductive cycle. It is therefore common practice to obtain blood in the follicular phase of the menstrual cycle and to obtain two to three samples several minutes apart.

Increased levels of gonadotropins. Increased gonadotropin levels are seen in primary ovarian failure such as Turner's syndrome or menopause. Increased levels are also seen following ovarian surgery. In polycystic ovary syndrome, LH levels may be elevated with normal or low levels of FSH. A ratio of LH to FSH of greater than 2:1 would support the diagnosis of polycystic ovary syndrome. Increased levels in the male may be attributable to primary gonadal failure.

Decreased levels. Low circulating levels of gonadotropins indicate that the primary defect is either in the hypothalamus or pituitary.

Androgens

Androgens measured are testosterone, androstenedione, DHEA-sulfate, and occasionally dihydrotestosterone. A screening test still in use is the measurement of 17-ketosteroids in urine.

Increased androgens. Very high levels are seen in the female with masculinizing tumors, while slightly increased concentrations are seen in patients with polycystic ovarian syndrome.

Decreased androgens. In males, decreased levels of androgens signify testicular hypofunction, either primary or secondary.

Estrogens

Increased levels of estrogen. Increased estrogen levels result from ovarian hyperfunction, whether from a feminizing tumor (pseudoprecocious puberty) or an activated hypothalamic-pituitary-gonadal axis. Very high levels are found during the induction of ovulation and in pregnancy. In the male, increased levels of circulating estradiol and estrone may reflect not only increased aromatization and secretion by the testis and adrenal, but also altered aromatization activity in other organs, as seen in liver disease. Increased estrogen levels in males manifest as gynecomastia.

Decreased levels of estrogen. In the female, low estrogen levels indicate ovarian hypofunction.

Progesterone

Progesterone levels in postpubertal premenopausal females vary with the phase of the reproductive cycle. Highest concentrations are achieved in the luteal phase following ovulation and in pregnancy. A single progesterone level of 4 to 10 ng/mL obtained between 4 and 10 days after ovulation is presumptive evidence of luteinization. A value of greater than 10 ng/mL is excellent evidence of good ovulation.

Increased progesterone. Increased progesterone levels are usually seen in pregnancy or the luteal phase of the menstrual cycle.

Decreased progesterone. Decreased protesterone levels

are of no diagnostic value. Measurement of progesterone in males is of little value.

BIBLIOGRAPHY

Corey, P: Polycystic ovarian syndrome—current concepts of pathophysiology and therapy, Fertil Steril **42**:667, 1984.

Edwards, RG: Conception in the human female, London, 1980, Academic Press.

Fritz, MA, and Sperhoff, L: The endocrinology of the menstrual cycle: the interaction of folliculogenesis and neuroendocrone mechanisms, Fertil Steril **38**;509, 1982.

Ganong, W: The gonads: development and function of the reproductive system. In Review of medical physiology, Los Altos, Calif, 1981, Lange Medical Publications.

Gibbons, WE, Battin, DA, and diZerega, GS: Mechanism of action of reproductive hormones. In Mishell, D, and Davajan, V, editors: Infertility, contraception, and reproductive endocrinology, Oradell, NJ, 1986, Medical Economics Books.

Goebelsmann, U: Steroid hormones. In Mishell, D, and Davajan, V, editors: Infertility, contraception, and reproductive endocrinology, Oradell, NJ, 1986, Medical Economics Books.

Hodgen, GD: The dominant ovarian follicle, Fertil Steril **38**:281, 1982.

Kirschner, MA, Zucker, IR, Jespersen, DL: Ovarian and adrenal vein catheterization studies in women with idiopathic hirsutism. In James, VHT, Serio, M, and Giusti, G, editors: The endocrine function of the human ovary, London, 1976, Academic Press.

Klopper, A: Steroids in pregnancy. In Sherman, RP: In clinical reproductive endocrinology, London, 1985, Churchill Livingstone.

Lachelin, G: The polycystic ovary syndrome. In Studd, J: Progress in obstetrics and gynaecology, 1986.

Lipsett, MG: Steroid hormones. In Yen, SSC, and Joffe, RB: Reproductive encodrinology: physiology, pathophysiology, and clinical management, Philadelphia, 1986, WB Saunders Co.

Mooradian, AP, Morley, JE, and Korenman, SG: Biological action of androgens, Endocr Rev 8:1, 1987.

Norman, RJ, and van der Spuy, ZM: Investigation and management of hirsutism, S Afr J Contin Med Educ **5**:37, 1987.

Ross, GT, Schreiber, JR: The ovary. In Yen, SSC, and Jaffe, RB: Reproductive endocrinology: physiology, pathophysiology, and clinical management, Philadelphia, 1986, WB Saunders Co.

Swerdloff, RS: Infertility in the male, Ann Intern Med **103**:906, 1985.

Tuomisto, J, and Mannisto, P: Neurotransmitter regulation of anterior pituitary hormones, Pharmacol Rev **37**:249, 1987.

CHAPTER 44 Parathyroid

RICHARD J. KOZERA

OBJECTIVES

- State the roles of ionized calcium, magnesium, cyclic AMP, and vitamin D metabolites in the regulation of PTH release.
- List the target organs and physiological actions of PTH.
- Describe expected serum levels of calcium, phosphorus, PTH, calcitonin, and urinary cyclic AMP in pathological conditions with elevated and decreased parathyroid function.

KEY TERMS

chief cells Portion of the parathyroid gland actively synthesizing and secreting parathyroid hormone (PTH).

hyperparathyroidism Condition defined by elevated serum PTH levels caused by hypersecretion of the parathyroid gland (primary hyperparathyroidism) or secondarily in response to decreased serum calcium levels.

hypoparathyroidism Condition defined by serum parathyroid hormone levels that are decreased from established reference ranges. Can be caused by primary parathyroid dysfunction or secondarily by elevated serum calcium levels.

ionized calcium The ionized, unbound, noncomplexed fraction of serum calcium that is biologically active.

multiple endocrine neoplasia Simultaneously occurring tumors of different endocrine glands.

osteoblasts Bone cells responsible for laying down new bone material.

osteoclast activating factor Factor released from certain tumors that enhances osteoclastic activity and results in increased levels of serum calcium. The factor is not parathyroid hormone.

osteoclasts Bone cells responsible for bone resorption.

pre-proparathyroid hormone (pre–pro-PTH) The initial translation product of the parathyroid hormone gene; this 13,000 molecular weight compound is degraded to a second precursor, pro-PTH.

pro-PTH An immediate precursor to active parathyroid hormone formed by cleavage of pre-pro-PTH.

pseudohypoparathyroidism Clinical conditions of hypocalcemia caused by cellular resistance to PTH. This includes pseudo-pseudohypoparathyroidism

PTH resistance Cellular dysfunction in which PTH molecules do not elicit a normal response from target cells.

tetany A muscular spasm that is characteristic of hypocalcemic conditions.

ANATOMY

Human parathyroid glands are small, ovoid, yellowish brown structures. There are usually four parathyroid glands, but there may be as few as two and as many as 12. They are paired structures usually embedded in the posterior capsule of the thyroid gland, but they may be located in other parts of the neck or upper chest.

The glands are populated with cells of epithelial origin. They are divided into two basic groups. In the first group are two subpopulations of chief cells, one that is lighter staining and considered to be inactive or resting and one that stains more darkly because it is densely packed with secretory granules and considered to be actively producing hormone. The second group of cells is the oxyphils, these cells are acidophilic, abundant in granules, and loaded with mitochondria. Their function is unknown.

PHYSIOLOGY

Biosynthesis of parathyroid hormone

The first precursor of parathyroid hormone (PTH) to be synthesized is pre-proparathyroid hormone (pre–pro-PTH), which has 115 amino acids and a molecular weight of 13,000 daltons. The first 25 amino acids are cleaved off, leaving a 90 amino acid product, pro-PTH, with a molecular weight of 10,200 daltons. This is transported through the endoplasmic reticulum to the Golgi apparatus. The first six amino acids of pro-PTH are cleaved, yielding the 84 amino acid PTH (Fig. 44-1), which is packaged in granules and either stored for subsequent secretion or degraded.

Secretion of parathyroid hormone

Biological activity of the PTH molecule resides in about the first 30 amino acids of its amino terminal end. PTH seems to be secreted intact but is quickly cleaved into an amino-terminal fragment with a half-life of only a few minutes and a carboxy-terminal fragment with a half-life of an hour or more. Most carboxy-terminal fragments are biologically inactive but constitute the largest pool of PTH fragments in the circulation. Other biologically inactive fragments of PTH may be found in the circulation in varying but small quantities. These fragments are mostly produced during the peripheral degradation of PTH and are not believed to be direct secretory products of the chief cell. There is some evidence that other precursors, intermediaries, and associates of PTH metabolism within the cell may be secreted, but the magnitude and importance of this is poorly understood. Unlike many other proteins that circulate in a free and bound (carrier) state, PTH exists as freely circulating intact molecules and fragments.

Regulation of parathyroid hormone secretion

The single most important determinant of the release of PTH from the parathyroid glands is the serum concentration of ionized calcium, which is usually adequately reflected in the concentration of total serum calcium. PTH secretion is stimulated as the calcium concentration falls and is suppressed as the calcium concentration rises. There is a sharp increase in the secretion rate of PTH when the total serum calcium concentration falls below 90 mg/L. The rate is maximum after an acute fall in serum calcium to 70 or 80 mg/L. The change in PTH secretion rate is linear, with changes of calcium concentrations between 90 and 105 mg/L. When calcium concentrations are greater than 105 mg/L, the PTH secretion rate is greatly diminished but not completely suppressed. Studies of subjects with hyperplastic or neoplastic parathyroid glands have shown similar relationships between PTH secretion or serum concentration and serum calcium concentration, but higher calcium concentrations were required to suppress PTH secretion in these abnormal glands.

Other substances are probably best regarded as modifiers of the calcium-PTH interaction. The most important of these is magnesium, which has an effect on PTH that both parallels and modulates the effect of calcium. In healthy persons and in those with hyperparathyroidism who have

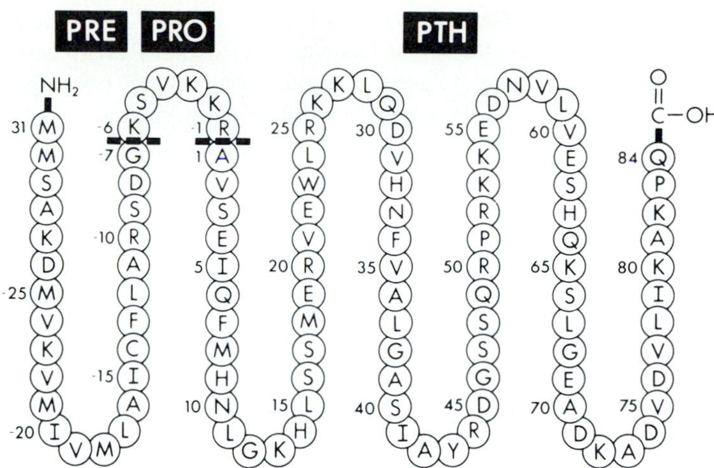

Fig. 44-1 Amino acid sequence of bovine pre-pro-PTH. Dashed lines separate "pre" and "pro" amino acid segments and native hormone (*PTH*, parathormone). *(From Cohn DV, and MacGregor RR: Endocrine Rev 2:1, 1981; copyright The Endocrine Society.)*

normal total body magnesium stores, an acute increase in the serum magnesium concentration will, if anything, decrease the amount of circulating PTH. In subjects who are both hypocalcemic and hypomagnesemic the PTH concentration increases with administration of magnesium regardless of whether the PTH was low, normal, or high to begin with. This suggests that magnesium at very low concentrations probably suppresses PTH secretion. The likely explanation for this is that the generation of cyclic AMP (cAMP) by adenylate cyclase, a magnesium-dependent process, is inhibited, blocking the cAMP-dependent release of PTH. In hypomagnesemic states PTH is quickly released with replenishment of magnesium. It is probable that this is a release from prepackaged stores and that the synthetic processes through the point of storage are not magnesium dependent.

Persons with high circulating levels of PTH who develop hypomagnesemia show a decreased effect of PTH at the target cell. In this circumstance the low magnesium level inhibits generation of cAMP by the target cell, thereby inducing a state of tissue resistance.

The role of biogenic amines in PTH secretion is somewhat unclear. Generally, substances that raise the level of cAMP in the parathyroid cells tend to enhance PTH secretion and those that lower cAMP tend to inhibit PTH secretion. cAMP levels may be raised by prostaglandin E_1, prostaglandin E_2, dopamine, cholera toxin, aminophylline, dibutyryl cAMP, and the beta agonists epinephrine, norepinephrine, and isoproterenol. cAMP levels may be lowered by beta antagonists (propranolol), alpha agonists (methoxamine), prostaglandin F_2, and nitroprusside. There may be different responses to different concentrations of the same drug. For example, at low concentrations isoproterenol has been shown to stimulate PTH release, but at higher concentrations it has had no effect on PTH release. Although there are several explanations for this phenomenon, including decreased blood flow and some masking by simultaneous alpha-agonist effects, it indicates the need to separate physiological from pharmacological effects of biologically active agents.

Although vitamin D is discussed extensively in Chapter 24, it is important to recognize here that there is probably direct interaction between vitamin D metabolites and parathyroid gland function. Parathyroid cell receptors for 1,25-dihydroxycholecaliciferol (calcitriol) have been demonstrated in several species. Some investigators have shown effects of vitamin D metabolites on serum calcium concentration, parathyroid gland size, and modulation of synthesis or secretion of PTH. The role of these interactions in humans is not yet clearly defined.

Physiological actions of parathyroid hormone

Mineral homeostasis is governed in large part through the action of PTH on its two major target organs, bone and kidney. The exact nature of the interaction between target cell and PTH is unknown, but biological activity is initiated by the interaction between PTH and cellular receptors and is almost certainly mediated through the generation of cAMP within the target cell. There are less data on bone, but in vitro studies of bone cells suggest similar PTH effects.

Although little is known of the molecular mechanisms of PTH action, the physiological effects of PTH have been well described. Although the control of calcium and phosphorus in the kidney depends on a complex interrelationship between PTH, vitamin D, calcium, and phosphorus, the net effect of PTH on the renal tubule is calcium and magnesium retention and phosphate and bicarbonate loss (Fig. 44-2). Specifically, PTH causes increased resorption of both calcium and magnesium in the ascending limb of the loop of Henle and in the distal tubule. It may cause a slight decrease in the proximal tubular resorption of isotonic fluid, which contains sodium, calcium, and magnesium. It causes decreased phosphate reabsorption in the proximal tubule and possibly in the distal tubule. It also causes a slight decrease in the glomerular filtration rate.

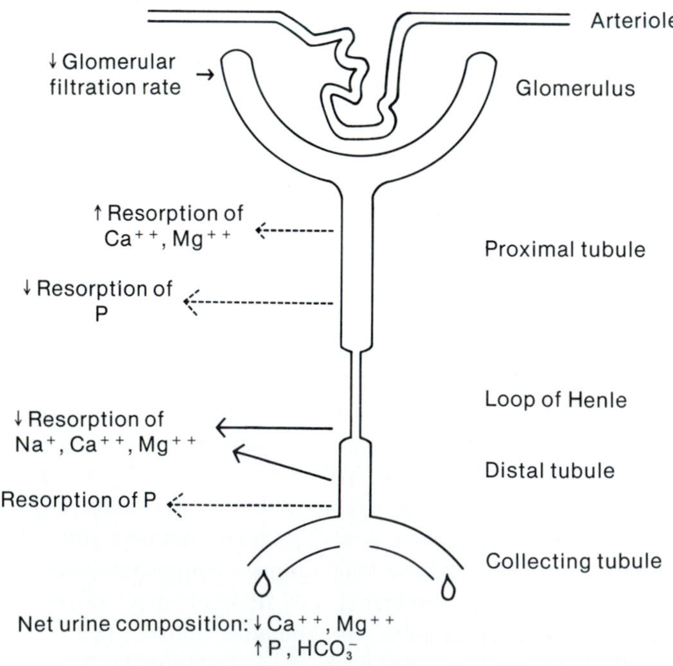

Fig. 44-2 Effect of PTH on renal tubules. *Arrows,* Effect of PTH on analytes at segments of tubule; *P,* inorganic phosphate.

The effect of PTH on bone is equally complex and again requires the presence of the hormone, an appropriate receptor system, vitamin D metabolites, calcium, and phosphorus. As mentioned in Chapter 24, there are three major types of cells found in bone: osteoblasts, responsible for the provision of new bone; osteoclasts, primarily responsible for bone resorption; and osteocytes, which have many complex functions. There are two primary effects of PTH. The first is mediated by the osteoclasts and over a period of time results in bone remodeling. The second is probably mediated through the osteocytes, and an important result is the rapid efflux of calcium from bone into the extracellular fluid. These processes are interdependent and help serve both the acute minute-to-minute maintenance of mineral homeostasis and the long-term demands of proper bone structure.

PATHOLOGICAL CONDITIONS

Calcium

States of abnormal parathyroid function most often are present with signs and symptoms of hypocalcemia or hypercalcemia. The measurement of serum calcium therefore assumes major importance. Remember that calcium circulates in both an ionized (''free'') fraction that is biologically active and a fraction that is chiefly ''bound'' to serum albumin. Each fraction makes up about half the total serum calcium. For most clinical purposes the total serum calcium varies in proportion to the ionized fraction and is considered a good indicator of the level of ionized calcium. In the following sections the terms *hypocalcemia* and *hypercalcemia* are used to indicate a decrease or increase in both total and ionized serum calcium concentrations.

Hypocalcemia

Tetany, or muscle spasm, is the most characteristic of the major signs and symptoms of acute or chronic hypocalcemia. It is the direct result of enhanced neuromuscular excitability. One may demonstrate latent tetany by (1) lightly tapping the seventh cranial (facial) nerve just in front of the ear and observing the muscle contractions around the eye, nose, and mouth (Chvostek's sign) or (2) inflating a blood pressure cuff applied around the upper arm to levels above the systolic pressure for 3 minutes, causing local ischemia to the nerve trunk and resulting in severe flexion at the wrists (carpal spasm, Trousseau's sign).

Other clinical findings are largely related to chronic hypocalcemia, which may arise from a variety of conditions including hypoparathyroidism. Emotional or behavioral symptoms can range from mild irritability to frank psychosis. Skin changes can include mild dryness, brittle nails, severe dermatitis, and psoriasis. The scalp hair may be coarse and brittle or fall out in scattered patches (alopecia areata). The lenses of the eyes may develop cataracts. If hypocalcemia is present early in childhood, the teeth may fail to erupt, be underdeveloped, or show poor enamel formation. The heart may develop both contractile and electrical abnormalities. Hypotension or heart failure may occur. Gastrointestinal manifestations include steatorrhea, gastric achlorhydria, and impaired absorption of vitamin B_{12}. Mucocutaneous infection by *Candida albicans* and the bony changes of osteomalacia are more related to PTH deficiency and vitamin D deficiency than they are to hypocalcemia itself.

Hypocalcemic disorders

Clinical disorders resulting in hypocalcemia may be grouped into those associated with deficient PTH concentration or activity and those that have a primary effect on calcium deposition or excretion that is largely independent of PTH activity. Vitamin D–deficiency states are of major importance in both the classification and pathogenesis of hypocalcemia but are discussed elsewhere in this book. A bewildering terminology has developed for the classification of hypoparathyroid disorders. It is reasonable to consider them in two large groups: (1) hyposecretion of PTH and (2) tissue resistance to PTH.

Primary hypoparathyroidism. Hyposecretion of PTH or primary hypoparathyroidism is most commonly seen after neck surgery. The clinical manifestations are those of hypocalcemia. The severity and duration of the disorder depends on the extent of surgery and the amount of parathyroid tissue injured or removed. The condition may be transient, usually the result of trauma, or permanent, caused either by extensive trauma or removal of excessive tissue. If transient, it may be asymptomatic and recognized only by a drop in the serum calcium concentration, or it may be more severe and cause latent or obvious tetany.

Radiation therapy has occasionally been implicated as a cause of hypoparathyroidism. The source of irradiation may be an external beam usually used for treatment of malignancy or radioactive iodine (^{131}I) commonly used for treatment of hyperthyroidism or thyroid malignancy.

Idiopathic hypoparathyroidism is an unusual condition that can be either sporadic or familial and occurs either independently or linked to other endocrine deficiency states. The serum calcium concentration tends to be quite low, averaging around 55 mg/L; the serum phosphorus concentration tends to be quite high, averaging around 75 mg/L. The disease manifests itself early in life, primarily with symptoms of hypocalcemia, which may be quite severe, even to the point of generalized seizures or asphyxia caused by laryngeal obstruction. There is often associated mental retardation. Most bases of polyendocrine failure (that is, hypofunctioning of several endocrine glands simultaneously) are familial and are believed to show an autosomal recessive inheritance. Associated disorders include failure of the adrenal glands, thyroid glands, or ovaries, mucocutaneous candidosis, steatorrhea, myasthenia gravis, cirrhosis, alopecia, and vitiligo.

Tissue resistance to PTH. Syndromes of PTH resistance are divided into two groups. Both show normal or increased circulating levels of PTH, and both involve multiple organ systems. In the first group, pseudohypoparathyroidism (PHP), there is lack of the expected renal responsiveness to PTH. The associated hypocalcemia is not caused by a decrease in PTH synthesis but by a lack of cellular response to PTH. The group is further subdivided into two types. PTH administered to subjects with PHP type I does not cause an increase in nephrogenous cAMP. PTH administered to subjects with type II PHP does increase nephrogenous cAMP, but this does not result in the expected increase in calcium reabsorption or phosphorus excretion. In both types the clinical presentation can vary considerably depending on the organ systems involved but most commonly includes a round face, short stature, stocky habitus, mental retardation, thickening of the calvarium, short metacarpals or metatarsals, ectopic calcification or ossification, and the signs and symptoms of hypocalcemia.

The second group, pseudo-pseudohypoparathyroidism (PPHP), demonstrates normal renal responsiveness to PTH, including stimulation of nephrogenous cAMP, increased calcium resorption, and increased phosphorus excretion. The clinical presentation excludes the signs and symptoms of hypocalcemia, but the physical habitus is that of PHP as described above. This is probably caused by resistance of selected tissues other than the kidney to the effects of PTH.

There is at least one report in the literature of a patient with manifestations of hypoparathyroidism that were caused by the release of PTH that was immunologically but not biologically active. The major signs and symptoms that developed were those of hypocalcemia and not parathyroid hormone resistance. The disorder has been termed *pseudoidiopathic* because of these features.

As noted previously, hypomagnesemic states will lead to ineffective generation of AMP by PTH at the target cell level and may also result in defective PTH production or release from the parathyroid gland. This can be seen in malabsorption states, renal losses of magnesium (use of diuretics), chronic alcoholism, diabetic ketoacidosis, and, occasionally, hyperthyroidism. The tissue resistance is reversible on replenishment of magnesium to serum levels above 10 mg/L.

Other causes of hypocalcemia. Hypocalcemia is a common finding in acute pancreatitis. The reason for this is unclear, but suggested mechanisms have included the formation of calcium soaps within the pancreas or in other soft tissues, inappropriate release of PTH, abnormal metabolism of PTH, or activation of other hormones including gastrin, glucagon, and calcitonin. The hypocalcemic effect of these hormones has been caused only by their administration in pharmacological rather than physiological doses. These hormones are therefore less likely causes of the hypocalcemia.

Persons with hyperparathyroidism resulting in bone and mineral loss who undergo curative surgery may have severe postoperative hypocalcemia, related not to PTH insufficiency but to an increased rate of deposition of calcium within the bone. This is usually a temporary problem that resolves in a matter of days or weeks with administration of adequate amounts of calcium.

Conditions resulting in hyperphosphatemia, including major surgery, cytotoxic therapy, or renal failure may cause hypocalcemia either by local effects on bone, suppression of formation of 1,25-dihydroxycholecalciferol, or precipitation of calcium phosphate complexes within soft tissues.

Drugs may cause hypocalcemia in a variety of ways: inhibiting resorption of calcium and bone matrix, inhibiting calcium resorption from bone but leaving the matrix undisturbed, promoting calcium deposition in bone and soft tissues, or chelating calcium within the blood. Some of the more important hypocalcemia-causing agents are estrogens, mithramycin, calcitonin, fluoride, CPDA1 (a citrate-phosphate-dextrose–containing preservative used in storing blood), and ethylene glycol. Diuretics that are active in the loop of Henle may cause calciuresis, resulting in hypocalcemia. States of vitamin D deficiency, rickets, or osteomalacia are commonly associated with hypocalcemia.

Hypocalcemia is commonly seen in the perinatal period. In the first few days of life it is most probably caused by transient developmental hypoparathyroidism. If it occurs 2 to 3 weeks after birth, it is usually caused by excess ingestion of phosphorus or immaturity of the renal tubule.

Hypercalcemia

Generally, a hypercalcemic disorder is suspected if the total serum calcium concentration is above 102 mg/L on repeated occasions, but it may also be suspected if on repeated determinations the serum calcium fails to fall below the level of 100 mg/L. Obtaining proper specimens for serum calcium determination is perhaps most important when a hypercalcemic disorder is suspected and the serum calcium is in the range of 100 to 110 mg/L. The sample should be collected from a subject, in a relaxed state for at least 30 minutes, who has been fasting for 4 hours. No tourniquet should be used. Collection under other conditions can increase the serum calcium level by as much as 5 to 10 mg/L.

The signs and symptoms of subjects with hypercalcemia may be related to the hypercalcemia itself, or to the underlying disease process causing the hypercalcemia, or both. Mild hypercalcemia may be entirely asymptomatic and may be seen as an unexpected finding on routine serum screening. The prevalence of asymptomatic hypercalcemia is unknown but has been estimated to be as high as 1 in 1000. More commonly hypercalcemia is seen in association with a chronic disease process and may be first manifested in several ways: as polyuria and polydipsia caused by a defect in renal concentrating ability; as progressive renal insufficiency; as nephrolithiasis; in association with hypertension, ulcers, or pancreatitis; or as an acid-base disorder. Subjects with hypercalcemia but without hyperparathyroidism may have a metabolic alkalosis caused by direct stimulation of hydrogen-ion secretion by the distal renal tubule or may show a distal renal tubular acidosis. If hypercalcemia develops acutely, the subject may show symptoms of anorexia, nausea, vomiting, confusion, stupor, and coma.

There are no specific physical signs. Laboratory studies may provide clues. The electrocardiogram may show abnormalities depending on the degree of hypercalcemia. There may be a metabolic acidosis or alkalosis as just discussed. The urine may show a lack of concentrating ability and may contain an excess of triple phosphate crystals. If renal stones are present, there may be hematuria.

A frequently difficult problem in clinical medicine is establishing the diagnosis of hyperparathyroidism in subjects with or without underlying disorders that may cause hypercalcemia. Determination of the serum PTH level assumes major importance in these cases.

Hypercalcemic disorders

Hypercalcemic disorders are broadly classified into five categories (see box on p. 669).

Hypercalcemia can be either a primary or secondary feature in the diseases listed below. The clinical presentation may be as subtle as an abnormal laboratory test in an asymptomatic subject or as severe as coma in a person with advanced malignancy. The disorders in the list that are responsible for most cases of hypercalcemia are malignancy and hyperparathyroidism.

Primary hyperparathyroidism. Primary forms may be subclassified as adenomas or hyperplasia. Adenomas tend to involve single glands, whereas hyperplasia tends to involve multiple glands. Many published series in the past indicated that the frequency of adenomas in subjects with hyperparathyroidism was approximately 80%. More recent

HYPERCALCEMIC DISORDERS

- Malignancy
  - Multiple myeloma
  - Leukemia
  - Lymphoma
  - Carcinoma
- Endocrine disorders
  - Hyperparathyroidism
  - Hyperthyroidism
  - Adrenal insufficiency
  - Acromegaly
  - Paget's disease
- Drug-induced disorders
  - Vitamin D
  - Vitamin A
  - Thiazide diuretics
  - Lithium
  - Milk and alkali
- Granulomatous disease
  - Sarcoidosis
  - Tuberculosis
  - Berylliosis
  - Systemic mycosis
- Miscellaneous causes
  - Familial hypocalciuria
  - After organ transplantation
  - Recovery from acute renal failure

experience suggests that this may be an overestimate and that hyperplasia accounts for an increasing proportion of this population.

Hyperparathyroidism can exist as an independent condition or can be associated with other endocrine disorders in a category termed *multiple endocrine neoplasia (MEN)*. Although MEN can be divided into specific groups, more overlap is being recognized among the groups. Type I MEN consists of lesions of the pituitary, pancreas, and parathyroid glands. Type II-A (Sipple's syndrome) consists of hyperparathyroidism, medullary carcinoma of the thyroid gland, and pheochromocytoma. Type II-B (also referred to as type III) consists of medullary carcinoma of the thyroid gland, pheochromocytoma, and mucosal neuromas. The histological appearance of the parathyroid glands in types I and II-A shows hyperplasia. The clinical manifestations are those of the underlying disease and hypercalcemia.

Symptomatic hyperparathyroidism is seen predominantly as bone disease or renal stones. The renal stones may cause pain and hematuria or obstruction, hydronephrosis, and renal failure. The bone changes include osteoporosis and subperiosteal resorption. In the past a severe disorder termed *osteitis fibrosa cystica* was seen and was characterized by pain, fractures, and skeletal deformities. This has largely vanished from the clinical scene, presumably because of the earlier recognition and greater ease of diagnosis of parathyroid disorders. Nonspecific features of hyperparathyroidism include vague neuromuscular symptoms, mild lethargy, altered behavior, probably peptic ulcer disease, acute or chronic pancreatitis, nonspecific abdominal pain, gallstones, and hypertension. Central nervous system abnormalities are almost always seen at serum calcium levels greater than 130 mg/L, with stupor and coma becoming prevalent as the serum calcium exceeds 170 mg/L.

Secondary hyperparathyroidism. Secondary hyperparathyroidism is regarded as a compensatory biological response to a hypocalcemic state. The most common example is seen in patients with renal failure whose kidneys become unable to excrete phosphorus. The serum phosphorus is increased leading to a secondary decrease in serum calcium. This stimulates the secretion of PTH in an attempt to restore the serum calcium concentration to normal, but this is done largely at the expense of demineralization of bone. This process can usually be kept in control with adequate treatment of renal failure, including dialysis and the use of substances that bind phosphorus within the gastrointestinal tract so that it cannot be absorbed. If the serum phosphorus can be maintained within the normal range, the serum calcium concentration will stay within the normal range. Some poorly controlled subjects may have a return of calcium to or above the normal range although the serum phosphorus is still elevated. This condition has been called *tertiary hyperparathyroidism* to signify an increased degree of autonomy of parathyroid gland function. It is suggested that the previously hyperplastic parathyroid tissue has now become autonomous, but such a pathogenetic mechanism is difficult to verify.

Other forms of secondary hyperparathyroidism are commonly seen in osteomalacic states, which are caused by lack of absorption of calcium through the gastrointestinal tract or excessive renal loss of calcium.

Hypercalcemia of malignancy. Hypercalcemia related to malignancy is most commonly seen in squamous cell carcinoma of the lung or of the head and neck but may also be seen in a variety of other solid tumors, including those of the breast, kidney, uterine cervix, or prostate. Of the "liquid" tumors, leukemia and lymphoma are infrequently associated with hypercalcemia, but multiple myeloma also has a frequent association. Factors responsible for tumor hypercalcemia include production of PTH or a PTH-like substance by the tumors, particularly those of squamous cell origin, production of prostaglandins, especially by breast tumors, and elaboration of a factor stimulating osteoclastic activity (OAF), which is seen in multiple myeloma. Demineralization of bone is caused by localized invasion by many tumors, both solid and liquid. Measurement of PTH in subjects with suspected ectopic PTH production may not show elevated levels, since the PTH fragments may be immunologically different from native PTH. This is unlike tumors that produce other polypeptide hormones because those polypeptides seem to be both biologically and immunologically similar to the naturally produced hormone.

The other conditions causing hypercalcemia listed in the box on p. 669 are generally diagnosed rather easily, and the hypercalcemia can be corrected by appropriate treatment of the underlying disorder. An interesting phenomenon is seen in some patients with sarcoidosis, a systemic granulomatous disease frequently accompanied by hypercalcemia. There is an increased rate of conversion of 25-hydroxycholecalciferol to the 1,25-dihydroxy form, which increases the transport of calcium from the gut to the extracellular fluid. It has been suggested but not proved that the granulomatous tissue itself is responsible for this biochemical alteration. Additionally, primary hyperparathyroidism and sarcoidosis can coexist, and measurement of serum PTH assumes major diagnostic importance because it is normal or suppressed in sarcoidosis alone and elevated in the combination of sarcoidosis and hyperparathyroidism.

CHANGE OF ANALYTE IN DISEASE (Table 44-1)

Calcium

In ideal circumstances the normal daily human diet contains about 1000 mg of calcium. Each day about 825 mg is excreted in the feces and 175 mg is excreted in the urine. This is a net result because an ongoing flux of serum calcium among the gut, extracellular fluid, bone, and kidney exists. Under normal circumstances there is considerable variation in urinary calcium excretion, which primarily depends on the amount of calcium ingested. It is usually less than 300 mg/day. More critical values may be obtained when the calcium excretion is expressed in terms of the amount of creatinine excreted or in terms of the glomerular filtration rate. Critical analyses must be made with the patient on a ''metabolic diet'' of known calcium content for a period of several days. Such determinations are not very practical in the diagnosis or management of hypercalcemic states but may be of some help in the diagnosis and management of hypercalciuric states unaccompanied by hypercalcemia.

In conditions resulting in decreased serum albumin concentration, the total serum calcium is not a good indicator of the metabolically important ionized fraction, which usually remains at normal levels. Albumin is the protein to which serum calcium is predominantly bound. Decreases in the serum albumin concentration result in proportional decreases in total serum calcium concentration. A variety of formulas have been suggested to estimate the ionized calcium from the total calcium in states of altered albumin concentrations. At physiological pH, each gram of albumin binds about 0.8 mg of calcium. The normal range of serum albumin is about 35 to 55 g/L. The measured albumin concentration may be subtracted from 46 (''normal'' concentration) and the difference multiplied by 0.8. The product is then added to the measured calcium concentration. This provides the ''corrected'' serum calcium concentration. In another method of calculation, the measured serum albumin (in grams per liter) may be subtracted from the measured serum calcium (in milligrams per liter) and 40 is added to the difference. A corrected value in the normal range for total serum calcium would indicate a normal circulating ionized calcium concentration. In states of acute acidosis or alkalosis the change in ionized calcium concentration is inversely proportional to the change in pH. In chronic states of acidosis or alkalosis, compensatory mechanisms usually allow for return of the ionized calcium to the normal range.

Phosphorus

Approximately 1400 mg of phosphorus is ingested each day, with a net excretion of 500 mg in the feces and 900 mg in urine. As with calcium there is a continuing flux among gut, extracellular fluid, bone, and kidney. Measurement of serum phosphorus is helpful in the diagnosis of hypoparathyroidism and may be helpful in discriminating among some hypercalcemic states. Measurement of

Table 44-1 Change of analyte with disease

Analyte	Primary hyperparathyroidism	Secondary hyperparathyroidism (osteomalacia)	Primary hypoparathyroidism	Pseudohypoparathyroidism type I	Pseudohypoparathyroidism type II	Sarcoidosis
PTH-N terminal	Frequently normal	Normal to ↑	↓	↑	↑	↓
PTH-C terminal	↑	↑	↓	↑	↑	↓
Calcium	↑	Low to normal	↓	↓	↓	↑
Inorganic phosphorus	Normal to ↓	Normal to high	↑	↑	↑	Normal to ↓
Nephrogenous cAMP (urine)	↑	↑	↓	↓	?	↓
Nephrogenous cAMP response to PTH	—	—	↑	↓	↑	—
Urinary calcium excretion	↑	↓	↓	—	—	↑

↑, Increased; ↓ decreased.

urinary phosphorus is now rarely used in the diagnosis or management of hypercalcemic states.

Parathyroid hormone

The serum PTH concentration should be interpreted in light of the serum calcium concentration. Subjects with hypercalcemia caused by hyperparathyroidism usually, but not always, have elevated levels of PTH. Subjects with hypercalcemia from other causes may have PTH levels that are measurable and within the "normal" range or levels that are below the limits of detectability of the assay. Subjects with secondary hyperparathyroidism usually have low serum calcium values and elevated circulating PTH. Subjects with primary hypoparathyroidism rarely, if ever, have detectable PTH.

These phenomena demand that caution be taken in the interpretation of "normal" PTH levels, especially in the evaluation of hypercalcemic states. Thus the PTH values cannot be properly interpreted unless the serum calcium concentration is also known.

Efforts to clarify ambiguous or confusing PTH levels by stimulation or suppression of PTH secretion through lowering or raising serum calcium levels have generally not provided diagnostic information that is more helpful than measurement of basal levels. Such studies have demonstrated, however, that either hyperplastic or adenomatous parathyroid tissue is responsive to changes in extracellular calcium concentrations but at a higher set point.

Nephrogenous cyclic AMP

About 40% of the cAMP that appears in the urine is produced by the kidney in response to PTH. The remainder is derived from the plasma as it is filtered by the glomerulus. It is primarily of adjunctive value in states of altered PTH production. In cases of suspected hyperparathyroidism when serum PTH levels are in the normal range, elevated urinary cAMP may more clearly define the diagnosis. In states of PTH resistance (pseudohypoparathyroidism), response of urinary cAMP to injected PTH can establish the diagnosis and type of pseudohypoparathyroidism.

Calcitonin

Calcitonin is secreted from the C cells of the thyroid gland. In animal species its behavior generally contrasts with that of PTH in response to changes in ionized calcium. A similar role in humans has not been clearly defined. The major diagnostic role of serum levels of calcitonin is in medullary carcinoma of the thyroid and not in states of abnormal parathyroid function.

BIBLIOGRAPHY

Physiology

Agus, ZS, Wasserstein, A, and Goldfarb, S: Disorders of calcium and magnesium homeostasis, Am J Med 72:473-488, 1982.

Cohn, DV, and MacGregor, RR: The biosynthesis, intracellular processing, and secretion of parathormone, Endocrinol Rev 2:1-26, 1981.

Epstein, FH: Calcium and the kidney, Am J Med 45:700-714, 1968.

Habener, JF: Responsiveness of neoplastic and hyperplastic parathyroid tissue to calcium in vitro, J Clin Invest 62:436-450, 1978.

Heath, H, III: Biogenic amines and the secretion of parathyroid hormone and calcitonin, Endocrinol Rev 1:319-338, 1980.

Lemann, J, Jr, Adams, H, and Gray, RW: Urinary calcium excretion in human beings, N Engl J Med 301:535-541, 1979.

Robinson, CJ: The physiology of parathyroid hormone, Clin Endocrinol Metab 3:389-417, 1974.

Stoff, JS, Phosphate homeostasis and hypophosphatemia, Am J Med 72:489-495, 1982.

Clinical pathology

Benson, RC, Jr, Riggs, BL, and Pickard, BM: Immunoreactive forms of circulating PTH in primary and ectopic hyperparathyroidism, J Clin Invest 54:175-181, 1974.

Berson, SA, and Yalow, RS: Immunochemical heterogeneity of parathyroid hormone in plasma, J Clin Endocrinol Metab 28:1037-1047, 1968.

Broadus, AE: Nephrogenous cyclic AMP as a parathyroid function test, Nephron 23:136-141, 1979.

Chen, IW, Park, HM, King, LR, et al: Radioimmunoassay of parathyroid hormone: peripheral plasma immunoreactive parathyroid hormone response to ethylenediaminetetraacetate, J Nucl Med 15:763-769, 1974.

Flueck, JA, DiBella, FP, Edis, AJ, et al: Immunoheterogeneity of parathyroid hormone in venous effluent serum from hyperfunctioning parathyroid glands, J Clin Invest 60:1367-1375, 1977.

Habener, JF, and Segre, GV: Parathyroid hormone radioimmunoassay, Ann Intern Med 91:782-784, 1979.

Clinical disorders

Bone, HG, II, Snyder, WH, III, and Pak, CYC: Diagnosis of hyperparathyroidism, Annu Rev Med 28:111-117, 1977.

Goldsmith, RE, Sizemorae, GW, and Chen, IW: Familial hyperparathyroidism: description of a large kindred with physiologic observations and a review of the literature, Ann Intern Med 84:36-43, 1976.

Mallette, LE, Bilezikian, JP, Heath, DA, and Aarback, DG: Primary hyperparathyroidism: clinical and biochemical features, Medicine 53:127-146, 1974.

Nusynowitz, ML, Frame, B, and Kolb, FO: The spectrum of the hypoparathyroid states, Medicine 55:105-119, 1976.

Scholz, DA, Purnell, DC, Edin, AJ, et al: Primary hyperparathyroidism with multiple parathyroid gland enlargement, Mayo Clin Proc 53:792-797, 1978.

Skrabanek, P, McPartlin, J, and Powell, D: Tumor hypercalcemia and ectopic hyperparathyroidism, Medicine 59:262-282, 1980.

CHAPTER 45 Adrenal hormones

MORRIS R. PUDEK

OBJECTIVES

- List the principal glucocorticoids, mineralocorticoids, and medullary hormones and state their physiological effects.
- Describe the synthesis, transport, catabolism, and regulation of glucocorticoids, mineralocorticoids, and medullary hormones.
- Describe each of the following pathological conditions, including diagnostic laboratory results: Cushing's syndrome, hyperaldosteronism, adrenogenital syndrome, Addison's disease, and pheochromocytoma.
- Describe each of the following adrenal function tests and the interpretation of results: clonidine suppression test, overnight and 2-day dexamethasone suppression tests, ACTH stimulation test, and metyrapone test.

KEY TERMS

Addison's disease Primary adrenal insufficiency most commonly the result of an autoimmune adrenalitis.

adrenal cortex The outer portion of the adrenal gland, which produces various steroid hormones.

adrenal medulla The inner portion of both adrenal glands, which produces catecholamines.

adrenocorticosteroids Refers to all steroids secreted by the adrenal cortex.

adrenocorticotropic hormone (ACTH) A polypeptide hormone secreted by the anterior pituitary gland, which primarily stimulates the synthesis and release of glucocorticoids from the adrenal cortex.

adrenocorticotropic hormone (ACTH) stimulation test An initial screening test used in the assessment of adrenal insufficiency.

catecholamines Epinephrine and norepinephrine, which are produced in the adrenal medulla and are responsible for maintenance of blood pressure.

chromaffin cells These cells are found in the adrenal medulla and other sites throughout the body and produce catecholamines.

clonidine-suppression test A function test used in the diagnosis of pheochromocytoma.

congenital adrenal hyperplasia Also known as adrenogenital syndrome. A group of hereditary diseases that result from enzyme deficiencies in the steroid hormone synthetic pathways leading to altered steroid hormone production patterns.

Conn's syndrome Another name used to denote primary hyperaldosteronism.

corticosteroid-binding globulin Also known as transcortin. This protein binds and transports the majority of cortisol in the circulation.

corticotropin-releasing hormone (CRH) A hypothalamic polypeptide that stimulates ACTH secretion.

Cushing's disease Elevation of glucocorticoid (cortisol) secretion as a result of oversecretion of ACTH.

Cushing's syndrome A spectrum of specific symptoms resulting from elevation of blood glucocorticoid levels from primary or secondary causes.

dexamethasone suppression test A function test that is used in the diagnosis and differentiation of various causes of Cushing's syndrome.

glucocorticoids A group of steroid hormones secreted by the adrenal cortex that have multiple physiological effects including regulation of carbohydrate metabolism. Cortisol is the major glucocorticoid in man.

hyperaldosteronism Increased secretion of aldosterone from the adrenal cortex either because of elevated blood renin levels or autonomous adrenocortical secretion (Conn's syndrome).

hypoadrenalism Adrenal insufficiency resulting in decreased output of steroid hormones from the adrenal cortex.

metyrapone test This adrenal function test can be used in the assessment of both hyper- and hypoadrenal function. Metyrapone blocks 11-hydroxylase activity and the response is usually determined by measuring 11-deoxycortisol in blood.

mineralocorticoids Steroid hormones secreted by the adrenal cortex that stimulate the resorption of sodium and the excretion of potassium in the distal tubules of the kidneys. Aldosterone is the major mineralocorticoid in man.

pheochromocytoma A tumor of the chromaffin cells, usually located in the adrenal medulla, which results in hypersecretion of epinephrine and norepinephrine.

renin-angiotensin system This system is responsible for the regulation of aldosterone secretion from the adrenal cortex.

zona fasciculata The middle portion of the adrenal cortex in which glucocorticoids and various sex hormones are produced.

zona glomerulosa The outer portion of the adrenal cortex in which the mineralocorticoids are produced.

zona reticularis The innermost portion of the adrenal cortex, next to the adrenal medulla, which acts in consort with the zona fasciculata.

ANATOMY

The human body contains a pair of adrenal glands situated at the upper pole of each kidney (Fig. 45-1). In the adult the adrenal cortex is made up of three distinct layers. The outer layer is called the zona glomerulosa. The wide middle layer and the inner layer are called the zona fasciculata and the zona reticularis, respectively. These three layers secrete steroid hormones that may have mineralocorticoid, glucocorticoid, or androgen functions.

The medulla consists of sheets of irregular cells with small nuclei. These cells synthesize the catecholamines and are known as chromaffin cells.

PHYSIOLOGY OF ADRENAL HORMONES

All adrenal steroids secreted by the adrenal cortex (adrenocorticosteroids) have the same basic steroid structure, consisting of four rings composed of 17 carbon atoms (Fig. 45-2). They differ only in the degree of carbon saturation and in the type of side chain. There are three major functional groups of steroids secreted by the adrenal cortex. These are the mineralocorticoids secreted by the zona glomerulosa, and the glucocorticoids and androgens secreted by the zona reticularis and zona fasciculata. Relatively minor differences in the chemical structure result in major differences in the physiological function of these steroid molecules. The regulation and biological function of the adrenal androgens are not well understood; therefore, these will not be discussed in any detail.

The adrenal medulla secretes the catecholamines. These molecules are not related in structure to the adrenal steroids and have very different physiological functions.

Glucocorticoids

The glucocorticoids, primarily cortisol in humans, are synthesized and secreted by the zona fasciculata and the zona reticularis. These steroid molecules are involved in the regulation of carbohydrate, protein, and lipid metabolism. Cortisol at high concentrations also demonstrates mineralocorticoid activity. Some of the more important physiological effects of the glucocorticoids are summarized in the box, p. 674. These hormones are essential for life, especially when the human body is subject to stress such as surgery, major illness, or severe trauma. Cortisol concentrations increase greatly during these stresses, with output of cortisol from the adrenal glands increasing from 25 mg/day to 200 to 300 mg/day. If steroids are not present in sufficient quantities in times of major stress, the patient may die of vascular collapse. Large quantities of glucocorticoids also have antiinflammatory and immunosuppressive properties.

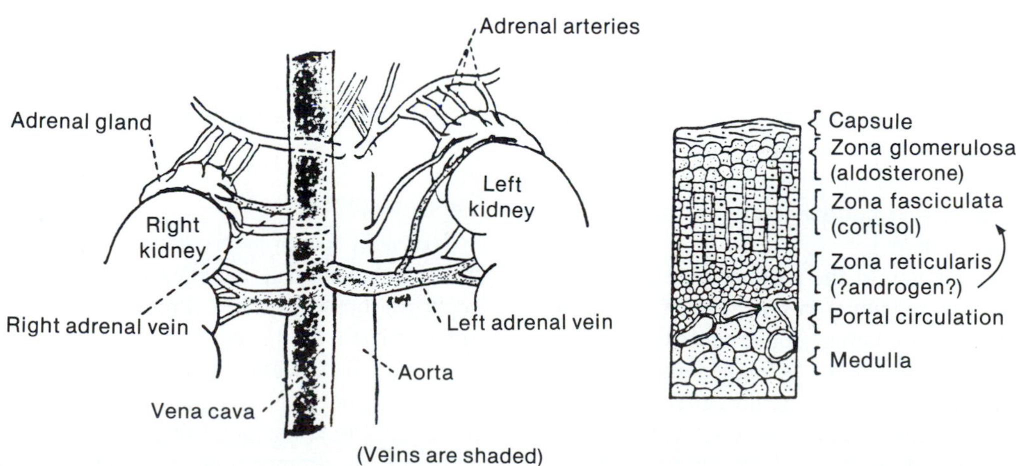

Fig. 45-1 Adrenal gland anatomy and histology. *(From Ryan W: Endocrine disorders, Chicago, 1980, Year Book Medical Publishing Co.)*

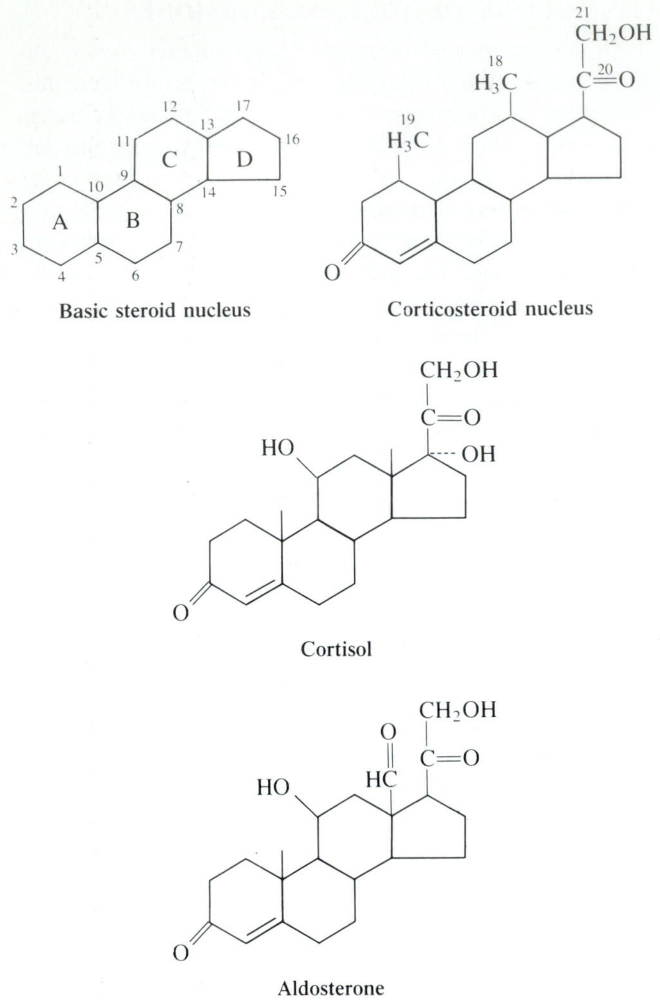

Fig. 45-2 Structures of adrenocortical hormones.

The overall action of glucocorticoids is catabolic, promoting protein and lipid breakdown and inhibiting protein synthesis. However, they have an anabolic effect on liver metabolism. The effects of cortisol are antagonistic to those of insulin, increasing the concentration of glucose by stimulating gluconeogenesis. The amino acids and glycerol released by the catabolic action of cortisol on protein and fat are used as gluconeogenic substrates. Cortisol increases the synthesis and activity of a number of enzymes in the liver involved in amino acid and glucose metabolism. The molecular basis for steroid hormone actions has been described in Chapter 40, Part I.

Cortisol also contributes to the maintenance of normal blood pressure through unknown mechanisms. Cortisol increases urine flow by stimulating glomerular filtration rate and decreasing water resorption. At high concentrations, however, cortisol can act like a mineralocorticoid promoting sodium and water retention and causing hypokalemia.

The hematologic effects of cortisol are multiple, stimulating erythropoiesis and causing leukocytosis, neutrophilia, lymphocytopenia, monocytopenia, and eosinopenia.

METABOLIC AND ANTIINFLAMMATORY ACTIONS OF GLUCOCORTICOIDS

- Carbohydrate metabolism
  - Increase in blood glucose
  - Decrease in glucose use
  - Increase in gluconeogenesis from body protein
  - Increase in liver glycogen formation
- Protein metabolism
  - Increase in protein catabolism
  - Decrease in protein synthesis
- Inhibition of allergic and inflammatory reactions
  - Decrease in antibody formation
  - Decrease in number of circulating lymphocytes and eosinophils
  - Decrease in mass of lymphatic tissue
  - Inhibition of activity of granulycotes and monocytes

From Feldkamp, CS, Powsner, ER, Zak, B, and Epstein, E: Adrenal cortex. In Sonnenwirth, AC, and Jarett, L, editors: Gradwohl's clinical laboratory methods and diagnosis, ed 8, St Louis, 1980, The CV Mosby Co.

Glucocorticoids also suppress the inflammatory and immune response by stabilizing lysozomes, interfering with leukocyte migration, and inhibiting phagocytosis.

Mineralocorticoids

Aldosterone is the primary product of the zona glomerulosa, with approximately 200 μg produced per day. This is about one one-hundredth the amount of cortisol synthesized daily by the adrenal cortex. The major physiological action of aldosterone is stimulation of sodium resorption in the distal convoluted tubules of the kidney in exchange for potassium or hydrogen.

Cortisol and other corticosteroids such as corticosterone and deoxycorticosterone have some mineralocorticoid activity that can become clinically significant when serum levels of these compounds are elevated. This can occur with the high cortisol levels seen in Cushing's syndrome. Aldosterone in turn has weak glucocorticoid activity, but its concentration is too low to have any physiological effect.

Catecholamines

The main secretory products of the adrenal medulla are the catecholamines, epinephrine (adrenaline) and norepinephrine (noradrenaline). Production of catecholamines is not restricted to the adrenal medulla. Synthesis of these hormones also occurs in the neurons of the sympathetic and central nervous systems (CNS) and in scattered groups of chromaffin cells found in other regions of the abdomen and neck. Norepinephrine is the principal product synthesized in the CNS and epinephrine is the principal catecholamine produced by the adrenal glands.

Physiological actions of the catecholamines are diverse. Norepinephrine functions primarily as the neurotransmitter. Both norepinephrine and epinephrine have influences on the vascular system, whereas epinephrine influences metabolic processes such as carbohydrate metabolism. The biological actions of the catecholamines are initiated through their interaction with two different types of specific plasma membrane receptors, alpha- and beta-adrenergic receptors. These receptors have different affinities for norepinephrine and epinephrine and cause opposing physiological effects. Norepinephrine primarily interacts with alpha receptors, whereas epinephrine interacts with both alpha and beta receptors.

Stimulation of alpha-adrenergic receptors results in vasoconstriction, decrease in insulin secretion, sweating, piloerection (hair standing on end), and stimulation of glycogenolysis in the liver and skeletal muscle, leading to an increase in blood glucose concentration. Stimulation of beta receptors, however, leads to vasodilation; stimulation of insulin release; increased cardiac contraction rate; relaxation of smooth muscle in the intestinal tract, and bronchodilation by relaxation of smooth muscles in bronchi; stimulation of renin release, which enhances sodium resorption from the kidney; and enhanced lipolysis.

BIOSYNTHESIS

Adrenocorticosteroids

All adrenal steroid synthesis begins with cholesterol. Cholesterol in the adrenal tissue may be synthesized in situ from acetate or may come from cholesterol made in the liver and transported to the adrenal glands by low-density lipoprotein.

The biosynthetic pathway leading to the three major groups of adrenal steroids is outlined in Fig. 45-3. The rate-limiting step in the synthesis of all steroids is the conversion of cholesterol to pregnenolone. This step is stimulated by ACTH in the zona fasiculata and zona reticularis and by angiotensin III in the zona glomerulosa. The pathway leading to progesterone is common to both aldosterone and cortisol synthesis. In the zona reticularis and fasciculata, progesterone is hydroxylated at the 17, 21, and 11 positions to form cortisol. Under normal circumstances, 10 to 30 mg of cortisol is synthesized per day. The zona glomerulosa does not contain 17 hydroxylase activity. Instead, hydroxylation occurs at positions 21, 11, and 18; finally, a dehydrogenase reaction forms aldosterone.

The androgens are derived from the major pathway of steroid biosynthesis, following cleavage of the side chain attached to carbon 17 to ring D. See Chapter 43 for a description of the synthesis of androgens. The adrenal gland synthesis of androgens is significant, producing 60% of the circulating testosterone levels in females.

Catecholamines

The biochemical pathway leading to the synthesis of the catecholamines is outlined in Fig. 45-4. The rate-limiting step is the hydroxylation of the amino acid tryosine leading to the formation of dehydroxyphenylalanine (dopa). This step is inhibited by both epinephrine and norepinephrine. Tyrosine comes from the diet or from hydroxylation of phenylalanine. Dopa is decarboxylated to form dopamine, which is a major end product in the central nervous system where it functions as a neurotransmitter. Dopamine is stored in granules that are present in both neurons and the

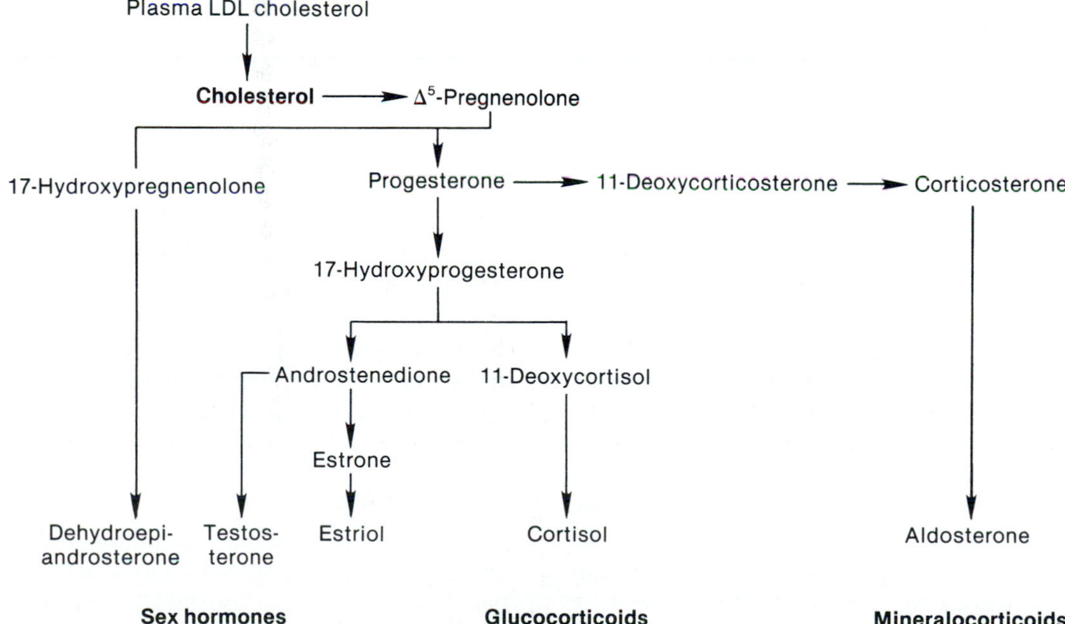

Fig. 45-3 Scheme of biosynthetic pathways of cortical hormones.

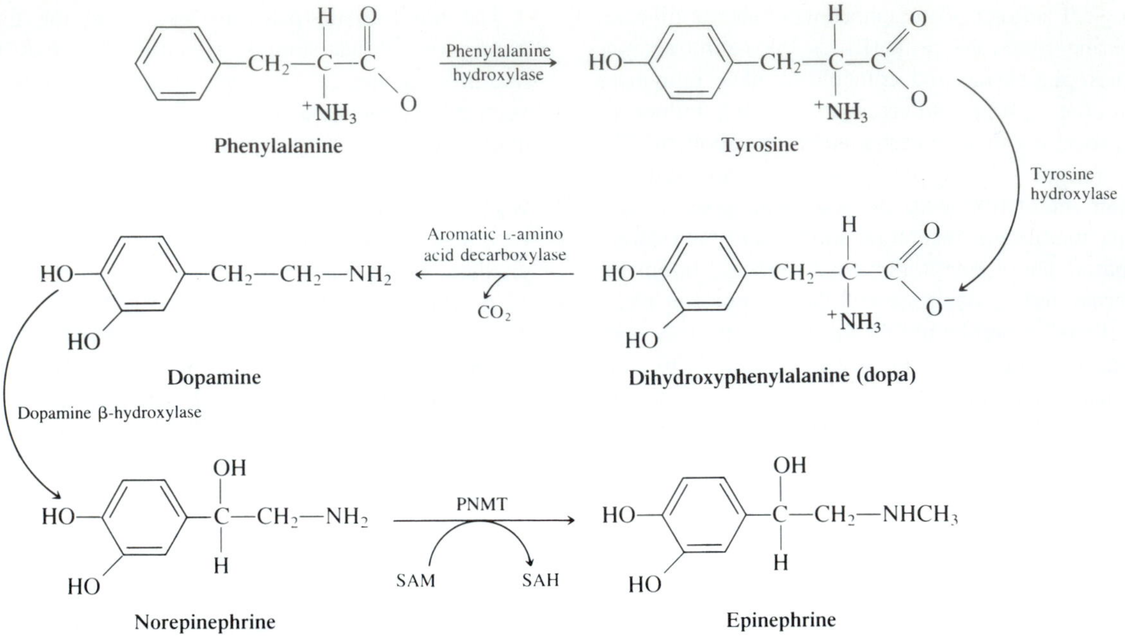

Fig. 45-4 Synthesis of medullary hormones. *PNMT,* Phenylethanolamine *N*-methyl transferase; *SAM, S*-adenosylmethionine; *SAH, S*-adenosyl homocysteine. *(From Orten JM, and Neuhaus OW: Human biochemistry, St Louis, 1982, The CV Mosby Co.)*

adrenal medulla. Within the granules, dopamine β-hydroxylase converts dopamine to norepinephrine. Finally in the adrenal medulla release from the storage granules, phenylethanolamine-*N*-methyl transferase (PNMT) converts norepinephrine to epinephrine. PNMT, which catalyzes the methylation of norepinephrine, is only found in the adrenal medulla. This enzyme is induced by glucocorticoids (cortisol). The hormones of the adrenal medulla are stored complexed with proteins (chromogranin A, dopamine-β-hydroxylase) and adenosine 5′-triphosphate (ATP) in chromaffin granules. Nerve stimulation results in release of the stored catecholamines from these vesicles by the process of exocytosis.

TRANSPORT AND CATABOLISM

Adrenocorticosteroids

The aldosterone and cortisol present in the plasma are not entirely free in solution but are bound in plasma proteins to different degrees. Aldosterone exists approximately 40% in the free state, whereas 4% of cortisol is free in solution. Albumin and corticosteroid-binding globulin (CBG) account for most of the binding of these two steroids. CBG is a high-affinity, low-capacity steroid binder, binding 90% of the cortisol under normal circumstances; whereas albumin is a low-affinity, high-capacity binding protein. The proportion of cortisol in the free state greatly increases as the concentration of cortisol exceeds the binding capacity of CBG (approximately 550 nmol/L). CBG levels are increased in hyperestrogenic states such as pregnancy and that engendered by women taking estrogen-containing birth control pills. The free cortisol levels remain normal under these circumstances owing to a compensatory increase in total cortisol. Aldosterone is much less affected by these hormonally induced changes.

Steroid metabolism is quite complex, and only a brief discussion of it is necessary for the understanding of the pathogenesis and laboratory investigation of adrenal disorders. Most steroids are catabolized by the liver and the kidneys. Examples of the types of reactions that are carried out include further hydroxylation of the steroid nucleus, conjugation with glucuronic acid, and reduction of the double bond in ring A. These transformations increase the water solubility of the steroids, allowing for their excretion into the urine. Only a small portion of aldosterone and cortisol is excreted unmetabolized into the urine.

The amount of cortisol directly secreted into the urine is related to the porportion of cortisol that circulates in the free form. CBG is saturated at high physiological concentrations of cortisol. Therefore any increase in cortisol above this level will result in a marked increase in the amount of cortisol excreted into the urine. Measurement of urinary free cortisol is a valuable test in the investigation of Cushing's syndrome, as will be discussed later.

Catecholamines

The catecholamines are stabilized within the storage granules of the adrenal medullary cells. However, when they are released they are rapidly degraded by two enzymes: catechol-*O*-methyltransferase (COMT) and monoamine oxidase (MAO) (Fig. 45-5). Only a small fraction

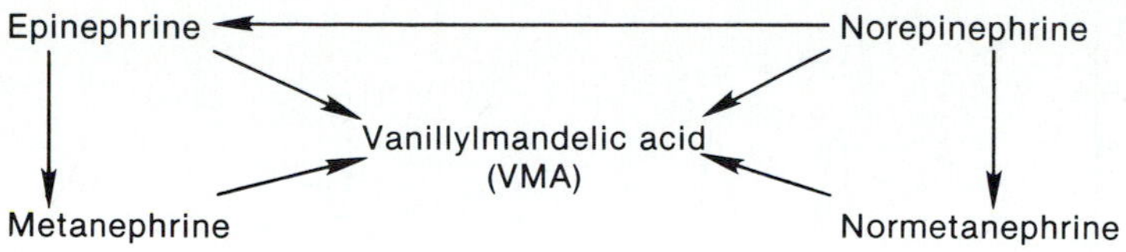

Epinephrine
Norepinephrine
Vanillylmandelic acid (VMA)
Metanephrine
Normetanephrine

Fig. 45-5 Metabolism of medullary hormones.

of catecholamines (less than 2%) is excreted unmetabolized as free catecholamines into the urine. COMT is present in many tissues, especially liver and kidney, and in erythrocytes. This enzyme methylates the C-3 hydroxyl group of norepinephrine and epinephrine, resulting in normetanephrine and metanephrine, respectively. Approximately 20% of catecholamines is excreted into the urine as metanephrines. The bulk of the catecholamines, however, is further converted to vanillylmandelic acid (VMA) by the combined action of COMT and MAO, a ubiquitous enzyme that deaminates these amines. The measurement of free catecholamines, metanephrine, and VMA in the urine all may be useful in the diagnosis of adrenal medullary disease.

CONTROL AND REGULATION

Glucocorticoids

Cortisol released from the adrenal cortex is regulated by the hypothalamic pituitary adrenal axis (Fig. 45-6 and Chapter 40). The hypothalamus synthesizes a 42 amino

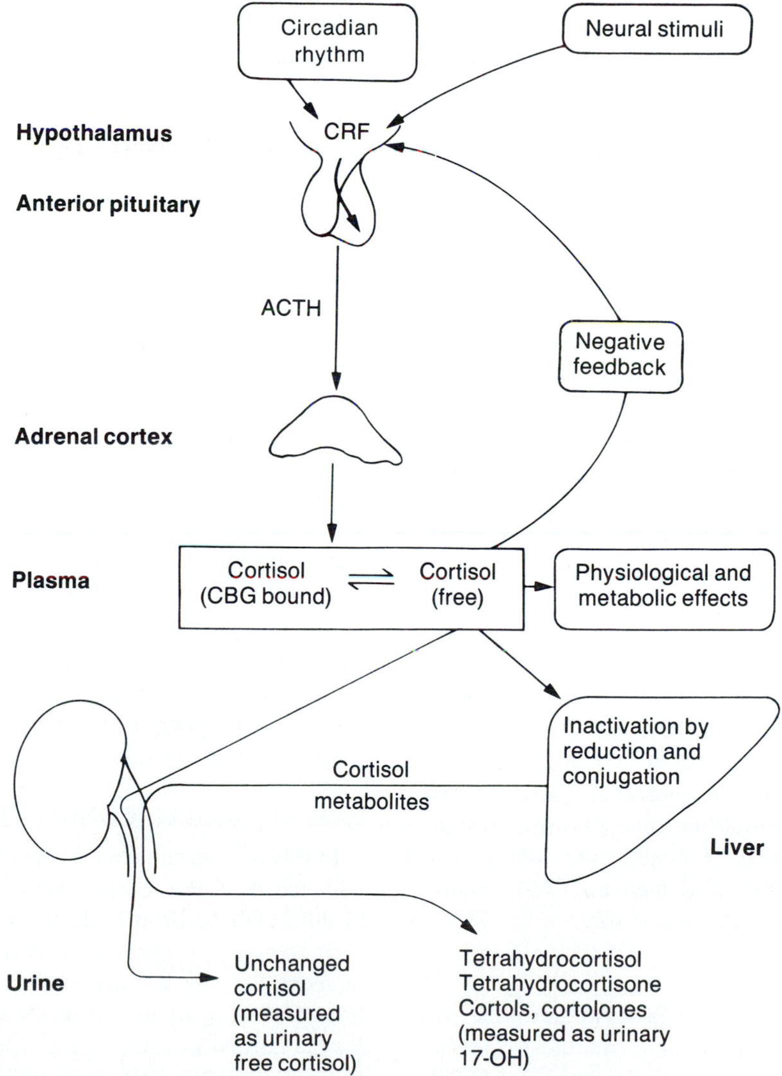

Fig. 45-6 Control and metabolism of glucocorticoids. *(From Toft A: Diagnosis and management of endocrine diseases, St Louis, 1981, Blackwell Scientific Publications.)*

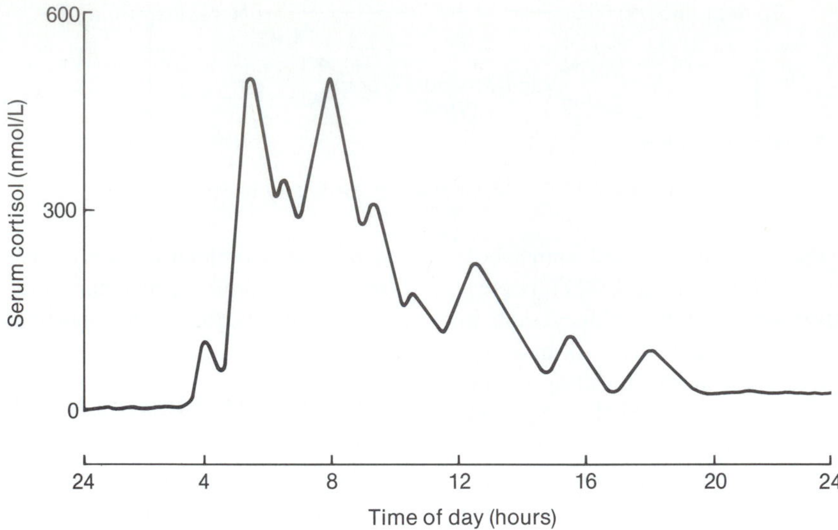

Fig. 45-7 Variation of serum cortisol concentration during 24-hour period in normal individual.

acid polypeptide, corticotropin releasing hormone (CRH), which is carried by the circulation to the anterior pituitary gland, where it causes the release of adrenocorticotropic hormone (ACTH), a 39 amino acid polypeptide. The ACTH molecule in turn interacts with membrane receptors of the cells of the adrenal cortex and through its second messenger, c-AMP, stimulates the rate-limiting step in steroidogenesis (the conversion of cholesterol to pregnenolone) leading to cortisol secretion. The free circulating cortisol in the circulation interacts with target tissues, where its effects are elicited (see above). Cortisol also acts in a negative feedback manner to control the release of ACTH from the pituitary gland. Overriding this system of negative feedback control are the higher centers of the brain, which establish the normal diurnal variation of cortisol (Fig. 45-7). Cortisol under normal circumstances is highest in the morning on waking and lowest in the late evening. The circadian pattern of ACTH release is caused by changes in CRH secretion from the hypothalamus. Short-term release of cortisol is episodic, following the pattern of ACTH pulses by about 2 to 3 minutes. Stress is another factor that can override the negative feedback of cortisol on ACTH release. Stress stimulates release of neurogenic amines, which in turn stimulate release of CRH. ACTH levels can increase up to tenfold in times of stress, resulting in high levels of cortisol. Hypoglycemia, which is a form of chemical stress, can also increase CRH release ultimately leading to an increase in cortisol.

Mineralocorticoids

In normal individuals three main factors control the secretion of aldosterone from the zona glomerulosa: (1) the renin-angiotensin system, (2) potassium, and (3) ACTH. Under normal circumstances, the renin-angiotensin system predominates. Fig. 45-8 outlines the normal regulation of aldosterone secretion. A more detailed description is found in Chapter 20.

Potassium stimulates aldosterone secretion directly at the adrenal level. Potassium may also act on the kidney. Hyperkalemia stimulates and hypokalemia inhibits renin release. This effect at the kidney level is mediated by prostaglandins and calcium. ACTH can also stimulate aldosterone secretion directly; however, this is only an acute phenomenon that is short lived.

There is normally a circadian rhythm in plasma aldosterone concentration and highest values occurring in the norming. In addition, there are alterations in aldosterone levels with postural changes. Approximately 150 μg of aldosterone are secreted per day under normal circumstances.

Catecholamines

The synthesis of epinephrine and norepinephrine is regulated by the intracellular concentrations of these hormones by negative feedback inhibition as stated previously. The catecholamines are released from the adrenal medulla in response to hypotension, hypoxia, exposure to cold, muscular exertion, pain, and emotional disturbances.

PATHOLOGICAL CONDITIONS

In this section the causes and clinical features associated with the disorders of the adrenal cortex and the adrenal medulla will be discussed. In general, the gland may hyperfunction and produce excess quantities of bioactive molecules or the gland may hypofunction and secrete too little or certain important molecules that may be essential for the normal maintenance of life. The pathological cause of these disorders may be neoplastic, hyperplastic, vascular, inflammatory, autoimmune, infectious, hereditary, or idiopathic.

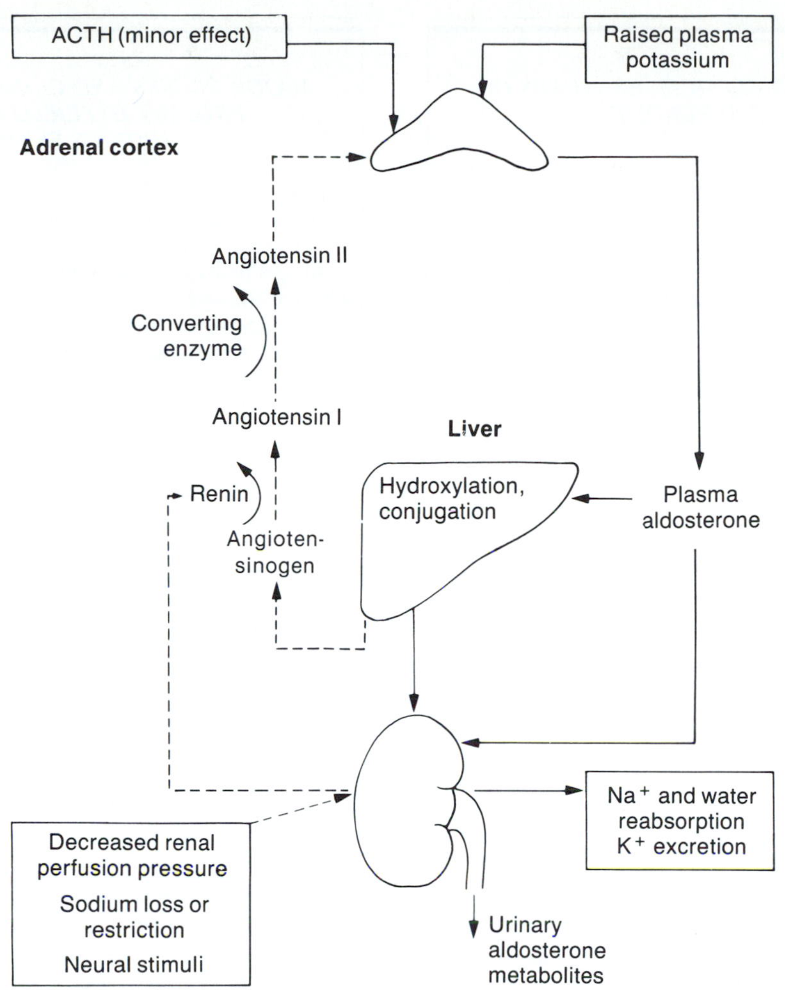

Fig. 45-8 Control and metabolism of aldosterone. *(From Toft A: Diagnosis and management of endocrine diseases, St Louis, 1981, Blackwell Scientific Publications.)*

Disorders of the adrenal cortex

Hyperadrenalism. There are three basic conditions associated with hyperadrenalism, each of which may have more than one cause. These are Cushing's syndrome, resulting from excess cortisol production; hyperaldosteronism; and congenital adrenal hyperplasia.

Cushing's syndrome. *Cushing's syndrome* is the term used to describe any condition resulting from an increased concentration of circulating glucocorticoid, usually cortisol. Excess cortisol can be caused by primary adrenal disorders (primary hyperadrenalism), for example, benign adrenal adenoma or carcinoma. These are autonomous secreting tumors that are independent of control by ACTH. Adrenal carcinoma has a particularly bad prognosis with a mean survival time following diagnosis of only 6 months. Another cause of excess cortisol may be bilateral adrenal hyperplasia resulting from autonomous secretion of ACTH by a benign pituitary adenoma (Cushing's disease) or from excess ectopic ACTH produced by a malignant tumor (secondary hyperadrenalism). The most common source of ectopic ACTH is bronchogenic carcinoma of the lung. The most common cause of Cushing's syndrome is, however, iatrogenic, the chronic administration of exogenous glucocorticoid used in the treatment or management of a wide variety of disorders (see box at left on p. 680).

The most common noniatrogenic causes are, in order: pituitary tumor (60%), ectopic ACTH (20%), and adrenal adenoma and adrenal carcinoma (combined 20%). The estimated prevalence of these disorders is approximately 1:10,000 women and 1:30,000 men. The most common cause of hyperadrenalism in children is adrenal carcinoma.

Cushing's syndrome in advanced stages may be easy to recognize. However, in early stages the patients can have a wide variety of clinical symptoms that may be confused with other common problems such as idiopathic hypertension, depression, and obesity. The laboratory also plays a major role is sorting out the diagnosis. In decreasing order of frequency, common clinical features include central obesity (90%), hypertension (85%), glucose intolerance (80%), plethoric facies (80%), purple striae (65%), hirsut-

MAJOR CAUSES AND CLINICAL FEATURES OF CUSHING'S SYNDROME

Causes
ACTH independent
- Adrenal adenoma
- Adrenal carcinoma

ACTH dependent
- Pituitary adrenal (Cushing's disease)
- Ectopic ACTH (bronchogenic carcinoma, carcinoid)
- Ectopic CRH

Iatrogenic
- Glucocorticoid therapy

Clinical features

Central obesity	Muscle weakness
Hypertension	Hirsutism
Glucose intolerance	Bruising
Plethoric facies	Osteoporosis
Purple striae	Personality change
Menstrual dysfunction	

MAJOR CAUSES AND CLINICAL FEATURES OF PRIMARY HYPERALDOSTERONISM (CONN'S SYNDROME)

Causes
- Adrenal aldosterone-producing adenoma (APA) (60% of cases)
- Idiopathic hyperaldosteronism (IHA) (40% of cases)

Clinical features
- Hypertension
- Symptoms resulting from hypokalemia
  - Muscle weakness
  - Polyuria and polydipsia
  - ECG changes
  - Glucose intolerance

ism (65%), menstrual dysfunction (60%), muscle weakness (60%), bruising (40%), and osteoporosis (40%). Less common findings include mental changes, pigmentation, acne, and hypokalemic alkalosis.

The catabolic effect of glucocorticoids on protein metabolism can account for the bruising, striae (stretch marks), osteoporosis, and muscle weakness. Hypertension and hypokalemia can be explained by the mineralocorticoid actions of excess cortisol. Hirsutism, acne, and menstrual dysfunction, a result of excess androgen production, may most dramatically be seen in some cases of adrenal carcinoma. Hyperpigmentation sometimes occurs in association with ectopic ACTH where the high levels of ACTH may have some melanocyte-stimulating activity causing generalized hyperproduction of melanin.

Hyperaldosteronism. Primary autonomous hypersecretion of aldosterone by the zona glomerulosa can be caused by two disorders: (1) an adrenal adenoma producing aldosterone, or Conn's syndrome (approximately 60% of cases); and (2) idiopathic hyperaldosteronism caused by bilateral hyperplasia (approximately 40% of cases). The major clinical feature of these disorders is hypertension. The estimation of prevalence of this disorder in the hypertensive population varies from 0.01% to as high a 2.2%. The hypertension associated with primary hyperaldosteronism can be explained by the actions of aldosterone that result in elevated plasma sodium levels and decreased plasma potassium levels (see Chapter 21). If the hypokalemia is significant, patients may have problems related to potassium depletion, including muscle weakness, fatigue, and cramping. Chronic hypokalemia can cause nephrogenic diabetes insipidus with polyuria and polydipsia, as well as cardiac dysrhythmias. Hypokalemia may also result in glucose intolerance because potassium is required for normal release of insulin from pancreatic cells.

Hyperaldosteronism may also result from secondary causes. In these situations the adrenal gland is not autonomously secreting aldosterone but is responding to enhanced production and release of renin from the kidney, which may be triggered by sodium loss, decreased renal perfusion, or volume depletion. Rarely the hypersecretion of renin may be inappropriate such as in *Bartter's syndrome,* a kidney defect in chloride resorption, or in patients with renin-secreting tumors. In contrast to primary aldosteronism where renin is decreased, plasma renin is elevated in secondary aldosteronism (see box above).

Congenital adrenal hyperplasia. Congenital adrenal hyperplasia or adrenogenital syndrome describes a group of inborn errors of metabolism involving deficiency of enzymes in the biosynthetic pathways leading to cortisol and aldosterone production. There are at least six distinct inheritable defects in this pathway, the most common of which is a 21-hydroxylase deficiency. These enzyme defects lead to diminished production of cortisol, which results in increased levels of ACTH; this in turn stimulates adrenal hyperplasia and steroid production in an attempt to overcome the enzyme deficiency. The block is usually partial, and the patient may be capable of maintaining normal levels of cortisol and aldosterone under normal circumstances at the expense of accumulation of steroid precursors that are diverted down other metabolic pathways. Commonly there is hypersecretion of various androgens, which may lead to precocious puberty in males and varying degrees of masculinization and sexual dysfunction in females (see Chapter 43). The clinical findings in these patients vary according to the site and the completeness of the enzyme deficiency. More severe deficiencies may be manifested in childhood. The less severe deficiencies may not become clinically apparent until after puberty. Some children with a 21-hydroxylase deficiency may have both glucocorticoid and mineralocorticoid deficiency, which may lead to problems early in life, including a life-threatening addisonian crisis (see later section). Others may simply have symptoms of excess androgen production.

A deficiency of 11-hydroxylase is the second most common cause of congenital adrenal hyperplasia. Again, depending on the severity of the deficiency, cortisol production may or may not be adequate. A unique feature of this enzyme deficiency is the accumulation of 11-deoxycorticosterone, a precursor in the aldosterone pathway. This steroid promotes sodium resorption and therefore can cause hypertension.

Hypoadrenalism (see box below)

Adrenal hypofunction or insufficiency can be caused by (1) primary disease at the adrenal level involving the entire adrenal cortex; (2) secondary adrenal insufficiency caused by decreased levels of CRH or ACTH, a result of pituitary or hypothalamic disease; or (3) long-term suppression of hypothalamic pituitary adrenal axis by glucocorticoids, which leads to adrenal atrophy. The causes of secondary adrenal insufficiency owing to decreased ACTH or CRH will not be discussed in this chapter.

Primary adrenal hypofuction or insufficiency, also known as *Addison's disease,* is relatively rare (estimated prevalence 1:50,000). The major cause today is autoimmune adrenalitis with circulating adrenal antibodies. This disorder, accounting for 70% of Addison's disease, may be associated with other autoimmune disorders such as Hashimoto's thyroiditis, hypoparathyroidism, diabetes mellitus, pernicious anemia, vitiligo, and primary ovarian failure. Other causes of primary adrenal failure include granulomatous diseases such as tuberculosis, histoplasmosis, or sarcoidosis; hemorrhagic disorders; neoplasms; amyloidosis; hemochromatosis; infections; and surgery. Tuberculosis was the leading cause of adrenal failure in the first half of this century. Symptoms of Addison's disease begin to appear after about 90% of the adrenal cortex has been destroyed. The disease usually develops slowly with progressive loss of cortisol and increasing ACTH levels, resulting in hyperpigmentation of the patient because of the melanocyte-stimulating hormone properties of ACTH. The patient may experience muscle weakness, fatigue, weight loss, and orthostatic hypotension. Because of coincident aldosterone deficiency, there is sodium loss and potassium retention. Hypoglycemia may be present because of cortisol deficiency. These symptoms may be vague and nonspecific. However, some patients may be seen with acute life-threatening disease following stress caused by illness, surgery, or trauma. *Addisonian crisis* is a result of an acute deficiency of both mineralocorticoids and glucocorticoids in these stressful states. The clinical presentation includes high fever, dehydration, nausea, vomiting, and hypotension, which rapidly evolve into circulating shock. Hyperkalemia and hyponatremia are common laboratory findings along with hemoconcenteration and elevated urea levels resulting from fluid loss.

Patients with secondary adrenal insufficiency do not normally experience symptoms related to hypoaldosteronism because its synthesis and secretion depend on the renin-angiotensin system rather than on ACTH. Hyperpigmentation is also not a feature of this disorder. However, patients with secondary adrenal insufficiency may show other signs of hypothalamic or pituitary disease including concomitant hypogonadism and hypothyroidism. Symptoms common to both primary and secondary disease of the adrenal cortex include weakness, hypoglycemia, weight loss, and gastrointestinal discomfort.

MAJOR CAUSES AND CLINICAL FEATURES OF PRIMARY ADRENAL INSUFFICIENCY

Causes
- Autoimmune adrenalitis
- Granulomatous disease
  - Tuberculosis
  - Histoplasmosis
  - Sarcoidosis
- Neoplastic infiltration
- Hemochromatosis
- Amyloidosis
- Bilateral adrenalectomy
- Infarction
- Infectious disease

Clinical features
- Muscle weakness
- Fatigue
- Weight loss
- Orthostatic hypotension
- Pigmentation
- Addisonian crisis
  - Fever
  - Dehydration
  - Nausea
  - Vomiting
  - Hypotension

Disorders of the adrenal medulla: pheochromocytoma

Pheochromocytoma, a relatively rare, usually benign tumor arising from chromaffin cells, results in hypersecretion of the catecholamines, epinephrine and norepinephrine. It is estimated that at least 0.1% of patients with persistent diastolic hypertension may have this tumor. Although it is a rare cause of hypertension, it is important to diagnose because it is surgically curable, and, even more significantly, it can cause death from acute hypertensive attacks. Approximately 90% of pheochromocytomas occur in the adrenal glands with the remainder occurring in extraadrenal chromaffin cells. Approximately 10% of pheochromocytomas are bilateral or multiple, and approximately 10% are malignant. Pheochromocytomas also occur as an inheritable disorder. In this case they may also be associated with neurofibromatosis and multiple endocrine neoplasia type II syndrome, which consists of pheochromocytoma, hyperparathyroidism, and medullary carcinoma of the thyroid.

Table 45-1 Change of analyte with disease

Disease	24-hour urinary-free cortisol	17-OHS	Plasma ACTH	Urinary aldosterone	Plasma aldosterone	Plasma cortisol	Serum renin	Plasma catecholamine	Vanillylmandelic acid and metanephrine
Hypercortical disease									
Primary Cushing's syndrome	↑	↑	↓			±			
Cushing's disease (secondary)	↑	↑	±			±			
Ectopic ACTH	↑	↑	↑			±			
Primary hyperaldosteronism				↑	↑		↓		
Secondary hyperaldosteronism				↑	↑		↑		
Hypocortical disease									
Primary	±	↓	↑			↓			
Secondary	±	↓	↓			↓			
Pheochromocytoma								↑	↑

↑, Elevated; ±, variable response; ↓ diminished.

MAJOR CAUSES AND CLINICAL FEATURES OF PHEOCHROMOCYTOMA

Causes
- Benign adrenal chromaffin cell tumor (80%)
- Malignant adrenal chromaffin cell tumor (10%)
- Extraadrenal chromaffin cell tumor (10%)

Clinical features
- Episodic or sustained hypertension
- Headache
- Sweating
- Palpitations with or without tachycardia
- Nervousness
- Weight loss
- Nausea
- Weakness or fatigue
- Less common
- Flushing
- Dyspnea
- Dizziness

These tumors may release their hormones in a sustained or episodic fashion. Clinically the most significant finding is persistent or paroxysmal hypertension. Other common findings include headache (70% to 90%), sweating (60% to 70%), and palpitations with or without tachycardia (50% to 70%). Other symptoms are summarized in the box above. Many of the symptoms may be persistent or episodic. Episodes can be as infrequent as once every few weeks or as frequent as 20 to 30 times daily, with attacks persisting for less than a minute or for as long as a week. The symptom pattern depends on the nature of the catecholamines secreted by the tumor.

CHANGE OF ANALYTE IN DISEASE (Table 45-1)

Hyperadrenalism

Cushing's syndrome. In the laboratory investigation of Cushing's syndrome, it is first established that the patient actually has autonomous cortisol production. Once this is established, the laboratory investigation is used to differentiate the cause of the Cushing's syndrome.

One of the simplest or most important initial tests to perform is the overnight dexamethasone suppression test. Dexamethasone is a synthetic glucocorticoid that is 30 times as potent as cortisol. The patient is given a tablet containing 1 mg of dexamethasone and is instructed to take this at 11 PM and come to the laboratory for plasma cortisol determination at 8 AM the following morning. A morning cortisol level less than 140 nmol/L usually excludes any cause of hypercortisolism. This is a normal response. Levels greater than 280 nmol/L indicate hypercortisolism. False-positive results (for example, patients who do not have Cushing's syndrome yet fail to suppress their 8 AM cortisol below 140 nmol/L following this overnight procedure) can be seen in patients with some forms of mental depression, stress-induced hypercortisolism, or pseudo-Cushing's syndrome resulting from chronic alcoholism. This test has a reported sensitivity of 98% for Cushing's syndrome. However, the specificity is not as good. However, it is an ideal test to screen for Cushing's syndrome.

Another useful initial investigative test is the estimation of a 24-hour urine-free cortisol level. This is in essence a direct measure of the amount of cortisol that is not bound to plasma protein that is excreted unmetabolized in the urine over a 24-hour period. This test also shows good sensitivity (95%) for Cushing's syndrome. A 24-hour urine cortisol determination is required along with a urine creatinine determination to ensure the adequacy of collection. The measurement of other steroid metabolites in urine, which reflect glucocorticoid output such as 17-hydroxycorticosteroids and 17-ketogenic steroids, is not considered as reliable.

Plasma cortisol values normally display diurnal variation with the highest levels occurring in the morning and the lowest levels in the early evening. Evening values are less than 50% of the early morning concentrations. Classically, samples are drawn at 8 AM (normal range, 140 to 660 nmol/L) and 4 PM (normal range, 50 to 330 nmol/L). Many patients with Cushing's syndrome will not show this diurnal variation and will have elevated concentrations at both times. However, the release of cortisol is episodic, and there is considerable overlap between normal patients and patients with Cushing's syndrome. Circadian variation is not always lost in Cushing's syndrome. To differentiate patients with Cushing's syndrome from normal patients it is best to take the sample at the time when cortisol is normally at its lowest concentration in the circulation. This time happens to be at midnight, and because it is not always practical to draw blood at this time, the more practical time of 8 PM is often chosen. Plasma cortisol levels greater than 420 nmol/L at this time are highly suggestive of hypercortisolism.

Another variation of the overnight dexamethasone suppression test is the low-dose dexamethasone suppression test, which can also be used to establish that the patient has Cushing's syndrome. This test is more time consuming. The patient is given a total of 2 mg of dexamethasone per day for 2 days. The dose is given in 0.5 mg doses every 6 hours. During the second day, a 24-hour urine specimen is collected for urine-free cortisol. Plasma cortisol measurements may also be performed. Patients with Cushing's syndrome generally will not show significant suppression of cortisol output with this dose of dexamethasone. Normal patients should show greater than 50% suppression.

There are several possible approaches that can be used to differentiate the cause of Cushing's syndrome. One can perform a high-dose dexamethasone suppression test. This can be done as an overnight procedure using 8 mg of dexamethasone, or it can be carried out over 2 days giving a total of 8 mg of dexamethasone per day divided into four aliquots of 2 mg every 6 hours. The response can be determined by measurements of plasma cortisol or 24-hour urine-free cortisol. Patients who have pituitary tumors secreting excess ACTH (Cushing's disease) will show greater than 50% suppression of their glucocorticoid output following the high-dose dexamethasone suppression test. Pituitary tumors remain responsive to negative feedback, however, requiring higher levels of corticosteroid than normal. Patients with adrenal tumors or ectopic sources of ACTH will fail to show suppression of glucocorticoids. Anomalous responses do occur.

Plasma ACTH levels, which can be measured by radioimmunoassay, are very useful in differentiating the cause of Cushing's syndrome. ACTH is a labile polypeptide hormone, and special precautions are required in its handling. Plasma samples should be collected on ice and the plasma separated in a refrigerated centrifuge and stored frozen. ACTH levels in the upper normal range (normal range, 0 to 22 pmol/L) and up to twice the upper limit of normal are consistent with a pituitary cause of Cushing's syndrome. Nondetectable levels of ACTH (less than 2 pmol/L) suggest an adrenal tumor. Very high levels of ACTH (greater than 50 pmol/L) are suggestive of an ectopic source such as a malignant tumor. There is some overlap in the plasma concentrations of ACTH associated with ectopic and pituitary causes of Cushing's syndrome.

The metyrapone stimulation test has also been used to delineate the cause of Cushing's syndrome. Metyrapone acts by inhibiting the enzyme 11-hydroxylase and therefore blocking the synthesis of cortisol. In normal patients, ACTH levels will increase because of loss of negative feedback inhibition from cortisol. The response to metyrapone can be determined by measuring either serum or plasma 11-deoxycortisol or urine 17-hydroxycorticosteroids. The following protocol is recommended. First measure the baseline serum cortisol and 11-deoxycortisol in serum at 8 A.M. Give the patient 750 mg of metyrapone every 4 hours for 24 hours, then repeat the measurement of serum cortisol and 11-deoxycortisol. In Cushing's syndrome caused by pituitary tumors the ACTH response remains intact and 11-deoxycortisol levels increase to levels greater than 200 nmol/L. Levels of 11-deoxycortisol that are less than this are consistent with an adrenal tumor or ectopic ACTH. The recent literature suggests that this test may be better than the high-dose dexamethasone suppression test in determining the cause of Cushing's syndrome.

Finally, since the recent discovery and availability of corticotropin-releasing hormone (CRH), there has been interest in using this peptide to diagnose the cause of Cushing's syndrome. Ovine CRH is administered intravenously. Patients with pituitary tumors will respond with an increase in the level of plasma ACTH and cortisol. Patients with ectopic ACTH and adrenal tumors should show

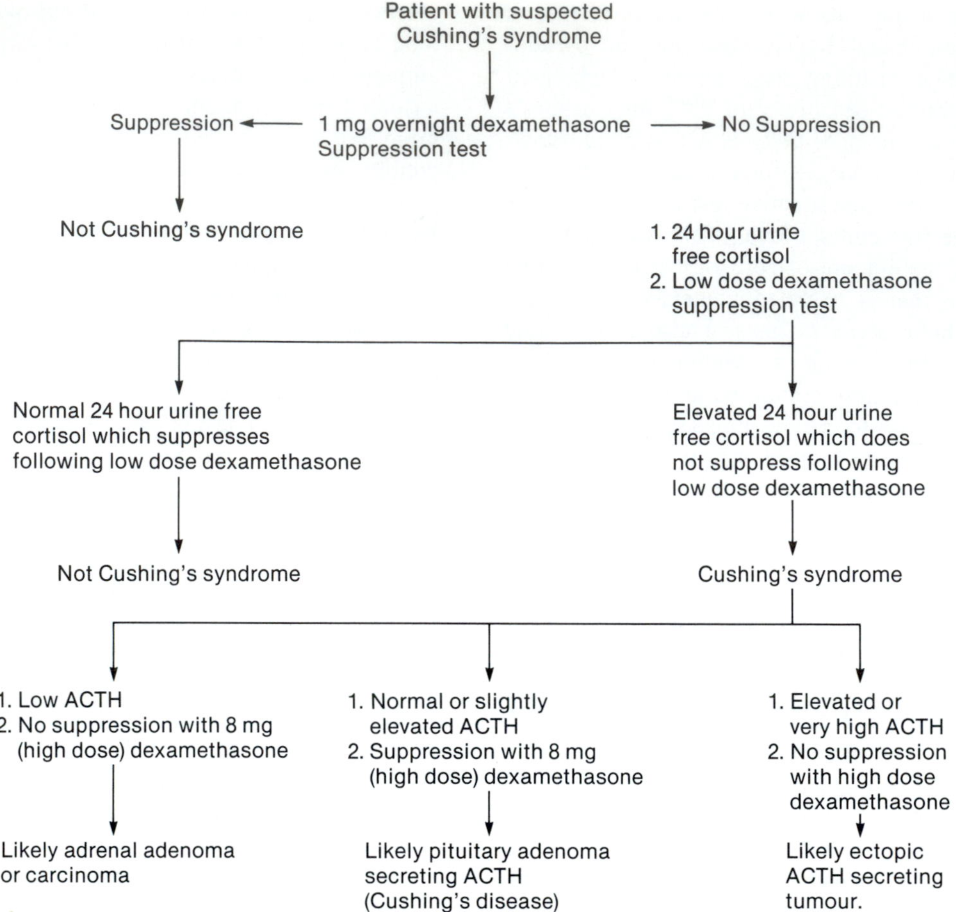

Fig. 45-9 Laboratory protocol for investigation of Cushing's syndrome.

no response. A major role for this test is the differentiation of pituitary causes of Cushing's syndrome from ectopic ACTH because basal ACTH levels are elevated in both cases. At the present time, this test is not widely used, but it may become more popular in the future if its diagnostic value is proven in the literature. It has the advantage of being a rapid test.

In conclusion, to establish the diagnosis of Cushing's syndrome, a recommended protocol should include the following laboratory tests. Screen the patient with an overnight dexamethasone suppression test and collect urine for 24-hour urine-free cortisol. If these tests are abnormal, a high-dose dexamethasone suppression test and a plasma ACTH level should allow delineation of the cause in most cases. A more complete approach is outlined in Fig. 45-9.

Primary hyperaldosteronism

Conn's syndrome or primary hyperaldosteronism is a rare cause of hypertension. The investigation for this disorder is costly and time consuming. Therefore not all patients with hypertension should be investigated for this disorder. The simplest method for screening for this disorder is the measurement of serum and urine potassium when the patient is not receiving diuretics. Values of serum potassium of less than 3.5 mmol along with a urine potassium excretion rate of greater than 30 mmol/24 hours are not usually seen with essential hypertension but are commonly seen with primary hyperaldosteronism. This screen is not entirely reliable because normal serum potassium may be seen in some patients with primary hyperaldosteronism. The test should be repeated on two or three occasions.

Plasma renin activity may be measured after stimulation with furosemide and with upright posture. This effectively causes hypovolemia and salt depletion. Patients with primary hyperaldosteronism will not respond, and renin activity will remain low. High autonomous secretion of aldosterone in this condition suppresses the release of renin. Normal patients and patients with most other causes of hypertension will respond to the hypovolemia and salt depletion protocol with an increase in renin activity. Low renin activity may also be seen with congenital adrenal hyperplasia caused by 11-hydroxylase deficiency, excessive licorice ingestion, and low-renin essential hypertension.

The definitive test for primary hyperaldosteronism is the measurement of serum or urine aldosterone following placement of the patient on saline infusion or a high-salt

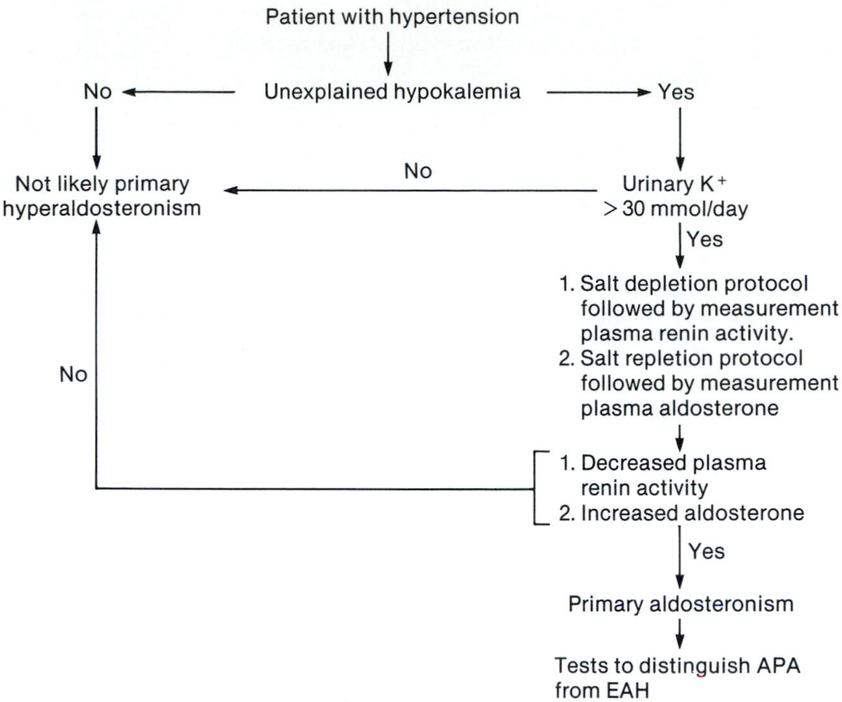

Fig. 45-10 Laboratory protocol for investigation of primary hyperaldosteronism.

diet. This results in volume expansion and suppression of the pretest aldosterone levels by 50% to 80% in normal individuals and in patients with other causes of hypertension. Nonsuppressibility of the aldosterone levels suggests autonomous aldosterone secretion caused by primary hyperaldosteronism.

To differentiate whether the primary hyperaldosteronism is due to an adenoma-producing aldosterone (APA) or to benign adrenal hyperplasia (BAH) is the next step in the investigation. This is important because APA responds favorably to surgery, whereas BAH does not, and these patients require medical therapy. Adrenal CT scans, nuclear scans with radioactive iodocholesterol, and adrenal vein catheterization studies may be useful to determine whether the aldosterone is being secreted from one adrenal gland or both. Another approach to the differentiation of APA from BAH is the measurement of aldosterone change following sodium depletion and assumption of upright posture. Patients with APA show no change, whereas patients with BAH will show a rise in plasma aldosterone.

In summary, a patient with diastolic hypertension, low serum potassium, and potassium loss in the urine exceeding 30 mmol/day may be investigated for primary hyperaldosteronism. Aldosterone should be measured following saline infusion or a high-salt diet. Renin should be measured following a low-salt diet or after administration of a diuretic. Nonsuppressibility of the aldosterone and lack of stimulation of renin activity are suggestive of primary hyperaldosteronism. A scheme for the investigation of primary hyperaldosteronism is outlined in Fig. 45-10.

Congenital adrenal hyperplasia

The incidence of congenital adrenal hyperplasia (CAH) reported in various studies ranges from 1:5000 to 1:62,000 births. The most common cause is 21-hydroxylase deficiency, accounting for 95% of all cases. A patient may have evidence of adrenal insufficiency as discussed earlier and may be investigated from this point of view as outlined in the next section of this chapter. However, the patient may also suffer the biochemical consequences of excess androgen, with females showing signs of virilization and males demonstrating precocious puberty. The definitive test for this condition is finding an elevated level of 17-hydroxyprogesterone, the immediate precursor to the metabolic block, in the serum.

Measurements of testosterone in serum and pregnanetriol in urine, a metabolite of 17-hydroxyprogesterone, may also be useful.

Hypoadrenalism

A patient with postural hypotension, low serum sodium, and increased serum potassium should suggest the possibility of primary adrenal insufficiency or Addison's disease. Samples should be drawn immediately for ACTH and baseline cortisol determinations. Aldosterone determinations may also be useful. A synthetic form of ACTH (cortrosyn), consisting of the first 24 amino acids of ACTH, can then be injected intravenously or intramuscularly. Blood samples for serum cortisol should then be drawn at 30 to 60 minutes following injection. A normal response to this cortrosyn (ACTH) stimulation test consists

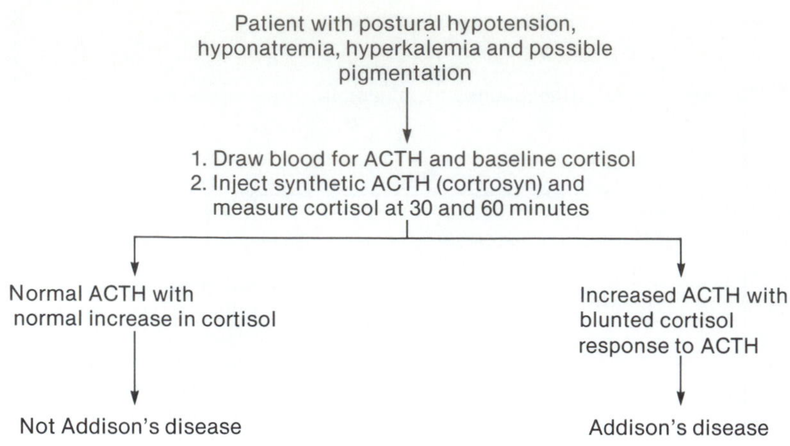

Fig. 45-11 Laboratory protocol for investigation of Addison's disease.

of a rise of the serum cortisol by at least 280 nmol/L unless the baseline is already above 550 nmol/L. A low baseline cortisol result with a failure to respond to ACTH may suggest primary adrenal failure or may be a result of atrophy caused by long-term steroid therapy or pituitary insufficiency. In primary adrenal insufficiency the ACTH levels will be greatly elevated (greater than 50 pmol/L) and in fact may be clinically apparent because of pigmentation of the patient. In secondary adrenal insufficiency or atrophy resulting from exogenous steroids, the ACTH levels will be suppressed (less than 10 pmol/L).

If the response to the cortrosyn stimulation test is abnormal and secondary adrenal insufficiency is suspected, a prolonged (3 to 5 day) ACTH stimulation test or an insulin-induced hypoglycemia stress test should be performed.

Intravenous administration of ACTH over several days generally results in a gradual increase in cortisol output if the adrenal insufficiency is a result of long-term deficiency of ACTH from the pituitary. Hypoglycemia normally stimulates release of ACTH from the pituitary. A failure to show increased ACTH or increased cortisol in response to hypoglycemia is suggestive of pituitary or hypothalamic disease.

The metyrapone stimulation test is sometimes used in the investigation of adrenal insufficiency. A protocol similar to that outlined earlier can be followed. A normal response usually results in increased levels of 11-deoxycortisol in the serum or increased output of 17-hydroxycorticosteroids in the urine. There is no response or an inadequate response in the case of both primary and secondary adrenal insufficiency. A simplified scheme for the investigation of adrenal insufficiency is shown in Fig. 45-11.

Pheochromocytoma

There are many possible routes to the investigation of pheochromocytoma. One can measure urinary VMA, metanephrines, total catecholamines, or fractionated catecholamines. In addition, one can measure plasma epinephrine and norepinephrine. All these analyses should be performed by HPLC. These methods are analytically specific, but many drugs such as MAO inhibitors and reserpine can alter the in vivo metabolism of the catecholamines and may interfere with the interpretation of the results. To be on the safe side, the patient should be off all medication, if possible, before initiating the laboratory investigation.

Most of the literature suggests that a 24-hour urine metanephrine (normal range, less than 5 μmol/day) determination is the most sensitive test for screening for pheochromocytoma. However, the recent HPLC methods for VMA analysis (normal range, 10 to 35 μmol/day) are probably not significantly different in their screening sensitivity for pheochromocytoma. Urine free catecholamine measurements are also considered very valuable, especially if the catecholamines are fractionated into norepinephrine (normal range, 60 to 470 nmol/day) and epinephrine (normal range, 0 to 160 nmol/day). An increase in urine norepinephrine is one of the more specific findings associated with pheochromocytoma. These measurements are especially useful if the patient has episodic hypertension of short duration. A random urine collected shortly after the attack may indicate abnormal catecholamine excretion, whereas the metabolic concentrations may be normal.

There has been considerable interest recently in plasma catecholamine measurements especially in combination with the clonidine suppression test. The analysis of plasma catecholamines is very difficult because the concentrations are very low and the levels very labile. A number of conditions can cause elevations of plasma catecholamines into the range seen in patients with pheochromocytoma, including volume depletion, anxiety, exercise, anoxia, smoking, renal failure, obesity, and several drugs such as L-dopa and methyldopa.

The clonidine suppression test may be of some use in difficult diagnoses. Plasma catecholamines are measured before and 3 hours after the administration of 0.3 mg of clonidine. Patients with pheochromocytoma show no suppression, whereas patients with essential hypertension suppress their catecholamine levels into the normal range.

To localize the tumor before surgery, CT scans, venous sampling for catecholamines, and, recently, radioisotope imaging with metaiodine-131-iodobenzylguanidine are all useful techniques.

To summarize, appropriate patients should be screened with 24-hour urinary VMA or metanephrine analyses preferably using the newer HPLC methods combined with electrochemical detection. The results should be confirmed by the analysis of fractionated free urinary catecholamines. In patients with episodic hypertension, an analysis of a random urine for fractionated free catecholamines may be useful. In very difficult cases, if available, a clonidine suppression test combined with plasma catecholamine measurements may be done.

BIBLIOGRAPHY

General

Felig, P, Baxter, J, Broadus, A, and Frohman, L: Endocrinology and metabolism, New York, 1987, McGraw-Hill Book Co, pp 511-692.

Mulrow, PJ, editor: The adrenal gland, New York, 1986, Elsevier Publishing Co.

Stern, N, and Tuck, ML: The adrenal cortex and mineralocorticoid hypertension. In Lavin, N, editor: Manual of endocrinology and metabolism, Little, Brown & Co, Boston, 1986, pp 107-130.

Glucocorticoids

Aron, DC, Tyrrell, JB, Fitzgerald, PC, et al: Cushing's syndrome: problems in diagnosis, Medicine 60:25-35, 1981.

Baxter, JC: Glucocorticoid hormone action. In Gill, GN, editor: Pharmacology of adrenal cortical hormones, Oxford, 1976, Pergamon Press, Ltd.

Besser, GM, and Rees, LH: The pituitary-adrenocortical axis, Clin Endocrinol Metab 14:765-1042, 1985.

Chrovsos, GP, Schulte, HM, Oldfield, EH, et al: The corticotropin-releasing factor stimulation test, N Engl J Med 310:622-626, 1984.

Crapo, L: Cushing's syndrome: a review of diagnostic tests, Metabolism 28:955, 1979.

Gold, PW, Loriaux, DL, Roy, A, et al: Responses to CRH in hypercortisolism of depression and Cushing's disease, N Engl J Med 314:1329-1335, 1986.

Hermus, AR, Pesman, GJ, Benraad, TJ, et al: The CRH test versus the high-dose dexamethasone test in the differential diagnosis of Cushing's syndrome, Lancet 2:540-544, 1986.

Howanitz, PJ, and Howanitz, JH: Hypercostisolism, Clin Lab Med 4:683-702, 1984.

Krieger, DT: Physiopathology of Cushing's disease, Endocrinol Rev 4:2-43, 1983.

Moore, A, Aitken, R, Burke, C, et al: Cortisol assays: guidelines for the provision of a clinical biochemistry service, Ann Clin Biochem 22:435-454, 1985.

Urbanic, RC, and George, JM: Cushing's diseases: 18 years' experience, Medicine 60:14-24, 1981.

Trecan, GV, Laudat, MH, Thomopoulos, P, et al: Urinary free corticoids: an evaluation of their usefulness in the diagnosis of Cushing's syndrome, Acta Endocrinol 103:110-115, 1983.

Wand, GS, and Ney, RL: Disorders of the hypothalamic pituitary-adrenal axis, Clin Endocrinol Metab 14:33-53, 1985.

Mineralocorticoids

Bravo, EL, Tarazi, RC, Dustan, HP, et al: The changing clinical spectrum of primary aldosteronism, Am J Med 74:641-651, 1983.

Drury, PL: Disorders of mineralocorticoid activity, Clin Endocrinol Metab 14:175-202, 1985.

Grangulay, A, Grim, CE, and Weinberger, MH: Primary aldosteronism: the etiologic spectrum of disorders and their clinical differentiation, Arch Intern Med 142:813-815, 1982.

Hiramatsu, K, Yamada, T, Zukimura, Y, et al: A screening test to identify aldosterone-producing adenomas by measuring plasma renin activity, Arch Intern Med 141:1589-1593, 1981.

Lyons, DF, Kem, DC, Brown, RD, et al: Single dose captopril as a diagnostic test for primary aldosteronism, J Clin Endocrinol Metab 57:892-896, 1983.

Marver, D, and Kokko, JP: Renal target sites and the mechanism of action of aldosterone, Mineral Electrolyte Metab 9:1-18, 1983.

Weinberger, MH, Grim, CE, Hollifield, JW, et al: Primary aldosteronism: diagnosis, localization, and treatment, Ann Intern Med 90:386, 1979.

Zipser, RD, Meidor, V, and Horton, R: Characteristics of aldosterone binding in human plasma, J Clin Endocrinol Metab 50:158-162, 1979.

Catecholamines

Bravo, EL, Tarazi, RC, Gifford, RW, et al: Circulatory and urinary catecholamines in pheochromocytoma: diagnosis and pathophysiologic implications, N Engl J Med 301:682, 1979.

Bravo, EL, Tarazi, RC, Fouad, RM, et al: Clonidine-suppression test: a useful aid in the diagnosis of pheochromocytoma, N Engl J Med 305:623, 1981.

Bravo, EL, and Tarazi, RC: Plasma catecholamines in clinical investigation: a useful index or a meaningless number? J Lab Clin Med 100:155-160, 1982.

Bravo, EL, and Gifford, RW: Pheochromocytoma: diagnosis, localization, and management, N Engl J Med 311:1298-1303, 1984.

Cryer, PE: Pheochromocytoma, Clin Endocrinol Metab 14:203-220, 1985.

Davidson, DF: Urinary free catecholamines: diagnostic application of an HPLC technique to the investigation of neural crest tumors, Ann Clin Biochem 24:494-499, 1987.

Eckfeldt, JH, and Engleman, K: Diagnosis of pheochromocytoma, Clin Lab Med 4:703-716, 1984.

Kaplan, NM, Kramer, NJ, Holland, OB, et al: Single-voided urine metanephrine assays in screening for pheochromocytoma, Arch Intern Med 137-190, 1977.

Plouin, PF, Duclos, MJ, Menard, J, et al: Biochemical tests for diagnosis of phaeochromocytoma: urinary versus plasma determinations, Br Med J 282:853-854, 1981.

St. John Sutton, MG, Sheps, SG, and Lie, JT: Prevalence of clinically unsuspected pheochromocytoma: review of a 50-year autopsy series, Mayo Clin Proc 56:354-360, 1981.

Sowers, JR: Pheochromocytoma. In Lavin, N, editor: Manual of endocrinology and metabolism, Little, Brown & Co, Boston, 1986, pp 131-142.

CHAPTER 46

Diseases of genetic origin

THADDEUS E. KELLY

OBJECTIVES

- Briefly describe the chromosome abnormality associated with the following pathological disorders: Down's syndrome, Turner's syndrome, Klinefelter's syndrome, chronic myelogenous leukemia, and Fanconi anemia.
- Briefly describe the metabolic disorder in each of the following and state the primary abnormal clinical chemistry results:
  - Gaucher's disease
  - Niemann-Pick disease
  - Tay-Sachs disease
  - Hurler's syndrome
  - Phenylketonuria
  - Maple-syrup urine disease
  - Galactosemia
  - Fanconi syndrome
  - von Gierke's disease
  - Vitamin D–resistant rickets
  - Wilson's disease
  - Familial hypercholesterolemia
  - Lesch-Nyhan syndrome
- Describe the role of genetic screening in the diagnosis of genetic disease.

KEY TERMS

allele One of various forms of a gene that may appear at a specific locus.

amniocentesis A transabdominal aspiration of the uterus by syringe to obtain amniotic fluid.

aneuploidy A chromosomal abnormality caused by the addition or absence of an entire chromosome.

autosomal Pertaining to any of the 22 chromosomes except the X and Y chromosomes.

Barr body The condensation of nuclear (genetic) material of the inactivated X chromosome.

chromosome Nuclear structure containing a linear array of genes. Humans have 23 pairs of chromosomes.

diploid The duplicate representation of each gene and chromosome.

dominant trait A trait that is expressed or determined by the heterozygous presence of an allele at the locus on the chromosome.

Down's syndrome A condition characterized by mental retardation and physical abnormalities caused by trisomy of chromosome 21.

dysmorphogenesis Physical defects caused by intrinsically altered embryonic development.

galactosemia A toxicity syndrome associated with intolerance to dietary galactose and characterized by deficiency of the enzyme galactose-1-phosphate uridyl transferase.

gene The smallest biological unit of heredity, located on specific sites of specific chromosomes. A gene contains the encoding for one specific protein.

Guthrie test A microbiologic assay for serum phenylalanine; used to screen for PKU.

heterozygous Pertaining to a state in which a pair of alleles are dissimilar at both positions of the same locus.

homozygous Pertaining to a state in which a pair of alleles are the same at both positions of the same locus.

karyotype The chromosomal makeup of a nucleated cell.

Lesch-Nyhan syndrome A rare X-linked error of purine metabolism characterized by mental retardation and self-mutilation. There is a deficiency of the enzyme hypoxanthine-guanine phosphoribosyl transferase.

locus The particular location on a given chromosome occupied by a structural gene.

lyonization A process of random inactivation of an X chromosome to compensate for the double-gene dosage of two X chromosomes in females.

meiosis Segregation of genes and chromosomes with reduction from a diploid to a haploid number in sperm and egg formation.

mitogen A chemical substance that induces cells in culture to divide.

monosomy Absence of one chromosome from an otherwise diploid cell ($2n - 1$). Only known disorder is absence of Y or X chromosome in Turner's syndrome.

phenylketonuria (PKU) An autosomal, recessively inherited disorder resulting from a defect in conversion of phenylalanine to tyrosine because of a phenylalanine hydroxylase deficiency.

recessive trait The expression of a trait that requires that both alleles at a locus are mutated.

Tay-Sachs disease Infantile form of a recessive hereditary disorder caused by a defect in lipid metabolism in which sphingolipids accumulate in the brain, resulting in progressive mental and physical degeneration. The disease occurs primarily in children of Jewish ancestry.

trait An observable feature of an organism that is visually apparent or laboratory derived.

translocation A term used in genetics to describe the movement of a portion of one chromosome into the structure of another.

trisomy Presence of one additional chromosome of a specific pair in an otherwise diploid cell ($2n + 1$ chromosomes). See *Down's syndrome*.

Turner's syndrome A condition consisting of absent ovarian function, short stature, and physical anomalies; most commonly caused by a 45,X karyotype.

X linked Pertaining to a trait carried on the X chromosome.

GENETIC BASIS OF INHERITANCE

In the 1860s, after a series of simple experiments, Gregor Mendel proposed two laws regarding the transmission of traits from generation to generation. In the first law Mendel stated that the segregation of factors determining a trait followed predictable patterns and that one could anticipate the proportion of offspring expressing various forms of a given trait. The second law, the law of independent assortment, stated that the inheritance of one trait had no effect on the inheritance of a second trait. Approximately 40 years later, with the capability of observing cellular division under the microscope, it was recognized that the behavior of chromosomes during cellular division was consistent with the predictions made by Mendel regarding the segregation of traits. We now recognize that the factors described by Mendel are *genes* and that these are carried by *chromosomes* whose segregation can be observed in cellular division.

Chemical basis

Genes must be described in functional terms. Genes are composed of DNA, and occasionally (in viruses) RNA, which are made up of nucleotide sequences. Genes of higher organisms occupy specific locations on chromosomes and code for the amino acid sequences of polypeptides that have specific physiological functions. Each polypeptide may function as a single protein or combine with other polypeptides to form a heteropolymeric functional unit. However, the basic premise remains that one structural gene encodes a unique polypeptide with a specific function (Fig. 46-1). A change in the DNA sequence of a structural gene, a mutation, will result in a structurally altered polypeptide. Different alterations in a gene, and thus in the resulting polypeptide, will produce different changes in the phenotype through a functional change in the gene product. A mutation in the gene for phenylalanine hydroxylase may have a moderate effect on enzyme activity and produce hyperphenylalaninemia, whereas a different mutation of the same gene may have a profound effect and result in phenylketonuria. The kinds of diseases that result from such mutations depend on the function of the polypeptide. Examples are illustrated in Table 46-1.

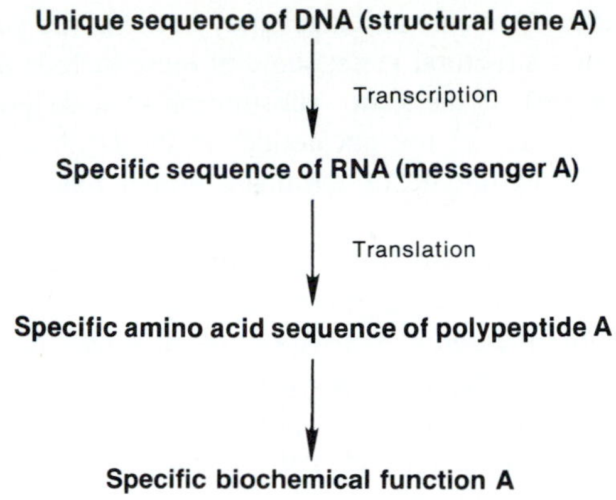

Fig. 46-1 Sequence of events from a structural gene leading to a specific, normal biochemical function.

The genetic material of human cells is *diploid* in that each gene is represented twice, one of each pair occupying a specific location, or a *locus,* on each of a pair of similar, or homologous, chromosomes. The chromosomes segregate during *meiosis,* reducing the chromosome number from a diploid to haploid number. Humans have a chromosomal or modal number of 46 in somatic cells and 23

Table 46-1 Diseases caused by various genetic mutations

Protein type	Consequences of mutation	Example
Enzyme	Loss of enzyme activity	Phenylketonuria
Hemoglobin	Altered oxygen affinity	Sickle cell disease
Structural protein (cartilage collagen)	Defective bone matrix	Osteogenesis imperfecta
Receptor	Altered metabolic regulation	Familial hypercholesterolemia
Membrane protein	Altered membrane transport	Cystinuria
Coagulation protein	Defective coagulation	Hemophilia
Carrier protein	Inability to transport compound	Hemochromatosis

in gametes. Numerous changes may occur in the DNA sequence of a structural gene; some of these include deletion of all or part of the gene, substitution of a different nucleotide for any of the nucleotides in the DNA sequence, and failure to initiate or terminate proper reading of the DNA code.

Virtually every structural gene locus has a series of *alleles* that may occur. In most instances these alleles were created by mutations consisting of a single nucleotide change in the structure of the gene and thus a single amino acid substitution in the gene product. Many of these gene products are equally functional but may be recognized by different electrophoretic or immunological properties of the gene product. On the long arm of chromosome 1 is the locus for the Duffy blood group. At that locus, various alternative forms of the gene, or alleles, may occur giving rise to different Duffy blood types. The two most common alleles at the Duffy blood group locus are designated *a* and *b*. The gene products for the *a* and *b* alleles of the Duffy blood group are recognized by standard blood-typing techniques. When the pair of alleles at a given locus are the same, the person is said to be *homozygous* at that locus. Thus a person with the *a* allele at both positions on the pair of 1 chromosomes for the Duffy blood group is homozygous *a*. When the pair of alleles are dissimilar, such as the Duffy allele *a* and *b*, the person is a heterozygote.

A series of allelic mutations may impart a range of functional consequences on the polypeptide gene product. Many different mutations occur in the structural gene for the beta chain of hemoglobin. This series of alleles results in functional changes in beta-hemoglobin chains that range from insignificant, no disease, to lack of the gene product and severe disease, beta thalassemia.

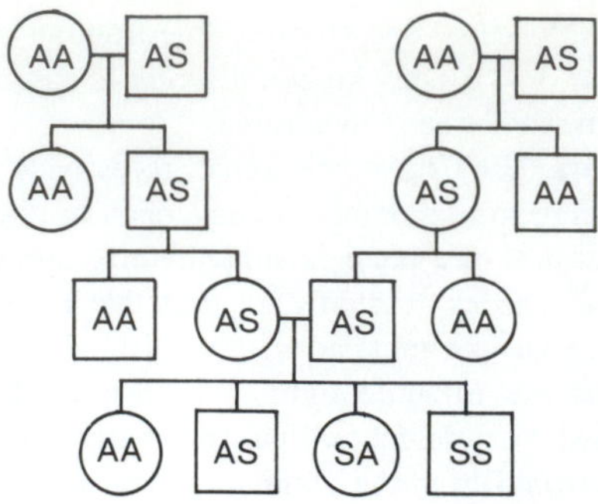

AA, Homozygous for normal hemoglobin A
AS, Heterozygous for hemoglobins A and S (sickle cell carrier)
SS, Homozygous for mutant hemoglobin S (sickle cell disease)

Fig. 46-2 Pedigree that demonstrates segregation of single-gene mutations. Heterozygous state (carrier for sickle-cell disease in this example) is inherited from a single parent, whereas homozygous state (a person affected with sickle cell disease in this example) requires inheritance of a mutant gene from each parent.

Single gene patterns of inheritance

The chromosomes in the pairs numbered 1 to 22 are called *autosomes,* and the chromosomes in the remaining pair are called the *sex chromosomes* (X and Y). When a *trait* is determined by the heterozygous presence of an allele at the locus on an autosome, that trait is inherited as an *autosomal dominant trait*. When the expression of the trait requires that both genes consist of the same mutant allele, the trait is said to be an *autosomal recessive trait*. The terms *dominant* and *recessive* refer to the trait in question and not to the genes determining that trait. For example, the presence of the S and A alleles for the beta chain of hemoglobin gives rise to the presence of normal hemoglobin A and sickle hemoglobin S in a person. Since that condition is present in the heterozygous state, it is dominantly inherited. Thus the presence of sickle hemoglobin by electrophoresis and a tendency for sickling of red blood cells to occur under reduced oxygen tension is inherited as an autosomal dominant trait. However, the disease *sickle cell disease* requires that both alleles code for S hemoglobin. The person with sickle cell disease is homozygous for the S allele and sickle cell disease is inherited as an autosomal recessive disorder. With this definition it becomes apparent that one parent may transmit an autosomal dominant disorder to his or her offspring as it is determined by the presence of a single allele, but that an autosomal recessively inherited disorder requires the inheritance of mutant alleles from both parents. It can be seen from Fig. 46-2 that one may inherit the carrier state for sickle cell disease from a single parent, but sickle cell disease itself must be inherited from both parents. In general, the disorders that are characterized by altered biochemical metabolism and are diagnosable through biochemical means are conditions that are autosomal recessively inherited.

Two sex chromosomes of the human male consist of an X chromosome and a Y chromosome. The human female has two X chromosomes. The enzyme glucose-6-phosphate dehydrogenase (G-6-PD) is coded for by a gene on the X chromosome. Because a female has two X chromosomes and a male has one X chromosome, it might be expected that a female would make twice as much G-6-PD as a male does. On the average, however, females and males make equal amounts of G-6-PD. There is gene dosage compensation that occurs by a mechanism called *lyonization,* or random inactivation of the X chromosome. During early embryonic development of the 46,XX female, one of the two X chromosomes is randomly inactivated; it no longer produces gene products. Because this is a random process, each female is a mosaic, or mixture, of cells in which one or the other, but not both, X chromosomes is functioning (Fig. 46-3). The presence of a Y chromosome is associated with maleness, and its absence is associated with femaleness. The Y chromosome carries

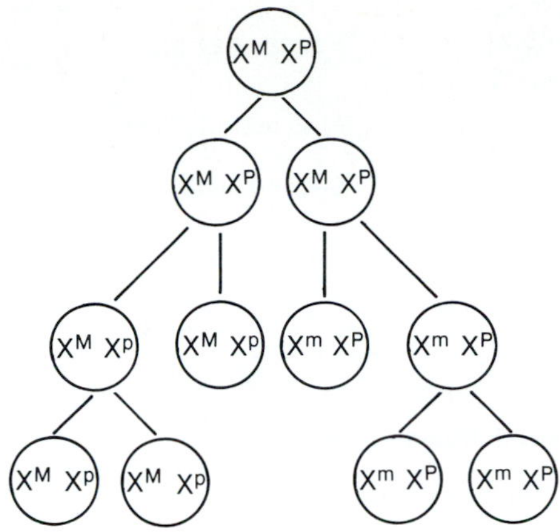

X^M, Active maternally derived X chromosome
X^P, Active paternally derived X chromosome
X^m, Inactive maternally derived X chromosome
X^p, Inactive paternally derived X chromosome

Fig. 46-3 Random inactivation of the X chromosome in the 46,XX female. Once an X chromosome is inactivated, that particular X is inactivated in all subsequent progeny of that cell.

a factor, called the testis determining factor (TDF), that induces the embryonic gonad to develop into a testis; in its absence the gonad develops as an ovary. There are no other genes residing on the Y chromosome that have been shown to code for proteins. The X chromosome carries many structural genes and codes for proteins similar to those coded for by autosomes.

The difference in the sex chromosome constitution of males and females is the basis for the particular pattern of inheritance that is known as X-linked. If there is a mutation on the single X chromosome of a male, that male will always transmit that mutant gene to all of his daughters and to none of his sons. A heterozygous female may give either of her X chromosomes, one of which has the normal gene and one of which has the mutant gene, to a son or daughter. If the mutation on the X chromosome results in a disease when present in the heterozygous state of the female, the trait is known as an *X-linked dominant disorder* and will be observed in both males and females. Because the single X of the male carries a mutation, the disease is usually more severe than in the female who has a mutant gene and a normal gene. If the mutation is not expressed as a recognizable trait in the heterozygous state of the female, the disorder is *X-linked recessively inherited*. Males will manifest the disorder and females will carry but not manifest the disorder. Females are carriers for hemophilia and Duchenne muscular dystrophy, but the disease occurs among males who have a single copy of the mutant gene. There are tests that will identify the carrier female for most X-linked recessive diseases. In general these tests demonstrate a partial expression in females of the defect seen in affected males.

A number of traits such as height and intelligence and isolated birth defects such as cleft lip and congenital heart disease are determined by the joint action of a number of genes and the environment. This mode of inheritance is called *multifactorial causation*. Such traits are familial and have a major genetic input but do not segregate in families in a manner similar to that seen for autosomal dominant and recessive traits. An expressed trait is the consequence of a number of genes received from each parent and a variable environmental influence that determines the expression of the trait. In addition to isolated birth defects seen in infants, multifactorial causation is responsible for a number of common diseases of adults. Examples include insulin-independent diabetes mellitus, osteoarthritis, gout, and certain forms of hypertension and coronary artery disease.

PATHOLOGICAL DISORDERS ASSOCIATED WITH ABNORMAL CHROMOSOMES

Aneuploidy

Disorders that occur as the result of a chromosome abnormality involve a change in the gene dosage for a large number of genes rather than a change in the gene structure. This produces a change in the blueprint for the structural development of the embryo. Such an abnormality can come about in two ways. First, there is the presence or absence of an entire chromosome, an *aneuploidy*. This is exemplified by the presence of an extra 21 chromosome in *trisomy 21,* which causes *Down's syndrome,* and by the presence of a single X chromosome resulting in monosomy, which occurs in *Turner's syndrome. Trisomy* means the presence of three copies of a chromosome rather than the normal two and *monosomy* means the presence of single copy. Second, there are structural alterations in chromosomes that result in the loss or addition of part but not all of a chromosome. A variety of structural abnormalities of chromosomes can lead to such a deviation from the normal amount of chromosomal material.

Chromosomal abnormalities involving the autosomes, whether they involve a partial or complete loss or addition of a chromosome, always result in altered morphogenesis expressed as major and minor birth defects and altered mental development in the form of mental retardation. The altered physical development in a chromosomally abnormal embryo is called *dysmorphogenesis;* the individual physical abnormalities are called *dysmorphic features*. In Down's syndrome this consists of such features as small ears, unusual creases on the palms, upward slant of the eyes, small head size, congenital heart disease, and short stature.

Down's syndrome is an easily recognized combination of major and minor physical abnormalities associated with mental retardation. It occurs as the specific result of the presence of a triple dose of chromosome 21 material. This occurs most commonly as trisomy 21, the presence of

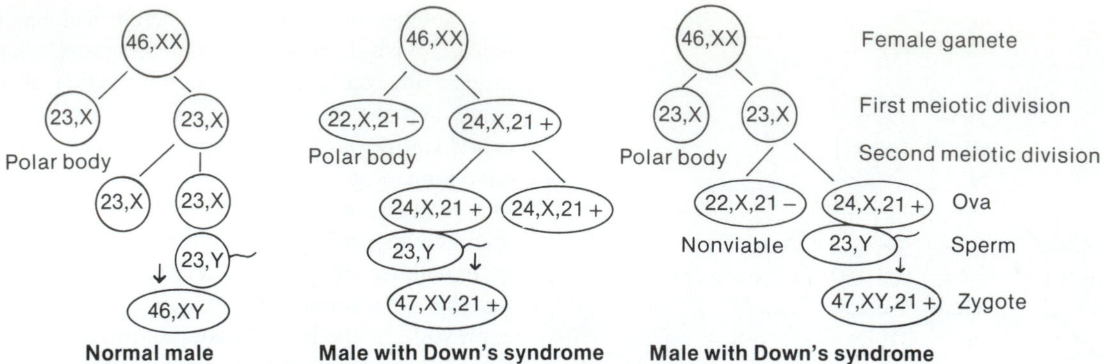

Fig. 46-4 Meiotic nondysjunction. *Left,* Normal meiosis of an ovum fertilized by a normal sperm yielding a normal zygote. *Center,* Error in first cell division of meiosis results in an aneuploid ovum that when fertilized yields a 47,XY,21 + male with Down's syndrome. *Right,* Error has occurred in second cell division of meiosis with same result.

three separate 21 chromosomes as the result of *meiotic nondysjunction*. *Meiosis* is the special form of cell division that occurs in germ cells that produce ova or sperm. This division reduces the chromosome number from 46 to 23, with one member of each pair of chromosomes being represented. *Nondysjunction* is the failure of one of a pair of chromosomes to go to each daughter cell during division; instead both go to one daughter cell (Fig. 46-4). The frequency of nondysjunction increases with maternal age. For that reason, prenatal diagnosis through *amniocentesis* with cytogenetic study is recommended for women 35 years of age or older. This form of Down's syndrome does not occur with an increased familial incidence.

Structural abnormalities of chromosomes

A familial form of Down's syndrome occurs with a structural abnormality known as a *robertsonian translocation*. This translocation is formed by the fusion of the centromeres, most commonly of chromosomes 14 and 21. A carrier of a 14/21 translocation will have a total chromosomal number of 45. However, the translocation chromosome contains the normal amount of chromosomal material of two separate chromosomes and the carrier is phenotypically normal. Meiosis in a translocation carrier can result in a variety of different gametes. Among the offspring of carriers of a 14/21 translocation, one may observe normal infants with 46 normal chromosomes, phenotypically normal infants with 45 chromosomes and a translocation similar to the parent, and infants with 46 chromosomes of which there are two normal 21 chromosomes and the material of a third 21 present in the translocation (Fig. 46-5). Such children have typical Down's syndrome. The translocation may occur in multiple members of a family, each of whom is a carrier at increased risk of having a child with Down's syndrome. One can differentiate the familial and sporadic forms of Down's syndrome by chromosomal

	14, 21 t14/21	45,XX,t14/21	Carrier
Gametes		**Zygotes**	**Phenotype**
Ovum Sperm	14, 21 14, 21	46,XX	Normal
Ovum Sperm	t14/21 14, 21	45,XX,t14/21	Normal (carrier)
Ovum Sperm	t14/21,21 14, 21	46,XX,t14/21	Down's syndrome

Fig. 46-5 Translocation 14/21. Carrier has 45 chromosomes with one normal 14, one 21, and genetic content of a 14 and 21 contained in translocation. A person with Down's syndrome as a result of such a translocation has two separate 21s and a third 21 as part of translocation. Such a person has a total of 46 chromosomes but genetic content of three number 21s, that is, trisomy 21.

studies. For that reason it is advisable that each person with Down's syndrome undergo chromosomal analysis to determine whether there is an increased risk for other family members of having children with Down's syndrome.

Sex chromosome abnormalities

Alterations in the number of sex chromosomes are common, and several types of aneuploidy can occur. These conditions were originally recognized through the use of *buccal smear* for *Barr body analysis*. Each X chromosome in excess of 1 in the cell nucleus is inactivated and is seen microscopically as the Barr body. The Barr body, named for its discoverer, is the nuclear condensation of the inactivated X chromosome and appears as a clump inside and adjacent to the nuclear membrane (Fig. 46-6). The cells of a normal male contain only one X chromosome; thus no Barr body is present. Cells of a normal female contain two X chromosomes and demonstrate a single Barr body.

Abnormalities of sex chromosomes in females occur most commonly as a 45,X and a 47,XXX karyotype. The former is the most common chromosomal finding in Turner's syndrome. This disorder has three major features: short stature, minor dysmorphic abnormalities and congenital malformations, and sexual infantilism. The sexual infantilism occurs as a result of fibrosis of the infantile ovary so that at the time of puberty there are neither follicles nor hormone-producing cells present. In its classic form this syndrome is diagnosed by buccal smear, which will show the absence of Barr bodies. However, a variety of X chromosome abnormalities may result in Turner's syndrome; for that reason a full lymphocyte karyotype should always be done regardless of the findings on buccal smear. The 47,XXX karyotype is not associated with significant clinical abnormalities.

The most common abnormalities of sex chromosomes in males include the 47,XXY and 47,XYY karyotypes. Klinefelter's syndrome (47,XXY) has three major clinical findings: minor to no dysmorphic abnormalities, normal to mild retardation in cognitive function, and failure of testicular development with secondary sexual infantilism and a eunuchoid body habitus. Because of the lack of testosterone production at the age of puberty, such males may respond to ovarian estrogens with breast enlargement or gynecomastia. Surgical correction of gynecomastia and testosterone replacement therapy will correct most of the problems associated with this disorder. There is considerable controversy regarding the type and frequency of clinical manifestations associated with the 47,XYY karyotype. Males with the 47,XYY karotype are physically and sexually normal, but there are data that suggest that a significant proportion of such men have problems with sociopathic behavior.

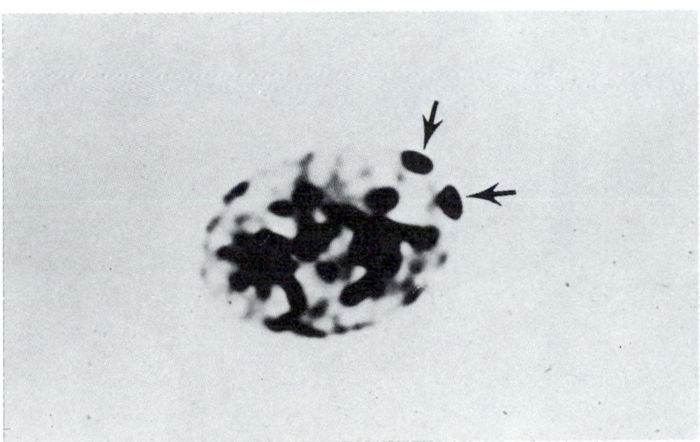

Fig. 46-6 Barr body. This cell has two Barr bodies seen as nuclear condensations just inside nuclear membrane at 3-o'clock position *(arrow)*. Two Barr bodies would be seen in a male with 48,XXXY and a female with 47,XXX.

Cytogenetic methods

The most common method of chromosome analysis involves the culture of peripheral lymphocytes. These are mature cells circulating in the peripheral blood that are not actively undergoing cell division. Such cells can be stimulated to divide by the addition of a *mitogen* to the culture medium. The most commonly used mitogen in human cytogenetics is phytohemagglutinin (PHA). Lymphocytes in the presence of PHA are commonly cultured for 72 hours. Toward the end of the culture period, colchicine is added to the culture. This agent acts as a microtubular poison and thereby disrupts the mitotic spindle during cell division. Thus cells can be arrested in metaphase, the only time during cell division when chromosomes can be easily visualized and studied (Fig. 46-7). At the end of the culture

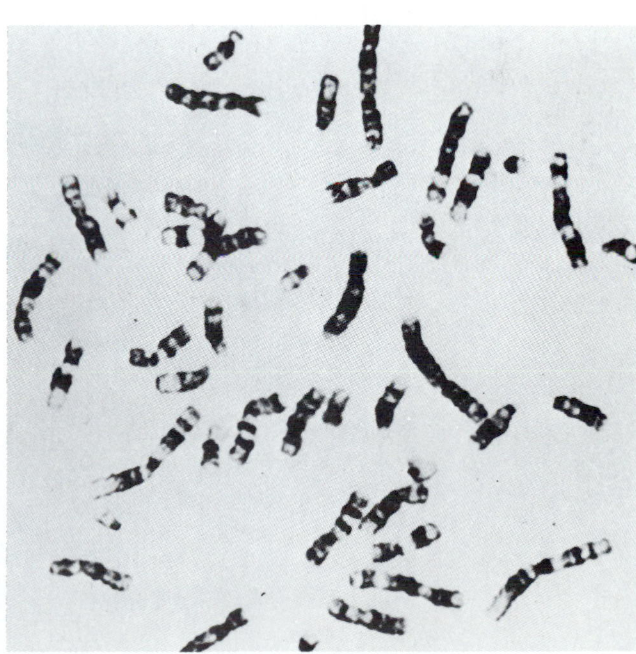

Fig. 46-7 Metaphase spread. Chromosomes as seen through microscope in a spread stained by G-banding technique.

Table 46-2 Buccal smear findings in sex chromosome abnormalities

Karyotype	Sex	Barr body	Y body
46,XX	F	+	0
46,XY	M	0	+
45,X	F	0	0
47,XXX	F	++	0
48,XXXX	F	+++	0
49,XXXXX	F	++++	0
47,XXY	M	+	+
47,XYY	M	0	++
48,XXYY	M	+	++
49,XXXXY	M	+++	+

period the cells are harvested, spread on a slide, and stained with a number of fluorogens or chromagens. When special fixative techniques are used, the chromosomes take on a banded appearance. These bands are specific for each chromosome and allow detailed analysis of the structure of each chromosome.

The normal Y chromosome contains a large fluorescent segment in the long arm. If a buccal smear from a 46,XY male is stained with a fluorescent dye, a bright fluorescent spot called the *F body* or *Y body* will be revealed. The use of Giemsa staining for Barr bodies and fluorescent staining for Y bodies allows determination of the sex chromosome constitution from a buccal smear. The recognition of both Barr bodies and Y bodies is sufficiently subjective that an experienced technician, using control cells from a normal male and female, is necessary for an accurate determination. The expected findings of Barr body–Y body determination on a buccal smear for normal and abnormal sex chromosomal constitutions are shown in Table 46-2.

Chromosomal markers of disease

The analysis of the chromosomes in dividing cells may reveal abnormalities of three types: first, a change in the amount of chromosomal material present as in aneuploidy or partial duplication or deletion; second, no alteration in the amount of chromosomal material but chromosomal changes that are diagnostic of specific disease states; third, nonspecific changes in chromosomal structure or number that reflect the consequences of environmental influences such as drugs or radiation or the abnormal chromosomal behavior seen in malignant cells.

Improvements in culture techniques and staining of chromosomes have revealed that malignant cells often demonstrate a consistent and specific chromosomal abnormality that is of major diagnostic assistance. Such an abnormality is best exemplified by the Philadelphia chromosome (Ph^1) found in the majority of persons with chronic myelogenous leukemia (CML). The Philadelphia chromosome represents a deletion of chromosome 22, which renders it a small acrocentric chromosome (Fig. 46-8). The deleted material is most often translocated to a 9 chromosome. Thus there is no loss or gain of chromosomal material, but there is a structural rearrangement. The Philadelphia chromosome is confined to the malignant cells of

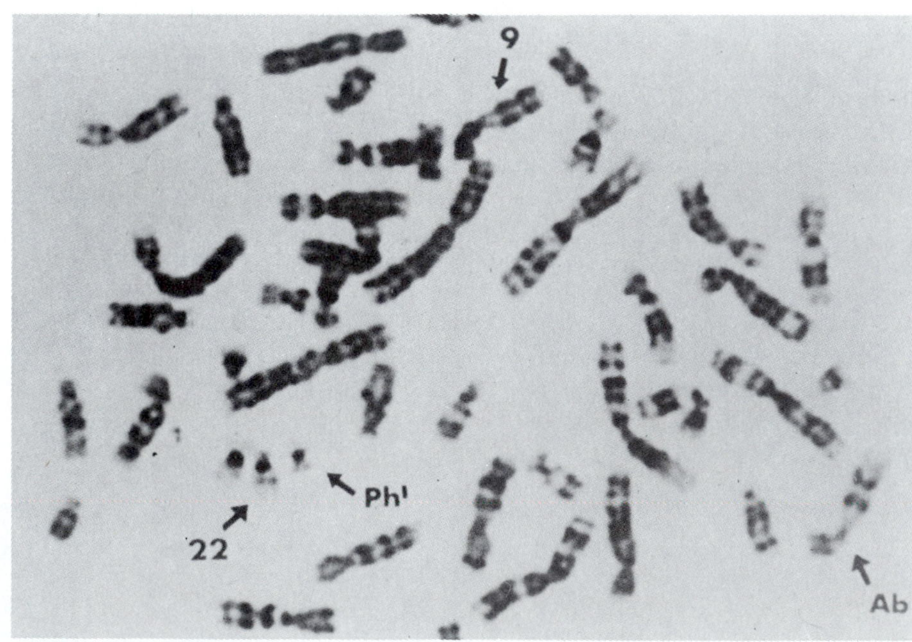

Fig. 46-8 Philadelphia chromosome. A metaphase spread with banded chromosomes shows small number 22, or Ph^1 chromosome, and normal 22 plus normal chromosome 9 and chromosome 9(Ab) involved in 9/22 translocation producing Ph^1 chromosome.

the bone marrow, and analysis for the Philadelphia chromosome is done on bone marrow aspirate.

A second type of chromosomal marker used diagnostically is that of chromosomal breakage. This is most dramatically illustrated in the Fanconi anemia. Persons with this form of autosomal recessively inherited aplastic anemia demonstrate a pronounced increase in the frequency of spontaneous chromosomal breaks and gaps in cultured peripheral lymphocytes. Aplastic anemia occurs as a result of failure of the bone marrow to produce blood cells. Fanconi anemia cells are quite sensitive to agents such as mitomycin C, which further increases the frequency of chromosomal breakage.

The autosomal recessively inherited disorder Bloom's syndrome is also characterized by a sharp increase in the frequency of sister chromatid exchanges in cultured lymphocytes. Bloom's syndrome is typically characterized by intrauterine growth retardation and short stature thereafter. Such patients have increased sensitivity of skin to ultraviolet light and a significant risk of malignancies including leukemia during late childhood and early adult years.

A fourth type of specific chromosomal marker is the fragile X chromosome recognized in one X-linked form of mental retardation. The characteristic findings in this disorder include postpubertal enlargement of the testes, large, simply formed ears, a jovial personality, and moderate to severe mental retardation. Approximately 10% of carrier females for this form of X-linked mental retardation show a mild form of mental retardation. The culture of lymphocytes with standard growth media used for cytogenetic studies will not demonstrate the fragile X chromosome. However, the use of media that are deficient in folic acid will show a fragile X chromosome in a significant percentage of the cells from an affected male. The designation *fragile X* is used because the X chromosome in these males appears to have a break at the end of the long arm.

PATHOLOGICAL DISORDERS ASSOCIATED WITH BIOCHEMICAL CHANGES

Lysosomal storage diseases

Lysosomal storage diseases are recessively inherited disorders, each of which is the result of a deficiency of a specific acid hydrolase. These acid hydrolases are located within membrane-bound cytoplasmic organelles called *lysosomes*. If there is a complete or nearly complete deficiency of a specific hydrolase, the macromolecular compound that it normally degrades will accumulate within tissue. The primary organs affected depends on the tissue distribution of the macromolecular compound. These include (1) the reticuloendothelial system, (2) the central nervous system (CNS), and (3) the skeleton and connective tissues plus other somatic tissues. In a given disorder the clinical manifestations may be apparent in only one or in any combination of these three organ systems. The box above lists a classification scheme for lysosomal storage diseases.

CLASSIFICATION OF LYSOSOMAL STORAGE DISEASE

- Mucopolysaccharidoses
- Mucolipidoses (ML)
  - (ML I—now called "sialidosis")
  - ML II or I-cell disease
  - ML III or pseudo-Hurler polydystrophy
  - ML IV
- Gangliosidoses
  - GM_1 gangliosidosis or generalized gangliosidosis
  - GM_2 gangliosidosis
    - Tay-Sachs disease
    - Sandhoff's disease
- Leukodystrophies
  - Metachromatic leukodystrophy
  - Krabbe's disease
  - Adrenoleukodystrophy
- Glycoproteinosis
  - Mannosidosis
  - Fucosidosis
  - Sialidosis
  - Aspartylglucosaminuria
- Others
  - Ceramidosis
  - Cholesterol ester storage disease
  - Pompe's disease or glycogen-storage disease II

Varying degrees of deficiency or altered activity of a hydrolase may result in several different clinical syndromes occurring as a result of the same enzyme deficiency. This is demonstrated by deficiency of the hydrolase α-L-iduronidase. Deficiency of this enzyme results in the accumulation of the glycosaminoglycans, dermatan sulfate, and heparan sulfate. In its most severe form with the greatest degree of enzyme deficiency, Hurler's syndrome occurs with severe skeletal, somatic, and central nervous system manifestations and death by 5 to 8 years of age. In its mildest form, deficiency of this enzyme results in the Scheie syndrome with normal height, life expectancy, and intelligence but with skeletal, ocular, and cardiac abnormalities. An intermediate form, known as the Hurler-Scheie syndrome, is intermediate in its clinical manifestations.

Reticuloendothelial system involvement. The two disorders in which involvement of the liver and spleen is most prominent and may represent the major manifestation are *Gaucher's* and *Niemann-Pick diseases*. Gaucher disease occurs as a result of the deficiency of the acid hydrolase β-glucosidase.

$$\text{Ceramide-}\beta\text{-glucose} \xrightarrow[\text{H}_2\text{O}]{\beta\text{-Glucosidase}} \text{Ceramide} + \text{Glucose}$$

Deficiency of this enzyme results in the lysosomal accumulation of the phospholipid glucosylceramide. This disorder occurs in several forms. The most severe form is known as neuronopathic Gaucher's disease in which, although liver and spleen enlargement occur, massive accumulation of phospholipids within the CNS predominates and affected infants die within the first year of life. The other extreme is illustrated by the adult form of Gaucher's disease recognized by hepatosplenomegaly, in which good general health and normal life expectancy occur. Before the use of enzyme assays for specific diagnoses, Gaucher's disease was most often recognized by the demonstration of Gaucher cells (large macrophages) in a bone marrow aspirate. The second disorder in this group, Niemann-Pick disease, also occurs in several forms ranging from a severe infantile to a mild adult disease. This disorder occurs as a result of a deficiency of sphingomyelinase.

$$\text{Sphingomyelin} \xrightarrow[H_2O]{\text{Sphingomyelinase}} \text{Ceramide} + \text{Phosphorylcholine}$$

It too is associated with enlargement of the liver and spleen and demonstration of storage cells on bone marrow aspirate that represent macrophages with lysosomes engorged with spingomyelin.

CNS predominance. The lysosomal storage diseases that affect primarily the CNS clinically involve gray matter or white matter initially. When the initial accumulation is within white matter, the disorder is known as a *leukodystrophy* and motor abnormalities are seen early in the course of the disease. The most commonly recognized leukodystrophy is metachromatic leukodystrophy, in which neural tissue demonstrates a particular metachromatic staining property resulting from the lysosomal accumulation of sulfated galactosylceramide. The deficient enzyme is called arylsulfatase A, or cerebroside sulfate sulfatase. Several screening tests are available for this disorder and include the examination of urine of metachromatic granules.

$$\text{Ceramide-}\beta\text{-galactose-3 sulfate} \xrightarrow[H_2O]{\text{Sulfate arylsulfatase A}} \text{Ceramide-}\beta\text{-galactose} + \text{Sulfate}$$

In normal urine adequate amounts of this enzyme are present for a simple colorimetric assay of arylsulfatase activity. The presence of an adequate amount of enzyme activity will exclude this diagnosis, but lack of enzyme activity is not diagnostic and suggests that more specific assay systems should be used. The specific diagnosis of metachromatic leukodystrophy requires the demonstration of arylsulfatase A deficiency in homogenized peripheral leukocytes or cultured skin fibroblasts. Arylsulfatase occurs in two lysosomal forms known as A and B. The use of cellulose acetate electrophoresis is a second diagnostic method. In metachromatic leukodystrophy a normal arylsulfatase B band will be observed and no A band will be recognized.

The best known lysosomal storage disease that results in macromolecular compound accumulation initially within gray matter is Tay-Sachs disease. This disorder occurs as a result of the accumulation of G_{M2} ganglioside (Fig. 46-9) within the CNS. Clinically it results in early regression of CNS function so that by 1 year of age children with this disease have delayed cognitive development, impaired hearing and sight, and a characteristic cherry-red spot found on funduscopic examination. The accumulation of G_{M2} ganglioside within the CNS occurs as a result of deficiency of the acid hydrolase *N*-acetyl-β-D-galactosaminidase. The activity of this enzyme is assayed with artificial substrates; either a glucosamine or galactosamine. Therefore the enzyme activity so measured is referred to as *hexosaminidase*. Hexosaminidase occurs as a heat-labile form, hexosaminidase A, and a heat-stabile form, hexosaminidase B. Tay-Sachs disease is characterized by the specific deficiency of hexosaminidase A. Measurement of enzyme activity with and without heat inactivation gives levels of hexosaminidase B and total hexosaminidase, respectively. The difference between the two determinations, or the heat-labile form is thus the calculated level of hexosaminidase A.

Because Tay-Sachs disease occurs most commonly among Ashkenazic Jewish infants, this target population can be screened for carrier detection. Determination of the percentage of hexosaminidase A present as a function of

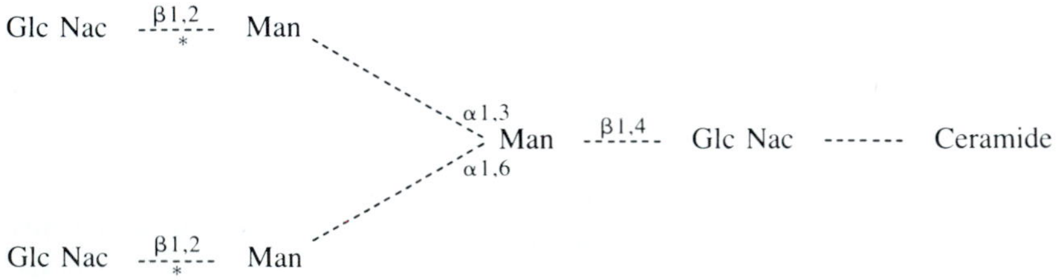

Glc Nac --β1,2*-- Man
α1,3 → Man --β1,4-- Glc Nac ------- Ceramide
Glc Nac --β1,2*-- Man
α1,6

G_{M2} ganglioside

*Cleaved by hexosaminidase A (*N*-acetyl-β-1,2-glucosaminidase)

Fig. 46-9 Tay-Sachs disease. G_{M_2} ganglioside accumulates in the gray matter of brain because of deficiency of *N*-acetyl-β-1,2-glucosaminidase (hexosaminidase A). *Glc Nac, N*-Acetylglucosamine; *Man,* mannose.

Table 46-3 Hurler-like disorders

Disorders	Screening test
Mucopolysaccharidoses	Urinary screening for mucopolysacchariduria
Mucolipidoses	
ML II and ML III	Serum levels of acid hydrolases Urinary bound sialic acid
Glycoproteinosis	Urinary oligosaccharide TLC Urinary bound sialic acid

total hexosaminidase activity represents the most accurate screening method for detecting carriers of Tay-Sachs disease. In establishing a screening assay system for Tay-Sachs disease carriers, it is necessary that each laboratory determine its reference values and values for Tay-Sachs disease carriers from obligate heterozygotes. Confirmation of the carrier state requires study of hexosaminidase A activity in peripheral leukocytes and additional family studies if necessary to resolve questionable results. The use of such carrier tests among Ashkenazic Jews and subsequent use of prenatal diagnosis has significantly reduced the incidence of Tay-Sachs disease in the United States.

Connective tissue and skeletal predominance. Lysosomal storage diseases in which involvement of connective tissue predominates, especially the skeletal system, are referred to as *Hurler-like disorders*. With skeletal system involvement there may also be CNS or reticuloendothelial involvement. Table 46-3 lists a classification of lysosomal disorders that are seen with this phenotype.

Hurler's syndrome represents the prototype of diseases involving glycosaminoglycan metabolism and resulting in a connective tissue disorder. This form of α-L-iduronidase deficiency is characterized by multisystem involvement that is usually apparent by 6 months of age (Fig. 46-10). It follows a rapidly progressive course thereafter with all organ systems involved and mental deterioration predominating late in the disorder with death ultimately occurring between 5 and 8 years of age. These diseases were classi-

IA—*Glc Nac—GA—Glc Nac—IA—*Glc Nac

Dermatan sulfate

IA—*Gal Nac—GA—Gal Nac—IA—*Gal Nac

Heparan sulfate

*Site of hydrolysis by α-L-iduronidase in oligosaccharide chain of dermatan and heparan sulfate

Fig. 46-10 Hurler's syndrome. Deficiency of lysosomal hydrolase α-L-iduronidase results in widespread accumulation of both heparan and dermatan sulfate. *GA,* Glucosamine; *Gal Nac, N*-acetylgalactose; *Glc Nac, N*-acetylglucosamine; *IA,* iduronic acid.

fied initially on the basis of the urinary pattern of mucopolysaccharide or glycosaminoglycan excretion. In Hurler's syndrome massive amounts of dermatan sulfate and heparan sulfate are excreted in the urine. Mucopolysacchariduria is detected by a variety of screening tests including a urine spot test, the toluidine blue test.

Recognition of the clinical phenotype and the presence of a positive urine screening test are followed by the study of mucopolysaccharide metabolism in cultured fibroblasts and the assay of specific hydrolases that can cause the mucopolysaccharidoses. Inorganic sulfate added to tissue culture media is exclusively incorporated into cultured fibroblasts as sulfated glycosaminoglycans. The use of radioactive sulfate allows the recognition of accumulation of these sulfated compounds within cultured fibroblasts and therefore provides direct evidence for tissue accumulation of glycosaminoglycans. A specific diagnosis can be pursued by the assay of the individual hydrolases that, when deficient, result in a mucopolysaccharidosis. The pattern of mucopolysacchariduria and the specific acid hydrolase deficient in the mucopolysaccharidoses are shown in Table 46-4.

Another disorder that results in a Hurler-like clinical picture is *mucolipidosis II,* or I-cell disease. Children with this disease are similar clinically to those with Hurler's

Table 46-4 Mucopolysaccharidoses

Eponym	Number	Mucopolysacchariduria*	Enzyme deficiency
Hurler	HPS I-H	DS, HS	α-L-Iduronidase
Scheie	MPS I-S	DS, HS	α-L-Iduronidase
Hurler-Scheie	MPS I-H/S	DS, HS	α-L-Iduronidase
Hunter	MPS II	DS, HS	Iduronide sulfate sulfatase
Sanfilippo	MPS III-A	HS	Heparan-*N*-sulfate sulfatase
Sanfilippo	MPS III-B	HS	α-Glucosaminidase
Sanfilippo	MPS III-C	HS	Acetyl-CoA:α-glucosaminide *N*-acetyltransferase
Morquio A	MPS IV-A	KS	Galactosaminyl-6-sulfate sulfatase
Morquio B	MPS IV-B	KS	β-Galactosidase
Maroteaux-Lamy	MPS VI	DS	Galactosaminyl-4-sulfate sulfatase (arylsulfatase B)
Sly	MPS VII	DS or HS or DS, HS	β-Glucuronidase

**DS,* Dermatan sulfate; *HS,* heparan sulfate; *KS,* keratan sulfate.

Table 46-5 Inborn errors of intermediary metabolism of amino aids

Condition	Defective enzyme	Biochemical features	Clinical features	Treatment
Alkaptonuria*	Homogentisate oxygenase	Urinary excretion of homogentisic acid	Urine darkens; ochronosis; arthritis in later life.	None known
Phenylketonuria†	Phenylalanine 4-hydroxylase	Phenylalanine accumulates in blood, CSF, etc.; urinary excretion of phenylpyruvic acid and related compounds.	Severe mental deficiency, epilepsy, abnormal EEG, eczema, behavioral disorders.	Diet low in phenylalanine beginning at early age
Albinism‡	*o*-Diphenol oxidase (tryosinase)	Lack of melanin in skin, hair, and eyes.	Photophobia, nystagmus, carcinomata of the skin.	None known
Goitrous cretinism (several types)	(1) Tryosine iodinase (2) Coupling enzyme (3) Deiodinase	Lack of thyroid hormone.	Cretinism, goiter.	Thyroid, thyroxine, or triiodothyronine
Maple-syrup urine disease (leucinosis)	Enzyme responsible for oxidative decarboxylation of α-ketoisocaproic, α-keto-β-methyl-*n*-valeric and α-ketoisovaleric acids	Leucine, isoleucine and valine accumulate in blood, CSF, etc.; urinary excretion of the 3 keto acids and related compounds.	Cerebral degeneration; usually early death, milder form with partial enzyme deficiency, symptomless except during infections, etc.	Diet low in leucine, isoleucine, and valine
Cystinosis	Cystine reductase (?)	Cystine is deposited in reticuloendothelial system; amino-aciduria, glucosuria, proteinuria, phosphaturia, dilute urine.	Dwarfism, photophobia, renal acidosis, hypokalemia, vitamin-resistant rickets; death before puberty. A benign (non renal?) variant occurs in adults.	Palliative: potassium salts, alkalis, vitamin D; diet low in cystine and methionine (efficacy doubtful)
Homocystinuria	L-Serine dehydratase	Urinary excretion of homocystine.	Mental retardation, retinal defects, dislocated lenses, malar flush, thromboses.	Diet low in methionine, high in cystine; pyridoxine
Hyperglycinemia (several types)	(Uncertain, depends on type)	Glycine accumulates in blood, etc.; urinary excretion of glycine and, in one type, methylmalonic acid.	Neonatal lethargy and ketosis, neutropenia, hypo-γ-globulinaemia; mental retardation.	Diet low in protein

Oxalosis	Excessive conversion of glycine to oxalic acid	Calcium oxalate accumulates in kidneys, heart, bone marrow and cartilages.	Nephrocalcinosis leading to progressive renal failure.	None known
Histidinemia	Histidine ammonialyase	Urinary excretion of β-imidazolylpyruvic acid and related compounds.	Speech defects; mental retardation in some.	Diet low in histidine
Familial tryosinemia	(Uncertain)	Tryosine levels in blood and urine raised; urinary excretion of phenolic acids related to tryosine; generalized amino-aciduria; glucosuria; fructosuria.	Rapidly enlarging liver; jaundice; hypoprothrombinemia; death common in infancy; survivors may have vitamin D–resistant rickets and acidosis.	Diet low in tyrosine and phenylalanine (efficacy doubtful)
Hyperprolinemia				
Type I	Pyrroline-5-carboxylate reductase	Hyperprolinemia; urinary excretion of proline, glycine, and hydroxyproline.	Mental retardation, convulsions, renal disease, deafness.	None known
Type II	Pyrroline-5-carboxylate dehydrogenase			
Hydroxyprolinemia	3-Hydroxypyrroline-5-carboxylate reductase (?)	High levels of hydroxyproline in blood and urine.	Mental retardation (?).	None known
Citrullinemia	Argininosuccinate synthetase	High blood and urinary levels of citrulline; blood ammonia increased; urea excretion normal.	Mental retardation, epilepsy, vomiting, ammonia intoxication.	Diet low in protein
Argininosuccinic aciduria	Argininosuccinate lyase	Urinary excretion of argininosuccinic acid; high blood and CSF ammonia levels; urea excretion normal.	Mental retardation, convulsions, hair abnormalities, ammonia intoxication.	Diet low in protein
Hyperammonemia				
Type I	Ornithine carbamoyltransferase	Blood ammonia about 10 mg/L; urea excretion normal.	Mental retardation, ammonia intoxication.	Diet low in protein(?)
Type II	Carbamoyl-phosphate synthase			

From Geigy scientific tables, ed 7, 1970, CIBA-Geigy Corp, Summit, NJ.

*Incidence 1 in 100,000. †Incidence varies from 1 in 3200 to 1 in 10^7 according to locality. ‡Incidence 1 in 13,000.

syndrome but are diagnosed at an earlier age and follow a more rapid downhill course with death by 3 to 5 years of age. There is no mucopolysacchariduria. When cultured skin fibroblasts from affected children were first analyzed under phase microscopy, a marked increase in cytoplasmic inclusions was noted; hence the name *I-cell,* or *inclusion cell disease*. Lysosomal hydrolases are glycoproteins. Each lysosomal enzyme has a unique structural gene that codes for the protein component of the glycoprotein. There is, however, commonality in the posttranslational modification of these proteins to form the oligosaccharide component of the glycoproteins. It has been recognized that mannose-6-phosphate sugars in the oligosaccharide chain of the glycoprotein structure of acid hydrolases are required for their appropriate intracellular localization. This is accomplished through a number of posttranslational enzymatic steps in the modification of the oligosaccharide chain. Deficiency of a transferase involved in such posttranslational modifications represents the primary molecular defect in I-cell disease. Cultured skin fibroblasts from affected children show deficient levels of numerous acid hydrolases in cultured cells. The ultimate defect resides in the abnormal intracellular localization of these hydrolases. The urine of patients with I-cell disease shows increased levels of lysosomal hydrolases and an increase in the amount of sialic acid oligosaccharides. The measurement of bound sialic acid in urine is a relatively simple procedure and represents an excellent screening test for a number of lysosomal storage diseases.

Disorders of intermediary metabolism

Disorders of intermediary metabolism occur as a result of three basic mechanisms: an enzyme deficiency with defective substrate conversion, a membrane transport defect resulting in failure of absorption or excessive excretion of a compound, and defects in receptors involved in mediating metabolism. The biochemical basis for the resulting disease occurs as a result of (1) accumulation of a substrate to levels that become toxic, (2) deficiency through lack of production or excessive loss of a needed compound, and (3) conversion of an elevated compound to an altered metabolite that itself is a toxic material (Fig. 46-11). In general, defects in intermediary metabolism result in alterations of lower molecular weight compounds than those seen in lysosomal storage diseases. As a result, they are more likely to result in more acute manifestations and are less likely to be associated with the striking physical features that dominate the lysosomal storage disorders. Defects in intermediary metabolism result in a disruption of a normal metabolic process and often produce an elevated urinary excretion of a normal metabolite or the urinary excretion of an abnormal metabolite. Therefore urine screening tests are the principal means of recognizing the presence of such a disorder, prompting more specific diagnostic assays.

Amino acidopathies. Table 46-5 is a list of the more common amino acid disorders and their clinical chemical characterizations. *Phenylketonuria* (PKU) is the best known and best understood of the amino acid disorders. This autosomal recessively inherited disorder occurs as a defect in phenylalanine conversion to tyrosine because of a deficiency of the enzyme phenylalanine hydrolase.

Phenylalanine —Defective phenylalanine hydrolase→ Tyrosine
↘ **Alternative metabolic pathway** ↘ Phenylpyruvic acid

The deficiency of phenylalanine hydrolase results in a shunting of phenylalanine metabolism to several ketoacids, the principal one being phenylpyruvic acid. This disease occurs in approximately one in 15,000 live-born infants. The enzymatic activity is restricted to the liver, and therefore diagnosis is accomplished through demonstration of alterations in phenylalanine metabolism.

Ferric chloride screening of the urine in the newborn is used both for diagnosis and dietary management from early infancy. The Guthrie test involves a bacterial inhibition assay to detect an elevated level of phenylalanine. In newborn screening a Guthrie test is considered positive when the phenylalanine levels exceed a range of 20 to 40 mg/L. Given a positive result, approximately 1 in 10 to 1 in 20 infants will in fact have phenylketonuria. The spe-

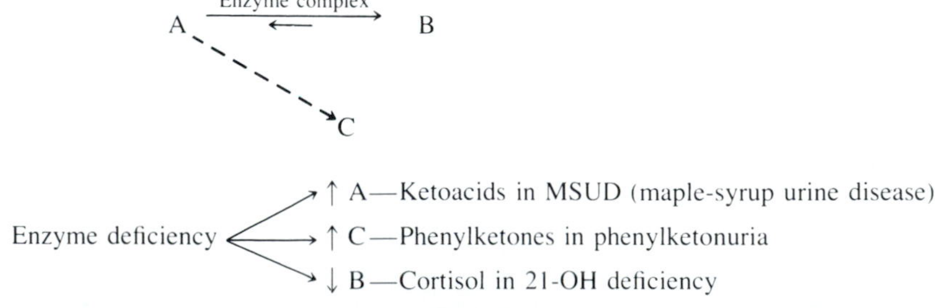

Fig. 46-11 Enzyme deficiency. Deficiency of enzyme in metabolic pathway may produce clinical symptoms because of accumulation of substrate, *A;* lack of production of product, *B;* or shunting of substrate to alternate pathway with production of a toxic compound, *C*.

cific diagnosis of phenylketonuria after a positive screening test requires the demonstration of plasma levels of phenylalanine greater than 200 mg/L on 2 consecutive days while normal feedings are given. After institution of dietary restriction in phenylalanine intake, urine screening tests will revert to normal. Subsequent monitoring of children on dietary management requires quantitative plasma determinations of phenylalanine levels.

Maple-syrup urine disease was so named because of the obvious odor of urine in affected infants. This disease occurs in roughly one in 250,000 live-born infants. Many disorders of amino acid metabolism are characterized by unusual odors that usually are the result of the excessive urinary excretion of organic acids that impart the peculiar odor. The three branched-chain amino acids, leucine, isoleucine, and valine, undergo transamination and transketolation and decarboxylation through a common pathway (Fig. 46-12). Maple-syrup urine disease occurs as a result of a deficiency in the multicomponent enzyme, branch-chain ketoacid decarboxylase. Deficiency of this enzyme results in the elevated urinary excretion of the ketoacid analogs of these three branch-chained amino acids and gives positive ferric chloride and dinitrophenol hydrazine screening tests. The plasma is characterized by elevations in the levels of leucine, isoleucine, and valine. This disorder is screened for in the newborn using a bacterial inhibition assay system, similar to the Guthrie test, which detects elevated levels of plasma leucine in the newborn. After the diagnosis, special diets that restrict the intake of isoleucine, leucine, and valine combined with careful monitoring can provide adequate control of this disease. Affected children who are not detected by newborn screening within the first month of life generally become acutely ill with hypoglycemia and ketoacidosis. Without early recognition and treatment death most commonly occurs, and in those in whom diet is instituted late, significant mental retardation results.

Homocystinuria is generally recognized in late childhood because of physical abnormalities with or without mental retardation. Homocystine is an intermediate amino acid in the metabolism of methionine to cystine. The most common form of homocystinuria occurs because of a deficiency of the enzyme cystathionine synthetase, which results in elevated plasma and urinary levels of homocystine. The elevation in plasma levels of homocystine results in conversion of most of this substance to methionine. Thus one can screen for homocystinuria in newborn blood by analyzing plasma levels of methionine. This condition results in clinical manifestations through two mechanisms: (1) the elevated levels of homocystine have been shown to be toxic to vascular endothelium and account for the thromboembolic phenomenon (plugging of brain or lung blood vessels by blood clots) of this disease, and (2) the failure of production of cystine results in a deficiency of this essential amino acid in connective tissue, specifically collagen, metabolism. The enzymatic activity of cystathionine synthetase requires as its cofactor the B vitamin, pyridoxine. Different mutations in the gene locus for cystathionine synthetase result in two distinct forms of homocystinuria. In approximately half the patients with cystathionine synthetase deficiency, a sufficient enhancement of residual enzyme activity can be achieved by the addition of therapeutic doses of pyridoxine to the diet to essentially cure the disease.

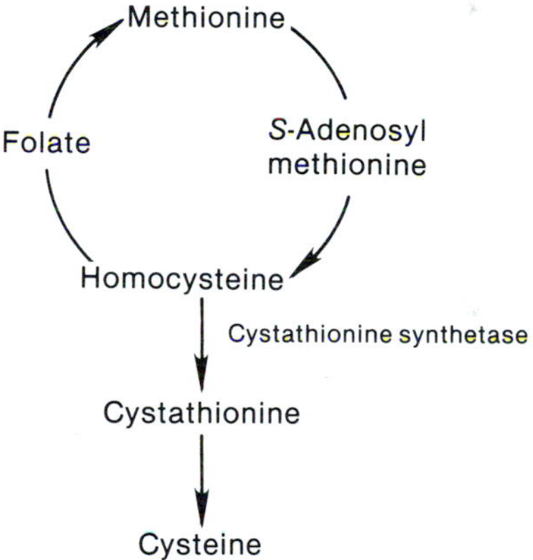

Cyanide-nitroprusside is used as a urine-screening test for this disease. Nitroprusside combines with sulfur-containing amino acids to produce a red color. The cyanide reduces disulfide bonds and gives a positive screening re-

$$\mathrm{R{-}\overset{\overset{\displaystyle O}{\|}}{C}{-}COOH + NAD^+ + CoA{-}SH \xrightarrow{BCKADH} R{-}\overset{\overset{\displaystyle O}{\|}}{C}{-}S{-}CoA + CO_2 + NADH + H^+}$$

$\mathrm{R{-}\overset{\overset{\displaystyle O}{\|}}{C}{-}COOH}$ = Branched-chain keto acid

BCKADH = Branched-chain keto acid dehydrogenase complex, a multisubunit enzyme complex involved in a five-step reaction

Fig. 46-12 Maple-syrup urine disease. Branched-chain keto acid decarboxylase enzyme complex when deficient results in elevation of branched-chain amino acids valine, isoleucine, and leucine, as well as their keto acid analogs.

$$\text{Propionyl CoA} + HCO_3^- \xleftrightarrow{1} \text{D-Methylmalonyl CoA} \xleftrightarrow{2} \text{L-Methylmalonyl CoA} \xleftrightarrow{3} \text{Succinyl CoA}$$

If enzyme 1 is missing → Propionic acid

If enzyme 2 is missing → Methylmalonic acid

1 = Propionyl/CoA carboxylase (biotin, ATP, Mg^{++})
2 = Methylmalonyl-CoA racemase
3 = Methylmalonyl-CoA mutase

Fig. 46-13 Propionic aciduria and methylmalonic aciduria. This sequence of metabolism of organic acids can be interrupted at several points, each of which results in a specific organic aciduria.

sult. Homocystinuria and cystinuria give a positive cyanide-nitroprusside reaction, whereas silver nitroprusside, which does not reduce the disulfide bond of homocystine, only gives a positive result when cystinuria is present.

Organic acidurias. The metabolism of amino acids involves a number of intermediate steps. During each step organic acids are produced. A defect in the subsequent metabolism of these compounds results in disorders that are characterized by a severe metabolic acidosis. Among this group of disorders, the ketotic hyperglycinemia syndrome is composed of several defects in propionic acid and methylmalonic acid metabolism. The amino acids leucine, isoleucine, valine, threonine, and methionine and odd-numbered fatty acids lead to the production of propionyl CoA, which is converted to methylmalonyl CoA. Methylmalonyl CoA is converted into succinyl CoA, which can then enter the tricarboxylic acid cycle (Fig. 46-13). Infants with these disorders usually are seen within the first few weeks of life with severe ketoacidosis, which may be accompanied by hypoglycemia. The urine yields a strongly positive dinitrophenyl hydrazine reaction and by amino acid analysis shows striking elevations in glycine. If methylmalonic acid is not found in the urine, one can make a presumptive diagnosis of propionic aciduria. A definitive diagnosis of this disorder can then be further pursued through the demonstration of the presence of methylcitrate in the urine, the gas-liquid chromatographic analysis of organic acids in the urine, or the assay of propionyl CoA carboxylase in cultured cells. The conversion of propionyl CoA to methymalonic CoA uses biotin as a cofactor, which may in therapeutic doses enhance residual enzyme activity. The conversion of methylmalonic CoA to succinyl CoA involves the metabolism of B_{12}, which in therapeutic doses can also result in significant biochemical improvement. In those instances in which vitamin therapy does not result in a significant therapeutic response, a low-protein diet is used.

Defects in carbohydrate metabolism. *Galactosemia* is a toxicity syndrome associated with an intolerance to dietary galactose. This recessively inherited disorder occurs as a result of a deficiency of the enzyme galactose-1-phosphate uridyl transferase.

$$\text{Galactose-1-phosphate} + \text{UDP-glucose} \xrightarrow{\text{Gal-1-P uridyl transferase}} \text{UDP-galactose} + \text{Glucose-1-phosphate}$$

Lactose, composed of galactose and glucose, is the major disaccharide in mammalian milk. Hydrolysis of lactose by the intestine results in the release of the monosaccharides, glucose and galactose. The main pathway of galactose metabolism in humans involves the conversion of galactose to glucose by epimerization of the hydroxyl group at the carbon-4 position. The reaction catalyzed by galactose-1-phosphate uridyl transferase involves galactose-1-phosphate plus UDP-glucose, yielding UDP-galactose plus glucose-1-phosphate. The UDP-galactose can by further conversion yield UDP plus glucose-1-phosphate. Humans are thus capable of metabolizing large amounts of galactose. However, with deficiency of the transferase enzyme, galactose is reduced to galactitol and oxidized to galactonate. It is the presence of these two intermediate products of galactose metabolism that has direct toxic effects and results in the clinical manifestations of galactosemia. The classic clinical presentation of galactosemia is one of failure to thrive in early infancy complicated by vomiting and diarrhea. In addition, these infants show deranged hepatic function with jaundice and hepatomegaly. Severe hemolysis can also occur, and cataracts may be noted shortly after birth. Without dietary therapy, retarded mental development can be apparent in the newborn after only a few months of age. A number of states have instituted newborn screening for galactosemia using the disk of blood collected for the Guthrie test to assay for galactose-1-phosphate uridyl transferase. Metabolic screening tests of the urine from acutely ill infants should include a test for reducing substances in the urine. Urine dip sticks impregnated with glucose oxidase are specific for glucose. Such a urine screening would be negative in most infants with galactosemia. An infant with the clinical manifestations described above who has a negative dip-stick urine test for glucose but a positive copper sulfate test (such as Clinitest) for reducing substances in the urine has strong presumptive evidence for galactosemia. Dietary control of the disease in the first few years of life can result in normal growth and development.

A second defect in galactose metabolism occurs as a result of deficiency of the enzyme *galactokinase*.

$$\text{Galactose} + \text{ATP} \xrightarrow{\text{Galactokinase}} \text{Galactose-1-phosphate} + \text{ADP}$$

This disorder is recognized most commonly following a

Table 46-6 Inborn errors of glycogen deposition or utilization

Condition	CORI type	Biochemical features	Clinical features
Glucose-6-phosphatase deficiency (von Gierke's disease)	1	Normal glycogen accumulates in liver and kidney	Hepatomegaly, hypoglycemia; stunted growth with retarded bone age, etc.
Idiopathic generalized glycogenosis (Pompe's disease)	2	Normal glycogen accumulates in all organs	Cardiac failure, muscle hypotonia, neurological disorders, death in infancy
Dextrin-1,6-glucosidase (debrancher) deficiency (limit dextrinosis; Forbes' disease)	3	Abnormal glycogen with short branches deposited in liver and, sometimes skeletal and cardiac muscle	Hepatomegaly, hypoglycemia; less severe than von Gierke's disease
α-Glucan-branching glycosyltransferase (brancher) deficiency (amylopectinosis; Andersen's disease)	4	Abnormal carbohydrate with long inner and outer branches deposited in liver, spleen, and lymph nodes	Hepatic cirrhosis; death within 2 years of birth
Glycogen phosphorylase (glycogen phosphorylase of the muscle deficiency (McArdle's syndrome)	5	Moderate accumulation of normal glycogen in skeletal muscles; lactate and pyruvate levels in blood fall during exercise	Generalized muscular fatigability and pain
Glycogen phosphorylase (hepatic glycogen phosphorylase) deficiency (Hers' disease)	6	Normal glycogen accumulates in liver; phosphorylase content of liver and leukocytes reduced	Hepatomegaly; relatively benign
Deficiency of UDP glucose glycogen glucosyltransferase (glycogen synthetase)	7	Liver glycogen almost completely absent	Severe fasting hypoglycemia

From Geigy's scientific tables, ed 7, 1970, CIBA-Geigy Corp, Summit, NJ. From Field, RA. In Stanbury et al, editors: The metabolic basis of inherited disease, ed 2, McGraw-Hill, New York, 1966, page 141; Hers, HG, Advanc Metab Disord 1:1, 1964; Control of glycogen metabolism, Ciba Foundation symposium, London, 1964, Churchill.

screening of urine metabolites of patients with cataracts. Deficiency of galactokinase results in the accumulation of galactitol as an end product in galactose metabolism. This compound accumulates within the lenses and is responsible for the cataracts. There are no hepatic or other systemic manifestations of this defect. One can assay the enzyme in red blood cells, and the excretion of galactose and galactitol in urine will give a positive reducing substance reaction. Early recognition of this disorder and the elimination of galactose from the diet will prevent the development of cataracts.

Lactic acidosis is a commonly observed complication of many diseases that result in increased anaerobic metabolism. Primary abnormalities in carbohydrate metabolism leading to lactic acidosis are rare. Anaerobic metabolism and defective carbohydrate metabolism may give rise to elevations in both pyruvic and lactic acid. However, as pyruvate is shunted rapidly to lactate and most clinical laboratories perform lactate assays but not pyruvate assays, lactic acidosis rather than pyruvic acidosis is more commonly recognized and referred to. A primary abnormality in pyruvate metabolism or in muscle mitochondrial metabolism of glucose will result in an increased pyruvate and lactate level after a glucose load. Patients with lactic acidosis in whom no abnormality in tissue blood perfusion or oxygenation is apparent may require additional tests to rule out a primary abnormality in carbohydrate or pyruvate metabolism.

Glycogen-storage diseases (see Chapter 29). The glycogen-storage diseases combine two different types of metabolic defects. First, because of the alterations in glycogen metabolism, inadequate glucose stores are available for metabolic needs, and symptoms of acute hypoglycemia occur. Second, the accumulation of glycogen results in the long-term, chronic effects of a storage disease. Table 46-6 lists a classification scheme of the glycogen-storage diseases.

The classic form of glycogen-storage disease is von Gierke's disease, which occurs as a result of a deficiency of glucose-6-phosphatase. All metabolic sources of blood glucose are channeled through the intrahepatic formation of glucose-6-phosphate (see Chapter 21). Glucose in this form cannot be transported outside the liver cell. Thus the formation of glucose from amino acids through gluconeogenesis or the conversion of other carbohydrates into glucose uses the intermediate of glucose-6-phosphate. With a deficiency of glucose-6-phosphatase, the only carbohydrate available to maintain blood glucose is the glucose metabolite glucose-1-phosphate. A simplified scheme of glycogen metabolism and the consequences of glucose-6-phosphatase deficiency are shown in Fig. 46-14. Infants with type 1 glycogen-storage disease, glucose-6-phosphatase deficiency, usually have recurring episodes of hypoglycemia within the first few days of life that are often accompanied by ketoacidosis and lactic acidosis. If there is no elevation in blood glucose after administration of gluca-

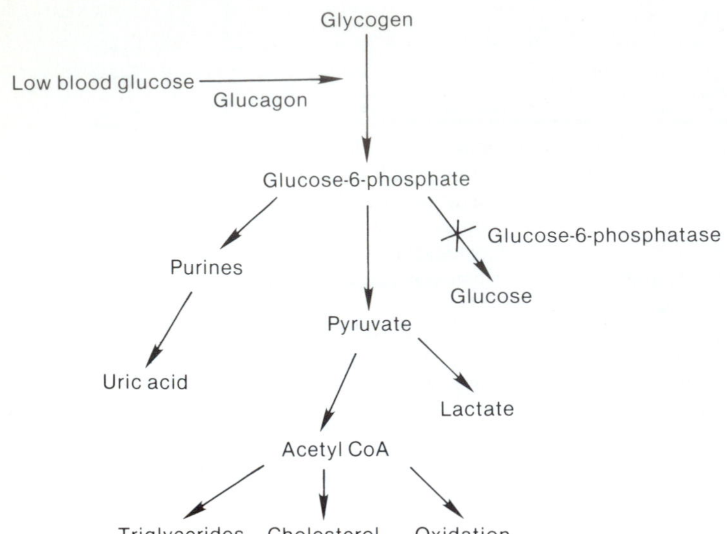

Fig. 46-14 Glycogen-storage disease type 1 (von Gierke's disease). Use of glycogen to form free glucose requires hepatic action of glucose-6-phosphatase to transport glucose extracellularly.

gon, a defect in glycogen metabolism must be assumed. Specific diagnosis of type 1 glycogen-storage disease requires a liver biopsy for the demonstration of glucose-6-phosphatase deficiency. Children with type 1 glycogen-storage disease are unable to maintain an adequate blood glucose for more than 2 to 2½ hours after a normal feeding. The management of infants with this disorder cannot be satisfactorily accomplished through the use of frequent oral feedings. For that reason, nasogastric tubes and feeding gastrostomies have been used to improve the management of this disease during the first few years of life.

Pompe's disease or type 2 glycogen-storage disease is a condition that is entirely different from the remainder of the glycogen-storage diseases. This disorder is the result of a defect in lysosomal degradation of glycogen. The deficient enzyme is α-glucosidase, an acid hydrolase similar to the deficient enzyme of other lysosomal storage diseases. Deficiency of this enzyme results in widespread, systematic lysosomal storage of glycogen and is seen in infancy with hepatomegaly, cardiomegaly, and central hypotonia. The disease follows a rapidly downhill course, with death by 1 to 2 years of age. This alteration of glycogen metabolism does not affect normal glucose homeostasis. Diagnosis of this form of glycogen-storage disease is indicated by lysosomal accumulation of glycogen demonstrated on muscle or liver biopsy. The enzyme α-glucosidase can be assayed in peripheral lymphocytes or in cultured skin fibroblasts to make a specific diagnosis. Because this enzyme is present in cultured cells, prenatal diagnosis exists for this form of glycogen-storage disease.

Transport defects. There are a number of conditions that lead to altered metabolic states as a result of a defect in membrane transport of metabolites rather than as a result of an enzyme deficiency in a metabolic pathway. After glomerular filtration of plasma, there is the selective reabsorption of metabolites through the renal tubules so that urine ultimately contains only waste products and nonwaste products are conserved. Single gene-determined components of the plasma membrane of the renal tubule are responsible for the selective reabsorption of individual compounds. One selective renal tubular reabsorption mechanism exists for the four amino acids cystine, lysine, ornithine, and arginine. Homozygotes for a mutation involving this reabsorptive system have a pronounced renal loss of these four amino acids. The condition is known as *cystinuria* because it is the renal loss of cystine that contributes to the clinical disorder. Cystine is relatively more insoluble in an acid urine than an alkaline urine. With an increased renal loss of cystine this amino acid precipitates out to form stones in renal papillae and the collecting system. Cystinuria is seen in early childhood as a form of chronic renal insufficiency usually first manifest as growth retardation. Metabolic screening of urine because of failure to thrive or short stature will detect cystinuria through a positive cyanide-nitroprusside reaction.

A number of conditions exist in which there is defective renal transport of amino acids; these are shown in Table 46-7.

Vitamin D–resistant rickets, or hypophosphatemic rickets, represents a defect in renal tubular reabsorption of phosphate. This condition is inherited as an X-linked dominant disorder. As a result, its effect is quite variable in heterozygous females who may show no clinical manifestations or who may exhibit moderate short stature and evidence of rickets. In affected males there is hypophosphatemia and hyperphosphaturia with short stature and bowing of the legs accompanying the roentgenographic evidence of rickets. Definitive diagnosis is made by demonstration of a greatly elevated 24-hour urine phosphate excretion. This disorder is called vitamin D–resistant rickets because massive doses of vitamin D will not correct the altered plasma and urine levels of calcium and phosphate. With regular intake of the maximum gastrointestinal tolerance of phosphate, males with this condition can have significant improvement of rickets and enhancement of their adult height. Phosphate is rapidly cleared from plasma when taken orally, and replacement therapy must include ingestion of phosphate every 4 hours.

Disorders of mineral metabolism. The most common and best known disorder involving mineral metabolism is *Wilson's disease,* or *hepatolenticular degeneration*. This autosomal recessively inherited disorder involving copper metabolism may be seen initially as cirrhosis in adolescents or as a psychiatric disorder in older teenagers or young adults. Although the exact genetic basis of the specific abnormality in copper metabolism remains unknown, derangements in tissue and urinary levels of copper are diagnostic for the disease. Clinically this disorder is char-

Table 46-7 Errors in renal amino acid transport

Disorder	Urine analysis	Findings
Cystinuria	Urine cyanide-nitroprusside	Renal failure resulting from cystine stones
Hyperdibasic aminoaciduria	Elevated urinary lysine, ornithine, arginine	Autosomal dominant asymptomatic
Fanconi syndrome	Elevated urinary amino acids, bicarbonate, phosphate, glucose, potassium, uric acid	Growth failure in primary form, occurs secondarily in Wilson's disease and galactosemia
Glucoglycinuria	Glucosuria, glycinuria	Autosomal dominant asymptomatic
Hartnup's disease	Elevated neutral and ring amino acids	Asymptomatic with good nutrition, especially nicotinic acid
Familial renal iminoglycinuria	Elevated urine proline, hydroxylproline, glycine	Autosomal recessive asymptomatic
Lowe's (oculocerebrorenal) syndrome	Renal aminoaciduria, proteinuria, aciduria, phosphaturia	X-linked recessive cataracts, mental retardation, and hypotonia

acterized by a triad of findings: a peculiar neurological syndrome, cirrhosis of the liver, and Kayser-Fleischer rings of the cornea. The neurological abnormalities take two forms: The first is lenticular degeneration (loss of cerebellar function), which is also known as the dystonic form (abnormal movement and tone of muscles) of the disease. The second neurological form is pseudosclerosis. This involves flapping tremors of the wrist and shoulders associated with rigidity and spasticity. The hepatic involvement of this disease is first manifest as an enlarged liver with associated splenomegaly. Thereafter the course is one similar to that of chronic hepatitis. Ultimately the disease progresses to the full-blown picture of cirrhosis. The Kayser-Fleischer ring is still considered as the single most important clinical diagnostic finding of this disease. It is a ring of a golden-brown or greenish discoloration that appears at the margin of the cornea near the limbus.

There are a number of laboratory approaches to the diagnosis of Wilson's disease. Most but not all patients with this disorder will have a depressed plasma level of ceruloplasmin. Ceruloplasmin is a metalloglycoprotein containing copper. It acts as an oxidase in the enzymatic oxidation of iron from the ferrous to the ferric state. Early in the course of the disease the 24-hour urine level of copper may be normal to slightly increased but is uniformly strikingly elevated in advanced stages of the disorder. Given a clinical suspicion of this disorder, one diagnostic approach is the 24-hour urine measurement of copper followed by a 24-hour urine measurement of copper during administration of penicillamine. With increased tissue levels of copper the chelating agent penicillamine will result in a striking increase in copper excretion. The most definitive and in some cases the only assured way of making a diagnosis is by liver biopsy and tissue analysis of copper content. Wilson's disease can be essentially controlled by the long-term administration of penicillamine.

A second disorder of copper metabolism is known as *Menkes' syndrome,* or *kinky-hair disease*. This disorder is an X-linked recessive neurological disease that is seen in early infancy with failure to thrive, lethargy, hypothermia, and myoclonic seizure activity. Affected male infants have pallid skin and a characteristic facial appearance. The hair may appear normal at birth, but thereafter a characteristic pattern develops. The hair lacks luster and is somewhat depigmented; it also has a steely feel. Children with Menkes' syndrome have low serum levels of copper and ceruloplasmin. However, attempts at treatment through copper administration have been unsuccessful. A defect in copper transport is demonstrated through the study of radioactive copper metabolism in cultured skin fibroblasts. This approach has also been successfully used in the prenatal diagnosis of this disease.

Disorders of urea cycle and hyperammonemias. The principal end product of nitrogen metabolism in humans is urea. Protein metabolism results in the production of ammonia, which enters the urea cycle (Fig. 46-15) in the synthesis of carbamyl phosphate and exits through urea in the conversion of arginine to ornithine. Within this cycle there have been recognized five distinct enzymatic deficiencies, each of which results in hyperammonemia. Disruption of the urea cycle does not produce acidosis, ketosis, or hypoglycemia but manifests itself clinically as the direct toxic effect of ammonia on the CNS. The clinical diagnosis should be suspected in infants with CNS depression after protein intake, a condition that should prompt an analysis of a blood ammonia level. The degree of hyperammonemia is massive and in itself does not indicate a particular disorder but rather an interruption in the urea cycle. Quantitative determination of plasma and urine amino acids may show elevations in amino acids suggestive of defects in earlier steps in the urea cycle or demonstrate the specific compound elevated in the latter steps of the urea cycle. Carbamyl phosphate synthetase and ornithine transcarbamylase deficiencies require liver biopsy for specific diagnosis using an enzymatic analysis. Citrullinemia, argininosuccinicaciduria, and argininemia are diagnosed by demonstrating an elevation of these substrates through two-dimensional paper chromatography of urine or

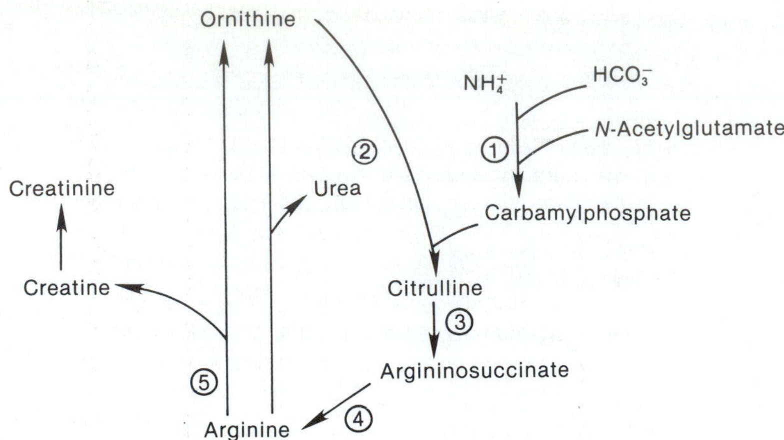

Fig. 46-15 Urea cycle defects. Five primary hyperammonemias include type 1, or carbamyl phosphate synthetase deficiency; type 2, or ornithine transcarbamyl transferase deficiency; type 3, or citrullinemia caused by argininosuccinate synthetase deficiency; type 4, or argininosuccinic aciduria caused by argininosuccinase deficiency; type 5, or hyperargininemia caused by arginase deficiency.

by column chromatography for amino acid analysis. The enzyme defect in these latter three conditions is demonstrated in cultured skin fibroblasts.

Receptor defects. Receptors are discrete gene products that function as membrane-bound or cytoplasmic agents required for the transport of specific compounds across membranes or as the intracellular regulators of metabolic activity. Knowledge about receptor function has been greatly enhanced through the study of single gene disorders that result in specific receptor defects. Such genetic disorders are illustrated by the conditions familial hypercholesterolemia and testosterone-resistant syndromes (testicular feminization syndrome).

Familial hypercholesterolemia is a dominantly inherited disorder that is the most common single gene cause of coronary artery disease and myocardial infarction in young adults. In humans, low-density lipoprotein (LDL) is the carrier for most of the cholesterol in plasma. After cholesterol is synthesized within the liver, it is transported within LDL particles to cells, where it is internalized through the binding of the LDL molecules to a specific cell surface receptor. After binding of LDL molecules to the cell surface receptors, an endocytotic vesicle is formed, which fuses with lysosomes. Within the lysosome the LDL molecules are degraded to free amino acids, eventually unesterified free cholesterol. As adequate amounts of cholesterol are internalized, there is suppression of 3-hydroxy-3-methylglutaryl coenzyme A reductase (HMG CoA reductase), which causes a reduction in cellular cholesterol synthesis. Conversely, an intracellular requirement for cholesterol stimulates HMG CoA reductase activity. The familial hypercholesterolemia disorder results from a mutation affecting the LDL receptor. With a reduced number of cell surface receptors for LDL cholesterol and thus reduced uptake of cholesterol, there is inadequate suppression of HMG CoA reductase activity. This results in stimulation of further cholesterol synthesis. The net result is familial or type IIa hypercholesterolemia. There are rare persons who are homozygous for a defect in the LDL receptor. During early childhood such persons have greatly elevated levels of plasma cholesterol and severe coronary artery disease, usually leading to death at adolescence. Study of cultured skin fibroblasts from homozygously affected persons allowed elucidation of the defect in LDL receptors and the concomitant effect on the regulation of HMG CoA reductase activity. As a result of this work, it was possible to show that the more common heterozygous form of autosomal dominant hypercholesterolemia represented a partial defect in LDL receptors and regulation of HMG CoA reductase activity.

The term *testicular feminization syndrome* is an older designation for a disorder involving an abnormality in testosterone metabolism. Affected individuals are phenotypic females who develop female secondary sexual characteristics at puberty but fail to undergo menarche. A laparotomy in such a person will show absence of a uterus and the presence of intra-abdominal testes. Chromosome analysis reveals a normal male 46,XY karyotype. These chromosomal males embryologically develop the external genitalia of a female because of a cellular resistance to the action of testosterone. There is normal testosterone generation by the testes, but there is an unresponsiveness of target tissues to testosterone. The action of testosterone (androgens) on target cells is mediated by specific cytoplasmic receptors. Dihydrotestosterone combines with a cytosol-binding protein (the receptor) to form a hormone-protein complex that is transported into the cell nucleus, where it exerts its action on chromatin. A defect in the receptor results in an inability of target cells to respond to the testosterone.

Disorders of purine metabolism. The best known disorder involving a defect in purine metabolism is the *Lesch-Nyhan syndrome*. Affected males are seen with a disorder of self-mutilation, mental retardation, and cerebellar dysfunction. Their urine contains excessive amounts of uric acid crystals. The hyperuricemia characteristic of this X-linked recessively inherited disorder is the result of a deficiency of hypoxanthine-guanine phosphoribosyltransferase (HGPRT). This enzyme is involved in the metabolism of the purine nucleotides guanylic acid (GMP) and inosinic acid (IMP). These two purines regulate the activity of a number of enzymes in purine metabolism including HGPRT. The deficiency of this enzyme results in an accelerated rate of purine biosynthesis, and because of the block, increased amounts of hypoxanthine and xanthine are synthesized, which are readily converted in the liver by xanthine oxidase to massive amounts of uric acid. Allopurinol is used to lower the serum and urinary levels of uric acid in these patients, but this does not alter the CNS abnormalities that are part of the disorder.

When cells deficient in HGPRT are cultured, they are resistant to the effects of purine analogs such as 8-azaguanine and 6-mercaptopurine, which inhibit the growth of normal cells. Because the female carrier for this X-linked condition by lyonization has cells with HGPRT and cells without HGPRT, fibroblasts will grow in both normal media and media with one of these purine analogs. Cells from a noncarrier female will grow in the normal media but not in media with the analog. This selective medium system has proved to be a powerful tool in the detection of carriers for the Lesch-Nyhan syndrome. As prenatal diagnosis is available, study of females in a family in which this condition has occurred is important for genetic counseling.

GENETIC SCREENING

Role of mass screening

Screening tests play a major role in the diagnosis and population study of genetic disease (Table 46-8). A screening test is not designed as a primary diagnostic tool. Rather, its purpose is to identify a subpopulation of persons on whom definitive diagnostic testing would be cost beneficial. In general, the more specific a particular screening test, the greater is its cost; thus its utility in screening large populations is reduced. A definitive diagnostic test for a given entity is one that contains close to 100% sensitivity and specificity. The sensitivity of a test is a measure of how frequently a test will detect a disorder under question. The specificity of a test is a measure of how well a test discriminates one disorder from other disorders. The sensitivity and specificity that are required of a screening test depend on the population to be studied and the purpose for which the screening test is undertaken. This is best illustrated when one considers the purposes for screening tests: (1) screening tests for the early detection of disease in the presymptomatic state when effective therapy exists (phenylketonuria), (2) carrier detection for inherited disease (Tay-Sachs disease), and (3) epidemiological studies of the frequency of a given disorder within the population (47,XXY or 47,XYY karyotype). If a disease can be effectively treated when detected early and if the consequences of nondetection are unacceptable, it is desirable to have a test with maximum sensitivity in order to

Table 46-8 Genetic screening tests

Nature of screening test	Condition(s) screened for	Definitive diagnostic test
Buccal smear for Barr bodies	Numerical abnormalities of the X chromosome	Peripheral blood lymphocyte karyotype
Serum assay for percent hexosaminidase A	Carriers for Tay-Sachs disease	Hexosaminidase A in peripheral leukocytes
Guthrie test for semiquantitative blood phenylalanine level	Phenylketonuria among newborn infants	Quantitative plasma level of phenylalanine
Maternal serum alpha-fetoprotein determination	In utero detection of neural tube defects and Down's syndrome	Amniocentesis for chromosomes, alpha-fetoprotein, and acetylcholinesterase
RBC mean corpuscular volume	Carriers for thalassemia	Hemoglobin electrophoresis
Stool for trypsin activity	Cystic fibrosis among newborns	Sweat test for 1 month of age
Serum CK levels in males with delayed walking	Duchenne muscular dystrophy	Microscopy of muscle biopsy
Fasting blood sugar or 2-hour postprandial blood sugar	Diabetes mellitus	Glucose tolerance test
Urine for reducing substances	Newborn with jaundice	Galactosemia
Hematocrit and RBC morphology among black infants	Sickle cell disease	Hemoglobin electrophoresis
Urine nitroprusside test among people with dislocated lenses	Homocystinuria	Quantitative plasma amino acids
Bleeding and clotting times in preoperative evaluation	Hemophilia A	Plasma factor VIII levels

detect all possibly affected persons. Those persons identified through a positive screening test can undergo additional testing with more specific assays that will establish a specific diagnosis. For a disease in which treatment is not readily available, successful, less-sensitive screening tests may be used. By lowering the sensitivity one may improve the specificity. The few persons with the disease who are missed by the screening test will not be unduly compromised.

Detection of heterozygote carriers

During the past 20 years considerable information has been amassed regarding the molecular abnormality in many genetically determined disorders. In recessively inherited disorders, knowledge of the molecular defect can be used for the detection of carriers through demonstration of a partial expression of the molecular defect in heterozygous individuals. The accuracy of these carrier detection tests generally is directly proportional to the accuracy of the assay system that measures the specific molecular defect of the disorder under question. A number of biochemical abnormalities have been noted in cystic fibrosis, most notably an elevation in the chloride content of sweat. However, the primary abnormality in cystic fibrosis remains unknown, and the application of numerous biochemical tests to known carriers of cystic fibrosis has to date failed to yield a means of carrier detection. It is therefore unlikely that an accurate carrier test for cystic fibrosis will become available until the specific molecular defect in cystic fibrosis is known. Recent advances in recombinant DNA analysis may contribute to the development of such a test (see below and Chapter 47). Some assay systems used for the detection of carriers of recessively inherited disorders are listed in Table 46-9.

Prenatal diagnosis

There are several criteria that a diagnostic method must meet if it is to be used successfully in prenatal diagnosis. First, there must be a means of identifying couples at sufficient risk for a particular condition to warrant the risks and costs of amniocentesis and other diagnostic procedures. A couple may be determined to be at risk for producing an offspring with genetic disease because of the knowledge of a previous birth of a child with a condition carrying a significant recurrence risk, because the parents are determined by previous carrier detection to be carriers for an autosomal recessively inherited disorder, or because of the age of mother.

Second, the disorder under question should by its nature warrant prenatal diagnosis. At present this is generally true in that practically all conditions for which prenatal diagnosis is available are severe disorders for which there is no effective therapy for affected infants.

Third, the accuracy of the diagnostic method used should be well established because there is little margin for error with these studies. For diagnoses of many cytogenetic disorders and biochemical disorders using cultured amniotic fluid cells, the laboratory receives a 20 to 30 mL sample for a one-time opportunity of establishing a culture and performing a limited number of assays. The results of these assays will be used to determine whether the pregnancy will be continued. One wants to have supreme confidence in the systems used under these circumstances.

The box on p. 709 lists the various diagnostic modalities used in prenatal diagnoses, with an example of each.

Table 46-9 Carrier detection for genetic diseases

Disease	Carrier test	Findings
Sickle cell disease	Hemoglobin electrophoresis	S and A hemoglobin bands
Tay-Sachs disease	Serum hexosaminidase (Hex) A and total assay	% Hex A < 45%
Hurler's syndrome	α-L-iduronidase assay	Level < 50% control
Duchenne muscular dystrophy	Serum CK assay	Elevated level
Hemophilia A	Factor VIII assay	Low level
	DNA genotyping	Genotype of carriers in family
Phenylketonuria	Phenylalanine (phe) tolerance test	Higher than normal phe level after load
	DNA genotyping	Genotype of one carrier patient

ALPHA-FETOPROTEIN TESTING

Maternal serum alpha-fetoprotein screening

Alpha-fetoprotein (AFP) is a plasma protein made by the fetal liver. In the nonpregnant state, a woman's plasma contains virtually no detectable AFP. During pregnancy, the maternal serum AFP level is related to the level of AFP in the amniotic fluid. Measurement of the level of AFP in maternal serum represents a screening test for fetal conditions associated with abnormal levels of amniotic fluid AFP. Any fetal condition that results in incresed passage of fetal plasma into amniotic fluid will increase the maternal serum level of AFP. Such conditions include open neural tube defects, such as anencephaly or meningomyelocele; abdominal wall defects, such as gastroschisis or oomphalocele, and renal loss of plasma proteins from congenital nephrotic syndrome. However, elevated levels of maternal serum AFP are also indicative of pregnancy complications in the absence of a fetal malformation; such complications include fetomaternal transfusion, increased risk of premature delivery, increased risk of fetal death, and the presence of twins.

From empirical experience it has been found that expression of the level of AFP in multiples of the median (MoM) lead to a better correlation with infant risk than

METHODS OF PRENATAL DIAGNOSIS

- Cultured amniotic fluid cells
  - Cytogenetic
    - Advanced maternal age for Down's syndrome
  - Biochemical
    - Hexosaminidase assays for Tay-Sachs disease
  - Morphological analysis
    - Electron microscopy for mucolipidosis IV
  - Substrate analysis
    - Radioactive copper in Menkes' syndrome
- Uncultured amniotic fluid cells
  - Restrictive endonuclease analysis
    - Beta-hemoglobin gene study in sickle cell disease
  - Genetic polymorphisms for linkage diagnosis
    - HLA determination and 21-hydroxylase deficiency
- Amniotic fluid analysis
  - Quantification of marker components
    - Alpha-fetoprotein for open neural tube defects
  - Analysis of abnormal metabolites
    - Methylmalonate in methylmalonicaciduria
  - Genetic polymorphisms for linkage diagnosis
    - Secretory status and myotonic dystrophy
- Imaging techniques
  - Ultrasonography
    - Autosomal recessive polycystic kidneys
  - Roentgenography
    - Skeletal dysplasia—achondrogenesis
  - Fetoscopy
    - Polydactyly in Ellis–van Creveld syndrome
- Fetal sampling
  - Fetal blood sampling
    - Factor VIII antigen assay in hemophilia A
  - Skin biopsy
    - Histological appearance in epidermolysis bullosa dystrophica

does use of standard deviations of the mean. Most screening programs use 2.5 MoM as the upper limit of normal for maternal serum; this is roughly equivalent to 5 SD above the mean. The amount of AFP produced by the fetus is relatively consistent from pregnancy to pregnancy, but the level of maternal serum AFP will vary with the maternal blood volume. It is necessary to adjust for this by correcting for maternal weight. Other factors that affect the maternal serum level are diabetes and race. The level of maternal serum AFP varies with gestational age, increasing gradually during the time appropriate for screening—14 to 24 weeks of pregnancy. The maternal serum level in MoM is based on the median determination for each gestational week of pregnancy and interpretation requires accurate dating of the pregnancy.

Retrospective analysis of maternal serum screening for elevated AFP levels showed a correlation between low levels of AFP and the presence of a fetus with Down's syndrome. This correlation was subsequently confirmed in a series of prospective studies so that, currently, low levels of maternal serum AFP are considered an indication for amniocentesis and determination of the fetal karyotype. An independent correlation also exists between maternal age and the risk for a fetus with Down's syndrome; a maternal age of 35 years or older is the most common indication for prenatal diagnostic testing by amniocentesis or chorionic villus sampling. Levels of maternal serum AFP can be expressed as a risk for Down's syndrome for specific maternal ages. Thus, if a woman's age and her low level of AFP combine to generate a risk equivalent to that of a 35-year-old woman, she is offered prenatal diagnosis.

Table 46-10 Inborn errors of purine and pyrimidine metabolism

Condition	Defective enzyme or system	Biochemical features	Clinical features	Incidence and genetics
Gout (hyperuricemia)	Excessive synthesis of uric acid from precursors	Concentration of uric acid increased in serum and often in urine.	Acute arthritic attacks, chronic arthritis with urate deposition in tissues; urinary urate calculi causing kidney damage; asymptomatic in 80% of cases	Hyperuricemia in 1%-2%, clinical gout in 2-4/1000; probably autosomal dominant with variable and sex modified expression
Xanthinuria	Deficiency of xanthine oxidase and defective renal tubular reabsorption of xanthine	Xanthine excreted in large amounts.	Xanthine calculi in urinary tract	Rare
Orotic-aciduria	Absence of orotidine-5′-phosphate pyrophosphorylase and/or decarboxylase	Orotic acid accumulates and is excreted in urine.	Severe megaloblastic anemia, orotic-acid crystalluria	Very rare; recessive
β-Aminoisobutyric-aciduria	Deficiency of a catabolic enzyme	High urinary excretion of β-amino-isobutyric acid.	Harmless	0%-46%, depending on ethnic group; recessive

From Scientific tables, ed 7, 1970, CIBA-Geigy Corp, Summit, NJ.

Table 46-11 Lipidoses

Condition	Lipid accumulating	Site	Clinical features	Age at which symptoms appear	Genetics
Gaucher's disease (a) 'Adult' (b) Acute infantile (c) Juvenile and adult neurological	Glucocerebroside	Spleen, liver, bone marrow, leucko-cytes. Brain in (b) and (c); lung in (b)	Splenomegaly, often gross; hepatomegaly; anemia; bone disorder; purpura; cerebral degeneration in (b) and (c)	(a) 1-60 years (b) 1st or 2nd half-year of life (c) 6-20 years	(a), (b), and (c) in different families; all recessive
Tay-Sachs disease (infantile amaurotic familial idiocy)	Ganglioside G_{M2} (G_0), amino glycolipid	White and gray matter of the brain	Cherry-red spot; progressive cerebral degeneration; death at age 1-5 years.	Usually 4-6 months, some times earlier	Recessive
Juvenile and adult amaurotic familial idiocy	Ganglioside G_{M1} (G_1)	Brain (moderate increase)	Progressive loss of vision and cerebral degeneration.	From 5 years onward	Probably recessive
Niemann-Pick disease (a) Acute infantile (b) Cerebral juvenile (c) Noncerebral	Mainly sphingomyelin	Spleen, bone marrow, liver; usually also brain and retina	Often cherry-red spot; hepatosplenomegaly; hepatic cirrhosis; usually cerebral degeneration and death in first 2½ years. Some adult cases are without neurological involvement.	(a) From birth (b) Childhood (c) Up to 30 years or later	(a) Recessive (b) Recessive (c) Uncertain
Metachromatic leucodystrophy (a) Infantile (b) Adult	Sulfhatides	Brain, kidney, urine, gallbladder			
Essential familial hyperlipemia	Triglycerides, lipoproteins	Blood plasma (chylomicrons)	(a) Cerebral and cerebellar degeneration; spasticity; dementia; death after 1-6 years. (b) Psychotic changes; blindness; aphasia; tetraplegia. Death after 3-12 years.	(a) 1-2 years (b) Late childhood or adulthood	(a) Recessive (b) Uncertain
Hypercholesterolemia	Cholesterol (free and esterified), phosphatides, sometimes triglycerides	Blood plasma (lipoproteins), tendons, skin, blood vessels			
			Hepatosplenomegaly; sometimes xanthomata.	Usually early childhood	Complex
			Relatively benign.		
			Cutaneous and tendinous xanthomata; atheroma of endocardium, coronary arteries, or great vessels.	From childhood onward	Usually dominant

*From Scientific tables, ed 7, 1970, CIBA-Geigy Corp, Summit NJ.

Amniotic fluid determination of alpha-fetoprotein

Maternal serum AFP determination is a screening test considered applicable to all pregnancies. Given an abnormal result, further testing in the form of ultrasonography and amniocentesis is offered. Normally, AFP determinations are done any time amniotic fluid samples are taken as with amniocentesis for advanced maternal age. A close correlation exists between the level of amniotic fluid AFP and the fetal conditions cited above; additionally, these structural defects can often be visualized by ultrasonography. If AFP is elevated in maternal serum and, upon amniocentesis, in amniotic fluid, but no fetal defect is seen upon ultrasonography, electrophoresis of amniotic fluid acetylcholinerase will reveal a specific band if an open neural tube defect is present. Amniotic fluid levels of AFP also vary with gestational age; levels are expressed as MoM with the median established for each gestational week of pregnancy between 16 and 26 weeks.

The recombinant DNA techniques that have proved so useful in the study of gene structure and function have provided means for accurate prenatal diagnosis of several conditions. The best known example is the test for the prenatal diagnosis of sickle cell disease. With sickle cell disease it has been possible to combine the use of restrictive endonucleases that recognize the specific nucleotide change of the mutant sickle cell gene with probes that bind to that portion of the beta chain of the hemoglobin gene. This technique requires neither the culturing of amniotic fluid cells nor special studies on the parents provided that they are known carriers for sickle cell disease. It is possible to imagine a time when probes and restrictive endonucleases will exist for the diagnosis of all genetic disorders. See Chapter 47 for a more detailed description of this technique.

CHANGE OF ANALYTE IN DISEASE

Each disease caused by an inborn genetic error results in a unique defect in a protein or enzyme function, or both. The consequence of this defect is a singular pattern of analyte change in biological fluids. Tables 46-4 to 46-11 list the pathognomonic analyte changes for a number of these genetic defects. In addition, the major clinical findings for each disease are also given.

BIBLIOGRAPHY

Kelly, TE: Clinical genetics and genetic counseling, ed 2, Chicago, 1986, Second edition, Yearbook Medical Publishers, Inc.

Grouchy, J de, and Turleau, C: Clinical atlas of human chromosomes, New York, 1977, John Wiley & Sons, Inc.

McKusick, VA: Mendelian inheritance in man, Baltimore, 1983, John Hopkins University Press.

Scriver, CR, and Rosenberg, LB: Amino acid metabolism and its disorders, Philadelphia, 1973, WB Saunders Co.

Simpson, JL: Disorders of sexual differentiation, New York, 1976, Academic Press, Inc.

Stanbury, JB, Wyngaarden, JB, Fredrickson, DS, et al, editors: Metabolic basis of inherited disease, ed 5, New York, 1983, McGraw-Hill Book Co.

Thomas, GH, and Howell, RR: Selected screening tests for genetic metabolic diseases, Chicago, 1973, Yearbook Medical Publishers, Inc.

CHAPTER 47

Use of nucleic acid probes in the clinical laboratory

LAWRENCE M. SILVERMAN
VICKY A. LEGRYS
NANCY C. PARKER

OBJECTIVES

- Review the structure of DNA, and describe how its properties of complementary base pairing and digestion by specific nucleases can be used to measure specific sequences of DNA.
- List the major clinical uses of techniques to measure specific DNA sequences.
- Describe restriction fragment length polymorphism, and explain how it can be used to detect the presence of specific genes.
- Describe the Southern blot technique.

KEY TERMS

base pairing The process by which purine and pyrimidine bases bind through hydrogen bonds. The bases pair in a specific complementary fashion: adenine with thymine (or uracil) and guanine with cytosine.

base sequence The exact order of purine bases (guanine and adenine) and pyrimidine bases (cytosine and thymine or uracil) found in nucleic acids. The order defines the primary sequence of the gene products, which are proteins.

complementary DNA (c-DNA) DNA that contains an exact sequence of bases that will pair to a strand of DNA or RNA through base pairing. c-DNA can be used as a probe for detecting specific sequences.

denaturation The process by which double-stranded nucleic acid separates to form single strands. Denaturation can be accomplished by heat, salts, or chemicals.

double-stranded DNA Two complementary strands of DNA that are bound together through base pairing.

hybridization The process by which complementary single strands of nucleic acid form double-stranded complexes of nucleic acid through base pairing.

linkage A term defining the association of genes on a chromosome to specific sequences of DNA.

Northern blot A process similar to the Southern blot, except RNA is the molecule transferred and analyzed.

polymerase chain reaction A process by which complementary DNA strands are enzymatically synthesized and amplified.

probe A sequence of complementary DNA or RNA that is labeled with a radioisotope, enzyme, or other marker. Probes are used to detect specific sequences of nucleic acid by hybridization.

restriction endonucleases Class of nucleases (usually bacterial) that act within a strand of DNA at specific nucleotide sequences to cleave the DNA.

restriction fragment length polymorphism (RFLP) The pattern of DNA fragments observed in a test population when analyzed by restriction endonucleases and specific DNA probes.

single-stranded DNA A length of DNA that is not paired to its complementary strand.

Southern blot A process by which electrophoretically separated, denatured DNA is transferred from the electrophoretic gel (usually agarose) onto a nitrocellulose filter or membrane for subsequent hybridization analysis.

Deoxyribonucleic acid (DNA) carries the specific information that determines all physiological processes. Some misinformation also contained in DNA can be responsible for certain disease states. Therefore clinical laboratories are currently becoming involved in the use of a new technology, *nucleic acid probes,* for the detection of specific sequences of DNA. These tools, often called *DNA probes,* are used (1) to detect viral and bacterial DNA, thus diagnosing microbial infections; (2) to characterize molecular rearrangements associated with various types of cancer; and

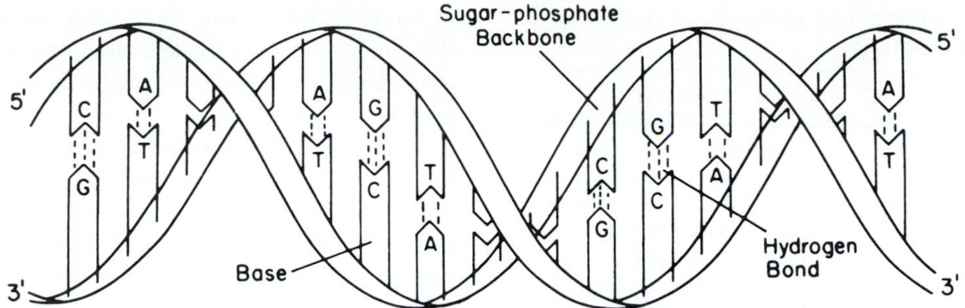

Fig. 47-1 Structure of DNA. DNA molecule is double helix that consists of two sugar-phosphate backbones with four bases cytosine *(C)*, guanine *(G)*, adenine *(A)*, and thymine *(T)* attached. *C* and *G* residues and *A* and *T* residues on opposite strands pair through hydrogen bonding. *(Reprinted with permission from LeGrys, V, Leinbach, SS, and Silverman, L: CRC Crit Rev Clin Lab Sci 25(4):255, 1987. Copyright CRC Press, Inc, Boca Raton, FL.)*

(3) to detect specific markers of genetic disorders. In the very near future laboratories will use similar approaches to study the specific portions of DNA that appear to be intimately related to the carcinogenic process in many cells, the so-called oncogenes. This chapter deals with the use of DNA probes in genetic diseases.

DNA STRUCTURE

Fig. 47-1 shows the unique arrangement of sugar, phosphate, and the purine and pyrimidine bases that form the double-helical structure known as DNA. Particular sequences of this structure form the *gene,* which codes for a specific protein (see Chapter 51, Part I).

DNA consists of two strands of base sequences that are bound to each other by hydrogen bonding between the bases of each single strand of DNA (Fig. 47-1). The bases bind to each other in a specific, or *complementary,* fashion. Adenine binds only to thymine, whereas guanine binds only to cytosine. Thus one chain of double-stranded DNA has a base sequence that is complementary to the other strand. A single strand of DNA will bind to another single DNA strand if the strands contain a high proportion of complementary sequences. For example, if a mixture consists of single strands of DNA-A and its complementary strand and an excess of other noncomplementary DNA strands, strand A will *only* form double-stranded complexes with its complementary strand and none other. In the laboratory the process of allowing complementary single strands of DNA to form double-stranded DNA is called *hybridization.* Hybridization can also be performed between single strands of DNA and complementary strands of ribonucleic acid (RNA). For analysis of a specific DNA base sequence, a known copy of that base sequence is prepared. This copy of complementary DNA (c-DNA) known as a DNA *probe,* is labeled in some fashion to allow monitoring of the hybridization reaction. The basis of the techniques described later in the chapter hinge on the hybridization properties of a specific sequence of bases that make up a single strand of the double-stranded helix of DNA.

A specific property of DNA (and RNA) is its ability to be acted on by enzymes called *nucleases.* Nucleases hydrolyze the phosphodiester bonds that connect bases within a nucleic acid strand, resulting in cleavage of the strand. Certain nucleases have a very high substrate specificity and will cleave a DNA strand only at specific base sequences, often as small as 6 to 8 bases in length. Because these nucleases are employed by bacteria to restrict entry of foreign DNA into their cells, they are called *restriction endonucleases.*

Restriction endonucleases are critical reagents in laboratories investigating DNA or RNA base sequences. After degrading DNA into a series of many smaller fragments, specific sequences can be more readily identified by the hybridization technique. To aid in the identification of a specific base sequence, the fragments can be first separated into molecules of differing molecular size. This is accomplished by agarose gel electrophoresis. The hybridization of the separated fragments is then achieved by the Southern or Northern blot techniques (see later discussion).

CLINICAL APPLICATION OF NUCLEIC ACID PROBE TECHNOLOGY

The nucleic acid probe technology can be used to detect specific DNA or RNA sequences in human samples. DNA sequences of interest to the clinician include:

- Normal gene sequences
- Abnormal gene sequences associated with specific disease states
- Abnormal gene sequences associated with cancer
- Exogenous base sequences associated with infectious organisms, such as bacteria and viruses

Specific gene sequences associated with cancer are called *oncogenes.* Oncogenes are normal cell genes that become expressed inappropriately (see Chapter 49). By detecting specific RNA transcripts of oncogenes, the laboratory may be able to help physicians classify cancers, establish a prognosis of certain cancers, or even predict the likelihood of an individual developing a cancer.

Table 47-1 Some nucleic acid probes available for diagnosis of infectious diseases

Agent	Supplier
Mycobacterium tuberculosis	Gen Probe (San Diego)
Mycoplasma pneumoniae	Gen Probe (San Diego)
Legionella	Gen Probe (San Diego)
Chlamydia trachomatis	Gen Probe (San Diego)
Salmonella	Integrated Genetics (Framingham, Mass.)
Neisseria gonorrhoeae	Ortho Diagnostic Systems (Westwood, Mass.)
Human papilloma virus	Life Technologies (Gaithersburg, Md.)
Human immunodeficiency virus	Specialty Laboratories Inc. (Los Angeles)

Table 47-1 lists some of the infectious diseases for which DNA probes are commercially available. DNA probe analysis for microbial genes can be used for very early diagnosis of infections. A common techniques used is called *in situ hydridization,* in which a labeled probe binds to bacterial or viral DNA on cells or tissue sections placed on microscopic slides. The labeled probe is used to produce a color reaction, which identifies the infectious DNA. These procedures are very sensitive and specific and may eventually replace conventional microbiological methods.

Abnormal variants of normal DNA sequences are associated with diseases of genetic origin (see Chapter 46). DNA probes can be used to make very early diagnoses of these diseases, giving prospective parents the opportunity to seek genetic counseling. The box lists genetic disorders for which DNA probe analysis is currently available. As research continues in this field, this list should rapidly expand.

GENETIC DISORDERS

In general, genetic disorders can be grouped into the following categories: chromosomal abnormalities, multifactorial disorders, and single gene disorders. The discussion in this chapter centers on single gene disorders because these are the most frequently encountered form of genetic disorders.

Single gene disorders are inherited in one of the following patterns: autosomal dominant, autosomal recessive, or X-linked. The terms *autosomal* and *X-linked* refer to the chromosomal location of the disease-causing gene, or *mutant* gene. Generally, in X-linked disorders, males with the mutant gene express the disorder, whereas females with the same gene are *carriers*. Examples of X-linked disorders include Duchenne's muscular dystrophy and hemophilia. *Dominant* disorders are expressed whether an individual is heterozygous or homozygous; *recessive* disorders are only expressed when an individual is homozygous for a mutant gene. Examples of the former are the autosomal dominant disorders achrondroplasia and Huntington's disease; examples of the latter are the autosomal recessive disorders cystic fibrosis, sickle cell anemia, and phenylketonuria. Inidividuals who are heterozygous for these recessive disorders are carriers.

COMMON CLINICAL APPLICATIONS OF DNA LINKAGE ANALYSIS

Cystic fibrosis
Duchenne's muscular dystrophy
Hemophilia A and B
Alpha$_1$-antitrypsin deficiency
Sickle cell anemia
Thalassemia
Phenylketonuria
Adult polycystic kidney disease
Huntington's disease

LABORATORY TESTS FOR GENETIC DISORDERS

Many laboratory tests used in the diagnosis of genetic disorders give results that may be characteristic of a disorder but are not entirely specific. These tests are generally *phenotypic* tests; that is, they reflect how a particular gene is expressed rather than the exact DNA sequence, or *genotype*. Therefore, phenotypic tests can be misleading, particularly in the absence of complete clinical data and family history. *Genotypic* tests could avoid these problems by detecting the specific alterations in DNA that result in the disorder. Unfortunately only a few such *direct* genotypic tests exist; sickle cell anemia can be detected in this manner because the specific mutation and DNA sequence are known. More often the specific mutation in DNA is unknown, the DNA sequence of the gene is undetermined, or a disorder is caused by one of several alterations in DNA. In these cases an *indirect* genotypic approach is necessary. This indirect approach, called *linkage analysis,* is the focus of the remainder of this chapter.

Linkage analysis can be used in following the inheritance of single gene disorders within a family. For example, color blindness is a phenotypic marker that can be followed within families. The most common form of color blindness is manifested only in males; this is an example of a genetic linkage. In this case the gene of interest is linked to the X chromosome.

This type of linkage analysis is at the level of the phenotypic expression of a specific gene on a specific chromosome. Monitoring the expression of abnormal genes at the molecular level is more difficult, especially when the gene product, and specific DNA sequence, may not be known. However, the presence of a closely linked DNA abnormal gene sequence can often affect the ability of restriction endonucleases to degrade the DNA. Thus the presence of an abnormal gene can be monitored by following the pattern of DNA fragments produced after restriction endonuclease digestion of closely linked DNA. The production of different patterns of DNA fragments from different individuals is called *restriction fragment length polymorphism (RFLP)*. The RFLP can be used to follow inheritance of the gene of interest as long as complementary DNA probes are available. Therefore the complementary DNA probe and the disease-carrying gene are linked, and the abnormal gene can be detected even though its gene product and its DNA sequence are not known.

To use linkage analysis, it is essential to compare various family members' RFLP patterns; thus key family members' DNA must be available. An important point to remember is that the DNAs complementary to the c-DNA probe do not cause the disorder; they merely allow one to follow the inheritance of a closely linked mutant gene throughout a family. A more detailed description of this technique follows.

Although this chapter emphasizes DNA techniques, phenotypic tests are also available that can be helpful in either prenatal diagnosis or carrier detection. The secretion of the intestinal isoenzyme of alkaline phosphatase into the amniotic fluid is decreased in fetuses affected with cystic fibrosis. After amniocentesis is performed, amniotic fluid can be tested for intestinal alkaline phosphatase activity, and fetal cells can be tested using DNA probe techniques. Frequently the results can be used to increase the accuracy of the DNA tests; occasionally the DNA tests are not applicable, and the intestinal enzyme activity is the only clinically useful prenatal test (see section on cystic fibrosis).

LINKAGE ANALYSIS

Nucleic acid probes and linkage analysis have been used successfully for the prenatal or presymptomatic diagnosis of genetic disorders. To perform prenatal tests, DNA must be obtained from the fetus either by *amniocentesis,* in which fetal cells are obtained by removing amniotic fluid (see Chapter 37), or by *chorionic villus sampling,* in which fetal cells are obtained when fetus is about 9 weeks old.

The use of linkage analysis to detect genetic disorders begins when an interested family seeks genetic counseling. The genetic counselor explains the procedures and limitations to the family (that is, amniocentesis versus chorionic villus sampling), obtains a *pedigree* (detailed family history), completes the appropriate consent forms, and answers any questions that the family may have. Peripheral blood samples are obtained from the *proband* (the family member affected with the disorder), the natural parents, and any interested family members. If prenatal diagnosis is requested, a fetal sample is obtained by either chorionic villus sampling (at 9 to 11 weeks of gestation) or amniocentesis (15 to 17 weeks).

Blood samples for linkage analysis are collected into acid-citrate-dextrose (ACD) anticoagulant tubes. The white cells are then isolated from each sample. DNA is extracted from the white cells and incubated with a restriction endonuclease to cleave the DNA into smaller fragments. The digested DNA sample is applied to an agarose gel and electrophoresed to separate the fragments according to size. The fragments are then treated so as to separate the double-stranded DNA into single strands; this process is termed *denaturation*. The separated, denatured fragments are then transferred from the gel onto another support medium, such as a nitrocellulose or nylon membrane (the *Southern blot* procedure). The fragment or fragments containing the DNA sequence of interest are identified by incubating the membrane with a labeled DNA probe that has

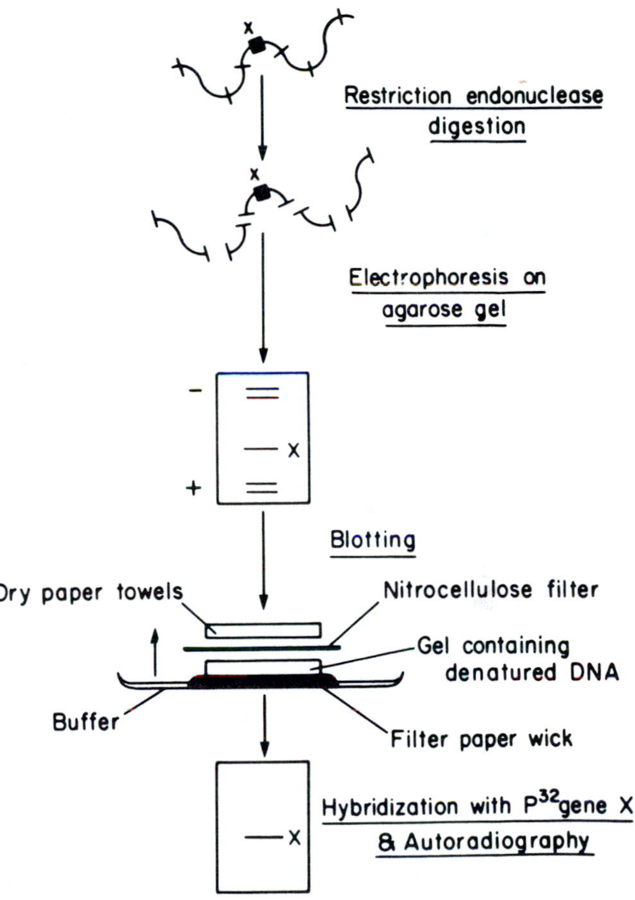

Fig. 47-2 Identification by Southern blot hybridization of DNA fragment containing gene X. DNA was digested with restriction endonuclease, and resulting fragments were fractionated according to size by electrophoresis in agarose gel. DNA fragments in gel were denatured and blotted to nitrocellulose filter as a result of flow of buffer (↑) through gel and nitrocellulose filter to dry paper towels. Subsequent hybridization of DNA on filter to ^{32}P-labeled gene X probe and autoradiography revealed single DNA fragment containing gene X. *(Reprinted with permission from LeGrys, V, Leinbach, SS, and Silverman, L: CRC Crit Rev Clin Lab Sci 25(4):255, 1987. Copyright CRC Press, Inc, Boca Raton, FL.)*

been associated with the sequence of interest. The label can be a radioisotope, an enzyme, or a fluorescent dye. The complementary sequence of the probe permits it to hybridize to the sample DNA containing the linked probe sequences. In the case of radiolabeled c-DNA, the membrane is then incubated with x-ray film to expose areas *(bands)* on the film where the probe has bound to the sample DNA, resulting in an *autoradiogram* (Fig. 47-2).

The procedure described requires 7 to 10 days from DNA extraction from whole blood to development of the autoradiogram. Usually DNA is extracted on the first day, digested with restriction endonucleases on day 2, electrophoresed overnight on day 3, and hybridized on days 4 and 5. The membrane is then placed in an x-ray cassette with x-ray film for 1 to 4 days and then developed. The resulting bands are interpreted. If the results can clearly determine the genetic status of all family members, the family is informed and counseled appropriately. If the status cannot be unequivocally determined, additional probes and restriction endonucleases are used in a repeat of the previous procedure. As a result, the entire process may require 2 to 3 weeks. In addition, DNA samples obtained by chorionic villus sampling and amniocentesis for prenatal diagnosis often require cell culturing to yield adequate amounts of DNA, which further lengthens the total turnaround time. Ideally prenatal diagnosis should follow preconception family studies that have determined the appropriate DNA probes and restriction endonucleases to be used on the fetal sample.

FACTORS AFFECTING ACCURACY AND AVAILABILITY OF GENE PROBE TESTING

Factors that can decrease test accuracy include:

1. Recombination that occurs between the DNA sequence recognized by the probe and the gene of interest can be minimized by employing probes that are tightly linked (very close) to the gene.
2. Incorrect identification of paternal relationships can cause misinterpretation of the results, since the procedure is based on comparing inheritance patterns from generation to generation.
3. Erroneous diagnosis of the proband will invalidate the results, since the basis of linkage analysis depends on comparing the restriction fragment patterns of the affected individual to those of others.
4. Genetic heterogeneity or disorders can be the result of more than one gene defect, sometimes on different chromosomes.
5. Laboratory errors can include specimen mislabeling, probe contamination, incomplete DNA digestion by the restriction endonuclease, or misinterpretation of the autoradiogram.

In addition, not all families can benefit from the application of linkage analysis to their genetic disorder for the following reasons:

1. Linkage analysis is an indirect test that requires comparing patterns of restriction fragments of the affected individual's DNA to those of other family members. The test cannot be done without a DNA sample from the affected individual and his or her natural parents. For this reason, the test cannot be used to screen the general population.
2. The DNA of some families may not yield informative RFLP patterns using the existing probes and restriction enzyme combinations.

CYSTIC FIBROSIS

Clinical background

Linkage analysis is used in the prenatal diagnosis and carrier detection of cystic fibrosis (CF). CF is inherited as an autosomal recessive disorder and appears to be caused by a single gene defect at a specific site on chromosome 7. The gene for CF has not yet been identified, and therefore no direct method is currently available for testing the general population for carriers.

CF is characterized by thick secretions obstructing the exocrine glands and leading to chronic pulmonary disease, pancreatic insufficiency, and abnormal sweat electrolytes. CF is diagnosed on the basis of clinical manifestations, a positive family history for the disease, and elevated sweat chloride concentration. Other laboratory tests that may be useful in documenting pancreatic insufficiency include the *p*-aminobenzoic acid (PABA) test and the pancreatic stimulation test. Pilot programs are screening newborns for CF on the basis of elevated immunoreactive trypsin (IRT) activity in blood spots collected shortly after birth. Decreased activity of microvillar enzymes (the intestinal isoenzyme of alkaline phosphatase, leukocyte alkaline phosphatase, and gamma-glutamyl transferase) is seen in amniotic fluid from fetuses with CF and can be a test for prenatal diagnosis when DNA testing is not available or useful.

Efforts to identify the gene responsible for CF have resulted in localizing the gene to the long arm of chromosome 7. Several clinically useful DNA probes are tightly linked to the CF gene, such as Met D, Met H, J3.11, XV2c, KM19, and CS.7. The use of these probes and their appropriate restriction endonucleases allows one to follow the inheritance of the CF gene in familiies with a previously affected child with CF and to determine carrier status or provide prenatal diagnosis. The accuracy of this method is approximately 98%.

Case report

Mr. and Mrs. K have one child, a 3-year-old girl with CF. The child was diagnosed at 18 months of age on the basis of recurrent pulmonary infections and elevated sweat chloride concentrations of 100 to 105 mEq/L. When Mrs. K was 16 weeks pregnant with another child, the K family asked their physician about prenatal diagnosis. Their phy-

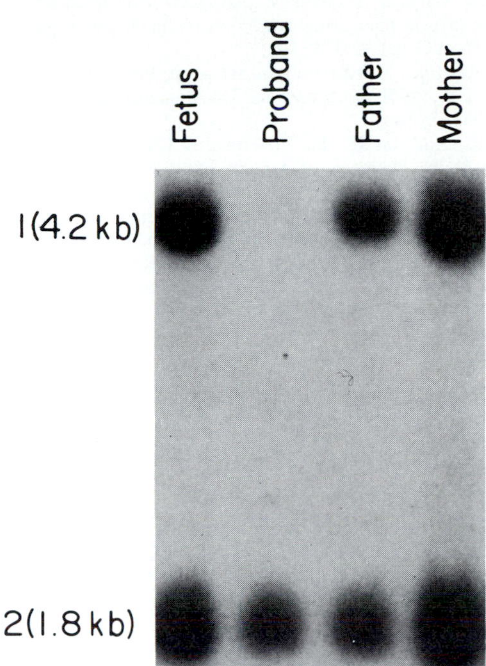

Fig. 47-3 Autoradiogram containing the J3.11/MspI polymorphisms of the K family. *(Reprinted with permission from LeGrys, V, Leinbach, SS, and Silverman, L: CRC Crit Rev Clin Lab Sci 25(4):255, 1987. Copyright CRC Press, Inc, Boca Raton, FL.)*

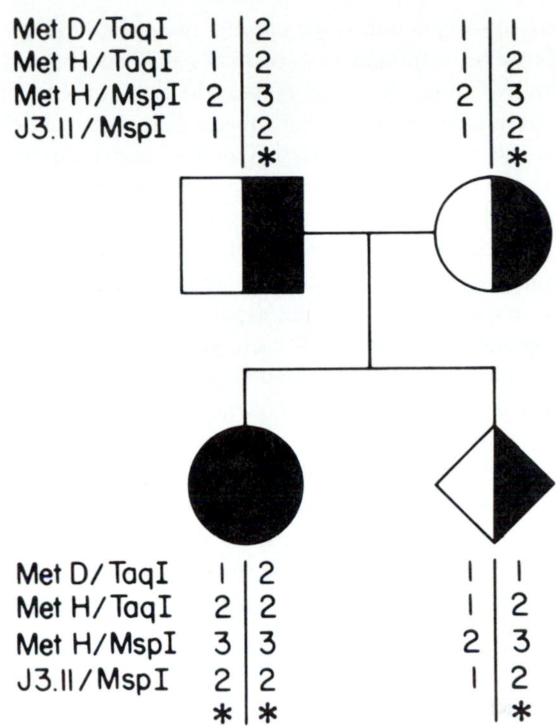

Fig. 47-4 Haplotypes for the K family. Asterisk denotes chromosome bearing mutant CF allele. *(Reprinted with permission from LeGrys, V, Leinbach, SS, and Silverman, L: CRC Crit Rev Clin Lab Sci 25(4):255, 1987. Copyright CRC Press, Inc, Boca Raton, FL.)*

sician referred them to a genetic counselor for information.

Following genetic counseling, blood was obtained on Mr. and Mrs. K and their affected child, and amniocentesis was performed to obtain fetal DNA. The DNA was analyzed using the Met D, Met H and J3.11 probes with the restriction endonucleases TaqI and MspI. It was possible to obtain the information with these probes and restriction endonucleases. Fig. 47-3 shows the autoradiogram for the J3.11/MspI analysis. With this probe and enzyme combination, the CF gene is associated with the *2* fragment (the smaller-sized fragment). The affected child (proband) has two alleles with the *2* fragment and the carrier parents are *1,2*. The fetus who has the same RFLP pattern as the parents appears to be carrier for CF. The result of this analysis, the pedigree of the K family for the CF allele, is shown in Fig. 47-4. Mr. and Mrs. K were informed of the results by the genetic counselor and decided to continue the pregnancy with plans to employ the sweat test for the baby a few months after birth. When the infant was 3 months old, sweat testing was performed with normal results.

FUTURE APPLICATIONS

The use of DNA probe technology has been limited by the difficulty in generating complementary probe DNA in sufficient quantities from very small amounts of DNA. A recent development called the *polymerase chain reaction (PCR)* was developed to increase the sensitivity and specificity of the DNA technique. The PCR is based on the ability of the DNA polymerase enzyme to generate many copies of a specific fragment of single-stranded DNA when an appropriate DNA template is available. The PCR uses the actual steps of in vivo DNA replication. Each cycle of PCR consists of three steps and results in a doubling of DNA molecules. In the first step double-stranded DNA is separated into single strands by denaturing at high temperatures. In the second step short sections of synthetic DNA are hybridized to the denatured DNA. These short pieces of DNA serve as primers or "start" signals for DNA replication in the next step. The third, or replication, step of the cycle uses an enzyme called a DNA polymerase, which allows replication of DNA, starting with the primer. The replication process forms a single strand by adding a series of single nucleotides that are complementary to the original DNA molecule. Newly synthesized DNA is separated from the original template by increasing the temperature, resulting in two molecules of DNA. This frees the original template to begin a new cycle.

PCR can result in amplification of the initial number of strands by a factor of at least 10^5 in less than 8 hours. Instrumentation is now commercially available to automate PCR. This procedure, however, cannot be used to identify genetic disorders unless the specific sequence of DNA that needs to be amplified is known. As this information becomes available, PCR may replace the traditional Southern blot technique because it has greater sensitivity and eliminates the need for radioactive probes and hybridization steps. In our laboratories cystic fibrosis testing can be completed in less than 48 hours using amplified DNA from peripheral blood or fetal cells obtained by chorionic villus sampling or amniocentesis. The resulting cost is significantly reduced, as one might expect.

Applications of the PCR technique are limited, at this time, to direct techniques involving disorders in which the normal and mutant allelic sequences are known, such as sickle cell anemia, and in which allelic heterogeneity is absent. In the last 3 years clinical applications of PCR have included (1) amplifying DNA extracted from tissue specimens obtained at autopsy or biopsy, (2) amplifying DNA for forensic purposes (such as semen stains for rape identification), and (3) amplifying DNA from filter blood spots.

As molecular research continues to describe other probes that are tightly linked to disease-causing genes, this list will continue to grow. Furthermore, researchers are now describing genetic risk factors for common, polygenic disorders such as heart disease, cancer, and diabetes. As we enter the 1990s we can expect this area to have a dramatic impact on laboratory medicine.

BIBLIOGRAPHY

Botstein D, White R, Scolnick M, and Davis R: Construction of a genetic linkage map in man using restriction fragment length polymorphisms, Am J Hum Genet 32:314, 1980.

Buffone GJ, Spence JE, Fernbach SD, et al: Prenatal diagnosis of cystic fibrosis: microvillar enzymes and DNA analysis compared, Clin Chem 34(5):933, 1988.

LeGrys V, Leinbach SS, and Silverman L: Clinical applications of DNA probes in the diagnosis of genetic diseases, CRC Crit Rev Clin Lab Sci 25(4):255, 1987.

Pearson PL: Restriction fragment length polymorphisms and their use in mapping the human genome. In Kare B, editor: Progress in Clinical Biology Research, New York, 1985, Alan R Liss Inc.

White R, Woodward S, Leppert M, et al: A closely linked genetic marker for cystic fibrosis, Nature 318:382, 1985.

CHAPTER 48

Psychiatric disorders

DAVID L. GARVER

OBJECTIVES

- Briefly describe schizophrenias and affective disorders.
- Interpret significant results of challenge or function tests used in the diagnosis of psychiatric disorders.
- Describe the action of the following types of drugs in treatment of psychiatric disorders:
  - Antipsychotic drugs
  - Tricyclic antidepressants
  - Monoamine oxidase inhibitors
  - Lithium

KEY TERMS

affective disorder A disorder of mood regulation manifest clinically by episodes or sustained periods of depression or mania or both.

antipsychotic drugs Drugs that are utilized for the reversal or attenuation of psychotic symptoms (hallucinations, delusions, and disorders of cognition). All current antipsychotic drugs act as a single mechanism: blockade of postsynaptic dopamine receptors by stereospecific attachment to the receptor.

bipolar affective disease Affective disorder in which both poles of mood disturbance are episodically present in the same patient: episodes of both depression and mania.

cognitive disorder (schizophrenia) Thinking disorder characterized by speech in which ideas shift from one subject to another that is completely unrelated or only obliquely related without the speaker's showing any awareness that the topics are unrelated; frank incoherence results from total lack of logical or meaningful connections between words, phrases, or sentences.

delusion False belief based on incorrect inference about external reality that is not held by other members of an individual's sociocultural milieu and is so firmly held that evidence to the contrary will not shake the false belief.

depression A mood disturbance often described as being sad, blue, hopeless, low, "down in the dumps," or irritable, accompanied by pervasive loss of interest or pleasure in almost all usual activities or pastimes.

genetic patterns of transmission Familial biological patterns of illness occurrence as manifested by increased density of particular or related illnesses both within families and in the adopted-away offspring (who share common genes with their biological as opposed to adopting families).

hallucination A sensory perception (such as voices, visions, tactile sensations) that has the immediate sense of reality of a true perception but without external stimulation of the relevant sensory organ.

limbic system A loosely defined group of brain structures, including the hippocampus and dentate gyrus with their archicortex, the cingulate gyrus, septal areas, amygdala, and parts of the hypothalamus, that are believed to regulate not only emotion, but also certain aspects of cognition (as well as olfaction and autonomic function).

mania A periodic disturbance of mood in which the mood is elevated, expansive, or irritable and is accompanied by hyperactivity, pressure of speech, flight of ideas, inflated self-esteem, decreased need for sleep, distractibility, and excessive involvement in activities that have a high potential for unrecognized painful consequences.

nerve receptor A protein complex embedded in the cell membrane that stereospecifically identifies a particullar neurotransmitter and responds to its attachment by initiating a series of events that alter the membrane's resting potential and rate of firing.

neurotransmitter A chemical substance released by one neuron onto a specific receptor or an adjacent cell, producing information transfer in the form of regulating or modulating the resting potential and discharge rates of postsynaptic cell.

neurotransmitter storage granules Membranous vesicles within the cell cytoplasm that both store the neurotransmitter for depolarization-induced release and protect the neurotransmitter from deamination by intracellular enzymes.

precursor loading A process whereby neurotransmitter levels are increased by an increase in the availability of precursors for enzymatic synthesis of the neurotransmitter.

reuptake An energy-dependent process by which neurons conserve their own neurotransmitter by recovering it from the synaptic cleft for storage and subsequent rerelease.

schizophrenia A deteriorating psychotic illness with characteristic delusions, hallucination, and disorders of cognition that has lasted more than 6 months and from which full recovery is not expected.

schizophreniform disorder An episodic psychotic illness with symptoms similar to those of schizophrenia, but the psychotic episodes of which last less than 6 months; recovery of function is full between illness episodes.

stereospecific binding Specific attachment of a neurotransmitter or drug to a receptor; the attachment depends on complementary configurations of the compound and its receptor; stereospecificity of the attachment indicates failure of a stereoisomer (a structural mirror image of the compound) to attach to the receptor.

synapse The structural junction of two neurons where chemical messages are carried from the presynaptic neuron to the postsynaptic receptor by neurotransmitters.

unipolar affective disease Affective disorders in which episodes of depression alone occur without episodes of mania.

CENTRAL NERVOUS SYSTEM: ANATOMY, BIOCHEMISTRY, AND PHYSIOLOGY

Biological psychiatry is concerned with abnormalities of the basic units of central nervous system function: neuronal systems. A brief review of the anatomy of the central nervous system has been presented in Chapter 39.

Neurons within the central nervous system process information arising from multiple internal and external sources. In maintaining physiological and psychobiological homeostasis, central nervous system neurons communicate both with one another and, eventually, with effectors outside the central nervous system by means of neurotransmitters released by each neuron onto specific receptors. Neurons are often characterized not only by their anatomical distribution (location of their cell bodies) and by the path of projection to their terminal areas, but also by the nature of the neurochemical hormone/transmitter that they synthesize and release.

Central norepinephrine systems

A norepinephrine (NE) neuron and its synaptic links are shown in Fig. 48-1. The presynaptic neuron contains machinery necessary for the synthesis, storage, release reuptake, and inactivation of NE. In the NE neuron, the precursor amino acid tyrosine is taken up and enzymatically converted to levodopa, dopamine, and then norepinephrine (see Fig. 45-4). Uptake of dopamine and NE into storage granules within the neuron protect the neurotransmitter from enzymatic degradation (by mitochondrial monoamine oxidase [MAO]) to inactive aldehydes and acids. With each electrical stimulation (depolarization) of the neuron, the granular packets migrate and fuse to the synaptic membrane, releasing their neurochemical contents into the synaptic cleft.

Once released into the synaptic cleft, NE acts on the postsynaptic neuron's NE receptor to alter the electrical resting potential of the postsynaptic cell membrane, either inhibiting or facilitating the rate of electrical and neurochemical discharge by the postsynaptic cell. In many cases, the changes in resting potential triggered by NE receptor activation are mediated by a second intracellular

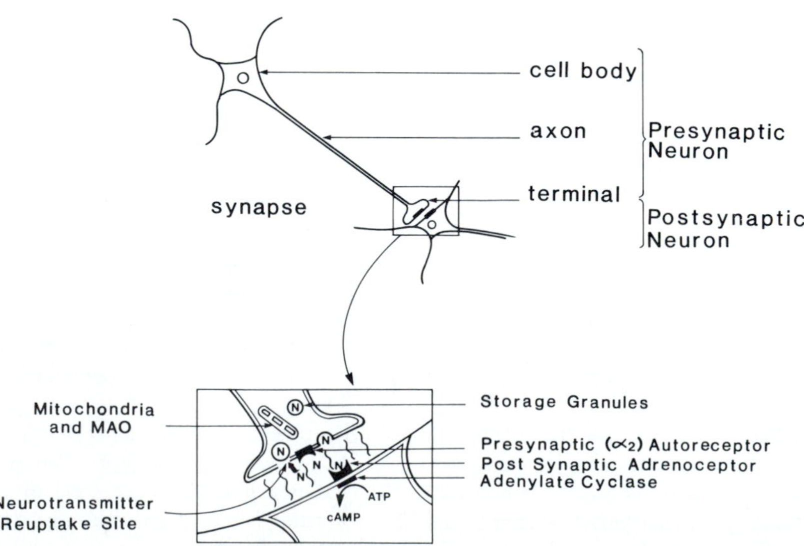

Fig. 48-1 Norepinephrine neuron, synapse, and postsynaptic connections.

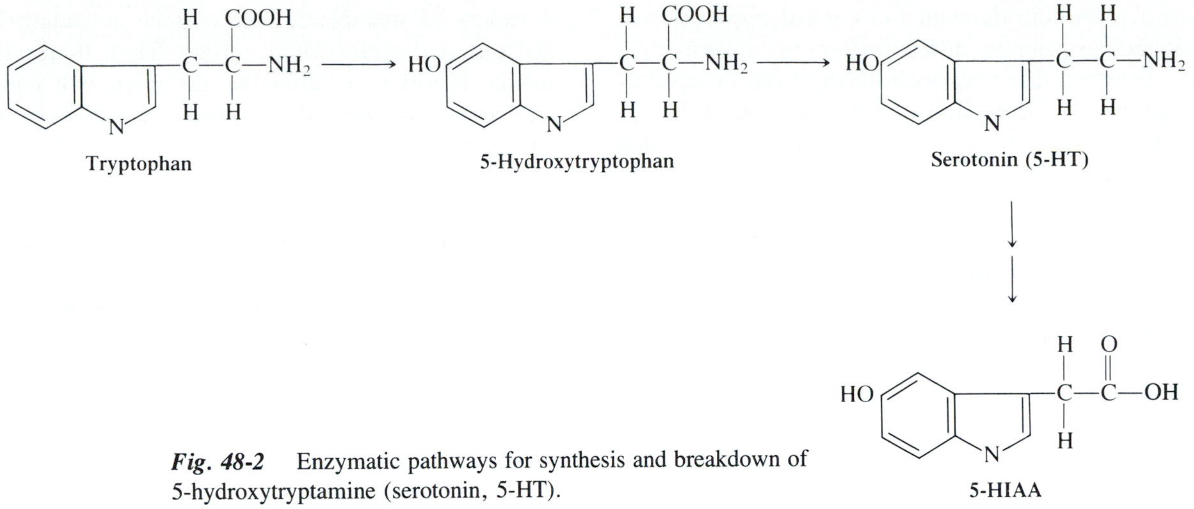

Fig. 48-2 Enzymatic pathways for synthesis and breakdown of 5-hydroxytryptamine (serotonin, 5-HT).

messenger, cyclic adenosine monophosophate (cAMP). NE released in the cleft also interacts with presynaptic ($alpha_2$) adrenoceptors, whose activation reduces the quantity of NE released with each electrical depolarization that arrives at the terminal area of the NE neuron.

The presynaptic NE neuron does not abandon the majority of released neurotransmitters. Rather, the neuron has a reuptake mechanism that actively recovers NE from the synaptic cleft and again stores the recovered NE for subsequent release. Some central NE, however, is metabolized to 3-methoxy-4-hydroxyphenylglycol (MHPG). The quantity of MHPG in a 24-hour urine sample can be used as an estimate of synthesis and breakdown of central nervous system NE.

Central dopamine systems

Dopamine (DA) neurons also utilize the precursor amino acid tyrosine from which they synthesize their neurotransmitter, but unlike NE neurons they lack the enzyme dopamine β-hydroxylase, which converts DA to NE. DA is stored in presynaptic granules, which are released into the DA synapse after electrical depolarization of the DA neuron. DA is released not only onto the postsynaptic receptor, but also onto a presynaptic DA autoreceptor, which, when activated, partially inhibits the activity of the DA neuron. Like the NE neuron, the DA neuron recovers synaptically released neurotransmitter by an energy-dependent process and stores the neurotransmitter for subsequent rerelease. Some DA is metabolized to homovanillic acid (HVA), which can be quantified in body fluids.

Central serotonin systems

Serotonin neurons within the central nervous system utilize the amino acid precursor tryptophan from which they synthesize serotonin (5-hydroxytryptamine, or 5-HT) (Fig. 48-2). The process involving release of 5-HT onto stereospecific 5-HT receptors followed by reuptake of the neurotransmitter is similar to that occurring in NE and DA neurons. 5-HT is metabolized to 5-hydroxy-indole acetic acid (5-HIAA), which, like HVA, can be measured in body fluids.

Other central neurotransmitter systems

NE, DA, and 5-HT neurons constitute a very small proportion of the neuronal systems within the central nervous system. Gamma-amino-butyric acid (GABA), glutamic acid, and acetylcholine systems are distributed through the central nervous system, as are neurons that use as their neurotransmitter complex polypeptides such as β-endorphin, the enkephalins, cholecystokinin, substance P, adrenocorticotropic hormone (ACTH), and somatostatin. The distribution of such neurotransmitter systems and their biochemistry and physiology are currently being studied. Their significance for psychiatric disorders is unknown.

PSYCHIATRIC DISORDERS

Schizophrenias

Description. The schizophrenias (syndrome diagnosis) are primarily disorders that manifest characteristic delusions, hallucinations, and disorders cognition.[1] During an active phase, patients may have the conviction that their thoughts, feelings, and behavior are not really their own, but forced on them by some alien force. Voices commenting on the patient's behavior are common; sometimes they command the patient to action. Thought processes lose common logical context; thoughts and speech move from one thought to another seemingly unrelated thought. Between such active phases or episodes of psychosis, residual social isolation, odd speech, and peculiar behavior persist to varying degrees.

A preliminary attempt has been made in the diagnostic nomenclature[1] to separate the larger disease group into distinct subentities. This is accomplished by isolation of a group of disorders with similar symptoms whose episodic course differs in both duration and clinical response. Additional findings support differentiation of at least three separate disorders that have traditionally been grouped as "schizophrenia." Such clinical differentiation is important, because it separates those patients who can be treated by certain drugs from those who cannot. The three subtypes to be discussed below include classical schizophrenias, organic psychoses, and atypical psychoses.

Classical schizophrenia. It has been observed that drugs that can prevent dopamine-mediated neurotransmission have an antipsychotic effect (reduce the numbers or severity of delusions, hallucinations, and disordered cognition of many psychotic patients). Such drugs bind stereospecifically to one form of the postsynaptic dopamine receptor; each drug binds with an affinity that is inversely related to the average dose of the particular antipsychotric drug required to reduce psychotic symptoms clinically.

At postmortem examination, the neurons of substantial numbers of chronic psychotic (schizophrenic) patients have been found to have an increased number of dopamine-related postsynaptic receptors. Such receptors can be identified and quantitated by measurement of the amount of stereospecific binding of a radiolabeled antipsychotic drug to the specific receptor. Binding of labeled psychotropic drugs to membrane protein has been found to be significantly greater in chronic psychotic patients than in age- and sex-matched controls.[2]

Organic psychoses. Although one group of chronic psychotic patients shows evidence of altered dopamine receptor number, this group must be discriminated from another group of psychotic patients with similar-appearing psychotic illness who do not exhibit dopamine hyperactivity. Many of this latter category of chronic psychotic patients fail to respond to antipsychotic drugs that block dopamine receptors. Many of these psychotic patients do have signs of at least mild neuropsychological deficits. Chronic psychotics of this separate group have evidence of enlarged cerebral ventricles by computerized axial tomography (CAT scan)[3] and perhaps of diminished size of the anterior cerebellar vermis[4] as well. It is presently not clear whether such "chronic organic" psychoses are the result of a chemical insult or trauma to brain substance with brain atrophy or whether such a group of chronic psychotic illnesses represents a developmental disorder affecting the differentiation of central nervous system tissues. The clinical history of poor premorbid adjustment (poor level of functioning before the first frankly psychotic episode) possibly indicates that abnormal developmental processes may be the most relevant in this subgroup of chronic psychotic illnesses.

Atypical psychoses (schizophreniform disorders). In a third group of psychotic patients with similar symptoms, the duration of a psychotic episode is shorter, usually less than 6 months, and patients exhibit essentially full recovery between episodes. These patients appear to respond to a variety of therapeutic interventions including lithium[5] and classical antipsychotic drugs. They frequently have family histories of affective disorders but not of the chronic deteriorating schizophrenias and the schizophrenic spectrum diseases.

The early diagnosis of such good-prognosis, schizophreniform, psychotic illnesses is important, and the discrimination of these patients from patients with the chronic progressively deteriorating schizophrenias is vital because of the necessity to anticipate the illness course in treatment planning. Most such good-prognosis patients with proper treatment can be expected to return to their homes and to employment at a level of functioning comparable to what was present before the psychotic episode. The use of lithium for both acute treatment and maintenance emphasizes the similarity of this group of psychotic disorders to the affective disorders, although, unlike the affective disorders, it presents with classical delusions, hallucinations, and disorders of cognition, which characterize the schizophrenic, psychotic syndrome.

Affective disorders

Description. The affective disorders are diseases of mood regulation and are accompanied by altered rates of cognition (slowed or speeded) and by physical signs and symptoms. Numerous investigators have observed that affective disorders may appear in either a unipolar or a bipolar form. In *unipolar disease,* only periods of depression occur. Specifically, major depressive episodes are characterized by appetite and sleep disturbance with substantial loss of ability to experience pleasure. Fatigue, agitation, feeling of worthlessness, and thoughts of suicide are recurrent. In *bipolar affective* disease both episodes of depression and episodes of mania occur. Manic episodes are characterized by increased activity, racing thoughts, and pressured speech together with inflated self-esteem, grandiosity, and an inability to anticipate negative consequences of actions.

The effective treatment of the affective disorders frequently requires careful attention to determine the underlying biological process that needs to be corrected or compensated for by pharmacological intervention. Therapeutic response to one type of medication but not to another has been observed in patients with affective disorders. The correct choice of drug for a patient with an affective disorder requires choosing a drug that compensates for specific underlying biological abnormalities. The identification of these abnormalities, or the clustering of the series of clinicobiological findings associated with the abnormalities is already beginning to be used to guide selection of psychopharmacological agents.

Several groups of investigators have explored the incidence of affective disorders in first-degree relatives of patients with affective disorders. Most investigators have found that bipolar illness breeds relatively true in families,[6] which clearly suggests that unipolar and bipolar affective illnesses are distinct entities. It follows that biological studies may reveal that different biological abnormalities exist in bipolar patients and unipolar depressed patients. Similarly, drugs that specifically correct or compensate for a primary biological abnormality in bipolar diseased patients may be ineffective in unipolar populations.

Unipolar affective disorders. After studying the occurrence of unipolar disease in families, Winokur et al.[7] suggested that unipolar disease may itself have at least two subtypes. They distinguished depressive spectrum disease from pure depressive disease. Depressive spectrum disease has an early onset (before 40 years of age); it is associated with a history of greater incidence of frankly affective disorder in first-degree female relatives and a prominent history of alcoholism or antisocial personality or both in first-degree male relatives. In contrast, pure depressive disease is characterized by a later onset, essentially equal incidence of affective disorder in male and female first-degree relatives, and little or no family history of alcoholism or antisocial personality.

Although such subtypes of unipolar disorders are indistinguishable symptomatically, family history findings suggest that the primary underlying biological abnormality of the two unipolar disorders might differ. One might expect that pharmacological agents with a high degree of specificity of action might be effective in one but probably not in both of the unipolar disorders. Indeed, there is growing evidence that this may be the case with commonly used, relatively specific antidepressant drugs.

Serotonin-related depressions. Evidence for the role of serotonin (5-HT) in depressive disease comes from numerous, often conflicting reports of diminished quantities of its metabolite 5-HIAA in the cerebrospinal fluid of depressed patients and from findings that 5-hydroxytryptophan, a precursor of 5-HT, is an effective antidepressant only in depressed patients with decreased cerebrospinal fluid 5-HIAA.

The routine evaluation of cerebrospinal fluid 5-HIAA in depressed patients might therefore identify a subgroup of depressive patients who have deficiencies in serotonin turnover, who might therefore benefit from precursor loading with tryptophan or 5-hydroxytryptophan.[8] Such 5-HT–deficient patients may also be candidates for treatment with more conventional antidepressant drugs that increase the 5-HT in the serotoninergic synapse by partially inhibiting the reuptake of 5-HT into the presynaptic neuron. Pharmacological agents such as amitriptyline and trazodone have been shown to inhibit selectively the neuronal reuptake of the 5-HT from the synapse.

Norepinephrine-related depressions. Much of our understanding of the biological basis of affective disorders comes from work that has pursued the hypothesis that an NE deficiency at critical central nervous system synapses underlies the cyclic emergence of depression in many patients with affective disorders.

The presence of large NE-containing cells in specific areas of the brain suggests the possibility of a critical regulating role of NE in a variety of activities that are impaired in depression: appetite, sexual function, sleep, cognition, and so forth.

MHPG IN NOREPINEPHRINE DEPRESSIONS. The early studies of depressive disease examined whether the turnover of central NE was abnormal in depressed patients. Urinary MHPG, half of which has its origin in central NE turnover, has been shown to be deficient in many depressed patients.[9] However, concentrations of the peripheral NE metabolites vanillylmandelic acid and normetanephrine were found to be similar in depressed patients and in age- and sex-matched controls. Such findings suggest that in some forms of depression, although the rates of synthesis and breakdown of NE in the periphery appear to be normal, turnover of NE in the central nervous system is diminished. Several investigators have demonstrated that patients who are deficient in urinary MHPG and therefore presumably have a deficit in central NE have a favorable therapeutic response to antidepressants that selectively inhibit the reuptake of NE from the synaptic cleft into the presynaptic neuron.[10,11] The consequence of such inhibition of reuptake is that more NE is available in the synaptic cleft, acting on the postsynaptic receptor. Such augmented activity of central NE may underlie the antidepressant response clinically. Antidepressant drugs that are relatively selective in inhibition of neuronal reuptake of NE include desipramine and maprotiline. Imipramine has also been shown to produce good antidepressant response in low-MHPG depressions, presumably because it is itself metabolized rapidly to desipramine, which has selective effects on NE reuptake.

ABNORMAL CORTISOL SECRETION. Considerable numbers of depressed patients have been shown to have an abnormality of cortisol secretion.[12] In some depressed patients, the usual diurnal rhythm of plasma cortisol is frequently obscured by continuous elevation of cortisol throughout the day. Such depressed patients are found to have a pronounced elevation of urinary free cortisol as well. The dexamethasone suppression test (DST) (in which dexamethasone suppresses plasma cortisol levels for 36 hours) is also abnormal in many depressed patients, showing either total nonsuppression or early escape of cortisol from dexamethasone suppression.[13] As discussed below, nonsuppression of cortisol by dexamethasone in depression is also consistent with a deficiency of hypothalamic NE, which indirectly suppresses cortisol release.

RECEPTOR SENSITIVITY. Receptors that ordinarily aid in the homeostatic feedback regulation of NE neurons may be abnormal in some depressive illnesses.

Direct evidence of abnormalities of alpha$_2$-adrenergic autoreceptors in some patients with major depressive disorders has come from studies of a peripheral tissue: the platelet. Like the NE neurons of the central nervous system, platelets have alpha$_2$-adrenergic receptors in the outer phospholipid bilayer of the cell membrane. Greater numbers of these receptors, identified by ^{3}H-clonidine binding, are found in depressed patients than in control subjects. Moreover, treatment with antidepressant drugs leads to a significant decrease (normalization) in the number of platelet alpha$_2$-adrenergic receptors.[14]

Alpha$_2$-adrenergic receptors within the central nervous system are part of the feedback loop that regulates the synthesis of NE and the firing rate of NE neurons. Activation of alpha$_2$-adrenergic receptors, which are located on NE neuron membranes, decreases NE synthesis and release, resulting in diminished plasma and urinary MHPG.[15] If increased numbers of alpha$_2$-adrenergic receptors occur on NE neurons of the central nervous system in certain depressed patients—similar to the increase of alpha$_2$-adrenergic receptors found in the peripheral platelet—the increased central alpha$_2$ sensitivity would be expected to cause both reduction of the synthesis of central NE (as demonstrated by diminished plasma and urinary MHPG) and a functional NE deficit at central NE synapses. The striking finding that chronic but not acute treatment with antidepressant drugs decreases the number of alpha$_2$ receptors in platelets[16] and diminishes the hypotensive effects of the presynaptic agonist clonidine indicates that similar events may be occurring at central alpha$_2$-adrenergic sites after chronic administration of antidepressant drugs.

Bipolar disorders

Clinical presentation and history. As noted previously, patients with a classical bipolar disease have a history of major depressive episodes and episodes of mania with affectively normal intervals interspersed. The diagnosis of bipolar illness is not always obvious from clinical presentation and clinical history. Many bipolar illnesses can present with a first or even second episode of depression without a previous episode of diagnosable mania. In addition, many patients with bipolar illness have psychotic symptoms suggestive of schizophrenia. Clinically, the history of mania in a first-degree relative should suggest the presence of a bipolar disorder in a patient who is in a major depressive episode of undergoing a brief psychotic decompensation. Treatments for a bipolar disorder and for unipolar disease and schizophrenia differ; the correct diagnosis early in the episode is important for therapeutic management.

Neurotransmitter dysregulation. The biology of the bipolar affective disorder indicates possible dysregulation of several neurotransmitter systems. During the depressed phase of bipolar illness, decreases in urinary MHPG have been clearly documented[17] and indicate a possible abnormality of central NE function. Manic episodes can be precipitated by dopamine-like drugs such as amphetamine. The episodes are at least partially resolved acutely by the use of dopamine receptor–blocking agents, such as antipsychotic drug.[18] However, it appears that central cholinergic systems may also have a role in bipolar illness. Mania can be attenuated not only by a blockade of dopamine receptors, but also by an increase in central cholinergic activity, such as that which occurs after the infusion of a cholinesterase inhibitor such as physostigmine.[19] The increased functional activity of cholinergic systems promptly relieves mania. As the effect of physostigmine wears off, mania recurs. Bipolar illness thus appears to be a dysregulation of balance between several central neurotransmitter systems. The restoration of balance by pharmacological intervention appears to be associated with cessation of the affective episode.

FUNCTION TESTS

Dexamethasone suppression test

In some depressed patients, plasma cortisol levels do not vary during the course of the day but tend to remain elevated; the neuroregulatory mechanism is altered. Ordinarily, the drug dexamethasone suppresses the secretion of cortisol into plasma. Some depressed patients do not suppress cortisol secretion after dexamethasone or fail to suppress it completely. The details of the dexamethasone suppression test (DST) are given in Table 48-1.

The dexamethasone suppression test is used to confirm a melancholic (unipolar) depression, which responds characteristically to NE drugs such as imipramine, desipramine, and maprotiline. Although early studies suggested that tricyclic drugs such as imipramine and desipramine were more effective in depressed dexamethasone nonsuppressors than drugs that block serotonin reuptake, such as

Table 48-1 Challenge or function test used in diagnosis of psychiatric disorders

Test	Sample	Procedure	Interpretation
DST (dexamethasone suppression)	Plasma	1 mg of dexamethasone at 11 PM followed by blood drawing for plasma cortisol at 4 PM and 11 PM the following day	4 PM or 11 PM cortisol greater than 50 μg/L suggestive of depression responsive to antidepressant drugs
Protirelin	Plasma	500 μg of protirelin infused over 30 min; basal, 15-, 30-, 60-, and 90-min plasma samples assayed for thyroid-stimulating hormone (TSH)	Blunted increase from baseline of TSH (less than 7 μU/mL) suggestive of affective disorder rather than schizophrenia

amitriptyline, further evaluations of this important issue are required before one draws any firm conclusions. Several investigators have suggested that dexamethasone nonsuppression predicts the response to all antidepressant agents, without discriminating well between NE and serotonin types of antidepressants.[20] Normalization of the DST following successful treatment of depression has been reported in many studies. When such normalization of nonsuppression does not occur following treatment, risk of early relapse is significant and continued observation and treatment, or even alternative antidepressant treatment or electroconvulsive therapy, may be required to induce or maintain and antidepressant response[21] (Table 48-1).

Protirelin test

The protirelin test measures the change in plasma thyroid-stimulating hormone (TSH) following a challenge with protirelin (thyrotropin-releasing hormone [TRH]). The protirelin test is useful in discriminating atypical psychoses from schizophrenia and in discriminating unipolar from bipolar depressions.

Atypical psychotics, who clinically resemble schizophrenics, have a blunted TSH response to TRH challenge; schizophrenics, on the other hand, have TSH responses that are similar to controls.[23] A blunted TSH response in a psychotic patient may therefore be an indication for lithium treatment. Both manic individuals and those with atypical psychoses have similar blunted responses to TRH, and both have a similar response to lithium.

The protirelin test can also aid in discriminating unipolar depression from bipolar depression. Unipolar depression is associated with a blunted TSH response to TRH challenge. Bipolar depression (unlike the manic phase of bipolar disease) shows an excessive TSH response (>13 μU/mL).[21,22] The distinction between bipolar and unipolar depressions is critical since with bipolar illness patients often develop a rapid cycling of mood states when treated with tricyclic and MAO inhibitor antidepressants and not additionally with lithium.

Interpretation of the protirelin test should be made cautiously because marked increases in TSH (>25 μU/mL) are also a sign of frank or impending hypothyroidism. Other signs of hypothyroidism include depression, anergy, and difficulty with memory and concentration. Thyroid testing (see Chapter 41) should be performed to rule out thyroid disease.

DRUG TREATMENT AND MONITORING

Treatment of schizophrenia: antipsychotic drugs

Mechanism of action. The classical antipsychotic drugs, those that successfully block postsynaptic dopamine receptors, have been shown since the mid 1950s to be moderately successful in the reversal of acute psychotic symptoms and in the prevention of subsequent episodes in the majority of schizophrenic patients. It is well known, however, that differing doses of the various antipsychotic drugs were required for treatment of the schizophrenic symptoms (Table 48-2). The dose of each antipsychotic drug required for treatment has been found to be inversely related to the affinity of the drugs for a specific form of the central dopamine receptor. Fluphenazine, which has a high affinity for the relevant receptor, can therefore be given at low doses; chlorpromazine, which has a comparatively low affinity for the receptor, reverses schizophrenic symptoms only at higher doses in the average patient.

Dose and blood level relationships for antipsychotic response. Although most clinical studies has suggested a sigmoid dose-response curve for antipsychotic drugs, with the optimum dose for the average schizophrenic patient above a range of 500 to 1000 mg of chlorpromazine equivalents per day,[23] studies relating blood levels of such drugs to therapeutic effects have indicated that for at least some patients antipsychotic patients may have an optimum therapeutic range, both below which and above which the antipsychotic effects of the drug are diminished.

There is wide variation in drug blood levels among patients receiving the same dose of the same antipsychotic drug. Between-patient variations of 100-fold have been reported for plasma chlorpromazine; 40-fold variations have been reported for butaperazine and fluphenazine. At average antipsychotic drug doses, substantial numbers of patients do appear to have extremely low or extremely high blood-and-tissue drug levels because of the pronounced

Table 48-2 Comparable doses and relative potencies of antipsychotic drugs defined from dose-ranging studies clinically

Generic name	Empirically defined relative potency	Average daily dose for acute treatment (mg)
Fluphenazine decanoate	163.9	—
Fluphenazine	83.3	9
Haloperidol	62.5	12
Trifluoperazine	35.7	21
Thiothixene	22.7	32
Molindone	16.7	44
Lexapine	11.4	64
Perphenazine	11.1	66
Butaperazine	11.1	66
Piperacetazine	9.1	81
Prochlorperazine	7.1	103
Acetophenazine	4.4	169
Carphenazine	4.0	184
Triflupromazine	3.6	206
Chlorprothixene	2.2	322
Mesoridazine	1.8	411
Thioridazine	1.3	712
Chlorpromazine	1.0	734

From Garver, D: In Hippus, H, and Winokur, G, editors: Psychopharmacology, 1. Part 2: Clinical psychopharmacology, Amsterdam, 1982, Excerpta Medica.

between-patient differences in drug absorption or metabolism.[24] Some chronic antipsychotic drug-resistant patients have such poor absorption or rapid metabolism of the drugs that they fail to respond to conventional doses because of their very low drug blood levels. On the other hand, patients who, by reason of slow metabolism, develop very high blood levels of antipsychotic drugs with conventional dosages also frequently fail to respond; such patients frequently show therapeutic response after the dose is lowered and consequently the drug blood level is reduced to a more therapeutic value.

Not all investigators have found clear-cut relationships between plasma antipsychotic drug levels and therapeutic response, perhaps because of the fact that many antipsychotic drugs also have active metabolites, which generally are not measured by the usual gas chromatographic, liquid chromatographic, or high-performance liquid chromatographic assays. Ignoring the active metabolites of a drug results in only a portion of the functional drug level actually being assayed. Relationships between serum levels of total active antipsychotic drug, including active metabolites, and antipsychotic effects may therefore be obscured.

Receptor-binding assay, a sophisticated bioassay in which the quantity of functional antipsychotic drug together with its functional, active metabolites is monitored, has recently been shown to have considerable potential for the monitoring of the quantity of functionally active drug and metabolites. Such a receptor-binding procedure measures the total dopamine receptor-blocking potency in a patient sample (plasma, cerebrospinal fluid, tissue). Using this technique, virtually all investigators[25-27] have found that it was possible to define a more accurate therapeutic range.

Treatment of affective disorders

The mainstays of the pharmacological treatment of the depressive disorders are the tricyclic antidepressant drugs and the MAO inhibitors. Lithium is used in the treatment of bipolar (manic-depressive) disease.

Reuptake inhibitors: tricyclic antidepressants

Choice of antidepressant drug. The most widely used drugs for treating depression are the tricyclic antidepressants.[28] Tricyclic antidepressants are generally well tolerated, cause minimum side effects, and are relatively safe (although drug overdose may cause severe cardiotoxicity). As yet, no clinical signs or epidemiological aspects of the patient's history have been definitively proven to be useful in the choice of the tricyclic antidepressant to be used for a particular patient. Previous response of the patient or of family members to a specific tricyclic antidepressant may be the most reliable method to use in the selection of a drug. The use of urinary MHPG measurements, cerebrospinal fluid 5-HIAA concentrations, or certain neuroendocrine tests such as the dexamethasone suppression test and the evaluation of receptor abnormalities do show some promise in aiding the clinician in the selection of an appropriate tricyclic drug for a particular patient. For now, these must be considered research tools that provide evidence for a theoretical framework or rationale for changing to other types of tricyclates when the first drug tried is not effective.

Table 48-3 Commonly used antidepressant drugs and their approximate effective dose

Drug	Effective dose range (mg/day)
Amine-reuptake inhibitors	
Amitriptyline	150-300
Desipramine	150-300
Doxepin	75-300
Imipramine	150-300
Maprotiline	150-300
Nortriptyline	50-100
Trazodone	200-600
Monoamine oxidase inhibitors	
Phenelzine	45-75
Tranylcypromine	20-40

In the physically healthy patient, the dosage of the tricyclic antidepressant should be advanced to the average mean dose as shown in Table 48-3 within 1 week. If the patient's depressive symptoms have not improved after 2 weeks at this average dose, the first question is whether the plasma levels of drug are too high or too low.

Use of antidepressant drug levels in adjustment of antidepressant drug dosage. The recognition of major differences in the absorption and metabolism of tricyclic drugs in individual patients has resulted in increased interest in relating plasma levels (rather than simply dosage) of antidepressant drugs to therapeutic response. For example, daily doses of 150 mg of amitriptyline given to a series of different patients result commonly in threefold differences in the quantity of plasma amitriptyline plus its active metabolite nortriptyline. Such differences in steady state plasma levels after comparable oral doses may account for the different doses of drug required by individual patients. Patients who absorb an antidepressant poorly or metabolize it rapidly need a higher oral dose in order to reach adequate therapeutic levels. On the other hand, patients who absorb the drug rapidly and metabolize it slowly require less drug to reach therapeutic drug levels.

The definition of therapeutic plasma levels of antidepressant drugs has been the center of considerable effort and some controversy over the past years. Relationships between plasma levels of the tricyclic antidepressant nortriptyline and therapeutic response have been most fully studied. The available evidence suggests an inverted U type of relationship between nortriptyline plasma levels and antidepressant effects in tricyclic-responsive, endogenously depressed inpatients. Maximum therapeutic efficacy is achieved with plasma nortriptyline levels between 50 and 175 ng/mL. Patients whose nortriptyline blood levels are below 50 ng/ml generally fail to respond to the drug; importantly, patients whose plasma nortriptyline levels are in excess of 175 ng/mL also fail to respond.[29]

For patients receiving the antidepressant drug imipramine, the relationship between plasma levels of imipramine plus its active metabolite desipramine and clinical response appears to be linear in nondelusional, endogenously depressed, tricylic drug–responsive patients.[30] That is, antidepressant response to imipramine occurs when a threshold plasma concentration of imipramine and desipramine is reached (approximately 200 ng/mL of plasma). Patients whose blood levels are significantly in excess of this threshold level continue to have an antidepressant response yet may be burdened by excessive side effects of the drug.

There are less clear relationsips between plasma levels of other antidepressant drugs and therapeutic response. The available evidence for amitriptyline suggests that a blood level in excess of 120 ng/mL of plasma amitriptyline plus nortriptyline produces a good antidepressant response in tricyclic-responsive, endogenously depressed patients.[29] Some reports suggest that an even better response occurs at higher blood concentrations of amitriptyline plus its metabolite, but other reports suggest that blood concentrations over 180 ng/mL reduce the therapeutic efficacy of the drug[29]; it is not clear whether the relationship of amitriptyline plus nortriptyline and their response is a linear or curvilinear one.

For other antidepressants such as protriptyline, desipramine, doxapine, maprotiline, and amoxapine, significant relationships, if any, await further elucidation by systematic studies.

The most important indications for obtaining antidepressant drug levels are as follows:

1. During use of an antidepressant such as imipramine or nortriptyline (which has known drug level–response relationships), routine drug blood levels should be performed on the fifth day following achievement of the targeted dose. Adjustment of drug dose up or down so as to achieve therapeutic levels in the blood should proceed immediately.
2. Drug levels should be obtained after treatment with another tricyclic drug for 2 to 3 weeks without response at an adequate dose. If low plasma concentration is present, the dosage should be increased to a generally accepted therapeutic range; antidepressant response may then occur.
3. Drug levels should be obtained when higher than usual doses of antidepressant drug are given. For example, doses of amitriptyline or imipramine of more than 300 mg/day may result in altered clearance with a nonlinear increase in plasma concentrations. Monitoring could help avoid problems with toxicity.
4. When treating very old or very young patients, plasma concentrations should be kept at about 50% of what one might wish in healthy patients in the middle years of life. Protein binding of these drugs is less at extremes of age, and more drug is present in pharmacologically active free form; the usual interpretations of plasma concentration may not hold under such circumstances.

Monamine oxidase inhibitors

Clinical use. MAO inhibitors have been generally found to be less effective than other antidepressant drugs, except in chronically depressed persons who are nonresponsive to other antidepressants. This generalization is somewhat surprising because MAO inhibitors raise the levels of both serotonin and NE and appear to have effects in altering receptor sensitivity similar to those of the more conventional antidepressant drugs. It is likely that previous trials of MAO inhibitors did not utilize adequate drug doses to achieve sufficient MAO inhibition for therapeutic response.

Platelet MAO inhibition and antidepressant response. During treatment with MAO inhibitors such as phenelzine, the degree of MAO inhibition can be related to therapeutic response. The inhibition of MAO activity in peripheral tissues, such as the platelet, is assumed to be proportional to the degree of inhibition produced by the MAO inhibitor in central nervous system tissue. Therapeutic response to phenelzine occurs only when MAO activity of the platelet can be reduced to between 10% and 20% of baseline activity.[31] Utilizing a platelet assay for MAO activity, one can titrate the oral dose of phenelzine to achieve optimum therapeutic response with minimum toxic side effects; such side effects generally appear when platelet MAO inhibition exceeds 95%. Laboratory monitoring of MAO activity at baseline and during inhibitory treatment may alleviate depression in patients who are otherwise resistant to conventional antidepressant drug therapies.

Lithium

Clinical use. Lithium is clearly the treatment of choice for bipolar illness. It appears to be a specific agent for the treatment of mania, dramatically converting an acute manic episode into a normal mood state without sedation. The major drawback with lithium therapy is the lag period between the beginning of lithium administration and its clinical effect. For this reason, acutely manic patients are frequently treated initially with antipsychotic drugs such as haloperidol, in addition to lithium, to achieve a rapid antimanic effect. After approximately 1 to 2 weeks, the neuroleptic drug can generally be gradually withdrawn, with lithium alone maintaining significant protection from both manic and depressive episodes.

Lithium is also effective acutely in the treatment of depression, especially the bipolar type, but again 3 to 4 weeks are generally required for lithium to produce an acute antidepressant effect. Generally, bipolar depressed patients are initially treated with a combination of a norepinephrine type of antidepressant, such as desipramine, and lithium. The antidepressant should be withdrawn after 3 to 4 weeks with continued maintenance on lithium. Addition of an antidepressant early in the course of treatment of bipolar depression generally shortens the acute depressive episode by about 2 weeks as compared to treatment with lithium alone.

Fifteen percent of depressed, bipolar patients treated with antidepressant drugs alone (rather than with lithium) switch into mania. Other bipolar patients appear to develop a more malignant, rapid cycling illness (rapid switching between depression and mania) when given antidepressant drugs for long periods of time. Such a rapid cycling illness is often resistant to further treatment. In general, bipolar depressed patients are therefore not given conventional antidepressants alone. Neither do bipolar patients with mania receive full therapeutic benefit from an antipsychotic alone. Such patients treated with antipsychotics feel constricted in physical movements (extrapyramidal effects) and often retain an antipsychotic agent–induced dysphoria. Long-term treatments of such patients with lithium is therefore clearly indicated.

Use of lithium plasma levels in adjustment of lithium dose. During both lithium treatment and lithium maintenance (prophylaxis) it is important for the clinician to adjust the lithium dose to achieve therapeutic plasma levels of lithium. Therapeutic plasma levels of lithium appear to vary widely from patient to patient, with some patients requiring as little as 0.5 mmole of lithium per liter of plasma and others requiring up to 1.5 mmole for acute antimanic or antidepressant effects. In general, for acute treatment lithium levels are maintained between 1 and 1.2 mmol/L. The dose of lithium required to achieve such levels varies between 600 and 2700 mg of lithium carbonate per day. Lithium toxicity including tremor and confusional states occurs at lithium levels above 1.5 mmol/L. Some patients appear more sensitive to toxic side effects and will have similar side effects at much lower levels. A combination of laboratory lithium values and clinical judgment is necessary to determine the optimum plasma concentration of lithium for acute therapeutic effects and for prophylaxis. Many elderly patients develop an energy-deficit syndrome when taking lithium at concentrations above 0.8 mmol/L; yet they receive therapeutic benefit at concentrations of lithium between 0.5 and 0.7 mmol/L.

Since lithium prevents antidiuretic hormone from having its full functional effect on the kidney, polyuria may occur in some lithium-treated patients. The response of the thyroid to thyroid-stimulating hormone is also reduced in the presence of lithium; hypothyroidism or goiter may occur. Basal and follow-up tests of thyroid function are often necessary during long-term lithium prophylaxis. A rare but possible effect of lithium on the kidney (sclerosis of both glomeruli and tubules)[32] generally requires periodic evaluation of kidney function (BUN, creatinine). Baseline creatinine clearance and urinary concentration ability should be determined in patients for whom long-term lithium treatment is warranted. In case of possible abnormality of kidney function during lithium treatment, creatinine clearance and urinary concentration ability determinations should be repeated as necessary and compared to baseline values.

CHANGE OF ANALYTE IN DISEASE

3-Methoxy-4-hydroxyphenylglycol

In 24-hour urine samples decreased MHPG (less than 1400 μg/24 hours in males or 1200 μg/24 hours in females) is suggestive of NE-related depressive disease.[11]

Many clinicians routinely collect 24-hour urine samples for the evaluation of urinary MHPG levels in depressed patients. However, most studies demonstrating clear relationships between low MHPG and response to NE type of antidepressants are based on studies in which patients were drug free for 2 or more weeks before MHPG determinations. Many drugs have prolonged effects on NE turnover and urinary MHPG. In particular, antidepressants themselves generally cause a suppression of NE turnover (lower MHPG), particularly in nonresponsive patients. After withdrawal of antidepressant drug therapy, MHPG precipitously rises (escapes) and remains elevated for up to 2 weeks. The use of MHPG as a means of identifying NE type of depression can often be impractical because the patient must be kept drug free for 2 or more weeks before a decision (based on MHPG) concerning the choice of antidepressant drug can be made. Urinary MHPG, although an intriguing research tool, may therefore have limited clinical practicality for patient subtype identification and the initiation of proper drug treatment.

5-Hydroxyindoleacetic acid

5-HIAA is collected from a single lumbar-puncture sample of cerebrospinal fluid. When 5-HIAA is decreased (less than 15 ng/mL), it is suggestive of serotonin-related depressive disease.

Not all unipolar depressed patients have evidence of decreased 5-HIAA in the cerebrospinal fluid. Several investigators have found that cerebrospinal fluid 5-HIAA is bimodally distributed in depressed patients.[33] A precursor of serotonin, 5-HT, has been shown to be an effective antidepressant only in patients with the lower distribution of 5-HIAA.[8] This observation supports the idea that loading with 5-HT precursors may be effective only in a subgroup of depressive patients who have diminished central serotonin turnover.

Alpha$_2$ adrenoceptors

Blood is collected and made into a platelet-rich fraction for assessment of the number of alpha$_2$ receptors per milligram of platelets. Stereospecific ^{3}H-clonidine binding to platelet alpha$_2$ adrenoceptors has been found to be significantly greater in a group of depressed patients than in a similar control population.[14] Although increased numbers of alpha$_2$ adrenoceptors may be related to NE type of depression, clinical studies have not yet demonstrated preferential antidepressant response after any particular antidepressant drug in depressed patients having the receptor abnormality.

REFERENCES

1. American Psychiatric Association: Diagnostic and statistical manual of mental disorders, ed 3, (DSM III), Washington, DC, 1980, The Association.
2. Lee, T, and Seeman, P: Brain dopamine receptors in schizophrenia. In Usdin, E, and Hanin, I, editors: Biological markers in psychiatry and neurology, New York, 1982, Pergamon Press, Inc.
3. Golden CJ, Moses, JA, Zelazowski, R, et al: Cerebral ventricular size and neuropsychological impairment in young chronic schizophrenics, Arch Gen Psychiatry 37:619-626, 1980.
4. Winberger, DR, and Wyatt, RJ: Structural pathology of the cerebellum in schizophrenia: CT and postmortem studies. In Jansson, B, Perris, C, and Struve, G, editors: Biological psychiatry 1981, New York, 1982, Elsevier/North Holland Inc.
5. Hirschowitz, J, Casper, R, Garver, DL, and Chang, S: Lithium response in good prognosis schizophrenia, Am J Psychiatry 137:916-920, 1980.
6. Perris, C: A study of bipolar (manic-depressive) and unipolar recurrent depressive psychoses. I. Genetic investigation, Acta Psychiatr Scand 194(suppl):15-44, 1966.
7. Winokur, G, Cadoret, K, Dorzab, J, and Baker, M: Depressive disease: a genetic study, Arch Gen Psychiatry 24:135-144, 1971.
8. Van Praag, HM, and Korf, J: Endogenous depression with and without disturbances in the 5-hydroxytryptamine metabolism: a biochemical classification? Psychopharmacology 19:148-152, 1971.
9. Maas, JW, Fawcett, J, and Dekirmenjian, H: 3-Methoxy-4-hydroxyphenyl-glycerol (MHPG) excretion in depressive states, Arch Gen Psychiatry 19:129-134, 1968.
10. Maas, JW, Fawcett, JA, and Dekirmenjian, H: Catecholamine metabolism, depressive illness and drug response, Arch Gen Psychiatry 26:252-262, 1972.
11. Beckmann, H, and Goodman, FK: Antidepressant response to tricyclics and urinary MHPG in unipolar patients, Arch Gen Psychiatry 32:17-21, 1975.
12. Carrol, BJ, Curtis, CG, and Mendels, J: Neuroendocrine regulation in depression. I. Limbic system-adrenocortical dysfunction, Arch Gen Psychiatry 33:1039-1058, 1976.
13. Carrol, BJ, Feinberg, M, Greden, JF, et al: Specific laboratory test for diagnosis of melancholia, Arch Gen Psychiatry 38:15-23, 1981.
14. Garcia-Sevilla, JA, Zis, AP, Hollingsworth, PJ, et al: Platelet alpha$_2$-adrenergic receptors in major depressive disorder, Arch Gen Psychiatry 38:1327-1333, 1981.
15. Charney, DS, Heninger, GR, Sternberg, ED, et al: Presynaptic adrenergic receptor sensitivity in depression, Arch Gen Psychiatry 38:1334-1343, 1981.
16. Siever, LJ, Cohen, RM, and Murphy, DL: Antidepressants and alpha$_2$-adrenergic autoreceptor desensitization, Am J Psychiatry 138:681-682, 1981.
17. Jones, FD, Maas, JW, Dekirmenjian, H, and Fawcett, JA: Urinary catecholamine metabolites during behavioral changes in a patient with manic-depressive cycles, Science 179:300-302, 1973.
18. Gerner, RH, Post, RM, and Bunney, WE: A dopaminergic mechanism in mania, Am J Psychiatry 133:1177-1180, 1976.
19. Janowsky, DW, El-Yousef, MK, Davis, JM, and Sekerke, JH: Parasympathetic suppression of manic symptoms by physostigmine, Arch Gen Psychiatry 28:542-547, 1973.
20. Greden, FJ: The dexamethasone suppression test: an established biological marker of melancholia. In Usdin, E, and Hanin, I, editors: Biological markers in psychiatry and neurology, New York, 1982, Pergamon Press, Inc.
21. Gold, MS, Pottash, AC, and Extein, I: The psychiatric laboratory. In Bernstein, JG, editor: Clinical psychopharmacology, Boston, 1984, John Wright-PSG, Inc.
22. Extein, I, Pottash, ALC, Gold MS, and Cowdry, RW: Using the protirelin test to distinguish mania from schizophrenia, Arch Gen Psychiatry 39:77-81, 1982.
23. Davis, JM, Schaffer, CB, Killian, GA, et al: Important issues in the drug treatment of schizophrenia. In Keith, SJ, and Mosher, LR, editors: Special report: schizophrenia 1981, Washington, DC, 1980, US Government Printing Office.
24. Garver, DL: Drug therapy of psychiatric disease: schizophrenias and related psychoses. In Graham-Smith, DG, Hippus, H, and Winokur, G, editors: Psychopharmacology. I. Clinical psychopharmacology, Amsterdam, 1982, Excerpta Medica.
25. Calil, HM, Avery, DH, Hollister, LE, et al: Serum levels of neuroleptics measured by dopamine radioreceptor assay and some clinical observations, Psychiatry Res 1:39-41, 1979.
26. Tune, LE, Creese, I, DePaulo, JR, et al: Clinical state and serum neuroleptic levels measured by radioreceptor assay in schizophrenia, Am J Psychiatry 137:187-190, 1980.
27. Cohen, BM, Lipinski, JF, Harris, PQ, et al: Clinical use of radioreceptor assay for neuroleptics, Psychiatry Res 1:173-177, 1980.
28. Kayton, W, and Roy-Byrne, P: Antidepressants in the medically ill: diagnosis and treatment in primary care, Clin Chem 34:829-836, 1988.
29. Risch, SC, Huey, LY, and Janowsky, DS: Plasma levels of tricyclic antidepressants and clinical efficacy: review of the literature, II, J Clin Psychiatry 40:4-16, 1979.
30. Risch, SC, Huey, LY, and Janowsky, DS: Plasma levels of tricyclic antidepressants and clinical efficacy: review of the literature. II, J Clin Psychiatry 40:59-69, 1979.
31. Robinson, DW, Nies, A, Ravaris, L, et al: Clinical pharmacology of phenelzine, Arch Gen Psychiatry 35:629-635, 1978.
32. Ramsey, AT, and Cox, M: Lithium and the kidney: a review, Am J Psychiatry 139:443-449, 1982.
33. Asberg, M, Thoren, P, Traskman, L, et al: Serotonin depression—a biochemical subgroup within affective disorders? Science 191:478-480, 1976.

CHAPTER 49

Neoplasia

BERNARD E. STATLAND
PER WINKEL

OBJECTIVES

- Briefly describe the biological factors that can result in cancer.
- List the roles of laboratory tests in the assessment of cancers.
- Define and describe an ideal tumor marker.
- List commonly used chemical and cellular markers and state their clinical significance in relation to cancers.

KEY TERMS

carcinoembryonic antigen A glycoprotein produced by or associated with cancer cells, which is also expressed by fetal cells. Small levels are detected in healthy individuals. Detection is by immunochemical analysis.

carcinogen An agent, usually a chemical, that transforms a cell from a normal to a cancerous state.

cocarcinogen An agent that, by itself, does not transform a normal cell into a cancerous state but in concert with another agent can effect the transformation.

confirmation Use of second test with very high specificity to verify the observation of a less specific test (for example, biopsy to verify a mass as a tumor).

dedifferentiation The process by which cells go from the more specific to the more general in nature. Usually such cells lose their morphological architecture and ability to synthesize specific cell components (for example, estrogen receptors).

dissemination The phase of cancer in which the cells spread to various parts of the body distant from the site of origin.

ectopic hormones Hormones that are produced in tissues that do not normally do so (for example, hormones produced by cancers).

estriol receptor The specific tissue membrane receptor in breast that binds the hormone estriol. Its presence in breast cancer signifies a differentiated tumor.

heterogeneity Variation in gene expression among cancer cells. Differences between cells exist, and not all cells within a tumor are positive for the same antigen or respond to the same drug.

induction phase The period during which a normal cell becomes transformed into a cancerous cell.

in situ A term used to show that cancer cells are localized at the place of origin.

invasion The process by which malignant cells move into deeper tissue and through the basement membrane and gain access to blood vessels and lymphatic channels.

metastases Cancer cells that have spread to other organs and have formed colonies that are growing and often invading the organ.

monitoring Measurement of a biochemical marker of cancer after a confirmed diagnosis (for example, the use of carcinoembryonic antigen to monitor colorectal cancer).

oncofetal protein A protein produced by or associated with cancer cells that is made by the fetus; however, it is normally produced in the child or adult at very low levels.

Pap (Papanicolaou) smear A common screening test for cancer in which cells from the cervix are examined for cytological abnormalities consistent with cancer.

staging A process of diagnosis in which the pathologist determines the position of cancer in the cycle of induction, in situ, invasion, or dissemination.

tumor marker A misused term applied to molecules that can be used to diagnose or monitor the presence or growth of a cancer. Usually such markers are not specific for cancer.

SCOPE OF THE PROBLEM
Incidence

Cancer and its complications rank second only to cardiovascular disease as a cause of death in the United States and in most Western countries. Approximately one in five persons in the United States will die of cancer. In addition, this disease accounts for a large proportion of the total money spent in health care in the United States. The various complications of cancer, the type of therapy used, and the long rehabilitation play a role in the costs. Finally, the psychological fear and dread of this disease and the long-term anxiety borne by cancer patients and their families represent an added burden caused by this disease.

Distribution of deaths by site

Table 49-1 presents the percentage of cancer deaths by site and sex as tabulated in 1987 in the United States. As noted in Table 49-1, the leading cancer sites associated with mortality in males of all ages are, in descending order, lung, colorectal area, prostate, pancreas, and stomach. In females of all ages the sites are breast, colorectal area, lung, uterus, and ovary. Five years ago 15% of cancer deaths in women were due to cancer of the lung. Today this figure is 20%. The incidence of smoking in women is probably related to the incidence of lung cancer. Obviously, if the survival of patients with cancer in general is to improve, it is imperative to attempt to find cures for the commonly seen cancers. It should be mentioned that the distribution of deaths by site depends on geographical considerations as well. For example, in Japan esophageal and stomach cancers are the most common types of malignancy, whereas this pattern is not true in the United States.

Table 49-1 Estimated cancer deaths in 1987 in the United States by site and sex

Site	Sex	
	Male (%)	**Female (%)**
Breast	—	18
Colon and rectum	11	14
Blood and lymphoid tissue (leukemia and lymphomas)	9	9
Lung	35	20
Oral cavity	2	1
Ovary	—	5
Pancreas	5	5
Prostate gland	11	—
Skin	2	1
Urinary tract	5	3
Uterus	—	4
Other	20	20

CANCER: NATURE OF THE DISEASE
Description

Clinical manifestations. The clinical manifestations of cancer vary widely, depending on the tissue affected. For example, cancer of the gastrointestinal tract is manifested by obstruction, hemoptysis, and bloody stools. Cancer of the lung is manifested by hypoxia, chest pain, and often various neurological symptoms. The clinical manifestations are related to the physiological function of the organ with the primary cancer and the effect of the cancer on other organs as well. For example, cancer of an endocrine gland can result in production of excess hormone with many systemic hormonal effects. New symptoms are evidenced with the spread (metastasis) of the cancer cells to other organs. Cancer spreads through the lymphatic system and the bloodstream, resulting in liver, bone, and pulmonary metastases.

Time as a factor

Cancer as a long-term process. Cancer is a long-term process and progresses through four obligatory phases: an induction phase, an in situ phase, an invasion phase, and a dissemination phase. During the *induction phase,* which can last up to 30 years or more, the cells are exposed to one or more carcinogens. These environmental carcinogens may include radiation or various toxins. It has been estimated that approximately three fourths of all human cancers may be caused by these environmental factors.

It is now believed that a period of many years following exposure may be necessary before a carcinogen is able to have its effect on the host. The histological changes begin with severe dysplasia, eventually leading to cancer. It should be obvious that not everyone who is exposed to the same carcinogen will develop cancer. Additional factors that play a role in deciding which individual may get cancer include individual (genetic) or tissue susceptibility, the presence of other carcinogens or cocarcinogens, the site at which the carcinogen may act, the duration of exposure, and obviously the nature, amount, and concentration of the carcinogen under question. Often the time between the induction phase and the clinically apparent cancer can be as long as 20 years.

After induction there is the *in situ phase.* The in situ phase represents that time during which the transformed cell actually develops into a cancer, but the cancer remains localized in the original site and does not invade other tissues.

The third phase is called the *invasion phase.* During the invasion phase the malignant cells multiply and invade into the deeper tissues through the basement membrane, thereby gaining access to blood vessels and lymphatic channels.

The fourth stage is that of dissemination. During the *dissemination phase,* which lasts 1 to 5 years, the invading cancer spreads to various parts of the body distant from the site of origin.

It is critical to detect cancer early, before metastatic

spread. Ideally it should be detected during the induction phase. However, this is impossible, because before the in situ phase one is *not* certain if cancer will actually develop in the individual. The next approach is to detect the cancer in the in situ phase. This has been done with great success in patients with cancer of the cervix. Here the Pap (Papanicolaou) smear technique has been of great benefit. When in situ cancer of the cervix is detected, the prognosis is excellent. Most cancers are detected during the invasion phase. If dissemination has not yet occurred, the prognosis is reasonable. Detection of local spreading with or without involvement of the lymph nodes often leads to a cure. However, if dissemination has already occurred, the prognosis is very poor.

Invasion by cancer cells of surrounding tissue. Several factors play a role in determining the cancer's ability to invade the surrounding tissue. Such factors include increased motility of the cells, increased pressure within the tumor caused by active multiplication of the cells, elaboration by the cancer of lytic substances, lack of intercellular bridges found between all normal cells, decreased cohesiveness between cells, and eventual spread of the tumor cells to the regional lymph nodes. Much work is now being done to assess these critical factors that play a role in the invasiveness of cancer cells. However, when the metastases are still microscopic (micrometastases), the clinician's ability to detect them is very poor. It has been estimated that approximately half the patients who appear to be clinically free of metastases do in fact have unrecognized distant micrometastases at the time of initial diagnosis and treatment.

Change in cell division. Cancer is often manifested by a change in cellular division rate. Although most cancers are associated with an increased rate of cell division, there are examples in which this is not always the case, such as nephroma.

Dedifferentiation of cells. A common phenomenon of cancer is dedifferentiation, in which cells go from a more specific cell type to a more general cell type. Thus it is not uncommon for cancer cells to synthesize various compounds that are normally present only in the embryonic or fetal stage. On the other hand, as cells dedifferentiate, they may lose certain specific cellular properties such as receptor activity or an enzyme activity. These phenotypic changes can be used as prognostic indicators.

Chromosomal changes in cancer. Chromosomal changes in cancer have been studied mainly in patients with leukemia. In fact, various types of leukemia can be confirmed on the basis of these chromosomal changes.

Etiology

Sager[1] presents an intriguing model of the origin of cancer. She discusses it as a multistage genetic process. The stages are of three types:

1. Initial DNA damage
2. Chromosome breakdown and rearrangement; gene replication
3. Selection of successfully growing mutant cells

The initial changes in cellular DNA can be caused by a variety of carcinogens, including radiation, chemicals, viruses, and unknown agents. This leads to faulty growth control and loss of chromosome stability.

The chromosome breakage and rearrangement occur in several continuous phases. There is first an initiation of cell division. This is later manifested in terms of aberrant chromosomal transpositions, which lead to genomic rearrangements, creating new phenotypes (see Chapter 47). The changes in the DNA and chromosomes result in a new pattern of gene expression, creating a new phenotype, in which previously quiescent genes are now expressed, and previously expressed genes are now quiescent. Among the genes that can be newly expressed are *oncogenes*. Oncogenes are viral genes that have the potential to induce neoplastic transformation in cells.[2] Genetic elements similar to viral oncogenes have been identified in eukaryotic cells as well. The cellular counterparts of the viral oncogenes are known as *protooncogenes*. These are believed to be activated in cancers as a result of mutational or chromosomal rearrangement events and, once activated, may play a role in the transformation of a normal cell to a cancer cell. Assays that detect oncogenes in human cancer tissues are rapidly becoming available (see Chapter 47). They may be potentially useful for the detection of oncogene-associated cancers in high-risk groups.[3]

Chemical carcinogens are present in the environment in increasing amounts. Although various carcinogens are known, there probably are many that still are unknown. Viruses also can cause certain cancers. Some of these, such as Burkitt's lymphoma, are seen in humans. There is experimental and epidemiological support for the presence of hepatic viruses and papilloma viruses that may cause hepatic and cervical carcinoma in humans. However, viruses as general cancer-causing agents are not very well substantiated at this point.

Diversity of cancer cells

Variation of gene expression. There is a broad spectrum of possible combinations of gene expression in the human cell. The spectrum varies from normal cells to the most atypical cancer cells. The phenotypic variation occurs not only from cancer cells to normal cells, or from cancer type to cancer type, but also within particular cancer types. For example, in patients with cancer of the breast there is a heterogeneity of genes expressed by various cells; that is, not all cells express the same genes.

Variable gene expression and its manifestations. Variable gene expression leads to biological and biochemical diversity of cancer cells; consequently various tumor-specific markers are not necessarily elaborated by all cancer cells of the same type or even of a single cancer over a period. This is very important clinically in trying to determine which analyte to follow in monitoring patients with known malignancy.

OVERVIEW OF ROLES OF LABORATORY TESTS

Detection (screening)

There are four major functions that laboratory tests can serve in the field of neoplasia. They are detection or screening, confirmation, classification (staging), and monitoring.

An essential feature of any screening program is the screening test. Table 49-2 lists a number of screening tests for early detection of cancer.[4] Note that no laboratory test is included in this list. The quality of a screening test is usually expressed by its sensitivity and specificity. The observations from the screening tests are divided into negative and positive results. Each person examined is classified as either a diseased or nondiseased person.

A rigid classification of test results into positive and negative results may sometimes be too simplistic. Outcomes of screening tests can usually be ordered from very negative to very positive. The latter approach allows for a more sophisticated test interpretation in actual screening programs. For example, patients whose results are not negative but also are not alarming enough to justify immediate diagnostic action can be scheduled for earlier repeat screenings.[5] Another example is a stepwise screening policy in which only individuals with positive results at the first screening test are subject to further diagnostic testing.[6]

Sometimes the use of more than one screening test may seem advantageous. However, assessment of the sensitivity and specificity of a combination of screening tests based on data available for the individual tests is complicated by the fact that usually the tests are not independent in a statistical sense. In general, it is more effective to combine two tests that are complementary (that is, directed at different anatomical or biochemical features of the tumor) than to combine tests directed at the same types of features.

Complementary tests include sputum cytology and chest x-ray examination for lung cancer screening.[7] Palpation and mammography in breast cancer screening are examples of two related tests. They both detect tumors largely on the basis of size. One study[8] showed that when mammography was performed, the physical examination proved to be almost completely redundant.

Table 49-2 Screening tests for early detection of cancer

Site	Test
Bladder	Cytological analysis of urine
Breast	Mammography, physical examination, self-examination
Cervix	Papanicolaou smear, pelvic examination
Colon and rectum	Testing stool for occult blood, sigmoidoscopy
Hodgkin's disease	Physical examination and roentgenography
Lung	X-ray, cytological analysis of sputum
Oral cavity	Visual examination
Prostate	Digital palpation per rectum, prostatic massage and cytological examination
Skin	Visual inspection
Stomach	Photofluorography, saline wash and cytological examination of gastric contents, examination of stool for occult blood

From Habbema, JDF, van Oortmarssen, GJ, and van der Maas, PJ. In Statland, BE, and Winkel, P, editors: Laboratory measurements in malignant disease, vol 2, Philadelphia, 1982, WB Saunders Co.

Confirmation

Additional tests are used to confirm the suspicion of cancer based on clinical symptoms or signs. Tests that tend to confirm the presence of a cancer include, for example, bone marrow examination for leukemia, urinary catecholamines for pheochromocytoma, and alpha-fetoprotein for testicular cancer. The confirmatory results must be above a certain decision level.[9] For a laboratory test result to be confirmatory, it should possess 100% diagnostic specificity, that is, contain no false-positive results. For example, all cases in which the catecholamine level is above a certain value should be associated with pheochromocytoma.

Classification and staging

Surgical pathologists have developed various staging approaches based on the size and extent of invasion of surrounding tissues by the tumor, the number of cancer cell–positive lymph nodes, and the presence or absence of metastases. This has been called the TNM (tumor, nodes, metastases) system.[10] The purpose of such staging is to give reasonable estimates of prognosis (that is, recurrence of cancer), appropriate response to therapy, or the likely course of the disease. In addition to staging based on gross or microscopic pathological data, it would be of great value to have biochemical tests that also classify cancers appropriately.

Monitoring

The most important function of laboratory tests in cancer is that of monitoring the course of the disease or its response to therapy. Winkel et al.[11] have developed various strategies to monitor patients known to have breast cancer. The problem addressed was that of predicting on the basis of sequential values whether a patient would have recurrence of this disease. Other approaches have been used to monitor patients on the basis of carcinoembryonic antigen (CEA) in colon cancer,[12] prostate-specific antigen (PSA) for prostate cancer,[13] and others. An increased CEA or PSA value is a signal to explore the patient surgically again to remove additional cancer. It is assumed that the CEA- or PSA-producing tumor has recurred when the serum values for PSA or CEA reach a certain threshold.

All four major functions—screening, confirming, classifying, and monitoring—are possible roles for laboratory tests for neoplasia.

DEFINITION OF IDEAL TUMOR MARKER

Coombes and Neville[14] have suggested that the *ideal* tumor marker should fulfill the following criteria:

1. Be easy and inexpensive to measure
2. Be specific to the tumor studied and commonly associated with it
3. Have a stoichiometric relationship between plasma level of the marker and tumor mass
4. Have an abnormal plasma level, urine level, or both in the presence of micrometastases, that is, at a stage at which no clinical or presently available diagnostic methods reveal their presence
5. Have plasma levels, urine levels, or both that are stable—not subject to wild fluctuations
6. If present in plasma of healthy individuals, exist at a much lower concentration than that found in association with all stages of cancer

Obviously much additional research must be done before such ideal tumor markers will be found. However, it is important to recognize that the evaluation of an ideal tumor marker should relate to the clinical setting. To do so, it has been suggested that all tumor markers should also comply with the following major criteria[15]:

1. They should prognosticate a higher or lower risk for eventual development of recurrence.
2. They should change as the current status of the tumor changes over time.
3. They should precede and predict recurrences before they are clinically detectable.

All tumor markers should be analyzed both according to the criteria that Coombes and Neville have presented and according to the considerations just mentioned. For a tumor marker to be of some value, it must give information beyond that readily seen on the basis of physical examination or history, and it must give this information with a reasonably long lead time to allow appropriate therapy to be given in a timely manner. Lead time is the time elapsed between the time a test result will be positive and the time that the disease will be clinically evident or advanced.

TYPES OF ANALYTES

Classes of biochemicals used as tumor markers (Table 49-3)

This section will review a number of biochemical tests that have been used either as primary tumor markers or as secondary tests to note invasion or dissemination of cancer. The types of analytes are listed in the box on p. 735 and are discussed in terms of their clinical usefulness and applications. This chapter will discuss assays that are commonly in use or that seem to have potential value. Furthermore, the analytical procedures are not dealt with in great detail. They are mentioned only if such procedures are critical in the interpretation of results. Reviews that cover a number of these assays in greater depth are recommended for further reading (for example, reference 16).

Oncofetal antigens

Many of the oncofetal antigens are measured in the laboratory using immunoassays; either competitive binding radioimmunoassays or solid-phase immunometric assays, employing either ^{125}I- or enzyme-labeled second antibodies.

Carcinoembryonic antigen. CEA is a glycoprotein present in colonic adenocarcinoma and fetal gut; it was first described by Gold and Freedman.[17] The detection of CEA in various tissues or serum is complicated by the presence in these tissues of CEA–cross-reacting antigens.

In general, CEA plasma levels increase with increasing age and smoking. This has prevented the use of CEA levels for the purpose of general screening.[18] The results of screening programs confined to subpopulations with higher-than average risk of developing cancer have been equally discouraging.[19]

Usually a CEA measurement is not a useful adjunct to cancer diagnosis. Neither the sensitivity nor the specificity of CEA justifies its use to support by itself a diagnosis of cancer.[20] In particular situations, however, CEA has proved of diagnostic value. CEA is useful, for example, for the detection of primary colorectal cancer[21] when used in combination with a barium enema and with radioiodide imaging for the detection of carcinoma metastatic to the

Table 49-3 Classes of biochemicals used as tumor markers

Class of biochemical	Examples	Use
Increased production of endogenous biochemicals	Hormones, enzymes, polyamines, and so on	Confirmation, diagnosis, monitoring
Synthesis of biochemicals of previously quiescent genes	Oncofetal proteins, cell surface antigens, enzymes	Monitoring, prognosis
Receptors	Estriol receptor (breast cancer), androgen receptor (prostate cancer)	Prognosis, treatment
Modification of usual cell or organ function	Gamma-glutamyl transferase (GGT) or 5′-nucleotidase	Diagnosis

TYPES OF ANALYTES

Oncofetal proteins
- Carcinoembryonic antigen (CEA)
- Alpha-fetoprotein (AFP) and human chorionic gonadotropin (HCG)
- Tissue polypeptide antigen (TPA)
- Tennessee antigen (TENAGEN)
- Fetal sulfoglycoprotein (FSA)
- Pancreatic oncofetal antigen (POA)
- Breast cancer–associated markers

Various proteins
- Glycoproteins
- Ferritin
- Casein

Enzymes
Collagen-breakdown products
Polyamines
Nucleosides
Cellular markers
Hemostasis-related factors

liver. According to the consensus statement of the National Cancer Institute,[18] only values five to ten times the upper normal reference limit in patients with symptoms should be considered strongly suggestive of the presence of cancer. In a number of cancers, including colorectal and breast cancer, the plasma level of CEA and the frequency of elevated values are positively correlated with the severity of the disease as assessed by clinical staging.* In general, CEA-producing tumors tend to be more aggressive after initial treatment than nonsecreting tumors are.[25] Alteration in CEA levels correlated well with change in disease status in patients with metastatic breast carcinoma and in patients with metastatic colorectal cancer.[10] However, because the relationship between clinical course and CEA plasma level is not completely consistent, it is difficult to assess the potential clinical usefulness of these observations.

Postoperative monitoring of plasma CEA levels for the detection of recurrence or metastases has proved valuable in colorectal cancer. Most clinicians now respond to consecutively increasing serum CEA values with second-look operations in patients with surgically treated colorectal cancer. Balz et al.[26] found that 19 of 22 patients with postoperatively rising CEA levels had otherwise undetected recurrent disease. Among these, the recurrent tumor was localized and could be resected in six patients. Five of these were asymptomatic with stable CEA values over the next 37 months.

Postoperative CEA levels are less frequently elevated in patients with breast cancer who eventually develop overt metastatic disease than in corresponding patients operated on for colorectal cancer. In only 10% to 15% of patients with breast cancer does the plasma CEA level rise to values above 10 μg/L.[27] On the average this happens 4 to 6 months before the appearance of overt metastatic disease.

Alpha-fetoprotein and human chorionic gonadotropin. Alpha-fetoprotein (AFP) is an oncofetal glycoprotein. In early embryonic life it is a predominant component of the serum proteins. It is first synthesized by the yolk sac and later by the fetal liver. Later in life it is mainly produced in the liver. AFP was first recognized as a tumor marker by Abele in 1963.[28]

Serum AFP values should be less than 10 μg/L in healthy subjects. In benign hepatic disorders, moderate (40 μg/L) elevations may be seen. Values above 400 μg/L are almost always associated with hepatocellular carcinoma, germ cell carcinoma (such as testicular carcinoma), chronic aggressive hepatitis, or subacute hepatic necrosis.

Human chorionic gonadotropin (HCG) is a glycoprotein hormone that shares indistinguishable biological activity and extensive structural homology with its pituitary counterpart, human luteinizing hormone (HLH).[29] Although the alpha subunits of HLH and HCG are essentially identical, the beta subunits can be differentiated on the basis of specific immunoassay techniques (see Chapter 37 and p. 938).

Tumors of the placenta and the testes that contain trophoblastic tissue secrete excessive amounts of HCG. Specific and sensitive assays have revealed that many cancers also secrete HCG. However, available data[30] clearly show that HCG determinations are of no value in screening for cancer.

The main clinical use of AFP and HCG is related to the diagnosis, therapy, and follow-up of germ cell tumors.[31] Table 49-4 presents the World Health Organization (WHO) classification of germ cell tumors and associated markers in tissue and serum. In general, AFP and HCG provide the most information about tumor status when they are persistently elevated. The absence of a marker does not preclude the presence of germ cell tumors.

Prostate specific antigen. PSA is a glycoprotein found in the epithelial cells of the prostatic duct and acini. PSA is elevated in all four stages of prostate cancer as well as in benign prostatic hypertrophy. It is a much more sensitive assay for a prostate cancer than is prostatic acid phosphatase.[13]

Carbohydrate antigen 19-9. Carbohydrate antigen 19-9 (CA 19-9) occurs in tissue as a monosialoganglioside and in serum as mucin, a high-molecular-weight, carbohydrate-rich glycoprotein. Results of clinical studies indicate that the CA 19-9 level in serum or plasma of patients with an intraabdominal carcinoma frequently is increased. It is correlated most strikingly with cancer of the pancreas, for which early studies have shown a sensitivity of 90% and a specificity of 85%. CA 19-9 also may be increased

*See references 10, 16-19, and 22-24.

Table 49-4 WHO classification of germ cell tumors and associated tumor markers

WHO classification	Immunohistochemistry		Serology		Comments
	AFP*	HCG†	AFP	HCG	
Seminoma (S)	–	±	No	± Yes	HCG in giant cells
Embryonal carcinoma (EC)	+	+	± Yes	± Yes	HCG in giant cells AFP controversial, may occur in undiagnosed yolk sac elements
Yolk sac tumor (YST)	+	–	± Yes	No	
Choriocarcinoma (CC)	–	+	No		
Teratoma (TT)	–	–	No?	No?	

From Norgaard-Pedersen, B, and Hangaard, J: In Statland, BE, and Winkel, P, editors: Laboratory measurements in malignant disease, Philadelphia, 1982, WB Saunders Co.
**AFP*, alpha-fetoprotein.
†*HCG*, human chorionic gonadotropin.

with other adenocarcinomas such as lung, gastric, biliary, and colonic.

Carbohydrate antigen-125. Serum Carbohydrate antigen-125 (CA-125), a glycoprotein antigen, is elevated in the serum of patients with ovarian cancer. Increased concentrations of CA-125 were found in many patients with epithelial ovarian cancer and in ovarian teratoma. Changes in CA-125 concentrations in serum during chemotherapy mirrored the progress of the disease as assessed by clinical and radiological evidence. It should be noted that CA-125 provides no real assistance for diagnosis; however, it does have value as a marker for monitoring responsiveness to chemotherapy.

Other proteins

Ferritin. Marcus and Zinberg[32] demonstrated that serum ferritin levels exceeded the upper reference limit in preoperative sera of 40% of women with breast cancer and in 67% of women with locally recurrent or metastatic breast cancer. It should be pointed out, however, that hepatic inflammatory disease also results in elevated serum ferritin values; for example, 431 of patients with hepatitis or cirrhosis had values above the upper reference limit. Furthermore, 13% of patients with ulcerative colitis or gastric duodenal ulcers had elevated values.

Enzymes

Schwartz[33] reviewed the use of enzyme tests in the management of patients with cancer. The box at right presents various enzymes that have been used for this purpose. As with many putative tumor markers discussed in this chapter, the use of enzyme markers is fraught with difficulties. Not all patients with a particular cancer type have elevations in an enzyme (poor sensitivity), and many noncancer diseases are associated with elevations of many of these enzymes. Thus the most frequent uses of these enzymes are as objective markers to give semiquantitative estimates of response to therapy or as prognostic indicators.

Acid phosphatase. There are two isoenzymes of the 13 known for acid phosphatase (ACP) that appear to have important roles in managing patients with cancer. The first is the prostatic isoenzyme, and the second is the bone isoenzyme. The former is inhibited by L-tartrate, whereas the bone isoenzyme is tartrate resistant.

The success in interpreting elevations of the prostatic acid phosphatase (PAP) isoenzyme for diagnostic purposes has been poor.[34,35] Most assays investigated have low sensitivities in the early, more treatable stages of prostatic carcinoma (A and B) (Table 49-5). The use of many of these assays to confirm the presence of prostatic carcinoma in men with urological complaints has also been limited.[36] The primary use of PAP is to monitor recurrence of cancer in men being treated for carcinoma of the prostate.

Pancreatic enzymes (amylase, lipase, ribonuclease, and trypsin). Serum amylase elevations have been found in approximately 25% of patients with pancreatic cancer. Unfortunately lipase assays are not more useful than are amylase assays. Various workers have examined ribonuclease as a marker of pancreatic cancer. Reddi and Holland[37] observed that this enzyme was elevated in 90% of patients with pancreatic cancer and in only 10% of pa-

ENZYMES USEFUL IN CANCER DETECTION

Acid phosphatase (ACP)
Alkaline phosphatase (ALP)
Creatine kinase BB (CK-BB)
Gamma-glutamyl transferase (GGT)
Glycosyl transferases
Lactate dehydrogenase (LD)
Lysozyme (muramidase)
5′-Nucleotidase
Pancreatic enzymes (amylase, lipase, ribonuclease, trypsin)
Phosphohexose isomerase (PHI)
Terminal deoxynucleotidyl transferase (TDT)

Table 49-5 Diagnostic sensitivities (percent positive) of several PAP assays for prostatic carcinoma*

Stage	Chemical†	Radioimmunoassay			
		†	‡	§	‖
A	12	33	13	0	8
B	15	79	26	20	21
C	29	71	30	33	40
D	60	92	94	79	86

*Sensitivities were calculated from data presented in the reference listed for each assay.
†Foti, AG, Cooper, JF, and Herschman, H, et al: N Engl J Med 297:1357-1361, 1977.
‡Mahan, DE, and Doctor, BP: Clin Biochem 12:10-17, 1979.
§Chu, TM, Wang, MC, Scott, WW, et al: Invest Urol 15:319-323, 1978.
‖New England Nuclear, April 1979, Study Report, Boston.

tients with pancreatitis. Fitzgerald et al.[38] repeated these studies and found that 50% of patients with pancreatic cancer and, unfortunately, 35% of patients with noncancerous pancreatic disease had elevated ribonuclease values.

Immunoreactive trypsin has been used to study patients with pancreatic cancer. In one study all of 17 patients with pancreatic cancer had elevated renal clearance of immunoreactive trypsin, whereas 67% of patients with acute pancreatitis had elevated values. This should be compared with the fact that all patients with chronic pancreatitis had values within the control range.[39]

Terminal deoxynucleotidyl transferase. Terminal deoxynucleotidyl transferase (TdT) is a polymerase that is found in high concentrations in normal thymus and in T and non-T, non-B acute lymphoblastic leukemia cells. High TdT activity in peripheral blood lymphocytes and in bone marrow is observed in most patients at initial diagnosis of acute lymphoblastic leukemia (see below).

Other enzymes (Table 49-6)

Additional enzyme tests that may be of value in patients with cancer include alkaline phosphatase that is elevated in the presence of bone or liver metastases; creatine kinase BB elevations in a small percentage of patients with lung cancer; gamma-glutamyl transferase and 5′-nucleotidase elevations in patients with liver metastases; galactosyl transferase II elevations in patients with pancreatic cancer; lactate dehydrogenase (LD) elevations as the LD 5 isoenzyme in patients with liver metastases and as the LD 2 or LD 3 isoenzyme in patients with lymphoma; and lysozyme (muramidase) elevations in patients with various types of leukemia.

Collagen-breakdown products

In patients with bone metastases of an osteolytic nature there is an increase in urinary excretion of both hydroxyproline and hydroxylysine. In fact, hydroxyproline excretion in urine is often the first sign of bony metastases in certain malignancies.

Polyamines and nucleosides

The polyamines (spermidine, spermine, and putrescine) are elevated in the urine of patients with cancer. Tormey et al.[40] noted that the excretion of the nucleoside 2-*N*-dimethylguanosine was increased in many patients with metastatic breast cancer and in patients with local lymph node involvement with the disease.

Cellular markers

A number of markers associated with the plasma membrane, cytoplasm, or nuclei of the lymphoid cell have been identified. Various techniques have been used. These techniques, which tend to be immunological in nature, include rosetting, immunofluorescence, and immunoenzymatic testing. The rosetting technique is based on a reaction between an indicator cell (usually an erythrocyte) and the lymphoid cell to form rosettes in cases in which the lymphoid cell carries a particular membrane marker. By such techniques the cells may be mixed directly, or the indicator cell may first be coded with antibody or complement to demonstrate receptors for the Fc part of immunoglobulin or complement components.

It appears that the various antigens demonstrated by these techniques are not tumor-specific antigens but rather are tumor-associated differentiation antigens that represent the expression of oncofetal antigens not normally expressed by differentiated cells.

Lymphocytic leukemias and non-Hodgkin lymphomas have been subdivided into clinically usable subgroups on the basis of biochemical cell markers. The most striking evidence of the value of typing lymphocytes with a panel

Table 49-6 Liver metastasis and serum enzyme activity in 95 patients

Enzyme	Patients with metastasis		Patients without metastasis		Number of patients*
	Normal	Abnormal	Normal	Abnormal	
Alkaline phosphatase (ALP)	9	31	36	18	94
5′-Nucleotidase (5′-NT)	13	24	50	4	91
Gamma-glutamyl transferase (GGT)	1	35	30	25	91
Glutamate dehydrogenase (GD)	11	26	36	19	92

From Kim, NK, Yasmineh, WG, Freier, EF, et al: Clin Chem 23:2034-2038, 1977.
*The total is less than 95 because some enzyme determinations were omitted.

Table 49-7 Phenotypic heterogeneity of ALL*

ALL	E	HTA	SmIg	Cyμ	CALLA	HLA-DR	TdT
Common ALL	−	−	−	−	+	+	+
Pre-B ALL	−	−	−	+	+	+	+
Null-ALL	−	−	−	−	−	+	+
T-ALL	+	+	−	−	−	−	+
B-ALL	−	−	+	−	−	+	−

From Plesner, T, Wilken, M, and Avenstrøm, S: In Statland, BE, and Winkel, P, editors: Laboratory measurements in malignant disease, Philadelphia, 1982, WB Saunders Co.

*Abbreviations: *E,* sheep erythrocyte receptor; *HTA,* human thymocyte antigen(s); *Sm,* surface membrane; *Cy,* cytoplasmic; μ, IgM heavy chain; *CALLA,* common ALL antigen; *HLA-DR,* human Ia-like antigen; *TdT,* terminal deoxynucleotidyltransferase; *B-ALL,* B-cell type of acute lymphoblastic leukemia; *T-ALL,* T-cell type of acute lymphoblastic leukemia.

of markers comes from studies of acute lymphocytic leukemia (ALL). Table 49-7 gives the five prognostically distinct groups of ALL and the relevant markers for each. The groups are ordered according to prognosis. The cells of the B-cell type are characterized by the presence of surface membrane immunoglobulin (SmIg), as are normal mature B cells. The cells of the T-cell type are characterized by the presence of sheep erythrocyte receptors and human thymocyte antigen, as are mature T cells. The cells of the pre-B cell type are characterized by a cytoplasmic IgM heavy chain but no SmIg, which corresponds to the characteristics of an early stage during the B-cell differentiation.

The terminal deoxynucleotidyl transferase and the common ALL antigen are of value not only for the classification of ALL (see Table 49-7), but also because they are very useful in distinguishing between acute lymphoblastic and myeloblastic leukemia.

ALL may be classified into B-cell leukemia (95%), which is characterized by low-density monoclonal surface membrane immunoglobulins, usually IgM or IgM and IgD with one light chain, and the more rare, but also more aggressive, T-ALL (5%). The cells of the last type form E rosettes and have T-antigens but lack SmIg.

Steroid receptor analysis

Both estrogen and progesterone receptor assays are useful in assessing the prognosis of patients with breast cancer.[41] These procedures evaluate the relative concentration of receptors for estrogen and progesterones in breast tumor excised during surgery. Individuals who are positive both for estrogen and progesterone receptors tend to have a longer survival time and thus a better prognosis than individuals who are deficient in these receptors. Estrogen receptor appears to be the most important of the two factors. This might also explain the value of antiestrogen therapy in treating patients with breast cancer. The presence or absence of steroid receptors can help determine the type of therapy; for those women devoid of receptors, a more aggressive therapy may be used.

Hemostasis-related factors

Plasma fibrinogen levels are generally elevated in patients with cancer. However, in patients with disseminated intravascular coagulation (DIC) hypofibrinogenemia has also been noted. As expected, DIC is also associated with decreased values for antithrombin III (AT III). Consequently, in patients with DIC associated with cancer, AT III values will be decreased.

Increased fibrinolytic or fibrinogenolytic activity has been reported in patients with cancer. Consequently the fibrinogen-degradation products are often elevated in the plasma or urine in patients with cancer.

It is interesting that plasminogen activators are often elevated in patients with cancer. Unfortunately techniques for the measurement of the plasminogen activators are still under development and are not readily available for routine use.

CONCLUSIONS

The scope of the problem is too large to be covered in one chapter in a textbook of clinical chemistry. More important than merely enumerating all the assays available is the fact that there is a challenge that must be presented for any candidate tumor marker. The challenge is that it be useful clinically in the management of patients with the disease or suspected of having the disease. Unfortunately the present scene is a very pessimistic one. It is hoped that with further research and clear thinking additional strategies and new markers will be made available.

REFERENCES

1. Sager, R: Explorations on the origin of cancer, Focus 2/3:1-3, 1983.
2. Huebner, RJ, and Todaro, GJ: Oncogenes of RNA tumor viruses as determinants of cancer, Proc Natl Acad Sci USA 64:1087-1094, 1969.
3. Niman, HL: Detection of oncogene-related proteins with site-directe monoclonal antibody probes, J Clin Lab Anal 1:28-41, 1987.
4. Habbema, JDF, and van Oortmarssen, GJ: Performance characteristics of screening tests. In Statland, BE, and Winkel, P, editors: Laboratory measurements in malignant disease, vol 2, Philadelphia, 1982, WB Saunders Co, pp 639-656.
5. EVAC: Rapport Eerste Screeningsronde, Leidschendam, 1980, Ministerie van Volksgezondheid en milieuhygiene.

6. Tabar, L, and Gad, A: Screening for breast cancer: the Swedish trial, Radiology 138:219-222, 1981.
7. Woolner, LB, Fontant, RS, Sanderson, DR, et al: Mayo Lung Project: evaluation of lung cancer screening through December 1979, Mayo Clin Proc 56:544-555, 1981.
8. Shapiro, S: Evidence on screening for breast cancer from a randomized trial, Cancer 39:2772-2782, 1977.
9. Statland, BE: Clinical decision levels for lab tests, Oradell, NJ, 1983, Medical Economics Co.
10. Rubin, P: Clinical oncology for medical students and physicians, ed 5, New York, 1978, American Cancer Society.
11. Winkel, P, Bentzon, MW, Statland, BE, et al: Predicting recurrence in patients with breast cancer from cumulative laboratory results: a new technique for the application of time series analysis, Clin Chem 28:2057-2067, 1982.
12. Ravry, M, Moertel, CG, Schutt, AJ, et al: Usefulness of serial serum carcinoembryonic antigen (CEA) determinations during anticancer therapy or long-term follow-up of gastrointestinal carcinoma, Cancer 34:1230-1234, 1974.
13. Chan, DW: PSA as a marker for prostatic cancer, Clin Chem 33:1916-1920, 1987.
14. Coombes, RC, and Neville, AM: Significance of tumor-index-substances in management. In Stoll, BA, editor: Secondary spread in breast cancer, Chicago, 1978, William Heinemann Medical Books.
15. Statland, BE: The challenge of cancer testing, Diagn Med 4:2-8, 1981.
16. Statland, BE, and Winkel, P, editors: Clinics in laboratory medicine: laboratory measurements in malignant disease, vol 2, Philadelphia, 1982, WB Saunders Co.
17. Gold, P, and Freedman, SO: Demonstration of tumor-specific antigens in human colonic carcinomata by immunological tolerance and absorption techniques, J Exp Med 121:439-462, 1965.
18. CEA as a cancer marker, vol 3, no 7, Bethesda, Md, 1981, National Institutes of Health, Consensus Development Conference Summary.
19. Holyoke, ED, Chu, TM, and Murphy, GP: CEA as a monitor of gastrointestinal malignancy, Cancer 35:830-836, 1975.
20. Costanza, ME, Das, S, Nathanson, et al: Carcinoembryonic antigen: report of a screening study, Cancer 33:583-590, 1974.
21. McCartney, WH, and Hoffer, PB: The value of carcinoembryonic antigen (CEA) as an adjunct to the radiological colon examination in the diagnosis of malignancy, Radiology 110:325-328, 1974.
22. McCartney, WH, and Hoffer, PB: Carcinoembryonic antigen assay in hepatic metastases detection, JAMA 236:1023-1027, 1976.
23. Wanebo, HJ, et al: Preoperative carcinoembryonic antigen level as a prognostic indicator in colorectal cancer, N Engl J Med 200:448-452, 1978.
24. Myers, RE, Sutherland, DJ, Meakin, JW, et al: Carcinoembryonic antigen in breast cancer, Cancer 42:1520-1525, 1978.
25. Chu, TM, Holyoke, ED, Cedermark, R, et al: A long-term follow-up of CEA in resective colorectal cancer. In Fishman, WH, and Sell, S, editors: Oncodevelopmental gene expression, New York, 1976, Academic Press, Inc, pp 427-431.
26. Balz, JB, Martin, EW, and Minton, JP: CEA as an early indicator for second-look procedure in colorectal carcinoma, Rev Surg 34:1-5, 1977.
27. Statland, BE, and Winkel, P: Usefulness of clinical chemistry measurements in classifying patients with breast cancer, CRC Crit Rev Clin Lab Sci 26:255-290, 1982.
28. Sell, S, and Becker, FF: Alpha-fetoprotein, J Natl Cancer Inst 60:19-26, 1978.
29. Vaitukaitis, JL: Secretion of human chorionic gonadotrophin by tumors. In Carcino-embryonic proteins, vol 1, New York, 1979, Elsevier/North Holland, Inc, pp 447-455.
30. Braunstein, GD: Human chorionic gonadotropin in nontrophoblastic tumors and tissues. In Talwar, GP, editor: Recent advances in reproduction and regulation of fertility, Amsterdam, 1979, Elsevier/North Holland Biomedical Press.
31. Anderson, CK, Jones, WG, and Ward, A: Germ cell tumors, London, 1981, Taylor and Francis, Ltd.
32. Marcus, D, and Zinberg, N: Measurement of serum ferritin by radioimmunoassay: results in normal individuals and patients with breast cancer, J Natl Cancer Inst 55:791-795, 1975.
33. Schwartz, MK: Enzyme tests in cancer. In Statland, BE, and Winkel, P, editors: Laboratory measurements in malignant disease, Philadelphia, 1982, WB Saunders Co, pp 479-491.
34. Chu, TM, Wang, MC, Scott, WW, et al: Immunological detection of serum prostatic acid phosphatase: methodology and clinical evaluation, Invest Urol 15:319-323, 1978.
35. New England Nuclear: Study report, Boston, April 1979 (research company publication).
36. Watson, RA, and Tang, DB: The predictive value of prostatic acid phosphatase as a screening test for prostatic cancer, N Engl J Med 303:497-499, 1980.
37. Reddi, K, and Holland, JF: Elevated serum ribonuclease in patients with pancreatic cancer, Proc Natl Acad Sci 73:2308, 1976.
38. Fitzgerald, PJ, Fortner, JG, Watson, RC, et al: The value of diagnostic aids in detecting pancreas cancer, Cancer 41:868-879, 1979.
39. Lake-Bakaar, G, McKavanaugh, S, and Summerfield, JA: Urinary immunoreactive trypsin excretion: a non-invasive screening test for pancreatic cancer, Lancet 2:878-880, 1979.
40. Tormey, DC, Waalkes, TP, Ahmann, D, et al: Biological markers in breast carcinoma. I. Incidence of abnormalities of CEA, HCA, three polyamines and three minor nucleosides, Cancer 35:1095-1101, 1975.
41. Pertschuk, LP, Eisenberg, KB, Carter, AC, and Feldmann, JG: Immunohistologic localization of estrogen receptors in breast cancer with monoclonal antibodies, Cancer 55:1513-1520, 1985.

CHAPTER 50

Toxicology

ALPHONSE POKLIS
AMADEO J. PESCE

OBJECTIVES

- Define toxicology, poison, toxicity, and LD_{50}.
- Describe the mechanisms of toxicity and factors that influence toxicity.
- List some general effects of toxic agents.
- List and describe classes of drugs frequently encountered in overdose situations.
- Define drug interaction.

KEY TERMS

abused substances Potentially addictive compounds such as alcohol, cocaine and marijuana that are taken to induce pleasure. These are often toxicants.

acute toxicity Usually refers to the harmful effect of a toxic agent that manifests itself in seconds, minutes, hours, or days after entering the patient.

analgesics A class of drugs that reduce pain.

antidote Any agent that counteracts the effects of a poison.

benzodiazepines A group of antianxiety sedative drugs.

biotransformation The chemical modifications of chemicals as they pass through the body.

chronic toxicity Usually refers to the long-term harmful effects of an agent weeks, months, and years after entering the patient.

drug interaction The observation that when two drugs are given, their actions or effects are not independent.

drug screen A test that qualitatively identifies the presence of any one of a class of drugs.

forensic toxicology A branch of the discipline of toxicology concerned with the medical and legal aspects of the harmful effects of chemicals or poisons.

hypnotics A class of drugs often used as sedatives.

LD_{50} The amount of a substance that will cause death in half the test animal population.

lethal dose The amount of a substance that, after being ingested, causes death.

mixed function oxidase A group of enzymes present in the microsomes of the liver that adds oxygen to a drug.

phenothiazines Drugs used to treat psychoses.

therapeutic dose The amount of a substance that will produce a desired pharmacological effect.

toxicant or poison A substance that, when taken in sufficient quantity, will cause sickness or death.

toxicokinetics The passage of a toxicant with time through an affected person. Similar to pharmacokinetics.

toxicology The study of poisons.

tricyclic antidepressants Mood-elevating drugs.

xenobiotics Compounds, usually drugs, that do not normally occur in the body.

DEFINITIONS

Toxicology is the study of poisons. More specifically, toxicology is concerned with the chemical and physical properties of poisons, their physiological or behavioral effects on living organisms, qualitative and quantitative methods for their analysis in biological and nonbiological materials, and the development of procedures for the treatment of poisoning. A poison (or toxicant) is regarded as any substance that, when taken in sufficient quantity, will cause sickness or death. The key phrase in this definition is ''sufficient quantity.'' As the sixteenth century physician Paracelsus observed, ''All substances are poisons; there is none which is not a poison.'' The right dose often differentiates a poison from a "remedy." For example, minute quantities of cyanide, arsenic, lead, and dichlorodiphenyltrichlorethane (DDT) are regularly ingested from food sources or inhaled as environmental contaminates and retained by the human body. However, the amounts of these toxicants are insufficient to cause obvious deleterious effects. On the other hand, a substance as apparently innocuous as pure water will, if ingested in sufficient quantity, cause incapacitating electrolyte imbalance or even death.

There is often no difference between the mechanism of action of a drug and a poison. A drug is administered in doses that alter physiological function to produce a desired therapeutic effect. If administered in greater than therapeutic quantities, a drug may produce toxic (harmful) effects. Thus toxicology is a quantitative discipline that seeks to identify the amount of a substance that, in particular exposure situations, will cause deleterious effects in a particular animal or patient.

Because of a diversity of concerns and applications, modern toxicology has developed into three specialized branches: environmental, clinical, and forensic. Environmental toxicology is concerned primarily with the harmful effects of chemicals that are encountered incidentally because they are in the atmosphere, food chain, or occupational or recreational environments. Clinical toxicology is concerned with the harmful effects of chemicals that are intentionally administered to living organisms for the purpose of achieving a specific effect. The desired effect achieved may be beneficial to the organ (therapeutic), in which case the toxicologist is interested in the adverse or side effects of the agent in question. Forensic toxicology is the branch of toxicology that is concerned with the medical and legal aspects of the harmful effects of chemicals or poisons.

DOSAGE RELATIONSHIPS

Toxicity is a relative term used to compare one substance with another. To state that one substance is more toxic than another is to make a quantitative comparison of dose. Such a comparison is only valid in a specific organism under identical exposure conditions. A toxicity rating of the lethal dose of a substance for a normal man (body weight 150 pounds or 70 kg) is presented in Table 50-1. In the common vernacular, *toxic substance* refers to substances with a toxicity defined in Table 50-1 as "extremely" or "super" toxic (arsenic, cyanide, strychnine).

Important parameters in toxicology include the amount of an agent introduced into a person (that is, a dose), the route of administration, the number of doses, and the time period over which the agent is administered. Time is an essential quantity, since the effect of many agents is time dependent. For convenience, terms such as *acute toxicity* and *chronic toxicity* are used. To some extent, these terms overlap. In general, the toxic reactions observed in the emergency room are caused by the acute ingestion, inhalation, or injection of some agent, whereas the reactions caused by environmentally toxic agents are more often observed in the chronic care setting. Thus toxic agents can have an acute effect when the action of the chemical is observed from seconds to days after dosage, or a chronic effect when the time scheme is in weeks, years, or decades.

Table 50-1 Toxicity rating chart

Toxicity rating or class	Probable lethal dose (human)	
	Amount	**For 70 kg (150 pound) man**
6 Super toxic	Less than 5 mg/kg	A taste (<7 drops)
5 Extremely toxic	5-50 mg/kg	Between 7 drops and 1 teaspoonful
4 Very toxic	50-500 mg/kg	Between 1 teaspoonful and 1 ounce
3 Moderately toxic	500-5 g/kg	Between 1 ounce and 1 pint (or 1 pound)
2 Slightly toxic	5-15 g/kg	Between 1 pint and 1 quart
1 Practically non-toxic	>15 g/kg	More than 1 quart

A dose-response relationship occurs when an increased toxic effect is observed as the dose is increased. However, the increased toxic effect may not be observed in all persons at the same dose. Only a portion of the population may be affected at a given dose, but as increasing amounts of the agent are given, the dose becomes great enough so the entire population is affected.

Fig. 50-1 illustrates the overlap between therapeutic and toxic effects on a population. The horizontal (logarithmic) axis depicts a drug dosage of 1 to 10 U, while the vertical

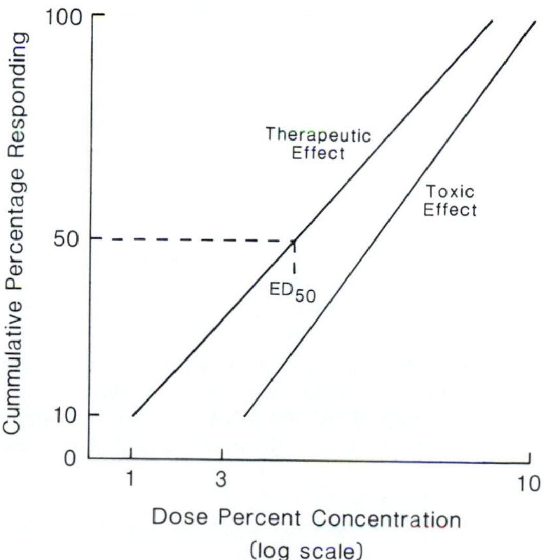

Fig. 50-1 Results of therapeutic and toxic effects on a population. ED_{50}, Effective dose for 50% of population.

axis shows the cumulative percentage of treated individuals who have a response, that is, the percent of the total population that is affected. The curve termed *therapeutic effect* applies to the designated drug effect, for example, prevention of seizures in the case of phenytoin. A point on the curve at which 50% of the people are effectively treated is the ED_{50}. The *toxic effect* curve describes an unwanted effect, such as nausea, dizziness, or in extreme cases, death. It is important to note that in this example, and indeed in many cases, the therapeutic and toxic dose ranges overlap. In some individuals a high dose is needed for treatment, whereas in other cases this same dosage can result in a toxic effect. Thus the therapeutic range usually describes the range of doses in which the desired effect occurs. However, in this range a few patients will not benefit from the drug, whereas others will have a toxic reaction. The reasons for such individual variability in response are many, including age, sex, race, health status, and genetic factors. All contribute to the distribution of the effect of an agent on a population.

One important quantitative parameter used in toxicology is the dose necessary to cause the death of an animal. If increasing doses are given, an increasing percentage of animals will die. The lethal dose for 50% of the population, termed the LD_{50}, is an important value that is used for many drugs or agents. For humans, such values are extrapolated from survivors or deaths reported in the literature. The agent does not have to affect the entire population. If only 1% of the animals died at a given dose, the term would become LD_1, the dose lethal for 1% of the animals. If 99% died, the term would then be LD_{99}.

MOLECULAR MECHANISMS OF TOXICITY*

- Interference with enzyme action and systems
  - Irreversible enzyme inhibition (organophosphates)
  - Reversible enzyme inhibition (atropine)
  - Uncoupling of oxidative phosphorylation (dinitrophenol)
  - Synthesis of lethal compounds (fluoroacetic acid)
  - Chelation of metals required for enzyme activity (dithiocarbamates)
- Blockage of oxygen usage and transport
  - Inhibition of cytochrome oxidase (cyanide)
  - Inhibition of oxygen transport (carbon monoxide)
  - Red cell lysis (arsine gas)
- Interference with cell function
  - Interaction with lipid phase of the cell membranes affecting its reaction to a depolarizing pulse (anesthetics)
  - Interference with DNA and RNA synthesis (chemotherapeutic agents such as 5-fluorouracil, and many carcinogens)
- Hypersensitivity reactions
  - Immune reaction to specific chemicals (diisocyanates)

*Compounds in parentheses are examples of each class of toxic agent.

MECHANISMS OF TOXICITY

There are many ways in which toxic agents can cause their effects. They can be subtle, such as *promoter* molecules, which affect cells only if a carcinogen has been given, or dramatic, such as that of a caustic agent, which destroys tissue on contact. The examples listed in the box at lower left illustrate only a few of the ways that toxic agents act.

FACTORS THAT INFLUENCE TOXICITY

Nature of toxicant

Solubility properties. One of the most important properties of a toxicant is its solubility in various tissues of the body. For example, toxicants that are water soluble are less likely to penetrate the skin. Those that are lipid soluble tend to be more readily absorbed; however, the ionic form of the chemical and its molecular size are also important chemical properties affecting the solubility.

Physical properties. One important property of a toxicant is the physical state of the compound, whether gas, liquid, or solid. Gases are readily inhaled and pass immediately into circulation. In addition, they pass through skin. In contrast, liquids are not inhaled unless in a vaporous state or aerosol form. Liquids can, however, pass through the epidermal barrier. Although particulates can be inhaled, they are not absorbed quickly through the skin. Solids usually must be ingested or dissolved before they have an effect. Compounds (for example, mercury) with a liquid form that have a significant gas phase under exposure conditions can have an adequate amount of the compound in the gas phase to induce toxicity by vapor inhalation.

Exposure variables

Dose dependency. For common pharmacological agents or drugs, increased dosage results in an increase in the percentage of a population that will be affected and also an increase in the severity of the toxicological reaction. In contrast, continued exposure in very low concentrations can be noncumulative and relatively innocuous. Such an effect of concentration is observed with the drug acetaminophen. At low concentrations this analgesic is commonly used to treat pain and is metabolized in the liver to nontoxic forms by a pathway involving glutathione. At high concentrations of acetaminophen, this pathway is depleted of glutathione, and free radical damage occurs. This can irreversibly damage liver cells, causing cell death.

Also keep in mind that the symptoms of toxicological response may not be obviously related to the clinical response to the drug. In the case of the tricyclic antidepressants, the desired effect is a mood elevation resulting from drug action on central nervous system receptors. In an overdose the drug may affect similar receptors in the heart, causing arrhythmia. As the amount of the tricyclic given increases, so do the arrhythmias that can cause heart failure and death.

Route, rate, and site of administration. The route of administration can be percutaneous, inhalation, oral, rectal, subcutaneous, or intravenous. Of these routes, the percutaneous route is usually not considered in a hospital environment because most agents are given by the other routes. However, the skin has a very large surface and is permeable to many chemicals. Some new systems that deliver medications using transcutaneous membranes are increasing the availability of the percutaneous absorption route.

Variation in the amounts of an agent causing death based on the route of administration can be quite large. For example, a 20-fold difference between the effect of procainamide given intravenously versus subcutaneously has been demonstrated. For other compounds, such as pentobarbital, there is less than a twofold difference in dosage between these two routes to achieve the same lethal effect. Thus route of administration can lead to considerable differences in the dose-response curve to a toxic agent (see Chapter 54).

Duration and frequency of exposure. Duration of exposure is the time in seconds, minutes, hours, days, months, or years that a person is exposed to a toxicant. For some agents, such as carbon monoxide, the duration can be a few seconds or minutes, such as an exposure in a room with a high concentration of carbon monoxide, or chronic as in cigarette smoking. One calculation often used is exposure time × dosage. The derived value is used to compare persons with different levels of toxicant, such as the blood lead levels of workers, and toxicological symptoms, such as renal disease. Frequent exposure in many cases leads to an accumulation of toxic effects, but this will vary, depending on the agent. An unusual case is that of a delayed hypersensitivity reaction, in which a minute amount of a substance can result in a toxic anaphylactic reaction, even though the amount may be subtoxic for the same person under presensitizing conditions.

Biological variables

Age affects the rate of metabolism of compounds. In general, younger subjects will metabolize a drug more rapidly than older ones. Older patients are often given the same dosage as younger ones; however, because their metabolism may be one half to one third as rapid as that of younger patients, the drug may build up to toxic levels. Part of the age difference is attributable to changes in organ function. For example, renal clearance in an aged population is 20% to 50% of that of young medical technology students.

Difference in genetic makeup can be one of the most dramatic factors influencing the toxic effect of an agent. Succinyl choline, a muscle relaxant, is hydrolyzed to succinic acid and choline. However, certain persons (1 in 3000) either do not have the enzyme that will hydrolyze this compound or have a defective form of the enzyme. Thus they have prolonged muscle relaxation and possible apnea (stoppage of breathing) and therefore are at risk when given this drug. It is possible to screen a population for this deficiency by measuring acetyl cholinesterase levels in the presence of the inhibitor dibucaine and thus find those persons with this enzyme deficiency.

Another example of genetic differences in drug metabolism involves the ability to *N*-acetylate certain compounds. One drug, procainamide, is *N*-acetylated as part of the normal metabolic detoxification pathway. It has been found that there are two groups of persons, those who are fast acetylators and those who are slow acetylators. The slow acetylators who form *N*-acetyl procainamide at a less rapid rate are more likely to have an immunological reaction to the drug. However, the *N*-acetylated drug is cleared more slowly, thus leading to higher blood levels at a given dosage for fast acetylators. Thus drug levels are dependent on the *N*-acetylation rate.

Toxicokinetics

The principles of toxicokinetics are similar to those of pharmacokinetics discussed in Chapter 54 on therapeutic drug monitoring. Most toxicants, such as gases or drugs, must pass through membranes or be absorbed and then pass through the circulation before they exert their effect on a particular target tissue. In addition, many toxicants must be transformed by metabolic processes before they can exert action on receptor or cell. Finally, most drugs or compounds must be metabolized before they are excreted. In general one can use the same equations presented for therapeutic monitoring to follow these agents. Other toxicants, such as caustic agents (lye, acid, corrosives), have a direct tissue-damaging effect and thus are not studied in this manner.

Biotransformation of xenobiotics

Most chemicals or agents that enter the body (xenobiotics) undergo a change in structure termed *biotransformation*. There are numerous pathways for this to occur. A list of such reactions is presented in the box on p. 744. Biotransformed molecules can have properties that make them biologically less active than, more active than, or as reactive as the parent compound. Atropine is hydrolyzed to the inactive moieties tropic acid and tropine. On the other hand, cyclophosphamide, an anticancer alkylating agent, must be activated by liver microsomes to be effective. Finally, the drug amitriptyline is demethylated to form nortriptyline, which is equipotent as a tricyclic antidepressant.

The microsomes of the liver contain a group of enzymes that oxidize various chemicals such as steroids, fatty acids,

BIOTRANSFORMATION REACTIONS OF XENOBIOTICS*

Azo reduction (azobenzene → aniline)
Nitro reduction (nitrobenzine → aniline)
Hydrolysis of esters (procaine → 2-diethylamino ethanol and *p*-aminobenzoic acid)
Acylation (procainamide → *N*-acetylprocainamide)
Conjugation with glucuronic acid (morphine → morphine glucuronide)
Conjugation with mercaptic acid (dichlorobenzene → dichlorobenzene mercaptic acid)
Oxidative deamination (amphetamine → phenylacetone)
Dealkylation (meperidine → normeperidine)
Aromatic ring oxidation (*n*-butylaniline → *n*-butyl-*p*-aminophenol)
N-oxidation (trimethylamine → trimethylamine oxide)
S-oxidation (chlorpromazine → chlorpromazine sulfoxide)
Oxidation of alcohols (benzyl alchol → benzoic acid)
Hydroxylation of alkyl side chain (*p*-nitrotoluene → *p*-nitrobenzyl alcohol)

*Examples of each reaction are given in parentheses.

drugs, pesticides, and carcinogens. These microsomal systems are termed *mixed-function oxidases*. The last oxidase in this chain of oxidases is cytochrome P-450, so termed because of its spectral absorption at 450 nm. Oxidation of a drug occurs by a reaction that involves oxygen, the drug or substrate, and the appropriate cytochrome P-450 enzyme:

$$\text{Substrate} + O_2 + \text{NADPH} + H^+ \xrightarrow{\text{Cytochrome P-450}} \text{Oxidized substrate} + H_2O + \text{NADP}^+$$

Many of the biotransformation reactions listed above are oxidations catalyzed by the microsomal cytochrome P-450 system; these include epoxidations, hydroxylations of aryl and alkyl hydrocarbons, oxidation of alcohols, removal of the alkyl group from ether-type linkages, *N*-hydroxylation, nitroxidation, *N*-dealkylations, oxidative deamination, sulfoxication, and dehalogenation. These enzyme reactions are induced by some drugs, such as phenobarbital or alcohol, and inhibited by others, such as cimetidine; thus administration of one of these drugs will change the rate of metabolism and effectively alter the steady-state level of the drug. In general, the drugs oxidized by the microsomal enzyme system are lipid soluble. After oxidation, they are more water soluble and thus more likely to be excreted.

DRUGS AND NONTHERAPEUTIC AGENTS ENCOUNTERED IN THE CLINICAL LABORATORY

One role of the clinical toxicology laboratory is to help the clinician determine whether a patient's pathological condition is the result of a toxic agent. Although a great many deaths are the result of poisonings, a far greater number of individuals are brought to the hospital with a diagnosis of potential poisoning. The physician must differentiate between a condition that is the result of poisoning and one that is not.

Table 50-2 Poisoning deaths in 1980[1]

Chemical class	Number
Carbon monoxide and other gases	3811
Analgesics, antipyretics, and so on	1148
Tranquilizers and psychotropic agents	1165
Other drugs, medicinal substances	1645
Unspecific drugs, medicinal substances	1217
Barbiturates	718
Sedatives and hypnotics	186
Other compounds	1034
TOTAL	10,924

Agents used in suicide

Table 50-2 gives the number of deaths from poisoning for the year 1980 by class of agent.[1] Most of these deaths are the result of suicides; the remainder are the result of accidental poisoning. It should be noted that a large number of these deaths are caused by gases, most frequently by carbon monoxide. Also note that most poisonings are caused by agents that are prescribed for medicinal purposes; these include tranquilizers, psychotropic agents, barbiturates, and others. The most frequent problem facing the laboratory is the determination of the agent employed in an attempted suicide. When depressed individuals are given access to potentially lethal agents, they often attempt suicide with these agents.

Drugs in emergency situations

In a hospital, particularly one involved in emergency care, the pattern of toxic drug ingestion or abuse observed in acutely ill patients is somewhat different from that seen in suicide. Table 50-3 lists the percentage of cases positive

Table 50-3 Frequency of positive results (by drug class) obtained in emergency room drug screens[2]

Drug or drug class	Frequency
Benzodiazepines	15.9
Opiates	12.8
Ethanol	12.1
Anticonvulsants	9.4
Tricyclic antidepressants	8.0
Stimulants	6.3
Phenothiazine	4.3
Antihistamines	3.8
Cardiac antiarrhythmics	3.4
Barbiturates	3.0
Hypnotics and sedatives	2.7
Other	1.0

Table 50-4 Some clinical features of severe drug intoxication encountered in overdose cases

Class	Agent	Pupils	Respiration	Other
Analgesic	Acetaminophen	—	—	Liver toxicity
Hypnotic	Barbiturates	Dilated, light reactive	↓	Flaccid paralysis, hypothermia, hypotension
Minor tranquilizer	Benzodiazepines	—	↓	CNS* depression, extrapyramidal system changes, autonomic nervous system changes, hypotension
Depressant	Ethanol	—	↓	CNS depression (in some grand mal seizures)
Analgesic	Narcotics and propoxyphene	Pinpoint	↓	CNS depression
	Salicylates	—	↓	Metabolic acidosis (later)
Antidepressant	Tricyclics	Constricted	—	Tachycardia, arrhythmia

**CNS,* Central nervous system.

for a particular class of drugs, from a survey conducted at several hospitals.[2] Note that the major drugs or drug classes observed were benzodiazepines, opiates, ethanol, anticonvulsants, and antidepressants. Again, prescription drugs make up the majority of the observed toxic agents. Most important is the observation that many patients will have ingested more than one drug. In one study about half the patients seen in an emergency room had ingested two or more drugs.[2] Drug screening tests are ordered for a variety of patients in emergency situations who may appear clinically hyperactive, hallucinating, or in coma. A description of the various grades of coma in toxic cases is given by Hanenson as follows[3]:

Coma gradation

Grade 1—Drowsy but responds to verbal commands
Grade 2—Unconscious but responds to mild painful stimulus
Grade 3—Unconscious and responds only to maximum painful stimulus
Grade 4—Unconscious without response to painful stimulus

When a comatose or bizarrely acting patient is seen in an emergency room, the possibility of a drug overdose is one of the diagnoses that must be considered by a physician. For the most efficient use of laboratory facilities, it is important for the physician to try and ascertain the nature or identity of the drug or drugs used. A physician's judgment about the presence of intoxicating or lethal drugs is based on a number of clinical observations (Table 50-4). A history obtained from an alert patient can be helpful, though in general, patient history is not a reliable index of what drugs may have been taken (less than 50% reliability).

General effects of toxic agents can include symptoms such as respiratory depression, shock, convulsions, hyperthermia, or hypothermia. Some clinical effects and symptoms observed in patients intoxicated with very high doses of these agents are summarized in Table 50-4. The size of pupils, respiration rate, cardiac status, and the condition of other organ systems are all important signs and can provide information about the nature of the toxic substance.

If the physician suspects the presence of one or more drugs, laboratory tests may be ordered to confirm the presence of the drug or drugs and to determine the amount present. These data can establish the therapy to be used. If the clinician is not certain if drugs are present, a general drug screen may be ordered. The drug screen will rapidly test for the presence of some of the major drug groups, and any positive result will be confirmed by an alternative procedure (see p. 1091). Electrolytes, BUN, and glucose may also be ordered to rule out renal or diabetic disease states.

ANTIDOTES AND TREATMENT

The history of poisons is intertwined with that of antidotes. Currently there are only a few antidotes specific for those agents commonly observed in a hospital setting. These are presented in Table 50-5. Usually most therapy is supportive, involving treatment for respiratory depression, aspiration, shock, convulsions, delirium, methemoglobinemia, hyperthermia, and hypothermia. Supportive therapy is used to maintain a patient's vital functions until the body is cleared of drug.

Removal of the toxic agent can often be effected by emesis (vomiting), gastric lavage, or skin decontamination. Other forms of treatment include agents that help remove the toxicant from the body. Dimercaprol is used to treat arsenic and mercury poisoning, and the salts of the divalent metals (lead, mercury, copper, nickel, zinc, cadmium, cobalt, beryllium, and manganese) are chelated by

Table 50-5 Specific antagonists and antidotes

Drug	Antagonist or antidote
Acetaminophen	Methionine, *N*-acetyl cysteine
Amphetamines	Chlorpromazine
Anticholinergic drugs (tricyclic antidepressants, atropine, scopolamine)	Physostigmine
Carbon monoxide	Oxygen
Cyanide	Sodium nitrite
Nitrates and nitrites	Methylene blue
Opiates	Naloxane
Organophosphates	Atropine and pralidoxime

Table 50-6 Examples of classes of drug interactions

Action	Example
Drugs with similar effects	Multiple nephrotoxic drugs such as gentamicin and cephalosporin yield increased nephrotoxicity.
Drugs with opposing effects	Hynotics and caffeine result in antagonism to the hypnotic effect.
Absorption interactions	Tetracycline and iron (Fe^{+2}) supplement result in decreased oral uptake of drug.
Drug displacement interactions	Theoretical displacement of bound drug from albumin yields increased free-drug level.
Drug metabolism interactions	
Enzyme induction	Barbiturates stimulate microsomal oxidation of drugs such as dicumarol, resulting in lower plasma levels.
Enzyme inhibition	Cimetidine blocks the P-450 oxidation pathway, slowing theophylline metabolism and resulting in a longer half-life and higher plasma levels.
Altered excretion interactions	
Changes in urine pH	Acid urine enhances basic drug excretion. The reverse is true for acidic drugs.
Competition for active tubular secretion	Probenecid decreases the active secretion of drugs such as penicillin, thus increasing serum levels.
Interactions at adrenergic neurons	Tricyclic antidepressants prevent the uptake of guanethidine into neurons, thus blocking its antihypertensive effect.

ethylenediaminotetraacetate (EDTA). The chemical deferoxamine is used to remove excess amounts of iron and aluminum. Charcoal may be swallowed or placed in the stomach to enhance the rate of removal of drug from the body.

A number of agents are removed more rapidly if urine flow is increased (forced diuresis). This is often accomplished by the intravenous administration of a diuretic such as mannitol. In addition, because of a lower concentration in the kidney under diuretic conditions, the potential for nephrotoxicity of some agents (such as cisplatin) is diminished. The pH of urine is also important. For basic drugs such as the amphetamines, excretion is enhanced by an acid urine, in which the drugs are ionized. Barbiturate excretion is increased by alkaline urine, in which the molecule is ionized and therefore more soluble.

DRUG INTERACTIONS

When two drugs are given to a patient, their actions may not be independent; that is, one drug may affect the pharmacological action of the other. This phenomenon is called *drug interaction*. The effects of drug interactions are not obvious, and much information about these phenomena has been gathered from case reports. Many types of interactions have been noted, and it is not possible to provide a comprehensive list here. Stockly[4] has listed the types of interactions, though he notes that "interactions which occur when drugs are given concurrently are often the result of not a single mechanism, but of two or more mechanisms acting in concert. . . ." Table 50-6 gives this listing with an example of each type of interaction.

Important drug interactions are observed between ethanol and other agents that affect the central nervous system. Ethanol is frequently encountered in drug interactions because of its wide usage and because ethanol has a wide range of depressant effects on the central nervous system. For example, the concomitant use of the barbiturates and alcohol can have deleterious effects. The lethal dose of barbiturates is almost 50% lower when combined with alcohol. This same effect occurs with concomitant use of alcohol and agents such as chloral hydrate, paraldehyde, glutethimide, meprobamate, and other tranquilizers.

MEDICOLEGAL ASPECTS

Technologists in the toxicology laboratory may become involved in the medicolegal system because of their need to document under oath the validity of analytical results. The technologist may also have to document the laboratory's procedures and produce a *chain of custody* document (see Chapter 2) to ensure that the specimen was not tampered with.

For some types of testing, such as screening for the presence of abused drugs, the methods of screening by immunoassay and confirmation by gas chromatography or mass spectroscopy may be specified by a federal agency such as the National Institute for Drug Abuse. The current testing for abused substances (alcohol, marijuana, cocaine, heroin, amphetamines) is based on the assumption that the use of certain drugs, particularly cocaine and marijuana, may be illegal, and that individuals will have impaired performance, affecting their job or resulting in an accident. In either case, litigation involving a civil or a criminal suit may result. The laboratory's role is to supply and document legally admissible and defendable data.

REFERENCES

1. Brancato, DJ, and Nelson, RC: Poisoning mortality in the United States 1980, Vet Hum Toxicol 26:273-275, 1984.
2. Merigian, KS, Schroeder, TJ, Tasset, JJ and Pesce, A: Toxicology screening pattern in Hamilton County Ohio: review of 1710 comprehensive drug screens in 5 area hospitals, J Clin Lab Anal 2:112-116, 1988.

3. Hanenson, IB, editor: Quick reference to clinical toxicology, Philadelphia, 1980, JB Lippincott Co.
4. Stockley, I: Drug interactions: a source book of adverse interactions, their clinical importance, mechanisms and management, Oxford, 1981, Blackwell Scientific Publications, Ltd.

BIBLIOGRAPHY

Bowman, WC, and Rand, MJ: Textbook of pharmacology, ed 2, Oxford, 1980, Blackwell Scientific Publications, Ltd.

Casarett and Doull's Toxicology: the basic science of poisons, New York, 1986, Macmillan Publishing Co, Inc.

Dreisbach, RH, and Robertson, WO: Handbook of poisoning, ed 12, Los Altos, Calif, 1987, Appleton & Lange.

Gilman, AG, Goodman, LS, and Gilman, A, editors: Goodman and Gilman's pharmacological basis of therapeutics, New York, 1980, Macmillan Publishing Co, Inc.

Lave, LH and Upton, AC, editors: Toxic chemicals, health, and the environment, Baltimore, 1987, Johns Hopkins University Press.

Loomis, TA: Essentials of toxicology, ed 3, Philadelphia, 1978, Lea & Febiger.

Stockley, I: Drug interactions: a source book of adverse interactions, their clinical importance, mechanisms and management, Oxford, 1981, Blackwell Scientific Publications, Ltd.

SECTION THREE

Methods of Analysis

CHAPTER 51

Classifications and description of proteins, lipids, and carbohydrates

KEY TERMS

aldose The chemical form of monosaccharides in which the carbonyl group is an aldehyde.

apoprotein Polypeptide chain not yet complexed to its specific prosthetic group.

carbohydrates Chemicals with the general formula of hydrated carbon, $(CH_2O)_n$, that are an aldehyde or ketone derivative of polyhydric alcohols. Commonly called *sugars*.

compound (conjugated) proteins Polypeptide chain complexed with other chemical classes such as lipids (lipoproteins), carbohydrates (glycoproteins), or nucleic acids (nucleoproteins).

conjugated lipids Esters of fatty acids and alcohols containing additional chemical moieties. Group includes phospholipids, sphingolipids, sterols, bile acids, and so on.

denaturation Unfolding the tertiary structure of a protein that renders it insoluble, causing it to precipitate out of solution.

derived lipids Lipids derived from the hydrolysis of simple and conjugated fats; these include the fatty acids.

disaccharide Two monosaccharides linked together by a $1 \rightarrow 4$ or $1 \rightarrow 6$ glycosidic linkage, such as sucrose, maltose, or lactose.

furanose Five-membered rings of monosaccharides formed by intramolecular reaction between the carbonyl group and a hydroxyl group; present in alpha or beta stereoisomeric forms.

isoelectric point pH at which a molecule containing many ionizable groups is electrically neutral; that is, the number of positively charged groups equals the number of negatively charged groups.

ketose The chemical form of a monosaccharide in which the carbonyl group is a ketone.

monosaccharide Basic monomeric carbohydrate unit in which n in the formula $(CH_2O)_n$ ranges from 3 to 8.

polypeptide bond The covalent amide bond between a primary amino group of one amino acid and the carboxylic acid group of a second amino acid.

polysaccharide Polymer usually containing more than 10 monosaccharides linked by glycosidic bonds; branched and unbranched chains up to many millions of molecular weight can be formed, as in cellulose, starch, or glycogen.

primary structure The linear sequence of amino acids in a protein, defined by the genetic code in cellular DNA.

prosthetic group A nonprotein chemical group that is bound to a protein and is responsible for the biological activity of the protein. The functional complex between protein and a prosthetic group is called a *holoprotein,* and the protein without the prosthetic group is called an *apoprotein*.

pyranose Six-membered rings of monosaccharides formed by intramolecular reaction between a carbonyl group and a hydroxylin group, present in an alpha or beta stereoisomeric form.

quaternary structure The three-dimensional spatial arrangement of polypeptide chains resulting from the combining of more than one polypeptide chain into a larger, stable complex.

Schiff's base Covalent complex between a primary amine and carbonyl function of an aldose.

secondary structure The spatial arrangement of a linear chain of amino acids in a polypeptide; the three groups are the beta-plated sheet, alpha-helix, and random coil.

sialic acids *N*-acetyl derivatives of neuraminic acid that are covalently linked to many proteins.

simple lipids Esters of fatty acids with various alcohols, including the triglycerides and some steroids.

simple proteins Polypeptide chain consisting only of amino acid groups.

steroids Lipids containing four six-membered rings and including many hormones, vitamins, and drugs.

tertiary structure The intramolecular folding of a polypeptide chain on itself resulting from interactions between side-chain groups of individual amino acids.

zwitterion Molecule containing two ionized groups of opposite charge. (Pronounced tsvit′-er-ión.)

The analytes in this section are grouped as much as possible by common chemical class. This grouping is important because the chemical nature of an analyte is often the single most important factor in determining which type of method will be used for analysis. Other important factors for consideration are the physiological concentration of the analyte, biological variation in concentration, and the clinical situations for which the analytes are measured. These chemical classes are now reviewed, along with the type of methods used for their analysis. This chapter is not intended to provide a complete biochemical review of the analytes measured in the chemistry laboratory. For this, readers are encouraged to use the excellent biochemistry texts listed in the bibliography. Instead, an emphasis has been placed on those properties of proteins, lipids, and carbohydrates that affect how the analytes may be measured.

Part I: Proteins

LAWRENCE A. KAPLAN

DEFINITION AND CLASSIFICATION

Proteins are linear polymers of alpha-amino acids. There are 20 natural amino acids with the general structure as shown in Fig. 51-1. These exist as the L-stereoisomeric form with the amino group placed on the alpha-carbon atom next to the carboxylic acid group. The pK_a of the carobxylic acid group is approximately 1.8 to 2.4, whereas the pK_a of the alpha-amino group is approximately 8.5 to 10.5. This means that at pH less than 2.5, the carboxylic acid will be in the nonionized form (COOH), whereas the alpha-amino group will remain ionized at pH values less than 9.5 (Fig. 51-2). At physiological pH (approximately 7.4), both groups are ionized. A compound, such as an amino acid, with two opposite charges is called a *zwitterion* (''hybrid ion'' or ''hermaphrodite ion'').

The side chain groups of the 20 amino acids are listed in Table 51-1, along with the pK_a's of all ionizable groups. These side-chain groups can interact with one another to determine the overall chemical, physical, and biological properties of the polypeptide chain.

The amino acids are covalently linked together by the protein-synthesizing machinery of cells. The actual *order* or sequence of amino acids in a protein chain is predetermined by the genetic code within the cell. The sequence of genetic information in DNA is transcribed into messenger RNA, which is translated in the cytoplasm into protein (Fig. 51-3). The specific sequence of amino acids for protein is called its *primary structure*.

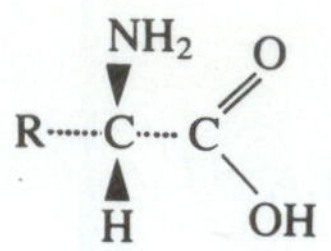

Fig. 51-1 General structure of amino acid of L-stereoisomeric form. *Heavy lines,* Bonds coming out of plane of page; *dotted lines,* bonds extending behind plane of paper.

The amino acids are linked together by the peptide bond. As shown in Fig. 51-4, this bond has a specific spatial arrangement in three-dimensional space. The linear polypeptide chain can exist in three possible conformations, alpha-helix, beta-plated sheet, and random coil (Fig. 51-5). These conformations are called the *secondary structure* of the protein.

When a polypeptide chain is in solution, it is flexible enough for the molecule to bend, allowing the side chain groups to interact with one another. The types of interactions are listed in Table 51-2. Although the interaction energy of each side group is small, the net energy of all these interactions is great enough to stabilize proteins in a folded, convoluted, three-dimensional spatial arrangement called the *tertiary structure* (Fig. 51-6). Each protein's unique tertiary structure confers on it specific biological properties.

The folded polypeptide chains are often organized as aggregates with identical or different polypeptides. The specific number and type of these polypeptide chains determines the specific properties of the entire complex. The spatial arrangement of these multichain proteins is called the *quaternary structure* of the protein. Usually the biological properties of such quaternary proteins consisting of subunit chains is the sum of each individual chain.

Proteins are generally classified into two major groups, simple and conjugated, with several subdivisions within each group. This classification scheme is based on the physical properties and chemical composition of the protein.

I. *Simple proteins*—generally not associated with other major chemical classes

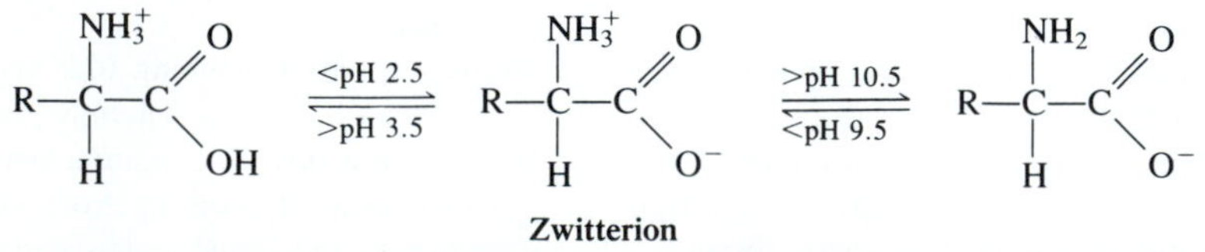

Zwitterion

Fig. 51-2 Various ionized and nonionized forms of amino acids present at various pH levels. When two opposite charges are present on same molecule, molecule is called a *zwitterion*.

Table 51-1 Classification and properties of side chain (R groups) for naturally occurring amino acids

R group (R—CH(NH$_2$)COOH)	L-Amino acid (symbol)	Amino acid molecular weight	pK$_a$* (25° C)		Secondary groups
			Primary —COOH	Primary —NH$_2$	
Nonpolar (hydrophobic)					
H–	Glycine (gly)	75.07	2.34	9.60	—
CH_3–	Alanine (ala)	89.09	2.34	9.69	—
$(CH_3)_2CH$–	Valine (val)	117.15	2.32	9.62	—
$(H_3C)_2CH–CH_2$–	Leucine (leu)	131.18	2.36	9.60	—
$CH_3CH_2–CH(CH_3)$–	Isoleucine (ile)	131.18	2.36	9.68	—
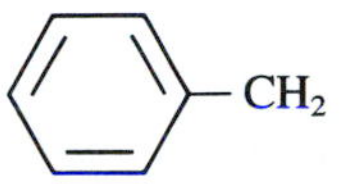 C_6H_5–CH_2	Phenylalanine (phe)	165.19	1.83	9.13	—
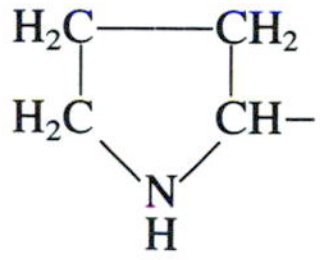 $H_2C–CH_2$, H_2C CH–, N, H (pyrrolidine ring)	Proline (pro)	115.13	1.99	10.60	—
$CH_3–S–CH_2CH_2$–	Methionine (met)	149.21	2.28	9.21	—
Neutral polar (hydrophilic)					
$OHCH_2$–	Serine (ser)	105.09	2.21	9.15	—
$CH_3CH(OH)$–	Threonine (thr)	119.12	—	—	—
$NH_2–C(=O)CH_2$	Asparagine (asp)	132.12	2.02	8.80	—
$NH_2–C(=O)CH_2CH_2$–	Glutamine (gln)	146.15	2.17	9.13	—
$HSCH_2$–	Cysteine (cys)	121.16	1.96 (30°)	10.28	8.18 (SH)
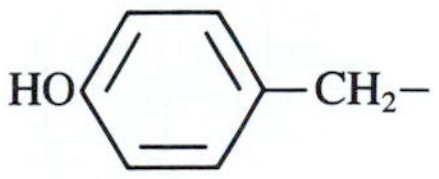 HO–C_6H_4–CH_2–	Tyrosine (tyr)	181.19	2.20	9.11	10.07 (OH)
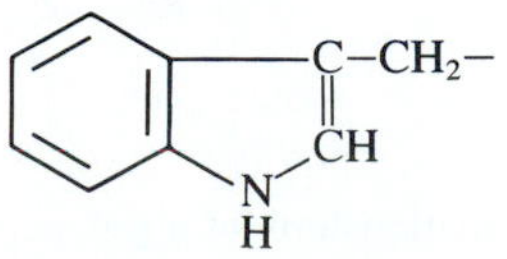 (indole ring) C–CH_2–, CH, N, H	Tryptophan (trp)	204.23	2.38	9.39	—

From Cohn, EJ, and Edsall, JT: Proteins, amino acids and peptides, New York, 1943, Reinhold Co.
*The pK$_a$ values will be slightly different in a protein molecule.

Continued.

Table 51-1 Classification and properties of side chain (R groups) for naturally occurring amino acids—cont'd

R group ($R—CH(NH_2)COOH$)	L-Amino acid (symbol)	Amino acid molecular weight	pK$_a$* (25° C) Primary —COOH	Primary $—NH_2$	Secondary groups
Acidic polar (hydrophilic)					
$HOOCCH_2-$	Aspartic acid (asp)	133.10	1.88	9.60	3.65 (COOH)
$HOOCCH_2CH_2-$	Glutamic acid (glu)	147.13	2.19	9.67	4.25 (COOH)
Basic polar (hydrophilic)					
$H_2NCH_2CH_2CH_2CH_2-$	Lysine (lys)	146.19	2.18	8.95	10.53 (E—NH_3)
$H_2N-C(=NH)-N(H)-CH_2CH_2CH_2-$	Arginine (arg)	174.20	2.17	9.04	12.48 (guanidinium)
HC═C–CH_2– (imidazole ring: N, NH, C–H)	Histidine (his)	155.16	1.82	9.17	6.00 (imidazolium)

A. Globular proteins—relatively symmetric, water-soluble or saline-soluble proteins
   1. Albumin—major serum protein
   2. Globulins—most other serum proteins
   3. Histones—basic proteins, found associated with nucleic acids
   4. Protamines—strongly basic proteins found associated with nucleic acids

B. Fibrous proteins—asymmetric proteins that are insoluble in water or dilute salts; highly resistant to most proteolytic enzymes
   1. Collagens—major proteins of connective tissue, high in hydroxyproline content
   2. Elastins—found in elastic tissue such as tendon and arteries
   3. Keratins—major proteins in animal hair, nails, hooves, and so on

II. *Conjugated proteins*—combined with other non–amino acid biochemicals; considered to consist of two components—the protein, called the *apoprotein,* and the nonprotein *prosthetic* group (The ability of the prosthetic group and apoprotein to dissociate varies from group to group.)

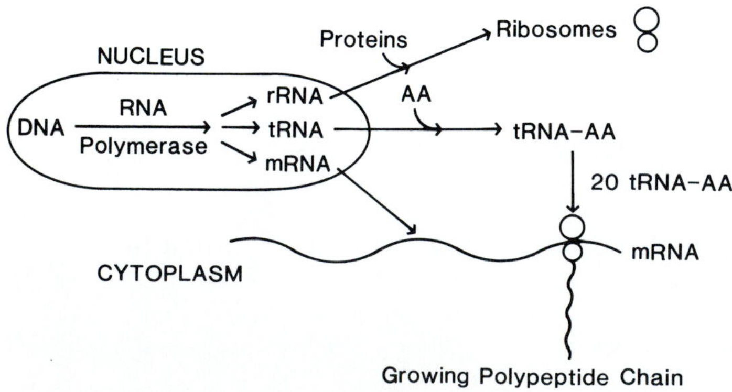

Fig. 51-3 Scheme of synthesis of proteins. *AA,* Amino acid; *DNA,* deoxyribonucleic acid; *tRNA,* transfer ribonucleic acid; *mRNA,* messenger ribonucleic acid; *rRNA,* ribosomal ribonucleic acid (18S and 28S forms); *tRNA-AA,* activated amino acid covalently bound to amino acid–specific tRNA.

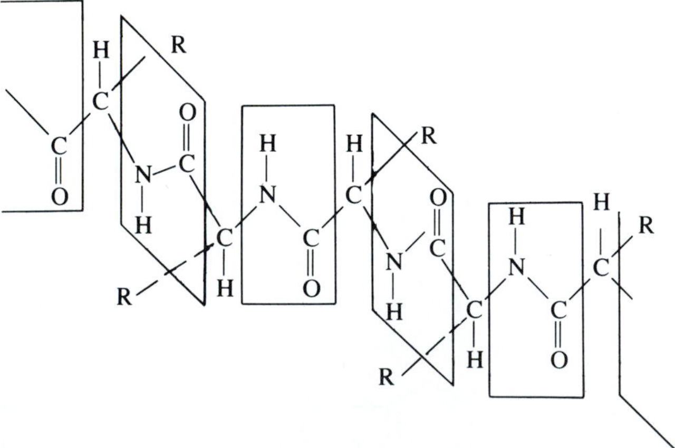

Fig. 51-4 Spatial relationships of a polypeptide bond. *C,* Carbon atom; *N,* nitrogen atom; *H,* hydrogen atom; *O,* oxygen atom. *(From Orten, JM, and Neuhaus, OW: Human biochemistry, ed 10, St Louis, 1982, The CV Mosby Co.)*

Table 51-2 Types of intramolecular side-chain interactions of protein R-groups

Type of bond	Schematic*
Covalent	
Disulfide (cystine)	—S—S—
Lysinonorleucine (in collagen)	—$(CH_2)_3$—CH_2—N(H)—$(CH_2)_4$—
Noncovalent	
Electrostatic	C(=O)—$O^{\ominus}$ $^{\oplus}NH_3$
Hydrogen	O—H···O=C—O^-
Hydrophobic	
Van der Waals	CH_2OH CH_2OH

**Wavy line,* Polypeptide chain.

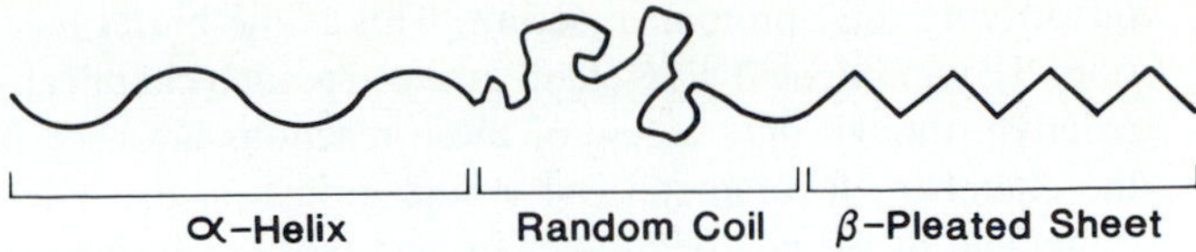

Fig. 51-5 Scheme of three possible polypeptide chain conformations defining secondary structure of proteins.

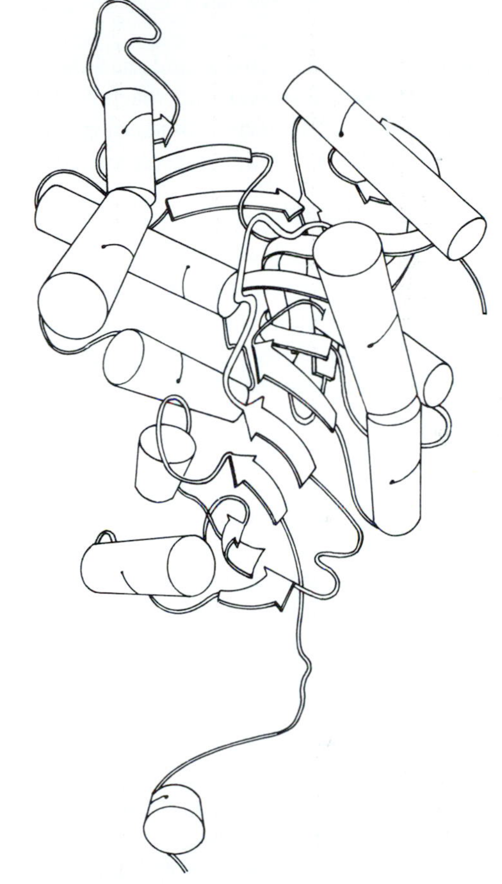

Fig. 51-6 Scheme of tertiary structure of a protein. *Cylinders,* α-helix; *flat arrows,* β-plated sheet; *lines,* random coil secondary structure of lactate dehydrogenase. *(From Orten, JM, and Neuhaus, OW: Human biochemistry, ed 10, St Louis, 1982, The CV Mosby Co.)*

A. Nucleoproteins—the prosthetic groups are the nucleic acids (DNA or RNA)
B. Mucoproteins—in which large amounts (more than 4% by weight) of complex carbohydrates are covalently linked to the protein
C. Glycoproteins—also contain covalently linked carbohydrate residues but usually less than 4% by weight.
D. Lipoproteins—contain cholesterol, triglycerides, and phospholipids associated with highly water-insoluble proteins.
E. Metalloproteins—include proteins that contain metals strongly bound to the protein, either as the ion, or as complex metals such as the flavoproteins and hemoproteins
F. Phosphoproteins—contain high concentrations of phosphate groups covalently linked to protein

The biological functions of the proteins are extraordinarily varied, but many functions (see p. 756) are specific for only one or two of these classes of proteins.

CHEMICAL PROPERTIES

The chemical properties of proteins are based on the sum of their parts, that is, the constituent amino acids and prosthetic groups. The peptide bond is chemically reactive and is the basis of the most popular, specific method for quantifying total protein in serum. This is the biuret reaction. The amino acid side-chain groups are also chemically reactive, though only a few of these reactions are used in the chemistry laboratory. The amino groups at the *N*-terminal end of the polypeptide chain and those of lysine and the guanidino groups of arginine can react with several compounds to produce an intense fluorescence. These same groups can react with ninhydrin to give a blue color. Both of these reactions have been used to quantify total protein.

The phenolic group of tyrosine and the indole group of tryptophan react with the oxidizing reagent of the Folin-Wu or Lowry reactions to form a blue color. This method is employed with dilute solutions or microanalysis.

PHYSICAL PROPERTIES

The aromatic amino acids (tryptophan, phenylalanine, and tyrosine) give most proteins an absorption spectrum with the unique absorption maximum at 278 to 280 nm. The absorption at 280 nm is used to estimate the concentration of proteins in solution. In addition, complex proteins such as hemoglobin, which have prosthetic groups with unique absorption properties, can be individually quantitated without extensive purification on the basis of their specific absorption spectra. The Soret absorption band of hemoglobin at 415 nm is used extensively to quantitate the concentration of hemoglobins.

Polypeptide chains vary widely in molecular size. The smallest polypeptides, such as the endorphins or the hypothalamic hormones, contain 5 to 25 amino acids, whereas the largest proteins, containing several subunits, can have molecular weights in the millions of daltons. Although separation of protein on the basis of different molecular weights can be done, it is a method rarely used in the chemistry laboratory.

The density of most proteins falls within a fairly narrow range of about 1.33 g/mL. Lipoproteins represent an important exception because the lipid content of these proteins gives them an unusually low density, allowing the various classes of lipoproteins to be separated from each other and from other proteins on the basis of their density. This technique is used mainly in specialized laboratories.

An important physical property of proteins is the net charge on the protein molecule. The net charge on a protein is the sum of all ionic charges of the amino acids and the carbohydrate and the prosthetic groups of the protein. Since the various chemical groups are ionized at different pH levels, the net charge of a protein varies with pH. The pH at which a protein carries no net charge is called the *pI*, or *isoelectric point;* the isoelectric point of a protein is the point at which the number of positively charged groups equals the number of negatively charged groups. At a pH greater than the pI, the protein will be negatively charged, whereas at a pH less than the pI, it will be more positively charged. At physiological pH most serum proteins are negatively charged.

Since proteins differ in the number and type of constituent amino acids, they also differ in their pIs. Therefore, at different pH levels, proteins will carry different net charges. This difference in net charge is the basis of many procedures for separating and quantifying classes of proteins or individual proteins. The most common procedures are electrophoresis, ion-exchange chromatography, and isoelectric focusing. After separation, individual proteins are detected by use of specific stains.

Proteins found in body fluids are readily water soluble but can become insoluble in the presence of a wide range of denaturing or precipitating agents. These include organic solvents (such as acetone and acetonitrile), heavy metals (such as tungstic acid), certain salts (such as zinc hydroxide and ammonium sulfate), and strong acids (such as sulfosalicylic acid, trichloroacetic acid, and mineral acids). The ability to precipitate proteins from solution by the use of one or more of these chemicals is the basis for a few routine clinical analyses. Cerebrospinal fluid and urine protein measurements are commonly performed by turbidimetric analysis. In addition, protein precipitation steps are often included as part of purification schemes for analytes before analysis.

BIOLOGICAL PROPERTIES

All proteins fulfill some physiological or biological function. The known functions of proteins cover a wide range of activities and are listed in Table 51-3. Often the known biological property of a protein is the basis of a method for its detection and quantification.

Of the important physiological functions of proteins, the most important to the clinical chemistry laboratory are the transport, receptor, and catalytic functions. Many serum proteins function as specific transporters of small molecules. Most transport proteins are globular proteins. Examples are thyroid-binding globulin (TBG), which binds thyroxine; transcortin, which binds cortisol; and albumin, which transports free fatty acids, unconjugated bilirubin, calcium, and many other endogenous and exogenous com-

Table 51-3 Biological functions of proteins

Function	Example
Transport of small molecules	Transcortin (cortisol), thyroxine (TBG)
Receptors	Estriol receptors (cytoplasmic), insulin receptors (surface)
Catalytic	All enzymes
Structural	Collagen
Nutritional (source of calories and amino acids)	Albumin
Oncotic pressure	Albumin
Host defense versus foreign antigens	Antibodies (all classes)
Hormonal	Thyroid-stimulating hormone (TSH)

pounds. The specific binding properties of these transport proteins have been used as the basis for procedures to measure the serum concentrations of cortisol, TBG saturation, and other analytes. The lipoproteins function as transporters of lipids in serum (see Chapter 30).

Many cellular proteins act as intermediary information processors for hormone molecules. Each protein, called a receptor, binds a specific hormone and then acts to transmit the hormonal message to the cell. Receptor proteins are usually glycoproteins. Assays for specific receptors, such as estriol and progesterone receptors, are valuable for the management of certain types of cancers.

An important biological property of some proteins is their ability to catalyze biochemical reactions. The serum concentrations of these proteins, called *enzymes,* are often important in determining the nature of a disease process. Most proteins exhibiting catalytic properties are globular or metalloproteins. An enzyme is most often measured by monitoring the specific biochemical reaction it catalyzes. The conditions of the enzyme assay are carefully defined so as to give maximum enzymatic activity and therefore sensitivity of analysis (see Chapter 52).

One of the most important biological properties of proteins is their ability to act as antigens (see Chapter 9). An antigen inserted into an immunologically competent host will stimulate the synthesis of antibodies by the immune system of the host. Antibodies are also globular proteins. An antibody raised against a specific antigen will be able to bind specifically to that antigen. This antibody-antigen interaction is the basis of many assays for the sensitive and specific measurement of proteins that are not able to be detected by other means, and it is used to increase both the sensitivity and specificity of older assays.

Proteins play a major role both intracellularly and in tissues. For example, connective tissue is composed primarily of collagen and mucoproteins. The proteins forming the cytoplasmic endoskeleton also fall into this group.

Part II: Lipids

HERBERT K. NAITO

DEFINITION AND CLASSIFICATION

Lipids (fats) comprise a wide spectrum of organic compounds that differ greatly in their chemical and physical properties and in their physiological roles. They include a variety of substances, such as fatty acids, sterols, triacylglycerides (more commonly called *triglycerides*), phosphorus-containing compounds (phospholipids), fat-soluble vitamins, bile acids, waxes, and other complex fats. As a consequence, it is difficult to provide a uniform and clear-cut definition of lipids that is broad enough to encompass all these diverse compounds. In general, however, one can say that lipids are substances that are insoluble in water but soluble in organic solvents such as alcohol, chloroform, ether, acetone, hexane, and benzene. Even with this general definition, there are some exceptions, such as phospholipids, which are rather insoluble in acetone. In addition, some phospholipids, such as phosphatidyl serine, phosphatidyl inositol, and phosphatidyl ethanolamine, have a limited but significant ability to dissolve in water.

There is no generally agreed on system for the classification of lipids, but for simplicity, the following commonly used classification of lipids can be used.

Simple lipids

Simple lipids are esters of fatty acids with various alcohols.

Neutral fats. Neutral fats are esters of fatty acids and glycerol (triglycerides). Because they are uncharged, cholesterol and cholesterol esters are also termed *neutral lipids*. However, structurally they are steroids and not neutral fats.

The neutral fats contain mixtures of triglycerides—esters of glycerol and fatty acids (such as stearic, palmitic, or oleic acid). The general formula for such a fat is:

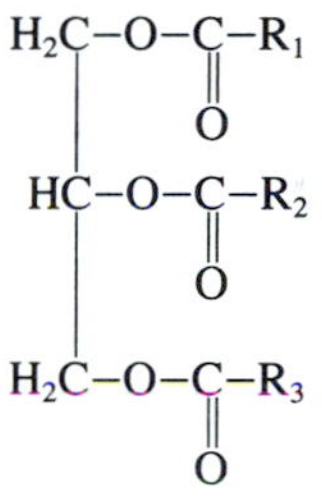

$$\begin{array}{l} H_2C-O-\underset{\underset{O}{\|}}{C}-R_1 \\ \;| \\ HC-O-\underset{\underset{O}{\|}}{C}-R_2 \\ \;| \\ H_2C-O-\underset{\underset{O}{\|}}{C}-R_3 \end{array}$$

If $R_1 = R_2 = R_3$ (where R = fatty acid), one has a simple triglyceride. If the Rs are not the same, one has a mixed triglyceride. Naturally ocurring fats usually exist as mixtures of mixed triglycerides.

The fats then are triesters of the trihydric alcohol (glycerol) and of certain, but not all, organic acids. Since all three glycerol alcohol radicals are esterified, they are termed *triacylglycerides* or more commonly called *triglycerides*. A simple ester would be formed by the combination of an acid and an alcohol:

$$CH_3COOH + C_2H_5OH \rightarrow CH_3COOC_2H_5 + H_2O$$

A fat is formed by the combination of a fatty acid (usually of relatively high molecular weight) with the alcohol glycerol.

Being esters, the fats are readily hydrolyzed:

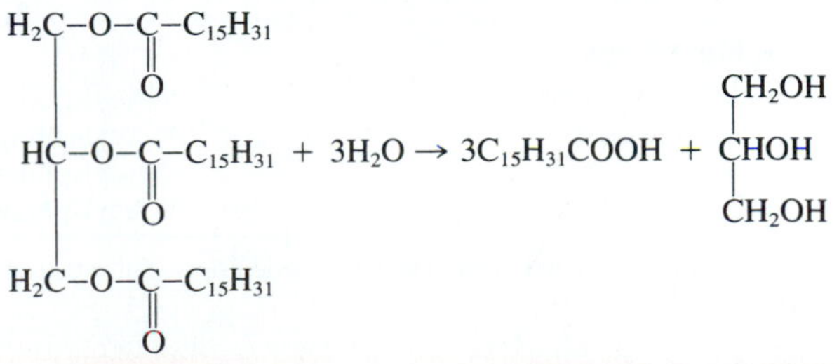

$$\underset{\text{Tripalmitin}}{\begin{array}{l} H_2C-O-\underset{\underset{O}{\|}}{C}-C_{15}H_{31} \\ \;| \\ HC-O-\underset{\underset{O}{\|}}{C}-C_{15}H_{31} \\ \;| \\ H_2C-O-\underset{\underset{O}{\|}}{C}-C_{15}H_{31} \end{array}} + 3H_2O \rightarrow \underset{\text{Palmitic acid}}{3C_{15}H_{31}COOH} + \underset{\text{Glycerol}}{\begin{array}{l} CH_2OH \\ \;| \\ CHOH \\ \;| \\ CH_2OH \end{array}}$$

This hydrolysis is accomplished by use of acid, alkali, superheated steam, or an appropriate enzyme (such as pancreatic lipase). In acid hydrolysis, the free fatty acid is liberated. When alkali is used, a soap is formed, and the process is called *saponification:*

$$\underset{\textbf{Stearin}}{C_3H_5(O{-}CO{-}C_{17}H_{35})_3} + 3NaOH \rightarrow \underset{\substack{\textbf{Sodium stearate}\\ \textbf{(a soap)}}}{3C_{17}H_{35}COONa} + \underset{\textbf{Glycerol}}{C_3H_5(OH)_3}$$

The fats we eat are mostly triglycerides that contain even-numbered fatty acids because of their mode of biosynthesis. These range from butyric (C_4) to lignoceric (C_{24}) and probably higher fatty acids (see Table 51-5). Odd-numbered fatty acids do occur naturally.

Waxes. Waxes are esters of fatty acids with higher molecular weight alcohols than glycerol. Examples are carnauba wax, wool wax, beeswax, and sperm oil. Industrially, they are used in the manufacture of lubricants (sperm oil), polishes (carnauba wax), ointments (lanolin, which contains wool wax), candles (spermaceti), and so on.

Aside from cholesterol, the common alcohols found in waxes are cetyl alcohol ($C_{16}H_{33}OH$), ceryl alcohol ($C_{26}H_{53}OH$), and myricyl alcohol ($C_{30}H_{61}OH$).

Conjugated lipids

Conjugated lipids are esters of fatty acids containing groups as well as an alcohol and a fatty acid (Table 51-4).

Phospholipids. Phospholipids are lipids having, in addition to fatty acids and glycerol, a phosphoric acid residue, nitrogen-containing bases, and other constituents. These lipids include phosphatidyl choline (lecithin), phosphatidyl ethanolamine, phosphatidyl inositol, phosphatidyl serine, sphingomyelins, and plasmalogens. Phosphatidyl ethanolamine, phosphatidyl serine, and phosphatidyl inositol (lipositol) are also known as cephalins.

This class of complex lipids is also called *glycerophosphatides, phosphoglycerides, glycerol phosphatides,* or more commonly *phospholipids*. Note that not all phosphorus-containing lipids are phosphoglycerides; that is, sphingomyelin is a phospholipid because it contains phosphorus, but it is better classified as a sphingolipid because of the nature of the backbone structure to which the fatty acid is attached. In phospholipids, one of the primary OH groups of glycerol is esterified to phosphoric acid; the other OH groups are esterified to fatty acids. The parent compound of the phospholipids is phosphatidic acid, which contains no polar alcohol head group. The phospholipids are constituents of all animal and vegetable cells. They are present in abundance in brain, heart, kidney, eggs, soybeans, and so on. The phospholipids have been commonly separated into five distinct groups of compounds: phosphatidic acid, lecithin, cephalin, plasmalogens, and sphingolipids. In addition to carbon, hydrogen, and oxygen, the compounds contain the elements nitrogen and phosphorus. In lecithin and cephalin, the nitrogen-phosphorus ratio is 1:1; in sphingomyelin it is 2:1.

$$\begin{array}{l} H_2{-}C{-}O{-}\overset{\overset{\large O}{\|}}{C}{-}R_1 \\ \quad\;\; | \\ H{-}C{-}O{-}\overset{\overset{\large O}{\|}}{C}{-}R_2 \\ \quad\;\; | \\ H_2{-}C{-}O{-}\overset{\overset{\large O}{\|}}{\underset{\underset{\large OH}{|}}{P}}{-}OH \end{array}$$

Phosphatidic acid

Phosphatidic acid. Phosphatidic acid is important as an intermediate in the synthesis of triglycerides and phospholipids, but it is not found in any quantity in tissues. Phos-

Table 51-4 Classification of phosphatides and glycolipids

Name	Main alcohol component	Other alcohol components
Glycerophosphatides		
Phosphatidic acid	Diglyceride (= glycerol diester)	
Lecithin	Diglyceride (= glycerol diester)	Choline
Cephalin	Diglyceride (= glycerol diester)	Ethanolamine, serine
Inositide	Diglyceride (= glycerol diester)	Inositol
Plasmalogens (acetal phosphatides)	Glycerol ester and enol ether	Ethanolamine, choline
Sphingolipids		
Sphingomyelins	*N*-Acylsphingosine	Choline
Cerebrosides	*N*-Acylsphingosine	Galactose,* glucose*
Sulfatides	*N*-Acylsphingosine	Galactose*
Gangliosides	*N*-Acylsphingosine	Hexoses,* hexosamine,* neuraminic acid*

*These components are not present as phosphoric esters, but rather in glycosidic linkage; for this reason, cerebrosides, sulfatides, and gangliosides are called *glycolipids*.

phatidic acid is the simplest type of phospholipid. Phosphatidic acid is derived from glycerophosphoric acid by esterification of the two remaining OH groups with fatty acids.

Lecithins. On hydrolysis, a typical lecithin forms glycerol, 2 mol of fatty acids, phosphoric acid, and the nitrogenous base, choline. Most lecithins have a saturated fatty acid in the C-1 position and an unsaturated fatty acid in the C-2 position. The structural formula may be written as follows:

```
         O
         ‖
H3–C–O–C–R1
   |
   |     O
   |     ‖
 H–C–O–C–R2
   |
   |     O
   |     ‖
 H–C–O–P–O⁻
         |
         |                  CH3
         |               + /
         O–CH2–CH2–N–CH3
                           \
                            CH3
```

Lecithin (phosphatidyl choline)

The lecithins, like cholesterol, are common cell constituents that occur principally in animal tissue, having both structural (as part of cell membranes) and metabolic functions. Although not found in depot fat, they make up a considerable proportion of the liver and brain lipids. They also occur in the plasma as part of the lipid-protein complexes called lipoproteins; thus they are important for the formation of these macromolecules, which play an important role in fat transport. Lecithins play an important role in the esterification of free cholesterol to form ester cholesterol. Lecithins are an important precursor for the formation of lung surfactant. Lecithins form colloidal solutions with water, but one can prepare an aqueous solution of lecithin by the addition of bile salts. It is believed that a bile salt–lecithin compound is formed that is water soluble. Lecithins are not soluble in acetone; this property is used to separate them from other phospholipids and lipids.

Cephalins. The cephalins resemble the lecithins in structure except for the component corresponding to choline. There are three main fractions: the ethanolamine cephalins, serine cephalins, and inositol cephalins.

The cephalins differ from lecithins in their insolubility in ethyl or methyl alcohol.

Phosphatidyl ethanolamine. Phosphatidyl ethanolamine differs from lecithins in that ethanolamine replaces choline. Both alpha and beta cephalins are known. This is one

```
         O
         ‖
H2–C–O–C–R1
   |
   |     O
   |     ‖
 H–C–O–C–R2
   |
   |     O
   |     ‖
H2–C–O–P–OH
         |
         O——[Ethanolamine]
                 or
             [Serine]
                 or
             [Inositol]
```

of the more abundant cephalins found in higher plants and animals.

Phosphatidyl serine. Phosphatidyl serine, which contains the amino acid serine rather than ethanolamine, has been found in tissues, like the brain.

Phosphatidyl inositol. Phosphatidyl inositol is found in phospholipids of brain tissue and of soybeans and in other plant phospholipids as well. The inositol is present as the stereoisomer myoinositol.

Plasmalogens. Plasmalogens constitute as much as 10% of the phospholipids of the membranes of nerves and muscles. They are also found in the liver and other organs. Their function is not known at present. Structurally, the plasmalogens resemble lecithins and cephalins but give a positive reaction when tested for aldehydes with Schiff's reagent (fuchsin–sulfurous acid) after pretreatment of the phospholipid with mercuric chloride. These phospholipids contain higher fatty aldehydes in place of fatty acids. Thus the basic units of this class of compounds include glycerol, phosphorus, fatty aldehyde, and ethanolamine.

Sphingolipids. The amino dialcohol group sphingosine characterizes all sphingolipids. It serves as a structural unit for substitution, just as the trihydroxyalcohol glycerol does in glycerides. Sphingosine is a long-chain C_{18} compound that contains a *trans*-double bond, an NH_2 group on C-2, and two OH groups (on C-1 and C-3). Sphingolipids are especially abundant in the brain. Some storage diseases are characterized biochemically by the accumulation of certain sphingolipids. There are four major categories (see Table 51-2): sphingomyelins, cerebrosides, sulfatides, and gangliosides.

Sphingomyelins. Sphingomyelins are found in the brain and other organs. Stearic, lignoceric, and nervonic acid are the sole fatty acids present in brain sphingomyelins, whereas palmitic and lignoceric acids are the fatty acids in lung and spleen sphingomyelins. A typical formula is shown at the top of p. 760.

Sphingomyelin

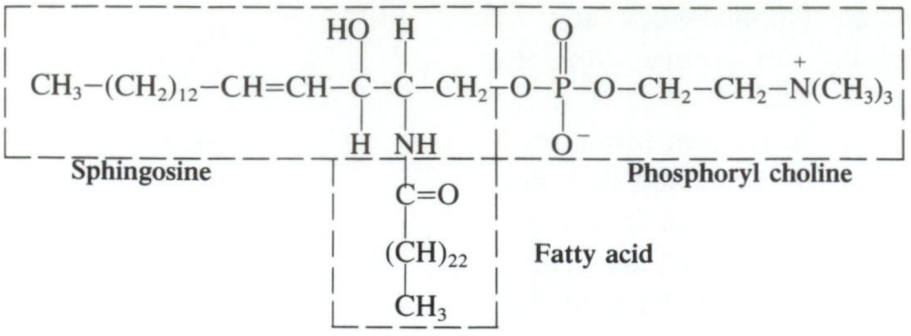

$$\underbrace{CH_3-(CH_2)_{12}-CH=CH-\overset{HO}{\underset{H}{C}}-\overset{H}{\underset{NH}{C}}-CH_2}_{\text{Sphingosine}}-\underbrace{O-\overset{O}{\overset{\|}{\underset{O^-}{P}}}-O-CH_2-CH_2-\overset{+}{N}(CH_3)_3}_{\text{Phosphoryl choline}}$$

$$\text{Fatty acid: } NH-C{=}O-(CH)_{22}-CH_3$$

and its two important constituents are:

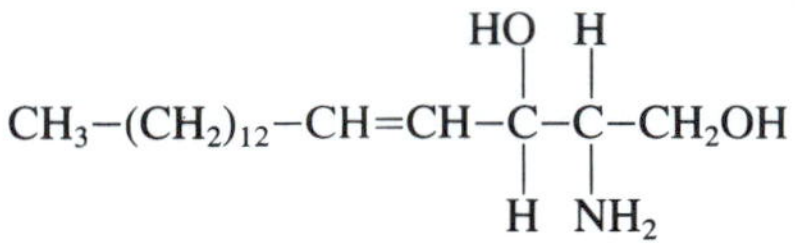

$$CH_3-(CH_2)_{12}-CH=CH-\overset{HO}{\underset{H}{C}}-\overset{H}{\underset{NH_2}{C}}-CH_2OH \qquad CH_3(CH_2)_{22}-COOH$$

Sphingosine **Lignoceric acid**

Cerebrosides. Cerebrosides contain galactose or glucose, a high-molecular-weight fatty acid, and sphingosine. Thus cerebrosides have the following basic structure:

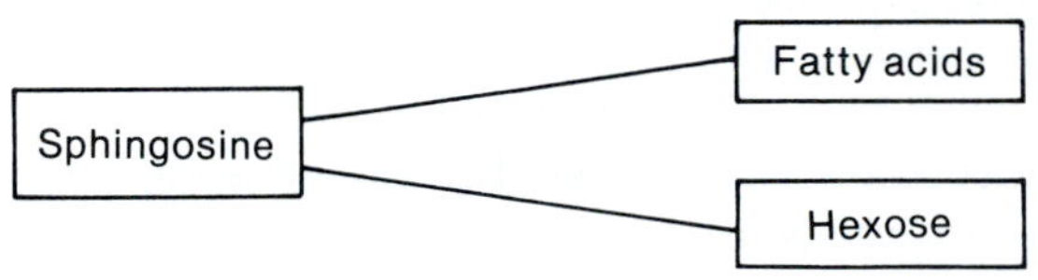

They are structurally similar to sphingomyelins. Cerebrosides may also be classified with the sphingomyelins as sphingolipids. Individual cerebrosides are differentiated by the type of fatty acid in the molecule: *kersins* contain lignoceric acid; *cerebrons* contain a hydroxylignoceric acid (cerebronic acid); *nervons* contain an unsaturated homolog of lignoceric acid called nervonic acid; and *oxynervons* apparently contain the hydroxyl derivative of nervonic acid as a constituent fatty acid.

$$CH_3-(CH_2)_{22}-COOH$$

Lignoceric acid

$$CH_3-(CH_2)_{21}-CH(OH)-COOH$$

Cerebronic acid

$$CH_3-(CH_2)_7-CH=CH-(CH_2)_{13}-COOH$$

Nervonic acid

$$CH_3-(CH_2)_7-CH=CH-(CH_2)_{12}-CH(OH)-COOH$$

Oxynervonic acid

Cerebrosides are found in many tissues besides the brain. In Gaucher's disease, the cerebroside content of the reticuloendothelial cells (as in the spleen) is very high. The cerebrosides are in much higher concentration in myelinated than in nonmyelinated nerve fibers.

Sulfatides. Sulfatides are sulfate derivatives of the galactosyl residue in cerebrosides.

Gangliosides. Gangliosides are glycolipids occurring in the brain (in ganglionic cells). The main components are sphingosine, fatty acids, and branched-chain carbohydrates with as many as seven sugar residues. The construction of gangliosides is similar to that of cerebrosides, but the carbohydrate moiety is far more complex. The various gangliosides differ primarily in the number of sugar residues.

Derived lipids

Derived lipids are compounds derived from the hydrolysis of simple and conjugated fats. These include the compounds described below.

Fatty acids. Fatty acids are straight-chain carboxylic acids (both saturated and unsaturated). More than 100 different kinds of fatty acids have been isolated from various lipids of animals, plants, and microorganisms. All possess a long hydrocarbon chain and a terminal carboxyl group. Fatty acids are obtained from the hydrolysis of fats or can be synthesized from two carbon units (acetyl radicals). Fatty acids that occur in naturally occurring fats usually contain an even number of carbon atoms (because they are synthesized from two carbon units) and are straight-chain derivatives. The chain may be saturated (containing no double bonds) or unsaturated (containing one or more double bonds).

Some generalizations may be made about the fatty acids present in lipids of higher plants and animals. Nearly all have an even number of carbon atoms and have chains that are between 14 and 22 carbon atoms long; those having 16 or 18 carbons are by far the most abundant. In general, unsaturated fatty acids predominate over the saturated type, particularly in the neutral fats and in cells of poikilothermic (cold-blooded) organisms living at lower temperatures. Unsaturated fatty acids have lower melting points than saturated fatty acids. Most neutral fats rich in unsaturated fatty acids are liquid down to 5° C or lower. In most unsaturated fatty acids in higher organisms, there is a dou-

Table 51-5 Common unsaturated fatty acids, number of double bonds, and length of carbon chain

Fatty acid	Number of double bonds	Number of carbons
Palmitoleic	1	16
Oleic	1	18
Linoleic	2	18
Linolenic	3	18
Arachidonic	4	20

ble bond between carbon atoms 9 and 10; additional double bonds usually occur between C-10 and the methyl end of the chain. In fatty acids containing two or more double bonds, the double bonds are never found in conjugation but are separated by one methylene group. The double bonds of nearly all the naturally occurring unsaturated fatty acids are in the *cis* configuration. The most abundant unsaturated fatty acids in higher organisms are oleic, linoleic, linolenic, and arachidonic acids (Table 51-5).

Alcohols. Straight-chain alcohols and cyclic alcohols (such as the sterols) are a subclass of derived lipids.

The steroids may be classified into the following groups:

Sterols
Bile acids
Substances obtained from cardiac glycosides
Substances obtained from saponins
Sex hormones
Adrenocorticosteroids
Vitamin D

These compounds are widely distributed in plant and animal tissues, either in the free state or in the form of esters (in combination with higher fatty acids). Chemically, they are known to be phenanthrene derivatives, or more correctly cyclopentanoperhydrophenanthrene derivatives.

Steroids. The best known steroid is cholesterol. It is present in all animal cells and is particularly abundant in nervous tissue and liver. Varying quantities of this steroid are found admixed in animal fats but not in vegetable fats. The structure of a cholesterol molecule is illustrated in Fig. 51-7.

Cholesterol is the precursor of many other steroids in animal tissues, including the bile acids, detergent-like compounds that aid in emulsification and absorption of lipids in the intestine; the androgens, or male sex hormones; the estrogens, or female sex hormones; the progestational hormones; and the adrenocortical hormones.

Cholesterol is a member of a large subgroup of steroids called the *sterols*. It is a steroid alcohol containing a hydroxyl group at carbon 3 of ring A and a branched aliphatic chain of eight or more carbon atoms at carbon 17. Sterols occur either as free alcohols or as long-chain fatty acid esters of the hydroxyl group at carbon 3; all are solids at room temperature. Cholesterol melts at 150° C and is insoluble in water but readily extracted from tissues with

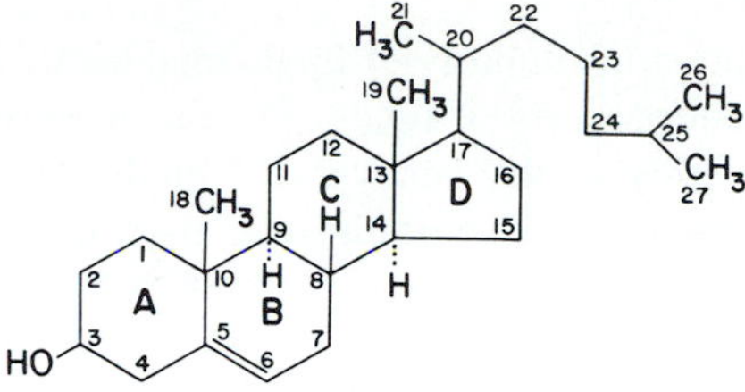

Fig. 51-7 Structure of cholesterol molecule, a C_{27} hydrocarbon sterol.

chloroform, ether, or hot alcohol. Cholesterol occurs in the plasma membranes of animal cells and in the lipoproteins of blood. Cholesterol is found only in animal tissues and fluids, never in plants.

Other similar steroids are phytosterols, which are steroids from plants. Among these are stigmasterol, campesterol, and sitosterol.

Fungi and yeasts contain still other types of sterols, the mycosterols. Among these is ergosterol, which is converted to vitamin D.

Bile acids. Bile acids (a C_{24} steroid) are digestion-promoting constituents of bile. They are surface-active agents. This means that they lower surface tension and thus can emulsify fats, an important step in the formation of micelles. Bile acids also activate gastrointestinal lipases. For these reasons, bile acids play an important physiological role in the digestion and absorption of fats.

The major primary bile acids are cholic acid and chenodeoxycholic acid (Fig. 51-2), which are made in the liver by the enzymatic cleavage of the terminal three carbons on the cholesterol molecule (a C_{27} hydrocarbon). Thus the bile acids are one of the end products of the metabolism of cholesterol; however, it should be noted that bile acid constitutes the acidic sterol fraction of the bile, which is about 50% to 60% of the total steroid excreted. The remainder of the steroid output in the bile is in the form of neutral steroids, such as cholesterol.

Hydrocarbons. The hydrocarbons are both aliphatic and cyclic compounds.

Vitamins. Since vitamins and hormones are not discussed in this chapter, their structural formulas are not presented. (See Chapters 36 and 66.)

Other compound lipids. Sulfolipids, aminolipids, and lipoproteins may also be placed in this category.

CHEMICAL AND PHYSICAL PROPERTIES

Melting point

The melting point of fatty acids is influenced by the chain length and degree of chain unsaturation. Increasing the chain length and decreasing the number of unsaturated double bonds will increase the melting point of fatty acids. The melting points of fatty acids and other lipids can be used to identify the compound, but this property is not routinely used in analysis.

Solubility

The relative insolubility of lipids in aqueous solutions is an important property of lipids. The major consequence of this insolubility is that analyses of lipids often require a prior treatment of the sample to extract the lipid into a more lipid-soluble medium, such as methanol, chloroform, or ether.

The insolubility of unesterified cholesterol has been used in the past for certain cholesterol assays. Cholesterol can be quantitatively separated from cholesterol esters by precipitation with digitonin as a 1:1 molecular complex of cholesterol digitonide. Measurement, gravimetrically, of the amount of digitonide complex formed has been the basis of cholesterol quantitation in the past.

Specific gravity

The specific gravity of all fat is less than 1 g/mL. Consequently, all fats float in water. This characteristic has enabled lipoproteins to be selectively separated from more dense proteins, and individual lipoproteins to be separated from each other on the basis of varying proportions of lipid content.

Alcohol groups of steroids

The chemically reactive alcohol group of steroids is the basis of many assays for quantitating cholesterol. The group can react with strong mineral acids, such as sulfuric acid, and salts to form a chromogen. The Burchard-Liebermann reaction is the most frequently used example of such a chemical reaction.

The hydroxyl group can be specifically oxidized by the enzyme cholesterol oxidase. Monitoring of this reaction is the basis for the increasingly used enzyme assays for cholesterol.

Triglyceride composition

The chemical composition of triglycerides (that is, the esterification of glycerol by three fatty acids) is the basis of all methods for quantitating triglycerides. These techniques are based on the quantitation of glycerol released from triglycerides after chemical or enzymatic hydrolysis of the fatty acid esters. The glycerol can be chemically or enzymatically oxidized to form measurable chromogens.

Chemical composition

The phospholipids are detectable by specific reactions for phosphorus after chemical reaction.

BIOLOGICAL PROPERTIES

The most important biological properties of lipids are structural, nutritional, and hormonal. Almost all classes of lipids are used as structural components of membranes. The triglycerides are essential components in the formation of the bimolecular protein-lipid–lipid-protein membranes. Cell membranes also contain varying amounts of steroids, phospholipids, and other complex lipids.

Triglycerides also function as an important source of calories and as a source of carbon atoms for the synthesis of other macromolecules.

The role of steroids as hormones has a large impact on a laboratory because the measurement of many of these hormones is required. These hormones are often measured by some of the chemical techniques discussed above. For example, one can measure cortisol and estriol by the chromophores formed after reaction with hot sulfuric acid.

Part III: Carbohydrates

LAWRENCE A. KAPLAN

DEFINITION AND CLASSIFICATION

The earliest carbohydrates were found to have the empirical formula of $(CH_2O)_n$. Thus these chemicals were simply defined as compounds consisting of hydrated (H_2O) carbon; hence the name *carbohydrate*. Subsequently the existence of complex carbohydrates containing other chemical moieties was noted. Thus carbohydrates can be covalently linked to proteins, lipids, and nucleic acids. The various classes of carbohydrates are discussed below.

Simple monomeric carbohydrates (saccharides)

Saccharides are also known as *sugars,* and their common names all end with the suffix *-ose,* meaning "sugar." The smallest sugar units are called *monosaccharides,* in which n in the above formula is from 3 to 8. If $n = 3$, the sugar is a triose; if $n = 4$, a tetrose; and so on. The monosaccharides are straight carbon chains in which each carbon atom except one carries a hydroxyl group (–OH); the one remaining carbon atom has a carbonyl group. If the carbonyl group is on the first or last carbon atom, the carbonyl group is an aldehyde and the monosaccharide is called an *aldose*. If the carbonyl group is on an internal carbon atom, it is a *ketone,* and the monosaccharide is called a *ketose* (Fig. 51-8). Thus a 4-carbon aldose is an *aldotetrose,* a 6-carbon ketose is a *ketohexose,* and so on.

The monosaccharides found in nature are all stereoisomers. Stereoisomerism is defined by the ability of a molecule to rotate the plane of incident polarized light. The physical and chemical properties of, for example, all the eight aldohexoses (6-carbon chain) are exactly the same

Aldose	Ketose
H–C=O	H_2COH
HCOH	C=O
HCOH	HCOH
⋮	⋮

Fig. 51-8 Structural differences between aldoses and ketoses, which are aldehydes and ketones, respectively.

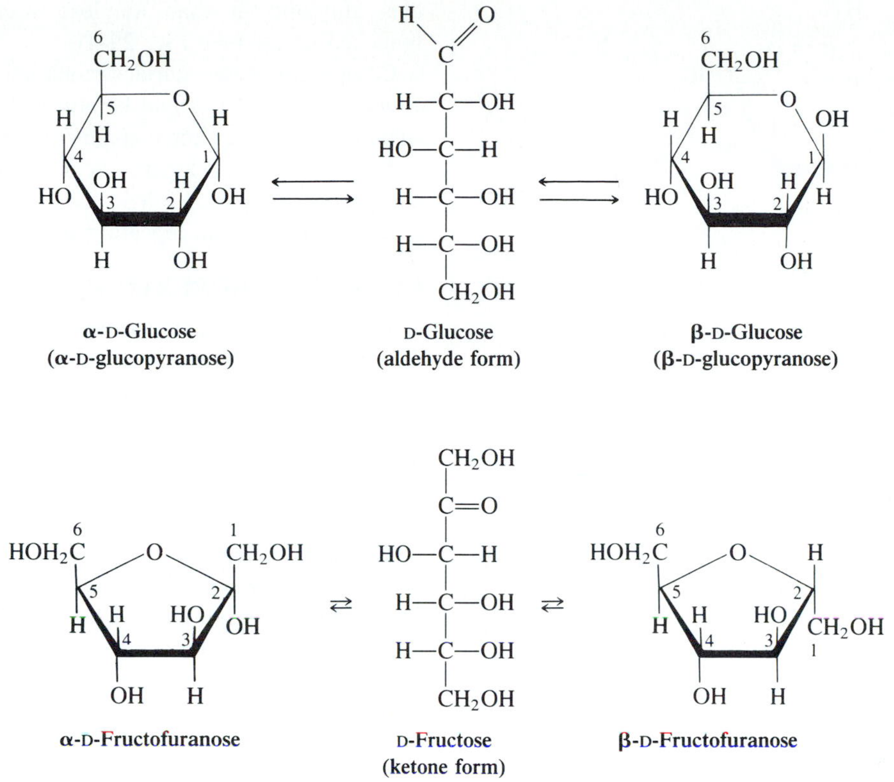

α-D-Glucose (α-D-glucopyranose) ⇌ D-Glucose (aldehyde form) ⇌ β-D-Glucose (β-D-glucopyranose)

α-D-Fructofuranose ⇌ D-Fructose (ketone form) ⇌ β-D-Fructofuranose

Fig. 51-9 Interrelationships between straight-chain and ring forms of D-glucose and D-fructose, which form pyranose and furanose rings. *(From Orten, JM, and Neuhaus, OW: Human biochemistry, ed 10, St Louis, 1982, The CV Mosby Co.)*

except for their different actions on polarized light. All the monosaccharides in human biochemistry are of the dextroisomeric (D) form. Examples of some monosaccharides are given in Fig. 51-9.

The pentose and hexose monosaccharides also have the ability to form ring structures by intramolecular reaction of the terminal hydroxyl group with the carbonyl function. The six-membered ring forms of the sugars are called *pyranoses,* whereas the five-membered rings are called *furanoses*. The aldohexoses, such as D-glucose, form six-membered rings, whereas an aldoketose, such as D-fructose, forms a five-membered ring (Fig. 51-9).

Glucose can form two types of six-membered rings. The rings differ in how the hydroxyl group at the number 1 carbon atom is positioned with respect to the plane of the ring. If the hydroxyl group is on the same side of the molecule as the ring oxygen (Fig. 51-9), the isomer is known as the α-D-glucose isomer, whereas if the hydroxyl group is on the opposite side of the ring oxygen, then this isomer is known as the β-D-glucose. Enzymes acting on carbohydrates usually have a specificity directed toward one of the isomers, usually the most common one found, such as β-D-fructose.

Derived monosaccharides

Derived monosaccharides are formed by reduction or oxidation of the carbonyl groups. The products of reductive reactions are polyols (polyalcohols), such as D-sorbitol or D-mannitol, whereas the products of oxidation are acids, such as D-glucuronic acid (from D-glucose). Many acid forms of monosaccharides are important constituents of more complex carbohydrates, such as mucopolysaccharides.

An important group of derived monosaccharides is the result of the replacement of a hydroxyl group by an amino group. The term *sialic acid* is used to describe the important *N*-acetyl derivatives of neuraminic acid, which are often found covalently linked to proteins (Fig. 51-10).

Complex carbohydrates

These molecules are formed by linking two or more monosaccharides by a glycosidic linkage (Fig. 51-11). The simplest disaccharides, important nutritionally, are maltose (two glucose), lactose (milk sugar, one galactose and one glucose), and sucrose (one fructose and one glucose). Oligosaccharides are often defined as carbohydrates containing two to 10 monosaccharide subunits. Polysaccharides are larger polymers of up to 100 million daltons. All three of the most important polysaccharides contain glucose as the monomeric subunit. Cellulose, a structural component of plant walls, consists of glucose units linked by a β-(1→4) glycosidic bond to form long, unbranched chains. Starch, a storage form of glucose in plants, consists of glucose residues connected by α-(1→4) glycosidic link-

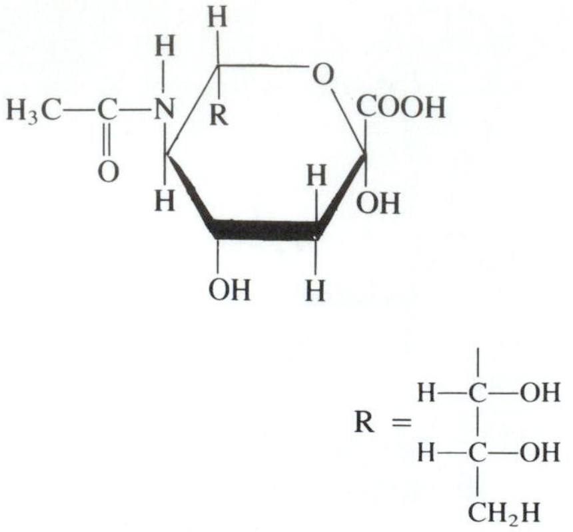

N-Acetylneuraminic acid

Fig. 51-10 Structure of *N*-acetylneuraminic acid ("sialic acid"). *(From Orten, JM, and Neuhaus, OW: Human biochemistry, ed 10, St Louis, 1982, The CV Mosby Co.)*

ages, which, unlike the β-(1→4) linkages of cellulose, are amenable to degradation by human hydrolytic enzymes (such as amylase). Starch also differs from cellulose in that it is a branched molecule. Branching points are scattered throughout the molecule formed by α-(1→6) bonds. The two forms of starch are therefore called *amylose* (the straight-chain fraction) and *amylopectin* (the highly branched fraction). Glycogen is the glucose-storage molecule found in animal cells. Glycogen more closely resembles amylopectin than amylose because of its highly branched nature (see Fig. 29-4).

Complex polysaccharides containing hyaluronic acid, chondroitin-4-sulfate, and keratin sulfates as the repeating subunits are important constituents of synovial fluid and connective tissue. Heparin is a complex polysaccharide containing D-glucuronic acid-2-sulfate *N*-acetyl-D-glucosamine-6-sulfate as the repeating subunit.

CHEMICAL PROPERTIES

The monosaccharides (pentoses and larger) can undergo dehydration in the presence of hot mineral acids to form the cyclic furfural derivatives. One can dehydrate glucose in this manner to form 3-hydroxymethylfurfural, a reaction that is the basis for a colorimetric assay for glycosylated proteins.

An important chemical property of the monosaccharides is the ability of these compounds to be oxidized or reduced and in turn to reduce or oxidize some other compounds. The ability of reducing aldoses, such as glucose, to be oxidized to the acid form has been the historic basis for chemical assays for glucose. The glucose in turn reduced such compounds as Cu^{+2} or $Fe(CN_6)^-$ with the formation of colored complexes of the reduced forms of these compounds (such as Cu^+ and Cu_2O).

The enzymatic oxidation of glucose by glucose oxidase is the basis of many of the current glucose assay procedures, whereas the oxidation of glucose-6-phosphate is the basis of the hexokinase assay for glucose.

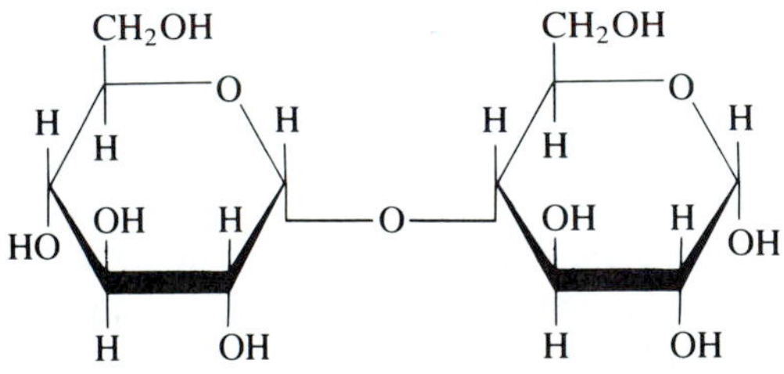

(D-glucose) (D-glucose)
(4-*O*-α-D-glucopyranosyl-D-glucopyranose)
Maltose (α-form)

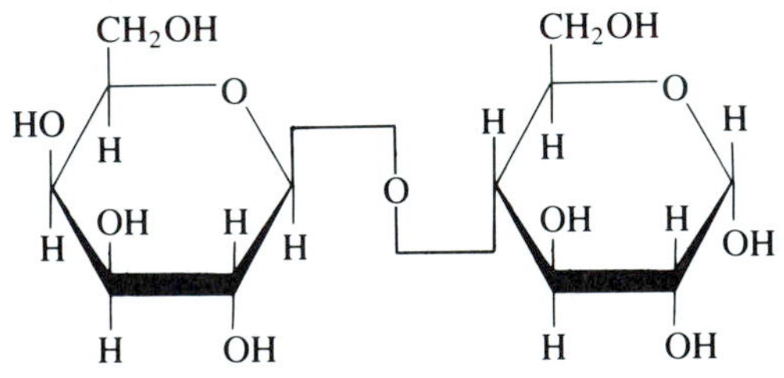

(D-galactose portion) (D-glucose portion)
(4-*O*-β-D-galactopyranosyl-α-D-glucopyranose)
Lactose (α-form)

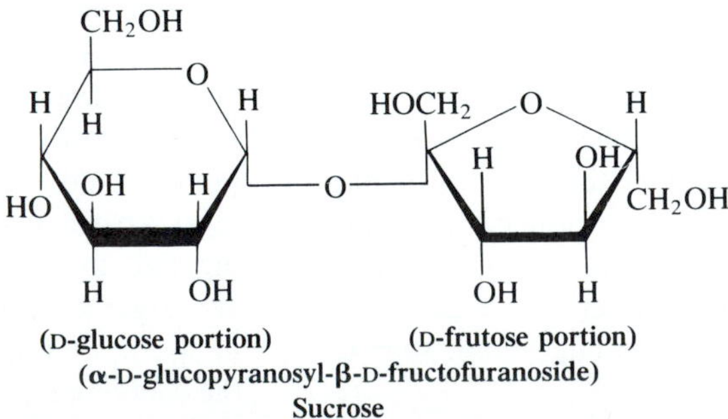

(D-glucose portion) (D-frutose portion)
(α-D-glucopyranosyl-β-D-fructofuranoside)
Sucrose

Fig. 51-11 Common disaccharides linked by β-glycosidic bonds. *(From Orten, JM, and Neuhaus, OW: Human biochemistry, ed 10, St Louis, 1982, The CV Mosby Co.)*

Aldoses, such as glucose, can react with primary amines to form a Schiff base. This nonenzymatic condensation is the mechanism for the formation of glycoproteins in blood. In addition, the reaction of glucose with aromatic primary amines, such as *o*-toluidine, is the basis of an important historic method for measuring blood glucose.

PHYSICAL PROPERTIES

The commonly measured monosaccharides and disaccharides are highly water-soluble compounds. Assays for these analytes thus do not require prior extraction or purification. Separation of the monosaccharides by adsorption chromatography is possible, though this is usually perfomed by specialized metabolic laboratories. The simple monosaccharides, dissaccharides, or polysaccharides are not readily distinguished by their spectral or electrophoretic properties.

BIOLOGICAL PROPERTIES

The monosaccharides and disaccharides are the major source of calories for the human body and as such serve as a primary form of nutrition. Polymeric forms of glucose, such as glycogen, serve as a storage for glucose in liver and muscle cells. Complex polysaccharides are found in body fluids and connective tissue.

BIBLIOGRAPHY

General

Frisell, WR: Human biochemistry, New York, 1982, Macmillan, Inc.

Orten, JM, and Neuhaus, OW: Human biochemistry, ed 10, St Louis, 1982, The CV Mosby Co.

Proteins and amino acids

Anfinsen, CB, Jr, Edsall, JT, and Richards, FM: Advances in protein chemistry, vols 1-26, New York, 1944-1972, Academic Press, Inc.

Bradshaw, R, and Tang, J, editors: Molecular architecture of proteins and enzymes, Baltimore, 1987, Johns Hopkins University Press.

Lipids

Christie, WW: Lipid analysis: isolation, separation, and structural analysis of lipids, ed 2, New York, 1987, Pergamon Press.

Mead, JF, et al, editors: Lipids: chemistry, biochemistry, and nutrition, New York, 1986, Plenum Press.

Carbohydrates

Whelan, WJ, editor: Biochemistry of carbohydrates, vol 5. In MTP International Review of Science, Baltimore, 1975, University Park Press.

Brinkley, RW: Modern carbohydrates chemistry, New York, 1988, Marcel Dekker.

CHAPTER 52

Enzymes

DAVID C. HOHNADEL

KEY TERMS

activation energy The energy required in a chemical reaction to convert reactants to activated or transition-state species that will spontaneously proceed to products.

activators Inorganic ions that are cofactors for an enzyme reaction.

active sites The specific areas on an enzyme where a substrate binds and catalysis takes place.

activity The amount of substrate for a particular enzymatic reaction that is converted to product per unit time under defined conditions.

allosteric sites, or regulatory sites The sites, other than the active site or sites, of an enzyme that bind regulatory molecules.

apoenzyme An enzyme without any associated cofactors or with less than the entire amount of cofactors or prosthetic groups.

auxiliary enzyme In a coupled assay system, an enzyme that links the enzyme being measured with an indicator enzyme.

binding sites The sites on the surface of the enzyme that serve to bind the substrate or product of the reaction.

bond specificity The nature of enzyme action that causes the disruption of only certain bonds between atoms.

catalyst A substance that increases the rate of a reaction without being changed by the reaction.

catalytic site Another name for active site.

coenzymes Organic cofactor compounds, such as thiamine pyrophosphate and pyridoxyl-5-phosphate.

cofactors Nonprotein substances associated with an enzyme that are needed for catalytic activity.

competitive inhibitor An inhibitor of an enzyme reaction that competes with the substrate by binding at the active site.

constitutive enzymes Enzymes that are present at a constant concentration during the life of a cell.

coupled assays Assays with several enzyme reactions leading to an indicator reaction that has an easily measured substance.

denaturation The loss of the biological properties of a protein, usually as a result of changes in tertiary or quaternary structure.

EC code The four-digit Enzyme Comission code for the systematic classification of enzyme reactions.

ELISA Enzyme-linked immunosorbent assay.

EMIT Enzyme-multiplied immunoassay technique.

end-point assays Assays in which a single measurement is made at a fixed time.

endopeptidases Protein-hydrolyzing enzymes that break bonds in the interior of a protein substrate.

enzyme kinetics The study of enzyme reaction rates and the factors that affect them.

enzyme specificity The degree to which an enzyme will catalyze one or more reactions.

enzyme substrate complex An intermediate active complex formed between the substrate and the enzyme during the reaction.

enzymes Biological materials (proteins) with catalytic properties.

equilibrium constant The ratio of the concentration of product to the concentration of substrate when the reaction is at equilibrium.

exopeptidases Protein-hydrolyzing enzymes that break bonds proceeding from one end of the protein substrate toward the center of the substrate.

first-order kinetics State occurring when the rate of an enzyme reaction is proportional to the concentration of the substrate.

holoenzymes The complete enzyme-cofactor complex that gives full catalytic activity.

hydrophilic amino acids Polar, water-loving amino acids.

hydrophobic amino acids Nonpolar, water-hating amino acids.

inactivation A reversible denaturation of a protein.

indicator enzymes Enzymes that produce (or consume) an easily measured substance.

inducible enzymes Enzymes whose cellular concentrations can be made to increase.

inhibitors Materials that reduce the catalytic activity of an enzyme.

initial rates Enzyme measurements made at the start of a reaction just after the lag phase.

international unit of enzyme activity The amount of enzyme that catalyzes the conversion of one micromole of substrate per minute under defined conditions, 1 U = 1.67×10^{-8} katal.

in vitro systems Those systems outside of a living organism, that is, in a test tube.

isoenzymes Different forms of an enzyme that catalyze the same reaction.

katal (kat, K) An enzyme unit in moles per second defined by the SI system: 1 K = 6.0×10^7 U.

K_m The symbol for the Michaelis-Menten constant.

kinetic assays Assays that form increasing amounts of product with time, usually monitored by multiple datum points.

labile enzymes Unstable or easily denatured proteins.

lag phase The early time in an assay when mixing occurs and temperature and kinetic equilibrium are becoming established.

linear phase Time when an assay is following zero-order kinetics producing a constant absorbance change per unit of time.

metalloenzymes Enzymes that contain very tightly bound metal ions.

Michaelis-Menten constant A constant related to the rate constants of an enzyme reaction and equal to the concentration of substrate that gives one half the maximal catalytic velocity.

noncompetitive inhibitor An inhibitor that binds to an allosteric site of an enzyme and does not compete with the substrate by binding at the active site.

optimal assay conditions Conditions for reaction concentrations of substrates, cofactors, activators, and buffer that produce the maximum rate of enzyme catalysis.

primary structure The sequence of amino acids of a protein.

prosthetic groups Cofactors that are so tightly bound that they are considered to be part of the enzyme structure.

quarternary structure The structural relationship of various enzyme subunits to one another.

reactivation The restoration of biological properties of a protein after a temporary loss.

regulatory sites See *allosteric sites*.

secondary structure of an enzyme The twisting of amino acids into a semifixed steric relationship in two dimensions.

specific activity The enzyme activity expressed as units per milligram of protein.

stereoisomeric specificity The specificity of an enzyme for one form of a DL pair of compounds with an asymmetric carbon atom.

substrate-depletion phase The time late in an enzyme assay when the substrate concentration is falling and the assay is not following zero-order kinetics.

substrates The materials enzymes act upon.

subunits Single protein chains from enzymes composed of two or more peptide chains in an active form.

Système International d'Unités An international system of rational and internally consistent units for all types of scientific quantities; SI units.

tertiary structure of an enzyme The folding of amino acid chains into a three-dimensional structure.

uncompetitive inhibitor An inhibitor that appears to bind only to the enzyme- substrate complex and not to free enzyme.

V_{max} The maximum rate of catalysis.

zero-order kinetics State occurring when the rate of an enzyme reaction is independent of the concentration of the substrate.

BIOCHEMICAL NATURE OF ENZYMES

Enzymes are biological materials with catalytic properties; that is, they increase the rate of chemical reactions in biological and in vitro systems that otherwise proceed very slowly. They are large organic compounds with molecular weights usually between 13,000 and 500,000 daltons. The study of these molecules and of the changes in enzyme activity that occur in body fluids over time has become a valuable diagnostic tool for the elucidation of various disease entities and for testing organ function.

Different tissues or cell types do not contain the same amounts or types of enzymes. The hundreds of different enzymes in each cell are attached to the cell walls and membranes and are also found in the cytoplasm, the nucleus, and many other specialized subcellular organelles (that is, microsomes, mitochondria, and lysosomes). Often the determination of one or several enzymes in plasma gives a pattern of results that is indicative of the tissue or cell type from which the enyzme or enzymes have been derived. Different cells or compartments within a single cell can even contain different forms of an enzyme (that is, isoenzymes) that catalyze the same chemical reaction. Assays for these different isoenzyme forms can sometimes be performed for specific determination of the tissue or compartment from which an enzyme has come. A few enzymes are found in plasma or other extracellular fluids where they seem to perform a physiological function, but most enzymes catalyze reactions inside cells or in the lumen of various organs.

Composition and structure

All enzymes are proteins: that is, they are compounds of high molecular weight; they contain amounts of carbon, hydrogen, oxygen, nitrogen, and sulfur that are similar to amounts found in other proteins; and hydrolysis with strong acid yields a mixture of amino acids and small peptides. Enzymes are distinguished from other proteins that are not enzymes by their catalytic action, which is usually quite specific for the materials they act on. The structure of proteins (and enzymes) is discussed in Chapter 51.

The catalytic behavior of an enzyme is dependent on the primary, secondary, tertiary, and quaternary structures of the protein molecule. Changes in any one of these structures can affect the enzymatic activity of the protein.

Apoenzymes and cofactors

An enzyme may have nonprotein substances associated with it that are needed for optimum activity. These other materials called *cofactors* may be either loosely or tightly bound to the protein portion of the enzyme. Those that are loosely bound can often be removed by dialysis. These materials may be organic compounds, such as thiamine pyrophosphate and pyridoxyl-5-phosphate, which are called *coenzymes,* or inorganic ions like chloride (Cl^-) and magnesium (Mg^{+2}), which are called *activators*. Cofactors like the heme portion of peroxidase that are so tightly bound that they are considered to be part of the enzyme structure are termed *prosthetic groups*. Enzymes that have metal ions (that is, activators) bound very tightly are called *metalloenzymes*. Ferroxidase (ceruloplasmin, EC 1.16.3.1), an enzyme containing a relatively large amount of tightly bound copper, and carbonate dehydratase (carbonic anhydrase, EC 4.2.1.1), an enzyme with a large amount of zinc, are two examples of metalloenzymes.

The term *coenzyme* is often loosely used when referring to NADH (or NADPH) for a reaction like the lactate dehydrogenase reaction.

$$\text{Pyruvate} + \text{NADH} + H^+ \overset{\text{LD}}{\rightleftharpoons} \text{L-Lactate} + \text{NAD}^+$$

In a formal kinetic sense, both pyruvate and NADH are substrates for the enzyme reaction, and lactate and NAD^+ are products. In this case, pyruvate and NADH react stoichiometrically with one another on a molar basis, and the enzyme reaction does not convert the NAD^+ that is produced back to NADH. Although NADH is still called a *coenzyme,* that is, a nonprotein organic material needed for maximal activity, perhaps for historic reasons, it should be more correctly called a *second substrate,* or *cosubstrate*.

Since it is possible to dialyze away loosely held cofactors from some enzymes and still retain some activity, an enzyme without the associated cofactors is referred to as an *apoenzyme,* and the complete enzyme-cofactor complex is termed a *holoenzyme*. In the clinical use of enzyme assays, the enzyme assay mixture must contain an excess of all the activators and coenzymes to ensure that the holoenzyme is the enzyme form being measured, rather than a mixture of apoenzyme and holoenzyme forms.

Catalysts

Enzymes function as biological catalysts. They are proteins that have the property of accelerating specific chemical reactions toward equilibrium without being consumed in the process. The material the enzyme acts on is termed the *substrate,* and a simple enzymatic reaction for one substrate and one product is listed below:

$$E + S \underset{k_{-1}}{\overset{k_{+1}}{\rightleftharpoons}} \{ES\} \underset{k_{-2}}{\overset{k_{+2}}{\rightleftharpoons}} P + E \qquad \textit{Eq. 52-1}$$

In this case the enzyme is represented by *E,* the substrate on which the enzyme acts by *S,* a postulated enzyme-substrate complex by $\{ES\}$, and the product of the reaction by *P*. The forward reaction rate constants are represented by k_{+1} and k_{+2}, whereas the reverse reaction rate constants are represented by k_{-1} and k_{-2}. An example of a single substrate enzyme reaction is the action of the enzyme urease (EC 3.5.1.5) on the substrate urea, though in this case two products are produced:

$$\underset{\textbf{Urea}}{H_2N{-}\overset{\overset{\large O}{\|}}{C}{-}NH_2} + E \underset{k_{-1}}{\overset{k_{+1}}{\rightleftharpoons}} \{\text{Urea-E}\} \underset{k_{-2}}{\overset{k_{+2}}{\rightleftharpoons}} \underset{\textbf{Ammonia}}{2NH_3} + \underset{\textbf{Carbon dioxide}}{CO_2} + E$$

Water also participates in the reaction, but for these purposes it is not a substrate. These biological catalysts are similar to other chemical catalysts in many respects, except that they function in biological systems. Enzyme catalysts, though they are unstable and easily destroyed, have catalytic properties which are similar to those of other chemical catalysts. These include the following: they are effective in small amounts; they are unchanged by the reaction; they affect the speed of attaining equilibrium but do *not* change the final concentrations of the substrates and products of the *equilibrium* state; and they demonstrate a much greater degree of specificity than the usual chemical catalysts for the reactions they accelerate.

It is the first property that makes enzymes such a valuable diagnostic tool. Since they are effective in such small amounts, measurement of changes in enzyme concentrations is a very sensitive way to follow changes that have occurred in various types of tissues.

The amount of enzyme involved in an enzyme assay is very much smaller than the amount of glucose, for instance, present in an assay for glucose, which makes conventional chemical assays for enzyme materials very difficult. Of the several thousand proteins in plasma, the measurement of the concentration of a single enzyme (protein), even if it is present at a very elevated value, is below the limit of detection for most chemical protein assays. What is easier to measure and is biologically related to many clinical conditions is the amount of catalytic activity of the enzyme and how it changes with time.

The activity of an enzyme is the amount of substrate for a particular enzyme reaction that is converted to product per unit time under defined conditions (see p. 778). The assumption that is made in the use of activity to express enzyme concentration is that a given amount (that is, weight) of enzyme has a given number of units of activity. That is, the specific enzyme activity (in units per milligram of protein) remains constant even when the increase in enzyme activity observed during a particular disorder may come from a different tissue. The increased enzyme activity is assumed to occur because of the presence of more enzyme with the same specific activity rather than because of the presence of another form of the enzyme with a different and perhaps higher specific activity. In practice, the analyst uses activity measurements of enzymes as if they were enzyme concentrations.

As shown in Equation 52-1, if the enzyme were acting as a catalyst, it would be unchanged by the reaction. Because of the unstable nature of most enzymes, this property is difficult to demonstrate. Using modern assay techniques, it has been demonstrated that enzymes do not participate in the reaction on a molar basis with the substrate; that is, they are not consumed during the reaction.

Biological catalysts accelerate the attainment of equilibrium but do not shift the final equilibrium state. One example of this is the effect of lactate dehydrogenase (LD, EC 1.1.1.27) on the conversion of pyruvate to lactate. In the presence of the enzyme LD and the coenzyme reduced nicotinamide adenine dinucleotide (NADH) the conversion of pyruvate to lactate occurs rapidly, but without the enzyme the process is so slow that it can hardly be demonstrated.

$$\text{Pyruvate} + \text{NADH} + \text{H}^+ \overset{\text{LD}}{\rightleftharpoons} \text{L-Lactate} + \text{NAD}^+$$

$$\text{Pyruvate} + \text{NADH} + \text{H}^+ \xrightarrow{\text{No enzyme}} \text{No detectable reaction}$$

This is not a one-way process but leads to an equilibrium between pyruvate and lactate, since the same enzyme converts lactate to pyruvate with the coenzyme nicotinamide adenine dinucleotide (NAD^+). The speed of the reaction and the conditions employed are not the same in both directions, since they are related to the equilibrium constant. It is possible to measure the conversion from either direction, and both methods are widely used to determine LD activity in the clinical laboratory.

Active sites and catalysis

The Gibbs free-energy change (ΔG) is the measure of the amount of work a chemical reaction can produce. All reactions that proceed from initial reactants to products have a net negative free energy ($-\Delta G$). However, the reactants do not become products directly but must absorb energy to pass through an activated or transition state. An energy diagram is given in Fig. 52-1 showing the effect of an enzyme on this process.

Enzymes lower the energy required for activation to the transition state. Without the enzyme present, even with a favorable negative free energy, the reaction may not proceed to any appreciable extent. The reactants must gain the energy to overcome an activation-energy barrier to enter the transition state (active state) and then pass on to products. Without a catalyst present, the reaction will occur only if enough heat (energy) can be added to the reaction system. With an enzyme catalyst, the reaction may easily

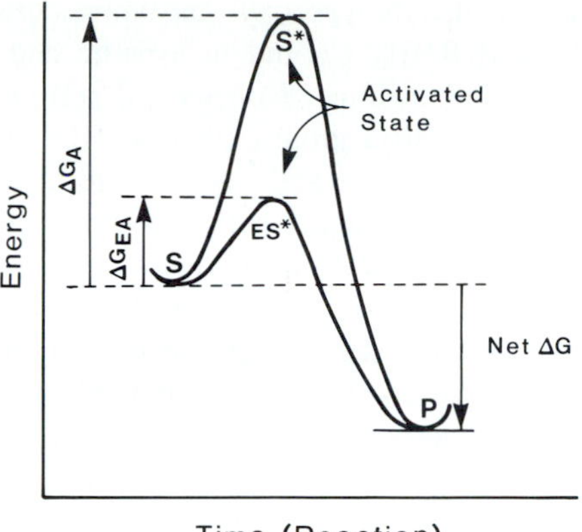

Fig. 52-1 Energy diagram showing reduction in activation energy $\Delta G_{EA} << \Delta G_A$ that occurs for same reaction with and without enzyme catalyst. *A*, Activated state; (also *); *EA*, enzyme activation; *ES*, enzyme substrate; *G*, energy; *P*, product; *S*, substrate.

proceed at normal physiological temperatures. Rewriting Equation 52-1 to account for this transition state in an enzyme-catalyzed reaction:

$$E + S \underset{k_{-1}}{\overset{k_{+1}}{\rightleftharpoons}} \{ES \rightarrow ES^* \rightarrow EP\} \underset{k_{-2}}{\overset{k_{+2}}{\rightleftharpoons}} P + E \qquad \textit{Eq. 52-2}$$

where *ES** is the transition state form of the reactant (substrate) and *ES* and *EP* are enzyme-substrate and enzyme-product intermediate forms in which substrate and product are bound but not activated. Substantial reductions in the activation energy requirements are often found when enzymes are used to catalyze the process. For example, the activation energy necessary for the decomposition of hydrogen peroxide is 18,000 cal/mol, but in the presence of the enzyme catalase less than 2000 cal/mol are required.

One of the most difficult problems facing enzymologists is to explain how an enzyme can reduce the activation energy and at the same time remain unchanged by the reaction. Equation 52-2 shows one general possibility but is not detailed enough to explain what happens on a molecular level. A further examination of some of the details of enzyme structure will serve to clarify the mechanism of enzyme catalysis.

A wide variety of nonpolar *hydrophobic* (water-hating) and polar *hydrophilic* (water-loving) amino acids are present in enzyme proteins. The external surface of the enzyme is believed to be composed of mostly polar but generally unreactive side chains of amino acids. The unreactive amino acid side chains may contain structures like the methyl and isopropyl groups (that is, R—CH_3 and R—CH—CH_3 with a CH_3 branch on the CH), found in alanine and leucine.

Some areas of the enzyme surface are believed to contain amino acids with reactive side chains as a part of their structure. The reactive amino acid side chains may contain charged groups like the carboxyl and amino groups (that is, $RCOO^-$ and RNH_3^+) found in aspartic and glutamic acids or lysine and arginine. Noncharged portions like the hydroxyl and sulfhydryl groups (that is, ROH and RSH) found in serine, tyrosine, and cysteine are also reactive. Also present in amino acids are other types of reactive groups, such as histidine, which has an active nitrogen in a ring structure. These active areas of the enzyme may be on the surface or may exist in clefts or folds in the enzyme surface. These areas can be involved in the binding of substrates, products, activators, and inhibitors or be involved in the catalytic process itself. Catalysis can take place in only a limited number of places on the enzyme; these areas are called *active sites* or *active centers* and may involve only five to ten amino acids out of a total of 200 to 300 amino acids in the entire enzyme. The active site that has catalytic properties also serves to bind the substrate in a specific way to facilitate the breaking and forming of new bonds. It is believed that certain areas of the catalytic site contain reactive amino acids that bind portions of the substrate through ionic and hydrogen bonds. The substrate is positioned so that other reactive amino acids at the active site cause conversion to product.

The sites on the surface of the enzyme that serve to bind the substrate or product of the reaction are termed *binding sites*. Enzymes, particularly those of complex structure with several subunits, often have other sites that are far removed from the active site but seem to affect enzyme activity. These sites are called *allosteric sites* or *regulatory sites,* and much information about enzyme mechanisms has been derived from studies with inhibitors that seem to affect these sites and consequently the enzyme activity.

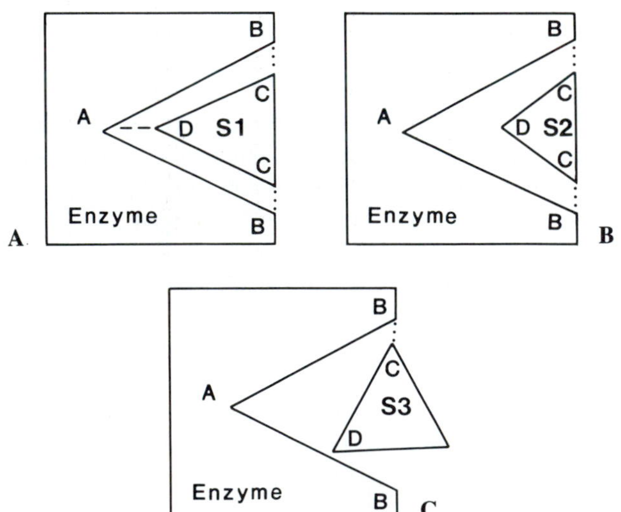

Fig. 52-2 Active site on enzyme is at point *A*, and binding sites are at points *B*. **A,** Correct substrate, *S1*, has complementary binding sites at *C*, and active site can react at point *D* on substrate. **B,** Substrate *S2* has complementary binding sites, but point *D* is too far away for catalysis to take place. If *S2* were present with *S1*, it could act as an inhibitor depending on relative binding constants by preventing *S1* from binding. **C,** Substrate *S3* has only one complementary binding site, *C*, and point *D* is not aligned correctly for catalysis.

A few schematic examples of different substrates with an enzyme are given in Fig. 52-2. Only the specific substrate S1 (in Fig. 52-2, *A*) properly fits (binds) or induces the fit at the active site and therefore enters into the reaction. Although one would expect great diversity among types of catalytic sites, a number of common features have been observed.

The substrate of the reaction binds to the active site and is oriented so that a particular bond is subject to attack. The reactive side-chain moieties of the enzyme interact with the group on the substrate so that the covalent bond to be altered becomes weakened. This bond weakening decreases the activation energy needed for chemical reaction. The weakened bond now undergoes a chemical reaction that breaks the covalent bond and allows new ones to form. The modified substrate, that is, product, no longer has the same affinity for the active site as the original substrate and is released from the enzyme.

Changes in the amino-acid sequence of a protein could produce different enzymes presumably with different active and binding sites, or even similar proteins without catalytic activity. Such changes, caused by genetic mutations, are often the cause of *inborn errors of metabolism* and other diseases of genetic origin (see Chapter 46).

The chemical reactions in which these reactive amino acids take part not only define the enzyme's specific catalytic activity but also determine the sensitivity of the enzyme to losses of activity by such factors as heavy metals, detergents, or even other reactive parts of the same protein molecule. Metals or detergents can bind to active groups and inactivate them. Changes in surface tension, that is, vigorous shaking, can cause unfolding of the protein, or denaturation. As a result, the spatial relationships of these reactive amino acids with each other are disrupted; thus the usual reaction is prevented from taking place.

Specificity of reaction

Differences in enzyme specificity are believed to be related to physical differences at the active site. Some enzymes will react with many related compounds and are said to have a broad specificity. Acid phosphatase (EC 3.1.3.2) is one of these enzymes. It exhibits a broad *bond specificity* for the hydrolysis (that is, addition of water to a bond) of many types of organic phosphate ester bonds, such as β-glycerol phosphate, thymolphthalein phosphate, *para*-nitrophenyl phosphate, and α-naphthyl phosphate. At an acid pH, the enzyme-catalyzed reaction

$$R{-}O{-}P + H_2O \xrightarrow{\text{Acid phosphatase}} R{-}O{-}H + P_i$$

produces an organic alcohol and inorganic phosphate (P_i).

Many proteases also exhibit a broad bond specificity,

hydrolyzing a large number and variety of peptide bonds within a protein substrate. If the peptide bonds that are hydrolyzed are located in the inside of the protein, the enzyme is called an *endopeptidase,* such as pepsin A (EC 3.4.23.1). Alternatively, carboxypeptidases are enzymes that act on protein substrates cleaving peptide bonds from the carboxyl terminus of the substrate toward the middle of the protein. These enzymes are termed *exopeptidases* because they hydrolyze proteins from one end of the peptide substrate, and they also demonstrate a broad substrate specificity.

In contrast to the broad specificity of many peptidases, other enzymes are more specific in their action and will catalyze only a definite reaction with a few substrates. In extreme cases, an almost absolute specificity is demonstrated and only a single compound will serve as a substrate for an enzyme reaction, such as phospho*enol*pyruvate for the pyruvate kinase (PK, EC 2.7.1.40) reaction.

$$\left.\begin{array}{c}\text{Phospho}\textit{enol}\text{pyruvate}\\ \text{(PEP)}\\ +\\ \text{Adenosine diphosphate}\\ \text{(ADP)}\end{array}\right\} \underset{}{\overset{\text{PK}}{\rightleftharpoons}} \left\{\begin{array}{c}\text{Pyruvate}\\ \text{(PYR)}\\ +\\ \text{Adenosine triphosphate}\\ \text{(ATP)}\end{array}\right.$$

Enzyme specificity should be described for each substrate involved in a reaction. In contrast to the absolute specificity shown for phospho*enol*pyruvate in the pyruvate kinase reaction, several natural and synthetic nucleoside diphosphates, such as UDP, IDP, GDP, and CDP will also serve as phosphate acceptors in the reaction in place of adenosine diphosphate (ADP).[1] Thus although an absolute specificity is shown for one substrate, an intermediate degree of specificity is shown for the other substrate.

An intermediate degree of specificity for each substrate is seen in the hexokinase (HK, EC 2.7.1.1) reaction, in which D-glucose and several other sugars may be phosphorylated, that is, D-mannose, 2-deoxy-D-glucose, and D-glucosamine. However, D-galactose and 5-carbon sugars like D-xylose are not substrates. The enzyme can also use a variety of nucleoside triphosphates as phosphate donors, such as inosine triphosphate (ITP) and guanosine triphosphate (GTP) as well as adenosine triphosphate (ATP).[2]

$$\left.\begin{array}{c}\text{D-Glucose}\\ +\\ \text{Adenosine triphosphate}\\ \text{(ATP)}\end{array}\right\} \xrightarrow{\text{HK}} \left\{\begin{array}{c}\text{Glucose-6-phosphate}\\ +\\ \text{Adenosine diphosphate}\\ \text{(ADP)}\end{array}\right.$$

Many enzymes demonstrate a *stereoisomeric specificity* for either the L-form or the D-form of a pair of compounds. Hexokinase is absolutely specific for the D-form of glucose; the L-form is not a substrate. Malate dehydrogenase (EC 1.1.1.37) acts only on the L-form of malate, not the D-form. Lactate dehydrogenase (EC 1.1.1.27) acts only on L-lactate, not D-lactate. However, stereoisomeric specificity does not necessarily mean that the enzyme is absolutely substrate specific, since some forms of lactate dehydrogenase act on hydroxybutyrate as well as on lactate, and, as mentioned earlier, hexokinase acts on several D substrates.

Subunits

Some enzymes have been shown to be composed of subunits, but only the multiple subunit form has been found to have significant enzyme activity. The enzymes whose subunit structures have been most widely studied have been creatine kinase, CK, and lactate dehydrogenase, LD. (See Isoenzyme sections in this chapter and Chapter 53.)

Anabolism and catabolism

The synthesis of enzymes is assumed to be carried out by intracellular protein synthetic pathways within the tissues that contain the enzymes. Extracellular enzymes like those involved in the coagulation process are synthesized in the liver and secreted into the plasma. In some cases other organs, that is, kidney, lung, and pancreas, also contribute to the extracellular enzyme pool.

The large size of enzymes and their complexity of structure result in molecular forms that are somewhat unstable and are therefore said to be *labile.* Many enzymes in vitro lose their catalytic activity with relatively slight changes in pH, temperature, or even salt concentration of the surrounding medium. It is presumed that similar processes occur intracellularly and that constant, though minimal, synthesis of enzymes occurs in a steady-state fashion to maintain the required amounts of intracellular enzymes needed for intermediary metabolism.

A loss of enzyme activity can be either reversible and temporary or irreversible and permanent. *Denaturation* is a process whereby biological properties are lost by a protein: that is, enzyme activity is lost. Experimental evidence suggests that the denaturation process is an unfolding or "melting" of tightly coiled peptide chains leading to a more disorganized structure. *Irreversible denaturation* can occur when the enzyme protein chains unfold and are unable to refold to their biologically active form, or when a heavy metal ion (such as mercury or lead) binds tightly at or near the active site. Many other factors and events can lead to denaturation and loss of activity including changes in temperature (heating or cooling), the addition of strong acids or bases, exposure to high pressure, treatment with ultraviolet light, repeated freezing, and the addition of detergents or organic solvents or the presence of high concentrations or urea or guanidine.

A *reversible denaturation* or loss of enzyme activity is called *inactivation.* For example, inactivation can occur if an enzyme solution is allowed to remain for some time at room temperature and the enzyme partially loses activity. This temporary activity loss can have several causes including heat instability with the breaking of hydrogen bonds or oxidation of sulfhydryl groups. In both of these cases, there is some loss of the natural structural form. With some enzymes, reducing the temperature of the solution or adding a sulfhydryl reducing agent like dithiothreitol may allow the enzyme to refold to the original active form, with reformation of hydrogen bonds or reduction of oxidized sulfhydryl groups, thus producing a *reactivation* of the enzyme and a restoration of the lost activity.

Little is known about the mechanism of removal of enzyme proteins from the extracellular fluid compartment. Presumably, extracellular proteases inactivate enzymes that are lost from cells to the extracellular compartment, and the inactive proteins are then removed by a mechanism utilizing one of several excretory routes, that is, excretion in bile or by the intestine, liver, kidney, or reticuloendothelial system. In addition, it is known that different enzymes have different half-lives, a finding that indicates that several mechanisms of removal may be present.

ENZYME NOMENCLATURE AND CLASSIFICATION

Many enzymes were first named for their function (such as lactate dehydrogenase), but some have also been named for the type of substrate on which they act: urease hydrolyzes urea, lipase hydrolyzes lipids, and phosphatases act on organic phosphates. Many clinically important enzymes are still known by these trivial names, which arose from historic circumstances and will continue to pervade the literature because of their simplicity. A more organized approach was needed because of the increasing complexity of names found in the literature and the increasing number of enzymes being discovered and studied. The Enzyme Commission (EC) of the International Union of Biochemistry (IUB) developed and proposed a systematic convention for the naming of enzymes.[3]

IUB names and codes

The IUB systematic name describes the reaction catalyzed. The IUB also recognized that trivial names were important and assigned practical names to many enzymes but no abbreviations. For each enzyme the system provides a numeric code designation consisting of four numbers separated by periods. The first number assigns the enzyme to one of six categories of reaction. The next two numbers are the subclass and the sub-subclass of the reaction, and the last is a serial number unique to the enzyme for the sub-subclass. A more complete description follows.

EC classification

Since the EC classification of the enzymes will be used increasingly in the future, a description of its basis is given. All enzymes are placed in one of six classes depending on the type of reaction they catalyze. A few clinically important enzymes are listed in Table 52-1 along with the EC code and systematic names of each.

The first class includes the *oxidoreductases,* those enzymes that catalyze electron transfer or oxidation-reduction reactions, which can be illustrated schematically as follows:

$$A_{red} + B_{ox} \rightleftharpoons A_{ox} + B_{red}$$

Some common names of enzymes in this category include dehydrogenases, reductases, oxidases, and peroxidases. If the reaction involves a direct participation of oxygen, the enzymes are termed *oxidases.* If the reaction involves the transfer of hydrogen, as in the change from L-malate to oxaloacetate, the enzymes are known as *dehydrogenases.*

The second group of enzymes contains the *transferases,* those enzymes that catalyze the transfer of a group, such as an amino, carboxyl, glucosyl, methyl, or phosphoryl group, from one molecule to another. These reactions can be illustrated as follows:

$$A{-}X + B \rightleftharpoons A + B{-}X$$

The term *aminotransferase* may be replaced by the term *transaminase.* Other common enzyme names in this category include kinases, transferases, and transcarboxylases.

A third group includes the *hydrolases,* which catalyze the cleavage of C—O, C—N, C—C and some other bonds with the addition of water. These hydrolysis reactions can be illustrated as follows:

$$A{-}B + H_2O \rightleftharpoons A{-}OH + B{-}H$$

Some common enzyme names in this category are acid phosphatase, amylase, urease, pepsin, trypsin, chymotrypsin, and various peptidases and esterases.

A fourth group contains the *lyases,* which hydrolyze C—C, C—O, and C—N bonds by elimination, with the formation of a double bond, or catalyze the reverse reaction, the addition of a group to a double bond. In cases where the reverse reaction is important the term *synthase* (not the group EC 6 synthetase) is used in the name. This type of reaction is illustrated as follows:

$$A + B \rightleftharpoons AB \text{ (synthase)}$$

or

$$AB \rightleftharpoons A + B \text{ (lyase)}$$

An examination of the EC listing shows that this and the subsequent groups contain relatively few enzymes that are used in clinical diagnosis.

The fifth group includes the *isomerases,* which catalyze structural or geometric changes in a molecule. They are called epimerases, isomerases, or mutases depending on the type of isomerism involved. This reaction can be illustrated as follows:

$$ABC \rightleftharpoons CAB$$

An example of this group is the enzyme glucose-phosphate isomerase (EC 5.3.1.9), which catalyzes the change of glucose-6-phosphate into fructose-6-phosphate; the two

Table 52-1 Examples of enzyme nomenclature

EC code	Recommended name (trivial)	Abbreviation*	Systematic name	Other name or abbreviation
Oxidoreductases				
1.1.1.27	Lactate dehydrogenase	LD	L-Lactate: NAD^+ oxidoreductase	LDH
1.1.1.37	Malate dehydrogenase	MD	L-Malate: NAD^+ oxidoreductase	MDH
1.1.1.42	Isocitrate dehydrogenase ($NADP^+$)	ICD	*threo*-Ds-Isocitrate: $NADP^+$ oxidoreductase (decarboxylating)	
1.1.1.49	Glucose-6-phosphate dehydrogenase	GPD	D-Glucose-6-phosphate: $NADP^+$ 1-oxidoreductase	G6PDH
1.4.1.2	Glutamate dehydrogenase	GMD	L-Glutamate: NAD^+ oxidoreductase (deaminating)	—
1.16.3.1	Ferroxidase	—	Iron(II): oxygen oxidoreductase	Ceruloplasmin
Transferases				
2.1.3.3	Ornithine carbamoyltransferase	OCT	Carbamoylphosphate: L-ornithine carbamoyltransferase	Ornithine carbamyltransferase
2.3.2.2	γ-Glutamyl transferase	GGT	(5-Glutamyl)-peptide: amino acid 5-glutamyl transferase	—
2.6.1.1	Asparate aminotransferase	AST	L-Aspartate: 2-oxoglutarate aminotransferase	Glutamic oxaloacetic transaminase/SGOT
2.6.1.2	Alanine aminotransferase	ALT	L-Alanine: 2-oxoglutarate aminotransferase	Glutamic pyruvic transaminase/SGPT
2.7.1.1	Hexokinase	HK†	ATP: D-hexose-6-phosphotransferase	—
2.7.1.40	Pyruvate kinase	PK	ATP: pyruvate 2-*O*-phosphotransferase	—
2.7.3.2	Creatine kinase	CK	ATP: creatine *N*-phosphotransferase	CPK
Hydrolases				
3.1.1.3	Triacylglycerol lipase	LPS	Triacylglycerol acyl hydrolase	Lipase
3.1.1.8	Cholinesterase	CHS	Acylcholine acyl hydrolase	Pseudocholinesterase
3.1.3.1	Alkaline phosphatase	ALP	Orthophosphoric-monoester phosphohydrolase (alkaline optimum)	—
3.1.3.2	Acid phosphatase	ACP	Orthophosphoric-monoester phosphohydrolase (acid optimum)	—
3.1.3.5	5′-Nucleotidase	NT	5′-Ribonucleotide phosphohydrolase	—
3.2.1.1	α-Amylase	AMS	1,4-α-D-Glucan glucanohydrolase	Diastase
3.4.11.1	Aminopeptidase (cytosol)	LAS‡	α-Aminoacyl-peptide hydrolase (cytosol)	Arylaminadase/LAP; leucine aminopeptidase
3.4.21.1	Chymotrypsin	—	None (preferred cleavage: Tyr, Trp, Phe, Leu)	Chymotrypsin A and B
3.4.21.4	Trypsin	TPS	None (preferred cleavage: Arg, Lys)	α- and β-trypsin
Lyase				
4.1.2.13	Fructose-bisphosphate aldolase	ALS	D-Fructose-1,6,-bisphosphate: D-glyceraldehyde-3-phosphatelyase	Aldolase
4.2.1.24	Porphobilinogen synthase	—	5-Aminolevulinate hydrolyase	—
Isomerases				
5.3.1.1	Triosephosphate isomerase	TPI	D-Glyceraldehyde-3-phosphate: ketol-isomerase	Triosephosphate mutase
5.3.1.9	Glucosephosphate isomerase	GPI	D-Glucose-6-phosphate: ketol-isomerase	Phosphohexose isomerase
Ligases				
6.3.1.2	Glutamine synthetase	—	L-Glutamate: ammonia ligase (ADP-forming)	—

*Baron, DN, et al: J Clin Pathol 24:656-657, 1971 (ref. 4) and Baron, DN, et al: J Clin Pathol 28:592-593, 1975 (ref. 5) are *not* recommended by the International Union of Biochemistry but are in common use.
†Not listed in references 4 and 5 but in common use in biochemistry laboratories.
‡Reference 5 incorrectly lists (EC 3.4.11.2) the microsomal form of this enzyme as "leucine aminopeptidase."

compounds mentioned have exactly the same number of atoms ($C_6H_{13}O_9P$) except that they are arranged into a slightly different structure in each compound.

A sixth and last group consists of the *ligases* (synthetases). In this reaction, the joining of two molecules is coupled with the hydrolysis of the pyrophosphate in ATP. Many of these enzymes are involved in DNA, RNA, and protein synthesis; some are currently used in clinical diagnosis. The synthetic reaction type is illustrated as follows:

$$A + B + ATP \rightleftharpoons AB + ADP + P_i$$

An example of this reaction is the enzyme glutamine synthetase (EC 6.3.2.1), which is an enzyme involved in an alternate pathway for the removal of ammonia.

In the EC code number, the second number denotes the subclass, which is often based on the type of group, such as amino group or hydroxyl group, that takes part in the reaction. The third number of the EC code indicates the sub-subclass of a reaction, often the acceptor group, and the last number is merely the serial number of the particular enzyme in this sub-subgroup. Thus, for the enzyme lactate dehydrogenase (EC 1.1.1.27) the first number, 1, indicates that the enzyme is an oxidoreductase; the second number, 1, indicates that the enzyme acts on the CH—OH group of donors; the third number, 1, indicates that the acceptor is NAD^+ or $NADP^+$; and the fourth number, 27, is merely the serial number of the enzyme in the EC 1.1.1.*x* group.

Isoenzymes

Multiple natural forms of an enzyme in a single species are known as *isozymes* or *isoenzymes*.[6,7] The IUB has designated that this term is to apply only to those forms of enzymes arising from genetically determined differences in primary (that is, amino acid) structure, though there is not complete agreement on this designation. Isoenzymes are usually distinguished on the basis of electrophoretic mobility and denoted by subscripts, with the first form having the mobility closest to the anode (+). Despite many reports in the literature to the contrary, isoenzymes are not to be labeled on the basis of tissue distribution (that is, heart type, brain type), since some confusion may arise because of differences in the predominant form found in various species. Three classes of multiple enzyme forms have been defined as isoenzymes by the IUB and are grouped as follows: genetically independent proteins, such as mitochondrial and cytosol forms of malate dehydrogenase (EC 1.1.1.37); heteropolymers of two or more different subunits, such as CK (EC 2.7.3.2) and LD (EC 1.1.1.27); genetic variants in protein structure, such as glucose-6-phosphate dehydrogenases (EC 1.1.1.49), with more than 50 varieties known in humans.

The polymeric forms of glutamate dehydrogenase (EC 1.4.1.2) are not isoenzymes by this definition, since they are polymers of a single subunit. Variations in the cleavage of a terminal segment of a protein chain can produce isoenzymes that do not fit the strict IUB definition. Hexokinase (EC 2.7.1.1) and carbonate dehydratase (EC 4.2.1.1) are examples of this type of isoenzyme. A more complete description of isoenzymes is given in Chapter 53.

Nonstandard abbreviations

Simple abbreviations containing four or fewer capital letters are also used to represent the enzymes in common usage. These abbreviations, suggested by Baron et al.[4,5] as a useful addition, are widely used in practice but are *not* part of the IUB system. These abbreviations are so convenient and have become so commonly used that it would be difficult to discard them completely. These nonstandard abbreviations are included in Table 52-1 along with the IUB systematic names.

MEASUREMENT OF ENZYMES

In most enzymatic procedures, the reaction rates are not constant with time. By observing the rate of change of absorbance at some wavelength, such as 340 nm, the amount of substrate converted to product can be followed. Initially, there is a *lag phase* with little change of absorbance per unit time when the reactants are mixed and reach thermal and kinetic equilibrium, then a *linear phase* with a constant absorbance change per unit time, and finally a *substrate depletion phase* with little change of absorbance per unit time.

Enzyme assays must be performed during the linear phase of absorbance change when a constant amount of activity can be determined for a period of time (Fig. 52-3, *A*). Thus measurements do not start at zero time but begin after the lag phase has occurred. Measurements can be made at any time during the linear phase and can continue up to the substrate depletion phase. The time course for the same assay with increasing amounts of enzyme present is shown in Fig. 52-3, *B*. There is a time period when all three enzyme activities can be measured with the same assay conditions, but the length of time of the linear phase is different for each amount of enzyme.

If the lag phase is very short and the linear phase is relatively long, the activity can be measured from zero time up to the substrate depletion phase with only a small error. Many older manual assays included the lag phase in the determination of enzyme activity because the measurement was started at zero time rather than after the lag phase was completed. The result of timing assays this way was that halving or doubling the time of assay only *sometimes* halved or doubled the amount of product produced. The activity thus measured would be either half or double that originally measured only if the lag was relatively insignificant. In an absolute sense, if the rate is taken from the linear phase, one can usually assay samples with high activity by halving the measurement time and calculating

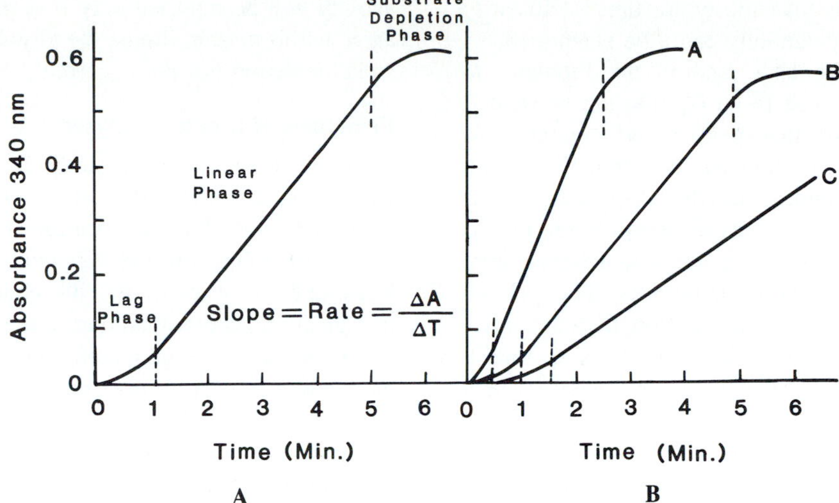

Fig. 52-3 **A,** Typical enzyme reaction with initial lag phase, linear change of absorbance, and final phase of substrate depletion. Enzyme activity is slope of linear phase. **B,** Time course of an enzyme reaction with three different amounts of enzyme present. Curve *A* has a high activity, *B* has a medium activity, and *C* has a low activity. As enzyme activity is increased in an assay system, lag phase decreases, linear phase decreases, and substrate depletion occurs sooner. ΔA, Change of absorbance; ΔT, change of time.

a new $\Delta A/\Delta T$. An alternate way to handle samples with high activity is to dilute them twofold or threefold with saline or water.[7] However, not all enzymes demonstrate linearity on dilution, particularly if the enzymes are active at a lipid-water interface, such as lipase (EC 3.1.1.3), or if there are inhibitors present in the sample, such as LD (EC 1.1.1.27) when measured in urine.

One of the more convenient methods of assaying enzyme activity is based on measurement of the absorbance of either the substrates or the products. Some enzyme systems involve the conversion of nicotinamide adenine dinucleotide (NAD^+) to its reduced form (NADH), or vice versa. The reduced form, NADH, has a much greater absorption at 340 nm than does the oxidized form, and consequently, reactions that convert one form to the other may be conveniently followed by measurement of the change in absorption at this wavelength. The difference in the absorption spectrum of the reduced and oxidized compounds is shown in Fig. 52-4.

In some instances, the NAD^+ may be replaced by $NADP^+$ (nicotinamide adenine dinucleotide phosphate), producing NADPH, which has properties similar to those of NADH. In addition, many other enzyme reactions that do not involve the NAD^+/NADH change directly can be coupled with another reaction involving NAD^+, so that the change in the NAD^+ becomes a measure of the enzyme reaction. Thus a number of enzyme reactions can be followed by measurement of the absorbance change at 340 nm.

Enzyme assays

Enzymes have been measured by several different techniques. The two most commonly employed methods are the one-point method at a fixed time, sometimes called the *end-point method,* and the multipoint fixed-time assay, called the *kinetic method.*[8]

In both the one-point and multiple-point methods, fixed time is a somewhat arbitrary term, since most current enzyme assays are performed for a fixed time of a few min-

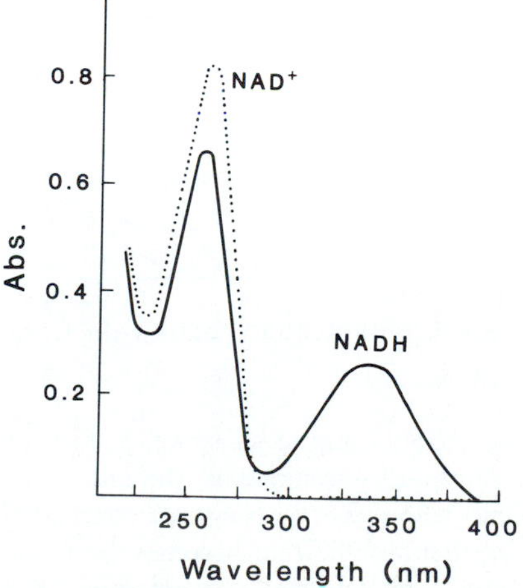

Fig. 52-4 Absorption spectrum, *Abs.*, of 5×10^{-5} M NAD^+ in 0.1 M Tris buffer, pH 7.5, and absorption spectrum of 4×10^{-5} M NADH in 0.1 M Tris buffer, pH 9.5.

utes, which is often less than the entire time of linearity during which the assay potentially could be performed.

End-point methods are also used in the literature to mean assays that have come to an equilibrium or steady-state point, that is, assays that measure the amount of an analyte after no further reaction has occurred. End-point assays are still used in some cases, but in general shorter time periods are employed. In these assays a reaction is started and allowed to incubate at a constant temperature for a fixed time period, such as 30 minutes. The reaction is then stopped, perhaps by the addition of another reagent, and the amount of absorbance change is measured. The assumption for this type of assay is that a constant amount of product is produced throughout the entire assay period.

If the rate of reaction is followed continuously or with many points as a function of time, the assay is termed a *kinetic assay*. Usually the reaction time is short, that is, some seconds to a few minutes, and there is little danger of enzyme degradation. A *kinetic method* is a term often used to mean *continuous* monitoring of the progress of the reaction. Although this terminology is not strictly correct, the continuous or multiple-point assays are superior to the single-point, fixed-time assays, since it is easier to demonstrate approximate linearity of the reaction over the entire measurement period.

Current instrumentation often permits multiple absorbance readings of the reaction to be made automatically for the determination of enzyme activity. With some instruments, one can take hundreds of readings and average them to determine the rate or to calculate the rate several times during the assay period. In some cases, the enzyme

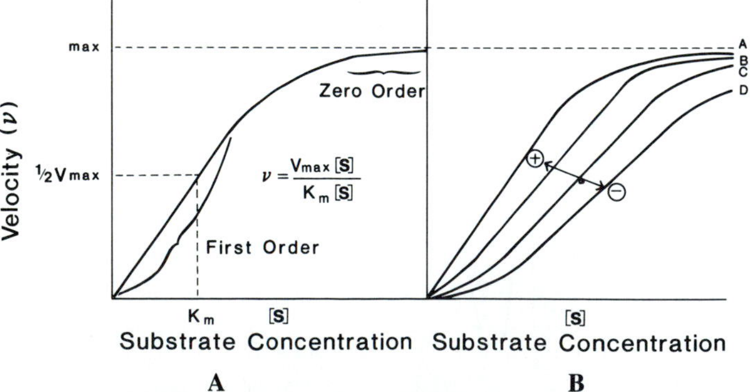

Fig. 52-5 **A,** Relationship of substrate, *S*, to velocity of reaction. At low substrate concentrations, the rate is first order (linearly dependent) with respect to substrate concentration. At high substrate concentrations the rate becomes zero order (independent) with respect to substrate concentration. K_m, Michaelis-Menten constant; V_{max}, maximal rate of reaction. **B,** Relationship between velocity and substrate concentration for an allosteric enzyme. Presence of positive or negative effectors shifts curve toward the + or − side respectively.

activity will be reported only if several rate measurements agree within certain limits, thereby demonstrating that substrate depletion has not occurred.

Principles of kinetic analysis

Enzyme kinetics is the study of enzyme reaction rates and the factors that affect them. Initially, many experiments are performed to examine the effects of different assay conditions on measurements of enzyme activity. Eventually, a series of specific conditions are established that give rise to the maximum rate of enzyme activity.

The general enzyme reaction given previously for a *single substrate reaction* may be rewritten slightly for *initial rates,* as shown in Equation 52-3. In this case, the amount of product is very small, and the reverse reaction of P combining with E and going back to form ES is ignored since initial rate measurements are to be made. The initial rate is the rate at the start of the reaction, after the lag phase and during the linear phase in Fig. 52-3, *A,* but before substantial product formation.

$$E + S \underset{k_{-1}}{\overset{k_{+1}}{\rightleftharpoons}} \{ES\} \xrightarrow{k_{+2}} P + E \qquad \textit{Eq. 52-3}$$

For a given quantity of enzyme, the rate of activity that is observed increases with increasing amounts of substrate as shown in Fig. 52-5, *A*. At low substrate concentrations the rate is linearly dependent on the amount of substrate, that is, *first order;* but at high substrate concentrations, the rate is essentially independent of substrate concentration, that is, *zero order*. A mathematical description of the reaction must explain how the reaction can be first order at low substrate concentrations and zero order at high substrate concentrations.

If the enzyme has a limited number of active sites, at a low substrate concentration the rate will be dependent on the amount of substrate present, since there will be a large effective concentration of unfilled active sites per substrate concentration. However, since the total number of sites, that is, enzyme, is limited, then as the amount of substrate is increased, the sites will become increasingly saturated with substrate until the reaction will appear to be independent of the substrate concentration. At these high substrate concentrations all the enzyme active sites are filled and the reaction proceeds at maximal velocity. Small changes in the substrate concentration will not affect the enzyme rate.

The second step, product formation, is assumed to be the rate-limiting step or the one that determines the overall activity. The equilibrium for the formation of the ES complex can be written as follows with the molar concentrations of all the reacting species expressed in brackets:

$$K_{eq} = \frac{k_{+1}}{k_{-1}} = \frac{[ES]}{[E]\ [S]}$$

The equilibrium constant, K_{eq}, is equal to the ratio of the forward over the reverse rate constants. From Equation 52-

3, the rate of formation of the product, P, is the amount of [ES] times the rate, k_{+2}, at which the enzyme complex is converted to P. Thus the rate of formation of product is expressed as follows:

$$\text{Velocity or rate} = [\text{ES}] \times k_{+2}$$

Since the rate is the amount of product formed for some period of time,

$$\text{Rate} = \frac{\Delta \text{P}}{\Delta \text{T}} = [\text{ES}] \times k_{+2}$$

substituting K_{eq} [E] [S] for [ES] and rearranging gives:

$$\Delta \text{P} = K_{eq} \times [\text{S}] \times [\text{E}] \times k_{+2} \times \Delta \text{T}$$

The amount of product formed is proportional to the amount of enzyme present, the time of the assay, and the amount of substrate present. When a proportionality constant is substituted for the rate constants, the equation becomes:

$$\Delta \text{P} = K1 \times [\text{S}] \times [\text{E}] \times \Delta \text{T}$$

where *ΔP* is the amount of product formed during the assay time, *[E]* is the amount of enzyme, *[S]* is the amount of substrate, *ΔT* is the assay time, and *K*1 is a proportionality constant. The enzyme activity or rate of product formation over time is then given by the following:

$$\text{Rate} = \frac{\Delta \text{P}}{\Delta \text{T}} = K1 \times [\text{S}] \times [\text{E}]$$

Usually enzyme assays are performed at a high substrate concentration for a short enough time period so that the substrate concentration can be assumed to be constant. The value of this constant substrate concentration can be combined with *K*1 to produce a second proportionality constant *K*2, which is the product of *K*1 times the substrate concentration. The rate can then be expressed so that it is dependent only on the amount of enzyme present, that is, a zero-order reaction, independent of substrate concentration.

$$\text{Rate} = \frac{\Delta \text{P}}{\Delta \text{T}} = K2 \times [\text{E}]$$

This rate of reaction or velocity is often listed as v, or V_i or V_o, in the enzyme kinetic literature.

K_m and V_{max}. The enzyme activity (that is, velocity) is dependent on the substrate concentration when the amount of substrate is low relative to the amount of enzyme present in an assay. This relationship for a single substrate reaction is shown graphically in Fig. 52-5, *A*, with the same enzyme concentration assayed at many different substrate concentrations.

At steady state, before much product is present, the rate of formation of the ES complex will equal the rate of breakdown. This can be described using the following rate equation:

Formation	**Breakdown**
k_{+1} [E] [S]	$= k_{-1}$ [ES] + k_{+2} [ES]

By collecting terms and rearranging, we can remove the rate constants and define a new constant, K_m.

$$\frac{[\text{E}]\,[\text{S}]}{[\text{ES}]} = \frac{k_{-1} + k_{+2}}{k_{+1}} = K_m \qquad \textit{Eq. 52-4}$$

The rate or velocity of product formation, v, at any time, and free enzyme concentration, [E], are related by the following:

$$v = k_{+2}\,[\text{ES}]$$

and

$$[\text{E}] = [\text{Et}] - [\text{ES}]$$

where [Et] is the total amount of enzyme and [ES] is the amount complexed with substrate. When all the enzyme is present in the form of ES (that is, at a very high [S] in a zero-order reaction), the maximum rate, V_{max}, is as follows:

$$V_{max} = k_{+2}\,[\text{Et}]$$

Combining the previous three equations gives:

$$[\text{E}] = \frac{V_{max}}{k_{+2}} - \frac{v}{k_{+2}} = \frac{V_{max} - v}{k_{+2}}$$

Since from Equation 52-4

$$[\text{E}] = \frac{K_m\,[\text{ES}]}{[\text{S}]}$$

and

$$[\text{ES}] = \frac{v}{k_{+2}}$$

then

$$\frac{K_m\,[\text{ES}]}{[\text{S}]} = \frac{V_{max} - v}{k_{+2}}$$

or

$$\frac{K_m \times v}{[\text{S}] \times k_{+2}} = \frac{V_{max} - v}{k_{+2}}$$

Rearranging gives:

$$K_m \times v = (V_{max} - v)\,[\text{S}]$$

or

$$v(K_m + [\text{S}]) = V_{max}\,[\text{S}]$$

When this equation is solved for v, it gives the *Michaelis-Menten equation,* which is the equation for the rectangular hyperbola shown in Fig. 52-5, *A*.

$$v = \frac{V_{max}\,[\text{S}]}{K_m + [\text{S}]} \qquad \textit{Eq. 52-5}$$

[S] is the concentration of substrate, v is the velocity or rate (that is, enzyme activity), V_{max} is the maximal rate of reaction when the enzyme is saturated with substrate (that is, when [S] is approximately a constant), and K_m, the Michaelis-Menten constant, is the substrate concentration that

Table 52-2 Enzyme activity as a function of substrate concentration (expressed as multiples of K_m)

Substrate × K_m	Activity (V_{max})
1	0.50
5	0.83
10	0.91
20	0.95
50	0.98
100	0.99

produces one half the maximal velocity (Fig. 52-5, *A*). Thus when K_m = [S]:

$$v = \frac{V_{max}\,[S]}{[S] + [S]} = \frac{1\cancel{[S]} \times V_{max}}{2\cancel{[S]}} = \frac{1}{2}V_{max}$$

At the fixed high substrate concentration found in the usual clinical laboratory assays, the velocity, v, approaches V_{max} and is proportional to the amount of enzyme present, since all other factors are constant. The reaction is said to be zero order with respect to substrate, that is, independent of the concentration of substrate. The usual condition used for assaying enzyme activity is a high substrate concentration where [S] ≅ 10 K_m or higher. The rate at a substrate concentration of 10 K_m is given by the following equation:

$$v = \frac{V_{max}\,(10\,K_m)}{K_m + (10\,K_m)} = V_{max}\frac{10\,K_m}{11\,K_m} = 0.91\,V_{max}$$

Thus at [S] = 10 K_m the rate produced is greater than 90% of V_{max}. The effect of various substrate concentrations on the enzyme activity is shown in Table 52-2.

If all other cofactors and activators are present in similar excess, the rate is said to follow zero-order kinetics and depends only on the amount of enzyme present. When this is so, changes in the enzyme activity are attributable to changes in the amount of enzyme present in a sample rather than to variations in the assay conditions. Another way to examine the Michaelis-Menten equation is to see if it is consistent with first-order kinetics at low substrate concentrations and zero-order kinetics at high substrate concentrations.

At low substrate concentrations where [S] << K_m, then

$$v = \frac{V_{max}\,[S]}{K_m + [S]} = \frac{V_{max}\,[S]}{K_m}$$

and since K_m and V_{max} are constants:

$$v = \mathrm{K1}[S]$$

showing that the rate is dependent only on the first power of the substrate concentration.

At high substrate concentrations where [S] >> K_m, then

$$v = \frac{V_{max}\,[S]}{K_m + [S]} \cong \frac{V_{max}\,\cancel{[S]}}{\cancel{[S]}} = V_{max}$$

showing that the rate (v) does not depend on substrate concentration.

As was shown in Fig. 52-5, *A*, the relationship of substrate concentration to enzyme activity is a curve that is often similar to a rectangular hyperbola. For multisubstrate enzyme reactions the kinetics are more complex and the reaction curves are similar to those shown in Fig. 52-5, *B*. The presence of activators and inhibitors acting at allosteric or regulatory sites tends to make the curves sigmoid because of the complex kinetics. A high concentration of activators, which may sometimes be the substrates themselves, and a low concentration of inhibitors (or products) tend to make these curves approach the case shown for a single substrate in Fig. 52-5, *A*. The presence of inhibitors will tend to make the curve more sigmoid, and thus they are to be avoided if possible in setting up enzyme assays. Reactions with two substrates must have the concentrations of both substrates optimized for maximal activity as well as any other activators needed for catalysis. For a more advanced treatment of kinetics see the text by Orten and Neuhaus.[9]

The accurate determination of K_m and V_{max} for each substrate or activator from Michaelis-Menten curves, such as those illustrated in Fig. 52-5, *A*, is very difficult. However, it is necessary to determine these constants so that assays may be established using optimal conditions to measure enzyme activity. If the curve is transformed into a straight line, the K_m and V_{max} can be determined with greater accuracy. One may transform the Michaelis-Menten equation mathematically and obtain the equation of a straight line in several ways. The K_m and V_{max} can then be graphically determined from the line slopes and intercepts by use of these transformed equations. Several common graphic presentations are shown in Fig. 52-6.

Calculation of enzyme activity

The result of an enzyme determination is expressed as an activity unit in terms of the amount of product formed per unit of time, under specified conditions, for a given amount (volume) of sample (usually serum). Thus one unit of enzyme activity might be the amount of enzyme that would, under certain specified conditions, cause the formation of 1 mg of the product, P, per minute when 1 mL of the sample was used. In older procedures arbitrary units like these were often employed.

Enzyme activity has been historically reported in the measuring units by the authors who developed the assay; for example, for alkaline phosphatase using β-glycerol phosphate as a substrate, 1 Bodansky unit equals 1 mg of phosphorus formed per deciliter of serum per hour; and using phenolphosphate as a substrate, 1 King-Armstrong unit equals 1 mg of phenol formed per deciliter of serum per 30 minutes. The units were expressed as so many King-Armstrong units or Bodansky units after those who developed the assay. Each time an assay was modified a

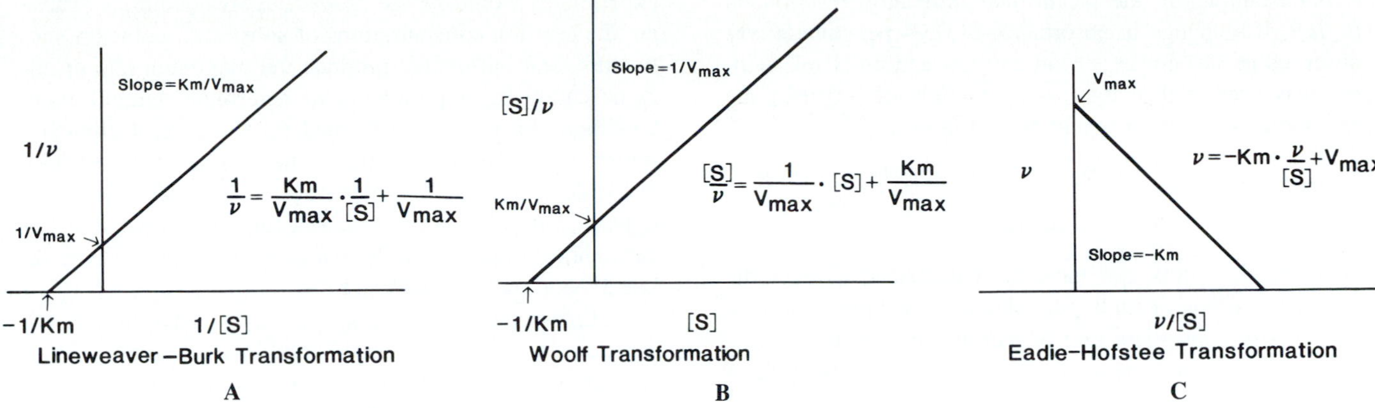

Fig. 52-6 Graphic representations of linear forms of Michaelis-Menten equation.

new unit would be created, which took on the name or names of the authors who proposed the modified assay. Such private units are no longer used, and all assays are now expressed in international units (U) or in Système International units (K, katal), which are described later.

In 1961 the Enzyme Commission (EC) recommended the adoption of an international unit of enzyme activity. This unit, U, was defined as the amount of enzyme that would convert 1 micromole of substrate per minute under standard conditions.

$$1 \text{ U} = 1 \text{ micromole/minute}$$

In those instances where one molecule of substrate is transformed into a number of molecules of product, the definition of activity is per molecule (micromole) of product formed.

This recommendation of an international unit was an attempt to standardize the units of assay and to reduce the large number of units in use. This unit has been widely adopted, and in some respects it has standardized assay units. It has not reduced the number of reference ranges because if the standard conditions change, the apparent enzyme activity changes. For example, if a new buffer were used in the assay, it might affect the enzyme rate and produce a different number of international units of activity, that is, a larger absorbance change per minute, and thus it would also change the reference range. Since the assay conditions are not usually included with the results, a clinician practicing at two different institutions might receive results from these institutions expressed in terms of two levels of international units for the same enzyme assay. To interpret the results properly, the physician must remember that the reference ranges are different.

The Système International d'Unités (SI), as originally adopted by the World Health Organization, established the unit of enzyme activity as the katal, K. This is defined as 1 mol/sec of substrate changed. This unit is too cumbersome to be useful clinically, and so it has met with little acceptance in the United States though it was recommended by the EC in 1972.

For conversion of international units to katals:

$$1 \text{ U} = \frac{\text{Micromole}}{\text{Minute}} \times \frac{10^{-6} \text{ mole}}{\text{Micromole}} \times \frac{1 \text{ min}}{60 \text{ sec}} = 1.67 \times 10^{-8} \text{ K}$$

Thus 1.0 U = 16.7 nK (nanokatals).

These international units have been adopted by most workers in the field of clinical enzymology, and professional journals such as *Clinical Chemistry* require the use of international units in papers presented for publication.

Pure human enzyme materials are not available. Thus enzyme assays cannot be standardized by calibration with pure materials. The alternative method that is used most widely for standardization requires an accurately calibrated spectrophotometer. Many enzyme assays are followed by making spectrophotometric measurements at a specific wavelength. With the spectrophotometric method usually one assumes that at 340 nm, NADH has a molar absorption coefficient of

$$A/(b \times c) = 6.22 \times 10^3 \text{ L} \cdot \text{mol}^{-1} \cdot \text{cm}^{-1}$$

where A is the actual absorbance of a solution, b is the light path in centimeters through the solution, and c is the concentration in moles per liter of the absorbing substance. Thus for a 1 cm light path, rearranging for c

$$c = \text{A} \times 10^{-3}/6.22 \text{ mol/L}$$

When the concentration is expressed in micromoles per liter (the international unit per liter) instead of moles per liter, the expression will be

$$c = A \times 10^3/6.22 \ \mu\text{mol/L}$$

Thus from the absorbance change that was measured and the volume of solution that was used, one can readily calculate the number of micromoles of NADH formed or used up during the enzyme measurement period.

$$c = \Delta A \times 10^3/6.22 \ \mu\text{mol/L}$$

For example, in the lactate dehydrogenase reaction on p. 769, if a change in absorbance of 0.04 per minute was observed at 340 nm in a 1 cm cuvette, and a 0.1 mL sample was used with a total assay volume of 3.0 mL, the calculation of activity would be as follows:

$$\text{International units/L} = \frac{0.04 \times 1000\ \mu\text{mol/mmol} \times 3.0\ \text{mL}}{6.22\ \text{mmol/L} \times 0.1\ \text{mL}}$$
$$= 193\ \text{U/L}$$

Both enzyme units that have been described express the activity in terms of units per volume of sample. This is a particularly convenient unit of measure in the clinical laboratory when one is assaying enzymes in biological fluids like serum and plasma. If one is measuring an enzyme found in red blood cells (RBC) or in white blood cells (WBC), another unit of measure is needed. In the case of RBC and WBC enzymes, the enzyme activity can be expressed as units per 10^{10} cells.

In biochemistry laboratories where enzyme purification is important, the activity might be expressed per milligram of protein or per dry weight of cells or per microgram of DNA, but these are not convenient units for the clinical laboratory.

Factors affecting enzyme measurements

The rate of reactions involving enzymes is greatly influenced by temperature, pH, concentration of substrate, and a number of other factors. Accordingly, all the details of a given procedure must be followed exactly so that precise and accurate results are produced.

Assays of enzyme activity should be performed under conditions of zero-order kinetics (that is, with the reaction rate dependent only on the amount of enzyme present) and with all reaction parameters optimized. If experiments are not performed to determine the substrate concentration that will result in maximal activity,[10,11] it may be that an assay will give different results if the substrate concentration is changed.

For optimization of an assay, such as the lactate dehydrogenase reaction given earlier, a series of reaction assays would be set up with increasing concentrations of lactate but with a high fixed NAD^+ concentration and a fixed amount of enzyme. The enzyme rates would then be measured, and a graph similar to Fig. 52-5, *A* would be constructed. A second series of assays would then be performed with increasing concentrations of NAD^+ but with the fixed high concentration of lactate determined from the first experiment, that is, $[S] \cong 10\ K_m$ for lactate, and with the same amount of enzyme present. The enzyme rates would again be determined and another figure created to determine the K_m for NAD^+. This same type of experiment would be performed for each item of the assay mixture (that is, metal ions, pH, buffer) until all the variables had been evaluated for the production of maximal enzyme activity. The final conditions determined from this set of experiments would be the *optimal assay* conditions. These are the reaction concentrations of substrates, cofactors, activators, and buffer that produce the maximum rate of enzyme catalysis. Experiments to determine optimal assay conditions have been performed for the current clinically important enzymes, and diagnostic kits are commercially available with all the materials at optimal concentrations.[10-12] It is important to check these concentrations because not all commercially available materials that are labeled "optimal" actually are.

At times optimal conditions cannot be used; that is, the substrate might have a limited solubility or might inhibit a secondary enzyme used in a coupled assay, and then compromises in the optimal assay conditions must be made.

pH. Changes in pH will greatly affect the enzyme reaction rate. For most enzymes there is a definite pH range where the enzyme is most active. A pH near the center of this range is usually specified for the measurement of that particular enzyme. The optimal pH is different for different enzymes. Reduced activity is observed at pH values greater or less than the optimal.

At pH values other than the optimal pH, the enzyme activity may be affected because of changes in the structure of the enzyme. These changes may occur at the active site or may be attributable to conformational changes affecting the three-dimensional structure. A change of pH might bring about an unfolding of the enzyme and loss of activity if the effect of pH change was to disrupt hydrogen bonds and other intramolecular forces holding the enzyme in an active conformation. There may also be changes in the charge of the enzyme or of the substrate. Since the active site of an enzyme often contains ionizable side chains of amino acids, such as $RCOO^-$ or RNH_3^+, a significant change in the pH can lead to the gain or loss of a proton. The result will be a substantial change in charge at the active site. The active site might therefore lose its ability to attract a substrate with a significant opposing charge. A similar loss of activity would occur if the change in charge were on the substrate molecule rather than on the enzyme.

Buffer. In many cases, as the enzyme reaction proceeds, the products tend to alter the pH. Most assays include a buffer to maintain the assay pH within the optimal pH range. The buffer chosen should have a pK_a within 1 pH unit of the optimal pH of the enzyme in order to exert effective pH control.

A typical pH curve of enzyme activity is given in Fig. 52-7. This bell-shaped curve showing maximal enzyme activity versus factor concentration (in this case pH) is seen for other assay factors, such as concentration of buffer, substrate, cofactors, and activators.

Buffers not only serve to regulate the pH of an assay, but they may also take part in the reaction. Alkaline phosphatase (ALP, EC 3.1.3.1) assays with *p*-nitrophenol phosphate as a substrate, as in the Bowers and McComb

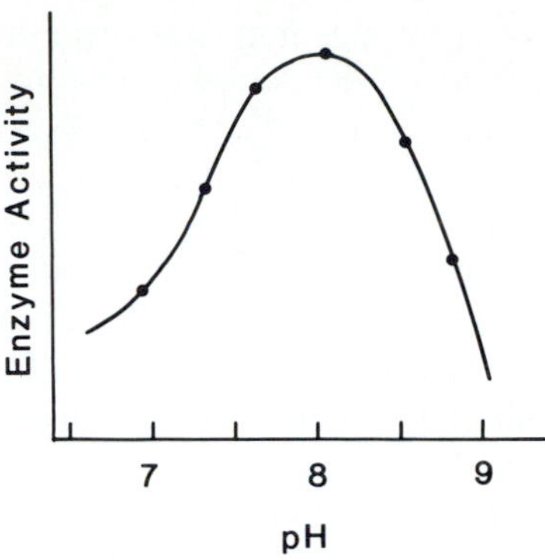

Fig. 52-7 Enzyme activity as a function of pH. Optimal pH range is 7.8 to 8.2; lower activities are observed at pH < 7.8 and pH > 8.2.

procedure,[13] use the buffer 2-amino-2-methyl-1-propanol (AMP) to maintain the pH at 10.2. The enzyme hydrolyzes the substrate into *p*-nitrophenol and inorganic phosphate in a multistep process, part of which involves a temporary phosphorylation of the enzyme. The final and rate-limiting step includes hydrolysis of the enzyme-phosphate bond to regenerate free enzyme. At similar pH values, buffers that are phosphate acceptors in a transphosphorylation process with the enzyme will produce rates of alkaline phosphatase activity that are higher than those seen with buffers that do not act as phosphate acceptors.[14] Thus AMP buffer produces higher rates of alkaline phosphatase activity at pH 10.2 than does glycylglycine buffer at pH 10.2 because glycine is not a phosphate acceptor. For any buffer, the concentration of buffer that gives maximal enzyme activity at the optimal pH must be experimentally determined.

It has been found that the buffer and certain salts may have an unusual effect on the K_m. When the buffer-to-substrate ratio is very large, the buffer may compete with the substrate for the enzyme and make the enzyme activity appear to be related to substrate concentration in a nonlinear way. This has been observed with NADH in the LD reaction.[15] Here the buffer-to-substrate molar ratio is $10^4:1$; and the rate of reaction is affected by Tris, phosphate, and NH_4HCO_3 buffers and certain salts, such as NaCl and $(NH_4)_2SO_4$, which are often found in the materials used to prepare enzyme assays. There seems to be no effect at buffer concentrations below 0.05 mol/L, a finding that is consistent with several recommendations for optimal assay conditions.[12,16,17] It would seem prudent to maintain a concentration of buffer as low as possible without compromising pH stability or enzyme rate.

Cofactors. Many enzymes require a nonprotein, often dialyzable material for maximal activity. Some of these materials are related to vitamin structures. For example, thiamin, or vitamin B_1, can be converted to thiamin pyrophosphate, a cofactor in many decarboxylation reactions. Niacin can be converted to nicotinamide adenine dinucleotide, and vitamin B_2, riboflavin, can be converted to flavin adenine dinucleotide. Both niacin and riboflavin are involved in many dehydrogenation reactions. Pyridoxine, vitamin B_6, is modified to pyridoxal phosphate, which is used in many transamination reactions.

In the lactate dehydrogenase reaction, NAD^+ is considered to be such a cofactor, but since it reacts on an equal molar basis with lactate, it is a substrate, though it is discussed as if it were a cofactor.

In analytical assays of transaminase activity, pyridoxyl-5-phosphate is an example of a cofactor that is not a substrate. The optimal concentration of a cofactor is determined in the same way as that of a substrate so that assay conditions can be established with a cofactor concentration of approximately 10 K_m or higher.

Activators and inhibitors. Many enzymes require specific ions for maximal activity. All phosphate-transferring enzymes, such as hexokinase, require magnesium ions (Mg^{+2}). Other common metal ion activators are manganese (Mn^{+2}), calcium (Ca^{+2}), zinc (Zn^{+2}), iron (Fe^{+2}), and potassium (K^+). Amylase requires chloride (Cl^-) for maximal activity, and there are enzymes that require several ions for maximal activity; for example, pyruvate kinase requires magnesium (Mg^{+2}) and potassium (K^+). In each case, the optimal concentration of the activator must be determined just as the optimal concentration of substrate is determined.

Inhibitors are materials that reduce the catalytic activity of an enzyme. There are many types of inhibitors and several classes of inhibition. Inhibitors may act by removing an activator by chelation; for example, Ca^{+2} and Mg^{+2} are removed by EDTA or oxalate in the inhibition of hexokinase. They may also act by binding to the active site to compete with the substrate, as shown in Fig. 52-2, or by forming a complex at a different site, that is, an allosteric site, which may affect the enzyme activity.

Inhibitors are classed into three main groups. *Competitive inhibitors* bind at the active site and compete with the substrate for binding sites. These materials demonstrate a reversible inhibition that can be reduced by use of a higher substrate concentration.

$$\begin{array}{l} \quad S \rightleftharpoons \{ES\} \rightarrow P + E \\ + \\ E \\ + \\ \quad I\ \{EI\} \end{array}$$

The maximum rate of reaction is not affected if enough substrate is present because of the reversibility of the reactions. The binding of the substrate is affected, and thus the apparent K_m will be higher while the V_{max} remains the same.

Noncompetitive inhibitors are a second group of materials that bind at an allosteric or regulatory site, which may be far removed from the active site. These inhibitors cannot be reversed by the addition of more substrate because they bind at a different location on the enzyme surface.

$$
\begin{array}{ccccc}
E & + S \rightleftharpoons & \{ES\} & \rightarrow P + E \\
+ & & + & \\
I & & I & \\
\downarrow\uparrow & & \downarrow\uparrow & \\
\{EI\} & + S \rightleftharpoons & \{ESI\} &
\end{array}
$$

Since the inhibitor does not compete with the substrate, the K_m will be unaffected, but the amount of E or ES that converts substrate to product will be reduced and the V_{max} will be lessened.

Uncompetitive inhibitors, a third group of inhibitors, are believed to bind to the enzyme substrate complex and not to the free enzyme. In this case, at low substrate concentrations, the addition of more substrate increases the inhibition, since it produces more enzyme-substrate complex to react with the inhibitor. The result of this type of inhibition is that the V_{max} is reduced and the K_m is decreased.

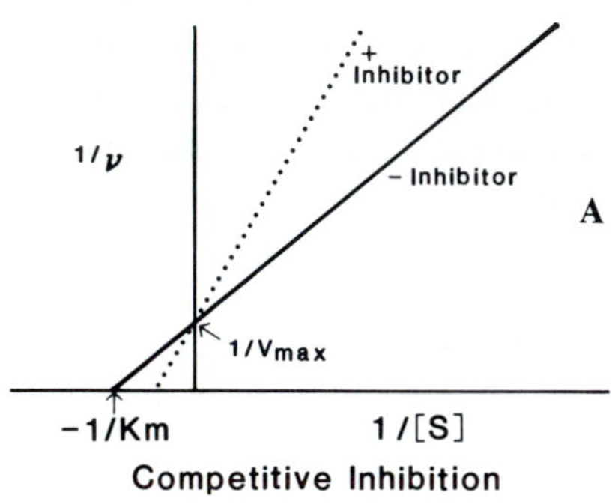

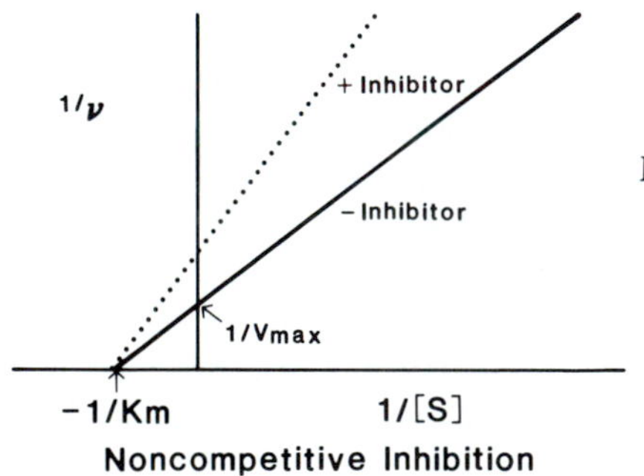

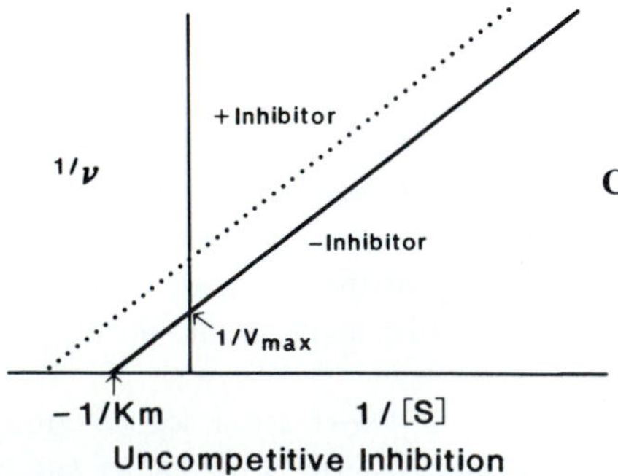

Fig. 52-8 The three types of inhibition are shown by use of Lineweaver-Burk graphic method to demonstrate effect of type of inhibition of K_m and V_{max}.

$$
\begin{array}{c}
E + S \rightleftharpoons \{ES\} \rightarrow P + E \\
+ \\
I \\
\downarrow\uparrow \\
\{ESI\}
\end{array}
$$

The type of inhibition a substance exerts can be determined when one examines the results of kinetic studies using a linear graph of enzyme activity. The studies are performed with and without inhibitors present, as shown in Fig. 52-8.

A brief summary of the effects of different types of inhibition is given in Table 52-3. The simple types of inhibition may be classified by examination of the kinetic effect of the inhibitor on the K_m and V_{max}.

Coupling enzymes. In some enzyme reactions of interest, such as alanine aminotransferase (ALT) and aspartate aminotransferase (AST), products that can be monitored directly are not formed. One may couple the initial enzyme reaction to a second *indicating* enzyme reaction that, for example, does contain the NAD^+/NADH conversion to make a convenient assay. The AST enzyme reaction can be coupled to the malate dehydrogenase reaction (MD, EC 1.1.1.37).

$$\text{L-Aspartate} + \alpha\text{-Ketoglutarate} \overset{AST}{\rightleftharpoons} \text{L-Glutamate} + \text{Oxaloacetate}$$

$$\text{Oxaloacetate} + \text{NADH} + H^+ \overset{MD}{\rightleftharpoons} \text{L-Malate} + NAD^+$$

This gives the following net reaction:

$$\text{L-Aspartate} + \alpha\text{-Ketoglutarate} + \text{NADH} + H^+ \rightleftharpoons \text{L-Glutamate} + \text{L-Malate} + NAD^+$$

In this case, the substrate for the second reaction, oxaloacetate, is supplied as the product of the first reaction. Oxaloacetate from the AST reaction serves as the substrate, along with the cofactor NADH, for the malate dehydrogenase reaction. This assay would have large excesses of L-aspartate, α-ketoglutarate, NADH, and the enzyme malate dehydrogenase (MD) present so that the rate-limiting item in the assay would be the amount of AST in the sample.

For other enzymes, such as creatine kinase (CK, EC 2.7.3.2), the measurement of the first enzyme requires an intermediate *auxiliary* enzyme reaction and then an *indicator* enzyme. In the measurement of CK, hexokinase (EC 2.7.1.1) is used as an *auxiliary* enzyme and glucose-6-phosphate dehydrogenase (EC 1.1.1.49) is used as an *indicating* enzyme. Large excesses of both of these enzymes

Table 52-3 Kinetic effects of inhibition

Type of inhibition	Change in K_m	Change in V_{max}
Competitive	Increased	No change
Noncompetitive	No change	Decreased
Uncompetitive	Decreased	Decreased

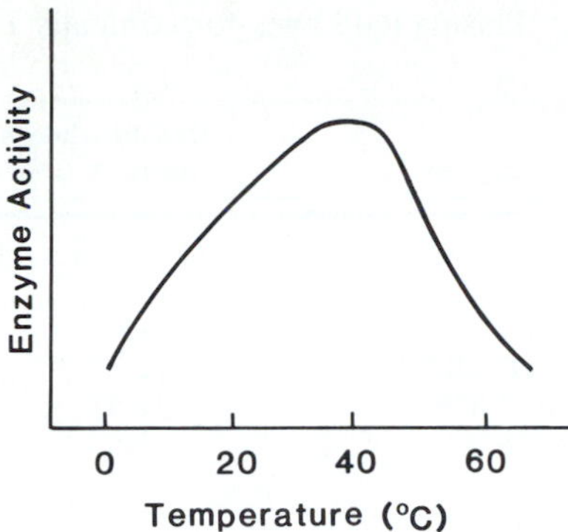

Fig. 52-9 Enzyme activity as a function of temperature of assay. At low temperatures activity decreases. As temperature is raised, activity increases until rate of denaturation is greater than increase in activity.

as well as their substrates, would have to be present for correct measurement of CK. It is difficult to establish optimal assays that have more than two coupled reactions because of the large number of components in the assay system and the problems with maximizing all the components without causing inhibition of the limiting reaction.

Temperature. There is no optimal temperature for enzyme assays. Most enzymes show an increasing amount of activity as the temperature is raised over a limited temperature range, such as, 10° to 40° C; an example is shown in Fig. 52-9.

To minimize any losses of activity if the enzyme cannot be assayed immediately after collection, one should store samples at refrigerator temperatures, 2° to 6° C, or frozen (Table 52-4). In a few cases, some forms of enzymes, such as LD_4 and LD_5, have been found to be more stable at room temperature than at refrigerator temperatures. The repeated freezing and thawing of a specimen will often cause denaturation and loss of activity. Above 40° C most enzymes are rapidly denatured and lose almost all activity after a short time. An exception to this general rule is amylase, which seems to be stable up to about 60° C before significant losses of activity occur.

There is an approximate doubling of the enzyme activity for a 10° C increase in temperature within the limits given above. To describe this phenomenon, a quantity was defined, Q10, which is the ratio of activity at the higher temperature over the activity at the lower temperature for a 10° C difference in temperature. For many enzymes a 1° C change in temperature would produce about a 10% change in activity. A tolerance of ± 0.1° C for temperature control of an enzyme analyzer is recommended, since this would produce approximately a ± 1% change in the activity that was measured. This amount of variation would be small enough to be ignored as an insignificant source of error for most clinical work. A recommendation of ± 0.05° C for temperature control has also been suggested.[19] This would reduce the change in activity ± 0.5%.

It has been determined that the Q10 for an enzyme is not fixed but varies from about 1.5 to 2.5, depending on the temperature range that is picked to compare enzyme activities. However, almost all clinical measurements of enzyme activity are performed within the range of 25° to 37° C. The apparent increase in activity with increasing temperature means that assays performed at higher temperatures, such as 37° C, will be more sensitive to slight changes in the amount of enzyme in a sample. The common enzymes employed for clinical diagnosis are less stable at this temperature than at 25° to 30° C, and therefore

Table 52-4 Enzyme stability under various storage conditions (less than 10% change in activity)

Enzyme	Room temperature (about 25° C)	Refrigeration (0° to about 4°C)	Frozen (−25° C)
Aldolase (ALS)	2 days	2 days	Unstable*
Alanine aminotransferase (ALT, GPT)	2 days	5 days	Unstable*
α-Amylase (AMS)	1 month	7 months	2 months
Aspartate aminotransferase (AST, GOT)	3 days	1 week	1 month
Ferroxidase I (ceruloplasmin)	1 day	2 weeks	2 weeks
Cholinesterase (CHS)	1 week	1 week	1 week
Creatine kinase (CK)	1 week	1 week	1 month
γ-Glutamyl transferase (GGT)	2 days	1 week	1 month
Isocitrate dehydrogenase (ICD)	1 day	2 days	1 day
Lactate dehydrogenase (LD)	1 week	1 to 3 days†	1 to 3 days†
Leucine aminopeptidase (LAP)	1 week	1 week	1 week
Lipase (LPS)	1 week	3 weeks	3 weeks
Phosphatase, acid (ACP)	4 hours‡	3 days§	3 days§
Phosphatase, alkaline (ALP)	2 to 3 days‖	2 to 3 days	1 month

*Enzyme does not tolerate thawing well.
†Depending on isoenzyme pattern in the serum.
‡Unacidified.
§With added citrate or acetate to pH ~ 5.
‖Activity may increase.

assays carried out at 37° C must be performed with relatively short assay times so that enzyme denaturation is minimized.

Arguments for the use of both higher, that is, 37° C, and lower temperatures, that is, 25° C, have appeared in the literature[18,19] based primarily on scientific and technical reasoning. A reasonable compromise seems to be measurement at 30° C, which is the recommendation of the International Federation on Clinical Chemistry (IFCC).[20] A very accurate gallium standard melting point cell is available to all laboratories from the National Bureau of Standards.[21] This material has a melting temperature plateau of 29.772° C and can be used to calibrate or check the assay temperature of a wide variety of instruments.

Enzymes as reagents

It is possible to determine substrate concentrations using the principles of enzyme kinetics applied in a slightly different way. Enzyme activity at low substrate concentrations is first order, that is, linear, with respect to substrate concentration (Fig. 52-5, *A*). To measure the concentration of pyruvate in a sample, for example, a special assay mixture is prepared. The samples containing the unknown amounts of pyruvate are diluted (or only a small amount used) so that pyruvate concentration in the assay is low, that is, less than the K_m. An assay mixture that contains an excess of lactate dehydrogenase (LD), an excess of coenzyme NADH, and a buffer is used. The decrease in the amount of NADH in this assay is related to the amount of pyruvate present. Alternatively, a series of pyruvate standards can be used to calibrate the assay. Other enzymatic assays of a similar nature that are commonly used in the clinical laboratory include the determination of glucose, urea, ethanol, cholesterol, tyriglycerides, and uric acid. There are many other uses of enzymes as reagents. They are often used as indicators in EMIT or in ELISA assays for drugs.

Storage of enzymes

Most enzymes used clinically are stable at refrigerator temperatures for 2 to 3 days to about 1 week and at room temperature for a shorter time. Table 52-4 summarizes data for three temperatures.[22,23] Additional information on enzyme stability can be found in Table 52-5.

Acid phosphatase activity is unstable at all temperatures unless the pH of the serum is reduced to about 5 to 6 with citrate or acetate. Alkaline phosphatase in human serum demonstrates a linear increase in activity dependent on temperature and time.[24] At 96 hours (4 days) there is a 6% increase at room temperature, a 4% increase at refrigerator temperature, and a 1% increase at −20° C. Enzymes in control materials are usually of nonhuman origin; some are more stable and some are less stable than enzymes in human serum.

Table 52-5 Plasma half-lives for clinically important enzymes

Enzymes	Half-life (hours) (mean ± 2 SD)
LD_1	53-173
LD_5	8-12
CK	15
AST (GOT)	12-22
ALT (GPT)	37-57
AMS	3-6
LPS	3-6
ALP	3-7 days
GGT	3-7 days

CLINICAL RATIONALE FOR ENZYME MEASUREMENTS

In 1954 after LaDue, Wróblewski, and Karmen[25] found a temporary increase in serum aspartate aminotransferase (EC 2.6.1.1) activity after an acute myocardial infarction, the measurement of changes in plasma enzyme activity gained importance as a means of following the course of a disease or of improving clinical diagnosis. Many investigators began to look for changes in enzyme activity that were specific for a disease state or organ. Changes in enzyme activity in the plasma (or serum) were followed, since it was known that intracellular enzymes were released after cell damage or cell death has taken place in a specific organ or tissue.

Although many disease-related enzyme changes have been found, the original hope of finding a simple marker for a specific organ or disease has not been realized. The changes that occur with many diseases or in a particular organ can now be understood, but only by examining the pattern of several enzyme or isoenzyme changes over a period of hours or days.

Extracellular versus cellular enzymes

The enzymes that are found in plasma can be categorized into two major groups, the plasma-specific enzymes and the nonplasma specific enzymes.[26]

The plasma-specific enzymes are those enzymes that have a definite and specific function in plasma. Plasma is their normal site of action, and they are present at higher concentrations in plasma than in most tissues. Among these are the enzymes involved in blood coagulation, and ferroxidase, pseudocholinesterase, and lipoprotein lipase. These enzymes are synthesized in the liver and are constantly liberated into the plasma to maintain a steady-state concentration. These enzymes are clinically of interest when their concentration decreases in plasma, and some have historically been used as estimates of liver function.

The nonplasma specific enzymes are those enzymes with no known function in plasma. Their concentrations are usually found to be lower in plasma than in most tissues. These enzymes can be divided into two groups, the enzymes of secretion and the enzymes of intermediary metabolism.

The enzymes of secretion are those enzymes secreted by exocrine glands, that is, the pancreas and prostate, and some enzymes from the gastric mucosa and the bones. Enzymes in this group are clinically important when their concentrations are either higher or lower than normal. Elevated values are found when the normal mode of excretion is blocked or when the amount of enzyme produced is increased. Decreases in the amount of enzyme are found when the tissue that normally produces the enzyme is damaged or necrotic. Common examples of this group are amylase, lipase, and acid and alkaline phosphatases.

The other major group of nonplasma specific enzymes are the enzymes of metabolism. The concentrations of these enzymes in tissues are very high, sometimes thousands of times higher than in the plasma. Cellular damage resulting in leakage or necrosis allows a fraction of these proteins to escape into the plasma, causing a sharp rise in the plasma concentration of the enzymes. Some common examples are creatine kinase (CK), lactate dehydrogenase (LD), alanine aminotransferase (ALT), and aspartate aminotransferase (AST).

Enzymes as tumor markers

Enzyme determinations can be used to estimate the amount of tumor mass. Excesses of specific proteins are sometimes elaborated into the plasma during tumor cell growth. In some cases, these specific proteins are enzymes, for example, PAP and CK, and they can be conveniently measured to monitor response to anticancer therapy.

SIGNIFICANT FACTORS AFFECTING ENZYME-REFERENCE VALUES

A number of important factors affect the reference ranges for enzyme determinations. If these factors are not accounted for in the interpretation of the results, a misdiagnosis is possible. In the following items a brief comment on the problem and an example will be given. Further details are covered in Section Three, Methods of Analysis (Chapters 53 and 59) and in the section on clinical interpretation (Section Two) and Chapter 17.

Age

There appear to be variations in the amounts of enzyme normally present in serum that are the result of differences in age between various subgroups in the population. There are perhaps three principle times to consider whether age will be important in the determination of a reference range: during the first year of life as various organs (such as the liver) mature, during puberty, and in late middle age when hormonal changes occur.

Perhaps some of the most dramatic changes are seen with the enzyme alkaline phosphatase.[27] Using the Bowers and McComb procedure, at 30° C the following values are found; 135 to 270 U/L for children 6 months to 10 years of age, 90 to 320 U/L for children 10 to 18 years of age, and 40 to 100 U/L for adults.

Sex

Differences in normal enzyme levels for male and female populations are seen with some enzymes. These differences are most probably related to muscle mass, exercise, or hormone concentration.

An example of these effects is seen with the enzyme creatine kinase; males are reported to have higher reference ranges than females, an effect that is most likely attributable to muscle mass.[28]

Race

Race may also be a factor for a limited number of enzymes, but data are sparse. Black populations are reported to have higher reference ranges than comparable white populations for creatine kinase,[29] but the effect may be an indirect result of several factors other than race in the two populations.

Exercise

Exercise and ambulation are important variables in the consideration of reference ranges for several enzymes. Patients who have had complete bed rest for several days are found to have 20% to 30% lower values for creatine kinase than ambulatory patients.[30] Normal amounts of exercise also elevate creatine kinase.[31] The additional creatine kinase attributable to normal exercise is of the MM type, CK_3. Thus the distinction between these elevations and those caused by an acute myocardial infarction, which is the MB type, CK_2, is easily accomplished by determination of the isoenzyme pattern. The increases seen after exercise usually disappear after 2 to 24 hours, unless the exercise is extremely strenuous. In ultralong-distance runners the normal CK-MB was found to be up to threefold higher than usual, and the total CK was found to be up to 40-fold higher than usual.[32]

Sampling time

Since enzyme levels do not undergo any significant circadian rhythm, sampling time is unimportant for the determination of enzyme reference ranges. On the other hand, the sampling time may be important for detecting a variety of acute and chronic conditions if the changes observed are sufficiently rapid. The classic mean time for maximum elevation for a series of enzymes in patients with myocardial infarctions was reported to be CK-MB, 6 hours; CK, 18 hours; AST, 24 hours; and LD, 48 hours.[33]

Assay type

Although there is an optimal set of conditions for the assay of an enzyme, it is clear that not all assays are optimal. At times the differences between assays may not appear to be significant and yet the results obtained will be substantially divergent. The effect of various components of an assay upon one another is even more significant when one considers a coupled enzyme assay. These assays must not only have adequate concentrations of substrates and activators for the primary reaction, but they must also have excesses of the auxiliary and indicating enzymes with their associated additional activators.

Alanine aminotransferase is often measured by addition of an excess of lactate dehydrogenase and NADH to a mixture containing L-alanine and α-ketoglutarate and buffer. The usual commercially available kits often specify that about 500 U/L of LD are present as an indicating enzyme and perhaps that the source of this enzyme is "animal." This is not sufficient to completely define the assay since the K_m of pyruvate varies with the isoenzyme type. About four times as many units of M_4-LD_5 would be required as H_4-LD_1, to achieve an equivalent reaction rate. A crude mixture of isoenzymes would be somewhere in between these extremes. Even if the units of LD added were the same, the measured enzyme rates might vary with each lot of a kit if the indicating enzyme were added without regard to the isoenzyme content. This same kind of variability would occur between manufacturers with the "same" concentrations of substrates, activators, and units of LD if a different source, such as bacterial, of the indicating enzyme was used. It is for this reason that a reference range should be checked by each laboratory, particularly when changing reagent manufacturers.

BIBLIOGRAPHY

Fersht, A: Enzyme structure and mechanism, ed 2, New York, 1985, WH Freeman & Co Publishers.

Page, MI: The chemistry of enzyme action, New York, 1984, Elsevier Science Publishing Co, Inc.

Zilva, JF, and Pannall, PR: Clinical chemistry in diagnosis and treatment, ed 4, Chicago, 1983, Year Book Medical Publishers, Inc., Chapter 15.

REFERENCES

1. Hohnadel, DC, and Cooper, C: The effect of structural alterations on the reactivity of the nucleotide substrate of rabbit muscle pyruvate kinase, FEBS Lett 30:18-20, 1973.
2. Hohnadel, DC, and Cooper, C: The effect of structural modifications of ATP on the yeast-hexokinase reaction, Eur J Biochem 31:180-185, 1972.
3. Nomenclature committee of the International Union of Biochemistry: Enzyme nomenclature 1984, Recommendations of the nomenclature committee of the International Union of Biochemistry on the nomenclature and classification of enzymes, (Webb, EC, editor), Orlando, 1984, Academic Press, Inc.
4. Baron, DN, Moss, DW, Walker, PG, and Wilkinson, JH: Abbreviations for names of enzymes of diagnostic importance: J Clin Pathol 24:656-657, 1971.
5. Baron, DN, Moss, DW, Walker, PG, and Wilkinson, JH: Revised list of abbreviations for names of enzymes of diagnostic importance: J Clin Pathol 28:582-593, 1975.
6. Commission of Editors of Biochemical Journals—International Union of Biochemistry: Biochemical Nomenclature and Related Documents, London, England, 1978, William Clowes & Sons, Limited.
7. Fendley, TW, Hochholzer, JM, and Frings, CS: Effect of various diluents on the activity of several enzymes present in serum, Clin Chem 19:1079-1080, 1973.
8. Pardue, HL: A comprehensive classification of kinetic methods of analysis used in clinical chemistry, Clin Chem 23:2189-2201, 1977.
9. Orten, JM, and Neuhaus, OW: Human biochemistry, St Louis, 1982, The CV Mosby Co, Chapter 5.
10. Wróblewski, F, and LaDue, JS: Serum glutamic pyruvic transaminase in cardiac and hepatic disease, Proc Soc Exp Biol Med 91:569-571, 1956.
11. Henry, RJ, Chiamori, N, Golub, OJ, and Berkman, S: Revised spectrophotometric methods for the determination of glutamic-oxalacetic transaminase, glutamic-pyruvic transaminase, and lactic acid dehydrogenase, Am J Clin Pathol 34:381-398, 1960.
12. Howell, BF, McCune, S, and Schaffer, R: Lactate-to-pyruvate or pyruvate-to-lactate assay for lactate dehydrogenase: a re-examination, Clin Chem 25:269-272, 1979.
13. Bowers, GN, Jr, and McComb, RB: A continuous spectrophotometric method for measuring the activity of serum alkaline phosphatase, Clin Chem 12:70-89, 1966.
14. McComb, RB, and Bowers, GN, Jr: Study of optimum buffer conditions for measuring alkaline phosphatase activity in human serum, Clin Chem 18:97-104, 1972.
15. Howell, BF, McCune, S, and Schaffer, R: Influence of salts on Michaelis-constant values for NADH, Clin Chem 23:2231-2237, 1977.
16. Bergmeyer, HU, Büttner, H, Hillman, G, et al: Standardization of methods for the estimation of enzyme activity in biological fluids, Z Klin Chem Klin Biochem 8:659-660, 1970.
17. Keiding, R, Hörder, M, Gerhardt, W, et al: Recommended methods for the determination of four enzymes in blood, Scand J Clin Lab Invest 33:291-306, 1974.
18. Bergmeyer, HU: Standardization of the reaction temperature for the determination of enzyme activity, Z Klin Chem Klin Biochem 11:39-45, 1973.
19. Duggan, PF: Activities of enzymes in plasma should be measured at 37° C, Clin Chem 25:348-352, 1979.
20. Committee on Standards—Expert Panel on Enzymes: Provisional recommendation (1974) of IFCC methods for the measurement of catalytic concentrations of enzymes, Clin Chem 22:384-391, 1976.
21. Bowers, GN, Jr, and Inman, SR: The gallium melting-point standard: its evaluation for temperature measurements in the clinical laboratory, Clin Chem 23:733-737, 1977.
22. Bergmeyer, HU: Standardization of enzyme assays, Clin Chem 18:1305-1311, 1972.
23. Labrosse, KR, Bixby, EK, and Lakatua, DJ: Stability of cardiac enzymes at different storage temperatures, Clin Chem 26:1026, 1980.
24. Massion, CG, and Frankenfeld, JD: Alkaline phosphatase: lability in fresh and frozen human serum and in lyophilized control material, Clin Chem 18:366-373, 1972.
25. LaDue, JS, Wróblewski, F, and Karmen, A: Serum glutamic oxaloacetic transaminase activity in human acute transmural myocardial infarction, Science 120:497-499, 1954.
26. Hess, B: Enzymes in blood plasma, New York, 1963, Academic Press, Inc.
27. McComb, RB, Bowers, GN, Jr, and Posen, S: Alkaline phosphatase, New York, 1979, Plenum Press, Chapter 9.
28. Garcia, W: Elevated creatine phosphokinase levels associated with large muscle mass, JAMA 228:1395-1396, 1974.
29. Blight, M, Wagman, E, Shastri, S, and Nevins, M: Race-related differences in reference intervals for creatine kinase, Clin Chem 26:1928-1929, 1980.
30. Tietz, NW, editor: Fundamentals of clinical chemistry, Philadelphia, 1976, WB Saunders Co.
31. LaPorta, MA, Linde, HW, Bruce, DL, and Fitzsimons, EJ: Elevations of creatine phosphokinase in young men after recreational exercise, JAMA 239:2685-2686, 1978.
32. Kielblock, AJ, Manjoo, M, Booyens, J, and Katzeff, IE: Creatine phosphokinase and lactate dehydrogenase levels after ultra long-distance running, S Afr Med J 55:1061-1064, 1979.
33. Blomberg, DJ, Kimber, WD, and Burke, MD: Creatine kinase isoenzymes: predictive value in the early diagnosis of acute myocardial infarction, Am J Med 59:464-469, 1975.

CHAPTER 53

Isoenzymes

LAWRENCE M. SILVERMAN
JOHN F. CHAPMAN
VICKI N. DAASCH

KEY TERMS

artifactual modification Change in protein structure caused by in vitro manipulation.

dimer A protein, such as creatine kinase, composed of two polypeptide subunits.

heteropolymer A polymeric compound that has more than one type of subunit.

homopolymer A polymeric compound in which all the subunits are the same.

immunoinhibition Inhibition of enzyme activity of a specific polypeptide subunit by its reaction with antibody.

isoforms Multiple forms of serum isoenzymes that result from enzymic modification of the parent form after its release from tissues.

isoenzymes Multiple forms of an enzyme family catalyzing the same biochemical reaction.

macroenvironment An environment specific to an organ or tissue that may be associated with a specific physiological function.

microenvironment A specific environment or location within a cell that may be associated with a specific physiological function.

Regan isoenzyme Name given to an alkaline phosphatase isoenzyme associated with some cancers.

subunit A polypeptide chain constituting a protein or enzyme.

tetramer A protein, such as lactate dehydrogenase, composed of four polypeptide subunits.

DEFINITIONS

Isoenzyme properties

Isoenzymes (isozymes) are multiple forms (isomers) of an enzyme family that catalyze the same biochemical reaction. Each isoenzyme of a family has a different affinity for substrates and cofactors. Thus the Michaelis constant (K_m) and specificity for different substrates may vary.[1] The various isoenzymes of a family may also differ in the ability to be inhibited by specific agents.[1] They may also differ in their physical properties, such as heat stability and charge, and their biochemical properties, such as amino acid composition and immunological reactivities. However, isoenzymes of a particular family usually do not differ in molecular size.

Most or all of these properties have been used to differentiate the various forms of an isoenzyme family. Assays based on these differences are discussed on p. 793.

Structural basis

The most commonly encountered isoenzyme families in clinical chemistry are proteins composed of two or more polypeptide chains or subunits. If the subunits are identical in primary, secondary, and tertiary structure, the resultant enzyme is a homopolymer. If the subunits are different, the enzyme is a heteropolymer. The final composition of an isoenzyme is dependent on the number of polypeptide subunits that compose the complete molecule. For example, if there are two subunits and two different subunit types (A and B), the three possible combinations are AA, AB, and BB. If there are three subunits with two different types, the four possible subunits are AAA, AAB, ABB, and BBB. With four subunits, five different combinations would result: AAAA, AAAB, AABB, ABBB, and BBBB. Therefore with these combinations the enzyme may be present as a homopolymer or a heteropolymer. An example of a dimeric isoenzyme family is creatine kinase (CK), which is formed by variable combinations of the subunits M and B. The subunits differ from each other in primary, secondary, and tertiary structure. The resulting isomeric forms are CK-MM, CK-MB, and CK-BB. The lactate dehydrogenase (LD) isoenzyme family is an example of different enzyme forms composed of combinations of four of the H and M subunits. The resulting tetrameric structures are HHHH, HHHM, HHMM, HMMM, and MMMM (H_4, H_3M_1, H_2M_2, H_1M_3, and M_4).

ORIGIN

Genetic basis

Since the early studies establishing the existence of isoenzymes, much progress has been made in elucidating the chemical bases for structural diversification. For example, tryptic digestion of the H and M subunit chains of the LD isoenzymes suggests that these subunits are the products of two related but separate genes.[2] Some amino acid sequences within the chains are identical, including the enzymatically active site, but there are differences in the amino acid composition and the antigenic properties of the H and M chains. These differences suggest that isoenzyme diversity may have arisen through gene duplication, followed by independent mutations of the two genes, resulting in different amino acid sequences. One could surmise that evolutionary selection influences the retention of genes for isoenzyme forms that impart some biological advantage to the organism.

Posttranslational and artifactual modifications

Isomeric forms of enzymes may be formed as a result of posttranslational modifications of the parent enzyme structure.[3] Variations in the patterns of aggregation of different monomer units of the enzyme can result in variation in enzymatic properties.[3] Oxidation or reduction of functional groups in the enzyme molecule may also result in identifiable isoenzyme forms.[3] Variability in modifications of the protein chain, such as the addition of sialic acid, can result in an apparently large number of isoenzymes with changes in net charge, which would allow separation and identification of the isoenzyme activities.[3] Addition of neutral sugars such as glucose and mannose is not likely to alter the overall electrical charge of the enzyme; therefore, these alterations cannot be detected by electrophoretic methods. Binding of a low-molecular-weight substance, such as NAD^+, may influence the electrophoretic pattern of an enzyme.

Isoforms

After isoenzymes are released from tissues, they are frequently degraded by serum enzymes, called peptidases. Of special interest is the family of CK-MM isoforms, referred to as CK-MM_3, CK-MM_2, and CK-MM_1. CK-MM_3 is the native form produced in tissues and is degraded first to MM_2 and then to MM_1 by a serum carboxypeptidase. Clinical interest in isoforms relates to their use in early diagnosis of acute myocardial infarction (AMI) and to monitoring response to antithrombolytic therapy, which can dissolve the clots that cause AMI's. Monitoring the serum levels of the CK-MM isoforms may provide earlier knowledge of the success of the antithrombolytic agents.

PURPOSE OR FUNCTION

Microenvironmental factors

The functional significance of isoenzymes remains an intriguing biological question. This question has been approached by the study of individual cells, of the organization of cells into tissues, and of the developmental process of the organism as a whole. Observations of the compartmentalization of isoenzymes within organelles of individual cells have led to theories related to subcellular interactions and metabolic requirements. Net charge of the isoenzyme may influence its interaction with other charged molecules within the cell. This may result in differential location of specific isoenzymes with respect to specialized parts of the cell. Charge probably plays a role in the preferential location of LD_1 in the mitochondrion.[4] This may result in improved metabolic activity, since the mitochondrion has a highly aerobic metabolism, and it has been suggested that the kinetic properties of the LD_1 isoenzyme are optimal under aerobic conditions. Aggregation of monomer enzymes may play a role in intracellular compartmentalization by hindering diffusion of the enzyme through the cell membrane.

Macroenvironmental factors

Differential location of isoenzymes has also been observed on a larger scale, that is, within different tissues. For example, tissues such as heart and brain show a predominance of LD_1, whereas other tissues, such as muscle, have a high content of LD_5. The tetramer of H chains that forms LD_1 has an affinity for pyruvate that is 10 times greater than the affinity of LD_5, a tetramer of M chains. Cahn et al.[5] suggested that because of its kinetic properties LD_1 isoenzyme predominates in tissues rich in oxygen supply that undergo oxidative metabolism and do not accumulate lactate or pyruvate but can use lactate as a fuel. The LD_5 isoenzyme is the major form in skeletal muscle, which undergoes anaerobic glycolysis with the accumulation of pyruvate. Pyruvate cannot be readily metabolized under anaerobic conditions. By having the LD_5 isoenzyme, muscle cells can force the conversion of pyruvate to lactate and regenerate NAD^+, permitting the energy producing reactions of the Embden-Meyerhof pathway (see Chapters 28 and 29) to continue. Thus it has been proposed that isoenzyme patterns vary according to the particular needs and metabolic environments of various tissues. In oxygen-rich tissues, such as the heart or brain, the predominant LD_1 favors the catalysis of lactate (picked up from blood and derived from muscles) to pyruvate in order to fully metabolize it to CO_2, H_2O, and energy (adenosine 5′triphosphate, ATP).

Because the actual tissue distribution does show this pattern, it must be assumed that the isoenzymes exist to fulfill different metabolic needs for the different conditions that exist within the body.

Developmental factors

If one considers the development of the organism itself, the pattern of isoenzymes in various tissues that evolves during ontogeny may reflect the tremendous changes in the interaction of the organism with its environment. As the environment changes during fetal development, the organism must adapt by optimizing enzymatic reactions. Dramatic changes in intracellular aerobic metabolism require major shifts in the isoenzyme composition. Although not a traditional enzyme, the switch from the fetal form to the adult form of hemoglobin (see Chapter 37) illustrates adaptation to the environmental conditions (true respiration) that begin at birth. The oxygen affinities of purified fetal and adult hemoglobin structures are about the same, but organic phosphate in the blood lowers the oxygen affinity of adult hemoglobin more than fetal hemoglobin. Maternal hemoglobin provides transport of oxygen to the fetal circulation, where oxygen exchange can occur as a result of the greater oxygen affinity of fetal hemoglobin. Thus different hemoglobins, as with certain isoenzymes, exist in response to different developmental biological needs.

Changes in isoenzyme patterns occur during early embryological development. For example, CK is initially present as the BB isoenzyme in all tissues. In developing myocardium MB and MM isoenzymes gradually appear and achieve adult levels (see below) as the M gene is expressed in association with myofibrillar contractile elements.[6] In a similar way, varying expression of the M and B genes in skeletal muscle tissue results in the appearance of CK-MB and CK-MM during development. However, the shift to greater expression of the M gene proceeds with fetal development to the point where almost 100% of the CK enzyme activity in skeletal muscle is of the CK-MM isoenzyme (adult) form. Changing patterns of isoenzyme expression, like that of hemoglobin, occur as a result of differential gene activation and repression in response to changing biological needs.

The LD isoenzymes of undifferentiated embryonal tissues have maximal activity in the hybrid isoenzymes LD_2, LD_3, and LD_4.[7] As tissues differentiate, the LD pattern appears to change to adapt to the type of energy production required for the tissues' specific function. Tissues such as heart, brain, and renal cortex have predominantly aerobic energy metabolism and show maximal enzyme activity in the LD_1 and LD_2 isoenzymes. Anaerobic energy production, such as occurs in skeletal muscle, liver, and renal medulla, is associated with increased activity of LD_4 and LD_5 isoenzymes.

TISSUE DISTRIBUTION OF MAJOR ISOENZYME FAMILIES

CK-MM is the predominant isoenzyme in adult skeletal muscle tissue, with only a small quantity of CK-MB present. Unlike skeletal muscle, adult myocardium contains 14% to 42% CK-MB (reports vary), with the remainder of activity contributed by CK-MM. Brain tissue primarily expresses the CK-BB isoenzyme (Table 53-1).

The heart is the organ richest in LD_1 isoenzyme (H_4), whereas LD_5 (M_4) is found predominately in liver and skeletal muscle.[8] Red blood cells are also rich in LD_1. The lung has large amounts of LD_2 and LD_3 (Table 53-2).

Another example of a family of isoenzymes is that of alkaline phosphatase. At birth, alkaline phosphatase in the serum appears to come entirely from bone, differing from the pattern observed in fetuses whose serum contains both bone and fetal intestinal forms. Serum from adults contains many alkaline phosphatase isoenzymes (Table 53-3), although the major forms released into serum are bone, liver, and kidney (all originating from a common gene product and representing postsynthetic modifications), and intestinal.[9]

Acid phosphatase is a ubiquitous enzyme, located in the lysosomes of all cells. Red blood cells, white blood cells,

Table 53-1 Creatine kinase activity in various human tissues

Tissue	Isoenzyme distribution in U/g of wet tissue (% total activity)		
	MM	MB	BB
Skeletal muscle	3281 (100)	0-623 (0-19)	0 (0)
Heart	313 (78)	56-169 (14-42)	0 (0)
Brain	0 (0)	0 (0)	157 (100)
Colon	4 (3)	1 (1)	143 (96)
Stomach	4 (3)	2 (2)	114 (95)
Uterus	1 (2)	1 (3)	45 (95)
Thyroid	7 (26)	0.3 (1)	21 (73)
Kidney	2 (8)	0 (0)	19 (92)
Lung	5 (35)	0.1 (1)	9 (64)
Prostate	0.3 (3)	0.4 (4)	9.3 (93)
Spleen	5 (74)	0 (0)	2 (26)
Liver	3.6 (90)	0.2 (6)	0.2 (4)
Pancreas	0.4 (14)	0 (1)	2.6 (85)
Placenta	1.4 (48)	0.2 (6)	1.4 (46)

From Chapman, J, and Silverman, L: Bull Lab Med (NCMH), no 60, pp 1-7, Jan 1982.

Table 53-2 Lactate dehydrogenase activity—percentage activity distributing

Organ	Isoenzyme distribution				
	H_4	H_3M_1	H_2M_2	H_1M_3	M_4
Heart	60	30	5	3	2
Kidney	28	34	21	11	6
Cerebrum	28	32	19	16	5
Liver	0.2	0.8	1	4	94
Skeletal muscle	3	4	8	9	76
Skin	0	0	4	17	79
Lung	10	18	28	23	21
Spleen	5	15	31	31	18

From Pfleiderer, G, et al: In Schmidt, E, et al, editors: Advances in clinical enzymology, Hanover, West Germany, 1979, S Karger AG.

Table 53-3 Alkaline phosphatase activity in human tissues*

Tissue	Activity (U/g of wet tissue)	
	MAP†	DEA‡
Adrenal	30	66
Placenta	36	—
Liver	12.6	27
Bone	7.5	18
Spleen	7.5	18
Lung	6.6	15
Intestine	4.8	9
Kidney	4.2	11
Prostate	3.3	6.6
Thyroid	2.1	5.1
Heart	1.8	3.6
Erythrocytes	0.02	—

Data calculated from Bowers, GN, et al: Clin Chem 21:1988-1995, 1975.

*Mean activity in two buffer systems of tissue specimens from human autopsies. Note the greater than twofold activity between the buffer systems.

†2-Methyl-2-amino-1-propanol buffer.

‡Diethylamine buffer.

platelets, and the prostate gland have particularly enriched levels of this enzyme. The molecules from the last three tissues listed are closely related immunologically. Most of the enzyme activity found in serum is probably derived from blood cells, whereas only a small fraction is derived from the prostate gland.

It is important to realize that many enzymes exist in isomeric or isoenzyme forms. However, except for a few of the isoenzyme families, such as the ones described here, there has been little demonstrable value for doing isoenzyme analyses. The clinical utility of isoenzyme analysis is described later.

CHANGE IN ISOENZYME PATTERNS SECONDARY TO PATHOLOGICAL PROCESSES

Changes in isoenzyme patterns may occur as a result of pathological processes such as ischemia, atherosclerosis, or cancer. Wilhelm[7] showed that LD activity in normal adult aortic tissue is found primarily in the LD_3 fraction. In atherosclerotic aortic tissue, maximal LD activity is present in the LD_5 fraction. Likewise, myocardial LD activity shifts from predominantly LD_1 to LD_3 during the progression of ischemic heart disease.[7] CK activity in ischemic myocardial tissue also shows greater expression of CK-MB activity compared with normal adult myocardium.[7]

Pathological conditions that result in regeneration of damaged tissue may result in changes in isoenzyme composition that resemble the patterns observed during embryological development. For example, many types of muscle disease, such as Duchenne's muscular dystrophy, are characterized by muscle fiber regeneration; that is, new muscle fibers are formed as a result of destruction of adult muscle fibers. Immature muscle fibers may contain isoenzymes that are normally expressed only during early development. Thus appearance of CK-MB isoenzyme in skeletal muscle may indicate an abnormality in that tissue. However one must use caution in the interpretation of isoenzymes, as the fetal isoenzyme form in one tissue may be the adult isoenzyme form in another tissue. For example, the CK-MB isoenzyme represents a significant proportion of the CK activity in both fetal and adult myocardium. However, increased amounts of CK-MB isoenzyme in the sera of normal adults would probably represent damage to the heart. For a more complete discussion of the interpretational problems associated with the presence of CK-MB in patient sera, refer to Chapter 28.

In addition to the processes of destruction and regeneration, the processes of transformation and dedifferentiation associated with the development of malignancy may change isoenzyme patterns. The isoenzymes associated with tumors are often referred to as *oncofetal tumor markers* because of the similarities with the isoenzyme expression observed during early embryological development. For example, although CK-BB is the predominant isoenzyme in all early embryonic tissue, its expression is more restricted in the adult and is associated primarily with the brain and some gut-associated tissues. In patients without malignancy, detection of CK-BB in the serum is often associated with a pathological condition affecting the nervous, pulmonary, or gastrointestinal systems.[10] However, during the process of cell transformation, some cells may express significant amounts of CK-BB, which may be detected in the serum.[10]

In lymphoid malignancies, the LD isoenzymes in serum are predominantly LD_2, LD_3, and LD_4.[11] This pattern reflects the presence of increased numbers of lymphoid cells resulting from malignant proliferation. The LD isoenzyme pattern in the malignant cells reflects the pattern in normal lymphoid cells.

A shift toward LD_5 expression is observed in many solid tumors, especially in carcinomas of the genitalia or the digestive tracts.[11] However, this shift does not always occur; in some tumors (such as germ cell tumors) there is a shift toward LD_1 expression. The inconsistency of the shift toward LD_5 expression argues against the relationship of isoenzyme expression and increased anaerobic glycolysis.[11] What seems clear is that the development of most malignancies is accompanied by an isoenzyme expression that differs from that of the normal adult tissue of origin.

Isoenzymes may provide metabolic flexibility to an organism, allowing the organism to interact with its environment most efficiently. Changes in the environment of the individual cells may occur during the process of ontogenesis, as a result of pathological destruction and the process of regeneration, or even as a result of the development of the malignant state.

CLINICAL SIGNIFICANCE OF SPECIFIC ISOENZYMES

For all practical purposes, serum has been the only clinical specimen examined for isoenzyme markers of specific tissue abnormalities. The release of intracellular isoenzymes after damage or disease can be demonstrated in tissue culture; however, the release of intracellular isoenzymes in vivo does not necessarily follow the same rules or have the same significance as in vitro situations. For instance, various mechanisms have been proposed to explain how isoenzymes, with molecular weights usually exceeding 15,000 daltons, can be released from living cells without irreversible damage to the cellular membrane. The fact that living cells apparently do release these molecules has been frequently demonstrated; this constitutes the baseline levels of activities of isoenzymes and enzymes that are found in serum. These levels define laboratory reference intervals (often referred to as ''normal'' values). Because these baseline levels exist for most enzymes and isoenzymes released from the cytoplasm of cells, reference intervals represent normal cell leakage and cellular turnover. Increases in the levels of enzymes and isoenzymes in excess of these reference ranges are associated with a variety of pathological abnormalities and are the basis of the clinical utility of enzyme and isoenzyme determinations. The purpose of the next section is to discuss the major abnormalities associated with increased levels of isoenzymes and to provide a basis for interpretation of abnormal serum isoenzyme values. Determination of isoenzymes in other body fluids is rare and is discussed briefly at the conclusion of this section.

Creatine kinase (CK)

As previously mentioned, CK is found primarily in skeletal muscle, cardiac tissue, and brain. Although the majority of the isoenzymes are cytoplasmic, there are reports describing CK in other subcellular locations, particularly in the mitochondrion. Significant increases in serum CK levels usually reflect either skeletal or cardiac muscle release. By fractioning CK into its isoenzymes, skeletal muscle release can usually be discriminated from cardiac tissue release. However, the complexities of enzyme release and the various clearance mechanisms by which the body recycles proteins require the interpretation of serum enzyme or isoenzyme levels to be made in the context of the clinical situation. Generally, multiple serum markers yield information that makes interpretation of enzyme values more practical. Thus the combined use of LD isoenzymes with CK isoenzymes yields the necessary information in most cases to evaluate patients for possible myocardial infarction. Nevertheless, multiple specimen analyses appropriately spaced in time provide even more useful information (see Chapter 27).

A more common interpretative problem exists when skeletal muscle abnormalities result in significant elevation of CK activity in serum. Because skeletal muscle contains 5 to 10 times more CK per gram than does cardiac tissue, small areas of muscle damage or disease can result in serum levels of CK consistent with substantial damage to the heart. The use of isoenzyme fractionation can usually differentiate the source of the elevated CK serum activity because skeletal muscle usually consists of more than 95% CK-MM isoenzyme. However, certain diseases of skeletal muscle result in an increased amount of CK-MB content frequently associated with muscle fiber regeneration. Thus diseases such as Duchenne's muscular dystrophy or polymyositis often result in serum elevations of total CK and an abnormal increase in serum CK-MB activity (usually elevated to 5% to 15% of the total CK activity). Because the majority of patients with these muscle diseases are not being evaluated for myocardial infarction, misinterpretation of these CK-MB elevations is infrequent. However, clinical information must be available to allow for the proper interpretation of isoenzyme values.

Further difficulty in CK isoenzyme interpretation may be encountered when evaluating patients undergoing thoracic surgery, particularly coronary artery bypass graft surgery. Surgical procedures involving the heart release myocardial enzymes and isoenzymes that reach levels consistent with myocardial infarction. In such patients the clinician is frequently concerned about infarction either during surgery or during recovery. Therefore isoenzyme levels are extremely difficult to interpret. Experience has led many laboratories to require multiple serum samples for CK isoenzymes during the recovery period (4, 12, 24, and 48 hours and 3, 5, and 7 days postoperatively). In an uncomplicated recovery the CK-MB levels return to normal within 24 hours or decrease to levels approaching normal. With serious complications involving extension of myocardial necrosis or reinfarction, CK-MB levels continue to rise during the recovery period or rise after an initial diminution. Additional information on this subject is found in Chapter 27.

Occasionally CK isoenzymes can be used to evaluate tissues other than the heart or muscle. For example, CK-BB activity can be elevated in the sera of patients with various conditions, including malignancy and prostatic, pulmonary, and neurological disorders. One cannot make the assumption that abnormal isoenzymes are associated only with a particular tissue, as evidenced here for CK or other isoenzymes used in the clinical laboratory.

Acid phosphatase

Although acid phosphatase isoenzymes are found in a variety of cells, including red blood cells, white blood cells, and platelets, the most common isoenzyme measured in the clinical laboratory is the so-called prostatic acid phosphatase. The method of measurement of prostatic acid phosphatase affects the clinical usefulness of the results and is discussed on p. 892.

For nearly 40 years prostatic acid phosphatase has been shown to be frequently elevated in patients with advanced stages of prostatic cancer. Thus levels of this isoenzyme are useful when monitoring patients receiving treatment to assess tumor burden (the relative size of the tumor that remains during or after treatment). This isoenzyme activity can be used to assess the success of various types of treatment, but is rarely useful for diagnosis of early stages of prostate cancer.

Prostatic acid phosphatase levels may be normal even when a patient has prostate cancer. Therefore false-negative rates are high. Interpretative problems also exist with false-positive elevations. The most common source of false-positive results is benign prostatic hypertrophy in which there is increased growth of the prostate leading to various urological problems. Unfortunately persons with this condition are frequently the very patients in whom prostate cancer is a strong possibility. Other false-positive elevations have been observed in patients with various malignancies other than prostate cancer, such as leukemia. Again, interpretation of isoenzyme levels depends on adequate clinical information. The increased use of a more sensitive prostate marker, called prostate specific antigen (PSA), may diminish the clinical usefulness of prostatic acid phosphatase.

Other clinical situations in which acid phosphate isoenzyme studies may be ordered include the diagnosis of Gaucher's disease (an inborn error of metabolism) and various leukemias in which leukocyte acid phosphatase can be measured (also leukocyte alkaline phosphatase). Total acid phosphatase and the prostatic isoenzyme associated with prostatic secretions have been used in rape cases as legal evidence of sexual assault.

Alkaline phosphatase

Alkaline phosphatase isoenzyme studies are frequently ordered for a variety of clinical situations. As in the case of acid phosphatase, the method of measurement plays a significant role in the clinical usefulness of the results. Although at least five different isoenzymes of alkaline phosphatase are reported in serum, many methods are able to discriminate only the bone isoenzyme from the liver form. In laboratories in which these methods are used, the only clinical situations in which fractionation of alkaline phosphatase isoenzymes is useful are those in which liver or bone is involved. With the advent of other tests to monitor liver involvement and dysfunction, for example, gamma-glutamyl transferase, the necessity to fractionate alkaline phosphatase isoenzymes has diminished. Even in those laboratories with methods to fractionate alkaline phosphatase into five or more components, there is controversy over the clinical usefulness of these additional fractions. For example, one isoenzyme (Regan isoenzyme) has been reported to be an effective tumor marker in certain malignancies (such as bronchogenic carcinoma) but is certainly no more useful for diagnosis than other tumor markers.

By far the most frequent uses of alkaline phosphatase isoenzymes involve diseases affecting the bone, liver, and intestine. In particular, alkaline phosphatase isoenzyme evaluations are often ordered in patients with cancer as a means of monitoring metastases. Because bone and liver are two of the more common metastatic sites, the ability to diagnose early metastases to either of these tissues is important. Unfortunately not all tumors that have metastasized to bone or liver are associated with increased alkaline phosphatase activity or an abnormal isoenzyme pattern. Again, the use of other markers for liver and bone involvement has decreased the usefulness of alkaline phosphatase isoenzyme determinations. In addition, the methodological variations in isoenzyme fractionation may result in errors in identification or the inability to totally resolve isoenzymes, leading to a high degree of uncertainty in interpretation.

Lactate dehydrogenase (LD)

Much discussion of LD has been covered in the previous section on CK and in Chapter 27 because LD isoenzyme levels are most frequently ordered in combination with CK isoenzymes for the diagnosis of myocardial infarction. Methodological considerations can also determine clinical usefulness of LD isoenzymes, particularly with the advent of newer, more sensitive immunological techniques. This discussion is confined to results obtained by use of conventional electrophoresis.

LD isoenzymes have been more widely investigated than most other isoenzyme families on which tests are currently performed routinely in clinical laboratories. Still the clinical usefulness of this isoenzyme family is somewhat limited to myocardial infarction, partly because of its tremendous variety of tissue sources. LD is found in virtually every tissue, though there is some relative tissue specificity for the various isoenzymes. Nevertheless, with only five isoenzyme forms commonly found in serum, there is considerable overlap in isoenzyme-tissue specificity. For example, although the LD_5 isoenzyme is frequently used to ascertain damage to skeletal muscle, it is also the predominant isoenzyme in liver. A similar situation exists for each of the other four isoenzymes, which leads most clinicians to depend on other more specific tests as their primary liver, muscle, or cardiac assessment tools. As mentioned, much of this lack of specificity may be method dependent, a situation that may be resolved with newer techniques.

Thus LD isoenzyme fractionation is most useful for ruling out myocardial damage and muscle injury and occasionally for monitoring progression of certain malignancies. For several decades an increase in LD_5 isoenzyme has been observed in the sera of patients with various types of cancer. Again, the use of isoenzymes as tumor markers is a nonspecific finding fraught with possible misinterpretation.

Other isoenzymes

Although various laboratories perform isoenzyme fractionation for other families besides those just described, the clinical usefulness of these other procedures is questionable. For completeness, isoenzymes of amylase, aspartate aminotransferase, and aldolase should be mentioned because they are reported to have some clinical applications. Fractionation of these and other isoenzyme families is performed by few laboratories, though the clinical usefulness of these data is not so well established as for CK and others previously discussed in this chapter.

Fluids other than serum

Various reports indicate isoenzyme fractionation of several enzymes has significance in cerebrospinal fluid, pleural effusions, urine, and so on. These fluids are rarely examined, and the conditions of isoenzyme fractionation are occasionally very different from serum. Because so little data exist, laboratories that are involved in analyzing these fluids generally determine their own guidelines for clinical interpretation.

MODES OF ISOENZYME ANALYSIS

Most of the physical and catalytic differences among individual isoenzymes of a family have been exploited, at one time or another, to determine the isoenzyme concentrations in serum (Table 53-4). All these methods depend on the differences of individual subunit polypeptide chains that impart differences to the complete isoenzyme molecule.

Many older methods, based on the catalytic properties and the chemical inhibition patterns of the isoenzymes, were in widespread use before the advent of more sophisticated methods. These latter methods differentiate between the isoenzymes based on physical or immunological differences of the subunit chains. Thus the use of substrate affinity and catalytic rates to differentiate between CK isoenzymes is rarely used today. Similarly, the use of β-hydroxybutyrate as a "specific" substrate for the measurement of LD_1 activity is now viewed as an inadequate and unsatisfactory method. Methods using L-tartrate to inhibit prostatic acid phosphatase activity are being replaced by immunoassays. Heat stability and catalytic inhibition methods for the differentiation of isoenzymes with alkaline phosphatase activity are also rarely used for routine clinical analysis. In contrast, methods using dibucaine to inhibit cholinesterase isoenzymes are still frequently used.

Electrophoretic techniques to separate CK and LD isoenzymes have become the reference method for CK-MB analysis. Electrophoretic methods for the separation and

Table 53-4 Modes of isoenzyme analysis

Technique	Principles of analysis	Isoenzyme family
Electrophoresis	Subunits have different charges; isoenzymes separated in an electrical field	All
Ion-exchange chromatography	Subunits have different charges; isoenzymes separated by differential affinity for ion-exchange resin	CK, LD
Immunoinhibition	Antibody reacts specifically with one subunit type; this property can be used to render an isoenzyme or isoenzymes catalytically inactive or be used to physically remove an isoenzyme or isoenzyme from solution	CK, LD, acid phosphatase
Immunoassay	Antibody reacts specifically with one subunit type; extent of reaction monitored by use of radioisotope, enzyme, or fluorescent tag	CK, LD, acid phosphatase
Heat stability	Individual isoenzyme subunits are rendered catalytically inactive at different temperatures	Alkaline phosphatase
Catalytic inhibition	Individual isoenzyme subunits bind low molecular weight inhibitors with different affinities; such binding results in different inhibition of each isoenzyme	Acid phosphatase (L-tartrate), alkaline phosphatase (urea and L-phenylalanine), cholinesterase (dibucaine)
Substrate specificity	Each isoenzyme subunit binds a substrate with different affinities (K_m), giving each isoenzyme various rates of activity	CK
	Also each isoenzyme subunit may bind various substrates with different affinities; different isoenzymes have increased catalytic rates with certain substrates, whereas others have very low activities.	LD_1

quantitation of amylase, gamma-glutamyl transferase, alkaline and acid phosphatases, and many other isoenzyme families have also been developed. These methods are based on differences in the net charge of each isoenzymatic member of a family. The popularity of electrophoretic methods is based on their relative ease of use and their good sample throughput.

Isoenzyme analysis based on immunological differences between isoenzymes has become an important and widespread technique. These methods are based on the different amino acid sequences and thus the different immunological characteristics of subunit polypeptides. The immunological methods include competitive radioimmunoassay (RIA) and immunometric techniques (see Chapters 10 and 11) and immunoinhibition techniques. Because of the ease and sensitivity of most enzymatic assays for isoenzymes, radioimmunoassays have not come into widespread use. The immunoinhibition assays for CK and LD_1 are used with more frequency. Solid phase immunometric "sandwich" assays for acid phosphatase have become viable replacements for the older tartrate-inhibition assays. Newer immunoassays are based on monoclonal antibodies derived against specific recognition sites (epitopes) on isoenzyme molecules. For instance, the sequential use of two monoclonal antibodies, each recognizing different sites, can add significant specificity to an immunoassay for prostatic acid phosphatase or for CK-MB. By attaching a different enzyme (such as alkaline phosphatase) to one of the monoclonal antibodies, one can measure the concentration of *only* the isoenzyme that binds to both monoclonals.

For additional information on isoenzyme analysis of the more commonly analyzed isoenzymes, refer to the individual methods listed on pp. 892, 902, 922, and 928.

BIBLIOGRAPHY

Foreback, CC, and Chu, JW: Creatine kinase isoenzymes: electrophoretic and quantitative measurements, CRC Crit Rev Clin Lab Sci 15: 187-230, 1981.

Moss, DW: Isoenzymes, New York, 1982, Chapman & Hall.

REFERENCES

1. Wilkinson, JH: Principles and practice of diagnostic enzymology, London, 1976, EJ Arnold & Son, Ltd.
2. Markert, CL: The molecular basis for isoenzymes, Ann NY Acad Sci 151:15-39, 1968.
3. Rothe, GM: A survey of the formation and localization of secondary isoenzymes in mammalia, Hum Genet 56:129-155, 1980.
4. Agostoni, A, Vergani, C, and Villa, L: Intracellular distribution of the different forms of lactic dehydrogenase, Nature 209:1024-1025, 1966.
5. Cahn, RD, Kaplan, NO, Levine, L, and Zwilling, E: Nature and development of lactic dehydrogenases, Science 136:962-969, 1962.
6. Chapman, J, and Silverman, L: Creatine kinase in the diagnosis of myocardial disease, Bull Lab Med (NCMH), no 60, pp 1-7, 1982.
7. Wilhelm, A: Topochemical variation of LDH and CK isoenzyme patterns in aorta, Artery 8:362-367, 1980.
8. Pfleiderer, G, et al: Tissue enzymes in phylogenetic and ontogenetic development. In Schmidt, E, et al, editors: Advances in clinical enzymology, Hanover, West Germany, 1979, S Karger AG.
9. Bowers, GN, McComb, RB, Statland, BE, et al: Measurement of total alkaline phosphatase activity in human serum (selected method), Clin Chem 21:1988-1995, 1975.
10. Lang, H, and Wurzburg, U: Creatine kinase, an enzyme of many forms, Clin Chem 28:1439-1447, 1982.
11. Schapira, F: Isoenzymes and cancer, Adv Cancer Res 18:77-153, 1973.

CHAPTER 54 Therapeutic drug monitoring (TDM)

WOLFGANG A. RITSCHEL

KEY TERMS

absorption Uptake of unchanged drug into circulation.

absorption rate constant Value describing how much drug is absorbed per unit of time.

active transport Movement of drug across a membrane by binding to a carrier molecule and delivery to the opposite side with expenditure of energy.

bioavailability The amount of drug in the formulation that the system of the patient can absorb.

biophase The site of interaction between the drug molecule and its receptor.

bound drug A pharmacological agent that exists in blood complexed with another molecule (usually protein or lipid).

C_{max} Maximum plasma level of drug.

C^{ss}_{av} Average steady-state concentration.

C^{ss}_{max} Maximum steady-state concentration (peak concentration).

C^{ss}_{min} Minimum steady-state concentration (trough concentration).

compartment A pharmacokinetic term consisting of the drug concentration, C, and a volume of distribution.

distribution Proportional division of drug into different compartments of the body, such as blood and extracellular fluid.

elimination Final excretion of an agent.

first-order kinetics The rate of change of plasma drug concentration that is dependent on the concentration itself; that is, a constant proportion of drug is removed with time, or $dC/dt = -k \cdot C$.

free drug A pharmacological agent that exists in biological fluids unbound by other molecules.

half-life ($t_{1/2}$) The amount of time required to reduce a drug level to one half its initial value. Usually it refers to time necessary to reduce the plasma value to one half of its initial value. The term is also applied to the disappearance of the total amount of drug from the body.

LADME An acronym for the time course of drug distribution: *l*iberation, *a*bsorption, *d*istribution, *m*etabolism, and *e*limination.

liberation The process of drug release from the dosage form.

limited fluctuation method of dosing A method of dosing in which the drug given is not to exceed or go below specified limits.

maintenance dose The amount of drug required to keep a desired mean steady-state concentration.

MEC, MIC The minimum effective concentration or the minimum inhibitory concentration for a drug to be active. A drug is effective at any level above this value.

metabolism The biotransformation of the parent drug into metabolites.

Michaelis-Menten kinetics A method of transforming drug plasma levels into a linear relationship using the parameters of drug concentration and a constant, K_m.

passive diffusion The transport of drug by a concentration gradient across the membrane.

peak concentration The highest concentration reached after a dosage (usually soon after the dose is given).

peak method of dosing A method whereby the drug must reach a specified maximum level to be effective.

pharmacokinetics The quantitative study of drug disposition in the body.

pharmacological effect The influence of a drug on a patient's biochemical or physiological state (such as lowering of blood pressure and bacteriostasis).

prodrug A parent compound that is usually not active and must be metabolized to the active form.

receptor The structure in the body with which the drug interacts, yielding its pharmacological effect. Most often it is located on a cell membrane or other cellular component.

slow release A dosage form of drug that allows the drug to be slowly placed into solution.

steady state A condition in which drug input and drug output are equal. This is obtained when, after multiple dosing, the peak concentration and the trough concentration after each dose oscillate within a certain range.

subtherapeutic A level of drug less than that necessary to have the desired clinical effect.

t_{max} The time of maximum drug concentration.

τ Dosing interval.

terminal disposition rate constant The overall elimination of drug from the body per unit time.

therapeutic index The ratio between the plasma concentrations yielding the desired and undesired effects of a drug.

therapeutic range The relationship between the desired clinical effect of a drug and the concentration of the drug in the plasma.

therapeutic window A term describing a bell-shaped response curve of drug level versus pharmacological response.

total clearance (Cl_{tot}) A term that describes how much of the volume of distribution of a drug is cleared per unit of time.

toxic Implies poisonous or deleterious, sometimes fatal, side effects from a therapeutic agents that is present at a level that is too high.

trough concentration The lowest drug concentration reached, usually before the next dose is given.

zero-order kinetics The rate of change of plasma concentration, independent of the plasma concentration. A constant amount is eliminated per unit of time, or $dC/dt = -k_0$.

zero-time blood level A hypothetical blood concentration obtained by extrapolating back to the initial or zero time period of administration. Usually this yields a maximal value.

FATE OF DRUG AND NEED FOR THERAPEUTIC DRUG MONITORING

Concept of therapeutic range

For many drugs a relationship has been established between the clinical effects and the drug concentration in plasma. In general, in order to achieve the desired pharmacological effect (such as lowering of blood pressure, pain relief, or bacteriostasis) a certain concentration must be reached at the site of interaction between the drug molecule and the receptor (cell membrane, cell component) to elicit the clinical effect.

Currently, the only qualitative measurements available to assess the drug interactions at the cellular level are measurements of serum drug levels. If the degree of clinical response to a drug is plotted against the logarithm of dose or blood concentration of that drug, one obtains a linear curve. One obtains the same semilogarithmic curve if the drug dose or blood concentration is plotted against the percentage of a given population that has a specific clinical response to the drug. Any dose or concentration that does not result in any measurable or quantifiable effect is subtherapeutic. Any dose or concentration larger than the minimum dose or concentration that gives 100% effectiveness is unwarranted and may be toxic.

Similar to the log dose-response curve, there usually is a log dose-toxicity curve (see Chapter 50). Often one finds an overlap between the upper portion of the log dose-response curve and the lower portion of the log dose-toxicity curve. For most drugs the therapeutic range is a concentration range somewhere in the lower third to middle portion of the log concentration-response curve. The steepness of the log concentration-response curve indicates the magnitude of the therapeutic range. Absolute toxicity is less important than the ratio between the average toxic dose and the average therapeutic dose or concentration. This ratio, called the *therapeutic index,* is narrow for some drugs (digoxin, lithium) and wide for others. Hence the therapeutic range may also be narrow or wide. It is particularly desirable to monitor drug concentrations for drugs that have a narrow therapeutic range and low therapeutic index (digoxin, lithium, gentamicin), have dose-dependent elimination kinetics (phenytoin), or show great individual variability in metabolism (tricyclic antidepressants). Thus it is important for the physician to know whether the drug is present in a concentration within the therapeutic range.

The purpose of this chapter is to describe the fate of drugs once administered, to provide some insight into the type of dose regimen and achieving the therapeutic range, and to describe some basic principles of pharmacokinetics.

LADME system to describe drug disposition

It is generally accepted that changes in drug concentrations in the body, which occur with time, are related to the course of the pharmacological effects. The change of drug concentration with time is described by the LADME system, in which the *l*iberation, *a*bsorption, *d*istribution, *m*etabolism, and *e*limination of a drug is considered in sequence.

Liberation or drug release from a dosage form. To be absorbed, a drug must be present in the form of a true solution at the site of absorption. Hence the active ingredient of any dosage form except those that are already true solutions (such as intravenous injection, peroral elixir, peroral syrup, rectal enema, eye drops, and nose drops) has to be released from the dosage form before the drug can be absorbed. The release or liberation is the process of the drug passing into solution. When given orally by tablets, capsules, or suspensions, the drug dissolves in gastric fluid. After intramuscular or subcutaneous injection of suspensions, the drug dissolves in tissue fluid. After rectal administration, suppositories melt in the rectum and the drug dissolves in rectal fluid. After application of ointments, the drug dissolves in the water of perspiration at the interface between the skin and the ointment. These are a few cases in which liberation is necessary for the drug to be absorbed.

Sustained or controlled-release dosage forms are preparations with slow release rates. These are designed for those drugs that do not remain in the body for a long time. Since the drug cannot be absorbed faster than it is released, the apparent absorption rate becomes a function of the release rate and the entire absorption process takes longer, resulting in a prolonged duration of clinical effect.

Absorption. Absorption is the process by which the drug molecule is taken up into systemic circulation. Systemic circulation is usually defined as the bloodstream. The process of absorption, must occur whenever a drug is administered *extravascularly;* that is, perorally, orally, intramuscularly, subcutaneously, rectally, topically, and so on. Whenever a drug is given *intravascularly* (intravenously, intra-arterially, intracardiacally), no absorption takes place because the drug is directly introduced into the bloodstream.

There are various mechanisms of absorption, including passive diffusion, active transport, facilitated transport, convective transport, and pinocytosis. *Passive diffusion,* applicable for about 95% of all drugs, depends on the concentration of nonionized drug being higher on one side of the membrane than on the other. As long as there is a concentration gradient across the membrane, the drug will be absorbed into the region of lower concentration. For weak electrolytes, the drug's pK_a and the pH at the absorption site (such as stomach pH 1.5 to 3, intestines pH 5 to 7, rectum pH 7.8, or skin pH 5) influence the degree of ionization. The pH of blood is 7.4 and rather constant. As a general rule, ionized drug species are passively absorbed much *less* readily than nonionized species. At two pH units below an acid drug's pK_a and 2 pH units above a basic drug's pK_a, the drugs will be 99% nonionized and have maximal rates of absorption.

The next important absorption mechanism is *active transport,* which requires binding of the drug molecule to a carrier (protein) in the membrane. The carrier delivers the drug to the opposite side of the membrane by an expenditure of energy. The process moves the drug (such as cardiac glycosides, hexoses, monosaccharides, amino acids, riboflavin) against a concentration gradient. *Facilitated transport* is a similar mechanism, but facilitated transport of a substance (such as vitamin B_{12}) follows the concentration gradient. *Convective transport* is the mechanism of absorption by which small molecules (such as urea) enter systemic circulation through water-filled pores in the membrane. For all these mechanisms the drug must be in true aqueous solution at the absorption site.

A unique absorption mechanism is that of *pinocytosis* of fats and solid particles. Engulfing vesicles form in the cellular membrane and open at the intracellular side, releasing the fat droplets or particles (such as vitamins A, K, D, and E; parasite eggs; fats; and starch).

Distribution. Once drug molecules are absorbed, they distribute within the bloodstream and can (1) be confined to the blood space, (2) leave the bloodstream and enter other extravascular fluids (such as interstitial fluid), or (3) migrate into various tissues and organs. The entire process of transfer of drug from the bloodstream to other compartments is called *distribution*. This process usually takes between 30 minutes and 2 hours but it may be completed within a few minutes or it may take much longer than 2 hours (distribution time for methotrexate is 15 hours).

Metabolism. Metabolism is the process of biotransformation of the parent drug molecule to one or more metabolites. The metabolites are usually more polar, that is, more water soluble, and can thus be more easily excreted by the kidney. Metabolism occurs primarily in the liver and the kidney but also takes place in plasma and muscle tissue. Usually but not always metabolites are less active and less toxic than their parent compounds. However, at this point a group of drugs, known as *prodrugs,* should be mentioned. The prodrug as parent compound is usually not active and must be metabolized to the active form (for example, the inactive cancer drug cyclophosphamide is biotransformed to the active compound 4-hydroxycyclophosphamide). The active form of prodrugs is either unstable, not readily soluble, or poorly absorbed.

Some drugs form metabolites that are also active. For example, the active drug procainamide is biotransformed to the equipotent metabolite acetylprocainamide. The knowledge of active metabolites is particularly important for therapeutic drug monitoring to correlate the total concentration of all active forms with pharmacological effects.

Elimination. The final excretion of the drug from the body either as unchanged parent compound or in the form of metabolites is called *elimination*. The major routes of excretion are through the kidney into urine and through the liver into bile and consequently into feces. Other pathways of elimination are through skin (sweat), lungs (expired air), mammary glands (milk), and salivary glands (saliva).

The elimination half-life is the time required to reduce the blood level concentration to one half after equilibrium is obtained. After the drug is absorbed and distributed, it takes one half-life to eliminate 50% of the drug, seven

Table 54-1 Elimination half-lives

Number of half-lives	Percentage of drug remaining in body	Percentage of drug eliminated
1	50	50
2	25	75
3	12.5	87.5
4	6.25	≅94
5	3.125	≅97
6	1.5625	≅98.5
7	0.78125	≅99.2
8	0.3906	≅99.6
9	0.195	≅99.8
10	0.097	≅99.9

half-lives to eliminate 99% of the drug, and 10 half-lives to eliminate 99.9% of the drug (Table 54-1).

Effects of biological variation on LADME. If a drug is given in identical amounts by the same route of administration at the same time of day to identical twins, the pharmacokinetic parameters will differ only very slightly. In fraternal twins there will be larger differences. Greater differences will occur within a population group, even if this group is homogeneous with regard to sex, age, body weight, and health. These differences are genetically based variations in drug handling, which may influence absorption, distribution, metabolism, elimination, and drug-receptor interactions. Hence pharmacokinetic parameters for healthy subjects reported in the literature are means with ranges. They are actually valid only for the group studied.

Physiological and pathological factors influencing drug disposition. Apart from biological variations caused by genetic differences, a number of physiological and pathological factors may alter considerably a drug's disposition.[1]

The most prominent physiological factors are body weight and composition, age, temperature (hyperthermia and hypothermia), gastric emptying time and gastrointestinal motility, bloodflow rates (during rest and exercise), environment (high altitude, mountain sickness), nutrition, pregnancy, and circadian rhythm.

Among the most important pathological factors are renal impairment, liver impairment, acute congestive heart failure, burns, shock, trauma, and gastrointestinal diseases.

Blood levels as indicators of clinical response

The rationale for the use of blood levels as indicators of clinical response is based on the concept that for those drugs that interact at a receptor site without being changed, the drug concentration at the site of action will determine the intensity and duration of the pharmacological effect. Since it is usually not possible to sample at the site of action or biophase (such as the cell membrane), the next alternative is to sample whole blood, plasma, or serum, which is the biological fluid in closest equilibrium with the receptor site that can be easily sampled. After the distribution phase is complete, the drug concentration in the central and peripheral compartments (that is, blood) will decline in parallel. At this point a pseudoequilibrium of distribution is obtained regardless of whether the site of action is in the central compartment or in any peripheral compartment. Although the total drug concentration may differ considerably between central and peripheral compartments, the concentration of *free* (unbound) drug will be the same. Hence, once the pseudoequilibrium of distribution is reached, a correlation should exist between pharmacological effect and drug concentration in blood. Usually only the total drug concentration is measured in plasma. This is quite acceptable under normal conditions because individual differences in plasma-protein binding seem to be small;[2] in some cases, however, this is not true.[3,4]

Blood levels after single dose of drug

Most graphical descriptions of a pharmacokinetic response are given as a plot of blood concentration versus time (Fig. 54-1). The shape and course of a blood level-time curve depend on the route of administration and the LADME system.

With rapid intravenous administration, each facet of the drug is instantly in the systemic circulation. If the drug is given by extravascular administration, none will be in systemic circulation at the moment of administration, that is, at time zero. After the drug is released from the dosage form, the blood level-time curve rises with continuous absorption. Once absorbed, a molecule is exposed to distribution, metabolism, and elimination. Since initially a greater proportion is absorbed than is distributed, metabolized, and eliminated, the blood level-time curve rises until input and output are equal. At this time (t_{max}) the peak concentration (C_{max}) is reached and the blood level-time curve declines as elimination exceeds absorption (Fig. 54-1).

Two liberation factors may change the shape of the curve: the rate and the extent of liberation. Drug products from different manufacturers may release the drug at various rates. A slow release may also be intentional, as in the case of slow-release (sustained-release) dosage forms. However, if all the drug is released, the areas under the blood level-time curves of different formulations will be the same. If the drug is not fully released, a so-called bioavailability problem might be present and the area under the curve will be reduced (Fig. 54-2). *Bioavailability* refers to the amount of drug systemically absorbed.

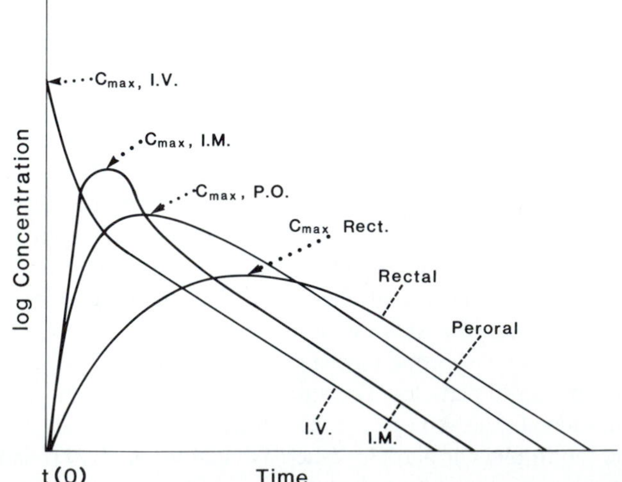

Fig. 54-1 Blood level-time curves of a hypothetical drug upon different routes of administration.

The absorption process can be influenced by many factors. Food (when the drug is given orally before, during, or after meals) may have no effect on the absorption, may accelerate or prolong the absorption, or may influence the extent of absorption. For instance, the blood level of griseofulvin is greatly enhanced when the drug is given with fat, whereas a tetracycline blood level decreases when the drug is ingested with milk.

The volume of distribution may change in various pathological conditions. If the volume of distribution increases, the blood level decreases and vice versa. In congestive heart failure the volume of distribution for certain drugs is reduced (digoxin, quinidine). The same dose will therefore result in a higher concentration (Fig. 54-3).

For drugs that are extensively metabolized, changes in the course of blood levels may result from impaired metabolism (liver damage) or other drugs given concomitantly that either compete for metabolic pathways (enzyme inhibition) or accelerate metabolism (enzyme induction) (Fig. 54-4).

Elimination of drugs, particularly of those predominantly eliminated through the kidney, may be tremendously prolonged in case of renal failure and in aged persons. The reduced elimination can result in a manyfold prolonged elimination half-life. Classic examples are the aminoglycosides. Gentamicin's normal half-life of 2 hours may easily be prolonged to 20 hours or more (Fig. 54-5).

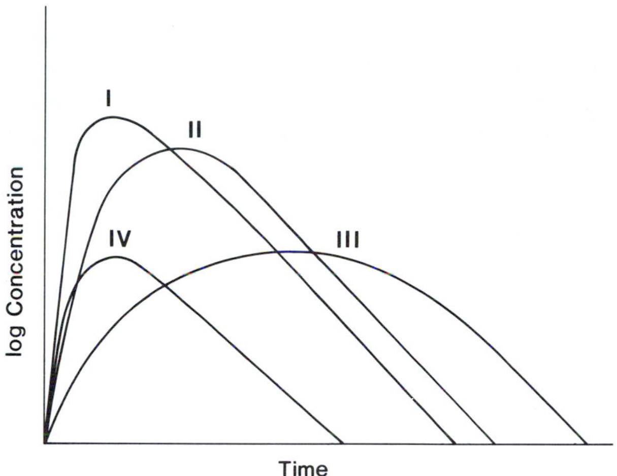

Fig. 54-2 Influence of liberation process on course of blood level-time curves. *I,* Fast-dissolving tablet; *II,* tablet with slower dissolution rate; *III,* sustained-release tablet; *IV,* tablet with poor bioavailability.

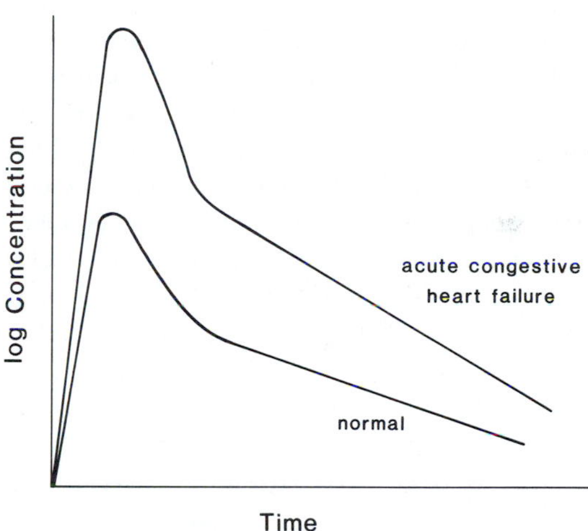

Fig. 54-3 Influence of distribution process on course of blood level-time curves of digoxin. In acute congestive heart failure a higher blood level is observed because of decreased volume of distribution.

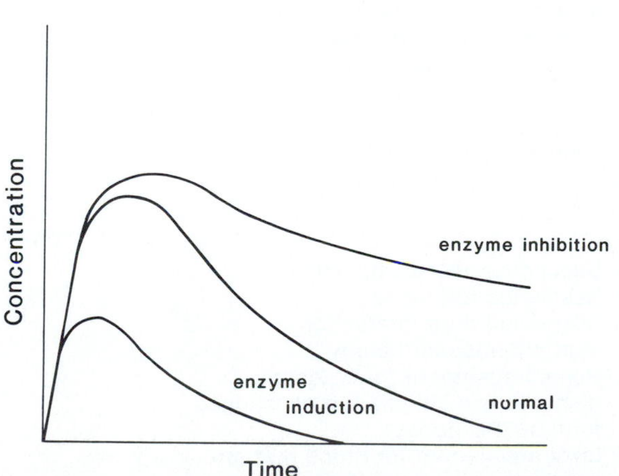

Fig. 54-4 Influence of metabolism processes on course of blood level-time curves. Enzyme inhibition and liver damage may greatly increase blood level, whereas enzyme induction may decrease it.

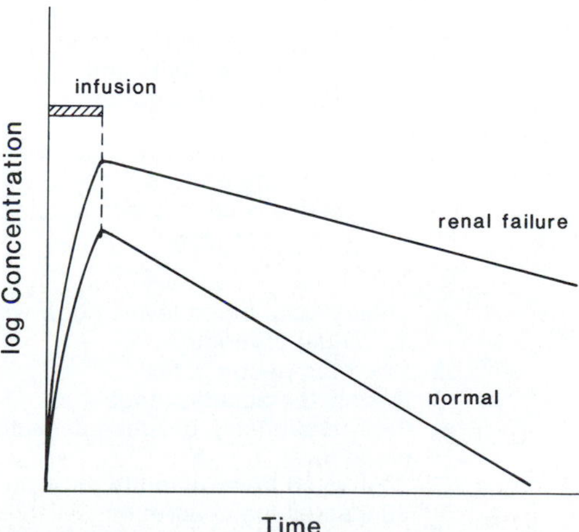

Fig. 54-5 Influence of elimination processes on course of blood level-time curve of gentamicin. In presence of renal failure, peak concentration after short-term infusion is higher and blood level remains elevated with a longer elimination half-life.

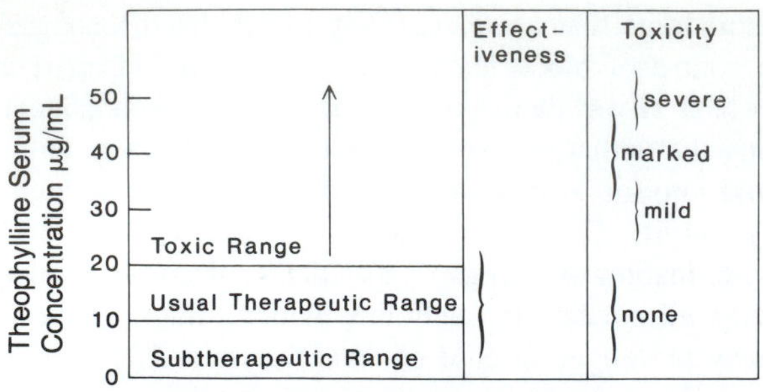

Fig. 54-6 Relationship between serum theophylline concentration and effectiveness and toxicity.

Effects of high levels of drugs

As stated earlier, for most drugs there is a relationship between drug concentration in the blood and the pharmacological and toxic response. Hence an increase in the blood level is usually associated with an increase not only in intensity of clinical effectiveness but also of toxicity. For most drugs it is desirable either to reach a therapeutic range with the peak concentration or to maintain the blood level throughout the dosage interval within the therapeutic range. A concentration below the therapeutic range is subtherapeutic or ineffective, and a concentration above the therapeutic range is likely to be toxic and cause side effects. However, the therapeutic and subtherapeutic range and the therapeutic and toxic range often overlap (Fig. 54-6). Additionally, an established therapeutic range may be applicable for the majority of patients but may be too low or too high for an individual patient. One such example is theophylline, for which the usual therapeutic range is between 10 and 20 µg/mL. However, some patients are perfectly controlled with levels as low as 5 µg/mL. On the other hand, whereas most patients do not experience theophylline side effects with levels of 23 µg/mL, some already show toxic signs at this level.

Need for monitoring of drug therapy

Many patients, regardless of whether they are hospitalized or ambulatory, receive more than one drug during any given day,[5] which increases the probability of drug-induced diseases, drug interactions, and side effects. One of the most widely used drugs, cimetidine, has been reported to interact with 21 different drugs.[6] A definite need for drug monitoring is also indicated by the finding that 30%

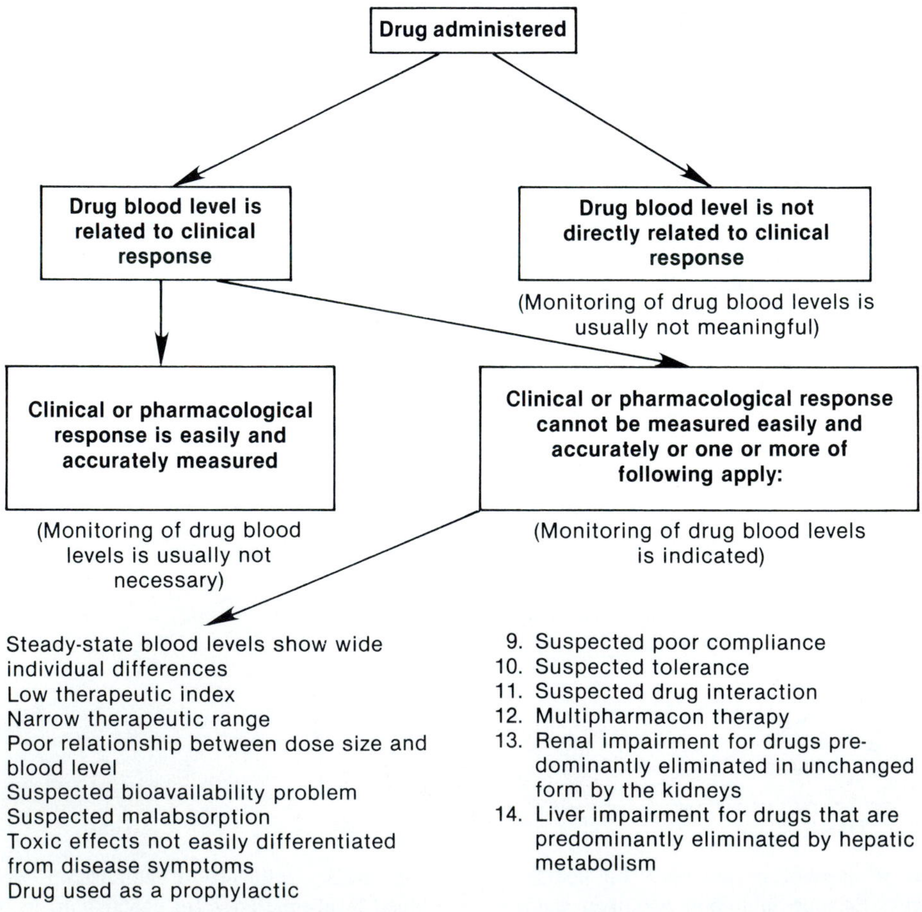

Fig. 54-7 Scheme to identify cases and situations when drug monitoring is indicated. *(Modified from Pippenger, CE: Ther Drug Monit 1:3-9. 1979.)*

to 50% of dosage administrations in hospitals and nursing homes were in error.[7]

Another reason for drug monitoring is noncompliance with a prescribed dosage regimen. One report states that the percentage of patients failing to take their medication as directed ranges between 20% and 82%.[8]

Drug monitoring for all drugs and all patients is neither possible nor feasible. Furthermore, total drug monitoring is, at least at present, not relevant for all drugs. For other drugs, for which a pharmacological response is easily, quickly, and accurately measured, it is clinically more relevant to directly monitor the clinical response (such as blood pressure, blood glucose, electrolyte excretion) instead of the blood level.

If a patient is responding well to the drug therapy without any signs of toxicity, this dosage regimen should be maintained even though the blood level might be outside the usual therapeutic range. One can answer the question "For which drugs is therapeutic drug monitoring indicated?" as follows. Monitoring is indicated for a number of drug groups such as antiepileptic drugs, antiarrhythmic agents, peroral anticoagulants, theophylline, tricyclic antidepressants, lithium, and aminoglycosides that either show large individual variation or are toxic above the therapeutic range. A flow chart showing the factors underlying the need for monitoring is given in Fig. 54-7.[9]

DOSAGE REGIMENS USED IN ACHIEVING THERAPEUTIC TARGET CONCENTRATION

Prediction of dosage for steady-state therapeutic levels

Most drugs are not administered as a single dose. Instead most drugs are administered in a series of doses given at specified intervals throughout the entire course of drug therapy. If the drug is administered repeatedly using dosing intervals shorter than the time required to eliminate the drug remaining in the body from the preceding dose, the drug will *accumulate* until a steady state is achieved, that is, one in which drug input and output are equal. Steady state is obtained when, with a specific regimen of dosage, the peak concentration (C^{ss}_{max}, or maximum steady-state concentration) and trough concentration (C^{ss}_{min}, or minimum steady-state concentration) after each dose oscillate within a certain range; the goal is to achieve the therapeutic range. By obtaining a blood level-time curve after a single dose, one can derive the necessary parameters to predict the steady state and in turn the dose required to achieve a desired steady state.

The *maintenance dose* required to maintain a desired mean steady- state concentration, C^{ss}_{av}, at a given dosage interval, τ, depends on the magnitude of C^{ss}_{av} (the required drug concentration in blood to elicit the pharmacological response), the pharmacokinetic parameters of drug disposition, and the patient's body weight. The generalized equation for determining the correct maintenance dose is given in the accompanying box.[10]

The dosing interval, τ, is freely chosen within a wide range, most often at times less than $t_{1/2}$ (half-life). It may have to be increased in renal or hepatic diseases because $t_{1/2}$ is often greatly extended in these cases. In general, at the end of four half-lives (if a dosing interval less than the half-life is chosen) a steady-state level is reached with multiple dosing (Fig. 54-8).

Dosing regimens

One can design the dosage regimen for multiple dosing maintenance therapy according to five different methods,

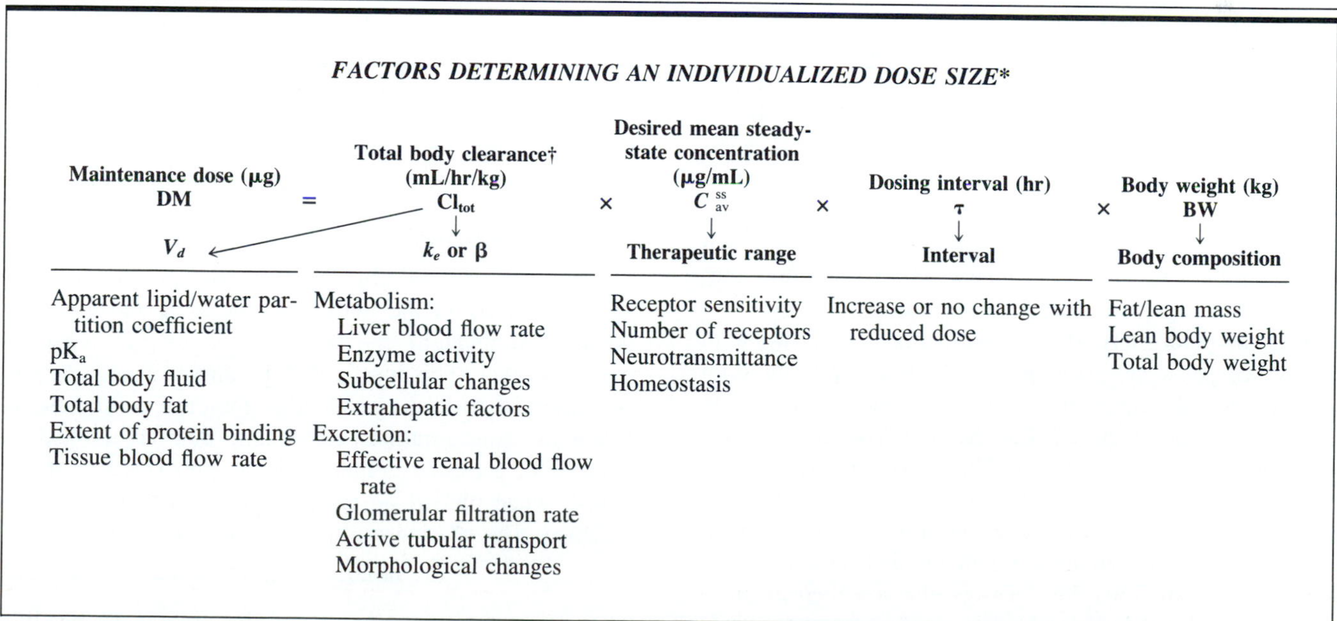

FACTORS DETERMINING AN INDIVIDUALIZED DOSE SIZE*

Maintenance dose (μg) DM	=	Total body clearance† (mL/hr/kg) Cl_{tot}	×	Desired mean steady-state concentration (μg/mL) C^{ss}_{av}	×	Dosing interval (hr) τ	×	Body weight (kg) BW
V_d ←		↓ k_e or β		↓ Therapeutic range		↓ Interval		↓ Body composition
Apparent lipid/water partition coefficient pK$_a$ Total body fluid Total body fat Extent of protein binding Tissue blood flow rate		Metabolism: Liver blood flow rate Enzyme activity Subcellular changes Extrahepatic factors Excretion: Effective renal blood flow rate Glomerular filtration rate Active tubular transport Morphological changes		Receptor sensitivity Number of receptors Neurotransmittance Homeostasis		Increase or no change with reduced dose		Fat/lean mass Lean body weight Total body weight

*For further information see Ritschel, WA: Contemp Pharmacy Pract 5:209-218, Washington, DC, 1982, American Pharmaceutical Association.
†V_d, Apparent volume of distribution (mL/kg); k_e or β, overall terminal disposition rate constant (hr^{-1}); $Cl_{tot} = V_d \cdot \beta$.

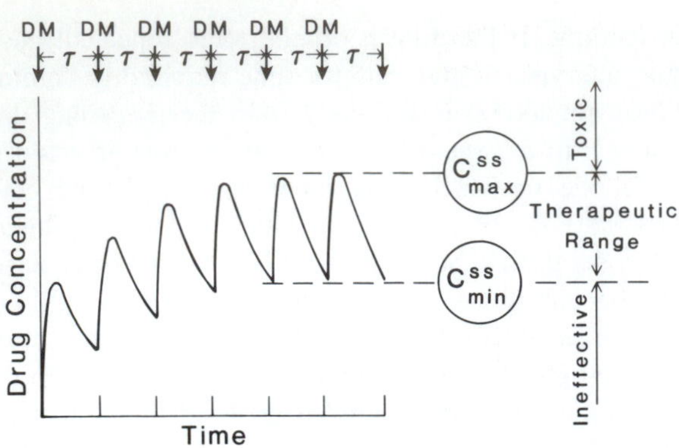

Fig. 54-8 Graph of blood-drug concentrations as function of dose and time. *DM*, Dose; τ, interval between doses; C^{SS}_{max}, concentration maximum at steady state; C^{SS}_{min}, concentration minimum at steady state. In this example τ is chosen to be equivalent to the half-life of elimination.

depending on the desired target concentration to be achieved or maintained throughout each dosing interval. For monitoring purposes it is necessary to know which method will be used because the optimum blood sampling protocol for laboratory analysis depends on the method in question. Five methods for dosage regimen design follow:

- Minimum effective concentration (MEC) or minimum inhibitory concentration (MIC) method
- C^{ss}_{max} or peak method
- C^{ss}_{max}-C^{ss}_{min} or limited-fluctuation method
- C^{ss}_{av} or log dose-response method
- TW or therapeutic window method

All methods refer to steady-state concentrations.

MEC or MIC method. For some drugs to be effective it is necessary to reach and maintain a minimum inhibitory concentration (MIC), or a minimum effective concentration (MEC), at steady state. Above the MIC or MEC the drug will be effective regardless of how high a peak level is reached as long as the entire steady-state blood level-time curve is above the required MIC or MEC. If the blood level-time curve falls below the MIC or MEC level, the drug will be ineffective as long as the concentration stays below this level. Drugs such as bacteriostatic antibiotics and other antimicrobial agents (sulfonamides) that have a relatively large *therapeutic index* are often prescribed at dosages calculated by this method.

C^{ss}_{max} or peak method. For some drugs it is desirable to reach a certain steady-state peak concentration during each dosing interval. However, for the remainder of the dosing interval it is not required that the drug concentration remain above a minimum level. This is particularly the case with bactericidal drugs, which act only on the proliferating microorganisms. In these cases one does not want to inhibit growth of those microorganisms that have not been killed by the previous dose. Drugs that are often given in dosage regimens based on the C^{ss}_{max} or peak method include penicillins, cephalosporins, gentamicin, and kanamycin.

C^{ss}_{max}-C^{ss}_{min} or limited-fluctuation method. For some drugs it might be desirable to maintain at steady state an MIC or MEC throughout the dosing interval but never exceed a certain peak value. This is particularly the case if the drug has a narrow therapeutic range. Drugs that might be administered using this method to compute dosage include gentamicin, kanamycin, streptomycin, isoniazid, and theophylline.

C^{ss}_{av} or log dose-response method. For drugs whose clinical effect follows a log dose-response curve, drug doses are selected to be in the lower portion of the log dose-response curve. The desired steady-state concentration is then usually in the lower third of the log dose-response curve. For drugs following a log dose response, the intensity of effect (and of toxicity) increases with increasing peak size. Drugs whose dosages are often based on this pattern are digoxin, lidocaine, procainamide, theophylline, quinidine, bactericidal antibiotics, analgesics, antipyretics, and hypoglycemic agents.

TW or therapeutic window method. With some drugs, such as antidepressants and antipsychotics, the clinical effect increases with dose size only up to a certain point and then actually diminishes as the dose size is further increased. Instead of a therapeutic range there exists a therapeutic window, showing a more or less bell-shaped log dose-response curve.

PHARMACOKINETICS

Pharmacokinetics is the quantitative study of drug disposition in the body. Pharmacokinetics permits one (1) to describe mathematically the fate of a drug after administration in a given dosage form by a given route of administration, (2) to compare one drug with others or one dosage form with other dosage forms, and (3) to predict blood levels of a drug with different dosage regimens or disease states.

Basically, in pharmacokinetics three types of kinetic processes are used to characterize the fate of drugs in the body: first-order or linear kinetics; zero-order or nonlinear kinetics; and Michaelis-Menten or saturation kinetics.

First-order kinetics

Most processes of drug uptake (absorption), diffusion and permeation in the body (distribution), and excretion (urinary elimination) can be described by first-order, or linear, kinetics. This means that the rate of change of concentration of drug is dependent on the drug concentration. When the concentration *versus* time data are plotted on numeric or cartesian graph paper, a concave curve is obtained; when plotted on semilog paper, a straight line is obtained. The relationship is expressed by Equation 54-1

$$dC/dt = -k \cdot C \qquad \textit{Eq. 54-1}$$

where C is the concentration of the drug, k is the first-order rate constant, and t is time. The minus sign indicates that the drug concentration decreases with time. Drugs are eliminated in a manner that can be described by first-order kinetics when a *constant percentage* of drug is eliminated per unit of time. Drugs exhibiting first-order elimination kinetics are antibiotics and sulfonamides, digoxin, lidocaine, procainamide, and theophylline. First-order kinetics describes the elimination of most drugs.

Zero-order kinetics

If the rate of elimination of a compound from the body is not proportional to the concentration of the drug taken, the elimination usually follows zero-order, or nonlinear, kinetics. This means that the rate of change of concentration is independent of the concentration of the particular drug. In other words, a *constant amount* of drug, rather than a constant proportion, is eliminated per unit of time (elimination depends on the amount per unit of time). When the concentration versus time data are plotted on numeric or cartesian graph paper, a straight line is obtained, whereas on semilog paper a convex curve is obtained. The classic example for zero-order kinetics is the disposition of alcohol (ethanol).

The relationship can be expressed by equation 54-2

$$dC/dt = -k_0 \qquad \textit{Eq. 54-2}$$

where the rate of change of concentration, dC/dt, is equal to the zero-order rate constant k_0, which has the units of amount per unit of time.

Michaelis-Menten kinetics

In metabolism nearly all biotransformation processes are catalyzed by specific enzyme systems with a limited capacity for the drug. Also in active transport of drugs across membranes the carriers have a limited capacity. Whenever the drug concentration present in a given system exceeds the capacity of the system, the rate of change of concentration is most precisely described by the Michaelis-Menten equation

$$dC/dt = -(V_m \cdot C)/(K_m + C) \qquad \textit{Eq. 54-3}$$

where C is the drug concentration, t is the time, V_m is a constant representing the maximum rate of the process, and K_m is the Michaelis constant, the drug concentration at which the process proceeds at exactly one half its maximal rate.

Examples of drugs that show saturation elimination kinetics are phenytoin, high doses of barbiturates, and glutethimide.

Compartment models

To describe the quantitative processes of a drug in the organism, pharmacokinetics uses the concept of compartments. A compartment is a unit characterized by two parameters: the drug concentration, C, and the volume, V_d. By multiplying the drug concentration by the apparent volume of distribution, one obtains the amount, A, of the drug in that compartment:

$$C \cdot V_d = A \qquad \textit{Eq. 54-4}$$

A given compartment model is not necessarily specific for a given drug. For instance, a drug given intravenously is often described by a two-compartment open model, whereas the same drug given orally or by any other extravascular route may be described by a one-compartment open model. *Open* means that there is input to and output from the compartment.

In reality the human body is a multimillion compartment system. However, usually one has in the intact organism easy access to only two kinds of biological fluids, blood (serum, plasma) and urine. Being restricted to blood or urine specimens, the drug has a fate in the body usually described by either a one-compartment or two-compartment open model. Clinically speaking, the concept of the one-compartment and two-compartment models is usually satisfactory for therapeutic use. The difference between a one-compartment and a two-compartment model is that in the former the distribution occurs instantly, whereas in the latter the distribution process needs a measurable time before pseudoequilibrium is obtained.

Terminal disposition rate constant

In the one-compartment open model, the last or terminal portion of the straight (monoexponential) slope of a semilog blood level-time curve gives the overall elimination rate constant, k_e (metabolism, renal excretion, and other pathways of elimination). In the two-compartment model it gives β, the slow-disposition rate constant (Figs. 54-9 and 54-10).

Zero-time blood level

Back extrapolation of the blood level-time curve after intravenous administration results in the zero-time blood

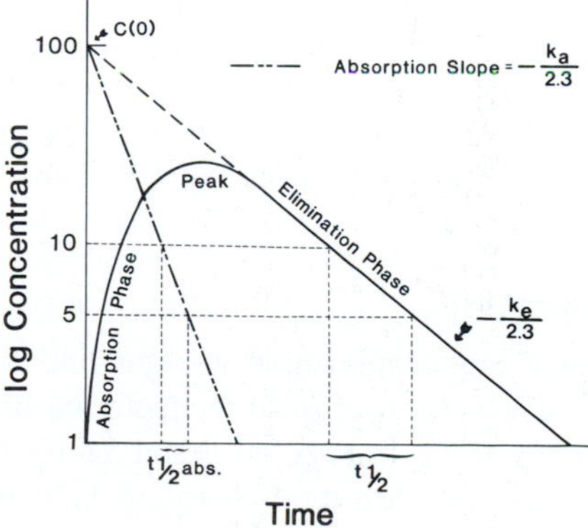

Fig. 54-9 One-compartment model blood level-time curve after extravascular administration, with monoexponential slopes for elimination, k_e, and absorption, k_a. *(From Ritschel, WA: Graphic approach to clinical pharmacokinetics, Barcelona, 1983, JR Prous.)*

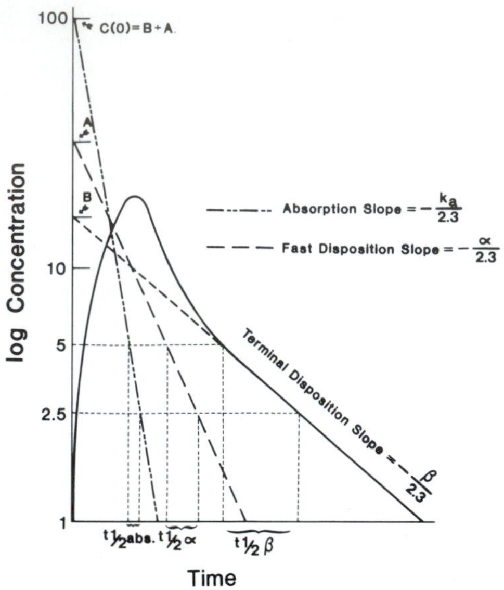

Fig. 54-10 Two-compartment model blood level-time curve after extravascular administration, with monoexponential slopes for slow disposition, β; fast disposition, α; and absorption, k_a. *(From Ritschel, WA: Graphic approach to clinical pharmacokinetics, Barcelona, 1983. JR Prous.)*

level C_0. After extravascular administration, the "fictitious" zero-time blood level, C_0, is the intercept of the k_e slope with the ordinate on a semilog plot in the one-compartment model, and the sum of the intercepts A + B of the α and β slopes in the two-compartment model (Figs. 54-9 and 54-10).

Absorption rate constant

The k_e slope is extrapolated back to time zero. This yields C_0, a theoretical concentration roughly equivalent to that obtained from an intravenous injection of the same amount of drug. By subtraction of the observed drug concentration during the absorption phase from the concentrations read from the back-extrapolated k_e slope, *residual* points are obtained. When plotted on semilog paper, they are described by a straight line, the slope of which is the absorption rate constant k_a (Fig. 54-10) in the one-compartment model.

Elimination half-life

Whenever a monoexponential straight line is obtained, one can calculate a drug's half-life. Note the line describing the terms k_e and β in Figs. 54-9 and 54-10. Other half-lives frequently used are the absorption half-life ($t_{1/2}$abs) and distribution half-life ($t_{1/2}$).

The terms *half-time, half-life, plasma half-life, elimination half-life,* and *biological half-life* are often used interchangeably. Half-life is equal to the time required for elimination of one half the total dose of drug from the body. The elimination half-life, or plasma half-life, is the time required for the elimination of one half the amount of drug that is in the blood (plasma or serum). In those instances in which the decline of drug concentrations in all tissues does not parallel the decline of drug concentration in plasma, blood, or serum, half- life and elimination half-life will be different. Most statements on drug disposition refer to the elimination half-life. In Fig. 54-10 the elimination half life ($t_{1/2}\beta$) is depicted graphically.

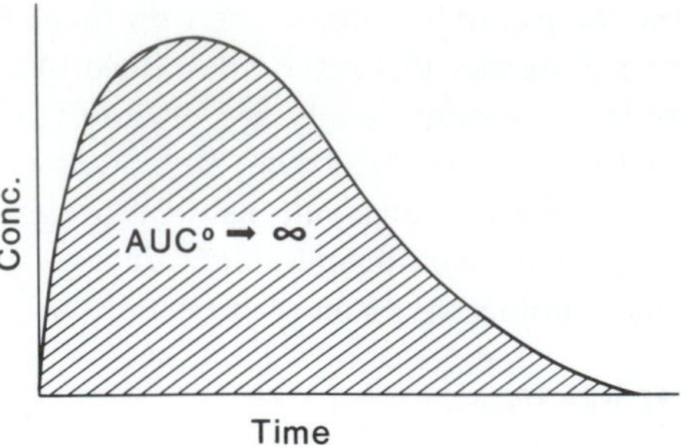

Fig. 54-11 Scheme of total area under the blood level-time curve. $AUC^{0\rightarrow\infty}$, Area under curve from time = 0 to time = ∞. *(From Ritschel, WA: Graphic approach to clinical pharmacokinetics, Barcelona, 1983, JR Prous.)*

Volume of distribution

The volume of distribution is not a real volume and usually has no relationship to any physiological space or body fluid volume. It is simply a term to make the mass-balance equation valid. On intravenous administration the amount of drug in the body is known; however, only the blood can be sampled. Because an amount of drug, *A,* equals the product of concentration and volume (μg/mL × mL), the volume of distribution is the hypothetical volume that would be required to dissolve the total amount of drug at the same concentration as that found in blood.

The volume of distribution is expressed in milliliters. If this value is divided by the patient's body weight, the distribution coefficient Δ′ is obtained in mL/g (or L/kg).

Area under blood level-time curve

The integral under a blood level-time curve is a measure of the total amount of drug in the body. One can calculate the area under the blood level-time curve from time zero to infinity, AUC, or one may approximate it by plotting the curve on graph paper and cutting out the area and weighing it. The AUC is shown in Fig. 54-11.

Total clearance

The total clearance in pharmacokinetics describes how much of the volume of distribution is cleared of the drug per unit of time, regardless of the pathway for the loss of drug from the body. In effect it is the sum of all clearances by different pathways. The total clearance is the product of the apparent volume of distribution and the terminal disposition rate constant.

Steady state

Steady state refers to the accumulation of drug in the body in multiple dosing when input and output are equal within a dosing interval. The magnitude of accumulation depends on the drug's elimination half-life and the dosing interval. The smaller the dosing interval for a given dosage, the greater the accumulation and the smaller the fluctuation around the mean serum value. At steady state the drug concentration oscillates around a mean steady-state concentration, C_{av}^{ss}, with a definite maximum steady-state concentration, C_{max}^{ss}, and a minimum steady-state concentration, C_{min}^{ss}. Only in the case of an intravenous constant rate infusion are C_{max}^{ss}, C_{min}^{ss}, and C_{av}^{ss} identical.

APPLICATION OF PHARMACOKINETICS TO TDM

Clinical assessment

Clinical (physician) estimation of patient response is the first and most important task in therapeutic monitoring. One should not forget that therapy requires an approach that considers all aspects of a patient's condition, including the disease symptoms; the disease itself; other diseases present; and the patient's physical condition, age, nutritional status, and psychological aspects. Furthermore, one should not forget that clinical pharmacokinetics is only a tool that can assist with but never substitute for clinical evaluation.

Application. The clinical evaluation of patient response comprises the evaluation of vital signs and change of symptoms in response to the drug therapy, such as blood pressure, pulse rate, electrocardiogram, measurement of edema, and urinary output. Furthermore, supportive laboratory analyses may be required, such as serum glucose and electrolytes. In all these cases the pharmacological response is evaluated *clinically,* either as direct measurement of pharmacological effect or as a measurement of body constituents, but *not* by the measurement of drug concentration in biological fluid.

Limitations. Sometimes the clinical evaluation might be difficult because of the presence of two or more disease states with similar or overlapping symptoms, polypharmacy (many drugs are given simultaneously), or *unexpected* results. Unexpected results that might occur are that (1) the patient is not responding as expected, showing either no effectiveness or limited effectiveness of therapy, or (2) the patient may exhibit unexpected toxicity or side effects.

A drug may be less effective than expected because (1) the drug has a low bioavailability, (2) the patient is not complying with the prescribed drug regimen, (3) a malabsorption syndrome exists and less of the drug than expected is being absorbed. A patient may exhibit unexpected toxicity or side effects because of drug interactions, enzyme induction, enzyme inhibition, renal or hepatic impairment, edema, dehydration, and so on. In these cases it is advisable, when possible, to request drug monitoring in biological samples (Fig. 54-7).

Assessment by drug analysis

When the therapeutic target concentration, therapeutic range, or toxic concentration is known, therapeutic drug monitoring can be used to support the clinical evaluation. When a patient is treated with drugs of low therapeutic index, such as aminoglycosides, digoxin, and lithium, or when unexpected side effects or toxicity are occurring (Fig. 54-7), therapeutic drug monitoring should be used.

Basis for monitoring. For monitoring, it is essential to fulfill certain requirements. Otherwise, any evaluation will be in error:[11]

1. The dose size, dosage form, and route of administration must be known.
2. The dosage regimen must be followed.
3. The time between the administration of the last dose and the drawing of the blood sample must be known.
4. The blood sampling time or times must be recorded exactly.
5. The sampling times must be appropriate.

Most of the requirements listed above are self-explanatory. To understand item 5, one should remember that any samples taken during the absorption or distribution phase are useless for monitoring. Samples taken at peak time allow only an approximation of pharmacokinetic data. Samples taken at peak time and during the terminal elimination phase will result in an overestimation of the elimination rate constant and an underestimation of the elimination half-life.

The optimal sampling times for the various dosage regimen methods are shown in Fig. 54-12. Sampling times of commonly monitored drugs are listed in Table 54-2.[12]

A review of the elimination half-lives and the therapeutic ranges of 16 commonly monitored drugs also is given in Table 54-2.[13]

Limitations. The ability to monitor a specific drug in a blood sample can sometimes be limited because (1) accurate information about the times of drug administration and blood drawing is not available, (2) a reliable assay method is not available, and (3) laboratory analysis time is not reasonable. Most difficulties experienced in monitoring are caused by limited or inaccurate information. Precision of the assay must also be considered. The coefficient of variation of the assay can be important when the drug concentration is found to be either at the lower or upper end of the therapeutic range. The therapeutic ranges reported in Table 54-2 are mean values applicable to the majority of patients; the therapeutic or toxic concentration may be different for a particular patient (see Fig. 54-6).

Assessment by pharmacokinetic calculations

To evaluate blood samples pharmacokinetically, one must know whether the drug concentration is at steady

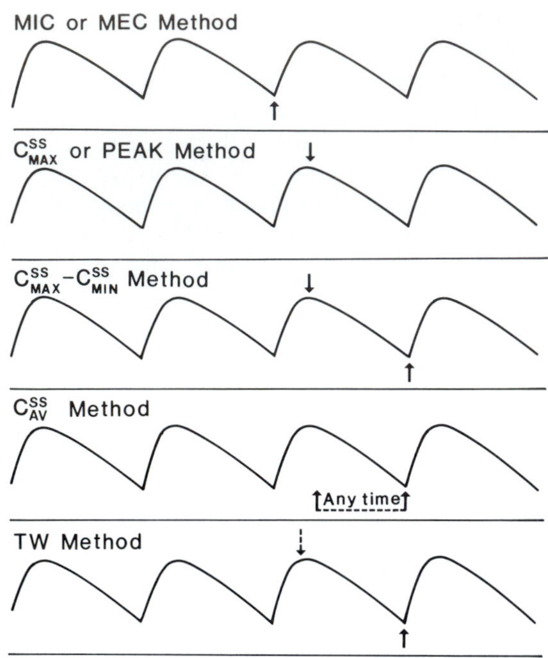

Fig. 54-12 Scheme showing optimal sampling times for monitoring for different methods used for dosage regimens.

state. To decide if a steady state has been achieved, one needs to know the dosage regimen (dose size and dosing interval) and how long this dosage regimen has been in effect. Usually it is assumed that a steady state is reached when the dosage regimen has been implemented for a time greater than four times the drug's elimination half-life. If the regimen has been in effect for less than $4t_{1/2}$, one should know the number of doses given before the sample was obtained. Using classic pharmacokinetic equations one can calculate what the drug concentration should be, based on mean pharmacokinetic parameters from the literature and patient information (age, body weight, sex, height, renal status, and so on). The purpose of drug monitoring is to compare the observed drug concentration to the expected or desired one. The observed concentration may be equal to, smaller than, or greater than the desired one. In the latter two cases an adjustment of the dosage regimen may be indicated and recommended.

If C_x (the measured drug concentration) differs from the desired concentration (such as $C^{ss}_{av\ des}$ or $C^{ss}_{max\ des}$), one should try to find the reason for the deviation. Some general and probable causes are presented here.

A concentration higher than the expected one could be associated with an increased bioavailability of a drug of generally low bioavailability (for example, cimetidine increases bioavailability of propranolol), patient noncompliance (for example, use of more drug or shorter dosage

Table 54-2 Recommended sampling times, half-lives, and therapeutic ranges for commonly monitored drugs[12]

Drug	Recommended sampling time	Normal $t_{1/2}$ (hours)	Therapeutic range (μg/mL)
Amikacin	0.5 to 1 hour after dose and end of dosing interval	2.5	10-25
Carbamazepine	End of dosing interval	10-60	5-12
Diazepam	3 to 5 hours after dose	33	0.1-1.0
Digitoxin	Any time at least 6 hours after dose	164	0.01-0.035
Digoxin	Any time at least 6 hours after dose	40	0.0005-0.002
Ethosuximide	2 to 5 hours after dose	24-72	40-100
Gentamicin	0.5 to 1 hour after dose and end of dosing interval	2.0	0.5-8
Lidocaine	24 hours after start of infusion	1.8	1.5-7
Lithium	End of dosing interval	19	0.6-1.4 (mmol/L)
Phenobarbital	End of dosing interval	84-108	10-40
Phenytoin	End of dosing interval	12-36	10-20
Primidone	End of dosing interval	6-18	5-12
Procainamide	End of dosing interval	2.4	20-100
Salicylate	1 to 3 hours after dose		10-20
Theophylline	Intravenous infusion: 24, 48, and 72 hours after start of infusion Oral or intravenous injection; 2 hours after dose or end of dosing interval, or both Oral sustained release; 6 hours after dose and end of dosing interval	6 (?)	
Tobramycin	0.5 to 1 hour after dose and end of dosing interval	2.0	2-10

intervals than prescribed), or decreased total clearance (for example, as occurs with renal or liver failure and acute congestive heart failure), or increased protein binding. A concentration lower than expected could result from decreased bioavailability (possibly drug interaction), insufficient drug dosing (longer dosing intervals, missed doses, or noncompliance), increase in the total clearance (drug interaction or enzyme induction), or decreased protein binding.

More complex pharmacokinetic analyses

The equations presented here are basic ones. Space and background information do not permit more detail. The purpose of this chapter is to present a general review and to transmit a general understanding of the pharmacokinetic principles involved, including dosage regimens.

A number of articles and books on pharmacokinetic assessment of drug monitoring can provide more information.[14-16]

REFERENCES

1. Ritschel, WA: Handbook of basic pharmacokinetics including clinical applications, Hamilton, IL, 1986, Drug Intelligence Publications.
2. Borga, O, Azarnoff, DL, Plym Forshell, G, and Sjoqvist, F: Plasma protein binding of tricyclic antidepressants in man, Biochem Pharmacol 18:2135-2143, 1969.
3. Reidenberg, MM, Odar-Cederlof, I, von Bahr, C, et al: Protein binding from diphenylhydantoin and desmethylimipramine in plasma from patients with poor renal function, N Engl J Med 285:264-267, 1971.
4. Levy, G: Relationship between pharmacological effects and plasma or tissue concentrations of drugs in man. In Davies, DS, and Pritchard, DNC, editors: Biological effects of drugs in relation to their plasma concentrations, Baltimore, 1973, University Park Press, pp 83-95.
5. Stewart, RB, Forgnone, M, and Cluff, LE: Drug utilization and reported adverse drug reactions in outpatients, Drugs in Health Care 2:231-243, 1975.
6. Ritschel, WA: Pharmacokinetics of H_2−receptor antagonists, Sci Pharm 50:250-259, 1982.
7. Barker, KN and McConnell, WE: How to detect medication errors, Mod Hosp 99:95-106, 1962.
8. Stewart, RB, and Cluff, LE: A review of medication errors and compliance in ambulant patients, Clin Pharmacol Ther 13:463-468, 1971.
9. Bochner, F, Carruthers, G, Kampmann, J, and Stiner, J: Handbook of clinical pharmacology, Boston, 1978, Little, Brown & Co, pp 22-25.
10. Ritschel, WA: The effect of aging on pharmacokinetics: a scientists' view of the future, Contemp Pharmacy Pract 5:209-218, 1982.
11. Hassan, FM, Pesce, AJ, and Ritschel, WA: Pitfalls and errors in drug monitoring: analytical aspects, Methods Find Exp Clin Pharmacol 5:567-573, 1983.
12. Slaughter, RL, and Koup, JR: Clinical pharmacokinetic service. In McLeod, DD, and Miller, WA, editors: The practice of pharmacy, Cincinnati, 1981, Harvey Whitney Books, pp 114-128.
13. Ritschel, WA: Handbook of basic pharmacokinetics, ed 2, Hamilton, IL, 1980, Drug Intelligence Publications, pp 412-427.
14. Garrett, ER, and Hirtz, JL: Drug fate and metabolism, New York, 1985, Marcel Dekker.
15. Welling, PG: Pharmacokinetics: processes and mathematics, Washington, DC, 1986 American Chemical Society.
16. Rowland, M, and Tucker, GT: Pharmacokinetics: theory and methodology, New York, 1986, Pergamon Press.

CHAPTER 55

Interferences in chemical analysis

LAWRENCE A. KAPLAN
AMADEO J. PESCE

KEY TERMS

Allen correction Multichromatic analysis of reaction to correct for background absorbance. Two wavelengths, in addition to the A_{max} of the chromophore, are monitored to subtract average background absorbance.

bichromatic analysis Spectrophotometric monitoring of a reaction at two wavelengths. Used to correct for background color.

chemical interferent A compound that either produces an endogenous color or interferes directly in the reaction or process being monitored.

end-point analysis Monitoring of a reaction after the reaction has been essentially completed.

hemolysis Breakage of red blood cells, either in vitro or in vivo. Hemolysis will give plasma specimen a red color.

icteria Pertaining to orange color imparted to a sample because of the presence of bilirubin.

interferent Any chemical or physical phenomenon that can interfere in or disrupt a reaction or process.

in vitro interferent An interferent that is not caused by any in situ physiological process.

in vivo interferent An interfering process resulting from physiological processes within the body.

kinetic analysis Analysis in which the *change* of the monitored parameter with time is related to concentration, such as change of absorbance per minute. Analysis made very early in reaction period.

lipemia Presence of lipid particles (usually very low density lipoprotein) in a sample, which gives the sample a turbid appearance.

reagent blank Reaction mixture *minus* the sample; used to subtract endogenous reagent color from the absorbance of the complete reaction (plus sample).

sample blank Sample plus diluent; used to correct absorbance of complete reaction mixture for endogenous sample color.

turbidity Scatter of light in a liquid containing suspended particles.

window Term used to denote a specific time during which reactions are monitored, a phenomenon can be observed, or a procedure can be initiated.

A chemistry laboratory uses a number of techniques for measuring the concentration of specific biochemicals. All these techniques are subject to interferences from a variety of sources. Specific interferences of one technique may not be important for another. There are, however, general concepts that, when understood, help to control and minimize the effects of interferences on method accuracy and precision.

There are four basic types of interferences in laboratory analysis: (1) those that arise from limitations of detectors; (2) chemical substances in the sample that directly interfere with the analytical method; (3) disease states or exogenous agents that modify certain physiological processes, thus changing the concentrations of an analyte in vivo; and (4) those that occur as a result of sample (blood) processing.

LIMITATIONS OF DETECTORS

Methods yielding quantitative answers usually employ a detector, such as a spectrophotometer. It is then possible to obtain a linear relationship between the detector response and the concentrations of analytes in various samples. In the case of absorption spectrophotometry, there is a complex logarithmic relationship between concentration and detector response (see Chapter 3). When fluorescence is used, there is a linear relationship between concentration and the fluorescence signal. Similarly, when other optical properties, such as refractive index, or electrochemical properties, such as ion current from oxidation, are used by detectors, the response is also linear. This knowledge of the type of relationship between concentration and detector response is important.

The first section describes errors in absorption spectrophotometry because the clinical chemistry laboratory quantifies most of its analytes by this technique. It is most important therefore to understand the interference problems associated with this mode of measurement. As the nature and sophistication of the laboratory change, other types of interferences will become important to consider in performing a laboratory analysis.

Absorption spectrophotometer

There are two interrelated types of error in spectrophotometric measurement. The first is caused by the nature of the mathematical relationship between absorbancy and percent transmission, and the second is related to limitations of the instrument.

Absorbance variance. As discussed in Chapter 3, there is a logarithmic relationship between the percentage of transmittance, which is the quantity actually measured, and the absorbance, which is calculated. Fig. 55-1 shows the relationship between the linear percent transmittance and the log absorbance scales. At very low percent transmittances, small changes in percent T result in large changes in absorbance. For example, a change in percent transmittance of 60% to 50% T represents an absorbance change of approximately 0.08 A. A change in percent transmittance from 15% to 5%, however, results in a change of absorbance of 0.65 A. Thus small changes of percent T at very low transmittances will result in disproportionately large changes in the calculated absorbance at this part of the scale and will lead to increased error of analysis.

One can consider the error of a spectrophotometric measurement as a function of the total or full-scale deflection of the detection meter or its electronics. When the absorbance scale is set at 0.000 or the transmittance scale is adjusted to 100%, the maximum electronic signal is obtained. For error analysis, the variation in this measurement is presumed to be constant throughout all readings on the scale. Because percent T directly reflects the electrical signal, some simple calculations using percent T can be done. At full-scale deflection (100% T), a 1% variation in percent T means an error of ±1% T. At half scale, a 1% variation of the 50% T value means a 2% absolute error (1/50), and at 10% T, this becomes a 10% error (1/10). However, it is not percent T that is directly proportional to concentration, it is absorbance. One can convert these percent T values into absorbance and calculate the error (Table 55-1).

Table 55-1 Absorbance error as a function of percent T

% T	Absorbance	Variation in absorbance	Percent error of absorbance measurement
4 ± 1	1.398	0.22	15.8
10 ± 1	1.000	0.041	8.60
25 ± 1	0.602	0.035	5.79
35 ± 1	0.456	0.025	5.44
50 ± 1	0.301	0.017	5.78
70 ± 1	0.155	0.012	8.03
90 ± 1	0.046	0.0097	21.2

Because the conversion of percent T to absorbance is a logarithmic function, both ends of the scale, 0.000 absorbance and high absorbance (more than 1.0), have the greatest error. In simple terms, when the solution has little color, it is difficult to tell the difference between no color and some color. The relative error can be huge, because this difference is so small. At high absorbance it is not easy to record accurately a small amount of light passing through the solution. Thus differences between the two values are minute compared to the total incident light used to calibrate 100% T, and it is difficult to measure these small changes.

The relative error of spectrophotometric measurements versus percent transmittance and absorbance is shown in Fig. 55-2. One should make most spectrophotometric measurements at absorbances between 0.1 and 1.1 to minimize this type of error.

Instrument limitations. Consider how a spectrophotometer functions. At 100% T or zero absorbance, all the light signal is converted to an electronic signal. Presume that this signal measures 1000 nanoamps (nA). If the absorbance changes by 0.010, there is a decrease to 990 nA, and the instrument must measure accurately 10/1000 nA or a 1% change in signal. To do this accurately (1%), it must measure the signal to ± 0.10 nA (1% of 10). Thus at zero absorbance there is the difficulty of accurate measurement of one part in 10^4 of signal. On the other hand, if the absorbance is 2.0, the signal to the photomultiplier is only 10 nA because only 1% of the light reaches the photodetector. To achieve the same degree of accuracy between a value of absorbance of 2 and 2.02, the instrument must measure the difference between 10 and 9.9 nA, or 0.1 nA. To do this accurately, it must measure to within 0.001 nA (1% of 0.1). Thus at high absorbance the limitations are caused by the inability of the detection system to measure small differences between high levels of absorbance accurately.

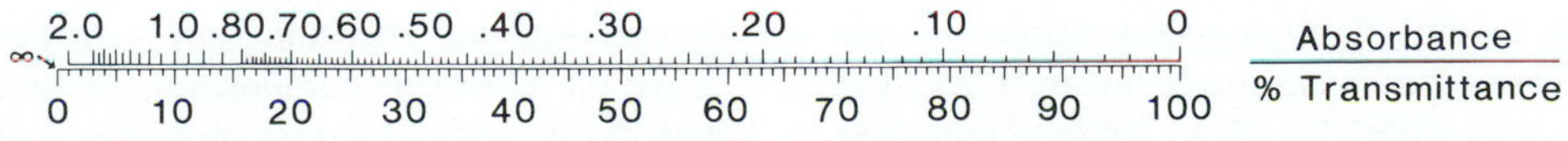

∞ 2.0 1.0 .80 .70 .60 .50 .40 .30 .20 .10 0 Absorbance

0 10 20 30 40 50 60 70 80 90 100 % Transmittance

Fig. 55-1 Absorbance and percentage of transmittance scales juxtaposed.

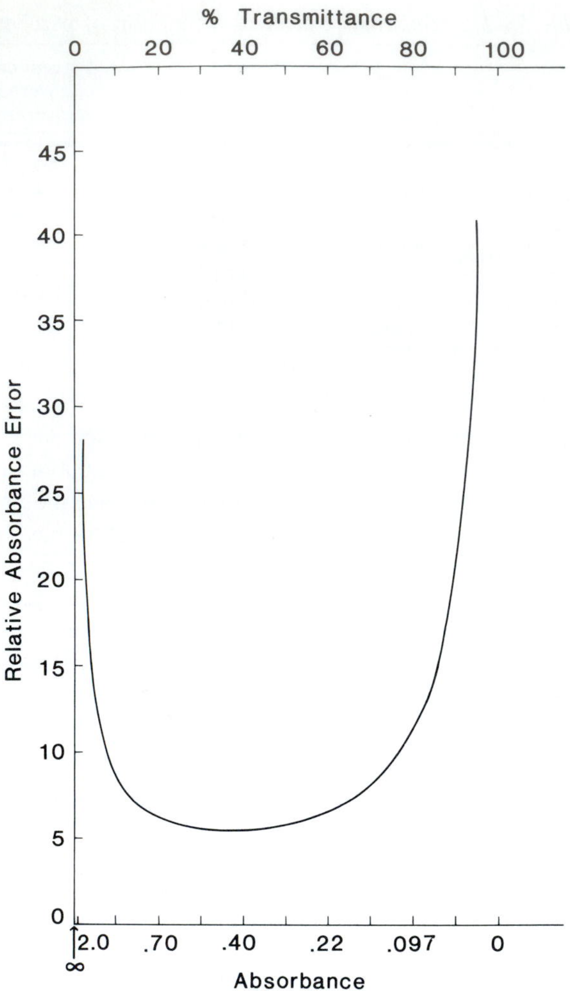

Fig. 55-2 Relative absorbance error versus absorbance *(A)* and percentage of transmittance *(% T)* for a $\pm 1\%$ error in measurement of transmitted light. Relative error is minimum at 36.8% *T*, and *A* is 0.434.

Thus in analyses with relatively high levels of absorbing compound or interferent there will be a large spectrophotometric error. An initial dilution to lower total absorption *plus* a sample blank may be needed to eliminate this problem (see the discussion of turbidity on p. 811 for an example).

Fluorescence spectrophotometer

Fluorescence measurements are different from those of transmission spectrophotometry in that the intensity of the fluorescence signal is linearly related to concentration. A doubling of the fluorescence signal is indicative of a twofold increase in concentration. This is true only if there is very little light absorbed by the sample. If a significant portion of the light passing through a fluorescence sample is absorbed, the relationship is no longer linear; it becomes a more complex mathematical function. Thus, to minimize error, fluorescence analysis should be performed with relatively dilute solutions, the absorbance of which is less than 0.1.

Scattering of light or the presence of stray light has pronounced effects on fluorescence measurements. If the sample scatters light, some may be observed by the detector as fluorescence. Similarly, if there is *stray light* (a term meaning that the light used to excite the sample was not pure, that is, not of a very narrow color band), this also may be recorded by the detector. Because only a small fraction of the incident light (less than 1% and often less than one part in a million of the total light input into the instrument) is detected as fluorescence, these extraneous signals have a disproportionate effect on the reading and thus on the error.

Fluorescence measurements, like absorption measurements, can be inaccurate or invalid because of high signals. Unlike absorption spectrophotometry, this usually occurs because of the blanking system. If the fluorometer is adjusted to read zero for the blank, the entire detector sensitivity is adjusted for this zero reading. Presume that there is a blank solution that the instrument records as 10 units on its most sensitive scale. The instrument is now set to read zero fluorescence units. If the instrument can record accurately this zero units to $\pm$ 1 unit and if it can also measure full scale of 100 units in the same way, it can accurately measure a signal of 100 $\pm$ 1 units. If a different blank solution is used and this records as 100 units, this value can be set at zero by use of an electronic manipulation (subtraction of the signal). At this new full-scale deflection of 100, the detector is really recording 200, of which 100 is subtracted as the blank. Because the instrument is accurate to 1%, it is now accurate to 1% of the new full scale of 200 units. The inaccuracy of measurement is now $\pm$ 2 units. The accuracy is therefore twice as poor as that for the first example. Similarly, if a blank records as 1000, the accuracy is one tenth as great. Therefore the blank limits the accuracy of fluorescence measurements.

This same line of argument also applies to other linear measurements such as in electrochemical analysis. The background can be blanked, but this must be considered in relation to the total signal.

IN VITRO INTERFERENCES

In vitro interferences arise from the fact that biochemical analyses are performed in the complex matrices that make up biological fluids (serum, plasma, urine, cerebrospinal fluid, and so on). These fluids contain hundreds of compounds that either have chemical groups that can react to some extent with the test reagents or can mimic the physical, chromatographic, or spectral properties of the desired analyte. This situation is further complicated because the chemical composition of body fluids can vary with the nature and extent of disease processes. This variability is increased by the possible presence of a large

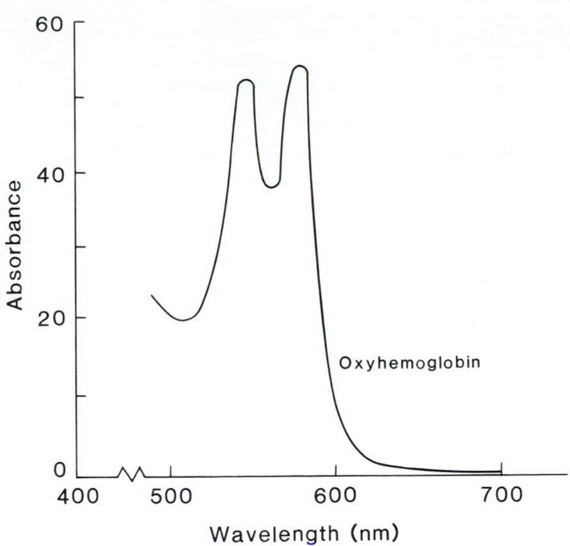

Fig. 55-3 Partial spectrum of oxyhemoglobin (HbO_2).

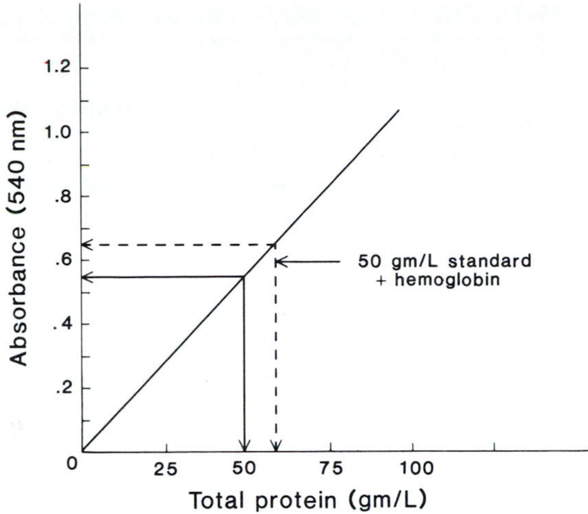

Fig. 55-4 Standard curve for measurement of total protein by the biuret reaction: A_{540} versus concentrations. *Solid arrow,* A_{540} for 50 g/L standard; *dotted arrow,* A_{540} for same standard containing hemoglobin.

number of drugs. Each of these factors, alone or in combination, can result in a possible interference.

The in vitro interferences can be subclassified into those of a spectral nature and those caused by competing chemical reactions. The most commonly observed interferences are hemolysis, icteria, and lipemia. From one quarter to one third of samples obtained from clinic patients[1a] or hospitalized patients[1b] are lipemic, icteric, or hemolyzed. A compendium listing the degree of interference by hemolysis, icteria, and lipemia on the analysis of 21 analytes on 22 different instruments is available.[2]

Spectral interferences

Absorbance. Spectral interferences are observed when a compound causes a response in the spectrophotometer similar to that of the analyte of interest, though the interferents themselves do not necessarily undergo any chemical change during the analytical reaction. The simplest and most common example is the effect of hemoglobin (Hb) on many analytical procedures. A partial spectrum of HbO_2 (Fig. 55-3) shows significant absorption in the 500 to 600 nm portion of the visible spectrum. If one were monitoring the reaction of a colorimetric procedure in this region of the visible spectrum, there would be significant positive interference whenever Hb was contaminating the specimen. Other molecules, such as bilirubin, cause a similar interference.

An example of hemoglobin interference can be seen when a serum total protein (TP) concentration is determined by monitoring the biuret reaction at 540 nm. A standard curve for this reaction is depicted in Fig. 55-4. If a significant concentration of hemoglobin is added to a sample, the absorption at 540 nm is increased, thus giving a falsely high total protein reading. In this example, the A_{540} of a 50 g/L standard is 0.550. If small amounts of hemoglobin are added to this standard, the A_{540} is 0.650. When this solution is read off the standard curve, a higher apparent concentration of protein is calculated *(dotted line)*. Most spectral interferences give falsely elevated results in this manner.

Turbidity. A common type of spectral interference is caused by the turbidity of the sample. Turbidity is caused by large lipoprotein molecules called very low density lipoproteins (VLDL), which are suspended in serum. When a turbid specimen is analyzed in a colorimetric reaction, the lipoproteins cause the incident light to scatter, much as in nephelometry (see Chapter 3).

Because spectrophotometric analysis normally measures transmitted light at 180 degrees to the incident light, any light scattering tends to decrease the transmitted light and therefore to increase the apparent absorbance of the specimen. This, of course, results in falsely elevated results. Sample blanks normally work poorly here, just as two-point kinetic analysis does (see below), because of the error resulting from the very high absorbances often encountered. The best method for eliminating the interference caused by turbidity is dilution of the specimen. The extent that the sample can be diluted to minimize turbidimetric interference is limited by the ability of the analytic procedure to measure the diluted analyte. If possible, several dilutions should be analyzed simultaneously to determine the best response. An example of the effect and elimination of turbidometric interference is presented in the analysis of equal amounts of lactate dehydrogenase (LD) activity in a turbid and a nonturbid specimen (Table 55-2). When the nonturbid specimen is diluted, all the corrected LD activities calculate out to the same approximate value. This indicates linearity of dilution. In contrast, when the

Table 55-2 Effect of turbidity on measurement of LD activity (U/L)

Dilution (with saline)	LD Activity (U/L)			
	Nonturbid sample		Turbid sample	
	Uncorrected	Corrected	Uncorrected	Corrected
Undiluted	440	—	28	—
1:2	245	450	32	64
1:4	136	444	30	120
1:8	62	496	26	208
1:16	30	480	25	400
1:32	14	450	13	416

turbid specimen is diluted, the calculated LD activity changes with dilution. Only at higher dilutions containing minimum turbidity do the calculated LD activities converge with the true values of the nonturbid specimen.

Fluorescence. Turbidity affects fluorescence measurements in a similar fashion. Here some scattered light will reach the detector set at 90 degrees to the incident light, thus giving an apparent increase in fluorescence and falsely elevated concentrations. Reducing problems of turbidity in fluorescence measurements is more difficult than for absorption spectroscopy. The best approach is the elimination of the source of light scattering by filtration or centrifugation.

Correction of spectral interferences

Sample blank. One can minimize spectral interferences by measuring the absorbance of the assay against a sample blank. The simplest sample blank is obtained by a mixture of the sample and diluent (instead of reagent). The correction for spectral interference is made by subtraction of the absorbance value of the blank from the absorbance of the complete reaction mixture. Any significant color inherent to the sample is eliminated by this calculation. In the example of the biuret reaction discussed previously, the absorbance of the sample plus hemoglobin diluted with saline is 0.100. If this is subtracted from the absorbance of the complete reaction mixture (0.650), the true absorbance of the standard (0.550) is obtained. Such a sample blank can usually work unless there are gross amounts of the interferent present. In these cases, the very large total absorption ($A_{\text{interferent}} + A_{\text{reaction}}$) results in large spectrophotometric and calculation errors. Random access analyses (see Chapter 14) easily permit the measurement of sample blank absorbances before the addition of reagent.

Reagent blanks (reagent plus diluent) are used in a similar fashion to correct for high absorbance of the reagent.

In the case of fluorescence, the sample blank allows for the correction of nonspecific fluorescence; however, the fluorescence of this blank cannot be a great portion of the total fluorescence signal (see above).

Kinetic measurements. One frequently used method for the correction of spectral interference is the measurement of a typical end-point reaction as a two-point kinetic reaction. If the absorbance of a noninstantaneous, colorimetric reaction is monitored versus time, a reaction curve as shown in Fig. 55-5 will be observed. An *end-point* reaction is monitored at a single time point when the reaction is mostly completed (Fig. 55-5, arrow 3). If no spectral interferences are present, the reaction curve should go through the origin. If such interferences are present, the curve will be *parallel* to the original curve but biased high because of the endogenous color in the sample. If a sample blank was used to subtract endogenous color, a line identical to the sample containing no interferences would be obtained.

In a *two-point kinetic* assay, the absorbance is measured at *two* different time points (Fig. 55-5, arrows 1 and 2) when (1) the final color development has not occurred and in fact may be small and (2) the absorbance versus time response is still linear.

The initial absorbance reading (Fig. 55-5, arrow 1) is actually taken when almost no color formation has occurred. Thus any absorbance at this time is primarily caused by endogenous spectral interferents. A second reading is taken a short time later when only a small

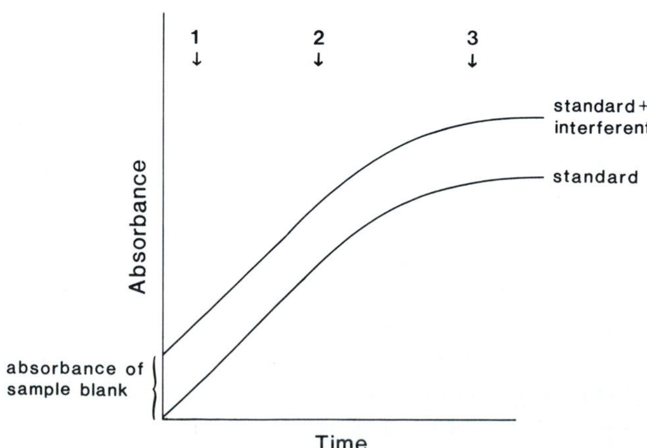

Fig. 55-5 Absorbance changes versus time for colorimetric reaction, with and without interferent present. *Arrows 1 and 2,* Time frame for kinetic analysis; *arrow 3,* end-point reading.

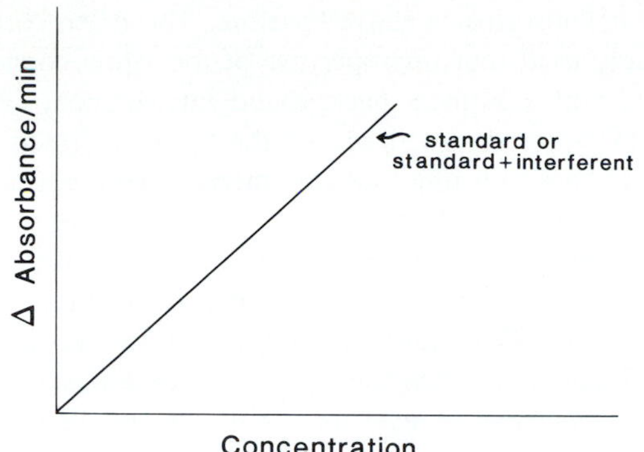

Fig. 55-6 Kinetic analysis of both reactions shown in Fig. 55-5. Change in absorbance (Δ absorbance) per minute versus concentration during linear portion of absorbance versus time curve of Fig. 55-5, between arrows 1 and 2.

amount of color has formed and the response of absorbance versus time is still linear (Fig. 55-5, arrow 2). This absorbance therefore includes both the original endogenous color plus color formed because of the specific analytical reaction. By subtraction of the first reading from the second, the calculated *delta absorbance* (ΔA) is caused only by the specific color formed by the analytical reaction. Standard curves based on kinetic analysis have the change in absorbance (ΔA) plotted versus concentration (Fig. 55-6). In this standard curve the presence of a *nonreacting,* endogenous, colored interferent has no effect. Thus no separate sample blank measurement needs to be made; a two-point kinetic reaction is self-blanking when there is no change in the nature of the interferent during the reaction. This is an important technique when one is performing automated chemical analysis on large numbers of specimens.

Biochromatic analysis.[3] Many instruments in current use employ a different technique for correction of spectral interferences. This technique involves measurement of the absorbance of the reaction mixture simultaneously at two different wavelengths. These are the primary wavelength (λ_1) and one other wavelength (λ_2) close by. As shown in Fig. 55-7, λ_1 is the wavelength at which the chromogen maximally absorbs. At λ_2 there is minimum absorbance of the chromogen. Because the reaction is monitored simultaneously at two wavelengths, this is known as *bichromatic analysis*. This technique is based on the premise that

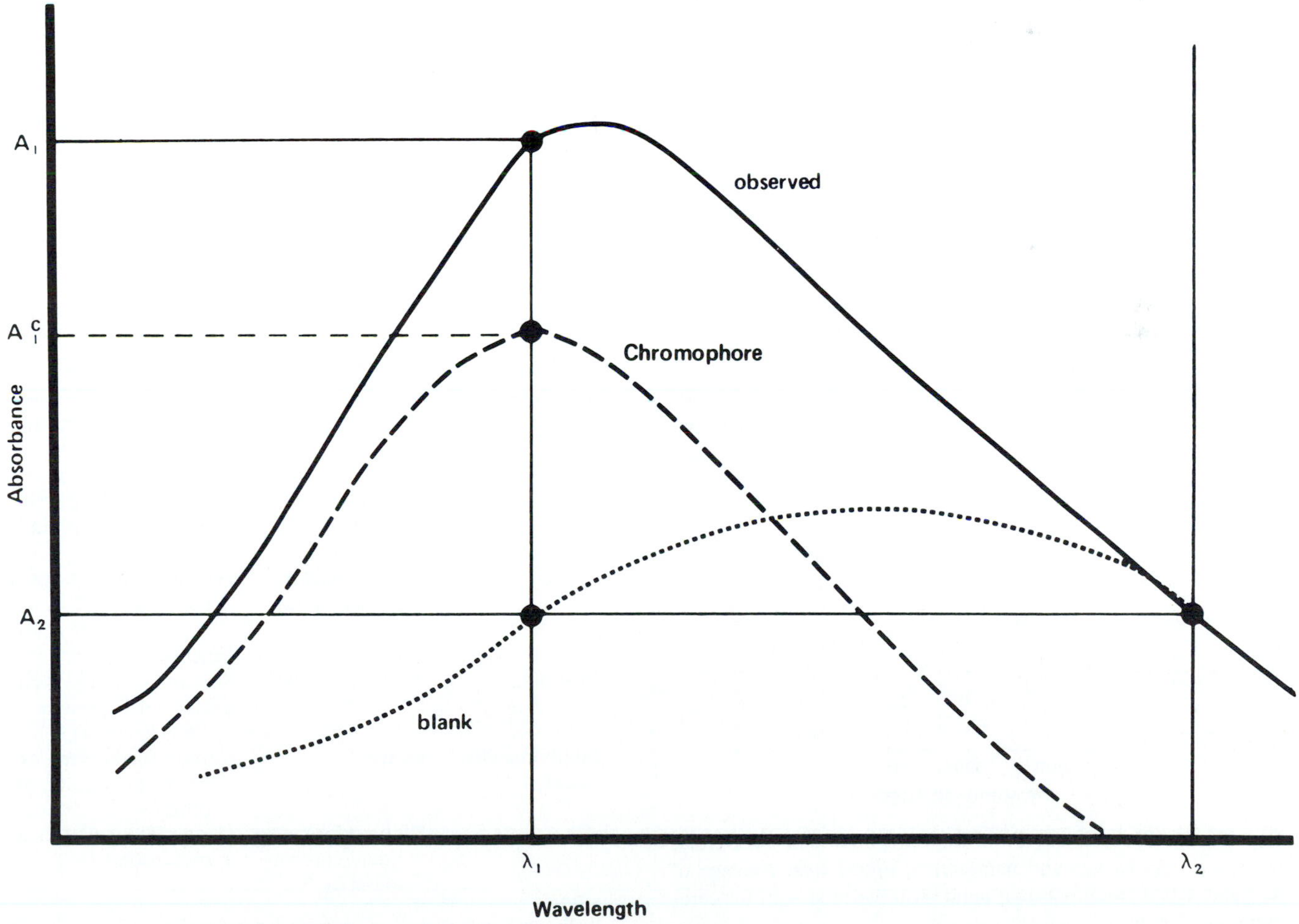

Fig. 55-7 Spectral curves for chromophore and nonreactive blank, where blank absorbance is equal at λ_1 and λ_2.

although a compound may give a spectral interference, the absorbance maxima of the interferent will differ from that of the actual analytical reaction. In addition, this procedure assumes that the absorption caused by the interfering compound is approximately the same at λ_1 as at λ_2. Although the measured absorbance at λ_1 will be caused by both the analytical reaction and the interferent, the absorbance at the second wavelength (λ_2) will be caused by only the interferent. This technique can also correct for instrument problems such as dirt on the cell, which causes light scattering or reflectance. Standard curves are then based on either $A_1 - A_2$ or the ratio of the two absorbances (A_1/A_2). Use of this procedure also allows each sample to serve as its own blank for endogenous color.

Another similar method for the correction of background interference is the measurement of absorbance at the primary wavelength A_{max} and at two additional wavelengths, usually equidistant from the peak, A_1 and A_2. The absorbance readings at these last two wavelengths are averaged to give the average background absorbance in the specimen. This technique for the correction of background absorbance from interfering substances is known as the *Allen correction.*[4]

The Allen correction is only valid if the background absorbance is approximately linear with wavelength over the region in which the measurements are being taken. Thus the shape of the absorption curve for both the analyte plus interferent *(solid line)* and the interferent or interferents *(dotted line)* must be obtained as shown in Fig. 55-8. Use of the Allen correction in this example, when the wavelengths used to correct for background are equidistant from the absorbance maxima, would give the following equation:

$$A_{300} \text{ corrected} = A_{300} - \frac{A_{320} + A_{280}}{2}$$

The "A_{300} corrected" has the *average* background absorbance subtracted from the absorbance maximum to give the actual absorbance above baseline. The Allen correction is widely used, but improper use of the Allen connection, that is, with nonlinear background interference, can lead to even larger errors. Because the final corrected absorbance is based on three measurements, there is a decrease in the precision of the assay.

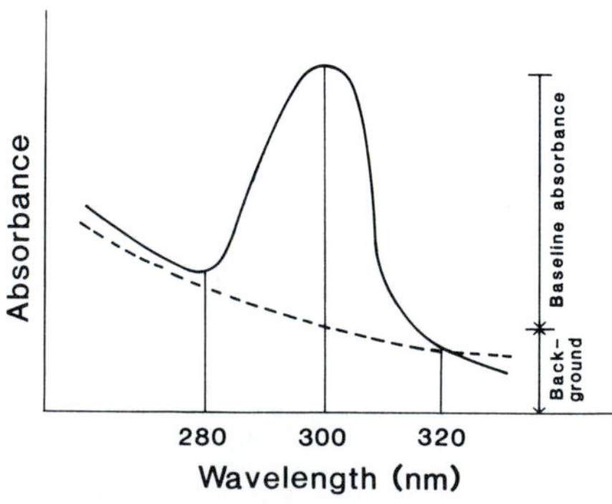

Fig. 55-8 Spectral curves of chromophore and interferent, *solid line,* and background interferents, *dotted line.* Average of A_{280} and A_{320} represents background absorbance at A_{max} for chromophore (300 nm).

Dilution. As discussed for turbidity, dilution of a sample containing a spectral interferent can sometimes reduce the problem. One must be careful not to overdilute the desired analyte or chromagen to a concentration below the minimum detectable level for a given assay. Several dilutions should be assayed simultaneously to determine the most effective dilution.

Chemical interferences

All the interferences discussed so far have been spectral interferences caused by compounds that do not react in the analytical chemical reaction. However, many interferents do react with the chemicals of the analytical reaction. The reaction products of these interferences usually result in positive interferences, although negative interferences are observed.

The types of nonspecific, chemically reacting interferents can vary greatly, as seen in the examples in Table 55-3. Uric acid produces a positive interference and bilirubin and ascorbic acid yield negative interferences in the glucose oxidase methods used for glucose measurement. The alkaline picrate reaction for the measurement of creatinine is known to have both positive (ketones, protein) and negative (bilirubin) interferences.

Table 55-3 Examples of chemically interfering biochemicals

Analyte	Method	Interferences
Glucose	Reducing sugar	Uric acid (+), creatinine (+), protein (+), glutathione (+)
	Glucose oxidase–horseradish peroxidase	Uric acid (+), ascorbic acid (−), bilirubin (−)
	Glucose oxidase consumption—O_2	Hemoglobin (−), ascorbic acid (+)
	Hexokinase	Fructose
Creatinine	Alkaline picrate	Ascorbic acid (+), glucose (+), protein (+), ketones (+)
Vanillylmandelic acid	Pisano	Certain foods (such as bananas), vanilla, aspirin (+)
	High-performance liquid chromatography	Certain drugs and their metabolites (+)

+, Positive interference; −, negative interference.

Correction of chemical interferences

Elimination of many nonspecific chemical interferents is often achieved by one or more of the following techniques:

1. Diluting the interferent
2. Increasing the specificity of the reaction
3. Removing the interferent
4. Monitoring an assay by kinetic measurement
5. Monitoring an assay by bichromatic measurement

Dilution of the sample is an effective method in the case of interferents that do not react at the same rate or produce the same color intensity as the analyte. Interference by protein is minimized in many automated analyzers by a large specimen dilution.

Increased specificity of an analytic reaction is often achieved by use of specific enzymes as reagents. Examples of this approach include the measurement of glucose by hexokinase or glucose oxidase, uric acid by uricase, and urea by urease. Immunochemical-based reactions are also used to increase the specificity of the analysis. An example would be the measurement of theophylline by enzyme immunoassay versus the older methods, which employed ultraviolet absorbance.

Separation of an interferent from the analyte may be achieved by the use of (1) a protein-free sample, (2) liquid-liquid extraction, or (3) adsorption or partition chromatography. Protein-free samples were originally prepared by precipitation of serum proteins and separation of the protein-free sample by filtration or centrifugation. Agents used to precipitate protein include tungstic acid (Folin-Wu procedure) and heavy-metal salts (such as barium and zinc; Somogyi-Nelson procedure). Protein-free specimens obtained by the dialysis technique are the basis of many Technicon Auto-Analyzer procedures.

Liquid-liquid extractions are used when the analyte and interferent or interferents can be separated into different liquid phases. Similarly, in adsorption and partition chromatography, the analyte and interferent are separated by their differential affinity for the stationary phase (see Chapters 4 to 6).

The basis for the elimination of nonspecific chemical reactants by use of a two-point kinetic reaction is that many interferents react at a different *rate* than the specific analyte of interest does. This is observed in the example of the Jaffe reaction with creatinine.[5]

Creatinine reacts with alkaline picrate at a finite rate (curve TC—true creatinine) (Fig. 55-9). Many of the nonspecific interferents (such as acetone) react at a faster kinetic rate (FR), whereas some (such as protein) react at a slower rate (SR), giving a complex change of absorbance with time for the reaction of a mixture of all three species (Fig. 55-10). Therefore, by properly choosing an optimal window of time for the two absorbance readings for the kinetic analysis *(arrows),* one can minimize the effect of the fast-reacting and slow-reacting nonspecific interferents and isolate the absorbance change caused primarily by creatinine. During the window time period, the reaction of the FR interferents is essentially complete, whereas that of the SR interferents is not yet occurring (Fig. 55-10). The ΔA during this time period is caused by the analyte, that is, creatinine. This concept has also been used for the glucose oxidase reaction and is, in fact, a popular technique for increasing the specificity of reactions in many instruments.

Chromatographic interferences

The third type of common in vitro, or methodological, interference is chromatographic interference. This often occurs when an interfering compound cochromatographs with the compound of interest to give falsely elevated results.

Currently, many analytical procedures use chromatography to separate the analyte to be measured from interfering compounds. A chromatographic method assumes that the desired analyte is completely isolated from other com-

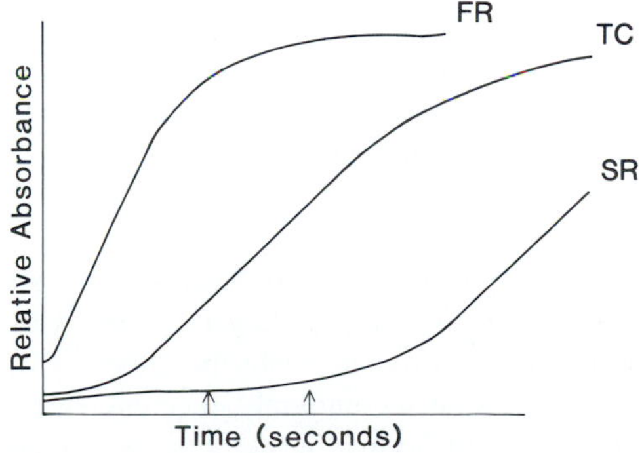

Fig. 55-9 Relative absorbance versus time curves for alkaline picrate reaction, for creatinine *(TC),* slow-reacting interferents *(SR),* and fast-reacting interferents *(FR). Arrows,* Time frame during which absorbance over time primarily reflects change attributable to the TC reaction.

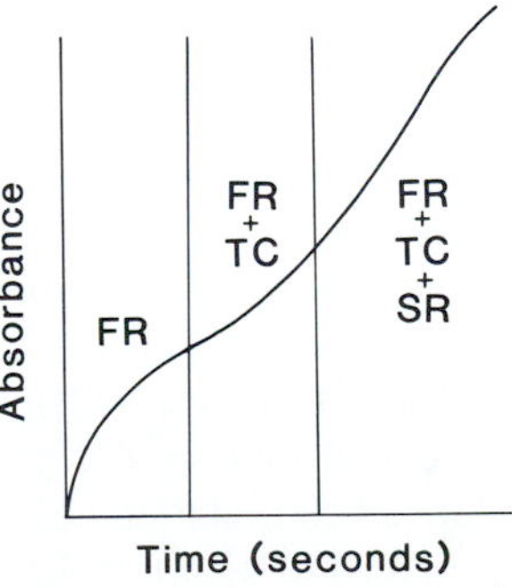

Fig. 55-10 Complex reaction of mixture containing fast-reacting *(FR)* and slow-reacting *(SR)* interferents plus creatinine *(TC).* Only by measurement of the change in absorbance over time can TC reaction be isolated and SR and FR interferences minimized.

pounds that may be recorded by the detection system. However, no single set of chromatography conditions can possibly prevent interferences from cochromatographing or closely chromatographing compounds, especially when the patient may be receiving a number of potentially interfering drugs.

An example of this type of interference and its correction can be seen in the high-performance liquid chromatographic (HPLC) separation of catecholamines and the drug, methyldopa (Aldomet), which is used in the treatment of hypertension. Methyldopa is eluted from the column just before norepinephrine. Because the pharmacological dose of Aldomet is much greater than the physiological concentrations of norepinephrine, it will obliterate or be confused with the norepinephrine peak. The only way to eliminate this type of interference is to remove the source of exogenous interferent. In the case of methyldopa, removal of the drug from the patient for 2 to 10 days is required. Drug or diet restrictions before biochemical analysis are often a necessity for many compounds.

There are two primary modes of minimizing chromatographic interferents: (1) increasing the specificity of the detector and (2) removing the interferent from the analyte. Detectors can measure compounds based on a variety of different principles. If the interferent has physical or chemical properties different from those of the analyte, it is possible to select a detector that will not respond significantly to a potential interferent. In liquid chromatography, refractive index detectors can detect almost any compound in the eluant; thus they are very nonspecific. Fluorescence detectors have much higher specificities, because not all compounds fluoresce. By appropriate setting of excitation and emission wavelengths, the detector can be made even more specific. Electrochemical detectors also have high specificities because not all compounds are electrochemically active. The specificity can be further increased by selection of an electrode voltage at which the interferent will not react.

As discussed earlier, separation of the analytes from potentially interfering compounds is achieved by several techniques. These techniques are based on differences in solubility or chromatographic behavior between the analyte and interferents. The techniques commonly used are single liquid-liquid extractions, multiple extractions including back-extractions, and adsorption and ion-exchange chromatography. The complexity of the procedure used depends on the nature of the interference and required sensitivity. HPLC methods for serum theophylline, present in relatively large amounts, usually employ only a simple liquid-liquid extraction. On the other hand, the HPLC analysis of the tricyclic antidepressants present in nanogram quantities requires several extraction steps, including back-extractions. Chapter 4 discusses these types of procedures in more detail.

A technique for detecting contaminating or cochromatographing compounds is dual-detection analysis. This technique uses two different types of detectors or parameters to monitor the column eluate. The response of the two different detectors (D_1 and D_2) is determined for a standard, and the ratio of the responses is calculated (D_1 standard/D_2 standard). The probability that another compound would have a similar characteristic ratio is quite small. Thus significant deviations of the ratio found in patient analyses would strongly suggest that the analyte peak contains a contaminating, coeluting compound.

Often the presence of a cochromatographing interferent is recognized because of the abnormal shape of the peak. The presence of a contaminant often causes a normally symmetric peak to be skewed. In these cases, rerunning the chromatogram at lower flow rates will sometimes separate or partially separate the analyte from the interfering compound, allowing quantitation.

IN VIVO INTERFERENCES

NATHAN A. PICKARD

Many in vivo factors may play a significant role in the proper interpretation of results obtained by clinicians. This is of special concern for those tests in which the reference range interval is narrow and a result may be improperly considered abnormal when in fact it may be normal for a specific person.

Age and sex

Some tests, such as alkaline phosphatase, may be both age and sex dependent.[6] Reference intervals for all laboratory tests should reflect both variables, if they occur. Consideration should also be given to the patient's state of health. For example, interpretation of glucose tolerance tests is performed under the assumption that the patient is otherwise healthy.[7]

Time of day

The time of day at which a sample is drawn may affect the interpretation of numerous tests. Cortisol and estriol levels in blood vary significantly during the course of the day, as do catecholamine and iron levels.[8-10]

Drugs

Virtually every drug affects some laboratory procedure, and any laboratory procedure may be affected by one or more drugs. The interference may be either in vivo or in vitro. An example of a commonly encountered drug is alcohol, whose ingestion may affect glucose, lactate, urate, bicarbonate, gamma-glutaryl transferase, and creatine phosphokinase levels.[11] Smoking may alter catecholamine, cortisol, and blood-gas results. A detailed compilation of the effects of drugs on laboratory tests is discussed on p. 817.

Diet

The patient's diet may considerably alter laboratory results. Fasting for 48 hours significantly elevates serum bilirubin, triglycerides, glycerol, nonesterified fatty acids, various amino acids, glucagon, secretin, insulin, and ketone bodies.[12] High-fat diets can result in elevated alkaline phosphatase levels of intestinal origin. Caffeine ingestion will increase plasma catecholamines and fatty acids and may interfere with certain theophylline procedures. Highly saturated fatty acid diets affect the cholesterol levels, and high-protein, meat diets can increase the urea, ammonia, and uric acid levels.[13]

Recent food ingestion produces alterations in bile acids, glucose, insulin, triglyceride, cortisol, and catecholamines. Therefore it is best for most procedures that samples be collected first thing in the morning after an overnight fast.[14]

Pregnancy and menses

Pregnancy has profound effects on the interpretation of laboratory tests of the mother because of, for example, hormonal changes and altered metabolism (see Chapter 37). Many of these changes are difficult to quantitate, and unessential testing should probably be delayed if possible until at least a couple of months after the birth of the child. Test results such as iron levels may be affected by the presence or absence of menses, and one should consider this when evaluating the results.[15]

Other conditions

Physical conditioning, jogging, and swimming have resulted in elevated enzyme results (including CK-MB) and in elevated lactate and glucose levels.[16] Physical and psychological stress affects laboratory results just as blood transfusions and other diagnostic manipulations such as surgery, physical examination, intravenous infusion, and intramuscular injection do.[17]

It would be an oversimplification to conclude that each variable will always produce a specific effect. The effect a variable has on a patient's test result often depends on the person, the duration of the variable, the time between initial stress and sample collection, and the degree of exposure. More important is a heightened awareness of the many factors occurring outside the laboratory in and around the patient that may affect the test result before the sample reaches the laboratory or even before the sample is collected.

The effect of in vivo interferences can be minimized when the clinician takes a good history and when there is good communication of such information between the clinician and laboratory.

SAMPLE-PROCESSING ERRORS

The final category of interferences is the result of improper phlebotomy or sample processing. The improper selection of the phlebotomy tube can lead to large laboratory errors. These errors most often arise from the use of anticoagulants that can inhibit the test procedure. Release of intracellular constituents, especially hemoglobin, can drastically affect test results.[18] Improper storage and stabilization of the specimen can also result in invalid laboratory results.

Laboratory personnel can prevent these problems of sample processing by fully understanding the precautionary steps required in general and for each analyte in particular. The laboratory supervisory staff should educate both the technologists and clinical staff and then implement the necessary precautions. A more complete discussion on the collection and handling of patient samples is found in Chapter 2.

SOURCE-REFERENCE MATERIAL

This chapter is only a brief description of the wide variety and types of interferences in chemistry laboratory testing. Many common interfering substances are well documented, and an alert laboratory staff can often eliminate these as a source of interference. For example, the development of the interferogram[2] allows the laboratory to estimate the effect of hemolysis, icteria, and lipemia on the analyses performed on frequently used instruments.

However, an ongoing problem for the laboratory is the rapid proliferation of drugs produced by pharmaceutical companies and consumed by the public. The difficulty is that these drugs can have both in vivo and in vitro effects on clinical laboratory analysis. The problem remaining for the laboratory is to know which drug is affecting a specific assay.

One of the few attempts to list the known effects of drugs and other interferences on chemical analysis was done by Donald Young. This compendium was published as an issue of *Clinical Chemistry* in April of 1975.[19] Although it was outdated when published, it remains the best and only single listing of drug effects on laboratory tests. One of the two major sections has the possible interferents (not only drugs) listed in alphabetical order. Listed under each interferent are those laboratory tests that may be affected by that interferent.

One section of Young's study is shown in Fig. 55-11. Barbiturates are the tested interferent. The initial capital letter indicates the type of body fluid in which the laboratory test was analyzed (U = urine, S = serum, and so on); the laboratory tests are then given in alphabetical order. After each test is the indication of how the tested interferent affects the laboratory test (*inc* = increases, *dec* = decreases, and z = no effect). The next capital letter indicates the type of interference (V = in vivo, M = methodological). A comment and a reference finishes the test listing.

This section is cross-indexed by another section that lists, in alphabetical order, laboratory tests. Each labora-

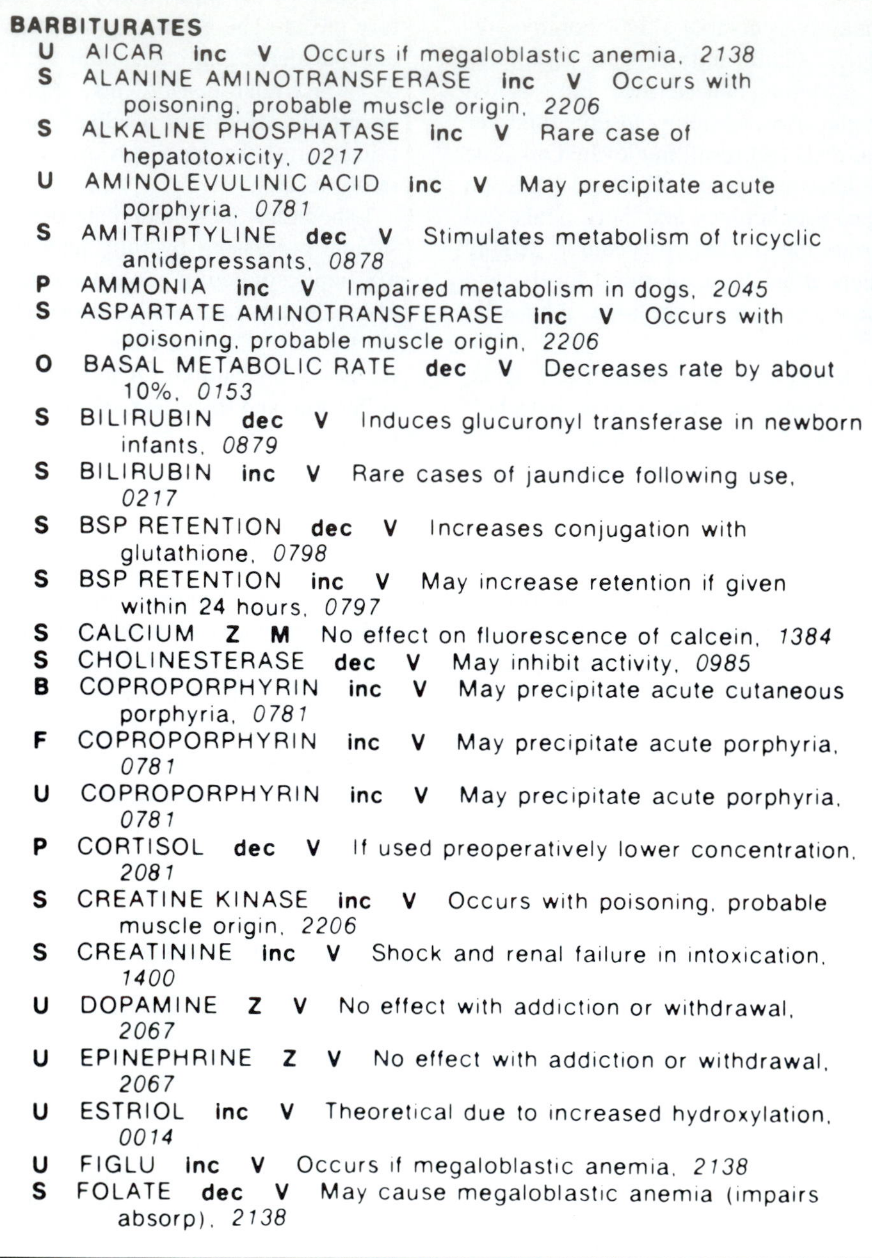

BARBITURATES
U AICAR **inc** **V** Occurs if megaloblastic anemia. *2138*
S ALANINE AMINOTRANSFERASE **inc** **V** Occurs with poisoning, probable muscle origin. *2206*
S ALKALINE PHOSPHATASE **inc** **V** Rare case of hepatotoxicity. *0217*
U AMINOLEVULINIC ACID **inc** **V** May precipitate acute porphyria. *0781*
S AMITRIPTYLINE **dec** **V** Stimulates metabolism of tricyclic antidepressants. *0878*
P AMMONIA **inc** **V** Impaired metabolism in dogs. *2045*
S ASPARTATE AMINOTRANSFERASE **inc** **V** Occurs with poisoning, probable muscle origin. *2206*
O BASAL METABOLIC RATE **dec** **V** Decreases rate by about 10%. *0153*
S BILIRUBIN **dec** **V** Induces glucuronyl transferase in newborn infants. *0879*
S BILIRUBIN **inc** **V** Rare cases of jaundice following use. *0217*
S BSP RETENTION **dec** **V** Increases conjugation with glutathione. *0798*
S BSP RETENTION **inc** **V** May increase retention if given within 24 hours. *0797*
S CALCIUM **Z** **M** No effect on fluorescence of calcein. *1384*
S CHOLINESTERASE **dec** **V** May inhibit activity. *0985*
B COPROPORPHYRIN **inc** **V** May precipitate acute cutaneous porphyria. *0781*
F COPROPORPHYRIN **inc** **V** May precipitate acute porphyria. *0781*
U COPROPORPHYRIN **inc** **V** May precipitate acute porphyria. *0781*
P CORTISOL **dec** **V** If used preoperatively lower concentration. *2081*
S CREATINE KINASE **inc** **V** Occurs with poisoning, probable muscle origin. *2206*
S CREATININE **inc** **V** Shock and renal failure in intoxication. *1400*
U DOPAMINE **Z** **V** No effect with addiction or withdrawal. *2067*
U EPINEPHRINE **Z** **V** No effect with addiction or withdrawal. *2067*
U ESTRIOL **inc** **V** Theoretical due to increased hydroxylation. *0014*
U FIGLU **inc** **V** Occurs if megaloblastic anemia. *2138*
S FOLATE **dec** **V** May cause megaloblastic anemia (impairs absorp). *2138*

Fig. 55-11 Example of format used in special issue of *Clinical Chemistry:* "Effects of drugs on clinical laboratory tests." *(From O'Kell, RT, and Elliott, JR: Clin Chem 16:161-165, 1970.)*

tory test is listed by subsections according to decreased *(dec)*, increased *(inc)*, or no effects *(z)* of the listed interferents on that test. The laboratory test is also listed as either a urinary *(U)* or serum *(S)* analyte. Under each laboratory test is listed the interferent for which the effect *(inc, dec,* or *z)* has been shown.

The complexity and extent of Young's work highlight the awareness laboratory personnel must have concerning interferences with clinical laboratory tests. A complementary issue to the 1975 listing was published in 1980.[20] This volume lists the effects of disease on clinical laboratory tests. The format for this volume is similar to the one just described. The first section lists each analyte and those disease states in which changes in the concentration of that analyte have been noted. The second section lists diseases and those analytes that change during the course of the disease.

REFERENCES

1a. Glick, MR: Ohio Valley Section Meeting on "Interferences," Cincinnati, Feb 27, 1988.

1b. Kaplan, LA: Ohio Valley Section Meeting on "Interferences," Cincinnati, Feb 27, 1988.

2. Glick, MR, and Ryder, KW: Interographs: user's guide to interferences in clinical chemistry instrument, Indianapolis, 1987, Science Enterprises, Inc.

3. Hahn, B, Vlastelica, DL, Snyder, LR, et al: Polychromatic analysis: new applications of an old technique, Clin Chem 25:951-959, 1979.

4. Allen, E, and Rieman, W: Determining only one compound in a mixture, short spectrophotometric method, Anal Chem 25:1325-1331, 1953.

5. Soldin, SJ, Henderson, L, and Hill, JG: The effect of bilirubin and ketones on reaction rate methods for the measurement of creatinine, Clin Biochem 11:82-86, 1978.

6. Leonard, PJ: The effect of age and sex on biochemical parameters in blood of healthy human subjects. In Proceedings of the Second International Colloquium, automation and prospective biology, Pont-a-Mousson, France, 1972, Basel, 1973, S Karger AG, Medical and Scientific Publishers.

7. Klimt, CR, Prout, JE, Bradley, RF, et al: Standardization of the oral glucose tolerance test: report of the Committee on Statistics of the American Diabetes Association, June 14, 1968, Diabetes 18:299-310, 1969.

8. Crane, JP, Sauvage, JD, and Arias, F: A high risk pregnancy management protocol, Am J Obstet Gyneol 125:227-235, 1976.

9. Jacobs, SL, Sobel, C, and Henry, RJ: Specificity of the trihydroxyindole method for the determination of urinary catecholamines, J Clin Endocrinol Metab 21:305-314, 1961.

10. Bowie, EJW, Tauxe, WN, Sjoberg, WE, Jr, and Yamaguchi, M: Daily variation in the concentration of iron in serum, Am J Clin Pathol 40:491-494, 1963.

11. Freer, DE, and Statland, BE: The effect of ethanol (.75 g/kg body weight) on the activities of selected enzymes in sera of healthy young adults. I. Intermediate-term effect, Clin Chem 23:830-834, 1977.

12. Stout, RW, Henry, RW, and Buchanan, KD: Triglyceride methabolism in acute starvation: the role of secretin and glucagon, Eur J Clin Invest 6:179-185, 1976.

13. Statland, BE, and Winkel, P: Effects of non-analytical factors on the intra-individual variation of analytes in the blood of healthy subjects: consideration of preparation of the subject and time of veni-puncture, CRC Crit Rev Clin Lab Sci 8:105-110, 1977.

14. Schwartz, MK: Interferences in diagnostic biochemical procedures, Adv Clin Chem 16:1-45, 1973 (see pp 17-20).

15. Latner, AL: Clinical biochemistry, Philadelphia, 1975, WB Saunders Co.

16. King, S, Statland, BE, and Savory, J: The effects of a short burst of exercise on activity values of enzymes in sera of healthy subjects, Clin Chem Acta 72:211-218, 1977.

17. Schwartz, MK: Interferences in diagnostic biochemical procedures, Adv Clin Chem 16:1-45, 1973 (see pp 21-26).

18. Frank, JJ, Bermes, EW, Bickel, MJ, and Watkins, BF: Effects of in vitro hemolysis on chemical vaues of serum, Clin Chem 24:1966-1970, 1978.

19. Young, DS, Pestaner, LC, and Gibberman, V: Effects of drugs on laboratory tests, Clin Chem 21:1D-432D, 1975.

20. Friedman, RB, Anderson, RE, Entine, SM, and Hirshberg, SB: Effect of diseases on clinical laboratory tests, Clin Chem 26:1D-476D, 1980.

CHAPTER 56

Examination of urine

G. BERRY SCHUMANN
SUSAN C. SCHWEITZER

KEY TERMS

acute tubular necrosis A disease that involves the destruction of renal tubular epithelial cells and is most commonly associated with reduced blood supply to the renal tubules (ischemia) or toxic exposures.

Addis count A quantitative urine sediment test in which the number of erythrocytes, leukocytes, and casts are quantified in a timed urine specimen.

albuminuria Increased albumin in urine.

aminoaciduria An excess of one or more amino acids in urine.

amorphous crystals The granular, noncrystalline precipitate of salts with no pathological importance.

bacteriuria The presence of bacteria in urine.

bilirubinuria The presence of bilirubin in urine.

calculus An abnormal concretion, usually composed of mineral salts, present in the urinary system or other tissues, renal stone.

cast A molded, cylindrical structure that is formed as a result of cell conglutination and protein precipitation in the lumen of the distal convoluted tubule or collecting duct of the nephron. It is extruded into the urinary sediment.

catheterization The passage of a thin, flexible, tubular instrument into the bladder or ureter for the withdrawal of urine.

crystalluria The presence of crystals in urine.

cylindruria The presence of casts in urine.

cystitis Inflammation of the bladder.

diuretic An agent that promotes the secretion of urine.

glitter cells Pale-staining, swollen, and degenerated neutrophils found in dilute urine, with cytoplasmic granules that exhibit a characteristic brownian movement.

glomeruli Coils of blood vessels projecting into the expanded end of the capsule of each of the uriniferous tubules of the kidney.

glycosuria The presence of glucose in urine.

hematuria The presence of blood in urine.

hemoglobinuria The presence of free hemoglobin in urine.

hyaline cast A transparent cast composed of mucoprotein.

hydrometer An instrument used for determining the specific gravity of a fluid.

ketone Any compound containing the carbonyl group, —CO—, and having hydrocarbon groups attached to the carbonyl carbon.

ketonuria The presence of ketones, which are intermediary products of fat metabolism, in urine, as in diabetes mellitus.

melanin The dark amorphous pigment of the skin, hair, and various tumors. It is produced by polymerization of oxidation products of tyrosine and dihydroxyphenol compounds and contains carbon, hydrogen, nitrogen, oxygen, and often sulfur.

myoglobinuria The presence of myoglobin, an oxygen-binding protein of muscle cells, in urine.

nephritis Inflammation of the kidney.

nephrosis A disease of the kidney.

osmolality The number of solute particles per unit mass of solvent (mOsm/kg).

Papanicolaou stain A differential stain that aids in the identification of nuclear chromatin, cytoplasmic properties (such as keratinization), noncellular entities (such as crystals and casts), and hematopoietic elements.

porphyrins A group of iron-free or magnesium-free pyrrole derivatives that occur universally in cells. They constitute the basis of the respiratory pigments in animals and plants.

proteinuria Increased protein in urine.

pyuria An abnormal number of leukocytes in urine.

reagent-strip testing The use of a chemical test strip to determine whether pathological concentrations of various substances are present in the urine.

specific gravity The weight of a substance compared with that of an equal volume of another substance taken as a standard.

Sternheimer-Malbin stain A crystal violet and safranin stain used in urinalysis. This stain provides additional contrast for some cells and casts.

urobilinogen A group of colorless compounds formed from the reduction of conjugated bilirubin by intestinal bacteria; about 1% of the total urobilinogen produced reaches the urine.

virus A self-replicating agent that consists of a core of nucleic acid enclosed by a protein coating. This microorganism can multiply only within living host cells.

yeast A unicellular nucleated microorganism that reproduces by budding.

INTRODUCTION AND CLINICAL UTILITY OF URINALYSIS

Urinalysis can provide a wide variety of clinical information on an individual's kidneys and the systemic diseases that may affect this excretory organ. Both structural (anatomical) and functional (physiological) disorders of the kidney and lower urinary tract may be elucidated, as well as sequential information about the disease, its cause, and prognosis. Careful laboratory examination of urine often narrows the clinical differential diagnosis of numerous urinary system diseases. Usually these laboratory data may be obtained without pain, danger, or distress. Therefore properly performed and interpreted urinalysis will always remain an essential test in clinical practice.

Urinalysis provides a front-line test for the detection of chemical and morphological abnormalities present in urine. Routine urinalysis is a standard part of almost every physical examination or hospital admission. Diabetic patients often monitor their own disease by the daily testing of urine specimens.[1] The purpose of examining urine is to screen asymptomatic, low-risk individuals, diagnose the symptomatic patient, and assist in the therapeutic monitoring of conditions that affect the urinary system.

Currently urinalysis procedures consist of two major components: (1) *macroscopic urinalysis,* or physicochemical testing (appearance, specific gravity, and reagent-strip measurements of several chemical constituents), and (2) *microscopic urinalysis,* or microscopic examination of urinary sediment.[2]

The purpose of this chapter is to describe briefly common methodologies employed by most conventional urinalysis laboratories.[3] Emphasis will be placed on the responsibilities of the urinalysis laboratory in the following areas: (1) common procedures and equipment; (2) quality reagents; (3) sensitivity, specificity, and limitations of each procedure; (4) confirmatory tests; (5) accurate identification of urine sediment entities; and (6) quality control. The technical and diagnostic significance of urinalysis results and the mechanism of diseases that produce urinalysis abnormalities are discussed. Readers are encouraged to supplement their knowledge by using the reference list.

SPECIMEN COLLECTION

Careful attention to collection of the urine specimen and its prompt delivery to the laboratory are crucial for optimum information to be derived from urinalysis. Urine should be collected in a clean, sterile container that has a tightly fitting lid to prevent spillage, evaporation, and contamination. Specimen containers should be marked with the patient's name, date, and time of collection.

The first voiding in the morning is usually the most desirable for urinalysis testing, since it provides the most concentrated urine. A clean-catch, midstream collection avoids contamination from the distal urethra. The sample must be free of vaginal secretions and other extraneous debris. Kunin[4] thoroughly describes a variety of collection techniques for urinary tract diseases.

If data from the urinalysis are to be accurate, it is essential that urine be either examined within 2 hours of collection or preserved in some manner, usually by refrigeration (2° to 8° C). Appropriate fixatives or preservatives may also be used as long as their effects on the urine and its testing are well understood. If urine is allowed to sit unpreserved at room temperature, it will begin to decompose. Preservatives work by modifying urine so that chemical changes associated with decomposition do not occur and by preventing growth and metabolism of microorganisms. Toluene, phenol, thymol, and acid preservatives are commonly used for urine chemistry determinations. Other forms of preservation include adjusting pH and excluding light. Ethanol (95%) or commercially available fixatives, such as Mucolexx and Saccomanno, may be used to preserve cell structure. Laboratories should be responsible for the correct type and amount of urine preservatives used for cellular preservation.

Timed urine specimens are frequently used to quantitate various aspects of renal function. The urine must reflect excretion over a precisely measured duration of time. Timed urine specimens must not include urine in the bladder before the timed test. One would therefore obtain a 24-hour urine specimen by discarding the first-voided morning urine on the first day and collecting all subsequent urine up to and including the first-voided morning urine on the second day. Recommended types of specimens for various urine tests are listed in Table 56-1.

When a specimen is received for urine testing, the laboratory workers performing that test must make a decision as to its acceptability. Specific criteria should be established in the form of written guidelines for the rejection of urine specimens and should include criteria such as visible signs of contamination, incorrect or incomplete labeling, inappropriate type of specimen or preservatives, and time lapse since collection.

Table 56-1 Recommended types of urine specimens

Tests	Types of collection					Types of storage		
	Random	**First morning**	**2-hour**	**12-hour**	**24-hour**	**Fresh**	**Refrigerated**	**Preserved**
Addis count				×			×	
Albumin								
Qualitative analysis	×					×		
Quantitative analysis					×		×	
Aldosterone					×		×	
Amino acid–nitrogen					×		×	×
Bence Jones		×				×		
Bilirubin	×					×		
Blood	×					×		
Calcium					×		×	
Catecholamine					×			×
Coproporphyrin								
Qualitative analysis	×					×		
Quantitative analysis					×			×
Creatinine				×	×		×	
Cytological		×				×	×	
Estrogen	×				×		×	
Glucose	×					×		
17-Hydroxycorticosteroid					×		×	
17-Ketosteroid					×		×	
Ketone	×					×		
Lead		×					×	
Microbiological		×				×		
Nitrite		×				×		
Osmolality	×					×		
pH	×					×		
Phosphorus					×		×	
Porphobilinogen	×					×		
Potassium					×		×	
Pregnanediol					×		×	
Pregnanetriol					×		×	
Protein						×		
Qualitative analysis	×						×	
Quantitative analysis					×	×		
Sediment		×					×	
Sodium					×	×		
Specific gravity	×					×		
Sugar	×						×	
Urea nitrogen					×		×	
Uric acid					×	×		
Urobilinogen			×					×
			(afternoon)					×
Uroporphyrin					×		×	
Vanillylmandelic acid					×			×
Volume					×			

PHYSICAL EXAMINATION OF URINE[5-8]

The physical examination of urine is the initial part of a routine urinalysis. This examination includes assessment of volume, odor, appearance (color and turbidity), and specific gravity or osmolality.

Volume

Urinary volume is influenced by the fluid intake; solutes to be excreted, primarily sodium and urea; loss of fluid by perspiration and respiration; and cardiovascular and renal status. Normally adults excrete 750 to 2000 mL/24 hours. Conditions that produce an increased or decreased amount of urine are discussed in Chapter 22. Although the volume of a random specimen is clinically insignificant, the volume of specimen received should be recorded for purposes of documentation and standardization.

Odor

Normal fresh urine has an inoffensive odor. An offensive odor can indicate that a specimen is too old for accurate analysis. A foul odor in a specimen that was collected (and not preserved or refrigerated) more than 2 hours earlier indicates an unacceptable specimen. Odor can also provide clues to certain urine abnormalities. An ammonia-like odor suggests urea-splitting bacteria, fruity odors indicate the presence of acetone (ketone), a sweet odor suggests the presence of glucose or other sugar, and a foul odor suggests pus or inflammation. Odor is important in the clinical detection of maple-syrup urine disease (a congenital metabolic defect).

Table 56-2 Common colors of urine

Colors	Potential causes				Clinical conditions	Commonly associated diseases
	Foodstuffs	Metabolites	Drugs	Organisms		
Yellow to colorless					Polyuria	Diabetes insipidus Diabetes mellitus Chronic renal failure
Yellow	Food color	Urochrome	Quinacrine (Atabrine) Sulfasalazine (Azulfidine) Phenacetin Nitrofurantoin Riboflavin		Normal	
Yellow-orange	Food color Carotene Rhubarb	Urobilin	Sulfisoxazole and phenazopyridine (Azo Gantrisin) Riboflavin Furazolidone (Furoxone)		Dehydration Jaundice	Liver disease
Yellow-green		Bilirubin-biliverdin	Methylene blue Indican Amitriptyline (Elavil)		Jaundice	Liver disease Biliary obstruction
Yellow-red-brown	Beets Food color Rhubarb	Hemoglobin Myoglobin Porphyrin	Sulfisoxazole and phenazopyridine (Azo Gantrisin) Phenyltoin (Dilantin) Pyrvinium pamoate (Povan) Phenazopyridine (Pyridium) Phenolsulfonphthalein Phenindione Amidopyrine	*Serratia marcescens*	Hematuria Hemoglobinuria Myoglobinuria Porphyrinuria Menstrual contamination	Hemolysis Transfusion Burns Renal disease Urological disease
Brown-black		Porphyrin Melanin Methemoglobin Homogentistic acid	Chloroquine (Aralen) Levodopa		Alkaptonuria	Melanoma Ochronosis Phenol poisoning
Blue-green		Indican	Methylene blue	*Pseudomonas* species	Dysuria	Urinary tract infection

Appearance (color and turbidity)

The color of urine is determined to a large degree by its degree of concentration. Normal urine varies widely from colorless to deep yellow. Interpretation of color is subjective and varies with each laboratory examiner. It may be helpful to the technologist to use standardized colored objects in the laboratory as reference points and to use defined colors to describe the urine, avoiding ambiguous terms such as *straw* or *bloody*. Common colors of urine are listed in Table 56-2.

Red urine is perhaps the most clinically important discoloration; it may be a result of urinary hemoglobin or myoglobin. Intact erythrocytes, hemolyzed erythrocytes, or free hemoglobin (hemolysis) can be responsible for the red color. Hemolysis may occur intravascularly, or after the urine has been formed. Myoglobinuria is rare and results from crushing damage to muscle. Urinary hemoglobin and myoglobin may not be differentiated by gross inspection or simple tests. Ingestion of beets or certain drugs may redden the urine, and thus such chromogens must be excluded as a source of urine coloration.

Urine characteristic of acute glomerulonephritis is brownish red. It is acidic and contains blood, a combination producing brownish-colored, acidic hematin. Blood in the urine also may assume a smoky appearance. One can observe this only by shaking the urine while holding it up to the light. The "smoke," or turbidity, is actually erythrocytes suspended in solution. Centrifugation will yield clear urine with a red plug at the bottom of the tube. The smoky appearance of erythrocytes should not be confused with the faint turbidity of other suspensions (such as crystals and cells).

Normally freshly voided urine is clear. When urine is allowed to stand, amorphous crystals, usually urates, may precipitate and cause urine to be cloudy. The turbidity of urine should always be recorded and microscopically explained. Table 56-3 lists common causes of cloudy urine.

Table 56-3 Common causes of cloudy or turbid urine

Causes	Methods of clearing	Comments
Chemical		
Urates	Soluble at 60° C or alkali	Pink sediment
Phosphates and carbonates	Soluble in dilute acid	
Crystals	See specific solubility characteristics (Table 54-4)	
Mucus	—	Sticky
X-ray contrast mediums	Soluble in 10% sodium hydroxide	
Lipids	Soluble in ether	Opalescent
Chyle	Soluble in ether	Milky
Cells		
Bacteria	Centrifugation	Foul-smelling odor
Fungi	Centrifugation	Sweet-smelling odor
Erythrocyte	Centrifugation	Red, smoky
Leukocyte	Centrifugation	
Epithelium	Centrifugation	
Spermatozoa	Centrifugation	

Specific gravity

Urinary specific gravity measurement serves as a partial assessment of the ability of the kidneys to concentrate urine. The normal range is 1.003 to 1.035 g/mL. A value of 1.020 or greater indicates good renal function and increased amounts of dissolved solutes excreted by the kidneys. A specific gravity greater than 1.035 represents the presence of extraneous solutes and should be investigated. Increased specific gravity greater than 1.030 is found in dehydration, diabetes mellitus, congestive heart failure, proteinuria, and adrenal insufficiency. Decreased specific gravity is found in patients with hypothermia and those using diuretics. In a patient with severe kidney disease, urine is produced with a fixed specific gravity that is identical to that of the glomerular filtrate, approximately 1.010. Sediment constituents are often poorly preserved in dilute urine. Therefore a low specific gravity indicates that microscopic examinations of urine may not yield optimally accurate results.

Large-molecular-weight substances will affect specific gravity to a greater extent than will simple crystalloids. This becomes significant when urine contains abnormal amounts of larger molecules such as glucose, protein, or x-ray contrast mediums. In cases of marked glucosuria or proteinuria correction factors can be used to adjust the specific gravity to a more representative value; 0.004 is subtracted for every 10 g/L glucose and 0.003 for every 10 g/L protein. The presence of x-ray contrast mediums or preservation is often associated with values of 1.040 or greater.

A hydrometer and a suitable container may be used to determine specific gravity. There are several limitations to the use of hydrometers: (1) they require a large volume of urine (10 to 15 mL); (2) they are calibrated to be used at 20° C (if urine is not tested at this temperature, a temperature correction is necessary); and (3) hydrometers cannot be recalibrated. For these reasons few laboratories employ hydrometers to measure specific gravity.

Most laboratories are now equipped with refractometers that can relate density of a solution to specific gravity (see Chapter 3). There are several advantages to the use of refractometers: (1) they require only a drop or two of urine, (2) they are temperature compensated, (3) readings are less affected by density than urinometers, and (4) refractometers have a zero set screw for calibrating the instrument. The biggest disadvantage of refractometers is their expense.

In specific gravity methodology calibration is essential. Instruments should be checked with distilled water on a regular basis. Standardized salt and sucrose solutions can be used for calibration and quality control (Table 56-4).

Table 56-4 Standardizing solutions for specific gravity

Solution	Specific gravity
Distilled water	1.000
3% NaCl	1.015 ± 0.001
5% NaCl	1.022 ± 0.001
9% Sucrose	1.034 ± 0.001

Refractometer method

1. Clean the surfaces of the cover with a damp cloth and then dry.
2. Using two applicator sticks or a capillary pipet, apply a drop of urine at the notched bottom of the cover so that it flows over the prism surface by capillary action.
3. Point the refractometer toward a light source and read the specific gravity value directly on the specific gravity scale at the sharp dividing line between light and dark contrast.
4. For specific gravity over 1.030, dilute the urine with equal parts of water and read. Multiply the last three digits by three.[3]

Reagent-strip method. A new indirect colorimetric method for estimating specific gravity is available on the Ames Multistix 10 SG (Miles Laboratories, Inc., Elkhart, IN) reagent strip. This method uses a strip that contains a pretreated electrolyte that elicits a pH change based on the ionic concentration of the urine. This test for specific gravity is fast and simple and requires no additional equipment. The manufacturer states there is no interference with glucose, protein, or radiographic dye. Sensitivity is poorer than that of the refractometer method (units of 0.005), and urines with a pH of 6.5 or greater require a correction factor.

Falling drop method. The falling drop method is a direct method of measuring specific gravity that is usually used with automated instruments, such as the Clinitek Auto 2000 (Ames Division, Miles Laboratories, Inc., Elkhart, IN). This instrument uses a silicone oil in a specially designed column. Specific gravity is related to the time it takes for a drop of urine to fall a distance defined by two optical gates. This procedure is more specific and accurate than using refractometry and more precise than using hydrometers.[9]

Osmolality

The normal kidney is capable of producing urine with a range of 50 to 1200 mOsm/kg (see Chapter 22). Urine ranges from one sixth to four times the osmolality of normal serum (280 to 290 mOsm/kg). Osmolality is measured by an osmometer (see Chapter 12).

Osmolality is determined by the number of particles per unit mass, whereas the specific gravity is a reflection of the density (size or weight) of the suspended particles. Generally, specific gravity and osmolality are directly related in a linear fashion, though there are important exceptions. For example, if iodinated dyes are administered to a patient for intravenous pyelography, the specific gravity may reach as high as 1.070 or 1.080, although the osmolality will remain within normal limits. The dye particles have a mass that is large enough to raise the specific gravity, but too few molecules are present to notably increase the osmolality.

CHEMICAL EXAMINATION OF URINE

Several chemical constituents are routinely analyzed by urinalysis. Both qualitative reagent-strip and tablet tests are used, as well as quantitative methods for protein, electrolytes, and porphyrins.

Normal constituents

Urine is composed of numerous chemical substances. Table 56-5 lists common chemical constituents measured by urinalysis laboratories.

Reagent-strip testing

Reagent-strip tests have enabled urinalysis laboratories to generate valuable semiquantitative chemical results in a rapid, accurate, and efficient manner. In general, properly performed urine test strips are sensitive, specific, and cost effective.

The urinalysis laboratory is responsible for selecting the most suitable type of reagent strip for its hospital or clinical setting. The following guidelines will ensure the best results:

1. Test urine promptly; use properly timed test readings only.
2. Beware of interfering substances.
3. Understand the advantages and limitations of the test.
4. Employ controls.

Strip tests should be performed on well-mixed urine equilibrated to room temperature. Each chemical parameter must be evaluated within a specific time interval, as suggested by the manufacturer's instructions. The correct number of strips needed for immediate analysis should be removed from the container and the lid tightly replaced. Reagent strips should be stored in a cool (not refrigerated), moisture-free environment. Outdated strips must never be used.

After dipping the reagent strip into the urine, remove excess urine by gently tapping the strip on the edge of the specimen container. Compare individual reagent-pad reactions with the correct color chart in a properly lighted area. Automated reagent-strip readers (reflectance photometers) are widely available. Table 56-6 lists the sensitivities of two commercially available multiple test strips. Reagent-strip results that are positive may require confirmation with chemical and microscopic methods. Manufacturers' inserts should be reviewed to identify sources of inhibitors and false-positive and false-negative results. Ascorbic acid in urine can interfere with reagent-strip reactions for glucose, hemoglobin, bilirubin, and nitrite. Manufacturers have been encouraged to minimize or eliminate this interference when possible because ingestion of vitamin C supplements is so common.[10] Quality control samples are essential to verify reagent-strip results.

Table 56-5 Composition of urine from healthy subjects

Constituent	Value
Calcium	100-240 mg/24 hr
Creatinine	1.2-1.8 g/24 hr
Glucose	<300 mg/L
Ketones	<50 mg/L
Osmolality	>600 mOsm/L
Phosphorus	0.9-1.3 g/24 hr
Potassium	30-100 mEq/24 hr
pH	4.7-7.8
Sodium	85-250 mEq/24 hr
Specific gravity	1.005-1.030
Total bilirubin	(Not detected)
Total protein	<150 mg/24 hr
Urea nitrogen	7-16 g/24 hr
Uric acid	300-800 mg/24 hr
Urobilinogen	<1 mg/L

Table 56-6 Practical sensitivities of commercial reagent strips

Urine parameters	Chemstrip*	N-Multistix†
pH	±1 pH unit	± 1.0 pH unit
Protein	60 mg/L	50-200 mg/L
Glucose	400 mg/L	750 mg/L
Ketones	50 mg/L acetoacetic acid	90 mg/L acetoacetic acid
	400-700 mg/L acetone	800-1400 mg/L acetone
Bilirubin	5 mg/L	4-8 mg/L
Blood	5 intact erythrocytes/μL or hemoglobin from 10 erythrocytes/μL	5-20 intact erythrocytes/μL or hemoglobin from 5 erythrocytes/μL
Urobilinogen	4 mg/L	2 mg/L
Nitrite	0.3 mg/L	0.6-1 mg/L
Esterase (neutrophils)	6-10 leukocytes/hpf	5-15 leukocytes/hpf

*Ames, Inc, Division of Miles Laboratories, Inc., Elkhart, IN.
†BMC/Biodynamics, Indianapolis, IN.

Confirmatory testing

Many laboratory workers equate a confirmatory test with rechecking for a given parameter. This confirms nothing; it only establishes the precision for that parameter. A confirmatory test is one that will establish the accuracy or correctness of another procedure. Examples of confirmatory tests are quantitative protein analysis, protein electrophoresis, bacterial cultures, and cytological appearance (Table 56-7). A confirmatory test should have either the same or better specificity, be based on a different principle, or have a sensitivity equal to or better than that of the original test.

Urinary pH

Although the standard method for pH measurements uses glass electrodes, urinary pH is usually measured with indicator paper, since small changes in pH are of little clinical significance. Most urinalysis laboratories use multitest reagent strips with two indicators—methyl red and bromthymol blue. These provide a pH range from 5.0 to 9.0 that is demonstrated by a color change from orange (acid) to green to blue (alkaline). The urinary pH range is 4.7 to 7.8. Extremely acidic or alkaline urine usually indicates a poorly collected specimen.

Table 56-8 lists common clinical causes of acidic and alkaline urine. The average American ingests an acid-residue diet high in protein that results in acidic urine (5.0 to 6.5). With alkaline urine (8.0 to 8.5) one should always suspect an unpreserved or old specimen with the presence of urease-producing ammonia bacteria, such as *Proteus* species. Patients with renal tubular acidosis, a clinical syndrome characterized by an inability to excrete acidic urine, may produce urine with a much higher pH than would be expected on the basis of the acidosis.

The urinary pH is important in the management of renal stones or crystals. Uric acid stones precipitate in acidic urine and are more soluble in basic urines. Alkaline urine will precipitate calcium or calcium phosphate stones, whereas acidic urine will tend to dissolve them. Alkaline urine is desirable during sulfonamide and streptomycin therapy to prevent precipitation of the drugs in the kidneys and the formation of uric acid, cystine, and oxalate stones. The alkaline pH is also maintained during treatment of transfusion reactions and salicylate intoxication. In cystitis, the pH is kept acidic to combat bacteriuria and to prevent formation of *alkaline stones,* such as calcium carbonate or calcium phosphate stones.[6] Technologists should be aware that alkaline urine interferes with the determination of proteins and may alter the urine sediment examination.

Proteins

A healthy person will have a daily protein excretion of about 100 mg/day, a very small fraction of the plasma protein content. The majority of the urine protein is albumin that has crossed the glomerular membrane, but smaller-molecular-weight proteins such as globulins may also be present. Once filtered, proteins are almost completely reabsorbed in the proximal tubule. Proteinuria, therefore, can be the result of either increased filtration or decreased reabsorption (tubular function).

Table 56-7 Useful confirmatory tests

Test	Confirmatory test	Reason
Protein	Sulfosalicylic acid (SSA) method	Increased specificity Increased sensitivity
	Protein electrophoresis	Increased specificity
Ketones	Acetest	Increased sensitivity Increased specificity Increased color stability
Bilirubin	Ictotest	Increased sensitivity
Bacteriuria	Culture	Identification Quantitation
Abnormal cells, casts	Cytological examination	Increased visualization Quantitation

Table 56-8 Common clinical conditions causing acidic and alkaline urine

Acidic urine	Alkaline urine
Protein diet	Vegetable diet
Starvation	Vomiting
Dehydration	Renal tubular acidosis
Diarrhea	Respiratory and metabolic alkalosis
Diabetic acidosis	
Metabolic and respiratory acidosis	Ammonia-producing, urea-splitting bacteria
Metabolism of fats	Acetazolamide therapy
Sleep	Low-carbohydrate diets
Acid-producing bacteria	Chronic renal failure

Reagent-strip tests represent a screening procedure for proteinuria. Since the specificity of strip tests is limited to the detection of albumin, it is highly recommended that the laboratory simultaneously perform both a reagent-strip test and an acid precipitation test for the detection of all types of protein. Reagent strips are pH sensitive and depend on the presence of protein for color generation (Sörensen's protein error). The presence of protein on the strip changes the pH environment of the dye embedded in the pad, resulting in a change in color. Highly buffered, alkaline urine can result in a false-positive test:

$$\text{Tetrabromphenol blue} \xrightarrow[\text{Protein}]{\text{pH 3}} \text{Positive results } (\textit{green-blue})$$

$$\text{Tetrabromphenol blue} \xrightarrow[\text{No Protein}]{\text{pH 3}} \text{Negative results } (\textit{yellow})$$

A positive or faintly positive result should be confirmed with a more specific test such as the trichloracetic acid (see p. 1060) or the sulfosalicylic acid tests. A mildly positive test-strip result and a grossly positive turbidity test result may indicate the presence of drugs or Bence Jones proteins.

Sulfosalicylic acid test and semiquantitative turbidity method

Principle. Protein, denatured by acid, precipitates and renders the urine specimens progressively more turbid as protein concentration increases.

Materials

Sulfosalicylic acid, 3% (30 g/L) aqueous solution
Test tubes, 13 x 100 mm

Procedure

1. Add 3 mL of reagent to 1 mL of centrifuged urine.
2. Mix and allow to stand for 5 minutes.
3. Observe the degree of turbidity and compare the test sample with the original urine specimen in a 13 × 100 mm test tube.

Reporting

Grade	Turbidity*	Protein range (mg/L)
None	Clear	≤75
Trace	Print can be easily seen and read when viewed through test tube	200
1+	Print can be easily seen and read with some difficulty when viewed through test tube	300-1000
2+	Print can be seen but not easily read when viewed through test tube	1000-2500
3+	Print cannot be seen when viewed through test tube	2500-4500
4+	Precipitate formed	4500

*Urine clarity when light is passed through specimen in a clear test tube.

In normal urine, turbidity should be absent and protein should represent less than 75 mg/L. Table 56-9 lists the sensitivities, types of proteins detected, and sources of false-negative and false-positive results for both the reagent-strip and the sulfosalicylic acid tests.

Table 56-9 Tests for proteinuria

	Reagent strip	Sulfosalicylic acid	Results
Sensitivity	50-200 mg/L (albumin)	100 mg/L (all proteins)	
Combined use	Negative result	Negative result	No protein
	Positive result	Negative result	Clinically insignificant protein level or false-positive reagent-strip test
	Negative result	Positive result	Bence Jones proteins
			Heavy-chain proteins (electrophoresis confirmation required)
			Drug interference (penicillin)
Common urine constituents causing false-positive or false-negative results			
Urine turbidity	—	False-positive and false negative result	
X-ray contrast mediums	—	False-positive result	
Tolbutamide (Orinase)	—	False-positive result	
Penicillin (massive dose)	—	False-positive result	
Sulfisoxazole (Gantrisin)	—	False-positive result	
Highly buffered alkaline urine	False-positive result	False-negative result	
Quaternary ammonium salts	False-positive result	—	
Tolmetin sodium (Tolectin)	—	False-positive result	

Sugars

Enzymatic tests. The reagent-strip test is an excellent test that is specific for glucose. It detects the oxidation of glucose to gluconic acid:

$$\text{Glucose} + \text{Oxygen in room air} \xrightarrow{\text{Glucose oxidase}} \text{Gluconic acid} + \text{Hydrogen peroxide}$$

$$\text{Hydrogen peroxide} + \text{Chromogen} \xrightarrow{\text{Peroxidase}} \text{Oxidized chromogen } (\textit{blue}) + H_2O$$

One reagent-strip product uses *o*-tolidine as the chromogen for the indicator reaction.

Copper reduction (Clinitest, Benedict's test)

$$\text{Cupric ions} + \text{Glucose (or reducing substances)} \xrightarrow[\text{Alkali}]{\text{Heat}} \text{Cuprous oxide } (\textit{red}) + \text{Cuprous hydroxide } (\textit{yellow})$$

The Clinitest tablet (Ames Division, Miles Laboratories, Inc., Elkhart, IN) provides another test for sugar. It is a copper reduction test that measures total reducing substances. In addition to glucose, Clinitest will detect sugars such as galactose, lactose, and pentoses. It also detects ascorbic acid and certain drugs (that is, nalidixic acid [NegGram]), used to treat urinary tract infections; probenecid, used to treat gout; and cephalosporin, an antibiotic.

Clinitest is an important test in pediatric screening. A negative test-strip result (specific for glucose) but a positive Clinitest result (sensitive for any reducing sugar) may indicate the presence of an inherited metabolic disorder in the newborn.

Clinitest will detect reducing substances at a concentration of 2000 mg/L (200 mg/dl) or greater. Reagent strips will detect glucose at a concentration of 400 to 750 mg/L. Because Clinitest is both less specific and less sensitive than the reagent strip, it cannot be used as a confirmatory test for a positive reagent-strip glucose test. Clinitest should be reserved for patient populations in whom nonglucose-reducing substances need to be detected.

Sugar will appear in the urine because of an increased filtered load, as in diabetes mellitus, or because of decreased tubular reabsorption, as in renal glucosuria. The presence of ascorbic acid may lead to erroneous low results.

Ketones

The term *ketone bodies* includes three discrete but related chemicals: acetoacetic acid, β-hydroxybutyric acid, and acetone (see p. 856 and Chapter 29). Reagent-strip testing for ketones uses a sodium nitroprusside reaction that detects acetone and acetoacetic acid but not β-hydroxybutyric acid, the primary ketone body. It is important to realize that the sodium nitroprusside reagent reacts primarily with acetoacetic acid; acetone has only a 20% reactivity compared with acetoacetate:

$$\text{Acetoacetic acid} + \text{Sodium nitroprusside} + \text{Glycine} \xrightarrow{\text{Alkaline pH}} \text{Purple color}$$

Ketone determinations are important in monitoring diabetes and ketoacidosis and should be performed whenever sugar determinations are made.

Blood and myoglobin

As previously discussed, red urine usually indicates the presence of erythrocytes, hemoglobin, or myoglobin in the urine. Hematuria most often represents a combination of intact erythrocytes (greater than 5 per high-power field), degenerated erythrocytes, and free hemoglobin. Gross hematuria implies hemorrhage or fresh bleeding and in acidic urine results in a red to brown, turbid or smoky appearance. The reagent-strip method for hemoglobin and myoglobin uses the peroxidase-like activity of these proteins:

$$\text{Hydrogen peroxide } (H_2O_2) + \text{Chromogen} \xrightarrow{\text{Myoglobin or hemoglobin}} \text{Oxidized chromogen } (\textit{blue}) + H_2O$$

A positive test indicates the presence of hematuria, hemoglobinuria, or myoglobinuria, and a microscopic urinalysis is needed to confirm the presence of erythrocytes. Oxidizing agents such as iodides and bromides in the urine may cause false-positive results; large quantities of ascorbic acid (used in some antibiotics) in the urine may produce false-negative results with some reagent strips.

Myoglobin is a ferrous porphyrin similar to hemoglobin that is commonly seen in urine after crush injuries and muscle trauma. When myoglobin is released in the circulatory system, it is rapidly excreted by the kidney. Like hemoglobin, its presence will produce pink to red urine. Myoglobinuria should always be confirmed with rapid immunodiffusion or radioimmunoassay procedures.

Bilirubin

Dark, yellow to brown, foamy urine suggests the presence of conjugated bilirubin. Normal urine does not contain bilirubin. Jaundiced patients with hepatocellular disease, such as hepatitis, or obstructive disease, such as biliary cirrhosis, may have conjugated bilirubin in the urine. The reagent-strip method for determining bilirubin involves a diazotization reaction:

$$\begin{matrix}\text{Bilirubin} \\ \text{glucuronide}\end{matrix} + \begin{matrix}\text{Diazonium} \\ \text{salt}\end{matrix} \xrightarrow{\text{Acid}} \text{Azobilirubin } (\textit{brown})$$

Negative results on suspicious urines and questionably positive results, such as on highly colored urines, should be confirmed by using Ictotest tablets (Ames Division, Miles Laboratories, Inc., Elkhart, IN). The Ictotest employs the same diazotization reaction as the reagent strip. False-negative results may occur if the urine is not fresh, because urinary bilirubin may hydrolyze or oxidize when exposed to light.

Urobilinogen

Urobilinogen, a colorless compound, is formed in the intestine by the bacterial reduction of bilirubin (see Chapter 23). Normal urine contains small amounts of urobilinogen. Decreased urobilinogen is found in infants, who lack reducing intestinal bacteria; in patients after administration of antibiotics that suppress intestinal flora; and in patients with obstructive liver disease. Increased urobilinogen is present in hemolytic anemia (increased bilirubin formation) and liver dysfunction.

The reagent-strip tests for urobilirubin differ with the manufacturer. Ames products use the Ehrlich reaction employing *p*-dimethylaminobenzaldehyde in a simple color reaction with porphobilinogen (see p. 1124). The reaction is not specific for urobilinogen, and false-positive findings may result from other Ehrlich reagent positive compounds (porphobilinogen, PAS). Boehringer Mannheim Diagnostics products employ a reaction that is specific for urobilinogen; urobilinogen reacts with a diazonium compound to form a red color. It should be noted that reagent strips will not detect the absence of urinary urobilinogen.

A fresh specimen is essential for the detection of urobilinogen, as it is a light-sensitive compound. The preferred specimen for detecting and/or quantitating urinary urobilinogen is a 2-hour early afternoon specimen. This collection takes into account the diurnal excretion pattern of urobilinogen.

Nitrites

The nitrite test is used in urinalysis laboratories to detect bacteriuria. The reagent-strip nitrite test depends on the reduction of nitrate to nitrite by the enzymatic action of certain bacteria in the urine. Under conditions of acid pH, nitrite reacts with *p*-arsanilic acid to form a diazonium compound, which in turn couples with *N*-1-naphthyl ethylenediamine to produce a pink color.[11] The nitrite test should be performed on the first morning specimen or on a urine sample that has been collected at least 4 hours or more after the last voiding to allow the organisms in the bladder time to metabolize the nitrate. Stale urine may have a positive nitrate test because of bacterial contamination after voiding. The nitrite test is specific for gram-negative organisms; however, some false-negative results will occur if organisms such as enterococci, streptococci, or staphylococci, which do not form nitrite, are present. The sensitivity of the nitrite test is about 60% when compared with microbiological procedures.[11] There are very few cases of false-positive nitrite results.

Nitrite testing is of dubious value in hospital clinical urinalysis because there is limited effective control over how and when the urine sample is collected. The nitrite test may have use in a clinic or physician's office because proper control over sampling can be better achieved.

Leukocytes

The presence of leukocytes (pyuria) is a clinically important indicator of inflammation. Reagent-strip tests for pyuria will detect both lysed and intact leukocytes and are based on the presence of intracellular esterases. These enzymes will catalyze the hydrolysis of esters, releasing components that are then used in a color reaction. The intensity of the color reaction is proportional to the number of leukocytes in the specimen. Sensitivities for the two reagent-strip manufacturers are listed in Table 56-6. False-positive results are seen with trichomonads and oxidizing agents. It has not yet been well established whether eosinophils and histiocytes will also give a positive reaction. False-negative values may be seen with high levels of protein and ascorbic acid. Several studies have documented the clinical utility of the leukocyte test as a screening test for pyuria, and many laboratory workers believe a microscopic examination for leukocytes need be performed only on urine specimens that are positive for esterase when tested with reagent strips.[12-14]

Porphyrins

Porphyrins are groups of intermediary products in the biosynthesis of heme and cytochromes, which are produced in the liver and bone marrow (see Chapter 23). Identification of various porphyrins and porphyrin precursors (especially porphobilinogen) is important in the clinical diagnosis of porphyrias, a group of genetic or metabolic disorders. Normal urinary excretion of porphyrins is approximately 2 mg/day. Increased quantities of excreted porphyrins (porphyrinuria) give urine a red or wine color. A screening urinalysis test for porphobilinogen is based on the Watson-Schwartz test (see p. 1124).[15] Confirmatory and quantitative tests are available.[7]

Melanin

Normal urine does not contain melanin. Melanin is found in the urine of patients with malignant melanoma. Patients with this malignancy will excrete a colorless precursor of melanin (melanogen) that, when exposed to air, will polymerize to form the dark pigment, melanin. Screening tests, using ferric chloride, oxidize melanogen to melanin, which turns urine brown-black (see p. 1122).

MICROSCOPIC EXAMINATION OF URINE

Methods

Accurate microscopic identification of urine sediment is important in the early recognition of infectious, inflammatory, and neoplastic conditions affecting the urinary system.[2] It is debatable whether all routine urine specimens require the more time-consuming microscopic analysis. Instead, most laboratory workers agree that a microscopic examination should be performed when the patient is symptomatic, when the physician specifically requests this examination, and when the macroscopic urinalysis is abnormal, that is, hematuria, proteinuria, or a positive result for pyuria (positive nitrate or esterase result).[16]

Several microscopic procedures are available for the sediment examination. Standardized bright-field microscopy is still the most common technique employed.[17] Supravital staining can be combined with bright-field microscopy to enhance cellular detail. Phase-contrast microscopy is probably the best method for rapid urine sediment evaluations without the use of stains. Commercially available standardized slide methods are far superior to conventional glass slide and coverslip methods and a practical alternative to hemocytometer chamber counts.[18]

Standardization

Standardization of the microscopic urinalysis is essential to reduce ambiguity and minimize subjectivity.[21] Aspects of the microscopic examination that should be standardized are:

1. Volume of urine analyzed
2. Length and force of centrifugation
3. Resuspending volume and concentration of sediment
4. Volume and amount of sediment examined
5. Terminology and reporting format

Bright-field microscopy of unstained urine. Unstained bright-field microscopy uses reduced light to delineate the more translucent formed elements of the urine, such as hyaline casts, crystals, and mucous threads.

Accurate identification of leukocytes, macrophages, renal tubular epithelial cells, and viral inclusion-bearing cells may be very difficult in unstained preparations. Cytological techniques, stained preparations, or both should be used to confirm results.[20,21]

Procedure. The urine specimen must be examined while fresh, since cells and casts begin to disintegrate within 1 to 3 hours. Refrigeration (2° to 8° C) for up to 48 hours usually prevents the disintegration of cells and pathological entities. Each specimen is concentrated tenfold or twentyfold for the purpose of standardization. The examination proceeds as follows:

1. Mix the specimen well.
2. Pour a fixed volume (10, 12, 15 mL) of urine into a graduated centrifuge tube.
3. Centrifuge at 1500 rpm or approximately 80 *g* for 5 minutes.
4. Remove the supernatant fluid by careful decantation or aspiration to a fixed volume; 1 mL and 0.4 mL are the most common. Resuspend the sediment by gently tapping the bottom of the tube.
5. Place a drop of resuspended sediment on one area of a standardized slide.
6. Examine with low power (100 ×) and subdued light. Vary the fine focus continuously while randomly scanning the area under the coverslip.

   During the scan, evaluate the specimen for the presence of squamous and transitional epithelial cells, crystals, mucus, bacteria, yeast, and artifacts. Report according to laboratory protocol. Further identification of casts, renal epithelial cells, erythrocytes and leukocytes can be accomplished using high power.
7. Examine at least 10 low-power fields using subdued light. Count and report the number of casts per low-power field. Be sure to examine the edges, as casts are often found along the edge of a coverslip. Abnormal crystals, when present, should also be counted on low power. Bacteriuria, visible at low power, should be reported as at least 2+.

8. Examine at least 10 high-power (440×) fields and report numerical values for erythrocytes, leukocytes, and renal tubular epithelial cells per high-power field.
9. Report all counts (average of 10 fields) and qualitative assessments according to standardized terminology.

Bright-field microscopy with supravital staining. Cellular detail is enhanced with stained sediments. A crystal violet–safranin O stain is often used in the rapid assessment of certain cellular elements.[22]

Reagent (Sternheimer-Malbin stain)

Solution 1	Crystal violet	3.0 g
	Ethyl alcohol (95%)	20.0 mL
	Ammonium oxalate	0.8 g
	Distilled water	80.0 mL
Solution 2	Safranin O	1 g
	Ethyl alcohol (95%)	40 mL
	Distilled water	400 mL

Three parts of solution 1 and 97 parts of solution 2 are mixed and filtered. The mixture should be clarified by filtration every 2 weeks and discarded after 3 months. Separately, solutions 1 and 2 keep indefinitely at room temperature. In highly alkaline urine, the stain will precipitate.

Procedure

1. Add one or two drops of crystal violet–safranin O stain to approximately 1 mL of precentrifuged, concentrated urine sediment.
2. Mix with a pipet and place a drop of this suspension on a slide.

Phase and interference microscopy. Many urinalysis laboratories recommend phase microscopy for the detection of more translucent formed elements of the urinary sediment. Sediment, notably hyaline casts, mucus, and bacteria, may escape detection using conventional, unstained, bright-field microscopy. Phase microscopy has the advantage of hardening the outlines of even the most ephemeral formed elements, making detection simple.[24] Even greater morphological detail of formed elements (notably casts and cells) is afforded by interference contrast microscopy.[25,26]

Standardized slide methods. The KOVA system (ICL Scientific, Fountain Valley, CA), Count-10 and Count-6 (V-Tech, Palm Desert, CA), and the Uri-System (Fisher Scientific, Pittsburgh) offer complete standardized procedures that are technically more precise, reproducible, and reliable than conventional bright-field microscopy.[18] A comparison of standardized microscopic urinalysis systems is shown in Table 56-10. All these standardized slides are superior to conventional glass slides.

In some laboratories, the hemocytometer continues to be used for quantifying urine sediment entities. Kesson et al.[27] provide evidence that chamber counts are more reliable in detecting sediment abnormalities than conventional methods of counting cells under high-power fields.

Semiautomated urinalysis work station. An automated urinalysis instrument called the Yellow IRIS (International Remote Imaging Systems, Chatsworth, CA) combines automated microscopy with a dipstick reader and specific gravity module. This technology may provide more accurate results over standardized manual systems when there are high volumes of test samples and very low concentrations of sediment elements.[28] Manual methods appear to be more sensitive in detecting casts.[29]

Combined cytocentrifugation and Papanicolaou stain. A technique that combines cytocentrifugation and Papanicolaou staining has been recommended for more accurate assessment of urine sediment. This more specialized method provides a simple, reproducible, and semiquantitative method for identifying urine sediment entities. Cellular casts, mononuclear cells, tissue fragments, and neoplastic cells may be clearly identified with this method. Although not found in routine laboratories, this technique's acceptance is growing.

Table 56-10 Physical features of urinalysis slide systems

	Conventional	Uni-Slide*	KOVA†	Count-10‡
Volume of urine sediment				
Mean amount	1 drop	16 µL	6 µL	6 µL
% Coefficient of variation		7%	6%	10%
Surface for microscopic examination				
Total area (mm^2)	484	90	32	36
Number of 100× fields	144	25	9	12
Number of 400× fields	2116	420	119	49
Number of focal planes before settling	1-2	1-2	1-2	3-4
Type of coverglass	Glass	Glass	Plastic	Plastic
Cost per test	7¢	9¢	9¢	6¢
Maximum number of tests or slides	2	4	10	10

*Uni-Slide, Uri-System Slide distributed by Fisher Scientific, Pittsburgh, as part of the Fisher brand Uri-System.
†Kova IV Slide, ICL Scientific, Fountain Valley, CA.
‡Count-10 Slide, V-Tech, Inc., Palm Desert, CA.

Crystals

Urinary crystals (Figs. 56-1 to 56-5) are commonly seen. Usually crystals are not present when urine is freshly voided, and in general, the formation of crystals should be regarded as an artifact of the system of collection. Crystal formation occurs when various chemical constituents become saturated or undergo altered solubilities when urine is stored at cooler temperatures. Certain chemical substances such as albumin prevent crystallization. When heated to 37° C, most crystals disappear.[6] Those still present might have some diagnostic significance when correlated with clinical symptoms.[5]

The types of urinary crystals formed depend on the pH of freshly voided urine. Table 56-11 lists the common types, properties, and clinical significance of various urinary crystals. Cystine, uric acid, leucine, and tyrosine crystals are the most diagnostically important crystals to recognize. Because of the limited clinical significance of the urinary crystals, most laboratory workers agree that time should not be wasted on their specific identification. Bradley et al.[5] illustrate and describe the clinical significance of various types of crystals.

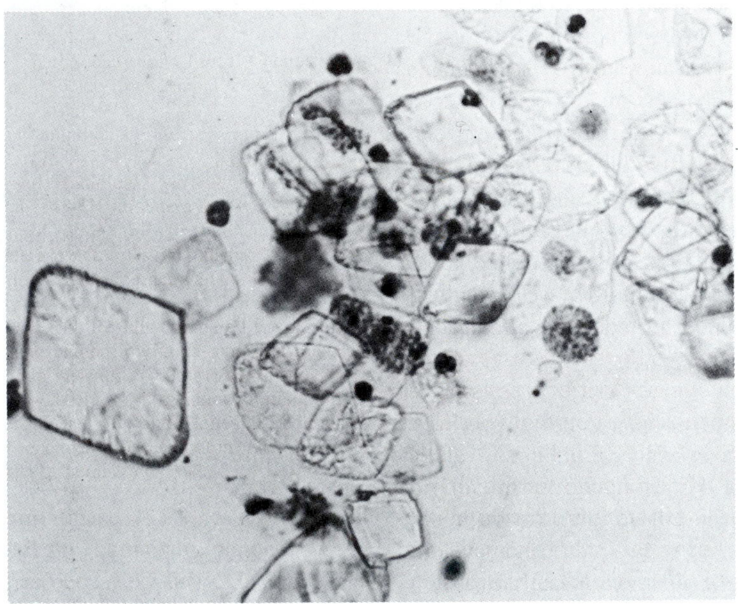

Fig. 56-1 Uric acid crystals. (Papanicolaou stain, 400×.)

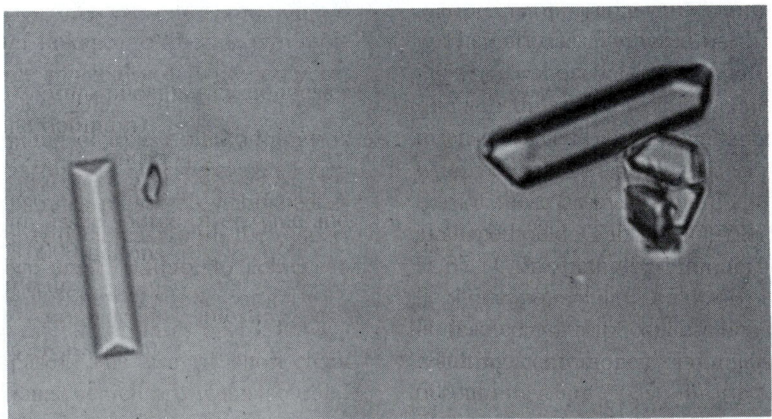

Fig. 56-2 Triple phosphate crystals. (Bright field, 400×.)

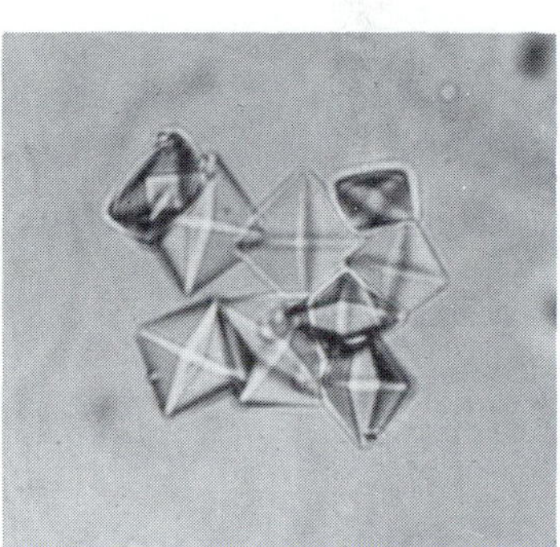

Fig. 56-3 Calcium oxalate crystals. (Bright field, 1000×.)

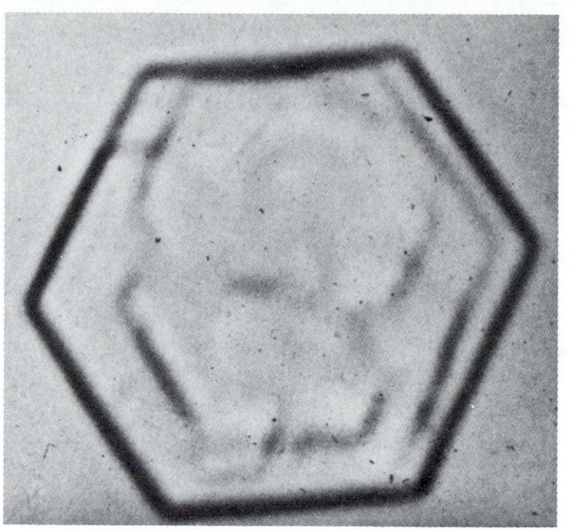

Fig. 56-4 Cystine crystals. (Bright field, 2500×.)

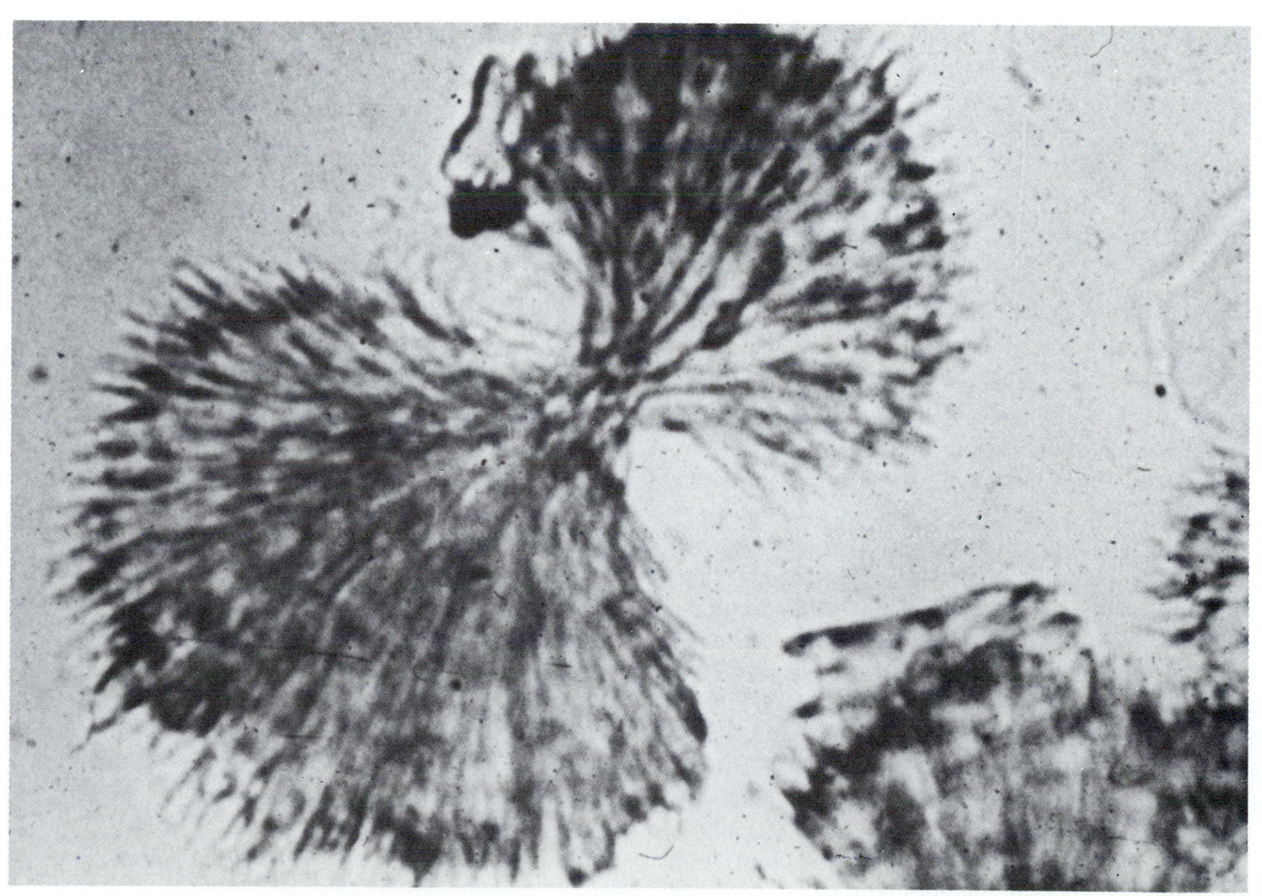

Fig. 56-5 Sulfonamide crystals. (Bright field, 2500×.)

Table 56-11 Common urinary crystals

Types	Urine pH				Diagnostic morphological appearance	Solubility characteristics	Clinical significance
	Acid	Alkaline	Neutral	Variable			
Urates	×				Colorless, amorphous, spherical, or needle shaped	Soluble in alkali or at 60°C	Normal
Uric acid	×				Colorless to yellow brown, pleomorphic, rhombic, four-sided plates or rosettes	Soluble in alkali	Normal, chemotherapy, gout
Calcium carbonate				×	Colorless dumbbells or spheres	Soluble in acetic acid	Normal
Phosphates (triple phosphates)		×			Colorless, three- to six-sided prisms; "coffin lids"	Soluble in acetic acid	Normal
Ammonium urates		×			Brown, "thorn apple"	Soluble in 60° C with acetic acid	Normal
Calcium oxalate				×	Colorless, octahedron, dumbbell, or envelope or internal "X" forms	Soluble in dilute hydrochloric acid	Normal, glycol poisoning
Cystine				×	Colorless, hexagonal, flat	Soluble in alkali	Cystinosis
Cholesterol				×	Colorless, flat plates with corner notched	Soluble in chloroform and ether	Renal damage
Tyrosine				×	Yellow to colorless, fine needles, in sheaves or rosettes	Soluble in alkali	Liver damage, aminoaciduria
Leucine				×	Yellow spheroids with striations	Soluble in hot alcohol	Liver damage, aminoaciduria
Bilirubin				×	Reddish brown, amorphous, needles, rhombic plates, or cubes	Soluble in alkali	Bilirubinuria
Sulfonamides				×	Cubes, globules, or sheaves	Soluble in acetone	Antibiotic therapy

Organisms

In a properly collected and processed urine specimen, the presence of organisms is clinically significant. Bacteria, fungi, parasites, and virally infected cells are frequently reported. Organisms seen in urine sediment are microscopically recognized as extracellular or intracellular structures.[20] Bright-field microscopy readily detects extracellular bacteria, fungi, and parasites. Detection of intracellular phagocytized bacteria and fungi, *Toxoplasma* organisms, and viral inclusion bodies usually requires cytological procedures.

Accurate identification of organisms aids in the clinical differential diagnosis of urinary system infections. Stained preparations are important in the evaluation of organisms, identification of associated inflammatory cells, and assessment of epithelial exfoliation and renal cast formation for purposes of localization.[2] Microbiological techniques should be used to confirm and fully classify various urinary organisms.

Bacteria. Normal urine is sterile and does not contain bacteria. Some bacteria may be present because of contamination during collection or prolonged storage. If bacteria are seen in centrifuged specimens but not in unspun urine specimens, less than 10^5 bacteria/mL are present. Bacteria in an unspun specimen indicate that greater than 10^5 bacteria are present. The presence of 10^5 bacteria or greater suggests a urinary tract infection. This number corresponds to 10 or more bacteria per high-power field. Identification of bacteria, cocci, or rods can readily be accomplished by bright-field or phase-contrast microscopy. Occasionally, there is difficulty in distinguishing bacteria from amorphous crystals.

Fungus. Urinary tract fungal infections are common in patients with diabetes, those taking birth control pills, or those undergoing intensive antibiotic or immunosuppressive therapy.

Candida albicans is the most common fungus and is identified as budding yeast or mycelia (Fig. 56-6). In general the budding yeast appearance indicates that the fungi are coexisting with the host, whereas the mycelial forms appear during tissue invasion. Yeasts of *Candida albicans* are oval and highly refractile and measure 3 to 5 μm. Often they can be misinterpreted as erythrocytes (7 μm). Unlike erythrocytes, yeasts are not lysed by acids.

Parasites. The presence of parasites in urine usually indicates vaginal or fecal contamination. *Trichomonas vaginalis,* a flagellate, is the most common parasite seen in urine. The incidence of this type of parasitism is very high in women and may be the cause of intense vaginitis. In men the parasite causes an asymptomatic urethritis. Because of the motility of this oval organism, bright-field microscopy is used for simple and rapid identification. Nonmotile trichomonads can easily be mistaken for leukocytes or epithelial cells.

Pinworm ova (Enterobius vermicularis) have been found in urine in children because of fecal contamination. Morphologically a pinworm ovum is surrounded by a thick two-layered transparent capsule, and a coiled embryo may be visible inside. Trematode ova, *Schistosoma haematobium* (found in North Africa), and *Schistosoma mansoni* (found in Central America) may be found in urine.

Virally infected cells. Virus-induced cellular changes have been recognized with increased frequency in urine sediment, especially in immunosuppressed patients. Cytological techniques should be used for the accurate identification of cytomegalovirus, herpes simplex, and *Polyomavirus,* which produce diagnostic intranuclear inclusion cells and are the most common types of urinary system viral infections. Viral inclusion cells must be distinguished from nonviral sources of inclusion cells such as heavy metal exposure (lead and cadmium) and nonspecific degenerative cellular changes. A detailed description of viral inclusion cells can be found in standard textbooks on urine cytology.

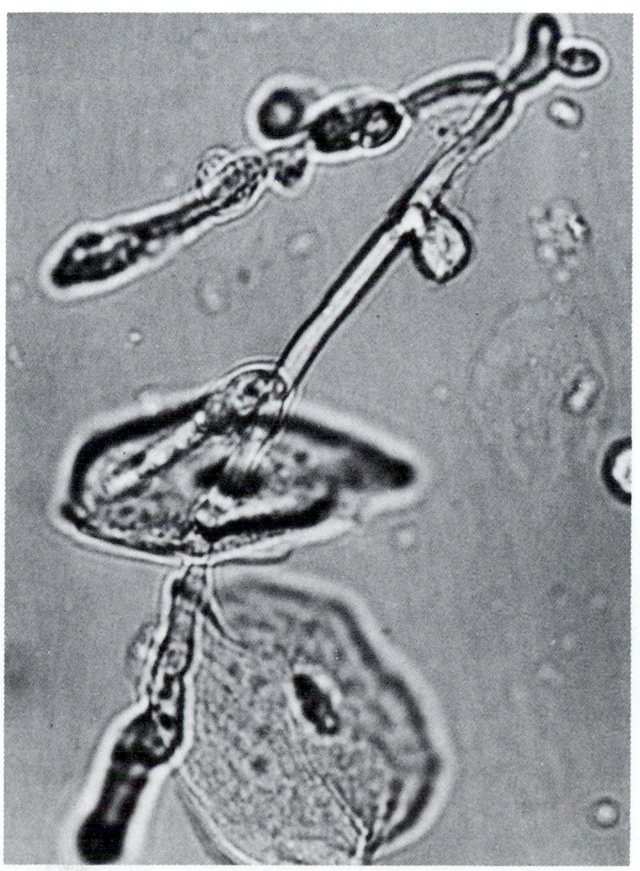

Fig. 56-6 Fungal yeast and mycelia, probable *Candida* species. (Bright field, 2500×.)

Cells

Microscopic identification and evaluation of cells is an important part of urinalysis. Common types of cells normally found in urine include a few erythrocytes, leukocytes, and epithelial cells of renal or lower urinary tract origin. All cells of pathological significance are usually quantified by high-power field examination.

Erythrocytes. Normal urine should never contain more than a few erythrocytes per high-power field. They appear in the urine stream after vascular injury or disorders in the kidney or lower urinary tract. Erythrocytes accompanied by blood casts or dysmorphic erythrocytes (Fig. 56-7) suggest renal parenchymal or glomerular bleeding.[31] Quantification aids in diagnosis and patient management. It is important that contamination from menstrual flow be avoided when urine from females is to be examined.

Erythrocytes measure approximately 7 μm in diameter, have a biconcave disk shape, and often appear pale yellow when bright-field microscopy is used (Fig. 56-8). In hypertonic urine they are smaller and crenated, whereas in hypotonic urine they are larger and swollen. When erythrocytes have been in urine for a considerable time, the hemoglobin may have leaked out of the cells.

On occasion hemoglobin is detected by reagent strip testing in urine in the absence of microscopic erythrocytes in the sediment. Possible explanations for this discrepancy include hypotonic urine or an alkaline urine, both of which can cause erythrocyte lysis. The absence of these conditions strongly suggests that the pigment that appears in the urine (which may be hemoglobin or myoglobin) originates from filtration of these products from the blood.

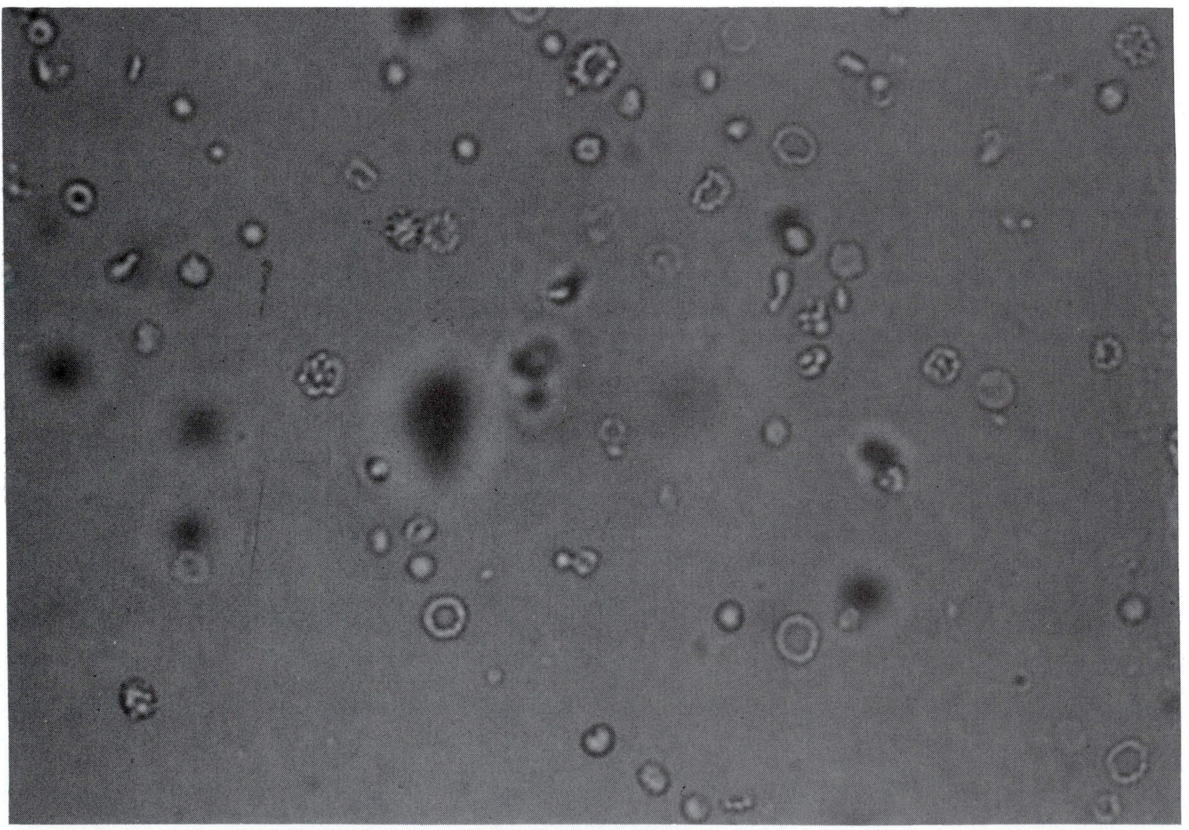

Fig. 56-7 Dysmorphic erythrocytes. (Bright field, 400×.)

Leukocytes (Fig. 56-8). The normal excretion rate for leukocytes in the urine is up to 1 leukocyte/3 high-power fields, 3000 cells/mL, or up to about 200,000 cells/hour. Elevated numbers of leukocytes (pyuria) are associated with numerous urinary tract inflammatory and infectious conditions. Most leukocytes recognized by bright-field microscopy are segmented neutrophils. The identification of lymphocytes, plasma cells, and eosinophils requires special stains.

Little[32] has shown that leukocyte excretion rates in excess of 400,000 cells/hour virtually always indicate urinary tract infection. This rate corresponds to more than 10 neutrophils per high-power field. Patients with active upper urinary tract infections frequently have more than 50 neutrophils per high-power field or have a leukocyte excretion rate in excess of 2 or even 3 million/hour.

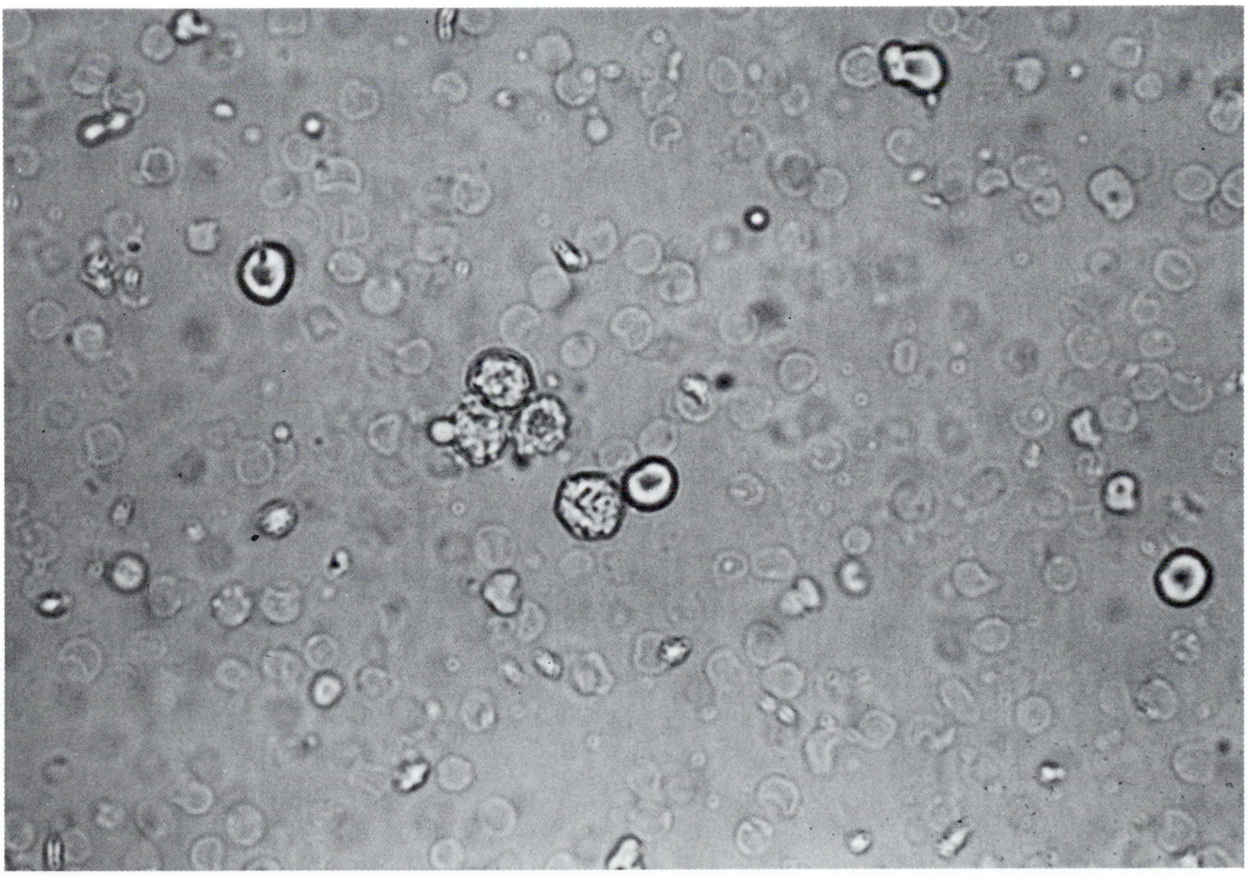

Fig. 56-8 Erythrocytes and leukocytes. (Bright field, 400 ×.)

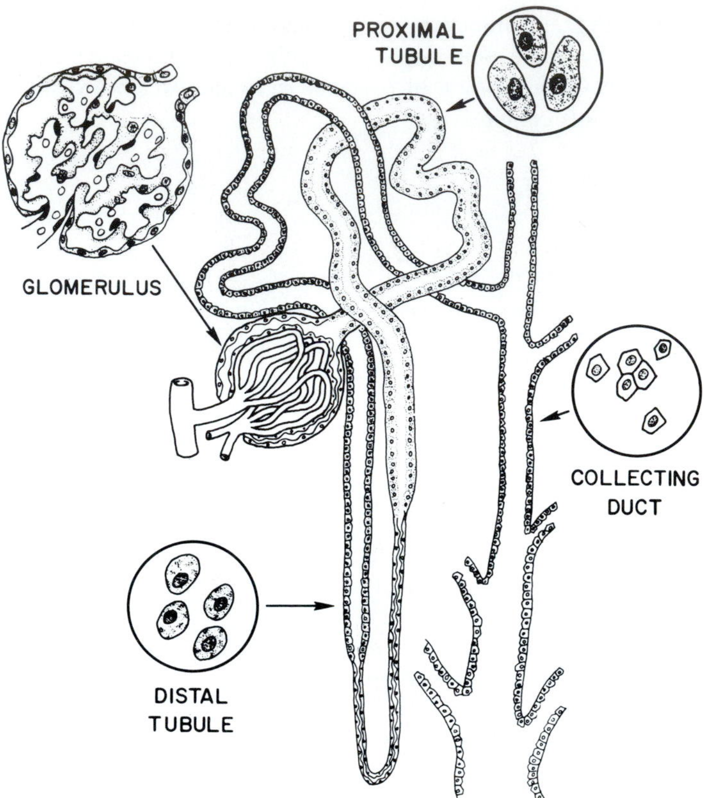

Fig. 56-9 Location of renal tubular epithelial cells.

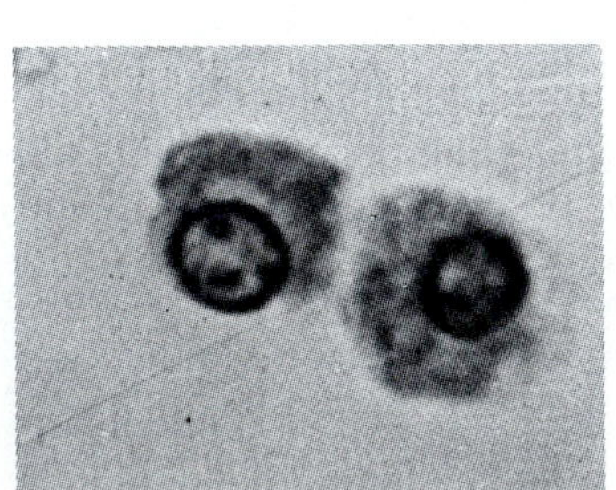

Fig. 56-10 Renal tubular epithelial cells. (Bright field, 1000×.)

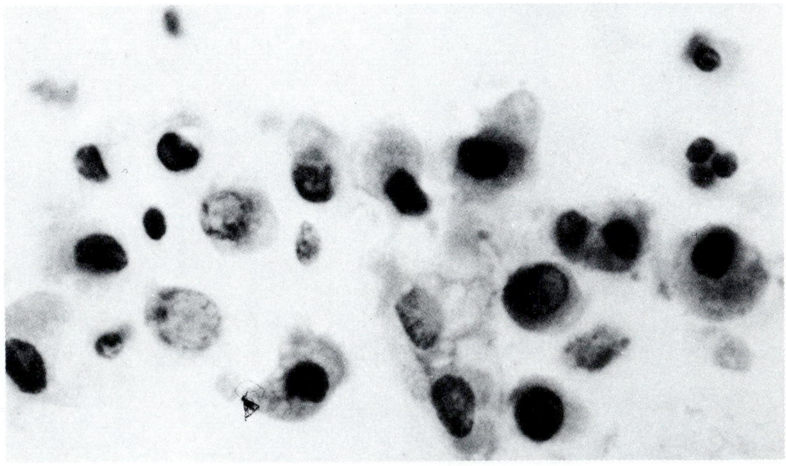

Fig. 56-11 Renal tubular epithelial cells. (Papanicolaou stain, 1000×.)

Renal tubular epithelial cells (Figs. 56-9 to 56-11). Various types of renal tubular epithelial cells line the nephron, and senescent or diseased cells are constantly being shed into the urine. Since they represent actual renal exfoliation, the presence of more than two renal tubular epithelial cells per high-power field indicates active renal tubular damage or injury.

Considerable difficulty exists in the accurate identification of renal tubular cells, particularly in distinguishing them from other mononuclear cells commonly found in urine. By bright-field microscopy they are polygonal and slightly larger than leukocytes. For accurate identification of various types of renal tubular cells (convoluted versus collecting duct) and sheets or fragments, cytological techniques are required.

Oval fat bodies (Fig. 56-12). Oval fat bodies are renal tubular epithelial cells that are filled with absorbed lipids or that have undergone degenerative cellular changes. Oval fat bodies are often associated with proteinuria and lipiduria and are characteristically seen in the nephrotic syndrome and diabetes mellitus.

Transitional epithelial cells (Fig. 56-13). A few transitional (urothelial) cells can be found in normal urine. Inflammatory conditions in the bladder, catheterization, or a pathological process such as malignancy may be indicated by large numbers of transitional cells.

By bright-field microscopy, transitional cells appear round to oval, measure 40 to 60 μm, and have a centrally located nucleus. The cytoplasm borders of these cells appear thickened and crisp. When the nuclei of transitional cells become enlarged or irregular, cytological techniques should be suggested for the purpose of detecting a urinary system malignancy.

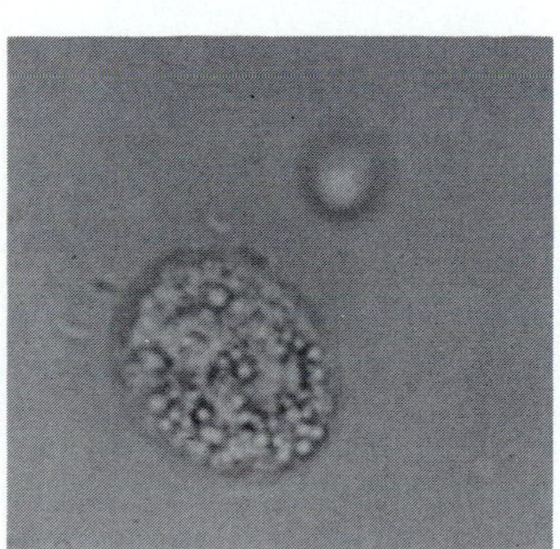

Fig. 56-12 Oval fat bodies. (Bright field, 1250×.)

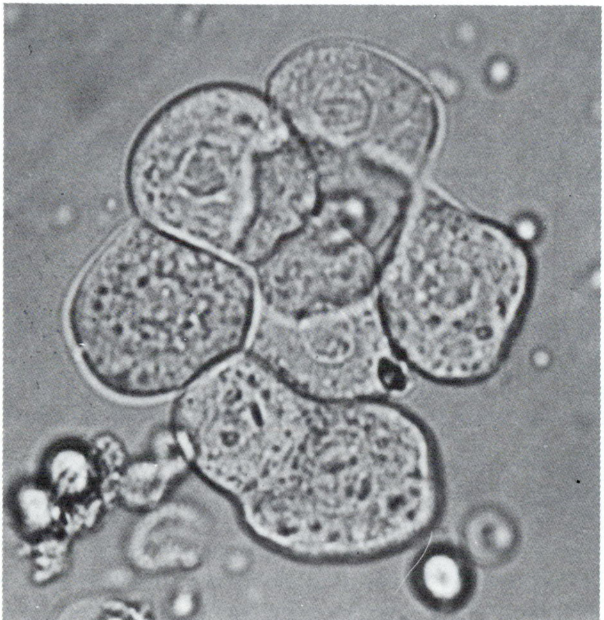

Fig. 56-13 Transitional epithelial cells. (Bright field, 1000×.)

Squamous epithelial cells (Fig. 56-14). Squamous epithelial cells line the distal portion of the lower urinary tract and the female genital tract. Squamous cells are the largest of the epithelial cells found in urine. They have abundant flat cytoplasm with small nuclei. Frequently one or more corners of these cells may be folded. Squamous cells in urine usually indicate contamination (vaginal in women, urethral in uncircumsized men) or squamous metaplasia of the bladder and represent the least significant type of epithelial cell found in urine.

Tissue fragments in urine (Fig. 56-15). Any clumps or chunks of solid-appearing material in urine specimens should be noted. This material is identified during the initial inspection of the specimen because of its large size. It is often white or tan. Precise identification of such material is important for an accurate diagnosis. This involves transfer of the observed material to an appropriate fixative to preserve it for a cytological or histological evaluation. Renal papillary necrosis and bladder tumors are most frequently responsible for shedding large tissue fragments into the urine.

Spermatozoa. Spermatozoa may be easily recognized in the urine of a man after ejaculation or in the urine of a woman as a vaginal contaminant after coitus. Their identification is of limited clinical significance, and the presence of spermatozoa is usually not reported.

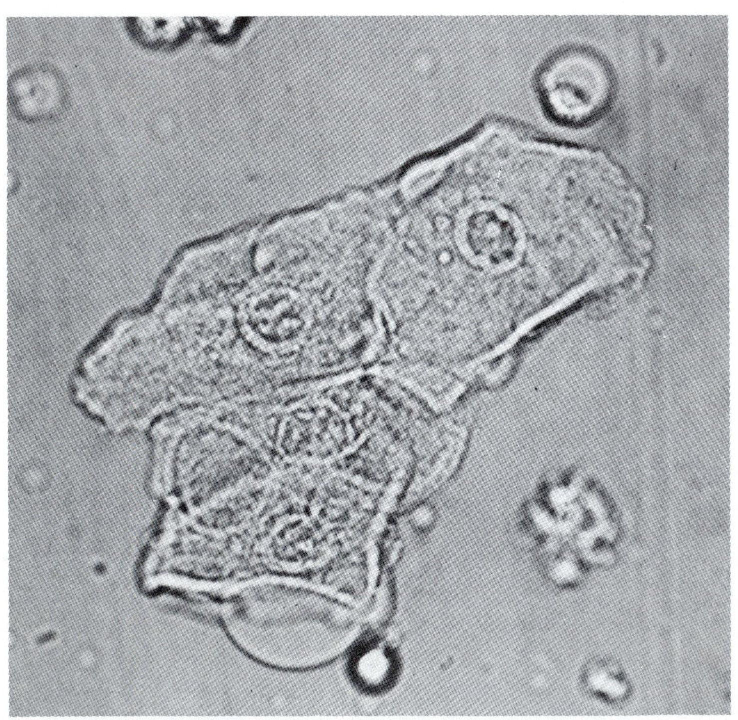

Fig. 56-14 Squamous epithelial cells. (Bright field, 1000×.)

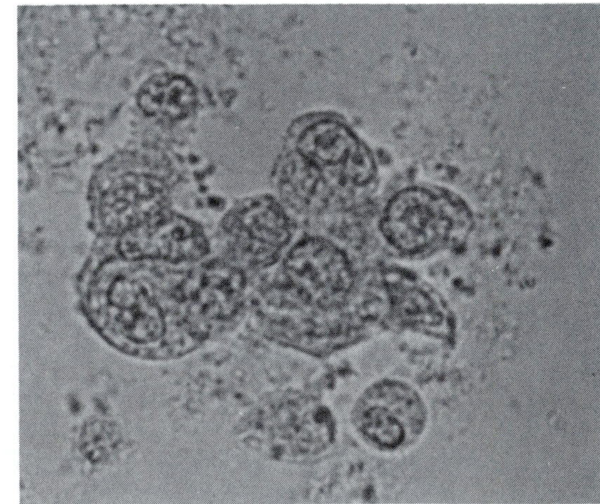

Fig. 56-15 Malignant tissue fragments in urine. (Bright field, 1000×.)

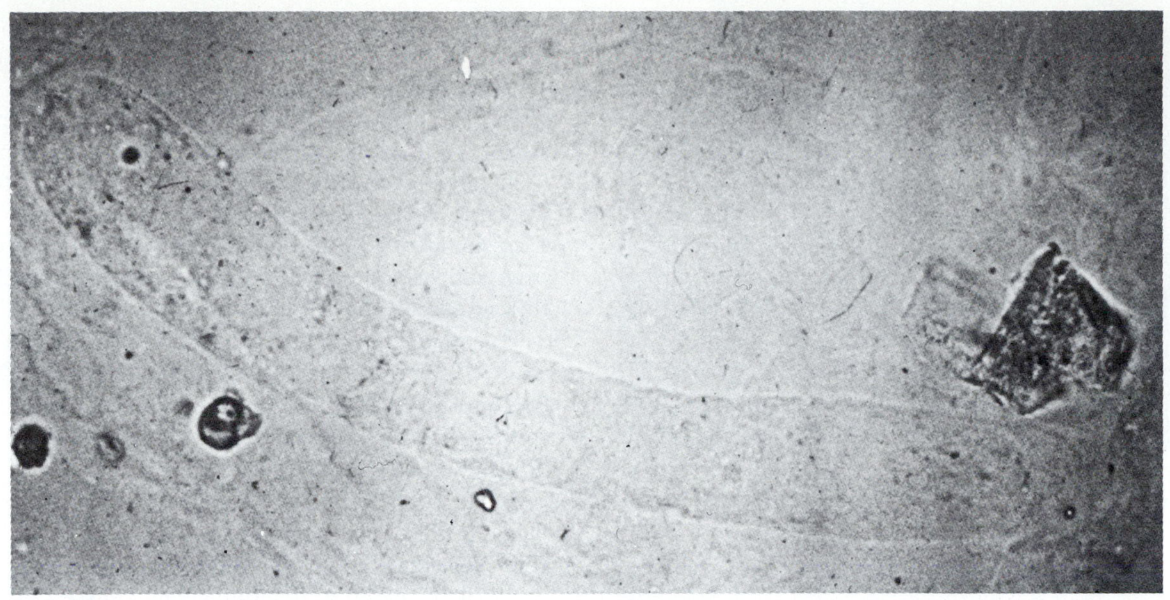

Fig. 56-16 Hyaline cast. (Bright field, 1000×.)

Renal casts (Figs. 56-16 to 56-21)

Renal (urinary) casts are formed cylindrical structures that are organized in the nephron. Casts are significant because of their localizing value. Casts are composed of uromucoid (Tamm-Horsfall mucoprotein), which is always present in urine, usually in solution. Uromucoid is produced by the renal tubular epithelial cells of the ascending limb of the loop of Henle. Casts are formed as a result of urine stasis allowing uromucoid to precipitate. Other factors contributing to cast formation include an increased concentration of protein, salts and a low urine pH. Since the precipitation of this protein depends on the concentration and composition of urine, casts are more likely to be formed at the distal portion of the nephron and in the collecting ducts of the kidney where the urine is more con-

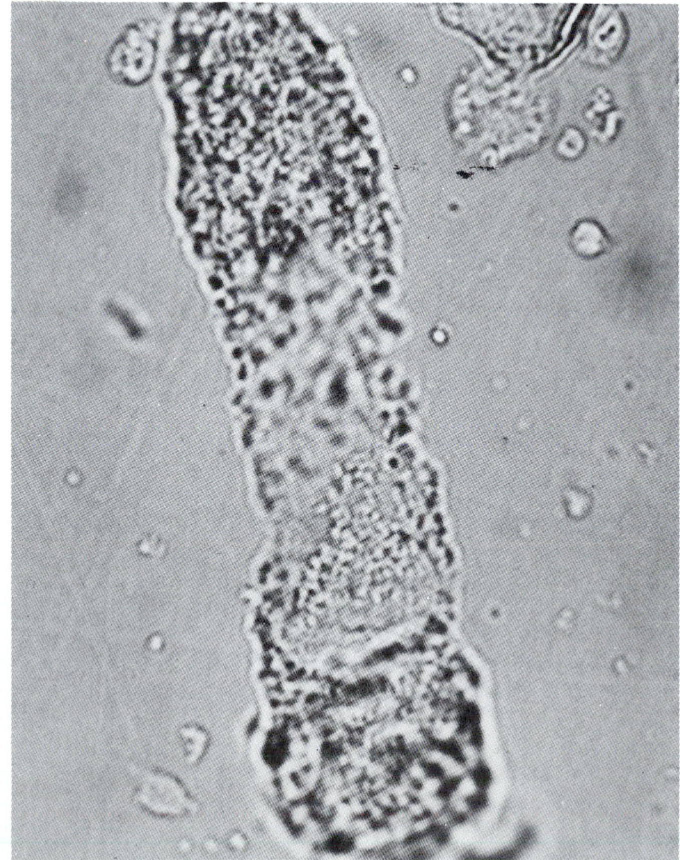

Fig. 56-17 Granular cast. (Bright field, 1000×.)

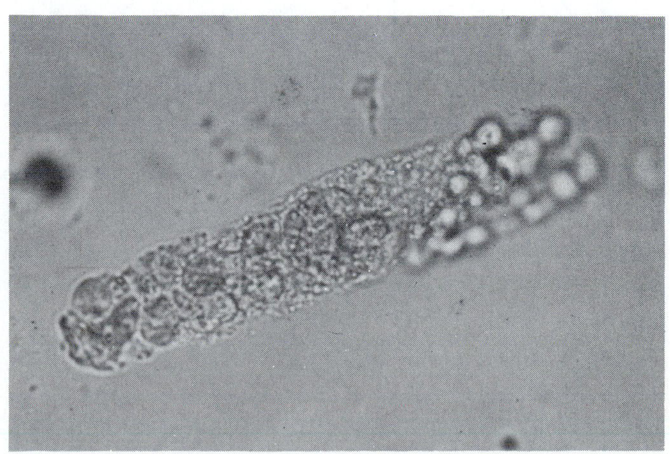

Fig. 56-18 Renal tubular cell cast. (Bright field, 400×.)

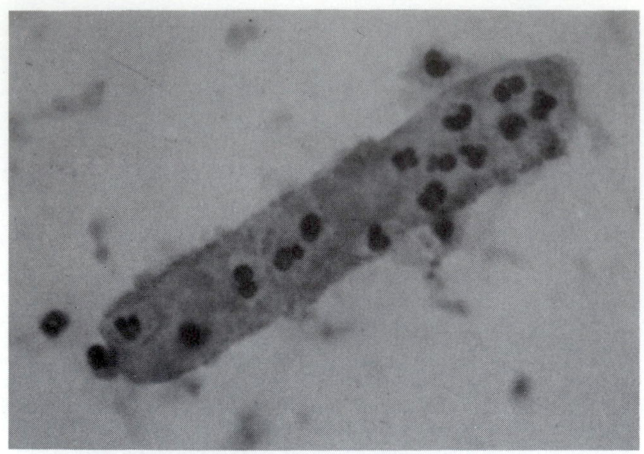

Fig. 56-19 White blood cell cast. (Bright field, 400×.)

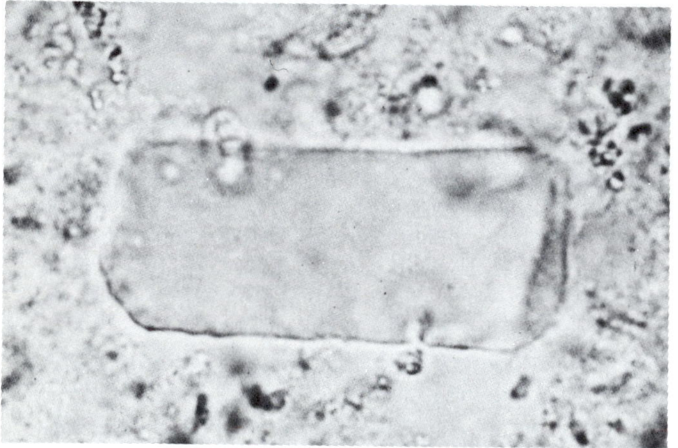

Fig. 56-20 Waxy cast. (Bright field, 1000×.)

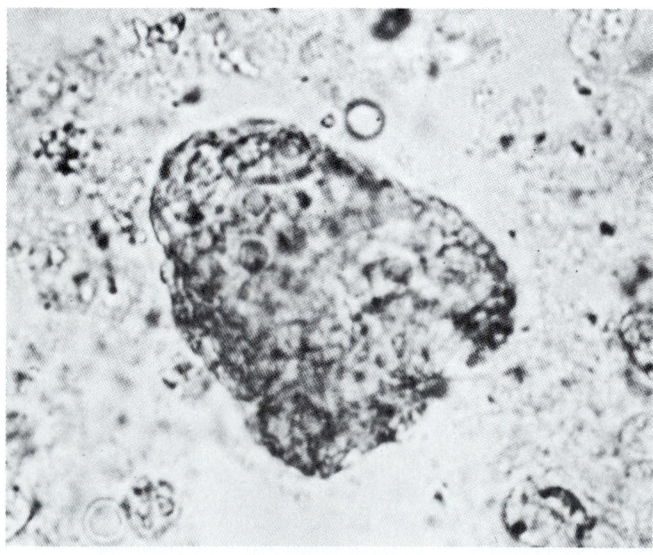

Fig. 56-21 Broad cast. (Bright field, 1000×.)

centrated. Casts may form in the proximal convoluted tubules in patients with Bence Jones proteins (multiple myeloma).

The appearance of these cylindrical formed elements in urine reflects the shapes (long versus short, straight versus convoluted) and diameter (thin versus broad) of the renal tubular lumens in which they originally were formed. Their number and measurable properties of size, form, and compositions are valuable clues to intrinsic renal parenchymal disease.

Microscopically casts are characterized by the appearance of the matrix (hyaline, granular, or waxy), the cellular constituents (erythrocytes, leukocytes, or renal tubular epithelial cells), or particulate matter embedded in the matrix (fine or coarse granules or fibrin).

Accurate identification of casts, especially the cellular types, is often difficult when unstained bright-field microscopy (wet preparation) is used. A skilled microscopist is essential to avoid misinterpretations. Phase contrast and interference contrast microscopy, as well as special stains, can be used to improve visualization. The cytocentrifugation-Papanicolaou technique is a superior method for the accurate identification of casts in urine specimens of patients with renal disease.

For diagnostic purposes, renal casts are classified as *physiological* or *pathological*. Table 56-12 lists the common types, properties, and clinical significance of casts.

Fats

Fats are found in the urine of patients who have fat emboli after bone-crush injuries, fatty degeneration of the kidney, or nephrotic syndrome. Fat will appear on top of urine and is the last part of voided urine. Vacuolated epithelial cells may be found in the urinary sediment. Oil-Red-O or Sudan III fat stains should be used for accurate identification of fat droplets in urine.

URINE CYTOLOGY

Urinalysis laboratory workers should be able to recognize abnormal mononuclear cells suggestive of malignancy. These should be referred for urine cytology. Holmquist[23] has advocated the use of supravital stains for malignant cell detection and suggests a role for the urinalysis laboratory in cancer screening.

Table 56-12 Common renal (urinary) casts

Types	Characteristic morphological appearance	Significance	Associated diseases
Physiological			
Hyaline	Transparent cylinder	Exercise, dehydration, and fevers	Nonspecific
Granular (fine)	Semitransparent cylinder containing fine refractile granules	Exercise, dehydration, and fever; accumulation of plasma proteins	Nonspecific
Pathological			
Cellular			
Erythrocytic	Semitransparent or granular cylinder containing distinct erythrocyte stroma	Renal parenchymal bleeding, glomerular leakage	Glomerular disease, interstitial hemorrhage (infarction)
Blood	Red-brown granular cylinder, but intact erythrocyte stroma *not* seen	Same as above	Same as above
Leukocytic	Transparent granular or waxy cylinder containing segmented neutrophils	Renal inflammation	Tubulointerstitial inflammation (pyelonephritis), glomerular disease
Renal tubular epithelial	Semitransparent granular or waxy cylinder containing intact or necrotic renal tubular epithelial cells	Renal tubular damage	Renal tubular injury, acute tubular necrosis, acute allograft rejections, tubulointerstitial disease
Bacterial	Semitransparent or granular cylinder containing bacteria	Renal infection	Acute pyelonephritis
Fungal	Semitransparent or granular cylinder containing fungi	Renal infection, probably sepsis	Acute pyelonephritis, papillary necrosis
Noncellular			
Granular (coarse)	Semitransparent cylinder containing coarse refractile granules	Cellular degeneration, accumulation of plasma proteins	Nonspecific
Waxy	Sharply defined, highly refractile, homogenous cylinder with broken-off borders and indentations	Cellular degeneration	Nonspecific
Fibrin	Transparent or granular cylinder containing thin long fibrils	Leakage of coagulation products	Glomerular disease, thrombosis
Fatty	Semitransparent or granular cylinder containing large highly refractile vacuoles or droplets	Lipiduria	Nephrotic syndrome
Bile	Deep yellow, transparent, granular waxy cylinder	Leakage of bile salts	Liver dysfunction, tubulointerstitial disease
Crystal	Crystalline inclusion in a semitransparent or granular cylinder	Cellular degeneration and malabsorption or excretion	Nonspecific
Other			
Broad	Width of cylinder two to six times that of other casts; waxy and granular most common types	Tubular dilatation and stasis	Advanced renal disease

CALCULI (LITHIASIS)

Urinary calculi are precipitates, concretions, or crystalloids embedded in a binding substance of mucus and protein. Bacteria and epithelial cells also may be included in the calculi.[6] Although the cause of calculus formation remains controversial, the detection and identification of calculi are important for diagnosis of urinary system conditions. Calculi formation in the kidney or lower urinary tract can be the source of gross hematuria and can cause serious anatomical damage and excruciating pain for the patient. Knowledge of the specific composition of a passed or surgically removed calculus may aid the physician in effectively treating lithiasis and preventing future stone formation.[6]

Calculus (stone) analysis for chemical constituents is a complex laboratory procedure. Urinary system calculi are usually composed of calcium oxalate, calcium oxalate mixed with calcium phosphate, ammonium magnesium phosphate, uric acid, or cystine. Most laboratories refer specimens for calculi analysis to more specialized laboratories.

URINE FINDINGS IN COMMON RENAL AND LOWER URINARY TRACT DISEASES

Numerous primary and secondary conditions and diseases occur in the kidney and lower urinary tract. An accurate diagnosis of urinary system disease requires correlation of abnormal urine findings with the patient history, physical examination, symptoms, signs, renal function tests, and other laboratory data. To be classified as abnormal, urine sediment must meet at least one of the following criteria:

1. More than five erythrocytes or leukocytes per high-power field (400×)
2. More than two renal tubular cells per high-power field (400×)
3. More than three hyaline casts, more than one granular cast, or the presence of any pathological cast per low-power field (100×)

Table 56-13 Urinalysis abnormalities found in common urinary system diseases

Conditions	Diagnostic physiochemical findings (macroscopic urinalysis)	Diagnostic urine sediment findings (microscopic urinalysis)
Renal		
Acute glomerulonephritis	Decreased urine volume, increased turbidity (smoky), proteinuria (<2 g/24 hr), hematuria (often gross)	Erythrocytic and blood casts, erythrocytes (often dysmorphic), neutrophils, mixed cellular casts, renal epithelial cells, occasional leukocytic or renal tubuloepithelial casts
Nephrotic syndrome	Lipiduria, significant proteinuria (>4.5 g/24 hr)	Fatty and waxy casts, doubly refractile oval fat bodies (Maltese cross with polarized light), lipid-laden renal tubuloepithelial cells
Chronic glomerulonephritis	Occasional lipiduria, decreased and fixed specific gravity, proteinuria (>2 g/24 hr), hematuria	Pathological casts, especially broad types
Acute tubular necrosis	Decreased urine volume, decreased specific gravity, minimum proteinuria, hematuria	Intact and necrotic renal epithelial cells, renal epithelial fragments, pathological casts
Acute pyelonephritis (tubulointerstitial inflammation)	Occasional odor, increased turbidity, minimum proteinuria, positive nitrite reaction	Leukocyte casts, neutrophils, especially in clumps; bacterial, granular, and waxy casts; renal tubuloepithelial cells
Diabetes mellitus	Proteinuria, glycosuria, ketonuria	Fatty and waxy casts, oval fat bodies, renal epithelial cells, leukocytes
Systemic lupus erythematosus	Proteinuria	Pathological casts, renal epithelial cells, neutrophils, erythrocytes
Cystinosis	Minimum proteinuria or hematuria	Cystine crystals
Acute allograft rejection	Decreased urine volume, minimum proteinuria, hematuria	Renal epithelial cells, renal epithelial casts, lymphocytes, pathological casts, especially renal epithelial casts
Viral nephropathy (cytomegalic inclusion disease)	Minimum proteinuria, hematuria	Mononuclear cells (occasional giant cell forms) with prominent intranuclear or cytoplasmic inclusions
Lower urinary tract		
Bacterial urinary tract infection	Occasional odor, increased turbidity, positive nitrite reaction, occasional hematuria	Bacteria, neutrophils, reactive transitional epithelial cells, absence of cast formation
Fungal urinary tract infection	Increased turbidity, occasional hematuria	Fungi, neutrophils and lymphocytes, reactive transitional epithelial cells
Viral urinary tract infection	Occasional hematuria	Viral inclusion bodies, neutrophils, transitional cells
Eosinophilic cystitis	Hematuria	Eosinophils (numerous), reactive transitional epithelial cells, absence of cast formation
Transitional cell carcinoma	Hematuria	Increased numbers of malignant transitional epithelial cells with high nuclear-to-cytoplasmic ratio, hyperchromasia, and chromatin clumping; cells occur singly and as tissue fragments

Modified from Schumann, GB: Urine sediment examination, Baltimore, 1980, The Williams & Wilkins Co.

4. More than 10 bacteria per high-power field (400×)
5. Presence of fungus, parasites, or viral inclusion cells
6. Presence of pathological crystals (such as cystine) or a large number of nonpathological crystals (such as uric acid)

A summary of urinalysis abnormalities found in common renal and lower urinary tract disease is shown in Table 56-13. Diagnostic findings of both the macroscopic and microscopic examination are also listed for quick review.

COORDINATED APPROACH TO URINALYSIS

Urinalysis continues to be one of the most commonly requested and demanding clinical laboratory procedures.[35] Laboratories involved in the examination of urine must define new responsibilities for both the rapid, routine assessment of urine and the more time-consuming, specialized interpretive tests.

Current, rapid dipstick technology and standardization of bright-field microscopy provide a quality program for the analysis of urine specimens from asymptomatic individuals. If symptomatic patients are to receive a more comprehensive urine examination, more definitive evaluation and confirmation procedures are required.

A coordinated approach to the examination of urine is depicted in Fig. 56-22. Proper use of this approach requires that the clinician and urine technologist understand the roles of both the routine laboratory, which offers basic urinalysis, and the specialized laboratory, which offers a more comprehensive sediment examination. After the collection of urine and macroscopic analysis, the physician or laboratory must differentiate between results from symptomatic and asymptomatic patients. Emphasis must be placed on the urinalysis technologist's ability to recognize the results that require additional follow-up testing. This coordinated approach represents a flexible system in which technical responsibilities can be shared, and additional laboratory confirmation procedures can be integrated into the system (Fig. 56-22). Communication among physicians,

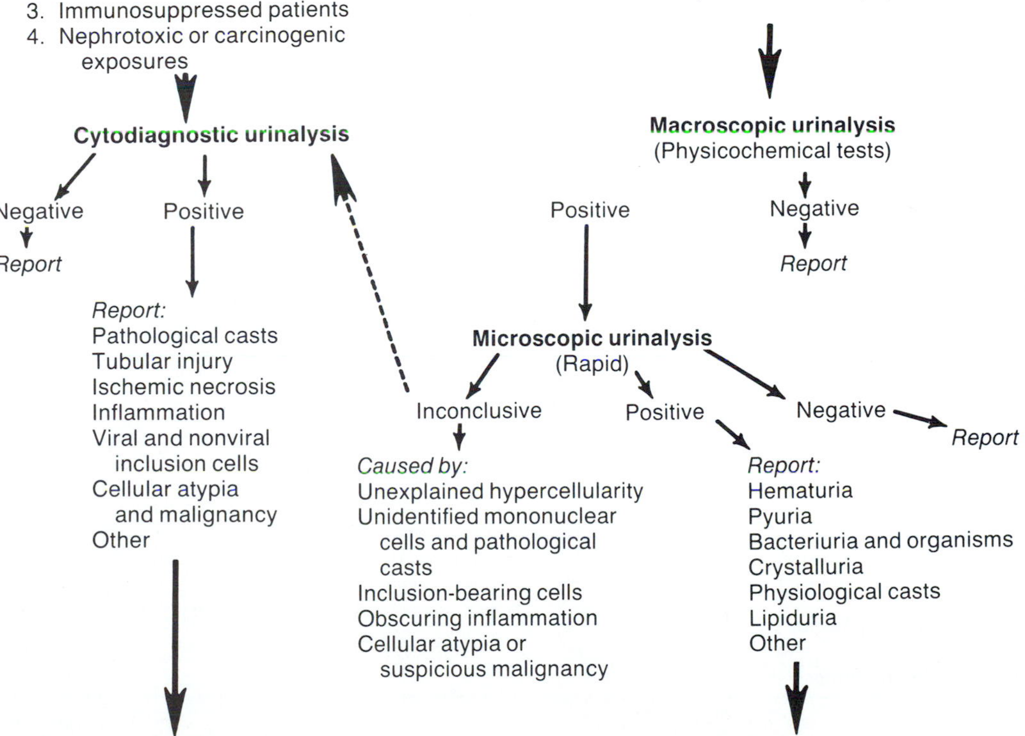

Fig. 56-22 Coordinated approach to urine sediment examination. *(From Schumann, GB, Schumann, JL, and Schweitzer, S: Lab Management 1:47, 1983.)*

the laboratory technologist, and the medical director is essential to resolve inconclusive results and reestablish credibility of the urine examination.

QUALITY CONTROL

An effective quality control program is essential to ensure accuracy in urinalysis. A program of quality assurance that covers all aspects of urinalysis and is similar to that used in other areas of the clinical laboratory must be implemented to achieve more reliable urinalysis results.[38]

Commercial quality control preparations are available for the assessment of specific gravity and reagent-strip testing. These preparations may be in tablet, strip, liquid, or lyophilized form. Hoeltge and Ersts[31] describe a 3-year experience using a synthetic-urine control prepared in their laboratory. Currently there is no ideal commercial preparation for urine sediment elements. Quality control of sediment evaluation should focus on standardized techniques and policies.

A suggested urinalysis quality control schedule is found in Table 56-14. It is important that reagents are dated when they are received by the laboratory and used before expiration. Urine control solutions should retain their utility if they are stored in a tightly stoppered container, refrigerated, and protected from light.[8] The new lot number should always be recorded. A laboratory manual containing operating instructions and documentation of equipment maintenance should be maintained and reviewed yearly. The importance of continuing education of technologists and the use of current references cannot be overemphasized.

Table 56-14 Suggested urinalysis quality control schedule

	Time				
Areas checked	**Daily**	**Weekly**	**Monthly**	**Semiannually or annually**	**As needed**
Reagents and supplies					
Reagent strip	×				
Reagent tablets	×				
Remaking protein standard				×	
Equipment					
Refrigerator temperature	×				
Freezer temperature	×				
Refractometer calibration		×			
Urinometer calibration		×			
Spectrophotometer calibration		×			
Microscope maintenance				×	
Thermometers		×			
Glassware			×		
Centrifuge maintenance				×	
Education					
Revise laboratory manual				×	
Technologist's proficiency testing			×		
Update library				×	
Clinicopathological correlations					×

REFERENCES

1. Free, AH, and Free, MA: Rapid convenience urine tests: their use and misuse, Lab Med 9:9-17, 1978.
2. Schumann, CB: Urine sediment examination, Baltimore, 1980, The Williams & Wilkins Co.
3. Schumann, GB, Schumann, JL, and Schweitzer, S: Coordinated approach to the urine sediment examination, Lab Management 1:45-48, 1983.
4. Kunin, CM: Detection, prevention and management of urinary tract infections, ed 2, Philadelphia, 1974, Lea & Febiger.
5. Bradley, M, Schumann, GB, and Ward, PCJ: Examination of urine. In Henry, JB, editor: Todd-Stanford clinical diagnosis by laboratory methods, ed 16, Philadelphia, 1979, WB Saunders Co.
6. Bauer, JD: Clinical laboratory methods, ed 9, St Louis, 1982, The CV Mosby Co.
7. Ross, DL and Neely, AE: Textbook of urinalysis and body fluids, East Norwalk, Conn, 1983, Appleton-Century-Crofts.
8. Free, AH, and Free, MA: Urinalysis in clinical laboratory practice, Boca Raton, Fla, 1975, CRC Press, Inc.
9. Free, MH: Modern urine chemistry, Elkhart, IN, 1987, Miles Laboratories, Inc.
10. Zweiss, MH and Jackson, A: Ascorbic acid interference in reagent-strip reactions for assay of urinary glucose and hemoglobin, Clin Chem 32:674-677, 1986.
11. Monte-Verde, D, and Nosanchuk, JS: The sensitivity and specificity of nitrite testing for bacteriuria, Lab Med 12:755-757, 1981.
12. Kusumi, RK, Grover, PJ, and Kunin, CM: Rapid detection of pyuria by leukocyte esterase activity, JAMA 245:1653-1655, 1981.
13. Gillenwater, NY: Detection of urinary leukocytes by Chemstrip-L, J Urol 125:383-384, 1981.
14. Avent, J, Schumann, GB, and Vars, L: Comparison of the Chemstrip leukocyte test with a standardized Papanicolaou-stained urine sediment evaluation, Lab Med 14:163-166, 1983.
15. Race, GJ, and White, MG: Basic urinalysis, Hagerstown, MD, 1979, Harper & Row, Publishers, Inc.
16. Schumann, GB, and Greenberg, NF: Usefulness of microscopic urinalysis as a screening procedure: a preliminary report, Am J Clin Pathol 71:452-456, 1979.
17. Ferris, JA: Comparison and standardization of the urine microscopic examination, Lab Med 14:659-662, 1983.
18. Schumann, GB, and Tebbs, RD: Comparison of slides used for standardized routine microscopic urinalysis, J Med Technol 3:54-58, 1986.
19. Schumann, GB, and Henry, JB: An improved technique for the evaluation of urine sediment, Lab Management 1:19-24, 1977.
20. Schumann, GB, and Weiss, MA: Atlas of renal and urinary tract cytology and its histopathologic bases, Philadelphia, 1981, JB Lippincott Co.
21. Winkel, P, Statland, B, and Jorgenson, K: Urine microscopy: an ill-defined method examined by a multifactorial technique, Clin Chem 20:436-439, 1974.
22. Sternheimer, R: A supravital cytodiagnostic stain for urinary sediment, JAMA 231:826-832, 1975.
23. Holmquist, N: Detection of cancer with urinary sediment, J Urol 123:188-189, 1980.
24. Brody, LH, Webster, MC, and Kark, RM: Identification of elements of urinary sediment with phase contrast, JAMA 206:1777-1781, 1969.
25. Haber, MH: Interference contrast microscopy for identification of urinary sediment, Am J Clin Pathol 57:316-319, 1972.
26. Haber, MH: Urinary sediment: a textbook atlas, Chicago, 1981, American Society of Clinical Pathologists.
27. Kesson, AM, Talbott, JM, and Gyory, AZ: Microscopic examination of urine, Lancet 2:809-812, 1978.
28. Roe, CE, Carlson, DA, Daigneault, RW, and Statland, BE: Evaluation of the Yellow IRIS®: an automated method for urinalysis, Am J Clin Pathol 86:661-665, 1986.
29. Elin, RJ, Hosseini, JM, Kestner, J, et al: Comparison of automated and manual methods for urinalysis, Am J Clin Pathol 86:731-737, 1986.
30. Schumann, GB, Burleson, RL, Henry, JB, et al: Urinary cytodiagnosis of acute renal allograft rejection using the cytocentrifuge, Am J Clin Pathol 67:134-140, 1977.
31. Thal, SM, DeBellis, CC, Iverson, SA, and Schumann, GB: Comparison of dysmorphic erythrocytes with other urinary sediment parameters of renal bleeding, Am J Clin Pathol 86:784-787, 1986.
32. Little, PJ: A comparison of the urinary white cell concentration with the white cell excretion rate, Br J Urol 36:360-363, 1964.
33. Voogt, HJ, Rathert, P, and Beyer-Boon, ME: Urinary cytology, New York, 1977, Springer-Verlag, Inc.
34. Mandel, N: Urinary tract calculi, Lab Med 17:449-458, 1986.
35. Schweitzer, SC, Schumann, JL, and Schumann, GB: Quality assurance guidelines for the urinalysis laboratory, J Med Technol 3:567-572, 1986.
36. Schweitzer, SC, Schumann, JL, and Schumann, GB: A model for educating future urine technologists, J Med Technol 2:251-255, 1985.

CHAPTER 57

Carbohydrates and metabolites

Glucose

LAWRENCE A. KAPLAN

Glucose, dextrose
Clinical significance: p. 436
Molecular formula: $C_6H_{12}O_6$
Molecular weight: 180.16 daltons
Merck Index: 4319
Chemical class: carbohydrate

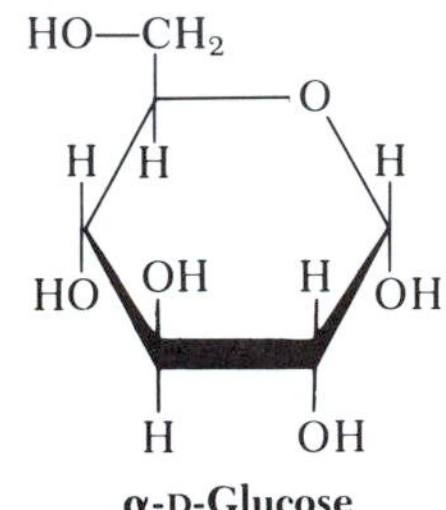

α-D-Glucose

PRINCIPLES OF ANALYSIS

Chemically, glucose is an aldohexose, with the aldohyde form in equilibrium with the glucopyranose form (which is shown above). The latter is the favored structure at physiological pH. The aldehyde/enediol equilibrium allows glucose to be reduced and oxidized easily.

Most older established methods for measurement of serum glucose were based on the ability of glucose to directly reduce cupric ions (Cu^{+2}) to monovalent cuprous ions (Cu^{+}). In the presence of heat, the reduced cuprous ions can form cuprous oxide (Cu_2O), which can be detected by a variety of methods. The most popular method has been the reduction of phosphomolybdate (Folin-Wu) or arsenomolybdate (Somogyi-Nelson) to form blue molybdenum compounds (methods 1 to 3, Table 57-1). Neocuproine (2,9-dimethyl-1,10-phenanthroline) can also be reduced by Cu^{+} ions to form a highly colored complex (method 4, Table 57-1). Although it is a more sensitive procedure than the other copper-reduction methods, this method also lacks specificity. However, in the 1960s and 1970s many millions of analyses were done using Neocuproine on Technicon Analyzers. Few laboratories use these procedures today, and they are listed in Table 57-1 only for historical interest.

Benedict's modification of the copper-reduction methods (method 3, Table 57-1) is still used but only as a semiquantitative method for estimation of urine glucose. This is marketed by the Ames Division of Miles Laboratories (Elkhart, IN) as Clinitest. This procedure, sensitive to total reducing compounds present in urine, yields red Cu_2O and yellow CuOH precipitates. The greater the concentration of glucose, the redder the final color. In combination with a more specific enzymatic glucose assay (see below), the Benedict reaction can be used to screen for genetic diseases of carbohydrate metabolism in newborns. A negative enzyme test and a positive Benedict reaction (Clinitest) are suggestive of such diseases.

The alkaline ferricyanide reaction (method 5, Table 57-1) involves the reduction of yellow ferricyanide, $Fe(CN)_6^{-3}$, to colorless ferrocyanide, $Fe(CN)_6^{-4}$, by glucose in alkaline conditions. When this method was automated on the Technicon AutoAnalyzer, the reaction was monitored by measurement of the decrease in yellow color. Later adaptations of this method in the 1960s and 1970s to AutoAnalyzer equipment utilized indirect measurement of ferrocyanide. The ferrocyanide was reacted with excess ferric ions to form ferric ferrocyanide (Prussian blue). The alkaline ferrocyanide reactions have largely been replaced by more specific methods.

The *o*-toluidine reaction is based on the ability of many aromatic amines in acid solutions to condense with the aldehyde group of glucose to form glycosamines. The initial reaction product is most likely an unstable *N*-glycoside that is in equilibrium with the stable Schiff base (method 6, Table 57-1). The most widely used aromatic amine, *o*-toluidine, is believed to be a carcinogen. The method is not widely used today, and, when it is, it is used as a manual procedure.

The most commonly used procedures for glucose analysis employ enzymes as reagents to increase analytical specificity. These are the glucose oxidase and hexokinase reactions. Both procedures have been automated with resulting high specificity and precision.

The hexokinase method (method 7, Table 57-1) involves two coupled reactions. The hexokinase reaction phosphorylates hexoses with ATP. Glucose-6-phosphate dehydrogenase (G6PD) reacts specifically with glucose-6-phosphate produced by the hexokinase reaction to yield 1 mole of NADH or NADPH for each mole of glucose that is oxidized. The earlier hexokinase procedures employed mammalian G6PD. Most present assays, however, use an enzyme derived from bacteria that employs NAD^{+} as a substrate rather than the $NADP^{+}$ used by mammalian enzymes. The advantage of using the bacterial enzyme is that red blood cell G6PD and 6-phosphogluconate dehydrogenase, which use NADP as a substrate, do not interere in

Table 57-1 Methods of glucose analysis*

Methods	Type of analysis	Principle	Usage	Comments
Copper reduction				
1. Phosphomolybdate (Folin-Wu) 2. Arsenomolybdate (Somogyi-Nelson)	Quantitative, EP	Cu^{+2} + Glucose $\xrightarrow[OH^-]{Heat}$ Cu_2O *(red)* Cu^+ + Molybdate (Mo^{+2}) → *blue* molybdenum complexes	Reactions 1 and 2 rarely used now; of historical interest	Large positive bias because of chemical interference by other sugars, creatinine, ascorbic acid, and other compounds
3. Benedict's	Qualitative, semiquantitative	Cu^{+2} + Glucose $\xrightarrow[OH]{Heat}$ $Cu_2O\downarrow$ *(red)* + $CuOH\downarrow$ *(yellow)*	Basis of semiquantitative tests for total reducing sugars in *urine*	Used in combination with more specific glucose oxidase/peroxidase urine screen to differentiate glucosuria from other sugars in urine, especially in neonates
4. Neocuproine	Quantitative, EP	Cu^{+2} + Glucose $\xrightarrow[OH^-]{Heat}$ Cu^+ Cu^+ + Neocuproine (2,9-dimethyl-1,10-phenanthroline) → *colored complex*	Historical interest	As per methods 1 and 2
Other reduction				
5. Alkaline ferricyanide	Quantitative, EP	$Fe(CN)_6^{-3}$ $\xrightarrow[Glucose]{Heat,\ OH^-}$ $Fe(CN)_6^{-4}$ (ferricyanide) (ferrocyanide) *(yellow)* *(colorless)* Decreased absorbance at 420 nm because of consumption of ferricyanide	Rarely used, sometimes seen in Technicon systems; of historical interest	1 mg of creatinine = 1 mg of glucose; 0.5 mg of uric acid = 1 mg of glucose; very poor specificity
6. *o*-Toluidine	Quantitative, EP	Increased absorbance at 630 nm **o-Toluidine** (CH_3–C_6H_4–NH_2) + **β-D-Glucose** (HCO–HCOH–HOCH–HCOH–HCOH–CH_2OH) → **Glycosamine** *(colored)* (CH_3–C_6H_4–NH–CH_2–C=O–HOCH–HCOH–HCOH–CH_2OH)	Serum or urine; rarely used in automated analysis	*o*-Toluidine is a suspected carcinogen; other sugars, especially mannose and galactose, give positive interferences; turbidity can cause a positive bias
Enzymatic				
7. Hexokinase (HK)	Quantitative, spectrophotometric EP	Glucose + ATP $\overset{HK}{\rightleftharpoons}$ Glucose-6-phosphate + ADP Glucose-6-phosphate + $NADP^+$ $\overset{G6PD}{\rightleftharpoons}$ 6-Phosphogluconate + NADPH + H^+ Increased absorbance at 340 nm related to glucose concentration	Serum, CSF, urine; automated; most commonly used method	Has been proposed as basis of reference method; very good accuracy and precision

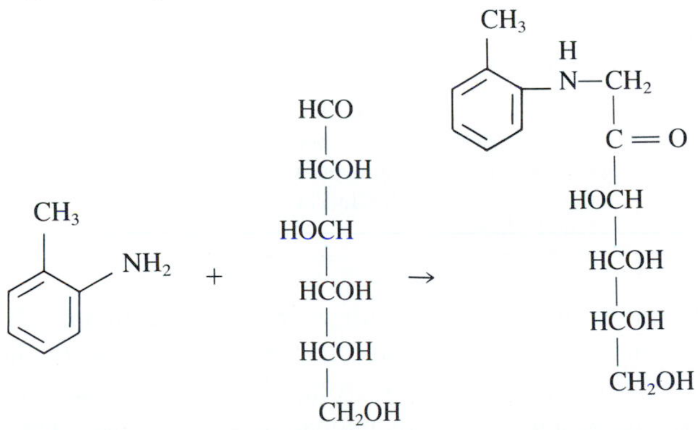

**EP*, End-point analysis mode; *CSF*, cerebrospinal fluid; *GDH*, glucose dehydrogenase; *K*, kinetic analysis mode.

Continued

Table 57-1 Methods of glucose analysis—cont'd

Methods	Type of analysis	Principle	Usage	Comments
8. Glucose oxidase coupled reaction ("Trinder")	Quantitative, using various types of dyes as final O_2 acceptor; K or EP	Glucose + O_2 $\xrightarrow{\text{Glucose oxidase}}$ Gluconic acid + H_2O_2 H_2O_2 + Reduced dye $\underset{\text{peroxidase}}{\overset{\text{Horseradish}}{\rightleftharpoons}}$ Oxidized dye + H_2O *(colored)* **Peroxidase indicator reaction** Increased absorbance related to glucose concentation	Serum, urine, CSF; easily and usually adapted to automated analysis	Second indicator reaction susceptible to false-positive interferences from a variety of compounds; good accuracy and precision
	Quantitative or semiquantitative in dipstick screen, visual or reflectance photometry		Used in all dipstick screens; serum, urine	
9. Glucose oxidase (GO) oxygen consumption	Quantitative, polarographic measurement using O_2 electrode; K	Glucose + O_2 $\xrightarrow{\text{Glucose oxidase}}$ Gluconic acid + H_2O_2 H_2O_2 consumed in side reactions O_2 consumption measured polarographically by oxygen electrode	Serum, CSF; semiautomated and fully automated systems	Correlates best with proposed reference method; very good accuracy and precision
10. Radiation energy attenuation	Quantitative, EP	Reaction described in method 8 is used; chromogen absorbs light used to excite fluor or absorbs fluorescent light itself; glucose concentration related to attenuation of fluorescence	Rare, serum	Used only on TDx automated assay
11. Glucose dehydrogenase	Quantitative, EP, K	Glucose + NAD^+ $\xrightarrow{\text{GDH}}$ D-Gluconolactone + NADH + H^+ Increased absorbance at 340 nm related to glucose concentration	Rare, serum, CSF	Can be adapted to automated instruments
Other				
12. Isotope dilution–mass fragmentography	Quantitative	Fragmentation of glucose into specific, detectable fragments; ratio of specific fragments for unlabeled and deuterated labeled glucose	Rare	Proposed definitive method

the analysis of glucose. This reduces the interference in the assay resulting from hemolysis. Although other hexoses can enter into the hexokinase reaction, normal serum concentrations of these sugars do not cause significant interference. The hexokinase assay is not commonly performed as a kinetic procedure because of the requirement for a rapid initial reading. However, it is often performed using bichromatic measurements.

One of the most frequently used specific glucose methods uses two coupled enzyme reactions (method 8, Table 57-1). In this case, the initial reaction is the specific one, and the indicator reaction is nonspecific. The first reaction employs glucose oxidase to oxidize glucose to glucuronic acid and hydrogen peroxide. Since glucose oxidase is highly specific for beta-D-glucose, most preparations of glucose oxidase contain the enzyme mutarotase to catalyze the conversion of alpha-D-glucose to the beta form. The hydrogen peroxide from the glucose oxidase reaction is consumed by a peroxidase-dye indicator reaction in which the oxidized dye is colored, allowing the reaction to be monitored photometrically. This reaction is catalyzed by horseradish peroxidase (HPO). The dyes employed as final oxygen acceptors can vary. The most widely used compound is 4-aminophenazone, although 3-methyl-2-benzothiazolinone hydrazone/*N,N*-dimethylaniline (MBTA/DMA) has been used as well. The glucose oxidase–peroxidase reaction with 4-aminophenazone as the final oxygen acceptor is usually monitored between 500 and 525 nm. The coupled glucose oxidase procedure has been adapted to a wide range of automated instruments, including those that employ dry chemistry reagents, either in a strip or film form. The Ames Division of Miles Laboratories (Elkhart, IN) employs the oxidation of iodine to monitor the peroxidase reaction of their urine dipstick:

$$\text{Glucose} + O_2 \xrightarrow{GD} \text{Glucuronic acid} + H_2O_2$$

$$H_2O_2 + 2\text{ KI} \xrightarrow{HPO} I_2\ (brown) + 2\text{ KOH}$$

Other manufacturers employ various compounds, such as 3,3′,5,5′-tetramethylbenzidine and 1,7-dihydroxynaphthalene, whose oxidized forms are colored. The reaction adapted to dipsticks has been widely used to screen for glucose in urine or serum to help monitor hyperglycemia in diabetics. These strips can be read visually or by reflectance photometry for a semiquantitative analysis of glucose.

The coupled glucose oxidase procedure has also been modified and adapted for use on the TDx (Abbott Laboratories, North Chicago, IL) using the technique termed *radiation energy attenuation* (method 10, Table 57-1). The peroxidase reaction catalyzes the reaction between H_2O_2, 3,5-dichloro-2-hydroxybenzene sulfonic acid, and 4-aminoantipyrene to form a chromogenic product. The reaction is not monitored by measurement of the color formed but by monitoring of the ability of the chromophore to attenuate the fluorescence of fluoroscein, which is also present in the reaction mixture. See Chapter 3 for a more detailed description.

A glucose method based solely on the highly specific enzymatic reaction (method 11, Table 57-1) of glucose dehydrogenase is popular in Europe[1]:

$$\text{Beta-D-glucose} + NAD^+ \overset{GDH}{\rightleftharpoons} \text{D-Gluconolactone} + NADH + H^+$$

The reaction is monitored at 340 nm, as either a kinetic or an end-point reaction. This method has been adapted for use on several automated instruments.[2]

Glucose can also be measured by mass fragmentography (method 12, Table 57-1). The sample is mixed with glucose labeled with deuterium and is subjected to an ion beam to break the glucose into specific fragments.[3] The ratio of the fragments from the unlabeled glucose to the fragments from the deuterated glucose can be used to measure the concentration of glucose in the sample. Automated and semiautomated procedures have also taken advantage of the unique specificity of the glucose oxidase reaction (method 9, Table 57-1) as a basis of highly specific and precise methods. These procedures follow only the specific glucose oxidase reaction by use of an oxygen electrode to monitor the consumption of oxygen kinetically. Hydrogen peroxide produced by the glucose oxidase reaction is removed by reaction either with iodides or with ethanol in the presence of catalase. Monitoring only the glucose oxidase reaction instead of the less specific peroxidase step greatly reduces the number of potential interferences. The method has been automated on the Astra system (Beckman Instruments, SmithKline-Beckman, Brea, CA) and is available with semiautomated devices, including the YSI analyzer (Yellow Springs Instrument, Yellow Springs, OH), which uses whole blood as the sample.

The use of semiquantitative dipsticks for assessment of blood glucose control in the management of diabetes is becoming increasingly widespread. Not only are they used in emergency units and at the hospital bedside, but also they are used by patients for home monitoring of blood or urine glucose levels. Thus these dipsticks probably represent one of the most frequently used forms of glucose analysis.

The hexokinase procedure, using a protein-free filtrate as the sample, has been proposed as the product class standard for the measurement of glucose.[4] All the copper reduction reactions including neocuproine have been shown to be nonspecific. Compounds such as creatinine and uric acid interfere even if a protein-free filtrate is used as the sample. The use of the Benedict reaction for measuring urinary glucose is also restricted by its poor specificity. Many reducing compounds in urine can produce a positive reaction.[5] The alkaline ferricyanide method is also fairly nonspecific because of significant positive interference by such compounds as creatinine and uric acid. Although the *o*-toluidine procedure has acceptable accuracy and precision, it has the disadvantages of the noxious nature of its reagents and the nonspecific reaction with urea and other hexoses, most notably mannose and galactose. It offers only an economic advantage over enzymatic methods, and its use has rapidly decreased over the past few years.

In one study, the authors concluded that the automated oxygen rate–glucose oxidase method compared best with the proposed reference method in terms of precision, accuracy, freedom from interferences, and sample carryover.[6]

In 1976, Björkhem et al.[3] suggested the use of a mass-fragmentation procedure as a reference method for serum glucose measurements. In their study, they compared their method with two hexokinase and two glucose oxidase assays. These assays were performed by manual analysis except for the oxygen rate–glucose oxidase procedure. Although all the reviewed methods had nonsignificant error from a clinical view, the oxygen rate–glucose oxidase method compared best with the mass spectrometer method from an analytical view (accuracy and precision).

Thus the best methods available for the routine automated measurement of glucose in body fluids appear to be the enzyme methods used in the kinetic analysis mode. A comparison of these methods is found in Table 57-2. The disadvantage of the coupled glucose oxidase assays is that many compounds present in serum or urine (such as bilirubin, ascorbic acid, and uric acid) can be oxidized by the hydrogen peroxide produced by the glucose oxidase reaction, resulting in a negative bias. The negative bias resulting from the reaction between the hydrogen peroxide and ascorbic acid[7] can often lead to strikingly low values in cerebrospinal fluid. This is because glucose levels are rel-

Table 57-2 Comparison of reaction conditions for glucose analysis

	Hexokinase (proposed reference)	Glucose oxidase–hydrogen peroxidase (proposed reference)	Glucose oxidase oxygen rate consumption*	Glucose oxidase–hydrogen peroxidase (modified Trinder)†
Temperature	25° C	37°C	37° C	37° C
pH	7.5	7.0	7.0	7.0
Final concentration of reagent components	Hexokinase (yeast): 83 U/L Glucose-6-phosphate dehydrogenase: 83 U/L *(Leuconostoc mesenteroides)* NADP: 1.0 mmol/L ATP: 1.1 mmol/L Tris buffer: 80 mmol/L Magnesium acetate: 3.3 mmol/L	Glucose oxidase *(Aspergillus niger)*: 2600 U/L Peroxidase (horseradish): 5000 U/L Chromogen (*o*-dianisidine): 0.324 mmol/L 0.036 mol/L phosphate buffer in 40% glycerol	Glucose oxidase *(A. niger)*: 140 U/L Methanol: 5% Potassium iodide: 10 mmol/L	Glucose oxidase *(A. niger)*: ≥12,000 U/L Peroxidase (horseradish): ≥ 1200 U/L Chromogen (4-aminophenazone): 1.48 mmol/L Phenol: 4.25 mmol/L Phosphate buffer: 0.01 mol/L
Fraction of sample volume	0.16	0.20	0.01	0.006
Sample	Protein-free supernatant according to Somogyi	Protein-free supernatant according to Somogyi	Serum, CSF, and urine	Serum, plasma, CSF, and urine
Linearity	6000 mg/L	10,000 mg/L	6000 mg/L	7500 mg/L
Time of reaction	End point at 10 minutes	End point at 30 minutes	Kinetic, 9.6 seconds	Kinetic, 16 seconds
Major interferences (concentration at which compound interfered at glucose concentration within normal range)	None	Ascorbic acid: 250 mg/L L-Cysteine: 1.5 g/L Citric acid: 15 g/L Uric acid: 150 mg/L L-Dopa: 100 mg/L	Hemoglobin: 10 g/L	Bilirubin: >100 mg/L Ascorbic acid: 250 mg/L L-Cysteine: 1.5 g/L Citric acid: 15 g/L
Precision $\bar{X}$‡ (% CV)	468 (5%-10-%)§ 921 (3%-7.9%) 3020 (2%-6.5%)	443 (6%-9.7%)‖ 911 (3.5%-6.1%) 4033 (2%-6.5%)	428 (4.4%-6.3%) 879 (3.4%) 2983 (2.8%)	396 (6.4%) 848 (3.6%) 2897 (1.7%)

*Beckman Instruments, Inc, Brea, CA.
†IL919, Instrumentation Laboratory, Lexington, MA.
‡Data from College of American Pathologists comprehensive chemistry surveys; $\bar{X}$, mean in mg/L; *CV*, coefficient of variation.
§Group mean for all hexokinase methods (without blank).
‖Group mean for all glucose oxidase–phenylaminophenazone methods.

atively low in cerebrospinal fluid, and ascorbate levels in cerebrospinal fluid are higher than in serum.[8] Since the extent of the interference is inversely related to the glucose concentration, cerebrospinal fluid samples from neonates are particularly vulnerable to this form of interference. The coupled glucose oxidase reactions performed as a kinetic analysis are especially sensitive to ascorbate interference.[7] Also, because ascorbate is very unstable and breaks down rapidly, the degree of interference will vary with time and temperature of storage before analysis. This interference can be avoided by analysis of all cerebrospinal fluid specimens by an alternative technique, one that does not use a glucose oxidase–hydrogen peroxidase generating method. These include the hexokinase, rate of oxygen consumption, and glucose dehydrogenase methods.[9]

In addition, there may be compounds that can react by oxidizing the indicator dye, resulting in positive biases. The small positive biases observed in these procedures indicate that the latter type of interferences predominate. The specificity and accuracy of the coupled glucose oxidase procedures have been increased by a choice of reaction conditions that minimize this bias. The most common approach is to employ kinetic analysis of the reaction. By selection of the most appropriate initial and subsequent times for spectrophotometric reading, a reduction in both direct photometric and chemical interferences can be made. Kinetic analysis–based methods will most likely increase in frequency for the analysis of glucose by this method. The reference hexokinase and glucose oxidase–hydrogen peroxidase methods minimize interference by protein and other serum constituents by means of a Somogyi protein precipitation step. Adaptations of these procedures for routine automated analysis employ large dilutions of the sample to minimize the effects of these interferences. Thus the fractions of sample volume noted for the last two methods listed in Table 57-2 are tenfold lower than for the reference methods. In addition, one can further minimize errors in accuracy resulting from the direct analysis of serum by performing the analysis as a kinetic reaction or by bichromatic analysis.

The semiquantitative dipsticks employing glucose oxidase and horseradish peroxidase are highly specific for glucose, with no other sugar producing a reaction. Strong oxidizing substances, such as hypochlorite and chlorine bleach, can produce a positive reaction, and ascorbic acid at high levels can interfere with the peroxidase step, reacting with the hydrogen peroxide to give erroneously low glucose levels (see previous discussion). The sensitivity of the urine dipstick ranges from 400 to 1000 mg/L, sufficient for clinical needs. The dipsticks measuring blood glucose do show varying degrees of accuracy when compared to automated glucose analyzers.[10-13] The most serious bias (a negative bias) is seen in the hypoglycemic range, that is, glucose levels less than a range of 500 to 700 mg/L.[10-13] Very often, nurses perform these analyses. Even with well-trained nurses, the degree of imprecision reported in one study was sufficiently large that hypoglycemia (less than 500 mg/L) in 12 of 72 infants was not detected when a reflectance meter with a dipstick was used.[13] Clinicians must be made aware of the potential for serious injury to patients, especially to neonates, because of the current limitations of this method. When dipsticks are used for blood glucose measurements, it is essential to train the users to calibrate and read the strips properly. The use of reflectance meters can minimize errors resulting from poor lighting or color bias, but only properly trained and motivated operators will minimize the errors resulting from variations in the amounts of blood used, in the time allowed for the reaction, in the washing and wiping of the strips, and in the calibration of reflectance readers.

SPECIMEN

Serum or plasma, free of hemolysis, is the specimen of choice. Other body fluids, such as cerebrospinal fluid and urine, can also be used. Common anticoagulants (oxalate, fluoride, ethylenediaminetetraacetic acid [EDTA], citrate, or heparin) do not cause interference. Since glucose in whole blood at room temperature can undergo glycolysis at a rate of appoximately 5% per hour, the sample should be centrifuged and removed from clot or cells as soon as possible. Glucose in serum or plasma separated from blood cells is stable for up to 3 days at 2° to 8° C. Fluoride and iodoacetate have been used as inhibitors of glycolysis to preserve blood that cannot be separated rapidly. Since iodoacetate-preserved blood yields a better serum specimen, this might be the preferred preservative.

Urine can be analyzed by enzymatic methods if a large dilution step and kinetic analysis are employed. Heparinized whole blood can be used on the YSI glucose analyzer.

REFERENCE RANGE

Reference ranges of serum glucose in healthy adults are listed next for several methods. There does not appear to be any significant difference in glucose levels between males and females or between races. There are some age-related differences. Below 5 years of age, normal glucose levels may be 10% to 15% below adult levels. Newborns can have blood glucose concentrations ranging from 200 to 800 mg/L (1.11 to 4.44 mmol/L), with blood glucose in premature infants even lower.[14] Glucose levels in cerebrospinal fluid are approximately 40% to 80% of serum or plasma levels. Reference ranges for individual methods can vary significantly and should be evaluated by each laboratory.

There is normally no glucose detectable in urine. Newborns have been reported to excrete sufficient galactose to give a positive result by the nonspecific copper-reduction methods.[15]

Method	Serum glucose, adult, mg/L (mmol/L)
1. Glucose oxidase, 4-aminophenazone, end point (Technicon)	700 to 1050 (3.89 to 5.83)
2. Glucose oxidase, 4-aminophenazone, kinetic reaction (IL919)	600 to 950 (3.33 to 5.27)
3. Oxygen rate–glucose oxidase (Beckman ASTRA)	650 to 1100 (3.61 to 6.11)
4. Hexokinase, end point (du Pont aca)	700 to 1100 (3.89 to 6.11)

REFERENCES

1. Banauch, D, Brummer, W, Ebling, W, et al: Z Klin Chem Klin Biochem 13:101-107, 1975.
2. Vormbrock, R: Clin Chem 29:1224(A), 1983.
3. Björkhem, I, Blomstrand, R, Falk, O, and Ohman, G: Clin Chem Acta 72:353-362, 1976.
4. Department of Health, Education and Welfare, Food and Drug Administration: Fed Reg 39(126):24136-24147, 1974.
5. Pileggi, VJ, and Szustkiewicz, C: Carbohydrates. In Henry, RJ, Cannon, DC, and Winkelman, JW, editors: Hagerstown, Md, 1974, Harper & Row, Publishers.
6. Passey, RB, Gillum, RL, Fuller, JB, et al: Clin Clem 23:1131-1139, 1977.
7. Maquire, GA, and Price, CP: Clin Chem 29:1810-1812, 1983.
8. Spector, R: N Engl J Med 296:1393-1398, 1977.
9. Price, CP, and Spencer, K: Ann Clin Biochem 16:100-105, 1979.
10. Gerson, B, and Figoni, M: Arch Pathol Lab Med 109:711-715, 1985.
11. Frantz, ID, III, Medina, G, and Tauensch, HW, Jr: J Pediatr 87:417-420, 1975.
12. Stewart, TC, and Kleyle, RM: Clin Chem 29:132-135, 1983.
13. Hay, WW, Jr, and Osberg, IM: Clin Chem 29:558-560, 1983.
14. Meites, S, editor-in-chief: Washington, DC, 1976, America Association for Clinical Chemistry.
15. Breusch, FL, and Tulus, R: Biochem Biophys Acta 1:77, 1947.

Ketones

LAWRENCE A. KAPLAN

Clinical significance: p. 436

Common name:	**Acetoacetic acid**	**Acetone**	**β-Hydroxybutyric acid**
Structure:	$CH_3C(=O)CH_2C(=O)OH$	$CH_3C(=O)CH_3$	$CH_3CH(OH)CH_2C(=O)OH$
Molecular formula:	$C_4H_6O_3$	C_3H_6O	$C_4H_8O_3$
Molecular weight:	102.09	58.08	104.10
Merck Index:	51	58	4727
Chemical class:	Ketocarboxylic acid	Ketone	Hydroxycarboxylic acid

PRINCIPLES OF ANALYSIS

Ketone bodies are biochemicals that either have structural ketone moieties $(R_1{-}\overset{O}{\overset{\|}{C}}{-}R_2)$ or are directly derived from ketones (that is, reduced ketones). Commonly found serum ketone bodies include the ketones, pyruvate, acetoacetic acid, and acetone and the reduced chemical forms, lactic acid and β-hydroxybutyric acid.

Analysis of ketones has been limited by the fact that the proportions of specific ketone bodies present during an illness vary considerably and by the fact that no single chemical or enzymatic method measures total ketone bodies. An assay that specifically measures one of these biochemicals will almost certainly lack the sensitivity to detect total ketone bodies produced. In addition, any clinically useful test for ketones should be able to be performed rapidly and should be available 24 hours a day (stat).

The reaction between ketones and nitroprusside (sodium nitroferricyanide) under alkaline conditions is a widely used procedure (method 1, Table 57-3).[1,2] Acetoacetic acid and acetone both form a purple color in this reaction, which is most frequently used as a semiquantitative measure of ketones in serum and urine.

$$\underset{\textbf{Sodium nitroferricyanide}}{Na_2[Fe(CN)_5NO]} + \underset{\textbf{Ketone}}{CH_3{-}\overset{O}{\overset{\|}{C}}{-}R} \xrightarrow[\text{Glycine}]{OH^-} \underset{\textit{(purple complex)}}{Na_2[Fe(CN)_5NOCH_2\overset{O}{\overset{\|}{C}}{-}R]} + H_2O \quad \textbf{\textit{Eq. 57-1}}$$

R is CH_3 (acetone) or $-CH_2{-}\overset{O}{\overset{\|}{C}}OH$ (acetoacetic acid)

Ames, Inc. (Division of Miles Laboratories, Elkhart, IN) markets this assay in the form of a paper impregnated with nitroprusside, glycine, and sodium phosphate (Ketostix). The phosphate provides a suitable buffered pH, enabling the reaction to occur. This technique is also incorporated into multitest urine chemistry dipsticks. The dipsticks are

Table 57-3 Methods for ketone analysis*

Method	Type of analysis	Principle	Usage	Comments
1. Colorimetric	Semiquantitative	$Na_2Fe(CN)_5NO$ + Acetone/acetoacetate → *Purple color*	Most common; frequently used as stat procedure	Sensitivity for acetoacetate five times greater than for acetone
2. Enzymatic	Quantitative	$NADH + H^+ + AcAc \underset{pH\ 8.5\text{-}9.5}{\overset{pH\ 7.0}{\rightleftharpoons}} \beta\text{-HB} + NAD^+$ (β-Hydroxybutyrate dehydrogenase)	Rare; not useful as stat procedure	By adjustment of pH and cofactor concentration (that is, NADH or NAD^+), reaction used to quantitate AcAc and β-HB
3. Gas chromatography	Quantitative	Acetone detected by flame ionization detector; AcAc converted to acetone by heating	Rare; not useful as stat test	Acetoacetate quantitated by subtraction of nonheated acetone from heated acetone

**AcAc,* Acetoacetic acid; *β-HB,* β-hydroxybutyric acid.

most accurate for urines with a specific gravity of 1.010 to 1.020. A similar reagent (including lactose as a color enhancer) is marketed as a spot test by Ames as Acetest. Both reagents can be used for serum or urine.

Several techniques have been recently developed that measure the individual ketone bodies with high specificity. Acetoacetic acid[3] and β-hydroxybutyric acid[4] can be quantitated by an enzyme assay (method 2, Table 57-3), which makes use of the following reversible reaction:

$$NADH + H^+ + \text{Acetoacetic acid} \underset{\text{dehydrogenase}}{\overset{\beta\text{-Hydroxybutyrate}}{\rightleftharpoons}} \beta\text{-Hydroxybutyrate} + NAD^+ \qquad \textit{Eq. 57-2}$$

The reaction can be monitored when one observes the absorbance of NADH at 340 nm.[5,6] At pH 8.5 to 9.5, the reaction proceeds to the left as written, and the concentration of β-hydroxybutyric acid is quantitated when the increase in absorbance is monitored at 340 nm. When performed at pH 7.0, the reaction proceeds to the right, and the amount of acetoacetic acid present is proportional to the decrease in absorbance at 340 nm.

Acetone and acetoacetic acid have been quantitated by gas chromatography (method 3, Table 57-3). Acetone is quantified by a flame ionization detector, and acetoacetic acid is estimated by measurement of acetone before and after heating to convert the acetoacetic acid to acetone. The difference represents the amount of acetoacetic acid present in the sample. Most of the methods listed in Table 57-3 are not useful in the most frequently encountered, potentially lethal clinical situation. This is the situation of diabetic ketoacidosis, which requires a rapid estimation of serum or urinary ketones to ensure a rapid diagnosis and prompt therapy. Because the analysis for ketones is often performed in a laboratory with limited facilities, as in an emergency room or a physician's office, the technique also must be simple to perform. For these reasons, the most frequently used test for ketones is the semiquantitative nitroprusside test in the form of a dipstick screen.

There are drawbacks to assays specific for ketones; the most important is that they detect only acetone and acetoacetate. Often these ketones represent only a very small proportion of the total ketone bodies present in serum or urine, and the test can only estimate the total ketone load. In diabetic ketoacidosis, β-hydroxybutyric acid is the predominant ketone body formed, and tests for acetoacetic acid may be negative or weakly positive, thereby underestimating the ketones. In fact, as the diabetic ketoacidosis is resolved, more acetoacetic acid can be seen than is found in the initial stages of the crisis.

The enzymatic procedures are specific for acetoacetate, pyruvate, or β-hydroxybutyrate and have been readily adapted to automated instruments, such as the Abbott ABA-100 and centrifugal analyzers.[5,6] The reagents for these assays can be easily prepared by most laboratories. Thus these methods are recommended for those laboratories interested in specifically quantitating the individual serum and urine ketone bodies.

For most analyses of ketones in serum and urine, the semiquantitative nitroprusside test appears to be the best method. The major drawback to this procedure is that it cannot detect β-hydroxybutyric acid at all and is fivefold to tenfold less sensitive for acetone than for acetoacetic acid. Acetone is rarely positive at serum concentrations less than 5 mmol/L. The primary advantage to the nitroprusside screen is its relative ease of use and reagent stability, making it well suited as a stat screen for ketones. As with any of the dry chemical processes, however, the reagent can deteriorate unless properly protected from air.

The reported sensitivity of the Ketostix is 50 to 100 mg/L of acetoacetate, whereas the sensitivity for Acetest is 25 to 50 mg/L. The dipstick methods are linear up to 1600 mg/L of acetoacetate. The Ketostix can give false-positive results in the presence of large amounts of levodopa.[7] Highly colored urine or hemolyzed serum samples can yield erroneously positive results.

SPECIMEN

Well-centrifuged urine, serum, or plasma can be used for the semiquantitative nitroprusside test. Serum or plasma can be used for the enzymatic assays as well.

Bacterial contamination of urine can hasten the disappearance of the ketones for urine. In addition, acetone is a volatile substance and can be lost by evaporation. Thus the sample should be refrigerated within 20 to 30 minutes of collection if not analyzed immediately.

REFERENCE RANGE

Drews[8] reports a reference range for serum acetoacetate of 5 to 30 mg/L, which compares well with results of 3 to 23 mg/L obtained by a laboratory using an enzymatic procedure. There do not appear to be any age-related differences in reference ranges for ketone bodies.[9] At these levels, serum and urine will be negative by the nitroprusside semiquantitative screening tests.

REFERENCES

1. Rothera, ACH: Note on the sodium nitroprusside reaction for acetone, J Physiol 37:491-494, 1908.
2. Free, AH, and Free, HM: Nature of nitroprusside reactive material in urine in ketosis, Am J Clin Pathol 30:7-10, 1958.
3. Kaplan, L: Acetoacetic acid. In Pesce, AJ, and Kaplan, LA: Methods in clinical chemistry, St Louis, 1987, The CV Mosby Co.
4. Kaplan, L: Beta-hydroxybutyric acid. In Pesce, AJ, and Kaplan, LA: Methods in clinical chemistry, St Louis, 1987, The CV Mosby Co.
5. Li, PL, Lee, JT, MacGilliray, MH, et al: Direct fixed-time kinetic assays for beta-hydroxybutyrate and acetoacetate with a centrifugal analyzer or a computer-backed spectrophotometer, Clin Chem 26:1713-1717, 1980.
6. Hansen, JL, and Frier, EF: Direct assays of lactate, pyruvate, beta-hydroxybutyrate and acetoacetate with a centrifugal analyzer, Clin Chem 24:475-479, 1978.
7. Ketostix package insert, 1988, Ames Division of Miles Laboratories, Inc, Elkhart, IN.
8. Drews, PA: Carboxyhydrate derivatives and metabolites. In Henry, RJ, Cannon, DC, and Winkleman, JW, editors: Clinical chemistry: principles and techniques, ed 2, Hagerstown, MD, 1974, Harper & Row, Publishers, Inc.
9. Peden, VH: Determination of individual serum ''ketone bodies,'' with normal values in infants and children, J Lab Clin Med 63:332-343, 1964.

CHAPTER 58 Electrolytes

Anion gap

F. PHILIP ANDERSON
W. GREGORY MILLER

Clinical significance: pp. 313, 332, and 436
Computation formula: $Na^+ - (Cl^- + CO_2)$

PRINCIPLES OF ANALYSIS

The anion gap, $Na^+ - (Cl^- + CO_2)$, is an estimate of the net number of anions not directly measured in serum. This parameter is usually calculated by subtraction of the *measured* anions from the *measured* cations to give a positive gap in milliequivalents per liter (mEq/L):

$$\text{Measured cations (Na)} - \text{Measured anions (Cl} + CO_2) = +\text{Anion gap}$$

The above equation is equivalent to subtraction of the unmeasured anions from the unmeasured cations, which results in the same absolute value as the above equation but with a negative value, as in the following equation:

$$\text{Unmeasured cations} - \text{Unmeasured anions} = -\text{Anion gap}$$

The major unmeasured cations are calcium (5 mEq/L) and magnesium (2 mEq/L), and the major unmeasured anions are protein (15 mEq/L), phosphate (2 mEq/L), sulfate (1 mEq/L), and organic acids, such as lactic acid (5 mEq/L). Since potassium (4 mEq/L) is usually not included in the computation of anion gap, the net "unmeasured anions," or anion gap, is approximately

$$(5 + 2 + 4) - (15 + 5 + 2 + 1) = -12 \text{ mEq/L}$$

The computation of the anion gap is useful clinically to alert the physician to the potential presence of metabolic disorders that alter electrolyte balance.[1] The accompanying box lists some causes of both increased and decreased anion gaps. A normal anion gap is observed in hyperchloremic metabolic acidosis such as that produced in renal tubular acidosis or diarrhea. In these cases the bicarbonate loss is ionically compensated by increased chloride, with no net change in the calculated anion gap.

A recent study of electrolyte measurements of healthy persons observed a range of anion gaps of 8 to 18 mEq/L, whereas a group of hospitalized patients had a range of 1 to 30 mEq/L, with 90% of the patients between 6 and 20 mEq/L.[2] Because the expected result for an anion gap calculation falls within a well-defined range, one can use it as a quality control criterion in individual patients to detect spurious electrolyte results. If the anion gap falls outside

ANION GAP CHANGES

Causes of increased anion gap

- Decreased unmeasured cations
  - Hypocalcemia
  - Hypomagnesemia
- Increased unmeasured anions
  - Associated with metabolic acidosis
    - Uremia (renal failure)
    - Ketoacidosis
    - Lactic acidosis
    - Salicylate poisoning
  - Not necessarily associated with metabolic acidosis
    - Hyperphosphatemia
    - Hypersulfatemia
    - Large doses of antibiotics (such as penicillin and carbenicillin)
    - Treatment with lactate, citrate, or acetate
- Increase in net protein charge as in alkalosis

Causes of decreased anion gap

- Decreased unmeasured anions
  - Hypoalbuminemia
  - Hypophosphatemia
- Increased unmeasured cations
  - Hypercalcemia
  - Hypermagnesemia
  - Paraproteins
  - Polyclonal gamma globulins
  - Drugs such as polymyxin B or lithium
- Underestimation of serum sodium
  - Hyperproteinemia
  - Hypertriglyceridemia (turbidity)
- Overestimation of serum chloride
  - Bromism
  - Turbidity (for ferric thiocyanate method)

the range of 6 to 20 mEq/L, it *may* indicate an analytical error. In this case it is necessary to repeat the electrolyte measurements to confirm the results. The imprecision of the anion gap (typical coefficient of variation [CV] of 9%) is much larger than the individual electrolyte results (typical CV for sodium, 1%; for chloride, 1.5%; and for CO_2, 4%) because it includes the imprecision from each term. Consequently the individual electrolyte results (not the anion gap) should agree within ±2 mEq/L of the original results for confirmation. If the repeated results do not agree with the original, it is necessary to repeat the analysis to determine which values are correct. If over a period of time all anion gaps are on the high side or on the low side, this trend may indicate the presence of a consistent analytical error and should be investigated.

Repeat analysis may include the use of alternative, confirmatory methods such as nondilutional, ion-selective electrodes or a chloridometer to minimize the presence of lipemia or bromism. Inspection of other laboratory findings, such as presence of hyperproteinemia, ketones, or hyperglycemia, may indicate a physiological reason for an abnormal anion gap.

A moderately increased anion gap (that is, 21 to 25) in the presence of greatly elevated BUN (more than 500 mg/L) or creatinine (more than 40 mg/L) or both is consistent with renal failure, and electrolyte results do not need to be confirmed. Similarly, a moderately increased anion gap in the presence of greatly increased glucose (more than 3000 mg/L) is consistent with diabetic ketoacidosis, and electrolyte results do not need to be confirmed. It is less common to observe gaps less than 6.

REFERENCE RANGE[2]

Ambulatory 7 to 18 mEq/L
Hospitalized 6 to 10 mEq/L

REFERENCES

1. Gar, AK, and Nanji, AA: The anion gap and other critical calculations, Diagn Med pp 32-43, March-April 1982.
2. Witte, DL, Rodgers, JL, and Barrett DA: The anion gap: its use in quality control, Clin Chem 22:643-646, 1976.

Blood gas analysis and oxygen saturation

BERNDT B. BRUEGGER
JOHN E. SHERWIN

Partial pressure of oxygen (Po_2), carbon dioxide (Pco_2), and the pH of whole blood

Clinical significance: pp. 313 and 332

PRINCIPLES OF ANALYSIS

Blood gas analyses are usually performed for the assessment of either the acid-base status or the respiratory oxygenation status of the patient. Blood gas analysis includes measurement of blood pH, Pco_2, and Po_2 and may include any or all of the calculated parameters of oxygen content, oxygen saturation, total CO_2, bicarbonate, and base excess. These calculated parameters are discussed in Chapter 21.

Analysis of pH

The components of the blood pH–measuring system are the glass electrode, the reference electrode, and the liquid junction between the two electrodes.[1] The hydrogen ions in the blood sample exchange with metallic ions in the glass membrane of the glass electrode, creating a potential difference across that membrane that is proportional to the hydrogen ion concentration in the blood specimen. A solution of constant pH inside the electrode is in contact with the glass membrane. A silver–silver chloride wire that is connected to a voltmeter lies inside the solution of constant pH and measures the potential difference created by the hydrogen ion concentration in the blood specimen (Fig. 58-1).

The potential difference measured by the glass electrode is assessed against the reference electrode (Fig. 58-1). The

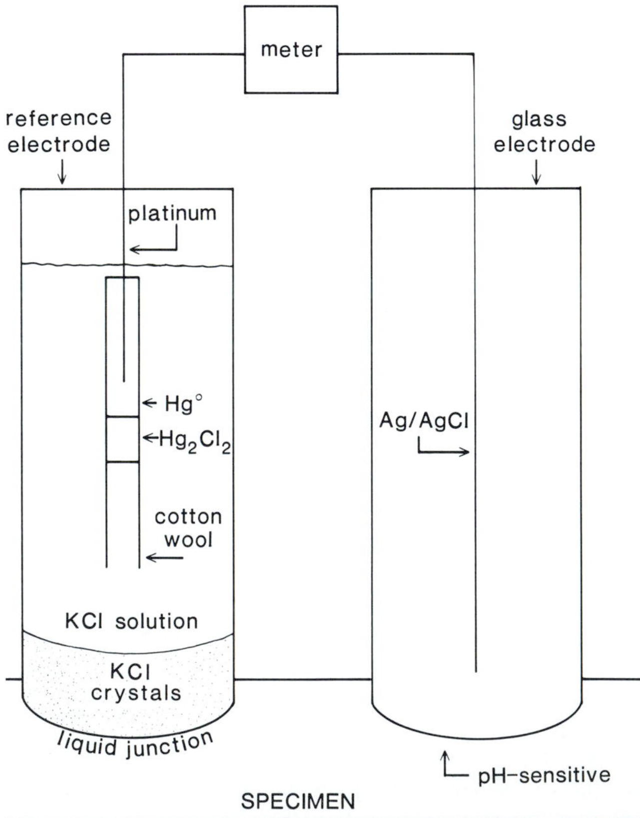

Fig. 58-1 Schema of pH electrode and calomel reference electrode. A potential develops across pH-sensitive glass membrane of pH electrode because of difference in pH of outer sample and inner electrode buffer solution. Calomel electrode is in electrical contact with sample and pH electrode through KCl salt bridge and acts as reference against potential developed at pH electrode.

reference electrode, which is also connected to the voltmeter, is usually a calomel electrode. The calomel electrode consists of a glass tube containing a platinum wire that extends into a calomel paste of mercurous chloride (Hg_2Cl_2) and mercury (Hg) and a solution of potassium chloride (KCl) at a constant concentration. One end of the glass tube usually contains a cotton wick to keep the calomel paste in contact with the platinum wire and to allow the system to be permeated with the KCl solution inside and surrounding the glass tube. The KCl solution is usually a saturated solution, although some pH electrode systems use lower concentrations of KCl; for example, the Radiometer ABL uses a 20% KCl solution.

A liquid junction is present between the reference electrode and blood specimen being analyzed for its pH. Generally the reference electrode has a permeable membrane through which the KCl solution inside the electrode escapes, coming into contact with the specimen and thus forming the junction. The KCl solution is at a high concentration so that the difference in ionic composition of the blood specimen does not change the constant potential of the reference electrode.

A representation of the entire pH-measuring system is shown in Fig. 58-2. The only variation in potential in the system is at the interface of the blood specimen to be analyzed and the glass electrode so that any voltage change measured by the voltmeter is attributable to the pH of that blood specimen.

Analysis of P_{CO_2}

One measures the P_{CO_2} by isolating a glass electrode in a weak bicarbonate buffer and separating this buffer from the blood sample by a membrane permeable to the carbon dioxide (CO_2) in the specimen. The CO_2 diffuses into a bicarbonate buffer solution inside the electrode, and the following reaction occurs:

$$CO_2 + H_2O \rightleftarrows H_2CO_3 \rightleftarrows H^+ + HCO_3^-$$

The change in hydrogen concentration in the buffer is then measured inside the P_{CO_2} electrode by the same type of components found in the pH electrode. A glass electrode inside the P_{CO_2} electrode develops a potential difference because of a change in concentration of H^+ ions, and a reference electrode also inside the P_{CO_2} electrode has a constant potential that is the standard to which the change in potential at the glass electrode is compared. The gas-permeable membrane of the P_{CO_2} electrode excludes hydrogen ions in solution so that the pH of the specimen will not affect the observed pH change inside the P_{O_2} electrode induced by the diffusion of the CO_2 from the specimen into the electrode.

Analysis of P_{O_2}

The measurement of P_{O_2} relies on the electrochemical measurement of the oxygen that diffuses across a gas-permeable membrane into an electrolyte buffer that is within a Clark electrode.[2] The gas-permeable membrane allows oxygen but not ions to cross into the electrode. Inside the electrode is an electrolyte solution consisting of potassium chloride and phosphate buffer, or buffered potassium hydroxide, a silver–silver chloride anode, and a platinum cathode. The cathode is held at a constant potential with respect to the anode so that when oxygen diffuses into the electrode the oxygen is reduced at the cathode as follows:

$$O_2 + 2H^+ + 4e^- \rightarrow H_2O_2 + 2e^- \rightarrow 2\ OH^-$$

The reduction of oxygen causes a current to flow between the cathode and anode. The current is measured and is

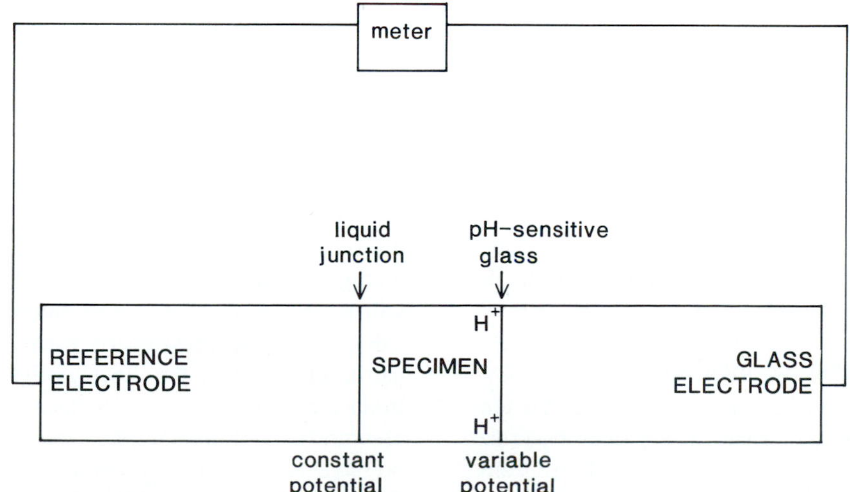

Fig. 58-2 Schema of contact sequence of specimen and glass electrode and reference electrode. One area of variable potential is at area of contact between specimen and pH-sensitive glass and depends on pH of specimen.

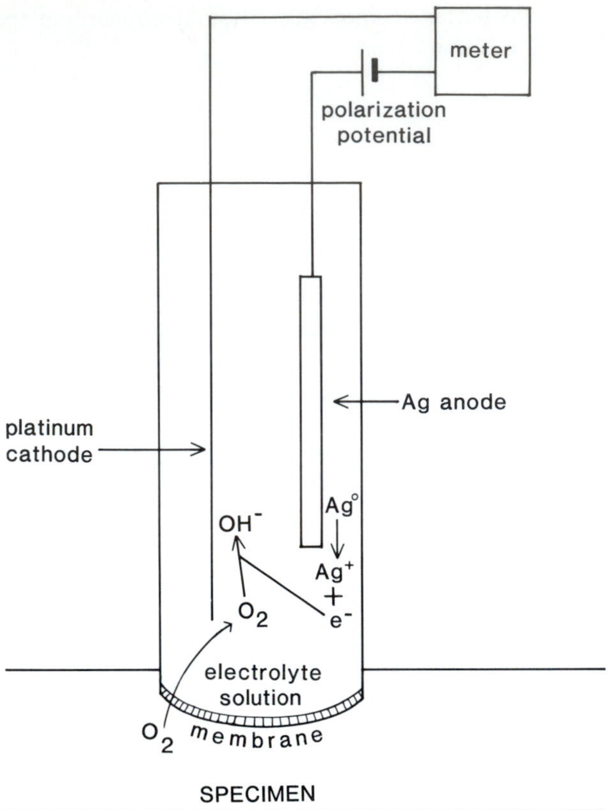

Fig. 58-3 Schema of Clark polarographic electrode. A constant voltage is generated between platinum cathode and silver anode. Oxygen molecules that diffuse past outer membrane into layer of electrolyte solution in contact with platinum cathode are reduced at cathode, producing current that can be measured and related to amount of diffusing oxygen.

proportional to the amount of oxygen diffusing into the electrode from the analyzed specimen. Fig. 58-3 illustrates the components of the Po_2 electrode. The electrolyte solution buffers against the accumulation of peroxide (H_2O_2) and hydroxide (OH^-) and serves as the conduction medium for the diffusing oxygen and the current.

The majority of available blood gas analyzers use certified gases saturated with water for calibration. This technique has the advantage of providing a constant and highly reproducible calibration mixture. Nonetheless, it is a gas rather than a liquid and therefore is not equivalent to the patient samples being analyzed.

Noninvasive in vivo blood gas analysis

The transcutaneous measurement of Pco_2 and Po_2 currently relies on miniature electrodes of the same basic design as the electrodes of blood gas analyzers.[3] These transcutaneous, noninvasive measurements are complicated by the presence of the skin, which acts as a second gas-permeable membrane.

A small amount of gas diffuses from the capillary bed in the skin toward the surface of the skin. If the skin is heated so that local blood flow is maximally increased, the Po_2 at the surface of the skin correlates well with the arterial Po_2. Temperatures of 42° to 43° C have been found to maximize blood flow while minimizing the risk of burns, so a major feature of the transcutaneous Po_2 ($tcPo_2$) electrode is a heating element maintaining this temperature. The $tcPo_2$ electrode uses the same concept for measuring Po_2 as the Clark polarographic Po_2 electrode described earlier. A membrane, through which Po_2 diffuses, lies between the skin and the electrode, and a heating device is present inside the electrode. A sealing ring around the outer edge of the electrode prevents nonregulated diffusion of atmospheric oxygen to the electrode. Because the Po_2 electrode consumes oxygen during measurement, the area of the cathode is kept as small as possible, minimizing O_2 consumption by the electrode so that measurements will correlate with arterial Po_2 values.

Analysis of oxygen saturation

Oxygen saturation, which is the percentage of total hemoglobin that has bound oxygen, can be calculated from the measured parameters Po_2 and pH, on the basis of standard oxygen-dissociation curves.[4] Most automated blood gas analyzers calculate oxygen saturation in such a manner.

One can also calculate oxygen saturation by using the difference between the absorption spectra of oxygenated hemoglobin and deoxygenated hemoglobin. Generally, a blood sample is analyzed at two wavelengths—one at which a large difference in absorbance occurs between deoxygenated hemoglobin and oxygenated hemoglobin, as at 600 and 577 nm, and the other, called an *isosbestic point,* at which the molar absorbance is identical for the two forms of hemoglobin, as at 506 and 548 nm. The measurement at the isosbestic point gives the total amount of hemoglobin present, and the absorption at the other wavelength depends on the difference in concentration of the two forms of hemoglobin.

Calibration of gas electrodes

Gases used to calibrate blood gas analyzers are bubbled through humidifiers at 37° C to saturate the gases with water vapor. The gases are saturated with water to eliminate the problem of maintaining a totally ''dry'' gas for calibrating the blood gas instruments. Some residual moisture will remain in the measuring chamber and contaminate a ''dry'' gas. The pressure of water vapor will reduce the partial pressures of the other gases in the mixture and introduce a variable error. For instance, at 37° C, the vapor pressure of a water-saturated gas will include the partial pressure of 47 mm Hg, contributed by the water vapor. Thus, to produce a reproducible gas mixture of known partial pressure of O_2 and CO_2, calibration requires a fixed partial pressure of water vapor. Since the specimen to be analyzed is aqueous, this is best achieved with a water-saturated gas.

Modern instruments

The majority of currently available blood gas analyzers require a whole blood sample of less than 100 μL and measure the pH, PCO_2, and PO_2 simultaneously. Many are now microprocessor controlled so that calibration occurs automatically at preset intervals, and both measured parameters and the calibrated parameters are displayed when analysis is complete.

SPECIMEN

The skill required to draw the arterial specimens and the potential for harm to the patient limit the personnel who can collect such specimens. Thus the capillary specimen is frequently used as a compromise, providing an arterialized specimen that is easily collected. A capillary specimen can be drawn by a component phlebotomist under conditions that produce pH, PCO_2, and PO_2 values closely paralleling those of arterial blood. To ''arterialize'' capillary blood, the limb is warmed for several minutes at about 45° C immediately before the arterialized specimen is collected. The warming dilates the capillaries, and the increased blood flow in that area decreases the accumulation of metabolic products of tissue respiration and ensures more blood of an arterial than a venous composition. Alternatively, a histamine cream can be used to produce local hyperemia so that an arterialized specimen can be collected.

Whole blood is collected in heparinized capillary tubes containing a small metal mixing wire, which is added to facilitate mixing with the magnet (see Fig. 2-2). These components are then sealed at one or both ends with clay. The capillaries are placed into a labeled test tube or plastic bag and delivered to the laboratory in crushed ice to prevent deterioration of the sample.

The time taken to complete the test procedure must not exceed 15 minutes. This includes drawing of sample, delivery to the laboratory, log-in, analysis, and reporting of results. If this time limit is adhered to, no medically significant difference is observed between specimens collected in plastic and glass syringes.

Anticoagulants

Blood gas measurements must be performed on whole blood, that is, unclotted and unseparated blood. An anticoagulant is used to inactivate the clotting mechanisms so that the sample remains unclotted in the syringe. Oxalates, EDTA, and citrates are not acceptable for blood gas samples because they significantly alter the blood sample.[5]

Heparin is the anticoagulant of choice; however, too much heparin may affect all parameters.

Anaerobic conditions

Room air contains a PCO_2 of essentially zero and a PO_2 of approximately 150 mm Hg. Air bubbles that mix with a blood sample will result in gas equilibrium between the air and the blood. Air bubbles may thus *significantly* lower the PCO_2 values of the blood sample and cause the PO_2 to approach 150 mm Hg. The greater the amount of air mixed with a blood sample, the greater the error. The syringe must be *immediately* sealed with a cork or a cap after the sample is obtained. *It is recommended that any sample obtained with more than minor air bubbles be discarded.*

Table 58-1 In vitro blood gas changes*

Parameter	37° C†	4° C
pH	0.01/10 min	0.001/10 min
PCO_2	1 mm Hg/10 min	0.1 mm Hg/10 min
PO_2	33 mm Hg/10 min	3 mm Hg/10 min

*Approximate changes with time and temperature after sample is drawn into syringe.
†A temperature of 37° C implies that the blood remains at body temperature in the syringe.

Temperature during transport and storage

Blood is living tissue in which oxygen continues to be consumed and carbon dioxide continues to be produced, even after the blood is drawn into a syringe. Table 58-1 shows the approximate rate of change of a sample that is held in a syringe at 37° C. If the sample is immediately placed in an ice slush, the temperature rapidly falls below 4° C, and the PCO_2 changes are insignificant over several minutes. Acceptable samples are those that have been properly iced immediately after being drawn and rapidly (less than 15 minutes) transported to the laboratory at 4° C.

Although a majority of pediatric blood gas specimens are capillary blood specimens, a variable percentage will be arterial. Arterial samples may be drawn into heparinized syringes either by a laboratory phlebotomist (from a catheter line only) or from the femoral or radial artery (by a physician only). The arterial sample is preserved in crushed ice and delivered to the laboratory.

Sample handling

For a specimen in a capillary tube

1. Mix the sample by moving the magnet gently back and forth along the capillary tube to cause the metal rod in the capillary tube to move through the sample.
2. Use a file to score the tube just above the sealing wax plug and break the tube. Expel one or two drops of blood onto a gauze pad to be sure blood is flowing freely.

For a specimen in a syringe

1. Rotate the syringe between your palms for about 30 seconds to mix and warm the specimen.
2. Remove the needle and expel one or two drops into a gauze pad to ensure that there are no blood clots present in the syringe.
3. Place a syringe adapter into the sample fill port, and then put the other end of the syringe adapter (the thin tube) into the syringe.

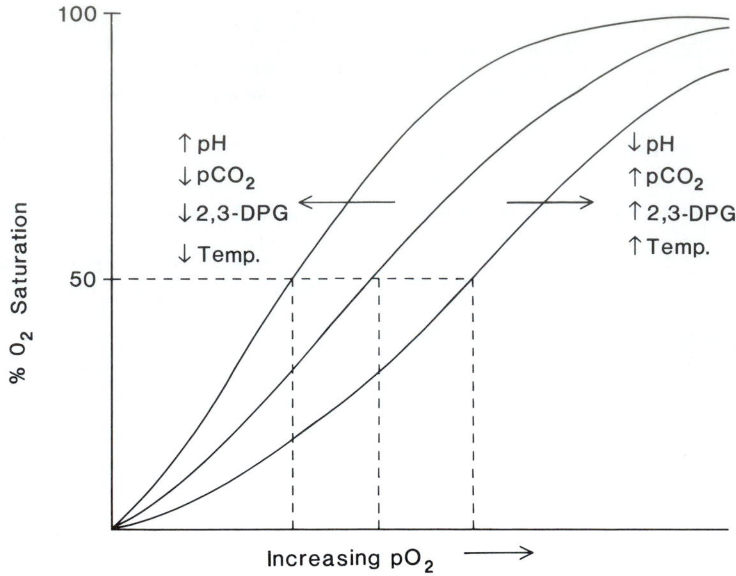

Fig. 58-4 Oxygen-dissociation curve in middle is developed under following conditions: pH is 7.40, P_{CO_2} is 40 mm Hg, and temperature is 37° C. P_{50} is derived from that curve. Also indicated are factors that shift curve to right and left.

Temperature corrections

If the blood sample is taken from a patient whose body temperature deviates from 37° C, the measured and calculated parameters for that temperature can be calculated by the instrument.

REFERENCE RANGE (at 37° C)

	Arterial	Venous
pH	7.35-7.45	7.35-7.43
P_{CO_2}	35-45 mm Hg	38-50 mm Hg
P_{O_2}	80-100 mm Hg	30-50 mm Hg
HCO_3^-	22-26 mmol/L	23-27 mmol/L
Total CO_2	23-27 mmol/L	24-28 mmol/L
O_2 saturation	95%-100%	60%-85%
Venous anion gap	5 to 14 mmol/L	
Base excess	−2 to +2 mEq/L	

The unit commonly used to express partial pressure of blood gases is millimeters of mercury (mm Hg) or its equivalent, the torr. However, in the International System of Units, the kilopascal (kPa) is the recommended unit of expression although it is not yet widely used. One millimeter of mercury, or 1 torr, is equal to 0.133 kPa.

The three types of blood specimens assayed for blood gas content are arterial, venous, and capillary. The preceding outline compares the normal ranges for arterial and venous blood gas parameters. Arterial blood is not as affected by the metabolic activity of the area being sampled as is venous blood. The slightly lower pH and P_{O_2} values and the slightly higher P_{CO_2} in the venous blood reflect the changes induced by tissue metabolic activity, which releases acids and uses oxygen. Arterial blood specimens reflect the state of pulmonary activity, which introduces oxygen to the blood and releases carbon dioxide from the blood to the air spaces of the lungs. Although the usually small differences in pH between arterial and venous blood specimens allow the venous specimens to be of value in the interpretation of the acid-base status of the patient, arterial blood is the specimen of choice when one wishes to understand the dynamics of the body's respiratory system.

P_{50} oxygen-dissociation curve

The term P_{50} is defined as the partial pressure of oxygen (P_{O_2}) at which hemoglobin is 50% saturated.[6] A graph of the percent oxygen saturation versus P_{O_2} is referred to as the Hill plot or the oxygen-dissociation curve (Fig. 58-4). A shift of the curve to the right represents a decrease in the affinity of hemoglobin for oxygen and an increase in P_{50}. The P_{50} value increases because a greater O_2 partial pressure is required to saturate 50% of the hemoglobin. The effect of pH and P_{CO_2} on the oxygen affinity of hemoglobin is known as the Bohr effect. A shift of the curve to the left represents an increase in the affinity of hemoglobin for oxygen and a decrease in P_{50}; that is, a smaller O_2 partial pressure is required to saturate 50% of the hemoglobin.

REFERENCES

1. Clark, LC: Monitor and control of blood and tissue oxygen tensions, Trans Am Soc Artif Intern Organs 2:41-48, 1956.
2. Evans, NTS, and Naylor PFD: The systemic oxygen supply to the surface of human skin, Respir Physiol 2:21-77, 1967.
3. Huch, R, and Huch, A: Continuous transcutaneous blood gas monitoring, New York, 1983, Marcel Dekker, Inc.

4. Rem, J, Siggaard-Andersen, O, Norgaard-Pedersen, B, and Sorensen, S: Hemoglobin pigments: photometer for oxygen saturation, carboxyhemoglobin, and methemoglobin in capillary blood, Clin Chim Acta 42:101, 1972.
5. Shapiro, BA, Harrison, RA, and Walton, JR, editors: In Clinical application of blood gases, ed 2, Chicago, 1977, Year Book Medical Publishers, Inc.
6. Fleisher, WR, and Gambino, SR: Blood pH, pCO_2, and oxygen saturation, Chicago, 1976, American Society of Clinical Pathologists.

Calcium

E. CHRISTIS FARRELL, Jr.

Clinical significance: pp. 373 and 664
Atomic symbol: Ca
Atomic weight: 40.08 daltons
Merck Index: 1613
Chemical class: alkaline earth element
Number of forms: free Ca^{++}, protein bound, or in inorganic complexes

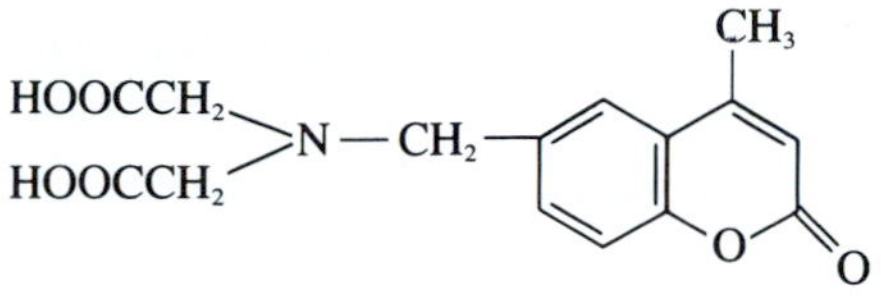

Calcein or calcein blue
4-methylumbelliferone-6-methyleneiminodiacetic acid

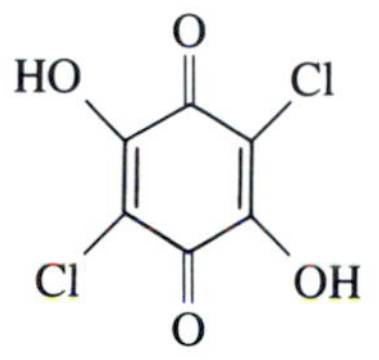

Chloranilic acid

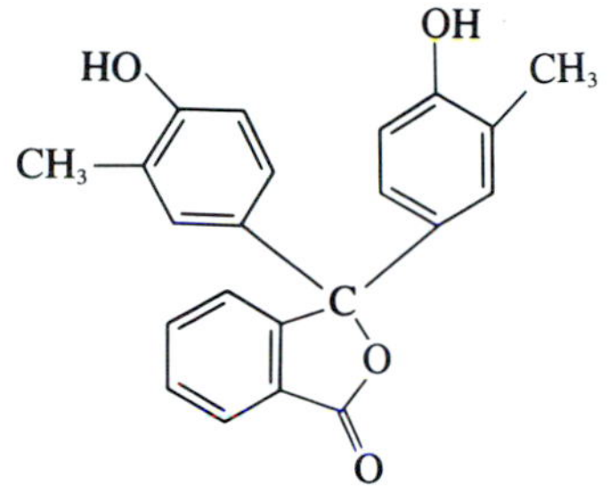

***o*-Cresolphthalein**

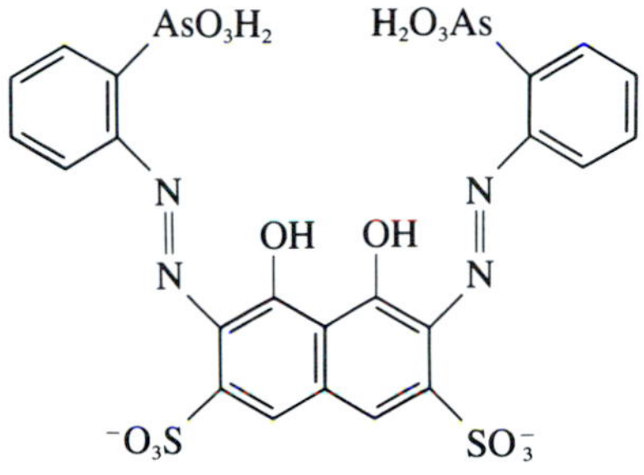

Arsenazo dye (on Ektachem)

PRINCIPLES OF ANALYSIS

Calcium is found 46% free, 32% bound to albumin, 8% bound to globulins, and 14% associated in freely diffusible calcium complex. All calcium is ionized no matter what it is bound to, but it may not be dialyzable or completely reactive with a chosen chromogen when complexed to another compound.

The oldest procedures used for the clinical determination of total calcium involved the quantitative precipitation of calcium with an excess of anions from oxalic, chloranilic, or naphthylhydroxamic acids.[1] In the case of oxalate, the precipitate was quantitated by a redox titration with potassium permanganate or ceric ions (method 1, Table 58-2). The most widely used oxalate methods were based on the modification introduced by Kramer and Tisdall and by Clark and Collips and were for years considered reference methods for calcium analysis.[2]

Methods using chloranilate or naphthylhydroxamate ions to precipitate calcium were based on quantification of the color of the precipitating ion. An excess of chloranilate was added to react with the serum calcium. The calcium-chloranilate complex was precipitated and redissolved in alkaline solutions of ethylenediaminetetraacetic acid (EDTA). The absorbance of the intense red-purple color of the liberated chloranilate acid (A_{max}, 520 nm) was proportional to the concentration of calcium in the specimen (method 2, Table 58-2). Similarly, naphthylhydroxamate has been used to precipitate calcium. The precipitate was dissolved in alkaline EDTA, and, when acid ferric nitrate was added, a red-orange color was produced. These precipitation methods are rarely used today because they are time consuming and insensitive.

A sensitive fluorescent method for the microanalysis of serum or urine is available.[3] In one method (Calcette), the calcium is added to an alkaline solution containing calcein (method 3, Table 58-2). The calcium combines with the calcein, which at high pH fluoresces at 520 nm (excitation at 490 nm). The fluorescent complex is titrated with ethylene glycol-bis(β-aminoethylether)-*N,N'*-tetraacetic acid (EGTA), which binds the calcium from the complex with a resulting decrease in fluorescence. When the baseline fluorescence is achieved, the volume of EGTA required to titrate the complex is directly proportional to the concentration of calcium. It has been reported that magnesium and phosphate do not interfere with this method.

Direct spectrophotometric measurements of calcium in serum or urine are also based on formation of color complexes between calcium and organic molecules. Examples

Table 58-2 Methods of calcium analysis

Method	Principle	Usage	Comments
1. Precipitation by oxalate and redox titration	Ca^{++} + Oxalate → Ca oxalate *(ppt)* Ca oxalate *(ppt)* + H_2SO_4 → Oxalate + $CaSO_4$ $2KMnO_4 + 5\text{ Oxalate} + 3H_2SO_4 \xrightarrow{70°\text{ C}} K_2SO_4 + 2MnSO_4 + 10CO_2 + 8H_2O$	Historical	Initial reference method
2. Precipitation by colored anions; spectrophotometric	Ca^{++} + Chloranilate → Ca-chloranilate *(ppt)* Ca-chloranilate *(ppt)* + EDTA $\xrightarrow{OH^-}$ Ca-EDTA + Chloranilic acid *(purple)*	Historical	Labor intensive; imprecise; many other dyes available
3. Titration of fluorescent Ca^{++} complex	Ca^{++} + Calcein → Ca-calcein *(fluorescent)* EGTA + Ca-calcein → Ca-EGTA + Calcein *(decreased fluorescence)*	Stat or small laboratories	Small sample size; dedicated instrument
4. Spectrophotometric measurement of Ca^{++} complexes			
a. Direct	Ca^{++} + *o*-Cresolphthalein $\xrightarrow{OH^-}$ *Red complex (570 nm)*	Most common	Early adapted to a variety of automated instruments; positive bias compared to atomic absorption
b. Dialysis	Ca^{++} complex + H^+ $\xrightarrow{\text{Dialysis}}$ Ca^{++} in recipient stream Ca^{++} detected as in (a)	On Technicon AutoAnalyzer	
5. Flame emission	$Ca^{++} \xrightarrow[2e^-]{\text{Heat}} Ca^0 \xrightarrow{\text{Heat}} Ca^* \rightarrow Ca^0$ + Photon	Historical	Poor sensitivity
6. Atomic absorption	$Ca^{++} \xrightarrow{2e^-} Ca^0$ Photon + $Ca^0 \rightarrow Ca^*$	Reference method	Excellent accuracy and sensitivity
7. Isotope-dilution mass spectrometry	Ca and known amount of Ca isotope; isolate Ca^{++}, and record ratio of two isotopes on mass spectrometer	Definitive method	Available in reference centers only

*Ca**, Calcium atom in excited state; *Ca^0*, calcium atom in ground state; *EGTA*, ethylene glycol-bis(β-aminoethylether)-*N,N'*-tetraacetic acid; *ppt*, precipitate.

of compounds that give colored reaction products with calcium are (1) glyoxal-bis(2-hydroxyanil), (2) alizarin, (3) chlorophosphonazo III, (4) methylthymol blue, (5) *o*-cresolphthalein complexone, and (6) arsenazo III.

Of these calcium-complexing, colorometric reagents, the *o*-cresolphthalein is certainly the most commonly used for routine calcium analysis (see figure of structure at beginning of section).

The reaction of calcium with *o*-cresolphthalein produces a red complex (quantitated at 570 to 575 nm) at pH 10 to 12 (method 4, Table 58-2). The reaction product is stabilized by the addition of KCN (potassium cyanide), which also acts to eliminate interference from heavy metals. Interference by magnesium ions is eliminated by the addition of 8-hydroxyquinoline.[4] To reduce interference from proteins, one can dialyze the sample with an acid solution to release bound calcium (AutoAnalyzers, Technicon, Inc., Tarrytown, NY) or make a large dilution of the sample (approximately fortyfold to a hundredfold) with an acid solution. Most methods used diethylamine to achieve a high buffered pH. The method of Moorehead and Biggs[5] involves the use of the nonirritating, less volatile 2-amino-2-methyl-1-propanol (AMP) buffer. A commercially available kit (Dow Diagnostics, Inc., Midland, MI) uses 2-ethylaminoethanol as the buffer. Another commercially available reagent eliminates the potentially hazardous cyanide and uses a sodium carbonate–bicarbonate buffer.

The Kodak Ektachem (Rochester, NY) uses the arsenazo III dye in a dry reagent multiple-layered system. When the patient fluid activates the system, the arsenazo dye migrates from the bottom layer to a mordant layer and is trapped. The calcium in the sample migrates to the mordant layer, complexing with the dye and changing its color. The absorption is recorded by reflectance photometry at 680 nm.[6]

Methods for measurement of calcium in serum and urine using atomic emission (method 5, Table 58-2) or atomic absorption (method 6, Table 58-2) have been suggested. Only the atomic absorption method, however, had the necessary specificity and sensitivity.[7,8] (See Chapter 3 for description.)

Calcium present in protein or inorganic complexes is detectable by flame atomic absorption only when steps are taken to dissociate the calcium from these complexes. Acid is used for the dissociation of protein-bound calcium, and lanthanum (La^{++}) or strontium (Sr^{++}) ions are added to displace Ca^{++} from phosphate, oxalate, citrate, and other complexes. The La^{++} or Sr^{++} is added to the sample diluent. Serum is diluted 1:50 with 1% La^{++}, whereas urine is diluted 1:50 with 5% La^{++} to overcome the higher phosphate concentration found in urine. Protein is precipitated by treatment of the sample with acid. Precipitation results in a 2% to 3% contraction in sample volume when the precipitate is removed. One can mathematically correct for the contracted volume. Acid treatment without protein precipitation requires increased maintenance for the burner head to prevent accumulation of debris.

In atomic absorption, interference by magnesium and other elements is reduced by the use of narrow-band pass, diffraction-grating spectrophotometers for specific isolation of the atomic absorption line of calcium (422.7 nm). Interference by sodium is eliminated by the addition of physiological concentrations of sodium ions to calcium standards. Atomic absorption is not frequently used for routine analysis, probably because it has rarely been automated for high sample throughput. However, it is considered a reference method.

The definitive method for calcium measurements is isotope-dilution mass spectroscopy (method 7, Table 58-2).[9] This method, available in only a few institutions, is the accuracy standard against which all methods must be compared.

The *o*-cresolphthalein methods have been adapted to a wide variety of automated analyzers with a resulting high precision. Most methods using direct analysis after dilution appear to be sensitive to interferences from lipemia, hemoglobin, and myeloma protein.[10-14]

The preferred clinical method is certainly atomic absorption on the basis of small sample size, precision, and degree of accuracy. The *o*-cresolphthalein method will suffice for most routine needs. The reaction conditions used for these two methods are summarized in Table 58-3. The Kodak Ektachem procedure also gives acceptable results.[6]

SPECIMEN

Serum or heparinized plasma is separated from cells as rapidly as possible. Blood anticoagulated with oxalate or EDTA is not acceptable because these chemicals will strongly chelate calcium. Venous stasis and erect posture can both elevate calcium by 4 to 6 mg/L.[15] Stasis changes concentration of protein-bound Ca^{++}, and concentrations of free Ca^{++} will be changed by pH shifts.

Urine calcium can be kept in solution by addition of 10 mL of 6 M HCl to the collection container before a 24-hour specimen is collected. The urine should be kept well mixed during the collection period.

PROCEDURE: CALCIUM BY ATOMIC ABSORPTION SPECTROPHOTOMETRY

Principle

Calcium determinations by atomic absorption spectroscopy are based on the fact that atoms of an element in the "ground," or unexcited, state absorb light of the same wavelength as that emitted by the element in the excited state. Each element has its own characteristic absorption or resonance lines, and no two elements are known to have an identical resonance line pattern.

Calcium is determined in serum or urine after the fluid is diluted sufficiently with lanthanum oxide solution to

Table 58-3 Reaction conditions for calcium analysis

Condition	*o*-Cresolphthalein spectrophotometric*	Atomic absorption
Temperature	25°-37° C	2300° C
Sample volume	20 μL	50 μL
Volume fraction	0.01	0.02
Final concentration	*o*-Cresolphthalein: 0.06 mmol/L 8-Hydroxyquinoline: 9.6 mmol/L Diethylamine: 193.4 mmol/L Potassium cyanide: 3.8 mmol/L	Lanthanum diluent 1 g/L (serum) 5 g/L (urine)
Wavelength	575 nm	422.7 nm
Reaction mode	End point	End point
Time of reaction	30 seconds	12 seconds
Precision† (mean, percent coefficient of variation)	Serum 78.0 mg/L, 3.7% CV 147.7 mg/L, 3.3% CV Urine 159 mg/L, 7.9% CV 286 mg/L, 7.7% CV	Serum 78.2 mg/L, 2.3% CV 146.1 mg/L, 3.6% CV Urine 158 mg/L, 5.5% CV 290 mg/L, 7.6% CV
Linearity	150 mg/L	Highest standard, usually 120 mg/L
Interferences	Gross lipemia, calcium chelators	Calcium chelators

*Baginski, ES, Marie, SS, Alcock, NW, et al: Calcium in biological fluids. In Faulkner, WR, and Meites, S, editors: Selected methods of clinical chemistry, vol 9, Washington, DC, 1982, American Association for Clinical Chemistry Press.

†From College of American Pathologists Quality Assurance Survey data. Cresolphthalein data are from all mutianalyzers. Individual means and coefficients of variation vary considerably from method to method.

avoid interference by chelating substances, especially phosphate ions.

Reagents

Use specially, cleaned atomic absorption glassware and deionized water (that is, free of calcium).

Stock lanthanum, 50 g/L (314 mmol/L). Weigh 58.64 g of lanthanum oxide (La_2O_3), place in a 1 L flask, and add about 50 mL of deionized water to wet the powder. Add 250 mL of concentrated HCl *slowly* and *cautiously* under hood while swirling the flask. Dilute to 1 L with deionized water. Store in dark at room temperature. Solution is stable for 6 months.

Working lanthanum diluent for serum, 1 g/L (6.28 mmol/L). Dilute 20 mL of stock lanthanum to 1 L with deionized water. Store in dark bottle at room temperature. Solution is stable for 6 months.

Working lanthanum diluent for urine, 5 g/L (31.4 mmol/L). Dilute 100 mL of stock lanthanum to 1 L with deionized water. Store in dark bottle at room temperature. Solution is stable for 6 months.

Stock calcium standard, 1000 mg/L (25 mmol/L). Dissolve 250 mg of reagent-grade $CaCO_3$ in 3 mL of 1 mol/L HCl. Dilute to 100 mL with deionized water. Solution is stable for 6 months at room temperature in a closed container. The calcium standard may be purchased; for example, there is calcium reference standard solution SC 191-500 (Fisher Scientific, Pittsburgh).

Calcium blank solution (140 mmol/L Na, 5 mmol/L K, 0.1 mmol/L phosphate). Place 8.2 g of NaCl, 0.373 g of KCl, and 0.014 g of Na_2HPO_4 into a 1 L volumetric flask. Dissolve salts in about 250 mL of deionized water, and dilute to mark with deionized water. Solution is stable for 1 month at refrigerator temperatures.

Working standards

Ca (mg/L)	Stock Ca (mL)	Dilute to volume with calcium blank solution (mL)
50	5	100
100	10	100
120	12	100
150	15	100
200	20	100

Store at room temperature in glass flasks. Do not pipet from flasks. Pour aliquots daily into separate containers.

Assay

Equipment: Perkin-Elmer Atomic Absorption Spectrophotometer Model 460 (or other similar instrument) with a single-slot burner head and calcium-magnesium hollow-cathode lamp. Follow each manufacturer's directions for maintenance of burner, optimization of lamp and flame by peaking the flame, and air-acetylene fuel mixture, as well as for the actual analytical procedure.

Notes

1. The dilution (1:50) of serum or urine prevents protein interference. The dilution can be performed accurately and conveniently with a semiautomatic dilutor (such as that of Micromedic Systems, Inc., Philadelphia).
2. Sera left overnight in plastic sample cups may give low calcium values.
3. There is no appreciable variation in the patient's calcium levels throughout the day if exercise is avoided, and there is no difference between fasting and nonfasting serum calcium levels.
4. There is no difference between venous and capillary serum calcium levels.
5. Serum or plasma should not be allowed to remain in prolonged contact with the erythrocytes, since with time the cells become permeable to calcium.
6. Cerebrospinal fluid (CSF) can be analyzed for calcium using the procedure for serum analysis. Since CSF concentrations may be lower than serum concentrations, use of a slightly lower standard is advisable.

REFERENCE RANGES

Many sources report serum calcium levels of 90 to 110 mg/L for disease-free persons (''normals''). Newer methods show lower ranges of about 80 to 105 mg/L.

Meites[16] reports that premature infants have serum calcium levels of 60 to 100 mg/L and that full-term infants have serum calcium concentrations of 73 to 120 mg/L. In the first 24 hours after birth, calcium drops from 100 to 80 mg/L. This is followed by a slight rise, a drop in 75 mg/L, and a return to 85 mg/L at 72 hours.

There is no clinically significant difference in serum calcium for men and women. Urine calcium values vary considerably and are only meaningful if the patient is kept on a low-calcium, neutral-ash diet for 3 days before collection. Low output is 50 to 100 mg/day; the average is 100 to 300 mg/day. Values are lower in persons above 70 years old.

Population	Serum Ca mg/L (mmol/L)	Urine Ca mg/24 hr (mmol/24 hr)
Adults		
Atomic absorption	80-105 (2.0-2.6)	Men, <275 (<6.87) Women, <250 (<6.25)
Cresolphthalein complexone[17]	80-105 (2.0-2.6)	Hypercalcemic, >300 (>7.50)
Pediatric (by atomic absorption)[16]		
Premature infants[16]	60-100 (1.5-2.5)	
Full-term infants	73-120 (1.8-3.0)	
1 to 2 years	100-120 (2.5-3.0)	

REFERENCES

1. Weissman, N, and Pileggi, VJ: Inorganic ions. In Henry, RJ, Cannon, DC, and Winkelman, JW, editors: Clinical chemistry principles and techniques, ed 2, New York, 1974, Harper & Row, Publishers, Inc.
2. Clark, EP, and Collips, JB: A study of the Tisdall method for the determination of blood serum calcium with suggested modification, J Biol Chem 63:461-464, 1925.
3. Jackson, JE, Breem, M, and Cheng, C: Fluorometric titration of calcium, J Lab Clin Med 60:700-708, 1962.
4. Walmsley, TA, and Fowler, RT: Optimum use of 8-hydroxyquinoline in plasma calcium determinations, Clin Chem 27:1782, 1981.
5. Moorehead, WR, and Biggs, HG: 2-Amino-2-methyl-1-propanol as the alkalinizing agent in an improved continuous flow cresolphthalein complexone procedure for calcium in serum, Clin Chem 20:1458-1460, 1974.
6. Shirey, TL: Development of a layered-coating technology for clinical chemistry, Clin Biochem 16:147-155, 1983.
7. Kaplan, LA, and Pesce, AJ: Clinical chemistry: theory, analysis, and correlation, St Louis, 1984, The CV Mosby Co.
8. Cali, JP, Bowers, GN, Young, DS, et al: A reference method for he determination of total calcium in serum. In Cooper, GR, editors: Selected methods of clinical chemistry, vol 8, Washington, DC, 1977, American Association for Clinical Chemistry, pp 3-8.
9. Cali, JP, Mandel, J, Moore, L, and Young, DS: A reference method for the determination of calcium in serum, NPS Spec Pub 260-36, Washington, DC, 1972, US Department of Commerce National Bureau of Standards, Superintendent of Documents, US Government Printing Office.
10. Brett, EM, and Hicks, JM: Total calcium measurement in serum from neonates: limitations of current methods, Clin Chem 27:1733-1737, 1981.
11. Leidtke, RJ, Kroon, G, and Batjer, JD: Centrifugal analysis with automated sequential reagent addition: measurement of serum calcium, Clin Chem 27:2025-2028, 1981.
12. Porter, WH, Carrol, JR, and Roberts, RE: Hemoglobin interference with the DuPont automatic clinical analyzer procedure for calcium, Clin Chem 23:2145-2147, 1977.
13. Hass, RG, and Mushel, S: Hemoglobin interference compensation using a two pack calcium method on the DuPont ACA, Clin Chem 25:1126, 1979.
14. Ladenson, JH, McDonald, JM, and Goren, M: Multiple myeloma and hypercalcemia, Clin Chem 25:1821-1825, 1979.
15. Steward, AF, Adler, M, Byers, CM, et al: Calcium homeostasis in immobilization: an example of resorptive hypercalciuria, N Engl J Med 306:1136-1140, 1982.
16. Meites, S: Normal values for pediatric clinical chemistry, Washington, DC, 1974, American Association for Clinical Chemistry.
17. Baginski, ES, Marie, SS, Alcock, NW, et al: Calcium in biological fluids. In Faulkner, WR, and Meites, S, editors: Selected methods of clinical chemistry, vol 9, Washington, DC, 1982, American Association for Clinical Chemistry, pp 125-129.

Carbon dioxide

WILLIAM J. KORZUN
W. GREGORY MILLER

CO_2, total CO_2 content
Clinical significance: pp. 313 and 332
Molecular formula: CO_2 or $O=C=O$
Molecular weight: 44.01 daltons
Merck Index: 1793
Chemical class: gas, organic acid in soluble phase

PRINCIPLES OF ANALYSIS

Total carbon dioxide in serum or plasma exists in three major chemical forms: dissolved CO_2 (3%), carbamino derivatives of plasma protein (33%), and bicarbonate (HCO_3^-) anion (64%).[1] Other quantitatively minor forms are carbonic acid (H_2CO_3) and carbonate ions ($CO_3^=$). The majority of the procedures to quantitate total carbon dioxide content in serum or plasma involve acidification of the sample to convert all carbon dioxide forms to CO_2 gas and measurement of the amount of gas formed. The established reference method uses the Natelson Microgasometer[2,3] to measure the liberated CO_2 gas manometrically (method 1, Table 58-4). The measured gas pressure is corrected for other gases dissolved in serum by addition of alkali to totally absorb the CO_2 and measurement of the residual gas pressure. In another procedure the pressure exerted by CO_2 gas, which is proportional to the number of moles of gas in the sample, is measured by an electronic pressure transducer.

A commonly employed method for carbon dioxide determination is the continuous-flow procedure used on Technicon AutoAnalyzers (Technicon Instruments Corp., Tarrytown, NY).[4] The CO_2 gas, released by acidification of the sample stream, diffuses across a silicon-rubber membrane into an alkaline bicarbonate buffer containing a pH-indicating dye (phenolphthalein or cresol red) (method 2, Table 58-4). The CO_2 gas is quantitatively converted to bicarbonate and hydrogen ions. A pH change is produced, resulting in a change in color intensity of the indicator, which is detected spectrophotometrically. Since other gases dissolved in the serum do not produce a pH change, they do not interfere. The cresol red Technicon method uses Tris buffer whose pH response is temperature sensitive. In addition, the relationship between absorbance and CO_2 concentration is nonlinear and requires correction. Phenolphthalein reagent in bicarbonate buffer, which does not have these deficiencies, is preferred.

A second commonly used method employs a P_{CO_2} electrode to quantitate the CO_2 gas produced by acidification of the sample. The CO_2 gas is allowed to diffuse across a silicon-rubber membrane, and this changes the pH of a bicarbonate electrode buffer. The pH change is detected by

Table 58-4 Methods of total CO_2 analysis

Method	Analysis	Principle	Usage	Comment
1. Gas pressure	Manometric	CO_2 is released from samples after addition of acid, and the total gas pressure is measured manometrically. CO_2 gas is absorbed by alkali, and residual gas pressure (other than CO_2) is measured. The difference is the P_{CO_2}. P_{CO_2} = Total gas pressure − Residual gas pressure	Rare	Reference method; used for calibration purposes
2. pH indicator	Spectrophotometric	CO_2 + Acid → CO_2 gas; CO_2 gas $\xrightarrow{\text{Silicone membrane}}$ CO_2 dissolved, ↓ pH; H^+ + pH indicator → Color change	Common	Used in continuous-flow analyzers
3. CO_2 electrode	Ion-selective (pH) electrode	After acidification of sample, CO_2 diffuses across a silicone membrane into dilute bicarbonate buffer. The resulting pH change, as measured by a pH electrode, is related to total CO_2 content.	Common	Same method by which P_{CO_2} is measured in blood-gas analyzers
	Ion-selective CO_2 electrode	CO_2 diffuses across an ion-selective membrane into potentiometric cell. Potential of this cell is compared to reference cell containing ions at fixed concentrations. Two solutions form a liquid junction and concentration cell.	Common	Uses disposable electrodes; specific for Ektachem
4. Enzymatic	Spectrophotometric	$HCO_3^- + PEP \xrightarrow{PEPC} OX + P_i$; $OX + NADH + H^+ \xrightarrow{MDH} MAL + NAD^+$ (↓ A, 340 nm)	Common	Used on discrete analyzers, such as du Pont aca
5. Calculation	Ion-selective (pH and P_{CO_2}) electrode	Total CO_2 calculated from data obtained in blood-gas analysis. Total $CO_2 = \alpha P_{CO_2} + [HCO_3^-]$	Common	Although only an estimate, compares well with measured values

MAL, Malate; *MDH,* malate dehydrogenase; *NADH,* dihydronicotinamide adenine dinucleotide; *OX,* oxaloacetate; P_i, phosphate; *PEP,* phospho*enol*pyruvate; *PEPC,* phospho*enol*pyruvate carboxylase.

a glass pH electrode within the P_{CO_2} electrode assembly (method 3, Table 58-4, and Fig. 58-5).

Disposable ion-selective electrodes are used in the Kodak Ektachem. This procedure uses a concentration cell approach. In effect, the difference in concentration of a specific component, such as CO_2, results in a potential difference between two electrochemical cells. In this system, two cells are prepared at the same time, one containing the test solution and the other the reference solution. The cells, which are constructed identically, consist of the silver–silver chloride (Ag/AgCl) reference electrode, an ion-selective membrane, and an ion-selective indicator electrode. A liquid junction formed by the test and reference solutions results in the formation of a concentration cell.

Many discrete analyzers measure total carbon dioxide by quantitatively converting all CO_2 forms to HCO_3^- by adding alkali to the serum. The bicarbonate is then enzymatically converted to oxaloacetic acid, which is measured by an NADH consumption reaction that is quantitated spectrophotometrically as follows (method 4, Table 58-4):

$$HCO_3^- + \text{Phospho}\textit{enol}\text{pyruvate} \xrightarrow{PEPC} \text{Oxaloacetate} + P_i$$

$$\text{Oxaloacetate} + NADH + H^+ \xrightarrow{MDH} \text{Malate} + NAD^+$$

PEPC is phospho*enol*pyruvate carboxylase, P_i is inorganic phosphate, MDH is malate dehydrogenase. The decrease in absorption at 340 nm is related to the serum concentration of CO_2.

In another method, total CO_2 is calculated during a blood-gas analysis (method 5, Table 58-4). Using the measured pH and P_{CO_2} values, one can estimate the total CO_2 using the following equation:

$$\text{Total } CO_2 \text{ (mEq/L)} = \alpha P_{CO_2} + [HCO_3^-]$$

Where α, the Bunson coefficient, is the solubility coefficient of CO_2, and the bicarbonate concentration can be estimated from the Henderson-Hasselbalch equation:

$$[HCO_3^-] = \alpha P_{CO_2}(\text{antilog}[pH - pK_a'])$$

Although this calculation can provide only an indirect, unmeasured estimate of total CO_2, it is frequently performed for many blood-gas analyses. The calculation correlates well with measured total CO_2 values.

Both of the methods based on gas diffusion/pH change and the enzymatic procedure must be calibrated either with primary aqueous standards of sodium bicarbonate or serum-based secondary standards that have had carbon dioxide calibration values assigned to them. The manometric microgasometer technique requires no standard because the result is calculated from physical properties of gases and remains the reference for total CO_2 measurement. Each microgasometer is supplied with a tempera-

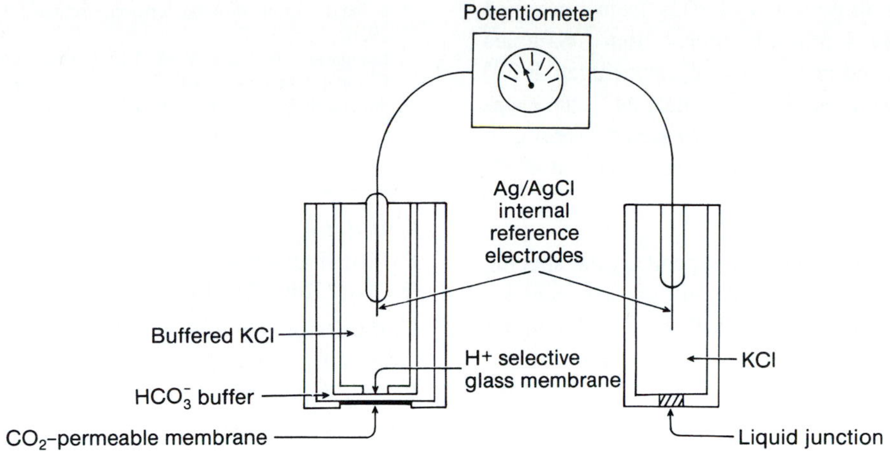

Fig. 58-5 Schema of a P_{CO_2} electrode assembly.

ture-correction chart that corrects for the gas pressure and the glass expansion for the temperature during the measurement. The technique is very accurate and specific and requires only 10 to 30 μL of sample. A drawback to the manometric procedure is that the instrument uses metallic mercury, which is toxic. Extra care therefore should be exercised when one is using this instrument. In addition, the analysis is time consuming and laborious. It is therefore only used as a reference procedure.

The enzymatic method reagents are all commercially available (such as from Sigma Chemical Co., St. Louis) and are reasonably stable once prepared if protected from absorbing atmospheric CO_2.

All these methods are capable of producing accurate and precise results (Table 58-5). For most laboratories with larger numbers of specimens, automated instruments using acidification and gas-diffusion methods or the enzymatic procedure are suitable. All provide clinically acceptable results.

SPECIMEN

One can use serum or heparinized plasma. Other anticoagulants cannot be used because they disturb the equilibrium between erythrocyte and plasma CO_2. They also lower the pH of the sample and accelerate the loss of CO_2 to ambient air. Whole blood cannot be used because the heme-bound CO_2 and carbamino-bound CO_2 vary with the hematocrit and oxygen saturation, producing variable results.

The major errors in total CO_2 measurement are associ-

Table 58-5 Comparison of reaction conditions for total CO_2 analysis

Parameter	Microgasometer*	pH indicator†	CO_2 electrode‡	Enzymatic§
Sample volume	30 μL	Approximately 8 μL	Approximately 175 μL	20 μL
Fraction of sample volume	0.25	0.013	0.05	0.004
Final concentration of reagents	Lactic acid: 833 mmol/L	H_2SO_4, 4 mmol/L Phenolphthalein, 63 μmol/L HCO_3^- buffer, 4.5 mmol/L	H_2SO_4, 80 mmol/L Surfactant	NADH: 320 μmol/L PEP-carboxylase, 160 U/L PEP, 16 mmol/L MDH, 3000 U/L
Wavelength	—	550 nm	—	340 nm
Reaction mode	End point	End point	End point	Kinetic
Temperature	Ambient	37° C	Approximately 40° C	37° C
Time of reaction	2-3 minutes	Approximately 9 minutes	45 seconds	56 seconds
Linearity		10-40 mmol/L	10-50 mmol/L	5 to 40 mmol/L
Interferences	None	None	None	None

*Natelson Microgasometer, Scientific Industries, Bohemia, NY.
†Technicon, AutoAnalyzer Method No. 5G4-0037PC6, Technicon Instruments Corp, Tarrytown, NY.
‡IL 446, Instrumentation Laboratory, Inc, Lexington, MA.
§Du Pont aca, EI du Pont de Nemours & Co, Wilmington, DE.

ated with sample handling. Loss of CO_2 from uncapped tubes is approximately 4 mmol/L after 1 hour.[5] Samples should ideally be handled anaerobically, centrifuged at 37° C, and stored tightly stoppered before analysis.[6] Once separated from erythrocytes and kept stoppered, serum or plasma total CO_2 content is stable for several days at 4° C.

Procedure

The details of total CO_2 methods depend on the specific design and operating parameters of various dedicated instruments. Procedures necessarily follow manufacturer's instructions.[7]

REFERENCE RANGE

Children	18-27 mmol/L
Adults	21-31 mmol/L

The reference ranges listed here are based on the manometric procedure. The values obtained by the other available methods range from 24 to 32 mmol/L, according to their respective manufacturer's product information.

REFERENCES

1. Davenport, HW: The ABC of acid-base chemistry, ed 6, Chicago, 1974, University of Chicago Press.
2. Natelson, S: Routine use of ultramicro methods in the clinical laboratory, Am J Clin Pathol 21:1153-1172, 1951.
3. Meites, S, and Faulkner, WR: Manual of practical micro and general procedures in clinical chemistry, Springfield, IL, 1961, Charles C Thomas.
4. Skeggs, LT, Jr: An automatic method for the determination of carbon dioxide in blood plasma, Am J Clin Pathol 33:181-185, 1960.
5. Gambino, SR, and Schreiber, H: The measurement of CO_2 content with the autoanalyzer, Am J Clin Pathol 45:406-411, 1966.
6. Astrup, P: A simple electrometric technique for the determination of carbon dioxide tension in blood and plasma, total content of carbon dioxide in plasma, and bicarbonate content in "separated" plasma at a fixed carbon dioxide tension, Scand J Clin Lab Invest 8:33-43, 1956.
7. Lustgarten, JA, and Wenk, RE: Carbon dioxide: review of methods, Clinical Chemistry, vol 27, no 7, check sample, Chicago, 1987, American Society of Clinical Pathologists.

Chloride

F. PHILIP ANDERSON
W. GREGORY MILLER

Clinical significance: pp. 313, 332, and 346
Atomic formula: Cl^-
Atomic weight: 35.453 daltons
Merck Index: 2062
Chemical class: Inorganic anion

PRINCIPLES OF ANALYSIS

Chloride is the major anion in extracellular body fluids and is measured in serum, plasma, urine, sweat, and occasionally other body fluids. Early methods to quantitate chloride depended on the very low solubility of silver chloride and mercury chloride salts. The most commonly used method today employs the quantitative displacement of thiocyanate by chloride from mercuric thiocyanate and subsequent formation of a red ferric thiocyanate complex, which is measured colorimetrically at 525 nm (method 2, Table 58-6).[1] Sensitivity and linear range of reaction are adjusted by adding excess mercuric ions as mercuric nitrate. Chloride first combines with free mercury ions (colorless) and then displaces any thiocyanate from mercuric thiocyanate. The free SCN^- reacts with Fe^{+3} to produce

Table 58-6 Methods of chloride measurement

Method	Type of analysis	Principle	Usage	Comments
1. Mercuric nitrate	Quantitative titration, end point	$2\,Cl^- + Hg(NO_3)_2 \rightarrow HgCl_2 + 2\,(NO_3)^-$ Excess Hg^{+2} + Diphenylcarbazone → Mercuric diphenylcarbazone *(blue)*	Serum, plasma, urine; manual	Relatively uncommon; poor precision
2. Mercuric/ferric thiocyanate	Quantitative, end point	$2\,Cl^- + Hg(SCN)_2 \rightarrow HgCl_2 + 2(SCN)^-$ $3\,(SCN)^- + Fe^{+3} \rightarrow Fe(SCN)_3$ *(red)* A_{max}, 525 nm	Serum, plasma, urine; manual, automated	Most frequently used; good accuracy and precision
3. Coulometric titration	Quantitative titration, end point	$Ag^+ + Cl^- \rightarrow AgCl(\downarrow)$	Serum, plasma, urine, fluids, sweat; manual, automated	Reference method; highly accurate
4. Ion-selective electrode	Quantitative, potentiometric end point or kinetic	$Ag \mid AgCl(s), Cl^- \mid AgCl, Ag_2S \mid$ Chloride-specific electrode (above) measures Cl^- in test solution versus reference electrode	Serum, plasma, urine, fluids, sweat; manual, automated	Increasingly used; good accuracy and precision
5. Isotope dilution	Mass spectrometer	Dilution of common isotopes with ^{37}Cl	Research or reference laboratories	Definitive method

a colored product measured at 525 nm. This approach will limit the minimum detectable chloride to 80 mmol/L but will greatly improve the sensitivity (absorbance change per millimole) in the clinically important serum range of 80 to 125 mmol/L. The ferric thiocyanate reaction is very temperature sensitive, and a constant temperature must be maintained to obtain accurate results.

A second, very common method quantitates chloride by coulometric titration with silver ions. Silver ions are generated at a constant voltage from a silver electrode, and they react with chloride ions to form insoluble silver chloride (method 3, Table 58-6).[2,3] The end point is detected amperometrically by a second pair of electrodes that specifically measure the free silver ions that result when all chloride ions are consumed. This procedure is accepted as the reference method for chloride. In principle one can determine the absolute amount of silver ions generated from the number of coulombs (current × time) produced and Faraday's constant (96,487 coulombs per equivalent). In practice, the time required to titrate a chloride standard solution or unknown sample with a constant current is measured. The unknown concentration is then calculated with the following proportion:

$$\frac{\text{Chloride concentration (standard)}}{\text{Titration time (standard)}} = \frac{\text{Chloride concentration (unknown)}}{\text{Titration time (unknown)}}$$

Methodological precision in coulometric titration is limited by the rate of mass transfer of silver ions from the generating electrode to the bulk solution. The time needed to detect the end point once all the Cl^- is consumed is called the *blank time* and depends on the rate of mixing in the vessel and the rate of Ag^+ formation. For maximum precision, the blank time should be a small fraction of the total titration time. Gelatin in the original procedure and polyvinyl pyrrolidone in current procedures prevent reduction of silver chloride at the indicating electrodes and promote uniform deposition of excess silver ions on the indicator cathode. It is necessary for the titration mixture to have an acid pH to prevent the formation of poorly soluble basic silver salts. Nitric acid in the reagent also provides ionic conductivity, and acetic acid sharpens the end point by reducing the slight solubility of silver chloride.[4]

Ion-selective electrodes, typically with a silver chloride–silver sulfide sensing element, are also used for the measurement of chloride (method 4, Table 58-6). Electrodes and measuring equipment are available from several manufacturers.

All the methods for chloride will show positive interference from other halides. The only clinically important interfering halide is bromide, which is administered in some drug preparations. Bromide and chloride do not react equivalently in all analytical systems. Bromide ion showed a reactivity equivalent to 2.3 chloride ions with one ion-selective electrode instrument, an equivalency to 1.6 chloride ions with the continuous-flow mercuric thiocyanate method, but an equivalency to 1 chloride ion with the coulometric titration method.[4]

The definitive method for chloride analysis is isotope dilution—mass spectrometry using samples spiked with ^{37}Cl.[5] Coulometric titration is both an AACC selected method[4] and the NBS reference method for serum chloride[6] because of its precision and its freedom from optical interferences. Because of the wide range of chloride concentrations possible in urine and sweat, chloride analyses are best performed with a coulometric method, although an ion-selective electrode method is also a satisfactory choice (Table 58-7).

SPECIMEN

Serum, heparinized plasma, urine, and other body fluids are acceptable. Serum, plasma, and other fluids should be promptly separated from cells (within 3 hours) to avoid shifts in the ionic equilibrium with metabolism and pH changes. Chloride in serum, plasma, urine, and other fluids is stable at least 1 week at room or refrigerator temperatures and at least 1 year frozen. Some ion-selective analyzers can measure chloride activity in whole blood, and such measurements should be made within 3 hours to avoid tonic shifts in electrolytes.

Sweat for chloride measurement is collected after iontophoretic delivery of pilocarpine to skin and sweat glands

Table 58-7 Comparison of reaction conditions for chloride measurement

Condition	Ferric thiocyanate*	Coulometric titration†
Temperature	~40° C	Ambient
pH	1.5-2.0	< 1
Final concentration of reagents	$Fe(NO_3)_3$, 50 mmol/L $Hg(SCN)_2$, 2 mmol/L $Hg(NO_3)_2$, 0.22 mmol/L	Acetic acid, 1.7 mol/L HNO_3, 0.1 mol/L Gelatin, 300 ng/L‡
Sample volume	175 µL	50 µL
Fraction sample volume	0.0025	0.012
Time of reaction	51 seconds	Variable
Precision (CAP survey data)	125 mmol/L, 2.2% CV 111 mmol/L, 1.9% CV 101 mmol/L, 1.6% CV	125 mmol/L, 2.5% CV 113 mmol/L, 2.4% CV 102 mmol/L, 2.0% CV
Interferences	Br^-, I^- Bilirubin (210 mg/L = 1 mmol/L of Cl^-) Lipemia (triglycerides, 6 g/L)	Br^-, I^-
Wavelength	525 nm	
Linearity	70-125 mmol/L	1-999 mmol/L

CV, Coefficient of variation.
*IL463 (Instrumentation Laboratory, Inc, Lexington, MA).
†Cotlove chloridometer (Buchler Instruments, Fort Lee, NJ).
‡Can be replaced by polyvinyl alcohol at a concentration of 1.8 g/L of acetic acid–nitric acid reagent.

to stimulate sweating. The area of skin that must be stimulated depends on the volume of sweat needed for analysis. Ion-selective electrode methods require less volume than coulometric titration methods. In any event, the rate of sweating must be greater than 1 $g/m^2/min$ to obtain clinically reliable results.[6] Children under 2 to 3 weeks of age may have unusually high sweat electrolyte concentrations. It is therefore advisable to delay the sweat test until after this time.[7] Sweat specimens should always be collected in duplicate to provide a check for contamination or evaporation during collection.

PROCEDURE: SERUM CHLORIDE BY COULOMETRY

Principle

This coulometric-amperometric titration is based on the method of Cotlove.[2]

Reagents

Nitric/acetic acid reagent (0.1 mol/L HNO_3 + 1.76 mol/L acetic acid). To 800 mL of distilled water, add 6.4 mL of concentrated nitric acid and 100 mL of glacial acetic acid. Mix. Solution is stable for 6 months at room temperature.

Polyvinyl pyrrolidone (PVA) solution, 18 g/L. Place 1.8 g of powdered PVA in 100 mL of cool distilled water. Heat just to boiling while stirring to dissolve PVA. Cool. Use immediately to prepare acid/PVA solution.

Acid/PVA solution. Add the 100 mL of PVA solution to the 900 mL of nitric/acetic acid reagent. Mix. This is stable at room temperature for 6 months.

Chloride standard, 100 mmol/L of Cl^-. Place 5.845 g of dry NaCl in a 1 L volumetric flask, and add approximately 900 mL of distilled water to dissolve the salt. Bring volume to 1 L with distilled water and mix well. Solution is stable for 6 months at room temperature.

Assay

Equipment: This procedure requires a Naake/Buchler Digital Chloridometer with concentrations in the range of 30 to 300 mmol/L. Lower and higher concentrations can be assayed by adjusting the conditions. Different instruments will require slightly different protocols, which will be found in the individual instrument operation manuals.

1. *Preparation of sample.* Dispense 4.0 mL of acid/PVA solution into vial. Pipet 0.1 mL of standard, sample, or control into appropriate vials, and rinse pipet in acid/PVA solution. Mix. For blanks, use 4.0 mL of acid/PVA solution.

   NOTE: Rinsing the inside of the sample pipet with reagent is absolutely necessary to avoid viscosity errors between aqueous standards and serum. Note that the exact volume of reagent in the vial is not important; however, it is critical that the reagent cover approximately 1 cm of the electrodes to ensure adequate mass transfer and conductivity.
2. *Daily setup.* Turn on chloridometer. Set test switch to low, titration switch to auto, and blank adjust to 00.0. Allow the instrument to warm up for 5 to 10 minutes. While waiting, polish the electrodes and check the silver generator wire. Clip off the thinned-out tip, and push down until it is almost flush with the stirrer. Tighten electrical connection, and polish. After cleaning, condition the electrodes by placing a vial of reagent in the instrument; raise the vial holder, and start the titration. A reading should appear within 30 seconds. If none appears, repeat this step with another vial of reagent.
3. *Reagent blank adjustment.* Set range to high (for 100 μL samples) or to low (for 10 μL samples); set titration to auto and blank adjust to 00.0. Sequentially place six vials of reagent in the instrument, titrate, and record the readings for the last four. Rinse electrodes with distilled water between each titration. Average the last four readings, and enter this value into the blank adjust thumbwheels.
4. *Calibration.* The instrument is electronically calibrated to give a correct result when either a 0.1 or 0.01 mL sample is assayed. Calibration should be checked by titrating a 0.1 mL standard sample (high setting) in triplicate. If the average value is not 100 mEq/L, then place the toggle switch to compensate and adjust the 10-turn potentiometer on the rear of the instrument one turn for each 2.5% change desired. Recheck the new calibration setting as described earlier.
5. *Sample titration.* Place the unknown sample vial in the instrument and titrate. The readout will give the result directly in milliequivalents per liter (mEq/L). Note that the blank, calibration, and patient samples must all be titrated with the range on the same setting.
6. Sweat Cl^- determinations may be made by placing the sweat collection filter paper directly into a sample vial and adding reagent. The filter paper must be pushed to the sides of the vial to avoid interfering with the mixing propeller, and the range switch (titration rate) must be set on low because of the small amount of Cl^- actually present.

REFERENCE RANGE

Serum or plasma, 101 to 111 mmol/L. These values were determined as the central 95% range obtained from analysis of 1205 (596 males, 609 females) healthy ambulatory persons 1 to 78 years old. No variation with sex or age was observed. An automated mercuric–ferric thiocyanate method, which was calibrated against the coulometric titration method using primary aqueous standards, was used.

Urine, 110 to 250 mmol/24 hr. Urine chloride levels are strongly dependent on dietary intake.

Sweat[6,7]

Normal	<35 mmol/L
Ambiguous	35 to 60 mmol/L
Cystic fibrosis	>60 mmol/L

Reliable sweat chloride measurements depend on an adequate rate of sweating being achieved, as discussed earlier. A repeat determination of sweat chloride (new collection and analysis) should always be made to confirm a positive finding. Intermediate values do not indicate heterozygote status. Adults have a more variable sweat composition than children and can have sweat chloride values up to 70 mmol/L.

REFERENCES

1. Levinson, SS: Chloride, colorimetric method. In Faulkner, WR, and Meites, S, editors: Selected methods of clinical chemistry, vol 9, Washington, DC, 1982, American Association for Clinical Chemistry.
2. Cotlove, E: Chloride: standard methods of clinical chemistry, vol 3, New York, 1961, Academic Press, Inc.
3. Dietz, AA, and Bond, EE: Chloride, coulometric-amperometric methods. In Faulkner, WR, and Meites, S, editors: Selected methods of clinical chemistry, vol 9, Washington, DC, 1982, American Association for Clinical Chemistry.
4. Elin, EJ, et al: Bromide interferes with determination of chloride by each of four methods, Clin Chem 27:778-779, 1981.
5. Velapoldi, RA, Paule, RC, Schaffer, R, et al: Standard reference materials: a reference method for the determination of chloride in serum, NBS Spec Publ 260-67, Washington, DC, 1979, US Government Printing Office.
6. Gibson, LE: The decline of the sweat test, Clin Pediatr 12:450-453, 1973.
7. Hammond, KB, and Johnston, EJ: Sweat test for cystic fibrosis. In Faulkner, WR, and Meites, S, editors: Selected methods of clinical chemistry, vol 9, Washington, DC, 1982, American Association for Clinical Chemistry.

Magnesium

E. CHRISTIS FARRELL, Jr.

Clinical significance: p. 373
Atomic formula: Mg
Atomic weight: 24.31 daltons
Merck Index: 5469
Chemical class: alkaline earth element
Forms: free Mg^{++}, protein bound or in inorganic complexes

PRINCIPLES OF ANALYSIS

Magnesium, like calcium, has offered a challenge to analysts because of the ease of contamination and problems with binding to serum proteins. Seventy-one percent of magnesium is free, 22% is bound to albumin, and 7% is bound to globulins.[1]

The first methods used for magnesium analysis were precipitation techniques,[2-4] in which the magnesium was precipitated from solution in the form of salts such as $MgNH_4PO_4$ (method 1, Table 58-8). The amount of precipitate, measured gravimetrically or by analysis of phosphorus in the precipitate, was directly related to the magnesium concentration. 8-Hydroxyquinoline was also used as a precipitate; the precipitate formed was quantified by a variety of techniques (method 2, Table 58-8). The manually performed precipitation methods are no longer in use.

Compleximetric techniques using EDTA[5] and other chelators[1,6] have been developed but are rarely used. Sensitive fluorometric assays employing calcein,[1] *o,o*-dihydroxyazobenzene,[7] and 8-hydroxyquinoline are also available.[8] In the latter method (method 3, Table 58-8), a chelated complex forms, has an excitation maximum at 420 nm, and fluoresces at 530 nm. This technique has been automated. Many fluorescence procedures employ chelating agents such as EGTA to inhibit interference from Ca^{++}.

One can measure magnesium by flame photometry at its band emission peaks at 370 or 383 nm, but this is also not a widely used procedure[1] (method 4, Table 58-8).

Several spectrophotometric methods that can be used directly on biological fluids after deproteinization have been developed. Several of these methods have been automated and are in widespread use today.

Gindler and Heth[9] introduced the use of calmagite (1-[1-hydroxy-4-methyl-2-phenylazo]-2-naphthol-4-sulfonic acid), a metallochromic dye, for direct determination of magnesium without deproteinization (method 5, Table 58-8). The alkaline, blue-colored working reagent forms a pink magnesium-calmagite complex with a resulting shift of the reagent color to reddish violet. This color is quantitated at 532 nm. The absorption spectra of the reaction are presented in Fig. 58-6. EGTA is used to prevent Ca^{++}

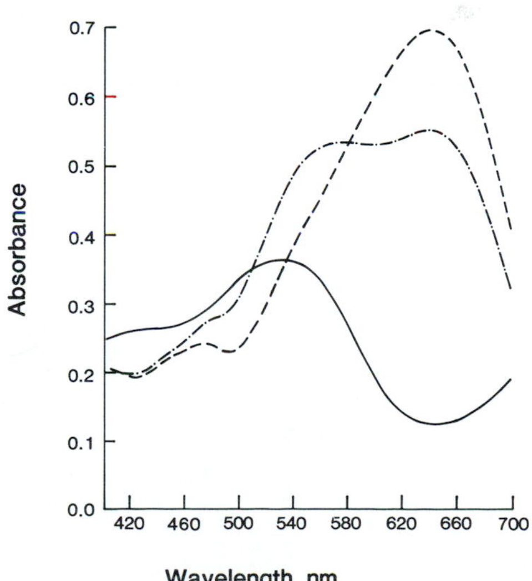

Fig. 58-6 Absorption spectra of calmagite and magnesium. *Dashed line,* reagent versus H_2O blank; *dotted-dashed line,* reagent and magnesium versus H_2O blank; *solid line,* reagent and magnesium versus reagent blank.

Table 58-8 Methods for analysis of magnesium

Method	Principle	Usage	Comments
1. Precipitation by ammonium phosphate with gravimetric analysis and with subsequent analysis for phosphate	Mg^{++} + Ammonium phosphate → Mg–ammonium phosphate *(precipitate)* Fisk-Subbarow phosphorus analysis of precipitate	Historical	One of first methods of analysis
2. Reaction with 8-hydroxyquinoline	Mg^{++} + 2 HO–(8-hydroxyquinoline) + 2 NH_4OH → (quinolyl)–O–Mg–O–(quinolyl) *(precipitate)* + 2 NH_4^+ + 2 H_2O Alternate analyses a. Quantitation with Folin's phenol reagent b. Blue-green color with ferric iron c. Bromination and titration of excess bromate	Historical	Labor intensive; supplanted by better methods
3. Titration of fluorescent complex	a. Mg^{++} + Calcein → Mg-calcein (*fluorescent;* λ_{ex}, 420 nm; λ_{em}, 530 nm)	Stat or small laboratories	Difficulty with fluorescence background noise as well as quenching
	b. Mg^{++} + *o,o'*-Dihydroxyazobenzene → Fluorescent complex	Rarely used	
4. Flame emission	$Mg^{++} \xrightarrow{2e^-} Mg^0 \xrightarrow{Heat} Mg^* \rightarrow$ Photon (λ_{em}; *370 nm, 383 mn)* + Mg^0	Historical	Poor sensitivity
5. Calmagite	Mg^{++} + (2-hydroxy-5-methylphenyl)–N=N–(1-hydroxy-naphthalene-4-SO_3H) $\xrightarrow{PVP}$ **Complex at 532 nm** *(violet)*	Widely used in laboratories without du Pont aca or atomic absorption; abapted to many automated analyzers	Slight negative bias; positive shift caused by lipemia
6. Methylthymol blue	Mg^{++} + Methylthymol blue → Complex at 510 and 600 nm	Du Pont aca, most commonly reported method	Good correlation with atomic absorption
7. Titan yellow	Mg^{++} + 2NaOH + Titan yellow $\xrightarrow{PVP}$ $Mg(OH)_2$–Titan yellow *(red lake; colloidal ppt)*	Infrequently used	Only one-eighth sensitivity of calmagite
8. Atomic absorption	$Mg^{++} \xrightarrow[+2e^-]{Heat} Mg^0 \xrightarrow[Photon]{} Mg^*$ (285.2 nm) (*indicates energized Mg)	Used in a number of routine laboratories	Excellent accuracy, reference method
9. Neutron activation isotope dilution	Uses magnesium isotope ^{27}Mg	Definitive method	
10. Enzyme colorimetric	Enzyme transfer of phosphate from Mg·ATP to substrate is dependent on Mg^{++} concentration in serum; reaction monitored by coupling to indicator system such as NADPH or peroxidase	Newly developed	Potential as new automated method

interference, and KCN is used to inhibit heavy metal complexes. Polyvinylpyrrolidone and its 9-ethyleneoxide adduct (Bion PVP, Bion Ne-9) have been employed to prevent serum proteins from shifting the absorbance maxima of the magnesium-calmagite complex.[10] The reaction is quite rapid and is completed in 60 seconds. This method has been adapted to a large number of automated analyzers, especially centrifugal analyzers, and is a widely used procedure.

Methylthymol blue, which has been used for calcium analysis,[1] has also been employed for the analysis of magnesium. Again, specific calcium chelators are used to increase the specificity of the analysis. The methylthymol blue (MTB) reaction has been automated on the du Pont aca; the color is quantitated bichromatically at 510 and 600 nm (method 6, Table 58-8).

Another direct spectrophotometric method is based on the reaction between magnesium and the dye Titan yellow (Clayton Yellow, Thiazol yellow, 2,2′[(di-azoamino)-di-*p*-phenylene]bis[6-methyl-7-benzothiazolesulfonic acid] disodium salt). Under the alkaline conditions of the reaction, the colloid $Mg(OH)_2$ forms.[1,11] The colloidal particles adsorb the reactive fraction of the Titan yellow dye to form a red complex (lake) measured at 540 nm (method 7, Table 58-8). To prevent precipitation of the dye-colloid complex, stabilizers, such as polyvinyl alcohol, have been added to the reaction mixture. The stabilizers also increase the intensity of the color.[1] This method has been adapted for automated analysis but is a rarely used procedure.

A widely used method for magnesium analysis is atomic absorption (method 8, Table 58-8), which is also the reference method. Magnesium has a strong spectral emission or absorption line at 285.2 nm, which can be readily isolated and used to measure magnesium concentration specifically in biological fluids. Although atomic absorption is most frequently employed as a manual procedure, several steps have been semiautomated.

The definitive method of magnesium analysis is neutron activation with ^{27}Mg (method 9, Table 58-8).

Recently enzyme assays for magnesium have been developed[12,13] (method 10, Table 58-8). These assays are based on the principle that an Mg-ATP complex must be present to enable the phosphorylating enzyme to transfer the high-energy phosphate from the ATP to another molecule. The original procedure of Tabata et al.[12] proposed the use of hexokinase in a coupled reaction, as follows:

$$\text{D-Glucose} + \text{Mg·ATP} \xrightarrow{\text{Hexokinase}} \text{D-Glucose 6-phosphate} + \text{Mg·ADP}$$

$$\text{D-Glucose 6-phosphate} + \text{NADP}^+ \xrightarrow{\text{G6PHD}} \text{D-Gluconolactone 6-phosphate} + \text{NADPH} + \text{H}^+$$

The reaction rate is followed by monitoring the rate of NADPH formation at 340 nm. This enzymatic approach has been modified by Wimmer et al.,[13] who used a glycerol kinase, glycerolphosphate oxidase, and peroxidase coupled sequence. In all these assays, it is presumed that the Mg^{+2} binds to the ATP in solution and remains bound to the ADP reaction product.

REFERENCE AND PREFERRED METHODS

When comparing methods or comparing the results of proficiency studies, one must be aware of the use of four different systems of units for the reporting of magnesium concentrations:

$$0.5 \text{ mmol/L} = 1.0 \text{ mEq/L} = 1.22 \text{ mg/dL} = 12.2 \text{ mg/L}$$

The direct spectrophotometric analyses, adapted to automated techniques, are fairly accurate and precise (Table 58-9).

On the basis of precision and accuracy, atomic absorption must be the preferred method. Since the methylthymol blue method correlates well with atomic absorption, I recommend it as a second choice. For those laboratories without atomic absorption instrumentation the automated calmagite method will provide adequate clinical data.

SPECIMEN

Serum must be separated from the clot as soon as possible, or the level of magnesium will increase because of its elution from the red blood cells.[1] The erythrocytes contain about three times the Mg^{+2} found in plasma, and the Mg^{+2} is free within the interior of the cell, not bound to the membrane like Ca^{+2}.[1] Thus, in general, hemolyzed samples are unacceptable for analysis. The presence of Ca^{+2} gluconate in serum because of intravenous administration can cause results of the Titan yellow method to be 35% low.[1] Blood anticoagulated with citrate, oxalate, or EDTA is unacceptable because these compounds can chelate the magnesium ions.

Urine should be acidified to pH 1 with concentrated HCl. If a precipitate forms, shake, mix, acidify, and warm to 60° C to redissolve it.[1]

PROCEDURE: MAGNESIUM BY ATOMIC ABSORPTION SPECTROPHOTOMETRY

Principle

The magnesium determinations by atomic absorption spectroscopy are based on the fact that atoms of an element in the "ground," or unexcited, state absorb light of the same wavelength as that emitted by the element in the excited state. Each element has its own characteristic absorption or resonance lines, and no two elements are known to have an identical resonance line.

Magnesium in serum or urine is determined after sufficient dilution with lanthanum oxide solution to avoid interference by phosphate ions.

Reagents

Use special atomic absorption glassware and water that are free of calcium and magnesium contamination.

Table 58-9 Reaction conditions for magnesium analysis

Condition	Calmagite*	Atomic absorption†	Methylthymol blue‡
Temperature	25°-37° C	2000° C	37° C
Sample volume	Variable	50 μL	20 μL
Fraction of sample volume	0.01	0.02	0.004
Final concentration of reagents	Calmagite: 0.06 g/L (10 parts dye plus 1 part base reagent) KCl: 0.34 mol/L Bion NE-9: 0.98 g/L Bion PVP: 0.91 g/L KCN: 0.280 mmol/L KOH: 25.6 mmol/L EGTA: 0.041 g/L	Lanthanum: 1 g/L for serum and CSF; 5 g/L for urine	EGTA§: 500 μmol/L Methylthymol blue: 60 μmol/L Sodium metaborate: pH > 11
Wavelength	532 nm	285.2 nm	600 and 510 nm (primary and secondary)
Reaction	End point	End point	End point
Time of reaction	30 seconds	30 seconds	261.5 seconds
Linearity	40 mg/L	50 mg/L	200 mg/L
Interferences (negative)	Chelators (EDTA, oxalate)	Chelators	Chelators
Precision ($\overline{X}$, *CV*)‖	39.4 mg/L, 11.7% CV 19.2 mg/L, 12.1% CV	40.9 mg/L, 7.0% CV 20.0 mg/L, 7.4% CV	40.2 mg/L, 4.8% CV 19.5 mg/L, 9.4% CV

*Performed on reagent obtained from Pierce Chemical Co (Rockford, IL), adaptable to any spectrophotometer.
†Performed as described in text.
‡As used on the du Pont aca.
§From College of American Pathologists (CAP) Quality Assurance Survey data; $\overline{X}$, mean; *CV*, coefficient of variation.
‖Ethylene glycol-bis(β-aminoethylether)*N,N'*-tetraacetic acid.

Stock lanthanum, 50 g/L (314 mmol/L) in 4 mol/L HCl. Weigh 58.64 g of lanthanum oxide (La_2O_3), and wet with about 50 mL of deionized water. Add 250 mL of concentrated HCl *slowly* and *cautiously* under hood. Mix to dissolve powder. Dilute to 1 L with deionized water. Store in dark at room temperature. Solution is stable for 6 months.

Working lanthanum diluent for serum, 1 g/L (6.28 mol/L). Dilute 20 mL of stock lanthanum to 1 L with deionized water. Store in dark bottle at room temperature. Solution is stable for 3 months.

Working lanthanum diluent for urine, 5 g/L (31.4 mmol/L). Dilute 100 mL of stock lanthanum to 1 L with deionized water. Store in dark bottle at room temperature. Solution is for 3 months.

Stock magnesium standard, 1000 mg/L (1 mg/mL, 82.2 mEq/L). Stock magnesium standard is commercially available from Fisher Scientific (50-M-51). Alternatively, dissolve 100 mg of magnesium metal (commercially available in powder or ribbon form) in 2 mL of 1 mol/L HCl in a 100 mL volumetric flask. After metal has completely dissolved, dilute to 100 mL with distilled, deionized water. Store in a polyethylene bottle at 4° C. Solution is stable for 12 months.

Working standards

Mg μg/mL (mEq/L)	Stock Mg (mL)	Dilute to volume with working lanthanum, 1 g/L (mL)
10 (0.82)	1	100
20 (1.64)	2	100
30 (2.47)	3	100
40 (3.28)	4	100
50 (4.11)	5	100

Assay

Equipment: Perkin-Elmer Atomic Absorption Spectrophotometer Model 460 with a single-slot burner head and calcium-magnesium cathode lamp. Follow manufacturer's instructions for proper analytical procedure and for establishing a proper air-acetylene fuel ratio.

A Micro-Medic dilutor, Model 25000, or its equivalent, is used to prepare 1:50 dilutions of samples with the appropriate lanthanum solution.

Note

The dilution (1:50) of serum or urine prevents protein interference.

REFERENCE RANGE

Urinary excretion of magnesium is diet dependent but normally equals about one third of daily intake.[1]

Method	Mg^{++}
Serum and cerebrospinal fluid	
Titan yellow[1] for serum	18.23-29.3 mg/L (1.5-2.4 mEq/L)
Average value[1] for CSF (no reported sex difference)	24.4-32.0 mg/L (2.0-2.7 mEq/L)
Atomic absorption	15.8-25.5 mg/L (1.3-2.6 mEq/L)
Urine	24-255 mg/24 hr (2-21 mEq/24 hr)

REFERENCES

1. Henry, RJ, Cannon, DC, and Winkelman, JW: Clinical chemistry principles and techniques, ed 2, New York, 1974, Harper & Row, Publishers, Inc.
2. Kramer, B, and Tisdall, FF: A single technique for the determination of calcium and magnesium in small amounts of serum, J Biol Chem 47:475-481, 1921.
3. Briggs, AP: A colorimetric method for the determination of small amounts of magnesium, J Biol Chem 52:349-355, 1922.
4. Denis, W: The determination of magnesium in blood, plasma and serum, J Biol Chem 52:411-415, 1922.
5. Schwartzenbach, G, Bidermann, W, and Banserter, F: Komplex-one VI: neue einfache Titriermethoden zur Bestimmung der Wasserhärte, Helv Chim Acta 29:811-818, 1946.
6. Thiers, RE: Magnesium (Fluorometric). In Meites, S, editor: Standard methods in clinical chemistry, vol 5, New York, 1965, Academic Press, Inc.
7. Brien, M, and Marshall, RT: An automated fluorometric method for the determination of magnesium in serum and urine using *o-o*-dihydrozyazobene: studies on normal and uremic subjects, J Lab Clin Med 68:701-712, 1966.
8. Schachter, D: The fluorometric estimation of magnesium in serum and urine, J Lab Clin Med 54:763-768, 1959.
9. Gindler, EM, and Heth, DA: Colorimetric determination with bound "Calmagite" of magnesium in human blood serum, Clin Chem 17:663, 1971 (abstract).
10. Pierce Magnesium Stat Kit, Pierce Chemical Co., Rockford, IL; insert revised Feb 1979.
11. Basinski, DH: Magnesium (Titan yellow). In Meites, S, editor: Standard methods in clinical chemistry, vol 5, New York, 1965, Academic Press, Inc.
12. Tabata, M, Kido, T, Totani, M, and Marachi, T: Direct spectrophotometry of magnesium in serum after reaction with hexokinase and glucose-6-phosphate dehydrogenase, Clin Chem 31:703-705, 1985.
13. Wimmer, MC, Artiss, JD, and Zak B: A kinetic colorimetric procedure for quantifying magnesium in serum, Clin Chem 31:629-632, 1986.

Osmolality

LAWRENCE A. KAPLAN

Clinical significance: p. 207

PRINCIPLES OF ANALYSIS

Osmolality is a colligative property of solutions that depends on the number of dissolved particles present in the solution. The three types of solutes most often encountered in biological fluids are electrolytes, organic molecules, and colloids. As the number of dissolved particles increases, the freezing point and vapor pressure of a solution are decreased and the osmotic pressure and boiling point are increased.[1] In dilute solutions there is a linear change in the colligative properties as the solute concentration increases. It is important to realize that it is not the mass concentration but the molal concentration (that is, moles per kilogram of solvent) that is the basis of colligative properties. For example, in biological fluids the concentration of salts and low-molecular-weight organic compounds (such as glucose and urea) affects osmolality much more than albumin, which is present in a large mass amount but, because of its large molecular weight, is present at low molar concentrations. Thus the molar concentration of serum NaCl, present at approximately 9 g/L, is approximately 150 mmol/L. The molar concentration of serum albumin, present at approximately 40 g/L, is approximately 0.58 mmol/L.

The most frequently measured colligative property used to estimate fluid osmolality is freezing-point depression. When 1 mol of solute is added to 1 kg of water, the freezing point is depressed by 1.858° C. The laboratory osmometer actually employs a sensitive heat thermistor to measure the heat released by the freezing fluid and relates it to the freezing point of the fluid.

A less frequently measured parameter is vapor-pressure depression. The vapor-pressure osmometer actually measures the dew-point depression of the vapor, that is, the vapor in equilibrium with the solution being measured. The more dissolved particles present (increased osmolality), the lower the vapor pressure of the aqueous component of the solution. An important exception to this is found when the solute itself is a volatile substance.

A stat procedure frequently used to estimate the osmolality of a fluid is the calculated osmolarity. To calculate serum osmolarity, one sums up the molar concentrations of the principal active osmolar solutes present in serum.[2] Although there are many equations used to calculate the osmolarity, the simplest and the most frequently used is as follows:

$$\text{Calculated mOsm/L } H_2O = 1.86\,[Na^+] + \frac{\text{Glucose (mg/L)}}{180\text{ mg/mmol}} + \frac{\text{BUN (mg/L)}}{28\text{ mg/mmol}} \qquad \textit{Eq. 58-1}$$

Table 58-10 Comparison of freezing-point and vapor-pressure osmometers

Conditions	Freezing point*	Vapor pressure†
Sample volume	20-2000 μL	8 μL
Time of analysis	< 60-70 sec	90 sec
Interferences	None	Volatile solutes (negative)
Precision‡ *($\bar{X}$, % CV)*		
Serum	277 mOsm/kg of H_2O, 1.8% CV	274 mOsm/kg of H_2O, 3.0% CV
	386 mOsm/kg of H_2O, 1.5% CV	377 mOsm/kg of H_2O, 2.8% CV
Urine	643 mOsm/kg of H_2O, 1.3% CV	628 mOsm/kg of H_2O, 2.3% CV
	882 mOsm/kg of H_2O, 1.4% CV	872 mOsm/kg of H_2O, 2.0% CV

$\bar{X}$, Mean; *CV*, coefficient of variation.
*Osmette A, Precision Systems, Inc, Sudbury, MA.
†Wescor model 5100C, Wescor, Inc, Logan, UT.
‡College of American Pathologists Quality Assurance Survey data.

where sodium concentration is expressed in milliequivalents per liter (mEq/L), and the concentrations of glucose and blood urea nitrogen (BUN) are divided by their respective formula equivalent weights, 180 and 28, to convert these mass concentrations into millimoles per liter (mmol/L).[3] It should be pointed out that most automated analyzers now determine a calculated serum osmolality. Thus the most frequently reported serum osmolality is a calculated serum osmolality.

It has been pointed out that the two types of osmometers differ in their ability to respond to volatile solutes that may be clinically encountered.[4,5] The contribution to serum osmolality of several commonly seen alcohols (such as ethanol, methanol, and isopropanol) is directly proportional to their molar concentrations when osmolality is measured by freezing-point depression osmometry. However, vapor-pressure osmometers do not detect these volatile substances because volatile solutes contribute to the total vapor pressure present above the solution, although they decrease the vapor pressure of the water portion of the solution.[6] Therefore the decreased vapor pressure of water is counterbalanced by the increase in vapor pressure caused by the presence of a volatile substance. As the concentration of the volatile substance increases in aqueous solutions, the vapor pressure above the solution actually increases, giving falsely low osmolality values. Table 58-10 compares freezing-point and vapor-pressure osmometers.

The calculated osmolality is not meant to provide a definitive result but only one sufficient for most acute emer-

Table 58-11 Osmolar gap associated with toxic substances present in serum at toxic concentrations

Substance	Osmolar gap (mOsm/kg of H_2O)
Ethanol	80
Isopropanol	60
Methanol	27
Acetone	10
Ethylene glycol	4

gency service work. Used this way, the calculated osmolality can provide clinicians with valuable information with no extra analysis time. The error introduced by calculation of the serum osmolality as mOsm/L instead of mOsm/kg can be clinically significant only if the percentage of water in the sample (usually 91% to 93%) drastically changes. Thus highly lipemic samples or samples with large amounts of protein (as from multiple myeloma patients) can have an inaccurate calculated osmolality. The laboratory might not wish to report a calculated osmolality on such specimens or alternatively might affix a comment to the results warning the physician of a possible error.

The calculated osmolality will give falsely low results in the presence of volatile solutes. The differences between measured and calculated serum osmolality, the ''osmolality gap,'' has been suggested as a rapid test for the presence of ingested volatiles, especially alcohols[7-9] (see also p. 1071). Alcohols have a significant effect on the serum osmolality because of their low molecular weights (Table 58-11).

SPECIMEN

Blood drawn without the use of any anticoagulant is the preferred sample. The serum should be removed from the clotted blood cells as rapidly as possible. Well-centrifuged urine or other body fluids are also acceptable samples. Random urine samples can vary considerably in the concentration of analytes and are rarely useful clinically. Therefore an aliquot of a 24-hour urine collection is the preferred specimen for the measurement of urine osmolality.

Any fluid left uncovered will have an increased osmolality because of evaporation and concentration of sample solutes. Serum osmolality has been reported[10,11] to be stable for 3 hours at room temperature, 10 hours at 4° C, and as long as 3 days at 4° C. Urine osmolality may be stable at 4° C for up to 24 hours.[11] Frozen samples may be stable for several weeks if the samples are sealed to prevent solvent loss by sublimation.

PROCEDURE: OSMOLALITY BY FREEZING-POINT DEPRESSION

Principle

The sample of standard is supercooled below its freezing point and then agitated by a rapidly vibrating probe. This induces rapid crystallization of the sample, and it begins to freeze. Heat of fusion is released by the crystallization and is accurately measured by a thermistor probe present in the solution. The released heat is related to the freezing-point depression (1 mOsm of solute depresses the freezing point by 0.001858° C), which is converted to osmolality.

Reagent

Osmolality standards can be purchased from a variety of commercial sources or can be provided by the instrument manufacturers. Calibrators are used to standardize the instrument for either serum or urine osmolalities.

A relatively simple procedure for preparing osmolality standards has been described.[11] Reagent-grade NaCl is heated to 200° C overnight to drive off any water present in the crystals. After it cools, the desired amount of NaCl is weighed out (see following) and added to 1 kg of class I distilled water. The most convenient way to do this is to fill a 1 L class A volumetric flask with water at 20° C to the mark. Add exactly 1.8 mL of additional water, and then add the salt to dissolve.

Desired osmolality (mOsm/kg)	Grams of NaCl per kilograms of H_2O
100	3.094
300	9.476
500	15.93
1000	32.12

The standard should be stored in clean polyethylene or borosilicate glass bottles. Always pour the standards into clean test tubes before use. When stored properly in closed containers, the standards will be stable for 3 to 6 months at room temperature.

Assay

Follow manufacturers' instructions for the calibration, use, and maintenance of osmometers.

REFERENCE RANGE

Serum. Most reported ranges for serum osmolality fall between 282 and 300 mOsm/kg. There are no sex or age differences for serum osmolality.

Urine. Urine osmolality will vary considerably with diet. Twenty-four–hour osmolalities will vary between 50 and 1200 mOsm/24 hr.

Cerebrospinal fluid. CSF and other true body fluids will have an osmolality that is essentially equal to that of serum drawn at the same time.

REFERENCES

1. Stevens, SC, Neumayer, F, and Gutch, CF: Serum osmolality as a routine test, Neb Med J 45:447, 1960.
2. Weisberg, HF: Osmolality—calculated "delta" and more formulas, Clin Chem 21:1182-1184, 1975.
3. Holmes, JH: Measurement of osmolality in serum, urine and other biologic fluids by the freezing point determination. In Preworkshop manual on urinalysis of renal function studies, Chicago, 1962, American Society of Clinical Pathologists, Commission on Continuing Education.
4. Rocco, RM: Volatiles and osmometry I, Clin Chem 22:399, 1976.
5. Barlow, WK: Volatiles and osmometry II, Clin Chem 22:1230-1232, 1976.
6. Mercier, DE, Feld, RD, and Wittell, DL: Comparison of dewpoint and freezing point osmolality, Am J Med Technol 44:1066-1069, 1978.
7. Pappas, AA, Gadsden, RH, Jr, Gadsden, RH, and Groves, WE: Computerized calculation of osmolality and its automatic comparison with observed serum ethanol concentration, Am J Clin Pathol 77:449-451, 1982.
8. Coakley, JC, Tobgui, S, and Dennis, PM: Screening for alcohol intoxication by the osmolar gap, Pathology 15:321-323, 1983.
9. Bhagat, CI, Gracia-Webb, P, Fletcher, E, and Beilby, JP: Calculated vs. measured osmolalities revisited, Clin Chem 30:1703-1705, 1984.
10. Johnson, RB, and Hoch, H: Osmolality. In Meites, S, editor: Standard methods of clinical chemistry, vol 5, New York, 1965, Academic Press, Inc.
11. Weissman, N, and Pilegg, VJ: Inorganic ions. In Henry, RJ, Cannon, DC, and Winkleman, JW, editors: Clinical chemistry principles and techniques, New York, 1974, Harper & Row, Publishers, Inc.

Phosphorus

E. CHRISTIS FARRELL, Jr.

Phosphorus, inorganic phosphorus, phosphate
Clinical significance: pp. 346, 373, and 664
Molecular weight: phosphate 97.00, 96.00 daltons (phosphorus at 30.98)
Chemical class: inorganic anions (P is nonmetallic group 5 element)
Forms: $H_2PO_4^{-1}$, HPO_4^{-2}

PRINCIPLES OF ANALYSIS

Because elemental phosphorus does not exist to any appreciable extent in the body, methods have been directed toward analysis of the two phosphate anions, which interchange rapidly, depending on the pH. The monovalent and divalent anion forms are present in serum at about equal concentrations in acidosis, in a ratio of 1:9 in alkalosis, in a ratio of 1:4 at pH 7.4, and in a ratio of 100:1 in a pH 4.5 urine, making it impossible to identify with any certainty the molecular weight of inorganic "phosphate." Therefore the units traditionally chosen have been milligrams or millimoles of phosphorus (in a volume) but never milliequivalents of phosphate, since that would change rapidly with the charge.

The oldest and still most commonly used methods are based on the reaction of phosphate ions with molybdate to form complex structures such as ammonium phosphomolybdate $(NH_4)_3[P(Mo_3O_{10})_4]$. One can measure phosphomolybdate complexes directly or convert them to molybdenum blue using a wide variety of reducing agents (methods 1 to 7, Table 58-12). Molybdenum blue is a complex heteropolymer of unknown structure whose absorbance is usually measured at 660 nm.

Table 58-12 Methods of phosphorus analysis

Method	Principle	Usage	Comments
1. Hydroquinone*	Reduction of phosphomolybdate to molybdenum blue	Historical	Irregular, rapid fading of color
2. ANS (1-amino-2-naphthol-4-sulfonic acid)*		Historical	Better than hydroquinone
3. ANS + 100° C, 5 min*		Historical	Intensified color
4. $SnCl_2$*		Early continuous flow	Replaced by improved methods because of reagent instability, poor linearity, and lack of precision
5. $SnCl_2$ + Hydrazine*		Continuous flow	Good accuracy, limited linearity, better reagent stability than method 4
6. NH_4FeSO_4*		Discrete analyzers	Not so sensitive as ultraviolet methods
7. *p*-Semidine*		Lipid phosphorus	Research applications
8. No reduction*	340 nm, monitoring	Discrete and continuous flow	Good accuracy, precision, sensitivity
9. No reduction*	Bichromatic	Some discrete analyzers	Bichromatic (340 and 380 nm) about one-half sensitivity of 340 nm; blanking problems with hemolysis and lipemia
10. Enzymatic†	HPO_4^{-2} + Inosine $\xrightarrow{PNP}$ Hypoxanthine + Ribose-1-phosphate; Hypoxanthine + $2H_2O$ + $2O_2$ $\xrightarrow{XOD}$ Uric acid + $2H_2O_2$; H_2O_2 + Chromogenic substrate $\xrightarrow{Peroxidase}$ *(reddish purple complex)* (A_{max}, 555 nm)	Discrete analyzers	New assay

*General principle: $PO_4 + H_2SO_4 + (NH_4)_6Mo_7O_{24}\cdot 4H_2O \rightarrow$ Mo-PO_4 complex.
Mo/PO_4 complex + Reducing agent $\rightarrow$ Heteropolymeric blue complex.
†*PNP*, Purine nucleoside phosphorylase (EC 2.4.2.1); *XOD*, xanthine oxidase (EC 1.2.3.2); chromogenic substrate is *N*-ethyl-*N*-(3-methylphenyl)-*N*-acetylethylenediamine.

The classical method of Bell and Doisy for phosphate was based on the reduction of phosphomolybdic acid by hydroquinone to molybdenum blue[1] (method 1, Table 58-12). It was limited by irregular and rapid fading of the color in alkaline solution. Fiske and Subbarow[2] demonstrated that 1-amino-2-naphthol-4-sulfonic acid (ANS) was a superior reducing reagent that could be used at room temperature (method 2, Table 58-12). Bartlett[3] intensified the color by heating the reaction to 100° C for 5 minutes (method 3, Table 58-12).

Kuttner and Cohen[4] showed that stannous chloride stock reagent was more stable than ANS and that it produced a more intense color with phosphomolybdic acid (method 4, Table 58-12). Hurst stabilized the dilute stannous chloride working reagent by combining it with hydrazine sulfate[5] (method 5, Table 58-12). This also permitted rapid development of the color at room temperature. Technicon[6] produced a method for the AutoAnalyzer and the Multichannel analyzers based on the work of Hurst[5] and Kraml,[7] in which the stannous concentration is doubled to increase linearity.

Taussky and Shorr[8] proposed the use of ferrous ammonium sulfate as a reducing agent; this procedure is still employed by several kit manufacturers (method 6, Table 58-12). The merits of several methods of phosphate analysis employing reducing reactions are discussed by Martinek in 1970 review.[9]

Simonsen et al.[10] reported that one can quantitate phosphate using the 340 nm absorbance of unreduced phosphorus-molybdate complex (methods 8 and 9, Table 58-12). This made the reaction faster and simpler and improved reagent stability. It also produced three to four times more absorbance change than the molybdenum blue reductions. Daly and Ertinghausen[11] adapted this technique to the CentrifiChem (Baker Instruments, Dallas), studied the reaction kinetics, and adjusted reagent concentrations so that only 2% of the reaction occurred before the 3-second initial read time on the CentrifiChem. Sera had half-reaction times of about 70 seconds, and aqueous standards had half-reaction times of 100 seconds. During the 12-minute total reaction time, 99% of the final color was produced. One interesting observation, which has been confirmed, is

Table 58-13 Reaction conditions for serum phosphate analysis

Condition	Nonreduction at 340 nm*	Reduction method†
Sample volume	3 μL	10 μL
Fraction of sample volume	0.013	0.014
Final concentration of reagents	Ammonium molybdate: 2.25 mmol/L Sulfuric acid: 750 mmol/L, pH 1.55 Tween 80: 0.45 mL/L	Sodium acetate: 55.7 mmol/L Acetic acid: 328 mmol/L Sodium metabisulfate: 37.2 mmol/L *p*-Methyammonium phenol sulfate: 11.0 mmol/L Ammonium paramolybdate · 7 H_2O: 587 μmol/L H_2SO_4: 16.2 mmol/L Cu^{+2}: as catalyst
Temperature	37° C	30° C
Wavelength	340 nm	620 nm
Reaction	End point with serum blank	Two-point kinetic
Time	200 sec	300 sec
Precision‡ ($\bar{X}$, % CV)	25 mg/L, 9.2% CV 78.7 mg/L, 5.3% CV	24.9 mg/L, 7.2% CV 77.3 mg/L, 4.8% CV
Linearity	80 mg/L	150 mg/L

$\bar{X}$, Mean; *CV,* coefficient of variation.
*As analyzed on COBAS analyzer (Roche Inc, Nutley, NJ), using reagent obtained from Gilford Diagnostics, Oberlin, OH.
†As analyzed on CentrifiChem Analyzer (Baker Instruments, Dallas), using reagent from Smith-Kline-Beckman (Sunnyvale, CA).
‡From College of American Pathologists Quality Assurance Survey data for nonreduction and reduction method groups (overall).

that the formation of the molybdate complex is slowed when the acidity is increased.

Monitoring of the unreduced phosphorus-molybdate complex has been adapted to a wide variety of automated and semiautomated analyzers, including those produced by Gilford, Abbott, Technicon, and du Pont. Bichromatic instruments use 340 and 380 nm filters to monitor the unreduced complex.

Direct measurement of inorganic phosphate without reduction has also been performed by using a mixed vanadium-molybdenum heteropoly acid.[12]

Enzymatic methods have been proposed to measure serum inorganic phosphate (method 10, Table 58-12). A cited advantage was the stability of organic phosphate compounds in the neutral range where most enzymes function. One early assay proposed by Schultz et al.[13] used a coupled enzyme system of glycogen phosphorylase, phosphoglucomutase, and glucose-6-phosphate dehydrogenase. The amount of phosphate present was determined by measurement of the NADPH formed. More recently a coupled enzymatic method has been marketed by Kyowa Medex, Kyoto, Japan.[14] The steps of the enzymatic reaction involve formation of ribose-1-phosphate and hypoxanthine from inorganic phosphate and inosine, oxidation of hypoxanthine to uric acid and H_2O_2, and formation of a chromogen-using peroxidase and a chromogenic substrate.

Any analyzer capable of good 340 nm photometry should be able to accomplish accurate, sensitive phosphate analysis employing direct analysis of the phosphorus-molybdate complex. Two factors complicate straightforward adaptation of this reaction to an automated system. The first factor is that aqueous standards react considerably more slowly than do protein-based standards, and they may not attain the end point at the same time as serum unknowns.[15] The second factor is the unusual influences of pH; acidity is necessary, but as the pH decreases to 1.0, the reaction is substantially slowed, and as it is increased beyond 2.0, spontaneous reduction of molybdate takes place (Table 58-13).[11,15]

SPECIMEN

Serum values will be lower after meals, and the patient should fast before the sample is drawn. Phosphate will also be lower during the menstrual period. Intravenous glucose and fructose also physiologically lower phosphate. Since phosphate is the major intracellular anion, serum samples should be removed promptly before phosphate loss to the serum can occur. Any observable hemolysis will render a sample unacceptable for phosphate analysis.

Urine may contain larger quantities of organic phosphates, which can decompose on exposure to elevated temperatures. When acidified with HCl, urine phosphate is stable for more than 6 months.[16] Urine samples are usually dilated 1:10 with 0.15 M saline before analysis.

REFERENCE RANGES

Method	Males and females reported as mg of phosphorus/L (mEq P/L)	Urine (g/24 hr)
Molybdenum blue methods		
Henry TCA filtrate[16]		
Children	40-70 (2.32-4.06)	0.5-0.8
Adults	25-48 (1.45-2.76)	0.3-1.3
Unreduced phosphomolybdate methods		
Gilford 340 nm rate—adults	25-48 (1.45-2.76)	
Abbott 340 nm kinetic—adults	20-46 (1.16-2.67)	

NOTE: There is a diurnal variation in urine phosphate excretion, with the highest output occurring in the afternoon. There is no significant sex difference reported, but serum and urine levels do vary with age.

REFERENCES

1. Bell, RD, and Doisy, EA: Rapid colorimetric methods for the determination of phosphorus in urine and blood, J Biol Chem 44:55-67, 1920.
2. Fiske, CH, and Subbarow, Y: The colorimetric determination of phosphorus, J Biol Chem 66:375-400, 1925.

3. Bartlett, GR: Phosphorus assay in column chromatography, J Biol Chem 234:466-468, 1959.
4. Kuttner, T, and Cohen, HR: Microcolorimetric studies. I. A molybdic acid stannous chloride reagent: the microestimation of phosphate and calcium in pus, plasma, and spinal fluid, J Biol Chem 75:517-531, 1927.
5. Hurst, RO: The determination of nucleotide phosphorus with stannous chloride-hydrazine sulphate reagent, Can J Biochem 42:287-292, 1964.
6. Method No SF4-0004FF5, Technicon Instruments Corp, Tarrytown, NY 10591; released June 1975.
7. Kraml, M: A semi-automated determination of phospholipids, Clin Chim Acta 13:442-448, 1966.
8. Taussky, HH, and Shorr, J: Microcolorimetric method for determination of inorganic phosphorus, J Biol Chem 202:675-685, 1953.
9. Martinek, RG: Revue of methods for determining inorganic phosphorus in biological fluid, Am J Med Technol 32:337, 1970.
10. Simonsen, DG, Wertman, M, Westover, LM, and Mehl, JW: The determination of serum phosphate by the molybdivanadate method, J Biol Chem 166:747-755, 1946.
11. Daly, JA, and Ertinghausen, G: Direct method for determining inorganic phosphorus in serum with the "Centrifichem," Clin Chem 18:263-265, 1972.
12. Quinlan, KP, and DeSesa, MA: Spectrophotometric determination of phosphorus as molybdovanadophosphoric acid, Anal Chem 27:1626-1629, 1955.
13. Schultz, DW, Passanneau, JV, and Lowry, OH: An enzymatic method for the measurement of inorganic phosphate, Anal Biochem 19:300-314, 1967.
14. Adam, A, Boulanger, J, Azzouzi, M, and Ers, P: Colorimetric versus enzymatic determination of serum phosphorus, Clin Chem 30:1724-1725, 1984 (letter).
15. Michelsen, OB: Photometric determination of phosphorus as molybdovanadophosphoric acid, Anal Chem 29:60-62, 1957.
16. Henry RJ: Clinical chemistry: principles and technics, ed 2, New York, 1974, Harper & Row, Publishers, Inc.

Sodium and potassium

WILLIAM J. KORZUN
W. GREGORY MILLER

Clinical significance: pp. 313 and 346
Atomic formula: Na, K
Atomic weight: Na, 22.990; K, 39.098 daltons
Merck Index: Na, 8398; K, 7473
Chemical class: inorganic cation

PRINCIPLES OF ANALYSIS

Sodium and potassium are usually quantitated simultaneously either by flame atomic emission spectroscopy (FAES) (method 1, Table 58-14; see Chapter 3) or by ion-selective electrode potentiometry (ISE) (method 2, Table 58-14; see Chapter 13). FAES methods typically employ a 1:100 or 1:200 dilution of sample with a diluent containing an internal standard. Lithium at 15 mmol/L has been the most widely used standard, but cesium at 1.5 mmol/L is being used by one manufacturer (Instrumentation Laboratory Inc., Lexington, MA).

The heat of the air-propane flame (about 1925° C) vaporizes the salt, which gains electrons from the reducing gases to form the ground-state atoms Na^o and K^o. These atoms are heated by the flame, resulting in the formation of electronically excited atoms, Na* and K*. The excited atoms instantly decay to the original ground state with the emission of light. Sodium emission is monitored at 589 nm, potassium at 766 nm, lithium internal standard at 671 nm, and cesium internal standard at 852 nm. The amount of emitted light is directly proportional to the concentration of sodium and potassium. Although only 1% to 5% of the

Table 58-14 Methods of sodium and potassium analysis

Method	Type of analysis	Principle	Usage	Comments
1. Flame atomic emission spectroscopy (FAES)	Quantitation of mass concentration	Excited atom emits photon	Serum, urine, CSF, other body fluids	Reference method as well as most commonly used; dilution error possible
2. Ion-selective electrode potentiometry (ISE)	Quantitation of chemical activity	Ion-selective electrode measures potentiometric change as function of ion concentration	Serum, urine, CSF, whole blood (direct procedure)	Dilution error possible by indirect procedure; urine analysis may have limited linear range
3. Atomic absorption	Quantitation of mass concentration	Ground state atoms absorb incident light from hollow cathode lamp	All biological fluids	Highest sensitivity but not useful for routine analysis
4. Spectrophotometry	Quantitation of mass concentration	Complex with macrocyclic iomophore	Serum, plasma	Physician office testing

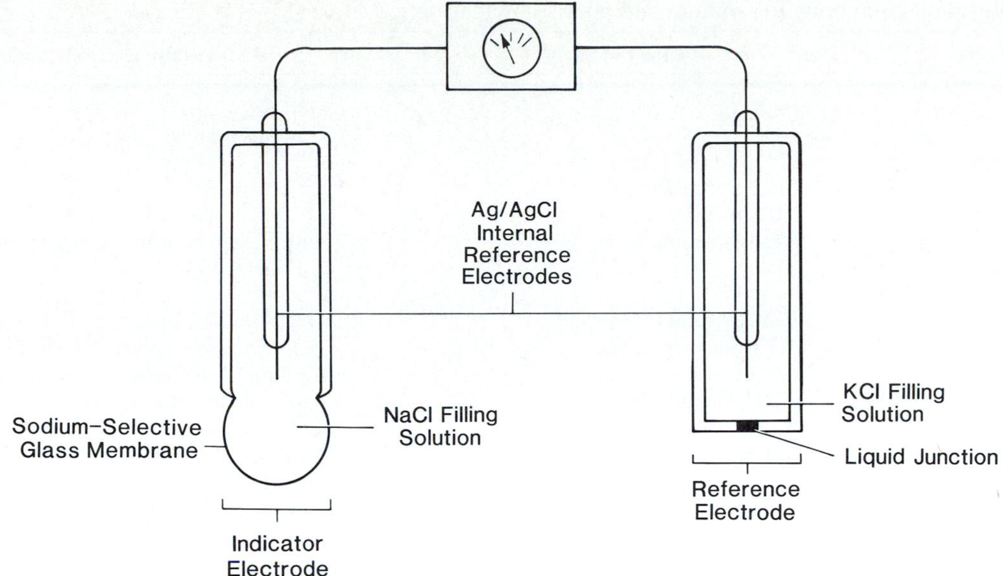

Fig. 58-7 Schema of ISE system with sodium-selective glass electrode. Potassium ISE system is similar, with a valinomycin-containing polymeric membrane replacing glass membrane and a KCl internal filling solution.

atoms in the flame are excited to emission, this is sufficient for accurate and precise quantitation.

The FAES method measures the concentration of ions in the total volume of specimen sampled. In the case of a serum or plasma specimen for a healthy person, the sample volume is composed of 93.5% water, 5.4% protein, 0.6% lipid, and 0.9% other molecules.[1] The physiological effect of ions is a function of their concentration (activity) in the volume of water in which they are dissolved. Consequently, in diseases that alter the volume fraction of water in serum, that is, those that have a different ratio

$$\frac{\text{Water volume}}{\text{Total serum volume}}$$

the physiological control mechanisms will strive to keep ionic activity per water volume at a constant level. However, if the water-volume fraction is reduced, as occurs in hyperproteinemia or hyperlipidemia, the actual concentration of ions in the *total* volume of the sample will be reduced, although the effective activity of ions in the *water* volume is approximately normal. This physiological interference in FAES methods is responsible for the *pseudohyponatremia* (and pseudohypochloremia) observed in such cases. When the interference is caused by lipemia, one can use an ultracentrifuge to remove the excess lipid volume from the serum and then analyze the sample by FAES to get a physiologically meaningful result. If the interference is caused by protein, the FAES method is inappropriate and a direct ISE method (see following discussion) must be used to obtain a clinically reliable result. This volume fraction error, when present, does not usually create a problem of clinical interpretation for potassium because the normal range is fairly large with respect to the magnitude of concentration of potassium in serum. For potassium, the size of the reference range (3.6 to 5 mmol/L) represents a 33% change in concentration, whereas for sodium the reference range (135 to 145 mmol/L) extends over only a 7% change in concentration. Hypoproteinemia and hypertriglyceridemia also produce an increased serum viscosity that may introduce error into the sample dilution step, depending on the design of the diluter, but that will not affect aspiration into the flame at the dilutions usually employed.

Another type of artifactual hyponatremia that occurs in hyperglycemia is a physiological response that shifts intracellular water to the extracellular space, increasing the blood volume, when elevated glucose in the circulation causes an increase in osmolality. The magnitude of this effect varies, depending on the state of hydration, from 1.2 to 2.0 mmol/L decrease in sodium per 5.56 mmol/L (100 mg/dL) increase in serum glucose.[2]

Ion-selective electrode (ISE) methods use a glass ion-exchange membrane for sodium and a valinomycin neutral-carrier membrane for potassium measurement. Fig. 58-7 shows a schematic ion-selective electrode. Typical sodium electrodes have a thousandfold greater selectivity for Na^+ than for K^+ and are insensitive to pH above 1. Although the Na^+/K^+ ratio in serum is approximately 30, the potassium electrode has a greater than 10^5 selectivity for K^+ than for Na^+ so that sodium does not interfere.

Table 58-15 Reaction conditions for sodium and potassium analysis

Condition	Flame emission	Ion-selective electrode (indirect/*direct*)
Temperature	1925° C	Ambient
pH	—	8.0/*(endogenous)*
Final reagent composition	Li: 15 mmol/L or Cs: 1.5 mmol/L	$MgSO_4$: 40 mmol/L Buffer: pH 8.0
Sample dilution factor	0.01 or 0.005	0.029/*1.0*
Sample	Plasma, serum, urine, CSF	Plasma, serum, urine, CSF/*plasma, serum, urine, whole blood*
Linearity		
Sodium	100-160 mmol/L (serum)	100-180 mmol/L (serum)/*100-180 mmol/L (serum)*
	0-200 mmol/L (urine)	25-150 mmol/L (urine)/*30-190 mmol/L (urine)*†
Potassium	2-8 mmol/L (serum)	1-9 mmol/L (serum)/*2-10 mmol/L (serum)*
	0-200 mmol/L (urine)	5-105 mmol/L (urine)/*12-200 mmol/L (urine)*†
Time of reaction	10-30 seconds	10-20 seconds/*10-60 seconds*
Major interferences	Dilution effect by lipids or proteins	Hemolysis (for potassium)
	Hemolysis (for potassium)	Sodium anticoagulants
	Sodium anticoagulants	Ammonium anticoagulants (for potassium)
Intralaboratory precision		
Sodium	0.7%-1.4% CV in normal range	0.7%-1.4% CV in normal range/*0.4%-1%*
Potassium	1.5%-2% CV in normal range	1.5%-2% CV in normal range/*0.5%-2%*
Interlaboratory precision*		
Sodium	1.5%-1.8% CV	1.2%-1.3% CV/*1.4%-2.1% CV*
Potassium	2.3%-3.5% CV	2.0%-3.2% CV/*2.3%-2.4% CV*

CV, Coefficient of variation.
*Interlaboratory precision data from College of American Pathologists survey for sodium and potassium values falling within the normal range.
†Some instruments require dilution of urine samples with sodium or potassium concentrations above 50 mmol/L.

Two general types of ISE measurements are made on clinical samples. *Direct* systems measure the ion activity in an undiluted sample. *Indirect* ISE systems measure the ion activity in a prediluted sample. Because ISE measurements determine the activity of an ion in the water-volume fraction in which it is dissolved, direct measurements are unaffected by conditions such as hyperproteinemia or hyperlipidemia, which alter the volume fraction of water in serum. However, indirect methods are usually sensitive to this physiological effect because the dilution step itself is based on total volumes, and after dilution the volume occupied by soluble serum molecules becomes insignificant with respect to the total diluent volume.

Colorimetric methods for sodium and potassium have been recently developed based on macrocyclic ionophores with binding sites specific for Na^+ or K^+. The ionophores can be covalently attached to chromophoric groups that undergo spectral changes on ion binding, or they can form ion pairs with chromogenic reagents in organic or micellar phases. These systems are found in some dry reagent strips and discrete chemical analyses. Table 58-15 lists reaction conditions for sodium and potassium analysis.

SPECIMEN

Serum, heparinized plasma, heparinized whole blood, urine, and other body fluids are acceptable specimens. Heparinized plasma, in which the heparin is present as the sodium salt, is unacceptable for sodium analysis unless the tube is completely filled with blood. Therefore heparinized phlebotomy tubes with Li^+ heparin are preferred. Ammonium heparin is unacceptable because ammonium ions in high concentrations give erroneously increased values for potassium when measured by ion-selective electrode or by chromogenic ionophore methods. Serum, plasma, and other fluids should be separated from cells within 3 hours to avoid shifts in ionic equilibria as a result of cell metabolism and pH changes. Similarly, whole blood should be analyzed within 3 hours. Plasma and serum sodium and potassium are stable for at least 1 week at room or refrigerator temperatures and for at least 1 year frozen. Whole blood analysis for sodium and potassium is only possible with certain direct ISE methods. Hemolysis will falsely elevate potassium values, and moderately to grossly hemolyzed specimens should be so indicated on the report form.

PROCEDURE: FLAME ATOMIC EMISSION SPECTROSCOPY

Principle

$$Na \xrightarrow{\text{Heat}} Na^o \xrightarrow{\text{Heat}} Na^* \rightarrow Na^o + 589 \text{ nm light}$$
$$K^+ \xrightarrow{\text{Heat}} K^o \xrightarrow{\text{Heat}} K^* \rightarrow K^o + 766 \text{ nm light}$$

Intensity of emission at each wavelength is proportional to the concentration of the respective ion in the sample.

Reagents

NBS-SRM 918: KCl and NBS-SRM 919: NaCl. These materials should be dried according to the NBS product insert.

Lithium carbonate, sodium chloride, potassium chloride, hydrochloric acid, nitric acid, chloroform, methanol, and 95% ethanol. All of these materials should meet ACS specifications.

Distilled or deionized water. Use water with a specific resistance of at least 0.01 megohm·meter at 23° C ± 5° C.

Assay

Equipment:

Volumetric glassware, of borosilicate material, meeting NBS class A specification

Pipettor-diluter with a maximum inaccuracy of 1% and a maximum imprecision of ±0.5% relative standard deviation

FAES instruments: this clinical laboratory method protocol was designed to be applicable to a typical flame atomic emission spectrophotometer capable of providing the stability and linearity of response required to achieve stated goals of maximum total imprecision and maximum total bias in the measurements. Interested readers should consult references 3 and 4 for details concerning the reference FAES method and verification of the necessary performance specifications of a specific instrument.

Analytical balance capable of weighing 0.1 mg.

1. Prepare stock standard solutions of Na and K from NBS-SRM reagents.
2. Prepare appropriate combined working standard solutions of 0, 120, 140, and 160 mmol Na/L and 0, 1, 2, 4, and 6 mmol K/L for serum measurements. For urine or other body fluid measurements, standards of 0, 50, 100, 150, and 200 mmol Na/L and 0, 25, 50, 100, and 150 mmol K/L will be necessary.
3. Dilute each working standard and unknown 1:100 or 1:200 with 15 mmol $LiCO_3$/L diluent or 1.5 mmol/L $CsCl_2$, depending on the internal standard used by the instrument. Various dilution procedures are used by different instruments, and the specific instructions provided by the manufacturer should be followed.
4. Set up the instrument and adjust the flame and aspiration rate according to the manufacturer's recommendations. Note that the zero standard dilution should be aspirated at all times to maintain a constant burner head temperature, which will decrease instrument drift.
5. Aspirate the standards into the instrument nebulizer and prepare a calibration graph of instrument response versus concentration. Most instruments will give a readout in concentration units. The response must be linear over the appropriate concentration range for serum or urine/fluid samples. If the response is not linear, consult the instrument manufacturer's literature for corrrective action.
6. Once the instrument is properly calibrated, aspirate patient samples and determine Na and K concentration. When analyzing large batches of samples, recheck instrument calibration every 10 samples with a single standard solution. It is generally not necessary to recheck linearity within a run of analyses.

NOTE: Using this method for Na and K analysis, it is possible to achieve a total imprecision less than CV = 1.0% for sodium and less than CV = 1.5% for potassium.

REFERENCE RANGE

Serum or plasma using FAES or indirect ISE methods

Sodium	135 to 145 mmol/L
Potassium	3.6 to 5.0 mmol/L

These values were determined as the central 95% range obtained from analysis of 1205 (596 males, 609 females) serum samples from healthy ambulatory persons between 1 and 78 years of age. No variation with sex or age was observed. Plasma potassium values may be 0.1 to 0.2 mmol/L lower than serum values because serum contains a small amount of potassium released from platelets during clotting.

Serum, plasma, or whole blood using direct ISE methods

Sodium	145 to 155 mmol/L
Potassium	3.9 to 5.3 mmol/L

These ranges were determined by adding 7% to the reference ranges for FAES or indirect ISE methods. Because of the 93% water-volume fraction in serum, plasma, or whole blood, these ranges for direct ISE methods represent the physiological activity of sodium and potassium ions in the sera of the population sampled earlier. However, it has recently been recommended that manufacturers of direct ISE instruments calibrate those instruments in such a manner that the reported concentration values will be the same as those obtained by flame photometry with normal serum or plasma.[5] This practice is intended to circumvent the difficult task of matching the activity coefficients and junction potentials of calibrating solutions and whole blood and will promote uniform calibration of these systems.

Urine[6]

Sodium	40 to 220 mmol/L
Potassium	2.5 to 125 mmol/L

Urinary excretion of sodium and potassium of healthy persons is highly dependent on dietary intake and state of hydration. In other persons hemodynamic status, acid-base balance, disease processes, and drug therapies can all additionally influence sodium and potassium excretion. Because of the selectivity coefficients of sodium-glass electrodes for hydrogen ions, urine collected in an acid preservative cannot be analyzed by ISE methods for sodium. Potassium-selective electrodes suffer significant liquid-junction potential errors with acidified urine samples.

REFERENCES

1. Czaban, JD, Cormier, AD, and Legg, KD: Establishing the direct-potentiometric ''normal'' range for Na/K: residual liquid junction potential and activity coefficient effects, Clin Chem 128:1936-1945, 1982.
2. Moran, SM, and Jamison, RL: The variable hyponatremic response to hyperglycemia, West J Med 142:49-53, 1985.
3. Velapoldi, RA, et al: A reference method for the determination of sodium in serum, NBS Special Publication 260-60, US Dept of Commerce, National Bureau of Standards, Washington, DC, 1978, US Government Printing Office.
4. Velapoldi, RA, et al: A reference method for the determination of potassium in serum, NBS Special Publication 260-63, US Dept of Commerce, National Bureau of Standards, Washington, DC, 1978, US Government Printing Office.
5. Maas, AHJ, Siggaard-Andersen, O, Weisberg, HF, and Zijlstra, WG: Ion-selective electrodes for sodium and potassium: a new problem of what is measured and what should be reported, Clin Chem 31:482-485, 1985.
6. Tietz, NW: In Wyngaarden, JB, and Smith, LH, Jr, editors: Cecil textbook of medicine, Philadelphia, 1982, WB Saunders Co, pp 2343 and 2346.

CHAPTER 59 Enzymes

Acid phosphatase, total

STEVEN A. NOEL
JOHN A. LOTT

Acid phosphatase, ACP, *ortho*-phosphoric monoester phosphohydrolase (acid optimum)
Clinical significance: p. 730
Enzyme number: EC 3.1.3.2
Molecular weight: approximately 100,000 daltons; varies with isoenzyme
Chemical class: enzyme, protein
Known isoenzymes: prostate, erythrocyte, platelet, liver, spleen, kidney, bone marrow
Biochemical reaction:

$$\underset{\textit{Organic phosphate ester}}{R_1—O—PO_3H_2} + \underset{\textit{Alcohol}}{R_2—OH} \xrightarrow[\text{pH 5 to 6}]{\text{ACP}} \underset{\textit{New alcohol}}{R_1—OH} + \underset{\textit{New phosphate ester}}{R_2—O—PO_3H_2}$$

PRINCIPLES OF ANALYSIS

The term *acid phosphatase* (ACP) refers to a group of nonspecific phosphatases that show maximum activity near pH 5.0 and catalyze the hydrolysis of an orthophosphoric monoester to yield an alcohol and a new phosphate ester, as shown in the biochemical reaction. ACP has various isoenzymes, which are distributed ubiquitously throughout the body. Three and possibly four ACP isoenzyme populations exist.[1] Erythrocytes have three isoenzymes capable of forming five phenotypical patterns. Intracellular ACPs, associated in part with lysozymes, are present in the reticuloendothelial system. Prostatic ACPs are formed primarily as secretory products with extracellular function, although lysosomal forms of these isoenzymes may also occur in the prostate. Finally, an intracellular, membrane-bound ACP not involved in lysosomal function contributes a likely fourth population.

Numerous procedures have been developed for the measurement of ACP activity in biological fluids. The methods vary with the type of substrate and buffer used, as well as temperature and time of incubation. Each modification represents an attempt to maximize the sensitivity for total ACP or specificity for a selected isoenzyme, most often for prostatic acid phosphatase (PAP). A representative summary of ACP methods, including those of historical interest and those in current use, can be found in Table 59-1.[2-10]

The natural substrate or substrates for ACP are not known, although the enzyme is capable of hydrolyzing many orthophosphoric monoesters. The selection of a particular substrate is influenced by the different reactivities of various substrates with the isoenzymes of ACP.[11,12] For example, PAP reacts more readily with thymolphthalein monophosphate than does ACP derived from platelets or erythrocytes,[5] but no substrate exhibits absolute specificity for a particular isoenzyme of ACP.[8,9] Michaelis-Menten constants (K_m) for each substrate are found in Table 59-1.[13,14]

Many early assays for ACP were adaptations of methods already in use for alkaline phosphatase. Identical substrates were used; the adaptation in many cases simply was to buffer the reaction medium at an acid pH.

As originally reported, all these methods were manual, end-point assays; they are of historical interest only. (See Table 59-1, first six methods.[2-7]) Some procedures have been adapted to automated methods of analysis and find limited current use in the clinical laboratory. They have been replaced by kinetic methods that employ more specific PAP substrates, such as thymolphthalein monophosphate, or by radioimmunoassay (RIA) for prostatic ACP.

Thymolphthalein monophosphate is a self-indicating substrate that was introduced for the quantitation of ACP by Roy et al.[8] Serum was incubated for 30 minutes at 37° C with the substrate in citrate buffer at pH 5.95. The reaction was stopped by the addition of NaOH. The liberated thymolphthalein produced a color in alkaline medium, which was measured at 590 nm. The method was subsequently modified to increase its sensitivity by Ewen,[9] who used an acetate buffer (pH 5.4), reduced the substrate concentration and reaction volume, and changed conditions for sample preservation.

A highly sensitive fluorometric assay for ACP was reported by Rietz and Guilbault.[10] ACP activity was determined from the rate of appearance of the fluorescent 4-

Table 59-1 Summary of methods for acid phosphatase (ACP) measurement in body fluids

Source of method	Type of analysis	Principle	Usage	Comments	K_m for PAP* (mmol/L)
Gutman and Gutman[2]	End-point spectrophotometric	Substrate: monophenyl phosphate Liberated phenol measured with Folin-Ciocalteu reagent	Serum, manual	Historical interest only; insensitive substrate	0.091
Jaffé and Bodansky[3]	End-point spectrophotometric	Substrate: β-glycerophosphate Measure liberated phosphate with color reaction	Serum, manual	Historical interest only; long incubation times; high background absorbance	2.0
Hudson et al.[4]	End-point or kinetic spectrophotometric	Substrate: *p*-nitrophenol phosphate Measure release of *p*-nitrophenol	Serum, manual or automated	Sensitive, self-indicating substrate; not specific for PAP	0.31
Huggins and Talalay[5]	End-point or kinetic spectrophotometric	Substrate: phenolphthalein phosphate Released phenolphthalein measured colorimetrically after addition of base	Serum, manual or automated	Self-indicating but relatively insensitive substrate	n.a.
Seligman et al.[6]	End-point or kinetic spectrophotometric	Substrate: β-naphthyl phosphate Liberated β-naphthol complexed with *o*-dianisidine to form dye	Serum, manual or automated	Multistep procedure; colored product requires extraction before reading	n.a.
Babson and Read[7]	End-point or kinetic spectrophotometric	Substrate: α-naphthyl phosphate Liberated α-naphthol complexed with *o*-dianisidine to form dye	Serum, manual or automated	Multistep procedure; precise timing required; no extraction involved	0.39
Roy et al.[8]	Kinetic spectrophotometric	Substrate: thymolphthalein monophosphate Measure liberated thymolphthalein after addition of base	All body fluids and tissues; manual or automated	Simple procedure; sensitive, self-indicating substrate; high specificity for PAP	1.2
Ewen[9]	End-point or kinetic spectrophotometric	Substrate: thymolphthalein monophosphate Measure liberated thymolphthalein after addition of base	All body fluids and tissues; manual or automated	Modification of Roy et al.[8]; enhanced sensitivity of substrate; preferred method	n.a.
Rietz and Guilbault[10]	Fluorometric kinetic	Substrate: 4-methylumbelliferone phosphate Measure rate of appearance of fluorescent 4-methylumbelliferone	All body fluids and semisolids; manual or automated	Highly sensitive substrate; equipment expense may be limitation	n.a.

**PAP*, Prostatic acid phosphatase; *n.a.*, not applicable.

methylumbelliferone liberated by the enzyme from 4-methylumbelliferone phosphate. One to 10 μL of serum was added to the substrate in citrate buffer (pH 4.9), and the reaction rate was measured in a fluorometer at 37° C (λ_{ex}, 365 nm; λ_{em}, 455 nm).

REFERENCE AND PREFERRED METHODS

Table 59-2 is a summary of the reaction conditions of the most frequently used method for ACP.[9] The method of Roy et al.,[8] as modified by Ewen,[9] the preferred method, is described in Table 59-2. It measures all isoenzymes of ACP but reacts preferentially with the prostatic form. The procedure, presented as a manual method, is simple and can easily be automated.[15] It has been adapted to several commercially available instruments and has been used by the National Bureau of Standards and Cooperating Laboratories Enzyme Study Group to assay human serum calibrators.[16]

SPECIMEN

Serum and heparinized plasma give similar results with thymolphthalein monophosphate as substrate. Plasma re-

Table 59-2 Enzymatic method for total acid phosphatase (ACP) analysis[9]

Condition	Measure
Temperature	37° C
Sample volume (μL)	50
Fraction of sample volume	0.0303
pH	5.4
Reagent concentration	Acetate buffer 0.15 mol/L Thymolphthalein monophosphate 1.1 mmol/L Brij-35 1.6 g/L
Reaction time	30 min
Detection method	Absorbance: 595 nm
Linearity	To 3.5 absorbance units at full scale
Precision (within-run) $\overline{X}$ (%CV)*	1.7 U/L* (2%) 5.0 U/L (1%)
Interferences	Tissue ACP reacts <0.25% that of PAP
Reference range	0.5-1.9 U/L

**U/L,* International units per liter; *$\overline{X}$ (%CV),* mean (percent coefficient of variation).

sults are typically slightly lower than serum because ACP is released by the platelets.[17] EDTA does not interfere, whereas oxalate is reported to inhibit the enzyme. Trace hemolysis is tolerable, but gross hemolysis results in increases in ACP activity ranging from 6% to 315%, depending on the particular method used.[18] Any drug causing hemolysis will increase ACP activity; detergent left on glassware inhibits the enzyme.

ACP is very unstable in separated serum at room temperature; the rise in pH caused by the loss of CO_2 from the serum results in the rapid and irreversible inactivation of the enzyme. Both temperature and pH play a role in the stability of serum ACP. Rapid inactivation occurs at room temperature and at a pH greater than 6.[19,20] An acid pH is maintained by addition of citric acid tablets to the serum or by storage of the serum in tubes containing an acetate buffer.[9] As soon as the serum or plasma is separated, add an acid stabilizer. The stabilizer tablets from Sigma Chemical Co. (St. Louis; catalog no. 104-9) are very convenient. Add 1 tablet per milliliter of serum. Control sera should be treated in the same way after reconstitution.

ACP is used for medicolegal determinations of seminal fluid in the vagina after coitus.[21] Obtaining vaginal specimens is a critical step in determining evidence of rape. The steps for vaginal swab technique are outlined as follows[21]:

1. Soak a cotton-tipped applicator in the posterior fornix.
2. Place the moist swab in 1.0 mL of preservative broth.[22]
3. Analyze immediately for ACP activity or store overnight at 4° C. (Swabs are stable for 1 month at 4° C in preservative broth or with acidification, as for serum specimens.)

REFERENCE RANGE[9]

The serum reference range for healthy persons is 0.5 to 1.9 U/L. Activities are similar for both sexes.

REFERENCES

1. Moncure, CW: Isoenzymes in prostatic carcinoma. In Tannenbaum, M, editor: Urologic pathology: the prostate, Philadelphia, 1977, Lea & Febiger.
2. Gutman, AB, and Gutman, EB: An "acid" phosphatase occurring in the serum of patients with metastasizing carcinoma of the prostate gland, J Clin Invest 17:473-478, 1938.
3. Jaffé, HL, and Bodansky, A: Diagnostic significance of serum alkaline and acid phosphatase values in relation to bone disease, Bull NY Acad Med 19:831-848, 1943.
4. Hudson, PB, Brendler, H, and Scott, WW: Simple method for the determination of serum acid phosphatase, J Urol 58:89-92, 1947.
5. Huggins, C, and Talalay, P: Sodium phenolphthalein phosphate as a substrate for phosphatase test, J Biol Chem 159:399-410, 1945.
6. Seligman, AM, Chauncey, HH, Nachlas, MM, et al: The colorimetric determination of phosphatases in human serum, J Biol Chem 190:7-15, 1951.
7. Babson, AL, and Read, PA: A new assay for prostatic acid phosphatase in serum, Am J Clin Pathol 32:88-91, 1959.
8. Roy, AV, Brower, ME, and Hayden, JE: Sodium thymolphthalein monophosphate: a new acid phosphatase substrate with greater specificity for the prostatic enzyme in serum, Clin Chem 17:1093-1102, 1971.
9. Ewen, LM: Acid phosphatase activity (thymolphthalein monophosphate substrate). In Faulkner, WH, and Meites, S, editors: Selected methods for the small clinical laboratory, Philadelphia, 1982, WB Saunders Co.
10. Rietz, B, and Guilbault, GG: Fluorometric assay of serum acid or alkaline phosphatase in solution or on a semisolid surface, Clin Chem 21:1791-1794, 1975.
11. Ladenson, JH, and McDonald, JM: Acid phosphatase and prostatic carcinoma, Clin Chem 24:120-134, 1978.
12. Henderson, AR, and Nealon, DA: Enzyme measurements by mass: an interim review of the clinical efficacy of some mass measurements of prostatic acid phosphatase and isoenzymes of creatine kinase, Clin Chim Acta 115:9-32, 1981.
13. Tsubai, KK, and Hudson, PB: Acid phosphatase. III. Specific kinetic properties of highly purified human prostatic phosphomonoesterase, Arch Biochem Biophys 55:191-205, 1955.
14. van Rijn, HJM, Boer, R, and Klosse, JA: The determination of acid phosphatase of prostatic origin with the Automatic Clinical Analyzer (aca, du Pont), J Clin Chem Clin Biochem 18:627-630, 1980.
15. Vihko, P, Sajanti, E, Janne, O, et al: Serum prostate-specific acid phosphatase: development and validation of a specific radioimmunoassay, Clin Chem 24:1915-1919, 1978.
16. Bowers, GN, Jr, Cali, JP, Elser, R, et al: Activity measurements for seven enzymes in lyophilized human serum SRM 909, Clin Chem 26:969, 1980 (abstract).
17. Lott, JA, and Wolf, PL: Clinical enzymology: a case oriented approach, Chicago, 1986, Year Book Medical Publishers.
18. Frank, JJ, Bermes, EW, Bickel, MJ, et al: Effect of in vitro hemolysis on chemical values for serum, Clin Chem 24:1966-1970, 1978.
19. Kahn, R, Turner, B, Edson, M, et al: Bone marrow acid phosphatase: another look, J Urol 117:79-80, 1977.
20. Ricci, LR, and Hoffman, SA: Prostatic acid phosphatase and sperm in the post-coital vagina, Ann Emerg Med 11:530-534, 1982.
21. Lantz, RK, and Eisenberg, RB: Preservation of acid phosphatase activity in medicolegal specimens, Clin Chem 24:486-488, 1978.
22. Bowers, GN, Jr, Onoroski, M, Schifreen, RS, et al: Spectrophotometric and liquid-chromatographic studies of thymolphthalein monophosphate: specifications for high-quality substrate for the measurement of prostatic acid phosphatase activity, Clin Chem 27:1372-1377, 1981.

Acid phosphatase, prostatic

JOHN F. CHAPMAN
LINDA L. WOODARD
LAWRENCE M. SILVERMAN

Acid phosphatase, ACP, *ortho*-phosphoric monoester phosphohydrolase (acid optimum)
Clinical significance: pp. 730 and 787
Enzyme number: EC 3.1.3.2
Molecular weight: <20,000-100,000 daltons (varies with tissue source)
Chemical class: enzyme, glycoprotein
Known isoenzyme forms: ACP_0, ACP_1, ACP_2, ACP_3, ACP_4, and ACP_5 are commonly recognized; at least 20 discrete electrophoretic isoenzymes have been reported[1,2]
Biochemical reaction: pH 5.0

$$R_1—O—PO_3H_2 + R_2—OH \xrightarrow[\text{pH 5.0}]{\text{ACP}} R_1—OH + R_2—O—PO_3H_2$$

Organic phosphate ester — *Alcohol* — *New alcohol* — *New phosphate ester*

Net reaction: hydrolysis if alcohol is H_2O

PRINCIPLES OF ANALYSIS

Acid phosphatase (ACP) is the name given a group of phosphohydrolases that hydrolyze phosphoric monoesters at an acidic pH. Several ACP isoenzymes are found in human tissues and cells, including liver, spleen, kidney, prostate, erythrocytes, platelets, osteoclast, and hairy cell leukemias.[1-4] Heterogeneity in the carbohydrate portion of the molecule may result in 20 or more isoenzyme forms in human tissues, although the clinical relevance of most of these remains to be established.[3] Of the ACP isoenzymes currently identified, only prostatic acid phosphatase (PAP), ACP isoenzyme 2 by electrophoresis, and isoenzymes 1 and 5 from human spleen in Gaucher's disease are associated with demonstrated clinical utility.

Historically, PAP enzymatic activity was first measured by use of substrates such as α-naphthyl phosphate, which were erroneously believed to be specific for this isoenzyme. After the discovery that ACP isoenzymes originating in leukocytes could also utilize this substrate,[5] the inhibitor L-tartrate was shown to be useful as a specific inhibitor of PAP (method 1, Table 59-3).[6] By this approach, the test is run in the absence and presence of L-tartrate, and the difference in activity between the assays is attributed to the prostatic isoenzyme. Although this technique is simple to perform and widely used, its sensitivity and specificity have been criticized.[7]

More recently, a variety of immunoassay techniques have been developed for the quantitation of PAP (methods 2 to 5, Table 59-3). Examples of these techniques include radioimmunoassay (RIA),[8] fluorescence immunoassay (FIA),[9] counterimmunoelectrophoresis (CIEP),[10] and enzyme immunoassays.[11] Specific (polyclonal or monoclonal) antibodies to PAP are used to separate this isoenzyme from serum. This can be accomplished by use of an antibody-coated support (beads or coated tubes) or by double-antibody precipitation. The RIA, FIA, and CIEP assays all quantify the mass amount of immunologically reactive PAP. Both RIA and FIA are competitive binding assays. In RIA, ^{125}I is used as the label, whereas in FIA a fluorescent ligand serves as the label. The immunoenzyme techniques employ either antibody coated to solid supports or double-antibody precipitation. In the first approach, anti–human PAP antibodies bound to polypropylene tubes selectively sequester the PAP during the initial incubation. The tubes are then emptied, washed, and incubated with substrate, and enzyme activity is measured as an ACP and

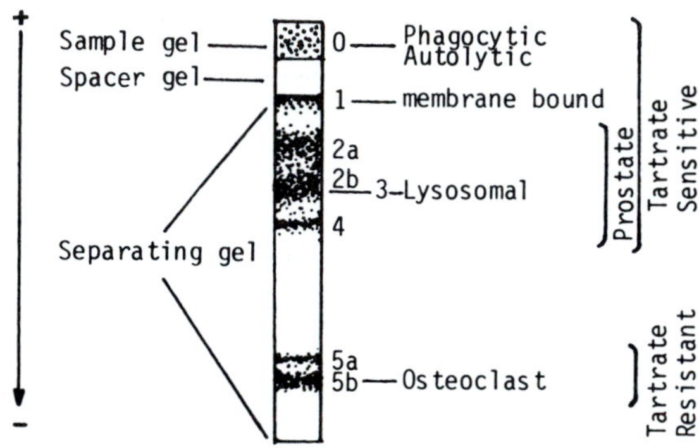

Fig. 59-1 Using electrophoresis and ion-exchange column chromatography, Lam et al. characterized tissue sources and locations of ACP (acid phosphatase) isoenzymes as well as their occurrence in normal persons and patients with variety of diseases. Band *O* is predominant in ascitic cells and is presumed to have role in autophagic function. Band *1* is membrane-bound isoenzyme present primarily in microsomal fraction. Band *3* is associated with lysosomal fraction of all nonprostatic cell types. Bands *2* and *4* are antigenically related and found almost exclusively in prostatic tissue. Band *2* is secretory enzyme normally secreted into seminal plasma and is found in exceptionally high concentration in prostate. Band *5* is characteristic isoenzyme found in osteoclasts and certain other cell types and has two subtypes: *a* and *b*. As with bands *2* and *4*, bands *5a* and *5b* are antigenically related, and their differential mobility is related to attachment of carbohydrate moieties. Band *5a* seems to predominate in Gaucher's cells and spleen, although band *5b* often predominates in serum of patients with Gaucher's disease. This apparent inconsistency may be attributable to release of large amounts of *5b* from increased osteoclastic activity in these patients or to posttranslational modifications of *5a* to *5b*, a phenomenon that has been observed in vitro. Band *5b* is often present in serum of normal adults and children and is associated with increased osteoclastic activity during normal bone growth in children. *5b* is also observed in patients with malignancies metastasized to bone. *(From Lam, KW, et al, in Shaw, LM, et al: Prostatic acid phosphatase measurements, Ann NY Acad Sci* ***390****:1-15, 1982.)*

Table 59-3 Methods for measurement of prostatic acid phosphatase (PAP)

Method	Type of analysis	Principle	Usage	Comments
1. Enzyme reactions and tartrate inhibition	Colorimetric, kinetic, or end point	Total ACP is measured by usual method* with and without L-tartrate, which inhibits PAP; subtraction of two results gives PAP activity.	Some usage	Poor sensitivity for prostatic cancer
2. Radioimmunoassay (RIA)	Competitive binding assay	^{125}I-labeled PAP competes with sample PAP for binding to anti-PAP antibodies.	Some usage	Higher sensitivity than method 1
3. Fluorescence immunoassay (FIA)	Competitive binding assay	Fluorescent ligand-labeled PAP competes with sample PAP for binding to anti-PAP antibodies.	Not frequently used	—
4. Counterimmunoelectrophoresis (CIEP)	Immunoelectrophoresis	Electrophoresis of PAP into countermigrating anti-PAP; measurement of height of precipitation line.	Not frequently used	—
5. Enzyme immunoassay	Immunoprecipitation of PAP followed by end-point colorimetric assay	Antibody-coated support or double-antibody precipitation with first Ab anti-PAP specific; enzyme assay by any ACP assay.	Some usage	Sensitivity for prostatic cancer between methods 1 and 2
6. Enzyme immunometric assay	Two-site solid-phase binding assay; enzyme label	Monoclonal antibody on solid phase binds PAP; a second monoclonal antibody linked to an enzyme binds to bound PAP; amount of enzyme activity bound to solid phase is proportional to PAP levels.	Most frequently used	Sensitivity similar to method 2
7. Electrophoresis	Electrophoresis	Separation of all isoenzymes on basis of charge.	Rarely used	Quantitates many ACP isoenzymes

*See previous section on total acid phosphatase for discussion on optimum assay.

end-point reaction. In the double-antibody precipitation technique, an anti-PAP antibody (Ab_1) binds to PAP molecules, and this complex is precipitated by a second antibody (anti-Ab_1). The precipitate is then resuspended in buffer and substrate, and the enzyme activity is measured as in the coated-tube approach. Antibody binding of PAP appears to have no inhibitory effect on catalytic activity and may actually impart enzymatic stability, since these methods have been performed successfully on specimens stored at $-20°$ C without acidification for up to 30 days with no significant loss of activity.

Solid-phase, sandwich immunoassays for PAP are also readily available (method 6, Table 59-3). These assays employ monoclonal or polyclonal antibodies to PAP affixed to a solid phase, usually a plastic bead. The bead is immersed in diluted sample for a set time, washed, and then incubated with a second antibody, specific for a different antigenic site (epitope) on the PAP molecule. This second antibody is conjugated to an enzyme, for example, horseradish peroxidase (HRP) or alkaline phosphatase (AP). The bead is washed again and incubated with substrate specific for HRP (*ortho*-phenylenediamine) or AP (*para*-nitrophenyl phosphate), and the absorbance of the product of the enzymatic reaction is measured (HRP, 492 nm; AP, 405 nm, *p*-nitrophenol). The amount of PAP present, in ng/mL, is determined by comparison to standards.

All these techniques are more sensitive than the so-called functional techniques such as tartrate inhibition, although specificity depends on the antiserum used, and elevated serum values have been reported in patients without apparent prostatic cancer.[12] Further, the diagnostic superiority of immunological methods over functional technique could not be demonstrated in at least one recent study.[13]

ACP isoenzymes have been separated by electrophoretic techniques (method 7, Table 59-3), but these methods have not been used for routine clinical analysis. Fig. 59-1 shows an example of the electrophoretic pattern. Using polyacrylamide electrophoresis at a pH of 4, the major prostatic acid phosphatase isoenzyme migrates closest to the origin, the osteocyte form migrates most anodally, and the lysosomal isoenzyme is in between. Generally, the red blood cell isoenzyme does not migrate on this system.

The solid-phase, sandwich immunoassays are commer-

Table 59-4 Comparison of several assays for prostatic acid phosphatase (PAP)

Condition	Enzymatic assay with L-tartrate inhibition*	Radioimmunoassay†	Enzyme immunoassay‡
Temperature	30° or 37° C	Ambient	Ambient—primary incubation and precipitation enzyme 37° C—reaction
Sample volume	40 μL each for total and L-tartrate inhibition	50 μL	100 μL each for reaction and blank
Fraction of sample volume	0.0625	0.01	Assay performed on precipitate or in antibody-coated tube
Final concentration of reagents	α-Naphthylphosphate · NaH_2O: 740.6 mg/L Fast Red TR: 412.5 mg/L Citric acid: 8.5 mmol/L Sodium citrate: 53.4 mmol/L pH 5.0 ± 0.2 L-Tartrate (only in inhibition reaction): 18.2 mmol/L	—	*p*-Nitrophenyl phosphate: 10 mmol/L Citric acid: 30 mmol/L Sodium citrate: 57 mmol/L pH: 5.0 ± 0.2 NaOH: 0.32 mol/L (for color development)
Time of reaction	10 minutes	18 hours with first antibody 3 hours with radioligand	Total procedure time approximately 3 to 4 hours Colorimetric reaction 60 min
Wavelength	405 nm	^{125}I counting by gamma-ray counter	405 nm
Linearity	45 U/mL	25 ng/mL	Approximately 20 U/L
Precision‡	1.0 U/L (69%)	3 ng/mL (4.7%-5.1%)	1.1 U/L (14.5%)
$\overline{X}$ (mean), % coefficient of variation	21.3 U/L (8.4%)	18 ng/mL (5.2%-5.9%)	19.4 U/L (8.2%)
Interferences	Observable hemolysis and lipemia pH <4.0 or >6.0	pH <4.0 or >6.0	Gross lipemia and hemolysis pH <4.0 or >6.0

*Smith-Kline, Inc, Division of Smith-Kline-Beckman, Inc, Sunnyvale, CA. Assay described here.
†New England Nuclear Corp, Boston. Modified method of Hillmann.
‡Unpublished data from University Hospital, University of Cincinnati, Departments of Pathology and Laboratory Medicine (chemistry laboratory) and Radiobiology, using reagents supplied by Smith-Kline-Beckman, Inc, Sunnyvale, CA.

cially available (Abbott Diagnostics, North Chicago; Hybritech, San Diego) either as manual, semiautomated, or automated procedures. The assays appear to be highly specific and reasonably reproducible. They have now become important alternatives to the enzymatic procedures and RIAs.

Although no established reference method exists, an L-tartrate inhibition method for PAP is often used as the established method against which all new methods are compared. (See the discussion of acid phosphatase, p. 889, on the use of other substrates in an L-tartrate inhibition assay.) However, for improved precision and specificity, either an RIA or enzyme immunoassay is the recommended method. Since the number of specimens requiring PAP analysis is not great, the lack of automation is not an important obstacle in the use of these methods.

SPECIMEN

Nonhemolyzed serum is preferred. *Grossly hemolyzed specimens are not acceptable,* since blood-cell acid phosphatases can result in falsely increased PAP results. Icteric specimens may cause a depression in the acid phosphatase activity. Samples collected in tubes containing anticoagulants that inhibit enzyme activity (oxalate, sodium fluoride) *cannot* be used. If plasma is used, heparin is the anticoagulant of choice. All specimens must be separated from cells immediately. One must then preserve the sample by adding 20 μL of 5 M acetate buffer or a citrate tablet (composed of 18 mg of disodium citrate in tablet form, Sigma Chemical Co., St. Louis) to 1 mL of serum. The pH should be between 5.0 and 6.0.[14] Specimens not preserved in this manner are unsuitable for analysis. ACP activity may decrease by up to 50% in 1 hour if unbuffered. ACP activity in preserved serum is stable for 2 days at 4° C.

REFERENCE RANGE

	Enzymatic	RIA, EIA
Male, total ACP	2.4 to 5.0 U/L	—
Male, PAP	<1.2 U/L	<2.5 ng/mL

Females have approximately the same levels of serum prostatic acid phosphatase activity as males. Females also excrete large amounts of PAP in their urine. The source of PAP in females may be glands lining the urethra.[18]

REFERENCES

1. Robinson, DG, and Glew, RH: Acid phosphatase in Gaucher's disease, Clin Chem 26:371-382, 1980.
2. Smith, JK, and Whitby, IG: The heterogeneity of prostatic acid phosphatase, Biochim Biophys Acta 151:607-613, 1968.
3. Lam, KW, Lee P, Eastlund, P, and Yam, LT: Antigenic and molecular relationship of human acid phosphatase isoenzymes, Invest Urol 18:209-211, 1980.
4. Bodansky, O: Acid phosphatase, Adv Clin Chem 15:43-147, 1972.
5. Amador, E, Price, JW, and Marshall, G: Serum acid α-naphthyl phosphatase activity, Am J Clin Pathol 51:202-206, 1969.
6. Fishman, WH, and Lerner, FA: A method for estimating serum acid phosphatase of prostatic origin, J Biol Chem 200:89-97, 1953.
7. Townsend, RM: Enzyme tests in diseases of the prostate, Ann Clin Lab Sci 7:254-261, 1977.
8. Vihko, P, Sanjanti, E, Janne, O, et al: Serum prostate-specific acid phosphatase: development and validation of a specific radioimmunoassay, Clin Chem 24:1915-1919, 1978.
9. Lee, CI, Wang, WC, Murphy, GP, and Chu, TM: A solid-phase fluorescent immunoassay for human prostatic acid phosphatase, Cancer Res 38:2871-2877, 1978.
10. Quinones, GR, Rohner, TJ, Jr, Drago, JR, and Demers, LM: Will prostatic acid phosphatase determination by radioimmunoassay increase the diagnosis of early prostatic cancer? J Urol 125:361-364, 1981.
11. Gericke, K, Kohse, KP, Pfleiderer, G, et al: Development and evaluation of a new solid-phase direct immunoenzyme assay for prostatic acid phosphatase, Clin Chem 28:596-602, 1982.
12. Foti, AG, Cooper, JF, Herschman, H, and Malvarz, RR: Detection of prostatic cancer by solid phase radioimmunoassay of serum prostatic acid phosphatase, N Engl J Med 297:1257-1261, 1977.
13. Gibb, I, Wilson, LA, Powell, PH, et al: Immunological and colorimetric determination of prostatic acid phosphatase: technical and clinical reappraisal in symptomatic patients, Clin Chem 32:1760-1766, 1986.
14. Chen, I-W, Sperling, MI, Maxon, HR, and Kaplan, LA: Stability of immunological activity of human prostatic acid phosphatase in serum, Clin Chem 28:1163-1166, 1982.
15. Sawtell, N, Weiss, M, and Kaplan, LA: Prostatic acid phosphatase positive paraurethral glands in human females, Am J Clin Pathol 83:263, 1985.

Alanine aminotransferase

ROBERT L. MURRAY

Alanine aminotransferase, ALT, L-alanine:2-oxoglutarate aminotransferase, serum glutamate pyruvate transaminase, SGPT

Clinical significance: pp. 359, 427, and 484

Enzyme number: EC 2.6.1.2

Molecular weight: approximately 101,000 daltons

Chemical class: enzyme, protein

Biochemical reaction: amino transfer catalyzed by ALT:

$$\underset{\text{L-Alanine}}{\begin{matrix}CO_2H\\|\\HC{-}NH_2\\|\\CH_3\end{matrix}} + \underset{\alpha\text{-Ketoglutarate}}{\begin{matrix}CO_2H\\|\\C{=}O\\|\\CH_2\\|\\CH_2\\|\\CO_2H\end{matrix}} \overset{ALT}{\rightleftharpoons} \underset{\text{Pyruvate}}{\begin{matrix}CO_2H\\|\\C{=}O\\|\\CH_3\end{matrix}} + \underset{\text{L-Glutamate}}{\begin{matrix}CO_2H\\|\\HC{-}NH_2\\|\\CH_2\\|\\CH_2\\|\\CO_2H\end{matrix}}$$

PRINCIPLES OF ANALYSIS

Alanine aminotransferase (ALT) catalyzes the transfer of an amino group between L-alanine and L-glutamate; the corresponding keto acids in this process are α-ketoglutarate and pyruvate. In vivo this reaction, shown in the reaction figure, goes to the right to provide a source of nitrogen for the urea cycle. The pyruvate thus generated is available for entry into the citric acid cycle, whereas the glutamate is deaminated (catalyzed by glutamate dehydrogenase), yielding ammonia and α-ketoglutarate.

The reaction is reversible; the chemical equilibrium favors the formation of alanine and α-ketoglutarate. Because these products are relatively difficult to assay, however, analytical techniques typically force the reverse reaction, allowing quantitation of pyruvate. Two methods of ALT analysis have enjoyed wide popularity for routine clinical use: the Reitman-Frankel method[1] (method 1, Table 59-5), involving the measurement of ALT activity by conversion of the reaction product, pyruvate, to its hydrazone[1]; and the Wróblewski method,[2] involving the coupling of the ALT reaction to a lactate dehydrogenase (LD) reaction, with measurement of that reaction's products (method 2, Table 59-5).[2]

In the DNPH procedure,[1] the serum is incubated with L-alanine and α-ketoglutarate; after a measured time the reaction is stopped, and the newly formed pyruvate is reacted with dinitrophenylhydrazine (DNPH), producing the corresponding hydrazone. This condensation is relatively rapid, even at room temperature. After condensation, the reaction mixture is alkalinized, producing a blue color caused by the anion form of the hydrazone. The absorbance is measured at 505 nm and compared to a standard curve. The dinitrophenylhydrazone of α-ketoglutarate produces negligible color, and little error is introduced by this source.[3]

In the other approach (methods 2 and 3, Table 59-5) reduced nicotinamide adenine dinucleotide (NADH) is the reaction product that is quantitated. Lactate dehydrogenase (LD) and its required cofactors are added, with allowance for the enzymatic conversion of pyruvate to lactate and with the simultaneous oxidation of NADH. The disappearance of NADH is followed spectrophotometrically (at 340 nm) or fluorometrically. Theoretically, one could follow this reaction by either continuous monitoring or by an end-point determination; in practice the continuous monitoring technique is more frequently encountered. Only under very unusual circumstances would the increased sensitivity afforded by fluorescence be needed; for routine clinical use, the absorbance technique is adequate.[4-7]

The coupled-enzyme technique, with continuous ultraviolet monitoring of NADH disappearance, is recommended as the preferred method for clinical analysis of ALT.

Table 59-6 gives various published recommendations for performance of the ALT assay using the coupled-enzyme

Table 59-5 Methods of alanine aminotransferase (ALT) analysis

Method	Type of analysis	Principle*	Usage
1. Dinitrophenylhydrazine (DNPH) coupling (colorimetric) (Reitman and Frankel[1])	Quantitative	*End-point absorbance measurement at 505 nm:* Ala + α-KG → Gl + Pyr Pyr + DNPH → Pyr-DNP-hydrazone	Serum, rarely performed assay
2. Enzymatic (ultraviolet monitoring) (Wróblewski and LaDue[2])	Quantitative	*UV monitoring of NADH disappearance at 340 nm:* Ala + α-KG → Gl + Pyr Pyr + NADH + $H^+ \xrightarrow{LD}$ Lac + NAD^+	Serum, most frequently employed procedure
3. Enzymatic (fluorescence)	Quantitative	*Fluorescence monitoring of NADH disappearance:* Same reactions as above	Serum, rarely used (high sensitivity)

**Ala,* Alanine; *α-KG,* α-ketoglutarate; *DNP,* dinitrophenyl; *DNPH,* dinitrophenylhydrazine; *Gl,* glutamate; *Lac,* lactate; *NAD^+,* nicotinamide adenine dinucleotide; *NADH,* reduced nicotinamide adenine dinucleotide; *Pyr,* pyruvate.

Table 59-6 Comparison of enzymatic alanine aminotransferase (ALT) methods

Condition	AACC[12]*	British[4]	IFCC[5]*	Scandinavian[6]	German[7]
Temperature	30° C	25° C	30° C	37° C	35° C
Fraction of serum volume	0.052	0.067	0.083	0.120	0.135
Final concentration of reagent					
L-Alanine	400 mmol/L	250 mmol/L	500 mmol/L	400 mmol/L	800 mmol/L
α-Ketoglutarate	15 mmol/L	6.7 mmol/L	15 mmol/L	12 mmol/L	18 mmol/L
pH	7.5	7.4	7.5	7.4	7.4
Tris buffer	86 mmol/L	—	100 mmol/L	20 mmol/L	—
Phosphate	—	90 mmol/L	—	—	80 mmol/L
NADH	0.18 mmol/L	0.25 mmol/L	0.18 mmol/L	0.15 mmol/L	0.18 mmol/L
Pyridoxal phosphate (PP)	—	—	0.1 mmol/L	—	—
Lactate dehydrogenase	2.4 U/mL	0.15 U/mL	1.2 U/mL	2.0 U/mL	1.2 U/mL
EDTA	—	—	—	5 mmol/L	—

**AACC,* American Association for Clinical Chemistry; *IFCC,* International Federation of Clinical Chemistry.

procedure; the International Federation of Clinical Chemistry (IFCC) method is recommended. A lag time, before initiation of the reaction with α-ketoglutarate, must be provided to allow for consumption of endogenous α-ketoglutarate and for any other NADH-consuming processes. Unless the absorbance change is monitored to ensure linearity, this lag phase should not be less than 90 seconds. A 10-minute preincubation period of serum with the cofactor PP is recommended.[8-11]

The American Association for Clinical Chemistry (AACC)[12] has proposed a method for the measurement of ALT in the small clinical laboratory that differs from the IFCC method in that (1) a single reagent is used to avoid a two-step addition procedure, (2) the reaction is read for 180 seconds after the reaction is allowed to proceed for 150 seconds, and (3) pyridoxal phosphate is not added.

SPECIMEN

Serum is the preferred specimen. Oxalate, heparin, and citrate do not inhibit the enzymatic activity but may introduce slight turbidity. Hemolyzed specimens should be avoided, since erythrocytes contain three to five times more ALT than serum does. ALT is stable in serum for 3 days at room temperature and for up to 1 week at 4° C. Urine has little or no activity and is not recommended for analysis.

PROCEDURE: KINETIC ANALYSIS OF ALT[5]

Principle

An amino group is transferred from L-alanine to 2-oxoglutarate to form pyruvate and L-glutamate, catalyzed by the action of ALT. The rate of the reaction is monitored by use of an indicator reaction. The pyruvate formed is converted to lactate, and the reaction is followed by monitoring the absorbance change at 339 nm of the consumed NADH.

Reagents

Tris, L-alanine buffer (110 mmol/L Tris, 630 mmol/L L-alanine, pH 7.5). Dissolve 1.32 g of Tris-

(hydroxymethyl)aminomethane and 5.62 g of L-alanine (free acid) in 80 mL of distilled water. Adjust to pH 7.5 at 30° C with 1 mol/L HCl (approximately 8.0 mL). Allow the solution to cool to the calibration temperature, and bring to a volume of 100 mL in a volumetric flask. This is solution 1, stable for 6 months at 0° to 4° C. Check for bacterial growth.

Tris/hydrochloric acid buffer (110 mmol/L Tris, pH 7.5). Dissolve 2.64 g of Tris(hydroxmethyl)aminomethane in 160 mL of distilled water. Adjust to pH 7.5 at 30° C with 1 mol/L HCl. Allow the solution to cool to the calibration temperature, and bring to volume of 200 mL in a volumetric flask. This is solution 2, stable for 6 months at 0° to 4° C. Check weekly for bacterial growth.

Pyridoxal phosphate solution (6.3 mmol/L pyridoxal phosphate). Dissolve 16.7 mg of pyridoxal phosphate in solution 2, and bring to 10 mL volume. This is solution 3, stable for 2 weeks at 0° to 4° C when stored in a dark bottle.

Reduced nicotinamide adenine dinucleotide (11.3 mmol/L NADH). Dissolve 16 mg of the sodium salt of NADH (or an equivalent amount correcting for water of hydration) in 2.0 mL of solution 2. This is solution 4, stable for 2 weeks at 0° to 4° C when stored in a dark bottle.

Lactate dehydrogenase (2.52×10^6 nkat/L or 150,000 U/L). Dilute the enzyme in a 50% (v/v) mixture of glycerol and solution 8 to give the desired catalytic concentration. This is solution 5, stable for 6 months at 0° to 4° C.

Working reagent for ALT reaction. Mix 1000 mL of solution 1 (Tris-alanine) with 2 mL of solution 3 (pyridoxal phosphate), 2 mL of solution 4 (NADH), and 1 mL of solution 5 (enzyme). This is solution 6. Make fresh daily; keep in a dark bottle.

Reagent mixture for sample blank (D-alanine, 630 mmol/L). Dissolve 5.62 g of D-alanine (free acid) in 50 mL of solution 2 (Tris-HCl). Adjust to pH 7.5 at 30° C with 1 mol/L HCl. Add 2 mL of solution 3 (pyridoxal phosphate), 2 mL of solution 4 (NADH), and 1 mL of solution 5 (enzyme). Bring to 100 mL with solution 2. This is solution 7. Make fresh daily, and keep it in a dark bottle.

Tris/2-oxoglutarate (110 mmol/L Tris, 180 mmol/L 2-oxoglutarate, pH 7.5). Dissolve 1.36 g of Tris(hydroxymethyl)aminomethane and 3.42 g of 2-oxoglutaric acid in 80 mL of distilled water. Adjust to pH 7.5 at 30° C. Allow the solution to cool to the flask calibration temperature, and make up to 100 mL with distilled water using a volumetric flask. This is solution 8, stable for 2 weeks at 0° to 4° C.

Sodium chloride (154 mol/L). Dissolve 0.9 g of sodium chloride in 100 mL of water. This is solution 9, stable for 6 months stored at 4° C. Check for bacterial growth.

Assay

Equipment: spectrophotometer with ≤10 nm pass capable of reading at 339 nm, with a constant temperature cuvette capable of maintaining less than 0.05° C fluctuation. A recording spectrophotometer is preferable.

1. Add 2.00 mL of solution 6 (working reagent) and 0.20 mL of serum to the cuvette.
2. Mix, and allow to stand for 10 minutes.
3. Monitor the absorbance to ensure that the values are stable.
4. Add 0.20 mL of solution 8.
5. Mix, and record the change in absorbance for at least 300 seconds. If no recorder is available, record every 15 seconds.

Calculations

If the values of the change in absorbance, ΔA, per second ($\Delta A/\Delta t$) are greater than 0.00025, or if they decrease during the step 3 monitoring period, dilute the sample five- to tenfold with solution 9 (0.154 M NaCl).

Correct for the blank using the following schema:

$(\Delta A/\Delta t)$A = Measured reaction of serum and solution 6
$(\Delta A/\Delta t)$B = Measured reaction of solution 9 and solution 6
$(\Delta A/\Delta t)$C = Measured reaction of serum and solution 7
$(\Delta A/\Delta t)$D = Measured reaction of solution 2 and solution 7

Then

$$(\Delta A/\Delta t)\text{ corrected} = [(\Delta A/\Delta t)\text{A} - (\Delta A/\Delta t)\text{B}] - [(\Delta A/\Delta t)\text{C} - (\Delta A/\Delta t)\text{D}]$$

Using the corrected $\Delta A/\Delta t$, if one uses a 1 cm path length and a value of 6.3×10^3 mol/L·cm path length, the catalytic concentration is:

$$1.905 \times 10^{-3} \times (\Delta A/\Delta t)\text{kat/L}$$

or

$$1.19 \times 10^5 \times (\Delta A/\Delta t)\text{U/L}$$

NOTE: To prevent growth of microorganisms, use sterilized containers. Sodium azide may be added to solutions 1, 2, and 5 to make a final concentration of 8 mmol/L.

REFERENCE RANGE

When analyzed at 37° C by methods employing activation with PP, the normal adult reference range of ALT is 0 to 55 U/L. Men have been reported to show slightly higher values than women. Because the levels of this enzyme are unusually sensitive to liver damage, increases in ALT may be a result of excessive use of alcohol or of exposure to a variety of hepatotoxic agents.

Normal newborns have been reported to show a reference range of up to double the adult upper level. These values decline to adult levels by approximately 3 months of age. This increased activity has been attributed to seepage from the neonate's immature hepatocytes, which have more permeable membranes.

REFERENCES

1. Reitman, S, and Frankel, S: A colorimetric method for the determination of serum glutamic oxalacetic and glutamic pyruvic transaminases, Am J Clin Pathol 28:56-63, 1957.
2. Wróblewski, F, and LaDue, JS: Serum glutamic-pyruvic transaminase in cardiac and hepatic disease, Proc Soc Exp Biol Med 91:569-571, 1956.
3. Brétaudière, JP, Burtis, C, Pasching, J, et al: Study of the alanine aminotransferase kinetic assay by response surface methodology, Clin Chem 26:1023, 1980 (abstract).
4. Wilkinson, JH, Baron, DN, Moss, DW, and Walker, PG: Standardization of clinical enzyme assays: a reference method for aspartate and alanine transaminases, J Clin Pathol 25:940-944, 1972.
5. Bergmeyer, HU, and Hørder, M: IFCC methods for the measurement of catalytic concentrations of enzymes. Part 3. IFCC method for alanine aminotransferase, J Clin Chem Clin Biochem 18:521-534, 1980.
6. Committee on Enzymes of the Scandinavian Society for Clinical Chemistry and Clinical Pathology: Recommended methods for the determination of four enzymes in blood, Scand J Clin Lab Invest 33:291-305, 1974.
7. Enzyme Commission of the German Society for Clinical Chemistry: Recommendations of the German Society for Clinical Chemistry, Z Klin Chem Klin Biochem 10:281-291, 1972.
8. Lustig, V: Activation of alanine aminotransferase in serum by pyridoxal phosphate, Clin Chem 23:175-177, 1977.
9. Bergmeyer, HU, Schiebe, P, and Wahlefeld, AW: Optimization of methods for aspartate aminotransferase, Clin Chem 24:58-73, 1978.
10. Siest, G, Schiele, F, Galteau, M-M, et al: Aspartate aminotransferase and alanine aminotransferase activities in plasma: statistical distributions, individual, variations, and reference values, Clin Chem 21:1077-1087, 1975.
11. Miller, DA, Glick, MR, and Oei, TO: Results compared for plasma aminotransferase activity with and without pyridoxal phosphate activation in infants and children, Clin Chem 27:1035, 1981 (abstract).
12. Butler, TJ, Klotzsch, SG, and Osberg, IM: Alanine aminotransferase, ALT provisional. In Faulkner, WR, and Meites, S, editors: Selected methods of clinical chemistry, Washington DC, 1982, American Association for Clinical Chemistry.

Alkaline phosphatase

STEVEN C. KAZMIERCZAK
JOHN A. LOTT

Alkaline phosphatase, ALP, *ortho*-phosphoric monoester phosphohydrolase (alkaline optimum)
Clinical significance: pp. 359, 373, and 664
Enzyme number: EC 3.1.3.1
Molecular weight: varies with tissue source of enzyme and ranges from 70,000 to 120,000 daltons
Chemical class: enzyme, protein
Known isoenzymes: bone, liver, placenta, intestine, kidney, Regan (fetal), Nagao, Kassahara
Biochemical reaction:

$$H_2O + R{-}O{-}\overset{\displaystyle O}{\overset{\|}{\underset{\displaystyle OH}{\underset{|}{P}}}}O^- \underset{pH>9}{\overset{ALP}{\rightleftharpoons}} R{-}OH + H_2PO_4^-$$

PRINCIPLES OF ANALYSIS

The group of nonspecific phosphatases that catalyze the reaction shown is known collectively as *alkaline phosphatase* (ALP). Phosphatases transfer a phosphate moiety from one group to a second, forming an alcohol and a second phosphate compound. When water is the phosphate acceptor, inorganic orthophosphate is formed. Optimum activity of these enzymes is exhibited at a pH of approximately 10.0. ALP requires Mg^{+2} and Zn^{+2} ions for stability and maximum activity; it is inhibited by Ca^{+2} and inorganic phosphate.

Reaction rates depend on such variables as the tissue source of the enzyme, the type of substrate and buffer used, and the incubation temperature. ALP is denatured slowly at 37° C; thus a reaction temperature of 25° to 30° C is often recommended for clinical assays. Some buffers enhance the enzyme rate by acting as phosphate-group acceptors in a process called *transphosphorylation*. Transphosphorylation buffers greatly increase the rate of reaction compared to the barbital, carbonate, or glycine buffers used in previous methods.[1-4] Examples of transphosphorylating buffers are 2-methyl-2-amino-1-propanol (MAP), diethanolamine (DEA), and Tris(hydroxy)aminomethyl methane (Tris). Mannitol, which also acts as a transphosphorylating phosphate acceptor, has been employed as an ''accelerator'' in some ALP reactions. Since the natural substrates of ALP are not known, almost all assays for ALP now employ *p*-nitrophenyl phosphate (*p*NPP) as substrate. At alkaline pH, *p*NPP is colorless; the reaction product *p*-nitrophenol (*p*NP) is intensely yellow with a molar absorptivity at 403 nm of approximately 18,450 $L \cdot mol^{-1} \cdot cm^{-1}$.

The earliest assays for total serum ALP measured the release of inorganic phosphate. Subsequently, ALP methods have been introduced in which refinements in the use of chromogenic substrates and rate-enhancing buffers have led to significant improvements in analytical sensitivity and precision. Table 59-7 summarizes these methods.

The early assays based on the measurement of liberated inorganic phosphate are now only of historical interest. The substrates and reaction conditions used were relatively insensitive, and long incubation times (1 to 1½ hours) were required to produce acceptable results. Representative of these assays is the classic method of Shinowara et al.[1] (method 1, Table 59-7). Serum is incubated for 1 hour at 37° C in diethylbarbiturate buffer (pH 9.3) containing β-glycerophosphate as substrate. Liberated phosphate ion is reacted with phosphomolybdic acid reagent, and the resulting chromogen is measured colorimetrically at 600 nm.

Another procedure of historical interest is that of King and Armstrong[2] (method 2, Table 59-7). Phenyl phosphate is used as substrate, and the liberated phenol is measured colorimetrically after the addition of Folin-Ciocalteu reagent (phosphotungstic-phosphomolybdic acid); this re-

Table 59-7 Methods of alkaline phosphatase (ALP) analysis

Type of analysis	Method source	Principle	Usage	Comments
1. Two-point spectrophotometric	Shinowara et al.[1]	Substrate: β-glycerophosphate Measure rate of release of inorganic phosphate; 1 hour incubation	All body fluids; manual	Requires long incubation time; high phosphate background in samples Considered obsolete
2. End-point spectrophotometric	King and Armstrong[2]	Substrate: phenyl phosphate Measure rate of release of phenol with Folin-Ciocalteu reagent; 30 min incubation	All body fluids; manual	Samples require deproteinization Considered obsolete
3. End-point spectrophotometric	Kind and King[3]	Substrate: phenyl phosphate Measure rate of release of phenol with 4-amino-antipyrine as chromogenic reagent; 15 min incubation	All body fluids; manual/automated	Faster rate than King and Armstrong[2] Requires no deproteinization
4. End-point or kinetic spectrophotometric	Bessey et al.[4]	Substrate: *p*-nitrophenyl phosphate (*p*NPP) Measure rate of formation of yellow *p*-nitrophenoxide ion	All body fluids; manual/automated	Rapid; linear change in absorbance with time
5. End-point or kinetic spectrophotometric	Moss[5]	Substrate: α-naphthol monophosphate Measure rate of formation of α-naphthol at 340 nm	All body fluids; manual/automated	Rapid; convenience of measuring at 340 nm
6. Fluorescent	Cornish et al.[6]	Substrate: 4-methylumbelliferyl phosphate	All body fluids; automated	Highly sensitive
7. Kinetic or end-point spectrophotometric	Bowers and McComb[7,8]	Substrate: *p*NPP Measure rate of release of *p*-nitrophenoxide in transphosphorylating buffer	Manual or automated	Proposed reference method; more sensitive than Bessey et al.[4]

agent forms a blue color with phenol in an alkaline solution that has an absorption maximum at 750 nm.

Kind and King[3] reported a similar but improved method that also used phenyl phosphate as substrate but measured the liberated phenol with 4-amino-antipyrine, which reacts with phenol to yield a red-colored quinone (method 3, Table 59-7). The quinone, which does not react with plasma proteins, can be measured colorimetrically at 500 nm. In this case protein precipitation and separation before analysis is not required, as is the case when the Folin-Ciocalteu reagent is used.

Bessey et al.[4] introduced the use of *p*NPP as substrate in ALP assays (method 4, Table 59-7). The phosphatase hydrolyzes the colorless substrate to yield the yellow salt of *p*NP, which has an absorption maximum at 400 nm and eliminates the need for additional color-producing reactions. The assay can be performed on as little as 5 μL of serum. In the original procedure, serum is added to 50 μL of 0.1 mol/L glycine buffer (pH 10.3) containing 2 g of *p*NPP/L. The samples are incubated at 38° C for 30 minutes, and the reaction is stopped by the addition of 0.5 mL of 20 mmol/L NaOH, followed by spectrophotometry at 400 to 420 nm. Background correction is performed by addition of acid to the samples, which converts the yellow sodium salt into colorless free nitrophenol, and a second absorbance measurement is made.

Moss[5] described a rapid continuous spectrophotometric assay that offers the convenience of measurement at 340 nm (method 5, Table 59-7). Assays are carried out in 0.1 mol/L sodium carbonate-bicarbonate buffer (pH 10.0 at 37° C) containing 5 mmol/L $MgCl_2$, with α-naphthyl phosphate acting as substrate, yielding α-napthol as the hydrolysis product. Reaction curves are recorded for 5 to 10 minutes, and rates are calculated from the initial, linear portions of the curves.

Cornish et al.[6] introduced a highly sensitive, automated fluorometric assay for ALP using 4-methylumbelliferyl phosphate as substrate, which is hydrolyzed to 4-methylumbelliferone, a highly fluorescent compound (method 6, Table 59-7). The excitation peak is at 360 nm, and the fluorescent light is isolated by a secondary cutoff filter at 465 nm. The procedure requires that 10 μL of serum be added to 0.1 mol/L of carbonate-bicarbonate buffer (pH 9.2 at 37° C) containing 1.0 mmol/L 4-methylumbelliferyl phosphate. Incubation time is approximately 8 minutes.

The *p*NPP method of Bowers and McComb,[7,8] widely used today because of its simplicity and sensitivity, is accepted by many as the reference method for ALP quantitation (method 7, Table 59-7). It is an advance over prior colorimetric assays because of its use of a transphosphorylating buffer (MAP) and because of carefully selected conditions of pH, temperature, and reagent concentrations. Specific parameters of the assay are summarized in Table 59-8.[9-12]

Although the methods just described have at one time or another been used in the clinical laboratory, recent ad-

Table 59-8 Comparison of reaction conditions for alkaline phosphatase (ALP) analysis*

Condition	Manual reference (Bowers and McComb)[7]	German[10]	Scandinavian[11]	AACC[12]
Temperature	30° C	25° C	37° C	30° C
pH	10.5	9.8	9.8	10.4
Final concentration of reagents	*p*NPP: 16 mmol/L MAP: 1.0 mol/L Mg^{+2}: 1.0 mmol/L	*p*NPP: 10 mmol/L DEA†: 1.0 mol/L Mg^{+2}: 0.5 mmol/L	*p*NPP: 10 mmol/L DEA†: 1.0 mol/L Mg^{+2}: 0.5 mmol/L	*p*NPP: 16 mmol/L MAP: 0.35 mol/L Mg^{+2}: 2.0 mmol/L
Fraction of sample volume	0.0164	0.009	0.009	0.0196
Linearity (approximate)	To 500 U/L	To 500 U/L	Linear for 10 minutes or up to 1500 U/L	To 900 U/L
Precision (in reference range)	3% to 6%	5%	5%	3% to 5%

AACC, American Association for Clinical Chemistry; *German,* German Society for Clinical Chemistry; *Scandinavian,* Scandinavian Society for Clinical Chemistry and Clinical Physiology.
DEA, Diethanolamine; *MAP,* 2-methyl-2-amino-1-propanol; p*NNP, p*-nitrophenyl phosphate.
*Major interferences: EDTA, citrate, oxalate, inorganic phosphate, calcium, and ammonium sulfate.
†It has been reported that a contaminant in some lots of DEA causes significant loss of ALP activity.[7]

vances in our knowledge of the nature and reactivity of ALP have led to modifications in assay conditions that render most earlier methods obsolete. Controversy over conditions for optimum ALP activity continues, and up to now a universally accepted reference method has not been established.

The method of Bowers and McComb[7,8] appears to be the most likely candidate for a reference method. It offers the convenience and sensitivity of a self-indicating, highly reactive substrate in a transphosphorylating buffer. The AACC reference method[12] optimizes all reaction conditions, including temperature, pH, reagent concentrations, and sample volume fraction. This method and slight modifications of it are under active consideration by various national and international organizations interested in establishing a reference method for ALP. The German Society for Clinical Chemistry[10] and the Scandinavian Society for Clinical Chemistry and Clinical Physiology[11] have proposed reference methods similar to that of Bowers and McComb. A summary of the four proposed reference methods is given in Table 59-8.

It is important to note that the measured activity of ALP depends on the volume fraction of serum. Increases in activity have been observed when the serum fraction was decreased from 1/26 to 1/51; no further increases were seen below 1/51.

The following procedure is adapted from the manual reference method used by the National Bureau of Standards.[8] It uses *p*NPP as substrate with MAP as the rate-enhancing buffer and involves rigorously controlled conditions of pH (10.5) and temperature (30° C).

SPECIMEN

ALP can be found in all body fluids and tissues. Blood samples should be drawn after a fast of at least 8 hours. Serum and heparinized plasma give similar results. Other anticoagulants, such as EDTA, oxalate, and citrate, inhibit the enzyme by complexing Mg^{+2} and should not be used. Slight hemolysis is tolerable, but gross hemolysis should be avoided. Bilirubin does not interfere with kinetic methods using *p*NPP as substrate. Most reports indicate that serum ALP activity increases slowly with storage at room or refrigerated temperatures. As a general rule, it is best to analyze ALP specimens the same day they are drawn.

PROCEDURE: *p*-NITROPHENYL PHOSPHATE KINETIC ASSAY

Principle

The colorless substrate is converted at alkaline pH to the yellow *p*-nitrophenoxide ion. The reaction is followed by measurement of the increase in absorbance at 403 nm.

Reagents

MAP buffer (1.5 mol/L). Liquefy 2-methyl-2-amino-1-propanol by warming to 30° to 35° C. Weigh 135 g of the liquid directly into a 1 L volumetric flask, add 500 mL of distilled water, and mix. Carefully add 190 mL of 1.0 mol/L HCl to the flask. When the solution has cooled to room temperature, dilute to 1 L with water. Confirm that the pH is 10.5 at 30° C. This buffer is stable for 1 month when stored in an airtight container at 25° C.

Magnesium acetate solution (3 mmol/L). Dissolve 650 mg of magnesium acetate · 4 H_2O in 1 L of water. This is stable indefinitely at 4° C.

***p*-Nitrophenyl phosphate solution (24.5 mmol/L).** Dissolve 91 mg of disodium 2-nitrophenyl phosphate · 6 H_2O in 10 mL of MAP buffer. Prepare fresh daily.

***p*-Nitrophenol standard solutions (1 mmol/L).** Dissolve 139.1 mg of *p*NP in 1 L of distilled water. This is stable for several months when stored in the dark. To prepare the working standard solution, add 25 mL of the 1 mmol/L solution to 900 mL of MAP buffer, and dilute to 1 L with distilled water. The working standard solution is stable for at least 2 months. This solution is used to standardize spectrophotometers at 403 nm.

Assay

Equipment: any recording spectrophotometer equipped with a temperature-controlled cell compartment is suitable for the procedure described. The temperature should be held constant at 30° ± 0.1° C.

1. Add 50 μL of specimen to 1.0 mL of magnesium acetate solution in a test tube. Thoroughly mix, and incubate for 5 minutes at 30° C.
2. Prewarm buffered *p*NPP solution to 30° C, and add 2.0 mL to the incubation mixture from step 1. Agitate thoroughly.
3. Transfer the reaction mixture to a cuvette with a 1 cm light path, and read the absorbance change versus time at 403 nm for 2 minutes. Readings can be taken immediately after mixing.

Calculations

International units (U) of activity are expressed as micromoles of *p*-nitrophenoxide formed per minute. Enzyme concentrations are expressed as international units per liter (U/L). U/L can be calculated from the change in absorbance by the following equation for a 1 cm light path:

$$\text{U/L} = \frac{(\Delta A \text{ for } t \text{ min}/t \text{ min})(\text{Total volume})}{\text{Sample volume}} \times 10^6/\epsilon$$

where ΔA = change in absorbance for time, t
ϵ = molar absorptivity for *p*-nitrophenoxide (18.8×10^3 L · mol^{-1} · cm^{-1}
10^6 = factor to convert the concentration to micromoles per liter (10^6 μmol/mol)

The values are as follows:

$$\text{U/L} = \Delta A/\text{min} \times (3050\ \mu\text{L}/50\ \mu\text{L}) \times 10^6/18{,}800$$
$$\text{U/L} = \Delta A/\text{min} \times 3245$$

The method is linear to approximately 500 U/L.

REFERENCE RANGE

The serum reference ranges for healthy persons, determined at 30° C by the method just outlined, are as follows[8,13]:

Group (age in years)	ALP (up to U/L)	Group (age in years)	ALP (up to U/L)
Females:		Males:	
Newborns	250	Newborns	250
1 to 12	350	1 to 12	350
10 to 14	280	10 to 14	275
15 to 19	150	15 to 19	155
20 to 24	85	20 to 24	90
25 to 34	85	25 to 34	95
35 to 44	95	35 to 44	105
45 to 54	100	45 to 54	120
55 to 64	110	55 to 64	135
65 to 74	145	65 to 74	140
75+	165	75+	190

Factors causing increased ALP activities in a normal population include exercise, periods of rapid bone growth in children, and pregnancy.

REFERENCES

1. Shinowara, G, Jones, LM, and Reinhart, HL: Estimation of serum inorganic phosphate and "acid" and "alkaline" phosphatase activity, J Biol Chem 142:921-933, 1942.
2. King, EJ, and Armstrong, AR: A convenient method for determining serum and bile phosphatase activity, Can Med Assoc J 31:376-381, 1934.
3. Kind, PRN, and King, EJ: Estimation of plasma phosphatase by determination of hydrolysed phenol with amino-antipyrine, J Clin Pathol 7:322-326, 1954.
4. Bessey, O, Lowry, OH, and Brock, MJ: Method for the determination of alkaline phosphatase with five cubic millimeters of serum, J Biol Chem 164:321-329, 1946.
5. Moss, DW: A note on the spectrophotometric estimation of alkaline phosphatase activity, Enzymologia 31:193-202, 1966.
6. Cornish, CJ, Neale, FC, and Posen, S: Automated fluorometric alkaline phosphatase microassay with 4-methylumbelliferyl-phosphate as a substrate, Am J Clin Pathol 53:68-76, 1970.
7. Bowers, GN, Jr, and McComb, RB: Measurement of total alkaline phosphatase activity in human serum, Clin Chem 21:1988-1995, 1975.
8. Bowers, GN, Jr, and McComb, RB: Alkaline phosphatase, total activity in human serum. In Faulkner, WH, and Meites, S, editors: Selected methods for the small clinical chemistry laboratory, Washington, DC, 1982, American Association for Clinical Chemistry.
9. Bowers, GN, Jr, McComb, RB, Christensen, RG, and Schaffer, R: High-purity 4-nitrophenol: purification, characterization, and specifications for use as a spectrophotometric reference material, Clin Chem 26:724-729, 1980.
10. German Society for Clinical Chemistry: Recommendations of the Enzyme Commission, Z Klin Chem Klin Biochem 10:281-291, 1972.
11. Scandinavian Society for Clinical Chemistry and Clinical Physiology: Recommended methods for the determination of four enzymes in blood, Scand J Clin Lab Invest 33:291-306, 1974.
12. Tietz, NW, Burtis, CA, Duncan, P, et al: A reference method for measurement of alkaline phosphatase activity in human serum, Clin Chem 29:751-761, 1983.
13. Munan, L, Kelly, A, Petitclerc, C, and Billon, B: Atlas of blood data. Prepared by the Epidemiology Laboratory and the Laboratory of Clinical Biochemistry, University of Sherbrooke, Sherbrooke, Quebec, 1980.

Alkaline phosphatase isoenzymes

JOHN F. CHAPMAN
LINDA L. WOODARD
LAWRENCE M. SILVERMAN

Alkaline phosphatase, ALP, *ortho*-phosphoric monoester phosphohydrolase (alkaline optimum)
Clinical significance: pp. 359, 373, and 664
Enzyme number: EC 3.1.3.1
Molecular weight: 120,000 daltons (varies considerably)
Chemical class: enzyme, protein
Known isoenzyme forms: bone, liver, fast liver, intestine, placenta, kidney, Regan, and Nagao
Biochemical reaction:

$$R_1—O—PO_3 + R_2—OH \xrightarrow[Mg^{+2},\ pH\ 9.8]{ALP} R_1—OH + R_2—O—PO_3H_2$$

Organic phosphate ester **+** ***Alcohol*** → ***New alcohol*** **+** ***New organic phosphate ester***

Net reaction: hydrolysis if alcohol is H_2O

PRINCIPLES OF ANALYSIS

Alkaline phosphatase (ALP) is an enzyme found in practically all body tissues, catalyzing the reaction shown at an alkaline pH. Although "optimum" reaction conditions for the assay of total ALP have been reported[1] (see p. 900), it is important to remember that these conditions were established using serum samples containing almost exclusively liver and bone isoenzymes. The optimum conditions for analysis of ALP isoenzymes from other tissue sources are known to vary widely with respect to substrate, buffer, and pH optima.[1-4]

Likewise, ALP isoenzymes from various tissue sources display considerable heterogeneity with respect to net molecular charge,[5] differential sensitivity to heat,[3] and inhibition by L-phenylalanine,[6] other amino acids,[7] and urea[8] (Table 59-9).

These unique molecular characteristics have provided the basis for the development of a variety of techniques for the separation and identification of ALP isoenzymes. Each technique possesses unique advantages and disadvantages, leading to uniformly recognized "preferred" methods. Electrophoresis, differential heat sensitivity, and differential chemical inhibition are currently the methods of analysis most often used.

Zone electrophoresis followed by staining for enzymatic activity is a frequently used technique for the qualitative analysis of ALP isoenzymes in serum (method 1, Table 59-10). Various supporting mediums, including agarose,[9] cellulose acetate,[10] paper,[11] polyacrylamide gel,[12] and starch gel,[11] have been used with varying degrees of success.[13] In polyacrylamide gels at a pH of approximately 9.0, the charge of the ALP isoenzymes results in the normal liver isoenzyme (liver I) migrating most rapidly toward the anode, with bone and intestinal isoenzymes migrating progressively more slowly. Fig. 59-2 demonstrates a typical polyacrylamide gel electrophoretic separation of serum ALP isoenzymes. Typically, only liver I and bone isoenzymes are detected by the isoenzyme procedures. With the polyacrylamide method shown, these two activities are clearly distinguishable (Fig. 59-2, *b* and *c*). The intestinal isoenzyme (Fig. 59-2, *a*) is not normally detected in serum, although some persons of blood types O and B have a small amount of intestinal isoenzyme in serum, particularly after meals. The intestinal isoenzyme may also be elevated with intestinal disease. The liver II isoenzyme (Fig. 59-2, *d* and *e*) is an intracellular enzyme that is not usually observed in serum. Its presence is often associated with severe parenchymal cell damage. The biliary isoenzyme is derived from the biliary tree, and its presence may indicate active cholestatic disease.

Differentiation of ALP isoenzymes on the basis of selective inactivation at 56° C remains one of the most widely used techniques (method 2, Table 59-10). The decrease in

Table 59-9 Properties of alkaline phosphatases

Property	Liver	Bone	Intestine	Placenta
L-Phenylalanine inhibition (%)	0-10	0-10	75	75
L-Homoarginine inhibition (%)	78	78	5	5
Thermal inactivation (%)	50-70	90-100	50-60	0
Starch gel anodal migration (cm)	4.4-5.0	4.0-6.0	3.0	3.8-4.2
Neuraminidase effect	+	+	0	+
Immunochemical effect with dilute antisera to:				
Placental form	0	0	0	+
Hepatic form	+	+	0	0
Intestinal form	0	0	+	0

0 indicates ALP is nonreactive; + indicates ALP is reactive.
From O'Carroll, D, et al: Am J Clin Pathol 63:564, 1975.

Table 59-10 Methods for measurement of alkaline phosphatase (ALP) isoenzymes

Method	Principle	Usage	Comment
1. Electrophoresis	Total charge on various isoenzymes differ, allowing differential migration in electric field	Most frequently used technique	Incomplete separation; bone and liver often not completely resolved
a. Polyacrylamide gel	Separation by both charge and molecular weight	Frequently used	Best separation, difficulty in densitometric scanning; qualitative estimate of isoenzymes
b. Cellulose	Separation by charge	Some usage	Bone and liver not completely resolved
2. Heat inactivation	Heat inactivation rates of isoenzymes differ. Rate of inactivation after incubation at 56° C is suggestive of presence of bone or liver isoenzymes	Frequently used	Calculation is made more difficult by presence of intestinal or other more heat-stable isoenzyme forms
3. Chemical inactivation	L-Phenylalanine and urea inactivate different isoenzymes at different rates	Some usage; can be automated on a centrifugal analyzer	Regan-like isoenzyme activities differentiated from bone and liver

ALP activity after exposure of serum samples to heat at 56° C can be resolved into two phases.[14-16] Each phase represents the exponential inactivation of one of the isoenzymes normally present in serum. The more rapidly decaying activity has an inactivation half-time ($t_{1/2}$) of 112 seconds, whereas the more stable activity has a $t_{1/2}$ of 456 seconds. These activities represent the bone and liver activities, respectively.

The original procedure[16] required up to seven sampling times during the inactivation period. The linear portion of the curve representing the liver isoenzyme could be extrapolated back to zero time to determine the percentage of liver activity in the sample (*y* intercept). However, it was observed that the portion of the decay curve representing bone activity included the time points up to 15 minutes at 56° C, whereas the time points greater than 15 minutes represented liver activity.[17] By extrapolating only the time points at 15 and 25 minutes, one can directly determine activity that can be attributed to the liver isoenzyme, which substantially reduces the labor required to obtain the results.

The heat inactivation assay requires careful control of assay conditions, including, most critically, a rigorous (±0.01° C) control of the inactivation temperature. To the extent that the assay variables can be controlled on a routine basis, selective inactivation at 56° C can provide qualitative and semiquantitative estimates of liver and bone ALP activity when these two isoenzyme forms predominate in serum.

The use of L-phenylalanine and urea together to inhibit ALP isoenzymes differentially (method 3, Table 59-10) was first introduced by Statland et al.[18] in 1972 for an assay that utilized a centrifugal fast analyzer. As mentioned previously, several chemicals and amino acids in-

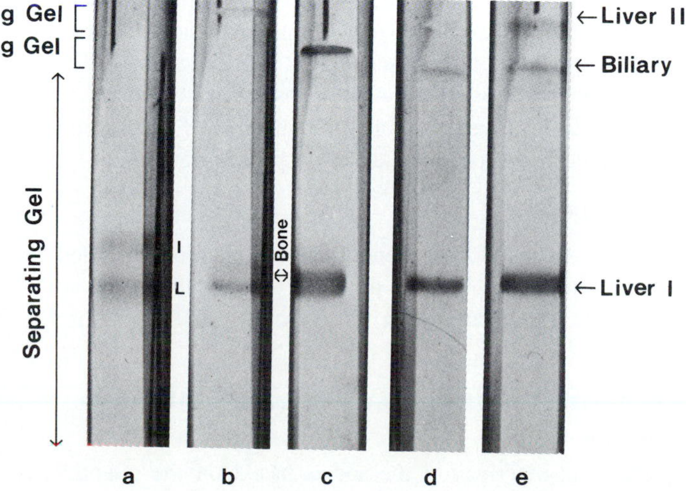

Fig. 59-2 Alkaline phosphatase isoenzymes separated by polyacrylamide disk electrophoresis. Normal blue stain of enzyme activity shown here as black bands. **a,** Control serum showing intestinal (I) and liver I isoenzymes. **b,** Heated patient sample of **c** showing residual bone and liver isoenzyme activities. **c,** Unheated patient sample showing elevated bone and liver isoenzyme activities. **d,** Heated patient sample of **e** showing residual liver I and biliary isoenzyme activities and *no* bone activity. **e,** Unheated patient sample showing large increase in liver I isoenzyme activity and presence of liver II and biliary isoenzyme activities.

hibit ALP isoenzymes to varying degrees. L-Phenylalanine, for example, at a concentration of 7 mmol/L, inhibits 91% ± 1.7% of intestinal ALP, 18% ± 5.6% of bone ALP, and 16% ± 3.1% of liver activity.[3] In addition, human placental ALP activity is inhibited by L-phenylalanine, L-tryptophan, and L-leucine, and human bone and liver ALP isoenzymes are inhibited by L-homoarginine, imidazole, and levamisole.[12] These inhibitors are stereospecific and uncompetitive, and inhibition most probably results from the formation of a poorly dissociable enzyme inhibitor–substrate complex.

ALP isoenzymes have also been denatured to varying degrees by urea. Generally, this inhibition proceeds in the following order: bone > liver > intestine > placenta. In the presence of 3.3 M urea, for example, approximately 93% of bone activity, 60% of liver activity, and 40% of intestinal ALP activity is inhibited. The use of both L-phenylalanine and urea in an assay allows the calculation of bone and liver isoenzyme activities and the determination of the presence of placental or Regan isoenzymes in serum. Although this technique does not differentiate between placental, intestinal, and Regan isoenzymes, it does allow the calculation of bone and liver activity in the presence of these other isoenzyme forms.

In general, electrophoretic methods yield only an incomplete separation of many of the ALP isoenzymes of interest. The best separation of ALP isoenzymes is obtained using polyacrylamide gel, whereas the poorest separation is obtained using cellulose acetate.

Although the concept of using selective inactivation by heat to identify ALP isoenzymes is relatively straightforward, the technical considerations and pitfalls are numerous. Thus, meticulous attention must be paid to details such as temperature control, exposure time, and thickness of glass tubes.

Although there is currently no one recognized reference method for ALP isoenzyme determination, both electrophoretic and heat-inactivation techniques are popular and relatively simple to perform and yield clinically useful results under most circumstances. Chemical inhibition methods, although not so widely used, may offer certain advantages over other techniques and should probably be considered by those laboratories possessing adequate instrumentation. Examples of a recommended electrophoretic method and a heat-inactivation procedure follow.

SPECIMEN

Serum from clotted venous blood is recommended. Plasma containing EDTA, fluoride, or oxalate should not be used. Samples are stable for 1 year at −20° C or for 1 week at 4° C.

PROCEDURE: HEAT-INACTIVATION METHOD

Principle

The heat-inactivation method is based on the quantitative differences in heat-stability characteristics of various ALP isoenzymes. Because the heat-stability characteristics of the same human ALP isoenzymes vary from one serum specimen to another, however, the calculation of one activity point after a single period of exposure can lead to inaccurate results. In this method, activity measurements are made after incubation for 15 and 25 minutes at 56° C. Using these activity points, the percentage of liver activity in the sample and the half-inactivation times for liver and bone ALP are calculated.[17]

Reagents

See total ALP activity method, p. 900.

Assay

Equipment: water bath set at 56° ± 0.01° C, spectrophotometer (band pass ≤ 10 nm) capable of reading at 505 nm.

1. Prepare incubation tubes. Adjust water bath to 56° C (± 0.01° C), and allow the appropriate number of small, thin-walled glass tubes (Dreyer tubes) to reach equilibrium at this temperature. Tubes should be immersed to a depth sufficient to allow 150 μL of serum sample to be at least 2 cm below the surface of the bath. A second set of Dreyer tubes is allowed to equilibrate in an ice bath during the first incubation.
2. At zero time, add approximately 150 μL of serum to the first tube. Start further specimens at exactly 90-second intervals thereafter. At exactly 15 minutes from the start of thermal inactivation for each sample, transfer 25 μL to a prechilled tube. At 25 minutes from the start of thermal inactivation, transfer the entire tube to the ice water bath.
3. Take measurements of ALP activity on (a) the unheated specimen, (b) the 15-minute specimen, and (c) the 25-minute specimen using the total ALP method (p. 900) or any suitable alternative method. Generally, the volume of serum taken for measurement is 10, 20, and 100 μL for a, b, and c, respectively. Absolute volume requirements may vary with the assay method used.

Calculations

After calculating enzyme activity (U/L) for each time point, calculate the percentage of residual activity:

$$\frac{\text{Tube } b \text{ or } c}{\text{Tube } a} \times 100\%$$

The percentage of residual activity values are plotted on semilog graph paper, and the straight line joining the 25- and 15-minute time points (linear axis) is extrapolated to zero time (Fig. 59-3). The percentage of liver ALP in the specimen is the *y*-intercept value obtained from the graph. The 15- and 25-minute time points correspond to the portion of the thermal inactivation curve associated with the more stable liver ALP isoenzyme. The slope of the thermal

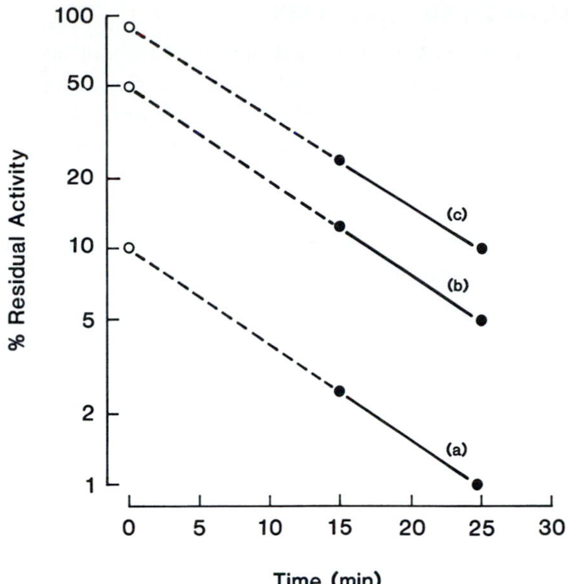

Fig. 59-3 Theoretical heat-inactivation plots for serum sample containing bone and liver alkaline phosphatase isoenzymes in ratios of, **a,** 90:10; **b,** 50:50; and, **c,** 10:90.

inactivation curve can be used to calculate the half-inactivation time for liver ALP.

Because the presence of heat-stable placental and Regan isoenzymes can influence the shape of the thermal inactivation curve during the measurement period just described, it is recommended that the apparent $t_{1/2}$ for liver isoenzyme be calculated for each sample.

The apparent $t_{1/2}$ can be conveniently determined from the semilog plot of residual activity in the following manner (Fig. 59-3). First, determine baseline liver ALP activity (y intercept) and divide this value by 2 to obtain one-half maximum liver activity. Locate the percentage of residual activity value corresponding to one-half maximum activity on the thermal inactivation curve, and by inspection determine the x intercept (time) for this point. This point represents the $t_{1/2}$ for the liver isoenzyme in minutes. Convert this value to seconds, and compare with the reference $t_{1/2}$ for liver isoenzyme using this method (451 sec ± 18 SD). Small between-batch temperature variations can affect the apparent half-life of liver ALP but have no effect on the calculated isoenzyme percentage, since each residual activity is altered proportionally. For example:

y intercept on curve c (in Fig. 59-3) = 90%

$$\frac{90\%}{2} = 45\%$$

x intercept for 45% point on curve c = 7.5 min

7.5 min × 60 sec/min = 450 sec = $t_{1/2}$ liver isoenzyme

Notes

1. An unusually high $t_{1/2}$ value would indicate a probable contribution from heat-stable isoenzyme forms. When the presence of placental or Regan isoenzyme forms is suspected, alternative tests such as L-phenylalanine inhibition[5] are advised to exclude any contribution of these isoenzymes to the liver isoenzyme–inactivation curve.
2. The heat-stability characteristics of intestinal ALP in serum are similar to those of liver ALP. The presence of intestinal ALP in serum therefore does not alter the shape of the thermal inactivation curve for liver ALP, and any contribution from this enzyme could erroneously be calculated as part of the liver isoenzyme activity. For this reason, it is recommended that samples be obtained from fasting patients and that alternative procedures such as electrophoresis or L-phenylalanine inhibition be available for the detection of intestinal ALP as required.
3. The percentages of liver and bone isoenzyme determined from a heat inactivation experiment are independent of the *absolute* value of the inactivation temperatures, provided that the temperature employed is kept *constant* during the analysis.

REFERENCE RANGE

The following ranges are based on the L-phenylalanine and urea inhibition method[17,19] but are similar to the results observed by polyacrylamide electrophoresis and heat-inactivation methods.

Age group	ALP isoenzyme		
	Bone	Liver	Other (intestinal and so on)
Child and adolescent	85%	5%	<10%
Adult	30%	60%	<10%
Elderly	30%	60%	<10%

REFERENCES

1. Bowers, GN, and McComb, RD: Measurement of total alkaline phosphate activity in human serum, Clin Chem 21(13):1988-1995, 1975.
2. Moss, DW, and King, EJ: Properties of alkaline phosphatase fractions separated by starch-gel electrophoresis, Biochem J 84:192-195, 1962.
3. Fishman, WH, and Ghosh, NK: Isoenzymes of human alkaline phosphatase, Adv Clin Chem 10:255-370, 1967.
4. Fernley, HN: Mammalian alkaline phosphatase. In Boyer, PD, editor: The enzymes, vol 4, New York, 1971, Academic Press, Inc.
5. Moss, DW: Scientific foundations of the estimation of isoenzymes in diagnosis. In Schmidt, E, and Schmidt, FW, editors: Multiple forms of enzymes, New York, 1982, S Karger Publishers, Inc.
6. Fernley, HN, and Walker, PG: Inhibition of alkaline phosphatase by L-phenylalanine, Biochem J 116:543, 1970.
7. O'Carroll, D, Statland, BE, Steele, BW, and Burke, MD: Chemical inhibition method for alkaline phosphatase isoenzymes in human serum, Am J Clin Pathol 63:564-572, 1975.
8. Bahr, M, and Wilkinson, JH: Urea as a selective inhibitor of human tissue alkaline phosphatases, Clin Chim Acta 17:367, 1967.
9. Sundblad, L, Wallin-Nilsson, M, and Brohult, J: Characterization of alkaline phosphatase isoenzymes in serum by agar gel electrophoresis, Clin Chim Acta 45:219, 1973.
10. Viot, M, Joulin, C, Cambon, P, et al: The value of serum alkaline phosphatase isoenzyme in the diagnosis of liver metastases, preliminary results, Biomedicine 31:74-77, 1979.

11. Kieding, NR: Differentiation into three fractions of the serum alkaline phosphatase and the behavior of the fractions in diseases of bone and liver, Scand J Clin Lab Invest 11:106-112, 1959.
12. Fishman, L: Acrylamide disc gel electrophoresis of alkaline phosphatase of human tissues, serum and ascites fluid using Triton X-100 in the sample and gel matrix, Biochem Med 9:309-315, 1974.
13. Moss, DW: Isoenzyme analysis, London, 1979, The Chemical Society.
14. Moss, DW: Alkaline phosphatase isoenzymes, Clin Chem 28(10):2007-2016, 1982.
15. Moss, DW, Shakespeare, MJ, and Thomas, DM: Observations on the heat stability of alkaline phosphatase isoenzyme in serum, Clin Chim Acta 40:35-41, 1972.
16. Whitby, LG, and Moss, DW: Analysis of heat inactivation curves of alkaline phosphatase isoenzymes activity in serum, Clin Chim Acta 59:361-367, 1975.
17. Moss, DW, and Whitby, LG: A simplified heat inactivation method for investigating alkaline phosphatase isoenzymes in serum, Clin Chim Acta 61:63-71, 1975.
18. Statland, BE, Nishi, NH, and Young, DS: Serum alkaline phosphatase: total activity and isoenzyme determinations made by use of the centrifugal fast analyzer, Clin Chem 18:1468-1474, 1972.
19. Gorman, L, and Statland, BE: Clinical usefulness of alkaline phosphatase isoenzyme determinations, Clin Biochem 10:171-174, 1977.

Amylase

MICHAEL D.D. McNEELY

Amylase, AMS, α-1,4-glucan 4-glucanhydrolase
Clinical significance: pp. 390, 398, and 484
Enzyme number: EC 3.2.1.1
Molecular weight: 40,000 to 50,000 daltons
Chemical class: enzyme, protein
Isoenzymes: at least seven in human tissue; salivary and pancreatic forms in serum

PRINCIPLES OF ANALYSIS

Amylase activity is increased in the serum and urine of patients with pancreatitis. Amylase measurements are used for the detection of this condition. Pancreatitis often is a medical emergency; thus amylase methods must be able to provide stat results.

Approximately 200 methods for amylase measurement have been devised. All begin with combining the patient's sample with a buffered solution of a polysaccharide. The assay conditions must be rigidly adjusted to suit the specific requirements of amylase. The pH optimum is 6.9 to 7.0, and calcium and chloride ions are absolutely required.

After incubation with the polysaccharide substrate, a variety of different detection techniques are available to measure the activity of amylase. The different approaches can be grouped into six categories.

Viscosimetric techniques (method 1, Table 59-11)

Viscosimetric techniques depend on the decrease in viscosity of the substrate that occurs after the action of amylase. Initially, the large polysaccharide substrate molecules form a highly viscid solution. After the action of amylase, many molecules have been broken into smaller fragments, and therefore the solution is less viscid. Viscosity measurements are performed before and after a timed incubation period, and the change in viscosity is proportional to the amylase activity. This technique has been abandoned because of the inconvenience of performing the viscosity measurements.

Table 59-11 Methods of amylase measurement

Method	Principle	Usage	Comments
1. Viscosimetric	Amylase degradation of starch decreases viscosity of solution; viscosity is inversely related to amylase concentration.	Historical	Neither precise nor accurate
2. Turbidimetric, nephelometric	Decrease in turbidity or scattered light (nephelometry) of starch solution is directly related to amylase concentration.	Rare	With consistent substrate, automated technique is acceptable
3. Iodometric	Degradation of starch by amylase reduces reaction of iodine with starch; reduction in iodine-starch product (A_{max} = 660 nm) is inversely related to amylase activity.	Rare	Can be easily adapted by most laboratories
4. Saccharogenic	Glucose released from substrate is quantitated, usually by enzymatic procedure, such as hexokinase or glucose oxidase.	Very frequent	Can be easily adapted to many automated, discrete analyzers
5. Chromolytic			
a.	Liberation of dye coupled to insoluble polysaccharide.	Frequent	Labor intensive; not suited for automation
b.	Liberation of chromogen from soluble, defined substrates.	Increasingly frequent	Suitable for automation
6. Fluorescence	Amylase degrades fluorescein-labeled starch into smaller fragments. More rapidly rotating labeled fragments cause polarized light to be emitted as depolarized fluorescence. Decrease in fluorescence polarization is related to amylase activity.	Rare	Available on a single instrument (TDx), suitable for stat usage

Turbidimetric and nephelometric techniques (method 2, Table 59-11)

In turbidimetric techniques, the unreacted substrate is designed to produce a turbid solution with a spectrophotometric absorbance of about 1.000. Amylase reduces the turbidity by fracturing the substrate. Turbidimetric and light-scattering measurements are carried out either kinetically or after a fixed time interval, and the change in turbidity or light scattering is proportional to amylase activity. These methods are difficult to standardize because of substrate variation, and, in general, they lack precision.[1]

Nephelometry has been used in a commercial instrument for amylase and lipase measurements (Model 91, Coleman Instruments Division, Perkin-Elmer Corp., Norwalk, CT).[2] This instrument uses a standardized substrate and carries out rate measurements over a short time. In this configuration the precision is acceptable, and the assay is rapidly completed.

Iodometric (amyloclastic) techniques (method 3, Table 59-11)

The iodometric methods are based on the ability of iodine to form a vivid blue color after reaction with starch.[3,4] The by-products of amylase action may also form colored substances with iodine, but their absorbance maxima are at different wavelengths from the characteristic starch-iodine complex. The methods are carried out by adding iodine color reagent to the substrate-sample mixture after an incubation period. The greater the amount of amylase activity, the lighter will be the color of the final solution.

Amyloclastic methods are no longer widely employed, partly because substrate variability makes standardization difficult.[5]

Saccharogenic techniques (method 4, Table 59-11)

Saccharogenic techniques depend on the measurement of monosaccharides or disaccharides liberated by amylase's reaction with the substrate. The classical amylase method of Somogyi[6] is a saccharogenic technique. Sample and substrate are incubated for 30 minutes; a serum blank and an incubated sample reaction tube then undergo measurement for reducing substances. Since the by-products of amylase action are reducing substances, the enzyme's action is directly proportional to the amount of reducing substrate produced.

More recently, the saccharogenic approach has been modified for automated kinetic analysis. In these techniques, the maltose split from the substrate is converted into glucose by maltase, which is included in the reagent mixture. One can measure the glucose liberated by this step by several enzymatic glucose techniques. These methods must be designed to account for glucose present in the patient's sample.

One can use a hexokinase assay to measure the glucose spectrophotometrically or fluorometrically.[7,8] One also can use glucose oxidase and measure glucose by monitoring oxygen consumption electrometrically.[9] Another simple approach employs a coupled color reagent instead of the O_2 electrode.[10] The Beckman method[10] employs a maltotetraose substrate.

The du Pont aca method[11] employs a maltopentaose substrate, which generates five molecules of glucose. Glucose interference is eliminated by gel filtration of the sample before analysis. The glucose released by the amylase reaction is measured by a hexokinase reaction.

Chromolytic techniques (methods 5a and 5b, Table 59-11)

In recent years, commercial manufacturers have produced a variety of intriguing and convenient amylase methods that depend on the liberation of a dye coupled to a complex, insoluble polysaccharide. In some of these methods, one incubates the sample with the synthetic substrate and after a fixed time measures the amount of color liberated as the dye is split from the original complex. The Pharmacia (Pharmacia, Inc., Piscataway, NJ) Phadebas method has been one of the most popular of this type.[12] These methods often require blanks, centrifugation steps, and decanting and are therefore difficult to automate.

An interesting version of this approach employs a substrate of reactone red 2 β-amylopectin in an agarose gel.[13] The amount of amylase activity is indicated by the size of the cleared diffusion ring around a central well containing the sample. The method has a large measurement range and is suitable for multiple samples.

A somewhat different tactic depends on the release of small, water-soluble fragments in such a way that the color can be measured continuously.[14,15] This can be done with *p*-nitrophenyl glucosides, such as maltoheptaosides, which produce *p*-nitrophenol by direct hydrolysis or by a coupled reaction involving α-glucosidase and β-glucosidase. The reaction is monitored by measurement of the liberated *p*-nitrophenol at 405 nm. These methods are readily automated and are commercially available.

Fluorescence polarization technique (method 6, Table 59-11)

This method is available on the commonly used Abbott TDx (Abbott Corp., North Chicago, Ill.). It employs a high molecular weight, fluorescein-labeled amylose molecule as the substrate. Since this large molecule rotates relatively slowly in solution, when polarized light interacts with the fluorescein label, polarized fluorescent light will be emitted. Amylase present in a sample will degrade the amylose molecule into smaller molecules. The smaller molecules will rotate more rapidly, causing the emitted fluorescent light to be depolarized. The amount of fluorescence depolarization is related to the amylase concentration.[16]

The Amylase Conference of the German Society for

Table 59-12 Comparison of methods for amylase analysis

Condition	Iodometric*	Saccharogenic†	Chromolytic‡
Temperature	37° C	37° C	37° C
pH	7.0	6.7	6.9
Final concentration of reagents	Starch: 392 mg/L Benzoic acid: 69 mmol/L Na_2HPO_4: 188 mmol/L Color reaction I^-: 1 mmol/L KIO_3: 167 μmol/L	Maltotetraose: 5 g/L NAD: 2.5 mmol/L Maltose phosphorylase: ≥3 U/mL β-Phosphoglucomutase: ≥1 U/mL Glucose-6-phosphate dehydrogenase: ≥6 U/mL	*p*-Nitrophenyl maltoheptaoside: 5 μmol/L α-Glucosidase: ≥37,990 U/L Phosphate buffer: 105 mmol/L
Sample volume	50 μL	10 μL	10 μL
Fraction of sample volume	0.0196	0.042	0.027
Time of reaction	7.5 min	11 min	8 min
Wavelength	660 nm	340 and 380 nm	415 and 660 nm
Linearity	8000 U/L	350 U/L	1500 U/L
Interference	Turbidity	Turbidity	Turbidity

*Method listed in this text.
†Boehringer Mannheim Diagnostics, Inc (Indianapolis): reagent employed on the ABA-100 (Abbott, North Chicago).
‡Boehringer Mannheim Diagnostics, Inc: reagent employed on the Hitachi 737.

Clinical Chemistry has proposed several characteristics:

1. A defined substrate with constant quality, reasonable cost, and well-defined reaction products
2. A continuous monitoring method that obeys zero-order kinetics and has no lag phase
3. Sensitive enough to read at 30° C
4. Lack of endogenous glucose interference

Several additional criteria are also obvious:

1. Assays must be able to be performed "stat" (by itself, minimum steps, and in less than 30 minutes).
2. Methods should be suitable for serum and urine.
3. Dilutions should be avoided or must be easily performed.
4. Viscosimetric techniques are not practical.

A method selected by the American Association for Clinical Chemistry for amylase screening is an iodometric method.[17] An iodometric assay is presented next.

The continuous chromolytic techniques satisfy all the criteria just enumerated and will probably become the standard approach in the future. Table 59-12 compares the reaction conditions for an iodometric, saccharogenic, and chromolytic assay.

SPECIMEN

Either serum or heparinized plasma can be used as the sample, since similar results are obtained with each.[18] Because amylase has an absolute requirement for calcium ions (as well as chloride), chelating anticoagulants, such as citrate, oxalate, and EDTA, cannot be used to collect plasma for amylase measurements. Nonacidified urines, with random or timed collections, are also valid specimens.

Serum and urine specimens free of bacterial contamination show no loss of amylase activity for 1 week at room temperature and for several months at 4° C.[19]

PROCEDURE: IODOMETRIC ANALYSIS FOR α-AMYLASE

Principle

Starch is hydrolyzed by amylase in the sample to liberate smaller molecules. An iodine reagent is added, and it forms a vivid blue color with the remaining starch. The amylase activity is inversely proportional to the amount of color in the final solution.

Reagents

Buffered starch substrate, pH 7.0. Dissolve 13.3 g of anhydrous disodium phosphate (Na_2PO_4) and 4.3 g of benzoic (C_6H_5COOH) in approximately 250 mL of distilled water. Heat to boiling. Add 200 mg of soluble starch in 5 mL of cold distilled water. Add this starch mixture to the boiling buffer, stirring and rinsing to ensure complete transfer. Allow the mixture to cool to room temperature, and dilute to 500 mL with distilled water. Store at 4° C. The solution should remain clear. One should assess its stability by measuring the absorbance of these mixed reagents (observing for significant decline in the reagent blank value from run to run).

Stock iodine solution (0.1 mol/L). Dissolve 3.567 g of potassium iodate (KIO_3) and 45 g of potassium iodide (KI) in 800 mL of distilled water. Add 9 mL of 12 M HCl slowly to the mixture, using ample mixing. Dilute to 1 L. Store in an amber bottle of 4° C. The solution is stable for 12 months.

Working iodine solution (0.01 mol/L). Dilute 50 mL of stock iodine solution to 500 mL. Store in amber bottle at 4° C. The solution is stable for 2 months.

Assay

Equipment: for each test, two 25 mL volumetric flasks (or graduated cylinders) are used. One is labeled "test" and the other "serum blank." Use a spectrophotometer with a ≤10 nm band pass capable of reading at 660 nm, and a 37° C water bath.

1. Pipet 2.5 mL of starch substrate into each container, and place in a 37° C water bath for 15 minutes.
2. Pipet 50 μL of sample into the test container. A series of flasks should be processed at timed intervals.
3. Incubate at 37° C for exactly 7½ minutes.
4. At timed intervals, remove each test container from the water bath, and pipet 1.5 mL of distilled water and 2.5 mL of working iodide solution into each tube. Dilute to 25 mL with distilled water. Mix each tube well, and read the absorbance against distilled water at 660 nm.
5. Remove the serum blank containers from the water bath, and add 50 μL of serum, approximately 15 mL of distilled water, and 2.5 mL of starch solution; dilute to 25 mL. This should be done in rapid sequence, and the absorbance read as in step 4.
6. If the amylase activity is more than 400 units, the test should be repeated with 25 μL of sample, since the reaction does not proceed linearly when more than half the substrate has been hydrolyzed.

Calculation

$$\frac{\text{Absorbance (serum blank)} - \text{Absorbance (test)}}{\text{Absorbance (serum blank)}} \times \text{Dilution factor} \times 8000 = \text{Amylase (U/L)}$$

$$\text{Dilution factor} = \frac{2.5\ \text{mL}}{0.05\ \text{mL}} = 50$$

REFERENCE RANGE

According to the previous assay, ranges for amylase in healthy adults are as follows:

Serum: less than 1800 U/L
Urine: less than 5000 U/24 hours

Newborns have serum amylase levels that are approximately 18% of adult levels.[18] Mean serum amylase levels increase from the neonatal period until adult levels are achieved at approximately 3 to 4 years of age. No significant differences exist in the serum activity of amylase between males and females.[18]

REFERENCES

1. Zinterhofer, L, Wardlaw, L, Jatlow, P, and Seligson, D: Nephelometric determination of pancreatic enzymes. I. Amylase, Clin Chim Acta 43:5-12, 1973.
2. Smeaton, JR, and Marquardt, HF: A reaction rate nephelometer for amylase determinations, Clin Chem 20:896, 1974 (abstract).
3. Wohlgemuth, J: A new method for the quantitative determination of amylolytic ferments, Biochemistry 91:9, 1908 (Title translated into English).
4. Caraway, WT: A stable starch substrate for the determination of amylase in serum and other body fluids, Am J Clin Pathol 32:97-99, 1959.
5. Alpha-amylase methodology survey I, Centers for Disease Control, US Public Health Service, Atlanta, Nov 1975.
6. Somogyi, M: Modifications of two methods for the assay of amylase, Clin Chem 6:23-35, 1960.
7. Guilbault, GG, and Rietz, EB: Enzymatic, fluorometric assay of α-amylase in serum, Clin Chem 22:1702-1704, 1976.
8. De Rijke, D, and Kretzer, HJH: Kinetic measurement of total amylase and isoamylase activities with a centrifugal analyzer, Clin Chem 29:1100-1104, 1983.
9. Niwa, M, Miyamoto, M, Osawa, H, and Kuroda, M: Elevation of Fuji amylase analyzer equipped with enzyme electrode, Clin Chem 29:1177, 1983 (abstract).
10. Kaufman, RA, and Tietz, NW: Recent advances in measurement of amylase activity: a comparative study, Clin Chem 26:846-853, 1980.
11. Balcom, RM, O'Donnell, CM, and Amano, E: Evaluation of the DuPont aca amylase method, Clin Chem 25:1831-1835, 1979.
12. Ceska, M, Birath, K, and Brown, B: A new and rapid method for the clinical determination of α-amylase in human serum and urine, optimal conditions, Clin Chim Acta 26:437-444, 1969.
13. Tauschel, HD, and Rudolph, C: A new sensitive radial diffusion method for microdetermination of α-amylase, Anal Biochem 120:262-266, 1982.
14. Lorentz, K: α-Amylase assay: current state and future development, J Clin Chem Clin Biochem 17:499-504, 1979.
15. Okabe, H, Uji, Y, Netsu, K, and Noma, A: Automated measurement of amylase with 4-nitrophenylmaltoheptaoside as a substrate and use of a selective amylase inhibitor, Clin Chem 30:1219-1222, 1984.
16. Hofman, M, and Shaffer, M: Fluorescence depolarization assay for quantitative α-amylase in serum and urine, Clin Chem 31:1478-1480, 1985.
17. Caraway, WT: Amylase screening test (starch-iodine method). In Faulkner, WR, and Meites, S, editors: Selected methods of clinical chemistry, vol 9, Washington, DC, 1982, American Association for Clinical Chemistry.
18. Gillard, BK, Simbala, JA, and Goodglick, L: Reference intervals for amylase isoenzymes in serum and plasma of infants and children, Clin Chem 29:1119-1123, 1983.
19. Henry, RJ: Clinical chemistry: principles and techniques, ed 4, New York, 1964, Harper & Row.

Aspartate aminotransferase

ROBERT L. MURRAY

Aspartate aminotransferase, AST, L-aspartate:2-oxoglutarate aminotransferase, formerly serum glutamate oxaloacetic transaminase (SGOT)
Clinical significance: pp. 359, 427, and 484
Enzyme number: EC 2.6.1.1
Molecular weight: 110,000 daltons
Chemical class: enzyme, protein
Biochemical reaction: amino transfer catalyzed by AST:

$$\begin{array}{c} CO_2H \\ | \\ HC{-}NH_2 \\ | \\ CH_2 \\ | \\ CO_2H \\ \text{L-Aspartate} \end{array} + \begin{array}{c} CO_2H \\ | \\ C{=}O \\ | \\ CH_2 \\ | \\ CH_2 \\ | \\ CO_2H \\ \alpha\text{-Ketoglutarate} \end{array} \underset{}{\overset{AST}{\rightleftharpoons}} \begin{array}{c} CO_2H \\ | \\ C{=}O \\ | \\ CH_2 \\ | \\ CO_2H \\ \text{Oxaloacetate} \end{array} + \begin{array}{c} CO_2H \\ | \\ HC{-}NH_2 \\ | \\ CH_2 \\ | \\ CH_2 \\ | \\ CO_2H \\ \text{L-Glutamate} \end{array}$$

PRINCIPLES OF ANALYSIS

Aspartate aminotransferase (AST) catalyzes the transfer of an amino group from specific amino acids (L-glutamate or L-aspartate) to specific keto acids (α-ketoglutarate or oxaloacetate) (see structural reaction figure). Although at physiological pH the reaction is energetically favored toward the formation of L-aspartate and α-ketoglutarate (to the left in the reaction figure), in vivo the reaction goes to the right to provide a source of nitrogen for the urea cycle. The glutamate thus produced is deaminated by glutamate dehydrogenase, resulting in ammonia and regeneration of α-ketoglutarate.

Two methods of AST analysis have enjoyed the widest popularity. The coupling of oxaloacetate with 2,4-dinitrophenylhydrazine (DNPH) to produce a blue hydrazone (method 1, Table 59-13) was introduced in 1957 and gained popularity because of its simplicity and relative accuracy.[1] In this method, as in all others, it is necessary to force the AST reaction toward the energetically less-favored products, of which only oxaloacetate is easily quantitated. Thus L-aspartate and α-ketoglutarate are added in excess. After incubation, the reaction is stopped with the addition of DNPH, which reacts with both α-ketoglutarate and oxaloacetate. After condensation, the pH is made alkaline, and the absorbance at 505 nm is measured. Because the absorptivity of the oxaloacetate hydrazone is much greater than the corresponding hydrazone of α-ketoglutarate (present in great excess when the patient specimen has low AST activity), little error is introduced.

In the second approach, first described by Karmen,[2] oxaloacetate is again the product quantitated. Malate dehydrogenase (MDH) is used to convert the oxaloacetate to malate (method 2, Table 59-13). The decrease in absorbance at 340 nm, caused by reduced nicotinamide adenine dinucleotide (NADH) consumption in the second reaction, is used to follow the course of the AST reaction. More than 95% of reporting College of American Pathologists (CAP) survey laboratories use the Karmen reaction.

Several photometric techniques have been described, all of which followed a similar principle—the coupling of an aryldiazonium salt to the active methylene of oxaloacetate[3] (method 3, Table 59-13). In all cases the aryl substitute (Ar) was a dye with absorbance characteristics that

Table 59-13 Methods of aspartate aminotransferase (AST) analysis

Method	Type of analysis	Principle*	Usage
1. Dinitrophenylhydrazone coupling (colorimetric) (Reitman and Frankel[1])	Quantitative	**Asp + α-KG $\xrightarrow{AST}$ Gl + Oa** 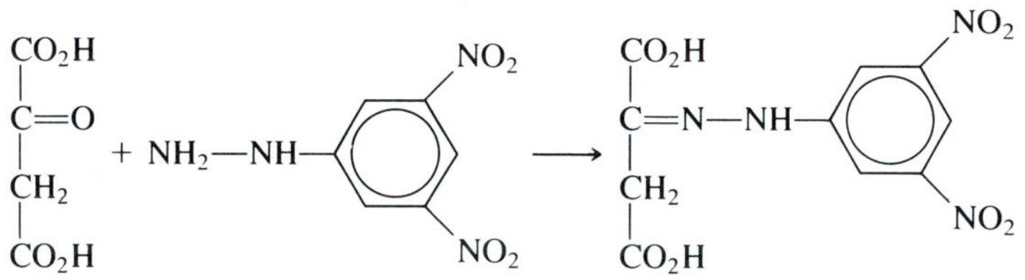 $HO_2C{-}CO{-}CH_2{-}CO_2H$ + $NH_2{-}NH{-}C_6H_3(NO_2)_2$ → $HO_2C{-}C(=N{-}NH{-}C_6H_3(NO_2)_2){-}CH_2{-}CO_2H$ **Oa** + **Dinitrophenylhydrazine** → **Oa-dinitrophenylhydrazone (absorbance at 505 nm)**	Serum†
2. Enzymatic (ultraviolet monitoring) (Karmen[2])	Quantitative	**Asp + α-KG $\xrightarrow{AST}$ Gl + Oa** 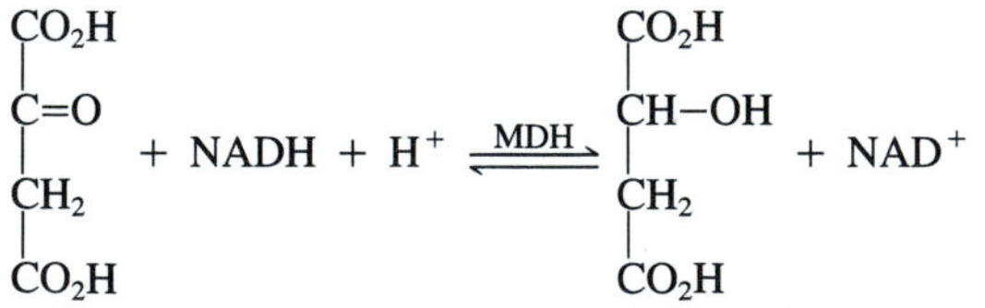 $HO_2C{-}CO{-}CH_2{-}CO_2H + NADH + H^+ \xrightleftharpoons{MDH} HO_2C{-}CH(OH){-}CH_2{-}CO_2H + NAD^+$ **Oa** → **Malic acid**	Serum‡
3. Diazonium dye coupling (colorimetric)	Quantitative	**Asp + α-KG $\xrightarrow{AST}$ Gl + Oa** 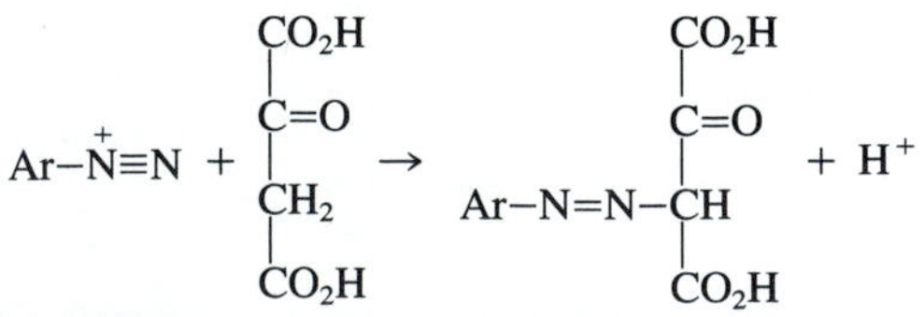 $Ar{-}\overset{+}{N}{\equiv}N + HO_2C{-}CO{-}CH_2{-}CO_2H \rightarrow HO_2C{-}CO{-}CH(N{=}N{-}Ar){-}CO_2H + H^+$ **Diazonium dye** + **Oa** → **Oa diazonium dye**	Serum†

*Asp, Aspartic acid; α-KG, α-ketoglutaric acid; Gl, glutamic acid; Oa, oxaloacetic acid. †Historical; ‡most frequently used.

Table 59-14 Recommended conditions for assay for aspartate aminotransferase activity

Organization or country recommendation	IFCC	France	Germany (FRG)	Germany (GDR)	Japan	Netherlands	Scandinavia	Switzerland	U.K.	U.S.	U.S.S.R.
L-Aspartate (mmol/L)	240	200	200	200	200	200	200	200	240	175	200
2-Oxoglutarate (mmol/L)	12	12	12	12	10	12	12	12	12	15	12
Buffer	Tris	Tris	Phosphate	Tris	Tris	Tris	Tris	Tris	Tris	Tris	Phosphate
Buffer concentration (mmol/L)	80	80	80	100	80	20*	20*	80	80*	60	80
pH	7.8	7.8	7.4	7.4	7.8	7.8	7.7	7.8	7.8	7.8	7.6
Pyridoxal phosphate (mmol/L)	0.1	0.1	0	0	0	0.1	0.03†	0.1	0.1	0.14	0
NADH (mmol/L)	0.18	0.18	0.18	0.13	0.16	0.15	0.15	0.18	0.18	0.15	0.18
MDH (U/L)‡	420	600	600	800	500	600	600	600	500	600	620
LDH (U/L)‡	600	600	1200	0	500	200	200	900	650	0	620
Volume fraction (v/v)	0.083	0.083	ns§	0.133	0.010	0.120	0.120	0.083	0.083	0.067	ns§
Temperature (°C)	30	30	25	37	30	30	37	37/30	30	30/37	30

Reprinted (modified) from Rej, R: Crit Rev Clin Lab Sci 21:99, 1984.
*Also includes EDTA at 5 mmol/L.
†Optional addition directly to serum at 0.25 mmol/L; final assay concentration is 0.03 mmol/L.
‡Activity in U/L under conditions defined by specific recommendation.
§*ns,* Not specified.

changed with formation of the azo bridge to oxaloacetate. The reaction was thus monitored by measurement of the increase in absorbance of the diazonium reaction product.

Identification and quantitation of AST isoenzymes has recently been reported.[4,5] The clinical utility and technical practicality of these have yet to be determined.

The reference method, the coupled enzymatic method (method 2, Table 59-13), has the characteristics of simplicity and speed, which make it the preferred method for routine operations as well.

There have been many minor modifications in the enzymatic technique since its introduction by Karmen.[2] The specifications listed by the International Federation of Clinical Chemistry (IFCC)[6] are recommended primarily for evaluation of preformulated reagent mixtures offered by various manufacturers. Although it may be impossible or impractical to modify such purchased mixtures, the product most closely matching the stated conditions can be identified. Other groups[7-11] have also published recommended conditions for measurement of AST (Table 59-14). Some differences among these groups reflect the span of almost 20 years during which these recommendations were developed.

A striking difference between the IFCC method and the others is the inclusion of pyridoxal phosphate. The need for addition of this cofactor has been widely debated. Although the essential nature of the cofactor has long been recognized, because it is usually present in human serum in adequate amounts, many investigators do not add this component to the reaction mixture. In the occasional patient with severe vitamin B_6 deficiency, this could lead to a serious underestimation of AST level. Addition of pyridoxal phosphate results in an increased activity, reportedly ranging up to 50%.[12]

SPECIMEN

Serum or plasma may be used; heparin, oxalate, EDTA, and citrate do not cause enzyme inhibiton.[13] (Anticoagulants with ammonium as cation should be avoided to reduce the possibility of error.) Because of the high levels of AST in red blood cells, hemolyzed samples are unacceptable.

PROCEDURE: IFCC RECOMMENDED PROCEDURE[14]

Principle

In the presence of aspartate aminotransferase, L-aspartate and 2-oxoglutarate exchange an amino group to form oxaloacetate and L-glutamate. The rate of this reaction is monitored by use of an indicator reaction in which the oxaloacetate formed is converted to malate by an excess of malate dehydrogenase. The change in absorbance at 339 nm is monitored as the NADH is consumed.

Reagents

Tris–L-aspartate buffer (97 mmol/L Tris, 302 mmol/L L-aspartate, pH 7.8). In a 100 mL volumetric flask, dissolve 1.18 g of Tris(hydroxymethyl)aminomethane and 4.02 g of L-aspartic acid in 80 mL of distilled water. Adjust to pH 7.8 with 0.5 mol/L NaOH. Make up to 100 mL with distilled water. It is stable for 6 months at 0° to 4° C. This is solution 1.

Tris-HCl buffer (97 mmol/L Tris, pH 7.8). In a 100 mL volumetric flask, dissolve 1.18 g of Tris(hydroxymethyl)aminomethane in 80 mL of water, adjust to pH 7.8 with 0.5 mol/L HCl, and bring to 100 mL volume with water. It is stable for 6 months at 0° to 4° C. This is solution 2.

Pyridoxal phosphate, 6.3 mmol/L. Dissolve 16.7 mg of pyridoxal-5-phosphoric acid monohydrate in 10.0 mL of the Tris-HCl buffer. Store in a dark bottle at 0° to 4° C. It is stable for 2 weeks. This is solution 3.

NADH, 11.3 mmol/L. Dissolve 16 mg of β-nicotinamide adenine dinucleotide reduced from disodium salt in 2.0 mL Tris-HCl buffer. It is stable for 2 weeks if stored at 0° to 4° C. This is solution 4.

Malate dehydrogenase/lactate dehydrogenase (0.88 mkat/L [528 × 10^5 U/L] and 1.36 mkat/L [816 × 10^5 U/L], respectively). Mix the enzyme solutions in glycerol according to their catalytic concentrations and adjust with 50% (v/v) glycerol to give indicated catalytic concentrations. It is stable for 6 months at 0° to 4° C. This is solution 5.

Working reagent mixture. Mix 100 mL of Tris-aspartate (solution 1) with 2.0 mL of pyridoxal phosphate (solution 3), 2.0 mL of NADH (solution 4), and 1.0 mL of the enzymes (solution 5). This is solution 6.

Reagent mixture for blank. Mix 100 mL of Tris-HCl buffer (solution 2) with 2.0 mL of pyridoxal phosphate (solution 3), 2.0 mL of NADH (solution 4), and 1.0 mL of the enzyme mixture (solution 5). This is solution 7.

2-Oxoglutarate, 0.144 mmol/L. Dissolve 273 mg of 2-oxoglutaric acid disodium salt in 10.0 mL of distilled water. Adjust to pH 7.8 (at 30° C) with 5 mmol/L HCl. This is solution 8.

Sodium chloride, 154 mmol/L. Dissolve 0.9 g of sodium chloride in 100 mL of water. This is stable for 6 months stored at 4° C. Check for bacterial growth. This is solution 9.

Assay

Equipment: spectrophotometer with ≤10 nm band pass at 339 nm with a constant temperature cuvette capable of maintaining a constant temperature with less than 0.05° C fluctuation. A recording spectrophotometer is preferable.

1. Add 2 mL of solution 6 (working reagent) and 0.2 mL of serum to the cuvette.
2. Mix, and allow to stand for 10 minutes.
3. Monitor the absorbance to ensure that the values are stable.
4. Add 0.2 mL of solution 8 (2-oxoglutarate).
5. Mix, and record the change in absorbance for 300 seconds. If no recorder is available, record every 15 seconds.
6. Repeat steps 1 to 5 with solution 7 for blank reaction.

Calculations

If the values of the change in absorbance (ΔA) per second ($\Delta A/\Delta t$) are greater than 0.0025 per second or if they decrease during the step 3 monitoring period, dilute the sample five- to tenfold with 154 mmol/L NaCl (solution 9) and repeat the measurement.

Correct for the blank reaction using the following schema:

$(\Delta A/\Delta t)$A = Measured reaction of serum and solution 6
$(\Delta A/\Delta t)$B = Measured reaction of solution 2 and solution 6
$(\Delta A/\Delta t)$C = Measured reaction of serum and solution 7
$(\Delta A/\Delta t)$D = Measured reaction of solution 2 and solution 7

Then

$$(\Delta A/\Delta t)\text{ corrected} = [(\Delta A/\Delta t)\text{A} - (\Delta A/\Delta t)\text{B}] - [(\Delta A/\Delta t)\text{C} - (\Delta A/\Delta t)\text{D}]$$

If a 1 cm path length is used and a value of 6.3 × 10^2 mol/L/mm path length at 339 nm is used for NADH, the catalytic concentration is as follows:

$$1.905 \times 10^{-3} \times \Delta A/\Delta t,\ \text{kat/L}$$
$$1.19 \times 10^{5} \times \Delta A/\Delta t,\ \text{U/L}$$

REFERENCE RANGE

Adults

5 to 34 U/L (37° C)
8 to 22 U/L (30° C)

Infants

Levels approximately twice the adult level are seen in neonates and in infants; these decline to adult levels by approximately 6 months of age.[15]

REFERENCES

1. Reitman, S, and Frankel, S: A colorimetric method for the determination of serum glutamic oxalacetic and glutamic pyruvic transaminase, Am J Clin Pathol 28:56-63, 1975.
2. Karmen, S: A note on the spectrophotometric assay of glutamic oxalacetic transaminase in human blood serum, J Clin Invest 34:131-135, 1955.
3. Sax, SM, and Moore, JJ: Determination of glutamic oxalacetic transaminase activity by coupling of oxalacetate with diazonium salts, Clin Chem 13:175-185, 1967.
4. Sampson, E, Hannon, W, McKneally, S, et al: Column chromatography and immunoassay compared for measuring the isoenzymes of aspartate aminotransferase in serum, Clin Chem 25:1691-1696, 1979.
5. Rej, R, Bretaudiere, J, and Graffunder, N: Measurement of aspartate aminotransferase isoenzymes: six procedures compared, Clin Chem 27:535-542, 1981.
6. Bergmeyer, H, Bowers, G, Jr, Horder, M, and Moss, DW: Provisional recommendations on IFCC methods for all measurements of catalytic concentrations of enzymes. Part 2. IFCC method for aspartate aminotransferase, Clin Chem 23:887-899, 1977.
7. Frieman, M, and Taylor, T: Transaminase. In American Association of Clinical Chemists: Standard method of clinical chemistry, vol 3, New York, 1961, Academic Press, Inc.
8. Henry, RJ, Chiamori, N, Golub, OJ, and Berkman, S: Revised spectrophotometric methods for the determination of glutamic-oxalacetic transaminase, glutamic-pyruvic transaminase and lactic acid dehydrogenase, Am J Clin Pathol 34:381-391, 1960.
9. Wilkinson, JH, Baron, DN, Moss, DW, and Walter, PG: Standardization of clinical enzyme assays: a reference method for aspartate and alanine aminotransferases, J Clin Pathol 25:940-948, 1972.

10. Recommendations of the German Society for Clinical Chemistry, Z Klin Chem Klin Biochem 6:37-45, 1972.
11. Committee on Enzymes of the Scandinavian Society for Clinical Chemistry and Clinical Physiology: Recommended methods for the determination of four enzymes in blood, Scand J Clin Lab Invest 33:291-296, 1974.
12. Bruns, D, Savory, J, Titheradge, A, et al: Evaluation of the IFCC-recommended procedure for serum aspartate aminotransferase as modified for use with the centrifugal analyzer, Clin Chem 27:156-159, 1981.
13. Demetriou, JA, Drewes, PA, and Gin, JN: Enzymes. In Henry, RJ, Cannon, DC, and Winkelman, JE, editors: Clinical chemistry: principles and technics, ed 2, Hagerstown, MD, 1974, Harper & Row, Publishers, Inc.
14. EPE/IFCC provisional recommendations on IFCC methods for the measurement of catalytic activity concentrations of enzymes. Part 2 (revised 1977). IFCC method for aspartate aminotransferase L-aspartate:2-oxoglutarate aminotransferase EC 2.6.1.1, Clin Chim Acta 80:F21-F22, 1977. Redefinition of conditions previously published as Part 2 in Clin Chim Acta 70:F19-F42, 1976.
15. Meites, S, editor: Pediatric clinical chemistry, Washington, DC, 1981, American Association for Clinical Chemistry.

Cholinesterase

MARY ELLEN KING

Cholinesterase, CHS, acylcholine acylhydrolase, pseudocholinesterase, butyrylcholinesterase, cholinesterase II
Clinical significance: pp. 594 and 740
Enzyme number: EC 3.1.1.8
Molecular weight: 360,000 daltons
Chemical class: enzyme, protein
Biochemical reaction:

$$CH_3{-}C({=}O){-}O{-}CH_2{-}CH_2{-}\overset{+}{N}(CH_3)_3 + H_2O \xrightarrow{\text{EC 3.1.1.8}} CH_3{-}C({=}O){-}O^- + H^+ + HO{-}CH_2{-}CH_2{-}\overset{+}{N}(CH_3)_3$$

Acetylcholine **Acetate** **Choline**

PRINCIPLES OF ANALYSIS

Cholinesterase (CHS, choline esterase), which is also referred to as plasma or serum cholinesterase, pseudocholinesterase, and butyrylcholinesterase, is synthesized by the liver and is present in plasma.[1-3] Its true physiological function is unknown. Although it can hydrolyze acetylcholine, CHS is less specific for this substrate than is the red blood cell enzyme acetylcholinesterase (true cholinesterase, AChE, EC 3.1.1.7); thus the function of CHS may be to hydrolyze other choline esters. CHS activity is usually measured for one of three reasons. The most common reason is the determination of a decrease in enzymatic activity as an indicator of exposure to organophosphorus compounds, including many pesticides. The second reason is to identify the presence of inherited abnormal variants of CHS. One may identify variants by assaying both total activity and the extent of inhibition by either dibucaine or fluoride; some of these variants lead to prolonged apnea in patients receiving the anesthetic succinylcholine. Since a patient with pesticide poisoning can be an emergency case, an assay for CHS must be simple enough to be performed "stat" but accurate enough for determination of CHS phenotyping. Finally, measurement of AChE activity is performed during amniotic fluid analyses for neural tube defects. The measurement of acetylcholinesterase is used to confirm an elevated amniotic fluid α-fetoprotein (AFP) level.

Most current assays for CHS are spectrophotometric and use propionylthiocholine or acetylthiocholine as substrate (method 1, Table 59-15). 5,5′-Dithiobis(2-nitrobenzoic acid) (DTNB) is added to react with the released thiocholine to form the yellow compound 5-thio-2-nitrobenzoic acid (absorption maxium, 410 nm).[4,5] The reactions may be followed as a rate or end-point procedure.

Alternative procedures include titrimetric[6-10] and electrometric[11-13] methods that follow the release of hydrogen ions from a choline ester. One titrimetric method (Michel method) directly measures hydrogen-ion release by monitoring pH changes over a 1- to 1½-hour period (method 2, Table 59-15).[6,7] A second titrimetric colorimetric method quantifies the amount of hydrogen ion released by the hydrolysis of acetylcholine by following the color change of a phenol red indicator over a 30-minute incubation time (method 3, Table 59-15). The color change is related to the moles of substrate converted.[8,9]

A special type of titrimetric method, usually referred to as an electrometric[11-13] procedure, uses highly automated titration instrumentation (pH Stat) (Radiometer, Copenhagen) to add base continuously to neutralize the hydrogen ions released from acetylcholine by CHS (method 4, Table 59-15). This instrument maintains the enzyme reaction at constant pH, and the amount of base added is directly related to the number of moles of acetylcholine hydrolyzed on a mole-per-mole basis.

The presence of true acetylcholinesterase (AChE) activity in amniotic fluid can be determined by a quantitative[14,15] or a qualitative assay.[16,17] In either case, an inhibitor of AChE activity is used to identify the activity specifically. The most frequently used inhibitor is 1,5-bis(4-allyldimethyl-ammoniumphenyl)pentan-3-one dibromide (BW 284C51). The activity remaining in the presence of this inhibitor can be quantitated by method 1 in Table 59-15. The qualitative assay for AChE employs polyacrylamide gel electrophoresis to separate isoenzymes of CHS activity.[16,17] The gels are then incubated with acetylthiocholine, with and without the inhibitor BW 284C51 present. The thiocholine formed by the reaction reacts with copper sulfate, also present in the incubation mixture, to

Table 59-15 Methods of quantitative cholinesterase (CHS) analysis

Method	Type of analysis	Principle	Usage	Comments
1. Colorimetric	Spectrophotometric, kinetic or end point	$C_2H_5{-}C(=O){-}S{-}CH_2CH_2\overset{+}{N}(CH_3)_3 + H_2O \xrightarrow{CHS}$ **Propionylthiocholine** $C_2H_5{-}COO^- + HS{-}CH_2CH_2{-}\overset{+}{N}(CH_3)_3$ **Propionate** **Thiocholine** Thiocholine + DTNB (O_2N, HO_2C ... S—S ... CO_2H, NO_2) → HS ... CO_2H, NO_2 + $(CH_3)_3\overset{+}{N}CH_2CH_2{-}S{-}S{-}$ ... CO_2H, NO_2 + Mixed disulfides **DTNB** **5-Thio-2-nitrobenzoic acid** (yellow, A_{max} = 410 nm)	Serum or plasma	Recommended method
2. pH change	pH measurement	Acetylcholine $\xrightarrow{CHS}$ Choline + Acetate + H^+	Serum or plasma	Wide interlaboratory variation of results
3. Colorimetric	pH dye color change, end point	$CH_3C(=O)O{-}CH_2CH_2\overset{+}{N}(CH_3)_3 + H_2O \xrightarrow{CHS}$ **Acetylcholine** $CH_3{-}C(=O){-}O^- + HO{-}CH_2CH_2{-}\overset{+}{N}(CH)_3 + H^+$ **Acetate** **Choline** Dye + H^+ → Color change	Serum or plasma	Wide interlaboratory variation of results
4. Electrometric	Titrate acid released, kinetic or end point	Acetylcholine $\xrightarrow{CHS}$ Choline + Acetate + H^+ H^+ + Added base → H_2O and constant pH	Serum or plasma	Requires automated pH titration

DTNB, 5,5′-Dithiobis(2-nitrobenzoic acid).

form a white cupric thiocholine precipitate. The zone of AChE activity is identified by its inhibition, in a separate gel, by BW 284C51. The qualitative assay is the most frequently used technique to determine the presence of AChE in amniotic fluid.

The following method (see Table 59-16 for conditions) is based on the procedure of Ellman and is recommended by the American Association for Clinical Chemistry.[4] However, it must be kept in mind that the Michel procedure[6] has served as a standard method, and it may be the procedure used by regulating agencies.

SPECIMEN

Either serum or heparinized plasma may be used. Moderate hemolysis will not interfere if the serum is well centrifuged to remove red blood cell ghosts.[4] CHS is stable for up to 80 days at room temperature and for 3 years if frozen at $-20°$ C.[1] Samples should be stored and shipped cold to avoid wide extremes of temperature.

Samples believed to contain a reversible inhibitor should be collected on ice, kept cold, and assayed as soon as possible to limit in vitro destruction of the inhibitor. Sample-to-reagent dilution and incubation times should be minimized in the assay to limit reactivation of the enzyme during assay.[2]

PROCEDURE: COLORIMETRIC ASSAY FOR CHOLINESTERASE

Principle

Propionylthiocholine is used as a substrate for CHS in the following coupled reactions:

$$\text{Propionylthiocholine} + H_2O \xrightarrow{\text{EC 3.1.1.8}} \text{Propionic acid} + \text{Thiocholine}$$

$$\text{Thiocholine} + \text{DTNB} \longrightarrow \text{Oxidized thiocholine} + \text{5-Thio-2-nitrobenzoic acid}$$

Phosphate buffer, pH 7.6, is used, and the rate of reaction is followed at 410 nm, an absorption maximum of 5-thio-2-nitrobenzoic acid. In addition to total activity, dibucaine and fluoride inhibitions are determined when CHS phenotyping is requested. Glass containers must be used, since inhibitors may be extracted from some plastics.

Reagents

Phosphate buffer, 0.036 mol/L, pH 7.6. Prepare a solution of 0.036 M Na_2HPO_4 by dissolving 4.73 g of the

Table 59-16 Conditions for cholinesterase analysis

Condition	Measure
Temperature	37° C
Wavelength	410 nm
pH	7.6
Final concentration of reagent	PCTI: 2 mmol/L DTNB: 0.253 mmol/L Phosphate: 25 mmol/L Dibucaine: 0.03 mmol/L Fluoride: 4 mmol/L
Sample volume	10 μL
Fraction of sample volume	0.002 (dilution corrected)
Reaction time	10 min
Coefficient of variation	3%[4]
Linearity	0 through normal range

PCTI, Propionylthiocholine, *DTNB,* 5,5′-dithiobis(2-nitrobenzoic acid).

salt in approximately 800 mL of distilled H_2O and diluting to 1 L. Prepare a 0.1 M solution of KH_2PO_4 by dissolving 13.6 g of the salt in a final volume of 1 L of distilled water. Mix 50 mL of KH_2PO_4 with 950 mL of Na_2HPO_4 to achieve a final pH of 7.6. This is stable for 3 months at 4° to 8° C. Check weekly for bacterial growth, pH, and stability.

Propionylthiocholine iodide solution (PCTI), 20 mmol/L. Place 606 mg of PCTI in a 100 mL volumetric flask, and dissolve with approximately 80 mL of distilled water. Bring volume to 100 mL with distilled water. This is stable for 1 day at ambient temperature.

5,5′-Dithiobis(2-nitrobenzoic acid) (DTNB), 167 mg/L (0.421 mmol/L). Weigh out 167 mg of DTNB, dissolve, and dilute to a final volume of 1 L with 0.036 M phosphate buffer. This is stable for 12 months when stored in a brown glass bottle at 4° C.

Dibucaine HCl (Nupercaine Hydrochloride, Ciba), 0.3 mmol/L. Place 57 mg into a 500 mL volumetric flask, dissolve in distilled water, and dilute to a final volume of 500 mL with distilled water. This is stable for 1 year at room temperature.

Sodium fluoride, 40 mmol/L. Weigh out 84 mg of NaF, dissolve in approximately 40 mL of distilled water, and dilute to a final volume of 50 mL. This is stable for 1 day at room temperature.

Quinidine sulfate, 14 mmol/L (5 g/L). Weigh out 500 mg, dissolve in approximately 90 mL of distilled water, and dilute to a final volume of 100 mL. This is stable for 1 year at room temperature.

Working substrate reagents. Prepare 3 mL of each solution per test sample.

1. Dilute PCTI with an equal volume of distilled water.
2. Dilute PCTI with an equal volume of dibucaine solution.
3. If NaF is used, dilute PCTI with an equal volume of sodium fluoride solution.

Assay

Equipment: spectrophotometer or photometer set at 410 nm, with band pass ≤10 nm, and 37° C water bath.

This procedure measures the activity of CHS and the effect of two inhibitors, dibucaine and fluoride.

Assay for total activity

1. Prepare two test tubes for each sample or control, one labeled "test" and the other "blank."
2. Add 3 mL of DTNB buffer to each test tube, and allow to equilibrate 5 minutes in a 37° C water bath.
3. Add 1 mL of working PCTI substrate diluted with water to each tube labeled "test."
4. Add 1 mL of a 1:100 dilution of serum. (Samples with low activity should be reanalyzed with a smaller dilution, such as ten- or fiftyfold.) Dilute the serum with distilled water shortly before the assay, since the CHS may slowly lose activity after dilution.
5. After exactly 3 minutes, add 1 mL of quinidine reagent to stop the reaction.
6. Prepare blank readings by reversing steps 2 and 5; that is, first add quinidine reagent to the "blank" tube and then serum, and then add the substrate solution.
7. Read absorbance of each "test" reaction against the corresponding blank within 5 minutes.

Assay for dibucaine or fluoride inhibition

1. Same as for previous step 1.
2. Add 3 ml of DTNB buffer to each test tube, and allow to incubate 5 minutes in a 37° C water bath.
3. Add 1 ml of working substrate-inhibitor solution, that is, PCTI diluted with either dibucaine or NaF.
4. Follow steps 4 to 7 of the assay for total activity.

Calculations

The CHS activity is expressed in international units per milliliter (U/mL = μmol/min/mL) at 37° C and calculated from the following equation:

$$\text{U/mL} = \frac{X \cdot \Delta A_{\text{unknown}}}{y \cdot 13.6 \cdot z} = 14.71 \cdot \Delta A_{\text{unknown}}$$

where X is the total reaction volume (in milliliters), y is the volume of undiluted sample used, $\Delta A_{\text{unknown}}$ is the reaction absorbance corrected for the blank, 13.6 is the millimolar absorptivity of the 5-thio-2-nitrobenzoate for a 1 cm cuvette, and z is the number of minutes of incubation.

Inhibition by dibucaine or fluoride is calculated as follows:

$$\text{Percent inhibition} = 100\% - \left(\frac{\text{U/mL}_{\text{with inhibitor}}}{\text{U/mL}_{\text{without inhibitor}}} \times 100\%\right)$$

REFERENCE RANGES

CHS values in neonates and infants less than 6 months are 40% to 50% of adult levels.[18] Young adult females,

18 to 35 years of age, have CHS levels 64% to 74% of adult male values. Older females and males (70 to 80 years of age) have values not significantly different from young adult females.[19] CHS decreases during pregnancy.[20]

AChE levels in neonates are 63% of adult levels[21] and rise gradually to adult levels by 1 year of age.[22] There are no male-female differences in AChE values in adults.[19]

The most common phenotype (U, *usual*) is not associated with prolonged response to succinylcholine and is inhibited approximately 84% by dibucaine and 80% by fluoride.[1-4] Other phenotypes that can occur include dibucaine-resistant A *(atypical)* and fluoride-resistant F *(fluoride)*. The so-called silent phenotypes, S_1 and S_2, display little or no CHS activity in the homozygous state. The homozygous variant forms are usually not inhibited more than 20% to 50% by the relevant inhibitor.

AChE is present in fetal serum but is not normally detected in amniotic fluid.

REFERENCES

1. Brown, SS, Kalow, W, Pilz, W, et al: The plasma cholinesterases: a new perspective, Adv Clin Chem 22:1-123, 1981.
2. Willis, JH: Blood cholinesterase: assay methods and considerations, Lab Management 20:53-64, 1982.
3. Lehmann, H, and Liddell, J: The cholinesterase variants. In Stanbury, JB, Wyngaarden, JB, and Fredrickson, DS, editors: The metabolic basis of inherited disease, ed 3, New York, 1972, McGraw-Hill Book Co.
4. Dietz, AA, Rubenstein, HM, and Lubrano, T: Colorimetric determination of serum cholinesterase and its genetic variants by the propionylthiocholine-dithiobis(nitrobenzoic acid) procedure. In Cooper, GR, and King, JS, editors: Selected methods for the small clinical chemistry laboratory, vol 8, Washington, DC, 1977, American Association for Clinical Chemistry.
5. Price, EM, and Brown, SS: Scope and limitations of propionylthiocholinesterase in the characterization of cholinesterase variants, Clin Biochem 8:384-390, 1975.
6. Michel, HO: An electrometric method for the determination of red blood cell and plasma cholinesterase activity, J Lab Clin Med 34:1564-1568, 1949.
7. Witter, RF, Grubbs, LM, and Farrior, WL: A simplified version of the Michel method for plasma or red cell cholinesterase, Clin Chim Acta 13:76-78, 1966.
8. Caraway, WT: Photometric determination of serum cholinesterase activity, Am J Clin Pathol 26:954, 1956.
9. Crane, CR, Sanders, DC, and Abbot, JN: Cholinesterase use and interpretation of cholinesterase measurements. In Sunshine, I, editor: Methodology for analytical toxicology, Cleveland, 1975, CRC Press, Inc.
10. Ellin, RI, and Vicario, P: A pH method for measuring blood cholinesterase, Arch Environ Health 30:263-265, 1975.
11. Nabb, DP, and Whitfield, F: Determination of cholinesterase by an automated pH stat method, Arch Environ Health 15:147-152, 1967.
12. Aldrich, FD, Walker, GF, and Patnow, CA: A micromodification of the pH-stat assay for human blood cholinesterasee, Arch Environ Health 19:617-620, 1969.
13. Serat, WE, Van Loon, AJ, and Lee, MK: Microsampling techniques for human blood cholinesterase analysis with the pH stat, Bull Environ Contam Toxicol 17:542-550, 1977.
14. Hay, DL, Ibrahim, GF, and Horacek, I: Rapid acetylcholinesterase screening test for neural tube defect, Clin Chem 29:1065-1069, 1983.
15. Hullin, DA, Laurence, KM, Elder, GH, et al: Amniotic fluid cholinesterase measurement as a rapid method for the exclusion of fetal neural-tube defects, Lancet 1:325-326, 1981.
16. Smith, AD, Wald, NJ, Cuckle, HS, et al: Amniotic fluid acetylcholinesterase as a possible diagnostic test for neural-tube defects in early pregnancy, Lancet 1:685-688, 1979.
17. Barlow, RD, Cuckle, HS, and Wald, NJ: A simple method for amniotic fluid gel-acetylcholinesterase determination, suitable for routine use in the antenatal diagnosis of open neural tube defects, Clin Chim Acta 119:137-142, 1982.
18. Zsigmond, EK, and Downs, JR: Plasma cholinesterase activity in newborns and infants, Can Anaesth Soc 18:278-283, 1971.
19. Shanor, S, Van Hess, GR, Baart, N, et al: The influence of age and sex on human plasma and red cell cholinesterase, Am J Med 242:357-361, 1961.
20. Howard, J, East, N, and Chaney, J: Plasma cholinesterase activity in early pregnancy, Arch Environ Health 33:277-279, 1978.
21. Kaplan, E, Herz, F, and Hsu, KS: Erythrocyte acetylcholinesterase activity in ABO hemolytic disease of the newborn, Pediatrics 33:205-211, 1964.
22. Kaplan, E, and Tildon, JT: Changes in red cell enzyme activity in relation to red cell survival in infancy, Pediatrics 32:371-375, 1963.

Creatine kinase

KORY M. WARD
JOHN A. LOTT

Creatine kinase (CK, CPK, adenosine triphosphate: creatine *N*-phosphotransferase, creatine phosphoryltransferase)

Clinical significance: pp. 415 and 427

Enzyme number: EC 2.7.3.2

Molecular weight: varies from 78,500 to 85,100 daltons; molecular weight of individual subunits is half that of intact molecule[1]

Chemical class: enzyme, protein

Known isoenzymes: CK_1 (CK-BB), CK_2 (CK-MB), CK_3 (CK-MM), CKm (mitochondrial—rare), CK-interband (rare, usually CK-BB complexed to IgG)

Biochemical reaction:

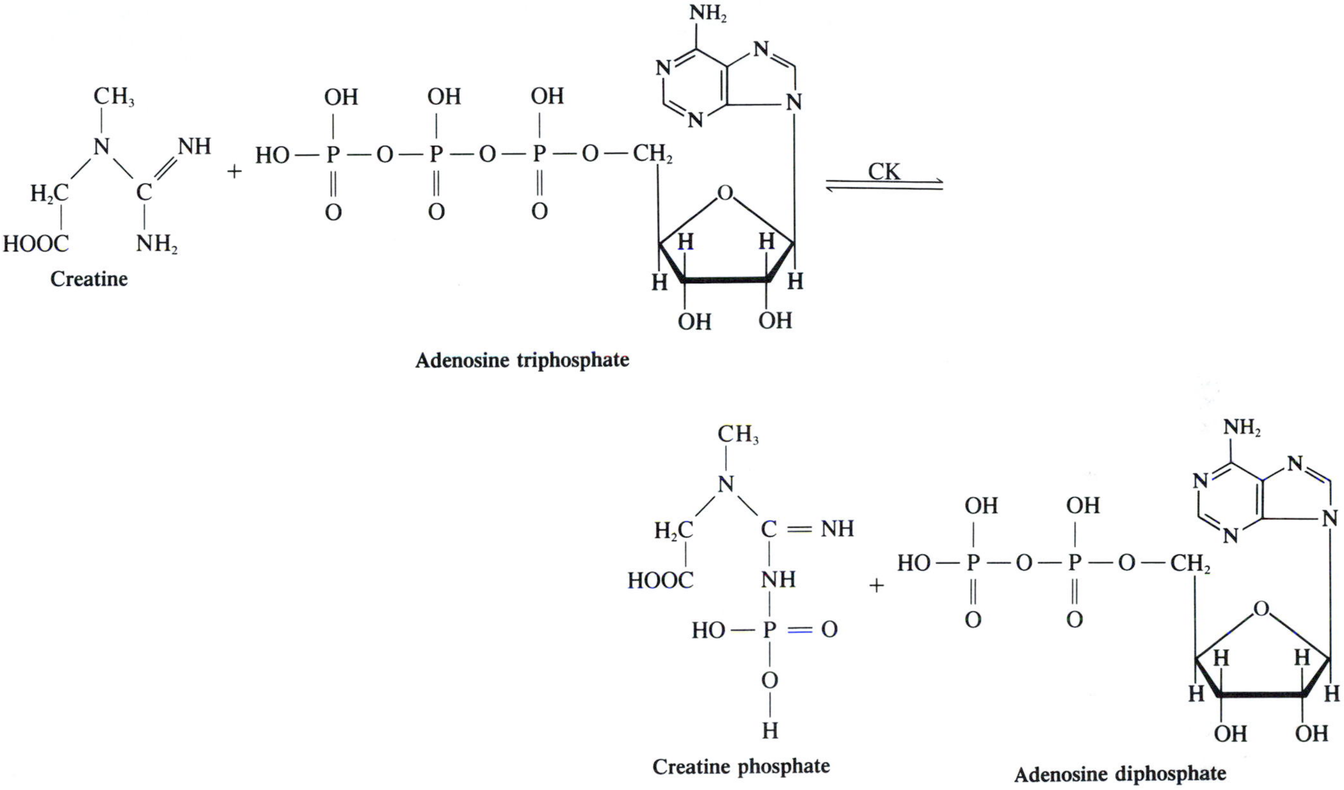

Creatine + Adenosine triphosphate $\overset{CK}{\rightleftharpoons}$ Creatine phosphate + Adenosine diphosphate

PRINCIPLES OF ANALYSIS

Creatine kinase (CK) is a cytoplasmic and mitochondrial enzyme with wide tissue distribution.[1,2] CK catalyzes the formation of adenosine triphosphate (ATP), a substance that is required for contractile or transport systems.[3] The enzyme also catalyzes the reversible phosphorylation of creatine with ATP as the donor of the phosphate group. The reaction to form creatine phosphate (the "forward" reaction) follows:

$$\text{Creatine} + \text{ATP} \underset{\text{pH 6.8}}{\overset{\text{pH 9.0}}{\underset{\text{CK}}{\rightleftharpoons}}} \text{Creatine phosphate} + \text{ADP}$$

At neutral pH, the formation of ATP is favored; a pH of 9.0 is optimum for the formation of creatine phosphate, another high-energy compound. CK combines with ATP or adenosine diphosphate (ADP) and a cation to form a metal-nucleotide complex; metalloactivators include Mg^{+2}, Mn^{+2}, Co^{+2}, Sr^{+2}, and Ba^{+2}. Cationic inhibitors are Be^{+2}, Ni^{+2}, Cr^{+2}, Ca^{+2}, Zn^{+2}, Ag^{+}, Hg^{+2}, and Cu^{+2}. Anionic inhibitors include malonate, iodoacetate, citrate, acetate, I^{-}, Cl^{-}, F^{-}, SO_4^{-2}, NO_3^{-}, Br^{-}, and S^{-2}.[4,5]

In analytical methods for serum CK, either the products formed in the forward or reverse reaction are measured. Historically and currently used methods are described in Table 59-17.[6-13] Creatine phosphate is more labile to acid hydrolysis than either ATP or ADP; under mild acid conditions, creatine phosphate is hydrolyzed to creatine and phosphate; the latter can be determined by use of the blue complex formed with molybdate (method 1, Table 59-17).

The ADP can be determined with the procedure described by Tanzer and Gilvarg.[7] The reaction sequence is described in method 2, Table 59-17. Because this reaction is carried out at an alkaline pH, both ATPase and alkaline phosphatase interfere. A serum blank is required to correct for endogenous serum pyruvate. Because of its low sensi-

Table 59-17 Methods of creatine kinase (CK) analysis*

Method	Direction of reaction	Type of analysis	Principle	Comments
1. Kuby et al.[6]	Forward	End point, manual modified for AutoAnalyzer	Acid hydrolysis of creatine phosphate to creatine and phosphate; phosphate then determined by Fiske-Subarow method	Requires serum blank because of endogenous phosphate; interference from alkaline phosphatase; very laborious
2. Tanzer and Gilvarg[7]	Forward	Kinetic	Adenosine diphosphate (ADP) produced,† followed by decrease in absorbance at 340 nm	Interference from ATPase and alkaline phosphatase; slow reaction; serum blank required to correct for endogenous pyruvate
3. Nuttal and Weddin[8]	Forward	End point	Pyruvate produced in method 2 reacts with 2,4-dinitrophenylhydrazine to form 2,4-dinitrophenylhydrazone, which is measured at 440 nm	Serum blank required to correct for endogenous pyruvate and other metabolites; not suited for clinical laboratory because of poor accuracy
4. Hughes[9]	Reverse	End point	Creatine produced complexes with diacetyl and α-naphthol by method of Voges and Proskauer; measured at 520 nm	Poor sensitivity; serum blank needed; reagent unstable; both diacetyl and α-naphthol must be prepared daily
5. Sax and Moore[10]	Reverse	Fluorometric	Creatine produced reacts with ninhydrin to produce a fluorophor	Needs serum blank; requires fixed incubation period
6. Witteveen et al.[11]	Reverse	Bioluminescence	ATP in reverse reaction determined with luciferin-luciferase method	Highly sensitive; expensive reagents; needs specialized instrumentation
7. Oliver[12]; Rosalki[13]	Reverse	Kinetic	ATP produced by reverse reaction is coupled to hexokinase and glucose-6-phosphate dehydrogenase reactions to produce NADPH	No serum blank needed; preferred method

*These methods can be used to analyze all body fluids.
†ADP produced reacts as follows:

$$\text{ADP} + \text{Phospho}\textit{enol}\text{pyruvate} \xrightleftharpoons{\text{Pyruvate kinase}} \text{ATP} + \text{Pyruvate}$$

$$\text{Pyruvate} + \text{NADH} + \text{H}^+ \xrightleftharpoons{\text{LD}} \text{Lactate} + \text{NAD}^+$$

ATP, Adenosine triphosphate; *LD,* lactate dehydrogenase; *NADH,* reduced nicotinamide adenine dinucleotide (NAD^+).

tivity, the method is obsolete. In a modification of the method, the resulting pyruvate was reacted with 2,4-dinitrophenylhydrazine to form a colored hydrazone derivative (method 3, Table 59-17).[8] The problem of interferences remained.

The reverse reaction, creatine phosphate to creatine, proceeds six times faster than the forward reaction. Hughes[9] reacted creatine with diacetyl and α-naphthol to form a chromophore (method 4, Table 59-17). Because compounds other than creatine also reacted, the method gave falsely increased values. Sax and Moore[10] reacted the creatine with ninhydrin in the presence of KOH to form a fluorescent product (method 5, Table 59-17). Both these methods have been adapted to fixed-incubation-time, continuous-flow analyzers[14-16] with fluorophotometers as detectors.

Bioluminescence has been applied to measure CK activity (method 6, Table 59-17). The reverse reaction is coupled with the luciferin-luciferase reaction:

$$\text{Creatine phosphate} + \text{ADP} \xrightleftharpoons[\text{Mg}^{+2}]{\text{CK}} \text{Creatine} + \text{ATP}$$

$$\text{ATP} + \text{Luciferin} + O_2 \xrightleftharpoons[\text{Mg}^{+2}]{\text{Luciferase}} \text{Adenosine monophosphate (AMP)} + \text{Oxyluciferin} + \text{Pyrophosphate} + CO_2 + \text{Light emission}$$

In the presence of ATP, luciferin, and oxygen, the light intensity is proportional to the CK activity. The luminescence assay is by far the most sensitive of any of the currently available CK methods but is rarely used because of the cost of the reagents and the complexity of the instrumentation. The currently used methods for CK and the preferred method described here have adequate sensitivity for serum specimens. For cerebrospinal fluid, where the activity is usually very low, the excellent sensitivity of bioluminescence methods is an advantage.[17]

Oliver[12] (method 7, Table 59-17) described a coupled reaction for the determination of CK as follows:

$$\text{Creatine phosphate} + \text{ADP} \xrightleftharpoons[\text{pH 6.8}]{\text{CK, Mg}^{+2}} \text{Creatine} + \text{ATP}$$

$$\text{ATP} + \text{Glucose} \xrightleftharpoons{\text{Hexokinase}} \text{Glucose-6-phosphate} + \text{ADP}$$

$$\text{Glucose-6-phosphate} + \text{NADP}^+ \xrightleftharpoons{\text{G6PD}} \text{6-Phosphogluconate} + \text{NADPH} + \text{H}^+$$

(Abbreviations: *NADP*$^+$, Nicotinamide adenine dinucleotide phosphate; *G6PD*, glucose-6-phosphate dehydrogenase.)

The rate of increase in absorbance at 340 nm is a measure of CK activity present in the specimen. The first improvements and modifications of the original Oliver method were introduced by Rosalki,[13] who included AMP to inhibit adenylate kinase and cysteine hydrochloride to reduce and reactivate CK (method 7, Table 59-17). Serum can be stored because thiols reduce the functional sulfhydryl groups of CK and restore activity.

Various thiols have been proposed as reactivators of CK. Among the most effective are beta-thioglycerol and beta-mercaptoethanol.[18] *N*-acetylcysteine is also effective and has the advantage of being a crystalline substance, whereas the former two are liquids and cannot be used in dry reagent formulations (Table 59-18).

The Oliver-Rosalki[12,13] reaction scheme is the procedure of choice. The Scandinavian,[19] French,[20] and German[21] clinical chemistry societies have published methods for the measurement of CK activity in serum using this scheme (Table 59-19).[13,19-24]

Table 59-18 Reactivators of serum CK activity

Activator	Comments
N-Acetylcysteine	At concentrations of 20 mmol/L, is reactivator of choice for dry reagents; it has good stability; is very soluble; and does not cause protein precipitation during assay.
Beta-thioglycerol	Causes gelling and turbidity of serum after 6 hours of storage at 37° C. Its decomposition products inhibit CK.
Beta-mercaptoethanol (BME)	Causes gelling and turbidity after 6 hours of storage at 37° C. Its decomposition products inhibit CK. When prepared as "reactivating solution" with EDTA, effectively reactivates CK-BB.[18]
Dithiothreitol (DTT) Dithioerythriotol (DTE)	Cause microprecipitation of albumin in sample and cause gelling on storage at 4° C.
Glutathione (GSH)	Oxidized glutathione reacts with NADPH and gives falsely low CK results.

Adapted from references 1, 5, 18, and 19.

SPECIMEN

Specimens for CK analysis include serum, cerebrospinal fluid,[17] and amniotic fluid.[25] Blood should be collected in plain evacuated tubes or syringes, centrifuged promptly after the blood clots, and the serum removed for analysis. Approximately 320 mg/L of hemoglobin (trace hemolysis) does not appreciably change the measured CK activity.[26]

Anticoagulants, other than heparin, should not be used in collection tubes because they inhibit CK activity.[27] Turbid and icteric samples can be analyzed; appropriate values are obtained if the starting absorbance is not too high.

The timing of blood collection must be considered because the total CK activity may be significantly increased after exercise, surgery, intramuscular injections, electrical defibrillation, or countershock.[28-30]

Serum should be stored in the dark at 4° or −20° C if analysis is delayed.[30] CK is inactivated by daylight, even at freezing temperatures.[5] Long-term storage of samples at −20° C has resulted in minimum loss of activity.[31]

Among the three cytoplasmic isoenzymes of CK, CK-BB is the least stable. Adding a thiol such as 2-mercaptoethanol and a chelating agent such as EDTA to the serum stabilizes CK-BB.[18] Nealon et al.[32] have demonstrated differences in the storage stability of the CK isoenzymes that depend on the pH of the serum.

Interferences

The major interferent in CK assays is adenylate kinase (AK), or myokinase, an enzyme abundant in erythrocytes. AK catalyzes the direct conversion of ADP to ATP as follows:

$$2\ \text{ADP} \xrightleftharpoons{\text{AK}} \text{ATP} + \text{AMP}$$

The production of ATP leads to falsely increased CK activities in the Oliver-Rosalki reaction sequence; 10 U/L of AK is equivalent to approximately 1 U/L of CK.[33] Several AK inhibitors have been employed to stop this reaction.[34-36] The most effective inhibitor of AK appears to be a combination of 5 mmol/L AMP and 10 μmol/L diadenosine-5-pentaphosphate[35] (Table 59-20).

PROCEDURE

Principle

Creatine kinase is measured by a coupled reaction sequence described earlier (Oliver-Rosalki method). After a

Table 59-19 Final reaction conditions for creatine kinase

Condition	CK-NAC†	Rosalki[13]	Szasz[22]	Morin[23]	Scandinavian[19,‡]	French[20,‡]	German[21,‡]
Temperature	30°, 37° C	37° C	25°, 30°, 37° C	30°, 37° C	37° C	30° C	25° C
Buffer, concentration (mmol/L)	Imidazole, 100	Tris, 50	Imidazole, 100	Bis-Tris, 200	Imidazole, 100	Imidazole, 100	Imidazole, 100
pH (at temperature)	6.7 (30° C)	6.8 (25° C)	6.7 (30° C)	6.7 (30° C)	6.5 (37° C)	6.8 (30° C)	6.7 (25° C)
Volume fraction	0.02	0.033	0.04	0.02	0.043	0.033	ca. 0.04
Final concentration of reagents in reaction mixture*							
Creatine phosphate	30	10	30	30	30	30	30
ADP	2	1	2	2	2	2	2
Mg^{+2}	10	30	10	20	10	10	10
Salt of Mg^{+2}	Acetate	Chloride	Acetate	Acetate	Acetate	Triethanolamine	Triethanolamine
D-Glucose	20	20	20	15	20	20	20
Pyridine nucleotide	2.0 NADP	0.8 NADP	2.0 NADP	2.0 NADP	2.0 NADP	0.8 NADP	0.6 NADP
Hexokinase (U/L)	2500	600	2500	3000	3000	3000	2500
G6PD (U/L)	1500	300	1500	2500	2000	2000	1500
Sulfhydryl reactivator	20 NAC	5 cysteine	20 NAC	20 monothioglycerol	20 NAC	20 NAC	9.0 glutathione
Adenylate kinase (AK) inhibitor	5 AMP, 10 DAPP	10 AMP	5 AMP	25 fluoride	5.0 AMP	5.0 AMP	5.0 AMP

*Concentrations in millimoles per liter in final reaction mixture.
†Boehringer Mannheim Diagnostics, Indianapolis, IN.
‡As recommended by the individual societies.
AMP, Adenosine monophosphate; *DAPP,* diadenosine-5-pentaphosphate; *G6PD,* glucose-6-phosphate dehydrogenase; *NAC, N*-acetylcysteine.

5-minute preincubation to allow the lag phase to go to completion, the change in absorbance at 340 nm is monitored at 37° C. The CK activity is directly related to the increase in absorbance of NADPH, obtained from the zero-order portion of the curve of absorbance versus time. The results are expressed in U/L.

Table 59-20 Inhibitors of adenylate kinase (AK)

AMP	CK is also inhibited. AMP:ADP ratio of 10:1 inhibits AK by 95% and CK by 15% to 25%; at AMP:ADP ratio of 5:2, AK is inhibited by 80% to 90% and CK by 5%.
Diadenosine 5-pentaphosphate (DAPP)	Potent AK inhibitor; selectively inhibits AK of muscle and erythrocytes. Combination of 10 μmol of DAPP and 5 mmol of AMP/L of assay mixture inhibits AK from erythrocytes and muscles by 97%; CK is inhibited by about 5%.
Sodium fluoride	Inhibits AK but not CK; tends to precipitate as MgF_2.
Elemental sulfur[36]	Does not appear to inhibit CK activity.

Adapted from references 5 and 32 to 35.

Reagents

Reagents may be purchased from several manufacturers.

Assay

Follow general instructions of instrument and reagent manufacturers.

Calculations

Average the change of absorbance (ΔA) per minute values. If the ΔA values are decreasing or appear inconsistent, repeat the assay with a diluted specimen.

CK activity:

$$\text{U/L} = \frac{\Delta A \text{ (average)}}{\text{Minute}} \times \frac{10^6\ \mu\text{mol}}{\text{Mol}} \times \frac{1}{\text{Molar absorptivity}} \times \frac{\text{Total volume}}{\text{Sample volume}}$$

$$\text{U/L} = \frac{\Delta A \text{ (average)}}{\text{Minute}} \times \frac{10^6}{6220} \times \frac{2.55}{0.05}$$

$$\text{U/L} = \frac{\Delta A \text{ (average)}}{\text{Minute}} \times 8220$$

NOTE: The molar absorptivity of NADPH is 6220 $L \cdot mol^{-1} \cdot cm^{-1}$ at 340 nm. The method should be linear to 1500 U/L.

REFERENCE RANGES

Although the manufacturer provides serum reference ranges, it is recommended that users confirm these at their location with persons in good health. Investigators have observed values exceeding these ranges in healthy populations of men and women.[37] Age-, sex-, and race-related differences for serum CK for healthy subpopulations have been found.[37-41] Newborns generally have increased CK activities resulting from skeletal muscle trauma during birth and the transient hypoxia that may cause enzyme release.[39] Serum CK in infants decreases to the adult reference range by 6 to 10 weeks. Males generally have higher serum CK activities than women. In both children and adults CK activities have been found to vary with age.[38] In females, the median CK values decrease during the first 2 decades of life, and there are no significant changes until at least 55 years of age.[39] In males, a fairly constant mean level of serum CK seems to exist, except for a peak near 15 to 18 years of age, which may be related to increased physical activity.

Muscle mass also appears to influence the serum CK activity. Males generally have a larger muscle mass, which results in higher serum CK activities than those in females. Race also has an effect on CK activities, and the mean activity in whites is 66% of the mean activity in blacks.[41]

The suggested serum reference range for the preferred method is as follows:

Males: up to 160 U/L
Females: up to 130 U/L
Newborns: 2 to 3 times adult values

REFERENCES

1. Tsung, SW: Creatine kinase isoenzyme patterns in human tissue obtained at surgery, Clin Chem 22:173-175, 1976.
2. Neumeier, D: Tissue specific and subcellular distribution of creatine kinase isoenzymes. In Lang, H, editor: Creatine kinase isoenzymes, Berlin, 1981, Springer-Verlag.
3. Watts, DC: Creatine kinase (adenosine-5′-triphosphate-creatine phosphotransferase). In Boyer, PD, editor: The enzymes, vol 8, ed 3, New York, 1973, Academic Press, Inc.
4. Helger, R: Methods for differentiation and quantitation of creatine kinase isoenzymes. In Lang, H, editor: Creatine kinase isoenzymes, Berlin, 1981, Springer-Verlag.
5. Bais, R, and Edwards, JB: Creatine kinase, Crit Rev Clin Lab Sci 16:291-335, 1982.
6. Kuby, SA, Noda, L, and Lardy HA: Adenosine triphosphate-creatine transphorylase, J Biol Chem 209:191-201, 1954.
7. Tanzer, ML, and Gilvarg, C: Creatine and creatine kinase measurement, J Biol Chem 234:3201-3204, 1966.
8. Nuttal, F, and Weddin, DJ: A simple rapid colorimetric method for determination of creatine kinase activity, J Lab Clin Med 68:324-332, 1966.
9. Hughes, BP: A method for estimation of serum creatine kinase and its use in comparing creatine kinase and aldolase activity in normal and pathological sera, Clin Chim Acta 7:597-603, 1965.
10. Sax, SM, and Moore, JS: Fluorometric measurement of creatine kinase activity, Clin Chem 11:951-958, 1965.
11. Witteveen, SAGJ, Sobel, BE, and DeLuca, M: Kinetic properties of the isoenzymes of human creatine phosphokinase, Proc Natl Acad Sci 71:1384-1387, 1974.
12. Oliver, IT: A spectrophotometric method for the determination of creatine phosphokinase and myokinase, Biochem J 61:115-122, 1955.
13. Rosalki, SB: An improved procedure for serum creatine phosphokinase determination, J Lab Clin Med 69:696-705, 1967.
14. Fleischer, GA: Fluorometric measurement of creatine kinase activity, Clin Chem 11:951-958, 1965.
15. Rokos, JAS, Rosalki, SB, and Tarlow, D: Automated fluorometric procedure for measurement of creatine phosphokinase activity, Clin Chem 18:193-198, 1972.
16. Armstrong, JB, Lowden, JA, and Sherwin, AL: Automated fluorometric creatine kinase assay: measurement of 100-fold normal activity without serum dilution, Clin Chem 20:560-565, 1974.
17. Savory, J, and Brody, JP: Measurement and diagnostic value of cerebrospinal fluid enzymes, Ann Clin Lab Sci 9:68-79, 1979.
18. Abbott, LB, and Lott, JA: Reactivation of serum creatine kinase isoenzyme BB in patients with malignancies, Clin Chem 30:1861-1863, 1984.
19. Committee on Enzymes, Scandinavian Society for Clinical Chemistry and Clinical Physiology: Recommended method for the determination of creatine kinase in blood, Scand J Clin Lab Invest 36:711-723, 1976.
20. Societe Francaise de Biologie Clinique Commission ''Enzymologie'': Recommendations pour la mesure de l'activite catalytique de la creatine kinase dans le serum human a 30° C, Ann Biol Clin 34:291-302, 1976.
21. The German Society for Clinical Chemistry: Standard method for the determination of creatine kinase activity, J Clin Chem Clin Biochem 15:255-260, 1977.
22. Szasz, G, Gruber, W, and Bernt, E: Creatine kinase in serum. 1. Determination of optimum reaction conditions, Clin Chem 22:650-656, 1976.
23. Morin, LG: Creatine kinase: re-examination of optimum reaction conditions, Clin Chem 23:1569-1575, 1977.
24. CK-NAC (activated), BioDynamics/BMC, ed 1/78, Indianapolis, IN.
25. Stempel, LE, and Lott, JA: Diagnosis of fetal death in utero with amniotic fluid creatine kinase, Am J Obstet Gynecol 15:1173-1176, 1980.
26. Frank, JJ, Bermes, EW, Bickel, MJ, and Watkins, BF: Effect of in vitro hemolysis on chemical values for serum, Clin Chem 24:1966-1970, 1978.
27. Lott, JA, and Heinz, JW: Creatine kinase in serum. In Faulkner, WR, and Meites, S, editors: Selected methods for the small clinical chemistry laboratory, Washington, DC, 1982, American Association for Clinical Chemistry.
28. Prellwitz, W: Clinical kinase isoenzymes in direct skeletal muscular damage. In Lang, H, editor: Creatine kinase isoenzymes, Berlin, 1981, Springer-Verlag.
29. Friedman, RB, Anderson, RE, Entine, SM, and Hirschberg, SB: Effects of disease on laboratory tests, Clin Chem 26:83D-84D, 1980.
30. Lott, JA, and Stang, JM: Serum enzymes and isoenzymes in the diagnosis and differential diagnosis of myocardial ischemia and necrosis, Clin Chem 25:1241-1250, 1980.
31. Bowie, LJ, Griffiths, JC, and Gochman, N: The preferred method for creatine kinase. In Griffiths, JC, editor: Clinical enzymology, New York, 1979, Masson Publishing USA, Inc.
32. Nealon, DA, Pettit, SM, and Henderson, AR: Activation of human creatine kinase isoenzymes by pH and various sulfhydryl and chelating agents, Clin Chem 27:402-404, 1981.
33. Szasz, G, Gerhardt, W, and Gruber, W: Creatine kinase in serum. 5. Effects of thiols on isoenzyme activity during storage at various temperatures, Clin Chem 24:1557-1563, 1978.
34. Szasz, G: Laboratory measurement of creatine kinase activity. In Tietz, NW, Weinstock, A, and Rodgerson, DO, editors: Proceedings of the Second International Symposium on Clinical Enzymology, Washington, DC, 1976, American Association for Clinical Chemistry.
35. Szasz, G, Gerhardt, W, and Gruber, W: Creatine kinase in serum. 3. Further study of adenylate kinase inhibitors, Clin Chem 23:1888-1892, 1977.
36. Connor, J, and Russel, PJ: Elemental sulfur: a novel inhibitor of adenylate kinase, Biochem Biophys Res Comm 113:348-352, 1983.
37. Miller, GW, Chinchilli, VM, Gruemer, H, and Nance, WE: Sampling from a skewed population distribution as exemplified by esti-

mation of the creatine kinase upper reference limit, Clin Chem 30:18-23, 1984.
38. Lott, JA: Serum enzyme determinations in the diagnosis of acute myocardial infarction: an update, Hum Pathol 15:706-716, 1984.
39. Lott, JA, and Landesman, P: The enzymology of skeletal muscle disorders, Crit Rev Clin Lab Sci 20:153-190, 1984.
40. Cherian, GA, and Hill, JG: Age dependence of serum enzymatic activities (alkaline phosphatase, aspartate aminotransferase, and creatine kinase) in healthy children and adolescents, Am Soc Clin Pathol 70:783-789, 1978.
41. Meltzer, HY: Factors affecting serum creatine phosphokinase levels in the general population: the role of race, activity, and age, Clin Chim Acta 33:165-172, 1971.

Creatine kinase isoenzymes

LAWRENCE M. SILVERMAN
JOHN F. CHAPMAN
LINDA L. WOODARD

Creatine kinase, CK, CPK, creatine phosphotransferase
Clinical significance: pp. 415 and 427
Enzyme number: EC 2.7.3.2
Molecular weight: approximately 80,000 daltons
Chemical class: enzyme, protein
Known isoenzymes 4: CK_1 (CK-BB), CK_2 (CK-MB), CK_3 (CK-MM), CKm (mitochondrial—rare)
Biochemical reaction: see section on creatine kinase

$$\text{Creatine} + \text{ATP} \underset{\text{pH 6.8}}{\overset{\text{EC 2.7.3.2 pH 9.0}}{\rightleftharpoons}} \text{ADP} + \text{Creatine phosphate}$$

PRINCIPLES OF ANALYSIS

Creatine kinase (CK) is an enzyme present in many human tissues (skeletal muscle, myocardium, brain, prostate, uterus, and others). It catalyzes the reversible reaction shown. The molecule is a dimer; its monomeric subunits are designated M and B. The subunits combine to form three isoenzymes, designated as CK_1 (BB), CK_2 (MB), and CK_3 (MM). Methods for separation and quantitation of the CK isoenzymes include electrophoresis, ion-exchange chromatography, immunoinhibition, radioimmunoassay (RIA), and immunochemiluminometry.

One can perform electrophoretic separation of CK isoenzymes by using agarose, cellulose acetate, or polyacrylamide as the support medium. Electrophoretic methods on agarose gels (method 1, Table 59-21) are currently the most frequently used techniques for separation and quantitation of CK isoenzymes. The net charge on the M and B subunits are distinct, imparting a different total charge to each of the isoenzymes. At a pH of 8.6, the mobility of the isoenzymes toward the anode is BB > MB > MM. At this pH, the MM remains at or slightly cathodic to the point of application. Fig. 59-4 shows a CK separation on a 1% agarose gel at pH 8.6, employing the tetrazolium colorimetric procedure. On agarose gels, fluorometric determination of the relative amounts of each isoenzyme present is accomplished through incubation of the gels with substrate mixtures for analysis of total CK.

Table 59-21 Methods for creatinine kinase (CK) isoenzyme analysis

Method	Principle	Usage	Comments
1. Agarose gel electrophoresis	At pH 8.6, major isoenzymes have different mobilities in 1% agarose; CK activities detected by coupled enzymatic assay (fluorescent or visible).	Most frequently used technique	Separates all three major isoenzymes Rapid, relatively simple Minute sample volume Unusual isoenzyme forms such as macro-CK; easily detectable
2. Ion-exchange chromatography	At pH 6.7, CK-MM isoenzyme does not absorb onto column and elutes; CK-MB is eluted from column with buffer of different ionic strength.	Historical usage; was most frequently used method for CK-MB activity quantitation	Reference method for quantitation of CK-MB activity Requires rigorous control of assay parameters for accurate performance
3. Immunoinhibition	M subunits of CK-MM and CK-MB are inactivated by reaction with anti-M antibody; measure remaining B subunit activity.	Frequently used	Can be subject to interference by CK-BB
4. Radioimmunoassay	Competitive binding assay for B or M subunits	Rarely used	Clinical utility of this assay not yet proved
5. Two-site (sandwich) immunoassay	Anti-M or -B Ab binds MM and MB or MB and BB. Labeled second Ab, anti-B or -M, used to detect MB.	Increased use	Requires dedicated analyzer

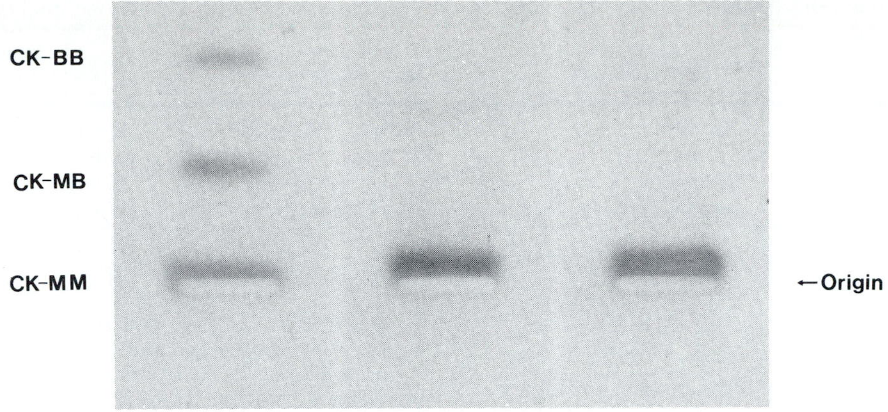

Fig. 59-4 Separation of creatine kinase (CK) isoenzymes on 1% agarose gel with tetrazolium dye staining. Specimen on left is control; other two, showing only CK-MM activity, are from a patient.

$$\text{Creatine phosphate} + \text{ADP} \underset{\text{MgCl}_2}{\overset{\text{CK}}{\rightleftharpoons}} \text{Creatine} + \text{ATP}$$

$$\text{ATP} + \text{Glucose} \xrightarrow{\text{Hexokinase}} \text{Glucose-6-phosphate} + \text{ADP}$$

$$\text{Glucose-6-phosphate} + \text{NADP}^+ \overset{\text{G6PD}}{\rightleftharpoons} \text{NADPH} + \text{H}^+ \text{ 6-Phosphogluconate}$$

The fluorescent NADPH produced in the agarose is considered to be directly proportional to the amount of CK present in the isoenzyme bands. Colorimetric determination of CK isoenzymes is also possible by the addition of phenazine methosulfate and nitroblue tetrazolium to the substrate, so that the following reaction can occur:

$$\text{NADPH} + \text{Phenazine methosulfate} + \text{Nitroblue tetrazolium} \rightarrow \text{Purple formazan}$$

Similar visualization of the isoenzymes is possible with other support media. Densitometric or fluorometric scanning of the resultant patterns yields relative percentages of the isoenzymes present.

Ion-exchange chromatography (method 2, Table 59-21) relies on separation of the isoenzymes through the use of minicolumns containing an anion-exchange material. The ion exchangers used are either diethylaminoethyl (DEAE)-Sephadex A-50 or DEAE-cellulose.[1-3] The isoenzymes are absorbed onto the ion-exchange resin at a pH of approximately 6.7. At this pH, the MM fraction is not charged and therefore is not absorbed and immediately is eluted from the column. The MB and BB isoenzymes are eluted from the column by stepwise increases of salt concentration in a Tris or imidazole-eluting buffer. The aliquots collected from the column may then be analyzed for CK activity by the usual procedures for total CK.

Immunoinhibition methods (method 3, Table 59-21) involve the use of specific antisera to selectively inhibit the B or, more frequently, the M subunits of CK; one then measures the remaining CK activity.[4] When anti-M antibodies are used, the residual B subunit activity is multiplied by 2 so that MB activity is calculated. This method assumes that there is no CK-BB activity present in the sample. Alternatively, a blank can be employed to correct for any CK-BB activity present.

A procedure to correct for CK-BB or adenylate kinase (AK) activity that can result in erroneously high levels of CK-MB was developed by Roche (Roche Diagnostics, Nutley, NJ). This method employs an antibody directed against the anti-M antibody, used to inhibit the M subunit. This second antibody is covalently attached to a large cellulose bead. When it is incubated in the blank tube, an insoluble complex is formed containing CK-MB and CK-MM, linked to the bead by the immunoglobulins. When the complex is removed by centrifugation, the ''background'' CK activity (CK-BB and AK) can be measured and subtracted from the tube measuring the ''total'' CK-MB activity (CK-MB, CK-BB, and AK) to yield a corrected CK-MB measurement.

RIA methods (method 4, Table 59-21) for CK isoenzymes have been reported[5,6] and may offer superior sensitivity in the diagnosis of myocardial infarction. Recently, a two-site chemiluminometric (sandwich) immunoassay specific for CK-MB was developed,[7] and preliminary evaluation indicate both high sensitivity and specificity and a relative independence from interferences and sample lability problems (method 5, Table 59-21).

Electrophoretic methods are usually considered to be simple to perform, although considerable time and skill may be required for proper interpretation of unusual results.

Table 59-22 compares methods for CK analysis.

There is no generally accepted reference method for CK isoenzyme fractionation. The Mercer ion-exchange column procedure under rigorously controlled conditions may be

Table 59-22 Comparison of conditions of several methods for creatine kinase (CK) isoenzyme analysis

Condition	Agarose gel electrophoresis	Ion-exchange chromatography	Immunoinhibition (Roche Diagnostics, Nutley, NJ)
Temperature (for separation)	Ambient	Ambient	Ambient
Sample volume	1-2 μL	500 μL	200 μL each for sample and blank
Assay time	20 min (separation)	30 min (separation)	25 min (separation)
	30 min (analysis)	5 min (quantitation)	5 min (quantitation)

accepted as a reference procedure for CK-MB quantitation. However, this is not widely accepted as a routine laboratory procedure. For ease of use and the clinical usefulness of the resulting information, the electrophoretic procedures in agarose gels are the recommended, preferred methods. An immunoinhibition assay employing a sample blank is certainly an equally satisfactory method.

SPECIMEN

Serum is the preferred specimen because CK activity is easily inhibited by EDTA, citrate, and fluoride. Because of the instability of CK at room temperature, all specimens should be separated from the cells as rapidly as possible and stored at 4° C for no longer than 24 hours. Specimens that are not to be assayed within 24 hours should be frozen at −20° C. As a result of the presence of adenylate kinase in red blood cells and its possible interference with electrophoretic methods, hemolyzed specimens are less than satisfactory. However, this interference may be reduced by the addition of an adenylate kinase inhibitor such as adenosine monophosphate (AMP), fluoride, or diadenosine-5-pentaphosphate[8,9] to the substrate reaction mixture.

REFERENCE RANGES

MM	MB	BB
96% to 100%	<4%	0%
	(or <5 U/L of MB)	

Adult levels of CK-MB are achieved by 1 month of life. Varying levels of CK-MB and CK-BB have been demonstrated in neonatal serum samples. Reference intervals have not been established. In general, however, the CK-MB and the CK-BB activities should disappear within the first month of life.

Two forms of macromolecular CK have been described. Macro-CK, type 1, a complex of CK-BB and IgG, migrates between MM and MB. Macro-CK, type 2, migrates to the cathodic side of MM and may be a polymerization product of mitochondrial CK.[10-12]

REFERENCES

1. Mercer, DW: Separation of tissues and serum creatine kinase isoenzymes by ion-exchange column chromatography, Clin Chem 20:36, 1974.
2. Nealon, DA, and Henderson, AR: Separation of creatine kinase isoenzymes by ion-exchange column chromatography (Mercer's method, modified to increase sensitivity), Clin Chem 21:393, 1975.
3. Mercer, DW, and Varat, MA: Detection of cardiac-specific creatine kinase isoenzyme in sera with normal or slightly increased total creatine kinase activity, Clin Chem 21:1088, 1975.
4. Wicks, R, Usategui-Gomez, M, Miller, M, and Warshaw, M: Immunochemical determination of CK-MB isoenzyme in human serum. II. An enzymic approach, Clin Chem 28:54-58, 1982.
5. Roberts, R, Sobel, BE, and Parker, CW: Radioimmunoassay for creatine kinase isoenzymes, Science 194:855-857, 1976.
6. Zweig, MH, Van Steirteghem, AC, and Schechter, AN: Radioimmunoassay for creatine kinase isoenzyme in human serum: isoenzyme BB, Clin Chem 24:422-428, 1978.
7. Piran, U, Kohn, DW, Uretsky, LS, et al: Immunochemiluminometric assay of creatine kinase MB with a monoclonal antibody to the MB isoenzyme, Clin Chem 33:1517-1520, 1987.
8. Galen, RS: Isoenzymes: which method to use, Diag Med 1:42-63, 1978.
9. Welch, SL, and Swanson, JR: Adenylate kinase interference in creatine kinase isoenzyme electrophoresis, Clin Chem 27:1026, 1981 (abstract).
10. Roe, CR, Limbird, LE, Wagner, GS, and Nerenberg, ST: Combined isoenzyme analysis in the diagnosis of myocardial injury: application of electrophoretic methods for the detection and quantitation of the creatine phosphokinase MB isoenzyme, J Lab Clin Med 80:4, 1972.
11. Wu, AHB, and Bowers, GN, Jr: Clinical correlation of macromolecular CK, type 2, Clin Chem 28:1565, 1982.
12. Stein, W, Bohner, J, Steinhart, R, and Eggstein, M: Macro creatine kinase: determination of two types by their activation energies, Clin Chem 28:19, 1982.

Lactate dehydrogenase

AMADEO J. PESCE

Lactate dehydrogenase, LD, LDH, L-lactate:NAD oxidoreductase
Clinical significance: p. 415
Enzyme number: EC 1.1.1.27
Molecular weight: 140,000 daltons
Chemical class: enzyme, protein
Known isoenzymes: LD_1, LD_2, LD_3, LD_4, LD_5
Biochemical reaction:

$$\begin{array}{c} CH_3 \\ | \\ C{=}O \\ | \\ COOH \end{array} + H^+ + NADH \xrightleftharpoons{\text{EC 1.1.1.27}} \begin{array}{c} CH_3 \\ | \\ CHOH \\ | \\ COOH \end{array} + NAD^+$$

Pyruvate **L-Lactate**

Table 59-23 Methods of lactate dehydrogenase (LD) analysis

Method	Type of analysis	Principle	Usage	Comments
1. Lactate to pyruvate (L → P)	End-point or kinetic spectrophotometric	Reaction buffered at alkaline pH to favor equilibrium to pyruvate	All body fluids Automated	Kinetic Suggested reference method
2. Pyruvate to lactate (P → L)	End-point or kinetic spectrophotometric	Reaction buffered at physiological pH	All body fluids Automated	Claimed to be less linear than method 1; however, faster rate than 1
3. Dinitrophenylhydrazine	End-point spectrophotometric	Pyruvate → Lactate Unreacted pyruvate forms colored product with reagent	Manual	For laboratories without ultraviolet spectrophotometer; imprecise
4. Tetrazolium	End-point spectrophotometric	Lactate → Pyruvate + NADH Reduced NADH forms colored product by reduction of tetrazolium dye	Manual	For laboratories without ultraviolet spectrophotometer Most often used for electrophoretic isoenzyme detection

PRINCIPLES OF ANALYSIS

Lactate dehydrogenase (LD) is an enzyme present in all cells of the body, catalyzing the reaction shown. The equilibrium of the reaction is pH dependent, with alkaline pH favoring the conversion of lactate to pyruvate and neutral pH favoring the reverse reaction.

Most current methods for LD quantitation use spectrophotometers to measure the interconversion of the coenzyme nicotinamide adenine dinucleotide (NAD) and reduced NAD (NADH) at 340 nm[1-4] (methods 1 and 2, Table 59-23). In contrast, most older reactions used an indicator dye after a set time to determine the amount of reactants consumed or formed.[5,6] The dinitrophenylhydrazine used in method 3 (Table 59-23) forms a colored adduct with pyruvate. For the reaction to be measured in the pyruvate-to-lactate direction, a significant amount of pyruvate must be converted. The tetrazolium reaction oxidizes NADH using a dye system to form a colored tetrazolium dye (method 4, Table 59-23).

Abbott Diagnostics (North Chicago) has introduced an assay for LD based on fluorescence attenuation and an inner filter effect.[7] The LD reaction is performed, with lactate being converted to pyruvate and NADH. The NADH in a coupled dye reaction reduces the dye iodonitrotetrazolium violet to form a colored product with an absorption maximum at 492 nm.

It has been shown that *initially* the rate of conversion of lactate-to-pyruvate (L-P) is nonlinear.[8] After 20 seconds, the reaction rate becomes linear, provided that not enough product is formed to inhibit the reaction. In both L-P and P-L directions, product inhibition can be observed. Therefore, there is a limit to the maximum rate that can be reached before the reaction becomes nonlinear, usually about 0.5 absorbance units per minute. The reaction is linear with dilution (zero-order kinetics), and high levels can be accurately quantified by use of smaller amounts of enzyme (dilution). In general, it is best to follow the reaction by measurement of the rate kinetically rather than by end-point analysis.

The reaction rate is temperature dependent, with significantly faster rates occurring at higher temperatures. Conversion factors have been established for comparison of reaction rates at 25°, 30°, and 37° C (for example, U/L at 25° C × Factor = U/L at 37° C). See Table 59-24.

The dry strip of film technology procedures as developed by Ames (Miles Laboratories, Elkhart, IN) and Kodak (Eastman Kodak, Rochester, NY) use dry reagents that are activated on contact with patient sample. The change in the concentration of NADH is determined by reflectance spectrophotometry. The Ames Seralyzer uses the pyruvate-to-lactate reaction (method 2, Table 59-23).

Selecting the best method for LD analysis is difficult because of the heterogeneity of the enzyme itself, the inability of various standardizing groups to agree on the correct temperature for analysis, and the disagreement about whether to follow the reaction in the L-P or the P-L direction.

In 1973 the Scandinavian Society for Clinical Chemistry and Clinical Physiology accepted a subcommittee report for a recommended method for LD in blood.[2] This method uses the P-L reaction. In contrast, the method most often used in the United States is the L-P reaction, as described by Henry.[3] Since there is no generally accepted LD method, assays based on either of these two procedures are adequate for routine use.

SPECIMEN

LD can be found in all body fluids and tissues. The most accurate blood levels are obtained from serum rather than plasma, since some anticoagulants such as oxalate interfere with the reaction. Hemolysis results in the release of LD from red cells into the body fluid, giving falsely high results. Enzyme stability is temperature dependent. Since different isoenzymes are stable at different temperatures,

Table 59-24 Comparison of reaction conditions for LD analysis*

Condition	Scandinavian[2]	Henry[3]	Optimum concentration[8]‡	
			L-P	P-L
Temperature†	37° C	32° C	37° C	37° C
pH	7.4	9.0	8.7	7.0
Final concentration of reagent components	Tris: 50 mmol/L EDTA: 5 mmol/L Pyruvate: 1.2 mmol/L NADH: 0.14 mmol/L	2-Amino-2-methyl-1-propanol: 0.65 mol/L Lactic acid: 91 mmol/L NAD: 5.4 mmol/L	2-Amino-2-methyl-1,3-propanediol: 200 mmol/L Lithium lactate: 70 mmol/L NAD: 7 mmol/L	Imidazole: 100 mmol/L Pyruvate: 1.5 mmol/L NADH: 0.22 mmol/L
Fraction of sample volume	0.022	0.033	0.05	0.05
Linearity	20% decrease at 1000 U/L	Better than P → L at same concentration	—	—
Precision	3%	3%	—	—

*Sources of analytical error: any specimen with visible hemolysis will give elevated results because of release of LD from red cells. *Interferences* include salicylate and inhibitors that may be present in uremic area.
†College of American Pathologists (CAP) surveys indicate that the most common temperatures of running LD assays are at 30° or 37° C.
‡The optimum concentrations for NAD + lactate and NADH + pyruvate are temperature dependent. The maximum reaction rate can vary depending on the buffer used. These conditions have been worked out by Buhl and Jackson.[8]

there is no absolute method of storing samples without some loss of activity. Consistency in storage technique will yield the most precise results. Freezing results in loss of activity of the enzyme.

PROCEDURE: SCANDINAVIAN COMMITTEE ON ENZYMES RECOMMENDED METHOD[2]

Principle

$$\text{Pyruvate} + \text{NADH} + \text{H}^+ \xrightarrow{\text{EC 1.1.1.27}} \text{Lactate} + \text{NAD}^+$$

The reaction of pyruvate and NADH is allowed to occur at a neutral pH. One follows the reaction rate by recording the rate of consumption of NADH at 340 nm.

Reagents

Working Tris buffer (0.056 mol/L Tris, 6.2 mmol/L EDTA). Tris(hydroxymethyl)aminomethane; ethylenediaminetetraacetate disodium solution, EDTA. To approximately 800 mL of distilled water, add 6.8 g of Tris and 2.1 g of EDTA. Allow to dissolve, and bring to pH 7.4 with 1 M HCl. Dilute to 1 L. This solution is stable for at least 6 weeks at 4° C.

Tris-NADH reagent. Dissolve 13 mg of NADH in 90 mL of working Tris buffer. Measure the absorption, and bring to an absorbance of 1.0 (161 μmol/L) by dilution with the same buffer. This is stable for 72 hours at 4° C.

Working pyruvate solution (13.5 mmol/L pyruvate). Dissolve 149 mg of sodium pyruvate in 100 mL of distilled water. This solution is stable for 20 days at 4° C.

Assay

Equipment: spectrophotometer with a ≤10 nm band pass and a temperature-controlled cuvette.

The procedure is described for any recording spectrophotometer, and the volume is calculated for a 3 mL cuvette. The temperature should be held constant at 37° C.

1. Place 2 mL of Tris-NADH reagent in a 3 mL cuvette, add 50 μL of serum, mix, and incubate 5 to 15 minutes at 37° C.
2. Initiate the reaction by adding 200 μL of the pyruvate working solution. Mix, and rapidly insert the cuvette into the spectrophotometer. Immediately start the recording of the change in absorbance at 340 nm.
3. If the resulting curve is linear (ΔA/min is constant over a period of analysis), proceed with the calculations described next. If the rate is too rapid to give a linear curve, dilute the sample with saline and repeat the assay.

Calculation

The international units of activity (U) are expressed as micromoles of NADH per minute, whereas enzyme concentrations are expressed as U/L. The change in absorbance is related to international units by the equation for a 1 cm light path:

$$\text{U/L} = \frac{\Delta A \text{ for } t \text{ min}}{t \text{ min}} \times \frac{\text{Total volume}}{\text{Sample volume}} \times \frac{1}{\epsilon} \times \frac{10^6\ \mu\text{mol}}{\text{mol}}$$

where ΔA = change in measured absorbance for time t
ϵ = molar extinction coefficient for NADH
10^6 = factor to convert concentration to μmol/L.

For the assay, the values are as follows:

$$\text{U/L} = \Delta A/\text{min} \times \frac{2250\ \mu\text{L}}{50\ \mu\text{L}} \times \frac{1}{6220} \times 10^6$$
$$= 7235 \times \Delta A/\text{min}$$

PROCEDURE: METHOD OF HENRY[3]

Principle

$$\text{Lactate} + \text{NAD}^+ \xrightarrow{\text{EC 1.1.1.27}} \text{Pyruvate} + \text{H}^+ + \text{NADH}$$

The reaction of lactate and NAD is allowed to occur at an alkaline pH. One follows the reaction by recording the rate of formation of NADH at 340 nm.

Reagents

2-Amino-2-methyl-1-propanol. This is a liquid and is stable for at least 1 year.

Lactic acid, 85%. This is a liquid and is stable for at least 1 year.

Working lactic acid solution (0.6 mol/L 2-amino-2-methyl-1-propanol, 0.094 mol/L lactic acid). To approximately 800 mL of distilled water, add 63.9 mL of 2-amino-2-methyl-1-propanol and 10 mL of 85% lactic acid. Mix, and adjust to pH 9.0 ± 0.05 with 5 M NaOH. Bring to 1 L volume. Store in a refrigerator. This solution is stable for 1 month.

Lactic acid-NAD reagent. For every 100 mL of working lactic acid solution to be used (enough for 30 samples), add 400 mg of NAD. Final concentration is 5.6 mmol/L. Prepare fresh each day a sufficient volume for the number of specimens to be assayed. Keep solution in refrigerator or on ice at all times.

Assay

Equipment: the original procedure is described for a Gilford 2400 recording spectrophotometer equipped with a temperature-controlled cuvette compartment held at 32° C by a circulating constant-temperature bath. I recommend either 30° or 37° C, since 32° C is no longer used in the United States.

1. Place 2.9 mL of lactic acid–NAD reagent into a test tube, and incubate in a water bath at 37° ± 0.5° C for 4 to 5 minutes. During this period, the cuvettes are kept in the instrument's cuvette compartment so that they reach temperature equilibrium (with spectrophotometer and circulating water bath turned on at least 2 hours before use).
2. Add 0.1 mL of serum using a to-contain pipet, mix, and transfer as quickly as possible to the prewarmed cuvette, which is rapidly reinserted into the cuvette compartment. The cuvette compartment lid is left open for as short a time as possible. Immediately start the automatic recording of the change in absorbance at 340 nm.
3. If the resulting recorded curve is linear for at least 6 minutes, proceed with calculations as described next. If the rate is too rapid to give a linear curve, dilute the sample with 0.15 M NaCl (dilution required if result is more than 300 units), and repeat the analysis.

Calculation

The units of enzyme activity are calculated according to the following relationship (see previous calculation section for definition of terms):

$$\text{U/L} = \Delta A/\text{min} \times \frac{3.0 \text{ mL}}{0.1 \text{ mL}} \times \frac{1}{6220} \times 10^6 = 4830 \times \Delta A/\text{min}$$

REFERENCE RANGE

	Method 1 (L-P)	Method 2 (P-L)
Temperature	37° C	37° C
Adult males	63 to 155 U/L	90 to 320 U/L
Adult females	62 to 131 U/L	90 to 320 U/L
Children	Approximately 10% to 15% higher	

Factors to convert LD activity from assayed temperature to 37° C are as follows:

Multiplication factor for 25° C	2.44
Multiplication factor for 30° C	1.68

A study was performed to establish intrapersonal and interpersonal variability of total LD levels of 24 healthy persons.[9] The analytical variability was only a small fraction of the intrapersonal and interpersonal variability, implying that current analytical techniques are adequate for long-term monitoring. Interpersonal variability was far greater than intrapersonal variability. Thus, determining the intrapersonal variance may be an appropriate method of monitoring changes in a person, since there is less change than from use of a reference range.

One of the major interferences in the measurement of LD is the release of enzyme from red blood cells. Rupture of red cells during collection, as well as by pathophysiology such as intravascular hemolysis, can raise the LD levels.[10-12]

REFERENCES

1. Howell, BF, McClure, S, and Schaffer, R: Lactate-to-pyruvate or pyruvate-to-lactate assay for lactate dehydrogenase: re-examination, Clin Chem 25:269-272, 1979.
2. Keiding, R, Horder, M, Gerhardt, W, et al: Recommended methods for determination of four enzymes in blood, Scand J Clin Lab Invest 33:291-306, 1974.
3. Demetriou, JA, Drewes, PA, and Gin, JB: Enzymes. In Henry, RJ, Cannon, DC, and Winkelman, JW, editors: Clinical chemistry: principles and techniques, New York, 1974, Harper & Row, Publishers, Inc.
4. Freer, DE, Statland, BE, Johnson, M, and Felton, H: Reference values for selected enzyme activities and protein concentrations in serum and plasma derived from cord-blood specimens, Clin Chem 25:565-569, 1979.
5. Natelson, S: Techniques of clinical chemistry, ed 3, Springfield, IL, 1971, Charles C Thomas, Publisher.
6. Spiegel, HE, Symington, JA, Hordynsky, WE, and Babson, AL: Colorimetric determination of lactate dehydrogenase (L-lactate:NAD oxidoreductase) activity, Stand Methods Clin Chem 7:43-46, New York, 1972, Academic Press, Inc.
7. Baugher, BW, Cohen-Giecek, CL, and Shaffar, M: REA LDH: the first enzyme assay for the Abbott TDx, Clin Chem 30:1001, 1984.
8. Buhl, SN, and Jackson, KY: Optimal conditions and comparison of lactate dehydrogenase catalysis of the lactate-to-pyruvate and pyru-

vate-to-lactate reactions in human serum at 25°, 30°, and 37° C, Clin Chem 24:828-831, 1978.

9. Moses, GC, and Henderson, AR: Biological variance of total lactate dehydrogenase and its isoenzymes in human serum, Clin Chem 30:1737-1741, 1984.
10. Frank, JJ, Bermes, EW, Bickel, MS, and Watkins, BF: Effect of in vitro hemolysis on chemical values for serum, Clin Chem 224:1966-1970, 1978.
11. Brydon, WG, and Roberts, LG: The effect of hemolysis on the determination of plasma constituents, Clin Chim Acta 41:435-438, 1972.
12. Blank, DW, Kroll, MH, Ruddel, ME, and Elin, RJ: Hemoglobin interference from in vivo hemolysis, Clin Chem 31:1566-1569, 1985.

Lactate dehydrogenase isoenzymes

JOHN F. CHAPMAN
LINDA L. WOODARD
LAWRENCE M. SILVERMAN

Lactate dehydrogenase, LD, LDH, L-lactate:NAD oxidoreductase
Clinical significance: p. 415
Enzyme number: EC 1.1.1.27
Molecular weight: 140,000 daltons
Chemical class: enzyme, protein
Known isoenzymes: LD_1, LD_2, LD_3, LD_4, LD_5
Biochemical reaction: see previous section on lactate dehydrogenase

PRINCIPLES OF ANALYSIS

Lactate dehydrogenase (LD) is an enzyme present in many tissues. Consequently, elevations of LD in serum have to be considered nonspecific for any single tissue or organ. Additional information can be obtained through the separation and quantitation of LD into the five known isoenzymes. There are two subunits, designated H (heart) and M (muscle), and each isoenzyme is composed of four of these subunits. They are LD_1 (HHHH, or H_4), LD_2 (HHHM, or H_3M), LD_3 (HHMM, or H_2M_2), LD_4 (HMMM, or HM_3), and LD_5 (MMMM or M_4). Methods used for separation and quantitation of these isoenzymes include electrophoresis, chromatography, and immunoinhibition (Table 59-25).

Support media for electrophoretic separation (method 1, Table 59-27) include agarose and cellulose acetate. Agarose gel electrophoresis is currently the most frequently used procedure for LD isoenzyme separation. The charge on each subunit differs; thus there is a different, total charge imparted to each isoenzyme. At a pH of about 8.6, LD_1 (H_4) has the highest negative charge and so moves farthest toward the anode. LD_5 migrates slightly toward the cathode, whereas LD_2, LD_3, and LD_4, in order of decreasing anodic mobilities, migrate toward the anode. Fig. 59-5 shows an LD isoenzyme separation in 1% agarose gel at pH 8.6. Colorimetric or fluorometric determination of the relative amounts of each isoenzyme present may be accomplished by the addition of substrate containing lactate, nicotinamide adenine dinucleotide (NAD), iodonitrotetrazolium chloride (INT), and phenazine methosulfate (PMS). LD catalyzes the following reaction:

$$\text{L-Lactate} + NAD^+ \xrightleftharpoons{LD} \text{Pyruvate} + \underset{\text{(fluorescent)}}{NADH} + H^+$$

The NADH (reduced NAD) generated is then used to reduce the tetrazolium salt to a colored formazan compound (INF):

$$NADH + INT + PMS \rightarrow NAD + INF$$

Densitometric scanning of the resultant pattern yields relative percentages of the isoenzymes present.

Anion-exchange methods (method 2, Table 59-25) for LD isoenzyme have been described[1,2] but are more tedious

Table 59-25 Methods of lactate dehydrogenase (LD) isoenzyme analysis

Method	Principle	Usage	Comments
1. Electrophoresis	At pH 8.6, major isoenzymes have different mobilities in 1% agarose; LD activity quantitatively determined by coupled dye reaction	Most frequently used	Separates all five major isoenzymes Rapid and relatively simple
2. Anion-exchange chromatography	At pH 8.0, LD_1 and LD_2 are absorbed to a diethylaminoethyl (DEAE) type of ion-exchange resin column LD_5 is eluted initially, with LD_1 being eluted at highest salt concentrations Either stepwise reactions or gradients elute isoenzyme fractions Fractions are quantitated by standard assay	Rare	Separation of all isoenzymes not readily accomplished
3. Immunoprecipitation	M subunits of H_3M, H_2M_2, HM_3, and M_4 (LD_{2-5}) removed by double-antibody, solid-phase capture reaction	Frequently used	Gives only LD_1 activity; rapid and relatively simple

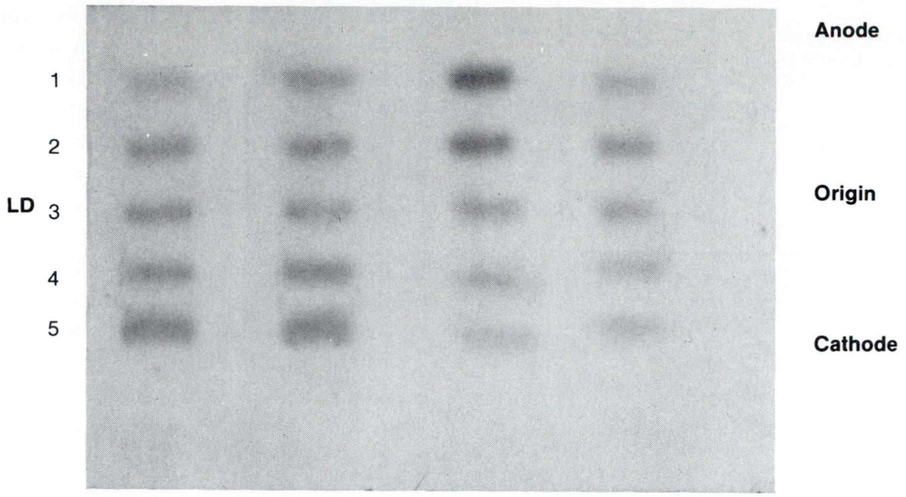

Fig. 59-5 Separation of five lactate dehydrogenase (LD) isoenzymes on 1% agarose gel with tetrazolium dye staining. Specimen on right is a control; others are three specimens from a patient. Notice LD ''flip'' on second specimen from right.

to perform and are inaccurate if special care is not taken to ensure proper separation.[3] Immunochemical methods for the determination of LD_1 use antibodies against the M subunit of LD. The anti-M antibody/M isoenzyme complex then reacts with a second antibody directly against the first to remove all the M-bearing isoenzymes, leaving only LD_1 in solution (method 3, Table 59-25). Alternately, the anti-M immunoglobulin can be covalently attached to an insoluble particle, and the bound M subunits—LD_2, LD_3, LD_4, and LD_5 isoenzymes—are removed from solution by centrifugation. In both cases, the remaining LD activity is measured by the usual LD procedure. This method is generally considered to be a more sensitive indicator of early elevations of LD_1 after acute myocardial infarction.

There has been no agreement on a reference method for LD isoenzyme analysis. Electrophoretic isoenzyme analysis is most often used in the clinical context of the diagnosis of acute myocardial infarction. Most frequently, a ''positive'' LD isoenzyme result occurs when the ratio of LD_1/LD_2 is greater than 1. Since the usual ratio is approximately 0.72 to 0.79, a ratio greater than 1 is called an ''LD flip.'' The presence of an LD flip, in the absence of hemolysis, has a very high specificity (approximately 100%) and can be considered very positive evidence of a myocardial infarction. Immunoprecipitation assays can certainly serve as an acceptable alternative to an electrophoretic procedure.

Table 59-26 compares conditions for LD isoenzyme analysis.

SPECIMEN

The preferred specimen is serum, which should be separated from the clot as quickly as possible, since red blood cells contain 100 times the amount of LD present in normal serum. If serum is allowed to stand on the clot, an elevation in LD_1 and LD_2 will be seen. For the same reason, hemolyzed specimens are unacceptable. LD_5 and to a smaller extent LD_4 are reported to be particularly labile at lower temperatures. Therefore, samples should be kept at room temperature to ensure maximum activity. There is some controversy about whether these forms are inactivated by the lower temperatures or by the rapid change in temperature when the samples are either thawed or warmed to ambient conditions.

REFERENCE RANGE

Reference intervals should be determined for the population of patients being tested. The following reference intervals are given only as a guide:

Isoenzyme fractions	Reference intervals (relative %)
LD_1	18 to 33
LD_2	28 to 40
LD_3	18 to 30
LD_4	6 to 16
LD_5	2 to 13

Table 59-26 Comparison of conditions for LD isoenzyme analysis

Condition	Electrophoresis	Immunoinhibition*
Temperature	Ambient	Ambient
Sample volume	1-2 μL	200 μL
Assay time	20 min (separation)	25 min (separation)
	30 min (analysis)	5 min (quantitation)

*Roche Diagnostics, Nutley, NJ.

REFERENCES

1. Mercer, DW: Simultaneous separation of serum creatine kinase and lactate dehydrogenase isoenzymes by ion-exchange column chromatography, Clin Chem 21:1102, 1975.
2. Mercer, DS: Improved column method for separating lactate dehydrogenase isoenzymes 1 and 2, Clin Chem 24:480, 1980.
3. Usategui-Gomez, M, Wicks, RW, and Warshaw, M: Immunochemical determination of the heart isoenzyme of lactate dehydrogenase (LDH 1) in human serum, Clin Chem 25:729, 1979.

Lipase

MICHAEL D.D. McNEELY

Lipase, LPS, triacylglycerol acylhydrolase
Clinical significance: pp. 390 and 454
Molecular weight: approximately 38,000 daltons
Enzyme number: EC 3.1.1.3
Chemical class: protein, enzyme
Biochemical reaction:

$$\begin{array}{l} \alpha\text{-}CH_2\text{-}O\text{-}\overset{O}{\overset{\|}{C}}\text{-}R_1 \\ \beta\text{-}CH\text{-}O\text{-}\overset{O}{\overset{\|}{C}}\text{-}R_2 \\ \alpha_1\text{-}CH_2\text{-}O\text{-}\overset{O}{\overset{\|}{C}}\text{-}R_3 \end{array} \xrightarrow{\text{Lipase}} \begin{array}{l} CH_2\text{-}OH \\ CH\text{-}O\text{-}\overset{O}{\overset{\|}{C}}\text{-}R_2 \\ CH_2\text{-}OH \end{array} + \begin{array}{l} R_1\text{-}\overset{O}{\overset{\|}{C}}\text{-}OH \\ R_3\text{-}\overset{O}{\overset{\|}{C}}\text{-}OH \end{array}$$

Triglyceride **β-Monoglyceride + 2 fatty acids**

PRINCIPLES OF ANALYSIS

The measurement of serum lipase (LPS) activity has long been recognized as useful for diagnosing pancreatitis. Unfortunately, the enzyme has been very difficult to measure accurately and precisely, and shortcut techniques have proved to be clinically unreliable. These factors have cast the diagnostic use of serum lipase into doubt.

Lipases are enzymes that hydrolyze the long-chain fatty acid (LCFA) triglyceride (triacylglycerol). They act on the glycerol (LCFA) bonds (ester linkage) to liberate two molecules of fatty acid and mole molecule of β-monoglyceride.

Essential to the understanding of lipase methodology is the fact that the enzyme acts only at an ester-water interface. Thus, lipase-assay substrates must be established as emulsions. The reaction rate will increase with the dispersion of the emulsion (surface area). Failure to achieve a suitable ester-water interface will permit the action of nonlipase enzymes (carboxylic ester hydrolase, aryl-ester hydrolase, and lipoprotein lipase). The use of short-chain fatty acid triglyceride substrates, which are water soluble, will also permit this false lipase reaction to occur.

Titration methods were the first practical assays. In the classical technique of Cherry and Crandall,[1] serum was allowed to incubate for 24 hours at 37° C, with a 50% emulsion of olive oil and 5% gum acacia in a pH 7.0 phosphate buffer. The liberated fatty acid was then quantified by titration against 0.05 M NaOH with a phenolphthalein indicator being used (method 1a, Table 59-27). Subsequent methods have attempted to improve on ease, analytical sensitivity, and precision of the assay. Some approaches have refined the assay conditions or employed more sensitive indicators to allow incubation periods as short as 1 hour.[2-5]

Alternative detection methods to replace titration methods have been developed. For example, Tietz and Fiereck[6] have done considerable work using pH-meter end-point measurements (method 1b, Table 59-27) and have a method using pH-stat apparatus.[7] A more recently developed variation of the pH technique involves measurement of the rate of change of pH as the fatty acids are liberated.[8] This method requires a very sensitive pH-measurement device, one that can measure pH changes of less than 0.02 for values in the normal range. This procedure employs olive oil as the substrate but does not use bile acids or colipase to activate lipase. The reaction is very rapid, with most analyses being completed in 3 minutes.

Gindler[9] introduced the Spectru Cationic Blue dye, which reacts directly with liberated fatty acids to form a blue complex that can be measured spectrophotometrically (method 2a, Table 59-27). A similar approach was used by Yang and Biggs,[10] who added Cu^{+2} to form colored complexes with fatty acids, which could be extracted into chloroform and measured in a photometer (method 2b, Table 59-27).

The method of Massion and Seligson[11] quantifies the liberated fatty acids after 1 hour of incubation by extracting them and allowing them to change the color of an appropriately buffered methyl red indicator (method 2c, Table 59-27).

Substrates other than LCFA triglycerides have been used to allow direct color production and produce a more rapid assay[12-14] or to permit the use of a highly sensitive fluorometric end point.[15] Unfortunately, these substrates react with nonlipase esterases and give confusing results. For this reason, natural oil substrates must be purified by alumina absorption.

Several immunochemically based assays have been reported for the measurement of lipase.[16-18] A radioimmunoassay (RIA) (method 3a, Table 59-27) is commercially available (NucLipase, NuClin Diagnostics, Northbrook, IL).[11,13] This is a sequential RIA that employs a 15- to 16-hour incubation with ^{125}I-labeled lipase and a second antibody to precipitate lipase-primary antibody complexes. In addition to measuring serum lipase,[16] this assay has been used to measure lipase in dried blood on filter paper.[17] A direct latex-agglutination procedure (method 3b, Table 59-27) is also commercially available (Behringwerke AG, Marburg, West Germany).[18] Antibodies to pancreatic lipase are bound to latex particles. These particles are mixed with the sample for 3 to 5 minutes. If lipase is present,

Table 59-27 Methods of lipase (LPS) analysis

Method	Principle	Usage	Comments
1. Titration of released fatty acids	Reaction *A:* Triglyceride $\xrightarrow{\text{LPS}}$ Monoglyceride + 2 fatty acids		
	(a) Reaction *A* with fatty acid + NaOH titration to neutrality using phenolphthalein indicator	Used	Cherry-Crandall method[1]
	(b) Reaction *A* with titration of released acids using pH meter to assess end point	Rarely used	Requires a pH meter accurate to ± 0.01 units
2. Colorimetric	(a) Reaction *A* with released fatty acids reacting with Spectru Cationic Blue dye to form blue complex	Rarely used	
	(b) Reaction *A* with released fatty acids reacting with Cu^{+2} ions	Rarely used	
	(c) Reaction *A* with released fatty acids changing the color of pH indicator, methyl red	Rarely used	
3. Immunochemical			
a. RIA	Competitive binding assay using radiolabeled lipase	Rarely used	Measures both active and inactive enzyme
b. Latex agglutination	Antibody-coated latex beads aggregate if lipase is present in sufficient quantity	Rarely used	Semiquantitative
4. Coupled enzymatic	1-Oleoyl-2,3-diacetoyl glycerol on TiO_2 particles $\xrightarrow{\text{Lipase}}$ Oleic acid + 2,3-Diacetoyl glycerol $\xrightarrow{\text{Esterase}}$ Glycerol + Acetate Glycerol quantitated by enzymatic reaction	Frequently used	Kodak Ektachem
5. Emulsion clearing	Turbidimetric or nephelometric monitoring of decrease in size of emulsion of substrate after action of lipase	Widely used	Very dependent on availability of stable, reproducible substrate

agglutinated clumps of particles appear on the reaction slide.

Several coupled enzymatic assays have been developed for the measurement of serum lipase.[19,20] An enzymatic approach has been adapted to the dry-film technology of the Kodak Ektachem (method 4, Table 59-27).[21] This method employs 1-oleoyl-2,3-diacetoyl glycerol absorbed onto TiO_2 particles as the substrate. The TiO_2 particles act to provide the surface needed for lipase activity. In addition, an emulsifying agent and porcine colipase are present to increase lipase activity. Lipase will act on the asymmetrical triglyceride to release the oleyl fatty acid and the water-soluble[2,3] diacetoyl glycerol. The latter molecule diffuses into another film layer of the reaction slide, where an esterase clears both acetate groups to release glycerol, which can then be quantitated by an enzymatic method. When one employs a lag period before measuring the glycerol formed, endogenous glycerol is consumed and will not interfere in the assay.

A completely different analytical approach that used clearing of an emulsion was introduced by Vogel and Zieve[22] and has been modified successfully by others.[23,24] In this method, the turbidity of the oil-water emulsion, monitored at 340 or 400 nm, is reduced as lipase hydrolyzes the triglyceride molecules. The technique requires a highly stable, reproducible substrate with an appropriate initial absorbance. The sample is added to the substrate, and kinetic measurements are made turbidimetrically or nephelometrically (method 5, Table 59-27). A special-purpose instrument has been developed for this method (Perkin-Elmer Corp., Norwalk, CT).

From 3% to 5% of specimens analyzed by emulsion-clearing methods will display an anomalous absorbance increase caused by molecular aggregation. This phenomenon occurs more frequently when specimens are frozen and thawed or when they contain increased concentrations of IgM, such as rheumatoid factor. By avoidance of freezing or by use of polyethylene glycol (PEG) to precipitate protein (mostly IgM), these problems will be reduced in frequency.[25]

Recently, colipase[26] was introduced to enhance lipase measurements by turbidity or nephelometry.[27,28] Colipase increases the reaction rate of lipase reactions by allowing the enzyme better access to the substrate.[26] Although variable amounts of colipase have been reported to be released into peripheral blood during acute pancreatitis,[29] colipase is added to ensure sufficient activation of the pancreatic isoenzyme.[30] The suggested "optimum" conditions for a lipase reaction containing colipase include use of NaCl (140 mmol/L) and a bile salt, sodium deoxycholate (18 mmol/L).

The most widely used lipase assay is most likely one that employs the turbidimetric or nephelometric approach.

Methods using substrates other than triglycerides must

be viewed with great suspicion. The use of purified olive oil substrate, extraction of LCFA, and quantitation against pure standards are recommended as a standard technique. The modified method of Massion and Seligson[11] detailed next has these characteristics carried out with microsamples in under 2 hours.

SPECIMEN

Serum or duodenal fluid should be stored at 4° C until analyzed. Repeated freezing and thawing should be avoided. Lipase activity is normally not found in urine, unless glomerular integrity has been compromised. Thus urine is not an appropriate or useful specimen.

PROCEDURE: TURBIDIMETRIC METHOD FOR SERUM LIPASE (MODIFIED)[31]

Principle

The clearing of an emulsion of olive oil is measured turbidimetrically at 340 nm.

Reagents

Tris-deoxycholate buffer. Dissolve 3.0 g (0.025 mol) of tris(hydroxymethyl)aminomethane and 6.0 g (0.014 mol) of sodium deoxycholate in distilled water and dilute to a final volume in 1 L. Adjust to pH 8.8 with concentrated HCl. This is stable for 6 months at 4° to 8° C.

Olive oil, reagent grade (no. 0-111, Fisher Scientific Co.). Separate the olive oil from the fatty acids by passing it though alumina (80-200 mesh, catalog no. A-540, Fisher Scientific Co.) contained in a column or separatory funnel plugged with glass wool. One volume of oil is purified by an equal volume of alumina. Check the fatty acid content of the oil by pipetting a 0.050 mL aliquot into 3 mL of methyl red reagent, shaking, and centrifuging. If the absorbance of the methyl red layer at 502 nm is more than 0.020 in a 10 mm cuvette, the oil must be passed through more alumina. The oil is stable for at least 1 week at 4° to 8° C.

Olive oil emulsion. Dissolve 1.0 g of purified olive oil in 100 mL of absolute ethanol. Add 4 mL of this solution slowly with stirring to 100 mL of Tris-deoxycholate buffer. The absorbance of this reagent at 340 nm should be adjusted until it is within the measurable range of the photometer by the addition of more Tris-deoxycholate buffer.

Standard. A lyophilized serum with known lipase activity is used.

Assay

Equipment: a $\leq$10 nm band pass spectrophotometer with temperature-controlled cuvette.

1. To 1 mL of emulsion, add 50 μL of sample, standard, or control. Mix immediately.
2. Read the absorbance of each tube at 340 nm at 1 minute ($A1'$) and again at 5 minutes ($A5'$).

Calculation

$$\frac{(A1' - A5')_{unknown} \times Value_{std}}{(A1' - A5')_{std}} = Value_{unknown}$$

$Value_{std}$ is the lipase activity of a known standard.

REFERENCE RANGE

Serum from healthy persons will have 2 to 7.5 U of lipase activity per milliliter. Serum values remain constant throughout most of the adult life, with no significant difference between male and female.[32] In persons over 60 years of age, however, the values may rise by 20%.

REFERENCES

1. Cherry, IS, and Crandall, IA, Jr: The specificity of pancreatic lipase: its appearance in the blood after pancreatic injury, Am J Physiol 100:266-273, 1932.
2. MacDonald, RP, and LeFave, RO: Serum lipase determination with an olive substrate using a three-hour incubation period, Clin Chem 8:509-519, 1962.
3. Roe, JH, and Byler, RE: Serum lipase determination using a one-hour period of hydrolysis, Anal Biochem 6:451-460, 1963.
4. Tietz, NW, Bordem, T, and Stepleton, JD: An improved method for the determination of lipase in serum, Am J Clin Pathol 31:148-154, 1959.
5. Vogel, WC, and Zieve, L: A rapid and sensitive turbidimetric method for serum lipase based on differences of normal and pancreatitis serum, Clin Chem 9:168-181, 1963.
6. Tietz, NW, and Fiereck, EA: Measurement of lipase in serum. In Cooper, GR, editor: Standard methods of clinical chemistry, vol 7, New York, 1972, Academic Press, Inc.
7. Tietz, NW, and Repique, EV: Proposed standard method for measuring lipase activity in serum by a continuous sampling technique, Clin Chem 19:1268-1275, 1973.
8. Cerolotti, F, Bonini, PA, Murone, M, et al: Measurement of lipase activity by a differential technique, Clin Chem 31:257-260, 1985.
9. Gindler, EM: Colorimetric estimation of lipase activity in blood serum with use of olive oil as substrate, Clin Chem 17:633, 1971.
10. Yang, JS, and Biggs, HG: Rapid, reliable method for measuring serum lipase activity, Clin Chem 17:512, 1971.
11. Massion, CG, and Seligson, D: Serum lipase: a rapid photometric method, Am J Clin Pathol 48:307-313, 1967.
12. Kramer, SP, Batatos, M, Karpa, JN, et al: Development of a clinically useful colorimetric method for serum lipase, J Surg Res 4:23-35, 1964.
13. Patt, HH, Kramer, SP, Woel, G, et al: Serum lipase determination in acute pancreatitis, Arch Surg 92:718-723, 1966.
14. Whitaker, JF: A rapid and specific method for the determination of pancreatic lipase in serum and urine, Clin Chim Acta 44:133-138, 1973.
15. Fleisher, M, and Schwartz, MK: An automated, fluorometric procedure for determining serum lipase, Clin Chem 17:417-422, 1971.
16. Royse, VL, Jensen, DA, Schaffney, JA, and Payne, JA: Laboratory and clinical evaluation of a sensitive radioimmunoassay for lipase, Clin Chem 30:1058, 1984 (abstract).
17. Hammond, KB, Ask, CG, and Watts, DC: Immunoresponsive pancreatic lipase measurements on dried blood spots: a simple neonatal screening test for cystic fibrosis, Clin Chem 30:1059, 1984 (abstract).
18. Moller-Petersen, J, Klaerke, M, Dati, F, and Toth, T: Immunochemical qualitative latex agglutination test for pancreatic lipase in serum evaluated for use in diagnosis of acute pancreatitis, Clin Chem 31:1207-1210, 1985.
19. Griebel, RJ, Knoblock, ED, and Koch, TR: Measurement of serum lipase activity with the oxygen electrode, Clin Chem 27:153-165, 1981.
20. Hoffmann, GE, Neumann, U, Hoffmann, S, et al: An enzymatic method for calibration of serum lipase assays, Clin Chem 32:545-547, 1986.

21. Mauck, JC, Weaver, MS, and Stanton, C: Development of a Kodak Ektachem clinical chemistry oxide for serum lipase, Clin Chem 30:1058, 1984 (abstract).
22. Vogel, WC, and Zieve, L: A rapid and sensitive turbidimetric method for serum lipase based upon differences between the lipases of normal and pancreatitis serum, Clin Chem 9:168-181, 1963.
23. Shipe, JR, and Savory, J: The simultaneous kinetic mesurement of amylase and lipase using an automatic sampling fluoronephelometer, Clin Chem 19:645, 1973.
24. Zinterhofer, L, Wardlaw, S, Jatlow, P, and Seligson, D: Nephelometric determination of pancreatic enzymes. II. Lipase, Clin Chem Acta 44:173-178, 1973.
25. Kannisto, J, Lalla, M, and Lukkari, E: Characterization and elimination of a factor in serum that interferes with turbidimetry and nephelometry of lipase, Clin Chem 29:96-99, 1983.
26. Borgstrom, B, Erlanson-Albertsson, C, and Weilock, T: Pancreatic colipase: chemistry and physiology, J Lipid Res 20:805-816, 1979.
27. Hoffman, GE, and Weiss, L: Specific serum pancreatic lipase determination with use of purified colipase, Clin Chem 26:1732-1733, 1980.
28. Arzoglou, PL, Ferard, G, Khalfa, K, and Metais, P: Conditions for assay of pancreatic lipase from human plasma using a nephelometric technique, Clin Chim Acta 119:329-335, 1982.
29. Junge, W, and Leybold, K: Detection of colipase in serum and urine of pancreatitis patients, Clin Chim Acta 123:293-302, 1982.
30. Arzoglou, PL, Lessinger, JM, and Ferard, G: Plasma lipase properties as related to pancreatic condition, Clin Chem 32:50-52, 1986.
31. Shihabi, ZK, and Bishop, C: Simplified turbidimetric assay for lipase activity, Clin Chem 17:1150-1153, 1971.
32. Mohiuddin, J, Katrak, A, Junglee, D, et al: Serum pancreatic enzymes in the elderly, Ann Clin Chem 21:101-104, 1984.

5′-Nucleotidase

GRAHAM ELLIS

5′-Nucleotidase, 5′-ribonucleotide phosphohydrolase, 5′-NT, 5′-Nase

Clinical interpretation: pp. 359, 484, and 730

Enzyme number: E.C. 3.1.3.5

Chemical class: protein, enzyme

Biochemical reaction catalyzed:

5′-Ribonucleotide phosphate → Ribonucleotide + P_i

PRINCIPLES OF ANALYSIS

In routine clinical chemistry practice, 5′-nucleotidase (5′-NT) has been measured in serum. The levels of 5′-NT activity are increased in hepatobiliary diseases and malignancy. Although 5′-NT is specific for ribonucleotide 5′-phosphates and has the same Michaelis-Menten constant K_m (3×10^{-5} mol/L) and V_{max} for at least two of the ribonucleotide phosphates, adenosine monophosphate (5′-AMP) and inosine monophosphate (5′-IMP),[1] it cannot simply be assayed by the addition of serum to substrate and the monitoring of product formation because nonspecific alkaline and acid phosphatases also hydrolyze this class of phosphate esters. This problem has stimulated the imagination of investigators and has led to the development of many methods with different approaches to assess 5′-NT activity in the presence of nonspecific phosphatases and to monitor the 5′-NT reaction. The frequency of publication of "new" methods attests to the fact that no 5′-NT method is ideal, and none has been adopted as a reference procedure.

Approaches to assessing or eliminating interferences

Potential interference by nonspecific phosphatases has been eliminated or assessed by various methods, which are essentially of three types: subtractive methods, selective activation or inhibition, and phosphatase diversion.

Subtractive methods. Nonspecific phosphatases are measured by use of substrates such as β-glycerophosphate (β-GP), or mixtures of 2′-AMP and 3′-AMP, under conditions where the rate of hydrolysis of these substrates approximates the rate of hydrolysis of 5′-AMP. Total 5′-adenosine monophosphatase (5′-AMPase) activity is measured independently, and "true" 5′-NT is calculated by subtraction of nonspecific phosphatase activity from total 5′-AMPase activity.[2-4] These methods fail to take into account any differences that may exist between various nonspecific phosphatases in their capacity to hydrolyze 5′-AMP or alternative substrate.

Methods based on selective activation or inhibition of 5′-NT or alkaline phosphatases. In the method of Young,[5] nonspecific alkaline phosphatase (APase) is inhibited by preincubation of serum with EDTA. The results of others using this method were mixed.[6-8]

According to Ahmed and Reis,[9] Ni^{+2} inhibited 5′-NT activity but was without effect on APase. This observation has formed the basis for several manual[10-14] and automated methods.[15,16] Technical difficulties arise because of the insolubility of some Ni^{+2} salts.[15] Incorrect estimates of 5′-NT activity are caused by effects of Ni^{+2} on APase itself. APase in peripheral nervous tissue is stimulated by Ni^{+2},[17] whereas APase in bone is inhibited by Ni^{+2}.[18]

As with Ni^{+2}, concanavalin A (con A) also inhibits 5′-NT but has no significant effect on APase, and a method has been proposed using this property.[19] Because some con A preparations contain ammonia and others may produce turbid solutions,[19] reagent suppliers should be chosen with care. Con A cannot be used universally in all assays because it may precipitate serum glycoproteins.[1]

Belfield and Goldberg[20] showed that Mg^{+2} stimulated the activity of 5′-NT at pH 7.9 but had negligible effect on APase activity at this pH. The increase of 5′-AMP hydrolysis on addition of Mg^{+2} was used as a measure of 5′-NT activity.

Methods based on alkaline phosphatase diversion. If any assay mixture contains several substrates, the rate of hydrolysis of each individual substrate by APase depends on the enzyme affinity for the substrate and also on the substrate concentration relative to that of the other substrates. If β-GP and 5′-AMP are present in an assay mixture and the β-GP concentration is 50- to 300-fold that of the 5′-AMP, nonspecific APases will hydrolyze the β-GP and their hydrolysis of 5′-AMP will be negligible. In contrast, 5′-NTs do not act on β-GP, and they are unaffected by β-GP at the concentrations at which it is used. Consequently, the rate of 5′-AMP hydrolysis in such a mixture represents the true 5′-NT activity. This approach has been termed *enzyme diversion* by its originators.[21,22]

Approaches to monitoring the 5′-NT reaction

5′-NT methods vary in the way the progress of the enzyme reaction is monitored and in the substrate used. Traditionally, 5′-AMP has been used as a substrate for serum 5′-NT, but in recent years alternative ribonucleotide phosphates have been tested and different indicator reactions used. There is little information on the relative merits of the various ribonucleotide phosphates. Serum 5′-NT shows similar K_m and V_{max} for 5′-IMP and 5′-AMP.[1] The relative rates of hydrolysis by 5′-NT from human liver are as follows: 5′-AMP, 100; 5′-guanosine monophosphate (5′-GMP), 52; 5′-cytidine monophosphate (5′-CMP), 105; and 5′-uridine monophosphate (5′-UMP), 104.[23] The methods may be divided into the following six groups.

Rate of phosphate release (Method 1a, Table 59-28). The rate of hydrolysis of any substrate may be monitored as phosphate release. Phosphate methods cannot be used when alkaline phosphatase interference is eliminated by enzyme diversion, that is, the provision of alternative substrates, because these also produce inorganic phosphate (P_i). Phosphate methods are susceptible to problems with contaminated glassware. Endogenous inorganic phosphate contributes to a high blank value.

Rate of substrate consumption monitored by ultraviolet spectrophotometry (method 2, Table 59-28). The rate of utilization of substrate may be monitored spectrophotometrically at 265 nm as 5′-AMP is being hydrolyzed to adenosine and inosine in the presence of excess adenosine deaminase (ADase).

$$5'\text{-AMP} \xrightarrow{5'\text{NT}} \text{Adenosine} + P_i \qquad \textit{Eq. 59-1}$$

$$\text{Adenosine} \xrightarrow{\text{ADase}} \text{Inosine} + NH_3 \qquad \textit{Eq. 59-2}$$

5′-AMP and adenosine have similar high absorbances at 265 nm, whereas inosine has low absorbance. This method[21] requires an ultraviolet spectrophotometer, and the high absorbance of both substrate and serum limits the concentration and volume fraction that may be used in the assay.

Rate of radioactive nucleoside formation. Several assays have been proposed in which ^{14}C- or ^{3}H-ribonucleotide phosphates are used as substrates. After an incubation with 5′-NT, substrate and products are separated by column chromatography, electrophoresis, or thin-layer chromatography and quantified by liquid scintillation counting. The methods are exquisitely sensitive,[24-27] but they have been applied to microanalysis of 5′-NT after cellular fractionation, rather than to the microanalysis of serum.

Rate of nucleoside formation assessed by deamination and assay of liberated ammonia (Method 1b, Table 59-28). In the presence of excess ADase, adenosine produced from 5′-AMP by 5′-NT is deaminated (Equation 59-2). The ammonia can be measured colorimetrically by the Berthelot reaction after a timed incubation of 5′-NT and substrate or by continuous kinetic assay at 340 nm in the presence of excess glutamate dehydrogenase (GLDH), 2-oxoglutarate, and reduced nicotinamide adenine dinucleotide (NADH).

$$\text{2-Oxoglutarate} + \text{NADH} + NH_4^+ \xrightleftharpoons{\text{GLDH}} \text{L-Glutamate} + \text{NAD}^+ \qquad \textit{Eq. 59-3}$$

A colorimetric assay of 5′-NT has also been described using a similar principle but with cytidine 5′-monophosphate as substrate and cytidine deaminase (CDase) as indicator enzyme.[28] The authors offer no explanation as to why this substrate is better than the 5′-AMP used in their previous 5′-NT methods; moreover, they chose to use 5′-AMP[29,30] or 5′-IMP[1] in their later studies.

Serum should be as fresh as possible for assays of this type; otherwise the serum ammonia may produce a high value for the blank. These decrease assay precision and

Table 59-28 Methods of 5′-nucleotidase analysis

Method	Type of analysis	Principle	Comments
1. Colorimetric	a. Phosphate release	Phosphate produced as product of substrate hydrolysis	Cannot be used when enzyme diversion methods used
	b. Ammonia formation	5′-AMP forms inosine, which can be deaminated to form NH_3	High serum ammonia causes increased values
	c. Inosine release	5′-AMP forms inosine, which can be converted to hypoxanthine and/or uric acid and H_2O_2	High absorbance
2. Spectrophotometric			
	a. Xanthine oxidase	Monitors hypoxanthine reaction but using 293 absorbency	High serum absorbance
	b. NADPH linked	As in Equations 59-4 to 59-6, but reaction of H_2O_2 coupled to NADP formation (Equations 59-7 and 59-8)	—

reduce the range of activity that can be measured without dilution.

Rate of inosine formation linked to xanthine oxidase (Method 2a, Table 59-28). With 5′-IMP as substrate, 5′-NT liberates inosine, which, in the presence of excess nucleoside-specific nucleoside phosphorylase and xanthine oxidase, is converted to hypoxanthine and urate. The urate formation may be monitored kinetically at 293 nm.

$$\text{5'-IMP} \xrightarrow{\text{5'-NT}} \text{Inosine} + P_i \qquad \textit{Eq. 59-4}$$

$$\text{Inosine} + P_i \xrightarrow[\text{phosphorylase}]{\text{Nucleoside}} \text{Hypoxanthine} + \text{Ribosine-1-phosphate} \qquad \textit{Eq. 59-5}$$

$$\text{Hypoxanthine} + 2H_2O + 2O_2 \xrightarrow[\text{oxidase}]{\text{Xanthine}} \text{Uric acid} + 2H_2O_2 \qquad \textit{Eq. 59-6}$$

The method[31] operates at an elevated absorbance (1.8 to 2.3), necessitating a high-quality ultraviolet instrument, and correlates quite poorly with an alternate technique that monitors the production of H_2O_2 in Equation 59-6.

The H_2O_2 produced by the oxidation of the hypoxanthine (Equation 59-6) may be channeled to $NADP^+$ reduction by catalase and aldehyde dehydrogenase,[1] as shown in the following:

$$H_2O_2 + \text{Ethanol} \xrightarrow{\text{Catalase}} \text{Acetaldehyde} + 2H_2O \qquad \textit{Eq. 59-7}$$

$$\text{Acetaldehyde} + NADP^+ + H_2O \xrightarrow[\text{dehydrogenase}]{\text{Aldehyde}} \text{Acetic acid} + NADPH + H^+ \qquad \textit{Eq. 59-8}$$

The reduction of the $NADP^+$ may be monitored at 334 or 340 nm (method 2b, Table 59-28).

In the presence of excess peroxidase and the acceptor system, 3,5-dichloro-2–hydroxybenzenesulfonic acid–4-aminophenazone, the H_2O_2 may instead be assayed by rate colorimetry at 510 nm.[31] The method is reported to be sensitive to bilirubin interference despite addition of ferrocyanide, but the extent of this interference is unclear.[32]

Assay of uridine production measured by HPLC at 254 nm. After a 1-hour incubation of erythrocytes or serum with 5′-UMP, protein is denatured by heat, the reaction mixture centrifuged, and the supernatant assayed for uridine. Uridine and unhydrolyzed 5′-UMP are separated by HPLC (run time 7 minutes per sample), and the uridine is quantified from its absorbance at 254 nm.[33]

Because there are several methods for monitoring the reaction and various ways of preventing interference by nonspecific APases, many combinations and permutations have been proposed.

No reference method is available for the enzyme 5′-NT.

A laboratory should choose the method best suited to its instrumentation and workload. My preference[34] is influenced by the following factors:

1. 5′-AMP is the substrate traditionally used for serum 5′-NT measurements.
2. Substrate diversion by β-GP is the preferred method for elimination of APase interference, because Ni^{+2} has some inhibitory effect on bone APase,[35] and phenylphosphate offers no advantage over β-GP as a diverter.[22]
3. Mn^{+2} is a better activator than Mg^{+2}.[34]
4. A triethanolamine buffer (pH 7.2) is best for 5′-NT activity in the presence of Mn^{+2}[42] and best for stabilization of GLDH.[36] This pH is within the pH optimum, 7.2 to 7.9, as determined by others.

SPECIMEN

Collect venous blood, and separate the serum from the clot within 1 hour of collection. There is no diurnal variation, but samples should not be taken soon after a fatty meal because of lipemia. Serum is stable at 4° C for up to 4 days or at −20° C for several months. Room temperature storage should be avoided because 5′-NT is less stable under these conditions and storage at room temperature increases the ammonia content of the serum. Hemolysis increases 5′-NT activity, but the effect is variable and depends on methodology, since red blood cell 5′-NTs are not necessarily affected to the same degree as serum 5′-NT by inhibitors, such as con A.[19] Red blood cells contain about four times the activity of serum with 5′-UMP as substrate.[33]

PROCEDURE: MEASUREMENT OF 5′-NT ACTIVITY USING COUPLED GLUTAMATE DEHYDROGENASE REACTION

Principle

In the presence of excess adenosine deaminase (ADase), the adenosine liberated by 5′-nucleotidase (5Nase) (Equation 59-1) is hydrolyzed to inosine and ammonia (Equation 59-2). This ammonia production is monitored as a fall in A_{340} resulting from NADH oxidation in the presence of excess GLDH and 2-oxoglutarate (Equation 59-3). The reaction mixture includes excess β-GP as an alternate substrate for nonspecific phosphatases.

The serum to be assayed is added to substrate mixture. After endogenous ammonia and adenosine are consumed and a blank reaction is measured, 5′-adenylic acid (5′-AMP) is added. After a short lag period, the rate of fall in A_{340} attributed to 5′-NT activity is monitored.

Reagents

Use reagents that are of the highest analytical grade. Water should be distilled and deionized.

Combined bulk substrate. Mix the following in about 800 mL of water:

22.9 g (123 mmol) of triethanolamine hydrochloride (Sigma, catalog no. T 1502).

29 g (496 mmol) of sodium chloride

361 mg (1.9 mmol) of 2-oxoglutarate
19.5 g (66.5 mmol) of β-GP sodium salt $5.5H_2O$ (Sigma, catalog no. G 6251)

Adjust the pH to 7.2 with NaOH (1 mol/L) and volume to 1 L. Store at −20° C in aliquots. This is stable 6 to 12 months.

$MnCl_2$ (100 mmol/L). Dissolve 9.895 g of the tetrahydrate in 500 mL of water. Store in aliquots at −20° C, or at 4° C for not more than 3 weeks.

GLDH in 50% glycerol. Use 10 mg/mL, approximately 450 U/mL (Boehringer Mannheim, BMC, Indianapolis). Store at 4° C. This is stable 6 to 12 months.

ADase in 50% glycerol. Use 2 mg/mL, approximately 220 U/mg (BMC). Store at 4° C. This is stable 6 to 12 months.

NADH (approximately 14 mmol/L). Each day, dissolve 10 mg of the disodium salt in 1 mL of water.

5′-AMP (20 mmol/L). Dissolve 998 mg of the disodium salt (BMC) in 100 mL of water. Store at −20° C in aliquots. This is stable for 6 months.

Working reagent. Prepare a suitable quantity fresh daily by mixing the reagents (see the following notes) in the indicated proportions:

2.415 mL of combined bulk substrate
30 μL of 100 mmol/L $MnCl_2$
50 μL of GLDH
5 μL of ADase
50 μL of NADH

Assay

Equipment: recording spectrophotometer with a constant-temperature cuvette block at 37° C.

1. Mix 2.55 mL of working reagent and 300 μL of serum or preparation to be assayed in a 10 mm light path cuvette, and incubate at 37° C for 20 to 30 minutes.
2. Measure ΔA_{340}/min of the blank reaction for 5 minutes.
3. Add 150 μL of 5′-AMP reagents, and mix.
4. Measure ΔA_{340}/min of the test, after a 3- to 4-minute lag phase, for 5 minutes.
5. Calculate the enzyme activity (U/L) from the following expression:

$$\text{Activity} = 1608(\Delta A_{340}/\text{min test} - \Delta A_{340}/\text{min blank})$$

Notes

1. If immediately before blank measurement the absorbance of the solution is less than 1.0, add more (50 μL) NADH to compensate. (Excess pyruvate, ammonia, or adenosine in serum, enzyme preparations, or reagents is responsible for this effect.)
2. Dilute samples with 5′-NT activity greater than 40 U/L with saline, and reassay.
3. Check the 5′-AMP reagent daily for excess ammonia or adenosine by adding 150 μL to 2.55 mL of working reagent with 330 μL of water. The resulting decrease in A_{340} should not exceed 0.08 and should be complete within 4 minutes, followed by a rate-of-absorbance decrease not greater than that of working reagent alone.
4. This method has been slightly modified by Arkesteijn,[29] and the modification forms the basis of the Sigma test kit no. 265-UV. Arkesteijn[29] increased the NADH concentration to 0.35 mmol/L (a modification not retained by Sigma). Without explanation, he omitted the 400 mmol/L NaCl advocated by Ellis and Goldberg[34] as a ''powerful stabilizer for GLDH,'' demonstrated reagent instability, and substituted L-leucine. The formulation of the Sigma kit with respect to the concentrations of β-GP, buffer, and ''activator'' is not stated. The correlation of the Sigma method against that of Arkesteijn gave a slope of 0.8, suggestive of a slightly suboptimum formulation.[37]
5. If the ammonia content of the serum submitted for analysis is consistently high, despite appropriate storage, it may be preferable to remove the ammonia from the serum by adding a suspension of a cation-exchange resin to the serum just before analysis.[38]
6. Grossly lipemic, icteric, or hemolyzed samples may increase the starting absorbance to greater than 1.8. These samples should be diluted and reanalyzed.
7. The method of Ellis and Goldberg[34] has been adapted for use on the LKB Reaction Rate Analyzer,[39] the GEMSAEC,[40] and the GeMENI analyzers,[41] although the modification of the GeMENI has not been properly validated and gives an inappropriately high reference range. Sigma describes applications of their kit using the ABA-100 Abbott VP and the Union Carbide CentrifiChem analyzers.[37]

REFERENCE VALUES

The method has a reference range for adults of 0 to 11 U/L at 37° C.[39] Similar 5′-nucleotidase ranges have been found by others.[13,16,19]

REFERENCES

1. Heinz, F, Pilz, R, and Reckel, S: A new spectrophotometric method for the determination of 5′-nucleotidase, J Clin Chem Clin Biochem 18:781-788, 1980.
2. Dixon, TF, and Purdom, M: Serum 5′-nucleotidase, J Clin Pathol 7:341-343, 1954.
3. Kowlessar, OD, Haeffner, LJ, Riley, EM, and Sleisenger, MH: Comparative study of serum leucine aminopeptidase, 5-nucleotidase and nonspecific alkaline phosphatase in diseases affecting the pancreas, hepatobiliary tree and bone, Am J Med 31:231-237, 1961.
4. Leybold, K, Beckman, J, and Weisbecker, L: Kinetische Bestimmung der 5′-nucleotidase in Serum, Z Klin Chem Klin Biochem 7:25-27, 1969.
5. Young, II: Serum 5-nucleotidase: characteristics and evaluation in disease states, Ann NY Acad Sci 75:357-362, 1958.
6. Walker, PG, and Bayliss, V: The determination of 5′-nucleotidase in serum, Sixth International Congress of Clinical Chemistry, Munich, 1966, abstract B13, p 180.
7. Reddi, KK: Serum 5′-nucleotidase of a breast cancer patient, Clin Biochem 24:51-54, 1980.
8. Von Beckmann, J, Leybold, K, and Weisbecker, L: Zur Bestim-

mung der 5′-Nucleotidase in Serum, Z Klin Chem Klin Biochem 7:18-24, 1969.
9. Ahmed, Z, and Reis, JL: The activation and inhibition of 5′-nucleotidase, Biochem J 69:386-387, 1958.
10. Campbell, DM: Determination of 5′-nucleotidase in blood serum, Biochem J 82:34P, 1962.
11. Schwartz, MK, and Bodansky, O: Serum 5′-nucleotidase activity in patients with cancer, Cancer 18:886-892, 1965.
12. Rieder, SV, and Otero, M: A simplified procedure for the assay of 5′-nucleotidase, Clin Chem 15:727-729, 1969.
13. Baginski, ES, Marie, SS, Epstein, E, and Zak, B: Simple, direct determination of serum 5′-nucleotidase without deproteinization, Ann Clin Lab Sci 7:469-478, 1977.
14. Wood, JW, and Williams, DG: Colorimetric determination of serum 5′-nucleotidase without deproteinization, Clin Chem 27:464-465, 1981.
15. Hill, PG, and Sammons, HG: An automated method for the determination of serum 5′-nucleotidase, Clin Chim Acta 13:739-745, 1966.
16. Ryan, ED: Serum 5′-nucleotidase: automation of a manual assay and brief observations on values in patients with breast cancer, Clin Biochem 16:249-253, 1983.
17. Pechan, I, and Tursky, T: Estimation of 5′-nucleotidase activity in nervous tissue, Enzymologia 37:73-82, 1969.
18. Schwartz, MK, and Bodansky, O: Properties and activity of 5′-nucleotidase and human serum, and applications in diagnosis, Am J Clin Pathol 42:572-580, 1964.
19. Zygowicz, ER, Sunderman, FW, Jr, Horak, E, and Dooley, JF: Inhibition by concanavalin A as the basis for a specific assay of serum 5′-nucleotidase activity, Clin Chem 23:2311-2323, 1977.
20. Belfield, A, and Goldberg, DM: Activation of serum 5′-nucleotidase by magnesium ions and its diagnostic applications, J Clin Pathol 22:144-151, 1969.
21. Belfield, A, and Goldberg, DM: Inhibition of the nucleotidase effect of alkaline phosphatase by β-glycerophosphate, Nature 219:73-75, 1968.
22. Belfield, A, and Goldberg, DM: Comparison of sodium β-glycerophosphate and disodium phenyl phosphate as inhibitors of alkaline phosphatase in determination of 5′-nucleotidase activity of human serum, Clin Biochem 3:105-111, 1970.
23. Song, CS, and Bodansky, O: Purification of 5′-nucleotidase from human liver, Biochem J 101:5C-6C, 1966.
24. Hurwitz, MY, and Edstrom, RD: A simple, sensitive radiomicroassay for 5′-nucleotidase, Anal Biochem 84:246-250, 1978.
25. Chatterjee, SK, Bhattacharya, M, and Barlow, JJ: A simple, specific radiometric assay for 5′-nucleotidase, Anal Biochem 95:497-506, 1979.
26. Klaushofer, K, Mayer, D, Hummel, W, and von Mayersback, H: A double-labelling radioassay for the determination of 5′-nucleotidase activity, Enzyme 24:77-84, 1979.
27. Kizaki, H, and Weber, G: Simple radioassay for uridine phosphorylase and 5′-nucleotidase, Anal Biochem 105:257-261, 1980.
28. van der Kooij, PJ, Hout, A, Persijn, JP, and van der Slik, W: Determination of serum nucleotidase with cytidine monophosphate as substrate, J Clin Chem Clin Biochem 14:469-473, 1976.
29. Arkesteijn, CLM: A kinetic method for serum 5′-nucleotidase using stabilised glutamate dehydrogenase, J Clin Chem Clin Biochem 14:155-158, 1976.
30. van Helden, WCH, van der Slik, W, Persijn, J-P, and Souverijn, JHM: Automated method for the determination of 5′-nucleotidase in serum by continuous flow analysis, J Clin Chem Clin Biochem 18:333-337, 1980.
31. Dooley, JF, and Racich, L: A new kinetic determination of serum 5′-nucleotidase activity, with modifications for a centrifugal analyzer, Clin Chem 26:1291-1297, 1980.
32. Bertrand, A, and Buret, J: A one-step determination of serum 5′-nucleotidase using a centrifugal analyzer, Clin Chim Acta 119:275-284, 1982.
33. Sakai, T, Yanagihara, S, and Ushio, K: Determination of 5′-nucleotidase activity in human erythrocytes and plasma using high-performance liquid chromatography, J Chromatogr 239:717-721, 1982.
34. Ellis, G, and Goldberg, DM: An improved kinetic 5′-nucleotidase assay, Anal Letters 5:65-73, 1972.
35. Bodansky, O, and Schwartz, MK: 5′-Nucleotidase, Adv Clin Chem 11:277-328, 1968.
36. Ellis, G, and Goldberg, DM: Activity and stability of glutamate dehydrogenase in various buffer systems, Clin Chim Acta 39:472-474, 1972.
37. Sigma Technical Bulletin, No. 265-UV (1-82).
38. Ismail, AAA, and Williams, DG: Scope and limitations of a kinetic assay for serum 5′-nucleotidase activity, Clin Chim Acta 55:211-216, 1974.
39. Goldberg, DM, and Ellis, G: Route determination of 5′-nucleotidase activity of human serum using the LKB 8600 reaction rate analyser, J Clin Pathol 25:907-909, 1972.
40. Harvey, MS, van der Stoel, AG, and Backer, ET: Determination of serum 5′-nucleotidase with a centrifugal analyzer, Clin Chem 25:918-923, 1979.
41. Hathaway, JA: Adaptation of a NADH-linked serum 5′-nucleotidase procedure to a small benchtop centrifugal analyzer, Clin Chem 27:1950-1951, 1981.

CHAPTER 60

Hormones and their metabolites

β-*hCG* (β-*human chorionic gonadotropin*)

DAVID C. HOHNADEL
LAWRENCE A. KAPLAN

Human chorionic gonadotropin, hCG, HCG, β-hCG
Clinical significance: p. 569
Molecular weight: 47,000 daltons
Chemical class: glycoprotein, hormone

PRINCIPLES OF ANALYSIS

Human chorionic gonadotropin (hCG) is a glycoprotein composed of two noncovalently linked polypeptides: the alpha and beta subunits. The individual subunits lack biological activity but become active when linked to form the intact complex. For clarity and for historical reasons, we will use hCG as the abbreviation for this hormone rather than HCG or β-hCG. The designation β-hCG is used to represent specifically the beta subunit of hCG, not the entire hormone. The term *human,* although usually superfluous in a clinical setting, is in common use with chorionic gonadotropin and with growth hormone and is indicated with a lowercase *h*. All other hormone references are in uppercase letters and do not include a species designation.

The polypeptide chains contain various amounts of carbohydrate moieties, including D-galactose, D-mannose, and *N*-acetylneuraminic acid, the last of which is essential for the biological activity of the hormone.[1,2] The α-hCG subunit is essentially identical to the alpha chain of several other pituitary polypeptide hormones, such as thyroid-stimulating hormone (TSH), follicle-stimulating hormone (FSH), and luteinizing hormone (LH). It is the individual beta subunit that gives each of these hormones its specific biological characteristics.

There is also a great degree of similarity between the beta subunit of hCG and the beta subunit of LH. This similarity of structure makes the hCG protein analytically difficult to distinguish from LH. Initial assays specific for the beta subunit of hCG often had significant amounts of cross-reactivity with LH, depending on the area of the β-hCG to which the antibody was directed. The 30 carboxy-terminal amino acids of the β-hCG chain are the most divergent from LH, and many current β-hCG antibodies are directed toward this amino acid region.[3]

Since it is the beta chain that specifies the biological activity of hCG, many current assays for hCG are designed to specifically detect this portion of the hCG molecule. Thus many commercial assays are known as β-hCG assays even though they may be specific for the intact hCG molecule, the β-hCG subunit, or both. Because of this difference in antibody specificity, it is possible to obtain different results from the same samples with different hCG kits, even if they are standardized with the same standard material.

hCG is synthesized by the placenta, and it appears in urine and serum relatively soon after implantation of the developing embryo. Thus the presence of this hormone serves as the basis for pregnancy testing. In an acute care setting, knowledge of pregnancy is needed within 5 to 15 minutes of sample receipt.

Early bioassays measured the response of gonadal tissue to hCG in various animals (mice, rabbits, frogs). These assays were possible because of the large amounts of hCG found in the urine of pregnant women. The assays were insensitive (2 to 5 U/mL), expensive, and tedious to perform. At present, these assays are available only in specialized research centers.

Qualitative assays

Previous typical methods developed for measurement of urinary hCG are commonly referred to as either slide tests or tube tests. More recently, concentration tests (methods 1 to 3, Table 60-1) have been developed.

Current tube and slide tests use the agglutination- (flocculation-) inhibition method. The agglutination-inhibition procedures use either latex particles or sheep red blood cells coated with the hCG molecules. These particles are mixed with the urine sample and then with a solution containing antibodies to the hCG subunit. In the absence of urinary hCG, the antibody reacts with the hCG-coated particles and causes agglutination. When hCG is present in the urine sample, it will react with and neutralize the antibody, thus inhibiting particle agglutination. In these tube tests a positive urine test thus results in the formation of a visible ring of particles at the bottom of the reaction tube. A negative urine test will not have the visible ring, but a button of agglutinated particles will settle at the bottom of the tube (Fig. 60-1). In the slide tests, urine positive for hCG will remain a homogeneous suspension, whereas negative urine shows a clumped, heterogeneous suspension of

Table 60-1 Methods for hCG measurements

Method	Type of analysis	Principle	Use	Comments
Qualitative assays				
1. Slide tests	Agglutination inhibition	Colored latex or other visible particles (red blood cells) coated with hCG, antibodies to hCG, and urine are mixed with particles. Negative urine results in visible agglutination; presence of hCG in urine inhibits agglutination (or protein flocculation).	Was frequently used as stat urinary pregnancy test; urine	Least sensitive of all hCG methods; most rapid (2-3 min)
2. Tube tests	Same as method 1	Same as method 1; reaction occurs in tube.	Was sometimes used for stat urine pregnancy tests; urine	More sensitive than slide; some approach upper limit of sensitivity of RIA methods; 45-120 min per assay
3. Immunoenzymatic concentration tests	Sandwich immunometric assay	Solid-phase, double-antibody sandwich ELISA in which hCG binds to antibody. Enzyme-labeled antibody added, and residual activity directly related to hCG concentration.	Has become assay of choice, has speed of "slide" and sensitivity of "tube" with a colored end point; urine and serum	Reported sensitivity 20-50 mU/ml, 5- to 15-min assay; many forms: membrane, bead, paddle, dipstick, coated tube
Quantitative assays—serum and urine				
4. Radioimmunoassay (RIA)	Competitive inhibition	Radiolabeled (radioactive iodine, ^{125}I) hCG competes with sample analyte for binding to anti-hCG. Increased hCG in sample, decreased bound radioactivity.	Infrequently used as stat procedure; serum or urine	Most sensitive hCG assay available; 40-60 min per assay
5. Enzyme-linked immunosorbent assay (ELISA)	Sandwich immunometric assay	Enzyme-labeled anti-hCG reacts with sample hCG bound to solid-phase anti-hCG. Amount of bound enzyme activity directly proportional to amount of hCG in sample.	Most frequently used assay; serum and urine	Reported sensitivity of 2-10 U/mL; assay time 1-3 hr
6. Radioreceptor assay (RRA)	Competitive inhibition	Radiolabeled hCG competes with sample analyte for binding to tissue-receptor sites. Increased hCG in sample, decreased bound radioactivity.	Infrequently used as stat procedure; serum	Not as sensitive as RIAs; more sensitive than tube tests

agglutinated particles (Fig. 60-2). The agglutination-inhibition tube tests typically require approximately 60 to 90 minutes for a negative result, although a positive result can be detected sooner. A few methods take as long as 120 minutes for a negative result. The agglutination-inhibition slide tests are complete within 2 to 3 minutes.

The newest qualitative tests for serum and urine hCG have been called concentration tests for want of a better term because any hCG in the sample is concentrated in a small area on a surface. These procedures, a widely divergent group of assays with many different physical forms, are all variants of the sandwich immunoenzymometric as-

Fig. 60-1 Hemagglutination "tube" test for urinary hCG. Tube on left shows results of urine positive for hCG. Tube on right shows results of urine hCG negative for hCG.

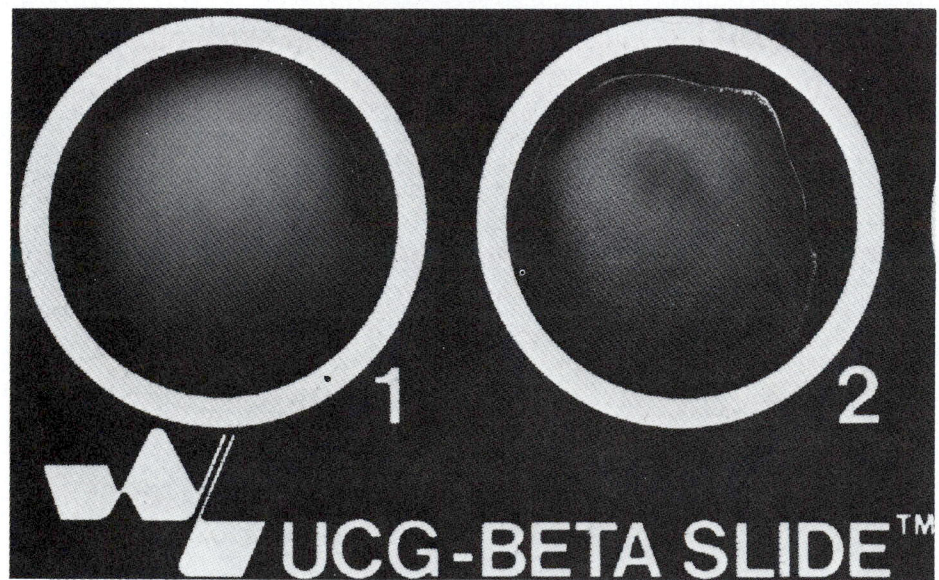

Fig. 60-2 Agglutination "slide" test for urinary hCG. *Left,* Test 1 shows results for positive urine. *Right,* Test 2 shows results for urine negative for hCG.

say. In virtually all cases, the assay (method 3, Table 60-1) can be described as follows: A monoclonal antibody to the α-hCG subunit (or the intact hCG or the β-hCG subunit) is bound to a solid surface. The surface can be a membrane (as in the Tandem ICON, Abbott Laboratories' TestPack, or Monoclonal Antibodies' RAMP), a plastic bead (Tandem V-hCG), a plastic dipstick (Monoclonal Antibodies' PregnaSTICK), a plastic paddle (Roche's Sensi-Chrome and Quidel's QUEST), or a coated tube (Monoclonal Antibodies' ModEL). Any hCG in the sample will be bound to the surface antibody. After incubation with the sample, the surface is washed, and then a second, enzyme-tagged monoclonal antibody to the β-hCG subunit is added. A substrate for the enzyme reaction is added, and development of a color (usually blue) is determined visually (see, for example, Fig. 60-3). Only samples with hCG

A

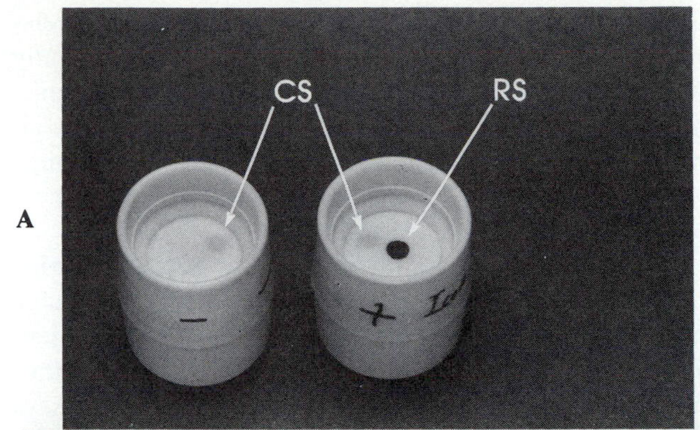

B

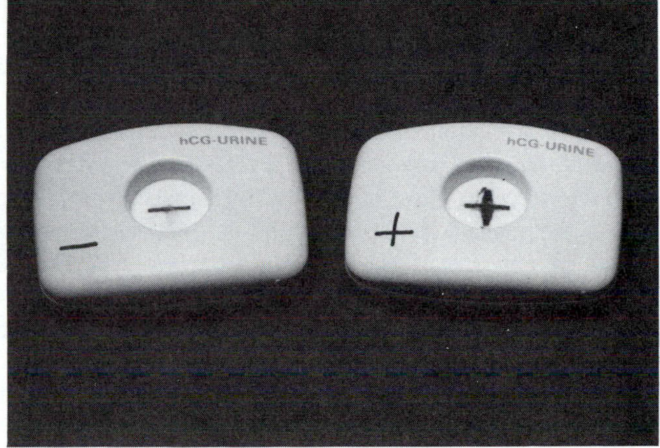

Fig. 60-3 Two examples of the recent concentration tests are shown. **A,** Tandem ICON hCG (Hybritech): *left,* negative urine; *right,* urine positive for hCG; *CS,* quality control spot; *RS,* positive reaction spot. **B,** Abbott Labs' TestPack hCG-Urine: *right,* urine positive for hCG; *left,* negative urine.

present can form the sandwich (first antibody–hCG–enzyme-linked second antibody) and ultimately any color. As with any qualitative assays that develop color, one must confirm that all the technologists performing the test can see a positive result. There is a male and female subpopulation with reduced blue visual acuity.

In most cases, these assays have been designed to be specific for intact hCG and to react within a few minutes at sensitivities of 20 to 50 mU/mL. This is an area of testing currently undergoing constant improvement and change. This type of sandwich assay has supplanted the stat pregnancy testing previously performed by slide and tube tests.

Quantitative assays

There are several radioimmunoassays (RIAs) available for measurement of serum and urine hCG levels (method 4, Table 60-1). The RIAs are typical competitive binding assays in which the hCG in the sample competes with hCG labeled with radioactive iodine for binding sites on an hCG antibody. Both solid-phase and double-antibody procedures are available. Some quantitative RIA procedures were modified to shorten the reaction time and allow the procedure to be used as a qualitative, stat pregnancy test. However, with the appearance of good qualitative immunoenzymometric sandwich assays with the appropriate specificity, sensitivity, and speed, the use of RIA for stat testing has become rare.

A second type of immunoassay uses solid-phase, and double-antibody, sandwich enzyme-linked immunosorbent assay (ELISA) technology (method 5, Table 60-1). One method (Abbott Laboratories, Inc., North Chicago) uses horseradish peroxidase coupled to anti-hCG. This reacts with the hCG in the serum sample, which, in an initial step, had reacted with anti-hCG antibodies coated on polystyrene beads. After the reaction is completed, the unbound anti-hCG–peroxidase complexes are removed from the tube, and the horseradish peroxidase bound to the bead is allowed to react with added substrate. The chromogen (*o*-phenylenediamine) is oxidized and measured at 492 nm. As the levels of hCG increase in the sample, the amount of chromogen formed increases.

Hybritech (San Diego) uses a bead coated with a monoclonal antibody that binds hCG present in the sample. After washing, a monoclonal antibody–alkaline phosphatase conjugate is added, and the amount of enzyme activity left after washing is related to the amount of hCG present in the sample. The sensitivity claimed for the ELISA assays (approximately 2 mU/mL) is similar to the sensitivity of RIAs.

Another type of assay system available for serum hCG measurements is a radioreceptor assay (RRA, method 6, Table 60-1). This method (Biocept-G, Wampole Laboratories, Cranbury, NJ) is based on the competition between the hCG in the serum sample and a ^{125}I-labeled hCG for binding to hCG receptors present in tissue (such as bovine corpus luteum cell membranes). The RRA technique is not widely used.

Standardization with this first International Reference Preparation (IRP) for chorionic gonadotropin has helped to produce more uniform results. Unfortunately, not all manufacturers have adopted this material as a standard. We hope that those manufacturers who have not adopted this standard can be encouraged to do so in the near future. There are problems associated with assays for hCG that a pure standard cannot address. The hCG is a complex protein with carbohydrate components that are necessary for biological activity. The different immunological assays detect different parts of the hCG molecule and may or may not be sensitive to the presence of free subunits in a biological sample. A variety of kits may give the same answer with a pure standard and quite divergent results with biological samples. In normal pregnancies, serum contains intact hCG and little, if any, free α- or β-hCG subunits. In urine, intact hCG is present and, depending on the gestational time, both free α- and β-hCG may appear, as well as desialated fragments of hCG. The relative concentrations of intact and subunit hCG have been found to vary greatly in patients with trophoblastic disease and ectopic pregnancies. Some patients seem to have a balanced production; some do not. Many tumors are characterized by an unbalanced production of subunits.[4] Thus it may be useful to have several assays available with different known specificities to help delineate various clinical problems.

It is desirable that hCG assays detect a pregnancy as soon after conception as possible. The clinical sensitivity required for this will depend on the levels of hCG present in early pregnancy. RIA has sufficient sensitivity to detect serum hCG levels (approximately over 0.02 U/mL) by the third to fourth week after the last menstrual period (LMP) (Fig. 60-4).[5] This is approximately 1 to 2½ weeks of gestation. The serum and urine hCG levels increase rapidly, peaking at approximately 8 weeks after the LMP, and then fall off for the remaining 32 weeks of pregnancy.

However, the serum levels present in an ectopic (tubal) pregnancy are often considerably lower than those present in a uterine pregnancy.[6] By 4 weeks of gestation the hCG levels are approximately the same for both types of pregnancies, but after 4 weeks the hCG levels fall off for the ectopic pregnancy, especially after a fallopian tube ruptures. Thus, after 4 to 8 weeks of ectopic gestation, serum hCG levels can range from less than 0.02 to 20 U/mL, whereas for normal pregnancies the serum hCG levels will be from two to two and one-half times higher.

Urine levels of hCG can range from 0.05 to 5 U/mL during the 4 to 8 weeks after the LMP. Urine hCG levels are normally not detected by slide or tube tests until the fourth to fifth weeks after the LMP. This level of sensitivity is unacceptable for best clinical practice.

Choosing a method for stat pregnancy testing is no longer difficult. Although the slide and tube tests are relatively inexpensive and use stable reagents, the currently available methods have poor sensitivity (100 to 500 mU/mL) and specificity (polyclonal antibodies). RIAs, although specific and sensitive, require expensive equipment and unstable reagents (iodinated tracers) that make them unsuitable for 24-hour stat testing.

The availability of ELISA assays for serum and urine testing has decided the issue. Many assays use double monoclonal antibodies[7-9] for specificity and have good sensitivity (20 to 50 mU/mL). Several assays designed to produce a result in 3 to 15 minutes are available. These assays are ''positive'' at hCG concentrations above the normal low concentrations of hCG found in nontrophoblastic tissue.[10] These assays require no specialized equipment, and their simplicity has made them widely accepted.

The confirmation of pregnancy is now possible within 24 to 48 hours of implantation, at a time several days before the expected onset of the next menstrual period. Thus the recommended procedure for stat pregnancy testing would be any one of the sensitive serum or urine ELISA procedures. An RIA or ELISA procedure should be available for quantitation of serum hCG measurements for those conditions that require such monitoring.

A comparison of the performance characteristics of qualitative assays in current use for serum and urine hCG analysis is found in Table 60-2.

SPECIMEN

Urine

Centrifuge random urine samples at 900 *g* (approximately 1500 to 2000 rpm with a tabletop centrifuge) for 10 minutes. Urinary hCG levels are highest in an early-morning specimen, approximating serum levels.

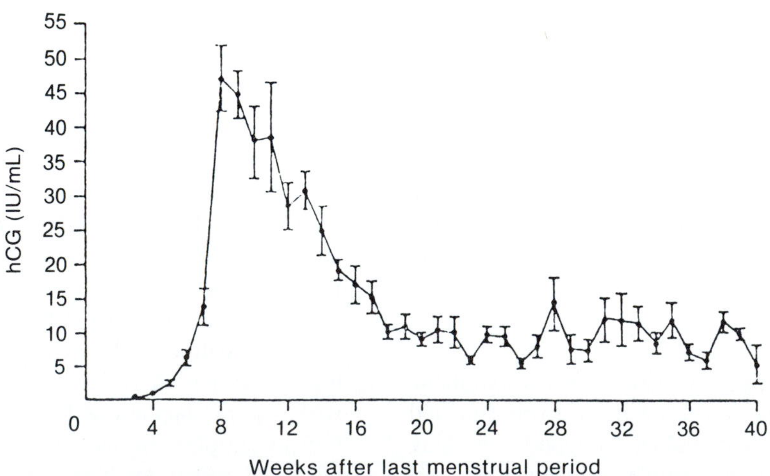

Fig. 60-4 Mean serum hCG levels throughout normal pregnancy. Arithmetical scale used on ordinate. Bars represent ±1 standard error of the mean. *(From Braunstein, GD, et al: Am J Obstet Gynecol 126:680, Nov 1976.)*

Table 60-2 Performance characteristics of 11 representative qualitative urine and serum hCG enzyme immunoassays*

Characteristic	1	2	3	4	5	6	7	8	9	10	11
Sample	Urine/serum	Urine	Urine/serum	Urine/serum	Urine/serum	Urine	Serum	Urine	Urine	Urine	Serum
Sample volume	0.1 mL	1 mL	2 drops	0.5 mL	2 drops	5 drops	50 μL	1 mL	5 drops	5 drops	10 drops
Sensitivity (mU/mL)	50/25	175	50	50	50	50	25	50	50	20	20
First antibody											
Type	Monoclonal	Monoclonal	Polyclonal	Monoclonal	Monoclonal	Monoclonal	Monoclonal	Monoclonal	Monoclonal	Monoclonal	Monoclonal
Specificity	i-hCG	α-hCG	α-hCG	β-hCG	α-hCG	β-hCG	α-hCG	β-hCG	β-hCG	α-hCG	α-hCG
Time of treatment	40 min	20 min	10 min	15 min	10 min	3 min	90 min	22 min	4 min	4 min	6 min
Second antibody											
Type	Monoclonal	Monoclonal	Monoclonal	Monoclonal	Monoclonal	Monoclonal	Monoclonal	Monoclonal	Monoclonal	Monoclonal	Monoclonal
Specificity	β-hCG	β-hCG	β-hCG	β-hCG	α- and β-hCG	β-hCG	β-hCG	β-hCG	α-hCG	β-hCG	β-hCG
Internal control	No	No	No	Yes	No	No	No	No	Yes	Yes	No
Indicator	Blue	Blue	Blue	Blue	Blue	Blue	Blue	Blue	Blue	Blue	Blue
Assay type	Tube	Dipstick	Coated tube	Dipstick	Coated tube	Membrane	Coated tube	Paddle	Membrane	Membrane	Membrane
Test name	PREGNO-LISA	Pregna-STICK	NIMBUS	QUEST	ECHO-Clonal	RAMP	ModEL	Sensi-Chrome	TestPack	ICON II	ICON
Reagent storage	RT	2°-8° C	2°-8° C	RT	2°-8° C	2°-8° C	2°-8° C	2°-8° C	2°-8° C	RT	2°-8° C
Standard	2nd IS	2nd IS	2nd IS	2nd IS	1st IRP	1st IRP	1st IRP	2nd IS	1st IRP	1st IRP	1st IRP

There are many quantitative assays available, including radioimmunoassay, radioreceptor assay, immunoradiometric assay, and immunoenzymometric assay.
i-hCG, Intact hCG; *RT*, room temperature; *IS*, International Standard; *IRP*, International Reference Preparation.
*Many other, less sensitive qualitative assays are available.

Serum

Obtain blood without the use of anticoagulants. Remove the serum from the clot as soon as possible.

Analyses for both urine and serum should be performed as soon as possible. For any undue delay, especially for serum analysis, the sample should be frozen at $-20°$ C.

REFERENCE RANGE

The absolute amount of hCG present in the serum during pregnancy varies greatly with gestational age and between patients. Following is a table of values for the first trimester:

Serum hCG levels with gestational age

Gestational age	hCG (mU/mL)
0.2-1 week	5-50
1-2 weeks	50-500
2-3 weeks	100-5,000
3-4 weeks	500-10,000
4-5 weeks	1,000-50,000
5-6 weeks	10,000-100,000
6-8 weeks	15,000-200,000
2-3 months	10,000-100,000

Within the first trimester, the hCG rises, doubling approximately every 2.2 days ($\pm$0.5 day is 1 SD).[4,] A slower rise may be associated with a higher risk of abortion. Values decline to 10% to 15% of the peak concentrations for the remainder of pregnancy with a smaller, second hCG surge in the third trimester.[5]

REFERENCES

1. Ross, GT: Clinical relevance of research on the structure of human chorionic gonadotropin, Am J Obstet Gynecol 129:795-805, 1977.
2. Orten, JM, and Neuhaus, OW: Human biochemistry, ed 10, St Louis, 1982, The CV Mosby Co, p 608.
3. Birken, S, Canfield, R, Agosto, G, and Lewis, J: Preparation and characterization of an improved immunogen for generation of specific and sensitive antisera to human chorionic gonadotropin, Endocrinology 110:1555-1563, 1982.
4. Schwarz, S, Berger, P, and Wick, G: Epitope-selective monoclonal antibody based immunoradiometric assay of predictable specificity for differential measurement of choriogonadotropin and its subunits, Clin Chem 31:1322-1328, 1985.
5. Braunstein, GD, Rasor, J, Adler, D, et al: Serum human chorionic gonadotropin levels throughout normal pregnancy, Am J Obstet Gynecol 126:678-681, 1976.
6. Romero, R, et al: The effect of different human chorionic gonadotropin assay sensitivity on screening for ectopic pregnancy, Am J Obstet Gynecol 153:72-74, 1985.
7. Demers, L: Pregnancy testing, Endocrinol Metab Cont Educ Prog (AACC) 4:1-5, 1986.
8. Wenk, RE: The perfect pregnancy test: special topics no. ST 85-6(ST-147), Check Sample Cont Educ Prog (ASCP) 23:1-6, 1985.
9. Valkirs, GE, and Barton, R: ImmunoConcentration™—a new format for solid-phase immunoassays, Clin Chem 31:1427-1431, 1985.
10. Vaitukaitis, JL: Human chorionic gonadotropin: a hormone secreted for many reasons, N Engl J Med 301:324-325, 1979.

Estriol

LAWRENCE A. KAPLAN

Estriol, 1,3,5,(10)-estratriene-3,16α, 17β-triol, E_3
Clinical significance: pp. 569 and 650
Molecular formula: $C_{18}H_{24}O_3$
Molecular weight: 288.37 daltons
Merck Index: 3654
Chemical class: steroids

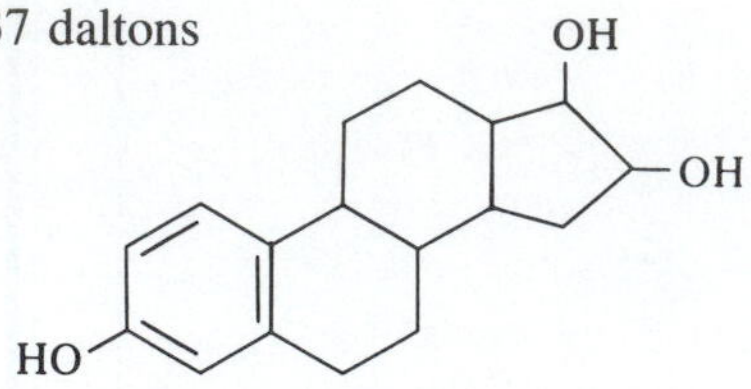

OH
OH
HO

PRINCIPLES OF ANALYSIS

Estriol is the major estrogen steroid produced during pregnancy. The concentration of plasma and urinary estriol rises a thousandfold during pregnancy, and by the third trimester estriol represents 85% to 90% of the C_{18} estrogenic steroids present in plasma and urine. The serum and urine levels of the estriol derivatives are compared in Table 60-3. Urinary estriol is present only in the conjugated forms because the protein-bound, unconjugated estriol is not filtered at the glomerulus.

Estriol is most frequently measured as part of the biochemical monitoring of high-risk pregnancies, and thus the sample analysis time required before a physician can obtain a result is an important consideration when one is choosing a method. The type of method used will depend most importantly on the decision to monitor estriol in either serum or urine. In either case, the monitoring is usually initiated in the beginning of the third trimester, when estriol levels rise to measurable amounts. The absolute measurement at any one time is usually less important than changes over time.

The urinary procedures were traditionally performed on a 24-hour sample. This is a difficult specimen to collect accurately and repeatedly. When reported as milligrams of estrogen per 24 hours, the collection error is added to the already significant biological and analytical errors. The introduction of reporting estrogen output normalized to creatinine output, the estrogen/creatinine (E/C) ratio, improved the assay by eliminating the need for an accurate,

Table 60-3 Estriol metabolites present in plasma and urine[1]

Estrogen	Urine (%)	Plasma (%)
Unconjugated estriol	0	9
Estriol-3-sulfate	4	15
Estriol-3-glucuronide	13	15
Estriol-16-glucuronide	73	21
Estriol-3-sulfate-16-glucuronide	10	40

timed collection. The premise of the E/C ratio yields results as valid as results obtained from the longer collection.[1]

Urinary estriol output depends not only on fetal and placental biochemical pathways that result in the synthesis of estriol but also on maternal hepatic conjugation and renal clearance of the estriol. It has therefore been suggested that monitoring urinary estrogen is less valid than measurement of serum estriol for the assessment of the status of the fetoplacental unit (fetus and placenta) because of the dependence of the urinary measurement on both maternal and placental function.[2] The measurement of plasma estriol has been recommended because it monitors only the fetoplacental unit. Unconjugated estriol, with a half-life of only 20 minutes, reflects most specifically the output of the fetoplacental unit at the time of phlebotomy.[3]

Few studies have demonstrated an overwhelming utility of one measurement (urine versus plasma) over the other. Many studies have demonstrated a good correlation between plasma estriol and urinary estriol measurements; included are studies both on entire populations and on individual, high-risk pregnancies.[3] Estriol output by the fetoplacental unit is variable, and circadian rhythms and hour-to-hour fluctuations have both been noted.[4,5] These variations appear to affect the concentration of plasma estriol more than urinary estriol. Urinary estrogens, the E/C ratio, and total serum estriol seem to have maximum values at around midday with minimum concentrations noted about midnight; the daytime concentrations for serum free estriol are greater than the nighttime values.[4] The daily fluctuations in total serum estriol appear greater than those of free serum estriol, although the percentage of fluctuation for both appears to be similar.[3] The individual diurnal variations of plasma estriol can vary from 15% to 45%, with total urinary estrogens varying from 30% to 40% during the day.[5] However, when the estrogen output is reported as an E/C ratio, the variation is reduced to 12% to 15%.

Urine

The estrogen composition of urine of a pregnant woman in the third trimester is approximately 80% to 95% estriol conjugates (sulfate and glucuronate) with the remainder consisting of conjugates of estrone and estridol.[6] The oldest, and still sometimes used, procedure for the analysis of estrogen in urine is the colorimetric reaction developed by Kober in 1931.[7] This method is based on the reaction of all estrogens with sulfuric acid to form a pinkish red Kober chromogen (method 1, Table 60-4). Since most of the urinary estrogen in the third trimester is estriol, this assay is effective in monitoring changes in excretion of estriol during pregnancy although it is not specific for this compound.

The Kober chromogen in aqueous solutions has an absorbance maximum at 514 nm, and the Allen correction measurements are made at 472 and 556 nm to enhance selectivity.[8]

An important modification of the Kober reaction was the extraction of the Kober chromophore into chloroform (containing 2% *p*-nitrophenol plus 1% ethanol), the so-called Ittrich extraction.[9] This extraction step was designed to remove the estrogen reaction product from other, more water-soluble, interfering chromogens. The Ittrich chromogen can be measured colorimetrically or fluorometrically. In chloroform, the chromophore has an absorbance maximum between 530 and 540 nm[8,9] (method 2, Table 60-4). The Allen correction has also been applied to the chloroform extract, with the secondary absorbance measurements made at 505 and 565 nm.[17] When the extracted chromophore is excited at 530 nm, an intense, yellow-green fluorescence, which can be monitored at 550 nm,[9,10] is produced.

Urinary estriol can be specifically quantitated by a variety of chromatographic techniques, all of which require a hydrolysis step to convert estriol to the unconjugated form. In addition, all chromatographic assays require an extraction and partial purification of the estrogens before analysis. Liquid-liquid partitioning, column extractions, and carbon black adsorption[11] have been used successfully. Gas-liquid chromatographic techniques usually require an extraction of the hydrolyzed estrogens before volatile derivatives, such as acetate, methyl, and tetramethylsialyl conjugates, are formed. Most separations have been performed on 3% OV-1 columns, with nitrogen gas as the mobile phase and flame ionization as the detection method.[12] Some methods use temperature programming (method 3, Table 60-4). High-performance liquid chromatography (HPLC) analyses of urinary estriol using reversed-phase columns have been reported. These methods use several different types of detection methods (method 4, Table 60-4), including measurement of absorbance at 280 nm,[13] measurement of native fluorescence at 308 nm after excitation at 220 nm,[14] and derivatization with dansyl chloride and measurement of the fluorescent derivative.[15]

Immunoassays developed for measurement of urine estriol include radioimmunoassay (RIA) and enzyme immunoassays (EIA) (methods 5 and 6, Table 60-4). These assays require a hydrolysis step to convert estriol conjugates to the free form. Several RIAs have been reported using tritium (^{3}H) or radioactive iodine (^{125}I) labeling and using a variety of techniques for separating bound from free label, including dextran-coated charcoal,[16] double-antibody precipitation,[17] and ammonium sulfate precipitation.[18] The EIA is a heterogeneous assay requiring polyethylene glycol to separate the antibody-bound estriol-enzyme conjugate from the free label.[19] The enzyme label used in these assays is horseradish peroxidase with 5-aminosalicylic acid as a substrate. The reaction is monitored as a 60-minute end-point analysis at 463 nm.

A fluorescence polarization assay (method 7, Table 60-

Table 60-4 Methods of estriol analysis

Method	Type of analysis	Principle of analysis	Use	Comment
Urine				
1. Kober reaction[7]—direct or indirect (prior extraction)	Spectrophotometric	Estriol heated in presence of sulfuric acid and hydroquinone forms pinkish red chromogen (absorbance maximum, 514 nm) Reaction can be performed with (indirect) or without (direct) prior extraction of estrogen	Historically most popular method Infrequently used today	Measures total estrogen Direct assay can have negative interference by glucose
2. Ittrich[9]	Fluorometric	Chromophore extracted and measured fluorometrically (excitation, 530 nm; emission, 550 nm) or colorimetrically	Rare	Measures total estrogen Automated on continuous-flow analyzers
3. Gas-liquid chromatography[12]	Flame ionization detector	Volatile derivatives chromatographed on OV-1 with N_2 mobile phase and flame ionization detector	Rare	Requires hydrolysis to free estriol Very specific
4. High-performance liquid chromatography (HPLC)[13-15]	Ultraviolet or fluorescence detection	Estriol chromatographed by reversed-phase (C_{18}) chromatography, detected at 280 nm (ultraviolet); fluorescence (excitation, 220 nm; emission, 608 nm) Dansyl derivatives detected by fluorescence	Rare	Requires hydrolysis to unconjugated estriol Very specific
5. Radioimmunoassay (RIA)[16-18]	Liquid scintillation or gamma-ray counting	Competitive binding assay, employing ^3H- or ^{125}I-labeled estriol* Free from bound label separated by second antibody, $(NH_4)_2SO_4$ or dextran-charcoal	Frequent	Requires hydrolysis to free estriol Specific
6. Enzyme immunoassay (EIA)[19]	Spectrophotometric, end point	Competitive binding assay; estriol-enzyme conjugate separated from free enzyme and enzyme reaction (peroxidase) monitored at 463 nm	Rare	Measures total estriol after hydrolysis
Plasma				
7. RIA[5,22,23]—free or total estriol	See method 5	See method 5 Extraction required for free estriol, hydrolysis for total	Most frequent	Very specific Automated
8. EIA[24,25]	See method 6	See method 6	Rare	See method 6
9. HPLC[26,27]	Electrochemical or fluorescence	Extracted unconjugated estriol chromatographed by reversed-phase chromatography and detected by electrochemical or fluorescence detection	Rare	Very specific
10. Fluorescence polarization immunoassay[28]	Fluorescence	Fluorescent-labeled estriol competes with sample estriol for binding sites; free label results in decreased polarized fluorescence	Rare	Measures total estriol in plasma and urine

*^3H, Tritium; ^{125}I, radioactive iodine.

4) for the measurement of total estriol in serum, plasma, or urine is available on the TDx (Abbott Diagnostics, North Chicago).[20] Like other fluorescence polarization assays on the TDx (see Chapter 11), this method uses a fluorescein conjugate that competes with the analyte (estriol) for binding sites on antibodies. This assay, which is partially automated on the TDx, requires no extraction of the sample. Hydrolysis of total (conjugate and free) estriol is achieved by the use of *Escherichia coli* glucuronidase to hydrolyze the glucuronide conjugate.

A luminescence immunoassay has been developed for the measurement of total urinary estrogens.[21] After enzymatic treatment of the urine to hydrolyze the conjugates, the estrogens are allowed to compete for antibody binding sites with 17β-estradiol conjugated to aminobutylated isoluminol. The amount of light emitted in the subsequent luminescence assay is inversely related to the concentration of total estrogens in the sample.

Plasma estriol

The thousandfold lower estriol concentration in serum as compared with urine has, for the most part, precluded the use of colorimetric or fluorometric assays[16] for analysis of blood estriol. Most of the assays for serum estriol, whether one is measuring total or only the unconjugated fraction, have been RIAs (method 8, Table 60-4). The RIAs use either ^{3}H- or ^{125}I-estriol as the tracer and use a variety of procedures[3,5,22,23] to separate bound from free label, including polyethylene glycol, ammonium sulfate, a second precipitating antibody, and solid-phase bound systems. The use of the last technique has led to the automation of the assay. Measurement of total serum estriol requires the hydrolysis of estriol conjugates to the free form before the analysis; either acid or enzyme hydrolysis is used.

Since there can be significant antibody cross-reactivity between unconjugated estriol and its conjugated forms (present in five to ten times higher concentrations), the analysis for free estriol often requires separation of the free estriol from its conjugates. This is most frequently accomplished by a liquid-liquid extraction step, in which the less water-soluble, unconjugated estriol is extracted into the organic phase with high efficiency. Alternatively, gel exclusion chromatography (such as Sephadex) has been used for this purpose. RIAs that directly measure serum unconjugated estriol without a prior extraction step are also available.[23] These methods use solid-phase procedures[23,29] and double-antibody techniques to separate bound from free label and can be automated as well.[23]

The measurement of total serum estrogens by EIA has been reported.[24] These assays are not specific for estriol but measure total serum estrogens, including conjugates of estriol. These assays use a horseradish peroxidase antibody label, and the reaction is monitored at 463 nm (method 8, Table 60-4). An EIA that allows direct measurement of serum unconjugated (as well as conjugated) estriol has been developed.[25] This assay uses a double-antibody precipitation technique to separate bound from free tracer. One measures the tracer, estriol conjugated to alkaline phosphatase, that is bound to the precipitated antiestriol antibody by following the hydrolysis of *p*-nitrophenyl phosphate at 400 nm.

HPLC assays for unconjugated serum estriol have also been reported. The estriol is extracted by liquid-liquid partitioning or adsorption chromatography and is separated from interfering compounds by reversed-phase chromatography (method 9, Table 60-4). Detection and quantitation of the estriol peak are either by electrochemical analysis at +0.75 volts using a glassy carbon electrode[26] or by fluorescence at 308 nm after excitation at 280 mn.[27]

American Association for Clinical Chemistry (AACC) and College of American Pathologists (CAP) quality assurance surveys from 1988 indicate that approximately 90% of the participating laboratories are measuring serum estriol rather than total urinary estriol. Approximately two thirds of those laboratories that report measuring serum estriol are measuring the unconjugated estriol form, whereas the remainder are measuring total estriol.

There is no reference method for the measurement of either urine or serum estriol. The lack of a reference method may be partly ascribed to the lack of a consensus as to how to best monitor estrogens during pregnancy. The Kober methods are not specific and should not be used.

The estriol assay on the TDx is reported to have good recovery (average recovery, 98%) and precision (approximately 3% to 7% between-run coefficient of variation for estriol concentrations expected in a normal pregnancy).[28] The Abbott Company reports insignificant cross-reactivity with other estrogenic steroids and no interference from lipemia (up to 10 g of triglycerides/L), hemolysis (up to 10 g of hemoglobin/L), or bilirubin (up to 100 mg/L). Because the TDx has a good history of precise, reliable, and convenient analysis, one might expect that this procedure will become more widely used. Its ability to provide 24-hour availability of results makes it especially attractive.

Table 60-5 compares assays for urine and serum estriol.

SPECIMEN

Either 24-hour collections or single voidings are appropriate urine specimens. Serum or plasma specimens are both valid for blood analysis. When one is monitoring patients with consecutive sample analyses, the random urine or serum specimens should be obtained at approximately the same time of the day to minimize any error caused by biological variation. It has been shown that bacterial contamination of urine left unpreserved leads to an apparent increase in estrogens measured by an automated fluorometric assay.[29] Preservation of urine during collection with thimerosal (about 25 mg/L) or storage at 4° C effectively prevents changes in measured estrogen.

Table 60-5 Comparison of representative assays for urine and serum estriol

Parameter	Urine			Serum		
	Kober[7]	**HPLC[13]**	**RIA***	**RIA (total)†**	**RIA (free)‡**	**HPLC[26]**
Reaction temperature	100° C	60° C (hydrolysis)	27° C (hydrolysis)	37° C (hydrolysis)	Ambient	Ambient
Sample volume	2 mL	5 mL	20 μL	25 μL	250 μL	2 mL
Label or detection system	Spectrophotometric 514 nm	Spectrophotometric 280 nm	^{125}I	^{125}I	^{125}I	Chromatographic, electrochemical detection
Linearity	100 mg/L	100 mg/L	400 mg/L	80 μg/L	800 ng/L	0.4-8 ng
Assay time						
Incubation/extraction	20 min	60 min	15 min	75 min	2.5 hr/batch	25 min
Operational	40 min	15 min	90 min	30 min	—	25 min
Sensitivity	—	2-3 mg/L	3 mg/L	1000 ng/L	100 ng/L	1000 ng/L
Final concentration of reagents	Hydroquinone: 180 mmol/L H_2SO_4: 24.5 mol/L	—	90 × 10^3 dpm/tube§	50-70 × 10^3 dpm/tube	76 × 10^3 dpm/tube	—
Reported precision (between run): $\overline{X}$, %CV§	—	15.9 mg/L, 3.7%	100 mg/L, 4.3%	150 μg/L, 8.4%	3900 ng/L, 8.8%	14.9 μg/L, 9.9%
Interferences	Ampicillin	Ampicillin	Ampicillin	Ampicillin	Estriol-3-sulfate, ampicillin	Ampicillin

*Amersham Corp., Arlington Heights, IL.
†Nuclear Medical Systems, Inc, Newport Beach, CA.
‡Becton-Dickinson, Automated RIA, Rutherford, NJ.
§*dpm,* Disintegrations per minute; $\overline{X}$, mean; *CV,* coefficient of variation.

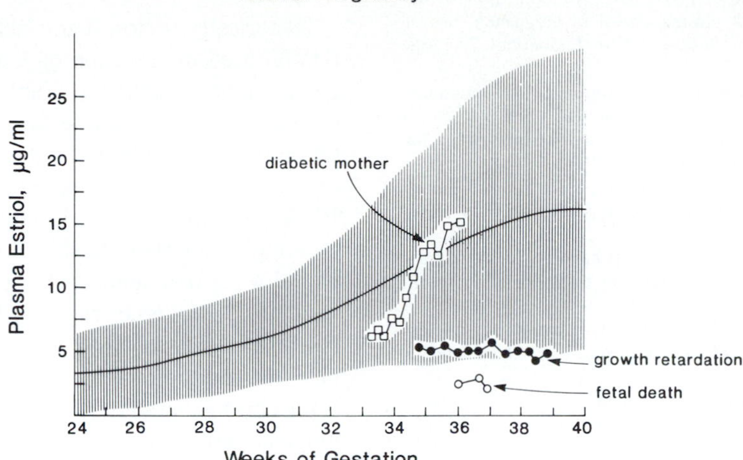

Fig. 60-5 Mean *(solid line)* and estimated 5th and 95th percentiles *(shaded area)* for plasma unconjugated estriol during normal pregnancy. Estriol patterns from three actual pregnancy conditions are shown.

Certain antibiotics, most notably ampicillin, have been shown to cause a decrease in the urinary excretion of estriol conjugates by reducing the intestinal bacteria that hydrolyze estriol conjugates.[3] This reduces the recirculation of estriol into serum and its eventual urinary excretion.

REFERENCE RANGE

The reference range for both urinary and serum and plasma estriol varies with gestational age (see Chapter 37). The biological variability of the fetoplacental unit (as the 95th percentile range) is very broad during the last month of pregnancy. This is illustrated in Fig. 60-5 for plasma estriol. The following reference ranges can serve to illustrate expected estriol values in the last trimester of pregnancy. Each laboratory must determine its own reference range. The normal ranges are shown in Table 60-6.

The levels of salivary estriol (all unconjugated) are approximately one tenth the levels of serum unconjugated estriol, with the plasma to salivary estriol concentrations ranging from 6.1 to 14.0 ng/mL.[30,31]

Table 60-6 Reported normal ranges for total estriol

Gestational week	Plasma, serum total estriol, ng/mL (nmol/L)	Urinary total estriol, mg/24 hr (µmol/24 hr)
28-30	38-140 (132-485)	5-18 (17-62)
30	31-140 (107-485)	5-18 (17-62)
32	35-330 (121-1144)	7-21 (24-73)
34	45-260 (156-902)	8-26 (28-90)
36	48-350 (166-1214)	10-30 (35-104)
38	59-570 (205-1977)	12-36 (42-125)
40	95-460 (329-1595)	13-42 (45-146)

REFERENCES

1. Dickey, RP, Grannis, GF, and Hanson, FW: Use of the estrogen/creatinine ratio and the "estrogen index" for screening of normal and "high-risk" pregnancy, Am J Obstet Gynecol 113:880-886, 1972.
2. Klopper, A: Criteria for the selection of steroid assays in the assessment of fetoplacental function. In Klopper, A, editor: Plasma hormone assays in the evaluation of fetal well-being, New York, 1976, Churchill Division of Longmans Press.
3. Kirkish, LS, Barclay, ML, Parra, JB, et al: Plasma estriol vs. estrogen assays in 24-hour urines as an index to fetal status, Clin Chem 24:1830-1832, 1978.
4. Townsley, JD, Dubin, NH, Grannis, GF, et al: Circadian rhythms of serum and urinary estrogens in pregnancy, J Clin Endocrinol Metab 36:289-295, 1973.
5. Katagiri, H, Distler, W, Freeman, RK, et al: Estriol in pregnancy. IV. Normal concentrations, diurnal and/or episodic variations, and day-to-day changes of unconjugated and total estriol in late pregnancy plasma, Am J Obstet Gynecol 24:272-280, 1976.
6. Goebelsmann, U: The uses of estriol as a monitoring tool, Clin Obstet Gynecol 6:223, 1979.
7. Kober, S: Ein kolorimetrische Bestimmung des Brunsthormons (Menformon), Biochem Z 239:209-212, 1931.
8. Grannis, GF, and Dickey, RP: Simplified procedure for determination of estrogen in pregnancy urine, Clin Chem 16:97-102, 1970.
9. Ittrich, G: Eine neue methode zur fhemischen Bestimmung der oestrogenen Hormone in Harn, Z Physiol Chem 312:1-14, 1958.
10. Adessi, G, and Jale, MF: Rapid fluorometric estimation of estrogens in urine after twenty weeks of pregnancy, Ann Biol Clin 30:127, 1972.
11. Andreslini, F, DiCorcia, A, Lagana, A, et al: Preliminary isolation of urinary placental estriol before gas or liquid chromatography, Clin Chem 29:2076-2078, 1983.
12. Gotelli, GR, Kabra, PM, and Marton, LJ: Determination of placental estriol in urine by gas-liquid chromatography with equilenin as internal standard, Clin Chem 23:165-168, 1977.
13. Gotelli, GR, Wall, JH, Kabra, PM, and Marton, LJ: Improved liquid chromatographic determination of placental estriol in urine, Clin Chem 24:2132-2134, 1978.
14. Taylor, JT, Krotts, JG, and Schmidt, GJ: Determination of urinary placental estriol by reversed-phase liquid chromatography with fluorescence detection, Clin Chem 26:130-132, 1980.
15. Schmidt, GJ, Vandmark, FL, and Slavin, W: Estrogen determination using liquid chromatography with precolumn fluorescent labeling, Anal Bioch 91:636-645, 1978.

16. Ertel, NH, Moskovitz, M, and Schiffer, MA: A modification of the rapid method for the assay of plasma estriol in pregnancy: use of unconjugated ^3H to correct for losses, J Clin Endocrinol 29:1266-1268, 1969.
17. Jawad, JJ, Wilson, EA, and Kincaid, HL: Improved radioimmunoassay for total urinary estriol, Clin Chem 25:99-102, 1979.
18. Anderson, DW, and Goebelsmann, U: Rapid radioimmunoassay for total urinary estriol, Clin Chem 22:611-615, 1976.
19. Korhonen, MK, Juntunen, KO, and Stenman, UH: Enzyme immunoassay of estriol in pregnancy urine, Clin Chem 26:1829-1831, 1980.
20. Vanderbilt, A, Spring, T, Fino, J, and Shipchandler, M: A fluorescence polarization immunoassay quantitating total estriol in serum and urine, Clin Chem 30:1043(A), 1984.
21. Messeri, G, Caldini, AL, Bolelli, GF, et al: Homogeneous luminescence immunoassay for total estrogens in urine, Clin Chem 30:653-657, 1984.
22. France, JT, Knox, BS, and Fisher, PR: Evaluation of a new commercial solid-phase direct radioimmunoassay for unconjugated estriol in pregnancy plasma, Clin Chem 28:2103-2105, 1982.
23. Jones, TB, and Hoffman, KL: An automated solid phase RIA for the direct measurement of unconjugated estriol in serum, Clin Chem 29:1268, 1983.
24. Osterman, TM, Juntunen, KO, and Gothoni, GD: Enzyme immunoassay of estrogen-like substances in plasma with polyethylene glycol as precipitant, Clin Chem 25:716-718, 1979.
25. Carter, JH, and DeBernardi, M: Double antibody enzyme immunoassays for unconjugated and total estriol in unextracted serum, Clin Chem 30:1043(A), 1984.
26. Kaplan, LA, and Hohnadel, DC: Measurement of unconjugated estriol in serum by liquid chromatography with electrochemical detection compared with radioimmunoassay, Clin Chem 29:1463-1466, 1982.
27. Andreolini, F, Borra, C, DiCorcia, A, et al: Improved assay of unconjugated estriol in maternal serum or plasma by adsorption and liquid chromatography with fluorometric detection, Clin Chem 30:742-744, 1984.
28. Vanderbilt, A, Spring, T, Fino, J, and Shipchandler, M: A fluorescence polarization immunoassay quantitating total estriol in serum and urine, Clin Chem 30:1043(A), 1984.
29. Simkins, A, and Crawley, M: Some chemical and bacterial contributions to analytical variations in urinary oestrogen quantitations during pregnancy, Med Lab Sci 35:325-334, 1978.
30. Truran, PL, and Read, GF: Automated continuous-flow radioimmunoassay for salivary estriol, Clin Chem 30:1678-1682, 1984.
31. Evan, JJ, Wilkinson, AR, and Aickin, DR: Salivary estriol concentrations during normal pregnancies and a comparison with plasma estriol, Clin Chem 30:120-121, 1984.

Homovanillic acid

STEVEN J. SOLDIN

Homovanillic acid, 3-methoxy-4-hydroxyphenylacetic acid, HVA

Clinical significance: p. 730

Molecular formula: $C_9H_{10}O_4$

Molecular weight: 182.17 daltons

Merck Index: 4638

Chemical class: substituted aromatic carboxylic acid (catecholamine metabolite)

H_3CO / HO — benzene ring — CH_2 — COOH (structure: H–C(–H)–COOH)

PRINCIPLES OF ANALYSIS

Historically, colorimetric analysis of homovanillic acid (HVA) used the reaction of 1-nitroso-2-naphthol with biogenic amines (method 1, Table 60-7). HVA formed a red chromogen, which was measured at 500 nm.[1] For preparation of a sample blank, potassium thiocyanate was added to remove the color that resulted from the reaction of HVA.[1] An improved colorimetric reaction was proposed by Knight and Hammond[2] who substituted 1-nitroso-2-naphthol-4-sulfonic acid for the previously used 1-nitroso-2-naphthol.[2] The assay can also be performed after the partial purification of HVA by anion-exchange chromatography. Another approach to quantitation was the use of thin-layer chromatography (TLC) performed on silica gel. The biogenic amines were measured by scanning reflectometry at 400 nm (method 2, Table 60-7). The compounds, which were allowed to photo-oxidize for 2 days, developed a red color.[3]

A gas chromatographic (GC) method was developed with *p*-hydroxyphenyl acetic acid used as an internal standard (method 3, Table 60-7). The sample was extracted with ether and derivatized with TRI-SIL/TBT (Pierce Chemical Co., Rockford, IL) forming the silyl derivative. The compounds were separated by gas chromatography with 3% OV-1 as the stationary phase, and the HVA was detected with a flame-ionization detector.[4] The HVA could also be measured by electron-capture detection.[5]

Gas chromatography–mass spectrometry was also used for quantitation of serum and urine levels of HVA (method 4, Table 60-7).[6] The HVA in the samples was first separated by use of Amberlite XAD-4 anion-exchange columns. The eluted compounds were converted to their trifluoroacetyl hexafluoroisopropanol ester derivatives, the derivatives were separated by GC, and the HVA derivative was detected by a mass spectrophotometer. The limit of detection was 2 ng/mL for plasma and 120 ng/mL for urine. HPLC has been used by many investigators for the measurement of HVA (method 5, Table 60-7). Usually HVA is first partially separated by anion-exchange chromatography on XAD-X4, DEAE-cellulose, or an equivalent resin, and then HPLC analysis is performed.[7-11] Final separation is accomplished by reversed-phase (C_{18}) chromatography, and detection is accomplished by ultraviolet spectrophotometry or amperometry. Many procedures allow the simultaneous measurement of HVA and vanillylmandelic acid (VMA).[7-10] HPLC methods are discussed in more detail in the section on VMA (p. 962).

The colorimetric methods are not specific for HVA and therefore cannot be recommended. The use of the 4-sulfonic acid derivative improves the stability of the chromogen formed on reaction of HVA.[2] The TLC method is slow and requires specialized equipment. The GC methods do work well for HVA, but the need for derivatization and the relatively poor sensitivity make them less likely to be adopted by many laboratories.

Table 60-7 Methods of homovanillic acid (HVA) analysis

Method	Principle	Usage	Comments
1. Colorimetric	Reaction of nitrosonaphthol to form blue compound with absorption maximum at 500 nm.	Historical	Lacks specificity
2. Thin-layer chromatography (TLC)	HVA is extracted and separated on silica gel; color developed by photo-oxidation, quantified by reflectance spectrophotometry.	Historical	Slow, not suitable for clinical laboratory
3. Gas chromatography (GC)	HVA is extracted and the silyl derivative formed; detected by flame ionization.	Rare	Slow, research method
4. Gas chromatography–mass spectroscopy (GC/MS)	HVA is extracted with an ion-exchange resin, converted to the trifluoroacetyl hexafluoroisopropanol ester derivatives, and separated by gas chromatography with detection by mass spectrometry.	Rare	Slow, research method
5. High-performance liquid chromatography (HPLC)	HVA is extracted with an ion-exchange resin and separated by reversed-phase chromatography.	Most common	Preferred method

The HPLC methods are reasonably rapid, allowing separation of the HVA within 20 to 30 minutes. The sensitivity is excellent, especially when one is using amperometric detection.[7,9-11] These methods have shown little interference from other endogenous compounds or exogenous drugs. The ability to simultaneously quantitate VMA and HVA may not have much practical advantage but are commonly done together to improve technical efficiency. The HPLC methods with amperometric detection seem to be the most sensitive, specific, and reproducible, and HPLC is the preferred method for HVA analysis.

SPECIMEN

The preferred sample for HVA analysis is a 24-hour urine sample collected with sufficient acid to maintain a pH of less than 3 during the collection period. Usually 15 mL of 6 M HCl added to the collection vessel before collection of urine is sufficient to maintain the proper pH. An investigation of the diurnal variation of urinary HVA secretion indicates that a random urine sample may be satisfactory for the diagnosis of neuroblastomas, especially when the excretion rate is normalized for creatinine excretion (mg of HVA/g of creatinine).[12] However, these results should be confirmed in each laboratory.

PROCEDURE: HPLC METHOD FOR URINARY HVA USING ELECTROCHEMICAL DETECTION

The procedure for VMA can be used for the quantitation of urinary HVA because this method allows for the simultaneous measurement of VMA and HVA in urine samples.

REFERENCE RANGE

The age-dependent percentile values for urinary HVA, as milligrams of HVA/24 hours or milligrams of HVA/g of creatinine, are shown in Table 60-8. There appears to be no difference in excretion rates between men and women. In several reports on diagnosis of neuroblastoma,[7,13] 27 of 30 cases of neuroblastoma had urinary HVA levels exceeding the 100th percentile; 29 of 30 exceeded the 95th percentile, and only one case had a value below the 95th percentile.

Table 60-8 Percentile reference values for urinary HVA excretion

		HVA			
		95th		100th	
Age	n	mg/24 hr (μmol/24 hr)*			
0-1	48	2.8	(15.4)	3.5	(19.2)
2-4	34	4.7	(25.8)	6.7	(36.8)
5-9	20	5.4	(29.6)	5.7	(31.3)
10-19	40	7.2	(39.5)	8.5	(46.7)
>19	56	8.3	(45.6)	10.3	(56.5)
		mg/g of creatinine (mmol/mol of creatinine)†			
0-1	37	32.6	(20.3)	76.9	(49.4)
2-4	49	22.0	(13.7)	58.8	(36.5)
5-9	79	15.1	(9.4)	33.2	(20.6)
10-19	55	12.8	(7.9)	44.2	(27.4)
>19	56	7.6	(4.7)	9.4	(5.8)

*Analyses performed on timed urine samples obtained from patients under investigation for hypertension.
†Analyses performed on untimed urine specimens obtained from hospital patients not suspected of having a neural crest tumor.

REFERENCES

1. Goldenberg, H: Specificity of the nitrosonapththol reaction: detection of metanephrine and other guaiacol derivatives, Clin Chem 13:698, 1967.
2. Knight, JA, and Hammond, RE: Improved colorimetry of urinary 3-methoxy-4-hydroxyphenylacetic acid (homovanillic acid), Clin Chem 23:2007-2010, 1977.
3. Huck, H, and Dworzak, E: Quantitative analysis of catecholamine and serotonin metabolites on thin-layer plates. II. Reflectance measurements after a specific photochemical reaction, J Chromatogr 74:303, 1972.
4. Brewster, MA, Berry, DH, and Moriarty, M: Urinary 3-methoxy-4-hydroxyphenylacetic (homovanillic) and 3-methoxy-4-hydroxymandelic (vanillylmandelic) acids: gas-liquid chromatographic methods and experience with 13 cases of neuroblastoma, Clin Chem 23:2247-2249, 1977.

5. Cahuhan, J, and Darbre, A: Determination of homovanillic, isohomovanillic and vanillylmandelic acids in human urine by means of glass capillary gas-liquid chromatography with temperature programmed electron-capture detection, J Chromatogr 183:391-401, 1980.
6. Takahashi, S, Yoshioka M, Yoshiue, S, and Tamura, Z: Mass fragmentographic determination of vanillylmandelic acid, homovanillic acid and isohomovanillic acid in human body fluids, J Chromatogr 145:1-9, 1978.
7. Soldin, SJ, and Hill, JG: Liquid-chromatographic analysis for urinary 4-hydroxy-3-methoxymandelic acid and 4-hydroxy-3-methoxyphenylacetic acid and its use in the investigation of neural crest tumors, Clin Chem 27:502-503, 1981.
8. Yoshida, A, Sakai, T, and Tamura, Z: Simple method for the determination of homovanillic acid and vanillylmandelic acid in urine by high-performance liquid chromatography, J Chromatogr 227:162-167, 1982.
9. Frattini, P, Santagostino, G, Schinelli, S, et al: Assay of urinary vanilmandelic, homovanillic and 5-hydroxyindole acetic acids by liquid chromatography with electrochemical detection, J Pharmacol Methods 10:193-198, 1983.
10. Baursfeld, W, Diener, U, Knoll, E, et al: Determination of urinary vanilmandelic acid and homovanillic acid by high performance liquid chromatography with amperometric detection, J Clin Chem Clin Biochem 20:217-220, 1982.
11. Morrisey, JL, and Shihabi, ZK: Assay of 4-hydroxy-3-methoxyphenylacetic (homovanillic) acid by liquid chromatography with electrochemical detection, Clin Chem 25:2045-2047, 1979.
12. Tuchman, R, Robison, LL, Maynard, RC, et al: Assessment of the diurnal variations in urinary homovanillic and vanillylmandelic acid excretion for the diagnosis and follow-up of patients with neuroblastoma, Clin Biochem 18:176-179, 1985.
13. Soldin, SJ: High-performance liquid chromatography: application in a children's hospital, Adv Chromatogr 20:139-163, 1982.

Thyroid-stimulating hormone

I-WEN CHEN
LINDA A. HEMINGER

Thyroid-stimulating hormone, TSH, thyrotropin
Clinical significance: p. 620
Molecular weight: 28,300 daltons
Chemical class: glycoprotein, hormone

PRINCIPLES OF ANALYSIS

Thyroid-stimulating hormone (TSH, thyrotropin) is a glycoprotein secreted by the anterior lobe of the pituitary gland that is capable of stimulating the normal thyroid gland to synthesize and secrete thyroxine (T_4) and triiodothyronine (T_3). TSH is structurally similar to two other pituitary glycoprotein hormones, follicle-stimulating hormone (FSH) and luteinizing hormone (LH), as well as to the placental glycoprotein hormone (human) chorionic gonadotropin (hCG), They are all composed of two dissimilar noncovalently bound subunits; the alpha subunit is hormone nonspecific, whereas the beta subunit is distinct for each hormone and confers biological and immunological specificity. The primary amino acid structure of both subunits has been established.[1] The alpha subunit of human TSH (MW 13,600 daltons) is a single-chain polypeptide consisting of 89 amino acids and two carbohydrate side chains; it is identical to the alpha subunit of human LH and differs from that of human FSH and hCG only by a tripeptide at the NH_2 terminal. The beta subunit of human TSH is also a single-chain polypeptide consisting of 112 amino acids and one carbohydrate side chain (MW 14,700 daltons); its primary amino acid structure is significantly different from those of FSH, LH, and hCG.

Initially, TSH was measured using bioassays, and the measurement of serum TSH did not become a routine laboratory test for clinical evaluation of the thyrometabolic state until purified human pituitary TSH became available. The development of antibodies was then possible. Hormone was also available for radiolabeling and for use as a standard in a radioimmunoassay (RIA) specific for human TSH.[2,3] The RIA for TSH (method 1, Table 60-9) is based on competition between a constant trace amount of radiolabeled TSH and TSH from patient serum samples for a fixed and limited number of TSH-antibody binding sites. The amount of labeled TSH bound to the antibody is inversely related to the amount of unlabeled TSH present in the sample.

Table 60-9 Methods of TSH analysis

Methods	Principle	Marker	Comments
1. Radioimmunoassay (RIA)	Competitive binding of radiolabeled TSH and nonlabeled TSH to limited binding sites on antibody	^{125}I	Not sensitive to measure subnormal TSH levels
2. Immunoradiometric assay (IRMA)	Binding of TSH to radiolabeled antibody	^{125}I	May be more sensitive and specific than RIAs; fully automated
3. Immunoenzymometric assay (IEMA)	Binding of TSH to enzyme-labeled antibody	Enzyme	May become an excellent alternative to radioassay as sensitivity is improved
4. Immunoluminometric assay (ILMA)	Binding of TSH to luminescent molecule–labeled antibody	Luminescent labels such as luminol and luciferase	Has potential of being very sensitive assay; available for routine use
5. Time-resolved immunofluorometric assay (TR-IFMA)	Binding of TSH to europium-labeled antibody	Europium chelate	Has potential of being very sensitive assay; available commercially

The RIA of TSH is complicated further by the variability of TSH preparations used as standards. Standard preparations of human TSH, calibrated by bioassay and expressed in international units (U, nonofficially IU), are available from several agencies, such as the World Health Organization, the Medical Research Council (Great Britain), and the National Institutes of Health (United States). The standard TSH used in RIAs should be calibrated against one of these standard preparations. However, it is important to remember, as with any immunoassay, that the TSH RIA measures the immunological activity, not the biological activity, of circulating TSH, although the results of TSH RIAs are customarily expressed as milli–international units (mU) of biological activity per liter (mU/L) of serum. It is possible for a TSH molecule to lose its biological activity without losing its immunological activity. However, all available data indicate that the TSH concentrations measured by RIA correlate well with the biological activity and are excellent biological indicators for the diagnosis of disorders of the hypothalamic-pituitary-thyroid axis.

In addition to the conventional RIA, a solid-phase, two-site immunoradiometric assay (IRMA) developed by Miles and Hales[4] has been used for the measurement of circulating TSH (method 2, Table 60-9). In this method TSH is assayed directly by reaction with excess radiolabeled specific antibodies rather than by competition with a radiolabeled TSH molecule for a fixed number of binding sites on a limited amount of antibody. In general, samples containing TSH are reacted with a radiolabeled antibody directed toward a unique site on the TSH molecule and with an immobilized solid-phase antibody directed against a different antigenic site on the same TSH molecule. The radiolabeled antibody/TSH/solid-phase antibody complex formed is separated from the reaction mixture, washed, and counted. The radioactivity measured is directly proportional to the concentration of TSH present in the test sample. An assay of this type is also known as a sandwich assay because the antigen TSH becomes sandwiched between the labeled antibodies and immobilized antibodies. Sereno Diagnostics Inc. (Randolph, MA) has marketed a TSH IRMA kit using three high-affinity monoclonal antibodies, each specific to different antigenic sites on the same TSH molecule, to achieve a level of sensitivity and specificity impossible with traditional methods.[5] Two monoclonal antibodies are labeled with ^{125}I, and the third monoclonal antibody is labeled with fluorescein isothiocyanate (FITC), which is a highly immunogenic group. Patient samples are incubated with a mixture of these three monoclonal antibodies in a liquid phase to form a ''sandwich.'' Attachment of two ^{125}I-labeled antibodies to a TSH molecule increases the specific activity of the labeled complex and thus the sensitivity of the assay. Anti-FITC linked to magnetizable particles is then added to the incubation mixture to achieve separation of excess ^{125}I-labeled antibodies from the labeled sandwich complex, which sediments rapidly in a magnetic field. In this assay method, since the interactions between TSH and antibodies proceed in a liquid phase, the diffusional and geometrical constraints encountered in the solid-phase, two-site IRMA are eliminated, resulting in a shortened reaction time. The minimum detectable level was claimed to be 0.1 mU/L. The assay turnaround time is 3 hours.

Immunoassays of TSH using enzymes as labels have also been recently developed. The major disadvantage of enzyme immunoassays (EIAs) has been the relatively low sensitivity, believed to be primarily the result of steric hindrance introduced into the antigen-antibody reaction by the presence of the enzyme macromolecule. The problem of steric hindrance appears to be partly eliminated by the sandwich technique, in which enzyme-labeled antibodies instead of antigens are used. In the sandwich, solid-phase, two-site enzyme-immunometric assay developed by Hybritech, Inc. (San Diego) (method 3, Table 60-9), the test serum is allowed to react with two different anti-TSH monoclonal antibodies. One antibody is affixed to a solid phase, and the other is labeled with bovine alkaline phosphatase (AP), each antibody directed against a distinctly different antigenic site on the TSH molecule. The AP-labeled antibody/TSH/solid-phase antibody complex formed is separated, washed, and reacted with a chromogenic substrate, *p*-nitrophenylphosphate. The yellow color developed is measured spectrophotometrically at 405 nm. The concentration of TSH is directly proportional to the color intensity (enzyme activity). Hybritech claims that there is no detectable cross-reactivity between their assay and hCG, FSH, or LH.

A somewhat different technique for labeling antibodies is used in the two-site ELISA method for TSH developed by Abbott Laboratories (North Chicago). In this method, biotin (a growth factor, vitamin H) is used to label antibodies (biotinylation). The biotinylated antibody/TSH/solid-phase antibody complex formed is separated, washed, and incubated with avidin-HRPO (horseradish peroxidase) conjugates. Avidin, a glycoprotein found in large amounts in raw egg white, is capable of avidly binding biotin. The resulting avidin-HRPO conjugate/biotinylated antibody/TSH/solid-phase antibody complex is separated, washed, and reacted with a chromogenic substrate for the determination of HRPO activity. The biotinylated antibody has several advantages over the HRPO-labeled antibody: (1) because biotin is a small molecule (244 daltons) compared with HRPO (40,000 daltons) and because HRPO is not incorporated into the sandwiched complex until after the antigen-antibody complex formation, the steric hindrance introduced into the antigen-antibody reaction by the lebel is minimum; (2) up to four biotin molecules have been incorporated into a molecule of antibody, resulting in an increase in assay sensitivity through signal enhancement; (3) biotinylation of antibodies is technically easier than labeling of antibodies with enzymes; and (4) the avidin-HRPO conjugate may be used in any ELISA technique using biotin as a label. Woodhead (Weeks et al.[7]) of the Welsh National School of Medicine reported development of a chemiluminescent immunoassay of TSH capable of detecting as little as 0.04 μU/L (method 4, Table 60-9). The extremely good sensitivity is said to be attributable to the high quantum efficiency of the luminescent compound used for labeling anti-TSH antibody.

The use of a fluoroimmunoassay technique for measurement of TSH has been described as well.[8] This fluoroimmunoassay used the TSH-antibody labeled with europium chelate as the tracer and time-resolved fluorescence as the detection method (method 5, Table 60-9). The limited sensitivity of most fluoroimmunoassay methods results from the high background signal encountered in conventional fluorometric determinations. To minimize the high background fluorescence, serum samples are usually pretreated before they are used in the fluoroimmunoassay. For example, in the fluorescence polorization immunoassay of digoxin developed by Abbott Laboratories (TDx, North Chicago), serum samples are pretreated with trichloroacetic acid for deproteinization to minimize the polarization of background fluorescence. The europium chelate used in the time-resolved fluoroimmunoassay of TSH has a much longer fluorescence lifetime (a half-life of decay in the range of 10^{-3} to 10^{-6} seconds) than that of fluorescence emitted from conventional fluorescent compounds including those causing the high background signal in serum (10^{-8} to 10^{-9} seconds), and thus its specific fluorescence can be measured after the background signal has decayed away. Two monoclonal antibodies directed against two separate antigenic determinants on the beta-subunit of the TSH molecule are used in this assay; one monoclonal antibody is labeled with europium as a nonfluorescent chelate, and the other antibody is immobilized on the surface of microtitration strip wells. After a 4-hour incubation of the serum sample with the europium-labeled antibody at room temperature in the antibody-coated microtitration well, the well is aspirated and washed with water to remove the unbound europium-labeled antibody. The fluorescence of the europium ion in the solid-phase antibody/TSH/europium-labeled antibody complex is developed after the ion is released by dissociation. The dissociation is carried out in the presence of an enhancement solution containing energy-absorbing ligands required for the formation of a fluorescent chelate (tri-*n*-octylphosphine oxide and 2-naphthoyltrifluoroacetone). The light emission of the europium ion is measured in a time-resolving (400 μsec) fluorometer. Europium concentrations as low as 5×10^{-14} M have been measured by this technique, and theoretically the assay sensitivity can be increased further be-

Table 60-10 Comparison of reaction conditions for TSH assays

Conditions	Radioimmunoassay (Nuclear Medical Laboratories)	Immunoradiometric assay		Immunoenzymometric assay (IEMA) (Hybritech)	Time-resolved immunofluorometric assay (TR-IRMA) (LKB Wallac)
		Manual IRMA (Sereno)	Automated IRMA (ARIA-HT)		
Temperature (C)	37°	25°	37°	25°-37°	25°
Reaction time (hr)	2 (without tracer) 3 (with tracer)	2	1	2-24	4
pH	7.8	7.4	8.0	8.0	7.4
Sample volume (μL)	200	200	240	100	50
Tracer	^{125}I-labeled TSH	^{125}I-labeled antibody	^{125}I-labeled antibody	Alkaline phosphatase	Europium ion
Antibodies	Rabbit primary Goat secondary	Three monoclonal antibodies: two ^{125}I-labeled and one fluorescein labeled	Rabbit antibodies: ^{125}I-labeled, polysaccharide supported	Two monoclonal antibodies: one alkaline phosphatase, one supported by beads	Two monoclonal antibodies: one europium chelate labeled, one immobilized on microtitration well
Enzyme substrate	Not required	Not required	Not required	*p*-Nitrophenylphosphate	Not required
Sensitivity* (mU/L)	1.0	0.1	1.2	0.5	0.03‡
Precision (%)	9.2-12.2†	6.5-7.0†	13.0-19.6†	11.5-22.0†	5-10‡
Interferences	Lipemic sera, radioactive drugs	Radioactive drugs	Hemolyzed or turbid sera, radioactive drugs	Enzyme inhibitors	

*Minimum detectable concentration.
†Interlaboratory coefficient of variation according to second quarter of 1988 College of American Pathologists survey with mean TSH concentration ranging from 1.05 to 439.16 mU/L.
‡Claimed by kit manufacturer.

cause a number of europium ions can be coupled per antibody molecule. A commercial kit using this technique has been marketed by LKB Wallac, Helsinki, Finland. The sensitivity of this assay kit is claimed to be 0.3 mU/L.

Most commercial TSH RIAs are designed to confirm a diagnosis of primary hypothyroidism, that is, to detect an elevated serum TSH level, and thus they usually place more emphasis on a shorter assay time than on the sensitivity sufficient to discern differences between normal circulating TSH levels and TSH levels in patients with a TSH deficiency (Table 60-10). However, it is also necessary for a TSH assay to be able to distinguish normal from low concentrations of TSH reliably for the assay to be useful in the diagnosis of both hypothyroidism and hyperthyroidism.

The avidity of the antiserum alone is the most important limiting factor in setting the sensitivity of the conventional RIA. However, the sandwich method requires two purified antibodies capable of recognizing two different antigenic sites on the same antigen molecule, which requires tedious and time-consuming antibody selection and purification processes. Such problems can be simplified through the use of monoclonal antibody technology, which can provide large quantities of pure, homogeneous antibody with precisely defined immunochemical properties. The sandwich technique has been successfully applied to nonisotopic immunoassays (Table 60-10). Certainly, the sensitivity and precision of these assays are similar to those achieved by RIA.

SPECIMEN

Since TSH secretion is relatively constant throughout the day, useful clinical information can be obtained from a blood specimen drawn at any convenient time. Serum or plasma samples can be used, but some kit manufacturers recommend use of serum samples only. Although TSH in serum is stable for at least 5 days at 4° C, if the test is not to be run within 24 hours, the serum should be kept frozen at −20° C. Frozen sample should be completely thawed and mixed before testing to ensure homogeneity. Repeated freezing and thawing of the sample should be avoided. The use of grossly hemolyzed or lipemic samples is not recommended.

In the thyroid-releasing hormone (TRH) stimulation test of pituitary TSH reserve, a baseline TSH sample is drawn before the test. Five hundred micrograms of TRH is given intravenously to the patient, and blood samples are drawn again 20, 30, and 40 minutes after TRH injection. Minor side effects caused by TRH injection include headaches, dizziness, nausea, and a momentary urge to urinate.

REFERENCE RANGE

Many older commercial kits for TSH were not sensitive enough to detect TSH in all healthy subjects. We estimate the reference range of serum TSH to be about 0.51 to 5.75 mU/L.[9,10]

Lipson et al.[11] found no difference in TSH values between men and women ranging from 20 to 60 years of age. Men from 60 to 70 years old continue to have stable TSH levels, but women older than 60 years show a significantly higher mean TSH level (3.4 ± 1.6 mU/L) than younger women (2.3 ± 1.3 mU/L).

In neonates serum TSH levels rise sharply within 10 minutes after delivery, reach a peak (10- to 24-fold increase) after 30 minutes, and then decline gradually and reach adult levels about 5 days after delivery.[12]

For the TRH stimulation test, the maximum increment of serum TSH from the baseline value (ΔTSH) after TRH injection is 2 to 20 mU/L in euthyroid subjects. Elderly men are less responsive to TRH stimulation than women and young men. ΔTSH is also normal in hypothalamic (tertiary) hypothyroidism, but the response is usually delayed (60 to 180 minutes versus 30 minutes for normal subjects). ΔTSH is less than 2 mU/L in pituitary (secondary) hypothyroidism and also in hyperthyroidism but is greater than 20 mU/L in thyroidal (primary) hypothyroidism (see p. 627).

REFERENCES

1. Rathnam, P: Structure-function relationship of pituitary hormones HFSH, HLH, and TTSH. In Abraham, GE, editor: Radioassay systems in clinical endocrinology, New York, 1981, Marcel Dekker, Inc.
2. Utiger, RD: Radioimmunoassay of human plasma thyrotropin, J Clin Invest 44:1277-1286, 1965.
3. Odell, WD, Wilver, JF, and Paul, WE: Radioimmunoassay of thyrotropin in human serum, J Clin Endocrinol Metab 15:1179-1188, 1965.
4. Miles, LEM, and Hales, CN: Labeled antibodies and immunological assay systems, Nature 219:186-189, 1068.
5. Rattle, SJ, Purnell, SR, Williams, PIM, et al: New separation method for monoclonal immunoradiometric assays and its application to assays for thyrotropin and human choriogonadotropin, Clin Chem 30:1457-1461, 1984.
6. Tandem™ R TSH Immunoradiometric Assay Publication 701081-013B, Jan 1983, Hybritech Inc, San Diego, CA.
7. Weeks, I, Sturgess, M, Siddle, K, et al: A high sensitivity immunochemiluminometric assay for human thyrotropin, Clin Endocrinol 20:489-495, 1984.
8. Lovgren, T, Hemmila, I, Petterson, K, et al: Determination of hormones by time-resolved fluoroimmunoassay, Talanta 31:909-916, 1984.
9. Chen, I-W, Heminger, LA, Barnes, EL, et al: A sensitive radioimmunoassay (RIA) for detection of serum thyrotropin (TSH) in healthy subjects and patients with suppressed pituitary function, J Nucl Med 24:114, 1983.
10. Tsay, JY, Chen, I-W, Maxon, HR, et al: A statistical method for determining normal ranges from laboratory data including values below the minimum detectable value, Clin Chem 25:2001-2014, 1979.
11. Lipson, A, Nickoloff, EL, Hsu, TH, et al: A study of age-dependent changes in thyroid function tests in adults, J Nucl Med 20:1124-1130, 1979.
12. Fisher, DA: Thyroid physiology and function tests in infancy and childhood. In Werner, SC, and Ingbard, SH, editors: The thyroid, New York, 1978, Harper & Row, Publishers, Inc.

Thyroxine

I-WEN CHEN
MATTHEW I. SPERLING

Thyroxine, T_4, 3,5,3′,5′-tetraiodothyronine
Clinical significance: p. 620
Molecular formula: $C_{15}H_{11}I_4NO_4$
Molecular weight: 776.93 daltons
Merck Index: 9262
Chemical class: amino acid (thyroid hormone)

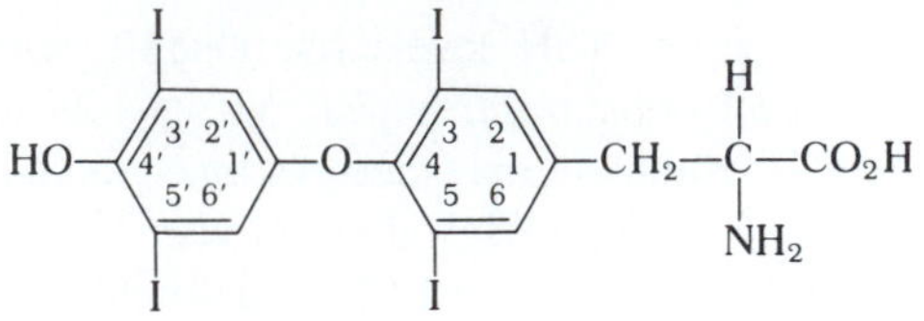

$$HO-C_6H_2I_2-O-C_6H_2I_2-CH_2-CH(NH_2)-CO_2H$$

PRINCIPLES OF ANALYSIS

Thyroxine (T_4) is an amino acid synthesized in and secreted from the thyroid gland. It plays an important role in the regulation of developmental and metabolic processes. The naturally occurring T_4 is the L-isomer. Although the D-isomer of T_4 has an affinity for nuclear binding proteins equal to that of the L-isomer, its biological activity is only 5% to 20% of the L-isomer. The differences in transport and clearance of the L- and D-isomers are probably responsible for differences in their metabolic activity.

Historically, serum T_4 concentration was estimated by measurement of the amount of iodine present in the partially purified T_4 fractions of serum samples. Such methods involved the measurement of the iodine content in a protein precipitate of serum (protein-bound iodine, PBI), in an alkali-washed butanol extract of a protein precipitate of serum (butanol-extractable iodine, BEI), and in an anion-exchange column–purified fraction of serum (T_4 by column). All these methods are relatively nonspecific and are subject to contamination by iodine-containing drugs and nonhormonal iodine. They are therefore used rarely in routine clinical chemistry laboratories at present. However, the test can be useful in a few clinical situations, such as for detecting the presence of iodoprotein abnormalities.

A more specific and sensitive method for measuring total T_4 concentration in serum, called the *competitive protein-binding assay* (CPBA), was developed in the early 1960s by Murphy and Pattee[1] (method 1, Table 60-11). This method is based on the competition between serum T_4 and added radioactive T_4 for the limited binding sites on specific T_4-binding proteins. The radioactivity (of T_4) bound to the specific binding protein is counted after the unbound radioactivity has been removed from the assay mixture. The bound radioactivity is inversely proportional to the amount of T_4 present in serum samples. The specific binding protein used in this assay is thyroxine-binding globulin (TBG) and is usually obtained by proper dilution of a human serum pool with a barbital buffer. The barbital buffer selectively inhibits any binding of T_4 to the other T_4-binding protein present in serum called *thyroxine-binding prealbumin*. Binding of T_4 to albumin is a low-affinity binding and can be eliminated merely by dilution of the serum pool with a barbital buffer. Since more than 99.9% of T_4 in serum is normally bound to TBG and the other tyroxine-binding proteins and since CPBAs require a constant quantity of TBG in each assay tube, it is necessary

Table 60-11 Methods of thyroxine analysis

Method	Tracer	Binder	Reaction phase	Use	Comment
1. Competitive protein-binding analysis (CPBA)	$^{125}I-T_4$	TBG	Liquid	Any biological specimens	Requires extraction of serum and phase separation
2. Radioimmunoassay (RIA)	$^{125}I-T_4$	Antibody	Liquid or solid	Serum only, fully automated	Requires phase separation; the most sensitive method
3. Enzyme-multiplied immunoassay (EMIT)	Enzyme-T_4	Antibody	Liquid	Serum only, fully automated	Requires pretreatment of samples and addition of enzyme substrate; requires no phase separation
4. Enzyme inhibitor immunoassay	Phosphonate-T_4	Antibody	Liquid	Serum only, fully automated	Requires no pretreatment of sample or phase separation; requires addition of enzyme substrate
5. Fluorescence immunoassay (FIA)	Fluorescein-T_4	Antibody	Liquid	Serum only, partially automated	Requires no pretreatment of serum; requires no phase separation
6. Isotope dilution–mass spectrometry	$^{2}H-T_4$	None	Requires separation and derivatization of T_4	Recommended as definitive method	Highly accurate but too laborious and complicated for routine clinical use

TBG, Thyroxine-binding globulin.

to extract T_4 and remove TBG from serum samples by an alcohol, usually ethanol. The dried T_4 extract is redissolved in the TBG solution and used in the CPBA for T_4.

Although CPBA for T_4 is a direct analysis of T_4 and thus is free from interference by nonhormonal iodine in the blood, it has now largely been replaced by a more sensitive and specific radioimmunoassay (RIA) (method 2, Table 60-11). The higher sensitivity and specificity of RIA is achieved by the use of high-affinity antisera. The other advantage of the T_4 RIA is avoidance of the time-consuming extraction step. This is achieved by the use of blocking agents, such as 8-anilinonaphthalenesulfonic acid.

Otherwise, the basic principles of RIA and CPBA are the same.

Many T_4 RIA kits are available commercially. Although all kits are based on the principle of RIA, many different techniques are used in the separation of antibody-bound and free radioligands. The most commonly used technique is the solid-phase separation procedure using the T_4 antibody complex chemically or physically bonded to a solid support such as glass beads, plastic tubes, cellulose, or magnetic particles. Other separation methods involve the use of second antibodies or polyethylene glycol to precipitate the antibody-bound ligand. Charcoal and ion-exchange resins have been used as the separating agent, but they are generally not used in T_4 RIA because they are relatively nonspecific.

In RIA a radionuclide (radioactive iodine, ^{125}I) is used as a marker to follow and measure the course of an immunological reaction. The radioactivity produced by the radionuclide is the end-point signal that is measured. The radioactivity emitted by the radionuclide through the radioactive decay process is not affected by the physicochemical environment. Therefore a separation step for bound and free radiolabeled ligands is always required in RIA.

Other types of labels have been developed as alternatives to radionuclides in immunoassays of T_4. Currently available nonisotopic methods are homogeneous immunoassays.[2]

The enzyme-multiplied immunoassay technique (EMIT) for T_4 developed by Syva (Palo Alto, CA) is a homogeneous immunoassay system that requires no separation step (method 3, Table 60-11). In this technique, malate dehydrogenase chemically bound to T_4 is used as the tracer. It is postulated that the enzyme with bound T_4 is inactive because the active site of the enzyme is blocked by T_4. The enzyme becomes active in the presence of T_4 antibody because T_4 binds to the antibody and is therefore displaced from the active site. Therefore the degree of binding of the enzyme-labeled T_4 to the T_4 antibody is directly proportional to the enzyme activity and can be measured without physical separation of bound and free T_4. One disadvantage of the homogeneous assay over the heterogeneous assay is that the serum sample has to be pretreated with alkali to eliminate serum effects on the enzyme activity measurement.

A homogeneous enzyme inhibitor immunoassay has been developed by Abbott Laboratories (North Chicago) (method 4, Table 60-11). This method uses a phosphonate-T_4 conjugate, which is a potent irreversible inhibitor of acetylcholine esterase.[3] When the conjugate is bound by the T_4 antibody, the inhibitory activity is blocked. Therefore the amount of bound phosphonate-T_4 conjugate is directly proportional to the acetylcholine esterase activity, which one can measure photometrically by monitoring the change in the absorbance. After the addition of appropriate enzyme substrate, the separation of bound and free T_4 conjugate is unnecessary. This enzyme inhibitor immunoassay is available only for use on the du Pont aca. Fluorescent probes are also used in place of the radiolabels in T_4 immunoassays. The homogeneous fluorescence immunoassay (method 5, Table 60-11) for T_4 developed by Abbott Laboratories uses fluorescence polarization (TDx system). Fluorescein-labeled T_4 and T_4 in serum samples are allowed to compete for the binding sites on the T_4 antibody. (See Chapter 3 for a description of fluorescence polarization and Chapter 11 for a description of the assay.)

To minimize the background interference, serum samples are usually treated with a denaturing agent before they are used in the immunoassay. In the TDx immunoassay, the serum sample is manually pretreated with a denaturing solution containing 8 M urea and 3% sodium dodecyl sulfate.

Isotope dilution–mass spectrometry has been developed and proposed as a reference method for serum T_4 measurement (method 6, Table 60-11).[4] In this technique, a fixed quantity of ^{2}H-labeled T_4 is added to a fixed quantity of serum. T_4 is then isolated from the serum by the extraction–solvent distribution method, derivatized to an *N*,O-bis(trifluoracetyl)methylester and subjected to combined gas chromatography–mass spectrometry with helium as the carrier gas. The ions at the mass-charge ratios M/799 and M/801 are traced simultaneously. The precision of the assay in terms of interassay coefficient of variation is less than 3.5%.

All three immunoassay methods—radioimmunoassay, enzyme immunoassay, and fluorescence immunoassay—have been either fully automated or semiautomated. Two fully automated RIA systems are available commercially in the United States at present. RIA is still the most commonly used method for thyroxine analysis.

The Concept 4 (Micromedic Systems, Horsham, PA) uses special 8 × 50 mm tubes coated with antibody to achieve separation of bound and free ligand. The system consists of a pipetting station for sample transfer, reagent addition and mixing, an incubator for incubating the mixture at ambient temperature or at 45° C for up to 17 hours, an aspirate-wash station for separation of bound and free ligands, a gamma ray–counting station capable of counting two tubes at a time, and a programmable calculator for online data reduction. The system is designed to use racks of

10 tubes each and is capable of processing up to 200 antibody-coated tubes per assay.

The separation of bound from free ligands in the ARIA II or ARIA-HT, introduced by Becton Dickinson (Salt Lake City), is achieved by sequential pumping of the assay mixtures through a reusable antibody chamber containing antibodies covalently bonded to a solid support medium. The system is composed of a sample carousel for dispensing samples and also for incubation, if needed; a flow-through system consisting of a flexible arrangement of pumps, valves, reservoirs, and tubing; an antibody chamber; a gamma-ray detector for counting both free and bound radioligands; and an online microcomputer for data reduction.

Methods involving the measurement of the iodine content in the partially purified T_4 fraction of serum samples are nonspecific and are seldom used for the evaluation of thyroid status though they are occasionally used for detection of the presence of an iodoprotein abnormality. CPBA requires the tedious extraction step for releasing T_4 bound to T_4-binding proteins, and the extraction efficiency may vary from sample to sample; CPBA kits for T_4 analysis are no longer commercially available. This method has now largely been replaced by more sensitive and specific T_4 RIA techniques.

The homogeneous enzyme immunoassay was developed as an alternative method for T_4.[5,6] This method has two major advantages over the RIA technique: the separation step for bound and free T_4 is not required and handling of radioactive materials is avoided. Although enzyme immunoassay is used extensively for therapeutic drug monitoring, it has not gained popularity in T_4 analysis, largely as a result of its relative insensitivity and imprecision, especially at low T_4 concentrations.

Fluorescent probes seem to be a more attractive alternative to the radioactive labels. Background interference is commonly encountered in fluoroimmunoassay and is the major cause of a loss of fluorometric sensitivity. It appears that RIA is still the best method available for the routine measurement of serum T_4 and is the most commonly used method at present because of its excellent sensitivity, precision, and accuracy. It has been shown that there is an excellent agreement between T_4 results obtained by RIA and by isotope dilution mass spectrometry.[4]

Table 60-12 compares reaction conditions for several methods of T_4 analysis.

SPECIMEN

Serum is preferred and should be collected using normal aseptic venipuncture techniques. Plasma may also be used but tends to form fibrin after freezing and thawing, which may mechanically interfere with the assay, especially in an automated system. T_4 in serum is quite stable; it has been shown that storage of serum samples at room temperature for up to 14 days resulted in no appreciable loss of T_4. However, it is recommended that serum samples be stored frozen if they will not be analyzed within 24 hours. Repeated freezing and thawing of the sample should be

Table 60-12 Comparison of reaction conditions for thyroxine

Conditions	Manual RIA (Corning)[7]	An automated RIA (ARIA II)[7]	Du Pont aca	Abbott TDx
Temperature (C)	25°	25°	37°	25°
pH	7.4	10.5	7.4	7.4
Sample volume (μL)	25	20	60	50
Sample pretreatment	Not required	Not required	Not required	Sodium clofibrate
Blocking agent	Thimerosal	8-Anilinonaphthalenesulfonic acid	8-Anilinonaphthalenesulfonic acid, sodium salicylate	Not required
Tracer	^{125}I-labeled T_4	^{125}I-labeled T_4	Phosphonate-acetylcholine–labeled T_4	Fluorescein-labeled T_4
Antibody, phase	Rabbit, bound to glass particles	Rabbit, bound to fiber particles	Sheep, liquid	Sheep, liquid
Enzyme substrate	Not required	Not required	Acetyl-β-(methylthio)choline iodide	Not required
Reaction time	60 min	Instantaneous	0.5 minute	5 minutes
Sensitivity* (μg/L)	20	10	20	6
Precision†	5.9%-8.6%	4.8%-5.4%	3.9%-8.2%	4.9%-7.5%
Interferences	Abnormal protein concentrations	Abnormal protein concentrations	Abnormal protein concentrations; enzyme inhibitors	Abnormal protein concentrations

*Minimum detectable concentration.
†Interlaboratory coefficient of variation (CV) according to second quarter of 1988 CAP survey, with mean T_4 concentration in the range of 54 to 166 μg/L.

avoided. Although hemolyzed specimens do not interfere with the assay, use of grossly hemolyzed samples should be avoided because hemolysis may be sufficient to have diluted the samples. Grossly lipemic specimens should not be used, especially in CPBA, since fatty acids are known to compete with T_4 for the binding sites on TBG.

As in the case of all radioassays, if a patient has received diagnostic or therapeutic radionuclides within the 2 weeks immediately before the T_4 determination, the radioactivity of the serum sample should be checked in a counter set for ^{125}I for determination as to whether the radioactivity contained in the sample will significantly affect assay results.

Most commercial T_4 assay kits are designed specifically for measurement of T_4 in serum or plasma samples and should not be used for other biological fluids, such as urine or cerebrospinal fluid, because of the extreme sensitivity of the antigen-antibody interaction to the matrix of samples to be measured. Antibodies to T_4 have been detected in some sera obtained from euthyroid subjects and patients with, for example, hypothyroidism, thyroid carcinoma, Hashimoto's thyroiditis, chronic lymphocytic thyroiditis, and Waldenström's macroglobulinemia, and it is important to be aware of their potential interference in immunoassays of T_4.[9]

REFERENCE RANGE

As with all diagnostic tests, it is recommended that each laboratory establish its own reference range, thereby allowing for variability resulting from such factors as geography and assay techniques. The reported reference ranges are, in general, within 41 to 120 μg/L (54.8 to 160 nmol/L). There are conflicting reports regarding the age and sex dependency of serum T_4 concentrations, but most researchers reported no age dependency.[10] Lipson et al.[11] found relatively constant T_4 levels in men of all ages and in women over 60 years of age but significantly high T_4 levels in women under 60 years of age (mean + standard deviation, 79 ± 13 μg/L [106 ± 17 nmol/L], n = 44) compared with men (72 ± 12 μg/L [96 ± 16 nmol/L], n = 120) and women over 60 years of age (74 ± 14 μg/L [99 ± 19 nmol/L]). However, these differences are small and can be ignored for routine testing.

Although the age-dependent differences in serum T_4 concentrations are of little significance in adults, they are important during childhood. Serum T_4 concentrations increase from cord blood values of about 127 μg/L (170 nmol/L) to a mean of 165 μg/L (221 nmol/L) by 26 hours after birth and fall gradually thereafter to a mean of 80 μg/L (107 nmol/L) at 17 years of age.[11]

REFERENCES

1. Murphy, B, and Pattee, CJ: Determination of thyroxine utilizing the property of protein binding, J Clin Endocrinol Metab 24:187, 1964.
2. Kaplan, LA, and Pesce, AJ, editors: Nonisotopic alternatives to radioimmunoassay, New York, 1980, Marcel Dekker.
3. Finlay, PR, William, RJ, Lichti, DA, et al: Evaluation of a new homogeneous enzyme inhibitor immunoassay of serum thyroxine with use of a biochromatic analyzer, Clin Chem 25:1723-1726, 1980.
4. Möller, B, Falk, O, and Björkhem, I: Isotope dilution-mass spectrometry of thyroxin proposed as a reference method, Clin Chem 29:2106-2110, 1983.
5. Yalow, RS: Radioimmunoassay: a probe for the fine structure of biologic systems, Science 200:236, 1978.
6. Kaplan LA, Chen, IW, Gau, N, et al: Evaluation and comparison of radio-, fluorescence, and enzyme-linked immunoassays for serum thyroxine, Clin Biochem 14:182, 1981.
7. Chen, I-W: Commercially available fully automated systems for radioligand assay. Part I. Overview, Ligand Rev 2(3):45, 1980.
8. Chen, I-W, Maxon, HR, Heminger, LA, et al: Evaluation and comparison of two fully automated radioassay systems with distinctly different modes of analysis, J Nucl Med 21:1162, 1980.
9. Neely, WE, and Alexander, NM: Polyclonal 3,4,3′-triiodothyronine (T_3) antibodies in a euthyroid woman and their effect on radioimmunoassays for T_3, J Clin Endocrinol Metab 58:851-854, 1983.
10. Caplan, RH, Wickus, G, Glasser, JE, et al: Serum concentrations of the iodothyronines in elderly subjects: decreased triiodothyronine (T_3) and free T_3 index, J Am Geriatr Soc 29:19, 1981.
11. Lipson, A, Nickoloff, EL, and Hsu, TH: A study of age dependent changes in thyroid function tests in adults, J Nucl Med 20:1124, 1979.

T_3 uptake

I-WEN CHEN
MATTHEW I. SPERLING

T_3 uptake, 3,5,3′-triiodothyronine uptake test; thyroxine-binding globulin (TBG) unsaturation
Clinical significance: p. 620

PRINCIPLES OF ANALYSIS

The triiodothyronine uptake test (T_3 uptake) is a test for estimating unoccupied binding sites on serum thyroxine-binding (T_4-binding) proteins. It was developed originally as an in vitro diagnostic test of thyroid function in 1959 by Hamolsky et al.[1] In the early T_3 uptake tests, a fixed amount of radioactive T_3 was incubated with whole blood, and the incorporation of radioactive T_3 by red blood cells was measured by counting the separated erythrocytes obtained after centrifugation and washing. The radioactivity taken up by red blood cells (the "T_3 uptake") was inversely proportional to the radioactivity bound to T_4-binding serum proteins, especially thyroxine-binding globulin (TBG); that is, a high T_3 uptake meant high TBG saturation and reduced available T_4-binding sites.

In the original T_3 uptake test[1] erythrocytes were used as the secondary binder for estimation of unsaturated T_4-binding capacity of serum proteins. This method, however, was found to be neither convenient nor reproducible because of the difficulty of uniform washing of erythrocytes and the effect of individual hematocrit variations on the test results. Since then, a variety of solid-phase materials have been used to improve the technique and the consistency in the measurement of T_3 uptake values. For example, ion-exchange resins, charcoal, solid-phase anti-T_3 an-

Table 60-13 Methods of T_3 uptake test

Methods (type of tracer used)	Principle	Assay system	Comments
1. ^{125}I-labeled T_3	Direct measurement of distribution of tracer between primary and secondary binder	Heterogeneous	Simplest method for manual estimation of TBG-binding capacity
2. Enzyme inhibitor–labeled T_4	Indirect measurement of tracer bound to serum T_4-binding proteins (primary binder)	Homogeneous	Requires no secondary binder; enzyme activity directly proportional to TBG-binding capacity
3. Fluorescein-labeled T_4	Tracer is used to measure bound T_4 by fluorescence polarization assay	Homogeneous	Requires no secondary binder; measures binding capacity of all T_4-binding proteins in serum
4. Fluorescein-labeled T_3 (fluor-labeled T_3)	Tracer is used to measure T_3 by fluorescent excitation transfer immunoassay technique	Homogeneous	Quencher-labeled T_3 antibody is used as specific binder in immunoassay of unbound T_3

tibody, organic polymers (Sephadex), silicate, talc, and macroaggregated albumin are used in a variety of commercially available T_3 uptake test kits. The 1988 College of American Pathologists (CAP) survey (first quarter) listed 26 commercial kits manufactured by 17 different companies. Although different secondary binders and variations in technique are used, the principle of the test remains the same: measurement of the distribution of ^{125}I-radiolabeled T_3 between endogenous serum binders (mainly TBG) and the secondary binder (method 1, Table 60-13).

Although radiolabeled T_4 can theoretically be used for assessment of the number of unoccupied binding sites on serum proteins, radiolabeled T_3 is preferred because of its relatively low affinity for TBG (affinity constant approximately 10^9 compared with 10^{10} mol/L for T_4). The radioactive T_3 will fill the unoccupied binding sites but will not displace bound T_4. Therefore differences between the test results obtained from patients with normal and abnormal numbers of unoccupied binding sites are greater with radioactive T_3 than with radioactive T_4. Further, T_4 binds to all T_4-binding proteins, including TBG, T_4-binding prealbumin, and albumin; thus the T_4 uptake values can be affected by the changes in any of the serum T_4-binding proteins. In the T_3 uptake test the T_3 binding is affected primarily by the TBG concentrations, since T_3 binds to neither prealbumin nor albumin under the conditions of most T_3-uptake tests.

In addition to radioactive T_3, nonradioactive labels are also used in the uptake test. In the A-gent Thyrozyme Uptake Diagnostic Kit developed by Abbott Laboratories (method 2, Table 60-13), phosphonate-T_4 conjugate, a potent inhibitor of acetylcholinesterase that still maintains the ability to bind TBG, is used as the label. When the T_4 conjugate is bound to the TBG, it loses the enzyme-inhibitory activity. Therefore the extent of the T_4 conjugate binding to TBG and hence the number of available T_4-binding sites on TBG can be assessed by measurement of the change in the enzymatic activity of acetylcholinesterase (AChE) by photometric monitoring of the change in the absorbance in the presence of a chromogenic enzyme substrate, acetyl(methylthio)choline iodide (AMTCI). The enzyme activity is proportional to the number of available T_4-binding sites. Since separation of free T_4 conjugate from TBG-bound T_4 conjugate is not required, the secondary binder used in the radiolabeled T_3-uptake test is not needed. Phosphonate-T_4 conjugate, rather than phosphonate-T_3 conjugate, is used in this method because the latter loses almost all of its ability to bind TBG.[2] The reaction sequence involved in this enzyme inhibitor assay is as follows:

Phosphonate-T_4 + TBG $\rightleftarrows$ Phosphonate-T_4-TBG + Phosphonate-T_4 *(residual)*

Phosphonate-T_4 *(residual)* + AChE (active) $\rightleftarrows$ AChE-phosphonate-T_4 *(inactive)* + AChE *(residual, active)*

AChE *(residual, active)* + Enzyme substrate *(AMTCI)* $\rightarrow$ β-(Methylthio)choline

β-(Methylthio)choline + DTNB* $\rightarrow$ Thionitrobenzoate *(absorbs at 405 nm)*

Fluorescent probes are also used in place of the radiolabels in the uptake test. In the TDx T-uptake test developed by Abbott Laboratories the extent of uptake of fluorescein-labeled T_4 by T_4-binding proteins in serum or plasma is estimated by the fluorescence polarization technique, the same basic technology used in the TDx thyroxine assay except that no T_4-antibody is used (see p. 957). An increase in the degree of polarization of fluorescent light emitted from the fluorescein-labeled T_4 resulting from its binding to the T_4-binding proteins in serum is proportional to the number of unoccupied T_4-binding sites on serum T_4-binding proteins. The condition of this assay is

*DTNB: The chromogenic reagent 5,5′-dithiobis(2-nitrobenzoic acid).

Table 60-14 Comparison of reaction conditions for T_3 uptake test

Conditions	Manual radioimmunoassay[3] (clinical assays)	Automated radioimmunoassay[4] (ARIA-HT)	Enzyme inhibitor assay (du Pont aca)	Fluorescence polarization assay (Abbott TDx)
Temperature (C)	18°-24°	30°	37°	20°-25°
pH	8.6	7.4	7.0	7.4
Sample volume (μL)	25	50	120	50
Tracer	$^{125}I-T_3$	$^{125}I-T_3$	Phosphonate-T_4	Fluorescein-T_4
Secondary binder	T_3-antibody coated tube	Hydrophobic resin column	Not required	Not required
Enzyme substrate	Not required	Not required	AMTCI*	Not required
Reaction time	60 min	Instantaneous	29 sec	5 min
Precision (coefficient of variation)	4.2%-4.9%†	2.9%-3.1%†	4.1%-5.5%†	8.4%-12.0%†
Interference	Radioactive drugs	Radioactive drugs	Hemolysis, enzyme inhibitors	Hemolyzed serum

FIA, Fluorescence immunoassay.
*Acetyl-β-(methylthio)choline iodide.
†Interlaboratory coefficient of variation according to second quarter of the 1988 College of American Pathologists survey.

such that it measures T_4-binding sites not only on TBG but also on prealbumin and albumin.

The Abbott TDx T-uptake uses a fluorescein thyroxine tracer and measures the binding of the tracer to the serum proteins using a surfactant organic base solution. Calibrators are standard amounts of thyroxine antibody. The amount of bound tracer is directly related to concentration by the degree of polarized fluorescence (method 3, Table 60-13). This in turn is related to percent uptake values by the following inverser transformation:

$$\%\ \text{Uptake} = \frac{\text{Mean normal value}}{\sqrt{0.8\ (\text{T uptake})^2 + 0.2}}$$

A somewhat different approach is taken on the Syva Advance T_3 Uptake assay using a fluorescent probe as an alternative to radionuclides (method 4, Table 60-13). Since the fluor-labeled T_3 loses the ability to bind serum proteins, it can no longer be used directly to determine the distribution of T_3 between the serum proteins and the secondary binder as in the radiolabeled T_3-uptake tests.[2] However, the fluor-labeled T_3 still maintains its reactivity toward T_3 antibody. A serum sample is incubated with a known amount of T_3, which binds to the available binding sites on the serum proteins. The amount of T_3 that remains unbound is determined by the fluorescence energy transfer immunoassay technique using fluor-labeled T_3 and quencher-labeled antibody to T_3. The binding of the fluor-labeled T_3 to the quencher-labeled antibody will result in the quenching of the fluorescence signal because of the transfer of the emission energy of the fluor to the quencher (method 4, Table 60-13). Since the fluor-labeled T_3 can compete with the unbound T_3 for antibody-binding sites, the amount of quenching is reduced in proportion to the amount of unbound T_3 and thus is indirectly related to T_3 uptake value. This is a homogeneous assay system requiring no separation of free and bound fluor-labeled T_3. A comparison of the reaction conditions for the T_3 uptake test is presented in Table 60-14.[3-5]

SPECIMEN

Most kit manufacturers recommend use of serum samples that are collected with use of the routine precautions required in venipuncture for clinical assays. Fresh serum samples may be stored at 2° to 8° C for up to 1 week without appreciable changes in the uptake values. However, the sample should be stored frozen at −20° C if the time between sample collection and analysis is longer than 6 days. Frozen samples should be mixed well by inversion after thawing and before assay. Repeated freezing and thawing and mixing by vortexing or vigorous agitation should be avoided because this may result in denaturation of proteins. The use of hemolyzed or lipemic samples is not recommended.

REFERENCE RANGE

When the radioactive T_3 uptake tests are used, uptake values are customarily expressed as the percentage of the total radioactivity that is taken up by the secondary binder or that is not bound to TBG. For most assays, the reference range is 25% to 35%; values below 25% are suggestive of hypothyroidism, and values greater than 35% are suggestive of hyperthyroidism. However, the actual value depends on various assay conditions used (the specific activity of the radiolabeled T_3, the avidity of the secondary binder for T_3, the incubation time and temperature, and so on). Variation in the uptake values resulting from imprecise assay conditions can be minimized by use of the T_3 uptake ratio, which is the T_3 uptake of the patient divided by the T_3 uptake of a pool of normal standard reference serum included in the same assay run. If the T_3 uptake value of the normal standard reference serum is 30%, the reference range of the T_3 uptake ratio will be 0.83 to 1.17. The T_3 uptake ratio can be calculated directly from the radioactivity taken up on the secondary binder (absorbent):

$$\text{Patient } T_3 \text{ uptake ratio} = \frac{\text{Absorbent counts for patient serum}}{\text{Absorbent counts for normal reference serum}}$$

The T_3 uptake ratio can then be converted to the percent T_3 uptake by multiplication of the patient T_3 uptake ratio by the percent T_3 uptake of the normal reference standard.

Almost all commercial T_3 uptake kits include a normal reference serum with a known percent T_3 uptake value for the determination of T_3 uptake ratio. However, it is strongly recommended that each laboratory prepare its own pool of normal standard reference serum, stored frozen at −20° C or preferably −70° C in small aliquots, and analyze it in every assay run. This will enable each laboratory to maintain the consistency in the reported T_3 uptake values in case of changes in the assay procedure or the kit components.

The Committee on Nomenclature of the American Thyroid Association (ATA) strongly recommends expression of the T_3 uptake as a ratio and considers it mandatory when one is calculating the free T_4 index (FTI).[6] The ATA committee also recommends that if the T_3 uptake is reported directly, it should always be reported with a serum total T_4 concentration so that the free T_4 index can be calculated. In other words, the T_3 uptake test is useful only as a means of calculating the free T_4 index:

$$\text{FTI} = \text{Total } T_4 \text{ concentration} \times T_3 \text{ uptake ratio}$$

Therefore, if the reference range of the T_3 uptake ratio is 0.83 to 1.17 and that of total T_4 concentration is 45 to 115 μg/L, the reference range of the free T_4 index would be 37.4 to 134 μg/L. The ATA committee recommends omission of the units from the free T_4 index to avoid confusion with measurements of T_4 itself. However, as with all diagnostic tests, each laboratory should establish its own reference range because differences may exist between laboratories and between the populations being studied.

In the case of uptake tests using enzyme- or fluorescein-labeled T_4 as the tracer, such as the Thyrozyme uptake test and the TDx-T uptake test, the end point of measurement is the enzyme activity expressed in absorbance units or degree of polarization, which may change in the direction opposite to the radioactivity measured in the T_3 uptake test. For example, in the Thyrozyme uptake test an enzyme inhibitor-T_4 conjugate is used as the tracer, and thus a higher enzyme activity or a higher absorbance reading means a lower free-inhibitor concentration or a greater number of available sites on serum proteins for binding the inhibitor. Therefore the Thyrozyme uptake ratio (absorbance of patient sample divided by absorbance of normal reference serum) used by the kit manufacturer is a reciprocal of the T_3.

$$\text{FTI} = T_3 \text{ uptake ratio} \times T_4 = \frac{T_4}{\text{Thyrozyme uptake ratio}}$$

REFERENCES

1. Hamolsky, MW, Goledtz, A, and Freedberg, AS: The plasma protein-thyroid complex in man. III. Further studies on the use of the in vitro red blood cell uptake of ^{131}I L-triiodothyronine as a diagnostic test of thyroid function, J Clin Endocrinol Metab 19:103-116, 1959.
2. Chen, I-W: Personal communication, Cincinnati, Ohio, 1983.
3. GammaCoat (^{125}I) T_3 Uptake Kit, catalog nos. CA-539 and 559, Clinical Assays, Cambridge, Mass.
4. Instruction manual for determination of T_3 uptake using the ARIA-HT, Becton Dickinson Immunodiagnostics, Salt Lake City, 1983.
5. Bellet, N, Winfrey L, Horton, A, et al: Development of homogeneous fluorescence rate immunoassays for thyroid function testing, Clin Chem 27:1071, 1982.
6. Solomon, DH, Denotti, J, DeGroot, LJ, et al: Revised nomenclature for tests of thyroid hormones in serum, J Clin Endocrinol Metab 42:595-598, 1976.

Vanillylmandelic acid

STEVEN J. SOLDIN

Vanillylmandelic acid, VMA, 3-methoxy-4-hydroxymandelic acid

Clinical significance: p. 422

Molecular formula: $C_9H_{10}O_5$

Molecular weight: 198.17 daltons

Merck Index: 4745

Chemical class: substituted carboxylic acid (catecholamine metabolite)

OH O
CH_3O C—C
H OH
HO

PRINCIPLES OF ANALYSIS

The methods for the measurement of vanillylmandelic acid (VMA) can be divided into those using spectrophotometric techniques and those using chromatographic techniques. Urine contains a large number of compounds, especially phenols, acid phenols, and other metabolites of aromatic ring compounds, that can interfere in either the colorimetric or chromatographic methods. Therefore almost all methods for the quantitation of urinary VMA use an extraction step to partially purify the analyte before analysis. In addition, an extraction followed by evaporation and reconstitution in small volume can be used to concentrate the VMA, improving the sensitivities of all the methods.

The usual procedure used for a liquid extraction of VMA first converts the VMA to a less soluble, nonionized form and then makes this form more insoluble by reducing the amount of water available for solvation. This is accomplished by adding concentrated HCl to decrease the pH to less than 2.0 and then saturating the solution with excess sodium chloride. The VMA can then be extracted by ethyl acetate with an 80% to 90% efficiency. The alternative methods used to partially purify VMA are anion-exchange or reversed-phase batch column chromatography.

The oldest quantitative method for VMA analysis is based on the direct method of Armstrong et al.,[1] in which extracted VMA reacts with diazotized *p*-nitroaniline to form a purple chromogen (method 1, Table 60-15). This chromogen is then spectrophotometrically quantitated at 520 nm. The other colorimetric methods are based on the conversion of VMA to vanillin. This can be accomplished by autoclaving the sample or by use of chemical oxidants such as sodium periodate, ferricyanide, Cu^{+2} ions, or aluminum oxide. The widely used procedure developed by Pisano et al.[2] measures vanillin directly at 348 to 360 nm (method 2, Table 60-15). The Pisano method uses sodium periodate to convert extracted VMA to vanillin. The vanillin is measured at 360 nm rather than at its peak at 348 nm because of the presence of other compounds that also absorb at about 348 nm. An alternative procedure was developed by Sunderman et al.[3] based on the observation of Sandler and Ruthven[4] that indole reacts with vanillin to produce a pinkish chromogen (method 3, Table 60-15). This chromogen, which is believed to be a carbonium ion complex, can be quantitated at 495 nm. The Sunderman method uses a number of extraction steps to increase the specificity and accuracy of the procedure: (1) the treatment of urine with magnesium silicate to remove interfering compounds, (2) oxidation of VMA to vanillin with ferricyanide, (3) extraction of vanillin with toluene and then back-extraction of the vanillin into an alkaline solution, and (4) formation of the pinkish carbonium complex of vanillin and indole. Wybenga and Pileggi[5] also used the indole condensation method for quantitation of VMA but improved on the specificity of the colorimetric reaction by partially purifying the VMA with anion-exchange chromatography before oxidizing it with sodium periodate.

All the colorimetric procedures have the potential for positive interference from either endogenous phenolic compounds, exogenous drugs, and dietary products. For example, the colorimetric methods that use vanillin have reported positive interference from dietary vanillin, as well as from a large number of phenolic acids found in certain foods, such as bananas, coffee, tea, and chocolate (for these methods, diets of patients must be restricted), or the VMA must be purified from urine. However, even with purification steps these methods may not have maximum specificity.

The nonspecificity of the colorimetric methods led to the use of chromatographic methods for separating the extracted VMA from interfering compounds. Whatman No. 1 and 3MM papers have been used for both unidirectional[6] and bidirectional[1,7] descending paper chromatography (method 4, Table 60-15). All these methods used diazotized *p*-nitroaniline to develop the chromatogram. The VMA spot was clearly distinguished from other compounds by both its purple color and its mobility (R_f). The amount of VMA present could be determined when the diameter of the spot is related to the diameters of standards or when the blue color is eluted and its absorbance is measured at 525 nm.

A thin-layer chromatographic (TLC) assay was proposed as a selected method in 1977 by the American Association for Clinical Chemistry (AACC) (method 5, Table 60-15).[8] An ethyl acetate extract was chromatographed on a silica stationary phase, and the chromatogram was developed with a toluene/acetic acid/ethyl acetate solvent phase. The VMA spot, developed with diazotized *p*-nitroaniline, was completely separated from other urinary constituents. The amount of VMA spotted could be determined either semiquantitatively by visual comparison with standards or quantitatively by densitometry at 505 nm using the method of standard addition to calibrate the mass of VMA chromatographed.

High-voltage electrophoresis on cellulose acetate has also been used for VMA analysis (method 6, Table 60-15). Ethyl acetate extracts are applied to cellulose acetate strips, and the components are separated by the influence of an electrical field of 200 to 250 V.[9] The spots are stained with diazotized *p*-nitroaniline, the purple VMA spot is eluted with an alkaline methanol solution, and VMA is quantitated at 520 nm. Methods 4 through 6 are now considered to be only of historical interest.

There have been many assays reported that use high-performance liquid chromatography (HPLC) to separate VMA from other constituents (method 7, Table 60-15). Although most of these assays use reversed-phase columns (RP-HPLC), the use of adsorption chromatography has also been reported.[112] The VMA can be quantitated by a variety of detectors, including ultraviolet light (254 to 260 nm),[10-13] amperometric detection,[14-17] and postcolumn reaction.[18,19]

One of the earliest RP-HPLC procedures described was a modification of the Pisano method. In this method, extracted VMA was converted to vanillin by sodium periodate. The vanillin was isolated by reversed-phase HPLC and quantitated by electrochemical detection.[12] Direct reversed-phase HPLC assays for VMA using electrochemical detection have also been reported. In some of these assays, the VMA is partially purified by liquid extraction (ethyl

Table 60-15 Methods of vanillylmandelic acid (VMA) analysis

Method	Principle	Use	Comment
1. Direct spectrophotometric	VMA + Diazotized *p*-nitroaniline → Purple color (520 nm)	Rare	Used most frequently to develop thin-layer and paper chromatography
2. Direct vanillin method (Pisano)	3-methoxy-4-hydroxymandelic acid (CH_3O, HO, CH(OH)COOH) $\xrightarrow{NaIO_4}$ Vanillin (CH_3O, HO, CHO) (360 nm)	Most common	Need anion-exchange chromatography to minimize interferences
3. Indole condensation	VMA (CH_3O, HO, CH(OH)COOH) $+ 2\ K_3Fe(CN)_6 + 2\ KCl + 3\ ZnCl_2 \rightarrow$ Vanillin $+ K_2Zn_3[Fe(CN)_6]_2\downarrow + 6\ KCl + 2\ HCl + CO_2\uparrow$; Vanillin + Indole $\xrightarrow{H_2SO_4,\ H_3PO_4}$ Carbonium salt (pink) (495 nm)	Common	Needs ion-exchange chromatography to minimize interferences
4. Paper chromatography	Separation of VMA from other compounds by differential absorption between paper and mobile phases; VMA spot developed using diazotized *p*-nitroaniline; amount of VMA determined from diameter of spot or by measurement of its absorption at 520-525 nm after extraction from paper	Rare	Requires 12-18 hr for analysis
5. Thin-layer chromatography (TLC)	Separation of VMA from other compounds by differential absorption between silica gel and mobile phases; VMA spot developed by diazotized *p*-nitroaniline; amount of VMA present determined by densitometer or reflectance spectroscopy	Rare	Requires 1-2 hr for analysis; an AACC selected method (1977)
6. High-voltage electrophoresis	Separation on cellulose acetate of VMA from other compounds having different mobilities in high-voltage electrical field (> 200 V); diazotized *p*-nitroaniline used to develop spot, which is quantitated spectrophotometrically (520 nm) after elution	Rare	Requires 4-5 hr for analysis
7. High-performance liquid chromatography (HPLC)	VMA separated from other compounds by differential partitioning between stationary (usually reversed) phase and mobile phase; VMA peak detected by variety of means	Infrequent, but use increasing	Most specific method, very few compounds known to interfere. Single analysis can take 15-20 min

Table 60-15 Methods of vanillylmandelic acid (VMA) analysis—cont'd

Method	Principle	Use	Comment
a. Ultraviolet	Absorbance at 260 nm		Many of these techniques can also be used to measure homovanillic acid (HVA)
b. Electrochemical	Oxidation of VMA to quinone forms; amperometric detection of electrons		
	CH_3O, HO, OH, CHCOOH (**VMA**) $\xrightarrow[-H^+]{-2e^-}$ CH_3O^+, O, OH, CHCOOH (**Quinone**) $\xrightarrow{-H_2O}$ O, O, OH, CHCOOH + CH_3OH + H^+ (**Diquinone**)		
	Or oxidation of VMA to vanillin and measurement of separated vanillin by its oxidation to its quinone form and amperometric detection of reaction:		
	CH_3O, HO, CHO $\xrightarrow{-2e^-}$ O, O, CHO + CH_3OH		
c. Postcolumn reaction	Eluate reacts with sodium periodate to form vanillin, which is measured spectrophotometrically at 360 nm		
8. Gas chromatography (GC)	Silyl and trimethylsilyl derivatives of VMA separated from other compounds by differential absorption on liquid stationary phase (phenylmethylsilicone) and use of a flame ionization detector	Infrequent	Highly specific Method can be adapted to measure HVA Analysis time short

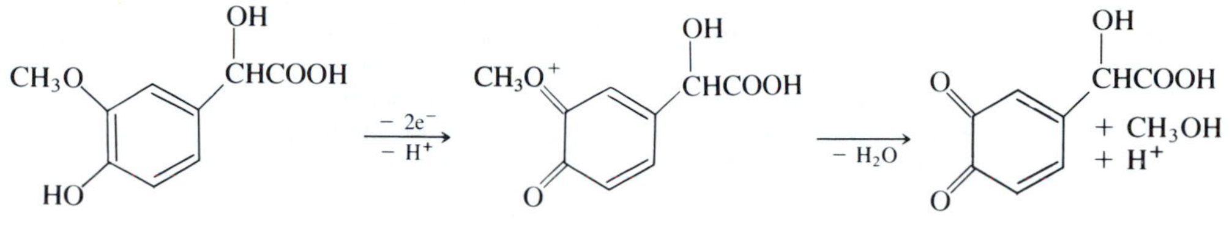

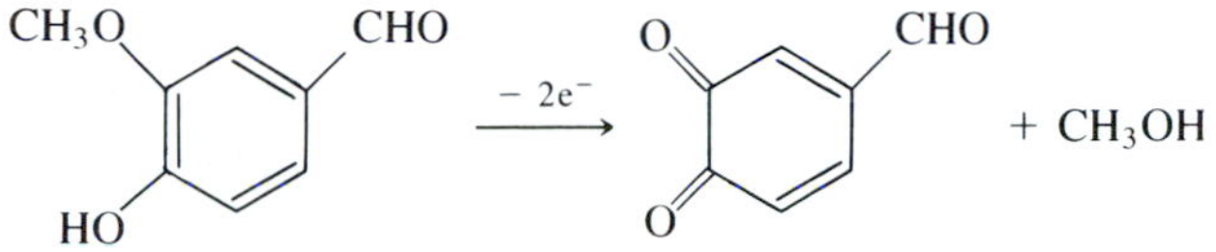

acetate)[16] or ion-exchange chromatography.[14] In others, the VMA is not purified at all.[17] In all these methods the VMA is oxidized to its quinone form by the positive potential (about +0.75 V) of the working electrode (usually, glassy carbon) electrochemical detector. (See reaction below.) The magnitude of the current generated by the electrochemical reaction is directly proportional to the amount of VMA eluted. Many of these reversed-phase HPLC assays can also be adapted for the simultaneous analysis of homovanillic acid (HVA).[11,13-15,17]

The postcolumn reaction used to monitor the HPLC eluate has been the sodium periodate oxidation of VMA to vanillin and the measurement of the vanillin at 360 nm.[11,12] One procedure uses direct injection of the urine,[12] whereas the other uses ion-exchange chromatography to purify the VMA partially.[11]

Many gas-chromatographic (GC) methods have been described for VMA quantitation (method 8, Table 60-15).[20,21] These methods require extraction of VMA and subsequent derivatization (trimethylsilyl) to form volatile conjugates. Silyl[20] and trimethylsilyl-ether conjugates[21] have been commonly used although other agents have been reported. A capillary GC method using 3,4-dihyroxybenzoic acid as an internal standard has been reported, with 5% phenylmethyl silicone as the stationary phase and a flame ionization detector.[21] The GC methods have also been used to quantitate HVA simultaneously with VMA. A GC method using mass fragmentography for detection has also been reported.[22]

For overall specificity and sensitivity of analysis, the HPLC procedures, especially with electrochemical detection, are the recommended methods of choice. Although

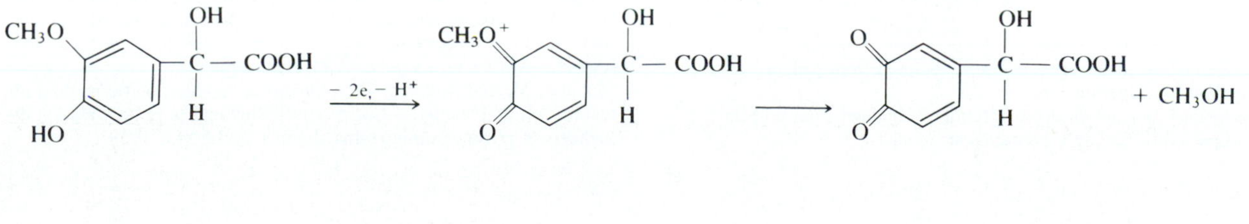

CH_3O, HO, OH, C—COOH, H $\xrightarrow{-2e,\,-H^+}$ CH_3O^+, O, OH, C—COOH, H $\longrightarrow$ O, O, OH, C—COOH, H + CH_3OH

Table 60-16 Comparison of methods of VMA analysis

Method parameters	Direct vanillin (Pisano)[2]	Indole condensation[5]	HPLC[15]
Reaction temperature	50° C	50° C	40° C
Time of analysis	1½-2 hr	2 hr	20 min
Sample volume (extracted)	Variable 0.2% of total 24-hr output	5 mL	5 mL
Final concentration of reagents	For formation of vanillin $NaIO_4$: 8.5 mmol/L (for extraction) $Na_2S_2O_5$: 45.8 mmol/L Phosphate buffer: 0.86 mmol/L (pH 7.5)	To 5 mL of column eluate For vanillin formation: $NaIO_4$: 7.2 mmol/L $Na_2S_2O_5$: 14.3 mmol/L For color formation: Indole: 0.39 mmol/L H_3PO_4: 0.16 mmol/L H_2SO_4: 4.91 mol/L	Mobile phase Phosphate buffer: 100 mmol/L, pH 6.7 Tetrabutylammonium phosphate: 0.5 mmol/L
Detector	Ultraviolet, 360 nm	Visible light: 495 nm	Electrochemical, +0.76
Precision ($\bar{X}$, % coefficient of variation)	—	4.8 mg/day, 5%	2.7 mg/day, 9.9%
Interferences	Dietary vanillin, aromatic phenols from various foods (bananas, tea, coffee, chocolate)	α-Methyldopa, *p*-hydroxymandelic acid	None known

few compounds have been reported to interfere in the HPLC assays, one must carefully review the chromatographs for closely eluting, interfering peaks.

For the small laboratory with a low volume of VMA requests, a modified Pisano or Sunderman procedure,[11] which uses anion-exchange chromatography to partially purify the VMA, may give adequate precision and specificity. Table 60-16 compares several methods of VMA analysis.

SPECIMEN

Because variations in VMA output occur throughout the day, a 24-hour urine collection is the recommended specimen of choice. It has been suggested that useful information can be obtained from specimens with shorter collection times, especially when the VMA output is expressed as micrograms of VMA per milligram of creatinine.[23] The urine should be collected in a container with acid preservative to prevent degradation of VMA. Usually 10 mL of 6 M HCl will suffice for a 24-hour collection, and the urine should be mixed throughout the collection period to ensure an adequately low pH of less than 2.0 throughout the entire specimen. Urine samples collected over shorter collection times should be kept on ice during the collection period and acidified to a pH below 3 immediately after the collection is completed. The specimen should be stored at 4° C until analyzed for VMA content.

Table 60-17 Percentile reference values for urinary VMA

Age (years)	Size of population (*n*)	VMA 95th	VMA 100th
Excretion*: mg/24 hr (μmol/24 hr)			
0-1	48	2.3 (11.6)	3.1 (15.6)
2-4	34	3.0 (15.1)	4.0 (20.2)
5-9	20	3.5 (17.7)	8.8 (44.4)
10-19	40	6.0 (30.2)	7.7 (38.9)
>19	56	6.8 (34.3)	8.1 (40.9)
Excretion†: mg/g of creatinine (μmol/mmol of creatinine)			
0-1	37	18.8 (10.7)	59.4 (33.9)
2-4	49	11.0 (6.3)	20.8 (11.9)
5-9	79	8.3 (4.7)	9.4 (5.4)
10-19	55	8.26 (4.7)	13.9 (7.9)
>19	56	6.0 (3.4)	8.3 (4.7)

*Analyses performed on timed urine samples obtained from patients under investigation for hypertension.

†Analyses performed on random urine specimens obtained from hospital patients not suspected of having a neural crest tumor.

REFERENCE RANGES

Age-related urinary excretion rates for VMA measured by HPLC are listed in Table 60-17. These ranges are given as the upper-percentile limits of excretion. The normal range for adults is 1.8 to 8.0 mg/24 hours (9.1 to 40.4 μmol/24 hours) for the modified Wybenga-Pileggi procedure.

REFERENCES

1. Armstrong, MD, McMillan A, and Shaw, KNF: 3-Methoxy-4-hydroxy-D-mandelic acid: a urinary metabolite of norepinephrine, Biochim Biophys Acta 25:422-423, 1957.
2. Pisano, JJ, Crout, JR, and Abraham, D: Determination of 3-methoxy-4-hydroxymandelic acid in urine, Clin Chim Acta 7:285-291, 1962.
3. Sunderman, FW, Jr, Cleveland, PD, Law, NC, and Sunderman, FW: A method for the determination of 3-methoxy-4-hydroxymandelic acid ("vanilmandelic acid") for the diagnosis of pheochromocytoma, Am J Clin Pathol 34:293-312, 1960.
4. Sandler, M, and Ruthven, CRJ: Quantitative colorimetric method for estimation of 3-methyl-4-hydroxymandelic acid in urine: value in diagnosis of phaeochromocytoma, Lancet 2:114-115, 1959.

5. Wybenga, D, and Pileggi, VJ: Quantitative determination of 3-methoxy-4-hydroxymandelc (VMA) in urine, Clin Chim Acta 16:147-154, 1967.
6. Vahidi, HR, Roberts, JS, San Filippo, J, Jr, and Sankar, DVS: Paper chromatographic quantitation of 4-hydroxy-3-methoxymandelic acid (VMA) in urine, Clin Chem 17:903-907, 1971.
7. Gitlow, SE, Mendlowitz, M, and Bertain, L: The biochemical techniques for detecting and establishing the presence of a pheochromocytoma, Am J Cardiol 26:270-279, 1970.
8. Badella, M, Routh, MW, Gump, BH, and Gigliotti, HJ: Thin-layer chromatographic method for urinary 4-hydroxy-3-methoxymandelic acid (vanilmandelic acid). In Cooper, GR, editor: Selected methods in clinical chemistry, Washington, DC, 1977, American Association for Clinical Chemistry, pp 139-145.
9. Hermann, GA: The determination of urinary 3-methoxy-4-hydroxymandelic (vanilmandelic) acid by means of electrophoresis with cellulose acetate membrane, Am J Clin Pathol 41:373-376, 1964.
10. Anderson, GM, Feibel, FC, and Cohen, DJ: Liquid-chromatographic determination of vanillymandelic acid in urine, Clin Chem 31:819-821, 1985.
11. Yoshida, A, Yoshioka, M, Tanimura, T, and Tamura, Z: Determination of vanilmandelic acid and homovanillic acid in urine by high speed liquid chromatography, J Chromatogr 116:240-243, 1976.
12. Felice, LJ, and Kissinger, PR: A modification of the Pisano method for vanilmandelic acid using high pressure liquid chromatography, Clin Chim Acta 76:317-320, 1977.
13. Bertani-Dziedzic, LM, Krstulovic, AM, Ciriello, S, and Gitlow, SE: Routine reversed-phase high performance liquid chromatographic measurement of urinary vanillylmandelic acid in patients with neural crest tumors, J Chromatogr 164:345-353, 1979.
14. Soldin, SJ, and Hill, JG: Simultaneous liquid chromatographic analysis for 4-hydroxy-3-methoxymandelic acid and 4-hydroxy-3-methoxyphenylacetic acid in urine, Clin Chem 26:291-294, 1980.
15. Soldin, SJ, and Hill, JG: Liquid chromatographic analysis for urinary 4-hydroxy-3-methoxymandelic acid and 4-hydroxy-3-methoxyphenylacetic acid and its use in investigation of neural crest tumors, Clin Chem 27:502-503, 1981.
16. Moleman, P, and Borstrok, JJM: Determination of urinary vanillylmandelic acid by liquid chromatography with electrochemical detection, Clin Chem 29:878-881, 1983.
17. Fujita, K, Maruta, K, Ito, S, and Nagatsu, T: Urinary 4-hydroxy-3-methoxymandelic (vanillymandelic) acid, 4-hydroxy-3-methoxyphenylacetic (homovanillic) acid, and 5-hydroxy-3-indoleacetic acid determined by liquid chromatography with electrochemical detection, Clin Chem 29:876-878, 1983.
18. Rosano, TG, and Brown, HH: Liquid chromatographic assay for urinary 3-methoxy-4-hydroxymandelic acid with use of a periodate oxidative monitor, Clin Chem 25:550-554, 1979.
19. Flood, JG, Granger, M, and McComb, RB: Urinary 3-methoxy-4-hydroxymandelic acid as measured by liquid chromatography, with on-line post-column reaction, Clin Chem 25:1234-1238, 1979.
20. Brewster, MA, Berry DH, and Moriarty, M: Urinary 3-methoxy-4-hydroxyphenylacetic (homovanillic) and 3-methoxy-4-mandelic (vanillylmandelic) acids: gas-liquid chromatographic methods and experience with 13 cases of neuroblastoma, Clin Chem 23:2247-2248, 1977.
21. Tuchman, M, Crippin, PJ, and Krivit, W: Capillary gas-chromatographic determination of urinary homovanillic acid and vanillylmandelic acid, Clin Chem 29:828-831, 1983.
22. Takahashi, S, Toshioka, M, Yoshiue, S, and Tamura, Z: Mass fragmentographic determination of vanilmandelic acid, homovanillic acid and isohomovanillic acid in human body fluids, J Chromatogr 145:1-9, 1978.
23. Tuchman, M, Robison, LL, Maynard, RC, et al: Assessment of the diurnal variations in urinary homovanillic and vanillylmandelic acid excretion for the diagnosis and follow-up of patients with neuroblastoma, Clin Biochem 18:176-179, 1985.

CHAPTER 61 Lipids

Amniotic fluid phospholipids: lecithin-to-sphingomyelin ratio and phosphatidyl glycerol

PAUL T. RUSSELL

Clinical significance: p. 569
Molecular formula: depends on fatty acid moieties
- Phosphatidyl choline (lecithin) (L) $C_{40}H_{82}O_9NP$ (dipalmitoyl-)
- Sphingomyelin (S) $C_{39}H_{81}O_7N_2P$ (palmitoyl-)
- Phosphatidyl inositol (PI) $C_{40}H_{80}O_{14}P$ (dipalmitoyl-)
- Phosphatidyl glycerol (PG) $C_{42}H_{83}O_{10}P$ (distearoyl-)

Molecular weight: depends on fatty acid moieties
- Phosphatidyl choline (lecithin) 752 (dipalmitoyl-)
- Sphingomyelin 721 (palmitoyl-)
- Phosphatidyl inositol about 883
- Phosphatidyl glycerol 779

Merck Index:
- L 5271
- S 8518

Chemical class: phospholipids

PRINCIPLES OF ANALYSIS

Tests for the estimation of pulmonary surfactant in amniotic fluid fall into three general categories: (1) those that measure the chemical constituents of surfactant, (2) those that measure the physical properties of lung surfactant, and (3) those that measure a variety of chemical agents that *correlate* with stages of fetal maturation.

The analysis of the chemical constituents of surfactant entails the separation of the individual components of the phospholipids present in amniotic fluid. Extraction and separation, followed by quantitation, permit the establishment of relationships such as the lecithin-to-sphingomyelin (L/S) ratio,[1] the phospholipid profile,[2] the fatty acid composition of amniotic fluid lecithin,[3] and many others that provide an index of fetal lung maturity.[4-6] Thin-layer chromatography (TLC) became the most widely used technique for most of these analyses after it was published as the method of choice by Gluck et al.[1,7,8] in their pioneering work on the L/S ratio.

The TLC systems (method 1, Table 61-1) most extensively used[5,6] employ silica gel stationary phases and mobile phases consisting of either chloroform-methanol-water or chloroform-methanol-ammonium hydroxide.[6] TLC separations are based on the differential adsorption of phospholipids with different polarities. Both one-dimensional and two-dimensional TLC systems have been used; the choice is made primarily by balancing the need to completely separate the phospholipids of interest (that is, lecithin, sphingomyelin, and phosphatidyl glycerol [PG]) from one another and the need to separate these compounds from the other phospholipid components of amniotic fluid.

A number of techniques can be used for the visualization of the separated phospholipids. One can char organic materials by spraying the TLC plates with sulfuric or phosphoric acid and then heating to 280° C; variations in charring temperatures can give variable results.[9] Alternatively, ammonium sulfate can be incorporated into the silica gel, eliminating the need for strong acid sprays during the charring procedure.[7] Lipids are visualized under ultraviolet radiation after spraying with rhodamine B or dichlorofluorescein. Bismuth subnitrate reacts with the choline moiety contained in lecithin and spingomyelin.[10] Other agents, such as cupric acetate and sulfuric acid with or without dichromate, react with the fatty acid double bonds, yielding colored products. The intensity of the colored spots depends on the degree of unsaturation of the phospholipid constituent fatty acids. Iodine vapor and molybdate ions also react with the unsaturated bonds in the fatty acid moieties of the phospholipids. After the phospholipids are separated by TLC and identified on the chromatogram by an indicator, the L/S ratio can be evaluated in a number of ways:

1. One can evaluate the chromatogram by transmission or reflection densitometry. Compounds that are separated and charred on thin-layer plates are quantitated densitometrically by use of a scanner. The L/S ratio is calculated from the recorder tracings, which are related proportionately to the plate concentrations of lecithin and sphingomyelin.
2. The phospholipids can be eluted from the chromatogram and the phospholipid quantitatively estimated by a phosphorus determination. Phosphorus determinations have been done by several methods.[11-15]

Table 61-1 Methods of amniotic fluid analysis

Method	Type of analysis	Principle	Usage	Comment
Chromatographic analysis of phospholipids				
1. Thin-layer	Semiquantitative	Silica gel adsorption chromatography; visualization of phospholipids by color-forming sprays or charring	Most common	Can be run as one dimension or two dimension
2. Gas-liquid chromatography	Quantitative	Analysis of fatty acids on polyester columns with flame ionization detector	Rare	Measured as methyl ester derivatives
3. High-performance liquid chromatography	Quantitative	Separation of L, S, PI, PE, PS, and PG by reversed-phase chromatography, detection at 203 nm	Rare	Ability to quantitate multiple components simultaneously may make this important method in future
Measurement of surfactant functional activity				
4. Foam stability (shake test, foam stability index [FSI])	Qualitative	Measures presence of surfactant by its ability to support foam bubbles generated by shaking amniotic fluid with ethanol	Common, as stat procedure	Problem of false-negative results
5. Surface tension	Quantitative	Presence of surfactant related to surface tension properties of amniotic fluid	Rare	Dedicated instrument, but relatively easy to perform
6. Fluorescence polarization	Quantitative	Surfactant increases microviscosity of fluid, which reduces rotation of dissolved fluorescent probe, resulting in increase in polarization of incident fluorescent light	Rare	Dedicated instrument, but relatively easy to perform

3. One can estimate the phospholipids on the chromatogram by planimetry. The area of each visualized spot is estimated either by taking the product of the length and width of the individual spots or by using a planimeter. The estimated area is proportional to the quantity of material present.

Gas-liquid chromatography (GLC) (method 2, Table 61-1) has provided a means to estimate pulmonary surfactant based on its unique fatty acid composition.[16] GLC involves quantification of the palmitic acid concentration of amniotic fluid, establishment of a ratio of palmitic acid to stearic acid, and determination of the relative content of palmitic acid contained in lecithin isolated from amniotic fluid. GLC for these fatty acid separations has been performed on polyester columns of diethyleneglycol succinate and ethyleneglycol succinate with flame ionization detectors. Phosphatidyl choline has been determined as the diacylglycerol trimethylsilyl ether derivative by GLC using a glass capillary column.[17] Preliminary TLC separation is required for this method and for some others (method 1, Table 61-1, also reference 3) before GLC analysis.

High-performance liquid chromatography (HPLC) (method 3, Table 61-1) has been used to separate phospholipid extracted from the amniotic fluid. One method employs diol bonded-phase columns with gradients (2% to 15% water) of acetonitrile to H_2O as the mobile phase.[18] Detection and quantitation are often based on ultraviolet absorption characteristics of double bonds (203 nm) in the fatty acid moieties of the phospholipids, though methods that measure saturated phospholipid palmitate have been published.[19]

The methods that have been proposed to measure specifically the desaturated phosphatidyl choline eliminate the unwanted unsaturated fatty acid–containing species with osmium tetroxide.[20] In general, some form of chromatography (column, TLC) is then necessary to isolate the unreacted lecithin.

All the procedures that measure phospholipid constituents of amniotic fluid require partial purification of the compounds before chromatography. Purification schemes usually involve one or more of the following steps: (1) centrifugation, (2) extraction, (3) acetone precipitation, (4) evaporation, and (5) reconstitution. However, there is considerable controversy about the efficacy of the first three steps. Thus a large number of extraction schemes using permutations and combinations of these techniques are in use. Various techniques have been developed for measurement of lecithin,[21-25] sphingomyelin,[21,25] and PG[26] using either spectrophotometry[22-25] or radiochemical techniques.[21,26] Radiochemical techniques in general have the inherent disadvantages of high cost and waste disposal.

Kits that can detect PG are commercially available (for

example, AmnioStat-FLM, Hana Biologics, Inc., Alameda, CA). These kits detect PG by means of an immunological agglutination test. This approach is sensitive (2 μg/mL) and specific. It requires about 15 minutes, and it is easy to perform. Results are reliable even with blood and meconium contaminate in the amniotic fluid sample. It has the theoretical limitation (as does any amniotic fluid test for concentration) that it would be subject to error in cases of extreme hydramnios or oligohydramnios.

Clinically these agglutination assays for PG are specific (high rate of true positives) but not very sensitive (high rate of false negatives). A positive result therefore can be used to indicate possible fetal maturity, but a negative result is not very reliable for indicating lack of fetal maturity. More recently, enzymatic methods for lecithin, sphingomyelin,[25] and PG[27] that have many attractive features have been published.

Methods that measure the physical properties of lung surfactant have recently attracted interest because they measure functional lung surfactant. Besides multiple phospholipid components, surfactant may also include a protein component for functional integrity. These tests measure surface tension, stable foam (bubbles), microviscosity, and other properties.[5,6] Many of these tests are simpler and quicker to perform than the chemical tests and thus have strong potential for routine laboratory use. To their disadvantage, some of them require expensive equipment and include techniques that are less familiar to most laboratory workers than the chromatographic methods.

One of the physical methods in most common use is the foam stability or *shake* test (method 4, Table 61-1). This test was devised by Clements et al.[28] and calls for serial dilutions of amniotic fluid mixed with equal volumes of 95% ethanol to exclude spurious surface-active compounds. The tubes are inspected for bubbles around the meniscus after being shaken. At the recommended final percentage of ethanol (47.5%), the formation of bubbles by other substances in amniotic fluid, such as proteins, bile salts, or salts of free fatty acids, is eliminated. Pulmonary surfactant forms stable surface films that support bubble stability. Therefore the test is positive if bubbles persist.

The foam stability index (FSI)[29] uses a mixture of different volumes of ethanol with 0.5 mL of amniotic fluid and provides semiquantitative evidence of surfactant content. The index is defined as the highest ethanol volume fraction of an amniotic fluid-ethanol mixture that will permit a stable ring of bubbles at the meniscus after vigorous shaking. An FSI value of 0.48 is comparable to an L/S ratio of 2.0 in correlating with fetal pulmonary maturity.[30]

One can also measure surface tension directly[29,31,32] using a surface balance or tensiometer (method 5, Table 61-1). These measurements give good correlations with other tests for lung surfactant.[33]

The fluorescence polarization assay measures the microviscosity of amniotic fluid lipids, which is related to surface tension (method 6, Table 61-1). A fluorescent probe, when mixed into amniotic fluid, dissolves in the hydrocarbon region. Its rotation in this hydrophobic environment depends on the microviscosity of the fluid. The greater the viscosity (that is, the more surfactant present), the more effectively opposed is the rotation of the probe. This results in an increased polarization of the emitted fluorescent light, and the extent of polarization can be measured by a specially designed instrument such as the Abbott TDx. The TDx assay employs a filtered or centrifuged sample and requires a simultaneous albumin measurement. More reliable results are obtained by relating the surfactant levels to amniotic fluid albumin levels.[34]

Nonchromatographic methods for assessing surfactant activity of amniotic fluid have only been recently developed and are rarely used. TLC remains the almost universal technique for amniotic fluid analysis, though there may be an increase in the use of alternative procedures as their clinical utility is demonstrated. The foam stability (shake) test is widely used as a stat procedure when TLC analysis is unavailable.

The reference method is the method of Gluck et al.[7,8,35] Either in its original form or in some modified form, this has been the most widely used method. Clinical interpretation and predictions have for the most part used Gluck's criteria.[7]

A large number of variations of the original procedure are in use today because virtually every step has more than one legitimate means for its execution. Other variations have developed because of differences of opinion over the need for specific steps in the procedure, such as the acetone precipitation step. Accurate clinical correlations are the only meaningful criteria for methodological considerations, and so it is incumbent on each laboratory to establish its own criteria with clinical experience.

SPECIMEN

Amniotic fluid obtained by amniocentesis is transported to the laboratory on ice and is promptly centrifuged at 500 *g* for 5 minutes. A 3 mL aliquot of the supernatant is needed for extractions.

Samples should be processed as soon as possible after collection. Untreated amniotic fluid held at room temperature can lose both lecithin and sphingomyelin over 48 hours.[36,37] According to Wagstaff et al.,[36] after only 4.5 hours lecithin levels are decreased by about 25%. Decreases of a small order occur with sphingomyelin over a similar period. This disproportionate loss of lecithin relative to sphingomyelin with time results in a consistent fall in the L/S ratio. However, if amniotic fluid is centrifuged at low speed, such as 500 *g* for 5 minutes, immediately after collection, it can be kept at room temperature for at least 4 days with no apparent change in the L/S ratio.[38] Centrifugation of amniotic fluid samples is necessary if storage is contemplated. Centrifuged samples stored at −20° C are stable for prolonged periods without noticeable alteration of the L/S ratio.

PROCEDURE: LECITHIN-SPHINGOMYELIN RATIO BY THIN-LAYER CHROMATOGRAPHY[39]

Principle

Phospholipids are extracted from the amniotic fluid into chloroform-methanol and then precipitated in cold acetone. The precipitated phospholipids are separated by TLC, visualized, and quantitated. An estimate is made of the amount of lecithin and sphingomyelin present in the amniotic fluid. This is expressed as the L/S ratio and is used to evaluate fetal lung maturity. Table 61-2 summarizes the important parameters of the procedure.

Reagents

Eastman Chromatographic Sheets, 6061 Silica gel without fluorescent indicator, 20 × 20 cm. Cut into 1 × 13 cm strips. Activate by storing in a desiccator over Drierite for 24 hours.

Developing solvent. Pipette into a 50 mL round-bottom capped centrifuge tube 5.0 mL of methanol, 0.8 mL of distilled water, and 13.0 mL of chloroform. Vortex mix. Keep solution capped when not in use. Prepare fresh daily.

Methanol. Spectral grade.

Chloroform. Spectral grade.

Acetone. Reagent grade.

Bromthymol blue (417 mg/L). A water-soluble indicator dye (3′,3′-dibromothymol sulfone phthalein, sodium salt) available from Aldrich Organic Chemicals (Milwaukee). Add 200 mg of bromthymol blue to 200 mL of distilled water and 32.0 mL of 1 M NaOH. Dilute to 480 mL with distilled water. Add 5.0 g of boric acid powder (reagent grade) and mix. This solution is stable for 3 months.

Table 61-2 Comparison of assay conditions for amniotic fluid analyses

Parameter	Thin-layer chromatography	Shake test
Assay temperature	Ambient	Ambient
Sample volume	1 mL	1.0, 0.75, 0.50, 0.25, and 0.20 mL
Fraction of sample volume	0.50 (extraction)	0.50
Final concentration of reagents	(Not applicable)	47.5% ethanol
Interferences	Blood, meconium, high-speed centrifugation before testing	Blood, meconium, soap, serum, biological fluids, surface evaporation from tubes, movement of tubes once they have been shaken, high-speed centrifugation of sample before testing

Nitrogen gas. Compressed, 100%.

1 M NaOH (40 g/L). Place 40 g of NaOH in a 1 L volumetric flask and add about 800 mL of distilled water slowly to dissolve. When the NaOH is fully dissolved and the temperature of the solution is back to room temperature, add distilled water to mark. Mix thoroughly. This is stable for 12 months at room temperature.

Standards. The lecithin, sphingomyelin, and PG standards can be purchased from Supelco, Inc. (Bellefonte, PA) as chloroform-methanol solutions. The contents of each ampule are quantitatively transferred to a volumetric flask (the ampule must be rinsed as well) and then diluted with chloroform-methanol (2:1, v/v) to a final concentration of 1 mg/mL.

Lecithin (catalog no. 4-6012). 50 mg is diluted to 50 mL with chloroform-methanol.

Sphingomyelin (catalog no. 4-6009). 25 mg is diluted to 25 mL with chloroform-methanol.

Phosphatidyl glycerol (catalog no. 4-6013). 10 mg is diluted to 10 mL with chloroform-methanol.

Assay

Equipment: centrifuge 20 × 50 mm test tubes with cork stoppers, forceps, 12 mL conical screw-capped centrifuge tubes, and 3 mL conical test tubes.

This procedure is best performed in duplicate:

1. Centrifuge 3 to 5 mL of amniotic fluid at 500 *g* for 5 minutes at ambient temperature.
2. Pipette 1 mL of the supernatant into duplicate 12 mL conical screw-capped test tubes. Pipette 1 mL into each of three tubes if PG is also to be determined.
3. To each 1 mL sample of amniotic fluid, add 1 mL of methanol. Vortex mix. Add 2 mL of chloroform, and vortex mix vigorously.
4. Centrifuge tubes at 500 *g* for 5 minutes to facilitate phase separations. Three layers should be seen:
   a. Aqueous layer (top)
   b. Protein fluff layer (middle)
   c. Chloroform layer (bottom)
5. Remove as much of the chloroform layer as possible using a disposable Pasteur transfer pipet (go through the top two layers carefully). Place this in a 3 mL conical tube. Evaporate to *moist dryness* under nitrogen. (NOTE: Water bath should be below 40° C.)
6. If PG is to be determined, repeat the extraction (steps 3 to 5) for the tube for PG.
7. Wash down the lipid adhering to the sides of each tube with 50 μL of chloroform. Take to moist dryness again.
8. Chill the 3 mL tubes with the lipid extract in ice for 1 minute. With a Pasteur pipet, add two drops of ice-cold acetone while swirling the tube in the ice. Add eight more drops of ice-cold acetone and ice

the tube for another minute. Decant the acetone. Wash the precipitate with another 10 drops of ice-cold acetone while the tube remains iced. Decant the supernatant. Dry the precipitate gently under a stream of nitrogen.

9. Dissolve the cold acetone precipitate in 6 μL of chloroform. The entire amount is then spotted 2 cm from the end of the activated TLC strip. (The PG tube is spotted on a plate to be described later.)
10. Using forceps to handle the strips, place each into 1.5 mL of the developing solvent in a 20 × 150 mm test tube. Stopper the tube with a cork. Chromatograph the strip to a distance of 2 cm from the top of the strip.
11. Remove the strips from the tubes with forceps, air dry, and then dip in the bromthymol blue indicator contained in a 20 × 150 mm test tube. Blot excess dye from the strip. The lecithin and sphingomyelin spots appear dark yellow (Fig. 61-1). Visualization is easier when the strip is briefly placed in a 20 × 150 mm test tube containing concentrated NH_4OH. The ammonia vapors turn the spots blue.

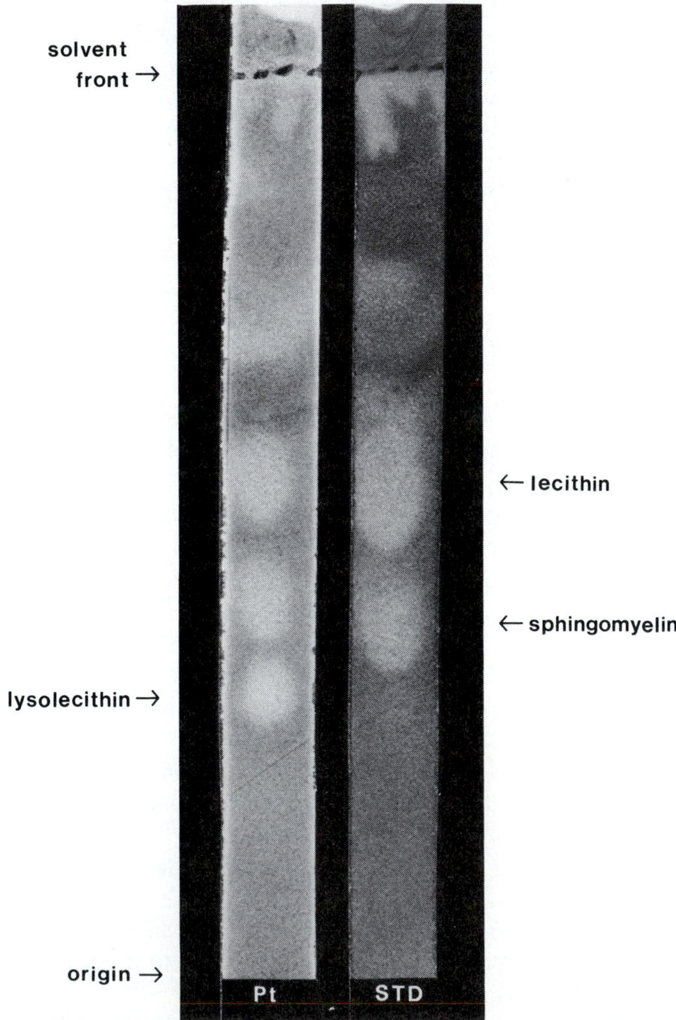

Fig. 61-1 Separation of lecithin and sphingomyelin on thin-layer strips as described for method to determine L/S ratio. *Pt,* Patient sample; *STD,* L and S standards.

Calculation

1. Circle each spot along the outermost edges with a pencil. Measure maximum width and length of each spot in millimeters. Use the following calculation:

$$\text{L/S ratio} = \frac{(\text{Length of lecithin}) \times (\text{Width of lecithin})}{(\text{Length of sphingomyelin}) \times (\text{Width of sphingomyelin})}$$

2. Average the values from the duplicate samples.
3. Apply L/S standards daily. Spot 30 mL of lecithin standard (10 mg/mL) on top of 15 μL of sphingomyelin standard (100 μg/mL) on a fresh strip and chromatograph.
4. Running time for L/S is 30 to 45 minutes.

PROCEDURE: PHOSPHATIDYL GLYCEROL BY THIN-LAYER CHROMATOGRAPHY

Reagents

Chloroform. Spectral grade.

Methanol. Spectral grade.

Glacial acetic acid. Reagent grade.

Solvent system. Chloroform-methanol–acetic acid–water (65:25:8:4, v/v).

2′2′-Dichlorofluorescein (0.2 g/L) in ethanol. Dissolve 2 g of dye into 100 mL of ethanol. This is the stock solution. Dilute 1:100 in ethanol for working solution. When stored at room temperature, the stock solution is stable for 1 year and the working solution for 3 months.

Assay

Equipment: silica gel 60 thin-layer plates (plates are activated at 100° C for 1 hour before use), 5 × 20 cm (E. Merck, Darmstad, West Germany) and filter paper–lined TLC tank.

1. Use the third tube prepared in step 2 of the L/S ratio procedure. Dissolve lipid contents in 10 μL of chloroform and spot 2 cm from the bottom and a third of the way in from the side of a silica gel (5 × 20 cm) TLC plate (activated).
2. Spot lecithin (20 μL of 1 μg/μL) and PG (20 μL of 1 μg/L) standards 2 cm from the bottom and a third of the way in from the opposite side of the same plate.
3. Place the plate in a filter paper–backed thin-layer tank previously equilibrated with the solvent system for at least 1 hour (replace solvent system weekly or more often if runs become atypical).
4. Permit the solvent system to run about 10 cm up the plate. Withdraw the plate from the tank and air dry for 15 to 20 minutes.

5. Spray the plate with dichlorofluorescein reagent, allow to dry, and view under ultraviolet radiation 15 minutes to 1 hour later (Fig. 61-2).
6. The visual identification of a PG spot confirms the presence of PG. This visualization is sensitive to 0.5 μg and indicates the presence of PG in amniotic fluid to be greater than or equal to 0.5 μg/mL.

Notes

Blood contamination of amniotic fluid has some effect on the L/S ratio, though there is no agreement about whether the blood increases or decreases the L/S ratio.[36,40] Samples contaminated with blood are still useful in the prediction of fetal lung maturity if PG is found. PG is virtually absent from blood, meconium, and vaginal secretions, all of which contain phospholipids and other components that cause inaccurate ratios.[41-44] Alternatively, one can isolate the 10,000 g pellet from the contaminated fluid and measure fetal pulmonary surfactant from the lamellar bodies.[41,44]

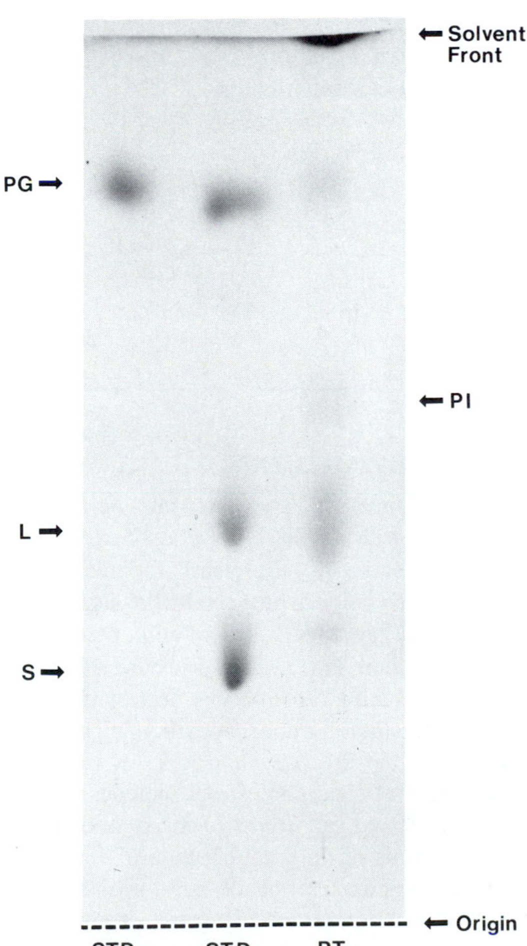

Fig. 61-2 PG spot on silica gel thin-layer plate according to method for PG. *L,* Lecithin; *PI,* phosphatidyl inositol; *PT,* patient; *S,* sphingomyelin; *STD,* standard *(left sample,* PG; *middle sample,* Pg, L, and S).

Contamination of amniotic fluid with meconium interferes with the interpretation of chromatograms,[45,46] though Gerbie et al.[47] did not consider that its presence interfered with the estimation of the L/S ratio. Wagstaff et al.[36] demonstrated a consistent rise in the L/S ratio with increasing contamination with meconium. Chromatograms from samples contaminated with meconium are characterized by a lysolecithin spot immediately after the sphingomyelin spot. If chromatographic separation is not good, the lysolecithin spot will come to lie so close to the sphingomyelin spot that the two may be confused, giving a falsely decreased L/S ratio.[42] Other authors report falsely high L/S ratios from fluid contamination with meconium.[37] Again, it seems prudent to rely on the presence of PG if samples are contaminated with meconium or to assess surfactant of the 10,000 g pellet.[41,44]

REFERENCE RANGE

An L/S ratio more than 2.0 is considered indicative of fetal maturity. If a spot is seen in the PG area, the test is considered positive. If no spot is seen, it is considered negative. (See Chapter 37 for details.)

REFERENCES

1. Gluck, L, Kulovich, MV, and Borer, RC, Jr: The diagnosis of the respiratory distress syndrome (RDS) by amniocentesis, Am J Obstet Gynecol 109:440-445, 1971.
2. Kulovich, MV, Hallman, MB, and Gluck, L: The lung profile. I. Normal pregnancy, Am J Obstet Gynecol 135:57-63, 1979.
3. Russell, PT, Miller WJ, and McLain, CR: Palmitic acid content of amniotic fluid lecithin as an index to fetal lung maturity, Clin Chem 20:1431-1434, 1974.
4. Wagstaff, TI: The estimation of pulmonary surfactant in amniotic fluid. In Fairweather, DVI, and Eskes, TKAB, editors: Amniotic fluid-research and clinical application, ed 2, Amsterdam, 1978, Excerpta Medica, pp 347-391.
5. Freer, DE, and Statland, BE: Measurement of amniotic fluid surfactant, Clin Chem 27:1629-1641, 1981.
6. Tsao FH, and Zachman, RD: Prenatal assessment of fetal lung maturations: a critical review of amniotic fluid phopholipid tests. In Farrell, PM, editor: Lung development: biological and clinical perspectives, vol 2, Neonatal respiratory distress, New York, 1982, Academic Press, Inc, pp 167-203.
7. Gluck, L, and Kulovich, MV: Lecithin-sphingomyelin ratios in amniotic fluid in normal and abnormal pregnancy, Am J Obstet Gynecol 115:539-546, 1973.
8. Gluck, L, Kulovich, MV, and Borer, RC Jr: Estimates of fetal lung maturity, Clin Perinatol 1:125-139, 1974.
9. Mueller, RG: Effect of charring temperature on observed L/S ratio, Clin Chim Acta 122:79-83, 1982.
10. Coch, EH, Kessler, G, and Meyer, JS: Rapid thin-layer chromatographic method for assessing the lecithin-sphingomyelin ratio in amniotic fluid, Selected Methods Clin Chem 8:63-70, 1977.
11. Fiske, CH, and Subbarow, Y: The colorimetric determination of phosphorus, J Biol Chem 66:375-400, 1925.
12. Chen, PS, Toribara, TY, and Warner, H: Microdetermination of phosphorus, Anal Chem 28:1756-1758, 1956.
13. Bartlett, GR: Phosphorus assay in column chromatography, J Biol Chem 234:466-468, 1959.
14. Badham, LP, and Worth, HGJ: Critical assessment of phospholipid measurement in amniotic fluid, Clin Chem 21:1441-1447, 1975.
15. Yee HY, Yee TML, and Jackson, B: Determination of phosphatidylcholine in amniotic fluid, Microchem J 25:61-71, 1980.
16. Adams, FH, Fujiwara, T, Emmanouilides, GC, and Raiha, N: Lung

phospholipids of human fetuses and infants with and without hyaline membrane disease, J Pediatr 77:833-841, 1970.
17. Lohinger, A, Salzer, H, Sumbruner, G, et al: Relationships among human amniotic fluid dipalmitoyl lecithin, postpartum respiratory compliance, and neonatal respiratory distress syndrome, Clin Chem 29:650-655, 1983.
18. Briand, RL: High-performance liquid chromatographic determination of the lecithin-sphingomyelin ratio in amniotic fluid, J Chromatogr 223:277-284, 1981.
19. Sax, SM, Moore, JJ, Oley, A, et al: Liquid-chromatographic estimation of saturated phospholipid palmitate in amniotic fluid compared with a thin-layer chromatographic method for acetoneprecipitated lecithin, Clin Chem 28:2264-2268, 1982.
20. Mason, RJ, Nellenbogen, J, and Clements, JA: Isolation of disaturated phosphatidylcholine with osmium tetroxide, J Lipid Res 17:281-284, 1976.
21. McDonald, L, Robin, NL, and Siegel, L: New method for determining lecithin and sphingomyelin in amniotic fluid, Clin Chem 27:410-416, 1981.
22. Artiss, JD, Draisey, TF, Thibert, RJ, et al: A procedure for the direct determination of micromolar quantities of lecithin employing enzymes as reagents, Microchem J 24:239-258, 1979.
23. Artiss, JD, Draisey, TF, Thibert, RJ, et al: The determination of lecithin and total choline-containing phospholipids in amniotic fluid employing enzymes as reagents, Microchem J 25:153-168, 1980.
24. Anaokar, S, Garry, PJ, and Standefer, JC: Enzymatic assay for lecithin in amniotic fluid, Clin Chem 25:103-107, 1978.
25. McGowan, MW, Artiss, JD, and Zak, B: Enzymatic colorimetry of lecithin and sphingomyelin in aqueous solution, Clin Chem 29:1513-1517, 1983.
26. Siegel, L, Walker, SI, and Robin, NI: An enzymatic radiochemical method for determining phosphatidylglycerol in amniotic fluid, Clin Chem 29:782-785, 1983.
27. Artiss, JD, McGowan, MW, Strandbergh, DR, et al: Enzymatic colorimetric determination of phosphatidylglycerol in amniotic fluid, Clin Chem 30:534-537, 1984.
28. Clements, JA, Platzker, ACG, and Tierney, DF: Assessment of risk of respiratory distress syndrome by a rapid test for surfactant in amniotic fluid, N Engl J Med 286:1077-1081, 1972.
29. Statland, BE, and Freer, DE: Evaluation of two assays of functional surfactant in amniotic fluid, surface tension lowering ability, and the foam stability index test, Clin Chem 25:1770-1773, 1979.
30. Sher, G, Statland, BE, and Freer, DE: Clinical evaluation of the quantitative foam stability index test, Obstet Gynecol 55:617-620, 1980.
31. Tiwary, CM, and Goldkrand, JW: Assessment of fetal pulmonary maturity by measurement of the surface tension of amniotic fluid lipid extract, Obstet Gynecol 48:191-194, 1976.
32. Goldkrand, JW, Varki, A, and McClurg, JE: Surface tension of amniotic fluid lipid extracts: prediction of pulmonary maturity, Am J Obstet 28:591-598, 1977.
33. Bichler, A, Daxenbichler, G, Oriner, A, et al: Amniotic fluid surface tension measurements versus various phospholipid determinations in the assessment of fetal lung maturity, Respiration 37:114-121, 1979.
34. Russell, JC: A calibrated fluorescence polarization assay for assessment of fetal lung maturity, Clin Chem 33:1177-1184, 1987.
35. Gluck, L, Landowne, RA, and Kulovich, MV: Biochemical development of surface activity in mammalian lung, Pediatr Res 4:352-364, 1970.
36. Wagstaff, TI, Whyley, GA, and Freedman, G: Factors influencing the measurement of the lecithin/sphingomyelin ratio in amniotic fluid, J Obstet Gynaecol Br Commonw 81:264-277, 1974.
37. Kulkarni, BD, Bieniarz, J, Burd, L, and Scommegna, A: Determination of lecithin/sphingomyelin ratio in amniotic fluid, Obstet Gynecol 40:173-179, 1972.
38. Whitfield, CR, and Sproule, WB: Fetal lung maturation, Br J Hosp Med 12:678-690, 1974.
39. Painter, PC: Simultaneous measurement of lecithin, sphingomyelin, phosphatidylglycerol, phosphatidylinositol, phosphatidylethanolamine, and phosphatidylserine in amniotic fluid, Clin Chem 26:1147-1151, 1980.
40. Wagstaff, TI, Whyley, GA, and Freedman, G: The measurement of the lecithin/sphingomyelin ratio of amniotic fluid after thin layer chromatography, Ann Clin Biochem 11:24-27, 1974.
41. Oulton, M: The role of centrifugation in the measurement of surfactant in amniotic fluid, Am J Obstet Gynecol 135:337-343, 1979.
42. Buhi, WC, and Spellacy, WN: Effects of blood or meconium on the determination of the amniotic fluid lecithin/sphingomyelin ratio, Am J Obstet Gynecol 121:321-323, 1975.
43. Torday, J, Carson, L, and Lawson, EE: Saturated phosphatidylcholine in amniotic fluid and prediction of the respiratory distress syndrome, N Engl J Med 301:1013-1018, 1979.
44. Oulton, M, Martin, TR, Faulkner, GT, et al: Development study of a lamellar body fraction isolated from the human amniotic fluid, Pediatr Res 14:722-728, 1980.
45. Bryson, MJ, Gabert, HA, and Stenchever, MA: Amniotic fluid lecithin/sphingomyelin ratio as an assessment of fetal pulmonary maturity, Am J Obstet 114:208-212, 1972.
46. Hobbins, JC, Brock, W, Speroff, L, et al: L/S ratio in predicting pulmonary maturity in utero, Obstet Gynecol 39:660-664, 1972.
47. Gerbie, MV, Gerbie, AB, and Boehm, J: Diagnosis of fetal maturity by amniotic fluid phospholipids, Am J Obstet Gynecol 114:1078-1082, 1977.

Cholesterol

HERBERT K. NAITO

Clinical significance: p. 454
Molecular formula: $C_{27}H_{46}O$
Molecular weight: 386.64 daltons
Merck Index: 2181
Chemical class: sterol, lipid

HO

PRINCIPLES OF ANALYSIS

Liebermann,[1] in 1885, first described the color reaction of sulfuric acid with a solution of cholesterol in acetic anhydride. Four years later, Burchard[2] reported that a more intense blue-green color is produced when acetic anhydride and sulfuric acid are added to a solution of cholesterol in chloroform (method 1, Table 61-3). Since that time, the *Liebermann-Burchard reaction* has been widely used as a colorimetric reaction for the estimation of cholesterol in biological fluids.

Measurement of total cholesterol includes measurement of both the ester and free forms of the steroid. In serum or plasma, two thirds of the total cholesterol exists in the esterified form, with the rest in the free form. This has some analytical implications, since in some chemical reactions the color development with the ester cholesterol is greater in intensity than that with free cholesterol. This in turn can lead to a large positive bias. In some enzymatic reactions, hydrolysis of longer-chain cholesterol esters, such as cholesterol arachidonate, is not complete, a condition that leads to a negative bias.

Table 61-3 Methods for serum cholesterol analysis

Method	Method classification	Principle	Usage	Comments
1. Liebermann-Burchard (L-B)	One-, two-, three-, or four-step	Cholesterol extracted and reacted with strong acid (sulfuric acid) and acetic anhydride to form colored cholestahexaene–sulfonic acid molecule (A_{max}, 410 nm); nonesterified cholesterol precipitated by digitoxin, and remaining cholesterol measured and free cholesterol calculated: Total − Esterified = Free	Infrequently used	Total cholesterol reaction overestimates concentration of esterified cholesterol Unstable color
2. Abell et al.[3]	Three-step	Cholesterol extracted with zeolite, esters chemically hydrolyzed (saponification), and total cholesterol measured by Liebermann-Burchard reaction	Considered current reference method	Laborious
3. Iron-salt-acid	Two-step	Similar to reaction conditions of method 2, except Fe^{+3} ions are added to yield tetraenylic cation (A_{max}, 563 nm)	Not frequently used	Sevenfold more sensitive than L-B method Free and esterified cholesterol give *same* color; no need to hydrolyze esters
4. *p*-Toluene-sulfonic acid (*p*-TSA)	Three-step	Similar to method 3; *p*-TSA reacts with cholesterol derivative to form chromophere (A_{max}, 550 nm)	Rarely used	Free and esterified cholesterol give same color Bilirubin causes large positive error
5. Enzymatic end point	One-step	a. Cholesterol esters $\xrightarrow{\text{Cholesterol esterase}}$ Cholesterol + Fatty acids b.* Cholesterol + O_2 $\xrightarrow{\text{Cholesterol oxidase}}$ Cholest-4-en-3-one + H_2O_2 c. H_2O_2 + 4-Aminophenazone (or other dye) $\xrightarrow{\text{Peroxidase}}$ Oxidized dye (A_{max}, 500 nm) + H_2O	Most common	Accurate and easily automated Future reference method

*Can monitor reaction by following O_2 consumption with oxygen electrode.

In single-step direct assays there is no sample preparation, that is, no isolation and purification of the steroid or steroids. Thus direct procedures are those carried out on serum or plasma samples without any prior solvent extraction steps.

In two-step assays on organic-phase extraction step is introduced before measurement of cholesterol and other chemically related steroids. This pretreatment step removes many nonspecific chromogens that might interfere with the assay.

Three-step procedures involve, in addition to extraction of cholesterol, a saponification step that hydrolyzes the fatty acid moiety from the cholesterol ester. Consequently, one measures only free cholesterol. The method of Abell et al.[3] (method 2, Table 61-3) belongs in this classification and is now considered a reference method.

Four-step methods go a step further than the three-step method of extraction, saponification, and color development. The total extractable steroids are purified for cholesterol determination by the addition of a saponin, digitonin. The reactive site on the C-3 position on the cyclopentanoperhydrolphenanthrene ring in cholesterol is a hydroxyl group that is esterified by the digitonin. This causes the complex to be precipitated. The addition of the digitonin step also eliminates the effect of interfering nonspecific chromogen constituents.

Methods of analysis

Literally hundreds of cholesterol methods have been published, usually as modifications of the following reactions: (1) Liebermann-Burchard, (2) iron-salt-acid, (3) *p*-toluene-sulfonic acid, or (4) enzymatic end point. Table 61-4 is a comparison of reaction conditions of the candidate reference method and two most frequently used methods for total cholesterol determinations.

Liebermann-Burchard reaction. Among the nonenzymatic colorimetric reactions for cholesterol, the Liebermann-Burchard (L-B) procedure (method 1, Table 61-3) is perhaps the most widely used. The L-B reaction generally is carried out in a strong acid medium—sulfuric acid, acetic acid, and acetic anhydride. In the chemical reaction cholesterol goes through a stepwise oxidation, with each

Table 61-4 Comparison of reaction conditions of methods for total cholesterol analysis

Condition	Abell-Levy-Brodie-Kendall*	Liebermann-Burchard (L-B)†	Enzymatic‡
Temperatures	50° C—hydrolysis 25° C—colorimetric reaction	Ambient—extraction, reaction	37° C
Sample volume	500 μL	500 μL	4 μL
Fraction of sample volume	0.091 for hydrolysis	0.1 for extraction	0.01
Final concentration of reagents	Hydrolysis step KOH: 0.322 mol/L Ethanol: 909 g/L Extraction step Hexane: 50% v/v Colorimetric reaction step Acetic anhydride: 64.5% (v/v) Sulfuric acid: 3.2% (v/v) Acetic acid: 32.3% (v/v)	Insoluble extraction mixture (in 99% isopropanol) 99% isopropanol: 990 g/L Zeolite: 320 g/L $CuSO_4 \cdot H_2O$: 16 g/L Kaolin: 32 g/L $Ca(OH)_2$: 32 g/L Reaction mixture Acetic anhydride: 60% (v/v) Sulfuric acid: 30% (v/v) Acetic acid: 10% (v/v)	Cholesterol esterase: 40 U/L Cholesterol oxidase: 56 U/L Peroxidase: 1 U/L Sodium cholate: 15 mmol/L 4-Aminophenazone: 0.4 mmol/L Sodium hydroxybenzoate: 10 mmol/L Phosphate buffer: 90 mmol/L (pH 6.75) EDTA: 8 mmol/L
Time of reaction	Hydrolysis, 60 min Reaction, 30 min	Extraction, 20 min AutoAnalyzer-II: 8 min	10 min
Wavelength	620 nm	630 nm	505 nm, 600 nm, secondary
Linearity	5 g/L (12.5 mmol/L)	4 g/L	5 g/L
Precision§	—	2879 mg/L, <3%	1845 mg/L, <3%
($\bar{X}$, % coefficient of variation)	—	3375 mg/L, <3%	2830 mg/L, <3%

*Candidate reference method from Abell, LL, et al: J Biol Chem 195:357-366, 1952. Modified by Centers for Disease Control Lipid Laboratory, Atlanta (by Duncan et al, unpublished data).
†Technicon AutoAnalyzer-II method. Modified from Lipid Research Clinics Program method.[50]
‡Sclavo Diagnostics reagents (Wayne, NJ). Analysis performed on Hitachi 705 analyzer (Boehringer Mannheim Diagnostics, Indianapolis).
§From College of American Pathologists Quality Assurance Survey data. L-B without extraction.

step yielding a cholestapolyene molecule, which has one more double bond than the compound from which it was derived. The initial L-B step involves protonation of the OH^- group in cholesterol and subsequent loss of water to give the carbonium ion 3,5-cholestadiene, which is the first step in color reaction. Sequential oxidation of this allylic carbonium ion by SO_3^- yields a cholestahexaene–sulfonic acid chromophoric compound with absorbance maxima (A_{max}) of 610 and 410 nm. This chemical reaction for cholesterol determination has undergone many modifications through the years and has included the measurement of either free or esterified cholesterol or both. Digitonin, a saponin, reacts with the free OH^- (at the C-3 position) group of the A ring of the sterol to form an insoluble complex, cholesterol digitonide. The measurement of the sterol content in the supernatant solution provides an estimate of esterified cholesterol. To obtain free cholesterol, one subtracts the esterified value from the total (free plus ester) cholesterol value.

Numerous investigators have attempted to apply the L-B reaction directly to serum without preliminary organic-solvent extraction of cholesterol. Simple, rapid, direct procedures for estimation of serum cholesterol were described by Richardson et al.,[4] Potsma and Stroes,[5] and Kim and Goldberg.[6]

Iron-salt-acid reaction. In 1953 Zlatkis et al.[7] proposed a new colorimetric procedure for cholesterol, dependent on the magenta color produced when a solution of ferric chloride in concentrated sulfuric acid is added to a solution of cholesterol in glacial acetic acid. The color developed in this reaction is more intense and more stable than that developed in the L-B reaction. This reaction (method 3, Table 61-3) involves acetic acid and sulfuric acid in the absence of acetic anhydride. In this reaction, however, Fe^{+3} must be added to obtain the desired chromogen. As in the L-B procedure, the initial step is the protonation of the OH^- group in the cholesterol molecule and subsequent loss of water to form the carbonium ion (3,5-cholestadiene). Serial oxidation of this allylic carbonium ion by Fe^{+3} yields a tetraenylic cation with an absorbance maximum of 563 nm. The iron-salt-acid procedures are about sevenfold more sensitive than the L-B methods.

Measurements of serum cholesterol by the $FeCl_3$-H_2SO_4 reaction were automated by Levine and Zak[8] and by Block et al.[9] The Technicon AutoAnalyzer technique was adapted for dual-chemical analysis of cholesterol and triglycerides by Kohring and Katterman[10] and by Wease et al.[11]

In 1969 a "micromethod" for serum cholesterol, which requires only 20 μL of serum, was described.[12] In this technique, cholesterol is precipitated as a dextran sulfate complex to avoid interference from bilirubin. The precipitate is dissolved in glacial acetic acid, and ferric chloride reagent is then added. Even greater sensitivity can be achieved by fluorometric detection of the ferric chloride–sulfuric acid chromogen.[13] Solow and Freeman[13] adapted Leffler's method to fluorometry and showed that the sensitivity permits quantitation of less than 1 μg of cholesterol. For example, they reported fluorometric measurements of cholesterol in 100 μL samples of spinal fluid. An automated fluorometric technique for serum cholesterol also was described.[14]

A ferric chloride reagent in acetic acid, followed by sulfuric acid, separately, gives the same intensity of color with cholesterol from serum, regardless of whether it is in the free form or esterified.[15] Therefore saponification is not necessary when one is using the ferric iron–salt-acid reaction for the determination of serum cholesterol. Methods using this reaction can then be simpler than those using the L-B reagent systems. A direct method for determining serum cholesterol uses ferric acetate–uranium acetate and sulfuric acid–ferrous sulfate reagents. In this method,[16,17] the ferric acetate–uranium acetate serves as a unique precipitating agent that removes bilirubin (as biliverdin) with proteins, clears the serum of lipids, and serves to extract total cholesterol without need of solvents. The acetate reagent extract, when mixed with the sulfuric acid–ferrous sulfate reagent, yields a purple color with an absorption maximum of 560 nm. The maximum color develops within 15 minutes and is stable for at least 1 hour. This method provides excellent correlation with analyses by the Abell et al.[3] technique and is reported to be practically free from interference.

Wybenga et al.[18] developed a stable reagent for direct manual determination of serum cholesterol. This reagent consists of ferric perchlorate in a mixture of sulfuric acid and ethyl acetate. Their method is subject to moderate interference from bilirubin, hemoglobin, and gamma-globulin.

In addition to the ferric methods, ferrous cholesterol assays have been reported. In 1960 Searcy and Bergquist[19,20] described a color reaction for cholesterol that is similar to that of Zlatkis et al.[7] except that ferrous sulfate is used instead of ferric chloride. The studies of Badzio[21] indicate that chromogenic property in the Searcy-Bergquist[19,20] reaction depends on the brand of acetic acid used and that this reaction is no more sensitive than the reaction of Zlatkis et al.[7] Zak and Epstein[22] reported the Searcy-Bergquist reagent to be unstable.

***p*-Toluenesulfonic acid reactions.** *p*-Toluenesulfonic acid methods are based on reactions of cholesterol with *para*-toluenesulfonic acid (*p*-TSA), acetic anhydride, glacial acetic acid, and H_2SO_4 (method 4, Table 61-3).

Enzymatic end-point reactions. Enzymatic techniques (method 5, Table 61-3) for determining cholesterol have emerged to compete with the classical L-B reaction and have become the most popular method for cholesterol analysis. The original work used preliminary alkaline saponification of the sample to produce only free cholesterol.[23-25] In the next step, cholesterol oxidase, an enzyme almost entirely specific for cholesterol, was added. This caused the breakdown of cholesterol to cholest-4-en-3-one and hydrogen peroxide, after which several different reaction systems have been used to produce a final chromogen.

Subsequent developments led to the technique of enzymatic hydrolysis of cholesterol esters by employing cholesterol esterase.[26-31]

The first chemical step in the enzymatic methods for cholesterol uses the enzyme cholesterol esterase to hydrolyze the cholesterol esters present in the serum to free cholesterol and free fatty acids:

$$\text{Cholesterol esters} + H_2O \xrightarrow{\text{Cholesterol esterase}} \text{Cholesterol} + \text{Free fatty acids} \quad \textit{Eq. 61-1}$$

As discussed previously, the second step uses the enzyme cholesterol oxidase in the presence of oxygen to oxidize the cholesterol (both the free cholesterol found in the serum and the free cholesterol generated in step 1) to cholest-4-en-3-one and hydrogen peroxide:

$$\text{Cholesterol} + O_2 \xrightarrow{\text{Cholesterol oxidase}} \text{Cholest-4-en-3-one} + H_2O_2 \quad \textit{Eq. 61-2}$$

In this reaction cholesterol concentration can be determined by amperometric measurement of the rate of oxygen depletion. Beckman Instruments (Brea, CA) produced the Cholesterol Analyzer-2, which measured cholesterol by this method.

Other assays make use of the ability of hydrogen peroxide to oxidize compounds to produce colored species that can be measured spectrophotometrically:

$$2H_2O_2 + \text{Phenol} + \text{4-Aminophenazone} \xrightarrow{\text{Peroxidase}} \text{Quinoneimine dye} + 4H_2O \quad \textit{Eq. 61-3}$$

or

$$H_2O_2 + \text{Methanol} \xrightarrow{\text{Catalase}} \text{Formaldehyde} + H_2O \quad \textit{Eq. 61-4}$$

$$\text{Formaldehyde} + \text{Acetylacetone} + \text{Ammonium acetate} \rightarrow \text{3,5-Diacetyl-1,4-dihydrolutidine} \quad \textit{Eq. 61-5}$$

or

$$H_2O_2 + \text{Ethanol} \xrightarrow{\text{Catalase}} \text{Acetaldehyde} + H_2O \quad \textit{Eq. 61-6}$$

$$\text{Acetaldehyde} + NADP^+ \xrightarrow[\text{dehydrogenase}]{\text{Aldehyde}} \text{Acetate} + NADPH + H^+ \quad \textit{Eq. 61-7}$$

Reaction 3 (Trinder's reaction), which forms the quinoneimine dye (absorbance maximum 500 to 525 nm), is the basis of the majority of the kits currently on the market.[26,27,32,29,33,34] Other indicator systems that have been used in Equation 61-3 include 2-hydroxy-3,5-dichlorobenzenesulfonic acid and 3,3′,5,5′-tetramethylbenzidine.

Reaction 5 (Equation 61-5), monitored at 405 nm,[31,35,36] has the advantage of not being subject to bilirubin interference and has been developed as a rate procedure to improve on the original time-consuming end-point assay.

Reaction 7 (Equation 61-7) is monitored at 340 nm as an end-point reaction.[37]

Additional methods that measure cholesterol by analysis of the H_2O_2 produced by cholesterol oxidase include polarography[23,37,38] employing electrodes specific for hydrogen peroxide, action of peroxidase to oxidize homovanillic acid to a fluorescent compound measured at 470 nm,[39] and the condensation with *o*-dianisidine in the presence of peroxidase to produce a colored compound measured at 450 nm.[40,41]

Most total cholesterol assays today have within-day precision of less than 4% (coefficient of variation [CV]), whereas day-to-day precision is slightly higher, that is, about 3% to 6% CV. With a great amount of care, one can achieve a within-day precision of less than 2% and a day-to-day precision of less than 3%.[42,43] The Centers for Disease Control (CDC) has proposed a modification of the procedure of Abell et al.[3] for use as a reference method for cholesterol.[44] This method includes hydrolysis of the cholesterol esters, extraction with petroleum ether, and color development with acetic acid–acetic anhydride–sulfuric acid reagent.[45,46] The CDC still uses this method for the assignment of *target values* on serum-calibrating materials and on reference quality control materials.

Interlaboratory comparisons of measurements of serum cholesterol by a Technicon AutoAnalyzer procedure that uses the L-B reagent on 2-propanol extracts were described by Lippel et al.[47] Their study clearly illustrated that interlaboratory performance can be significantly improved when a common standardization material is used.

The major cause of method-specific biases that lead to inaccuracy of cholesterol measurements and poor interlaboratory comparability is inaccurate assignment of cholesterol target values to the calibrating materials. The manufacturers must standardize the method of assigning accurate values to the calibrating materials that should be traceable to the National Committee for Clinical Laboratory Standards National Reference System for Cholesterol. This approved system includes the use of the National Bureau of Standards (NBS) definitive method (isotope dilution mass spectroscopy) and certified reference materials and the CDC reference method (Abell-Kendall) and certified reference materials. Thus the user should select analytical systems (investments and reagents) from manufacturers that have adopted this procedure of providing reliable serum calibrators that will help ensure more accurate cholesterol measurements. To verify the manufacturer's claims, the laboratory should perform cholesterol analysis on fresh patient materials on the laboratory analytical system in question and have the samples analyzed by a CDC Standardized and Certified National Reference Method Laboratory. If the values agree within 3% of the true values generated by the reference laboratory, one can be assured that the analytical system is standardized for accuracy.

A properly calibrated enzymatic method is certainly the preferred method for routine laboratory analysis. Enzymatic methods can be as accurate and precise as the Abell-Kendall reference method. Since these methods are readily available as kits, this technique is not presented in detail.

SPECIMEN

Patient preparation

Determinations of lipid constituents in plasma or serum are normally done on blood drawn from patients fasting for 12 to 16 hours. Patient preparation is an integral part of ensuring proper interpretation of lipid analysis. The optimum patient condition at the time of blood drawing is achieved by (1) no change in dietary habits for at least 3 weeks, (2) stable body weight, and (3) fasting for at least 12 hours (preferred but not necessary for cholesterol determinations). Proper fasting means no food of any sort, except for water and possibly black coffee (without sugar) in the morning. The fasting sample is essential for triglyceride analysis, since triglyceride levels increase as soon as 2 hours postprandially and reach a maximum at 4 to 6 hours. Nonfasting samples are not suitable for triglyceride analysis, since elevated results caused by normal assimilation of food cannot be distinguished from elevated results resulting from abnormal lipid metabolism or inborn errors of metabolism. Keep in mind that a sample from a nonfasting patient can be hyperchylomicronemic and hyperprebeta-lipoproteinemic, and this excess can in itself interfere with a given cholesterol assay. This is especially true for greatly hypertriglyceridemic samples (greater than 20,000 mg/dL), which have chylomicrons whose particle size is greater than 80 nm; these can interfere with colorimetric assays. Ethanol consumption causes acute but transient elevations in serum triglyceride concentrations. This is especially evident in carbohydrate-sensitive hypertriglyceridemic persons. Therefore it would be advisable to request that the patient refrain from drinking alcohol for 72 hours before the day of blood drawing. Thus for cholesterol and phospholipid determinations the fasting sample is not so important, although the fasting time may make a different in cholesterol levels for a small percentage of people.

Blood-drawing techniques

Setting up a standardized procedure for drawing blood is important, especially when long-term comparison studies are being considered. Posture, emotional and physical stress, selected vein, tourniquet use, tube selection, and fasting time are some important factors to be considered. In one convenient method for the drawing of blood from an adult patient, the patient is seated, blood is drawn from the antecubital vein, and the patient is subjected to a minimum amount of stress.

It is well recognized that plasma volume increases and the concentrations of nondiffusible plasma components decrease when a standing subject assumes a recumbent position, a result of redistribution of water between the vascular and extravascular compartments. A significant reduction in total plasma cholesterol has been measured after 5 minutes, and decreases of as much as 10% to 15% have been recorded 20 minutes after a recumbent position is assumed. The effect on cholesterol concentration when the subject changes from a standing position to a sitting position is also significant, though somewhat smaller—about 6% after 10 to 20 minutes. Although these changes should probably not be regarded as causing errors in the lipid analyses, since they occur physiologically, they can nonetheless complicate the interpretation of the measurements made, especially when one is monitoring the patient.

If a tourniquet is used, it should be removed before blood sampling, since prolonged use of a tourniquet has been reported to increase lipid values. Serum cholesterol concentrations were found to increase an average of 10% to 15% after 5 minutes of occlusion. From a practical standpoint, errors of this magnitude would not normally be encountered, since the tourniquet is usually removed within 30 to 60 seconds and the changes occurring during this period are insignificant. Increases of 2% to 5% have been observed after about 2 minutes; these increases may be seen if difficulties arise during sampling or if blood is taken by an inexperienced technician.

Choice of plasma or serum

If phlebotomy is properly performed, either plasma or serum is usually suitable for total cholesterol, triglyceride, or phospholipid determinations. Fibrinogen in improperly prepared serum can cause difficulties with continuous-flow systems (such as AutoAnalyzers) and other newer generation micro-sample-reagent analytical systems by plugging or coating the tubing or joint connectors or by causing a change in flow rates. On the other hand, if plasma is used, spontaneous hydrolysis of triglycerides can occur if the tubes are left at ambient temperature too long. Serum or plasma left at room temperature too long can have increased lecithin-cholesterol acyltransferase activity. This can alter blood lipid or lipoprotein composition, changing the free cholesterol-to-cholesterol ester and lecithin-to-lysolecithin ratios. Plasma is usually preferred when lipids and lipoproteins are being chemically analyzed. However, the new cholesterol cut points being recommended by the National Institutes of Health (NIH) are based on serum cholesterol measurements. To use these new cut points, serum is the preferred sample; if plasma is used, the value is multiplied by 1.03 to convert the plasma value to a serum-equivalent value. If plasma is chosen, the suggested anticoagulant is solid EDTA, 1 mg/mL of blood, and the blood cells should be separated as soon as possible (within 2 hours).

Certain anticoagulants, such as fluoride, citrate, and oxalate, cause rather large shifts of water from the red blood cells to the plasma, which result in the dilution of plasma components. For example, the apparent total cholesterol concentration of plasma containing 2.5 mg/mL of sodium oxalate was found to be 10% lower than that of serum obtained concurrently without anticoagulant. A recent study by Kuba et al.[48] possibly indicates a similar finding between plasma and serum, except that the difference between the two was about 2% (at levels of 1220 and 2440 mg/L) for cholesterol and 5% (at levels of 380 to 3740 mg/L) for triglycerides. Heparin, on the other hand, causes no detectable change in red blood cell volume and decreases total cholesterol concentration by only 1% or less. EDTA, which is commonly used to prepare plasma for lipoprotein analysis, causes a slightly greater decrease in lipid concentration. We have found that the cholesterol and triglyceride concentrations of plasma containing 1 mg/mL of EDTA are about 3% lower than those of serum. Although EDTA causes slightly greater changes than heparin does in cholesterol concentration, it is generally preferred for lipoprotein analysis for several reasons. EDTA retards the autooxidation of unsaturated fatty acids and cholesterol by chelating heavy metal ions such as Cu^{+2}. Oxidation leads to alterations in the physical properties of the lipoproteins. In addition, EDTA can reduce changes that can occur as a result of contamination by phospholipase C–producing bacteria. Such contamination is associated with spurious increases in total glyceride concentration that are caused by the production of partial glycerides from some phospholipids. These changes are minimized by the addition of EDTA, which is an antioxidant and an inhibitor of phospholipase C activity.

Sodium fluoride may be used when one is collecting plasma for glucose determinations. Its major function is to prevent red blood cell glycolysis, which will alter glucose results. It has been reported by Martinek[49] that colorimetric determination of total cholesterol on samples containing fluoride lowers values by 300 to 500 mg/L. Our data on enzymatic determination of total cholesterol indicate that fluoride may lower values by about 52%.[42]

For newer instruments, whole blood can often be used as the sample. For these samples, EDTA is the anticoagulant of choice.

Plasma and serum preparation

For serum, good technique dictates centrifuging the blood samples after an adequate clotting time but without excessive delay after the sample is drawn; alterations in lipoprotein composition can result from the exchange of cholesterol esters and triglycerides between high-density lipoproteins (HDLs) and the other lipoproteins. Several of these processes can occur simultaneously and influence results. Their cumulative effects can be either additive (they may increase the error of the analysis) or compensatory (the processes may have opposite effects on the results minimizing the error), depending on the exact procedure.

Storage of samples

Two of the common variables of sample handling that influence measured lipid and lipoprotein values are the length of time and the conditions under which samples are stored before analysis. It is generally recommended that plasma be stored in the liquid state when it is to be used for lipid and particularly lipoprotein analysis or for lipoprotein electrophoresis studies; in addition, the determination should be performed promptly. In practice, however, delays can occur for a variety of reasons, and it may even be necessary to analyze frozen samples. We have observed that the process of repeated freezing and thawing has a noticeable effect on lipoprotein electrophoretic patterns—that is, the presence of lipid-stainable material at the application point—which is suggestive of lipoprotein degradation. Several groups of workers have measured lipoprotein levels in samples that have been stored at different temperatures for varying periods of time, frozen and unfrozen.[50] The measurements were made with the analytical ultracentrifuge, and the results generally indicated that the levels of all lipoproteins may decrease with storage. No definitive studies have been made determining the optimum freezing temperature and the length of storage permissible at this temperature. Cooper[51] reported that when specimens are stored in the liquid state at 4° C for longer than 1 week, HDL cholesterol values, reproducibility of the test, and resolution of lipoprotein patterns decrease. The HDL cholesterol values were found to be more stable when the specimens were stored at −60° C than when the specimens were stored at −20° or −5° C over 10 months. At present, the general consensus is that freezing at −60° C provides the longest stable storage and may allow for reproducible results even after a year or more. Freezing at −20° C will keep samples stable for a few months if self-defrosting freezer units are avoided, since samples stored in these units periodically undergo thawing and freezing, causing a breakdown of the lipid and lipoprotein components. If the samples are to be used in a few days, refrigeration at 4° C is sufficient. However, even at refrigerated temperatures, spontaneous hydrolysis of the triglycerides to free glycerol and fatty acids takes place and effectively reduces the true triglyceride concentration. In addition, the sharpness of the lipoprotein bands decreases with time when electrophoresis is performed. Therefore, if more than short-term storage is anticipated, serum and plasma samples (but not whole blood) should be frozen to prevent the spontaneous hydrolysis and oxidation of the unsaturated lipids, which occurs during storage, especially at warmer temperatures.

Stored serum or plasma samples must be adequately mixed before use. Lipids will layer by density within the sample when stored and give different results, depending on the layer from which an aliquot is obtained. Simple inversion of the tube that contains the serum or plasma is not sufficient, particularly if the sample was frozen. Instead, the sample should be carefully mixed on a vortex mixer for 5 to 10 seconds. Particular attention should be given to those samples that have appreciable amounts of "standing" chylomicrons. One can obtain accurate lipid results only when the chylomicrons are dispersed evenly through the sample before taking an aliquot for lipid analysis. If pronounced chylomicronemia exists, the sample should be diluted with saline solution so that positive-displacement errors do not occur with the excessive macromolecules.

If an organic solvent extraction method is used, the extract can be stored at 4° C for short periods (a few days) if one is concerned only with total cholesterol, triglyceride, or phospholipid analyses. On the other hand, if one is interested in fatty acid profiles, the extract should be frozen immediately to inhibit autooxidation. Care must be taken when one is storing the purified lipid extracts. Cork or rubber stoppers or caps with paper, cork, or rubber liners should be avoided when one is storing lipid extraction organic solvents. Screw-capped tubes with Teflon-lined caps are good storage vials if a good seal is formed. Evaporation of the solvent can be a problem if care is not taken to ensure the selection of a good storage container. A well-sealed container or vial also helps to keep oxygen out of the sample and to prevent oxidation of the unsaturated lipids. If the samples are to be stored in this form for a long period, it is best to overlay the sample extract with nitrogen before the vial is sealed.

REFERENCE RANGE AND INTERPRETATION

The 95th percentile limits for normal values, once advocated by Fredrickson et al.[52] for lipoprotein phenotyping, are shown in Table 61-5.

Although it is customary to use normal limits or reference ranges to determine abnormal values for most blood constituents, the determination of abnormal cholesterol levels may be an exception to the rule. In 1980 this issue was addressed.[53] A problem arises in defining what levels of plasma lipid will separate persons with elevated blood fats from the rest of the "normal" population. Cut-off values for plasma lipids are assumed to be the 95th percentile of their distribution in a given population. Thus before addressing the question of what is adequate dietary or drug therapy for the control of serum lipid and lipoprotein levels, one must first consider the degree to which blood lip-

Table 61-5 Historical normal ranges for total serum cholesterol

Age (years)	Men*	Women*	Suggested normal limits
0-19	1720 ± 340	1790 ± 330	1200-2300 (3.10-5.95)
20-29	1830 ± 370	1790 ± 350	1200-2400 (3.10-6.21)
30-39	2100 ± 330	2040 ± 370	1400-2700 (3.62-6.98)
40-49	2300 ± 550	2170 ± 350	1500-3100 (3.88-8.02)
50-59	2400 ± 480	2510 ± 490	1600-3300 (4.14-8.53)

From Fredrickson, DS, et al: N Engl J Med 276:148-156, 1967.
*Total cholesterol in mg/L in mean ± standard deviation (values in parentheses are in mmol/L).

Table 61-6 Classifications based on total cholesterol and low-density lipoprotein (LDL) cholesterol

Classification	Total cholesterol (mg/dL)*	LDL cholesterol (mg/dL)
Desirable	<200	<130
Borderline to high risk	200-239	130-159
High risk	≥240	≥160

From National Cholesterol Education Program: Adult Treatment Panel Report.
*Milligrams of cholesterol per deciliter of blood; to convert mg/dL cholesterol to mmol/L, divide by 38.7 or multiply by 0.02586.

Table 61-7 National Institutes of Health 1984 Consensus Development Conference Statement on Lowering Cholesterol*

Age (years)	Moderate risk† mg/L (mmol/L)	High risk‡ mg/L (mmol/L)
2-19	>1700 mg/L (4.40) and >1850 mg/L (4.78)	Still being studied as per approximate cutoff value
20-29	>2000 mg/L (5.17)	>2200 mg/L (5.69)
30-39	>2200 mg/L (5.19)	>2400 mg/L (6.21)
≥40	>2400 mg/L (6.21)	>2600 mg/L (6.72)

*Proposed values for selecting men and women at moderate and high risk requiring treatment.
†75th percentile.
‡90th percentile.

The consensus values for persons 2 to 19 years of age should be considered preliminary estimates. Further studies are being performed to determine the appropriate cutoff range for this age group. After carefully examining epidemiological, clinical, and genetic data, the expert panel recommended the adoption of these new cholesterol limits for adults as the most prudent means of having a meaningful impact on reducing the risk for coronary heart disease.[59]

ids should be lowered; that is, at what blood lipid level should the clinician consider a sample ''hyperlipidemic''? Many clinical laboratories and practicing physicians still use normal ranges in evaluating whether a sample can be considered normal or abnormal. These nomograms are based on sampling apparently normal persons and arbitrarily defining hyperlipidemia as being present when the plasma cholesterol, triglycerides, or both are above the 95th-percentile value for the population to which the persons belong.

Unfortunately, because of the way we have defined ''normal'' in the past, many test results do not correlate very well with disease states or health-risk conditions on an individual basis. Thus the ''normal'' range may not be a ''healthy'' range. Although a cholesterol value greater than 3100 mg/L may be within the 95th percentile of the distribution of an apparently normal male population between 51 and 59 years of age in the United States, about 50% of these persons eventually develop coronary heart disease (CHD). Epidemiological studies[54] show that CHD risk is linear down to a cholesterol concentration of at least 1800 mg/L. To rely on these ''normal'' ranges, however, one needs to redefine the word ''normal.''

In 1975 Galen and Gambino[55] stated that the concept of normality is itself inadequate for the proper interpretation of test results. They indicated that normality is a matter of semantics, and they used the example of total cholesterol. In the United States the mean value for cholesterol for men between 40 and 60 years of age is 2250 mg/L.[55] However, there are populations relatively free of atherosclerotic heart disease in which the mean cholesterol level for the same age group is 1500 mg/L.[56,57] Therefore the mean value of 2250 mg/L is a normal value only in a statistical sense. With regard to morbidity and mortality, a value of 2250 mg/L may be quite ''abnormal.'' It is perhaps more appropriate now if we work with a reference range based on the risk for developing disease, as in CHD.

The goal of the Adult Treatment Panel (National Cholesterol Education Program) was to develop practical and detailed guidelines for clinicians to use in measuring, assessing, and treating high blood cholesterol in adult patients.[58] The panel has developed recommendations for total cholesterol and low-density lipoprotein (LDL) cholesterol cut points that slightly modify the NIH Consensus Development Conference cut points (Table 61-6).

The American Heart Association recommends that the upper limits for total cholesterol levels for children and adults should be 2000 and 2200 mg/L, respectively. The NIH in 1984 addressed this issue and recommended the adoption of new upper limits (Table 61-7).[59] The NIH Consensus Development Conference on Lowering Blood Cholesterol was charged in 1984 with reviewing the evidence relating cholesterol levels to CHD. The panel unanimously concluded that elevated blood cholesterol is a major cause of coronary artery disease and that lowering elevated blood cholesterol levels (specifically blood levels of LDL) will reduce the risk of heart attacks attributable to CHD, as established by the findings of the Lipid Research Clinics–CPPT and related studies. After careful review of genetic, experimental, epidemiological, and clinical trial evidence, the NIH panel recommended classifying and treating substantially higher-risk adults, that is, those with blood cholesterol above the 75th percentile (borderline high risk) and 90th percentile (high risk) (Table 61-6).

In November 1985 the National Cholesterol Education program was initiated by the National Heart, Lung, and Blood Institute with the goal of decreasing the prevalence of elevated blood cholesterol in the United States to help reduce CHD morbidity and mortality. A major objective of the program is to identify and treat the one in four American adults at significantly increased risk for the development of CHD because of high cholesterol levels. As part of the overall cholesterol education program, two panels of experts have been created to date: the Panel on Detection, Evaluation, and Treatment of High Blood Cholesterol in Adults (Adult Treatment Panel) and the Laboratory Standardization Panel on Blood Cholesterol Measurement.

REFERENCES

1. Liebermann, C: Ueber das Oxychinoterpen, Dtsch Chem Geselsch 18:1803-1809, 1885.
2. Burchard, H: Beitraĝe zur Kenntnis des Cholesterins, Chem Zentralbl 61:25-27, 1890.
3. Abell, LL, Levy, BB, Brodie, BB, and Kendall, FE: Simplified methods for the estimation of the total cholesterol in serum and demonstration of its specificity, J Biol Chem 195:357-366, 1952.
4. Richardson, RW, Setchell, KD, and Woodman, DD: An improved procedure for the estimation of serum cholesterol, Clin Chim Acta 31:403-407, 1971.
5. Potsma, T, and Stroes, JAP: Lipid screening in clinical chemistry, Clin Chim Acta 22:569-578, 1968.
6. Kim, E, and Goldberg, M: Serum cholesterol assay using a stable Liebermann-Burchard reagent, Clin Chem 15:1171-1179, 1969.
7. Zlatkis, A, Zak, B, and Boyle, AJ: A new method for the direct determination of serum cholesterol, J Lab Clin Med 41:486-492, 1953.
8. Levine, JB, and Zak, B: Automated determination of serum total cholesterol, Clin Chim Acta 10:381-384, 1964.
9. Block, WD, Jarett, KJ Jr, and Levine, JB: An improved automated determination of serum total cholesterol with a single color reagent, Clin Chem 12:681-689, 1966.
10. Kohring, B, and Katterman, R: Simultaneous colorimetric determination of cholesterol and triglycerides in serum with a dual-channel AutoAnalyzer, Z Klin Chem Klin Biochem 12:282-286, 1974.
11. Wease, DF, Espinosa, ES, and Anderson, YJ: New system for automated extraction of simultaneous determination of serum cholesterol and triglycerides, Clin Chem 21:1430-1436, 1975.
12. Jordan, WJ Jr, and Knoblock, EC: A micromethod for total cholesterol eliminating the effect of high bilirubin levels, Clin Chem 15:807-808, 1969.
13. Solow, EB, and Freeman, LW: A fluorometric ferric chloride method for determining cholesterol in cerebrospinal fluid and serum, Clin Chem 16:472-476, 1970.
14. Robertson, G, and Cramp, DG: An evaluation of cholesterol determination in serum and serum lipoprotein fractions by a semiautomated fluorimetric method, J Clin Pathol 23:243-245, 1970.
15. Webster, D: The chromatographic separation of ester and free cholesterol in blood, Clin Chim Acta 8:19-25, 1963.
16. Jung, DH, and Parekh, AC: A new color reaction for cholesterol assay, Clin Chim Acta 35:73-78, 1971.
17. Parekh, AC, and Jung, DH: Cholesterol determination with ferric acetate-uranium acetate and sulfuric acid-ferrous sulfate reagents, Anal Chem 42:1423, 1970.
18. Wybenga, DR, Pillegi, VJ, and Dirstine, PH: Direct manual determination of serum total cholesterol with a single stable reagent, Clin Chem 16:980-984, 1970.
19. Searcy, RL, and Bergquist, LM: Use of ferric ammonium sulfate in serum cholesterol determination, Clin Chim Acta 5:192-199, 1960.
20. Searcy, RL, Bergquist, LM, Jung, RC, et al: Rapid Ultramicro estimation of serum total cholesterol, J Lipid Res 1:349-351, 1960.
21. Badzio, T: The possibility of errors in the determination of cholesterol in blood with $FeCl_2$-reagent, Clin Chim Acta 11:53-56, 1965.
22. Zak, B, and Epstein, E: New cholesterol reagent, Clin Chem 7:268-270, 1961.
23. Noma, A, and Nakayama, K: Polarographic method for rapid microdetermination of cholesterol with cholesterol esterase and cholesterol oxidase, Clin Chem 22:336-340, 1976.
24. Zak, B, and Ressler, N: Methodology in determination of cholesterol, Am J Clin Pathol 2:433-443, 1955.
25. Carr, JJ, and Drekter IJ: Simplified rapid technic for the extraction and determination of serum cholesterol without saponification, Clin Chem 2:353-368, 1956.
26. Borner, K, and Klose, S: Enzymatische Bestimmung des Gesamtcholesterins mit dem Greiner Selective Analyzer (GSA-11), J Clin Chem Biochem 15:121-130, 1977.
27. Heuck, CC, and Wirth, A: Simultane Bestimmung von freiem cholesterin und Gesamtcholesterin auf dem Technicon AA1, J Clin Chem Clin Biochem 15:213-216, 1977.
28. Huang, HS, Kuan, SS, Guilbault, GG, et al: Amperometric determination of total cholesterol in serum with use of immobilized cholesterol ester hydrolase and cholesterol oxidase, Clin Chem 23:671-676, 1977.
29. Pesce, MA, and Bodowrian, SH: Enzymatic measurement of cholesterol in serum with the CentrifiChem centrifugal analyzer, Clin Chem 23:280-282, 1977.
30. Trocha, PJ: Improved continuous flow enzymatic determination of total serum cholesterol and triglycerides, Clin Chem 23:146-147, 1977.
31. Ziegenhorn, J: Enzymatische Bestimmung des Gesamt-cholesterins im Serum mit Analysenautomaten, Z Klin Chem Klin Biochem 13:109-115, 1975.
32. Witte, DL, Barrett, DA, and Wycoff, DA: Evaluation of an enzymatic procedure for determination of serum cholesterol with the Abbott ABA-100, Clin Chem 20:1282-1286, 1974.
33. Allain, CC, Poon, LS, Chan, CSG, et al: Enzymatic determination of total serum cholesterol, Clin Chem 20:470-475, 1974.
34. Nobbs, BT, Smith, JM, and Wallar, AW: Enzymic determination of plasma cholesterol on discrete automatic analyzers, Clin Chim Acta 79:391-398, 1977.
35. Knob, M, and Rosenmund, H: Enzymatische Bestimmung des Gesamtcholesterins im Serum mit Zentrifugalanalyzern, Z Klin Chem 13:493-498, 1975.
36. Pesce, MA, and Bodowrian, SH: Enzymatic rate method for measuring cholesterol in serum, Clin Chem 22:2042-2045, 1976.
37. Carlson, SE, and Goldfarb, S: A sensitive enzymatic method for the determination of free and esterified tissue cholesterol, Clin Chim Acta 79:575-582, 1977.
38. Yao, T, Sato, M, Kobayashi, Y, and Wasa, T: Amperometric assays of total and free cholesterols in serum by the combined use of immobilized cholesterol esterase and cholesterol oxidase reactors and peroxidase electrode in a flow injection system, Anal Biochem 149:387-391, 1985.
39. Huang, HS, Kuan, JC, and Guilbault, GG: Fluorometric enzymatic determination of total cholesterol in serum, Clin Chem 21:1605-1608, 1975.
40. Tarbutton, PN, and Gunter CR: Enzymatic determination of total cholesterol in serum, Clin Chem 20:724-725, 1974.
41. Sale, FO, Marchesini, S, Fishman, PH, and Berra, B: A sensitive enzymatic assay for determination of cholesterol in lipid extracts, Anal Biochem 142:347-350, 1984.
42. Naito, HK, and David, JA: Laboratory considerations: determination of cholesterol, triglyceride, phospholipid, and other lipids in blood and tissues. In Story, JA, editor: Lipid research methodology, New York, 1984, Alan R Liss, Inc, pp 1-76.
43. Naito, HK: Reliability of lipid and lipoprotein testing, Am J Cardiol 56(suppl):65-95, 1985.
44. Duncan, IW, Mather, A, and Cooper, GR: The procedure for the proposed cholesterol reference method, Atlanta, 1982, Centers for Disease Control.

45. Cooper, GR, Gill, JB, Biegeleisen, JJ, et al: CBC point of reference for total cholesterol measurements, Clin Chem 26:966, 1980 (abstract).
46. Technicon Methods Nos. SF40040F66 and SE0040FC6, Tarrytown, NY, March 1986, Technicon Instruments Corp.
47. Lippel, K, Ahmed, S, Albers, JJ, et al: Analytical performance and comparability of the determination of cholesterol by 12 lipid-research clinics, Clin Chem 23:1744-1752, 1977.
48. Kuba, K, Folsom, AR, Frantz, ID Jr, et al: Lipid concentrations in fasting and non-fasting serum and plasma, Clin Chem 28:1607, 1982.
49. Martinek, RG: Review of methods for determining cholesterol and cholesterol esters in serum, J Am Med Technol 32:64-86, 1970.
50. Manual of laboratory operations, Lipid Research Clinics Program, vol 1, Lipid and lipoprotein analysis, DHEW Pub No NIH 75-628, Washington, DC, May 1974, US Government Printing Office.
51. Cooper GR: High density lipoprotein reference materials. In Lipel, K, editor: Report of the high density lipoprotein methodology workshop, DHEW Pub No NIH 79-1661, Washington, DC, 1979, US Government Printing Office, p 178.
52. Fredrickson, DS, Levy, RI, and Lees, RS: Fat transport in lipoproteins: an integrated approach to mechanisms and disorders (continued), N Engl J Med 276:148-156, 1967.
53. Naito, HK: Role of specific nutritional components on plasma lipids, lipoproteins, and coronary heart disease. In Brewster, MA, and Naito, HK, editors: Nutritional elements and clinical biochemistry, New York, 1980, Plenum Press, pp 277-316.
54. Kannel, WB, Castelli, WP, Gordon, T, et al: Serum cholesterol, lipoproteins, and the risk of coronary heart disease: The Framingham Study, Ann Intern Med 74:1, 1971.
55. Galen, RS, and Gambino, SR: Beyond normal: the predictive value and efficiency of medical diagnosis, New York, 1975, John Wiley & Sons, Inc.
56. McGill, H Jr, editor: The geographic pathology of atherosclerosis, Baltimore, 1968, Williams & Wilkins.
57. Roberts, WC: Extreme hypercholesterolemia = malignant atherosclerosis, Am J Cardiol 54:242-243, 1984.
58. National Cholesterol Education Program: Report of the Panel on Detection, Evaluation, and Treatment of High Blood Cholesterol in Adults, Arch Intern Med 148:36-39, 1988.
59. National Institutes of Health Consensus Development Conference: Lowering blood cholesterol to prevent heart disease, Bethesda, MD, 1986, US Dept of Health and Human Services.

High-density lipoprotein (HDL) cholesterol

HERBERT K. NAITO

Clinical significance: p. 454
Molecular weight: 170,000 to 360,000 daltons
Chemical class: lipoprotein

PRINCIPLES OF ANALYSIS

The widespread use of high-density lipoprotein (HDL) cholesterol values in medicine warrants a critical analysis of the methods currently available. Although HDL cholesterol determinations are fairly simple, it is well established that interlaboratory differences are high, with a 10% to 25% coefficient of variation (CV). The coronary heart disease (CHD) risk tables, which are used by many clinical laboratories for data interpretation, are primarily based on the epidemiological studies in Framingham.[1-3] The methods used in the Framingham studies were rigorously standardized for accuracy by the Centers for Disease Control (CDC). The mean difference in plasma HDL cholesterol

Table 61-8 LDL cholesterol–HDL cholesterol risk ratio*

Ratio	Risk
Males	
1.00	One-half average
3.55	Average
6.25	Twice average
7.99	Three times average
Females	
1.47	One-half average
3.22	Average
5.03	Twice average
6.14	Three times average

*This risk ratio is not valid if (1) chylomicrons are present and (2) serum triglyceride is greater than 4000 mg/L. Therefore the serum should be centrifuged in the preparative ultracentrifuge at a density of 1.006 for 20 hours at 30,000 rpm, and the supernatant solution can be used to determine the cholesterol and triglyceride concentration of the *top* fraction. In addition, the *bottom* fraction can be analyzed and LDL cholesterol calculated:

LDL cholesterol = *Bottom* fraction cholesterol − HDL cholesterol

concentration between subjects with coronary artery disease and subjects who were apparently normal was about 40 mg/L.[3] This small difference between the two populations highlights the importance of ensuring that the HDL cholesterol method is highly precise and accurate. Table 61-8 presents an example of the risk tables based on the low-density lipoprotein (LDL) cholesterol–HDL cholesterol ratio. These tables, now being used by physicians, are based on the Framingham study. It is necessary to use the same standards for accuracy that were derived from CDC methods to employ the Framingham data base.

By definition, HDL is the fraction of plasma lipoproteins with a hydrated density of 1.063 to 1.21 g/mL isolated in a preparative ultracentrifuge. Electrophoretically, it has a mobility of alpha$_1$-globulins. The densest lipoprotein, it is composed of about 50% protein, 25% phospholipid, 20% cholesterol, and 5% triglyceride. The major lipid in HDL is phospholipid, of which lecithin is the major fraction, followed by sphingomyelin and lysophosphatides.[4] Cholesterol is the next major lipid found in HDL. The ratio of esterified to free cholesterol is about 3:1. HDL is a heterogeneous group of compounds that have been divided into two major density classes: HDL_2 (d = 1.063 to 1.125 g/mL) and HDL_3 (d = 1.125 to 1.21 g/mL).

There are two direct ways to measure HDL: (1) by analytical ultracentrifugation and (2) by isolation of HDL and measurement of the particles gravimetrically. The former method is more accurate and is considered the classic method for quantitation of HDL and other lipoprotein fractions as well.

These techniques are laborious and tedious and require a highly skilled, trained individual. Because of this, the isolated HDL fraction is not measured in its entirety. Instead, either a protein moiety or a lipid moiety is measured as an indirect means of quantitating HDL. Although the phospholipid component is the largest lipid constituent by mass, since cholesterol is simpler to measure than phospholipids are, HDL cholesterol has prevailed as an indirect means of determining HDL concentrations.

Numerous techniques are available for HDL cholesterol quantitation. They are all, however, based on two steps: (1) isolation of the HDL and (2) quantitation of the cholesterol in the isolated HDL. The various methods differ primarily in how the HDL fraction is isolated: the cholesterol analyses are usually done by one of the many acceptable cholesterol methods available with sufficient linearity and sensitivity at the low end of the concentration range. The Albers et al. method[5] can be recommended, though certain compatible enzymatic procedures would be preferable (see previous section on total cholesterol analysis). The following HDL isolation procedures are used today in the clinical laboratory.[6-17]

Preparative ultracentrifugation (method 1, Table 61-9)

For the simultaneous separation of both very-low-density lipoproteins (VLDL) and LDL, the plasma or serum is adjusted to a density of 1.063 g/mL by overlayering of the sample with the potassium bromide solution (415 mg of KBr per 5 mL) and centrifugation of the sample at 105,000 *g* for 24 hours at 16° C. After the supernatant solution containing the VLDL and LDL is removed, the infranatant solution can be analyzed for the cholesterol concentration. This method of isolating the HDL and measuring the cholesterol content for HDL cholesterol estimation is a classical procedure and is often regarded as a reference procedure.

Column chromatography (method 2, Table 61-9)

Ion-exchange chromatography and gel-permeation chromatography have both been used in the isolation of the major lipoprotein groups and the subpopulations within each group. These procedures separate HDL subclasses based on differences in charge or molecular size, respec-

Table 61-9 Methods for high-density lipoprotein (HDL) isolation

Method	Principle	Usage	Comments
1. Ultracentrifugation	Plasma or serum adjusted to density of 1.063 g/mL with potassium bromide and centrifuged at high speeds for 24 hours; all lipoproteins separated by density, with HDL fraction in 1.063 to 1.21 g/mL range	Specialized research laboratories	Reference method Laborious and time consuming
2. Column chromatography	HDL isolated and separated into subclasses on basis of charge (ion-exchange) or molecular size (gel permeation)	Specialized research laboratories Rarely used	Difficult to properly control chromatographic conditions
3. Starch block electrophoresis	HDL separated from other lipoproteins on basis of charge and size	Rarely used	Used for isolation of large amounts of HDL, not for quantitative purposes
4. Agarose gel electrophoresis	HDL separated from other lipoproteins on basis of charge and size	Infrequently used	Precision not adequate for clinical use
5. Precipitation	Polyanions (heparin, dextran sulfate), phosphotungstate, polyethylene glycol, in presence of divalent cations, used to precipitate larger, less dense lipoproteins; HDL quantitated in supernatant as HDL cholesterol	Most frequently used	
a. Heparin–manganese chloride		Infrequently used	Current method of choice Not compatible with all cholesterol procedures
b. Dextran sulfate–magnesium chloride		Frequently used	Use of lower molecular weight dextran can produce biased results; higher molecular weight dextrans produce excellent results Compatible with enzymatic procedures
c. Phosphotungstate		Most frequently used	Underestimates HDL Sensitive to temperature fluctuations
d. Polyethylene glycol		Used infrequently	Poorest accuracy and precision; not recommended
6. Polyacrylamide gel electrophoresis	HDL separated from other lipoproteins on basis of charge and size	Used infrequently	Probably underestimates HDL levels

tively. These methods are not widely used in the routine clinical laboratory but may be available in specialized laboratories.

Preparative block electrophoresis (method 3, Table 61-9)

Both starch block and Geon-Pevikon block electrophoresis methods are employed in the isolation of major lipoprotein fractions and their subfractions. These electrophoresis procedures separate the lipoprotein classes on the basis of their net charge and size. The smaller HDL molecules have the highest mobility toward the anode. These methods are mainly used as preparative techniques, when large sample size is required.

Agarose gel electrophoresis (method 4, Table 61-9)

Agarose gel electrophoresis uses the standard lipoprotein electrophoresis procedure on agarose medium, followed by the overlaying of the electrophoresed sample with enzymatic cholesterol reagent. A densitometer with an automated integrator is used to scan the agarose strip after color development and quantitate each lipoprotein fraction (see p. 991).

Precipitation with polyanion solutions (method 5, Table 61-9)

Differential precipitation of lipoproteins with various polyanion solutions is common practice and is suitable for the clinical laboratory because of its simplicity, elimination of expensive instrumentation, speed, and low cost. These techniques are based on the ability of various agents to selectively precipitate the major lipoprotein fractions, except HDL. HDL is left in the supernatant solution to be quantitated. The agents most frequently employed include heparin–manganese chloride (Lipid Research Clinics method), dextran sulfate–magnesium chloride, sodium phosphotungstate, and polyethylene glycol. Of these, the phosphotungstate and dextran sulfate–magnesium chloride are the two most frequently used precipitating reagents.

A detailed description of the heparin–manganese chloride procedure is included in this chapter. Since the mean difference in HDL cholesterol concentration between persons at normal risk for coronary artery disease and persons at high risk is small (about 40 to 50 mg/L), it is suggested that the minimum precision for this method should be less than 30 mg/L for within-day and day-to-day variance.

Although no ideal method for measuring serum HDL cholesterol is available, it is important to recognize the limitations of all procedures. Particular caution is urged for those setting up this test for clinical evaluation of persons at risk for heart disease. Since the only available data for assigning risk are based on tables generated from the Framingham study, the assay developed in the laboratory must be as accurate as the assay used in the Framingham study. The CDC can play an important role in helping ensure test accuracy by providing a common benchmark for method standardization. Precision is also important; remember that the mean difference in HDL cholesterol concentration in the Framingham Study populations with and without myocardial infarction is only 40 mg/L.

In conclusion, although no absolute reference or definitive method has been designated for HDL cholesterol determinations, precipitation techniques have been favored, with the heparin–manganese chloride procedure generally considered the method of choice. Included here is a modified CDC-proposed HDL cholesterol procedure[7] for cholesterol analysis by enzymatic or Liebermann-Burchard cholesterol methods (see previous section).

SPECIMEN

For a detailed review, see p. 978. Briefly, the patient must be fasting for at least 12 hours before the blood is drawn. It should be emphasized that although nonfasting conditions do not appear to influence the blood HDL cholesterol levels in most persons, postprandial lipemia has the potential of interfering with many of the analytical methods. To minimize this analytical problem, it is always good laboratory practice to request fasting specimens. Serum or plasma can be used as the sample. EDTA is again the anticoagulant of choice if plasma is used, and the final concentration should be 1 mg of EDTA per milliliter of blood.

If serum is used, the sample should be removed from the blood clot within 2 hours and stored at 4° C, unless analyzed immediately.

PROCEDURE: ISOLATION OF HIGH-DENSITY LIPOPROTEINS BY HEPARIN–MANGANESE CHLORIDE

Principle

The larger, less dense apo B–containing lipoproteins (those with low, very low, and intermediate density) are precipitated by heparin–manganese chloride. After centrifugation to separate the precipitated lipoproteins, HDL cholesterol in the supernatant solution is quantitated (Fig. 61-3).

Reagents

Manganese chloride solution 1 mol/L. Dissolve 19.791 g of $MnCl_2 \cdot 4H_2O$ in distilled water, and bring volume to 100 mL. This is stable for 3 months at 4° C.

Heparin solution, 4000 U/mL. Dilute 1 mL of heparin (A.H. Robins; 10,000 USP U/mL) with 1.5 mL of 0.9% saline solution. This is stable for 1 week at 4° C.

Assay

Equipment: refrigerated centrifuge capable of maintaining temperature of 4° C at 2000 *g*.

1. Pipette 1 mL of serum into a 13 × 100 mm disposable culture tube.

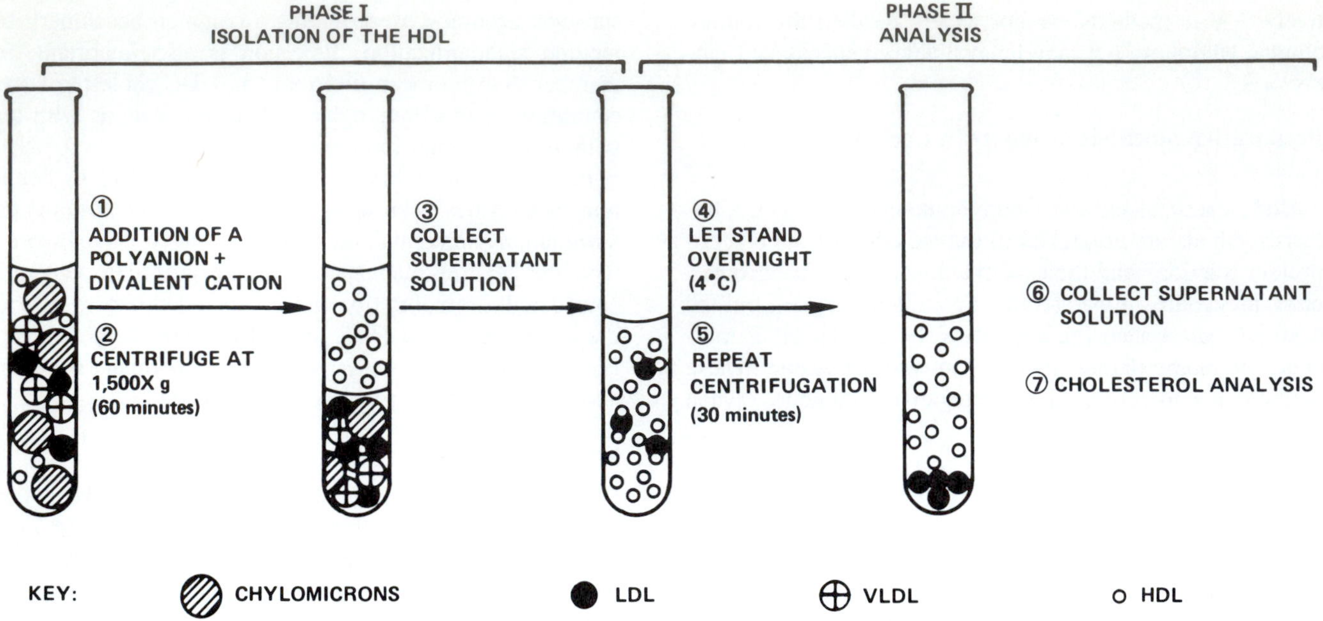

Fig. 61-3 Scheme of polyanion precipitation method (heparin-$MnCl_2$) for determination of HDL cholesterol.

2. Cap tubes with polyethylene stoppers, and place in refrigerator for 30 minutes at 4° C.
3. Add 50 μL of heparin solution; mix on a vortex mixer for 1 minute.
4. Add 50 μL of manganese chloride solution; mix on a vortex mixer.
5. Cap tubes with polyethylene stoppers, and place in refrigerator for 30 minutes at 4° C.
6. Centrifuge the heparin-lipoprotein precipitate at 2000 *g* for 1 hour at 4° C.
7. Using a transfer pipet, transfer supernatant solution into a clean test tube. Store in a refrigerator overnight to ensure complete precipitation of all apo B–containing lipoproteins.
8. The next morning, sediment any remaining precipitate at 2000 *g* for 30 minutes at 4° C.
9. Using a transfer pipet, transfer the clear supernatant solution into a clean test tube for measurement of cholesterol using a standard method compatible with heparin–manganese chloride. The Liebermann-Burchard method is recommended.

Notes

1. Biological sample. If one uses plasma samples, the 92 mM Mn^{+2} salt concentration should be selected. However, if one uses serum, the 46 mM Mn^{+2} concentration is more ideally suited for HDL cholesterol determinations. If EDTA plasma samples are used with the heparin 46 mM Mn^{+2} method, slightly higher HDL cholesterol values will be obtained because of incomplete precipitation of the apo B–containing lipoproteins. If one uses the 92 mM Mn^{+2} salt preparation on serum samples, the HDL cholesterol values will be slightly lower than results obtained using the ultracentrifugal procedure (the reference procedure).

2. Purity check. If the sample shows an unusual value (HDL concentrations less than 200 or greater than 800 mg/L) or the alpha-lipoprotein bands on the electrophoresis medium do not agree with the serum or plasma HDL cholesterol value, perform a purity check, repeat the analysis, or both.

A purity check on the supernatant solution should be done if there is the possibility of incomplete precipitation of apo B–containing lipoproteins. To do this, electrophorese the supernatant solution and stain with a lipid dye to show whether only one band exists. If there is more than one lipid-staining band and the second band is in the beta-globulin area or alpha$_2$-globulin area, an apo B–containing lipoprotein or Lp(a) is probably present. One can confirm this by doing immunological studies using apo B or Lp(a) antisera.

3. Double-concentration precipitation. If the serum is turbid or hypertriglyceridemic (greater than 4000 mg/L), a double-strength solution should be used to precipitate the beta-containing lipoproteins. To do this, use 1 mL of serum, 100 μL of heparin, and 100 μL of manganese chloride solution. Multiply the result by 1.2 to account for the dilution. A double-strength solution should also be used for a serum sample if, after conventional manganese chloride–heparin precipitation, a floating lipid precipitate is present and the purity check of the supernatant shows contamination.

4. Pooled human sera for quality control. Reference ma-

terials are needed for calibration, standardization, and quality control of HDL cholesterol analyses. The calibration and control materials must be suitable for determining cholesterol in the HDL fraction when specimens contain relatively low concentrations of cholesterol and when nonspecific and interfering substances may be present. Calibration materials applicable to low cholesterol level analyses can be standards containing about 250 and 600 mg/L of cholesterol. Serum control materials for these cholesterol analyses can be sera diluted to contain about 250 and 600 mg/L of cholesterol. These levels cover the range of values of most HDL specimens. An alternative procedure is to use a normal-range cholesterol serum pool for calibration and to use both primary standard solutions and serum pools of low concentration levels for monitoring. Since a major cause of discrepancies between HDL results among laboratories can be attributed to lack of appropriate standardization of the total cholesterol method, an effort should be made first to standardize the cholesterol method. According to Cooper,[18] reference materials that appear suitable for calibrating and monitoring the determination of cholesterol in HDL isolated fraction are (1) primary cholesterol standards, (2) serum calibrator pools, (3) serum quality control pools, and (4) fractions of HDL isolated by ultracentrifugation of precipitation methods and labeled accurately with total cholesterol content.

The calibration and control materials produced for use in the first step—preparation of the HDL sample for cholesterol analysis—are often unsuitable because of an unpredictable instability of HDL particles, varying composition and flux of lipids and apolipoproteins among lipoproteins,[2] or apolipoprotein B from HDL particles.[3] Therefore compromises have to be accepted. Under these circumstances, it appears practical to seek HDL reference materials that are homogeneous, are stable up to 2 years, and possess a suitable similar-to-serum matrix. The HDL reference materials that might meet these criteria for control of the HDL isolation procedure are (1) serum with low triglyceride or VLDL levels, (2) bottom fraction 1.006 g/mL density prepared by ultracentrifugation, and (3) bottom fraction 1.063 g/mL density prepared by ultracentrifugation.

Studies by Cooper[18] suggest that for long-term storage of HDL cholesterol pools a temperature less than −60° C, rather than −20° or −5° C, is necessary.

At present most commercially available quality control materials for HDL cholesterol are not suitable or stable. However, we have had success with the Omega Lipid

Table 61-10 Plasma LDL cholesterol in white males, mg/L (mmol/L)

Age (years)	*N*	Overall mean ± SE	Percentiles						
			5	10	25	50	75	90	95
5-9	131	925 ± 18	630	690	800	900	1030	1170	1290
		(2.39)	(1.63)	(1.78)	(2.07)	(2.33)	(2.66)	(3.03)	(3.34)
10-14	284	965 ± 14	640	720	810	940	1090	1330	1320
		(2.50)	(1.66)	(1.86)	(2.09)	(2.43)	(2.82)	(3.16)	(3.41)
15-19	298	944 ± 13	620	680	800	930	1090	1230	1300
		(2.44)	(1.60)	(1.76)	(2.09)	(2.41)	(2.82)	(3.18)	(3.36)
20-24	118	1033 ± 24	660	730	850	1010	1180	1380	1470
		(2.67)	(1.71)	(1.89)	(2.20)	(2.61)	(3.05)	(3.57)	(3.80)
24-29	253	1167 ± 19	700	750	960	1160	1380	1570	1650
		(3.02)	(1.81)	(1.94)	(2.48)	(3.00)	(3.57)	(4.06)	(4.27)
30-34	403	1264 ± 16	780	880	1070	1240	1440	1660	1850
		(3.27)	(2.02)	(2.28)	(2.77)	(3.21)	(3.72)	(4.29)	(4.78)
35-39	371	1332 ± 17	810	920	1100	1310	1540	1760	1890
		(3.44)	(2.09)	(2.38)	(2.84)	(3.39)	(3.98)	(4.55)	(4.89)
40-44	385	1356 ± 16	870	980	1150	1350	1570	1730	1860
		(3.51)	(2.25)	(2.53)	(2.97)	(3.49)	(4.06)	(4.47)	(4.81)
45-49	325	1439 ± 18	980	1060	1200	1410	1630	1860	2020
		(3.72)	(2.53)	(2.74)	(3.10)	(3.65)	(4.22)	(4.81)	(5.22)
50-54	340	1423 ± 17	890	1020	1180	1430	1620	1850	1970
		(3.68)	(2.30)	(2.64)	(2.05)	(3.70)	(4.19)	(4.78)	(5.10)
55-59	261	1458 ± 21	880	1030	1230	1450	1680	1910	2030
		(3.77)	(2.28)	(2.66)	(3.18)	(3.75)	(4.35)	(4.94)	(5.25)
60-64	131	1463 ± 31	830	1060	1210	1430	1650	1880	2100
		(3.78)	(2.15)	(2.74)	(3.13)	(3.70)	(4.27)	(4.86)	(5.43)
65-69	105	1504 ± 35	980	1040	1250	1460	1700	1990	2100
		(3.89)	(2.53)	(2.69)	(3.13)	(3.78)	(4.40)	(5.15)	(5.43)
70+	119	1429 ± 29	880	1000	1190	1420	1640	1820	1860
		(3.70)	(2.28)	(2.59)	(3.08)	(3.67)	(4.24)	(4.71)	(4.81)

Data derived from the LRC Prevalence Study of North America, 1980.[20,21]
SE, Standard error of the mean.

Fraction Control Serum (Cooper Biomedical, Malvern, PA). Since the matrix of commercial materials may be different from human samples, we also make our own quality control materials by pooling human serum that has no chylomicrons and only low amounts of VLDL and triglyceride (less than 1500 mg/L). This quality control material can be used if it is free of hepatitis antigens and antibodies to human immunodeficiency virus. The pooled sample (about 2 L) should be aliquoted into 12 × 75 mm tubes and stored in a freezer at −60° C.

5. Calculation of VLDL and LDL cholesterol. Although one can isolate VLDL (density less than 1.006 g/mL) and determine the cholesterol content, VLDL cholesterol can be estimated with a fair amount of accuracy by the following calculations[19]:

$$\frac{\text{Triglyceride (mg/L)}}{5} = \text{VLDL cholesterol (mg/L)}$$

This holds true only if serum triglyceride concentration is less than 4000 mg/L. Furthermore

$$\text{LDL cholesterol} = \text{Total cholesterol} - (\text{VLDL cholesterol} + \text{HDL cholesterol})$$

REFERENCE RATE

To define a reference value based on a percentile, one needs to establish the distribution of the relevant lipid or lipoprotein levels through study of a well-defined population sample. Up to this time, the existing data for lipid distributions in the North American population, from which reference values have been derived, have been limited. The Prevalence Study of the Lipid Research Clinics (LRC) Program has provided detailed information on the distribution of levels of plasma lipids and lipoproteins determined in a random sample of a variety of well-defined North American populations.[20,21] The reference values derived from this study can be used to screen and to diagnose hyperlipidemia (and hypolipidemia). In addition to 95th percentiles, 90th and 75th percentiles are provided for cholesterol and LDL cholesterol to permit the identification of a larger group of persons who are at risk for CHD.[22]

In 1985 the National Institutes of Health (NIH) recommended new guidelines; the consensus panel recommended treatment of persons with blood total cholesterol and LDL cholesterol levels above the 75th percentile (upper 25% of values) rather than the customary 90th or 95th percen-

Table 61-11 Plasma LDL cholesterol in white females, mg/L (mmol/L)

Age (years)	*N*	Overall mean ± SE	Percentiles						
			5	10	25	50	75	90	95
5-9	114	1004 ± 21 (2.60)	680 (1.76)	730 (1.89)	880 (2.28)	980 (2.53)	1150 (2.97)	1250 (3.23)	1400 (3.62)
10-14	244	973 ± 13 (2.52)	680 (1.76)	730 (1.89)	810 (2.09)	940 (2.43)	1100 (2.84)	1260 (2.56)	1360 (3.52)
15-19	294	958 ± 15 (2.48)	590 (1.53)	650 (1.68)	780 (2.02)	930 (2.40)	1110 (2.87)	1290 (3.34)	1370 (3.54)
20-24	119	1037 ± 22 (2.68)	570 (1.47)	650 (1.68)	820 (2.12)	1020 (2.64)	1180 (3.05)	1410 (3.65)	1590 (4.11)
24-29	314	1102 ± 16 (2.85)	710 (1.84)	750 (1.94)	900 (2.33)	1080 (2.79)	1260 (3.26)	1480 (3.83)	1640 (4.24)
30-34	332	1112 ± 15 (2.88)	700 (1.81)	770 (1.99)	1910 (2.35)	1090 (2.82)	1280 (3.31)	1470 (3.80)	1560 (4.03)
35-39	299	1197 ± 20 (3.09)	750 (1.94)	810 (2.09)	960 (2.48)	1160 (3.00)	1390 (4.16)	1610 (4.45)	1720 (4.89)
40-44	318	1250 ± 18 (3.23)	740 (1.91)	840 (2.17)	1040 (2.69)	1220 (3.16)	1460 (3.78)	1650 (4.27)	1720 (4.50)
45-49	326	1294 ± 19 (3.34)	790 (2.04)	890 (2.30)	1050 (2.71)	1270 (3.28)	1500 (3.88)	1730 (4.47)	1860 (4.81)
50-54	256	1381 ± 23 (3.57)	880 (2.28)	940 (2.43)	1110 (2.87)	1340 (3.47)	1600 (4.14)	1860 (4.81)	2010 (5.20)
55-59	250	1461 ± 24 (3.78)	890 (2.30)	970 (2.51)	1200 (3.10)	1450 (3.75)	1680 (4.35)	1990 (5.15)	2100 (5.43)
60-64	145	1520 ± 36 (3.93)	1000 (2.59)	1050 (2.71)	1260 (3.86)	1490 (3.85)	1680 (4.35)	1910 (5.22)	2240 (5.79)
65-69	130	1538 ± 41 (3.98)	920 (2.38)	990 (2.56)	1250 (3.23)	1510 (3.90)	1840 (4.76)	2050 (5.30)	2210 (5.72)
70+	143	1486 ± 27 (3.84)	960 (2.48)	1080 (2.79)	1270 (3.78)	1470 (3.80)	1700 (4.40)	1890 (4.81)	2060 (5.33)

Data derived from the LRC Prevalence Study of North America, 1980.[20,21]
SE, Standard error of the mean.

tiles.[23] In 1988 the NIH Adult Treatment Panel of the National Cholesterol Education Program[24] recommended new cut points for cholesterol and LDL cholesterol (see p. 981) that should have a greater impact on reducing the morbidity and mortality CHD rates in the United States, where it is still a major disease. To simplify the classification system and to make it more convenient to remember the cut points, the panel eliminated the age and sex stratification so that there would be substantially fewer numbers to remember. These new cut points apply to all adults 20 years and older. This philosophy was also extended to HDL cholesterol. The Adult Treatment Panel recommended that HDL cholesterol values below 350 mg/L constitute an elevated CHD risk factor and that individuals with such levels should be treated. The panel did not advocate the use of ratios, that is, HDL cholesterol–to–total cholesterol or HDL cholesterol–to–LDL cholesterol ratios. For HDL cholesterol, the 5th and 10th percentiles are given for screening of persons for HDL deficiency syndromes. It is also provided to identify persons at greater risk for developing CHD in view of the inverse relationship between HDL cholesterol levels and CHD. The 95th percentile is given to identify persons with so-called hyperalphalipoproteinemia, who seem to be at decreased risk for CHD.

Tables 61-10 to 61-13 provide LDL cholesterol and HDL cholesterol values for white males and females according to age.[21] Similar data for black males and females have not yet been completed.

USE OF RISK TABLES FOR DATA INTERPRETATION

In the past it was customary to use the values for LDL and HDL cholesterol simultaneously as a ratio and to compare them to CHD risk assessment tables generated from the Framingham Study. The 1988 Adult Treatment Panel guidelines do not, at the moment, advocate this strategy. In the past, these tables allowed one to assess only one of the risk factors (for example, lipids for heart disease) and therefore were not designed to take into account the other primary and secondary risk factors, which ultimately should be included in the overall assessment of a person's risk. Now there is a unified system that simultaneously accounts for all the major CHD risk factors that ultimately will determine if aggressive or passive treatment goals for

Table 61-12 Plasma HDL cholesterol in white males, mg/L (mmol/L)

Age (years)	*N*	Overall mean ± SE	Percentiles						
			5	10	25	50	75	90	95
5-9	142	555 ± 10 (1.44)	380 (0.98)	420 (1.09)	490 (1.27)	540 (1.40)	630 (1.63)	700 (1.81)	740 (1.91)
10-14	296	549 ± 7 (1.42)	370 (0.96)	400 (1.03)	460 (1.19)	550 (1.42)	610 (1.58)	710 (1.84)	740 (1.91)
15-19	299	461 ± 6 (1.19)	300 (0.78)	340 (0.88)	390 (1.01)	460 (1.19)	520 (1.34)	590 (1.53)	630 (1.63)
20-24	118	454 ± 10 (1.17)	300 (0.78)	320 (0.83)	380 (0.98)	450 (1.16)	510 (1.32)	570 (1.47)	630 (1.63)
25-29	253	447 ± 7 (1.16)	310 (0.80)	320 (0.83)	370 (0.96)	440 (1.14)	500 (1.29)	580 (1.50)	630 (1.63)
30-34	403	446 ± 6 (1.18)	280 (0.72)	320 (0.83)	380 (0.98)	450 (1.16)	520 (1.34)	590 (1.53)	630 (1.63)
35-39	371	433 ± 6 (1.12)	290 (0.75)	310 (0.80)	360 (0.93)	430 (1.11)	490 (1.27)	580 (1.50)	620 (1.60)
40-44	383	443 ± 6 (1.15)	270 (0.70)	310 (0.80)	360 (0.93)	430 (1.11)	510 (1.32)	600 (1.55)	670 (1.73)
45-49	325	454 ± 6 (1.17)	300 (0.78)	330 (0.85)	380 (0.98)	450 (1.16)	520 (1.34)	600 (1.55)	640 (1.66)
50-54	340	441 ± 6 (1.14)	280 (1.72)	310 (0.80)	360 (0.93)	440 (1.14)	510 (1.32)	580 (1.50)	630 (1.63)
55-59	261	476 ± 9 (1.23)	280 (0.72)	310 (0.80)	380 (0.98)	460 (1.19)	550 (1.42)	640 (1.66)	710 (1.80)
60-64	131	515 ± 13 (1.33)	300 (0.78)	340 (0.88)	410 (1.06)	490 (1.27)	610 (1.58)	690 (1.78)	740 (1.91)
65-69	105	511 ± 15 (1.32)	300 (0.78)	330 (0.85)	390 (1.01)	490 (1.27)	620 (1.60)	740 (1.91)	780 (2.02)
70+	119	505 ± 17 (1.31)	310 (0.80)	330 (0.85)	400 (1.03)	480 (1.24)	560 (1.45)	700 (1.81)	750 (1.94)

Data derived from the LRC Prevalence Study of North America, 1980.[20,21]
SE, Standard error of the mean.

Table 61-13 Plasma HDL cholesterol in white females, mg/L (mmol/L)

Age (years)	*N*	Overall mean ± SE	Percentiles						
			5	10	25	50	75	90	95
5-9	124	532 ± 11 (1.38)	360 (0.93)	380 (0.98)	470 (1.22)	520 (1.34)	610 (1.58)	670 (1.73)	730 (1.89)
10-14	247	522 ± 7 (1.35)	370 (0.96)	400 (1.03)	450 (1.16)	520 (1.34)	580 (1.50)	640 (1.16)	700 (1.81)
15-19	295	522 ± 7 (1.35)	350 (0.91)	380 (0.98)	430 (1.11)	510 (1.32)	610 (1.58)	680 (1.76)	740 (1.91)
20-24	199	533 ± 10 (1.38)	330 (0.85)	370 (0.96)	440 (1.14)	510 (1.32)	620 (1.60)	720 (1.86)	790 (2.04)
25-29	314	560 ± 8 (1.45)	370 (0.96)	390 (1.01)	470 (1.22)	550 (1.42)	630 (1.63)	740 (1.91)	830 (2.15)
30-34	335	561 ± 7 (1.45)	360 (0.93)	400 (1.03)	460 (1.19)	550 (1.42)	640 (1.66)	730 (1.89)	770 (1.99)
35-39	298	550 ± 8 (1.42)	340 (0.88)	380 (0.98)	440 (1.14)	530 (1.37)	640 (1.66)	740 (1.91)	820 (2.12)
40-44	318	578 ± 9 (1.49)	340 (0.88)	390 (1.01)	480 (1.24)	560 (1.45)	650 (1.68)	790 (2.04)	880 (2.28)
45-49	328	594 ± 10 (1.54)	340 (0.88)	410 (1.06)	470 (1.22)	580 (1.50)	680 (1.76)	820 (2.12)	870 (2.25)
50-54	256	620 ± 10 (1.60)	370 (0.96)	410 (1.06)	500 (1.29)	620 (1.60)	710 (1.84)	840 (2.17)	920 (2.38)
55-59	250	620 ± 11 (1.60)	370 (0.96)	410 (1.06)	500 (1.29)	600 (1.55)	730 (1.89)	850 (2.20)	910 (2.35)
60-64	145	622 ± 14 (1.61)	380 (0.98)	440 (1.14)	510 (1.32)	610 (1.58)	750 (1.94)	870 (2.25)	920 (3.28)
65-69	130	633 ± 18 (1.64)	350 (0.91)	380 (0.98)	490 (1.27)	620 (1.60)	730 (1.89)	850 (2.20)	980 (2.53)
70+	143	607 ± 14 (1.57)	330 (0.85)	380 (0.98)	480 (1.24)	600 (1.55)	710 (1.84)	820 (2.12)	920 (2.38)

Data derived from the LRC Prevalence Study of North America, 1980.[20,21]
SE, Standard error of the mean.

lowering total cholesterol and LDL cholesterol will be used. This comprehensive approach is expected to save 300,000 American lives annually. It should be remembered that the success of this national program hinges on accurate laboratory measurements of lipids and lipoproteins. The pressure to perform is now even greater.

REFERENCES

1. Gordon, T, Castelli, WP, Hjortland, MC, et al: High-density lipoprotein as a protective factor against coronary heart disease: the Framingham Study, Am J Med 62:707-714, 1977.
2. Heiss, G, Johnson, N, Reiland, S, et al: The epidemiology of plasma high-density lipoprotein cholesterol levels, Circulation 62(suppl):116-136, 1980.
3. Castelli, WP, Doyle, JT, Gordon, T, et al: HDL cholesterol and other lipids in coronary heart disease: the Cooperative Lipoprotein Phenotyping Study, Circulation 55:767-772, 1977.
4. Scanu, AM, Lim, CT, and Edelstein, C: On the subunit structure of the protein of human serum high density lipoprotein. II. A study of Sephadex fraction IV, J Biol Chem 247:5850-5855, 1972.
5. Albers, JJ, Warnick, GR, Wieve, D, et al: Multi-laboratory comparison of three heparin-Mn precipitation procedures for estimating cholesterol in high-density lipoprotein, Clin Chem 24:853-856, 1978.
6. Manual of laboratory operations, Lipid Research Clinics Program, Lipid and Lipoprotein Analysis, DHEW Pub No NIH 751-628, Washington, DC, May 1974, US Government Printing Office.
7. Bachorik, PS, Wood, PD, Albers, JJ, et al: Plasma high-density lipoprotein cholesterol concentrations determined after removal of other lipoproteins by heparin-manganese precipitation or by ultracentrifugation, Clin Chem 22:1828-1834, 1976.
8. Burstein, M, and Scholnick, MR: Lipoprotein-polyanion-metal interactions, Adv Lipid Res 11:67-108, 1973.
9. Burstein, M, Scholnick, HR, and Morgin, R: Rapid method for the isolation of lipoproteins from human serum by precipitation with polyanions, J Lipid Res 11:583-595, 1970.
10. Warnick, GR, and Albers, JJ: A comprehensive evaluation of the heparin-manganese precipitation procedure for estimating high density lipoprotein cholesterol, J Lipid Res 19:65-76, 1978.
11. Finley, PR, Schifman, RB, Williams, RJ, and Lichti, DA: Cholesterol in high-density lipoprotein: use of Mg^{2+}/dextran sulfate in its enzymatic measurement, Clin Chem 24:931-933, 1978.
12. Kostner, GM: Enzymatic determination of cholesterol in high-density lipoprotein fractions prepared by polyanion precipitation, Clin Chem 22:695, 1976 (letter).
13. Hatch, FT, Lindgren, FT, Adamson, G, et al: Quantitative agarose gel electrophoresis of plasma lipoproteins: a sample technique and two methods for standardization, J Lab Clin Med 81:946-960, 1973.
14. Vilkari, J: Precipitation of plasma lipoproteins by PEG-6000 and its evaluation with electrophoresis and ultracentrifugation, Scand J Clin Lab Invest 36:265-268, 1976.
15. Allen, JK, Hensley, WJ, Nicholis, AV, and Whitfield, JB: An enzymic and centrifugal method for estimating high-density lipoprotein cholesterol, Clin Chem 25:325-327, 1979.
16. Lopes-Virella, MF, Stone, P, Ellis, S, and Colwell, JA: Cholesterol determination in high-density lipoproteins separated by three different methods, Clin Chem 23:882-884, 1977.
17. Abell, LL, Levy, BB, Brodie, BB, and Kendall, FE: A simplified method for the estimation of total cholesterol in serum and demonstration of its specificity, J Biol Chem 195:357-366, 1952.

18. Cooper, GR: High-density lipoprotein reference materials. In Lippel, K, editor: Reports of the High Density Lipoprotein Methodology Workshop, DHEW Pub No NIH 79-1661, Washington, DC, 1979, US Government Printing Office, pp 178-188.
19. Friedewald, WT, Levy, RJ, and Fredrickson, DS: Estimation of the concentration of low-density lipoprotein cholesterol in plasma without use of the preparative ultracentrifuge, Clin Chem 18:499-509, 1972.
20. Heiss, G, Tamir, I, Davis, CG, et al: Lipoprotein-cholesterol distributions in selected North American populations, the Lipid Research Clinics Program Prevalence Study, Circulation 61:302-315, 1980.
21. Lipid Research Clinics Program: The Prevalence Study, vol 1. In The Lipid Research Clinics Population Studies Data Book, Pub No NIH 80-1527, Bethesda, MD, 1980, US Dept of Health and Human Services.
22. Rifkind, BM, and Segal, P: Lipid Research Clinics Program references values for hyperlipidemia and hypolipidemia, JAMA 250:1869-1872, 1983.
23. National Institutes of Health Consensus Development Conference: Lowering blood cholesterol to prevent heart disease, Bethesda, MD, 1985, US Dept of Health and Human Services.
24. National Cholesterol Education Program: Report of the Panel on Detection, Evaluation, and Treatment of High Blood Cholesterol in Adults, Arch Intern Med 148:36-39, 1988.

Lipoprotein electrophoresis

HERBERT K. NAITO

Clinical significance: p. 454
Chemical class: lipoprotein

PRINCIPLES OF ANALYSIS

During the latter half of the 1960s and through the 1970s, lipoprotein electrophoresis was in common use in the clinical laboratory to aid in the classification of the different forms of dyslipoproteinemias. In recent years, the use of lipoprotein electrophoresis has dwindled in the laboratory, and many laboratories have discontinued the use of this technique for the characterization of lipoprotein disorders. Approximately 80% of the hyperlipidemias can be accurately and inexpensively categorized by simple interpretation of the serum total cholesterol and triglyceride levels and the examination of the serum or plasma by the overnight refrigeration test. About 10% to 15% of the dyslipoproteinemias observed do not fit into the classical Fredrickson lipoprotein classification system.

The goal of lipoprotein electrophoresis is to separate the major lipoprotein fractions into relatively homogeneous bands of lipoproteins so that one can evaluate, semiquantitatively and qualitatively, the complete lipoprotein profile. A review of electrophoretic theory and technique can be found elsewhere[1] and in Chapter 8.

The major factors generally affecting the migration of molecules in an electric field must also be controlled for accurate and reproducible lipoprotein electrophoresis. Briefly, these are as follows:

1. Net charge on molecule. The pH has to be accurately and reproducibly controlled to obtain consistent electrophoretic patterns with optimum resolution of the lipoprotein bands.
2. Ionic strength and viscosity. The best and most reproducible results require a properly prepared electrophoretic buffer medium. Although high-ionic-strength buffers yield sharper bands and better separations, the resolution can be adversely affected by increased heating, which can denature heat-labile solutes.
3. Strength of electric field. The electric field strength must be reproducibly controlled. Too high a current can raise the temperature of the support medium, causing protein denaturation and poor migration and band resolution.
4. Type of support medium. The type of medium is one of the most important factors in lipoprotein separations. The charge on the support medium can cause an endosmotic effect; on support mediums where endosmosis is strong, as in agar gel, endosmotic effects cause the slowly migrating lipoprotein-X to be swept cathodically behind the application point.

Since the most critical element in lipoprotein electrophoresis is the support medium used, much of the following discussion is a comparison of the various support mediums.

Paper

The paper method, once widely used in the clinical chemistry laboratory for the separation of serum lipoproteins and proteins, has largely been replaced by agarose gel electrophoresis.

Paper electrophoresis is still considered the conventional and classic method for lipoprotein studies. Two basic types of apparatus for paper electrophoresis have proved most useful. In the first type, the amount of moisture on the filter paper is kept uniform throughout the period of electrophoresis. The paper moistened with buffer is laid on a siliconized glass plate, the sample is applied, and a second siliconized glass plate is laid on top. This system has been referred to as the *closed horizontal method*.

The second method applies the so-called open strip or inverted V principle. The Beckman-Durrum electrophoretic cell (Beckman Instruments, Inc., Brea, CA) is one of the most popular and widely used cells for this purpose.

In 1953 Swahn introduced the use of filter paper for lipoprotein electrophoresis. Most paper electrophoretic methods use either Whatman No. 1 or No. 3 paper. The original method of Lees and Hatch[2] was modified to increase the speed of the procedure.[3] In contrast to the method of Jencks and Durrum,[4] Lees and Hatch introduced the use of 1% albumin in barbital buffer and showed that the presence of the albumin provided a sharper delineation of the major lipoprotein components and a resolution or partial resolution of the very-low-density lipoprotein (VLDL). The method of Naito and Lewis[3] used 1%

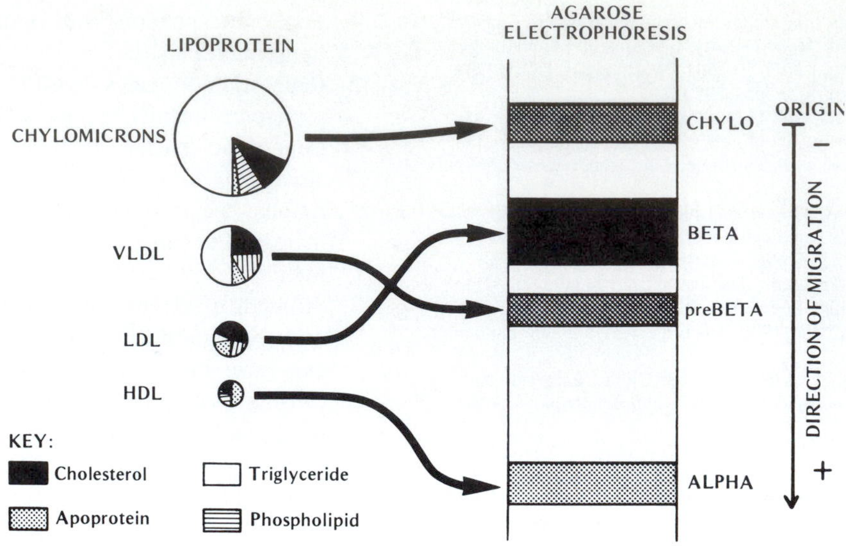

Fig. 61-4 Schema of lipoprotein electrophoretic pattern on agarose support medium. *HDL,* High-density lipoprotein; *LDL,* low-density lipoprotein; *VLDL,* very-low-density lipoprotein.

albuminated buffer for the paper strips only. This buffer was not used in the chambers or wells, which contained barbital without albumin. This minimized the microbial growth in the buffer, which, in previous methods, altered the pH of the buffer and caused unsatisfactory resolution of the lipoprotein bands. In addition, the method increased the ionic strength of the buffer from 0.075 to 0.1 mol/L, resulting in sharper delineation of the lipoprotein bands.

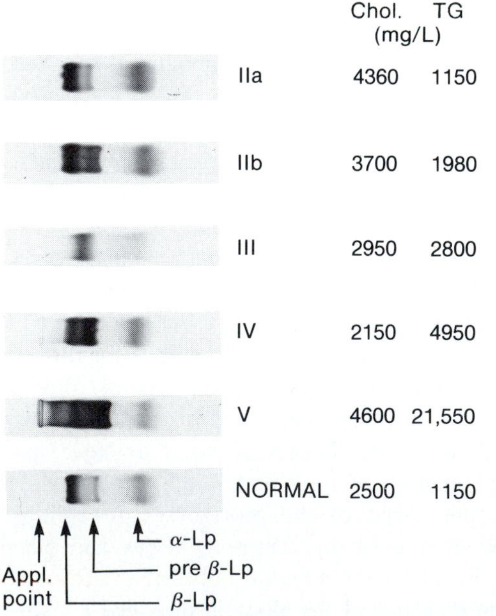

Fig. 61-5 Examples of agarose gel electrophoretic patterns depicting different forms of hyperlipoproteinemia. *Chol,* Cholesterol; *TG,* triglycerides.

Agarose

Fifteen years after Swahn reported the use of paper as a support medium for electrophoresis of serum lipoproteins, agarose gel was introduced. Agar is composed of at least two major fractions—agaropectin and agarose. Agaropectin contains sulfate and carboxylic acid groups and is responsible for the considerable endosmosis and background color observed with most unfractionated agar. Purified agarose is essentially neutral and exhibits few of the problems of unpurified agar. With the advent of premade gel plates produced by commercial companies, the agarose method came into widespread use. Although the basic knowledge of electrophoretic patterns in health and disease has been acquired from electrophoresis on filter paper, the agarose method was developed as an analytical, research, and clinical method for studying serum lipoproteins qualitatively and semiquantitatively (Fig. 61-4). Fig. 61-5 illustrates an agarose gel electrophoretic pattern of human sera, depicting the different forms of hyperlipoproteinemia. As illustrated in Fig. 61-6, hypobetalipoproteinemia and hyperbetalipoproteinemia can be readily detected by agarose gel electrophoresis. As can be seen from the electrophoretic patterns (Fig. 61-7), the agarose gel system separates the pre-beta-lipoprotein fraction from the beta-lipoprotein fraction with better resolution than other electrophoretic systems can, and, in addition, it can often show multiple pre-beta-lipoprotein bands that are generally not demonstrated with other support mediums, except for polyacrylamide gels.

Polyacrylamide gel

The technique of disk electrophoresis, the most recent addition to the many variations of electrophoresis, is based

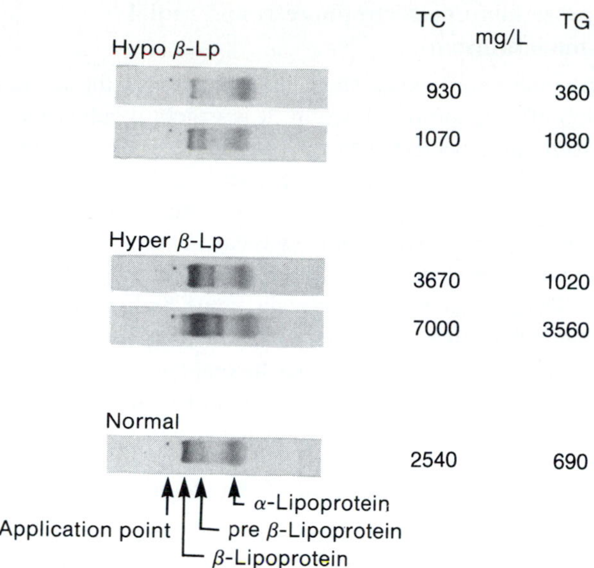

Fig. 61-6 Variations in beta-lipoprotein concentration. Agarose gel electrophoretic pattern of sera from human subjects with hypobetalipoproteinemia and hyperbetalipoproteinemia. Semiquantitation of staining intensity by visual comparison with normobetalipoproteinemic pattern can serve as means of classifying dyslipoproteinemic patterns. *TC,* Total cholesterol. (Agarose gel, lipid stain.)

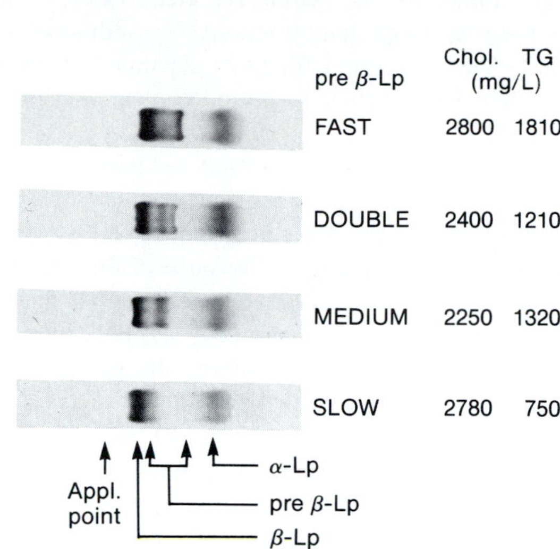

Fig. 61-7 Human pre-beta-lipoprotein bands with different mobilities (slow, medium, fast). It is not uncommon to see multiple pre-beta-lipoprotein bands as demonstrated in the second pattern from the top. *Chol,* Cholesterol; *TG,* trigylcerides.

on the bands or zones that stack up as a series of concentrated disks at the beginning of the electrophoretic run. The support medium is polyacrylamide gel.

The migration and separation of the components in a given sample is based mainly on (1) the electrophoretic charge of the particle and (2) the molecular size and configuration of the particle. Thus the polyacrylamide support medium produces a molecular sieving effect. As a result of this phenomenon, polyacrylamide gel electrophoresis tends to separate protein and lipoprotein fractions more discretely than other support mediums do (Fig. 61-8).

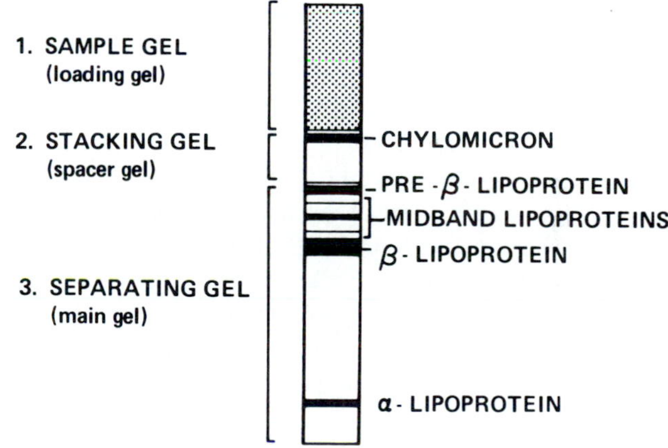

Fig. 61-8 Scheme of a lipoprotein electrophoretic pattern on a polyacrylamide-gel support medium.

Cellulose acetate

Before the introduction of agarose electrophoresis, cellulose acetate support medium was widely used in the clinical laboratory for lipoprotein studies. It is used less frequently now.

Starch gel

For some unknown reason, starch gel electrophoresis has not been used in the clinical laboratory as much as other support mediums. Starch gel has unique properties that permit better electrophoretic separation of proteins and lipoproteins than is achieved using other support mediums. The resolved fractions can be determined by staining, immunodiffusion, or radioisotopic or other techniques.

The original starch gel procedure described by Smithies[5] used a block of starch gel in which a slit was made for insertion of a paper strip moistened with the serum to be tested. The gels were pressed firmly against the filter paper, and electrophoresis was carried out using much the same technique as in horizontal paper electrophoresis. After electrophoresis, the gels were sliced into thin sheets, which could be stained for protein and lipoprotein; a third sheet was cut into appropriate sections as determined by the stained sheets and subjected to immunodiffusion studies. The major use of starch gel electrophoresis today is to prepare lipoprotein fractions for additional investigations. The lipoprotein fractions separated in starch gels can be eluted for chemical, immunological, and physical studies.

In a modification of the horizontal technique, subsequently developed, the starch gel was cast in a vertical holder. This technique was specifically designed to permit the application of larger samples. Because of this arrangement, the filter paper was no longer necessary; at the beginning of electrophoresis, a sharpening and concentration of the proteins occurred as they entered the gel. This resulted in sharper resolution of the fractions.

Another modification of the starch gel procedure involves the preparation of a thin-layer starch gel on an inert support film. The sheet coated with gel is placed in a Durrum inverted V type of electrophoresis cell, and the samples are applied to small depressions made in the gel at the level of the top support rod. After electrophoresis, the sheets are stained with oil red O for lipoprotein, amido black for protein, or other appropriate stains. After staining, they can be scanned in a densitometer to determine the relative percentages of the fractions resolved.

The lipoprotein patterns of young healthy adults obtained on serum using starch gel electrophoresis show the usual components resolved by other techniques such as paper or agarose electrophoresis. Alpha-lipoprotein is occasionally resolved as a double or triple fraction, with one or more of the subfractions electrophoresing more slowly than normal. Starch gel electrophoresis has proved valuable in the past as a guide in the study, isolation, and purification of lipoproteins and apolipoproteins. This electrophoretic system has also been used for special lipoprotein studies of patients with unusual lipoprotein-immunoglobulin complexes. An example of this is the evaluation of serum from multiple myeloma patients. Since the development of acrylamide gel electrophoresis, the use of starch gel in the evaluation of apolipoproteins and peptides has decreased.

Agar gel

One of the known lipid disturbances associated with liver disease involves the presence of hypercholesterolemia and an abnormal serum lipoprotein called lipoprotein-X (Lp-X) in patients with cholestasis. Lp-X is characterized by a high content of phospholipid and free cholesterol and a very low content of protein. Lp-X migrates in the beta-lipoprotein area on most support mediums. However, in agar gels at the usual pH of 8.4 to 8.6, the Lp-X moves in the opposite direction (cathodically). One can use this peculiar behavior to identify Lp-X. The serum is electrophoresed on 1% agar support medium, and the Lp-X is visualized by polyanionic precipitation or by lipid stain. For further details, refer to Seidel et al.[6]

Agarose immunoelectrophoresis and radial immunodiffusion

If some doubt exists about the accuracy of the identification of a lipoprotein fraction, it is generally advisable to perform immunoelectrophoresis and identify the fraction in question by using appropriate antisera, such as anti-alpha-lipoprotein antiserum, anti-beta-lipoprotein antiserum, or anti-whole-serum antiserum. One can further evaluate the lipoproteins by performing double immunodiffusion studies. These studies will provide some indication as to whether there has been alteration in the apoprotein component or structure, that is, an incomplete or partial line of identity as compared with a complete line of identity with known apolipoprotein standards.

Stains used for visualization of electrophoretic patterns

Many stains are available for visualization of the separated lipoprotein fractions. Sudan black-B was initially used for staining the lipoproteins. Since then, oil red O, fat red 7B, and other specific lipid stains have been used.

REFERENCE AND PREFERRED METHODS

The various electrophoretic methods just listed have a variety of advantages and disadvantages, and the support mediums chosen must suit the particular needs of the laboratory.

There is little information available on the comparative accuracy and reproducibility of the various support mediums. However, on the basis of speed and convenience, the agarose gel technique can certainly be recommended for most laboratory situations and is the method described in this text.

SPECIMEN

Before one performs any accurate quantitation of lipids and lipoproteins, a series of sampling conditions must be met to allow for a representative analysis. All too often the classification and diagnosis of a hyperlipoproteinemia is based on a single sample that is flawed, or nonrepresentative, because one or more of the following sampling conditions is not met (see p. 978 for details):

1. Fasting for at least 12 hours before blood is drawn
2. Habitual diet for at least 3 weeks before sampling; substantial changes in caloric, alcoholic, or type of food intake discouraged in advance so that the sample is representative
3. Stable weight or at least no sampling during weight changes, particularly during weight loss
4. No acute illness, trauma, or recent surgery
5. No sampling from patients after myocardial infarction for a minimum of 3 months
6. Sampling deferred in subjects taking drugs that affect lipid or lipoprotein levels (thyroxine, estrogen, corticosteroids, androgenic compounds, lipid-lowering drugs, and antihypertensive drugs)

	Electrophoretic Pattern	24 hr Standing Plasma (4°c)	Choles-terol	Triglye-cerides	Dietary Management	Drugs
TYPE I (very rare)	CHYLOMICRONS↑↑↑ chylo β pre β α – migration → +	Creamy layer over clear plasma	↔↑	↑↑↑	1. Restriction of fat to about 35 gm/day 2. Supplementation with medium-chain triglycerides 3. Restriction of alcohol intake	None effective at present
TYPE II-A (common)	LDL↑↑↑ β pre β α	Clear	↑↑↑	↔	1. Low-cholesterol diet (less than 300 mg/day) 2. Decreased intake of saturated fats (S) 3. Increased intake of polyunsaturated fats (P) 4. P/S Ratio: 1.0-1.2	1. Cholestyramine, 16-24 gm/day 2. Nicotinic acid, 3 gm/day 3. Probucol, 1 gm/day 4. Colestipol, 4-6 sachets/day
TYPE II-B (common)	LDL↑↑↑ VLDL↑ β pre β α	Clear to Slightly cloudy	↑↑↑	↑	1. Reduction to ideal body weight 2. Low-cholesterol diet (less than 300 mg/day) 3. Decreased intake of saturated fats 4. Increased intake of polyunsaturated fats 5. P/S Ratio 1.0-1.2	1. Cholestyramine, 16-24 gm/day 2. Nicotinic acid, 3 gm/day 3. Probucol 1 gm/day 4. Colestipol, 4-6 sachets/day
TYPE III (very rare)	B-VLDL,LDL OF ABNORMAL COMPOSITION β pre β α	Slightly cloudy to cloudy	↑↑	↑↑	1. Reduction to ideal body weight 2. Low-cholesterol diet (less than 300mg/day) 3. Decreased intake of saturated fat 4. Restriction of alcohol intake	1. Clofibrate, 2 gm/day
TYPE IV (very common)	VLDL↑↑↑ β pre β α	Clear cloudy or milky	↔↑	↑↑↑	1. Reduction to ideal body weight 2. Carbohydrate restricted to 40% of calories 3. Substitute polyunsaturated fats for saturated fats 4. Restrict alcohol intake 5. Reduce cholesterol intake to 300-400 mg/day	1. Clofibrate, 2 gm/day 2. Nicotinic acid, 3 gm/day
TYPE V (rare)	CHYLOMICRONS↑↑ VLDL↑↑↑ chylo β pre β α	Creamy layer over milky plasma	↑	↑↑	1. Reduction to ideal body weight 2. Reduction to fat to 30% of daily calories 3. Alcohol not recommended 4. Reduce cholesterol intake to 300-400 mg/day	1. Nicotinic acid, 3 gm/day 2. Clofibrate, 2 gm/day 3. Oxandrolone, 2.5 mg t.i.d.

Fig. 61-9 Summary of six types of hyperlipoproteinemias. *Chylo,* Chylomicrons; β, beta-lipoprotein; *pre* β, pre-beta-lipoprotein *(VLDL,* very-low-density lipoprotein); α, alpha-lipoprotein.

PROCEDURE: AGAROSE GEL ELECTROPHORESIS OF LIPOPROTEINS

Principle

Lipoproteins are separated on thin agarose film, stained with fat red 7B, destained, and scanned.

Reagents

The reagents can be readily obtained from Corning. Destaining solution is prepared when 2 volumes of methanol are mixed with 1 volume of distilled water. Prepare daily.

Assay

Equipment: electrophoresis apparatus, incubation, and drying ovens.

Lipoprotein electrophoresis on agarose gel (Corning) is performed as follows:

1. Prepare the buffer according to the manufacturer's directions.
2. Apply samples to the gel plates according to manufacturer's directions.
3. Electrophorese specimens for 30 minutes at 90 volts. The amperage should remain between 12 and 21 mA throughout the run.
4. Lift cell cover from cell base, remove agarose film carefully from cell cover, and wipe off *all* buffer condensation from the plastic backing of the agarose film.
5. Dry agarose film by placing it *gel side up* in the oven compartment of the incubator oven (Corning no. 470041), preheated to 60° C. The film should rest flat on the dryer shelf, and the rubber bands should *not* touch the gel.
6. Plug dryer shelf into the oven and set timer for *30 minutes*. NOTE: *Failure to dry the gel adequately will result in excessive background staining and blurring.*
7. Dispense 10 mL of fat red 7B stain and 2.2 mL of type 3 distilled water into a 50 mL beaker. Swirl the solution gently until the formed sediment dissolves. The ratio of stain to distilled water is flexible, depending on the age of the stock stain solution.
8. Pour the stain onto the agarose-gel surface and allow the uniformly flooded agarose film to be stained for 4 to 5 minutes. Staining time is flexible, depending on the age of the stock stain solution.
9. Transfer the agarose film into the destaining solution and agitate *gently* until the background clears up.
10. Rinse the agarose film in distilled water for a few seconds.
11. Dry the back of the agarose film with facial tissue; avoid scratching.
12. Place the agarose film in the incubator oven for *15 minutes*.
13. The electrophoretogram can now be scanned by a densitometer.

Table 61-14 Percent distribution of serum lipoproteins

Age (years)	Sex	Lipoproteins (%) $\overline{X}$ ± SD		
		Beta	Pre-beta	Alpha
Children (4-14)	M+F	55 ± 5	12 ± 3	33 ± 3
Adults (18-65)	M	65 ± 8	12 ± 4	23 ± 4
Adults (18-65)	F	60 ± 6	8 ± 4	32 ± 5

Naito, HK, unpublished data.

REFERENCE RANGE

Fig. 61-9 depicts a typical lipoprotein electrophoresis diagrammatic pattern of a patient with hyperlipoproteinemia. The percentage of each lipoprotein class present in a fasting, normal person is listed in Table 61-14.

REFERENCES

1. Smith I: In Chromatographic and electrophoretic techniques, New York, 1968, John Wiley & Sons, Inc, pp 1-524.
2. Lees, RS, and Hatch FT: Sharper separation of lipoprotein species by paper electrophoresis in albumin containing buffer, J Clin Med 6:518-528, 1962.
3. Naito HK, and Lewis, LA: Rapid lipid-staining procedures for paper electrophoresis, Clin Chem 21:1454-1456, 1973.
4. Jenks, WP, and Durrum, EL: Paper electrophoresis as a quantitative method: the staining of serum lipoproteins, J Clin Invest 34:1437-1447, 1955.
5. Smithies, O: Zone electrophoresis in starch gel: group variations in the serum proteins of normal human adults, Biochem J 61:629-641, 1951.
6. Seidel, D, Wieland, H, and Ruppert, C: Improved techniques for assessment of plasma lipoprotein patterns. I(1). Precipitation in gels after electrophoresis with polyanionic compounds, Clin Chem 19:737-739, 1973.

Triglycerides

HERBERT K. NAITO

Triglycerides, triacylglycerol
Clinical significance: p. 454
Molecular formula: varies with R group of fatty acids
Molecular weight: 800 to 900 daltons (varies with R group of fatty acid); average molecular weight for human samples, 885 daltons
Chemical class: neutral lipid

$$
\begin{array}{l}
H_2C-O-\overset{\overset{\Large O}{\|}}{C}-R_1 \\
\;\;| \\
HC-O-\overset{\overset{\Large O}{\|}}{C}-R_2 \\
\;\;| \\
H_2C-O-\overset{\overset{\Large O}{\|}}{C}-R_3
\end{array}
$$

PRINCIPLES OF ANALYSIS

Before the 1950s blood triglyceride levels were estimated by the subtraction method:

Triglycerides = Total lipids − (Cholesterol + Phospholipids)

This indirect method was widely used until 1957, when Van Handel and Zilversmit[1] published a direct manual method in which the phospholipids were removed from the lipid extract by an adsorbent and the triglycerides were determined by measurement of the amount of glycerol released by saponification with potassium hydroxide (KOH). This method has been widely adopted and has been modified by many investigators. In 1964 a semiautomated adaptation of this procedure for serum triglycerides was established on the Technicon AutoAnalyzer-I.[2] In 1966 Kessler and Lederer[3] improved the method by automating the potassium hydroxide saponification step; their method was once the most widely used automated procedure and has been regarded as a potential reference method. This method was also used by the Lipid Research Clinics Population Studies supported by the National Heart, Lung, and Blood Institute. The Lipid Research Clinics Laboratories' triglyceride method was standardized by the Centers for Disease Control (CDC) Lipid Standardization Section.

Most current methods use chemical or enzymatic procedures to determine the glyceride glycerol concentration, which is then converted to the equivalent mass concentration of an average triglyceride (in milligrams per liter). Alternatively, concentration may be expressed on a molar basis.

Recently, triglyceride analysis has been simplified by the introduction of enzymatic methods, which have been automated to provide the analyst with quick, easy, and direct procedures.

Since the automated method employing the Hantzsch condensation[3] is still considered a classical and is possibly a candidate reference method, the procedure is described in some detail in Pesce and Kaplan's *Methods in Clinical Chemistry*. This method can be used to measure cholesterol and triglycerides simultaneously. The classical manual extraction method of Van Handel and Zilversmit[1] is provided under the procedure section. Although triglyceride procedures based on thin-layer, gas-liquid, or column chromatography and infrared spectrometry have been published, these are not discussed: they have very specific and sometimes limited use and are not suitable for routine use.

Chemical determination of glyceride glycerol

In reactions based on chemical determination of glyceride glycerol (reactions A to C, Table 61-15) the *first step* is the extraction of triglycerides and the removal of interfering substances. The extraction is accomplished by use of solvents such as methanol, ethanol, isopropanol, or chloroform. These solvents cause the denaturation of the lipoproteins and therefore the dissociation of the bound triglycerides. Interfering substances are removed by either (1) solvent partition with organic solvents such as nonane or hexane or (2) use of adsorbents such as zeolite, silicic acid, or Florisil (activated magnesium silicate). The principal interfering substances removed are phospholipids and glucose, along with certain chromogens and sometimes free glycerol.

The *second step* in these procedures hydrolyzes triglycerides to glycerol and fatty acids and is usually carried out in ethanolic potassium hydroxide at elevated temperatures (saponification):

Tryglycerides → Glycerol + 3 fatty acids

An alternative approach is the use of an alkoxide transesterification reaction to form glycerol and fatty acids from the triglycerides.

The *third step* oxidizes glycerol to formaldehyde and is usually accomplished by the following reaction:

Glycerol + Periodate → Formaldehyde + Formic acid + Iodate + Water

In the *fourth step* the formaldehyde formed is quantitatively measured by the following various reactions:

1. Eegriwe's reaction: formaldehyde reaction with a chromotropic acid/sulfuric acid mixture; read at 570 nm (method 1, Table 61-15).
2. Schryver's reaction: formaldehyde reaction with a phenylhydrazine/ferricyanide/hydrochloric acid mixture; read at 540 nm (method 2, Table 61-15).
3. Pay's reaction: formaldehyde reaction with 3-methyl-2-benzothiazolinone/ferric chloride mixture; read at 620 nm (method 3, Table 61-15).
4. Hantzsch's reaction: formaldehyde reaction with am-

Table 61-15 Methods of serum triglyceride analysis

Method	Principle	Usage	Comments
Chemical determinations of glycerol	The following three reactions are common to methods 1 to 4: A. Triglycerides extracted with organic solvents or purified with adsorbents such as zeolite B. Triglycerides $\xrightarrow[OH^-]{ROH}$ Glycerol + 3 fatty acids C. Glycerol + IO_4^- → HCHO (Formaldehyde) + HCOOH (Formic acid) + H_2O + IO_3^-		
1. Eegriwe's reaction	HCHO 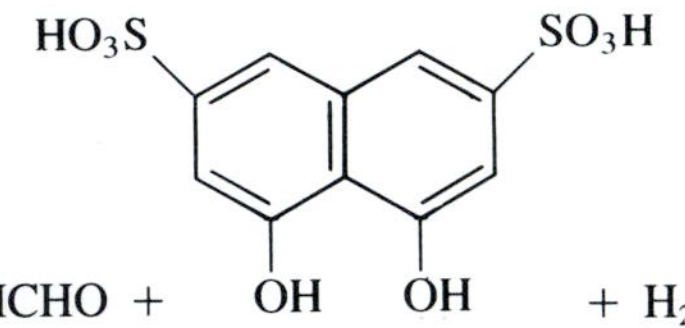 + Chromotropic acid + H_2SO_4 → Chromophore (570 nm)	Rarely used in routine laboratories	Widely used as comparative method
2. Schryver's reaction	HCHO 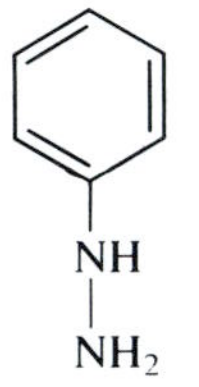 + Phenylhydrazine $\xrightarrow[HCl]{Ferricyanide}$ Chromophore (540 nm)	Rarely used	
3. Pay's reaction	HCHO + 3-Methyl-benzothiazolin-2-one $\xrightarrow{FeCl_3}$ Chromophore (620 nm)	Rarely used	
4. Hantzsch's reaction	HCHO + NH_4^+ + $2CH_3{-}CO{-}CH_2{-}CO{-}CH_3$ (Acetylacetone) 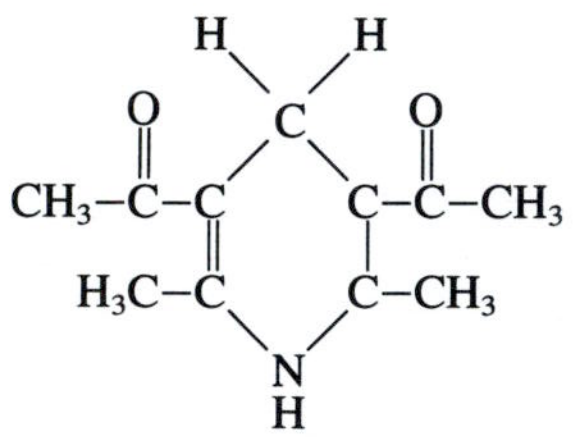 → 3,5-Diacetyl-1,4-dihydrolutidine (A_{max}, 412 nm; $\lambda_{excitation}$, 400 nm; $\lambda_{emission}$, 485 nm)	Most frequently used chemical method in routine laboratories	Fluorescent assay used as reference method of CDC

Enzymatic determinations of glycerol		Most commonly used
5. NADH consumption (decreased A_{340})	a. Triglycerides $\xrightarrow[\text{Protease}]{\text{Lipase}}$ Glycerol + 3 fatty acids	Very frequently used
or	b. Glycerol + ATP $\xrightarrow{\text{Glycerol kinase}}$ Glycerol-3 phosphate + ADP	
(decreased fluorescence: λ_{ex}, 355 nm; λ_{em}, 460 nm)	c. ADP + Phospho*enol*pyruvate $\xrightarrow{\text{Pyruvate kinase}}$ ATP + Pyruvate	
	d. Pyruvate + NADH + H^+ $\xrightarrow{\text{Lactate dehydrogenase}}$ Lactate + NAD^+	
6. Formazan colorimetric	*Reactions 5a and 5b plus:*	Very frequently used
	Glycerol-3-phosphate + NAD^+ $\xrightarrow{\text{Glycerol phosphate dehydrogenase}}$ Dihydroxyacetone phosphate + NADH + H^+	
	NADH + Oxidized tetrazolium $\xrightarrow{\text{Diaphorase}}$ Reduced tetrazolium (500-590 nm)	
7. Fluorescent	*Reaction 5a plus the following:*	Not frequently used
	Glycerol + NAD^+ $\xrightarrow{\text{Glycerol dehydrogenase}}$ Dihydroxyacetone + NADH + H^+	
	NADH + H^+ + Resazurin $\xrightarrow{\text{Diaphorase}}$ Resorufin + NAD^+ λ_{ex}, 548 λ_{em}, 580 (fluorescent)	

monium acetate and acetylacetone; read colorimetrically or fluorometrically (method 4, Table 61-15).

In the Hantzsch condensation reaction the following occurs:

Acetylacetone + Formaldehyde + Ammonium acetate →
3,5-Diacetyl-1,4-dihydrolutidine

The yellow end product can be measured colorimetrically at 412 nm or fluorometrically (primary filter at 400 nm, secondary filter at 485 nm). The fluorometric method is more widely used and has been applied to semiautomated methods, one of which is discussed later in this section. The described semiautomated method for the triglyceride determinations has been standardized and certified by the CDC for use as a reference method in our laboratory.

In summary, the following four steps are usually used in the chemical determination of triglycerides:

1. Extraction of triglycerides; removal of interfering substances
2. Saponification of the triglycerides to glycerol and fatty acids
3. Oxidation of the glycerol to formaldehyde
4. Measurement of the formaldehyde

Enzymatic determination of glyceride glycerol

The enzymatic methods are based on the determination of the glycerol portion of the triglyceride molecules after hydrolysis (chemical or enzymatic) to remove the fatty acids (see Table 61-16). Methods employing the enzymatic determination of glycerol have been used for many years; however, in the early years an alkaline hydrolysis of the triglycerides was used in these methods. The recent development of employing enzymes (lipase, usually combined with a protease) to catalyze hydrolysis has made possible methods that are direct, rapid, and specific. These enzyme methods are now used by virtually all laboratories. The completely enzymatic systems eliminate the use of caustic reagents, extraction solvents, high-temperature baths, and adsorption mixtures for phospholipid removal. Recent studies have focused on developing lipase reagents that completely hydrolyze triglycerides. The role played by the protease is as yet unknown, since the protease enzymes

Table 61-16 Comparison of methods for triglyceride analysis

Parameter	Egriwe's reaction*	Hantzsch's reaction†	Enzymatic‡
Temperature	Extraction: ambient Saponification: 65° C Colorimetric reaction: 100° C	50° C	37° C
Sample volume	0.5 mL	0.2 mL	0.004 mL
Fraction of sample volume	0.3 (extraction)	Approximately 0.01	0.01
Final concentration of reagents	KOH: 356.5 mmol/L (saponification) $NaIO_4$: 4.5 mmol/L (formaldehyde formation) Colorimetric reaction Chromotropic acid: 1.61 g/L H_2SO_4: 8 mmol/L	KOH: 22.5 mmol/L (saponification) $NaIO_4$: 1.1 mmol/L (formaldehyde formation) Ethanol: 57.5% (v/v) Colorimetric reaction Acetylacetone: 36.6 mmol/L	Glycerol kinase: 25 × 10^3 U/L Lipase: 300 × 10^3 U/L Glycerol kinase: 250 U/L L-α-Glycerophosphate oxidase: 100 × 10^3 U/L Horseradish peroxidase: 10 × 10^3 U/L ATP: 0.5 mmol/L $MgCl_2$: 5.0 mmol/L Triton X-100: 0.1 g/L 3,5-Dichloro-2-hydroxybenzene sulfonic acid: 1.5 mmol/L 4-Aminoantipyrine: 1.0 mmol/L Tris-HCl, pH 7.6: 50 mmol/L
Time of reaction	Extraction: 30 min Saponification: 15 min Colorimetric reaction: 30 min	Extraction: 20 min AutoAnalyzer-II: 8 min	10 min
Wavelength	570 nm	Excitation: 400 nm Emission: 485 nm	505 nm (primary) 600 nm (secondary)
Linearity	—	5000 mg/L	5000 mg/L
Interferences		Free glycerol	Free glycerol, gross lipemia, bilirubin (40 mg/L decrease in result for every 10 mg/L bilirubin)
Precision ($\overline{X}$, % coefficient of variation)§	Free glycerol	310 mg/L (11.9%)	1665 mg/L (1.6%) 2550 mg/L (2.1%)

*Modification of Van Handel and Zilversmit.[1]
†Technicon AutoAnalyzer-II method.
‡Reagents adapted to Hitachi 705 (Boehringer Mannheim Diagnostics, Indianapolis).
§College of American Pathologists Quality Assurance Survey data.

alone do not hydrolyze triglycerides; however, many methods find a protease enzyme necessary to attain complete hydrolysis. One of the most commonly added proteases is alpha-chymotrypsin.[4] The critical point of whether phospholipids were hydrolyzed by these enzymes has also been investigated; it was found that any glycerol from this source was insignificant.

The method of Bucolo and David[4] in common use today is based on the following sequence of reactions (method 5, Table 61-15):

1. Triglyceride $\xrightarrow[\text{Alpha-chymotrypsin}]{\text{Lipase}}$ Glycerol + Fatty acids
2. Glycerol + ATP $\underset{\text{Mg}^{++}}{\overset{\text{Glycerol kinase}}{\rightleftharpoons}}$ Glycerol-3-phosphate + ADP
3. ADP + Phospho*enol*pyruvate $\overset{\text{Pyruvate kinase}}{\rightleftharpoons}$ ATP + Pyruvate
4. Pyruvate + NADH + H^+ $\overset{\text{Lactate dehydrogenase}}{\rightleftharpoons}$ Lactate + NAD^+

The decrease in absorbance of reduced nicotinamide adenine dinucleotide (NADH) is measured at 340 nm. To provide a blank for this method, the procedure is repeated with a buffer used in place of the lipase reagent.

Rietz and Guilbault[5] modified the procedure using fluorometric measurement. The disappearance of NADH fluorescence is read at 460 nm after excitation at 355 nm.

A procedure by Megraw et al.[6] used reaction steps 1 and 2 and then steps 5 and 6 to form formazan, a highly colored compound that is measured in the 500 to 590 nm range (method 6, Table 61-15), as follows:

5. Glycerol-3-phosphate + NAD^+ $\xrightarrow{\text{Glycerol-3-phosphate dehydrogenase}}$ Dihydroxyacetone phosphate + NADH + H^+
6. NADH + H^+ + 2-*p*-Iodophenyl-3-nitrophenyl-5-phenyltetrazolium *(oxidized)* $\xrightarrow{\text{Diaphorase}}$ 2-*p*-Iodophenyl-3-*p*-nitrophenyl-5-phenyltetrazolium *(reduced)* (a formazan dye) + NAD^+

On the other hand, the method of Winartasaputra et al.[7] involves glycerol dehydrogenase and diaphorase to form a fluorescent compound from glycerol released by enzymatic hydrolysis (method 7, Table 61-15):

7. Glycerol + NAD^+ $\xrightarrow{\text{Glycerol dehydrogenase}}$ Dihydroxyacetone + NADH + H^+
8. NADH + H^+ + Resazurin $\xrightarrow{\text{Diaphorase}}$ Resorufin + NAD^+ *(fluorescent)*

The fluorescence is measured at 580 nm after excitation at 548 nm.

Recently, Nagele et al.[8] described a triglyceride method that is based on the action of L-alpha-glycerol phosphate oxidase (GPO) on the glycerol-3-phosphate released by reactions 1 and 2 above:

9. Glycerol-3-phosphate + O_2 $\xrightarrow{\text{GPO}}$ Dihydroxyacetone phosphate + H_2O_2

In the presence of horseradish peroxidase (EC 1.11.1.2), hydrogen peroxide oxidizes the chromogen, consisting of 4-aminophenazone and 4-chlorophenol, to form a red-colored quinonemonoimine dye:

10. H_2O_2 + 4-Chlorophenol + 4-Aminophenazone + K hexacyanoferrate (II) $\xrightarrow{\text{Peroxidase}}$ 4-*p*-Benzoquinone-monoiminophenazone + K hexacyanoferrate (III) + HCl + $2H_2O_2$

In this Trinder type of reaction, the absorbance (500 nm) of the quinone-monoimine dye is proportional to the concentration of triglycerides *and* free glycerol in the sample. L-Alpha-glycerol phosphate oxidase is highly specific for L-alpha-glycerol phosphate, which is generated in a specific reaction from glycerol with glycerol kinase. The oxidative coupling in the last step under the catalytic influence of peroxidase is less specific. For example, bilirubin is known to interfere with the quantitation of hydrogen peroxide and decrease the amount of chromophore formed. For this reason, results obtained by determination of serum triglycerides tend to be lower than true triglyceride values. Nagele et al.[8] used potassium hexacyanoferrate (II) to minimize bilirubin interferences. Naito et al.[9] adapted this assay to the Hitachi 705 bichromatic analyzer. In this assay, total glycerol (triacylglycerol + free glycerol) is measured by the GPO method, and free glycerol is measured by the ultraviolet absorption method, which monitors the decreased absorbance at 340 nm.

Several semiautomated analyzers using the enzymatic methods are now available for use in the physician's office for the testing of triglyceride levels. These include the BMD Reflotron (Boehringer Mannheim Diagnostics, Indianapolis), the Ames Seralyzer (Ames Co., Elkhart, IN), the Kodak DT60 (Kodak Co., Rochester, NY), and the Abbott Vision (Abbott Corp., North Chicago). The Reflotron, Seralyzer, and DT60 instruments employ an enzymatic procedure with the dry-reagent technology, whereas the Vision employs a more traditional wet-chemistry approach.

At present there is no officially recognized reference method for triglyceride determinations. The method of Van Handel and Zilversmit[1] and the Hantzsch condensation method are probably the most widely used comparative methods.

Triglyceride methods employing enzyme reactions that produce a decrease in NADH concentration and absorbance at 340 nm provide good precision, sensitivity, and specificity.

A major factor in the accuracy of an enzymatic determination of serum triglycerides is the complete hydrolysis of triglycerides by triacylglycerol acylhydrolase. Furthermore, there must be no interfering side reactions during phosphorylation of the liberated glycerol by glycerol kinase and oxidation of glycerol-3-phosphate by L-alpha-glycerol phosphate oxidase to dehydroxyacetone phosphate with the equimolar consumption of oxygen and the concomitant formation of hydrogen peroxide. The major problem one faces when setting up one of the completely enzymatic methods is the inconsistent quality of the enzymes available. Finally, there is a need to determine the free glycerol content in the blood for accurate triacylglycerol measurements.

For an accurate triglyceride level, the free glycerol of the sample should be subtracted from the total glycerol (both free and that produced from saponification of the triglycerides) by a free glycerol blanking procedure.

There are instances in which intravenous infusates may contain glycerol or glycerol-like products that will cause a high blank value. Lindblad et al.[10] reported increases in the blood glycerol level during total parenteral nutrition administration. It has also been found that mannitol (Osmitrol) infusion will cause spuriously high glycerol blank values.[11] Conditions known to contribute to high glycerol levels in blood include the following:

1. Stress (epinephrine effect)
2. Mannitol infusion
3. Treatment with nitroglycerin
4. Diabetes mellitus
5. Glycerol-coated stopper used in Vacutainer phlebotomy tubes
6. Certain liver diseases
7. Hemodialysis for kidney disease

Few laboratories in the United States correct for free glycerol concentration. This has important implications in determining the concentration of low-density lipoprotein cholesterol (LDL-C), which is often calculated from estimates of very-low-density lipoprotein cholesterol (VLDL-C) levels. Most clinical laboratories, for convenience, indirectly estimate VLDL-C and LDL-C levels by using the Friedewald formula, which is based on the assumption that VLDl-C approximates *plasma* true triglyceride divided by 5. From this calculation, LDL-C can be calculated as follows:

$$\text{LDL-C} = \text{Total cholesterol} - (\text{VLDL-C} + \text{HDL-C})$$

where high-density lipoprotein cholesterol (HDL-C) is directly quantitated. There are several technical problems with this technique. First, the Friedewald formula does not apply to all persons. Second, the formula is based on true triglyceride determination, which is often not performed, and makes the false assumption that the free glycerol level in the blood is small and relatively constant. Although most unstressed healthy persons may have free serum glycerol concentrations of 80 to 200 mg/L, it is not uncommon to find values greater than 750 mg/L.[12] Thus it is easy to see that when inaccurate triglyceride values are obtained the estimated VLDL cholesterol value is incorrect, leading to inaccurate estimates of LDL cholesterol. It should also be remembered that the Friedewald formula does not apply when the triglyceride values are greater than 4000 mg/L.

Standards

A pure triglyceride standard is essential to obtain the most accurate results. Highly purified triglycerides are available from commercial sources; however, one should check for impurities by thin-layer chromatography until reliability is verified.

Currently, reference or calibration sera are being used to standardize many of the enzymatic procedures. Determining the correct set point for a reference or calibration serum is difficult but critical for attaining accurate and reliable results. The assigned target value should be traceable to a reference method that has been standardized for accuracy through the Centers for Disease Control (CDC) Lipid Standardization Program. It is also desirable to compare the results obtained with CDC standardized reference laboratory results or to use calibrated sera from CDC to evaluate the accuracy of the method.

Error is also introduced when results are reported in mass per volume units, such as milligrams per liter. The molecular weights of the triglycerides vary; for example, tripalmitin's molecular weight is 807 daltons, whereas triolein's is 885 daltons. In theory nearly 10% disparity would exist in the same method using the two different triglycerides as standards. This disparity indicates that one should report triglyceride values in molar units. The official units designated for triglycerides using the International System of Units *(Système International d'Unités,* SI units) are millimoles per liter.

A properly calibrated enzymatic method is certainly the preferred method for routine laboratory analysis. Since these methods are readily available in kits, this technique is not presented here in detail. Instead, a method based on the potential reference method is presented, since this may be used for the initial calibration of a method. A comparison of several methods for triglyceride analysis is found in Table 61-16.

The enzymatic method currently used in our laboratory is primarily based on the method of Bucolo and David[4] (method 5, Table 61-15). It has proved to be as accurate and precise as the chemical method of Kessler and Lederer,[3] which has been considered a reference method for serum triglyceride determinations.

SPECIMEN

To understand blood collection, review the specimen discussion on cholesterol methods in this section. Included in that section is a discussion of serum versus plasma and stability of triglycerides.

It should be reemphasized that a fasting sample (from 12 to 24 hours) is essential for triglyceride analysis, since triglyceride levels increase as soon as 2 hours postprandially and reach a maximum at 4 to 6 hours. Samples drawn from nonfasting patients are *not* suitable for analysis, since elevated triglyceride levels caused by normal assimilation of food cannot be distinguished from elevated triglyceride levels resulting from abnormal lipid metabolism or inborn errors of metabolism.

Plasma should be the sample of choice if one uses the Lipids Research Clinics data base, which is based on plasma triglyceride determinations. Plasma values are about 2% to 5% lower than serum because of the dilution effect of the efflux of water from the red blood cells caused by the anticoagulant. Certain anticoagulants (such as fluoride, citrate, and oxalate) cause rather large shifts of water from the red cells to the plasma,[13] which can result in a large decrease in triglyceride values (as much as 10%). Thus it is always wise to centrifuge the sample and remove the plasma from the blood cells as soon as possible, certainly within 2 hours after the blood is drawn. If plasma is chosen, the suggested anticoagulant is solid EDTA, 1 mg/mL whole blood. Ethanol consumption causes acute but transient elevations in serum triglyceride concentrations. This is especially evident in carbohydrate-sensitive hypertriglyceridemics. Therefore it would be advisable to request that the patient refrain from drinking alcohol for 36 hours before the day of blood drawing.

If Becton-Dickinson Vacutainer tubes are used, the stoppers should be silicon coated, not glycerin coated. The blood should be rapidly centrifuged to minimize the spontaneous hydrolysis of triglycerides to glycerol and fatty acids in the blood. If the blood sample cannot be analyzed for triglycerides within 24 hours, freezing the samples at −20° C, preferably at −40° to −60° C or colder (such as −80° C), is recommended.

REFERENCE RANGE

The most reliable U.S. data base to use for plasma triglyceride levels is that of the Lipid Research Clinics (LRC) population studies, which is based on 60,502 participants of 10 separate, well-defined North American populations.[14] The method used to measure triglyceride employed the Technicon AutoAnalyzer I or II system, adapted and modified for the program. The assay method was fluorometric (method 4, Table 61-15) and was corrected for free glyceride concentration. All participating centers were standardized by the CDC Lipid Standardization Program. The reference ranges are age- and sex-strat-

Table 61-17 Reference values for plasma triglycerides for white males (mg/L)

Age (years)	Percentiles			
	5	50	90	95
0-9	300	550	850	1000
10-14	300	650	1000	1250
15-19	350	800	1200	1500
20-24	450	1000	1650	2000
25-29	450	1150	2000	2500
30-34	500	1300	2150	2650
35-39	550	1450	2500	3200
40-54	550	1500	2500	3200
55-64	600	1400	2350	2900
65+	550	1350	2100	2600

From the LRC Prevalence Study (North America).

Table 61-18 Reference values for plasma triglycerides for white females (mg/L)

Age (years)	Percentiles			
	5	50	90	95
0-9	350	600	950	1100
10-19	400	750	1150	1300
20-34	400	900	1450	1700
35-39	400	950	1600	1950
40-44	450	1050	1700	2100
45-49	450	1100	1850	2300
50-54	550	1200	1900	2400
55-64	550	1250	2000	2500
65+	600	1300	2050	2400

From the LRC Prevalence Study (North America).

Table 61-19 Reference values for plasma triglycerides for black males (mg/L)

Age (years)	Percentiles*			
	5	50	90	95
0-9	310	470	750	880
10-19	310	530	880	1020
20-29	—	710	1250	—
30-39	420	910	1660	2240
40-49	520	970	2150	2940
50-59	—	1050	—	—
60+	—	960	—	—

From the LRC Prevalence Study (North America).
*5th and 95th percentiles not given if $n < 100$; 90th percentile not given if $n < 75$.

Table 61-20 Reference values for plasma triglycerides for black females (mg/L)

Age (years)	Percentiles*			
	5	50	90	95
0-9	330	500	830	940
10-19	360	600	950	1100
20-29	380	680	1180	1370
30-39	380	740	1290	1500
40-49	430	840	1530	1880
50-59	—	940	1760	—
60+	—	1060	—	—

From the LRC Prevalence Study (North America).
*5th and 95th percentiles not given if $n < 100$; 90th percentile not given if $n < 75$.

ified for a white population (Tables 61-17 and 61-18) and a black population (Tables 61-19 and 61-20) and are provided in percentile ranks (such as 5, 50, 90, 95). The 90th percentile is given because it is customarily used to define hypertriglyceridemia, especially type IV hyperlipoproteinemia. The 5th percentile is provided to screen for persons with lipoprotein deficiencies. For reference ranges that are more detailed, the *LRC Population Studies Data Book*[15] should be consulted.

It is important for laboratory workers and clinicians to remember that if they wish to use the LRC data for their reference ranges, they must also standardize their method by using CDC calibrators or calibrators that have been standardized by a CDC method. It is, in fact, not acceptable and even dangerous to interpret a patient's triglyceride results using a data base obtained from a method that has been standardized differently from the method used by the laboratory reporting the patient's results. Naito[16] has clearly demonstrated the clinical importance of stricter laboratory goals for accurate and precise triglyceride measurements.

REFERENCES

1. Van Handel, E, and Zilversmit, BD: Micromethod for the direct determination of triglycerides, J Lab Clin Med 50:152-157, 1957.
2. Technicon Methods, Nos. SF40040F66 and SE0040FC6, Tarrytown, NY, March 1976, Technicon Instruments Corp.
3. Kessler, G, and Lederer, H: Fluorometric measurement of triglycerides. In Skeggs, LT Jr, et al, editors: Automation in analytical chemistry, Technicon Symposium, Tarrytown, NY, 1965, 1966, pp 341-344.
4. Bucolo, G, and David, H: Quantitative determination of serum triglycerides by the use of enzymes, Clin Chem 19:476-482, 1973.
5. Rietz, EB, and Guilbault, GG: Fluorometric estimation of triglycerides in serum by a modification of the method of Bucolo and David, part 1, Clin Chem 23:286-288, 1977.
6. Megraw, RE, Dunn, DE, and Biggs, HG: Manual and continuous-flow colorimetry of triglycerides by a fully enzymatic method, Clin Chem 25:273-278, 1979.
7. Winartasaputra, H, Mallet, B, Kuan, S, and Guilbault, G: Fluorometric and colorimetric enzymic determination of triglycerides (triglycerols) in serum, Clin Chem 26:613-617, 1980.
8. Nagele, U, Hagele EO, Sauer, G, et al: Reagent for the enzymatic determinations of serum total triglycerides with improved lipolytic efficiency, J Clin Chem Clin Biochem 22:165-174, 1984.
9. Naito, HK, Gatautis, VJ, and Galen, RS: The measurement of serum triacylglycerol on the Hitachi 705 chemistry analyzer, Clin Chem 31:948, 1985.
10. Lindblad, BS, Settergren, G, Feychting, H, and Persson, B: Total parenteral nutrition in infants: blood levels of glucose lactate, pyruvate, free fatty acids, glycerol, D-beta-hydroxybutyrate, triglycerides, free amino acids and insulin, Acta Paediatr Scand 66:409-419, 1977.
11. Naito, HK, Gatautis, VJ, and Popowniak, KL: Effect of mannitol (Osmitrol) intoxication on serum "triglyceride" values, Clin Chem 22:935-936, 1976 (letter).
12. Naito, HK, and David, JA: Laboratory considerations: determination of cholesterol, triglyceride, phospholipid, and other lipids in blood and tissues. In Story, JA, editor: Lipid research methodology, New York, 1984, Alan R Liss, Inc, pp 1-76.
13. Alper, C: Specimen collection and preservation. In Henry, RJ, Cannon, DC, and Winkelman, JW, editors: Clinical chemistry: principles and techniques, Hagerstown, MD, 1974, Harper & Row, Publishers, Inc, p 373.
14. Rifkind, BM, and Segal, P: Lipid Research Clinics Program reference values for hyperlipidemia and hypolipidemia, JAMA 250:1869-1872, 1983.
15. Lipid Research Clinics Program: The prevalence study, vol 1. In the Lipid Research Clinics population studies data book, NIH Pub No 80-1527, Bethesda, MD, 1980, US Dept of Health and Human Services.
16. Naito, HK: Reliability of lipid and lipoprotein testing (I), Am J Cardiol 56:63-93, 1985.

CHAPTER 62

Nonprotein nitrogenous compounds

Amniotic fluid bilirubin

PAUL T. RUSSELL

Clinical significance: p. 569
Molecular formula: $C_{33}H_{36}H_4O_6$
Molecular weight: 584.65 daltons
Merck Index: 1222
Chemical class: tetrapyrrole, bile pigment

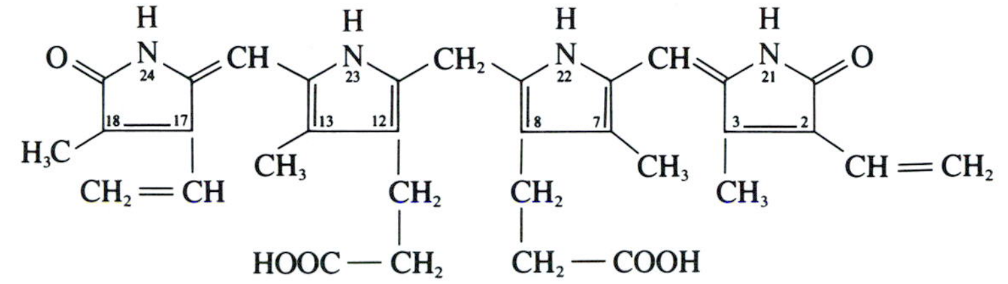

PRINCIPLES OF ANALYSIS

Assessment of unconjugated bilirubin in amniotic fluid is important in the evaluation of Rh-immunized patients.[1] Spectral absorption curves of amniotic fluid in cases of erythroblastosis fetalis show a peak at 450 nm, which is not present in fluid of normal pregnancies, an observation first made by Bevis[2] and subsequently refined by Liley,[3] Walker,[4] and others for clinical application. The 450 nm peak is the result of free bilirubin, and the extent of elevation of the peak has been correlated with the severity of erythroblastosis fetalis.[2,3,5] From these pioneering studies the magnitude of the absorbance difference at 450 nm (ΔA_{450}) became the basis for prognostications[3,6-9] of delivery of an affected fetus (method 1, Table 62-1).

Because bilirubin-free amniotic fluid has a high absorbance at 350 nm, which declines steadily toward the longer wavelengths, Liley[3] connected points on the absorption curve between 365 and 550 nm and established an arbitrary line that represented the baseline value for a sample with no bilirubin present (Fig. 62-1). The difference at 450 nm between the peak and this baseline represented the absorption in optical density (or absorbance, expressed as ΔOD at 450 nm), a value that was proportional to the bilirubin concentration. Because serial samples were taken, absolute quantitation was not necessary. It was soon realized, however, that it is not always easy to establish satisfactory baseline values,[2,6] particularly if the fluid contains either blood or meconium.

In the earlier studies heavy contamination with blood rendered the fluid samples worthless for analysis because oxyhemoglobin peaks at 575, 540, and 412 nm obscured the bilirubin reading at 450 nm, and the increase in absorbance from 350 to 400 nm noticeably affected how the baseline value was established. Blood-contaminated samples gave falsely low ΔOD_{450} values for these reasons. Likewise, because meconium absorbs maximally from 350 to 400 nm and with decreasing absorption up to 550 nm, misleading bilirubin values were associated with the presence of this material.

Modifications of the original spectrophotometric methods attempted to avoid false values resulting from contamination of fluid with blood or meconium. These methods and the direct chemical methods are comprehensibly discussed in recent monographs.[7-10] The most practical of the

Table 62-1 Methods of amniotic fluid bilirubin analysis

Method	Type of analysis	Principle	Usage	Comments
1. Spectrophotometry	Differential absorbance	Bilirubin absorbs at 450 nm; background absorption corrected by extrapolation	Common	Accurate except in cases of hemolysis
2. Extraction spectrophotometry	Differential absorbance	Bilirubin extracted with chloroform; absorbance at 450 nm measured	Less common	Minimizes effects of interferences

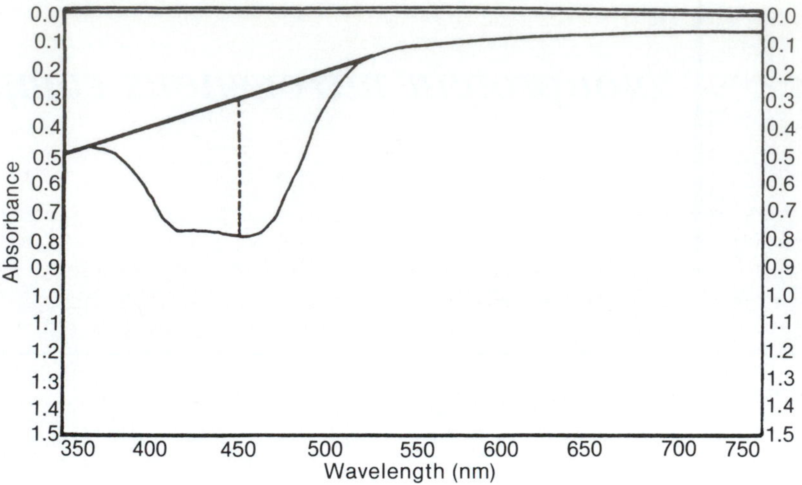

Fig. 62-1 Amniotic fluid scan showing method for determining absorbance at 450 nm. Arbitrary baseline drawn from 375 to 525 nm shows where scan would have traced if no bilirubin pigment were present. *Broken line,* Absorbance recorded (ΔOD when ordinate is in optical density units).

proposals suggested that contaminated samples be extracted by chloroform before spectroscopic assay to separate bilirubin from other interfering pigments.[11-13] Bilirubin is readily soluble in organic solvent and thus extracts efficiently into chloroform. Potential contaminants from meconium and blood are more water soluble and therefore remain partitioned with the water. Biliverdin, for example, a green oxidation product of bilirubin, is quite water soluble and is separated from bilirubin by chloroform extraction. The same is true of hemoglobin and its degradation products.

Simkins and Worth[14] compared the method of Liley,[3] a modified version of the method of Knox et al.,[15] and the method of Mallikarjuneswara et al.[12] for the estimation of bilirubin in amniotic fluid and for potential effects of various concentrations of hemoglobin, methemoglobin, and methemalbumin on the methods. They found that the methods were in good agreement unless hemoglobin degradation products were present. To circumvent this circumstance, chloroform extraction was necessary (method 2, Table 62-1).

Many samples of amniotic fluid from affected pregnancies contain oxyhemoglobin as well as methemoglobin and methemalbumin. These pigments most often arise from lysis of erythrocytes that have entered the amniotic fluid from intrauterine hemorrhage associated with partial placental separations or from the amniocentesis itself ("bloody tap"). Intrauterine (fetal) transfusions introduce considerable blood contamination into the fluid, which can take nearly 2 weeks to clear. Oxyhemoglobin is converted first into methemoglobin and then into methemalbumin. The last two compounds may also enter the amniotic fluid directly when the fetus has died or in some cases of fetal distress or impending fetal death. The methemoglobin peak at 620 nm is considered a bad prognostic sign for fetal well-being.

As mentioned previously, oxyhemoglobin has a large spectral peak at 412 nm (Soret band) that can overlap with absorption at 450 nm if oxyhemoglobin contamination is present. When the ferrous ion of hemoglobin is oxidized to the ferric state for methemoglobin, the absorption spectrum changes. Peaks for oxyhemoglobin at 540 and 575 nm disappear, and a peak appears at 630 nm. This last peak is shifted toward 620 nm when the globin of the compound is replaced by albumin. Alvey[16] found the extinction coefficients for oxyhemoglobin at 574 and 545 nm to be identical. By subtracting the optical density at 574 nm from that at 454 nm, he obtained a "bilirubin index" used for prediction.

Biliverdin, the oxidation product of bilirubin, can also contaminate amniotic fluid and thereby contribute to its spectral character. In the fetus the small proportion of the bilirubin that is conjugated is excreted into the intestines, where it undergoes conversion into biliverdin. With excretion of meconium the amniotic fluid becomes stained by biliverdin. The absorption spectrum of biliverdin is ill defined, but the amount of biliverdin in amniotic fluid can be estimated by the difference in absorption at 480 and 500 nm.[7] When certain methods[15,17] are used, biliverdin can be mistaken for bilirubin. These methods cannot be used therefore when even small amounts of meconium contaminate the amniotic fluid.

The spectrophotometric method was the first method reported for amniotic fluid bilirubin,[2,3] and it prevails as the standard for amniotic fluid bilirubin because it is simple, sensitive, quantitative, and qualitative.

Direct spectrophotometric analysis of bilirubin in amniotic fluid is adversely affected by blood contamination. This is directly related to high absorptions that make it difficult or impossible to interpret a scan. If contamination is caused by a single insult of blood (as in intrauterine transfusion), it can take 2 or more weeks for the amniotic fluid to be fully free of contamination. In these cases, a method such as that of Brazie et al.,[11] Mallikarjuneswara et al.,[12] or Hochberg et al.,[13] using a chloroform extraction, is required. A single extraction with chloroform recovers about 90% of the bilirubin present. A spectrophotometer can then be used for the determination of the ΔA_{450} and with the method of Liley for the interpretation.[3]

SPECIMEN

The amniotic fluid specimen should be placed in an opaque container to protect it from light. Ultraviolet energy causes a change in the molecules of bilirubin, which can lead to a false low reading.

At the laboratory, the specimen is centrifuged at 1000 *g* for 10 minutes in a clinical centrifuge to remove cellular material and vernix. If the fluid still remains turbid, it should be filtered through Whatman No. 42 filter paper. The filtration should be done out of direct light. As long as the sample is protected from light, the spectrophotometric analysis can be performed within the next 24 hours. The sample should be frozen if it is necessary to store it for longer than 24 hours. Centrifuged amniotic fluid can be stored frozen for months without significant change in spectrophotometric properties.[18]

If blood contaminates the amniotic fluid, the fluid should be centrifuged immediately to remove the red blood cells before they hemolyze. If the amount of blood is small and the fluid is centrifuged immediately, the spectrophotometric absorption will not be appreciably affected. If the amniotic fluid is contaminated with blood and is not centrifuged immediately, the red blood cells may hemolyze and the specimen will become contaminated with oxyhemoglobin, which cannot be removed by centrifugation.

Amniotic fluid is always more or less turbid, depending on the duration of pregnancy. This turbidity is caused by cells and debris, derived from the membranes and from meconium and vernix, which are suspended in the amniotic fluid. Centrifugation of the fluid is the most appropriate way to eliminate at least part of the turbidity. Depending on the degree of turbidity, time for centrifugation can vary from 5 to 15 minutes or more at speeds at or near the maximum for a clinical centrifuge. A further decrease of turbidity can be achieved by means of filtration under pressure through filters with small-sized pores. A disadvantage of filtration, however, is the unwanted loss of sample. A comprehensive discussion of turbidity is given by Van Kessel.[7]

PROCEDURE: DIRECT SPECTROPHOTOMETRY[3]

Principle

Amniotic fluid is centrifuged to remove particulate matter and to clear turbidity. The supernatant is then transferred to a cuvette and scanned between 350 and 750 nm. Absorbance in the 450 nm range is recorded.

Assay

Equipment: if a recording spectrophotometer is available, the centrifuged amniotic fluid specimen is scanned between the wavelengths of 350 and 750 nm; 1 cm lightpath glass or silica cuvettes are used.

To establish the baseline value, draw a line connecting the optical-density points at 365 and 550 nm. The difference in either absorbance or optical density (ΔA or ΔOD) is determined by measurement of the deviation beween the spectrophotometric tracing at 450 nm and the baseline as it intersects at 450 nm (Fig. 62-1).

For spectrophotometers that do not provide the option for a continuous scan, place the centrifuged sample in a cuvette, with distilled water used as a blank, and measure the absorbance at 365, 415, 450, and 550 nm.

Calculation

A line is constructed between the points at 365 and 550 nm on semilogarithmic graph paper. The intercepts at 450 and 415 nm on the baseline are determined, and these values are subtracted from the absorbance readings at 450 and 415 mn, respectively. The ΔA or ΔOD at 415 indicates possible oxyhemoglobin contamination of the fluid.

If, from the character of the spectrum, contamination of the amniotic fluid is suspected, chloroform extraction is recommended.

Notes

1. Oxyhemoglobin contamination causes peaks at 412, 540, and 575 nm. If the packed cell volume of erythrocytes is greater than 5%, the scan will be noticeably altered even if immediately centrifuged,[4,6,19] because plasma components of the contaminating blood can also distort the spectrophotometric reading.
2. Meconium in amniotic fluid will give a noticeably abnormal tracing that is easy to identify. There is significant absorbance from 350 to 400 nm, with decreasing absorption up to 550 nm. Small bands in the 412 nm range (Soret bands) may also be seen.
3. The spectral scan is normally done on undiluted fluid, but occasionally pigmentation will require dilution of the sample with distilled water.

PROCEDURE: CHLOROFORM EXTRACTION[11]

Principle

Centrifuged amniotic fluid is extracted with chloroform to separate bilirubin from interfering hemoglobin. A spectral scan is obtained for the chloroform extract as described previously.

Reagents

Chloroform. Chloroform is equilibrated with 0.05 M phosphate buffer, pH 6, and stored in a separatory funnel. This is stable for 2 weeks in the dark.

Assay

Equipment: spectrophotometer ($\leq$10 nm band pass), glass-stoppered centrifuge tubes (12 mL), and laboratory centrifuge in cold room.

1. Mix equal quantities of amniotic fluid and chloroform for 30 seconds in a tight-fitting glass-stoppered tube.
2. Centrifuge the tube in the cold to facilitate separation of the phases. If a protein mesh persists in the chloroform layer after centrifugation, it may be broken up with a glass stirring rod and recentrifuged briefly.
3. Warm the tubes to room temperature, and pipette the chloroform (lower layer) into a 1 cm cuvette.
4. Read the absorbance at 450 nm; use chloroform for the blank.

Calculation

The spectral scan of the chloroform extract is treated the same as the scan obtained by direct spectrophotometry.

Note

In a small study on the use of the chloroform extraction technique[13] the ΔA_{450} for the chloroform extract differed from the ΔA_{450} by direct spectral scanning in 50% of the cases. However, in all the cases, the analysis of the chloroform extract of a bloody amniotic fluid provided a more accurate prediction of fetal outcome than the direct spectrophotometric scan did.

INTERFERENCES

1. Sunlight (ultraviolet light) will cause a false low reading for bilirubin.
2. Oxyhemoglobin interferes because it affects the establishment of the baseline value. It increases absorption from 350 to 400 nm, and large amounts of oxyhemoglobin can magnify the absorption at 450 nm. Oxyhemoglobin interference thus inaccurately lowers bilirubin results. A spectrophotometric method called the *corrected bilirubin method* has been used to correct for blood contamination.[18]
3. Meconium gives an abnormal tracing that includes noticeable absorbance from 350 to 400 nm. An inaccurately low bilirubin value can be expected in the presence of meconium.
4. When the amniotic fluid volume is considerably abnormal,[20] as in hydramnios, inaccurately lower levels of amniotic fluid bilirubin will occur by dilution and pose the problem of incorrect prediction of the clinical status of the fetus.
5. If, during the course of amniocentesis, urine is inadvertently aspirated, it will be readily detectable with the spectrophotometer. The absorption pattern of urine shows an increased absorption beginning at 365 nm and decreasing to 550 nm.[18]

Because many of the inaccuracies of amniotic bilirubin measurements result in a qualitatively abnormal spectral curve (see 2, 3, and 5 above), it is very important for technologists to have a firm understanding of what a *normal* spectral pattern looks like. Deviations from a normal pattern can then be more readily detected, and the clinician can be made aware of potential problems with the analysis.

REFERENCE RANGE

Values over 0.01 ΔA (ΔOD) indicate possible fetal distress at 36 to 37 weeks of gestation.

REFERENCES

1. Brazie, JV, Ibbott, FA, and Bowes, WA: Identification of the pigment in amniotic fluid of erythroblastosis as bilirubin, J Pediatr 69:354-358, 1966.
2. Bevis, DCA: Blood pigments in haemolytic disease of the newborn, J Obstet Gynaecol Br Emp 63:68-75, 1956.
3. Liley, AW: Liquor amnii analysis in the management of the pregnancy complicated by rhesus sensitization, Am J Obstet Gynecol 82:1359-1370, 1961.
4. Walker, AHC: Liquor amnii studies in prediction of haemolytic disease of the newborn, Br Med J 2:376-378, 1957.
5. Eberlein, WR: A simple solvent partition method for measurement of free and conjugated bilirubin is serum, Pediatrics 25:878-885, 1960.
6. Liley, AW: Errors in the assessment of hemolytic disease from amniotic fluid, Am J Obstet 86:485-494, 1963.
7. Van Kessel, H: Spectrophotometry of amniotic fluid. In Sandler, M, editor: Amniotic fluid and its clinical significance, New York, 1981, Marcel Dekker, Inc, pp 131-164.
8. Robertson, JG: Clinical value and application of measurements of bilirubin and protein levels in amniotic fluid in Rh-isoimmunization. In Sandler, M, editor: Amniotic fluid and its clinical significance, New York, 1981, Marcel Dekker, Inc, pp 165-186.
9. Koch, TR: Bilirubin measurements in neonates, Clin Lab Med 1:311-327, 1981.
10. Queenan, JT: Amniotic fluid analysis in Rh and other blood group immunizations. In Sandler, M, editor: Amniotic fluid and its clinical significance, New York, 1981, Marcel Dekker, Inc, pp 277-290.
11. Brazie, JV, Bowes, WA, and Ibbott, FA: An improved, rapid procedure for the determination of amniotic fluid bilirubin and its use in the prediction of the course of Rh-sensitized pregnancies, Am J Obstet Gynecol 104:80-86, 1969.
12. Mallikarjuneswara, VR, Clemetson, CAB, and Carr, JJ: Determination of bilirubin in amniotic fluid, Clin Chem 16:180-184, 1970.
13. Hochberg, CJ, Witheiler, AP, and Cook, H: Accurate amniotic fluid bilirubin analysis from the bloody tap, Am J Obstet Gynecol 126:531-534, 1976.
14. Simkins, A, and Worth, HGJ: Determination of bilirubin in amniotic fluid: a comparison of some current methods, Ann Clin Biochem 13:510-515, 1976.
15. Knox, EG, Fairweather, DVI, and Walker, W: Spectrophotometric measurements on liquor amnii in relation to the severity of haemolytic disease of the newborn, Clin Sci 28:147-156, 1965.
16. Alvey, JP: Obstetrical management of Rh-incompatibility based on liquor amnii studies, Am J Obstet Gynecol 90:769-775, 1964.
17. Ovenstone, JA, and Connon, AF: Optical density differencing: a new method for the direct measurement of bilirubin in liquor amnii, Clin Chim Acta 20:397-402, 1968.

18. Queenan, JT: Amniotic fluid analysis, Clin Obstet Gynecol 14:505-536, 1971.
19. Queenan, JT, and Goetschel, E: Amniotic fluid analysis for erythroblastosis fetalis, Obstet Gynecol 32:120-133, 1968.
20. Whitfield, CR: Effect of amniotic fluid volume on prediction, Clin Obstet Gynecol 14:537-547, 1971.

Bilirubin

JOHN E. SHERWIN
RONALD OBERNOLTE

Clinical significance: pp. 359 and 496
Molecular formula: $C_{33}H_{36}N_4O_6$
Molecular weight: 584.65 daltons
Merck Index: 1222
Chemical class: tetrapyrrole, bile pigment
Isomeric forms: unconjugated bilirubin present in several isomeric forms; in serum is circulated as monoglucuronic and diglucuronic acid derivatives

PRINCIPLES OF ANALYSIS

Bilirubin was first demonstrated to be present in normal serum by van den Bergh and Snapper.[1] They found that bilirubin in normal serum reacted with Ehrlich's diazo reagent (diazotized sulfanilic acid) only when alcohol was added. It was later observed that pigment in human bile reacted with the diazo reagent without the addition of alcohol.[2] The pigment that reacted in the absence of alcohol was termed *direct*. The pigment that required the presence of alcohol was termed the *indirect* bilirubin fraction. The response of serum to van den Bergh's test, with or without the presence of alcohol, has been the basis for several classifications of jaundice.

It is now clear that indirect bilirubin is unconjugated bilirubin bound to albumin en route to the liver from the reticuloendothelial system, where it is formed. The unconjugated bilirubin is a nonpolar molecule and is not soluble in water. Consequently, it will react with diazo reagent only in the presence of an agent (historically called an *accelerator*), such as alcohol, in which both bilirubin and the diazo reagent are soluble. (Alcohol also enhances the intensity of the color formed.) Because of the nonpolar nature of the unconjugated bilirubin and its strong affinity to albumin, unconjugated bilirubin is not normally found in urine in more than trace amounts. Unconjugated bilirubin is bound so tightly to albumin that it cannot be filtered at the glomerulus, and there appears to be no tubular excretion of bilirubin.

In contrast, bilirubin conjugated with glucuronate is a polar and water-soluble compound that exists in plasma not bound to any protein. Therefore conjugated bilirubin reacts *directly* with the diazo reagents to form azobilirubin. When blood levels of conjugated bilirubin are high, it is filtered at the glomerulus and excreted in the urine.

Thus conjugated and unconjugated bilirubin fractions have historically been differentiated by the time of the reaction and solubility of the fractions. Bilirubin that reacts quickly in the absence of solvent is called *direct,* or *conjugated,* bilirubin; in the presence of a solvent both forms of bilirubin readily react. However, it has been known for some time that a portion of the unconjugated bilirubin is available for reaction in the absence of solvent and that the extent of the reaction depends on the time and temperature of the reaction and on the final concentration of reagents.[3] Decreasing the time of reaction to minimize reactivity of the unconjugated bilirubin in the absence of solvent can increase the specificity of the conjugated (direct) bilirubin assay. The extent of reaction of unconjugated bilirubin in the presence of solvent is also highly dependent on the final concentration of the solvent.

More recently, a third bilirubin fraction has been identified.[4-7] This fraction, called *delta* (δ) bilirubin, is a covalent conjugate between protein and unconjugated bilirubin resulting from a nonenzymatic reaction between albumin and bilirubin. The δ-bilirubin fraction was initially identified by HPLC. The δ-bilirubin fraction reacts like conjugated bilirubin and is usually present in very low amounts in blood. However, when the levels of unconjugated bilirubin increase, so does the concentration of δ-bilirubin, resulting in an apparent increase in the conjugated fraction in most methods. The δ-bilirubin fraction does not react quantitatively in most current diazo reactions.[4]

Qualitative analysis of serum bilirubin as indirect or direct, according to the type of van den Bergh reaction, has long been replaced by quantitative determinations of the amount of conjugated bilirubin and total bilirubin, in which the difference between the total and direct reaction is assumed to represent the indirect or unconjugated bilirubin.

Commonly used procedures for measurement of bilirubin and its fractions are modifications of the method of Malloy and Evelyn.[8,9] All these methods employ some variation of the reaction of bilirubin with diazotized sulfanilic acid to form a colored chromophore (Fig. 62-2). The diazotized sulfanilic acid reacts at the central methylene carbon of bilirubin to split the molecule, forming two molecules of azobilirubin. Modifications of the Malloy-Evelyn procedure primarily differ in the pH at which the reaction is carried out and the reagent used to solubilize the unconjugated (indirect) bilirubin fraction. The Malloy-Evelyn method is typically performed at pH 1.2. At this pH, the azobilirubin is red purple in color with an absorption maximum of approximately 560 nm (method 1, Table 62-2). The Malloy-Evelyn method most typically uses methanol as a solubilizer of the unconjugated fraction.

The Jendrassik-Grof modification is carried out at a pH near 6.5, but the absorbance of the reaction is measured after alkalinization of the reaction solution to pH 13.[10,11]

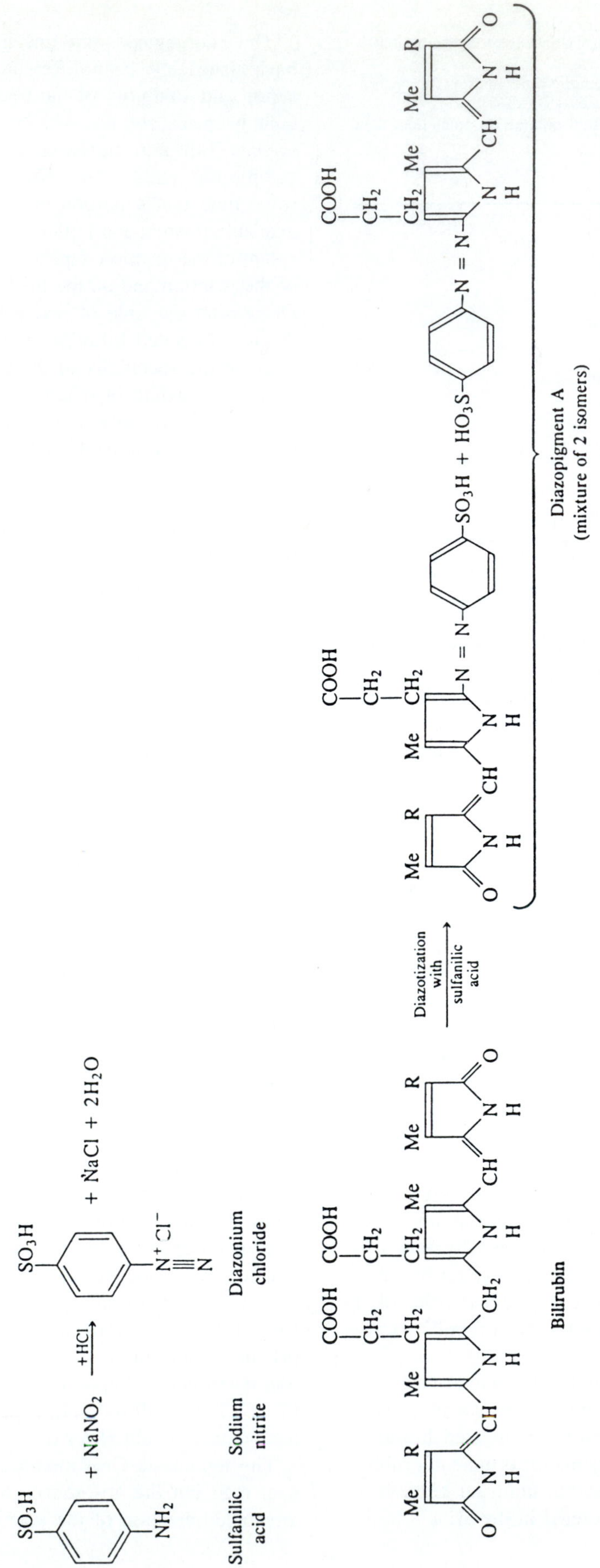

Sulfanilic acid + $NaNO_2$ (Sodium nitrite) $\xrightarrow{+HCl}$ Diazonium chloride + NaCl + $2H_2O$

Bilirubin $\xrightarrow{\text{Diazotization with sulfanilic acid}}$ Diazopigment A (mixture of 2 isomers)

Fig. 62-2 Formation of diazotized sulfanilic acid and its reaction with esterified and nonesterified forms of bilirubin to form azobilirubin derivatives. *Me*,—CH_3 group; *R*,—CH=CH_2 group.

Table 62-2 Methods of bilirubin analysis

Method	Type of analysis	Principle	Usages	Comments
1. Malloy-Evelyn	Kinetic, end point, with or without blank	Reaction shown in Fig. 62-2 is performed at pH 1.2; azobilirubin measured at 560 nm.	Very frequently used, especially as automated procedure	Susceptible to significant hemoglobin interference
2. Jendrassik-Grof	Kinetic, end point, with or without blank	Reaction shown in Fig. 62-2 is performed near neutral pH, but chromophore is measured at alkaline pH (approximately 13) at 600 nm.	Most frequently used method	Has higher molar absorptivity and thus is more sensitive and precise at low bilirubin concentrations than Malloy-Evelyn method
3. Bilirubinometer	Direct spectrophotometric	Bilirubin concentration is directly determined by its absorbance at 454 nm; HbO_2 interference is corrected by subtraction of absorbance at a second wavelength (540 nm).	Not frequently used; primarily for neonatal analysis	Very simple to perform but very strong interference from carotenoids
4. High-performance liquid chromatography	Chromatographic separation	Methyl esters of conjugated and unconjugated bilirubin are detected at 430 nm.	Research use only	May become future reference method
5. Bilirubin oxidase	Kinetic, end point	Enzymatic oxidation of bilirubin to biliverdin and water; reaction is monitored at 405 to 460 nm.	Newly available	May become future reference method; no hemoglobin interference
6. Spectral shift	End point	Binding of bilirubin by hydrophobic cationic polymer causes shift in spectrum of bilirubin. Magnitude of change, as measured by reflectance photometry, is related to bilirubin concentration.	Available on Kodak Ektachem only	No hemoglobin interference; can measure δ-bilirubin concentration

HbO_2, Oxyhemoglobin.

At this pH the absorption spectrum of the azobilirubin is shifted to a more intense blue color measured at 600 nm (method 2, Table 62-2). The original Jendrassik-Grof method employed sodium benzoate–caffeine to solubilize unconjugated bilirubin for the measurement of total bilirubin.

Modifications of these two methods vary primarily in the solvent used to solubilize the unconjugated bilirubin. Solvents used for the measurement of both the unconjugated and conjugated bilirubin fractions include antipyrine, methanol, urea, and dimethyl sulfoxide (DMSO).[3] Both the Malloy-Evelyn and Jendrassik-Grof procedures have been successfully automated and are currently the most frequently used methods for bilirubin analysis. Table 62-3 compares the two methods.

Some laboratories measure total bilirubin by a direct spectrophotometric technique[12] (method 3, Table 62-2). Because oxyhemoglobin (HbO_2) absorbs strongly at 454 nm, the absorption maximum of bilirubin, it is necessary to correct for the contribution of HbO_2. This is accomplished by subtraction of the absorbance at 540 nm, because HbO_2 absorbs equally at each of these wavelengths.[13] Thus either a spectrophotometer or a biochromatic photometer (''bilirubinometer'') capable of measurements at 454 and 540 nm is required. This method does not, however, correct for the presence of carotenoids, which also absorb in the 454 nm region of the spectrum and can cause falsely elevated results. Consequently direct spectrophotometric methods are rarely used for measurement of bilirubin in adults. However, neonates up to 3 months of age do not have a sufficient intake of carotene to interfere with this direct spectrophotometric assay. Its rapidity and small sample volume make the direct spectrophotometric assay appealing for use in an immediate re-

Table 62-3 Comparison of reaction conditions for analysis of total and direct bilirubin in serum

Condition	Malloy-Evelyn* (total and direct)		Jendrassik-Grof† (total)	
Temperature	30° or 37° C		Room temperature	
Sample volume	0.4 mL of 1:20 diluted sample; uses 100 μL of sample initially		500 μL	
Fraction of sample volume	0.4 in reaction mixture, but 0.02 in final dilution		0.0588	
Final concentration of reagents	**Total**	**Direct**		
	Methanol: 50% (v/v)	$NaNO_3$: 0.77 mmol/L	Sodium acetate: 496 mmol/L	In reaction mix
	$NaNO_3$: 0.77 mmol/L		Sodium benzoate: 283 mmol/L	In reaction mix
	Sulfanilic acid: 2.5 mmol/L	Sulfanilic acid: 2.5 mmol/L	Na_2EDTA: 2.5 mmol/L	In reaction mix
			Caffeine: 140 mmol/L	In reaction mix
	HCl: 0.07 mol/L	HCl: 0.07 mol/L	$NaNO_3$: 12.7 mmol/L	In reaction mix
			Sulfanilic acid: 0.2 mmol/L	In reaction mix
			HCl: 1.2 mmol/L	
			NaOH: 662 mmol/L (pH 13)	In final colored solution
			Potassium-sodium tartrate: 400 mmol/L	In final colored solution
Time of reaction	10 minutes	1 minute	10 minutes	
Wavelength	560 nm		598 nm	
Linearity	To 300 mg/L		To 250 mg/L	
Precision‡ (interlaboratory, $\bar{X}$, % CV)	7.2 mg/L, 15.3%		8.1 mg/L, 3.7%	
	28 mg/L, 6.9%		57.7 mg/L, 0.5%	
	93 mg/L, 4.9%		102.1 mg/L, 0.7%	
Interference	Hemoglobin		No interference from hemoglobin at concentrations as high as 2000 mg of Hb per liter	

*Method as described by Meites et al.[9]
†Method as described Doumas et al: Clin Chem 31:1779, 1985.
‡Taken from College of American Pathologists (CAP) Quality Assurance Survey data.

sponse laboratory. Direct spectroanalysis is also employed for the estimation of bilirubin to amniotic fluid as an indicator of a hemolytic disease process in fetuses.

A high-performance liquid chromatography (HPLC) procedure (method 4, Table 62-2) that can separate the methyl esters of unconjugated and conjugated bilirubin isomers has been developed.[14] This technique uses normal-phase (silica) chromatography, and the methylated derivatives are detected when the column eluate is monitored at 430 nm. However, this technique is used only as a research tool at this time.

Oxidation of bilirubin to colorless biliverdin by bilirubin oxidase, an enzyme obtained from *Myrothecium* species (xerophytic South African shrubs), has also been reported as a feasible method of analysis[15] (method 5, Table 62-2). The reaction is monitored by measurement of the decrease in absorbance at 405 to 460 nm. This technique is not in current use, but this is expected to change in the near future.

Transcutaneous measurement of bilirubin has been accomplished by means of reflectance photometers (Minolta Jaundice Meter 101, Narco Scientific Air Shields Division, Hatboro, PA), which measure the absorbance at 460 and 550 nm.

The method employed in the Kodak Ektachem analyzers separates bilirubin from the protein matrix, using film technology (method 6, Table 62-2). The bilirubin diffuses into a layer containing a complex, hydrophobic cationic polymer that complexes with the bilirubin, resulting in a shift in the bilirubin spectrum from about 440 to 460 nm. The reaction is monitored by reflectance spectrophotometry at 455 mn.[16] Because the δ-bilirubin fraction is a protein complex, it does not react in the Kodak direct bilirubin reaction. The original Kodak procedure was modified to monitor the reaction at lower wavelengths, that is, between 400 and 420 nm. At this portion of the spectrum, both conjugated and unconjugated bilirubin have equal molar absorptivities, making the total bilirubin assay more accurate.[17]

Urine bilirubin measurements are most often made using dipsticks impregnated with direct-reacting diazo reagent, such as 2,4-dichloroaniline diazonium salts (Ames Co., Elkart, IN). The Ictotest (Ames Co.) uses *p*-nitrobenzene diazonium *p*-toluenesulfonate as the active reagent. With a sensitivity of about 10 mg/L, it is two to four times more sensitive than the dipstick procedure. Both methods are subject to frequent interference from endogenous color.

The recent development of an HPLC procedure[11] that can physically separate and quantitate bilirubin fractions has led to a better understanding of the composition of bilirubin in normal blood. The HPLC results indicate that little (less than 5%) conjugated bilirubin may actually exist in the plasma of healthy persons and that most diazo methods greatly overestimate the concentration of this fraction.

In addition, the HPLC analysis indicates the presence in serum of nonbilirubin compounds that are also diazo reacting, such as mesobilifuscin and uroerythrin.

These studies indicate that chemical assays that give a low proportion (15% or less) of direct-reacting bilirubin in healthy persons probably give the most accurate results.

The American Association for Clinical Chemistry (AACC) and the National Bureau of Standards (NBS) have published a candidate reference method for total bilirubin.[18] This method is a modified Jendrassik-Grof procedure that employs caffeine-benzoate reagent to solubilize all the various bilirubin fractions with excellent recovery. Only zinc had any significant interference in this procedure.

SPECIMEN

Total bilirubin determinations using a diazo method require either serum or plasma. Serum is preferred for the Malloy-Evelyn procedure, because the addition of alcohol in the analysis can precipitate proteins from plasma that interfere with subsequent analysis. The presence of hemolysis falsely depresses the bilirubin result in most assays because of an increase in absorbance of the blank; therefore samples used for analysis should not have any visible hemolysis. Red blood cell contamination should be removed by centrifugation before analysis. Turbid serum creates artifacts in the performance of spectrophotometers and should not be directly analyzed. Bilirubin is readily destroyed by light and heat; therefore the analysis should be performed promptly in subdued light to avoid falsely low results.

Conjugated bilirubin may be determined in either serum or plasma, but serum is the usual sample because total bilirubin can be determined concomitantly. Spinal fluid samples can be analyzed for both total and direct bilirubin. Urine samples can also be analyzed by direct diazo methods, because the polar conjugated bilirubin is, in large part, not protein bound and is filtered at the glomerulus and excreted into urine.

PROCEDURE: MALLOY-EVELYN METHOD FOR TOTAL BILIRUBIN

Principle

The method is a variation of the Malloy-Evelyn procedure as described in *Selected Methods of Clinical Chemistry*.[9] The specimen is added to a solution of methanol and diazotized sulfanilic acid in an acid solution. The methanol accelerates the coupling of bilirubin with the diazotized sulfanilic acid. The purple azobilirubin color is measured at 560 nm.

Reagents

Sulfanilic acid (26.2 mmol/L in 0.702 mol/L HCl). Add 5.0 g of sulfanilic acid to 60 mL of concentrated hydrochloric acid (approximately 12 mol/L), mix, and dilute to 1 L with distilled water. The solution is stable for up to 6 months at ambient temperature.

Sodium nitrite stock (2.9 mol/L). Place 20 g of sodium nitrite ($NaNO_2$) in a 100 mL volumetric flask, dissolve in 80 mL of distilled water, and dilute to 100 mL. Store at 4° to 8° C in a brown glass-stoppered bottle. Discard when the reagent becomes discolored with yellow nitrate, after several weeks.

Working sodium nitrite solution (290 mmol/L). Dilute the stock sodium nitrite tenfold with distilled water to prepare a 0.29 mol/L solution. This reagent is unstable and should be prepared daily.

Diazo blank. Dilute 60 mL of concentrated HCl to 1 L with distilled water. This reagent can be kept at room temperature for at least 1 year.

Diazo reagent. Add 0.3 mL of working sodium nitrite solution to 10 mL of sulfanilic acid reagent, and mix. This reagent is unstable and should be used within a few hours of preparation.

Methanol (absolute). Follow American Chemical Society specifications for purity, that is, at least 99.5% pure. This must be stored in a glass container.

Serum-based calibrating standard. Bilirubin concentration should be approximately 30 to 50 mg/L. Also, serum-based controls, preferably one at elevated levels, are available commercially.

Assay

Equipment: spectrophotometer, with band pass ≤10 nm, capable of reading at 560 nm. For accurate results, the analysis should be performed in subdued light.

1. Dilute serum samples, standards, and control specimens 1:20 by pipetting 1.9 mL of distilled water into a 12 × 75 mm test tube and adding 100 μL of sample. Mix. This can be accurately done with an automatic pipettor.
2. Add 0.5 mL of methanol to cuvettes labeled *T* (for total) and *B* (for blank) for each patient, standard, and control.
3. Add 100 μL of fresh diazo reagent to the cuvette labeled *T*, and mix.
4. Add 100 μL of diazo blank reagent to the cuvette labeled *B*, and mix.
5. Add 0.5 mL of diluted serum, standard or control, to cuvettes *T* and *B*. Mix thoroughly.
6. Allow reaction to proceed for 10 minutes in covered tubes at room temperature. Read the absorbance of cuvette *T* at 560 nm against water set at zero absorbance. Read *B* reaction versus water, and subtract this reading from that of the sample to obtain corrected absorbance.

Calculation

The blank-corrected absorbance of cuvette *T* is compared to the absorbance of the bilirubin calibrator standard to determine the concentration of total bilirubin in the sample. The calculation is as follows:

$$\frac{A_s}{C_s} = \frac{A_u}{C_u} \qquad C_u = \frac{A_u}{A_s} \times C_s$$

where A_s = absorbance of calibrator standard
A_u = absorbance of unknown
C_s = concentration of standard (mg/L)
C_u = concentration of unknown (mg/L)

Notes

1. It is critical to add reagents and diluted samples in the given order to avoid turbidity.
2. This procedure was originally established for children. This method can be modified to eliminate the initial dilution step by increasing the volumes as follows:
   a. Methanol: 1.0 mL
   b. Diazo reagent: 0.2 mL (or 0.2 mL of diazo blank)
   c. Water: 0.75 mL
   d. Serum: 0.04 mL (using micropipettor, such as SMI or Eppendorf) (Pipette in given order.)
3. This procedure is linear up to 300 mg/L. Color is stable for at least 2 hours within this range.
4. Hemolysis will decrease bilirubin values.
5. To determine the concentration of the unconjugated bilirubin fraction, subtract the concentration of the conjugated bilirubin value (see following procedure) from the value for total bilirubin.

PROCEDURE: MALLOY-EVELYN METHOD FOR CONJUGATED BILIRUBIN

The conjugated bilirubin method, also from *Selected Methods in Clinical Chemistry,*[9] is based on the Malloy-Evelyn procedure. The azobilirubin pigment produced by bilirubin glucuronides and diazotized sulfanilic acid is formed in 1 minute, as opposed to the 10 minutes required in the total bilirubin procedure. The methanol is not added, and the time has been shortened to minimize the reaction of unconjugated bilirubin. To prevent turbidity, it is critical that the order of addition of reagents and diluted specimen be followed exactly as listed.

Reagents

Same as for total bilirubin.

Assay

Equipment: see total bilirubin assay.

1. Add 0.5 mL of distilled water to a cuvette labeled *B* (for blank) and 0.5 mL of water to a cuvette labeled *C* (for conjugated).
2. Add 100µL of diazo reagent to each *C* cuvette and 100 µL of diazo blank reagent to each *B* cuvette.
3. Add 400 µL of diluted serum, standard, or control specimen (see total bilirubin procedure) to the respective *B* (blank) and *C* (conjugated) cuvettes. Mix well.
4. Exactly 1 minute after the addition of diazo reagent in step 3, read the absorbance of cuvettes *B* and *C* at 450 nm against water set at zero absorbance. Subtract the blank absorbance for each sample from the test absorbance to obtain the net reaction absorbance. Add the samples at 30-second intervals so that absorbance of each reaction tube (*B* and *C*) can be measured within 60 seconds of sample addition.

Table 62-4 Mean serum total bilirubin values in healthy subjects[19]

Race	Serum bilirubin (mg/L)	
	Women	Men
Black	4.6 ± 2.2	6.0 ± 3.0
White	5.5 ± 3.2	6.5 ± 3.1
Latin American	5.0 ± 2.0	6.7 ± 4.5
Asian	5.6 ± 2.3	6.9 ± 2.7

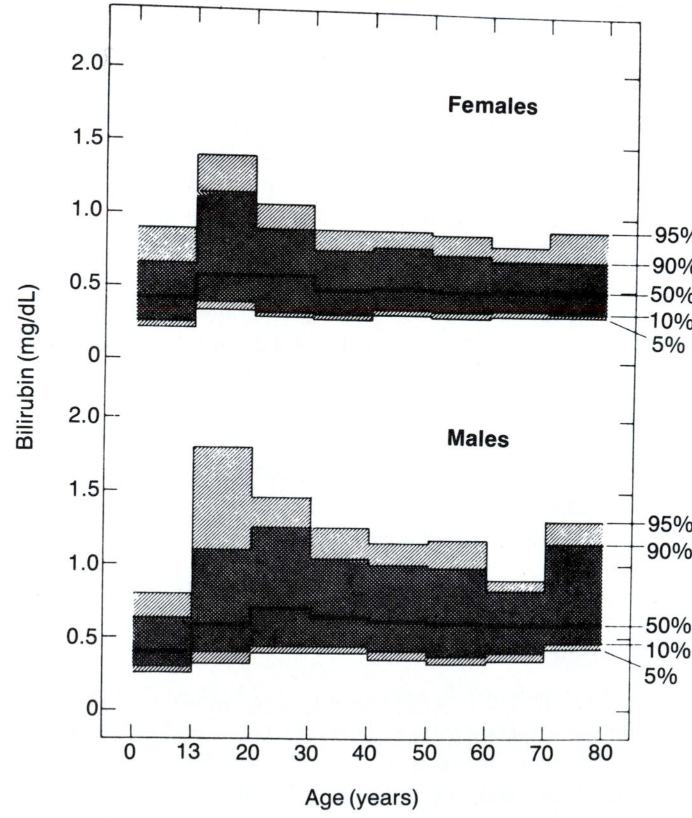

Fig. 62-3 Serum total bilirubin concentrations in percentiles. In both sexes, median concentration rises after puberty, falls during third decade, and thereafter remains stable. Notice that similar shifts are more pronounced at 90th and 95th percentile levels and that 5th and 10th percentile levels hardly change. After puberty, males have both higher median values and more pronounced skewness at high values than females do. (From Werner, M, et al: Z Klin Chem 8:105-115, 1970.)

Note

A commercially available conjugate of bilirubin (Ultimate D; SmithKline-Beckman, Sunnyvale, CA) can be used to standardize this reaction.

Calculation

The blank-corrected absorbance of cuvette *C*, the absorbance of the standard, and the concentration of the standard are used to determine the concentration of conjugated bilirubin in the sample.

REFERENCE RANGE (Table 62-4)

Total bilirubin levels in healthy adults are up to 15 mg/L (25.7 μmol/L), with a median of 7.0 mg/L (12 μmol/L). In both sexes the median concentration rises after puberty, falls during the third decade, and thereafter remains stable. After puberty males have both higher median values and a more pronounced skewness to high levels than females do.[18,19] The observation that men have a higher serum value than women has been confirmed, though the difference is probably not clinically significant. In addition, significant racial differences have been reported.[19] The racial differences were more pronounced between the female groups than the male.

Conjugated bilirubin levels up to 2 mg/L are found in infants by 1 month of age, and conjugated bilirubin remains at this level thereafter. Fig. 62-3 summarizes serum bilirubin concentrations as a function of age and sex.[20]

REFERENCES

1. van den Bergh, AAH, and Snapper, J: Die Farbenstoffe des Blutserums, Dtsch Arch Klin Med 110:540-541, 1913.
2. van den Bergh, AAH, and Müller, P: Uber eine direkte und eine indirekte Diazoreaktion auf Bilirubin, Biochem Z 77:90-103, 1916.
3. Winkelman J, Cannon, DC, and Jacobs, SL: Bilirubin. In Henry, RJ, et al, editors: Clinical chemistry: principles and techniques, ed 2, New York, 1974, Harper & Row, Publishers, Inc, pp 1042-1079.
4. Doumas, BT, Wu, TW, and Jendrzejczak, B: The reaction of bilirubin firmly bound to protein (δ-bilirubin) with the diazo reagent, Clin Chem 30:971, 1984.
5. Brett, EM, Hicks, JM, Powers, DM, and Rand, RN: Delta bilirubin in pediatric patients: correlations with age and disease, Clin Chem 30:1561-1564, 1984.
6. Ou, CN, Gilman, GE, and Buffone, GJ: Evaluation of the Kodak Ektachem clinical chemistry slide for the measurement of bilirubin in newborns, Clin Chim Acta 140:167-172, 1984.
7. Lauff, JJ, Kasper, ME, Wu, TW, and Ambrose, RT: Isolation and preliminary characterization of a serum bilirubin fraction firmly bound to protein, Clin Chem 28:629-637, 1982.
8. Malloy, HT, and Evelyn, KA: The determination of bilirubin with the photoelectric colorimeter, J Biol Chem 119:481-490, 1937.
9. Meites, S, Cheng, MH, Arnold LH, et al: Bilirubin, direct reacting and total, modified Malloy-Evelyn method. In Faulkner, WR, and Meites, S, editors: Selected methods of clinical chemistry, vol 9, Washington, DC, 1982, American Association of Clinical Chemists Press, pp 119-125.
10. Jendrassik, L, and Grof, P: Vereinfachte photometrische Methoden zur Bestimmung des Blutbilirubin, Biochem Z 297:81-89, 1938.
11. Koch, TR, Doumas, DT, Elser, RC, et al: Bilirubin, total and conjugated, modified Jendrassik-Grof method. In Faulkner, WR, and Meites, S, editors: Selected methods of clinical chemistry, vol 9, Washington, DC, 1982, American Association for Clinical Chemistry Press, pp 113-118.
12. Evans, RT, and Holton, JB: An assessment of a bilirubinometer, Ann Clin Biochem 7:104-106, 1970.
13. Leukoff, AH, Westphal, MC, and Finkler, JF: Evaluation of a direct reacting spectrophotometer for neonatal bilirubinometry, Am J Clin Pathol 54:562-565, 1970.
14. Blanckaert, N, Kabra, PM, Fraina, FA, et al: Measurement of bilirubin and its monoconjugates and diconjugates in human serum by alkaline methanalysis and high performance liquid chromatography, J Lab Clin Med 96:198-212, 1980.
15. Osaki, S, and Anderson, S: Enzymatic determination of bilirubin, Clin Chem 30:971, 1984.
16. Spayd, RW, Bruschi, B, Burdick BA, et al: Multilayer film elements for clinical analysis: applications to representative chemical determinations, Clin Chem 24:1343-1350, 1978.
17. Wu, TW, Dapper, GM, Powers, DM, et al: The Kodak Ektachem clinical chemistry slide for measurement of bilirubin in newborns: principles and performance, Clin Chem 28:2366-2372, 1982.
18. Doumas, BT, Kwok-Cheung, PP, Perry, BW, et al: Candidate reference method for determination of total bilirubin in serum: development and validation, Clin Chem 31:1779-1789, 1985.
19. Werner, M, Toos, RE, Hultin, JV, and Mellecker, J: Influence of sex and age on the normal range of eleven serum constituents, Z Clin Chem 8:105-115, 1970.
20. Carmel, R, Wong, ET, Weiner, JM, and Johnson, CS: Racial differences in serum total bilirubin levels in health and disease (pernicious anemia), JAMA 253:3416-3418, 1985.

Creatinine

ROBERT L. MURRAY

Creatinine, Cr
Clinical significance: pp. 346 and 427
Molecular formula: $C_4H_7N_3O$
Molecular weight: 113.12 daltons
Merck Index: 2552
Chemical class: creatinine end-product metabolite

H
N
HN=C C=O
N—CH_2
CH_3

PRINCIPLES OF ANALYSIS

The Jaffé method for creatinine analysis (method 1, Table 62-5), first described in 1886,[1] has the distinction of being the oldest clinical chemistry method still in common use. This assay is based on the reaction of creatinine (I) with an alkaline solution of sodium picrate (II) to form a red Janovski complex (III). The absorbance of III is measured between 510 and 520 nm, though its maximum is reported as 485 nm.[2] Picrate ion (II, present in excess in the reaction solution) absorbs significantly at wavelengths below 500 nm.

Table 62-5 Methods of creatinine analysis

Method	Type of analysis	Principle	Usage	Comments
1. Jaffé	Spectrophotometric (520 nm), end point, quantitative	Creatinine + Picrate $\xrightarrow{OH^-}$ Janovski complex *(red)*	Serum, plasma, diluted urine	Described by Jaffé, 1886
2. Jaffé/fuller's earth	Same as 1; creatinine isolated before analysis; can be removed with buffer or picrate reagent added directly to creatinine adsorbent suspension	As above	Serum, plasma, diluted urine	Reference method; alternatively one can use cation exchanges as adsorbent
3. Jaffé, kinetic	Spectrophotometric, quantitative, kinetic analysis during early color formation	As above	Serum, plasma, diluted urine	Requires automated equipment for accurate, precise absorbance measurements
4. 3,5-Dinitrobenzoic acid (DNBA)	Spectrophotometric, quantitative, end point	Creatinine + DNBA $\xrightarrow{OH-}$ Purplish rose color	Serum, plasma	Dry strip technology
5. Creatinine amidohydrolase	Enzymatic hydrolysis to creatine, which reacts in indicator reactions monitored spectrophotometrically at 340 nm	Creatinine + H_2O $\xrightarrow[\text{amidohydrolase}]{\text{Creatinine}}$ Creatine Creatine + ATP $\xrightarrow{CK}$ Creatine phosphate + ADP ADP + Phospho*enol*pyruvate $\xrightarrow{\text{Pyruvate kinase}}$ ATP + Pyruvate Pyruvate + NADH + H^+ $\xrightarrow{LD}$ NAD^+ + Lactate (*CK*, Creatine kinase; *LD*, lactate dehydrogenase)	Serum	Not widely used or automated
6. Creatinine iminohydrolase	Enzymatic hydrolysis of creatinine with formation of ammonia, which can be quantitated spectrophotometrically or electrometrically	Creatinine $\xrightarrow[\text{iminohydrolase}]{\text{Creatinine}}$ *N*-methylhydantoin + NH_3 NH_3 measured by: GLDH reaction, ammonia electrode, or colorimetrically	Serum	Requires high enzyme purity, rarely used, may be possible reference method
7. High-performance liquid chromatography (HPLC)	Cation-exchange or reversed-phase chromatographic separation of creatinine from other compounds	Creatinine quantitated by method 1 or absorption at 200 nm	Serum, plasma, urine	Highly specific, not useful for routine analysis, possible reference method
8. HPLC and gas chromatography/mass spectrometry (GC/MS)	Isotope dilution GC/MS	N^{15}-labeled creatinine is added to sample; creatinine is separated by HPLC and quantified by GC/MS by use of isotope ratio	Serum, urine	Highly specific, proposed reference/definitive method
9. Creatinine amidohydrolase	Coupled enzymatic reactions leading to peroxidase indicator reaction	Creatinine + H_2O $\xrightarrow[\text{amidohydrolase}]{\text{Creatinine}}$ Creatine Creatine + H_2O $\xrightarrow[\text{amidohydrolase}]{\text{Creatinine}}$ Sarcosine + Urea Sarcosine + O_2 $\xrightarrow[\text{oxidase}]{\text{Sarcosine}}$ Glycine + Formaldehyde + H_2O_2 H_2O_2 + Dye $\xrightarrow{\text{Peroxidase}}$ Colored dye + H_2O	Serum, urine	Automated on Kodak Ektachem, minimal interference from lidocaine metabolites

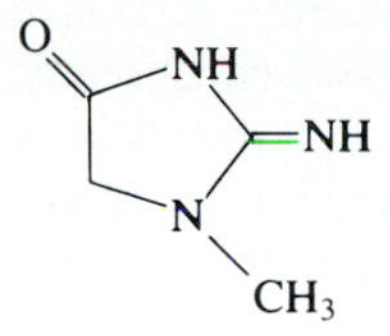

I

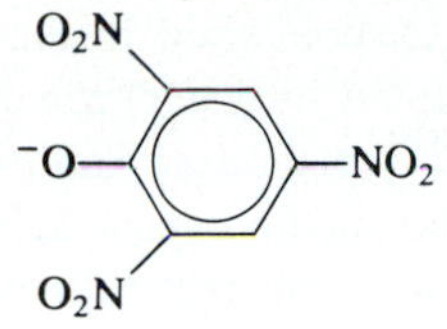

II

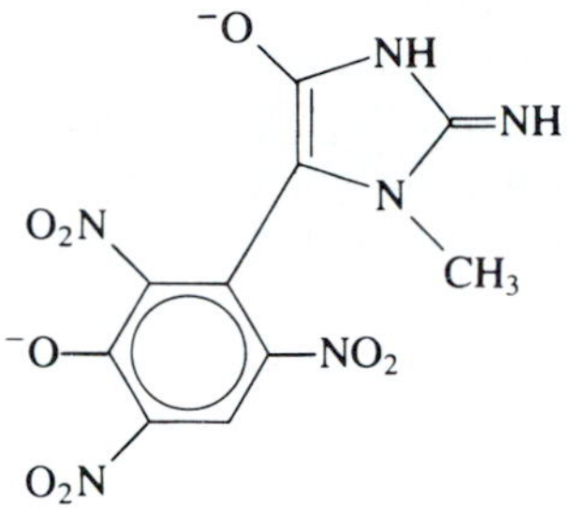

III

The reaction is run at a constant temperature of less than 30° C; at higher temperatures glucose, uric acid, and ascorbic acid can have an unacceptably high reductive reactivity toward picrate, resulting in formation of picramate, which has a maximum absorbance at 482 nm and causes an overestimation of creatinine. *Constant* temperature is also important because both the absorbance of the picrate ion and the creatinine-picrate reaction product increase with increasing temperature. The reaction is usually not buffered but is carried out in about 0.1 mol/L NaOH.

Fuller's earth (floridin) has been used to increase the specificity of the Jaffé reaction by absorbing the creatinine present in a protein-free filtrate, thus isolating it from potential interferents (method 2, Table 62-5). Other materials with comparable properties are Lloyd's reagent and bentonite. All of these materials are porous aluminum magnesium silicate clays that form colloidal suspensions with high natural adsorptive power. A small amount of the adsorbent is added to a protein-free filtrate at room temperature; adsorption is essentially complete after 1 minute. Creatinine is adsorbed with about 92% efficiency, and after centrifugation and decanting of the interferent-containing supernatant, one can add the alkaline picrate directly to the creatinine-adsorbent pellet. Most potential interferents are not adsorbed: only pyruvate in excess of 0.9 mmol/L and 2-oxoglutarate in excess of 0.5 mmol/L are adsorbed and can thus cause interference.[3]

The kinetic method for determination of creatinine (method 3, Table 62-5) gained popularity as instruments capable of making accurate absorbance readings at precise, highly reproducible intervals became available.[4-6] Sequential readings are made at precise points in the reaction sequence and compared to standards.

In one modification of the kinetic method, after reactants are mixed, there is a 10- to 60-second delay before the picrate-creatinine complex formation is monitored; this allows fast-reacting interferents such as acetoacetate to be largely eliminated before the initial absorbance reading is made. Additional measurements are made during or at the end of a period that ranges from 16 to 120 seconds, before the creatinine reaction has gone to completion and before the more slowly reacting interferents have been able to react significantly. Deproteinization is not necessary, because the reaction between alkaline picrate and protein is slow and does not significantly occur during the usual kinetic reaction interval.

Another method of creatinine analysis involves reaction of creatinine with 3,5-dinitrobenzoic acid (DNBA)[6a,6b,7] (method 4, Table 62-5) or its derivatives,[8,9] with 1,4-naphthoquinone-2-sulfonate,[10] and with *o*-nitrobenzaldehyde.[11] A purplish rose-colored product formed when the reagent was reacted in an alkaline solution. The reaction was sensitive to 2 mg/L and was thus recommended for clinical use. The DNBA procedure has been adapted for use with the dry strip technology. The reaction is monitored by reflectance spectrophotometry at 560 nm.

Two creatinine-degrading enzymes have been investigated for use in creatinine analysis. *Creatinine amidohydrolase* (EC 3.5.3.10), also referred to as *creatinase* or *creatinine hydrolase,* has been used to convert creatinine to creatine, coupled to an indicator reaction as shown below (method 5, Table 62-5). This allows the continuous monitoring of NADH disappearance.[12]

$$\text{Creatinine} + H_2O \xrightarrow{\text{Creatinine amidohydrolase}} \text{Creatine}$$

$$\text{Creatine} + \text{ATP} \xrightarrow{\text{Creatine kinase}} \text{Creatine phosphate} + \text{ADP}$$

$$\text{ADP} + \text{PEP} \xrightarrow{\text{Pyruvate kinase}} \text{ATP} + \text{Pyruvate}$$

$$\text{Pyruvate} + \text{NADH} + H^+ \xrightarrow{\text{Lactate dehydrogenase}} \text{Lactate} + \text{NAD}^+$$

The second enzyme system used for creatinine quantitation is *creatinine deiminase* (EC 3.5.4.21), also known as *creatinine iminohydrolase* (method 6, Table 62-5). This enzyme is responsible for conversion of creatinine to *N*-methylhydantoin and ammonia.

$$\text{Creatinine} + H_2O \xrightarrow{\text{Creatinine iminohydrolase}} N\text{-Methylhydantoin} + NH_3$$

$$\text{NADPH} + NH_4^+ + \alpha\text{-Ketoglutarate} \xrightarrow{\text{GLDH}} \text{L-Glutamate} + \text{NADP}^+$$

In this method one can quantitate the ammonia directly by following the colorimetric reaction of NADPH with α-ketoglutarate in the presence of glutamate dehydrogenase[13] or by use of an ammonia electrode.[14] In all of these procedures the reaction mixture must be free of ammonia-producing and ammonia-consuming materials. In addition, a correction for endogenous ammonia must be made. Although the concentration of physiologically produced ammonia is relatively low, substantial amounts of ammonia can be formed by protein deamination if blood is left at room temperature.

The creatinine iminohydrolase reaction was used in the Kodak Ektachem clinical chemistry slide technology.[15,16] In this technology, the enzyme is immobilized on one of the layers of the slide. The serum is ultrafiltered by the first slide layer so that the creatinine is delivered as an ultrafiltrate to the lower layers. The immobilized iminohydrolase converts the creatinine to ammonia and *N*-methylhydantoin. The ammonia passes through a barrier layer to remove organic bases and then reacts with a bromphenol blue dye complex, which is measured by reflectance spectrophotometry at 670 nm (A_{max}, 593 nm). The technique measures both serum creatinine and ammonia; therefore it is necessary to subtract the contribution of endogenous serum ammonia. To determine serum ammonia values, a similar assay is performed on a slide that does not contain the iminohydrolase. The second-generation Kodak slides employ method 9, Table 62-5.

Potential reference methods for measurement of creatinine include separation by high-performance liquid chromatography (HPLC) using a cation exchange resin. The creatinine can be monitored after elution from the column either by reaction with the Jaffé reagent or by use of ultraviolet spectrophotometry[17,18] (method 7, Table 62-5).

Mass spectrometry has been proposed as a definitive method for creatinine measurement. In this procedure $^{51}N_2$-labeled creatinine is used as the internal standard. Creatinine is separated from the lyophilized serum by HPLC, converted to the ditrifluoroacetate derivative, and separated by gas chromatography/mass spectrometry (GC/MS)[19] (method 8, Table 62-5).

The method least subject to interferences is the Jaffé reaction with pretreatment by fuller's earth. The disadvantages of this procedure are that the introduction of fuller's earth necessarily involves a manual procedure and the sample requirement (typically 100 μL) is higher than in fully automated methods. This method, described next, must be considered the reference method at this time.[20] However, for the reasons just discussed, it is not used in routine automated analysis.

The major disadvantages of the original Jaffé reaction (without fuller's earth) are related to lack of specificity. Among substances reported to give a positive reaction are ascorbic acid, pyruvate, acetone, acetoacetic acid, levulose, glucose, aminohippurate, uric acid, protein, and cephalosporin antibiotics.[21-23] Bilirubin or other hemoglobin degradation products cause a *negative* bias, probably by their oxidation in strong base to colorless compounds.[24] The resulting decrease in background decreases the measured absorbance and is interpreted as a lower creatinine concentration. This negative bias is frequently seen in kinetic analyses for creatinine.

Table 62-6 Comparison of reaction conditions for creatinine

Condition	Reference method Jaffé + fuller's earth*	Kinetic Jaffé†	End-point Jaffé‡
Temperature	Ambient, 30° or 37° C	37° C	37° C
pH	12.4	≥12	≥12
Final concentration of reagents	Fuller's earth: 3.7 g/L Picric acid: 12 mmol/L NaOH 155 mmol/L	Picric acid: 12 mmol/L LiOH: 96 mmol/L	Picric acid: 15 mmol/L NaOH: 88 mmol/L
Fraction sample volume	0.3 (in protein-free filtrate)	0.056 (serum), 0.0059 (urine)	0.002 after dialysis
Volume of sample	100 μL (serum) (urine diluted 1:100 before use)	40 μL serum 4 μL urine	90 μL
Sample	Serum, plasma, urine	Serum, plasma, urine	Serum, plasma
Time of reaction	30 minutes	16 seconds	3 to 6 minutes
Wavelength	509 nm	525 nm	505 nm
Linearity	≤100 mg/L serum ≤2000 mg/L urine	≤150 mg/L serum ≤2000 mg/L urine	≤200 mg/L
Interferences	None	Hemoglobin—negative; interference at 1 g/L Hb Bilirubin—negative; interference of 4 mg/L at creatinine concentration, of 14 mg/L (bilirubin, 55 mg/L) Ketones—positive; interference (4 mg/L) at 400 mg/L acetoacetate	Acetoacetic acid, ascorbic acid, glucose, acetone, methyldopa, L-dopa, pyruvate, and cephalosporins can give elevated creatinine values
Precision (from CAP quality control surveys)	$\bar{X}$, 10.7 mg/L; CV, 7%-15% $\bar{X}$, 51.1 mg/L; CV, 3%-10%	$\bar{X}$, 10.1 mg/L; CV, 12% $\bar{X}$, 47.6 mg/L; CV, 3%	$\bar{X}$, 11.7 mg/L; CV, 8.6% $\bar{X}$, 53.4 mg/L; CV, 2.8%

*Reference method described in text.
†IL919, Instrumentation Laboratory, Inc, Lexington, MA.
‡Technicon SMAC Method No SG-40011 FH9, Technicon Instruments Corp, Tarrytown, NY.

Without modifications to the Jaffé reaction one might expect approximately 20% of the "creatinine" measured in serum or plasma and 5% of the "creatinine" measured in urine to be, in reality, noncreatinine materials.

The kinetic modification of the Jaffé reaction is also easily adapted to automation. The initial delay reduces error caused by fast-reacting materials, and the final measurement made before the reaction has gone to completion reduces error caused by slow-reacting materials, but no correction is made for noncreatinine reactants that have reaction rates comparable to that of creatinine. Pyruvate, α-ketoglutarate, and oxaloacetate have reaction rates that fall into this range. Many kinetic methods still have significant interference from acetoacetic acid. This can be a significant problem when patients with end-stage renal disease are monitored, because approximately 40% of such patients are diabetic. The major advantages of the kinetic method are the small sample required (less than 25 μL), the speed of the analysis, and the adaptability to automation.

The enzymatic procedures show promise of replacing the chemical creatinine assays now in common use.[25,26]

The 3,5-dinitrobenzoic acid reagent is useful in reagent strip formulations because, in contrast to picric acid, the compound is stable in the dry state. In addition, this reaction may be less subject to interferents. However, the color is not as stable as that produced by the Jaffé reaction.

Table 62-6 compares several methods of creatinine analysis.

SPECIMEN

One can analyze serum, plasma, or diluted urine. Urine should be diluted to a final creatinine concentration of approximately 300 to 600 μmol/L (34 to 68 mg/L); a 1:100 dilution will usually accomplish this. The common anticoagulants (fluoride and heparin) do not cause interference, though heparin, which can be formulated as the ammonium salt, must be avoided in enzymatic methods that measure ammonia production. Such methods also require prompt removal of serum from red cells and prompt analysis to minimize in vitro ammonia production; both of these precautions are necessary because of the imprecision that results from an elevated ammonia background. If analyzed by the Jaffé reaction, specimens are stable for at least 7 days at 4° C.

PROCEDURE: REFERENCE METHOD—ALKALINE PICRATE REACTION WITH FULLER'S EARTH

Principle

Creatinine purified with fuller's earth reacts with picrate ions in a highly alkaline solution to form a red complex measured at 509 nm. This method is essentially the selected method of the American Association for Clinical Chemistry.[20]

Reagents

HCl, 1.0 mol/L. Dilute 85.5 mL of concentrated HCl to 1 L with distilled water. This is stable for 2 years at room temperature.

Tungstate reagent, 0.15 mol/L. Dissolve 50 g of $Na_2WO_4 \cdot 2H_2O$ in distilled water; dilute to 1 L. This is stable for 1 year at room temperature.

Sulfuric acid, 666 mmol/L. Dilute 18.5 mL of concentrated sulfuric acid to 1 L with distilled water. This is stable for 2 years at room temperature.

Fuller's earth suspension, 6 g/L. Suspend 6 g of fuller's earth (30 to 60 mesh) in 10 mL of HCl (1.0 mol/L), and dilute to 1 L with distilled water. This is stable for 2 years at room temperature.

Picrate solution. Dissolve 2.7 g of picric acid and 6.2 g of NaOH in 1 L of distilled water. This solution is approximately 12 mmol/L in picrate and 155 mmol/L in NaOH and is stable for at least 3 months at room temperature if stored in a capped, dark bottle. The absolute concentration is not critical; various authors recommend concentrations from 50% to 200% of the strengths given here.[4] There are simple alternative methods of preparing bulk quantities of this solution, many of which are acceptable depending on the volumes of solution that are needed and the storage conditions. The solutions can also be purchased.

Creatinine stock solution, 8.84 mmol/(1 g/L). Dissolve 500 mg of creatinine in 5 mL of HCl (1.0 mol/L), and dilute to 500 mL with HCl. (One can obtain a pure creatinine preparation from the National Bureau of Standards, Washington, DC 20234, as Standard Reference Material No. 914).

Working standard solution 17.8 μmol/L (20 mg/L). Dilute 2.0 mL of creatinine stock solution to 100 mL with distilled water. This is stable for 1 week at room temperature.

Assay

Equipment: spectrophotometer with a ≤10 nm band pass capable of reading at 509 nm and a 30° ± 0.5° C incubator.

1. Pipette into labeled tubes 200 μL of serum, diluted urine, standard, or water (reagent blank). Add 100 μL of tungstate reagent and 100 μL of sulfuric acid to each tube.
2. Mix well; after 3 minutes at room temperature, centrifuge for 5 minutes at 12,000 *g*, and transfer 300 μL of the supernatant into a second tube. To this add 500 μL of fuller's earth from a vigorously stirred suspension. Mix for 1 minute, centrifuge for 2.5 minutes at 12,000 *g*, and aspirate the supernatant fluid completely. Add to the pellet 500 μL of picrate solution.
3. Mix vigorously until the pellet is thoroughly resuspended. Let the mixture stand in a 30° C incubator

Table 62-7 Reference ranges for creatinine and creatinine clearance

Age	Serum creatinine[3,27] mg/L (μmol/L)	Urine creatinine[28,29] g/day (mmol/day)	Creatinine clearance, height/weight adjusted, mL/min
<12 years	2.5-8.5 (22-75)	0.057 g (0.5 mmol/L)/kg of muscle	50-90
Adult male	6.4-10.4 (57-92)	1.0-2.0 (8.8-17.7)	97-137
Adult female	5.7-9.2 (50-81)	0.8-1.8 (7.1-15.9)	88-128

for 30 minutes, centrifuge for 1 minute, and measure the absorbance of the supernatant fluid versus the reagent blank at 509 nm (490 to 520 nm). Precision is improved by the use of a constant temperature cuvette (30° or 37° C).

Calculation

The standard solution and the reagent blank are both taken through the same procedure as the unknown samples. The concentration, *C*, of creatinine in the unknown samples is calculated as follows:

$$C = A \times F$$

where *A* is the absorbance of the unknown sample, and *F* is a factor determined in each series as follows:

$$F = \frac{C_{std}}{A_{std}} \times D$$

where A_{std} is the absorbance of the standard, *C* is the concentration of the standard 20 mg/L (176.8 μmol/L), and *D* is the dilution factor (to be used for urines or elevated serum specimens only).

The reference ranges given in Table 62-7 were determined with the fuller's earth modification of the Jaffé reaction (Table 62-6). A less specific method would be expected to result in slightly increased serum levels but would have little effect on the urine levels. Because of this nonuniform effect, a less specific method would result in a lower calculated creatinine-clearance rate.

Creatinine clearance

To do the creatinine clearance test, one needs a precisely timed urine collection and a blood sample taken during the collection period. Best results are obtained from a 24-hour urine collection. The test is initiated by having patients empty their bladder at the beginning of the timed period. Urine is collected throughout the period, the bladder is again emptied at the end of the period, and a blood sample is obtained.

Creatinine determinations are performed on both samples. The creatinine clearance is calculated from the following formula:

$$\text{Creatinine clearance (mL/min)} = \frac{UV}{P} \times \frac{1.73}{S}$$

where *U* is urinary creatinine (mg/L), *V* is volume of urine (mL/min), *P* is plasma creatinine (mg/L), *S* is the calculated surface area of the patient, and 1.73 is the surface area (m^2) of a standard 70 kg person.

The range of creatinine clearance in healthy persons corrected to a surface area of 1.73 m^2 is 88 to 137 mL/min. Appendixes H and I have nomograms for correcting body weight and height to surface area.

REFERENCES

1. Jaffé, M: Ueber den Niederschlag welchen Pikrinsaure in normalen Harn erzeugt und uber eine neue Reaction des Kreatinins, Z Physiol Chem 10:391-400, 1886.
2. Narayanan, S, and Appleton, HD: Creatinine: a review, Clin Chem 26:1119-1126, 1980.
3. Haeckel, R: Assay of creatinine in serum with use of fuller's earth to remove interferents, Clin Chem 27:179-183, 1981.
4. Spencer, K, and Price, CP: A review of non-enzyme mediated reactions and their application to centrifugal analysers. In Price, CP, and Spencer, K, editors: Centrifugal analysers in clinical chemistry, New York, 1980, Praeger Publishers, pp 231-253.
5. Bowers, LD: Kinetic serum creatinine assays. I. The role of various factors in determining specificity, Clin Chem 26:551-554, 1980.
6a. Benedict, SR, and Behre, JA: Some applications of a new color reaction for creatinine, J Biol Chem 114:515-532, 1936.
6b. Bowers, LD, and Wong, ET: Kinetic serum creatinine assays. II. A critical evaluation and review, Clin Chem 26:555-561, 1980.
7. Langley, WD, and Evans, M: The determination of creatinine with sodium 3,5-dinitrobenzoate, J Biol Chem 115:333-341, 1936.
8. Parekh, AC, Cook, S, Sims, C, and Jung, DH: A new method for the determination of serum creatinine based on reactions with 3,5-dinitrobenzoyl chloride in an organic medium, Clin Chim Acta 73:221-231, 1976.
9. Sims, C, and Parekh, AC: Determination of serum creatinine by reaction with methyl-3,5-dinitrobenzoate in methyl sulfoxide, Ann Clin Biochem 14:227-232, 1977.
10. Sullivan, MS, and Irreverre, F: A highly specific test for creatinine, J Biol Chem 223:530-533, 1958.
11. Van Pilsum, JF, Martin, RP, Kito E, and Hess, J: Determination of creatine, creatinine, arginine, guanidinoacetic acid, guanidine, and methyguanidine in biological fluids, J Biol Chem 222:225-236, 1956.
12. Moss, GA, Bondar, RJL, and Buzzelli, DM: Kinetic enzymatic method for determining serum creatinine, Clin Chem 21:1422-1426, 1975.
13. Tanganelli, E, Prencipe, L, Bassi, D, et al: Enzymatic assay of creatinine in serum and urine with creatinine iminohydrolase and glutamate dehydrogenase, Clin Chem 28:1461-1464, 1982.
14. Thompson, H, and Rechnitz, GA: Ion electrode based enzymatic analysis of creatinine, Anal Chem 46:246-249, 1974.
15. Slickers, K, Fame, N, Powers, D, and Rand, R: Performance of Kodak Ektachem clinical chemistry slides for creatinine and ammonia, Clin Chem 28:1570, 1982.
16. Shirey, TL: Development of a layered-coating technology for clinical chemistry, Clin Biochem 16:147-155, 1983.
17. Brown, ND, Sing, HC, Neeley, WE, and Koetitz, ES: Determination of "true" serum creatinine by high-performance liquid chromatography combined with a continuous-flow microanalyzer, Clin Chem 23:1281-1283, 1977.
18. Soldin, SJ, and Hill, GJ: Micromethod for determination of creati-

nine in biological fluids by high-performance liquid chromatography, Clin Chem 24:747-750, 1978.
19. Bjorkhem, I, Blomstrand, R, and Ohman, G: Mass fragmentography of creatinine proposed as a reference method, Clin Chem 23:2114-2121, 1977.
20. Haickel, R, Godsden, RH, Sherwin, JE, et al: Assay of creatinine in serum with use of fuller's earth to remove interferents. In Cooper, GR, editor: Selected methods of clinical chemistry, vol 10, Washington, DC, 1983, American Association for Clinical Chemistry, pp 225-229.
21. Swain, RR, and Briggs, SL: Positive interference with the Jaffé reaction by cephalosporin antibiotics, Clin Chem 23:1340-1342, 1977.
22. Young, DS, Pestaner, LC, and Gibberman, V: Effects of drugs on clinical laboratory tests, Clin Chem 21:1D-432D, 1975.
23. Soldin, SJ, Henderson, L, and Hill, JG: The effect of bilirubin and ketones on reaction rate methods for the measurement of creatinine, Clin Biochem 11:82-86, 1978.
24. Henry, RJ, Cannon, DC, and Winkelman, JW: Clinical chemistry: principles and technics, New York, 1964, Harper & Row, Publishers, Inc, pp 287-292.
25. Jaynes, PK, Field, RD, and Johnson, GF: An enzymatic reaction-rate assay for serum creatinine with a centrifuged analyzer, Clin Chem 28:114-117, 1982.
26. Landesman, PW, and Lott, JA: Determination of creatinine in serum by using creatinase: enhancement of a commercial method, Clin Chem 29:2002-2003, 1983.
27. Meites, S, editor: Pediatric clinical chemistry, Washington, DC, 1981, American Association for Clinical Chemistry, pp 171-177.
28. Newkirk, RE, and Rawnsley, HM: Creatinine clearance, ASCP Check Sample Clinical Chemistry no CC-110, Chicago, 1978, American Society of Clinical Pathologists.
29. Faulkner, WR, and King, JW: Renal function. In Tietz, NW, editor: Fundamentals of clinical chemistry, ed 2, Philadelphia, 1976, WB Saunders Co, pp 975-1014.

Urea

LAWRENCE A. KAPLAN

Urea, BUN (blood urea nitrogen), carbamide
Clinical significance: p. 356
Molecular weight: 60.06 daltons
Merck Index: 9671
Chemical class: amino acid metabolite

$$NH_2-\overset{\overset{\displaystyle O}{\|}}{C}-NH_2$$

PRINCIPLES OF ANALYSIS

Urea has traditionally been quantitated either directly by chemical analysis or indirectly by conversion of urea to ammonia and subsequent analysis of the ammonia (NH_3). In most historical methods, the ammonia nitrogen is measured after samples are either autoclaved at 125° C or acted on by the enzyme urease:

$$2\ H_2O + O = C\begin{matrix} \diagup NH_2 \\ \diagdown NH_2 \end{matrix} \xrightarrow[\text{or urease}]{125°\ C} (NH_4)_2CO_3 \rightarrow 2\ NH_4^+ + CO_3^=$$

The extraordinarily high specificity of urease for urea makes the enzymatic reaction the preferred method for converting urea into ammonia. Because the results of blood urea analyses were originally reported in terms of released nitrogen, urea concentration was expressed in terms of milligrams of blood urea nitrogen (BUN) per volume (such as deciliters). This term, unfortunately, has survived the years and changes in methodology and is still in current use. BUN values can be converted to urea concentrations as follows:

1. Atomic weight of nitrogen = 14 g/mol; molecular weight of urea = 60.06 g/mol.
2. Urea contains two nitrogen atoms per molecule.
3. Urea nitrogen (urea N) is 46.6% by weight of urea (28 divided by 60.06).
4. Therefore:

$$10 \text{ mg/L of BUN divided by } 0.466 = 21.46 \text{ mg/L of urea} = 0.36 \text{ mmol/L of urea}$$

or

$$\text{mg of urea N/L} \times 2.146 = \text{mg of urea/L}$$
$$\text{mg of urea N/L} \times 0.036 = \text{mmol of urea/L}$$

Although it is certainly preferable to report urea concentration in body fluids as mass of moles of urea per liter, the usual convention of reporting BUN values is used in this text.

The older methods for quantitating ammonia released from urea used either acid titration or the nesslerization (method 1, Table 62-8) or Berthelot reaction (method 2, Table 62-8) to form a colored product.[1-4] The titration method titrated the released gaseous ammonia with dilute sulfuric acid in the presence of a colored pH indicator solution.[1]

The most common method for determining urea concentrations in serum or urine is measurement of the NH_3 formed by the urease reaction, which uses a coupled enzyme system employing an NAD/NADH indicator reaction[5,6] (method 3, Table 62-8). These reactions are monitored at 340 nm. Other endogenous enzymes can compete with the indicator reaction (GLDH) to oxidize the NADH, decreasing the accuracy of the coupled enzyme system. In addition, exogenous ammonia from the reagents may give falsely high values. The urease GLDH–coupled reaction performed in the kinetic analysis mode can be used to measure urine urea at normal levels of endogenous ammonia.

A method for quantitating urea by measurement of the change in conductivity of a sample after the action of urease[7] (method 4, Table 62-8). The CO_2 and ammonia formed by the urease reaction form ammonium carbonate

Table 62-8 Methods of urea analysis

Method	Type of analysis	Principle	Usage	Comments
1. Nesslerization	Quantitative, end-point, spectrophotometric	$\text{Urea} \xrightarrow{\text{Urease}} (NH_4)_2CO_3$ $2HgI_2 + 4KI + NH_4^+ + NaOH \rightarrow NH_2Hg_2I_3 + H_3O^+ + 4KI + NaI$ *(yellow-orange colloid)*	Rarely used, of historical interest	Nonspecific, long reaction times
2. Berthelot	Quantitative, end-point, spectrophotometric	$\text{Urea} \xrightarrow{\text{Urease}} (NH_4)_2CO_3$ $NH_4^+ + HOCl \xrightarrow{\text{pH 10.5}} H_2NCl + H_3O^+$ (hypochlorous acid) (chloramine) H_2NCl + Phenol $\xrightarrow[NP]{OH^-}$ $ClN{=}C_6H_4{=}O$ (Quinonechloramine) $ClN{=}C_6H_4{=}O$ + Phenol $\xrightarrow{OH^-}$ $^-O{-}C_6H_4{-}N{=}C_6H_4{=}O$ (Indolphenol (blue)) (*NP*, Nitroprusside, used as a catalyst)	Rarely used	Nonspecific, relatively long reaction times
3. Coupled enzymatic (urease/glutamate dehydrogenase [GLDH])	Quantitative, end-point, kinetic, spectrophotometric	$\text{Urea} \xrightarrow{\text{Urease}} (NH_4)_2CO_3$ $NH_4^+ + \alpha\text{-Ketoglutaric acid} + NADH \xrightarrow[\text{ADP, } H^+]{\text{GLDH}} NAD^+ + \text{Glutamic acid}$	Most frequently used procedure	Very specific, rapid
4. Conductimetric	Quantitative, kinetic	$\text{Urea} \xrightarrow{\text{Urease}} (NH_4)_2CO_3 \rightarrow 2NH_4^+ + CO_3^{-2}$ Increased ions change conductivity	Frequently used	Very specific, rapid
5. Diacetyl monoxime	Quantitative, end-point, colorimetric	$CH_3{-}C({=}O){-}C({=}NOH){-}CH_3 \xrightarrow[H^+]{H_2O} CH_3{-}C({=}O){-}C({=}O){-}CH_3 + HNO_2$ Diacetyl monoxime; Diacetyl; Hydroxylamine $NH_2{-}C({=}O){-}NH_2 + CH_3{-}C({=}O){-}C({=}O){-}CH_3 \xrightarrow{H^+} CH_3{-}C{-}C{-}CH_3 + 2\,H_2O$ Urea; Diacetyl; Diazine *(yellow)* (ring: C=N–C(=O)–N=C)	Less frequently used	Some nonspecificity of reaction, uses noxious, dangerous reagents
6. *o*-Phthalaldehyde	Colorimetric, quantitative, end-point	$\text{Urea} + o\text{-Phthalaldehyde} \xrightarrow{H^+} \text{Isoindoline}$ $\text{Isoindoline} + \text{8-(4-amino-1-methylamino)-6-methoxyquinoline} \xrightarrow{H^+} \text{Chromophore (510 nm)}$	Rarely used	Interferences, often primary amines
7. Indicator dye	Quantitative end point	$\text{Urea} \xrightarrow{\text{Urease}} (NH_4)_2CO_3 \rightarrow 2NH_4^+ + CO_3^{-2}$ NH_3 + pH indicator dye → Change in absorbance spectrum of dye	Rarely used now, but will be used more frequently	Very specific, used with dry chemistry technology

($[NH_4]_2CO_2$), which increases the conductivity of the reaction mixture. When it is performed in a kinetic analysis mode to correct for the endogenous conductivity, one can analyze both serum and urine samples by the conductivity method. This method is employed by the Beckman Astra system[8] (Beckman Instruments, Inc., Brea, CA).

The only direct chemical analysis for urea has been the diacetyl monoxime reaction[1,2,9] (method 5, Table 62-8). The diacetyl monoxime does not directly react with urea but is first hydrolyzed to form diacetyl and hydroxylamine. The diacetyl condenses with urea in an acid solution to form a yellow diazine product. This chromogen is monitored at 550 nm. The chromogen can also be monitored by its fluorescence at 415 nm.[10] Several chemicals have been used to enhance and stabilize the color produced in the diacetyl monoxime reaction, either directly (such as thiosemicarbazide, ferric ions, or glucuronolactone) or indirectly, by elimination of the hydroxylamine formed in the initial hydrolysis (by compounds such as potassium persulfate).[2] The diacetyl monoxime reaction has been automated for analysis of both serum and urine samples.

Several instrument manufacturers have employed the reaction of *o*-phthalaldehyde with primary amines to quantitate urea (method 6, Table 62-8). The isoindoline product of the reaction is coupled to a complex quinoline to form a chromogen that is monitored at 510 nm.[11] The Ames Seralyzer (Ames Division, Miles Laboratories, Inc., Elkhart, IN) combines the *o*-phthalaldehyde reaction with their dry paper strip technology. A cation-exchange matrix is added to the reagent layer to catalyze the reaction between the isoindoline and the quinoline to form the chromogen. The reaction is monitored by reflectance spectrophotometry.

A more recent approach to serum urea analysis has employed the reaction between NH_3 (produced by the action of urease) and a pH indicator dye to produce a color change[12-14] (method 7, Table 62-8). This approach has used dry reagent technology, either film[12,13] or strip,[14] for the reaction and reflectance photometry for monitoring the reaction. All these methods have urease, immobilized in an initial sample spreading layer, reacting with urea to produce NH_4^+. Under alkaline conditions, NH_3 passes through a semipermeable layer and reacts with the pH indicator dye. The Kodak Ektachem (Eastman Kodak Co., Rochester, NY) uses merocyamine as the dye A_{max} at 520 nm (monitoring at 670 nm),[12] the Fuji Drichem-1000 (Fuji Film Co., Saitama, Japan) uses bromcresol green (monitoring the change in color from green to blue at 600 nm, A_{max} at 620 nm)[13] whereas the BMD Reflotron (Boehringer Mannheim Diagnostics, Indianapolis, IN) uses *n*-[bis(dinitrophenol)methyl]-4-*t*-butylpyridinium chloride (monitoring the change from colorless to blue at 642 nm).[14] The Reflotron is semiautomated, whereas the other two instruments are automated. The Drichem-1000 and the

Table 62-9 Comparison of reaction conditions for urea analysis

Reaction component	Coupled enzymatic urease/GLDH*	Conductimetric†	Diacetyl monoxime‡
Temperature	30° C	37° C	37° C
pH	7.8	7.3	Highly acidic
Final concentration of reagents	Somogyi supernatant: $Ba(OH)_2$: 36.7 mmol/L $ZnSO_4$: 36.7 mmol/L Urease reaction: Urease: 4 × 10^3 U/L GLDH: 30 × 10^3 U/L ADP: 2.0 mmol/L NADH: 200 μmol/L α-Ketoglutaric acid: 6.0 mmol/L Tris buffer: 100 mmol/L EDTA: 4.0 mmol/L	Urease: 2 × 10^5 U/L Tris buffer EDTA	Diacetyl monoxime: 9.4 mmol/L Thiosemicarbazide: 2.1 mmol/L $FeCl_3$: 5.3 mmol/L H_3PO_4: 2.2 mmol/L H_2SO_4: 5.4 mmol/L
Fraction of sample volume	0.048 for both Somogyi precipitation and urease reaction	0.009	0.111 (before dialysis) (about 0.002 after dialysis)
Linearity (urea nitrogen per liter)	800 mg/L	1500 mg/L	1500 mg/L
Reaction time	30-minute end point	11.5-second kinetic	9-minute end point at 520 nm
Interferences	Fluoride (F^-)	Fluoride (F^-)	Nitrogen-containing compounds
Precision (at interlaboratory precision from CAP surveys)	220 mg of BUN/L—1.5% CV for overall precision	220 mg of BUN/L—3.4% CV	220 mg of BUN/L—4.1% CV

*Candidate reference method, reference 15 (Sampson, EJ, et al: Clin Chem 26:816, 1980).
†Beckman ASTRA, Beckman Instruments, Division of Beckman-SmithKline, Inc, Brea, CA. Actual reagent concentrations are proprietary information of Beckman Instruments.
‡Technicon, Inc, Tarrytown, NY, Method no SE4000 IFD4.

Reflotron can use whole blood, plasma, and serum, whereas the Ektachem can use only serum.

The diacetyl monoxime reaction has been replaced by other procedures primarily those using GLDH coupled reaction. The dry reagent methods are used in physician office testing and in those laboratories with Kodak Ektachem instruments. The conductivity method also gives suitable results.

SPECIMEN

Serum and heparinized plasma can be used for the diacetyl monoxime, urease/GLDH, and conductivity methods. Fluoride will inhibit the urease reaction; therefore methods employing urease cannot use serum preserved with fluoride.[3] Ammonium heparin also cannot be used as an anticoagulant for urease methods. One can analyze urine by all three methods after a 1:20 to 1:50 sample dilution, depending on the method and instrument employed.

Because of urea's susceptibility to bacterial degradation, serum and urine samples should be kept at 4° to 8° C until analysis. One can also preserve urine samples by maintaining the pH at less than 4.

Table 62-9 compares reaction conditions for urea analysis.

REFERENCE RANGE

The reference ranges for serum BUN will vary with the method; several are listed here. Statistically higher values are seen in men and in older age groups. For adults, these differences are not clinically significant, and one can combine reference-range data for these groups. Young children have slightly lower serum urea values than older children and adults have. Individual variations in BUN values will also depend on the dietary habits of the person; less affluent persons or those consuming less protein have lower serum urea concentrations. Similarly, urine output of urea will vary with diet.

Method (Table 62-9)	**Adult reference range (serum)**
Urease/GLDH	50 to 170 mg BUN/L (107 to 365 mg of urea/L, 1.8 to 6.1 mmol/L)
Urease conductivity	60 to 200 mg BUN/L (129 to 429 mg of urea/L, 2.2 to 7.2 mmol/L)
Diacetyl monoxime	80 to 260 mg BUN/L (172 to 558 mg of urea/L, 2.9 to 9.4 mmol/L)
	Urine urea output (average diet)
Urease/GLDH	7 to 16 g of BUN/24 hours (0.25 to 0.57 mol of urea/24 hours)

REFERENCES

1. Natelson, S: Techniques of clinical chemistry, ed 3, Springfield, Ill, 1971, Charles C Thomas, Publisher, pp 728-745.
2. Henry, RJ, Cannon, DC, and Wilkelmann, JW: Clinical chemistry principles and technics, ed 2, New York, 1974, Harper & Row, Publishers, Inc, pp 504-506.
3. Patton, CJ, and Crouch, SR: Spectrophotometric and kinetics investigation of the Berthelot reaction for determination of ammonia, Anal Chem 49:464-469, 1977.
4. Faulkner, WR, and Meites, S, editors: Selected methods of clinical chemistry, vol 9, Washington, DC, 1982, American Association for Clinical Chemistry, pp 357-363.
5. Talke, H, and Schubert, GE: Enzymatische Harnstoffbestimmung in Blut und Serum im optischen Test nach Warburg, Klin Wocheschr 43:174-175, 1965.
6. Tiffany, TO, Jansen, JM, Burtis, CA, et al: Enzymatic kinetic rate and end-point analysis of substrate, by use of a GeMSAEC fast analyzer, Clin Chem 18:829-840, 1972.
7. Chin, WT, and Kroontje, W: Conductivity method for determination of urea, Anal Chem 33:1757-1760, 1961.
8. Paulson, G, Ray, R, and Sternberg, J: A rate sensing approach to urea measurement, Clin Chem 17:644, 1971.
9. Faulkner, WR, and Meites, S, editors: Selected methods of clinical chemistry, vol 9, Washington, DC, 1982, American Association for Clinical Chemistry, pp 365-373.
10. McCleskey, JE: Fluorometric method for the determination of urea in blood, Anal Chem 36:1646-1648, 1964.
11. KDA Application notes, American Monitor Corporation, Indianapolis, IN, 1983.
12. Spayd, RW, Bruschi, B, Burdick, BA, et al: Multilayer film elements for clinical analysis: applications to representative chemical determinations, Clin Chem 24:1343-1350, 1978.
13. Ohkubo, A, Kamei, S, Yamanaka, M, et al: Multilayer-film analysis for urea nitrogen in blood, serum, or plasma, Clin Chem 30:1222-1225, 1984.
14. Hammond, BR, and Lester, E: Evaluation of a reflectancce photometric method for determination of urea in blood, plasma, or serum, Clin Chem 30:596-597, 1984.
15. Sampson, EJ, Baird, MA, Burtis, CA, et al: A coupled-enzyme equilibrium method for measuring urea in serum: optimization and evaluation of the AACC Study Group on Urea candidate reference method, Clin Chem 26:816-826, 1980.

Uric acid

ARNOLD L. SCHULTZ

Clinical significance: pp. 346 and 587
Molecular formula: $C_5H_4H_4O_3$
Molecular weight:f 168.11 daltons
Merck Index: 9683
Chemical class: purine

PRINCIPLES OF ANALYSIS

The reduction of phosphotungstic acid to tungsten blue by alkaline solutions of uric acid was first used for the analysis of uric acid in blood in 1912.[1] This early method used protein precipitation and isolation of the uric acid from the filtrate as the silver salt before reaction with phosphotungstic acid in sodium carbonate solution (method 1, Table 62-10). Urea-cyanide was later used as the alkaline reagent.[2,3] This modification did not require isolation of the uric acid from the filtrate. Many other alkaline reagents have been used to enhance the color of the tungsten blue produced (A_{max}, 700 nm). Proteins have been removed by precipitation with tungstic acid, trichloroacetic acid, phosphotungstic acid, heat coagulation, and

Table 62-10 Methods of uric acid analysis

Method	Type of analysis	Principle	Usage	Comments
1. Phosphotungstic acid	Spectrophotometric	Oxidation of uric acid to allantoin and carbon dioxide with reduction of phosphotungstic acid to tungsten blue (A_{max}, 700 nm)	Serum, urine	Nonspecific, but widely used
2. Uricase	Enzymatic	Oxidation of uric acid to allantoin, hydrogen peroxide, and carbon dioxide	Serum, urine	
	a. Differential absorption	Uric acid absorbs in the 290 to 293 nm (at pH ≥7) and 283 nm (at pH <7) region of ultraviolet spectrum, but allantoin does not		Basis for a candidate reference method, increased specificity
	b. Colorimetric	Quantitation of hydrogen peroxide produced, especially when coupled to NAD/NADH indicator reaction		Specificity varies from method to method, NADH reaction widely used
	c. Polarographic	Rate of oxygen consumption measured		Not widely used, some interferences
	d. Coulometric	Titration with iodine		Instrumentation not readily available
3. High-performance liquid chromatography	Chromatographic			
	a. Spectrophotometric	Reversed-phase chromatography	Serum, urine	Increased specificity and sensitivity
	b. Electrochemical	Ion-exchange separation	Serum, urine	Proposed selected method

membrane filtration. Other oxidizing reagents have included arsenotungstic acid, arsenophosphotungstic acid, arsenomolybdic acid, potassium ferricyanide, and uranyl acetate.

Uricase has been used extensively to increase the specificity of the uric acid assay (method 2, Table 62-10). The uricase methods are based on the specificity of the uricase-catalyzed oxidation of uric acid to allantoin and hydrogen peroxide. Allantoin, unlike uric acid, does not have an absorption peak in the 290 to 293 nm region of the ultraviolet spectrum. Absorption measurements at these wavelengths before and after incubation of uric acid with uricase have been used to quantitate uric acid in serum, plasma, and urine.[4-6]

The hydrogen peroxide produced when uric acid is oxidized in the uricase-catalyzed reaction has been quantified by measurement of the chromogenic response at 530 nm when *o*-dianisidine is oxidized by hydrogen peroxide.[7] Hydrogen peroxide oxidatively reacts with 3-methyl-2-benzothiazolinone and *N,N*-dimethylaniline, in the presence of peroxidase, to produce a blue indamine dye, the absorbance of which is measured at 600 nm.[8] It also reacts with 3,5-dichloro-2-hydroxybenzenesulfonic acid and 4-aminophenazone to form a red quinoneimine dye.[9] The absorbance of the quinoneimine dye is measured at 520 nm to avoid spectral interference from hemolysis, bilirubin, and turbidity, though the maximum absorbance occurs at 512 nm and is 3% more intense. 2,4,6-Tribromophenol has also been used in the reaction with 4-aminophenazone.[10] The absorbance of the product from this reaction is read at 492 nm. Similarly, the reaction of formaldehyde (formed in the catalase-catalyzed reaction of hydrogen peroxide and methanol) with acetylacetone and ammonia to produce the yellow dye 3,5-diacetyl-1,4-dihydrolutidine has been measured at 410 nm to quantitate the oxidation of uric acid in the presence of uricase.[11] The hydrogen peroxide produced by the uricase-catalyzed oxidation has been detected through the catalase-catalyzed oxidation of ethanol to acetaldehyde, coupled to the oxidation of acetaldehyde to acetate in the presence of aldehyde dehydrogenase and NAD^+. The change in absorbance at 340 nm is related to uric acid levels.[12]

The rate of oxygen consumption, which is a measure of the rate of the uricase-catalyzed oxidation of uric acid, is proportional to the uric acid concentration and has been used to measure uric acid concentration in serum and urine.[13] A polarographic oxygen sensor is used.

Several high-performance liquid chromatographic (HPLC) procedures for the quantitation of uric acid in serum[14,15] and urine[16] have been introduced (method 3, Table 62-10). These HPLC methods use either reversed-phase chromatography with spectrophotometric detection at 280[14] or 235 nm[16] or ion-exchange separation followed by amperometric detection in a thin-layer flowthrough electrochemical cell.[15]

The preferred method for routine uric acid measurement

is an automated coupled-enzymatic procedure available in kit form from many commercial sources. This method is both accurate and precise and is readily adapted to most automated instruments. As future automated spectrophotometric analyzers improve in their ability to monitor reactions in the near-ultraviolet range, the ultraviolet (290 to 293 nm) method may be the method of choice. The method presented here is based on the manual uricase candidate reference method as given by the Centers for Disease Control.[17-19]

SPECIMEN

Serum or plasma may be used. EDTA and sodium fluoride should be avoided as an anticoagulant and a preservative, respectively, because they contribute a positive interference to the method described here. Uric acid is stable at 2° to 6° C for 3 to 5 days and for at least 6 months at −20° C.

Aliquots of a 24-hour urine collection are also useful for uric acid determinations (Table 62-11). To prevent urate precipitation, add 10 mL of 500 g/L sodium hydroxide to the collection bottle *before* collection of the specimen. Uric acid in urine is usually stable for approximately 3 days at room temperature, provided that there is no bacterial growth to destroy it.

PROCEDURE: URIC ACID ANALYSIS BY ULTRAVIOLET URICASE METHOD

Principle

Incubation of uric acid with uricase in Tris buffer, pH 8.5 at 37° C, results in the production of one mole of allantoin for every mole of uric acid oxidized. The reaction is quantified by measurement of the decrease in absorbance at 283 nm (see following note 1). Uric acid exhibits absorbance at this ultraviolet wavelength, but allantoin does not. Before one takes absorbance readings, proteins are precipitated with trichloroacetic acid to eliminate high serum background absorbances. A blank is also used for each sample to correct for nonprotein, endogenous ultraviolet-absorbing substances that vary from specimen to specimen.

Table 62-11 Reaction conditions for uric acid analysis

Condition	Candidate reference method
Sample	0.5 mL of serum or plasma or 0.5 mL of a 1:10 dilution of an aliquot of a 24-hour urine collection
Protein precipitant	100 g/L trichloroacetic acid
Reaction temperature	37° C
Reaction time	95 minutes
Linearity	200 mg/L
Precision	2.6% at 30 mg/L, 1.5% at 99 mg/L
Final concentration of reagents	Tris buffer, 0.041 mol/L Uricase, 17 U/L Trichloroacetic acid, 33.3 g/L

Reagents

Tris hydrochloride, 15.8 g/L (0.1 mol/L). This is stable for 2 months at 2° to 6° C if free of bacterial contamination.

Tris base, 12.1 g/L (0.1 mol/L). This is stable for 2 months at 2° to 6° C if free of bacterial contamination.

Tris buffer, 0.1 mol/L, pH 8.5 at 37° C. Mix 940 mL of 12.1 g/L Tris base and 224 ml of 15.8 g/L Tris-HCl. Check the pH with a pH meter at 37° C ± 1° C. If necessary, adjust with either Tris-HCl or Tris base solution to pH 8.5 ± 0.1. Filter the buffer through a sterile, 0.45 μm membrane filter, and transfer to a sterilized, borosilicate glass, screw-capped bottle. This is stable for 6 months at 2° to 6° C if protected from bacterial contamination.

Trichloroacetic acid, 100 g/L (612 mmol/L). Place 10 g of trichloroacetic acid in a 100 mL volumetric flask, and add 50 mL of deionized water to dissolve. Bring volume to mark with water, and mix. Filter the solution through a 0.45 μm membrane filter into a sterilized borosilicate glass, screw-capped storage bottle. This is stable for 3 months at 2° to 6° C.

Uricase, 0.1 U/ml. Weigh or measure volumetrically an amount of uricase (from *Bacillus fastidiosus* or *Candida utilis*) estimated to have a total activity of 1000 International Units (U) at 37° C (see following note 2). Transfer this amount to a 1000 mL volumetric flask, and dilute to volume with Tris buffer, pH 8.5. One can aliquot the resulting solution in 100 mL portions in borosilicate screw-capped bottles. This is stable for 1 month at −20° C.

"Blank" enzyme reagent. Add 50.0 mL of 100 g/L trichloroacetic acid solution to 25.0 mL of 0.1 U/mL uricase solution. This is enough to prepare approximately 25 blanks. Prepare fresh 1 hour before use; prepare an appropriate amount for the number of samples to be analyzed. The solution must be allowed to stand at least 30 minutes to ensure that the uricase has been inactivated.

Uric acid stock standard, 1000 mg/L (5.95 mmol/L). Weigh and transfer 500 mg of anhydrous NBS Certified Uric Acid, Standard Reference Material No. 913, which has been stored in its original capped bottle over anhydrous calcium sulfate in a vacuum desiccator, and 375 mg of ACS-grade lithium carbonate to a 500 mL volumetric flask. Add 125 mL of sterile, distilled, deionized water that has been heated to 50° C. Mix the uric acid and lithium carbonate into complete solution with swirling. The water should not be heated higher than 50° C. After the solids have completely dissolved, allow the solution to cool to room temperature. Dilute to volume with sterile, distilled, deionized water. Mix well. Divide into five 100 mL portions, and transfer each to a screw-capped borosilicate bottle. This is stable for 2 months at −20° C. One bottle is used to prepare a complete set of standards.

Uric acid working standards, 20, 40, 60, 80, 100, 150, and 200 mg/L. Dilute 2, 4, 6, 8, 10, 15, and 20 mL of the 1000 mg/L stock uric acid standard to volume with sterile, distilled, deionized water in 100 mL volumetric flasks. Working standards should be dispensed into screw-capped borosilicate vials of about 15 mL capacity and stored tightly capped. This is stable for 2 months at −20° C. When a vial standard is thawed, it is used only that day and discarded.

"Zero" standard, lithium carbonate, 150 mg/L (2 mmol/L). Dissolve and dilute 15 mg of ACS-grade lithium carbonate in sterile, distilled, deionized water to volume in a 100 mL volumetric flask. This is stable for 6 to 12 months at 4° C.

Assay

Equipment: ultraviolet spectrophotometer with a ≤10 nm band pass; a 37° C ± 0.5° C water bath.

1. Warm all reagents to room temperature.
2. Measure and record the volume in milliliters of 24-hour urine collections.
3. If the urine is cloudy, warm the specimen to 60° C for 10 minutes to dissolve precipitated urates and uric acid. Cool the specimen to room temperature, and centrifuge an aliquot.
4. Prepare a 1:10 dilution of a centrifuged aliquot of urine with deionized water.
5. Label a sufficient number of 16 × 100 nm borosilicate, cappable tubes. Each standard, unknown, and control requires a "test" and a "blank" tube.
6. Transfer 0.5 mL of sample and 2.5 mL of Tris buffer into each "test" and "blank" of the appropriately labeled tubes.
7. Add 1.0 mL of uricase solution 0.1 U/mL to the tubes marked "test," mix well, and cap.
8. Immerse all capped "test" and "blank" tubes into a 37° C ± 1° C water bath for 60 minutes.
9. After 60 minutes, remove the tubes from the water bath and add 2.0 mL of 100 g/L trichloroacetic acid to all the tubes marked "test." Add 3.0 mL of the "blank" enzyme reagent to all tubes marked "blank." Mix well by inversion, being careful to avoid foaming. (Do *not* shake vigorously, and do *not* use a vortex mixer). Let stand 45 minutes, mixing several times by gentle inversion.
10. Centrifuge all the "test" and "blank" tubes at approximately 1200 *g* for 10 minutes. Remove the tubes from the centrifuge; mix again by gentle inversion. (It is not necessary to remix the precipitate at the bottom of the test tube. This step is helpful in bringing down the fine precipitate that remains at the meniscus.) Centrifuge the tubes for an additional 10 minutes at 1200 *g*. Transfer the supernatants by decanting to clean, dry, borosilicate tubes, and centrifuge again for 20 minutes. With Pasteur pipets, transfer the supernatants to clean, dry, borosilicate tubes. (Visually inspect all the supernatants, and if any particulate matter is noticed, centrifuge the tubes again for 20 minutres, and transfer the supernatant to clean, dry, borosilicate tubes with Pasteur pipets.) NOTE: *Do not filter*. Uric acid has been shown to adhere to filter paper and to filter membranes.[20]
11. Zero the spectrophotometer with deionized water. Obtain the absorbance readings at 283 nm for all "tests" and "blanks" samples. Read the absorbance of the standards, unknowns, and controls. Timing is not critical, but spectrophotometry is performed in the order of addition, so that each sample is read after approximately the same period of analysis time.

Calculation

1. Subtract the "test" absorbance (A_{test}) from the "blank" absorbance (A_{blank}) for each standard, unknown, and control to obtain the "corrected reading" (A_{corr}).
$$A_{corr} = A_{blank} - A_{test}$$
2. Draw a calibration curve relating the corrected reading to the quantity of uric acid in the standards in mg/L.
3. Interpolate the concentration of uric acid in mg/L in each of the samples and controls from the calibration curve using the corrected absorbance readings.
4. Calculate the amount of uric acid in the 24-hour specimen from the equation:
$$\text{mg of uric acid/24 hours} = \text{mg/L} \times \frac{\text{Total urine volume (mL)}}{1000 \text{ mL/L}} \times 10$$
where 10 is the dilution factor of the aliquot of urine, and 1000 converts mL to L.
5. Samples with absorbance readings that are more than 10% higher than the highest standard accepted for the run, or lying beyond the previously demonstrated linear range, are diluted with equal volumes of deionized water and reanalyzed. Multiply the result obtained from the calibration curve by the dilution factor.

Notes

1. Uric acid in lithium carbonate solution, pH 9.8, exhibits an absorbance maximum at 293 nm. However, in trichloroacetic acid solution, pH 1.1, the absorbance maximum is shifted to 283 nm.
2. It is not absolutely necessary to measure exactly 1000 U of uricase, but the measurement should be made as accurately as possible. If it is impossible to estimate enzyme activity, prepare only 10 mL of an approximately 1 U/mL uricase solution in Tris buffer, pH 8.5, and assay it for uricase activity on the day of preparation. Store at −20° C. This is stable for 1 month.

REFERENCE RANGES

Serum or plasma

36 to 77 mg/L (214 to 458 μmol/L) for males
25 to 68 mg/L (149 to 405 μmol/L) for females

Urine

250 to 750 mg (1.49 mmol to 4.46 mmol)/24 hours for average diet
Up to 450 mg (2.68 mmol)/24 hours for low-purine diet
Up to 1 g (5.95 mmol)/24 hours for high-purine diet

REFERENCES

1. Folin, O, and Denis, W: A new (colorimetric) method for the determination of uric acid in blood, J Biol Chem 13:469-475, 1912-1913.
2. Folin, O: An improved method for the determination of uric acid in blood, J Biol Chem 86:179-187, 1930.
3. Brown, H: The determination of uric acid in human blood, J Biol Chem 158:601-608, 1945.
4. Kalckar, HM: Differential spectrophotometry of purine compounds by means of specific enzymes. I. Determination of hydroxypurine compounds, J Biol Chem 167:429-443, 1947.
5. Praetorius, E, and Poulsen, H: Enzymatic determination of uric acid with detailed directions, Scand J Clin Lab Invest 5:271-280, 1953.
6. Feichtmeier, TV, and Wrenn, HT: Direct determination of uric acid using uricase, Am J Clin Pathol 25:833-839, 1955.
7. Marymount, JH, Jr, and London, M: Analyses performed with heat-coagulated blood and serum. VI. Direct determination of urates by means of *o*-dianisidine oxidation, Am J Clin Pathol 42:630-633, 1964.
8. Gochman, N, and Schmitz, JM: Automated determination of uric acid, with use of a uricase-peroxidase system, Clin Chem 17:1154-1159, 1971.
9. Fossati, P, Prencipe, L, and Berti, G: Use of 3,5-dichloro-2-hydroxy-benzenesulfonic acid/4-aminophenazone chromogenic system in direct enzymatic assay of uric acid in serum and urine, Clin Chem 26:227-231, 1980.
10. Kabasakalian, P, Kalliney, S, and Westcott, A: Determination of uric acid in serum, with use of uricase and a tribromophenol aminoantipyrine chromogen, Clin Chem 19:522-524, 1973.
11. Kageyama, N: A direct colorimetric determination of uric acid in serum and urine with uricase-catalase system, Clin Chim Acta 31:421-426, 1971.
12. Haeckel, R: The use of aldehyde dehydrogenase to determine H_2O_2 producing reactions. 1. The determination of the uric acid concentration, J Clin Chem Clin Biochem 14:101-107, 1976.
13. Bell, R, and Ray, RA: A rate-sensing approach to the measurement of uric acid in serum and urine, Clin Chem 17:644, 1971.
14. Kiser, EJ, Johnson, GF, and Witte, DL: Serum uric acid determined by reversed-phase liquid chromatography with spectrophotometric determination, Clin Chem 24:536-540, 1978.
15. Pachla, LA, and Kissinger, PT: Measurement of serum uric acid by liquid chromatography, Clin Chem 25:1847-1852, 1979.
16. Hausen, A, Fuchs, D, König, K and Wachter, H: Quantitation of urinary uric acid by reversed-phase liquid chromatography, Clin Chem 27:1455-1456, 1981.
17. Duncan, P, Gochman, N, Cooper, T, et al: Development and evaluation of a candidate reference method for uric acid in serum, 1979, US Dept of Health, Education and Welfare, Public Health Service, Atlanta, 1980, US Centers for Disease Control.
18. Duncan, PH, Gochman, N, Cooper, T, et al: A candidate reference method for uric acid in serum. I. Optimization and evaluation, Clin Chem 28:284-290, 1982.
19. Duncan, P, Gochman, N, Bayse, D, et al: A candidate reference method for uric acid in serum. II. Interlaboratory testing, Clin Chem 28:291-293, 1983.
20. Fales, FW: Recovery of uric acid from serum, Clin Chem 14:449-455, 1968.

CHAPTER 63 Proteins

Albumin

STEPHEN M. GENDLER

Clinical significance: pp. 346 and 359
Molecular weight: 66,248 daltons
Merck Index: 203
Chemical class: protein

PRINCIPLES OF ANALYSIS

The earliest techniques for albumin analysis are based on acid or salt precipitation of the protein (method 1, Table 63-1).[1] In the late nineteenth century albumin was defined as the serum protein that remained in solution at 2.05 mol/L ammonium sulfate at 25° C.[2,3] The albumin in the supernatant was measured by total nitrogen analysis or by biuret reaction. Modern salt fractionation techniques are cumbersome because of the many steps required in the procedure and not very specific because of protein interactions.[4] Since they are not easily automated, the precipitation methods are largely of historical interest.

The measurement of albumin can be performed by a direct determination of globulin based on tryptophan content and calculation of the albumin content by subtraction of globulin from total protein (method 2, Table 63-1). The tryptophan content of serum albumin is 7% to 10% that of globulin on a weight basis. A procedure proposed by Goldenberg and Drews[5] takes advantage of this large difference in tryptophan content between albumin and globulin or globulins. In this one-reagent system, glyoxylic acid, in the presence of Ca^{+2} in an acid medium, condenses with the tryptophan residues in globulins to produce a purple color measured at 540 nm. The method needs to be standardized with serum to compensate for the albumin interference in the reaction. Savory et al.[6] have adapted this globulin technique and a total protein analysis to the AutoAnalyzer (Technicon Instruments Corp., Tarrytown, NY). Tryptophan content methods have never come into common use because of the ease and specificity of the dye-binding methods for albumin.

Serum albumin can be quantitated by electrophoretic techniques (method 3, Table 63-1). The major classes of serum protein are separated by a serum protein electrophoresis method (such as cellulose acetate or agarose; see p. 1054). The separated fractions are stained, and the percentage of each fraction present in the sample is determined by densitometric analysis. The concentration of albumin is calculated by multiplication of the concentration of total protein in the sample by the percentage of albumin.

Immunological methods (method 4a and 4b, Table 63-1) used for serum albumin quantitation include radial immunodiffusion (RID) and electroimmunodiffusion (EID) in which albumin either passively diffuses (RID) or is electrophoresed (EID) into a stationary phase (such as agarose) that contains antibodies to albumin. See Chapters 9 and 10 for a description of these techniques.

The reaction between albumin and an antialbumin antibody can be monitored by turbidimetric or nephelometric means (method 4c and 4d, Table 63-1).[7-9] See Chapters 3 and 10.

The most widely used methods for the analysis of serum albumin are dye-binding procedures (method 5, Table 63-1). Albumin has the ability to bind a wide variety of organic anions, including complex dye molecules. The dye-binding techniques are based on a shift in the absorption maximum of the dye when bound to albumin. The shift in the absorption maximum allows the resulting color to be measured in the presence of excess dye, which, in concert with the high-binding affinity to albumin, allows all the albumin molecules to take part in the reaction. A variety of dyes has been employed for the measurement of albumin including methyl orange, 2-(4′-hydroxyazobenzene)-benzoic acid (HABA), bromcresol green (3,3′,5′-tetrabromo-*m*-cresolsulfonphthalein, or BCG), and bromcresol purple (5,5′-dibromo-*o*-cresolsulfonphthalein, or BCP). Introduced by Rodkey in 1965,[10] the bromcresol green reaction is usually performed at pH 4.2 to 4.5 and is monitored at 620 to 630 nm. A bromcresol green method was recommended by the American Association for Clinical Chemistry (AACC) in 1972 and 1982.[11] In this procedure, the absorbance of bromcresol green when albumin binds

Table 63-1 Methods of albumin analysis

Method	Type of analysis	Principle	Usage	Comment
1. Precipitation a. Salt fractionation b. Solvent fractionation c. Acid fractionation	Quantitative	Changes of net charge of protein result in precipitation.	Serum, manual	Historical; still used in manufacturing albumin
2. Tryptophan content	Quantitative	Glyoxylic acid + Tryptophan in globulin → *purple chromogen* (A_{max}, 540 nm) Total protein − Globulin = Albumin	Serum, manual, and automated	Correlates well with electrophoresis but requires total protein measurement
3. Electrophoresis a. Moving boundary b. Cellulose acetate c. Cellulose acetate with elution of peak	Quantitative	Albumin is separated from other proteins in electrical field. Percent staining of albumin fraction multiplied by total protein value.	Serum, manual, and automated	Very labor intensive, but if albumin is eluted for measurement, very accurate
4. Immunochemical				
a. Electroimmunoassay	Quantitative	Protein migrates in electrical field through medium containing a specific antibody.	Serum, manual	Reference method; somewhat labor intensive
b. Radial immunodiffusion	Quantitative	Protein diffuses through medium containing specific antibody.	Serum, CSF, manual	Reference method; very long incubation time
c. Turbidimetry	Quantitative	Antigen-antibody complexes decrease light transmission more than free antigen.	Serum, manual or automated	Reagent cost high
d. Nephelometry	Quantitative	Antigen-antibody complexes scatter light more than free antigen.	Serum, CSF, automated	Reagent cost high
e. Radioimmunoassay	Quantitative	Radiolabeled albumin competes with test albumin for limited amount of antibody.	Serum, CSF, urine	Primarily used for urine
f. Enzyme immunoassay	Quantitative	Sandwich assay uses antibody bound to surface and peroxidase-labeled antibody.	Serum, CSF, urine	New
5. Dye binding				
a. Methyl orange	Quantitative	Albumin binds to dye and changes spectral profile of dye.	Serum, manual, or automated	Nonspecific for albumin
b. HABA(2-[4′-hydroxyazobenzene]benzoic acid)	Quantitative	Same as above.	Serum, manual, or automated	Specific for albumin; poor sensitivity; many drug interferences
c. BCG (bromcresol green)	Quantitative	Same as above (A_{max}, 628 nm).	Serum, manual, or automated (most often used method)	Nonspecific for albumin if absorbance reading taken after 30 seconds
d. BCP (bromcresol purple)	Quantitative	Same as above (A_{max}, 603 nm).	Serum, manual, or automated	Specific for albumin; albumins from animal sources do not bind equivalently to human albumin
e. Bromphenol blue	Semiquantitative	Bromphenol blue in test strip changes color from yellow to be in presence of albumin.	Urine	Nonspecific; most sensitive with albumin; most commonly used test for urine protein

CSF, Cerebrospinal fluid.

the dye is measured at 628 nm in a 0.075 mol/L succinate buffer at pH 4.20. Brij-35, a nonionic detergent, is added to buffer and bromcresol green solutions to reduce the blank absorbance, prevent turbidity, and provide linearity. Human serum albumin Cohn fraction V (HSA-V) is used for standardization in the method published by the AACC, but bovine serum albumin Cohn fraction V has also been used. Bromcresol purple reacts with albumin at pH 5.2 and shows a color shift measured at 603 nm. One can monitor these dye reactions as end-point reactions or as blanked, fast (60-second) reactions.

Cerebrospinal fluid albumin is readily measured directly by electroimmunodiffusion, RID, or immunonephelometry. All these products are available from commercial vendors. These assay kits function in the range of about 200 mg/L.

The Albustix method was developed by the Ames Company (Division of Miles Laboratories, Elkhart, IN) as a semiquantitative measurement of urine protein. This method is based on the change in color of the dye bromphenol blue in the presence of a binding protein such as albumin (method 5e, Table 63-1). At pH 3, the protein is in the unionized (yellow) form. As albumin is added, the ionized form of the dye (blue color) preferentially binds; thus it is removed from solution. As more protein is added, more of the ionized (blue-colored) form is bound and a significant portion of the dye becomes ionized, so that the color of the solution is changed from yellow to green and then to blue. In some ways, this test is similar to the dye-binding methods reported on p. 1029, but in this circumstance the term "pH error" is used to describe the binding of the ionized form because the dye behaves as if it were at a significantly higher pH in which a greater portion is in the ionized form.[12]

Historically, urine albumin measurements have been made when one first performs a concentration step and then measures the albumin by either electrophoretic or immunochemical procedures. The procedures were subject to severe protein loss, which was dependent on the type of concentrating apparatus used.[13] More recently urine albumin has been directly measured by radioimmunoassay and enzyme immunoassay techniques. A commercial kit (Diagnostic Products Corp., Los Angeles) that uses ^{125}I-labeled human albumin has been introduced. The labeled albumin competes for a limiting amount of antibody. The free and bound radioactive albumin are separated by precipitation with a second antibody using polyethylene glycol to accelerate the precipitation. The ELISA immunoassays have not been standardized for urine measurement as a commercial product. The simplest procedure appears to be a sandwich assay in which the affinity purified antibody is bound to a solid support.[14,15] The albumin in a highly diluted sample is allowed to react. This is followed by reaction with a second albumin-specific antibody labeled with peroxidase.

The electroimmunoassay has been presented as a proposed reference procedure[16] because (1) it has an acceptable coefficient of variation of 6.2% and 3.7% at concentrations of 13 and 37 g/L respectively, (2) its analytical recovery is 99% to 100%, and (3) it has negligible interference from substances, such as bilirubin and salicylate, that bind to albumin and little interference from hemoglobin, which can interfere with colorimetric determinations. Given the availability of automated nephelometers and adaptations of turbidimetric techniques to automated analyzers,[7-9] these two techniques may become more popular.

The dye-binding techniques have variable specificities for albumin because of the ability of other serum proteins to also bind the dyes and cause a color shift. Specificity of the bromcresol green method is good as shown by correlation with salt fractionation when sera from normal patients were analyzed.[11]

The question of specificity has become an issue with the bromcresol green method when used for certain populations such as those with nephrotic syndrome or end-stage renal disease.[17] Bromcresol purple seems to have overcome most disadvantages seen in other dye-binding methods. The reaction conditions for this assay are given in Table 63-2. Given the specificity of the reaction without fast absorbance readings, the lack of many interferences, the excellent correlation with reference methods, and the precision and ease of use, bromcresol purple is the method of choice.

Albustix is probably the major method of albumin determination in any body fluid. This technique serves as a good initial screening for the presence of possible leakage of albumin into urine. False-positive results may be obtained with highly buffered or alkaline urines. Urines containing proteins such as Bence Jones proteins may not give positive results. The techniques of radioimmunoassay or ELISA assay appear to be the only current approaches suitable for the measurement of microalbuminuria.

Table 63-2 Reaction conditions for albumin analysis

Condition	Requirement
Temperature	Room temperature
pH	5.20 ± 0.03
Final concentration of reagent components in cuvette	Bromcresol purple: 63 μmol/L Brij-35: 196 mg/L Polybrene: 39 mg/L Acetate: 78 mmol/L
Fraction of sample volume	0.039
Sample	Serum, heparinized plasma
Linearity	5 to 50 g/L
Reaction time	Immediate
Major interferences	Gross hemolysis, gross lipemia, bilirubin greater than 100 mg/L
Precision (between day)	Coefficient of variation (CV): 2.2% at 30 g/L

Table 63-3 Reference ranges for albumin

Men			Women		
Age	Albumin, g/L	mmol/L	Age	Albumin, g/L	mmol/L
21-44	33.3-61.2	0.50-0.92	20-44	27.8-56.5	0.42-0.85
45-54	29.1-61.2	0.44-0.92	45-54	24.6-54.4	0.37-0.82
55-93	31.6-54.6	0.48-0.82	55-81	32.0-52.8	0.48-0.80

SPECIMEN

Serum is the specimen of choice, but heparinized plasma can also be used if precautions are taken to prevent heparin interference (see preceding discussion).

PROCEDURE: BROMCRESOL PURPLE

Principle

Bromcresol purple (BCP) complexes with albumin, resulting in the dye having a spectral shift. The presence of albumin increases the absorbance at 603 nm (Table 63-2).

Reagents[18,19]

Stock BCP solution (80 mmol/L). Dissolve 1.08 g of BCP (pH indicator grade) in 15 mL of absolute ethanol. When a clear orange solution is obtained, dilute to 25 mL with absolute ethanol. The reagent is stable at 4° C for at least 3 months.

Brij-35 solution (250 g/L). Dissolve 25 g of "Brij-35" (polyoxyethylene lauryl ether; Sigma Chemical Co., St. Louis) in distilled water, with warming, and dilute to 100 mL. Stable for 1 year at room temperature.

Stock acetic acid solution (2.5 mol/L). Dilute 150 mL of glacial acetic acid (AR grade) to 1 L with distilled water. Stable for 1 year at room temperature.

Working BCP reagent (BCP, 80 μmol/L; acetate buffer, 0.1 mol/L, pH 5.2). Dissolve 10 g of sodium acetate trihydrate (analytical reagent grade) in about 800 mL of distilled water. Add 10 mL of stock acetic acid solution, 1 mL of Brij-35 solution, 1 mL of stock BCP solution, and 50 mg of hexadimethrine bromide (Polybrene; Aldrich Chemical Co., Milwaukee), and dilute to 1 L with distilled water. Check pH and adjust to 5.20 ± 0.03 with stock acetic acid solution or 0.5 mol/L sodium hydroxide solution. The reagent is stable for at least 1 week at room temperature.

Sodium chloride solution (0.154 mol/L). Dissolve 9 g of sodium chloride in a final volume of 1 L of distilled water. Stable for 1 year at room temperature.

Standardization. At present, there is no adequate human-based albumin standard. Any native serum can be used, as long as it is calibrated against a monometric albumin solution standardized by a reference method such as EID. For standardization that might be less than optimal, highly purified human serum albumin fraction V might be used, if confirmed for purity by electrophoresis.

Assay

Equipment: GEMSAEC Centrifugal Analyzer (Electro-Nucleonics, Inc., Fairfield, NJ) is used with the following settings:

Sample volume	20 μL
Flush volume (saline)	90 μL
Reagent volume	400 μL
Wavelength	603 nm
Filter	560 to 700 nm
Number of readings	1
Starting absorbance	1.25

Linearity. The method is linear from 5 to 50 g/L.[20]

Interference.[19] For every 150 mg/L salicylate, 100 mg/L bilirubin, and 4.56 g/L hemoglobin, albumin is decreased by 1 g/L. For severely lipemic sera (10 g/L neutral fat) the "apparent albumin" value is increased by 2 g/L.

REFERENCE RANGE

The reference range by age and sex is shown in Table 63-3.[21]

The normal ranges for children and neonates have been reported to be slightly lower than the ranges seen in adults (newborns, 29 to 55 g/L [0.44 to 0.83 mmol/L]; children, 38 to 55 g/L [0.57 to 0.80 mmol/L]).[22]

Serum albumin values are lower in the recumbent (lying-down) patient compared to ambulant ones. Some laboratories take note of the difference between inpatients and outpatients. For example, Walmsley and White[23] use a reference range of 30 to 50 g/L for inpatients and 37 to 52 g/L for outpatients. A difference of 5 g/L between recumbent and ambulant values has been used.[24]

REFERENCES

1. Peters, T, Jr: Serum albumin, Adv Clin Chem 13:37-111, 1970.
2. Peters, T, Jr: Serum albumin. In Putnam, FW, editor, The plasma proteins, vol 1, New York, 1975, Academic Press, Inc.
3. Watson, D: Albumin and "total globulin" fractions of blood, Adv Clin Chem 8:237-303, 1965.
4. Henry, RJ: Proteins. In Henry, RJ: Clinical chemistry: principles and technics, New York, 1964, Harper & Row, Publishers, Inc.
5. Goldenberg, H, and Drewes, PA: Direct photometric determination of globulin in serum, Clin Chem 17:358-362, 1971.
6. Savory, J, Heintges, MG, and Sobel, RE: Automated procedure for simultaneously measuring total globulin and total protein in serum, Clin Chem 17:301-306, 1971.
7. Blom, M, and Hjørne, N: Immunochemical determination of serum albumin with a centrifugal analyzer, Clin Chem 21:195-198, 1975.
8. Kamp, HH, Luderer, TKJ, Muller, HJ, and Sopjes-Kruk, A: Rapid immunoturbidimetric assay of albumin and immunoglobulin G in

serum and cerebrospinal fluid with an automatic discrete analyzer, Clin Chim Acta 114:195-205, 1981.
9. Dito, WR: Rapid immunonephelometric quantitation of eleven serum proteins by centrifugal fast analyzer, Am J Clin Pathol 71:301-308, 1979.
10. Rodkey, FL: Direct spectrophotometric determination of albumin in human serum, Clin Chem 1:478-487, 1965.
11. Doumas, BT, and Biggs, HG: Determination of serum albumin, Standard Methods of Clinical Chemistry 7:175-188, 1972.
12. Peters, T, Jr, Biamonte, GT, Durnan, SM, et al: Protein (total protein) in serum, urine and cerebrospinal fluid: albumin in serum. In Faulkner, WP, and Meites, S, editors: Selected methods of clinical chemistry, vol 9, Washington, DC, 1982, American Association for Clinical Chemistry, Inc.
13. Pesce, AJ, and First, MR: Proteinuria an integrated review, New York, 1979, Marcel Dekker, Inc.
14. Puri, A, Casburn-Budd, R, Eisen, V, and Slater, JDH: Simpler measurement of albumin in urine or plasma, Clin Chem 31:1241-1242, 1985.
15. Feldt-Rasmussen, B, Dinesen, B, and Deekert, M: Enzyme immunoassay: an improved determination of urinary albumin in diabetes with incipient nephropathy, Scand J Clin Lab Invest 45:539-544, 1985.
16. Pascucci, MW, Grisley, DW, Jr, and Rand, RN: Electroimmunoassay of albumin in human serum: accuracy and long term precision, Clin Chem 29:1787-1790, 1983.
17. Webster, DA: A study of the interaction of bromocresol green with isolated serum globulin fractions, Clin Chim Acta 53:109-115, 1974.
18. Pinnel, AE, and Northam, BE: New automated dye-binding method for serum albumin determination with bromcresol purple, Clin Chem 24:80-86, 1978.
19. Duggan, J, and Duggan, PF: Albumin by bromcresol green: a case of laboratory conservatism, Clin Chem 28:1407-1408, 1982 (letter).
20. Cederblad, G, Hickey, BE, Hollender, A, and Akerlund, G: Improved continuous-flow (SMAC) determination of serum albumin, Clin Chem 24:1191-1193, 1978.
21. Denko, CW, and Gabriel, P: Age- and sex-related levels of albumin, ceruloplasmin, a_1-antitrypsin, a_1-acid glycoprotein, and transferrin, Ann Clin Lab Sci 11:63-68, 1981.
22. Meites, S, editor: Pediatric clinical chemistry, Washington, DC, 1977, American Association for Clinical Chemistry.
23. Walmsley, RN, and White, GH: A guide to diagnostic clinical chemistry, Victoria, Australia, 1983, Blackwell Scientific Publications.
24. Jacobs, DS, Kasten, BL, Jr, Demott, WR, and Wolfson, WL: Laboratory test handbook with DRG index, St Louis, 1984, The CV Mosby Co.

Carcinoembryonic antigens (CEA)

ALLYN H. RULE

Clinical significance: pp. 398, 587, and 730
Molecular weights: Colon CEA, 160,000 to 180,000 daltons
Ovarian CEA, 200,000 daltons
Breast CEA, 90,000 to 105,000 daltons
Chemical class: glycoprotein

PRINCIPLES OF ANALYSIS

Cancer-specific antigens of the digestive tract were first discovered by Gold and Freedman in 1965.[1,2] When these antigens were also found in fetal gut, they were collectively called carcinoembryonic antigens (CEA).[2] The first CEA-reacting molecule that was extensively characterized by this group was obtained from an ovarian carcinoma.[3] Later reports indicated that colon-CEA showed a lower molecular weight range (160,000 to 180,000 daltons).[4]

Although the collective term for these antigens, CEA, has been used from the beginning, these cancer-associated molecules have always existed in different forms.[5] Molecular weight determinations, isoelectric focusing, electron microscopy after concanavalin A affinity chromatography, and amino acid sequencing have all indicated major differences in CEA-reacting molecules.[6] For this reason, not all CEA-containing sera from cancer patients will produce linear dilution curves in parallel with CEA standard curves,[7] nor will the results obtained from one CEA-ligand assay be identical with another. The most recent data reveal organ-specific CEA differences in molecular weight and subunit specificities that result in reproducible molecular heterogeneity.[8]

In 1969, a radioimmunoassay (RIA) was developed to measure PCA-extracted CEA (method 1, Table 63-4) in sera obtained from patients with adenocarcinoma of the colon.[9] RIA used to investigate CEA-reacting molecules soon showed that many CEA-reacting molecules could be identified from fetal gut extracts obtained at different developmental stages in utero. RIA of saline extracts from patients with 20 primary adenocarcinomas of the colon showed 8 to 12 CEA peaks by isoelectric focusing.[6] It is now known that three of these bands are cancer specific, four belong to a family of normal colon cross-reacting antigens (CEA), which may also be produced in excess during carcinogenesis, whereas other CEA-reacting molecules are bound to serum lipoproteins, membranes, and immune complexes.

A number of assays have used perchloric acid (PCA) (method 1, Table 63-4), heat (method 2, Table 63-4), or both, to extract serum glycoproteins, including CEA, into the fluid phase while precipitating 90% to 98% of the other serum proteins (globulins and albumins).[10,11] Both methods inactivate proteolytic enzymes, cause antigenic loss of different CEA molecules, and remove proteins that would otherwise cause nonspecific protein binding.

Although PCA releases CEA from immune complexes and allows a more sensitive assay for CEA, a number of approaches have been used to resolve the problem of removing *all* the acid before assay at an ion-free neutral pH (6.5 to 6.8). Dialysis is the least expensive but most universally disliked method (method 1a, Table 63-4).

Gel exclusion chromatography can be used for PCA removal (method 1b, Table 63-4). Amicon ultrafiltration cuvettes can be used to remove acid from CEA extracts (method 1c, Table 63-4).

Heat extraction procedures (method 2, Table 63-4) entail heating the sample to 70° C. This precipitates most proteins, while allowing free CEA to remain in the supernatant.

Direct assay of CEA (because plasma extraction is not

Table 63-4 Methods for carcinoembryonic antigens (CEA) analysis

Immunoassay	Type of analysis	Principle	Usage	Comment
1. Indirect RIA assay (acid extraction)	Sequential binding assay	Plasma is extracted to remove ≥95% of all other proteins and to release CEA from immune complexes. After centrifugation and removal of all acid, antibody first reacts with "cold" CEA for 30 minutes at 45° C and then with ^{125}I-CEA for 30 minutes at 45° C. Heat speeds the reaction; sequential addition gives better low-end sensitivity.	Plasma or serum	More CEA is available to measure. Assay is sensitive to pH, ions, and nonspecific protein quench.
Acid removal				
a. Dialysis		Dialysis allows acid to diffuse totally out of cellophane pores after many changes of water and a final buffer step.		Method is cheap but tedious and labor intensive.
b. Column chromatography		CEA is separated from acid based on molecular weight differences; CEA comes out in first fraction.		Columns are moderately expensive and must be carefully washed or they can become contaminated. CEA values may be lower.
c. Ultrafiltration		CEA is retained on top of filter or inside of bag while acid is removed. Serum CEA "sticks" to filter.		Expensive equipment; CEA values are significantly lower.
2. Indirect RIA assay (heat extraction)	Sequential binding	Heat (70° C) causes most serum proteins to precipitate out; after centrifugation, assay performed on supernatant as in method 1.	Plasma or serum	Immune-complex CEA is not released to supernatant.
3. Direct RIA assay	Sequential binding	Plasma is directly placed into distilled water with antibody, and assay proceeds similarly to that in method 1.	Plasma or serum	Used to monitor patients with CEA ≥20 ng/mL because of lack of sensitivity in low range; sensitive to ions, pH, and protein effects.
4. Solid-phase extraction	"Sandwich" immunoassay	Antigen binds to a monoclonal anti-CEA fixed on a bead. A second anti-CEA labeled with ^{125}I or enzyme attaches to a second site in either a single or sequential incubation. For an enzyme immunoassay dye-substrate is added after 3 wash steps. Reaction is stopped and read. For RIA, tube is counted in gamma-ray scintillation counter.	Plasma or serum	Washing is cumbersome; dye substrate may be carcinogenic. Vapors from washing machine must be trapped for health safety.

needed) is employed for samples containing ≥20 ng CEA/mL (method 3, Table 63-4). A sequential radioimmunoassay uses two 30-minute incubations at 45° C—the first with antibody and unlabeled CEA, the second with ^{125}I-CEA. A final incubation with a second antibody is employed to precipitate antibody complexes and separate bound from free CEA.

Solid-phase assay of CEA (method 4, Table 63-4) employs antibodies to CEA that are irreversibly bound to beads or an inner surface of a test tube. These bound antibodies selectively remove CEA from the fluid phase. Nonspecific protein binding must be eliminated by a series of washes before addition of a second anti-CEA, which contains a label. The second anti-CEA binds to additional antigenic determinants on CEA to form a ''sandwich.'' In some assays, both CEA and the labeled antibody can be added simultaneously in the first step before washing. At this point, tubes can be counted for RIA or color developed for enzyme immunoassay (EIA) (method 4, Table 63-4). In the latter procedure, a dye substrate is added to the tube in a second or third step before the spectrophotometric determination of color. The color developed is directly proportional to CEA content. The solid phase is thus an ''extracting agent'' though serum or plasma is ''directly'' added to the assay tube.

No two kits or experimental assays obtain the same quantitative results for serum CEA. Since cancer patients are monitored over long periods of time (5 to 10 years), considerable thought must be given before a laboratory changes to an alternative kit or method. The newer assay may not be suitable for the same patient population or indeed the same patient.

The importance of the ability of an assay to measure CEA found in immune complexes cannot be overemphasized. In the region of 2.5 to 10.0 ng of CEA/mL, where changing CEA values most clearly indicate early tumor progression, the CEA assay has the greatest technical difficulties. Yet at this point the CEA-ligand assay is more cost effective than the CT scan, ultrasound, proctoscopy, or clinical examination.

Differences in quantitative results between the various assays on the market can be explained by several experimental facts. An assay that uses either an acid- or heat-extraction step, and a ''direct'' (nonextraction) assay that uses either monoclonal or polyclonal antibodies may all give different CEA values with the same CEA-containing serum. The different forms of CEA-reacting molecules have different heat or acid stabilities. Additionally, acid will extract CEA from an immune precipitate, whereas heat will not. Most direct assays (including the solid-phase assays) will not be so sensitive below 20 ng of CEA/mL because CEA complexed to immunoglobulins, lipoproteins, or membrane proteins will have lost antigenic sites for assay. When comparing indirect and direct assays, a direct assay for CEA will yield higher CEA values 60% of the time, will yield the same results 25% of the time, and will yield lower CEA values 15% of the time.[12]

The Roche (Roche Diagnostics, Nutley, NJ) and Abbott (Abbott Corporation, North Chicago) EIA's employ solid-phase monoclonal ''sandwich'' assays using anti-CEA coated beads in an EIA form.

An analysis of quality control data illustrates the problems inherent in the various CEA kit methodologies: (1) the CEA values do not agree for all tests, (2) the coefficients of variation are not widely divergent, but (3) the ranges of values vary widely. These data indicate that (1) a laboratory cannot casually change from one CEA method to another without making it difficult for the physician to interpret long-range results and that (2) the EIA procedure does not clearly resolve CEA values less than 2.5 ng/mL. Clear resolution of values in this range is important because they are used to distinguish between normal values and early malignant progression.

STANDARDIZATION

The direct assay of CEA is exceptionally sensitive to ions, pH, colloidal metals, bacterial products, and organic molecules. The water should be filtered before use because bacteria can grow in distilled water systems. Additionally, improperly deionized tanks may leak resins. In troubleshooting problems with the CEA assay, one must thoroughly check the water as a possible cause of problems.

Many problems arise in the control of CEA values. Calibrators must be independent of the current laboratory methods whether they are internally devised or obtained from an external commercial source. Additionally, because cancer is considered to be cured after a 5-year interval, standards must be available in sufficient quantity to assure long-term quality control. This is particularly true if one contemplates changing CEA methodologies, manufacturers, or source materials. Because of the heterogeneity of CEA-reacting molecules, it is well known that CEA values obtained from one method cannot be obtained with a second method and a correction factor cannot be applied among different standards or different assays. Above all, one should not obtain reference CEA from the same company from which one obtains reagents because the ranges of the reference material will be established by the same assay used to evaluate the reagents. It will be impossible to determine if assay results have been ''drifting'' over the years.

How should one proceed? Obtain your own hospital or laboratory standards. First realize that CEA has traditionally been quantitated in immunochemical reactivity in nanograms of CEA per milliliter. This has no reality in terms of nanograms of CEA per milligrams of protein. The carbohydrate-to-protein ratio may vary from 1:1 to 1:4 in different preparations. CEA purified from different tumors will have different immunological reactivities because of the different number of CEA-reacting antigenic sites: that is, ovarian CEA > colon CEA > breast CEA. However, in terms of serum CEA: colon CEA > breast CEA > ovarian CEA. Additionally, spiking commercially obtained CEA into CEA-free plasma will give spurious effects because of the variability of these preparations, lyophilization aggregation, nonspecific heterologous reactions, and deleterious matrix effects.[8]

To create a CEA standard, first obtain a fluid containing an exceptionally high amount of CEA from a patient with colon cancer. Dilute the fluid in distilled water to achieve a final concentration that falls below the high end of the standard curve with the addition of the appropriate amount of CEA-free plasma required for that assay. This latter addition controls nonspecific protein quench. CEA at 0.5, 2.0, 4.0, and 8.0 ng/mL should be used for assessment of low-end sensitivity. These CEA values should parallel the standard curve. Our group uses an ascites fluid that contains 16,000 ng of CEA/mL, as measured by perchloric acid extraction. Aliquots of 0.5 mL of 1:100 (perchloric acid assay) and 1:200 (direct assay) dilutions are frozen and used monthly for monitoring of CEA assays, along with five previously analyzed CEA-containing plasma samples. Then 100, 50, 20, and 6 μL of these standards are delivered into the appropriate assay tubes along with the appropriate amounts of CEA-free plasma normally required for that assay, that is, 0.5 mL of plasma for perchloric acid extraction or 50 to 200 μL for various "direct" or unextracted CEA ligand assays. The small amount of extra volume (water) does not alter the CEA measured by any of the currently available commercial CEA assays, nor does it affect the protein matrix (ions, pH, or salt) if deionized distilled water is used as diluent.

Second, the reference CEA prepared in the laboratory should be experimentally tested by a reference laboratory whose CEA standard has been experimentally calibrated against early Phil Gold or Roche-CEA preparations from the past 17 years. If the laboratory chooses to purchase reference controls in quantity, these also should be sent out to a well-standardized, qualified laboratory.

Finally, standardized CEA reference serum should be run once monthly with the assay of your choice or at any time when CEA values (or curves) seem to drift. Sufficient standards should be obtained to last for 6 years. However, in the fifth year new reference control materials should be calibrated against old standards so that continuity in monitoring patient samples is assured.

SPECIMEN

EDTA plasma is the preferred sample for CEA analysis because calcium, which is a necessary cofactor in the activation of a serum proteolytic enzyme, plasmin, is chelated by this anticoagulant. CEA has been known to undergo proteolytic degradation in serum samples that have not been carefully centrifuged and immediately frozen at −20° C. Serum samples must also be kept in 0° to 4° C water baths during analysis so that if a repeat CEA value

Table 63-5 Quality control of carcinoembryonic antigens (CEA): a comparison of EIA and RIA methods currently used*

Method	Median or mean CEA (ng/mL)	Standard deviation	Coefficient of variation	Range (ng/mL)
Specimen K-7				
Abbott-EIA	43.7	4.94	11.3	28.6-57.8
Abbott-RIA	39.7	3.55	9.0	31.0-50.9
Roche/Indirect (Clinetics)	8.4	1.25	15.0	5.9-12.0
Roche/Indirect (dialysis)	8.9	—	—	8.2-12.4
Specimen K-8				
Abbott-EIA	56.1	6.41	11.4	38.4-80.0
Abbott-RIA	51.4	4.97	9.7	36.5-65.3
Roche/Indirect (Clinetics)	10.3	1.81	17.2	5.6-15.7
Roche/Indirect (dialysis)	11.1	—	—	9.7-15.3
Specimen K-9				
Abbott-EIA	2.07	0.38	18.3	0.7-3.7
Abbott-RIA	2.66	0.47	17.6	1.6-4.0
Roche/Indirect (Clinetics)	2.04	0.58	28.6	0.3-3.5
Roche/Indirect (dialysis)	2.6	—	—	1.6-3.7
Specimen K-9 (less than 2.5 ng/mL values)				
Abbott-EIA	2.2	—	—	0.5-4.0
Abbott-RIA	3.0	—	—	1.5-5.0
Roche/Indirect (Clinetics)	2.5	—	—	0.5-2.5

*Modified after College of American Pathologists 1984 survey for CEA.

is required or future dilutions are necessary, the samples have been carefully preserved. Plasma CEA values and serum CEA values are not identical. Because the laboratory cannot always control the manner in which blood samples are obtained or handled, plasma is preferred.

REFERENCE RANGE

Data from the first clinical study that used an RIA to measure CEA in cancer patients[9] indicated that a specific, sensitive, precise, and accurate assay with a normal cutoff value of 2.5 ng of CEA/mL might be used to distinguish healthy people,[13] or those with other tumors, from patients with gastrointestinal cancers. Another study[14] indicated that a 2.5 ng of CEA/mL cutoff value would be appropriate in nonsmokers but would not distinguish patients with gastrointestinal cancers from those with cancers of other sites. Results shown in the CAP survey (Table 63-5) indicate that a 4.0 ng of CEA/mL cutoff value might be more appropriate for CEA-EIA. However, since the rising or falling values of CEA predict exacerbation or regression for an individual patient, it is critical that laboratories maintain the low-end sensitivity of their operating CEA assay.

The regular and sequential use of plasma CEA is the best noninvasive technique currently available for postoperative surveillance of patients to detect disseminated recurrence of colorectal cancer. As a monitor of colorectal cancer, CEA has been found to be raised when residual disease is present or clinically progressing. After complete surgical removal of a colorectal malignancy, a raised plasma CEA value should usually return to normal by 6 weeks. Failure to observe a reduction in a previously raised preoperative CEA titer strongly indicates the presence of residual tumor. In a substantial number of patients, CEA values also become significantly raised before metastatic disease can be detected by clinical or other diagnostic measures. This information can best be achieved by obtaining plasma samples for CEA before operation, 4 to 6 weeks after surgery, and thereafter at regular intervals as an integral component of overall patient follow-up. Whereas slowly rising values may be more indicative of local recurrence, rapidly rising values reaching very high concentrations, usually over 20 ng/mL, are found most often with hepatic and osseous metastases.

REFERENCES

1. Gold, P, and Freedman, SO: Demonstration of tumor specific antigens in human colonic carcinomata by immunologic and absorption techniques, J Exp Med 121:439, 1965.
2. Gold, P, and Freedman, SO: Specific carcinoembryonic antigens of the human digestive system, J Exp Med 122:467, 1965.
3. Krupey, J, Gold, P, and Freedman, SO: Physicochemical studies of the carcinoembryonic antigens of the digestive tract, J Exp Med 128:387, 1968.
4. Krupey, J, Wilson, T, Freedman, SO, and Gold, P: The preparation of purified carcinoembryonic antigen of the human digestive system from large quantities of tumor tissue, Immunochemistry 9:617, 1972.
5. Rule, AH: Carcinoembryonic antigen (CEA): activity of meconium and normal colon extracts, Immunol Commun 2:15, 1973.
6. Rule, AH, and Kirch, ME: Gene activation of molecules with carcinoembryonic antigen determinants in fetal development and in adenocarcinoma of the colon, Cancer Res 36:3503, 1976.
7. Vrba, R, Alpert, E, and Isselbacher, KJ: Immunological heterogeneity of serum carcinoembryonic antigens, Immunochemistry 13:87, 1976.
8. Rule, AH, Ball, HO, Rajarajan, B, and Diaz, S: An immunochemical comparison of colon, ovarian, and breast carcinoembryonic antigens (CEA), Presented at the XII annual meeting of the International Society for Oncodevelopmental Biology and Medicine, Houston, Texas, Oct 4, 1984.
9. Thompson, DMP, Krupey, J, Freedman, SO, and Gold, P: The radioimmunoassay of circulating carcinoembryonic antigens of the human digestive system, Proc Soc Natl Acad Sci USA 64:161, 1969.
10. Lo Gerfo, P, Krupey, J, and Hansen, JH: Demonstration of an antigen common to several varieties of neoplasia, N Engl J Med 285:138, 1971.
11. Tomita, JT, Kim, YD, Schenck, JR, et al: Ligand assay of CEA: solid phase immunoassays for CEA, J Clin Immunoassay 7:107, 1984.
12. Rule, AH, Nathanson, L, Binder, S, and Miller, HH: Comparison of CEA-Roche "indirect" and "direct" radioimmunoassays: a second look, N Engl J Med 295:1079, 1976.
13. Hansen, HJ, Snyder, JJ, Miller, E, et al: Carcinoembryonic antigen (CEA) assay: a laboratory adjunct in the diagnosis and management of cancer, Hum Pathol 5:139, 1974.
14. Von Kleist, S, and Burtin, P: Antigens cross-reacting with CEA. In Herberman, RB, and McIntire, KR, editors: Immunodiagnosis of cancer, vol 9, part 1, New York, 1979, Marcel Dekker, Inc.

Cerebrospinal fluid proteins—quantitation

GAYLE B. JACKSON

Clinical significance: pp. 587 and 594
Chemical class: protein

PRINCIPLES OF ANALYSIS

Cerebrospinal fluid (CSF) proteins originate primarily from the ultrafiltration of plasma across the choroidal capillary wall, although some proteins are peculiar to CSF and are synthesized in the central nervous system. The ultrafiltration process removes most plasma proteins so that the total protein concentration of CSF (150 to 450 mg/L) is much lower than that of serum (60 to 78 g/L).[1]

Total protein in CSF is most frequently measured by one of the turbidimetric procedures[1-3] (methods 1 and 2, Table 63-6). These procedures employ either sulfosalicyclic acid with or without sodium sulfate or trichloroacetic acid to form a protein precipitate in the sample. The turbidity of the precipitate is measured spectrophotometrically. The amount of turbidity produced by albumin differs from that produced by an equal mass of other globular proteins. However, it has been reported that the differences are less pronounced when trichloroacetic acid is used. Recently, the reagent benzethonium chloride has been suggested as a precipitating agent.[4,5]

Although the biuret method (method 3, Table 63-6) has a very low color yield in the range of CSF protein concentration, a Beckman Astra method, which uses a kinetic

Table 63-6 Methods of cerebrospinal fluid protein analysis

Method	Type of analysis	Principle	Usage	Comment
1. Sulfosalicylic: 3% sulfosalicylic and 7% sodium sulfate	Total protein, turbidimetric	Precipitation of protein with resultant turbidity measured in spectrophotometer	CSF (500 μL), commonly used	Sulfosalicylic acid alone results in greater turbidity with albumin than with globulin. Same phenomenon is seen with sulfosalicylic acid–sodium sulfate in combination
2. Trichloroacetic acid	Total protein, turbidimetric	Precipitation of protein with resultant turbidity measured in spectrophotometer	CSF (500 μL), commonly used	Less sensitivity and poorer reproducibility than sulfosalicylic acid–sodium sulfate
3. Biuret	Total protein	Protein reacts with copper in alkaline solution	Rare	Low color yield restricts usage to very sensitive instruments with rate-measurement capability
4. Lowry method	Total protein, chemical	Uses Folin phenol reagent; (1) protein reacts with copper in alkaline solution; (2) copper-protein complex and any tyrosine and tryptophan present reduce phosphotungstic-phosphomolybdic acids (A_{max}, 750 nm)	CSF (200 μL), rarely used	Time consuming; influenced by endogenous phenols and drugs
5. Coomassie brilliant blue dye binding	Total protein, chemical, spectrophotometric	When Coomassie brilliant blue G-250 binds to protein, color changes from brownish orange to intense blue color, absorbance of dye shifts from 465 to 595 nm	CSF (25-100 μL), infrequently used	Problem with standardization; different proteins show variability in sensitivity and variation in standard curves

method and very sensitive optics, has been described for CSF proteins.[5]

The Lowry method[1,6] is used in Europe but is infrequently used in the United States (method 4, Table 63-6). The method uses Folin-Ciocalteu reagent and involves the following two steps:

1. Protein reacts with copper in an alkaline solution.
2. Copper-protein complex and any tyrosine and tryptophan present reduce phosphotungstic-phosphomolybdic acids to a colored product, with an A_{max} at 750 nm.

Dye-binding methods have also been reported (method 5, Table 63-6). A shift in the absorption maximum from 465 to 595 nm is observed when protein is bound to the dye Coomassie brilliant blue G-250. The change in absorbance at 595 nm is used to measure the amount of protein present.[7,8] A number of modifications have been made to decrease the variability of the color yield with different proteins. These include the addition of sodium dodecylsulfate or methanol and the use of different dyes such as Serva Blue G.[9]

Table 63-7 compares the reaction conditions for two widely performed turbidimetric assays for CSF protein.

SPECIMEN

Cerebrospinal fluid can be stored between 2° and 8° C for at least 5 days if protected from evaporation. Specimens that will not be tested within 5 days should be frozen at −20° C immediately after collection.

PROCEDURE: SULFOSALICYLIC ACID TURBIDIMETRIC ASSAY

Principle

Protein is precipitated as a fine white precipitate by the addition of sulfosalicylic acid. The resulting turbidity is determined spectrophotometrically at 430 nm.

Reagents

Sulfosalicylic acid, 30 g/L (199 mmol/L). Dissolve 30 g of sulfosalicylic acid in 800 mL of distilled water, and bring to 1 L. Store in a brown bottle. Stable for 1 year at room temperature.

Control. Ortho Diagnostics CSF control (Ortho Diagnostics, Raritan, NJ).

Assay

Equipment: spectrophotometer capable of reading at 430 nm.

Table 63-7 Comparison of reaction conditions for cerebrospinal fluid protein*

Condition	Sulfosalicylic acid	ACA†
Temperature	Ambient	37°
pH	Acid	Acid
Final concentration of reagents	Sulfosalicylic acid, 98.4 mmol/L	Trichloroacetic acid, 0.14 mol/L
Wavelength (nm)	430	340 and 540
Fraction of sample volume	0.166	0.06
Volume of sample	0.2 mL	0.3 mL
Linearity	0-2000 mg/L	0-2000 mg/L
Time of reaction	10 minutes	40 sec
Interferences	Xanthochromia (−) Turbidity (+) Hemolysis (−) Methotrexate (+)	Xanthochromia (−) Turbidity (+) Hemolysis (−)
Precision ($\bar{X}$, % CV)‡	501 mg/L, 7.3%	500 mg/L, 4%† 200 mg/L, 2%

*(−), Negative interferent; (+), positive interferent.
†From du Pont Instruments, Inc, Wilmington, DE.
‡$\bar{X}$, Mean; *CV*, coefficient of variation.

1. Add 1.0 mL of 3% sulfosalicylic acid to three "test" tubes—one for an instrument blank, one for the Ortho standard, and one for the patient.
2. Add 0.2 mL of Ortho Diagnostics CSF control to the standard tube (see note 1).
3. Add 0.2 mL of CSF to the patient's tube. Add 0.2 mL of saline solution to blank tube.
4. If the spinal fluid is strongly colored, prepare a patient blank using 0.2 mL of CSF and 1.0 mL of saline.
5. Cover all test tubes with plastic cups or Parafilm, and gently invert several times.
6. Let stand 10 minutes at room temperature.
7. Set absorbance at zero on the spectrophotometer at 430 nm using the instrument blank of sulfosalicylic acid plus saline.
8. Invert test tube gently to mix, introduce sample into cuvette, and read absorbance *(A)* of all standards and patient samples.
9. Dilute samples 1:2 with saline, and rerun if concentration is greater than 2000 mg/L.

Calculations

1. Correct absorbance of samples with blank:

$$A_{\text{patient}} - A_{\text{patient blank}} = A_{\text{corrected patient}}$$

2. Read the corrected absorbance from the standard curve (see next section) to obtain the protein concentration.

Controls and standards

1. Because sulfosalicylic acid produces different degrees of turbidity with equal concentrations of different types of CSF proteins, it is necessary to construct a curve using a standard that has the same albumin-globulin ratio as the spinal fluid being tested. Usually spinal fluid and serum have a very similar albumin-globulin ratio so that a normal serum control with the proper dilutions can be used to construct the curve. The albumin-globulin ratio of the standard should be between 1.0 and 1.5.
2. To prepare the standard curve, dilute a control serum of 70 g/L of total protein with saline so that 5 dilutions with protein concentration between 100 and 1500 mg/L are prepared as follows:

Stock volume	Total volume (mL)	Concentration (mg/L)
2.0	100	1400
1.5	100	1050
1.0	100	700
0.5	100	350
0.2	100	140
0	100	0

Do not use a serum control with elevated lipid or bilirubin content.

3. Construct the standard curve, plotting absorbance at 430 nm versus protein concentration. This should be a straight line that passes through the origin.
4. The standard curve needs to be run only when there is a change in the spectrophotometer (that is, bulb change) or that lot of quality control material or the quality control sample does not give the expected value.

Notes

1. The first drop of spinal fluid should be added slowly to the sulfosalicylic acid reagent. If a dense cloud forms immediately, the protein concentration is very high and dilution will be necessary. In this case, the rest of the 0.2 mL sample should not be added to the acid but instead returned to the original tube. The spinal fluid should be diluted with saline.
2. If red blood cells are present in the spinal fluid, the fluid must be centrifuged before protein analysis.

REFERENCE RANGE

CSF protein levels in adults usually range between 150 and 450 mg/L. CSF protein concentration of newborns (0 to 1 month) has been reported to be much higher, 600 to 1200 mg/L.[10,11]

REFERENCES

1. Henry, JB: Clinical diagnosis and management of laboratory methods, Philadelphia, 1979, The WB Saunders Co.
2. Henry, RJ, et al: Turbidimetric determination of proteins with sulfosalicylic and trichloroacetic acids, Proc Soc Exp Biol Med 92:748-751, 1956.
3. Schriever, H, and Gamblino, SR: Protein turbidity produced by trichloroacetic acid and sulfosalicylic acid at varying temperatures and varying ratios of albumin and globulin, Am J Clin Pathol 44:667-672, 1965.
4. Flachaire, E, Damour, O, Bienvenu, J, et al: Assessment of the ben-

zethonium chloride method for routine determination of protein in cerebrospinal fluid and urine, Clin Chem 29:343-345, 1983.
5. Finley, PR, and Williams, RJ: Assay of cerebrospinal fluid protein: a rate biuret method evaluated, Clin Chem 29:126-129, 1983.
6. Lowry, OH, et al: Protein measurement with Folin phenol reagent, J Biol Chem 193:265, 1951.
7. McIntosh, JC: Application of a dye-binding method to the determination of protein in urine and cerebrospinal fluid, Clin Chem 23:1939-1940, 1977.
8. Spector, T: Refinement of the Coomassie blue method of protein quantitation, Anal Biochem 86:142-146, 1978.
9. Stahl, M: Improved dye-binding procedure for determining microgram quantities of protein in cerebrospinal fluid—once again, Clin Chem 30:1878, 1984.
10. Friedman, HS: Total proteins in cerebrospinal fluid (colorimetric), Standard Methods of Clinical Chemistry 5:223-230, 1965.
11. Rice, EW: Total protein in cerebrospinal fluid (turbidimetric), Standard Methods of Clinical Chemistry 5:231-236, 1965.

Glycosylated hemoglobin

MARY ELLEN KING

Glycosylated hemoglobin, Ghb, Hb A_1
Clinical significance: p. 436
Molecular weight: 64,500 daltons
Merck Index: 4538 (hemoglobin)
Chemical class: glycosylated protein

HbA—Val—NH_2 + HCO—HCOH—HOCH—HCOH—HCOH—CH_2OH ⇌ HbA—Val—N=HC—HCOH—HOCH—HCOH—HCOH—CH_2OH —Amadori rearrangement→ HbA—Val—NH—CH_2—C=O—HOCH—HCOH—HCOH—CH_2OH

Glucose + Hb A | **Unstable Schiff's base** *Labile* | **Ketoamine** *Stable as Hb A_{1c}*

PRINCIPLES OF ANALYSIS

A wide variety of analytical techniques have been applied to the separation and quantitation of A_1 glycohemoglobins. Two basic approaches are used, either quantitation of the combined A_1 fraction (A_{1a} + A_{1b} + A_{1c} + A_{1d}) after its separation from Hb A, or separation and quantitation of the subcomponent A_{1c}. For separations of A_1, one can use either short cation-exchange columns[1,2] (method 1, Table 63-8) or electrophoresis[3,4] (method 2, Table 63-8). Typically, the sample is applied to a cation-exchange column such as BioRex 70 and the glycosylated subcomponents eluted with a low ionic strength phosphate buffer. The remaining hemoglobins are then eluted with a high ionic strength phosphate buffer. Each hemoglobin fraction is quantitated by direct spectrophotometry at 415 nm. The basis for separation by electrophoresis seems to lie in the ability of the free *N*-terminus of nonglycosylated hemoglobin to interact with sulfate groups, changing the molecule's electrophoretic mobility. A_1 hemoglobins cannot interact with these groups, and their mobility is unchanged. In these procedures the hemoglobins are exposed to sulfate groups in an agar-based support medium[3] or to dextran sulfate in the buffer used with a cellulose acetate support[4] and separated by short electrophoretic runs (20 to 45 minutes). Hemoglobin A_1 is separated as one band and measured by densitometry.

To separate A_{1c}, however, the additional resolution of high-performance liquid chromatography (HPLC)[5,6] (method 3, Table 63-8) or isoelectric focusing[7,8] is required. These A_{1c} procedures require more time and more sophisticated skills and equipment and are not generally used as routine procedures.

Separation of a fraction more closely approximating total glycosylated hemoglobin, including alpha-chain and lysine-glycosylated forms, is achieved by use of boronic acid affinity chromatography columns (method 4, Table 63-8).[9] The glycosylated hemoglobins bind by forming a strong

Table 63-8 Methods of hemoglobin analysis

Method	Basis of Hb separation	Basis of analysis	Type of glycohemoglobin measured
1. Minicolumn	Ion-exchange chromatography	Spectral absorbance of separated Hb components (415 nm)	A_1 + Labile A_1
2. Electrophoresis	Charge differences	Spectral absorbance of separated Hb components (415 nm)	A_1 + Labile A_1 S_1 and C_1 can also be measured
3. High-performance liquid chromatography	Ion-exchange chromatography	Spectral absorbance of separated Hb components (415 nm)	A_{1c} + Labile A_{1c}
4. Minicolumn	Affinity chromatography	Spectral absorbance of separated Hb components (415 nm)	A_1 + Non-A_1 glycohemoglobins Labile glycohemoglobin will not bind
5. Colorimetric	No Hb separation required, but interfering sugars must be removed before assay	Hb-bound glucose hydrolyzed, reacted with TBA, reaction monitored at 443 nm	A_1 + Non-A_1 glycohemoglobins Labile glycohemoglobin will not react

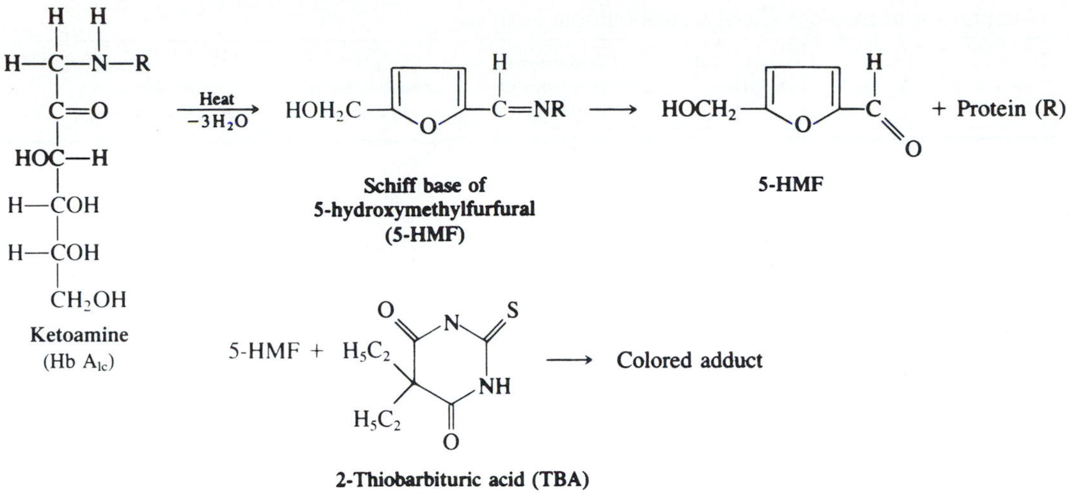

Ketoamine ($Hb\ A_{1c}$) $\xrightarrow[-3H_2O]{Heat}$ HOH_2C–(furan)–$\overset{H}{C}{=}NR$ ⟶ $HOCH_2$–(furan)–$\overset{H}{C}{=}O$ + Protein (R)

Ketoamine ($Hb\ A_{1c}$)

Schiff base of 5-hydroxymethylfurfural (5-HMF)

5-HMF

5-HMF + 2-Thiobarbituric acid (TBA) ⟶ **Colored adduct**

2-Thiobarbituric acid (TBA)

but reversible five-membered ring complex through their *cis*-glycol groups to immobilized boronic acid. The nonglycosylated hemoglobins are eluted first, and the glycosylated fraction is then eluted from the column by a sorbitol-containing buffer that competes with the bound glycosylated protein for the boronic acid-binding sites. A_{1c} and some glycosylated non-A_1 hemoglobins bind to the column, but A_{1a} and A_{1b} do not.

Quantitation of separated hemoglobin fractions, either by spectrophotometric techniques (chromatography) or densitometry (electrophoresis, isoelectric focusing), is based on the absorbance of hemoglobin at 415 to 420 nm.

Another analytical approach to determine total glycosylated hemoglobin is based on the release and quantitation of the bound sugar moieties rather than the separation and quantitation of the glycosylated hemoglobins.[10,11] The sample is dialyzed to remove free glucose, fructose, and sucrose, and the hemoglobin-bound sugars are hydrolyzed, converted to 5-hydroxymethylfurfural, and quantitated by a chemical reaction with 2-thiobarbituric acid (TBA), which is monitored at 443 nm (method 5, Table 63-8). The reaction is shown in the accompanying figure. This is the most frequently used nonchromatographic method for the analysis of glycosylated hemoglobin.

An alternative to measuring HbA_1 is to estimate total glycosylated serum protein using the fructosamine assay. This assay is based on the ability of the bound glucose to reduce nitroblue tetrazolium chloride at pH 10.35. This method has been adapted for use on the ABA-100 (Abbott Diagnostics, North Chicago). The results of the fructosamine assay compare well with methods that measure glycosylated hemoglobin by cation-exchange chromatography.[14]

An assay has been developed by Abbott Diagnostics based on the ability of inositol hexaphosphate (phytic acid, IHP) to bind at a specific binding point near the *N*-terminal region of the beta-chain.[14,15] When IHP binds to nonglycosylated hemoglobin to form a stable complex, the spectrum of hemoglobin shifts, with a resulting absorbance maximum at 577 nm. Glucose bound to the *N*-terminal amino group of hemoglobin will block the binding of IHP. The difference in absorbance at 577 nm before and after the addition of IHP is proportional to the concentration of Hb A_{1c}. This assay does not measure the other glycosylated hemoglobins or the labile form of Hb A_1. A difference in absorbance at 577 nm of 0.006 represents a 1% concentration of Hb A_{1c}. This assay, adapted to the ABA-100, requires glycosylated standards provided by Abbott and is not based on a percentage measurement. Because carboxyhemoglobin Hb CO causes a positive bias, the method requires the simultaneous measurement of Hb CO to correct for levels of this compound. This method is not commercially available.

The most generally accepted reference method is HPLC, which allows separation and quantitation of A_1 components.[5,6] Hb A_{1c} is measured separately or summed with a fraction containing A_{1a} + A_{1b} to give a total A_1 value (Table 63-9).

SPECIMEN

Blood should be collected with EDTA as anticoagulant. A hemolysate of saline-washed red cells is used. Either washed whole blood or the hemolysate may be stored for 4 to 7 days at 4° C and up to 30 days at −70° C. If a sample is not to be analyzed within 48 hours after it is received in the laboratory, the plasma should be removed. The cells should be washed once in 0.15 M (0.9%) saline and resuspended to the original blood volume with 0.15 M saline. Although much attention has been given to the presence of labile glycosylated hemoglobin, the amount present is generally very low. Typically, increases of 1% to 2% in apparent A_1 (expressed here as percent total hemoglobin) were seen both after glucose tolerance tests in normal volunteers and diabetics and after in vitro incuba-

Table 63-9 Comparison of methods for glycohemoglobin analysis

Method	Coefficient of variation (%)	Temperature dependence of separation	pH dependence of separation	Assay time of one sample	Interferences
HPLC (ion exchange)	3	Significant	Significant	10-20 min*	Hb F
Minicolumn ion exchange	2-16	Significant	Significant	20-40 min†	Hb F
Minicolumn affinity	1-3	Negligible	Minor	15-30 min†	None
Electrophoresis	4-10	Negligible	Minor	20-45 min†	Hb F
Colorimetric	4-18	‡	‡	hours†	None

*Samples cannot be batched.
†Samples can be batched.
‡No hemoglobin separation step.

tion of glucose with red blood cells.[16-19] Blood glucose values ranged up to 5000 mg/L in these studies. Similarly, small decreases in A_1 values are seen if the labile component is removed by dialysis from samples from diabetic patients and normal persons and the samples reassayed. One report does show a near doubling of normal A_1 levels after incubation with glucose, but only when levels of 10,000 mg/L are used.[20] If removal of the labile component is desired, one can use incubation or dialysis of the erythrocytes with physiological saline[16-19] or a much shorter incubation with semicarbizide pluse aniline at pH 5.[20]

REFERENCE RANGE

Reference ranges for glycosylated hemoglobin will vary greatly with the type of procedure employed. The amount of Hb A_1 present in nondiabetic adults, as measured by the temperature-controlled minicolumn assay, is 4.5% to 8.5%.[21] By electrophoretic separation the range is 5% to 7.6%.[3] Hemoglobin A_{1c} levels in nondiabetic persons have been reported to be 4.5% to 5.7% and 4% to 5.2% by HPLC[11] and isoelectric focusing[8] methods, respectively.

Total glycosylated hemoglobin as measured by affinity chromatography is reported to be 5.3% to 7.5%, whereas the colorimetric thiobarbituric acid procedure has a reference range of 20 to 22 nmol per 10 mg of hemoglobin.[9]

Since the percent Hb A_1 values by the minicolumn method are reported to be age dependent,[21] this may also be true for other methods. No sex or race dependence for glycosylated hemoglobin has been noted.

INTERFERENCES

Hemoglobin F will cause false-positive results for many methods, as listed in Table 63-9. Any increase in turnover of red blood cells, as in certain hemoglobinopathies or anemias, will give low values for glycosylated hemoglobin. Hemoglobin S and C can cause false-negative results by the minicolumn cation-exchange procedure.[22] Other chemical modifications of the *N*-terminal amino group of the beta-chain, as by salicylate, carbamate, and galactose, form products that can interfere in the minicolumn assay.[23]

Lactescence (from lipemia) will cause falsely elevated results in the minicolumn ion-exchange procedure. The red blood cells in these specimens should be washed with saline to remove the interfering lipemia.

REFERENCES

1. Jones, MB, Koler, RD, and Jones, RT: Micro-column method for the determination of hemoglobin minor fractions A_{1a+b} and A_{1c}, Hemoglobin 2:53-58, 1978.
2. Abraham, EC, Huiff, TA, Cope, JD, et al: Determination of the glycosylated (HB A_1) hemoglobins with a new microcolumn procedure, Diabetes 27:931-937, 1978.
3. Menard, L, Dempsey, ME, Blankstein, LA, et al: Quantitative determination of glycosylated hemoglobin A_1 by agar gel electrophoresis, Clin Chem 26:1598-1602, 1980.
4. Ambler, J: The importance of measuring hemoglobin A_1 in diabetes mellitus, Electrophoresis Today 3:1-2, 1982.
5. Davis, JE, McDonald, JM, and Jarett, L: A high performance liquid chromatography method for hemoglobin A_{1c}, Diabetes 27:102-107, 1978.
6. Dunn, PJ, Cole, RA, and Soeldner, JS: Further development and automation of a high pressure liquid chromatography method for the determination of glycosylated hemoglobins, Metabolism 28:777-779, 1979.
7. Simon, M, and Cuan, J: Hemoglobin A_{1c} by isoelectric focusing, Clin Chem 28:9-12, 1982.
8. Spicer, KM, Allen, RC, and Buse, MG: A simplified assay of hemoglobin A_{1c} in diabetic patients by use of isoelectric focusing and quantitative microdensitometry, Diabetes 27:384-388, 1978.
9. Klenk, DC, Hermanson, GT, Krohn, RT, et al: Determination of glycosylated hemoglobin by affinity chromatography: comparison with colorimetric and ion-exchange methods and effects of common interferences, Clin Chem 28:2088-2094, 1982.
10. Parker, KM, England JD, DaCosta, J, et al: Improved colorimetric assay for glycosylated hemoglobin, Clin Chem 27:669-672, 1981.
11. Burrin, J, Worth, R, Ashworth, L, et al: Automated colorimetric estimation of glycosylated hemoglobins, Clin Chim Acta 106:45-50, 1980.
12. Baker, JR, Metcalf, PA, Hollaway, IM, and Johnson, RN: Serum fructosamine concentration as measure of blood glucose control in type I (insulin dependent) diabetes mellitus, Br Med J 290:352-355, 1985.
13. Baker, JR, Metcalf, PA, Johnson, RN, et al: Use of protein-based standards in automated colorimetric determinations of fructosamine in serum, Clin Chem 31:1550-1554, 1985.
14. Gau, N, Kaplan, L, and Stein, E: Evaluation of a spectrophotometric assay for glycosylated hemoglobin, Clin Chem 30:977-978, 1984.
15. Deeg, R, Schmitt, U, Rollinger, U, and Ziegenhorn, J: A new approach to photometry of glycated hemoglobin in human blood, Clin Chem 30:790-893, 1984.
16. Svendsen, PA, Christiansen, JS, Welinder, B, and Nerup, J: Fast glycosylation of haemoglobin, Lancet 1:603, 1979.

17. Botterman, P: Rapid fluctuations in glycosylated haemoglobin concentration, Diabetologia 20:159, 1981.
18. Goldstein, DE, Peth, SB, England, JD, et al: Effects of acute changes in blood glucose on Hb A_{1c}, Diabetes 29:623-628, 1980.
19. Svendsen, PA, Christiansen, JS, Soegaard, U, et al: Rapid changes in chromatographically determined A_{1c} induced by short-term changes in glucose concentration, Diabetologia 19:130-136, 1980.
20. Nathan, DM, Avezzano, E, and Palmer, JL: Rapid method for eliminating labile glycosylated hemoglobin from the assay for hemoglobin A_1, Clin Chem 28:512-515, 1982.
21. Aleyassine, H: Low proportions of glycosylated hemoglobin associated with hemoglobin S and hemoglobin C, Clin Chem 25:1484-1486, 1979.
23. Gluckinger, R, Harmons, W, Meier, W, et al: Hemoglobin carbamylation in uremia, N Engl J Med 304:823-827, 1981.

Hemoglobin separation and quantitation

GERARDO PERROTTA

Clinical significance: pp. 512 and 688
Molecular weight: 64,000 daltons
Merck Index: 4538
Chemical class: heme protein

PRINCIPLES OF ANALYSIS

Three types of hemoglobins are normally found in erythrocytes: hemoglobins A, A_2, and F. They are all composed of two alpha and two nonalpha chains. Hemoglobin A, the most prevalent, is made of two alpha and two beta chains ($\alpha_2\beta_2$); HB A_2 constitutes up to 3.5% of erythrocyte hemoglobin and is made up of two alpha and two delta chains ($\alpha_2\delta_2$); Hb F normally constitutes up to 2% of the hemoglobin and is made of two alpha and two gamma chains ($\alpha_2\gamma_2$).

This section presents methods that will detect abnormal hemoglobins that are the result of an amino acid substitution, such as sickle cell hemoglobin, or hemoglobinopathies that are the result of aberrant globin-chain synthesis, as in the thalassemias. A combination of methods is usually necessary to separate and identify the presence of normal, abnormal, or variant hemoglobin. These include electrophoretic methods, nonelectrophoretic methods, structural analysis, and others.

Electrophoretic procedures

Electrophoretic methods separate hemoglobins based on differences in electrical charge. Specific methods may vary the strength of the electric field, pH, type of buffer, or the supporting medium used (methods 1 to 3, Tables 63-10 and 63-11).

The two most common electrophoretic techniques use alkaline and citrate buffer. Cellulose acetate[1] represents by far the most commonly used support medium for the alkaline-buffer procedure, whereas agar gel is preferred for the acid buffer. At alkaline pH, electrophoresis with cellulose acetate, agarose, or starch gel as the support phase separates HbA, A_2, and F. However, these conditions do not differentiate Hb S from hemoglobins D and G because all three have the same mobility in this system. Likewise, Hb C, E, and O have a mobility similar to the A_2 band at alkaline pH. Citrate agar (acid pH) electrophoresis is used to separate the hemoglobins that are not discriminated from each other in the alkaline cellulose acetate procedure. Thus citrate agar confirms the presence of hemoglobins D or G and differentiates hemoglobinopathy SD from SS and hemoglobin C from E.

After the actual electrophoresis, most procedures require fixation (denaturation) of the hemoglobin, followed by staining with such protein stains as Ponceau S, Paragon blue, or amido black 10 to render more visible the separated fractions. The position (relative mobility) and intensity of resultant hemoglobin bands are then compared to known controls. Quantitation is accomplished by use of a densitometer.

Confirmatory tests for specific hemoglobins

Nonelectrophoretic methods are used to quantitate specific hemoglobin fractions (methods 4 to 6, Table 63-10). The solubility test, column chromatography, and alkali denaturation are the ones most frequently used for hemoglobins S, A_2, and F, respectively.

Hemoglobin S. In 1953, Itano (Nalbrandian et al.[2]) reported that deoxy-Hb S is less soluble than normal Hb A when reduced by dithionite in phosphate buffer solution. In this reduced state, a turbid suspension of protein crystals is formed. These crystals disperse transmitted light, and the solution appears turbid. Therefore, when Hb S is present, the lines on a sheet of paper placed behind this solution are invisible. Hb A solutions, on the other hand, remain transparent, and one can easily see the lines. When this test is coupled with alkaline and acid electrophoresis, it confirms or rules out definitively the presence of hemoglobin S. The solubility test does not distinguish among the hemozygous SS, the heterozygous AS, SC, and S thalassemia states.

Hemoglobin A_2. Column chromatography provides an accurate method for quantitating hemoglobin A_2. This hemoglobin is separated on a small diethylaminoethyl (DEAE)-cellulose anion-exchange column. Hemoglobin solution is adsorbed onto the column, but Hb A_2, which has a different charge from the other hemoglobins, is not adsorbed under the chromatographic conditions employed and is eluted in the first fraction. A higher ionic strength buffer is then used to elute the remaining hemoglobins from the column. The percentage of Hb A_2 is determined from the absorbances of the eluates at 415 nm (Soret band). Gottfried et al.[3] reported that A_2 can also be quantitated by densitometry of the electrophoretic patterns when carefully controlled techniques and properly calibrated instruments are used.

Table 63-10 Methods of hemoglobin analysis

Method	Type of analysis	Principle	Usage	Comments
Electrophoretic				
1. Cellulose acetate	Quantitative-qualitative alkaline pH (8.5)	Negatively charged Hb migrate anodally: **Cathode** − | CA_2 (E,O) S (D,G Lepore) F A | + **Anode**	Whole blood prepared as hemolysate	Simple, fast, and most frequently used
2. Citrate agar	Qualitative acid pH (6.0)	Hb move from anode to cathode: − | F A E,D,G (Lepore) S C | +	Whole blood prepared as hemolysate	Not a good screening test; best used to distinguish S from D, A from F, and C from E or O
3. Globin chain	Qualitative alkaline or acid pH	Under reducing conditions in the presence of 6 mol/L urea, globin chains dissociate and are separated depending on pH of buffer: pH 8.9 − | αA βS γF βA | + pH 6.0 | αA βS βA γF |	Whole blood prepared as hemolysate	Separates alpha and nonalpha globin chain variants
Nonelectrophoretic				
4. Column chromatography	Quantitative for Hb A_2	Hb A binds more strongly to DEAE cellulose than A_2 does, resulting in a faster separation and elution of A_2.	Whole blood prepared as hemolysate	For thalassemia states
5. Alkali denaturation	Quantitative for Hb F	Hb F + Hb A + 1/12 mol/L NaOH → Precipitate (Hb A) Nondenatured Hb F is measured in supernatant at 415 nm.	Whole blood prepared as hemolysate	For thalassemia and other hemoglobinopathies
6. Solubility	Qualitative	Hb + $\xrightarrow[\text{Buffer, saponin}]{\text{Dithionite } Na_2S_2O_4}$ Turbid solution if Hb S is present	Whole blood	For Hb S

Table 63-11 Comparison of conditions for hemoglobin analysis by electrophoresis

Technique	CDC*	Helena	Beckman	Corning	Gelman
Alkaline					
pH	8.4	8.6	8.6	8.6	Not specified
Support media	Cellulose acetate	Cellulose acetate	Agarose	Agarose	Cellulose acetate
Application (volt/time)	450 V/20 min	450 V/15 min	200 V/25 min	240 V/20 min	400-425 V/20-30 min (membrane dependent)
Stain	Ponceau S	Ponceau S	Paragon blue stain	Amido black 10 colorimetric stain	Ponceau S
Acid					
pH	6.0-6.2	6.3	Not available	6.3	Not available
Support media	Citrate agar	Citrate agar	Not available	Citrate agar	Not available
Application (volt/time)	70-90 V/70 min	70 V/60 min	Not available	90 V/35 min	Not available
Stain	Benzidine diluted with distilled H_2O	*o*-Dianisidine	Not available	Amido black 10B colorimetric stain	Not available

**CDC,* Centers for Disease Control, Washington, DC.

Hemoglobin F. Alkali denaturation methods are most frequently used for the determination of Hb F. This technique is based on the principle that fetal hemoglobin is more resistant to denaturation under alkaline conditions than are other hemoglobins. After a strong alkali, such as KOH or NaOH, is added to a hemolysate, the denaturation is stopped by the addition of saturated or half-saturated ammonium sulfate, which results in a lower pH and a precipitated denatured hemoglobin. The denatured hemoglobin is removed by centrifugation, and the amount of unaltered hemoglobin is measured spectrophotometrically and is expressed as the percentage of alkali-resistant fetal hemoglobin.

Other confirmatory tests. Globin-chain electrophoresis, isoelectric focusing (IEF), and amino acid sequencing (fingerprinting) provide additional information for characterizing hemoglobins. These are performed in more specialized centers and are not part of a routine approach to hemoglobin analysis but are used to confirm the presence of rare hemoglobins. Globin-chain electrophoresis and IEF are the most frequently used methods. Globin-chain electrophoresis separates alpha and nonalpha globin chains by mixing a cell membrane-free hemolysate with 6 M urea in the presence of 0.05 M 2-mercaptoethanol. Fractionation takes place on cellulose acetate plates in Tris-EDTA-borate (TEB) buffer at pH 6 and 8.9. By comparing the electrophoretic pattern of the unknowns to some known Hb variant, one can make a characterization of the variant globin chain.

IEF is fast becoming a useful tool for identifying Hb mutants.[4] It is based on the principle that all proteins have a zero net charge in solution at a specific pH, called the "isoelectric point" (pI). The hemoglobin or hemoglobins being analyzed migrate toward their isoelectric points in a pH-gradient gel or support medium (most commonly a polyacrylamide gel). The band formed at this point is referred to as the "focused band." IEF has greater resolving power than either of the two electrophoresis techniques currently in use in most laboratories. The greatest advantage of the IEF technique seems to be in the detection and separation of those mutant hemoglobins that migrate with Hb A in cellulose acetate.

Other useful methods employed to differentiate rare abnormal hemoglobins include heat-stability tests, the isopropanol test, oxygen-affinity determination, and high-performance liquid chromatography.

Finally, erythrocyte size (mean corpuscular volume, MCV) and shape (poikilocytes), serum-iron studies, reticulocyte counts, and family studies are necessary and helpful to complete a definitive genetic picture of a suspected hemoglobinopathy or variant.

In studies comparing three methods for quantitating variant Hb fractions, Pearce[5] showed that cellulose acetate electrophoresis with densitometry was better than cellulose acetate with elution or microchromatography.

In 1980, the National Committee for Clinical Laboratory Standards (NCCLS) published a tentative standard for the quantitation of Hb A_2.[6] They considered the column chromatography technique as simple, precise, and accurate. However, their procedure did not address the issue of measuring Hb A_2 in the presence of Hb S. The NCCLS report did warn that trying to quantitate Hb A_2 from a sample containing hemoglobin S, C, or other slow variants could introduce error. It remains to be seen, in light of the recent

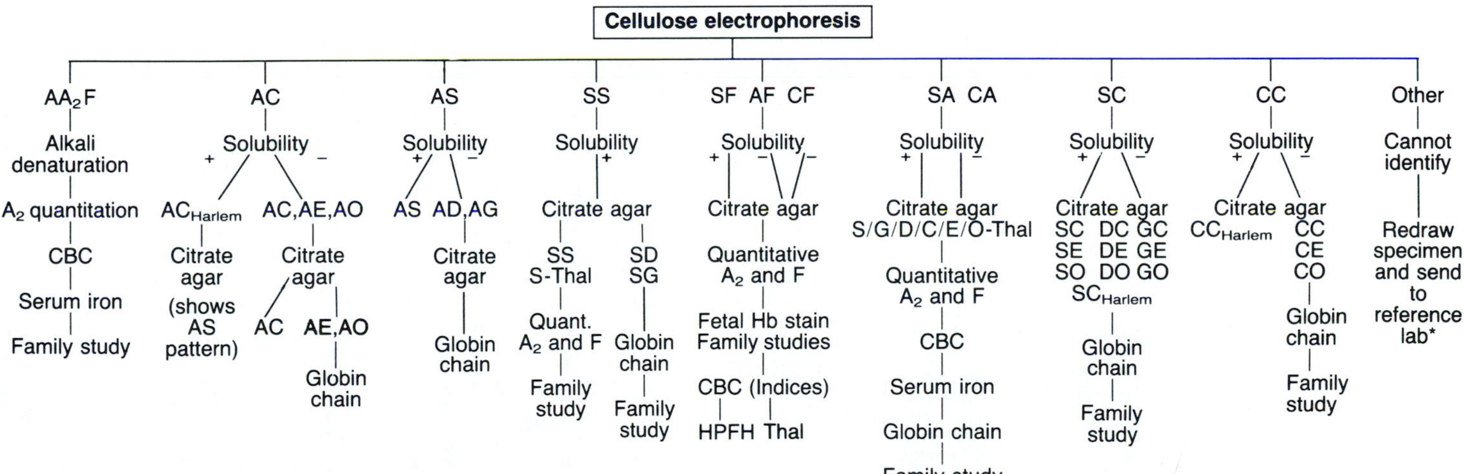

Fig. 63-1 Flow sheet for laboratory diagnosis of common hemoglobinopathies based on cellulose acetate electrophoresis as initial test. *CBC,* Complete blood count; *HPFH,* hereditary persistance of fetal hemoglobin; *Thal,* thalassemia. *(Modified from Centers for Disease Control: Hemoglobinopathy manual, HEW pub no (CDC) 77-8266, Washington, DC, 1976, US Government Printing Office.)*

developments by Helena and Isolab, if this tentative proposal will be revised to reflect the advancements made by the two commercially available column procedures.

Hemoglobin S

Given the ease with which any laboratory can prepare its own reagents for the dithionate test, it is the preferred method for use. The procedure described here is essentially that recommended by the Centers for Disease Control (CDC). The reagents used are concentrated phosphate buffer mixed with saponin and sodium dithionite. Regardless of the procedure used, it is recommended that the solubility test be used to confirm the presence of Hb S after presumptive identification by electrophoresis. It should definitely not be used to differentiate the genotype of the sickle cell state because of possible misleading results, as with Hb S thal.

Given this background, the NCCLS has proposed a tentative standard procedure for the detection of abnormal hemoglobins. The committee report suggests that cellulose acetate can provide accurate and reproducible results as an initial test.[6] The method described here is consistent with these recommendations but uses essentially the procedure outlined by Helena Laboratories.

In summary, Fig. 63-1 describes one of the most established approaches for the detection of hemoglobin variants. This does not mean that every request for Hb electrophoresis goes through this entire process; rather, this sequence of tests is used for those unknowns that cannot be identified at any one particular stage.

Hemoglobin electrophoresis

College of American Pathologists (CAP) and CDC[7,8] surveys indicate that if this algorithm is applied, there will be fewer misidentified or incompletely identified unknowns. The CDC survey clearly demonstrates that errors of interpretation are made more frequently when cellulose acetate is used alone. Some errors could have significant impact on proper patient counseling.

Similarly, the CAP survey shows quite conclusively that laboratories that include alkaline and acid electrophoresis and the hemoglobin solubility test have the best chance of correctly identifying hemoglobinopathies.

SPECIMEN

Cellulose acetate

Whole blood is collected in EDTA. Other anticoagulants are also suitable. A hemolysate is required for this method and is prepared by centrifugation of an aliquot (1 mL) of whole blood at 2500 rpm (900 *g*) for approximately 5 minutes. The plasma is discarded, and the packed red blood cells are washed with 4 mL of 0.9% NaCl. After mixing, centrifuging, and discarding the supernatant, two drops of the washed red blood cells are added to 12 drops of the hemolysate reagent, consisting of 5 mmol/L EDTA and hemoglobin preservatives. Anticoagulated samples are stable for 2 weeks at 4° C.

Citrate agar

One drop of this hemolysate is mixed with six drops of distilled water.

Hemoglobin A_2 by column chromatography

The specimen is freshly drawn blood using EDTA or heparin as an anticoagulant. Blood should be refrigerated at 4° C and can be stored as whole blood (not as a hemolysate) up to 1 week.

Before testing, one can prepare the hemolysate in either of two ways:

1. Mix 50 μL of whole blood with 200 μl of distilled water. Allow to stand 5 minutes, and mix again.
2. Wash red blood cells three times with saline solution (0.9% NaCl). The cells are then lysed; 50 μL of this hemolysate are mixed with 200 μL of distilled water.

Solubility test

The specimen is 20 μL of whole blood, with EDTA or heparin being used as an anticoagulant.

REFERENCE RANGE

Hemoglobin electrophoresis. See Figs. 63-2 and 63-3 for the usual hemoglobin patterns.

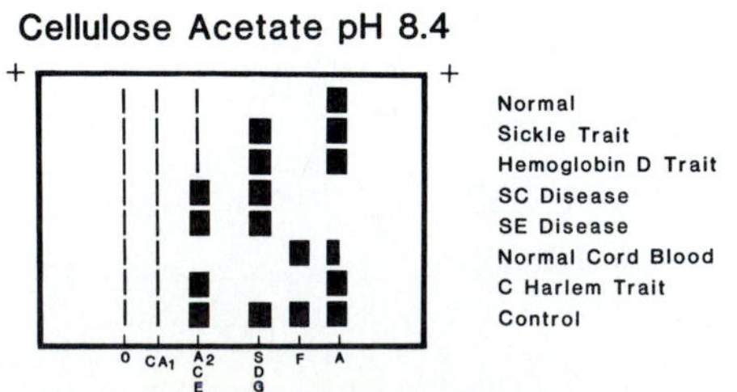

Fig. 63-2 Representation of hemoglobin electrophoresis on cellulose acetate strips, pH 8.4. CA_1, Carbonic anhydrase; others are hemoglobin variants discussed in text.

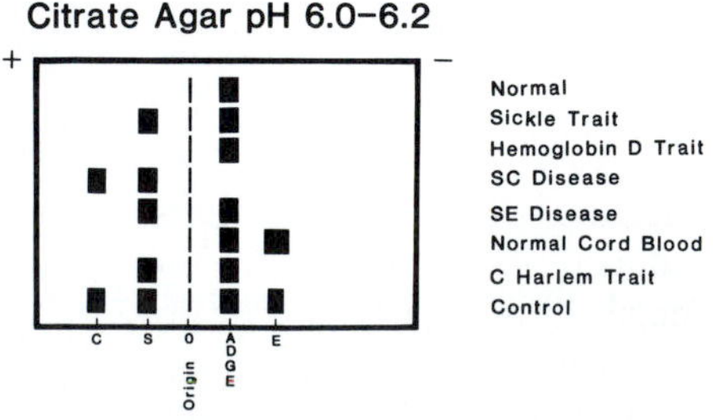

Fig. 63-3 Representation of hemoglobin electrophoresis on citrate agar, pH 6. On some plates the origin may not be as visible or as separated from Hb S as shown here.

Hemoglobin A_2. Reference ranges of Hb A_2 will vary between laboratories. Each laboratory must determine its own range, which must be reported with each test to facilitate interpretation of results. The Isolab Company claims the range of Hb A_2 to be 1.5% to 3.0%. My own experience indicates a range of 1.5% to 4.0%. Values greater than 4.0% are presumptive of the beta-thalassemia trait.

Hemoglobin S. Hemoglobin S is not present in normal blood but will be detected in patients with sickle cell trait or sickle cell anemia.

REFERENCES

1. Brosious, EM, et al: Hemoglobinopathy testing, a report of the 1976 and 1977 College of American Pathologists surveys, Am J Clin Pathol 70:563-566, 1978.
2. Nalbandian, RM, et al: Dithionite tube test—a rapid inexpensive technique for the detection of hemoglobin S and non-S sickling hemoglobin, Clin Chem 17:1028-1032, 1971.
3. Gottfried, EL, et al: Reliable estimation of hemoglobin A_2 concentration by electrophoresis with densitometry, Am J Clin Pathol 72:415-420, 1979.
4. Hedland, B: Isoelectric focusing as a tool for identifying variant hemoglobins in hemoglobinopathies and thalassimas. In Fairbanks, VF, editor: Laboratory methods and clinical cases, New York, 1980, B.C. Decker.
5. Pearce, C: A comparison of three methods for quantitation of variant hemoglobin fractions, Am J Med Technol 46:698-703, 1980.
6. National Committee for Clinical Laboratory Standards (NCCLS): Standard for abnormal hemoglobin detection by cellulose acetate electrophoresis, NCCLS Tentative Standard: TSH-8, 1980.
7. Mahoney, MA: Centers for disease control CDC Performance Evaluation Program Hematology Detection, 1986, II, Atlanta, 1986, Centers for Disease Control.
8. Rabinovitel, A: Hemoglobinopathy survey Set HGA College of American Pathologists, Skokie, IL, 1988.

BIBLIOGRAPHY

Lubin, BH, et al: Sickle cell disease and the thalassemias: diagnostic assays, Lab Management 18:38-47, 1980.

Schmidt, RM: Laboratory diagnosis of hemoglobinopathies, JAMA 224:1276-1280, 1973.

Schneider, RG, et al: Laboratory identification of the hemoglobinopathies, Lab Management 19:29-43, 1981.

Sonnenwirth, AC, and Jarett, L: Gradwohl's clinical laboratory methods and diagnosis, ed 8, vol 1, St Louis, 1980, The CV Mosby Co.

Immunoelectrophoresis

GAYLE B. JACKSON

Clinical significance: p. 730

PRINCIPLES OF ANALYSIS

Zone electrophoresis and immunodiffusion are commonly used methods in clinical laboratories. Zone electrophoresis employs an inert solid carrier medium, such as cellulose acetate or agarose, for the separation of proteins. The technique is routinely used to screen for serum protein abnormalities, hemoglobin variants, and isoenzymes. Specific bands are compared to a standard and are identified by their mobility and by their staining characteristics. Immunoelectrophoresis is a combination of zone electrophoresis and an antigen-antibody interaction. The immunoprecipitation step enhances the ability to identify specific proteins after electrophoretic separation.

Immunoelectrophoresis may be divided into the following methods: electroimmunodiffusion, crossed immunoelectrophoresis, counterimmunoelectrophoresis, immunoelectrophoresis (IEP), and immunofixation.

Electroimmunodiffusion or the Laurell rocket technique[1] (method 1, Table 63-12) uses electrophoresis to shorten the diffusion time of the antigen through agarose. The sample is added to a well in an agarose plate with antibody incorporated throughout the agarose. The plate is then placed in an electrophoresis chamber. As each antigen migrates through the agarose under the force of the electric field, precipitin peaks (rockets) form in the agarose. The method is quantitative, since the height of the "rocket" is proportional to the immunoglobulin concentration (see methods for immunoglobulin quantitation, p. 1050).

Crossed immunoelectrophoresis (method 2, Table 63-12) may also be used for quantitation purposes. Crossed immunoelectrophoresis or two-dimensional electrophoresis[2] is performed in two steps. In the first step, the sample is placed in a well cut in agarose. The proteins in the sample are separated by conventional zone electrophoresis. The excess agarose parallel to the separated proteins is removed and replaced by an antibody-containing gel. The second-dimensional electrophoresis then causes the separated proteins to migrate perpendicularly to the first separation into the antibody-containing agarose. Antigen-antibody precipitation peaks then form in the second gel. Quantitation is achieved by measurement of the area under the peaks by planimetry or by measurement of the area in square millimeters and comparison of these results to those obtained with standards.[3] Two-dimensional immunoelectrophoresis is very similar to one-dimensional immunoelectrophoresis or electroimmunodiffusion. Both are quantitative; however, only one antibody is used, and consequently one antigen is quantitated at a time with electroimmunodiffusion. A polyspecific antiserum is used with two-dimensional immunoelectrophoresis, thus allowing for the quantitation of multiple proteins with each run.

Counterimmunoelectrophoresis[4] (method 3, Table 63-12) again uses electrophoresis to shorten diffusion time. Various terms used for this technique are *countercurrent electrophoresis, electro-osmodiffusion,* and *crossover electrophoresis*. One performs counterimmunoelectrophoresis by placing an anodically migrating antigen in a well in agarose and a cathodically migrating antibody in another well nearby. When current is applied, the antigen and the antibody migrate toward each other and form a precipitin line where they meet. The method is much faster than standard diffusion techniques (that is, Ouchterlony plates) and can be used qualitatively and semiquantitatively (by use of a titer).

Table 63-12 Immunoelectrophoresis

Method	Type of analysis	Principle	Usage	Comments
1. Electroimmunodiffusion (EID) (Laurell rocket)	Quantitation by size of immunoprecipitate in gel	Electrophoresis of antigens into antibody-containing agarose gel; height of gel pattern (rocket) proportioned to antigen concentration	Serum	Used for special tests such as factor VIII–related antigen
2. Two-dimensional or crossed immunoelectrophoresis	Quantitation of multiple proteins in one gel by use of polyspecific antiserum; quantitation based on area under precipitin peak	Proteins separated by zone electrophoresis in agarose gel; second electrophoresis causes separated proteins to migrate perpendicularly to first separation into antibody-containing agarose	Serum, fluids	Too time consuming for the clinical laboratory
3. Counterimmunoelectrophoresis	Qualitative identification of protein by formation of a precipitin band with monospecific antiserum	Anodically migrating antigen and cathodically migrating antibody move through agarose toward each other in an electrophoretic field; precipitin band forms where they meet	Serum, CSF, other fluids	Used primarily by microbiology laboratories to identify bacterial antigens
4. Immunoelectrophoresis (IEP)	Identification of monoclonal proteins through precipitin arc formation	Proteins separated in agarose by zone electrophoresis; antiserum is placed in trough horizontal to electrophoretogram and allowed to diffuse for 24 hours; precipitin arcs form where antigen and antibody meet; monoclonality can be determined from shape and mobility of arc	Serum, urine, body fluids	Should be used as the confirmatory test for the presence and identification of monoclonal proteins
5. Immunofixation	Identification of protein bands on zone electrophoresis by formation of precipitin bands with antisera	Proteins separated by zone electrophoresis; area containing protein to be identified is overlayed with monospecific antibody and its corresponding antigen	Serum, urine, other body fluids	Can be used with the IEP to help identify monoclonal proteins; more sensitive than IEP
	Western blot	Proteins separated by zone electrophoresis and transferred to nitrocellulose sheet by electrophoresis; labeled antibody layered on sheet; Ab-antigen reaction detected by label reaction	Serum, other body fluids	Commonly used as confirmation test for HIV

Immunoelectrophoresis (method 4, Table 63-12) combines zone electrophoresis and simple antigen-antibody diffusion reactions. This method requires a longer diffusion time than that described in the previous methods but permits the accurate identification of various proteins by use of specific antibodies. Grabar and Williams[5] first described immunoelectrophoresis in 1953. Scheidegger[6] modified the original technique for use as a micromethod. Immunoelectrophoresis is usually performed on an agarose plate with wells and troughs precut in the agarose. The plates are available commercially or may be prepared by the laboratory. The patient sample is placed in a well and separated into albumin, $alpha_1$, $alpha_2$, beta, and gamma globulin components by zone electrophoresis. Either a monospecific or a polyspecific antiserum is placed in a trough that is cut parallel to the zone electrophoresis separation. The plate is then incubated in a moist chamber for 24 hours, during which time the antibody and antigen diffuse toward each other and form precipitin arcs where they meet. The arcs can be identified by both the shape and the migration rate when compared to a control. Immunoelectrophoresis is both a qualitative and a semiquantitative technique.

Immunofixation[7] (method 5, Table 63-12) is also a com-

Table 63-13 Comparison of reaction conditions for immunoelectrophoresis

	Electroimmunodiffusion	Counterimmunodiffusion	Immunoelectrophoresis (IEP)	Immunofixation
Temperature	Ambient	Ambient	Ambient	Ambient
pH	Usually pH 8.2	Varies (pH 5 to 10)	Usually pH 8.6	pH 8.6
Buffer	Barbital, 0.2 M	Barbital, 0.05 M	Barbital, 0.2 M	Barbital, 0.2 M
Sample volume	1 μL	20 μL	0.8 μL	100 μL
Time of reaction	2-4 hours	2-4 hours	18-48 hours	8-10 hours

bination of zone electrophoresis and an antibody-antigen interaction. Either cellulose acetate or agarose is used as the supporting medium.[8,9] Immunofixation, unlike immunoelectrophoresis, eliminates the long diffusion time for antigen-antibody interaction. The proteins are first separated by zone electrophoresis. Strips of cellulose acetate or filter paper, soaked in a monospecific antiserum, are then laid over the zone occupied by the protein to be identified, and after approximately a 1-hour incubation at room temperature, the strips are removed. The excess protein is washed from the plate, and the remaining precipitation band is stained with a protein stain such as amido black.

Immunoblotting and Western blot are described in Chapter 10.

REFERENCE AND PREFERRED METHODS

Crossed or two-dimensional immunoelectrophoresis is not used routinely in clinical laboratories. Electroimmunodiffusion (EID) (Table 63-13) or one-dimensional immunoelectrophoresis is used on a limited basis in some clinical laboratories for the quantitation of factor VIII-related antigen, an important test for the diagnosis of von Willebrand's disease.[10]

Counterimmunoelectrophoresis (Table 63-13) was used by clinical laboratories in the early 1970s to detect hepatitis B surface antigen in both patient and donor blood samples. However, radioimmunoassay and immunoenzyme techniques have replaced counterimmunoelectrophoresis for hepatitis testing.

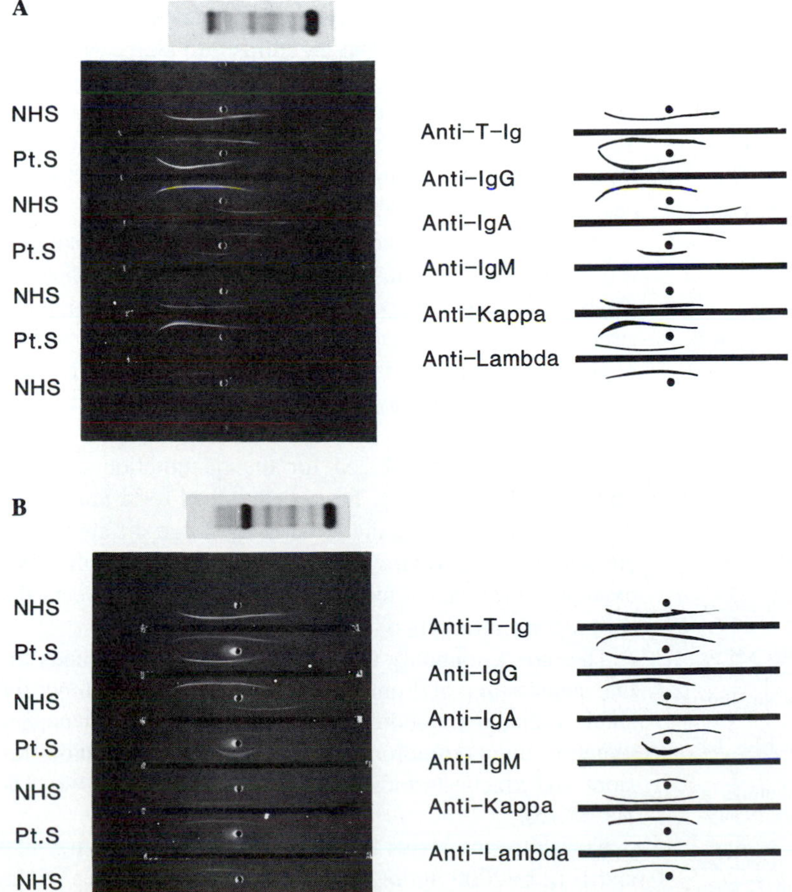

Fig. 63-4 **A,** Serum protein and immunoelectrophoresis patterns. *Upper pattern* is formed by protein electrophoresis of serum from patients with multiple myeloma. Notice strongly staining monoclonal band in gamma region of patient serum. Immunoelectrophoretic patterns of same patient's serum. *NHS,* Normal human serum; *Pt.S,* patient serum; *anti-T-Ig,* anti-immunoglobulins; *anti-IgG, anti-IgA, anti-IgM, anti-kappa, anti-kappa, anti-lambda,* antisera made to specific components of human immunoglobulins. *Left pattern,* Actual photograph of the agarose gel; *right pattern,* artist's tracing. Notice gull-wing pattern formed by the anti-IgG and anti-kappa antisera compared to the control. **B,** Serum from another patient with multiple myeloma. Notice sharp precipitation band formed by the anti-IgM and the gull-wing pattern of the anti-lambda.

Immunofixation has been used to study protein polymorphism[7] and to identify specific proteins.[8] It was used in the genetic typing of alpha$_1$-antitrypsin.[11]

Immunoelectrophoresis (IEP) (Table 63-13) is the most commonly used method in clinical laboratories for the identification of monoclonal *(M)* proteins. However, the detection and identification of monoclonal proteins is accomplished most effectively by a combination of the following tests: protein electrophoresis combined with total protein quantitation, immunoelectrophoresis, immunoglobulin quantitation, and in some cases immunofixation. The first screening tests for serum or urine should include the quantitation of the total serum proteins and protein electrophoresis. If the total protein and protein electrophoresis are normal, no further tests need to be performed. However, if a homogeneous band or a suspicious heterogeneous band is detected by protein electrophoresis, the sample should be evaluated by immunoelectrophoresis to determine whether a monoclonal gammopathy is present.

When a monoclonal protein is detected by immunoelectrophoresis, immunoglobulin quantitation should be performed. Immunofixation can be used for additional information when the immunoelectrophoresis results are unclear or when more than one monoclonal protein is present.

SPECIMEN

The immunoelectrophoresis techniques may be performed on serum, urine, or other body fluids. Plasma is not recommended. The fibrinogen peak or band may interfere with the detection of a monoclonal protein. Urine and body fluids are concentrated 30 to 50 times before analysis by protein electrophoresis, immunoelectrophoresis, or immunofixation. Samples should be frozen if not analyzed within 1 to 2 days of collection.

REFERENCE RANGE

See Fig. 63-4 for a typical immunoelectrophoresis pattern of a healthy person.

REFERENCES

1. Laurell, C: Quantitative estimation of proteins by electrophoresis in agarose gel containing antibodies, Anal Biochem 15:45-52, 1966.
2. Laurell, CB: Antigen-antibody crossed electrophoresis, Anal Biochem 10:358-361, 1965.
3. Versey, JMB, et al: Semi-automated two-dimensional immunoelectrophoresis, J Immunol Methods 3:63, 1973.
4. Goecke, DJ, and Howe, C: Rapid detection of Australia antigen by counterimmunoelectrophoresis, J Immunol 104:1031, 1970.
5. Grabar, P, and Williams, LA, Jr: Method permettant l'etude conjuguee des proprietes electrophoretiques et immunochemiques d'un melange de proteines: application au serum sanguin, Biochim Biophys Acta 10:193, 1953.
6. Scheidegger, JJ: Une micromethode de l'immunoelectrophorese, Int Arch Allergy Appl Immunol 7:1-3, 1955.
7. Alper, CA, and Johnson, AM: Immunofixation electrophoresis: a technique for the study of protein polymorphism, Vox Sang 17:445-452, 1969.
8. Ritchie, RF, and Smith, R: Immunofixation. I. General principles and application to agarose gel electrophoresis, Clin Chem 22:497-499, 1976.
9. Chang, C, and Inglis, N: Convenient immunofixation electrophoresis on cellulose acetate membrane, Clin Chim Acta 65:91-97, 1975.
10. Moehring, C: A rapid factor VIII-related antigen electroimmunoassay, Am J Clin Pathol 72:821-828, 1979.
11. Ritchie, RF, and Smith, R: Immunofixation. II. Application to typing of alpha 1-antitrypsin at acid pH, Clin Chem 22:497-499, 1976.
12. Grey, HM, et al: A subclass of IgA globulins (IgM_2), which lacks the disulfide bonds linking heavy and light chains, J Exp Med 128:1223-1236, 1968.
13. Levin, AS, and Perin, GM: Clinical immunology no. CL-5, alpha heavy chain disease, Chicago, 1977, American Society of Clinical Pathologists.
14. Cawley, LP, et al: Electrophoresis and immunochemical reactions in gels, ed 2, Chicago, 1978, American Society of Clinical Pathologists.

Immunoglobulin quantitation

GAYLE B. JACKSON

Clinical significance: p. 594
Molecular formula and weight: see Table 63-14
Chemical class: protein

PRINCIPLES OF ANALYSIS

Historically, immunoglobulins were measured only as part of the total globulins present in human serum. Globulins were isolated by salt or gold precipitation and quantitated by the biuret assay (methods 1 and 2, Table 63-15). The introduction of electrophoretic separation techniques allowed fractionation of the globulins into alpha, beta, and gamma components (method 3, Table 63-15). The gamma fraction consists almost completely of immunoglobulins. Today electrophoretic separation offers an excellent mode for examining serum for total immunoglobulin and for the detection of monoclonal antibodies.

Qualitative and semiquantitative estimates of total immunoglobulin levels may be obtained from immunoelectrophoresis respectively. Knowledge of immunoglobulin structure and the availability of antibodies to each immunoglobulin class, that is, specific for the constant regions of the heavy chains, have made it possible to measure each immunoglobulin class. Different immunochemical procedures have been developed for the quantitation of these immunoglobulin classes. In part, the need for a number of different procedures is dictated by the large differences in concentration between each class of immunoglobulin. For example, there is a millionfold difference between the serum concentrations of IgG and IgE.

The most commonly used immunochemical methods for the quantitation of immunoglobulins (G, M, and A) are single radial immunodiffusion, rate and end-point nephelometry, immunofluorometric assay, electroimmunodiffusion, and kinetic turbidimetric systems (methods 4-8, Table 63-15).

Normal serum levels for immunoglobulin D (30 μg/mL) and E (2 to 2000 ng/mL) are below the sensitivities of the above immunochemical systems. Serum levels of IgD gen-

Table 63-14 Physical properties of human immunoglobulins

Immunoglobulin class	IgG	IgM	IgA	IgD	IgE
Molecular weight (daltons)	150,000	900,000	160,000 (monomer) 320,000 (dimer)	185,000	200,000
Sedimentation coefficient, S	6.6	18.0-19.0	6.25-10.9	6.2-7.0	7.86-7.92
Heavy chains	γ	μ	α	δ	ϵ
Heavy-chain subclasses	$\gamma_1, \gamma_2, \gamma_3, \gamma_4$	μ_1, μ_2	α_1, α_2	—	—
Light chains	κ or λ	κ or λ	κ or λ	κ or λ	κ or λ
Molecular formula	$IgG(\kappa)2\gamma2\kappa$ $IgG(\lambda)2\gamma2\lambda$	$IgM(\kappa)(2\mu2\kappa)_5$ $IgM(\gamma)(2\mu2\lambda)_5$	$IgA(\kappa)(2\alpha2\kappa)_{1\text{-}3}$ $IgA(\lambda)(2\alpha2\lambda)_{1\text{-}3}$	$IgD(\kappa)2\delta2\kappa$ $IgD(\lambda)2\delta2\lambda$	$IgE(\kappa)2\epsilon2\kappa$ $IgE(\lambda)2\epsilon2\lambda$
Normal serum concentrations (mg/mL) (by age)					
Cord specimen	7.66-16.93	0.04-0.26	0.0004-0.09		
0.5-3 months	2.99-8.52	0.15-1.49	0.03-0.66		
3-6 months	1.42-9.88	0.18-1.18	0.04-0.90		
6-12 months	4.18-11.42	0.43-2.23	0.014-0.95		
1-2 years	3.56-12.04	0.37-2.39	0.13-1.18		116-122 ng/mL
2-3 years	4.92-12.69	0.49-2.04	0.23-1.37		80-122 ng/mL
3-6 years	5.64-13.81	0.51-2.14	0.35-2.09		
4-7 years					140-442 ng/mL
6-9 years	6.58-15.35	0.50-2.28	0.29-3.84		
10-14 years					374-674 ng/mL
12-16 years	6.80-15.48	0.45-2.56	0.81-2.52		
Adult	8.00-16.00	0.50-2.00	1.40-3.50	0-0.14	2-2000 ng/mL

Data from Seligson, O, editor: Handbook series in clinical laboratory science. Section F. Immunology, vol 1, part 1, Boca Raton, FL, 1978, CRC Press, Inc; and Meites, S, editor: Pediatric clinical chemistry, ed 2, Chicago, 1981, American Association for Clinical Chemistry.

erally are not of clinical importance, unless a plasma cell malignancy is present with the production of a monoclonal IgD protein. Single radial immunodiffusion may be used to quantitate IgD in this instance, since the serum IgD level is well above normal.

Radioimmunoassay, radioimmunometric, and enzyme immunoassay (ELISA) techniques are used to quantitate serum IgE levels (methods 9-11, Table 63-15).

The single radial immunodiffusion (RID) test consists of an agar plate with antibody incorporated throughout the agar. See Chapter 9.

Two radial immunodiffusion methods are commonly used in clinical laboratories. The Mancini method[1] is based on an end-point (equivalence-zone) reading. The Fahey-McKelvey method[2] measures the ring diameter before equivalence (such as 18 ± 2 hours). The Fahey technique has the advantage of producing results more rapidly; however, the accuracy and precision is somewhat less than that obtained with the end-point technique.[3,4] Commercial plates are readily available for radial immunodiffusion quantitation. Many plates can be used with either the Fahey or Mancini technique.

The quantitation of the immunoglobulins by nephelometry is a more recent development (method 5, Table 63-15). Several commercial nephelometers, with appropriate reagent test kits, are available.

Since light scattering produced by contaminating particles in the antiserum can interfere with quantitation, antiserum that is free of contamination must be used. Commercial test kits produced for these systems are readily available. All manufacturers offer IgG, IgM, IgA, C3, and C4 kits as well as kits for a number of other serum proteins. Turbid patient samples may have to be filtered before use on the nephelometer.

A commercial immunoflourometric system for the quantitation of IgG, IgM, and IgA by clinical laboratories is the FIAX system (International Diagnostic Technology, Santa Clara, CA) (method 6, Table 63-15). This is the indirect fluoroimmunoassay described in Chapter 10.

In electroimmunodiffusion or in the Laurell rocket technique (method 7, Table 63-15), the sample is added to a well in an agarose plate with antibody incoporated throughout the agarose. The plate is then placed in an electrophoresis chamber. As the immunoglobulin migrates through the agarose under the force of the electric field, precipitin peaks (rockets) form in the agarose. The height of the rocket can be plotted as a standard curve.[5] Commercial kits are not readily available for this technique.

Immunoglobulin quantitation can be performed by a turbidimetric method with a centrifugal fast analyzer (method 8, Table 63-15). With this method, the patient's sample and appropriate reagent antiserum are mixed by the use of centrifugal force, and the solution containing an antigen-antibody precipitate is forced into a cuvette at the end of the rotor arm. Light passes perpendicularly through the rotor arm to a photodetector. The increase in absorbance resulting from the presence of the precipitate is used to calculate the antigen concentration. The absorbance reading

Table 63-15 Methods of immunoglobulin quantitation

Method	Type of analysis	Principle	Usage	Comments
1. Salt precipitation	Isolation of total immunoglobulin by salt precipitation	Immunoglobulins are less soluble in certain salt solutions and precipitate; amount of immunoglobulins quantified by total protein assay (such as biuret)	Serum	Historical, nonspecific
2. Gold precipitation	Isolation of total immunoglobulin by gold precipitation	Similar to salt precipitation; gold used as precipitant	Serum	Historical, nonspecific
3. Electrophoresis	Estimation of total immunoglobulin by physical separation	Proteins separate based on class (albumin, gamma globulin); calculation of each class as percentage of total protein	Serum	Good screening method for detection of monoclonal immunoglobulins; commonly used
4. Radial immunodiffusion (RID)	Quantitation by immunoprecipitation in gel	Immunoglobulin diffuses into gel containing antibody, forming a ring-shaped immunoprecipitate; diameter of ring proportional to concentration	Serum, body fluids	Accurate, slow, commonly used
5. Nephelometry	Quantitation by immunoprecipitation in solution	Reaction of immunoglobulin with its specific antibody results in immunoprecipitate, which has light-scattering properties; amount of light scatter proportional to immunoglobulin concentration	Serum, body fluids	Accurate, rapid, commonly used
6. Immunofluorometry	Quantitation by competition for labeled antibody on a solid phase	Immunoglobulin adsorbed onto solid surface competes for fluorescent-labeled antibody with immunoglobulin sample; fluorescence signal is inversely proportional to immunoglobulin concentration	Serum, some body fluids	Precision slightly less than nephelometer, RID
7. Electroimmunodiffusion (Laurell rocket)	Quantitation by size of immunoprecipitate in gel	Immunoglobulin electrophoresis into antibody-containing agarose gel; height of gel pattern (rocket) proportional to immunoglobulin concentration	Serum	Not readily available commercially
8. Turbidimetry	Quantitation by light absorbance of immune precipitate	Immunoglobulin reacts with antibody or other precipitation agent; turbidity proportional to immunoglobulin concentration	Serum	Requires centrifugal analyzer and high level of expertise to establish procedure
9. Radioimmunoassay (RIA)	Quantitated by radioisotope	Immunoglobulin reaction with antibody displaces radiolabeled immunoglobulin	Any body fluid	Very sensitive, used for IgE
10. Radioimmunometric assay (IRMA)	Quantitated by radioisotope	Immunoglobulin reacts with antibody on solid surface and reaction quantitated by second radiolabeled antibody	Any body fluid	Very sensitive, used for IgE (PRIST)
11. Enzyme immunoassay (EIA)	Quantitated by enzyme	Similar to above radioisotope procedures; substitute enzyme label for radioisotopic ones	Any body fluid	Very sensitive, used for IgE equivalent to paper radioimmunosorbent test (PRIST)

can be taken at equilibrium or on a kinetic basis (at selected time intervals during the reaction). A standard curve is generated and entered into a computer. The calculation of antigen in the patient sample is generated by use of a computer program.

Finley et al.[6] described a two-point kinetic technique that measured changes in absorbance at 340 nm between 10 and 225 seconds using a 36-place centrifugal analyzer.

Although radioimmunoassay (RIA) procedures (method 9, Table 63-15) can be used to quantify IGE, the two techniques used most often to quantitate total IgE are the PRIST (paper-radioimmunosorbent test) (method 10, Table 63-15) and the ELISA (enzyme-linked immunosorbent assay) (method 11, Table 63-15). Commercial kits are available for both techniques (Pharmacia Diagnostics, Piscataway, NJ; Calbiochem-Behring Corporation, La Jolla, CA).

The PRIST technique is a sandwich technique using ^{125}I anti-IgE. Immobilized antibody is reacted with the IgE in the patient's serum. Excess serum is washed away, and labeled antibody is added. During incubation, the labeled antibody reacts with the immobilized IgE. Excess labeled antibody is washed away, and the amount of IgE initially in the patient's serum can be estimated by detection of the isotopic label in a gamma-ray scintillation counter.

With the ELISA technique (Enzygnost IgE, Calbiochem-Behring Corporation, La Jolla, CA), the test specimens are incubated in tubes coated with goat anti-human IgE. Peroxidase-labeled rabbit anti-human IgE, enzyme substrate, and sulfuric acid are added successively. The tubes are read in a spectrophotometer at a wavelength of 492 nm.

Another kit known as RAST (radioallergosorbent test) is available for the detection of specific IgE (that part of the total IgE that is directed against a certain allergen). In this test, patient's serum is incubated with an allergen-coated paper disk. ^{125}I-labeled rabbit anti-human scintillation counter is used to detect the amount of bound, labeled antibody. Specific IgE may also be quantitated by the ELISA technique. An allergen-coated disk is incubated with serum. Excess serum is washed away. Beta-galactosidase–labeled rabbit anti-human IgE is detected on a spectrophotometer at 420 nm.

STANDARDIZATION AND QUANTITATION

The main problem in the accurate quantitation of immunoglobulins is the extreme heterogeneity of the group.[7] The immunoglobulins are divided into five major classes with four immunologically characterized subclasses of IgG, two of IgA, and several probable subclasses of IgM. The ratio of these subclasses may vary with some infectious and autoimmune diseases and will certainly vary with malignant plasma cell dyscrasias. In addition to the immunologically characterized subclasses, the immunoglobulins also vary in size. For instance, IgM is usually present as the stable 19S pentamer of 7S subunits, but occasionally may be found in free 7S form. IgA is 7S in the monomeric form but 11S in the dimeric form. In addition, specificity and binding affinity of the reagent antisera will affect immunochemical systems.

Accurate quantitation of monoclonal proteins is difficult because the reagent antiserum is raised against normal human serum containing a mixture of subclasses. Therefore, in the case of a monoclonal IgG_3 protein, only a portion of the antibodies in the antiserum will react with the IgG_3. In addition, the standard curve is constructed by use of a mixture of all subclasses. Also, variations in size and idiotypic antigens may result in a different reaction between a monoclonal protein and a reagent antiserum.

Daniels et al.[8] found that serum protein electrophoresis gives a more accurate quantitation of a monoclonal protein (once identified) than immunochemical methods.

The main determining factor in choosing between radial immunodiffusion and the nephelometer is the quantity of test specimens analyzed. The nephelometer is the method of choice if large batches of specimens are to be run.

Most commercial manufacturers offer standard-level and low-level radial immunodiffusion plates. Kallestad (Austin, TX) standard plate ranges are approximately 4 to 25 g/L for IgG, 600 to 4000 mg/L for IgA, and 300 to 3000 mg/L for IgM. Sera with concentrations above these limits should be diluted. Concentrations below these limits should be reported as less than the lower limit, or the actual concentration should be determined by use of the low-level plates.

The low-level plates are used for fluids, for children's sera, and for adult sera with low concentrations if determination of the actual concentration is desired. The low-level plates have sensitivity levels equivalent to the nephelometer. The sensitivity of the Kallestad low-level plates for IgG, IgM, and IgA are 18.8, 25.2, and 18.8 mg/L, respectively.

IgE may be quantitated by either RIA technique or by solid-phase enzyme immunoassay. The RIA method has been used most frequently in the past because of ready availability of commercial kits (Kallestad, Austin, TX; Pharmacia, Piscataway, NJ). However, commercial kits for the enzyme immunoassay technique are now readily available (Calbiochem-Behring, La Jolla, CA).

The enzyme immunoassay technique offers the same precision and correlates well with radioimmunoassay. It may offer a wider assay range and a faster turnaround time, depending on the commercial kit used. The enzyme immunoassay does not use radioactive material and requires only a spectrophotometer rather than a gamma-ray scintillation counter as with radioimmunoassay. Therefore the enzyme immunoassay technique may become the method of choice.

SPECIMEN

Immunoglobulin quantitation for routine diagnostic purposes is performed most commonly on serum specimens, IgG is frequently measured in cerebrospinal fluid and combined with the albumin quantitation for a ratio. Normal levels for immunoglobulin in thoracentesis fluid, synovial fluid, and other serous exudates are not well defined; and the clinical usefulness of these fluid quantitations is controversial. Urine specimens generally are concentrated 20 to 50 times before they are evaluated by electrophoresis or by immunoelectrophoresis coupled with the total urine protein.

Samples for quantitation may be stored for up to 5 days at 2° to 8° C if they are protected from contamination and evaporation. For sample shipment or longer storage periods, samples should be frozen at −20° C or colder. Once thawed, samples generally should not be refrozen, as repeated freezing and thawing may damage the proteins.

Samples for RAST (radioallergosorbent test) assays should not be repeatedly frozen and thawed, for the IgE molecule is unstable under these conditions.

REFERENCE RANGE

See Table 63-14.

REFERENCES

1. Mancini, G, Carbonara, AO, and Heremans, JF: Immunochemical quantitation of antigens by single radial immunodiffusion, Immunochemistry 2:235-254, 1965.
2. Fahey, JL, and McKelvey, EM: Quantitative determination of serum immunoglobulins in antibody-agar plates, J Immunol 94:84-90, 1965.
3. Ritzman, SE, and Daniels, JC: Serum protein abnormalities, diagnostic and clinical aspects, Boston, 1975, Little, Brown & Co.
4. Berne, BH: Differing methodology and equations used in quantitating immunoglobulins by radial immunodiffusion in a comparative evaluation of reported and commercial techniques, Clin Chem 20:61-68, 1974.
5. Laurell, C: Quantitative estimation of proteins by electrophoresis in agarose gel containing antibodies, Anal Biochem 15:45-52, 1966.
6. Finley, PR, et al: Immunochemical determination of human immunoglobulins: use of kinetic turbidimetry and a 36-place centrifugal analyzer, Clin Chem 25:526-530, 1979.
7. Reimer, CB, and Madison, SE: Standardization of human immunoglobulin quantitation: a review of current status and problems, Clin Chem 22:577-582, 1976.
8. Daniels, JC, et al: Methodologic differences in values for M-proteins in serum, as measured by three techniques, Clin Chem 21:243-248, 1975.

Serum protein electrophoresis

TIMOTHY G. McMANAMON
JOHN A. LOTT

Clinical significance: p. 359
Chemical class: protein

PRINCIPLES OF ANALYSIS

Human serum contains more than 125 identified proteins, which perform a number of functions. The protein components of plasma and serum constitute almost all the mass of the serum solutes (approximately 80 g/L). Present are carrier proteins, antibodies, enzyme inhibitors, and clotting factors. Plasma contains approximately 3 g/L of fibrinogen, a protein that is absent from serum. The total serum protein concentrations and the proportions of the individual protein fractions change during a variety of diseases. Thus quantitation of total serum protein and individual fractions is of considerable value in clinical diagnosis.

One of the simplest techniques for separating serum proteins is serum protein electrophoresis (SPE). When an electric field is applied to a medium containing charged particles, the negatively charged particles or molecules migrate toward the positive electrode (anode), and the positively charged particles migrate toward the negative electrode (cathode). This principle is applied to separate the protein fractions in serum. After separation, permanent fixation of the fractions at the position in the medium to which they migrated is possible.

Support

Supporting media fall into two main classes. Class I supports separate molecules solely on the basis of net molecular charge. These include paper, cellulose acetate, thin-layer materials, and agarose gel. Class II supports separate molecules on the basis of size as well as charge. These include starch gel and acrylamide gel (Table 63-16).

Class II support gels are porous; the pore size is about the same as the size of the protein molecules. The result is that a molecular sieving effect is observed in addition to the electrostatic separation observed with class I supports. Class II supports can separate different-sized molecules that have the same size-to-charge ratio and that cannot be separated on class I supports. Starch and acrylamide gels will act as traps for larger molecules, impeding their movement through the support, whereas smaller molecules will migrate without hindrance. Thus the resolution power of these gels is much greater than that of class I supports. With normal serum, five protein bands are observed on class I supports, whereas about 25 bands or more are seen on class II supports.

Paper as a support for serum protein electrophoresis is largely of historical interest. Cellulose acetate and agarose gel electrophoresis are now in common use in clinical laboratories, and both material (supports) and equipment for electrophoresis assays are commercially available. Another electrophoretic method called *high-resolution electrophoresis* is also commercially available. High-resolution electrophoresis resolves serum proteins into 13 bands, and the technique is superior to agarose gel electrophoresis for certain diagnostic procedures, such as identification of oligoclonal bands in cerebrospinal fluid (CSF) for the diagnosis of multiple sclerosis.[1,2]

Buffers and pH for serum protein electrophoresis

Although barbital buffer at pH 8.6 is frequently used for serum protein electrophoresis, more recently other buffers have been used to improve the separation. The serum protein electrophoresis system of LKB (LKB Instruments, Rockville, MD) uses a Tris-barbital buffer with calcium lactate, pH 8.6 (LKB application Note 310). A buffer solution composed of boric acid, Tris, and EDTA allows the separation of prealbumin, three alpha globulins, three beta globulins, and gamma globulin.[3] High-resolution capabilities of systems such as the Panagel (Worthington Diagnostics, Freehold, NJ) and Gelman (Gelman Sciences, Inc., Ann Arbor, MI) systems are believed to result from minor changes in the components of the buffer system and efficient heat dissipation during electrophoresis.

Another technique, which makes use of the differing pI values of proteins, is isoelectric focusing. The gel contains a pH gradient from approximately 4 to 8. This method is

Table 63-16 Methods of analysis for serum protein electrophoresis

Method	Usages	Comment
Class I supports		
Paper*	Historical	Has slight denaturing properties causing tailing; difficult to do densitometric analysis
Cellulose acetate†	Most widely used	Rapid analysis, separates five major classes of serum proteins, densitometric analysis readily performed; endosmosis effects present; high-resolution electrophoresis resolves serum proteins into 10 to 13 bands
Agarose‡	Widely used	Rapid analysis, separates five major classes of serum proteins plus prealbumin; densitometry readily performed; endosmosis effects present; high-resolution electrophoresis resolves serum proteins into 10 to 13 bands
Class II supports		
Starch	Rarely used	Sieving properties result in separation based on molecular size and charges: additional protein bands (25) can be separated; opaqueness makes densitometry difficult
Polyacrylamide	Less widely used	Protein separation based on molecular size and charge; up to 100 protein bands can be resolved though of unproved clinical utility; densitometry difficult

*From Aronsson, T, and Grönwall, A: Scand J Clin Lab Invest 9:338, 1957.
†From Ojala, K, and Weber, TH: Clin Chem 26:1754, 1980.
‡From Wieme, RJ: Agar gel electrophoresis, New York, 1965, Elsevier/North Holland, Inc.

reported to be the most sensitive available for the demonstration of slight differences in proteins. Isoelectric focusing has been used to detect and confirm the presence of paraproteins in serum[4] and oligoclonal bands in CSF.[5]

Voltage, current, temperature, and time

In practice, it is desirable to separate the proteins rapidly to preserve the sharpness of the bands. The longer the electrophoresis is allowed to proceed, the greater the radial diffusion and the broader the bands. One can decrease the time for a given separation by either decreasing the length of the support or increasing the voltage.

With an increase in voltage, there will be a corresponding current increase, and more heat will be generated. The increased heat may become a limiting factor because it leads to protein denaturation. One may reduce heating by lowering the ionic strength of the buffer or by using a cooling chamber.

Use of constant voltage is a standard practice in serum protein electrophoresis. Because the electrophoretic separation is achieved within about 30 minutes at about 90 to 100 volts, the change in current and temperature that occurs during the analysis does not adversely affect the results. However, controlling current and temperature may improve the resolution.

Development of electrophoretogram

Zones of proteins in an electrophoretogram can be fixed in the support medium by evaporation of the solvent and thus precipitation of the protein. Staining the proteins after electrophoretic separation and subsequent densitometric scanning is the most popular method for quantitation of individual protein fractions. Some serum specimens will show diffuse bands between the major electrophoretic fractions. Scanning the pattern gives quantitative information on well-resolved protein fractions only. Observation with an experienced eye is helpful in the detection of minor fractions and unsatisfactory separations. Analysis of a control serum on each electrophoretic plate is essential.

Staining solutions should be applied to the dry film rather than to the wet gel. Amido black 10B is widely used for staining agarose gels because it leaves the background clear, gives distinct zones, and has adequate sensitivity. Amido black 10B has a slightly greater affinity for albumin and transferin than for the globulins. Ponceau S stain is preferable for cellulose acetate; however, it gives more color to the albumin band than it does to the globulin bands. Bromphenol blue has been used for staining protein bands on agarose, but this stain also gives more color to albumin than to the globulins. A combination of amido black 10B and Coomassie brillant blue R250 has been used to stain proteins that were separated on cellulose acetate.[6] Coomassie brilliant blue R250 gives an intense stain, making it suitable as a stain for proteins after agarose gel electrophoresis of urine and CSF. Because of its high sensitivity, silver stain has been used for dilute fluids that have not been concentrated before analysis.

It is important to realize that, because of the unequal stain characteristics noted earlier, SPE is a semiquantitative technique.

Cellulose acetate and agarose give similar results (Table 63-17); however, agarose gel electrophoresis has found increased use because of more consistent performance and ease of handling, and thus it is considered to be superior to cellulose acetate for electrophoresis.[7]

Serum protein electrophoresis with agarose as the supporting medium is recommended as the preferred method.[8]

SPECIMENS

Serum is the specimen of choice. Plasma can be used; however, an extra band produced by fibrinogen will be observed. Urine and CSF can also be analyzed if their pro-

Table 63-17 Comparison of results (percent composition) of serum protein electrophoresis

Method	Albumin %	Alpha$_1$ %	Alpha$_2$ %	Beta %	Gamma %
Cellulose acetate electrophoresis*	69.64	3.22	7.86	8.93	10.34
	±1.43 SD	±0.37 SD	±0.59 SD	±0.86 SD	±0.82 SD
	2.2 CV	11.4 CV	7.7 CV	9.7 CV	7.9 CV
Agarose gel electrophoresis†	69.4	3.78	9.56	6.87	10.23
	±0.76 SD	±0.35 SD	±0.61 SD	±0.74 SD	±0.66 SD
	1.1 CV	9.1 CV	6.4 CV	10.7 CV	6.4 CV

*Fisher Pool Lot No. 133-101, n = 107. Ponceau S stain, Beckman Microzone System, Beckman Densitometer at 520 nm.
†Same pool, n = 79. Amido black 10B stain, Corning Electrophoresis System, Beckman Densitometer at 520 nm.
CV, Coefficient of variation; *SD*, standard deviation.

tein content is increased by ultracentrifugation, dialysis, or other concentration techniques.

PROCEDURE: AGAROSE GEL ELECTROPHORESIS

Principle

At pH 8.6, most serum proteins will have a negative charge and will separate when placed in an electric field. The separation takes place on an agarose solid support. After fixing, the separated bands are visualized with amido black 10B stain.

Reagents

Sodium hydroxide, 180 g/L (4.5 mol/L). Dissolve 180 g of NaOH in 700 mL of distilled water. Make up to 1 L with distilled water. Store in a plastic bottle. This is stable for at least 1 year at room temperature.

Tricine buffer, 0.05 mol/L, pH 8.6. Place 8.9 g of Tricine in a 1 L flask, and add 900 mL of distilled water to dissolve. Titrate the pH to 8.6 on a pH meter with 4.5 mol/L NaOH. Make up to 1 L with distilled water. This is stable for at least 2 months in a glass bottle at 4° C.

Amido black 10B stain, 2 g/L (3.2 mmol/L) in 5% (v/v) (0.88 mol/L) acetic acid. Transfer 2 g of amido black 10B to a 1 L volumetric flask. Add 5% acetic acid to dissolve the dye, and make up to the 1 L mark with 5% acetic acid. Store the stain solution in an airtight glass container. This is stable for at least 3 months at room temperature.

Five percent acetic acid clearing solution (0.88 mol/L). Add 50 mL of glacial acetic acid to 800 mL of distilled water in a volumetric flask. Allow the solution to cool, and then bring to 1 L with distilled water. Stable for at least 1 year at room temperature.

Assay

Equipment:

1. Cassette electrophoresis cell and power supply; several versions of this type of electrophoresis cell are available. They are convenient to use and do not require paper wicks to connect the electrophoresis support medium with the buffer.
2. Incubation oven set at 65° C.
3. Agarose gel film. As with the electrophoresis cell, several variations of agarose films are available. The agarose film manufactured by Corning Medical comes with sample wells (Agarose Universal Electrophoresis Film). Others use a template for sample applications, such as Paragon (Beckman Instruments, Inc., Carlsbad, CA) and Panagel.
4. Quantitative microliter sample dispenser, such as Hamilton.
5. Disposable sample tips.
6. Stir-stain dishes and humidity chambers.
7. Disposable liners for stir-stain dishes.
8. Densitometer with 520 nm capability.

Follow manufacturer's instructions for the electrophoresis procedure.

1. Stain the agarose film in the amido black stain solution for 15 minutes.
2. Transfer the agarose film to the first 5% acetic acid clearing solution, and leave there for 30 seconds. Use forceps and disposable gloves to handle the gels.
3. Wipe the moisture from the back of the film, and place it in the oven. Dry at 65° C for 15 to 20 minutes or until dry.
4. Remove the agarose film from the oven, and allow it to cool.
5. Rinse agarose film in the first 5% acetic acid for 1 minute.
6. Transfer the films to the second 5% acetic acid, and rinse until clear.
7. Wipe moisture from the back of the film, and dry for 10 minutes at 65° C.
8. Scan the plate at 520 nm, and obtain the percent value of each fraction.

An example of such patterns is presented in Fig. 63-5.

Calculations

To quantitate each well-defined fraction in g/L, use the following formula:

$$\text{g/L} = \frac{\text{Fraction (in \%)} \times \text{Total protein (g/L)}}{100}$$

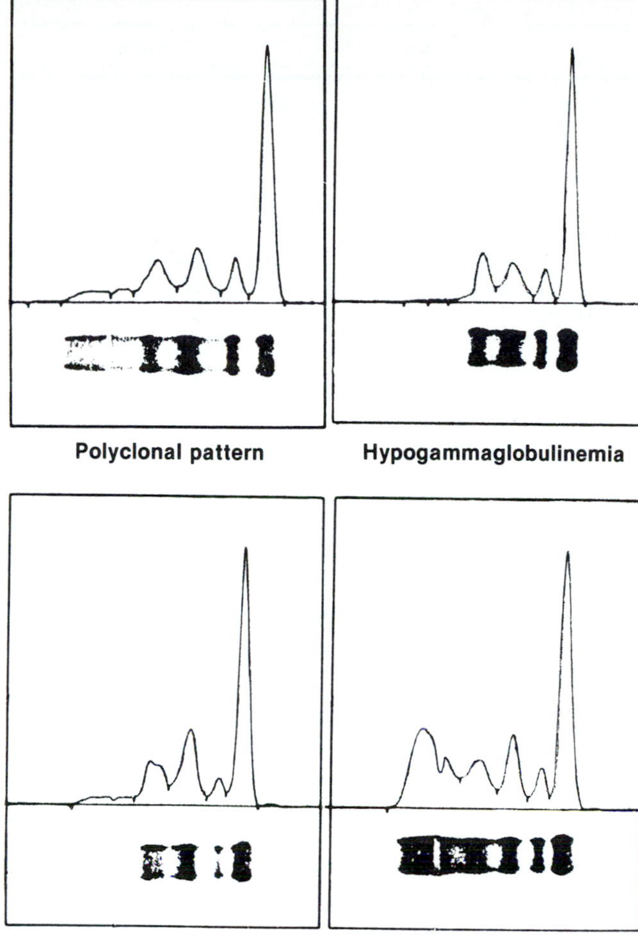

Fig. 63-5 Serum electrophoretic patterns in disease.

REFERENCE RANGE

Each laboratory performing serum protein electrophoresis should establish reference ranges on the populations it serves. In general, reference values for the five well-defined fractions are as follows:

Protein	g/L
Total protein	60-80
Albumin	32-50
$Alpha_1$	10-40
$Alpha_2$	6-10
Beta	6-13
Gamma	7-15

The percent composition is given in Table 63-17.

REFERENCES

1. Laurell, CB, Jeppsson, JO, and Tejler L: Plasma protein analysis, Bedford, MA, 1978, Millipore Corp.
2. Sun, T, Lien, YY, and Gross, S: Clinical applications of a high-resolution electrophoresis system, Ann Clin Lab Sci 8:219-227, 1978.
3. Aronsson, T, and Grönwall, A: Improve separation of serum proteins in paper electrophoresis: a new electrophoresis buffer, Scand J Clin Lab Invest 9:338-345, 1957.
4. Sinclair, D, Kumeraratre, DS, Forrester, JB, et al: The application of isoelectric focusing to routine screening for paraproteinemia, J Immunol Methods 64:147-156, 1983.
5. Roos, RP, and Lichter, M: Silver staining of cerebrospinal fluid IgG in isoelectric focusing gels, J Neurosci Methods 8:375-380, 1983.
6. Ojala K, and Weber, TH: Some alternatives to the proposed selected method for "agar gel electrophoresis," Clin Chem 26:1754-1755, 1980.
7. Alper, CA: Plasma protein measurements as a diagnostic aid, N Engl J Med 291:287-290, 1974.
8. Wieme, RJ: Agar gel electrophoresis, New York, 1965, Elsevier/North Holland, Inc.

Total serum protein

ANTHONY KOLLER
LAWRENCE A. KAPLAN

Clinical significance: p. 359
Chemical class: protein

PRINCIPLES OF ANALYSIS

The earliest approach to the determination of total protein in serum is the determination of protein nitrogen. The Kjeldahl procedure[1] is a method for determining the total nitrogen content in biological material (method 1, Table 63-18). The nitrogen-containing compounds in serum are converted to NH_4^+ by oxidation in a digestion mixture of concentrated sulfuric acid, a catalyst, and a salt to increase the boiling point of the mixture. The NH_4^+ is best analyzed by conversion to NH_3 by the addition of alkali. After steam distillation into a boric acid solution, the NH_3 is then titrated with a standardized solution of HCl. The NH_4^+ can also be quantitated photometrically with Nessler's reagent. A correction for nonprotein nitrogen is performed by use of a protein-free filtrate of serum. Based on the assumption that proteins from biological sources contain 16% nitrogen by weight, the total nitrogen content (in grams per liter) of a sample minus the nonprotein nitrogen is multiplied by 6.25 to obtain the protein content in grams per liter. The Kjeldahl procedure is rarely used in routine analysis.

Instruments that can automatically perform total nitrogen determinations are now available (Antek Instruments, Houston, TX). Dilutions of liquids or digested solid materials are introduced into the combustion chamber set at approximately 1000° C. In the presence of flowing oxygen gas, the material is oxidized, with the nitrogen forming nitric oxide (NO). The nitric oxide reacts with ozone (O_3) to form an excited form of nitrogen dioxide (NO_2^*), which rapidly decays to form the ground state NO_2 and light.

$$\text{Nitrogenous material} + O_2 \xrightarrow{1000^\circ\text{ C}} NO + \text{Combustion products}$$

$$NO + O_3 \rightarrow NO_2^* + O_2$$

$$NO_2^* \rightarrow NO_2 + \text{light (650 to 900 nm)}$$

Table 63-18 Methods of total protein analysis

Method	Type of analysis*	Principle	Usage	Comment
1. Kjeldahl	Quantitative, protein nitrogen determination	Oxidation of N-containing compounds to NH_4^+; conversion to NH_3 with alkali; steam distillation into boric acid and titration with standard HCl; correction for nonprotein nitrogen	Historical	Cumbersome and time consuming; good accuracy and precision; used as reference method in past
2. Biuret	Quantitative, increased absorption at 540 nm; EP or K	Formation of violet-colored complex between Cu^{+2} ions and peptide bonds in alkaline medium	Usually adapted to automated analysis	Good specificity, accuracy, and precision; has been proposed as basis for reference method
3. Lowry	Quantitative, increased absorption at 745 to 750 nm; EP	Pretreatment with alkaline copper solution followed by addition of Folin and Ciocalteu phenol reagent; reduction of phosphotungstic and phosphomolybdic acids produces color	Historical for serum, useful for other more dilute biological fluids	Good sensitivity but poor specificity and accuracy
4. Ultraviolet absorption	Quantitative, absorption at 210 nm	Light absorption by peptide bonds	Manual or semiautomated; not used routinely	Good sensitivity; acceptable accuracy and specificity; rapid
5. Refractometry	Quantitative	Measurement of refractive index of dissolved solids	Manual	Acceptable accuracy and precision; rapid; susceptible to false-positive interferences from variety of compounds

**EP*, End point; *K*, kinetic.

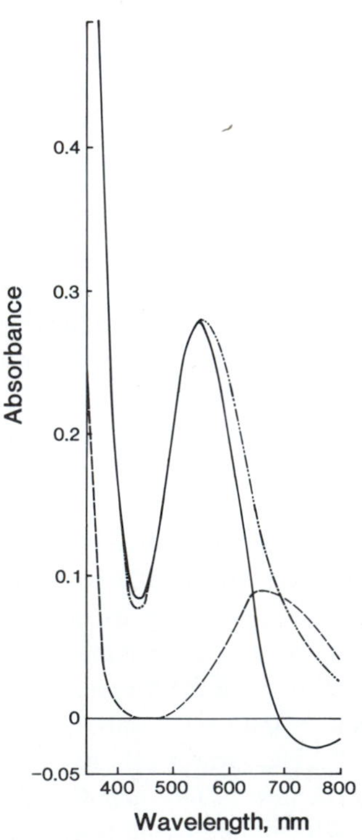

Fig. 63-6 Absorption spectra of biuret reagent and 50 g/L albumin. *Dashed line,* Biuret reagent versus water blank; *dotted-dashed line,* biuret reagent and protein versus water blank; *solid line,* biuret reagent and protein versus biuret reagent blank.

The amount of nitrogen present in the injection volume is determined by comparison of the chemiluminescence signal to the signal of a standard. The instrument is sensitive to 20 μg/L.

The most frequently used method for determining total protein in serum is the biuret reaction. In this reaction, cupric ion complexes with the nitrogen atoms of the peptide linkages of protein. These complexes react with the biuret (carbamylurea, $NH_2CONHCONH_2$) reagent in a highly alkaline (usually NaOH) solution to form a violet-colored condensation product with an absorption maximum of 540 nm (Fig. 63-6 and method 2, Table 63-18).

The method of Lowry et al.,[2] a widely used procedure for the quantitative determination of protein, has also been applied to serum protein analysis (method 3, Table 63-18). In this technique, the protein is pretreated with an alkaline copper solution. On addition of the phenol reagent of Folin and Ciocalteu,[3] the color produced (A_{max}, 745 to 750 nm) results from the reduction of the phosphotungstic and phosphomolybdic acids to molybdenum blue and tungsten blue by the copper-peptide bond complex and by the tyrosine and tryptophan of the protein. Cystine, cysteine, and histidine are also reactive under these conditions but to a lesser extent.[4]

Serum proteins have also been estimated by use of ultraviolet absorption (method 4, Table 63-18). Protein solutions show strong absorption in the 270 to 290 nm region and in the 200 to 225 nm region. Virtually all the ultraviolet absorption in serum is attributable to protein. The absorption at the higher wavelengths is attributable to the aromatic rings of tyrosine, tryptophan, and phenylalanine. The absorption at the lower wavelengths is much more intense (about 20-fold more intense than at 280 nm) and is primarily attributable to the peptide bond.

Another direct approach to the estimation of total protein relies on the measurement of refractive index (method 5, Table 63-18). This method is based on the refraction of incident light by total dissolved solids; for serum this reflects the mass of protein present. Since serum contains a substantial mass of nonprotein, dissolved solids (electrolytes, glucose, and so on), the refractometer must be specifically calibrated with serum of a known protein concentration. Refractometers are temperature calibrated as well and should not be used at a temperature significantly different from 25° C. To use the instrument, one simply allows the sample to fill the space between the closed cover and the glass plate by capillary action. Alternatively, one can place a drop of serum on the plate and carefully close the cover. One can view the refractometer's scale by pointing the instrument at a light source or using a built-in light source. The sharp line dividing the dark and light fields is used as an indicator to read the protein concentration (in g/dL) off the scale. The ease of use and rapidity of measurement makes refractometry a useful method for stat analyses of serum protein.

Almost all routine total serum protein determinations are performed by use of the biuret reaction. The only other approach used routinely to any extent is the measurement of the refractive index.

The Kjeldahl procedure is still used in method evaluation work, since it has historically been the standard to which other methods were compared.

The Lowry procedure, though a hundred times more sensitive than the biuret method, is not routinely used for serum protein determination. The method lacks specificity, and many substances are known to interfere with it.[4]

Reinhold[5] described a reagent that was compatible with protein-precipitating salts such as sodium sulfate and sodium sulfite. Doumas[6] introduced a modification of the Reinhold reagent that had a linearity up to 120 g/L. This proposed reagent was incorporated in a candidate reference method for the determination of total protein in serum (Table 63-19).[7,8]

In 1972 the National Committee for Clinical Laboratory Standards (NCCLS) adopted the use of bovine serum albumin as a reference material to be used primarily for assays for total protein by spectrophotometric procedures such as the biuret and the Lowry method. In 1977, the National Bureau of Standards (NBS) released a commercially prepared bovine serum albumin that met NCCLS specifications.[9] The standard is available in lyophilized form (Standard Reference Material No. 926) and as a 70 g/L (1.06 mmol/L) solution (SRM No. 927). This certified albumin is recommended as the most suitable standard for the biuret procedure.

Table 63-19 Candidate reference method reaction conditions for total serum protein

Conditions	Requirements
Temperature	25° C
Final concentration of reagent components	$CuSO_4 \cdot 5H_2O$: 12 mmol/L KNa tartrate ($KNaC_4H_4O_6 \cdot 4H_2O$): 30 mmol/L KI: 30 mmol/L NaOH: 0.6 mol/L
Fraction of sample volume	0.02
Sample	Serum
Linearity	140 g/L
Time of reaction	End point at 60 minutes
Major interferences	Dextran
Precision	Within-run ($\bar{X}$, 67 g/L), CV, 0.15% Day-to-day ($\bar{X}$, 67 g/L), CV, 0.60%

From Doumas, BT, Bayse, DD, Carter, RJ, et al: Clin Chem 27:1642-1650, 1981.
$\bar{X}$, Mean; *CV*, coefficient of variation.

SPECIMEN

Serum, exudates, and plasma may be used; all usually yield comparable results,[10] although because of the presence of fibrinogen, plasma levels for total protein are 2 to 4 g/L higher than serum levels. The refractometer is not calibrated for the matrices of other fluids and can be used only with serum and plasma. Total protein is stable in serum and plasma for 1 week at room temperature, for at least 1 month when refrigerated, and for up to 2 months at −20° C.[11,12] Lipemic samples may have to be treated for accurate results[12]; if not, such samples should be rejected for analysis. Hemolysis may cause a 3% increase in total protein values for each 1 g/L of hemoglobin present in the sample.

REFERENCE RANGE

The combined male (134 subjects) and female (97 subjects) range established by the Doumas method was 66.6 to 81.4 g/L.[7] Results for males are approximately 1 g/L higher than results for females; this difference is probably not of clinical significance. The subjects were healthy adults who had fasted for 10 to 12 hours (although no differences in serum protein concentrations have been noted after meals) and had been in the upright position for at least 2 hours before blood was collected without anticoagulant. This range is similar to the one established by Reed et al.,[13] which was 66.0 to 83.0 g/L based on 1419 subjects.

There is a slight and probably insignificant decrease in serum protein concentration with age.[14] In newborns, the mean serum protein concentration is 57 g/L, increasing to

60 g/L (±4 g/L) by 6 months[15] and to adult levels by about 3 years of age. Serum protein levels of premature infants can be much lower than that of term infants, ranging from 36 to 60 g/L.[11]

Shifts in body fluid between the vascular bed and the interstitial spaces can bring about significant changes in serum protein concentration. For example, total serum protein is lower by 4 to 8 g/L with the subject supine than with the subject ambulatory or in the upright position.[11] In addition, for several hours after vigorous exercise, an increase in serum protein concentration of 4 to 8 g/L may be noted.[12] In pregnancy, serum protein concentration has been noted to decrease from 69 to 61 g/L by parturition.[16]

REFERENCES

1. Archibald, RM: Nitrogen by the Kjeldahl method, Stand Methods Clin Chem 2:91-99, 1958.
2. Lowry, OH, Rosebrough, NJ, Farr, AL, and Randall, RJ: Protein measurement with the Folin reagent, J Biol Chem 193:265-275, 1951.
3. Folin, O, and Ciocalteu, V: On tyrosine and tryptophane determinations in proteins, J Biol Chem 73:627-650, 1927.
4. Peterson, GL: Review of the Folin phenol protein quantitation method of Lowry, Rosebrough, Farr, and Randall, Anal Biochem 100:201-220, 1979.
5. Reinhold, JG: Total protein albumin and globulin, Stand Methods Clin Chem 1:88-97, 1953.
6. Doumas, BT: Standards for total serum protein assays—a collaborative study, Clin Chem 21:1159-1166, 1975.
7. Doumas, BT, Bayse, DD, Carter, RJ, et al: A candidate reference method for determination of total protein in serum. I. Development and validation, Clin Chem 27:1642-1650, 1981.
8. Doumas, BT (chairman), Bayse, DD, Borner, K, et al: A candidate reference method for determination of total protein in serum. II. Test for transferability, Clin Chem 27:1651-1654, 1981.
9. Reeder, DJ, and Schaffer, R: Standard reference material (SRM) for total protein determination—bovine serum albumin, Clin Chem 23:1136, 1977.
10. Chorine, V: Influence des anticoagulants sur le dosage des elements due sang, Ann Inst Pasteur Immunol 63:213-256, 1939.
11. Cannon, DC, Olitzky, I, and Inkpen, JA: Proteins. In Henry, RJ, Canno, DC, and Winkelman, JA, editors: Clinical chemistry: principles and technics, Hagerstown, MD, 1974, Harper & Row, Publishers, Inc.
12. Peters, T, Jr, Biamonte, GT, and Doumas, BT: Protein (total protein) in serum, urine, and cerebrospinal fluid; albumin in serum. In Faulkner, WR, and Meites, S, editors: Selected methods of clinical chemistry, Washington, DC, 1982, American Association for Clinical Chemistry.
13. Reed, AH, Cannon, DC, Winkelman, JW, et al: Estimation of normal ranges from a controlled sample survey. I. Sex and age-related influence on the SMA 12/60 screening group of tests, Clin Chem 18:57-66, 1972.
14. Keating, FR, Jr, Jones, JD, Elveback, JR, and Randall, RV: The relationship of age and sex to distribution values in healthy adults of serum calcium, inorganic phosphorus, magnesium, alkaline phosphatase, total proteins, albumin, and blood urea, J Lab Clin Med 73:825-834, 1969.
15. Trevorrow, V, Klaser, M, Patterson, JP, and Hill, RM: Plasma albumin, globulin, and fibrinogen in healthy individuals from birth to adulthood, J Lab Clin Med 27:471-486, 1941.
16. Elliott, JR, and O'Kell, RT: Normal clinical chemistry values for pregnant women at term, Clin Chem 17:156-157, 1971.

Urine protein, total

DAVID C. HOHNADEL

Clinical significance: p. 346
Chemical class: protein, glycoprotein

PRINCIPLES OF ANALYSIS

The quantitative analysis of total protein in urine can be classified into three main approaches: turbidimetric, dye binding, and chemical.

The reagents most frequently used for turbidimetric estimation of protein are trichloroacetic acid, sulfosalicylic acid, and more recently benzethonium chloride in alkali.[1] For dye-binding protein estimation, Coomassie brilliant blue and Ponceau S are used most often. For chemical determination of protein, the biuret reaction, the Folin-Lowry reaction, and the reaction of ferric chloride with tannic acid protein precipitates have been used.

In turbidimetry, a protein precipitant, such as trichloroacetic acid (TCA), is added to the sample, and the denatured protein precipitates in a fine suspension that is quantitated turbidimetrically (method 1, Table 63-20). The turbidity varies appreciably with the chemical nature of the acid precipitant, the type of protein, and the concentration of the acid, the temperature, and the time elapsed between addition of the acid and turbidimetric measurement. Any wavelength may be used, but light dispersion will increase as the wavelength is decreased.

In 1976, Bradford[2] proposed the use of the dye Coomassie brilliant blue G-250 for the estimation of protein at low concentration (method 2, Table 63-20). The binding of the dye to protein causes a shift of the absorption maximum from 465 nm (red form) to 595 nm (blue form). The increase in absorbance at 595 nm is used to monitor the extent of the reaction. The binding is complete in about 2 minutes, and the color is stable for 1 hour. The method has been adapted for automated analysis.[3,4]

In 1973, Pesce and Strande[5] developed a procedure for the determination of protein in urine based on the formation of a protein-dye complex with Ponceau S (method 3, Table 63-20).

The biuret reaction involves the reaction of biuret reagent (Cu^{+2} ions) in alkaline solution with the peptide bonds in proteins. Since the biuret reaction is relatively insensitive and sometimes suffers from interferences when applied directly to urine, the protein is usually concentrated before analysis. This is accomplished by precipitation of the protein with either trichloroacetic acid or ethanolic hydrochloric-phosphotungstic acid. The precipitated protein is concentrated by centrifugation. The dissolved protein is then reacted with the biuret reagent (method 4, Table 63-20).

The Folin-Lowry (method 5, Table 63-20) reaction,

Table 63-20 Methods for total urine protein analysis

Method	Sensitivity (mg/L)	Principle	Usage	Comments
1. Turbidimetric		Protein-denaturing agent precipitates proteins; resulting turbidity is measured photometrically at either 450 or 620 nm	Most common technique	Technically simple, rapid, fairly accurate
a. Sulfosalicylic acid (SSA)	10 to 25		Frequently used method	Overestimates albumin, which produces 4 × greater turbidity than for gamma globulins
b. Trichloroacetic acid (TCA)	20		Frequently used method	Estimates albumin and gamma globulins equally
c. Benzethonium chloride	10		Most frequently used method	Most sensitive of turbidimetric techniques
2. Coomassie brilliant blue	2.5	Dye binds to NH_3^+ residues in proteins with resulting absorption at 595 nm	Second most frequently used method	Rapid, highly sensitive; overestimation of albumin
3. Ponceau S	20	Precipitation of dye-protein complex, which is redissolved in alkali; color intensity is measured at 560 nm	Infrequently used	Reacts with albumin and gamma globulins equally; aminoglycosides can interfere
4. Biuret (modified)	5 to 17	Proteins are concentrated by precipitation with TCA or ethanolic-HCl-phosphotungstic acid (Tsuchiya's reagent) and redissolved in biuret reagent (alkaline-Cu^{+2}); the Cu^{+2} reagent forms colored complex with peptide bonds, which is measured at 540 nm	Used by small percentage of laboratories	With Tsuchiya's reagent this method is very sensitive with a good linear range
5. Folin-Lowry	10	Folin reagent (mixture of molybdic and phosphotungstic acids and alkaline copper) reacts with peptide bonds, tyrosine, and tryptophan residues to produce blue color monitored at 650 nm	Infrequently used	Very sensitive but color varies with amino acid composition of protein; urate can interfere
6. Tannic acid precipitation	5	Tannic acid is used to precipitate proteins that are redissolved in triethanolamine-$FeCl_3$ to produce a purple color (510 nm)	Infrequently used	Overestimates albumin; erratic results for urines with pH >6
7. pH indicator	100	Protein (principally albumin) binds to pH indicator dye causing color change	Semiquantitative method; most frequently used	Reacts primarily with albumin; false positive if urine pH >8

which is 100 times more sensitive than the unmodified biuret reaction, has also been applied to urine protein estimations after a purification step to remove interfering materials present in urine. The Folin reagent consists of phosphotungstic and phosphomolybdic acids dissolved in phosphoric and hydrochloric acids. The initial step in this assay, similar to the Cu^{+2} interaction seen in the biuret reaction, is the binding of Cu^{+2} to peptide bonds and to the amino acids tyrosine and tryptophan under alkaline conditions. The Folin reagent is then added, oxidizing the

Cu^{+2} protein complex, the copper-tyrosine, and the tryptophan complexes and reducing the phosphotungstic and phosphomolybdic acids to the chromogens molybdenum blue and tungsten blue. The blue color has a broad absorption maximum but is most frequently measured at 660 nm.

A very widely used, semiquantitative technique for estimating urinary proteins is a dye binding-pH indicator technique (method 6, Table 63-20). This technique is the one employed by urine dipsticks. The reagent strips use a pH indicator, such as tetrabromo blue, embedded in the pad. When protein binds to the dye, there is a change in the pH environment (equivalent to a pH less than 3) of the dye, resulting in a change in the color of the dye.

$$\text{Tetrabromol blue} \xrightarrow{\text{Protein}} \text{Green-blue color}$$

Because of the widespread use of the urine dipsticks, this method may be the technique used most often for estimating urine protein levels. The dipstick method is also used to screen urines submitted for a quantitative urine protein determination. The result of the semiquantitative dipstick analysis helps the technologist choose the proper sample dilution for the quantitative procedure.

Total protein determination is much more difficult in urine than in serum. The concentration of urine protein is normally low (100 to 200 mg/L); large sample-to-sample variation in the amount and composition of proteins is common; the concentration of nonprotein interfering substances is high relative to the protein concentration and very variable; and the inorganic ion content is high. All these factors affect the precision and accuracy of the various methods.

Of the methods and techniques proposed over the years for total protein determination in urine, the turbidimetric procedures have gained the widest acceptance. This is attributable, for the most part, to their sensitivity and simplicity. A serious drawback encountered with the sulfosalicylic acid procedure is that it produces four times more turbidity with albumin than it does with gamma globulin.[6]

Some drugs have been shown to cause a positive interference with turbidimetric methods. Miconazole interfered with the trichloroacetic acid (TCA) procedure in urine when the protein content was above 0.5 g/L.[7] Benzylpenicillin and cloxacillin both interfered with the TCA procedure adapted to a centrifugal analyzer.[8] Changes in the blank and other modifications have been suggested to overcome problems of the turbidimetric procedures.

The three turbidimetric methods have been compared for sensitivity and bias with purified fractions of human serum proteins.[9] Nishi and Elin concluded that these methods will give results of varying accuracy for a urine sample that contains proteins different from those in the standard material. The protein-Coomassie brilliant blue dye complex has a high extinction coefficient and has four times the sensitivity of the Folin-Lowry method. However, as is true for the Folin-Lowry method, the amount of color development varies with the nature of the protein standard used and the nature of the protein in the sample.[10] Bradford[2] found that interfering color was caused by relatively large amounts of detergents, such as dodecyl sulfate. Thymol, often used as a urine preservative, and salicylates were shown to interfere with this procedure.[11,12]

McIntosh proposed the application of Coomassie blue to the estimation of protein in urine.[12] The correlation of the Coomassie blue method with the Sephadex-biuret method of Doetsch and Gadsden[13] was acceptable ($r = 0.91$), but the comparison with a turbidimetric methods was poor ($r = 0.65$). Slight differences in reagent composition, standard, method, or dye lot can produce significant differences in the response of the assay.

The quantitative turbidimetric methods are the methods of choice because of their simplicity and sensitivity. Although the sulfosalicylic acid methods are the most sensitive, the trichloroacetic acid method is preferred because it measures both albumin and gamma globulins without unacceptable bias for either class of protein. The trichloracetic acid turbidimetric procedure, described in the next section, can be performed using equipment readily available in any clinical chemistry laboratory. For analyses that require more accurate values at low protein concentrations, the method using Tsuchiya's reagent followed by the biuret reaction is recommended.[14]

A summary of reaction conditions for common methods of analysis is given in Table 63-21.

SPECIMEN

A 24-hour or 12-hour urine specimen with no preservatives is preferred. Other timed samples are acceptable, but as the time shortens, the possibility of volume errors with discarded samples increases. The sample should be well mixed and an aliquot frozen if the sample cannot be analyzed within 48 hours after collection. Tightly stoppered frozen specimens are stable for at least 1 year.

The use of random urine samples instead of timed urine samples is advocated by some authors[15] as a convenient index of proteinuria. However, the great variability of individual excretion and the nonuniform elimination of fluid and protein argue against this practice, and it should be discouraged.

PROCEDURE: TURBIDIMETRIC ASSAY WITH TRICHLOROACETIC ACID[16]

Principle

Trichloroacetic acid, 12.5%, added to a solution of protein in low concentration precipitates the protein as a fine suspension, which is quantitated photometrically. Trichloroacetic acid (TCA) produces approximately equal turbidities with both albumin and globulins when one uses the procedure outlined here.

Table 63-21 Reaction conditions for analysis of total protein in urine

Condition	TCA*	Biuret†	Coomassie brilliant blue‡
Temperature	20° to 25° C	0° C, precipitation; ambient, reaction	Ambient
Sample volume	800 μL	20 mL	100 μL
Fraction of sample volume	0.80	0.5 (for initial precipitation)	0.02
Final concentration of reagents	TCA: 153 mmol/L (25 g/L) TCA in blank: 153 mmol/L	Precipitation step HCl: 0.7 mol/L Ethanol: 895 ml/L Phosphotungstic acid: 17.4 g/L Colorimetric reaction (biuret) Potassium iodide: 4 mmol/L Potassium-sodium tartrate: 22 mmol/L Cupric sulfate: 6.4 mmol/L Sodium hydroxide: 600 mmol/L Blank contains all but $CuSO_4$	Coomassie brilliant blue: 100 mg/L Ethanol: 47 g/L Phosphoric acid: 85 g/L
Time of reaction	35 minutes	15-minute precipitation 20-minute reaction	2 minutes
Wavelength	420 nm	540 nm	595 nm
Linearity	500 mg/L	300 mg/L	1000 mg/L
Precision§ mean (%CV)	1284 mg/L (15%)	1432 mg/L (5%)	1467 mg/L (16%)
Interferences	Miconazole, penicillin derivatives, radiocontrast material	None	Detergents, thymol, and salicylates

*Trichloroacetic acid, turbidimetric assay.
†Biuret using Tsuchiya's reagent to precipitate and concentrate protein. From Savory, J, et al: Clin Chem 14:1160, 1968.
‡From Bradford, MM: Anal Biochem 72:248, 1976.
§*CV*, Coefficient of variation. From College of American Pathologists Quality Assurance Survey data, 1985.

Reagents

Albustix (Ames, Inc., Division of Miles Co., Elkhart, IN)

Stock trichloroacetic acid, 100% (1 g/mL, 6.12 mol/L). PRECAUTION: TCA is extremely corrosive and must be handled with caution. The person preparing this solution must wear gloves and safety glasses during the process. Extreme care must be taken to avoid splashing, and any spills must be carefully cleaned up immediately.

Add 175 mL of deionized water directly into one 500 g bottle of TCA (Fisher-certified ACS, No. A-322). Close tightly, and allow to dissolve overnight, swirling occasionally. Transfer the solution quantitatively with rinsings to a 500 mL volumetric flask, and dilute to 500 mL with deionized water. Mix carefully by inversion 10 times. The 100% TCA should be stored in a brown glass bottle (avoid metal bottle-cap liner), labeled ''CORROSIVE,'' at room temperature in a hood. The solution is stable for 12 months at room temperature.

Working trichloroacetic acid, 12.5% (765 mmol/L). PRECAUTION: Because of the corrosive nature of TCA, gloves and safety glasses must be worn during preparation of this solution. All spills must be carefully cleaned up.

With a 25 mL volumetric pipet, carefully transfer 25 mL of stock 100% TCA into a 200 mL volumetric flask. Dilute to 200 mL with deionized water, and mix by inversion 10 times. Store in a brown glass bottle (no metal cap liner), labeled ''CORROSIVE,'' at room temperature. The solution is stable for 1 month at room temperature.

Controls. Use commercially available lyophilized serum pools.

Stock standard. Use pooled serum from routine daily analysis. Pool clear, nonlipemic, nonicteric, nonhemolyzed specimens. Analyze the total protein concentration by a biuret procedure. The pooled serum must be diluted to achieve a protein concentration similar to that found in urine, as demonstrated in the following paragraph.

Working standards. *Assume* a total protein pool value of approximately 60 g/L was obtained by measurement. Dilute 8.3 mL of 60 g/L stock to 1000 mL with saline to give a working standard of 498 mg/L, a nominal 500 mg/L standard. The following saline dilutions can be used to plot a curve. All dilutions are prepared volumetrically to 10 mL with saline. Refrigerated working standards are stable for several days.

Milliliters of stock (498 mg/L)	Milliliters of saline	Working standard (mg/L)
2.0	8.0	100
4.0	6.0	200
6.0	4.0	300
8.0	2.0	400
10.0	—	500

Analyze standards in same manner as urines. Plot absorbance versus milligrams per liter.

Assay

Equipment: spectrophotometer or photometer capable of reading at 420 nm; centrifuge capable of 1500 *g*.

1. Centrifuge 10 mL of urine. Check urine with Albustix, and record results. If the dipstick estimate of urine protein concentration is +1 or greater, dilute urine with saline before performing analysis.
2. Set up 2 rows of 1.5 mL plastic centrifuge tubes; use a test and blank tube for each urine specimen.
3. Pipet 0.8 mL of urine into both test and blank tubes.
4. Add 0.2 mL of 12.5% TCA to both sets of tubes, and mix each gently by vortexing at a slow speed or inverting after sealing the tubes with Parafilm. Mix immediately after addition of TCA. Temperature control is critical. Ambient temperature should be between 23° and 27° C for proper results. All reagents and samples must also be at room temperature.
5. Twenty minutes after TCA addition to the blank tubes, centrifuge for 10 minutes at 1500 *g*.
6. Let test tubes stand for 35 minutes, and invert gently. Avoid excessive mixing to prevent trapping air bubbles and thus increasing the apparent turbidity. Immediately read absorbance at 420 nm against blank.

Calculations

Read mg/L off the absorbance-versus-mg/L standard curve. Do not extend standard curve beyond highest value employed to prepare the curve in present use.

Random urines. Report in mg/L.

24-hour urines. To calculate total protein excretion (TV, total volume) per 24 hours:

$$\text{mg/L} \times \text{TV (liters)/24 hours} = \text{mg/24 hours}$$

Dilutions. To calculate the dilutions, use the following:

$$\text{mg/L} \times \text{TV/24 hours} \times \text{Dilution factor} = \text{mg/24 hours}$$

Notes

1. With TCA being used as the precipitating agent, similar turbidity values are obtained for both albumin and globulin.
2. Any wavelength in the short end of the visible spectrum can be used. The choice of 420 nm is entirely arbitrary. At 600 nm, the method is linear to 1000 mg/L but is less sensitive to the amounts of protein in normal urine.
3. In regard to interfering colors, urine color is usually corrected for by use of the urine blank.
4. The validity of the preliminary results obtained with the reagent strip tests depends on the pH of the urine. In heavily buffered alkaline urine, the buffer on the indicator stick is insufficient to achieve the pH required for adequate indicator function, thus leading to false-positive results. Deeply pigmented urines also tend to give false-positive results.
5. Excretion of protein in the urine is not constant and varies considerably over a 24-hour period with no particular pattern. Testing random samples therefore can be misleading.
6. There can be discrepancies between results obtained using the Albustix (which measures albumin) and those obtained using the TCA turbidimetric procedures. A negative or low Albustix result and a high TCA result could be caused by the following:
   a. Presence of myeloma protein (gamma globulin, Bence Jones proteins) in urine.
   b. Nonhomogeneous suspensions caused by the presence of certain drugs (such as tolmetin, a nonsteroid anti-inflammatory drug used in treatment of rheumatoid arthritis). Metabolites of this drug can give false-positive results with acid precipitation procedures.

REFERENCE RANGE

Less than 150 mg/24 hours.

REFERENCES

1. Iwata, I, and Nishikaze, O: New micro-turbidimetric method for determination of protein in cerebrospinal fluid and urine, Clin Chem 25:1317-1319, 1979.
2. Bradford, MM: A rapid and sensitive method for the quantitation of microgram quantities of protein utilizing the principle of protein-dye binding, Anal Biochem 72:248-254, 1976.
3. Heick, HMC, Begin-Heick, N, Acharya, A, and Mohammed, A: Automated determinations of urine and cerebrospinal fluid proteins with Coomassie brilliant blue and the Abbott ABA-100, Clin Biochem 13:81-85, 1980.
4. Helmer, GR, Borer, WZ, and Palmer, JJ: Urine protein concentration measured by selected method and multistat microprotein method, Clin Chem 28:1628, 1982.
5. Pesce, MA, and Strande, CS: A new micromethod for determination of protein in cerebrospinal fluid and urine, Clin Chem 19:1265-1267, 1973.
6. Schriever, H, and Gambino, SR: Protein turbidity produced by trichloroacetic acid and sulfosalicylic acid at varying temperatures and varying ratios of albumin and globulin, Am J Clin Pathol 44:667-672, 1965.
7. Lievens, MM, and Celis, PJ: Drug interference in turbidimetry and colorimetry of proteins in urine, Clin Chem 28:2328, 1982 (Letter).
8. Muir, A, and Hensley, WJ: Pseudoproteinuria due to penicillins, in the turbidimetric measurement of proteins with trichloroacetic acid, Clin Chem 25:1662-1663, 1979.
9. Nishi, HH, and Elin, RJ: Three tubidimetric methods for determining total protein compared, Clin Chem 31:1377-1380, 1985.
10. Sedmak, JJ, and Grossberg, SE: A rapid, sensitive, and versatile assay for protein using Coomassie Brilliant Blue G-250, Anal Biochem 79:544-552, 1977.
11. McElderry, LA, Tarbit, IF, and Cassells-Smith, AJ: Six methods for urinary protein compared, Clin Chem 28:356-360, 1982.

12. McIntosh, JC: Application of a dye-binding method for the determination of protein in urine and cerebrospinal fluid, Clin Chem 23:1939-1940, 1977 (Letter).
13. Doetsch, K, and Gadsden, RH: Determination of total urinary protein, combining Lowry sensitivity and biuret specificity, Clin Chem 19:1170-1178, 1973.
14. Savory, J, Pu, PH, and Sunderman, FW, Jr: A biuret method for determination of protein in normal urine, Clin Chem 14:1160-1171, 1968.
15. Houser, M: Assessment of proteinuria using random urine samples, J Pediatr 104:845-848, 1984.
16. Shahangian, S, Brown, PI, and Ash, KO: Turbidimetric measurement of total urinary proteins: a revised method, Am J Clin Pathol 81:651-654, 1984.

CHAPTER 64

Toxicology and therapeutic drug monitoring (TDM)

Acetaminophen

JOSEPH SVIRBELY

Acetaminophen, *N*-acetyl-*p*-aminophenol, paracetamol
Clinical significance: p. 740
Molecular formula: $C_8H_9NO_2$
Molecular weight: 151.16 daltons
Merck Index: 39
Chemical class: *p*-aminophenol derivative

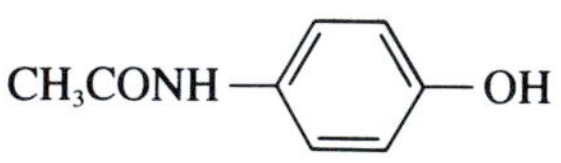

CH_3CONH—⟨benzene ring⟩—OH

PRINCIPLES OF ANALYSIS

Qualitative analysis

A spot test for the detection of acetaminophen in urine is available for stat laboratories (method 1, Table 64-1). This test requires the hydrolysis of the conjugated metabolites of acetaminophen, which then are allowed to react with *o*-cresol in ammonium hydroxide. The indigo color of the reaction product (indophenol blue) indicates the presence of acetaminophen. Acetaminophen can be detected by silica gel thin-layer chromatography (TLC) as a quenching spot under long ultraviolet radiation. Using the TLC screen described on p. 1100, one can find that acetaminophen has an R_f of 0.41.

Quantitative analysis

The first methods for analysis of serum acetaminophen were spectrophotometric.[1-3] Current methods include spectrophotometric, gas-liquid chromatographic (GLC), and high-performance liquid chromatographic (HPLC) procedures.

Glynn and Kendal[4] modified previous spectrophotometric methods (methods 2 and 3, Table 64-1). The hydrolysis step was eliminated, and instead plasma was deproteinated with trichloracetic acid solution followed by centrifugation. An aliquot of the supernatant was mixed with 6 M hydrochloric acid and sodium nitrite to form the nitrous acid derivative. Sulfamic acid was added, followed by sodium hydroxide; the ensuing yellow color was read 430 nm. The entire reaction took 15 minutes (method 3, Table 64-1). The Folin-Ciocalteu reagent has also been used (method 4, Table 64-1), as well as differential ultraviolet spectroscopy (method 5, Table 64-1). GLC has been used to quantitate acetaminophen (method 6, Table 64-1). In 1971 Prescott[5] published a GLC method utilizing flame ionization detection. The sample was extracted into ethyl acetate, and the extract was then evaporated. After reconstitution in pyridine and derivativization to the trimethylsilyl derivative, the extract was injected onto an OV-17 column at 200° C.

Thomas and Coldwell[6] modified Prescott's method by using a more powerful silylating reagent (Regisil, bis[trimethylsilyl]trifluoroacetamide) and a more selective extractant (diethyl ether). They also substituted an OV-1 column for Prescott's OV-17 column. Dechtiaruk et al.[7] formed the *o*-heptyl-*N*-methyl derivative and utilized temperature programming from 150° to 260° C at 16° C per minute with an OV-17 column.

HPLC has also been used to quantitate acetaminophen with several different chromatographic approaches being used. Howie et al.[8] have published a method employing reversed-phase HPLC (method 7, Table 64-1). The column was packed with 10-octadecylsilane-coated silica, and the mobile phase was water/acetic acid/ethyl acetate. Detection was at 254 nm. Plasma was deproteinated with trichloroacetic acid containing the internal standard 4-fluorophenol, and the supernatant was injected directly onto the column.

The methods developed by Horvitz and Jatlow,[9] Gotelli et al.,[10] and Lo and Bye[11] also used 10-octadecylsilane-coated silica columns, with spectrophotometric detection at 254 nm but with different mobile phases. All extracted the sample in ethyl acetate, evaporated the extract, reconstituted it, and injected it onto the column.

Immunoassays are, at present, among the most frequently used methods for measuring acetaminophen. The Syva EMIT Acetaminophen (Syva Corp., Palo Alto, CA) assay is a drug-enzyme–linked immunoassay in which drug in the sample competes with enzyme-linked drug for an antibody to acetaminophen (method 8, Table 64-1) (see Chapter 11). In this case, the enzyme is glucose-6-phos-

Table 64-1 Methods for acetaminophen analysis

Method	Type of analysis	Principle	Usage	Comments
1. Spot test	Qualitative, colorimetric	o-Cresol + Acetaminophen $\xrightarrow{NH_4OH}$ Indo–phenol blue	Screening test, urine	Highly sensitive and relatively specific
2. Azo dye formation (Lester and Greenberg[1])	Spectrophotometry	$CH_3CONH{-}C_6H_4{-}OH \xrightarrow[H^+]{\Delta} NH_2{-}C_6H_4{-}OH$ **Acetaminophen** → **Aminophenol** $NH_2{-}C_6H_4{-}OH + NaNO_2 \rightarrow HN{=}N{-}C_6H_4{-}OH$ **Sodium nitrite** → **Diazophenol** $HN{=}N{-}C_6H_4{-}OH$ + Ammonium sulfamate $\xrightarrow[5.5\ M\ NaOH]{\alpha\text{-Naphthol}}$ Azo dye	Of historical interest, replaced by immunoassay and HPLC	Hydrolysis step is time consuming
3. Nitro dye formation (Glynn and Kendal[4])	Spectrophotometry	$CH_3CONH{-}C_6H_4{-}OH \xrightarrow[HCl]{NaNO_2} CH_3CONH{-}C_6H_3(NO_2){-}OH$ $CH_3CONH{-}C_6H_3(NO_2){-}OH \xrightarrow{OH^-} CH_3CONH{-}C_6H_3(NO_2){-}O^-$ A_{max} = 430 nm	Most efficient spectrophotometric method	Linear in range of 100-500 μg/mL; HPLC and GLC more sensitive; no known interferences
4. Folin-Ciocalteu reagent	Spectrophotometry	Indophenol dye complex formed by reaction of acetaminophen with phenol reagent of Folin-Ciocalteu at pH 11.0	Not commonly used because of interferences	Tryptophan, tyrosine, uric acid, and salicylate react to give falsely high readings
5. Spectrophotometry	Differential ultraviolet spectrophotometry	Measures acetaminophen at differential absorbance peak of 266 nm, avoiding interference with salicylate	Replaced by HPLC and immunoassay	Reading at isosbestic point of salicylate removes interference from this compound; requires narrow band pass spectrophotometer
6. Gas-liquid chromatography (GLC)	Chromatographic separation	Extracted into ethyl acetate; converted to trimethylsilyl derivative; injected isothermically onto gas-liquid chromatogram with OV-17 column and flame ionization detector	Adequate; replaced by HPLC and immunoassay	Sensitive to 1 μg/mL; no known interferences; only drawback is time needed for ethyl acetate to evaporate

Continued.

Table 64-1 Methods for acetaminophen analysis—cont'd

Method	Type of analysis	Principle	Usage	Comments
7. High-performance liquid chromatography (HPLC)				
a. Ion exchange	Chromatographic separation	Separation on cation-exchange column; run time 50 min	Rarely used	Time-consuming procedure; not to be favored over reversed-phase HPLC
b. Silica adsorption	Chromatographic separation	Separation on Zorbax Sil silica column using mobile phase of chloroform/heptane/ethanol/glacial acetic acid	Alternative to reversed-phase HPLC	Sensitive to 5 μg/mL; recovery of 97%; most drugs do not interfere
c. Reversed-phase	Chromatographic separation	Packing octadecylsilane-bonded silica; mobile phase is dilute acetic acid/methanol/ethyl acetate; detection at 254 nm	Possible reference method	Most rapid and sensitive (to 1 μg/mL); some interferences with internal standard; future improvements may include use of electrochemical detector
8. Enzyme-multiplied immunoassay technique (EMIT)	Competitive binding	Drug in patient sample competes with drug-enzyme complex for limited amount of antibody; enzyme activity related to drug level	Commercially available; widespread usage	Adaptable to semiautomation; available on stat basis
9. Fluorescence polarization	Competitive binding	Drug in patient sample competes with drug-fluorescein tracer for limited amount of antibody; polarization inversely related to drug level	Commercially available; widespread usage	Adaptable to semiautomation; available on stat basis

phate dehydrogenase, and the substrate for the enzyme and NAD is included in the reaction mixture. Conversion of NAD to NADH is measured at 340 nm on a spectrophotometer, and the increase in the absorbance (ΔA_{340}) is compared to absorbance changes obtained for standards that are analyzed along with the sample.

The TDx acetaminophen assay from Abbott (Irving, TX) is based on the method of fluorescence polarization (method 9, Table 64-1) (see Chapter 11).

REFERENCE AND PREFERRED METHODS

Qualitative methods

The spot test is a very sensitive, inexpensive screen for acetaminophen in urine. A strong positive result can be observed in urine samples obtained 24 hours after a 1 g dose of acetaminophen. The spot test will also be positive in patients with greater than 2 μg/mL in the serum. However, since acetaminophen and *p*-aminophenol are metabolites of phenacetin, a positive spot test can indicate phenacetin ingestion.

Quantitative methods

The serum spectrophotometric assays are sufficiently specific in the overdose situation. In our experience, the Syva EMIT Acetaminophen assay is quick and accurate but requires expensive reagents and exhibits relatively large intertechnologist variability, requiring that the standard curves must be run often. This assay would be favored in the small laboratory without HPLC or GLC, since it requires only a spectrophotometer for instrumentation. The fluorescence polarization immunoassay (FPIA) method is more precise than Syva EMIT but, in our hands, gives lower values on spiked serum than HPLC or Syva EMIT. Like the Syva EMIT assay, the reagents for the Abbott TDx method are expensive, and the method offers no analytical advantages over HPLC, but in the laboratory equipped with an Abbott TDx analyzer, the method is attractive from a logistical standpoint.

A comparison of the conditions for several of these assays is presented in Table 64-2.

Table 64-2 Comparison of reaction conditions for acetaminophen assays

Condition	EMIT*	GLC	HPLC	FPIA
Temperature (C)	30°	220°	Ambient	35°
pH	8.0	—	—	7.0 (100 mmol of phosphate)
Sample volume	50 μL	1 mL	0.1 mL	10 μL
Fraction of sample volume	0.01	1.0	0.33	0.05
Precision (as %CB)	10% (20-200 μg/mL)	<10%	5%	6%
Linearity (μg/mL)	10-200	20-200	5-300	10-200
Interferences	None	None	Cefoxitin, loxapine HCl, thioridazine (all three interfere with internal standard)	None

EMIT, Enzyme-multiplied immunoassay technique (Syva Co, Inc, Palo Alto, CA); *FPIA,* fluorescence polarization immunoassay (Abbott, Inc, Irving, TX); *GLC,* gas-liquid chromatography; *HPLC,* high-performance liquid chromatography.
*Methods described in text.

SPECIMEN

Analysis for acetaminophen using HPLC or GLC techniques can be performed on serum or plasma. No special handling of specimens is required.

REFERENCE RANGE

Optimum therapeutic concentration of acetaminophen in serum is in the range of 10 to 20 μg/mL (66 to 132 μmol/L).

Interpretation of whether or not a patient has ingested a toxic amount of acetaminophen requiring overdose therapy is complex. In the severe toxic state, which occurs when the 4-hour blood levels (after ingestion) are greater than 200 μg/mL, the liver is exposed to attack by toxic minor metabolites of acetaminophen oxidation as well as free radical products from the metabolism of other chemicals. In these situations, treatment with acetylcysteine (Mucomyst), which contains sulfhydryl groups, provides protection for the liver, preventing toxic hepatitis. A nomogram of plasma acetaminophen levels versus hours after ingestion is used to estimate probable liver damage (Fig. 64-1).

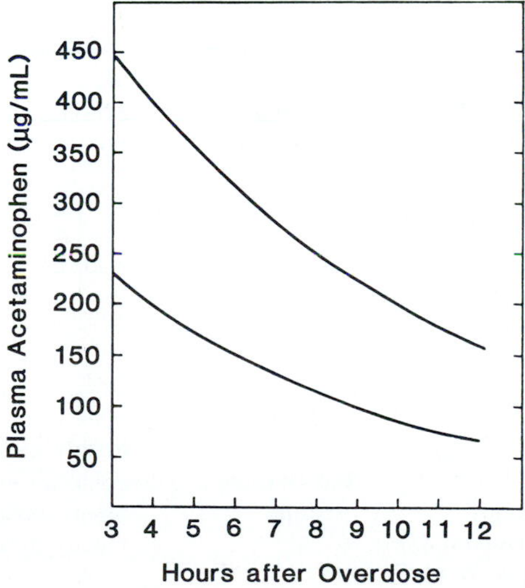

Fig. 64-1 Plasma acetaminophen concentration in relation to time after an acute overdose. Liver damage is likely to be severe above upper line, severe to mild between lines, and clinically insignificant under lower line. *(From Prescott, LF, et al: Lancet 2:109-113, 1976.)*

REFERENCES

1. Lester, L, and Greenberg, LA: The metabolic rate of acetanilid and other aniline derivatives, J Pharmacol Exp Ther 90:68-75, 1947.
2. Brodie, BB, and Axelrod, J: The estimation of acetanilid and its metabolic products, aniline, acetyl-p-aminophenol and p-aminophenol (free and total conjugated), in biological fluids and tissues, J Pharmacol Exp Ther 94:22-28, 1948.
3. Heirwegh, KPM, and Fevery, J: Determination of unconjugated and total *N*-acetyl-p-aminophenol (NAPA) in urine and serum, Clin Chem 13:215-219, 1967.
4. Glynn, JP, and Kendal, SE: Paracetamol measurement, Lancet 1:1147-1148, 1975.
5. Prescott, LF: The gas-liquid chromatographic estimation of phenacetin and paracetamol in plasma and urine, J Pharm Pharmacol 23:111-115, 1971.
6. Thomas, BH, and Coldwell, BB: Estimation of phenacetin and paracetamol in plasma and urine by gas-liquid chromatography, J Pharm Pharmacol 24:243, 1972.
7. Dechtiaruk, WA, Johnson, FG, and Solomon, HM: Gas-chromatographic method for acetaminophen (*N*-acetyl-p-aminophenol) based on sequential alkylation, Clin Chem 22:879-883, 1976.
8. Howie, D, Adriaenssens, PI, and Prescott, LF: Paracetamol metabolism following overdosage: application of high performance liquid chromatography, J Pharm Pharmacol 29:235-237, 1977.
9. Horvitz, RA, and Jatlow, PI: Determination of acetaminophen concentrations in serum by high-pressure liquid chromatography, Clin Chem 23:1596-1598, 1977.
10. Gotelli, GR, Kabra, PM, and Marton, LJ: Determination of acetaminophen and phenacetin in plasma by high-pressure liquid chromatography, Clin Chem 23:957-959, 1977.
11. Lo, LY, and Bye, A: Rapid determination of paracetamol in plasma by reversed phase high-performance liquid chromatography, J Chromatogr 173:198-201, 1979.

Alcohol

TIMOTHY J. SCHROEDER

Clinical significance: pp. 484 and 740

	Ethanol	Methanol	Isopropanol
Molecular formula:	C_2H_5OH	CH_3OH	C_3H_7OH
Molecular weight (daltons):	46.07	32.04	60.09

Merck Index: 210, 211, 212, 213, 5816, 5057

Chemical class: alcohol

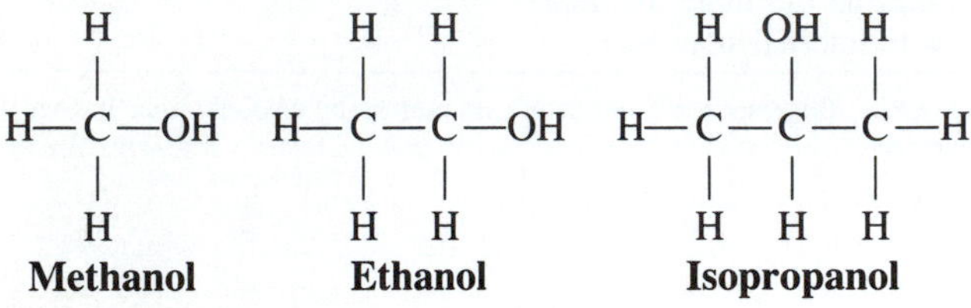

Methanol **Ethanol** **Isopropanol**

PRINCIPLES OF ANALYSIS

An alcohol analysis is frequently requested by the hospital's emergency department to aid in the differential diagnosis of central nervous system depression (for example, coma) that can be caused by this drug class. In most cases, the analyses must be rapid, and because of medicolegal requirements they must also be accurate. Not only must the physician in many instances know the type of alcohol ingested (usually ethanol, methanol, isopropanol, or ethylene glycol) but also the amount. This information is required on a stat basis since it can affect patient care. Techniques for the determination of alcohol include semiquantitative technique (such as osmometry, and diffusion or distillation, both followed by oxidation of the alcohol) and quantitative methods (such as enzymatic and gas-chromatographic procedures).

Early techniques for blood alcohol determination used distillation, aeration, or diffusion to separate the alcohol from the plasma matrix. The distilled alcohol was measured by oxidation of the alcohol by strong oxidizing agents such as dichromate, permanganate, or osmic acid.[1,2] The concomitant reduction of the oxidizing agent resulted in a color change that was used to monitor the reaction (method 1, Table 64-3). For example, it was observed that when alcohol was placed in a strongly acidic potassium dichromate solution, which was yellow-orange, the dichromate was reduced to a blue-green chromous ion solution, whereas the alcohol was oxidized to products such as acetaldehyde, acetic acid, and carbon dioxide and water. The extent of oxidation depended on the reaction conditions.

Williams et al.[3] employed the Conway diffusion dish for distillation of the alcohol. In this procedure, a solution of 18 N sulfuric acid–potassium dichromate was placed in the center compartment of the Conway dish (Fig. 64-2). The sample to be tested was placed in the outer ring. The diffusion of alcohol into the acid oxidizing solution was allowed to proceed for 1 hour at 55° to 60° C. The center compartment fluid was brought to a specified volume with a 1% brucine buffer solution in 10% sulfuric acid. The

Table 64-3 Methods for alcohol analysis

Method	Type of analysis	Principle	Usage	Comments
1. Distillation-oxidation	Colorimetric	Alcohol diffuses into gas phase and reacts with oxidizing agent, changing its color $2\ K_2Cr_2O_7 + 10\ H_2SO_4 + 3\ C_2H_5OH \rightarrow$ *(yellow-orange)* $2\ Cr_2(SO_4)_4 + 2\ K_2SO_4 + 3\ CH_3COOH + 11\ H_2O + 4\ H^+$ *(blue-green)*	Stat or routine; all body fluids; tissue	Nonspecific; gives reaction with all volatiles
2. Osmometry	Freezing point depression	Alcohol in high concentration increases serum osmolality; difference between normal and measured value is proportional to alcohol levels	Stat; serum	Nonspecific
3. Enzymatic	a. Spectrophotometric	$NAD^+ + \text{Alcohol} \xrightarrow{\text{Alcohol dehydrogenase}} NADH + H^+ + \text{Acetaldehyde}$	Stat or routine; serum	Specific for ethanol; other alcohols not readily measured
	b. Fluorometric	Above reaction (3a) plus $NADH \rightarrow NAD^+ + \text{Reduced monotetrazolium dye}$ *(fluorescein fluorescence decreased)*	Stat or routine; serum	Specific for ethanol; other alcohols not readily measured
	c. Colorimetric, dipstick	Above reaction (3a) plus $NADH \xrightarrow{\text{Diaphorase}} NAD^+ + \text{Reduced iodonitrotetrazolium dye}$	Stat or routine; serum; urine	Specific for ethanol; other alcohols not readily measured; new assay
4. Gas chromatography	Flame ionization	Alcohol separates on chromatographic column	Stat or routine; all body fluids	Specific for all alcohols

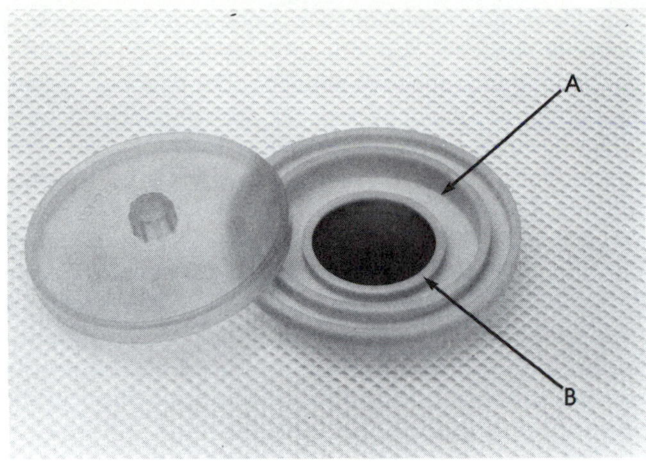

Fig. 64-2 Photograph of Conway diffusions dish showing sample well, *A*, and inner well containing reacted dischromate solution (blue green), *B*.

blue color of the reduced dichromate was read in a spectrophotometer at 425 nm. The coefficient of variation was reported to be ±2%. The Conway diffusion dish can also be used as a stat procedure for the detection of volatile substances.

An osmometric method of analysis for alcohols was first introduced by Redetzki et al. in 1972.[4] In this method, the increase of blood osmolality caused by the presence of alcohol was measured (see the discussion of osmolality, p. 879). Serum osmolality was determined on 2 mL samples by the freezing-point depression technique (method 2, Table 64-3). The contribution of alcohol to the serum osmolality was found to be directly related to its concentration in the blood. These authors found a correlation between enzymatically determined alcohol levels and those computed from a measurement of the osmolality. The alcohol concentration was observed to be proportional to the difference between the osmolality in the presence of alcohol (patient's sample) and the normal serum osmolality determined in the absence of alcohol, rather than proportional to the absolute osmolality alone. The average normal osmolality of serum (no alcohol) was considered to be 290 mOsm/kg of body water.[5] Serum osmolality was converted into ethanol concentration by use of the following equation:

$$(\text{Measured osmolality} - 290\ \text{mOsm/L}) \times 46.07\ \text{mg/mmol} \times 0.93 = \text{Milligrams of ethanol per liter of serum}$$

where 46.07 mg/mmol is the molecular weight of ethanol, 290 is the normal serum osmolality, and 0.93 is the correction for the volume fraction of water in serum.

The equation can be reduced to yield

$$\Delta\ \text{Osmolality} \times 42.4 = \text{Ethanol (mg/L)}$$

Rather than use a value of 290 mOsm/L, the serum osmolality may be calculated by use of the following formula:

$$\text{Osmolality (mOsm/kg)} = \frac{1.86 \times \text{Na (mmol/L)} + \dfrac{\text{mg of glucose/L}}{180} + \dfrac{\text{mg of serum urea nitrogen/L}}{28}}{0.93}$$

where 0.93 represents the fraction of serum volume occupied by water.

Since the alcohols are volatile and contribute significantly to the vapor pressure above a solution, vapor pressure osmometers cannot be used to estimate blood alcohol.

Enzymatic methods employ alcohol dehydrogenase (ADH), which reacts with ethanol but not with methanol or acetone. The enzyme does have some reactivity with propanol (6% for 2-propanol and 1% for 1-propanol) and butanol (17.5%). The reaction results in the oxidation of alcohol to acetaldehyde, with consequent reduction of the coenzyme nicotinamide adenine dinucleotide (NAD) to reduced NAD (NADH) (method 3, Table 64-3). The reduced coenzyme may be measured directly at 340 nm or indirectly by linking the NADH-generating reaction to the reduction of a tetrazolium dye to generate a visible color. Roos[6] reported the use of diaphorase to react NADH enzymatically with the tetrazolium salt iodonitrotetrazolium (INT) and phenazine methosulfate to form a colored formazan derivative that is stable between pH 4.0 and 5.0. The formazan derivative has an absorption maximum at 495 to 500 nm.

A fluorescence inhibition technique for serum alcohol has been developed for use on the Abbott TDx.[7] In this assay, the alcohol dehydrogenase reaction is coupled with the diaphorase reaction.

$$NAD^+ + \text{Ethanol} \xrightarrow{ADH} \text{Acetaldehyde} + NADH + H^+$$

$$NADH + INT \xrightarrow{\text{Diaphorase}} NAD^+ + \text{Pink product}$$

The NADH is oxidized in the presence of an INT dye to generate a colored product, which interferes with the intensity of the fluorescence of a dye, fluorescein, in the solution. Both the intensity of the light reacting with the fluorescein tracer in solution and the emitted fluorescent light are attenuated by the colored tetrazolium product formed by the diaphorase reaction. Thus there is less fluorescence if the sample contains ethanol. This is essentially an end-point enzymatic reaction. The decrease in fluorescence in a sample is proportional to ethanol concentration, as determined from a computer-calculated curve.

There has been considerable effort to develop a dipstick method for rapid determination of alcohol in body fluids. One such method involves the use of the coupled alcohol dehydrogenase–diaphorase reaction on cellulose strips.[8] The NADH generated by the ADH in the presence of ethanol reduces the dye INT chloride, also incorporated into the strip, to a pink form. One can quantify the color by stopping the reaction with pyrazole and measuring the color generated with a reflectance photometer. Another dipstick method uses alcohol dehydrogenase and an MBTH-DMAB indicator system to produce a blue color (λ_{max}, 590).[9]

Gas-chromatographic analyses (method 4, Table 64-3) of alcohol have used extraction, distillation, and head-space procedures to obtain samples free of the blood matrix. Methods also exist for the direct injection of blood into the chromatograph system.[10] In 1958 Cadman and Johns[11] reported a procedure for the separation and quantitation of alcohol and alcohol-related compounds using a gas chromatograph equipped with a thermal conductivity detector. The alcohol was extracted from blood into *n*-propyl acetate.

In 1964 Goldbaum et al.,[12] using a gas chromatograph equipped with a beta-particle ionization microdetector, introduced the first head-space gas chromatographic procedure. This method utilized an air sample removed from a confined space above blood, or other biological sample, in a closed container. The samples were collected in containers stoppered with a puncture type of rubber cap. Both the unknowns and standards were placed in a water bath at 25% C until the samples were ready. An airtight syringe was filled with 1 mL of the air on top of the blood sample; the needle of the syringe was inserted through the rubber cap and moved up and down several times to obtain an equilibrated air sample. The air sample was then introduced into the chromatograph. Several modifications of this method have been introduced.

In 1962, Machata[10] used a Perkin-Elmer chromatograph equipped with a flame ionization detector for analysis using direct sample injection. In this procedure, quantitation was accomplished by analysis of a known amount of acetone, which served as an internal standard, with the alcohol. A modification of the method of Jain[13] is considered to be the most widely used GLC procedure for alcohol analysis. This procedure includes the use of a flame ionization detector and a 20% Carbowax column. Injection temperature is 100° C, and the injection port is kept at 160° C.

The choice of the best method for the determination of ethanol is complicated by the medicolegal problems involved in accurately quantitating ethanol levels above certain legal amounts. Therefore in many states there are legal specifications defining the methods that must be used to determine ethanol levels. Many laboratories engaged in medicolegal analysis must follow these specified methods.

The distillation or diffusion methods for measurement of the commonly ingested ethanols take advantage of the volatility of this compound. These methods lack specificity, since other volatile oxidizable compounds, including other alcohols and ketones, may also be distilled into and react with the indicator solution. The Conway diffusion dish is a simple apparatus that can give accurate, reproducible, quantitative or qualitative answers. The Conway system, because of its simplicity and low cost, is preferred for a rapid, qualitative stat screening procedure.

One should note that the rapid osmolality method provides only an estimate of blood alcohol levels in the range of 1000 to 5000 mg/L.[14] This method is nonspecific; any condition that changes the osmolality affects the estimation of alcohol. Significant error can occur in the results for patients with diabetic hyperglycemia, uremia, or hypernatremia. Our opinion is that this method is not analytically specific and that it should be used only to rule out the presence of ethanol or other volatiles.

SPECIMEN

Alcohol swabs should not be used to clean the site of the venipuncture. Specimens must be kept well stoppered and preferably refrigerated to prevent loss of ethanol. For the distillation or diffusion procedure, any body fluid or tissue may be tested. For the enzymatic procedure, plasma, serum, or urine may be used, although serum is preferred.[15] Anticoagulants do not interfere with the enzyme or gas-chromatographic procedures. For gas chromatography any tissue or body fluid may be used.[16]

The problem of storage of ethanol specimens is a critical one because of the medicolegal implications of the results. It has been shown that sealed samples of whole blood, or whole blood plus fluoride, can be stored at 0° to 3° C, or at room temperature (22° to 29° C) without any significant loss of ethanol content over a 14-day period.[17]

PROCEDURE: GAS CHROMATOGRAPHIC ANALYSIS OF VOLATILE ALCOHOLS (Table 64-4)

Principle

Volatile compounds such as ethanol, methanol, isopropanol, and acetone separate from serum in the heated injection port. They are separated by chromatography and are quantitated by use of a flame ionization detector and comparison of the peak height ratio of the sample to the peak height ratio of a standard.

Reagents

Ethanol stock standards. Dilute 100, 200, 300, and 400 µL of 100% ethanol to 100 mL with distilled water. This yields stock standards of 789 µg/mL (17.15 mmol/L), 1578 µg/mL (34.30 mmol/L), 2367 µg/mL (51.45 mmol/L), and 3156 µg/mL (68.60 mmol/L). These are stable for 1 month at room temperature in a stoppered container.

***n*-Propanol internal standard, 3120 µg/mL (51.92 mmol/L).** Dilute 400 µL of *n*-propanol to 100 mL with distilled water. This is stable for 1 year at room temperature in a stoppered container.

Standard serum, 789 µg/mL (17.15 mmol/L). Add

Table 64-4 Comparison of assay conditions for ethanol

Parameter	Enzyme reaction	Gas chromatography
Temperature	37° C	100° C
pH	8.8	7.35
Final concentration of reagent components	Reduced nicotinamide adenine dinucleotide: 0.6 μmol/mL Alcohol dehydrogenase: 50 U/mL Tetrasodium pyrophosphate: 0.075 mol/L Sodium semicarbazide: 0.075 mol/L Glycine: 0.022 mol/L	*n*-Propanol: 1559 μg/mL (25.96 mmol/L)
Fraction of sample volume	Approximately 0.008	0.5
Sample volume	0.5 mL	0.2 mL (0.5 μL injection)
Linearity	0-3000 μg/mL (65.1 mmol/L)	0-3000 μg/mL (65.1 mmol/L)
Precision*	5%	9%
Time of reaction	20 min	8 min
Interferences	Other alcohols	None known

*Coefficient of variation; obtained from College of American Pathologists (CAP) toxicology survey data.

100 μL of 100% ethanol to 100 mL of pooled serum, and mix well. Store 1 mL aliquots at −20° C. This is stable for 1 year.

Assay

Equipment: gas-liquid chromatograph with flame ionization detector; 6-foot column containing 0.2% Carbowax 1500 on 60/80 Carbopack.

1. Sample preparation. Pipette 200 μL of internal standard into 12 × 75 mm glass tube. Add 200 μL of serum, and mix.
2. Inject 0.5 μL of sample or standard into chromatograph. Chromatographic run time is approximately 6 minutes.
3. Chromatographic parameters: injection temperature, 210° C; column temperature, 100° C; detector temperature, 260° C; gas flow at 60 mL/min with the proportions of nitrogen/air/hydrogen at 40:10:20.

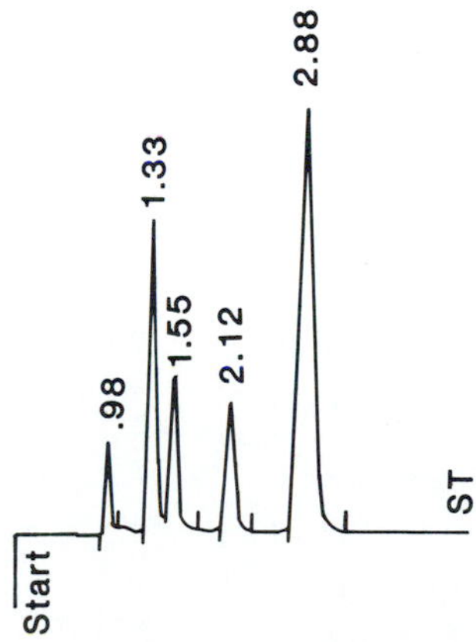

Fig. 64-3 Gas chromatographic tracing of a series of alcohol standards. Retention time in minutes is recorded next to peak height. Alcohols are eluted in following time sequence in minutes: 0.98, methanol; 1.33, ethanol; 1.55, acetone; 2.12, isopropanol; 2.88, *n*-propanol (internal standard). *ST,* Stop.

Calculation

1. Determine the peak height (in millimeters) for the ethanol (E) and internal standard (S) peaks on each chromatogram. (See Fig. 64-3 for sample chromatogram.)
2. Calculate peak-height ratio (E/S) for each specimen.
3. Calculate ethanol in serum samples as follows:

$$\text{Ethanol } (\mu g/mL) = \frac{\text{E/S (sample)}}{\text{E/S (standard serum)}} \times 789\ \mu g/mL$$

4. Alternatively, a standard curve may be generated using the series of working standards and the results interpolated from the peak-height ratio.

Notes

1. Anhydrous alcohols rapidly absorb water. Therefore, once opened and used, an anhydrous bottle of alcohol cannot be considered a stable reagent. For this reason, some laboratories use 95% alcohol, which is stable.
2. Isopropanol, methanol, and acetone can also be quantified by this procedure. The same peak-height ratio method with *n*-propanol as the internal standard is used. The same chromatographic conditions can be used with the appropriate standard added to the standard serum (see note 3).
3. Standards for the measurement of other volatiles.
   a. Isopropanol, 1570 μg/mL (26.12 mmol/L). Add 200 μl of 100% isopropanol to a 100 mL flask, and dilute to volume with distilled water.
   b. Acetone, 1576 μg/mL. Prepare in the same manner as isopropanol, using pure acetone.
   c. Methanol, 1582 μg/mL (46.4 mmol/L). Prepare in the same manner as isopropanol, using pure methanol.
4. To calculate serum values, use the following formula:

$$\text{Concentration} = \frac{\text{Peak-height ratio (unknown)}}{\text{Peak-height ratio (standard)}} \times \text{Concentration of standard}$$

REFERENCE RANGE

A patient with a blood ethanol level of 1000 μg/mL (21.7 mmol/L) or higher is considered legally intoxicated in many states.

A value of 3000 μg/mL (65.1 mmol/L) is usually associated with coma. The other volatiles, methanol and isopropanol, are not usually present.

Medicolegal interpretation of ethanol levels are usually based on state laws. However, unless there is a chain of custody to show that the sample actually came from a particular patient, the data cannot be used in court as evidence.

REFERENCES

1. Widmark, EMP: Concentration of alcohol in blood and urine under various conditions, Scand Arch Physiol 32:85-96, 1915.
2. Widmark, EMP: Modification of the Niclocex method for estimating ethyl alcohol, Scand Arch Physiol 35:125-130, 1916.
3. Williams, LA, Linn, RA, and Zak, B: Determinations of ethanol in fingertip quantities of blood, Clin Chim Acta 3:169, 1958.
4. Redetzki, HM, Koerner, TA, Hughes, JR, and Smith, AG: Osmometry in the evaluation of alcohol intoxicants, Clin Toxicol 5(3):343, 1972.
5. Cravey, RH, and Jain, NC: Current status of blood alcohol methods, J Chromatogr Sci 12:209, 1974.
6. Roos, KJ: Rapid, sensitive and inexpensive method for estimating blood and urine alcohol concentrations, Clin Chim Acta 31:285, 1971.
7. Yost, DA, Boehnlein, L, and Shaffer, M: A novel assay to determine ethanol in whole blood on the Abbott TDX, Clin Chem 30:1029A, 1984.
8. Kapur, BM, and Israel, Y: A dipstick methodology for rapid determination of alcohol in body fluids, Clin Chem 29:1178A, 1983.
9. Matzinger, D, Envin, KR, and Phillips, R: A solid state approach to alcohol testing, Clin Chem 30:1029, 1984.
10. Machata, G: Diphorase method of blood alcohol determination, Mikrochim Acta 6:91, 1962.
11. Cadman, WJ, and Johns, T: Paper presented at ninth annual Conference on Analytical and Chemical Applications in Spectrometry, Pittsburgh, 1958.
12. Goldbaum, LR, Domanski, TJ, and Schloegel, EL: Analysis of biological specimens for volatile compounds by gas chromatography, J Forensic Sci 9:63, 1964.
13. Jain, NC: Direct blood-injection method for gas chromatographic determination of alcohols and other volatile compounds, Clin Chem 17:82, 1971.
14. Pappas, AA, Godsden, RH, and Taylor, EH: Serum osmolality in acute intoxication: a prospective clinical study, Am J Clin Pathol 84:74-79, 1985.
15. Kane, K, and Fontaine, RB: Enzymatic measurement of ethyl alcohol. In Sunderman, FW: Seminar on clinical and analytical toxicology, Philadelphia, 1987, Institute for Clinical Science, Inc.
16. Kahn, SE, Holmes, EW, and Bermes, EW: Quantitative gas chromatographic volatile screen for serum acetone, ethanol, isopropanol, and methanol. In Sunderman, FW: Seminar on clinical and analytical toxicology, Philadelphia, 1987, Institute for Clinical Science, Inc.
17. Winek, CL, and Paul, LJ: Effect of short-term storage conditions on alcohol concentrations in blood from living human subjects, Clin Chem 29:1959-1960, 1983.

Anticonvulsant drugs

STEVEN J. SOLDIN

Clinical significance: pp. 594, 740, and 795
Molecular structure:

Drug:	**Valproic acid**	**Phenobarbital**	**Phenytoin**
Molecular formula:	$C_8H_{16}O_2$	$C_{12}H_2N_2O_3$	$C_{15}H_{12}N_2O_2$
Molecular weight (daltons):	144.21	232.23	252.26
Merck Index:	9574	7032	7130
Chemical class:	Carboxylic acid	Barbiturate	Hydantoin

Drug:	**Carbamazepine**	**Primidone**	**Ethosuximide**
Molecular formula:	$C_{15}H_{12}N_2O$	$C_{12}H_{14}N_2O_2$	$C_7H_{11}NO_2$
Molecular weight (daltons):	236.26	218.25	141.17
Merck Index:	1781	7542	3674
Chemical class:	Dibenzazepine-5-carboxamide	Pyrimidinedione	Succinimide

PRINCIPLES OF ANALYSIS

Most anticonvulsant drugs have specific ultraviolet spectral characteristics that were exploited in the initial attempts to monitor the therapeutic levels of these drugs.[1,2] The drugs were extracted into an organic solvent and reextracted into an aqueous solution, and the absorbance of the solution was recorded (method 1, Table 64-5). By use of appropriate solvents, the drugs could be extracted with relative specificity. By recording the difference spectra of the extract at two pH values, one could also increase the specificity. The difference spectra were obtained by use of a dual-beam spectrophotometer and placement of one solution into the reference compartment and the other into the usual test compartment.

A second method for measuring anticonvulsant drugs involved extraction and separation by thin-layer chromatography.[3-5] Quantitation was achieved by either ultraviolet scanning of the plate or by elution of the drug from the plate and recording of its ultraviolet absorption (method 2, Table 64-5).

Gas chromatographic techniques (method 3, Table 64-5) were for many years the primary techniques used to analyze antiepileptic drugs in biological fluids.[6-11] MacGee[6] developed an on-column methylation technique that resulted in thermally stable and easily volatilized derivatives with quantitative recovery. Initially, flame ionization detection was the system of choice for quantification of the eluted drug. Current gas chromatographic techniques prepare derivatives of anticonvulsant drugs. Methylation is the most commonly used procedure,[6,12] and in the method described by Solow and Green,[12] seven antiepileptic drugs may be analyzed simultaneously. Almost all antiepileptics

Table 64-5 Methods for anticonvulsant analysis

Method	Principle	Usage	Comments
1. Extraction and ultraviolet spectroscopy	Anticonvulsant extracted free of other ultraviolet compounds; ultraviolet spectrum or the ultraviolet difference spectra are specific for the drug.	Serum	Interference with drugs and endogenous compounds
2. Thin-layer chromatography (TLC)	Extracted drug separated by TLC and its R_f noted; drug quantitatively eluted and quantified by ultraviolet spectroscopy.	Serum	Difficult to separate all anticonvulsants Technically difficult to achieve reproducibility
3. Gas chromatography (GC)	Extracted drugs converted to derivatives: $(CH_3)_4N^+OH^-$ + Anticonvulsants → Methyl derivatives Retention time identifies specific drug.	Blood, plasma, serum, saliva	Can resolve all anticonvulsants simultaneously; requires extraction and derivatization; therefore technically difficult
4. High-performance liquid chromatography (HPLC)	After proteins are precipitated from solution, anticonvulsants are separated by reversed-phase chromatography and monitored by ultraviolet spectroscopy or electrochemical detection.	Blood, plasma, serum, saliva	Can resolve all anticonvulsants except valproate simultaneously; commonly used
5. Competitive-binding assays			
a. Enzyme-multiplied immunoassay technique (EMIT)	Competitive binding with drug attached to enzyme	Plasma, serum, saliva	Can measure all drugs, but each must be done separately; commonly used
b. Radioimmunoassay (RIA)	Competitive binding of radioactive ligand; radiolabeled hapten competes with unknown for antibody-binding site.	Plasma, serum, saliva	All problems of radioactive usage; slower than other immunoassays
c. Substrate-labeled fluorescence immunoassay (SLFIA)	Competitive binding substrate label; substrate competes with unknown antigen for antibody-binding site.	Plasma, serum, saliva	Fluorescence interferes; not widely used
d. Fluorescence polarization immunoassay (FPIA)	Competitive binding with fluorescein-labeled drug; fluorescent drug competes with unknown for antibody-binding site.	Plasma, serum, saliva	Each drug assayed individually; requires specialized instrument; most commonly used
e. Ames test strip	Antibody binding to drug labeled with prosthetic group prevents glucose oxidase activity. Competition with endogenous drug allows prosthetic group to activate enzyme. Glucose oxidase is coupled to colorimetric reaction.	Serum	Designed for inexpensive instrument; reflectance spectrophotometer; rapid test
f. Rate nephelometric inhibition immunoassay	Competitive binding for hapten; hapten-protein competes with unknown for antibody-binding site.	Plasma, serum	Each drug individually assayed; requires specialized instrument

except valproic acid consist of compounds containing nitrogen. Considerable interest has therefore been shown for use of the nitrogen flame ionization detection system, which appears to offer increased specificity and sensitivity.

High-performance liquid chromatography (HPLC) assays (method 4, Table 64-5) require protein precipitation of the sample before separation of the drug on a column.[13-21] Detection and quantitation are usually by ultraviolet absorption spectrophotometry, but electrochemical detection may also be used. Most methods use reversed-phase liquid chromatography on a C_{18} type of column.

Immunological procedures (methods 5a to 5e, Table 64-5) developed for measurement of the anticonvulsants include radioimmunoassay (RIA), enzyme immunoassay (EIA), fluorescence immunoassay (FIA), fluorescence polarization immunoassay (FPIA), and nephelometric inhibition immunoassay (NIA). (See Chapter 11 for descriptions.) The RIA procedures involve the standard competitive binding procedure in which there is competition between the labeled drug with that from the sample for a limited amount of antibody.[22]

The introduction of the enzyme-multiplied immunoassay technique (EMIT, Syva Co., Inc., Palo Alto, CA) into the clinical chemistry laboratory in the 1970s made possible the routine monitoring of all the anticonvulsant drugs (method 5a, Table 64-5). See Chapter 11 for a description of the assay.

The Ames (Elkhart, IN) fluorescent immunoassay for anticonvulsant drugs also uses the principles of competitive protein binding to measure levels of these drugs in serum or plasma[23] (method 5c, Table 64-5) (see Chapter 11). Fluorescence polarization immunoassay (FPIA) TDx (Abbott Laboratories, North Chicago, IL) uses the principle of a competitive binding assay and measures binding of tracer directly by fluorescence polarization[24] (see Chapter 11). The fluorescent tracer consists of fluorescein covalently attached to the drug being analyzed. The Abbott TDx system is highly automated, and the standard curves are automatically calculated by the instrument. This has become the most widely used procedure based on TDM surveys (method 5d, Table 64-5).

The Ames Seralyzer has methods for the measurement of phenytoin and phenobarbital that employ a labeled prosthetic group assay[25] (see Chapter 11). The enzyme converts glucose and O_2 to gluconolactone and H_2O_2. The color reaction is generated by use of peroxidase and a dye substrate 3,3′,5,5′-tetramethylbenzidine (TMB), which, when oxidized, forms a blue-colored product. This method has been incorporated into the test strip technology employed on the Ames Seralyzer. The strip is read in a solid-phase reflectance photometer. The strip color is at 740 nm, and the results are calculated from a calibration curve.

A rate nephelometric inhibition immunoassay is the procedure employed by the immunochemistry Systems (ICS) of Beckman Instruments, Inc. (Fullerton, CA) (method 5f, Table 64-5)[26] (see Chapter 11).

The ideal analytical procedure for anticonvulsant drugs should be as follows:

1. Both accurate and precise, affording the reliable quantitation of the antiepileptic drugs.
2. Technically easy, so that little effort is required in training technologists to perform the task.
3. Rapid, to minimize delay in reporting the results to the clinician. Many requests for drug quantitation arise as a result of problems in patient management. In our institution, results for approximately half the antiepileptic drugs quantitated fall outside the therapeutic range and necessitate alterations in dose and dosing schedule. Thus it is important that the interval between specimen receipt and report of the result be short.
4. Cost effective. A key issue here is laboratory work load. It would be inappropriate to spend $30,000 for a piece of equipment (such as an HPLC apparatus with automated sample injector and a data-handling facilities) if the laboratory work load is only 10 samples per week. On the other hand, if the laboratory work load is large (greater than 40 samples per week), the purchase of such equipment would provide a cheaper service than comparable radioimmunoassay and homogeneous enzyme immunoassay procedures, where reagent costs are considerable.
5. Sensitive, to allow analysis for drugs in a microscale sample and permit the simultaneous analysis of the most commonly used drugs. Many patients with epilepsy (about half at our institution) are on multiple-drug regimens.

Unfortunately, no method meets all these requirements, and the decision on which procedure to adopt depends largely on the environment and on the personal preferences of those involved. Table 64-6 compares methods for anticonvulsant analysis.

Gas-liquid chromatography (GLC) affords reliable analysis for anticonvulsant drugs on microscale samples. Disadvantages of GLC include the requirements of specialized equipment and considerable expertise. GLC is one of the methods of choice for the analysis of valproic acid[27-30] primarily because the poor ultraviolet absorption characteristics of this compound prevent its analysis by HPLC.

Some advantages of HPLC over most other current analytical techniques include the following:

1. Analyte volatility and thermal stability—so essential for GLC—are not required for this type of chromatography.
2. Relatively little sample workup is required before analysis.
3. Characteristically, methods involving HPLC require only a short analysis time.
4. There is no need for derivatization of the analyte.
5. For laboratories with a large work load, the cost per analysis is very low because reagent costs are low.
6. HPLC affords good sensitivity. The sensitivity varies

Table 64-6 Comparison of methods for anticonvulsant analysis

Parameter	High-performance liquid chromatography (HPLC)	Gas-liquid chromatography (GLC)	Enzyme-multiplied immunoassay technique (EMIT)	Fluorescence polarization immunoassay (FPIA)
Temperature	Ambient	Above 200° C	30° C	35° C
Sample volume	25 μL	1 mL	50 μL for each determination	20 μL for each determination
Fraction of sample volume	0.50	—	0.01	0.1
Linearity (μg/mL)				
Phenytoin	1-50	1-40	2.5-4.0	2.5-4.0
Phenobarbital	1-100	2-80	5-80	5-80
Primidone	1-40	2-40	2.5-20	2-24
Ethosuximide	1-250	—	10-150	150
Carbamazepine	1-30	—	1-20	2-20
Valproic acid	—	5-180	10-150	12.5-150
Precision (CV)*	7%-12%	5%-9%	6%-12%	4%-8%

*Average coefficient of variation (CV) for anticonvulsant drugs within therapeutic range. From Therapeutic Drug Monitoring Survey of the American Association for Clinical Chemistry, Washington, DC.

with the drug, but reliable quantitation at drug concentrations as low as 1 μg/mL can often be achieved.

7. HPLC methods are readily automated. Routine analyses therefore require a minimum of technician time, a major factor in arriving at the low cost per analysis.
8. HPLC makes possible the simultaneous analysis of most anticonvulsant drugs in a microsample.

Some limitations of HPLC in drug analysis are as follows:

1. Spectrophotometric, fluorometric, and amperometric detectors are commonly used. Therefore the analyte in question must either absorb light, fluoresce, or be electrochemically active.
2. Equipment is expensive. Investment of capital (a fully automated system currently costs about $30,000) is only warranted if the laboratory has a substantial work load (more than 40 samples per week).
3. Although HPLC analysis of valproic acid has been described,[31,32] these procedures require either post-column reactions[30] or precolumn derivatization.[32] The methods of choice for the analysis of valproic acid are GLC, EIA, or FPIA.

Several HPLC methods that allow simultaneous analysis for phenobarbital, phenytoin, primidone, ethosuximide, and carbamazepine have been used routinely in hospital laboratories.[20,21]

Enzyme immunoassay (EIA), in particular the enzyme-multiplied immunoassay technique (EMIT), is very popular. The advantages of the technique are that no separation steps are involved and simple equipment readily available in clinical laboratories is used. Such approaches to analysis considerably reduce both technician time and reagent cost. The fluorescence polarization procedures have the best between-laboratory precision of any of the methods presented. Disadvantages include costly reagents and the fact that EIA does not permit the simultaneous analysis of anticonvulsant drugs.

In general, the chromatographic procedures require greater technical skill to perform than the procedures based on immunological principles. In particular, the Ames Seralizer technique is designed for a physician's office, and performance of the test by minimally trained nurses and other workers in the physician's office is possible.

SPECIMEN

Plasma or serum may be used for the GLC or HPLC methods. Serum is the preferred specimen for the immunoassays. Icteric or hemolyzed specimens have less effect on the chromatographic methods than on immunoassays. Sera containing the anticonvulsants discussed in this chapter can be stored at room temperature for several hours. If frozen at −20° C, samples containing these drugs are stable for at least 1 year. Saliva may also be used for the measurement of some anticonvulsants. The saliva concentration provides a reliable estimate of the free drug concentration for phenytoin and carbamazepine.[33]

PROCEDURE: HPLC ANALYSIS OF ANTICONVULSANT DRUGS

Principle

Five anticonvulsant drugs (primidone, phenobarbital, ethosuximide, carbamazepine, and phenytoin), as well as the primidone metabolite phenylethylmalonamide and the carbamazepine metabolite carbamazepine-10,11-epoxide, are separated by reversed-phase liquid chromatography and quantitated by measurement of peak areas relative to the peak area of the internal standard, dihydrocarbamazepine. The compounds are identified by their retention times and also by their absorbance ratio at 200 (or 214) and 254 nm.

Each drug has a known absorbance ratio. Any skewing of the ratio alerts the operator to the presence of an interfering compound.

Reagents

Acetonitrile, methanol, and dimethyl sulfoxide. These reagents may be obtained from Burdick and Jackson (Muskegon, MI).

Standards. The pure drugs can be purchased from Applied Science Laboratories, State College, PA (phenobarbital), Parke-Davis, Morris Plains, NJ (phenytoin and ethosuximide), Ayerst Laboratories, New York (primidone), Aldrich Chemical Co., Milwaukee (10,11-dihydrocarbamazepine), and CIBA-Geigy, Ardsley, NY (carbamazepine). The National Bureau of Standards provides serum-based standards for all five drugs.

Potassium phosphate buffer, 10 mmol/L, pH 7.3. This is stable for 3 months at 4° C. Check monthly for microbial growth.

Stock internal standard, 200 mg/mL (839 μmol/L). Dissolve 20 mg of 10,11-dihydrocarbamazepine in 100 mL of acetonitrile. This is stable for 1 month when stored in a dark bottle.

Mobile-phase buffer, acetonitrile-phosphate, 37:63 (v/v). Pass 370 mL of acetonitrile through a nylon 66, 0.2 μm filter (D and L Filter Co., Woburn, MA) or equivalent member filter into a 2 L Erylenmeyer filtering flask. Replace the filter, and pass 630 mL of phosphate buffer into the flask. Degas the solution for 15 minutes at room temperature by applying a vacuum. This is stable for 1 week at room temperature.

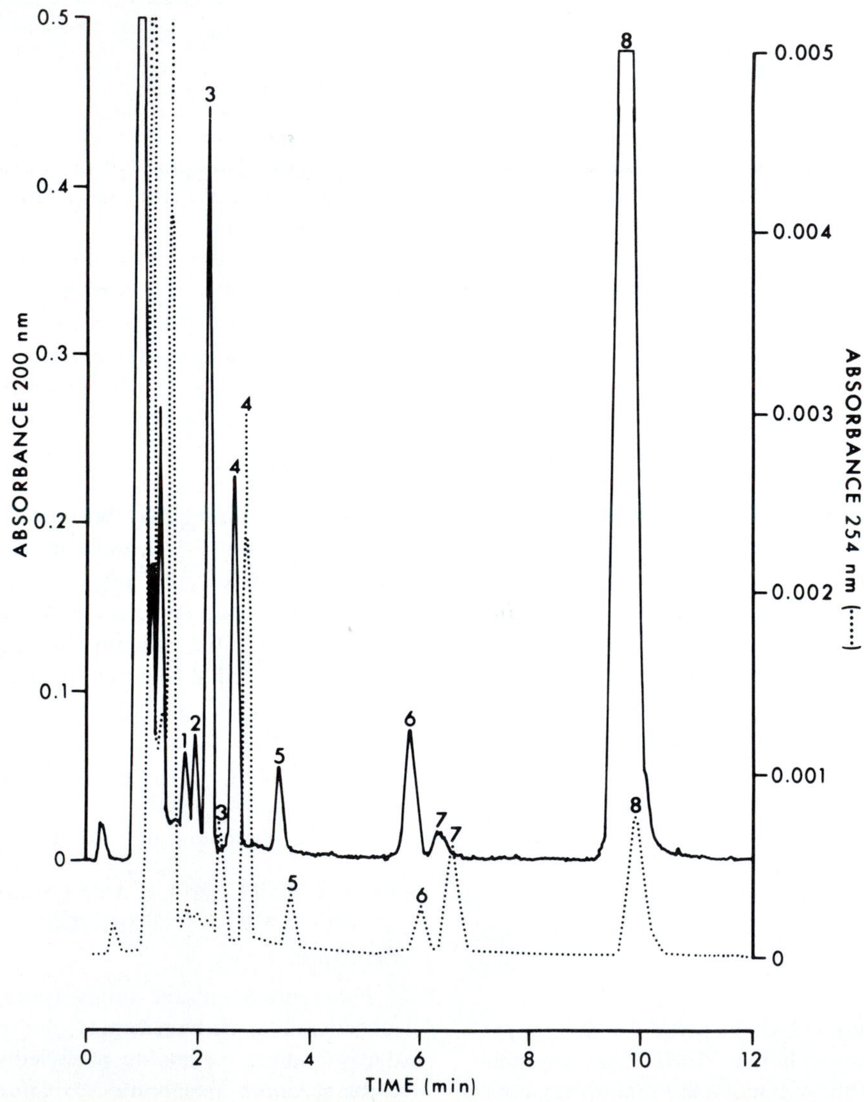

Fig. 64-4 Chromatogram obtained for a plasma specimen containing therapeutic concentrations of primidone, *2;* ethosuximide, *3;* phenobarbital, *4;* phenytoin, *6;* and carbamazepine, *7.* Peaks *1, 5,* and *8* represent phenylethylmalonamide, carbamazepine-10,11-epoxide, and the internal standard dihydrocarbamazepine, respectively. *Solid tracing,* 200 nm; *dotted tracing,* 254 nm.

Assay

Equipment: high-performance liquid chromatograph, model ALC/GPC/6000A, with an autosampler (WISP, Model 710A), a radial compression system (RC-100), and 10 × 0.8 Radial Pak A columns, all purchased from Waters Associates, Inc. (Milford, MA). Detectors used were the Waters Model 440 with 254 nm kit and the Model LC55 Variable Wavelength Spectrophotometer (Perkin-Elmer Corp., Norwalk, CT). Alternatively, the Waters QA1 with both zinc or mercury lines can be employed. In this configuration the 214 nm zinc line can be used to obtain a 214/254 ratio. Under these conditions a variable wavelength spectrophotometer is not needed. The recorder was a Honeywell Electronic, Model 196 (Honeywell, Inc., Fort Washington, PA). A Sigma 10 data-handling system (Perkin-Elmer) was employed to receive and compute the spectrophotometer signals.

1. Sample preparation
   a. Add 25 μL of acetonitrile containing approximately 200 mg/L of 10,11-dihydrocarbamazepine to 25 μL of serum or plasma.
   b. Vortex mix the sample for 30 seconds. Centrifuge for 5 minutes, and save the supernatant fraction for analysis.
2. Chromatographic parameters
   a. Chart speed, 1.0 cm/min
   b. Ultraviolet sensitivity, 254 nm, 0.01 AUFS (absorbance units at full scale)
   c. Ultraviolet sensitivity, 200 nm, 0.05 AUFS
   d. Flow rate, 2.0 mL/min at ambient temperature
3. Inject 4 μL of sample or standard. Sample analysis time is approximately 10 minutes. A typical chromatogram is shown in Fig. 64-4.

Calculations

Standards purchased from the National Bureau of Standards are diluted according to the described directions. The secondary standards (UTAK Laboratories, Inc., Canyon County, CA) are calibrated against the primary NBS standard and are then used for daily instrument calibration. Quality control materials are from Ortho Diagnostics. In each sample, peak heights of each drug are compared to the internal standard, and peak-height ratios are calculated. The calculation is as follows:

$$\text{Anticonvulsant drug (mg/L)} = C_{\text{cal}}\ \text{(mg/L)} \times \frac{\text{Peak height of sample/Peak height of internal standard}}{\text{Peak height of calibrator/Peak height of internal standard}}$$

where C_{cal} is the concentration of the calibrators.

Notes

1. No interferences are known. The recovery of all the anticonvulsants is greater than 80%.[16]
2. Linearity of this material for primidone, ethosuximide, phenobarbital, phenytoin, and carbamazepine is up to concentrations of at least 40, 250, 100, 50, and 30 mg/L, respectively.
3. The procedure described provides for the simultaneous analysis of the five anticonvulsant drugs, the primidone metabolite phenylethylmalonamide, and the carbamazepine metabolite carbamazepine-10,11-epoxide. It is not recommended for valproic acid.

PROCEDURE: VALPROIC ACID BY GAS CHROMATOGRAPHY[34]

Principle

The drug is extracted from an aqueous, acidic medium into an organic solvent. The valproic acid is then back-extracted into a dilute NaOH solution. The NaOH phase is made acidic and re-extracted with a second organic solvent. The final organic phase is injected into the gas chromatograph.

Reagents

Valproic acid (Abbott Laboratories)

Stock standard solution 1. Add 0.5 mL of valproic acid to a 50 mL volumetric flask and make up to volume with absolute methanol. (Concentration = 905.4 μg/0.1 mL or 62.9 mmol/L.)

Stock standard solution 2. Add 10 mL of stock solution 1 to a 50 mL volumetric flask and make up to volume with H_2O. (Concentration = 181.1 μg/0.1 mL or 12.6 mmol/L.)

The solutions are stable for 2 months when stored tightly sealed at 4° C.

NOTE: See below for working standard instructions.

Cyclohexane carboxylic acid internal standard (Eastman Kodak, Rochester, NY)

Working internal standard, 1 mg/mL (7.8 mmol/L). Weigh out one drop of the acid in a small beaker and record weight. Dissolve the drop in 1 mL of methanol, and then bring to volume with H_2O such that the total volume in milliliters is equal to the amount of drug in milligrams. (For example, if drop weighs 18 mg, add 1 mL of methanol and 17 mL of H_2O. Total volume = 18 mL.)

10% isopropanol/dichloromethane solution. Add 1 part isopropanol (Chromatoquality) to 9 parts dichloromethane (Chromatoquality).

Chloroform. Use Chromatoquality chloroform.

HCl, 1.0 mol/L. Dilute 8.35 mL of concentrated HCl to 100 mL with H_2O.

NaOH, 0.5 mol/L. Add 20 g of NaOH to 800 mL of water, and bring to 1 L with H_2O.

Working valproic acid (VA) standards. Prepare these by mixing the following portions of stock solution 2 with water and then adding 0.1 mL of the appropriate working solution to the designated tube containing 0.4 mL of pooled serum with an Eppendorf pipet. The total volume is 0.5 mL. These working standards are stable for 2 months at 4° C.

VA concentration (μg/0.1 mL)		VA stock solution 2 (mL)	H_2O (mL)
VA 1	181	10	10
VA 2	126.8	7	13
VA 3	90.6	5	15
VA 4	54.4	3	17
VA 5	18.2	1	19

Frozen controls. Prepare a valproic acid standard by weighing out one drop of valproic acid in a small beaker and recording the weight. Dissolve the drop in 1 mL of methanol, and then bring to volume with H_2O such that the volume in milliliters is equal to the amount of drug in micrograms. (For example, if drop weighs 18 mg, add 1 mL of methanol and 17 mL of H_2O. Total volume = 18 mL.)

Add 10 mL of the above standard to a 100 mL volumetric flask and dilute the volume with pooled serum. These controls are aliquoted into 1 mL portions and stored at −20° C. They are stable for 6 months.

Assay

Equipment: a gas chromatograph equipped with a 10% Carbowax 20 MTPA column and a flame ionization detector.

1. To a 25 mL screw-capped tube, add:
   0.5 mL of serum (run patient samples in duplicate)
   0.5 mL of HCl, 1.0 mol/L
   100 μL of cyclohexane carboxylic acid internal standard
   10 mL of 10% isopropanol/dichloromethane solution
2. Shake for 3 minutes, and then centrifuge for 3 minutes.
3. Discard top serum layer, and filter solvent layer through filter paper into a 10 mL screw-capped tube.
4. Add 2 mL of 0.5 mol/L NaOH.
5. Shake for 3 minutes, and then centrifuge for 3 minutes.
6. Transfer top NaOH layer into a 5 mL conical tube. Add 1 mL of 1 mol/L HCl and mix.
7. Add 0.1 mL of chloroform with a serological pipet. Vortex mix for 10 seconds, and then shake by hand for 30 seconds. Centrifuge for 3 minutes.
8. Aspirate off top layer. Inject 2 μL of the chloroform layer into the gas chromatograph.
9. Chromatographic parameters: injection temperature 280° C, column temperature 210° C, detector temperature 300° C, nitrogen flow 60 mL/min.

Calculation

1. Process a frozen control (10 μg/mL), and read from curve.
2. The following equations can be used, or peak heights can be read directly from the standard curve.

$$\text{Ratio } X = \frac{\text{Peak height}_x}{\text{Peak}_{\text{internal standard}}}$$

$$\text{Concentration } X = \frac{\text{Ratio}_x}{\text{Ratio}_{\text{std}}} \times \text{Concentration}_{\text{std}}$$

Notes

1. Process a new working curve if the internal standard or gas chromatograph conditions change.
2. Because of the nature of the valproic acid compound, sample should not be subjected to temperatures greater than 25° C.
3. Sample requirements: Approximately 1.5 mL of serum should be sent to allow for duplicate analyses. Minimum sample is 0.5 mL of serum.

REFERENCE RANGE

Generally accepted therapeutic ranges for these drugs are as follows:

Drug	μg/mL	μmol/L
Carbamazepine	4-12	17-51
Ethosuximide	40-100	284-459
Phenobarbital	15-40	64-172
Phenytoin	10-20	40-80
Primidone	5-12	23-55
Valproic acid	50-120	347-833

REFERENCES

1. Plaa, GL, and Hine, CH: A method for the simultaneous determination of phenobarbital and diphenylhydantoin in blood, J Lab Clin Med 47:649-657, 1956.
2. Svensmark, O, and Kristensen, P: Determination of diphenylhydantoin and phenobarbital in small amounts of serum, J Lab Clin Med 61:501-507, 1963.
3. Gardner-Thorpe, Parsonage, MJH, Smethurst, PF, and Toothill, C: A comprehensive gas chromatographic scheme for the estimation of antiepileptic drugs, Clin Chim Acta 36:223-250, 1971.
4. Pippenger, CE, Scott, JE, and Gillen, HW: Thin-layer chromatography of anticonvulsant drugs, Clin Chem 15:255-260, 1969.
5. Huisman, JW: The estimation of some important anticonvulsant drugs in serum, Clin Chim Acta 13:323-328, 1966.
6. MacGee, J: Rapid determination of diphenylhydantoin in blood plasma by gas-liquid chromatography, Anal Chem 42:421-442, 1970.
7. Vandemark, FL, and Adams, RF: Ultramicro gas-chromatographic analysis for anticonvulsants, with use of a nitrogen-selective detector, Clin Chem 22:1062-1065, 1976.
8. Sengupta, A, and Peat, MA: Gas-liquid chromatography of eight anticonvulsant drugs in plasma, J Chromatogr 137:206-209, 1977.
9. Toseland, PA: Gas chromatographic analysis of anticonvulsant drugs, Lablore 7:433-436, 1977.
10. Least, CJ, Jr, Johnson, GF, and Solomon, HM: Therapeutic monitoring of anticonvulsant drugs: gas-chromatographic simultaneous determination of primidone, phenylethylmalonamide, carbamazepine, and diphenylhydantoin, Clin Chem 21:1658-1662, 1975.
11. Abraham, CV, and Joslin, HD: Simultaneous gas-chromatographic analysis for phenobarbital, diphenylhydantoin, carbamazepine, and primidone in serum, Clin Chem 22:769-771, 1976.
12. Solow, EB, and Green, JB: The simultaneous determination of multiple anticonvulsant drug levels by gas-liquid chromatography, Neurology 22:540-550, 1972.
13. Adams, RJ, and Vandemark, FL: Simultaneous high-pressure liquid-chromatographic determination of some anticonvulsants in serum, Clin Chem 22:25-31, 1976.
14. Kabra, PM, Stafford, BE, and Marton, LJ: Simultaneous measurement of phenobarbital, phenytoin, primidone, ethosuximide, and carbamazepine in serum by high-pressure liquid chromatography, Clin Chem 23:1284-1288, 1971.

15. Kabra, PM, McDonald, DM, and Marton, LJ: A simultaneous high-performance liquid chromatographic analysis of the most common-anticonvulants and their metabolites, J Anal Toxicol 2:127-133, 1978.
16. Soldin, SJ, and Hill, JG: Rapid micromethod for measuring anticonvulsant drugs in serum by high-performance liquid chromatography, Clin Chem 22:856-859, 1976.
17. Soldin, SJ, and Hill, JG: Interference with column-chromatographic measurement of primidone, Clin Chem 23:782, 1977.
18. Soldin, SJ, and Hill, JG: Routine dual-wavelength analysis of anticonvulsant drugs by high-performance liquid chromatography, Clin Chem 23:2352-2353, 1977.
19. Soldin, SJ, and Hill, JG: The therapeutic monitoring of anticonvulsant drugs in a 650-bed children's hospital. In Hawk, GL, editor: Biological/biomedical application of liquid chromatography, New York, 1979, Marcel Dekker, Inc.
20. Soldin, SJ: High performance liquid chromatographic analysis of anticonvulsant drugs using radial compression columns, Clin Biochem 13:99-101, 1980.
21. Soldin, SJ, and Walter, M: Improved HPLC analysis for anticonvulsant drugs employing radial compression columns, Clin Ciochem 14:161, 1981.
22. Cook, CE, Christensen, HD, Amerson, EW, et al: Radioimmunoassay of anticonvulsant drugs: phenytoin, phenobarbital, and primidone. In Kellaway, P, and Petersen, IS, editors: Quantitative analytical studies in epilepsy, New York, 1976, Raven Press.
23. Wong, RC, Burd, JF, Carrico, RJ, et al: Substrate-labeled fluorescent immunoassay for phenytoin in human serum, Clin Chem 25:686-691, 1979.
24. Frings, CS, and Phillips, G: Therapeutic drug monitoring of anticonvulsant drugs using fluorescence polarization immunoassay, Clin Chem 28:1611, 1982.
25. Tybach, R: Adaptation of prosthetic-group label homogenous immunoassay to reagent strip format, Clin Chem 27:1499-1504, 1981.
26. Nishikawa, T, Kubo, H, and Saito, M: Competitive nephelometric immunoassay method for antiepileptic drugs in patient blood, J Immunol Methods 29:85-89, 1979.
27. Willox, S, and Foote, SE: Simple method for measuring valproate (Epilim) in biological fluids, J Chromatogr 151:67-70, 1978.
28. Jakobs, C, Bohasch, M., and Hanefeld, F: New direct micromethod for determination of valproic acid in serum gas chromatography, J Chromatogr 146:494-497, 1978.
29. Berry, DJ, and Clarke, LA: Determination of valproic acid (dipropylacetic acid) in plasma by gas-liquid chromatography, J Chromatogr 156:301-307, 1978.
30. Freeman, DJ, and Rawall, N: Extraction of underivatized valproic acid from serum before gas chromatography, Clin Chem 26:674-675, 1980.
31. Fairinotte, R, Pfaff, MC, and Mahuzier, G: Simultaneous determination of penobarbital and valproic acid in plasma using high performance liquid chromatography, Ann Biol Clin 36:347-353, 1978.
32. Gupta, RN, Keane, PM, and Gupta, ML: Valproic acid in plasma, as determined by liquid chromatography, Clin Chem 25:1984-1985, 1979.
33. Mucklow, JC: The use of saliva in therapeutic drug monitoring, Ther Drug Monit 4:229-247, 1982.
34. Dusci, LJ, and Hatchett, LP: Gas chromatographic determination of valproic acid in human plasma, J Chromatogr 132:145-147, 1977.

Barbiturates

AMADEO J. PESCE

Clinical significance pp. 594, 740, and 795

Molecular structure:	**Amobarbital**	**Butabarbital sodium**	**Hexobarbital**
Molecular formula:	$C_{11}H_{18}N_2O_3$	$C_{10}H_{15}N_2NaCO_3$	$C_{12}H_{16}N_2O_2$
Molecular weight (daltons):	226.27	234.23	236.26
Merck Index:	592	1471	4598

Molecular structure:	**Pentobarbital sodium**	**Phenobarbital**	**Secobarbital sodium**
Molecular formula:	$C_{11}H_{17}N_2NaO_3$	$C_{12}H_{12}N_2O_3$	$C_{12}H_{17}N_2NaO_3$
Molecular weight (daltons):	248.6	232.23	260.27
Merck Index:	6998	7109	8268

Chemical class: barbiturate, organic heterocyclic

PRINCIPLES OF ANALYSIS

Barbiturates, classified as hypnotics, are a group of drugs that are prescribed for a number of conditions. They act as sedatives or analgesics (painkillers) and are used to help control seizure disorders. Most commonly, phenobarbital is taken for the control of epilepsy, and its determination, as part of therapeutic drug monitoring, is usually the most frequently performed analysis for this group of compounds. Recently, the use of high-dosage pentobarbital has become important in the therapy of head trauma. However, barbiturates are also abused substances; thus the barbiturate determination in the overdose patient is one of the major components of the toxicological drug screen. Both qualitative and quantitative tests are performed by the laboratory. The qualitative tests in the drug screen are selected for their rapid performance in drug screening and usually react with any drug of the barbiturate class. Most qualitative analyses are performed on urine or gastric contents because these body fluids are best for drug screening. In contrast, quantitative analyses are performed on blood because blood levels correspond most closely with pharmacological effect.

Qualitative tests

One performs a qualitative color test for barbiturate by extracting the drug from a slightly acidic solution into an organic solvent such as chloroform or dichloromethane and complexing the barbiturate with mercury. The barbiturate mercury complex is then allowed to react with diphenylthiocarbazone, and an orange-colored product is formed (method 1, Table 64-7).[1]

One can use ultraviolet absorption spectroscopy both to identify and to quantify barbiturates, since the absorption spectrum is fairly specific for this class of drugs (method 2, Table 64-7). Extraction of the barbiturate is necessary to free it from other analytes that have similar absorption spectrums.[2] The presumptive barbiturate-containing body fluid is usually acidified and extracted into an organic solvent such as chloroform. It is then re-extracted into an alkaline aqueous solution. One scans the spectrum from

Table 64-7 Methods for barbiturate analysis

Method	Principle	Usage	Comments
Qualitative			
1. Extraction and colorimetric reaction	Barbiturate is extracted from slightly acidic solution and complexed with Hg^{+2}; reacts with diphenylthiocarbazone to form an orange-colored product.	Blood, plasma, serum, urine, gastric juice	Nonspecific for barbiturates
2. Extraction and ultraviolet spectroscopy	Barbiturate is extracted free of other ultraviolet-absorbing compounds; ultraviolet spectrum is characteristic and specific for each drug.	Blood, plasma, serum, urine, gastric contents, tissue	Original specific method Some interferences Can be quantitative *Not* recommended for mixtures
3. Ultraviolet difference spectroscopy	As in method 2, extracted barbiturate is divided into aliquots and difference between spectra at pH greater than 13 and at pH 10.5 recorded.	Blood, serum	More specific than method 2 Requires recording dual-beam spectrophotometer
4. Thin-layer chromatography (TLC)	Extracted barbiturate is chromatographed and its R_f factor* and reaction with stain recorded; R_f factor and color reactions presumptive for barbiturates.	Blood, plasma, serum, urine, gastric contents, tissue	Used as initial qualitative test or confirmatory Can separate mixtures
5. Gas chromatography	Extracted barbiturate is chromatographed and its retention time recorded; parent and possible degradation products have specific retention times.	Blood, plasma, serum, urine, gastric contents, tissue	Not quantitative Degradation of some barbiturates, not a gaussian elution pattern Can separate mixtures
6. Gas chromatography–mass spectrometry	Steps identical to method 5; mass spectrophotometric detector can identify specific barbiturates and help resolve mixtures.	Same as for method 5	Improves identification of method 5
7. EMIT (enzyme-multiplied immunoassay technique)	Competitive binding assay using EMIT technology; antibody has broad specificity and reacts with most barbiturates.	Urine	Extent of reaction varies with amount and type of barbiturate

*A relative mobility factor.

Table 64-7 Methods for barbiturate analysis—cont'd

Method	Principle	Usage	Comments
Quantitative—all barbiturates			
Ultraviolet spectroscopy (similar to method 2)	Extracted barbiturate has characteristic absorbance spectrum, with amount of absorbance proportional to concentration; some identification depends on spectral characteristics.	Blood, plasma, serum, urine, gastric contents, tissue	Original method, but subject to interference by endogenous compounds and other drugs Does not work well on mixtures
Differential ultraviolet spectroscopy (similar to method 3)	Extracted barbiturate has characteristic differential spectrum when solution at pH > 13 and pH 10.5 are compared.	Blood, serum	Less interference than ultraviolet spectroscopy, but specific barbiturate must be known to achieve quantitation
Gas chromatography (variation of method 5)	Extracted barbiturate(s) is converted to alkylated derivative(s), such as dimethylphenobarbital: $(CH_3)_4N^+OH^-$ + Barbiturates → Dimethylated barbiturates Retention time identifies which barbiturate present.	Blood, plasma, serum, urine, gastric contents, tissue	Only a few drugs may interfere Recommended procedure for mixtures
8. High-performance liquid chromatography (HPLC)	Extracted barbiturates are chromatographed and quantified by refractive index or absorbance.	Blood, plasma, serum, urine, gastric contents, tissue	Should be as efficient as gas chromatography, but fewer laboratories have used this technique
Quantitative—phenobarbital in plasma or serum			
Ultraviolet spectroscopy (same as methods 2 or 3)	As for methods 2 and 3	As for methods 2 and 3	More subject to interference than other procedures
Enzyme immunoassay (EIA; variation of method 7)	Competitive binding assay (EMIT nephelometry)	Serum	Common technique
Gas-liquid chromatography derivatized or underivatized (variation of method 5)	As for method 5	Serum, plasma	Requires technical skill Can quantitate several antiepileptic drugs simultaneously
High-performance liquid chromatography (HPLC; same as method 8)	As for method 8	Serum, plasma	Requires skill Can quantitate several antiepileptic drugs simultaneously
9. Radioimmunoassay	Competitive binding assay		Less used
10. Fluorescence polarization	Competitive binding with fluorescence polarization; bound labeled drug has higher polarized fluorescence than free; amount of polarized fluorescence inversely proportional to the concentration of the drug.	Serum, plasma	Commonly used procedure Lowest coefficient of variation
11. Ames strip test	Competitive binding with enzymatic procedure; antibody binding to drug labeled with prosthetic group prevents glucose oxidase activity; competition with endogenous drug allows prosthetic group to activate enzyme; coupled to colorimetric reaction.	Serum	Designed for inexpensive instrument Reflectance spectrophotometer Rapid test, useful for small laboratories or physician's office
12. Nephelometric inhibition	Competitive inhibition of precipitation; reaction is measured by nephelometry.	Serum	Restricted to laboratories with nephelometers

230 to 340 nm and records the resulting absorbance.[3] It is compared to the characteristic spectrum of each of the barbiturates. The amount of barbiturate present is proportional to the absorption at specific wavelengths. A refinement of this approach is the use of differential spectroscopy of the ultraviolet spectrum for the quantitation of blood levels (method 3, Table 64-7). In this test one divides the extracted drug into two aliquots, one with a strongly alkaline pH greater than 13 and the other with a pH of 10.5. The spectral difference between these solutions is obtained by use of a dual-beam recording spectrophotometer.[4] The difference spectrum is definitive for barbiturate, and the amount of the absorbance difference is proportional to the concentration (Fig. 64-5).[5]

Chromatographic procedures are also useful for the identification of barbiturates; these include thin-layer chromatography (TLC) and gas chromatography. TLC (method 4, Table 64-7) is usually performed on urine that has been acidified and extracted into an organic phase. One places the sample on a TLC plate and identifies the position of the barbiturate after chromatography by developing a color reaction with one of several sprays, such as potassium permanganate, mercuric sulfate, or diphenylcarbazone.[6] The R_f value and the color are the specific criteria for identification.

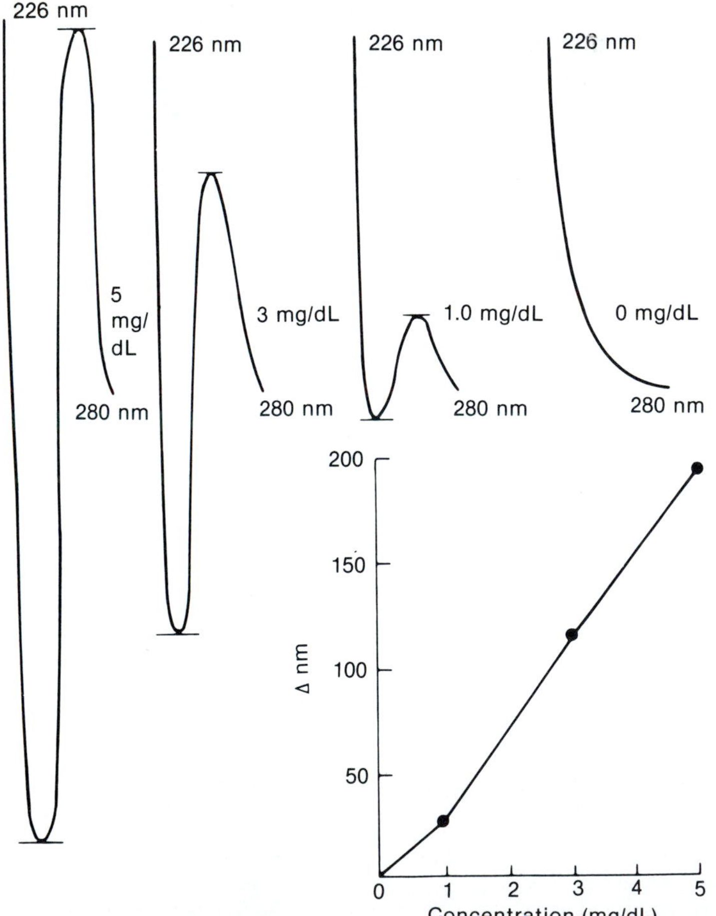

Fig. 64-5 Differential spectra of various concentrations of phenobarbital and a standard calibration curve. Absorbance of sample of pH 13 (sample cuvette) minus absorbance of sample at pH 10 (reference cuvette) is recorded. Difference in absorbance between 240 and 260 nm (Δnm) is plotted versus drug concentration. *(From Schumann, GB, et al: Am J Clin Pathol 66:823-830, 1976.)*

Gas chromatography detection (method 5, Table 64-7) is performed by the extraction of the acidified urine specimen with toluene or other selected organic solvents and the separation of the extracted analyte.[7] An OV-series column with a flame ionization detector is the detection system most commonly used as part of an overall screening procedure for drugs, including barbiturates (see discussion of drug screen, p. 1094). The relative retention time is specific for each barbital derivative and is used for presumptive identification. Alternatively, the extract can be analyzed with a gas chromatograph/mass spectrometer (GC/MS) and identified by the pattern of mass fragmentation (method 6, Table 64-7).

The enzyme-multiplied immunoassay technique (EMIT) for barbiturate in urine (method 7, Table 64-7) is a homogeneous enzyme immunoassay (EIA). (See Chapter 11.) In the original assay for urine, lysozyme acted to clear a bacterial substrate. A variation of this assay is currently used for the detection of barbiturates in serum and urine. In this circumstance the enzyme glucose-6-phosphate dehydrogenase is used instead of lysozyme.

Quantitative tests

Techniques of quantitation include gas chromatography, high-performance liquid chromatography (HPLC), EMIT, fluorescence polarization immunoassay (FPIA), prosthetic group ligand immunoassay, and other immunoassay procedures. In general, immunoassays are used only when the specific drug is known, such as phenobarbital. These procedures are all competitive binding assays; the drug in the unknown sample competes with radiolabeled drug (radioimmunoassay, RIA) (method 9, Table 64-7), enzyme-bound drug (EMIT), fluorescein-labeled drug (TDx) (method 10, Table 64-7), prosthetic group–labeled drug (ARIS) (method 11, Table 64-7), or protein-bound drug (nephelometric inhibition assays) (method 12, Table 64-7). (See Chapter 11.) To quantitate several different types of barbiturates, such as phenobarbital and secobarbital, one must establish the extent of cross-reactivity for each barbiturate.

The FPIA developed by Abbott Laboratories has become very popular (method 10, Table 64-7).[8,9]

The Ames Seralyzer method for phenobarbital (method 11, Table 64-7) employs a labeled prosthetic-group assay.[10] This method has been incorporated into the solid-phase strip technology employed on the Ames Seralyzer. In this assay, the patient serum or plasma sample is diluted with distilled water, and a specific amount of the prepared specimen is placed on the reagent area of the strip. The strip is then placed in the instrument, which is a solid-phase reflectance photometer. The color of the strip is read at 740 nm, and the results are calculated from a calibration curve.

Gas chromatographic analysis was the first method to isolate and quantify a mixture of barbiturates effectively with a minimum of possible interferences.[11-13] Initial methods required derivatization (methylation) of the barbiturates with trimethyl anilinium hydroxide. Current variations use tetramethyl ammonium hydroxide as the derivatizing reagent. Once derivatized, the barbiturate is more stable and is eluted with more symmetrically defined peaks than that of the underivatized parent. Improved column packings, such as the SP (stationary phase support, Supelco, Inc., Bellefonte, PA) series, enable analysis of the underivatized drugs without noticeable column effects.

Competitive binding immunoassays have focused primarily on the measurement of phenobarbital. Currently, the EMIT, TDx, and ARIS procedures are the most widely used immunoassays. In addition, the nephelometric inhibition assay has been developed for measurement of this specific drug. In these procedures for the measurement of phenobarbital one must keep in mind that there can be cross-reactivity with several or all of the other barbitals.

HPLC has been adapted for the measurement of the barbiturates, but the most common procedure quantifies phenobarbital in the presence of other antiepileptic drugs, such as primidone and phenytoin.[14] The chromatographic system uses a 30 × 0.4 cm μPorasil column (Waters Associates, Inc., Milford, MA). The drug is detected with an ultraviolet spectrophotometer, which monitors the absorbance at 254 nm.

The ultraviolet absorption analyses (method 2, Table 64-7) have acceptable accuracy and precision, but to quantify mixtures, one must make several calculations and approximations.

PREFERRED METHODS

Qualitative methods

The identification of barbiturates requires two procedures, one of which must be the confirming procedure. In the case of an overdose, urine is the fluid of choice for testing. In this circumstance, the enzyme-multiplied (EMIT) drug abuse test offers the best choice for a rapid screen for the presence of barbiturates. However, one must keep in mind that there is a cross-reactive response between the barbiturates; this response will vary depending on the nature of the individual barbiturates taken. Following are the lower limits of detection:

Secobarbital	2 μg/mL
Pentobarbital	5 μg/mL
Butabarbital	10 μg/mL
Phenobarbital	3 μg/mL
Amobarbital	30 μg/mL

The presence of barbiturates must be confirmed by TLC, ultraviolet spectroscopy, or another method.[7,15] If TLC is used, one can identify individual drugs by observing the R_f values and the color produced by the staining reagent.

Alternatively, EMIT can be used for the qualitative assay of barbiturate in serum. However, it is more difficult to obtain a confirmatory test on serum.

The ultraviolet procedure of Goldbaum, with extraction and measurement of the drug at pH 10 and at a pH greater than 13, also offers a valid assay for analysis; however, it does not work well if there is more than one barbital present. This procedure has been the standard against which others have been compared.[16]

Gas chromatography is the procedure of choice for qualitative analysis of barbiturates. HPLC has not been used as extensively for qualitative estimation of barbiturates, but it should yield data comparable to gas chromatography. For confirmation in legal cases, the recommended procedure is GC/MS.

Quantitative methods

For analysis of mixtures of barbiturates, only the gas and HPLC procedures can be recommended because the individual barbiturates are separated before quantitation.[17] For pentobarbital, both gas chromatography and HPLC are acceptable. An alternative method using an EMIT kit designed for serum barbiturates may be used, but great care is required. A comparison of the colorimetric, GLC, ultraviolet absorption, FPIA, and EMIT procedures is given in Table 64-8. This table shows that the GLC method has the highest variability.

The Ames ARIS technology is designed for physician office testing or for the small laboratory. It meets the criteria for this type of testing: minimum technical skill, stable reagents, low-cost instrumentation, and a reproducibility of approximately 10%.

SPECIMEN

The specimen can be blood, plasma, serum, urine, or tissue extract. The drug is stable in any of these specimens for 1 week. However, most laboratories refrigerate or freeze samples until analysis.

PROCEDURE: HIGH-PERFORMANCE LIQUID CHROMATOGRAPHY[18]

Principle

HPLC is used for the specific quantitative analysis of any of the following barbiturates and drugs that are extracted together.

Barbiturates	Other drugs
Amobarbital	Glutethimide
Butabarbital	Methaqualone
Butalbital	Methyprylon
Pentobarbital	
Phenobarbital	
Secobarbital	

Table 64-8 Comparison of assay conditions for serum phenobarbital

Factor	Gas-liquid chromatography	Ultraviolet spectrophotometry (difference)	Enzyme-multiplied immunoassay technique	Fluorescence polarization immunoassay
Sample size (mL)	1	3	0.05	0.02
Sensitivity (mg/L)	2	2	5.0	0.34
Coefficient of variation (%)	8.5	5.4	8.2	5.0
Interference	Phenobarbital and mephobarbital not separated	Endogenous absorption for *p*-aminophenol and sulfonamide drugs	Some barbitals	Some barbiturates
Time	1 hour	40 minutes	15 to 30 minutes	20 minutes

These compounds are isolated by use of C_{18} bonded-phase columns and analyzed by HPLC with spectrophotometric detection of 214 nm, with comparison of retention times and peak-height ratios of unknowns against known standards and controls.

Reagents

Methanol (HPLC grade). Methanol is stable for 2 years at room temperature.

Sodium acetate buffer (pH 4.5, 0.1 mol/L). Dissolve 8.2 g of sodium acetate is distilled water, and bring to 1000 mL with distilled water. Adjust the pH to 4.5 with acetic acid if necessary. Store at room temperature in brown bottle. Solution is stable for 2 years; check monthly for growth.

Phosphate buffer (pH 4.4, 0.15 mmol/L). Prepare KH_2PO_4, 1 mol/L, by dissolving 13.6 g of KH_2PO_4 in 100 mL of distilled water. Prepare 0.9 mol/L phosphoric acid by adding 6.125 mL of phosphoric acid to distilled water. Bring the volume to 100 mL.

Add 300 μL of 1 mol/L KH_2PO_4 to 1800 mL of distilled water. Using a pH meter, adjust to pH 4.4 with the 0.9 mol/L phosphoric acid (approximately 50 μL).

Acetonitrile: phosphate buffer (mobile phase). Add 190 mL of acetonitrile to 810 mL of the phosphate buffer; filter through a 0.2 μm pore filter and degas under vacuum for 5 minutes.

Internal standard (nitrazepam). Stock solution is 1 mg/mL. Dissolve 10 mg of nitrazepam in methanol, and bring to 10 mL with methanol. Store at 4° C. Solution is stable for 1 year.

Working internal standard solution, 0.01 mg/mL. Dilute 1 mL of working standard to 100 mL with sodium acetate buffer. Store at 4° C. Solution is stable for 6 months.

Calibration standards. Dissolve the desired barbiturate in methanol, and dilute to a concentration of 5 μg/mL in serum. For methyprylon and phenobarbital, dilute to a final concentration of 20 μg/mL. Solution is stable for 3 months at 4° C.

Assay

Equipment: HPLC equipment, such as Waters HPLC System, including injector, column heater, pump, detector, recorder, and a 15 cm C_{18} μBondapak Column; Vac-Elut extraction apparatus (Analytichem International); and 1 mL C_{18} Bond-Elut bonded-phse extraction columns (catalog no. 607101, Analytichem International).

1. Sample preparation
   a. Set up a Bond-Elut column for each sample to be analyzed. With vacuum on, prime columns twice with methanol and twice with distilled water.
   b. Turn off vacuum.
   c. To the respective columns, add 0.5 mL of working internal standard followed by 0.5 mL of appropriate sample.
   d. Turn on vacuum; samples should be pulled through the columns.
   e. Immediately wash columns twice with distilled water.
   f. Turn off vacuum.
   g. Wipe needle tips inside Vac-Elut, and position collection tubes in rack.
   h. Add 100 μL of methanol; wait 20 seconds.
   i. Turn on vacuum; then turn off after methanol passes through.
   j. Repeat with another 100 μL of methanol.
   k. Remove collection tubes, and with a disposable glass pipet, transfer the contents of a collection tube to a larger 12 × 75 mm glass tube.
   l. Inject 20 μL into the HPLC system.
2. Chromatographic parameters
   a. Flow rate, 1.5 mL/min
   b. Column temperature, 60° C
   c. Detector AUFS, 0.1 (absorbance units at full scale)
   d. Detector wavelength, 214 nm
   e. Chart speed, 0.5 cm/min

Calculations

A sample chromatogram of a series of barbiturate standards is presented in Fig. 64-6.

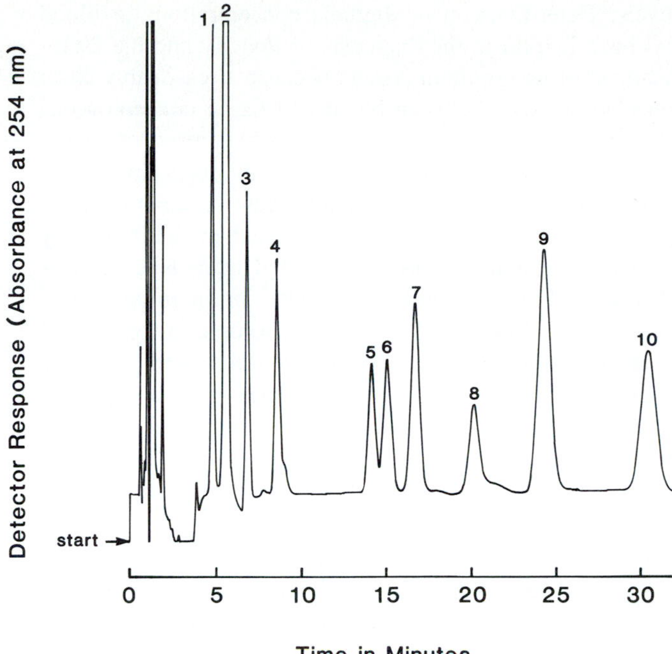

Fig. 64-6 Elution pattern of barbiturate standards after HPLC. *1,* Methyprylon; *2,* phenobarbital; *3,* butabarbital, *4,* butalbital; *5,* pentobarbital; *6,* amobarbital; *7,* glutethimide; *8,* secobarbital; *9,* nitrazepam internal standard; *10,* methaqualone.

Calculate the peak-height ratio of drug of interest to internal standard:

$$\frac{\text{Peak height (mm) of drug}}{\text{Peak height (mm) of internal standard}} = \text{Peak-height ratio}$$

Compare elution times of standards to the elution times of all unknown patient peaks of interest. See Fig 64-6 for elution order of the various drugs.

Concentration unknown (μg/mL) =

$$\frac{\text{Peak-height ratio in patient or control}}{\text{Peak-height ratio of calibration standard}} \times \text{Concentration of appropriate drug in standard } (\mu g/mL)$$

Notes

1. The following limits of sensitivity for each of the barbiturates have been observed:

Compound	Limit (μg/mL)
Amobarbital	1
Butabarbital	1
Butalbital	1
Phenobarbital	5
Pentobarbital	1
Secobarbital	1
Glutethimide	1
Methaqualone	1
Methyprylon	5

Specific negatives are reported as "none detected," "less than," and "approximately less than" value.

2. Specificity
   a. Phenytoin is known to have the same retention time as amobarbital.
   b. Clonazepam is known to have the same retention time as methaqualone.
   c. Carbamazepine has a similar retention time to secobarbital.
   d. A glutethimide metabolite has the same retention time as butalbital.
   e. Hemolyzed, lipemic, and icteric samples do not interfere with the analysis.

REFERENCE RANGES

Therapeutic concentrations in serum are as follows:

Amobarbital	Up to 8 μg/mL (<35 μmol/L)
Butabarbital	Up to 8 μg/mL (<34 μmol/L)
Pentobarbital	Up to 4 μg/mL (<16 μmol/L)
Phenobarbital	15 to 40 μg/mL (64-172 μmol/L)
Secobarbital	Up to 6 μg/mL (<23 μmol/L)

REFERENCES

1. Curry, AS: Rapid quantitative barbiturate estimation, Br Med J 1:354-355, 1965.
2. Walker, JT, Fisher, RS, and McHugh, JJ: Qualitative estimation of barbiturates in blood by ultraviolet spectrophotometry, Am J Clin Pathol 18:451-461, 1948.
3. Broughton, PMG: A rapid ultraviolet spectrophotometric method for the detection, estimation, and identification of barbiturates in biological material, Biochemistry 63:207-213, 1956.
4. Goldbaum, LR: Determination of barbiturates: ultraviolet spectrophotometric method with differentiation of several barbiturates, Anal Chem 24:1604-1607, 1952.
5. Schumann, GB, Leuenstein, K, LeFever, D, and Henry, JB: Ultraviolet spectrophotometric analysis of barbiturates, Am J Clin Pathol 66:823-830, 1976.
6. Ganshirt, H: Pharmaceutical products. In Stahl, E, editor: Thin-layer chromatography: a laboratory handbook, New York, 1962, Academic Press, Inc.
7. Sunshine, I, Maes, R, and Finkle, B: An evaluation of methods for the determination of barbiturates in biological materials, Clin Toxicol 1:281-296, 1968.
8. Popelka, SR, Miller, DM, Holen, JT, and Kelso, DM: Fluorescence polarization immunoassay. II. Analyzer for rapid precise measurement of fluorescence polarization with use of disposable cuvettes, Clin Chem 27:1198-1201, 1981.
9. Loomis, KF, and Frye, RM: Evaluation of the Abbott TDM for the stat. measurement of phenobarbital, phenytoin, carbamazepine, and theophylline, Am J Clin Pathol 80:686-691, 1983.
10. Tybach, R: Adaptation of prosthetic-group label homogeneous immunoassay to reagent strip format, Clin Chem 27:1499-1504, 1981.
11. MacGee, J: Rapid determination of diphenylhydantoin in blood plasma by gas liquid chromatography, Anal Chem 42:421-422, 1970.
12. MacGee, J: Rapid identification in quantitative determination of barbiturates in glutethimide in blood by gas liquid chromatography, Clin Chem 17:587-591, 1971.
13. Solow, EB, and Green, JB: The simultaneous determination of multiple anticonvulsant drug levels by gas liquid chromatography, Neurology 22:540-550, 1972.
14. Atwell, S, Green, V, and Honey, W: Development and evaluation of method for simultaneous determination of phenobarbital and diphenylhydantoin in plasma by HPLC, J Pharm Sci 64:806-809, 1975.
15. Sunshine I: Methodology for analytical toxicology, Cleveland, 1975, CRC Press, Inc.
16. Spiegel, HE, Chamberlain, RT, Dubowski, KM, et al: Barbiturates, ultraviolet spectrophotometric method. In Faulkner, WR, and Meites, S, editors: Selected methods of clinical chemistry, vol 9.

Selected methods for the small clinical laboratory, Washington, DC, 1982, American Association for Clinical Chemistry.

17. Spiehler, V, Sun, L, Miyada, DS, et al: Radioimmunoassay, enzyme immunoassay, spectrophotometry and gas-liquid chromatography compared to determination of phenobarbital and diphenylhydantoin, Clin Chem 22:749-753, 1976.
18. Kabra, PM: Simultaneous liquid-chromatographic determination of 12 common sedatives and hypnotics in serum, Clin Chem 24:657-662, 1978.

Digoxin and digitoxin

I-WEN CHEN
LINDA A. HEMINGER

Clinical significance: p. 415
Molecular formula: $C_{41}H_{64}O_{14}$ (α- and β-acetyldigitoxin, digoxin), $C_{41}H_{64}O_{13}$ (digitoxin)
Molecular weight: 780.97 and 764.92 daltons
Merck Index: 3148, 3143
Chemical class: cardiac glycoside

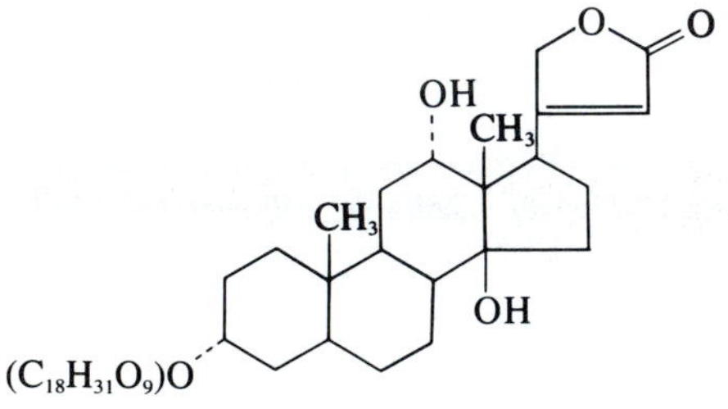

OH CH3 CH3 OH O O $(C_{18}H_{31}O_9)O$

Digoxin

O O CH3 CH3 OH $(C_{18}H_{31}O_9)O$

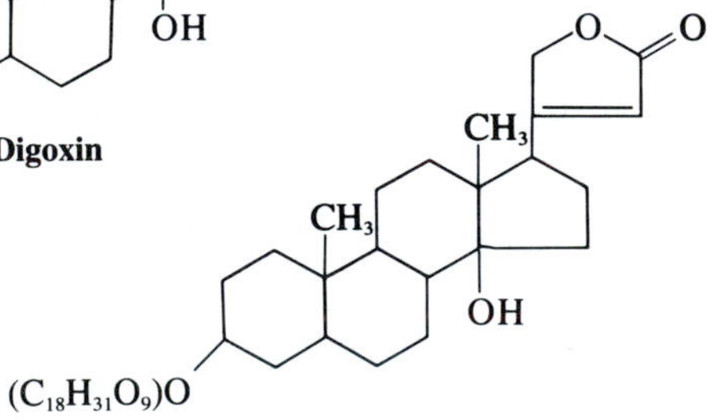

Digitoxin

PRINCIPLES OF ANALYSIS

The term *digitalis* is used to represent all steroid glycoside compounds sharing common features of chemical structure and common effects on cardiac activity. Digoxin and digitoxin are the two digitalis preparations most frequently used in the United States.

Digoxin and digitoxin are obtained in the crystalling form from the leaves of *Digitalis lantana* and *Digitalis purpurea,* respectively. Digoxin is sparingly soluble in water, chloroform, ether, ethyl acetate, and acetone but soluble in dilute (50%) alcohol or pyridine, whereas digitoxin, a nonpolar molecule (structure is that of digoxin without the C_{12}-OH group), is relatively more soluble in organic solvents. Digoxin and digitoxin are used to control cardiac arrhythmias through improvement of the strength of the myocardial contraction. However, intoxication is a frequent complication of digitalis therapy and can be life threatening, frequently requiring stat analysis for digitalis levels. Determination of digitalis concentrations in blood has been helpful in the diagnosis of toxicity and the establishment of an optimum dosage because a reasonably close correlation exists between blood and tissue concentrations of digitalis.

Many techniques have been used to determine blood levels of digoxin. They include gas chromatography[1] (method 1, Table 64-9), methods based on inhibition by digoxin in sodium-potassium-activated adenosine triphosphatase (ATPase) in red blood cells[2] or in microsomes[3] (method 2, Table 64-9), and an isotopic displacement method using a sodium-potassium-activated ATPase (method 3, Table 64-9).[4] However, these methods are relatively tedious and have been largely replaced by more practical and sensitive radioimmunoassay (RIA) techniques in routine clinical chemistry laboratories (method 4, Table 64-9).

Digoxin was one of the first haptenlike small molecules to be analyzed by the RIA technique.[5] Digoxin antisera were successfully raised in animals immunized with digoxin coupled to a variety of macromolecules. Unlike thyroid hormones, a digoxin molecule contains neither iodine nor any functional groups that can be iodinated. Tritiated digoxin preparations were initially the only radioligands available for digoxin RIA. Tritiated radioligands are beta-particle emitters and must be counted by means of liquid scinitillation counters. This is expensive and inconvenient in comparison to the counting of gamma-ray emitters such as iodine (^{125}I) in a crystal scintillation well counter. Techniques were developed that attach a functional group that can be easily iodinated to a digoxin molecule such as tyrosine, histamine, and tyramine. Nearly every separation technique used in RIAs has been applied successfully to the RIA of digoxin. The dextran-coated charcoal method seems to be the most widely used method, although the solid-phase method is popular among commercial digoxin kits because of its technical simplicity and rapidity.

Other types of labels have been developed as alternatives to radionuclides. In the homogeneous enzyme-multiplied immunoassay technique (EMIT, Syva Co., Inc., Palo Alto, CA) (method 5, Table 64-9), an enzyme, glucose-6-phosphate dehydrogenase, chemically coupled to digoxin is used as a label. The assay is described in Chapter 11.

A sensitive affinity column–mediated immunometric assay (method 6, Table 64-9) has been developed by E.I. du Pont de Nemours & Co. (Wilmington, DE).[6] In this assay, the $F(ab')_2$ fragment of the digoxin antibody is labeled with β-galactosidase. An excess amount of this enzyme-antibody conjugate is incubated with the sample so that digoxin contained in the sample can be rapidly and quantitatively bound to the antibody. To accomplish this, equal volumes of serum sample and reagent containing β-galactosidase–antibody conjugate are mixed and incubated manually for 10 minutes to 2 hours at room temperature. The sample cup is then introduced into the instrument (du Pont

Table 64-9 Methods for digoxin analysis

Method	Principle	Usage	Comment
1. Gas chromatography	Physicochemical separation with ^{3}H-digoxin as internal standard	Plasma; may be applied to other biological specimen	Requires prepurification by thin-layer chromatography and derivatization; tedious and time consuming
2. ATPase inhibition	Dose-dependent inhibition of Na- and K-activated ATPase	Plasma; may be applied to other biological specimens	Requires extraction; tedious and time consuming; less precise and less specific than RIA
3. Isotope displacement	Displacement of ^{3}H-ouabain from purified Na-K ATPase preparation by digoxin	Plasma; may be applied to other biological specimen	Requires extraction and beta-counting; tedious, nonspecific
4. Radioimmunoassay (RIA)	Competitive binding of ^{3}H- or ^{125}I-labeled and nonlabeled digoxin with digoxin antibody	Plasma or serum	Requires phase separation and beta-particle or gamma-ray counting; fully automated; most sensitive and precise
5. Enzyme-multiplied immunoassay techniques (EMIT)	Competitive binding for antibody between endogenous digoxin and enzyme-labeled digoxin; antibody binding to enzyme-digoxin complex inhibits enzymatic activity	Serum	Lacks sensitivity; requires sample pretreatment
6. Enzyme immunometric assay (EIA)	Quantitative binding of digoxin by excess amount of enzyme-labeled antibody	Plasma or serum	Requires no sample pretreatment but requires phase separation by affinity column; system semiautomated
7. Fluorescence polarization immunoassay (FPIA)	Competitive binding of fluorescein-labeled and nonlabeled digoxin with digoxin antibody	Plasma or serum	Requires sample pretreatment but requires no phase separation; quantitation depends on degree of polarization of emitted fluorescence; system semiautomated

aca) along with a digoxin text pack containing an ouabain (digoxin analog)–bovine serum albumin–Sephadex G-10 affinity column positioned in the pack head and a substrate tablet. The system than automatically pumps the sample through the column, allowing only the enzyme-antibody conjugate with the bound digoxin to be eluted from the column. The column eluate is mixed with the substrate tablet (containing *o*-nitrophenyl-β-D-galactopyranoside), and the enzyme activity is quantified by measurement of the rate of changes of absorbance at 405 nm. The β-galactosidase activity present in this fraction is proportional to the digoxin concentration in the serum. The capacity of the system is approximately 50 test packs per assay.

A fluorescent probe has also been used as an alternative to radioactive labels in the determination of digoxin. In the homogeneous fluorescence polarization immunoassay (FPIA) developed for digoxin by Abbott Laboratories (TDx, North Chicago, IL) the label on digoxin is a fluorescein (method 7, Table 64-9). See Chapter 11 for the principles of this method.

In the Abbott TDx system, serum samples are pretreated with trichloroacetic acid to precipitate proteins, and a clear supernatant obtained after centrifugation is assayed in the TDx system in a fully automated fashion.

All methods described in Table 64-9 are also applicable to, and some of them have actually been used for, the measurement of circulating digitoxin.[7-9] As with digoxin, digitoxin is determined in the clinical laboratory almost exclusively by immunoassay. However, only a few commercial immunoassay kits are currently available because digitoxin analyses are ordered much less frequently than digoxin analyses.

Although the RIA technique is still the most widely used, the enzyme and fluorescence immunoassays have been used more frequently in recent years.

Table 64-10 lists reaction conditions for digoxin analysis.

SPECIMEN

In most instances either serum or plasma may be used for the assay although some commercial manufacturers specifically indicate that plasma is not to be used with their assay kits. Digoxin in serum is quite stable; no appreciable loss of digoxin is observed in serum stored at room temperature for up to 2 weeks. However, if the specimen is not to be tested within 24 hours, it should preferably be stored frozen.

Time of blood collection is an important factor in determining digoxin toxicity. Serum digoxin levels will rise sharply during the first hour after an oral dose, indicating postabsorption uptake. This is followed by a sharp decrease as digoxin is taken up by the myocardium and other tissues. Usually 4 to 6 hours are needed for serum and tissue stores of digoxin to reach equilibrium. Thereafter, serum digoxin levels tend to stabilize and decrease very slowly over the next 24 to 40 hours. One obtains the great-

Table 64-10 Reaction conditions for digoxin analysis

Conditions	Manual RIA (Becton Dickinson)	Automated RIA (ARIA II)	Du Pont aca	Abbott TDx
Temperature	25° C	25° C	25° C	25° C
pH	7.0	7.2	7.0	7.5
Sample volume (μL)	50	50	300	200
Sample pretreatment	Not required	Not required	Not required	Protein precipitation
Tracer	^{125}I	^{125}I	β-Galactosidase	Fluorescein
Antibody	Rabbit	Rabbit	Rabbit labeled with β-galactosidase	Rabbit
Enzyme substrate	Not required	Not required	NGP (*o*-nitrophenyl-β-D-galactopyranoside)	Not required
Reaction time (min)	18	45	10	20
Sensitivity* (μg/L)	0.4	0.1	0.5	0.2
Precision (CV)†	8.2%-15%	4.0%-6.9%	6.8%-13%	5.4%-17.0%
Interferences	Digitoxin, digoxin metabolites	Digitoxin, digoxin metabolites	Digitoxin, digoxin metabolites, enzyme inhibitors	Digitoxin, digoxin metabolites, abnormal protein concentrations

*Minimum detectable concentration.
†Interlaboratory coefficient of variation according to the first quarter of 1985 College of American Pathologists survey, with mean digoxin concentrations in the range of 0.72 to 3.6 ng/mL.

est diagnostic accuracy by drawing the blood sample at a standard time during the stable phase, when serum digoxin levels reflect the average cardiac concentrations. In our laboratory, blood samples are routinely drawn between 5 and 30 hours after the last digoxin dose.

Digoxin levels cannot be accurately determined in a patient who is being switched from digitoxin to digoxin therapy because both drugs may be present in the serum. The digoxin antisera used in most commercial digoxin immunoassay kits cross-react significantly with digitoxin (2% to 6% cross-reactivity, with some as high as 40%). The problem of cross-reaction is further exacerbated by the fact that therapeutic levels of digitoxin are about 10 times higher than those of digoxin. The digitoxin antisera also cross-react with digoxin (2% to 4% cross-reactivity), but, because of the higher therapeutic levels, the interference by digoxin is less significant in the digitoxin immunoassay.

INTERFERENCES

A number of potential interferences can occur in immunoassays. Most commonly observed are those caused by natural metabolites, patient-derived antibody, or, more recently, exogenously administered antibody. Over the past few years, there has been considerable interest in digoxin-like immunoreactive factors (DLIFs), which have been observed in certain naturally occurring and pathophysiological states.[10,11] These compounds were originally detected because serum or plasma from patients with conditions such as uremia gave positive results for digoxin even though they were not receiving the drug. Further studies showed that the amount of these compounds observed in plasma was dependent on the assay method used. Characterization of these molecules shows them to be low-molecular-weight (200-1000 daltons) molecules bound to proteins in serum. They are water-soluble, heat-stable, neutral molecules that do not possess carboxylic or primary amine groups. In addition to being present in plasma, they are present in urine and amniotic fluids. These molecules can give values of immunoreactivity amounts as great as 3 ng/mL in some assays (1 to 2 ng/mL therapeutic range). They are commonly found in conditions such as renal failure or hepatic failure, and significant amounts are observed in infants and third-trimester pregnant women. It is argued that the presence of these molecules is detected because of either increased synthesis or decreased elimination in these conditions. DLIFs are observed because of the cross-reactivity of the assay antibody, and thus the observed values are antibody dependent. DLIFs are extractable from adrenal tissue, and it has been argued that the reason for their presence is that they probably react on certain receptors of the body. These receptors may or may not be similar to those from which digoxin normally reacts. However, such digoxin-like substances have been postulated for many years as being involved in the regulation of sodium and potassium transfer. The argument is that the antibody reactivity may reflect receptor site commonality. Thus the values found describe a class of molecules with potentially the same or similar type of function.

The recently introduced hydrophobic chromatography as a pretreatment step appears to overcome the problem of DLIF influence on digoxin assays.[12] Thus, in these specialized cases, it is most appropriate to use pretreatment to obtain values free of these interferences. In addition to assay pretreatment, the DLIF can be eliminated by the use of an antibody with better specificity for the digoxin.[13]

REFERENCE RANGE

As with all diagnostic tests, it is advisable for the individual laboratory to accumulate and analyze its own data to establish a range of toxic and nontoxic values; differences in assay techniques may affect assay results. In our laboratory we found mean digoxin levels of 1.1 ± 0.1 ng/mL (1.41 pmol/mL) for 108 nontoxic patients and 3.8 ± 0.5 ng/mL (4.87 pmol/mL) for 21 toxic patients. Ninety percent of the nontoxic patients had serum digoxin levels below 2 ng/mL (2.56 pmol/L), whereas 86% of the toxic patients had levels greater than 2 ng/mL.[14] A mean serum digitoxin concentration of 20 ± 11 ng/mL (2.61 pmol/mL) with a range of 9 to 32 mg/mL (11.8 to 41.8 pmol/mL) for nontoxic patients receiving oral doses of 0.07 to 0.15 mg/day (0.89 to 1.96 μmol/day) and a mean concentration of 34 ± 18 ng/mL (44.4 pmol/mL) with a range of 16 to 52 ng/mL (20.9 to 67.9 pmol/mL) for toxic patients receiving doses of 0.10 mg/day (1.31 μmol/day) were reported by Beller et al.[15]

It is generally believed that infants are more resistant to cardiac glycosides than adults are. O'Malley et al.[16] found that a mean nontoxic digoxin level of 3.8 ± 1.2 ng/mL (4.87 pmol/mL) for eight infants younger than 1 month was significantly higher than 1.4 ± 1.1 ng/mL (1.79 pmol/mL) for 16 adults, but the digoxin level in infants older than 1 month (1.4 ± 0.5 ng/mL; n = 5) was no higher on the average than that of the adult controls. The higher values of the infants may be caused by DLIF.

REFERENCES

1. Watson, E, and Kalman, SM: Assay of digoxin in plasma by gas chromatography, J Chromatogr 56:209, 1971.
2. Bertler, A, and Redfors, A: An improved method of estimating digoxin in human plasma, Clin Pharmacol Ther 11:665, 1970.
3. Burnett, GH, and Conklin, RL: The enzymatic assays of plasma digoxin, J Lab Clin Med 78:799, 1971.
4. Brooker, G, and Jelliffe, RW: Serum cardiac glycoside assay based upon displacement of ^{3}H-ouabain from Na-K-ATPase, Circulation 45:20, 1972.
5. Butler, VP, Jr, and Chen, JP: Digoxin specific antibody, Proc Natl Acad Sci 57:71, 1967.
6. Leflar, CC, Freytag, JW, Powell, LM, et al: An automated, affinity-column mediated, enzyme-linked immunometric assay for digoxin on the DuPont *aca* discrete clinical analyzer, Clin Chem 30:1809-1811, 1984.
7. Burnett, GH, and Conklin, RL: The enzymatic assay of plasma digitoxin levels, J Lab Clin Med 71:1040-1044, 1968.
8. Lukas, DS, and Peterson, RE: Double isotope dilution derivative assay of digitoxin in plasma, urine, and stool of patients maintained on the drug, J Clin Invest 45:782-795, 1966.
9. Oliver, GC, Jr, Parker, BM, Brasfield, DL, et al: The measurement of digitoxin in human serum by radioimmunoassay, J Clin Invest 47:1035-1042, 1968.
10. Valdes, R, Jr: Endogenous digoxin-like immunoreactive factor: impact on digoxin measurement and potential physiological implications, Clin Chem 31:1525-1532, 1985.
11. Solden, SJ: Digoxin—issues and controversies (review), Clin Chem 32:5-12, 1986.
12. Skogen, WF, Rea, MR, and Valdes, R, Jr: Endogenous digoxin-like immunoreactive factors eliminated from serum samples by hydrophobic silica gel extraction and enzyme immunoassay, Clin Chem 33:401-404, 1987.
13. Withespoon, L, Shuler, S, Neely, N, et al: Digoxin radioimmunoassay that does not detect digoxin-like substance in serum of newborns, infants or patients with renal failure, Clin Chem 33:420, 1987.
14. Park, HM, Chen, IW, Manitasas, GT, et al: Clinical evaluation of radioimmunoassay of digoxin, J Nucl Med 14:531-533, 1973.
15. Beller, GA, Smith, TW, Abelmann, WH, et al: Digitalis intoxication, N Engl J Med 284:989-997, 1971.
16. O'Malley, K, Coleman, EN, Doig, WB, et al: Plasma digoxin levels in infants, Arch Dis Child 48:55, 1973.

Drug screen

F. MICHAEL HASSAN

Clinical significance: p. 740

PRINCIPLES OF ANALYSIS AND CURRENT USAGE

"Coma panel, toxic screen, complete drug screen, comprehensive screen, drug abuse screen, stimulant panel, overdose panel, specific screen for drug identification," these test names used by various laboratories indicate that the term *drug screen* is nebulous at best. Each laboratory defines an analysis for possible toxic substances in a way that enables its personnel to obtain the information required by the physician ordering the test. There are many different types of qualitative drug screens, each developed with that particular laboratory's interests, finances, and personnel expertise in mind.

To describe the specific drug screen, one must consider the reasons for screening most often encountered by the clinical chemist. Areas in which drug screening is often indicated include (1) clinical toxicology, (2) forensic-medicolegal toxicology, (3) employee-screening programs, (4) drug trials, and (5) drug abuse programs. Although screening for forensic purposes is still largely the domain of the coroner's laboratory, other drug-screening programs have been adopted by the clinical laboratories, with special emphasis on clinical toxicology (the overdose patient) and drug-abuse screening programs.

Defining the test

Qualitative analysis for a drug or a group of drugs can generally be divided into three basic steps: (1) isolation or separation of the compounds of interest from the biological specimen, (2) analysis of the drug content of the partially purified sample by a "screening" method, and (3) employment of one or more "confirmatory" methods, different from the method used initially to detect the drug, to validate the findings (Table 64-11). Screening methods use analytical techniques that give either a positive or negative finding. Numerous possible analytical permutations exist for the screening-confirmation format.

Most drug screens fall into one of two classes: (1) a specific screen for a (particular) known or suspected drug

Table 64-11 Suggested confirmation procedures for screening tests

Screen	Confirmation
SPOT TEST ☞	UV
SPOT TEST ☞	TLC
TLC ☞	UV
TLC ☞	GC
EMIT ☞	TLC
EMIT ☞	GC
EMIT ☞	UV
RIA ☞	TLC
GC ☞	GC/MS

''Hand'' indicates that test on right is the suggested follow-up test to confirm a result obtained by a screen. See text for full test names indicated by abbreviations.

(or class of drug) or (2) a complete screen, with analysis for as many drugs as is technically and economically feasible for the laboratory to measure. Of course, searching for one particular compound is much easier than searching for any one or more of a large number of possible compounds. If a single known or suspected drug is of interest, the analyst can immediately optimize the screening test to ensure adequate specificity and sensitivity. If the screen is broadened to include several compounds, a series of specific analyses for each compound can be combined together to form a limited or ''panel'' type of screen. In this format, one tests each compound by a specific procedure that optimizes sensitivity. In screening for drugs of abuse a series of assays for six or seven compounds may be specified by an agency such as the National Institute for Drug Abuse. It is evident that if the number of compounds is increased further the time required for performing specific individual assays for a large group of drugs will eventually limit the number of drugs that can be included in the screen. When it is necessary to expand the screen to include many substances, less specific assays that enable the analyst to look for a vast array of compounds at a single glance must be employed. The actual number of drugs included in this expanding screening process varies from laboratory to laboratory and depends heavily on the experience of the analyst performing the assay.

Once the laboratory has decided which drugs it is feasible to include in its qualitative drug identification test list, it must gather the appropriate analytical data on these drugs and their metabolites. The necessity of understanding metabolic pathways of drugs of interest cannot be overemphasized. Many drugs are present in urine only as metabolites; thus it is essential to have standards of known metabolites as well as the parent drugs for accurate identification and confirmation.

General analytical procedures

Excellent general references detail not only general screening procedures, but also specific individual assay techniques.[1-3] Selected general and confirmation procedures utilized in drug screening are presented here. Table 64-11 presents suggested confirmation procedures for several different classes of tests. For analyses used in the forensic cases, confirmation is best done by GC/MS. Table 64-12 presents an overview of each type of test.

Extraction

Except for spot tests, most procedures used by the analyst require preparation of the specimen before actual analysis. For the drug screen, isolation is usually accomplished by solvent extraction procedures or by column chromatography employing exchange resins or solid bonded-phase stationary phases. Solvent extraction of urine samples involves a two-phase partitioning system in which the drug is separated from an aqueous phase and redistributed into a second, usually organic, phase. This results in the removal of many of the interfering compounds present in the specimen.

Spot tests

The spot tests are among the simplest and quickest methods of performing an initial screening process or a confirmatory test. Usually the unprocessed specimen (urine or serum) is added directly to test reagents to assay for the presence of a specific drug or class of drugs, which is presumptively indicated by the formation of a colored reaction product. Table 64-13 lists commonly used spot tests routinely employed by the laboratory.[4-11] Although spot tests offer inexpensive testing with good turnaround times and provide an early lead to the analyst, they should be used selectively and with care because they are relatively nonspecific and often require subjective interpretation.

Ultraviolet spectroscopy

As with the spot tests, ultraviolet analysis is best employed in testing for a specific drug or in confirming compounds detected by other techniques. The ultraviolet scan has relatively poor specificity unless the unknown drug is isolated by other test procedures, such as thin-layer chromatography (TLC), or extracted by procedures that minimize spectral interferences from the body fluids and from other drugs. Fig. 64-7 emphasizes that compounds may have overlapping or similar ultraviolet spectra, necessitating further confirmatory techniques. Excellent sources have compiled spectra of most drugs encountered by the analyst.[12,13] Ultraviolet spectra are simple to obtain, but the sample preparation time can be lengthy.

Table 64-12 Types of assays used in screening procedures

Assay	Principle	Advantages	Disadvantages
Spot test	Drug or agent in specimen reacts, giving specific color (see Table 64-13)	Very rapid test Indicates possible drug group present	Not specific Different spot test required for each group of drugs Limited to a few drug groups
Enzyme immunoassay (EIA)	Competitive binding assay between nonlabeled and enzyme-labeled drug	Rapid for single or several assays Most convenient method for detection of certain drugs	Class specific Assays must be done separately for each class of drugs or individual drug Costly
Gas chromatography	Extracted drug chromatographed and identified by retention time	Can separate and detect large variety of drugs	Not possible to detect all drugs Extraction and confirmation necessary
Thin-layer chromatography (TLC)	Extracted drug chromatographed and detected by specific chemical reaction or properties	Can separate and detect more compounds than any other method	Not possible to detect all drugs with one extraction or one chromatography system Detection is subjective
Ultraviolet spectroscopy	Extracted drug identified by its specific absorbance spectrum	Spectrum often gives specific identification and quantitation of compound	Not all drugs have suitable absorbance spectra Limited sensitivity Some degree of ambiguity among drugs of same class

Table 64-13 Spot tests

Drug	Specimen required	Reaction	Test time (min)	Comments	Reference
Acetaminophen	Urine	*o*-Cresol + Acetaminophen $\xrightarrow{NH_4OH}$ Blue color	15	Highly sensitive and relatively specific	4
Ethanol	Urine, serum	Microdiffusion into dichromate $2K_2Cr_2O_7 + 10H_2SO_4 + 3C_2H_5OH \rightarrow 2Cr_2(SO_4)_4 + 2K_2SO_4 + 3CH_3COOH + 11H_2O + 4H^+$ (green to blue color)	15-30	Good sensitivity Nonspecific for ethanol	5
Salicylate	Urine, serum	Trinder's solution Salicylate + $FeCl_3 \rightarrow$ Violet-colored complex	2	Good specificity if serum used Good sensitivity	6
Carbamates (meprobamate)	Urine	Furfural + Meprobamate + Antimony trichloride → Black color on thin-layer chromatography plate	5	Not specific for meprobamate Good sensitivity	7
Imipramine/desipramine	Urine	Forrest reagent K_2CrO_3 (acidic) + Imipramine → Green-colored complex	2	Phenothiazines may interfere	8
Ethchlorvynol	Urine, serum	Diphenylamine + Ethchlorvynol in acid → Red color	10	Good sensitivity and specificity	9
Phenothiazines	Urine	FPN reagent (ferric chloride/perchloric acid/nitric acid) $FeCl_3$ (acidic oxidizing agent) → Red- to violet-colored complex	2	Nonspecific Poor sensitivity for some phenothiazines	10
Iron	Serum	Bathophenanthroline color change with iron (blue)	15	Will not react at normal serum concentrations	11

Immunoassay

Both the fluorescence polarization (Abbott Laboratories, Chicago) and the EMIT systems (Syva Co., Inc., Palo Alto, CA) offer commercially available kit assays for individual drugs or drug classes. In the EMIT system, the sample is added to an antibody-substrate reagent, and then an enzyme-labeled drug reagent is introduced, allowing competitive binding of the patient and reagent drug. Assays, especially those performed on serum, offer good specificity and adequate sensitivities for detection of drugs

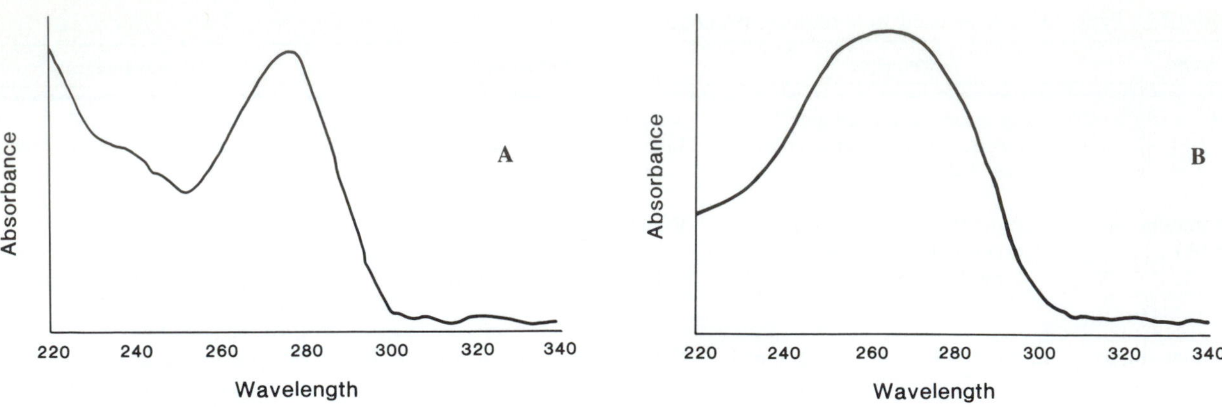

Fig. 64-7 **A,** Theophylline in 0.1 N NaOH. **B,** Sulfamethoxazole in 0.1 N NaOH.

in both overdose and drug-abuse situations. Following are test assays currently available:

Urine	Serum
Amphetamine	Acetaminophen
Barbiturates	Barbiturates
Benzodiazepines	Benzodiazepines
Cannabinoids	Methotrexate
Cocaine	Selected anticonvulsants
Ethanol	Selected antimicrobials
Methadone	Selected cardiac drugs
Methaqualone	Theophylline
Opiates	Tricyclic antidepressants
Phencyclidine	
Propoxyphene	

Each of these tests must be performed with appropriate controls. Thus the greater the number of standards and controls tested, the more time required for the drug screen analysis.

As with enzyme immunoassays both the FPIA and RIA procedures provide the analyst with commercially available assays for a select number of individual drugs.[14] Specificity of the individual procedures depends completely on the specificity of the assay's antibody. Test sensitivities are excellent, and most assays are quantitative, enabling detection of drug concentrations found in most clinical applications. Most assays require more than 30 minutes of incubation time, and few laboratories can offer immediate results for more than one or two of these assays. The cost of RIA reagents is comparatively high. The FPIA has good test sensitivity for drug abuse screening with potential for rapid turnaround time.

Thin-layer chromatography

Thin-layer chromatography (TLC) serves as an excellent screening or confirmation technique for both large laboratories and those with more limited capabilities. The method requires no instrumentation and is relatively inexpensive.[15-18] Additional advantages include the flexibility to perform simultaneous analyses of a larger number of drugs on a larger number of specimens. Sensitivities of most TLC screening procedures are adequate for both overdose and drug-abuse cases, and specificity is often acceptable. Commercially available methods have assay times in the range of 1 to 3 hours.

Gas chromatography and gas chromatography–mass spectrometry

Gas chromatography (GC) offers one of the most powerful analytical tools for the separation of complex mixtures, including drug mixtures, in biological fluids. As the TLC, GC enables the analyst to look for a large number of possible constituents simultaneously. Unlike TLC, however, GC techniques cannot screen a large number of patient specimens simultaneously; therefore individual samples must be analyzed consecutively.

Specimens must generally be extracted before GC can be performed,[19-23] and the drug must either be volatile or derivatized to make it volatile. For screening purposes, samples can be chromatographed on a single column or injected into various compound-selective columns to enhance specificity and sensitivity.[24] The compounds are identified by their respective retention times relative to an internal standard (Fig. 64-8, *A*). In addition, specific detectors are available (for nitrogen and phosphorus, or halogens), and they can give more specific information. Fig. 64-8, *B,* shows the same sample chromatographed as Fig. 64-8, *A,* but analyzed with a nitrogen detector.

When a gas chromatograph is interfaced to a mass spectrometer detector (gas chromatography/mass spectrometry, GC/MS), the combination provides the most objective analytical data of existing methods for structural elucidation of an unknown compound. The GC/MS system has been utilized both as a confirmatory procedure and as a primary screening method for drug screens when improved turnaround times are necessitated (immediate overdose screening).[25,26] Both conventional GC and GC/MS require considerable technical expertise.

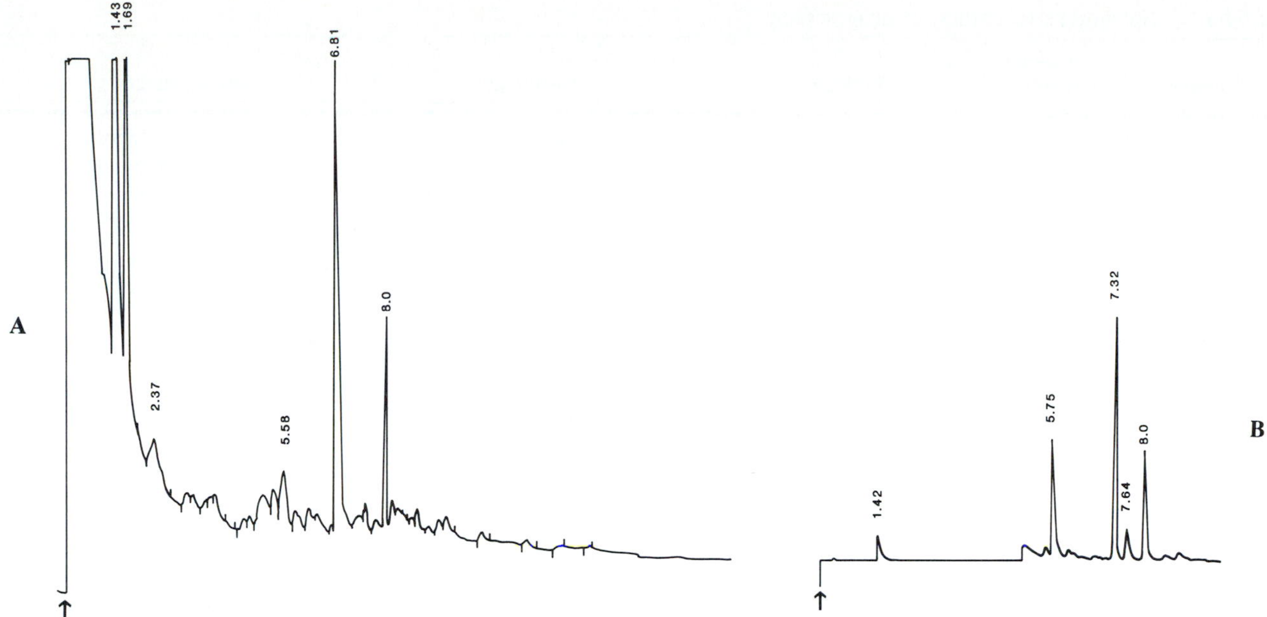

Fig. 64-8 Gas chromatography. **A,** Nonselective detector (FID, flame ionization detector). **B,** Selective detector (NPD, nitrogen-phosphorus detector).

High-performance liquid chromatography

High-performance liquid chromatography (HPLC) is analogous to GC and TLC in its ability to separate complex mixtures of drugs extracted from biological fluids. Like GC, HPLC is limited in the number of patient samples that can be processed, and individual ones must be processed consecutively.

Although specimens are usually extracted before HPLC is performed, the advantage of HPLC over GC is that the sample does not have to be volatile or derivatized. For screening purposes, there are currently no columns and buffer systems available for separating many different classes of drugs. Usually columns such as C_{18} are used with specificity for classes of drugs such as benzodiazepines. In addition, it is possible to connect several detectors, spectrophotometers, fluorometers, and electrochemical detectors in series to characterize the compounds partially as they are eluted from the column. Currently, this technique has been applied to only a few drug classes, but it may one day be as useful as TLC. It is particularly useful for the detection of drugs that are not volatile.

Quality control

As with all laboratory tests, one must continuously monitor the performance and reliability of the drug-screening procedure, documenting adequate sensitivity, accuracy, and specificity. One can accomplish this by employing an internal quality control program and by participating in outside proficiency surveys.

Proficiency surveys offer a check of screening accuracy and should allow evaluation of both the sensitivity and specificity of the methods utilized. Currently, the programs offered by the American Association for Clinical Chemistry (Washington, DC) or the College of American Pathologists (Traverse City, MI) are suitable.

The internal program involves daily analysis of known control materials spiked with an assortment of drugs, which can be provided commercially or can be prepared in-house if the laboratory possesses the proper licenses for handling controlled substances.* Control samples should be processed in the same manner as patient specimens and should be submitted to the same rigorous analyses as unknowns. It is important to monitor all three stages of the procedure (extraction or preparation, screening, and confirmation).

REFERENCE AND PREFERRED METHODS

As is so often found with esoteric laboratory tests, no reference procedure is widely accepted for the qualitative drug screen. However, GC/MS analysis has become the confirmatory method required by federal agencies. The optimum reference method would, in theory, be difficult to present because it would entail (1) defining which compounds should be included in various screens, which is impossible, since the drugs of interest are invariably changing; (2) establishing a preferred separation or extrac-

*For information regarding regulations concerning licensing for handling controlled drugs, contact the Drug Enforcement Administration, PO Box 28083, Central Station, Washington, DC 20005.

Table 64-14 Specimens of choice in drug screen

Specimen	Volume required	Indication	Advantages	Disadvantages
Urine	30 mL	Drug abuse screening Overdose screening Employment screening	Generally easy to obtain in high volume Most drugs found in sufficient concentration to enable identification	Contains many metabolic products that may interfere with identification Parent drug may not be present Quantitation offers little correlation with clinical effects
Gastric lavage or emesis	30 mL	Overdose screening	Parent drug present	Matrix problems (interference from foodstuffs) Drugs quickly absorbed may be missed by "gastric screen" Drugs not orally ingested will not be detected
Blood, serum, and plasma	10 mL	Overdose screening Therapeutic drug monitoring	Parent drug present Quantitative level may assist with patient management (therapeutic and toxic reference levels often known)	Limitation of sample volume Concentration of select drugs often too low to enable detection (especially in non-overdose situations)

tion scheme; (3) describing the best screening method; and (4) suggesting the best confirmation procedures. The following suggested methods include procedures that will best suit analysts of varying expertises and budgets and include the simplest to the most complex assays. One should note that several analytical techniques (nuclear magnetic resonance, infrared spectroscopy, and HPLC) have not been considered in the drug-screening assays because these procedures, with the possible exception of HPLC, have not yet been used extensively and routinely for qualitative analyses usually encountered by the clinical chemistry laboratory.

The decision about which type of screen to perform depends on both economic and clinical needs. If there is good clinical-laboratory interaction, it is possible to offer more efficient specific testing. For example, in the acute care setting, an agitated or hyperactive patient does not need to be screened for morphine, tricyclic antidepressants, and benzodiazepines because drugs such as the amphetamines, phencyclidine, and the cocaine agents are more probable causes of this altered mental status. Therefore the laboratory can offer faster, more accurate service if the drug screen is limited to a few agents. The final decision, of course, on which tests will be performed is dictated by the clinician ordering the test. Similarly, most drug abuse screening related to employee testing programs or drug rehabilitation programs requires only the testing of a limited number of drugs. We have presented in the procedures section examples of the comprehensive and drug abuse screens.

SPECIMEN

The submission of proper and timely specimens to the analyst is critical for obtaining useful results. Samples should be collected immediately if there is an indication that drugs may be involved in a particular case. Depending on the drug screen request, various biological samples are made available to the laboratory (Table 64-14). At present, urine is generally considered to be the optimum specimen for drug screening.

PROCEDURE: COMPREHENSIVE DRUG SCREEN

Principle

The box shows a flow diagram of the suggested method for a comprehensive drug screen.

Urine or gastric specimens are extracted with organic solvents at a selected pH, followed by analysis by TLC.[15] Selected ancillary spot tests and enzyme immunoassays are simultaneously performed on urine. Positive findings are confirmed by use of appropriately chosen techniques. A check sheet showing how the data can be recorded is presented in Fig. 64-9.

Indications

The comprehensive drug screen is ordered to establish information regarding usage as it may pertain to cases of a toxicological nature, including accidental or suicidal overdose with coma, or for evaluation of potential drug involvement in other medical emergencies, such as seizures, jittery baby syndrome, or drug interactions.

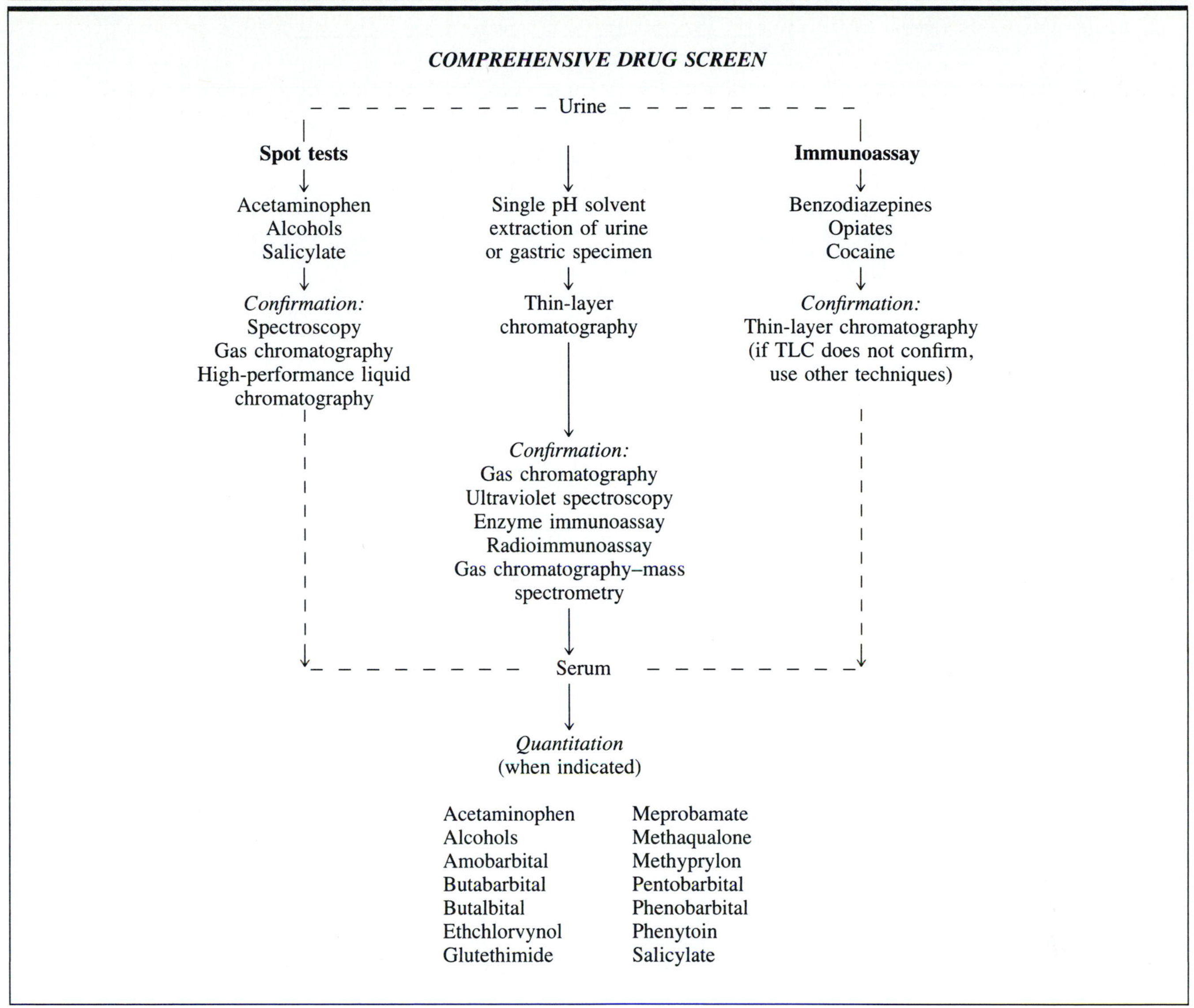

Reagents

All reagents should be chromatographic grade.

Trinder's solution: ($Fe[NO_3]_3$, 100 mmol/L; $HgCl_2$, 14.7 mmol/L; HCl, 120 mmol/L). Add 4 g of ferric nitrate and 4 g of mercuric chloride to a 100 mL volumetric flask. Dissolve salts in 12 mL of 1 M hydrochloric acid. Dilute to mark with distilled water. Store at room temperature. This is stable for 6 months.

Potassium dichromate (3.4 mmol/L). Add 1 g of potassium dichromate to a 1 L volumetric flask. Dissolve salt in 500 mL of distilled water. Add 0.1 g of silver nitrate to the flask, and then slowly, with stirring, add 500 mL of concentrated sulfuric acid. Store in a brown bottle at room temperature. This is stable for 1 year at room temperature.

***o*-Cresol reagent (0.096 mol/L).** Add 10 mL of *o*-cresol to a 1 L volumetric flask, and dilute to mark with distilled water. Store at room temperature. This is stable for 1 year at room temperature.

Forrest reagent

Nitric acid (7.8 mol/L). Add 50 mL of concentrated nitric acid to 50 mL of distilled water.

Perchloric acid (1.85 mol/L). Add 20 mL of 70% perchloric acid to 50 mL of distilled water. Allow to cool, and bring volume to 100 mL.

Sulfuric acid, 12 N (6 mol/L). Add 30 mL of concentrated sulfuric acid slowly, with constant stirring, to 50 mL of distilled water. Allow to cool, and bring volume to 100 mL.

Potassium dichromate, 2 g/L (6.8 mmol/L). Add 200 mg of potassium dichromate to 80 mL of distilled water. Dissolve, and bring to a final volume of 100 mL.

TOXICOLOGY LABORATORY DATE:______ TECH:______ CHECKED BY______
STAT DRUG SCREEN PATIENT #:______

(1) SPOT TESTS:

"Salicylates"

Serum
Urine (use only if serum is not available)

Sample	Pos.	Neg.
Std.		
Patient		

"Acetaminophen"

Use urine only

Sample	Pos.	Neg.
Std.		
Patient		

(2) ALCOHOLS: Microdiffusion

Serum
Urine (use only if serum is not available)

Sample	Pos.	Neg.
Std.		
Patient		

Gas Chromatography Confirmation

Alcohol detected Ethanol
Methanol
Isopropanol

Quantitation (if serum is available)

______ μg/mL

(3) EMIT DAUS: Use urine only

"Opiates"

Sample	OD	Pos.	Neg.
Neg. Cal.	___		
Low Cal.	___		
Patient	___		
Pat. Blk.			

"Benzodiazepines"

Sample	OD	Pos.	Neg.
Neg. Cal.	___		
Low Cal.	___		
Patient	___		
Pat. Blk.			

SCREEN: Analyze urine and gastric (if available)

(5) Gas chromatography (confirmation). List relative retention times of note.

(6) GC/MS (Confirmation)
– Utilize GC/MS analysis sheets. List all compounds detected with their respective scan numbers.
– Search all peaks in the "MAP" mode and perform appropriate limited mass analyses in the "CHRO" mode.

(4) TLC

6 Ref.
7 ___
8 ___
9 Std. I
10 Ref.
11 Std. II
12 Std. III
13 ___
14 ___
15 Ref.

COMMENTS:

(7) QUANTITATIONS:
(The following are routinely quantitated if detected)

Drug	Concentration (μg/mL)
Acetaminophen	___
Alcohols	(see above)
Barbiturates Specify	___
Ethchlorvynol	___
Glutethimide	___
Meprobamate	___
Methaqualone	___
Methyprylon	___
Phenobarb	___
Phenytoin	___
Salicylate	___

Fig. 64-9 Laboratory check sheet for recording data from a comprehensive drug screen (see box).

Mix 100 mL of each of these reagents just listed together. This solution is stable for 1 year at room temperature when stored in an opaque (amber) glass container.

FPN (ferric chloride/perchloric acid/nitric acid) reagent

Nitric acid (7.8 mol/L). See under Forrest reagent.

Perchloric acid (1.85 mol/L). See under Forrest reagent.

Ferric chloride (0.308 mol/L). Dissolve 5 g of ferric chloride in 50 mL of distilled water in a volumetric flask. Bring to final volume of 100 mL.

Mix these solutions just listed in the following proportions: one part ferric chloride, nine parts perchloric acid, and 10 parts nitric acid. This is stable for 1 year at room temperature when stored in an opaque (amber) glass container.

Concentrated hydrochloric acid. This is stable for 2 years at room temperature.

Ammonium hydroxide, 4 mol/L. Dilute 284 mL of concentrated ammonium hydroxide to 1 L with distilled water. This is stable for 2 years at room temperature.

EMIT d.a.u. (drug of abuse urine) opiate and benzodiazepine assay kits (Syva Corp., Palo Alto, CA). These kits are stable for 1 year before opening. Follow manufacturer's recommendations for storage and use.

Extraction buffer. Prepare a saturated ammonium chloride solution by dissolving 600 g in 1 L of hot, distilled water. Cool, and add concentrated ammonium hydroxide to adjust the pH to 9.5. Store at room temperature. This is stable for 6 months.

Extraction solvent. Prepare a mixture of dichloromethane and isopropanol (90:10 by volume). Store at room temperature. Prepare weekly.

Methanolic 0.1 M hydrochloric acid. Add 0.9 mL of concentrated hydrochloric acid to a 100 mL volumetric flask. Dilute to mark with methanol. Store at room temperature. This is stable for 1 year.

Reconstitution solvent. Prepare a 1:1 (v/v) mixture of dicholormethane and methanol. Store at room temperature. This is stable for 1 year.

Thin-layer chromatographic developing solution. Mix 85 mL of ethyl acetate, 10 mL of methanol, and 5 mL of ammonium hydroxide. Prepare fresh for each development.

Thin-layer chromatographic sprays

Ninhydrin (5.61 mmol/L). In a 100 mL of volumetric flask, dissolve 100 mg of 1,2,3-indantrione in 5 mL of acetone, and dilute to mark with acetone. Prepare fresh daily.

Diphenylcarbazone (2 mmol/L). In a 100 mL volumetric flask, dissolve 10 mg of diphenylcarbazone in 5 mL of acetone, and dilute to mark with a 1:1 acetone to water solution. This is stable for 1 month at room temperature.

Mercuric sulfate (11.5 mmol/L). Add 0.5 m of mercuric oxide to 20 mL of concentrated sulfuric acid. Add this acid solution slowly to 150 mL of distilled water, and bring to a final volume of 200 mL. This is stable for 6 months at room temperature.

Iodoplatinate

Stock solution (0.38 mol/L). Dissolve 10 g of platinum chloride in 100 mL of distilled water. Refrigerate at 4° C. This is stable for 1 year.

Working solution. Add 5 mL of stock solution to 3 g of potassium iodide in 100 mL of distilled water. Dilute this mixture to 125 mL with distilled water, and then dilute this with an equal volume of methanol (final volume 250 mL). Refrigerate at 4° C. This is stable for 6 months.

Dragendorf reagent. Dissolve 1.3 g of bismuth subnitrate in a solution composed of 60 mL of distilled water and 15 mL of glacial acetic acid. Dissolve 12 g of potassium iodide in 30 mL of distilled water. Combine the two solutions. Dilute this mixture with 100 mL of distilled water and 25 mL of glacial acetic acid. Refrigerate at 4° C. This is stable for 6 months.

Standards

Ethanol (789 mg/L, 17.15 mmol/L). Pipette 0.1 mL of absolute ethanol into a 100 mL volumetric flask. Dilute to mark with distilled water. Store at 4° C. This is stable for 2 months.

Salicylate standard (100 mg/L, 724 µmol/L). Dissolve 10 mg of salicylic acid (USP grade) in 0.5 mL of methanol in a 100 mL volumetric flask, and dilute to 100 mL with distilled water. Store at 4° C. This is stable for 6 months.

Acetaminophen urine control (100 µg/mL, 0.66 µmol/mL). Add 10 mg of acetaminophen to a 100 mL volumetric flask. Dissolve in 1 mL of methanol. Dilute to mark with urine previously noted to be free of the drug. Refrigerate at 4° C. This is stable for 6 months.

Urine drug control. Select a suitable commercially available control, or prepare one using the following procedure:

1. Collect approximately 2 L of drug-free urine (urine previously noted to be negative for all drugs through testing). To clarify, filter through Whatman No. 1 filter paper.
2. Prepare a stock drug standard by adding 4 mg each of phenobarbital, secobarbital, morphine, codeine, meperidine, methyprylon, and oxazepam and 10 mg each of D-amphetamine and methamphetamine to a 10 mL volumetric flask. Dissolve the drugs in methanol, and dilute to mark with methanol.
3. Transfer the entire contents to the drug standard flask to a 2 L volumetric flask. Rinse the flask several times with distilled water, and add rinses to the 2 L flask. Dilute to the 2 L mark with the urine pool. Mix well.
4. Divide into 20 mL aliquots, and store in plastic vials.
5. Store at −20° C until used. This is stable for 1 year.
6. Control contains 2 µg/mL of each drug except the amphetamines, which have concentrations of 5 µg/mL.

Thin-layer chromatographic reference standard (1 µg/µL). Add 10 mg each of propoxyphene, methadone, chlorpromazine, meperidine, quinine, phenobarbital, phenylpropanolamine, and morphine to a 10 mL volumetric flask. Dissolve in methanol, and dilute to mark with methanol. Refrigerate at 4° C. This is stable for 6 months.

Individual drug standards (1 µg/µL). Prepare in the same manner as previsions standard with drug choice. This is stable for 6 months at 4° C.

Assay

Equipment:

Porcelain spot test plates
Conway microdiffusion dish
Teflon 125 mL volume separatory funnels (Fisher Scientific, Pittsburgh)
60 mL glass centrifuge tubes
Sample concentrator/evaporator (Brinkmann, Buchii Rotavap, or other convenient evaporation system)
Thin-layer chromatographic apparatus
- Spotting platen
- 20 × 20 cm glass silica gel 60 F-254 plates (EM Science, Gibbstown, NJ)
- Developing tank to accommodate 20 × 20 cm plates
- Chromatographic spray cans
- Air blower
- Short (254 nm)/long (366 nm) combination ultraviolet lamp

Ancillary tests

Salicylate spot test. Add 3 drops of salicylate control or patient specimen to a porcelain dish. Serum is the first specimen of choice, followed by urine. Gastric lavage or emesis is unacceptable without special treatment. (If urine is used, boil the specimen first to remove diacetic acid, a potentially interfering substance). Add 3 drops of Trinder's reagent. A violet color indicates the presence of salicylate.

Alcohol spot test. To the center well of a microdiffusion dish, add 0.5 mL of potassium dichromate reagent. To the outer ring, add 0.5 mL of alcohol control or patient specimen (urine or serum). Lightly grease the outer ring of the microdiffusion dish lid with vacuum grease. Cover, and allow to stand at room temperature. The dichromate solution will change from a yellow to a light green or blue color in the presence of methanol, ethanol, or isopropanol.

Acetaminophen spot test. To 1 mL of patient or control urine, add 1 mL of concentrated hydrochloric acid. Heat in a boiling water bath for 10 minutes. Dilute 0.1 mL of this sample with 0.9 mL of *o*-cresol reagent. Add 2 mL of 4 M ammonium hydroxide. Blue color indicates presence of acetaminophen.

EMIT d.a.u. opiate, cocaine, and benzodiazepine assays (Syva Corp., Palo Alto, CA). Use as described by manufacturer.

Extraction of urine

1. Process the drug screen control along with patient specimen.
2. To a 125 mL separatory funnel, add 20 mL of urine and 2 mL of ammonium chloride–ammonium hydroxide buffer, and mix.
3. Add 30 mL of extraction solvent, and extract for 10 minutes.
4. Allow layers to separate. Collect bottom solvent layer in a 50 mL glass centrifuge tube, and centrifuge for 5 minutes at 500 *g* at room temperature.
5. With a glass stirring rod, break emulsions that may have formed by aggregation, and recentrifuge. Aspirate off top aqueous layer, and discard. Filter solvent through filter paper (P5, Fisher Scientific, Pittsburgh, qualitative) into evaporation flask.
6. Add 2 drops of methanolic 0.1 M hydrochloric acid, and evaporate to dryness at 50° C under nitrogen.
7. Using a Pasteur pipet, reconstitute with 10 drops (approximately 150 μL) of methylene dichloride–methanol solution, being sure to rinse down the sides of the evaporation flask.

Thin-layer chromatography

Thin-layer chromatography spotting. Use a conditioned (preheated for 1 hour at 70° C) TLC plate. Using a spotting platen, mark the spot origin at 2 cm from the bottom of the plate. From this mark measure 15 cm upward, and draw the end line. Apply approximately 1 μL of reference standard in the 3, 10, and 17 positions on the plate. Use positions 9, 11, and 12 to spot drug standards of interest (drugs indicated by history). Apply extracted control and patient specimens in open positions. Apply approximately 3 μL at a time, and spot 10 times, allowing the applied spots to dry between applications.

Plate development. Prepare the TLC development solvent, and pour into tank lined with 20 × 20 cm filter paper on both sides. Position the plate or plates in the development tank, and allow time for migration to proceed to the end line (approximately 40 minutes).

Detection. Air dry the plate. Observe the plate under a long-wave (366 nm) ultraviolet lamp. Mark any fluorescing compound by lightly tracing the spot with a pencil. Mark subsequent positive spots with a pencil. Place under short (254 nm) light, and look for any absorbing spots. Spray with ninhydrin, irradiate with long-wave ultraviolet radiation for 5 minutes, and then heat plate at 70° C for 5 minutes. Amines, such as phenylpropanolamine and D-amphetamine, will stain pink. Spray with diphenylcarbazone, followed by the mercuric sulfate spray. Observe the blue-violet colors of barbiturates, phenytoin, glutethimide, and ethchlorvynol. Respray with diphenylcarbazone, and heat plate at 70° C for 10 minutes. The pink, red, and violet colors of metabolites of phenothiazine drugs will appear. Observe under long-wave ultraviolet lamp. Observe the yellow fluorescence of the benzodiazepine drugs and their metabolites. Spray with iodoplatinate and Dragendorf's reagents. Most nitrogenous basic drugs will stain dark. Soak the finished plate in water for 30 seconds, remove, and dry with a jet of hot air. Observe the chalk-white colors of methyprylon and carbamates, such as meprobamate. R_f values (distance of spot migration divided by distance of developing solvent migration), spray reactions, and ultraviolet detections of unknowns should be compared to known standards.

Confirmation

The presence of drugs indicated as being positive by the TLC or ancillary test screening methods must be confirmed by a second analytical procedure different from that used to screen for the compound (see Table 64-11). Occasionally, when serum is provided and when indicated, as in overdose cases, the quantitation of the suspected drug in serum is the confirmatory step in the method. At other times, additional tests must be performed on the urine to confirm the presence of the drug.

PROCEDURE: DRUG ABUSE SCREEN

Principle

Drugs of interest are first screened by enzyme immunoassay, with presumptive positive cases being confirmed by TLC or GC/MS.

Indications and comments

The screen is used to rule out the presence of select drugs of abuse as part of employee screening programs, to monitor clients in drug rehabilitation programs, or for any clinical situation requiring a knowledge of drug usage.

As with the comprehensive drug screen, the drugs chosen to be included in this screen will be selected based on the current prevalence of their abuse, along with technical and economic considerations. The constituents of the screen will be continuously changing, just as drug-usage patterns are. When a limited approach to drug-abuse screening is the goal, as presented here, the participation by commercial suppliers of drug assay kits is essential for a successful program.

Reagents and supplies

EMIT d.a.u. assay kits. Use the kits available from Syva Corp. (Palo Alto, CA).

Toxi-Lab A and B drug detection systems and cannabinoid (THC) assay kits. Use the kits available from Analytical Systems, Inc. (Laguna Hills, CA).

Assay

1. Perform the following EMIT d.a.u. enzyme immunoassays as described by the manufacturer: amphetamines, barbiturates, benzodiazepines, cannabinoids (THC), cocaine, opiates, phencyclidine (PCP), methadone, methaqualone, and propoxyphene.
2. If found to be presumptively positive by immunoassay, confirm with the appropriate Toxi-Lab procedure as described by the manufacturer.
   a. *Positive amphetamines*. Confirm using the Toxi-Lab "Sympathomimetic Amines (Amphetamine) Differentiation" procedure using Toxi-Lab A system.
   b. *Positive barbiturates*. Confirm using the Toxi-Lab B system.
   c. *Benzodiazepines*. Confirm using the Toxi-Lab "Benzodiazepines: Hydrolysis" special procedure using Toxi-Tubes B and Toxi-Lab A system.
   d. *Cocaine*. Confirm using conventional Toxi-Lab A system for basic drugs and the "Benzoylecgonine: Extraction and Detection" special procedure.
   e. *Phencyclidine*. Confirm using the Toxi-Lab "Phencyclidine: Confirmation by Remigration" special procedure.
   f. *Methaqualone, methadone, and propoxyphene*. Confirm using the Toxi-Lab A system. For methaqualone, also use the "Methaqualone: Confirmation with Sodium Borohydride" special procedure.
   g. *Opiates*. Confirm using the conventional Toxi-Lab A system and the "Morphine and Codeine: Confirmation using Sodium Hydroxide" special procedure. In addition, the special procedure "Morphine: Hydrolysis of the Glucuronide Conjugate" may be required.
   h. *Cannabinoids*. Confirm using the Toxi-Lab cannabinoid (THC) assay system.

INTERPRETATION

A finding is generally termed positive when a positive immunoassay is observed along with an appropriate spot on the Toxi-Lab system at the expected R_f value. In fact, if potential medicolegal specimens are routinely analyzed (that is, employee, athletic, or forensic screening), a second confirmatory analytical procedure such as gas chromatography or mass spectrometry may be a necessity.

A discussion of drug abuse screening would not be complete without some comment on the proper processing of specimens, including collection, transportation, storage, and analytical phases of the process.

Again, since the ramifications of this type of screening are of concern, the analyst must be certain of the accuracy not only of the testing process but also of the specimen integrity. Generally, all urine collections should be witnessed, followed by a foolproof sample identification method.

Specimens should be properly stored (refrigerated at 4° C) until time of transportation and must be adequately sealed to assure their integrity. This is best done with standard "legal-specimen" tape. The number of persons handling the specimen during its transportation should be kept to a minimum, and the sample should be delivered directly to laboratory personnel. The laboratory's assignment of an accession number must be straightforward and easy to follow through all phases of the testing procedure (such as instrument logs, labeling of chromatogram data sheets, and final reports).

Fig. 2-3 shows a typical chain-of-evidence or custody form most often employed for the purposes of documenting sample continuity throughout.

CANNABINOIDS

Detection of the use of *Cannabis sativa* plant as an intoxicant is very important in drug abuse screening. The major psychoactive agent is Δ^9-tetrahydrocannabinol (Δ^9-THC), which is metabolized to 11-nor-Δ^9-tetrahydrocannabinol-9-carboxylic acid (Δ^9-THC-COOH) and excreted in urine as the free and conjugated forms. The plasma level of THC is too low to measure routinely, and most screening methods use urine in the search for the Δ^9-THC-COOH metabolite.

Because of the frequent medicolegal implications of this testing, the detection of Δ^9-THC-COOH requires a screening and a confirmation method unless measured by GC/MS. The preferred GC/MS procedures use deuterated or tritiated Δ^9-THC-COOH internal standards such as $5'$-d_3-11-nor-Δ^9-tetrahydrocannabinol-9-carboxilic acid.[27] Other procedures for screening and confirmation include homogeneous enzyme immunoassay (EMIT), RIA, TLC, gas chromatography, and HPLC. The EMIT kit formulation is in two forms: the EMIT d.a.u. and the EMIT s.t. (single test), which differ in their positive cutoff values of 20 and 100 ng/mL, respectively.

The most commonly used TLC system is that of Toxi-Lab, in which urine is hydrolyzed with KOH and neutralized with maleic acid. The Δ^9-THC-COOH is extracted into ethyl acetate–hexane, concentrated, and chromatographed. One can detect the THC metabolite by dipping the chromatogram in 0.1% (w/v) fast blue BB in dichloromethane, followed by drying and incubation in a container saturated with diethylamine vapors and then HCl vapors. The Δ^9-THC-COOH gives a rose red–orange color in the diethylamine and a purple color in the HCl standoff jars, respectively.

The GC/MS assays using the deuterated derivatives are clearly the optimum procedures for detection and confirmation of urinary cannabinoids. These are recommended by the National Institute for Drug Abuse (NIDA).

REFERENCES

1. Thoma, JJ, editor: Guidelines for analytical toxicology programs, vols I and II, Cleveland, 1977, CRC Press, Inc.
2. Sunshine, I, editor: Methodology for analytical toxicology, ed 2, Cleveland, 1975, CRC Press, Inc.
3. Clarke, EGC: Isolation and identification of drugs, London, 1969, Pharmaceutical Press.
4. Berry, DJ, and Grove, J: Emergency toxicological screening for drugs commonly taken in overdose, J Chromatogr 80:205-220, 1973.
5. Sunshine, I, editor: Methodology for analytical toxicology, ed 2, Cleveland, 1975, CRC Press, Inc.
6. Natelson, S: Techniques in clinical chemistry, ed 3, Springfield, IL, 1971, Charles C Thomas, Publisher.
7. Curry, A: Poison detection in human organs, ed 2, Springfield, IL, 1969, Charles C Thomas, Publisher.
8. Sunshine, I, editor: Methodology for analytical toxicology, ed 2, Cleveland, 1975, CRC Press, Inc.
9. Selected methods of emergency toxicology. In Frings, CS, and Faulkner, WR, editors: Selected methods of clinical chemistry, vol 11, Washington, DC, 1986, American Association for Clinical Chemistry.
10. Sunshine, I, editor: Methodology for analytical toxicology, ed 2, Cleveland, 1975, CRC Press, Inc.
11. Fisher, DS: A method for rapid detection of acute iron toxicity, Clin Chem 13:6-11, 1967.
12. Jatlow, P: UV spectrophotometry for sedative drugs frequently involved in overdose emergencies. In Sunshine, I, editor: Methodology for analytical toxicology, ed 2, Cleveland, 1975, CRC Press, Inc.
13. Sunshine, I: Spectroscopic analysis of drugs including atlas of spectra, Springfield, IL, 1963, Charles C Thomas, Publisher.
14. Byers, BJ: Radioimmunoassays. In Thoma, JJ, editor: Guidelines for analytical toxicology programs. VII, Cleveland, 1977, CRC Press, Inc.
15. Davidow, B: A thin layer chromatographic screening procedure for detecting drug abuse, Am J Clin Pathol 50:714-7419, 1968.
16. Blass, KG: A rapid simple thin-layer chromatography drug screening procedure, J Chromatogr 95:75-79, 1974.
17. Toxi-Lab Applications note, Analytical Systems, Inc, 23162 La-Cadena Dr, Laguna Hills, CA 92653.
18. Michaud, JD: Thin layer chromatography for broad spectrum drug detection, Am Lab 12:104, 1980.
19. Mule, SJ: Routine identification of drugs of abuse in human urine. II. Development and application of XAD-2 resin column method, J Chromatogr 63:289-301, 1971.
20. Fujimoto, JM: A method of identifying narcotic analgesics in human urine after therapeutic doses, Toxicol Appl Pharmacol 16:186-193, 1970.
21. Weisman, N, and Lowe, ML: Screening method for detection of drugs of abuse in human urine, Clin Chem 17:875-881, 1971.
22. Tox Elut drugs of abuse in urine, applications note, Analytichem International, Inc, 15620 S Ingelwood Ave, Lawndale, CA 90260.
23. Breiter, J: Evaluation of column extraction: a new procedure for the analysis of drugs in body fluids, Forensic Sci 7:131-140,1976.
24. Finkle, B.: A GLC based system for the detection of poisons, drugs and human metabolites encountered in forensic toxicology, J Chromatogr Sci 9:393-419, 1971.
25. Castello, CE: Routine use of a flexible gas chromatograph–mass spectrometer–computer system to identify drugs and their metabolites in body fluids of overdose victims, Clin Chem 20:255-265, 1974.
26. Ullucci, PA: A comprehensive GC/MS drug screening procedure, J Anal Toxicol 2:33-38, 1978.
27. Folz, RL: Analysis of cannabinoids in physiological specimens by gas chromatography/mass spectrometry. In Baselt, RC, editor: Advances in analytical toxicology, Davis, CA, 1984, Biomedical Publications.

Gentamicin and other aminoglycosides

JOSEPH R. DIPERSIO

Clinical significance: p. 795

Isomeric forms, Molecular formula, Molecular weight, Merck Index: See Table 64-15

Chemical class: aminoglycoside

PRINCIPLES OF ANALYSIS

Gentamicin is a member of the aminoglycoside group of antibiotics and includes a number of structurally related polycationic compounds composed of two or more amino

Table 64-15 Commonly used aminoglycosides

Name	Discovery date	Molecular formula	Molecular weight	Merck Index number	Structure	Route of adminis-tration*
Streptomycin	1944	$C_{21}H_{39}N_7O_{12}$	581.58	8685	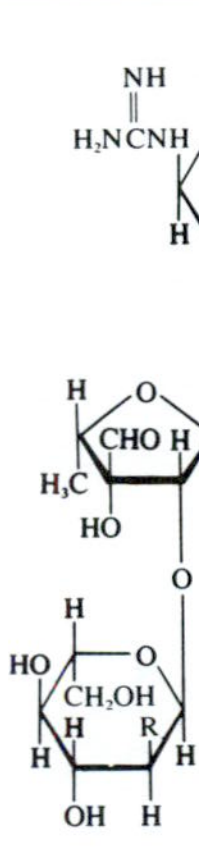 R = CH_3NH	IM
Neomycin	1949			6300	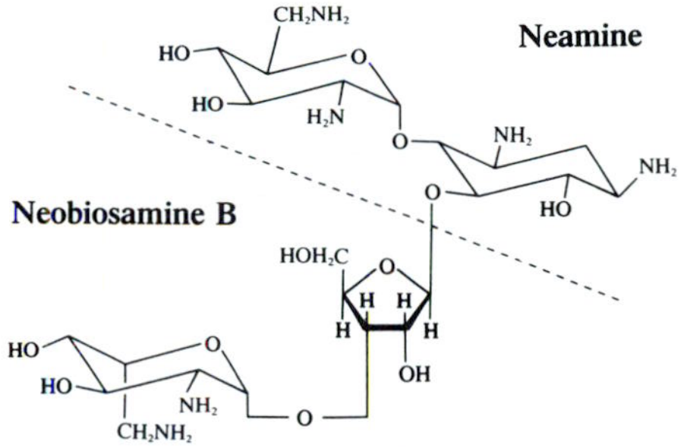 Neamine; Neobiosamine B; Neomycin B	PO†
Neomycin A (neamine)		$C_{12}H_{26}N_4O_6$	322.36	6283		
Neomycin B		$C_{23}H_{46}N_6O_{13}$	614.65	6300		
Neomycin C		$C_{23}H_{46}N_6O_{13}$	614.65	6300		
Kanamycin	1957			5118	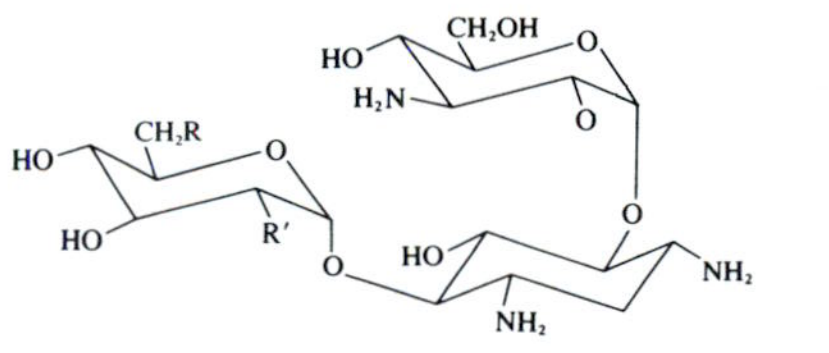	IM, IV
Kanamycin A		$C_{18}H_{36}N_4O_{11}$	484.50	5118		
Kanamycin B		$C_{18}H_{37}N_5O_{10}$	483.52	5118		
Kanamycin C		$C_{18}H_{36}N_4O_{11}$	484.50	5118		

**IM*, Intramuscular; *PO*, by mouth; *IV*, intravenous.
†Neomycin is no longer used parenterally because of toxicity.

Continued.

Table 64-15 Commonly used aminoglycosides—cont'd

Name	Discovery date	Molecular formula	Molecular weight	Merck Index number	Structure	Route of adminis-tration*
Gentamicin	1963			4251	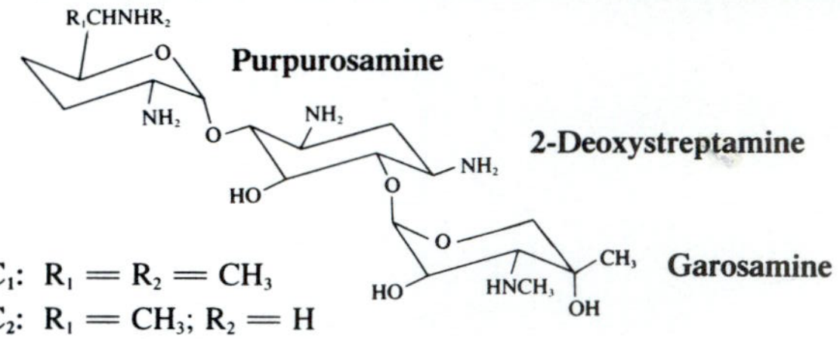 Purpurosamine; 2-Deoxystreptamine; Garosamine Gentamicin C_1: $R_1 = R_2 = CH_3$ C_2: $R_1 = CH_3$; $R_2 = H$ C_{1a}: $R_1 = R_2 = H$	IM, IV
Gentamicin A		$C_{18}H_{36}N_4O_{11}$	484.50	4251		
Gentamicin C_1		$C_{21}H_{43}N_5O_7$	477.60	4251		
Gentamicin C_2		$C_{20}H_{41}N_5O_7$	463.57	4251		
Gentamicin C_{1a}		$C_{19}H_{39}N_5O_7$	449.55	4251		
Tobramycin	1967	$C_{18}H_{37}N_5O_9$	467.54	9318	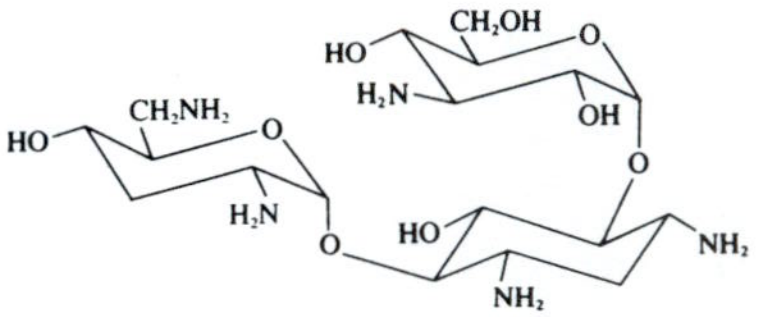	IM, IV
Amikacin	1972	$C_{22}H_{43}N_5O_{13}$	585.62	405	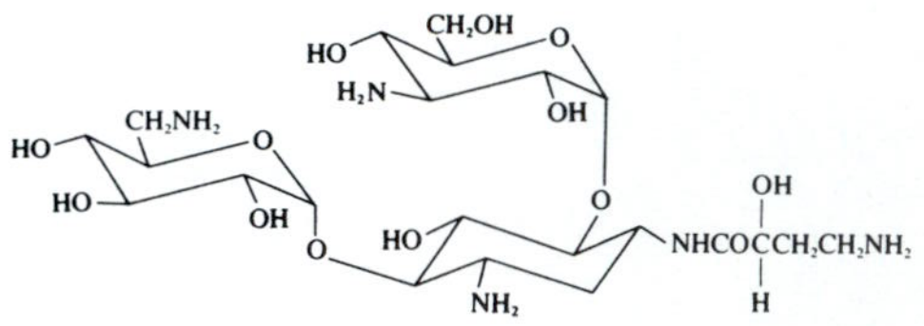	IM, IV
Netilmicin	1976	$C_{21}H_{41}N_5O_7$	475.60	6322	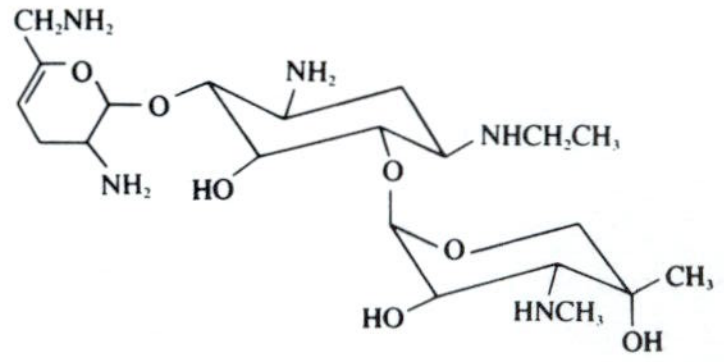	IM, IV

sugars connected by a glycosidic linkage to a central hexose or aminocyclitol nucleus. Other important members of the group include streptomycin, kanamycin, tobramycin, amikacin, and netilmicin (Table 64-15). All these except netilmicin are produced by microorganisms. Netilmicin is a semisynthetic derivative of gentamicin. The aminoglycoside antibiotics are used primarily to treat serious infections caused by aerobic gram-negative bacilli including *Pseudomonas aeruginosa*. Aminoglycoside antibiotics in general have a narrow margin between effective and toxic concentrations; thus laboratory measurements are often required to ensure adequate drug levels in body fluids and to adjust potentially toxic accumulations.

The measurement of gentamicin serves as a prototype for the analysis of the other aminoglycosides. Most of the assays described for the measurement of gentamicin are therefore applicable to the other compounds.

Before 1970 microbiological assays were commonly used to quantitate aminoglycosides. The method employed most frequently was the agar plate diffusion method (method 1, Table 64-16).[1] In this procedure one adds antibiotic standards (diluted in serum) to blank paper disks in measured amounts, bracketing the expected concentration in the unknown serum. Measured amounts of the unknown serum are likewise added to blank disks. All disks are placed on the surface of an agar plate that has been previously seeded with an indicator bacterium susceptible to the antibiotic being measured. The plate is incubated 4 to 18 hours, depending on the antibiotic assayed and the indicator organism used. During this period the indicator organism will grow throughout the agar medium, producing a visual haze of growth. However, zones of growth inhibition will form around the antibiotic-containing disks because of diffusion of drug into the agar. One measures the diameter of each zone and plots it against the appropriate standard disk concentration on semilogarithmic graph paper. The zone of inhibition produced by the unknown serum is also measured, and the concentration of antibiotic is determined from the standard curve. A variation of this method consists of punching small wells in the seeded agar and adding the various dilutions of standards and unknowns to the wells.

Table 64-16 Methods of gentamicin analysis

Method	Type of analysis	Principle	Advantages	Disadvantages
1. Microbiological (agar plate diffusion)	Growth inhibition	Zones of bacterial inhibition proportional to antibiotic concentration	Inexpensive, versatile; no special equipment needed	Slow (4 to 18 hours); variable accuracy
2. Radioenzymatic assay	Radiometry	Enzymatic transfer of radiolabel to antibiotic	Sensitive, specific, accurate	Requires radioisotopes, expensive equipment, and labor-intensive protocol; not appropriate for stat tests or low work load
3. Radioimmunoassay (RIA)	Competitive binding assay of radioactive drug	Antibody binding to radiolabeled hapten competes for unknown	Sensitive, specific, accurate	Requires radioisotopes and expensive equipment; not appropriate for stat tests or low work load
4. Enzyme-multiplied immunoassay technique (EMIT)	Competitive binding assay, enzyme label	Enzyme-hapten bound to antibody is inhibited	Sensitive, specific, accurate, rapid; can handle small work loads and stat procedures	Moderate to expensive equipment; may have some problems establishing and maintaining standard curve
5. Fluorescence polarization immunoassay (FPIA)	Competitive binding assay, fluorescent-labeled drug	Antibody-bound drug has higher fluorescence polarization than free drug	Sensitive, specific, accurate, extremely rapid; excellent for stat procedures; greater curve stability, low coefficients of variation; totally automated procedure	Specialized, expensive equipment
6. Nephelometric inhibition immunoassay (NIA)	Competitive binding assay, drug-protein conjugate label	Antibody bound to label has higher light scattering than antibody bound to drug	Sensitive, specific instrumentation can also perform immunoglobulin quantitation	Expensive equipment not appropriate for stat tests or low work load
7. High-performance liquid chromatography (HPLC)	Chromatography, fluorometry	Chromatographic separation followed by derivatization to fluorescent product	Sensitive, highly specific, accurate, versatile; can handle small work loads	Not appropriate for large work loads; expensive equipment

If a serum contains other antimicrobial agents in addition to an aminoglycoside such as gentamicin, they must be accounted for. To minimize the effects of other antimicrobial drugs, an indicator organism resistant to these antibiotics may be used, or techniques that inactivate the interfering drugs can be employed. The various shortcomings of the agar plate method have prompted the development of alternative procedures.

Radioenzyme procedures[2,3] were among the first rapid tests to be adapted for the assay of gentamicin and other aminoglycosides (method 2, Table 64-16). In these procedures, naturally occurring bacterial enzymes known to inactivate gentamicin are used to form a radioactive derivative by enzymatic transfer of radiolabeled substrate to the gentamicin. Knowing the specific activity (counts per minute per mole of label) of the labeled substrate, one can determine the gentamicin concentration by counting the amounts of radioactivity incorporated into the antibiotic, as follows:

$$\text{Labeled group} + \text{Gentamicin} \xrightarrow{\text{Bacterial enzyme}} \text{Labeled gentamicin}$$

Immunoassays are currently among the most popular methods for measuring aminoglycoside levels. All methods use the principle of competitive ligand binding. Radioimmunoassay (RIA), which was originally employed for the detection of cardiac glycosides,[4] was first used to quantitate gentamicin in 1972 (method 3, Table 64-16).[5]

Homogeneous immunoassays that do not require a separation step before measurement were developed in the late 1970s.[6] The enzyme-multiplied immunoassay technique (EMIT, Syva Co., Inc., Pal Alto, CA), which is based on the linkage of the enzyme glucose-6-phosphate dehydrogenase to gentamicin, was the first homogeneous immunoassay to be developed for clinical use (method 4, Table 64-16) (see Chapter 11). When enzyme-labeled gentamicin becomes bound to its specific antibody, the activity of the enzyme is reduced. Active enzymes converts NAD^+ to reduced NAD (NADH), resulting in an absorbance change at 340 nm, which is measured spectrophotometrically. Syva has also developed a dry reagent procedure called the QST system for performing EMIT assays. All the necessary reagents are together in dry powder form. The reaction occurs when patient sample and diluent are added to the tube.

The fluorescence polarization immunoassay (FPIA) (method 5, Table 64-16)[7,8] provides a direct measure of bound and free labeled drug in a homogeneous competitive binding immunoassay. In this test system, fluorescent-labeled gentamicin and unlabeled (sample) gentamicin compete for binding sites on a specific antibody[9] (see Chapter 11). A bench-top FPIA is available to perform these assays[10] (TDx, Abbott Diagnostics, North Chicago).

The nephelometric inhibition immunoassay (NIA)[11] can also be used to measure gentamicin levels in serum (method 6, Table 64-16) (see Chapter 10).

Table 64-17 Available methods for aminoglycoside determinations

	Aminoglycoside			
Method*	**Gentamicin**	**Tobramycin**	**Amikacin**	**Netilmicin**
RIA	+	+	+	?
EMIT	+	+	+	+
FPIA	+	+	+	+
NIA	+	+	−	−
HPLC	+	+	+	+

**EMIT,* Enzyme-multiplied immunoassay technique; *FPIA,* fluorescence polarization immunoassay; *HPLC,* high-performance liquid chromatography; *NIA,* nephelometric inhibition immunoassay; *RIA,* radioimmunoassay.
"+" indicates that a technique has been reported used to measure an aminoglycoside; "?" indicates no known report of an aminoglycoside being measured by a technique; "−" indicates technique not currently done.

Thin-layer chromatography and ion-exchange chromatography have also been used to assay gentamicin. These techniques are generally used in research applications but have not proved practical for routine clinical use. High-performance liquid chromatography (HPLC), however, can be used for the rapid and accurate assay of aminoglycoside antibiotics (method 7, Table 64-16).[12] The basic steps of HPLC for the assay of gentamicin include (1) extraction of the antibiotic from serum using a CM-Sephadex column, (2) analysis by reversed-phase ion-pair chromatography, (3) continuous-flow, postcolumn derivatization with *o*-phthalaldehyde to form a fluorescent product, and (4) fluorescent detection followed by quantitation using peak-height or peak-area analysis.

At present there is no one established reference method for the quantitation of gentamicin in body fluids. The various immunoassays are considered as a group to be reference methods. Microbiological assays are still widely used to measure antibiotic levels, but their use for aminoglycoside determinations has greatly diminished with the development of new immunoassays.

Immunoassays are now the most widely used method for aminoglycoside quantitation. Commercial kits are available for the commonly used aminoglycosides (Table 64-17). Immunoassays are sufficiently rapid, sensitive, accurate, and precise. The coefficients of variation for most methods are similar. However, no specific data comparing the amount of immunological cross-reactivity in these methods are available.

Table 64-18 compares reaction conditions for the gentamicin assays. The final choice of which method to use for gentamicin analysis will depend on many factors, including laboratory work loads, test cost, turnaround time, and available instrumentation. Since most instrumentation described can be used to measure a variety of chemical analytes, versatility may be an important determining factor. Much of the new instrumentation designed for the

Table 64-18 Comparison of reaction conditions for gentamicin assays

Parameter	Microbiological assay	Enzyme-multiplied immunoassay technique	Fluorescence polarization immunoassay
Temperature	35° C	30° C	35° C
pH	7.9	8.0	7.3 to 7.7
Sample (μL)	20	50	50
Linearity (μg/mL)	1 to 20	1-10	0.5 to 10
Percent coefficient of variation* (gentamicin concentration)	5% to 50%	≤10%	2.6% (1.0 μg/mL) 1.7% (4.0 μg/mL) 2.66 (8.0 μg/mL)
Known interferences	Other antimicrobials	Sisomicin,† netilmicin, heparin; hemolytic, lipemic, or icteric samples	Sisomicin, netilmicin (25% cross-reactivity)

*CV data supplied by manufacturer.
†Sisomicin is a semisynthetic aminoglycoside not approved by the FDA for clinical use.

Table 64-19 Usual doses and desired serum concentrations (μg/mL) for aminoglycosides

Antibiotic	Usual dose*	Peak†	Trough‡	Toxic range
1. Gentamicin Tobramycin	3 to 5 mg/kg/day divided q8h	4-8	1-2	>10-12
2. Netilmicin	4.0 to 6.5 mg/kg/day divided q8h	6-10	0.5-2	>10-20
3. Amikacin Kanamycin	15 mg/kg/day divided q12h	20-25	5-10	>30-55
4. Streptomycin	15 mg/kg/day divided q12h	5-20	<5	>40-50

*For adults with normal renal function.
†Peak concentration determined 1 hour after intramuscular dose or 30 minutes after intravenous dose.
‡Trough determined just before next dose.

clinical chemistry laboratory has the capability of performing homogeneous enzyme assays.

SPECIMEN

The most common sample tested is serum or plasma. However, the microbiological assay may be used to assay aminoglycosides in other body fluids, including cerebrospinal fluid, synovia, or any other nonviscous fluid. The antibiotic standards should be prepared in normal fluid from patients not on antibiotics. The immunoassays can be performed on serum or plasma but not on whole blood. There are no concrete data on the general use of these methods with other body fluids.

INTERFERENCES

As stated previously, the microbiological assay will be affected by any additional antibiotic or antibiotics present in the sample to which the indicator organism is sensitive. Unless this activity is counteracted, the results may be invalid. Netilmicin, which is structurally related to gentamicin, will cross-react with the antibodies to gentamicin in most immunoassays. However, netilmicin would not usually be administered with gentamicin. The immunoassays may also be affected to varying degrees by hemolytic, lipemic, or icteric serum although specific data on the effects of each on the individual assays are not available.

REFERENCE RANGE

The therapeutic range of gentamicin is approximately 2 to 9 μg/mL (4.3 to 19.4 μmol/L). See Table 64-19 for the other aminoglycosides.

REFERENCES

1. Sabath, LD, Casey, JI, Ruch, RA, et al: Rapid microassay of gentamicin, kanamycin, neomycin, streptomycin, and vancomycin in serum or plasma, J Lab Clin Med 78:457, 1971.
2. Butcher, RH: Rapid serum gentamicin assay by enzymatic adenylation, Am J Clin Pathol 68:566, 1977.
3. Stevens, P, Young, LS, and Hewitt, WL: Improved acetylating radioenzymatic assay of amikacin, tobramycin, and sisomicin in serum, Antimicrob Agents Chemother 7:374, 1975.
4. Oliver, GC, Parker, BM, Brasfeld, DL, and Parker, CW: The measurement of digitoxin in human serum by radioimmunoassay, J Clin Invest 47:1035, 1968.
5. Lewis, JE, Nelson, JC, and Wilson, TN: Radioimmunoassay of an antibiotic: gentamicin, Nature (New Biol) 239:214, 1972.
6. Bastiani, RJ: The EMIT system: a commercially successful innovation, Antibiot Chemother 26:89, 1979.
7. Dandliker, WB, Kelly, RJ, Dandliker, J, et al: Fluorescence polarization immunoassay: theory and experimental method, Immunochemistry 10:219, 1973.

8. Dandliker, WB: Investigation of immunochemical reactions by fluorescence polarization. In Atassi, MA, editor: Immunochemistry of proteins, New York, 1977, Plenum Publishing Co.
9. Jolly, MD, Stroupe, SD, Wang, CJ, et al: Fluorescence polarization immunoassay. I. Monitoring aminoglycoside antibiotics in serum and plasma, Clin Chem 17:1190, 1981.
10. Popelka, SR, Miller, DM, Holen, JT, and Kelso, DM: Fluorescence polarization immunoassay. II. Analyzer for rapid, precise measurement of fluorescence polarization with use of disposable cuvettes, Clin Chem 27:1198, 1981.
11. Finley, PR: Nephelometry: principles and clinical laboratory applications, Lab Management 20:34, 1982.
12. Arnhalt, JP: Assay of gentamicin in serum by high-pressure liquid chromatography, Antimicrob Agents Chemother 11:651, 1977.

Lithium

ROBERT L. MURRAY

Clinical significance: pp. 719 and 795
Atomic weight: 6.94
Atomic symbol: Li
Merck Index: 5343
Chemical class: metal

PRINCIPLES OF ANALYSIS

Two techniques have been used most frequently with almost equal success for the measurement of lithium: flame atomic emission spectroscopy (FAES)[1] and flame atomic absorption spectroscopy (FAAS) (methods 1 and 2, Table 64-20).

In the flame emission technique a 1:50 sample dilution is typically used for serum or plasma. With an exception to be noted later, the diluent is 1.5 millimoles of potassium ion per liter (mmol/L K^+). The diluent, which also serves as the internal standard, is used to prepare standards and the blank as well. Since the serum or plasma adds a small amount of potassium, the diluted sample would contain slightly more potassium than the blank or the standard. To correct for this, potassium chloride, 5.0 mmol/L, is added to both blank and standard to approximate the K^+ concentration of the sample. An aqueous standard of lithium carbonate, 1.0 mmol/L, is used for calibration. A blank is prepared in the same fashion as the standards, except that water is added in place of the 1.0 mmol/L lithium carbonate.

The principles of flame emission are discussed in detail in Chapter 3. The heat of the air-propane flame (approximately 1925° C) vaporizes the lithium chloride (boiling point, 1350° C). In the presence of the heat and the reducing gases carbon monoxide, hydrogen, and carbon dioxide, the lithium chloride dissociates into uncharged ground-state atoms, Li^0 and Cl^0. The heat further causes the 2*s* electron to be excited to the 2*p* state; the excited electron instantly decays to the original 2*s* ground state, with emission of light at 670.8 nm. Although only a very small fraction of the lithium is ultimately converted to the excited state with subsequent emission, this is sufficient for accurate quantitation.

The potassium in the internal standard and in the sample follows a similar pathway, entering the vapor state at a higher temperature by sublimation at approximately 1500° C. The 4*s* electron of potassium is raised to the 4*p* excited state and immediately decays to the ground state, resulting in emission of a doublet at 767 and 769 nm. (This doublet is not resolved by the interference filter commonly used with the flame photometer.) The emission intensity depends on conditions that are subject to minor fluctuations, such as the fuel/oxidant ratio, the sample aspiration rate, the reducing gas composition and concentration, and the flame temperature; these fluctuations are all potential sources of error. To circumvent this, one measures the ratio of the emission intensity of potassium to the emission intensity of lithium. In this way instrument fluctuations that affect both elements are compensated for and error is minimized.

A modification of the flame emission technique is also widely used.[2] This small but significant change allows the use of cesium (at 1.5 mmol/L) instead of potassium as the internal standard. Cesium functions in the same fashion as previously described for potassium, except that it is undetectable in serum or plasma, and therefore no correction or compensation for endogenous cesium is necessary. In addition, sodium, potassium, and lithium can all be analyzed using the cesium internal standard, with minimum instrumental modifications. Cesium chloride vaporizes at 2390° C. When thermally excited, its 6*s* electron is raised to the 6*p* orbital; decay of this electron results in light emission at 852 nm.

The theoretical basis of FAAS bears a certain similarity to that of FAES (see Chapter 3). As in FAES, uncharged ground-state lithium atoms are produced in the approxi-

Table 64-20 Methods of lithium analysis

Method	Type of analysis	Principle	Usage
1. Flame emission	Quantitative	Emission of light at 670.8 nm by Li^0	Serum, plasma
2. Atomic absorption	Quantitative	Absorption of light at 670.8 nm by Li^0	Serum, plasma, urine, RBCs
3. Ion-selective electrode	Quantitative	Measurement of voltage difference generated by Li^+ in contact with lithium ionophore	Serum, plasma, whole blood

mately 2200° C atomic absorption spectrometry (AAS) flame generated by burning acetylene and air. Neutral lithium atoms absorb light at 670.8 nm. Most AAS instruments employ a hollow cathode lamp to produce monochromatic light whose nominal peak wavelength is at 670.8 nm.[3] Because lithium ionizes only to a very small extent (approximately 1%) under the flame conditions described, little advantage is gained by the use of an electron-donating, ionization-suppression compound, such as tin chloride ($SnCl_2$). In atomic absorption analysis of lithium, an aqueous dilution of the serum, plasma, urine, or hemolysate is prepared. For serum, plasma, or hemolysate the typical dilution employed is 1:20 in physiological solution. For urine, the necessary dilution may vary from 1:100 to 1:1000, depending on the lithium concentration in the urine. The required dilution may be calculated based on the patient's dosage and the urine volume, or a range of dilutions may be used, and the appropriate dilution selected for calculation and reporting.

In 1986 automated analysis of lithium by direct potentiometry was introduced (method 3, Table 64-20). In this method a Li^+ ion-selective ionophore is embedded into a glass electrode. In the presence of Li^+ a voltage that is proportional to the concentration of Li^+ in the sample is generated.

The usual criteria of inherent analytical accuracy, precision, cost, and ease of performance do not affect the choice of a method for lithium analysis in serum or plasma.[3,4] The two flame techniques compare remarkably well for each of these criteria (Table 64-21).

Flame emission photometers are much more frequently found in clinical laboratories than are atomic absorption instruments. There can be clinical situations when a lithium analysis is needed on a stat basis. Thus, because of the availability and ease of use of flame photometers, FAES is probably the method of choice for the analysis of lithium.

Direct potentiometric methods using ion-selective electrodes have only recently been introduced, but these may become very useful stat methods in the future.

Table 64-21 Comparisons of methods for lithium analysis

	Flame emission*		Atomic absorption†	
Sample	Plasma or serum		Plasma, serum, urine, or hemolysate	
Dilution (plasma, serum)	1:100		1:20, 1:25	
Diluent	Cs^+ 1.5 mmol/L		Water	
Specimen required (μL)	16		100	
Precision‡ (mmol/L)	SD	Mean	SD	Mean
	0.08	0.55	0.06	0.54
	0.08	1.20	0.08	1.20
	0.12	2.39	0.15	2.39
Detection limit (mmol/L)	0.01		0.01	
Linearity (mmol/L)	3.0		5.0	

*IL 643, Instrumentation Laboratory, Lexington, MA.
†Perkin-Elmer 460, Perkin-Elmer Corp, Norwalk, CT.
‡1985 College of American Pathologists Quality Assurance Comprehensive Chemistry report.
SD, Standard deviation.

SPECIMEN

Serum or plasma drawn 8 to 10 hours after an oral dose of lithium is the specimen of choice for routine monitoring; if repeated levels are drawn, the dose-to-phlebotomy interval should be kept identical for each patient. Some heparin anticoagulants are formulated with lithium as the cation; if plasma is used, such tubes should be avoided.

The plasma or serum should be separated from the cells if storage of more than 4 hours is anticipated. In the compliant patient this would not be critical; it is advised only in the event of an unusually large plasma–to–red blood cell concentration disparity. Unseparated, anticoagulated blood stored at 4° C shows an increase in red blood cell lithium concentration; storage at 37° C results in a decrease in red blood cell lithium. Once the serum is separated, lithium is stable for at least 24 hours at room temperature, for 7 days at 4° C, and indefinitely if frozen.

PROCEDURE: ATOMIC ABSORPTION

Principle

Diluted sample is aspirated into flame where Li^0 absorbs light emitted from a hollow cathode lamp. Absorbance at 670.8 nm is proportional to Li^+ concentration.

Reagents

Deionized water should be of such a quality as to give 10 megaohms/cm of specific resistance at 25° C.

Stock blank (sodium chloride, 140 mmol/L; potassium chloride, 5.0 mmol/L). Prepare as in flame emission method.

Stock 1.0 mmol/L lithium standard (lithium carbonate, 0.5 mmol/L). Prepare as in flame emission method.

Stock 2.0 mmol/L lithium standard (lithium carbonate, 1.0 mmol/L). See note 1 before preparing. Dissolve 73.89 mg of the dried lithium carbonate in 50 mL of deionized water containing 10 mL of 0.1 M hydrochloric acid. Add water to the mark in a 1 L volumetric flask. This is stable for 1 year at room temperature.

Assay

Equipment: any modern atomic absorption spectrophotometer.

1. Dilute 0.50 mL of blank, standard, and each unknown serum, plasma, urine (see note 2), or hemolysate (see note 3) with 9.50 mL of deionized water.
2. Follow the instrument conditions specified by the manufacturer.

3. Zero the instrument with the diluted blank, and calibrate with 1.0 mmol/L lithium standard. Analyze the unknown specimens, checking the blank and calibration readings periodically.
4. Specimens containing more than 2.0 mmol/L lithium should be diluted with diluent until the measured concentration falls below 2.0 mmol/L. The appropriate correction is then made for the dilution.

Notes

1. The 2.0 mmol/L lithium standard is used only if curve correction is not available on the instrument. Since most atomic absorption spectrophotometers can automatically correct for nonlinearity, this standard should be run only initially for verification of the accuracy of the correction.
2. Urine samples are prepared by pipetting 10 mL aliquots of urine to each of the following volumetric flasks: 100 mL, 500 mL, and 1 L. Add deionized water to the mark in each, and then process as described in step 1 of the procedure. Make the appropriate corrections (times 10, times 50, and times 100, respectively), and report the value that reads closest to the 1.0 standard.
3. Hemolysates are made by centrifugation of anticoagulated blood for not less than 15 minutes at 2000 *g* to exclude plasma from the cellular fraction as much as possible. Aspirate the plasma completely and a small portion of the packed red blood cells. Washing the cells is avoided to eliminate the possibility of exchange of lithium with the wash solution. Hemolyze the cells either by freezing and thawing or by immersion of the tube into an ultrasonic bath for 60 seconds. Centrifuge the hemolysate at 2000 *g* again before removal of the sample.
4. Using a Perkin-Elmer 460 atomic absorption instrument, between-day precision (as percent coefficient of variation) of 2% and 1.5% was achieved at lithium levels of 0.97 and 2.43 mmol/L, respectively.[5]

REFERENCE RANGE

Maintenance therapy, serum	0.6-1.3 mmol/L
Acute therapy, serum	0.9-1.4 mmol/L
Red blood cell concentration	0.2-0.8 mmol/L
Urine content	95%-99% of daily intake (after steady state)

Lithium levels in cerebrospinal fluid are 40% to 50% of the levels in plasma at steady state.[6]

REFERENCES

1. Barrow, GRJ: The estimation of lithium in blood, J Med Lab Technol 17:236-240, 1960.
2. Bergkuist, C, Kelley, TF, and Moran, BL: Cesium as the internal standard in a new four-element flame photometer, Clin Chem 24:1061, 1978 (abstract).
3. Brybus, J, and Bowers, GN, Jr: Serum lithium determination by atomic absorption spectroscopy. In MacDonald, RP, editor: Standard methods of clinical chemistry, 6:189-192, New York, 1970, Academic Press, Inc.
4. Velapoldi, RA, Paule, RC, Schaffer, R, et al: Standard reference materials: a reference method for the determination of lithium in serum, National Bureau of Standards Special Pub No 260-69, Washington, DC, 1980, US Government Printing Office.
5. University of Cincinnati Medical Center, Division of Chemical Pathology, unpublished results.
6. Platman, SR, and Fieve, RR: Biochemical aspects of lithium in affective disorders, Arch Gen Psychiatry 19:659-663, 1968.

Procainamide and N*-acetylprocainamide*

JOHN E. SHERWIN

Clinical significance: pp. 415 and 795

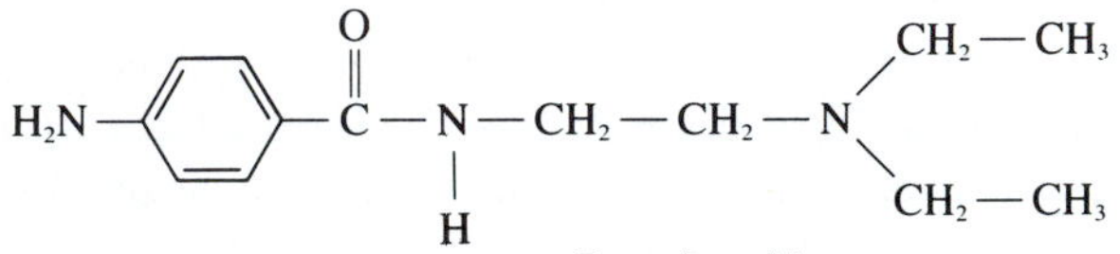

	Procainamide	*N*-Acetylprocainamide
Molecular formula:	$C_{13}H_{21}N_3O$	$C_{15}H_{23}N_3O_2$
Molecular weight:	235.54 daltons	277.37 daltons
Merck Index:	7553	14

Chemical class: amines

Structure:

$H_2N-C_6H_4-C(=O)-N(H)-CH_2-CH_2-N(CH_2-CH_3)_2$

Procainamide

$CH_3C(=O)-N(H)-C_6H_4-C(=O)-N(H)-CH_2-CH_2-N(CH_2-CH_3)_2$

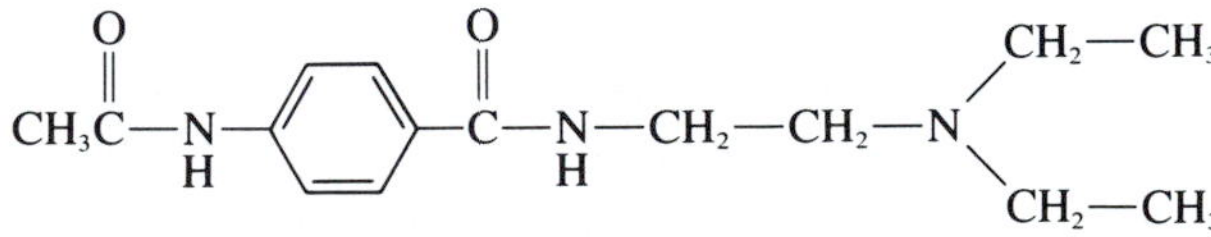

***N*-Acetylprocainamide**

PRINCIPLES OF ANALYSIS

Procainamide and its metabolite, *N*-acetylprocainamide (NAPA), are generally prescribed for cardiac arrhythmias.[1] Analysis for blood concentrations is most commonly requested for monitoring of the therapeutic efficacy of the drug and patient compliance with the medication regimen. Ordinarily, analysis of both procainamide and NAPA is requested because NAPA is also pharmacologically active, and both contribute to the therapeutic and toxic manifestations of drug therapy.

It is rare that procainamide and NAPA analyses are required in an emergency toxicology setting. Therefore, although it is possible to identify these substances qualitatively, only quantitative analysis is ordinarily performed.

A variety of colorimetric[2] (method 1, Table 64-22) and fluorometric (method 2, Table 64-22) procedures have been used to measure procainamide and NAPA.[2-4] Gas chromatography (GC; method 3, Table 64-22) has been used to separate and quantitate procainamide and NAPA.[5] However, high-performance liquid chromatography (HPLC) and immunoassay methods have replaced all other methods for analysis of these drugs in the clinical laboratory.

Table 64-22 Methods of procainamide and NAPA analysis

Method	Type	Principle	Usage	Comments
1. Colorimetric	Spectrophotometry	Extraction, diazotization measurement at 550 nm	Historical	Not specific
2. Fluorescence	Extraction; analysis by fluorescence	Extraction, fluorescence at 354 nm; excitation wavelengths, 298 nm for procainamide and 288 nm for NAPA	Historical	Coefficient of variation generally higher than that of other methods
3. Gas chromatography (GC)	Chromatography; flame ionization detection	Extraction; chromatography in gas-liquid phase	Limited	Both procainamide and NAPA determined with one assay
4. High-performance liquid chromatography (HPLC)	Chromatography; spectrophotometric analysis at 254 nm or lower	Extraction, chromatography	Limited	Both procainamide and NAPA determined with one assay
5. Fluorescence polarization immunoassay (FPIA)	Competitive binding of fluorescent-labeled drug	Fluorescence polarization of fluorescein drug derivative with competitive immunoassay	Common	Separate assay for procainamide and NAPA
6. Enzyme-multiplied immunoassay technique (EMIT)	Competitive binding of enzyme-labeled drug	Enzyme-multiplied competitive immunoassay using NADH	Common	Separate assay for procainamide and NAPA

Procainamide and NAPA can be analyzed simultaneously by use of HPLC (method 4, Table 64-22) after extraction of the drug into methylene chloride or chloroform.[6,7] Evaporation of the solvent and resuspension of the drug in a small amount of methanol concentrate the drugs sufficiently for accurate analysis.

The Abbott TDx therapeutic drug monitoring instrument (Abbott Instruments, Chicago, IL) uses a fluorescence polarization immunoassay (FPIA) (method 5, Table 64-22) to determine concentrations of procainamide or NAPA separately in serum or plasma samples.

In the enzyme-multiplied immunoassay technique (EMIT) assay systems (method 6, Table 64-22), one separately assays procainamide and its metabolite, NAPA, utilizing the EMIT method with separate assay kits. Patients procainamide/NAPA levels are obtained by comparison to a calibration curve constructed from samples containing known concentrations of procainamide/NAPA.

The immunoassays are the preferred procedures. The FPIA has less variability than the EMIT system. Table 64-23 compares assay conditions.

SPECIMEN

Either plasma or serum can be used. Acceptable anticoagulants for EMIT are EDTA, heparin, or oxalate; heparin is not recommended for FPIA. Store serum or plasma refrigerated (2° to 8° C) if the analysis is delayed beyond 8 hours. Procainamide and NAPA are stable for 2 weeks

Table 64-23 Comparison of assay conditions for procainamide and NAPA analysis

Condition	HPLC	FIPA	EMIT
Temperature	70° C	35° C	37° C
pH	—	7.0 (100 mmol phosphate)	8.0 (55 mmol Tris)
Sample volume	0.5 mL	20 μL	50 μL
Fraction of sample volume	0.25 (extraction)	0.0976	0.01
Linearity (μg/mL)	0.5 to 20	0.2 to 20	1.0 to 16
Sensitivity (μg/mL)	0.1	0.2	0.5
Time of analysis (minutes)	8	20	2
Interferences	None known	NAPA cross-reacts with procainamide 4.7%; desethyl-NAPA cross-reacts 1.7% with procainamide and 18.4% with NAPA.	Severe hemolysis, lipemia, or icterus will interfere; procaine cross-reacts.

at this temperature. Severely hemolyzed, lipemic, or icteric samples are unacceptable for all methods except HPLC and GC because of photometric interference.

PROCEDURE: HPLC ASSAY FOR PROCAINAMIDE AND *N*-ACETYLPROCAINAMIDE

Principle

Reversed-phase HPLC of serum or plasma is performed after alkaline extraction of the analytes into chloroform. Detection is by ultraviolet spectroscopy at 215 nm.

Reagents

Stock procainamide/*N*-acetylprocainamide (each 0.5 mg/mL; 2.12 mmol/L and 1.80 mmol/L, respectively). To a 10 mL volumetric flask, add 5.0 mg each of procainamide and *N*-acetylprocainamide (free base). Add 2 mL of methanol, and dilute to 10.0 mL with distilled water. Solution is stable frozen in glass vial for 2 years.

Procainamide/*N*-acetylprocainamide standard (5.0 μg/mL, 21.2 μmol/L, and 18.0 μmol/L, respectively). To a 200 mL volumetric flask, add 2.0 mL of stock procainamide/*N*-acetylprocainamide. Dilute to 200.0 mL with drug-free plasma. Store frozen in 1 mL aliquots. Solution is stable for 1 year.

Stock internal standard *N*-proprionylprocainamide (*N*-PPA; 1.0 mg/mL; 3.43 mmol/L). Add 10.0 mg of *N*-PPA to a 10 mL volumetric flask. Dissolve in approximately 8 mL of methanol, and bring to volume with additional methanol. Solution is stable refrigerated for 2 years.

International standard/extractant (0.5 μg/mL, 1.72 μmol/L). To a 1 L volumetric flask, add 0.5 mL of stock *N*-PPA, and bring to a final volume of 1 L with chloroform or methylene chloride. Store in 10 mL repipettor at room temperature. Solution is stable refrigerated for 2 years.

Mobile-phase buffer, pH 6.0. To 3.0 L of distilled water, add 1.5 mL of glacial acetic acid and 3.0 mL of triethylamine. Mix well. Check pH. Adjust to 6.0 ± 0.10 with triethylamine or glacial acetic acid, as necessary. Solution is stable at room temperature for 3 months.

Mobile phase (acetonitrile/buffer, 12:88 by volume). Add 410 mL of acetonitrile to 3.0 L of mobile-phase buffer. Mix well. Solution is stable stored at room temperature for 3 months. Filter through a 0.22 μm filter before use, and degas immediately before use.

Concentrated ammonium hydroxide, reagent grade

Anhydrous sodium sulfate

Assay

Equipment: 15 mL screw-capped tubes; 15 mL conical centrifuge tubes; a centrifuge capable of generating 3000 *g* force; 1 mL disposable pipets; 10 mL repipettor; an HPLC system with pump, oven, spectrophotometer, deuterium lamp source, strip-chart recorder, and a reversed-phase C_{18} column; 25 μL Hamilton syringe; funnels and phase-separating or Whatman No. 1 filter paper.

1. To separate screw-capped tubes, add 0.5 mL of patient samples (in duplicate), standards (in duplicate), and controls.
2. Add 4 drops of NH_4OH and 5 mL of internal standard/extractant.
3. Cap, shake, centrifuge for 5 minutes at 3000 *g*, and aspirate aqueous (top) layer.
4. Filter through anhydrous sodium sulfate, and evaporate chloroform to dryness in 15 mL conical centrifuge tubes.
5. Chromatographic parameters:
   Chart speed: 8 mm/min
   Ultraviolet sensitivity: 0.02 AUFS (absorbance units at full scale) at 215 nm
   Flow rate: 2.5 mL/min at 70° C
6. Reconstitute with 50 μL of methanol, and inject 20 μL into the HPLC system. Run time is approximately 8 minutes.

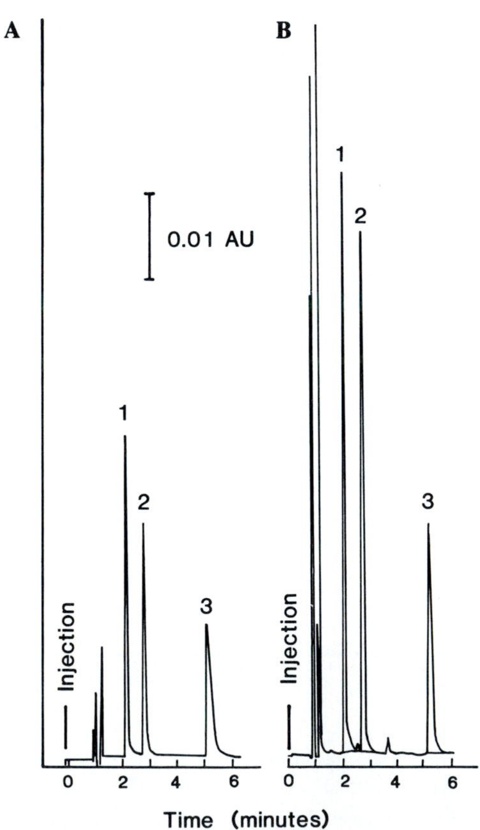

Fig. 64-10 Chromatogram depicting elution profile of pure drugs, **A,** and a standard extracted in drug-free plasma, **B,** containing 10 μg of procainamide per milliliter and NAPA (*N*-acetylprocainamide, acecainide) according to protocol described in text. *1,* Procainamide; *2,* NAPA; *3,* internal standard, *N*-proprionylprocainamide. *AU,* Absorbance unit.

7. The elution times for the analytes and internal standard are as follows:

	Approximate retention (minutes)	Relative retention to internal standard (IS)
Procainamide	2.8	0.47
N-Acetylprocainamide	3.6	0.60
N-Proprionylprocainamide (IS)	6.0	1.0

The typical chromatograms of the elution profile of pure drugs, standard, and patient are presented in Fig. 64-10.

Calculations

1. Identify peaks of the drugs and internal standard (IS).
2. Measure peak heights for procainamide, NAPA, and the internal standards.
3. Calculate the peak-height ratios.
4. Calculate the procainamide and NAPA concentrations from a standard curve or as follows:

$$\frac{\text{Peak height of drug}}{\text{Peak height of IS}} = R$$

$$\frac{R_{\text{unknown}}}{R_{\text{standard}}} \times \text{Concentration of standard} = \text{Concentration of unknown}$$

REFERENCE RANGE

Procainamide

Therapeutic	4 to 10 μg/mL (15 to 37 μmol/L)
Potentially toxic	Over 20 μg/mL

REFERENCES

1. Lima, JJ, and Lewis, RP: Procainamide therapeutic use and serum concentration monitoring. In Taylor, WJ, and Finn, AL, editors: Individualizing drug therapy: practical applications of drug monitoring, vol 3, New York, 1918, Gross Townsend Frank, Inc.
2. Koch-Weser, J, and Klein, SW: Procainamide dosage schedules, plasma concentrations, and clinical effects, JAMA 215:1454-1460, 1971.
3. Ambler, PK, and Masarei, JRL: A new fluorometric method for procainamide, Clin Chim Acta 70:379-383, 1976.
4. Matusik, E, and Gibson, TP: Fluorometric assay for *N*-acetylprocainamide, Clin Chem 21:1899-1902, 1975.
5. Karbson, E, Molin, L, Norlander, B, and Sjoqvist, B: Acetylation of procainamide in man studied with a new gas chromatographic method, Br J Clin Pharmacol 1:457-467, 1974.
6. Baselt, R: Analytical procedures for therapeutic drug monitoring and emergency toxicology, Davis, CA, 1980, Biomedical Publications.
7. Mackiohan, JJ, Coyle, JD, Shields, BJ, et al: Fluoroimmunoassay for procainamide and *N*-acetylprocainamide compared with a liquid-chromatographic method, Clin Chem 30:768-773, 1984.

Salicylates

JOSEPH SVIRBELY

Salicylate, acetylsalicyclic acid, aspirin
Clinical significance: p. 740

	Acetylsalicylic acid	Salicylic acid
Molecular formula:	$C_9H_8O_4$	$C_7H_6O_3$
Molecular weight:	180.15 daltons	138.12 daltons
Merck Index:	863	8190
Chemical class:	Benzoic acid derivative	Hydroxybenzoid acid

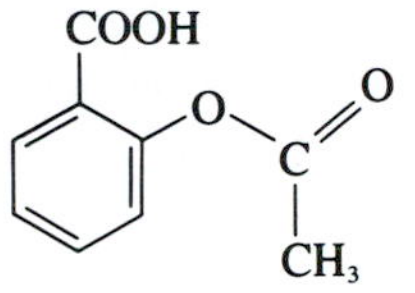

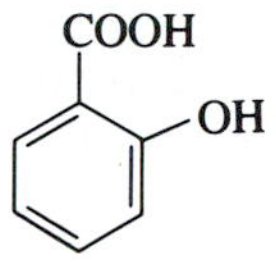

PRINCIPLES OF ANALYSIS

Acetylsalicylic acid (aspirin) is rapidly hydrolyzed to salicylic acid and acetic acid in the stomach; salicylic acid is rapidly absorbed into the blood. There are a variety of methods for determination of salicylate in serum or plasma. The most widely used methods in the clinical laboratory employ colorimetric techniques, but other methods include gas-liquid chromatography (GLC), ultraviolet spectroscopy, and fluorescence spectrophotometry. The earliest and still most widely used method for measurement of salicylates is the spectrophotometric method, which measures the absorbance of the complex produced by the mixture of salicylates with ferric ion (method 1, Table 64-24). The first widely accepted method using this technique was described by Brodie et al.[1] (method 1a, Table 64-24). In this procedure salicylate is separated from plasma by extraction into acidified ethylene dichloride, followed by back-extraction into an aqueous phase as the ferric complex, which is then measured in a colorimeter at 540 nm.

Keller[2] published a modification of Brodie's method, which eliminated the extraction steps (method 1b, Table 64-24). Instead, total salicylate was measured as the sum of acetylsalicylate and its metabolites. In another method, published by Trinder,[3] protein is precipitated with mercuric chloride and hydrochloric acid (method 1c, Table 64-24). Color is produced when salicylate is complexed with ferric ion supplied as ferric nitrate.

A second type of spectrophotometric method involves reducing salicylate with a mixture of phosphotungstic and phosphomolybdic acids (Folin-Ciocalteu reagent) (method 2, Table 64-24).[4] Weichselbaum and Shapiro[5] published a method in which serum was mixed with this reagent, and the resulting blue color was read at 660 nm.

Ultraviolet spectrophotometry has also been employed

Table 64-24 Methods for salicylate analysis

Method	Type of analysis	Principle	Usage	Comments
1. Ferric nitrate complex	Spectrophotometry	Salicylates + Fe^{+3} → Ferric complex (A_{max}, 540 nm)	Most commonly used clinical salicylate procedure for urine screening	Sensitive to 200 mg/L Linear to 1000 mg/L Only interference is salicylate metabolites
a. Brodie et al.		Extracts salicylate into ethylene dichloride		Method 1a uses toxic ethylene dichloride
b. Keller		Eliminates extraction step		Method 1b less complicated but specific enough for clinical needs
c. Trinder		Deproteinates with mercuric chloride		Method 1c uses toxic mercuric chloride
2. Folin-Ciocalteu reagent reduction	Spectrophotometry	Phosphotungstic acid + phosphomolybdic acid reduce salicylates to form blue complex (A_{max}, 660 nm)	Not commonly used because of interferences	Tryptophan, tyrosine, and uric acid react with reagent to give falsely high readings
3. Spectrophotometry	Differential ultraviolet spectrophotometry	Measures salicylate + acetylsalicylate by differences in absorbance at pH 9.0 and pH 13.5	Still in use	Has advantage of measuring acetylsalicylate Requires narrow band pass ultraviolet spectrophotometer; sensitive to 5 mg/L
4. Fluorometry	Spectrofluorometry	Protein can be removed by precipitation Alkali intensifies fluorescence of salicylate	Not commonly used	Has advantage of greater sensitivity Requires smaller sample than spectrophotometry
5. Gas-liquid chromatography	Chromatographic separation	Acetylsalicylate and major metabolites extracted with ether; derivation with BSTFA (see text); analysis with gas-liquid chromatograph using temperature programming and OV-17 column monitored by flame ionization detector	Used in research: best candidate for reference method	Sensitive to 1 mg/L Linear to 2000 mg/L No interfering compounds
6. Enzymatic	Spectrophotometric	Salicylate hydroxylase converts salicylate to catechol with the oxidation of NADPH to NADP; reaction rate, monitored by measurement of decrease in A_{340}, is proportional to salicylate concentration	Used only in research	Reagents not commercially available, highly specific
7. Enzyme-multiplied immunoassay technique (EMIT)	Competitive binding, enzyme label	Antibody binding to enzyme-hapten inhibits catalytic reactivity; competition by analyte in sample prevents inhibition	Serum, plasma	Some interferences caused by sample matrix
8. Fluorescence polarization immunoassay (FPIA)	Competitive binding with fluorescence polarization	Bound labeled drug has higher polarized fluorescence than free; amount of polarized fluorescence is inversely proportional to amount of unlabeled drug	Serum, plasma; commonly used for serum quantification	Fluorescence interferences; rapid test

as an analytical technique to measure acetylsalicylate plus salicylate (method 3, Table 64-24). Routh et al.[6] extracted the acetylsalicylate using the Brodie procedure and quantified the drugs using differential spectrophotometry by measuring the differences in absorbance at pH 9 and 13.5.

A fourth analytical technique historically employed for measurement of salicylate is spectrofluorometry (method 4, Table 64-24). In 1948 Saltzman[7] published a spectrofluorometric method in which the salicylate was separated from protein by precipitation with dilute tungstic acid. Strong alkali was added to convert the salicylate into a more fluorescent form. The ion was excited at 370 nm, and fluorescent emission was monitored at 460 nm. Variations include separation by gel filtration and use of enhancing agents.

Gas-liquid chromatography (GLC) is another attractive analytical technique for measuring salicylates, especially since acetylsalicylate and its metabolites can be separated and quantitated independently by this technique (method 5, Table 64-24). Two papers report GLC procedures[8,9] that incorporate a silylation step after extraction of the salicylates from the biological sample. Thomas et al.[8] use a GLC equipped with a flame ionization detector (FID), fitted with a 5% OV-17 column, and run isothermally at a column temperature of 150° C. Plasma is extracted into ether along with *p*-toluic acid, the internal standard, and the ether is evaporated. Bis(trimethylsilyl)trifluoroacetamide (BSTFA) is added to the residue and heated, and an aliquot is analyzed by GLC. Acetylsalicylate and salicylate are both analyzed.

Rance et al.[9] used a similar procedure, but by using temperature programming they were able to measure salicylamide in addition to salicylate and acetylsalicylate.

You and Bittikofer[10] reported an enzymatic assay for salicylate that seems promising for use in pharmacokinetic studies because of good precision at low levels of salicylate (less than 100 μg/ml). The assay utilizes salicylate hydroxylase, derived from *Pseudomonas cepacia* and purified by affinity chromatography (method 6, Table 64-24). Salicylate hydroxylase catalyzes the stoichiometric, unidirectional conversion of salicylate and NAD(P)H to catechol and NAD(P) in the presence of molecular oxygen. The change in absorbance at 340 nm is monitored and compared to a standard.

Recently, competitive immunoassays have been used to measure salicylate in serum. Both the EMIT enzyme and the TDx polarized fluorescence methods of measurement have been developed.

REFERENCE AND PREFERRED METHODS

The simplest method for measuring salicylates on a routine basis is the spectrophotometric method of Keller, which yields a measure of total salicylate and its metabolites.

The main advantage of spectrofluorometric methods is that they have better sensitivity than the colorimetric methods. Spectrofluorometric methods do not measure acetylsalicylate directly, since acetylsalicylate does not fluoresce. Instead, they derive the value indirectly by subtracting free salicylate from the total salicylate measured after the sample is hydrolyzed.

At present GLC is the method of choice for separating and quantitating acetylsalicylate and its major metabolites; its use is recommended in studies requiring separate determination of these compounds.

Spectrophotometric assay remains the method of choice for routine measurement of salicylates because it is relatively free from interferences, has accuracy and precision comparable to other methods, and is inexpensive.

SPECIMEN

Either serum or plasma is an acceptable sample for analysis. There is no known interference by commonly used anticoagulants. Urine specimens can be analyzed by either method described below.

PROCEDURE: SPECTROPHOTOMETRIC METHOD

Principle

The following spectrophotometric method is based on salicylates complexing with ferric ion, which is supplied by ferric nitrate in dilute nitric acid. The resulting complex is water soluble and absorbs at 540 nm against a serum blank pipetted into dilute nitric acid (Table 64-25).

Reagents

Ferric nitrate solution (6 g/L, 24.8 mmol/L). Dissolve 2.5 of $Fe(NO_3)_3 \cdot 9H_2O$ in 1.0 mL of concentrated nitric

Table 64-25 Comparison of reaction conditions for salicylate analysis

Parameter	Spectrophotometry	Gas-liquid chromatography
Fraction of sample volume	0.077	1.0 mL (sample volume)
Sample	Serum	Serum
Linearity	1000 mg/L	Acetylsalicylate: 100 mg/L Sodium salicylate: 2000 mg/L
Major interferences	Background color caused by iron Interference from salicylate metabolites	None
Precision (coefficient of variation)	9.6%	—

acid, and dilute to 250 mL with distilled water. Filter through Whatman No. 1 filter paper into a glass bottle. Solution is stable for 1 year at room temperature.

Dilute nitric acid (62 mmol/L). Dilute 1.0 mL of concentrated nitric acid to 250 mL with distilled water. This is stable for 2 years at room temperature.

Stock salicylate standard (1 g/L, 7.24 mmol/L). Dissolve 100 mg of salicylic acid in 100 mL of water. This is stable for 1 year at 4° to 8° C.

Working salicylate standard (100 mg/L, 724 μmol/L). In a 10 mL volumetric flask, dilute 1.0 mL of stock standard to 10 mL with distilled water. This is stable for 3 months at 4° to 8° C.

Control. A serum sample containing known amounts of salicylate is used as the control.

Assay

Equipment: ≤10 nm band pass spectrophotometer.

1. Label six test tubes: standard, standard blank, control, control blank, unknown(s), and unknown(s) blank.
2. To the standard, control, and unknown tubes, add 3.6 mL of ferric nitrate solution.
3. To the unknown, standard, and control blank tubes, add 3.6 mL of dilute nitric acid.
4. Measure 0.3 mL of standard into standard blank tubes. Mix.
5. Measure 0.3 mL of control into control and control blank tubes. Mix.
6. Measure 0.3 mL of unknown into unknown and unknown blank tubes. Mix.
7. Read the standard, control, and unknown against their respective blanks in the spectrophotometer at a wavelength of 540 nm.

Calculation

$$\frac{\text{Absorbance}_{\text{unknown (or control)}}}{\text{Absorbance}_{\text{standard}}} \times 100\ \mu g/mL = \text{Salicylate } (\mu g/mL)$$

Notes

1. Because of the presence of iron in the unknown, a low absorbance reading corresponding to approximately 20 μg/mL will be obtained in the absence of salicylate. A value of 20 μg/mL or less should not be considered significant.[11]
2. The colorimetric reaction is linear from 0 to 1000 μg/mL.
3. This procedure can be used for a urine specimen after a heating step to remove volatile interferences (especially ketones). Place a test tube containing 5 mL of centrifuged urine in a 100° C heating block or boiling water for 15 minutes. After cooling, use as a specimen in the ferric nitrate procedure. Phenothiazine, homogentisic acid, and thiocyanates will interfere and cause false-positive results. Salicylate cannot be confirmed in urine in the presence of these compounds.

REFERENCE RANGE

A serum concentration of total salicylates greater than 200 μg/mL is considered above the accepted therapeutic range, except for arthritic patients, for whom the therapeutic range is extended to 300 μg/mL.

REFERENCES

1. Brodie, BB, Udenfriend, S, and Coburn, AF: The determination of salicylic acid in plasma, J Pharmacol 80:114-117, 1944.
2. Keller, WJ: A rapid method for the determination of salicylates in serum or plasma, Am J Clin Pathol 17:415-417, 1947.
3. Trinder, P: Rapid determination of salicylate in biological fluid, Biochem J 57:301-303, 1954.
4. Folin, O, and Ciocalteu, V: On tyrosine and tryptophan determinations in proteins, J Biol Chem 73:627-650, 1927.
5. Weichselbaum, TE, and Shapiro, I: A rapid and simple method for the determination of salicylic acid in small amounts in blood plasma, Am J Clin Pathol 9:42-44, 1945.
6. Routh, JI, Shane, NA, Arredondo, EG, and Paul, WD: Method for the determination of acetylsalicylic acid in the blood, Clin Chem 13:734-743, 1967.
7. Saltzman, A: Fluorophotometric method for the estimation of salicylate in blood, J Biol Chem 174:399-404, 1948.
8. Thomas, BJ, Solomonraj, G, and Coldwell, BB: The estimation of acetylsalicylic acid and salicylate in biological fluids by gas-liquid chromatography, J Pharm Pharmacol 25:201-204, 1973.
9. Rance, ML, Jordan, BI, and Nichols, JD: A simultaneous determination of acetylsalicylic acid, salicylic cid, and salicylamide in plasma by gas-liquid chromatography, J Pharm Pharmacol 27:425-429, 1975.
10. You, K, and Bittikofer, JA: Quantification of salicylate in serum by use of salicylate hydroxylase, Clin Chem 30:1549-1551, 1984.
11. Natelson, S: Techniques of clinical chemistry, ed 3, Springfield, IL, 1971, Charles C Thomas, Publisher.

Theophylline and caffeine

AMADEO J. PESCE
GORDON L. BILLS

Clinical significance: p. 795

	Theophylline	Caffeine
Molecular formula:	$C_7H_8N_4O_2$	$C_8H_{10}N_4O_2$
Molecular formula:	180.17 daltons	194.19 daltons
Merck Index:	9114	1606

Chemical class: purine derivatives

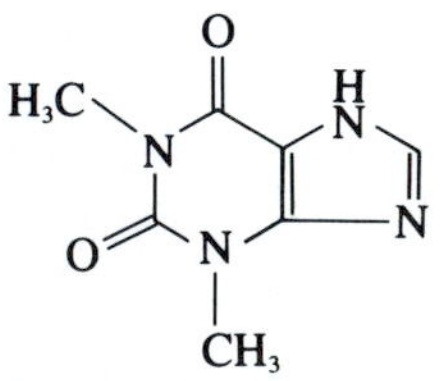

Theophylline 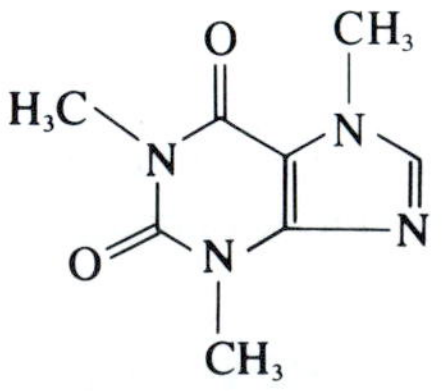 **Caffeine**

PRINCIPLES OF ANALYSIS

The first method used for routine determinations of serum theophylline was the ultraviolet spectrophotometric procedure described by Schack and Waxler in 1949 (method 1, Table 64-26).[1] It involved an initial extraction of serum with a mixture of chloroform and isopropanol,

Table 64-26 Methods for theophylline analysis

Method	Type of analysis	Principle	Usage	Comments
1. Ultraviolet	Spectrophotometric	Extraction of theophylline, absorption measurement at two wavelengths	Serum, plasma	Historical, nonspecific
2. Gas chromatography	Chromatographic, flame ionization detection	Extraction of theophylline and derivatization, chromatography in gas-liquid phase	Serum, plasma	Derivatization cumbersome
3. High-performance liquid chromatography (HPLC)	Chromatographic, spectrophotometric detection	Extraction of theophylline and analysis by reversed-phase chromatography	Serum, plasma	Some interferences; specialized equipment
4. Enzyme-multiplied immunoassay technique (EMIT)	Competitive binding, enzyme label	Antibody binding to enzyme-hapten inhibits catalytic reactivity, competition by analyte in sample prevents inhibition	Serum, plasma	Some interferences caused by sample matrix; proposed selected method
5. Radioimmunoassay (RIA)	Competitive binding, radioactive label	Radiolabeled hapten competes with unknown for antibody-binding sites	Serum, plasma	All problems of radioactive usage
6. Substrate-labeled fluorescence immunoassay (SLFIA)	Competitive binding, substrate label	Unbound drug–fluorogenic substrate converted by enzyme to fluorescent product	Serum	Fluorescence interferents
7. Fluorescence polarization immunoassay (FPIA)	Competitive binding with fluorescence polarization	Bound labeled drug has higher polarized fluorescence than free; amount of polarized fluorescence is inversely proportional to amount of unlabeled drug	Serum, plasma	Fluorescence interferents; rapid test
8. Ames test strip	Competitive binding with enzymatic procedure	Antibody binding to drug labeled with prosthetic group prevents glucose oxidase activity; competition with endogenous drug allows prosthetic group to activate enzyme; coupled to colorimetric reaction	Serum	Designed for inexpensive instrument; reflectance spectrophotometer; rapid test
9. Syntex test strip	Immunochromatography, spatial distribution	Drug and drug-labeled enzyme migrate through strip impregnated with antibody; length of migration proportional to concentration of theophylline; coupled reaction forms visible product	Whole blood	No instrument necessary; whole blood can be used

followed by reextraction into dilute sodium hydroxide. The absorbance of the final aqueous extract was measured at 277 and 310 nm, and concentrations were obtained from a standard curve. This procedure was modified by Jatlow in 1975 to eliminate barbiturate interference,[2] a significant problem in the original procedure.

Several procedures that used gas-liquid chromatography (GLC) to measure serum theophylline have been described (method 2, Table 64-26). Although there are many differences in the specific details of these procedures, most are bascially similar in principle. Most available GLC procedures utilize 3-isobutyl-1-methylxanthine as an internal standard and determine unknown theophylline concentrations from a standard curve. Sample preparation always involves at least one extraction of the serum with an organic solvent, usually chloroform or a mixture of isopropanol and chloroform, and the theophylline is usually derivatized by alkylation. Use of a 3% OV-17 column is fairly standard, although various carrier gases and detection systems have been utilized.[3-6] Initial serum volumes of at least 1 mL are usually required with most GLC procedures; however, use of an organic nitrogen detector has been shown to reduce the volume of serum needed to perform the assay to only 50 μL,[5] and a procedure using a flame ionization detector and involving on-column alkylation has been described as requiring only 100 μL of serum.[7] The GLC methods have not been widely used for routine analysis.

In contrast, high-performance liquid chromatography (HPLC) is a widely accepted method for analysis of serum theophylline (method 3, Table 64-26). A variety of procedures are available, essentially all of which utilize reversed-phase chromatography with a C_{18} column and an acetonitrile mobile phase.[8-13] Differences between procedures are found primarily in the initial steps of sample preparation, which are basically designed to remove serum proteins from the sample. This may be accomplished by either filtration or extraction with acetonitrile. Failure to remove serum proteins before sample injection results in protein binding to the column, leading to the development of significant back pressure after only a short period of use.[8] Most HPLC procedures use either 8-chlorotheophylline or 8-hydroxyethyltheophylline as an internal standard and calculate theophylline concentration by comparison of the peak-height ratio of the specimen to that of a plasma standard.[9,11] HPLC, however, was only used by about 10% of the respondents in the 1986 TDM survey. The HPLC procedures are also suitable for the measurement of caffeine. The same buffers are used, and caffeine elutes later than theophylline.

The homogeneous enzyme immunoassay (EIA) marketed by Syva Corp. (Palo Alto, CA) as the EMIT theophylline assay is the secona most widely employed method of theophylline determination (method 4, Table 64-26). It is a competitive binding immunoassay in which theophylline in the serum sample competes with glucose-6-phosphate dehydrogenase (G6PD)–labeled theophylline to bind with a theophylline-specific antibody. Enzyme activity is measured spectrophotometrically at 340 nm as the rate of conversion of nicotinamide adenine dinucleotide to reduced NAD.[14] The Syva assays were at one time used by 80% or more of the laboratories participating in TDM surveys. The high proportion of reporting laboratories using this method was decreased as other assay systems have been developed. This method is also available for the measurement of caffeine.

Radioimmunoassays (RIAs) have also been reported for theophylline analysis (method 5, Table 64-26). These have used the typical competitive binding procedure with radiolabeled hapten and antibody.[15]

The fluorescence polarization immunoassay (FPIA) developed by Abbott Laboratories (North Chicago) has recently become the most popular (method 7, Table 64-26). The FPIA method is a competitive binding assay in which fluorescein-labeled theophylline competes with endogenous theophylline from a patient sample for a limited amount of antibody[16-18] (see Chapter 11).

The Ames Seralyzer has a method for theophylline that employs a labeled prosthetic group assay[19] using the prosthetic group FAD and the enzyme glucose oxidase (see Chapter 11). This method has been incorporated into the test strip technology employed on the Ames Seralyzer. In this assay, the patient serum or plasma sample is first diluted with distilled water, and a specific amount of the prepared specimen is placed on the reagent area of the strip. The strip is then placed in the instrument, which is a solid-phase reflectance photometer. The color of the strip is read at 740 nm, and the results are calculated from a calibration curve. In all the above procedures, instruments are used to quantify the amount of drug present in the patient sample. Syntex Corp. (Palo Alto, CA) has developed a theophylline assay, termed *enzyme immunochromatography,* that uses a dry chemistry test strip technology.[20] The quantitation is based on the spatial distribution of the enzyme label rather than on quantitation of enzyme activity as in the EMIT or Ames assays (method 9, Table 64-26). The components of the assay system are an enzyme labeled with the drug to be measured (in this case, peroxidase-theophylline), a second enzyme with substrate for the coupled reaction (glucose oxidase), a substrate that turns color and is immobilized on a test strip surface when converted (4-chloro-1-naphthol), and a test strip with capillary properties that has immobilized antibody homogeneously dispersed throughout.

The colorimetric reaction is as follows:

$$\text{Glucose} + O_2 \xrightarrow{\text{Glucose oxidase}} \text{Gluconolactone} + H_2O_2$$

$$H_2O_2 + \text{4-chloro-1-naphthol} \xrightarrow{\text{Peroxidase}} \text{Insoluble blue-gray product} + H_2O$$

In the marketed reagent strip, a capillary tube of patient blood from a finger stick is mixed with the coupled enzymes (horseradish peroxidase–theophylline and glucose oxidase), and the solution is allowed to migrate by capillary action. This migration is complete in 10 minutes.

During the migration step, a specific amount of liquid from the patient sample–enzyme mixture is absorbed by the paper. This is essentially an aliquot of the patient sample. The antibody is eveny dispersed throughout the capillary strip, and the drug is bound by the antibody as it migrates through the matrix. Since one may assume that there is little or no difference in the retention of the drug and the enzyme-labeled drug by the immobilized antibody and since the amount of antibody in the matrix is limited, the more drug that is present in the sample, the greater the distance through the matrix the drug and the enzyme-labeled drug will travel. Essentially, it can be imagined that the drug migrates up the strip until it is all bound to antibody. The assay strip is then immersed in a developer solution of glucose and the chromogenic substrate. The blue-gray reaction product forms a colored product with a sharp front. The height of the migration front is proportional to the concentration of drug in the patient sample. The amount of drug present is determined from a calibration curve. This method is currently used in physician office testing.

Only recently has it been important to monitor the level of caffeine. Because this is usually requested on neonatal samples, micromethods are necessary. The most commonly used current techniques are EMIT and HPLC with ultraviolet absorption.

Currently there is no established or accepted reference method for the determination of serum theophylline concentration. However, an EMIT method[21] and an HPLC method[13] have both been suggested as the proposed selected method.

Ultraviolet spectrophotometry has been largely replaced in recent years by HPLC, EIA, FPIA, and other techniques. These methods have better specificity, lower sample volume requirements, and higher throughput than ultraviolet spectrophotometry.[2,9,13]

The choice of a preferred method will, in a large part, depend on the purpose of the analysis, equipment available, training of the staff, and cost of the assay. The EMIT

Table 64-27 Comparison of reaction conditions for theophylline assays

Condition	EMIT*	FPIA†	HPLC‡
Temperature	30° C	35° C	Ambient
pH	8.0 (55 mmol of Tris)	7.0 (100 mmol of phosphate)	—
Sample volume	50 μL	20 μL	100 μL
Fraction of sample volume	0.01	0.0976	0.5 (extraction)
Precision ($\bar{X}$, %CV)	13.7 μg/mL, 7%	13.7 μg/mL, 4.0%	13.7 μg/mL, 7%
Linearity	2.5-40 μg/mL	2.5-40 μg/mL	2.5-25 μg/mL
Interferences	Caffeine Renal failure§	Caffeine, 9.7% Other xanthine compounds (2%-10%) Renal failure§	Aminoglycosides Cefoxitin Corticosteroids

*Enzyme-multiplied immunoassay technique, Syva Co, Inc, Palo Alto, CA.
†Fluorescence polarization immunoassay, Abbott Diagnostics, North Chicago.
‡Assay described in text.
§Data from Elin, RJ, Ruddel, M, Korn, WR, et al: Clin Chem 29:1275, 1983.

procedure is adaptable to a wide variety of commonly available instruments, such as the ABA, COBAS, and du Pont aca, and is thus practical for many laboratories. The FPIA procedure has a smaller coefficient of variation than any of the other procedures (approximately 4% versus 8% to 10% for the others) and is probably better for pharmacokinetic computation of patient dosages because of its greater precision. HPLC is considerably more economical to perform than any of the other common used procedures,[22] since it uses reagents that not only are considerably less expensive, but also are available from a number of suppliers. These reagents also have a long shelf life. However, HPLC is more susceptible to interference by drugs that may be coadministered to the patient, particularly in an acute care hospital setting.

The Ames test strip and the Syntex immunochromatography system are specifically designed for physician office, small laboratory, and emergency unit use. They are adequate for that purpose. The Syntex system also uses capillary samples of whole blood and is less subject to icteric, hyperlipidemic, or hemolyzed samples. Several of the more common assays are compared in Table 64-27.

The EMIT immunochemical method is currently available for the measurement of caffeine. Both EMIT and the HPLC procedures provide adequate assay results.

SPECIMEN

Analyses using enzyme immunoassay (EIA), FPIA, test strips, or HPLC may be performed on either plasma or serum. Heparin should be used for anticoagulation in procedures that call for plasma. Otherwise, no special handling of specimens is required. HPLC can also be used for analysis of saliva. The saliva, collected while the patient chews on paraffin wax, is centrifuged to remove sediment. The Syntex immunochromatographic procedure uses whole blood. Whole blood has 82% of the theophylline levels found in serum or plasma.

PROCEDURE: HPLC ANALYSIS OF THEOPHYLLINE AND CAFFEINE

Principle

After deproteinization with acetonitrile, reversed-phase HPLC of serum is employed to separate the theophylline or caffeine from other compounds. Detection is by ultraviolet spectroscopy at 254 nm.

Reagents

Acetic acid reagent (875 mmol/L). Dilute 5 mL of glacial acetic acid to 1 L with glass-distilled water.

Acetonitrile, ultraviolet grade

Stock internal standard (100 μg/mL, 480 μmol/L). Place 10 mg to β-8-hydroxyethyltheophylline in a 100 mL volumetric flask, and dilute to 100 mL with methanol. This is stable for 6 months at 4° C.

Working internal standard (10 μg/mL, 48 μmol/L). Dilute 10 mL of stock internal standard to 100 mL with acetonitrile. This is stable for 3 months at 4° C.

Theophylline serum standard (10 μg/mL, 55.5 μmol/L). Dissolve 10 mg of theophylline in 10 mL of ethanol. This stock solution (1 mg/mL) is stable for 3 months at 4° C. To prepare working serum standard, dilute 100 μL of the stock solution with 9.9 mL of pooled human serum. This solution is stable for 1 week at 4° C and for 6 months at −20° C.

Caffeine serum standard (10 μg/mL, 51.5 μmol/L). Prepare in the same manner as the theophylline stock and working standards.

Mobile phase (acetonitrile–acetic acid, 10:90 by volume). Filter 450 mL of acetic acid reagent using a nitrocellulose filter (such as Millipore) of 0.47 μm pore size. Into the acetic acid, filter 50 mL of acetonitrile using an organic filter of the same dimension. Degas solution for 15 minutes at room temperature by applying a vacuum.

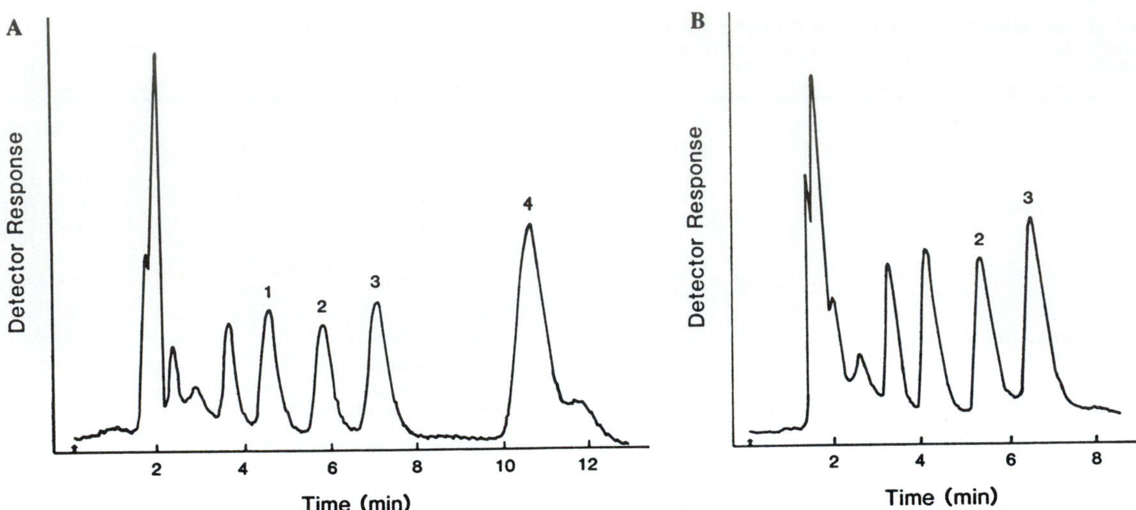

Fig. 64-11 HPLC chromatograms. **A,** Separation of standard solution; *1,* theobromine; *2,* theophylline; *3,* 8-hydroxyethyltheophylline (internal standard); *4,* caffeine. **B,** Separation of patient sample: *2,* theophylline; *3,* internal standard.

Assay

Equipment: HPLC equipment (recorder, pumps, injector), including an ultraviolet spectrophotometer capable of reading at 254 nm. Column is a μBondapak C_{18} column from Waters Associates (Milford, MA). Centrifuge capable of 2000 *g*.

1. Sample preparation
   a. Pipette 0.10 mL of working internal standard solution into a 12 × 75 mm glass tube.
   b. Add 0.10 mL of serum.
   c. Vortex mix, and centrifuge at 1000 *g* for 2 minutes at ambient temperature. Save supernatant as sample for HPLC analysis.
   d. Prepare calibrator by mixing 0.10 mL of internal-standard working solution with 0.10 mL of theophylline or caffeine serum standard (10 μg/mL).
2. Chromatographic parameters
   a. Chart speed, 1.0 cm/min
   b. Ultraviolet sensitivity, 0.01 AUFS (absorbance units at full scale) at 254 nm
   c. Flow rate, 2.0 mL/min at ambient temperature
3. Inject 10 μL of sample or standard. Sample run time is approximately 9 minutes.
4. Fig. 64-11, *A,* shows the chromatogram of the standards, theobromine, theophylline, internal standard, and caffeine. Fig. 64-11, *B,* shows a typical chromatogram of a patient treated with theophylline.

Calculation

1. Determine the peak height (in millimeters) for the theophylline (or caffeine), T, and internal standard, S, peaks on each chromatogram.
2. Calculate peak-height ratios, T/S.
3. Theophylline (or caffeine) concentrations in the serum samples can be calculated by direct proportions as follows:

$$\text{Theophylline } (\mu\text{g/mL}) = \frac{\text{T/S (sample)}}{\text{T/S (calibrator)}} \times 10\ \mu\text{g/mL}$$

Alternatively, standard curves can be used.

INTERFERENCES

The chromatographic techniques for theophylline are subject to interferences primarily from antibiotics.[23,24] The immunoassay procedures for theophylline occasionally use antibodies with cross-reactivity to caffeine.

REFERENCE RANGE

The therapeutic range for steady-state levels of serum theophylline in children and adults treated prophylactically for asthma has been established as 10 to 20 μg/mL (55.5 to 111 μmol/L).[25] The range of neonates treated for apnea is 5 to 20 μg/mL.[26] Concentrations greater than 20 μg/mL are associated with signs of toxicity in most patients, regardless of age.[27,28] The range of caffeine in neonates with apnea is 8 to 20 μg/mL.[29]

REFERENCES

1. Schack, JA, and Waxler, SH: An ultraviolet spectrophotometric method for the determination of theophylline and theobromine in blood and tissues, J Pharmacol Exp Ther 97:283, 1949.
2. Jatlow, P: Ultraviolet spectrophotometry of theophylline in plasma in the presence of barbiturates, Clin Chem 21:1518, 1975.
3. Bailey, DG, Dais, HL, and Johnson, GE: Improved theophylline serum analysis by an appropriate internal standard for gas chromatography, J Chromatogr 121:263, 1976.
4. Reid, R, Fareed, J, Bermes, EW, et al: A rapid gas chromatographic (GLC) procedure for the determination of theophylline in biological fluids using a simple double extraction method, Clin Chem 22:1166, 1976.

5. Lowry, JD, Williamson, LJ, and Raisys, VA: Micromethod for the gas chromatographic determination of serum theophylline utilizing an organic nitrogen sensitive detector, J Chromatogr 143:83, 1977.
6. Dechtiaruk, W, Johnson, GF, and Solomon, HM: Faster gas chromatographic procedure for theophylline, Clin Chem 21:1038, 1975.
7. Perrier, D, and Lear, E: Gas-chromatographic quantitation of theophylline in small volumes of plasma, Clin Chem 22:898, 1976.
8. Adams, RF, Vandemark, FL, and Schmidt, GJ: More sensitive high pressure liquid chromatography of theophylline in serum, Clin Chem 22:1903, 1976.
9. Orcutt, JJ, Kozak, PP, Sherwin, AG, and Cummins, LH: Microscale method for theophylline in body fluids by reversed phase, high pressure liquid chromatography, Clin Chem 23:599, 1977.
10. Franconi, LC, Hawk, GL, Sandmann, BJ, and Haney, WB: Determination of theophylline in plasma ultrafiltrate by reversed phase high pressure liquid chromatography, Anal Chem 48:372, 1976.
11. Weidner, N, Dietzler, DN, Ladenson, JH, et al: A clinically applicable high pressure liquid chromatograhic method for measurement of serum theophylline, with detailed evaluation of interferences, Am J Clin Pathol 73:79, 1980.
12. Soldin, SJ, and Hill, JG: A rapid micromethod for measuring theophylline and reverse phase high pressure liquid chromatography, Clin Biochem 10:74, 1977.
13. Broussard, LA, Stearns, FM, Tulley, R, and Frings, CS: Theophylline determination by "high pressure" liquid chromatography. In Cooper, GR, editor: Selected methods of clinical chemistry, vol 10, Washington, DC, 1983, American Association for Clinical Chemistry.
14. Chang, J, Gotcher, S, and Gashaw, JB: Homogeneous enzyme immunoassay for theophylline in serum and plasma, Clin Chem 28:361, 1982.
15. Cook, CE, Twine, ME, Meyers, M, et al.: Theophylline radioimmunoassay: synthesis of antigen with characterization of antiserum, Res Comm Chem Pathol Pharmacol 13:497-501, 1976.
16. Li, TM, Benovic, LJ, Buckler, RT, and Burd, JF: Homogeneous substrate-labeled fluorescent immunoassay for theophylline in serum, Clin Chem 27:22, 1981.
17. Popelka, SR, Miller, DM, Holen, JT, and Kelso, DM: Fluorescence polarization immunoassay. II. Analyzer for rapid precise measurement of fluorescence polarization with use of disposable cuvettes, Clin Chem 27:1198-1201, 1981.
18. Loomis, KF, and Frye, RM: Evaluation of the Abbott TDM for the stat. measurement of phenobarbital, phenytoin, carbamazepine, and theophylline, Am J Clin Pathol 80:686-691, 1983.
19. Tybach, R: Adaptation of prosthetic-group label homogeneous immunoassay to reagent strip format, Clin Chem 27:1499-1504, 1981.
20. Zuk, RF, Ginsberg, VK, Houts, T, et al: Enzyme immunochromatography: a quantitative immunoassay requiring no instrumentation, Clin Chem 31:1144-1150, 1985.
21. Chang, J, Gotcher, S, Gushaw, JB, et al: Homogeneous enzyme immunoassay for theophylline in serum and plasma. In Cooper, GR, editor: Selected methods of clinical chemistry, vol 10, Washington, DC, 1983, American Association for Clinical Chemistry.
22. Kampa, IS, Dunikoski, LK, Jarzabek, JJ, and Grubesich, D: Comparison of three assay procedures for theophylline determination, Ther Drug Monitor 1:249, 1979.
23. Weidner, N, McDonald, JM, Tieber, VJ, et al: Assay of theophylline: comparison of EMIT on the ABA-100 to HPLC, GLC and UV procedures with detailed evaluation of interferences, Clin Chim Acta 97:9, 1979.
24. Kelly, RC, Prentice, DE, and Hearne, GM: Cephalosporin antibiotics interfere with the analysis for theophylline by high pressure liquid chromatography, Clin Chem 24:838, 1978.
25. Jenne, JW, Wyze, E, Rood, FS, and McDonald, FM: Pharmacokinetics of theophylline: application to adjustment of the clinical dose of aminophylline, Clin Pharmacol Ther 13:349, 1972.
26. Shannon, DC, Gotay, F, Stein, IM, et al: Prevention of apnea and bradycardia in low birth weight infants, Pediatrics 55:589, 1975.
27. Zwillich, CW, Sutton, F, Neff, TA, et al: Theophylline induced seizures in adults: correlation with serum concentration, Ann Intern Med 82:784, 1975.
28. Neese, CL, and Soyka, LF: Development of a radioimmunassay for theophylline, Clin Pharmacol Ther 21:633, 1977.
29. Aranda, JV, et al: Effect of caffeine on control of breathing in infantile apnea, J Pediatr 103:975-978, 1983.

CHAPTER 65 Urine analysis

Melanins

ARNOLD L. SCHULTZ

Clinical significance: p. 730
Merck Index: 5629
Chemical class: dihydroxyphenylalanine derivatives

PRINCIPLES OF ANALYSIS

A number of simple tests[1,2] have been devised to aid in the diagnosis of melanuria and in the differentiation of melanuria from alkaptonuria and indicanuria. Alkaptonuria is associated with an abnormally high concentration of homogentisic acid. Indican is indoxyl sulfate, a product of tryptophan that accumulates under conditions of intestinal putrefaction.

In one approach to melanin analysis, melanogens are oxidized to melanins by agents such as ferric chloride, nitric acid, bromine water, and potassium chlorate in hydrochloric acid (methods 1 to 5, Table 65-1). A second approach takes advantage of the reducing properties of the melanogens. In the Thormählen reaction (method 6, Table 65-1), sodium nitroprusside is reduced to ferric ferrocyanide (Prussian blue). In another method, ammoniacal silver nitrate is reduced to colloidal silver by urine that contains melanogens (method 7, Table 65-1). Reactions of melanogens with *p*-dimethylaminobenzaldehyde, sodium nitrate in hydrochloric acid, and diazonium salts have also been reported.

Urine that contains melanogens or homogentisic acid darkens when exposed to air and sunlight. If the urine is tested with ammoniacal silver nitrate, the presence of homogentisic acid will be indicated if the solution darkens rapidly. If a brown color forms slowly, the presence of melanogens is indicated. If indican is present, the urine will also have a brown appearance and thus may be mistaken for melanotic urine. However, one can use the Obermayer test for indican to readily differentiate melanins from indican[3] (method 8, Table 65-1).

Methods that employ the oxidation of melanogens to melanins by nitric acid, bromine water, or potassium chlorate in hydrochloric acid do not have sufficient sensitivity to detect small concentrations of melanogens. The method that employs the reduction of ammoniacal silver

Table 65-1 Methods of urinary melanin analysis

Method	Type of analysis	Principle	Comments
1. Ferric chloride in water	Qualitative	Oxidation of melanogens to melanins	Lacks sensitivity
2. Ferric chloride in hydrochloric acid	Qualitative	Oxidation of melanogens to melanins	Best sensitivity
3. Nitric acid	Qualitative	Oxidation of melanogens to melanins	Lacks sensitivity
4. Bromine water	Qualitative	Oxidation of melanogens to melanins	Lacks sensitivity
5. Potassium chlorate in hydrochloric acid	Qualitative	Oxidation of melanogens to melanins	Lacks sensitivity
6. Thormählen reaction	Qualitative	Reduction of sodium nitroprusside to ferric ferrocyanide (Prussian blue)	Best sensitivity
7. Ammoniacal silver nitrate	Qualitative	Reduction of silver	Homogentisic acid reacts rapidly; melanogens react slowly Differentiates homogentisic acid from melanins
8. Obermayer test	Qualitative	Conversion of indican to indigo blue	Differentiates indican from melanins Indican reacts; melanins do not

nitrate by urine that contains melanogens also lacks sensitivity. The most sensitive tests are the ferric chloride and the Thormählen tests. The ferric chloride test may be performed either in water or 10% hydrochloric acid. When the procedure is done in water, the melanins are entrapped on the precipitate of ferric phosphate that is formed, resulting in a gray or black precipitate. If hydrochloric acid is used as the solvent, the precipitate of ferric phosphate does not form, and only the color change that results from the oxidation of the melanogens is observed. It has been claimed that the use of hydrochloric acid in the ferric chloride test increases the sensitivity of the reaction.

SPECIMEN

Freshly voided urine must be used. The tests for melanuria are actually reactions involving the presence of melanogens. If the urine is allowed to stand, the melanogens will be converted to melanins, less chromogen will be present, and false-negative results may be obtained.

PROCEDURE: SCREENING TESTS

Principle

Melanogens are oxidized to dark brown or black melanins in the presence of ferric chloride in hydrochloric acid.

Prussian blue (ferric ferrocyanide) is formed in the reduction of sodium nitroprusside by urine containing melanogens.

Colloidal silver is *rapidly* formed by the reduction of silver nitrate by homogentisic acid. Melanogens will also reduce silver nitrate; however, this reaction requires the presence of ammonium hydroxide and proceeds at a *slower* rate.

The reaction of indican with ferric chloride in the Obermayer test produces indigo blue. The blue pigment is extracted into chloroform.

A summary of the reactions of melanogens, homogentisic acid, and indican are given in Table 65-2.

Reagents (see Table 65-3)

Ferric chloride test

1.2 M hydrochloric acid (1.2 mol/L). To approximately 50 mL of deionized water in a 100 mL volumetric flask, add 10 mL of concentrated hydrochloric acid. Mix, and dilute to volume with deionized water.

Working ferric chloride reagent (100 g/L, 616 mmol/L). Dissolve 10 g of ferric chloride hexahydrate in 1.2 M hydrochloric acid, and dilute to 100 mL with 1.2 M hydrochloric acid. This is stable for 1 year at room temperature.

Thormählen test

Working sodium nitroferricyanide reagent (50 g/L, 191 mmol/L). Dissolve 0.5 g of sodium nitroferricyanide in 10 mL of deionized water. Prepare fresh.

10 M sodium hydroxide (10 mol/L). Cautiously dissolve 40 g of sodium hydroxide in deionized water, and dilute to 100 mL with deionized water. This is stable in a plastic bottle for 1 year at room temperature.

33% acetic acid (5.61 mol/L). To about 40 mL of deionized water in a 100 mL volumetric flask, add 33 mL of glacial acetic acid. Mix, and dilute to volume with deionized water. This is stable for 1 year at room temperature.

Ammoniacal silver nitrate test

3% silver nitrate (30 g/L, 176.6 mmol/L). Dissolve 0.3 g of silver nitrate in 10 mL of deionized water. This is stable for 6 months at room temperature.

2% ammonium hydroxide (0.3 mol/L). Dilute 0.2 mL of concentrated ammonium hydroxide to 10 mL with deionized water. This is stable for 6 months at room temperature.

Obermayer test

Working ferric chloride reagent (4 g/L, 24.6 mmol/L). Dissolve 0.1 g of ferric chloride hexahydrate in 25 mL of concentrated hydrochloric acid. This is stable for 6 months at room temperature.

Assay

Ferric chloride test

1. To 5.0 mL of urine, add 1 mL of 10% ferric chloride in 1.2 M hydrochloric acid.
2. If the urine color rapidly changes to brown, dark brown, or black, melanogens are present.

Table 65-2 Summary of test reactions with melanogen, homogentisic acid, and indican

	Oxidizing reagent $FeCl_3$	Thormählen test nitroferricyanide	Ammoniacal silver nitrate	Obermayer test	Air oxidation
Melanogen	Rapid formation of dark brown or black color	Green-blue to blue-black color	Slowly darkening solution	No color in chloroform layer	Dark brown or black melanin
Homogentisic acid	Transparent blue color	—	Rapidly darkening solution	No color in chloroform layer	Dark-colored urine
Indican	—	—	—	Colored chloroform layer	—

Table 65-3 Reaction conditions for melanin analysis

Conditions	Ferric chloride	Thormählen reaction	Ammoniacal silver nitrate	Obermayer reaction
Temperature	Ambient	Ambient	Ambient	Ambient
Final concentration of reagent components	Ferric chloride: 1.7% Hydrochloric acid: 0.2 M	Sodium nitroferricyanide: 0.4% Sodium hydroxide: 0.2 M Acetic acid: 2.8%	Silver nitrate: 2.5% Ammonium hydroxide: 0.2%	Ferric chloride: 0.2%
Fraction of sample volume	0.83	0.77	0.83	0.5
Reaction color	Brown, dark brown, black	Greenish blue to bluish black	Darkens slowly for melanins, rapidly for homogentisic acid	Clear for melanins, blue for indican

Thormählen test

1. Add 5 or 6 drops of freshly prepared 5% sodium nitroferricyanide solution to 5.0 mL of urine in a test tube.
2. Add 0.5 mL of 10 M sodium hydroxide, and mix vigorously.
3. Rapidly cool the tube in cold tap water, and using litmus paper, acidify with 33% acetic acid.
4. With normal urine an amber or pale brown color is obtained; melanogens give a color that may vary from greenish blue to bluish black, depending on the quantity present. In a positive test, the color also depends on the color of the urine. The deeper the yellow of the urine, the greener the final color. The less pigmented the urine, the bluer the final color.

Ammoniacal silver nitrate test

1. To 0.5 mL of urine, add 5 mL of 3% silver nitrate solution, and mix. A precipitate will form.
2. Add 2% ammonium hydroxide until the silver chloride precipitate is almost dissolved.
3. The solution will darken as a result of the formation of both melanins and colloidal silver. The reaction develops slowly. In contrast, homogentisic acid darkens the silver solution rapidly, even before the addition of ammonia.

Obermayer test

1. Mix 6 mL of urine with 6 mL of 0.4% ferric chloride in concentrated hydrochloric acid in an 18 × 200 mm test tube.
2. Add 3 to 4 mL of chloroform.
3. Stopper the tube, and mix by inverting the tube vigorously 10 to 12 times.
4. If indican is present, the chloroform (lower layer) will be colored blue (indigo blue).

Note

For quality control a normal urine with about the same pigmentation as the sample to be tested and freshly prepared solutions (1 mg/10 mL of deionized water) of indoxyl sulfate and homogentisic acid (both available from Sigma Chemical Co., St. Louis) should be processed with each test.

REFERENCE RANGE

Normal urines produce a negative reaction for melanin.

REFERENCES

1. Harrison, GA: Chemical methods in clinical medicine: their applications and interpretation with techniques of simple tests, ed 4, New York, 1957, Grune & Stratton, Inc, pp 252-255.
2. Varley, H: Practical clinical biochemistry, ed 2, New York, 1958, Interscience Publishers, Inc, pp 563-564.
3. Kachmar, JF: Proteins and amino acids. In Tietz, NW, editor: Fundamentals of clinical chemistry, Philadelphia, 1970, WB Saunders Co, pp 257-259.
4. Beeler, MF, and Henry, JB: Melanogenuria: evaluation of several commonly used laboratory procedures, JAMA 176:52-54, 1961.
5. Scott, RE, Ward, VL, Grinstead, GF, et al: Melanogenuria: laboratory evaluation of the qualitative Thormählen and ferric chloride tests and their clinical utility, Clin Chem 34:582-585, 1988.
6. Cohn, RM, and Jezyk, P: Metabolic screening. In Hicks, JM, and Boeck, RL, editors: Pediatric clinical chemistry, Philadelphia, 1984, WB Saunders Co, pp 657-689.

Porphobilinogen

MICHAEL D.D. McNEELY

Clinical significance: pp. 359 and 496
Molecular formula: $C_{10}H_{14}N_2O_4$
Molecular weight: 226.23 daltons
Merck Index: 7469
Chemical class: pyrrole

$HOOCCH_2$ CH_2CH_2COOH
H_2NCH_2 N

PRINCIPLES OF ANALYSIS

Porphobilinogen is a monopyrrole precursor of heme that is found in high concentrations in the urine of persons suffering from acute intermittent porphyria.

PBG + **p-Dimethylamino-benzaldehyde** $\xrightarrow{\text{strong acid}}$ **Pyrrole condensation product** ***(red)***

The simplest method for porphobilinogen detection is the "windowsill" test, in which fresh urine is acidified and exposed to sunlight. Under these conditions colorless porphobilinogen polymerizes to form porphyrins and other pigments that darken the urine. This crude test is neither sensitive nor specific.

The standard test for clinical detection of porphobilinogen was devised by Watson and Schwartz[1] in 1941. The method, which is qualitative, is designed to indicate the presence or absence of porphobilinogen. The basis of the procedure is the reaction between porphobilinogen and Ehrlich's aldehyde reagent (p-dimethylaminobenzaldehyde in strong acid) to form a red condensation compound, as shown above. The color forms quickly and then fades slowly as the colored product reacts with a second molecule of porphobilinogen to form colorless dipyrrylphenylmethane.[2,3] The color production is not specific for porphobilinogen and is caused by urobilinogen, indole, and indican as well.

Sodium acetate is added to the reagent to increase the pH. This maximizes the color produced by urobilinogen and renders it preferentially soluble in chloroform. Thus, if color is produced during the first step of the reaction, a chloroform extraction is carried out. The colored products of urobilinogen and other compounds are then extracted into the chloroform layer. Color that remains in the supernatant aqueous phase is usually caused by porphobilinogen.

This test was evaluated by testing the urine from 1000 hospitalized patients. There were no false-positive results.[4] However, when patients with severe systemic disease were studied, positive reactions were noted in those suffering from Hodgkin's disease, tetanus, hepatic cirrhosis, pancreatic carcinoma, poliomyelitis, carcinomatosis, and intraabdominal hemorrhage. This study did not confirm the exact nature of the positive pigment.[5]

To increase specificity, the aqueous phase color remaining after chloroform extraction can be further extracted with butanol.[6] Any color *not* passing into the butanol phase is virtually certain to be porphobilinogen. In a study of 1000 hospitalized patients, 59 urines produced weakly positive reactions when the original Watson-Schwartz method was used. None of these tests remained positive after butanol extraction.[7]

A somewhat simpler method is the Hoesch test,[8,9] which uses Ehrlich's aldehyde reagent without sodium acetate. No extractions are required. Test results have been observed to become positive with indoles, indole-3-acetic acid, alpha-methyldopa, phenazopyridine hydrochloride, and end-stage alcoholic malnutrition.[10]

Several variations of the Watson-Schwartz test have been suggested. These include direct application of Ehrlich's aldehyde reagent to urine on a talc pad,[11] preliminary ion-exchange resin isolation before Ehrlich's aldehyde reaction,[12] and dual-wavelength spectrometry.[13]

A popular misconception is that strip tests for urinary urobilinogen that employ Ehrlich's aldehyde reagent will routinely change color in the presence of porphobilinogen. Thus it has been suggested that a negative strip test will rule out acute intermittent porphyria. This is inappropriate, since porphobilinogen will not reliably change the test strip's color.[14]

The method of Mauzerall and Granick[2] is suitable for the quantitative analysis of porphobilinogen after chromatographic isolation.

SPECIMEN

A freshly voided urine specimen collected during or immediately after an attack of acute abdominal pain is recommended. The urine should be cooled to room temperature before being tested to prevent the formation of a "warm aldehyde" reaction. The assay should be conducted within several hours to prevent the degeneration of porphobilinogen or the formation of an inhibitor to the reaction with Ehrlich's reagent.[15]

PROCEDURE: WATSON-SCHWARTZ ASSAY FOR PORPHOBILINOGEN

Principle

As described earlier, porphobilinogen in urine reacts with Ehrlich's aldehyde reagent to form a colored chromogen. Sodium acetate is added to adjust the pH, and nonporphobilinogen color is removed by extraction into chloroform. Further specificity is gained by extraction of nonspecific color into a butanol extract.

Reagents

Ehrlich's aldehyde reagent (18.8 mmol/L in 7.2 M HCl). Combine 700 mg of reagent-grad p-dimethylaminobenzaldehyde, 150 mL of concentrated HCl, and 100 mL of deionized water. The solution should be colorless or very light yellow. Brown or brownish red material should not be used. Store in amber bottle at room temperature. This is stable for at least 1 to 2 months. Discard if discolored.

Sodium acetate (saturated). Use about 200g/L. This is stable at room temperature for at least 1 year.

Chloroform
***n*-Butanol**

Control urine. Run a fresh, otherwise normal, urine. Commercially available urine controls usually contain no detectable porphobilinogen and can be used as negative controls. A positive control can be prepared by addition of porphobilinogen (Sigma Chemical Co., St. Louis, no. P1134) to a final concentration of 10 mg/L. The spiked pool should then be frozen at −20° C in 2 mL aliquots. It is stable for up to 6 months.

Assay

1. Label three tubes: test, negative control, and positive control.
2. To the appropriate tubes, add 1 mL control or test urine.
3. Measure 1 mL of Ehrlich's reagent into each tube and shake for 30 seconds.
4. Measure 2 mL of saturated sodium acetate into each tube and shake for 30 seconds.
5. Check the pH with an indicator paper. The pH should be 4 to 5. If necessary, further sodium acetate should be added. Observe the color of the solutions. If the solution is colorless, the test is negative.
6. If color is seen in step 5, measure 2 mL of chloroform into each tube and shake for 30 seconds.
7. Observe the tubes. If the aqueous supernatant is colorless or has only a faint trace of color, the test is deemed negative. If color is seen in the supernatant, the test is presumed positive, and step 8 must be undertaken.
8. Transfer the colored aqueous-phase supernatant to a separate tube. Add a half volume of *n*-butanol and shake for 30 seconds.
9. Observe the tube. If the aqueous (lower) phase retains color, the specimen has porphobilinogen.

Notes

1. Color identification may be difficult. Usually when acute intermittent porphyria is present, the test color is intense.
2. Particular caution must be exercised if any intense color develops. The color may not be fully extracted into the organic solvent and may lead to a false-positive test. Thus, if the color appears in both phases, a second extraction of the aqueous phase is warranted.
3. Color formation can be inhibited by thiols, urea, and possibly certain indolic compounds that can be present in urine.[16]

REFERENCE RANGE

The range is less than 2 mg/L of urine, which gives a negative result in the screening test.

REFERENCES

1. Watson, CJ, and Schwartz, S: Simple test for urinary porphobilinogen, Proc Soc Exp Biol Med 47:393-394, 1941.
2. Mauzerall, D, and Granick, S: Occurrence and determination of amino-levulinic acid and porphobilinogen in urine, J Biol Chem 219:435-446, 1956.
3. Rimington, C, Krol, S, and Tooth, B: Detection and determination of porphobilinogen in urine, Scand J Clin Lab Invest 8:251-262, 1956.
4. Hammond, RL, and Welcker, ML: Porphobilinogen tests on 1000 miscellaneous patients in search for false positive reactions, J Lab Clin Med 33:1254-1257, 1948.
5. Watson, CJ: Some studies of nature and clinical significance of porphobilinogen, AMA Arch Intern Med 93:643-657, 1954.
6. Schwartz, S, Keprios, M, and Schmid, R: Experimental porphyria: type produced by lead, phenylhydrazine and light, Proc Soc Exp Biol Med 79:463-468, 1952.
7. Townsend, JD: Evaluation of recent modification of Watson-Schwartz test for porphobilinogen, Ann Intern Med 60:306-307, 1964.
8. Hoesch, K: Über die Pantothensäurebehandlung der Porphyrie, Dtsch Med Wochenschr 72:252-254, 1947.
9. Lamon, J, With, TK, and Redeker, AG: The Hoesch test: bedside screening for urinary porphobilinogen in patients with suspected porphyria, Clin Chem 20:1438-1440, 1974.
10. Pierach, CA, Cardinal, R, Bossenmaier, I, and Watson, CJ: Comparison of the Hoesch and the Watson-Schwartz tests for urinary porphobilinogen, Clin Chem 23:1666-1668, 1977.
11. With, TK: Screening for acute porphyria, Lancet 2:1187-1188, 1970.
12. Doss, MO: Tests for porphyria, Lancet 2:983-984, 1971.
13. Moore, DJ, and Labbe, RF: A quantitative assay for urinary porphobilinogen, Clin Chem 10:1105-1111, 1964.
14. Kanis, JA: Detection of urinary porphobilinogen, Lancet 1:1511, 1973.
15. Watson, CJ, Bossenmaier, I, and Cardinal, R: Acute intermittent porphyria, JAMA 175:1087-1091, 1961.
16. Jacobs, SL: Porphyrins and their precursors. In Henry, RJ, Cannon, DC, and Winkleman, JW, editors: Clinical chemistry: principles and technics, ed 2, Hagerstown, MD, 1974, Harper & Row, Publishers, pp 1229-1233.

CHAPTER 66 Vitamins

Vitamin B_{12}

I-WEN CHEN
MATTHEW I. SPERLING
LINDA A. HEMINGER

B_{12}, cyanocobalamin, cobalamin
Clinical significance: p. 543
Molecular formula: $C_{63}H_{88}CoN_{14}O_{14}P$
Molecular weight: 1355.42 daltons
Merck Index: 9822
Chemical class: corrinoid (complexes pyrrole rings)

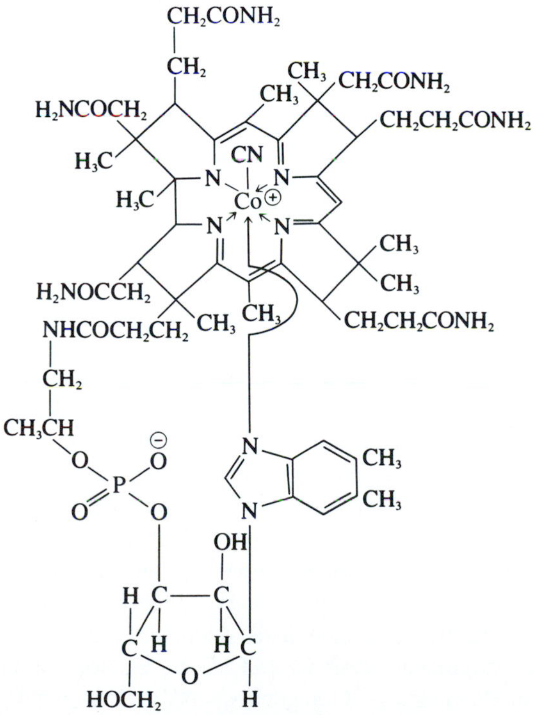

Cyanocobalamin (vitamin B_{12})

PRINCIPLES OF ANALYSIS

Vitamin B_{12} is also known as *cyanocobalamin* because cyanide can artificially occupy one of the coordination positions of the cobalt atom. A vitamin B_{12} molecule without the cyanide group is called *cobalamin*. Substitution of the cyanide group with a hydroxy group forms hydroxycobalamin; with a methyl group, methylcobalamin; and with a 5-deoxyadenosyl group, 5-deoxyadenosylcobalamin. All three cobalamins are present in plasma bound to plasma-binding proteins called transcobalamins, but methylcobalamin is the major form of vitamin B_{12} in the plasma.

Growth-dependent microbiological assays were established in the 1950s for the measurement of vitamin B_{12} concentrations in biological specimens (method 1, Table 66-1). Several microorganisms have been used in such assays, but *Euglena gracilis* and *Lactobacillus leichmannii* are the two most commonly used at present. Because of strict requirements regarding temperature and exposure to light that require special and involved apparatus for the growth of *E. gracilis, L. leichmannii* appears to be the current organism of choice.[1,2] The growth of the bacteria, as measured by the absorbance at 640 to 700 nm, is related to the concentration of vitamin B_{12} in the specimen.

The microbiological assay may be considered the reference method because it measures only the biologically active vitamin B_{12} derivatives. However, because of the specialized techniques and equipment required, it is used primarily as a research tool. The National Committee for Clinical Laboratory Standards (NCCLS) recommends that all commercial radioassay kits be verified by bioassays using either *E. gracilis* or *L. leichmannii* or by a radioassay that has been previously verified by one of the bioassays.

Radioligand assays for serum vitamin B_{12} (method 2, Table 66-1) were developed in the early 1960s.[3] Like other competitive inhibition radioligand assays, the measurement of vitamin B_{12} concentrations is based on the competition between endogenous serum vitamin B_{12} and radiolabeled vitamin B_{12} for the limited number of binding sites on specific vitamin B_{12}-binding proteins. Since vitamin B_{12} is bound to plasma proteins, serum is heated in a boiling water bath under acidic or alkaline conditions in the presence of cyanide before the radioassay to denature the endogenous B_{12} binder and convert vitamin B_{12} to the more stable cyanocobalamin derivative.

Commercially available crude hog intrinsic factor was used as the binder in B_{12} radioassays in the past but was largely replaced by purified intrinsic factor because the crude intrinsic factor preparation was found to contain

Table 66-1 Methods of vitamin B_{12} analysis

Method	Type of analysis	Principle	Usage	Comments
1. Microbiological	Turbidimetric	Microorganism growth depends on amount of vitamin B_{12} in serum sample.	Original procedure now reference	Tedious, time consuming; not suitable for clinical laboratories
2. Radioligand	Competitive binding	Competitive binding of radiolabeled (^{57}Co or ^{125}I) and nonlabeled vitamin B_{12} to limited binding sites on specific binders.	Most common	Sensitive and precise; suitable for clinical laboratories; often done with folate assay

more than 50% of rapidly migrating (R) proteins, which could also bind biologically inactive vitamin B_{12} analogs to give spuriously high serum vitamin B_{12} concentrations.[4,5] The presence of vitamin B_{12} analogs in serum has been demonstrated by Kolhouse et al.[4] The NCCLS carried out a thorough investigation of this problem and established the guidelines for the intrinsic factor preparations used as the binder in vitamin B_{12} radioassays. First, the intrinsic factor preparations should not measure cobinamide (a vitamin B_{12} analog) up to concentrations of 10 ng/mL in serum. Second, vitamin B_{12} binding to the binder should be inhibited more than 95% by specific antiintrinsic factor–blocking antibody, or the intrinsic factor–vitamin B_{12} complex should be more than 95% precipitated by specific intrinsic factor–precipitating antibody.[6] Affinity chromatography has been used in the purification of intrinsic factor to eliminate essentially all the R protein-binding activity.[7]

Since vitamin B_{12} contains a cobalt atom as an integral part of its molecule, ^{57}Co, a gamma-ray emitting nuclide, is used almost exclusively for labeling vitamin B_{12} for use in the radioassay. A ^{125}I-labeled vitamin B_{12} was recently developed and is claimed to offer greater sensitivity than the ^{57}Co-labeled vitamin B_{12}.[8] It would also allow B_{12} analysis in laboratories that have gamma-ray scintillation counters that are not capable of measuring ^{57}Co. Dextran-coated charcoal is a widely used agent for separating bound and free ligands in the liquid-phase radioassay of vitamin B_{12}. However, several commercial vitamin B_{12} radioassay kits using the solid-phase technique have become available and have gained popularity because of their technical simplicity and rapidity.

Radioassays are by far the most commonly used methods for the determination of serum vitamin B_{12} levels in routine clinical laboratories mainly because of their technical simplicity and the commercial availability of assay kits.

Most commercial radioassay kits for vitamin B_{12} are designed for the simultaneous measurement of serum folate levels, because assays for both vitamins are frequently ordered on the same serum sample. The reason for this is that the hematological manifestations of vitamin B_{12} and folate deficiencies are indistinguishable because of the close metabolic interrelationships of these two vitamins, and it is important to differentiate between the two deficiencies so that therapy will be appropriate. The simultaneous assay is possible because of the difference in the emission-energy spectrum of ^{57}Co-labeled vitamin B_{12} and ^{125}I-labeled folate used as tracers in the radioassay. This allows a certain portion of each radionuclide's energy to be detected almost without interference from the other. A scintillation detector equipped with at least two pulse-height analyzers is needed for simultaneous counting of two or more radionuclides. All modern gamma-ray scintillation counters are equipped with multichannel analyzers and are suitable for the simultaneous assay.

In most commercial kits, a purified intrinsic factor preparation essentially free of R protein is used as the specific binder to determine only the so-called true vitamin B_{12} (biologically active vitamin B_{12} with affinity to intrinsic factor). However, a crude intrinsic factor preparation contaminated with R protein has been used in at least one commercial kit (SimulTRAC Radioassay Kit, Becton Dickinson Immunodiagnostics, Orangeburg, NY). This kit provides the option of measuring either the "true" or the "total" vitamin B_{12} levels (vitamin B_{12} plus its analog) in serum. For the measurement of true vitamin B_{12} the assay is carried out in the presence of cobinamide to block binding of endogenous vitamin B_{12} to the R protein, whereas total vitamin B_{12} is measured in the absence of cobinamide. Clinical significances of true and total vitamin B_{12} levels in blood still remain controversial. It has been reported that measurement of total vitamin B_{12} with crude intrinsic factor may give an equally good or even better separation between healthy and vitamin B_{12}-deficient patients.[9,10]

Two automated systems (ARIA II, Becton Dickinson Immunodiagnostics, Orangeburg, NY; Concept 4, Micromedic Systems, Inc., Horsham, PA) are capable of assaying serum vitamin B_{12} (Table 66-2). In the ARIA II,[11] vitamin B_{12} and folate are assayed simultaneously employing purified porcine intrinsic factor and folate binder from bovine milk as the specific binders. The Concept 4 instrument measures vitamin B_{12} only and uses a special tube coated with an antibody to vitamin B_{12}. Both systems require off-line denaturation of endogenous serum binders by boiling, which requires manual pipetting of the serum

Table 66-2 Comparison of assay conditions for vitamin B_{12} radioassays

	Manual			**Automated**
Condition	**Dualcount (Boil) (Diagnostic Products)**	**Immo Phase (Corning Medical)**	**No-Boil Combo Stat II (RIA Products, Inc.)**	**ARIA II (Becton Dickinson)**
Binders	Purified hog IF*	Purified hog IF	Purified hog IF	Purified hog IF
Folate assay	Simultaneous or separate	Simultaneous	Simultaneous	Simultaneous
Sample pretreatment	Boil	Boil	Alkali denaturation	Boil
Separating agent	Charcoal suspension	Solid phase (glass beads)	Solid phase (cellulose particles)	Hydrophobic resin chamber
Sample volume	200 μL	100 μL	200 μL	100 μL
Incubation time	60 min	60 min	60 min	30 min
Incubation temperature	25° C	25° C	25° C	25° C
Incubation pH	9.4	9.2	9.3	9.3
Sensitivity	50 pg/mL[19]	17 pg/mL†	24 pg/mL†	100 pg/mL[19]
Precision (coefficient of variation)	3.8% to 8.5%[19]	3% to 12%†	2.5% to 10.6%†	4.6% to 8.2%[19]
Interferences	Ascorbic acid, F^-, radioactive drugs	Ascorbic acid, F^-, radioactive drugs	Ascorbic acid, F^-, radioactive drugs	Ascorbic acid, F^-, radioactive drugs

*Intrinsic factors.
†Reported by manufacturers.

sample and the denaturation buffer, mixing, boiling in a boiling water bath, and cooling. For radioassays of vitamin B_{12} to be carried out in a fully automated fashion, the denaturation of endogenous binders by processes other than boiling has to be developed. The use of an alkali treatment has been suggested as an alternative to the heat-denaturation procedure.[12] The alkali denaturation technique has already been used by commercial kit manufacturers in the so-called no-boil assay kits (Table 66-2).[13,14] The no-boil kits still require incubation and an additional pipetting step for the neutralization of the alkaline extractant before the competitive binding reaction and therefore offer no real advantages in saving technician time. Moreover, studies by Zucker et al.[15] indicate that the alkaline denaturation procedures specified by the kit manufacturers fail to denature cobalamin-binding proteins completely in the sera of patients with chronic myelogenous leukemia and fail to denature antiintrinsic factor–blocking antibodies completely in the sera of some patients with vitamin B_{12} deficiency.

SPECIMEN

Samples should be collected from fasting persons, since recent food intake may increase vitamin levels in the blood. For patients who are to be tested for malabsorption of B_{12}, blood samples should be drawn *before* the Schilling test because such patients will receive not only radioactive vitamin B_{12}, but also an injection of a large dose (1 mg) of nonradioactive vitamin B_{12} to block all potential tissue sites of vitamin B_{12} binding.

Either serum or plasma may be used for the assay, but heparin should not be used as an anticoagulant because it shows a modest B_{12}-binding ability. Large amounts of fluoride or ascorbic acid are added in vitro to blood samples for certain procedures; such blood samples should not be used for vitamin B_{12} assay because either of these two agents in high concentrations appears to destroy vitamin B_{12}.

Vitamin B_{12} is susceptible to photolytic degradation; therefore excessive exposure to light should be avoided. Serum samples can be stored overnight in a refrigerator (4° C) or several months in a freezer (−20° C) without appreciable loss of vitamin B_{12}. Repeated freezing and thawing of samples should be avoided.

REFERENCE RANGE

As with all diagnostic tests, the laboratory should establish its own reference ranges to conform with the characteristics of the populations that are being tested and to allow for the variation in assay techniques used. The following reference ranges in picograms per milliliter (in parentheses, picomoles per liter) were obtained in our laboratory using three different intrinsic factor (IF) preparations.[5]

B_{12} status	**Crude IF**	**Pure IF**	**Crude IF and blocking agent**
Deficient	<200 (<147)	<100 (<73.8)	<145 (<107)
Borderline	200-290 (147-214)	100-180 (73.8-133)	145-225 (107-166)
Normal	290-900 (214-664)	180-960 (133-708)	225-830 (166-613)
Above normal	>900 (>664)	>960 (>708)	>830 (>613)

A statistically significant negative correlation was found between age and serum levels of vitamin B_{12} measured by microbiological assays.[16] Elderly persons in geriatric homes had lower B_{12} values when assayed with pure IF than when assayed with crude IF in radioassay, an indication that biologically inactive vitamin B_{12} analogs were more frequently present in elderly persons than in the reference sample group.[17] Mean serum vitamin B_{12} levels have been found to be higher in blacks (546 ± 198 pg/mL, n = 49) than in whites (382 ± 131 pg/mL, n = 49) and suggest that clinical laboratories should establish their own separate reference values of serum vitamin B_{12} for blacks and whites to prevent misinterpretation of test results.[18]

High serum vitamin B_{12} levels in blacks appear to be attributable to the elevation of serum transcobalamins seen in this population. Serum vitamin B_{12} and unsaturated B_{12} binding capacity (UBBC) were measured in 59 black and 65 white blood donors. Serum levels for both analytes were higher among blacks (504 ± 205 and 702 ± 198 pg/mL, respectively) than whites (385 ± 146 and 520 ± 175 pg/mL, respectively).[19] Although higher vitamin B_{12} levels in blacks are accompanied by higher UBBC, there was no statistically significant correlation between these two values, suggesting that elevated vitamin B_{12} levels among blacks are attributable to causes other than elevated UBBC.

INTERFERENCE

It has been reported that ingestion of megadoses of ascorbic acid may be harmful to human vitamin B_{12} metabolism and may produce artificially low serum vitamin B_{12} levels by radioassay because of destruction of vitamin B_{12} by ascorbic acid.[20] However, we did not find a statistical difference between the mean serum vitamin B_{12} levels of 20 children with myelomeningocele receiving daily mean doses of 1.65 g of supplemental ascorbic acid to acidify the urine for prevention of possible urinary tract infection (768 ± 60 pg/mL) and those of 20 children with myelomeningocele receiving no ascorbic acid therapy (815 ± 55 pg/mL). Their mean serum ascorbic acid levels were 15.1 ± 0.7 and 7.4 ± 0.7 mg/L, respectively.[21]

Drugs capable of interfering with B_{12} absorption may cause a depression of B_{12} levels in blood. These include methotrexate, phenytoin, barbiturates, and oral contraceptives.

REFERENCES

1. Matthews, DM: Observations on the estimation of serum vitamin B_{12} using *Lactobacillus leichmannii,* Clin Sci 22:101-111, 1962.
2. Powell, DEB, Thomas, JH, Mandal, AR, et al: Effect of drugs on vitamin B_{12} levels obtained using the *Lactobacillus leichmannii* method, J Clin Pathol 22:672, 1969.
3. Lau, KS, Gottlieb, C, Wasserman, LR, et al: Measurement of serum vitamin B_{12} level using radioisotope dilution and coated charcoal, Blood 26:202-214, 1965.
4. Kolhouse, JF, Kondo, H, Allen, NC, et al: Cobalamin analogues are present in human plasma and can mask cobalamin deficiency because current radioisotope dilution assays are not specific for true cobalamin, N Engl J Med 299:785-792, 1978.
5. Chen, I-W, Silberstein, EB, Maxon, HR, et al: Clinical significance of serum vitamin B_{12} measured by radioassay using pure intrinsic factor, J Nucl Med 22:447-451, 1981.
6. National Committee for Clinical Laboratory Standards: Guidelines for evaluating a B_{12} (cobalamin) assay, Villanova, PA, 1980, National Committee on Clinical Laboratory Standards.
7. Kolhouse, JF, and Allen, RH: Isolation of cobalamin and cobalamin analogs by reverse affinity chromatography, Anal Biochem 84:486-490, 1978.
8. Hoffman, KL, Thomas,C, and Ebert, E: An automated no-boil combination B_{12}/folate assay using solid-phase binding proteins and radioiodinated B_{12} and folate, Clin Chem 31:903, 1985.
9. Herbert, V, and Colman, N: Evidence humans may use some analogues of B_{12} as cobalamins (B_{12}): pure intrinsic factor (IF) radioassay may diagnose B_{12} deficiency where it does not exist, Clin Res 29:571A, 1981.
10. Schilling, RF, Fairbanks, VF, Miller, R, et al: ''Improved'' vitamin B_{12} assays: a report on two commercial kits, Clin Chem 29:582-583, 1983.
11. Chen, I-W, Silberstein, EB, Maxon, HR, et al: Semiautomated system for simultaneous assays of serum vitamin B_{12} and folic acid in serum evaluated, Clin Chem 28:2161-2165, 1982.
12. Ithakissios, DS, Kubiatowicz, DO, and Wickes, JH: Room temperature radioassay for B_{12} with oyster toadfish *(Opsanus tau)* serum as binder, Clin Chem 26:323-329, 1980.
13. Sturgeon, MF, Hogle, DM, Luddy, BJ, et al: Advantages of solid phase technology over charcoal separation in the determination of B_{12} concentrations by radioassay, Ligand Q 4:52-56, 1981.
14. Chen, I-W, Sperling, MI, Heminger, LA, et al: Denaturation of interfering serum proteins under the assay conditions of a solid-phase no-boil vitamin B_{12} (B_{12}) assay kit, Clin Chem 29:1241, 1983.
15. Zucker, RM, Podell, ER, and Allen, RH: Multiple problems with current no-boil assays for serum cobalamin, Ligand Q 4:52-58, 1981.
16. Boger, WP, Wright, LD, Strickland, SC, et al: Vitamin B_{12}: correlation of serum concentrations and age, Proc Soc Exp Biol Med 89:375-378, 1955.
17. Magnus, EM, Bache-Wiig, JE, Anderson, TR, et al: Folate and vitamin B_{12} (cobalamin) blood levels in elderly persons in geriatric homes, Scand J Haematol 28:360-366, 1982.
18. Kwee, HG, Bowman, HS, and Wells, LW: A racial difference in serum vitamin B_{12} levels, J Nucl Med 26:790, 1985.
19. Chen, I-W, Silberstein, EB, Heminger, LA, et al: Comparison of serum vitamin B_{12} (B_{12}), unsaturated B_{12} binding capacity (UBBC), and folate levels in white and black subjects, Clin Chem 32:1186, 1986.
20. Herbert, V: Vitamin B_{12}, Am J Clin Nutr 34:971-972, 1981.
21. Ekvall, S, Chen, I-W, and Bozian, R: The effect of supplemental ascorbic acid on serum vitamin levels in myelomeningocele patients, Am J Clin Nutr 34:1356-1361, 1981.

Folic acid

MICHAEL D.D. McNEELY

Clinical significance: p. 543

Folates	Molecular formula	Merck Index	Molecular weight (daltons)
Pteroylglutamic acid (PGA, folic acid)	$C_{19}H_{19}N_7O_6$	4110	441.40
Tetrahydrofolic acid (THFA)	$C_{19}H_{23}N_7O_6$	—	445.40
N-5-Methyltetrahydrofolic acid (MTHFA)	$C_{20}H_{25}N_7O_6$	—	459.40

Chemical class: pteroylglutamic acid

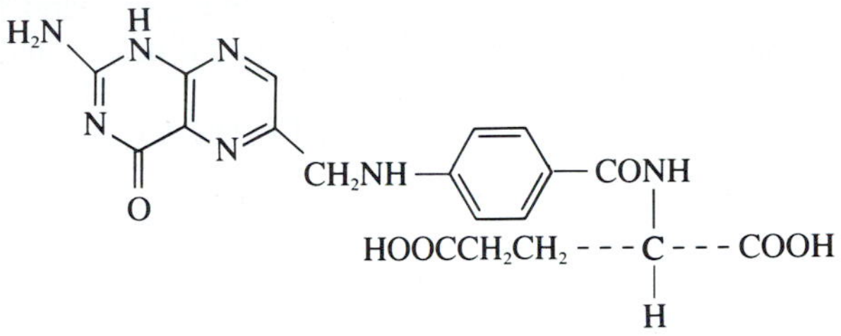

Folic acid

PRINCIPLES OF ANALYSIS

Folates are a group of chemically and biologically related compounds[1] that function as coenzymes in the metabolism of one-carbon compounds, in purine and pyrimidine synthesis, and in the degradation of histidine to glutamic acid. The measurement of folate in serum and whole blood is important in the differential diagnosis of megaloblastic anemia.

There are two fundamental approaches to folate measurement: microbiological assays and radiometric competitive protein-binding (CPB) techniques. Microbiological assays were developed first, but CPB methods are now used as the standard assay procedure in most laboratories.

The microbiological assay (method 1, Table 66-3) depends on the fact that the growth of *Lactobacillus casei* requires methyltetrahydrofolic acid (MTHFA).[2] The amount of *L. casei* growth permitted by a patient's serum is compared with the growth permitted by standard solutions of MTHFA. The greater the concentration of folates in the patient's sample, the greater the growth of the indicator organism. Growth is monitored by measurement of the turbidity of the test solutions at 640 to 700 nm, and the absorbance is directly related to the logarithm of folate concentration. This assay requires a 16- to 18-hour incubation period.

Radiometric CPB assays are based on the high affinity of B_{12} or folate for proteins that specifically bind these molecules (method 2, Table 66-3). Using a limited amount of binding protein and radiolabeled vitamin (ligand), one can quantitate the concentration of the molecules (see Chapter 11). The CPB assays are the most widely used methods because they are more efficient, faster, and more precise. Such assays may be arranged to allow the simultaneous measurement of vitamin B_{12} and folate.[3]

The most common binding agent for the CPB assays is derived from milk.[4] Unpurified milk powder or partially purified β-lactoglobulin may be used. Hog kidney[5] and porcine serum[6] contain folate-binding proteins, but these offer no advantage over milk. It is important to recognize that milk binders have different binding affinities for dif-

Table 66-3 Methods of folate analysis

Method	Type of analysis	Principle	Usage	Comments
1. Microbiological	Bioassay	Growth of folate-dependent bacteria related to specimen folate concentration; growth monitored by turbidimetry	Rare	False-negative results because of presence of antibiotics or antibodies to target organism; false-positive results because of turbidity
2. Radiometric	Competitive protein binding assay (CPB)	Folates in sample compete with ^3H- or ^{125}I-labeled ligand for binding to milk proteins (such as lactoglobulin); separation of bound and free labeled ligand by dextran-coated charcoal	Most frequent	Must destroy endogenous folate binders

ferent forms of folate, and this binding is pH dependent.[7]

The CPB assay is usually carried out at pH 7.4 to 8.2 as a single- or two-stage assay. The one-stage assay is the conventional competitive assay approach in which sample, binder, and tracer are combined simultaneously.

The two-stage assay is noncompetitive and involves sequential incubation. The binder is first incubated with sample (or standard). The temperature is then lowered to 4° C to minimize dissociation. In the second step, tracer is added and occupies the unoccupied binding sites. This approach is more sensitive.

For both assays the separation of bound and free tracer is almost invariably performed using dextran-coated charcoal with centrifugation in the cold.

Red blood cell folate is measured by the same analytical techniques. Folate is much more concentrated in red blood cells than in serum. Measurement of folate in red blood cells more closely reflects tissue folate stores and is less variable as a consequence of recent folate ingestion.[8,9]

To liberate the intracellular folate, freezing and thawing are considered more complete than incubation with ascorbic acid.[10] A hematocrit should be determined on the sample to express the results as a function of red blood cell mass.

Patient samples contain substances that bind folate and produce a variable blank, which causes falsely low results.[11] For destruction of these materials, the samples may be subjected to severe conditions that inactivate the endogenous binding proteins. The most common approach has been boiling at a high pH. A popular version uses a lysine buffer at pH 10.5 with 2-mercaptoethanol.[12] Since these extraction steps may destroy endogenous folate,[13] a serum blank may be used instead of an aqueous one.

Standardization of the assays must be done with care. Impure assay material containing a variety of folate derivatives is a major source of variation between assays.

Tritiated pteroylglutamic acid (PGA) has been widely employed as the assay tracer. Iodinated tracers in which ^{125}I is linked to PGA by tyrosine offer much greater counting activity and the added ease of gamma-ray counting.

The microbiological assay is not suitable for routine work but is appropriate as a reference procedure.

Radiometric CPB methods using milk binder at pH 8, tritiated PGA tracer, MTHFA standard, and dextran-coated charcoal separation provide the most reliable assays. A preliminary step should be undertaken to eliminate serum folate binding. Methods that do not use preliminary steps ("no boil") must thoroughly justify themselves.

The recent trend to develop commercial kit assays capable of measuring both vitamin B_{12} and folate must be evaluated with caution. Such methods are certainly more convenient than two individual assays but tend to be optimized for vitamin B_{12}, possibly at the expense of the folate assay.

SPECIMEN

Samples for serum folate should be collected in the fasting state, since blood folate levels vary with food intake. Stability is maintained for 24 hours at 4° C; thereafter, the serum should be kept frozen at −10° C. Ascorbic acid has been used to stabilize MTHFA but is inconvenient and unnecessary.[14]

Blood can be collected into oxalate or heparin anticoagulants and must be carefully maintained at 4° C; it is stable at this temperature for 3 days. Since 95% of whole blood folate is found in red blood cells, hemolyzed samples should be rejected for analysis.

REFERENCE RANGE

Serum folate	1.9 to 14 ng/mL (4.3 to 31.7 pmol/mL)
Red blood cell folate	200 to 1000 ng/mL (453.1 to 2266 pmol/mL) packed cells

REFERENCES

1. Usdin, E: Blood folic acid studies. VI. Chromatographic resolution of folic acid-active substances obtained from blood, J Biol Chem 234:2373-2376, 1959.
2. Herbert, V: Aseptic addition method for *Lactobacillus casei* assay of folate activity in human serum, J Clin Pathol 19:12-16, 1966.
3. Gutcho, S, and Mansbach, L: Simultaneous radioassay of serum vitamin B_{12} and folic acid, Clin Chem 23:1606-1614, 1977.
4. Chitis, J: The folate binding in milk, Am J Clin Nutr 20:1-4, 1967.
5. Kamen, BA, and Caston, JD: Direct radiochemical assay for serum folate: competition between ^{3}H-folic acid and 5-methyltetrahydrofolic acid for a folate binder, J Lab Clin Med 83:164-174, 1974.
6. Mantzos, J: Radioassay of serum folate with use of pig plasma folate binders, Acta Haematol 54:289-296, 1975.
7. Givas, JK, and Gutcho, S: pH dependence of the binding of folates to milk binder in radioassay of folates, Clin Chem 21:427-428, 1975.
8. Mortensen, E: Determination of erythrocyte folate by competitive protein binding assay preceded by extraction, Clin Chem 22:982-992, 1976.
9. Mortensen, E: Effect of storage on the apparent concentration of folate in erythrocytes, as measured by competitive protein binding radioassay, Clin Chem 24:663-668, 1978.
10. Netteland, B, and Bakke, OM: Inadequate sample-preparation technique as a source of error in determination of erythrocyte folate by competitive binding radioassay, Clin Chem 23:1505-1506, 1977.
11. Zettner, A, and Duly, PE: New evidence for a binding principle specific for folates as a normal constituent of human serum, Clin Chem 20:1313-1319, 1974.
12. Dunn, RT, and Foster, LB: Radioassay of serum folate, Clin Chem 19:1101-1105, 1973.
13. Mitchell, GA, Pochron, SP, Smutny, PV, and Guity, R: Decreased radioassay values for folate after serum extraction when pteroylglutamic acid standards are used, Clin Chem 22:647-649, 1976.
14. Waddell, CC, Domstad, PA, Pircher, FJ, et al: Serum folate levels: comparison of microbiologic assay and radioisotope kit methods, Am J Clin Pathol 66:746-752, 1976.

Schilling test

MICHAEL D.D. McNEELY

Clinical significance: pp. 398 and 543

PRINCIPLES OF ANALYSIS

The Schilling test is a method of quantifying the intestinal absorption of vitamin B_{12}. It is used to diagnose the cause of vitamin B_{12} deficiency.

Ingested vitamin B_{12} is separated from food matter in the stomach, where parietal cells secrete intrinsic factor (IF), a 55,000-dalton glycoprotein. Two moles of IF dimerize with 1 M of vitamin B_{12} at a pH greater than 6 in the presence of ionic calcium. Once coupled, the IF-B_{12} complex is resistant to separation by the usual intestinal environment. The IF-B_{12} complex is carried to the terminal portion of the ileum where it is captured by a specific epithelial receptor. The vitamin B_{12} is then conveyed across the epithelium into portal blood by a specific transport mechanism. Vitamin B_{12} is readily stored in the liver, and when storage sites are saturated, the excess B_{12} is excreted unmodified into the urine after glomerular filtration.

The Schilling test has been developed not only to quantify vitamin B_{12} absorption but also to distinguish whether the source of an absorptive defect is in the upper gastrointestinal tract (IF coupling) or lower gastrointestinal tract (complex absorption).

The test is considered to be the definitive test of vitamin B_{12} absorption. The original test was described by Schilling in 1953.[1,2] Several modifications have been proposed to avoid some of the diagnostic inconsistencies of the method or to make it easier to perform.

In the classical test, the patient is first given a parenteral injection of 1 g of nonlabeled vitamin B_{12} (referred to as a *flushing dose*). This fills the body storage sites completely. The patient is then fed a physiological dose of radioactive vitamin B_{12}. If absorption mechanisms are functioning normally, the vitamin B_{12} will enter the patient's circulation. However, because tissue storage sites have been saturated, the radioactive B_{12} will not be stored and will spill into the urine. Thus the radioactivity of a 24-hour, post-ingestion urine sample will be directly proportional to the amount of vitamin B_{12} absorbed.

If absorption is judged adequate by this first stage, the entire vitamin B_{12}-absorption mechanism is judged to be intact.

If absorption is observed to be inadequate during the first-stage test, a second-stage test may be conducted. In this variation, the radioactive vitamin B_{12} is administered orally with hog IF. If vitamin B_{12} is absorbed during the second stage but not during the first, the absorptive defect is judged to be the result of a deficiency of normal intrinsic factor.

If vitamin B_{12} is still not absorbed after the second test stage, a third-stage test may be performed. This test is carried out to determine whether microorganism overgrowth of the ileum is the cause of the malabsorption. The patient is given a full course of an antihelminthic (antitapeworm) or antibiotic therapy. Then the first-stage test is repeated. If absorption is restored to normal, the microorganism is judged to be the cause of the vitamin B_{12} malabsorption. A 1.5 to 2.0 μg load of B_{12} is recommended. This balances maximum challenge and absorption by nonspecific diffusion.[3]

It is recommended that the test be administered after an overnight fast and that the fast be continued for an additional 4 hours after the oral administration of the B_{12}.

The flushing dose of parenterally administered nonlabeled B_{12} must not be given more than 12 hours before the test, since a greater delay may allow some storage sites to open up. Orally administered B_{12} usually does not begin to cross the ileal wall for 6 hours after ingestion, with the peak absorption occurring at 12 hours. If the flushing dose is administered parenterally more than 6 hours after the administration of the oral dose, the tissue sites may not be occupied until after absorption of the orally administered B_{12}. On the other hand, if given before oral administration, the parenterally administered B_{12} may have time to enter the intestine through the biliary tract and compete with the oral dose.

Optimally, the flushing dose should be given 6 hours after the oral dose, but this is generally inconvenient. Thus the flushing dose is often given at the same time as the oral dose or as an intramuscular injection 1 hour later.

There are other factors that must be considered in the interpretation of the test. The flushing dose will obscure other tests of B_{12} metabolism and therefore should be the last test done in a B_{12} workup study. If the patient is already receiving B_{12} therapy, the Schilling test is still valid. However, regular therapy must be avoided for at least 1 week before the test.

A very practical approach has been the simultaneous administration of free radioactive vitamin B_{12} and radioactive vitamin B_{12} bound to IF.[4] Although administered simultaneously, two different radioisotopes, cobalt 58 and cobalt 57 (^{58}Co, ^{57}Co), are used to label the test preparations. This allows simultaneous absorption studies to be carried out (that is, a test for intrinsic factor deficiency and for malabsorption).

The dual isotope is commercially available as Dicopac (Amersham, Arlington Heights, IL), in which a preparation of ^{58}Co cyanocobalamin and ^{57}Co cyanocobalamin–IF complex are given simultaneously.[5] This is the procedure described next.

PROCEDURE: SCHILLING TEST

Specimen and patient protocol

1. The patient is requested to fast overnight before the test.

2. The patient arrives in the laboratory and passes a urine sample, which is discarded. Blood is collected for vitamin B_{12} measurement.
3. For the dual-isotope test, the patient ingests two capsules (Dicopac). For individual tests employing single-labeled B_{12}, the patient ingests the specific dose. IF is available from ER Squibb and Sons (Princeton, NJ) or Mallinckrodt, Inc. (St. Louis); vitamin B_{12} is obtained as cyanocobalamin (0.5 to 0.7 μg) capsules from Mallinckrodt, Inc.
4. A 1000 mg flushing dose is administered intramuscularly 1 hour after ingestion of the labeled B_{12}.
5. Urine is collected for 24 hours or 48 hours if the creatinine clearance is below 60 mL/min.

Assay

1. The activity of the oral preparation must be determined.
   a. For the dual-isotope test, standards are provided. One milliliter of each should be diluted with 3 mL of water to make working standards. These are placed in the same counting tubes used for the patient sample. Count both ^{58}Co and ^{57}Co working standards with the gamma-ray scintillation counter optimized for ^{58}Co and ^{57}Co counting.
   b. For the single isotope test, a suitable dilution of the oral preparation is prepared in water. Four milliliters of this working standard should be counted.
2. Four milliliters of the 24-hour urine is placed into a counting tube.
   a. For the dual isotope test, counts are made at the ^{57}Co and ^{58}Co settings.
   b. For the single isotope method, counts are made at the appropriate isotope setting.

Calculation

Dual-isotope setting

1. Corrected standard counts per minute (cpm)
   ^{58}Co cpm in ^{58}Co setting ______ A
   ^{58}Co cpm in ^{57}Co setting ______ B
   ^{57}Co cpm in ^{57}Co setting ______ C
2. Correction factor (CF) = B:A = ______ CF
3. Corrected urine cpm
   Urine cpm in ^{58}Co setting ______ E
   Urine cpm in ^{57}Co setting ______ F
4. Compute

$$\%^{58}\text{Co(E) in 24 hours} = \frac{E \times \text{24-hr urine vol (mL)}}{A \times 2} = \text{Excretion of test stage I}$$

$$^{57}\text{Co (F) in 24 hours} = \frac{[F - (E \times CF)] \times \text{24-hr urine vol (mL)}}{C \times 2} = \text{Excretion of stage H}$$

Single isotope test

$$\%\ \text{Excretion} = \frac{\text{Sample cpm} \times \text{24-hr urine vol (mL)}}{\text{Standard cpm} \times \text{Dilution}}$$

Note

It is essential that the gamma-ray scintillation counter be optimized for reading the two isotopes. It is also possible to set the counter so that only those emissions larger than the upper discriminator setting for ^{57}Co are measured as ^{58}Co. It is important to subtract a background count obtained using water in a vial from each count.

REFERENCE RANGE AND INTERPRETATION

1. *Absorption of the radioactively labeled vitamin B_{12} (test stage I) or the labeled vitamin B_{12} plus IF (test stage II) is greater than 15% of the administered dose.* Vitamin B_{12} absorption is normal. Any deficiency of B_{12} must be attributed to low dietary intake.
2. *Absorption in test stage II is greater than 10% with absorption in test stage I less than 10%.* A problem with IF coupling must exist. Causes are pernicious anemia, failure to digest the IF capsule, defective commercial IF, biologically inactive IF, defective IF-B_{12} coupling (partial gastrectomy, vagotomy, gastric ulcer, cimetidine therapy).
3. *Absorption in both test stages I and II less than 15%.* A problem with ileal absorption—terminal ileitis, bacterial overgrowth, blind-loop syndrome, severe pernicious anemia—that has caused generalized epithelial deficiency, incomplete urine collection, renal failure, tapeworm *(Diphyllobothrium latum),* or transport defect (Immerslund syndrome).[6] The dual-isotope test has been interpreted using a ratio of the two isotopes of 1:1.7, beyond which pernicious anemia is diagnosed. This is too simplistic and may be misleading. An analysis of actual values as described in 1 and 2 is recommended. The test has 83% sensitivity and 98% specificity for pernicious anemia.[7]

REFERENCES

1. Schilling, RF: Intrinsic factor studies. II. The effect of gastric juice on the urinary excretion of radioactivity after the administration of vitamin B_{12}, J Lab Clin Med 42:860-866, 1953.
2. Schilling, RF, and Herbert, V: The megaloblastic anemias, New York, 1979, Grune & Stratton, Inc.
3. Herbert, V: Detection of malabsorption of vitamin B_{12} due to gastric or intestinal dysfunction, Semin Nucl Med 2:220-234, 1972.
4. Katz, JH, DiMase, J, and Donaldson, RM Jr: Simultaneous administration of gastric juice-bound and free radioactive cyanocobalamin, J Lab Clin Med 82:266, 1963.
5. England JM, Snashall, EA, and DeSilva, AN: Comparison of the Dicopac with the conventional Schilling test, J Clin Pathol 34:1191-1192, 1981.
6. Streeter, AM, Bathur, FA, Arnold, BJ, et al: Limitations of the Schilling test, Lancet 1:39-40, Jan 3, 1981 (letter).
7. Domstad, PA, Choy, YC, Kim, EE, and DeLand, FH: Reliability of the dual-isotope Schilling test for the diagnosis of pernicious anemia or malabsorption syndrome, Am J Clin Pathol 75:723-726, 1981.

Appendixes

APPENDIX A Buffer solutions*

Buffer solutions (or buffers) are solutions whose pH value is to a large degree insensitive to the addition of other substances. It is important to realize, however, that the pH value of a buffer solution does not change only when acids or bases are added or on dilution but also when the temperature changes or neutral salts are added. In accurate work, therefore, it is important to check the pH value electrometrically after all the ingredients have been added. The extent to which the pH values of buffer solutions vary when acids or bases are added or the temperature changes is shown in the tables that follow. In general, dilution to half the concentration changes the pH value by only some hundredths of a unit (Buffer No. 1 in the table is an exception in that the change amounts to ca. pH 0.15); addition of 0.1-molar neutral salt solution may change the pH value of ca. 0.1.

In the table opposite the solutions are classified into general buffers (mostly in use for the last 50 years), universal buffers with a low buffering capacity but a wide pH range, and buffers for biological media with a moderate pH range but containing stable ingredients (phosphate and borate, for example, often undergo side reactions with biological media). An important property is often the transparency to ultraviolet light. Occasionally it is desirable to have a volatile buffer, which can be readily removed[1] (examples are buffers Nos. 20 and 21), but the use of very volatile systems makes a close control of the pH essential. Most of the pH data to be found in the literature relate to the Sørensen scale, and it should be noted that the values given in the following table of buffers are on the conventional pH scale.

Both stock and buffer solutions should be made up with distilled water free of CO_2. Only standard reagents should be used. If there is any doubt as to the purity or water content of solutions, their molarity must be checked by titration. The amounts x of stock solutions required to make up a buffer solution of the desired pH value are given in the second table in Appendix B.

*This appendix has been compiled by F. Kohler, Department of Physical Chemistry, University of Vienna, and taken from CIBA-Geigy AG, Basel, Switzerland.

REFERENCE

1. For a list of volatile buffers see Michl, H: In Heftmann, E, editor: Chromatography, part 1, New York, 1961, Reinhold, p 250.

No.	Name	pH range	Temperature	pH change per °C
	General Buffers			
1	KCl/HCl (Clark and Lubs)[1]	1.0- 2.2	Room	0
2	Glycine/HCl (Sørensen)[2]	1.2- 3.4	Room	0
3	Na citrate/HCl (Sørensen)[2]	1.2- 5.0	Room	0
4	K biphthalate/HCl (Clark and Lubs)[1]	2.4- 4.0	20° C	+0.001
5	K biphthalate/NaOH (Clark and Lubs)[1]	4.2- 6.2	20° C	
6	Na citrate/NaOH (Sørensen)[2]	5.2- 6.6	20° C	+0.004
7	Phosphate (Sørensen)[2]	5.0- 8.0	20° C	−0.003
8	Barbital-Na/HCl (Michaelis)[3]	7.0- 9.0	18° C	
9	Na borate/HCl (Sørensen)[2]	7.8- 9.2	20° C	−0.005
10	Glycine/NaOH (Sørensen)[2]	8.6-12.8	20° C	−0.025
11	Na borate/NaOH (Sørensen)[2]	9.4-10.6	20° C	−0.01
	Universal Buffers			
12	Citric acid/phosphate (McIlvaine)[4]	2.2- 7.8	21° C	
13	Citrate-phosphate-borate/HCl (Teorell and Stenhagen)[5]...........................	2.0-12.0	20° C	
14	Britton-Robinson[6]	2.6-11.8	25° C	at low pH 0 at high pH −0.02
	Buffers for Biological Media			
15	Acetate (Walpole)[7-9]	3.8- 5.6	25° C	
16	Dimethylglutaric acid/NaOH[10]	3.2- 7.6	21° C	
17	Piperazine/HCl[11,12]	4.6- 6.4 8.8-10.6	20° C	
18	Tetraethylethylenediamine*[12]	5.0- 6.8 8.2-10.0	20° C	
19	Trismaleate[7,13]	5.2- 8.6	23° C	
20	Dimethylaminoethylamine*[12]	5.6- 7.4 8.6-10.4	20° C	
21	Imidazole/HCl[14]	6.2- 7.8	25° C	
22	Triethanolamine/HCl[15]	7.0- 8.8	25° C	
23	*N*-Dimethylaminoleucylglycine/NaOH[16]	7.0- 8.8	23° C	−0.015
24	Tris/HCl[7]	7.2- 9.0	23° C	−0.02
25	2-Amino-2-methylpropane-1,3-diol/HCl[7,13]	7.8-10.0	23° C	
26	Carbonate (Delory and King)[7,17]	9.2-10.8	20° C	

From Geigy scientific tables, ed 8, Basel, Switzerland, 1981, CIBA-Geigy AG.
*Can be combined with tris buffer to give a cationic universal buffer (see reference 12).

REFERENCES

1. Clark and Lubs: J Bact 2:1, 1917.
2. Sørensen, SPL: Biochem Z 21:131, 1909, 22:352, 1909; Ergebn Physiol 12:393, 1912; and Walbum, LE: Biochem Z 107:219, 1920.
3. Michaelis, L: J Biol Chem 87:33, 1930.
4. McIlvaine, TC: J Biol Chem 49:183, 1921.
5. Teorell and Stenhagen: Biochem Z 299:416, 1938.
6. Britton and Welford: J Chem Soc, 1937, 1848.
7. Gomori, G: In Colowick and Kaplan, editors: Methods in Enzymology, vol 1, New York, 1955, Academic Press, p 138.
8. Walpole, GS: J Chem Soc 105:2501, 1914.
9. Green, AA: J Am Chem Soc 55:2331, 1933.
10. Stafford et al: Biochim Biophys Acta 18:319, 1955; Krebs, HA, unpublished, 1957.
11. Smith and Smith: Biol Bull 96:233, 1949.
12. Semenza et al: Helv Chim Acta 45:2306, 1962.
13. Gomori, G: Proc Soc Exp Biol (NY) 68:354, 1948.
14. Mertz and Owen: Proc Soc Exp Biol (NY) 43:204, 1940, quoted by Rauen, HM, editor: Biochemisches Taschenbuch, ed 2, part 2, Berlin, 1964, Springer, p 90.
15. Beisenherz et al: Z Naturforsch 8b:555, 1953.
16. Leonis, J: CR Lab Carlsberg, Sér Chim 26:357, 1948.
17. Delory and King: Biochem J 39:245, 1945.

APPENDIX B Preparation of buffer solutions

When not otherwise specified, both stock and buffer solutions should be made up with distilled water free of CO_2. Only standard reagents should be used. If there is any doubt as to the purity or water content of solutions, their molarity must be checked by titration. The amounts *x* of stock solutions required to make up a buffer solution of the desired pH value are given in the table in the second part of Appendix B.

Buffer No.	Stock solutions A	Stock solutions B	Composition of the buffer
1	KCl 0.2-N (14.91 g/l)	HCl 0.2-N	25 ml A + *x* ml B made up to 100 ml
2	Glycine 0.1-molar in NaCl 0.1-N (7.507 g glycine + 5.844 g NaCl/l)	HCl 0.1-N	*x* ml A + (100 − *x*) ml B
3	Disodium citrate 0.1-molar (21.01 g $C_6H_8O_7 \cdot 1H_2O$ + 200 ml NaOH 1-N per litre)	HCl 0.1-N	*x* ml A + (100 − *x*) ml B
4	Potassium biphthalate 0.1-molar (20.42 g $KHC_8H_4O_4$/l)	HCl 0.1-N	50 ml A + *x* ml B made up to 100 ml
5	As No. 4	NaOH 0.1-N	50 ml A + *x* ml B made up to 100 ml
6	As No. 3	NaOH 0.1-N	*x* ml A + (100 − *x*) ml B
7	Monopotassium phosphate 1/15-molar (9.073 g KH_2PO_4/l)	Disodium phosphate 1/15-molar (11.87 g $Na_2HPO_4 \cdot 2H_2O$/l)	*x* ml A + (100 − *x*) ml B
8	Barbital sodium 0.1-molar (20.62 g/l)	HCl 0.1-N	*x* ml A + (100 − *x*) ml B
9	Boric acid, half-neutralized, 0.2-molar (corr. to 0.05-molar borax: 12.37 g boric acid + 100 ml NaOH 1-N per litre)	HCl 0.1-N	*x* ml A + (100 − *x*) ml B
10	As No. 2	NaOH 0.1-N	*x* ml A + (100 − *x*) ml B
11	As No. 9	NaOH 0.1-N	*x* ml A + (100 − *x*) ml B
12	Citric acid 0.1-molar (21.01 g $C_6H_8O_7 \cdot 1H_2O$/l)	Disodium phosphate 0.2-molar (35.60 g $Na_2HPO_4 \cdot 2H_2O$/l)	*x* ml A + (100 − *x*) ml B
13	To citric acid and phosphoric acid solutions (ca. 100 ml), each equivalent to 100 ml NaOH 1-N, add 3.54 cryst. orthoboric acid and 343 ml NaOH 1-N, and make up the mixture to 1 litre	HCl 0.1-N	20 ml A + *x* ml B made up to 100 ml
14	Citric acid, monopotassium phosphate, barbital, boric acid, all 0.02857-molar (6.004 g $C_6H_8O_7 \cdot 1H_2O$, 3.888 g KH_2PO_4, 5.263 g barbital, 1.767 g H_3BO_3/l)	NaOH 0.2-N	100 ml A + *x* ml B
15	Sodium acetate 0.1-N (8.204 g $C_2H_3O_2Na$ or 13.61 g $C_2H_3O_2Na \cdot 3H_2O$/l)	Acetic acid 0.1-N (6.005 g/l)	*x* ml A + (100 − *x*) ml B

From Geigy scientific tables, ed 8, Basel, Switzerland, 1981, CIBA-Geigy AG.

Buffer No.	Stock solutions		Composition of the buffer
	A	B	
16	ββ-Dimethylglutaric acid 0.1-molar (16.02 g/l)	NaOH 0.2-N	(a) 100 ml A + x ml B made up to 1000 ml (b) 100 ml A + x ml B + 5.844 g NaCl made up to 1000 ml (NaCl ≙ 0.1-molar)
17	Piperazine 1-molar (86.14 g/l)	HCl 0.1-N	5 ml A + x ml B made up to 100 ml
18	Tetraethylethylenediamine 1-molar (172.32 g/l)	HCl 0.1-N	5 ml A + x ml B made up to 100 ml
19	Tris acid maleate 0.2-molar (24.23 g tris[hydroxymethyl]aminomethane + 23.21 g maleic acid or 19.61 g maleic anhydride/l)	NaOH 0.2-N	25 ml A + x ml B made up to 100 ml
20	Dimethylaminoethylamine 1-molar (88 g/l)	HCl 0.1-N	5 ml A + x ml B made up to 100 ml
21	Imidazole 0.2-molar (13.62 g/l)	HCl 0.1-N	25 ml A + x ml B made up to 100 ml
22	Triethanolamine 0.5-molar (76.11 g/l) containing 20 g/l ethylenediaminetetraacetic acid disodium salt ($C_{10}H_{14}O_8N_2Na_2 \cdot 2H_2O$)	HCl 0.05-N	10 ml A + x ml B made up to 100 ml
23	*N*-Dimethylaminoleucylglycine 0.1-molar (24.33 g $C_{10}H_{20}O_3N_2 \cdot \frac{3}{2}H_2O$/l) containing NaCl 0.2-N (11.69 g/l)	NaOH 1-N 100 ml made up to 1 liter with A	x ml A + (100 − x) ml B
24	Tris 0.2-molar (24.23 g tris[hydroxymethyl]aminomethane/l)	HCl 0.1-N	25 ml A + x ml B made up to 100 ml
25	2-Amino-2-methylpropane-1,3-diol 0.1-molar (10.51 g/l)	HCl 0.1-N	50 ml A + x ml B made up to 100 ml
26	Sodium carbonate anhydrous 0.1-molar (10.60 g/l)	Sodium bicarbonate 0.1-molar (8.401 g/l)	x ml A + (100 − x) ml B

The table gives the amounts (x ml) of the stock solutions listed in the first part of Appendix B required to make up a buffer solution of the desired pH value.

pH	1	2	3	4	5	6	7	8	9	10	11	12	13	14	15	16a	16b	17	18	19	20	21	22	23	24	25	26	pH
1.0	54.2	...	...	...	...	...	...	...	...	...	...	...	...	...	...	...	...	...	...	...	...	...	...	...	...	...	...	1.0
1.2	36.0	11.1	9.0	...	...	...	...	...	...	...	...	...	...	...	...	...	...	...	...	...	...	...	...	...	...	...	...	1.2
1.4	23.2	26.4	17.9	...	...	...	...	...	...	...	...	...	...	...	...	...	...	...	...	...	...	...	...	...	...	...	...	1.4
1.6	14.7	36.2	23.6	...	...	...	...	...	...	...	...	...	...	...	...	...	...	...	...	...	...	...	...	...	...	...	...	1.6
1.8	9.3	43.9	27.6	...	...	...	...	...	...	...	...	...	...	...	...	...	...	...	...	...	...	...	...	...	...	...	...	1.8
2.0	5.9	50.7	30.2	...	...	...	...	...	...	...	...	...	74.4	...	...	...	...	...	...	...	...	...	...	...	...	...	...	2.0
2.2	3.8	56.5	32.2	...	...	...	...	...	...	...	...	98.8	68.8	...	...	...	...	...	...	...	...	...	...	...	...	...	...	2.2
2.4	...	62.3	34.1	41.0	...	...	...	...	...	...	...	94.5	64.6	...	...	...	...	...	...	...	...	...	...	...	...	...	...	2.4
2.6	...	68.4	36.0	34.3	...	...	...	...	...	...	...	90.0	61.3	1.6	...	...	...	...	...	...	...	...	...	...	...	...	...	2.6
2.8	...	74.7	37.9	27.8	...	...	...	...	...	...	...	85.1	58.9	3.6	...	...	...	...	...	...	...	...	...	...	...	...	...	2.8
3.0	...	81.0	39.9	21.6	...	...	...	...	...	...	...	80.3	56.9	5.7	...	...	...	...	...	...	...	...	...	...	...	...	...	3.0
3.2	...	86.2	42.1	15.9	...	...	...	...	...	...	...	76.0	55.2	7.8	...	7.0	14.4	...	...	...	...	...	...	...	...	...	...	3.2
3.4	...	90.3	44.8	10.9	...	...	...	...	...	...	...	72.0	53.9	9.9	...	13.3	20.9	...	...	...	...	...	...	...	...	...	...	3.4
3.6	...	...	47.8	6.7	...	...	...	...	...	...	...	68.4	52.9	11.7	...	20.7	26.8	...	...	...	...	...	...	...	...	...	...	3.6
3.8	...	...	51.2	3.3	...	...	...	...	...	...	...	65.1	51.8	13.5	10.9	26.3	32.4	...	...	...	...	...	...	...	...	...	...	3.8
4.0	...	...	55.1	0.0	...	...	...	...	...	...	...	62.0	50.7	15.3	16.6	32.4	36.6	...	...	...	...	...	...	...	...	...	...	4.0
4.2	...	...	60.0	...	3.0	...	...	...	...	...	...	59.1	49.7	17.5	23.9	36.2	40.3	...	...	...	...	...	...	...	...	...	...	4.2
4.4	...	...	66.4	...	6.7	...	...	...	...	...	...	56.4	48.6	19.7	33.5	39.3	43.1	...	...	...	...	...	...	...	...	...	...	4.4
4.6	...	...	74.9	...	11.1	...	...	...	...	...	...	53.7	47.5	21.9	44.9	41.3	45.7	94.3	...	...	...	...	...	...	...	...	...	4.6
4.8	...	...	85.6	...	16.5	...	...	...	...	...	...	51.2	46.4	24.1	56.6	43.5	48.3	91.5	...	...	...	...	...	...	...	...	...	4.8
5.0	...	...	100.0	...	22.6	...	99.2	...	...	...	...	49.0	45.4	26.3	67.8	45.7	51.5	87.8	94.3	...	...	...	...	...	...	...	...	5.0
5.2	...	...	...	...	28.8	87.1	98.4	...	...	...	...	46.9	44.3	28.6	76.8	48.4	53.6	83.6	91.5	3.2	...	...	...	...	...	...	...	5.2
5.4	...	...	...	...	34.4	78.0	97.3	...	...	...	...	44.7	43.2	31.0	84.0	51.3	58.2	77.6	87.8	5.0	...	...	...	...	...	...	...	5.4
5.6	...	...	...	...	39.1	70.3	95.5	...	...	...	...	42.4	42.0	33.4	89.3	55.0	63.6	71.8	83.1	7.3	94.3	...	...	...	...	...	...	5.6
5.8	...	...	...	...	42.4	64.5	92.8	...	...	...	...	40.0	40.8	35.8	...	58.8	68.7	66.5	77.6	9.7	91.7	...	...	...	...	...	...	5.8
6.0	...	...	...	...	45.0	60.3	88.9	...	...	...	...	37.4	39.7	38.3	...	63.9	73.6	61.8	71.7	12.4	88.0	...	...	...	...	...	...	6.0
6.2	...	...	...	...	46.7	57.2	83.0	...	...	...	...	34.5	38.4	40.8	...	69.5	78.5	58.2	66.4	15.2	83.3	43.4	...	...	...	...	...	6.2
6.4	...	...	...	...	...	54.8	75.4	...	...	...	...	31.4	37.0	43.3	...	74.1	83.3	55.5	61.7	17.9	77.9	40.4	...	...	...	...	...	6.4
6.6	...	...	...	...	...	53.2	65.3	...	...	...	...	27.9	35.6	45.8	...	83.5	87.4	...	58.0	20.8	72.0	36.5	...	...	...	...	...	6.6
6.8	...	...	...	...	...	...	53.4	...	...	...	...	23.5	34.2	48.3	...	87.4	91.0	...	55.3	22.2	66.6	31.4	...	...	...	...	...	6.8
7.0	...	...	...	...	...	...	41.3	53.3	...	...	...	19.0	32.9	50.9	...	90.0	93.2	...	...	23.7	61.9	25.4	86.2	86.4	...	...	...	7.0
7.2	...	...	...	...	...	...	29.6	55.0	...	...	...	13.8	31.7	53.4	...	91.8	94.9	...	...	25.2	58.1	19.6	79.6	80.6	44.7	...	...	7.2
7.4	...	...	...	...	...	...	19.7	57.6	...	...	...	9.8	30.6	55.8	...	93.0	95.8	...	...	26.7	55.3	14.6	71.3	72.8	42.0	...	...	7.4
7.6	...	...	...	...	...	...	12.8	60.8	...	...	...	6.8	29.6	58.2	...	93.8	96.8	...	...	28.6	...	10.2	62.0	63.2	39.3	...	...	7.6
7.8	...	...	...	...	...	...	7.4	65.2	53.0	...	...	4.6	28.8	60.5	...	...	...	...	...	31.2	...	6.6	52.0	52.1	33.7	43.9	...	7.8
8.0	...	...	...	...	...	...	3.7	70.6	55.4	...	...	...	28.1	62.8	...	...	...	...	...	33.9	...	...	42.0	41.1	27.9	41.6	...	8.0
8.2	...	...	...	...	...	...	...	75.9	58.0	...	...	...	27.6	65.0	...	...	...	...	46.4	36.9	...	...	31.9	31.4	22.9	38.4	...	8.2
8.4	...	...	...	...	...	...	...	81.2	62.1	...	...	...	27.0	67.2	...	...	...	...	43.9	39.9	...	...	22.5	23.0	17.3	34.8	...	8.4
8.6	...	...	...	...	...	...	...	86.2	66.9	94.7	...	...	26.3	69.3	...	...	...	...	40.9	42.7	45.4	...	16.0	15.9	13.0	30.7	...	8.6
8.8	...	...	...	...	...	...	...	90.1	73.6	92.0	...	...	25.2	71.3	...	...	...	45.5	36.8	...	42.8	...	11.7	10.3	8.8	23.3	...	8.8
9.0	...	...	...	...	...	...	...	93.2	83.5	88.4	...	...	24.0	73.2	...	...	...	43.2	31.8	...	39.2	...	...	...	5.3	17.7	...	9.0
9.2	...	...	...	...	...	...	...	...	95.6	84.0	...	...	22.6	75.1	...	...	...	40.0	26.2	...	34.7	...	...	...	...	13.3	10.0	9.2
9.4	...	...	...	...	...	...	...	...	...	78.9	87.0	...	21.4	77.0	...	...	...	35.8	20.4	...	29.3	...	...	...	...	9.2	18.4	9.4
9.6	...	...	...	...	...	...	...	...	...	73.2	75.5	...	20.2	78.8	...	...	...	30.8	15.2	...	23.6	...	...	...	...	5.2	29.3	9.6
9.8	...	...	...	...	...	...	...	...	...	67.2	65.1	...	19.0	80.4	...	...	...	25.0	10.8	...	19.0	...	...	...	...	4.1	42.0	9.8
10.0	...	...	...	...	...	...	...	...	...	62.5	59.6	...	18.1	81.8	...	...	...	19.4	7.4	...	13.1	...	...	...	...	2.3	53.4	10.0
10.2	...	...	...	...	...	...	...	...	...	58.8	56.4	...	17.1	83.1	...	...	...	14.3	...	...	9.2	...	...	...	...	...	63.7	10.2
10.4	...	...	...	...	...	...	...	...	...	55.7	54.1	...	16.5	84.3	...	...	...	10.0	...	...	6.2	...	...	...	...	...	73.1	10.4
10.6	...	...	...	...	...	...	...	...	...	53.6	52.3	...	16.0	85.4	...	...	...	6.9	...	...	...	...	...	...	...	...	81.2	10.6
10.8	...	...	...	...	...	...	...	...	...	52.2	...	...	15.5	86.5	...	...	...	...	...	...	...	...	...	...	...	...	87.9	10.8
11.0	...	...	...	...	...	...	...	...	...	51.2	...	...	14.7	87.8	...	...	...	...	...	...	...	...	...	...	...	...	...	11.0
11.2	...	...	...	...	...	...	...	...	...	50.4	...	...	13.5	89.3	...	...	...	...	...	...	...	...	...	...	...	...	...	11.2
11.4	...	...	...	...	...	...	...	...	...	49.5	...	...	11.7	91.3	...	...	...	...	...	...	...	...	...	...	...	...	...	11.4
11.6	...	...	...	...	...	...	...	...	...	48.7	...	...	9.1	94.5	...	...	...	...	...	...	...	...	...	...	...	...	...	11.6
11.8	...	...	...	...	...	...	...	...	...	47.6	...	...	5.5	99.0	...	...	...	...	...	...	...	...	...	...	...	...	...	11.8
12.0	...	...	...	...	...	...	...	...	...	46.0	...	...	1.3	...	...	...	...	...	...	...	...	...	...	...	...	...	...	12.0
12.2	...	...	...	...	...	...	...	...	...	43.2	...	...	...	...	...	...	...	...	...	...	...	...	...	...	...	...	...	12.2
12.4	...	...	...	...	...	...	...	...	...	39.1	...	...	...	...	...	...	...	...	...	...	...	...	...	...	...	...	...	12.4
12.6	...	...	...	...	...	...	...	...	...	31.8	...	...	...	...	...	...	...	...	...	...	...	...	...	...	...	...	...	12.6
12.8	...	...	...	...	...	...	...	...	...	21.4	...	...	...	...	...	...	...	...	...	...	...	...	...	...	...	...	...	12.8

From Geigy scientific tables, ed 8, Basel, Switzerland, 1981, CIBA-Geigy AG.

APPENDIX C

Concentrations of common acids and bases

Compound	Molecular weight	Specific gravity	Percent	Normality	mL/liter for 1N* solution
HCl	36.46	1.19	36.0	11.7	85.5
HNO_3	63.02	1.42	69.5	15.6	64.0
H_2SO_4	98.08	1.84	96.0	35.9	28.4
CH_3COOH	60.03	1.06	99.5	17.6	56.9
NH_4OH	35.04	0.90	58.6	15.1	66.5
H_3PO_4	98.00	1.69	85.0	44.1	22.7
Thioglycolic acid	92.12	1.26	80.0	10.9	91.3
HCOOH	46.03	1.21	97.0	25.5	39.2
	46.03	1.19	88.0	22.7	44.1
$HClO_4$	100.5	1.67	70.0	11.65	85.7
Pyridine	79.10	0.98	100.0	12.4	80.6
2-Mercaptoethanol	78.13	1.14	100.0	14.6	68.5

To calculate concentration *(c)* from the weight percent *(w)* of a compound, use the formula:

$$\frac{10\ ws}{M} = c$$

M is the molecular weight, and *s* the specific gravity.

From Brewer, JM, Pesce, AJ, and Ashworth, H: Experimental techniques in biochemistry, Engelwood Cliffs, NJ, 1974, Prentice-Hall, Inc.
*Remember, the normality *(N)* is not the same as the molarity *(M)* for sulfuric and phosphoric acid.

APPENDIX D

Gases in common laboratory use—technical information

Product	Formula	State	CGA no. valve outlet	Highest purity grade (%)	Cylinder size (cubic feet)	Approximate cylinder pressure (psi)	Molecular weight	Vapor pressure at 21.1° C (psig)	Specific gravity at 21.1° C (1 atm)
		Cylinder specifications					**Thermophysical properties**		
Acetylene	C_2H_2	Dissolved gas (in acetone)	510/300	99.6	3-5	250	26.04	635	0.095
Air		Compressed gas	346/677	Mixture	200-500	2200-6000	28.96		1
Carbon dioxide	CO_2	Liquefied gas	320	99.999	2-6	274-838	44.01	839	1.53
Carbon monoxide	CO	Compressed gas	350	99.99	2-6	315-1602	28.01	*	0.97
Helium	He	Compressed gas	580/677	99.9999	200-300	2200-2640	4.003	*	0.138
Hydrogen	H_2	Compressed gas	350	99.9995	2-200	323-2200	2.02	*	0.0695
Methane	CH_4	Compressed gas	350	99.992	12	780-2000	16.04	*	0.555
Nitrogen	N_2	Compressed gas	580/677	99.999	30-200	2200-2640	28.01	*	0.967
Oxygen	O_2	Compressed gas	540	99.995	30-200	2200-2640	32.0	*	1.105
Propane	C_3H_8	Liquefied gas	510	99.98	12	109	44.1	109	1.55

Data derived from the Airco Industrial Cases catalogue, Murray Hill, NJ, 1977.
*Above critical temperature at 21.1° C.
SA, Simple asphyxiant.

Thermophysical properties			Hazardous properties			
			Flammability			
Critical temperature (°C)	Critical pressure (psia)	Specific volume (cf/lb)	Flammable limits in air (vol%)	Ignition temperature (°C)	Physiological properties	Threshold limit value (ppm)
35.1	890	14.7	2.3-100	305		SA
		13.3			Oxidant	
31.0	1071	8.74			Inert	5000
−140.0	507.4	13.8	12.5-74	651.1	Toxic	50
−267.8	33.2	96.7				SA
−239.98	190.8	192	4-75	585		SA
−82.1	673	23.7	5-15	538		SA
−146.9	492.9	13.8			Inert	SA
−118.4	736.9	12.1			Oxidant	
96.8	617.4	8.5	2.1-9.5	468		SA

APPENDIX E

Major plasma proteins

Protein	Molecular weight	Concentration, mg/100 mL	Electrophoretic† mobility	Biological function
Prealbumins				
Thyroxine binding (TBPA)	55,000	10-40	7.6	Thyroxine transport
Retinol binding (RBP)	21,000	3-6		Vitamin A transport
Albumin	66,300	3500-5500	5.92	Maintain osmotic pressure, transport of bilirubin, free fatty acids, anions, and cations, cell nutrition
α_1 Globulins				
α_1 Acid glycoprotein (α_1S)	40,000	55-140	5.7	Unknown, inactivates progesterone
α_1 Antitrypsin (α_1AT)	54,000	200-400	5.42	Antiserine type of protease
α_1 Glycoprotein (9.5S, α_1M)	308,000	3-8	α_1	Unknown
α_1 Glycoprotein B (α_1B)	50,000	15-30	α_1	Unknown
α_1 Glycoprotein T (α_1T)	60,000	5-12	α_1	Unknown, tryptophan poor
α_1 Antichymotrypsin (α_1X)	68,000	30-60	α_1	Chymotrypsin inhibitor
α_1 Lipoproteins, high density (HDL)	28,000	254-387	α_1	Lipid transport
α_2 Globulins				
G_0 Globulin (Gc)	51,000	40-70	α_2	Vitamin D transport
Ceruloplasmin (Cp)	134,000	15-60	4.6	Copper transport, peroxidase activity
α_2 Glycoprotein, histidine rich (HRG)	58,000	5-15	α_2	Unknown
Zn-α_2-glycoprotein (Znα_2)	41,000	2-15	4.2	Unknown, binds Zn^{2+}
α_2 HS-glycoprotein (α_2HS)	49,000	40-85	4.2	Unknown, binds Ba^{2+}
α_2 Macroglobulin (α_2M)	725,000	150-420	4.2	Inhibitor of thrombin, trypsin, and pepsin
Transcortin (TC)	49,500	<7	α_2	Cortisol transport
Haptoglobins (Hp)				
Type 1-1	100,000	100-200	4.1	Binds hemoglobin, prevents loss of iron
Type 2-1	200,000	160-300	α_2	
Type 2-2	400,000	120-260	α_2	
α_2 Lipoproteins (VLDL)	250,000	150-230	Pre-β	Lipid transport
Thyroxine-binding protein (TBG)	58,000	1-2	α_2	Thyroxine transport
β Globulins				
Hemopexin (Hpx)	57,000	50-100	3.1	Binds heme
Transferrin (Tf)	76,500	200-320	3.1	Iron transport
β Lipoproteins (LDL)	250,000	280-440	3.1	Lipid transport
C4 Complement component (C4)	206,000	40-80	β_1	Complement system
β_2 Microglobulin ($\beta_2\mu$)	11,818	Trace	β_2	Common portion of the HLA transplantation antigen
β_2 Glycoprotein I (β_2 I)	40,000	15-30	1.6	Unknown
β_2 Glycoprotein II (GGG)	63,000	12-30	β_2	C3 activator (activates properidin)
β_2 Glycoprotein III (β_2 III)	35,000	5-15	β_2	Unknown
C-Reactive protein (CRP)	118,000	1	β_2	Opsonin, motivates phagocytosis in inflammatory disease
C3 Complement component (C3)	180,000	55-180	β_2	Complement system
Fibrinogen (ϕ, Fib.)	341,000	200-600	2.1	Blood clotting
γ Globulins				
Immunoglobulin M (IgM)	950,000	60-250	2.1	Antibodies, early response
Immunoglobulin E (IgE)	190,000	0.06	2.1	Reagin of the allergy system
Immunoglobulin A (IgA)	160,000	90-450	2.1	Tissue antibodies
Immunoglobulin D (IgD)	160,000	15	1.9	Cell surface and plasma antibodies
Immunoglobulin G (IgG)	160,000	800-1800	1.2	Antibodies, long range

From Natelson, S, and Natelson, EA: Principles of applied chemistry, vol 3, New York, 1980, Plenum Publishing Corp.

*Does not include clotting factors, complement factors, or enzymes except fibrinogen, C3 and C4 of complement, which occur in substantial concentrations.

†Tiselius moving boundary electrophoresis in Tiselius units ($cm^2 V^{-1} sec^{-1} \times 10^5$, at 0° C, pH 8.6, and ionic strength 0.15).

APPENDIX F

Conversions between conventional and SI units

	Conventional units	× Factor	= SI units
Gram	g/mL	$\frac{10^{15}}{mw}$	pmol/L
	g/100 mL	10	g/L
	g/100 mL	$\frac{10}{mw}$	mol/L
	g/100 mL	$\frac{10^4}{mw}$	mmol/L
	g/d	$\frac{1}{mw}$	mol/d
	g/d	$\frac{10^3}{mw}$	mmol/d
	g/d	$\frac{10^9}{mw}$	nmol/d
Microgram	μg/100 mL	$\frac{10}{mw}$	μmol/L
	μg/d	$\frac{1}{mw}$	μmol/d
	μg/d	$\frac{10^3}{mw}$	nmol/d
Picogram	pg	$\frac{10^3}{mw}$	fmol
	pg/mL	$\frac{10^3}{mw}$	pmol/L
Milliequivalent	mEq/L	$\frac{1}{valence}$	mmol/L
	mEq/kg	$\frac{1}{valence}$	mmol/kg
	mEq/d	$\frac{1}{valence}$	mmol/d
Milligram	mg/100 mL	10^{-2}	g/L
	mg/100 mL	$\frac{10^{-2}}{mw}$	mol/L
	mg/100 mL	$\frac{10}{mw}$	mmol/L
	mg/100 mL	$\frac{10^4}{mw}$	μmol/L
	mg/100 g	10	mg/kg
	mg/100 g	$\frac{10}{mw}$	mmol/kg
	mg/d	$\frac{1}{mw}$	mmol/d
	mg/d	$\frac{10^3}{mw}$	μmol/d
Milliliter	mL/100 g	10	mL/kg
	mL/min	1.667×10^{-2}	mL/s
Millimeters of mercury	mm Hg	1.333	mbar
	mm Hg	0.133	kPa
Minute	min	60	s
	min	0.06	ks
Percent	%	10^{-2}	1 (unit)
	% (g/100 g)	10	g/kg
	% (g/100 g)	10^{-2}	kg/kg
	% (g/100 mL)	10	g/L
	% (g/100 mL)	$\frac{10}{mw}$	mol/L
	% (g/100 mL)	$\frac{10^4}{mw}$	mmol/L
	% (mL/100 mL)	10^{-2}	L/L

Modified from Campbell, JM, and Campbell, JB: Laboratory mathematics, ed 3, St Louis, 1983, The CV Mosby Co.
d, Day; *Eq,* equivalent; *g,* gram; *L,* liter; *min,* minute; *mw,* molecular weight; *Pa,* pascal; *s,* second.
f, Femto (10^{-15}); *p,* pico (10^{-12}); *n,* nano (10^{-9}); *μ,* micro (10^{-6}); *m,* milli (10^{-3}); *k,* kilo (10^3).

APPENDIX G

Conversions between conventional and SI units for specific analytes

Analyte	Conventional units	Multiply by: Conventional to SI	Multiply by: SI to conventional	SI unit
Acetominophen	μg/mL	6.61	0.151	μmol/L
Albumin	g/100 mL	144.9	0.0069	μmol/L
Ammonia	μg/100 mL	0.59	1.7	μmol/L
Anticonvulsant drugs				
Carbamazepine	μg/mL	4.32	0.23	μmol/L
Ethosuximide	μg/mL	7.08	0.14	μmol/L
Phenobarbital	μg/mL	4.31	0.23	μmol/L
Phenytoin	μg/mL	3.96	0.25	μmol/L
Primidone	μg/mL	4.58	0.22	μmol/L
Valproic acid	μg/mL	6.93	0.14	μmol/L
Bilirubin	mg/100 mL	17.1	0.059	μmol/L
Bromide	μg/mL	0.0125	80	mmol/L
Calcium	mg/100 mL	0.25	4	mmol/L
Chloride	mEq/L	1	1	mmol/L
Cholesterol	mg/100 mL	0.026	38.7	mmol/L
Cortisol	μg/100 mL	0.0276	36.2	μmol/L
Creatinine	mg/100 mL	88.4	0.0113	μmol/L
Digoxin	ng/mL	1.28	0.781	nmol/L
Estriol	μg/L	3.47	0.288	nmol/L
Ferritin	μg/L	2.2	0.445	pmol/L
Folic acid	μg/100 mL	22.7	0.044	nmol/L
Gentamicin	μg/mL	2.22	0.450	μmol/L
Glucose	mg/100 mL	0.055	18.0	mmol/L
Haptoglobin	mg/100 mL	0.118	8.47	μmol/L
HDL cholesterol	mg/100 mL	0.026	38.7	mmol/L
HCG	U/L	—	—	—
5-HIAA	mg	5.23	0.19	μmol
Ig A	mg/100 mL	0.0625	16	μmol/L
D	mg/100 mL	0.054	18.5	μmol/L
E	ng/mL	0.005	200	nmol/L
G	mg/100 mL	0.067	15	μmol/L
M	mg/100 mL	0.011	91	μmol/L
Insulin	pg/mL	0.174	5.74	nmol/L
	μU/mL	7.25	0.138	nmol/L
Iron	μg/100 mL	0.179	5.58	μmol/L
Ketones (acetoacetate)	mg/L	0.111	9.01	mmol/L
Lead	μg/L	4.83	0.207	nmol/L
Lithium	mEq/L	1	1	mmol/L
LDL cholesterol	mg/100 mL	0.026	38.7	mmol/L
Magnesium	mg/100 mL	0.41	2.43	mmol/L
Phosphorus	mg/100 mL	0.323	3.1	mmol/L
Phenylalanine	mg/L	6.05	0.165	μmol/L
Potassium	mEq/L	1	1	mmol/L

Analyte	Conventional units	Multiply by: Conventional to SI	Multiply by: SI to conventional	SI unit
Quinidine	μg/mL	3.09	0.324	μmol/L
Salicylate	mg/100 mL	0.0724	13.8	mmol/L
Sodium	mEq/L	1	1	mmol/L
TIBC	μg/100 mL	0.179	5.58	μmol/L
Theophylline	μg/mL	5.55	0.180	μmol/L
Thyroid-stimulating hormone	mU/L	—	—	—
Thyroxine	μg/100 mL	12.9	0.078	nmol/L
Transferrin	mg/100 mL	0.11	9.09	μmol/L
Triglycerides	mg/100 mL	0.0114	87.5	mmol/L
Urea	mg/100 mL	0.166	6.01	mmol/L
Urea N	mg/100 mL	0.356	2.81	mmol/L
Uric acid	mg/100 mL	59.5	0.0168	μmol/L
Vanillylmandelic acid	mg	5.03	0.20	μmol
VLDL cholesterol	mg/100 mL	0.026	38.7	mmol/L
Vitamin B_{12}	pg/mL	0.738	1.36	pmol/L
Gases	mm Hg	0.133	7.51	kPa
Enzymes	U/L	1.67×10^{-8}	0.6×10^{8}	katal/L

APPENDIX H Body surface of children

Nomogram for determination of body surface from height and mass*

Height	Body surface	Mass

Height — cm: 120, 115, 110, 105, 100, 95, 90, 85, 80, 75, 70, 65, 60, 55, 50, 45, 40, 35, 30, cm 25; in: 47 in, 46, 45, 44, 43, 42, 41, 40, 39, 38, 37, 36, 35, 34, 33, 32, 31, 30, 29, 28, 27, 26, 25, 24, 23, 22, 21, 20, 19, 18, 17, 16, 15, 14, 13, 12, 11, 10 in

Body surface — 1.10 m^2, 1.05, 1.00, 0.95, 0.90, 0.85, 0.80, 0.75, 0.70, 0.65, 0.60, 0.55, 0.50, 0.45, 0.40, 0.35, 0.30, 0.25, 0.20, 0.19, 0.18, 0.17, 0.16, 0.15, 0.14, 0.13, 0.12, 0.11, 0.10, 0.09, 0.08, 0.074 m^2

Mass — kg: 40.0, 35.0, 30.0, 25.0, 20.0, 15.0, 10.0, 9.0, 8.0, 7.0, 6.0, 5.0, 4.5, 4.0, 3.5, 3.0, 2.5, 2.0, 1.5, kg 1.0; lb: 90 lb, 85, 80, 75, 70, 65, 60, 55, 50, 45, 40, 35, 30, 25, 20, 15, 10, 9, 8, 7, 6, 5, 4, 3, 2.2 lb

From Geigy scientific tables, ed 8, Basel, Switzerland, 1981, CIBA-Geigy AG.

*From the formula of Du Bois and Du Bois: Arch Intern Med 17:863, 1916: $S = M^{0.425} \times H^{0.725} \times 71.84$, or $\log S = \log M \times 0.425 + \log H \times 0.725 + 1.8564$ (S: body surface in cm^2, M: mass in kg, H: height in cm).

APPENDIX I

Body surface of adults

Nomogram for determination of body surface from height and mass*

Height	Body surface	Mass

Height

cm 200 — 79 in; 78; 77; 195; 76; 75; 190; 74; 73; 185; 72; 71; 180; 70; 69; 175; 68; 67; 170; 66; 65; 165; 64; 63; 160; 62; 61; 155; 60; 59; 150; 58; 57; 145; 56; 55; 140; 54; 53; 135; 52; 51; 130; 50; 49; 125; 48; 47; 120; 46; 45; 115; 44; 43; 110; 42; 41; 105; 40; cm 100 — 39 in

Body surface

2.80 m²; 2.70; 2.60; 2.50; 2.40; 2.30; 2.20; 2.10; 2.00; 1.95; 1.90; 1.85; 1.80; 1.75; 1.70; 1.65; 1.60; 1.55; 1.50; 1.45; 1.40; 1.35; 1.30; 1.25; 1.20; 1.15; 1.10; 1.05; 1.00; 0.95; 0.90; 0.86 m²

Mass

kg 150 — 330 lb; 145; 320; 140; 310; 135; 300; 290; 130; 280; 125; 270; 120; 260; 115; 250; 110; 240; 105; 230; 100; 220; 95; 210; 90; 200; 190; 85; 180; 80; 170; 75; 160; 70; 150; 65; 140; 60; 130; 55; 120; 50; 110; 105; 45; 100; 95; 90; 40; 85; 80; 35; 75; 70; kg 30 — 66 lb

From Geigy scientific tables, ed 8, Basel, Switzerland, 1981, CIBA-Geigy AG.

*From the formula of Du Bois and Du Bois: Arch Intern Med 17:863, 1916: $S = M^{0.425} \times H^{0.725} \times 71.84$, or $\log S = \log M \times 0.425 + \log H \times 0.725 + 1.8564$ (S: body surface in cm^2, M: mass in kg, H: height in cm).

Index

Page numbers in **boldface** indicate structural formulas.
Page numbers in *italics* indicate illustrations and boxed material.
Page numbers followed by *t* indicate tables.

E

H

ANALYTE REFERENCE CHART—cont'd

Analyte*	Expected adult reference or therapeutic range		Clinical correlation (page)	Method of analysis (page)
	Conventional units	SI units		
Insulin (S)	<1042 pg/mL (<25 µU/mL)	<181 nmol/L	436	*(M-124)*
Iron (S)	400-1600 µg/L	7.16-28.6 µmol/L	496	*(M-1258)*
Iron-binding capacity, total (TIBC)	2.6-4.3 mg/L	46-77 µmol/L	496	
Saturation	20%-50%			
Ketones, total (S)	<50 mg/L	<500 µmol/L	436	856
See listing for individual ketone bodies, acetoacetate, β-hydroxybutyrate				
17-Ketosteroids (U)	5-15 mg/24 hr	—	650	
Lactate dehydrogenase (LD) (S)	90-320 (P-L)U/L	1.5-5.3 × 10^{-6} katal/L	415	924
Lactate dehydrogenase isoenzymes (S)	LD_1 18%-33%; LD_2 28%-40%; LD_3 18%-30%; LD_4 6%-16%; LD_5 2%-13%		415, 787	928
Lactic acid (S)	<180 mg/L	<2 mmol/L	332, 436	78
Lead (B)	100-200 µg/L	482-965 nmol/L	740	394
Lecithin-sphingomyelin (LS) ratio (amniotic fluid)	>2 indicates fetal maturity	—	569	968
Lidocaine (S)	1.2-5 µg/mL	5.1-21.3 µmol/L		415
Lipase (S)	2-7.5 lipase units/mL (66-248 U/L)	1.1-4.1 × 10^{-6} katal/L	390, 454	930
Lipoprotein electrophoresis (S or P)	—	—	454	991
Lithium (S, P)	0.4-1.0 mEq/L	0.4-1.0 mmol/L	719	1108
Low-density lipoprotein (LDL) cholesterol; *see* Cholesterol				
Lysozyme (S)	4-15.6 µg/mL	0.28-1.1 µmol/L	730	988
(U)	<1.4 µg/mL	<0.1 µmol/L		
Magnesium (S)	15.8-25.5 mg/L	0.65-1.05 mmol/L	373	875
(U)	24-255 mg/24 hr	1-10.5 mmol/24 hr		
Melanins (U)	Not detectable	—	730	1122
Mercury (U)	<20 µg/24 hr	<0.1 µmol/24 hr	740	405
Metanephrines (U)	Age dependent; see chart on p. *M-969*		415	964
Methotrexate (S)	<4.5 mg/L 24 hr after dose	<10 µmol/L 24 hr after dose	730	752
Methylmalonic acid (U)	2.7-8.6 mg/L	22-73 µmol/L	688	183
β_2-Microglobulin (S)	0.97-2.64 mg/L	82-224 nmol/L	346, 730	5
(U)	<320 µg/L	<27 nmol/L		
Mucopolysaccharides (U)	Age and sex dependent; see Table 29-3 (p. *M-199*)		688	189
Myoglobin (S, U)	Age, sex, and race dependent; see Table 120-3 (p. *M-919*)		415, 427	917
5′-Nucelotidase (S)	<11 U/L	<18.4 × 10^{-8} katal/L	359	933
Opiates—morphine (S)	>50 ng/mL	>175 nmol/L	740	409
Organic acid screen (U)	negative	—	688	201
Osmolality (S)	282-300 mOsm/kg	—	313, 346, 436	897
(U)	50-1200 mOsm/kg	—		
pH (arterial blood)	7.35-7.45	—	313, 332	860
P_{CO_2} (arterial blood)	35-45 mm Hg	4.6-6.0 kPa	313, 332	860
P_{O_2} (arterial blood)	80-100 mm Hg	10.7-13.9 kPa	313, 332	860
Parathyroid hormone (parathyrin) (PTH) (S)	<1 ngEq hPTH/mL	<105 pmol/L	373, 664	1027
Phenylalanine (S)	12-18 mg/L adults	73-109 µmol/L	688	205
	8-16 mg/L children	49-97 µmol/L		
Phosphatidyl glycerol (amniotic fluid)	Presence indicates fetal maturity	—	569	968
Phosphorus (S)	25-48 mg of P/L	0.81-1.55 mmol of P/L	373, 664	881
(U)	0.3-1.3 g of P/24 hr	9.7-42.0 mmol of P/24 hr		
Porphobilinogen screen (U)	Not detectable (<2 mg/L)	—	359, 496	1124
Porphobilinogen quantitation (U)	0.5-2.1 mg/day	2.25-9.25 µmol/day		*(M-1265)*
Porphyrin fractionation (U):			496	1270
Coproporphyrin	<200 µg/24 hr	<305 nmol/24 hr		
Uroporphyrin	<40 µg/24 hr	<48 nmol/24 hr		
Potassium (S)	3.6-5.0 mEq/L	3.6-5.0 mmol/L	313, 346	884
(U)	25-150 mEq/24 hr	25-150 mmol/24 hr		
Procainamide (S)	4-10 µg/mL	17-42 µmol/L	415	1110
N-acetulprocainamide and procainamide (S)	5-20 µg/mL	—		
Progesterone (S)	<0.4 ng/mL males	<1.27 pmol/mL	569, 650	253
	0.1-1.5 ng/mL females (pp. *M-255-256*)	0.3-4.8 pmol/mL		
Prolactin (S)	<20 µg/L males and nonpregnant females	<0.91 nmol/L	569	258
Propranolol (S)	60-400 ng/mL	0.23-1.5 µmol/L	415	971
Prostacyclin (P): 6-keto-$PGF_1\alpha$	<2.8 ng/L	<8 pmol/L	346	1208
Protein electrophoresis (S)	See interpretation	—	730	1054

**B*, Whole blood; *P*, plasma; *S*, serum; *U*, urine. Page numbers in parentheses and preceded by *M-* are cross-references to Pesce, AJ, and Kaplan, LA: *Methods in Clinical Chemistry*, St Louis, 1987, CV Mosby Co.

Continued on next page.

ANALYTE REFERENCE CHART—cont'd

Analyte*	Expected adult reference or therapeutic range			Clinical correlation (page)	Method of analysis (page)
	Conventional units		SI units		
Protein, total:					
Cerebrospinal fluid	150-450 mg/L		—	594	1037
(S)	66.6-81.4 g/L		—	359	1057
(U)	40-150 mg/24 hr		—	346	1060
Pyruvic acid (S)	<8.8 mg/L		<0.1 mmol/L	436	*(M-83)*
Quinidine (S)	2-5 μg/mL		6-15 μmol/L	415	*(M-927)*
Renin (P)	<4.5 ng/mL/hr	normal diet	—	415	*(M-979)*
Salicylate (S)	<200 μg/mL		<1.44 mmol/L	740	1113
	<300 μg/mL	arthritics	<2.17 mmol/L		
Schilling's test (U)	>15% absorption of dose			398	1133
Screens; *see* Amino acids; Carbohydrates; Drugs; Organic acids					
Selenium (B)	103-190 μg/L		1.3-2.4 μmol/L	533	*(M-543)*
Serotonin (B)	50-200 ng/mL		0.28-1.3 nmol/mL	719	*(M-796)*
SGOT (*see* AST)					
SGPT (*see* ALT)					
Sodium (S)	135-145 mEq/L		135-145 mmol/L	313, 346	884
(U)	40-220 mEq/24 hr		40-220 mmol/24 hr		
Steroid hormone receptors	<10 fmol/mg of cytosol protein		—	650	*(M-767)*
T_3 uptake (S)	25%-35%		—	620	95
Testosterone (S, P)	3-10 ng/mL	adult male	10.4-34.7 pmol/mL	650	*(M-266)*
	Depends on woman's menstrual cycle; see p. *M-268*				
Theophylline (S, P)	10-20 μg/mL		55-111 umol/L	795	1146
Thromboxane (P)	25-44 ng/L		71-125 pmol/L	346	*(M-1208)*
Thyroid-stimulating hormone (TSH) (S, P)	0.51-5.75 mU/L		—	620	952
Thyroxine (T_4) (S, P)	41-120 μg/L		0.053-0.154 μmol/L	620	956
TIBC; *see* Iron-binding capacity, total (S)					
Transferrin (S)	1.87-3.12 g/L		21-35 μmol/L	594	*(M-1279)*
Trazodone (S)	0.5-2 mg/L		1.2-4.9 μmol/L	594	*(M-661)*
Tricyclic antidepressants with their metabolites (S):				719	*(M-667)*
Amitriptyline and nortriptyline	125-150 μg/L		450-540 nmol/L		
Amoxapine, 8-hydroxy- and 7-hydroxy-forms	200-500 μg/L		638-1597 nmol/L		
Desipramine	150-300 μg/L		563-1126 nmol/L		
Doxepine and desmethyldoxepine	75-200 μg/L		269-716 nmol/L		
Imipramine and desipramine	150-300 μg/L		535-1071 nmol/L		
Nortriptyline	50-125 μg/L		190-475 nmol/L		
Triglycerides (P)	Age and sex dependent, p. *M-1225*			454	997
Triiodothyronine, free (S)	2.6-4.2 ng/L		4-6.5 pmol/L	620	*(M-277)*
Trypsin, duodenal fluid	>4 mEq H^+/mL		—	390, 398	*(M-857)*
Urea (S)	50-170 mg of urea nitrogen/L		3.57-12.1 mmol of urea nitrogen/L	346	1021
	107-365 mg of urea/L		1.78-6.08 mmol of urea/L		
(U)	7-16 g of urea nitrogen/24 hr		0.5-1.14 mol of urea nitrogen/24 hr		
	15-34 g of urea/24 hr		0.25-0.57 mol of urea/24 hr		
Uric acid (S)	36-77 mg/L	males	214-458 μmol/L	346	1024
	25-68 mg/L	females	149-405 μmol/L		
(U)	250-750 mg/24 hr	average diet	1.49-4.46 mmol/24 hr		
Urine protein, total	40-150 mg/24 hr		—	346	1060
Vanillylmandelic acid (VMA) (U)		Age dependent		415	962
	<6.0 mg/g of creatinine	adults	<3.4 μmol/μmol of creatinine		
Vitamin A (S)	0.45-0.80 mg/L	males	1.57-2.79 μmol/L	526, 543	*(M-551)*
	0.35-0.75 mg/L	females	1.22-2.62 μmol/L		
Vitamin B_6 (S)	>8.64 μg/L		>35 nmol/L	526, 543	*(M-558)*
Vitamin B_{12} (S)	180-960 pg/mL		133-708 pmol/L	526, 543	1127
Vitamin C (S)	6-20 mg/L		0.034-0.114 mmol/L	543	*(M-574)*
Vitamin D metabolites				604	*(M-1043)*
25-Hydroxyvitamin D (S)	4-60 ng/mL	seasonal average	10-150 pmol/mL		
1,25-$(OH)_2$-vitamin D (S)	20-65 pg/mL		48-156 fmol/mL		
Vitamin E (S)	5-20 mg/L		11.6-46.4 μmol/L	543	*(M-582)*
Vitamin K (S)	0.5-2.0 ng/mL		1.1-4.4 nmol/L	543	*(M-591)*
D-Xylose (S)	>300 mg/L 1-2 hr after dose		>2 mmol/L 1-2 hr after dose	398	*(M-862)*
(U)	>6.25 g excreted/5 hr		>41 mmol excreted/5 hr		
Zinc (S)	654-1150 μg/L		10-17.6 μmol/L	533	*(M-596)*

**B*, Whole blood; *P*, plasma; *S*, serum; *U*, urine. Page numbers in parentheses and preceded by *M-* are cross-references to Pesce, AJ, and Kaplan, LA: *Methods in Clinical Chemistry*, St Louis, 1987, CV Mosby Co.